Prefixes and Suffixes of Anatomical and Medical Terminology

Element	Definition and Example
labi-	lip: *levator labii superioris*
lacri-	tears: *nasolacrimal*
later-	side: *lateral*
leuk-	white: *leukocyte*
lip-	fat: *lipid*
-logy	science of: *morphology*
-lysis	solution, dissolve: *hemolysis*
macro-	large, great: *macrophage*
mal-	bad, abnormal, disorder: *malignant*
medi-	middle: *medial*
mega-	great, large: *megakaryocyte*
meso-	middle or moderate: *mesoderm*
meta-	after, beyond: *metatarsal*
micro-	small: *microtome*
mito-	thread: *mitosis*
mono-	alone, one, single: *monocyte*
mons	mountain: *mons pubis*
morph-	form, shape: *morphology*
multi-	many, much: *multinuclear*
myo-	muscle: *myology*
narc-	numbness, stupor: *narcotic*
necro-	corpse, dead: *necrosis*
neo-	new, young: *neonatal*
nephro-	kidney: *nephritis*
neuro-	nerve: *neurolemma*
noto-	back: *notochord*
ob-	against, toward, in front of: *obturator*
oc-	against: *occlusion*
-oid	resembling, likeness: *sigmoid*
oligo-	few, small: *oligodendrocyte*
-oma	tumor: *lymphoma*
oo-	egg: *oocyte*
or-	mouth: *oral*
orchi-	testicles: *cryptorchidism*
ortho-	straight, normal: *orthopnea*
-ory	pertaining to: *sensory*
-ose	full of: *adipose*
osteo-	bone: *osteoblast*
oto-	ear: *otolith*
ovo-	egg: *ovum*
par-	give birth to, bear: *parturition*
para-	near, beyond, beside: *paranasal*
path-	disease, that which undergoes sickness: *pathology*
-pathy	abnormality, disease: *neuropathy*
ped-	children: *pediatrician*
pen-	need, lack: *penicillin*
-penia	deficiency: *thrombocytopenia*
per-	through: *percutaneous*
peri-	near, around: *pericardium*
phag-	to eat: *phagocyte*
-phil	have an affinity for: *neutrophil*
phlebo-	vein: *phlebitis*
-phobe	abnormal fear, dread: *hydrophobia*
-plasty	reconstruction of: *rhinoplasty*
platy-	flat, side: *platysma*
-plegia	stroke, paralysis: *paraplegia*
-pnea	to breathe: *apnea*
pneumato-	breathing: *pneumonia*

Element	Definition and Example
pod-	foot: *podiatry*
-poiesis	formation of: *hemopoiesis*
poly-	many, much: *polyploid*
post-	after, behind: *postnatal*
pre-	before in time or place: *prenatal*
prim-	first: *primitive*
pro-	before in time or place: *prosect*
proct-	anus: *proctology*
pseudo-	false: *pseudostratified*
psycho-	mental: *psychology*
pyo-	pus: *pyoculture*
quad-	fourfold: *quadriceps femoris*
re-	back, again: *repolarization*
rect-	straight: *lateral rectus*
reno-	kidney: *renal*
rete-	network: *retina*
retro-	backward, behind: *retroperitoneal*
rhin-	nose: *rhinitis*
-rrhage	excessive flow: *hemorrhage*
-rrhea	flow or discharge: *diarrhea*
sanguin-	blood: *sanguiferous*
sarc-	flesh: *sarcoplasm*
-scope	instrument for examining a part: *stethoscope*
-sect	cut: *dissect*
semi-	half: *semilunar*
serrate-	saw-edged: *serratus anterior*
-sis	process or action: *dialysis*
steno-	narrow: *stenohaline*
-stomy	surgical opening: *tracheostomy*
sub-	under, beneath, below: *subcutaneous*
super-	above, beyond, upper: *superficial*
supra-	above, over: *suprarenal*
syn- (sym-)	together, joined, with: *synapse*
tachy-	swift, rapid: *tachycardia*
tele-	far: *telencephalon*
tens-	stretch: *tensor fascia lata*
tetra-	four: *tetrad*
therm-	heat: *thermogram*
thorac-	chest: *thoracic cavity*
thrombo-	lump, clot: *thrombocyte*
-tomy	cut: *appendectomy*
tox-	poison: *toxic*
tract-	draw, drag: *traction*
trans-	across, over: *transfuse*
tri-	three: *trigone*
trich-	hair: *trichology*
-trophy	a state relating to nutrition: *hypertrophy*
-tropic	turning toward, changing: *gonadotropic*
ultra-	beyond, excess: *ultrasonic*
uni-	one: *unicellular*
-uria	urine: *polyuria*
uro-	urine, urinary organs or tract: *uroscope*
vas-	vessel: *vasoconstriction*
vermi-	worm: *vermiform*
viscer-	organ: *visceral*
vit-	life: *vitamin*
zoo-	animal: *zoology*
zygo-	union, join: *zygote*

Human Anatomy

Human Anatomy

Third Edition

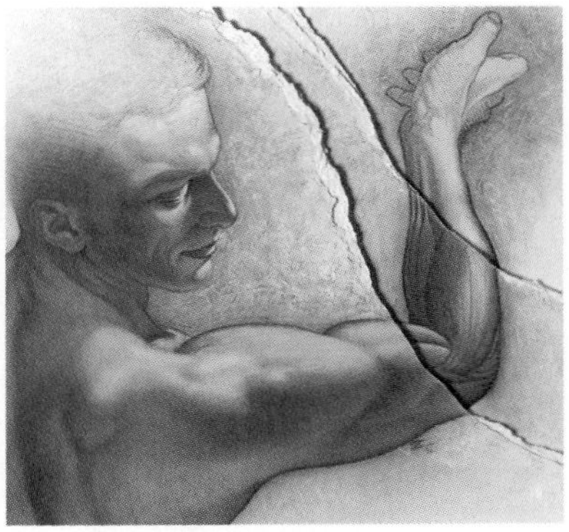

Kent M. Van De Graaff
Brigham Young University

WCB Wm. C. Brown Publishers

Book Team

Editor *Colin H. Wheatley*
Developmental Editor *Elizabeth M. Sievers*
Production Editor *Anne E. Scroggin*
Designer *K. Wayne Harms*
Art Editor *Mary E. Swift*
Permissions Editor *Karen L. Storlie*
Visuals Processor *Joseph P. O'Connell*

Wm. C. Brown Publishers

President *G. Franklin Lewis*
Vice President, Publisher *George Wm. Bergquist*
Vice President, Operations and Production *Beverly Kolz*
National Sales Manager *Virginia S. Moffat*
Group Sales Manager *Vincent R. Di Blasi*
Vice President, Editor in Chief *Edward G. Jaffe*
Executive Editor *Earl McPeek*
Marketing Manager *John W. Calhoun*
Advertising Manager *Amy Schmitz*
Managing Editor, Production *Colleen A. Yonda*
Manager of Visuals and Design *Faye M. Schilling*
Production Editorial Manager *Julie A. Kennedy*
Production Editorial Manager *Ann Fuerste*
Publishing Services Manager *Karen J. Slaght*

WCB Group

President and Chief Executive Officer *Mark C. Falb*
Chairman of the Board *Wm. C. Brown*

Cover illustration by Craig Zuckerman

The credits section for this book begins on page 775, and is considered an extension of the copyright page.

Photo research by Kathy Husemann

Library of Congress Catalog Card Number: 90-85080

ISBN 0-697-07892-2 (case)
0-697-14988-9 (paper)

Printed in the United States of America by Wm. C. Brown Publishers, 2460 Kerper Boulevard, Dubuque, IA 52001

10 9 8 7 6 5 4 3 2

To the users and reviewers of the previous editions of this text who have so thoughtfully suggested ways for its improvement.

Brief Contents

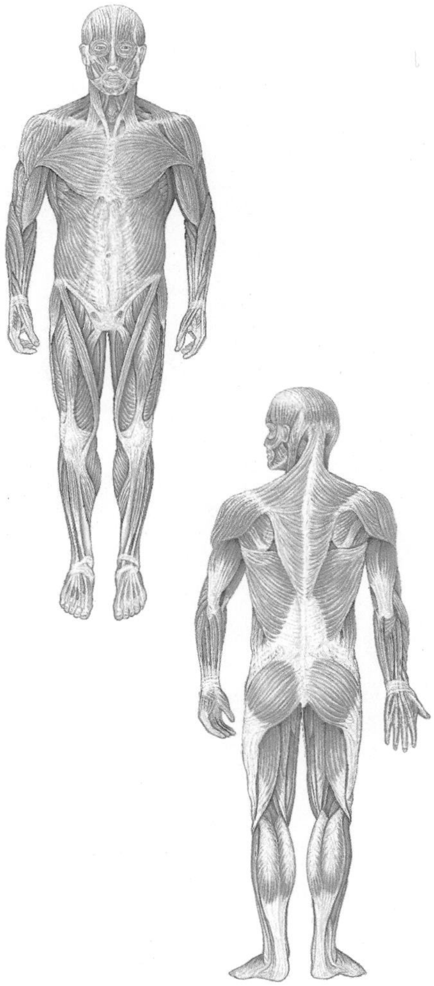

Contents

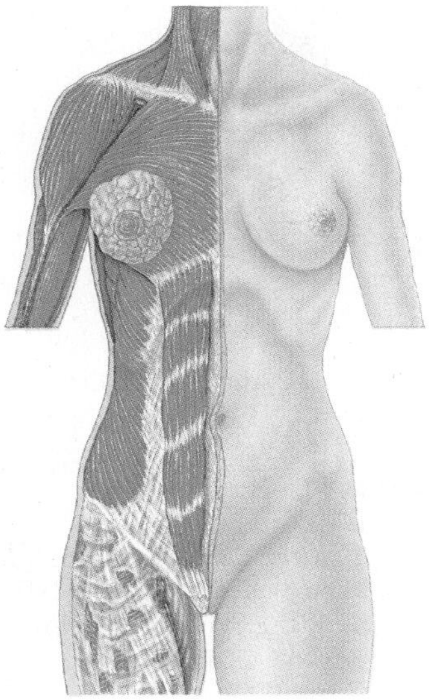

UNIT IV

Support and Movement 114

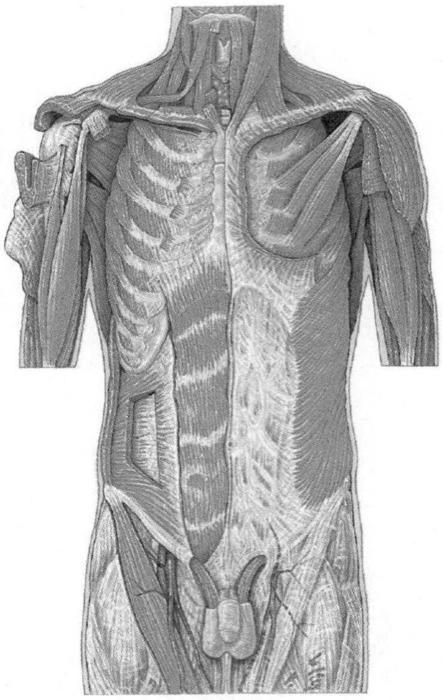

5 Integumentary System 117

6 Skeletal System: Introduction and the Axial Skeleton 141

7 Skeletal System: The Appendicular Skeleton 175

8 Articulations 193

9 Muscular System 221

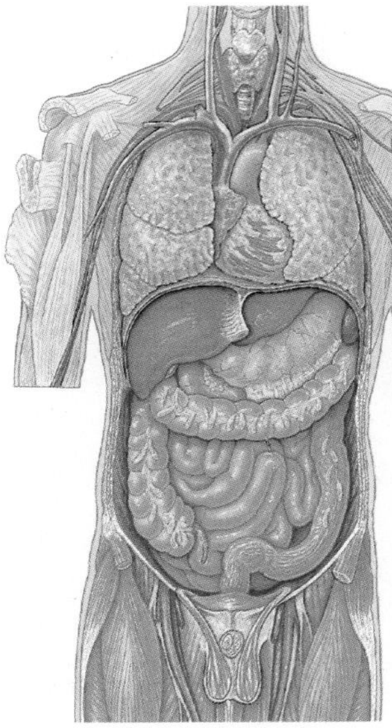

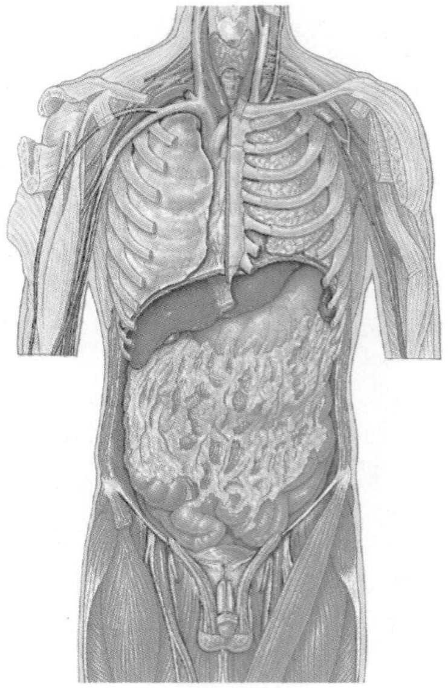

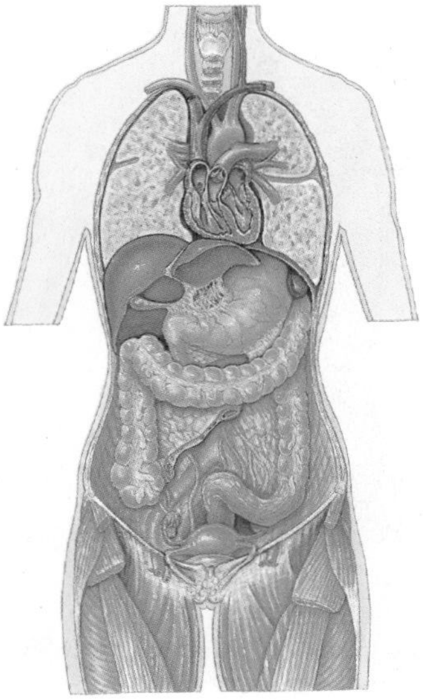

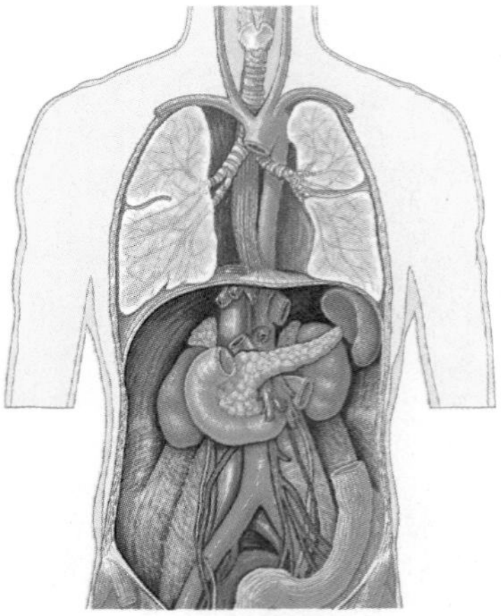

Preface

The third edition of *Human Anatomy* is written to be a basic introduction to human anatomy for students majoring in biological, medical, allied-health, and physical education programs. The book achieves balance by presenting embryological development, histology, gross anatomical structure, functions, and possible dysfunctions for each principal organ system. A synopsis of the clinical aspects of human anatomy is presented as a summary chapter at the end of the book. The information in this book is presented in an interesting and concise manner to maintain readability. Application is emphasized, since anatomy is a foundation for many careers, including nursing, paramedicine, dentistry, medical technology, physical therapy, medicine, and athletic training.

Objectives

With the needs of students and instructors in mind, these six objectives of *Human Anatomy* were formulated.

1. To provide an accurate and current text that is visually appealing, interesting, and readable—a text that will entice students to study the material sufficiently to gain an appreciation of life and a better understanding of the structure, function, and magnificence of the human body.
2. To identify and state clearly the basic concepts and accompanying learning objectives of the discipline of human anatomy.
3. To provide a correlated and balanced presentation of anatomy at the developmental, cellular, histological, clinical, and gross levels.
4. To develop in the student a sufficient anatomical vocabulary so that he or she understands and is conversant with medical terminology.
5. To encourage proper care of the body so that a more healthy and productive life can be enjoyed.
6. To acquaint the student with the history of the science of anatomy, from its primitive beginnings to recent advances.

Organization

The twenty-four chapters in this text are grouped into eight units.

Unit I In this unit, the stage is set for studying human anatomy by providing a historical perspective of how this science has developed over a long and arduous period. Anatomy is an exciting and dynamic science that has had and continues to have many contributors. An attempt is made in this unit to invite the student to feel a part of the heritage of anatomy.

Unit 2 In this unit, the human is described as a vertebrate animal. The various levels of body organization are also described, and the basic terminology is defined, which is necessary for understanding the structure and functioning of the body.

Unit 3 Body organization is discussed at the cellular and histological levels in this unit. Cellular chemistry is also introduced and emphasized as an integral aspect of understanding bodily functions.

Unit 4 Support, protection, and movement of the human body are the topics of discussion in this unit. The body's external support and protection are provided by the integumentary system, and the internal support and protection of certain bodily organs are provided by the skeletal system. Movement is possible at the joints of the skeleton as the associated skeletal muscles are contracted.

Unit 5 This unit includes the nervous system, endocrine system, and sensory organs. The concepts identified and discussed in these chapters are concerned with the integration and coordination of bodily functions and the perception of environmental stimuli.

Unit 6 In this unit, the structure and function of the circulatory, respiratory, digestive, and urinary systems are discussed as they regulate and maintain bodily processes. The concepts and learning objectives presented in this unit emphasize those structures that directly relate to maintaining normal body functions and health.

Unit 7 The male and female reproductive system are described in this unit and the continuance of the human species through sexual reproduction is discussed. An overview of the entire sequence of human life, including prenatal development and postnatal growth, development, and aging, is presented.

Unit 8 In this unit, the body systems are presented as they are functionally integrated in the principal regions of the body. This unit also includes photographs of excellent dissections of human cadavers. The text portion of this unit provides a synopsis of the developmental, traumatic, and disease conditions that commonly afflict the organ systems within each of the body regions.

Pedagogical Aids

Each of the twenty-four chapters of this text contains a complete system of pedagogical aids, which organizes, stimulates an interest in, and greatly facilitates the study of human anatomy. A student who utilizes these aids will find the study of this material effective and more enjoyable.

Chapter Introductions. The chapter introductions contain an overview of the *contents of each chapter* in outline form, which will help the students locate sections dealing with particular topics. All of the *concepts* for each chapter are stated in the chapter introductions so that the students gain a perspective of the salient points that will be discussed. In addition, most of the chapters contain a *clinical investigation* in the form of a *case study in anatomy.* These hypothetical medical situations set the stage for the type of application information that will be presented in the chapter. The solution to the case study is presented at the end of the chapter following the review questions.

Understanding Terminology. Where each technical term first appears in the narrative, it is set off in *boldface* or *italic type* and often followed by a bracketed *phonetic pronunciation.* Many of the terms are further elucidated in footnotes explaining their *derivations.* In addition, the roots of each term can be identified by referring to the *glossary of prefixes and suffixes* on the inside of the front and back covers. Knowing the pronunciation, derivation, and roots of a scientific term will help the student retain its meaning.

Chapter Sections. Each chapter is divided into several principal sections, and each section is prefaced by a *concept statement* and a list of *learning objectives.* A concept is a carefully worded sentence that expresses the essence, or synthesis, of the voluminous information presented in a chapter section; it is the formulation of a unifying thought pertinent to the detail that follows. The learning objectives for a chapter section indicate the level of competency the student should attain in order to understand the concept thoroughly and be able to make application in practical situations. *Review questions* at the end of each chapter section test the student's understanding of the concept and mastery of the learning objectives.

Clinical Information. Short paragraphs set off in boxes of colored type occur throughout each chapter. These *boxed clinical asides* contain information on the practical application of learning the preceding facts.

Selected developmental disorders, clinical procedures, and diseases or dysfunctions of the specific organ systems are described in the *clinical considerations sections* at the end of each chapter. Pathological changes are illustrated in many of the discussions.

Chapter Summaries. A summary, in outline form, at the end of each chapter presents the salient thoughts and facts presented in the chapter. Reading these *comprehensive summaries* is an excellent way for students to review the chapters.

Review Activities. At the end of each chapter the review activities, comprised of objective and essay questions, provide the students with feedback as to the depth of their understanding and learning.

Illustrations and Tables. The third edition of *Human Anatomy* is designed to have the most effective illustration program of any anatomy text. Because anatomy is a visual science, great care has been taken to construct completely labeled, informative figures that are rendered clearly and accurately. Each figure is integrated with the narrative of the text. The figures in each chapter include many original *full-color art* and two-color anatomical drawings. Chapters 9, 11, and 16 on the muscular, nervous, and circulatory systems respectively, have been extensively illustrated with new color art. *Color graphics* are used to depict functional processes where complex ideas need to be clarified. Microscopic anatomy is illustrated by *photomicrographs* at the light microscope level and electron micrographs from scanning and transmission electron microscopy. Carefully selected *photographs* are used throughout the text to provide a balanced perspective of the gross anatomy. *Color-coding* is used in certain places within the text as a pedagogical aid in learning. The bones of the skull, in chapter 6 for example, are color-coded so that each bone can be identified readily in the many renderings that are displayed. A set of *anatomical reference plates* follows chapter 3, where it will be of most value for review purposes. Actual photographs of *dissected human cadavers* are displayed in chapter 24 to show the complexity and structural relationships of the body that can be appreciated only when seen in a human specimen.

Numerous *tables* have been specifically constructed for each chapter to summarize information and clarify complex data. These tables are placed near their related textual material or at the end of the chapter sections.

Appendixes, Glossary, and Index. The appendixes, following chapter 24, contain valuable reference information. *Appendix A* contains the answers to the objective questions that appear at the end of each chapter (see page 750). *Appendix B* lists the important scientific journals that publish current research relative to anatomy (see page 750). *Appendix C* presents suggestions for the preparation of laboratory demonstrations in anatomy (see page 751). *Appendix D* lists the medical and pharmacological abbreviations commonly used in clinical practice (see page 751). *Appendix E* lists the various units of measurements and their equivalents together with a description of how to convert one unit into another (see page 752).

Provided in the *Glossary* are pronunciations and definitions of the more important scientific and clinical terms used in the text. If a term has a commonly used synonym or eponym, it is also indicated in the glossary.

The comprehensive *Index* is complete and has numerous cross-references.

Supplementary Materials

The supplementary materials that accompany the third edition of *Human Anatomy* are designed to aid the students in their learning activities and help the instructors plan course work and presentations.

Since the laboratory is an integral aspect of instruction in human anatomy, two laboratory manuals are available that provide for the range of a laboratory's diverse needs. The third edition of *Human Anatomy Laboratory Textbook* by Kent M. Van De Graaff accommodates the use of human cadavers, as well as selected cat dissections. Clinical application is emphasized. *Human Anatomy Laboratory Manual* by Kent M. Van De Graaff is specifically designed to accompany the third edition of *Human Anatomy* textbook. This laboratory manual contains cat dissections and selected organ dissections. It emphasizes learning anatomical structures through visual observation, palpation, and knowing the functional relationship of one body system to another.

A *Student Study Guide to Accompany Human Anatomy* by Kent M. Van De Graaff contains chapter concepts and objectives, focus questions, mastery quizzes, study activities, and answer keys that correspond to the chapters of the text.

A *Computer Review of Human Anatomy and Physiology* by S. Scott Zimmerman, Thomas V. Davis, and Kent M. Van De Graaff is contained in a thirteen-disk set designed for Apple II®- or IBM®-compatible computers. This computer disk set provides a graphic and innovative way to learn and review human anatomy.

A *set of sixty-four color slides* that depict important clinical aspects of human anatomy is also available. These slides correlate with the clinical information presented in the text and have a descriptive narrative explaining the conditions or diseases visualized on the slides.

A *set of fifty-two selected acetate transparencies* is available to instructors who adopt this text. These transparencies, which are taken from select illustrations in the text, are ideal supplements to classroom lectures or may be used as quiz material.

An *Instructor's Manual and Test Item File* by Kerry Openshaw contains chapter overviews, possible student assignments and discussion of chapter concepts, lists of supplemental films, and directories of suppliers of audiovisual and laboratory materials. It also contains over fifty test items from each chapter of the text, designed to evaluate student comprehension of factual data, concepts, clinical situations, and applications of this knowledge.

WCB *QuizPak,* a student self-testing program that operates on an Apple II*e* or IIc or IBM PC microcomputer, is available to instructors who adopt this text.

WCB *TestPak,* a free, computerized testing service for generating examinations, is available to instructors who adopt this text.

Acknowledgments

The third edition of *Human Anatomy* could not have been prepared without the support and encouragement of colleagues and students. My sincere appreciation is extended to professors Duane H. Smith, R. Ward Rhees, Kerry Openshaw, Stuart I. Fox, LaMont W. Smith, Ferron L. Andersen, Richard A. Heckmann, and Wilford M. Hess for their professional counsel and example. Connie J. Erdmann, Sidney L. Palmer, and Karin B. Anderson provided valuable input from a student perspective.

I wish to acknowledge friends who have assisted in specific ways: W. Geoff Williams, M.D., J. Phillip Freestone, M.D., Douglas W. Hacking, M.D., and Charles H. Stewart, M.D., provided professional advice and reviewed the clinical portions of the text. Brent C. Chandler, M.D. and Jason Kerr, provided many of the X rays used in the text. The photographs of surface anatomy were taken by Glenn L. Anderson and Dr. Sheril D. Burton.

Quality illustrations for the third edition of this text were a top priority. The talented artists assembled for this immense project were Chris Creek, Thomas Waldrop, Ruth Krabach, William Loechel, Steven Moon, Diane Nelson, Michael P. Schenk, and Thomas Sims. It has been a pleasure working with these professionals, and I appreciate their tremendous contribution.

The editorial staff and production people at Wm. C. Brown Publishers have been superb to work with. Associate Acquisitions Editor Colin H. Wheatley and Developmental Editor Elizabeth M. Sievers worked diligently throughout the rewriting and production of this text providing continual guidance and insight. K. Wayne Harms designed the text and Anne E. Scroggin orchestrated its production by coordinating the efforts of many.

Wm. C. Brown Publishers assembled a review panel of competent anatomists to review the manuscript and illustrations. These individuals aided my work immeasurably, and I am especially grateful for their frank criticisms, comments, and encouragements: Dr. David H. Carr, McMaster University, Hamilton, Ontario; Mary Jo Fourier, Johnson County Community College; Larry Ganion, Ball State University; John H. Green, Nicholls State University; Mark Nielsen, University of Utah; Eric L. Sun, Macon College; Eric Wise, Santa Barbara City College.

Finally, but most importantly, I wish to extend appreciation to my family for their patience, encouragement, and love.

Kent Marshall Van De Graaff

Human Anatomy

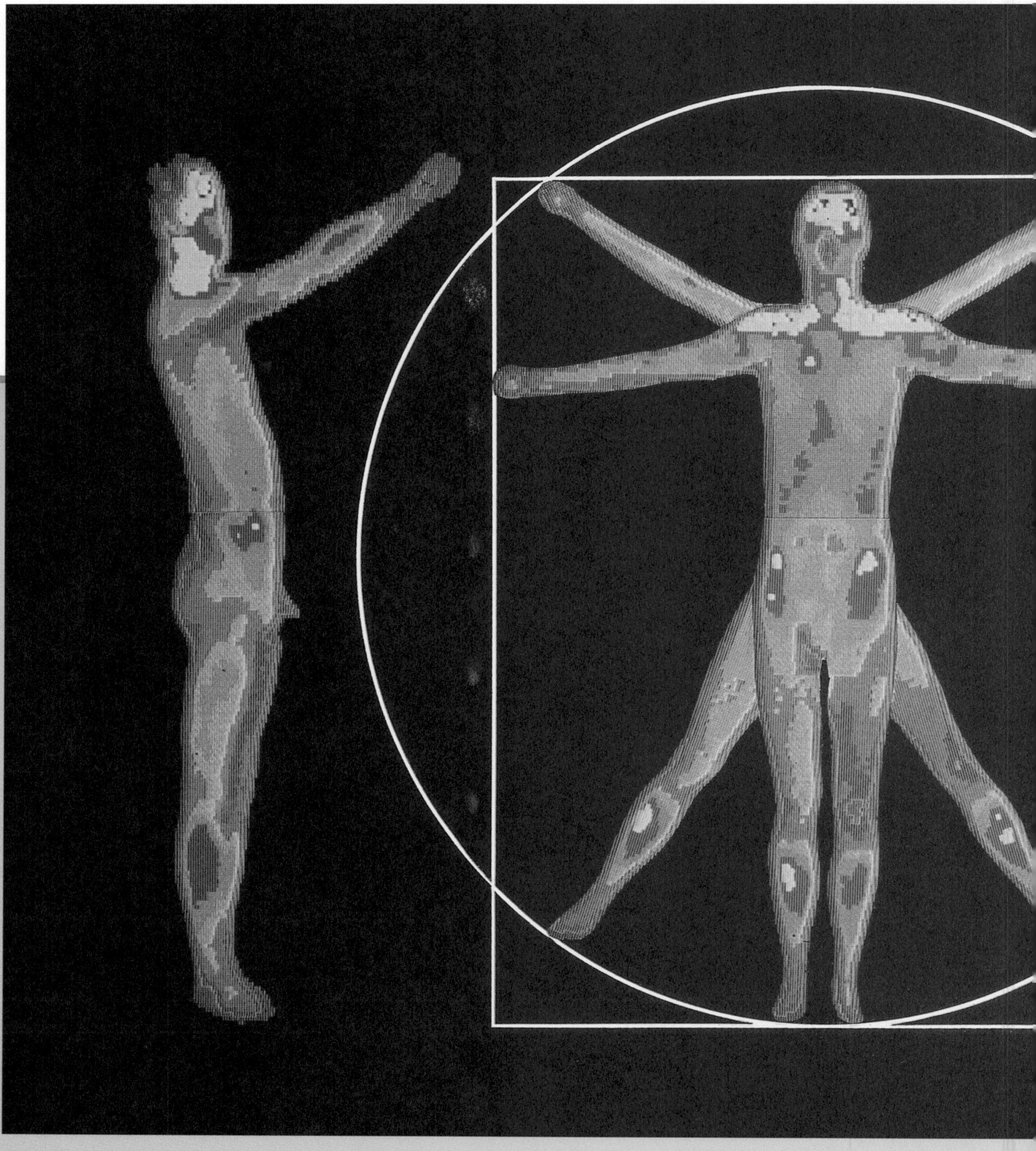

A modern thermograph rendition of the famous image of man by Leonardo da Vinci. The colors represent the range in skin temperature with red and yellow being the warmest and blue and green being the coolest. Thermograms may be used to determine relative rate of metabolism and even the presence of tumors.

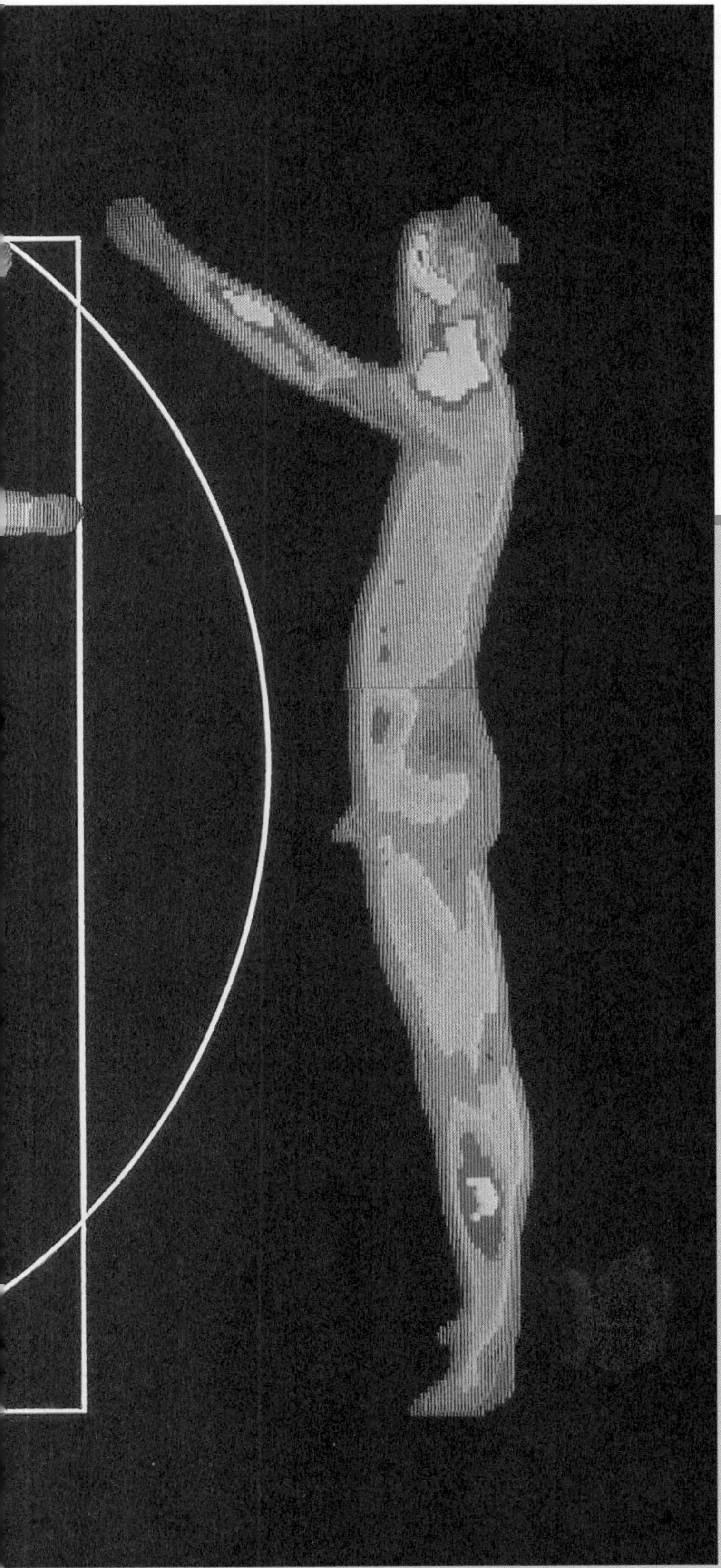

Historical Perspective

A brief historical review of anatomy is presented in the chapter in unit 1. The chapter is concerned with the various cultures, events, and people that made contributions to or had an impact on the science of anatomy. The stage is set for a modern day study of anatomy by providing a historical view of anatomy's ancient, exciting, but frequently troubled heritage. After studying this unit, a beginning student of anatomy should understand and appreciate the years of effort that earlier researchers have dedicated to gaining knowledge of the structure of the human body.

This unit includes:

1

History of Anatomy

CHAPTER 1

History of Anatomy

Outline and Concepts

Definition of the Science

The science of human anatomy is concerned with the structural organization of the human body. The descriptive anatomical terminology is principally of Greek and Latin derivation.

Prescientific Period

Evidence indicates that prehistoric anatomy was of survival value and provided the foundation for medicine.

Scientific Period

Human anatomy is a dynamic science that has had a long, exciting heritage and is currently providing the foundation for medical, biochemical, developmental, cytogenetic, and biomechanical research.

- Mesopotamia and Egypt
- Chinese and Japanese
- Grecian Period
- Alexandrian Era
- Roman Era
- Middle Ages
- Contributions of Islam
- Renaissance
- Seventeenth and Eighteenth Centuries
- Nineteenth Century
- Twentieth Century

Chapter Summary

Review Activities

Definition of the Science

The science of human anatomy is concerned with the structural organization of the human body. The descriptive anatomical terminology is principally of Greek and Latin derivation.

Objective 1. Define *anatomy.*
Objective 2. Distinguish between anatomy, physiology, and biology.
Objective 3. Explain why Greek and Latin have been used to describe structures of the human body.

Human anatomy is the science concerned with the structure of the human body. The term **anatomy** *(an-nat'o-me)* is derived from a Greek word meaning "to cut up," and in the past, the word *anatomize* was more commonly used than the word *dissect.* **Physiology** *(fiz''e-ol'o-je)* is an extension beyond anatomy in that, as a science, it is concerned with the function of the body. The term *physiology* is also derived from a Greek word meaning "the study of nature"—the "nature" of an organism is its function. Anatomy and physiology are both subdivisions of the science of **biology,** the study of living organisms. The anatomy of every structure of the body is adapted for performing a function or perhaps several functions. Natural selection eliminates organisms with inappropriate structures and functions and determines which favorable structures will be passed from one generation to the next.

The dissection of human **cadavers** *(kah-dav'erz)* has been the basis for understanding the structure and function of the human body for a long time. Each beginning student can discover and learn firsthand as the structures of the body are systematically studied. The anatomical terms that a student learns while becoming acquainted with a structure represent the work of hundreds of dedicated past anatomists who have dissected, diagramed, described, and named the multitude of body parts.

The majority of the terms that form the language of anatomy are of Greek or Latin derivation. Latin was the language of the Roman Empire, during which time an interest in scientific description was cultivated. With the decline of the Roman Empire, Latin became a "dead language," but it retained its value in nomenclature because it remained unchanged throughout history. As a consequence, anatomy is a descriptive science in which many of the terms can be understood through a knowledge of the basic prefixes and suffixes (see inside of the front and back covers). Although the Greeks and Romans made significant contributions to anatomical terminology, over the centuries many great persons from several cultures have contributed to the science of human anatomy.

The science of human anatomy has had a rich, long, and frequently troubled heritage. The history of the science of human anatomy parallels that of medicine. In fact, interest in the structure of the body often results from the desire of the medical profession to explain a body dysfunction. Various religions, on the other hand, have stifled the study of human anatomy through their restrictions on human dissections and religious explanations of diseases and debilitations.

anatomy: Gk. *ana,* up; *tome,* a cutting
physiology: Gk. *physis,* nature; *logos,* study
biology: Gk. *bios,* life; *logos,* study
cadaver: L. *cadere,* to fall

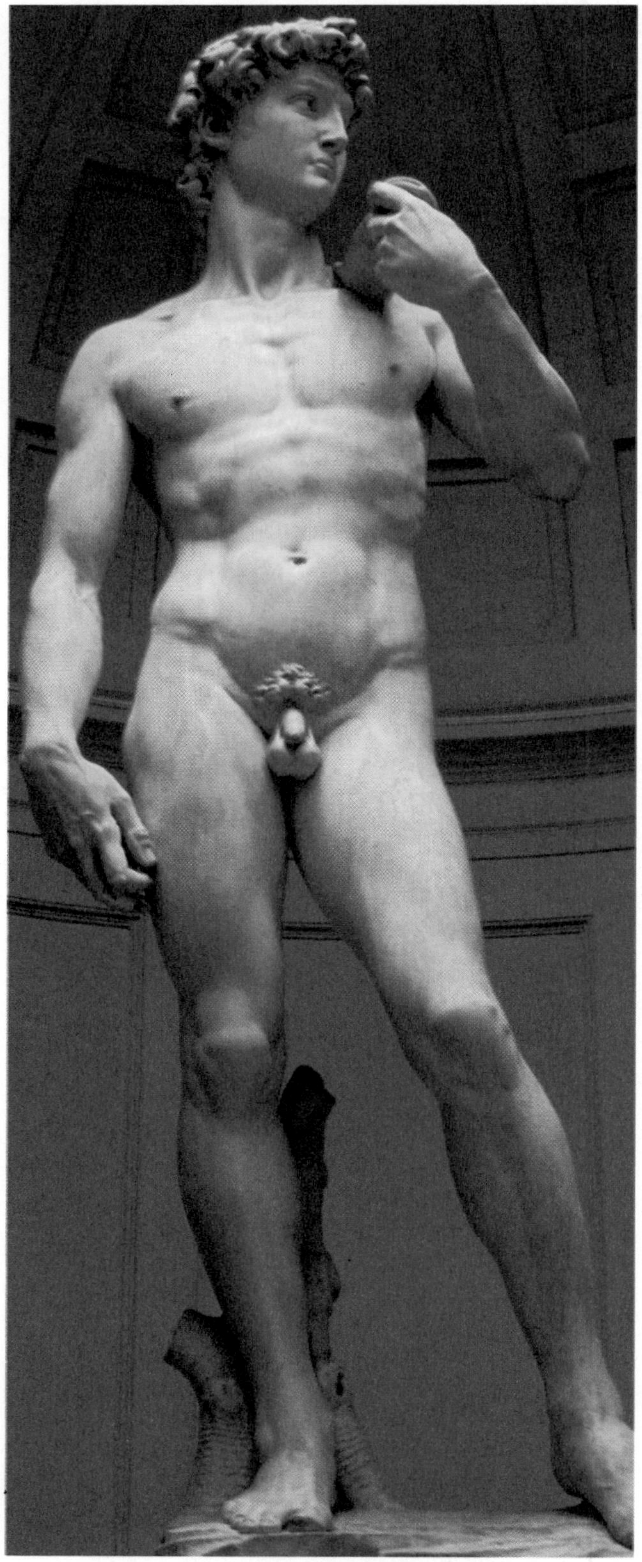

Figure 1.1 Michelangelo completed the seventeen-foot-tall *David* in 1504, sculptured from a single block of white, unflawed Carrara marble. This masterpiece captures the physical beauty of the human body in an expression of art.

People have always had an innate interest in their own structure and physical capabilities. The Greeks esteemed physical athletic competition and expressed the beauty of the body in their sculptures. Many of the great masters of the Renaissance portrayed human figures in their art. Indeed, many of these artists were excellent anatomists because of their attention to

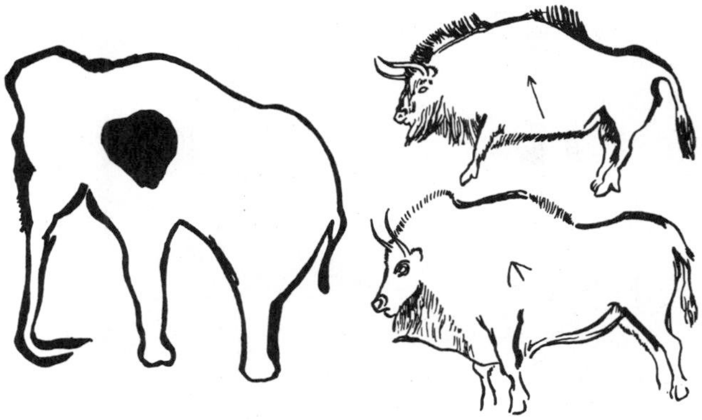

Figure 1.2 Contemporary redrawings of large game mammals that were depicted on caves occupied by prehistoric man in western Europe. Presumably the location of the heart is drawn on the mammoth, and vulnerable anatomical sites are shown on the two bison. Prehistoric people needed a practical knowledge of anatomy simply for survival.

detail. Such an artistic genius was Michelangelo, who was preoccupied with the detail and beauty of the human form and captured it in sculpture with the *David* (fig. 1.1) and in paintings such as those in the Sistine Chapel.

Even Shakespeare was awed by the structure of the human body as he wrote, "What a piece of work is a man! How noble in reason! how infinite in faculty! in form, moving, how express and admirable! in action how like an angel! in apprehension how like a god! The beauty of the world! The paragon of animals" (*Hamlet* 2.2.315–319).

In the past, human anatomy was an academic, descriptive science primarily concerned with identifying and naming body structures. Although dissection and description form the basis of anatomy, the importance of human anatomy today is in its functional approach and clinical applications. Human anatomy is a practical, applied science forming the foundation of an understanding of physical performance and body health. Studying the history of anatomy provides a perspective and appreciation of the relevant science that it is today.

1. What is the derivation and meaning of *anatomy?*
2. Explain the statement that anatomy is a science of observation and physiology is a science of experimentation and observation.
3. Why is knowing the anatomy and physiology of an organism essential to understanding its biology?
4. Discuss the value of naming a newly described structure using established Greek or Latin prefixes and suffixes.

Prescientific Period

Evidence indicates that prehistoric anatomy was of survival value and provided the foundation for medicine.

Objective 4. Discuss why understanding human anatomy is essential to medicine.

Objective 5. Define *trepanation* and *paleopathology.*

It is likely that a type of practical comparative anatomy is the oldest science. Certainly humans have always been aware of some of their anatomical structures and how they function. Our prehistoric ancestors undoubtedly knew their own functional abilities and limitations as compared to other animals. Through the trial and error of hunting, they discovered the "vital organs" of an animal, which, if penetrated with an object, would cause death (fig. 1.2). Likewise, they knew the vulnerable areas of their own bodies.

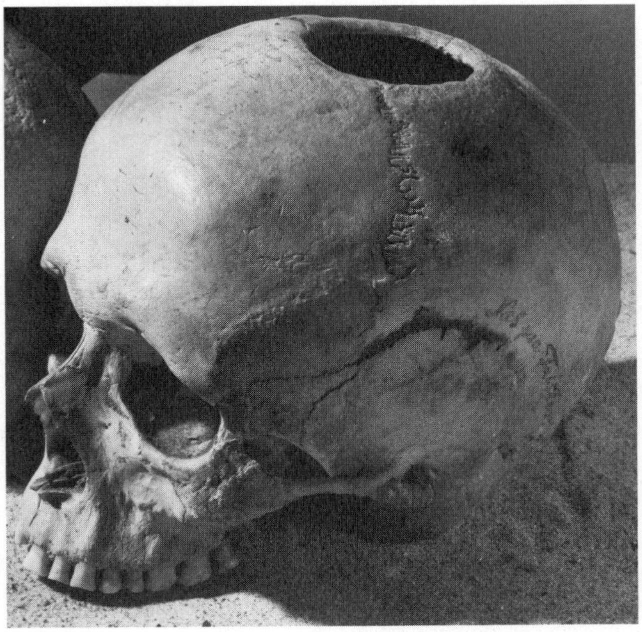

Figure 1.3 The surgical art of trepanation was practiced by several prehistoric cultures. Ossification around the bony edges of the wound indicates recovery from the operation.

Prehistoric people knew which parts of an animal's body were useful as food, clothing, or implements. Undoubtedly, they knew that the muscles functioned in locomotion as well as provided a major source of food. The skin from mammals with its associated fur served as a protective covering for the sparsely haired and vulnerable human. These people knew that the skeletal system formed a durable framework within themselves and other vertebrates. They used the bones from the animals on which they fed to make a variety of tools and weapons. They knew that their own bones could be broken through accidents and that improper healing would result in permanent disability. They knew that if an animal was wounded, it would bleed, and that excessive loss of this vital fluid would cause death. Perhaps they also realized that a severe blow on the head could cause deep sleep and debilitate an animal without killing it. Sexual differences were anatomically apparent even though basic reproductive functions could not have been understood. Their knowledge was the basic, practical type, necessary for survival.

Certain surgical skills are also ancient. **Trepanation,** the drilling of a hole in the skull or removal of a portion of a cranial bone, seems to have been practiced by several groups of prehistoric people. Trepanation was probably used as a ritualistic procedure to release evil spirits or perhaps, on some patients, to relieve cranial pressure resulting from a head wound. Trepanated skulls have repeatedly been found in archaeological sites (fig. 1.3). Judging from the partial reossification in some of these skulls, apparently the patients occasionally survived.

trepanation: Gk. *trypanon,* a borer

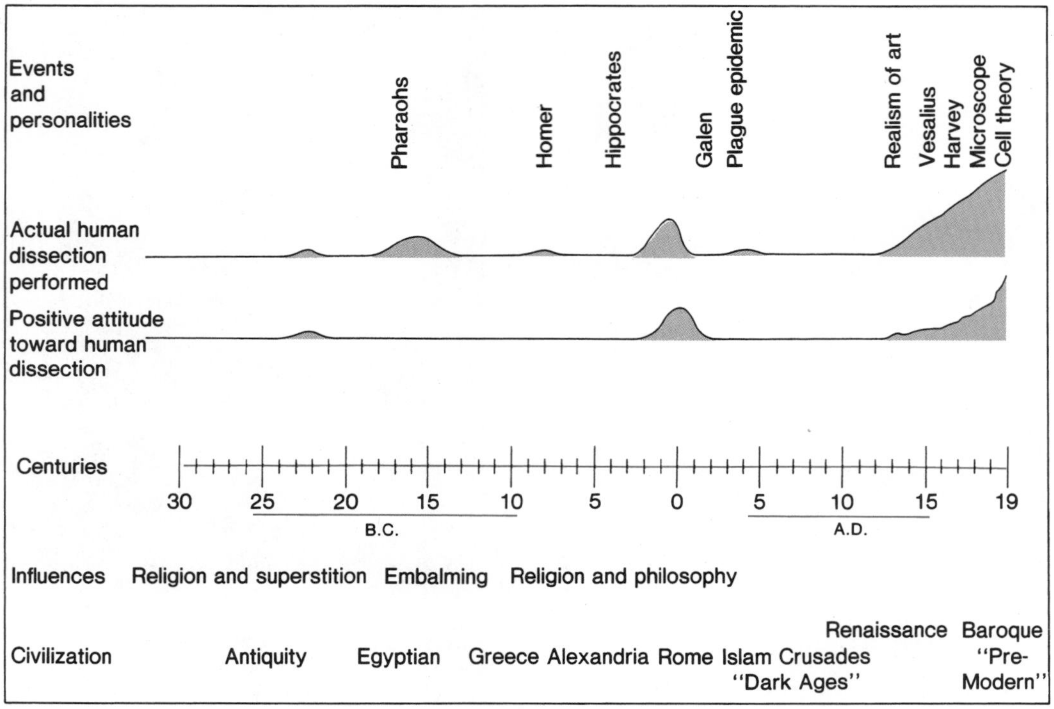

Figure 1.4 An anatomical timeline depicting the story of cadaver dissections.

What is known about prehistoric humans is conjectured through information derived from the cave drawings and fossils that contain paleopathological information. **Paleopathology** is the science concerned with studying diseases and the causes of death in prehistoric humans. Certain injuries, ages, and diseases, including nutritional deficiencies, can be determined from skeletal remains. Diets and dental conditions are indicated by fossil teeth. What cannot be determined, however, is the amount of anatomical information and knowledge that was transmitted orally until humans invented symbols to record their thoughts, experiences, and history.

---

1. Using the example of trepanation, discuss the importance of knowing the anatomy of the skull and brain in order for the procedure to be successful.
2. List the types of data that a paleopathologist would be interested in obtaining from an Egyptian mummy.

Scientific Period

Human anatomy is a dynamic science that has had a long, exciting heritage and is currently providing the foundation for medical, biochemical, developmental, cytogenetic, and biomechanical research.

Objective 6. Discuss key historical events in the science of human anatomy.

Objective 7. List the periods of time when cadavers have been used in studying human anatomy.

Objective 8. Explain why an understanding of human anatomy is important and personally applicable.

Objective 9. List ways a person may remain informed of anatomical research, and discuss why this is important.

The scientific period begins with recorded anatomical observations made in early Mesopotamia on clay tablets in *cuneiform script* over three thousand years ago and continues to the present day. Obviously, all of the past contributions to the science of anatomy cannot be mentioned; however, certain individuals and cultures had a tremendous impact and will be briefly commented upon in this section.

The history of anatomy has an interesting parallel with the history of the dissection of human cadavers, as is depicted in figure 1.4. The anatomists who made significant contributions are presented in table 1.1. Some of these contributions were in the form of books (table 1.2) that describe and illustrate the structure of the body and in some cases explain various body functions.

Mesopotamia and Egypt

Mesopotamia was the name given to the long, narrow wedge of land between the Tigris and Euphrates rivers, which is now a large part of present-day Iraq. Archaeological excavations and ancient records show that this area was settled prior to 4000 B.C. Because of the recorded information and the culture of the people, Mesopotamia is frequently called the cradle of civilization.

---

paleopathology: Gk. *palaios,* ancient; *pathos,* suffering disease; *logos,* to study

cuneiform: L. *cuneus,* wedge; *forma,* shape

Table 1.1 Historical perspective of contributors to the science of human anatomy

Person	Civilization	Lifetime or date of contribution	Contribution
Menes	Egyptian	About 3400 B.C.	Wrote first anatomy manual
Homer	Ancient Greece	About 800 B.C.	Described the anatomy of wounds in the *Iliad*
Hippocrates	Ancient Greece	About 460–377 B.C.	Father of medicine; Hippocratic oath
Aristotle	Ancient Greece	384–322 B.C.	Comparative anatomist; first recorded anatomical illustrations
Herophilus	Alexandria	About 325 B.C.	Remarkable work on aspects of the nervous system
Erasistratus	Alexandria	About 300 B.C.	Sometimes called father of physiology; concerned with body dysfunction and physiological changes
Celsus	Roman	30 B.C.–A.D. 30	Compiled information from Alexandrian school
Galen	Roman	130–201	Influential anatomical and medical writer
dé Luzzi	Renaissance	1487	Dissection guide of anatomy
da Vinci	Renaissance	1452–1519	Outstanding anatomical sketches and innovative anatomical preparations
Vesalius	Renaissance	1514–64	Replaced past misconceptions by direct observation and experiment; often called father of anatomy
Harvey	Premodern (European)	1578–1657	Demonstrated function of circulatory system; applied experimental method to anatomy
van Leeuwenhoek	Premodern (European)	1632–1723	Refinement of microscope; described various cells and tissues
Malpighi	Premodern (European)	1628–94	Regarded as father of histology
Sugita	Premodern (Japanese)	1774	Compiled a five-volume book of anatomy
Schleiden and Schwann	Modern (European)	1838–39	Formulation of cell theory
Roentgen	Modern (European)	1895	Discovery of X rays

Many early investigations of the body were an attempt to describe basic life forces. For example, people wondered what specific organ constituted the soul or the governing structure of the body. Some cuneiform writings from this era depict and describe body organs that were thought to serve this function. The liver, which was extensively studied in sacrificial animals (fig. 1.5), was thought to be the "guardianship of the soul and of the sentiments that make us men." This was a logical assumption because of the size of the liver and its close association with blood, which was observed to be vital for life. Even today, several European cultures attribute the liver with various emotions.

The warm blood and arrangement of blood vessels are obviously a governing system within the body, and this influenced the search for the soul. When excessive blood is lost, the body dies. Therefore, some concluded, blood must contain a vital, life-giving force. The scholars of Mesopotamia were influenced by this idea as was Aristotle, the Grecian scientist, who lived centuries later. Aristotle believed that the seat of the soul was the heart and that the brain functioned in cooling the blood that flowed from the heart. The association of the heart, in song and poetry, with the emotions of love and caring has its basis in Aristotelian thought.

The ancient Egyptian culture neighbored Mesopotamia to the west. Here the sophisticated science of embalming the dead in the form of mummies was perfected (fig. 1.6). No attempts were made to perform anatomical or pathological studies on the corpses, however, because embalming was strictly a religious ritual, reserved for royalty and the wealthy to prepare them for a life after death.

Several written works concerning anatomy have been discovered from ancient Egypt, but none of these influenced succeeding cultures. Menes, a king-physician during the first Egyptian dynasty in about 3400 B.C. (even before the pyramids were built), wrote what is thought to be the first manual on anatomy. Later writings (2300–1250 B.C.) attempted a systematization of the body, beginning with the head and progressing downward.

The techniques of embalming could have been a major contribution to the science of anatomy had these procedures been recorded and shared with future cultures. Apparently, however, embalmings were not well accepted by the people in ancient Egypt. In fact, the persons that performed embalmings were looked down upon and were frequently persecuted and even stoned. Embalming was a mystic art related more to religion than to science, and since its practice required a certain amount of mutilation of the dead body, it was regarded as demonism and antireligious. Consequently, embalming techniques that would provide embalmed cadavers as dissection specimens were not rediscovered until centuries later.

embalm: L. *in,* in; *balsamum,* balsam

Table 1.2 Influential anatomical books and publications

Aristotle. 384–322 B.C. *Historia animalium (History of animals), De partibus animalium (On the parts of animals),* and *De generatione animalium (On the generation of animals).* These classic works of the great Greek philosopher exerted a profound influence on biological thinking for centuries.

Celsus, Cornelius. 30 B.C.–A.D. 30. *De re medicina.* This eight-volume work was primarily a compilation of the medical data that was available from the Alexandrian school.

Galen, Claudius. 130–201. Nearly 500 medical papers on descriptive anatomy. Although Galen's writings contained numerous errors, his authoritative explanations of the structure and function of the body remained unchallenged and influenced anatomists and physicians for nearly fifteen hundred years.

dé Luzzi, Mondino. 1487. *Anathomia.* This book was used as a dissection guide for over 225 years, during which time it underwent forty editions.

Vesalius, Andreas. 1543. *De humani corporis fabrica (On the structure of the human body).* The beautifully illustrated *Fabrica* boldly challenged many of the errors that had been perpetuated by Galen. In spite of the controversies this book produced, it was well accepted and established a new standard of excellence in anatomy texts.

Fabricius ab Aquapendente, Hieronymus. 1600–1621. *De formato foetu (On the formation of the fetus)* and *De formatione ovi pulli (On the formation of the eggs of birds).* This book marked the beginning of embryological study.

Harvey, William. 1628. *Exercitatio anatomica de motu cordis et sanguinis in animalibus (On the motion of the heart and blood in animals).* Harvey advanced knowledge of circulation, and his experimental methods are still regarded as classic examples of method in science.

Descartes, René. 1637. *Discourse on method.* This philosophic thesis stimulated a mechanistic interpretation of biological data.

Linnaeus, Carolus. 1758. *Systema naturae.* The basis for the classification of living organisms was presented in this monumental work. Its anatomical value is in comparative anatomy, where the anatomy of different species is compared.

Haller, Albrecht von. 1760. *Elementa physiologiae (Physiological elements).* Some basic concepts of physiology are presented in this book, including a summary of what was known of the functioning of the nervous system to that date.

Sugita, Genpaku. 1774. *Kaitai shinsho (A new book of anatomy).* This book adopted the European concepts of the body and ushered in a new era of anatomy for the Japanese people.

Cuvier, Georges. 1817. *Le regne animal (The animal kingdom).* This comprehensive comparative vertebrate anatomy book had enormous influence on contemporary zoological thought.

Baer, Karl Ernst von. 1828–37. *Über entwickelungsgeschichtė der thiere (On the development of animals).* This book provided the foundation for modern embryology by discussing germ layer formation.

Beaumont, William. 1833. *Experiments and observations on the gastric juice and the physiology of digestion.* Basic digestive functions are accurately described in this classic work.

Müller, Johannes. 1834–40. *Handbuch der physiologie des menschen für vorlesungen (Elements of physiology).* This book established physiology as a science concerned with the functioning of the body.

Schwann, Theodor. 1839. *Mikroskopische untersuchungen über die übereinstimmung in der struktur and dem wachstum der thiere and pflanzen (Microscopic researches into accordance in the structure and growth of animals and plants).* The concept of cellular organization of the body is presented in this classic study.

Kölliker, Albrecht von. 1852. *Mikroskopische anatomia. (Microscopic anatomy).* This was the first textbook in histology. It became the foundation for the new science of histology.

Gray, Henry. 1858. *Anatomy, descriptive and surgical.* This masterpiece, better known as *Gray's anatomy,* is still in print and still contains over two hundred of the original illustrations. Thousands of physicians have used this book to learn gross human anatomy.

Virchow, Rudolf. 1858. *Die cellularpathologie. (Cell pathology).* Descriptions of normal and diseased tissues are presented in this book.

Darwin, Charles. 1859. *On the origin of species.* This was one of the most influential books ever published in biology. Its importance to anatomy is that it provided an explanation for the anatomical variation seen among different species.

Mendel, Gregor. 1866. *Versuche über pflanzenhybriden (Experiments with plant hybrids).* In this work, Mendel demonstrated the basic principles of heredity.

Owen, Richard. 1866. *Anatomy and physiology of the vertebrates.* Some basic concepts of structure and function, such as homologue and analogue, are presented in this book.

Balfour, Francis M. 1880. *Comparative embryology.* This book is considered the source of information for the science of modern experimental embryology.

Weismann, August. 1892. *Das keimplasma (The germplasm).* Weismann postulated the theory of meiosis, which states that a reduction in the chromosome number is necessary in the gametes of both the male and female for fertilization to occur.

Hertwig, Oskar. 1893. *Zelle and gewebe (Cell and tissue).* An important distinction between the sciences of cytology and histology is presented in this book.

Wilson, Edmund B. 1896. *The cell in development and heredity.* This book had a profound influence on the development of cytogenetics.

Pavlov, Ivan. 1897. *Le travail des glandes digestives (The work of the digestive glands).* The physiological functioning of the digestive system is presented in this classic experimental work.

Sherrington, Charles. 1906. *The integrative action of the nervous system.* The basic concepts of neurophysiology were established in this book.

Garrod, Archibald. 1909. *Inborn errors of metabolism.* Genetic defects were discussed in this pioneer book and were shown to be caused by defective genes.

Bayliss, William M. 1915. *Principles of general physiology.* This book provided a much needed synthesis of a newly emerging science.

Spemann, Hans. 1938. *Embryonic development and induction.* This masterful book provided the foundation of the science of experimental embryology.

Crick, Francis H. C., and James D. Watson. 1953. *Genetic implications of the structure of deoxyribonucleic acid.* This remarkable work provided explanations of genetic replication and control of cellular functions.

Like the people of Mesopotamia and Greece, the ancient Egyptians were concerned with a controlling spirit of the body. In fact, they even had a name for this life force, the Ba spirit, and believed that it was associated with the bowels and the heart. For this reason, food was placed in the tomb of a mummy to feed the Ba spirit during the journey to Osiris, god of the underworld.

Chinese and Japanese

Chinese The Chinese have always had an interest in the structure and function of the human body, but their explanations were based on philosophy rather than investigation. The Chinese revered the body and abhorred its mutilation. Their

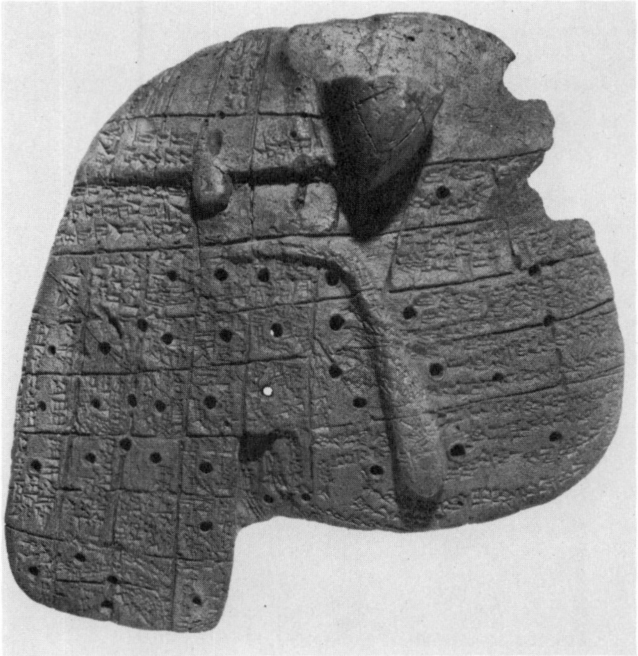

Figure 1.5 An inscribed clay model of a sheep's liver from the eighteenth or nineteenth century B.C. The people of ancient Mesopotamia regarded the liver as the source of human emotions.

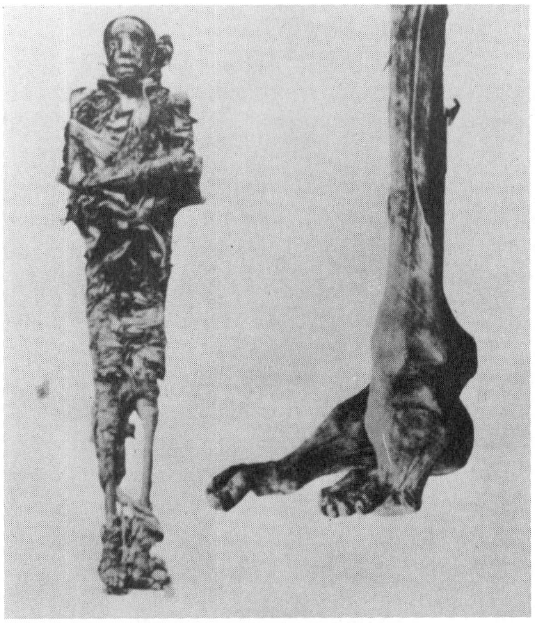

Figure 1.6 Perhaps the greatest contribution of the ancient Egyptian era to anatomy and medicine is the information obtained from the mummies. Certain diseases, injuries, deformities, and occasionally the cause of death can be determined from paleopathogenic examination of the mummified specimens. Shown on the right is a congenital clubfoot from a mummy of a person who lived during the Nineteenth Dynasty (about 1300 B.C.).

only knowledge of the internal organs was from wounds and injuries. Only in recent times have the dissections of cadavers been permitted.

The ancient Chinese had an abiding belief that everything within the universe depended on the balance of the two opposing principles of *yin* and *yang*. They believed that yin and

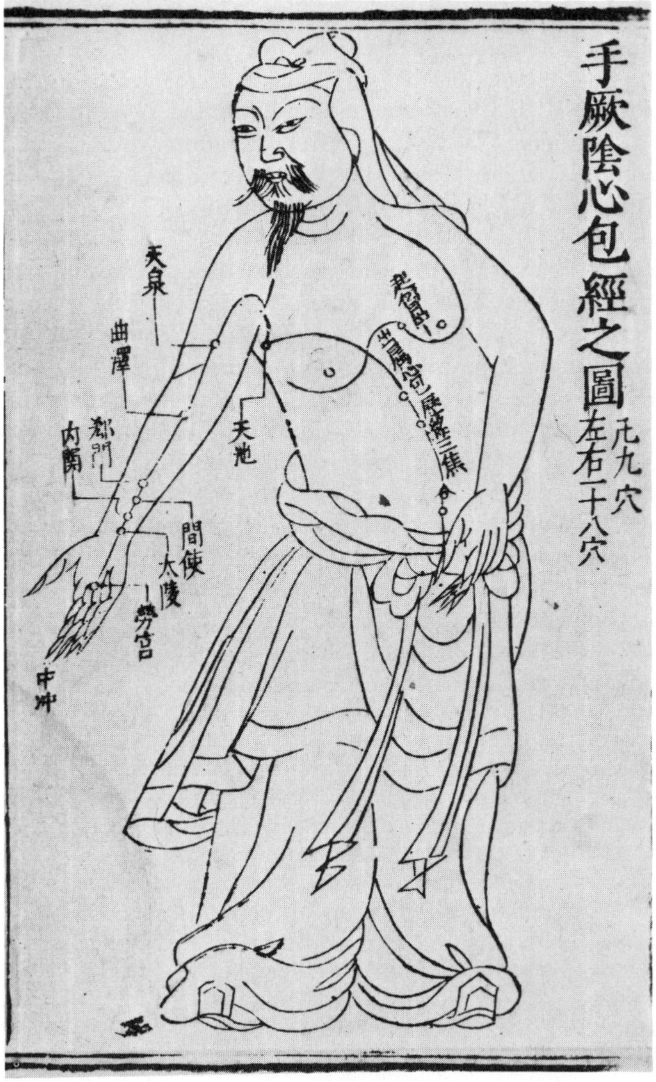

Figure 1.7 An acupuncture chart from the Ming dynasty of ancient China.

yang in the body functioned as the circulatory system. The blood was the conveyor of the yang, and the heart and vessels represented the yin. Other structures of the body were composed of lesser forces termed *zō* and *fū*.

The Chinese were great herbalists. In writings over five thousands years old, various herbal concoctions and potions were described to alleviate numerous ailments such as diarrhea and constipation. Opium was described as an excellent pain killer.

Until recently, the Chinese have been possessive of their beliefs, and, consequently, Western cultures were not influenced by Chinese thoughts or writings. Perhaps the best known but least understood of the Chinese contributions to human anatomy and medicine is acupuncture.

Acupuncture is an ancient practice established to maintain a balance between the yin and the yang. Three hundred sixty-five precise meridian sites, or vital points, that correspond to the number of days in a year, were identified on the body (fig. 1.7). Needles inserted into the various sites were believed to release bad secretions and rid the tissues of obstructions. Acupuncture is still practiced in China and has gained acceptance with some medical specialists in the United States as a technique of anesthesia and a cure for certain ailments. The effectiveness of acupuncture has been documented and is more than psychological. Acupuncture sites have been identified on domestic animals and have been used to a limited degree in veterinary medicine. Why acupuncture is effective remains a mystery, although it has recently been correlated with endorphin production within the brain (see chap. 11).

Japanese The advancement of anatomy in Japan was largely due to the influence of the Chinese and Dutch. The earliest records of anatomical interest in Japan date back to the sixth century. Buddhist monks from Japan were trained in China where they were exposed to Chinese philosophy, and so the Chinese beliefs concerning the body were prevalent in Japan as well. By the eighteenth century, the influences from the Western nations, especially the Dutch, were such that the Japanese themselves desired to sort out which version of anatomy was correct. In 1774, a five-volume book called *Kaitai Shinsho (A New Book of Anatomy)* that was published by a Japanese physician, Genpaku Sugita, totally adopted the Dutch concepts of the body. This book marked the beginning of a modern era in anatomy and medicine for the Japanese people.

For several hundred years, Western nations were welcome in Japan. In 1603, however, the Japanese government banned all contact with the Western world because they feared the influences of Christianity on their society. Although this ban was strictly enforced and Japan became isolated, Japanese scholars continued to circulate Western anatomical and medical books. These books led the Japanese physicians to reassess what they had been taught concerning the structure of the body.

Grecian Period

Anatomy was first widely accepted as a science in ancient Greece. The writings of several Greek philosophers had a tremendous impact on future scientific thinking. During this period, the Greeks were obsessed with the physical beauty of the human body, which they expressed through exquisite sculptures.

The youth of Greece were urged to be athletic and develop their physical abilities, but at about age eighteen they were directed to intellectual pursuits of science, literature, and philosophy. With this degree of cultural emphasis, it was only natural that great strides were made in the sciences.

Perhaps the first written reference to the anatomy of wounds sustained in battle is contained in the *Iliad,* written by Homer in about 800 B.C. Homer's detailed descriptions of the anatomy of wounds were exceedingly accurate and clean—hardly what would be observed from a traumatic battle wound. This has led to speculation that human dissections were conducted during this period and that anatomical structure was well understood. Victims of human sacrifice may have served as subjects for anatomical study and demonstration.

Hippocrates Hippocrates (460–377 B.C.) was the famous Grecian physician who is regarded as the father of medicine because of the sound principles of medical practice that he established (fig. 1.8). His name is memorialized in the Hippocratic oath, which many graduating medical students repeat as a promise of professional stewardship and duty to mankind (table 1.3).

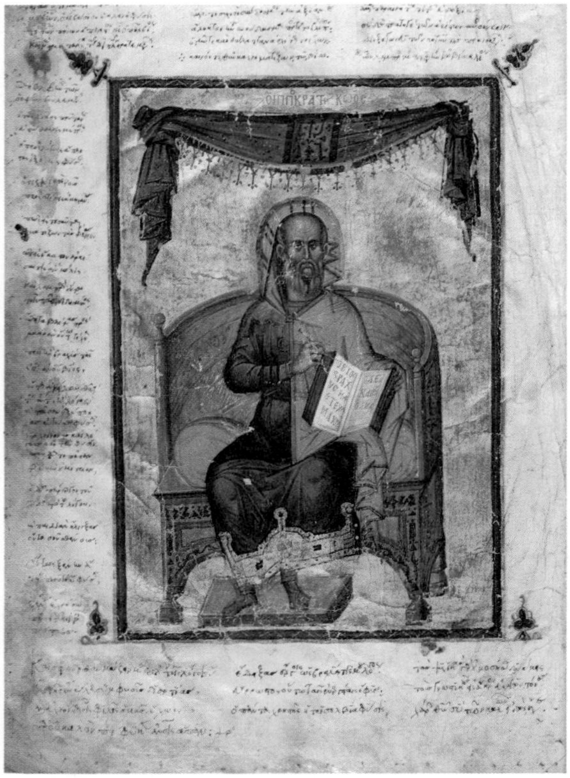

Figure 1.8 A fourteenth-century painting of the famous Greek physician Hippocrates. Hippocrates is referred to as the father of medicine, and his creed is immortalized as the Hippocratic oath.

Hippocrates probably had limited exposure to human dissections but was well disciplined in the popular *humoral theory* of body organization. Four body humors were recognized, and each was associated with a particular body organ: sanguine with the liver; choler, or yellow bile, with the gallbladder; phlegm with the lungs; and melancholy, or black bile, with the spleen. A healthy person was thought to have a balance of the four humors. The concept of humors has long since been discarded, but it dominated medical thought for over two thousand years.

Perhaps the greatest contribution of Hippocrates was that he attributed diseases to natural causes rather than to the displeasure of the gods. His application of logic and reason to medicine was the beginning of observational medicine.

The four humors are a part of our language and medical practice even today. *Melancholy* is a term used to describe a depressed or despondent temperament in a person, while *melanous* refers to a black or sallow complexion. The prefix *melano* means black. *Cholera* is an infectious intestinal disease that causes diarrhea and vomiting. *Phlegm* (pronounced *flem*) within the upper respiratory system is symptomatic of several respiratory disorders. *Sanguine,* which originally referred to blood, is used as an adjective to describe an ardent temperament.

physician: Gk. *physikos,* natural

sanguine: L. *sanguis,* bloody
choler: Gk. *chole,* bile
phlegm: Gk. *phlegm,* inflammation
melancholy: Gk. *melan,* black; *chole,* bile
humor: L. *humor,* fluid

Table 1.3 The Hippocratic oath

I swear by Apollo Physician and Aesculapius and Hygeia and Panacea and all the gods and goddesses, making them my witnesses, that I will fulfill according to my ability and judgment this oath and this covenant:

To hold him who has taught me this art as equal to my parents and to live my life in partnership with him, and if he is in need of money to give him a share of mine, and to regard his offspring as equal to my brothers in male lineage and to teach them this art—if they desire to learn it—without fee and covenant; to give a share of precepts and oral instruction and all the other learning to my sons and to the sons of him who has instructed me and to pupils who have signed the covenant and have taken an oath according to the medical law, but to no one else.

I will apply dietetic measures for the benefit of the sick according to my ability and judgment; I will keep them from harm and injustice.

I will neither give a deadly drug to anybody if asked for it, nor will I make a suggestion to this effect. Similarly I will not give to a woman an abortive remedy. In purity and holiness I will guard my life and my art.

I will not use the knife, not even on sufferers from stone, but will withdraw in favor of such men as are engaged in this work.

Whatever houses I may visit, I will come for the benefit of the sick, remaining free of all intentional injustice, of all mischief, and in particular of sexual relations with both female and male persons, be they free or slaves.

What I may see or hear in the course of the treatment or even outside of the treatment in regard to the life of men, which on no account one must spread abroad, I will keep to myself, holding such things shameful to be spoken about.

If I fulfill this oath and do not violate it, may it be granted to me to enjoy life and art, being honored with fame among all men for all time to come; if I transgress it and swear falsely, may the opposite of all this be my lot.

Aristotle Aristotle (384–322 B.C.), a pupil of Plato, was an accomplished writer, philosopher, and zoologist (fig. 1.9). He was also a renowned teacher and was hired by King Philip of Macedonia to tutor his son, Alexander, who later became known as Alexander the Great.

Aristotle made careful investigations of all kinds of animals, including references to man, and established a type of scientific method in obtaining data. He wrote the first known account of embryology, in which he described the development of the heart in a chick embryo. He named the aorta and contrasted the arteries and veins. Aristotle's best known zoological works are *History of Animals, On the Parts of Animals,* and *On the Generation of Animals.*

In spite of his tremendous accomplishments, Aristotle perpetuated some erroneous views of anatomy. For example, the doctrine of the humors formed the boundaries of Aristotle's thought. Plato had described the brain as the seat of feeling and thought, but Aristotle disagreed. He stated that the heart was the seat of intelligence and that the function of the brain, which was bathed in fluid, was to cool the blood that was pumped from the heart.

Alexandrian Era

Alexander the Great founded Alexandria in 332 B.C. and established it as the capital of Egypt and a center of learning. A great library as well as a school of medicine existed in Alexandria.

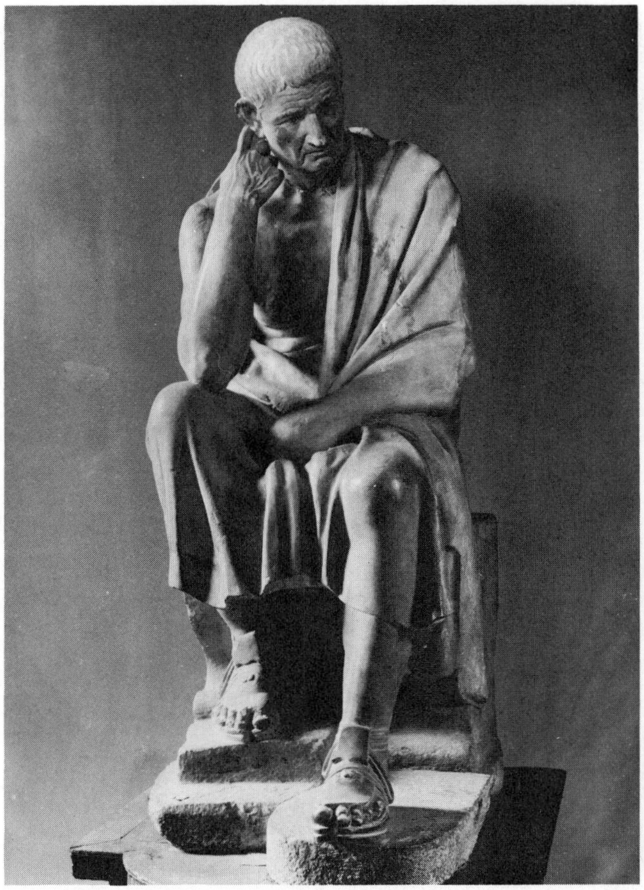

Figure 1.9 This Roman copy of a Greek sculpture is believed to be of Aristotle, the famous Grecian philosopher.

The study of anatomy flourished because of the acceptance of dissections of human cadavers and human **vivisections** (dissection of a living thing). This seemingly barbaric practice was commonly performed on condemned criminals. People reasoned that to best understand the functions of the body, it should be studied while the subject was alive, and that a condemned human could best repay society through the use of his body for a scientific vivisection.

Unfortunately, the scholarly contributions and scientific momentum of Alexandria did not endure. Most of the written works were destroyed when the great library was burned by the Romans as they conquered the city in 30 B.C. What is known about Alexandria was obtained from the writings of later scientists, philosophers, and historians such as Pliny, Celsus, Galen, and Tertullian. Two men of Alexandria, Herophilus and Erasistratus, made lasting contributions to the study of anatomy.

Herophilus Herophilus (about 325 B.C.) was trained in the Hippocratic school but became a great teacher of anatomy in Alexandria. Through the use of vivisections and human cadavers, Herophilus made excellent descriptions of the skull, eye,

Aesculapius: Gk. (mythology) son of Apollo and god of medicine
Hygeia: Gk. (mythology) daughter of Aesculapius; personification of health; *hygies,* healthy
Panacea: Gk. (mythology) also a daughter of Aesculapius; assisted in temple rites and tended sacred serpents; *pan,* all, every; *akos,* remedy

vivisection: L. *vivus,* alive; *sectio,* a cutting

various visceral organs and organ relationships, and the functional relationship of the spinal cord to the brain. Two monumental works of Herophilus were the books *On Anatomy* and *Of the Eyes.* Herophilus regarded the brain as the seat of intelligence and described many of its structures, such as the meninges, cerebrum, cerebellum, and fourth ventricle. He was the first to identify nerves as either sensory or motor.

Erasistratus Erasistratus (about 300 B.C.) was more interested in body functions than structure and is therefore frequently referred to as the father of physiology. Erasistratus authored a book on the causes of diseases, in which he included observations on the heart, vessels, brain, and cranial nerves. He noted the toxic effects of snake venom on various visceral organs and described changes in the liver resulting from various diseases. Although some of the writings of Erasistratus were very scientific, other concepts of his were primitive and mystical. He thought that the cranial nerves carried animal spirits and that muscles contracted because of distention by spirits. He believed that the left ventricle of the heart was filled with a vital air spirit (pneuma) that came in from the lungs and that the vessels between the lungs and heart contained pneuma rather than blood. Erasistratus thought that veins contained blood and that arteries transported pneuma but filled with blood that drained from veins when an artery was cut.

Both Herophilus and Erasistratus were greatly criticized later in history for the vivisections they performed. Celsus (about 30 B.C.) and Tertullian (about A.D. 200) were particularly critical of the practice of vivisection. Herophilus was described as a butcher of men who had dissected as many as six hundred living persons, some of them as public demonstrations.

Roman Era

In many respects, the Roman Empire stifled scientific advancements and set the stage for the Dark Ages. The interest and emphasis of science shifted from theoretical to practical during this time. Few dissections of cadavers were performed other than in attempts to determine the cause of death in criminal cases. Medicine was not preventive but was almost solely limited to the treatment of soldiers injured in battle. Later in Roman history, laws were established that indicated the influence of the church on medical practice. According to Roman law, for example, no deceased pregnant women could be buried without prior removal of the fetus from the womb so that it could be baptized.

The scientific documents that are preserved from the Roman Empire are mostly compilations of information obtained from the Grecian and Alexandrian scholars. New anatomical information was scant and for the most part consisted of speculation or was derived from dissections of animals. Two important anatomists from the Roman era were Celsus and Galen.

visceral: L. *viscus,* internal organ of torso
pneuma: Gk. *pneuma,* air

Celsus The major contribution of Cornelius Celsus (30 B.C.–A.D. 30) was the preservation of a portion of the medical information from the Alexandrian school. This information was compiled into an eight-volume work called *De re medicina.* Some pathological information is included in this record.

Galen Claudius Galen (A.D. 130–201) was perhaps the best physician since Hippocrates. He was certainly the most influential writer of all times on medical subjects. For nearly fifteen hundred years, the writings of Galen were the unquestionable authority on anatomy and medical treatment. Galen probably dissected no more than two or three human cadavers during his career, limiting his anatomical descriptions to animal dissections. Galen compiled nearly five hundred medical papers (of which eighty-three have been preserved) from earlier works as well as his personal studies. He believed in the humors of the body and perpetuated this concept and gave authoritative explanations of nearly all body functions.

Galen's works contain many errors, primarily because he desired to present definitive answers and because of his interpretation of data from nonhuman animals. He did, however, provide some astute and accurate anatomical details that are still regarded as classics. He proved to be an experimentalist, demonstrating that the heart of a pig would continue to beat when spinal nerves were transected so that nerve impulses could not reach the heart. He showed that the squealing of a pig stopped when the particular nerve that innervated the vocal cords was cut. He also proved that arteries contained blood rather than pneuma.

Galen compiled a list of many medicinal plants and used medications to treat many illnesses. Although he extensively used bloodletting to balance the four humors, he cautioned against removing too much blood. He accumulated a wide variety of medical instruments and suggested their use as forceps, retractors, scissors, and splints (fig. 1.10). He was also a strong advocate of helping nature heal through hygiene, diet, rest, and exercise.

Middle Ages

The Middle Ages, frequently referred to as the Dark Ages, began with the fall of Rome to the Goths in A.D. 476 and lasted nearly one thousand years until Constantinople was conquered by the Turks in 1453. The totalitarian Christian church suppressed science, and Europe sank into stagnation. Human existence continued to be miserably precarious, and people no longer felt capable of learning from personal observation but rather accepted life "on faith."

Dissections of cadavers were totally prohibited, and molesting a corpse was a criminal act frequently punishable by burning at the stake. If mysterious deaths occurred, examinations by inspection and palpation were acceptable. During the plague epidemic in the sixth century, however, a few necropsies and dissections were performed in hopes of determining the cause of this dreaded disease (see fig. 1.4).

epidemic: Gk. *epi,* upon; *demos,* people
necropsy: Gk. *nekros,* corpse; *opsy,* view

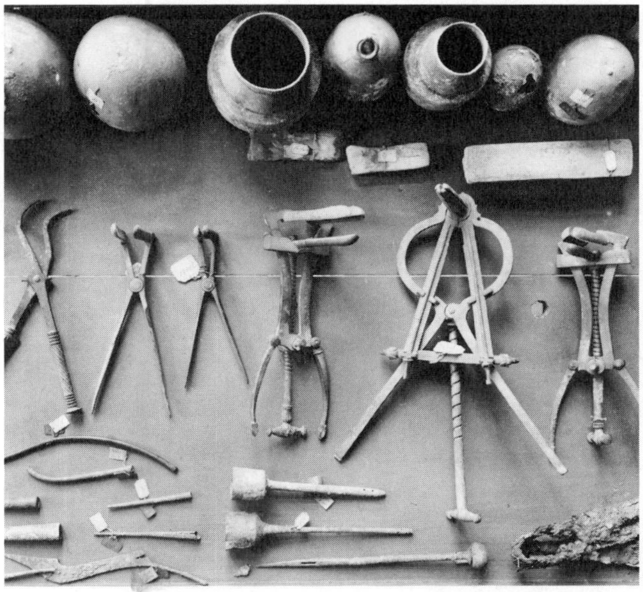

Figure 1.10 The surgical and gynecological instruments shown were found in the House of the Surgeon (about A.D. 62–79) in Pompeii and are representative of the medical equipment used throughout the Roman Empire during this time.

During the crusades soldiers cooked the bones of their dead comrades so that they could be returned home for proper burial. Even this act, however, was eventually considered sacrilegious and was strongly condemned by the Church. One of the ironies of the Middle Ages was that the peasants had far less respect and fewer rights when they were alive than when they were dead.

Contributions of Islam

The Arabic-speaking people made a profound contribution to the history of anatomy in a most unusual way. On several occasions during the Middle Ages, Arab armies conquered sections of the Middle East. As part of the spoils of their conquest, most of the anatomical writings were seized from the libraries of Alexandria. These materials were then carried back to the Arab countries and translated from Greek to Arabic.

As the Dark Ages stagnated Europe, the Christian church attempted to destroy any scholarship of worldly knowledge that was not acceptable within Christian dogma. The study of the human body was considered heretical, and the Church forbade the existence of any written anatomical material. Had it not been for the Arab invasions, the early writings of anatomy from Western civilization would have never survived. It wasn't until the thirteenth century that the Arabic translations were returned to Europe and, in turn, translated to Latin. During the translation process, any Arabic terminology that had been introduced was systematically removed so that today we find few anatomical terms of Arabic origin.

Renaissance

The period of time known as the Renaissance was characterized by a rebirth of science. It lasted from the fourteenth through the sixteenth century and was a transitional period from the Middle Ages to the modern age of science.

The Renaissance was ushered in by the great European universities established in Bologna, Salerno, Padua, Montpellier, and Paris. The first recorded human dissections at these newly established universities were the work of the surgeon William of Saliceto (1215–80), from the University of Bologna. The study of anatomy quickly spread to other universities, and by the year 1300 human dissections became an integral part of the medical curriculum. However, there persisted the Galenic dogma that normal human anatomy was sufficiently understood, so interest at this time centered on methods and techniques of dissection rather than the study of the human body.

The development of movable type in about 1450 revolutionized the production of printed books. One of the first anatomy books to be printed in this manner was that of Jacopo Berengario of Carpi, who was a professor of surgery at Bologna. He described many anatomical structures, including the appendix, thymus, and larynx.

The most influential text of this period was written by Mondino dé Luzzi, also of the University of Bologna. This book, first published in 1487, was more of a dissection guide than a study of gross anatomy, and in spite of its numerous Galenic errors, it underwent forty editions until the time of Vesalius.

Because of the rapid putrefaction of an unembalmed corpse, the anatomy textbooks of the early Renaissance were organized so that the more perishable portions of the body were dissected first. Dissection began with the abdominal cavity, then the chest, followed by the head, and finally the appendages. Dissection was a marathon event, frequently continuing for perhaps four days.

With the increased interest in anatomy during the Renaissance, obtaining cadavers for dissection became a serious problem. Medical students regularly practiced grave robbing until finally an official decree was issued that permitted the bodies of executed criminals to be used as specimens.

Corpses were embalmed to prevent deterioration, but it was not especially effective and the stench from cadavers was apparently a continual problem. Anatomy professors removed themselves from the immediate area of the cadaver to a thronelike chair where they could lecture to their students and perhaps be away from the smell (fig. 1.11). The phrase "I would not touch that with a ten foot pole" probably originated during this time in reference to the smell of a decomposed cadaver.

The major advancements in anatomy during the Renaissance came from the artistic and scientific abilities of Leonardo da Vinci and Andreas Vesalius. Working in the fifteenth and sixteenth centuries, each produced monumental studies of the human form.

Da Vinci The great Renaissance Italian Leonardo da Vinci (1452–1519) is best known for his art (e.g., *Mona Lisa*) and his scientific ability. He displayed genius as a painter, sculptor, architect, musician, and anatomist—although his contributions to anatomy were lost until the twentieth century. As a young

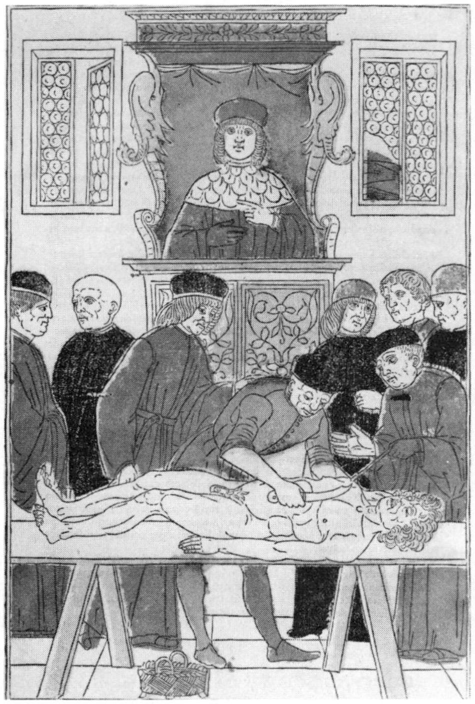

Figure 1.11 A scene of a cadaver dissection, 1491, from *Fasciculus Medicinae* by Johannes de Ketham. The anatomy professor removed himself from the scene of the cadaver to a thronelike chair overlooking the proceedings. Hired assistants did the dissections, while another, called the *ostensor*, pointed to the internal structures with a wand as the professor lectured.

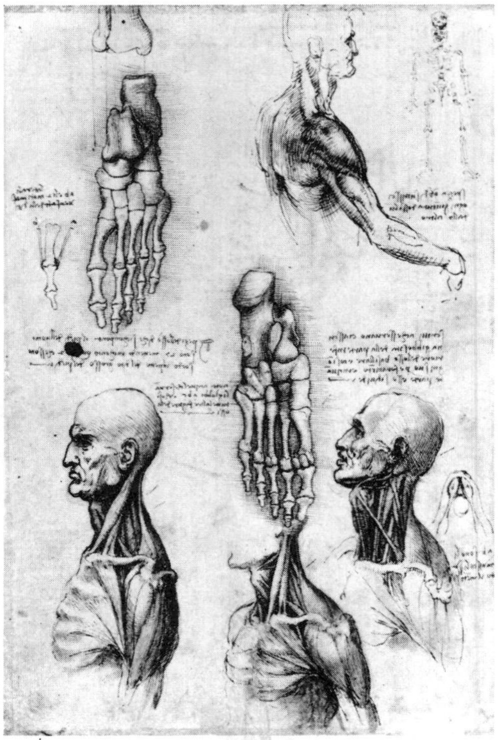

Figure 1.12 The anatomical sketches of Leonardo da Vinci have the detail and accuracy that only a master artist could achieve. "The painter who has acquired a knowledge of the nature of the sinews, muscles and tendons," da Vinci wrote, "will know exactly in the movement of any limb how many and which of the sinews are the cause of it, and which muscle by its swelling is the cause of the sinew's contracting."

man, da Vinci regularly observed cadaver dissections and intended to publish a textbook on anatomy with the Pavian professor Marcantonio della Torre. The untimely death of della Torre at the age of thirty-one halted their plans. When da Vinci died, his anatomical sketches were lost and were not discovered for over four centuries. The advancement of anatomy would have been accelerated by many years if da Vinci's notebooks had been available to the world at the time of his death.

Da Vinci presented his illustrations in a totally new perspective that showed the structural relationships of the various body organs. He was intent on accuracy, and his sketches are incredibly detailed (fig. 1.12). He experimentally determined the structure of complex body organs such as the brain and the heart. He made wax casts of the ventricles of the brain to study its structure. He constructed models of the heart valves to demonstrate their action.

Vesalius The contribution of Andreas Vesalius (1514–64) to the science of human anatomy and to modern medicine is immeasurable. Vesalius was born in Brussels to a family of physicians. He trained in Paris prior to accepting a position at the University of Padua in Italy. It was at Padua that he earned recognition as an anatomist and an artist. Vesalius participated in human dissections and initiated the use of live models to determine surface landmarks for internal structures (fig. 1.13).

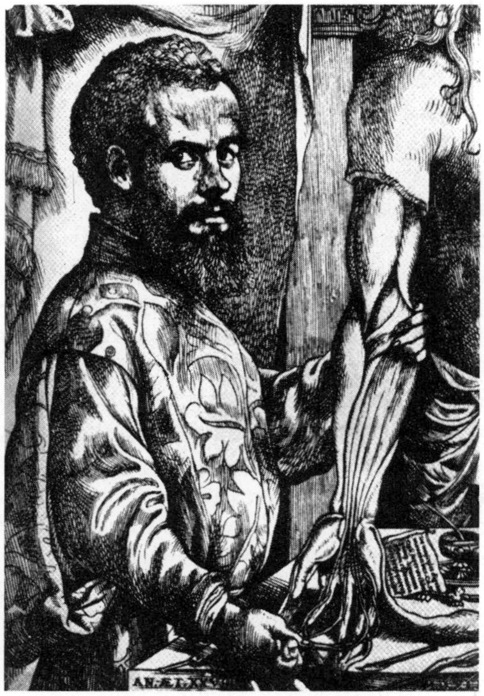

Figure 1.13 A painting of the great anatomist Andreas Vesalius, as he dissects a cadaver. From his masterpiece, *De Humani Corporis Fabrica.*

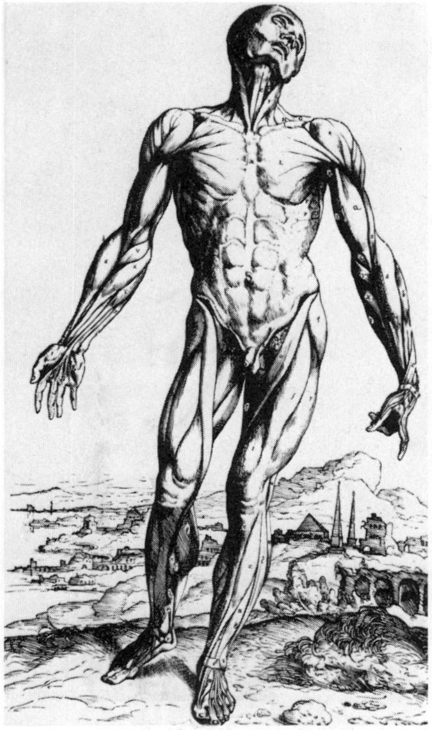

Figure 1.14 A plate from *De Humani Corporis Fabrica,* which Vesalius completed at the age of twenty-eight. This book, published in 1543, revolutionized the science of anatomy.

Vesalius apparently had enormous energy and ambitions. He completed the masterpiece of his life, *De Humani Corporis Fabrica,* by the time he was twenty-eight years old. The various body systems and individual organs were beautifully illustrated and described in the *Fabrica* (fig. 1.14). Vesalius was a devoutly religious person who considered the human body to be God's most beautiful creation. His book was especially important in that it boldly challenged hundreds of Galen's erroneous teachings. Vesalius wrote of his surprise at finding numerous anatomical errors that were taught as fact, and he refused to accept Galen's explanations on faith. Bitter controversies ensued between Vesalius and the traditional Galenic anatomists, including Vesalius's former teacher Sylvius. Vesalius became so incensed by the relentless attack that he destroyed much of his later unpublished work and ceased his dissections. Another point to his credit is that unlike Sylvius, Fallopius, Eustachius, and other anatomists of his time, Vesalius chose not to have his name attached to parts of the body that he described.

Although Vesalius was the greatest anatomist of his epoch, others made significant contributions and to an extent paved the way for Vesalius. Michelangelo pursued anatomy in 1495, being supplied with corpses by the friar of a local monastery. Mondino de' Luzzi and the surgeon Jacopo Berengario of Carpi also corrected many errors of Galen. Fallopius (1523–62) and Eustachius (1524–74) completed detailed dissections of specific body regions.

Seventeenth and Eighteenth Centuries

During the seventeenth and eighteenth centuries, the science of anatomy attained an unparalleled acceptance and theatrical-like status. Elaborate amphitheaters were established in various parts of Europe for public demonstrations of human dissections (fig. 1.15). Exorbitantly priced tickets for the public anatomies were sold to the wealthy, who witnessed the dissection of a cadaver by elegantly robed anatomists who were splendid orators. The subject was chosen from condemned criminals, and the performance was scheduled during cold weather because of the perishable nature of the body.

Fortunately there were also serious, scientific-minded anatomists during this period who made several significant contributions. The two most outstanding contributions were the explanation of blood flow and the development and use of the microscope.

Harvey In 1628, the English anatomist William Harvey (1578–1657) published his outstanding work *On the Motion of the Heart and Blood in Animals.* This important research established brilliant proof of the continuous circulation of blood within contained vessels, and the technique of investigation presented in this publication is still regarded as a classic example of the scientific method for conducting research (fig. 1.16). Like Vesalius, Harvey was severely criticized for his departure from Galenic philosophy. The controversy over circulation of the blood raged for twenty years until other anatomists finally repeated Harvey's experiments and added information.

Van Leeuwenhoek Antonie van Leeuwenhoek (1632–1723) was a Dutch lens grinder who improved the microscope to where he achieved a magnification of 270 times. His many contributions included developing techniques for tissue examination and describing blood cells, spermatozoa, and the striped appearance of skeletal muscle. Van Leeuwenhoek did not understand the role of sperm in fertilization, but thought that a human sperm contained a miniature human being called a *homunculus.*

The development of the microscope added an entirely new dimension to anatomy and eventually led to explanations of basic body functions. In addition, the improved microscope was invaluable for understanding the etiologies of many diseases and discovering cures for many of them. Although van Leeuwenhoek improved the microscope, credit for its invention is usually given to the Dutch maker of spectacles, Zacharius Janssen. The first scientific investigation using a microscope was done by Francisco Stelluti in 1625 on the structure of a bee.

Malpighi and Others Marcello Malpighi (1628–94), an Italian anatomist, has been referred to as the father of histology. He discovered the capillary blood vessels that Harvey had postulated, described the alveoli of lungs, and discussed the histological structure of the spleen and kidneys.

homunculus: L. diminutive form of *homo,* man

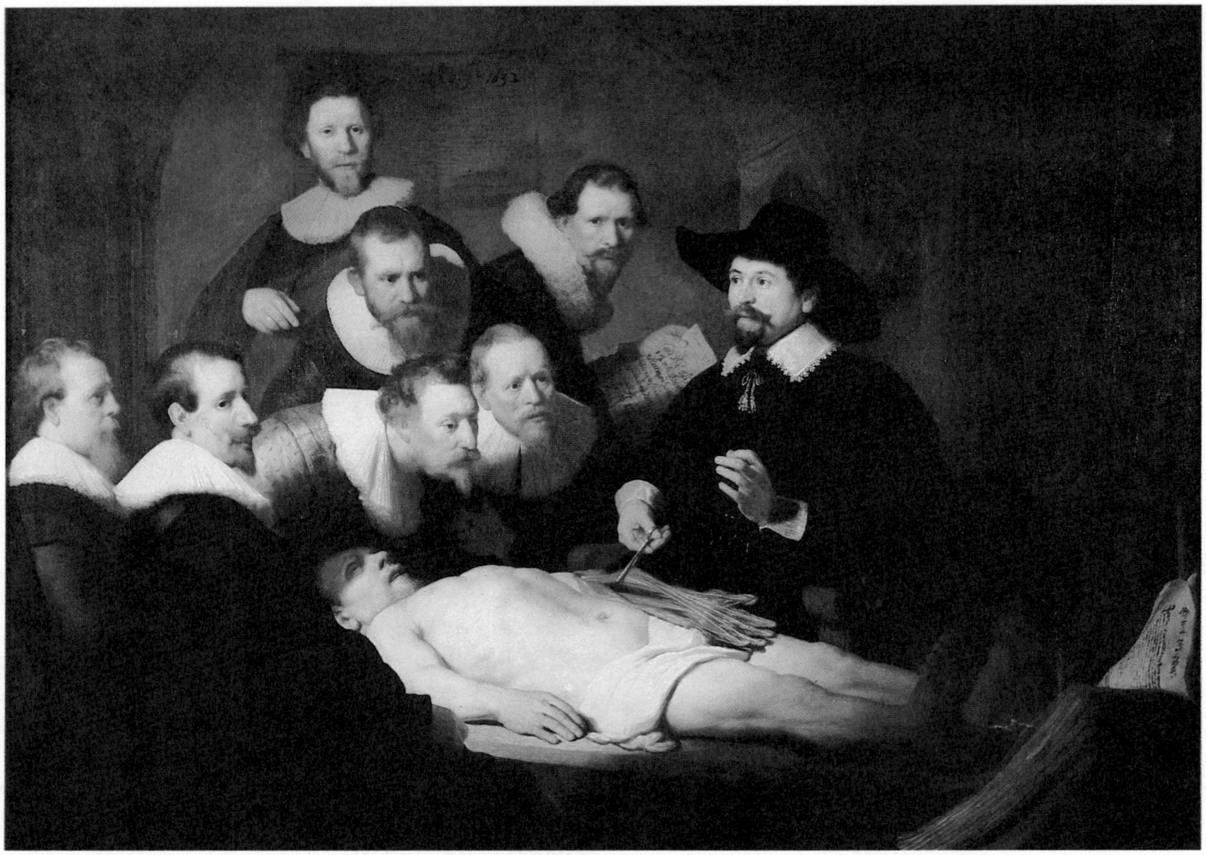

Figure 1.15 *The Anatomy Lesson of Dr. Tulp,* a famous Rembrandt painting completed in 1632, depicts one of the public anatomies that were popular during this period.

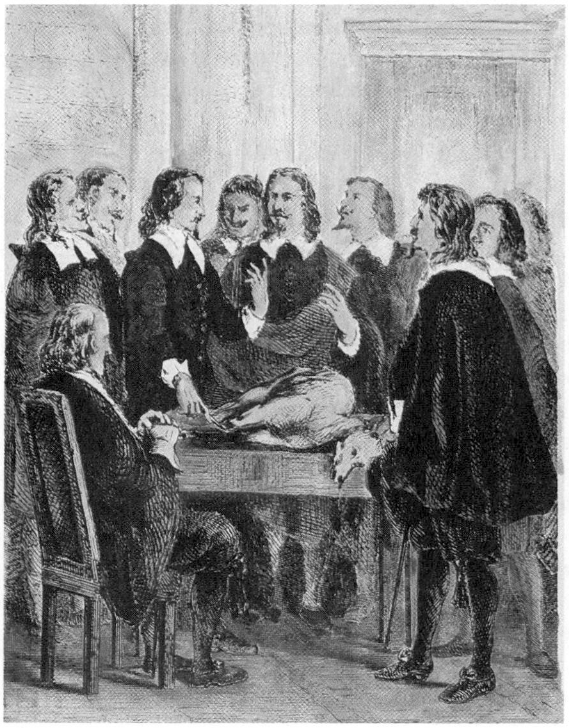

Figure 1.16 The English physician, William Harvey, demonstrated with experiments in 1628 that blood circulates and does not flow back and forth through the same vessels.

Many other anatomists made contributions during this two-hundred-year period. In 1672, the Dutch anatomist de Graaf described the ovaries of the female reproductive system, and in 1775 Spallanzani showed that both the ovum and the sperm were necessary for conception. Francis Glisson (1597–1677) described the liver and associated the gallbladder, the stomach, and the intestines. Thomas Wharton (1614–73) and Niels Stensen (1638–86) separately contributed to knowledge of the salivary glands and lymph nodes within the neck and facial regions. In 1664, Thomas Willis published a summary of what was known about the nervous system.

A number of anatomical structures throughout the body are named in honor of these early anatomists. Thus we have *graafian follicles, Stensen's* and *Wharton's ducts, Fallopian tubes, Bartholin's glands,* the *circle of Willis,* and many others. These terms are more meaningful when something is known of their origin.

Nineteenth Century

The major contribution in the nineteenth century was the formulation of the cell theory and the implications it had for a clearer understanding of the structure and functioning of the body. Once the microscope was invented, it was merely a matter of time before cells were discovered and described.

The term *cell* was coined in 1665 by an English physician, Robert Hooke, as he examined the structure of cork in an attempt to explain its buoyancy. What Hooke actually observed were the rigid walls that surrounded the empty cavities of dead cells. The significance of cellular structure did not become apparent until approximately one hundred fifty years after Hooke's work.

With improved microscopes, finer details were observed. In 1809, a French zoologist, Jean Lamarck, observed the jellylike substance within a living cell and speculated that this material was more important than the outside structure of a cell. Fifteen years later, René H. Dutrochet described the differences between plant and animal cells.

Two German scientists, Schleiden and Schwann, are credited with the biological principle referred to as the *cell theory*. In 1838 the botanist Matthias Schleiden suggested that each plant cell leads a double life. He believed that in some respects the cell was an independent organism, but at the same time it cooperated with the other cells that made up the whole plant. A year later Theodor Schwann, working with animal cells, concluded that all organisms were composed of cells that were essentially alike. This was followed nineteen years later by a further biological principle that seemed to complete the explanation of cells. In 1858 the German pathologist Rudolf Virchow wrote a book entitled *Cell Pathology* in which he proposed that cells come only from other cells. The mechanism of cellular replication, however, was not understood for several more decades.

Johannes Müller (1801–58) is noted for applying the sciences of physics, chemistry, and psychology to the study of the human body. With this increased dimensional breadth to the subject, anatomy became a comparative science.

Twentieth Century

The contributions to the science of anatomy during the twentieth century have not been as astounding as they were when little was known about the structure of the body. The study of anatomy within the twentieth century became specialized, and research became more detailed and complex.

One innovation that gained momentum early in the twentieth century was the simplification and standardization of nomenclature. Because of the proliferation of scientific literature toward the end of the nineteenth century, over thirty thousand terms for structures in the human body were reported, many of which were synonymous and referred to similar structures. In 1895, in an attempt to clarify the confusion, the German Anatomical Society met in Basel, Switzerland, and an official list of approximately five hundred terms was approved, which was called the *Basle Nomina Anatomica (BNA)*. The terms on this list were universally accepted for use in teaching and publications.

Other conferences on nomenclature have been held throughout the century, with increased interest. The meetings have become known as the *International Congress of Anatomists.* During the meetings of the Seventh International Congress held in New York City in 1960, a resolution was passed to eliminate all proper names from anatomical terminology and use descriptive names in their place. Structures like Stensen's duct and Wharton's duct, for example, are now correctly referred to respectively as the parotid duct and submandibular duct. Because of popular usage, however, it will be extremely difficult to eliminate all the proper names from anatomical terminology. But at least there is a trend toward descriptive simplification.

In response to the increased technology and depth of understanding in the twentieth century, new disciplines and specialities have appeared in the science of human anatomy in an attempt to categorize and use the new knowledge. The techniques of cognate disciplines such as chemistry, physics, electronics, mathematics, and computer science have been incorporated in research efforts.

There are several well-established divisions of human anatomy. The oldest, of course, is **gross anatomy,** which is the study of the structures of a cadaver that can be observed with the unaided eye. Stringent courses in gross anatomy in professional schools provide the foundation for the student's entire medical or paramedical training. Gross anatomy also forms the basis for the other specialities within anatomy. **Surface anatomy** (see chap. 10) deals with surface features of the body that can be observed or palpated (felt firmly).

Microscopic Anatomy Structures smaller than 0.1 mm (100 μm) can be seen only with the aid of a microscope. The sciences of **cytology** (study of cells), or **cellular biology,** and **histology** (study of tissues) are specialities of anatomy that have provided additional understanding of the structure and function of the human body. One can observe greater detail with the *electron microscope* than with the *light microscope* (fig. 1.17). New techniques in staining and histochemistry have aided study using the electron microscope—sometimes referred to as *ultra structure,* or *fine structure.*

Radiological Anatomy Radiological anatomy, or *radiology,* provides a way of observing structures within the living body. Radiology is based on the principle that substances of different densities absorb different amounts of X rays and therefore cause a differential exposure on film. Radiopaque substances such as barium can be ingested (swallowed) or injected into the body to produce even greater contrasts (fig. 1.18). **Angiography** involves making an X ray after injecting a dye into the bloodstream. In *angiocardiography* the heart and its associated vessels are x-rayed. *Cineradiography* permits the study of certain body systems through the use of motion picture X rays. Traditional X rays have had limitations as diagnostic tools for understanding human anatomy because of the vertical two-dimensional plane that is photographed. Since X rays compress the body image with an overlap of organs and tissues, diagnosis is often difficult.

X rays were discovered in 1895 by Wilhelm Konrad Roentgen. The x-ray image that is produced on film is frequently referred to as a roentgenograph. The recent development of the computer tomographic technique has been hailed as the greatest advancement in diagnostic medicine since the discovery of X rays themselves.

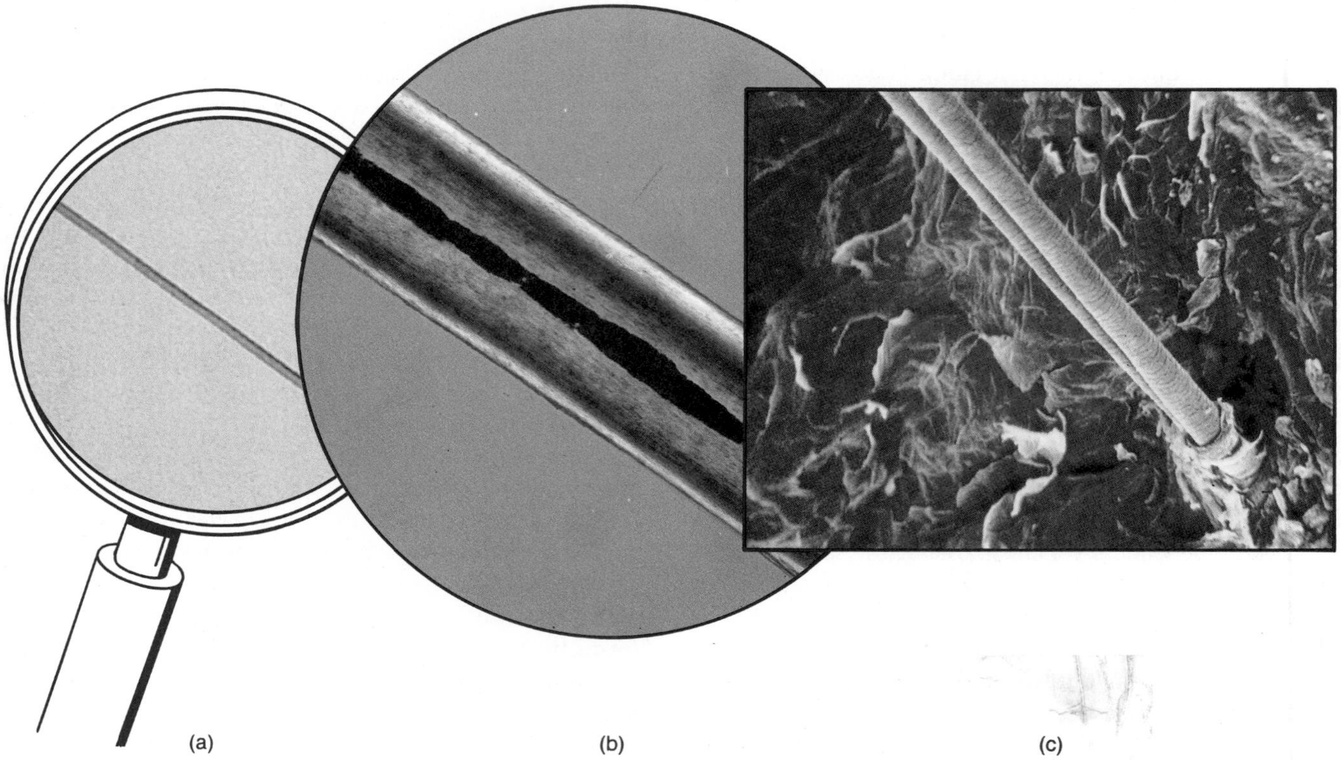

Figure 1.17 Different techniques for viewing microscopic anatomy have greatly enhanced our understanding of the structure and function of the human body. (*a*) The appearance of hair under a simple magnifying glass, (*b*) an observation of a stained section of a hair and skin through a light microscope, and (*c*) a hair as viewed through an electron microscope (340× at 35mm size).

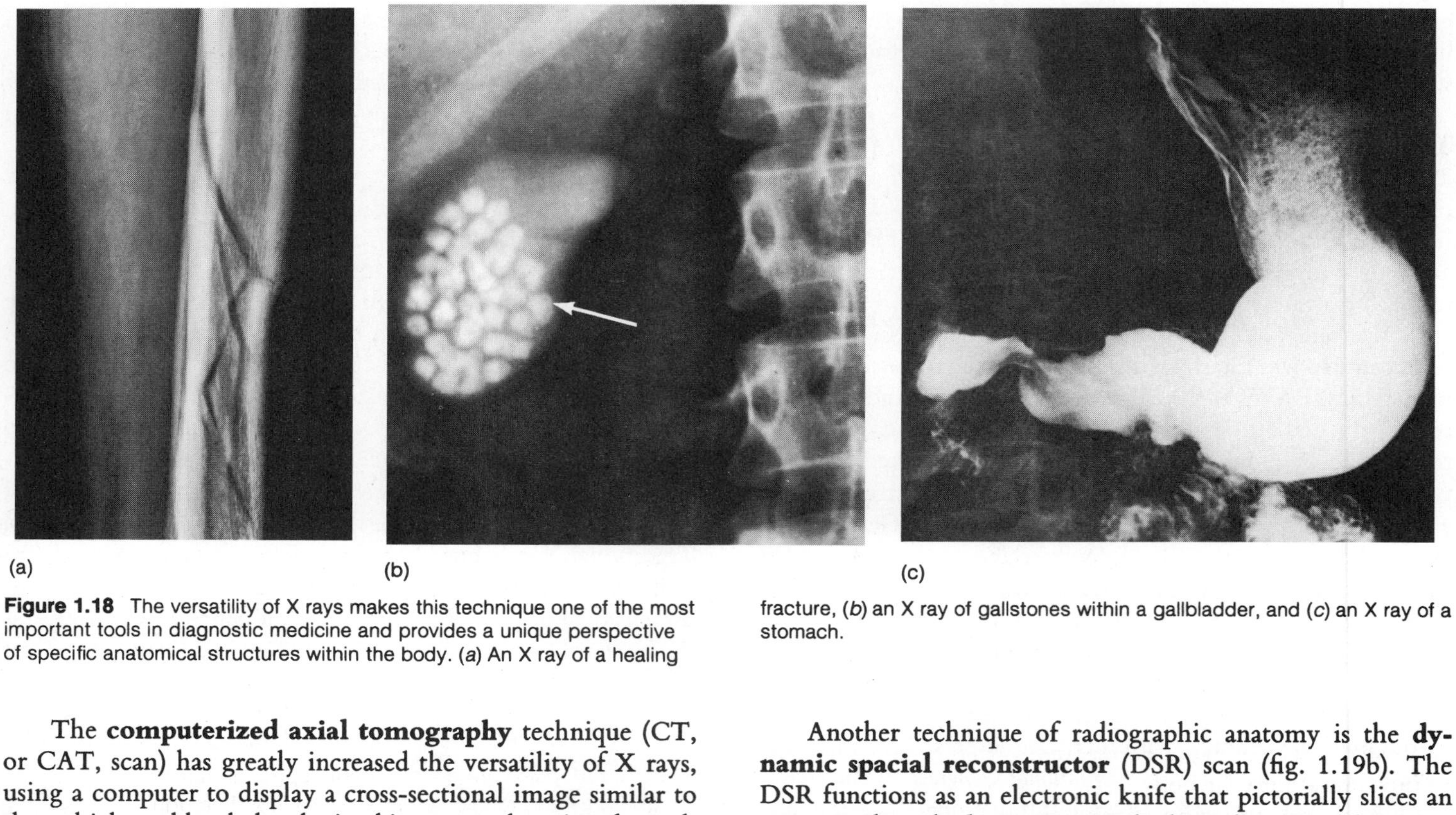

Figure 1.18 The versatility of X rays makes this technique one of the most important tools in diagnostic medicine and provides a unique perspective of specific anatomical structures within the body. (*a*) An X ray of a healing fracture, (*b*) an X ray of gallstones within a gallbladder, and (*c*) an X ray of a stomach.

The **computerized axial tomography** technique (CT, or CAT, scan) has greatly increased the versatility of X rays, using a computer to display a cross-sectional image similar to that which could only be obtained in an actual section through the body (fig. 1.19a).

Another technique of radiographic anatomy is the **dynamic spacial reconstructor** (DSR) scan (fig. 1.19b). The DSR functions as an electronic knife that pictorially slices an organ, such as the heart, to provide three-dimensional images. The DSR can be used to observe movements of organs, detect

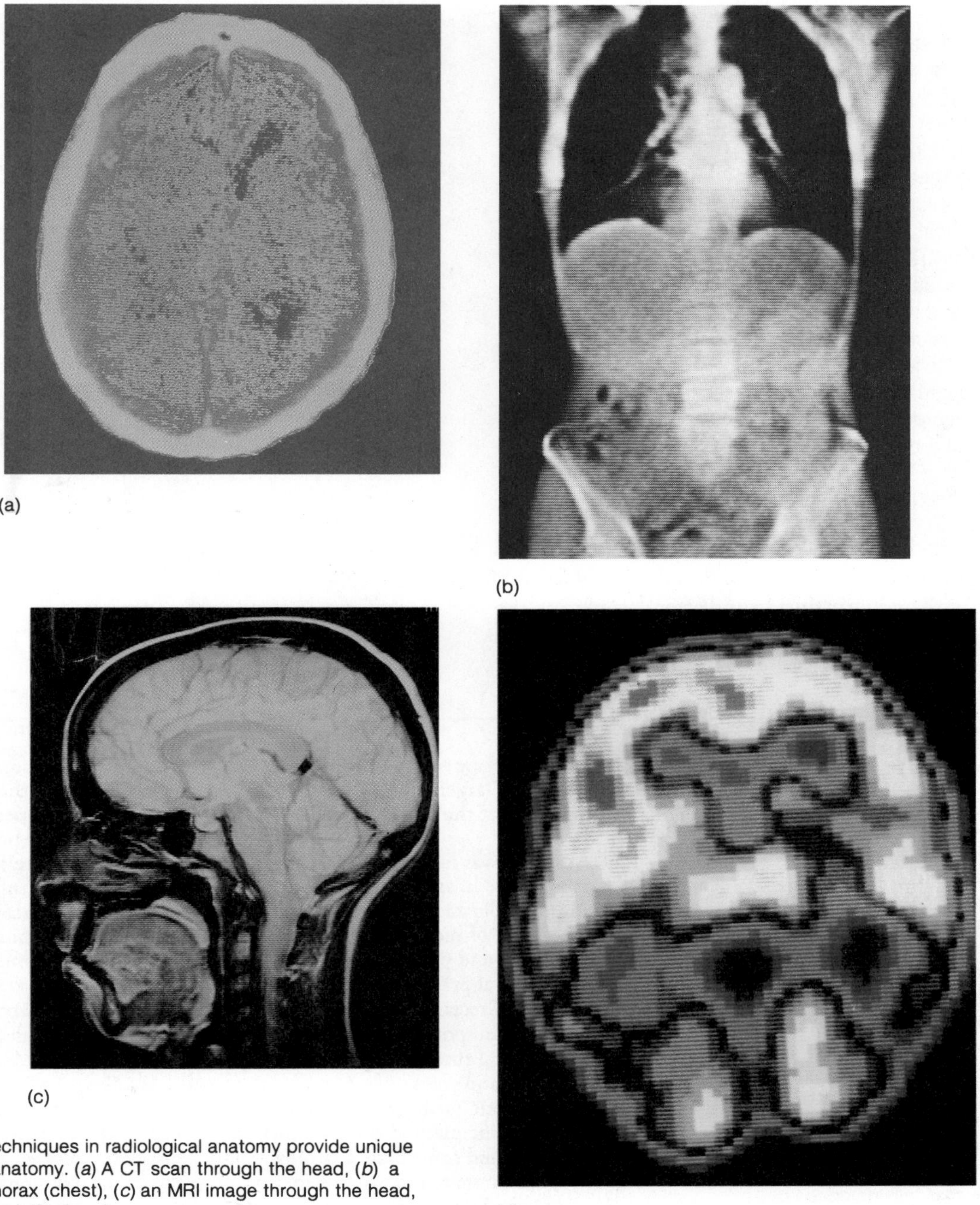

(a) (b) (c) (d)

Figure 1.19 Different techniques in radiological anatomy provide unique perspectives of human anatomy. (*a*) A CT scan through the head, (*b*) a DSR scan through the thorax (chest), (*c*) an MRI image through the head, and (*d*) a PET scan through the head.

defects, assess the extent of a disease such as cancer, or determine the extent of trauma to tissues after a stroke or a heart attack.

Magnetic resonance imaging (MRI), also called **nuclear magnetic resonance** (NMR), provides a new technique for diagnosing diseases and following the response of a disease to chemical treatment (fig. 1.19c). An MRI image is created rapidly as individual molecules of cells are focused upon simultaneously to determine their response to magnetism. MRI has the advantage of being noninvasive—that is, no chemicals are introduced into the body.

A **positron emission tomography** (PET) scan is a radiological technique used to observe the metabolic activity in organs (fig. 1.19d) after the injection of a radioactive substance, such as treated glucose, into the bloodstream.

Human anatomy will always be a relevant science. It is important to a personal understanding of body functioning as well as to the medical profession, which is concerned with the body's dysfunctions. Human anatomy is no longer confined to the isolated observation and description of structures, but has widened to include the complexities of how the body functions. The science of anatomy is dynamic and has remained vital because anatomists have dared to explore the intricate territory of the human body.

Autopsies are an important aspect of human anatomy and medicine. An autopsy is a thorough postmortem examination of all of the organs and tissues of a body. Autopsies were routinely performed in the early part of the twentieth century but their numbers have declined significantly in the last three decades. Currently, only 15% of corpses are autopsied in the United States, which is down from over 50% thirty years ago. Autopsies are of value because: 1) they reveal the cause of death, which may confirm preliminary death statements; 2) they frequently uncover diseases or structural defects that were undetected in life; 3) they check the value of a particular drug therapy for a patient or the success of a particular surgery; and 4) they are important in the training of medical students. One interesting note regarding the value of an autopsy in confirming the cause of death was revealed in a study of 2,557 autopsies conducted over a thirty-year period to determine the accuracy of physicians' diagnoses of deaths (see Autopsy, S. A. Geller, *Scientific American*, March 1983). In this study, the causes of death had been improperly or inaccurately recorded in 42% of the cases. This means that because of a decline in the number of autopsies, there may be over 1 million death certificates filed in the United States each year that are in error.

An objective of this text is to enable students to become educated and conversant in anatomy. An excellent way for students to keep up with anatomy during and after completing the formal course is to subscribe to and read scientific magazines such as *Science, Scientific American, Discover,* or *Science Digest.* These publications and others have many articles on scientific discoveries understandably written. There are many specialities of anatomy that are undergoing a rapid proliferation of exciting knowledge. It is most important and interesting to become and stay informed so that you may be an educated contributor to society.

1. Discuss the impact that each of the following has had upon the science of human anatomy: the humoral theory of body organization, vivisections, the Middle Ages, human dissections, movable type, embalming cadavers, the invention of the microscope, and the development of x-ray techniques.
2. Give some examples of how culture and religion influenced, positively or negatively, the science of anatomy.
3. Explain how a knowledge of anatomy can be of practical value to a person.
4. How can a student stay current in the science of human anatomy? Why is it important to do so?

Chapter Summary

I. Definition of the Science
   A. Human anatomy is the science concerned with the structure of the human body.
   B. The terms of anatomy are descriptive and are generally of Greek or Latin derivation.
   C. The history of human anatomy parallels that of medicine and has also been greatly influenced by various religions.

II. Prescientific Period
   A. Prehistoric interest in anatomy was undoubtedly limited to practical information necessary for survival.
   B. Trepanation was a surgical technique that was practiced by several cultures.
   C. Paleopathology is the science concerned with diseases of prehistoric man.

III. Scientific Period
   A. A few anatomical descriptions were inscribed upon clay cuneiform tablets by people who lived in Mesopotamia in about 4000 B.C.
   B. Egyptians of about 3400 B.C. developed an embalming technique. It was not recorded, however, and therefore was not of value to further the study of anatomy.
   C. The belief in a balance between yin and yang was a compelling influence in Chinese philosophy and even formulated the basis of acupuncture.
   D. The advancement of anatomy in Japan was largely due to the influence of the Chinese and Dutch.
   E. Anatomy was first widely accepted as a science in ancient Greece.
      1. Hippocrates is regarded as the father of medicine because of the sound principles of medical practice he established.
      2. The Grecian philosophy of body humors dominated medical thought for over two thousand years.
      3. Aristotle established a type of scientific method for obtaining data, and some basic anatomy was presented in the classic books he wrote.
   F. Alexandria was a center of scientific learning from 300 to 30 B.C.
      1. Human dissections and vivisections were performed in Alexandria.
      2. Erasistratus is referred to as the father of physiology because of his interpretations of various body functions.
   G. A de-emphasis of theoretical data occurred during the Roman era.
      1. Celsus's eight-volume work was a compilation of a portion of the medical data from the Alexandrian school.
      2. Galen was an influential medical writer who established some sound principles as well as introduced serious erroneous ideas that were adhered to for centuries.
      3. The Middle Ages suppressed science for nearly one thousand years, and dissections of human cadavers were prohibited.
      4. Anatomical writings were taken from Alexandria by Arab armies and thus saved from destruction during the Dark Ages in Europe.
   H. Great European universities were established during the Renaissance.
      1. Vesalius and da Vinci were renowned Renaissance men who produced monumental studies of the human form.
      2. *De Humani Corporis Fabrica,* written by Vesalius, made a tremendous impact on the advancement of human anatomy.
   I. The two major contributions during the seventeenth and eighteenth centuries were an explanation of blood flow and the development and utilization of the microscope.
      1. Harvey, in 1628, correctly described the circulation of blood.
      2. The microscope was perfected by van Leeuwenhoek, and shortly after many investigators added new discoveries to the rapidly changing specialty of microscopic anatomy.

J. The cell theory was formulated during the nineteenth century by Schleiden and Schwann, and cellular biology became established as a science separate from anatomy.

K. A trend toward simplification and standardization of anatomical nomenclature began during the twentieth century, and many specialties within anatomy have developed, including cytology, histology, electron microscopy, and radiology.

Review Activities

Objective Questions

1. *Anatomy* is derived from a Greek word meaning
   (a) to cut up.
   (b) to analyze.
   (c) functioning part.
   (d) to observe death.
2. The most important contribution of William Harvey was his research on the
   (a) continuous circulation of blood.
   (b) microscopic structure of spermatozoa.
   (c) detailed structure of kidney.
   (d) striped appearance of skeletal muscle.
3. Which of the following men would be most likely to disagree with the concept of body humors?
   (a) Galen (c) Vesalius
   (b) Hippocrates (d) Aristotle
4. Anatomy was first widely accepted as a science in ancient
   (a) Rome. (c) China.
   (b) Egypt. (d) Greece.
5. The establishment of sound principles of medical practice earned this man the title of father of medicine.
   (a) Hippocrates (c) Erasistratus
   (b) Aristotle (d) Galen
6. Which of the four body humors was believed by Hippocrates to be associated with the lungs?
   (a) black bile (c) phlegm
   (b) yellow bile (d) sanguine
7. The anatomical masterpiece *De Humani Corporis Fabrica* was written and illustrated by
   (a) da Vinci.
   (b) Harvey.
   (c) Vesalius.
   (d) van Leeuwenhoek.
8. What event of about 1450 helped usher in the Renaissance?
   (a) development of the microscope
   (b) acceptance of the scientific method
   (c) positive attitude toward dissection of human cadavers
   (d) development of movable type
9. The body organ that Aristotle attributed to being the seat of intelligence was the
   (a) liver. (c) brain.
   (b) heart. (d) intestine.
10. X rays were discovered during the late nineteenth century by
    (a) Roentgen. (c) Schleiden.
    (b) Hooke. (d) Müller.

Essay Questions

1. Define the terms *anatomise, trepanation, paleopathology, vivisection,* and *cadaver.*
2. Why were the techniques of embalming a corpse, which were perfected in ancient Egypt, not shared with other cultures or recorded for future generations?
3. What is acupuncture? What are some of its uses today?
4. Why is Latin an ideal language for the derivation of anatomical terms? What is the current trend in anatomical nomenclature regarding the use of proper names in referring to anatomical structures?
5. Why do you suppose the Hippocratic oath has survived for over two thousand years as a creed for medical practice? What aspects of the oath are difficult to conform to in today's society?
6. What is meant by the humoral theory of body organization? Which great anatomists were influenced by this theory? When did the humoral theory cease to be an influence upon anatomical investigation and interpretation?
7. Discuss the impact that Galen had on the advancement of anatomy and medicine. What ideological circumstances permitted the philosophies of Galen to survive for such a long period?
8. Briefly discuss the establishment of anatomy as a science during the Renaissance.
9. Who invented the microscope? What part did it play in the advancement of anatomy? What specialities of anatomical study have arisen since the introduction of the microscope?
10. Discuss the impact that Andreas Vesalius had on the science of anatomy.
11. Give some examples of how culture and religion influenced, positively or negatively, the science of anatomy.
12. List some techniques currently used to study anatomy and identify the specialities with which these techniques are utilized.

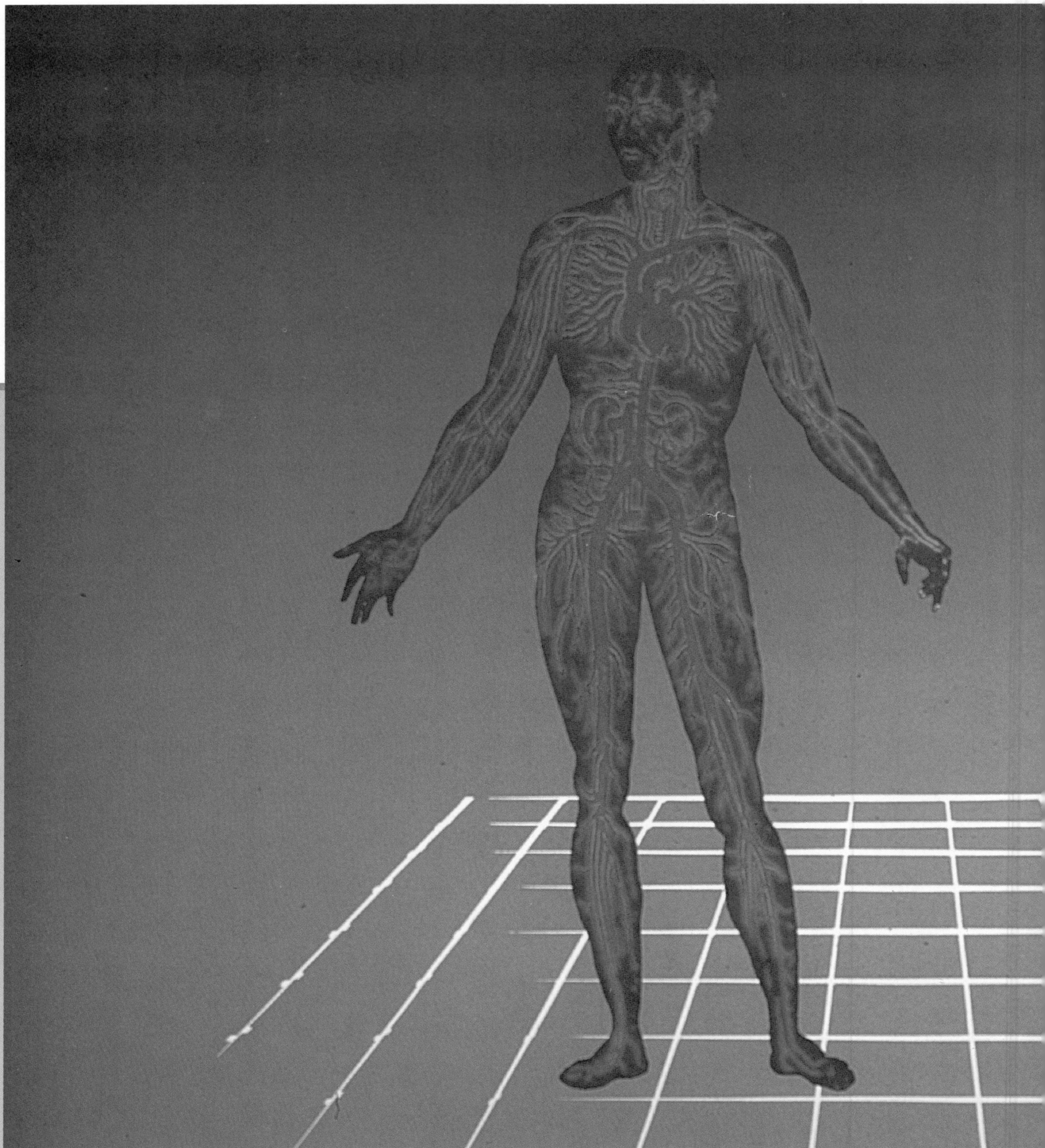

Techniques are available to focus on a particular body system, such as the circulatory system depicted in the photograph, specific organs, or particular body regions. Human anatomy is based on descriptive terminology that reflects the orientation of one body structure to another, the appearance of an organ, or an organ's function.

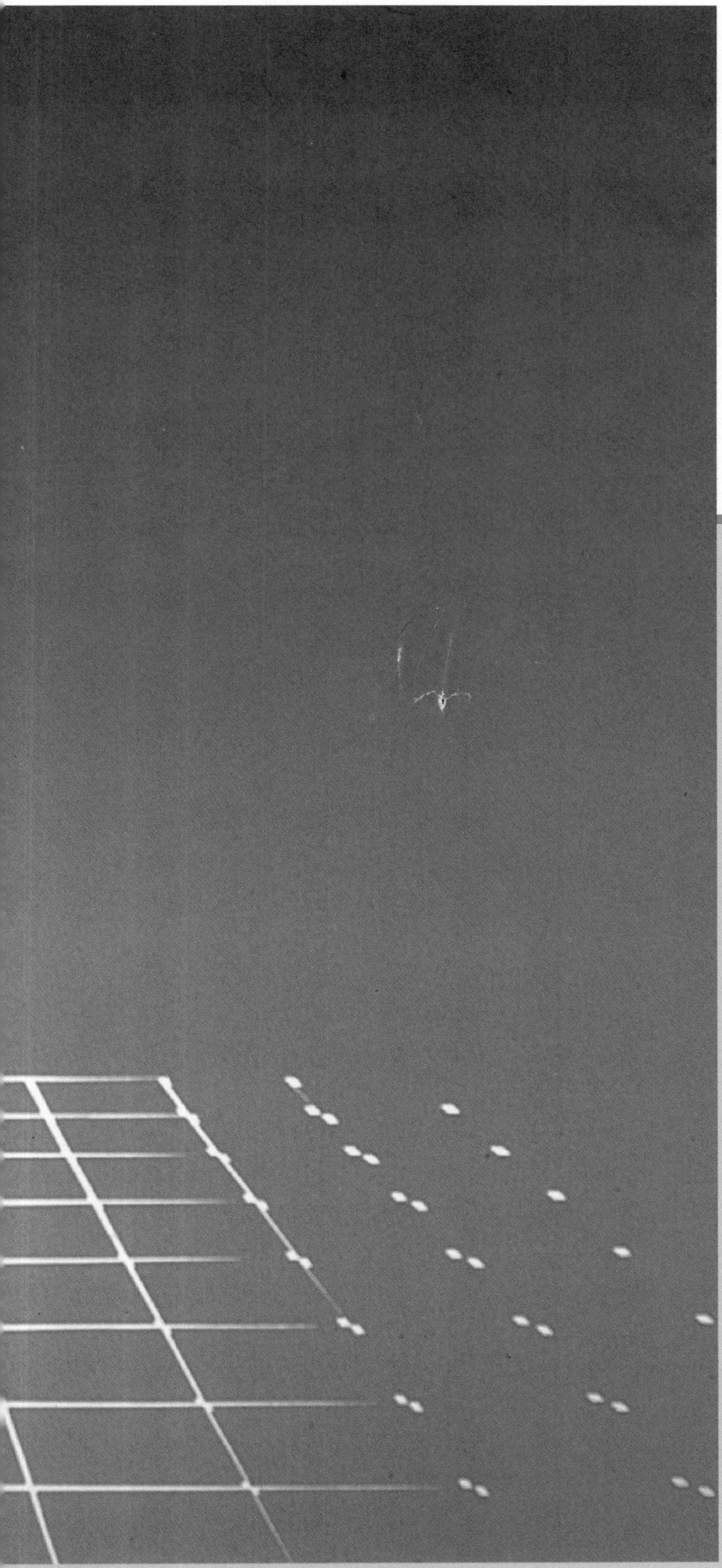

U N I T

2

Terminology, Organization, and the Human Organism

The general characteristics of humans as vertebrate animals and humans as excellent models from which to study anatomy are described in the chapter in unit 2. The various levels of body organization, from the cells to the tissues (histological) to the organs are defined and described. The fundamental terminology for describing the structure and function of the body is presented.

This unit includes:

CHAPTER

2

Body Organization and Anatomical Terminology

Outline and Concepts

Clinical Case Study

Classification and Characteristics of Humans

Humans are biological organisms belonging to the phylum Chordata within the kingdom Animalia and to the family Hominidae within the class Mammalia and the order Primates.

Phylum Chordata
Class Mammalia
Order Primates
Family Hominidae
Characteristics of Humans

Body Organization

Structural and functional levels of organization exist in the human body, and each of its parts contributes to the total organism.

Cellular Level
Tissue Level
Organ Level
System Level

Anatomical Terminology

As a science, anatomy has descriptive terminology that must be mastered by students who would understand and be conversant in the subject.

Body Regions

The human body is divided into regions and specific local areas, which can be identified on the surface. Each region contains internal organs, the location of which are anatomically and clinically important.

Head
Neck
Thorax
Abdomen
Upper Extremity
Lower Extremity

Body Cavities and Membranes

For functional and protective purposes, the viscera are compartmental and supported in specific body cavities by connective and epithelial membranes.

Body Cavities
Body Membranes

Planes of Reference and Descriptive Terminology

All of the descriptive planes of reference and terms of direction used in anatomy are standardized because of their reference to the body in anatomical position.

Planes of Reference
Descriptive Terminology

Clinical Case Study Answer

Chapter Summary

Review Activities

CLINICAL CASE STUDY

A twenty-year-old female was hit by a car while crossing a street. Upon arrival at the scene, paramedics find the patient to be a bit dazed but reasonably lucid, complaining of pain in her abdomen and the left side of her chest. Otherwise, her vital signs are within normal limits. Initial evaluation in the emergency room reveals a very tender abdomen and left chest. The chest X ray demonstrates a left-sided pneumothorax (collapsing of the lung due to air in the pleural space). The emergency room (E.R.) physician places a drainage tube into the left chest (into the pleural space) in order to treat the pneumothorax. Attention is then turned to the abdomen. Due to the finding of tenderness, a peritoneal lavage is performed. This maneuver involves penetration of the abdominal wall and insertion of a tube into the peritoneal cavity. Clear fluid is then instilled to the abdomen and siphoned out again. Fluid used in this procedure is called lavage fluid. A return of lavage fluid containing blood, fecal matter, or bile indicates injury to an abdominal organ that requires surgery. The return of lavage fluid from our patient is clear. However, the nurse states that lavage fluid is draining out of the chest tube.

Could this phenomenon be explained based on normal anatomy? Explain your answer using knowledge of the organization of, and relationships between, the various body cavities. Explain what could account for this finding in our patient. Does the absence of bile, blood, etc., in the peritoneal lavage fluid guarantee that rupture of an organ(s) is not present. If not, explain why in terms of the various organs' relationship to the membranes within the abdomen.

Classification and Characteristics of Humans

Humans are biological organisms belonging to the phylum Chordata within the kingdom Animalia and to the family Hominidae within the class Mammalia and the order Primates.

Objective 1. List the taxonomic classification of humans.
Objective 2. List the characteristics that identify humans as chordates and as mammals.
Objective 3. Describe the characteristics that humans have but other primates do not.

The human organism, or ***Homo sapiens,*** as we have named ourselves, is unique in many ways. Our scientific name translates from Latin to mean man the intelligent, and indeed this is our most distinguishing feature. Through our intelligence we have built civilizations, conquered many diseases, and established cultures. We have devised a means of communicating through written symbols. We can record our own history as well as that of other animals and speculate on our future. We are social organisms with well-established patterns of behavioral interactions. Most of us have become so intellectually specialized that we are not self-sufficient. We need one another as much as we need the recorded knowledge of the past.

As we seek knowledge, we are challenged to learn more about ourselves. It becomes intriguing to study our structure and function and to realize our close relationship to other living organisms. Often it is sobering to realize our biological imperfections and limitations.

We share many characteristics with all living animals. As human organisms, we breathe, eat and digest food, excrete bodily wastes, locomote, and reproduce our own kind. We are subject to disease, injury, pain, aging, mutations, and death. Being composed of organic materials, we will decompose after death as microorganisms consume our flesh as food. The processes by which our bodies produce, store, and utilize energy are similar to those found in all living organisms. The genetic code that regulates our development is found throughout nature. The fundamental patterns of development of many animals are also found in the formation of the human embryo.

The classification, or taxonomic, scheme has been established by biologists to organize the structural and evolutionary relationships of living organisms. Each category of classification is referred to as a *taxon.* The highest taxon is the kingdom and the most specific taxon is the species. Humans are species belonging to the **animal kingdom.** *Phylogeny (fi-loj'ĕny)* is the origin and evolutionary development of animal species.

Phylum Chordata

Human beings belong to the phylum Chordata *(fi'lum kor-dah'tah)* along with fishes, amphibians, reptiles, birds, and other mammals. All chordates have three structures in common: a **notochord** *(no'to-kord),* a **dorsal hollow nerve cord,** and **pharyngeal** *(fah-rin'je-al)* **pouches** (fig. 2.1). These chordate characteristics are well expressed during the embryonic stage of development and, to a certain extent, are present in an adult. The notochord is a flexible rod of tissue that extends the length of the back of an embryo. A portion of the notochord persists in the adult as the **nucleus pulposus,** located within each **intervertebral disc** (fig. 2.2). The dorsal hollow nerve cord is positioned above the notochord and develops into the **brain** and **spinal cord,** which are highly functional as the **central nervous system** in the adult. Pharyngeal pouches form gill openings in fishes and some amphibians. In other chordates, such as humans, embryonic pharyngeal pouches develop, but only one of the pouches persists, becoming the **auditory (eustachian)** *(u-sta'ke-an)* **canal,** a connection between the middle ear and **pharynx** *(far'ingks)* (throat area).

The function of the nucleus pulposus and the intervertebral discs is to allow flexibility between vertebrae for movement of the entire spinal column while preventing compression. Spinal nerves exit between vertebrae, and the discs maintain the spacing to avoid nerve damage. A "slipped disc," resulting from straining the back, is a misnomer. What actually occurs is a herniation, or rupture, because of a weakened wall of the nucleus pulposus, causing severe pain as a nerve is compressed.

Class Mammalia

Mammals are chordate animals with hair and mammary glands. Hair is a thermoregulatory protective covering for most mammals, and mammary glands serve for suckling the young (fig. 2.3). Other characteristics of mammals include three ear ossi-

taxon: Gk. *taxis,* order
phylogeny: L. *phylum,* tribe; Gk. *logos,* study.

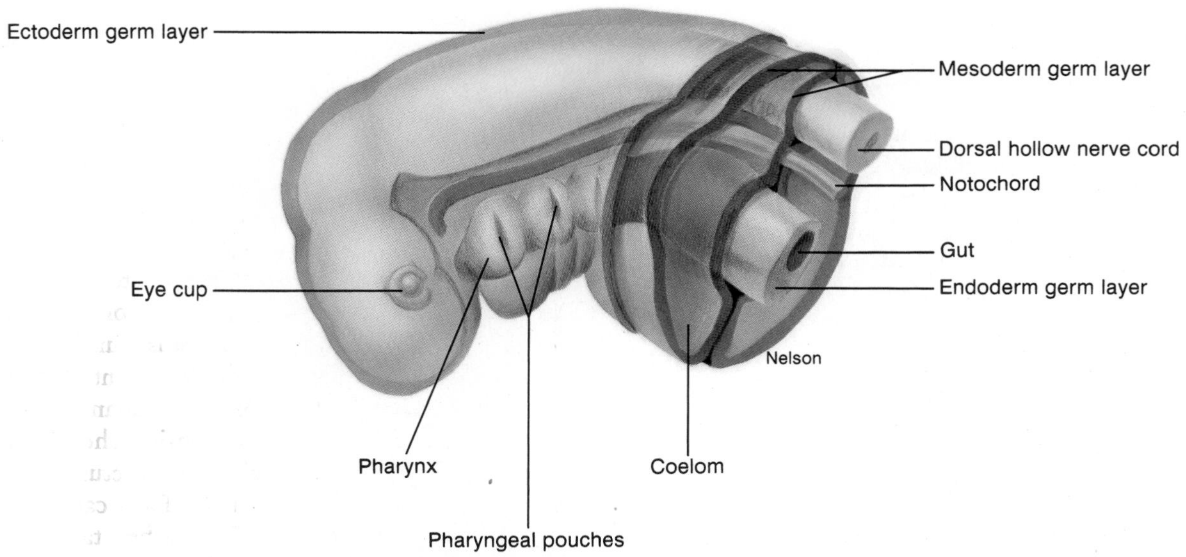

Figure 2.1 A schematic diagram of the front part of a chordate embryo.

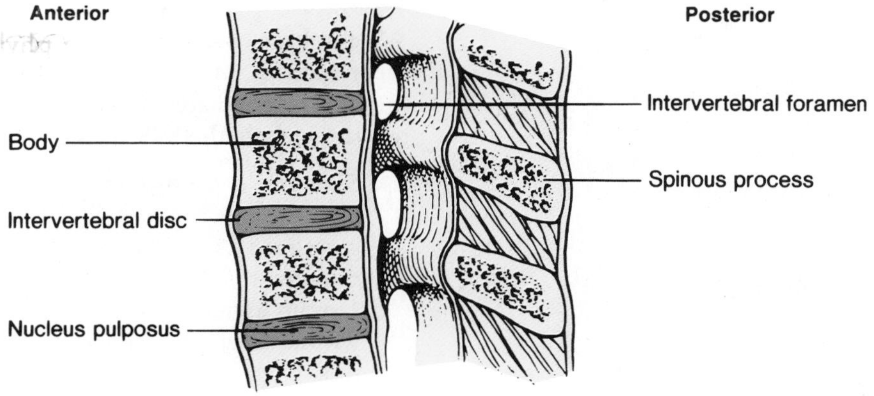

Figure 2.2 A midsagittal section through vertebrae to show an intervertebral disc and nucleus pulposus.

cles (bones), heterodont dentition (differently shaped teeth), squamosal-dentary jaw articulation (a joint between the lower jaw and skull), an attached placenta *(plah-sen'tah),* well-developed facial muscles, a muscular diaphragm, and a four-chambered heart with a left aortic arch (fig. 2.4).

Order Primates

There are several subdivisions of closely related groupings of mammals, called orders. Humans, along with lemurs, monkeys, and great apes, belong to the order called Primates. Members of this order have prehensile hands (fig. 2.5), digits modified for grasping, and relatively large, well-developed brains (fig. 2.6).

Family Hominidae

Humans are the sole living members of the family **Hominidae.** *Homo sapiens* is included within this family to which all the varieties or ethnic groups of humans belong (fig. 2.7).

Table 2.1 Classification scheme of human beings

Taxon	Designated grouping	Characteristics
Kingdom	Animalia	Eucaryotic cells that lack walls; plastids; and photosynthetic pigments
Phylum	Chordata	Dorsal hollow nerve cord; notochord; pharyngeal pouches
Subphylum	Vertebrata	Vertebral column
Class	Mammalia	Mammary glands; hair
Order	Primates	Well-developed brain; prehensile hands
Family	Hominidae	Large cerebrum; bipedal locomotion
Genus	*Homo*	Flattened face; prominent chin and nose with inferiorly positioned nostrils
Species	*sapiens*	Largest cerebrum

Each "racial group" has distinguishing features that have been established in isolated populations over thousands of years. Our classification pedigree is presented in table 2.1.

heterodont: Gk. *heteros,* other; *odontos,* tooth
placenta: L. *placenta,* flat cake
Primates: L. *primas,* first
prehensile: L. *prehensus,* to grasp

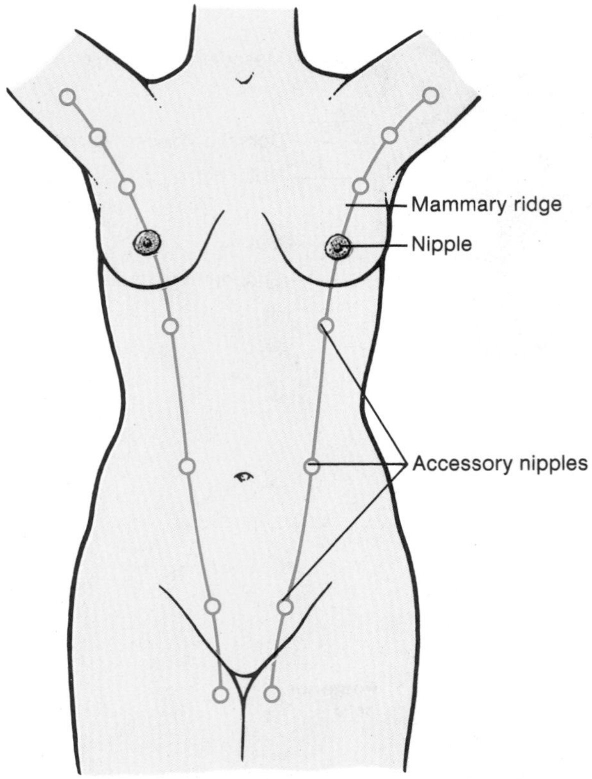

(a)

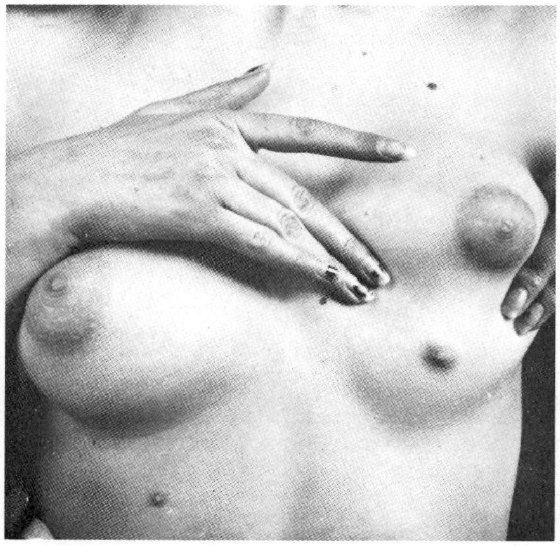

(b)

Figure 2.3 The mammary ridge and accessory nipples. (*a*) Mammary glands are positioned along a mammary ridge; (*b*) occasionally in humans, additional nipples (polythelia) develop elsewhere along the mammary ridge.

With greater ease of travel and communication, many traditional cultural barriers no longer inhibit interracial marriage. This may lead to a mixing of the "gene pool" until distinct racial groups are less evident. Perhaps multiple ethnic ties in everyone's pedigree is what is needed to lessen cultural hostility and strife.

Characteristics of Humans

As human beings, we have certain anatomical characteristics that are so specialized that they are diagnostic in separating us from other animals and even from other closely related mammals.

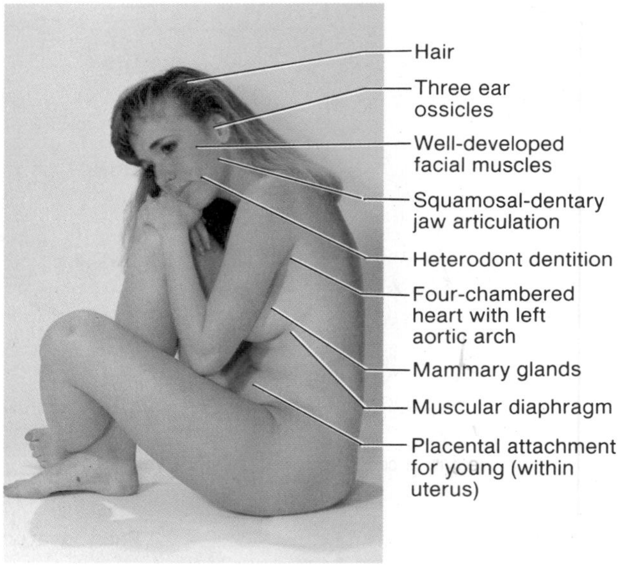

Figure 2.4 Mammals have several distinguishing characteristics, some of which are listed in the photo with their approximate location within the body.

We also have characteristics that are equally well developed in other animals, but when these function with the human brain, they provide remarkable capabilities. Our anatomical characteristics include the following:

1. **Size and development of brain.** The average human brain weighs between 1,350 and 1,400 g (3 lbs). This gives man a large brain-to-body-weight ratio. But more important is the development of portions of the brain. Certain extremely specialized regions and structures within the brain account for emotion, thought, reasoning, memory, and even precise, coordinated movement.
2. **Style of locomotion.** Because man stands and walks on two appendages, our style of locomotion is said to be *bipedal.* Upright posture imposes certain other diagnostic structural features such as the sigmoid (S-shaped) curvature of the spine, the anatomy of the hip and thighs, and arched feet. Some of these features may cause clinical problems in older individuals.
3. **Opposable thumb.** The human thumb joint is structurally adapted for tremendous versatility in grasping objects. Most primates have opposable thumbs.
4. **Vocal structures.** Humans, like no other animals, have developed articulated speech. The anatomical structure of the vocal organs and the well-developed brain have made this possible.
5. **Stereoscopic vision.** Although this characteristic is well developed in several other animals, it is also keen in humans. Our eyes are directed forward so that when we focus upon an object, we view it from two angles. Stereoscopic vision gives us depth perception, or a three-dimensional image.

---
bipedal: L. *bi,* two; *pedis,* foot
sigmoid: Gk. *sigma,* shaped like the letter *s*
stereoscopic: Gk. *stereos,* solid; *skopein,* to view

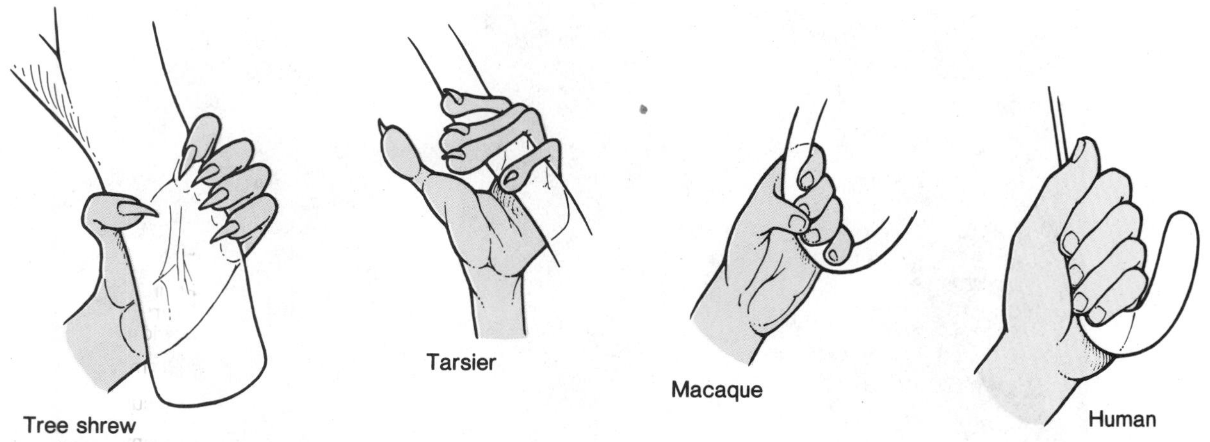

Figure 2.5 A prehensile grip is a characteristic of primates.

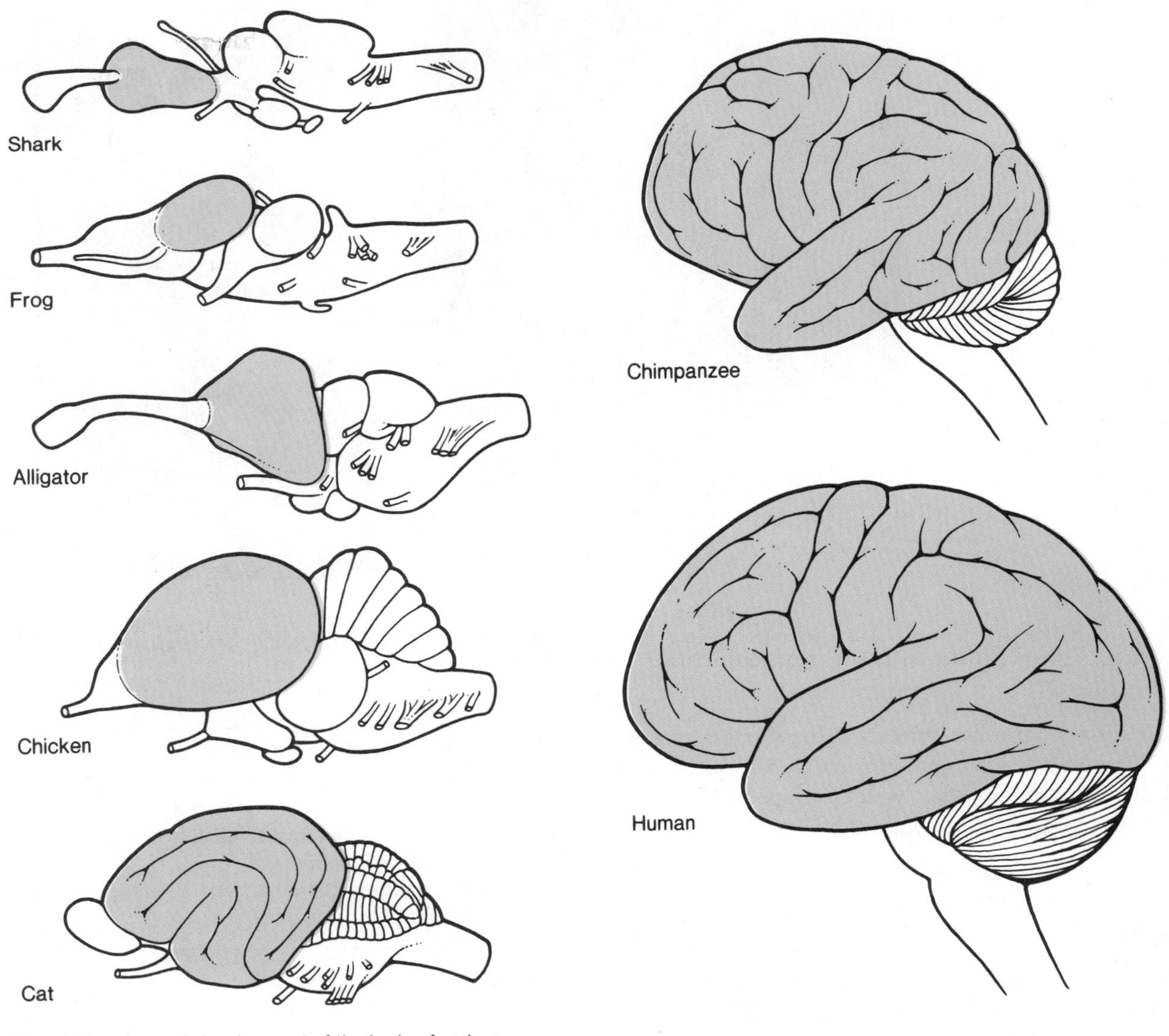

Figure 2.6 The relative size and development of the brain of various vertebrates. The cerebrum in each representative is shaded. (Note that only mammals have a convoluted cerebrum.)

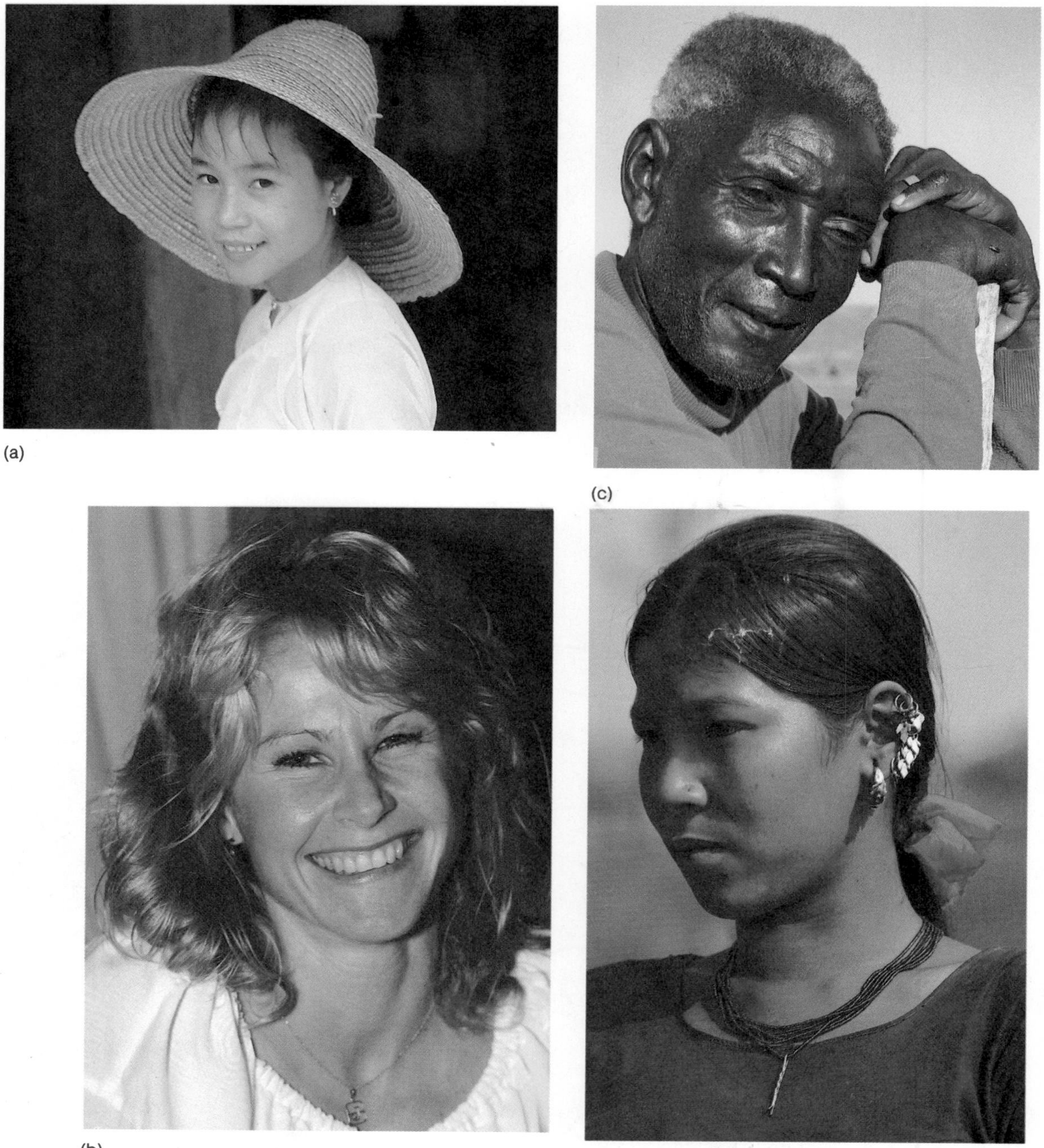

Figure 2.7 The principal races of humans: (*a*) Mongoloid (Thailand), (*b*) Caucasoid (U.S.), (*c*) Negroid (Cameroon, Africa), (*d*) people of Indian subcontinent (Nepal), (*e*) Capoid (Kalahari bushman), and (*f*) Australoid (Ngatatjara man, West Australia).

We also differ from other animals in the number and arrangement of vertebrae (vertebral formula), the kinds and number of teeth (tooth formula), well-developed facial muscles, and the structural design of various body organs.

The characteristics just described account for the success of humans in acquiring culture. As bipedal animals, we have our hands free to grasp and manipulate objects with our opposable thumb. We can assimilate experiences into our highly developed brain, apply this knowledge at a later time, and even share this learning through vocal or written communication.

---

1. What is a chordate? Why are humans considered members of the phylum Chordata?
2. Why are humans designated as mammals and primates? What characteristics distinguish humans from other primates?
3. Which of the characteristics of humans are adaptive for social organization?

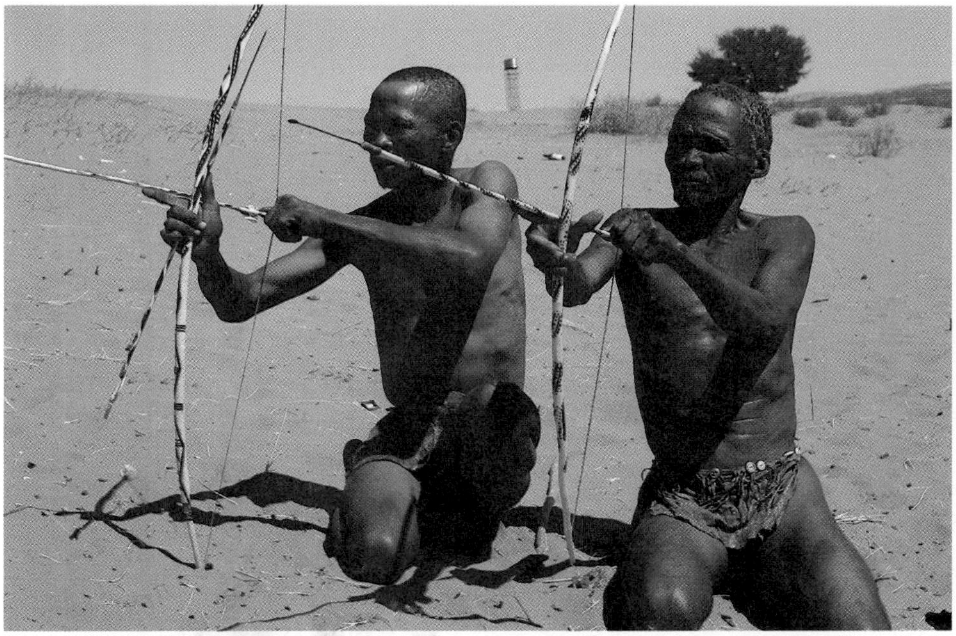

(e)

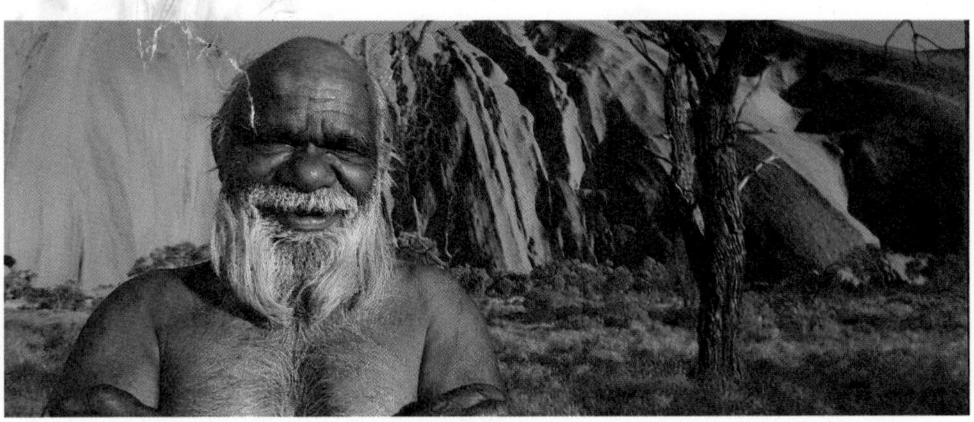

(f)

Body Organization

Structural and functional levels of organization exist in the human body, and each of its parts contributes to the total organism.

Objective 4. Identify the structures of a cell, tissue, organ, and system, and explain the relationships among these structures as they constitute an organism.

Objective 5. Explain the general function of each system.

Cellular Level

The **cell** is the basic structural and functional component of life. Humans are multicellular organisms composed of between 60 and 100 trillion cells. It is at the cellular level that the vital functions of life are carried out, such as metabolism, growth, irritability and adaptability, repair, and reproduction.

Cells are composed of minute particles called **atoms,** which are bound together to form larger particles called **molecules** (fig. 2.8). Certain molecules are arranged into small functional structures called **organelles** *(or''gah-nelz)*. Each organelle carries out a specific function within the cell. The nucleus, mitochondria, and endoplasmic reticulum are organelles. The structure of cells and the function of the organelles will be examined in more detail in chapter 3.

The human body contains many distinct kinds of cells, each specialized to perform specific functions. Examples of specialized cells are bone cells, muscle cells, fat cells, blood cells, and nerve cells. Each of these cell types has a unique structure directly related to its function.

Tissue Level

Tissues are layers or aggregations of similar cells that perform specific functions (fig. 2.8). An example of a tissue is the muscle within the heart, which functions to contract and pump the blood through the body. The outer layer of skin is a tissue because it is composed of similar cells that are bound together and function to protect and contain body contents.

cell: L. *cella,* small room

tissue: Fr. *tissu,* woven; from L. *texo,* to weave

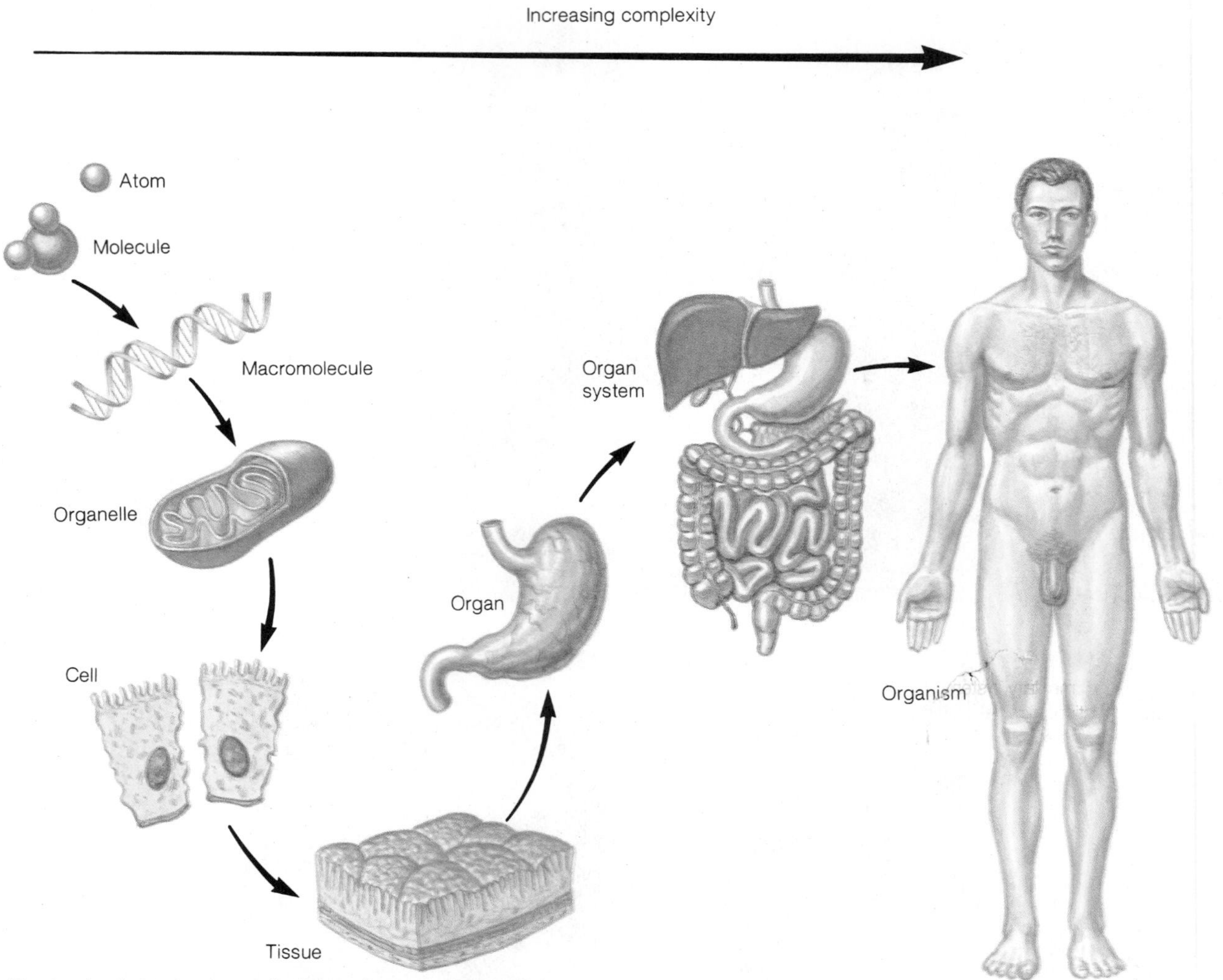

Figure 2.8 The levels of structural organization and complexity within the human body.

Organ Level

An **organ** is an aggregate of two or more tissues integrated to perform a particular function. Organs occur throughout the body and vary greatly in size and function. Examples of organs are the heart, spleen, pancreas, ovary, skin, and even any bone within the body. Each organ usually has one or more primary tissues and several secondary tissues. In the stomach, for example, the inside epithelial lining is considered the primary tissue because the basic functions of secretion and absorption occur within this layer. Secondary tissues of the stomach are the supporting connective tissue and vascular, nervous, and muscle tissues.

System Level

The **systems** of the body constitute the next level of structural organization. A body system consists of various organs that have similar or related functions. Examples of systems are the circulatory system, nervous system, digestive system, and endocrine system. Certain organs may serve several systems. The pancreas, for example, functions with both the endocrine and digestive systems. All the systems of the body are interrelated and function together, constituting the total organism.

In a *systematic (systemic) approach* to studying anatomy the functional relationships of various organs within a system are emphasized. For example, the functional role of the digestive system can be better understood if all of the organs within that system are studied together. Another approach to anatomy, the *regional approach,* has merit in professional schools because the structural relationships of portions of several systems are observed simultaneously in this approach. Dissections of cadavers are usually conducted on a regional basis. Trauma or injury usually affects a region of the body, whereas a disease that affects a region may also involve an entire system.

This text uses a systematic approach to anatomy. In the chapters that follow, you will become acquainted system by system with the functional anatomy of the entire body. An overview of the structure and function of each of the body systems is presented in figure 2.9.

1. Construct a diagram to illustrate the levels of structural organization that characterize the body. Which of these levels are microscopic?
2. Why is the skin considered an organ?
3. Which body systems control the functioning of the others; which are supportive of the organism; and which serve a transportive role?

organ: Gk. *organon,* instrument
system: Gk. *systema,* being together

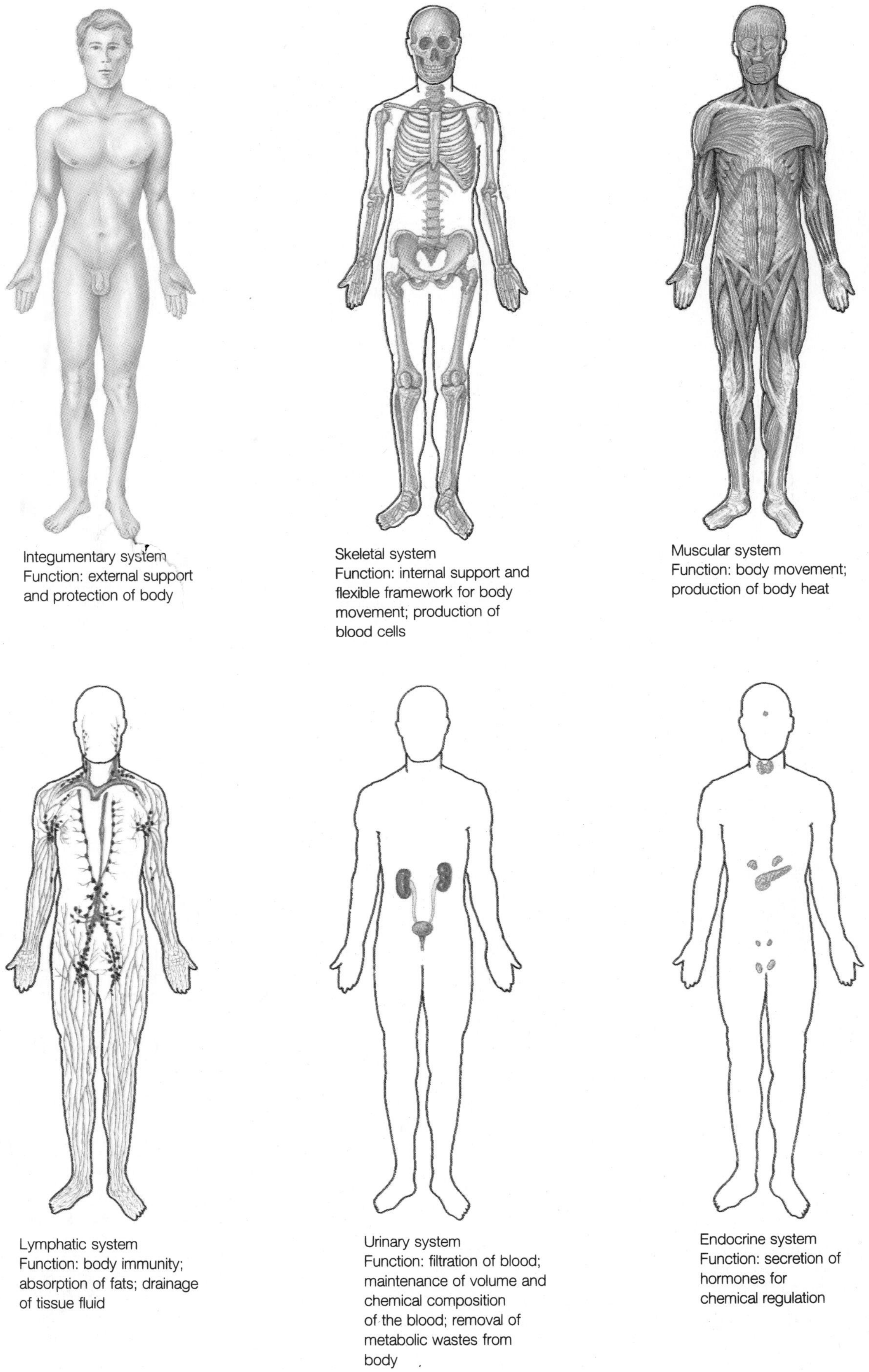
Integumentary system
Function: external support and protection of body
Skeletal system
Function: internal support and flexible framework for body movement; production of blood cells
Muscular system
Function: body movement; production of body heat
Lymphatic system
Function: body immunity; absorption of fats; drainage of tissue fluid
Urinary system
Function: filtration of blood; maintenance of volume and chemical composition of the blood; removal of metabolic wastes from body
Endocrine system
Function: secretion of hormones for chemical regulation

Figure 2.9 Various systems of the human body.

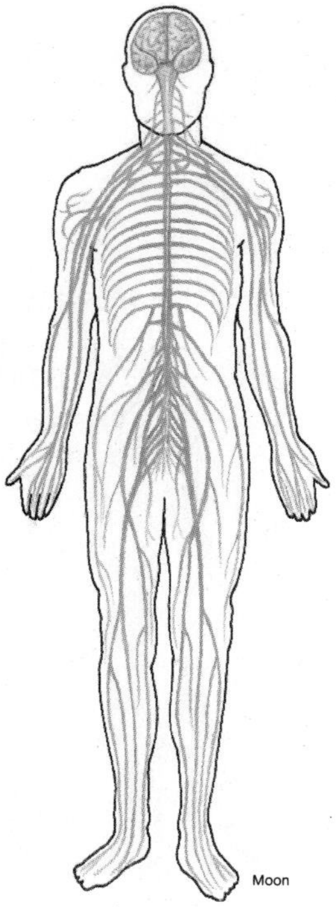

Nervous system
Function: regulation of all body activities: learning and memory

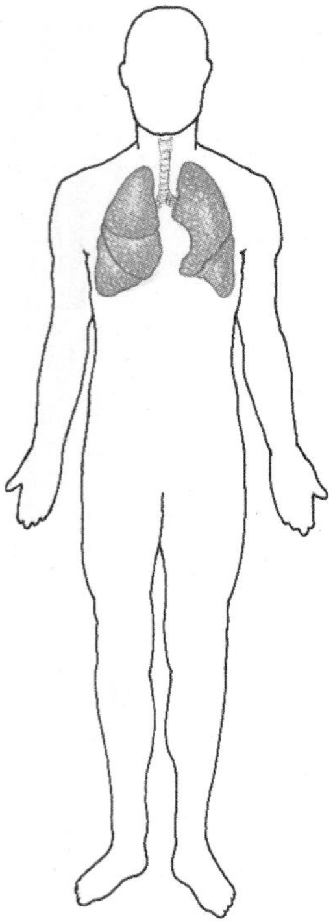

Respiratory system
Function: gaseous exchange between external environment and blood

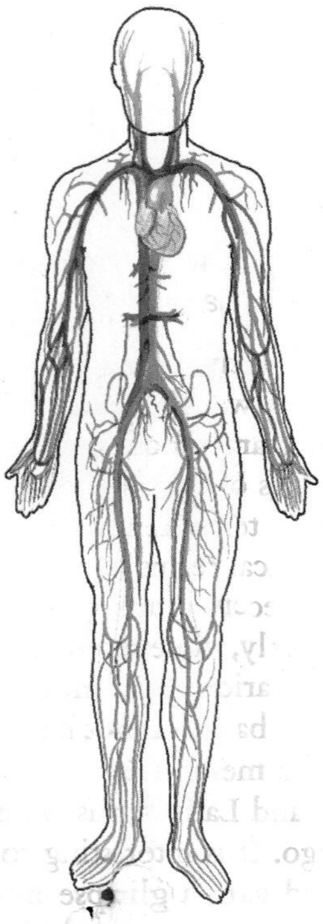

Circulatory system
Function: transport of life-sustaining materials to body cells; removal of metabolic wastes from cells

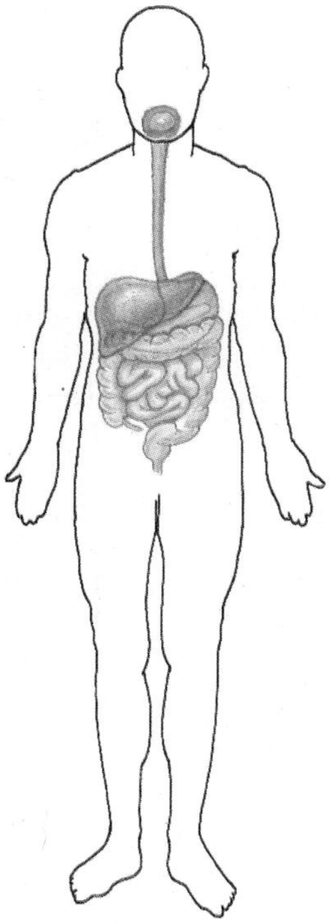

Digestive system
Function: breakdown and absorption of food materials

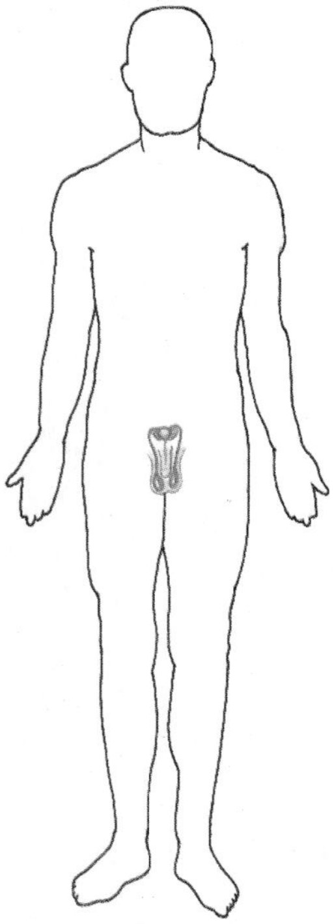

Male reproductive system
Function: production of male sex cells (sperm); transfer of sperm to reproductive system of female

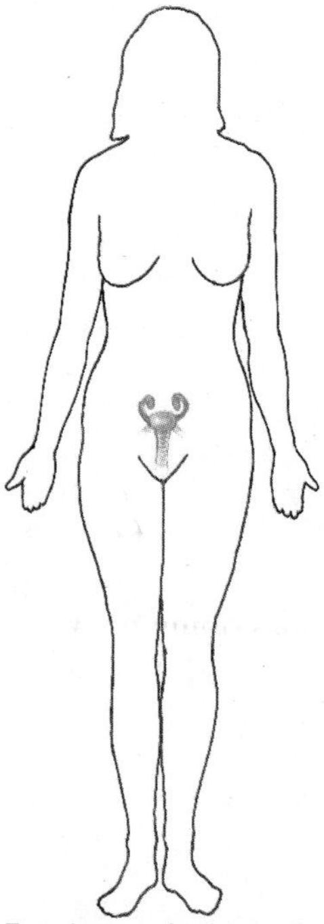

Female reproductive system
Function: production of female sex cells (ova); receptacle of sperm from male; site for fertilization of ovum, implantation, and development of embryo and fetus; delivery of fetus

Figure 2.9 *continued*

Anatomical Terminology

As a science, anatomy has descriptive terminology that must be mastered by students who would understand and be conversant in the subject.

Objective 6. Explain how anatomical terms are derived.
Objective 7. Define what is meant by prefixes and suffixes.

Anatomy is a descriptive science. Analyzing anatomical terminology can be a rewarding experience in itself as one learns something of the character of antiquity. Not only is an understanding of the roots of words of academic interest, but familiarity with technical terms reinforces the learning process. The majority of anatomical terms have Greek or Latin derivatives. Some of the more recent terms are of German and French derivation. Unfortunately, more recent anatomical terms have been coined in honor of various anatomists or physicians. Such terms have no descriptive basis, cannot be associated with anything, and must simply be memorized.

Many Greek and Latin terms were coined more than two thousand years ago. It is interesting to decipher the meanings of these terms and gain a glimpse into our medical heritage. Many terms referred to common plants or animals. Thus, the term *vermis* means worm, *cochlea (kok'le-ah)* means snail shell, *cancer* refers to crab, *uvula* means little grape, and even *muscle* comes from the Latin *musculus,* which means mouse. Other terms reveal the warlike environment of the Greek and Latin era. *Thyroid,* for example, means shield, *xiphos (zi'fos)* means sword, and *thorax* breastplate. *Sella* means saddle and *stapes (sta'pēz)* the stirrup. Various tools or instruments were referred to in early anatomy. The malleus and anvil resemble miniatures of a blacksmith's implements, and tympanum refers to a drum.

You will encounter many new terms throughout your study of anatomy. You can learn these terms more easily by understanding the prefixes and suffixes of the new words. Use the glossary of prefixes and suffixes (on the inside of the front and back covers) as an aid in learning new terms. Pronouncing these terms as you learn them will also aid your retention.

Learning the material presented in the remainder of this chapter establishes a basic foundation for anatomy as well as for all medical and paramedical fields. Anatomy is a very precise science because there is a universally accepted and used reference language for describing body parts and locations.

---

1. Explain the statement that "anatomy is a descriptive science."
2. Refer to the glossary of prefixes and suffixes listed on the inside of the front and back covers to decipher the terms: *blastocoel, hypodermic, dermatitis,* and *orchiectomy.*

Body Regions

The human body is divided into regions and specific local areas, which can be identified on the surface. Each region contains internal organs, the location of which are anatomically and clinically important.

Objective 8. List the regions of the body and the principal local areas comprising each region.
Objective 9. Explain why it is important to be able to describe the body areas and regions where each major internal organ is located.

The human body is divided into several regions that can be identified on the surface of the body. Learning the terminology used in reference to these regions now will help one learn the names of underlying structures later. The major body regions are the **head, neck, trunk, upper extremity,** and **lower extremity** (fig. 2.10). The trunk is frequently divided into the thorax or the pectoral region, and abdomen.

Head

The **head,** or **caput** *(kap'ut),* is divided into a **facial region** that includes the eyes, nose, and mouth, and a **cranium** *(kra'ne-um),* or **cranial region,** that covers and supports the brain. Further regionalization of the head is important for identifying specific glands, muscles, nerves, or vessels located under the skin. The identifying names for detailed surface regions are based on associated organs—such as the orbital (eye), nasal (nose), oral (mouth), mental (chin), and auricular (ear) regions.

Neck

The neck, referred to as the **cervix** *(ser'viks),* or **cervical region,** supports the head and permits it to move. As with the head, detailed subdivisions of the neck can be identified. Additional information concerning the neck region can be found in chapter 10.

Thorax

The thorax, or **thoracic** *(tho-ras'ik)* **region,** is commonly referred to as the chest. The **mammary region** of the thorax surrounds the nipple and in sexually mature females is enlarged as the breast. Between the mammary regions is the **sternal region.** The armpit is called the **axillary fossa,** or simply **axilla,** and the surrounding area the **axillary region.** The **vertebral region,** following the vertebral column, extends the length of the back.

The heart and lungs are contained within the thoracic cavity. Easily identified surface landmarks greatly facilitate a physical examination of them. A physician must know, for example, where the valves of the heart can best be detected and where to listen for respiratory sounds. The axilla becomes important when examining for infected lymph nodes. When fitting a patient for crutches, a physician will instruct the patient to avoid supporting the weight of the body on the axillary region because of the possibility of damaging the underlying nerves and vessels.

---

thorax: L. *thorax,* chest
mammary: L. *mamma,* breast
axillary: L. *axilla,* armpit

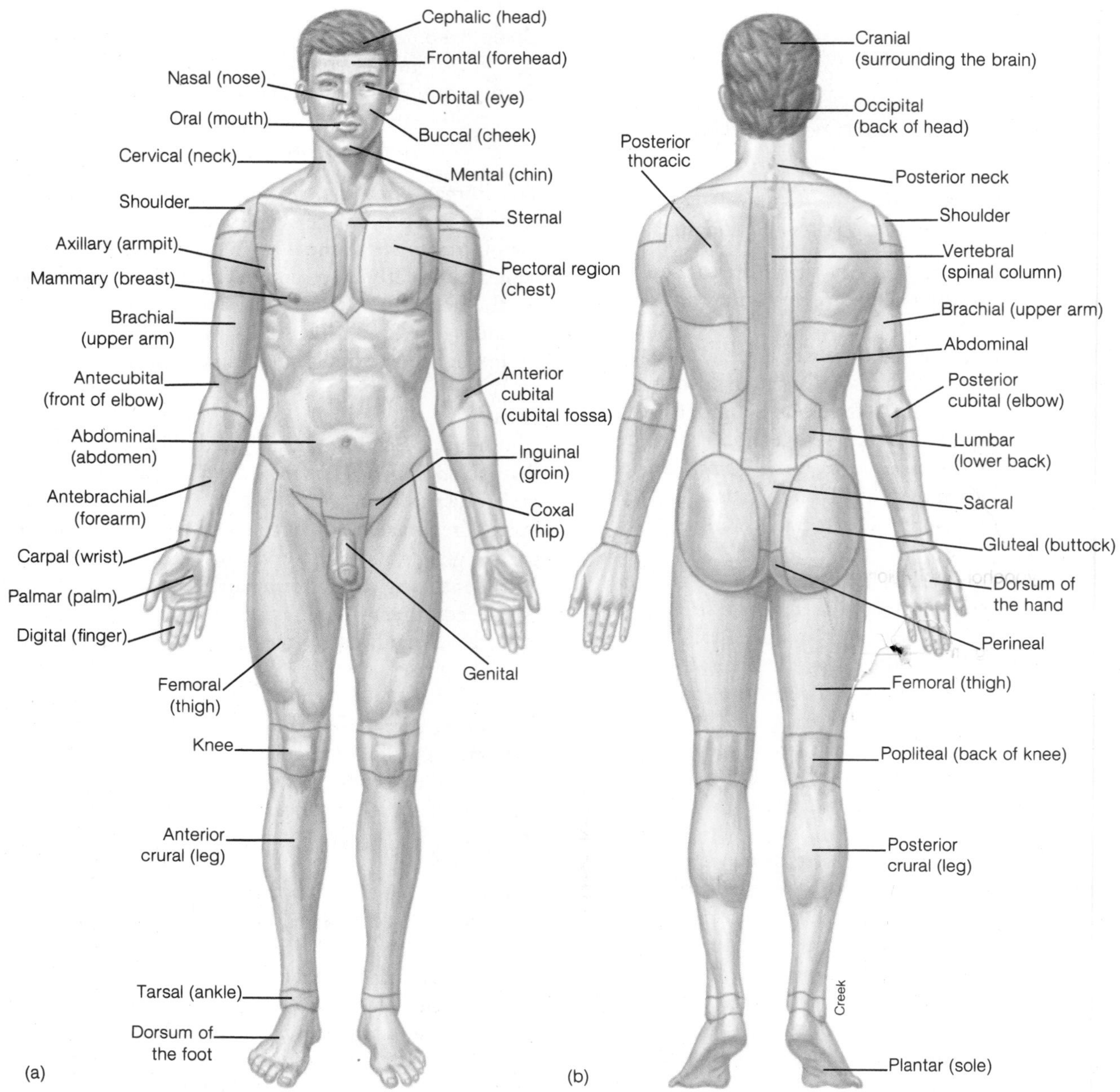

Figure 2.10 Body regions. (*a*) Anterior view, (*b*) posterior view.

Abdomen

The abdomen is located below the thorax. The **navel,** or **umbilicus,** is an obvious landmark on the front and center of the abdomen. The abdomen has been divided into nine regions in order to describe the location of internal organs. In figure 2.11 the subdivisions of the abdomen are diagramed, and the internal organs located within these regions are identified in table 2.2. Subdividing the abdomen into four quadrants (fig. 2.12) is a common clinical practice so that various pains and conditions can be more easily located.

The **pelvic region** forms the lower portion of the abdomen. Within the pelvic region is the **pubic area,** which is covered with pubic hair in sexually mature persons. The **perineum** *(per''ĭ-ne'um)* (fig. 2.13) is the region containing the external sex organs and the anal opening. The center of the back side of the abdomen, commonly called the small of the back, is the **lumbar region.** The **sacral region** is located further down at the point where the vertebral column terminates. The large hip muscles form the **buttock,** or **gluteal region.** This region is a common injection site of hypodermic needles.

Upper Extremity

The upper extremity is anatomically divided into the **shoulder, brachium** *(bra'ke-um)* (upper arm), **antebrachium** (forearm), and **manus** (hand) (see fig. 2.10). The shoulder is the region between the pectoral girdle and the brachium that contains the shoulder joint. The shoulder is referred to as the **omos,** or **deltoid region.** Between the arm and forearm is a flexible joint called the **elbow.** The surface area of the elbow is known as the **cubital region.** The front surface of the elbow, or anterior

cubital: L. *cubitis,* elbow

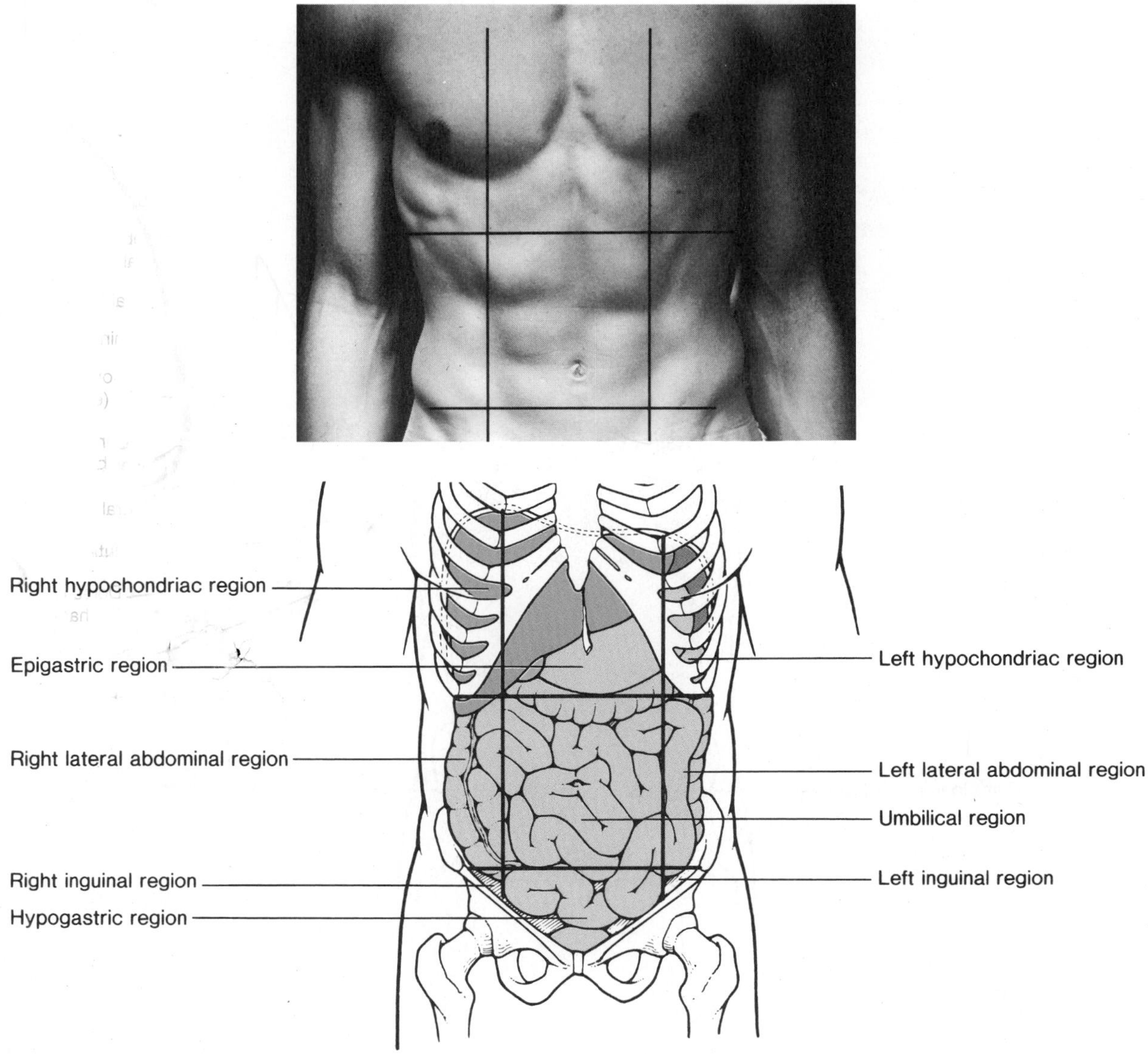

Figure 2.11 The abdomen is subdivided into nine regions. The vertical planes are positioned just medial to the nipples; the upper horizontal plane is positioned at the level of the rib cage; and the lower horizontal plane is even with the upper border of the hipbones.

Table 2.2 Regions of the abdomen and pelvis

Region	Location	Internal organs
Right hypochondriac	Right, upper one-third of abdomen	Gallbladder; portions of liver and right kidney
Epigastric	Upper, median abdomen	Portions of liver, stomach, pancreas, and duodenum
Left hypochondriac	Left, upper one-third of abdomen	Spleen; splenic flexure of colon; portions of left kidney and small intestine
Right lateral	Right, lateral one-third of abdomen	Cecum; ascending colon; hepatic flexure; portions of right kidney and small intestine
Umbilical	Center of abdomen	Jejunum; ileum; portions of duodenum, colon, kidneys, and major abdominal vessels
Left lateral	Left, lateral one-third of abdomen	Descending colon; portions of left kidney and small intestine
Right inguinal	Right, lower one-third of abdomen	Appendix; portions of cecum and small intestine
Pubic (hypogastric)	Lower, center one-third of abdomen	Urinary bladder; portions of small intestine and sigmoid colon
Left inguinal	Left, lower one-third of abdomen	Portions of small intestine and descending and sigmoid colon

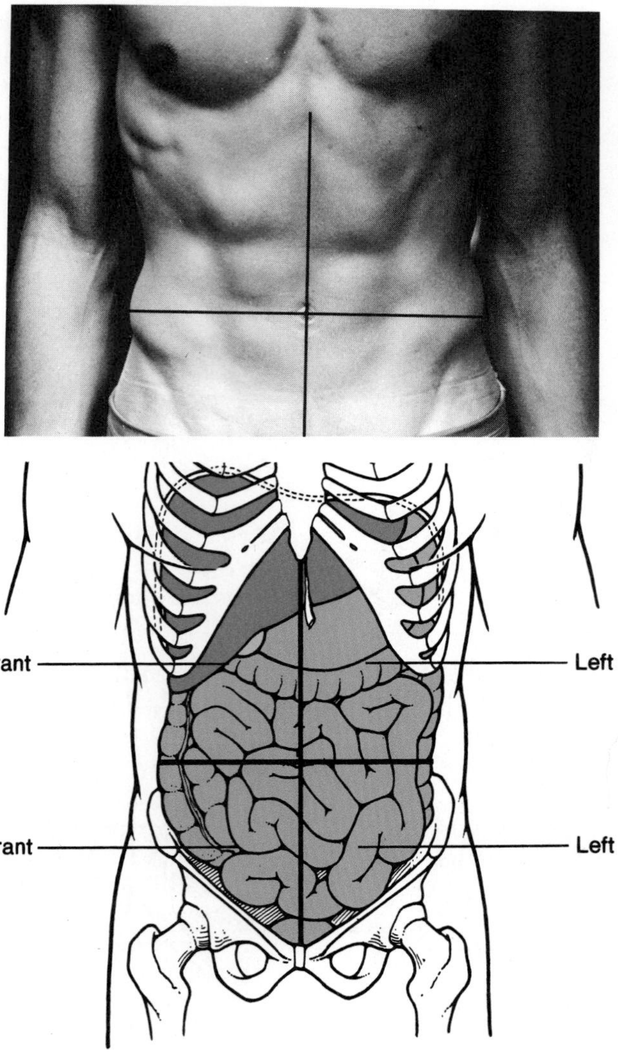

Figure 2.12 A clinical subdivision of the abdomen into four quadrants.

antebrachial region, is known as the **cubital fossa,** an important site for intravenous injections or the withdrawal of blood. The wrist is the flexible junction between the forearm and the hand. The front of the hand is referred to as the **palm,** or **palmar region,** and the back of the hand is called the **dorsum of the hand.** Each hand has five **digits,** or fingers.

Lower Extremity

The lower extremity consists of the **thigh, knee, leg,** and **foot.** The thigh is commonly called the **upper leg** or **femoral region.** The knee joint has two surfaces: the front surface is the **patellar region,** or kneecap; the back of the knee is called the **popliteal** *(pop''lĭ-te'al)* **fossa.** The leg has anterior and posterior crural regions (see fig. 2.10). The **shin** is a prominent bony ridge extending longitudinally along the anterior crural region, and the **calf** is the thickened muscular mass of the posterior crural region. The **ankle** is the junction between the leg and the foot. The **heel** is the back of the foot, and the **sole** of the foot is referred to as the **plantar surface.** The **dorsum of the foot** is the top surface. Each foot has five **digits,** or toes.

popliteal: L. *poples,* ham (hamstring muscles) of the knee

1. Using yourself as a model, identify the various body regions that are depicted in figure 2.10. Which of these regions have surface landmarks that help distinguish their boundaries?
2. In which region of the body are intravenous injections made?
3. What is the distinction between the pelvic, pubic, and perineal regions?
4. Identify the joint between the following regions: the brachium and antebrachium, the pectoral girdle and brachium, the leg and foot, the antebrachium and hand, and the thigh and leg.
5. Explain how a knowledge of the body regions has practical medical application.

Body Cavities and Membranes

For functional and protective purposes, the viscera are compartmental and supported in specific body cavities by connective and epithelial membranes.

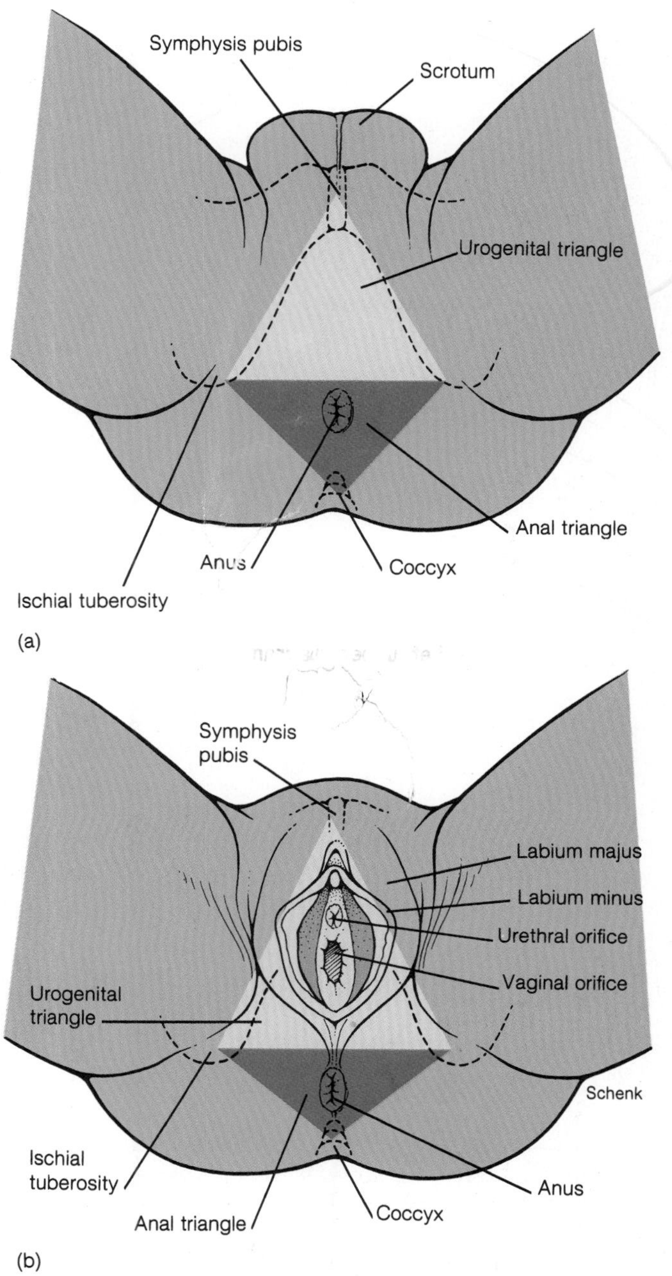

Figure 2.13 A superficial view of the (*a*) male perineum and the (*b*) female perineum. The perineum consists of a urogenital triangle and an anal triangle.

Objective 10. Identify the various body cavities and the organs found in each.

Objective 11. Discuss the types and functions of the various body membranes.

Body Cavities

Body cavities are confined spaces within the body. They contain organs that are protected, compartmentalized, and supported by associated membranes. There are two principal body cavities: the **dorsal body cavity** and the larger **ventral body cavity.** The dorsal body cavity contains the brain and the spinal cord.

During development, the ventral cavity forms from a cavity called the **coelom** *(see'lom)*. The coelom is a body cavity within the trunk, which is lined with a membrane that secretes a lubricating fluid. As development progresses, the coelom is partitioned by the muscular **diaphragm** into an upper **thoracic cavity,** or chest cavity, and a lower **abdominopelvic cavity** (figs. 2.14 and 2.15). Organs within the coelom are collectively called **viscera,** or **visceral** *(vis'er-al)* **organs** (fig. 2.16). Within the thoracic cavity are two **pleural** *(ploo'ral)* **cavities** for the right and left lungs and a **pericardial** *(per''ĭ-kar'de-al)* **cavity** containing the heart. The area between the two lungs is known as the **mediastinum** *(me''de-ah-sti'num)*.

The abdominopelvic cavity consists of an upper **abdominal cavity** and a lower **pelvic cavity.** The abdominal cavity contains the stomach, small intestine, large intestine, liver, gallbladder, pancreas, spleen, and kidneys. The pelvic cavity is occupied by the terminal portion of the large intestine, the urinary bladder, and certain reproductive organs (uterus, uterine tubes, and ovaries in the female; seminal vesicles and prostate gland in the male).

Body cavities serve to confine organs and systems that have related functions. The major portion of the nervous system occupies the dorsal cavity; the principal organs of the respiratory and circulatory systems are in the thoracic cavity; the primary organs of digestion are in the abdominal cavity; and the reproductive organs are in the pelvic cavity. Not only do these cavities house and support various body organs, they also effectively compartmentalize them so that infections and diseases cannot spread from one compartment to another. For example, pleurisy of one lung membrane does not usually spread to the other, and an injury to the thoracic cavity will usually cause only one lung to collapse rather than both.

In addition to the large ventral and dorsal cavities, there are several smaller cavities within the head. The **oral,** or **buccal** *(buk'al)* **cavity** functions primarily with digestion and secondarily with respiration. It contains the teeth and tongue. The **nasal cavity,** which is part of the respiratory system, has two chambers created by a nasal septum. There are two **orbital cavities,** each of which houses an eyeball and its associated muscles, vessels, and nerves. Likewise there are two **middle ear cavities,** which contain ear ossicles (bones) and function with hearing. The location of the cavities within the head is shown in figure 2.17.

Body Membranes

Body membranes are composed of thin layers of connective and epithelial tissue that serve to cover, separate, and support visceral organs and line body cavities. There are two basic types of body membranes: **mucous** *(mu'kus)* **membranes** and **serous** *(se'rus)* **membranes.**

---

buccal: L. *bucca,* cheek or mouth
coelom: Gk. *koiloma,* a cavity
orbital: L. *orbis,* circle

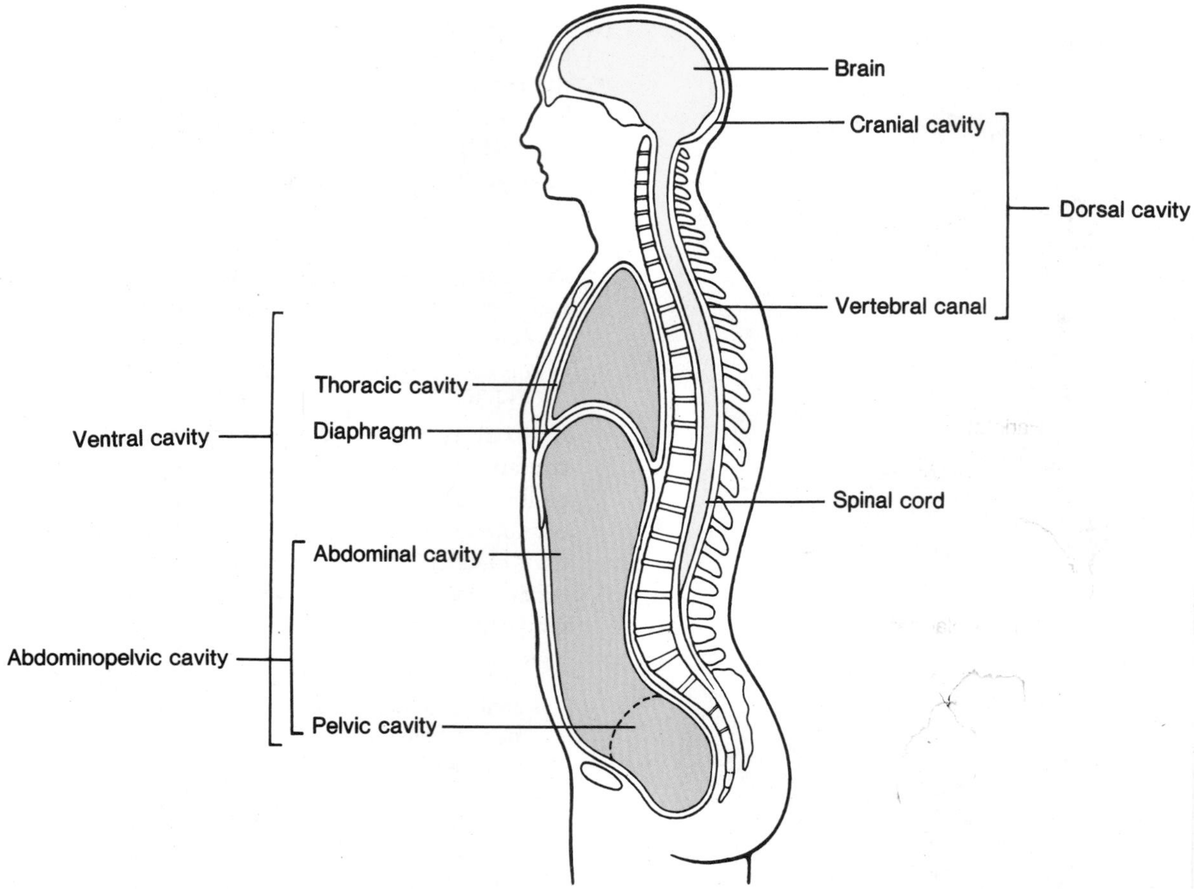

Figure 2.14 A midsagittal (median) section showing the body cavities.

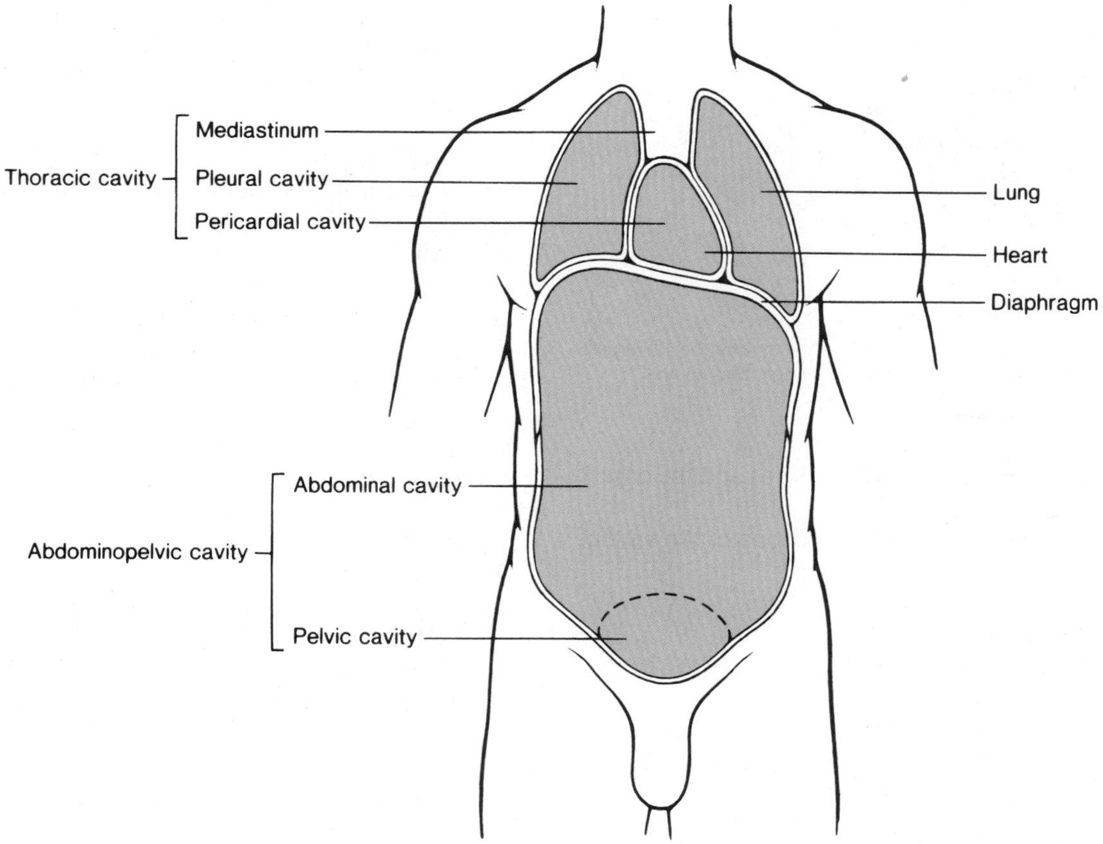

Figure 2.15 An anterior view of body cavities.

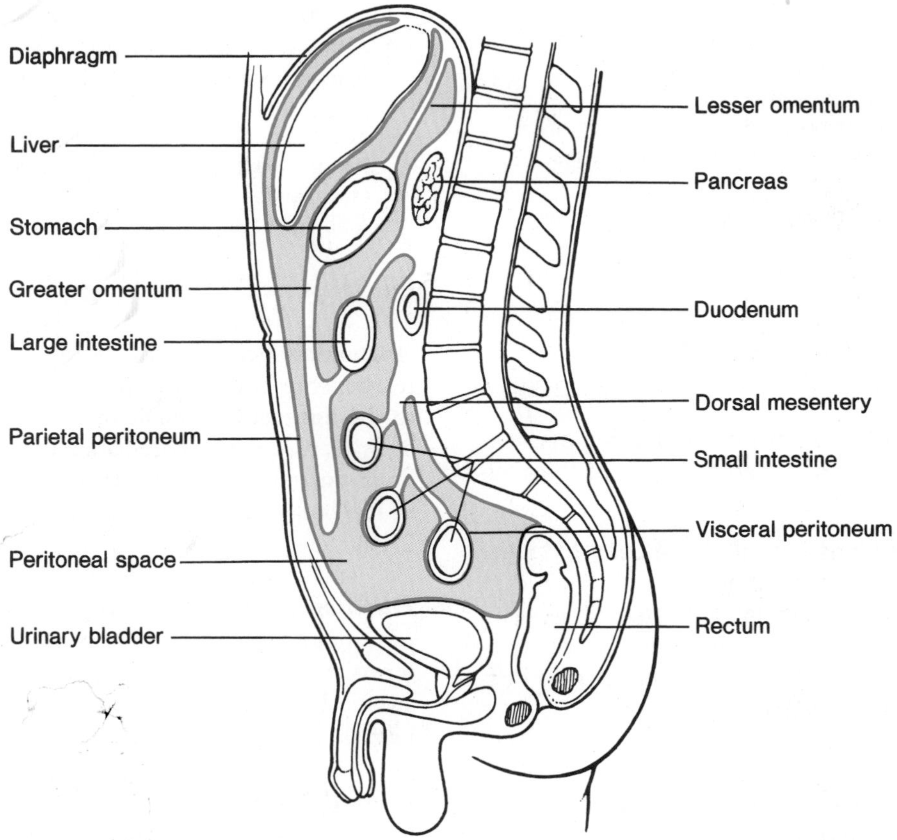

Figure 2.16 Visceral organs of the abdominal cavity and supporting serous membranes.

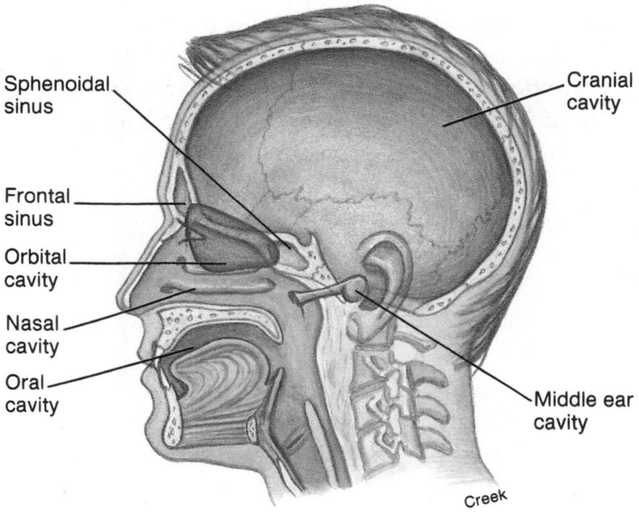

Figure 2.17 Cavities within the head.

Mucous membranes secrete a thick, viscid substance called *mucus.* Generally, mucus lubricates or protects the associated organs where it is secreted. Mucous membranes line various cavities and tubes that enter or exit from the body, such as the oral and nasal cavities and the tubes of the respiratory, reproductive, urinary, and digestive systems.

Serous membranes line the thoracic and abdominopelvic cavities and cover visceral organs, secreting a watery lubricant called *serous fluid.* **Pleurae** are serous membranes associated with the lungs. Each pleura (pleura of right lung and pleura of left lung) has two parts. The **visceral pleura** adheres to the outer surface of the lung, while the **parietal** *(pah-ri'ĕ-tal)* **pleura** lines the thoracic walls and the thoracic surface of the diaphragm. The moistened space between the two pleurae is known as the **pleural cavity.**

Pericardial membranes are the serous membranes of the heart. A thin **visceral pericardium** covers the surface of the heart, and a thicker **parietal pericardium** is the durable covering that surrounds the heart. The space between these two membranes is called the **pericardial cavity.**

Serous membranes of the abdominal cavity are called **peritoneal** *(per''ĭ-to-ne'al)* **membranes.** The **parietal peritoneum** lines the abdominal wall, and the **visceral peritoneum** covers the visceral organs. The **peritoneal cavity** is the potential space within the abdominopelvic cavity between the parietal and visceral peritoneal membranes. The *lesser omentum* and the *greater omentum* are folds of the peritoneum that extend from the stomach to store fat, cushion, and protect visceral organs of the abdominal cavity. Certain organs, such as the kidneys, adrenal glands, and a portion of the pancreas, which are within the abdominal cavity, are positioned behind the parietal peritoneum and are, therefore, said to be *retroperitoneal.* **Mesenteries** *(mes'en-ter''ez)* are double folds of peritoneum that connect the parietal to the visceral peritoneum (see fig. 2.16).

---

peritoneum: Gk. *peritonaion,* stretched over

1. Describe the divisions and boundaries of the ventral body cavity, and list the major organs contained within each division of the ventral body cavity.
2. Distinguish between mucous and serous membranes. List the specific serous membranes of the thoracic and abdominopelvic cavities.
3. Explain the importance of separate and distinct body cavities.

Planes of Reference and Descriptive Terminology

All of the descriptive planes of reference and terms of direction used in anatomy are standardized because of their reference to the body in anatomical position.

Objective 12. Identify the planes of reference used to locate structures within the body.
Objective 13. Describe the anatomical position.
Objective 14. Define and be able to properly use the descriptive and directional terms that have been designated to refer to body structures.

Planes of Reference

In order to visualize and study the structural arrangements of various organs, the body may be sectioned and diagramed according to planes of reference. Three fundamental planes, **midsagittal** *(mid-saj'i-tal),* **coronal,** and **transverse,** are frequently used to depict structural arrangement (fig. 2.18).

A *midsagittal* plane passes lengthwise through the midplane of the body, dividing it into right and left halves. *Sagittal* planes also extend vertically and divide the body into unequal right and left portions. *Coronal,* or *frontal* planes also pass lengthwise and divide the body into front and back portions. *Transverse* planes, also called *horizontal,* or *cross-sectional,* planes, divide the body into superior (upper) and inferior (lower) portions.

The value of the computerized tomographic X-ray (CT) scan is that it displays an image along a transverse plane similar to that which could otherwise be obtained only in an actual section through the body. Prior to the development of this X-ray technique, conventional X-ray images were on a vertical plane, and dimensions of body irregularities were difficult, if not impossible, to ascertain (see fig. 1.19).

Descriptive Terminology

Anatomical Position All terms of direction that describe the relationship of one body part to another are made in reference to the **anatomical position.** In the anatomical position, the body is erect; the feet are parallel to one another and flat on the floor, the eyes are directed forward, and the arms are at the sides of the body with the palms of the hands turned forward (fig. 2.19).

Directional Terms Directional terms are used to locate the position of structures, surfaces, and regions of the body. These terms are always relative to the specimen positioned in the anatomical position. A summary of directional terms is presented in table 2.3.

Clinical Procedures Certain clinical procedures are important in determining anatomical structure and function in a living individual. The more common of these are as follows:

1. **Observation.** Visual inspection for any revealing clinical symptoms such as abnormal skin color, swelling, or rashes. Other observations may include needle marks on the skin, irregular breathing rates, or abnormal behavior.
2. **Palpation.** Applying the fingers with firm pressure to the surface of the body to detect surface landmarks, lumps, tender spots, or pulsations.

Table 2.3 Directional terms for the human body

Term	Definition	Example
Superior (cranial, cephalic)	Toward the head; toward the top	The thorax is superior to the abdomen.
Inferior (caudal)	Away from the head; toward the bottom	The legs are inferior to the trunk.
Anterior (ventral)	Toward the front	The navel is on the anterior side of the body.
Posterior (dorsal)	Toward the back	The kidneys are posterior to the intestine.
Medial	Toward the midline of the body	The heart is medial to the lungs.
Lateral	Away from the midline of the body	The ears are on the lateral sides of the head.
Internal (deep)	Away from the surface of the body	The brain is internal to the cranium.
External (superficial)	Toward the surface of the body	The skin is external to the muscles.
Proximal	Toward the trunk of the body	The knee is proximal to the foot.
Distal	Away from the trunk of the body	The hand is distal to the elbow.
Visceral	Related to internal organs	The lungs are covered by a thin membrane called the visceral pleura.
Parietal	Related to the body walls	The parietal pleura is the inside lining of the thoracic cavity.

coronal: L. *corona,* crown

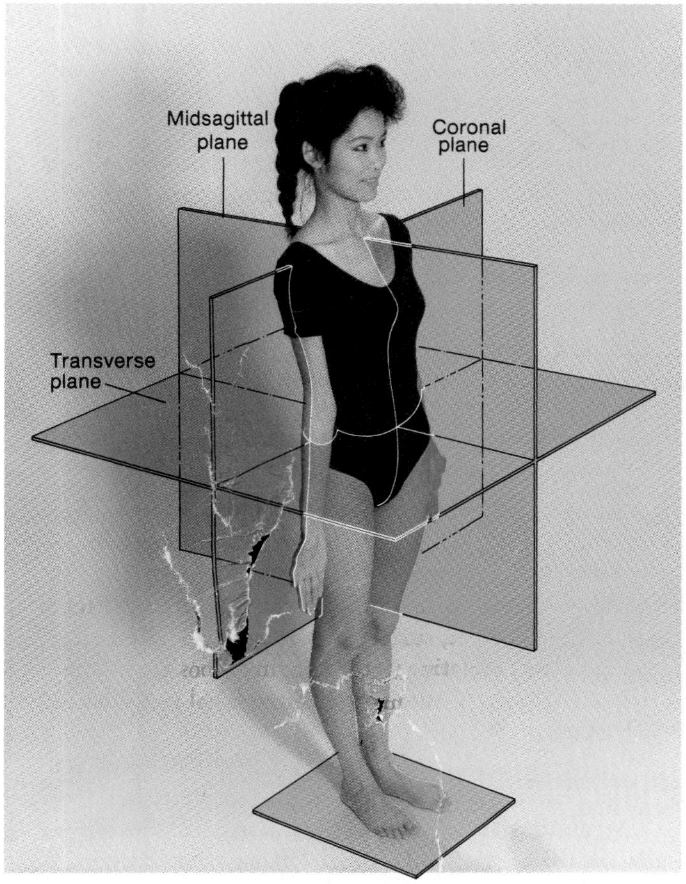

Figure 2.18 Planes of reference through the body.

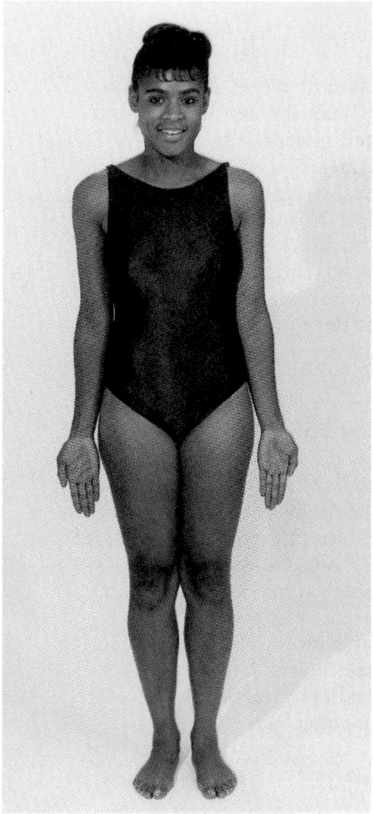

Figure 2.19 In the anatomical position, the body is erect, the feet parallel, the eyes directed forward, and the arms to the sides with the palms directed forward.

3. **Percussion.** Tapping sharply on various locations on the thorax or abdomen to detect resonating vibrations in determining fluid concentrations and organ densities.
4. **Auscultation.** Listening to the sounds that various organs make as they perform their functions (breathing sounds, heartbeats, digestive sounds, etc.).
5. **Reflex response.** Reflex responses are used to determine the condition of parts of the nervous system and some associated organs. One test of a reflex mechanism involves tapping a predetermined tendon with a reflex hammer and observing the response.

---

1. Explain why a transverse plane through the body is important in studying regional anatomy, whereas a transverse plane through an organ is more important in studying specific systems.
2. What does it mean that directional terms are relative and must be used in reference to a body structure or a body in anatomical position?
3. Write a list of statements, similar to the examples in table 2.3, that correctly expresses the directional terms used to describe the relative positions of various body structures.

CLINICAL CASE STUDY ANSWER

In normal anatomy, the thoracic cavity is effectively partitioned from the abdominal cavity by the diaphragm, peritoneum, and pleura. The phenomenon of peritoneal lavage fluid draining out of a tube properly placed in the chest can only be explained by the presence of a defect in the diaphragm. The defect in our patient is likely a traumatic rupture or laceration of the diaphragm caused by a blow to the abdomen. The blow produces sudden, upward pressure against the diaphragm causing its rupture. The absence of bile, blood, etc., in peritoneal lavage fluid does not guarantee the absence of trauma to organs such as the duodenum and pancreas. These organs do not exist within the peritoneal cavity but rather are retroperitoneal, or fully posterior to the peritoneal membrane. This keeps hemorrhage or the leaking of enzymes from these two organs out of the peritoneal space and away from lavage fluid. Other signs must therefore be relied upon, and a high index of suspicion maintained, in order to not miss injury to these organs. Evidence of injury to any of the intra-abdominal organs or the presence of diaphragmatic rupture are indications for emergency laparotomy for repair of the structures involved.

Chapter Summary

I. Classification and Characteristics of Humans
   A. Our scientific name, *Homo sapiens,* means "man the intelligent," and our intelligence is the most distinguishing feature.
   B. Humans belong to the phylum Chordata because of the presence of a notochord, a dorsal hollow nerve cord, and pharyngeal pouches during the embryonic stage of development.
   C. Humans are mammals and as such have mammalian characteristics.
   D. Humans are also classified within the order Primates.
   E. Humans are the sole members of the family Hominidae.
   F. Some of the characteristics of humans are size and development of brain, bipedal locomotion, opposable thumb, vocal structures, and stereoscopic vision.

II. Body Organization
   A. Cells are the fundamental structural and functional components of life.
   B. Tissues are aggregations of similar cells that perform specific functions.
   C. An organ is a structure consisting of two or more tissues, which performs a specific function.
   D. A body system is composed of a group of organs that function together.

III. Anatomical Terminology
   A. The majority of anatomical terms are descriptive and are of Greek or Latin derivation.
   B. Learning the basic prefixes and suffixes greatly facilitates the understanding and retention of the material.
   C. Anatomy is a foundation science for all of the medical and paramedical fields.

IV. Body Regions
   A. The head is divided into a facial region including the eyes, nose, and mouth, and a cranial region that covers and supports the brain.
   B. The neck is called the cervical region and functions to support the head and permit movement.
   C. The front of the thorax is subdivided into two mammary regions and one sternal region.
   D. On either side of the thorax is an axillary fossa and a lateral pectoral region.
   E. The abdomen may be divided into nine anatomical regions or into four quadrants.
   F. Regional names are given to the upper extremity and include the shoulder, brachium, antebrachium, and manus.
   G. Regional names for the lower extremity include the thigh, leg, and foot.

V. Body Cavities and Membranes
   A. The dorsal cavity contains the cranial and spinal cavities.
   B. The ventral cavity is composed of thoracic and abdominopelvic cavities.
   C. Other body cavities include the oral, orbital, nasal, and middle ear cavities.
   D. The body has two principal types of membranes: mucous membranes, which secrete protective mucus, and serous membranes, which line the ventral cavities and cover visceral organs.
   E. There are three categories of serous membranes: plural membranes, pericardial membranes, and peritoneal membranes.

VI. Planes of Reference and Descriptive Terminology
   A. Midsagittal, coronal, and transverse planes divide the body into regions.
   B. In the anatomical position, the subject stands erect and faces forward with arms at the sides and palms turned forward.
   C. Directional terms are used to describe the location of one body part with respect to another part in anatomical position.
   D. Clinical procedures include observation, palpation, percussion, auscultation, and reflex response.

Review Activities

Objective Questions

1. Which of the following is *not* a principal chordate characteristic?
   (a) dorsal hollow nerve cord
   (b) distinct head, thorax, and abdomen
   (c) notochord
   (d) pharyngeal pouches
2. Prehensile hands, digits modified for grasping, and large, well-developed brains are structural characteristics of the grouping of animals referred to as
   (a) primates. (c) mammals.
   (b) humans. (d) chordates.
3. Layers or aggregations of similar cells that perform specific functions are called
   (a) organelles. (c) organs.
   (b) tissues. (d) glands.
4. Filtration and maintenance of the volume and the chemical composition of the blood are functions of the
   (a) urinary system.
   (b) lymphatic system.
   (c) circulatory system.
   (d) endocrine system.
5. The cubital fossa is located in the
   (a) thorax.
   (b) upper extremity.
   (c) abdomen.
   (d) lower extremity.
6. Being composed of more than one tissue type, the skin is considered a(an)
   (a) composite tissue. (c) organ.
   (b) system. (d) organism.
7. Which of the following is *not* a fundamental plane?
   (a) coronal (c) vertical
   (b) transverse (d) midsagittal
8. The external genitalia (reproductive organs) are located in the
   (a) popliteal fossa.
   (b) perineum.
   (c) hypogastric region.
   (d) epigastric region.
9. The region of the thoracic cavity between the two pleural cavities is called the
   (a) midventral space.
   (b) mediastinum.
   (c) ventral cavity.
   (d) median cavity.
10. The abdominal region superior to the umbilical region that contains most of the stomach is the
    (a) hypochondriac region.
    (b) epigastric region.
    (c) diaphragmatic region.
    (d) inguinal region.
11. Regarding serous membranes, which of the following word pairs is *incorrect?*
    (a) visceral pleura—lung
    (b) parietal peritoneum—body wall
    (c) mesentery—heart
    (d) parietal pleura—body wall
    (e) visceral peritoneum—intestines
12. The sectional plane that divides the body into anterior and posterior portions is
    (a) sagittal.
    (b) transverse.
    (c) coronal.
    (d) cross-sectional.
13. In the anatomical position the
    (a) arms are extended away from the body.
    (b) palms of the hands face posteriorly.
    (c) body is erect and palms face anteriorly.
    (d) body is in a fetal position.
14. Listening to sounds that functioning visceral organs make is
    (a) percussion. (c) audiotation.
    (b) palpation. (d) auscultation.

Essay Questions

1. Discuss the characteristics an animal must possess to be classified as a chordate; a mammal; and a human.

2. Describe the relationship of the notochord to the vertebral column, the pharyngeal pouches to the ear, and the dorsal hollow nerve cord to the central nervous system.
3. Which group of animals are our "closest relatives"? What anatomical characteristics do we have in common with them?
4. Identify the four levels of complexity that constitute the human body.
5. Outline the systems of the body, and identify the major organs composing each.
6. Identify which major region of the body contains each of the following structures or minor regions: scapular region, carotid triangle, zygomatic region, brachium, popliteal fossa, lumbar region, cubital fossa, hypochondriac region, perineum, axillary fossa.
7. Diagram the thoracic cavity showing the relative position of the pericardial cavity, pleural cavities, and mediastinum.
8. What is a serous membrane? Explain how the names of the serous membranes differ with each body cavity.
9. What is meant by "anatomical position"? Why is the anatomical position important in studying anatomy?
10. Define the terms *palpation, percussion,* and *auscultation.*

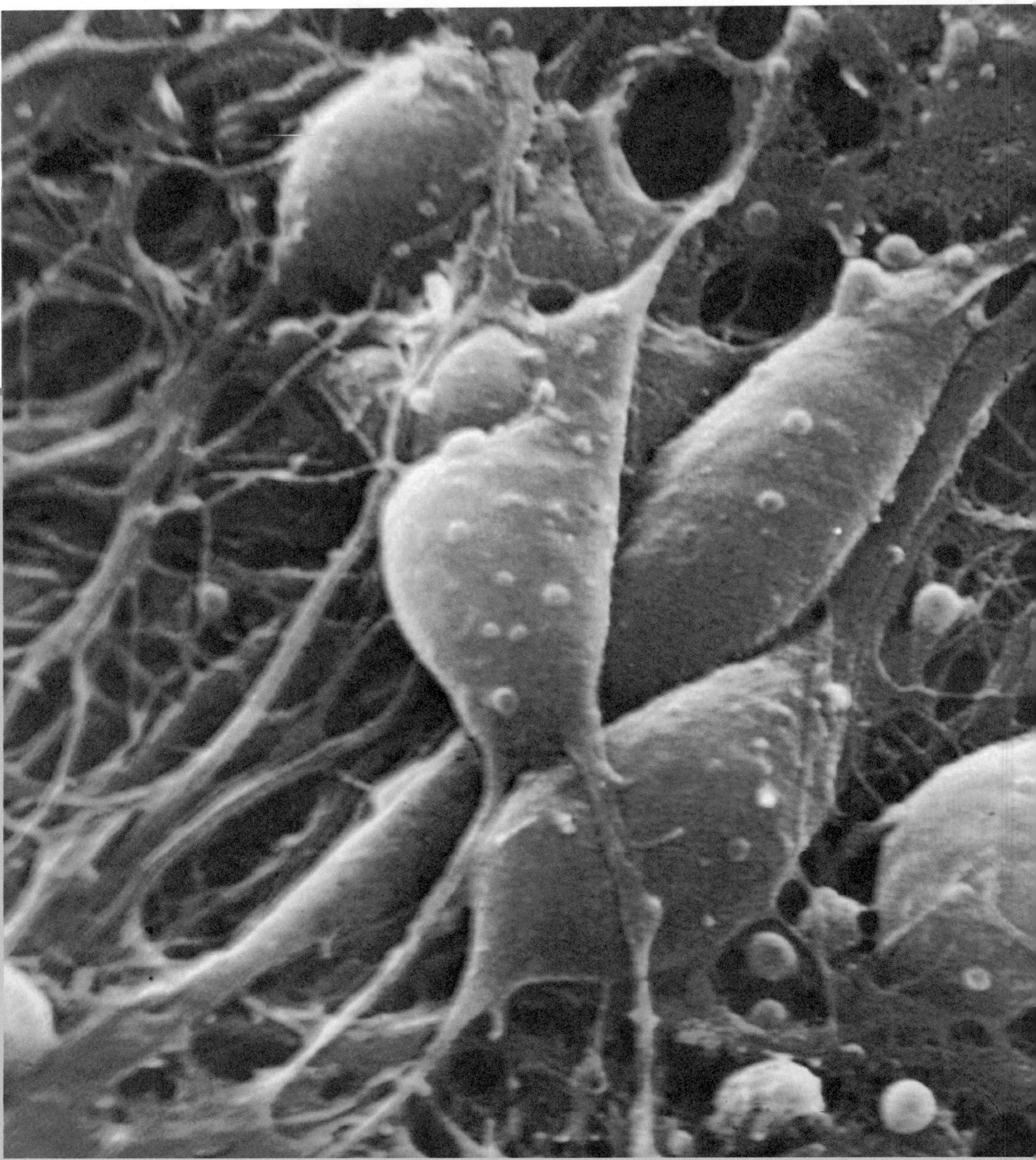

A color-enhanced scanning electron micrograph (SEM) of neurons from nervous tissue within the cerebral cortex. The abundance of neuron cell bodies within the cerebrum account for the gray color of the convoluted cerebral cortex.

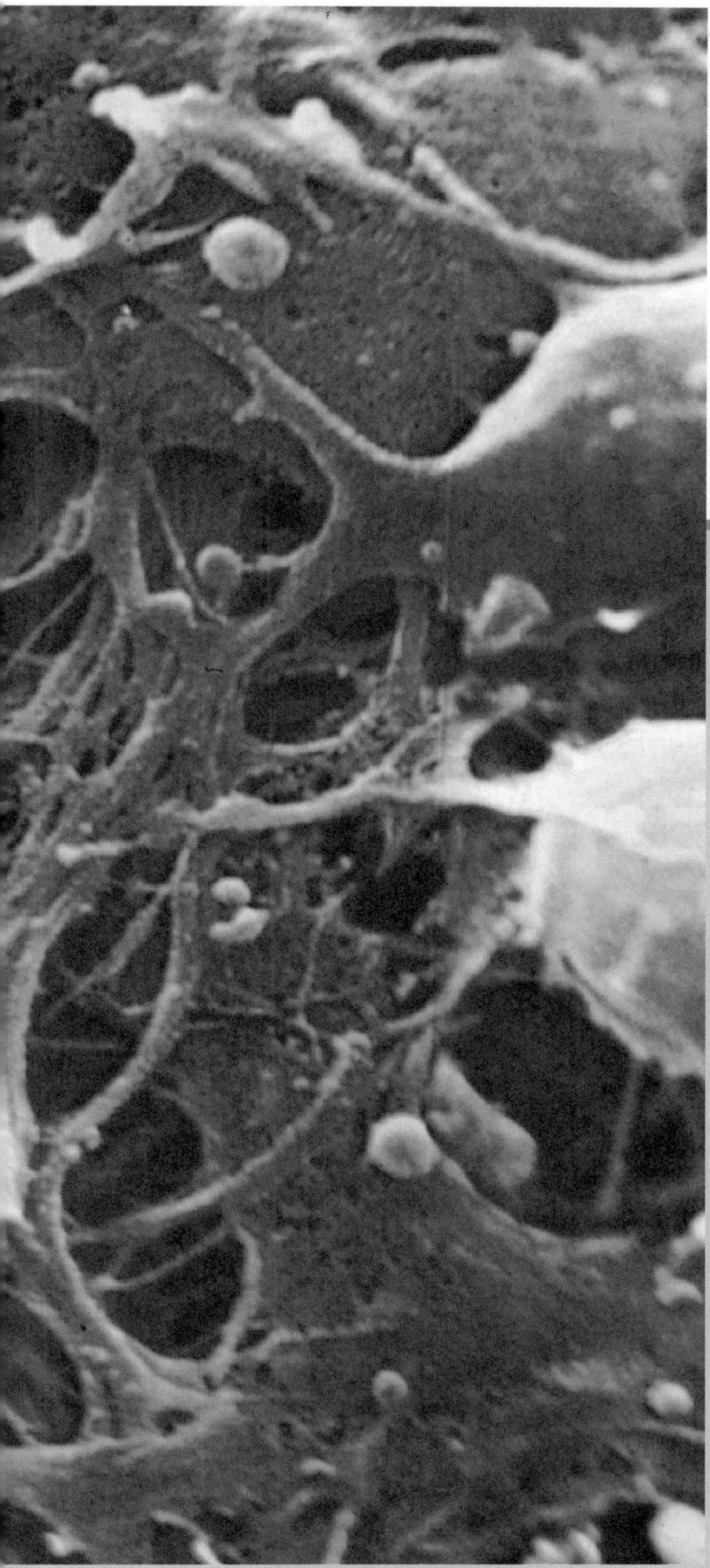

Microscopic Structure of the Body

T he microscopic organization of the human body at the cellular and the histological levels is discussed in the chapters in unit 3. The structure and function of the body depend on interactions between its parts, and hence a thorough understanding of the cellular and histological organization of the body is vital in learning anatomy.

This unit includes:

3

Cellular Anatomy

4

Histology

CHAPTER 3

Cellular Anatomy

Outline and Concepts

Introduction to Cellular Anatomy

The cell is the fundamental structural and functional unit of the body. Although diverse, cells have basic structural similarities and each metabolizes in order to stay alive.

Objective 1. Define the terms *cell, metabolism,* and *cytology.*
Objective 2. Using examples, explain how cells vary from one another and how the structure of a cell determines its function.

The Study of Cells

Human anatomy is concerned with the structure of the human body and the relationship of its parts. The body is a masterpiece of organization. The **cell** is the basis of this organization and, as such, it is called the functional unit. As discussed in chapter 2, cellular organization forms tissues, whose organization in turn forms the organs, which in turn form systems. In order for the organs and systems to function properly, cells must function properly. Cellular function is referred to as **metabolism.** In order for cells to remain alive and metabolize, certain requirements must be met: Each cell must have access to nutrients and oxygen and be able to eliminate wastes. In addition, a constant, protective environment must be maintained. All of these requirements are achieved through organization.

Cells were first observed more than three hundred years ago by the English scientist Robert Hooke. He observed the cell walls and the boxlike cavities in slices of cork and leaves and called them "little boxes or cells." Improvements in the microscope enabled further investigations and the formulation of the **cell theory,** which states that all living organisms are composed of one or more cells. The cell theory was unified in 1838 and 1839 by two German biologists, Matthias Schleiden and Theodor Schwann. Their work laid the foundation for a new science called *cytology,* which is concerned with the structure and function of cells.

A knowledge of the cellular level of organization is important for understanding basic body processes such as cellular respiration, protein synthesis, mitosis, and meiosis. An understanding of cellular structure enables one to grasp the concept of tissue, organ, and system levels of functional body organization. Furthermore, many body dysfunctions and diseases originate in the cells. Although cellular structure and function have been investigated for many years, there remains a vast amount that is not known about cells. There are several complex diseases whose *etiologies,* or causes, are unknown. Scientists are seeking why and how the body ages. Answers to these questions will come only through a better understanding of cellular structure and function.

Advancements in microscopic technology have immensely aided the science of cytology. In a new process called *microtomography,* the capabilities of electron microscopy are combined with those of computerized tomography (CT) scanning to produce high-magnification, three-dimensional, microtomographic images of living cells. With this technology, living cells can be observed as they move, grow, and divide. The clinical applications of this are immense as scientists can observe living, diseased (including cancerous) cells and how they respond to various drug treatments.

Cellular Diversity

It is amazing that from a single cell, the fertilized egg, hundreds of kinds of cells arise, composing the estimated 60 trillion to 100 trillion cells that make up an adult human. Cells vary greatly in size and shape, and their basic structure is often indicative of their function. The smallest cells are visible only through a high-powered microscope, whereas the largest, an egg cell (ovum), is barely visible to the unaided eye. The sizes of cells are measured in micrometers (μm)—one micrometer equals 1/1000th of a millimeter. Using this basis of comparison, an ovum is about 140 μm in diameter and a red blood cell is about 7.5 μm in diameter. The most common type of white blood cell varies in size from 10–12 μm in diameter. Although still microscopic, some cells can be extremely long, such as a nerve cell (neuron), which may traverse the entire length of an appendage and be about a meter in length.

Although a typical diagram of a cell depicts it as round or cube-shaped, the shapes of cells are actually highly variable. They can be flattened, oval, elongate, stellate, columnar, and so on (fig. 3.1). The shape of a cell is frequently an indication of its function. An oval red blood cell is adapted to transport oxygen and carbon dioxide. Thin, flattened cells may be bound together to form selectively permeable membranes. An irregularly shaped cell, such as a neuron, has a tremendous ratio of surface area to volume, which is ideal for receiving and transmitting stimuli.

Some cells have smooth surfaces that permit easy movement of substances across their surface, whereas others have marked depressions or grooves to facilitate absorption. Some support structures like cilia, flagella, microvilli, or gelatinous coats, which assist movement or provide adhesion. Regardless of the sizes or shapes of cells, they all have structural modifications that serve functional purposes.

---

1. Why is the cell considered the basic structural and functional unit of the body? List some aspects of cells that would be of concern to a cytologist.
2. What conditions are necessary for metabolism to occur?
3. Why is structural organization essential to a multicellular organism?
4. Construct a table to compare the structure and function of several kinds of cells.

Chemistry of a Cell

All tissues and organs are composed of cellular structures of basically the same chemicals. The most important inorganic substances in the body include water, acids, bases, and salts. The most important organic substances in the body include proteins, carbohydrates, and lipids.

---

metabolism: Gk. *metabole,* change
cytology: L. *cella,* small room; Gk. *logos,* study of
etiology: L. *aitia,* cause; Gk. *logos,* study of

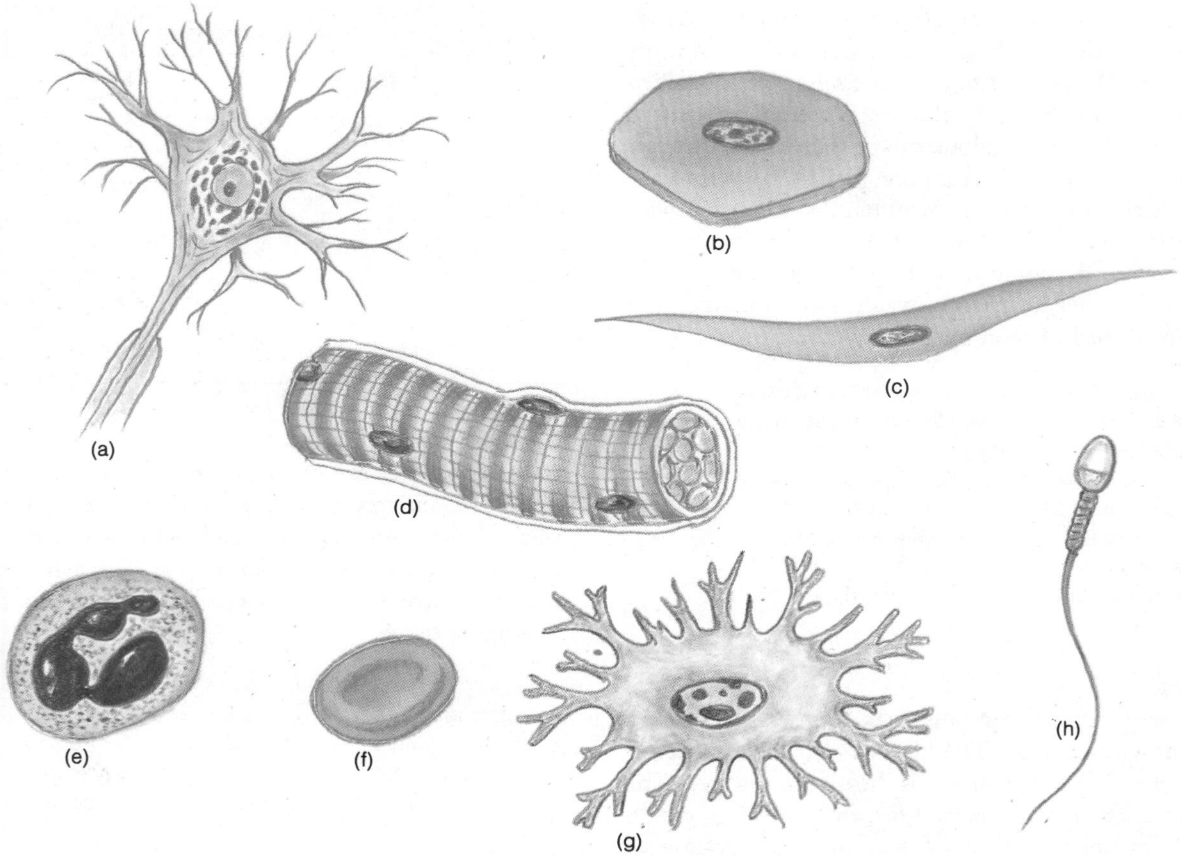

Figure 3.1 Samples of the various shapes of cells within the body. The cells are not drawn to scale. (*a*) A nerve cell, showing a cell body surrounded by numerous dendritic extensions and a portion of the axon extending below; (*b*) an epithelial cell from the lining of the lumen (hollow tube) of a blood vessel; (*c*) a smooth muscle cell from the intestinal wall; (*d*) a skeletal muscle cell; (*e*) a white blood cell; (*f*) a red blood cell; (*g*) a bone cell; and (*h*) a human sperm cell.

Objective 3. List the common chemical elements found within cells.
Objective 4. Differentiate between inorganic and organic compounds, and give examples of each.
Objective 5. Explain the importance of water in maintaining body homeostasis.
Objective 6. Differentiate between proteins, carbohydrates, and lipids.

To understand cellular structure and function, one must have a knowledge of basic cellular and general body chemistry. All of the processes that occur in the body comply with principles of chemistry. Furthermore, many of the dysfunctions of the body have a chemical basis.

Elements and Compounds

An *element* is a basic chemical substance. Four elements compose over 95% of the body. These elements and their percent of body weight are oxygen (O) 65%, carbon (C) 18%, hydrogen (H) 10%, and nitrogen (N) 3%. Additional common elements within the body include calcium (Ca), potassium (K), sodium (Na), phosphorus (P), magnesium (Mg), and sulfur (S). These various elements are linked together to form molecules and compounds.

A few elements exist separately in the body, but most are linked with others to form either *inorganic* or *organic compounds.*

Table 3.1 A comparison of the compounds within adult males and females (expressed as the percentage of body weight)

Substance	Male	Female
Water	62	59
Protein	18	15
Lipid	14	20
Carbohydrates	1	1
Other (electrolytes, nucleic acids)	5	5

A *compound* is a substance composed of two or more elements joined by chemical bonds. Organic compounds contain carbon and are formed by living organisms. Inorganic compounds generally lack carbon and are not formed by living organisms. Water and electrolytes (acids, bases, and salts) are examples of inorganic compounds. Proteins, carbohydrates, and lipids are organic compounds. Table 3.1 compares the percentage of inorganic and organic compounds within adult males and females.

Water

Water is the most abundant compound within cells and in the extracellular environment. Water generally occurs within the body as a homogeneous mixture of two or more compounds called a *solution.* In this condition, the water is the *solvent,* or the liquid portion of the solution, and the *solutes* are dissolved substances suspended within the solution. Water is an excellent solvent through which many solutes can be transported through the cell membrane of a cell or from one part of the cell to another. Water is also important in maintaining a constant cellular temperature and body temperature since it absorbs and releases heat slowly. Evaporative cooling through the skin also involves water. Another function of water is as a *reactant* in the breakdown *(hydrolysis)* of food material in digestion.

Dehydration is a condition in which fluid loss exceeds fluid intake with a resultant decrease in the volume of cellular and intracellular fluids. Rapid dehydration through vomiting, diarrhea, or excessive sweating can cause serious medical problems by impairing cellular function. Infants are especially vulnerable because their fluid volume is so small. They can die from dehydration caused by diarrhea within a matter of hours.

Electrolytes

Electrolytes are inorganic compounds that, when dissolved in water, break down into ions. The kind of electrolyte is determined by the ions it yields when it is dissolved in water. The three classes of electrolytes are *acids, bases,* and *salts,* all of which are important for normal cellular function. The functions of ions include the control of water movement through cells; the maintenance of normal acid-base *(pH)* balance; the conduction of electrical currents; the development of certain kinds of tissues, such as bones and teeth; and the normal functioning of blood, muscles, and nerves.

The three kinds of electrolytes are summarized in table 3.2.

Proteins

Proteins are nitrogen-containing, organic compounds composed of amino acid molecules. An amino acid is an organic compound that contains an amino group ($-NH_2$) and a carboxyl group ($-COOH$). Proteins are the most abundant of the organic compounds and may exist by themselves or may be *conjugated* (joined) with other compounds, such as with nucleic acids (RNA or DNA) to form nucleoproteins, with carbohydrates to form glycoproteins, or with lipids to form lipoproteins.

Proteins form the framework of the cell. The cell (plasma) membrane is composed primarily of protein and lipid molecules. Cellular growth, repair, and division depend upon the availability of proteins. *Enzymes* are a type of protein that serve to catalyze chemical reactions within cells. Many *hormones* are a specialized group of messenger and regulator proteins produced by endocrine glands. Proteins, under certain conditions, may even be metabolized to supply cellular energy.

Table 3.2 Kinds of electrolytes

	Characteristic	Examples
Acid	Ionizes to release hydrogen ions (H^+)	Carbonic acid, hydrochloric acid, acetic acid, phosphoric acid
Base	Ionizes to release hydroxyl ions (OH^-) that combine with hydrogen ions	Sodium hydroxide, potassium hydroxide, magnesium hydroxide, aluminum hydroxide
Salt	Substance formed by the reaction between an acid and a base	Sodium chloride, aluminum chloride, magnesium sulfate

Carbohydrates

Carbohydrates are organic compounds that contain carbon, hydrogen, and oxygen, with a 2:1 ratio of hydrogen to oxygen. Carbohydrates include the starches and sugars and are the most abundant source of cellular energy. Excessive carbohydrate intake is converted to *glycogen* (animal starch) or to *fat* for storage in adipose tissue.

If a person is deprived of food, the body uses the glycogen and fat reserves first and then metabolizes the protein within the cells. The gradual destruction of the cellular protein accounts for the lethargy, extreme emaciation, and ultimate death of starvation victims.

Lipids

Lipids are a third group of important organic compounds found in cells. Lipids include both fats and fat-related substances such as phospholipids and cholesterol. Fats are important in building cell parts and supplying metabolic energy. Fats also serve to protect and insulate various parts of the body. Phospholipids and protein molecules make up the cell membrane and play an important role in regulating which substances enter or leave a cell.

Lipids, like carbohydrates, are composed of carbon, hydrogen, and oxygen atoms. Lipids, however, contain a smaller proportion of oxygen than do carbohydrates.

Table 3.3 summarizes the locations and functions of inorganic and organic substances within cells.

1. List the four most abundant elements in the body, and state their relative percentages.
2. What is a compound? What two kinds of compounds exist in the body and what distinguishes each?
3. List the functions of water as related to cells, and define *solvent* and *solute.*
4. Explain how electrolytes are important in maintaining homeostasis within a cell.
5. Describe a protein, and discuss the functions of proteins within a cell. Discuss how proteins differ from carbohydrates and lipids.

hydrolysis: Gk. *hydor,* water; *lysis,* dissolution
electrolyte: L. *electrum,* amber; Gk. *lysis,* a loosening
acid: L. *acidus,* sour
solution: L. *solvere,* loosen or dissolve
protein: Gk. *proteios,* of the first quality
hormone: Gk. *hormon,* setting in motion

lipid: Gk. *lipos,* fat

Table 3.3 Summary of the chemical substances of cells

Substance	Location in cell	Function
Water	Throughout	Dissolves, suspends, and ionizes materials; helps to regulate temperature
Electrolytes	Throughout	Establish osmotic gradients, pH, and membrane potentials
Proteins	Membranes, cytoskeleton, ribosomes, enzymes	Give structure, strength, contractility; catalyze, buffer
Lipids	Membranes, Golgi apparatus, inclusions	Reserve energy source; shape, protect, insulate
Carbohydrates	Inclusions	Preferred fuel for metabolic activity
Nucleic acids		
DNA	Nucleus, in chromosomes and genes	Controls cell activity
RNA	Nucleolus, cytoplasm	Transfers instruction; transports amino acids
Trace materials		
Vitamins	Cytoplasm, nucleus	Work with enzymes in metabolism
Hormones	Cytoplasm, nucleus	Work with enzymes; activate or deactivate genes
Metals	Cytoplasm, nucleus	Synthesize various substances, such as insulin; maturation of blood cells

Structure of a Cell

The cell membrane separates the interior of the cell from the extracellular environment. Substances that enter or leave the cell are moved across the cell membrane. Most of the metabolic activities occur within the cytoplasmic organelles. The nucleus functions in protein synthesis and cell reproduction.

Objective 7. Describe the components of a cell.
Objective 8. Describe the composition and structure of the cell membrane, and relate the structure to the functions of the membrane.
Objective 9. Distinguish between passive and active transport, and describe the different ways whereby each is accomplished.
Objective 10. Describe the structure and function of the endoplasmic reticulum, ribosomes, Golgi apparatus, lysosomes, and mitochondria.
Objective 11. Describe the structure and function of the nucleus.

As the basic functional unit of the body, each cell is a highly organized molecular factory. As previously discussed, cells come in a great variety of shapes and sizes. This variation, which is also apparent in the subcellular structures within different cells, reflects the variation of the functions of different cells in the body. All cells, however, share certain characteristics—such as, they are surrounded by a cell membrane—and cells possess most of the structures listed in table 3.4. Thus, although no single cell can be considered "typical," the general structure of cells can be shown by a single illustration (fig. 3.2).

For descriptive purposes, a cell can be divided into three principal parts:

1. **Cell (plasma) membrane.** The outer, selectively permeable cell membrane gives form to the cell and separates the cell's internal structures from the extracellular environment.
2. **Cytoplasm** and **organelles** *(or''gah-nelz')*. The cytoplasm is the content of a cell between the nucleus and the cell membrane. Organelles are minute membrane-bound structures within the cytoplasm of a cell that are concerned with specific functions.
3. **Nucleus.** The nucleus is a large, generally spheroid body within a cell that contains the DNA, or genetic material, of a cell.

Cell Membrane

The extremely thin **cell (plasma) membrane** is composed of phospholipid and protein molecules. Its thickness ranges from 65 to 100 angstroms (Å), or less than a millionth of an inch. The structure of the cell membrane is not fully understood, but most cytologists believe that it is composed primarily of phospholipid molecules with larger globular proteins scattered within the lipid layer (fig. 3.3). Such a structure suggests that the membrane is not static but has considerable movement. Minute openings, or pores, ranging between 7 and 10 Å in diameter extend through the membrane.

The two most important functions of the cell membrane are to enclose the components of the cell and to regulate the passage of substances into and out of the cell. A highly selective exchange of substances occurs across the membrane boundary, involving several types of passive or active processes. The kinds of movement across a cell membrane are summarized in table 3.5 and illustrated in figure 3.4.

The permeability of the cell membrane depends on the following factors:

1. **Structure of cell membrane.** Although cell membranes on all cells are composed of phospholipids, there is evidence that their thickness and structural arrangement vary considerably—both of which could affect permeability.
2. **Size of molecules.** Macromolecules, such as certain proteins, are not allowed into the cell. Water and amino acids are small molecules and can readily pass through the cell membrane.
3. **Ionic charge.** The protein portion of the cell membrane may have an ionic charge. Ions with an opposite charge

plasma: Gk. *plasma,* to form or mold

nucleus: L. *nucleus,* kernel or nut

Table 3.4 Structure and function of cellular components

Component	Structure	Function
Cell (plasma) membrane	Membrane composed of phospholipid and protein molecules	Gives form to cell and controls passage of materials in and out of cell
Cytoplasm	Fluid, jellylike substance in which organelles are suspended	Serves as matrix substance in which chemical reactions occur
Endoplasmic reticulum	System of interconnected membrane-forming canals and tubules	Supporting framework within cytoplasm; transports materials and provides attachment for ribosomes
Ribosomes	Granular particles composed of protein and RNA	Synthesize proteins
Golgi apparatus	Cluster of flattened, membranous sacs	Synthesizes carbohydrates and packages molecules for secretion; secretes lipids and glycoproteins
Mitochondria	Membranous sacs with folded inner partitions	Release energy from food molecules and transform energy into usable ATP
Lysosomes	Membranous sacs	Digest foreign molecules and worn and damaged cells
Peroxisomes	Spherical membranous vesicles	Contain certain enzymes; form hydrogen peroxide

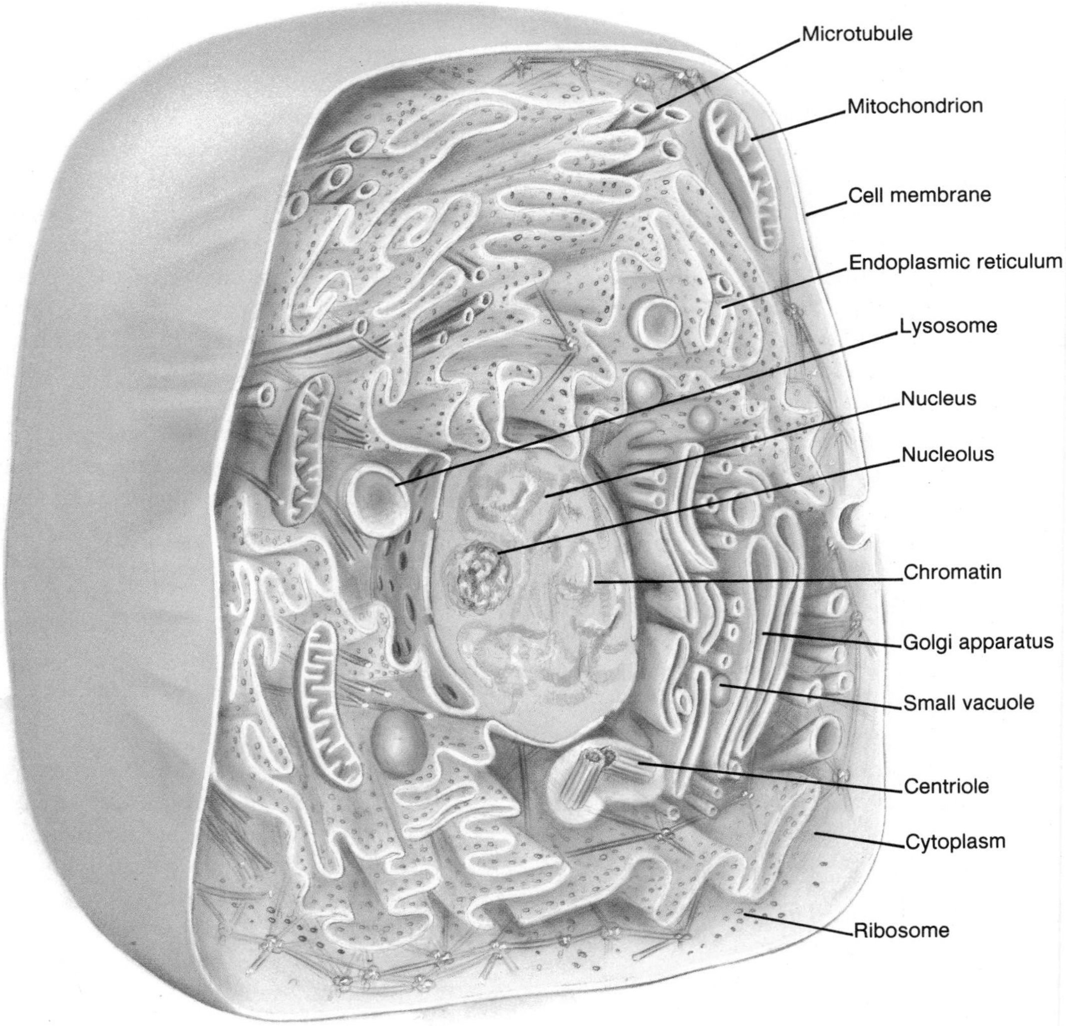

Figure 3.2 A generalized cell and the principal organelles.

Table 3.4 *Continued*

Component	Structure	Function
Centrosome	Nonmembranous mass of two rodlike centrioles	Helps organize spindle fibers and distribute chromosomes during mitosis
Vacuoles	Membranous sacs	Store and excrete various substances within the cytoplasm
Fibrils and microtubules	Thin, hollow tubes	Support cytoplasm and transport materials within the cytoplasm
Cilia and flagella	Minute cytoplasmic extensions from cell	Move particles along surface of cell or move cell
Nuclear membrane	Membrane surrounding nucleus, composed of protein and lipid molecules	Supports nucleus and controls passage of materials between nucleus and cytoplasm
Nucleolus	Dense, nonmembranous mass composed of protein and RNA molecules	Forms ribosomes
Chromatin	Fibrous strands composed of protein and DNA molecules	Controls cellular activity for carrying on life processes

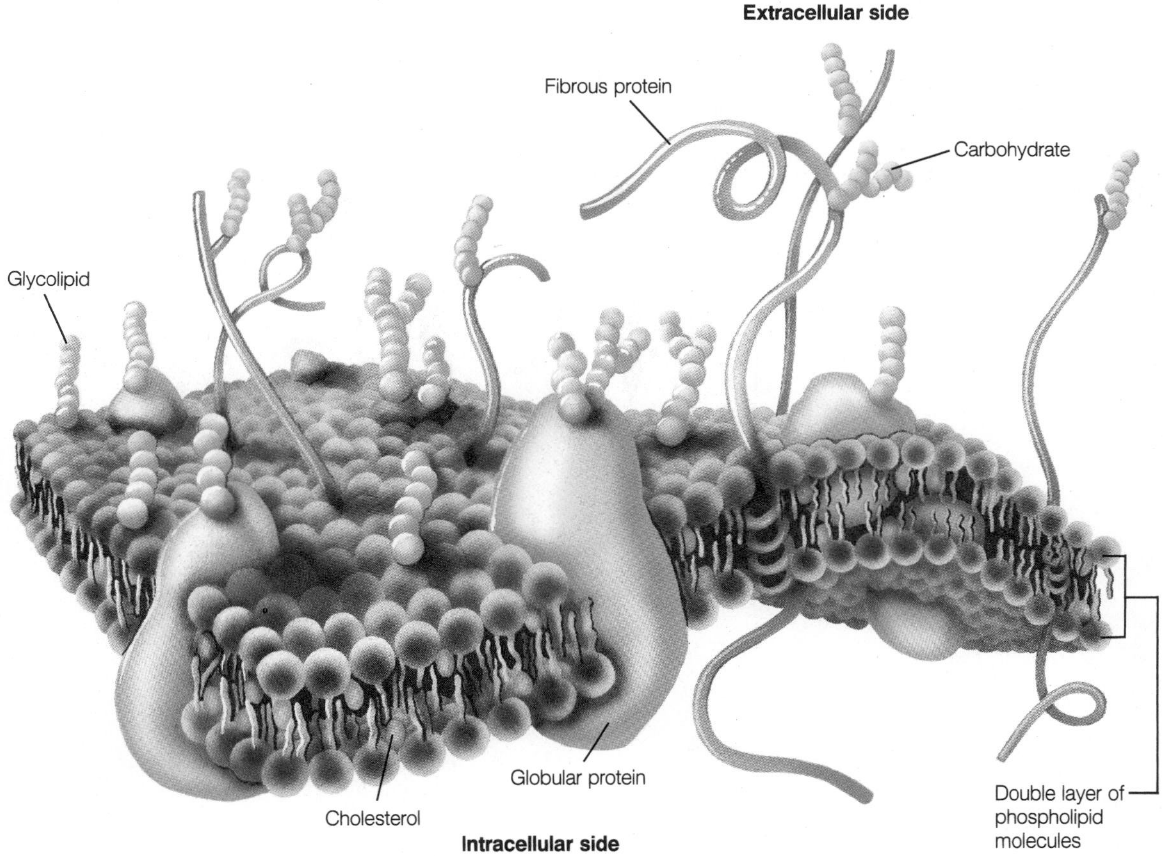

Figure 3.3 The cell membrane consists of a double layer of phospholipids. Proteins may completely or partially span the membrane. Carbohydrates are attached to the outer surface.

are attracted to and readily pass through the membrane, whereas those with a similar charge are repelled.

4. **Lipid solubility.** Substances that are easily dissolved in lipids pass into the cell with no problem since a portion of the cell membrane is composed of lipid material.
5. **Presence of carrier molecules.** Specialized carrier molecules positioned within the cell membrane are capable of attracting and transporting substances across the membrane regardless of size, ionic charge, or lipid solubility.

Cell membranes of certain cells are highly specialized to facilitate specific functions (fig. 3.5). The columnar cells lining the lumen of the intestinal tract, for example, have numerous fine projections, called **microvilli,** that aid in the absorptive process of digestion. A single columnar cell may have as many as 3,000 microvilli on the exposed portion of the cell membrane, and a square millimeter of surface area may contain over 200 million microvilli.

---

microvilli: Gk. *mikros,* small; *villus,* tuft of hair

Table 3.5 Kinds of movement through cell membranes

Processes	Characteristics	Energy source	Example
Diffusion	Passive movement of molecules from regions of higher concentration toward regions of lower concentration	Molecular motion	Exchange of respiratory gases in lungs
Facilitated diffusion	Carrier substances are used to speed process	Carrier energy and molecular motion	Glucose entering cell
Osmosis	Passive movement of solvent molecules through semipermeable membrane due to a concentration difference for water	Molecular motion	Water movement through cell wall to maintain constant turgidity of cell
Filtration	Molecules are forced by hydrostatic pressure from regions of higher pressures to regions of lower pressures	Blood pressure	Removal of wastes within kidneys

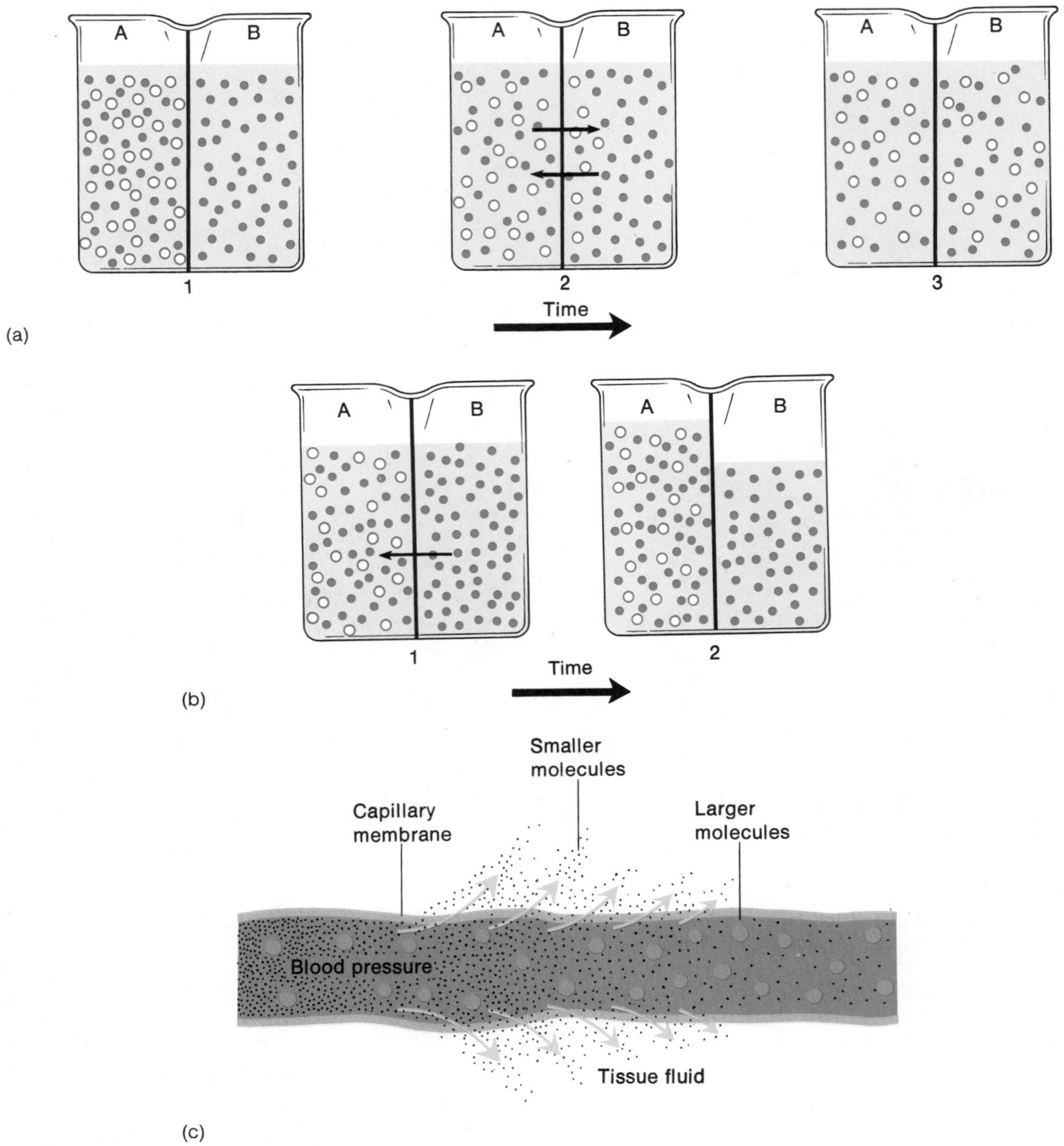

Figure 3.4 Examples of various kinds of movements through membranes. (*a*) Diffusion of sugar molecules from compartment A to compartment B until equilibrium is achieved in 3. (*b*) Osmosis is shown, as a selectively permeable membrane allows only water to diffuse through the membrane between compartments A and B, causing the level of the liquid to rise in A. (*c*) Filtration occurs as small molecules are forced through a membrane by blood pressure, leaving the larger molecules behind.

Table 3.5 *Continued*

Processes	Characteristics	Energy source	Example
Active transport	Molecules or ions are transported through cell membrane by other molecules	Cellular energy (ATP)	Movement of glucose and amino acids through membranes
Pinocytosis	Membrane engulfs minute droplets of fluid from surroundings	Cellular energy	Membrane forms vacuoles containing solute and solvent
Phagocytosis	Membrane engulfs solid particles from surroundings	Cellular energy	White blood cell membrane engulfs bacterial cell

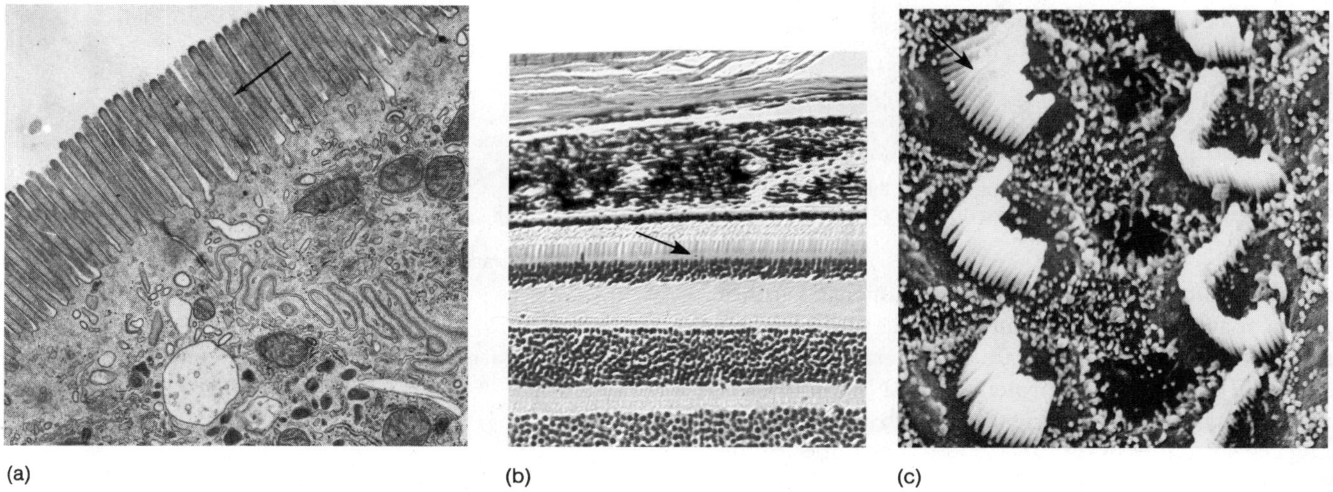

Figure 3.5 Specialized cell membranes indicated by arrows. (*a*) Microvilli of the cells that line the lumen of the small intestine (640×), (*b*) rod sacs within the eye for photoreception (160×), and (*c*) hair cells within the organ of Corti of the inner ear for tactile stimulation.

Cells composing certain sensory organs have specialized cell membranes. The photoreceptors, or light-responding rods and cones of the eye, have double-layered, disc-shaped membranes, called **sacs,** that contain pigments associated with vision. Within the organ of Corti in the inner ear are **hair cells** that are tactile (touch) receptors stimulated through mechanical vibration. Hair cells are so named because of the fine, cilialike processes that extend from the cell membrane.

Cytoplasm and Organelles

Cytoplasm refers to the material within the cell membrane but outside of the nucleus. The material within the nucleus is frequently called the **nucleoplasm. Protoplasm** refers collectively to both the cytoplasm and nucleoplasm.

When observed through an electron microscope (fig. 3.6), the cytoplasm is seen to be highly structural and contains distinct cellular components called **organelles.** The matrix of the cytoplasm is a jellylike substance that is 80% to 90% water. The organelles and inorganic *colloid substances* (suspended particles) are dispersed throughout the cytoplasm. Colloid substances have similar ionic charges that space them uniformly.

Metabolic activity occurs within the organelles of the cytoplasm. Specific roles, such as heat production, cellular maintenance, repair, storage, and protein synthesis, are carried out within the organelles.

The structure and function of each of the major organelles are discussed in the following paragraphs, and a summary of this information is presented in table 3.4.

Endoplasmic Reticulum The endoplasmic reticulum, or **ER,** is widely distributed throughout the cytoplasm as complex networks of interconnected membranes (fig. 3.7). Between the membranes are minute spaces, or **tubules,** that are connected at one end to the cell membranes. The tubules may also be connected to other organelles or to the nuclear envelope.

There are two types of endoplasmic reticulum. **Rough endoplasmic reticulum** is characterized by the presence of numerous small granules, called **ribosomes,** attached to the outer surface of the membranous wall. **Smooth endoplasmic reticulum** lacks ribosomes. Endoplasmic reticulum provides a pathway for transportation of substances within the cell and a storage area for synthesized molecules. Smooth endoplasmic reticulum manufactures certain lipid molecules. The membranous wall of rough endoplasmic reticulum provides a site for protein synthesis within ribosomes.

A person who repeatedly uses certain drugs, such as alcohol or phenobarbital, develops a tolerance to them, so that greater quantities are required to achieve the same effect. The cytological explanation of this is that repeated use causes the smooth endoplasmic reticulum to proliferate in an effort to detoxify these drugs and protect the cell. With increased amounts of smooth endoplasmic reticulum, cells can handle an increased volume of drugs.

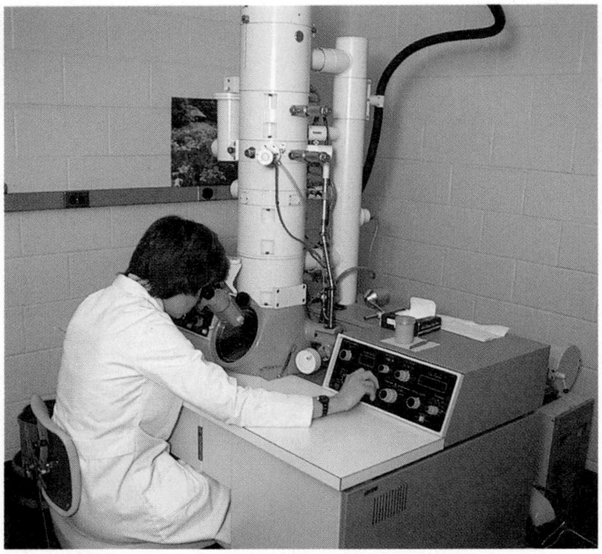

Figure 3.6 A transmission electron microscope such as this is used to observe and photograph organelles within the cytoplasm of a cell.

Ribosomes Ribosomes occur as free particles suspended within the cytoplasm or are attached to the membranous wall of rough endoplasmic reticulum. Ribosomes are small, granularlike organelles (fig. 3.7) composed of protein and RNA molecules. They function to synthesize protein molecules that may be used for building cell structures or for functioning as enzymes. Some of the proteins synthesized by ribosomes are secreted outside the cell to be used elsewhere in the body.

Golgi Apparatus The Golgi apparatus (Golgi complex) consists of several tiny membranous sacs, or vesicles, located near the nucleus (fig. 3.8a), which are continuous with the endoplasmic reticulum.

The Golgi apparatus is involved in the synthesis of carbohydrates and cellular secretions. As large carbohydrate molecules are synthesized, they combine with proteins to form compounds called glycoproteins. The glycoproteins accumulate in the channels of the Golgi complex, and after a critical volume is reached the vesicles break off from the complex and are carried to the cell membrane and released as a secretion (fig. 3.8b). Once the vesicle is outside the cell membrane, it ruptures to release its contents to complete the process known as *exocytosis.*

The Golgi apparatus is prominent in cells of certain secretory organs of the digestive system, such as the pancreas and the salivary glands. Pancreatic cells, for example, produce digestive enzymes that are packaged in the Golgi apparatus and secreted as droplets that flow into the pancreatic duct and are transported to the digestive tract.

Mitochondria Mitochondria are double-membraned, saclike organelles. The outer membrane is smooth, whereas the inner membrane is arranged into intricate folds called **cristae** (fig. 3.9). The folded cristae create a tremendous surface area for chemical reactions.

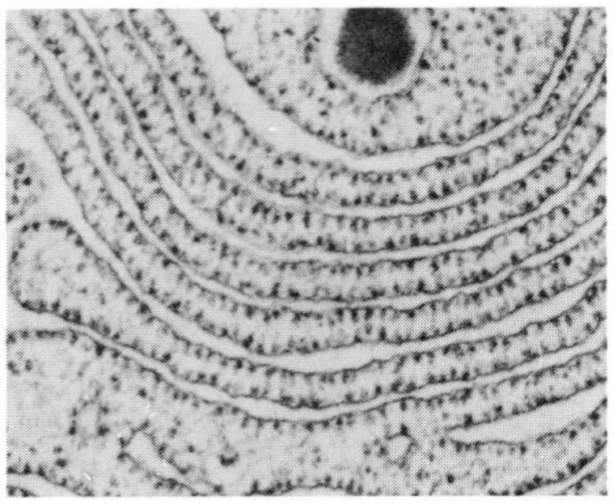

(a)

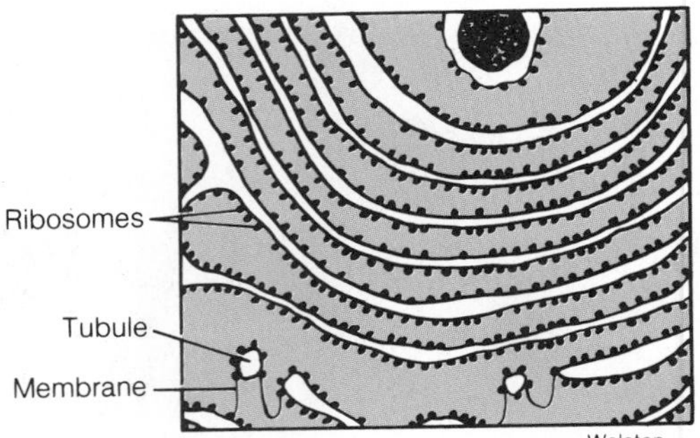

(b)

Figure 3.7 (*a*) An electron micrograph of rough endoplasmic reticulum magnified about 100,000 times. (*b*) Rough endoplasmic reticulum has ribosomes attached to its surface.

Mitochondria vary in size and shape. They can migrate through the cytoplasm and have the ability to reproduce themselves by budding or cleavage. Mitochondria are often called the powerhouses of cells because of their role in producing metabolic energy. Enzymes connected to the cristae control the chemical reactions that form ATP. Metabolically active cells, such as muscle cells, liver cells, and kidney cells, have a large number of mitochondria because they use energy at a high rate.

> The darker color of red meat (the thigh of a chicken, for example) is due to the larger amounts of myoglobin, a pigmented compound in muscle tissue that acts to store oxygen. Mitochondria are likewise more abundant in red meat. Both mitochondria and myoglobin are important for the greater metabolic activity in red muscle tissue.

Lysosomes Lysosomes vary in appearance from granularlike bodies to small vesicles to membranous spheres (fig. 3.10). They are scattered throughout the cytoplasm. Powerful enzymes enclosed within lysosomes are capable of breaking down protein and carbohydrate molecules. White blood cells contain large numbers of lysosomes and are said to be *phagocytic,* meaning that they will ingest, kill, and digest bacteria through the enzymatic activity of their lysosomes.

Golgi: from Camillo Golgi, Italian histologist, 1843–1926
cristae: L. *crista,* crest

phagocytic: Gk. *phagein,* to eat; *kytos,* a cell
lysosome: Gk. *lysis,* a loosening; *somo,* body
mitochondrion: Gk. *mitos,* a thread; *chondros,* lump, grain

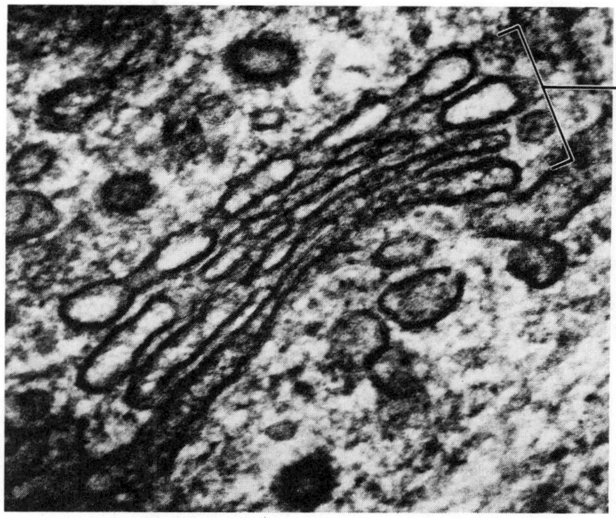

(a)

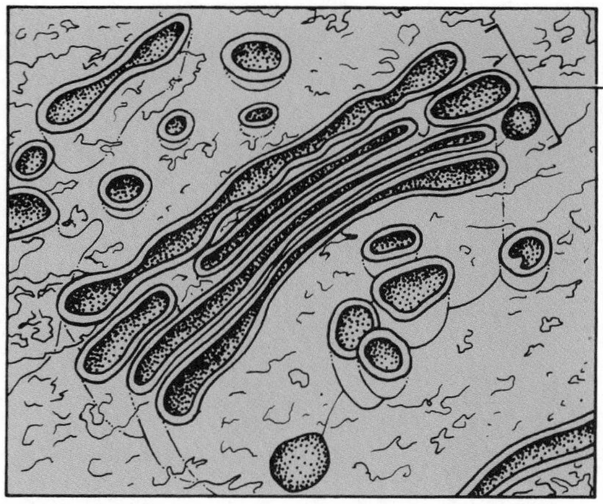

(b)

Figure 3.8 (*a*) An electron micrograph of a Golgi apparatus. Notice the formation of vesicles at the ends of some of the flattened sacs. (*b*) An illustration of the secretion of these vesicles by exocytosis. Portions of the Golgi apparatus function in packaging glycoproteins, which are carried to the cell membrane and released as a secretion.

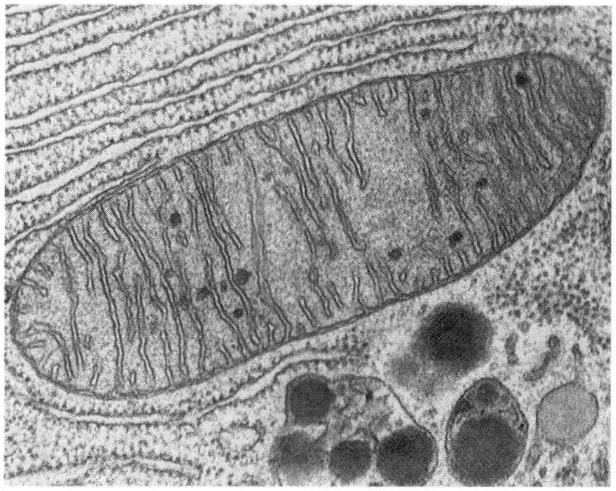

(a)

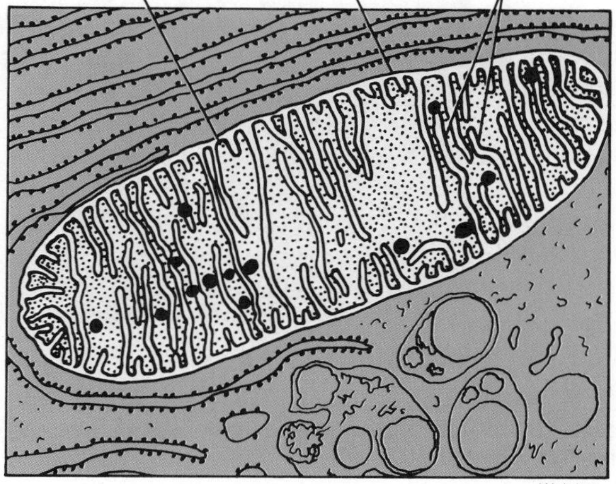

(b)

Figure 3.9 (*a*) An electron micrograph of a mitochondrion magnified about 40,000 times. (*b*) A diagram showing how membranous partitions form the cristae within the organelle.

The normal atrophy, or decrease in size, of the uterus following the birth of a baby is due to lysosomal digestive activity. Likewise, the secretions of lysosomes are responsible for the regression of the mammary tissue of the breasts after the weaning of an infant.

Lysosomes also function to destroy their own worn cell parts or the entire cell within which they reside if the cell is injured. For this reason, lysosomes are frequently called suicide packets.

Several diseases arise from abnormalities in lysosome function. The painful inflammation of *rheumatoid arthritis,* for example, occurs when enzymes from lysosomes are released into the joint capsule and initiate digestion of the surrounding tissue.

Peroxisomes Peroxisomes are membranous sacs that resemble lysosomes in structure and, like lysosomes, contain certain enzymes. Peroxisomes occur in most cells but are particularly abundant in the kidney and the liver. All peroxisomes contain one or more enzymes that produce hydrogen peroxide, a highly toxic substance. Hydrogen peroxide is an important compound in the white blood cells, which phagocytize diseased or worn-out cells. These organelles also contain the enzyme catalase, which breaks down excess hydrogen peroxide into water and oxygen so that there is no toxic effect on other organelles within the cytoplasm.

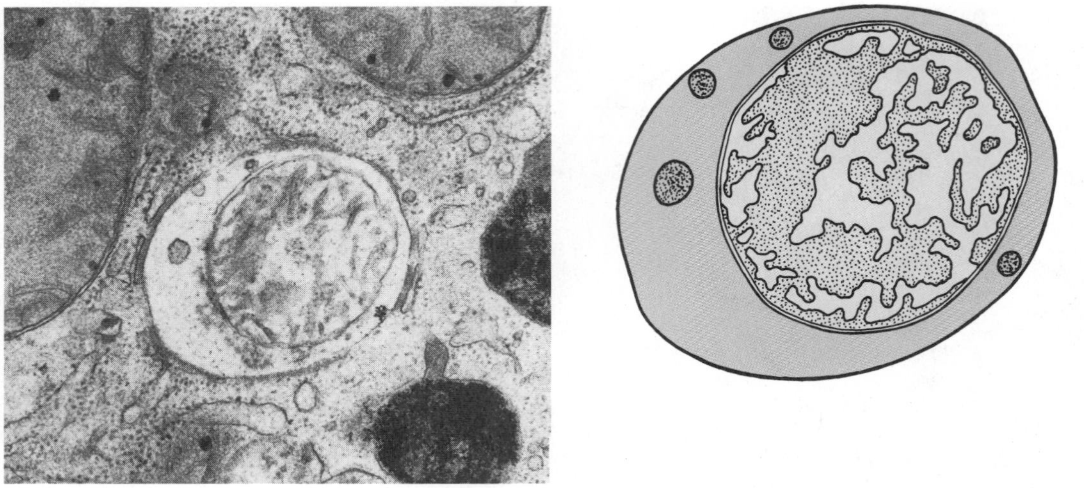

(a) (b)

Figure 3.10 (*a*) An electron micrograph of a lysosome magnified about 30,000 times. (*b*) A saclike lysosome with a diagram showing digestive phagocytic enzymes.

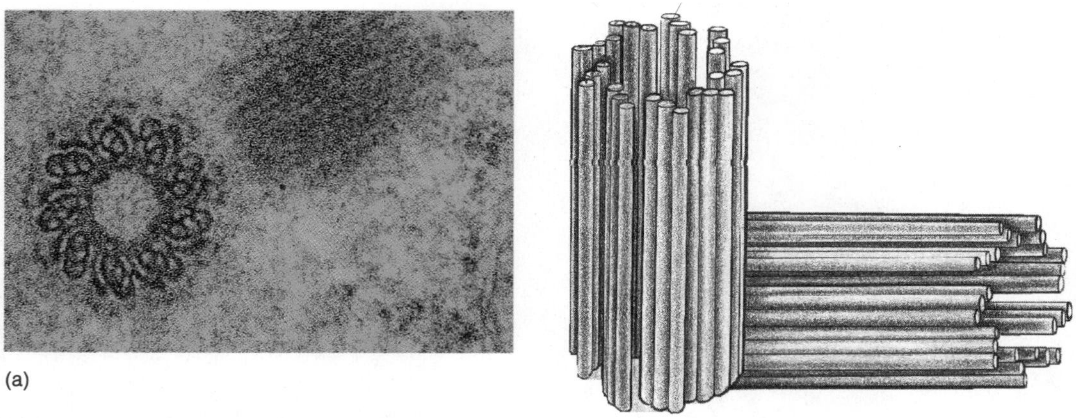

(a) (b)

Figure 3.11 The centrioles. (*a*) A micrograph of the two centrioles in a centrosome. (*b*) A diagram showing that the centrioles are positioned at right angles to one another.

Centrosome and Centrioles The **centrosome** (central body) is a nonmembranous, spherical mass positioned near the nucleus. Within the centrosome is a pair of rodlike structures called **centrioles** (fig. 3.11), which are positioned at right angles to each other. The wall of each centriole is composed of nine evenly spaced bundles, and each bundle contains three microtubules.

Centrosomes are found only in those cells that can divide. During the mitotic (reproductive) process, the centrioles migrate from one another and take positions on either side of the nucleus. They are then involved in the distribution of the chromosomes during cellular reproduction. Mature muscle and nerve cells lack centrosomes and thus cannot divide.

Vacuoles Vacuoles are membranous sacs that usually function as storage chambers. They vary in size and are formed when a portion of the cell membrane invaginates and pinches off during a process called *vacuolation.* Vacuolation is initiated either by *phagocytosis* or *pinocytosis.* In phagocytosis, the cell membrane engulfs solid particles; whereas in pinocytosis, cells take in minute droplets of liquid through the cell membrane (fig. 3.12). Vacuoles may contain liquid or solid materials that were previously outside the cell.

Fibrils and Microtubules Both fibrils and microtubules are found throughout the cytoplasm. The fibrils are minute, rodlike structures, whereas the microtubules are fine, threadlike tubular structures of varying lengths (fig. 3.13). Both provide support to the cell by forming a type of cytoskeleton. Specialized fibrils called myofibrils are particularly abundant in muscle cells, where they aid in the contraction of these cells. Microtubules are also involved in the transportation of macromolecules throughout the cytoplasm. They are especially abundant in the cells of endocrine organs, where they aid the movement of hormones to be secreted into the blood. Microtubules in certain cells provide flexible support to cilia and flagella.

---

vacuole: L. *vacuus,* empty

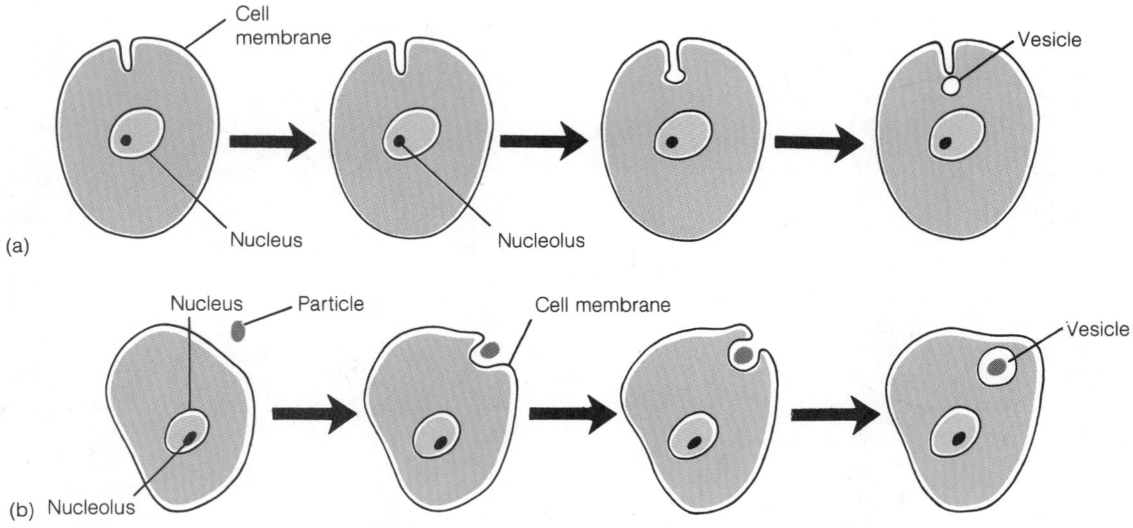

Figure 3.12 Pinocytosis compared to phagocytosis. (*a*) During pinocytosis, the cell takes in a minute droplet of fluid from its surroundings. (*b*) During phagocytosis, a solid particle is engulfed and ingested through the cell membrane.

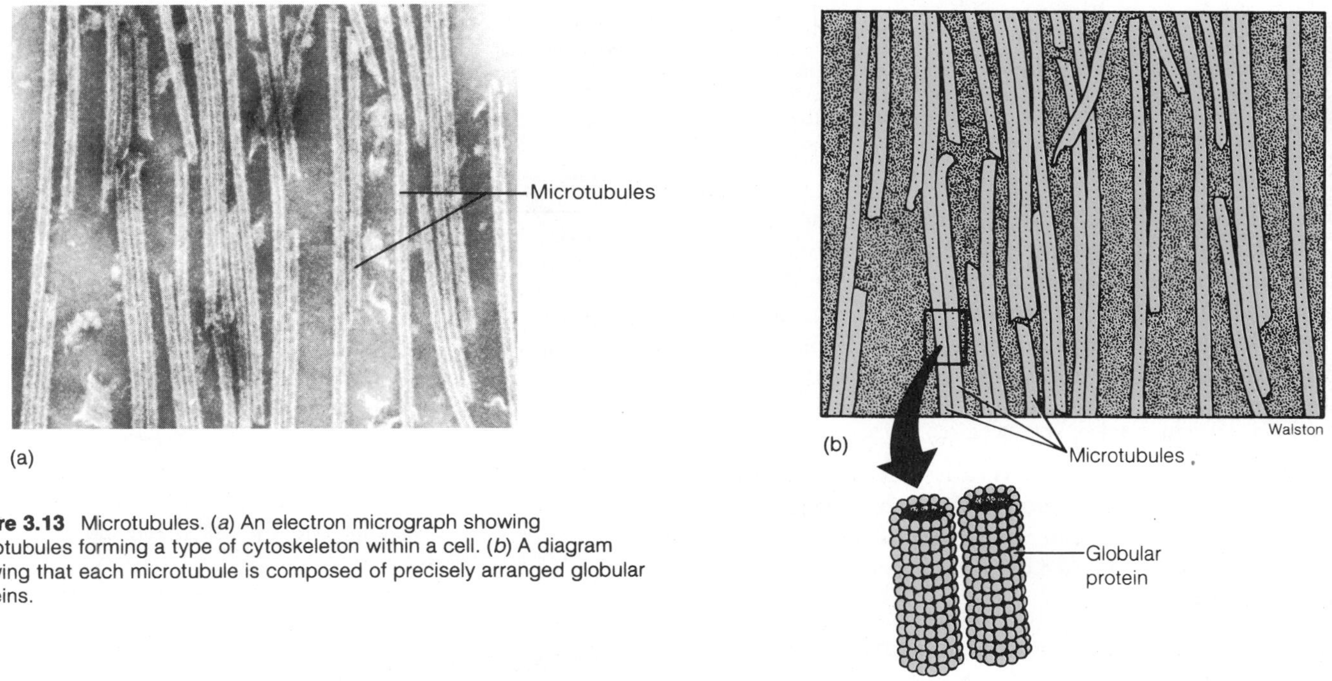

Figure 3.13 Microtubules. (*a*) An electron micrograph showing microtubules forming a type of cytoskeleton within a cell. (*b*) A diagram showing that each microtubule is composed of precisely arranged globular proteins.

Cilia and Flagella Although cilia and flagella appear to extend from the cell membrane, they are actually cytoplasmic projections from within the cell. These projections contain cytoplasm and supportive microtubules bounded by the cell membrane (fig. 3.14). Cilia and flagella should not be confused with the previously discussed microvilli, hair cells, or stereocilia, which are specializations of cell membranes.

Cilia are numerous, short projections from the exposed border of certain cells (fig. 3.15). Ciliated cells are interspersed with mucus-secreting **goblet cells.** There is always a film of mucus on the free surface of ciliated cells. Ciliated cells line the lumina (hollow portions) of sections of the respiratory and reproductive systems. The function of cilia is to move the mucus and any adherent material toward the exterior of the body.

Flagella are similar to cilia in basic microtubular structure (fig. 3.14). They are somewhat longer than cilia, however, and there are only one or two flagella per cell. The only type of cell in humans that has a flagellum is a spermatozoa cell, which uses the single structure for locomotion.

Cell Nucleus

The spherical **nucleus** is usually located near the center of the cell (fig. 3.16). It is the largest structure of the cell and contains the genetic material that determines cellular structure and controls cellular activity.

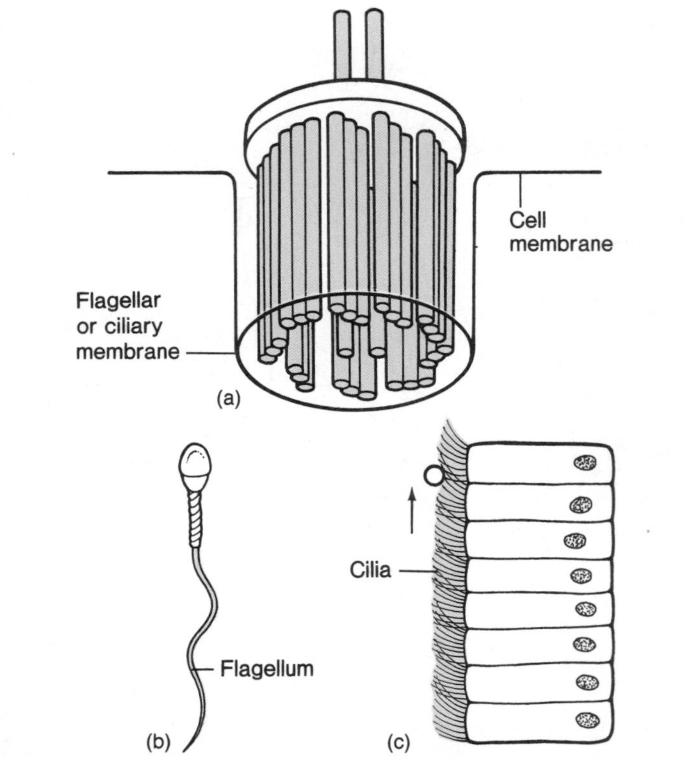

Figure 3.14 (*a*) Cilia and flagella are similar in the structural arrangement of their microtubules. (*b*) A spermatozoa cell has a single flagellum for propulsion. (*c*) Cilia produce a wavelike motion to move particles toward the outside of the body.

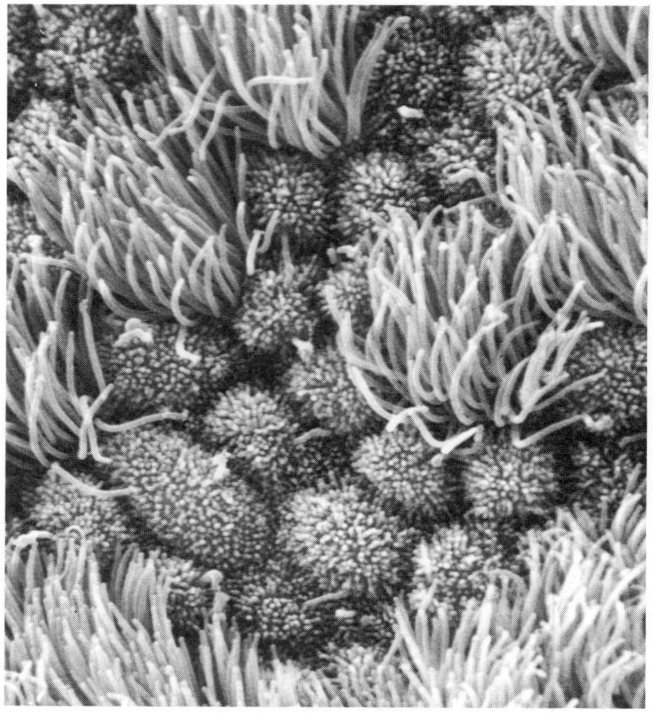

Figure 3.15 A scanning electron micrograph of ciliated cells that line the lumen of the uterine tube.

Most cells contain a single nucleus. Certain cells, however, such as skeletal muscle cells, are multinucleated. The long skeletal muscle fibers contain so much cytoplasm that several governing centers are necessary. Other cells, such as mature red blood cells, lack nuclei. These cells are limited to certain types of chemical activities and are not capable of reproduction.

The nucleus is enclosed by a double membrane called the **nuclear membrane** (fig. 3.16). The narrow space between the two walls of the nuclear membrane is called the **perinuclear cisterna.** Minute pores located along the nuclear membrane allow substances to move between the nucleoplasm and the cytoplasm. A cell is governed by two important structures within the nucleoplasm of the nucleus:

1. **Nucleoli.** The nucleoli are small, nonmembranous, spherical bodies composed largely of protein and RNA. It is thought that they function in the production of ribosomes. As ribosomes are formed, they migrate through the nuclear membrane into the cytoplasm.
2. **Chromatin.** Chromatin is a coiled, threadlike mass. It is the genetic material of the cell and consists principally of protein and DNA molecules. When the cell begins to undergo a division, the chromatin shortens and thickens into rod-shaped chromosomes (fig. 3.17). Each chromosome has thousands of genes distributed along its threadlike structure. The genes determine the structure and function of a cell.

---

1. Describe the composition and specializations of the cell membrane. Discuss why it is important that the cell membrane is selectively permeable.
2. Discuss the various kinds of movements across the cell membrane. Which are passive and which are active?
3. Describe the structure and function of the following cytoplasmic organelles: rough endoplasmic reticulum, a Golgi apparatus, lysosomes, and mitochondria.
4. Explain the distinction between nucleus and nucleoli.
5. Explain the distinction between chromatin and chromosomes.

The Cell Cycle

A cell cycle consists of growth, synthesis, and mitosis. Growth is the increase in cellular mass as the result of metabolism; synthesis is the production of DNA and RNA to regulate cellular activity; mitosis is the division of the nucleus and cytoplasm of a cell resulting in the formation of two daughter cells.

Objective 12. Describe the structure of a DNA molecule.
Objective 13. List the stages of mitosis, and discuss the events of each stage.
Objective 14. Discuss the significance of mitosis.

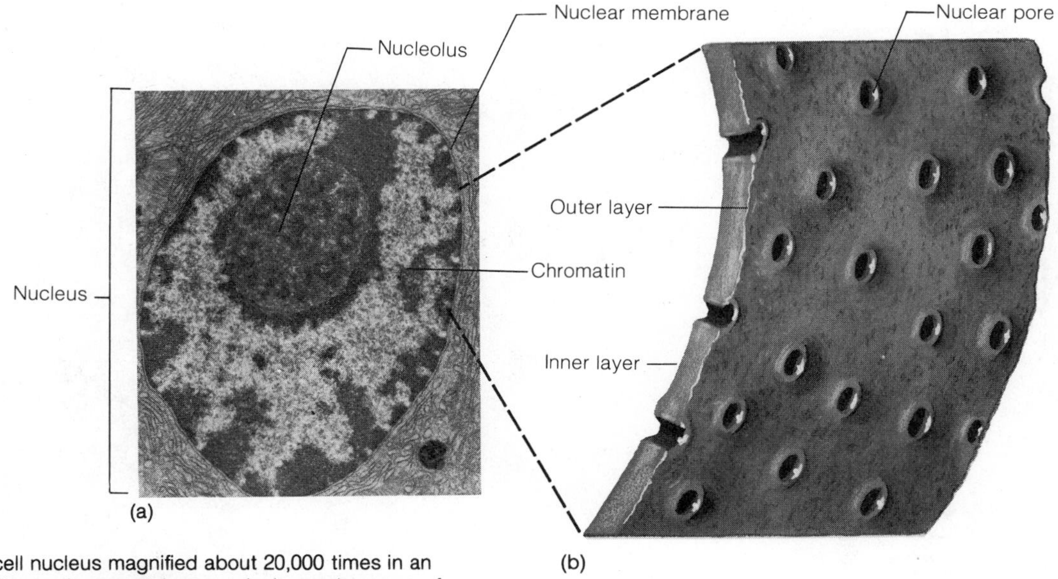

Figure 3.16 (*a*) The cell nucleus magnified about 20,000 times in an electron micrograph. The nucleus contains a nucleolus and masses of chromatin. (*b*) Notice the double-layered nuclear envelope with pores to permit substances to pass between the nucleus and the cytoplasm.

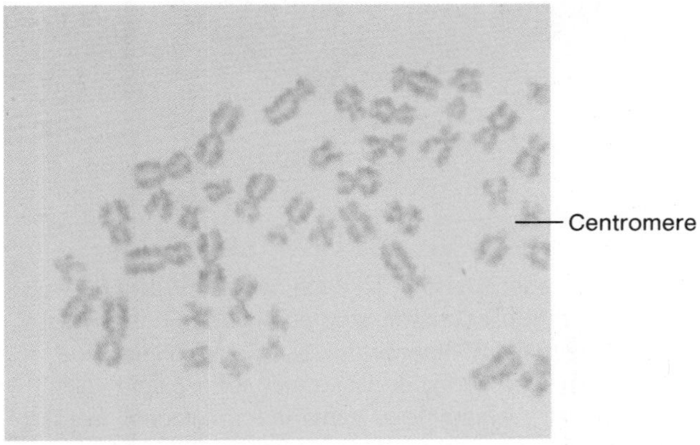

Figure 3.17 A chromosome smear showing the twenty-three pairs within a typical human cell.

Cellular replication is one of the principal concepts of biology. Through the process of cellular division, called **mitosis,** a multicellular organism can develop and be maintained. Mitosis makes possible body growth and the replacement of damaged, diseased, or worn-out cells. The process of mitosis ensures that each reproduced cell has the same number and kind of chromosomes as the original parent cell.

In an average, healthy adult, over 100 billion cells will die and be mitotically replaced during a twenty-four hour period. This represents a replacement of about 2% of the body mass each day. Some of the most mitotically active sites are the outer layer of skin, the internal lining of the digestive tract, and the liver.

Before a cell can divide, it must first duplicate its chromosomes so that the genetic traits can be passed to the succeeding generations of cells. A chromosome consists of a coiled **deoxyribonucleic acid (DNA) molecule** and is not visible until the cell begins to divide. There are twenty-three pairs of chromosomes in each human body (somatic) cell, and approximately twenty thousand genes are positioned on each chromosome.

Chromosomes are of varied lengths and shapes—some twisted, some rodlike. During mitosis, the chromosomes shorten and condense so that each pair assumes a characteristic shape (fig. 3.17). On the chromosome is a **centromere** to which are attached the **spindle fibers** that direct the chromosome toward the pole during mitosis.

Structure of DNA

The DNA molecule is frequently called a *double helix* because of its resemblance to a spiral ladder (fig. 3.18). The sides of the DNA molecule are formed by alternating units of sugar, called *deoxyribose,* and a phosphoric acid called the *phosphate group.* The rungs of the molecule are composed of pairs of *nitrogenous bases.* The ends of each nitrogenous base are attached to the deoxyribose-phosphate units. There are only four types of nitrogenous bases in a DNA molecule: adenine (A), thymine (T), cytosine (C), and guanine (G).

The basic structural units of the DNA molecule are called **nucleotides.** Each nucleotide consists of a molecule of deoxyribose sugar, a phosphate group, and one of the four nitrogenous bases. Thus, there is a nucleotide type for each of the four bases.

The pairing of the nitrogenous bases of the nucleotides is highly specific. The molecular configuration of each base is such that adenine always pairs with thymine and cytosine always pairs with guanine. The hydrogen bonds between these bases are relatively weak and can be easily split during cellular division (fig. 3.19). During division, the sequence of bases along the sides of the DNA molecule serves as a template that determines the sequence along each new strand.

Cell Cycle and Cell Division

A **cell cycle** is the series of changes that a cell undergoes from the time it is formed until it has completed a division and re-

mitosis: Gk. *mitos,* thread

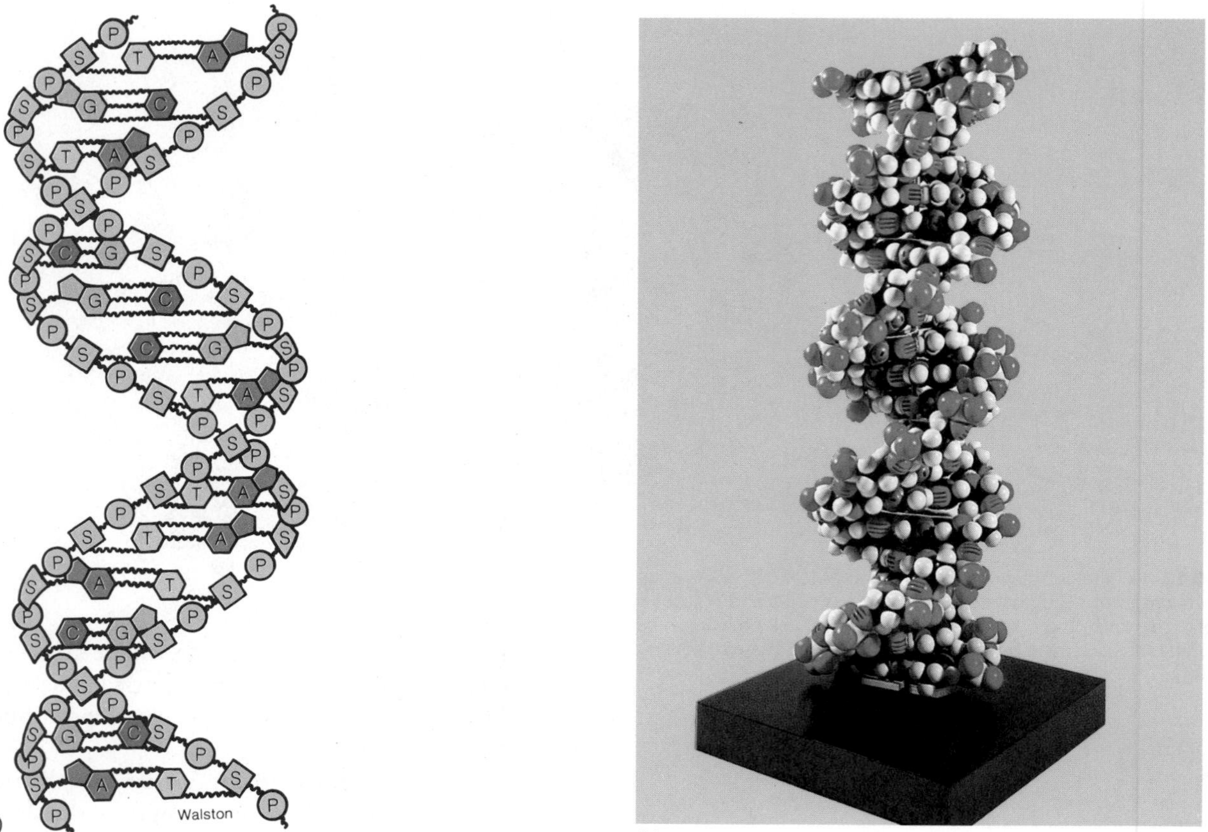

(a) (b)

Figure 3.18 (*a*) The molecular double helix of a DNA molecule. Each strand of the helix contains only four kinds of organic bases (*A, T, C,* and *G*). (*b*) A space-filling model of a portion of a DNA molecule.

produced itself. **Interphase** is the first period of the cycle from cell formation to the start of cell division (fig. 3.20). During interphase, a cell grows, carries on metabolic activities, and prepares itself for division.

Interphase is divided into **G1, S1,** and **G2 phases.** During the G1 (first growth) phase, a cell grows rapidly and is metabolically active. The length of time cells are within G1 varies considerably. It may be only hours in cells that have rapid division rates or even days or years for other cells. At the end of G1, the centrioles replicate in preparation for their role in cell division. During the S1 (synthetic) phase, the DNA in the nucleus of a cell replicates so that the two future cells will receive identical copies of the genetic material. During the G2 (second growth) phase, the enzymes and other proteins needed for the division process are synthesized and the cell continues to grow.

The actual division of a cell is referred to as the **mitotic phase** or simply **M phase** (fig. 3.20). The mitotic phase is further divided into mitosis and cytokinesis. **Mitosis** is the period of a cell cycle during which there is nuclear division and the duplicated chromosomes separate to form two genetically identical daughter nuclei. The process of mitosis takes place in four successive stages, each stage passing into the next without sharp structural distinctions. These stages are *prophase, metaphase, anaphase,* and *telophase* (fig. 3.21). Mitosis is followed by **cytokinesis** during which there is division of the cytoplasm.

Highly specialized cells, such as muscle and nerve cells, do not replicate after a person is born. If these cells die, as the result of disease, injury, or even disuse, they are not replaced and scar tissue may form. Nerve cells are especially vulnerable to damage from alcohol, oxygen deprivation, and various drugs, which can jeopardize the health of these types of cells.

1. Explain why the DNA molecule is described as a double helix.
2. List the phases in the life cycle of a cell, and describe the principal events that occur during each phase.
3. Explain why mitosis is such an important biological process.

Clinical Considerations

Cellular Adaptations

Except for cells located on exposed surfaces, most cells of the body are positioned in a rather homogeneous environment where continual adaptation to change is not necessary for survival. Cells do, however, have tremendous adaptability and resilience to ensure survival against conditions that might otherwise be lethal. Prolonged exposure to sunlight, for example, stimulates the synthesis of melanin and tanning of the

cytokinesis: Gk. *kytos,* a hollow; *kinesis,* movement

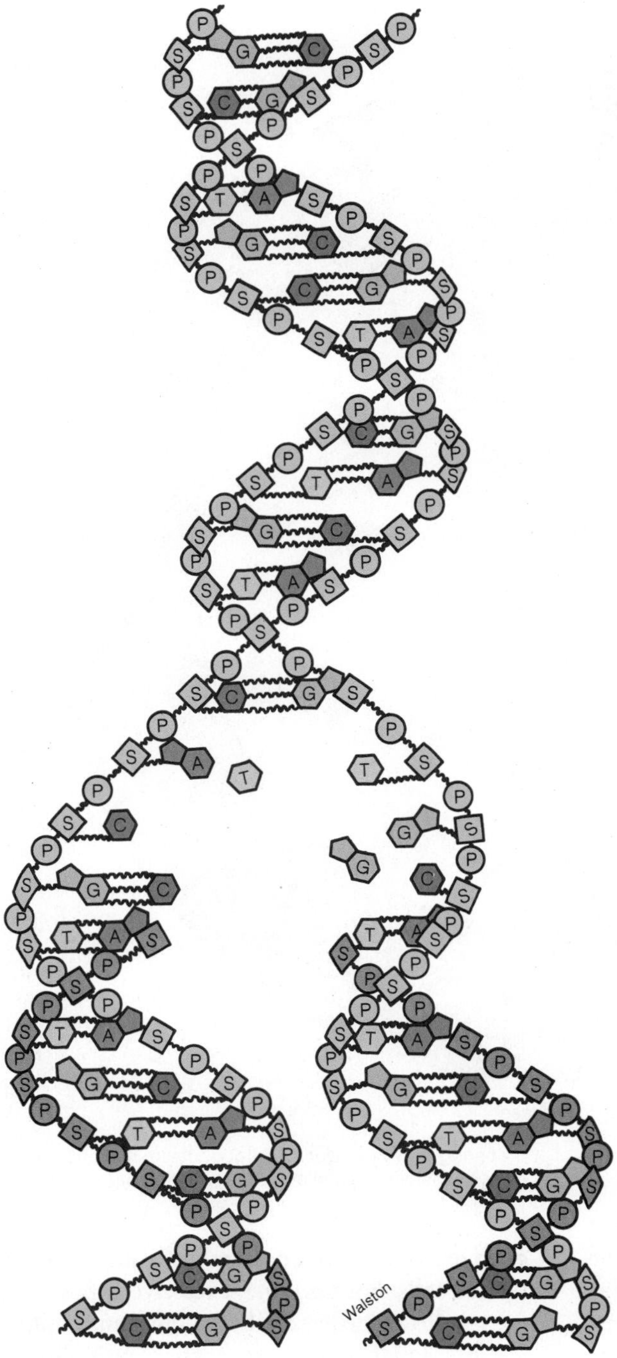

Region of parental DNA helix. (Both backbones are tan.)
Region of replication. Parental DNA is unzipped and new nucleotides are pairing with those in parental strands.
Region of completed replication. Each double helix is composed of an old parental strand (tan) and a new daughter strand (reddish). Notice that each double helix is exactly like the other one.
Walston

Figure 3.19 The replication of DNA. Each new double helix is composed of one old and one new strand. The base sequences of each of the new molecules is identical to that of the parent DNA because of complementary base pairing.

skin. Likewise, mechanical friction to the skin increases mitotic activity and the synthesis of a fibrous protein, *keratin,* which results in the formation of a protective callus.

Cells adapt to potentially injurious stimuli by several specific mechanisms. **Hypertrophy** refers to an increase in the size of cells caused by an increased synthesis of protein, nucleic acids, and lipids. Cellular hypertrophy can be either compensatory or hormonal. **Compensatory hypertrophy** occurs when increased metabolic demands on particular cells cause an increase in cellular mass. Examples of compensatory hypertrophy include skeletal muscle fibers enlarging as a result of exercise, and cardiac (heart) muscle fibers or kidney cells enlarging because of an increased work demand. Hypertension (high blood pressure) causes cardiac cells to hypertrophy to pump blood against raised pressures. After the removal of a diseased kidney, there is a compensatory increase in the size of the cells of the remaining kidney. The remaining kidney usually increases from a normal weight of 180 gm to about 350 gm. Examples of **hormonal hypertrophy** are the increase in the size of the breasts and the increase in the size of the smooth muscles of the uterus in a pregnant woman.

hypertrophy: Gk. *hyper,* over; *trophe,* nourishment

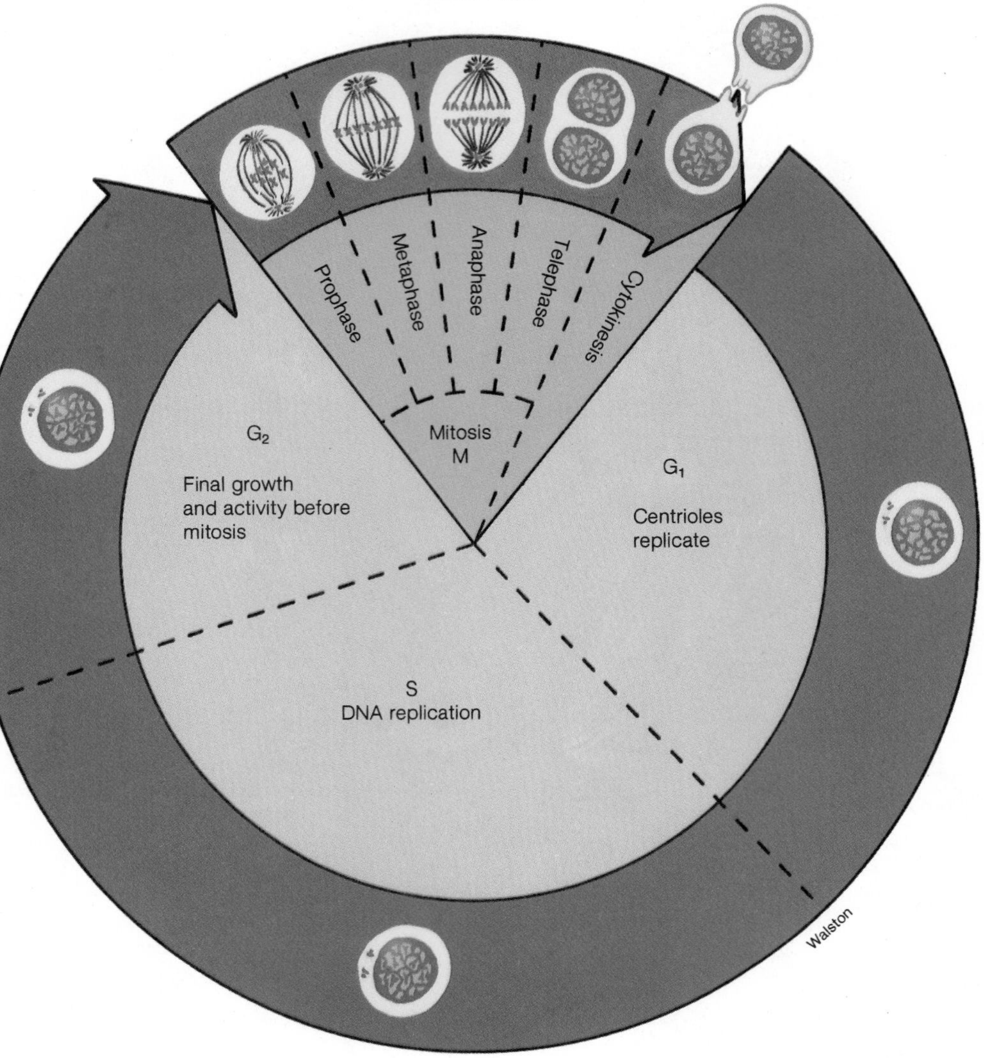

Figure 3.20 Interphase and mitotic phase are the two principal divisions of the cell cycle. During interphase, a cell is growing, metabolizing and preparing to divide. During mitotic phase, nuclear division occurs followed by cytoplasmic division and the formation of two daughter cells.

Hyperplasia refers to an increase in the mitotic activity of cells that causes a greater number of cells to be formed. The stimuli to induce hyperplasia are not well understood. The removal of a portion of the liver, for example, leads to regeneration, or hyperplasia, of the remaining liver cells to restore the loss. But the triggering mechanism for this is not known. A type of hormonally induced hyperplasia occurs in cells of the endometrium of a woman after menstruation, which restores this reproductive layer to a suitable state for implantation.

Atrophy refers to a decrease in the size of cells and a corresponding decrease in the size of the affected organ. Atrophy can occur in the cells of any organ and may be classified as **disuse atrophy, disease atrophy,** or **aging (senile) atrophy.**

Metaplasia is a specialized cellular change in which one type of cell transforms into another. Generally, it involves the change of highly specialized cells into more generalized, protective cells. For example, excessive exposure to inhaled smoke causes the specialized ciliated columnar epithelial cells lining the bronchial airways to change into squamous epithelium, which is more resistant to injury from smoke.

Cell Trauma

As adaptable as cells are to environmental changes, many agents and factors are injurious to cells. If the trauma causes extensive cellular death, the condition may become life threatening. A person dies when a vital organ can no longer perform its necessary metabolic role in sustaining the body.

Energy deficit means that more energy is required than is available. Cells can tolerate certain mild deficits because of various reserves stored within the cytoplasm, but a severe or prolonged deficit will cause cells to die. An energy deficit occurs if the cells have insufficient glucose or insufficient oxygen to cause glucose combustion. Examples of energy deficits are low levels of blood sugar (hypoglycemia) and the impermeability

---

hyperplasia: Gk. *hyper,* over; *plasis,* a molding
atrophy: Gk. *a,* without; *trophe,* nourishment
metaplasia: Gk. *meta,* between; *plasis,* a molding

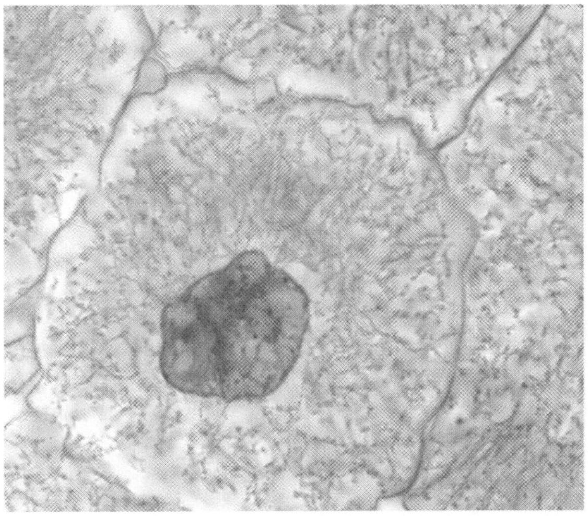

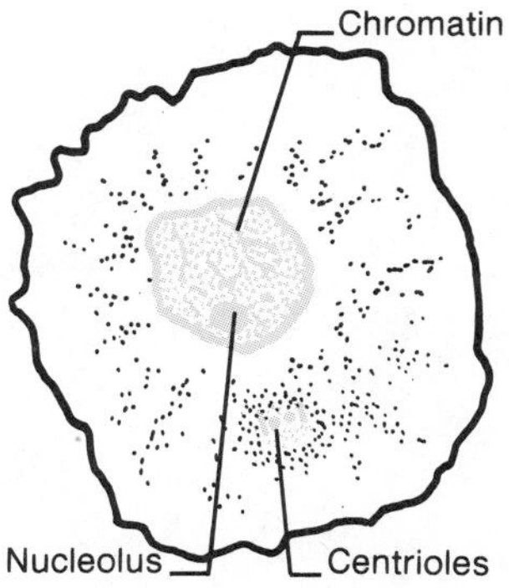

(a) Interphase

The chromosomes are in an extended form and seen as chromatin in the electron microscope.
The nucleus is visible.

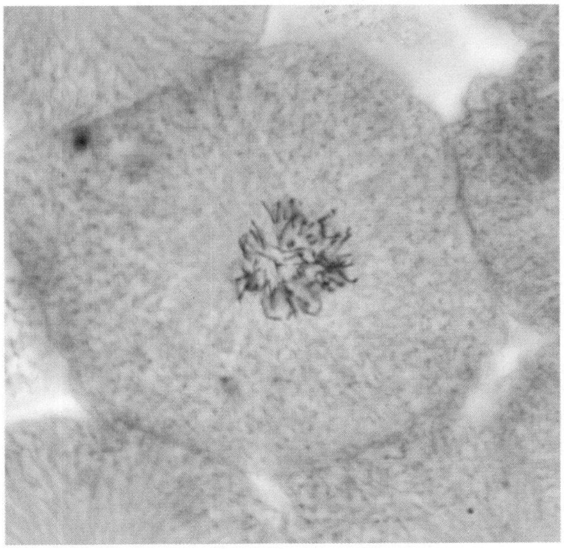

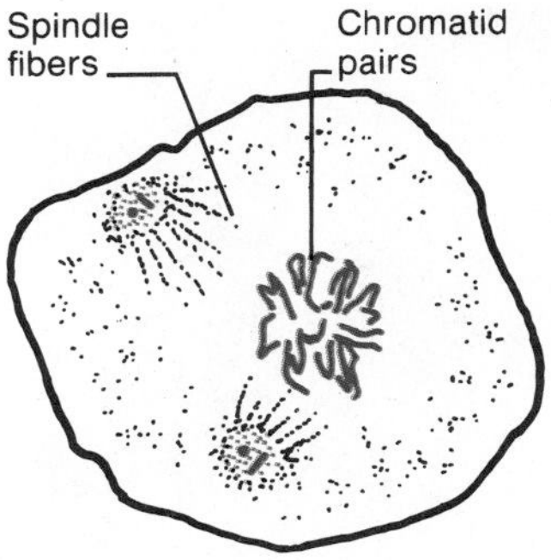

(b) Prophase

The chromosomes are seen and observed to consist of two chromatids joined together by a centromere.
The centrioles move apart towards opposite poles of the cell.
Spindle fibers are produced and extend from each centriole.
The nuclear membrane starts to disappear.
The nucleolus is no longer visible.

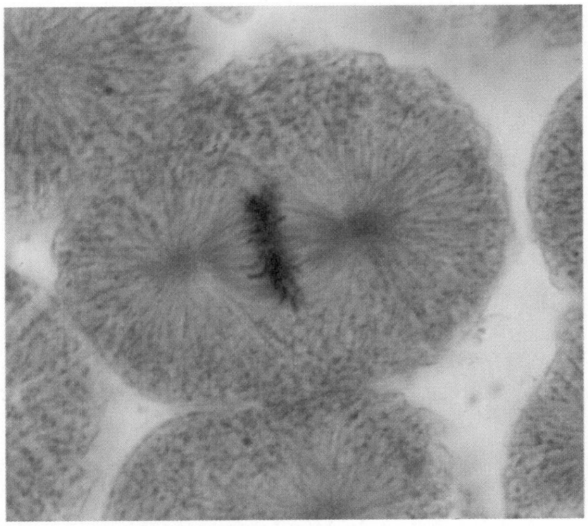

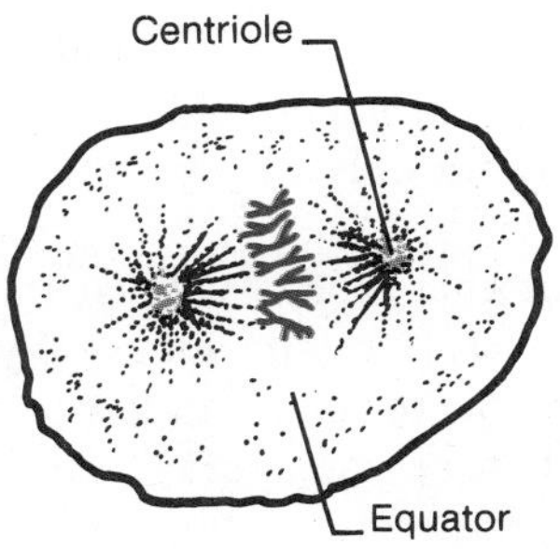

(c) Metaphase

The chromosomes are lined up at the equator of the cell.
The spindle fibers from each centriole are attached to the centromeres of the chromosomes.
The nuclear membrane has disappeared.

Figure 3.21 The stages of mitosis.

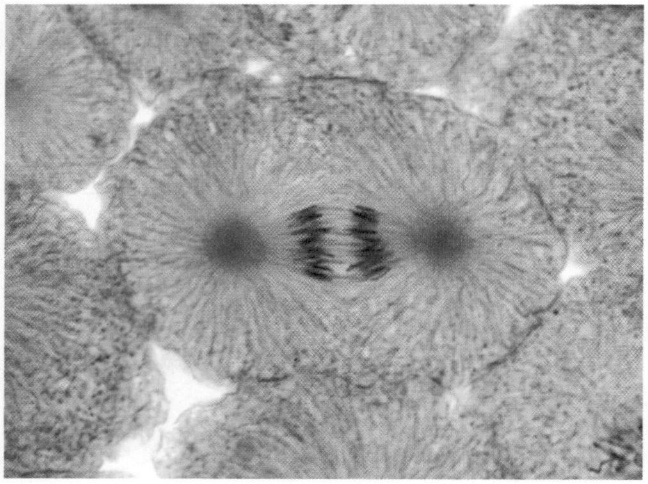

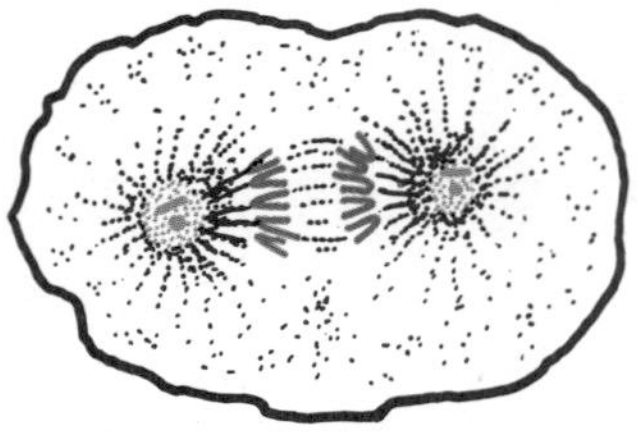

(d) Anaphase

The centromeres split, and the sister chromatids separate as each is pulled to an opposite pole.

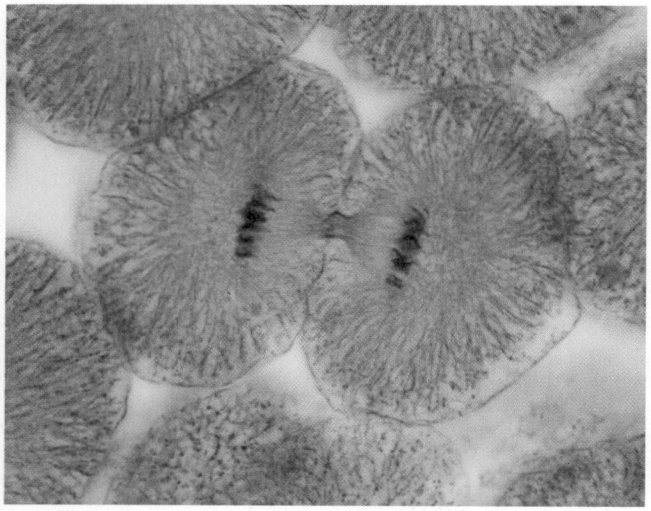

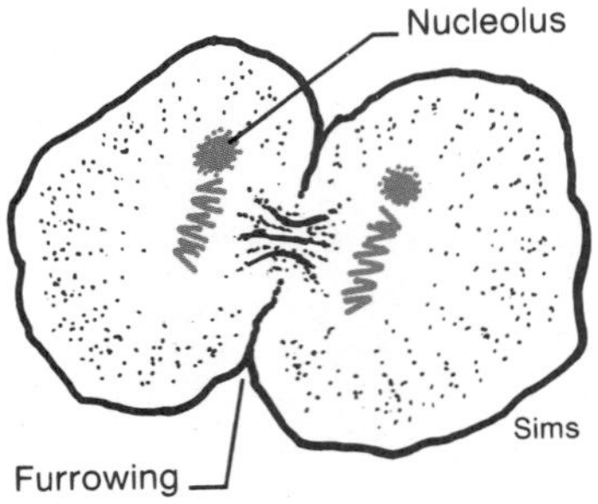

(e) Telophase

The chromosomes become longer, thinner, and less distinct.
New nuclear membranes form.
The nucleolus reappears.
Cell division (cytokinesis) is nearly complete.

Figure 3.21 *continued*

of the cell membrane to the entry of glucose (as in diabetes mellitus). Malnutrition also causes energy deficit. Few cells can tolerate an interruption in oxygen supply. Cells of the brain and the heart have tremendous oxygen demands, and an interruption of the supply to these organs for only a matter of minutes can cause death.

Physical injury to cells, another type of cellular trauma, occurs in a variety of ways. High temperatures **(hyperthermia)** are generally less tolerable to cells than low temperatures **(hypothermia).** Respiratory rate, heart rate, and metabolism accelerate with hyperthermia. Continued hyperthermia causes protein coagulation within cells and eventually cellular death. In frostbite, rapid or prolonged chilling causes cellular injury. In severe frostbite, ice crystals form and burst the cells.

Burns are a problem especially if they cause damage to the deeper skin layers, which interferes with the mitotic activity of cells (see chap. 5). Of immediate concern with burns, however, is the devastating effect of fluid loss and infections through traumatized cell membranes.

Accidental poisoning and suicide through drug overdose cause numerous deaths in the United States and in other countries. **Drugs** and **poisons** cause cellular dysfunction by disrupting DNA replication, RNA transcription, active transport, or cell membrane activity.

Radiation is a type of cellular trauma that is cumulative in effect. When X rays are given for therapeutic purposes (radiotherapy), small doses are focused on a tumorous area for many days to prevent widespread cellular injury. Some cells are more sensitive to radiation than others. Immature or mitotically active cells are highly sensitive, whereas cells that are no longer growing, such as neurons and muscle cells, are not as vulnerable to radiation injury.

Infectious agents, or **pathogens,** invade and inhabit cells, causing cellular dysfunction. Viruses and bacteria are the most common pathogens. Viruses usually destroy cells as they reproduce themselves. Bacteria, on the other hand, will frequently poison cells with their toxic metabolic wastes.

Cellular Organelles

Lysosomes are important organelles in the function of phagocytic cells. Organs such as the liver, spleen, and lymph nodes have phagocytes that are immobile, or "fixed," within the organs. These are sometimes referred to as the *reticuloendothelial* *system.* Among other functions, the reticuloendothelial system helps to remove old red blood cells and bacteria from blood and to remove and inactivate drugs from the blood. Lysosomes also participate in *autophagy* and by this process help to prevent the accumulation of damaged organelles and stored molecules. Genetic diseases that result in the defective production of lysosomal enzymes may produce abnormal accumulations of glycogen or lipids within particular cells. Examples of such diseases include Tay-Sachs disease and Gaucher's disease.

The **smooth endoplasmic reticulum** in liver cells and other cells contains enzymes used for the inactivation of steroid hormones and many drugs. This inactivation is generally achieved by reactions that convert these compounds to more water-soluble and less active forms, which can be more easily excreted by the kidneys. When people take certain drugs, such as alcohol and phenobarbital for a long period of time, a larger dose of these compounds is required to produce a given effect. This phenomenon, called *tolerance,* is accompanied by an increase in the smooth endoplasmic reticulum and thus an increase in the enzymes charged with inactivation of these drugs.

Medical Genetics

Medical genetics is a branch of medicine concerned with diseases that have a genetic origin. Genetic factors include abnormalities in chromosome number or structure and mutant genes. Genetic diseases are a diverse group of disorders, including malformed blood cells (sickle-cell anemia), defective blood clotting (hemophilia), and mental retardation (Down's syndrome).

Chromosomal abnormalities occur in approximately 0.6% of live-birth infants. The majority of them (70%) are subtle, cause no problems, and are usually undetected. Structural changes in the genetic DNA that is passed from parent to offspring by means of sex cells are called **mutations.** Mutations either occur naturally or are environmentally induced through chemicals or radiation. Natural mutations are not well understood. About 12% of all congenital malformations are caused by mutations and probably come about through an interaction of genetic and environmental factors. Many of these problems can be predicted by knowing the genetic pedigree of prospective parents and prevented through genetic counseling. **Teratology** is the science concerned with developmental defects and the diagnosis, treatment, and prevention of malformations.

Genetic problems are occasionally caused by having too few or too many chromosomes. The absence of an entire chromosome is termed **monosomy.** Embryos with monosomy usually die. Persons with **Turner's syndrome** have only one X chromosome and have a better chance of survival than persons missing one of the other chromosomes. **Trisomy,** a genetic condition in which an extra chromosome is present, occurs more frequently than monosomy. The best known among the trisomies is **Down's syndrome.**

autophagy: Gk. *autos,* self; *phagein,* to eat
mutation: L. *mutare,* to change
teratology: Gk. *teras,* monster; *logos,* study
Turner's syndrome: from Henry H. Turner, American endocrinologist, 1892–1970
Down's syndrome: from John L. H. Down, English physician, 1828–96

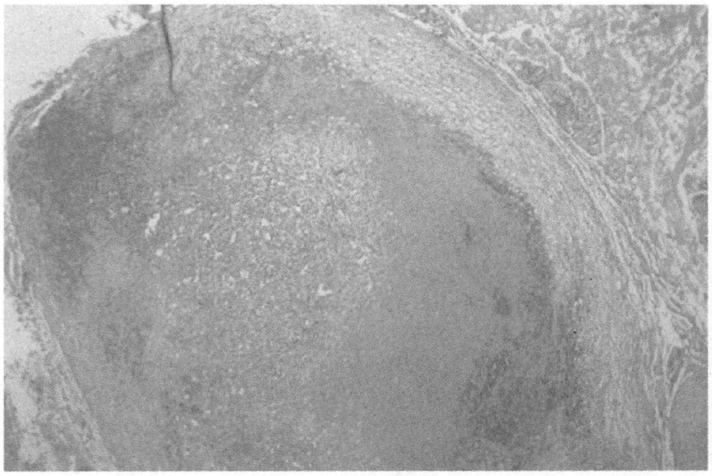

(a)

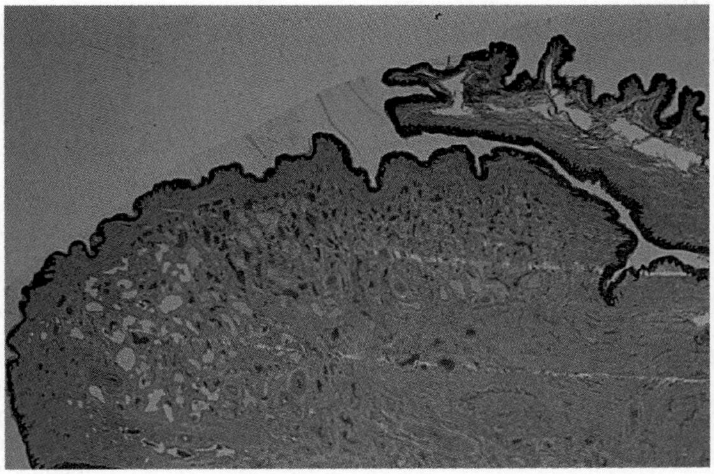

(b)

Figure 3.22 A light micrograph of (*a*) cancerous tissue of the penis and (*b*) normal tissue of the penis (0.9×). (Note the abnormal epithelial growth in the cancerous tissue.)

Cancer

Cancer is a complex group of diseases characterized by uncontrolled cell reproduction. The rapid proliferation of cells results in a **neoplasm,** or new cellular mass, being formed. Neoplasms, frequently called tumors, are classified as benign or malignant based on their cytological and histological features. **Benign neoplasms** usually grow slowly and are confined to a particular area. These types are usually not life threatening unless they grow to large sizes on vital organs such as the brain. **Malignant neoplasms** (fig. 3.22) grow rapidly and **metastasize** (fragment and spread) easily through lymphatic or blood vessels. The origin of a malignant neoplasm is a *primary growth,* and the locations to which it metastasizes are called *secondary growths.*

Cancer cells resemble undifferentiated or primordial cell types. Generally they do not mature before they divide and are not capable of maintaining normal cell function. Cancer causes death when a vital organ regresses because of competition from cancer cells for space and nutrients. The pain associated with cancer develops when the growing neoplasm affects sensory neurons.

The various types of cancers are classified on the basis of the tissue in which they develop. Lymphoma, for example, is a cancer of lymphoid tissue; osteogenic cancer is a type of bone cancer; myeloma is cancer of the bone marrow; and sarcoma is a general term for any cancer arising from cells of connective tissue.

The etiology (cause) of cancers is largely unknown. However, initiating factors, or **carcinogens,** such as viruses, chemicals, or irradiation may provoke cancer to develop. Cigarette smoking, for example, causes various respiratory cancers to develop. The tendency to develop other types of cancers has a genetic basis. Some researchers even think that physiological stress can promote certain types of cancerous activity. Because the causes of cancer are not well understood, early detection with prompt treatment is the best course of action.

Aging

As discussed in chapter 23, there are obvious external indicators of aging: graying and loss of hair, wrinkling of skin, loss of teeth, and decreased muscle mass. Changes due to aging within cells are not as apparent and not well understood. Certain organelles alter with age. The mitochondria, for example, may change in structure and number, and the Golgi apparatus may fragment. Also, lipid vacuoles tend to accumulate in the cytoplasm, whereas the inclusions that contain glycogen decrease.

The chromatin and chromosomes within the nucleus show changes with aging such as clumping, shrinking, or fragmenting with repeated mitotic divisions. There is strong evidence that certain cell types have a predetermined number of mitotic divisions that are genetically controlled, thus determining the overall vitality and longevity of an organ. If this is true, identifying and genetically manipulating the "aging gene" might be possible.

Extracellular substances also change with age. Protein strands of *collagen* and *elastin* change in quality and number in aged tissues. Elastin is important in the walls of arteries, and its deterioration is thought to be associated with vascular diseases such as arteriosclerosis in aged persons.

carcinogen: Gk, *karkinos,* cancer

Chapter Summary

I. Introduction to Cellular Anatomy
   A. Cells are the structural and functional units of the body. Cellular function is referred to as metabolism, and the study of cells is referred to as cytology.
   B. The structure and function of cells depend upon the specific membranes and organelles characteristic of each type of cell.

II. Chemistry of a Cell
   A. Four elements (O, C, H, N) compose over 95% of the body and are linked together to form either inorganic or organic compounds.
   B. Water is the most abundant inorganic compound in cells and is an excellent solvent.
      1. Water is important in temperature control and hydrolysis.
      2. Dehydration is a condition in which fluid loss exceeds fluid intake and may be a serious problem especially in infants.
   C. Electrolytes are inorganic compounds that form ions.
      1. The three classes of electrolytes are acids, bases, and salts.
      2. Electrolytes are important in maintaining pH, in conducting electrical currents, and in the development and functioning of certain tissues.
   D. Proteins are organic compounds that exist by themselves or are conjugated.
      1. They provide the framework of cells and are necessary for growth, repair, and division.
      2. Enzymes and hormones are examples of specialized proteins.
   E. Carbohydrates are organic compounds that include starches and sugars.
      1. The composition of carbohydrates is always in a 2:1 ratio of hydrogen to oxygen.
      2. Carbohydrates form the most abundant source of cellular energy.
   F. Lipids are organic fats and fat-related substances.
      1. Lipids are composed of carbon, hydrogen, and oxygen.
      2. Lipids are an important source of energy and serve to protect and insulate various parts of the body.

III. Structure of a Cell
   A. A cell is composed of a cell membrane, cytoplasm containing organelles, and a nucleus.
   B. The cell membrane, composed of phospholipid and proteins, encloses the contents of the cell and regulates the passage of substances.
      1. The permeability of the cell membrane is dependent upon its structure, size of molecules, ionic charge, lipid solubility, and the presence of carrier molecules.
      2. Cell membranes may be specialized with microvilli, sacs, hair cells, or stereocilia.
   C. Cytoplasm is the material between the cell membrane and the nucleus. Nucleoplasm is the material within the nucleus, and protoplasm includes both the cytoplasm and nucleoplasm.
   D. Organelles are highly structural and specialized components within the cytoplasm.
      1. Endoplasmic reticulum provides a framework within the cytoplasm and forms a site for the attachment of ribosomes. It functions in the synthesis of lipids and proteins and in cellular transport.
      2. Ribosomes are particles of protein and RNA that function in protein synthesis. The protein particles may be used within the cell or secreted.
      3. The Golgi apparatus consists of membranous vesicles that synthesize glycoproteins and secrete lipids. The Golgi apparatus is prominent in secretory cells such as those of the pancreas and salivary glands.
      4. Mitochondria are membranous sacs that consist of an outer smooth layer and an inner folded crista. The mitochondria produce ATP and are called the powerhouses of a cell.

5. Lysosomes are spherical bodies that contain digestive enzymes. They are abundant in the phagocytic white blood cells.
6. Peroxisomes are membranous sacs that contain specific enzymes. They are abundant in the kidneys and liver and are important in producing hydrogen peroxide.
7. The centrosome is the dense area of cytoplasm near the nucleus that contains the centrioles. The paired centrioles play an important role in cell division.
8. Vacuoles are membranous sacs that function as storage chambers.
9. Fibrils and microtubules provide support in the formation of a cytoskeleton.
10. Cilia and flagella are projections of the cell that have the same basic structure and function in producing movement.

E. The cell nucleus is enclosed in a nuclear membrane that controls the movement of substances between the nucleoplasm and the cytoplasm.
   1. The nucleoli are small bodies of protein and RNA that produce ribosomes.
   2. Chromatin are coiled fibers of protein and DNA that form chromosomes during cell reproduction.

IV. The Cell Cycle

A. A cell cycle consists of growth, synthesis, and mitosis.
   1. Growth is the increase in cellular mass as the result of metabolism, synthesis is the production of DNA and RNA to regulate cellular activity, and mitosis is the splitting of the nucleus and cytoplasm of a cell that results in two diploid cells being formed.
   2. Mitosis permits an increase in the number of cells (body growth) and allows for the replacement of damaged, diseased, or worn out cells.

B. A DNA molecule is in the shape of a double helix. The structural unit of the molecule is a nucleotide, which consists of a deoxyribose sugar, a phosphate, and a nitrogenous base.

C. Cell division consists of a division of the chromosomes, called mitosis, and a division of the cytoplasm, called cytokinesis. The stages of mitosis include prophase, metaphase, anaphase, and telophase.

Review Activities

Objective Questions

1. Inorganic compounds that form ions when dissociated in water are
   (a) hydrolites. (d) ionizers.
   (b) metabolites. (e) nucleic acids.
   (c) electrolytes.
2. The four elements that compose over 95% of the body are
   (a) oxygen, potassium, hydrogen, carbon.
   (b) carbon, sodium, nitrogen, oxygen.
   (c) potassium, sodium, magnesium, oxygen.
   (d) carbon, oxygen, nitrogen, hydrogen.
   (e) oxygen, carbon, hydrogen, sulfur.
3. Which organelle contains strong hydrolytic enzymes? The
   (a) lysosome.
   (b) Golgi apparatus.
   (c) ribosome.
   (d) vacuole.
   (e) mitochondrion.
4. Ciliated cells occur in the
   (a) trachea.
   (b) ductus deferens.
   (c) bronchioles.
   (d) uterine tubes.
   (e) All of the above.
5. Osmosis deals with the movement of
   (a) gases. (c) only oxygen.
   (b) only water. (d) Both a and c.
6. The phase of mitosis in which the chromosomes line up at the equator (equatorial plane) of the cell is called
   (a) interphase. (d) anaphase.
   (b) prophase. (e) telophase.
   (c) metaphase.
7. The phase of mitosis in which the chromatids separate is called
   (a) interphase. (d) anaphase.
   (b) prophase. (e) telophase.
   (c) metaphase.
8. The organelle that combines protein with carbohydrates and packages them within vesicles for secretion is the
   (a) Golgi apparatus.
   (b) rough endoplasmic reticulum.
   (c) smooth endoplasmic reticulum.
   (d) ribosome.
9. The larger skeletal muscle fibers that result from an increased work demand are an example of
   (a) disuse atrophy.
   (b) compensatory hypertrophy.
   (c) metaplasia.
   (d) inertia.
10. Regeneration of liver cells is an example of
   (a) compensatory hypertrophy.
   (b) hyperplasia.
   (c) metaplasia.
   (d) hypertrophy.

Essay Questions

1. Explain why understanding cellular anatomy is necessary for understanding tissue and organ functioning within the body.
2. Why is water a good fluid medium of the cell?
3. How are proteins, carbohydrates, and lipids similar? How are they different? What are enzymes and hormones?
4. Describe the cell membrane. List the kinds of movement through the cell membrane, and give an example of each.
5. Describe, diagram, and list the functions of the following: (*a*) endoplasmic reticulum; (*b*) ribosome; (*c*) mitochondrion; (*d*) Golgi apparatus; (*e*) centrioles; (*f*) cilia.
6. Define *inorganic compound* and *organic compound,* and give examples of them.
7. Define the terms *protoplasm, cytoplasm,* and *nucleoplasm.* Describe the position of the membranes associated with each of these substances.
8. Describe the structure of the nucleus and the functions of its parts.
9. What is a nucleotide? How does it relate to the overall structure of a DNA molecule?
10. Explain the relationship between DNA, chromosomes, chromatids, and genes.
11. What are the differences between mitosis and cytokinesis? Describe the major events of mitosis, and discuss the significance of the mitotic process.
12. Give examples of factors that bring about cellular hypertrophy, hyperplasia, atrophy, or metaplasia.
13. Explain how cells respond to the following:
   (a) energy deficit (d) radiation
   (b) hyperthermia (e) pathogens
   (c) burns
14. Define the following genetic terms: *teratology, monosomy, trisomy,* and *mutations.*
15. In what ways do neoplasms differ from normal cells? How may a malignant neoplasm cause death?

Reference Figures

The Human Organism

Figure 1.
An anterior view of the human torso with the superficial muscles exposed.

Figure 2.
The torso with the deep muscles exposed.

Figure 3.
The torso with the anterior abdominal wall removed to expose the abdominal viscera.

Figure 4.
The torso with the anterior thoracic wall removed to expose the thoracic viscera.

Figure 5.
The torso as viewed with the thoracic viscera sectioned in a coronal plane, and the abdominal viscera as viewed with most of the small intestine removed.

Figure 6.
The torso as viewed with the heart, liver, stomach, and portions of the small and large intestines removed.

Figure 7.
The torso with the anterior thoracic and abdominal walls removed, along with the viscera, to expose the posterior walls and body cavities.

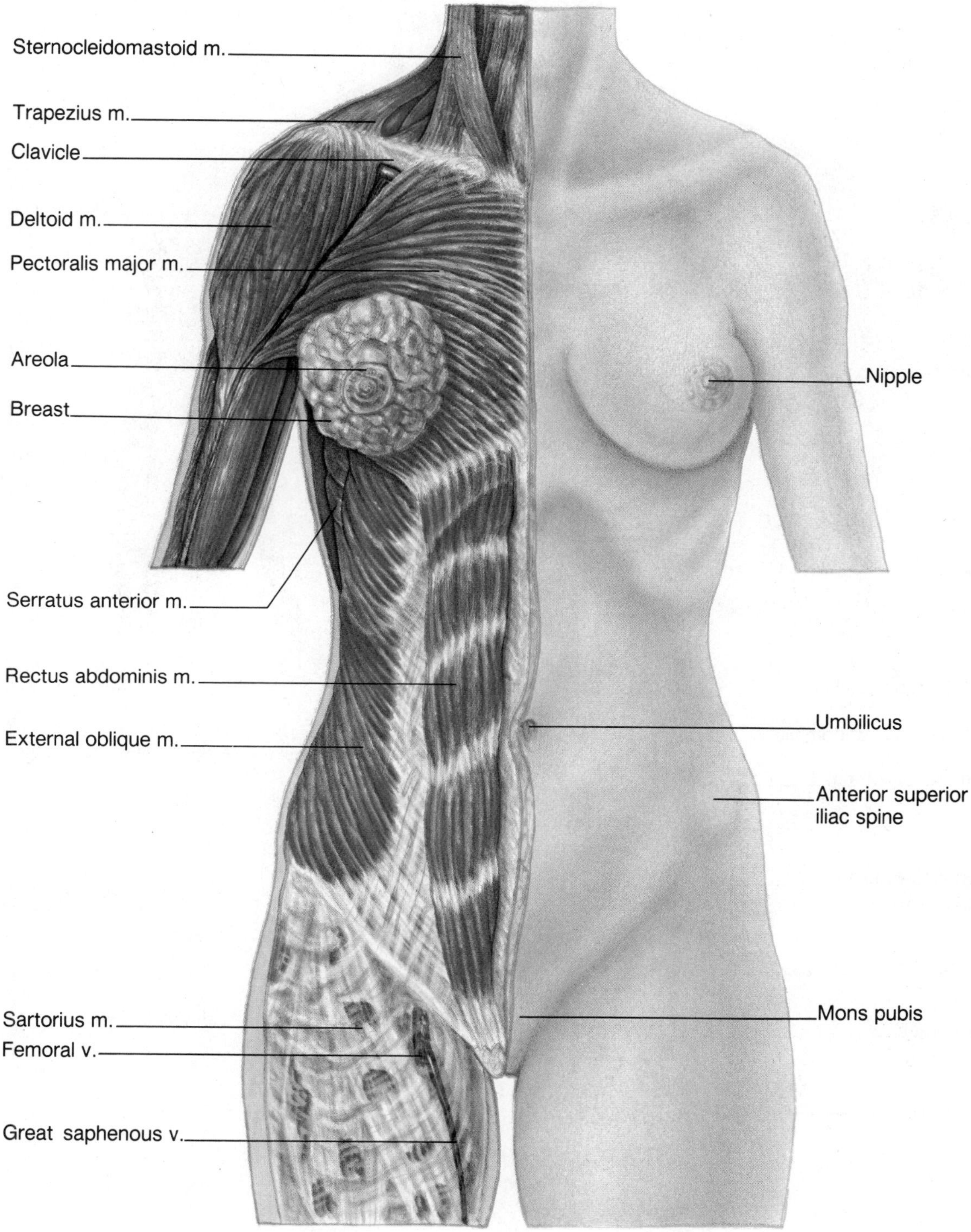

Figure 1 An anterior view of the human torso with the superficial muscles exposed. (m = muscle; v = vein)

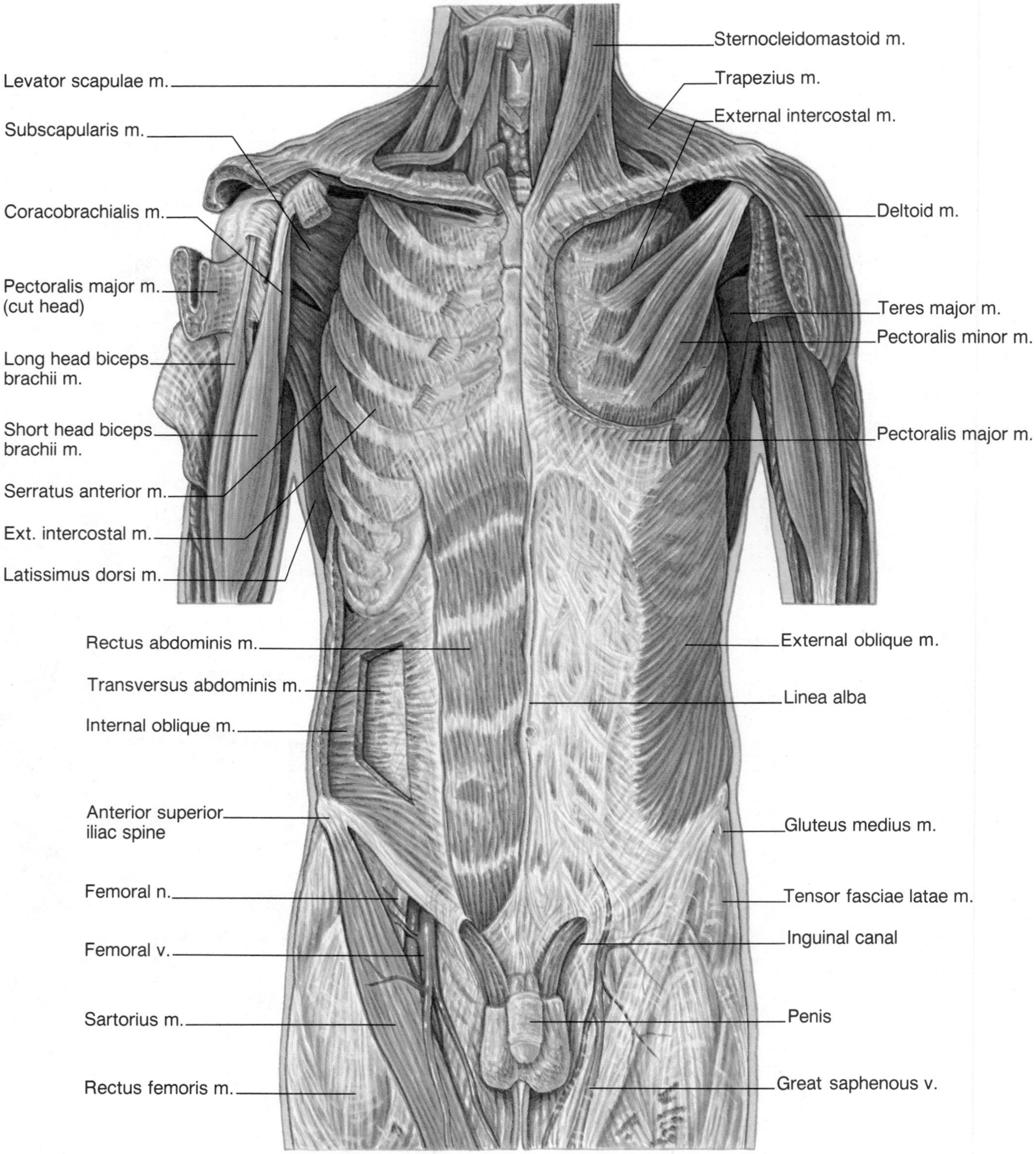

Levator scapulae m.
Subscapularis m.
Coracobrachialis m.
Pectoralis major m. (cut head)
Long head biceps brachii m.
Short head biceps brachii m.
Serratus anterior m.
Ext. intercostal m.
Latissimus dorsi m.
Rectus abdominis m.
Transversus abdominis m.
Internal oblique m.
Anterior superior iliac spine
Femoral n.
Femoral v.
Sartorius m.
Rectus femoris m.
Sternocleidomastoid m.
Trapezius m.
External intercostal m.
Deltoid m.
Teres major m.
Pectoralis minor m.
Pectoralis major m.
External oblique m.
Linea alba
Gluteus medius m.
Tensor fasciae latae m.
Inguinal canal
Penis
Great saphenous v.

Figure 2 The torso with the deep muscles exposed. (n = nerve)

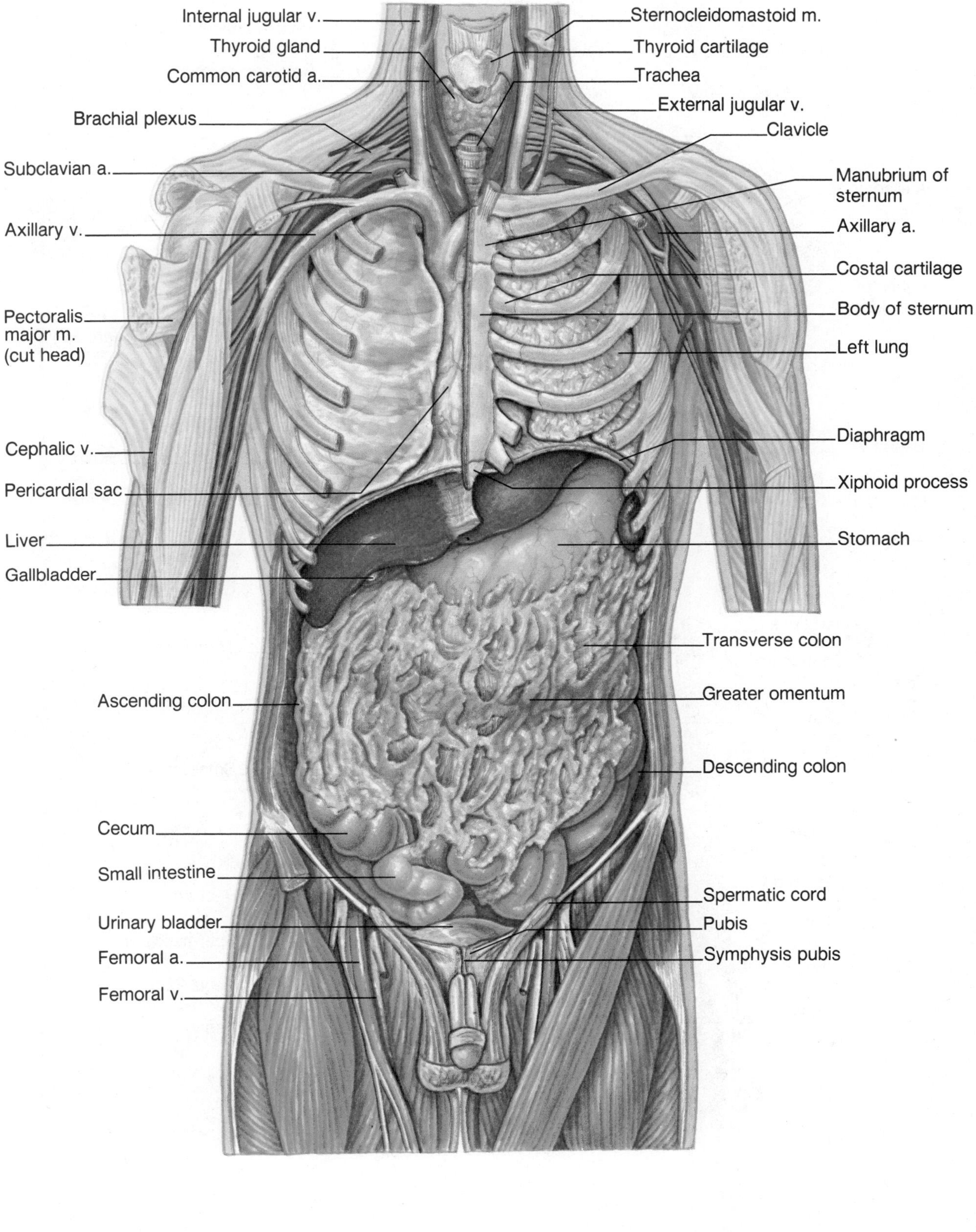

Figure 3 The torso with the anterior abdominal wall removed to expose the abdominal viscera. (a = artery)

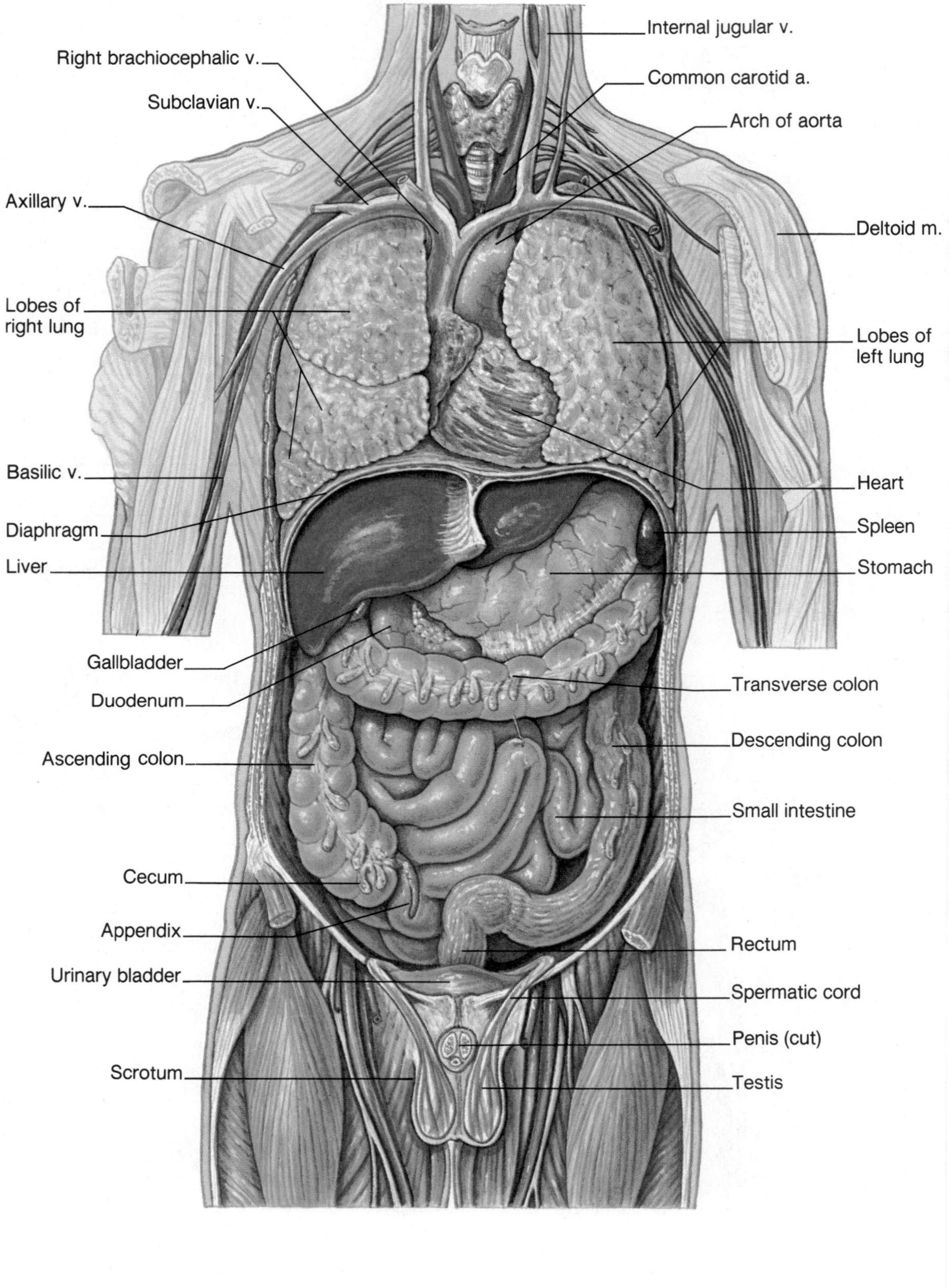

Figure 4 The torso with the anterior thoracic wall removed to expose the thoracic viscera.

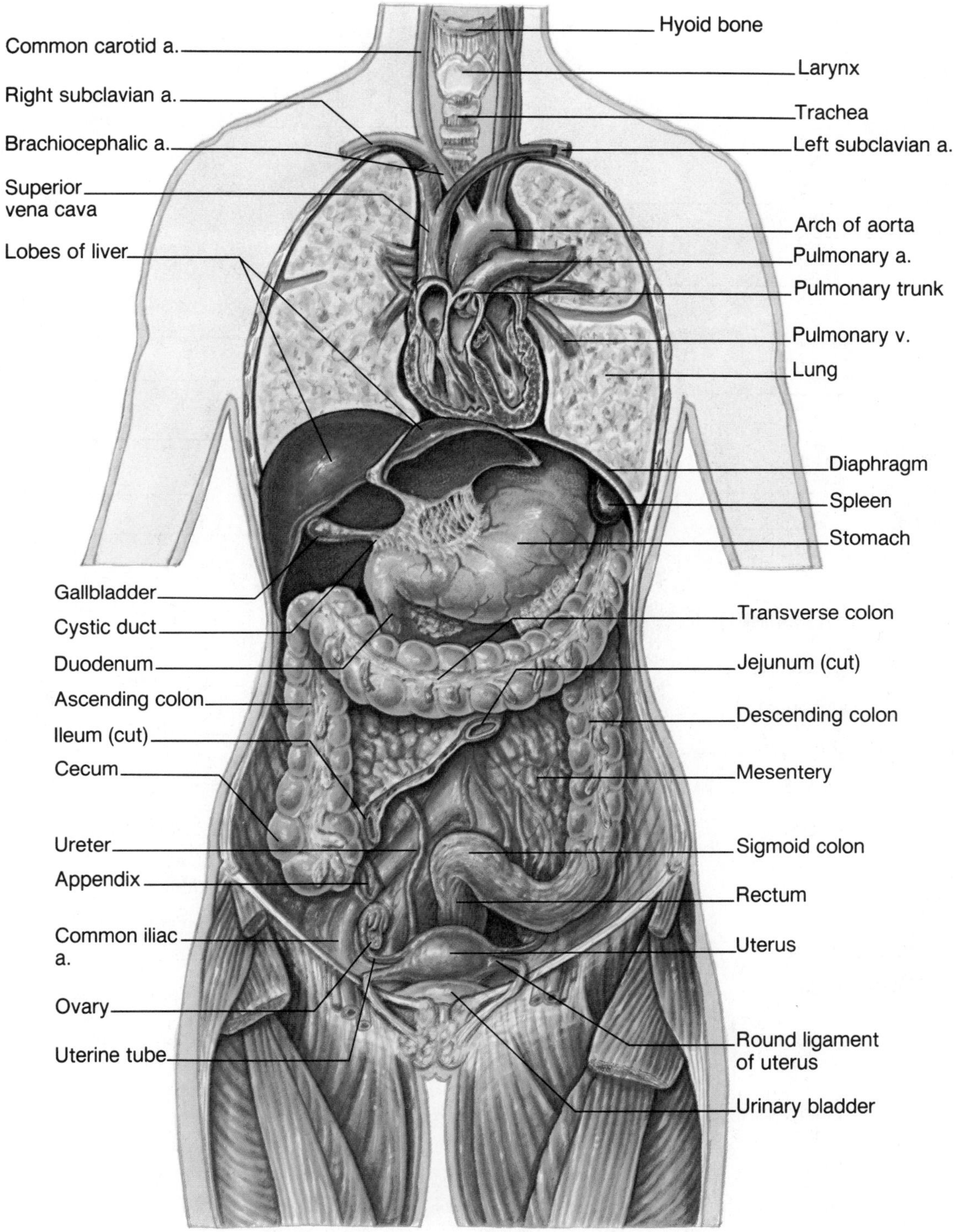

Figure 5 The torso as viewed with the thoracic viscera sectioned in a coronal plane, and the abdominal viscera as viewed with most of the small intestine removed.

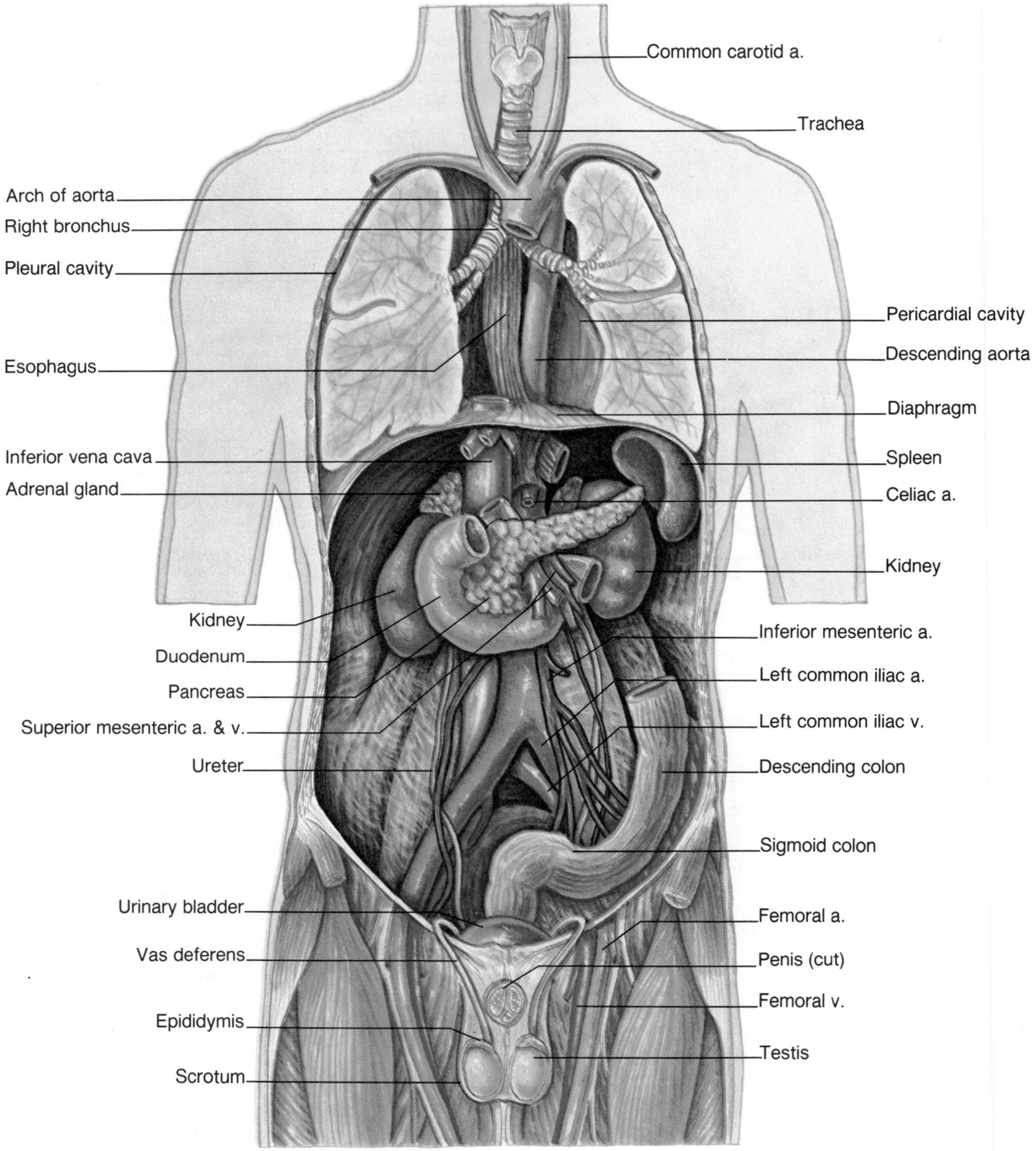

Figure 6 The torso as viewed with the heart, liver, stomach, and portions of the small and large intestines removed.

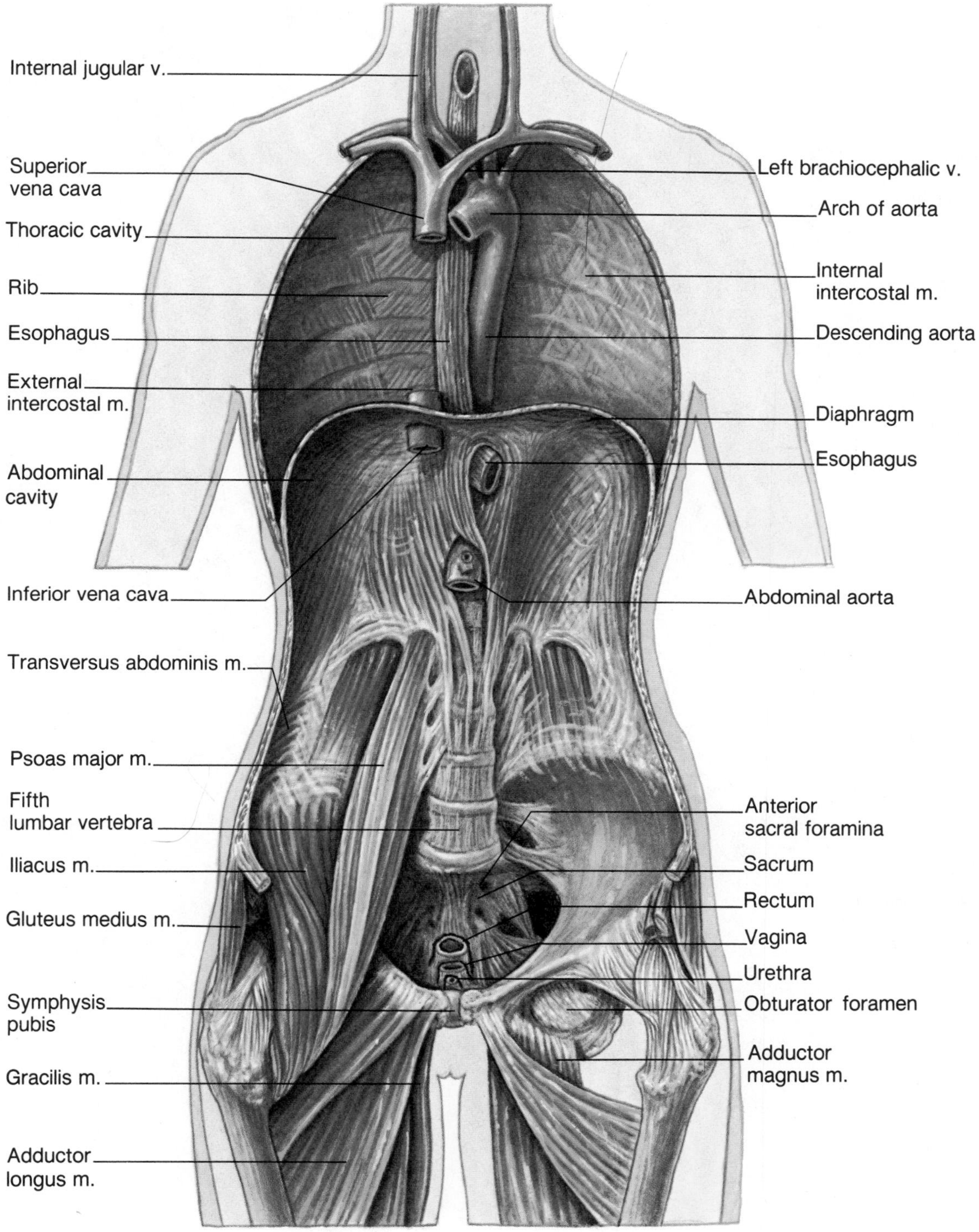

Figure 7 The torso with the anterior thoracic and abdominal walls removed, along with the viscera, to expose the posterior walls and body cavities.

CHAPTER

4

Histology

Outline and Concepts

Definition and Classification of Tissues

Histology is an integral part of anatomy because it imparts an understanding of the structure and function of organs at the tissue level of study. Tissues are classified into four principal kinds on the basis of their cellular composition and histological appearance.

Development of Tissues

The various types of tissues are established during embryonic development, and as differentiation continues, organs form, each of which is composed of a specific arrangement of tissues.

Epithelial Tissue

Epithelia are classified according to the physical features of the tightly packed cells that comprise these tissues. Epithelia line all body surfaces, cavities, and lumina and are adapted for protection, absorption, and secretion.

Characteristics of Epithelia
Simple Epithelia
Stratified Epithelia
Glandular Epithelia

Connective Tissue

Connective tissues are classified according to the characteristics of the matrix that binds the cells. Connective tissues provide structural and metabolic support for other tissues and organs of the body.

Characteristics and Classification of Connective Tissue
Embryonic Connective Tissue
Connective Tissue Proper
Cartilage
Bone
Vascular Connective Tissue

Muscle Tissue

Muscle tissues are responsible for the movement of materials through the body, the movement of one part of the body with respect to another, and for locomotion. Fibers in the three kinds of muscle tissue are adapted to contract in response to stimuli.

Nervous Tissue

Nervous tissue is composed of neurons, which respond to stimuli and conduct impulses to and from all body organs, and neuroglia, which functionally support and physically bind neurons.

Clinical Considerations

Changes in Tissue Composition
Tissue Analysis
Tissue Transplantation

Chapter Summary

Review Activities

Definition and Classification of Tissues

Histology is an integral part of anatomy because it imparts an understanding of the structure and function of organs at the tissue level of study. Tissues are classified into four principal kinds on the basis of their cellular composition and histological appearance.

Objective 1. Define *tissue,* and discuss the importance of histology.
Objective 2. Explain the functional relationship between cells and tissues.
Objective 3. Classify the tissues of the body into four major types, and give the distinguishing characteristics of each type.

Although cells are the structural and functional units of the body, the cells of a multicellular organism are so specialized that they do not function independently. *Tissues* are aggregations of similar cells that perform specific functions. The study of tissues is referred to as **histology** *(his-tol'o-je)*. The various types of tissues are established during embryonic development, and as differentiation continues, organs form, each composed of a specific arrangement of tissues. Many adult organs, such as the heart and muscles, contain the original cells and tissues, which were not replaced through mitotic activity during further growth and development. Some functional changes may occur, however, as the tissues of an organ are acted upon by hormones or as their effectiveness diminishes with age.

Although histology is actually microscopic anatomy, it is an essential part of anatomy because it imparts an understanding of the structure and function of organs at the tissue level. Many diseases profoundly alter the tissues within an affected organ; therefore, by knowing the normal tissue structure, a medical person can recognize the abnormal. Histology in medical schools is usually followed by a course in *pathology,* which is primarily concerned with identifying diseased tissues.

Although histologists utilize many different techniques for preparing, staining, and sectioning tissues, basically only two kinds of microscopes are used to view the prepared tissues. *Light microscopy (mi-kros'ko-pe)* is used for the general observation of tissue structure (fig. 4.1), and *electron microscopy* permits observation of the fine details of tissue and cellular structure. Most of the histological photomicrographs in this text will be at the light microscopic level to present an overview of tissue structure. A few electron micrographs from an electron microscope are used to depict the fine structural detail needed to understand a particular function.

Tissue cells are separated and bound together by a nonliving intercellular **matrix** *(ma'triks)* that the cells secrete. Matrix varies in composition from one tissue to another and may take the form of a liquid, semisolid, or solid. Blood, for example, has a liquid matrix, permitting this tissue to flow through vessels, whereas bone cells are separated by a solid matrix, permitting this tissue to support the body.

The tissues of the body are classified into four principal types, determined by structure and function: (1) *epithelial tissues* cover body and organ surfaces, line body and lumen cavities, and form various glands; (2) *connective tissues* bind, support, and protect body parts; (3) *muscle tissues* contract to produce movement; and (4) *nervous tissues* initiate and transmit nerve impulses from one body part to another.

1. Define the term *tissue,* and explain why histology is important to the study of anatomy and medicine.
2. Cells are the functional units of the body. Explain how the matrix permits specific kinds of cells to be even more effective and functional as tissues.
3. List and give the distinguishing characteristics of the four principal types of tissues in the body, and explain their basic functions.

Development of Tissues

The various types of tissues are established during embryonic development, and as differentiation continues, organs form, each of which is composed of a specific arrangement of tissues.

Objective 4. Describe the location of the three primary germ layers during early embryonic development.
Objective 5. List the tissues and body organs derived from each of the three primary germ layers.

Human prenatal development is initiated with the fertilization of an ovulated ovum (egg) from a female by a sperm cell from a male. The chromosomes within the nucleus of a **zygote** *(zi'gōt)* (fertilized egg) contain all the genetic information necessary for the differentiation and development of all body structures.

Within thirty hours after fertilization, the zygote undergoes a mitotic division as it moves through the uterine tube toward the uterus (chap. 22). After several more cellular divisions, the embryonic mass consists of sixteen or more cells and is called a **morula** *(mor'u-lah)* (fig. 4.2). Three or four days after conception, the morula enters the uterine cavity where it remains unattached for about three days. During this time, the center of the morula fills with fluid passing in from the uterine cavity. As the fluid-filled space develops inside the morula, two distinct groups of cells form. The single layer of cells, called *trophoblast cells,* forming the outer wall becomes the **trophoblast,** and the small, inner aggregation of cells becomes the **embryoblast,** or **inner cell mass.** After further development the trophoblast becomes a portion of the placenta, and the embryoblast becomes the embryo. With the establishment of these two groups of cells, the morula becomes known as a **blastocyst** *(blas'to-sist)*. Implantation of the blastocyst begins between the fifth and seventh day (chap. 22).

As the blastocyst completes implantation during the second week of development, the embryoblast undergoes marked differentiation. A slitlike space called the **amniotic** *(am'ne-ot-ic)* **cavity** forms within the embryoblast adjacent to the trophoblast (see fig. 4.2). The embryoblast now consists of two layers: an upper **ectoderm,** which is closer to the amniotic cavity, and

histology: Gk. *histos,* web (tissue); *logos,* study
pathology: Gk. *pathos,* suffering, disease; *logos,* study
matrix: L. *matris,* mother

zygote: Gk. *zygotos,* yolked
morula: Gk. *morus,* mulberry
trophoblast: Gk. *trophe,* nourishment; *blastos,* germ
embryoblast: Gk. *embryon,* to be full, swell; *blastos,* germ
ectoderm: Gk. *ecto,* outside; *derm,* skin

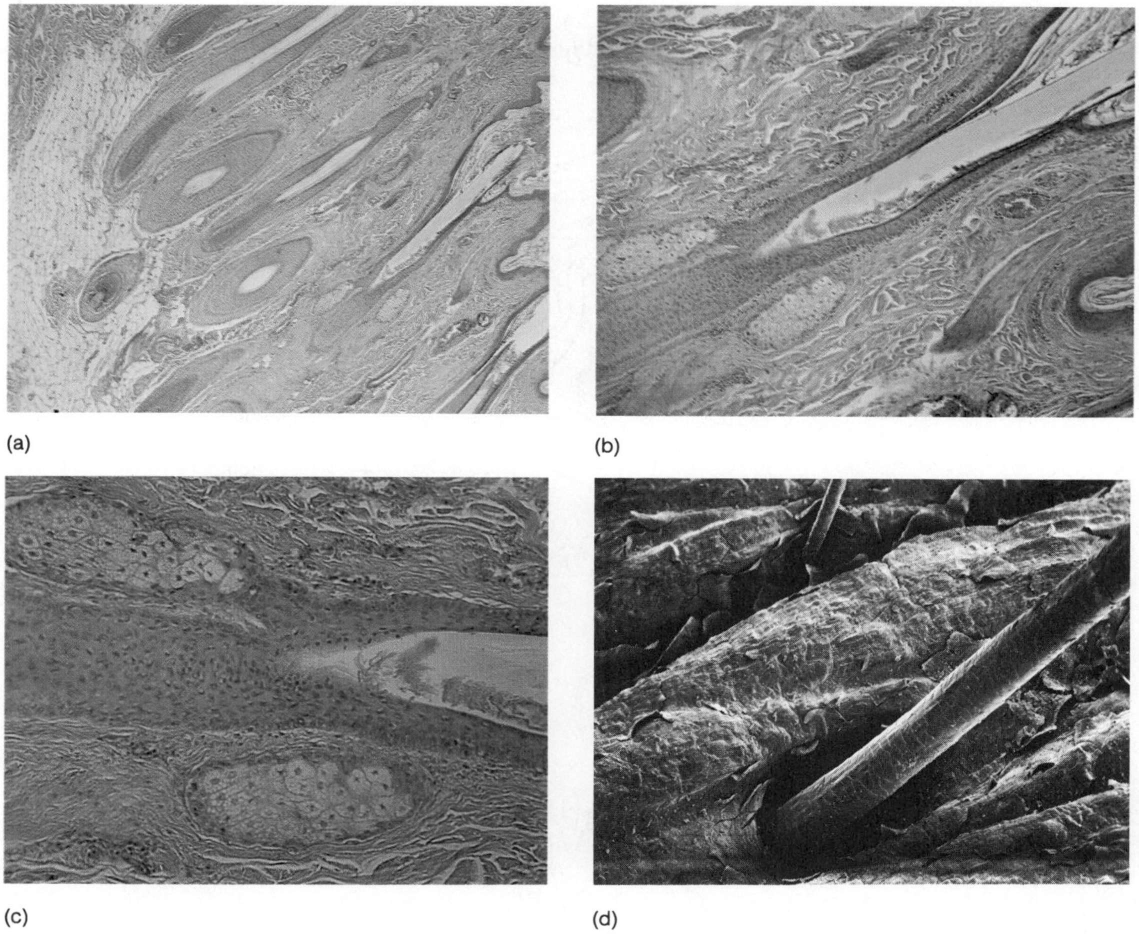

Figure 4.1 The appearance of skin at various magnifications: (*a*) = 10× (*b*) = 25×; (*c*) = 50× through a compound light microscope; and (*d*) a hair emerging from a follicle as seen through a scanning electron microscope (SEM) at 280×.

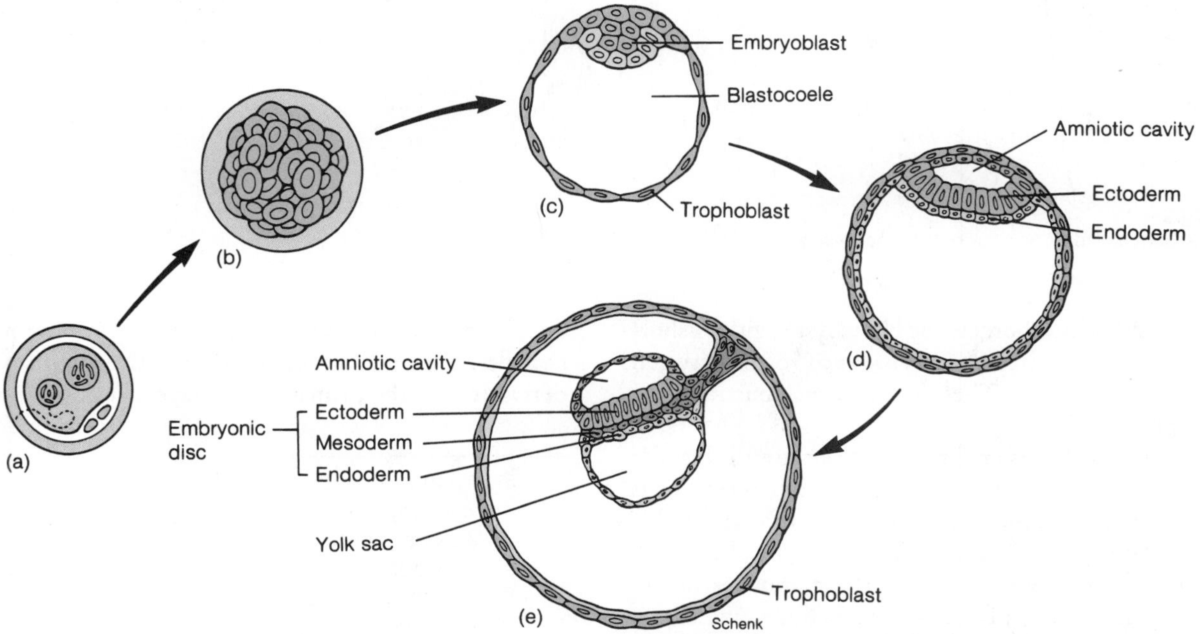

Figure 4.2 The early stages of embryonic development. (*a*) The zygote immediately following conception. (*b*) The morula at about the third day as it enters the uterine cavity. (*c*) The early blastocyst at the time of implantation between the fifth and seventh day. (*d*) A cross section of an implanted blastocyst at two weeks. (*e*) A cross section of a blastocyst at three weeks showing the three primary germ layers, which constitute the embryonic disc.

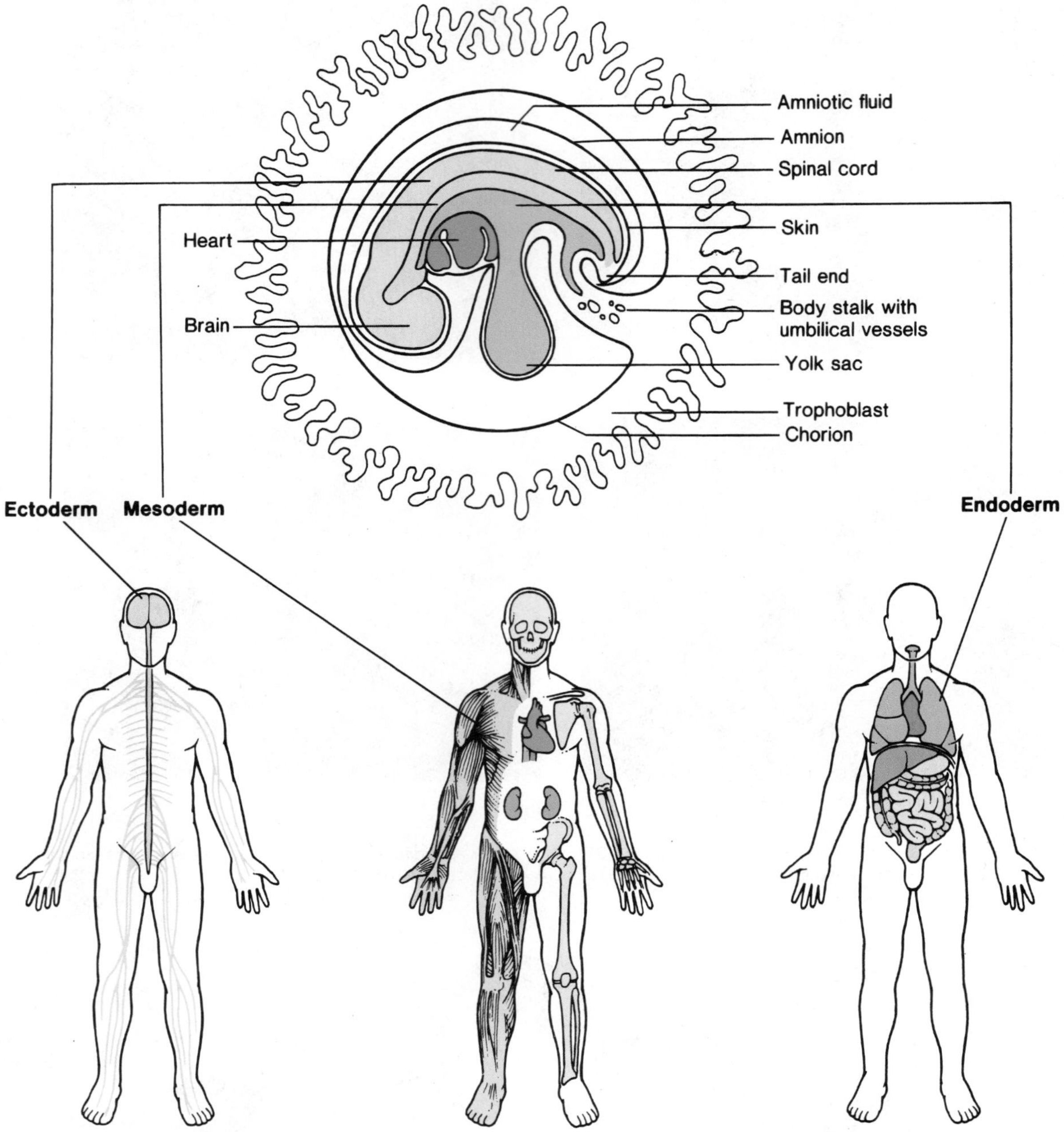

Figure 4.3 The body systems and the primary germ layers from which they develop.

a lower **endoderm,** which borders the blastocyst cavity. A short time later, a third layer called the **mesoderm** forms between the endoderm and ectoderm. These three layers constitute the **primary germ layers.**

The primary germ layers are important because all the cells and tissues of the body are derived from them. Ectodermal cells form the nervous system; the outer layer of skin (epidermis), including hair, nails, and skin glands; and portions of the sensory organs. Mesodermal cells form the skeleton, muscles, blood, reproductive organs, dermis of the skin, and connective tissue. Endodermal cells produce the lining of the digestive tract, the digestive organs, the respiratory tract and lungs, and the urinary bladder and urethra.

Illustrated in figure 4.3 are the organs and body systems that derive from each of the three primary germ layers. The derivatives of the primary germ layers are listed in table 4.1.

---

1. What is a morula? How does it differ from a blastocyst?
2. At what developmental age (weeks following conception) are the germ layers present, and where are they located?
3. Referring to chapter 2, list the ten body systems, and indicate the principal germ layers from which the organs of each system derive.

---

endoderm: Gk. *endo,* within; *derm,* skin
mesoderm: Gk. *meso,* middle; *derm,* skin

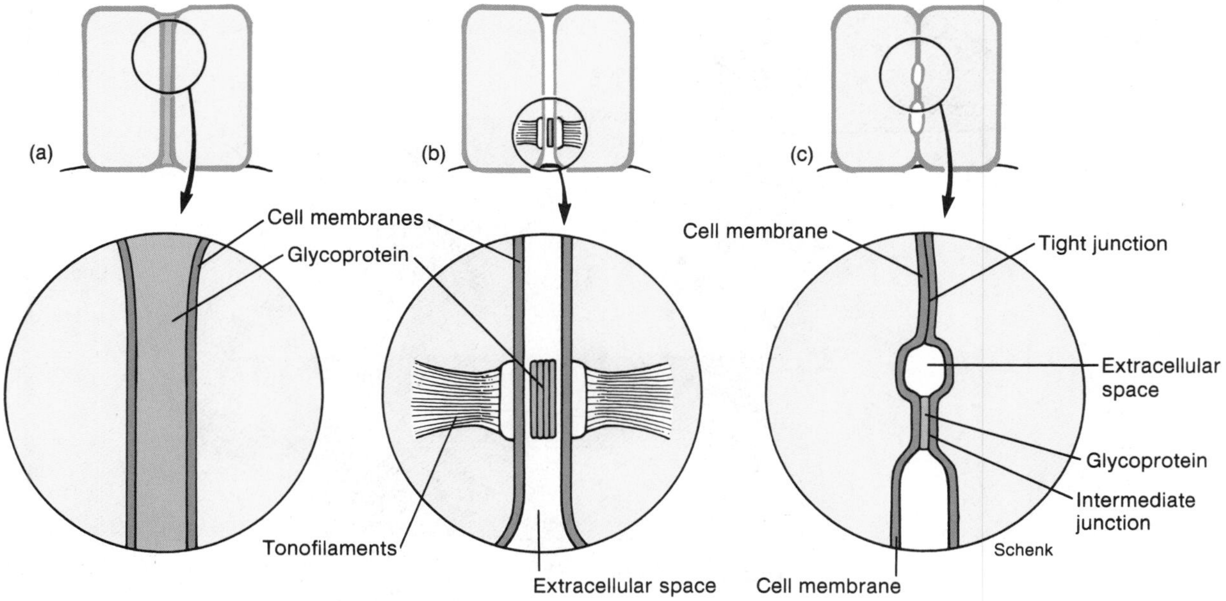

Figure 4.4 Types of cellular bonding in epithelia. (*a*) Glycoprotein deposits, (*b*) desmosomes, and (*c*) tight and intermediate junctions.

Table 4.1 Derivatives of the germ layers

Ectoderm	Mesoderm	Endoderm
Epidermis of skin and epidermal derivatives: hair, nails, glands of the skin; linings of oral, nasal, anal, and vaginal cavities	Muscle: smooth, cardiac, and skeletal	Epithelium of pharynx, auditory canal, tonsils, thyroid, parathyroid, thymus, larynx, trachea, lungs, digestive tract, urinary bladder and urethra, and vagina
Nervous tissue; sense organs	Connective tissue: embryonic, connective tissue proper, cartilage, bone, blood	Liver and pancreas
Lens of eye; enamel of teeth	Dermis of skin; dentin of teeth	
Pituitary gland	Epithelium of blood vessels, lymphatic vessels, body cavities, joint cavities	
Adrenal medulla	Internal reproductive organs	
	Kidneys and ureters	
	Adrenal cortex	

Epithelial Tissue

Epithelia are classified according to the physical features of the tightly packed cells that comprise these tissues. Epithelia line all body surfaces, cavities, and lumina and are adapted for protection, absorption, and secretion.

Objective 6. Differentiate between the kinds of epithelia.
Objective 7. Describe how epithelial cells are held together.
Objective 8. Define *gland,* and distinguish between the various types of glands in the body.

Characteristics of Epithelia

Epithelia *(ep''i-the'le-ah)* are located throughout the body and form such structures as the outer layer of the skin, the inner lining of body cavities and lumina (hollow portions of body tubes), the covering of visceral organs, and the secretory portion of glands.

There is always one side of epithelia exposed to a body cavity, lumen, or skin surface. Some epithelia are derived from ectoderm, such as the outer layer of the skin and integumentary glands; some from mesoderm, such as the inside lining of blood vessels; and others from endoderm, such as the inside lining of the digestive tract.

Epithelia may be one layer or several layers thick. The upper surface of epithelia may be exposed to gases, as in the case of epithelium in the integumentary and respiratory systems; to liquids, as in the circulatory and urinary systems; or to semisolids, as in the digestive system. The deep surface of epithelia is bound to underlying supportive tissue by a **basement membrane,** consisting of glycoprotein from the epithelial cells and a meshwork of collagenous and reticular fibers from the underlying connective tissue. With few exceptions, epithelia are avascular (without blood vessels) and must be nourished by diffusion from underlying connective tissues. Cells comprising epithelia are tightly packed together, and there is little intercellular matrix between them.

The cells of epithelia are tightly bonded in one of three ways (fig. 4.4): (1) Some epithelial cells have a bonding matrix of *glycoprotein deposits.* Glycoprotein is a combination of polysaccharides and protein secreted by the cells to bind them to other like cells in a process similar to that which binds epithelia

epithelium: Gk. *epi,* upon; *thelium,* to cover

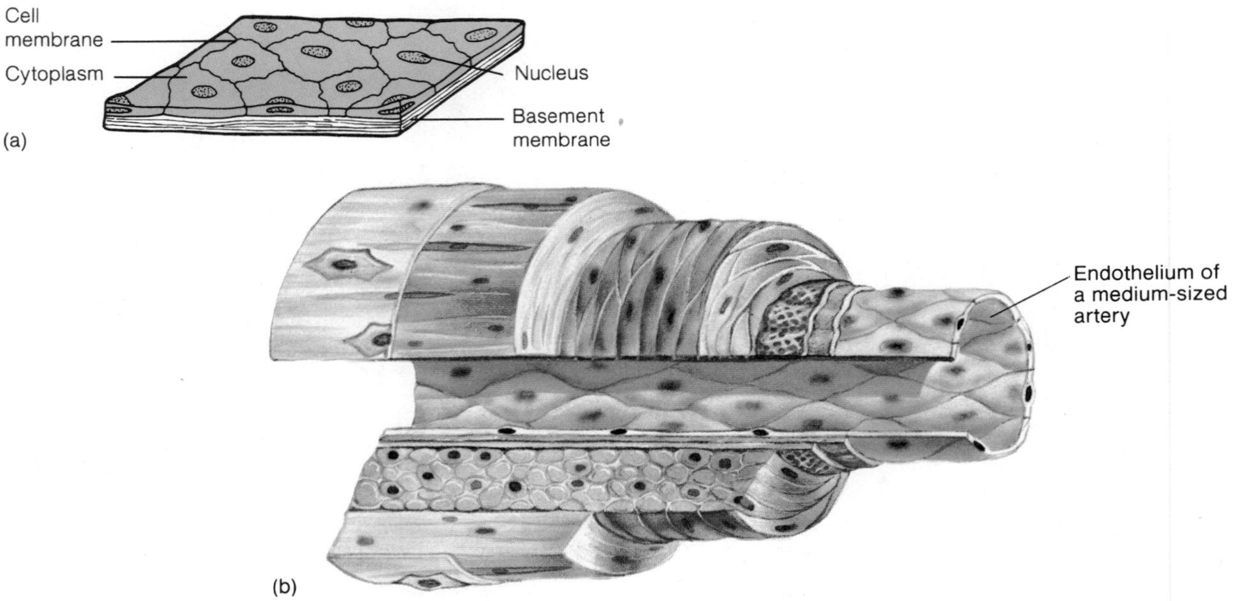

Figure 4.5 Simple squamous epithelium. (*a*) As the name implies, this type of epithelium consists of a single layer of flattened cells and is highly adapted to diffusion, osmosis, and filtration. (*b*) Where it lines the lumina of blood and lymphatic vessels, it is referred to as endothelium.

to the basement membrane. (2) Certain epithelial cells are bonded with *desmosomes.* Desmosomes are characterized by V-shaped **tonofilaments** *(ton''o-fil'ah-mentz),* reinforcing the cell membrane where glycoprotein is secreted to bond adjacent cells. (3) *Tight junctions* and *intermediate junctions* occur in the same type of epithelium. A tight junction exists where the cell membranes of adjacent cells form a serrated pattern. In an intermediate junction, the cell membranes of adjacent cells are bonded with glycoprotein.

Some of the functions of epithelial tissues are quite specific, but certain generalities can be made. Epithelia that cover or line surfaces provide *protection* from pathogens, physical injury, toxins, and desiccation. Epithelia lining the lumen of the digestive tract function in *absorption* and *secretion.* Glandular epithelia elsewhere in the body are also secretory. The epithelium of the kidneys provides *filtration,* whereas the epithelium within the air sacs of the lungs allows *diffusion.* Highly specialized *neuroepithelium* in the taste buds and in the nasal region functions as *chemoreceptors.*

Many epithelial tissues are exposed and, therefore, subject to trauma and destruction. For this reason, epithelial tissues have remarkable regenerative abilities. The mitotic replacement of the outer layer of skin and the lining of the digestive tract, for example, is a continuous process.

Epithelial tissues are histologically classified by the number of layers of cells and the shape of the cells along the exposed surface. Epithelial tissues that are composed of a single layer of cells are called *simple,* and those that are layered are said to be *stratified. Squamous (skwa'mus)* cells are flattened, *cuboidal* cells are cube-shaped, and *columnar* cells are taller than they are wide. Glandular epithelia are specialized types that have secretory functions.

squamous: L. *squamosus,* scaly

Simple Epithelia

Simple epithelial tissues are a single layer thick and are located where diffusion, filtration, and secretion occur. The cells that constitute simple epithelia range in size from thin, flattened cells to tall, columnar cells, depending on function. These cells may also exhibit surface specializations, such as cilia and microvilli, which facilitate specific surface functions.

Simple Squamous Epithelium Simple squamous epithelium is composed of flattened, irregularly shaped cells that are tightly bound together in a mosaic pattern (fig. 4.5a). Each cell contains an oval, centrally located nucleus. Simple squamous epithelium is adapted for diffusion and filtration and occurs in such places as the lining of air sacs within the lungs (where gaseous exchange occurs), portions of the kidney (where blood is filtered), the inside lining of the walls of blood vessels, the lining of body cavities, and in the covering of the viscera. The simple squamous epithelium lining the lumina of blood and lymphatic vessels is sometimes referred to as **endothelium** (fig. 4.5b). **Mesothelium** covers visceral organs and lines body cavities.

Simple Cuboidal Epithelium Simple cuboidal epithelium is composed of a single layer of tightly fitted, hexagonal cells (fig. 4.6). This type of epithelium is found lining small ducts and tubules that may have excretory, secretory, or absorptive functions. It occurs on the surface of the ovaries, forms a portion of the tubules within the kidney, and lines the ducts of the salivary glands and pancreas.

endothelium: Gk. *endon,* within; *thelium,* to cover
mesothelium: Gk. *meso,* middle; *thelium,* to cover

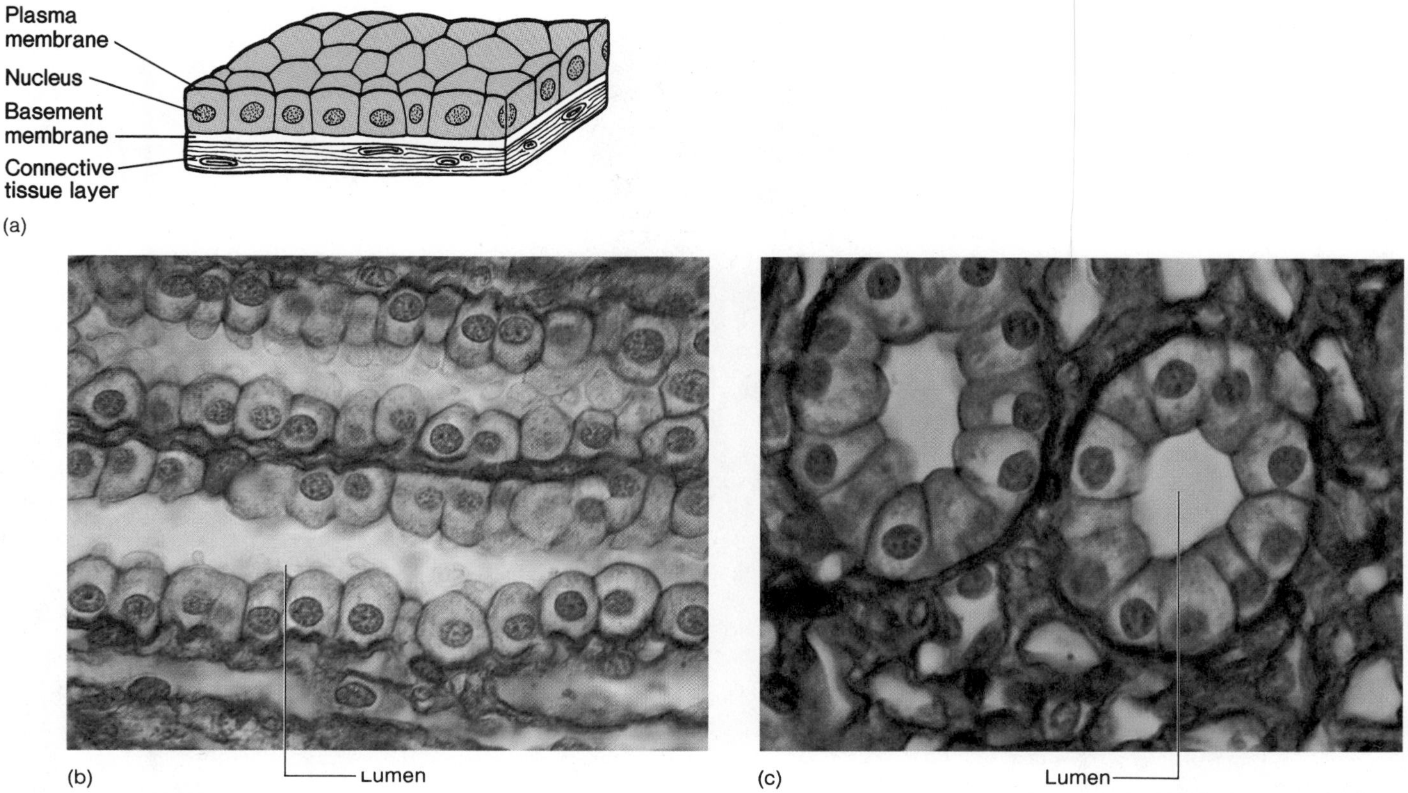

Figure 4.6 Simple cuboidal epithelium lines the lumina of ducts, as seen in the photomicrographs of kidney tubules. (*a*) A diagrammatic drawing of this type of epithelium that consists of a single layer of tightly packed, cube-shaped cells, each with a large, centrally located nucleus; (*b*) a longitudinal view of a tubule; and (*c*) a transverse view of several tubules.

Simple Columnar Epithelium Simple columnar epithelium is composed of tall, columnar cells (fig. 4.7). The height of the cells varies depending on the site and function of the tissue. Each cell contains a single nucleus usually located near the basement membrane. Specialized unicellular glands, called **goblet cells,** are dispersed throughout this tissue and secrete a lubricative and protective mucus along the free surfaces of the cells. This type of epithelium is found lining the lumen of the stomach and intestine. In the digestive system, simple columnar epithelium forms a highly absorptive surface and also secretes certain digestive substances. Within the stomach, simple columnar epithelium has a tremendous rate of mitotic activity—replacing itself every two or three days.

Simple Ciliated Columnar Epithelium Simple ciliated columnar epithelium differs from the simple columnar type by the presence of cilia along the free surface (fig. 4.8). Cilia produce wavelike movements that transport materials through tubes or passageways. This type of epithelium occurs in the uterine tubes of the female where the currents generated by the cilia propel the ovum toward the uterus.

Not only do cilia of the columnar epithelium move the ovum, but recent evidence indicates that sperm introduced during sexual intercourse may be moved along the return currents, or eddies, produced by ciliary movement. This gives fertilization a much greater chance of occurring.

Pseudostratified Ciliated Columnar Epithelium As the name implies, this type of epithelium appears stratified but is actually simple since each cell is in contact with the basement membrane, though not all cells are exposed to the surface (fig. 4.9). The epithelium has a stratified appearance because the nuclei of these cells are located at different levels. Numerous goblet cells and a ciliated, exposed surface are characteristic of this epithelium. The lumina of the trachea and the bronchial tubes are lined with this tissue; hence, it is frequently called respiratory epithelium. Its function is to remove foreign dust and bacteria entrapped in mucus from the lower respiratory system.

Coughing, sneezing, or simply clearing the throat are protective reflex mechanisms for clearing the respiratory passages of obstruction or of inhaled particles that have been trapped in the mucus along the ciliated lining. The material that is coughed up consists of the mucus-entrapped particles. For the health and proper functioning of the air sacs within the lungs, foreign material must not be permitted to enter this level.

Stratified Epithelia

Stratified epithelia are tissues consisting of two or more layers of cells. In contrast to simple epithelia, stratified epithelia are poorly suited for absorption and secretion because of their thickness. Stratified epithelia have a primarily protective function that is enhanced by a characteristic rapid mitotic activity.

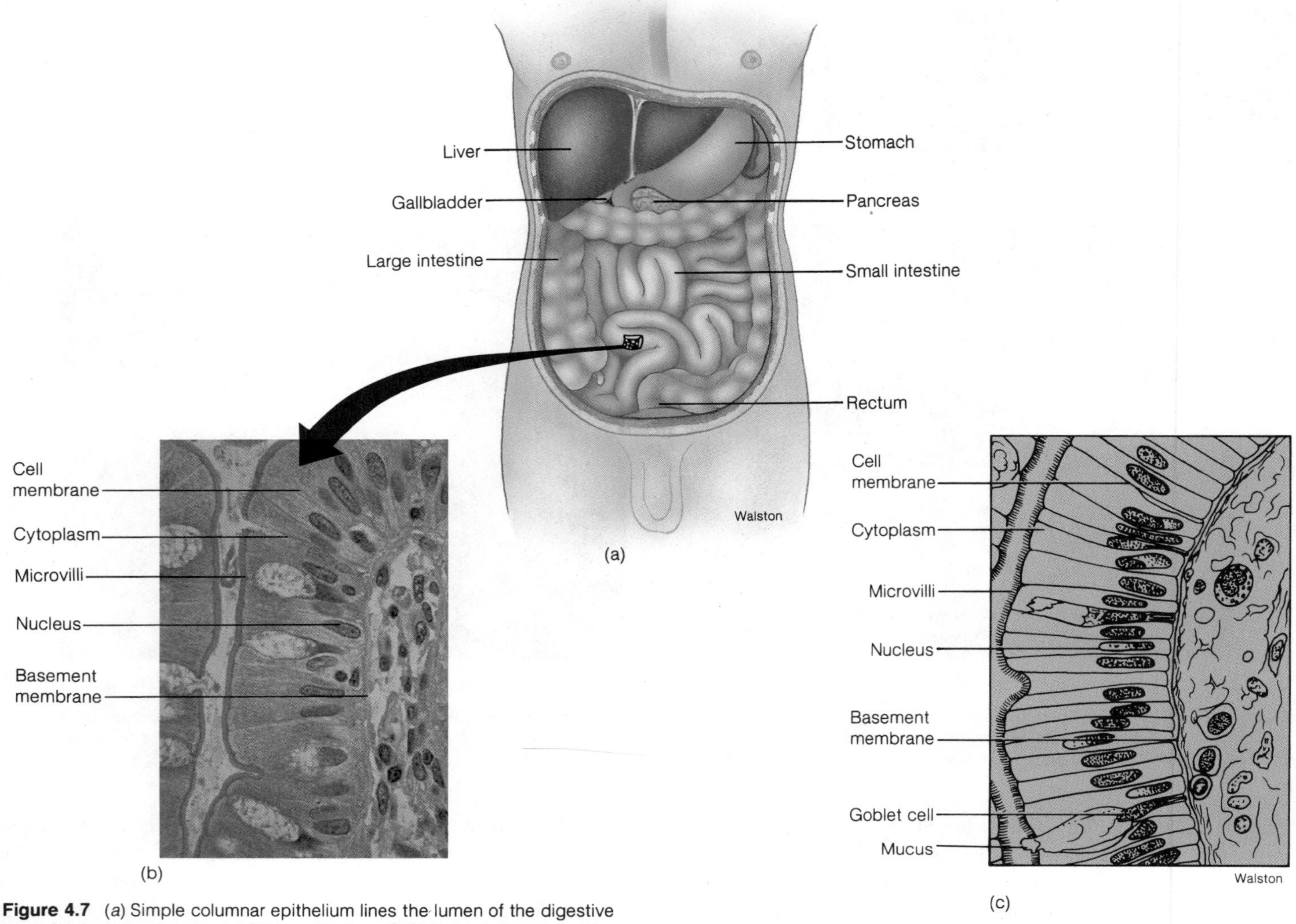

Figure 4.7 (*a*) Simple columnar epithelium lines the lumen of the digestive tract. (*b*) A photomicrograph of this epithelium (250×). (*c*) A diagrammatic drawing of the tall, columnar cells and the associated goblet cells.

Stratified epithelia are classified according to the shape of the surface layer of cells, since the layer in contact with the basement membrane is cuboidal or columnar in shape.

Stratified Squamous Epithelium Stratified squamous epithelium is composed of a variable number of cell layers that tend to flatten near the surface (fig. 4.10). Only at the deepest layer, called the **stratum basale,** does mitosis occur. The mitotic rate approximates the rate that cells are sloughed off at the surface. As the newly produced cells grow in size, they are pushed toward the surface where they will replace the cells that are sloughed off. Movement of the epithelial cells away from the supportive basement membrane is accompanied by the production of **keratin** *(ker'ah-tin),* progressive dehydration, and flattening.

There are two types of stratified squamous epithelia: *keratinized* and *nonkeratinized.* Stratified squamous epithelium that is keratinized forms the outer layer, or *epidermis,* of the skin (see chap. 5). Keratin is a protein that strengthens the tissue. This type of epithelium is especially durable and can generally withstand physical abrasion, desiccation, and bacterial invasion. The outer layers of the stratified squamous epithelium of the skin are dead but are kept pliable by local glandular secretions.

Nonkeratinized stratified squamous epithelium lines the oral cavity and pharynx, nasal cavity, vagina, and anal canal. This type of epithelium, called *mucosa,* is well adapted to withstand moderate abrasion but not fluid loss. The cells on the free surface of this tissue remain alive and are always moistened.

Stratified squamous epithelium is the first line of defense against the entry of living organisms into the body. Stratification as well as rapid mitotic activity and keratinization within the epidermis of the skin are important protective features. An acidic pH along the surfaces of this tissue also helps prevent disease. The pH of the skin is between 4 and 6.8. The pH in the oral cavity ranges between 5.8 and 7.1, which tends to retard the growth of microorganisms. The pH of the anal region is about 6, and the pH along the surface of the vagina is 4 or lower.

Stratified Cuboidal Epithelium Stratified cuboidal epithelium usually consists of only two or three layers of cuboidal cells forming the lining around a lumen (fig. 4.11). This type of epithelium is confined to the linings of the larger ducts of sweat

keratin: Gk. *keras,* horn

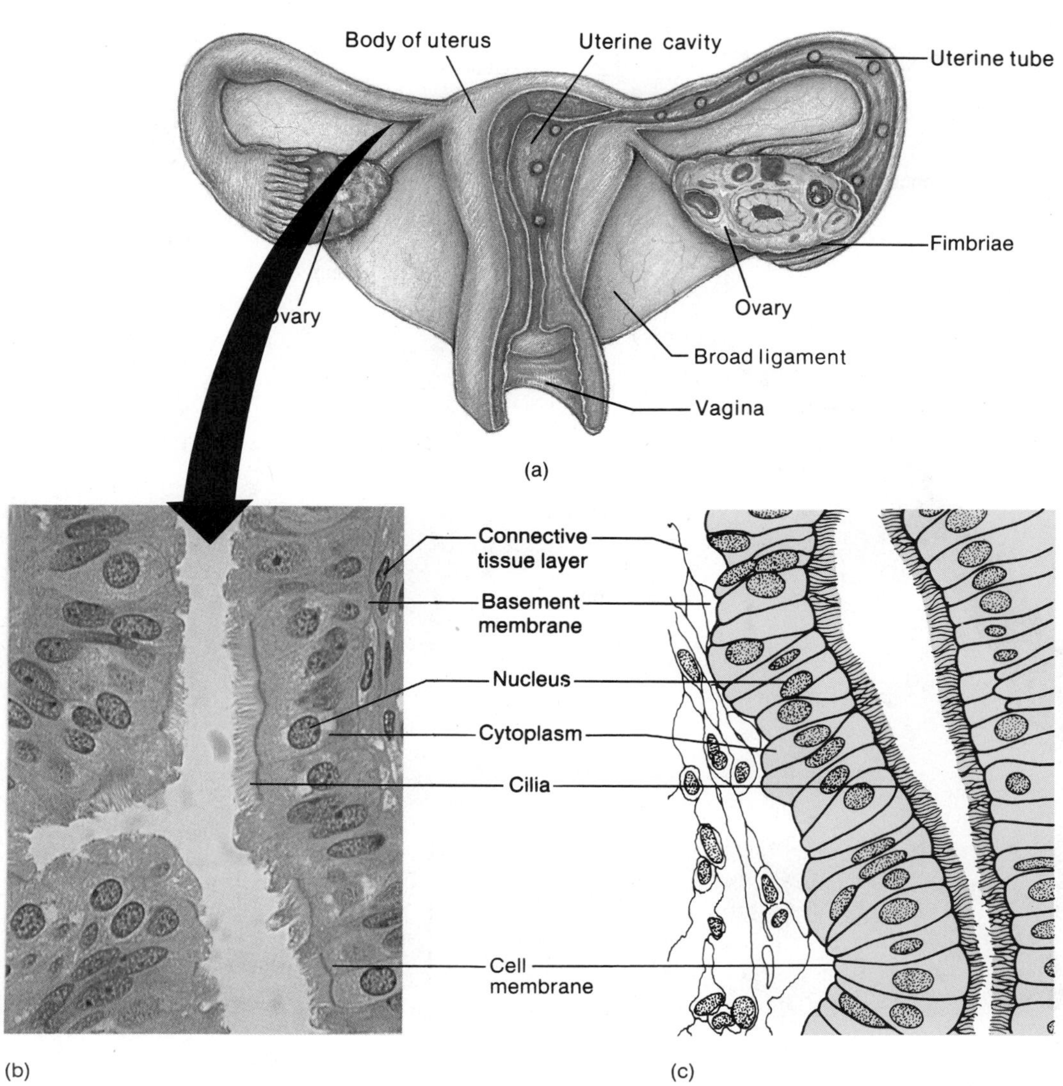

Figure 4.8 (*a*) Simple ciliated columnar epithelium occurs within the uterine tubes of the female reproductive system, where it moves the ovum toward the uterine cavity. (*b*) A photomicrograph showing the cilia (250×), and (*c*) a diagram.

glands, salivary glands, and the pancreas. The stratification of this tissue probably provides a more robust lining than would be afforded by simple epithelium.

Transitional Epithelium Transitional epithelium is similar to nonkeratinized stratified squamous epithelium except that the surface cells are large and round rather than flat and some may have two nuclei (fig. 4.12). Transitional epithelium is located only within the urinary system, particularly in the luminal surface of the urinary bladder and the walls of the ureters. This tissue is specialized to permit distension (stretching) of the urinary bladder.

The epithelial tissues are summarized in table 4.2.

Glandular Epithelia

During the prenatal development of epithelial tissue, certain epithelial cells invade the underlying connective tissue, forming specialized secretory accumulations called *exocrine (ek'so-krin) glands.* Exocrine glands retain a connection to the surface in the form of a duct through which secretions flow. Exocrine glands within the integumentary system include sebaceous (oil) glands, sweat glands, and mammary glands. Within the digestive system, they include the salivary and pancreatic glands.

Exocrine glands are derived from epithelial tissue and secrete a substance through ducts to the surface of the skin or the lumen of a body cavity. Exocrine glands should not be confused with endocrine glands, which are ductless and secrete hormones into the blood.

Exocrine glands are classified according to the structure of the gland and its means of discharging the secretory product. Structurally, there are two types, unicellular and multicellular.

Unicellular Glands Unicellular glands are single-celled glands interspersed with the various columnar epithelia. A mucus-secreting goblet cell is a good example of a unicellular gland. Goblet cells are found in the epithelial linings of the respiratory, digestive, urinary, and reproductive systems, where the mucus secretion lubricates and protects the surface linings (fig. 4.13).

exocrine: Gk. *exo,* outside; *krinein,* to separate

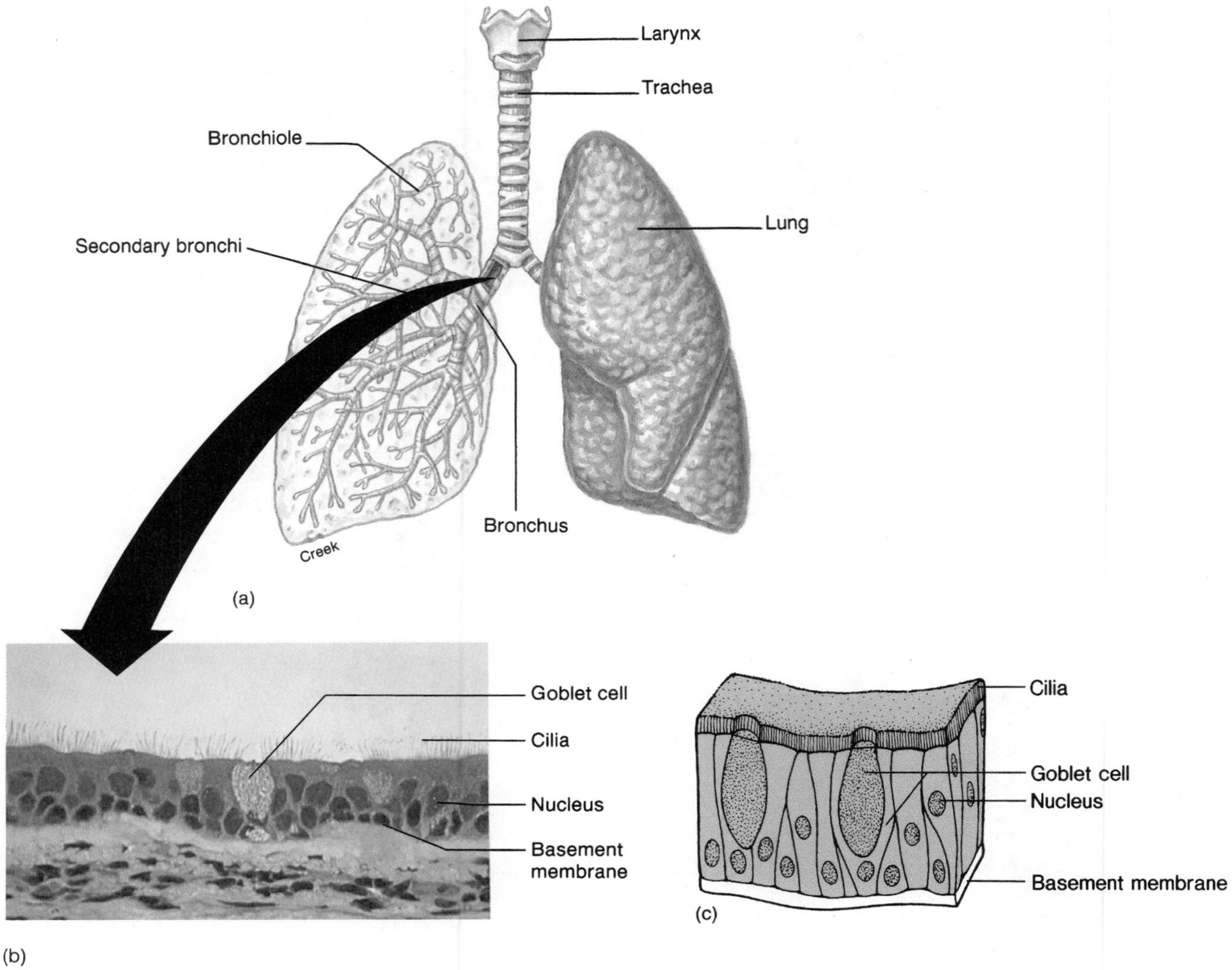

Figure 4.9 (*a*) Pseudostratified ciliated columnar epithelium lines most of the respiratory tract. The action of the cilia and the mucus secreted by goblet cells trap foreign material and move it away from the alveoli of the lungs. (*b*) This type of tissue appears stratified (250×), but (*c*) all the cells are in contact with the basement membrane.

Multicellular Glands Multicellular glands, as their name implies, are composed of numerous secretory cells as well as of cells forming the walls of the ducts. Multicellular glands are divided into *simple* and *compound* glands. The ducts of the simple gland do not branch, whereas those of the compound type do (fig. 4.14). Multicellular glands are further classified according to the shape of the secretory portion. They are identified as *tubular* if the secretory portion resembles the ductule portion and as *acinar* if the secretory portion is flasklike. Multicellular glands with a secretory portion that resembles both a tube and a flask are termed *tubuloacinar.*

Multicellular glands are also classified according to the means by which they discharge the secretory product (fig. 4.15). Glands that secrete a watery substance through the cell membrane of the secretory cells are called **merocrine** *(mer'o-krīn)* **glands.** Salivary glands, pancreatic glands, and certain sweat glands are of this type. In **apocrine** *(ap'o-krīn)* **glands** the secretion accumulates on the surface of the secretory cell, and then a portion of the cell along with the secretion is pinched off to be discharged. Mammary glands and certain sweat glands are apocrine glands. In a **holocrine** *(hol'o-krīn)* **gland,** the entire secretory cell is discharged along with the secretory product. An example of a holocrine gland is a sebaceous, or oil-secreting gland of the skin.

Glandular epithelia is summarized in table 4.3.

---

1. List the functions of simple squamous epithelia.
2. What are the three types of columnarlike epithelia? What do they have in common? How are they different?
3. What are the two types of stratified squamous epithelia, and how do they differ?
4. Describe the three ways that epithelial cells are bonded together.
5. Distinguish between unicellular and multicellular glands. Explain how multicellular glands are classified according to their mechanism of secretion.
6. In what ways are mammary glands and certain sweat glands similar?

---

merocrine: Gk. *meros,* part; *krinein,* to separate
apocrine: Gk. *apo,* off; *krinein,* to separate
holocrine: Gk. *holos,* whole; *krinein,* to separate

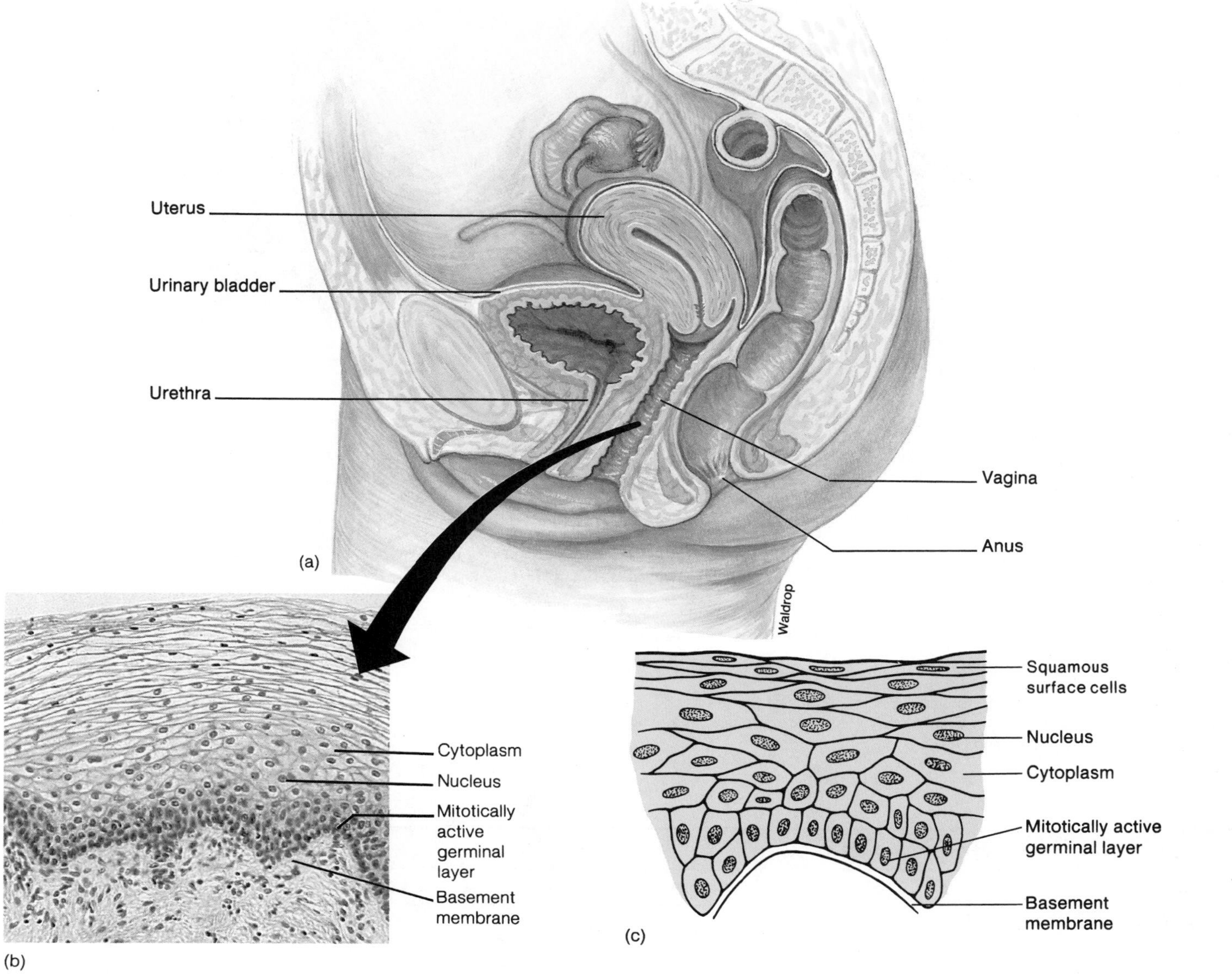

Figure 4.10 Stratified squamous epithelium is a multilayered, protective tissue that forms the outer layer of skin and the lining of body openings. In the moistened areas such as in the vagina (*a*), it is nonkeratinized, whereas in the epidermis of the skin it is keratinized. The lower layers are cube-shaped as can be seen in (*b*) the photomicrograph and (*c*) the diagram, whereas the outer layers are flattened and scalelike.

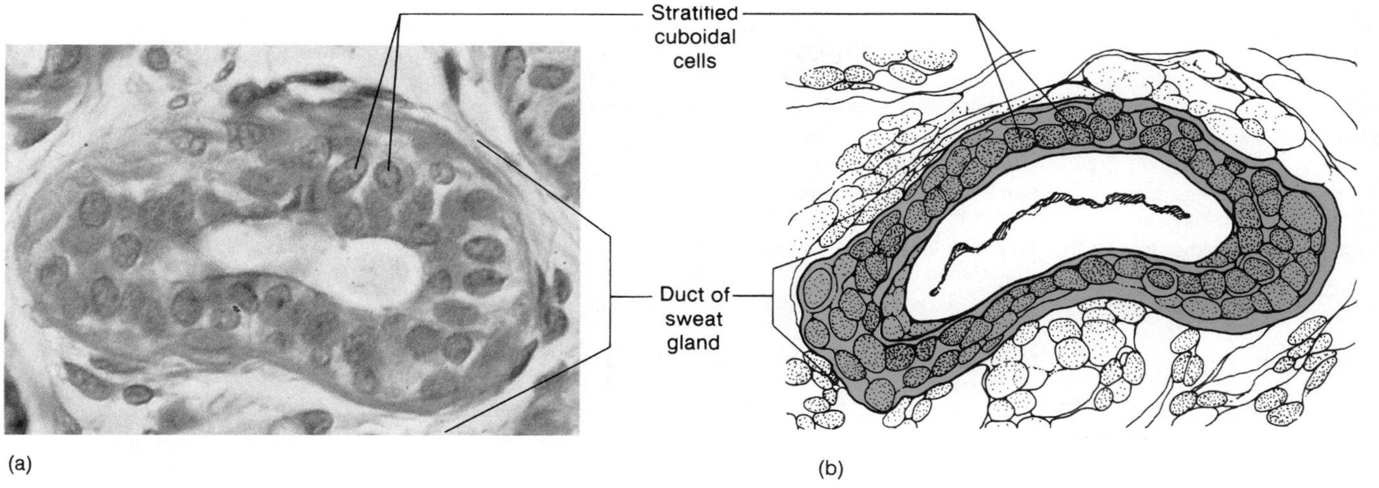

Figure 4.11 Stratified cuboidal epithelium consists of two or more layers of cube-shaped cells surrounding a lumen. (*a*) A photomicrograph of this tissue (100×) and (*b*) a diagram.

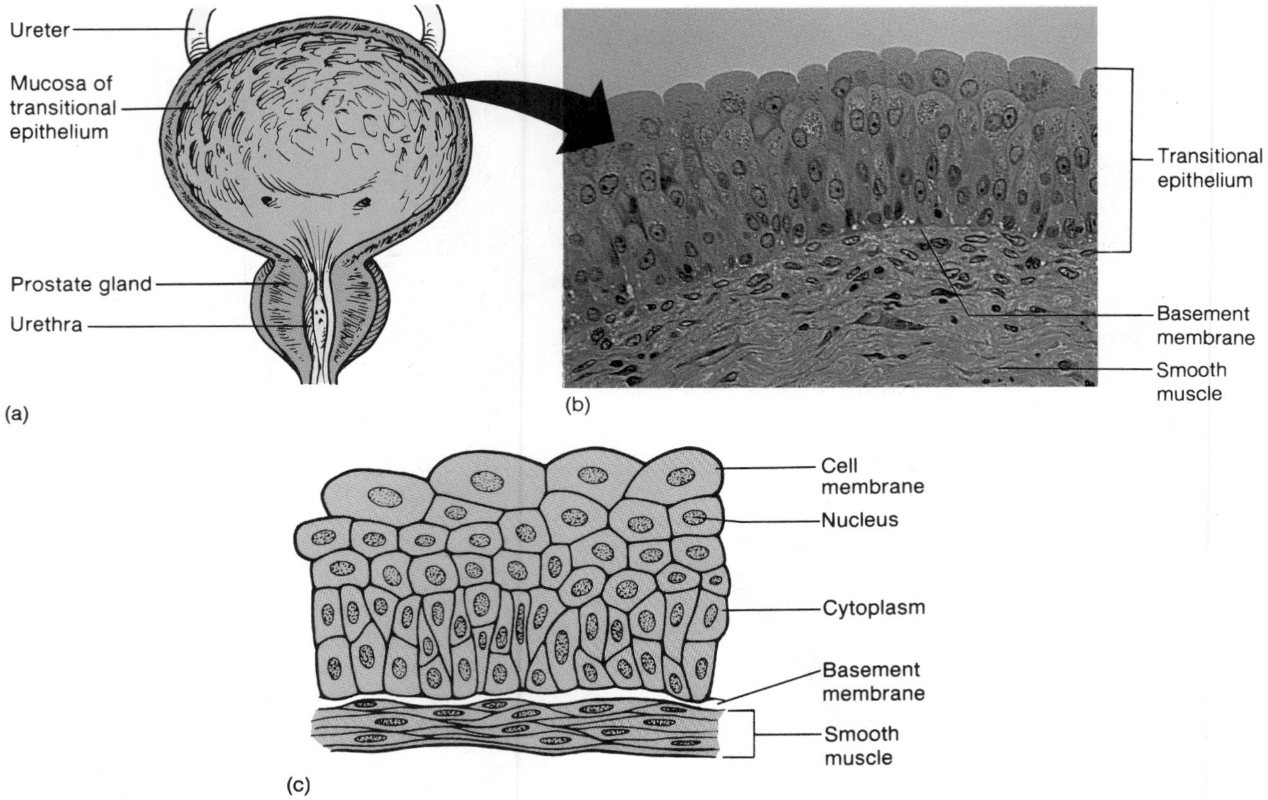

Figure 4.12 (*a*) Transitional epithelium lines the lumina of the urinary bladder and the ureters. (*b, c*) The cells of this tissue are stratified in a unique way to permit distension (100×).

Table 4.2 Summary of epithelial tissues

Type	Structure and function	Location
Simple epithelia	Single layer of cells; diffusion and filtration	Covering visceral organs, linings of lumina and body cavities
Simple squamous epithelium	Single layer of flattened, tightly bound cells; diffusion and filtration	Capillary walls, air sacs of lungs, covering visceral organs, linings of body cavities
Simple cuboidal epithelium	Single layer of cube-shaped cells; excretion, secretion, or absorption	Surface of ovaries; linings of kidney tubules, salivary ducts, and pancreatic ducts
Simple columnar epithelium	Single, nonciliated layer of tall, columnar-shaped cells; protection, secretion, and absorption	Lining of digestive tract
Simple ciliated columnar epithelium	Single, ciliated layer of columnar-shaped cells; transportive role through ciliary motion	Lining the lumen of the uterine tubes
Pseudostratified ciliated columnar epithelium	Single layer of ciliated, irregularly shaped cells, many goblet cells; protection, secretion, ciliary movement	Lining of respiratory passageways
Stratified epithelia	Two or more layers of cells; protection, strengthening, or distension	Epidermal layer of skin; linings of body openings, ducts, urinary bladder
Stratified squamous epithelium (keratinized)	Numerous layers, contains keratin, outer layers flattened and dead; protection	Epidermis of skin
Stratified squamous epithelium (nonkeratinized)	Numerous layers, lacks keratin, outer layers moistened and alive; protection and pliability	Linings of oral and nasal cavities, vagina, and anal canal
Stratified cuboidal epithelium	Usually two layers of cube-shaped cells; strengthen luminal walls	Larger ducts of sweat glands, salivary glands, and pancreas
Transitional epithelium	Numerous layers of rounded, nonkeratinized cells; distension	Luminal walls of ureters and urinary bladder

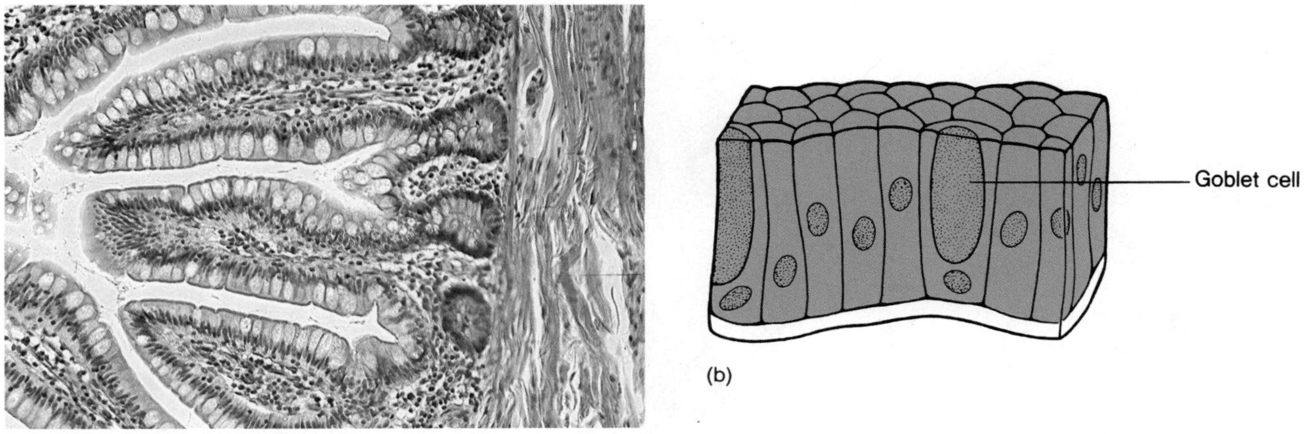

Figure 4.13 The goblet cell is a unicellular gland that secretes mucus to lubricate and protect surface linings. (*a*) A photomicrograph through the ileum of the small intestine shows the numerous goblet cells. (*b*) Columnar epithelia always contain mucus-secreting goblet cells.

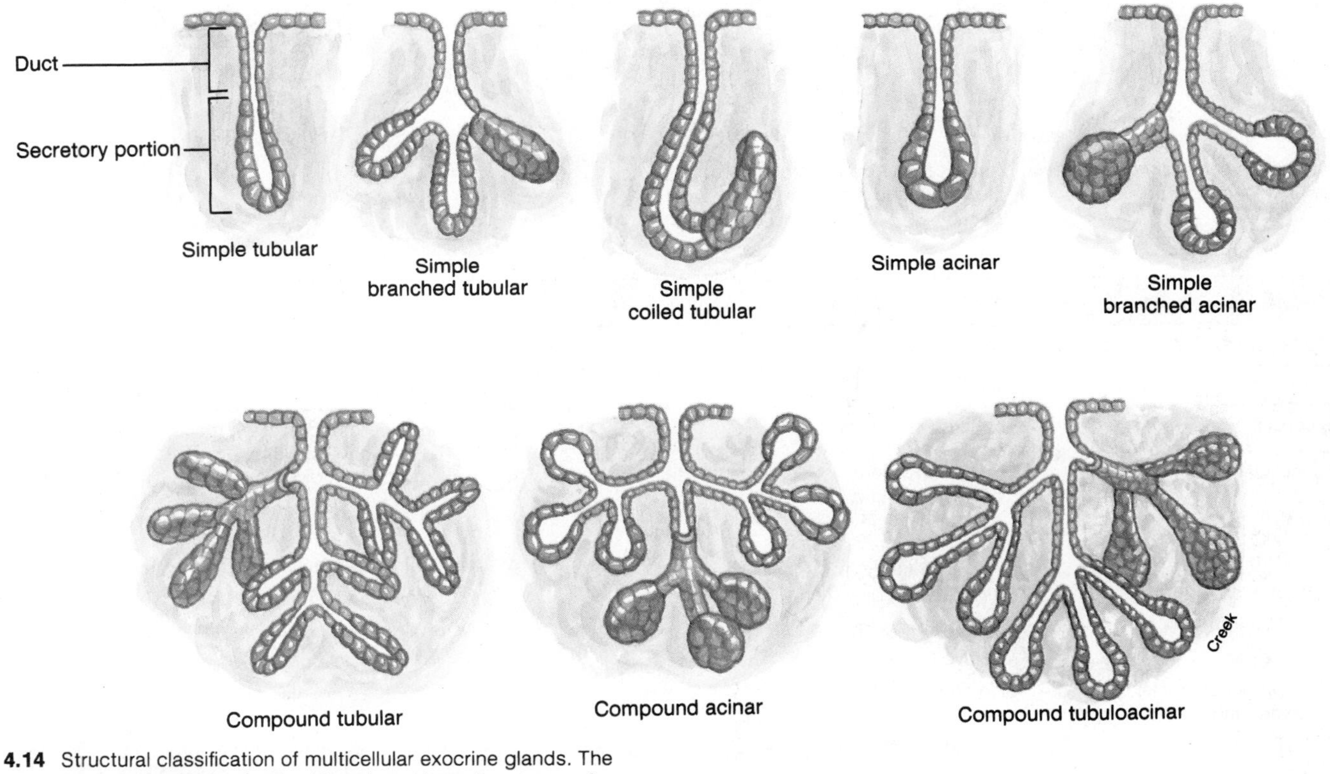

Figure 4.14 Structural classification of multicellular exocrine glands. The secretory portions of the simple glands either do not branch or have a few branches, whereas those of the compound type have multiple branches.

Connective Tissue

Connective tissues are classified according to the characteristics of the matrix that binds the cells. Connective tissues provide structural and metabolic support for other tissues and organs of the body.

Objective 9. Describe the general characteristics, locations, and functions of connective tissue.

Objective 10. Explain the functional relationship between embryonic and adult connective tissue.

Objective 11. List the various ground substances, fiber types, and cells that constitute connective tissue, and explain their functions.

Characteristics and Classification of Connective Tissue

Connective tissue is found throughout the body and, as the name indicates, supports or binds other tissues and provides for the metabolic needs of all body organs. Certain types of connective tissue store nutritional substances, whereas other types manufacture protective and regulatory materials.

Although connective tissue varies tremendously in structure and function, all connective tissues have similarities. With the exception of cartilage, connective tissues are highly vascular and well nourished. They are able to replicate and, by so

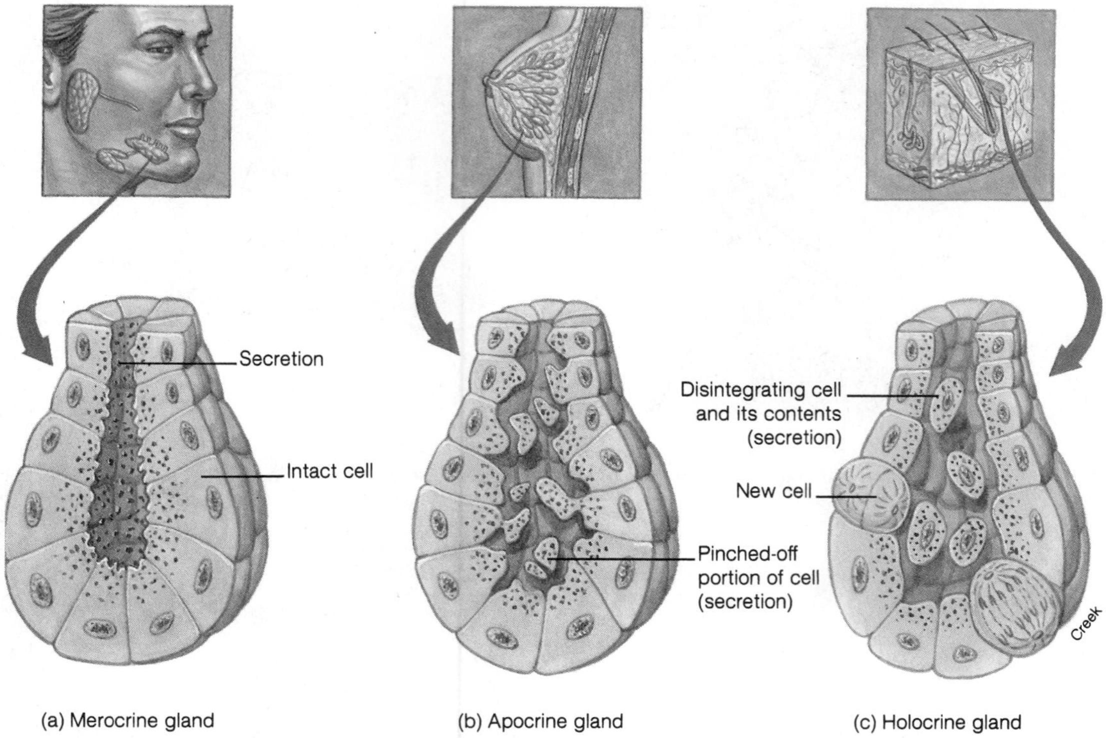

Figure 4.15 The secretory classification of multicellular exocrine glands: (*a*) merocrine gland, (*b*) apocrine gland, and (*c*) holocrine gland.

Table 4.3 Summary of glandular epithelia

Structural classification of exocrine glands

Type	Function	Example
I. Unicellular	Lubricate and protect	Goblet cells of digestive, respiratory, urinary, and reproductive systems
II. Multicellular	Protect, cool body, lubricate, aid in digestion, maintain body homeostasis	Sweat glands, digestive glands, mammary glands, sebaceous glands
A. Simple		
1. Tubular	Aid in digestion	Intestinal glands
2. Branched tubular	Protect, aid in digestion	Uterine glands, gastric glands
3. Coiled tubular	Regulate temperature	Certain sweat glands
4. Acinar	Additive to spermatozoa	Seminal vesicle of male reproductive system
5. Branched acinar	Skin conditioner	Sebaceous skin glands
B. Compound		
1. Tubular	Lubricate urethra of male, assist body digestion	Bulbourethral gland of male reproductive system, liver
2. Acinar	Nourishment to infant, aid in digestion	Mammary gland, salivary gland (sublingual and submandibular)
3. Tubuloacinar	Aid in digestion	Salivary gland (parotid), pancreas

Secretory classification of exocrine glands

Type	Description of secretion	Example
Merocrine glands	Watery secretion for regulating temperature or enzymes that promote digestion	Salivary and pancreatic glands, certain sweat glands
Apocrine glands	Portion of secretory cell and secretion are discharged; provides nourishment to infant, assists in regulating temperature	Mammary glands, certain sweat glands
Holocrine glands	Entire secretory cell with enclosed secretion is discharged; skin conditioner	Sebaceous glands of the skin

doing, are responsible for the repair of body organs. Unlike epithelial tissues, which are composed of tightly fitted cells, connective tissues contain considerably more matrix than cells. Connective tissues do not occur on free surfaces of body cavities or on the surface of the body as do epithelial tissues. Furthermore, connective tissues are embryonically derived from mesoderm, whereas epithelial tissues derive from ectoderm, mesoderm, and endoderm.

The classification of connective tissues is not exact, and several schemes have been devised. In general, however, they are named according to the kind and arrangement of the matrix. The following are the basic kinds of connective tissues.

A. Embryonic connective tissue
B. Connective tissue proper
   1. Loose (areolar)
   2. Dense regular
   3. Dense irregular
   4. Elastic
   5. Reticular
   6. Adipose
C. Cartilage
   1. Hyaline
   2. Fibrocartilage
   3. Elastic
D. Bone
E. Vascular (blood) tissue

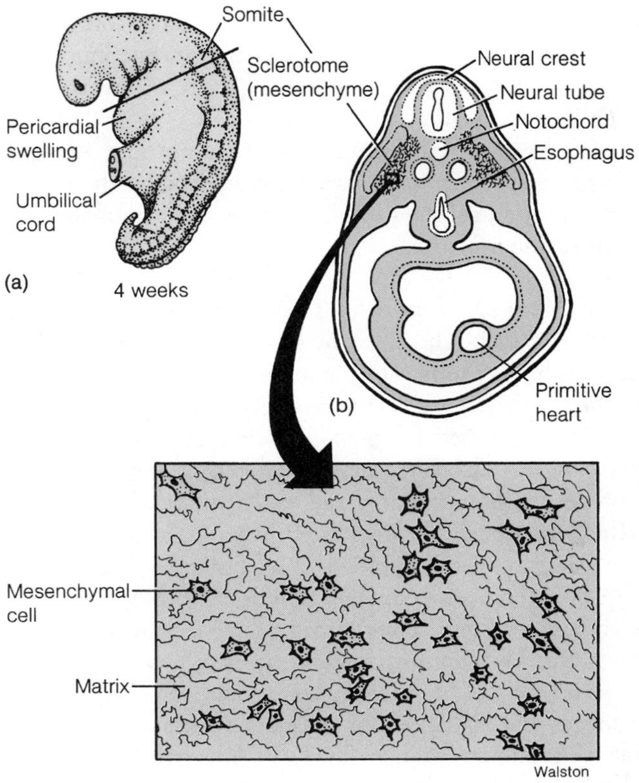

Figure 4.16 Mesenchyme is undifferentiated embryonic mesodermal connective tissue that can migrate and give rise to all other kinds of connective tissue. (*a*) It is found within an early developing embryo and (*b*) consists of irregularly shaped cells lying in a homogeneous, jellylike matrix.

Embryonic Connective Tissue

The embryonic period of development, which lasts six weeks (from the beginning of the third to the end of the eighth week), is characterized by a tremendous amount of tissue differentiation and organ formation. At the beginning of the embryonic period, all connective tissue appears the same and is referred to as **mesenchyme** *(mes'en-kīm)*. Mesenchyme is undifferentiated embryonic connective tissue, which is derived from mesoderm and consists of irregularly shaped cells lying in large amounts of a homogeneous, jellylike matrix (fig. 4.16). In certain areas of development, mesenchyme migrates to predisposed sites where it interacts with other tissues to form organs. Before the end of the embryonic period, once mesenchyme is in the appropriate position, it differentiates and from it all other kinds of connective tissues are formed. The causes of tissue differentiation are not fully understood.

Some mesenchymal-like tissue persists past the embryonic period in certain sites within the body. Good examples are the undifferentiated cells that surround blood vessels and form fibroblasts if the vessels are traumatized. Fibroblasts assist in healing wounds (chap. 5).

Another kind of prenatal connective tissue exists only in the fetus (the fetal period is from nine weeks to birth) and is called *mucous connective tissue,* or *Wharton's jelly.* It provides a turgid consistency to the umbilical cord.

Connective Tissue Proper

Connective tissue proper has a loose, flexible matrix, frequently called **ground substance.** The most common cell within connective tissue proper is called a **fibroblast.** Fibroblasts are large, star-shaped cells that produce collagenous, elastic, and reticular *(rĕ-tik'u-lar)* fibers. **Collagenous** *(kol-laj'ĕ-nus)* **fibers,** are composed of a protein called *collagen (kol'ah-jen);* they are flexible, yet have tremendous strength. **Elastic fibers** are composed of a protein called *elastin,* which gives elasticity to certain tissues. Collagenous and elastic fibers may be either sparse and irregularly arranged, as in loose connective tissue, or tightly packed, as in dense connective tissue. Tissues with loosely arranged fibers generally form packing material that cushions and protects various organs, whereas those that are tightly arranged form the binding and supportive connective tissues of the body.

Resilience in tissues that contain elastic fibers is extremely important for several physical functions of the body. Consider, for example, that elastic fibers are found in the walls of arteries and in the walls of the lower respiratory passageways. As these walls are expanded by blood moving through vessels or by inspired air, the elastic fibers must first stretch and then recoil. This maintains the pressures of the fluid or air moving through the lumina, which provides adequate flow rates and, thus, adequate rates of diffusion through capillary and lung surfaces.

Wharton's jelly: from Thomas Wharton, English anatomist, 1614–73

collagen: Gk. *kolla,* glue
elastin: Gk. *elasticus,* to drive

Reticular fibers reinforce by forming thin, short threads that branch and join to form a delicate lattice or reticulum. Reticular fibers are common in lymphatic glands where they form a meshlike **stroma.**

Six basic types of connective tissue proper are generally recognized. These tissues are distinguished by the consistency of the ground substance and the type and arrangement of the reinforcement fibers.

Loose Connective (Areolar) Tissue Loose connective tissue is distributed throughout the body as a binding and packing material. It binds the skin to the underlying muscles and is highly vascular, providing nutrients to the skin. Loose connective tissue surrounding muscle fibers and muscle groups is known as **fascia** *(fash'e-ah)*. It also surrounds blood vessels and nerves where it provides both protection and nourishment. Specialized cells called **mast cells** are dispersed throughout the loose connective tissue surrounding blood vessels. Mast cells produce *heparin (hep'ah-rin)*, an anticoagulant that prevents blood from clotting within the vessels. They also produce *histamine,* which is released during inflammation and acts as a powerful vasodilator.

The cells of loose connective tissue are predominantly fibroblasts, with collagenous and elastic fibers dispersed throughout the ground substance (fig. 4.17). The irregular arrangement of this tissue provides flexibility and yet strength in any direction. This tissue layer, for example, permits the skin to move when a part of the body is rubbed.

Much of the fluid of the body is found within loose connective tissue and is called *tissue fluid.* Sometimes excessive tissue fluid accumulates, causing a swelled condition called *edema (ĕ-de'mah).* Edema is a symptom of a variety of dysfunctions or disease processes.

Dense Regular Connective Tissue Dense regular connective tissue is characterized by large amounts of densely packed collagenous fibers that are arranged parallel to the direction of force placed on this tissue during body movement. Because this tissue is silvery white in appearance, it is sometimes called dense or white fibrous connective tissue

Dense regular connective tissue occurs where strong flexible support is necessary (fig. 4.18). **Tendons,** which attach muscles to bones and transfer the forces of muscle contractions, and **ligaments,** which connect bone to bone across articulations, are composed of this type of tissue.

Trauma to ligaments, tendons, and muscles are common sports-related injuries. A *strain* is an excessive stretch of the tissue comprising the tendon or muscle with no serious damage. A *sprain* is a tearing of the tissue of a ligament and may be slight, moderate, or complete. A complete tear of a major ligament is especially painful and disabling. Ligamentous tissue does not heal itself well because it has a poor blood supply. Surgical reconstruction is generally a necessary treatment of a severed ligament.

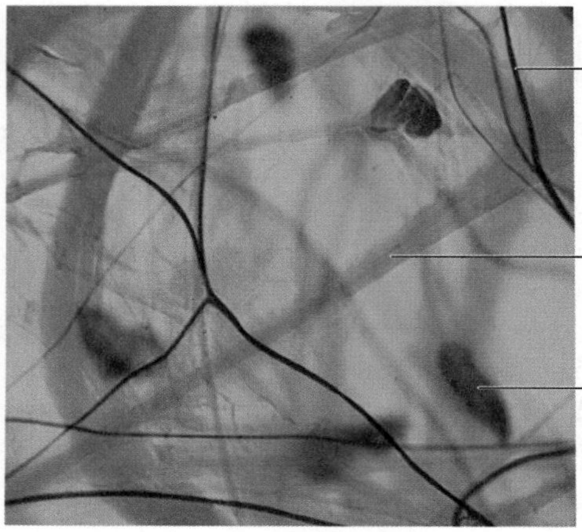

(a)

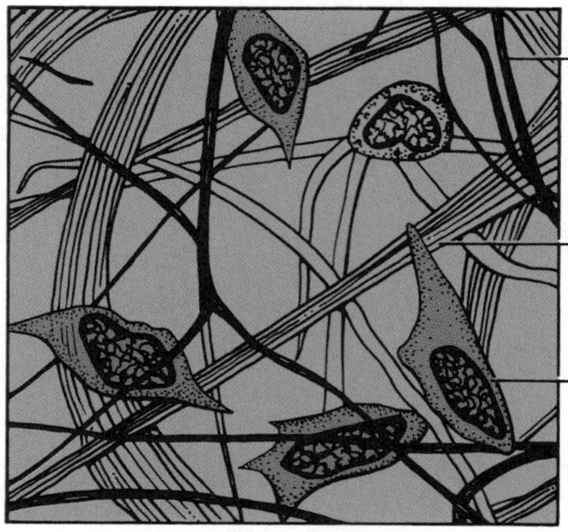

(b)

Figure 4.17 Loose connective tissue consists of collagenous and elastic fibers as well as fibroblasts and mast cells. This type of connective tissue is an important packing and binding material. It surrounds muscle, nerves, and vessels and binds the skin to the underlying muscles. (*a*) A photomicrograph and (*b*) a diagram.

Dense Irregular Connective Tissue Dense irregular connective tissue is characterized by large amounts of densely packed collagenous fibers that are interwoven to provide tensile strength in any direction. This tissue is found in the dermis of the skin, submucosa of the digestive tract, and comprising the fibrous capsules of organs and of joints (fig. 4.19).

Elastic Connective Tissue Elastic connective tissue has a predominance of elastic fibers that are irregularly arranged and yellowish in color (fig. 4.20). They can be stretched to one and a half times their original lengths and will snap back to their former size. Elastic connective tissue is found in the walls of large arteries, in portions of the larynx, and in the trachea and bronchial tubes of the lungs. It is also present between the arches of vertebrae, which make up the vertebral column.

reticular: L. *rete,* net or netlike
stroma: Gk. *stroma,* a couch or bed
fascia: L. *fascia,* band or girdle
heparin: Gk. *hepatos,* the liver
tendon: L. *tendere,* to stretch
ligament: L. *ligare,* bind

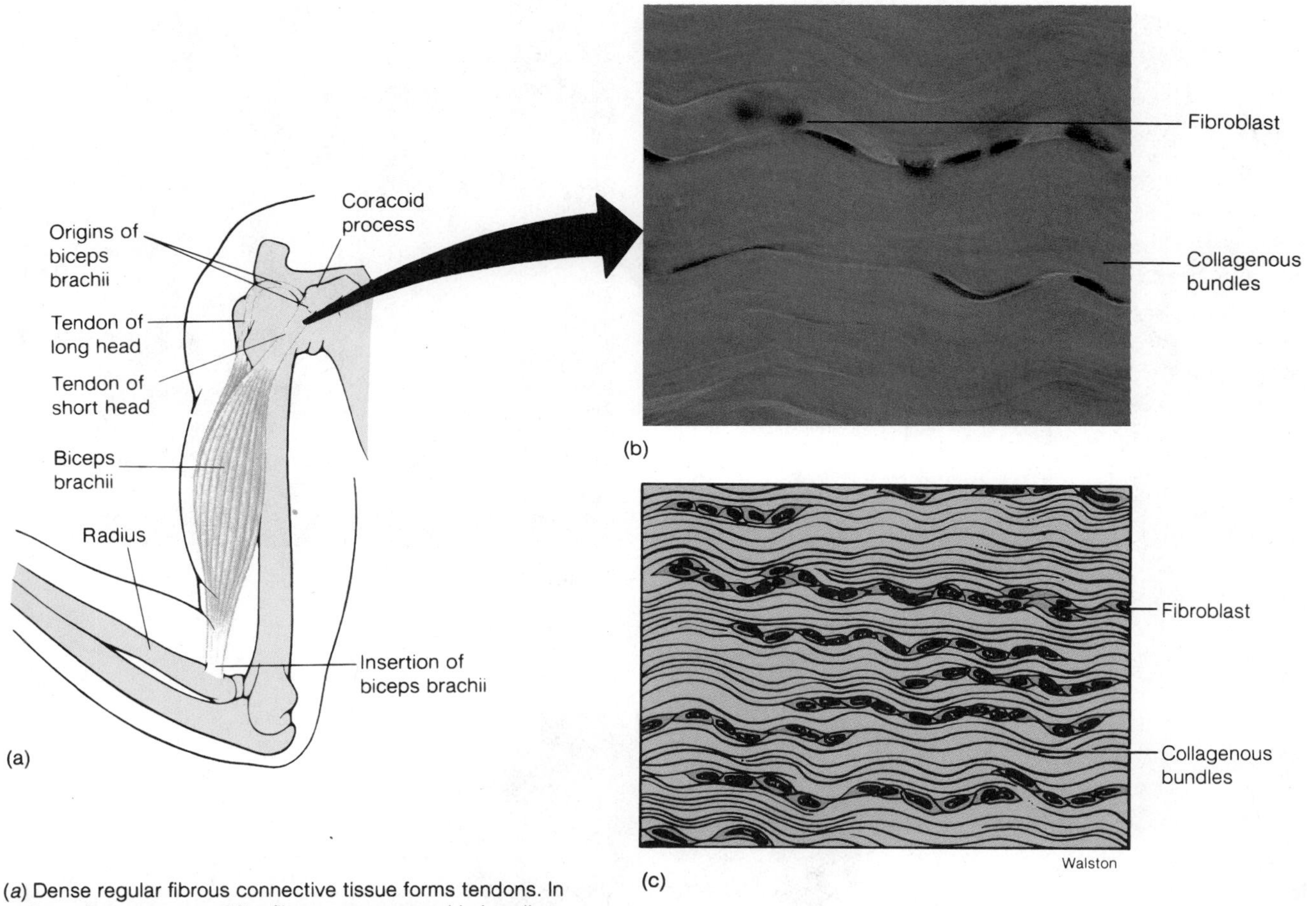

Figure 4.18 (*a*) Dense regular fibrous connective tissue forms tendons. In this tissue type, the collagenous, or white, fibers, are arranged in bundles, and fibroblasts are positioned between the bundles. The structure of this flexible, strong tissue is seen in (*b*) a photomicrograph (250×) and (*c*) a diagram.

Reticular Connective Tissue Reticular connective tissue is characterized by a network of reticular fibers woven through a jellylike matrix (fig. 4.21). Certain specialized cells within reticular tissue are *phagocytic (fag''o-sit'ik)* and therefore ingest foreign materials. The liver, spleen, lymph nodes, and bone marrow contain reticular connective tissue.

Adipose Tissue Adipose tissue is a specialized type of loose connective tissue that contains large quantities of **adipose cells** or **adipocytes.** Adipose cells form from mesenchymal cells and are, for the most part, formed prenatally and during the first year of life. Adipose cells store droplets of fat within their cytoplasm, causing them to swell and forcing their nuclei to one side (fig. 4.22).

Feeding an infant an excessive amount during its first year, when adipose cells are forming, may cause a greater amount of adipose tissue to develop. A person with large amounts of adipose tissue is more susceptible to developing obesity later in life than a person with a lesser amount of adipose tissue. Dieting eliminates the fat stored within the tissue but not the tissue itself.

Adipose tissue is found throughout the body but is concentrated around the kidney, in the hypodermis of the skin, on the surface of the heart, surrounding joints, and in the breasts of sexually mature females. Fat functions not only as a food reserve but supports and protects various organs. Fat is a good insulator against cold because it is a poor conductor of heat.

Excessive fat can be both unsightly and unhealthy, placing a strain on the heart and perhaps causing early death. For these reasons, good exercise programs and sensible diets are extremely important. Adipose tissue can also retain environmental pollutants that are ingested or absorbed through the skin.

The surgical procedure of *suction lipectomy* may be used to remove small amounts of adipose tissue from certain localized body areas such as the breasts, abdomen, buttocks, or thighs. Suction lipectomy is used for cosmetic purposes rather than as a treatment for obesity. Potential candidates should have good skin elasticity, be only about 15 to 20 pounds overweight, and be between thirty and forty years old.

The characteristics, functions, and locations of connective tissue proper are summarized in table 4.4.

adipose: L. *adiposus,* fat

lipectomy: Gk. *lipos,* fat; *ektome,* excision

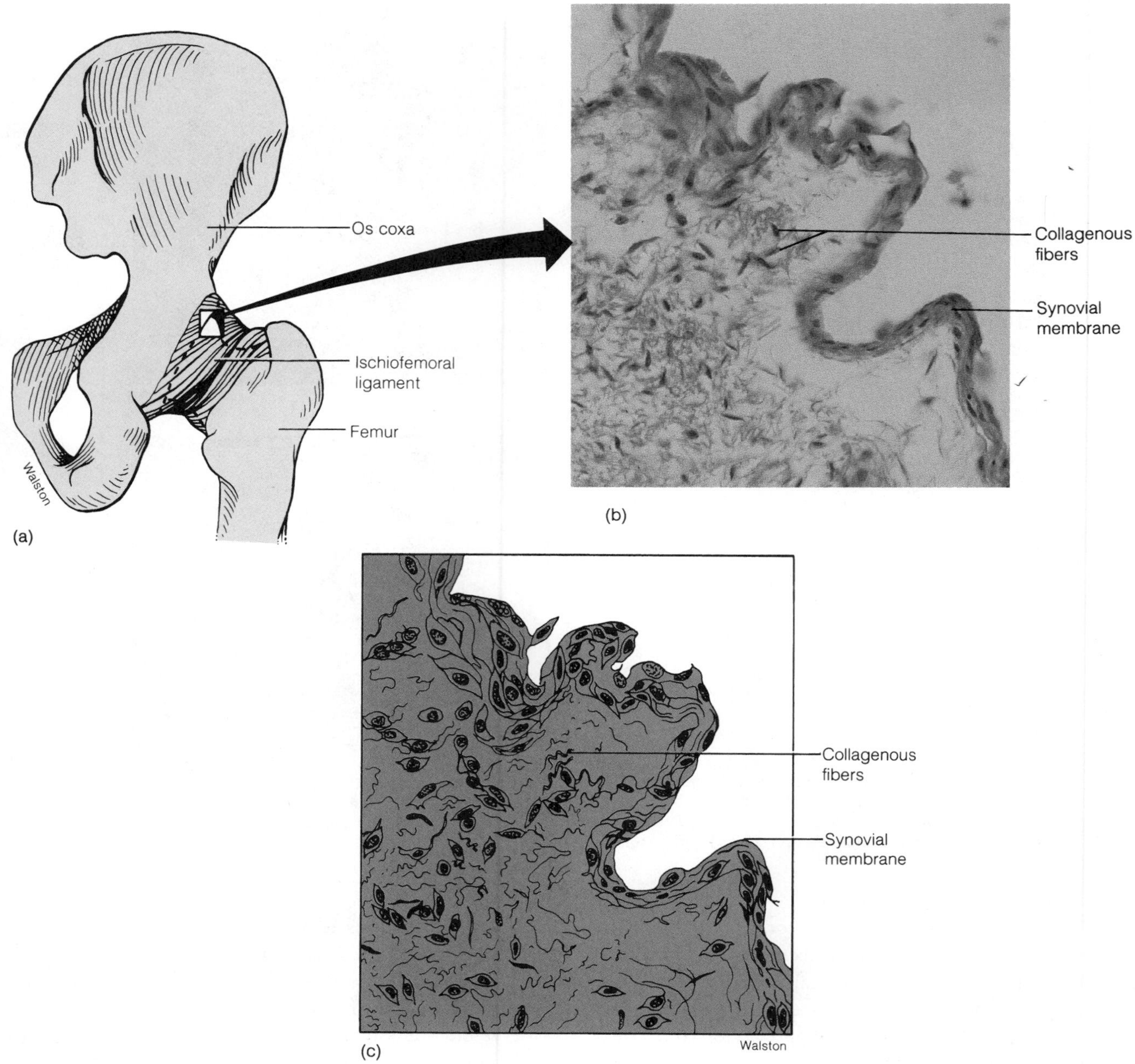

Figure 4.19 (*a*) Dense irregular connective tissue forms joint capsules that contain synovial fluid for lubricating movable joints. In this tissue type, the collagenous are irregularly arranged to withstand tension exerted in any direction. The structure of this flexible, strong tissue is seen in (*b*) a photomicrograph (400×) and (*c*) a diagram.

Cartilage

Cartilage tissue consists of cartilage cells, called **chondrocytes** *(kon'dro-sītz),* and a semisolid matrix that imparts marked elastic properties to the tissue. Cartilage is a type of supportive and protective connective tissue commonly called gristle. Cartilage is frequently associated with bone. It forms a precursor to one type of bone and persists at the articular surfaces on the bones of all movable joints.

The chondrocytes within cartilage may occur singly but are frequently clustered. Chondrocytes occupy spaces, called **lacunae** *(lah-ku'ne),* within the matrix. Most cartilage tissue is surrounded by a dense fibrous connective tissue called **perichondrium** *(per''ĭ-kon'dre-um).* Cartilage at the articular surfaces of bones (articular cartilage) lacks perichondria. Because cartilage is avascular, it must receive nutrients through diffusion from the surrounding tissue. For this reason, cartilaginous tissue has a slow rate of mitotic activity and, if damaged, heals with difficulty.

There are three kinds of cartilage, each of which is distinguished by the types of fibers embedded within the matrix: hyaline cartilage, fibrocartilage, and elastic cartilage.

Hyaline Cartilage Hyaline *(hi'ah-lın)* cartilage has a homogeneous, bluish-staining matrix in which the collagenous fibers are so fine that they can be observed only with an electron microscope. When viewed through a microscope, hyaline cartilage has a clear, glassy appearance (fig. 4.23).

---

lacuna: L. *lacuna,* hole or pit

hyaline: Gk. *hyalos,* glass

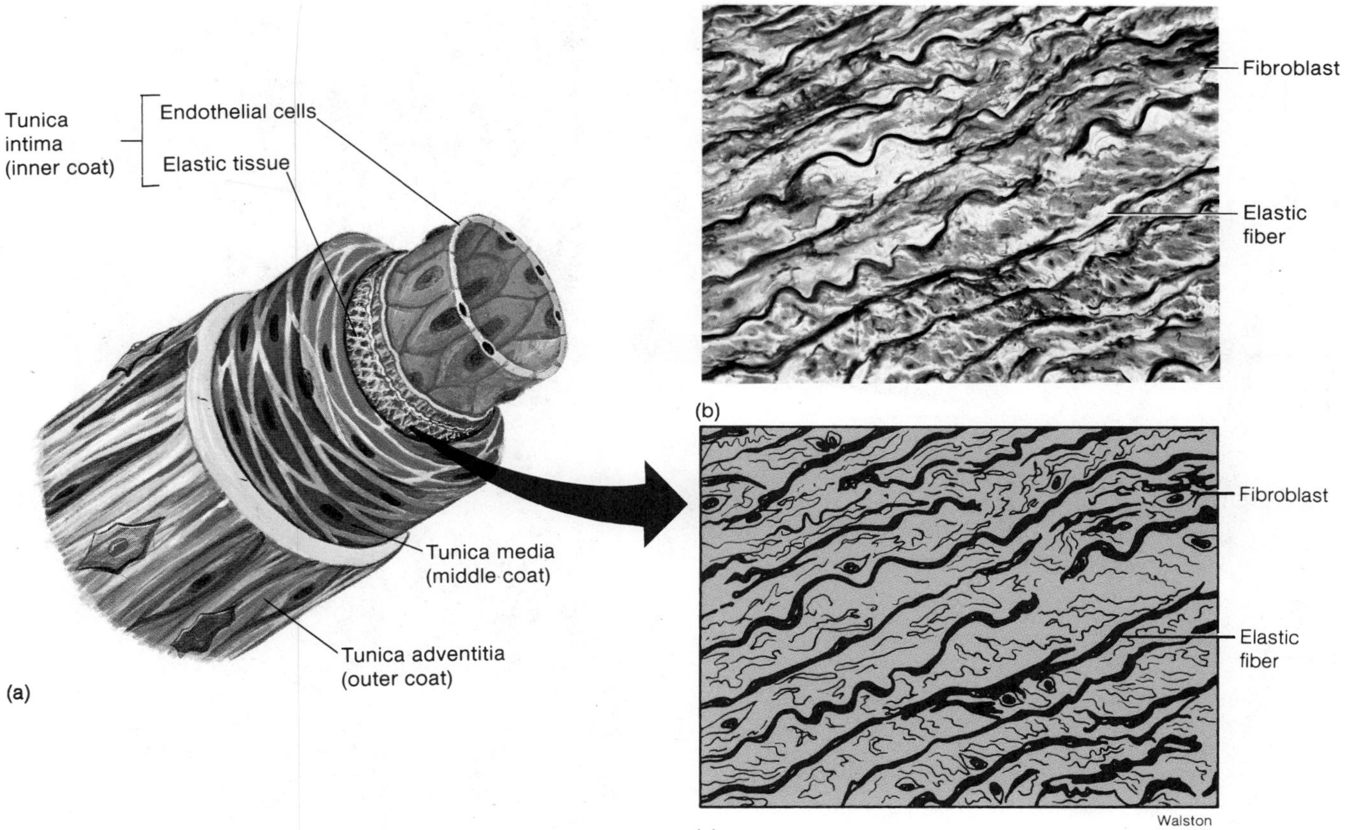

Figure 4.20 (*a*) The wall of an artery is composed of several tissue types. (*b*) The cellular structure. (*c*) Elastic connective tissue is found within the inner coat. This type of connective tissue allows stretching as the blood flows through. The outer coat consists of loose fibrous connective tissue, and the middle coat contains smooth muscle.

Hyaline cartilage is the most abundant cartilage within the body. It covers the articular surfaces of bones, supports the tubular trachea and bronchi of the respiratory system, reinforces the nose, and forms the flexible bridge, called **costal cartilage,** between the ventral end of each of the first ten ribs and the sternum. Most of the bones of the body form first as hyaline cartilage and later become bone (ossification).

Fibrocartilage Fibrocartilage has its matrix reinforced with numerous collagenous fibers (fig. 4.24). It is a durable tissue adapted to withstand tension and compression. It is found at the symphysis pubis where the two pelvic bones articulate and between the vertebrae as intervertebral discs. It also forms the cartilaginous wedges, called *menisci,* within the knee joint.

By the end of the day the intervertebral discs of the vertebral column are somewhat compacted. So a person is actually slightly shorter in the evening than in the morning following a recuperative rest. Aging brings a gradual compression of the fibrocartilaginous discs that is irreversible.

Elastic Cartilage Elastic cartilage is similar to hyaline except for the presence of abundant elastic fibers, which make it very flexible while maintaining its strength (fig. 4.25). The numerous elastic fibers also give it a yellowish appearance. This tissue is found in the outer ear, portions of the larynx, and the auditory canal.

The structure, function, and location of cartilage within the body are summarized in table 4.5.

Bone

Bone, or **osseous connective tissue,** is the most rigid of the connective tissues. Unlike cartilage, bone has a rich vascular supply and is the site of considerable metabolic activity. The hardness of bone is largely due to the inorganic calcium phosphate (calcium hydroxyappatite) deposited within the intercellular matrix. Numerous collagenous fibers, also embedded within the matrix, give some flexibility to bone.

In healthy bone tissue, a balance exists between the organic living and the inorganic nonliving materials. As a person ages, the proportion of organic material decreases and the bones become brittle. The bones of elderly persons not only fracture easily but do not heal readily.

When bone is placed in a weak acid the calcium salts dissolve away and the bone becomes pliable. It retains its basic shape but can be easily bent and twisted. In calcium deficiency diseases, such as rickets, the bone tissue becomes pliable and bends under the weight of the body.

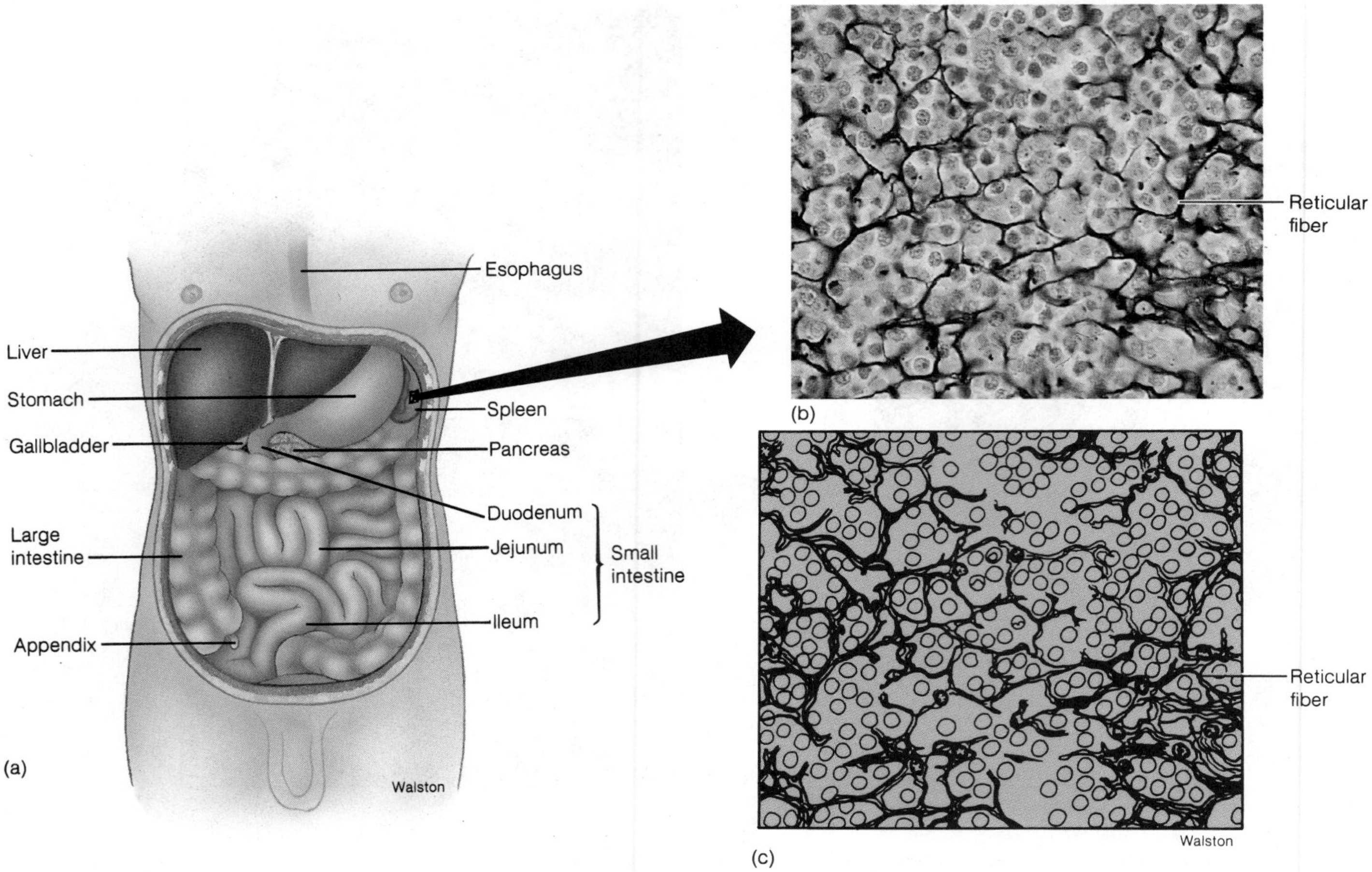

Figure 4.21 Reticular tissue forms the stroma, or framework, of organs such as the spleen and liver (*a*), thymus, and lymph nodes. It consists of a network of woven reticular fibers, as seen in (*b*) the photomicrograph (250×) and (*c*) the diagram.

Table 4.4 Summary of connective tissue proper

Type	Structure and function	Location
Loose connective (areolar) tissue	Predominantly fibroblast cells with lesser amounts of collagen and elastin proteins; binds organs, holds tissue fluids, diffusion	Surrounding nerves and vessels, between muscles, beneath the skin
Dense regular connective tissue	Densely packed collagenous fibers arranged parallel to the direction of force; provides strong, flexible support	Tendons, ligaments
Dense irregular connective tissue	Densely packed collagenous fibers arranged in a tight interwoven pattern; provides tensile strength in any direction	Dermis of skin, fibrous capsules of organs and joints
Elastic connective tissue	Predominantly irregularly arranged elastic fibers; supports, provides framework	Large arteries, lower respiratory tract, between vertebrae
Reticular connective tissue	Reticular fibers forming supportive network; stores, phagocytic	Lymph nodes, liver, spleen, thymus, bone marrow
Adipose connective tissue	Adipose cells; protects, stores fat, insulates	Hypodermis of skin, surface of heart, omentum, around kidneys, back of eyeball, surrounding joints

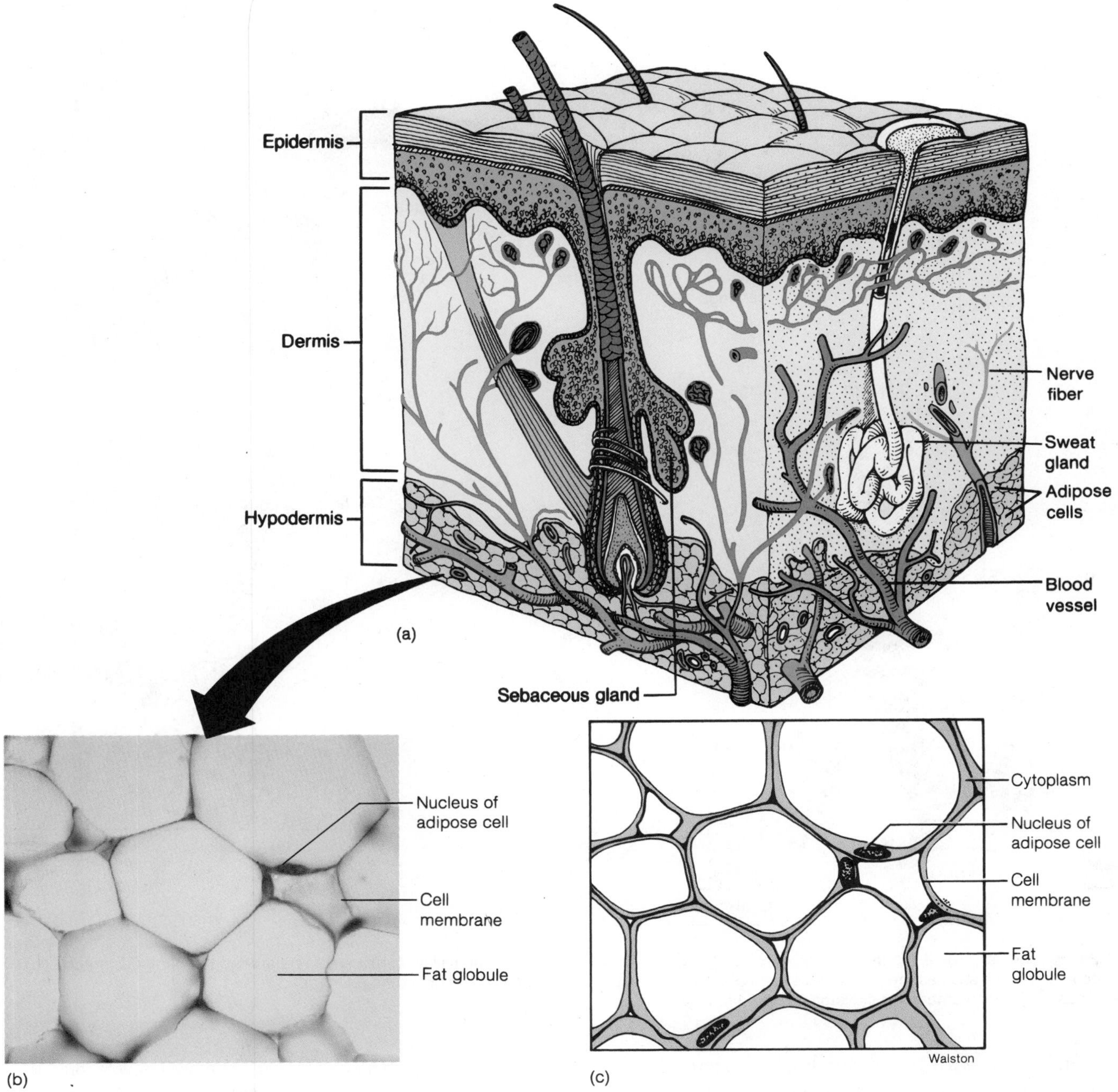

Figure 4.22 Adipose connective tissue is abundant in (*a*) the hypodermis of the skin and around various internal organs. It is protective and serves to store fat. The fat globules (*b, c*) are stored in the adipose cells (250×).

There are two kinds of bone tissue, based on porosity, and most bones have both types (fig. 4.26). **Compact,** or **dense,** bone tissue is the hard, outer layer, whereas **spongy,** or **cancellous,** bone tissue is the porous, highly vascular, inner portion. Compact bone tissue is covered by the periosteum, serves for attachment of muscles, provides protection, and gives durable strength to the bone. Spongy bone tissue makes the bone lighter and provides a space for bone marrow where blood cells are produced.

In compact bone tissue, the bone cells, called **osteocytes,** are arranged in concentric layers around a **central** (Haversian) **canal,** which contains a vascular and nerve supply (fig. 4.27). Each osteocyte occupies a space called a **lacuna.** Radiating from each lacuna are numerous minute canals, called **canaliculi,** that traverse the dense matrix of the bone tissue to adjacent lacunae. Nutrients diffuse through the canaliculi to reach each osteocyte. The inorganic matrix is deposited in concentric layers called **lamellae.**

Haversian canal: from Clopton Havers, English anatomist, 1650–1702

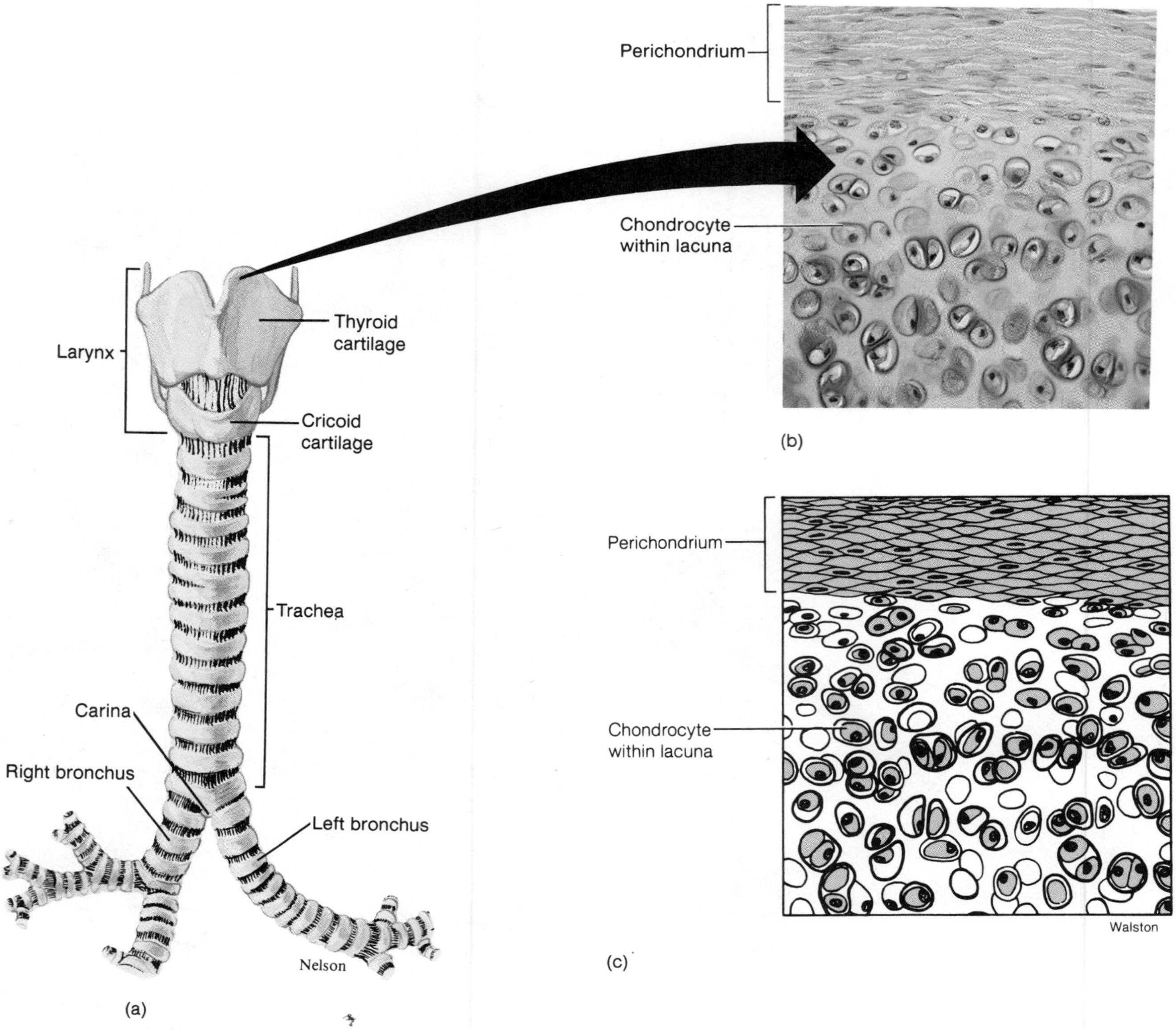

Figure 4.23 Hyaline cartilage is the most abundant cartilage within the body. It occurs in places such as the larynx and trachea (*a*), articular ends of bones, ends of ribs, and embryonic skeleton. (*b*) The homogeneous matrix of hyaline cartilage can be seen in the photomicrograph (100×). Its structure is diagrammed in (*c*).

A central canal, with its surrounding osteocytes, lacunae, canaliculi, and concentric lamellae, constitutes an **osteon,** or Haversian system (fig. 4.26). Metabolic activity within bone tissue occurs at the osteon level. Areas between the osteons contain **interstitial** *(in″ter-stish′al)* **lamellae.** These areas possess osteocytes within lacunae and associated canaliculi but are arranged irregularly. Transverse channels, called **perforating** (Volkmann's) **canals** (see fig. 6.8), penetrate compact bone and connect various osteons with blood vessels and nerves. Bone tissue is further described in chapter 6.

Electron microscopy is a relatively new technique for examining tissues. The structure of the central canal, for example, shown in figure 4.27, can be more accurately described using a scanning electron micrograph than when using a conventional photomicrograph. Electron microscopy is an extension of light microscopy and should be used to demonstrate the continuity of structural organization within tissues.

Vascular Connective Tissue

Blood is a highly specialized, viscous connective tissue. **Formed elements** (red blood cells, white blood cells, and platelets) are suspended in the liquid plasma matrix. Blood plays a vital role in maintaining internal body homeostasis. The three types of formed elements found within the blood are erythrocytes, leukocytes, and thrombocytes (fig. 4.28). Blood is further described in chapter 16.

osteon: Gk. *osteon,* bone
Volkmann's canals: from Alfred W. Volkmann, German physiologist, 1800–1877

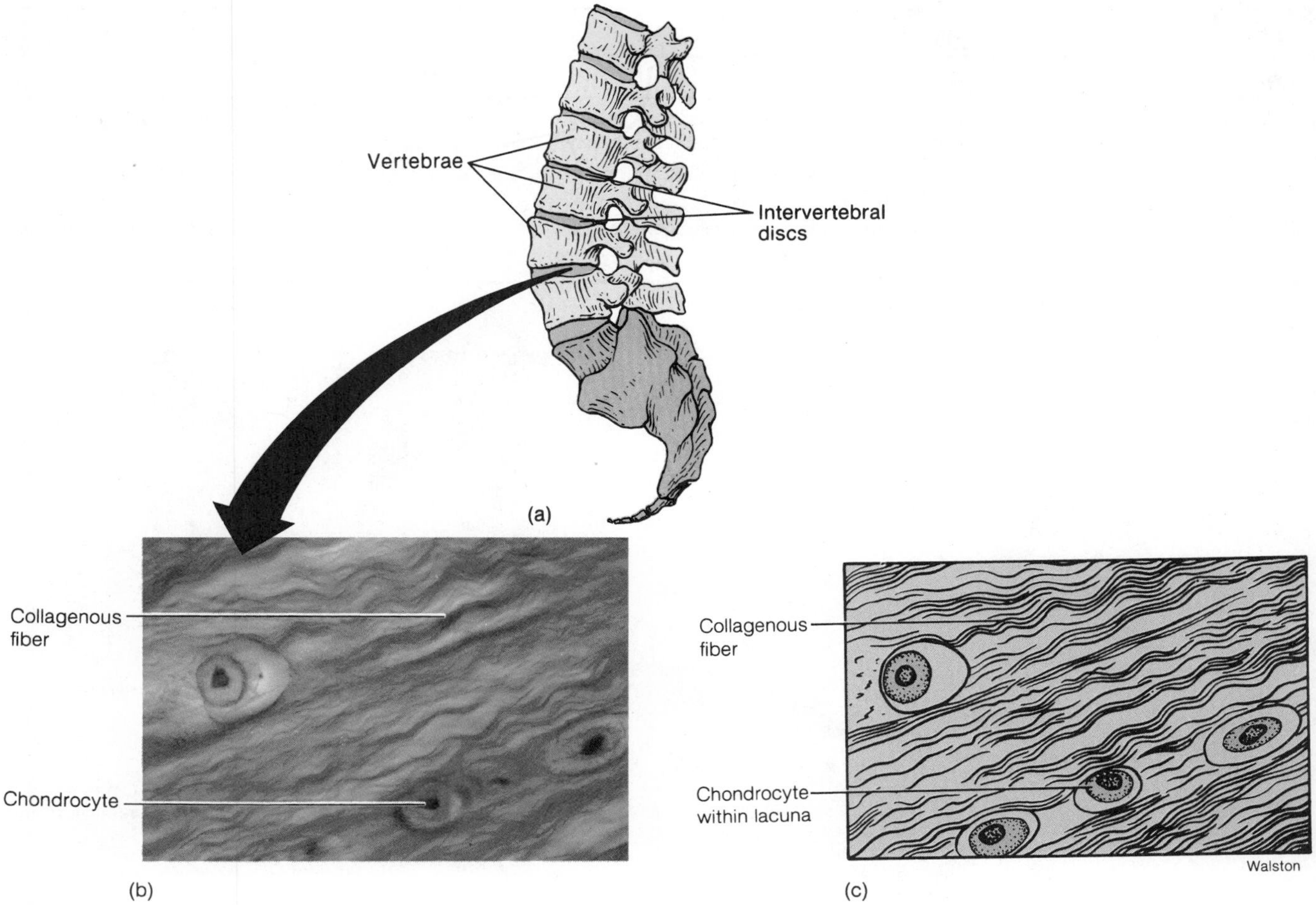

Figure 4.24 Fibrocartilage is located at the symphysis pubis, within the knee joint, and between the vertebrae as the intervertebral discs (*a*). It consists of chondrocytes within a matrix that is reinforced with collagenous fibers (*b, c*) to withstand stress and compression as it provides durable support (250×).

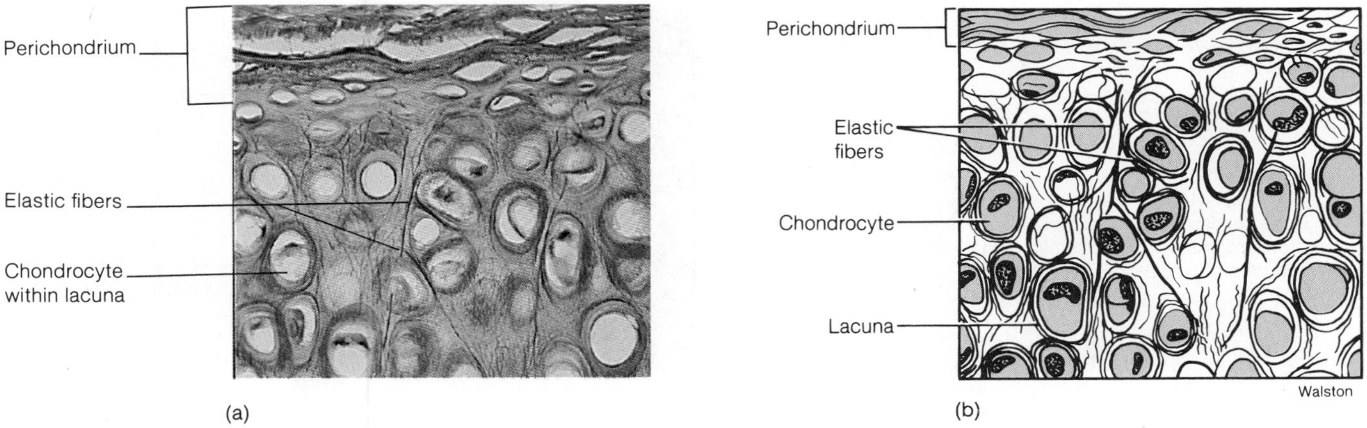

Figure 4.25 Elastic cartilage gives support to the outer ear, auditory canal, and parts of the larynx. (*a*) A photomicrograph of elastic cartilage (100×). (*b*) The matrix of elastic tissue is laced with elastic fibers to provide flexibility while maintaining shape.

Table 4.5 Summary of cartilage tissue

Type	Structure and function	Location
Hyaline cartilage	Homogeneous matrix with extremely fine collagenous fibers; flexible support, protection, precursor to bone	Articular surfaces of bones, nose, walls of respiratory passages, fetal skeleton
Fibrocartilage	Abundant collagenous fibers within matrix; support, withstand compression	Symphysis pubis, intervertebral discs, knee joint
Elastic cartilage	Abundant elastic fibers within matrix; support, flexibility	Framework of outer ear, auditory canal, portions of larynx

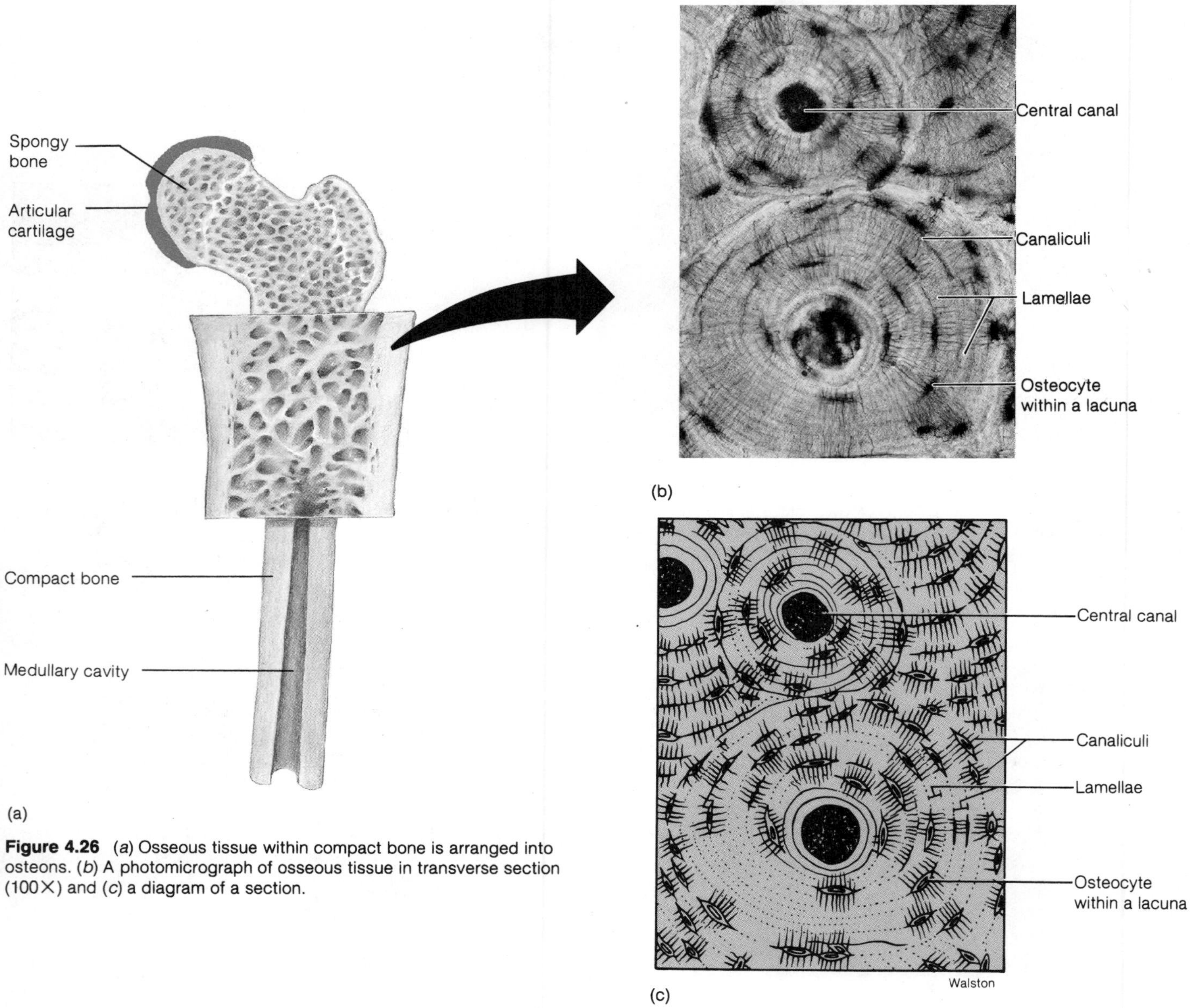

Figure 4.26 (*a*) Osseous tissue within compact bone is arranged into osteons. (*b*) A photomicrograph of osseous tissue in transverse section (100×) and (*c*) a diagram of a section.

Erythrocytes Erythrocytes *(ĕ-rith'ro-sītz),* or **red blood cells** (RBC) are minute, nonnucleated, biconcave discs that transport respiratory gases. The red color results from the pigment *hemoglobin.* In an infant, erythrocytes are produced in the spleen and bone marrow, but in a mature person, the bone marrow is the only production site. Erythrocytes live between 90 and 120 days and are disposed of in the liver and spleen.

Leukocytes Leukocytes, *(loo'ko-sītz)* or **white blood cells** (WBC), are ameboid in movement and slightly larger than erythrocytes. Leukocytes primarily defend the body against invasions by microorganisms. They are produced in bone marrow and lymphatic tissue and have a life span that ranges from 3 to 300 days. There are five kinds of leukocytes: **neutrophils, eosinophils** *(e''o-sin'o-filz),* **basophils** *(ba'so-filz),* **lymphocytes**

erythrocyte: Gk. *erythros,* red; *kytos,* hollow (cell)
hemoglobin: Gk. *haima,* blood; *globus,* globe

leukocyte: Gk. *leukos,* white; *kytos,* hollow (cell)

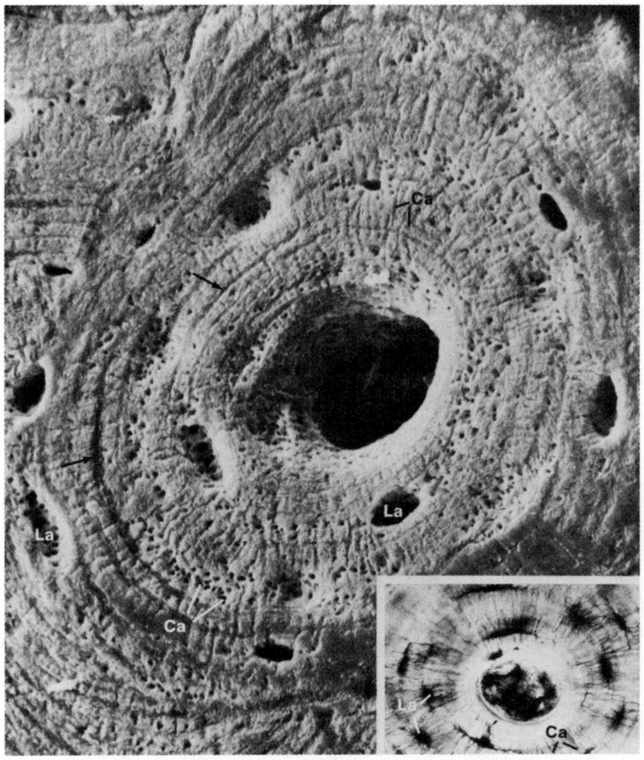

Figure 4.27 A scanning electron micrograph of a central canal of osseous tissue. The lacunae (*La*) provide spaces for the osteocytes, which are connected to one another by canaliculi (*Ca*). (Note the divisions between concentric lamellae [*arrows*].) (From: *Tissues and Organs: A Text-Atlas of Scanning Electron Microscopy* by R. G. Kessel and R. H. Kardon. W. H. Freeman and Company. © 1979.)

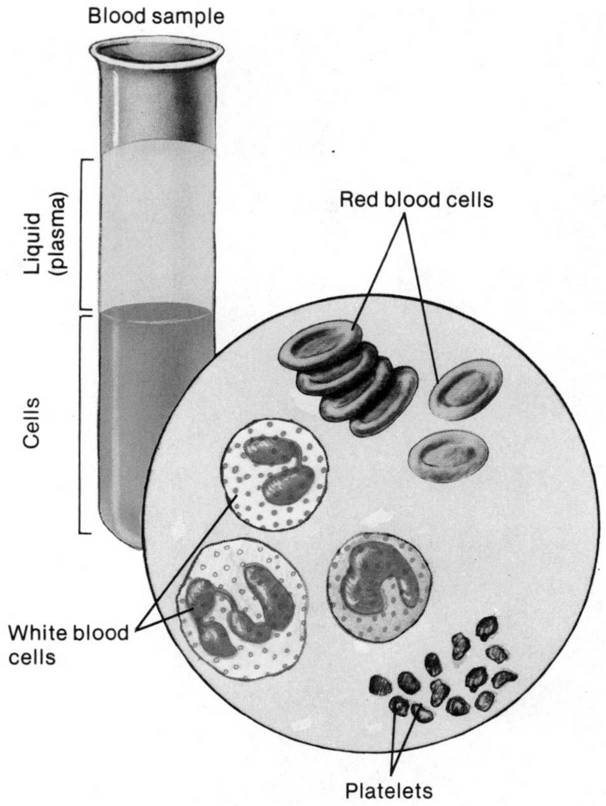

Figure 4.28 Vascular connective tissue, or blood, is composed of a fluid matrix, called plasma, and formed elements of which there are three types: erythrocytes (red blood cells), thrombocytes (platelets), and leukocytes (white blood cells).

(lim'fo-sītz), and **monocytes.** The first three can be further classified as **granular leukocytes** and the remaining two as **nongranulocytes.**

Thrombocytes Thrombocytes are the **platelets** of the blood; along with the protein *fibrinogen* of the plasma, they play a role in blood clotting. Platelets form from the surface of large, multinucleated cells in the bone marrow, which are called **megakaryocytes** *(meg"ah-kar'e-o-sītz)*. Platelets, like leukocytes, have ameboid movement.

An injury to a portion of the body may stimulate tissue repair activity, usually involving connective tissue. A minor scrape or cut results in platelet and plasma activity of the exposed blood and the formation of a scab. The epidermis of the skin regenerates beneath the scab. A severe open wound brings about connective tissue granulation, in which collagenous fibers form from surrounding fibroblasts to strengthen the traumatized area. The healed area is known as a *scar.*

thrombocyte: Gk. *thrombos,* a clot; *kytos,* hollow (cell)
megakaryocyte: Gk. *megas,* great; *karyon,* nut or kernel; *kytos,* hollow (cell)

1. List the kinds of connective tissues, and describe the structure, function, and location of each.
2. Which of the previously discussed connective tissues function to protect body organs? Which type is phagocytic? Which types bind and support various structures? Which types are associated in some way with the skin?
3. What is the developmental significance of mesenchyme, and how does it functionally differ from adult connective tissue?
4. Define or describe ground substance, reticular fibers, fibroblasts, collagenous fibers, elastic fibers, and mast cells.

Muscle Tissue

Muscle tissues are responsible for the movement of materials through the body, the movement of one part of the body with respect to another, and for locomotion. Fibers in the three kinds of muscle tissue are adapted to contract in response to stimuli.

Objective 12. Describe the structure, location, and function of the three types of muscle tissue.

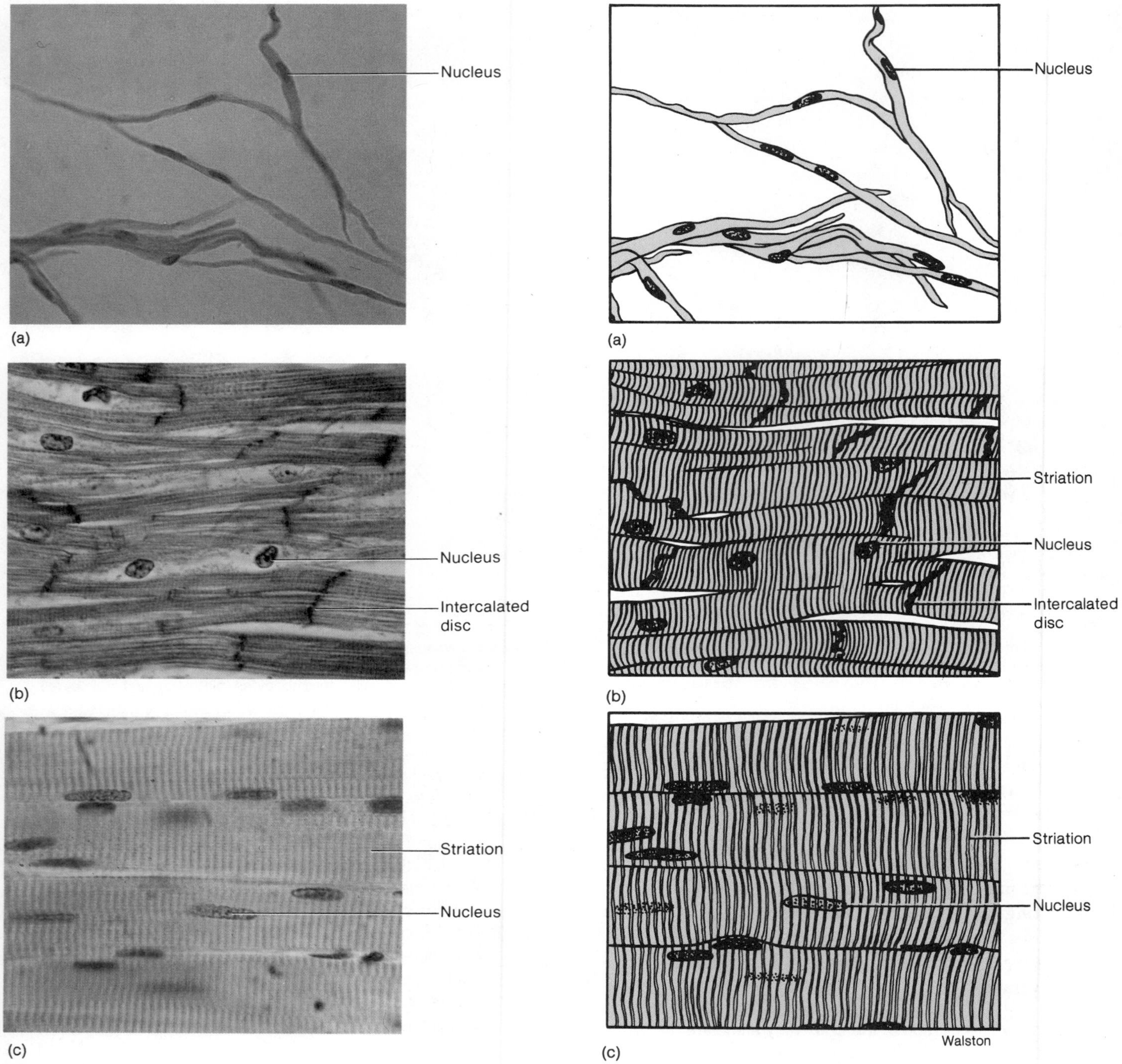

Figure 4.29 The three types of muscle tissue: (*a*) smooth muscle, fibers teased apart; (*b*) cardiac muscle; and (*c*) skeletal muscle.

Muscle tissues are unique in possessing the property of *contractility*. The muscle cells, or *fibers,* are elongated in the direction of contraction. Movement is accomplished through the shortening of the fibers in response to a stimulus. Muscle tissue is derived from mesoderm. There are three types of muscle tissue in the body: smooth, cardiac, and skeletal (fig. 4.29).

Smooth Muscle Smooth muscle tissue is common throughout the body, occurring in many of the systems. For example, in the wall of the digestive tract it provides the motive power for mechanically digesting food and for the peristaltic movements. Smooth muscle is also found in the walls of arteries, the walls of respiratory passages, and in the urinary and reproductive ducts. The contraction of smooth muscle is under autonomic (involuntary) nervous control.

Smooth muscle fibers are long, spindle-shaped cells that contain a single nucleus and lack striations. These cells are usually grouped together in flattened sheets, forming the muscular portion of a wall around a lumen.

Cardiac Muscle Cardiac muscle tissue makes up most of the wall of the heart. This tissue is characterized by bifurcating (branching) fibers, a centrally positioned nucleus, and transversely positioned **intercalated** *(in-ter'kah-lāt-ed)* **discs.** Intercalated discs help to hold adjacent cells together and transmit the force of contraction from cell to cell. Like skeletal muscle, cardiac muscle is striated, but unlike skeletal muscle it experiences rhythmical involuntary contractions. Cardiac muscle is further discussed in chapter 16.

Skeletal Muscle Skeletal muscle tissue attaches to the skeleton and is responsible for voluntary body movements. Each fiber is elongated, multinucleated, and has distinct transverse striations. Fibers of this muscle tissue are grouped into parallel fasciculi (bundles), which can be seen without a microscope in fresh muscle. Both cardiac and skeletal muscle fibers are striated and cannot replicate once tissue formation is completed shortly after birth. Skeletal muscle tissue is further discussed in chapter 9.

The three types of muscle tissue are summarized in table 4.6.

1. Identify the general characteristics of muscle tissue. What is meant by a voluntary or involuntary muscle?
2. Distinguish between smooth, cardiac, and skeletal muscle tissue on the bases of structure, location, and function.

Table 4.6 Summary of muscle tissue

Type	Structure and function	Location
Smooth	Elongated, spindle-shaped fiber with single nucleus; involuntary movements of internal organs	Walls of hollow internal organs
Cardiac	Branched, striated fiber with single nucleus and intercalated discs; involuntary rhythmic contraction	Heart muscle
Skeletal	Multinucleated, striated, cylindrical fiber that occurs in fasciculi; voluntary movement of skeletal parts	Associated with skeleton; spans joints of skeleton via tendons

Nervous Tissue

Nervous tissue is composed of neurons, which respond to stimuli and conduct impulses to and from all body organs, and neuroglia, which functionally support and physically bind neurons.

Objective 13. Give the basic characteristics and functions of nervous tissue.

Objective 14. Distinguish between neurons and neuroglia.

Neurons Although there are several kinds of neurons *(nu'rons)* in nervous tissue, all have three principal components: (1) a cell body, or perikaryon; (2) dendrites; and (3) an axon (fig. 4.30). **Dendrites** function to receive a stimulus and conduct the impulse to the cell body. The **cell body,** or **perikaryon** *(per''ĭ-kar'e-on),* contains the nucleus and specialized organelles and microtubules. The **axon** is a cytoplasmic extension that conducts an impulse away from the cell body. The term **nerve fiber** refers to an axon and its associated myelin sheath.

Neurons derive from ectoderm and are the basic structural and functional units of the nervous system. They are specialized to respond to physical and chemical stimuli, conduct nerve impulses, and perform other functions such as storing memory, thinking, and regulating the activity of other organs or glands. Of all the body's cells, neurons are probably the most specialized. As with muscle cells, the number of neurons is established shortly after birth, and thereafter they lack the ability to undergo mitosis, although under certain circumstances a severed portion can regenerate.

Neuroglia In addition to neurons, nervous tissue is composed of neuroglia *(nu'rog'le-ah)* (fig. 4.31). Neuroglial cells are about five times as abundant as neurons and have limited mitotic abilities. They do not transmit impulses but support and bind neurons together. Certain neuroglial cells are phagocytic, and others assist in providing sustenance to the neurons.

1. Compare the structure, function, and location of neurons and neuroglia, which comprise nervous tissue.
2. List the structures of a neuron in the sequence that a nerve impulse would pass through the cell.

Clinical Considerations

As discussed in the beginning of this chapter, understanding histology is extremely important in determining organ and system structure and function. Histology has immense clinical importance as well. Many diseases are diagnosed through microscopic examination of tissue sections. Even in performing an autopsy, an examination of various tissues is vital for establishing the cause of death.

Several sciences are concerned with specific aspects of tissues. *Histopathology* is the study of diseased tissues. *Histochemistry* is concerned with the biochemical physiology of tissues as they function to maintain body homeostasis. *Histotechnology* studies the various ways tissues can be better stained and observed. In all of these disciplines, a thorough understanding of normal, or healthy, tissues is imperative for recognizing altered, or abnormal, tissues.

Changes in Tissue Composition

Most diseases alter tissue structure *locally* where the disease is prevalent, though some diseases, called *general conditions,* cause changes remote from the location of the disease. **Atrophy**

neuron: Gk. *neuron,* sinew or nerve
perikaryon: Gk. *peri,* around; *karyon,* nut or kernel

atrophy: Gk. *a,* without; *trophe,* nourishment

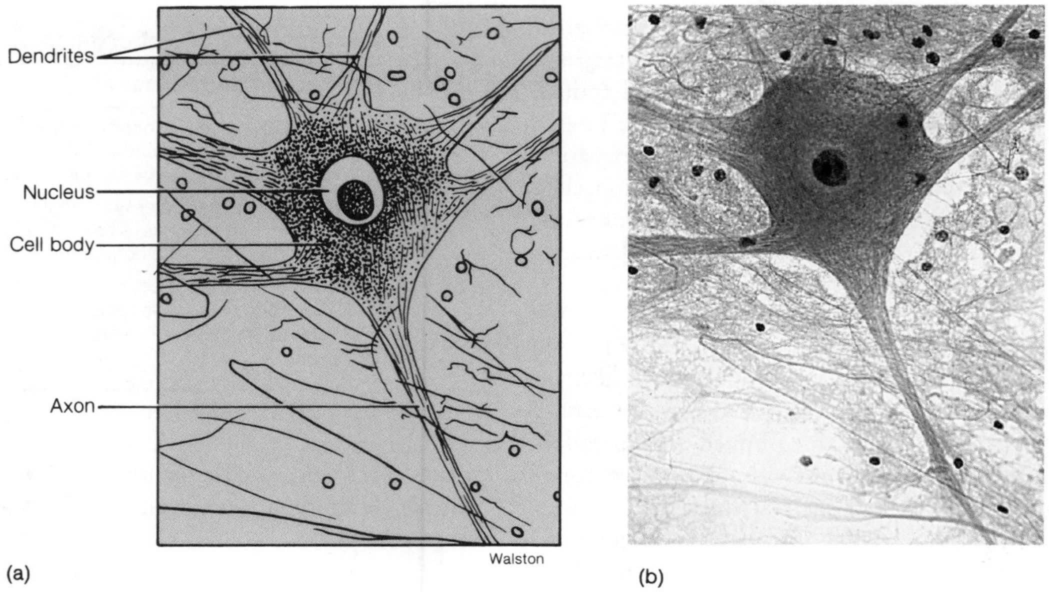

Figure 4.30 The neuron. (*a*) A simplified diagram of a neuron and its principal parts. The arrows indicate the direction of the nerve impulse. (*b*) A photomicrograph of nervous tissue (100×).

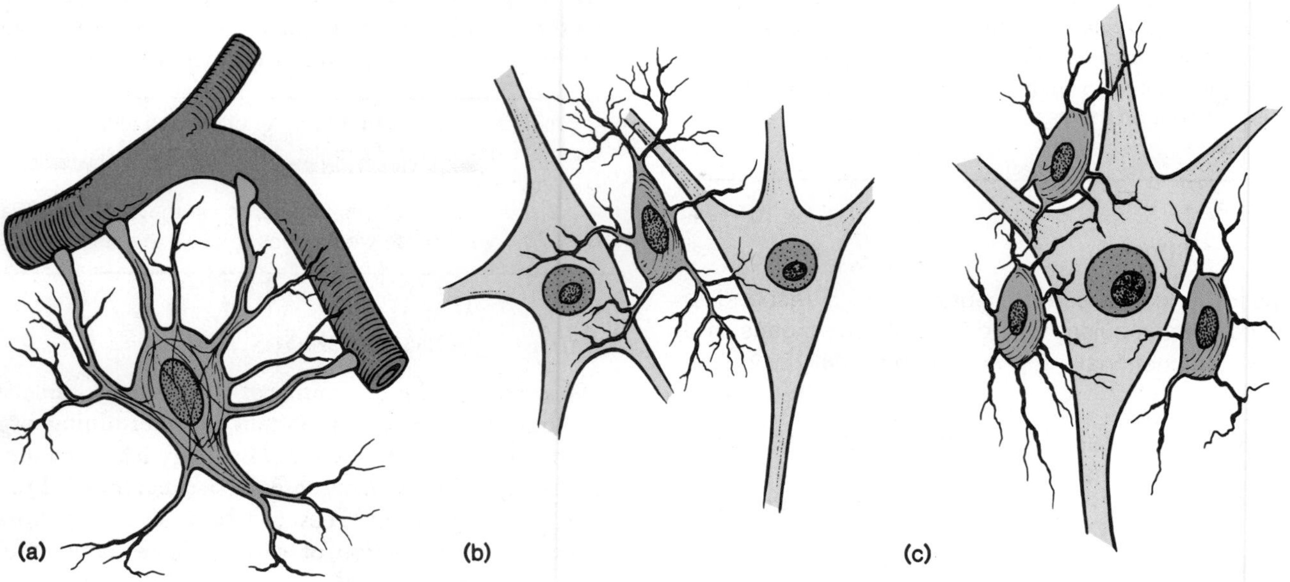

Figure 4.31 Various types of neuroglial cells that connect, support, and nourish neurons. (*a*) An astrocyte in contact with a capillary of a blood vessel serving the brain, (*b*) a microglia with processes extending to two nerve cell bodies, (*c*) oligodendrocytes near a nerve cell body.

(wasting of body tissue), for example, may be confined to a particular organ where the disease interferes with the metabolism of that organ; or it may involve an entire limb if nourishment or nerve impulses are decreased or prohibited. *Muscle atrophy,* for example, can be caused by a disease of the nervous system like polio or can be the result of a diminished blood supply to a muscle. *Senescence (se-nes'ens) atrophy,* or simply *senescence,* is the natural aging of tissues and organs within the body. *Disuse atrophy* is a local atrophy that results from the inactivity of a tissue or organ.

Necrosis *(ne-kro'sis)* is cellular or tissue death within the living body. It can be recognized by changes in the dead tissues. Necrosis can be caused by a number of factors, such as severe injury, physical agents (trauma, heat, radiant energy, chemical poisons), or interference with the nutrition of tissues. When histologically examined, the necrotic tissue usually appears opaque, and a whitish or yellow color is assumed. **Gangrene** is a massive necrosis of tissue accompanied by an invasion of microorganisms that live on decaying tissues.

Somatic death is the death of the body as a whole. Following somatic death, tissues undergo irreversible changes such as **rigor mortis** (muscular rigidity), clotting of the blood, and

necrosis: Gk. *nekros,* corpse

gangrene: Gk. *gangraina,* gnaw or eat

cooling of the body. Postmortem changes occur under varying conditions at predictable rates of time, which help in estimating the approximate time of death.

Tissue Analysis

In diagnosing a disease, it is frequently important to histologically examine tissues from a living person. When this is necessary, a **biopsy,** the removal of a section of living tissue, is taken. There are several techniques for obtaining a biopsy. *Surgical removal* is usually done on large masses or tumors. *Curettage* involves cutting and scraping tissue, as may be done in examining for uterine cancer. In a *percutaneous needle biopsy,* a biopsy needle is inserted through a small skin incision and tissue samples are aspirated. Both normal and diseased tissues are removed for purposes of comparison.

There are several steps to preparing tissues for examination. **Fixation** is fundamental for all histological preparation. It is the rapid killing, hardening, and preservation of tissue to maintain its existing structure. **Embedding** the tissue in a supporting medium such as paraffin wax usually follows fixation. The next step is **sectioning** the tissue into extremely thin slices, followed by **mounting** the specimen on a slide. Some tissues are fixed by rapid freezing and then sectioned while frozen, making embedding unnecessary. Frozen sections enable the pathologist to make a quick diagnosis during a surgical operation. These are done frequently, for example, in cases of suspected breast cancer. **Staining** is the next step. The hematoxylin and eosin (H & E) stains are routinely used on all tissue specimens. They give a differential blue and red color to the basic and acidic structures within the tissue. Other dyes may be needed to stain for specific structures.

Examination is first done with the unaided eye and then with the use of a microscope. Practically all histological conditions can be diagnosed with low magnification (100×). Higher magnification is used to clarify specific details. Further examination may be performed with an electron microscope, which makes visible the cellular structures that are the morphological bases of metabolic processes. Histological observation is the foundation of subsequent diagnosis, prognosis, treatment, and reevaluation.

Tissue Transplantation

In the last two decades, medical science has made tremendous advancements in tissue transplants. Tissue transplants are necessary for replacing nonfunctional, damaged, or lost body parts. The most successful transplant is one where tissue is taken from one place on a person's body and moved to another place, such as a skin graft from the thigh to replace burned tissue of the hand. This type of transplant is termed an **autotransplant. Isotransplants** are transplants between genetically closely related persons. Identical twins have the best acceptance success in this type of transplant. **Homotransplants** (between individuals of the same species) and **heterotransplants** (between two different species) have low acceptance percentages because of a *tissue rejection reaction.* When this occurs, the recipient's immune mechanisms are triggered, and the donor's tissue is identified as foreign and is destroyed. The reaction can be minimized by "matching" recipient and donor tissue. Immunosuppressive drugs also may lessen the rejection rate. These drugs act by interfering with the recipient's immune mechanisms. Unfortunately, immunosuppressive drugs may lower the recipient's resistance to infections as well. New techniques involving blood transfusions from donor to recipient before transplant are proving successful. In any event, tissue transplants are an important aspect of medical research, and significant breakthroughs are on the horizon.

Chapter Summary

I. Definition and Classification of Tissues
   A. Tissues are aggregations of similar cells that perform specific functions. The study of tissues is histology.
   B. Cells are separated and bound together by an intercellular matrix; its composition varies from solids to liquids.
   C. The four principal types of tissues are epithelial tissue, connective tissue, muscle tissue, and nervous tissue.

II. Development of Tissues
   A. Prenatal development following fertilization results in a blastocyst, which becomes implanted in the uterine wall between the fifth and seventh day.
   B. The three primary germ layers, from which all body tissues and organs eventually derive, form shortly after implantation.
   C. The various types of tissues are established during embryonic development, and as differentiation continues, organs form, each composed of a specific arrangement of tissues.

III. Epithelial Tissue
   A. Epithelia derive from all three germ layers and may be one or several layers thick; the lower surface is supported by a basement membrane.
   B. Epithelial cells are bonded by glycoprotein deposits, desmosomes, or tight and intermediate junctions.
   C. Simple epithelia vary in shape and surface characteristics; they are located where diffusion, filtration, and secretion occur.
   D. Stratified epithelia consist of two or more layers of cells and are adaptive for protection.
   E. Transitional epithelium lines the urinary bladder and is adapted for distension.
   F. Glandular epithelia derive from developing epithelial tissue and function as secretory exocrine glands.

IV. Connective Tissue
   A. Connective tissues derive from mesenchymal mesoderm and, with the exception of cartilage, are highly vascular.
   B. Connective tissue proper contains fibroblasts, collagenous fibers, and elastic fibers within a flexible ground substance.
   C. Cartilage provides a flexible framework for many organs and consists of a semisolid matrix of chondrocytes and various fibers.
   D. Bone (osseous) tissue consists of osteocytes, collagenous fibers, and a durable matrix of mineral salts.
   E. Vascular (blood) tissue consists of formed cellular elements (erythrocytes, leukocytes, and thrombocytes) suspended in a fluid plasma matrix.

V. Muscle Tissue
   A. Muscle tissues (smooth, cardiac, and skeletal) are responsible for the movement of materials through the body, the movement of one part of the body with respect to another, and for locomotion.

B. Fibers in the three kinds of muscle tissue are adapted to contract in response to stimuli.

VI. Nervous Tissue

A. Neurons are the functional units of the nervous system; they respond to stimuli and conduct impulses to and from all body organs.

B. Neuroglia support and bind neurons; some are phagocytic, and others provide sustenance to neurons.

Review Activities

Objective Questions

1. Which of the following is *not* a principal type of body tissue?
   (a) nervous (d) muscular
   (b) integumentary (e) epithelial
   (c) connective
2. The correct sequence of the prenatal stages of development is
   (a) ovum, zygote, blastocyst, morula.
   (b) ovum, morula, blastocyst, zygote.
   (c) ovum, zygote, morula, blastocyst.
   (d) zygote, ovum, blastocyst, morula.
3. Connective tissue, muscle, and the dermis of the skin derive from embryonic
   (a) mesoderm. (c) ectoderm.
   (b) endoderm.
4. Which statement is *false* regarding epithelia?
   (a) They are derived from both ectoderm and endoderm.
   (b) They are strengthened by elastic and collagenous fibers.
   (c) One side is exposed to the lumen, cavity, or external environment.
   (d) They have very few extracellular matrix-binding cells.
5. A gastric ulcer of the stomach would involve
   (a) simple cuboidal epithelium.
   (b) transitional epithelium.
   (c) simple ciliated columnar epithelium.
   (d) simple columnar epithelium.
6. Which structural and secretory designation describes mammary glands?
   (a) acinar, apocrine
   (b) tubular, holocrine
   (c) tubular, merocrine
   (d) acinar, holocrine
7. Dense regular connective tissue is in
   (a) blood vessels.
   (b) the spleen.
   (c) tendons.
   (d) the wall of the uterus.
8. The phagocytic connective tissue found in the lymph nodes, liver, spleen, and bone marrow is
   (a) reticular. (c) mesenchyme.
   (b) areolar. (d) elastic.
9. Cartilage is slow in healing following an injury because
   (a) it is located in body areas that are under constant physical strain.
   (b) it is avascular.
   (c) its chondrocytes cannot reproduce.
   (d) it has a semisolid matrix.
10. Cardiac muscle tissue has
    (a) striations.
    (b) intercalated discs.
    (c) rhythmical involuntary contractions.
    (d) All of the above.

Essay Questions

1. Define *tissue.* What are the differences between cells, tissues, glands, and organs?
2. What are the four principal types of tissues? What are the general functions of each?
3. What physiological functions are epithelial tissues adapted to perform?
4. Identify the epithelial tissue found in each of the following structures or organs, and give the function of the tissue in each case.
   (a) in the alveoli of the lungs
   (b) lining the lumen of the digestive tract
   (c) in the outer layer of skin
   (d) in the urinary bladder
   (e) in the uterine tube
   (f) lining the lumina of the lower respiratory tract
5. Why are both keratinized and nonkeratinized epithelia found within the body?
6. Why is it important that epithelial tissues, but not connective tissues, have tremendous mitotic potential?
7. Explain the possible consequences of excessive smoking to the linings of the lower respiratory tract and to the alveoli (air sacs) of the lungs.
8. Characterize and list the locations of connective tissue.
9. Describe how epithelial glands are classified according to structural complexity and secretory function.
10. Identify the connective tissue found in each of the following structures or organs, and give the function of the tissue in each case.
    (a) on the surface of the heart and surrounding the kidneys
    (b) within the lumen of the aorta
    (c) forming the symphysis pubis
    (d) supporting the outer ear
    (e) lymph node
    (f) tendo calcaneous
11. Distinguish among the following: reticular fibers, collagenous fibers, elastin, fibroblasts, and mast cells.
12. What is the relationship between adipose cells and fat? Discuss the function of fat, and explain the potential danger of excessive fat.
13. Identify the types of cartilage and describe their similarities and differences.
14. Discuss the mitotic abilities of each of the four principal types of tissues.
15. Distinguish among the following terms: atrophy, necrosis, gangrene, and somatic death.

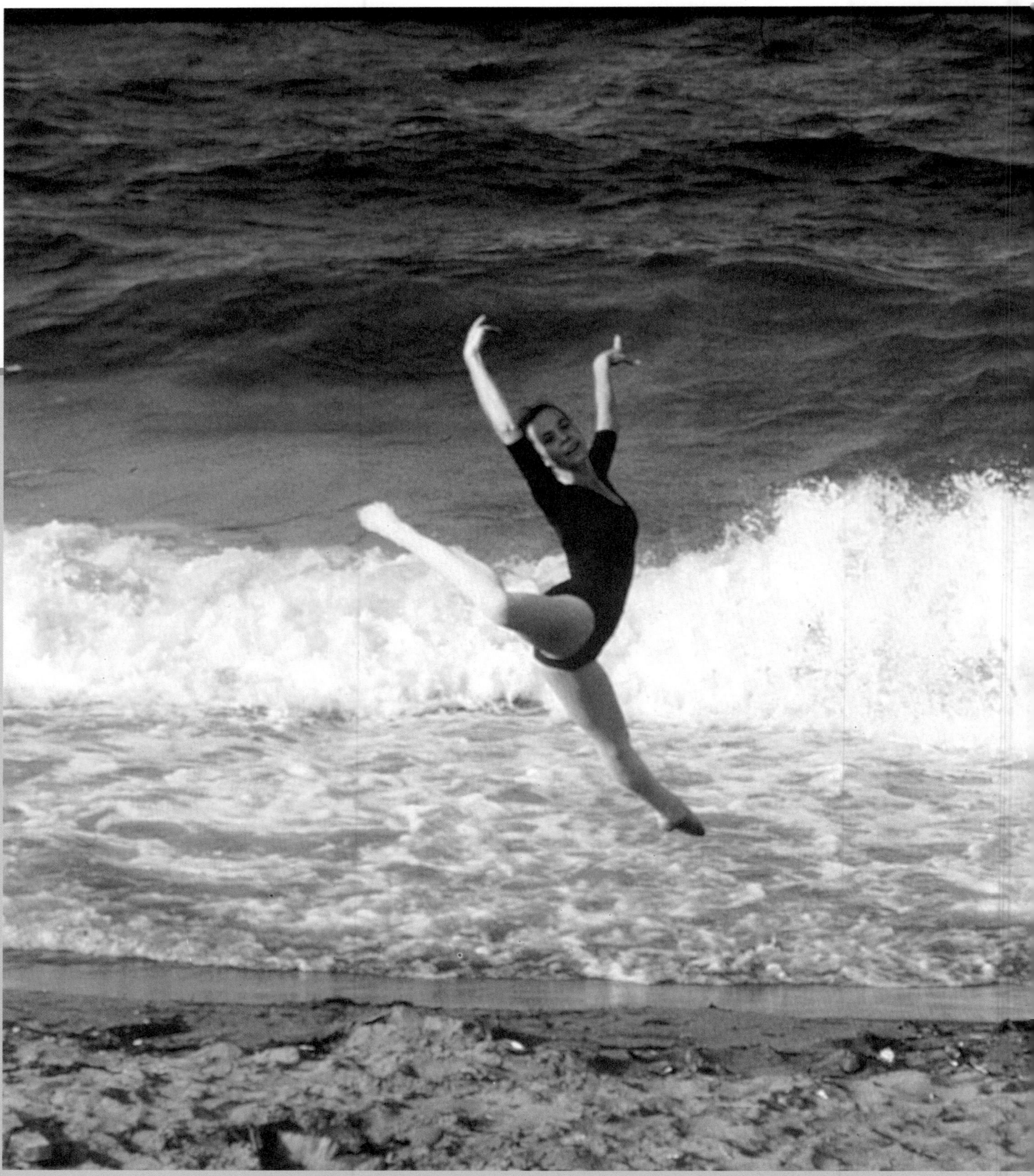

The dynamics of body movement directly involve the highly coordinated interaction of several body systems including the skeletal system, articulations, and the muscle system. Improved coordination results from intense physical conditioning of these systems.

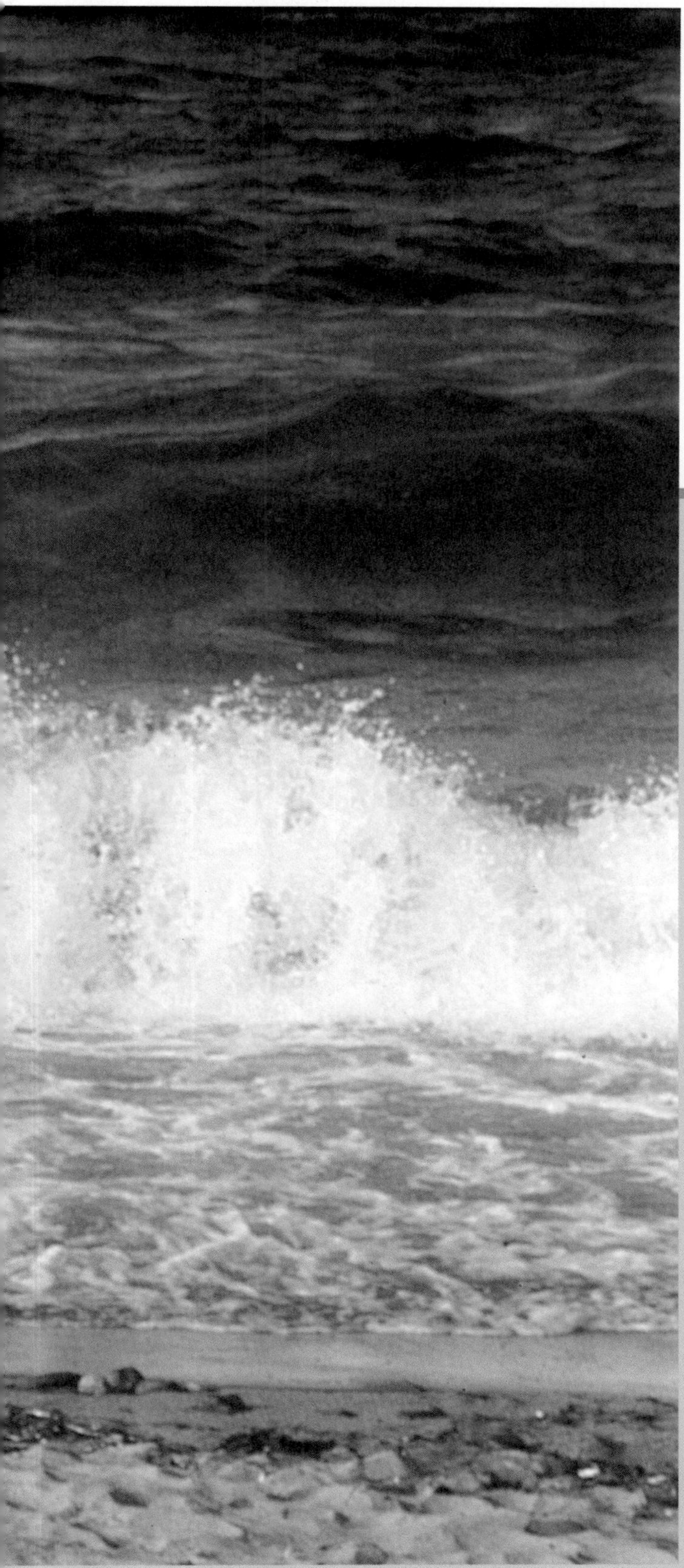

UNIT

4

Support and Movement

Body support and movement are the principal concerns of the chapters in unit 4. A description is given of how the integumentary and skeletal systems provide support as well as protection to the body. The structural and functional relationships between the bones, articulations, and muscles are discussed with respect to body movement. The basic anatomical terminology for each of these body systems is presented, as is a discussion about how the specific organs within each of these systems functions to maintain body homeostasis. Much of the organization of the body can be visualized from the surface of the body, and many of the internal structures can be observed or palpated. The important surface landmarks are identified and discussed in this unit.

This unit includes:

CHAPTER 5

Integumentary System

Outline and Concepts

Clinical Case Study

The Integument as an Organ

The integument is the largest organ of the body, and together with its epidermal structures (hair, glands, and nails), it constitutes the integumentary system. It has adaptive modifications in certain body areas that accommodate protective or metabolic functions. The integument is a dynamic interface between the continually changing external environment and our internal environment and aids in maintaining homeostasis.

Development of the Integumentary System

All of the structures of the integumentary system are derived from ectodermal and mesodermal germ layers. Although integumentary formation is an ongoing process throughout prenatal development, the integument as a body covering is established early during the embryonic period.

Development of the Layers of the Integument
Development of the Associated Structures of the Integument

Layers of the Integument

The integument consists of three principal layers. The outer epidermis is stratified into four or five structural and functional layers, and the thick and deeper dermis consists of two layers. The hypodermis, a single layer, binds the skin to underlying tissues or organs.

Epidermis
Dermis
Hypodermis

Functions of the Integument

The integument is a highly dynamic organ, which not only protects the body from pathogens and external injury, but is also extremely important for body homeostasis.

Protection of the Body from Disease and External Injury
Regulation of Body Fluids and Temperature
Cutaneous Absorption
Synthesis
Sensory Reception
Communication

Epidermal Derivatives

Hair, nails, and integumentary glands form from the epidermal layer and are therefore of ectodermal derivation. Hair and nails are structural features of the integument and have a limited functional role, whereas integumentary glands are extremely important in body defense and maintaining homeostasis.

Hair
Nails
Glands

Clinical Considerations

Inflammatory (Dermatitis) Conditions
Neoplasms
Burns
Frostbite
Skin Grafts
Wounds and Wound Healing

Clinical Case Study Answer

Important Clinical Terminology

Chapter Summary

Review Activities

CLINICAL CASE STUDY

A twenty-seven-year-old male is involved in a gasoline explosion, sustaining burns to the face, neck, chest, and arms. Upon arrival at the emergency room, he complains of pain in the face and neck, which exhibit extensive blistering and erythema (redness). These findings are all curiously absent, however, on the burned chest and arms, which have a pale, waxy appearance.

Examination reveals the skin to be leathery and without sensation, including pain. The emergency room (E.R.) physician comments to the medical student that third-degree burns are present on the chest and arms and that excision of the burn eschar (traumatized tissue) with subsequent skin grafting will be required.

Why is the area of second-degree burn red, blistered, and painful, while the third-degree burn is pale and insensate (without sensation, including pain)? Why will the chest and arms require skin grafting while the face and neck probably will not? Hint: Think in terms of (1) function of the skin, and (2) survival of cells for regeneration of functional skin.

The Integument as an Organ

The integument is the largest organ of the body, and together with its epidermal structures (hair, glands, and nails), it constitutes the integumentary system. It has adaptive modifications in certain body areas that accommodate protective or metabolic functions. The integument is a dynamic interface between the continually changing external environment and our internal environment and aids in maintaining homeostasis.

Objective 1. Explain why the integument is considered an organ and a component of the integumentary system.

Objective 2. Describe some common clinical conditions of the integument that result from nutritional deficiencies or body dysfunctions.

We are more aware of and concerned with our integumentary system than perhaps any other system of our body. One of our first actions in the morning is to examine ourselves in a mirror and assess how to make our appearance acceptable. Periodically, we examine our skin for wrinkles or our scalp for gray hairs as signs of aging. We recognize other persons to a large degree by features of the skin.

Unfortunately, the appearance of the skin frequently determines first impressions and social acceptance. For example, social rejection during teenage years, imagined or real, can be directly associated with skin problems such as acne. One's image of oneself and consequent personality may be closely associated with physical appearance.

Even clothing styles are somewhat determined by how much skin we, or the designers, want to expose. But our skin is much more than a showpiece; it protects and regulates structures within the body.

The skin, or *integument (in-teg'u-ment)*, and its associated structures (hair, glands, and nails) constitute the integumentary system. Included in this system are the millions of sensory receptors and the vascular network of the skin. The skin is a dynamic interface between the body and the external environment. It protects the body from the environment and at the same time allows communication with the environment.

The skin is considered an organ since it consists of several kinds of tissues that are structurally arranged to function together. It is the largest organ of the body, covering over 7,600 sq cm (3,000 sq in.) in the average adult and accounts for approximately 7% of a person's body weight. The skin is of variable thickness, averaging between 1.0 and 2.0 mm. It is thickest on the parts of the body exposed to wear and abrasion, such as the soles of the feet and palms of the hands, where it is about 6 mm. It is thinnest on the eyelids, external genitalia, and tympanum, where it is approximately 0.5 mm. Even the texture varies from the rough, callous skin covering the elbow and knuckle joints to the soft, sensitive areas of the eyelids, nipples, and genitalia.

The general appearance and condition of the skin are clinically important because they provide clues to certain body conditions or dysfunctions. Pale skin may indicate shock, whereas red, flushed, overly warm skin may indicate fever and infection. A rash may indicate allergies or local infections. Abnormal textures of the skin may be the result of glandular or nutritional problems (table 5.1). Even chewed fingernails may be a clue to emotional problems.

1. Explain why the skin is considered an organ and why the skin plus the integumentary derivatives are considered a system.
2. Which vitamins and minerals are important for healthy skin?
3. Describe the appearance of the skin that may accompany each of the following conditions: allergy; shock; infection; hyperpigmentation; and general dermatitis.

Development of the Integumentary System

All of the structures of the integumentary system are derived from ectodermal and mesodermal germ layers. Although integumentary formation is an ongoing process throughout prenatal development, the integument as a body covering is established early during the embryonic period.

Objective 3. Describe the development of the ectodermal and mesodermal components of the integument.

The embryonic germ layers involved in forming the various body tissues were introduced in chapter 4 (see figs. 4.2, 4.3). Just as the germ layers contribute to tissue formation, they also contribute to the formation of body systems.

integument: L. *integumentum*, a covering

Table 5.1 Variations of the skin and associated structures indicating nutritional deficiencies or body dysfunctions

Condition	Deficiency	Comments
General dermatitis	Zinc	Redness and itching
Scrotal or vulval dermatitis	Riboflavin	Inflammation in genital region
Hyperpigmentation	Vitamin B_{12}, folic acid, or starvation	Dark pigmentation occurs on backs of hands and feet
Dry, stiff, brittle hair	Protein, calories, and other nutrients	Usually in young children or infants
Follicular hyperkeratosis	Vitamin A, unsaturated fatty acids	Rough skin due to keratotic plugs from hair follicles
Pellagrous dermatitis	Niacin and tryptophan	Lesions develop on areas exposed to sun
Thickened skin at pressure points	Niacin	Noted at belt area at the hips
Spoon nails	Iron	Nails are thin and concave or spoon-shaped
Dry skin	Water or thyroid hormone	Dehydration, hypothyroidism, rough skin
Oily skin (acne)		Hyperactivity of sebaceous glands

The skin is composed of three structural and functional layers (see fig. 5.5). The **epidermis** *(ep''ĭ-der'mis)* is the thinnest, outermost layer and is derived from the embryonic *ectoderm.* The **dermis** underlies the epidermis, and the **hypodermis** (subcutaneous layer) affixes the dermis of the skin to underlying tissues or organs. Both the dermis and hypodermis derive from embryonic *mesoderm.*

Development of the Layers of the Integument

By the fourth week following conception, the embryo is surrounded by a single-cell layer of ectoderm (fig. 5.1). Just beneath the ectoderm is a thicker layer of undifferentiated mesoderm called *mesenchyme (mez'en-kīm'').*

The ectodermal layer differentiates by six weeks into an outer flattened **periderm** and an inner, cuboidal **germinal (basal) layer** in contact with the mesenchyme. The periderm eventually sloughs off forming the **vernix caseosa** *(ka''se-o'sah),* a cheeselike protective coat that covers the skin of the fetus. The germinal layer gives rise to the entire epidermis as well as all the glands, nails, hair, and hair follicles of the integument.

By eleven weeks, the mesenchymal cells below the germinal cells have differentiated into the distinct collagenous and elastic connective tissue fibers of the dermis. The tensile properties of these fibers cause a buckling of the epidermis and the formation of **dermal papillae** (fig. 5.1c). During the early fetal period (about ten weeks), specialized *neural crest cells* called **melanoblasts** *(mel'ah-no-blastz'')* migrate into the developing dermis and differentiate into **melanocytes** *(mel'ah-no-sītz'').* The melanocytes soon migrate to the germinal layer of the epidermis (fig. 5.1d), where they produce *melanin (mel'ah-nin),* which colors the epidermis.

Development of the Associated Structures of the Integument

Hair, glands, and nails develop from the germinal layer of the epidermis and are therefore ectodermal in origin. Before hair can form, a hair follicle must be present. Each **hair follicle** *(fol'ĭ-k'l)* begins to develop at about twelve weeks as a mass of germinal cells, called a **hair bud,** and proliferates into the underlying mesenchyme (fig. 5.2). As the hair bud becomes club-shaped, it is known as a **hair bulb.** The hair follicle, which physically supports and provides nourishment to the hair, is derived from specialized mesenchyme, called the **hair papilla** *(pah-pil'ah),* which is localized around the hair bulb, and from the epithelial cells of the hair bulb, called the **germinal matrix.** The wall of the hair follicle, known as the **epithelial root sheath,** is supported by the specialized, mesenchymal **dermal root sheath.** Continuous mitotic activity in the epithelial cells of the hair bulb results in the growth of the hair.

Sebaceous *(sĕ-ba'shus)* **glands** and **sudoriferous** *(soo'dor-if'er-us),* or **sweat glands,** are the two principal types of integumentary glands. Both develop from the germinal layer of the epidermis. Sebaceous glands develop as proliferations from the sides of the developing hair follicle (fig. 5.2). In a mature sebaceous gland, the oily secretions empty onto the shaft of a hair in a hair follicle. Sweat glands (fig. 5.2) become coiled as the secretory portion of the developing gland proliferates into the dermal mesenchyme.

Mammary glands are modified sweat glands, which develop like sweat glands as downward proliferations of the germinal epithelium (fig. 5.3). The site of development is specific, however, along a **mammary ridge** (see fig. 2.3). In humans, generally only one pair of pectoral (chest) mammary glands develop. As the mammary bud extends into the dermis, it branches (rather than coils as sweat glands) to form the secretory and ductal portions of the mammary glands (see fig. 5.3).

The mammary glands of newborns (males and females) are frequently swollen, and small amounts of secretion, called witch's milk, may be discharged. Stimulation of the infant's mammary glands is caused by maternal hormones passing into the fetal circulation via the placenta.

Nails begin developing at about ten weeks along the dorsal aspect of each digit (fig. 5.4). The appearance of a thickened area of epidermis, called the **nail field,** is the first indication of nail development. The proximal and lateral borders of the nail field become thickened as the **nail folds.** Continued mitotic activity at the proximal nail fold causes a toughened **nail plate (nail)** to grow over the nail bed.

epidermis: Gk. *epi,* on; *derma,* skin
hypodermis: Gk. *hypo,* under; *derma,* skin
mesenchyme: Gk. *mesos,* middle; *enchyma,* infusion
periderm: Gk. *peri,* around; *derm,* skin
vernix caseosa: L. *vernix,* varnish; *caseus,* cheese
melanin: Gk. *melas,* black

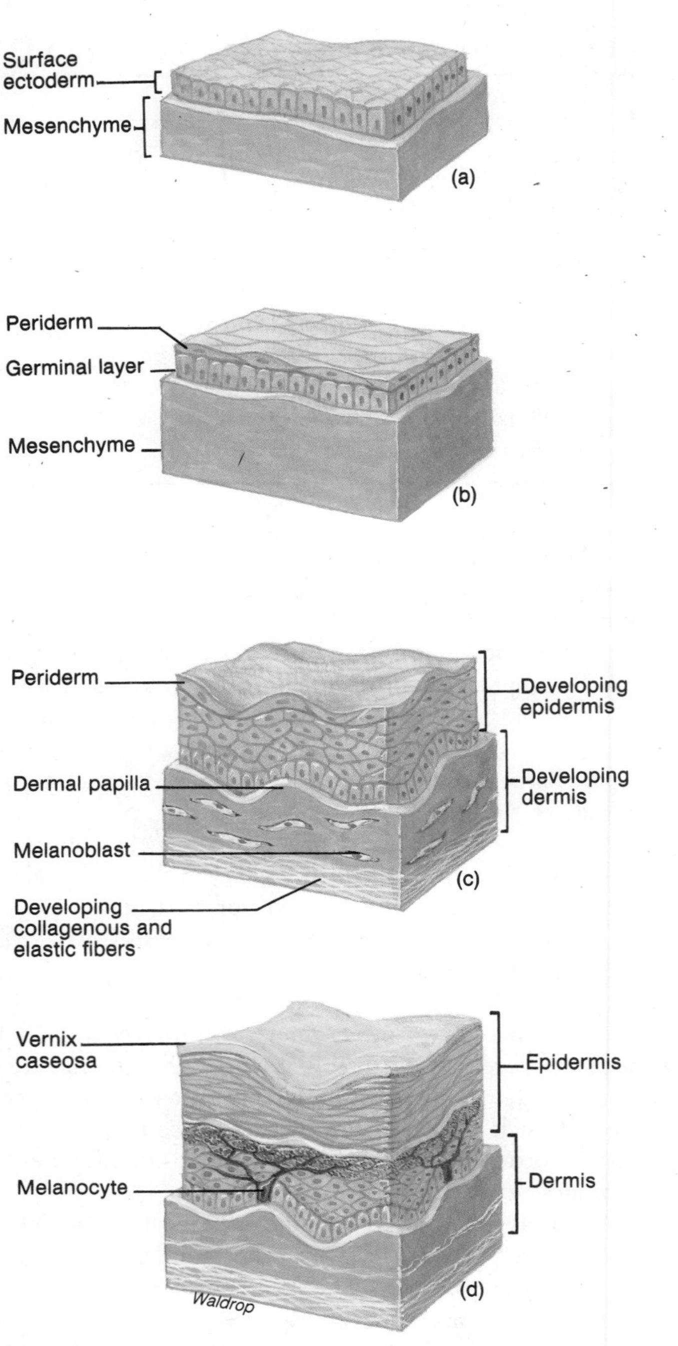

Figure 5.1 The development of the integument. (*a*) At four weeks only the surface ectoderm and mesenchyme of the mesoderm are apparent. (*b*) At seven weeks the surface ectoderm has differentiated into periderm and germinal layers. (*c*) By the eleventh week, the epidermis is stratified (layered) and the dermis is forming. (*d*) The integument at birth has distinct layers in both the epidermis and dermis. (Note the protective vernix caseosa over the epidermis and the developing melanocytes within the deeper epidermal layers.)

1. Discuss the role of the ectodermal germinal layer in the formation of the epidermis and associated structures of the skin, as well as the role of the mesodermal mesenchymal layer in the formation of the dermis and hypodermis of the skin.
2. Describe the sequence involved in establishing melanocytes in the epidermal germinal layer.

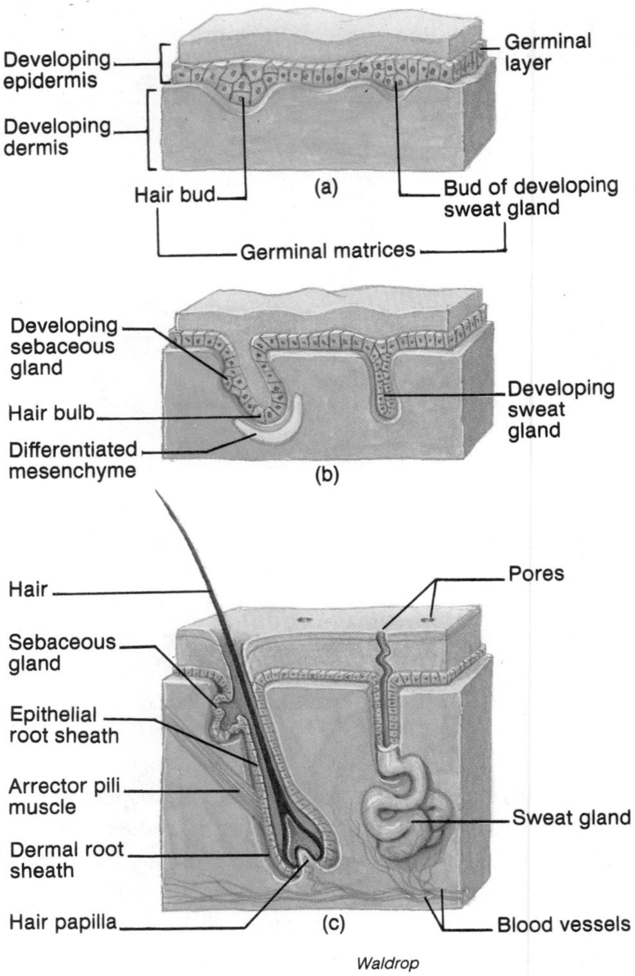

Figure 5.2 The development of hair and the integumentary glands. (*a*) Twelve weeks, (*b*) fifteen weeks, and (*c*) at birth.

Layers of the Integument

The integument consists of three principal layers. The outer epidermis is stratified into four or five structural and functional layers, and the thick and deeper dermis consists of two layers. The hypodermis, a single layer, binds the skin to underlying tissues or organs.

Objective 4. Describe the histological characteristics of each layer of the integument.

Objective 5. Summarize the process of cellular replacement within the epidermis.

Epidermis

The **epidermis** is the superficial protective layer of the skin and is composed of stratified squamous epithelium. Over most of the body, this layer is between thirty and fifty cell layers thick, or 0.007–0.12 mm. All but the deepest layers of the epidermis are dead cells that contain a protein called **keratin** *(ker'ah-tin),* which toughens and waterproofs the skin. The epidermis is composed of either four or five layers, depending on its location within the body (figs. 5.5, 5.6). The epidermis of the palms and soles has five layers because these areas are exposed to greater friction. The epidermis of all other areas of the body has only four layers. The names and characteristics of the epidermal layers are as follows.

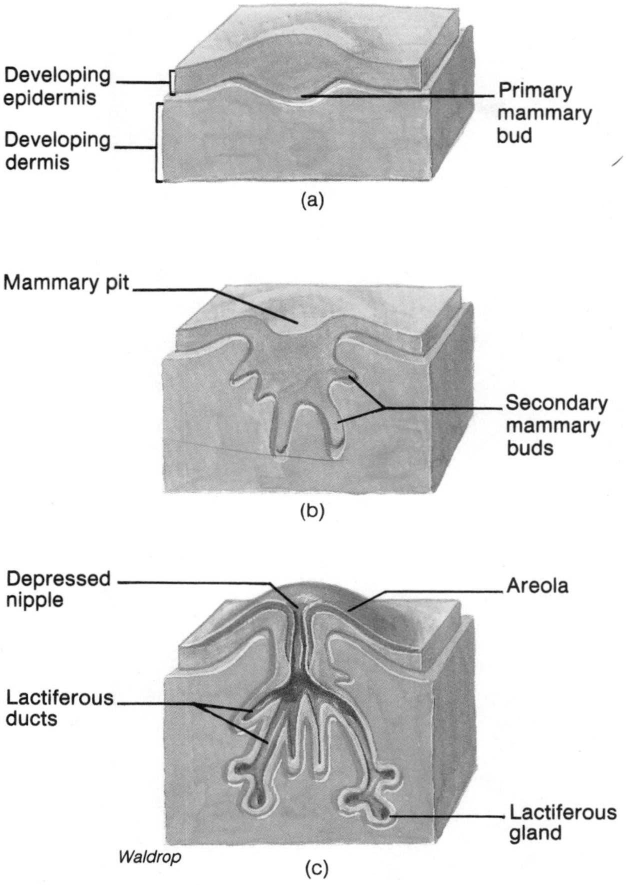

Figure 5.3 The development of mammary glands. (*a*) Twelve weeks, (*b*) sixteen weeks, and (*c*) about twenty-eight weeks.

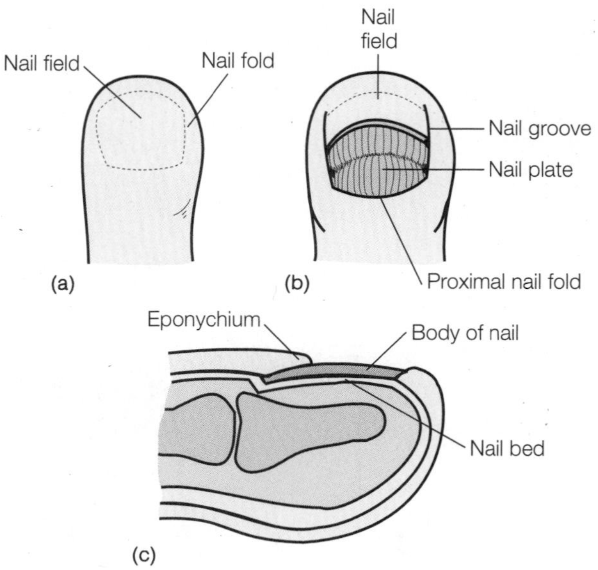

Figure 5.4 The development of nails. (*a*) Ten weeks, (*b*) about fourteen weeks, and (*c*) a sagittal view of distal digit and nail at birth.

1. **Stratum basale.** The stratum basale is composed of a single layer of cells in contact with the dermis. These cells are constantly dividing mitotically and moving outward to renew the epidermis. As the cells push away from the dermis and move away from the vascular nutrient and oxygen supply, their nuclei degenerate and the cells die. In approximately six to eight weeks, the cells are sloughed off from the top layer of the epidermis.
2. **Stratum spinosum** (stratum malpighii). The stratum spinosum contains several stratified layers of polygonal (many-sided) cells, which are tightly attached by spinelike projections. Since there is limited mitosis in this layer, it along with the stratum basale, these two layers are collectively referred to as the **stratum germinativum.**
3. **Stratum granulosum.** The cells of the stratum granulosum are flattened and contain dark-staining granules (the source of this layer's name). The process of *keratinization (ker''ah-tin''i-za'shun),* which is associated with cellular death, is initiated in this layer.
4. **Stratum lucidum.** The nuclei, organelles, and cell membranes are no longer visible in the cells of the stratum lucidum, and so histologically this layer appears clear. This layer exists only in the thickened skin of the soles and palms.
5. **Stratum corneum** *(kor'ne-um).* The stratum corneum is composed of twenty-five to thirty layers of flattened, scalelike cells, which are continuously shed as flakelike residues of cells. This surface layer is cornified and is the real protective layer of the skin (fig. 5.7). *Cornification* is brought on by keratinization, and the hardening, flattening process takes place as the cells migrate to the surface.

Friction at the surface of the skin stimulates additional mitotic activity of the stratum basale, resulting in the formation of a *callus* for additional protection. Table 5.2 describes the specific characteristics of each epidermal layer.

Tattooing colors the skin permanently because pigmented dyes are injected below the mitotic basale layer into the dermis. Because of frequent nonsterile conditions, those who administer the tatoo must take care not to introduce infections along with the dye.

Coloration of the Skin Normal skin color is caused by the expression of a combination of three pigments: **melanin** *(mel'ah-nin),* **carotene,** and **hemoglobin** *(he''mo-glo'bin).* Melanin is a brown-black pigment formed in the pigment cells, called **melanocytes.** Melanocytes are found throughout the strata basale and spinosum (fig. 5.8). All races have virtually the same number of melanocytes, but the amount of melanin produced and the degree of granular aggregation of the melanin

stratum: L. *stratum,* something spread out
basale: Gk. *basis,* base
spinosum: L. *spina,* thorn
germinativum: L. *germinare,* sprout or growth
granulosum: L. *granum,* grain

lucidum: L. *lucidus,* light
corneum: L. *corneus,* horny
carotene: L. *carota,* carrot (referring to orange coloration)
hemoglobin: Gk. *haima,* blood; *globus,* globe

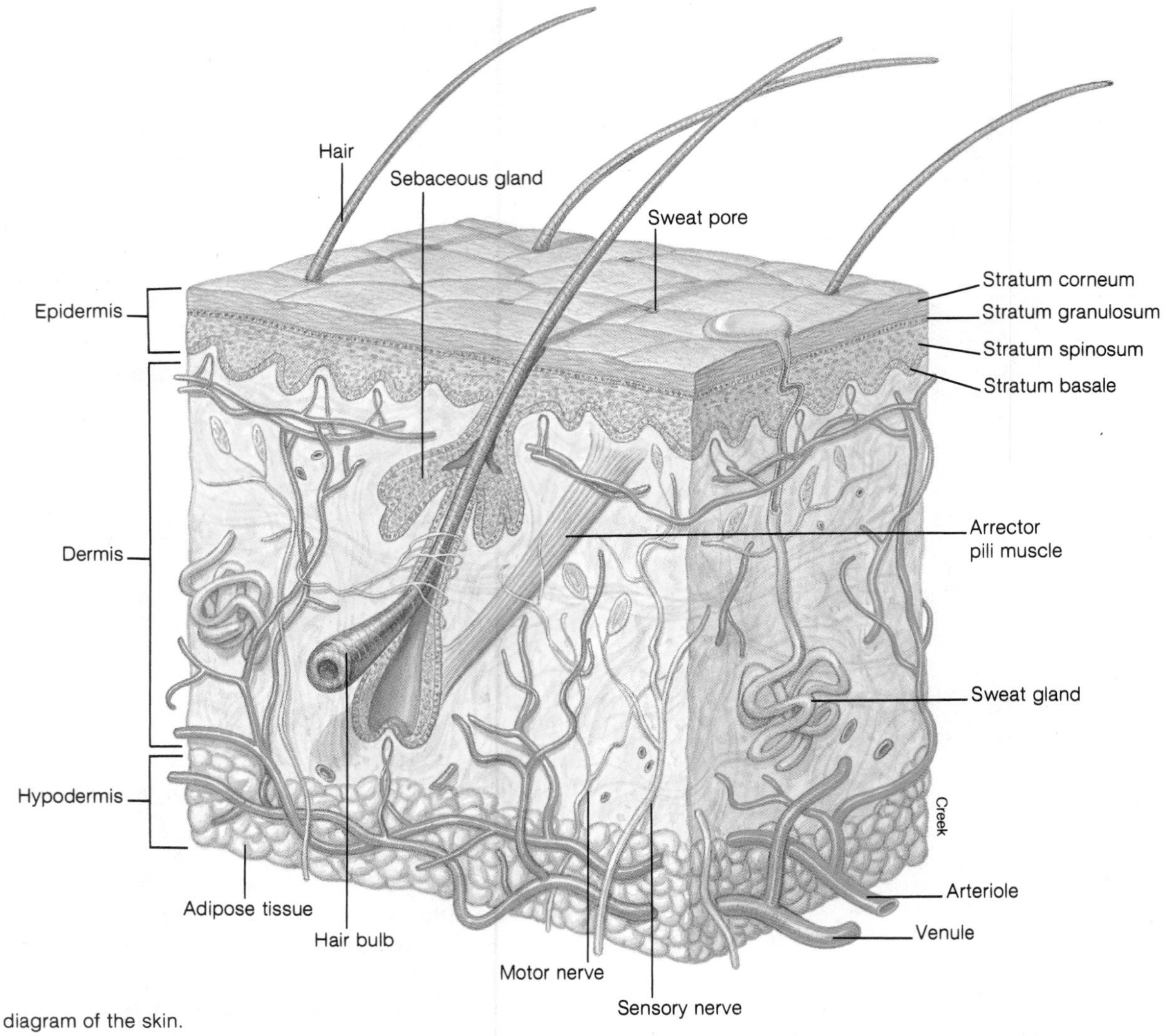

Figure 5.5 A diagram of the skin.

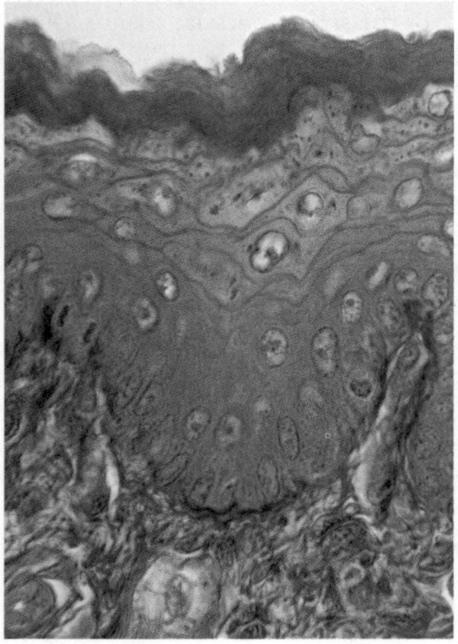

Figure 5.6 A photomicrograph of the epidermis (250×).

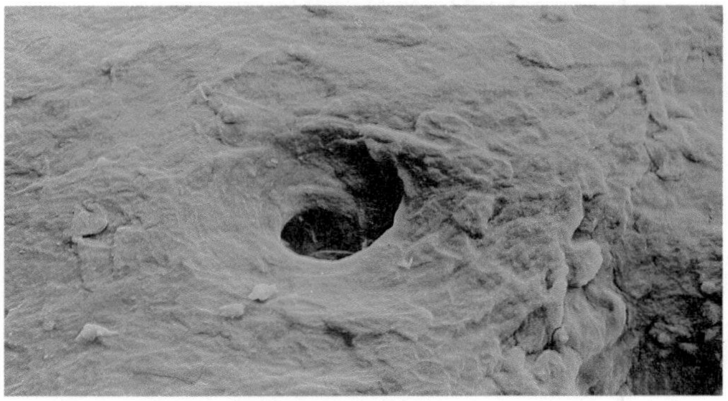

Figure 5.7 A scanning electron micrograph of the surface of the skin showing the opening of a sweat gland. The fragmented appearing particles are bacteria, which are present throughout the body on the surface of the skin.

Table 5.2 Layers of the epidermis

Layer	Location	Characteristics
Stratum basale	Deepest layer	Consists of a single layer of cells that undergo mitosis; contains pigment-producing melanocytes
Stratum spinosum	Above the stratum basale	Composed of several layers of cells with centrally located, large, oval nuclei and spinelike processes; limited mitosis
Stratum granulosum	Above the stratum spinosum	Composed of one or more layers of granular cells that contain fibers of keratin and shriveled nuclei
Stratum lucidum	Above the stratum granulosum	A thin, clear layer found only in the epidermis of the palms and soles
Stratum corneum	Outermost layer	Consists of many layers of keratinized, dead cells that are flattened and nonnucleated; cornified

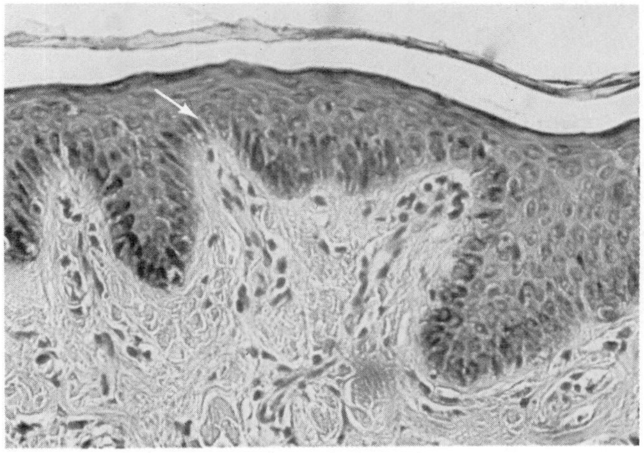

Figure 5.8 Melanocytes throughout the strata basale and spinosum (see arrow) produce melanin.

determine whether an individual's skin color is black, brown, red, tan, or white. Melanin is a protective device that guards against the damaging effect of the ultraviolet rays in sunlight. A gradual exposure to the sunlight promotes the increased production of melanin within the melanocytes and hence tanning of the skin. The skin of a genetically determined albino has the normal complement of melanocytes in the epidermis but lacks the enzyme tyrosinase, which converts the amino acid tyrosine to melanin.

There are other genetical expressions of melanocytes that are more common than albinism. *Freckles,* for example, are caused by aggregated patches of melanin. A lack of melanocytes in localized areas of the skin causes distinct white spots in the condition called *vitiligo (vit-i-li'gō).*

Excessive exposure to the sunlight can cause *basal cell carcinoma,* or *malignant melanoma.* While in sunlight, one's skin absorbs two wavelengths of ultraviolet rays known as UV-A and UV-B. The DNA within the basal skin cells may be damaged as the sun's more dangerous UV-B rays penetrate the skin. Although it was once believed that UV-A rays were harmless, findings now indicate that excessive exposure to UV-A rays may inhibit the DNA repair process that follows exposure to UV-B. Therefore, individuals who are exposed solely to UV-A rays in tanning salons are still in danger of basal cell carcinoma since they will later be exposed to UV-B rays of sunlight when they are out-of-doors.

Carotene is a yellowish pigment found in the epidermal cells and fatty parts of the dermis. Carotene is abundant in the skin of Oriental people and, together with melanin, accounts for their yellow-tan skin.

Hemoglobin is not a pigment of the skin but the oxygen-binding pigment found in red blood cells. Oxygenated blood flowing through the dermis gives the skin its pinkish tones.

Certain physical conditions or diseases cause symptomatic discoloration of the skin. Cyanosis is a bluish discoloration of the skin that appears in people with certain cardiovascular or respiratory diseases. People also become cyanotic during an interruption of breathing. In jaundice, the skin appears yellowish because of an excess of bile pigment in the bloodstream. Jaundice is usually symptomatic of liver dysfunction and sometimes of liver immaturity, as in a jaundiced newborn.

Surface Patterns The exposed surface of the skin has recognizable patterns that are either congenital or acquired. Congenital patterns called **fingerprints,** or **friction ridges,** are present on the finger and toe pads as well as on the palms and soles. The designs formed by these lines have basic similarities but are not identical in any two individuals (fig. 5.9). They are formed by the pull of elastic fibers within the dermis and are prenatally well established. As the name implies, friction ridges function to prevent slippage when grasping objects. Because they are precise and easy to reproduce, fingerprints are customarily used for identifying individuals.

Acquired lines include the deep **flexion creases** on the palms and the shallow **flexion lines** that can be seen on the knuckles and on the surface of other joints. Furrows on the forehead and face are acquired from continual contraction of facial muscles, such as from smiling or squinting in bright light or against the wind. Facial lines become more strongly delineated as a person ages.

vitiligo: L. *vitiatio,* blemish

cyanosis: Gk. *kyanosis,* dark-blue color
jaundice: L. *galbus,* yellow

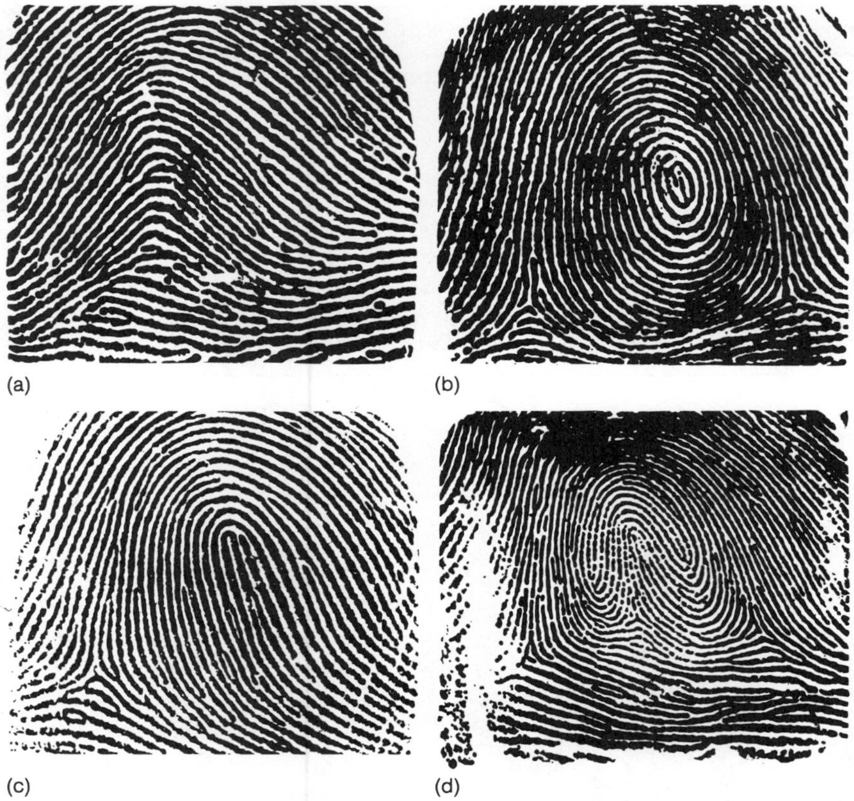

Figure 5.9 Basic dermatoglyphic patterns in the digits: (*a*) arch, (*b*) whorl, (*c*) loop, and (*d*) combination. Individual variations are commonly understood, but it is neither known why basic differences exist between men and women nor why the prints of mentally handicapped persons consistently deviate from normal patterns.

The science known as dermatoglyphics is concerned with the classification and identification of fingerprints. As seen in figure 5.9, there are four basic dermatoglyphic patterns. Every individual's prints are unique, including those of identical twins. Fingerprints, however, are not exclusive to humans. All other primates have fingerprints, and even dogs have a characteristic nose print that is used for identification in the military canine corps and in certain dog kennels.

Dermis

The **dermis** is deeper and thicker than the epidermis (fig. 5.5). Vessels within the dermis nourish the living portion of the epidermis, and numerous collagenous, elastic, and reticular fibers give support to the skin. The fibers within the dermis radiate in definite directions and are responsible for producing lines of tension on the surface of the skin (fig. 5.10). Elastic fibers are more superficial and provide skin tone. There are considerably more elastic fibers in the dermis of a young person than in that of an elderly one. The decreasing amount of elastic fiber is apparently directly associated with aging. The dermis is highly vascular and glandular and contains many nerve endings and hair follicles.

Layers of the Dermis The dermis is composed of two layers. The upper layer, called the **papillary layer,** is in contact with the epidermis and accounts for about one-fifth of the entire dermis. Numerous projections, called papillae, extend from the upper portion of the dermis into the epidermis. Papillae form the base for the friction ridges on the fingers and toes. The papillae also provide vascular concentrations for the avascular epidermis.

The deeper and thicker layer of the dermis is called the **reticular layer.** Fibers within this layer are more dense and regularly arranged to form a tough, flexible meshwork. It is quite distensible, as is evident in pregnant women or obese persons, but it can be stretched too far, causing "tearing" of the dermis. The repair of a strained dermal area leaves a white streak called a stretch mark, or **linea albicans.** Lineae albicantes are frequently found on the buttocks, thighs, abdomen, and breasts (fig. 5.11).

It is the strong, resilient reticular layer of domestic mammals that is used in making leather and suede. In the tanning process, the hide of an animal is treated with various chemicals that cause the epidermis with its hair and the papillary layer of the dermis to separate from the underlying reticular layer. The reticular layer is then softened and treated with protective chemicals to make it usable.

Innervation of the Skin The dermis of the skin has extensive innervation. Specialized integumentary *effectors* consist of muscles or glands within the dermis that respond to motor impulses transmitted from the central nervous system to the skin by autonomic nerve fibers.

Several types of **sensory (afferent) receptors** respond to various tactile (touch), pressure, temperature, tickle, or pain

papilla: L. *papula,* swelling or pimple

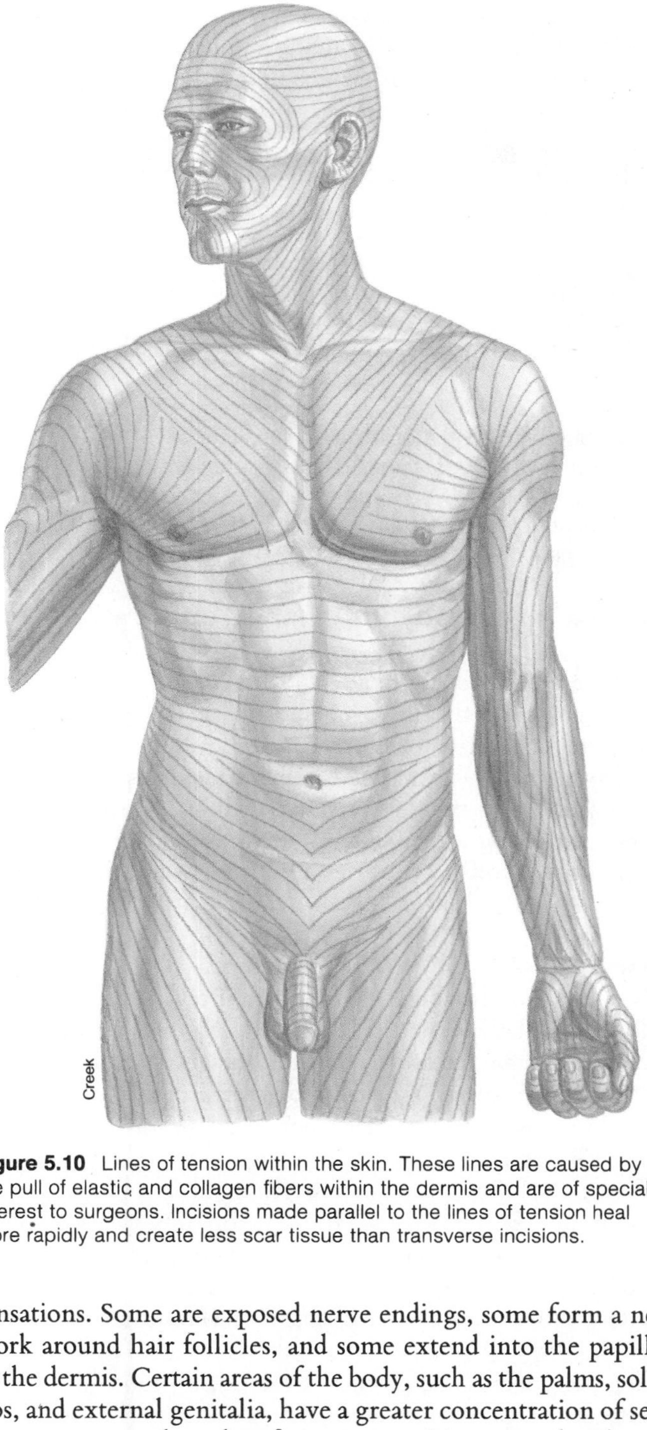

Figure 5.10 Lines of tension within the skin. These lines are caused by the pull of elastic and collagen fibers within the dermis and are of special interest to surgeons. Incisions made parallel to the lines of tension heal more rapidly and create less scar tissue than transverse incisions.

sensations. Some are exposed nerve endings, some form a network around hair follicles, and some extend into the papillae of the dermis. Certain areas of the body, such as the palms, soles, lips, and external genitalia, have a greater concentration of sensory receptors and are therefore more sensitive to touch. Chapter 15 contains a detailed structural and functional account of the various sensory receptors.

Vascular Supply of the Skin Blood vessels within the dermis supply nutrients to the mitotically active strata basale and spinosum of the epidermis and to the cellular structures of the dermis such as glands and hair follicles. Dermal blood vessels play an important role in regulating body temperature and even blood pressure. Autonomic vasoconstriction or vasodilation responses can either shunt the blood away from the superficial dermal arterioles or permit it to flow freely throughout dermal

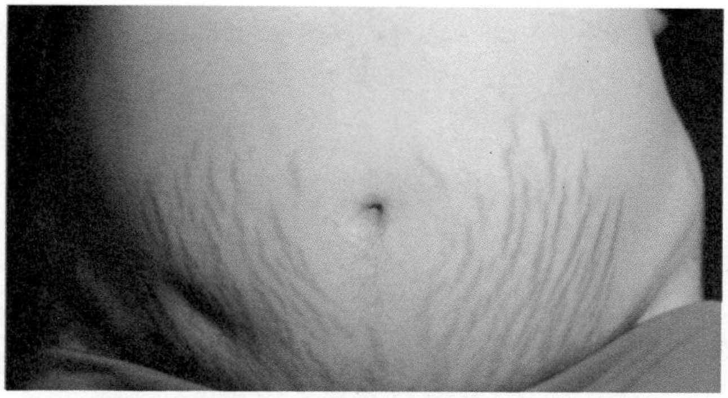

Figure 5.11 Stretch marks (lineae albicantes) on the abdomen of a pregnant woman. Stretch marks generally fade with time but frequently leave permanent integumentary markings.

vessels. Fever or shock can be detected by the color and temperature of the skin. Blushing is the result of involuntary vasodilation of dermal blood vessels.

A healthy circulating blood flow in debilitated bed patients is important for the prevention of bedsores, or *decubitus ulcers.* When a person lies in one position for an extended period, the dermal blood flow is restricted where the body presses against the bed. As a consequence, cells die and open wounds develop (fig. 5.12). Changing the position of the patient frequently and periodically massaging the skin to stimulate blood flow are good preventive measures against decubitus ulcers.

Hypodermis

The deepest layer of the integument is called the **hypodermis,** or subcutaneous layer. It binds the dermis to underlying organs. This layer is composed primarily of loose connective tissue and adipose cells interlaced with blood vessels (see fig. 5.5). Collagenous and elastic fibers reinforce the hypodermis, particularly on the palms and soles where the skin is firmly attached to underlying structures. The amount of adipose in the hypodermis varies with the sex, age, region of the body, and nutritional state of the individual. Females generally have about an 8% thicker hypodermis than do males. This layer also functions to store lipids, insulate and cushion the body, and regulate temperature.

The hypodermis is of clinical importance as an injection site for subcutaneous medication. Subcutaneous injections may be administered to patients who are unconscious or uncooperative, and when oral medications are not practical. Experimental subcutaneous medications are being tested that are slow release, low dosage, and long duration. These types of medications for diabetics regulate metabolism or mitigate pain. Subcutaneous devises, which have a continuous release of small dosages of insulin, are currently available.

decubitus: L. *decumbere,* lie down
ulcer: L. *ulcus,* sore

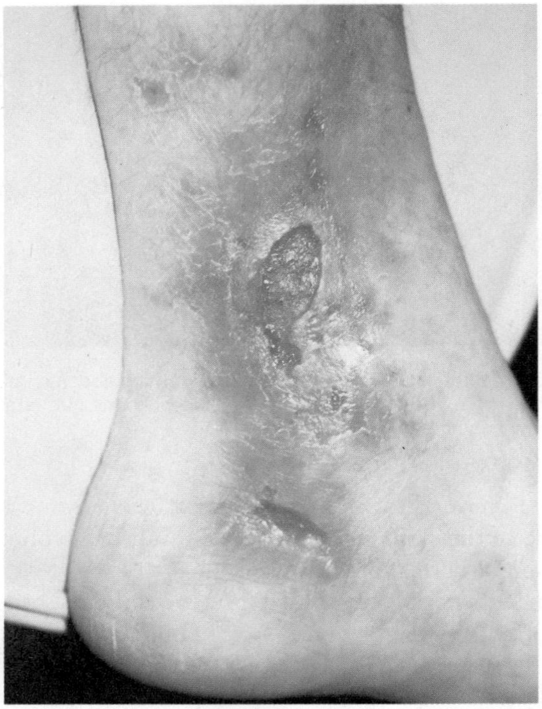

Figure 5.12 A decubitus ulcer on the medial surface of the ankle of the right leg. The most frequent sites for decubitus ulcers are in the skin overlying a bony projection, such as on the hip, ankle, heel, shoulder, or elbow.

---

1. List the layers of the epidermis and dermis, and explain how they differ in structure and function.
2. Describe the sequence of cellular replacement within the epidermis and the processes of keratinization and cornification.
3. How do both the dermis and hypodermis function in thermoregulation?
4. What two basic types of innervation are found within the dermis?

Functions of the Integument

The integument is a highly dynamic organ, which not only protects the body from pathogens and external injury, but is also extremely important for body homeostasis.

Objective 6. Discuss the role of the integument in the protection of the body from disease and external injury; the regulation of body fluids and temperature; absorption; synthesis; sensory reception; and communication.

Protection of the Body from Disease and External Injury

The integument is a physical barrier to most microorganisms, water, and excessive ultraviolet (UV) light. Oily secretions onto the surface of the skin form an acidic (pH 4.0–6.8) protective film, which waterproofs the body and retards the growth of most pathogens. The protein keratin in the epidermis also waterproofs the skin. Cornification toughens the mostly dead cells to withstand abrasion and the penetration of microorganisms. Upon exposure to UV light, the melanocytes are stimulated to synthesize melanin, which in turn absorbs and disperses sunlight. Surface friction causes the epidermis to thicken, as a protective response, by increasing mitosis in the cells of the strata basale and spinosum, which results in the formation of a protective *callus.*

Regardless of skin pigmentation, everyone is susceptible to skin cancer if his or her exposure to sunlight is sufficiently intense and continuous. There are an estimated eight hundred thousand new cases of skin cancer yearly in the United States, and approximately ninety-three hundred of these are diagnosed as the potentially life-threatening *melanoma* (cancer of melanocytes). Melanomas (fig. 5.13) are usually termed malignant as they may spread rapidly. Sunscreens are advised for persons who must be in direct sunlight for long periods of time.

Regulation of Body Fluids and Temperature

Fluid Loss The epidermis of the skin is adapted for continuous exposure to the air by being thickened, keratinized, and cornified. In addition, the outer layers are dead and scalelike, and a protein-polysaccharide basement membrane adheres the stratum basale to the dermis. Human skin is virtually waterproof, protecting the body from desiccation (dehydration) on dry land and even from water absorption when immersed in water.

Temperature Regulation The integument plays a crucial role in the regulation of body temperature. Body heat comes from cellular metabolism, particularly in muscle cells as they maintain tone or a degree of tension. A normal body temperature of 37° C is maintained by the antagonistic effects of sweating and shivering involving feedback mechanisms. Excess heat is actually lost from the body in three ways, all involving the skin: (1) through radiation from dilated blood vessels, (2) through excretion and the evaporation of sweat; and (3) through convection and the conduction of heat directly through the skin (fig. 5.14). Sweat excretion increases approximately 100–150 mL/day for each 1° C elevation in body temperature. Up to 10 L of sweat may be excreted to cool the body of a person doing hard physical work out-of-doors in the summertime.

A serious danger of continued exposure to heat and excessive water and salt loss is *heat exhaustion,* characterized by nausea, weakness, dizziness, headache, and a decreased blood pressure. *Heat stroke* is similar to heat exhaustion, with the exception that sweating is inhibited (for reasons that are not clear) and body temperature rises. Convulsions, brain damage, and death may follow.

Excessive heat loss triggers a shivering response in muscles, which increases cellular metabolism. Not only do skeletal muscles contract, but tiny smooth muscles, called **arrector pili** *(ah-rek'tor pi'li),* attached to hair follicles, contract involuntarily, causing goose bumps.

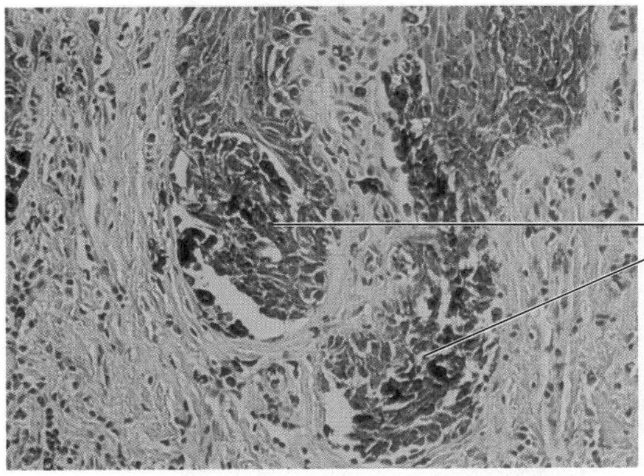

Figure 5.13 Skin cancer.

When the body's heat-producing mechanisms cannot keep pace with heat loss, *hypothermia* results. A lengthy exposure to temperatures below 20° C and dampness may lead to this condition. This is why it is so important that a hiker, for example, dress appropriately for the weather conditions, especially on cool, rainy spring or fall days. The initial symptoms of hypothermia are numbness, paleness, delirium, and uncontrolled shivering. If the core temperature falls below 32° C (90° F), the heart loses its ability to pump blood and will go into fibrillation (erratic contractions). If the victim is not warmed, extreme drowsiness, coma, and death follow.

Cutaneous Absorption

Because of the effective protective barriers of the integument already described, cutaneous (through the skin) absorption is limited. Some gases, such as oxygen and carbon dioxide, may pass through the skin and enter the bloodstream. Small amounts of UV light, necessary for synthesis of vitamin D, are absorbed readily. Of clinical consideration is the fact that certain toxins and pesticides enter the body through cutaneous absorption.

Synthesis

The integumentary system synthesizes melanin and keratin, which remain in the skin, and vitamin D, which is used elsewhere in the body. The integumentary cells contain a compound called *dehydrocholesterol (de-hi"dro-ko-les'tah-rol)* from which they synthesize vitamin D in the presence of UV light. Only small amounts of UV light are necessary for vitamin D synthesis, but these amounts are very important to a growing child (fig. 5.15). Synthesized vitamin D enters the blood and helps regulate the metabolism of calcium and phosphorus, which are important in the development of strong and healthy bones. *Rickets* is a disease caused by vitamin D deficiency.

Sensory Reception

Highly specialized sensory receptors (see chap. 15) that respond to the precise stimuli of heat, cold, pressure, touch, vibration, and pain are located abundantly throughout the dermis and hypodermis of the integument. These receptors, referred to as **cutaneous receptors,** are abundant in the skin in parts of the face, palms and fingers of the hands, soles of the feet, and the genitalia. They are less abundant along the back and on the back

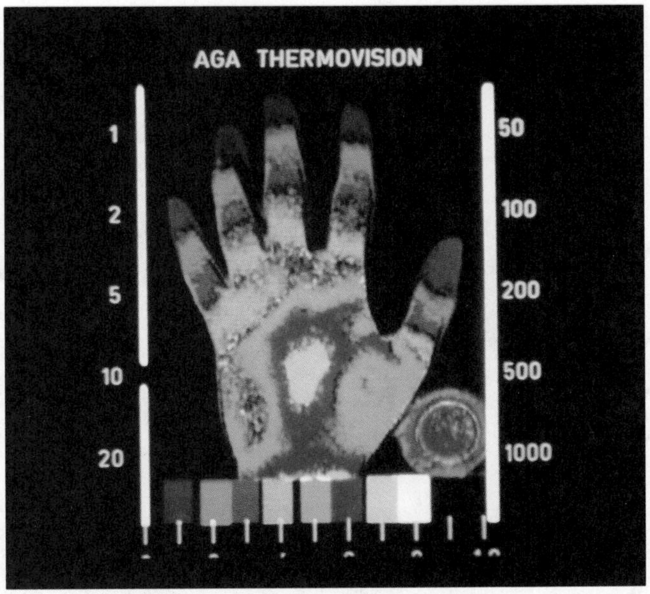

Figure 5.14 A thermogram of the hand showing differential heat radiation. Hair and body fat are good insulators. Red and yellow indicate the warmest portions of the body, whereas blue, green, and white indicate the coolest.

of the neck and are sparse in the skin over joints, especially the elbow. Generally speaking, the thinner the skin, the greater the sensitivity.

Communication

Humans are highly social animals, and the integument plays an important role in communication. Various emotions, such as anger or embarrassment, may be reflected in changes of skin color. Contracting specific facial muscles causes facial expressions, which convey an array of emotions, including love, surprise, happiness, sadness, or despair. Secretions from certain integumentary glands frequently elicit subconscious responses from those that detect the odors.

---

1. List five modifications of the integument that are structurally or functionally protective.
2. Explain how the integument participates in regulating body fluids and temperature.
3. What substances are synthesized in the integument?

Epidermal Derivatives

Hair, nails, and integumentary glands form from the epidermal layer and are therefore of ectodermal derivation. Hair and nails are structural features of the integument and have a limited functional role, whereas integumentary glands are extremely important in body defense and maintaining homeostasis.

Objective 7. Describe the structure of hair, and list the three principal types.

Objective 8. Discuss the structure and function of nails.

Objective 9. Compare the structure and function of the three principal kinds of integumentary glands.

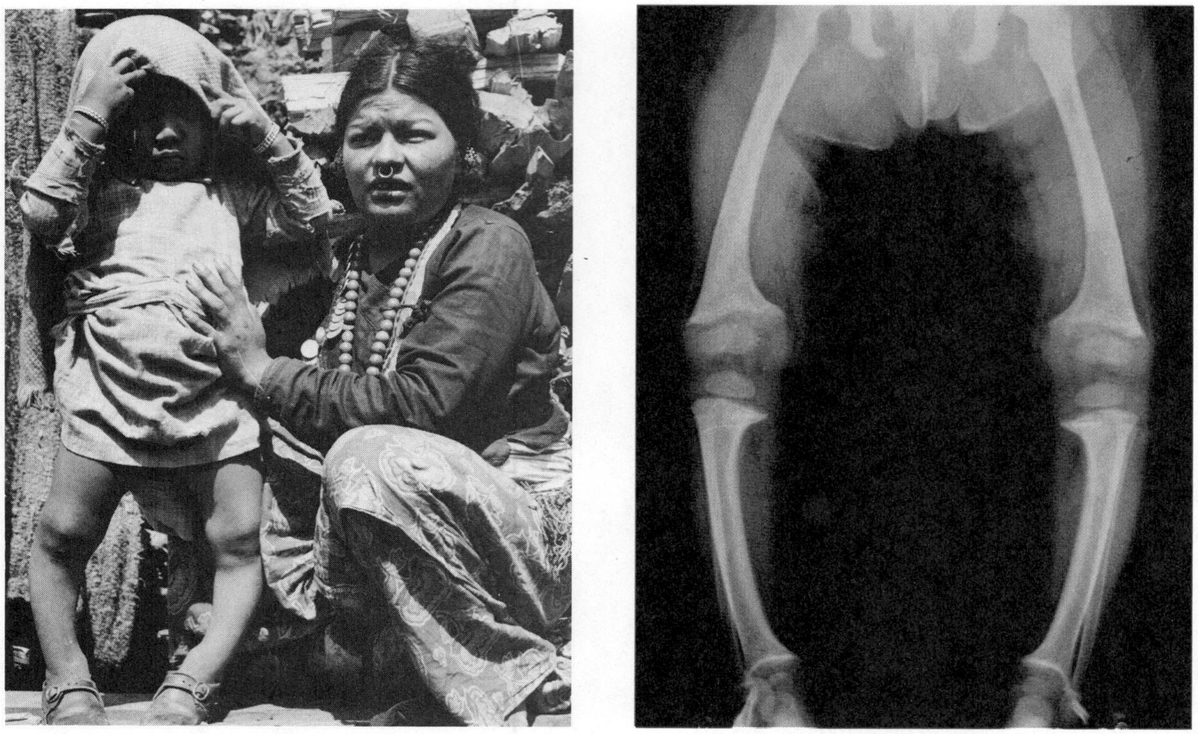

(a) (b)

Figure 5.15 (*a*) A case of rickets in a child who lives in a village in Nepal, where the people reside in windowless huts. During the five-to six-month rainy season, the children are kept indoors. (*b*) An X ray of rickets in a ten-month-old child. Rickets develop from improper diets and also from lack of UV light in the sunlight necessary to synthesize vitamin D.

Hair

Hair, or *pili (pi'li),* is characteristic of all mammals but its distribution, function, density, and texture varies among different species of mammals. Humans are relatively hairless, with only the scalp, face, pubis, and axilla being densely haired. Men and women have about the same density of hair on their bodies, but it is generally more obvious on men due to male hormones (fig. 5.16). Certain structures and regions of the body are hairless, such as the palms of the hands, soles of the feet, lips, nipples, penis, and labia minora.

Hirsutism (her'sut-ism) is a condition of excessive bodily and facial hair, especially in women. It may be a genetic expression as in certain ethnic groups or as the result of a metabolic disorder, usually endocrine. Hirsutism occurs in some women as they experience hormonal changes during menopause. There are various treatments for hirsutism including hormonal injections or undergoing electrolysis to permanently destroy selected hair follicles.

The primary function of hair is protection even though its effectiveness is limited. Hair on the scalp and eyebrows protect against sunlight. The eyelashes and the hair in the nostrils protect against airborne particles. Hair on the scalp may also protect against mechanical injury. Some secondary functions of hair are to distinguish individuals and to serve as an ornamental sexual attractant.

pili: L. *pilus,* hair
hirsutism: L. *hirsutus,* shaggy

Each hair consists of a diagonally positioned **shaft, root,** and **bulb** (fig. 5.17). The shaft is the visible, but dead, portion of the hair projecting above the surface of the skin. The bulb is the enlarged base of the root within the **hair follicle.** Each hair develops from stratum basale cells within the bulb of the hair, where nutrients are received from dermal blood vessels. As the cells divide, they are pushed away from the nutrient supply toward the surface, and cellular death and keratinization occur. In a healthy person, hair grows at the rate of approximately one millimeter every three days. As the hair becomes longer, however, it goes through a resting period during which it is anchored in its follicle.

The life span of a hair varies from three to four months for an eyelash to three to four years for a scalp hair. Each hair lost is replaced by a new hair that grows from the base of the follicle and pushes the old hair out. Between 10 and 100 hairs are lost each day through replacement. Baldness results when hair is lost and not replaced. This condition may be disease-related, but it is generally inherited and most frequently occurs in males because of genetic influences combined with the action of the male sex hormone *testosterone (tes-tos'tĕ-rōn).* No treatment is available for genetic baldness, other than grafting flaps or plugs of skin containing healthy follicles from hairy parts of the body to hairless regions.

Three layers can be observed in hair that is cut in cross section. An inner **medulla** *(mĕ-dul'ah)* is composed of loosely arranged cells separated by many air cells. The thick median layer,

medulla: L. *medulla,* marrow

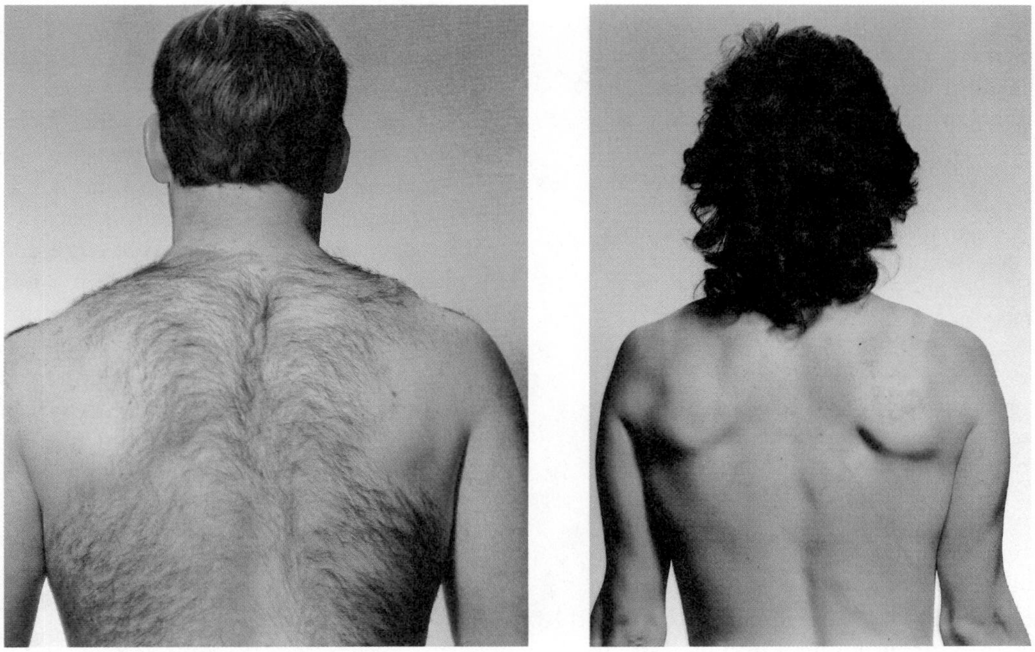

Figure 5.16 The difference between male and female in the expression of hair on the body.

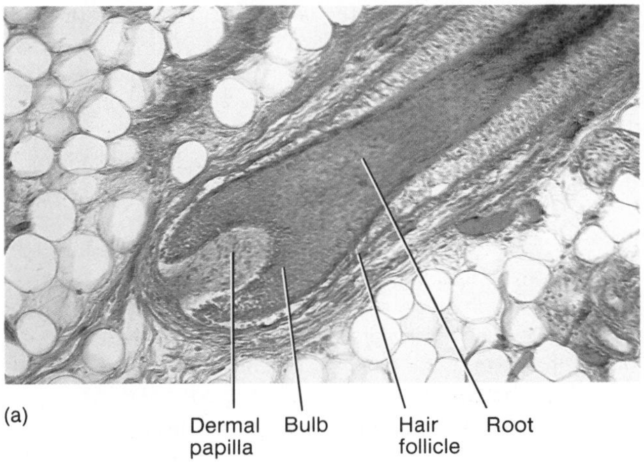

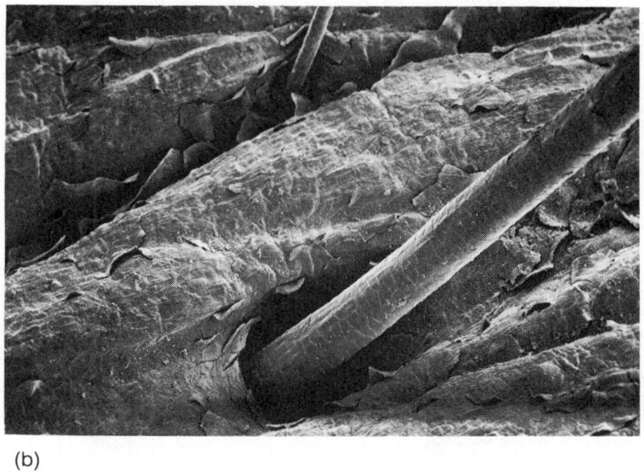

Figure 5.17 The structure of hair and the hair follicle. (*a*) A photomicrograph of the bulb and root of a hair within a hair follicle. (*b*) A scanning electron micrograph (280×) of a hair as it extends from a follicle.

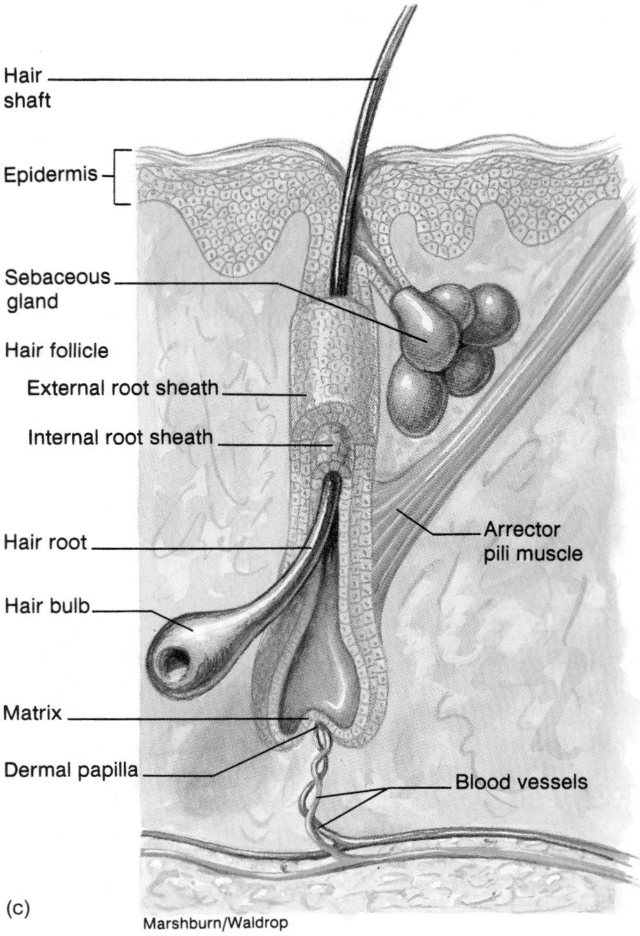

(*c*) A diagram of hair, a hair follicle, a sebaceous gland, and an arrector pili muscle.

called the **cortex,** consists of hardened, tightly packed cells. A **cuticle** layer covers the cortex and forms the toughened outer portion of the hair. Cells of the cuticle have serrated edges that give a hair a scaly appearance when observed under a dissecting scope.

People exposed to heavy metals such as lead, mercury, arsenic, or cadmium will have concentrations in the hair ten times as great as that found in their blood or urine. Because of this, hair samples may be extremely important in certain diagnostic tests.

Even certain metabolic diseases or nutritional deficiencies may be detected in hair samples. For example, the hair of children with cystic fibrosis will be deficient of calcium and have excessive sodium. There is a deficiency of zinc in the hair of malnourished individuals.

Hair color is determined by the type and amount of pigment produced in the stratum basale at the base of the hair follicle. Varying amounts of melanin produce hair ranging in color from blond to brunette to black; the more abundant the melanin, the darker the hair. A pigment with an iron base (trichosiderin) causes red hair. Gray and white hair is caused by lack of pigment production and by air spaces within the layers of the shaft of the hair and generally accompanies aging. The texture of hair is determined by the cross-sectional shape; straight hair is round in cross section, wavy hair is oval, and kinky hair is flat.

Sebaceous glands and specialized smooth muscles, called arrector pili, are attached to the hair follicle (fig. 5.17). Arrector pili muscles are involuntary, responding to thermal or psychological stimuli. When these muscles contract, the hair is pulled into a more vertical position, causing goose bumps.

Humans have three distinct kinds of hair.

1. *Lanugo.* Lanugo is a fine, silky, fetal hair that appears during the last trimester of development. Lanugo is usually not evident on the baby at birth unless the baby is born prematurely.
2. *Angora.* Angora hair grows continuously in length, as on the scalp of males and females, and on the face of males.
3. *Definitive.* Definitive hair grows to a certain length and then ceases growing. It is the most common type of hair. Eyelashes, eyebrows, pubic, and axillary hair are examples.

Anthropologists have referred to humans as the naked apes because of our relative hairlessness. The clothing that we wear over the exposed surface areas of our body functions to insulate and protect us as hair does in other mammals. The nakedness of our skin does lead to some problems. *Skin cancer* occurs frequently in humans, particularly in regions of the skin exposed to the sun. Acne, another problem unique to humans, is partly related to the fact that hair is not present to dissipate the oily secretion from the sebaceous glands.

Nails

Nails are found on the distal dorsum of each of the fingers and toes. Both fingernails and toenails serve to protect the digits, and fingernails also aid in grasping and picking up small objects. Nails form from a hardened, transparent stratum corneum of the epidermis. The hardness of the nail is due to a dense, parallel arrangement of keratin fibrils between the cells.

Each nail consists of a **body, free edge,** and **root** (fig. 5.18). The platelike body of the nail rests on a **nail bed,** which is actually the stratum spinosum of the epidermis. The body and nail bed appear pinkish because of the underlying vascular tissue. The sides of the body of the nail are protected by a **nail fold,** and the furrow between the two is the **nail groove.** The free edge is the distal exposed border, which is attached to the undersurface by the **hyponychium** *(hi''po-ni'kē-um)* (quick). The root of the nail is attached proximally.

An **eponychium** (cuticle) covers above the root of the nail. The eponychium frequently splits, causing a hangnail. The growth areas of the nail are the **matrix** and the **lunula** *(loon'ū-la),* which is the white half-moon-shaped area at the base of the nail. The nail grows by the transformation of the superficial cells of the matrix into nail cells. These harder, transparent cells are then pushed forward over the strata basale and spinosum of the nail bed. Fingernails grow at the rate of approximately one millimeter per week. The growth rate of toenails is somewhat slower.

The condition of nails can be an indication of the general health and personality of a person. Nails should appear pinkish, showing the rich vascular capillaries beneath the translucent nail. A yellowish hue may indicate certain glandular dysfunctions or nutritional deficiencies. Split nails may also be caused by nutritional deficiencies. A prominent bluish tint may indicate improper oxygenation of the blood. Spoon nails (concave body) may be the result of iron-deficiency anemia, and clubbing at the base of the nail may be caused by lung cancer. Dirty or ragged nails may indicate poor personal hygiene, and chewed nails may suggest emotional problems.

Glands

Although they originate in the epidermal layer, all of the glands of the skin are located in the dermis, where they receive physical support and nutritive sustenance. Glands of the skin are referred to as *exocrine* since they either secrete or excrete substances through ducts. The glands of the skin are of three basic types: sebaceous, sudoriferous *(soo'dor-if'er-us),* and ceruminous *(sĕ-roo'mĭ-nus).*

Sebaceous Sebaceous, or oil, glands are associated with a hair follicle since they develop from the follicular epithelium of the hair. They are simple, branched glands that are connected to hair follicles, where they secrete **sebum** onto the shaft of the hair (fig. 5.17). Sebum, which consists mainly of lipids, is dispersed along the shaft of the hair to the surface of the skin, where it lubricates and waterproofs the stratum corneum layer, and also prevents the hair from becoming brittle. If the drainage

cortex: L. *cortex,* bark
cuticle: L. *cuticula,* small skin
lanugo: L. *lana,* wool

hyponychium: Gk. *hypo,* under; *onyx,* nail
lunula: L. *lunula,* small moon
sebum: L. *sebum,* tallow or grease

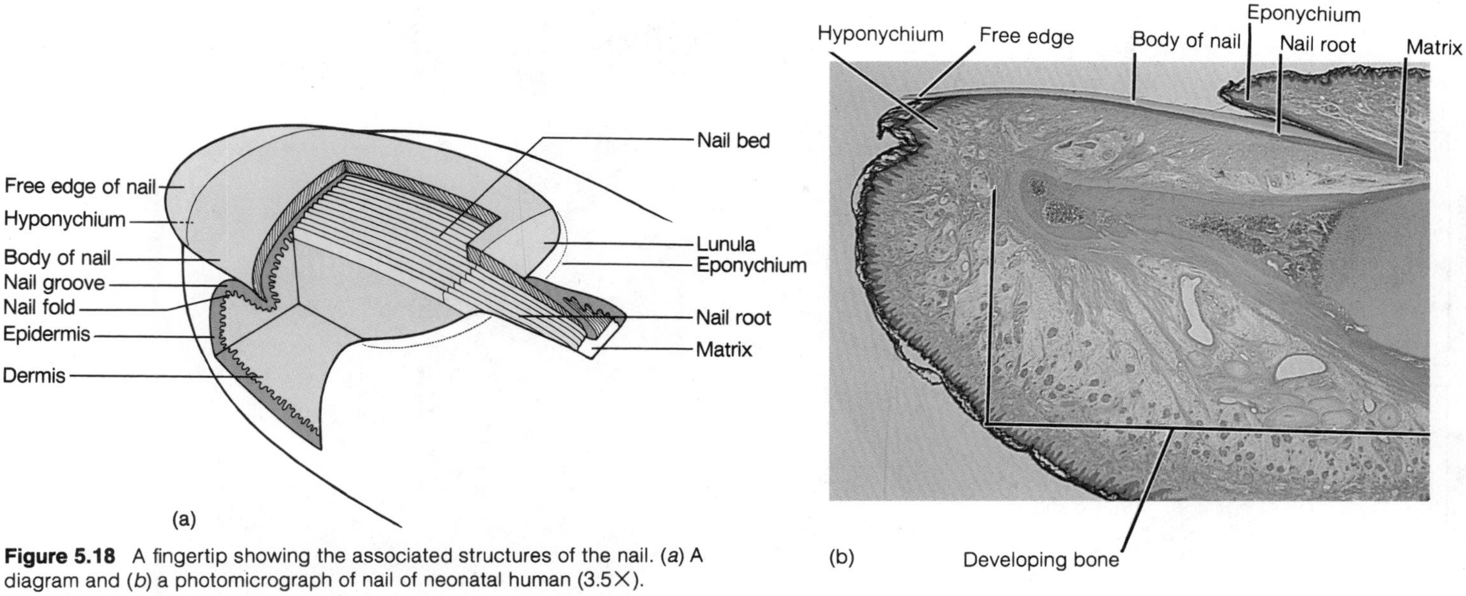

Figure 5.18 A fingertip showing the associated structures of the nail. (*a*) A diagram and (*b*) a photomicrograph of nail of neonatal human (3.5×).

pathway for sebaceous glands becomes blocked for some reason, the glands may become infected, resulting in acne. Sex hormones regulate the production and secretion of sebum, and hyperactivity of sebaceous glands can result in serious acne problems, particularly during teenage years.

Sudoriferous Sudoriferous, or **sweat glands,** excrete perspiration, or sweat, onto the surface of the skin. Sweat glands are most numerous on the palms, soles, axillary and pubic regions, and the forehead. They are coiled and tubular shaped (fig. 5.19) and are of two types:

1. **Eccrine** *(ek'rin)* **sweat glands** are widely distributed over the body, especially on the forehead, back, palms, and soles. These glands are formed totally before birth and function in evaporative cooling (figs. 5.19, 5.20).
2. **Apocrine** *(ap'o-krin)* **sweat glands** are much larger, localized glands found in the axillary and pubic regions and secrete into hair follicles. Apocrine glands are not functional until puberty, and their odoriferous secretion is thought to act as a sexual attractant.

Perspiration is composed of water, salts, urea, uric acid, and traces of other elements. Perspiration is therefore valuable not only for evaporative cooling, but also for the excretion of certain wastes.

Mammary glands, found within the breasts, are specialized sudoriferous, or sweat, glands that secrete milk during lactation periods (see fig. 5.21). The breasts of the human female reach their greatest development during the childbearing years under the stimulus of pituitary and ovarian hormones.

Ceruminous Glands These highly specialized glands are found only in the external auditory canal (ear canal). They secrete **cerumen** *(sĕ-roo'men),* or earwax, which is a water and insect repellent and also keeps the tympanum (eardrum) from drying out. Excessive amounts of cerumen may interfere with hearing.

Good routine hygiene is very important for health and social reasons. Washing away the dried residue of perspiration and sebum eliminates dirt. Excessive bathing, however, can wash off the natural sebum and dry the skin, causing it to itch or crack. The commercial lotions used for dry skin are, for the most part, refined and perfumed lanolin, which is sebum from sheep.

---

1. Draw and label a hair. Indicate which portion is alive, and discuss what causes the cells in a hair to die. What are the three principal types of hair?
2. Describe the structure and function of nails.
3. List the types of integumentary glands, and discuss the structure and function of each.
4. Are skin glands mesodermal or ectodermal in derivation? Are they epidermal or dermal in functional position?

Clinical Considerations

The skin is a buffer against the external environment and is therefore subject to a variety of disease-causing microorganisms and physical assaults. A few of the many diseases and disorders of the integumentary system are briefly discussed here.

Inflammatory (Dermatitis) Conditions

Inflammatory skin disorders are caused by immunologic hypersensitivity or infectious agents. Some persons are allergic to certain foreign proteins and, because of this inherited predisposition, experience hypersensitive reactions, such as asthma, hay fever, hives, drug and food allergies, and eczema. Eczema is characterized by redness, itching, and swollen vascular lesions that become dry, scaly, and crusted. **Lesions,** as applied to inflammatory conditions, are defined as more or less circumscribed pathologic changes in the tissue. Depicted in figure 5.22 are some common inflammatory skin disorders and their usual sites.

---

sudoriferous: L. *sudorifer,* sweat; *ferre,* to bear
cerumen: L. *cera,* wax

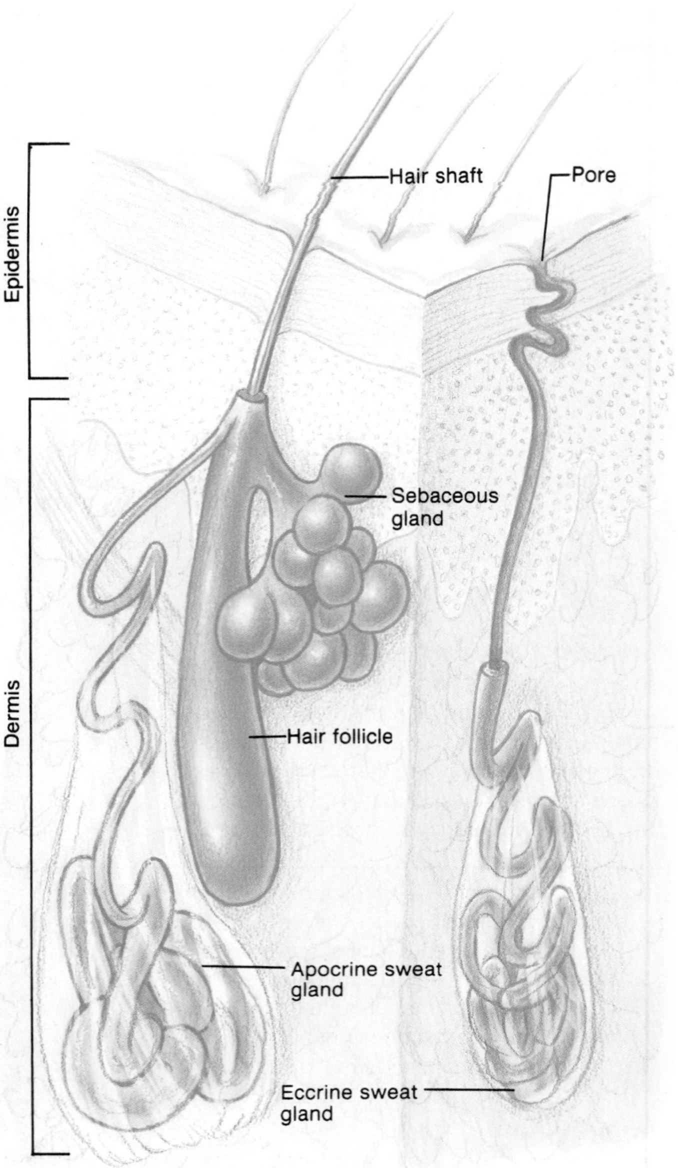

Figure 5.19 Types of skin glands.

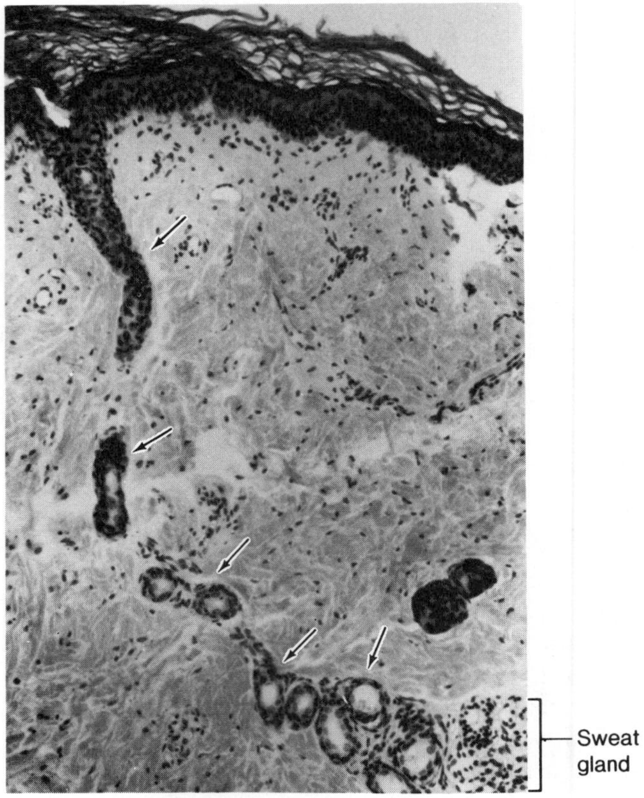

Figure 5.20 A photomicrograph of an eccrine sweat gland (100×). The coiled structure of the ductule portion of the gland (see arrows) accounts for its irregular appearance.

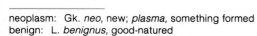

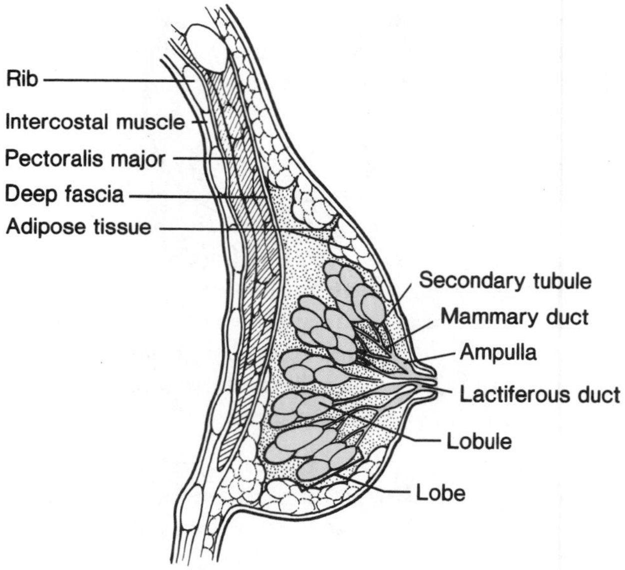

Figure 5.21 A mammary gland within the breast of a human female.

There are a number of *infectious diseases* of the skin, which is not surprising considering the highly social and communal animals that we are. Effective prevention and treatment are now available for most of these diseases, but too frequently persons do not avail themselves of safeguards or treatments. Infectious diseases of the skin include childhood viral infections (measles and chicken pox), bacteria such as staphylococcus (impetigo), sexually transmitted diseases, leprosy, fungi (ringworm, athlete's foot, candida), and mites (scabies).

Neoplasms

Both benign and malignant neoplastic conditions or diseases are common in the skin. Pigmented moles (nevi), for example, are a type of benign neoplastic growth of melanocytes. Dermal cysts and benign viral infections are also common. Warts are virally caused abnormal growths of tissue that occur frequently on the hands and feet. A different type of wart, called a venereal wart, occurs in the anogenital region of sexual partners. Both types of warts are easy to treat by excision or various drugs. Aging spots on elderly persons, which appear as pigmented patches on the surface of the skin, are a benign growth of melanin pigmented basale cells. Usually no treatment is required, unless for cosmetic purposes.

Skin cancer is the most common malignancy in the United States. Except for malignant melanomas (see fig. 5.8), which arise from melanocytes, skin cancer is generally not life threatening. Excessive exposure to ultraviolet light from the sun is a

neoplasm: Gk. *neo*, new; *plasma*, something formed
benign: L. *benignus*, good-natured
malignant: L. *malignus*, acting from malice

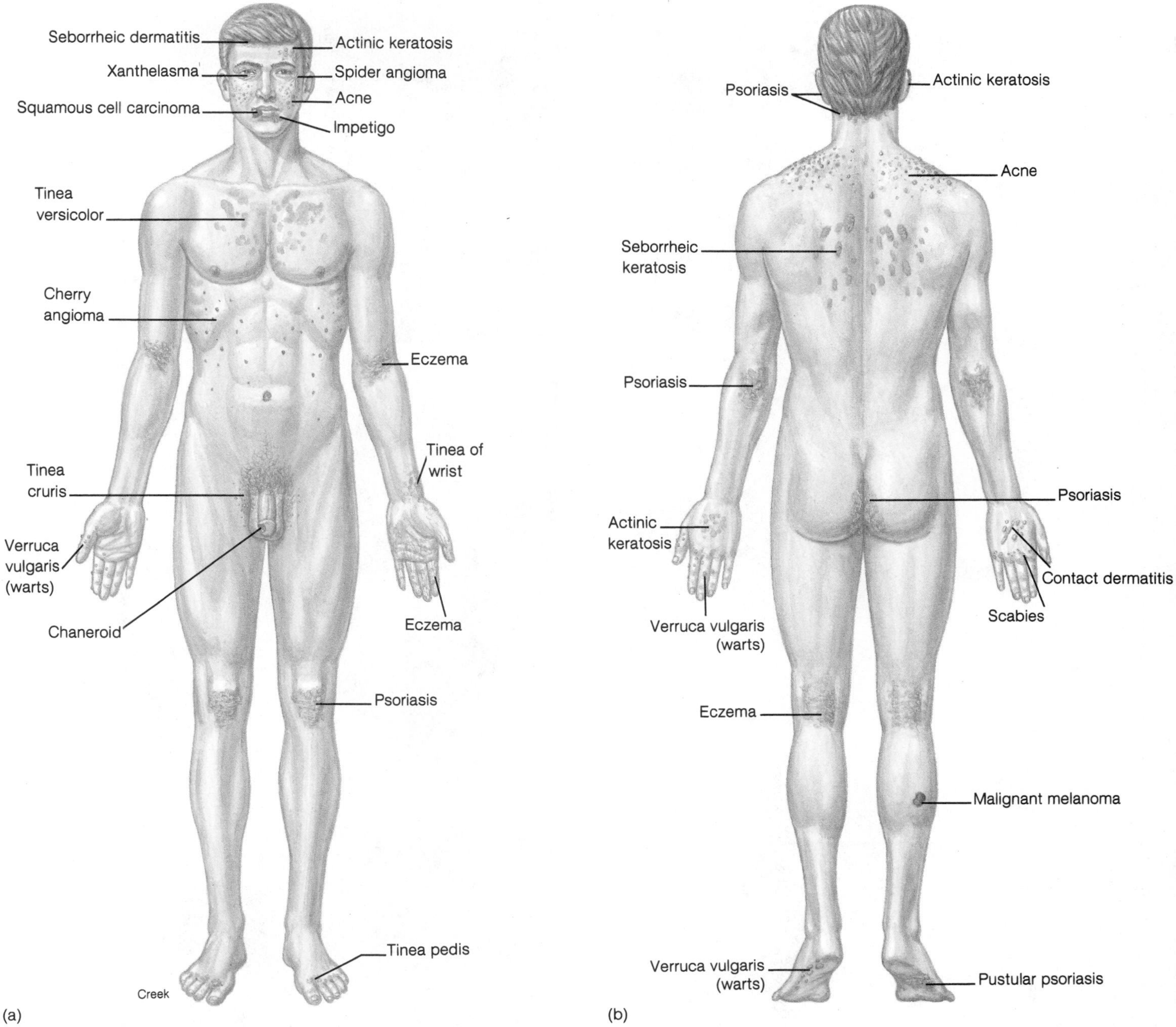

Figure 5.22 Common inflammatory skin disorders and their usual site of occurrence.

known cause of skin cancer. The preferred treatment for this disease is complete surgical excision of the cancerous portion.

Burns

A burn is an epithelial injury caused by contact with thermal, radioactive, chemical, or electrical agents. Burns generally occur on the skin, but burns can involve the linings of the respiratory and digestive tracts. The extent and location of a burn is frequently less important than the degree to which it disrupts body homeostasis. Burns that have a **local effect** (local tissue destruction) are not as serious as those that have a **systemic effect.** Systemic effects directly or indirectly involve the entire body and are a threat to life. Possible systemic effects include body dehydration, shock, reduced circulation and urine production, and bacterial infections.

Burns are classified into three types according to their severity: first-degree, second-degree, and third-degree (fig. 5.23). In **first-degree burns,** the epidermal layers of the skin are damaged and symptoms are restricted to local effects such as redness, pain, and edema (swelling). A shedding of the surface layers (desquamation) generally follows in a few days. A sunburn is an example. **Second-degree burns** involve both the epidermis and dermis. Blisters appear and recovery is usually complete although slow. **Third-degree burns** destroy the entire thickness of the skin and frequently some of the underlying muscle. The skin appears waxy or charred and is insensitive to touch. As a result, ulcerating wounds develop and the body attempts to heal itself by forming scar tissue. Skin grafts are frequently used to assist recovery.

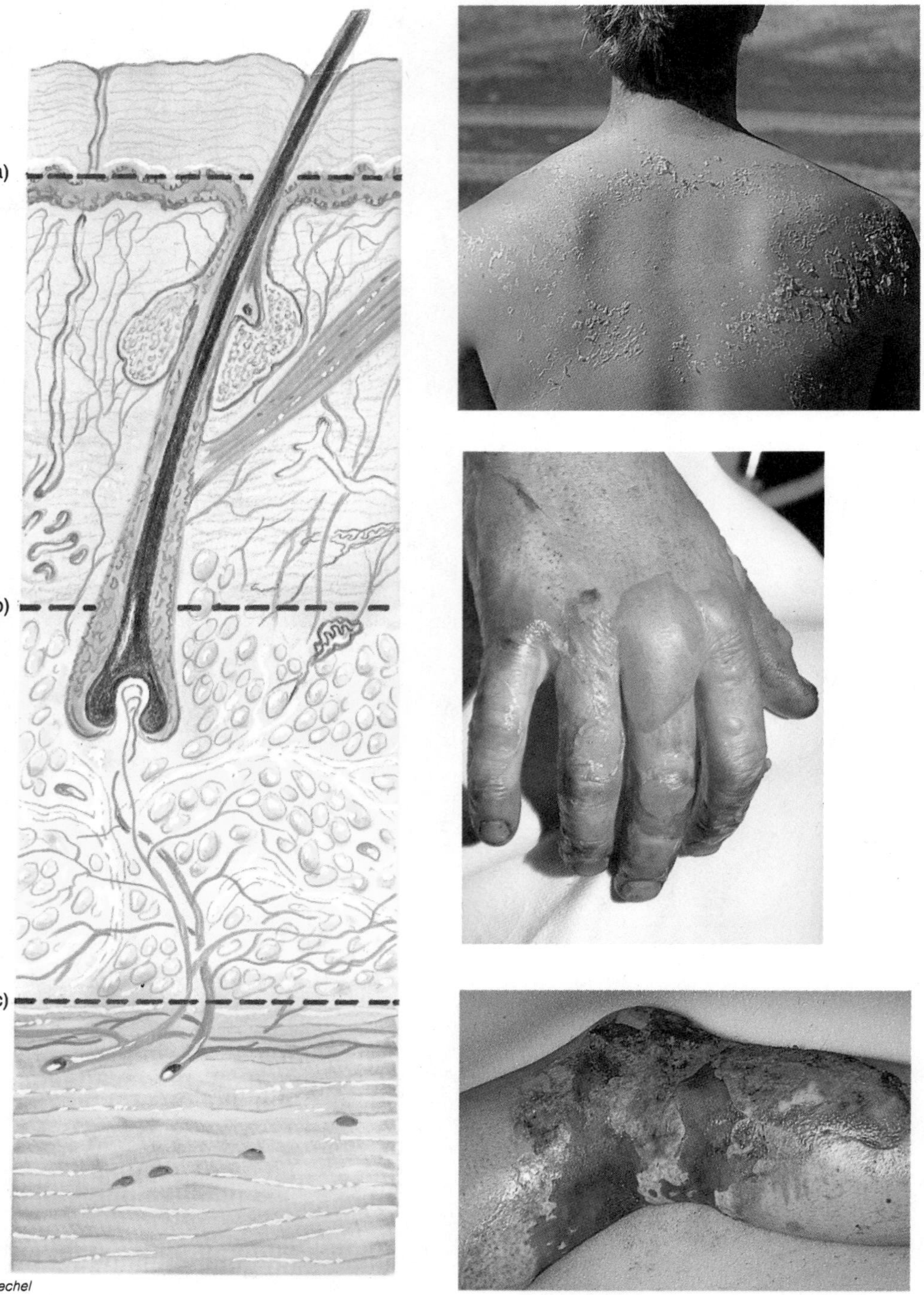

Figure 5.23 The classification of burns. (*a*) First-degree burns involve the epidermis and are characterized by redness, pain, and edema—such as with a sunburn; (*b*) second-degree burns involve the epidermis and dermis and are characterized by intense pain, redness, and blistering; and (*c*) third-degree burns destroy the entire skin and frequently expose the underlying organs. The skin is charred and numb and does not protect against fluid loss.

A procedure for estimating the extent of damaged integument suffered in burned patients is to apply the *rule of nines* (fig. 5.24), in which the surface area of the body is divided into regions accounting for about 9% of the area (or a multiple of 9%). Collectively, these areas account for 100% of the integumentary (body) surface. An estimation of the percentage of surface area damaged is important in treating with intravenous fluid, which replaces the fluids lost from tissue damage.

Frostbite

Frostbite is a local destruction of the skin resulting from freezing. Like burns, frostbite is classified by its degree of severity: first-degree, second-degree, and third-degree. In **first-degree frostbite** the skin will appear cyanotic (bluish) and swollen. Vesicle formation and hyperemia (swollen with blood) are symptoms of **second-degree frostbite.** As the affected area is warmed, there will be further swelling, and the skin will

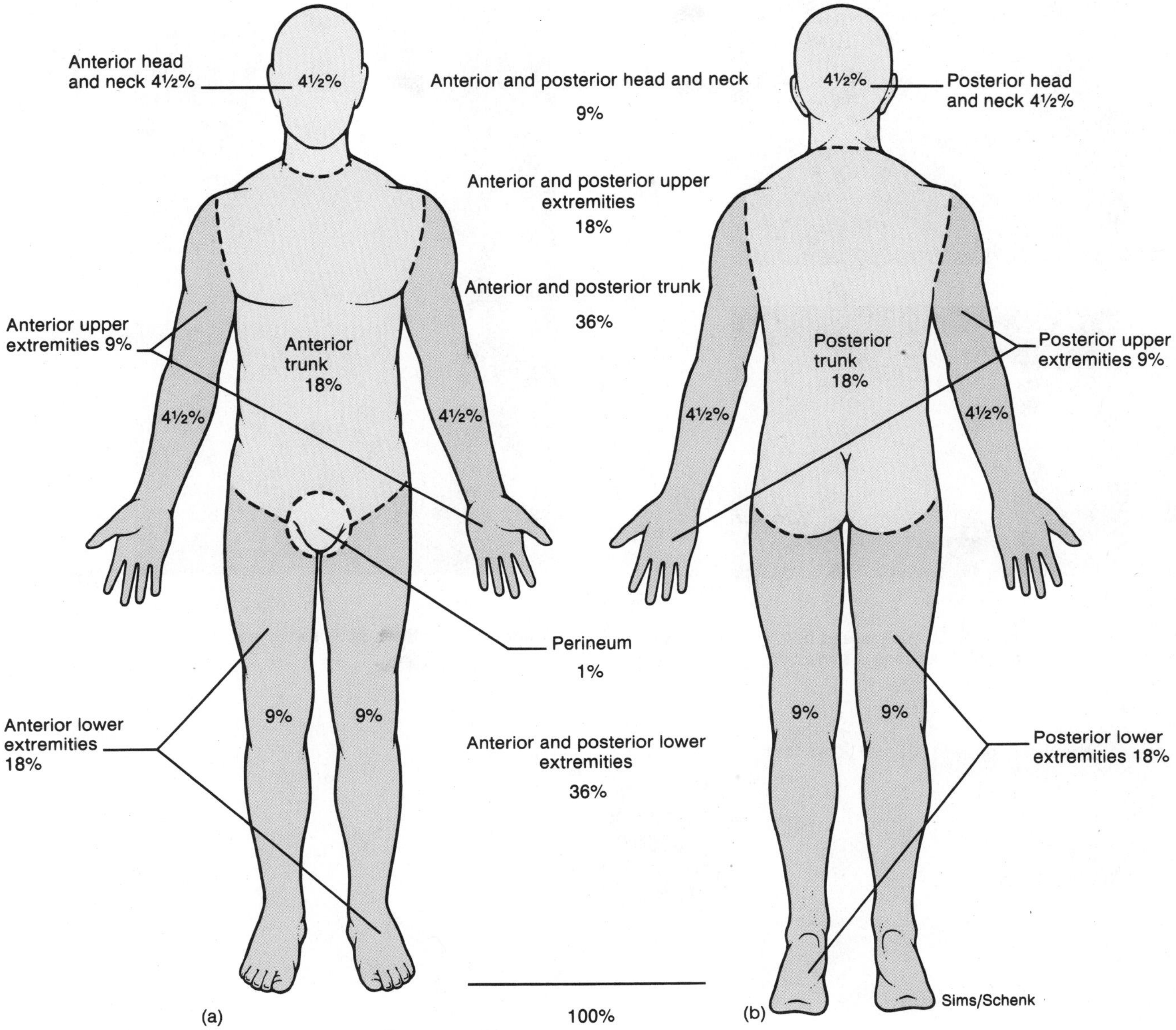

Figure 5.24 The extent of burns, as estimated by the rule of nines. (*a*) Anterior and (*b*) posterior.

redden and blister. In **third-degree frostbite,** there will be severe edema, some bleeding, and numbness followed by intense throbbing pain and necrosis of the affected tissue. Gangrene will follow untreated third-degree frostbite.

Skin Grafts

If extensive areas of the stratum basale of the epidermis are destroyed in second-degree or third-degree burns or frostbite, new skin cannot grow back. In order for this type of wound to heal, a skin graft must be performed.

A **skin graft** is a segment of skin that has been excised from a *donor site* and transplanted to the *recipient site,* or *graft bed.* An *autotransplant* is the most successful graft. This type of tissue transplant involves taking a thin sheet of healthy epidermis from a donor site of the burn or frostbite patient and moving it to the recipient site (fig. 5.25). A *heterotransplant* (between two different species) can be a temporary treatment to prevent infection and fluid loss.

Synthetic skin fabricated from animal tissue bonded to a silicone film (fig. 5.26) may be used on a patient who is extensively burned. The completed process includes seeding the synthetic skin with basal skin cells obtained from healthy locations on the patient. This technique and treatment eliminates the problems of skin grafting, such as additional trauma, widespread scarring, and rejection as in the case of skin obtained from a cadaver.

Wounds and Wound Healing

The skin effectively protects against many abrasions, but if a wound does happen (fig. 5.27) a sequential chain of events promotes rapid healing. The process of wound healing depends on the extent and severity of the injury. Trauma to the epidermal layers stimulates an increased mitotic activity in the stratum basale, whereas injuries that extend to the dermis or subcutaneous layer elicit activity throughout the body as well as within the wound itself. General body responses include a temporary elevation of temperature and pulse rate.

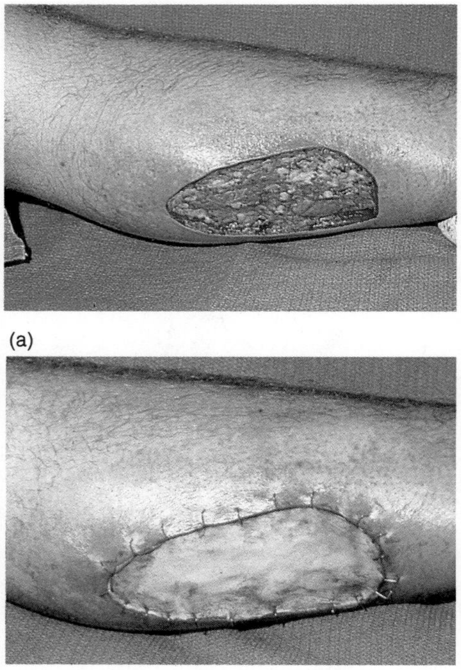

Figure 5.25 Skin graft to the arm. (*a*) Traumatized tissue is excised and (*b*) healthy tissue from another body location is transplanted.

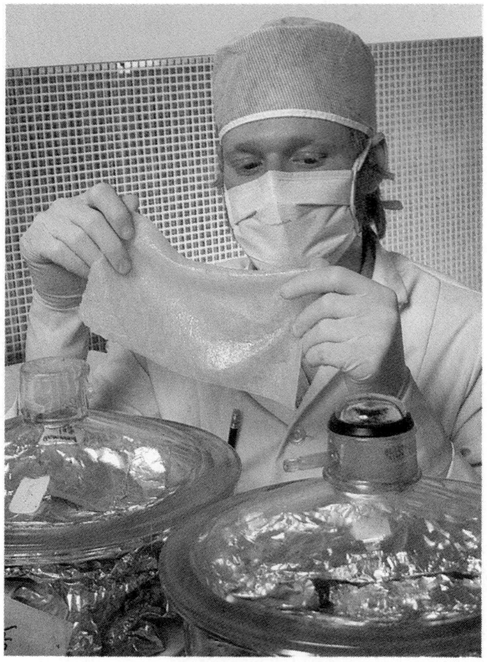

Figure 5.26 Synthetic skin used in grafting.

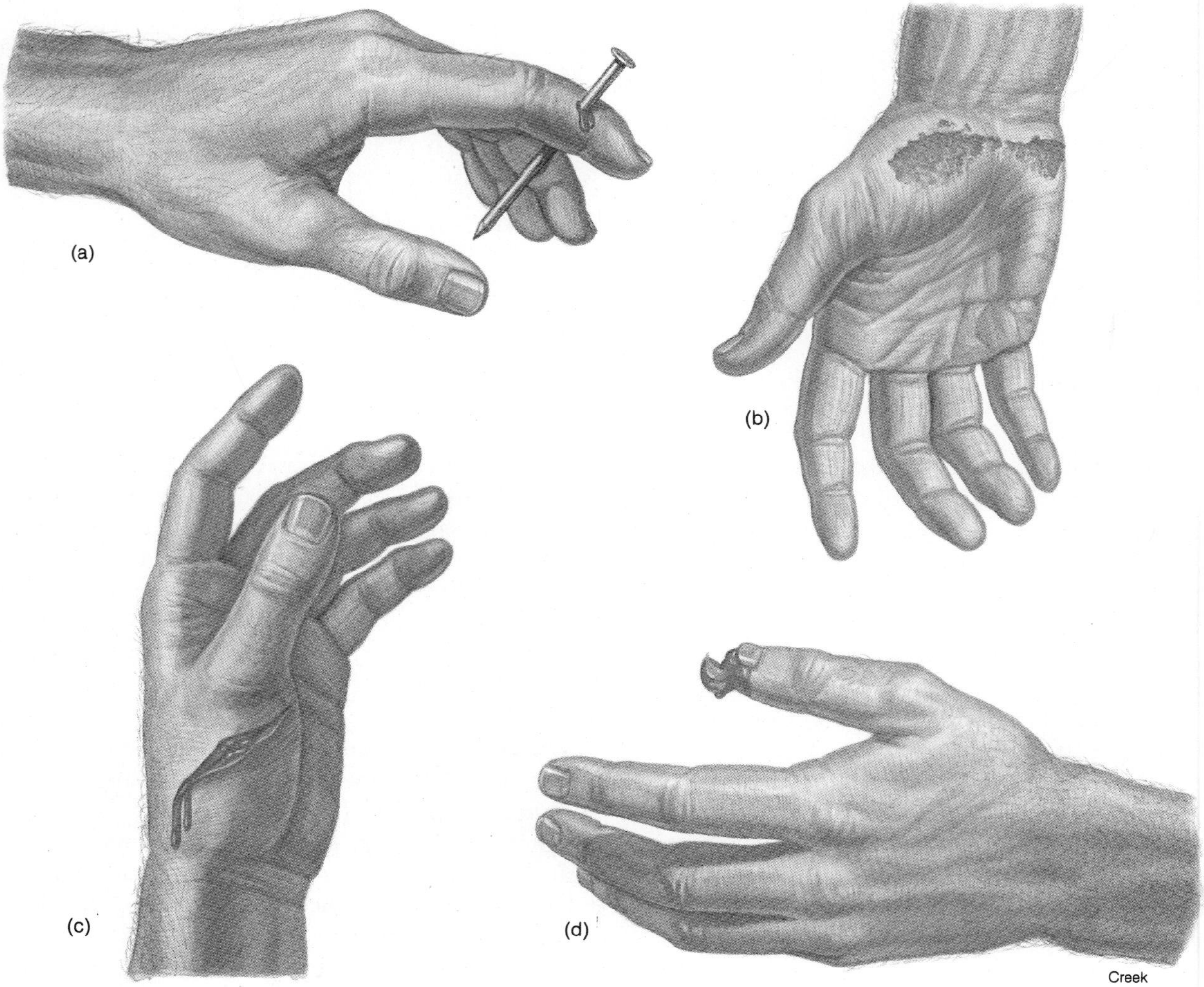

Figure 5.27 Various kinds of wounds. (*a*) Puncture, (*b*) abrasion, (*c*) laceration, and (*d*) avulsion.

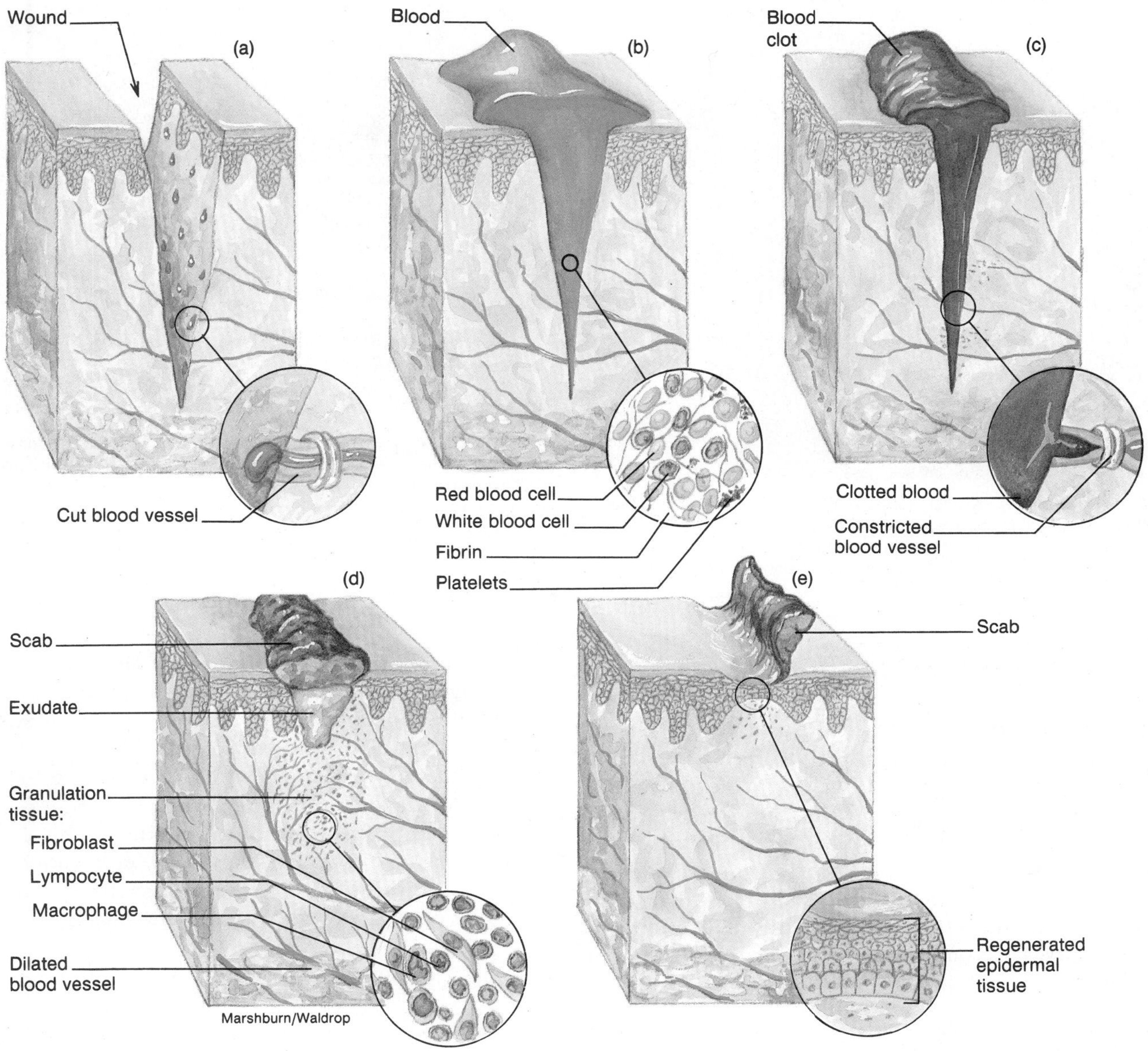

Figure 5.28 The process of wound healing. (*a*) A penetrating wound into the dermis ruptures blood vessels. (*b*) Blood cells, fibrinogen, and fibrin flow out of the wound. (*c*) Vessels constrict and a clot blocks the flow of blood. (*d*) A protective scab is formed from the clot, and granulation forms within the site of the wound. (*e*) The scab sloughs off as the epidermal layers are regenerated.

In an open wound (fig. 5.28), blood vessels are broken and bleeding occurs. Through the action of **blood platelets** *(plat'letz)* and protein molecules, called **fibrinogen** *(fi-brin'o-jen),* a clot forms and soon blocks the flow of blood. A scab forms and covers and protects the damaged area. Mechanisms are activated to destroy bacteria, dispose of dead or injured cells, and isolate the injured area. These responses are collectively referred to as *inflammation* and are characterized by redness, heat, edema, and pain. Inflammation is a response that confines the injury and promotes healing.

The next step in healing is the differentiation of binding **fibroblasts** from connective tissue at the wound margins. Together with new branches from surrounding blood vessels, **granulation tissue** is formed. Phagocytic cells migrate into the wound and ingest dead cells and foreign debris. Eventually the damaged area is repaired, and the protective scab is sloughed off.

If the wound is severe enough, the granulation tissue may develop into **scar tissue** (fig. 5.29). Scar tissue differs from normal skin in that its collagen fibers are more dense and it has no stratified squamous or epidermal layer. Scar tissue has fewer blood vessels and may lack hair, glands, and sensory receptors. The closer the edges of a wound, the less granulation tissue develops and the less obvious a scar. This is one reason for suturing a large break in the skin.

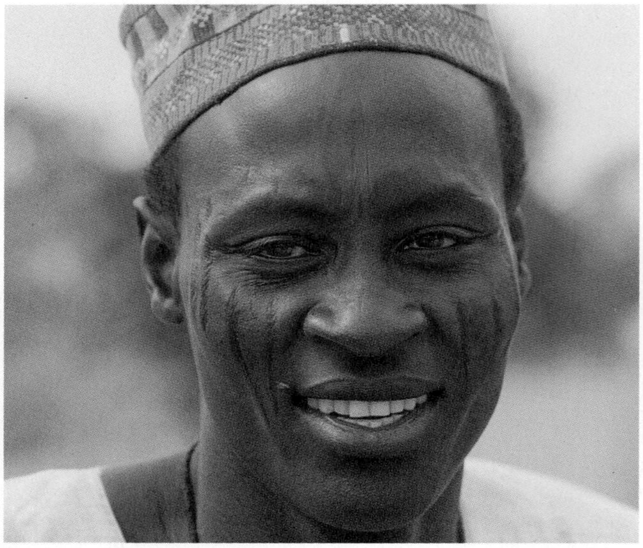

Figure 5.29 Scars for body adornment on the face of this Buduma man from the islands of Lake Chad, are created by instruments that make crescent-shaped incisions into the skin in beadlike patterns. Special ointments are applied to the cuts to retard healing and promote the forming of scars.

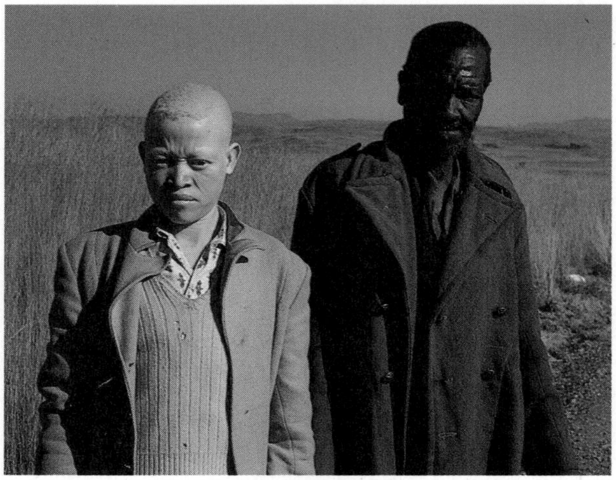

Figure 5.30 The albino individual in this photograph has melanocytes within his skin, but due to a mutant gene he lacks the ability to synthesize melanin.

CLINICAL CASE STUDY ANSWER

The blistering and erythema characteristic of second-degree burns is a manifestation of intact and functioning blood vessels, which exist in abundance within the spared dermis. In third-degree burns the entire dermis and its vasculature is destroyed, thus explaining the absence of these findings. Likewise, in third-degree burns, nerve endings and other nerve end organs that reside in the dermis are destroyed, resulting in a painless injury. In second-degree burns, however, significant numbers of these structures are spared and functional, thus preserving sensation, including pain. The third-degree burn areas will all require skin grafting in order to prevent infection, one of the skin's most vital functions. In second-degree burns, the spared dermis serves somewhat as a barrier to bacteria. Consequently, skin grafting is usually unnecessary, especially if sufficient numbers of skin adnexa, in other words, hair follicles, sweat glands, etc., which generally lie deep within the dermis, are spared. These structures serve as starting points for regeneration of surface epithelium and skin organs.

Important Clinical Terminology

acne An inflammatory condition of sebaceous glands. Acne is effected by gonadal hormones and is therefore more common during puberty and adolescence. Pimples and blackheads on the face, chest, and back are expressions of this condition.

albinism A congenital, genetic deficiency of the pigment of the skin, hair, and eyes due to a metabolic block in the synthesis of melanin (fig. 5.30).

alopecia Loss of hair, baldness. Baldness is usually due to genetic factors and cannot be treated. Baldness may signify anatomical maturity.

athlete's foot *(tinea pedis)* A skin fungus disease of the foot.

blister A collection of fluid between the epidermis and dermis, caused by excessive friction or a burn.

boil (furuncle) A localized bacterial infection originating in a hair follicle or skin gland.

carbuncle A bacterial infection similar to a boil, except that it infects the subcutaneous tissues.

cold sore *(fever blister)* A lesion on the lip or oral mucous membrane, caused by type I herpes simplex virus (HSV), transmitted by oral or respiratory exposure.

comedo A plug of sebum and epithelial debris in the hair follicle and excretory duct of the sebaceous gland. Also called a blackhead or whitehead.

corn A type of callus that is localized on the foot, usually over toe joints.

dandruff Common dandruff is the continual shedding of epidermal cells of the scalp, which can be removed by normal washing and brushing of the hair. Abnormal dandruff may be caused by certain skin diseases such as seborrhea or psoriasis.

decubitus ulcer A bedsore. An exposed ulcer caused by a continual pressure that restricts dermal blood flow to a localized portion of the skin (see fig. 5.12).

dermabrasion A procedure for removing tattoos or acne scars by high speed sanding or scrubbing.

dermatitis An inflammation of the skin.

dermatology A specialty of medicine concerned with the study of the skin, its anatomy, physiology, histopathology, and the relationship of cutaneous lesions to systemic disease.

eczema *(eg'ze-mah)* A noncontagious inflammatory condition of the skin producing red, itching, vesicular lesions, which may be crusty or scaly.

erythema *(er''ĭ-the'mah)* Redness of the skin caused by vasodilation from skin injury, infection, or inflammation.

furuncle A boil; a localized abscess resulting from an infected hair follicle.

gangrene Necrosis of tissue due to the obstruction of blood flow. It may be localized or extensive and may secondarily be infected with anaerobic microorganisms.

impetigo A contagious skin infection that results in lesions followed by scaly patches. It generally occurs on the face and is caused by staphylococci or streptococci.

keratosis Any abnormal growth and hardening of the stratum corneum layer of the skin.

melanoma A cancerous tumor originating from proliferating melanocytes within the epidermis of the skin.

nevus *(ne'vus)* A mole or birthmark; congenital pigmentation of a certain area of the skin.

papilloma A benign epithelial neoplasm, such as a wart or a corn.

papule A small inflamed elevation of the skin, such as a pimple.

pruritus Itching. It may be symptomatic of systemic disorders but is generally due to dry skin.

psoriasis *(so-ri'ah-sis)* An inherited inflammatory skin disease usually expressed as circular scaly patches of skin.

pustule A small, localized elevation of the skin containing pus.

seborrhea *(seb''o-re'ah)* A disease characterized by an excessive activity of the sebaceous glands and accompanied by oily skin and dandruff.

ulcer An open sore. A decubitus ulcer denotes a bedsore (see fig. 5.12).

urticaria (hives) A skin eruption of reddish weals, usually with extreme itching. It may be caused by an allergic reaction, stress, or contact with some external or internal precipitating factor.

wart A roughened projection of epidermal cells caused by a virus.

Chapter Summary

I. The Integument as an Organ
   A. The skin is considered an organ because it consists of several kinds of tissues that are structurally arranged to function together.
   B. The appearance of the skin is clinically important because it provides clues to certain body conditions or dysfunctions.

II. Development of the Integumentary System
   A. Epidermal layers, hair, nails, and cutaneous glands, including mammary glands, derive from ectoderm that specializes into a mitotically active germinal layer.
      1. Hair and sebaceous glands develop together during the formation of a hair follicle.
      2. Sweat glands develop independently from the germinal layer of the epidermis.
      3. Mammary glands are modified sweat glands that develop along a mammary ridge.
      4. Nails begin development at ten weeks at the nail fields.
      5. During early fetal development, melanoblasts migrate to the germinal layer and give rise to melanocytes, which produce melanin.
   B. The dermis and hypodermis derive from mesoderm. Specialized mesoderm, called mesenchyme, gives rise to specific dermal structures.

III. Layers of the Integument
   A. The stratified squamous epithelium of the epidermis is divisible into five structural and functional layers: stratum basale, stratum spinosum, stratum granulosum, stratum lucidum, and stratum corneum.
      1. Normal skin color is due to a combination of melanin and carotene in the epidermis and hemoglobin in the blood of the dermis and hypodermis.
      2. Fingerprints on the surface of the epidermis are individually unique; flexion creases and flexion lines are acquired.
   B. The thick dermis of the skin is composed of fibrous connective tissue interlaced with elastic fibers. The two layers of the dermis are the upper papillary layer and the deeper reticular layer.
   C. The hypodermis, composed of adipose and fibrous connective tissue, binds the dermis to underlying organs.

IV. Functions of the Integument
   A. Structural features of the skin provide protection of the body from disease and external injury.
      1. Keratin and an acidic oily secretion on the surface protect the skin from water and microorganisms.
      2. Cornification of the skin protects against abrasion.
      3. Melanin is a barrier to UV light.
   B. The skin regulates body fluids and temperatures.
      1. Fluid loss is minimal due to keratinization and cornification.
      2. Temperature regulation is maintained by radiation, convection, and the antagonistic effects of sweating and shivering.
   C. The skin permits the absorption of UV light, respiratory gases, steroids, fat-soluble vitamins, and certain toxins and pesticides.
   D. The integument synthesizes melanin and keratin, which remain in the skin, and vitamin D, which is used elsewhere in the body.
   E. Sensory reception in the skin is provided through cutaneous receptors throughout the dermis and hypodermis. Cutaneous receptors respond to precise sensory stimuli and are more sensitive in thin skin.
   F. Certain emotions are communicated through the integument.

V. Epidermal Derivatives
   A. Hair is characteristic of all mammals, but its distribution, function, density, and texture varies among species of different mammals.
      1. Each hair consists of a shaft, root, and bulb, which is positioned in a hair follicle.
      2. The three hair layers are the medulla, cortex, and cuticle.
      3. Lanugo, angora, and definitive are the three kinds of human hair.
   B. Hardened, keratinized nails are found on the distal dorsum of each digit, where they protect the digits; fingernails aid in grasping and picking up small objects.
      1. Each nail consists of a body, free edge, and a root.
      2. The hyponychium, eponychium, and nail fold support the nail on the nail bed.
   C. Integumentary glands are exocrine since they either secrete or excrete substances through ducts.
      1. Sebaceous glands secrete sebum onto the shaft of the hair.
      2. The two types of sudoriferous (sweat) glands are eccrine and apocrine.
      3. Mammary glands are specialized sudoriferous glands that secrete milk during lactation.
      4. Ceruminous glands secrete cerumen (earwax).

Review Activities

Objective Questions

1. Hair, nails, integumentary glands, and the epidermis of the skin are derived from embryonic
   (a) ectoderm. (c) endoderm.
   (b) mesoderm. (d) mesenchyme.
2. Spoon-shaped nails may result when a person has a dietary deficiency of
   (a) zinc. (c) niacin.
   (b) iron. (d) vitamin B_{12}.
3. The epidermal layer *not* present in the thin skin of the face is the stratum
   (a) granulosum. (c) spinosum.
   (b) lucidum. (d) corneum.

4. Which of the following does *not* contribute to skin color?
   (a) dermal papillae
   (b) melanin
   (c) carotene
   (d) hemoglobin
5. Integumentary cells synthesize vitamin D in the presence of UV light and
   (a) keratin.
   (b) cortisol.
   (c) trichosiderin.
   (d) dehydrocholesterol.
6. Integumentary glands that empty their secretions into hair follicles are
   (a) sebaceous glands.
   (b) sudoriferous glands.
   (c) eccrine glands.
   (d) ceruminous glands.
7. Fetal hair that is present during the last trimester of development is referred to as
   (a) angora. (c) lanugo.
   (b) definitive. (d) replacement.
8. Which of these conditions is potentially life threatening?
   (a) acne (c) eczema
   (b) melanoma (d) seborrhea
9. The skin of a burn victim has been severely damaged through the epidermis and into the dermis; integumentary regeneration will be slow with some scarring but complete. What degree burn is it?
   (a) first (c) third
   (b) second
10. The technical name for a blackhead or whitehead is
    (a) carbuncle. (c) nevus.
    (b) melanoma. (d) comedo.

Essay Questions

1. Discuss the development of the skin and its associated hair, glands, and nails. What role do the ectoderm and mesoderm play in integumentary development?
2. List the functions of the skin. Which of these functions occurs passively due to the structure of the skin, and which occurs dynamically due to physiological processes?
3. What are the types of tissues found in each of the three layers of skin?
4. Discuss the growth process and regeneration of the epidermis.
5. What are some physical and chemical features of the skin that make it an effective protective organ?
6. Of what practical value is it for the outer layers of the epidermis and hair to be composed of dead cells?
7. Define the following: *lines of tension, friction ridges, flexion lines.* What causes each of these to develop?
8. Distinguish between a hair follicle and a hair. What other accessory structures are associated with a hair follicle and a hair?
9. Contrast the structure and function of sebaceous, sudoriferous, mammary, and ceruminous glands.
10. Discuss what is meant by an inflammatory lesion. What are some frequent causes of skin lesions?

CHAPTER 6

Skeletal System: *Introduction and the Axial Skeleton*

Outline and Concepts

Clinical Case Study

Organization of the Skeletal System

The axial and appendicular components of the skeletal system of an adult human consist of 206 individual bones arranged into a strong, flexible body framework.

Functions of the Skeletal System

The bones of the skeleton perform the mechanical functions of support, protection, and body movement and the metabolic functions of hemopoiesis and mineral storage.

Development of the Skeletal System

All bones derive from specialized mesenchymal mesoderm through endochondral ossification or through intramembranous ossification. Ossification begins by the fourth week of prenatal development and is not completed in certain bones until about age twenty-five or thirty.

Bone Structure and Growth

Spongy bone and compact bone differ in histological structure, gross appearance, location, and function. Each bone has a characteristic pattern of ossification and growth, a characteristic shape, and diagnostic surface features that indicate its functional relationship to other bones, muscles, and to the body structure as a whole.

Bone Structure
Bone Growth

Skull

The human skull, consisting of eight cranial and fourteen facial bones, contains several cavities that house the brain and sensory organs. Each bone of the skull articulates with the adjacent bones and has diagnostic and functional processes, surface features, and foramina.

Cranium
Facial Bones

Vertebral Column

The supporting vertebral column consists of vertebrae, separated by fibrocartilaginous intervertebral discs, which lend flexibility and absorb the stress of movement. Vertebrae enclose and protect the spinal cord, support and permit movement of the skull, articulate with the rib cage, and provide for the attachment of trunk muscles.

General Structure of Vertebrae
Regional Characteristics of Vertebrae

Rib Cage

The cone-shaped and flexible rib cage consists of the sternum, costal cartilages, and the twelve paired ribs attached to the thoracic vertebrae. It encloses and protects the thoracic viscera and is directly involved in the mechanics of breathing.

Clinical Considerations

Developmental Disorders
Nutritional and Hormonal Disorders
Trauma and Injury
Neoplasms of Bone

Clinical Case Study Answer

Important Clinical Terminology

Chapter Summary

Review Activities

CLINICAL CASE STUDY

A sixty-eight-year-old man visits his family doctor for his first physical examination in thirty years. Upon initial greeting, the doctor senses that the patient is a little disgruntled. He gently probes the situation to which the patient responds, "The nurse that measured my height is incompetent! I know for a fact I used to be six feet even when I was in the navy, and she tells me I am 5′10″!" The doctor then performs the measurement himself, noting that although the patient's posture is excellent, he is indeed 5′10″. He explains that the spine contains some nonbony tissue, which shrivels up a bit over the years. The patient impatiently interrupts and states indignantly that he is aware of anatomic terms and principles and would like a detailed explanation. How would you explain the anatomy of the vertebral column and the changes that accompany it during the aging process? ■

Organization of the Skeletal System

The axial and appendicular components of the skeletal system of an adult human consist of 206 individual bones arranged into a strong, flexible body framework.

Objective 1. Describe the structural organization of the skeletal system, and list the bones of the axial and appendicular portions.

Osteology is the science concerned with the study of bones. Each bone is an organ that plays a part in the total functioning of the skeletal system. The skeletal system of an adult human is composed of approximately 206 bones. Actually, the number of bones differs from person to person depending on age and genetic variations. At birth, the skeleton consists of approximately 270 bones. As further bone development (ossification) occurs during infancy, the number increases. During adolescence, however, the number of bones decreases, as there is a gradual fusion of separate bones.

Some adults have extra bones within the sutures (joints) of the skull called **sutural** *(soo′cher-ahl)*, or **wormian, bones.** Additional bones may develop in tendons in response to stress as the tendons repeatedly move across a joint. Bones formed this way are called **sesamoid** *(ses′ah-moid)* **bones.** Sesamoid bones, like the sutural bones, vary in number. The patellae (kneecaps) are two sesamoid bones present in all people.

For the convenience of study, the skeleton is divided into the axial and appendicular portions. Anterior and posterior views of the skeleton are shown in figure 6.1. Listed in table 6.1 are the divisions of the skeleton and the number of bones in each portion.

The **axial skeleton** consists of the bones that form the axis of the body and that support and protect the organs of the head, neck, and torso:

1. **Skull.** The skull consists of two sets of bones: the cranial bones that form the cranium, or braincase, and the facial bones that support the eyes, nose, and jaws.

Table 6.1 Classification of the bones of the adult skeleton

I. Axial skeleton	
a. Skull	22 bones
8 cranial bones:	
frontal 1	
parietal 2	
occipital 1	
temporal 2	
sphenoid 1	
ethmoid 1	
13 facial bones:	
maxilla 2	
palatine 2	
zygomatic 2	
lacrimal 2	
nasal 2	
vomer 1	
inferior nasal concha 2	
1 mandible (frequently considered a facial bone)	
b. Middle ear ossicles	6 bones
malleus 2	
incus 2	
stapes 2	
c. Hyoid	1 bone
hyoid bone 1	
d. Vertebral column	26 bones
cervical vertebra 7	
thoracic vertebra 12	
lumbar vertebra 5	
sacrum 1 (4 or 5 fused bones)	
coccyx 1 (4 or 5 fused bones)	
e. Rib cage	25 bones
rib 24	
sternum 1	
II. Appendicular skeleton	
a. Pectoral girdle	4 bones
scapula 2	
clavicle 2	
b. Upper extremities	60 bones
humerus 2	
radius 2	
ulna 2	
carpal 16	
metacarpal 10	
phalanx 28	
c. Pelvic girdle	2 bones
os coxa 2 (each os coxa contains 3 fused bones)	
d. Lower extremities	60 bones
femur 2	
tibia 2	
fibula 2	
patella 2	
tarsal 14	
metatarsal 10	
phalanx 28	
Total	206 bones

2. **Ear ossicles.** Three ear ossicles are present in the middle ear chamber of each ear and serve to transmit sound impulses.
3. **Hyoid** *(hi′oid)* **bone.** The hyoid bone is located above the larynx and below the lower jaw. It supports the tongue and assists in swallowing.

sesamoid: Gk. *sesamon,* like a sesame seed

ossicle: L. *ossiculum,* little bone

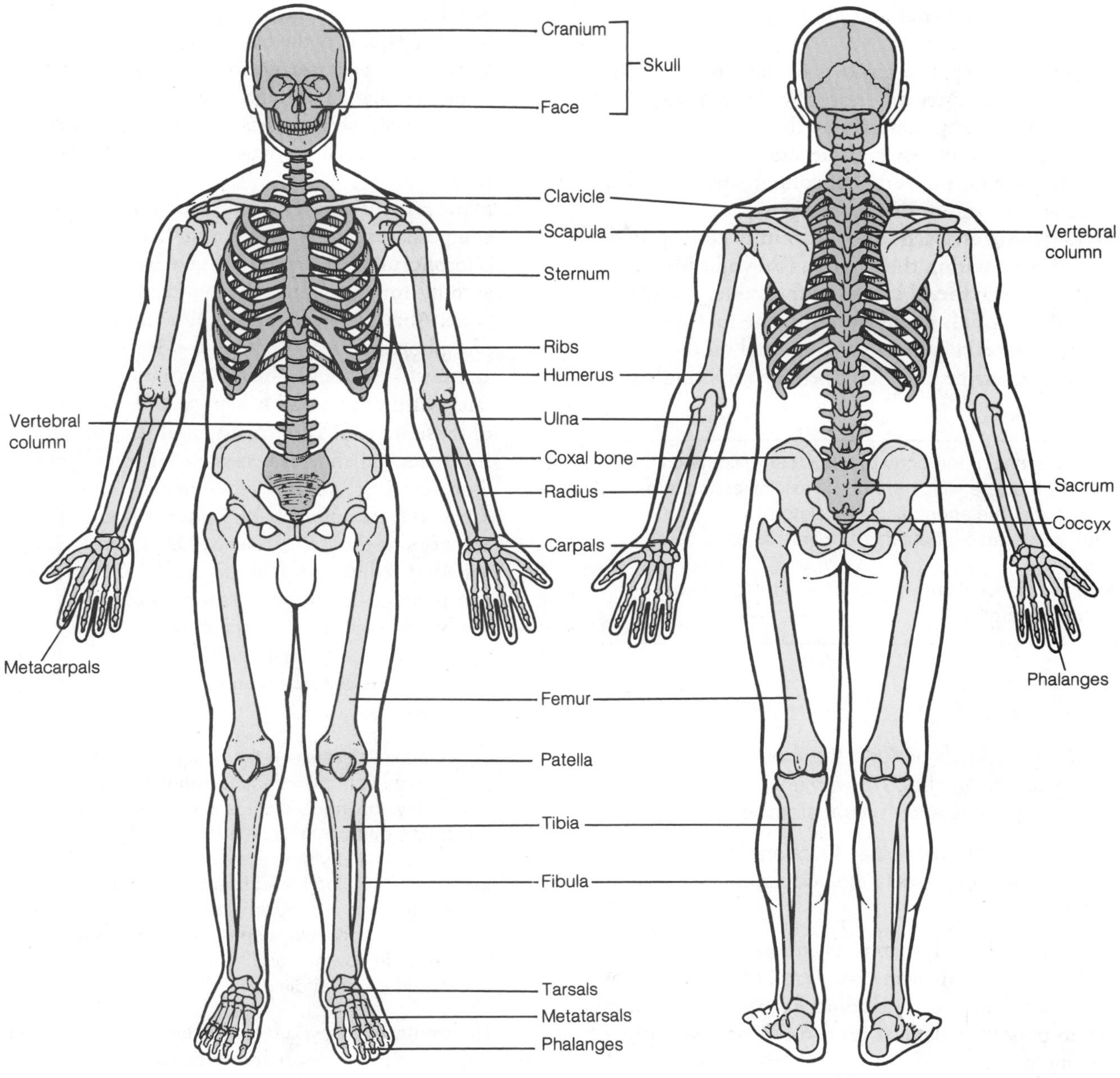

Figure 6.1 The human skeleton. (*a*) An anterior view. (*b*) A posterior view. The bones of the axial skeleton are colored differently to distinguish them from the bones of the appendicular skeleton.

4. **Vertebral column.** The vertebral column (backbone) consists of twenty-six individual bones separated by cartilaginous intervertebral discs. In the pelvic region several vertebrae are fused to form the sacrum, which provides attachment for the pelvic girdle. A few terminal vertebrae are fused to form the coccyx, the so-called tailbone.
5. **Rib cage.** The rib cage, or thoracic cage, forms the bony and cartilaginous framework of the thorax. The rib cage articulates posteriorly with the thoracic vertebrae and includes the twelve pairs of **ribs,** the flattened **sternum,** and the **costal cartilages,** which connect the ribs to the sternum on the anterior side.

The **appendicular skeleton** is composed of the bones of the upper and lower extremities and the bony girdles, which anchor the appendages to the axial skeleton:

1. **Pectoral girdle.** The paired **scapulae** and **clavicles** constitute the pectoral girdle. It is not a complete girdle, having only an anterior attachment to the axial skeleton at the sternum via the clavicles. The primary function of the pectoral girdle is to provide attachment for the muscles that move the brachium and forearm.
2. **Upper extremities.** Each upper extremity consists of a proximal **humerus** within the brachium, an **ulna** and **radius** within the forearm, the **carpal** bones of the

wrist, and the **metacarpal** and **phalangeal** *(fah-lan'je-al)* **bones** of the hand.

3. **Pelvic girdle.** The pelvic girdle is formed by two **ossa coxae** (hipbones), united anteriorly by the **symphysis** *(sim'fi-sis)* **pubis** and posteriorly by the sacrum of the vertebral column. The pelvic girdle supports the weight of the body through the vertebral column and protects the lower viscera within the pelvic cavity.
4. **Lower extremities.** Each lower extremity consists of a proximal **femur** within the thigh, a **tibia** and **fibula** within the leg, the **tarsal bones** of the ankle, and the **metatarsal** and **phalangeal bones** of the foot. In addition, the **patella** *(pah-tel'ah)* is located on the anterior surface of the knee joint between the thigh and leg.

---

1. List the bones of the body that you can palpate on yourself. Indicate which are of the axial skeleton and which are of the appendicular skeleton.
2. What are sesamoid bones, and where are they found?
3. Describe the locations and functions of the pectoral and pelvic girdles.

Functions of the Skeletal System

The bones of the skeleton perform the mechanical functions of support, protection, and body movement and the metabolic functions of hemopoiesis and mineral storage.

Objective 2. Discuss the principal functions of the skeletal system, and identify the body systems that these functions serve.

The strength of bone comes from its inorganic components, which resist decomposition even after death. Much of what we know of prehistoric animals, including humans, has been determined from preserved skeletal remains. When we think of bone, we frequently think of a hard, dry structure. In fact, the term *skeleton* comes from a Greek word meaning "dried up." Living bone is not inert material, however; it is dynamic and adaptable in performing many body functions, including support, protection, body movement, hemopoiesis, and mineral storage.

1. **Support.** The skeleton forms a rigid framework to which the softer tissues and organs of the body are attached.
2. **Protection.** The skull and vertebral column enclose the central nervous system; the rib cage protects the heart, lungs, great vessels, liver, and spleen; and the pelvic cavity supports and protects the pelvic viscera. Even the site where red blood cells are produced is protected within the central, hollow portion of the bone.
3. **Body movement.** Bones serve as anchoring attachments for most skeletal muscles. In this capacity, the bones act as levers with the joints functioning as pivots when muscles contract to cause body movement.
4. **Hemopoiesis** *(he''mo-poi-e'sis)*. The red bone marrow of an adult produces white blood cells, red blood cells, and platelets. In an infant, the spleen and liver produce red blood cells, but as the bones mature, the bone marrow assumes the performance of this formidable task. It is estimated that an average of one million red blood cells are produced every second by the bone marrow to replace those that are worn out and destroyed by the liver.
5. **Mineral storage.** The inorganic matrix of bone is composed primarily of the minerals calcium and phosphorus. These minerals give bone its rigidity and account for approximately two-thirds of the weight of bone. About 95% of the calcium and 90% of the phosphorus within the body are deposited in the bones and teeth. Although the concentration of these organic salts within the blood is kept within narrow limits, both of these mineral salts are essential for other body functions. Calcium is necessary for muscle contraction, blood clotting, and the movement of ions and nutrients across cell membranes. Phosphorus is required for the activities of the nucleic acids, DNA and RNA, as well as for ATP utilization. If mineral salts are not present in the diet in sufficient amounts, they may be withdrawn from the bones until they are replenished through proper nutrition. In addition to calcium and phosphorus, lesser amounts of magnesium and sodium salts are stored in bone tissue.

Vitamin D assists in the absorption of calcium and phosphorus from the intestine into the blood. As the bones develop in a child, it is immensely important that the child's diet contains an adequate amount of these two minerals and vitamin D. If the diet is deficient in these essentials, the blood level falls below that necessary for calcification and a condition known as *rickets* develops (see fig. 5.15). Rickets is characterized by soft bones that may result in bowlegs and malformation of the head, chest, and pelvic girdle.

In summary, the skeletal system is not an isolated body system. It is closely associated with the muscle system since it stores the calcium needed for muscular contraction and provides an attachment for muscles as they span the movable joints. The skeletal system serves the circulatory system by producing blood cells in protected sites. Also, many of the vessels of the circulatory system are named according to the bones they parallel. The skeletal system supports and protects, to varying degrees, all of the systems of the body.

---

1. List the functions of the skeletal system.
2. Discuss two ways the skeletal system serves the circulatory system in the production of blood. In which two ways does it serve the muscular system?

Development of the Skeletal System

All bones derive from specialized mesenchymal mesoderm through endochondral ossification or through intramembranous ossification. Ossification begins by the fourth week of prenatal development and is not completed in certain bones until about age twenty-five or thirty.

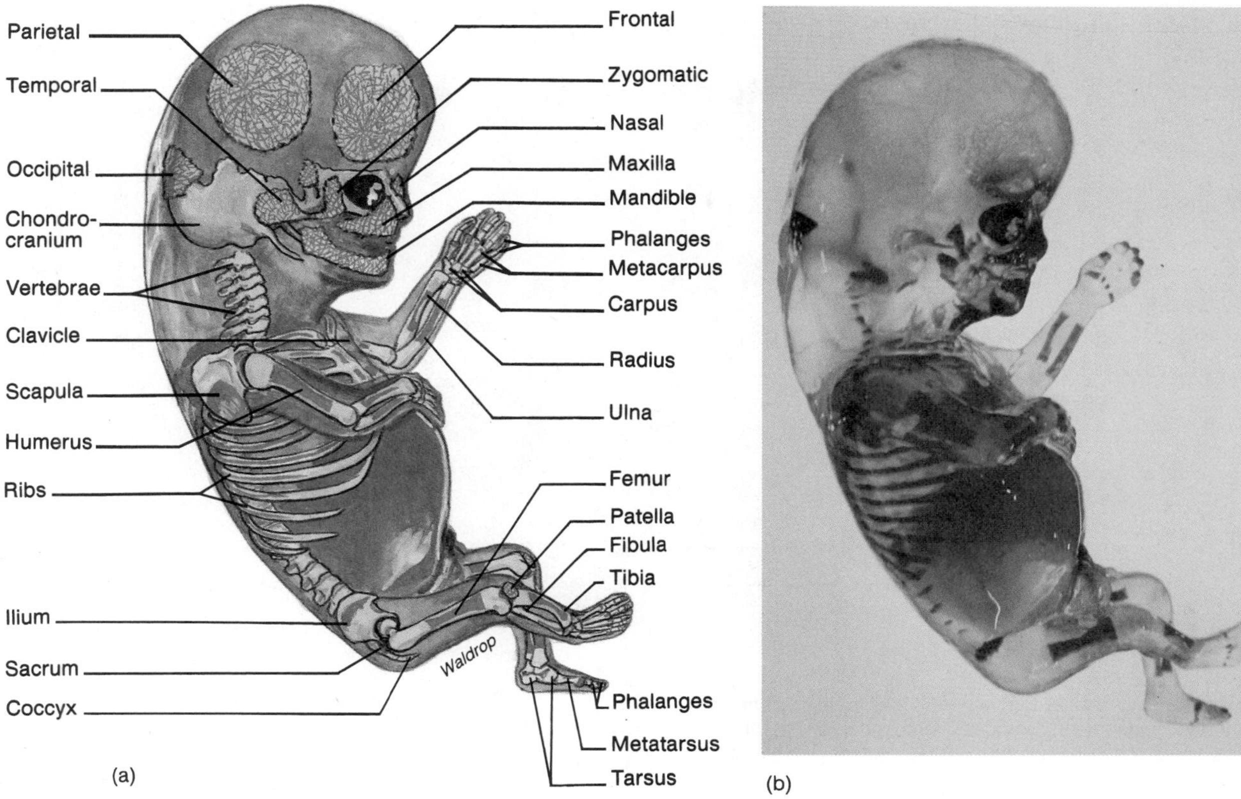

Figure 6.2 Ossification centers of the skeleton of a ten-week-old fetus. (*a*) The diagram depicts endochondral ossification in red and intramembranous ossification in a stippled pattern. The cartilaginous portions of the skeleton are shown in gray. (*b*) The photograph shows the ossification centers stained with a red indicator dye.

Objective 3. Distinguish between endochondral ossification and intramembranous ossification.

Objective 4. Describe the fontanels of a fetal skull, and explain their importance.

Bone formation, or ossification, begins about the fourth week of embryonic development, but ossification centers cannot be observed until about the eighth week (fig. 6.2). Bone tissue derives from specialized migratory cells known as *mesenchyme.* Some of the embryonic mesenchymal cells will transform into *chondroblasts (kon'dro-blastz)* and develop a cartilage matrix that is later replaced by bone in a process known as **endochondral** *(en''do-kon'dral)* **ossification.** Most of the skeleton is formed in this fashion—first it goes through a hyaline cartilage stage and then ossifies as bone (see fig. 6.9).

A smaller number of mesenchymal cells develop directly into bone without first going through a cartilage stage. This type of bone formation process is referred to as **intramembranous** *(in''trah-mem'brah-nus)* **ossification.** Facial bones and certain bones of the cranium are formed this way. **Sesamoid bones** are specialized intramembranous bones that develop in tendons.

The formation of the skull is a complex process that begins during the fourth week of embryonic development and continues well beyond the birth of the baby. Three factions are involved in the formation of the skull: the chondrocranium, the dermatocranium, and the viscerocranium (fig. 6.3). The **chondrocranium** is the portion of the skull that undergoes endochondral ossification to form the bones supporting the brain. The **dermatocranium** is the portion of the skull that develops through membranous ossification to form the bones covering the brain and facial region. The **viscerocranium** (splanchnocranium) is the portion that develops from the embryonic visceral arches and forms the ear ossicles, the hyoid bone, and specific processes of the skull.

During fetal development and infancy, the bones of the dermatocranium covering the brain are separated by fibrous sutures. There are also six large membranous areas of the skull that provide spaces between the developing bones (fig. 6.4). These areas are called **fontanels** *(fon''tah-nel)* (soft spots) and permit the skull to undergo changes of shape, called *molding,* during parturition (childbirth). The fontanels also allow for rapid growth of the brain during infancy. Ossification of the fontanels is normally complete by twenty to twenty-four months of age. A description of the six fontanels follows:

1. **Anterior** (frontal) **fontanel.** The anterior fontanel is diamond-shaped and is the most prominent of the six. It is located on the anteromedian portion of the skull.
2. **Posterior** (occipital) **fontanel.** The posterior fontanel is positioned at the back of the skull on the median line.
3. **Anterolateral** (sphenoidal) **fontanels.** The paired anterolateral fontanels are found on both sides of the skull directly below the anterior fontanel.

chondrocranium: Gk. *chondros,* cartilage; *kranion,* skull

dermatocranium: Gk. *derma,* skin; *kranion,* skull
viscerocranium: L. *viscera,* soft parts; Gk. *kranion,* skull
fontanel: Fr. *fontaine,* little fountain

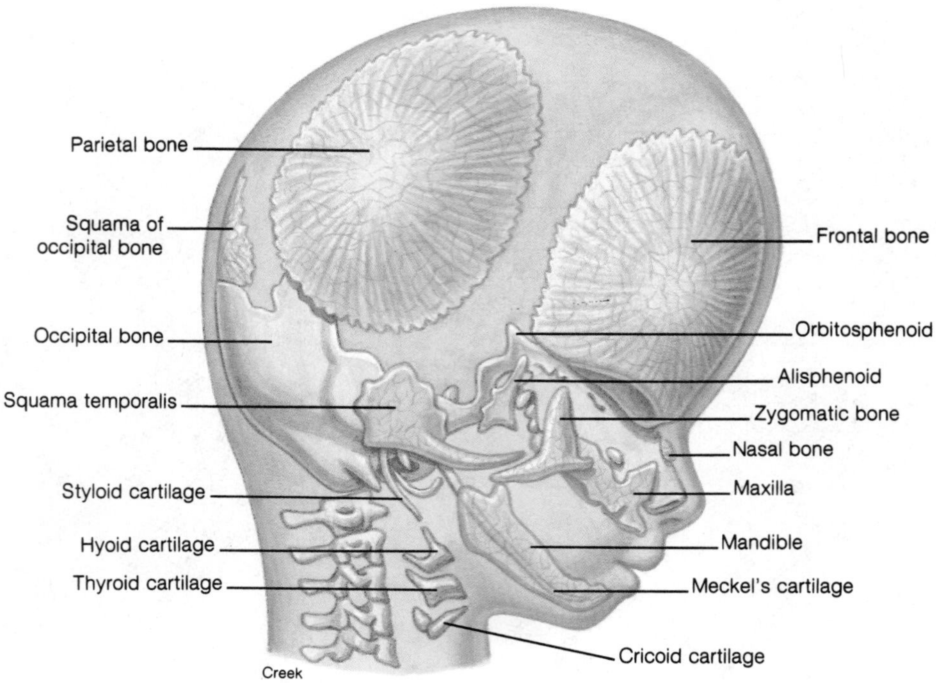

Figure 6.3 The embryonic skull at twelve weeks is composed of bony elements from three developmental sources: the chondocranium (blue-gray), the dermatocranium (yellowish), and the viscerocranium (salmon).

4. **Posterolateral** (mastoid) **fontanels.** The paired posterolateral fontanels are located on the posterolateral sides of the skull.

A prominent **sagittal suture** extends the anteroposterior median length of the skull between the anterior and posterior fontanels. A **coronal suture** extends from the anterior fontanel to the anterolateral fontanel. A **lambdoidal suture** extends from the posterior fontanel to the posterolateral fontanel. A **squamosal suture** connects the posterolateral fontanel to the anterolateral fontanel.

During normal parturition, the molding of the fetal skull is such that the occipital bone is usually pressed under the two parietal bones. In addition, one parietal bone overlaps the other, with the depressed one against the promontory of the mother's sacrum. If a baby is born breech (buttocks first), molding does not occur and delivery is more difficult.

1. Discuss how endochondral ossification differs from intramembranous ossification.
2. Describe the fontanels, and list their two functions.

Bone Structure and Growth

Spongy bone and compact bone differ in histological structure, gross appearance, location, and function. Each bone has a characteristic pattern of ossification and growth, a characteristic shape, and diagnostic surface features that indicate its functional relationship to other bones, muscles, and to the body structure as a whole.

Objective 5. Classify bones according to their shapes, and give an example of each type.
Objective 6. Describe the various markings on the surfaces of bones.
Objective 7. Describe the histological structure and gross features of a typical long bone, and list the functions of each.
Objective 8. Describe the process of endochondral ossification as it relates to bone growth.

Bone Structure

Each bone of the skeleton is an organ since it consists of several types of tissue. Bone tissue is the principal tissue, but nervous, vascular, and cartilaginous tissues also contribute to the structure and function of bone.

The shape and surface features of each bone indicate its functional role in the skeleton. Bones that are long, for example, function as levers during body movement. Bones that support the body are massive and have large articular surfaces and processes for muscle attachment. Roughened areas on these bones may serve for the attachment of ligaments, tendons, or muscles. A flattened surface provides placement for a large muscle or may serve for protection. Grooves around an articular end of a bone indicate where a tendon or nerve passes, and openings through a bone permit the passage of nerves or vessels.

Shapes of Bones The bones of the skeleton are classified into four principal types on the basis of shape rather than size. The

lambdoidal: Gk. *lambda,* letter (λ) in Greek alphabet

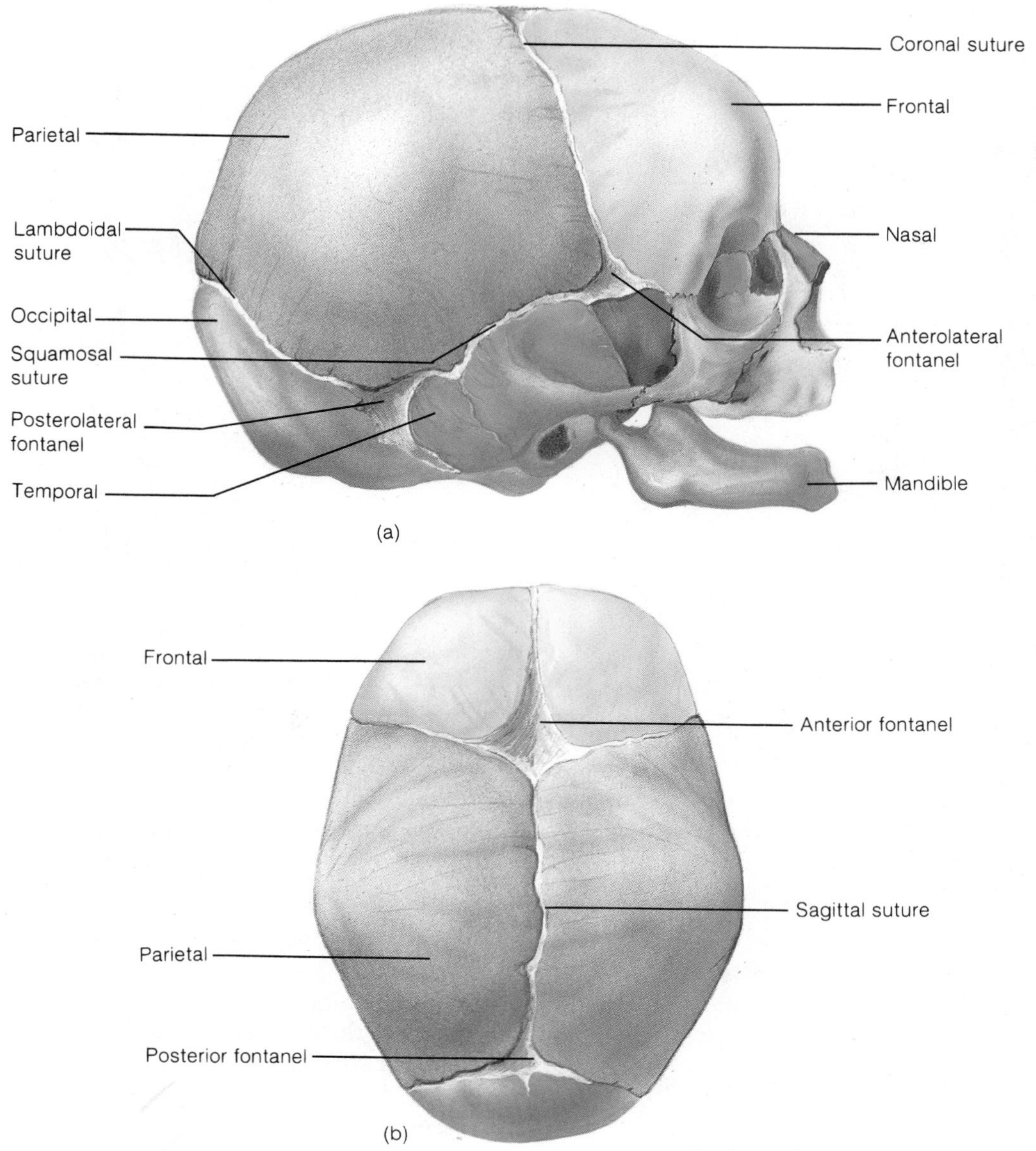

Figure 6.4 The fetal skull showing the six fontanels and the sutures. (*a*) A right lateral view and (*b*) a superior view.

four classes are long bones, short bones, flat bones, and irregular bones (fig. 6.5).

1. **Long bones.** Long bones are longer than they are wide and function as levers. Most of the bones of the upper and lower extremities are of this type (e.g., the humerus, radius, ulna, metacarpals, femur, tibia, fibula, metatarsals, and phalanges).
2. **Short bones.** Short bones are somewhat cube-shaped and are found in confined spaces where they transfer forces (e.g., the carpals and tarsals).
3. **Flat bones.** Flat bones have a broad, dense surface for muscle attachment or protection of underlying organs (e.g., the cranium, ribs, and bones of the pectoral girdle).
4. **Irregular bones.** Irregular bones have varied shapes and have many surface markings for muscle attachment or articulation (e.g., the vertebrae and certain bones of the skull).

Surface Features of Bone The following are the various descriptive terms (with examples) used to identify the surface features of bone:

Articulating surfaces

condyle *(kon'dīl)* A large, rounded, articulating knob (the occipital condyle of the occipital bone)

facet *(fas'et)* A flattened or shallow articulating surface (the costal facet of a thoracic vertebra)

head A prominent, rounded, articulating proximal end of a bone (the head of the femur)

Nonarticulating prominences

crest A narrow, ridgelike projection (the iliac crest of the os coxa)

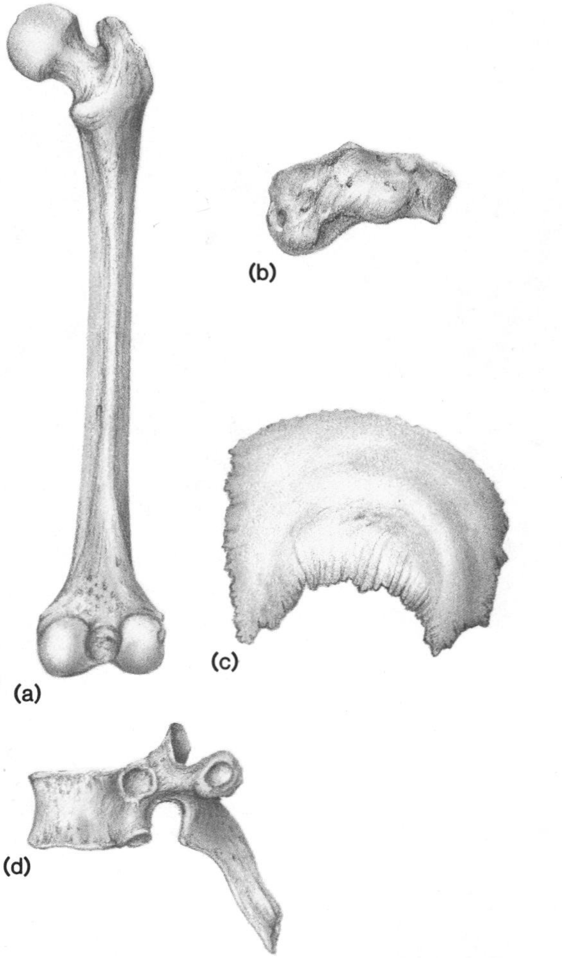

Figure 6.5 Shapes of bones. (*a*) A long bone, (*b*) a short bone, (*c*) a flat bone, and (*d*) an irregular bone.

epicondyle A projection above a condyle (the medial epicondyle of the femur)
process Any marked bony prominence (the mastoid process of the temporal bone)
spine A sharp, slender process (the spine of the scapula)
trochanter *(tro-kan'ter)* A massive process found only on the femur (the greater trochanter of the femur)
tubercle *(tu'ber-kl)* A small rounded process (the greater tubercle of the humerus)
tuberosity A large roughened process (the radial tuberosity of the radius)

Depressions and openings

alveolus *(al-ve'o-lus)* A deep pit or socket (the alveoli for teeth in the maxilla)
fissure A narrow, slitlike opening (the superior orbital fissure of the sphenoid bone)
foramen *(fo-ra'men)*, pl. **foramina** A rounded opening through a bone (the foramen magnum of the occipital bone)
fossa A flattened or shallow surface (the mandibular fossa of the temporal bone)

trochanter: Gk. *trochanter*, runner
tuberosity: L. *tuberosus*, lump

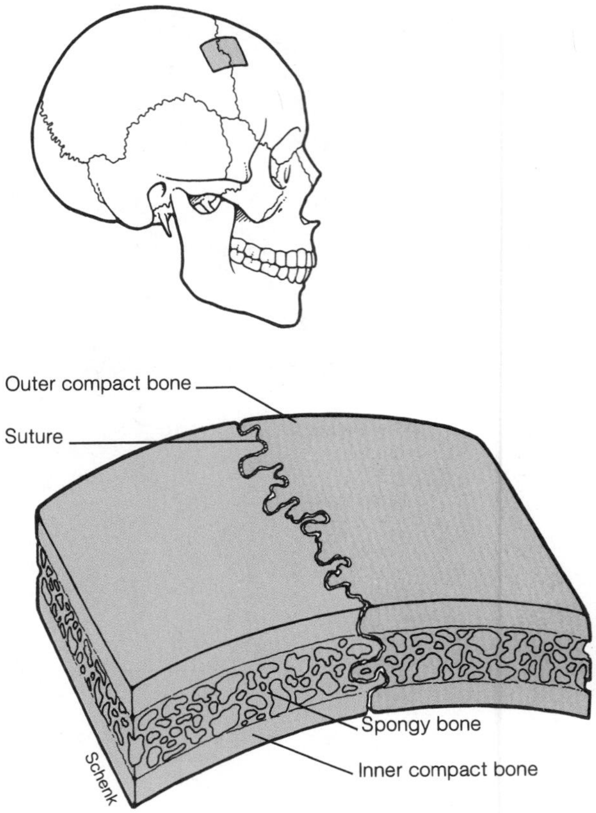

Figure 6.6 A section through the skull showing diploe. Diploe is a layer of spongy bone sandwiched between two surface layers of compact bone.

fovea A small pit or depression (the fovea capitis of the femur)
meatus *(me-a'tus)*, or **canal** A tubelike passageway through a bone (the external auditory meatus of the temporal bone)
sinus A cavity or hollow space in a bone (the frontal sinus of the frontal bone)
sulcus *(sul'kus)* A groove that accommodates a vessel, nerve, or tendon (the intertubercular sulcus of the humerus)

Organization of Bone Bone (osseous) tissue is organized as **spongy** (cancellous) **bone** or **compact** (dense) **bone,** and most bones have both types. In a flat bone of the skull, for example, the spongy bone is sandwiched between the compact bone and is called a **diploe** *(dip'lo-e)* (fig. 6.6). Because of this protective layering, a blow to the head may fracture the outer compact bone layer without harming the inner compact bone layer and the brain.

In a long bone from an appendage, the bone shaft, or **diaphysis** *(di-af'ĭ-sis)*, consists of compact bone forming a cylinder that surrounds a central cavity called the **medullary cavity** (fig. 6.7). The medullary cavity is lined with a thin layer of connective tissue called the **endosteum** *(en-dos'te-um)* and in the adult contains **yellow bone marrow,** so named because of the large amounts of fat it contains. On each end of the diaphysis is an **epiphysis** *(ĕ-pif'ĭ-sis)*, consisting of spongy bone

diploe: Gk. *diplous*, double
diaphysis: Gk. *dia*, throughout; *physis*, growth
epiphysis: Gk. *epi*, upon; *physis*, growth

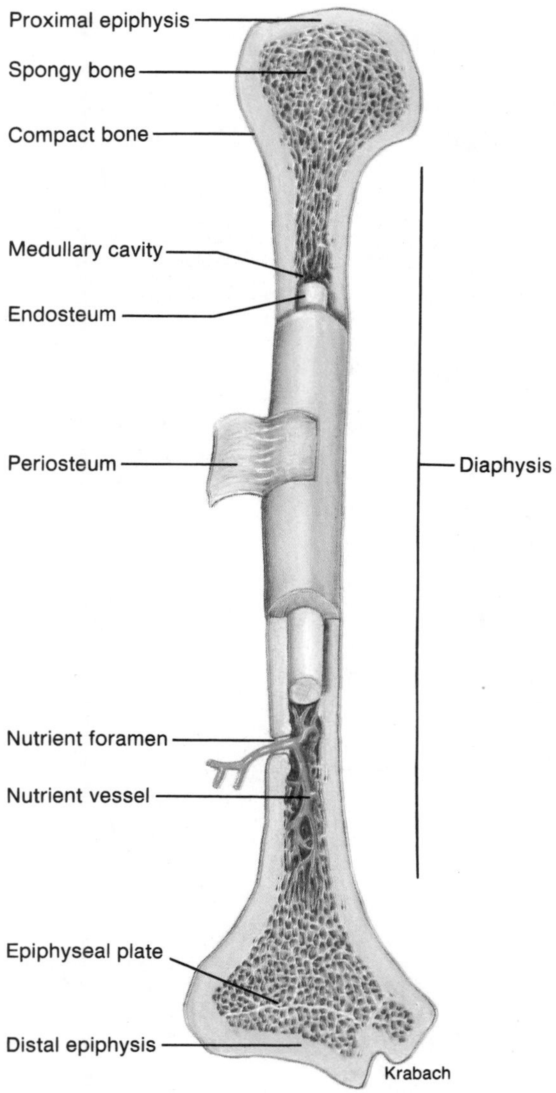

Figure 6.7 A diagram of a long bone shown in longitudinal section.

surrounded by a layer of compact bone. **Red bone marrow** is found within the porous chambers of spongy bone. In an adult, erythropoiesis, the production of red blood cells, occurs in the red bone marrow, especially that of the sternum, vertebrae, portions of the ossa coxae, and the proximal epiphyses of the femora and humeri. The red bone marrow is also responsible for the formation of certain white blood cells and platelets and for the phagocytosis of worn-out red blood cells. **Articular cartilage,** which is composed of thin hyaline cartilage, caps each epiphysis and facilitates joint movement (see fig. 6.9). Along the diaphysis are **nutrient foramina,** small openings into the bone that allow passage of nutrient vessels into the bone for nourishment of the living tissue.

Between the diaphysis and epiphysis is an **epiphyseal** *(ep''ĭ-fiz'e-al)* **plate,** a region of mitotic activity that is responsible for elongation of bone. As bone growth is completed, an **epiphyseal line** replaces the plates and ossification occurs between the epiphysis and the diaphysis. A **periosteum** *(per''e-os'te-um)* of dense, white fibrous tissue covers the surface of the bone, except at the articulating surfaces. This highly vascular layer serves as a place for a tendon-muscle attachment and is responsible for diametric (width) bone growth. The periosteum is secured to the bone by **perforating** (Sharpey's) **fibers** (fig. 6.8) composed of bundles of collagenous fibers.

Fracture of a long bone in a young person may be especially serious if it results in displacement of an epiphyseal plate. If such an injury is untreated, or treated improperly, longitudinal growth of the bone may be arrested or retarded, resulting in permanent shortening of the limb.

Compact bone consists of precise arrangements of microscopic cylindrical structures parallel to the long axis of the bone (fig. 6.8). These column-like structures are the **osteons** (Haversian systems). The matrix of an osteon is laid down in concentric rings, called **lamellae,** around a **central** (Haversian) **canal** that contains minute nutrient vessels and a nerve. A bone cell, or **osteocyte,** is within a space called a **lacuna,** which is connected to other lacunae by **canaliculi** through which nutrients diffuse. Metabolic activity within bone tissue occurs at the osteon level. Between osteons there are incomplete remnants of osteons, called **interstitial systems. Perforating** (Volkmann's) **canals** penetrate compact bone connecting osteons with blood vessels and nerves.

Bone Growth

The development of bone from embryonic to adult size depends on the orderly processes of mitotic divisions, growth, and the structural remodeling determined by genetics, hormonal secretions, and nutritional supply. In most bone development, a cartilaginous model is gradually replaced by bone tissue during endochondral bone formation (fig. 6.9). As the cartilage model grows, the chondrocytes (cartilage cells) in the center of the shaft hypertrophy, and minerals are deposited within the matrix in a process called *calcification.* Calcification restricts the passage of nutrients to the chondrocytes, causing them to die. At the same time, some cells of the perichondrium (dense fibrous connective tissue surrounding cartilage) differentiate into primordial bone cells called **osteoblasts,** which secrete **osteoid,** the organic component of bone. As the perichondrium calcifies, it gives rise to a thin plate of compact bone called the **periosteal bone collar,** which is surrounded by the periosteum.

A **periosteal bud,** consisting of osteoblasts and blood vessels, invades the disintegrating center of the cartilage model from the periosteum. Once in the center, the osteoblasts secrete osteoid, and a **primary ossification center** is established from which ossification expands into deteriorating cartilage. This process is repeated in both the proximal and distal epiphyses, forming **secondary ossification centers.**

Bone growth continues as long as the cartilage cells at the epiphyseal plate between the two ossification centers continue

periosteum: Gk. *peri,* around; *osteon,* bone

Sharpey's fibers: from William Sharpey, Scottish physiologist and histologist, 1802–1880
osteoblast: Gk. *osteon,* bone; *blastos,* offspring or germ

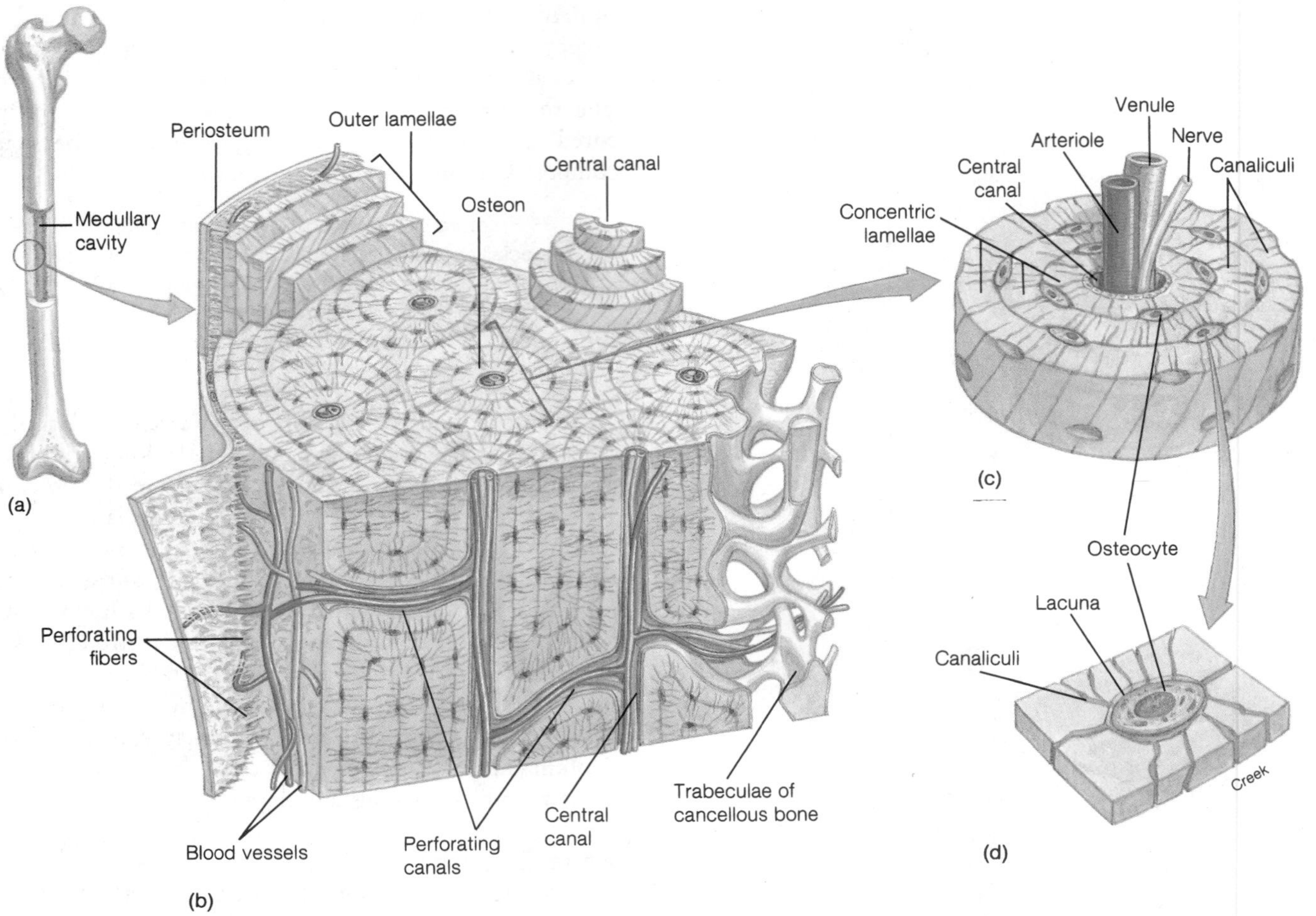

Figure 6.8 Compact bone tissue. (*a*) A diagram of the femur showing a cut through the compact bone into the medullary cavity. (*b*) The arrangement of the osteons within the diaphysis of the bone. (*c*) An enlarged view of an osteon showing the osteocytes within lacunae and the concentric lamellae. (*d*) An osteocyte within a lacuna.

to divide (fig. 6.10). When mitosis of the chondrocytes gradually decreases, ossification finally occurs within the plate, thus prohibiting the further lengthwise growth of the bone. The time when epiphyseal plates ossify varies greatly from bone to bone, but generally it occurs between the ages of eighteen and twenty within the long bones of most persons (table 6.2).

Radiologists can determine the ages of persons who are still growing by examining x-ray pictures of their bones. Ossification at the epiphyseal plates occurs at predictable periods of development and is therefore a reliable indicator of age. Large discrepancies between bone age and chronological age may indicate a genetic or endocrine dysfunction.

Bone remodeling is a continual process throughout a person's life. The diagnostic processes on the surface of bones develop as stress is applied to the periosteum, resulting in the osteoblastic secretion of osteoid and the formation of new bone tissue. These processes may continue to change somewhat in persons who are athletically active even though they have stopped growing in height.

Specialized bone cells called **osteoclasts** can enzymatically cause bone resorption. As new bone layers are deposited on the outside surface of the bone, osteoclasts dissolve bone tissue adjacent to the medullary cavity. In this way, the size of the cavity keeps pace with the increased growth of the bone.

The movement of teeth in orthodontics involves bone remodeling. The teeth sockets (alveoli) are reshaped through the activity of osteoclast and osteoblast cells as strain is applied through the application of braces. The use of traction in treating certain skeletal disorders has a similar effect.

The size, shape, and processes of bones depend on the physical stresses and pressures applied. Initial bone formation, bone growth, and the adaptation of mature bone to units of applied force is reflected in the **stress lines,** or the structure of ossification within a bone (fig. 6.11). Stress lines provide maximum strength to the bone in the direction of greatest applied pressure. The diagnostic processes on some bones, such as the greater trochanter of the femur, develop in response to forces of strain applied to the periosteum of the bone where the tendons of muscles attach. In other words, as a muscle is contracted, a pulling strain is applied to the periosteum of the bone where the tendon is attached. This mechanical tension causes an increase in osteoblastic activity and bone deposition, and a bony process is gradually formed.

---
osteoclast: Gk. *osteon,* bone; *klastos,* broken

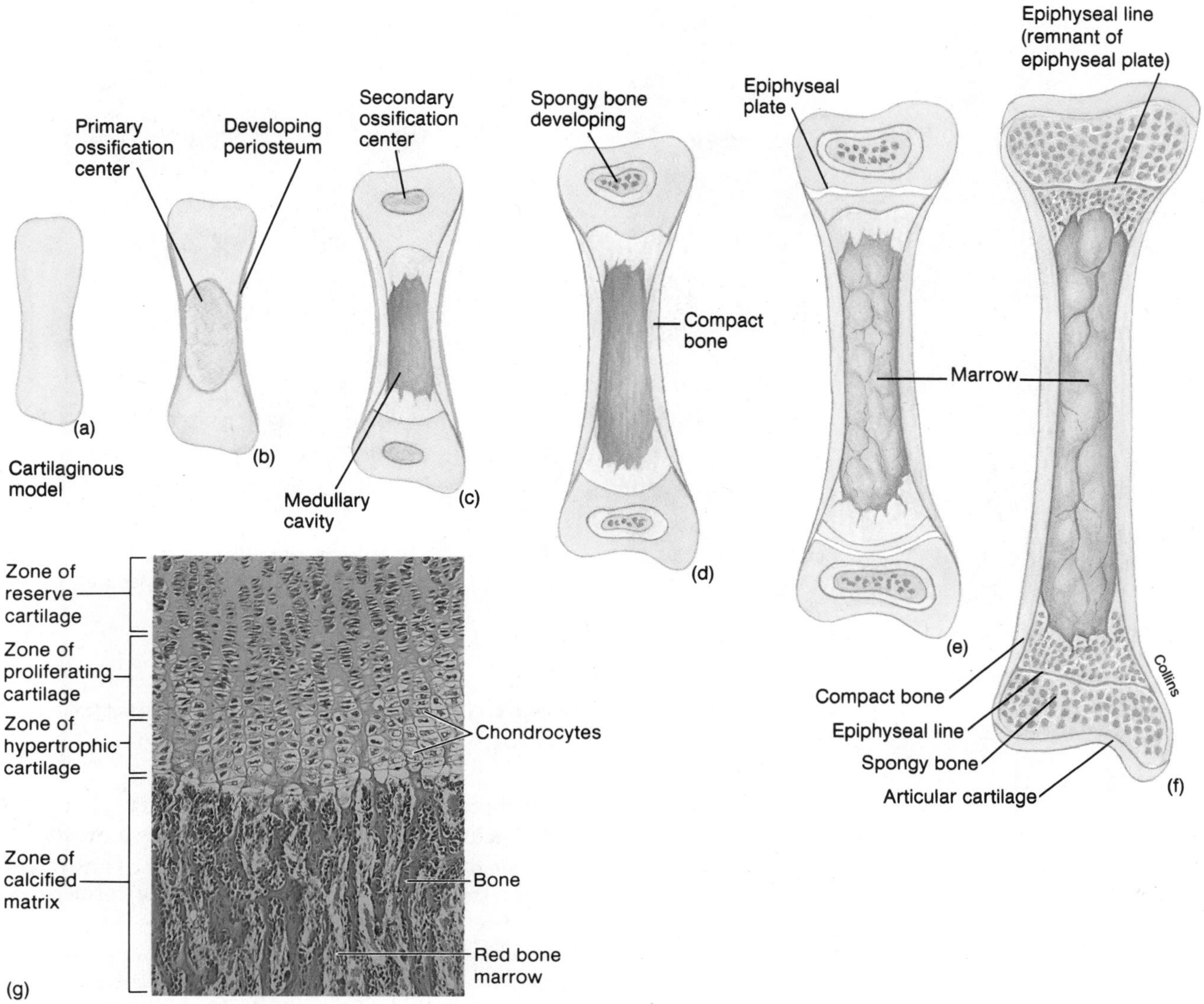

Figure 6.9 The process of endochondral ossification, beginning with (*a*) the cartilaginous model as it occurs in a fetus. The bone develops (*b–e*) through intermediate stages to (*f*) adult bone, and (*g*) the activity of ossification is shown in a photomicrograph from an epiphyseal region (63×).

Bone is highly dynamic and is continually being remodeled in response to mechanical strain or even the absence of strain. The effect of the absence of strain can best be seen in the bones of inactive patients confined to bed or the bones of persons who are paralyzed. X-ray examination shows the loss of bone mass or even *osteoporosis.* The absence of gravity that accompanies space flight may result in mineral loss from bones if an exercise program is not maintained.

1. Using examples, discuss the function of each of the four kinds of bones as determined by shape.
2. Define each of the following surface markings on bones: condyle, head, facet, process, crest, epicondyle, fossa, alveolus, foramen, and sinus.
3. Diagram a sagittal view of a typical long bone, and label the following: diaphysis, medullary cavity, epiphyses, articular cartilages, nutrient foramen, periosteum, and epiphyseal plates. Explain the function of each of these structures.
4. Define *osteocyte, osteoblast,* and *osteoclast,* and explain the function of each as related to the processes of endochondral ossification and bone growth.

Skull

The human skull, consisting of eight cranial and fourteen facial bones, contains several cavities that house the brain and sensory organs. Each bone of the skull articulates with the adjacent bones and has diagnostic and functional processes, surface features, and foramina.

Objective 9. Identify the cranial and facial bones of the skull, and describe their structural characteristics.

Objective 10. Describe the location of each of the bones of the skull, and identify the articulations that affix them together.

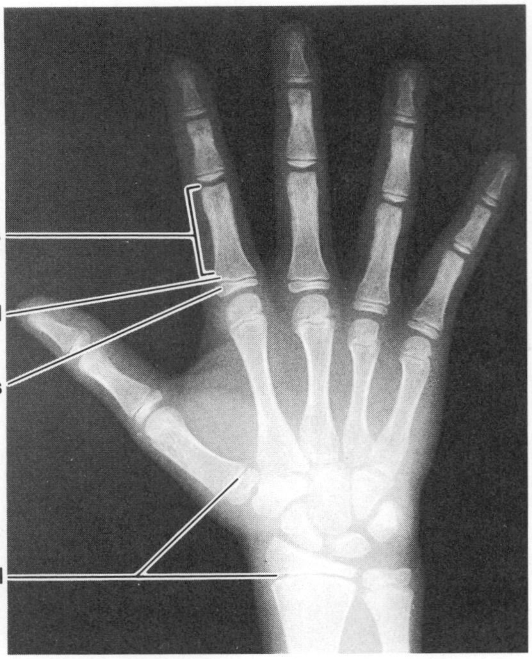

Figure 6.10 The presence of epiphyseal plates, as seen in an X ray of a child's hand, indicates that bones are still growing in length.

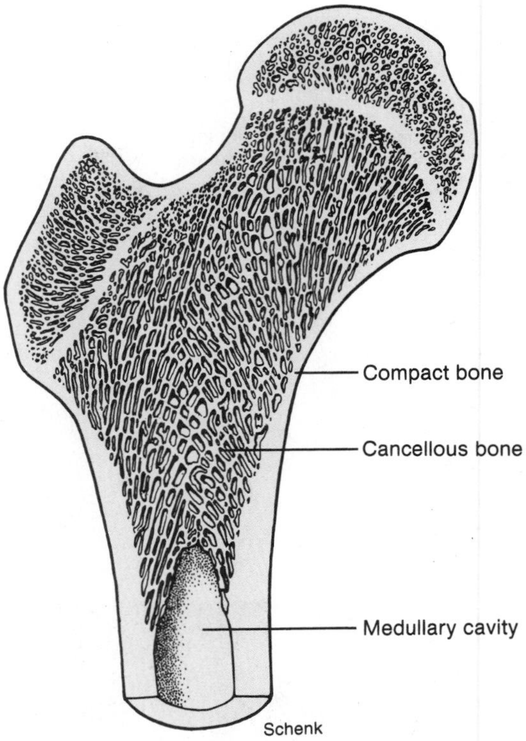

Figure 6.11 A longitudinal section of proximal end of femur showing stress lines.

Table 6.2 Average age of completion of bone ossification

Bone	Chronological age of fusion
Scapula	18–20
Clavicle	23–31
Bones of upper extremity (brachium, forearm, hand)	17–20
Os coxa	18–23
Bones of lower extremity (thigh, leg, foot)	18–22
Vertebra	25
Sacrum	23–25
Sternum (body)	23
Sternum (manubrium, xiphoid)	after 30

The skull consists of **cranial bones** and **facial bones.** The eight bones of the cranium articulate firmly with one another to enclose and protect the brain and sensory organs. The fourteen facial bones form the framework for the facial region and support the teeth. A variation in shape and density of the facial bones is a major contributor to the individuality of each human face. The facial bones, with the exception of the bone within the lower jaw, are also firmly interlocked with one another and the cranial bones.

The skull has several cavities. The **cranial cavity** is the largest, with a capacity of about 1,300–1,350 cc. The **nasal cavity** is formed by both cranial and facial bones and is partitioned into two chambers, or **nasal fossae,** by a **nasal septum** of bone and cartilage. Four sets of **paranasal sinuses** are located within the bones surrounding the nasal area and communicate via ducts into the nasal cavity. **Middle** and **inner ear chambers** are positioned inferior to the cranial cavity and house the organs of hearing and balance. The two **orbits** for the eyeballs are formed by facial and cranial bones. The **oral,** or **buccal, cavity** (mouth), which is only partially formed by bone, is completely within the facial region.

The bones of the skull contain numerous foramina to accommodate nerves, vessels, and other structures. Summarized in table 6.3 are the foramina of the skull. Various views of the skull are shown in figures 6.12 through 6.19. The diagrams of the bones of the skull are color-coded to facilitate learning their relative positions. X rays (roentgenograms) of the skull are shown in figure 6.20.

Although the hyoid bone and the three paired ear ossicles are not considered part of the skull, they are within the axial skeleton and will be described immediately following the discussion of the skull.

Cranium

The cranial bones enclose the brain and consist of one frontal, two parietals, two temporals, one occipital, one sphenoid, and one ethmoid.

Frontal The frontal bone forms the anterior roof of the cranium, the forehead, the roof of the nasal cavity, and the superior arch of the **bony orbits,** which contain the eyeballs. The bones of the orbit are summarized in table 6.4. The frontal bone develops in two halves that grow together and are usually completely fused by age five or six. Occasionally a complete suture

roentgenogram: from Wilhelm K. Roentgen, German physicist, 1845–1923
cranium: Gk. *kranion,* skull

Table 6.3 Summary of major foramina of the skull

Foramen	Location	Structures transmitted
Carotid canal	Petrous portion of temporal bone	Internal carotid artery and sympathetic nerves
Greater palatine	Palatine bone of hard palate	Greater palatine nerve and descending palatine vessels
Hypoglossal foramen/canal	Anterolateral edge of the occipital condyle	Hypoglossal nerve and branch of ascending pharyngeal artery
Incisive	Anterior region of hard palate, posterior to the incisor teeth	Branches of descending palatine vessels and nasopalatine nerve
Inferior orbital fissure	Between maxilla and greater wing of sphenoid	Maxillary branch of trigeminal nerve, zygomatic nerve, and infraorbital vessels
Infraorbital	Inferior to orbit in maxilla	Infraorbital nerve and artery
Jugular	Between petrous portion of temporal and occipital, posterior to carotid canal	Internal jugular vein; vagus, glossopharyngeal, and accessory nerves
Lacerum	Between petrous portion of temporal and sphenoid	Branch of ascending pharyngeal artery, and internal carotid artery
Lesser palatine	Posterior to greater palatine foramen in hard palate	Lesser palatine nerves
Magnum	Occipital bone	Union of medulla oblongata and spinal cord, meningeal membranes, accessory nerves; vertebral and spinal arteries
Mandibular	Medial surface of ramus of mandible	Inferior alveolar nerve and vessels
Mental	Below the first molar on the lateral side of mandible	Mental nerve and vessels
Nasolacrimal canal	Lacrimal bone	Nasolacrimal (tear) duct
Olfactory	Cribriform plate of the ethmoid	Olfactory nerves
Optic	Back of orbit in lesser wing of sphenoid	Optic nerve and ophthalmic artery
Ovale	Greater wing of sphenoid	Mandibular branch of trigeminal nerve
Rotundum	Within body of sphenoid	Maxillary branch of trigeminal nerve
Spinosum	Posterior angle of sphenoid	Middle meningeal vessels
Stylomastoid	Between styloid and mastoid processes of temporal	Facial nerve and stylomastoid artery
Superior orbital fissure	Between greater and lesser wings of sphenoid	Four cranial nerves (oculomotor, trochlear, ophthalmic branch of trigeminal and abducens)
Supraorbital	Supraorbital ridge of orbit	Supraorbital nerve and artery
Zygomaticofacial	Anteriolateral surface of zygomatic bone	Zygomaticofacial nerve and vessels

persists between these two portions beyond age six and is referred to as a **metopic** *(me-top'ik)* **suture.** The **supraorbital margin** is a prominent bony ridge over the orbit. Openings along this ridge, called supraorbital foramina, allow passage of small nerves and vessels.

The frontal bone contains a **frontal sinus,** which is connected to the nasal cavity (fig 6.20). This sinus, along with the other paranasal sinuses, lessens the weight of the skull and acts as a sound chamber for voice resonance.

Parietal The **coronal suture** separates the frontal bone from the parietals, and the **sagittal suture** along the dorsal midline separates the right and left parietals. The parietal bones form the upper sides and roof of the cranium. The inner concave surface of the parietal bone, as well as the inner concave surfaces of other cranial bones, is marked by shallow impressions from convolutions of the brain and vessels serving the brain.

Temporal The two temporal bones form the lower sides of the cranium (figs. 6.13, 6.14, 6.21). Each temporal is joined to its adjacent parietal bone by the **squamosal** *(skwa-mo'sal)* **suture.** Structurally, each temporal has four parts:

1. **Squamous portion.** The squamous portion is the flattened plate of bone at the sides of the skull. Projecting forward is a **zygomatic process** that forms the posterior portion of the **zygomatic arch.** On the inferior surface of the squamous portion is the **mandibular fossa,** which receives the articular condyle of the mandible. This articulation is referred to as the **temporomandibular** *(tem''po-ro-man-dib'u-lar)* **joint.**
2. **Tympanic portion.** The tympanic portion of the temporal bone contains the **external auditory meatus,** or ear canal, located immediately posterior to the mandibular fossa. A thin, pointed **styloid process** (figs. 6.13, 6.21) projects downward from the tympanic portion.
3. **Mastoid portion.** The **mastoid process,** a rounded projection posterior to the external auditory meatus, accounts for the mass of the mastoid portion. The **mastoid foramen** (fig. 6.14) is directly posterior to the mastoid process. The **stylomastoid foramen,** located between the mastoid and styloid processes, is the passage for part of the facial nerve.

metopic suture: Gk. *metopon,* forehead; L. *sutura,* sew

zygomatic: Gk. *zygoma,* yolk
styloid: Gk. *stylos,* pillar
mastoid: Gk. *mastos,* breast

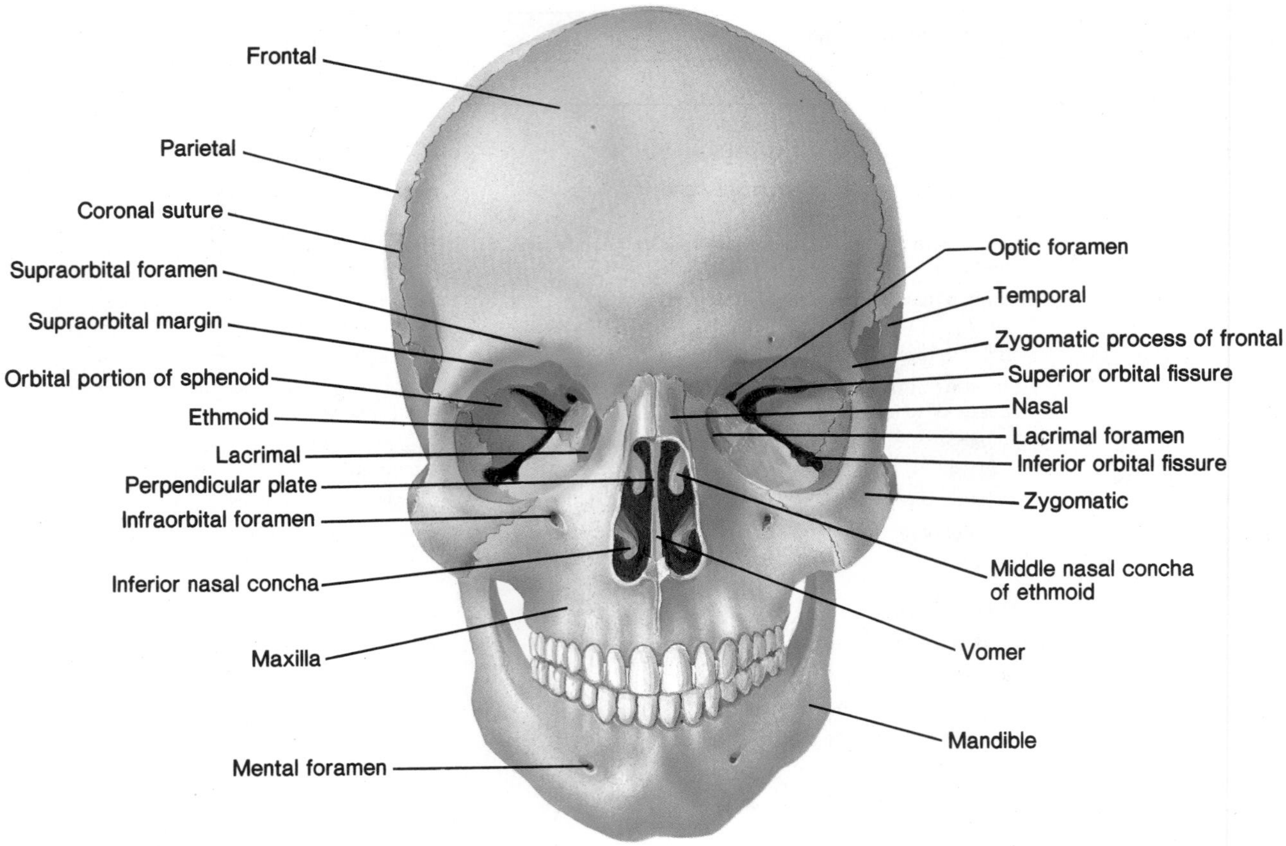

Figure 6.12 An anterior view of the skull.

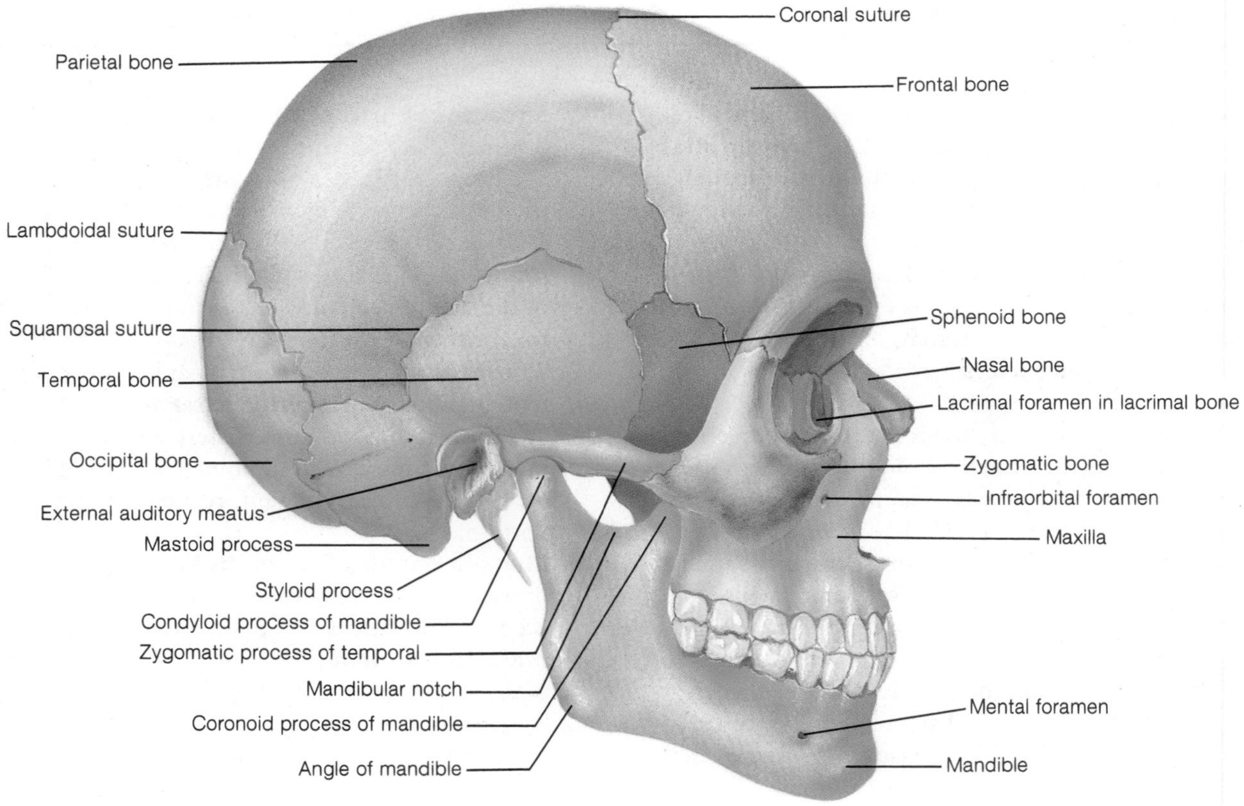

Figure 6.13 A lateral view of the skull.

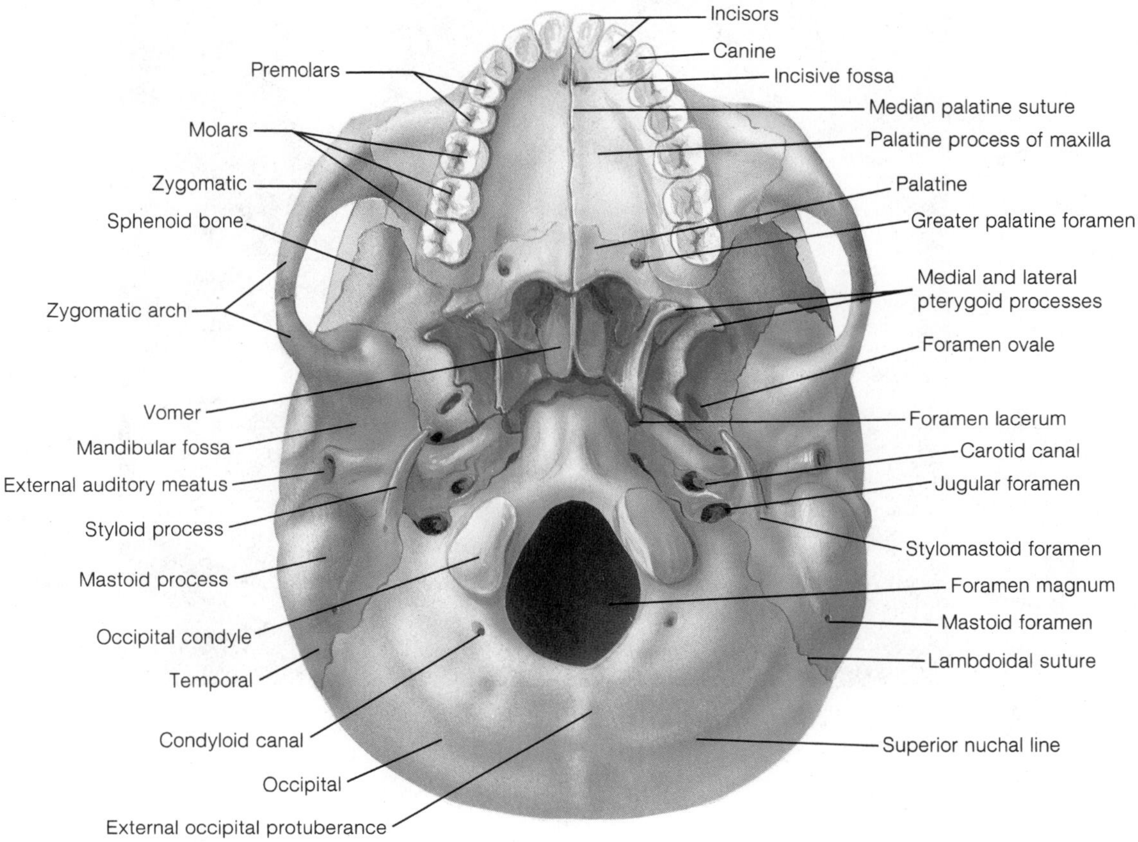

Figure 6.14 An inferior view of the skull.

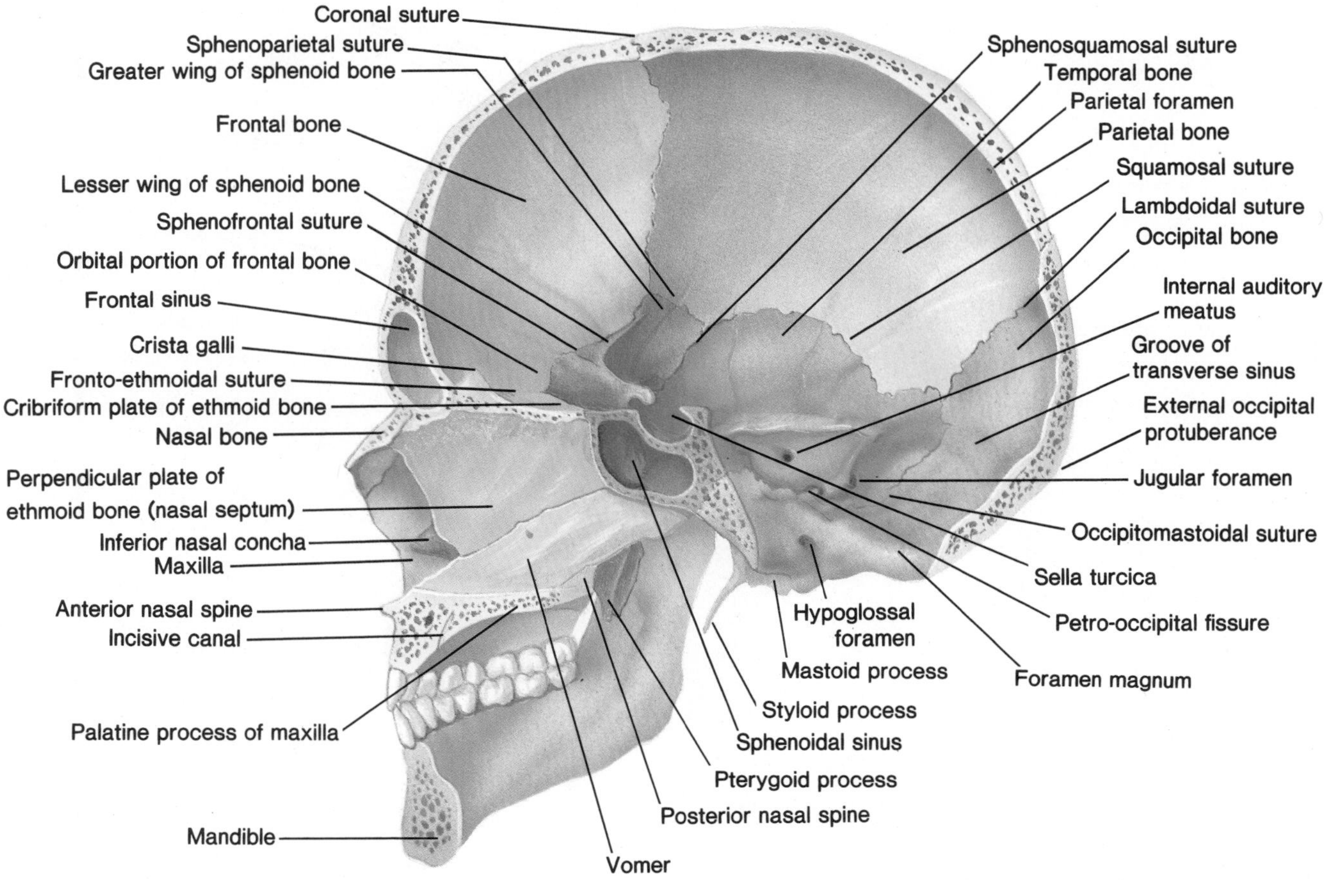

Figure 6.15 A midsagittal view of the skull.

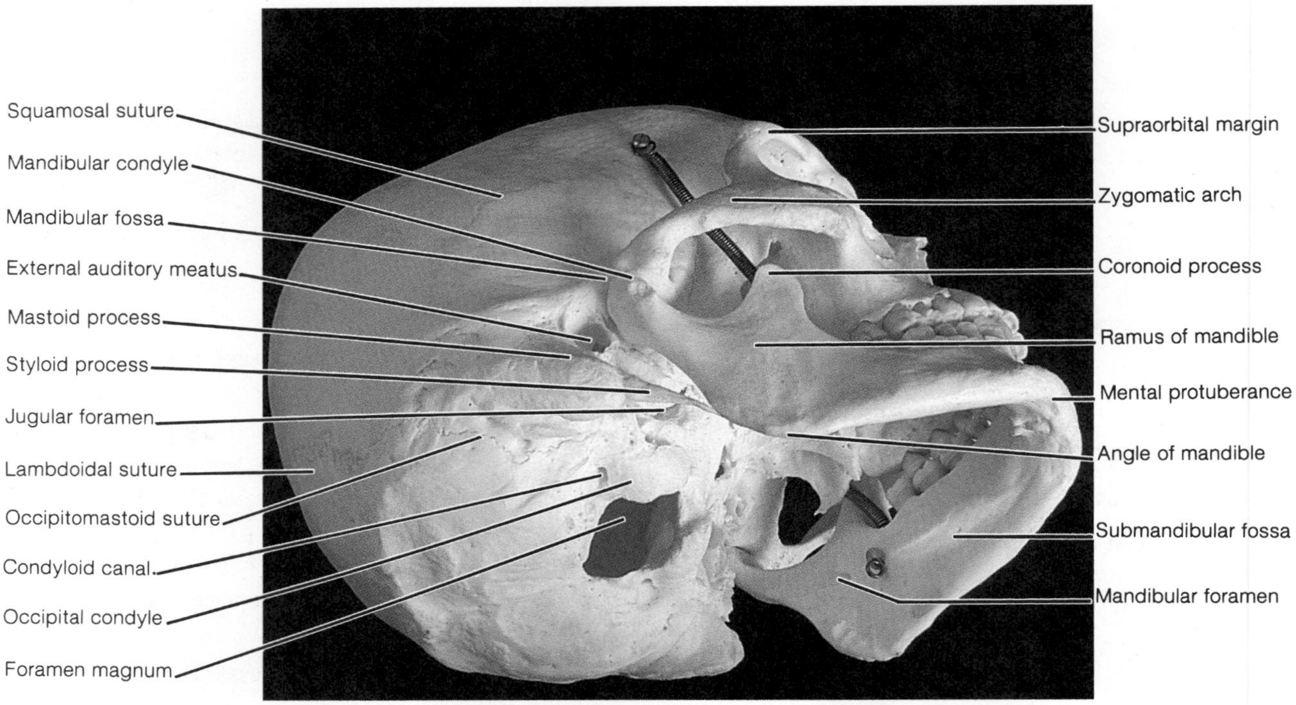

Figure 6.16 An inferolateral view of the skull.

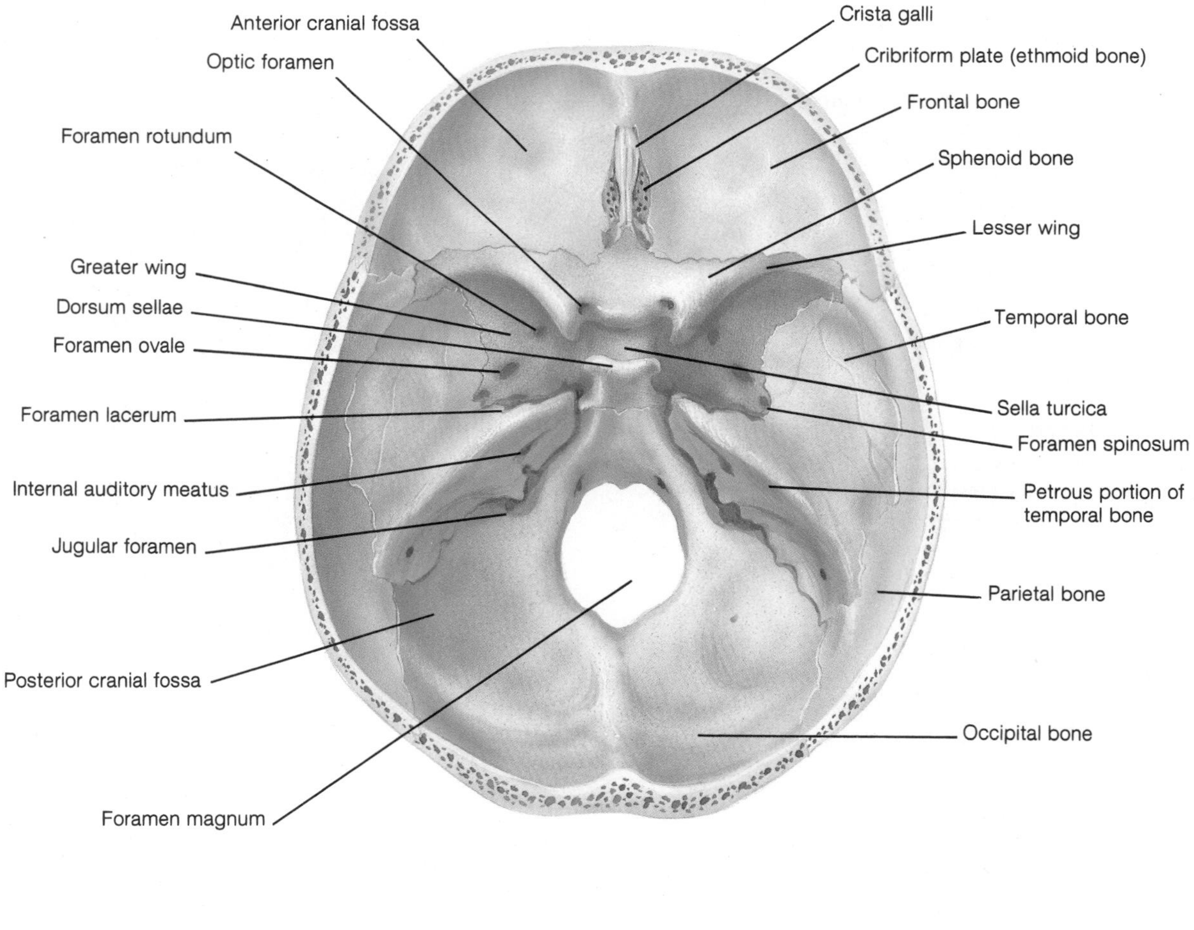

Figure 6.17 The floor of the cranial cavity.

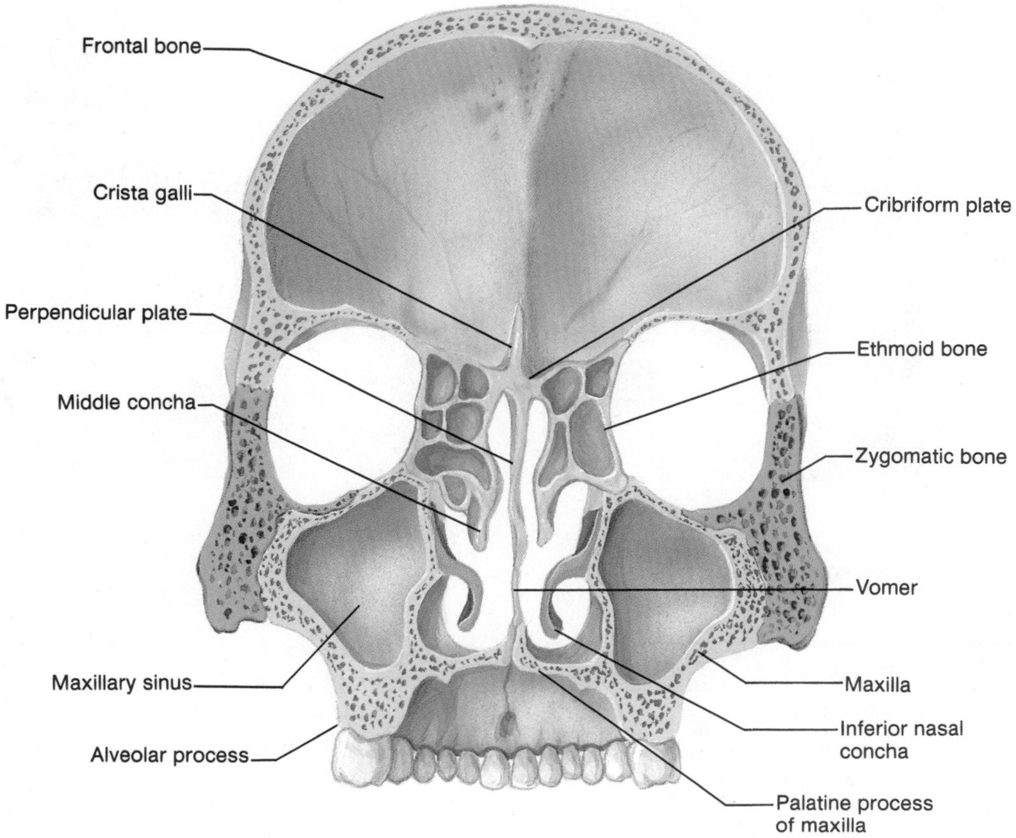

Figure 6.18 A posterior view of a frontal (coronal) section of the skull.

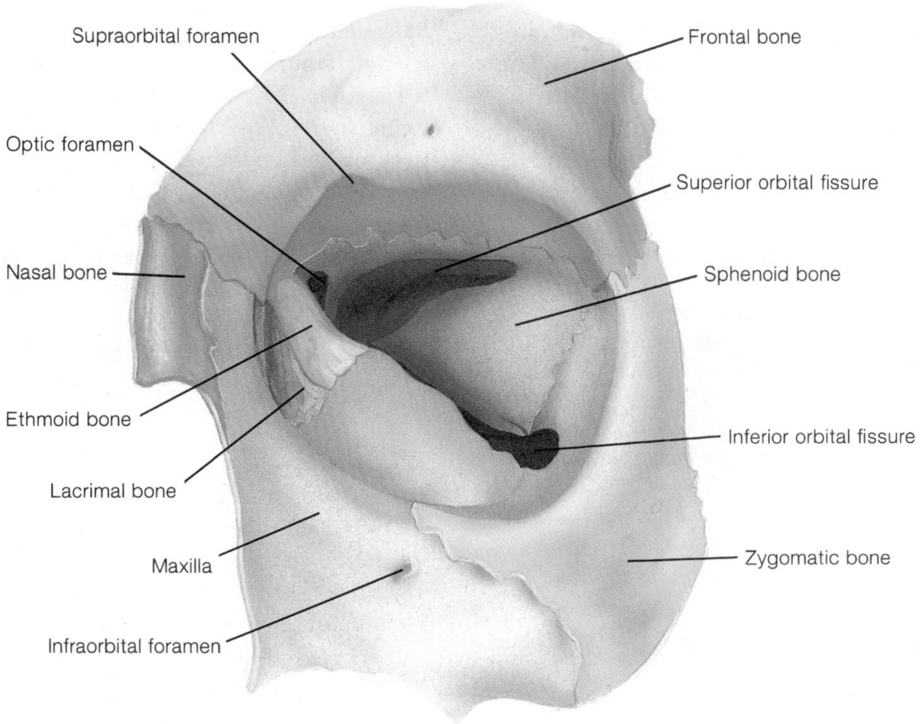

Figure 6.19 Bones of the orbit of the eye.

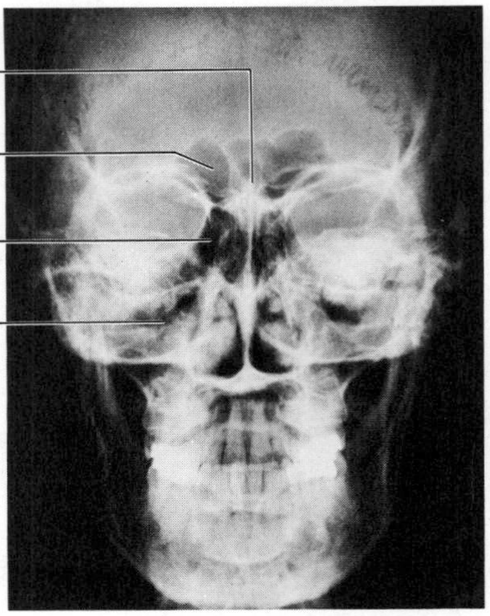

(a)

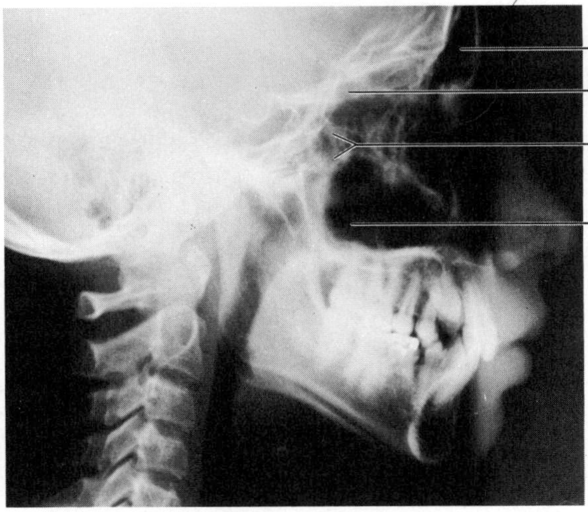

(b)

Figure 6.20 X rays of the skull showing paranasal sinuses. (*a*) An anteroposterior view and (*b*) a right lateral view.

Table 6.4 Bones forming the orbit

Region of the orbit	Contributing bones
Roof (superior)	Frontal; lesser wing of sphenoid
Floor (inferior)	Maxilla; zygomatic; palatine
Lateral wall	Zygomatic
Posterior wall	Greater wing of sphenoid
Medial wall	Maxilla; lacrimal; ethmoid
Superior rim	Frontal
Lateral rim	Zygomatic
Medial rim	Maxilla

4. **Petrous** *(pet'rus)* **portion.** The petrous portion is viewed inferiorly (figs. 6.14, 6.17). The structures of the middle and inner ear are housed in this dense, bony portion. The **carotid** *(kah-rot'id)* **canal** and the **jugular foramen** border on the medial side of the petrous portion. The carotid canal allows blood into the brain via the internal carotid artery, and the jugular foramen lets blood drain from the brain via the internal jugular vein. Three cranial nerves also pass through the jugular foramen.

The mastoid process of the temporal bone can be easily palpated as a bony knob immediately behind the earlobe. This process contains a number of small sinuses that are clinically important because they can become infected in *mastoiditis.* A tubular communication from the mastoid sinuses to the middle ear chamber may permit prolonged ear infections to spread to this region.

Occipital The occipital bone forms the back and much of the base of the skull. It is fastened to the parietal bones by the **lambdoidal suture.** The **foramen magnum** is the large hole in the occipital bone through which the spinal cord attaches to the brain. On each side of the foramen magnum are the **occipital condyles** (fig. 6.14), which articulate with the atlas of the vertebral column. At the anterolateral edge of the occipital condyle is the **hypoglossal foramen** (fig. 6.15), through which the hypoglossal nerve passes. A **condyloid** *(kon'dĭ-loid)* **canal** lies posterior to the occipital condyle (figs. 6.14, 6.16). The **external occipital protuberance** is a prominent posterior projection on the occipital bone that can be felt as a definite bump just under the skin. The **superior nuchal** *(nu'kal)* **line** is a ridge of bone extending laterally from the occipital protuberance to the mastoid portion of the temporal bone. **Sutural bones** are small clusters of irregularly shaped bones that may be found between the joints of certain cranial bones but generally occur along the lambdoidal suture.

Sphenoid The sphenoid *(sfe'noid)* bone forms the anterior base of the cranium and can be viewed laterally and inferiorly (figs. 6.13, 6.17). This bone resembles a butterfly with outstretched wings (fig. 6.22). It consists of a **body** with laterally projecting **greater** and **lesser wings,** which form part of the orbit. The body is a wedgelike central portion that contains the **sphenoidal** *(sfe-noi'dal)* **sinuses** and a prominent depression called the **sella turcica** *(sel'ah tur'si-kah),* which supports the pituitary gland. The sella turcica (meaning Turk's saddle) is seen on the floor of the cranium (fig. 6.17). A pair of **pterygoid** *(ter'ĭ-goid)* **processes** project inferiorly from the sphenoid bone to help form the lateral walls of the nasal cavity. Several foramina (figs. 6.14, 6.17, 6.22) are located within the sphenoid bone:

1. **Optic foramen.** A large opening through the lesser wing into the back of the orbit for passage of the optic nerve and the ophthalmic artery.
2. **Superior orbital fissure.** A triangular opening between the wings of the sphenoid for passage of four cranial nerves (oculomotor, trochlear, ophthalmic branch of the trigeminal, and abducens).

petrous: Gk. *petra,* rock
magnum: L. *magnum,* great

nuchal: Fr. *nuque,* nape of neck
sphenoid: Gk. *sphenoeides,* wedgelike

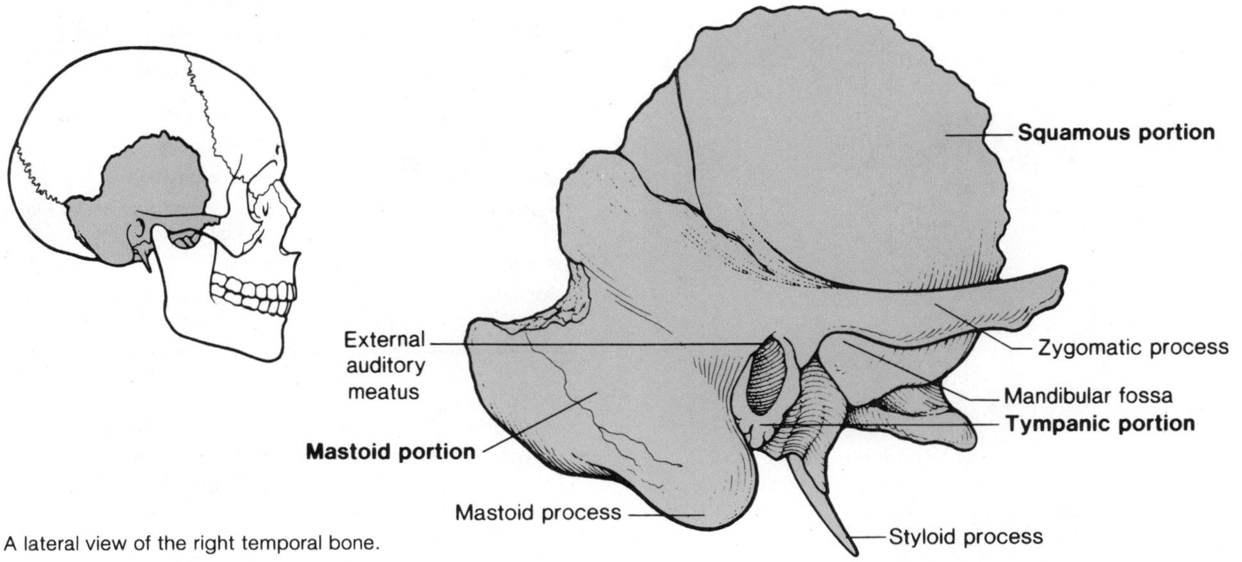

Figure 6.21 A lateral view of the right temporal bone.

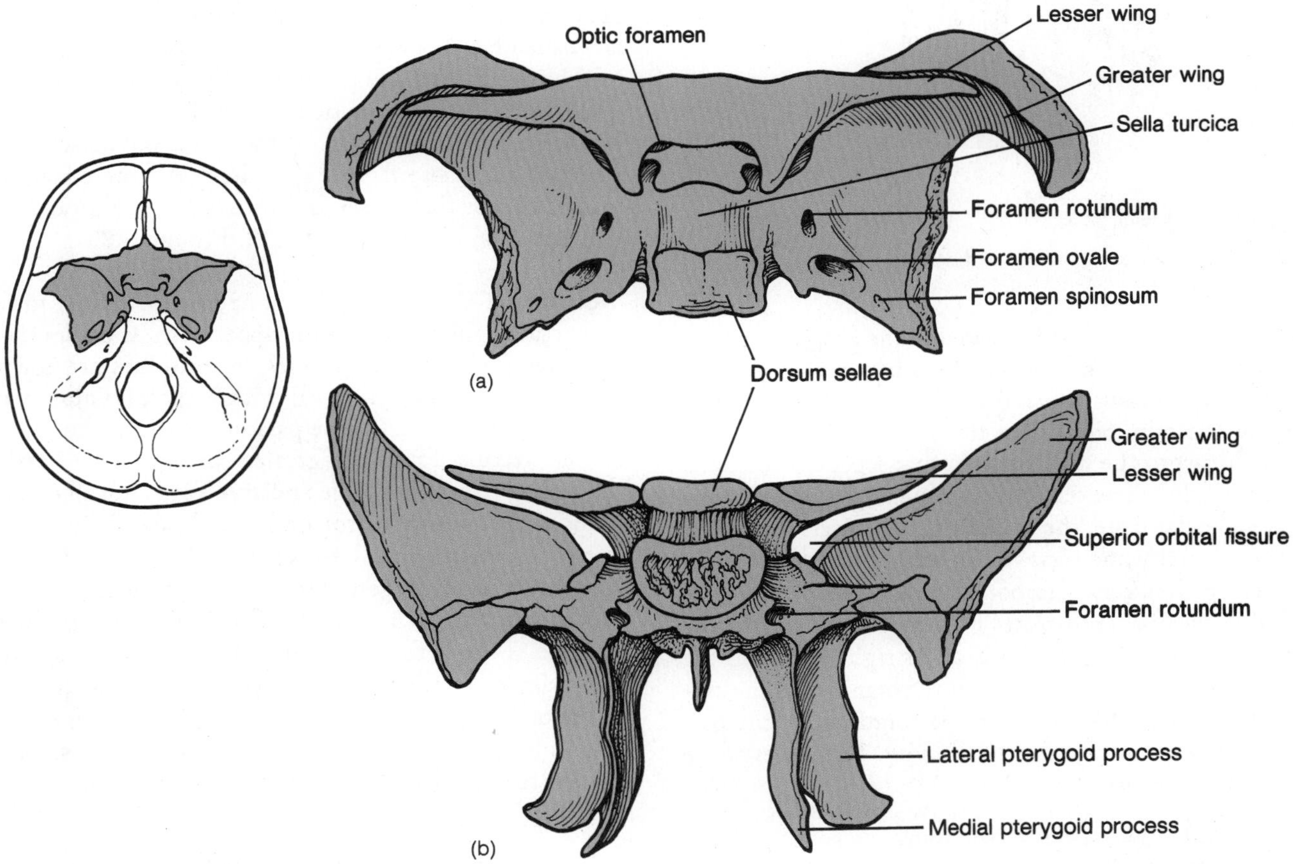

Figure 6.22 The sphenoid bone. (*a*) A superior view and (*b*) a posterior view.

3. **Foramen ovale.** An opening at the base of the pterygoid process, through which passes the mandibular branch of the trigeminal nerve.
4. **Foramen spinosum.** A small opening at the posterior angle of the sphenoid for passage of the middle meningeal vessels.
5. **Foramen lacerum** *(las'er-um)*. An opening between the sphenoid and the petrous portion of the temporal bone, through which pass the internal carotid artery and the meningeal branch of the ascending pharyngeal artery.
6. **Foramen rotundum.** An opening located just posterior to the superior orbital fissure at the junction of the anterior and medial portions of the sphenoid bone. The maxillary branch of the trigeminal nerve passes through this foramen.

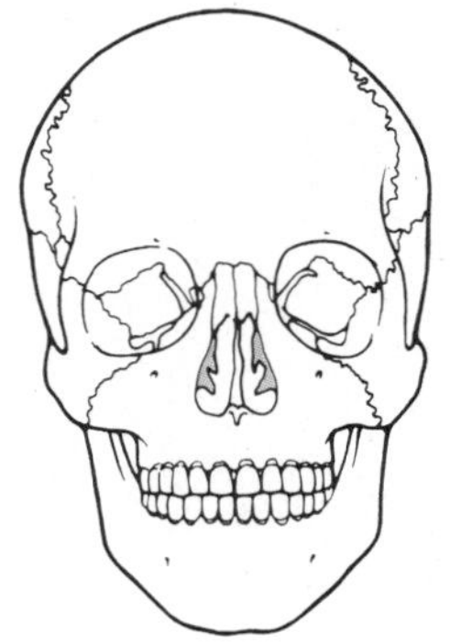

Table 6.5 Bones forming the nasal cavity	
Region of nasal cavity	**Contributing bones**
Roof (superior)	Ethmoid (cribriform plate); frontal
Floor (inferior)	Maxilla; palatine
Lateral wall	Maxilla; palatine
Nasal septum (medial)	Ethmoid (perpendicular plate); vomer; nasal
Bridge	Nasal
Conchae	Ethmoid (superior and middle conchae); inferior nasal concha

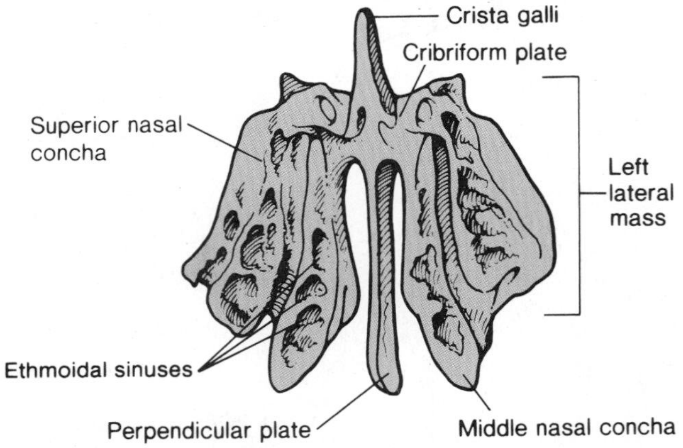

Figure 6.23 An anterior view of the ethmoid bone.

Ethmoid The ethmoid bone is located in the anterior portion of the floor of the cranium between the orbits, where it forms the roof of the nasal cavity (figs. 6.15, 6.17, 6.23). An inferior projection of the ethmoid, called the **perpendicular plate,** contributes in part to the nasal septum that separates the nasal cavity into two chambers referred to as **nasal fossae.** A spine of the perpendicular plate, the **crista galli,** projects superiorly into the cranial cavity and serves as an attachment for the meninges covering the brain. On both lateral walls of the nasal cavity are two scroll-shaped plates of the ethmoid, called the **superior** and **middle nasal conchae.** At right angles to the perpendicular plate, within the floor of the cranium, is the **cribriform plate,** which has numerous perforations for the passage of olfactory nerves from the nasal cavity. The bones of the nasal cavity are summarized in table 6.5.

The moist, warm vascular lining within the nasal cavity is susceptible to infections, particularly if a person is not in good health. Infections of the nasal cavity can spread to several surrounding areas. The paranasal sinuses connect to the nasal cavity and are especially prone to infection. The eyes may become reddened and swollen during a nasal infection because of the connection of the nasolacrimal duct, through which tears drain from the orbit to the nasal cavity. Organisms may spread via the auditory canal from the nasopharynx to the middle ear. With prolonged nasal infections, organisms may even ascend to the meninges covering the brain, along the sheaths of the olfactory nerves and through the cribriform plate to produce *meningitis.*

Facial Bones

The fourteen bones of the skull not in contact with the brain are called facial bones. These bones, together with certain cranial bones (frontal and portions of the ethmoid and temporals), provide the basic shape of the face. Facial bones also support the teeth and provide attachments for various muscles that move the jaw and cause facial expressions. All the facial bones are paired except the vomer and mandible. The articulated facial bones can be seen in figures 6.12 through 6.20.

Maxilla The two maxillae unite at the midline to form the upper jaw, which supports the upper teeth. **Incisors** *(in-si'zorz),* **canines** *(ka'nīnz),* **premolars,** and **molars** are contained in sockets, or **alveoli** *(al-ve-li),* within the **alveolar process** of the maxilla (fig. 6.24). The palatine process, a horizontal plate of the maxilla, forms the greater portion of the **hard palate,** or roof of the mouth. The **incisive foramen** (fig. 6.14) is located in the anterior region of the hard palate behind the incisor teeth. An **infraorbital foramen** is located under each orbit and serves as a passageway for the infraorbital nerve and artery to the nose (figs. 6.12, 6.13, 6.24). A final opening within the maxilla is the **inferior orbital fissure.** It is located between the maxilla and the greater wing of the sphenoid (fig. 6.12) and is the opening for the maxillary branch of the trigeminal nerve and infraorbital vessels. The large **maxillary sinus** located within the maxilla is one of the four paranasal sinuses (figs. 6.18, 6.20).

If the two palatine processes fail to join during early prenatal development (about twelve weeks), a *cleft palate* results. A cleft palate may be accompanied by a *cleft lip* (harelip) lateral to the midline. These conditions can be surgically and cosmetically treated with excellent success. A more immediate problem, however, is that a newborn with a cleft palate may have a difficult time swallowing while nursing because it is unable to create the necessary suction within the oral cavity.

ethmoid: Gk. *ethmos,* sieve
crista galli: L. *crista,* crest; *galli,* cock's comb
cribriform: L. *cribrum,* sieve; *forma,* like

incisor: L. *incidere,* to cut
canine: L. *canis,* dog
molar: L. *mola,* millstone
alveolus: L. *alveus,* little cavity

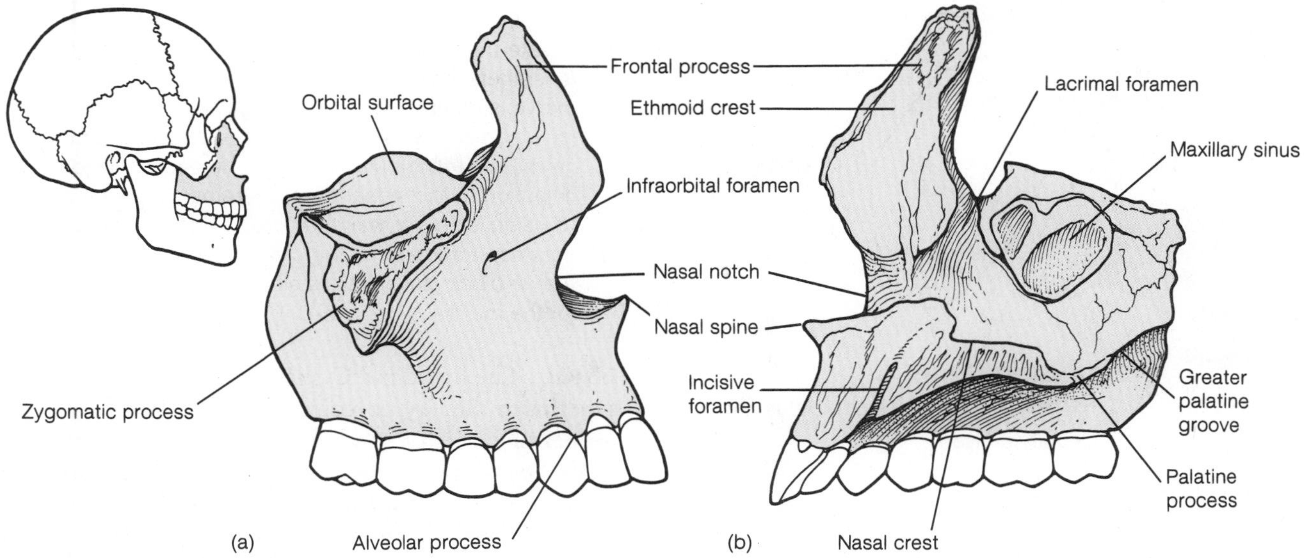

Figure 6.24 The right maxilla in (*a*) lateral view and (*b*) medial view.

Palatine The L-shaped palatine bones form the posterior third of the hard palate, a portion of the orbits, and a part of the nasal cavity. The **horizontal plates** of the palatines contribute to the formation of the hard palate (fig. 6.25). On the hard palate of each palatine bone is a large **greater palatine foramen,** which permits the passage of the greater palatine nerve and descending palatine vessels (fig. 6.14). Two or more smaller **lesser palatine foramina** are positioned posterior to the greater palatine foramina. Branches of the lesser palatine nerve pass through these openings.

Zygomatic (Malar) The two zygomatic bones form the cheekbones of the face. A posteriorly extending **temporal process** of this bone unites with the zygomatic process of the temporal bone to form the **zygomatic arch** (fig. 6.14). The zygomatic bone also forms the lateral margin of the orbit. A small **zygomatico-facial** (*zi''go-mat''i-ko-fa'shal*) **foramen** (not illustrated), located on the anterolateral surface of this bone, allows passage of the zygomatic nerves and vessels.

Lacrimal The small lacrimals *(lak'rĭ-mals)* are thin bones that form the anterior part of the medial wall of each orbit (figs. 6.13, 6.19). Each has a **lacrimal sulcus,** a groove that helps to form the **lacrimal foramen.** This opening permits the tears of the eye to drain into the nasal cavity (fig. 6.13).

Nasal The small, rectangular nasal bones join in the midline to form the bridge of the nose (fig. 6.26). The nasal bones support the flexible cartilaginous plates, which are a part of the framework of the nose. Common facial injuries include fractures of the nasal bones or fragmentation of the associated cartilage.

Inferior Nasal Concha The two inferior nasal conchae are fragile, scroll-like bones that project horizontally and medially from the lateral walls of the nasal cavity (fig. 6.12). They extend into the nasal cavity just below the superior and middle nasal

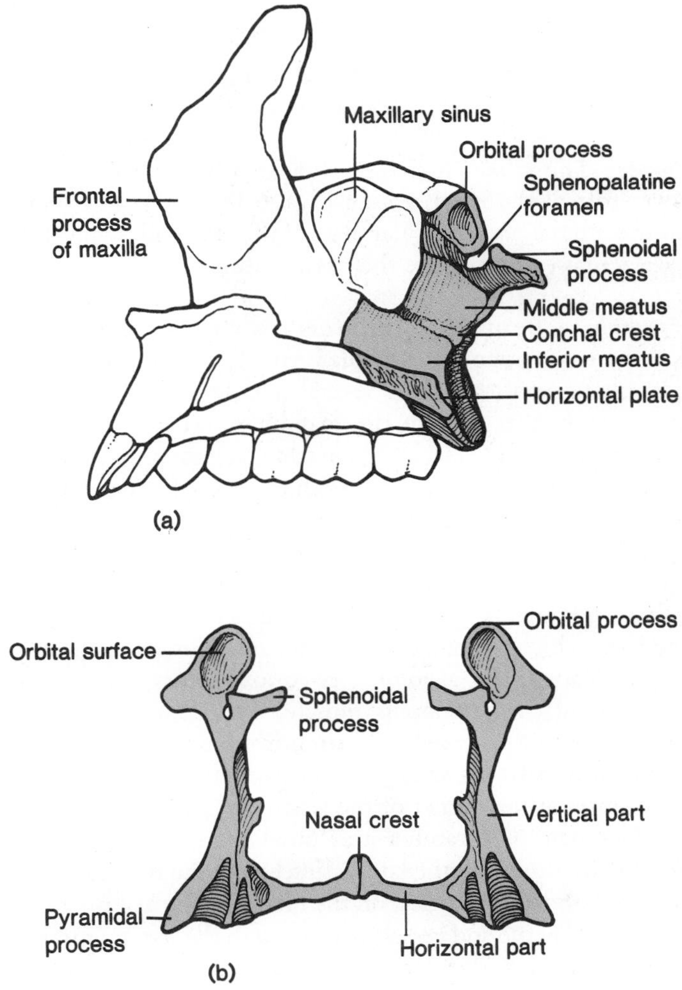

Figure 6.25 (*a*) The right palatine bone as it articulates anteriorly with the maxilla. (*b*) The two palatine bones viewed posteriorly. The two palatine bones form the posterior portion of the hard palate where they articulate at the nasal crest.

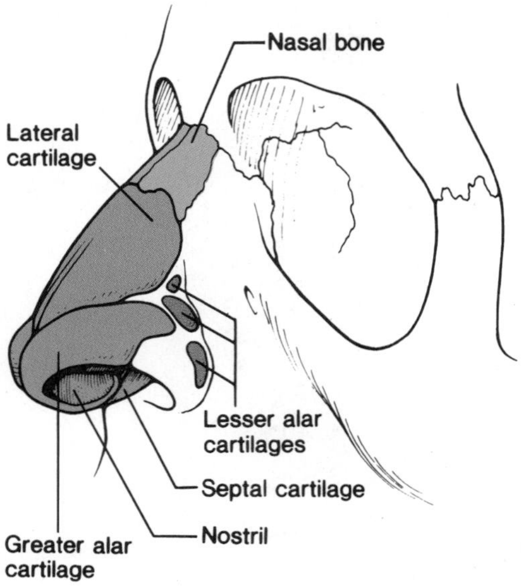

Figure 6.26 A lateral view of the bony and cartilaginous structure of the external nose.

conchae, which are part of the ethmoid bone (fig. 6.23). The inferior nasal conchae are the largest of the three conchae, and, like the other two, are covered with mucous membranes to warm, moisten, and cleanse inhaled air.

Vomer The vomer is a thin, elongated bone that forms the lower part of the nasal septum (figs. 6.14, 6.27). The vomer, along with the perpendicular plate of the ethmoid, supports the *septal cartilage* to complete the nasal septum.

Mandible The mandible, or lower jawbone, is attached to the skull by two temporomandibular articulations and is the only movable bone of the skull. Several muscles that close the jaw extend from the skull to the mandible and are discussed in chapter 9. The mandible of an adult supports sixteen teeth within alveoli, which occlude with those of the maxilla.

The horseshoe-shaped front and horizontal lateral sides of the mandible are referred to as the **body** (fig. 6.28). Extending vertically from the posterior portion of the body are two **rami** *(ra'mi)* (singular, *ramus*). Each ramus has a knoblike **condyloid process,** which articulates with the mandibular fossa of the temporal bone, and a pointed **coronoid process** for the attachment of the temporalis muscle. The depressed area between these two processes is the **mandibular notch.** The **angle of the mandible** is where the horizontal body and vertical ramus meet at the corner of the jaw.

Two sets of foramina are found on the mandible: the **mental foramen** on the lateral side below the first molar, and the **mandibular foramen** on the medial surface of the ramus. The mental nerve and vessels pass through the mental foramen, and the inferior alveolar nerve and vessels are transmitted through the mandibular foramen.

vomer: L. *vomer,* plowshare
mandible: L. *mandere,* to chew
ramus: L. *ramus,* branch
condyloid: L. *condylus,* knucklelike
coronoid: Gk. *korone,* like a crow's beak

Dentists use bony landmarks of the facial region to locate the nerves that traverse the foramina so that anesthetics can be injected. For example, the trigeminal nerve is composed of three large branches, the lower two of which convey sensations from the teeth, gums, and jaws. The mandibular teeth can be desensitized by an injection near the mandibular foramen, called a *third division,* or *lower, nerve block.* An injection near the foramen rotundum of the skull, called a *second division nerve block,* desensitizes all the upper teeth on one side of the maxilla.

Hyoid The hyoid is a U-shaped bone located in the neck just superior to the larynx (voice box). The hyoid is unique in that it does not attach directly to any other bone but is suspended from the styloid processes of the skull by the stylohyoid *(sti''lo-hi'oid)* muscles and ligaments. The hyoid has a **body,** two **lesser cornua** extending anteriorly, and two **greater cornua,** which project posteriorly to the stylohyoid ligaments (fig. 6.29). Several neck and tongue muscles attach to the hyoid bone. The hyoid may be palpated by placing a thumb and a finger on either side of the upper neck under the lateral portions of the mandible and firmly squeezing medially.

Ear Ossicles Three small, paired ear ossicles, the **malleus, incus,** and **stapes,** are located within the middle ear chambers in the petrous portion of the temporal bones (fig. 6.30). These bones transfer and amplify sound impulses through the middle ear.

---

1. Indicate which facial and cranial bones of the skull are paired and which are unpaired. Also, list at least two structural features of each of the bones of the skull.
2. Describe the location of each of the bones of the skull, and name the sutures that join them.
3. Describe the location and give the function of the sella turcica, foramen magnum, petrous bone, crista galli, and nasal conchae.
4. Which facial bones support the teeth?

Vertebral Column

The supporting vertebral column consists of vertebrae separated by fibrocartilaginous intervertebral discs, which lend flexibility and absorb the stress of movement. Vertebrae enclose and protect the spinal cord, support and permit movement of the skull, articulate with the rib cage, and provide for the attachment of trunk muscles.

Objective 11. Identify the bones of the five regions of the vertebral column, and describe the characteristic curves of each region.

Objective 12. Describe the structure of a typical vertebra.

cornu: L. *cornu,* horn
malleus: L. *malleus,* hammer
incus: L. *incus,* anvil
stapes: L. *stapes,* stirrup

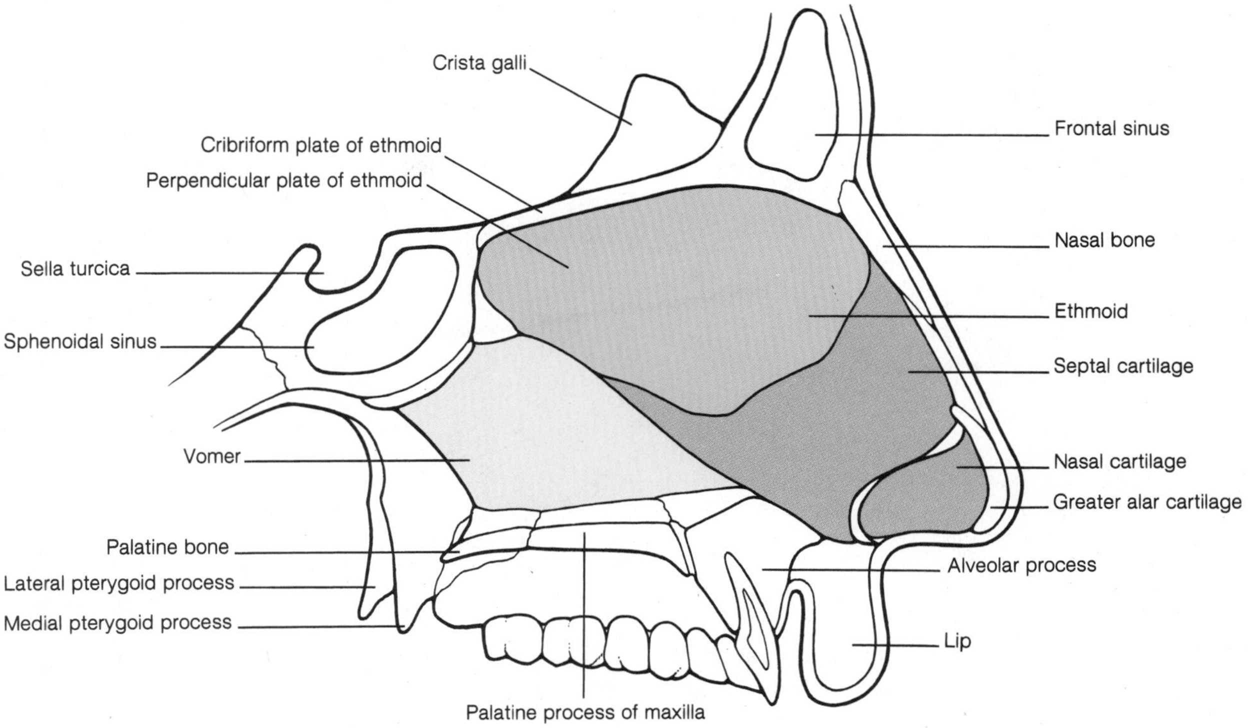

Figure 6.27 The bones and cartilages that comprise the nasal septum.

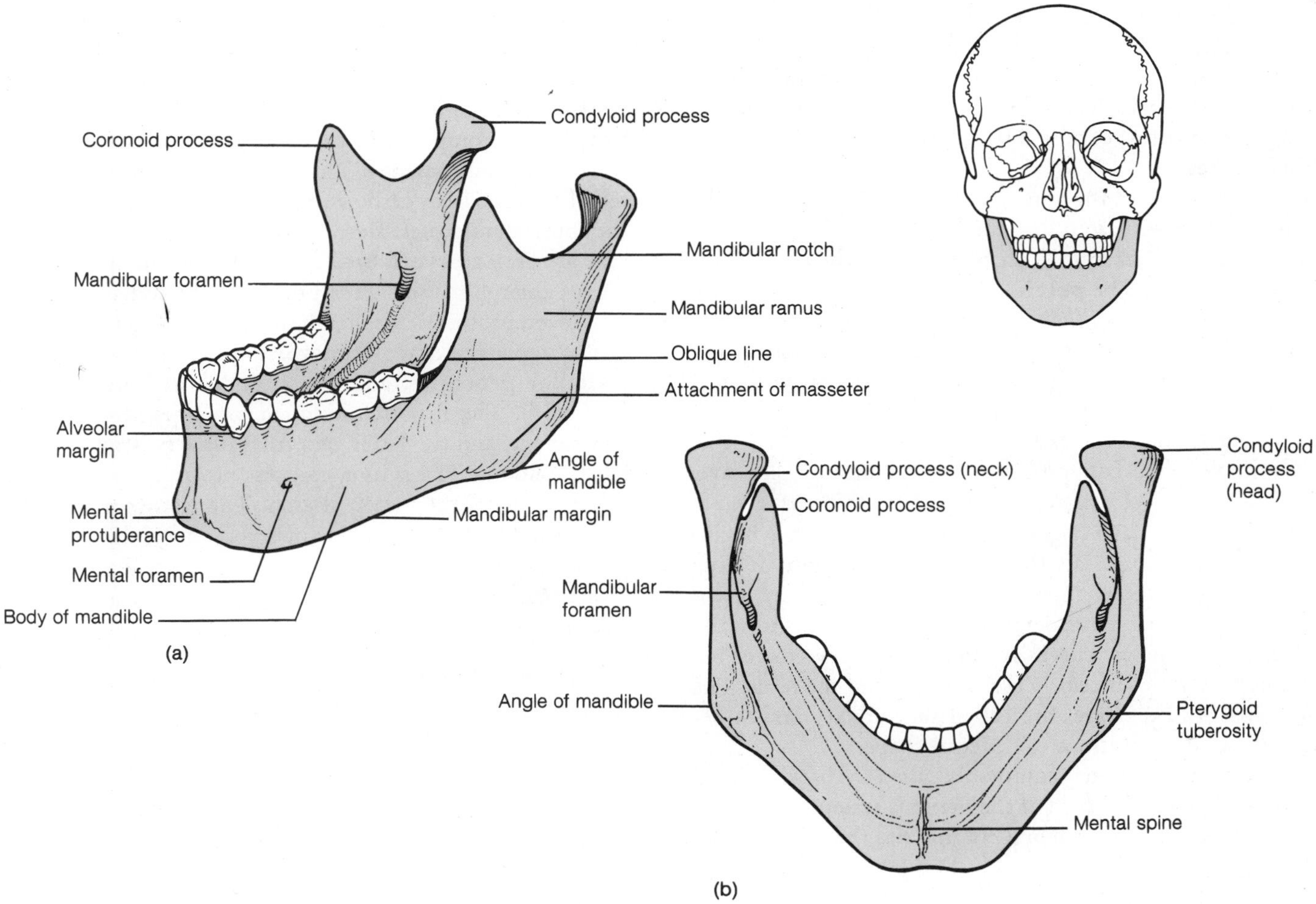

Figure 6.28 The mandible. (*a*) Lateral view and (*b*) posterior view.

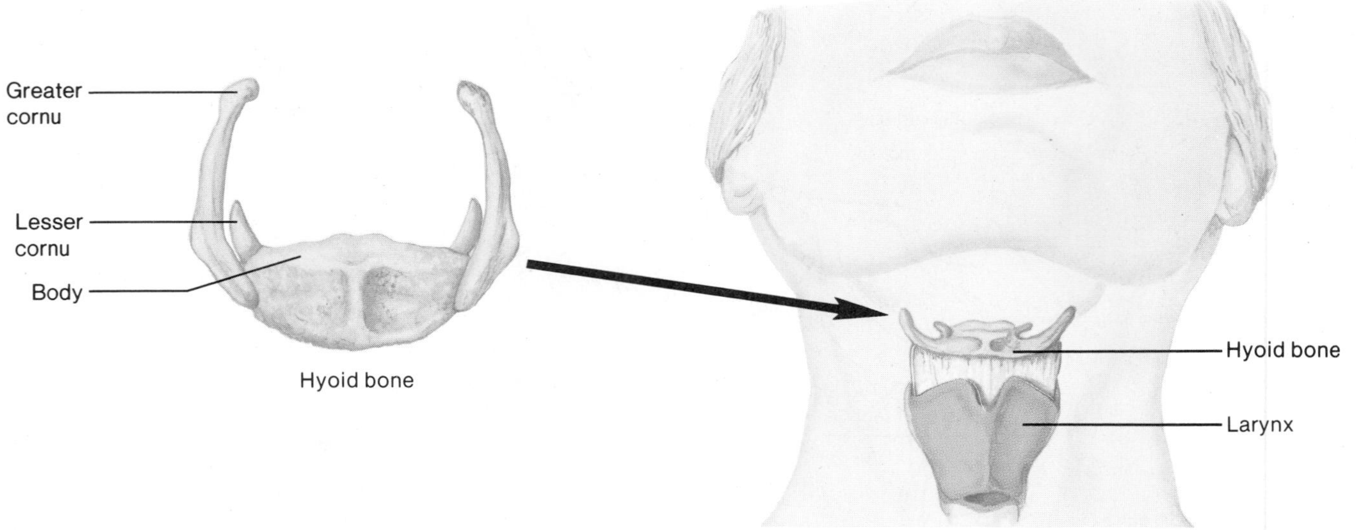

Figure 6.29 An anterior view of the hyoid bone.

The vertebral column is composed of thirty-three individual vertebrae. There are seven **cervical,** twelve **thoracic** *(tho-ras'ik),* five **lumbar,** four or five fused **sacral,** and four or five fused **coccygeal** *(kok-sij'e-al)* **vertebrae;** thus, the vertebral column is composed of a total of twenty-six movable parts. Vertebrae are separated by fibrocartilaginous intervertebral discs and are secured to one another by interlocking processes and binding ligaments. This structural arrangement provides limited movements between vertebrae but extensive movements for the entire vertebral column. Between the vertebrae are openings called **intervertebral foramina** that permit passage of spinal nerves.

Four curvatures of the vertebral column of an adult can be identified and viewed from the side (fig. 6.31). The **cervical, thoracic,** and **lumbar curves** are identified by the type of vertebrae they include. The **pelvic curve** is formed by the shape of the sacrum and coccyx *(kok'siks).* The curves of the vertebral column play an important functional role in increasing the strength and maintaining the balance of the upper portion of the body and also make possible a bipedal stance.

The four vertebral curves are not present in a newborn baby. Instead, the vertebral column is somewhat anteriorly concave, and except for the cervical region, it remains this way even as an infant learns to crawl (fig. 6.32). The cervical curve begins to develop at about three months of age as a baby begins holding up its head and becomes more pronounced as the baby learns to sit up. The lumbar curve develops as a child begins to walk. The thoracic and pelvic curves are called **primary curves** because they retain the anteriorly concave shape of the fetus. The cervical and lumbar curves are called **secondary curves** because they are modifications of the fetal shape.

The vertebral column is commonly called the backbone and together with the spinal cord of the nervous system constitutes the spinal column. The vertebral column has three basic functions:

1. To support the head and upper extremities while permitting freedom of movement
2. To provide attachment for various muscles, ribs, and visceral structures
3. To protect the spinal cord and permit passage of the spinal nerves

General Structure of Vertebrae

Vertebrae show similarities in their general structure from one region to another. A typical vertebra is illustrated in figure 6.33. A vertebra is usually composed of an anterior drum-shaped **body (centrum)** adapted to withstand compression. The body is in contact with **intervertebral discs** on each end. The **vertebral,** or **neural, arch** is affixed to the posterior surface of the body and is composed of two supporting **pedicles** *(ped'ĭ-k'l)* and two arched **laminae.** The hollow space formed by the vertebral arch and body is the **vertebral foramen,** through which the spinal cord passes. Between the pedicles of adjacent vertebrae are the **intervertebral foramina,** through which spinal nerves emerge as they branch off from the spinal cord.

Seven processes arise from the vertebral arch: the **spinous** *(spi'nus)* **process,** two **transverse processes,** two **superior articular processes,** and two **inferior articular processes** (fig. 6.34). The first two kinds of processes serve for muscle attachment, and the latter two pairs limit twisting of the vertebral column. The spinous process protrudes posteriorly and inferiorly from the vertebral arch. The transverse process laterally extends from each side of a vertebra at the point where the lamina and pedicle join. The superior articular processes of a vertebra have interlocking articulations with the inferior articular processes of the adjacent anterior vertebra.

A *laminectomy* is the surgical removal of the spinous processes and their supporting vertebral laminae in a particular region of the vertebral column. A laminectomy may be performed to relieve pressure on the spinal cord or nerve root caused by a blood clot, a tumor, or a herniated disc. It may also be performed on a cadaver to expose the spinal cord and its surrounding meninges.

pedicle: L. *pediculus,* small foot
lamina: L. *lamina,* thin layer

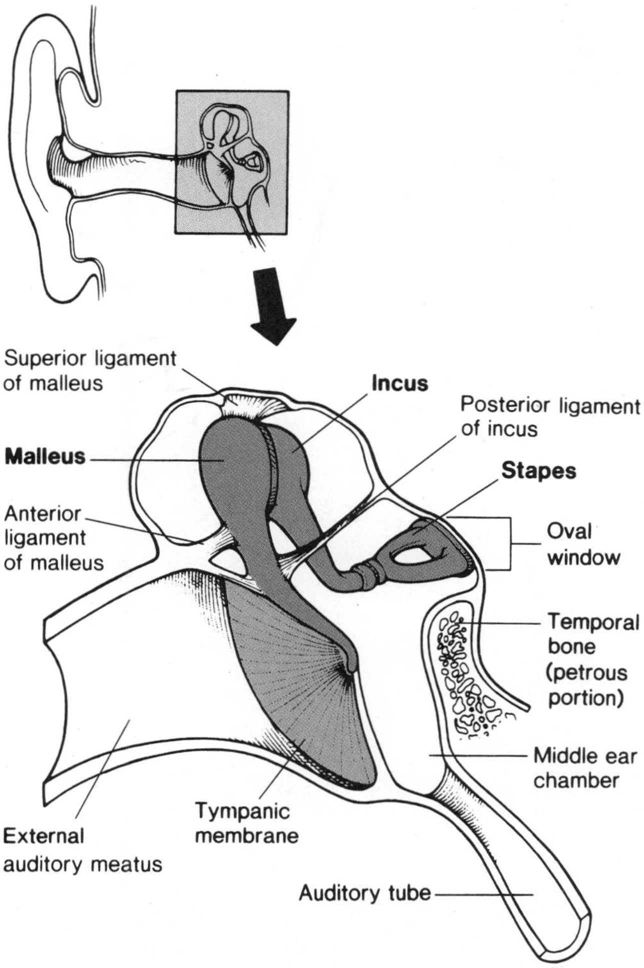

Figure 6.30 The three ear ossicles within the middle ear cavity.

Regional Characteristics of Vertebrae

Cervical Vertebrae The seven cervical vertebrae form the flexible framework of the neck region and support the head. The osseous tissue of cervical vertebrae is more dense than that of the other vertebral regions, and, except for the coccygeal region, the cervical vertebrae are smallest. Cervical vertebrae are distinguished by the presence of a **transverse foramen** in the transverse process (fig. 6.33). The vertebral vessels pass through this opening as they transfer blood to and from the brain. The spinous processes of the second through the sixth cervical vertebrae are **bifid** *(bi'fid)*, or notched, for the attachment of the strong **nuchal ligament,** which attaches to the back of the skull for added support.

The first cervical vertebra, the **atlas,** is adapted to articulate with the occipital condyles of the skull while supporting the head. The atlas has concave **superior articular surfaces** to articulate with the oval-shaped occipital condyles. This joint permits the nodding of the head in a yes movement. The atlas lacks a body. It has a short, rounded spinous process.

The second cervical vertebra is called the **axis** and is easily identified by the presence of a peglike projection called the

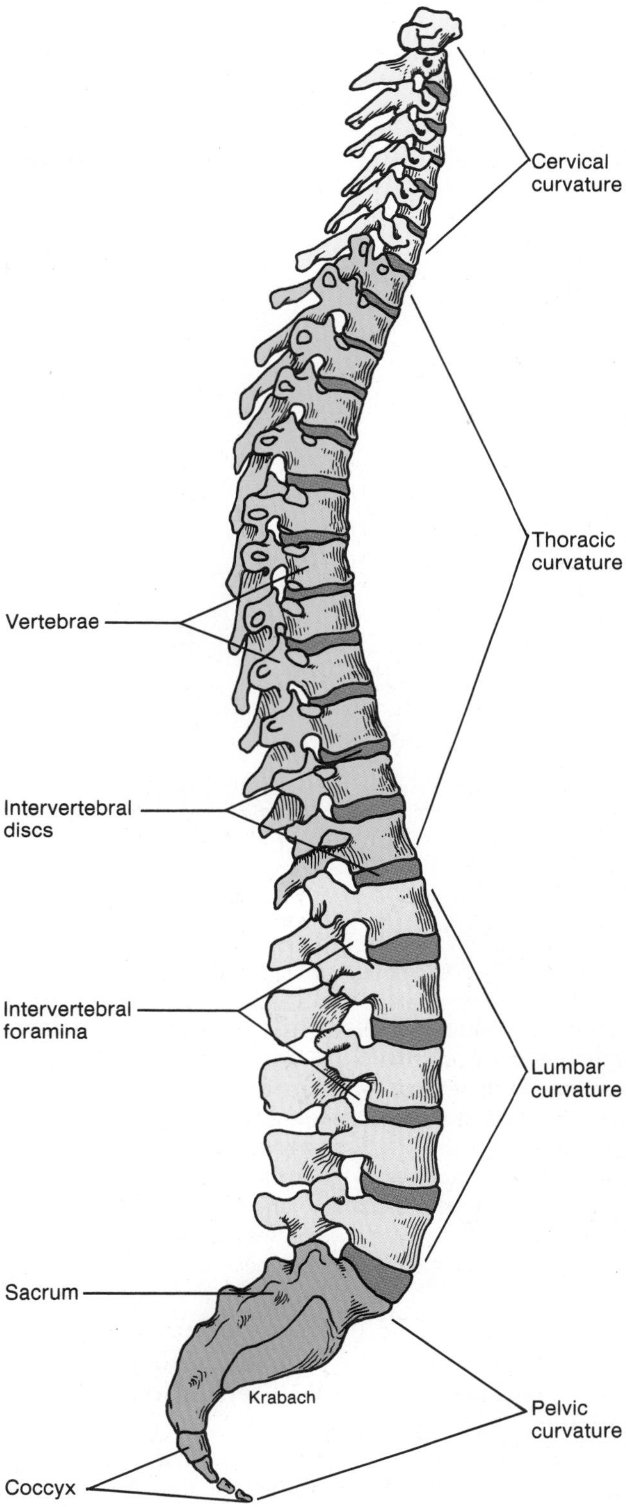

Figure 6.31 The vertebral column of an adult has four curves named according to the region in which they occur. The vertebrae are separated by intervertebral discs, which allow flexibility.

odontoid *(o-don'toid)* **process,** or **dens.** This process extends superiorly to provide a pivot for rotation with the atlas. This joint permits rotation, or the turning of the head to the side, as in a no movement.

atlas: from Gk. mythology, Atlas—the Titan who supported the heavens
axis: L. *axis,* axle

odontoid: Gk. *odontos,* tooth

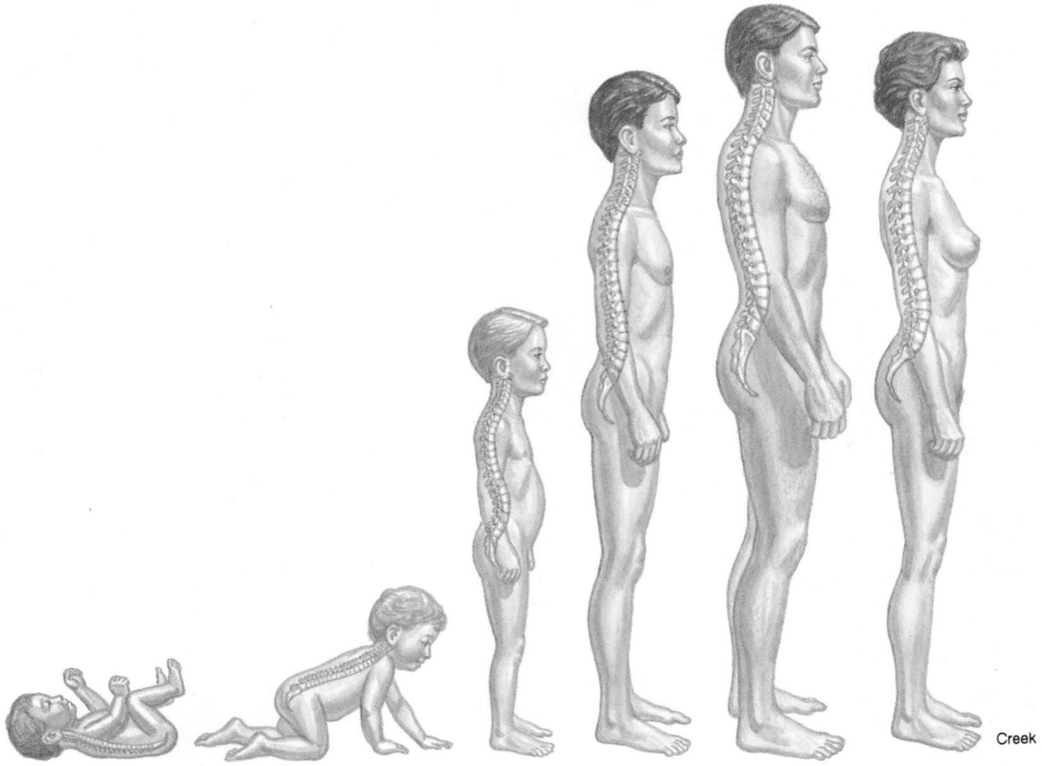

Figure 6.32 The development of the vertebral curvatures. An infant is born with the two primary curves and does not develop the secondary curves until it begins sitting upright and walking. (Note the differences in curvature between the sexes.)

Whiplash of the neck is a common, general term for injury to the cervical region. Muscle, bone, or ligament injury in this portion of the spinal column is relatively common in individuals involved in automobile wrecks and sports accidents. Joint dislocation without vertebral fracture occurs commonly between the fourth and fifth or fifth and sixth cervical vertebrae, where mobility is greatest. Bilateral dislocations are particularly dangerous because of the probability of spinal cord injury. Compression fractures of the bodies of the first three cervical vertebrae are common and follow abrupt forced flexion of the neck. Fractures of this type may be extremely painful because of pinched spinal nerves.

Thoracic Vertebrae The thoracic vertebrae serve as attachments of the ribs to form the posterior anchor of the rib cage. Thoracic vertebrae are larger than cervical vertebrae and increase in size from superior (T1) to inferior (T12). Each thoracic vertebra has a long spinous process that slopes obliquely downward and **facets** or **demifacets** for articulation with the ribs (fig. 6.34). The first thoracic vertebra has a superior facet and an inferior demifacet on both sides. The second through eighth have two demifacets on each side so that each head of a rib articulates with two vertebrae. The ninth vertebra has a single demifacet on each side, and the tenth through the twelfth have facets. The first ten vertebrae also have facets on their transverse processes for articulation with the tubercles of the ribs.

Lumbar Vertebrae The five lumbar vertebrae are easily identified by their heavy bodies and thick, blunt spinous processes (fig. 6.35) for attachment of powerful back muscles. They are the largest vertebrae of the vertebral column. Their articular processes are also distinctive in that the superior articular processes are directed medially instead of superiorly, and the inferior articular processes are directed laterally instead of inferiorly.

Sacrum The wedge-shaped sacrum (fig. 6.36) consists of four or five sacral vertebrae, which become fused after age twenty-six. The sacrum is functionally adapted to provide a strong foundation for the pelvic girdle. The sacrum has an extensive **auricular surface** on each side for the formation of an immovable **sacroiliac** *(sa''kro-il'e-ak)* **joint** with the os coxa. A **median sacral crest** is formed along the dorsal surface by the fusion of the spinous processes. **Dorsal sacral foramina** on both sides of the crest allow the passage of nerves from the spinal cord. The **sacral canal** is the tubular cavity within the sacrum that is continuous with the vertebral canal. Paired **superior articular processes,** which articulate with the fifth lumbar vertebra, arise from the roughened **sacral tuberosity** along the dorsal surface.

---

lumbar: L. *lumbus,* loin
sacrum: L. *sacris,* sacred

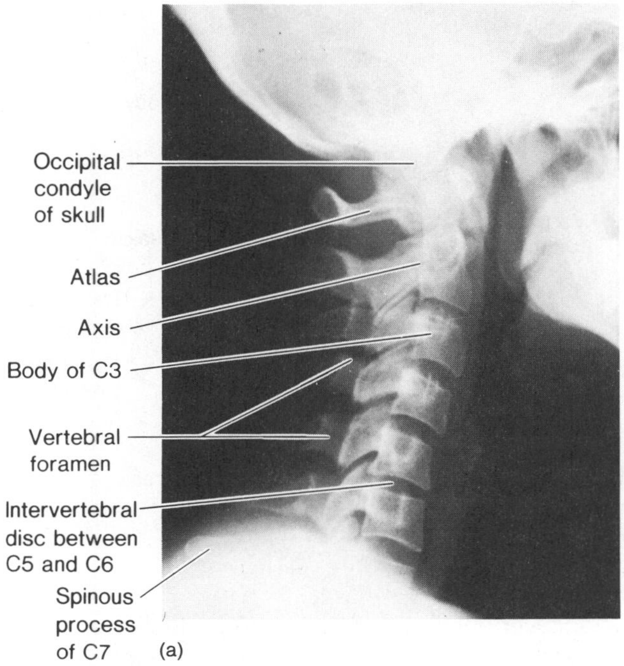

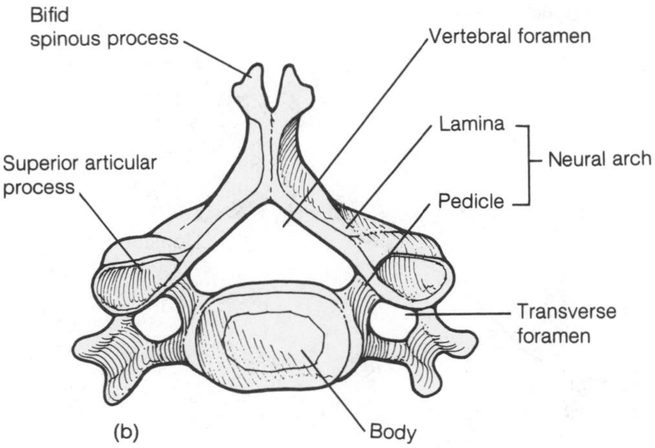

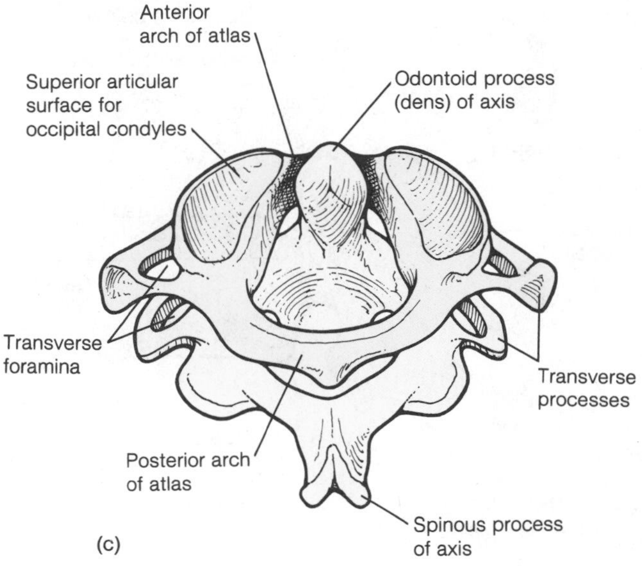

Figure 6.33 Cervical vertebrae. (*a*) An X ray of the cervical region. (*b*) A typical cervical vertebra. (*c*) The atlas and axis as they articulate.

The smooth anterior surface of the sacrum forms the posterior surface of the pelvic cavity. It has four **transverse lines** denoting the fusion of the vertebral bodies. On both sides of the transverse lines are the paired **pelvic foramina (anterior sacral foramina).** The superior border of the anterior surface of the sacrum, called the **sacral promontory** *(prom'on-to''re)*, is an important obstetrical landmark for pelvic measurements.

Coccyx The coccyx is the so-called tailbone. It is composed of four or five fused coccygeal vertebrae, which form a triangular-shaped structure. The first vertebra of the fused coccyx has two long **coccygeal cornua,** which are attached by ligaments to the sacrum (fig. 6.36). Lateral to the cornua are the transverse processes.

The regions of the vertebral column are summarized in table 6.6.

Table 6.6 Regions of the vertebral column

Region	Number of bones	Diagnostic features
Cervical	7	Transverse foramina; superior facets of atlas articulate with occipital condyle; odontoid process of axis; spinous processes of second through sixth vertebrae are bifid
Thoracic	12	Long spinous processes that slope obliquely inferiorly; facets and demifacets for articulation with ribs
Lumbar	5	Large bodies; prominent transverse processes; short, thick spinous processes
Sacrum	4 or 5 fused vertebrae	Extensive auricular surface; median sacral crest; dorsal sacral foramina; sacral promontory; sacral canal
Coccyx	4 or 5 fused vertebrae	Small, triangular; coccygeal cornua

coccyx: Gk. *kokkyx,* like a cuckoo's beak

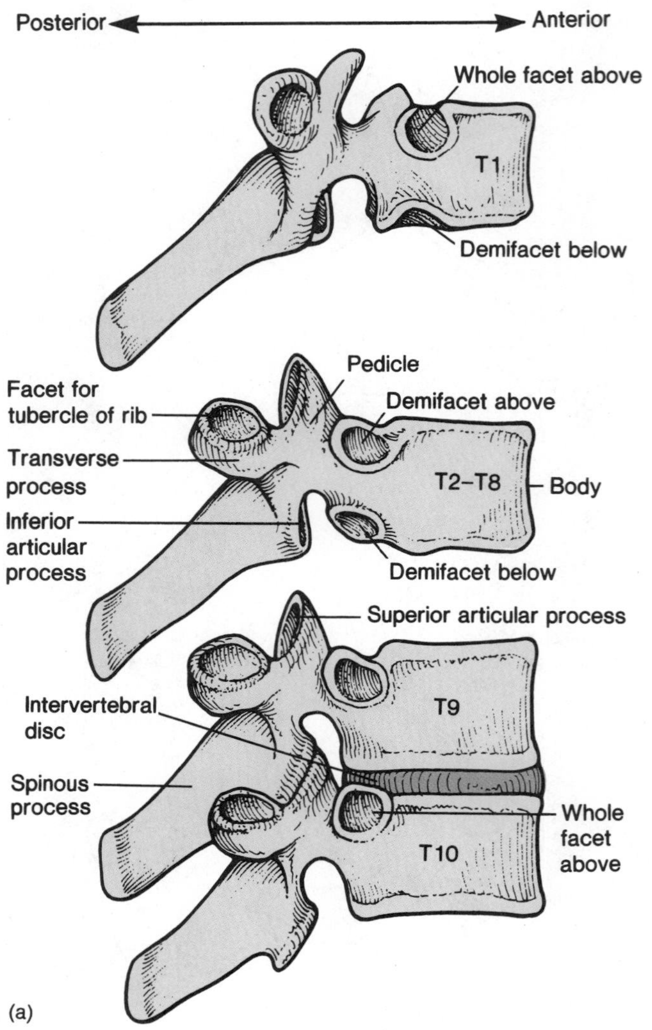

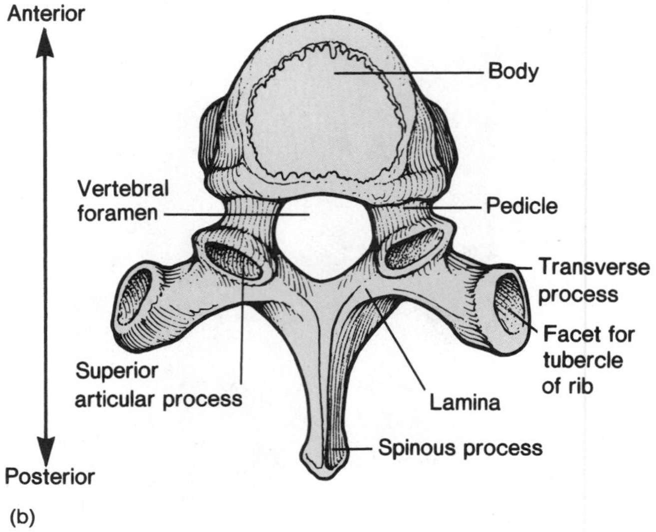

Figure 6.34 Thoracic vertebrae. (*a*) Representative vertebrae in a right lateral view and (*b*) a superior view.

When a person sits, the coccyx flexes anteriorly somewhat, acting as a shock absorber. An abrupt fall on the coccyx, however, may cause a painful subperiosteal bruising, fracture, or fracture-dislocation of the sacrococcygeal joint. An especially difficult childbirth can even injure the coccyx of the mother. Coccygeal trauma is painful and may require months to heal.

---

1. Which are the primary curves of the vertebral column, and which are the secondary curves? Describe the characteristic curves of each region.
2. What is the function of the transverse foramina of the cervical vertebrae?
3. Describe the diagnostic differences between a cervical, a thoracic, and a lumbar vertebra. Which structures are similar and could therefore be characteristic of a typical vertebra?

Rib Cage

The cone-shaped and flexible rib cage consists of the sternum, costal cartilages, and the twelve paired ribs attached to the thoracic vertebrae. It encloses and protects the thoracic viscera and is directly involved in the mechanics of breathing.

Objective 13. Identify the parts of the rib cage, and distinguish between the various types of ribs.

The **sternum, ribs, costal cartilages,** and the previously described thoracic vertebrae form the rib cage, or **thoracic cage,** of the thorax (fig. 6.37). The rib cage is anteroposteriorly compressed and more narrow superiorly than inferiorly. It supports the pectoral girdle and upper extremities, protects and supports the thoracic and upper abdominal viscera, and plays a major role in breathing. Certain bones of the rib cage contain active sites for blood cell production in the red bone marrow.

Sternum The sternum (breastbone) is an elongated, flattened bony plate consisting of three separate bones: the upper **manubrium,** the central **body,** and the lower **xiphoid** *(zi'foid)* **process.** On the lateral sides of the sternum are **costal notches** where the costal cartilages attach. A **sternal notch** is formed at the superior end of the manubrium, and a **clavicular** *(klah-vik'u-lar)* **notch** for articulation with the clavicle is present on both sides of the sternal notch. The manubrium articulates with the costal cartilages of the first and second ribs. The body of the sternum attaches to the costal cartilages of the second through the seventh ribs. The xiphoid process does not attach to ribs but is an attachment for abdominal muscles. The costal cartilages of the eighth, ninth, and tenth ribs fuse to form the **costal margin** (see fig. 11.15). A **costal angle** is formed where

---

sternum: Gk. *sternon,* chest
manubrium: L. *manubrium,* a handle
xiphoid: Gk. *xiphos,* sword
costal: L. *costa,* rib

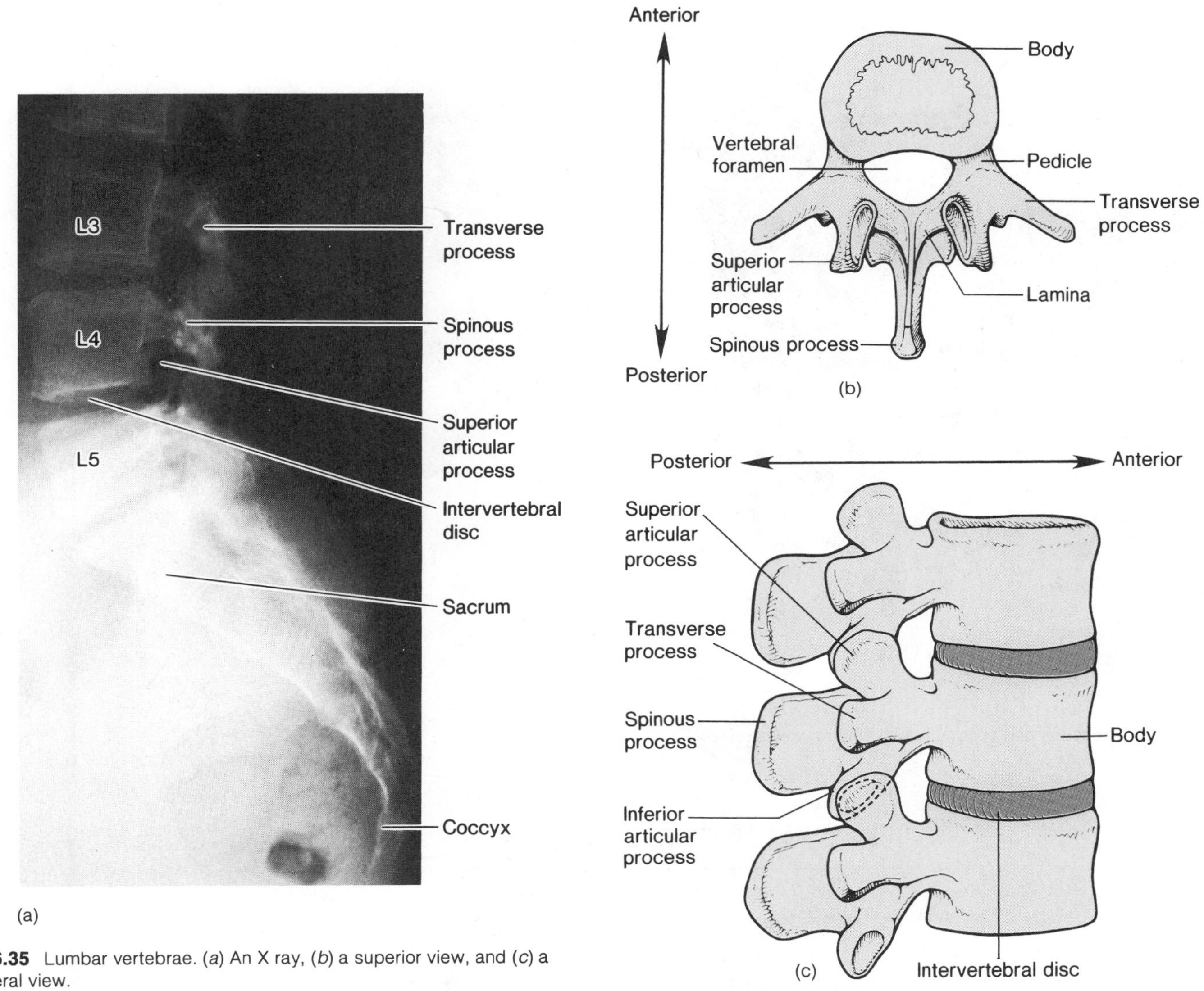

Figure 6.35 Lumbar vertebrae. (*a*) An X ray, (*b*) a superior view, and (*c*) a right lateral view.

the two costal margins come together at the xiphoid process. The **sternal angle** (angle of Louis) may be palpated as an elevation between the manubrium and body of the sternum at the level of the second rib (see fig. 6.37). The costal angle, costal margins, and sternal angle are important surface landmarks of the chest and abdomen.

Ribs There are twelve pairs of ribs, each pair being attached posteriorly to a thoracic vertebra. Anteriorly, the first seven pairs are anchored to the sternum by individual costal cartilages and are called **true ribs.** Ribs 8, 9, and 10 are attached to the costal cartilage of rib 7 and are termed **false ribs.** The remaining two paired ribs do not attach to the sternum and are referred to as **floating ribs.** They are attached to the muscles of the body wall.

Although the structure of ribs varies, each of the first ten pairs has a **head** and a **tubercle** for articulation with a vertebra. The last two have a head but no tubercle. In addition, each of the twelve pairs has a **neck, angle,** and **shaft** (fig. 6.38). The head of a rib projects posteriorly and articulates with the body of a thoracic vertebra (fig. 6.39). The tubercle is a knoblike process just lateral to the head. It articulates with the facet of the transverse process. The neck is the constricted area between the head and the tubercle. The **shaft,** or **body,** is the main, curved part of the rib. Along the inner surface of the shaft is a depressed canal called the **costal groove,** which protects the costal vessels and nerve. Spaces between the ribs are called **intercostal spaces** and are occupied by the intercostal muscles.

Fractures of the ribs are relatively common injuries, and most frequently occur between ribs 3 and 10. The first two pairs of ribs are protected by the clavicles, and the last two pairs move freely and will give with an impact. Little can be done to assist the healing of broken ribs other than binding them tightly to restrict movement.

---

1. Describe the rib cage, and list its functions. What determines if a rib is designated as true, false, or floating?
2. Define the costal margin and the costal angle.

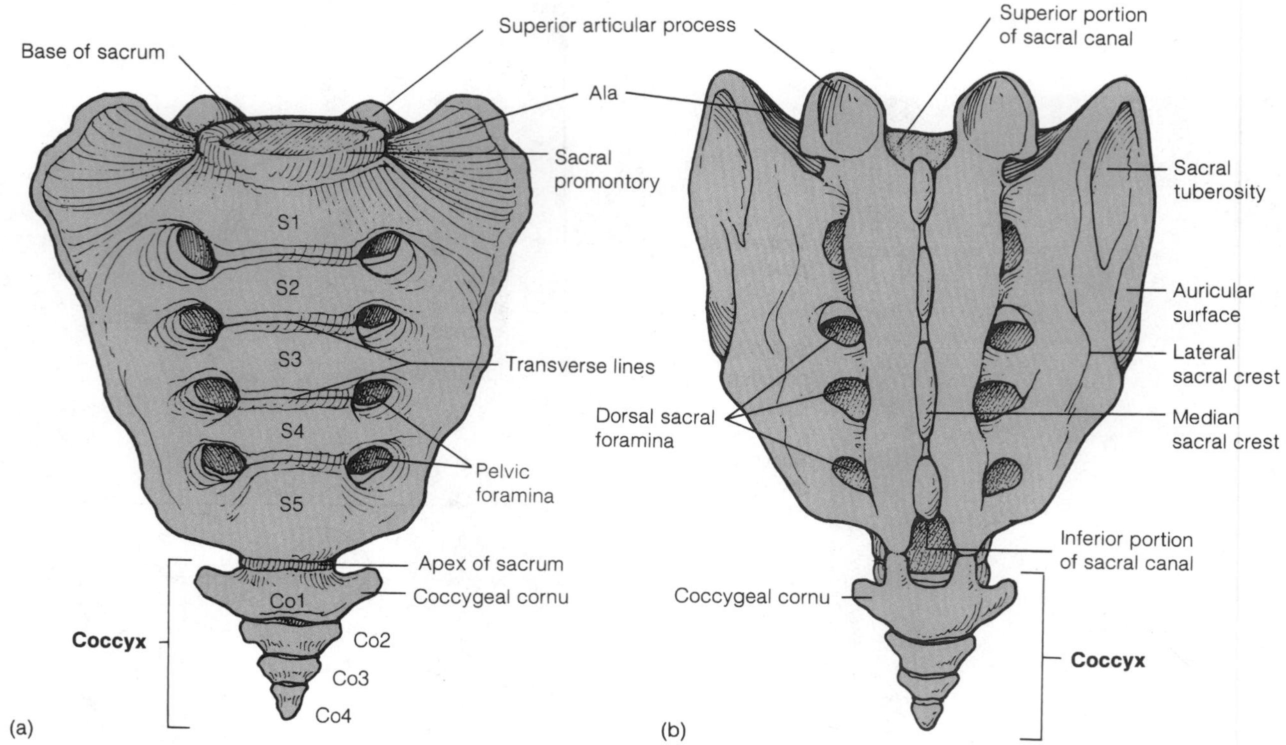

Figure 6.36 The sacrum and coccyx. (*a*) An anterior view and (*b*) a posterior view.

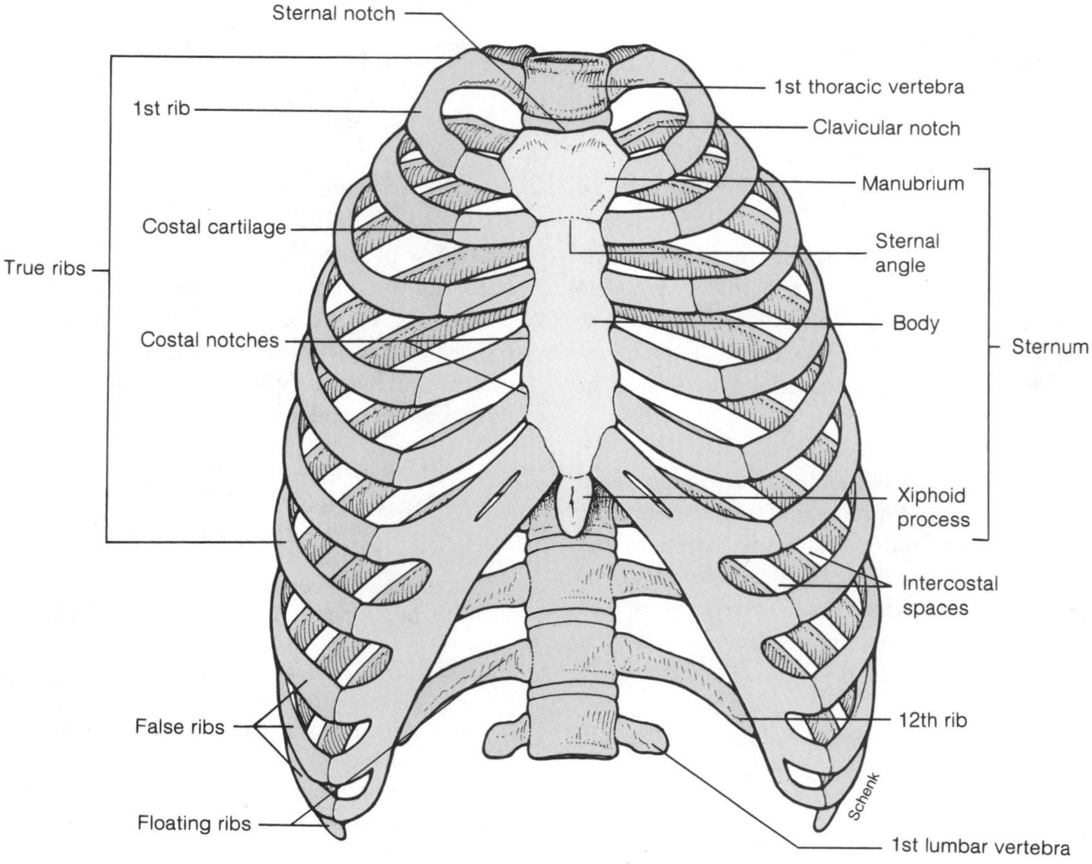

Figure 6.37 The rib cage.

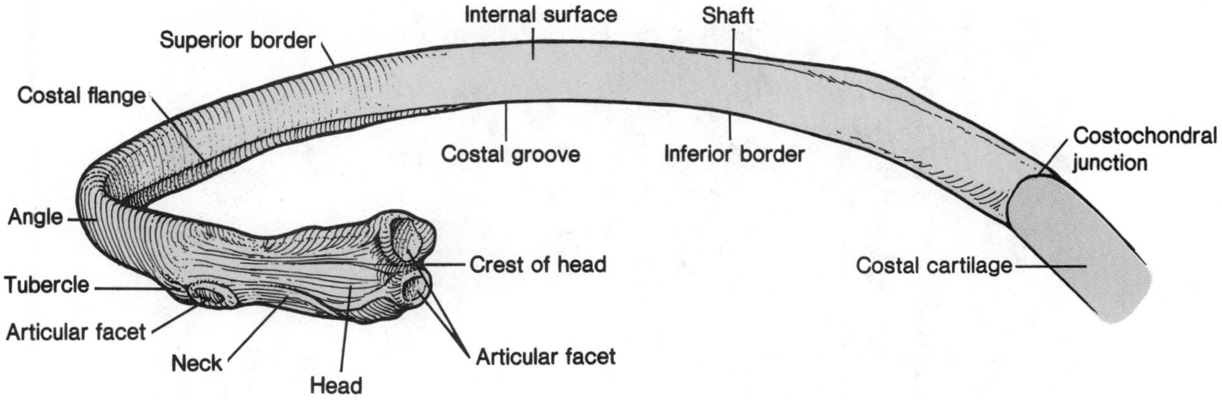

Figure 6.38 The structure of a rib.

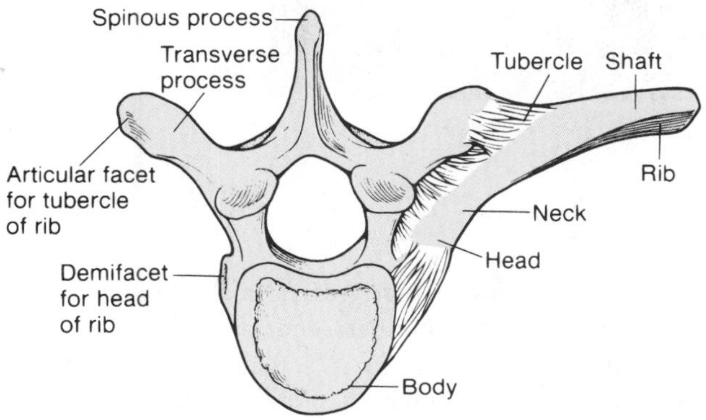

Figure 6.39 Articulation of a rib with a thoracic vertebra as seen in a superior view.

Clinical Considerations

Bone is a dynamic living tissue that is susceptible to hormonal or nutritional deficiency, diseases, and changes brought on by age. Since the development of bone is genetically governed, congenital conditions can occur. The hardness of bone gives it strength, yet it lacks the resiliency to avoid fracture if it undergoes excessive trauma. All of these aspects of bone provide some important and interesting clinical considerations.

Developmental Disorders

Congenital malformations account for several types of skeletal deformities. Certain bones may fail to form during osteogenesis, or they may form abnormally. **Cleft palate** and **cleft lip** are malformations of the palate and face. Cleft palates vary in severity and seem to have a mixed genetic and environmental cause. **Spina bifida** *(spi'nah bif'ĭ-dah)* is a congenital defect of the vertebral column resulting from a failure of the laminae of the vertebrae to fuse, exposing the spinal cord (fig. 6.40). The lumbar area is mainly affected, and frequently only a single vertebra is involved.

Nutritional and Hormonal Disorders

Several bone disorders result from nutritional deficiencies or from excessive or deficient amounts of the hormones that regulate bone development and growth. Vitamin D has a tremendous influence on proper bone structure and function. When there is a deficiency of this vitamin, the body is unable to metabolize calcium and phosphorus. Vitamin D deficiency in children causes **rickets.** The bones of a child with rickets remain soft and are deformed from the weight of the body (see fig. 5.15).

A deficiency in vitamin D in the adult causes the bones to demineralize, or to give up stored calcium and phosphorus. This condition is called **osteomalacia** *(os''te-o-mah-la'she-ah)*. Osteomalacia is prevalent in women who have repeated pregnancies and poor diets and are seldom exposed to the sun.

The consequences of endocrine disorders are described in chapter 14. Since the impact of hormones on bone development is great, however, a few endocrine disorders will be briefly mentioned. Hypersecretion of the growth hormone from the pituitary gland leads to **gigantism** in young persons if it starts before ossification of their epiphyseal plates and to **acromegaly** *(ak''ro-meg'ah-le)* in adults. Acromegaly is characterized by hypertrophy of the bones of the face, hands, and feet. In contrast, in a child, a hyposecretion of the growth hormone can lead to **dwarfism.**

Osteoporosis is a bone disorder characterized by marked demineralization, which weakens bone. The causes of osteoporosis include aging, prolonged inactivity, malnutrition, and an unbalanced secretion of hormones. It is most common in postmenopausal women. People with osteoporosis are prone to bone fracture, particularly at the pelvic girdle and vertebrae, as the bones become too brittle to support the body. Although there is no known cure for osteoporosis, it can be somewhat prevented in younger adults through proper diet, exercise, and good general health habits. Treatment of the disease in women through dietary calcium, exercise, and estrogens has limited positive results.

Paget's disease is a disease of disorganized metabolic processes within bone. The activity of osteoblasts and osteoclasts becomes irregular, producing areas with thickened osseous deposits and other areas where too much bone is removed. The etiology of the disease is unknown, but it is a relatively common affliction in persons over age fifty, and it occurs more frequently in males than in females.

Paget's disease: from Sir James Paget, English surgeon, 1814–1899

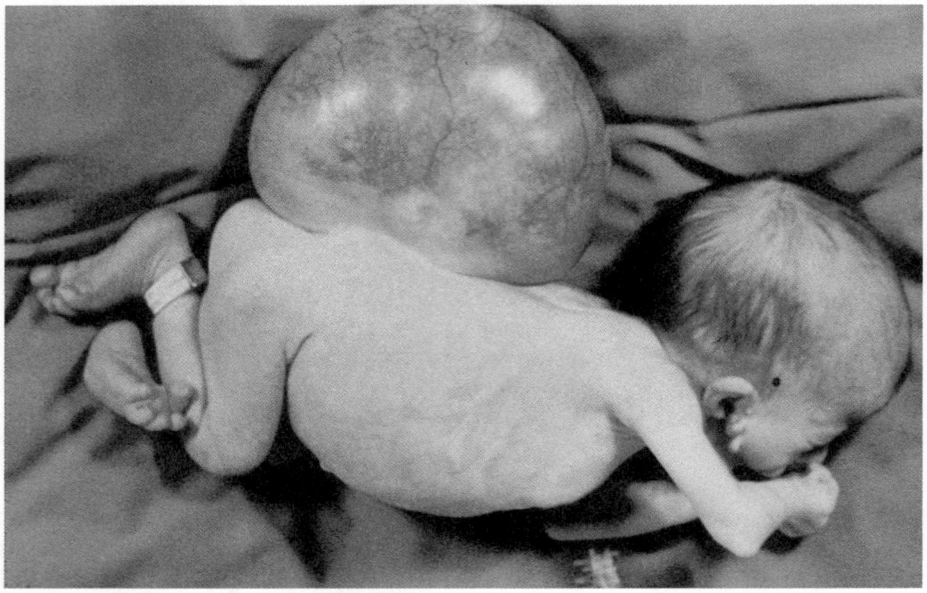

Figure 6.40 In spina bifida failure of the vertebral arches to fuse permits a herniation of the meninges that cover the spinal cord through the vertebral column, resulting in a condition called meningomyelocele.

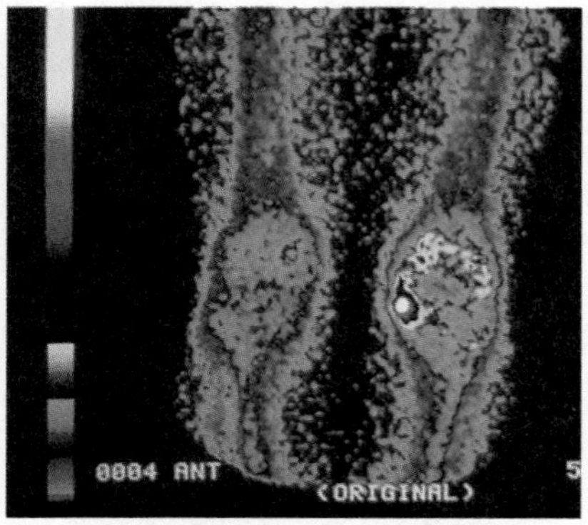

Figure 6.41 Bone scan of the legs of a patient suffering from arthritis in the left knee joint. In a bone scan, arthritis is depicted as a lighter image than a normal joint.

Trauma and Injury

There are a variety of types of trauma to the skeletal system, ranging from injury to the bone itself to damage of the joints in the form of sprains or dislocations. Fractures and the healing of fractures will be discussed in chapter 7, and joint injuries will be discussed in chapter 8.

Neoplasms of Bone

Malignant bone tumors are three times more common than benign tumors. Pain is the usual symptom of either type of osseous neoplasm, although benign tumors may not have accompanying pain.

Two types of benign bone tumors are **osteoma,** which is the more frequent and often involves the skull, and **osteoid osteoma,** which is a painful neoplasm of the long bones, usually in children.

Osteogenic sarcoma is the most virulent type of bone cancer and frequently metastasizes through the blood to the lungs. This disease usually originates in the long bones and is accompanied by aching and persistent pain.

A **bone scan** (fig. 6.41) is a diagnostic procedure frequently done on a person who has had a malignancy elsewhere in the body that may have metastasized to the bone. The patient receiving a bone scan may be injected with a radioactive substance that accumulates more rapidly in malignant tissue than normal tissue. Entire body radiographs show malignant bone areas as intensely dark dots.

CLINICAL CASE STUDY ANSWER

The height (overall length) of the vertebral column is a combination of the thickness of each vertebra plus the thickness of each intervertebral disc. The body of a vertebra consists of outer compact bone and inner spongy bone. An intervertebral disc consists of a fibrocartilage sheath called the annulus fibrosus and a mucoid center portion called the nucleus pulposus. The intervertebral discs generally change their anatomical configuration as one ages. In early adulthood, the nucleus pulposus is spongy and moist. With advanced age, however, the nucleus pulposus desiccates resulting in a flattening of the intervertebral disc. The collective effect of this occurring in all of the discs causes the vertebral bodies to lie closer together and an overall decrease in a person's height. Another cause of height loss may be the presence of undetectable compression fractures of the vertebral bodies, which are common in elderly people. This phenomenon, however, is considered pathological and not an aspect of the normal aging process. If a person has osteoporosis, there can be marked decrease in height and perhaps more serious clinical problems as well, such as compression of spinal nerves. ■

Important Clinical Terminology

achondroplasia *(ah-kon''dro-pla'ze-ah)* A genetic defect resulting in the retarded formation of cartilaginous bone during fetal development.

craniotomy Surgical cutting into the cranium to provide access to the brain.

epiphysiolysis *(ep''i-fiz''e-ol'i-sis)* A weakening and separation of the epiphysis from the diaphysis of a long bone.

laminectomy The surgical removal of the posterior arch of a vertebra, usually to repair a herniated intervertebral disc.

orthopedics A branch of medicine concerned with the diagnosis and treatment of trauma, diseases, and abnormalities involving the skeletal and muscular systems.

osteitis An inflammation of bone tissue.

osteoblastoma A benign tumor produced from bone-forming cells most frequently in the vertebrae of young children.

osteochondritis An inflammation of bone and cartilage tissues.

osteomyelitis An inflammation of bone marrow caused by bacteria or fungi.

osteonecrosis The death of bone tissue, usually caused by obstructed arteries.

osteopathology The study of bone diseases.

osteosarcoma A malignant tumor of bone tissue.

osteotomy The cutting of a bone, usually by means of a saw or a chisel.

Chapter Summary

I. Organization of the Skeletal System
   A. The axial skeleton consists of the skull, ear ossicles, hyoid bone, vertebral column, and rib cage.
   B. The appendicular skeleton consists of the pectoral girdle, upper extremities, pelvic girdle, and lower extremities.

II. Functions of the Skeletal System
   A. The mechanical functions of bones include the support and protection of softer body tissues and organs, and certain bones function as levers during body movement.
   B. The metabolic functions of bones include hemopoiesis and mineral storage.

III. Development of the Skeletal System
   A. Bones derive from mesenchymal mesoderm through endochondral ossification or through intramembranous ossification.
   B. The development of the skull involves the chondrocranium, dermatocranium, and viscerocranium.

IV. Bone Structure and Growth
   A. Bone structure includes the shape and surface features of each bone as well as gross internal components.
      1. Structurally speaking, bones may be classified as long, short, flat, or irregular.
      2. The surface features of bones can be broadly classified into articulating surfaces, nonarticulating prominences, and depressions and openings.
      3. A typical long bone has a diaphysis filled with marrow in the medullary cavity, epiphyses, epiphyseal plates for linear growth, and a covering of periosteum for diametric growth and the attachment of ligaments and tendons.
      4. Compact bone is arranged into osteons that contain osteocytes and lamellae.
   B. Bone growth from embryonic to adult size is an orderly process determined by genetics, hormonal secretions, and nutritional supply.
      1. Most bones develop through endochondral ossification.
      2. Bone remodeling is a continual process involving osteoclasts in bone resorption and osteoblasts in the formation of new osseous tissue.

V. Skull
   A. The cranium encloses and protects the brain and provides for the attachment of muscles.
      1. Sutures are immovable joints between cranial bones.
      2. The eight cranial bones include the frontal, parietals, temporals, occipital, sphenoid, and ethmoid.
   B. Facial bones form the basic shape of the face, support the teeth, and provide for the attachment of the facial muscles.
      1. The fourteen facial bones are the nasals, maxillae, zygomatics, mandible, lacrimals, palatines, inferior nasal conchae, and vomer.
      2. The hyoid bone is located in the neck between the mandible and the larynx.
      3. The three paired ear ossicles are located within the middle ear chambers of the petrous portion of the temporal bones.

VI. Vertebral Column
   A. The vertebral column consists of seven cervical, twelve thoracic, five lumbar, four or five fused sacral, and four or five fused coccygeal vertebrae.
   B. Cervical vertebrae have transverse foramina; thoracic vertebrae have facets and demifacets for articulation with ribs; lumbar vertebrae have heavy bodies; sacral vertebrae are triangularly fused and articulate with the pelvic girdle; the coccygeal vertebrae form a small triangular bone.

VII. Rib Cage
   A. The sternum consists of a manubrium, body, and xiphoid process.
   B. There are seven pairs of true ribs, three pairs of false ribs, and two pairs of floating ribs.

Review Activities

Objective Questions

1. A bone is considered to be a(n)
   (a) tissue. (c) organ.
   (b) cell. (d) system.
2. Which of the following statements is *false?*
   (a) Bones are important in the synthesis of vitamin D.
   (b) Bones and teeth contain about 99% of the body's calcium.
   (c) Red bone marrow is the primary site for hemopoiesis.
   (d) Most bones develop through endochondral ossification.
3. Bone tissue derives from specialized migratory mesodermal cells called
   (a) dermatomes. (c) myotomes.
   (b) mesenchyme. (d) somites.
4. Small portions of the embryonic notochord persist in the adult as the
   (a) primary curves.
   (b) nucleus pulposus.
   (c) pedicles.
   (d) centrum.
5. Which of the following is *not* a long bone? The
   (a) talus.
   (b) proximal phalanx.
   (c) metatarsal.
   (d) fibula.
6. Specialized bone cells that enzymatically reabsorb bone tissue are
   (a) osteoblasts. (c) osteons.
   (b) osteocytes. (d) osteoclasts.
7. The mandibular fossa is located in which structural portion of the temporal bone? The
   (a) squamous. (c) mastoid.
   (b) tympanic. (d) petrous.

8. The crista galli is a structural feature of which bone? The
   (a) sphenoid. (c) palatine.
   (b) ethmoid. (d) temporal.
9. Transverse foramina are characteristic of
   (a) lumbar vertebrae.
   (b) sacral vertebrae.
   (c) thoracic vertebrae.
   (d) cervical vertebrae.
10. The bone disorder common in aged persons particularly if they have prolonged inactivity, malnutrition, or an unbalanced secretion of hormones is
   (a) osteitis. (c) osteoporosis.
   (b) osteonecrosis. (d) osteomalacia.

Essay Questions

1. What are the functions of the skeletal system? Does each of the bones of the skeleton carry out these functions equally? Explain.
2. Explain why there are approximately 270 bones in an infant but 206 bones in a mature adult.
3. List the bones of the skull that are paired. Which are unpaired? Identify the bones of the skull that can be palpated.
4. Describe the development of the skull. What are the fontanels, where are they located, and what are their functions?
5. Which facial bones contain foramina? What structures traverse these openings?
6. Distinguish between the axial and appendicular skeletons. Describe where these two components articulate.
7. List four types of bones based on shape and give an example of each group.
8. Diagram a typical long bone, and label the epiphyses, diaphysis, epiphyseal plates, medullary cavity, nutrient foramina, periosteum, and articular cartilages.
9. List the bones that form the cranial cavity, the orbit, and the nasal cavity. Describe the location of the paranasal sinuses, the mastoidal sinus, and the inner-ear cavity.
10. Describe how bones grow in length and in circumference. How are these processes similar, and how do they differ? Explain how X rays can be used to determine normal bone growth.
11. Explain the process of endochondral ossification of a long bone. Why is it important that a balance be maintained between osteoblast activity and osteoclast activity?
12. Describe the curvature of the vertebral column. What is meant by primary curves as compared to secondary curves?
13. List two or more characteristics by which vertebrae from each of the five regions of the vertebral column can be identified.
14. Identify the bones that form the rib cage. What functional role do the bones and the costal cartilages have in respiration?
15. Explain why a proper balance of vitamins, hormones, and minerals is essential in maintaining healthy osseous tissue. Give examples of diseases or skeletal conditions that may occur if there is an imbalance of any of these three essential substances.

CHAPTER 7

Skeletal System: *The Appendicular Skeleton*

Outline and Concepts

CLINICAL CASE STUDY

A twelve-year-old boy is hit by a car while crossing a street. He is brought in stable condition to the emergency room complaining of severe pain in his right leg. X rays reveal a longitudinal fracture beginning on the surface of the tibial plateau, coursing 4 in. inferiorly into the anterior cortex of the tibia. The fragment of bone created by the fracture is moderately displaced. The orthopedic surgeon takes the X rays into the waiting room and confers with the boy's parents. He explains that this particular injury is more serious in children and growing adolescents than in adults. He elaborates by stating that future growth of the bone may be jeopardized, and that surgery will be required but cannot guarantee normal growth. The parents ask, "What is it about the fracture that threatens future growth?"

How would you respond?

Development of the Appendicular Skeleton

The development of the appendicular skeleton begins at the end of the fourth week and progresses through the embryonic period as the limb buds follow the growth of the apical ectodermal ridges.

Objective 1. Explain the process of appendicular skeletal development.

The development of the upper and lower extremities is initiated toward the end of the fourth week as four small elevations called **limb buds** appear (fig. 7.1). The anterior pair are the arm buds, which precede the development of the posterior pair of leg buds by a few days. Each limb bud consists of a mass of undifferentiated mesoderm partially covered with a layer of ectoderm, called the **apical ectodermal ridge,** which promotes bone and muscle development.

As the limb buds elongate, migrating mesenchymal tissues differentiate into specific cartilaginous bones. Soon primary ossification centers form in each bone, and the hyaline cartilage tissue is gradually replaced by bony tissue in the process of endochondral ossification.

Initially, the developing limbs are directed caudally, but later there is a lateral rotation in the upper extremity and a medial rotation in the lower extremity. As a result, the elbows are directed backward and the knees directed forward.

Digital rays that will form the hands and feet are apparent by the fifth week, and the individual digits separate by the end of the sixth week.

A large number of limb deformities occurred between 1957 and 1962 as a result of mothers ingesting thalidomide during early pregnancy to relieve morning sickness. It is estimated that 7,000 infants were malformed by thalidomide. The malformations ranged from *micromelia* (short limbs) to *amelia* (absence of limbs).

micromelia: Gk. *mikros,* small; *melos,* limb
amelia: Gk. *a,* without; *melos,* limb

1. Define limb buds, apical ectodermal ridge, and digital rays.
2. During which weeks of pregnancy is an embryo most susceptible to deformities of the appendages from the maternal ingestion of teratogenic drugs?

Pectoral Girdle and Upper Extremity

The structure of the pectoral girdle and upper extremities is adaptive for freedom of movement and extensive muscle attachment.

Objective 2. Describe the bones of the pectoral girdle and the positions of articulations.

Objective 3. Identify the bones of the upper extremity, and list the diagnostic features of each bone.

Pectoral Girdle

The two scapulae and two clavicles make up the pectoral (shoulder) girdle. It is not a complete girdle, having only an anterior attachment to the axial skeleton at the sternum. The primary function of the pectoral girdle is to provide attachment for the numerous muscles that move the brachium and forearm. The pectoral girdle is not weight-bearing and is therefore more delicate in structure than the pelvic girdle.

Clavicle The slender S-shaped clavicle (collarbone) binds the shoulder to the axial skeleton and positions the shoulder joint away from the trunk for freedom of movement. The articulation of the medial **sternal end** of the clavicle to the manubrium is referred to as the **sternoclavicular joint** (fig. 7.2). The lateral **acromial** *(ah-kro'me-al)* **end** of the clavicle articulates with the acromion process of the scapula. This articulation is referred to as the **acromioclavicular joint.** A **conoid tubercle** is present on the inferior surface of the lateral end, and a **costal tuberosity** is present on the inner surface of the medial end. Both processes serve as points of attachment for ligaments.

The long, delicate clavicle is the most commonly fractured bone in the body. Blows to the shoulder or an attempt to break a fall with an outstretched hand causes the force to be displaced to the clavicle. The most vulnerable area for a fracture of this bone is through its center immediately proximal to the conoid tubercle. Because the clavicle is subcutaneous and not covered with muscle, a fracture can easily be palpated.

Scapula The scapula (shoulder blade) is a large triangular flat bone positioned on the posterior aspect of the rib cage, overlying ribs 2 to 7. The **spine** of the scapula is a prominent diagonal bony ridge seen on the posterior surface (fig. 7.3). The spine increases the strength to the scapula for resistance to

clavicle: L. *clavicula,* a small key
acromial end: Gk. *akros,* peak; *omos,* shoulder
conoid tubercle: Gk. *konos,* cone; L. *tuberculum,* a small swelling
costal tuberosity: L. *costa,* rib; *tuberosus,* a knob
scapula: L. *scapula,* shoulder

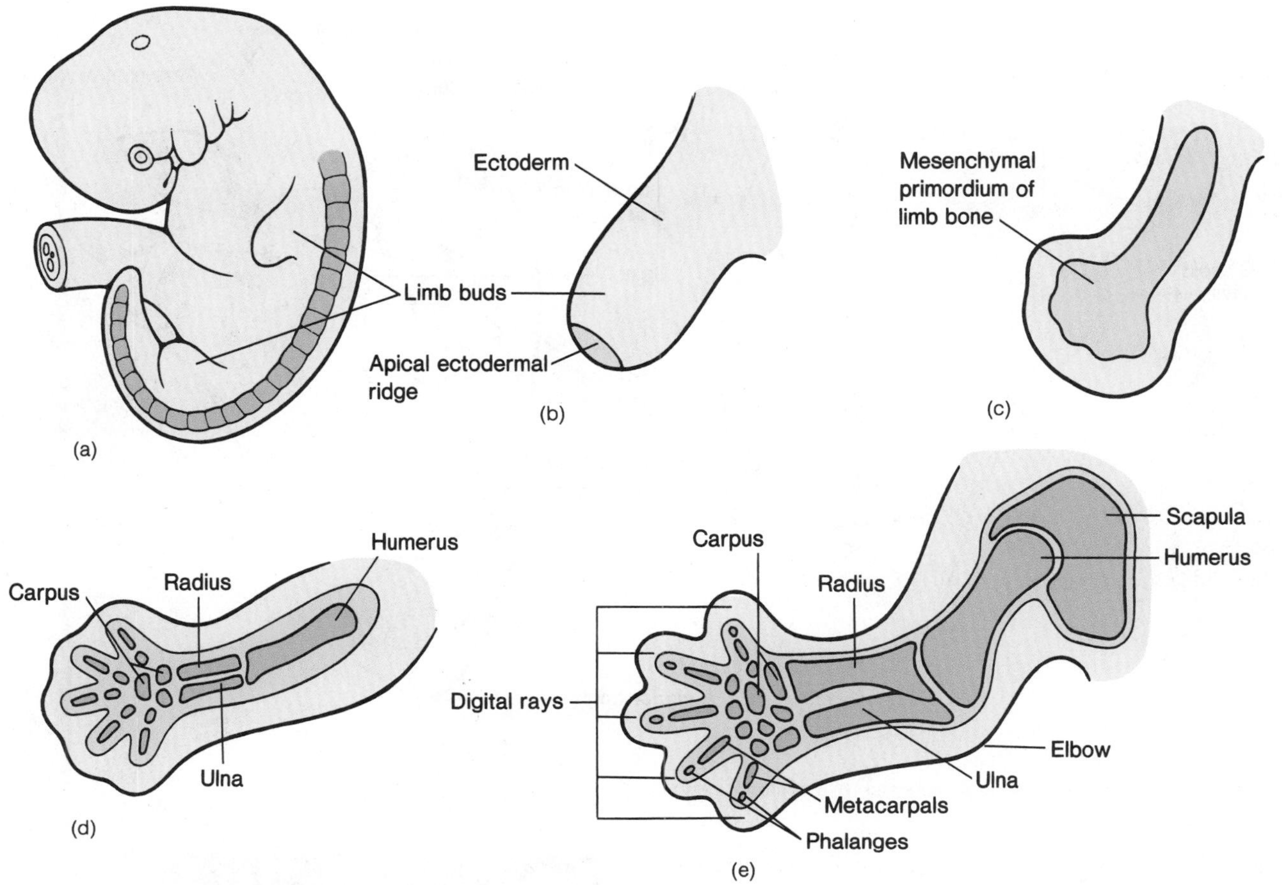

Figure 7.1 The development of the appendicular skeleton. (*a*) Limb buds are apparent in an embryo by twenty-eight days, and (*b*) an ectodermal ridge is the precursor of the skeletal and muscular structures. (*c*) Mesenchymal primordial cells are present at thirty-three days. (*d*) Hyaline cartilaginous models of individual bones develop early in the sixth week. (*e*) Later in the sixth week, the cartilaginous skeleton of the upper extremity is well formed.

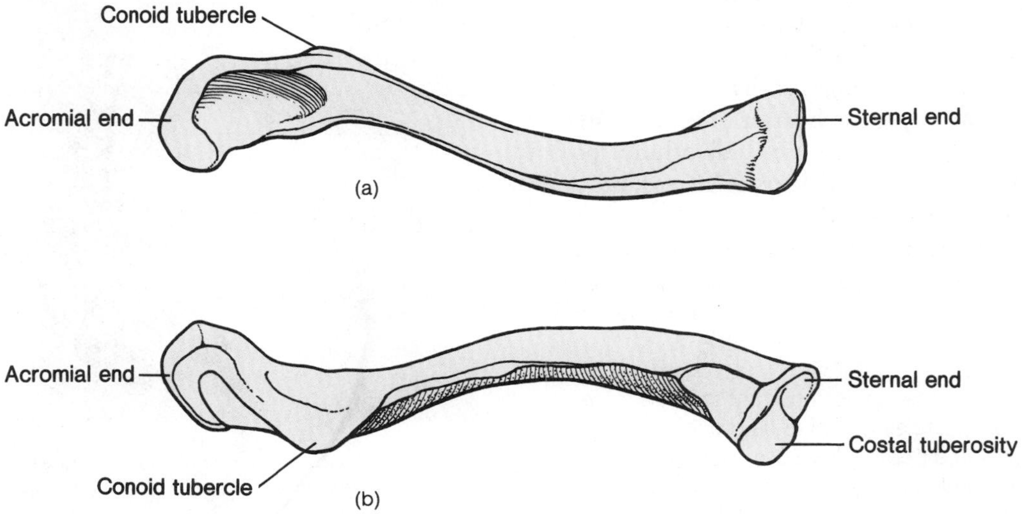

Figure 7.2 The right clavicle. (*a*) A superior view and (*b*) an inferior view.

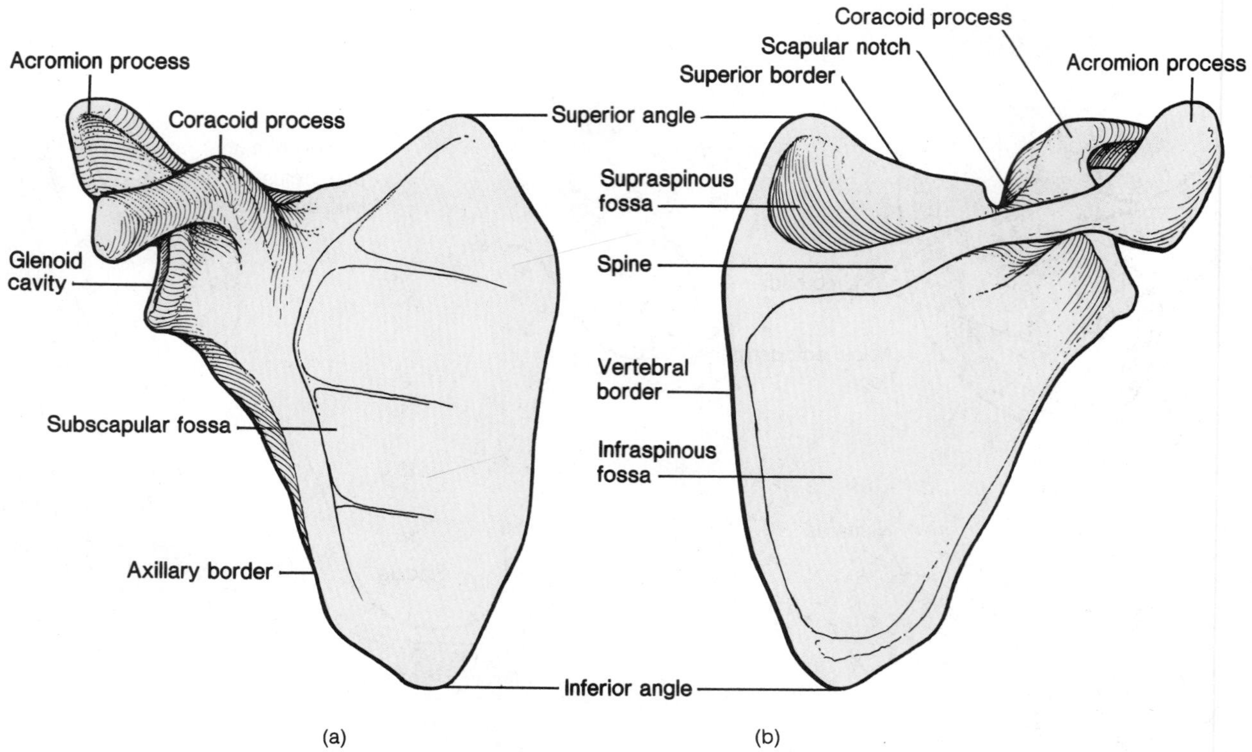

Figure 7.3 The right scapula. (*a*) An anterior view and (*b*) a posterior view.

bending. Above the spine is the **supraspinous** *(su''pra-spi'nus)* **fossa,** and below the spine is the **infraspinous fossa.** The spine broadens toward the shoulder as the **acromion process** (figs. 7.3, 7.4). This process serves for the attachment of several muscles as well as articulation with the clavicle. Inferior to the acromion is a shallow depression, the **glenoid cavity,** into which the head of the humerus fits. The **coracoid process** is a thick, upward projection that lies superior and anterior to the glenoid cavity. On the anterior surface of the scapula is a slightly concave area known as the **subscapular fossa.**

The scapula has three borders separated by two angles. The superior edge is called the **superior border.** The **vertebral,** or **medial, border** is nearest to the vertebral column positioned about 5 cm (2 in.) away. The **axillary, or lateral, border** is directed toward the arm. The **superior angle** is between the superior and vertebral borders, and the **inferior angle** is between the vertebral and axillary borders. Along the superior border is a distinct depression called the **scapular notch,** which serves as a passageway for a nerve.

The anatomy of the scapula is important to know because some fifteen muscles attach to its processes and fossae. Clinically, the pectoral girdle is significant because the clavicle and acromion process of the scapula are frequently fractured in trying to break a fall. The acromion process is palpated when locating the proper site for an intramuscular injection of the arm. This site is chosen because of its thick musculature, and it is away from the presence of a nerve.

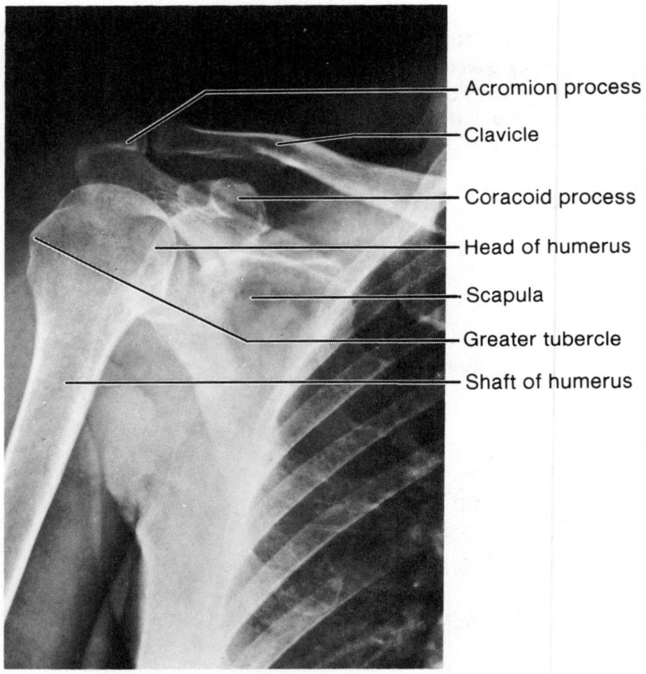

Figure 7.4 An X ray of the right shoulder shows the articulation of the clavicle, scapula, and humerus forming the shoulder joint.

Brachium

The brachium is the upper arm and contains a single bone, the humerus.

Humerus The humerus (fig. 7.5) is the longest bone of the upper extremity. It consists of a proximal **head,** which artic-

glenoid: Gk. *glenoeides,* shallow form
coracoid process: Gk. *korakodes,* like a crow's beak

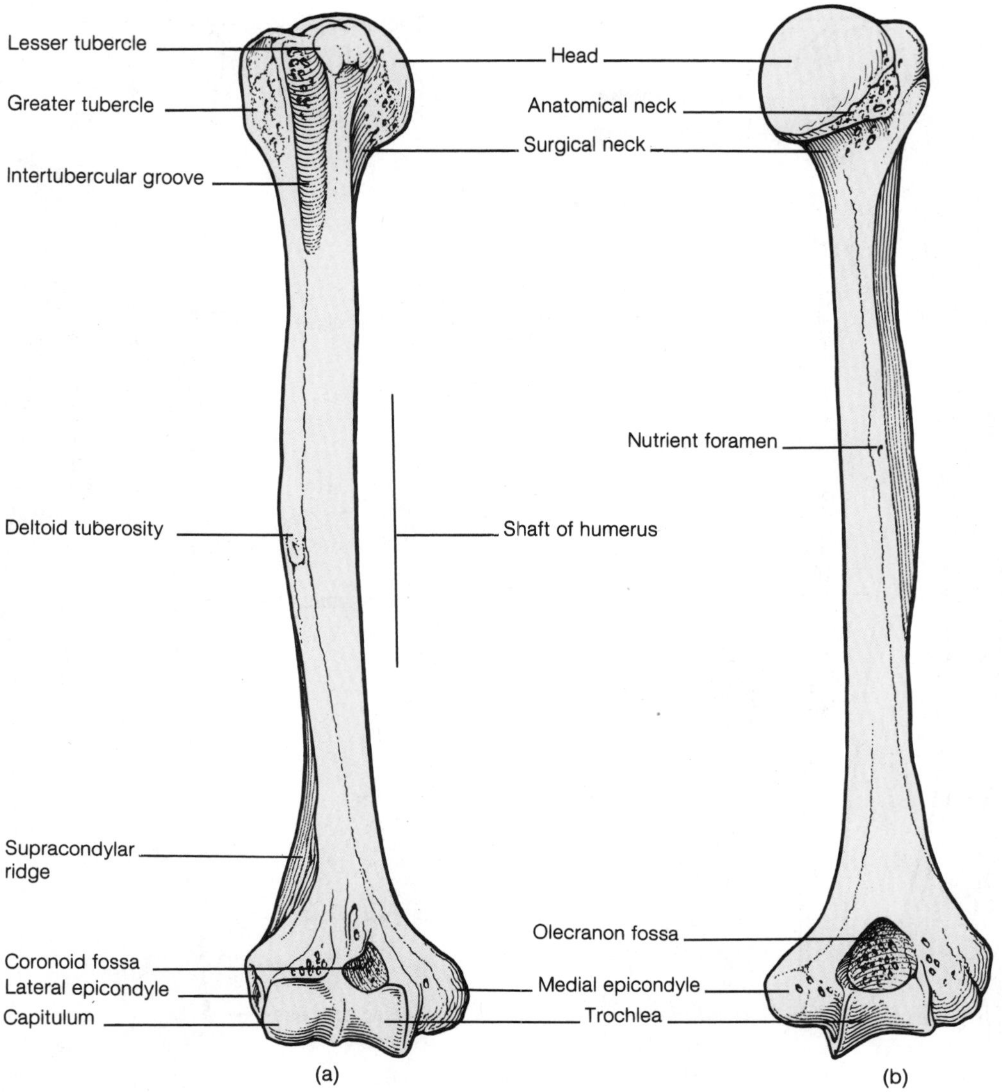

Figure 7.5 The right humerus. (*a*) An anterior view and (*b*) a posterior view.

ulates with the glenoid cavity of the scapula, a **shaft** (body), and a distal end, which is modified to receive the two bones of the forearm. Surrounding the margin of the head is a slightly indented groove denoting the **anatomical neck.** The region where the shaft begins to taper is referred to as the **surgical neck,** a frequent site of fractures. Lateral to the head is a large eminence, the **greater tubercle.** The **lesser tubercle** is slightly anterior to the greater and is separated from the greater by an **intertubercular (bicipital) groove,** through which passes the tendon from the biceps brachii muscle.

Along the lateral midregion of the shaft is a roughened area, the **deltoid tuberosity** for the attachment of the deltoid muscle, which elevates the arm horizontally to the side.

The distal end of the humerus has two rounded condyloid articular surfaces. The **capitulum** *(kah-pit'u-lum)* is the lateral rounded condyle that receives the radius. The **trochlea** *(trok'le-ah)* is the pulleylike medial surface that articulates with the ulna. On either side above the condyles are the **lateral** and **medial epicondyles.** The large medial epicondyle protects the ulnar nerve that passes posteriorly through the ulnar sulcus (see fig. 10.22). The **coronoid fossa** is a depression above the trochlea on the anterior surface. The **olecranon** *(o-lek'rah-non)* **fossa** is a depression on the distal posterior surface. Both fossae are adapted to receive parts of the ulna during movement of the forearm.

The medical term for tennis elbow is *lateral epicondylitis,* which means an inflammation of the tissues surrounding the lateral epicondyle of the humerus. At least six muscles that control backward (extension) movement of the hand and fingers originate on the lateral epicondyle. Repeated strenuous contractions of these muscles, as in stroking with a tennis racket, may cause a strain on the periosteum and tendinous muscle attachments resulting in tenderness and pain around the epicondyle. Binding usually eases the pain, but only rest can eliminate the causative factor, and recovery generally follows.

deltoid tuberosity: Gk. *deltoeides,* shaped like the letter (Δ)
capitulum: L. *caput,* little head
trochlea: Gk. *trochilia,* a pulley
olecranon: Gk. *olene,* ulna; *kranion,* head

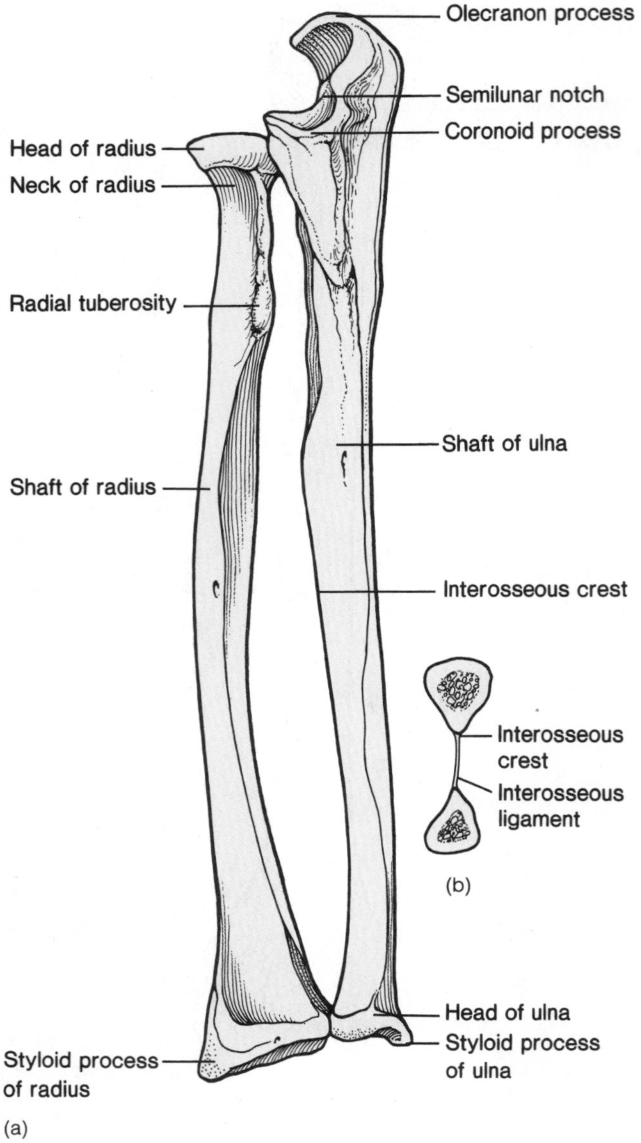

Figure 7.6 (*a*) An anterior view of the right radius and ulna. (*b*) A transverse section shows the binding interosseous ligament.

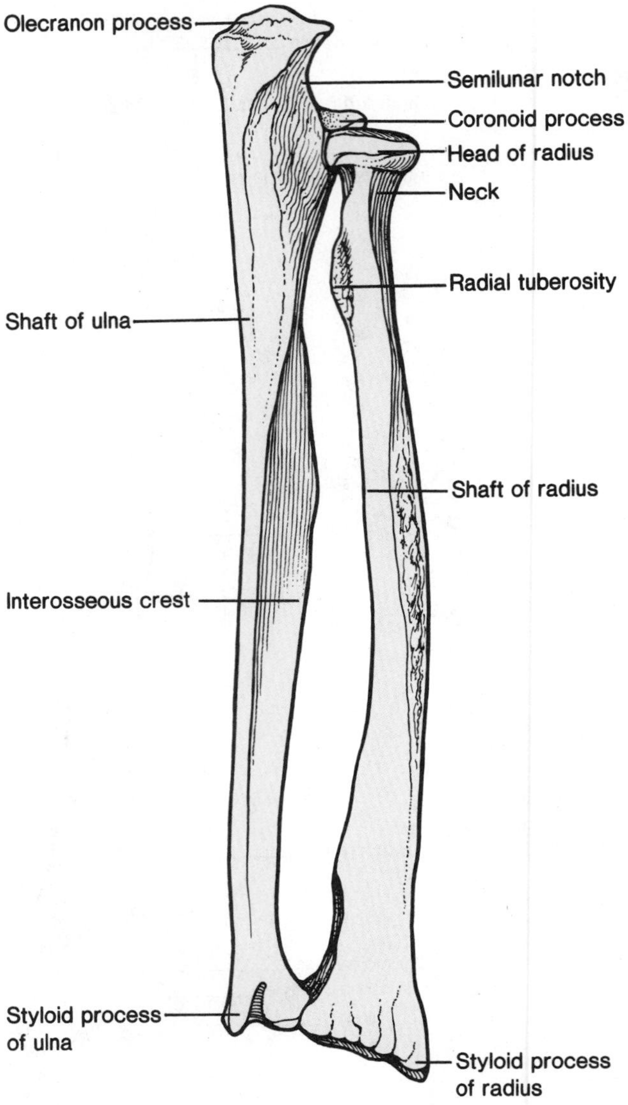

Figure 7.7 A posterior view of the right radius and ulna.

Forearm

The skeletal structures of the forearm are the ulna on the medial side and the radius on the lateral (thumb) side (figs. 7.6, 7.7). The ulna is more firmly connected to the humerus than the radius and is longer. The radius, however, contributes more significantly at the wrist joint than does the ulna.

Ulna The proximal end of the ulna articulates with the humerus and radius. A distinct depression, the **semilunar (trochlear) notch,** articulates with the trochlea of the humerus. The **coronoid process** forms the anterior lip of the semilunar notch, and the **olecranon process** forms the posterior portion, or elbow. Lateral and inferior to the coronoid process is the **radial notch,** which accommodates the head of the radius.

Along the anterolateral surface of the shaft is a sharp ridge called the **interosseous crest.** The **interosseous ligament** extends from this crest to the radius to bind these two bones together (fig. 7.6).

The tapered distal end of the ulna has a knobbed portion, the **head,** and a posteromedial projection, the **styloid process.** The ulna articulates distally and proximally with the radius.

Radius The radius consists of a **shaft** that has a small proximal end and a large distal end. A proximal disc-shaped **head** articulates with the capitulum of the humerus and the radial notch of the ulna. The prominent **radial tuberosity,** for attachment of the biceps muscle, is on the medial side of the shaft just below the head. The distal end of the radius has a double-faceted surface for articulation with the proximal carpal bones: a lateral **navicular facet** and a medial **lunate facet.** The distal end of the radius also has the **styloid process** on the lateral tip and the **ulnar notch** on the medial side for receiving the distal end of the ulna. The styloid processes on the ulna and radius provide lateral and medial support for articulation at the wrist.

---

styloid process: Gk. *stylos,* pillar; *eidos,* resemblance

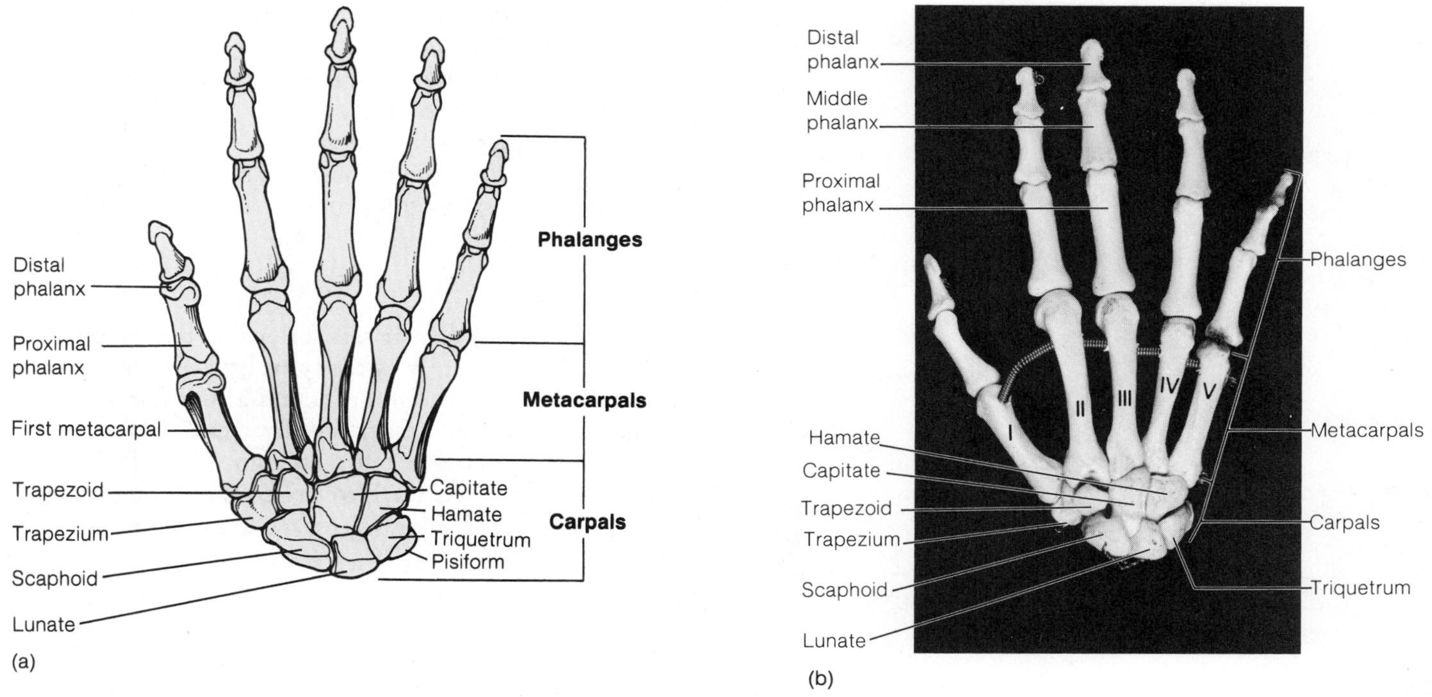

Figure 7.8 A posterior view of the right wrist and hand. (*a*) A drawing and (*b*) a photograph.

When a person falls, the natural tendency is to extend the hand to break the fall. This reflexive movement frequently results in fractured bones. Common fractures of the radius include a fracture of the head as it is driven forcefully against the capitulum, a fracture of the neck, or a fracture of the distal end *(Colles' fracture)* caused by landing on an outstretched hand.

When falling, it is less traumatic to the body to withdraw the appendages, bend the knees, and let the entire body hit the surface. Athletes learn that this is the safe way to fall.

Wrist (Carpus) and Hand (Manus)

The wrist and hand contain twenty-seven bones arranged into the carpus, metacarpus, and phalanges (figs. 7.8, 7.9, 7.10).

Carpus The carpus, or wrist, consists of eight carpal bones arranged into two transverse rows of four bones each. The proximal row, naming from the lateral (thumb) to medial side, consists of the **scaphoid** (navicular), **lunate, triquetrum** *(tri-kwe'trum)* (triangular), and **pisiform** *(pi'sĭ-form)*. The pisiform forms in a tendon as a sesamoid bone. The distal row, from lateral to medial, consists of the **trapezium** (greater multangular), **trapezoid** (lesser multangular), **capitate,** and **hamate** *(ham'at)*. The scaphoid and lunate of the proximal row articulate with the distal end of the radius.

carpus: Gk. *karpos,* wrist
navicular: L. *navicula,* small ship
lunate: L. *lunare,* crescent or moon-shaped
triquetrum: L. *triquetrus,* three-cornered
pisiform: Gk. *pisos,* pea
trapezium: Gk. *trapesion,* small table
capitate: L. *capitatus,* head
hamate: L. *hamatus,* hook

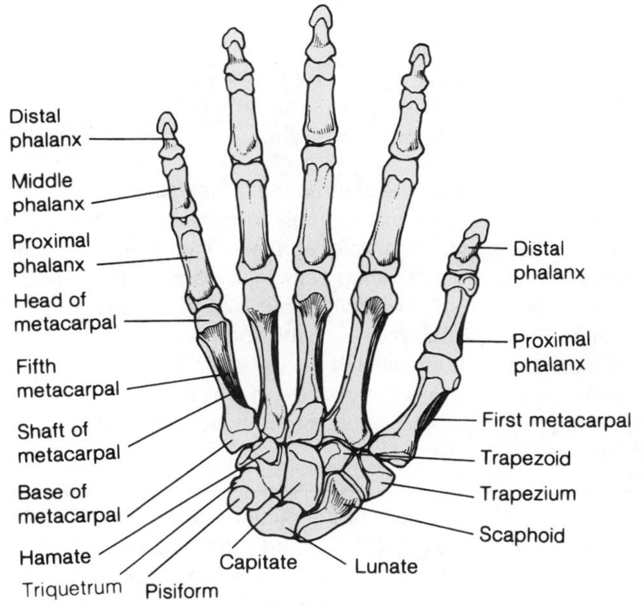

Figure 7.9 An anterior view of the right wrist and hand.

Metacarpus The metacarpus, or palm of the hand, is composed of five metacarpal bones. Each metacarpal bone consists of a proximal **base,** a **shaft,** and a distal **head,** which is rounded for articulation with the base of each proximal phalanx. The heads of the metacarpals are distally located and form the knuckles of a clenched fist. The metacarpal bones are numbered from one to five; the lateral, or thumb, side being one.

Phalanges The fourteen phalanges are the skeletal elements of the fingers. A single finger bone is called a **phalanx** *(fa'lanks)*. The phalanges of the fingers are arranged into a proximal row,

phalanx: Gk. *phalanx,* finger bone or toe bone

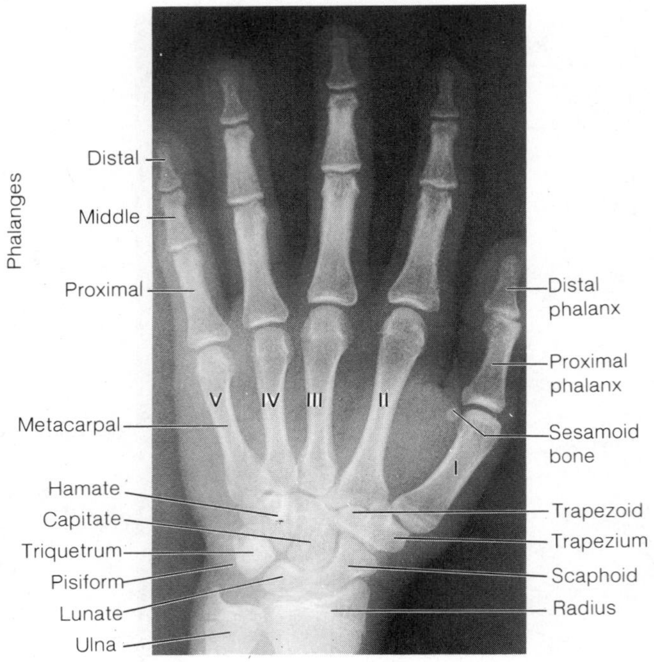

Figure 7.10 An X ray of the left wrist and hand shown in an anteroposterior projection. (Note the presence of a sesamoid bone at the thumb joint.)

a middle row, and a distal row. The thumb (pollex) has only a proximal and a distal phalanx. A summary of the bones of the upper extremities is presented in table 7.1.

The hand is a marvel of structural organization, which despite its complexity, is able to sustain considerable abuse. Other than sprained fingers and dislocation, the most common bone injury is a fracture to the scaphoid bone of the wrist (about 70% of carpal fractures occur here). When immobilizing the wrist joint with a plaster cast, the wrist is positioned in the plane of relaxed function. This is the position in which the hand is about to grasp an object between the thumb and index finger.

1. Describe the structure of the pectoral girdle. Why is the pectoral girdle considered an incomplete girdle?
2. Identify the fossae and processes of the scapula.
3. Describe each of the long bones of the upper extremity.
4. Where are the styloid processes of the wrist area? What are their functions?
5. Name the bones in the proximal row of the carpus. Which of these bones articulate with the radius?

Pelvic Girdle and Lower Extremity

The structure of the pelvic girdle and lower extremities is adaptive for support and locomotion. There are extensive processes and surface features on certain bones of the pelvic girdle and lower extremities that accommodate massive muscles for posture and locomotion.

Objective 4. Describe the structure of the pelvic girdle, and list its functions.

Objective 5. Describe the structural differences between the male and female pelvis.

Objective 6. Identify the bones of the lower extremity, and list the diagnostic features of each bone.

Objective 7. Describe the structural features and functions of the arches of the foot.

Pelvic Girdle

The pelvic girdle, or pelvis, is formed by two **os coxae** (hipbones) united anteriorly by the **symphysis pubis** (fig. 7.11). It is attached posteriorly to the sacrum of the vertebral column. The pelvic girdle and its associated ligaments support the weight of the body from the vertebral column. The pelvic girdle also supports and protects the lower viscera, including the urinary bladder, the reproductive organs, and in a pregnant woman, the developing fetus.

Clinically speaking, the basinlike pelvis is frequently divided into a **greater,** or **false, pelvis** and a **lesser,** or **true, pelvis** (see fig. 7.14). These two components are divided by the **pelvic brim,** a curved bony rim passing inferiorly from the sacral promontory to the upper margin of the symphysis pubis. The greater pelvis is the expanded portion of the pelvis superior to the pelvic brim. The pelvic brim not only divides the two portions but surrounds the **pelvic inlet** of the lesser pelvis. The lower circumference of the lesser pelvis bounds the pelvic outlet. During parturition, a child must pass through its mother's lesser pelvis for a natural delivery. *Pelvimetry* measures the dimension of the lesser pelvis to determine if a cesarean delivery might be necessary. Diameters may be determined by vaginal palpation or by x-ray measurements (fig. 7.12).

Each os coxa actually consists of three separate bones: the **ilium,** the **ischium** *(is'ke-um),* and the **pubis** (fig. 7.13). These bones are fused together in the adult. On the lateral surface of the os coxa where the three bones ossify together is a large circular depression, the **acetabulum** *(as''ĕ-tab'u-lum),* which receives the head of the femur. Although in the adult both ossa coxae are single bones, the three components are considered separately for descriptive purposes.

Ilium The ilium is the largest and uppermost of the three pelvic bones. The ilium presents a crest and four angles, or spines, which serve for muscle attachment and are important surface landmarks. The **iliac crest** forms the prominence of the hip. This crest terminates anteriorly as the **anterior superior iliac spine.** Just below this spine is the **anterior inferior iliac spine.** The posterior termination of the iliac crest is the **posterior superior iliac spine,** and just below this is the **posterior inferior iliac spine.**

coxae: L. *coxae,* hips
ilium: L. *ilia,* loin
ischium: Gk. *ischion,* hip joint
pubis: L. *pubis,* genital area
acetabulum: L. *acetabulum,* vinegar cup

Table 7.1 Bones of the pectoral girdle and upper extremities

Name and number	Location	Diagnostic features
Clavicle (2)	Anterior base of neck, between sternum and scapula	S-shaped; sternal and acromial ends
Scapula (2)	Upper back forming part of the shoulder	Triangular-shaped; spine; acromion and coracoid processes
Humerus (2)	Brachium between scapula and elbow	Longest bone of upper extremity; greater and lesser tubercles; surgical neck; deltoid tuberosity; capitulum; trochlea; coronoid and olecranon fossae
Ulna (2)	Medial side of forearm	Semilunar notch; olecranon and styloid processes
Radius (2)	Lateral side of forearm	Head; radial tuberosity; styloid process
Carpal (16)	Wrist	Short bones arranged in two rows of four bones each
Metacarpal (10)	Palm of hand	Long bones arranged one in line with each digit
Phalanx (28)	Digits	Three in each finger, except two in thumb

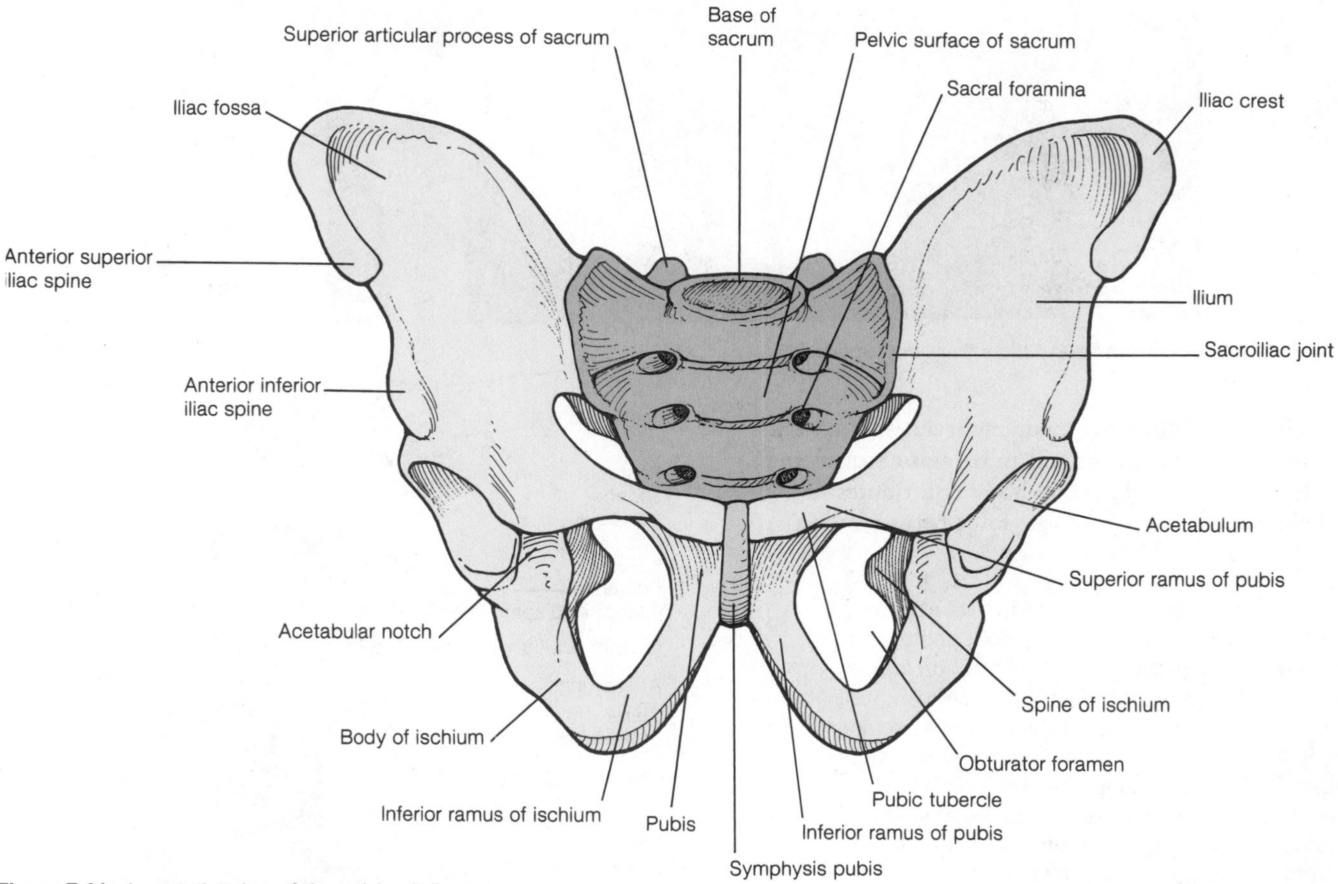

Figure 7.11 An anterior view of the pelvic girdle.

Below the posterior inferior iliac spine is the **greater sciatic** *(si-at'ik)* **notch.** On the medial surface of the ilium is the roughened **auricular surface** that articulates with the sacrum. The **iliac fossa** is the smooth, concave surface on the anterior portion of the ilium. The iliacus muscle originates from this fossa. The **iliac tuberosity,** for the attachment of the sacroiliac ligament, is positioned posterior to the iliac fossa. Three roughened ridges are present on the **gluteal surface** of the posterior aspect of the ilium. These ridges serve to attach the gluteal muscles and are the **inferior, anterior,** and **posterior gluteal lines.**

Ischium The ischium is the posterior-inferior component of the os coxa. This bone has several significant features. The **spine of the ischium** is the projection immediately posterior and inferior to the greater sciatic notch of the ilium and ischium. Inferior to this spine is the **lesser sciatic notch** of the ischium. The **ischial tuberosity** is the bony projection that supports the weight of the body in the sitting position. A deep **acetabular notch** is present on the inferior portion of the acetabulum. The large **obturator foramen** is formed by the **ramus** of the ischium together with the pubis. The obturator foramen is covered by the obturator membrane to which several muscles attach.

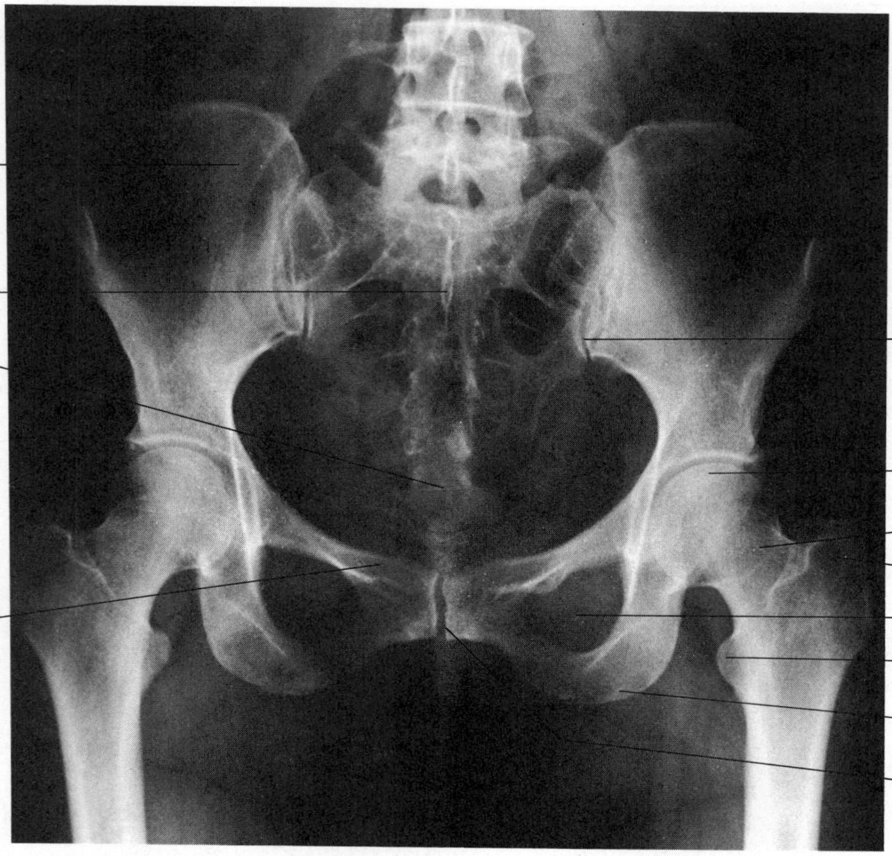

Figure 7.12 An X ray of the pelvic girdle and the articulating femora.

Pubis The pubis is the anterior component of the os coxa. This bone consists of a **superior ramus** and an **inferior ramus** that supports the **body** of the pubis. The body contributes to the formation of the symphysis pubis—the joint between the two ossa coxae.

The structure of the human pelvis, in its attachment to the vertebral column, permits an upright posture and locomotion on two legs (bipedal) rather than on four legs like other mammals. Although these structures are well adapted for *bipedal locomotion,* an upright posture may cause problems. The sacroiliac joint may weaken with age, causing lower back pains. The weight of the viscera may weaken the walls of the lower abdominal area and cause hernias. Some of the problems of childbirth are related to the structure of the mother's pelvis. The hip joint tends to deteriorate with age, and many older persons suffer from fractured hips.

Sex Differences in the Pelvis There are structural differences between the pelvis of an adult male and that of an adult female (fig. 7.14, and table 7.2) that reflect the female's role in pregnancy and childbirth.

In addition to the differences listed in table 7.2, the symphysis pubis and sacroiliac joints stretch during pregnancy and parturition.

Thigh

The patella will be considered along with the femur in the discussion of the thigh, even though the femur is the only bone of the thigh.

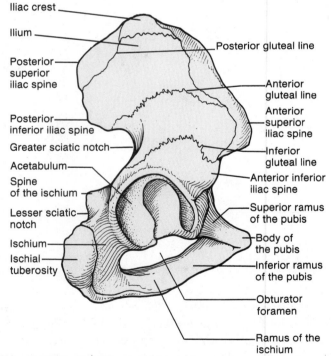

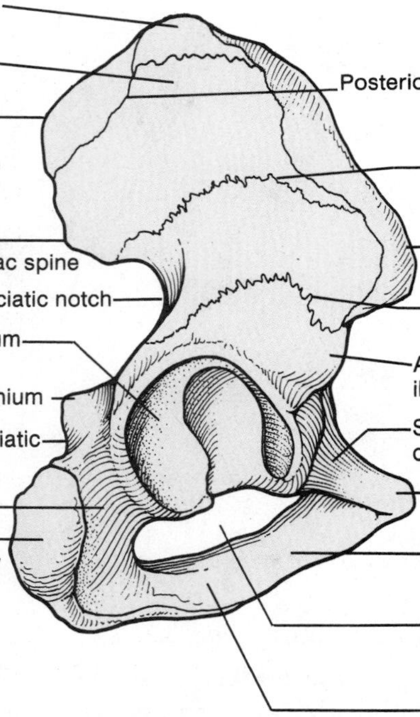

Figure 7.13 The lateral aspect of the right os coxa.

Femur The femur (thighbone) is the longest, heaviest, and strongest bone in the body (fig. 7.15). The proximal rounded **head** of the femur articulates with the acetabulum of the os coxa. A roughened, shallow pit, called the **fovea capitis,** is in

femur: L. *femur,* thigh

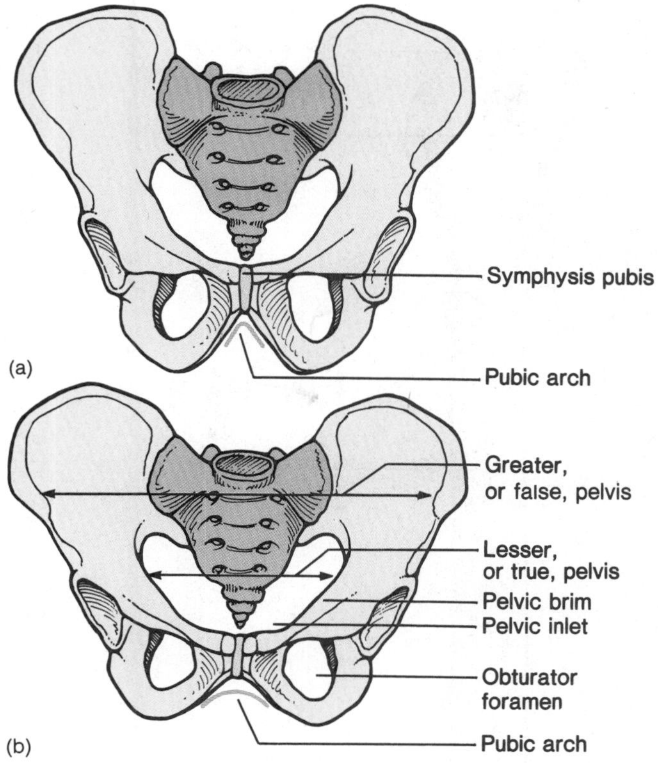

Figure 7.14 A comparison of the (*a*) male and (*b*) female pelvic girdle.

the lower center of the head of the femur. The fovea capitis provides the point of attachment for the ligamentum teres, which helps to support the head of the femur against the acetabulum. The constricted region supporting the head is called the **neck** and is a common site for fractures in aged persons.

The **shaft** of the femur has a slight medial bow so that it converges with the femur of the opposite thigh and brings the knee joints more in line with the body's plane of gravity. The degree of convergence is even greater in the female because of the wide pelvis. The shaft has several important structures for muscle attachment. On the proximo-lateral side of the shaft is the **greater trochanter,** and on the medial side is the **lesser trochanter.** On the anterior side between the trochanters is the **intertrochanteric line.** On the posterior side between the trochanters is the **intertrochanteric crest.** The **linea aspera** is a vertical ridge on the posterior surface of the shaft. Just proximal to the linea aspera is the roughened **gluteal tubercle.**

The distal end of the femur is expanded for articulation with the tibia. The **medial** and **lateral condyles** are the articular processes for this joint. The depression between the condyles on the posterior aspect is called the **intercondylar fossa.** The **patellar surface** is between the condyles on the anterior side. Above the condyles on the lateral and medial sides are the **epicondyles** for ligament and tendon attachment.

Patella The patella, or kneecap, is a sesamoid bone positioned on the anterior side of the knee joint (figs. 7.16, 7.17). It develops in response to strain in the tendon of the quadriceps femoris muscle. The patella is a triangular bone with a broad **base** and an inferiorly pointed **apex. Articular facets** on the posterior surface of this bone articulate with the medial and lateral condyles of the femur.

linea aspera: L. *linea,* line; *asperare,* rough
patella: L. *patina,* small plate

Table 7.2 Sexual differences of the pelvic girdle

Characteristics	Male pelvis	Female pelvis
General structure	More massive; prominent processes	More delicate; processes not so prominent
Pelvic inlet	Heart shaped	Round or oval
Pelvic outlet	Narrower	Wider
Anterior superior iliac spines	Less wide apart	More wide apart
Obturator foramen	Oval	Triangular
Acetabulum	Faces laterally	Faces more anteriorly
Symphysis pubis	Deeper, longer	Shallower, shorter
Pubic arch	Acute (less than 90°)	Obtuse (greater than 90°)

The functions of the patella are to protect the knee joint and to strengthen the quadriceps tendon. It also increases the leverage of the quadriceps femoris muscle as it straightens (extends) the leg.

The patella can be fractured by a direct blow. It usually does not fragment, however, because it is confined within the tendon. Dislocations of the patella may result from injury or may be congenital due to underdevelopment of the lateral condyle of the femur.

Leg

The tibia and fibula are the skeletal elements of the leg. The tibia is the larger and more medial of the two bones. The skeletal structure of the leg is illustrated in figure 7.17.

Tibia The tibia (shinbone) articulates proximally with the femur at the knee joint to bear the weight of the body. On the distal end, the tibia articulates with the talus of the ankle. Two slightly concave surfaces on the proximal end of the tibia, the **medial** and **lateral condyles,** articulate with the condyles of the femur. Between the condyles is a slight upward projection called the **intercondylar eminence.** The **tibial tuberosity,** for attachment of the patellar ligament, is located on the proximal-anterior portion of the shaft. There is a sharp ridge along the anterior surface of the shaft called the **anterior crest.**

The **medial malleolus** *(mal-le'o-lus)* is a prominent medial knob of bone located on the distal end of the tibia. A **fibular notch,** for articulation with the fibula, is located on the distal-lateral end.

Fibula The fibula is a long, narrow bone that is more important for muscle attachment than for support. The **head** of the fibula articulates with the lateral-proximal end of the tibia. The distal end has a prominent knob called the **lateral malleolus.**

tibia: L. *tibia,* shinbone, pipe, flute
fibula: L. *fibula,* clasp or brooch
malleolus: L. *malleolus,* small hammer

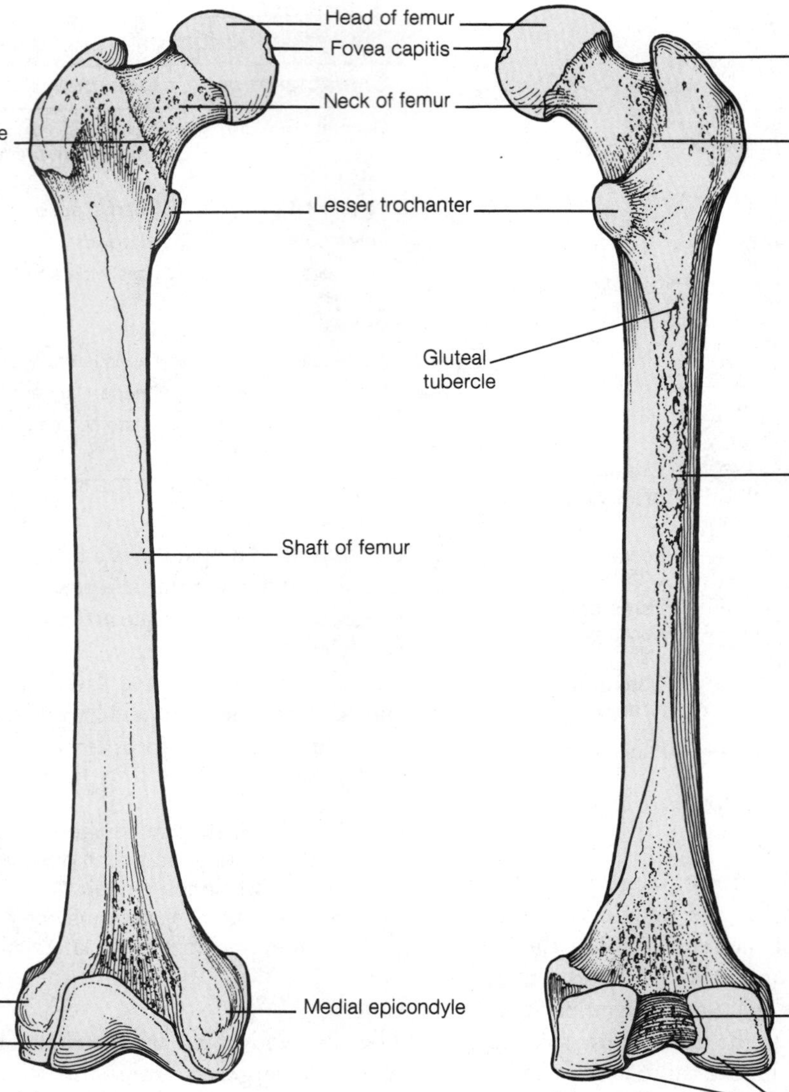

Figure 7.15 The right femur. (*a*) An anterior view and (*b*) a posterior view.

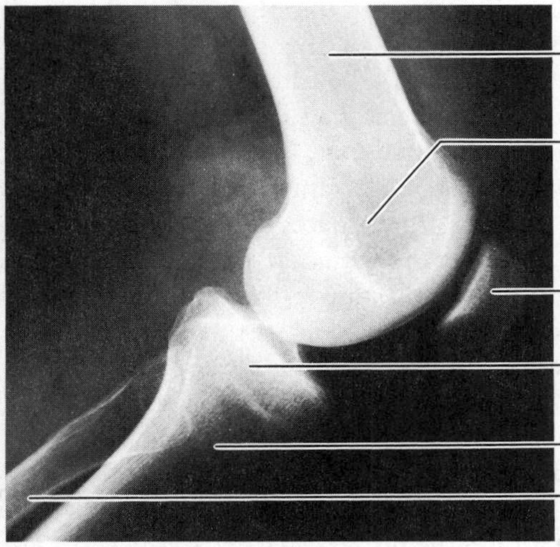

Figure 7.16 A lateral X ray of the right knee region.

The lateral and medial malleoli are positioned on either side of the talus and help stabilize the ankle joint. Both processes can be seen as prominent surface features and are easily palpated. Fractures to either or both malleoli are common in skiers. These fractures, clinically referred to as *Pott's fractures,* result from a shearing force occurring at a vulnerable spot on the leg.

Ankle and Foot (Pes)

The ankle and foot contain twenty-six bones, arranged into the tarsus, metatarsus, and phalanges (figs. 7.18, 7.19). The bones of the ankle and foot are basically similar to those of the wrist and hand. They do, however, have distinct structural differences in order to provide weight support and leverage during walking.

Tarsus There are seven tarsal bones. The **talus** is the tarsal bone that articulates with the tibia to form the ankle joint. The

tarsus: G. *tarsos,* flat of the foot
talus: L. *talus,* ankle

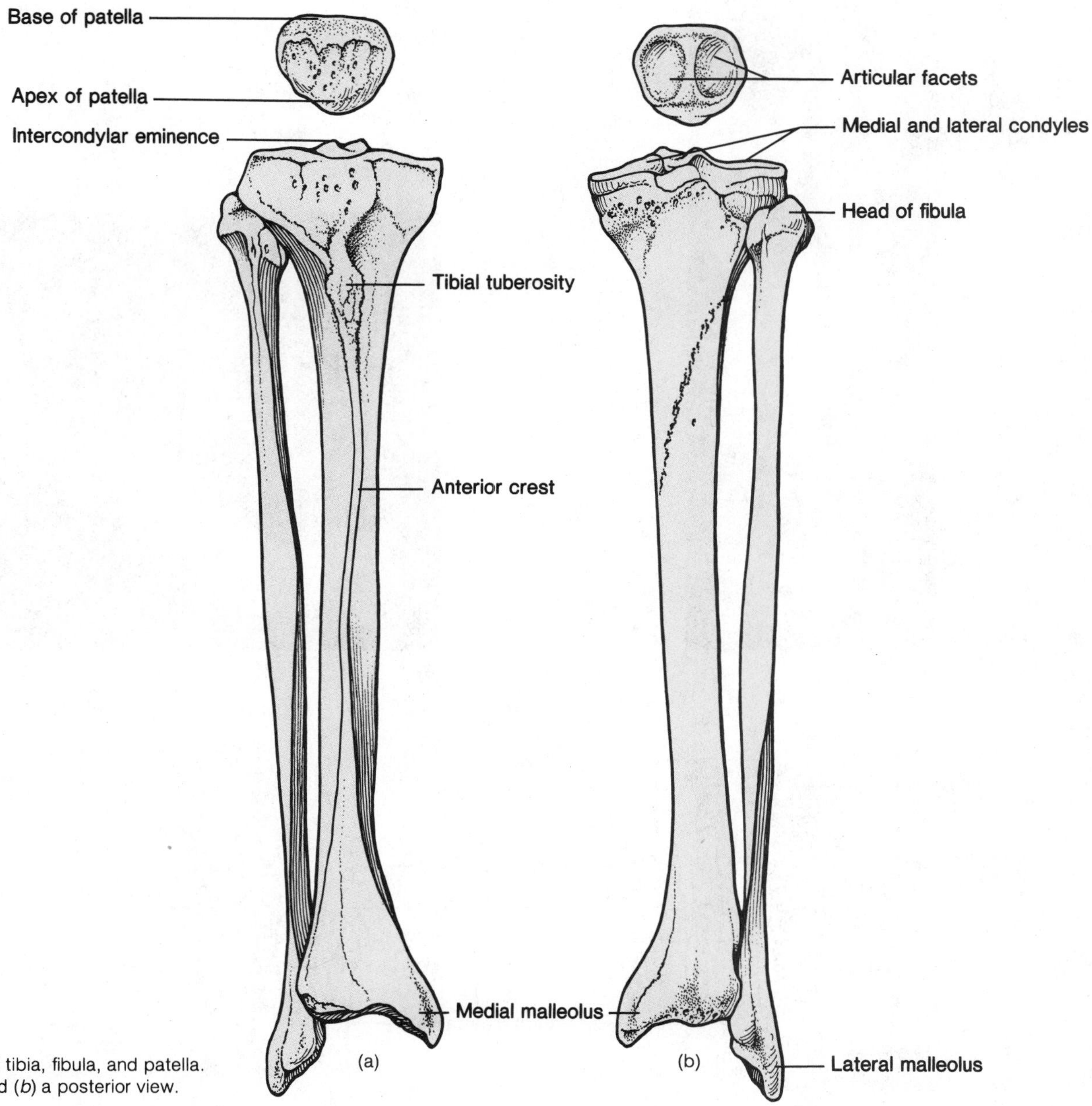

Figure 7.17 The right tibia, fibula, and patella. (*a*) An anterior view and (*b*) a posterior view.

calcaneus *(kal-ka'ne-us)* is the largest of the tarsal bones and forms the heel of the foot. It has a large posterior extension, called the **tuberosity of the calcaneus,** for the attachment of the calf muscles. Anterior to the talus is the block-shaped **navicular** *(nah-vik'u-lar)* bone. The remaining four tarsal bones form a distal series that articulate with the metatarsals. They are, from the medial to lateral side, the **first, second,** and **third cuneiforms** *(ku-ne'i-formz)* and the **cuboid.**

Metatarsus The metatarsus is composed of five metatarsal bones. The metatarsals are numbered from one to five, with the medial, or big toe, side being one. The first metatarsal is larger than the others because of its weight-bearing function.

The metatarsals are long bones, each having a **base, shaft,** and **head.** The proximal bases of the first, second, and third metatarsals articulate proximally with the cuneiforms. The heads of the metatarsals articulate distally with the proximal phalanges. The proximal joints are called **tarsometatarsal joints,** and the distal joints are called **metatarsophalangeal joints.** The ball of the foot is formed by the heads of the first two metatarsal bones.

Phalanges The fourteen phalanges are the skeletal elements of the toes. As with the fingers of the hand, the phalanges of the toes are arranged into a proximal row, a middle row, and a distal row. The great toe (hallux) has only a proximal and a distal phalanx.

Arches of the Foot The foot has two arches that support the weight of the body and provide leverage when walking. These arches are formed by the structure and arrangement of the bones held in place by ligaments and tendons. The arches are not rigid; they yield when weight is placed on the foot and spring back as the weight is lifted (fig. 7.19).

---

calcaneus: L. *calcis,* heel

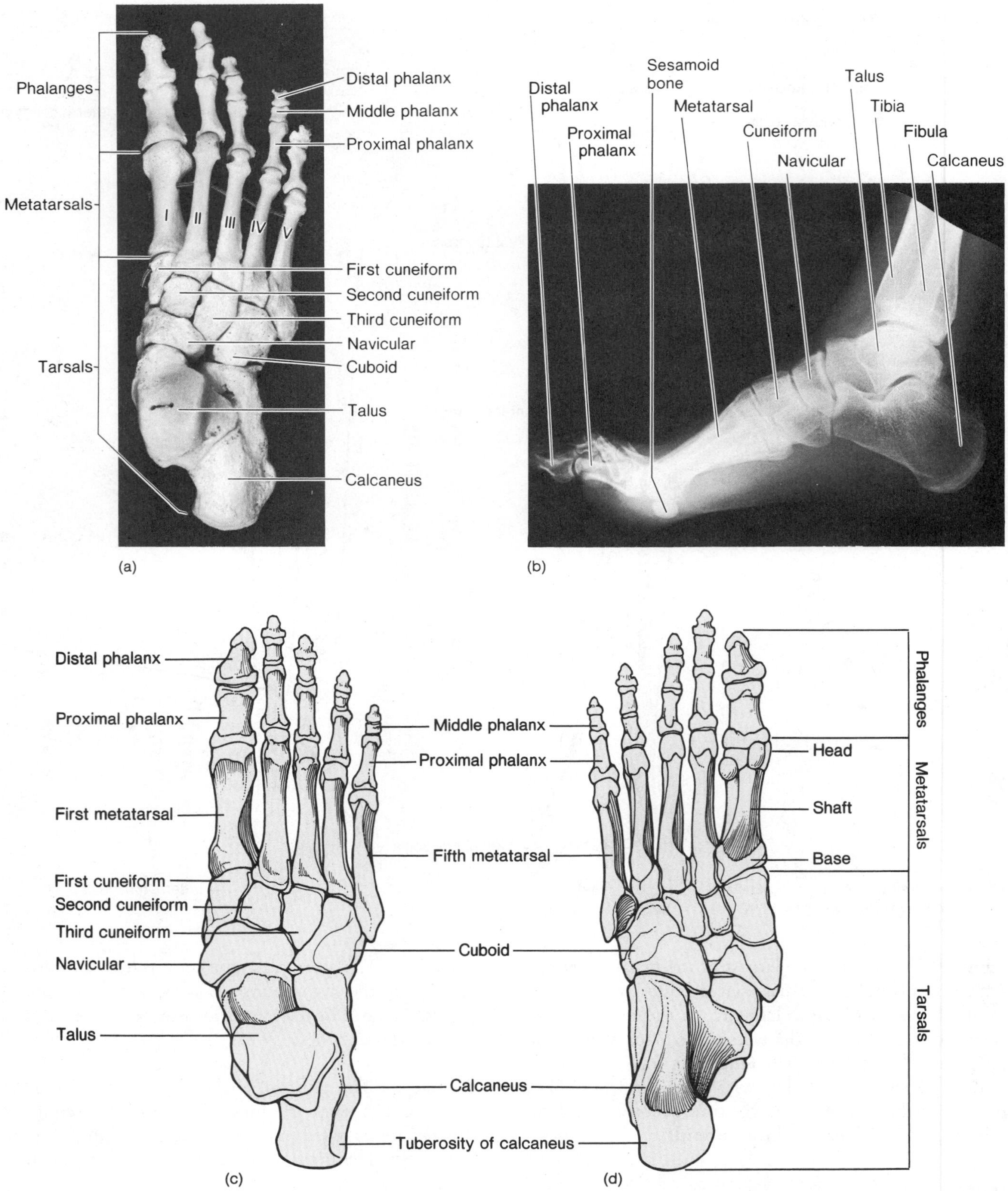

Figure 7.18 (*a*) A superior view of the right foot, (*b*) a medial X ray of the right foot and ankle, (*c*) a superior view of the bones of the right ankle and foot, and (*d*) an inferior view of the bones of the right ankle and foot. (Note the presence of a sesamoid bone in (*b*) at the base of the big toe.)

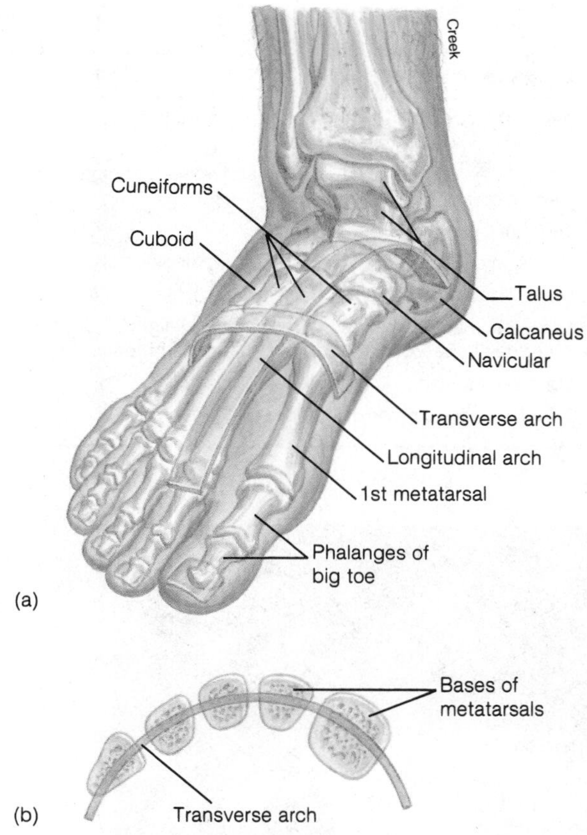

Figure 7.19 The arches of the foot. (*a*) A medial view of the left foot showing both arches. (*b*) A transverse view through the bases of the metatarsals showing a portion of the transverse arch.

The **longitudinal arch** is divided into medial and lateral portions. The medial portion is the larger of the two. It is supported by the calcaneus proximally and by the heads of the first three metatarsals distally. The wedge, or keystone, of this portion of the longitudinal arch is the talus. The shallower lateral portion consists of the calcaneus, cuboid, and fourth and fifth metatarsals. The cuboid is the keystone of this portion.

The **transverse arch** extends across the width of the foot and is formed by the distal portion of the calcaneus, navicular, cuboid, and the proximal portions of all five metatarsals.

A weakening of the ligaments and tendons of the foot decreases the height of the longitudinal arch in a condition called *pes planus,* or flatfoot.

The bones of the lower extremities are summarized in table 7.3.

---

1. Describe the structure of the pelvic girdle that reflects its weight-bearing role.
2. How can female and male pelves be distinguished? What is the clinical importance of the lesser pelvis in females?
3. Describe the structure of each of the long bones of the lower extremity and the position of each of the tarsal bones.
4. Which bones of the foot participate in the formation of the arches of the foot? What are the functions of the arches?

Clinical Considerations

Developmental Disorders

Minor defects of the extremities are relatively common malformations. Extra digits, a condition called **polydactyly** *(pol''e-dak'tĭ-le)* (fig. 7.20), is the most common limb deformity. Usually an extra digit is incompletely formed and does not function. **Syndactyly** *(sin-dak'tĭ-le),* or webbed digits, is likewise a relatively common limb malformation. Polydactyly is inherited as a dominant trait, whereas syndactyly is a recessive trait.

Talipes *(tal'ĭ-pēz),* or clubfoot (fig. 7.21), is a congenital malformation in which the sole of the foot is twisted medially. It is not certain if abnormal positioning or restricted movement *in utero* causes this condition, but both genetics and environmental conditions are involved in most cases.

Trauma and Injury

The most common type of bone injury is a **fracture.** A fracture is the cracking or breaking of a bone. In *radiology,* X rays are used to diagnose the position and extent of a fracture. Fractures may be classified in several ways, and the type and severity of the fracture varies with the age and the general health of the body. **Spontaneous,** or **pathologic, fractures,** for example, result from diseases that weaken the bones. Most fractures, however, are called **traumatic fractures** because they are caused by injuries. The following are descriptions of several kinds of traumatic fractures (fig. 7.22).

1. **Simple,** or **closed.** The fractured bone does not break through the skin.
2. **Compound,** or **open.** The fractured bone is exposed to the outside through an opening in the skin.
3. **Partial (fissured).** The bone is incompletely broken.
4. **Complete.** The fracture has separated the bone into two portions.
5. **Capillary.** A hairlike crack occurs within the bone.
6. **Comminuted** *(kom'ĭ-nut''ed)***.** The bone is splintered into small fragments.
7. **Spiral.** The bone is twisted as it is broken.
8. **Greenstick.** In this incomplete break one side of the bone is broken, and the other side is bowed.
9. **Impacted.** One broken end of a bone is driven into the other.
10. **Transverse.** A fracture occurs across the bone at right angles to the shaft.
11. **Oblique.** A fracture occurs across the bone at an oblique angle to the axis of the bone.
12. **Colles'.** A fracture of the distal portion of the radius.
13. **Pott's.** A fracture of either or both of the distal ends of the tibia and fibula at the level of the malleoli.
14. **Avulsion.** In this fracture a portion of a bone is torn off.
15. **Depressed.** The broken portion of the bone is driven inwards, as in certain skull fractures.

---

polydactyly: Gk. *polys,* many; *daktylos,* finger
syndactyly: Gk. *syn,* together; *daktylos,* finger
talipes: L. *talus,* heel; *pes,* foot

Table 7.3 Bones of the pelvic girdle and the lower extremities

Name and number	Location	Diagnostic features
Os coxa (2)	Hip, part of the pelvic girdle; composed of 3 fused bones	Iliac crest, acetabulum, anterior superior iliac spine, ischial tuberosity, obturator foramen
Femur (2)	Bone of the thigh between hip and knee	Head, fovea capitis, neck, greater and lesser trochanters, lateral and medial condyles
Patella (2)	Anterior surface of knee	Triangular sesamoid bone
Tibia (2)	Medial side of leg between knee and ankle	Medial and lateral condyles, tibial crest, medial malleolus
Fibula (2)	Lateral side of leg between knee and ankle	Head, lateral malleolus
Tarsal (14)	Ankle	Large talus and calcaneus to receive the weight of leg; five other wedge-shaped bones to help form arches of foot
Metatarsal (10)	Sole of foot	Long bones, one in line with each digit
Phalanx (28)	Digits	Three in each toe, two in big toe

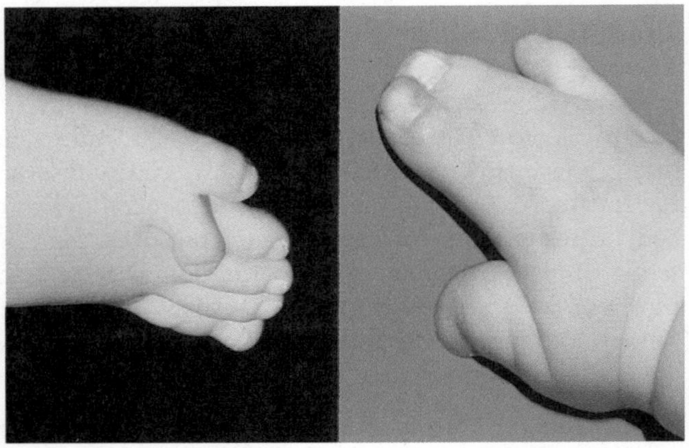

Figure 7.20 Polydactyly is having extra digits. It is the most common congenital deformity of the foot, although it also occurs in the hand. Syndactyly is when two or more digits are webbed together. It is a common congenital deformity of the hand, although it also occurs in the foot. Both conditions can be surgically corrected.

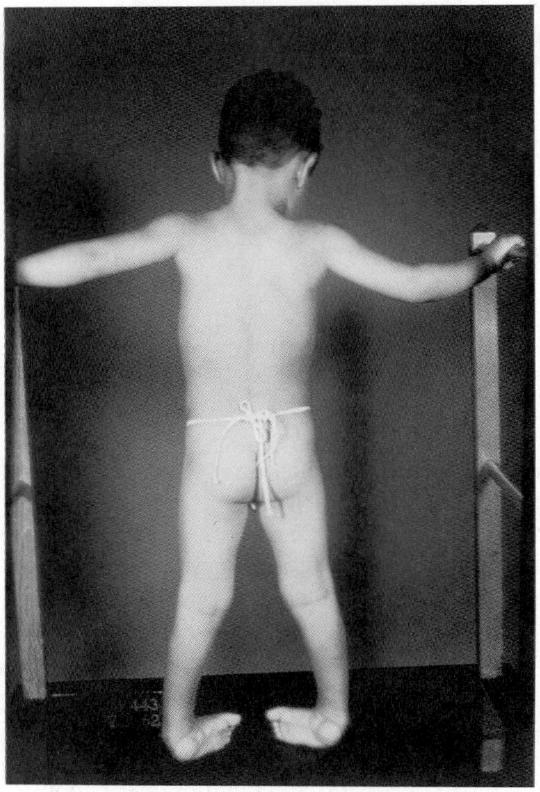

Figure 7.21 Talipes, or clubfoot, is a congenital malformation of a foot or both feet. The condition can be effectively surgically treated if done at a young age.

16. **Displaced.** In this fracture the bone fragments are not in anatomical alignment.
17. **Nondisplaced.** In this fracture the bone fragments are in anatomical alignment.

When a bone fractures, medical treatment involves realigning the broken ends and then immobilizing them until new bone tissue is formed and the fracture is healed. The site and severity of the fracture and the age of the patient will determine the type of immobilization. The methods of immobilization include tape, splints, casts, straps, wires, and steel pins. Even with these various methods of treatment, certain fractures heal poorly. Ongoing research into these problems is producing promising results. It has been found, for example, that applying weak electrical currents to fractured bones promotes healing and reduces the time of immobilization by half.

Physicians can realign and immobilize a fracture, but the ultimate repair of the bone occurs naturally within the bone itself. Several steps occur in the repair of a fracture (fig. 7.23).

1. When a bone is fractured, the surrounding periosteum is usually torn and blood vessels in both tissues are ruptured. A blood clot called a **fracture hematoma** *(hem-ah-to'mah)* soon forms throughout the damaged area. A disrupted blood supply to osteocytes and periosteal cells at the fracture site causes localized cellular death. This is followed by swelling and inflammation.
2. The traumatized area is "cleaned up" by the activity of phagocytic cells within the blood and osteoclasts that resorb bone fragments. As the debris is removed, fibrocartilage fills the gap within the fragmented bone, and a cartilaginous mass called a **callus** is formed. The callus becomes the precursor to the formation of bone in

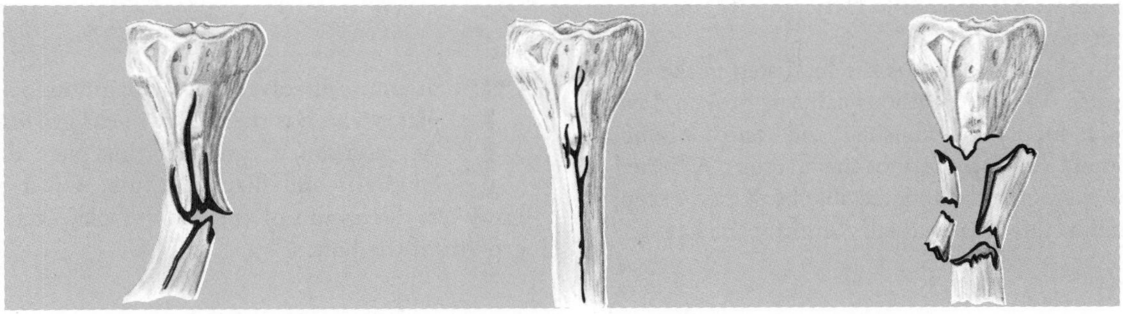

A *greenstick* fracture is incomplete, and the break occurs on the convex surface of the bend in the bone

A *partial* (*fissured*) fracture involves an incomplete longitudinal break

A *comminuted* fracture is complete and results in several bony fragments

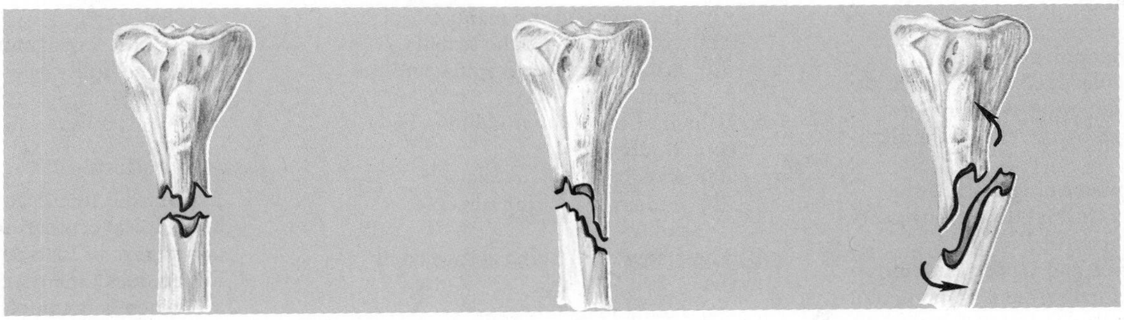

A *transverse* fracture is complete, and the break occurs at a right angle to the axis of the bone

An *oblique* fracture occurs at an angle other than a right angle to the axis of the bone

A *spiral* fracture is caused by twisting a bone excessively

Figure 7.22 Examples of types of traumatic fractures.

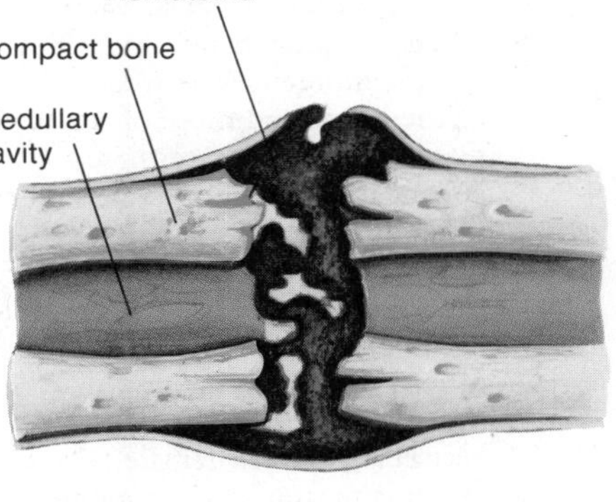

(a) Blood escapes from ruptured blood vessels and forms a hematoma

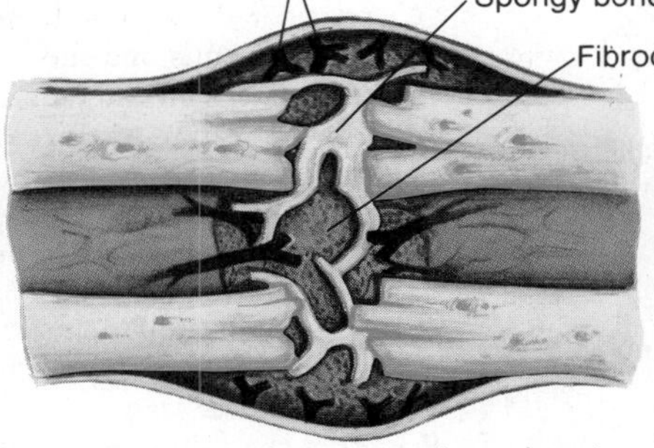

(b) Spongy bone forms in regions close to developing blood vessels, and fibrocartilage forms in more distant regions

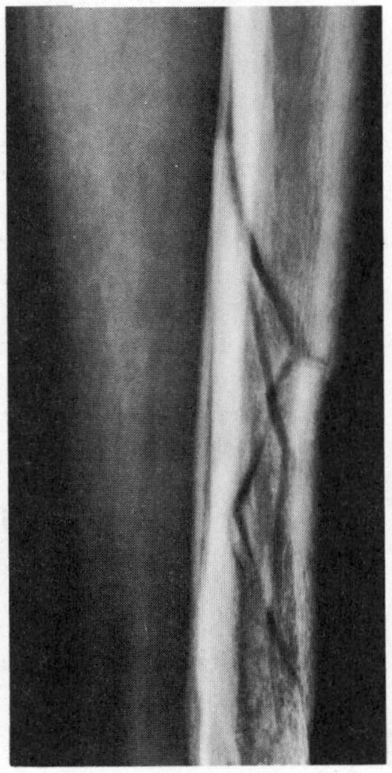

(e)

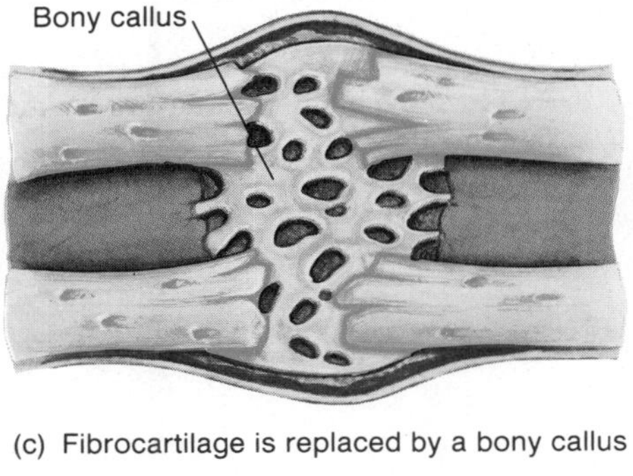

(c) Fibrocartilage is replaced by a bony callus

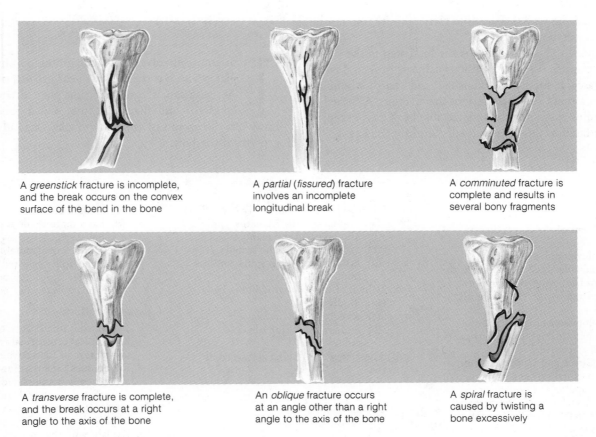

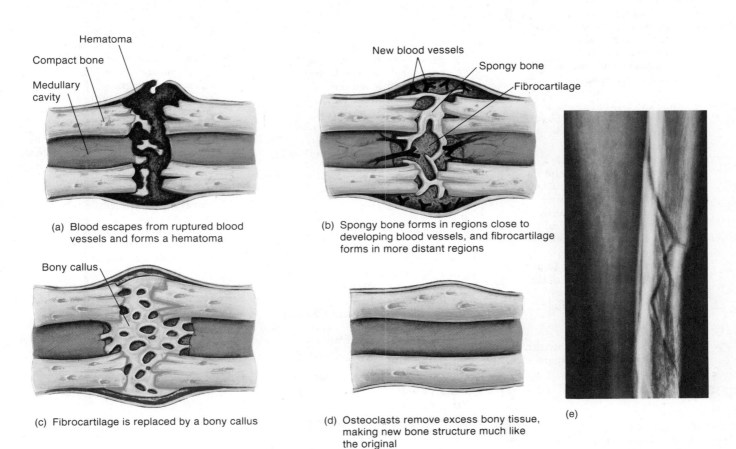

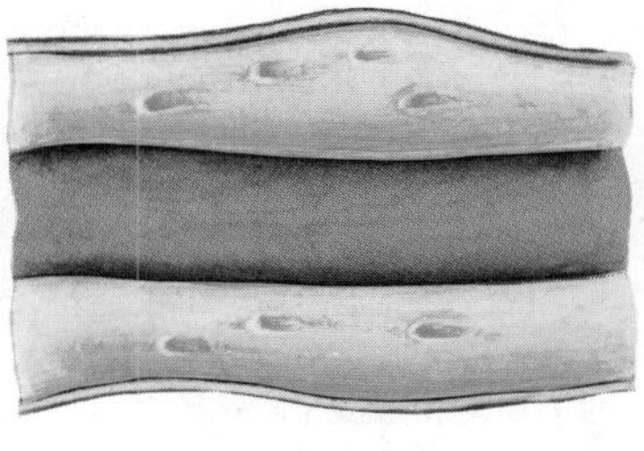

(d) Osteoclasts remove excess bony tissue, making new bone structure much like the original

Figure 7.23 (*a–d*) Stages of the repair of a fracture. (*e*) An X ray of a healing fracture.

much the same way that hyaline cartilage is the precursor to developing bone.

3. The remodeling of the callus is the final step in the healing process. The cartilaginous callus is broken down, a new vascular supply is established, and compact bone develops around the periphery of the fracture. A healed fracture line is frequently undetectable by X ray, except that the bone in this area is usually slightly thicker.

CLINICAL CASE STUDY ANSWER

The injury involves the cartilaginous epiphyseal growth plate, which is the site of linear growth of long bones. At cessation of growth, this plate disappears as the epiphysis and diaphysis fuse. Until this occurrence, however, disruption of the growth plate can adversely affect growth of the bone. ■

Chapter Summary

I. Development of the Appendicular Skeleton
   A. The development of the appendicular skeleton begins at the end of the fourth week and is completed by the end of the sixth week.
   B. Development progresses as the limb buds follow the growth of apical ectodermal ridges.

II. Pectoral Girdle and Upper Extremity
   A. The pectoral girdle is composed of two scapulae and two clavicles. The clavicles attach the pectoral girdle to the axial skeleton at the sternum.
   B. The brachium contains the humerus, which extends from the scapula to the elbow.
   C. The forearm contains the medial ulna and the lateral radius.
   D. The wrist and hand contain twenty-seven bones arranged into the carpus, metacarpus, and phalanges.

III. Pelvic Girdle and Lower Extremity
   A. The pelvic girdle is formed by two ossa coxae united anteriorly by the symphysis pubis.
      1. The pelvis is divided into a greater pelvis, which helps to support the pelvic viscera, and a lesser pelvis, which forms the walls of the birth canal.
      2. Each os coxa consists of an ilium, ischium, and pubis.
   B. The thigh contains the femur, which extends from the hip to the knee, where it articulates with the tibia and the patella.
   C. The leg contains the medial tibia and the lateral fibula.
   D. The ankle and foot contain twenty-six bones arranged into the tarsus, metatarsus, and phalanges.

Review Activities

Objective Questions

1. The cartilaginous bones of the appendicular skeleton derive from
   (a) mesoderm. (c) endoderm.
   (b) ectoderm. (d) chondroderm.
2. The clavicle articulates with the
   (a) scapula and the humerus.
   (b) humerus and the manubrium.
   (c) manubrium and the scapula.
   (d) manubrium, the scapula, and the humerus.
3. Which of the following bones has a conoid tubercle?
   (a) scapula (d) clavicle
   (b) humerus (e) ulna
   (c) radius
4. The elbow of the ulna is formed by the
   (a) lateral epicondyle.
   (b) olecranon process.
   (c) coronoid process.
   (d) styloid process.
   (e) medial epicondyle.
5. Which of the following statements concerning the carpus is *false*?
   (a) There are eight carpal bones arranged into two transverse rows of four bones each.
   (b) All of the carpal bones are considered sesamoid bones.
   (c) The navicular and the lunate articulate with the radius.
   (d) The trapezium, trapezoid, capitate, and hamate articulate with the metacarpal bones.
6. Pelvimetry is a measurement of the
   (a) os coxa.
   (b) symphysis pubis.
   (c) pelvic brim.
   (d) lesser pelvis.
7. Which of the following is *not* a structural feature of the os coxa?
   (a) obturator foramen
   (b) acetabulum
   (c) auricular surface
   (d) greater sciatic notch
   (e) linea aspera
8. A fracture across the intertrochanteric line would involve the
   (a) ilium. (d) fibula.
   (b) femur. (e) patella.
   (c) tibia.
9. As compared to the male pelvis, the female pelvis
   (a) is more massive.
   (b) is more narrow at the pelvic outlet.
   (c) is tilted backward.
   (d) has a more shallow symphysis pubis.
10. Clubfoot is a congenital malformation that is medically referred to as
   (a) talipes. (c) pes planus.
   (b) syndactyly. (d) polydactyly.

Essay Questions

1. Explain the significance of the limb buds, apical ectodermal ridges, and digital rays in limb development. When does limb development begin, and when is it completed?
2. Compare the pectoral and pelvic girdles in structure, articulation to the axial skeleton, and function.
3. Explain why the clavicle is more frequently fractured than the scapula.
4. List the processes of the bones of the upper and lower extremities that can be palpated. Why are these bony landmarks important to know?
5. There are some basic similarities and specific differences between the bones of the hands and those of the feet. Contrast and compare these appendages, taking into account the functional role of each.
6. Define *bipedal locomotion,* and discuss the adaptations of the pelvic girdle and lower extremities that permit this type of movement.
7. What are the structural differences between male and female pelves?
8. What is meant by a congenital skeletal malformation? Give two examples of such abnormalities that occur within the appendicular skeleton.
9. What are the differences between spontaneous and traumatic fractures? List some kinds of traumatic fractures.
10. How does a fractured bone repair itself? Why is it important that the fracture be immobilized?

CHAPTER 8

Articulations

Outline and Concepts

Clinical Case Study

Classification of Joints

The articulations between the bones of the skeleton are classified by arthrologists into three types according to structure or into three types according to function, or the degree of movement permitted.

Development of Freely Movable Joints

Diarthroses form by the third month as the epiphyses of adjacent endochondral bones develop distinct configurations and as the muscles that move the joints undergo contractions.

Synarthroses

Articulating bones in synarthroses are tightly connected by either fibrous tissue or cartilage. Synarthrotic joints are rigid and immovable and are of two types: sutures and synchondroses.

Sutures
Synchondroses

Amphiarthroses

Amphiarthroses allow limited motion in response to twisting, compression, or stress. The two types of amphiarthroses are symphyses and syndesmoses.

Symphyses
Syndesmoses

Diarthroses

Diarthroses are freely movable joints enclosed by joint capsules that contain synovial fluid. Types of diarthroses include gliding, hinge, pivot, condyloid, saddle, and ball-and-socket.

Structure of a Diarthrotic Joint
Kinds of Diarthroses

Movements at Diarthroses

Movements at diarthroses are produced by the contraction of skeletal muscles spanning the joints and attaching to or near the bones forming the articulations. In these actions, the bones act as levers, the muscles provide the force, and the joints are the fulcra, or pivots.

Angular
Circular
Special Movements
Biomechanics of Body Movement

Specific Joints of the Body

Of the numerous joints in the body, some have special structural features that enable them to perform particular functions. Furthermore, these joints are somewhat vulnerable to trauma and are therefore clinically important.

Temporomandibular Joint
Sternoclavicular Joint
Humeroscapular (Shoulder) Joint
Elbow Joint
Metacarpophalangeal Joints and Interphalangeal Joints
Coxal (Hip) Joint
Tibiofemoral (Knee) Joint
Talocrural (Ankle) Joint

Clinical Considerations

Trauma to Joints
Diseases of Joints
Treatment of Joint Disorders

Clinical Case Study Answer

Important Clinical Terminology

Chapter Summary

Review Activities

CLINICAL CASE STUDY

A 20-year-old male sustained injury to his right knee during a college football game. Because of rapid swelling and intense pain, he was taken to the emergency room at the local hospital. Upon questioning by the EM physician, the athlete provided this explanation to how the injury occurred. "I was carrying the football on an end-run left, on third down and two. As I planted my right foot to make a cut, I was hit in the knee from the side. I felt my knee give-way and this was followed by severe pain on the inside of my knee."

Close examination by the EM physician reveals marked swelling on the medial portion of the knee. The doctor determines that *valgus stress* (an inward bowing stress on the knee) has caused the medial aspect of the joint to open. What stabilizing structure is most likely injured? What cartilaginous structure is frequently injured in association with the above? Is there an anatomical explanation? What other stabilizing structures within the knee are subject to athletic injury? ■

Classification of Joints

The articulations between the bones of the skeleton are classified by arthrologists into three types according to structure or into three types according to function, or the degree of movement permitted.

Objective 1. Define the terms *arthrology* and *kinesiology.*
Objective 2. Distinguish between a structural and a functional classification of joints.

One of the functions of the skeletal system is to permit body movement. It is not the rigid bones that allow movement but the **articulations,** or **joints,** between the bones. The structure of a joint determines the range of movement it permits. Not all joints are flexible, however, and as one part of the body moves, other joints remain rigid to stabilize the body and maintain balance. The coordinated activity of all of the joints permits the sinuous, elegant movements of a gymnast or ballet dancer, as well as the mundane activities of walking, eating, writing, and speaking.

Arthrology is the science concerned with the study of joints. Generally speaking, an arthrologist is interested in the structure, classification, and function of joints as well as any dysfunction that may develop. **Kinesiology** *(ki-ne''se-ol'o-je)* is a more practical and dynamic science concerned with the functional relationship, or biomechanics, of the skeleton, joints, muscles, and innervation as they work together to produce coordinated movement.

To study the joints, one should adopt a kinetic approach and be able to demonstrate the various movements permitted at each of the movable joints and understand the adaptive advantage as well as the limitations of each type of movement.

arthrology: Gk. *arthron*, joint; *logos*, study
kinesiology: Gk. *kinesis*, movement; *logos*, study

The joints of the body may be classified according to structure or function. Structural classification is based on the presence or absence of a joint cavity and the kind of supportive connective tissue surrounding the joint. In classification by structure, there are three types of joints:

1. **Fibrous** *(fi'brus)*. A fibrous joint lacks a joint cavity, and fibrous connective tissue connects articulating bones.
2. **Cartilaginous** *(kar''tĭ-laj'ĭ-nus)*. A cartilaginous joint lacks a joint cavity, and cartilage binds articulating bones.
3. **Synovial** *(sĭ-no've-al)*. A synovial joint has a joint cavity, and ligaments help to support the articulating bones.

The functional classification of joints is based on the degree of movement permitted within the joint. Using this type of classification, there are three kinds of articulations:

1. **Synarthroses** *(sin''ar-thro'sēz)*. Immovable joints.
2. **Amphiarthroses** *(am''fe-ar-thor'sēz)*. Slightly movable joints.
3. **Diarthroses** *(di''ar-thro'sēz)*. Freely movable joints.

This book uses the functional classification scheme in discussing the various kinds of joints of the body. But continual reference will be made to the structural classification as the specific anatomy of a particular joint is presented.

---

1. Explain the statement that kinesiology is applied arthrology.
2. According to both structural and functional classifications, how would a joint be classified that is tightly bound with hyaline cartilage so that only slight movement is possible?
3. List the three types of functional joints, and speculate as to which provides the greatest support and which is the most vulnerable to trauma.

Development of Freely Movable Joints

Diarthroses form by the third month as the epiphyses of adjacent endochondral bones develop distinct configurations and as the muscles that move the joints undergo contractions.

Objective 3. Describe the development of diarthroses, and discuss the importance of fetal movement to the normal development of freely movable joints.

The sites of developing diarthroses (freely movable joints) are discernible at six weeks as mesenchyme becomes concentrated in the areas where precartilage cells differentiate (fig. 8.1). The future joints at this stage appear as intervals of less concentrated mesenchymal cells. As cartilage cells develop within a forming bone, a thin flattened sheet of cells forms around the cartilaginous model to become the perichondrium. These same cells are continuous across the gap between the adjacent developing bone. Surrounding the gap, the flattened mesenchymal cells differentiate to become the **joint capsule.**

During the early part of the third month of development, the mesenchymal cells still remaining within the joint capsule begin migrating toward the epiphyses of the adjacent developing bones. The cleft eventually enlarges to become the **joint**

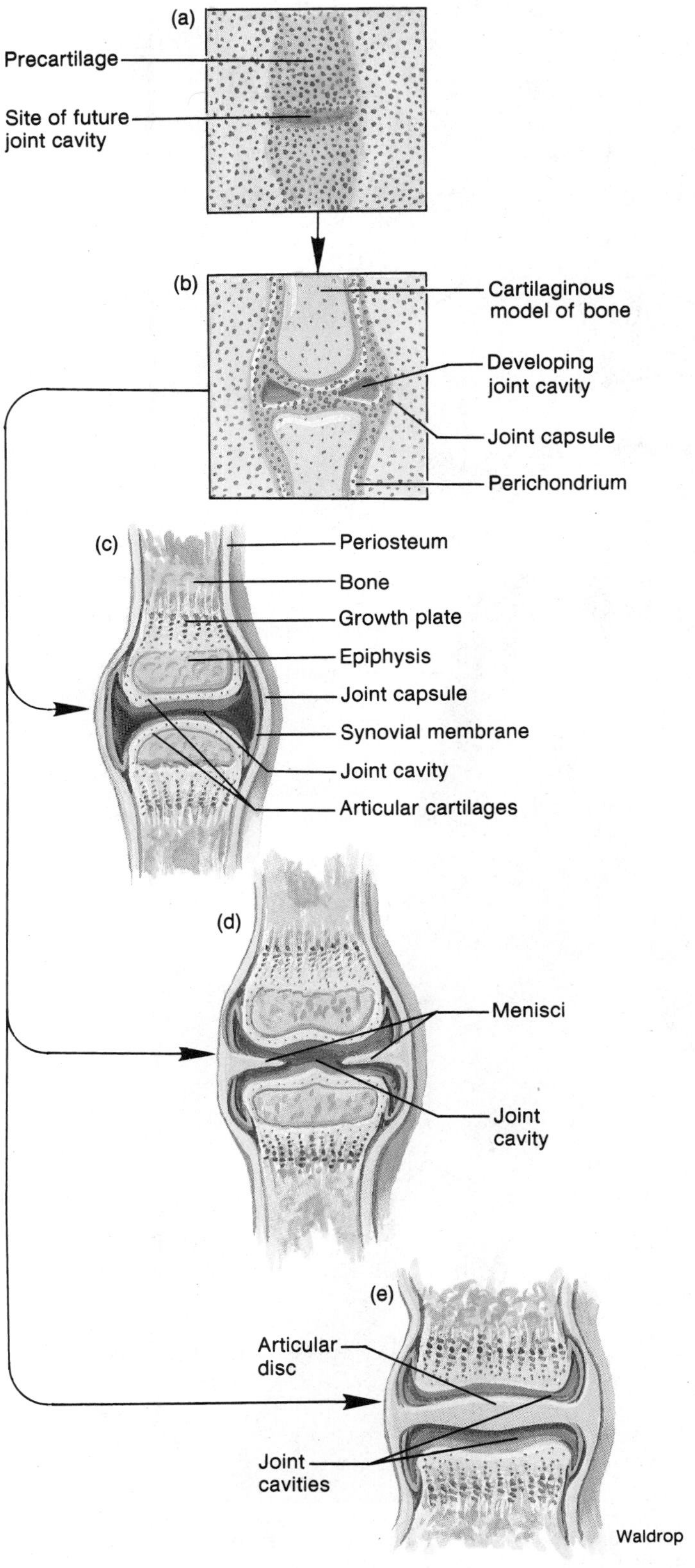

Figure 8.1 The development of diarthroses. (*a*) At six weeks, different densities of mesenchyme denote where the bones and joints will form. (*b*) At nine weeks, a basic diarthrotic model is present. At twelve weeks, the diarthroses are formed and have either (*c*) a free joint cavity (e.g., interphalangeal joint); (*d*) a cavity containing menisci (e.g., knee joint); or (*e*) a cavity with a complete articular disc (e.g., sternoclavicular joint).

cavity. Thin pads of hyaline cartilage develop on the surfaces of the epiphyses that contact the joint cavity. These pads become the **articular cartilages** of the functional joint. As the joint continues to develop, a highly vascular **synovial membrane** forms on the inside of the joint capsule and begins secreting a watery *synovial fluid* into the joint cavity.

In certain developing diarthroses, the mesenchymal cells do not migrate away from the center of the joint cavity but, rather, give rise to cartilaginous wedges, called **menisci,** as in the knee joint (fig. 8.1), or to complete cartilaginous pads, called **articular discs,** as in the sternoclavicular joint.

The formation of most diarthroses is completed by the end of the third month. Shortly after this time, there are fetal muscle contractions, known as *quickening,* that cause movement at these joints. Joint movement enhances the nutrition of the articular cartilage and prevents the fusion of connective tissues within the joint.

---

1. Describe the development of a diarthrotic joint, and give the approximate prenatal date of each event.
2. Describe the variation of cartilaginous separations between the articular cartilages of diarthroses.
3. Explain the importance of quickening to the health of a newly formed diarthrotic joint.

Synarthroses

Articulating bones in synarthroses are tightly connected by either fibrous tissue or cartilage. Synarthrotic joints are rigid and immovable and are of two types: sutures and synchondroses.

Objective 4. Describe the structure of a suture, and indicate where sutures are located.

Objective 5. Describe the structure of a synchondrosis; indicate where synchondroses are located and how they change as a person ages.

Sutures

Sutures, one type of synarthroses, are found only within the skull and are characterized by a thin layer of dense fibrous connective tissue that binds the articulating bones (fig. 8.2). Sutures form at about eighteen months of age and replace the pliable fontanels of an infant's skull (see fig. 6.4).

There are several types of sutures, which can be distinguished on the basis of the appearance of the articulating margin of bone. A **serrate suture** is characterized by interlocking, sawlike articulations. This is the most common type of suture, an example being the sagittal suture between the two parietal bones. In a **lap** (squamous) **suture,** the margin of one bone overlaps that of the articulating bone. The squamous suture formed between the temporal and parietal bones is an example. In a **plane** (butt) **suture,** the margins of the articulating bones are fairly smooth. An example is the maxillary suture, where the two maxillary bones articulate to form the hard palate. A **synostosis** *(sin″os-to′sis),* a unique type of suture, is present

suture: L. *sutura,* sew
synostosis: Gk. *syn,* together; *osteon,* bone

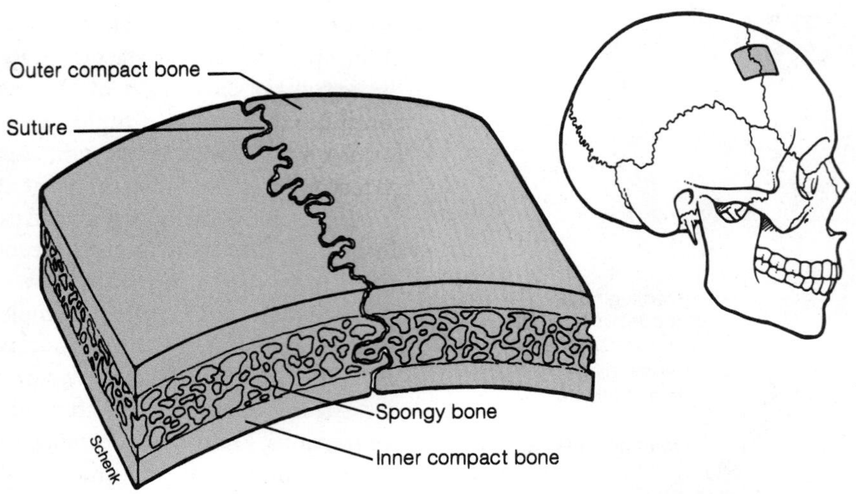

Figure 8.2 A section across the skull showing a suture.

during growth of the skull but becomes totally ossified in the adult. The union between the right and left portions of the frontal bone is an example of synostosis.

Fractures of the skull are fairly common in an adult but much less so in a child. The skull of a child is resilient to blows because of the nature of the bone and the layer of fibrous connective tissue within the sutures. The skull of an adult is much like an eggshell in its lack of resilience and will frequently splinter on impact.

Synchondroses

Synchondroses *(sin''kon-dro'sēs)* are synarthrotic joints that have hyaline cartilage between the bone segments. These joints are typically temporary, forming the growth lines (epiphyseal plates) between the diaphyses and epiphyses in the long bones of children. When growth is complete, the synchondrotic joints ossify. A totally ossified synchondrosis can also be referred to as a synostosis. A synchondrosis can be clearly seen in the X ray of a long bone of a child in figure 8.3.

A fracture of a long bone in a child may be extremely serious if it involves the mitotically active epiphyseal plate of a synchondrotic joint. If such an injury is untreated, there likely will be retarded or arrested bone growth resulting in permanent shortening of the appendage.

1. Distinguish between the three principal kinds of sutures, and give an example and the location of each. What does each have in common? How do synchondroses change with age?
2. Explain why synchondroses are found on bones that have epiphyses and a diaphysis.

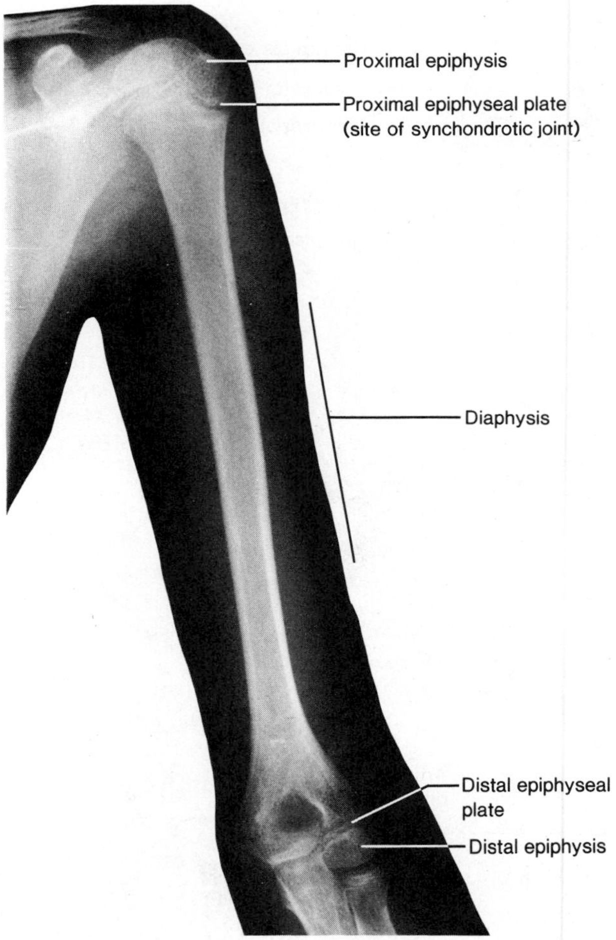

Figure 8.3 An X ray of the left humerus of a ten-year-old child, showing a synchondrotic joint. In a long bone, this type of joint occurs at both the proximal and distal epiphyseal plates. The mitotic activity at a synchondrotic joint is responsible for bone growth in length.

Amphiarthroses

Amphiarthroses allow limited motion in response to twisting, compression, or stress. The two types of amphiarthroses are symphyses and syndesmoses.

Objective 6. Describe the structure of a symphysis, and indicate where symphyses occur.

Objective 7. Describe the structure of a syndesmosis, and indicate where syndesmoses occur.

---

synchondrosis: Gk. *syn,* together; *chondros,* cartilage

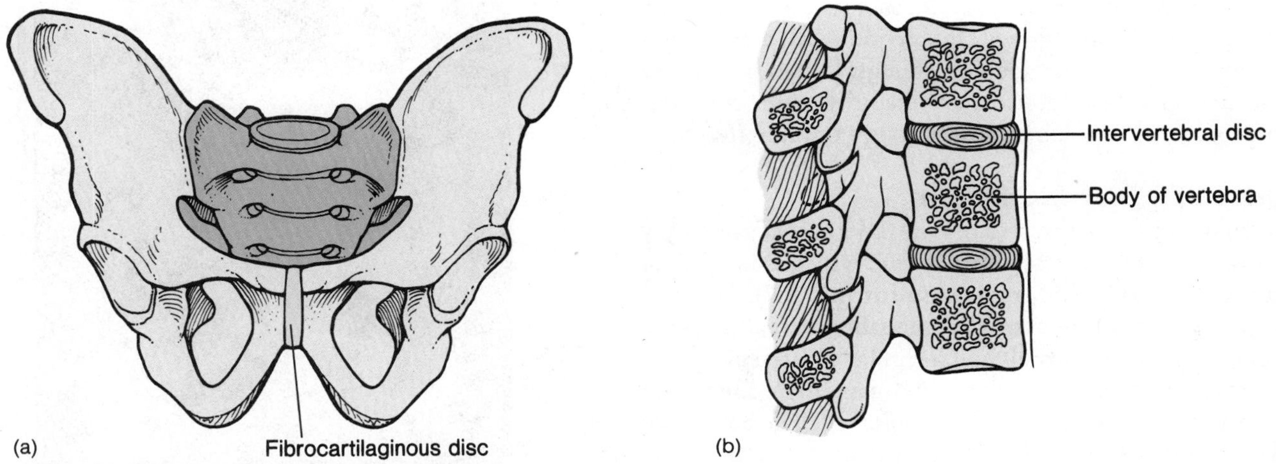

Figure 8.4 Amphiarthrotic joints. (*a*) The symphysis pubis and (*b*) the intervertebral joints between vertebral bodies.

Symphyses

The adjoining bones of a symphysis *(sim'fi-sis)* joint are separated by a pad of fibrocartilage. This pad cushions the joint and allows limited movement. The symphysis pubis and the intervertebral discs (fig. 8.4) are examples of symphyses. Although only limited motion is possible at each intervertebral joint, the combined movement of all the joints of the vertebral column results in extensive spinal action.

Hormonal action in a pregnant woman causes the symphysis pubis, sacroiliac, and sacrococcygeal joints to soften and become more flexible. This relaxation of the pelvic joints increases the potential capacity of the pelvic cavity and reduces trauma during childbirth by increasing the diameter of the pelvic outlet. These joints will strengthen postpartum, but following the birth of a mother's first child, the diameter of her pelvis will generally be slightly wider than before.

Syndesmoses

In a syndesmotic joint, adjacent bones are held together by collagenous fibers or interosseous ligaments. A syndesmosis *(sin''des-mo'sis)* is characteristic of the distal ends of the tibia-fibula and the radius-ulna (fig. 8.5).

---

1. Discuss the function of the pad of fibrocartilage in a symphysis joint, and give two examples of symphyses.
2. What structural feature is characteristic to all syndesmoses? Give two examples of syndesmoses.

Diarthroses

Diarthroses are freely movable joints enclosed by joint capsules that contain synovial fluid. Types of diarthroses include gliding, hinge, pivot, condyloid, saddle, and ball-and-socket.

symphysis: Gk. *symphysis,* growing together
syndesmosis: Gk. *syndesmos,* binding together

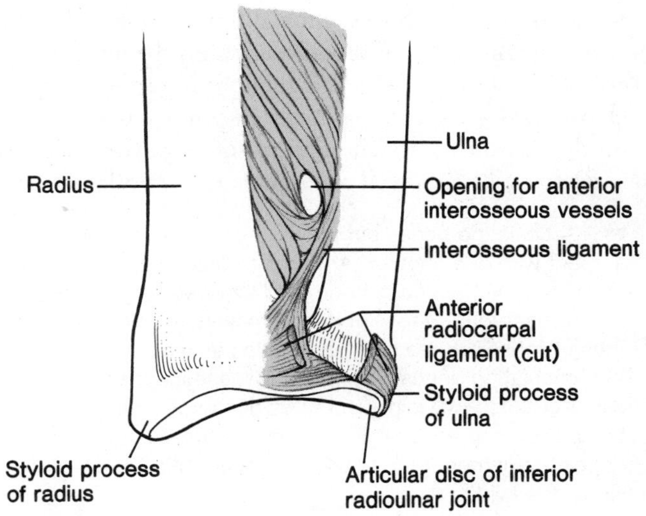

Figure 8.5 The distal articulation of the ulna and radius forms a syndesmotic joint. Interosseous ligaments tightly bind these bones and permit only slight movement between them.

Objective 8. Describe the structure of a diarthrotic joint, and indicate where diarthroses occur.

Objective 9. List and discuss the various kinds of diarthroses, where they occur, and the movements they permit.

The most obvious type of articulation in the body is the freely movable diarthrosis. The dual function of diarthrotic joints is to provide a wide range of precise, smooth movements and yet maintain stability, strength, and, in certain aspects, rigidity in the body.

Diarthroses are the most complex and varied of the three major types of joints. A diarthrotic joint's range of movement is limited by three factors: (1) the structure of the bones participating in the articulation (certain processes on bones, for example, the olecranon process of the ulna, actually limit the range of motion and "lock" the articulation to prevent overextension of the joint); (2) the strength and tautness of the associated ligaments, tendons, and joint capsule; and (3) the size, arrangement, and action of the muscles that span the joint. There is tremendous individual variation in joint motility, most of which is related to body conditioning. Double-jointed is a misnomer because such a joint is not double although it does permit extreme maneuverability. Joint maneuverability varies considerably in people and can be increased through training (fig. 8.6).

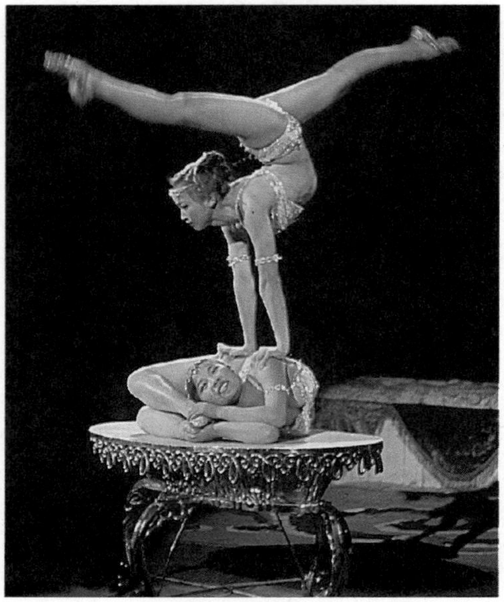

Figure 8.6 Although joint flexibility is structurally determined and limited, the range of movement can become extraordinary through extensive training as in gymnasts and acrobats.

Structure of a Diarthrotic Joint

In structural terms, diarthroses are synovial joints because they are enclosed by a fibroelastic joint capsule, which is filled with lubricating **synovial fluid,** or **synovium** (fig. 8.7). Synovial fluid is secreted by a thin **synovial membrane** that lines the inside of the capsule. Synovial fluid is similar to interstitial fluid (fluid surrounding cells of a tissue), and has a high concentration of hyaluronic acid, a lubricating substance. The bones that articulate in a diarthrosis are capped with a smooth **articular cartilage.** The avascular articular cartilage is only about 2 mm thick and depends upon the alternating compression and decompression during joint activity for the exchange of nutrients and waste products with the synovial fluid. Ligaments help to bind a diarthrosis and may be located within the joint cavity or on the outside of the capsule. Tough, fibrous, cartilaginous pads called **menisci**—singular, *meniscus (mĕ-nis'kus)*—are located within the capsule of certain diarthroses (e.g., the knee joint) and serve to cushion as well as to guide the articulating bones.

Articulating bones of diarthroses do not come in contact with one another. Articular cartilage caps the articular surface of each bone, and synovial fluid circulates through the joint during movement. Both of these joint structures minimize friction as well as cushion the articulating bones. Trauma or disease may render either of these two joint structures nonfunctional, and the two articulating bones will come in contact. Bony deposits will then form, and a type of arthritis will develop within the joint.

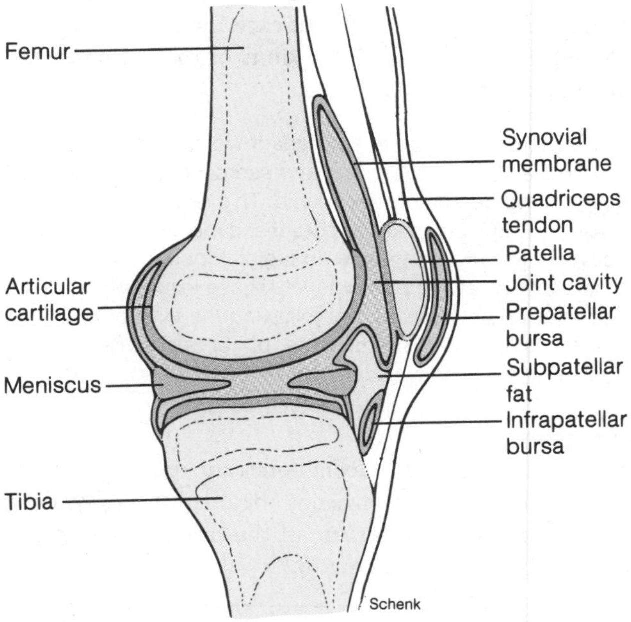

Figure 8.7 A diarthrotic joint is represented by this diagrammatic lateral view of the knee joint.

Located near certain diarthrotic joints are flattened, pouchlike sacs called **bursae**—singular, *bursa (ber'sah)*—which are filled with synovial fluid (fig. 8.7). These closed sacs are commonly located between muscles or within an area where a tendon passes over a bone. The functions of bursae are to cushion certain muscles and to facilitate the movement of tendons or muscles over bony or ligamentous surfaces.

A **tendon sheath** is a modified bursa that surrounds and lubricates the tendons of certain muscles, particularly those that cross the wrist and ankle joints (fig. 8.8). The tendons that are surrounded by sheaths are strong and cordlike and are usually attached to important flexor muscles.

Improperly fitted shoes or inappropriate shoes can cause joint related problems to develop. People who wear high-heeled shoes frequently have perpetual back aches and leg aches because their posture has to counteract the forward tilt of their body that they experience when standing or walking. Their knees are excessively bent and their spine is thrust forward at the lumbar curvature in order to maintain balance. Tightly fitted shoes, especially those with pointed toes, may cause *hallux valgus* to develop. This condition is a deviation of the hallux (great toe) laterally toward the other toes. Hallux valgus is generally accompanied by a bunion forming at the medial base of the proximal phalanx of the hallux. A *bunion* is an inflammation and an accompanying callus that develops because of pressure and rubbing of a shoe.

Kinds of Diarthroses

Diarthroses are classified into six main types according to their structure and the kinds of motion they permit. The six types are gliding, hinge, pivot, condyloid, saddle, and ball-and-socket.

meniscus: Gk. *meniskos,* small moon
bursa: Gk. *byrsa,* bag or purse

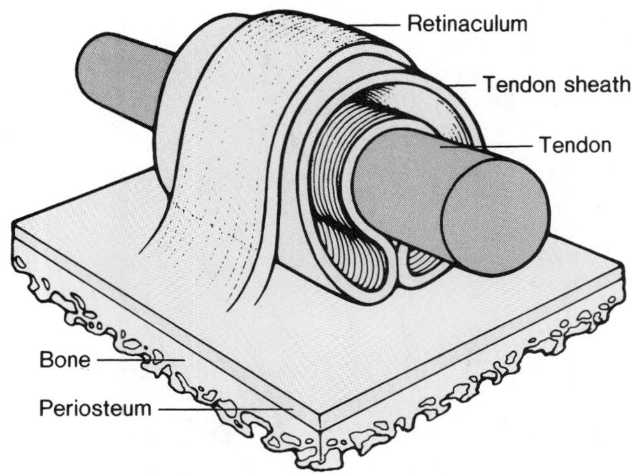

Figure 8.8 A tendon sheath facilitates the gliding of a tendon as it traverses a fibrous or bony tunnel. Tendon sheaths are closed sacs, one layer of the synovial membrane lining the tunnel, the other folding over the surface of the tendon.

Gliding Gliding joints allow only side-to-side and back-and-forth movements with some slight rotation. This is the simplest type of joint movement. The articulating surfaces can be nearly flat, or one may be slightly concave and the other slightly convex. The intercarpal and intertarsal joints, as well as the sternoclavicular joint and the joint between the articular processes of adjacent vertebrae, are examples (fig. 8.9).

Hinge The structure of a hinge joint permits bending in only one plane, much like the hinge of a door. In this type of articulation the surface of one bone is always concave, the other convex (fig. 8.10). Hinge joints are the most common type of diarthroses and include such specific joints as the knee, the elbow (humeroulnar articulation), and the interphangeal joints.

Pivot The movement in a pivot joint is limited to rotation about a central axis. In this type of articulation the articular surface on one bone is conical or rounded and fits into a depression on another bone. Examples are the proximal articulation of the radius and ulna for rotation of the forearm, as in turning a doorknob, and the articulation between the atlas and axis that makes rotational movement of the head possible (fig. 8.11).

Condyloid (Ellipsoid) A condyloid *(kon'dĭ-loid)* articulation is structured so that an oval, convex articular surface of one bone fits into an elliptical, concave depression on another bone. This permits angular movement in two directions (biaxial), as in an up-and-down and side-to-side motion. A condyloid joint does not permit rotational movement. The radiocarpal joint of the wrist is an example (see fig. 8.9).

Saddle Each articular process of a saddle-shaped joint has a concave surface in one direction and a convex surface in another. This unique articulation is a modified condyloid joint that allows a wide range of movement and is found only at the base of the thumb at the articulation of the trapezium of the carpus with the first metacarpal bone (fig. 8.12).

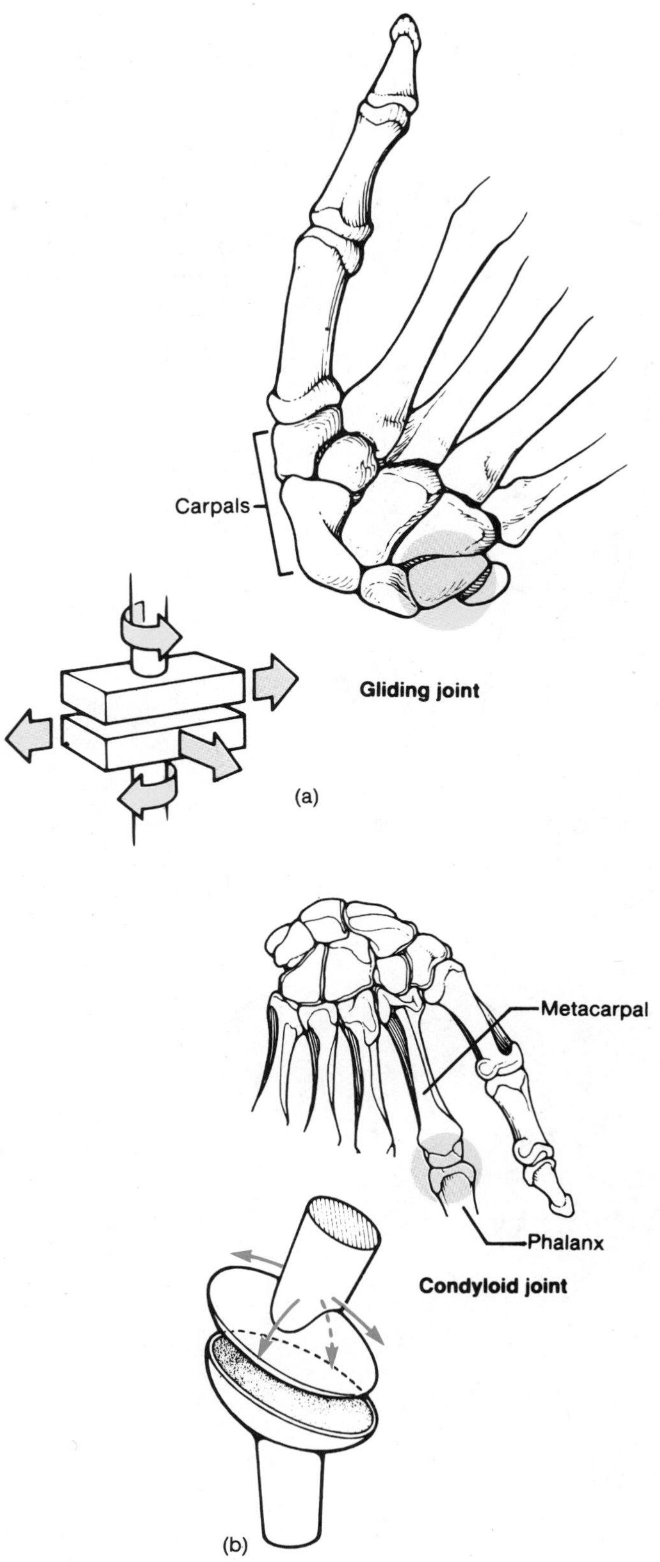

Figure 8.9 The locations of gliding and condyloid joints. (*a*) Gliding joints are located between the proximal and distal carpals and between the individual carpals. (*b*) Condyloid joints are located between the metacarpal bones and the phalanges. (Note the diagrammatic representation of these joints showing the directions of possible movements.)

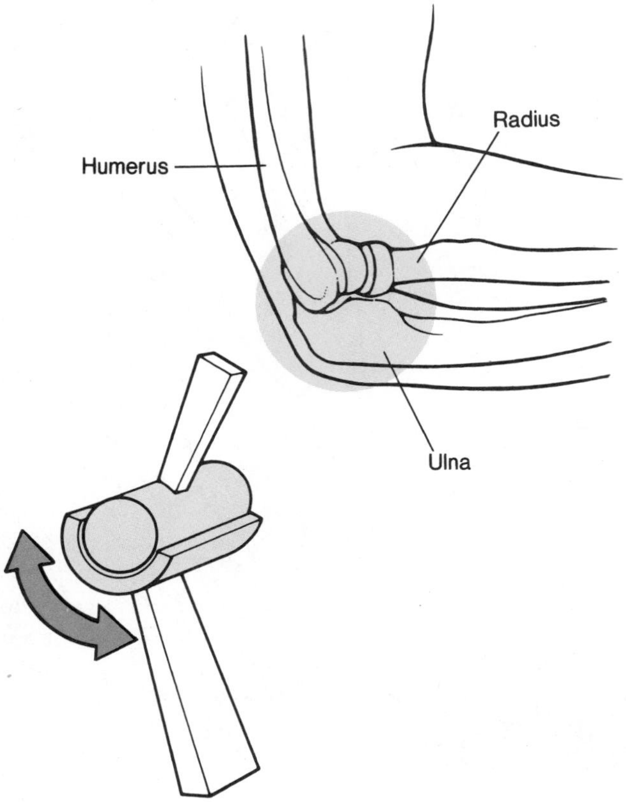

Figure 8.10 A hinge joint permits only a bending movement (flexion and extension). The hinge joint of the elbow involves the distal end of the humerus articulating with the proximal end of the ulna. (Note the diagrammatic representation of this joint showing the direction of possible movement.)

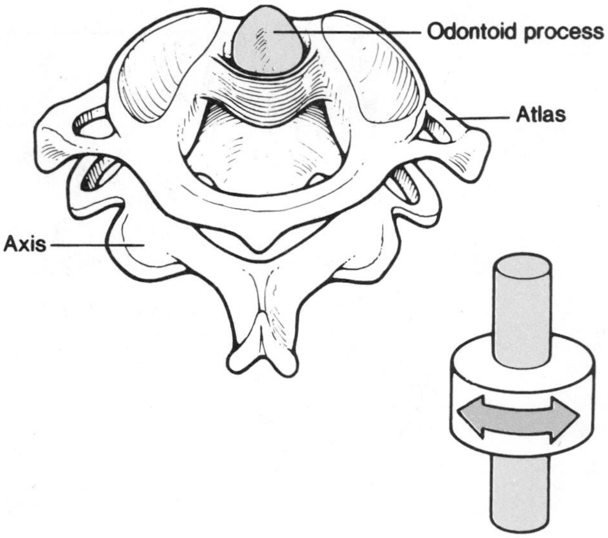

Figure 8.11 The atlas articulating with the axis forms a pivot joint that permits a rotational movement in one axis. (Note the diagrammatic representation showing the direction of possible movement.) Refer to figure 8.10 and determine which articulating bones of the elbow region form a pivot joint.

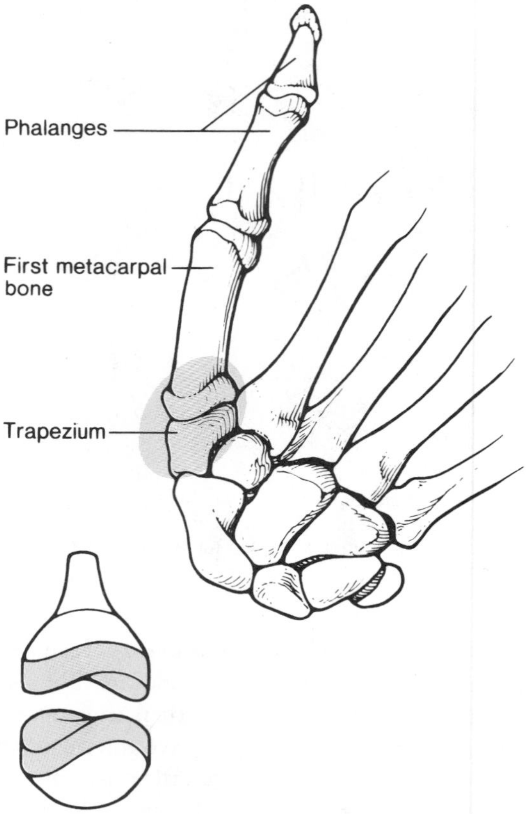

Figure 8.12 A saddle joint is formed as the trapezium articulates with the base of the first metacarpal bone. (Note the diagrammatic representation of this joint.)

Ball-and-Socket Ball-and-socket joints are formed by the articulation of a rounded convex surface with a cuplike cavity (fig. 8.13). This type of articulation provides the greatest range of movement of all of the diarthrotic joints. Examples are the hip and shoulder joints.

A summary of the various types of joints within the body is presented in table 8.1.

Synovial fluid within the diarthrotic joints serves to lubricate the articular surfaces and provide nourishment to the articular cartilage. Trauma to the joint causes the excessive production of synovial fluid in an attempt to cushion and immobilize the joint. This leads to swelling and discomfort to the joint. The most frequent type of joint injury is a sprain, in which the supporting ligaments or joint capsule are damaged to varying degrees.

---

1. List the structures of a diarthrosis, and explain the function of each.
2. What three factors limit the range of movement within diarthroses?
3. Give an example of each of the six kinds of diarthroses, and describe the range of movement possible at each.

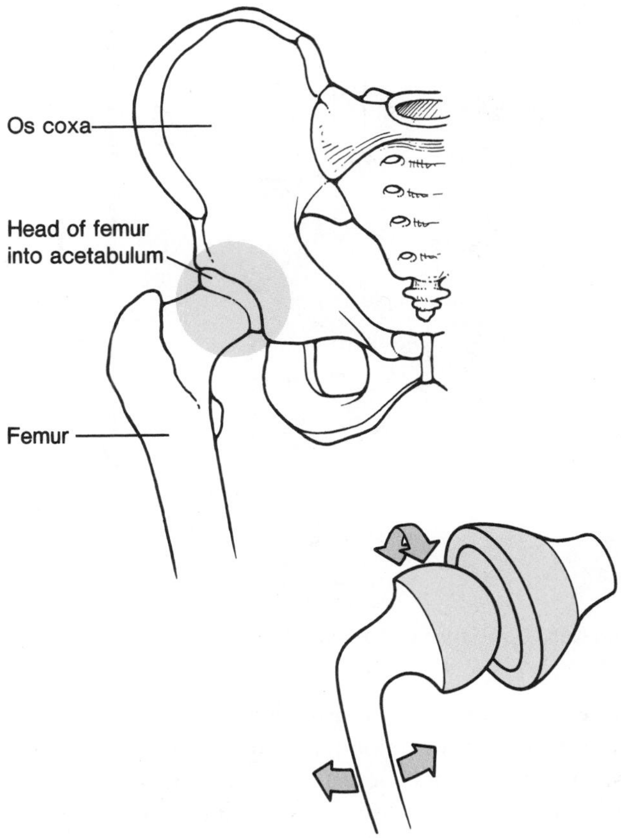

Figure 8.13 A ball-and-socket articulation illustrated by the hip joint. (Note the diagrammatic representation showing the directions of possible movements.)

Movements at Diarthroses

Movements at diarthroses are produced by the contraction of skeletal muscles spanning the joints and attaching to or near the bones forming the articulations. In these actions, the bones act as levers, the muscles provide the force, and the joints are the fulcra, or pivots.

Objective 10. List and discuss the various kinds of movements possible within diarthrotic joints.

Objective 11. Describe the components of a lever, and explain the role of diarthrotic joints in lever systems.

Objective 12. Compare the structure of first-, second-, and third-class levers.

As previously mentioned, the range of movement is determined by the structure of the individual joint and the arrangement of the associated muscle and bone. The movement at a hinge joint, for example, occurs in only one plane, whereas the structure of a ball-and-socket joint permits movement around many axes. Movements within joints are broadly classified as **angular** and **circular.** Within each of these categories are specific types of movements, and certain special movements may involve several of the specific types.

Angular

Angular movements increase or decrease the joint angle produced by the articulating bones. The four types of angular movements are flexion, extension, abduction, and adduction.

Table 8.1 Types of articulations

Type	Structure	Movements	Example
Synarthroses			
Suture	Frequently serrated edges of articulating bones; separated by thin layer of fibrous tissue	None	Sutures between bones of the skull
Synchondroses	Mitotically active hyaline cartilage between bony segments	None	Epiphyseal plates within long bones
Amphiarthroses			
Symphyses	Articulating bones separated by pad of fibrocartilage	Slightly movable	Intervertebral joints; symphysis pubis and sacroiliac joint
Syndesmoses	Articulating bones bound by interosseous ligament	Slightly movable	Joints between tibia-fibula and radius-ulna
Diarthroses	Joint capsule containing synovial membrane and synovial fluid	Freely movable	
Gliding	Flattened or slightly curved articulating surfaces	Sliding	Intercarpal and intertarsal joints
Hinge	Concave surface of one bone articulates with convex surface of another	Bending motion in one plane	Knee; elbow; joints of phalanges
Pivot	Conical surface of one bone articulates with a depression of another	Rotation about a central axis	Atlantoaxial joint; proximal radioulnar joint
Condyloid	Oval condyle of one bone articulates with elliptical cavity of another	Biaxial movement	Radiocarpal joint
Saddle	Concave and convex surface on each articulating bone	Wide range of movements	Carpometacarpal joint of the thumb
Ball-and-Socket	Rounded convex surface of one bone articulates with cuplike socket of another	Movement in all planes and rotation	Shoulder and hip joints

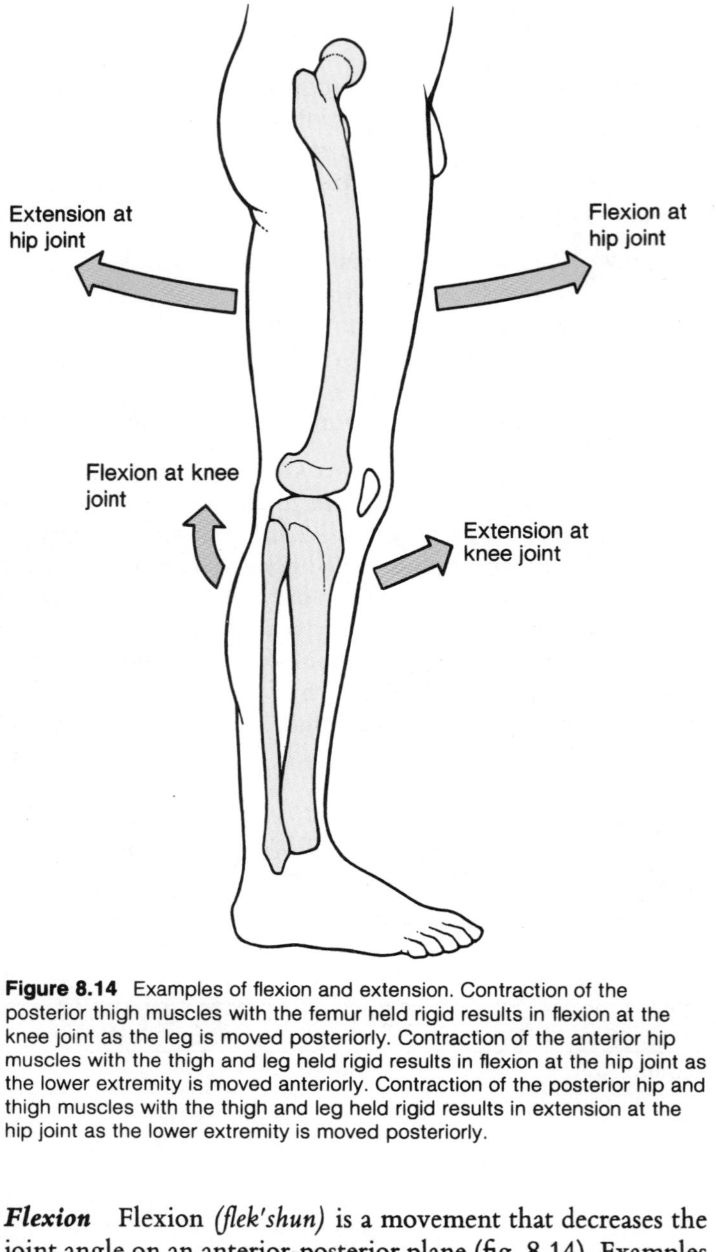

Figure 8.14 Examples of flexion and extension. Contraction of the posterior thigh muscles with the femur held rigid results in flexion at the knee joint as the leg is moved posteriorly. Contraction of the anterior hip muscles with the thigh and leg held rigid results in flexion at the hip joint as the lower extremity is moved anteriorly. Contraction of the posterior hip and thigh muscles with the thigh and leg held rigid results in extension at the hip joint as the lower extremity is moved posteriorly.

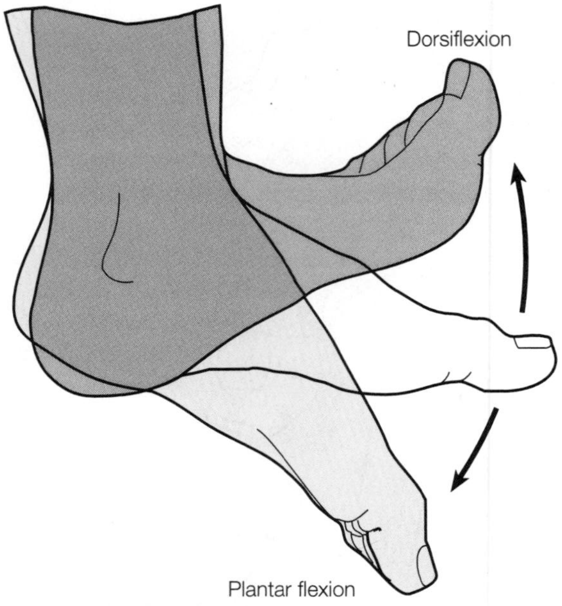

Figure 8.15 Dorsiflexion and plantar flexion of the foot at the ankle joint.

Flexion Flexion *(flek'shun)* is a movement that decreases the joint angle on an anterior-posterior plane (fig. 8.14). Examples of flexion are the bending of the elbow or knee. Flexion of the elbow is a forward movement, whereas flexion of the knee is a backward movement. Flexion in most joints is simple to understand, such as flexion of the head as it bends forward or the flexing of a digit, but flexion of the ankle and shoulder joints needs further explanation. In the ankle joint, flexion occurs as the dorsum of the foot is elevated. This movement is frequently called **dorsiflexion** (fig. 8.15). Pressing the foot downward (as in rising on the toes) is **plantar flexion.** The shoulder joint is flexed when the arm is brought forward, thus decreasing the joint angle.

Extension In extension *(ek-sten'shun)*, which is the reverse of flexion, the joint angle is increased (fig. 8.14). Extension returns a body part to the anatomical position. In an extended joint the angle between the articulating bones is 180°. The exception to this is the ankle joint, in which a 90° angle exists between the foot and the leg in the anatomical position. Examples of extension are the straightening of the elbow or knee joints from flexion positions. **Hyperextension** occurs when a portion of the body is extended beyond the anatomical position so that the joint angle is greater than 180°. An example of hyperextension is bending the head backward.

Abduction Abduction is the movement of a body part away from the main axis of the body, or away from the midsagittal plane, in a lateral direction (fig. 8.16). This term usually applies to the arm or leg but can also apply to the fingers or toes, in which case the line of reference is the longitudinal axis of the limb. Examples of abduction are moving the arms sideward and away from the body or spreading the fingers apart.

Adduction Adduction, the opposite of abduction, is the movement of a body part toward the main axis of the body (fig. 8.16). In the anatomical position, the arms and legs have been adducted toward the midplane of the body.

Circular

Joints that permit circular movement are composed of a bone with a rounded or oval surface articulating with a corresponding cup or depression on another bone. The two basic types of circular movements are rotation and circumduction.

Rotation Rotation is the movement of a bone around its own axis (see fig. 8.11). There is no lateral displacement during this movement. Examples are turning the head from side to side in a "no" motion and moving the forearm from a palm-up position to a palm-down position.

---

flexion: L. *flectere,* to bend
extension: L. *ex,* out, away from; *tendere,* stretch

abduction: L. *abducere,* lead away
adduction: L. *adductus,* bring to
rotation: L. *rotare,* a wheel

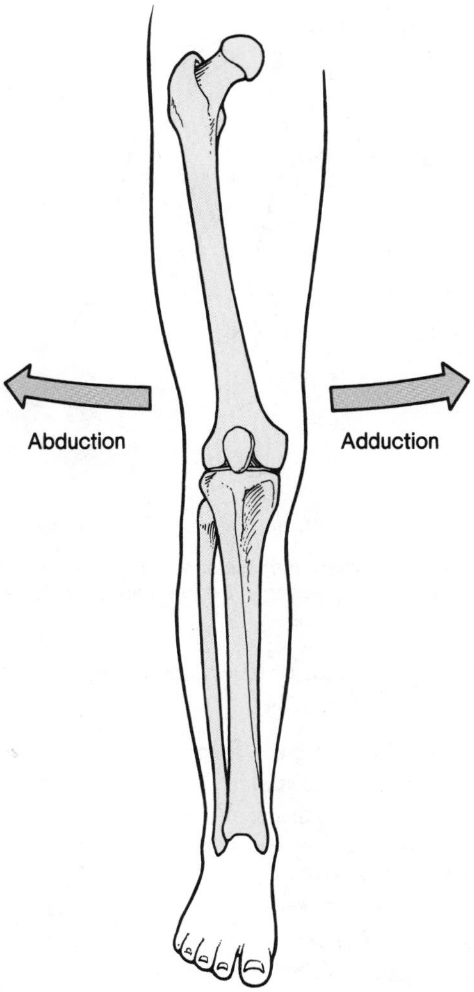

Figure 8.16 Abduction and adduction of the lower extremities in reference to the main axis of the body.

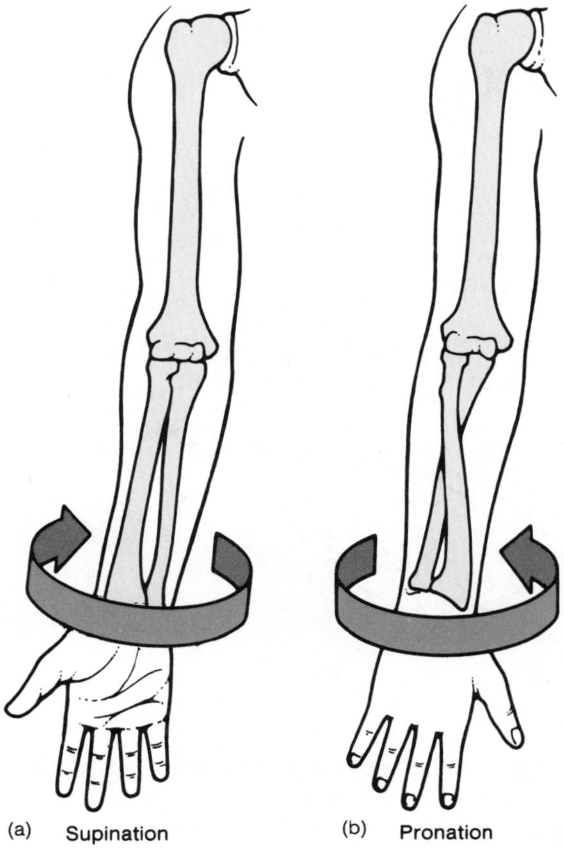

Figure 8.17 (*a*) Supination and (*b*) pronation of the hand. (Note the relative position of the ulna and radius in both positions.) Pronation requires medial rotation of the forearm relative to the anatomical position.

Supination *(su''pĭ-na'shun)* is a specialized rotation of the forearm that results in the palm of the hand being turned forward (anteriorly). In the supine position, the ulna and radius of the forearm are parallel and in the anatomical position. **Pronation** *(pro-na'shun)* is the opposite of supination (fig. 8.17). It is a rotational movement of the forearm that results in the palm of the hand being directed backward (posteriorly).

Circumduction Circumduction is the circular, conelike movement of a body segment. The distal extremity forms the circular movement, and the proximal attachment forms the pivot (fig. 8.18). This type of motion is possible at the shoulder, wrist, trunk, hip, and ankle joints.

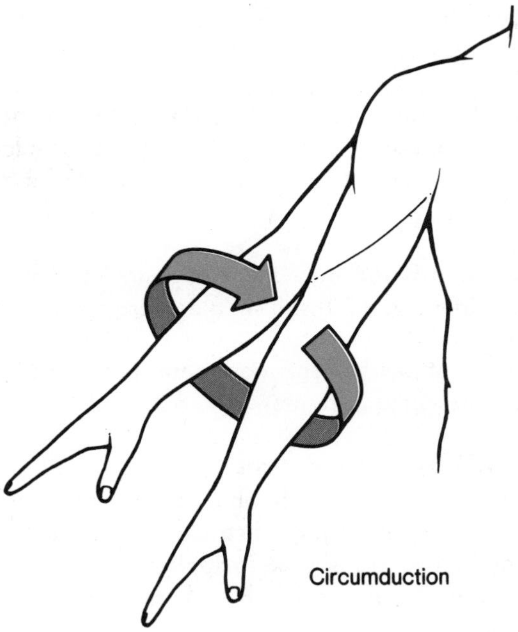

Figure 8.18 Circumduction of the shoulder joint.

Special Movements

Because the terms used to describe generalized movements around axes do not apply to the structure of certain joints, other terms must be used to describe the motion of such joints.

Inversion Inversion is the movement of the sole of the foot inward or medially (fig. 8.19). The pivot axis is at the ankle and intertarsal joints.

Eversion Eversion is the opposite of inversion and is the movement of the sole of the foot outward or laterally (fig. 8.19). Both inversion and eversion are clinical terms usually used to describe developmental abnormalities.

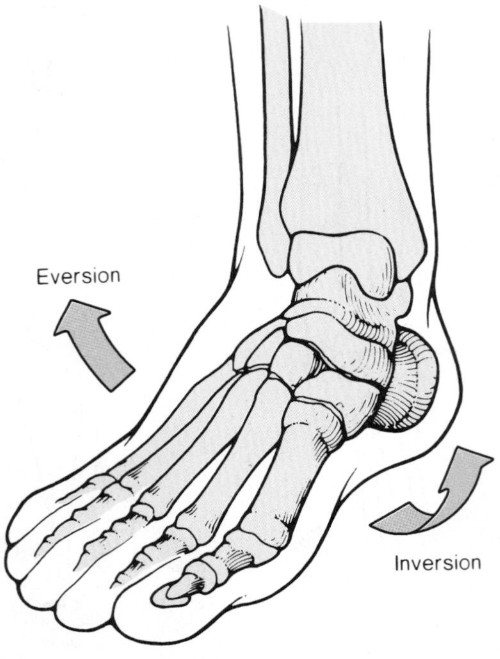

Figure 8.19 Inversion and eversion of the foot on the intertarsal joints.

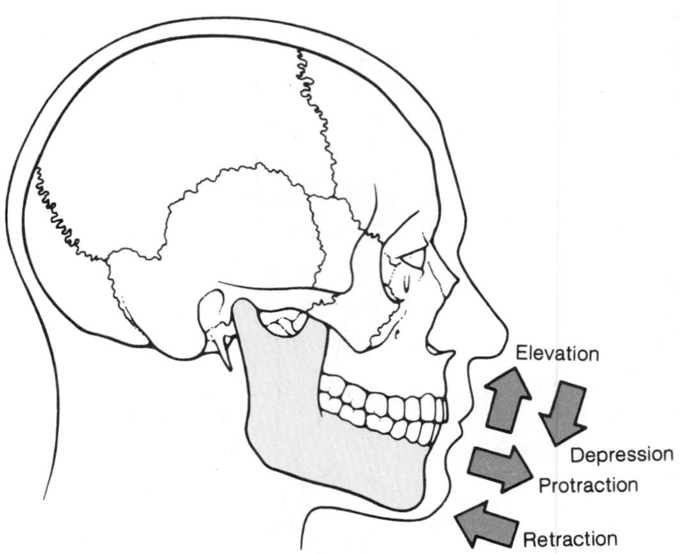

Figure 8.20 The antagonictic (opposite) movements of protraction-retraction and elevation-depression of the lower jaw as indicated by arrows.

Protraction Protraction is the movement of part of the body forward on a plane parallel to the ground. Examples are thrusting out the lower jaw (fig. 8.20) or movement of the pectoral girdle forward.

Retraction Retraction is the pulling back of a protracted part of the body on a plane parallel to the ground (fig. 8.20). Retraction of the mandible brings the lower jaw back in alignment with the upper jaw so that the teeth occlude.

Elevation Elevation is a movement that results in a portion of the body being lifted upward. Examples of elevation include elevating the mandible to close the mouth (fig. 8.20) or lifting the shoulders to shrug.

Depression Depression is the movement opposite of elevation. Both the mandible (fig. 8.20) and shoulders are depressed when moved downward.

A visual summary of many of the movements permitted at diarthrotic joints is presented in figures 8.21, 8.22, 8.23.

Biomechanics of Body Movement

A lever is any rigid structure that turns about a fulcrum when force is applied. Because bones are rigid structures that can be moved at diarthrotic joints in response to applied forces, they fit the criteria of levers. There are four basic components to a lever: (1) a rigid bar or other such structure, (2) a pivot or fulcrum, (3) an object or resistance that is moved, and (4) a force that is applied to one portion of the rigid structure.

Levers are generally associated with machines but can equally apply to other mechanical structures such as the human body. Diarthrotic joints are usually the fulcra (*F*), the muscles provide the force, or effort (*E*), and the bones are the rigid structures that move the resisting object (*R*).

There are three kinds of levers, determined by the arrangement of their parts (fig. 8.24).

1. In a **first-class lever,** the fulcrum is positioned between the effort and the resistance. The parts in a first-class lever are much like those of a seesaw—a sequence of resistance-pivot-effort. Scissors and hemostats are mechanical examples of the first-class levers. In the body, the head at the atlantooccipital *(at-lan''to-ok-sip'ĭ-tal)* joint is a first-class lever. The weight of the skull and facial portion of the head is the resistance, and the posterior neck muscles that contract to maintain the balance of the head on the joint are the effort.
2. In a **second-class lever,** the resistance is positioned between the fulcrum and the effort. The sequence of arrangement is pivot-resistance-effort, as in a wheelbarrow or the action of a crowbar when one end is placed under a rock and the other end lifted. Contracting the calf muscles (*E*) to elevate the body (*R*) on the toes, with the ball of the foot acting as the fulcrum, is another example.
3. In a **third-class lever,** the effort lies between the fulcrum and the resistance. The sequence of the parts is pivot-effort-resistance. An example is using a pair of forceps to grasp an object. The third-class lever is the most common type within the body. The flexion of the forearm at the elbow is an example. The effort occurs as the biceps muscle is contracted to move the resistance of the forearm with the elbow joint forming the fulcrum.

Each skeletal-muscular interaction at a diarthrotic joint forms some kind of lever system. The specific kind of lever is not always easy to identify. Certain joints are adapted for power at the expense of speed, whereas others are clearly adapted for speed (fig. 8.25). The specific attachment of muscles that span a joint plays an extremely important role in determining the mechanical advantage. The position of the insertion of a muscle relative to the joint is an important factor in the biomechanics

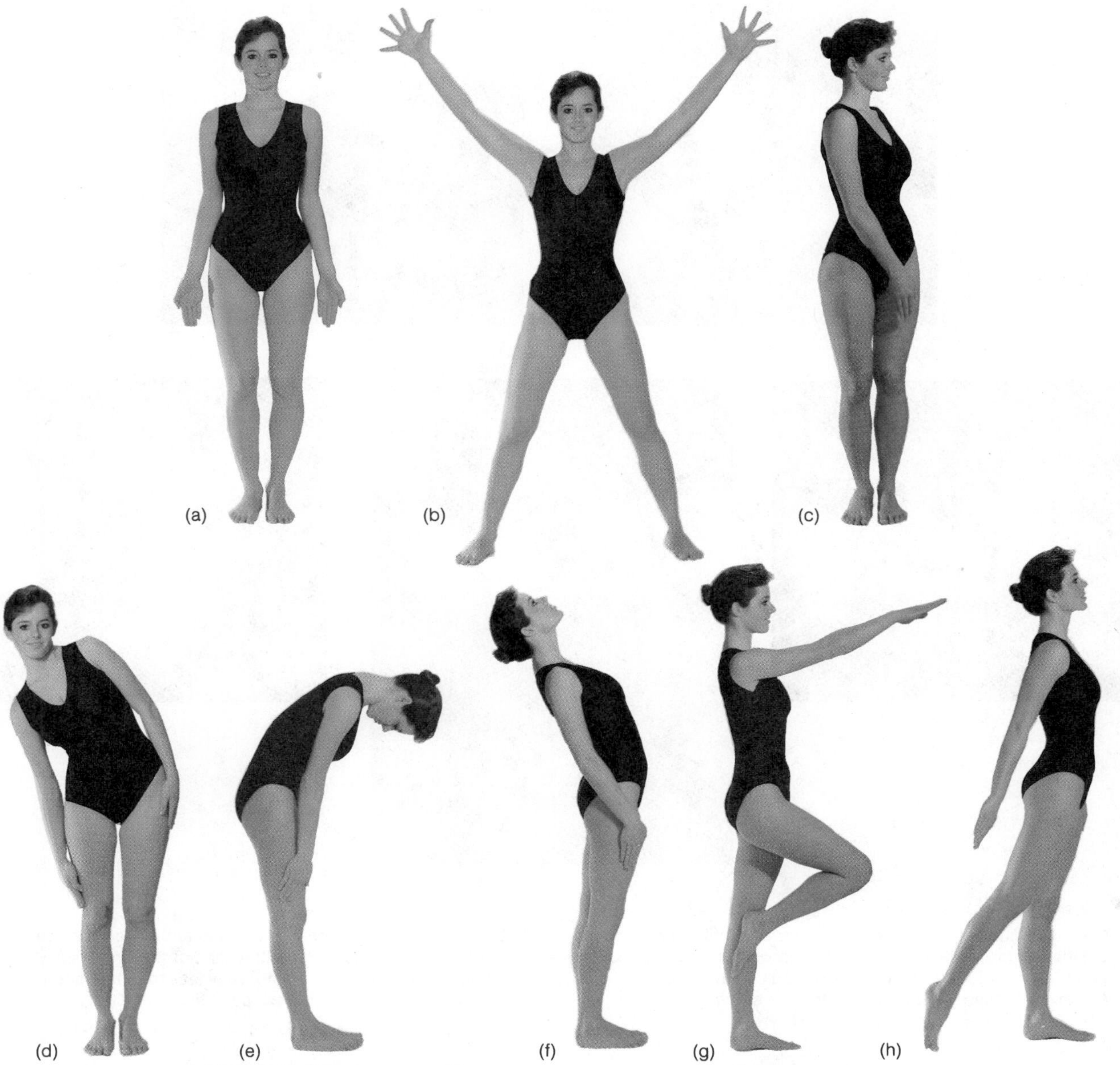

Figure 8.21 A photographic summary of joint movements. (*a*) Adduction of shoulder, hip, and carpophalangeal joints; (*b*) abduction of shoulder, hip, and carpophalangeal joints; (*c*) lateral rotation of vertebral column; (*d*) lateral flexion of vertebral column; (*e*) flexion of vertebral column; (*f*) hyperextension of vertebral column; (*g*) flexion of shoulder, hip, and knee joints of right side of body; extension of elbow and wrist joints; and (*h*) extension of shoulder and hip joints on right side of body; plantar flexion of right ankle joint.

of the contraction. An insertion close to the joint (fulcrum), for example, will produce a faster and greater range of movement than an insertion far away from the joint. An attachment far away from the joint takes advantage of the lever arm of the bone and increases power at the sacrifice of speed and range of movement.

1. Describe the structure of a joint that permits rotational movement.
2. What types of joints are involved in lever systems within the body?
3. Which is the most common type of lever in the body?
4. Considering the type of diarthrosis at the hip joint and the location of the gluteal muscles of the buttock, explain why this type of lever system is adapted for rapid, wide-ranging movements.

Specific Joints of the Body

Of the numerous joints in the body, some have special structural features that enable them to perform particular functions. Furthermore, these joints are somewhat vulnerable to trauma and are therefore clinically important.

Objective 13. Describe the structure, function, and possible clinical importance of the following joints: temporomandibular, sternoclavicular, humeroscapular, elbow, metacarpophalangeal, interphalangeal, coxal, tibio-femoral, and talocrural.

Temporomandibular Joint

The temporomandibular *(tem"po-ro-man-dib'u-lar)* joint is the only diarthrotic joint in the skull and is a unique combination of a hinge joint and a gliding joint (fig. 8.26). It is formed by

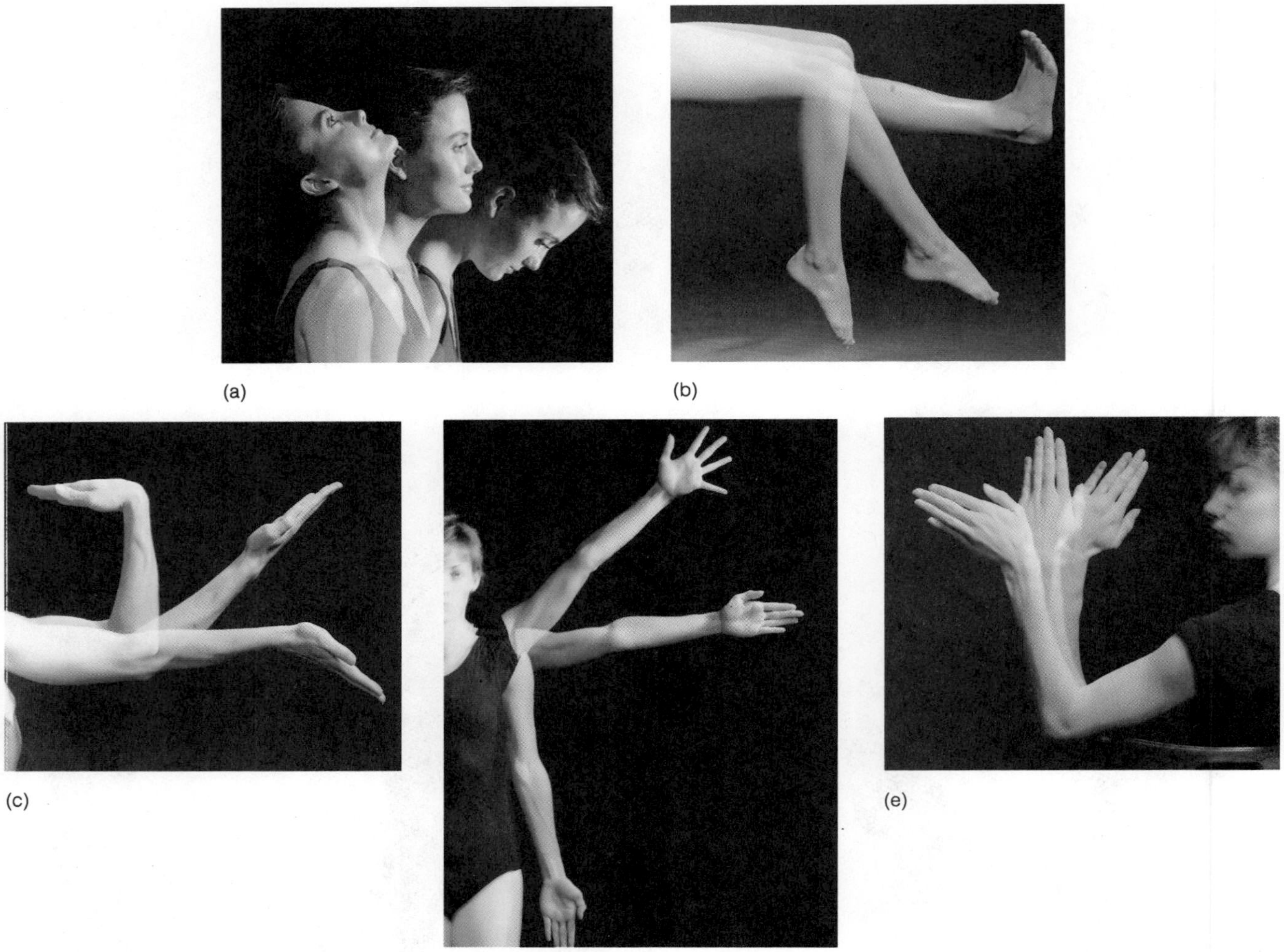

(a) (b) (c) (d) (e)

Figure 8.22 A visual summary of some angular movements at diarthroses. (*a*) Flexion, extension, and hyperextension in the cervical region; (*b*) flexion and extension at the knee joint, and dorsiflexion and plantar flexion at the ankle joint; (*c*) flexion, extension, and hyperextension at the wrist joint; (*d*) adduction and abduction of the arm and fingers; and (*e*) posterior view of abduction and adduction of the hand at the wrist joint. (Note that the range of abduction at the wrist joint is less than the range of adduction due to the length of the styloid process of the radius.)

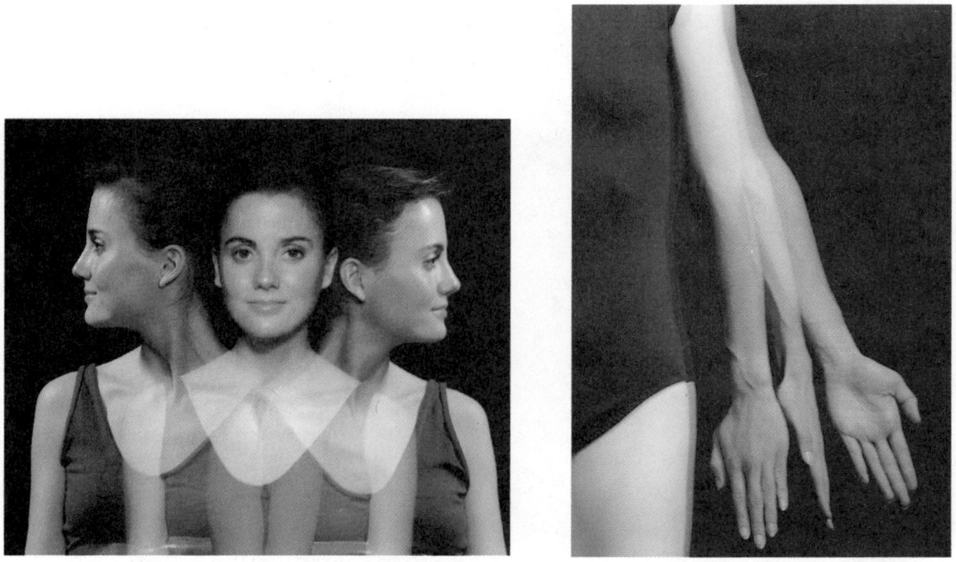

(a) (b)

Figure 8.23 A visual summary of some rotational movements at diarthroses. (*a*) Rotation of the head at the cervical vertebrae—especially at the atlantoaxial joint and (*b*) rotation of the forearm (antebrachium) at the proximal radioulnar joint.

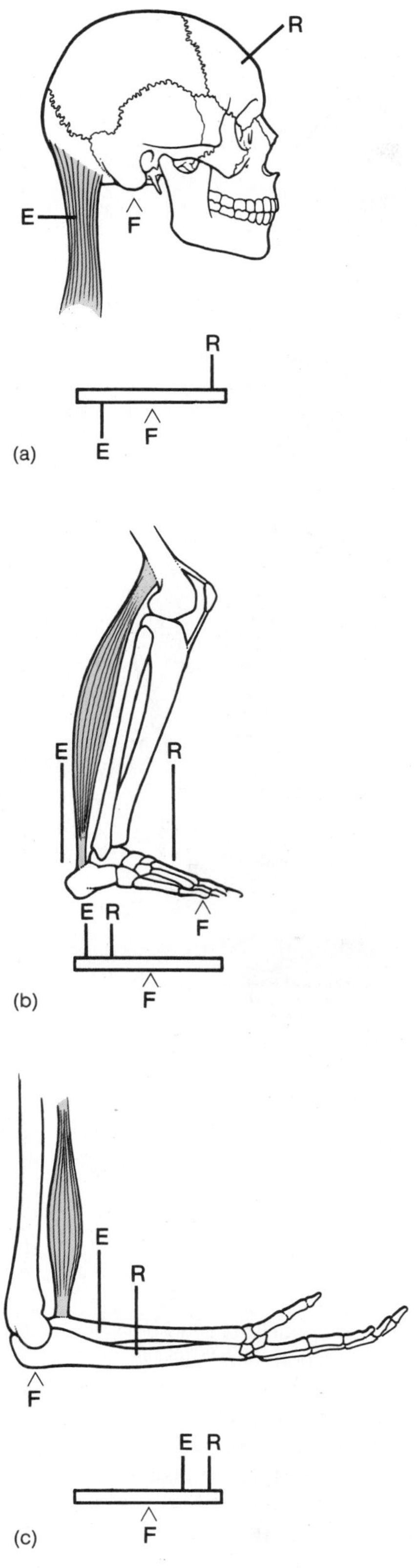

Figure 8.24 The three classes of levers. (*a*) In a first-class lever, the fulcrum (F) is positioned between the resistance (R) and the effort (E). (*b*) In a second-class lever, the resistance is between the fulcrum and the effort. (*c*) In a third-class lever, the effort is between the fulcrum and the resistance.

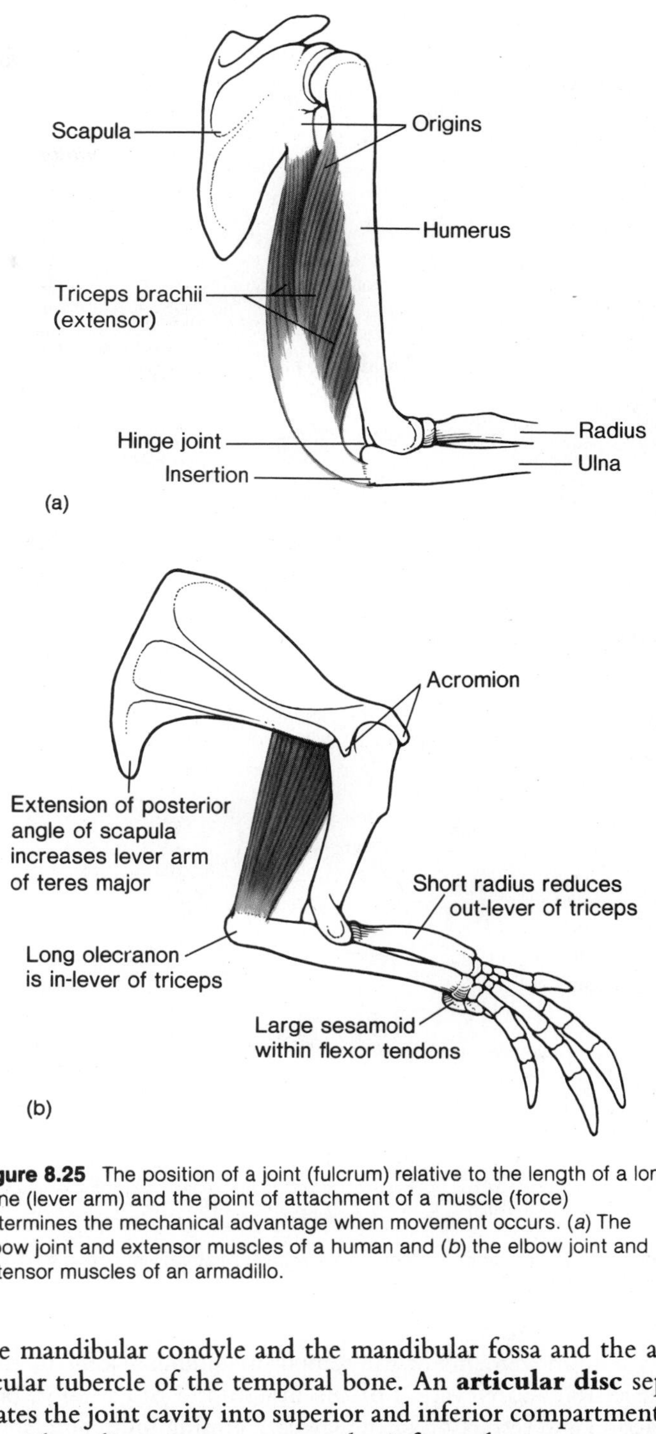

Figure 8.25 The position of a joint (fulcrum) relative to the length of a long bone (lever arm) and the point of attachment of a muscle (force) determines the mechanical advantage when movement occurs. (*a*) The elbow joint and extensor muscles of a human and (*b*) the elbow joint and extensor muscles of an armadillo.

the mandibular condyle and the mandibular fossa and the articular tubercle of the temporal bone. An **articular disc** separates the joint cavity into superior and inferior compartments.

Three ligaments support and reinforce the temporomandibular joint. The **temporomandibular ligament** is positioned on the lateral side of the joint capsule extending from the zygomatic arch to the neck of the mandible, and is covered by the parotid salivary gland. This ligament prevents the head of the mandible from being displaced posteriorly and fracturing the tympanic plate when the chin suffers a severe blow. The **stylomandibular ligament** does not directly touch the joint but extends inferiorly and anteriorly from the styloid process to the posterior border of the ramus of the mandible. A **sphenomandibular** *(sfe''no-man-dib'u-lar)* **ligament** crosses on the medial side of the joint from the spine of the sphenoid bone to the ramus of the mandible.

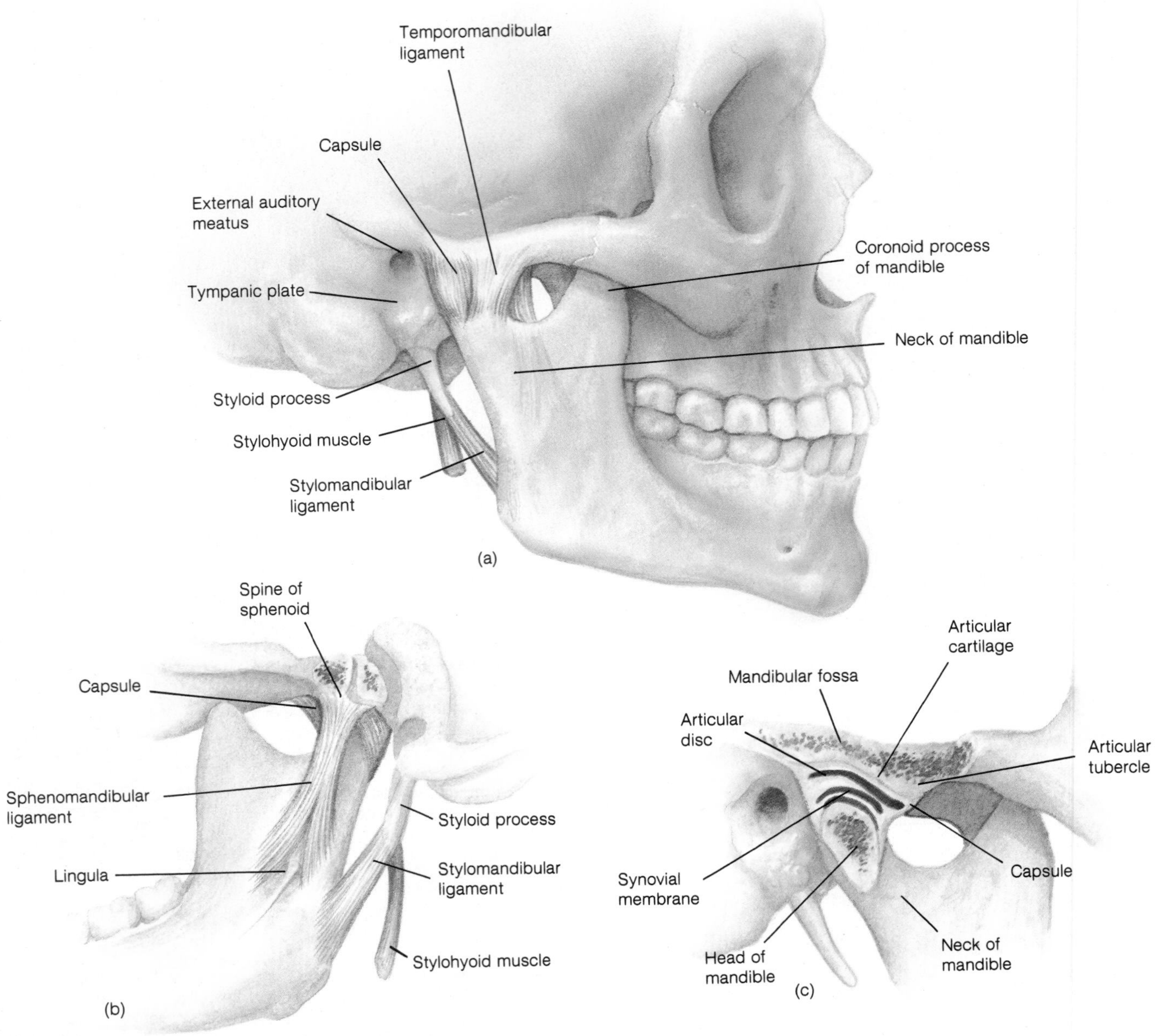

Figure 8.26 The temporomandibular joint. (*a*) A lateral view, (*b*) a medial view, and (*c*) a parasagittal section.

The movements of the temporomandibular joint include depression and elevation of the mouth as a hinge joint, protraction and retraction of the jaw as a gliding joint, and lateral rotatory movements. The lateral motion is made possible by the articular disc.

The temporomandibular joint can be easily palpated by applying firm pressure to the area in front of your auricle and opening and closing your mouth. This joint is most vulnerable to dislocation when the mandible is completely depressed, as in yawning. Relocating the jaw is usually a simple task, however, and is accomplished by pressing downward on the molar teeth while pushing the jaw backward.

Temporomandibular joint (TMJ) syndrome is an ailment that may afflict an estimated 75 million Americans. The apparent cause of TMJ syndrome is a malalignment of one or both temporomandibular joints. The symptoms of TMJ vary from moderate and intermittent pain to intense and continuous pain of the head, neck, shoulders, or back. Some vertigo (disorientation of coordination) and tinnitus (ringing in ear) may be experienced.

Sternoclavicular Joint

The sternoclavicular *(ster''no-klah-vik'u-lar)* joint is formed where the clavicle articulates with the manubrium of the sternum (fig. 8.27). Although a gliding joint, the sternoclavicular joint has a good range of movement because of the presence of an articular disc within the joint capsule.

Four ligaments support and give flexibility to the sternoclavicular joint. An **anterior sternoclavicular ligament**

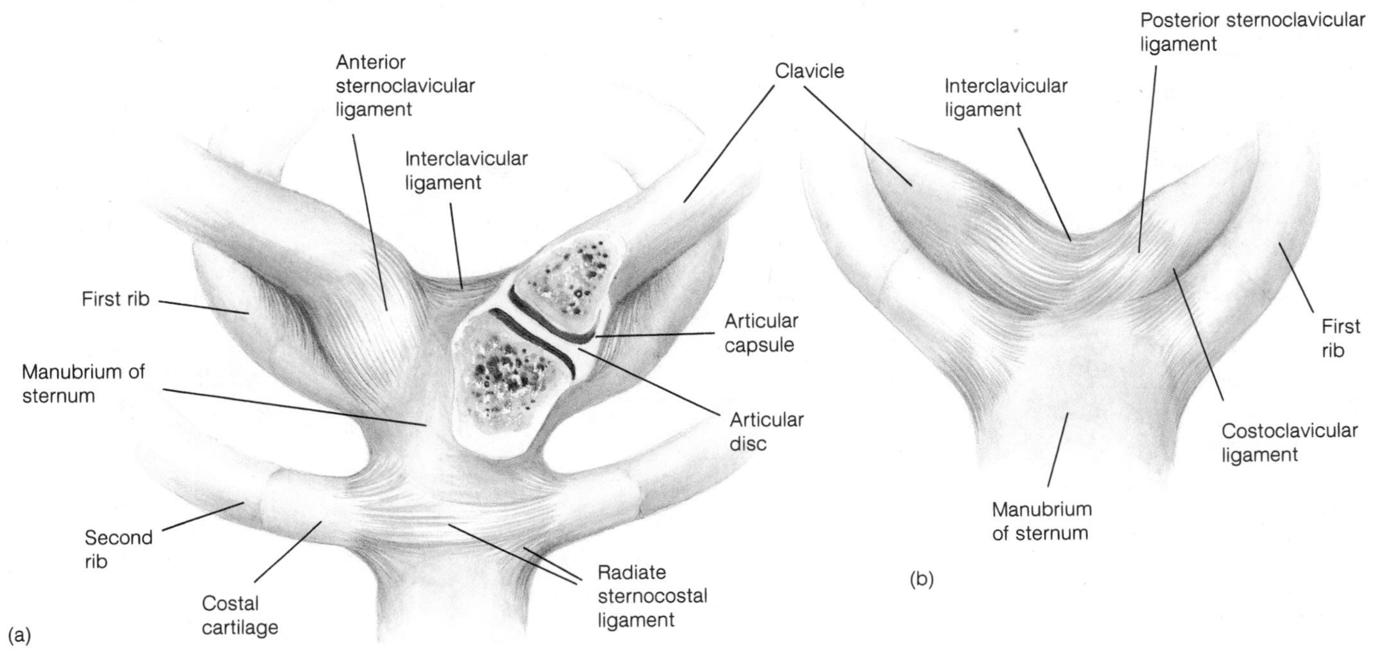

Figure 8.27 The sternoclavicular joint and associated ligaments. (*a*) An anterior view showing a frontal section and (*b*) a posterior view.

covers the anterior surface of the joint, and a **posterior sternoclavicular ligament** covers the posterior surface. Both ligaments extend from the sternal end of the clavicle to the manubrium. An **interclavicular ligament** extends between the sternal ends of both clavicles, binding them together. The **costoclavicular ligament** extends from the costal cartilage of the first rib to the costal tuberosity of the clavicle.

Of all the joints associated with the rib cage, the sternoclavicular joint is most frequently dislocated. Excessive force along the long axis of the clavicle may displace the clavicle forward and inferiorly. Injury to the costal cartilages is painful and is caused most frequently by a forceful, direct blow onto the costal cartilages.

Humeroscapular (Shoulder) Joint

The shoulder joint is formed by the articulation of the head of the humerus with the glenoid cavity of the scapula (fig. 8.28). It is a ball-and-socket joint and the most freely movable joint in the body. A circular band of fibrocartilage called the **glenoid** *(gle'noid)* **labrum** passes around the rim of the shoulder joint to deepen the concavity of the glenoid cavity (fig. 8.28). The shoulder joint is protected from above by an arch formed by the coracoid and acromion processes of the scapula and by the clavicle.

Although three ligaments surround and support the shoulder joint, most of the stability of this joint depends on the powerful muscles and tendons that cross over it. Thus it is an extremely mobile joint, in which stability has been sacrificed for mobility. The **coracohumeral** *(kor''ah-ko-hu'mer-al)* **ligament** extends from the coracoid process of the scapula to the greater tubercle of the humerus. The joint capsule is reinforced with three ligamentous bands called the **glenohumeral ligaments.** The final ligament of the shoulder joint is the **transverse humeral ligament,** a thin band that extends from the greater tubercle to the lesser tubercle of the humerus.

The stability of the shoulder joint is principally provided by the tendons of the subscapularis, supraspinatus, infraspinatus, and teres minor muscles together as they form the *musculotendinous (rotator) cuff.* The cuff is fused to the underlying capsule except in its inferior aspect. Because of the lack of inferior stability, most dislocations (subluxations) occur in this direction. The shoulder is most vulnerable to trauma when the joint is fully abducted and a sudden force from the superior direction is applied to the appendage. Degenerative changes in the musculotendinous cuff produce an inflamed, painful condition known as *pericapsulitis.*

Two major and two minor bursae are associated with the shoulder joint. The larger bursae are the **subdeltoid bursa** between the deltoid muscle and the joint capsule and the **subacromial bursa** between the acromion and joint capsule. The **subcoracoid bursa** lies between the coracoid process and the joint capsule and is frequently considered an extension of the subacromial bursa. A small **subscapular bursa** is located between the tendon of the subscapularis muscle and the joint capsule.

The shoulder joint is vulnerable to dislocations from sudden jerks of the arm, especially in children before strong shoulder muscles have developed. Because of the weakness of this joint in children, parents should be careful not to force a child to follow by yanking on the arm. Dislocation of the shoulder is extremely painful and may cause permanent damage or perhaps muscle atrophy due to disuse.

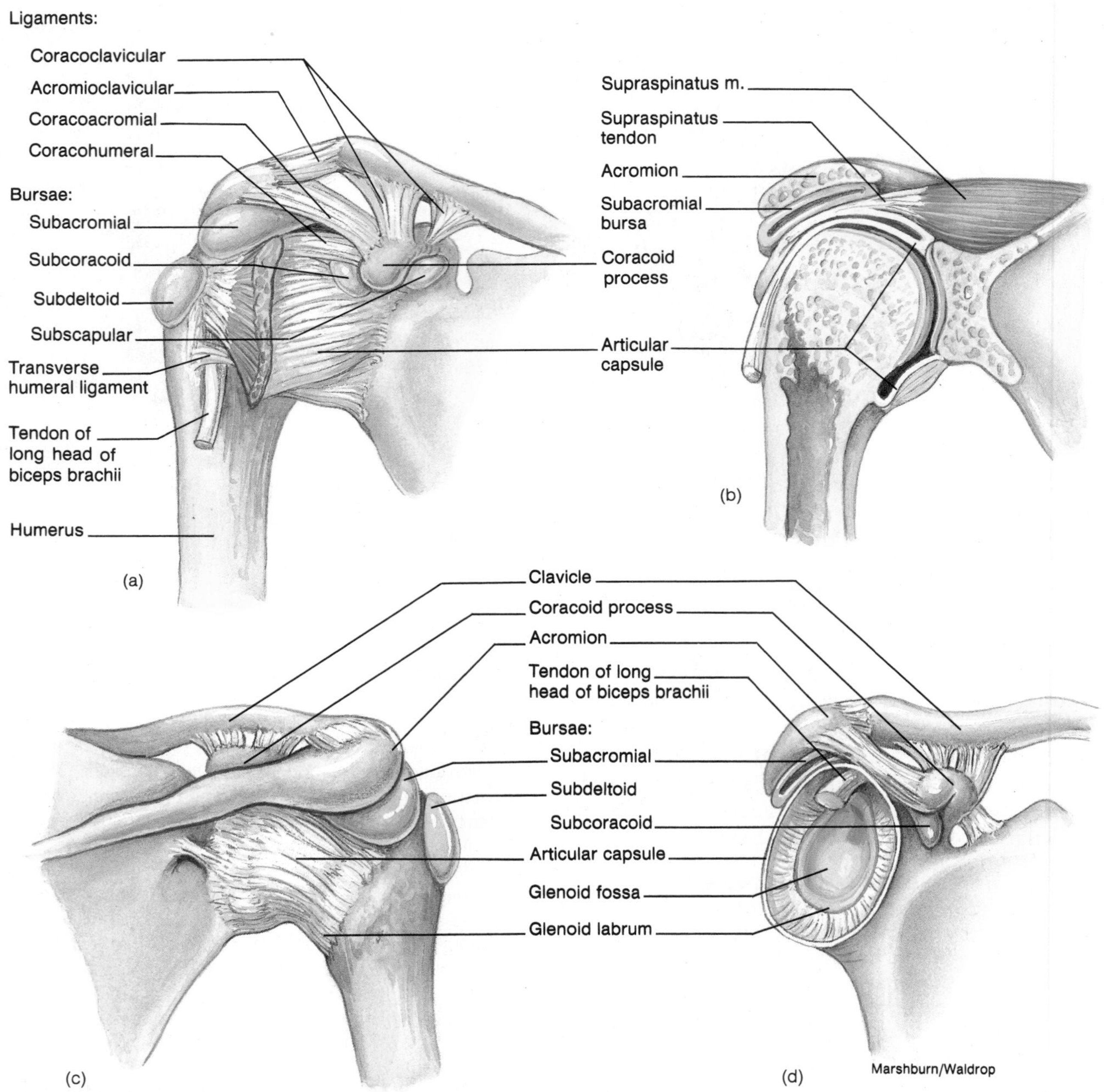

Figure 8.28 The humeroscapular joint. (*a*) An anterior view, (*b*) a coronally sectioned anterior view, (*c*) a posterior view, and (*d*) a lateral view with the humerus removed.

Elbow Joint

There are three specific articulations in the elbow region, two of which constitute the **elbow joint** (fig. 8.29). The elbow joint is a hinge joint, formed by the trochlea of the humerus articulating with the trochlear notch of the ulna and the capitulum of the humerus articulating with the head of the radius. Although there are two sets of articulations at the elbow joint, there is only one joint capsule and a large **olecranon bursa** to lubricate this area. A **radial** (lateral) **collateral ligament** reinforces the elbow joint on the lateral side, and an **ulnar** (medial) **collateral ligament** strengthens the medial side.

Because so many muscles originate or insert near the elbow, it is a common site of localized tenderness, inflammation, and pain. *Tennis elbow* is a general term for musculotendinous soreness in this area. The structures most generally strained are the tendons attached to the lateral epicondyle of the humerus. The strain is caused by repeated extension of the wrist against some force, as occurs during the backhand stroke in tennis.

Metacarpophalangeal Joints and Interphalangeal Joints

The **metacarpophalangeal joints** are condyloid diarthroses and the **interphalangeal joints** are hinge diarthroses. These joints are formed as the heads of the metacarpals articulate with

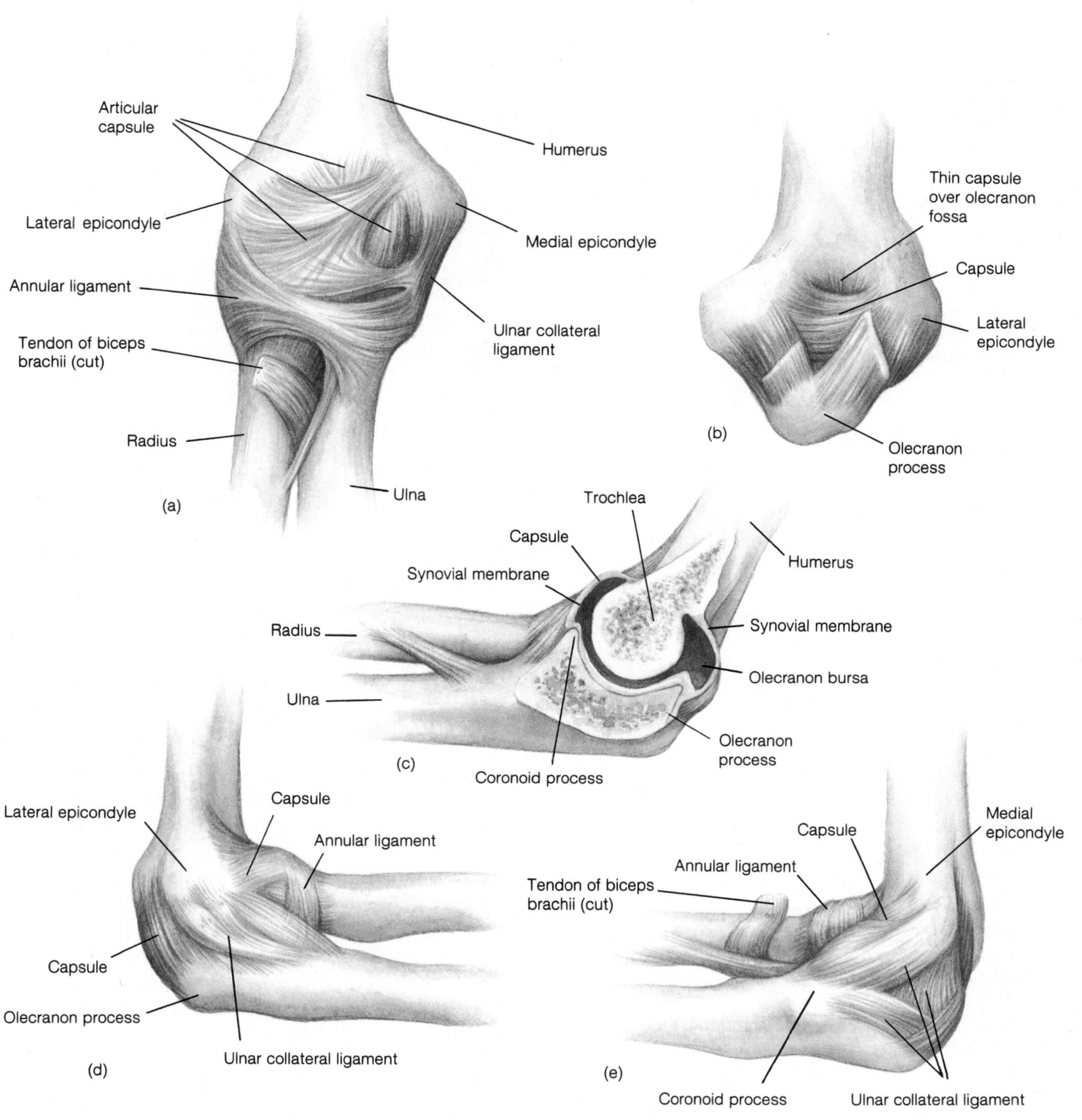

Figure 8.29 The right elbow region. (*a*) An anterior view, (*b*) a posterior view, (*c*) a midsagittal section, (*d*) a lateral view, and (*e*) a medial view.

the proximal phalanges and as the phalanges articulate with one another (fig. 8.30). Each joint in both joint types has three ligaments. A **palmar ligament** spans each joint on the palmar, or anterior, side of the joint capsule. Each joint also has two **collateral ligaments,** one on the lateral side and one on the medial side, to further reinforce the joint capsule. There are no supporting ligaments on the posterior side.

Athletes frequently jam a finger. It occurs when a ball forcefully strikes a distal phalanx as the fingers are extended, causing a sharp flexion at the joint between the middle and distal phalanges. No ligaments support the joint on the posterior side, but there is a tendon from the digital extensor muscles of the forearm. It is this tendon that is damaged when the finger is jammed. Treatment involves splinting the finger for a period of time. If splinting is not effective, however, surgery is necessary or the person will suffer a permanent crook in the finger.

Coxal (Hip) Joint

The ball-and-socket **hip joint** is formed by the head of the femur articulating with the acetabulum of the os coxa (fig. 8.31). It bears the weight of the body and is therefore much stronger and more stable than the shoulder joint. The hip joint is secured

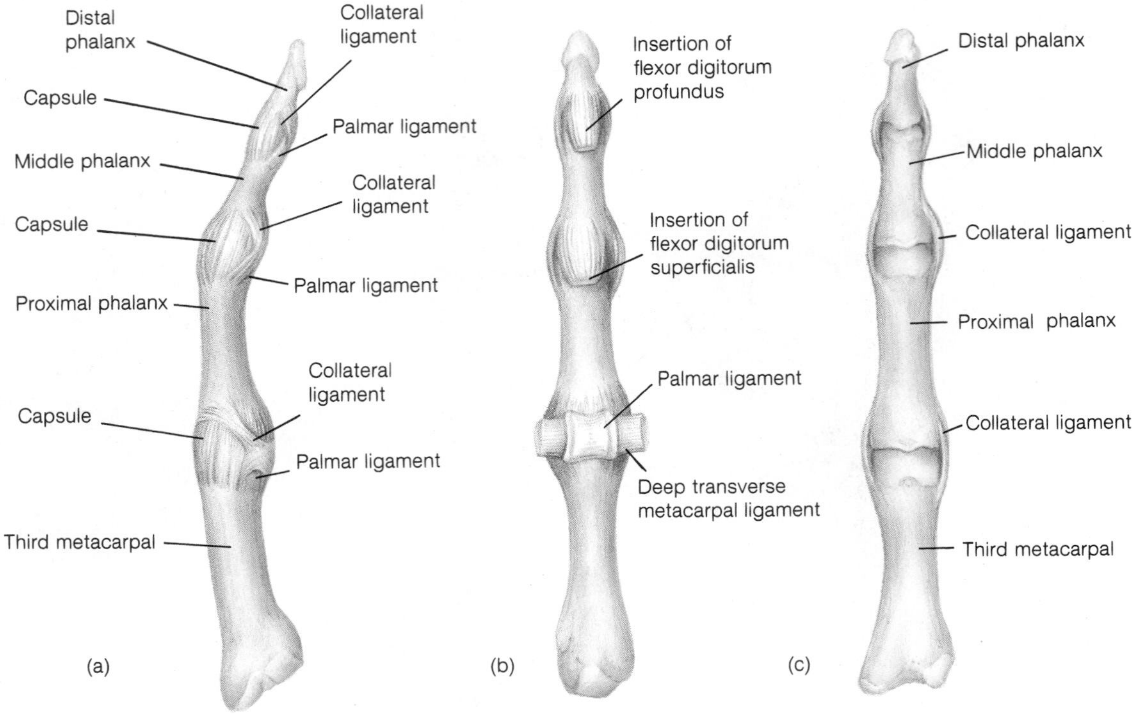

Figure 8.30 Metacarpophalangeal and interphalangeal joints. (*a*) A lateral view, (*b*) an anterior (palmar) view, and (*c*) a posterior view.

by a strong fibrous joint capsule, several ligaments, and a number of powerful muscles.

The primary ligaments of the hip joint are the anterior **iliofemoral** *(il″e-o-fem′or-al)* and **pubofemoral ligaments** and the posterior **ischiofemoral** *(is″ke-o-fem′or-al)* **ligament.** The **ligamentum teres** is located within the articular capsule and attaches the head of the femur to the acetabulum. The **transverse acetabular** *(as″ĕ-tab′u-lar)* **ligament** crosses the acetabular notch and connects to the joint capsule and the ligamentum teres. The **acetabular labrum,** a fibrocartilaginous rim that rings the head of the femur as it articulates with the acetabulum, is attached to the margin of the acetabulum.

Osteoarthritis is a degenerative disease of the articular cartilage of diarthrotic joints accompanied by the formation of bony spurs in the joint cavities. Immobility results if the hip joint is severely afflicted with this disease. Fortunately, the entire hip joint can be replaced in a procedure called *hip arthroplasty.* During this surgery, the acetabulum is replaced by a low-friction polyethylene socket, fit into the os coxa using bone cement. The head of the femur is replaced by a stainless steel, ball-shaped prosthesis (see fig. 8.36).

Tibiofemoral (Knee) Joint

The **knee joint** is the largest, most complex, and probably the most vulnerable joint in the body. The knee joint is formed as the femur and the tibia articulate. It is a complex hinge diarthrosis, which in addition to flexion and extension, permits limited rolling and gliding movements. On the anterior side, the knee joint is stabilized and protected by the patella and the **patellar ligament,** which forms a gliding **patellofemoral joint.**

Because of the complexity of the knee joint, only the relative positions of the ligaments, menisci, and bursae will be presented. Although the detailed attachments and functions will not be discussed, the locations of these structures can be seen in figure 8.32.

In addition to the patella and the patellar ligament on the anterior surface, the tendinous insertion of the quadriceps femoris muscle forms two supportive bands called the **lateral** and **medial patellar retinacula** *(ret″ĭ-nak′u-lah).* Four bursae are present on the anterior aspect of the knee: the **superficial infrapatellar** *(in″frah-pah-tel′ar)* **bursa,** the **suprapatellar bursa,** the **prepatellar bursa,** and the **deep infrapatellar bursa.**

The posterior aspect of the knee is referred to as the **popliteal region.** The broad **oblique popliteal ligament** and the **arcuate** *(ar′cu-āt)* **popliteal ligament** are superficial in position, whereas the **anterior** and **posterior cruciate** *(kroo′she-āt)* **ligaments** are deep within the joint. The **popliteal bursa** and the **semimembranosus bursa** (not illustrated) are the two bursae present on the back of the knee.

Strong **collateral ligaments** support both the medial and lateral sides of the knee joint. Two fibrocartilaginous discs called the **lateral** and **medial menisci** are located within the knee joint interposed between the distal femoral and proximal tibial condyles. The two menisci are connected by a **transverse ligament.** Several other bursae (not described) are associated with

cruciate: L. *crucis,* cross

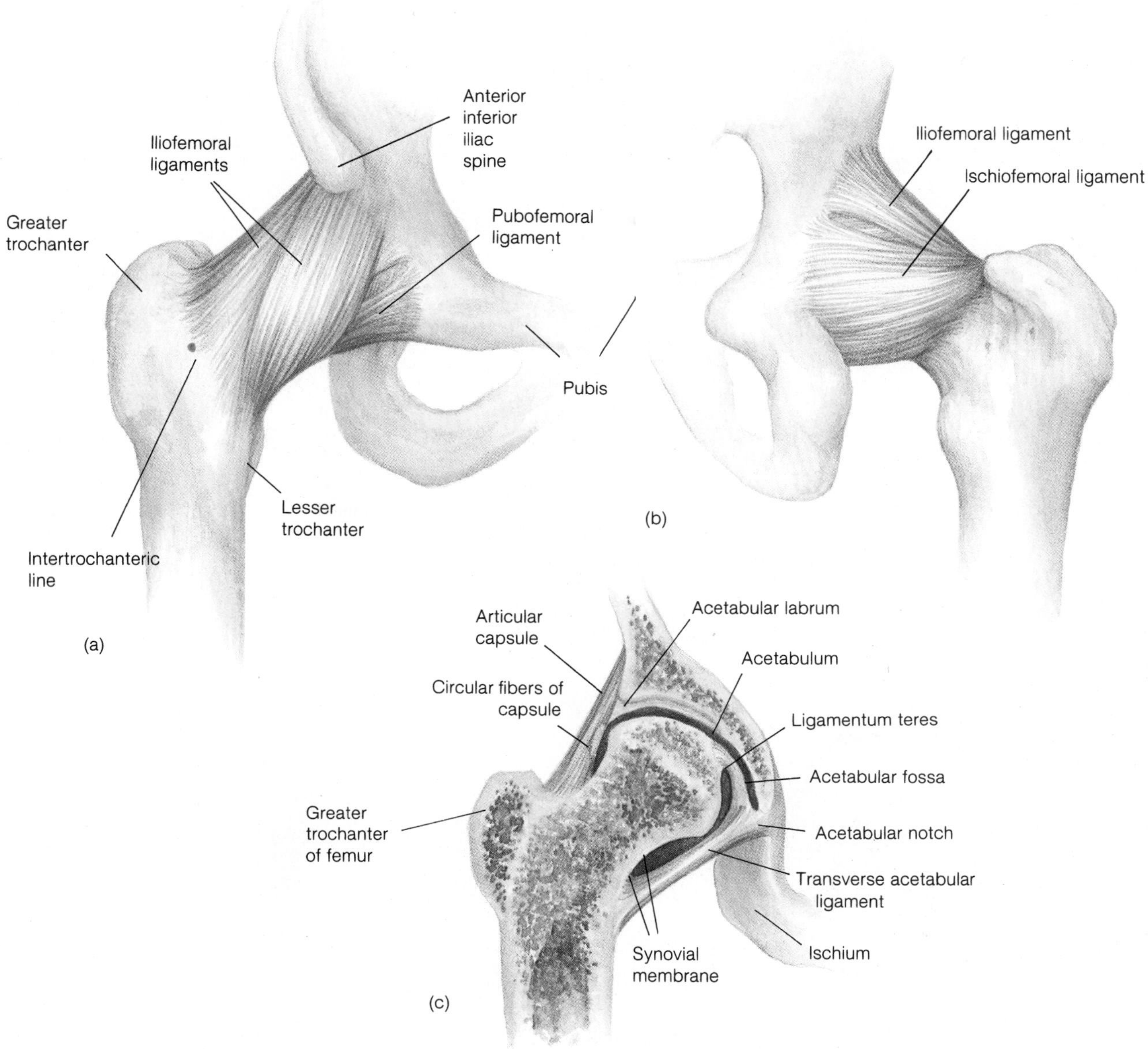

Figure 8.31 The right coxal (hip) joint. (*a*) An anterior view, (*b*) a posterior view, and (*c*) a frontal section.

the knee joint. In addition to the four on the anterior side and the two on the posterior side, there are seven bursae on the lateral and medial sides, making a total of thirteen.

During normal walking, running, and supporting of the body, the knee joint functions superbly. It can tolerate even moderate stress without tissue damage. However, the knee lacks bony support to withstand sudden forceful stresses, such as commonly occur among athletes. Knee injuries frequently require surgery and heal with difficulty due to the avascularity of the cartilaginous tissue. Because this joint is potentially so vulnerable, it is important to realize its limitations by understanding its anatomy.

Talocrural (Ankle) Joint

There are actually two articulations within the ankle joint, both of which are hinge diarthroses. One articulation is formed as the distal end of the tibia and its medial malleolus articulate with the talus, and the other is formed as the lateral malleolus of the fibula articulates with the talus (fig. 8.33).

One joint capsule surrounds the articulations of the three bones, and four ligaments support the ankle joint on the outside of the capsule. The strong **deltoid ligament** is associated with the tibia, whereas the **anterior talofibular** *(ta″lo-fib′u-lar)* **ligament, posterior talofibular ligament,** and **calcaneofibular** *(kal-ka″ne-o-fib′u-lar)* **ligament** are associated with the fibula.

The malleoli form a cap over the upper surface of the talus that prohibits side-to-side movement at the ankle joint. Unlike the condyloid joint at the wrist, the movements of the ankle are limited to flexion and extension. Dorsiflexion of the ankle is checked primarily by the tendo calcaneous, whereas plantar flexion, or ankle extension, is checked by the tension of the

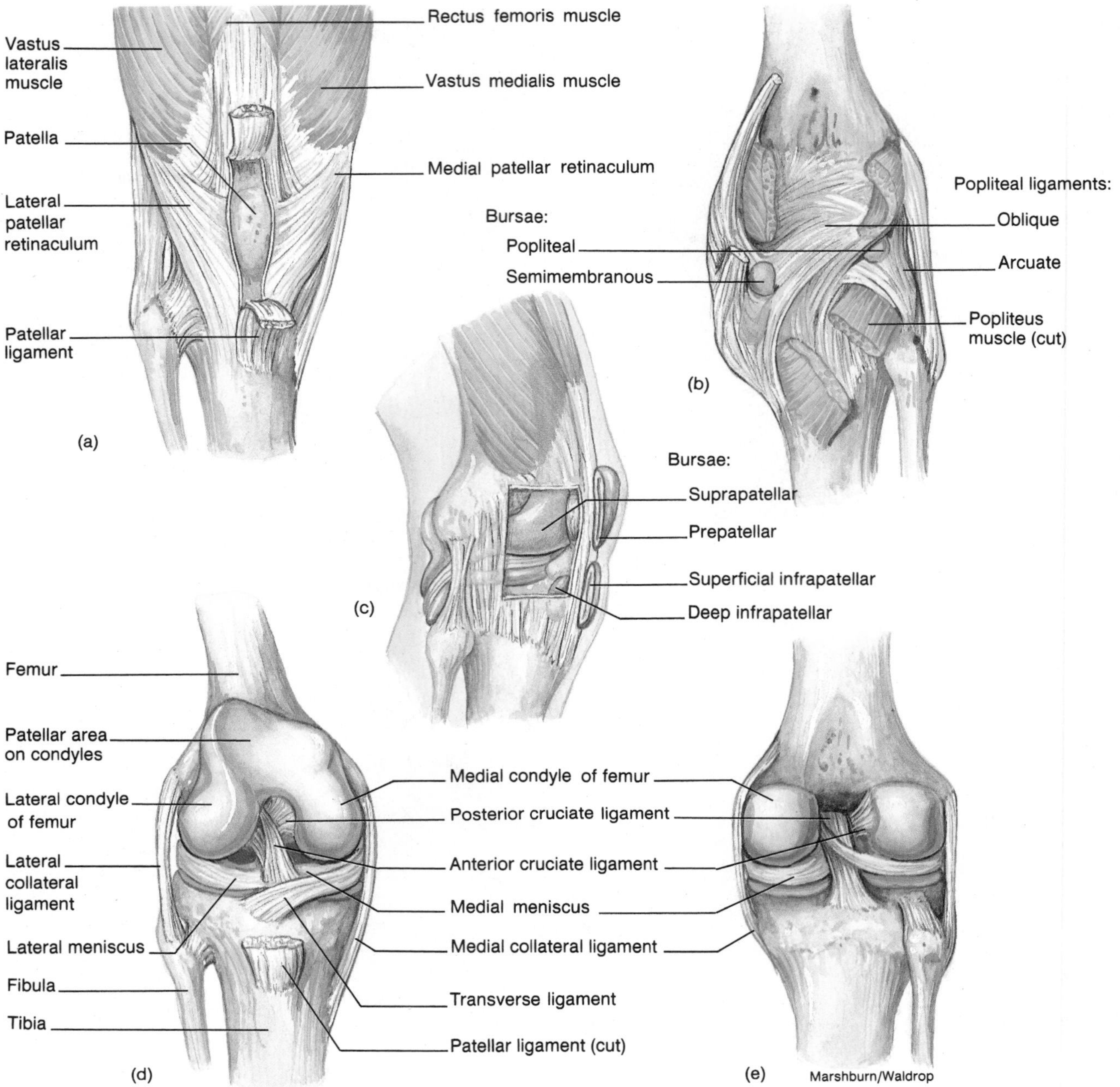

Figure 8.32 The right tibiofemoral (knee) joint. (*a*) An anterior view, (*b*) a superficial posterior view, (*c*) a lateral view showing the bursae, (*d*) an anterior view with the knee slightly flexed and the patella removed, and (*e*) a deep posterior view.

extensor tendons on the front of the joint and the anterior portion of the joint capsule.

Ankle sprains are a common type of locomotor injury. They vary extensively in seriousness but generally occur in certain locations. The most common cause of ankle sprain is excessive inversion of the foot, resulting in partial tearing of the anterior talofibular ligament and the calcaneofibular ligament. Less commonly, the deltoid ligament is injured by excessive eversion of the foot. Torn ligaments are extremely painful and are accompanied by immediate local swelling. Reducing the swelling and immobilizing the joint are about the only treatments for moderate sprains. Extreme sprains may require surgery and casting of the joint to facilitate healing.

A summary of the principal joints of the body and their movement is presented in table 8.2.

1. Which diarthrotic joints have menisci?
2. What are the two types of joints found in the shoulder region? Why is the shoulder joint so vulnerable?
3. Which joints are reinforced with muscles that span the joint?
4. Describe the structure of the knee joint, and indicate which structures protect and reinforce its anterior surface.

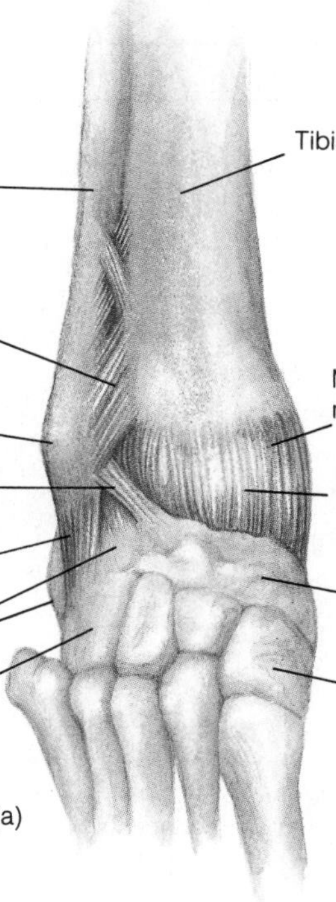

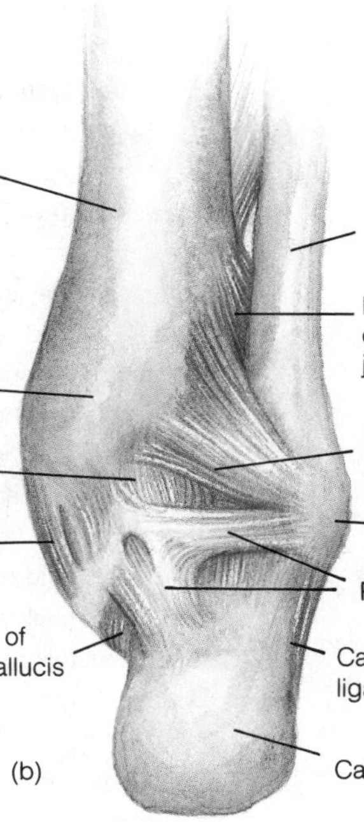

Figure 8.33 The right talocrural (ankle) joint. (*a*) An anterior view and (*b*) a posterior view.

Clinical Considerations

A diarthrotic joint is a remarkable biologic system, which acts as a self-lubricating bearing surface, able to move with almost frictionless precision under tremendous loads and impacts. Under normal circumstances and in most people, the many joints of the body perform without problems throughout life. Joints are not indestructible, however, and are subject to various forms of trauma and disease. Although not all of the diseases of joints are fully understood, medical science has made remarkable progress in the treatment of arthrological problems.

Trauma to Joints

Joints are well adapted to withstand compression and tension forces. Torsion or sudden impact to the side of a joint, however, can be devastating. These types of injuries frequently occur in athletes.

In a **strained joint,** unusual or excessive exertion stretches the tendons spanning a joint or the surrounding muscles but causes no serious damage. Strains frequently result from not warming up, or activating joints and muscles, prior to strenuous use. A **sprain** is a tearing of the ligaments or tendons that surround a joint. There are various grades of sprains, and the severity will determine the treatment. Severe sprains of the knee joint are frequently accompanied by damage to articular cartilages and menisci, which generally requires surgery. Sprains are usually accompanied by **synovitis** *(sin''o-vi'tis),* an inflammation of the joint capsule.

Luxation, or **joint dislocation,** is derangement of the articulating bones that compose the joint. Joint dislocation is more serious than a sprain and is usually accompanied by sprains. The shoulder and knee joints are the most vulnerable to dislocation. Self-healing of a dislocated joint, like the knee joint, may be incomplete, leaving the person with a trick knee that may unexpectedly give way.

Subluxation is partial dislocation of a joint. Subluxation of the hip joint is a common type of birth defect, which can be treated by bracing or casting the hip joints to promote suitable bone development.

Bursitis *(ber-si'tis)* is an inflammation of a bursa. Because of its close proximity to the joint, bursitis may affect the joint capsule as well. Bursitis may be caused by excessive stress on the bursa from overexertion, or it may be a local or systemic inflammatory process. As the bursa swells, the surrounding muscles become sore and stiff. **Tendonitis** involves the tendon, in or out of a tendon sheath, and usually has the same causes as bursitis.

The flexible vertebral column is a marvel of mechanical engineering. Not only do the individual vertebrae articulate one with another, but together they form the portion of the axial skeleton with which the head, ribs, and pelvic girdle articulate. The vertebral column also encloses the spinal cord and provides exits for thirty-one pairs of spinal nerves. With all the articulations within the vertebral column and the physical abuse it receives, it is no wonder that back ailments are second only to headaches as the most common physical complaint. Our way of life causes many of the problems associated with the vertebral column. Improper shoes, athletic exertion, sudden stops in vehicles, or improper lifting can all cause the back to go awry. Body weight, age, and general body condition influence a person's susceptibility to back problems.

luxation: L. *luxus,* out of place

Table 8.2 Principal articulation

Joint	Type	Movement
Most skull joints	Synarthrosis (suture)	Immovable
Temporomandibular	Diarthrosis (hinge; gliding)	Elevation, depression; protraction, retraction
Atlantooccipital	Diarthrosis (condyloid)	Flexion, extension, circumduction
Atlantoaxial	Diarthrosis (pivot)	Rotation
Intervertebral		
bodies of vertebrae	Amphiarthrosis (symphysis)	Slight movement
articular processes	Diarthrosis (gliding)	Flexion, extension, slight rotation
Sacroiliac	Diarthrosis (gliding)	Slight gliding movement; may fuse in adults
Costovertebral	Diarthrosis (gliding)	Slight movement during breathing
Sternocostal	Amphiarthrosis diarthrosis (gliding)	Slight movement during breathing
Sternoclavicular	Diarthrosis (gliding)	Slight movement when shrugging shoulders
Sternal	Amphiarthrosis (symphysis)	Slight movement during breathing
Acromioclavicular	Diarthrosis (gliding)	Protraction, retraction; elevation, depression
Humeroscapular (shoulder)	Diarthrosis (ball-and-socket)	Flexion, extension; adduction, abduction; rotation, circumduction
Elbow	Diarthrosis (hinge)	Flexion, extension
Proximal radioulnar	Diarthrosis (pivot)	Rotation
Distal radioulnar	Amphiarthrosis (syndesmosis)	Slight movement
Radiocarpal (wrist)	Diarthrosis (condyloid)	Flexion, extension; adduction, abduction; circumduction
Intercarpal	Diarthrosis (gliding)	Slight movement
Carpometacarpal		
fingers	Diarthrosis (condyloid)	Flexion, extension; adduction, abduction
thumb	Diarthrosis (saddle)	Flexion, extension; adduction, abduction
Metacarpophalangeal	Diarthrosis (condyloid)	Flexion, extension; adduction, abduction
Interphalangeal	Diarthrosis (hinge)	Flexion, extension
Symphysis pubis	Amphiarthrosis (symphysis)	Slight movement
Coxal (hip)	Diarthrosis (ball-and-socket)	Flexion, extension; adduction, abduction; rotation; circumduction
Tibiofemoral (knee)	Diarthrosis (hinge)	Flexion, extension; slight rotation when flexed
Proximal tibiofibular	Diarthrosis (gliding)	Slight movement
Distal tibiofibular	Amphiarthrosis (syndesmosis)	Slight movement
Talocrural (ankle)	Diarthrosis (hinge)	Dorsiflexion, plantar flexion; slight circumduction
Intertarsal	Diarthrosis (gliding)	Inversion, eversion
Tarsometatarsal	Diarthrosis (gliding)	Flexion, extension; adduction, abduction

The most common cause of back pain is *strained muscles,* generally the result of overexertion. The second most frequent back ailment is a *herniated disc.* The dislodged nucleus pulposus of a disc may push against a spinal nerve and cause excruciating pain. The third most frequent back problem is a *dislocated articular facet* between two vertebrae, caused by sudden torsion of the vertebral column. The treatment of back ailments varies from bed rest to spinal manipulation to extensive surgery.

Curvature disorders are another problem of the vertebral column. **Kyphosis** *(ki-fo'sis)* (hunchback) is an exaggeration of the thoracic curve. **Lordosis** (swayback) is an abnormal anterior convexity of the lumbar curve. **Scoliosis** (crookedness) is an abnormal lateral curvature of the vertebral column (fig. 8.34), which may be caused by one leg being longer than the other and uneven muscular development on the two sides of the vertebral column.

kyphosis: Gk. *kyphos,* hunched
lordosis: Gk. *lordos,* curving forward
scoliosis: Gk. *skoliosis,* crookedness

Diseases of Joints

Arthritis is a generalized term for over fifty different joint diseases, all of which have the symptoms of edema, inflammation, and pain. The causes of arthritis are unknown, but certain types follow joint trauma or bacterial infections. There is evidence that some types of arthritis are the result of hormonal or metabolic disorders. The most common forms are rheumatoid arthritis, osteoarthritis, and gouty arthritis.

In **rheumatoid** *(roo'mah-toid)* **arthritis,** the synovial membrane thickens and becomes tender and synovial fluid accumulates. This change is usually followed by an invasion of fibrous tissue and deterioration of the articular cartilage. When the cartilage is destroyed, the exposed bone tissue is joined by the fibrous tissue and instigates ossification of the joint. It is the joint ossification that causes the crippling effect of rheumatoid arthritis. This disease affects females slightly more than males and occurs most commonly between the ages of thirty and fifty.

Osteoarthritis is a degenerative joint disease that results from aging and irritation of the joints. Although osteoarthritis

rheumatoid: Gk. *rheuma,* a flowing

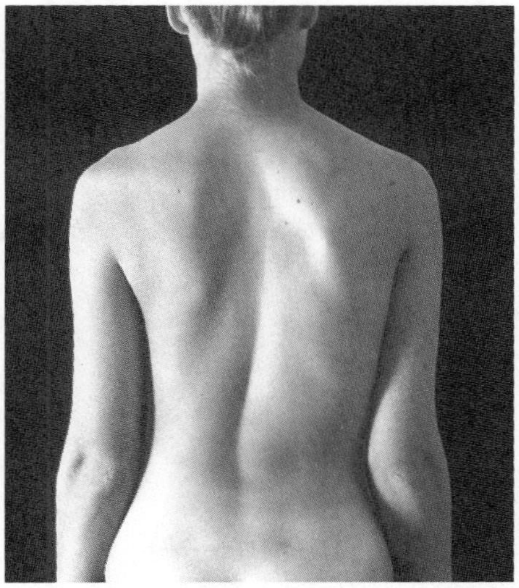

Figure 8.34 Scoliosis is a lateral curvature of the spine usually in the thoracic region. It may be congenital, acquired, or disease related.

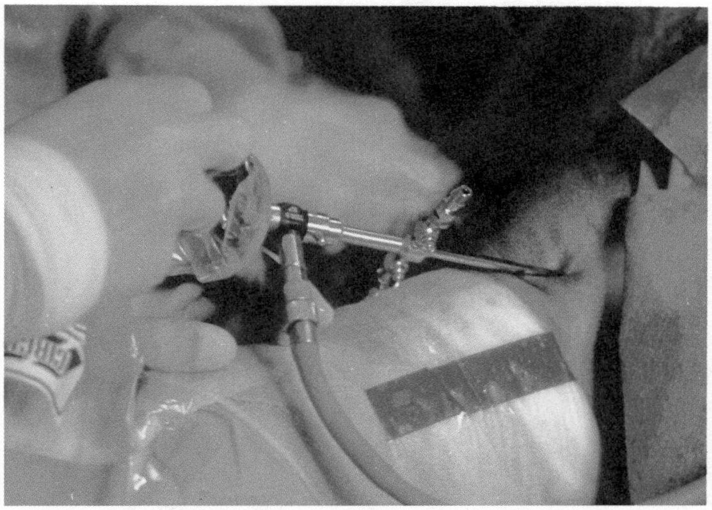

Figure 8.35 Arthroscopy. In this technique, a needlelike viewing arthroscope is threaded into the joint capsule through a tiny incision. The arthroscope has a fiberoptic light source, which illuminates the interior of a joint and the position of the surgical instruments that may be inserted through other small incisions.

is far more common than rheumatoid arthritis, it is usually less damaging. Osteoarthritis is a slow, progressive disease in which the articular cartilages gradually soften and disintegrate. The affected joints seldom swell, and the synovial membrane is rarely damaged. As the articular cartilage deteriorates, ossified spurs are deposited on the exposed bone, causing pain and restricting the movement of articulating bones. Osteoarthritis most frequently affects the knee, hip, and intervertebral joints.

Gouty arthritis results from a metabolic disorder in which an abnormal amount of uric acid is retained in the blood and sodium urate crystals are deposited in the joints. The salt crystals irritate the articular cartilage and synovial membrane, causing swelling, tissue deterioration, and pain. If gout is not treated, the affected joint fuses. Males have a greater incidence of gout than females, and apparently the disease is genetically determined. About 85% of gout cases affect the joints of the foot and legs. The most common joint affected is the metatarsophalangeal joint of the hallux (great toe).

Treatment of Joint Disorders

Arthroscopy *(ar-thros'ko-pe)* is widely used in diagnosing and, to a limited extent, treating joint disorders. Arthroscopic inspection involves making a local incision through the skin and into the joint capsule through which the tubelike arthroscopic instrument is threaded (fig. 8.35). In arthroscopy of the knee, the articular cartilage, synovial membrane, menisci, and cruciate ligaments can be observed. Samples can be extracted, and pictures taken for further evaluation.

Remarkable advancements have been made in the last fifteen years in **joint prostheses** *(pros-the'sēz)* (fig. 8.36). Joint prostheses, or artificial articulations, do not take the place of normal, healthy joints, but they are a valuable option for chronically disabled arthritis patients.

CLINICAL CASE STUDY ANSWER

The mechanism of knee injury, the location of the pain, along with the exam findings indicate a complete or near-complete tear of the medial collateral ligament. Because the medial meniscus is attached to the medial collateral ligament, it is frequently torn as well in an injury of this sort. Other ligaments susceptible to athletic injury are the anterior cruciate (most common), lateral collateral, and posterior cruciate ligaments. Complete tears of these ligaments usually require surgical repair. Incomplete tears can often be managed without surgery. ■

gout: L. *gutta,* a drop (thought to be caused by drops of viscous humors)

prosthesis: Gk. *pros,* in addition to; *thesis,* a setting down

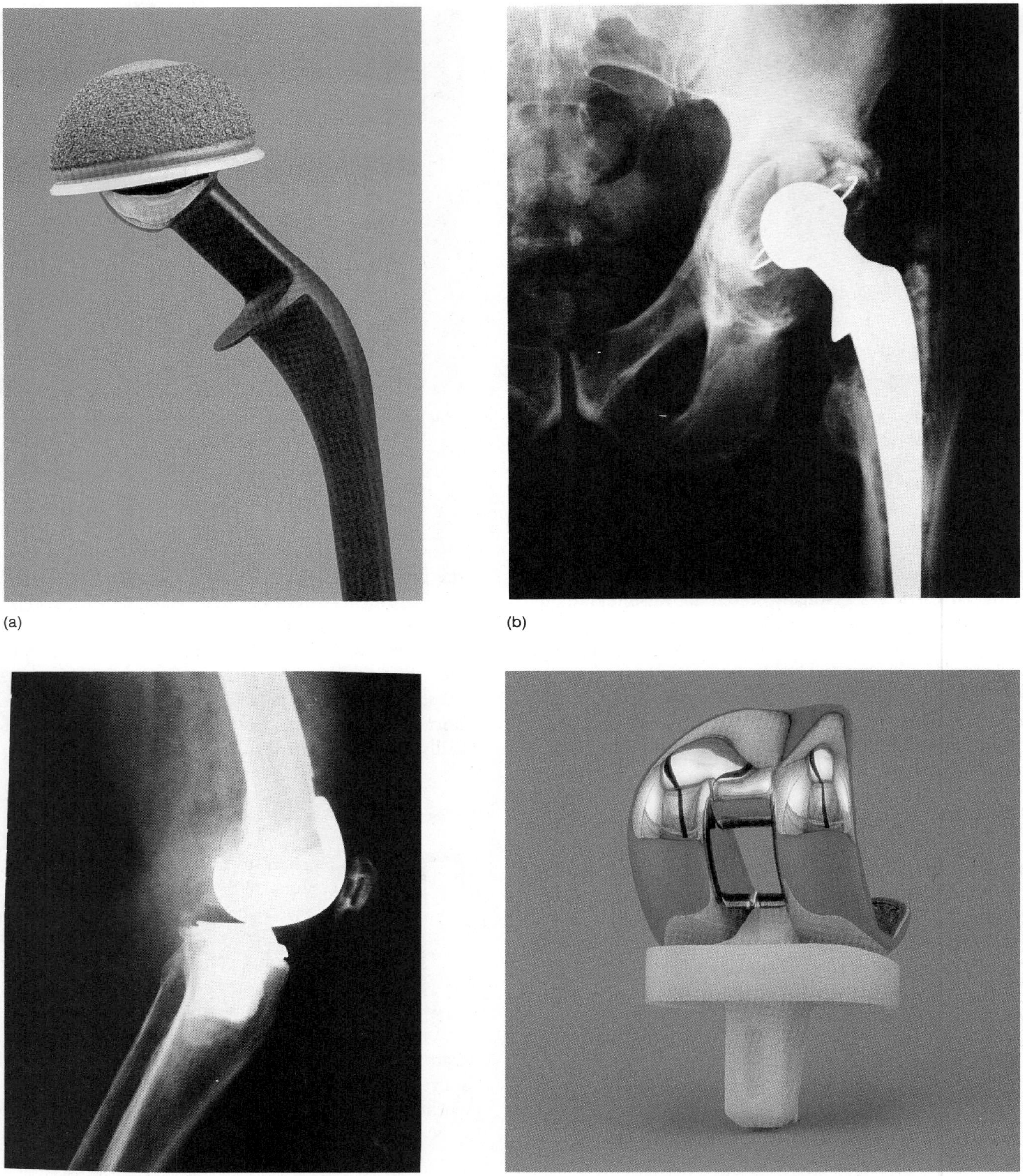

(a) (b)

(c) (d)

Figure 8.36 Two examples of joint prostheses. (*a, b*) The hip joint and (*c, d*) the knee joint.

Important Clinical Terminology

ankylosis Stiffening of a joint resulting in severe or complete loss of movement.
arthralgia (also *arthrodynia*) Severe pain within a joint.
arthrolith A gouty deposit in a joint.
arthrometry The measurement of the range of movement in a joint.
arthroncus Swelling of a joint due to trauma or disease.
arthropathy Any disease affecting a joint.
arthroplasty The surgical repair of a joint.
arthrosis A joint or an articulation; also, a degenerative condition of a joint.
arthrosteitis An inflammation of the bony structure of a joint.
chondritis An inflammation of the articular cartilage of a joint.
coxarthrosis A degenerative condition of the hip joint.
hemarthrosis An accumulation of blood in a joint cavity.
rheumatology The medical speciality concerned with the diagnosis and treatment of arthritis.
spondylitis An inflammation of one or several vertebrae.
synovitis The inflammation of the synovial membrane lining the inside of a joint capsule.

Chapter Summary

I. Classification of Joints
   A. Joints are formed as adjacent bones articulate. Arthrology is the science concerned with the study of joints, and kinesiology is the study of the biomechanics of joint movement.
   B. Joints can be classified according to structure or function.
      1. The structural classification divides joints into fibrous, cartilaginous, and synovial types.
      2. The functional classification divides joints into synarthroses, amphiarthroses, and diarthroses.

II. Development of Freely Movable Joints
   A. The development of each movable joint is initiated during the sixth week as the perichondrium from adjacent developing bones spans the future joint cavity and gives rise to the joint capsule.
   B. Diarthroses form by the third month as the epiphyses of adjacent endochondral bones develop distinct configurations and as the muscles that move the joints undergo contractions.
   C. Articular cartilages form on the epiphyses, and the synovial membrane secretes synovial fluid into the joint cavity.

III. Synarthroses
   A. Articulating bones in immovable joints are connected by either fibrous tissue or cartilage.
   B. Sutures are found only in the skull and are classified as serrate, lap, plane, or synostosis.
   C. Synchondroses are temporary joints formed in the growth lines (epiphyseal plates) between the diaphyses and epiphyses in the long bones of children.

IV. Amphiarthroses
   A. Amphiarthroses allow limited motion in response to twisting, compression, or stress.
   B. The articulating bones of amphiarthroses are connected by ligaments or fibrocartilaginous pads.
   C. The symphysis pubis and the intervertebral disc joints are examples of symphyses; the distal articulations of the tibia-fibula and the ulna-radius are examples of syndesmoses.

V. Diarthroses
   A. Diarthroses are freely movable joints enclosed by joint capsules that contain synovial fluid. They also contain joint cavities, synovial membranes, and articular cartilages.
   B. The range of movement of a diarthrosis is determined by the structure of the articulating bones, the ligaments and tendons, and the muscles that act on the joint.

VI. Movements at Diarthroses
   A. Movements at diarthroses are produced by the contraction of the skeletal muscles spanning the joints and attaching to or near the bones forming the articulations. In these actions, the bones act as levers, the muscles provide the force, and the joints are the fulcra, or pivots.
   B. Angular movements increase or decrease the joint angle produced by the articulating bones.
      1. Flexion decreases the joint angle on an anterior-posterior plane; extension increases the same joint angle.
      2. Abduction is the movement of a body part away from the main axis of the body; adduction is the movement of a body part toward the main axis of the body.
   C. Circular movements can occur only in joints that are composed of a bone with a rounded surface articulating with a corresponding depression on another bone.
      1. Rotation is the movement of a bone around its own axis.
      2. Circumduction is a conelike movement of a body segment.
   D. Special joint movements include inversion and eversion, protraction and retraction, and elevation and depression.
   E. Diarthroses can be classified as first-, second-, or third-class levers.

VII. Specific Joints of the Body
   A. The temporomandibular joint, a combined hinge and gliding joint, is of clinical importance because of temporomandibular joint (TMJ) syndrome.
   B. The sternoclavicular joint, a gliding joint, is frequently dislocated in injuries to the rib cage.
   C. The humeroscapular (shoulder) joint, a ball-and-socket joint, is vulnerable to dislocations from sudden jerks of the arm, especially in children before strong shoulder muscles have developed.
   D. There are two diarthroses at the elbow as the distal end of the humerus articulates with the proximal ends of the ulna and radius. Strain on the elbow joint is common during certain sports.
   E. Both metacarpophalangeal and interphalangeal joints are condyloid and hinge diarthroses, respectively, which may be traumatized when an athlete jams a finger.
   F. The ball-and-socket coxal (hip) joint is especially subject to osteoarthritis in elderly people.
   G. The hinged tibiofemoral (knee) joint is the largest, most complex, and most vulnerable joint in the body.
   H. There are two hinged articulations within the talocrural (ankle) joint. Ankle sprains are common injuries of this joint.

Review Activities

Objective Questions

1. Which statement regarding joints is *false*?
   (a) Joints are the locations where two or more bones articulate.

(b) The structural classification of joints includes fibrous, membranous, and cartilaginous types.
(c) Arthrology is the study of joints; kinesiology is the study of the biomechanics of joint movement.

2. Synchondroses are a type of
   (a) synarthrosis. (c) amphiarthrosis.
   (b) diarthrosis. (d) syndesmosis.
3. A synostosis is a unique type of
   (a) synchondrosis. (c) symphysis.
   (b) syndesmosis. (d) suture.
4. Which of the following joint type-function word pairs is *incorrect*?
   (a) synchondrosis—growth at the epiphyseal plate
   (b) symphysis—movement at the intervertebral joint
   (c) suture—strength and stability in the skull
   (d) syndesmosis—movement of the jaw
5. Structurally, diarthroses are
   (a) synovial joints.
   (b) fibrous joints.
   (c) membranous joints.
   (d) cartilaginous joints.
6. Which of the following is *not* characteristic of all diarthroses?
   (a) articular cartilage
   (b) synovial fluid
   (c) joint capsule
   (d) meniscus
7. The atlantoaxial and the proximal radioulnar diarthrotic joints are specifically classified as
   (a) hinge. (c) pivot.
   (b) gliding. (d) condyloid.
8. Which of the following joints can be readily and comfortably hyperextended?
   (a) interphalangeal
   (b) coxal
   (c) tibiofemoral
   (d) sternocostal
9. Which of the following is most vulnerable to luxation? The
   (a) elbow.
   (b) humeroscapular.
   (c) coxal.
   (d) tibiofemoral.
10. A thickening and tenderness of the synovial membrane and the accumulation of synovial fluid are signs of the development of
    (a) arthroscopitis.
    (b) gouty arthritis.
    (c) osteoarthritis.
    (d) rheumatoid arthritis.

Essay Questions

1. What is meant by a structural classification of joints, as compared to a functional classification?
2. Why is the anatomical position so important in explaining the movements that are possible at joints?
3. Differentiate between the two types of amphiarthroses.
4. Describe the structure of a typical diarthrotic joint.
5. What are the structural components of a diarthrotic joint that determine or limit the range of movement at that joint?
6. What are the advantages of a hinge joint over a ball-and-socket type? If ball-and-socket joints allow a greater range of movement, why are not all the diarthrotic joints of this type?
7. What is synovial fluid? Where is it produced, and what are its functions?
8. What is a bursa? What is the difference between a bursa and a tendon sheath? What are the functions of each?
9. Identify four types of diarthrotic joints found in the wrist and hand region, and state the types of movement permitted by each.
10. Discuss the articulations of the pectoral and pelvic regions to the axial skeleton with regard to range of movement, ligamentous attachments, and potential clinical problems.
11. List and describe the joints of the upper and lower extremities, giving the bones involved in each articulation, the functional classification, and the anatomical components.
12. Why is the knee joint so susceptible to injury?
13. What is meant by a sprained ankle? How does a sprain differ from a strain or a luxation?
14. Discuss the three common causes of back pain. What can cause a herniated intervertebral disc? Why is it so painful and what may be done to help the problem?
15. What occurs within the joint capsule in rheumatoid arthritis? How does rheumatoid arthritis differ from osteoarthritis?

CHAPTER 9

Muscular System

Outline and Concepts

CLINICAL CASE STUDY

A 66-year-old man comes to his doctor for a routine physical exam. The man's medical history reveals that he was treated surgically for cancer of the oropharynx, six years prior. The patient states that the cancer spread to the lymph nodes in the left side of his neck. He elaborates by pointing to the involved area and states that lymph nodes, a vein, and a muscle, among other things, were removed. On the right side, only lymph nodes were removed but contained no cancer. The patient adds, in a bewildered tone, that he can only turn his head to the left. "It seems to me Doc, that if they took the muscle out of the left side of my neck that I would only be able to turn my head to the right."

Does the patient have a valid point? If not, how would you explain the reason for his disability in terms of neck musculature? ■

Organization and General Functions

The muscles of the skeletal system are adapted to contract in order to carry out the functions of motion, heat production, and posture and body support.

Objective 1. Define the term *myology*, and describe the three principal functions of muscles.

Objective 2. Explain how muscles are described according to their anatomical location and cooperative function.

Myology is the study of muscles. Myology by itself does not have much meaning, however, except as it relates to the skeletal system and joints in the performance of body movements and to the nervous system in the performance of motor control. Muscles are dynamic, and an understanding of them requires an applied approach.

Functions of Muscles

Muscle cells (fibers) are adapted to contract when stimulated by nerve impulses. The stimulation of several fibers is not enough to cause a noticeable effect within a muscle, but these types of isolated fiber contractions are important and occur continuously within a muscle. When many fibers of a skeletal muscle are activated, the muscle contracts and causes body movement. Muscles perform three general functions: (1) motion, (2) heat production, and (3) posture and body support.

1. **Motion.** The most obvious type of motion performed by the skeletal muscles is to move the body and/or its appendages, as in walking, running, writing, chewing, and swallowing. The contraction of skeletal muscle is equally important in breathing and in moving body fluids. The stimulation of isolated fibers within the muscles maintains muscle tonus and is important in the movement of venous blood and lymphatic fluid. The eye and even the ear ossicles have associated skeletal muscles responsible for various movements.

   The primary impetus for blood flow is the contraction of cardiac muscles within the heart. All of the involuntary body systems (urinary, digestive, respiratory, circulatory, etc.) contain smooth muscles for the involuntary movement of materials through the body. The act of speaking involves several different muscle groups that control air volume, tense the vocal cords, move the tongue, and purse the lips. This is a highly coordinated effort of several muscle groups controlled by several different nerve centers.
2. **Heat production.** Body temperature is remarkably consistent. Metabolism within the cells releases heat as an end product. Since muscles constitute approximately 40 percent of the body weight and are in a continuous state of fiber activity, they are very important in the production of heat. The rate of heat production increases immensely as a person exercises strenuously.
3. **Posture and body support.** The skeletal system gives form and stability to the body, but skeletal muscles maintain posture and support around the flexible joints. Certain muscles are active postural muscles whose primary function is to work in opposition to gravity. Some postural muscles are working even when you think you are relaxed. As you are sitting, for example, the weight of your head is balanced at the atlantooccipital joint through the efforts of the muscles located at the back of the neck. If you start to get sleepy, your head will suddenly nod forward as the postural muscles relax and the weight (resistance) overcomes the effort.

There are three types of muscle tissue in the body: smooth, cardiac, and skeletal. Although these three types differ in structure and function, all muscle tissue possesses some basic characteristic properties:

1. **Irritability.** A muscle tissue receives and responds to a stimulus from a nerve impulse.
2. **Contractility.** A muscle tissue responds to a stimulus by contracting, or shortening, its length.
3. **Extensibility.** Once a stimulus has subsided and the fibers within muscle tissue are shortened and relaxed, they may be stretched back to their original length by the contraction of an opposing muscle. The fibers are then stretched in readiness for another contraction.
4. **Elasticity.** A muscle tissue has an innate tension that causes it to assume a desired shape regardless of how it might be stretched.

A histological description of each of the three muscle types was presented in chapter 4 and should be reviewed at this time. Cardiac muscle is discussed further with the heart in chapter 16. Smooth muscle functions with various involuntary organs throughout the body and is discussed, when appropriate, with these organs. The remaining information presented in this chapter pertains only to skeletal muscle and the skeletal muscular system of the body.

Skeletal muscle constitutes a body system by itself and accounts for approximately 40% of body weight. Over six

muscle: L. *mus*, mouse (from the appearance of certain muscles)
myology: Gk. *myos*, muscle; *logos*, study of

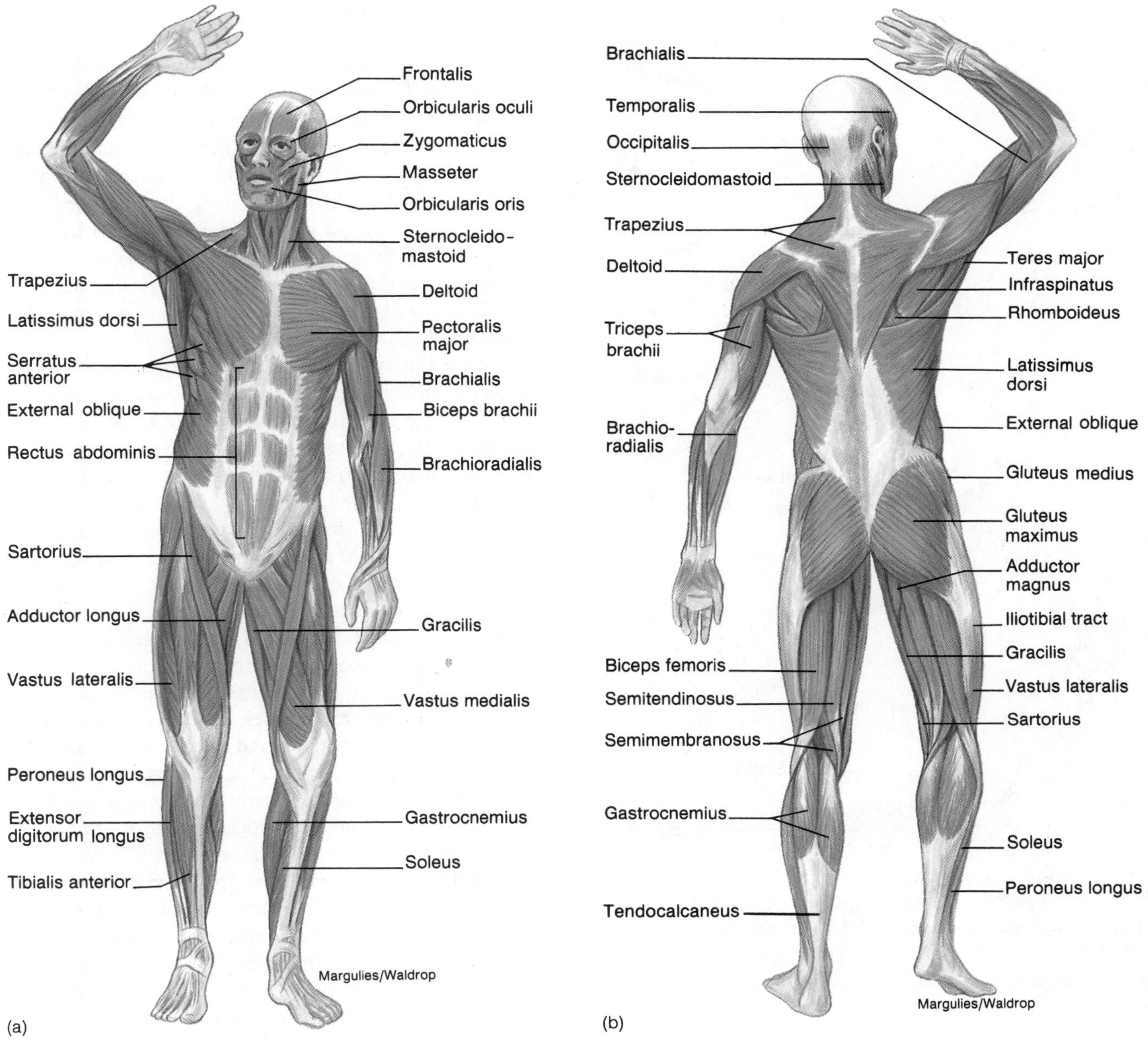

Figure 9.1 (*a*) An anterior view and (*b*) a posterior view of some of the superficial skeletal muscles.

hundred individual muscles comprise the skeletal muscular system. The principal superficial muscles are shown in figure 9.1. Muscles are usually described in groups according to their anatomical location and cooperative function. The *muscles of the axial skeleton* have their attachments to the bones of the axial skeleton and include facial muscles, neck muscles, and anterior and posterior trunk muscles. The *muscles of the appendicular skeleton* include those that act on the pectoral and pelvic girdles and those that move the segments of the appendages.

1. Discuss how the functions of muscles aid in maintaining body homeostasis.
2. What is meant by a postural muscle?
3. Distinguish between the axial and the appendicular muscles.

Development of Skeletal Muscles

Skeletal muscles begin development in the fourth week of embryonic life from blocks of mesoderm, called myotomes, in the trunk area, and from loosely organized mesenchyme in the head and appendage areas. Each muscle fiber is derived from several embryonic myoblast cells.

Objective 3. Describe the formation of a muscle fiber from myoblast cells.

Objective 4. Describe the development of skeletal muscles from myotomes.

The formation of skeletal muscle tissue begins during the fourth week of embryonic development as specialized mesodermal cells, called **myoblasts,** begin rapid mitotic division (fig. 9.2).

myoblast: Gk. *myos,* muscle; *blastos,* germ

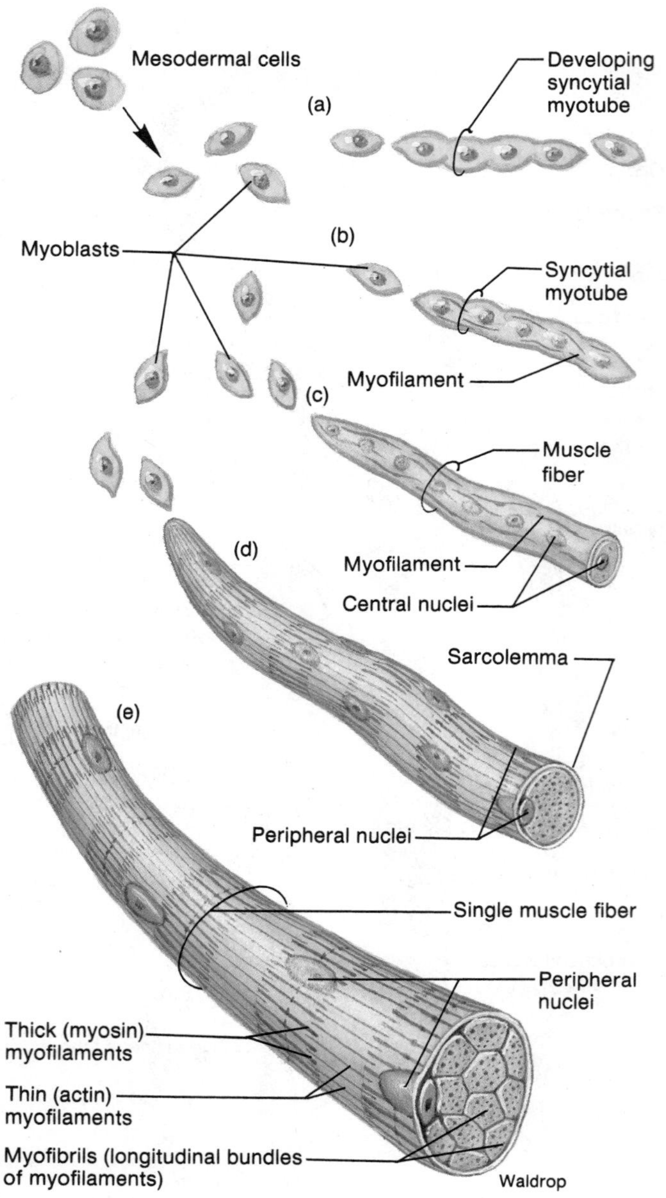

Figure 9.2 The development of skeletal muscle fibers. (*a*) At four weeks, mesodermal cells differentiate into myoblasts that aggregate into a developing syncytial tube. (*b*) At five weeks, the syncytial myotube is formed as individual cell membranes are broken down and longitudinal myofilaments appear. Myotubes grow in length by incorporating additional myoblasts; each adds an additional nucleus. (*c*) Muscle fibers are distinct at nine weeks, but the nuclei are still centrally located, and growth in length continues through the addition of myoblasts. (*d*) At five months, thin (actin) and thick (myosin) myofilaments are present, and moderate growth in length still continues. (*e*) By birth, the striated myofilaments have aggregated into bundles, the fiber has thickened, and the nuclei have shifted to the periphery. Myoblast activity ceases, and all the muscle fibers a person will have are formed.

The proliferation of new cells continues while the myoblast cells fuse together into **syncytial** *(sin-sish'al)* **myotubes.** A syncytium is a multinucleated protoplasmic mass formed by the secondary union of originally separate cells. At nine weeks, primitive myofilaments course through the myotubes and the nuclei of the contributing myoblasts become centrally located. Growth in length continues through the addition of myoblasts.

It is not certain when skeletal muscle is sufficiently developed to sustain contractions, but by the seventeenth week the fetal movements known as **quickening** are strong enough to be recognized by the mother. At this time, the individual muscle fibers have thickened, the nuclei have moved peripherally, and the filaments (myofilaments) have differentiated into alternating thin *(actin)* and thick *(myosin)* bands. Growth in length still continues through addition of myoblasts.

Shortly before a baby is born, the formation of myoblast cells ceases, and all of the muscle cells of a person have been determined. Differences in strength, endurance, and coordination are somewhat genetically determined but are primarily the result of individual body conditioning. Muscle coordination is an ongoing process of achieving a fine neural control of muscle fibers. A mastery of muscle movement is comparatively slow in humans. It is several months before a newborn has the coordination to crawl, and about a year before it achieves bipedal posture and locomotion. By contrast, most mammals can walk and run within a few hours after they are born. Refined, precise movements of the hands for delicate work, such as playing a violin or performing surgery, may be learned only after many hours of disciplined practice, and some persons may not be able to develop these movements at all.

The process of muscle fiber development occurs within specialized mesodermal masses, called **myotomes** *(mi'o-tōms)*, in the embryonic trunk area and from loosely organized masses of mesoderm in the head and appendage areas. At six weeks, the trunk of an embryo is segmented into distinct myotomes (fig. 9.3) that are associated dorsally with specific sclerotomes. Sclerotomes give rise to vertebrae. As will be explained in chapter 12, spinal nerves arise from the spinal cord and exit between vertebrae to innervate (serve with motor and sensory nerve branches) developing muscles in the adjacent myotomes. As myotomes develop, they elongate ventrally toward the midline of the body, or distally into the developing limbs. The muscles of the entire muscular system are differentiated and in correct position by the eighth week (fig. 9.3). The orientation of the developing muscles is preceded and influenced by cartilaginous models of bones.

1. Describe the process by which muscle fibers become multinucleated.
2. Distinguish between the terms *myoblasts, myotubes,* and *muscle fibers.*
3. Briefly describe the embryonic events that occur between weeks six and eight in the formation of specific muscles.

Structure of Skeletal Muscles

Skeletal muscle tissue and its binding connective tissue are arranged in a highly organized pattern so that the forces of the contracting muscle fibers are united and directed onto the structure being moved.

---

syncytial: Gk. *syn,* with; *cyto,* cell

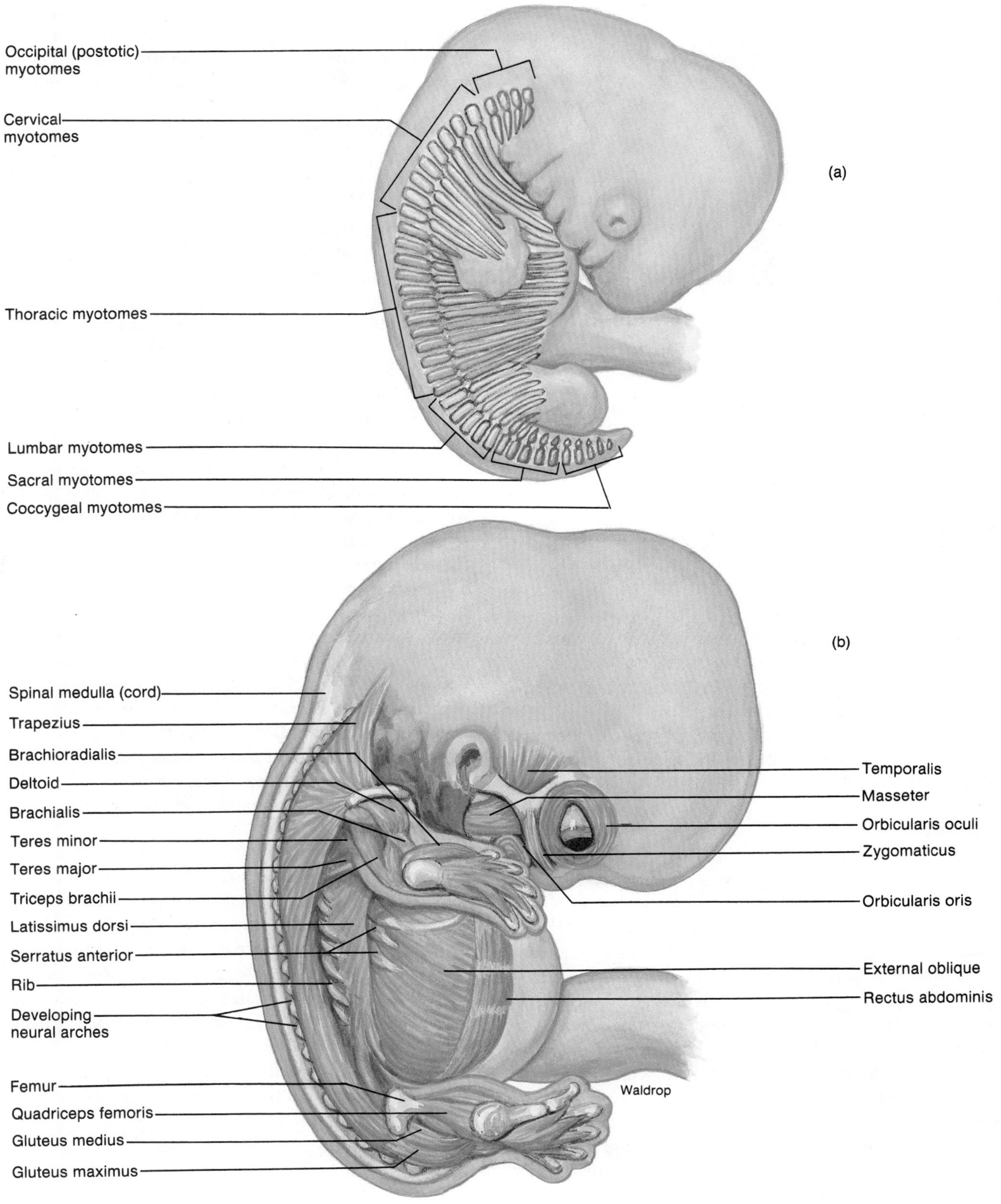

Figure 9.3 The development of skeletal muscles. (*a*) The distribution of embryonic myotomes at six weeks. Segmental myotomes give rise to the muscles of the trunk area and girdles. Loosely organized masses of mesoderm form the muscles of the head and extremities. (*b*) The arrangement of skeletal muscles at eight weeks. The development of muscles is influenced by the preceding cartilaginous models of bones. The innervation of muscles corresponds to the development of spinal nerves and dermatome arrangement.

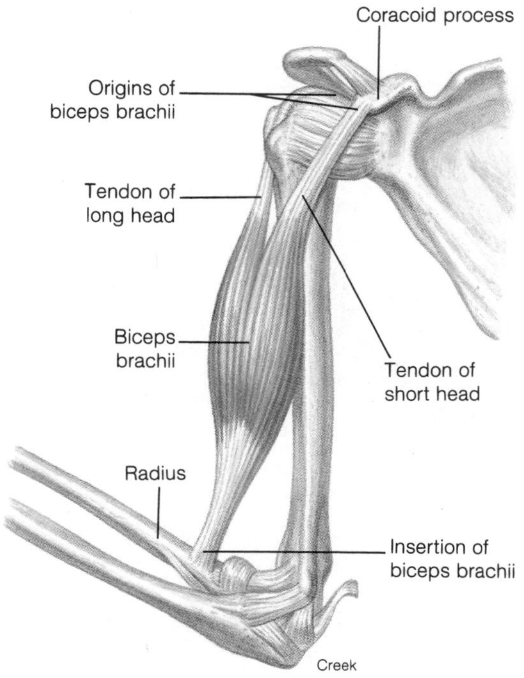

Figure 9.4 The skeletomuscular relationship. The more proximal and fixed point of muscle attachment is the origin, whereas the distal, maneuverable point of attachment is the insertion. The contraction of muscle fibers causes one bone to move relative to another around a joint.

Objective 5. Distinguish between the various binding connective tissues associated with skeletal muscles.
Objective 6. Explain what is meant by synergistic and antagonistic muscle groups.
Objective 7. Describe the various types of muscle fiber architecture, and discuss the biomechanical advantage of each.

Attachments

Skeletal muscles are usually long and narrow, span a joint, and are attached to a bone at either end by a tendon (fig. 9.4). As the muscle contracts, one of the bones moves relative to the other at the joint. The more fixed, or stationary, attachment is designated as the **origin** of a muscle, whereas the movable end is its **insertion.** In muscles associated with the girdles and appendages, the origin is generally the proximal attachment and the insertion the distal attachment. The fleshy, thickened portion of a muscle is referred to as its **belly,** or **gaster.** Generally the belly of a muscle is located on the bone proximal to the bone that is to be moved. The joint is spanned by a **tendon** from the muscle. A tendon is toughened, dense fibrous connective tissue that connects a muscle to the periosteum of a bone. It functions to transfer the force of contraction from the muscle across the joint and onto the bone that is to be moved. Flattened, sheetlike tendons are called **aponeuroses** *(ap''o-nu-ro'sēz)* and occur where attachment is over a broad line. In certain places, especially in the wrist and ankle, the tendons are not only enclosed by protective **tendon sheaths,** which lubricate the tendons with synovial fluid, but also the entire group of tendons is covered by a thin, but strong, band of connective tissue called a **retinaculum** *(ret''ĭ-nak'u-lum)* (see fig. 9.33).

aponeurosis: Gk. *aponeurosis,* change into a tendon

Associated Connective Tissue

Contracting muscle fibers would not be effective if they worked as isolated units. Each fiber is bound to adjacent fibers to form bundles, and the bundles in turn are bound to other bundles. With this arrangement, the contraction in one area of a muscle works in conjunction with contracting fibers elsewhere in the muscle. The binding substances within muscles are the associated connective tissues.

Connective tissue is structurally arranged within muscle to protect, strengthen, and bind muscle fibers into bundles and bind the bundles together (fig. 9.5). The individual fibers of skeletal muscles are surrounded by fine sheaths of connective tissue called **endomysium** *(en''do-mis'e-um).* The endomysium binds adjacent fibers and supports capillaries and nerve endings serving the muscle. Another connective tissue, called **perimysium,** binds groups of fibers together into bundles called **fasciculi** *(fah-sik'u-li).* The perimysium supports blood vessels and nerve fibers serving the various fasciculi. The entire muscle is covered by the **epimysium,** which in turn is continuous with a tendon.

Fascia *(fash'e-ah)* is a fibrous connective tissue that covers muscle and attaches to the skin. *Superficial fascia* secures the skin to the underlying muscles. It varies in thickness throughout the body. For example, superficial fascia over the buttocks and abdominal wall is thick and laced with adipose tissue. The superficial fascia under the skin of the dorsum of the hand and facial region is thin. *Deep fascia* is an inward extension of the superficial fascia. It lacks adipose tissue and blends with the epimysium of muscle. Deep fascia surrounds adjacent muscles compartmentalizing and binding them into functional groups.

Muscle Groups

Just as individual muscle fibers seldom contract independently, muscles generally do not contract separately but work as functional groups. Muscles that contract together and are coordinated in accomplishing a particular movement are said to be *synergistic* (fig. 9.6). *Antagonistic* muscles perform opposite functions and are generally located on the opposite sides of the limb. The two heads of the biceps brachii muscle along with the brachialis muscle, for example, contract together to *flex* the elbow joint. The triceps brachii muscle, the antagonist to the biceps brachii and brachialis muscles, *extends* the elbow as it is contracted.

retinaculum: L. *retinere,* to hold back (retain)
endomysium: Gk. *endon,* within; *myos,* muscle
perimysium: Gk. *peri,* around; *myos,* muscle
fasciculus: L. *fascis,* bundle
epimysium: Gk. *epi,* upon; *myos,* muscle
fascia: L. *fascia,* a band or girdle
synergistic: Gk. *synergein,* cooperate
antagonistic: Gk. *antagonistes,* struggle against

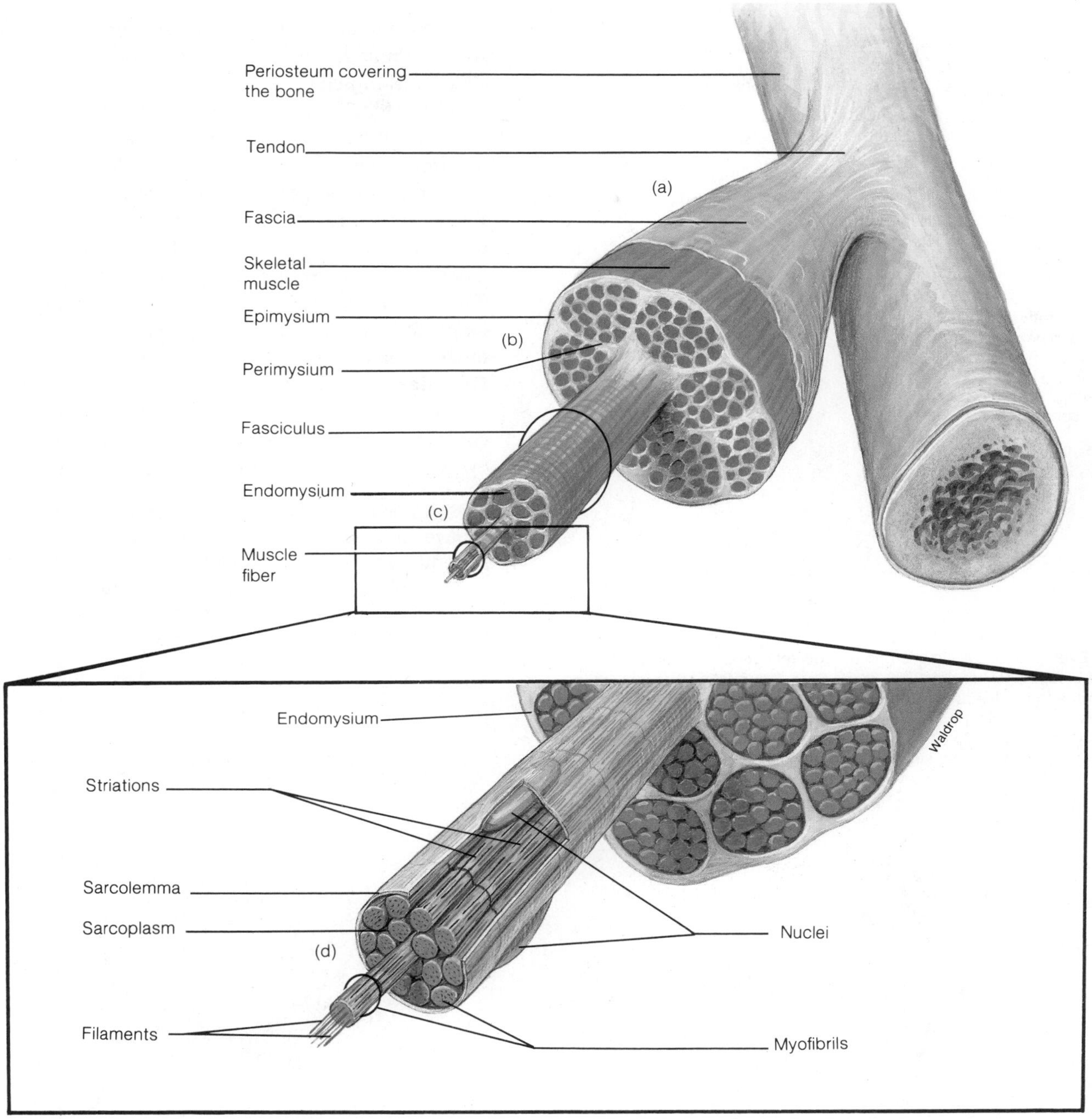

Figure 9.5 The relationship between muscle tissue and connective tissue. (*a*) The fascia and tendon attaches a muscle to the periosteum of a bone. (*b*) The epimysium surrounds the entire muscle, and the perimysium separates and binds the fasciculi (muscle bundles). (*c*) The endomysium surrounds and binds individual muscle fibers. (*d*) An individual muscle fiber is composed of myofibrils consisting of filaments of actin and myosin.

Seldom does a single muscle cause a movement at a joint. A division of labor is achieved by having several synergistic muscles rather than one massive muscle. One muscle may be an important postural muscle, for example, whereas another may be adapted for rapid, powerful contraction. When total output of the synergistic muscles is required, the contraction of several smaller muscles provides more strength than the contraction of one large one.

Muscle Architecture

Skeletal muscles may be classified on the basis of fiber arrangement (fig. 9.7). Certain advantages are inherent in each of the types of muscles. The following are the major types of fiber arrangement:

1. **Parallel,** or **fusiform.** Parallel muscles have relatively long excursions (contract over a great distance) and good

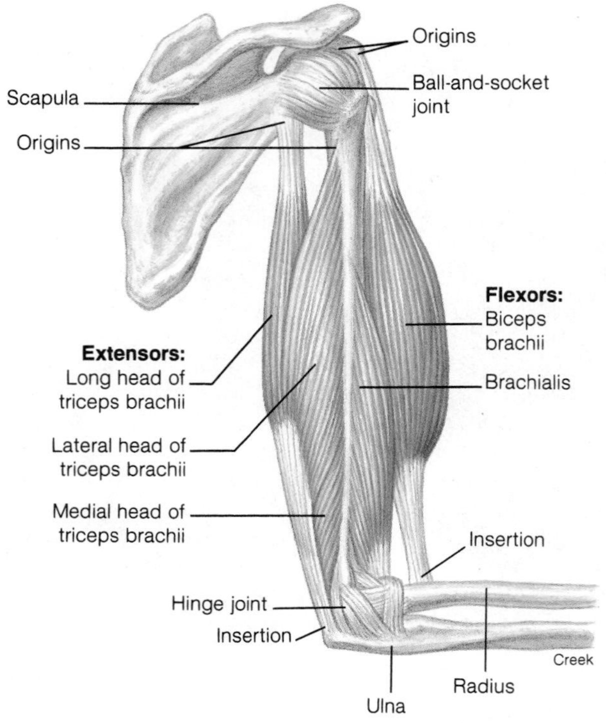

Figure 9.6 Examples of synergistic and antagonistic muscles. The two heads of the biceps and the brachialis muscle are synergistic to each other, as are the three heads of the triceps. The biceps and the brachialis are antagonistic to the triceps, and the triceps are antagonistic to the biceps and the brachialis muscle. When one antagonistic group contracts, the other one must relax, or movement does not occur.

endurance but are not especially strong. They are long, straplike muscles such as the sartorius and rectus abdominis muscles.

2. **Convergent.** Convergent-fibered muscles are so named because the fibers converge at the insertion point to maximize contraction. They are fan-shaped muscles such as the deltoid and pectoralis major muscles.
3. **Pennate.** Pennate-fibered muscles have many fibers per unit area and hence are strong. They provide dexterity. They have short excursions but generally tire quickly. There are three kinds of pennate-fibered muscles: *unipennate, bipennate,* and *multipennate.* Examples of each of the pennate types can be found in the forearm.
4. **Circular.** Circular-fibered muscles surround a body opening, or **orifice,** and act as a sphincter when contracted. Examples are the orbicularis oculi around the eye and orbicularis oris surrounding the mouth.

Muscle-fiber architecture can be observed on a cadaver or other dissection specimen. If you have the opportunity to learn the muscles of the body from a cadaver, observe the fiber architecture of specific muscles, and try to determine the advantages afforded to each muscle by its location and action.

pennate: L. *pennatus,* feather
orifice: L. *orificium,* mouth; *facere,* to make

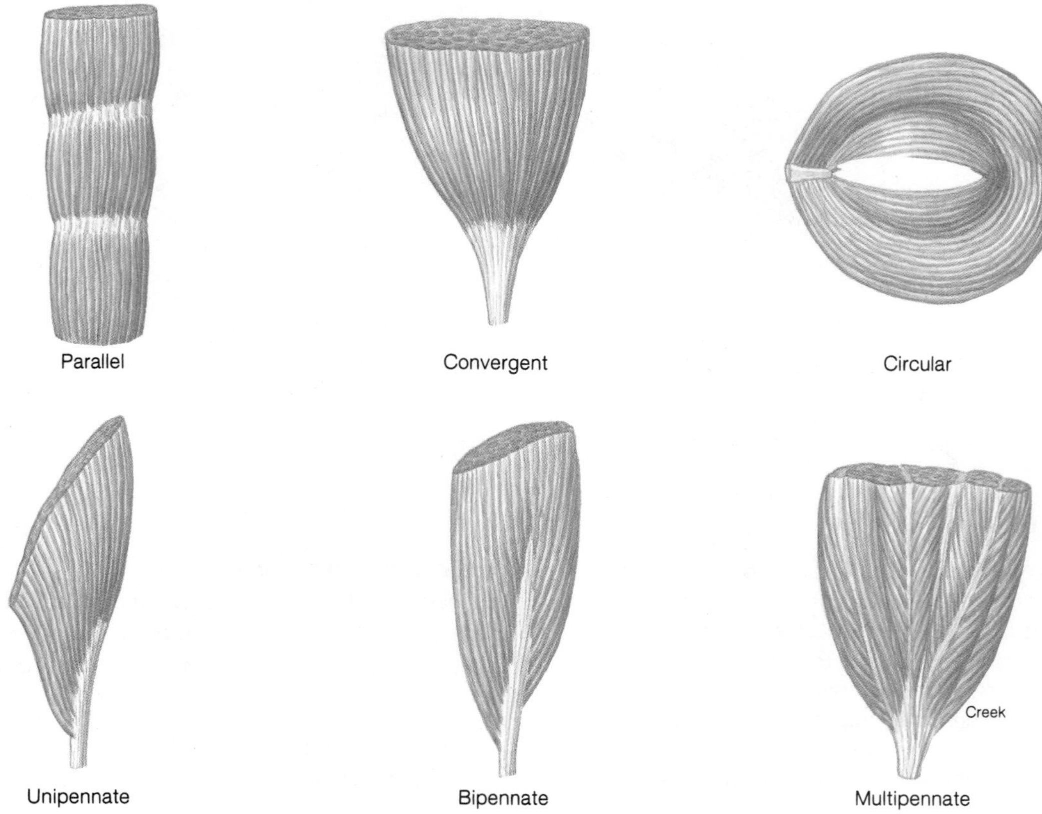

Figure 9.7 Skeletal muscle architecture.

Blood and Nerve Supply to Skeletal Muscle

Muscle cells have a high rate of metabolic activity and therefore require extensive vascularity to receive nutrients and oxygen and to eliminate waste products. Smaller muscles generally have a single artery supplying blood and perhaps two veins returning blood (fig. 9.8). Large muscles may have several arteries and veins. The microscopic capillary exchange between arteries and veins occurs throughout the endomysium that surrounds individual fibers.

A skeletal muscle fiber cannot contract unless it is stimulated by a nerve impulse. This means that there must be extensive innervation to a muscle to ensure the connection of each muscle fiber to a nerve cell. Actually there are two nerve pathways for each muscle. A **motor (efferent) neuron** is a nerve cell that conducts nerve impulses to the muscle fiber to stimulate it to contract. A **sensory (afferent) neuron** conducts nerve impulses away from the muscle fiber to the central nervous system. Muscle fibers will atrophy if they are not periodically stimulated to contract. The names of both the nerves and vessels usually refer to the muscle they serve.

The feeling of stiffness the day after strenuous exercise is due to trauma to muscle fibers and the accumulation of lactic acid within the muscle. Lactic acid is a waste product of metabolism and is removed by the venous flow from the muscle. If a person is in good physical condition, the vascularity can readily remove the metabolic wastes, and stiffness is not as apt to occur. Being in good physical condition not only improves vascularity, but enlarges muscle fibers and allows them to work more efficiently over a longer duration.

---

1. Contrast the following terms: *endomysium* and *epimysium; fascia* and *tendon; aponeurosis* and *retinaculum.*
2. Discuss the biomechanical advantage of having synergistic muscles. Give some examples of synergistic muscles, and state which muscles are antagonistic.
3. Which type of architecture provides dexterity and strength?

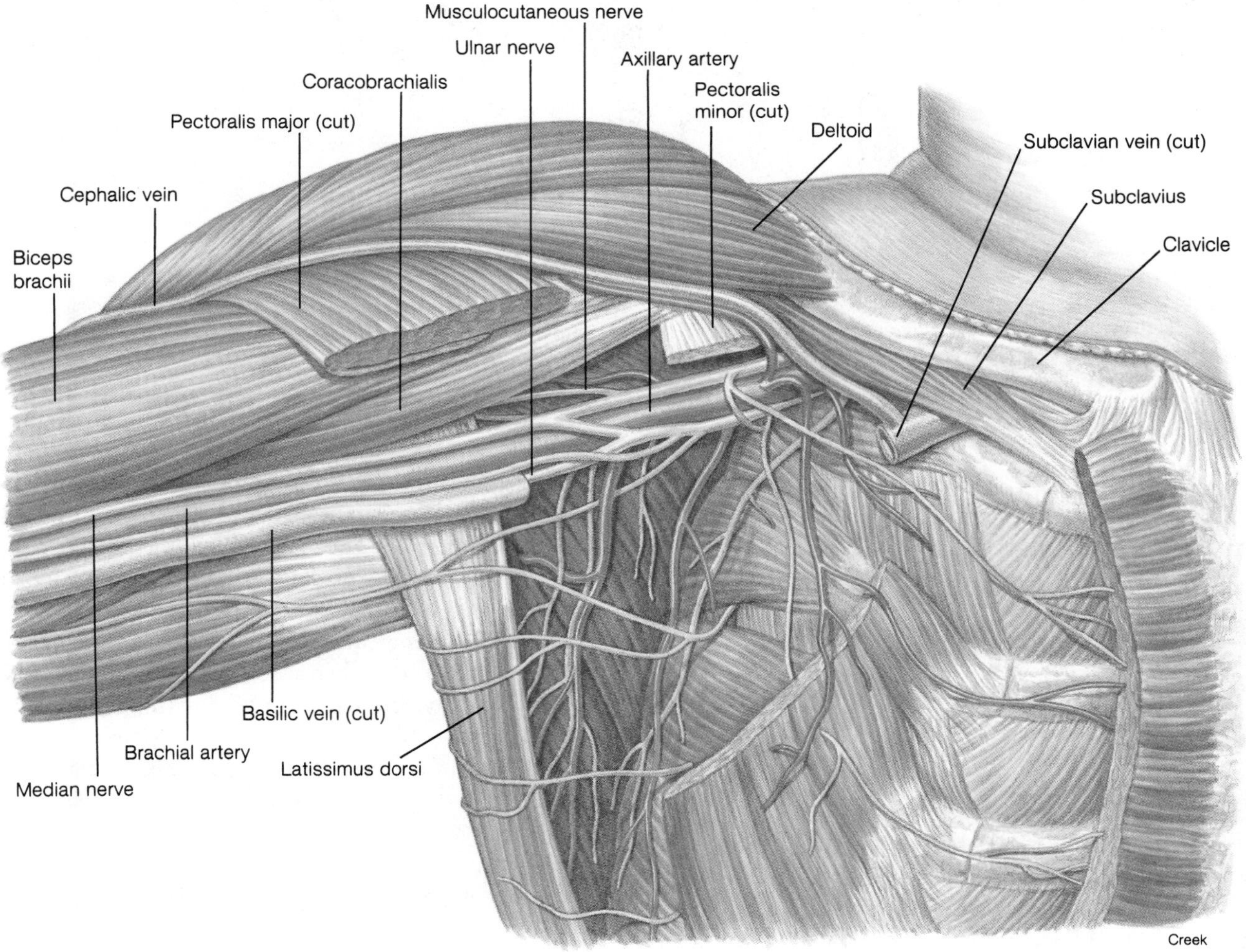

Figure 9.8 The relationship of blood vessels and nerves to skeletal muscles of the axillary region. (Note the close proximity of the nerves and vessels as they pass between muscle masses.)

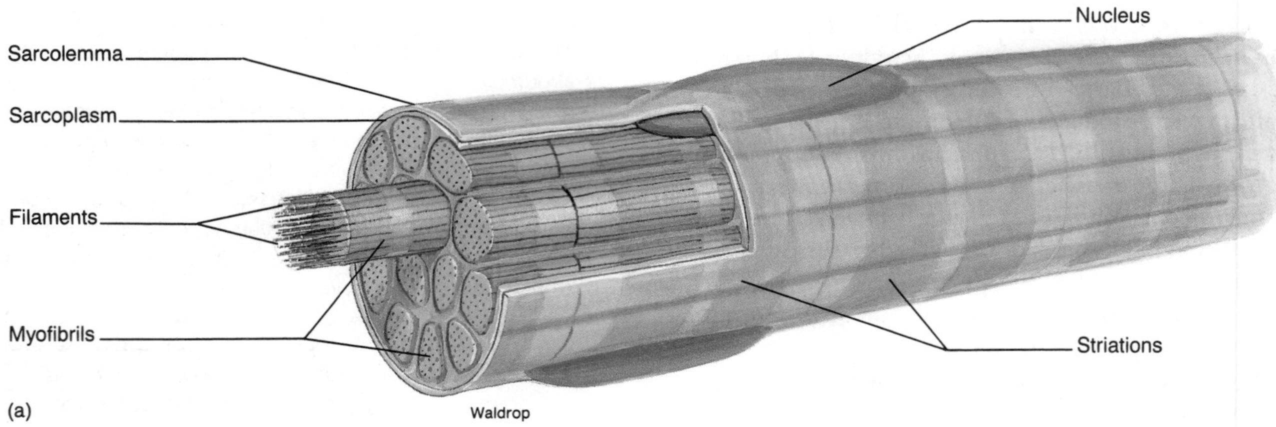

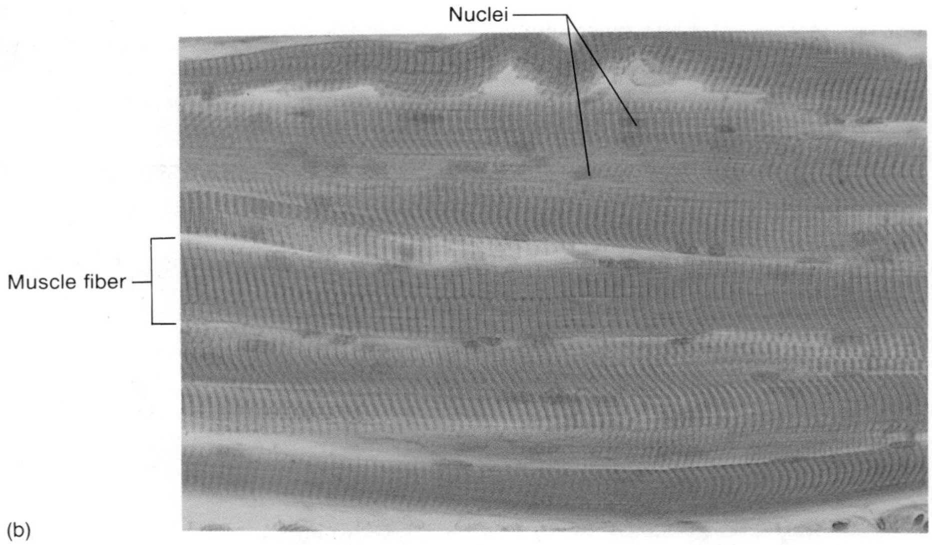

Figure 9.9 (*a*) A skeletal muscle fiber is composed of numerous threadlike strands of myofibrils that contain the filaments of actin and myosin. A skeletal muscle fiber is striated and multinucleated. (*b*) An electron micrograph of skeletal muscle fibers showing the striations and the peripheral location of the nuclei.

Skeletal Muscle Fibers and Mode of Contraction

Muscle fiber contraction in response to a motor impulse results from a sliding movement within the myofibrils in which the length of the sarcomeres is reduced.

Objective 8. Identify the major components of a muscle fiber, and discuss the function of each part.

Objective 9. Contrast isotonic and isometric contractions.

Objective 10. Define *motor unit,* and discuss the role of motor units in muscular contraction.

Skeletal Muscle Fibers

A skeletal muscle fiber is an elongated, cylindrical cell that may reach a length of 30 cm (12 in.) and have a diameter of 10 to 100 μm. Despite their unusual shape, muscle fibers have the same organelles that are present in other cells: mitochondria, intracellular membranes, glycogen granules, and others. Unlike most other cells in the body, however, skeletal muscle fibers are multinucleated and striated (fig. 9.9). Each fiber is surrounded by a cell membrane, called a **sarcolemma,** and the cytoplasm within the cell is called **sarcoplasm.** A network of membranous channels, called the **sarcoplasmic reticulum,** extends throughout the sarcoplasm (fig. 9.10). A system of **transverse tubules** (T-tubules) runs perpendicular to the sarcoplasmic reticulum and opens to the outside through the sarcolemma. Also embedded in the sarcoplasm and extending the entire length of the fiber are many parallel, threadlike structures called **myofibrils** *(mi''o-fi'brils)* (fig. 9.11). Each myofibril is composed of even smaller protein strands called **filaments.** *Thin filaments* are about 6nm in diameter and are composed of the protein **actin** *(ak'tin)*. *Thick filaments* are about 16nm in diameter and are composed of the protein **myosin** *(mi'o-sin)*.

The characteristic dark and light striations of skeletal muscle myofibrils are due to the arrangement of these filaments. The

actin: L. *actus,* motion, doing
myosin: L. *myosin,* within muscle

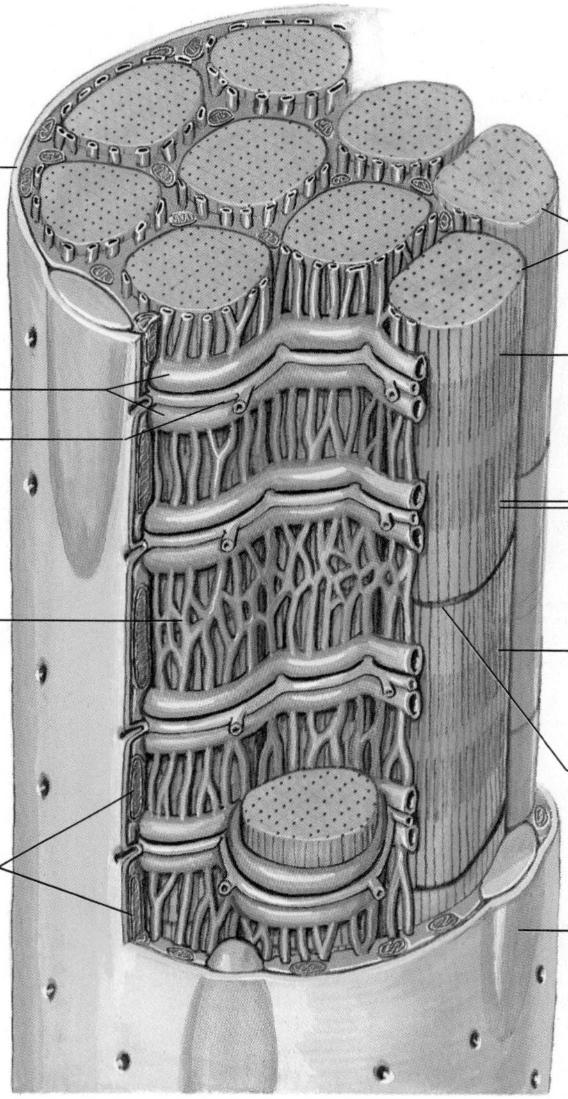

Figure 9.10 The structural relationship of the myofibrils of a muscle fiber to the sarcolemma, transverse tubules, and sarcoplasmic reticulum. (Note the position of the mitochondria.)

dark crossbands, produced by overlapping myosin filaments, are able to polarize visible light (anisotropic) and have been named *A bands.* The light crossbands, composed of thin actin filaments, do not polarize light (isotropic) and are called *I bands* (see figs. 9.10, 9.11).

The I bands within a myofibril extend from the edge of one stack of thick myosin filaments to the edge of the next stack of thick filaments. They are light bands because they contain only thin actin filaments. Each thin filament, however, extends partway into the A bands on each side (between the thick filaments of each stack). Since thick and thin filaments overlap at the edges of each A band, the edges are darker in appearance than the central region of the A band. These central lighter regions of the A bands are called the *H bands* (for *helle,* a German word for bright). The central H bands thus contain only myosin that is not overlapped with thin filaments.

At high magnification, thin dark lines can be seen in the middle of the I bands. These are labeled *Z lines* (for *Zwischenscheibe,* a German word meaning "between disc"). The arrangement of thick and thin filaments between a pair of Z lines forms a repeating pattern that serves as the basic subunit of skeletal muscle contraction. These subunits, from Z line to Z line, are known as **sarcomeres** *(sar'ko-mērs).* A longitudinal section of a myofibril thus presents a side view of successive sarcomeres (fig. 9.12a, b).

This side view is, in a sense, misleading; there are numerous sarcomeres within each myofibril that are out of the plane of the section (and out of the picture). A better appreciation of the three-dimensional structure of a myofibril can be obtained by viewing the myofibril in transverse section. In this view, it can be seen that the Z lines are in reality disc shaped and that the thin filaments that penetrate these Z discs surround

anisotropic: Gk. *anisos,* uneven; *tropos,* a turning
isotropic: Gk. *isos,* equal; *tropos,* a turning

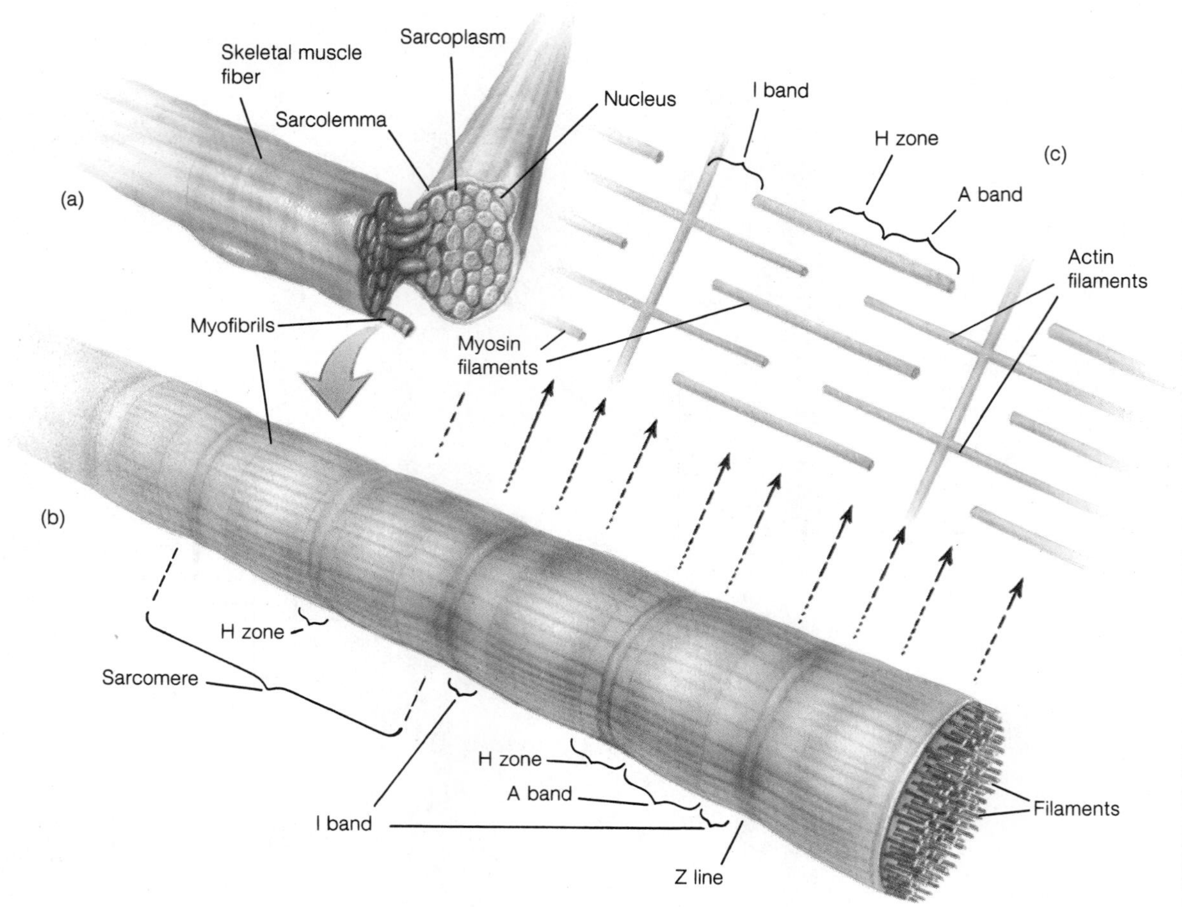

Figure 9.11 (*a*) A skeletal muscle fiber contains many threadlike structures called myofibrils, each arranged (*b*) into compartments called sarcomeres. (*c*) The characteristic striations of a sarcomere are due to the arrangement of thin and thick filaments composed of actin and myosin, respectively.

the thick filaments in a hexagonal arrangement (see fig. 9.12c). If one concentrates on a single row of dark thick filaments in this transverse section, the alternating pattern of thick and thin filaments seen in longitudinal section becomes apparent.

When a muscle is stimulated to contract, it decreases in length as a result of the shortening of its individual fibers. Shortening of the muscle fibers, in turn, is produced by shortening of their myofibrils, which occurs as a result of the shortening of the distance from Z line to Z line (fig. 9.13). As the sarcomeres shorten in length, however, the A bands do *not* shorten but instead appear closer together. The I bands—which represent the distance between A bands of successive sarcomeres—decrease in length.

Close examination reveals that the thick myosin and thin actin filaments remain the same length during muscle contraction. Shortening of the sarcomeres is produced, not by shortening of the filaments, but rather by the *sliding* of thin filaments over thick filaments. In the process of contraction, the thin filaments on either side of the A bands extend deeper and deeper toward the center, producing increasing amounts of overlap with the thick filaments. The central H bands thus get shorter and shorter during contraction.

Isotonic and Isometric Contractions

In order for muscle fibers to shorten when they contract, they must generate a force that is greater than the opposing forces that act to prevent movement of the muscle's insertion. Flexion of the forearm at the elbow, for example, occurs against the force of gravity and the weight of the objects being lifted. The tension produced by the contraction of each muscle fiber separately is insufficient to overcome these opposing forces, but the combined contractions of large numbers of muscle fibers may be sufficient to overcome the opposing force and flex the forearm as the muscle fibers shorten in length.

The contraction that results in muscle shortening is called *isotonic contraction,* so called because the force of contraction remains relatively constant throughout the shortening process. In an *isometric contraction,* the length of a muscle remains constant because the antagonist force equals the force in the muscle being

isotonic: Gk. *isos,* equal; *tonos,* tension
isometric: Gk. *isos,* equal; *metron,* measure

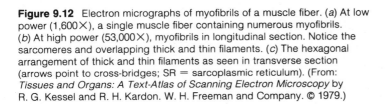

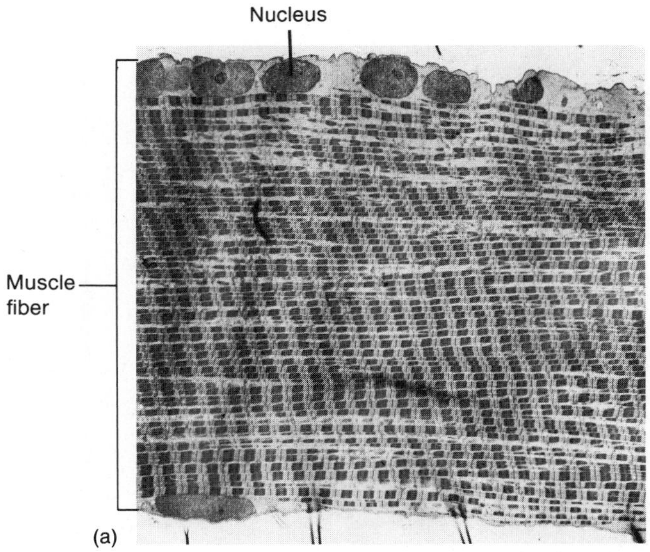

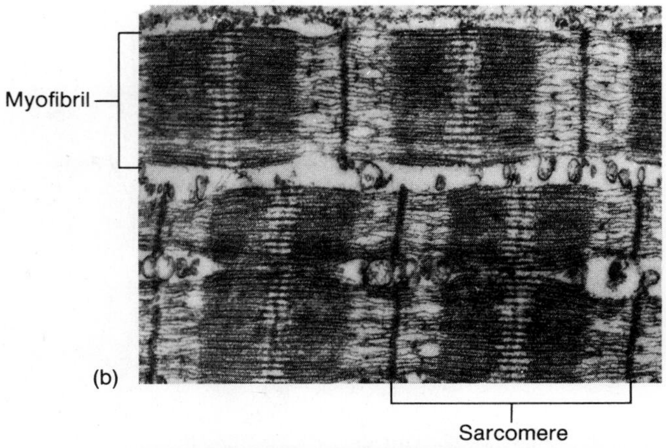

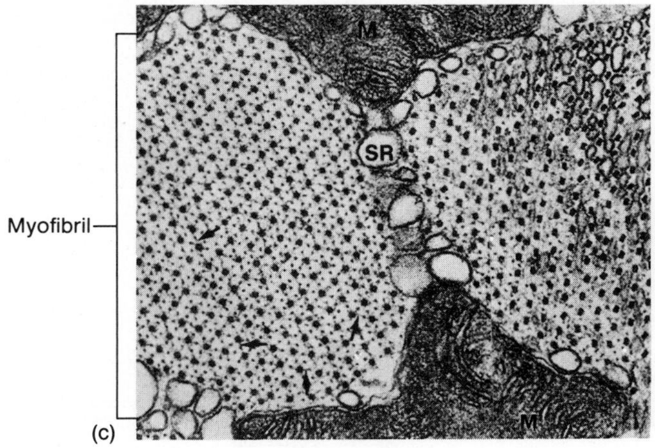

Figure 9.12 Electron micrographs of myofibrils of a muscle fiber. (*a*) At low power (1,600×), a single muscle fiber containing numerous myofibrils. (*b*) At high power (53,000×), myofibrils in longitudinal section. Notice the sarcomeres and overlapping thick and thin filaments. (*c*) The hexagonal arrangement of thick and thin filaments as seen in transverse section (arrows point to cross-bridges; SR = sarcoplasmic reticulum). (From: *Tissues and Organs: A Text-Atlas of Scanning Electron Microscopy* by R. G. Kessel and R. H. Kardon. W. H. Freeman and Company. © 1979.)

contracted. Isometric contractions occur when a person supports an object in a fixed position. An isometric contraction is converted to an isotonic contraction when increased force is generated within the muscle overcoming the resistance and resulting in muscle movement.

Neuromuscular Junction

A nerve serving a muscle is composed of both motor and sensory neurons. As the motor portion of the nerve penetrates a muscle, it splays into a number of branching neuron processes called **axons.** The terminal ends of axons contact the sarcolemma of muscle fibers by means of **motor end plates** (fig. 9.14). Each axon may split into enough branches to serve dozens of muscle fibers. The area consisting of the motor end plate and the sarcolemma of a muscle fiber is known as the **neuromuscular** (myoneural) **junction.**

Acetylcholine *(as''ĕ-til-ko'lēn)* is a *neurotransmitter chemical* stored in **synaptic vesicles** at the terminal ends of the axons. A nerve impulse reaching the terminal end of an axon causes the release of acetylcholine into the *neuromuscular cleft* of the neuromuscular junction. As this chemical mediator contacts the sarcolemma, it initiates physiological activity within the muscle fiber resulting in contraction.

Motor Unit

A **motor unit** consists of a single motor neuron and the aggregation of muscle fibers innervated by the motor neuron (fig. 9.14b). When a nerve impulse travels through a motor unit, all of the fibers served by it contract simultaneously to their maximum. Most muscles have an innervation ratio of one motor neuron per 100–150 muscle fibers. Muscles that are capable of precise, dexterous movements, such as an eye muscle, may have an innervation ratio of 1:10. Massive muscles that are responsible for gross body movements, like those of the thigh, may have an innervation ratio exceeding 1:500.

All of the motor units controlling a particular muscle, however, are not the same size. Innervation ratios in a large thigh muscle may vary from 1:100 to 1:2,000. Neurons that innervate smaller numbers of muscle fibers have smaller cell bodies and axon diameters than neurons that have larger innervation ratios. The smaller neurons also are stimulated by lower levels of excitatory input. The small motor units, as a result, are the ones that are used most often. The larger motor units are only activated when very forceful contractions are required.

Skeletal muscles are voluntary in that they can be consciously contracted. The magnitude of the task determines the number of motor units that are activated. Performing a light

---

axon: Gk. *axon,* axis

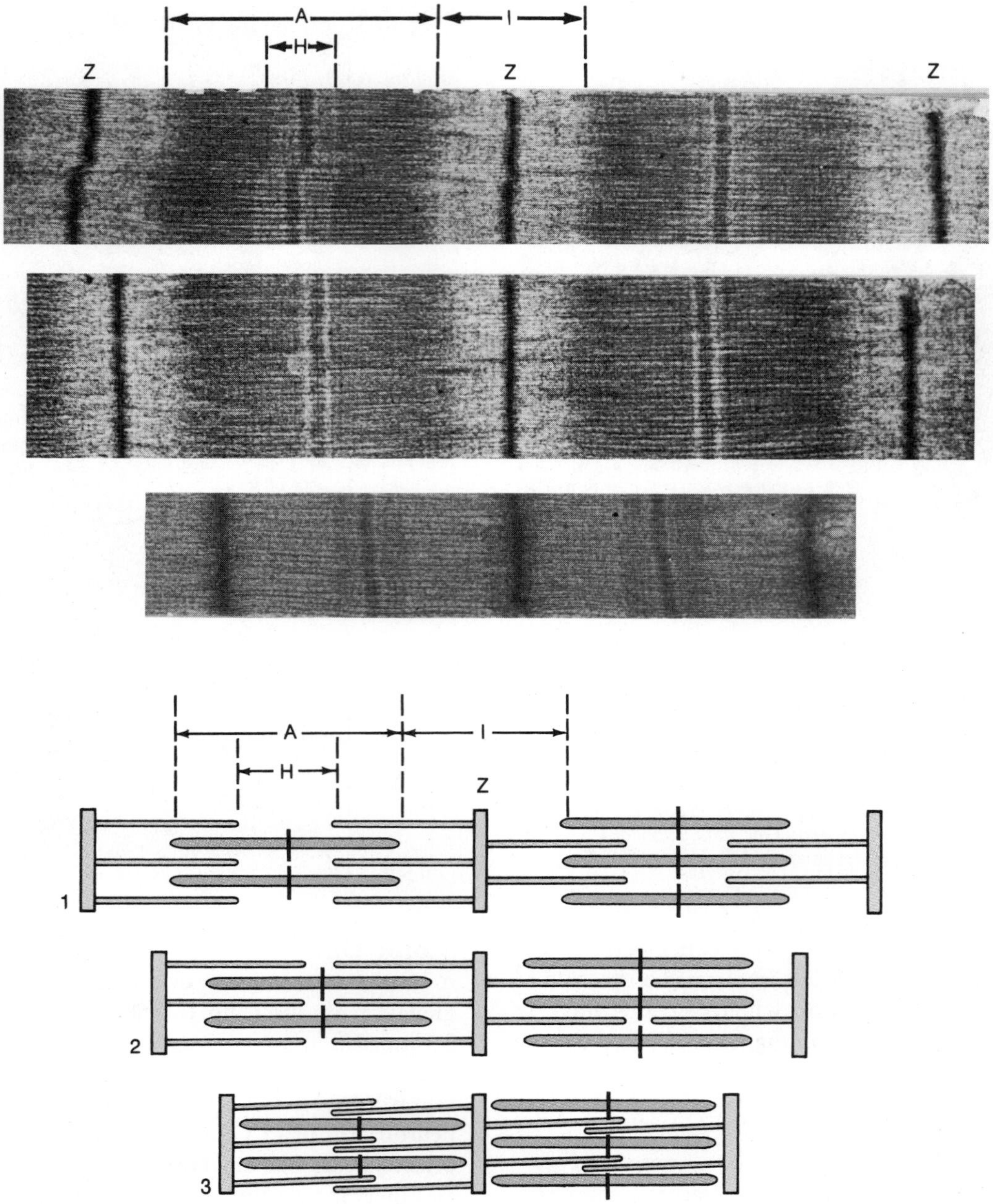

Figure 9.13 The sliding filament model of contraction. As the filaments slide, the Z lines are brought closer together. The A bands remain the same length during contraction, but the I and H bands get progressively narrower and may eventually become obliterated.

task, such as lifting a book, requires few motor units, whereas lifting a table requires many. Muscles with pennate architecture have many motor units, are strong and dexterous, but generally fatigue more readily than muscles with fewer motor units. Being mentally "psyched up" to accomplish an athletic feat involves voluntary activation of more motor units within the muscles. Seldom does a person utilize all the motor units within a muscle, but the secretion of *epinephrine (ep''ĭ-nef'rin)* from the adrenal gland does promote an increase in the force that can be produced when a given number of motor units are activated.

Steroids are hormones produced by the adrenal glands, testes, and ovaries. Because they are soluble in lipids, they readily pass through cell membranes and into the cytoplasm where they combine with proteins to form steroid-protein complexes, necessary for the syntheses of specific kinds of messenger RNA molecules. Synthetic steroids were originally developed to promote weight gain in cancer and anorexic patients. It was soon determined, however, that steroids taken by body builders and athletes would provide them with increased muscle mass, strength, and aggressiveness. The use of steroids is now considered illegal by most athletic associations because they provide unfair advantages in physical competition and because they can cause serious side effects, such as gonadal atrophy, induction of malignant tumors of the liver, and excessive aggressive behavior.

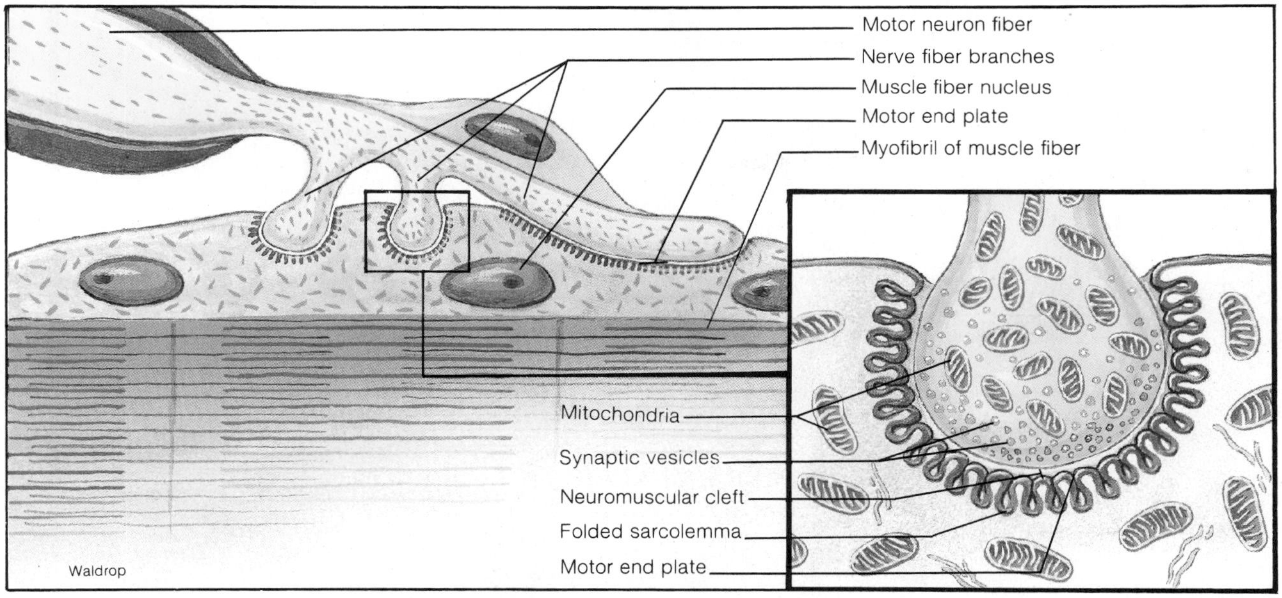

(a)

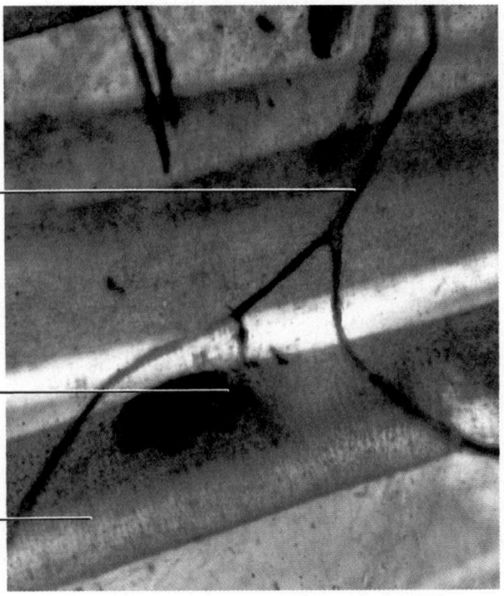

(b)

Figure 9.14 A motor end plate at the neuromuscular junction. (*a*) A neuromuscular junction is the site where the nerve fiber and muscle fiber meet. The motor end plate is the specialized portion of the sarcolemma of a muscle fiber surrounding the terminal end of the axon. (Note the slight gap between the membrane of the axon and that of the muscle fiber.) (*b*) A photomicrograph of muscle fibers and motor end plates. A motor neuron and the muscle fibers it innervates constitute a motor unit.

1. Draw three successive sarcomeres in a myofibril of a resting muscle fiber. Label the myofibril, sarcomeres, A bands, I bands, H bands, and Z lines.
2. Why do the A bands appear darker than the I bands?
3. Draw three successive sarcomeres in a myofibril of a contracted fiber. Indicate which bands get shorter during contraction, and explain how this occurs.
4. Describe how the antagonistic muscles in the brachium can be exercised through both isotonic and isometric contractions.
5. Explain why motor units are considered the basic functional units of muscle contraction

Naming of Muscles

Skeletal muscles are named on the basis of shape, location, attachment, orientation of fibers, relative position, or function.

Objective 11. Describe, using examples, the various ways in which muscles are named.

One of the tasks of a myologist is to learn the names of the more than six hundred skeletal muscles within the body. This task is somewhat simplified by the fact that most of the muscles are paired; that is, the right side is the mirror image of the left. Further simplifying the task are the descriptive names of muscles. Only the principal muscles of the body are described in the remaining pages of this chapter.

As you learn the muscles of the body, understand how they are named and be able to identify them on yourself. Learn them as functional groups so that you will better understand their actions. If you can identify a muscle on yourself, you will be able to contract the muscle and describe its action. Learning anatomy this way will be easier, more meaningful, and improve retention.

The following are some ways in which the names of muscles have been logically derived, with examples of each given:

1. **Shape:** rhomboideus (like a rhomboid); trapezius (like a trapezoid); or denoting the number of heads of origin: triceps (three heads), biceps (two heads)
2. **Location:** pectoralis (in the chest, or pectus); intercostal (between ribs); brachii (upper arm)
3. **Attachment:** many facial muscles (zygomaticus, temporalis, nasalis); sternocleidomastoid (sternum, clavicle, and mastoid process of the skull)
4. **Size:** maximus, (larger, largest); minimus (smaller, smallest); longus (long); brevis (short)
5. **Orientation of fibers:** rectus (straight); transverse (across); oblique (in an oblique direction)
6. **Relative position:** lateral, medial, internal, and external
7. **Function:** adductor, flexor, extensor, pronator, and levator (lifter)

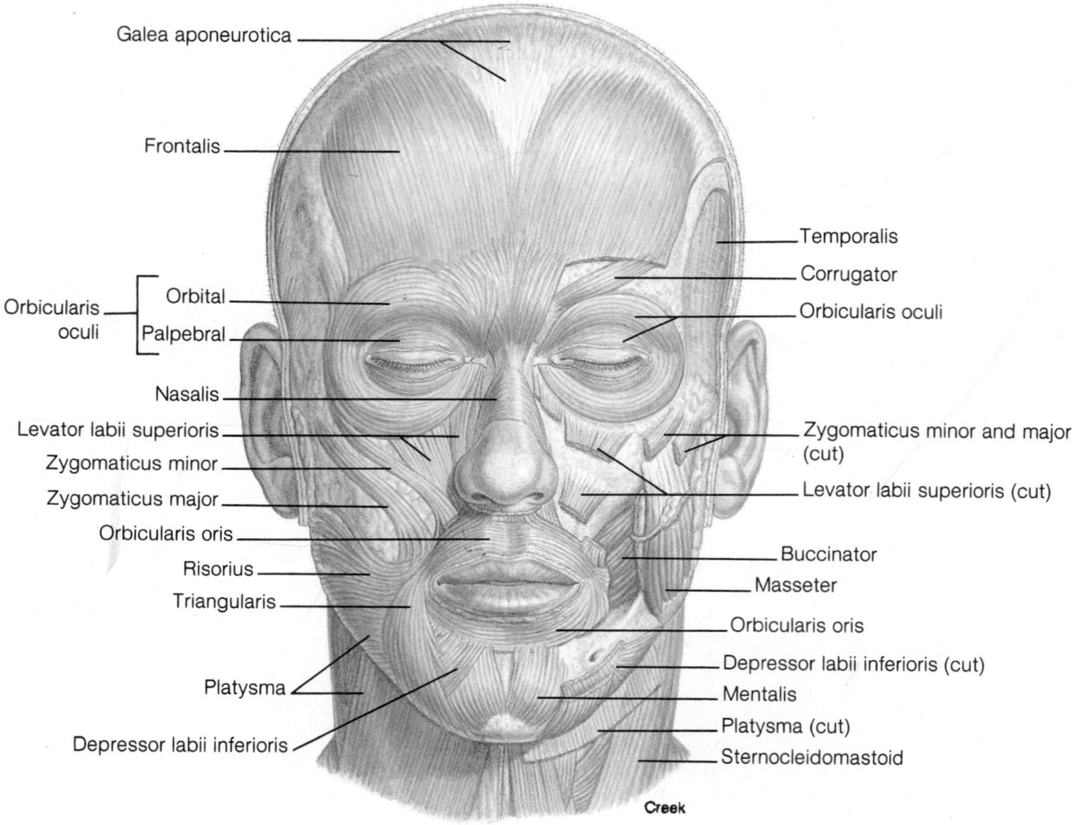

Figure 9.15 Anterior view of the superficial facial muscles involved in facial expression.

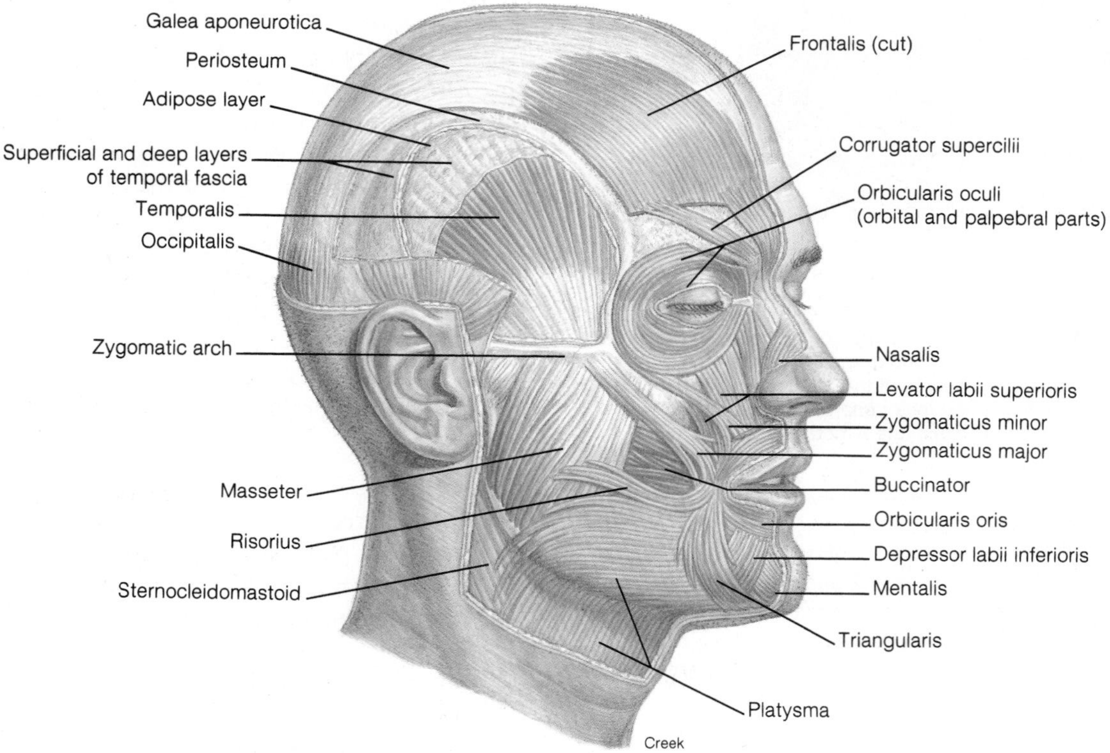

Figure 9.16 Lateral view of the superficial facial muscles involved in facial expression.

1. Refer to chapter 2, and review the location of the following body regions: cervical, pectoral, abdominal, gluteal, perineal, brachial, antebrachial, inguinal, thigh, and popliteal.
2. Refer to chapter 8, and review the movements permitted at diarthrotic joints.

Muscles of the Axial Skeleton

Muscles of the axial skeleton include those of facial expression, mastication, eye movement, tongue movement, neck movement, respiration, the abdominal wall, the pelvic outlet, and movement of the vertebral column.

Objective 12. Locate the major muscles of the axial skeleton; identify synergistic and antagonistic muscles, and describe the action of each.

Muscles of Facial Expression

Humans have a well-developed facial musculature (figs. 9.15, 9.16) that provides complex facial expression as a means of social communication. Indeed, many messages can be conveyed by a person without speaking a word.

The muscles of facial expression are in a superficial position located on the scalp, face, and neck. Although highly variable in size and strength, these muscles all originate from the bones of the skull or flat sheet-like tendons and insert onto the hypodermis of the skin (table 9.1). They are all innervated by one of the two facial cranial nerves. The locations and points of attachments of most of the facial muscles are such that when contracted they cause movements around the eyes, nostrils, or mouth (fig. 9.17).

The muscles of facial expression are of clinical concern for several reasons, each of which involves the facial nerve. Located right under the skin, the many branches of the facial cranial nerve are vulnerable to trauma. Facial lacerations and fractures of the skull frequently damage branches of this nerve. The extensive pattern of motor innervation (see chapter 12) becomes apparent in stroke victims and persons suffering from *Bell's palsy.* The facial muscles on one side of the face are affected in these people and that portion of the face appears to sag as there is loss of tonus in the muscles.

Muscles of Mastication

The large **temporalis** and **masseter** *(mas-se'ter)* muscles (fig. 9.18) are powerful elevators of the mandible in synergism with the **medial pterygoid** *(ter'i-goyd)*. The primary function of the medial and lateral pterygoid muscles is to provide grinding movements of the teeth. The **lateral pterygoid** also protracts the mandible (table 9.2).

Table 9.1 Muscles of facial expression

Muscle	Derivation	Origin	Insertion	Action	Innervation
Epicranius	*epi* = above; *crani* = skull	Galea aponeurotica and occipital bone	Skin of eyebrow and galea aponeurotica	Wrinkles forehead and moves scalp	Facial
Frontalis	Referring to frontal bone	Galea aponeurotica	Skin of eyebrow	Wrinkles forehead and elevates eyebrow	Facial
Occipitalis	Referring to occipital bone	Occipital bone and mastoid process	Galea aponeurotica	Moves scalp backward	Facial
Corrugator	*ruga* = a wrinkle	Fascia above eyebrow	Root of nose	Draws eyebrows toward midline	Facial
Orbicularis oculi	*orbis* = orbit, circular; *oculus* = eye	Bones of medial orbit	Tissue of eyelid	Closes eye	Facial
Nasalis	*nasus* = nose	Maxilla and nasal cartilage	Aponeurosis of nose	Compresses nostrils	Facial
Orbicularis oris	*orb* = circular; *or* = mouth	Fascia surrounding lips	Mucosa of lips	Closes and purses lips	Facial
Levator labii superioris	*levator* = lifter; *labii* = lip; *superior* = upper	Upper maxilla and zygomatic bone	Orbicularis oris and skin above lips	Elevates upper lip	Facial
Zygomaticus	Referring to zygomatic bone	Zygomatic bone	Superior corner of orbicularis oris	Elevates corner of mouth	Facial
Risorius	*risor* = laughter	Fascia of cheek	Oribcularis oris at corner of mouth	Draws angle of mouth laterally	Facial
Triangularis	Referring to triangular shape	Mandible	Inferior corner of orbicularis oris	Depresses corner of mouth	Facial
Depressor labii inferioris	*depressor* = depress; *labii* = lip; *inferior* = lower	Mandible	Orbicularis oris and skin of lower lip	Depresses lower lip	Facial
Mentalis	*mentum* = chin	Mandible (chin)	Orbicularis oris	Elevates and protrudes lower lip	Facial
Platysma	*platy* = broad	Fascia of neck and chest	Inferior border of mandible	Depresses lower lip	Facial
Buccinator	*bucca* = cheek	Maxilla and mandible	Orbicularis oris	Compresses cheek	Facial

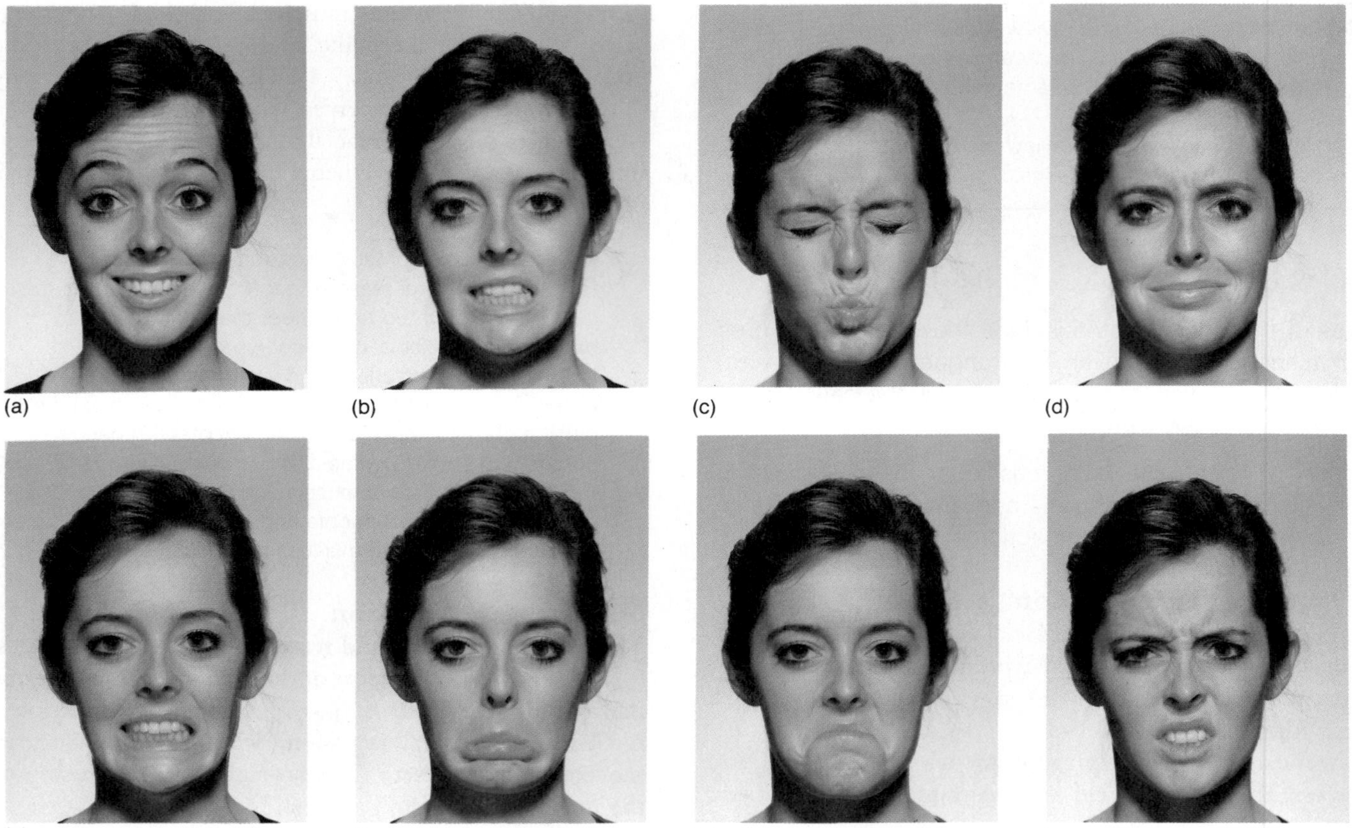

(a) (b) (c) (d) (e) (f) (g) (h)

Figure 9.17 Muscles of facial expression. The actions of each of the muscles of facial expression can be easily identified as a person contracts them. In each of these photographs, identify the muscles that are being contracted.

Table 9.2 Muscles of mastication

Muscle	Derivation	Origin	Insertion	Action	Innervation
Temporalis	*templus* = time, temple	Temporal fossa	Coronoid process of mandible	Elevates mandible	Trigeminal
Masseter	*maseter* = chew	Zygomatic arch	Lateral ramus of mandible	Elevates mandible	Trigeminal
Medial pterygoid	*medial* = toward midline; *pteryg* = winglike (referring to process of sphenoid bone)	Sphenoid bone	Medial ramus of mandible	Elevates mandible and moves mandible laterally	Trigeminal
Lateral pterygoid	*latus* = side	Sphenoid bone	Anterior side of mandibular condyle	Protracts mandible	Trigeminal

Tetanus is a bacterial disease caused by anaerobic *Clostridium tetani* being introduced into the body, usually from a puncture wound. The bacteria produce a neurotoxin that is carried to the spinal cord by sensory nerves. The motor impulses relayed back cause certain muscles to contract continuously (tetany). The muscles that move the mandible are affected first, which has given the disease the common name of *lockjaw.*

Ocular Muscles

The movements of the eyeball are controlled by six extrinsic eye muscles called the extrinsic ocular muscles (fig. 9.19). Five of these muscles arise from the margin of the optic foramen at the back of the orbital cavity and insert on the outer layer (sclera) of the eyeball. Four **rectus muscles** maneuver the eyeball in the direction indicated by their names (**superior, inferior, lateral,** and **medial**), and two **oblique muscles** (**superior** and **inferior**) rotate the eyeball on its axis. The medial rectus on one side contracts with the medial rectus of the opposite eye when focusing on close objects. When looking to the side, the lateral rectus of one eyeball works with the medial rectus of the opposite eyeball to keep both eyes focused together. The superior oblique muscle passes through a pulleylike cartilaginous loop, the *trochlea,* before attaching to the eyeball. The ocular muscles are innervated by three cranial nerves (table 9.3).

Another muscle, the **levator palpebrae** *(le-va'tor pal'pĕ-bre)* **superioris** (fig. 9.19), is located in the ocular region but is not attached to the eyeball. It extends into the upper eyelid and raises the eyelid when contracted.

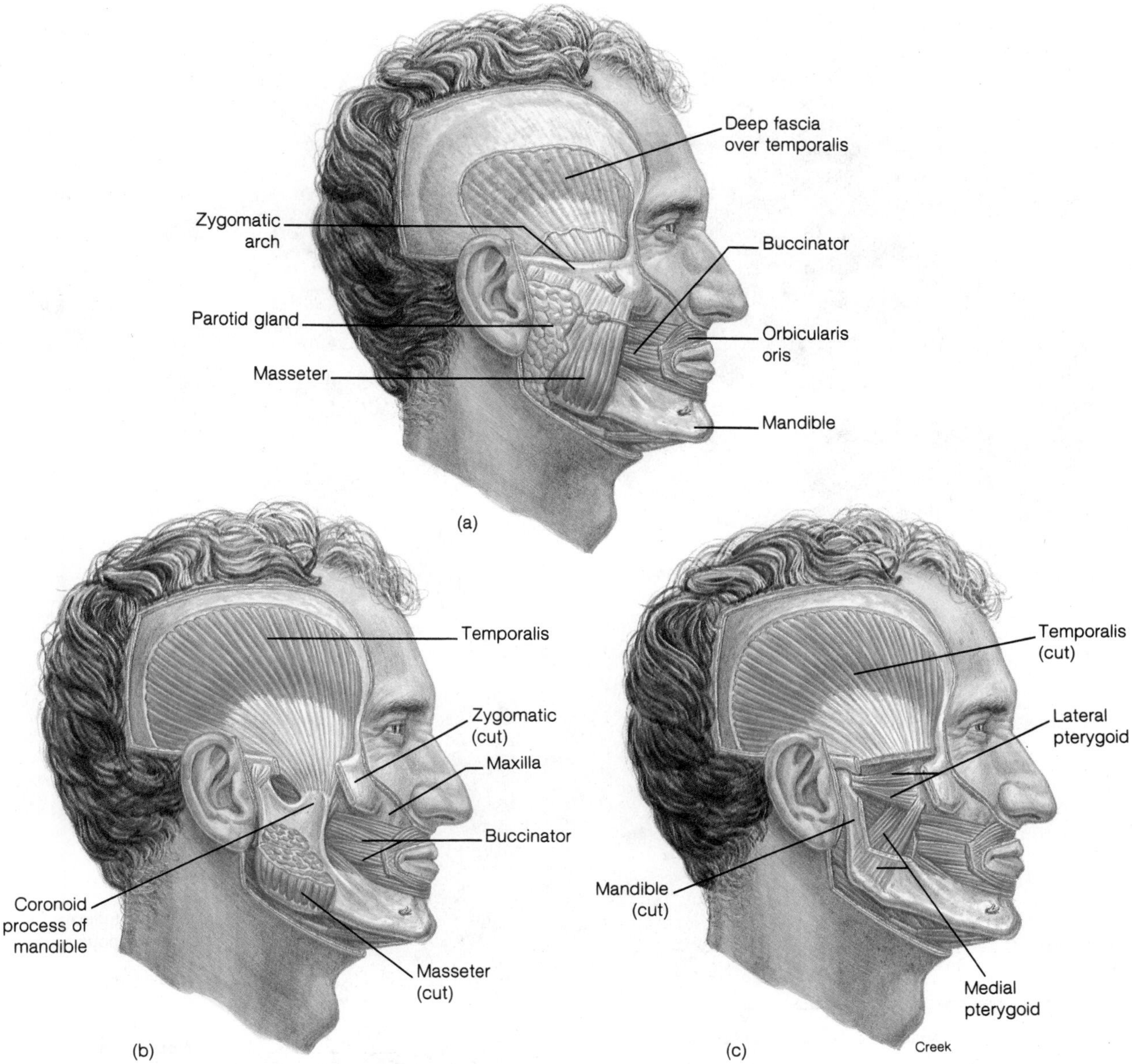

Figure 9.18 Muscles of mastication. (*a*) A superficial view, (*b*) a deep view, and (*c*) the deepest view showing the pterygoid muscles.

Table 9.3 Ocular muscles

Muscle	Cranial nerve innervation	Movement of eyeball
Lateral rectus	Abducens	Lateral
Medial rectus	Oculomotor	Medial
Superior rectus	Oculomotor	Superior and medial
Inferior rectus	Oculomotor	Inferior and medial
Inferior oblique	Oculomotor	Superior and lateral
Superior oblique	Trochlear	Inferior and lateral

Muscles That Move the Tongue

The tongue is a highly specialized muscular organ that functions in speaking, manipulating food during mastication, cleansing the teeth, and swallowing. Two groups of muscles are responsible for tongue movement: intrinsic and extrinsic. The *intrinsic muscles* are confined within the tongue and are responsible for its mobility and changes of shape. *Extrinsic tongue muscles* are those that originate on structures away from the tongue and insert onto it to cause gross tongue movement (see fig. 9.20 and table 9.4). The three paired extrinsic muscles are the **genioglossus, styloglossus,** and **hyoglossus.** When the

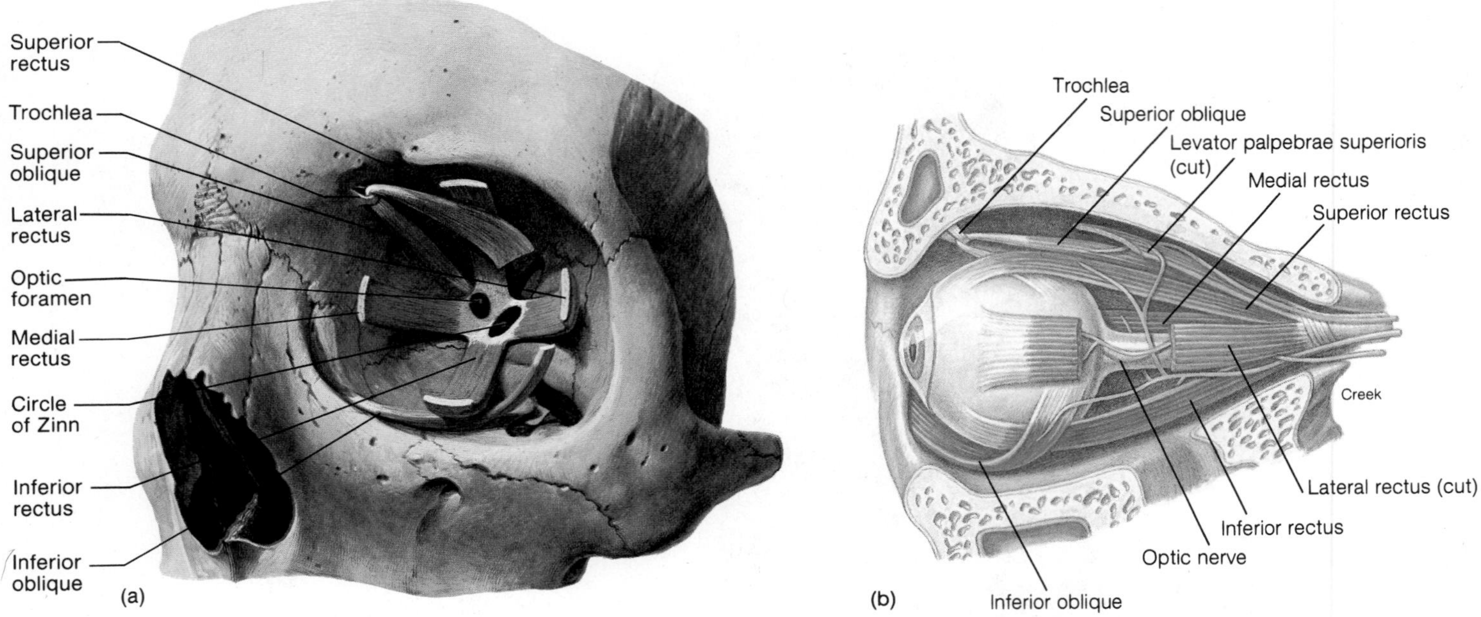

Figure 9.19 Extrinsic ocular muscles of the left eyeball. (*a*) An anterior view and (*b*) a lateral view.

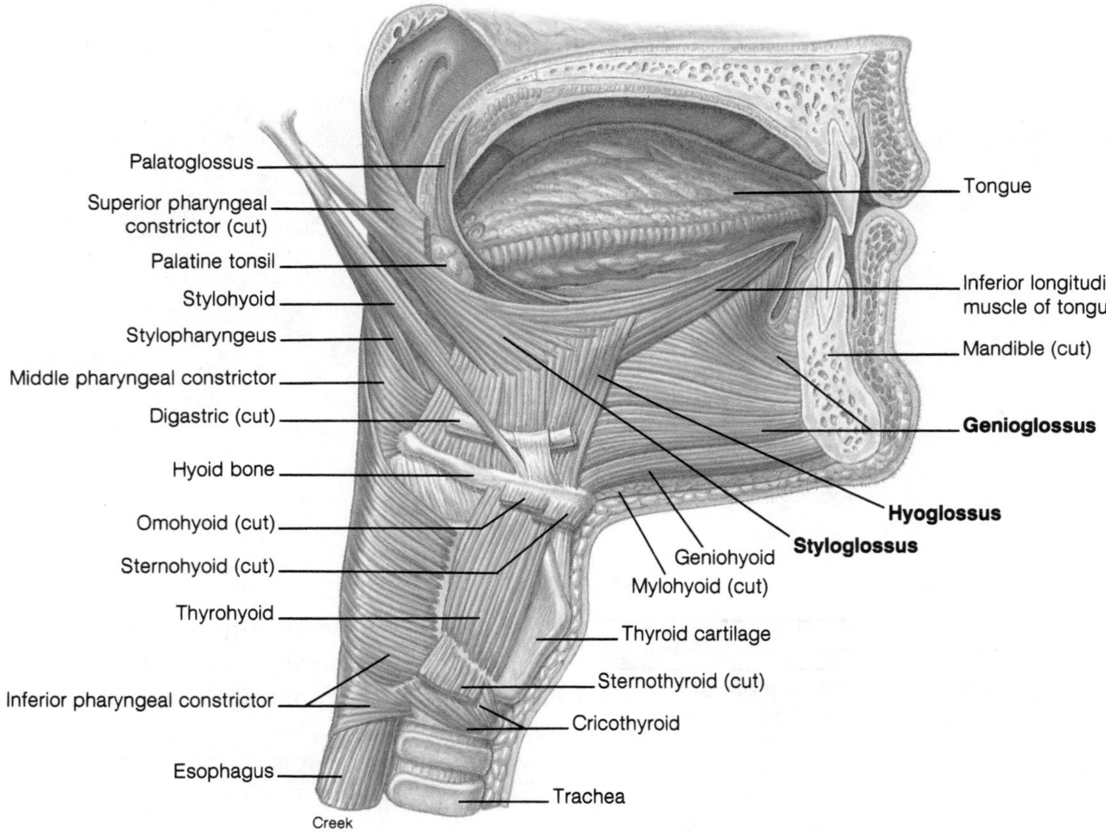

Figure 9.20 Extrinsic muscles of the tongue.

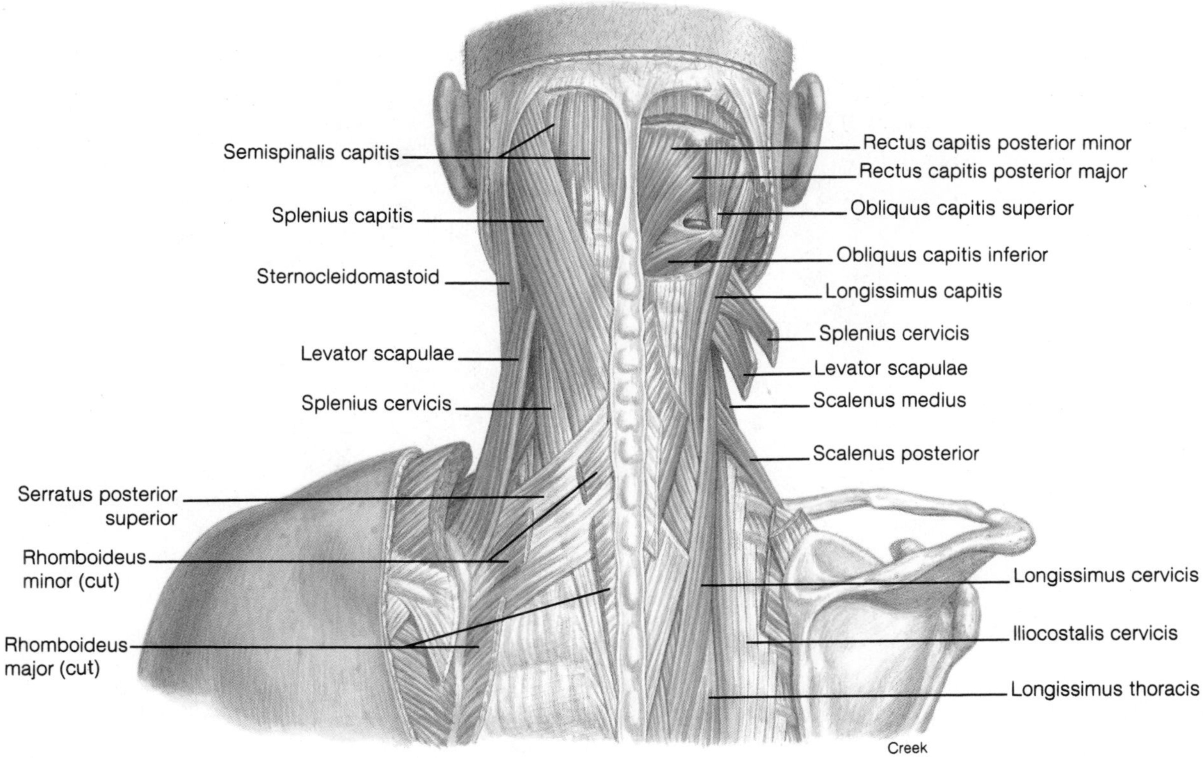

Figure 9.21 Deep muscles of the posterior neck and upper back regions.

Table 9.4 Extrinsic tongue muscles

Muscle	Derivation	Origin	Insertion	Action	Innervation
Genioglossus	*geneion* = chin; *glossus* = tongue	Mental spine of mandible	Undersurface of tongue	Depresses and protracts tongue	Hypoglossal
Styloglossus	*stylo* = referring to styloid process of skull	Styloid process of temporal bone	Lateral side and undersurface of tongue	Elevates and retracts tongue	Hypoglossal
Hyoglossus	*hyo* = referring to hyoid bone	Body of hyoid	Side of tongue	Depresses sides of tongue	Hypoglossal

anterior portion of the genioglossus is contracted, the tongue is depressed and thrust forward. If both genioglossus muscles are contracted together along their entire lengths, the dorsal surface of the tongue becomes transversely concave. This muscle is extremely important in infants for sucking; the tongue is positioned around the nipple with a concave groove channeled toward the pharynx.

Muscles of the Neck

Muscles of the neck either support and move the head or are attached to structures within the neck region, such as the hyoid bone and larynx. Only the more obvious neck muscles will be considered in this chapter.

Several of the muscles in this section and the sections to follow can be observed on yourself. Refer to chapter 10 to determine which muscles form important surface landmarks. These muscles are illustrated in figures 9.21 and 9.22 and are summarized in table 9.5.

Posterior Muscles The posterior muscles include the sternocleidomastoid (originates anteriorly), trapezius, splenius capitis, semispinalis capitis, and longissimus capitis.

As the name implies, the **sternocleidomastoid** *(ster''no-kli''do-mas'toid)* muscle originates on the sternum and clavicle and inserts on the mastoid process of the skull (fig. 9.22). When contracted on one side, it turns the head sideways in a direction opposite the side on which the muscle is located. If both sternocleidomastoid muscles are contracted, the head is pulled forward and down. The sternocleidomastoid is covered by the platysma, (see fig. 9.16 and table 9.5).

Although a portion of the **trapezius** muscle extends over the posterior neck region, it is primarily a superficial muscle of the back and will be described later.

The **splenius capitis** is a broad muscle positioned deep to the trapezius (fig. 9.21). It originates on the ligamentum nuchae and the spinous processes of the seventh cervical and first three thoracic vertebrae. It inserts on the back of the skull below the

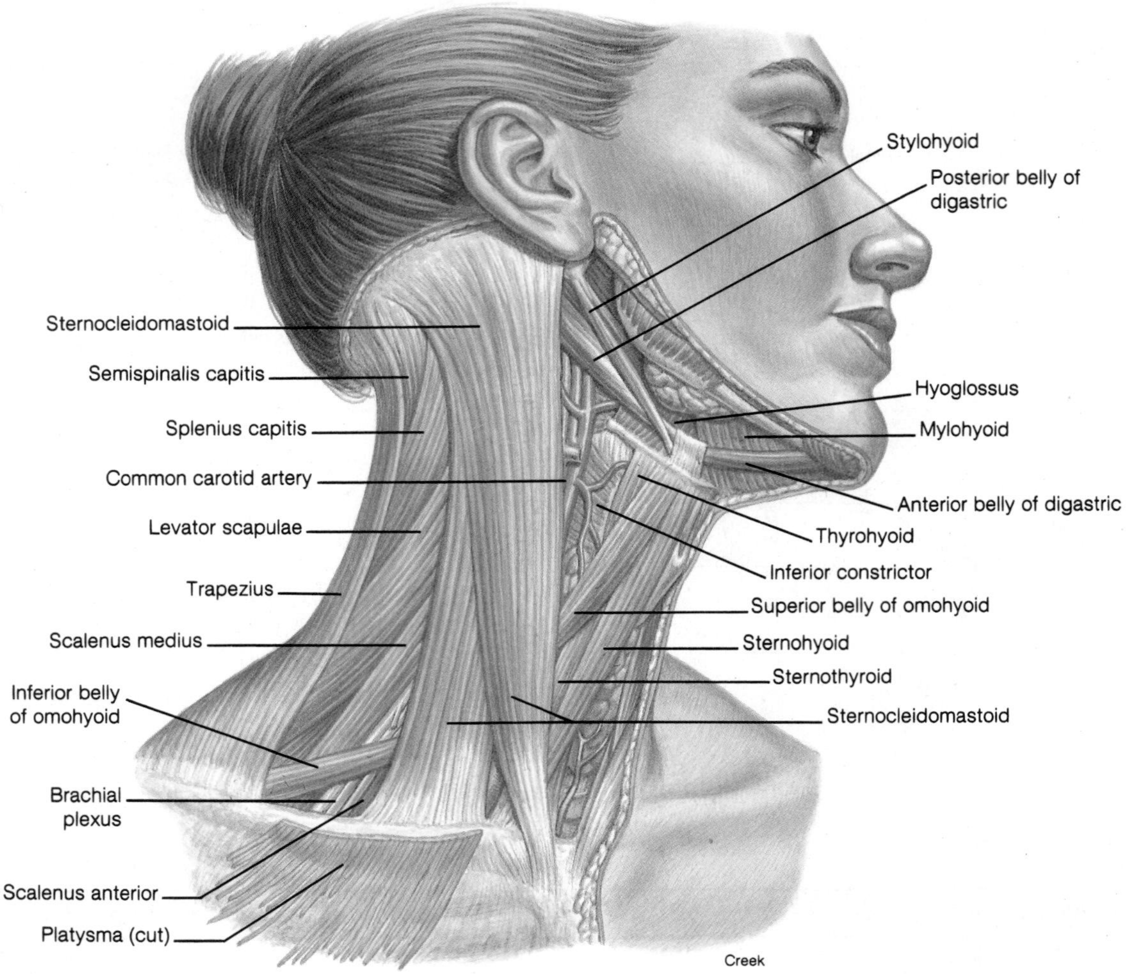

Figure 9.22 Muscles of the anterior and lateral neck regions.

Table 9.5 Muscles of the neck

Muscle	Derivation	Origin	Insertion	Action	Innervation
Sternocleidomastoid	*sternum* = referring to sternal bone; *cleido* = clavicle; *mastoid* = mastoid process of skull	Sternum; clavicle	Mastoid process of temporal bone	Turns head to side; flexes neck and head	Accessory
Digastric	*di* = two; *gaster* = belly	Inferior border of mandible; mastoid groove	Hyoid bone	Opens mouth; elevates hyoid	Trigeminal (ant. belly); facial (post. belly)
Mylohyoid	*mylos* = akin to; *hyo*=hyoid bone	Inferior border of mandible	Body of hyoid and median raphe	Elevates hyoid bone and floor of mouth	Trigeminal
Stylohyoid	*stylo* = referring to styloid process	Styloid process of temporal bone	Body of hyoid	Elevates and retracts tongue	Facial
Hyoglossus	*hyo* = hyoid bone; *glossus*=tongue	Body of hyoid bone	Side of tongue	Depresses side of tongue	Hypoglossal
Sternohyoid		Manubrium	Body of hyoid	Depresses hyoid	Hypoglossal
Sternothyroid	*thyro* = shield shaped	Manubrium	Thyroid cartilage	Depresses thyroid cartilage	Hypoglossal
Thyrohyoid		Thyroid cartilage	Great cornu of hyoid	Depresses hyoid; elevates thyroid	Hypoglossal
Omohyoid	*omo* = shoulder	Superior border of scapula	Clavicle; body of hyoid	Depresses hyoid	Hypoglossal

superior nuchal line and on the mastoid process of the temporal bone. When the splenius capitis contracts on one side, the head rotates and extends to one side. Contracted together, these muscles extend the head at the neck. Further contraction causes hyperextension of the neck and head.

The broad, sheetlike **semispinalis capitis** muscle extends upward from the seventh cervical and first six thoracic vertebrae to insert on the occipital bone (fig. 9.21). When the two semispinalis capitis muscles contract together, they extend the head at the neck along with the splenius capitis muscle. If one of the muscles acts alone, the head is rotated to the side.

The narrow, straplike **longissimus capitis** muscle ascends from processes of the lower four cervical and upper five thoracic vertebrae and inserts on the mastoid process of the temporal bone (fig. 9.21). This muscle extends the head at the neck, bends it to the one side, or rotates it slightly.

Suprahyoid Muscles The group of suprahyoid muscles located above the hyoid bone includes the digastric, mylohyoid, stylohyoid, and hyoglossus (fig. 9.22).

The **digastric** is a two-bellied muscle of double origin that inserts on the hyoid bone. The anterior origin is on the mandible at the point of the chin, and the posterior origin is near the mastoid process of the skull. The digastric can open the mouth or elevate the hyoid.

The **mylohyoid** forms the floor of the mouth. It originates on the inferior border of the mandible and inserts on the median raphe and body of the hyoid. As this muscle contracts, the floor of the mouth is elevated. It aids swallowing by forcing the food toward the back of the mouth.

The slender **stylohyoid** muscle extends from the styloid process of the skull to the hyoid bone, which it elevates as it contracts. The secondary effect of this muscle on tongue movement has already been described.

Infrahyoid Muscles Infrahyoid muscles are paired, thin, straplike muscles located below the hyoid bone. They are individually named on the basis of their origin and insertion and include the sternohyoid, sternothyroid, thyrohyoid, and omohyoid (fig. 9.22).

The **sternohyoid** muscle originates on the manubrium of the sternum and inserts on the hyoid. It depresses the hyoid bone as it contracts.

The **sternothyroid** muscle also originates on the manubrium but inserts on the thyroid cartilage of the larynx. When this muscle contracts, the larynx is pulled downward.

The short **thyrohyoid** muscle extends from the thyroid cartilage to the hyoid bone. It elevates the larynx and lowers the hyoid bone.

The long, thin **omohyoid** muscle originates on the superior border of the scapula and inserts on the clavicle bone and on the hyoid bone. It acts to depress the hyoid bone.

The coordinated movements of the hyoid bone and the larynx are impressive. The hyoid bone does not articulate with any other bone, yet it has eight paired muscles attached to it. Two paired muscles involve tongue movement, one paired muscle lowers the jaw, one muscle elevates the floor of the mouth, and four paired muscles depress the hyoid or elevate the thyroid cartilage of the larynx.

Muscles of Respiration

The muscles of respiration are skeletal muscles that continually and rhythmically contract, usually involuntarily. Breathing, or *pulmonary ventilation,* is divided into two phases: *inspiration (inhalation)* and *expiration (exhalation).*

During normal, relaxed inspiration, the important muscles are the **diaphragm** *(di'ah-fram),* the **external intercostal** muscles, and the interchondral portion of the **internal intercostal** muscles (fig. 9.23). A downward contraction of the dome-shaped diaphragm causes a vertical increase in thoracic dimension. A simultaneous contraction of the external intercostals and the interchondral portion of the internal intercostals produces an increase in the lateral dimension of the thorax. In addition, the **sternocleidomastoid** and **scalenes** muscles may assist in inspiration through elevation of the first and second ribs respectively. The intercostal muscles are innervated by the intercostal nerves, and the diaphragm receives its stimuli through the phrenic nerves.

Expiration is primarily a passive process, occurring as the muscles of inspiration are relaxed and the rib cage recoils to its original position. During forced expiration, the interosseous portion of the **internal intercostal** muscles contract, causing the rib cage to be depressed. This portion of the internal intercostal muscles lie under the external intercostals, and their fibers are directed downward and backward. The **abdominal** muscles may also contract during forced expiration, which increases pressure within the abdominal cavity and forces the diaphragm superiorly, squeezing additional air out of the lungs.

Muscles of the Abdominal Wall

The anterolateral abdominal wall is composed of four pairs of flat, sheetlike muscles: the external oblique, internal oblique, transversus abdominis, and rectus abdominis (fig. 9.24). These muscles support and protect the organs of the abdominal cavity and aid in breathing. When they contract, the pressure in the abdominal cavity increases, which can aid in defecation and in stabilizing the spine during heavy lifting.

The **external oblique** muscle is the strongest and most superficial of the three layered muscles of the lateral abdominal wall (figs. 9.24, 9.25). Its fibers are directed inferiorly and medially. The **internal oblique** muscle lies deep to the external oblique, and its fibers are directed at right angles to the external oblique. The **transversus abdominis** is the deepest of the abdominal muscles, and its fibers pass directly medially around the abdominal wall. The long, straplike **rectus abdominis** muscle is entirely enclosed in a fibrous sheath formed from the aponeuroses of the other three abdominal muscles. The *linea alba* is a band of connective tissue on the midline of the abdomen that separates the two rectus abdominis muscles. *Tendinous inscriptions* transect the rectus abdominis muscles at

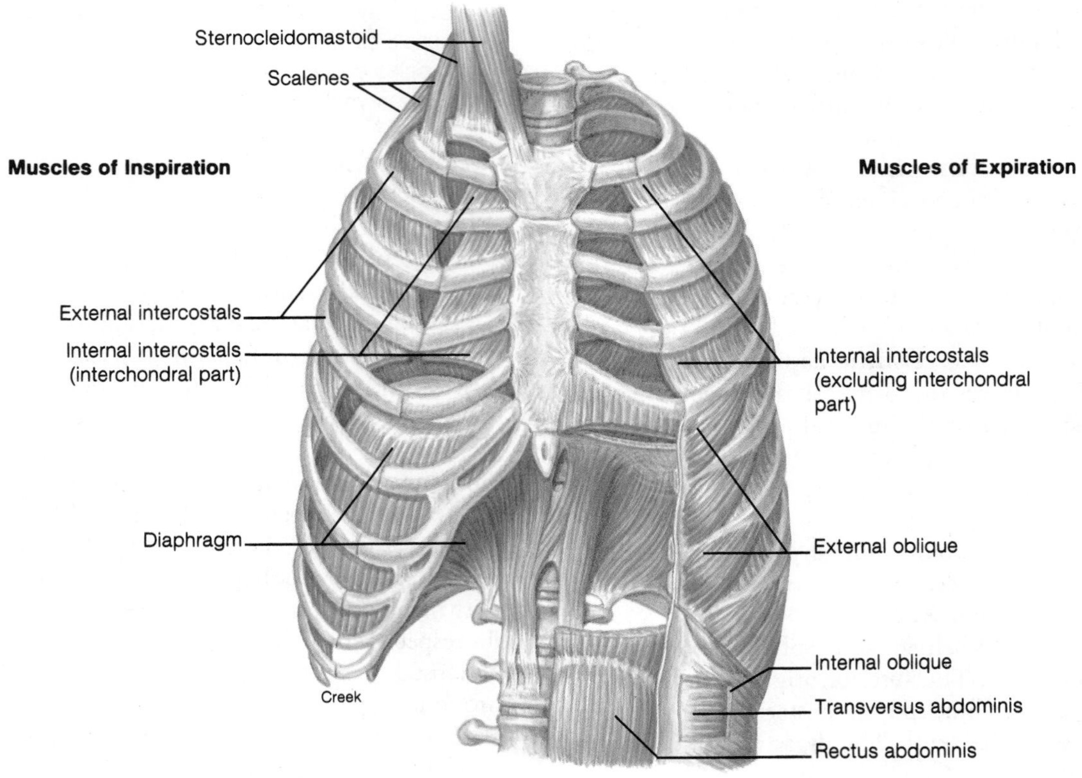

Figure 9.23 Muscles of respiration.

Table 9.6 Muscles of the abdominal wall

Muscle	Derivation	Origin	Insertion	Action	Innervation
External oblique	*external* = closer to surface; *oblique* = diagonal	Lower eight ribs	Iliac crest, linea alba	Compresses abdomen; lateral rotation	Intercostals, iliohypogastric, and ilioinguinal
Internal oblique	*internal* = deep	Iliac crest, inguinal ligament, and lumbodorsal fascia	Linea alba, costal cartilage of last three or four ribs	Compresses abdomen; lateral rotation	Intercostals, iliohypogastric, and ilioinguinal
Transversus abdominis	*transverse* = horizontal; *abdomino* = belly	Iliac crest, inguinal ligament, lumbar fascia, and costal cartilage of last six ribs	Xiphoid process, linea alba, and pubis	Compresses abdomen	Intercostals, iliohypogastric, and ilioinguinal
Rectus abdominis	*rectus* = straplike; straight	Pubic crest and symphysis pubis	Costal cartilage of fifth to seventh ribs and xiphoid process	Flexes vertebral column	Intercostals

several points, causing the surface abdominal anatomy of a well-muscled person to appear segmented.

Refer to table 9.6 for a summary of the muscles of the abdominal wall.

Muscles of the Pelvic Outlet

The **pelvic outlet,** or **floor of the pelvis,** is the inferiorly positioned muscular wall that supports the pelvic viscera. It consists of the levator ani and coccygeus muscles that collectively are called the *pelvic diaphragm.* The openings for the urethra and the rectum pass through the pelvic diaphragm, as well as the vaginal opening in the female.

The **levator ani** muscle forms a thin sheet of muscle that helps support the pelvic viscera and constrict the lower part of the rectum, pulling it forward and aiding defecation. The fan-shaped **coccygeus** muscle aids the levator ani in its functions. Either or both of these muscles are occasionally stretched and torn during parturition.

The *perineal muscles* are positioned inferior to the pelvic diaphragm and are arranged into a superficial layer consisting of the bulbocavernosus, ischiocavernosus, and the transversus perinei superficialis muscles, and a deep layer consisting of the deep transversus perinei muscle and the external anal sphincter (fig. 9.26). The deep layer, with a fascia, constitute the *urogen-*

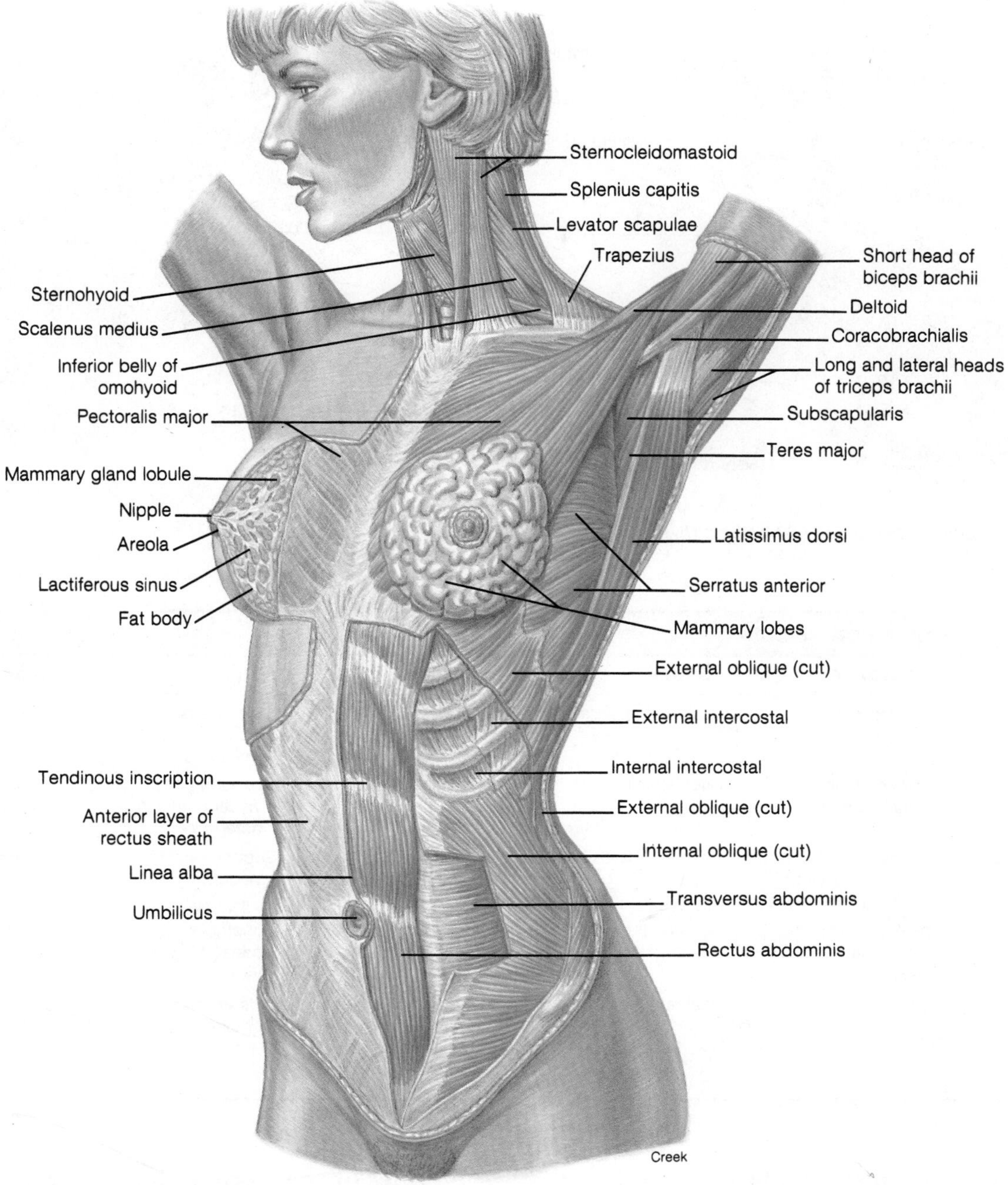

Figure 9.24 Muscles of the anteriolateral neck, shoulder, and torso regions. The mammary gland is an integumentary structure positioned over the pectoralis major muscle.

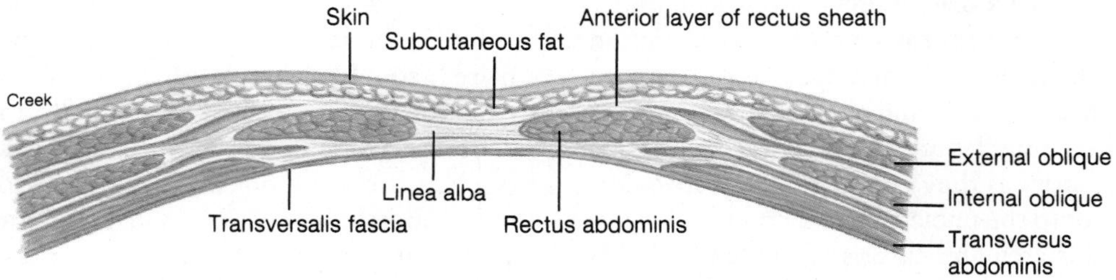

Figure 9.25 Muscles of the anterior abdominal wall shown in a transverse view.

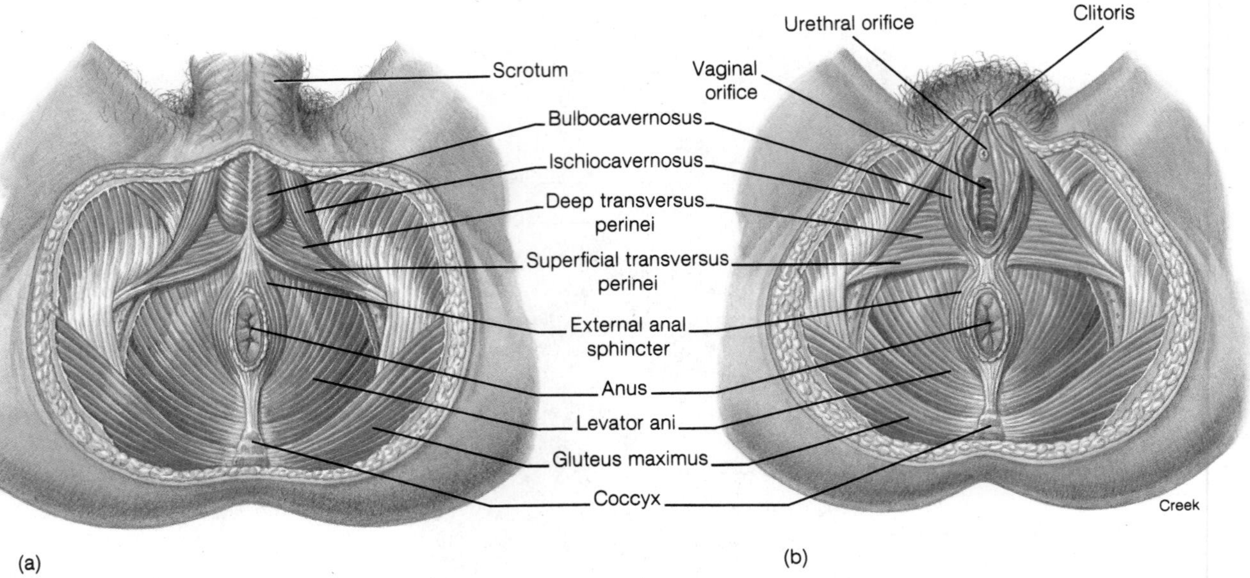

Figure 9.26 Muscles of the pelvic outlet. (*a*) Male and (*b*) female.

Table 9.7 Muscles of the pelvic outlet

Muscle	Derivation	Origin	Insertion	Action	Innervation
Levator ani	*anus* = ring	Spine of ischium and pubic bone	Coccyx	Supports pelvic viscera; aids in defecation	Pudendal plexus
Coccygeus	*coccyx* = like a cuckoo's bill	Ischial spine	Sacrum and coccyx	Supports pelvic viscera; aids in defecation	Perineal branch of pudendal nerve
Transversus perinei	*perineum* = referring to area between genitals and anus	Ischial tuberosity	Central tendon	Supports pelvic viscera	Pudendal plexus
Bulbocavernosus	*bulbus* = little bulb; *caverna* = hollow spaces	Central tendon	Males: base of penis; females: root of clitoris	Constricts urethral canal; constricts vagina	Perineal branch of pudendal nerve
Ischiocavernosus	*ischio* = ischium	Ischial tuberosity	Males: pubic arch and crus of the penis Females: pubic arch and clitoris	Aids erection of penis or clitoris	Perineal branch of pudendal nerve

ital diaphragm. The **external anal sphincter** of the deep layer is a funnel-shaped constrictor muscle that surrounds the anal canal. The muscles of the pelvic diaphragm and the urogenital diaphragm are similar in the male and female, but the perineal muscles of each sex markedly differs.

In males, the **bulbocavernosus** muscle of one side unites with that of the opposite side to form a muscular constriction surrounding the base of the penis. When contracted, the two muscles constrict the urethral canal and assist in emptying the urethra. In females, these muscles are separated by the vaginal orifice, which they constrict as they contract. The **ischiocavernosus** muscle inserts onto the pubic arch and crus of the penis in the male and the pubic arch and clitoris of the female. This muscle aids the erection of the penis in the male and the clitoris in the female.

The muscles of the pelvic outlet are illustrated in figure 9.26 and summarized in table 9.7.

Muscles of the Vertebral Column

The muscles that move the vertebral column are strong and complex because they have to provide support and movement in resistance to the effect of gravity.

The vertebral column can be flexed, extended, abducted, adducted, and rotated. The muscle that flexes the vertebral column, the rectus abdominis, has already been described as a paired, straplike muscle of the anterior abdominal wall. The extensor muscles located on the posterior side of the vertebral column have to be stronger than the flexors because extension (such as lifting an object) is in opposition to gravity. The extensor muscles consist of a superficial group and a deep group. Only some of the muscles of the vertebral column will be described.

The **erector spinae** is a massive superficial muscle group that extends from the sacrum to the skull. It actually consists of three groups of muscles: **iliocostalis, longissimus,** and

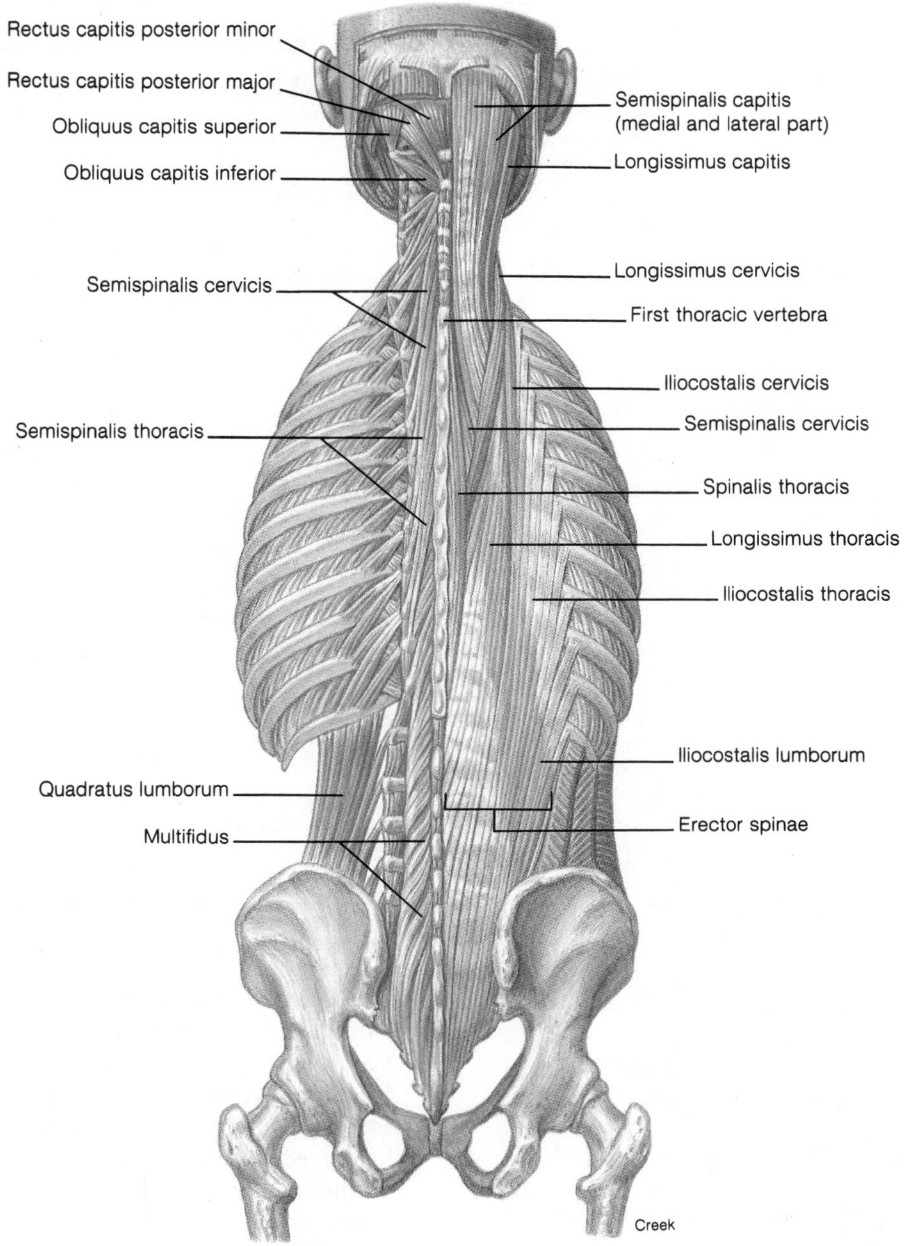

Figure 9.27 Muscles of the vertebral column. The superficial neck muscles and erector spinae group of muscles are illustrated on the right, and the deep neck and back muscles are illustrated on the left.

spinalis (fig. 9.27). Each of these groups, in turn, consists of overlapping slips of muscle. The iliocostalis is the most lateral group, the longissimus is intermediate in position, and the spinalis, in medial position, is positioned in contact with the spinous processes of the vertebrae.

The erector spinae muscles are frequently strained through improper lifting. A heavy object should not be lifted with the vertebral column flexed; instead, the thighs and knees should be flexed so that the pelvic and leg muscles can aid in the task.

The erector spinae muscles are also frequently strained in women during pregnancy. Pregnant women will try to counterbalance the effect of the protruding abdomen by hyperextending the vertebral column. This causes an exaggerated lumbar curvature, strained muscles, and a peculiar gait.

The **quadratus lumborum** is a deep muscle that originates on the iliac crest and the lower three lumbar vertebrae. It inserts on the transverse processes of the first four lumbar vertebrae and the inferior margin of the twelfth rib. When the right and left quadratus lumborum contract together, the vertebral column in the lumbar region extends. Separate contraction causes lateral flexion of the spine.

A summary of the major muscles of the back is presented in table 9.8.

Table 9.8 Muscles of the vertebral column

Muscle	Derivation	Origin	Insertion	Action	Innervation
Quadratus lumborum	*quad* = four; *lumb* = referring to lumbar region	Iliac crest and lower three lumbar vertebrae	Twelfth rib and upper four lumbar vertebrae	Extends lumbar region; lateral flexion of vertebral column	T12, L1–L4
Erector spinae	Consists of three groups of muscles: iliocostalis, longissimus, and spinalis. The iliocostalis and longissimus are further subdivided into three groups on the basis of location along the vertebral column.				
Iliocostalis lumborum	*ilium* = flank	Crest of ilium	Lower six ribs	Extends lumbar region	Dorsal rami of lumbar nerves
Iliocostalis thoracis	*thorax* = chest	Lower six ribs	Upper six ribs	Extends thoracic region	Dorsal rami of thoracic nerves
Iliocostalis cervicis	*cervix* = neck	Angles of third to sixth rib	Transverse processes of fourth to sixth cervical vertebrae	Extends cervical region	Dorsal rami of cervical nerves
Longissimus thoracis		Transverse processes of lumbar vertebrae	Transverse processes of all the thoracic vertebrae and lower nine ribs	Extends thoracic region	Dorsal rami of spinal nerves
Longissimus cervicis		Transverse processes of upper four or five thoracic vertebrae	Transverse processes of second to sixth cervical vertebrae	Extends cervical region and lateral flexion	Dorsal rami of spinal nerves
Longissimus capitis		Transverse processes of upper five thoracic vertebrae and articular processes of lower three cervical vertebrae	Posterior margin of cranium and mastoid processes	Extends head	Dorsal rami of middle and lower cervicals
Spinalis thoracis		Spinous processes of upper lumbar and lower thoracic vertebrae	Spinous processes of upper thoracic vertebrae	Extends vertebral column	Dorsal rami of spinal nerves

1. Identify the facial muscles that do the following: (a) wrinkle the forehead; (b) purse the lips; (c) protrude the lower lip; (d) produce frown lines; (e) smile; (f) frown; (g) wink; and (h) elevate the upper lip to show the teeth.
2. Describe the actions of the extrinsic muscles that move the tongue.
3. Which muscles of the neck either originate from or insert on the hyoid bone?
4. Describe the actions of the muscles of inspiration. Which muscles participate in forced expiration?
5. Which muscles of the pelvic outlet support the floor of the pelvic cavity, and which are associated with the genitalia?
6. What is the extent and the compartmentalization of the erector spinae muscle?

Muscles of the Appendicular Skeleton

The muscles of the appendicular skeleton include those of the pectoral girdle, arm, forearm, wrist, hand, and fingers, and those of the pelvic girdle, thigh, leg, ankle, foot, and toes.

Objective 13. Locate the major muscles of the appendicular skeleton; identify synergistic and antagonistic muscles, and describe the action of each.

Muscles That Act on the Pectoral Girdle

The shoulder is attached to the axial skeleton only at the sternoclavicular joint, so strong, straplike muscles are necessary. Furthermore, muscles that move the brachium originate on the scapula, and during brachial movement the scapula has to be held fixed. The muscles that act on the pectoral girdle originate on the axial skeleton and can be divided into anterior and posterior groups.

The anterior group of muscles that act on the pectoral girdle includes the **serratus anterior, pectoralis minor,** and **subclavius** (fig. 9.28). The posterior group includes the **trapezius, levator scapulae,** and **rhomboideus** (fig. 9.29). The position of these muscles is such that one muscle does not cause an action on its own, but rather several muscles contract synergistically to result in any movement of the girdle.

Treatment of advanced stages of *breast cancer* requires the surgical removal of both pectoralis major and pectoralis minor muscles in a procedure called a *radical mastectomy.* Postoperative physical therapy is primarily geared toward strengthening the synergistic muscles of this area. As the muscles that act on the brachium are learned, determine which are synergists with the pectoralis major.

The muscles that act on the pectoral girdle are summarized in table 9.9.

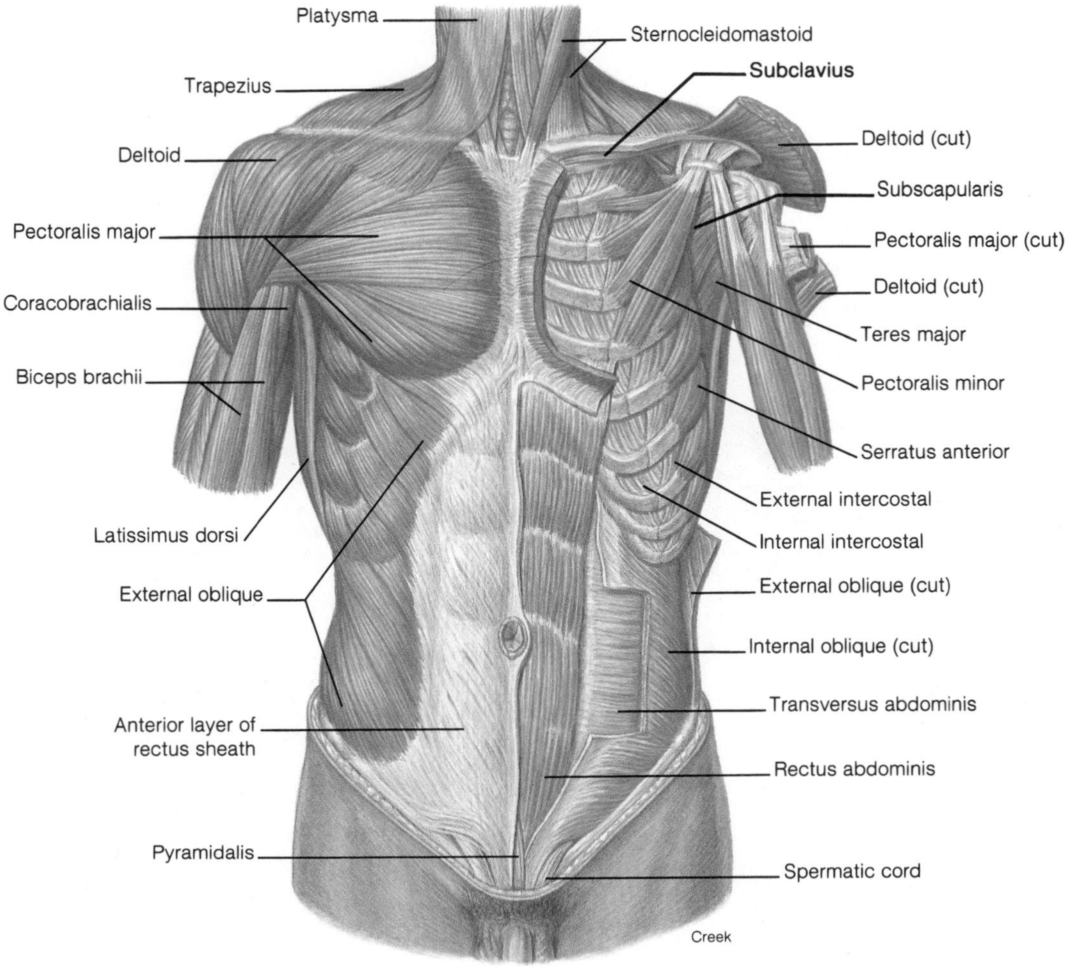

Figure 9.28 Muscles of the anterior torso and shoulder regions. The superficial muscles are illustrated on the right and the deep muscles are illustrated on the left.

Table 9.9 Muscles that act on the pectoral girdle

Muscle	Derivation	Origin	Insertion	Action	Innervation
Serratus anterior	*serratus* = serrated; *anterior* = front	Upper eight or nine ribs	Anterior vertebral border of scapula	Pulls scapula forward and downward	Long thoracic
Pectoralis minor	*pectus* = chest; *minor* = lesser	Sternal ends of third, fourth, and fifth ribs	Coracoid process of scapula	Pulls scapula forward and downward	Medial pectoral
Subclavius	*sub* = below, under	First rib	Subclavian groove of clavicle	Draws clavicle downward	C5, C6
Trapezius	*trapezoeides* = trapezoid shaped	Occipital bone and spines of seventh cervical and all thoracic vertebrae	Clavicle, spine of scapula, and acromion process	Elevates scapula, draws head back, adducts scapula, braces shoulder	Accessory, C3, C4
Levator scapulae	*levator* = elevator; *scapulae* = scapula	First to fourth cervical vertebrae	Vertebral border of scapula	Elevates scapula	Dorsal scapular
Rhomboideus major	*rhomboides* = rhomboid shaped	Spines of second to fifth thoracic vertebrae	Vertebral border of scapula	Elevates and adducts scapula	Dorsal scapular
Rhomboideus minor		Seventh cervical and first thoracic vertebrae	Vertebral border of scapula	Elevates and adducts scapula	Dorsal scapular

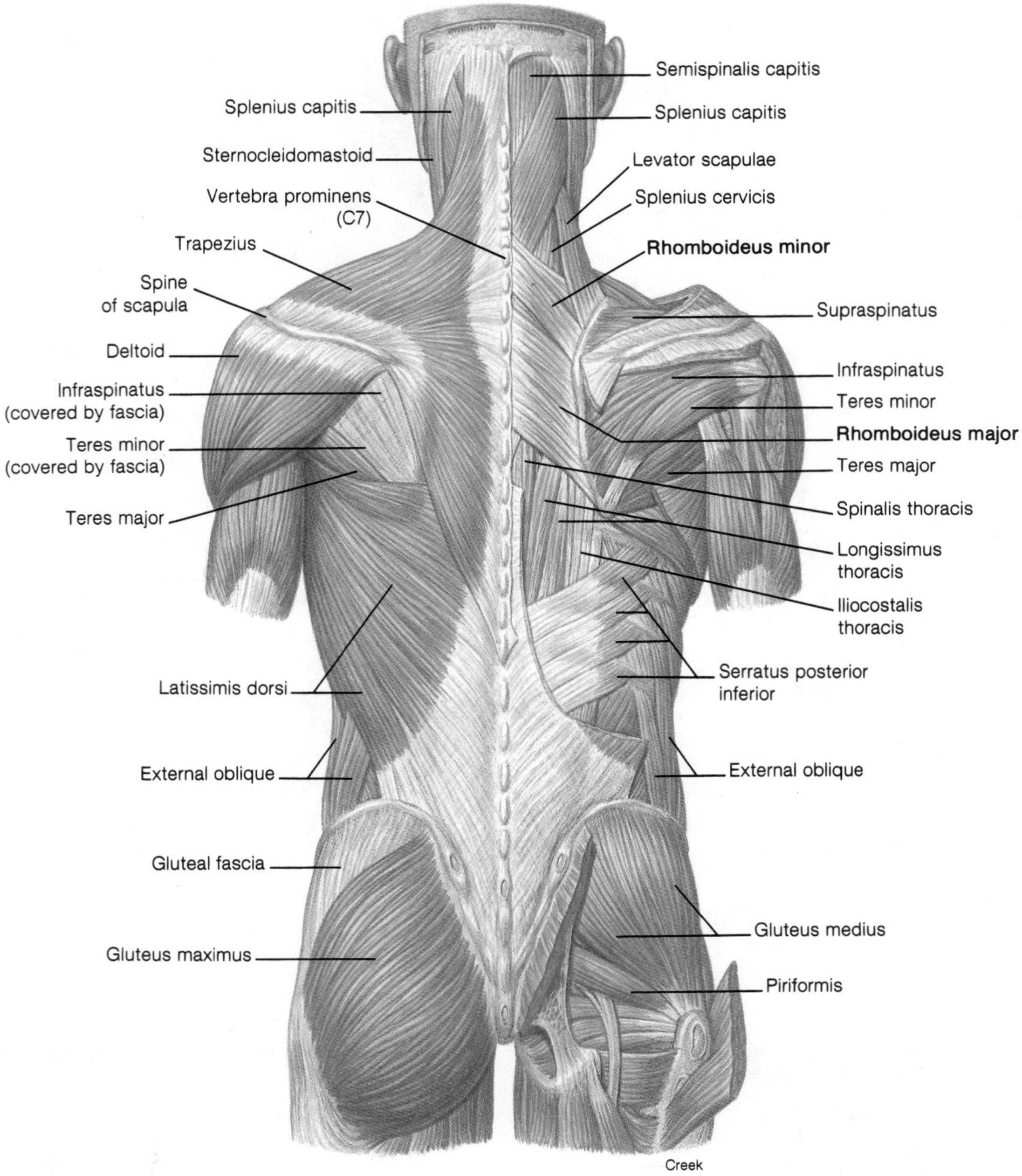

Figure 9.29 Muscles of the posterior neck, shoulder, torso, and gluteal regions. The superficial muscles are illustrated on the left and the deep muscles are illustrated on the right.

Muscles That Move the Humerus

Of the nine muscles that span the shoulder joint to insert on the humerus, only two of them, the pectoralis major and latissimus dorsi, do not originate on the scapula. These two are designated as axial muscles, whereas the remaining seven are scapular muscles. The muscles of this region are shown in figures 9.28 and 9.29, and the attachments of all the muscles that either originate or insert on the scapula are shown in figure 9.30.

In terms of their development, the pectoralis major and the latissimus dorsi muscles are not axial muscles at all. They develop in the forelimb and extend to the trunk secondarily. They are considered axial muscles only because their origins are on the axial skeleton.

Axial Muscles The axial muscles include the pectoralis major and latissimus dorsi.

The **pectoralis major** is a large chest muscle (fig. 9.28) that binds the humerus to the scapula and is the principal flexor muscle of the arm. The large, flat, triangular **latissimus dorsi** muscle covers the lower half of the thoracic region of the back (see fig. 9.29), and is the antagonist to the pectoralis major. The latissimus dorsi is frequently called the "swimmer's muscle" because it powerfully extends the arm, drawing it downward and backward while it rotates medially. Extension of the arm is in reference to anatomical position and is therefore a backwards, retracting (increasing the shoulder joint angle) movement of the arm.

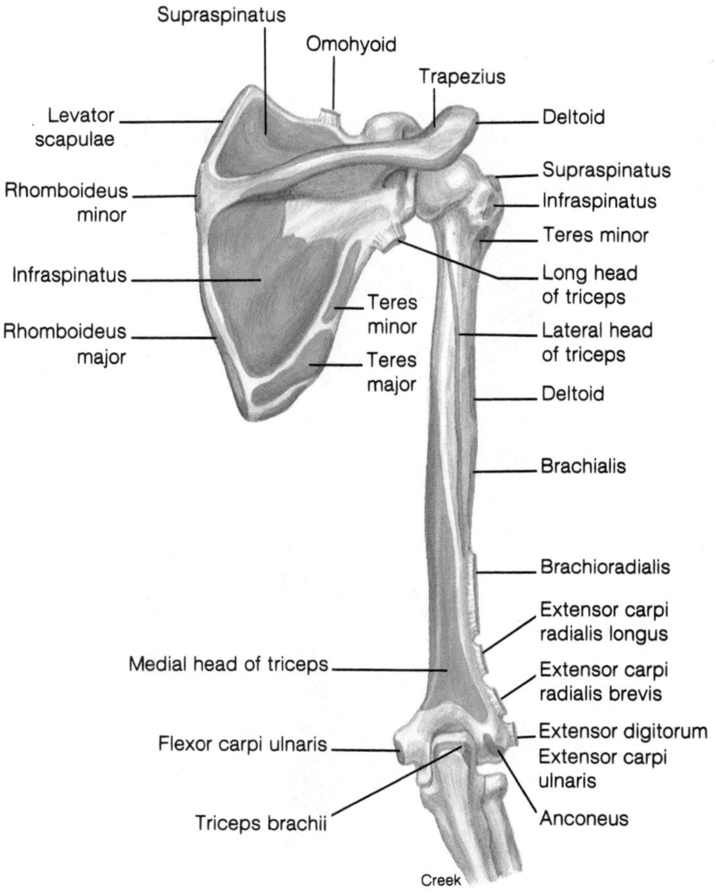

Figure 9.30 A posterior view of the scapula and humerus, showing the areas of attachment of the associated muscles. Points of origin are color coded red, and points of insertion are color coded blue.

A latissimus dorsi muscle, conditioned with pulsated electrical impulses, will in time come to resemble cardiac muscle tissue in that it is nonfatigable and uses oxygen at a steady rate. Following conditioning, the muscle may be used in an autotransplant to repair a surgically removed portion of a patient's diseased heart. The procedure involves detaching the latissimus dorsi from its vertebral origin, leaving the blood supply and innervation intact, and slipping it into the pericardial cavity where it is wrapped around the heart like a towel. A pacemaker is required to provide the continuous, rhythmic contractions.

Scapular Muscles The nonaxial, scapular muscles include the deltoid, supraspinatus, infraspinatus, teres major, teres minor, subscapularis, and coracobrachialis.

The **deltoid** is a thick, powerful muscle that caps the shoulder joint (figs. 9.31, 9.32). Although it has several functions (see table 9.10), the principal action of the deltoid is abduction of the arm at the shoulder joint. Functioning together, both the pectoralis major and the latissimus dorsi muscles are antagonists to the deltoid in that they cause adduction of the arm at the shoulder joint. The deltoid is a common site for intramuscular injections.

The remaining six scapular muscles also help stabilize the shoulder, and have specific actions at the shoulder joint (see table 9.10). The **supraspinatus** laterally rotates the humerus and is synergistic with the deltoid in abducting the arm at the shoulder joint. The **infraspinatus** rotates the humerus laterally. The action of the **teres major** is similar to that of the latissimus dorsi, adducting and medially rotating the humerus. The **teres minor** works with the infraspinatus in laterally rotating the humerus. The **subscapularis** is a strong stabilizer of the shoulder and also aids in medially rotating the humerus. The **coracobrachialis** is a synergist to the pectoralis major in flexing and adducting the arm at the shoulder joint.

Four of the nine muscles that cross the shoulder joint, the supraspinatus, infraspinatus, teres minor, and subscapularis, are commonly called the *musculotendinous cuff,* or *rotator cuff,* muscles. Their distal tendons blend with and reinforce the fibrous capsule of the shoulder joint en route to their points of insertion on the humerus. This structural arrangement plays a major role in stabilizing the shoulder joint. Musculotendinous cuff injuries are common in baseball players. When throwing a baseball, there is complete abduction of the shoulder followed by a rapid and forceful rotation and flexion of the shoulder, which may strain the musculotendinous cuff.

Muscles That Act on the Forearm

The powerful muscles of the brachium are responsible for movement of the forearm at the elbow and at the radioulnar joint. These muscles are the biceps brachii, brachialis, brachioradialis, and triceps brachii (figs. 9.31, 9.32). In addition, a short triangular muscle, called the anconeus, is positioned over the distal end of the triceps brachii near the elbow. A transverse section through the brachium in figure 9.33 provides a different perspective of the brachial region.

The powerful **biceps brachii** muscle, positioned on the anterior surface of the humerus, is the most familiar muscle of the arm; yet it has no attachments on the humerus. The biceps has a dual origin: a medial tendinous head, the **short head,** arises from the coracoid process of the scapula, and the **long head** originates on the superior tuberosity of the glenoid fossa and passes through the shoulder joint and descends in the intertubercular groove on the humerus. Both heads of the biceps insert on the radial tuberosity. The **brachialis** is located on the distal anterior half of the humerus, deep to the biceps brachii. It is synergistic to the biceps in flexing the forearm at the elbow joint.

The **brachioradialis** is the prominent muscle positioned along the lateral (radial) surface of the forearm. It too flexes the forearm at the elbow joint.

The **triceps brachii** muscle, located on the posterior surface of the brachium, is the muscle antagonistic to the biceps brachii. It has three heads, or origins. Two of the three, the **lateral head** and **medial head,** arise from the humerus, whereas the **long head** arises from the infraglenoid tuberosity of the scapula. A common tendinous insertion attaches the triceps brachii muscle to the olecranon process of the ulna. The triceps and the small **anconeus** muscles extend the forearm at the elbow joint.

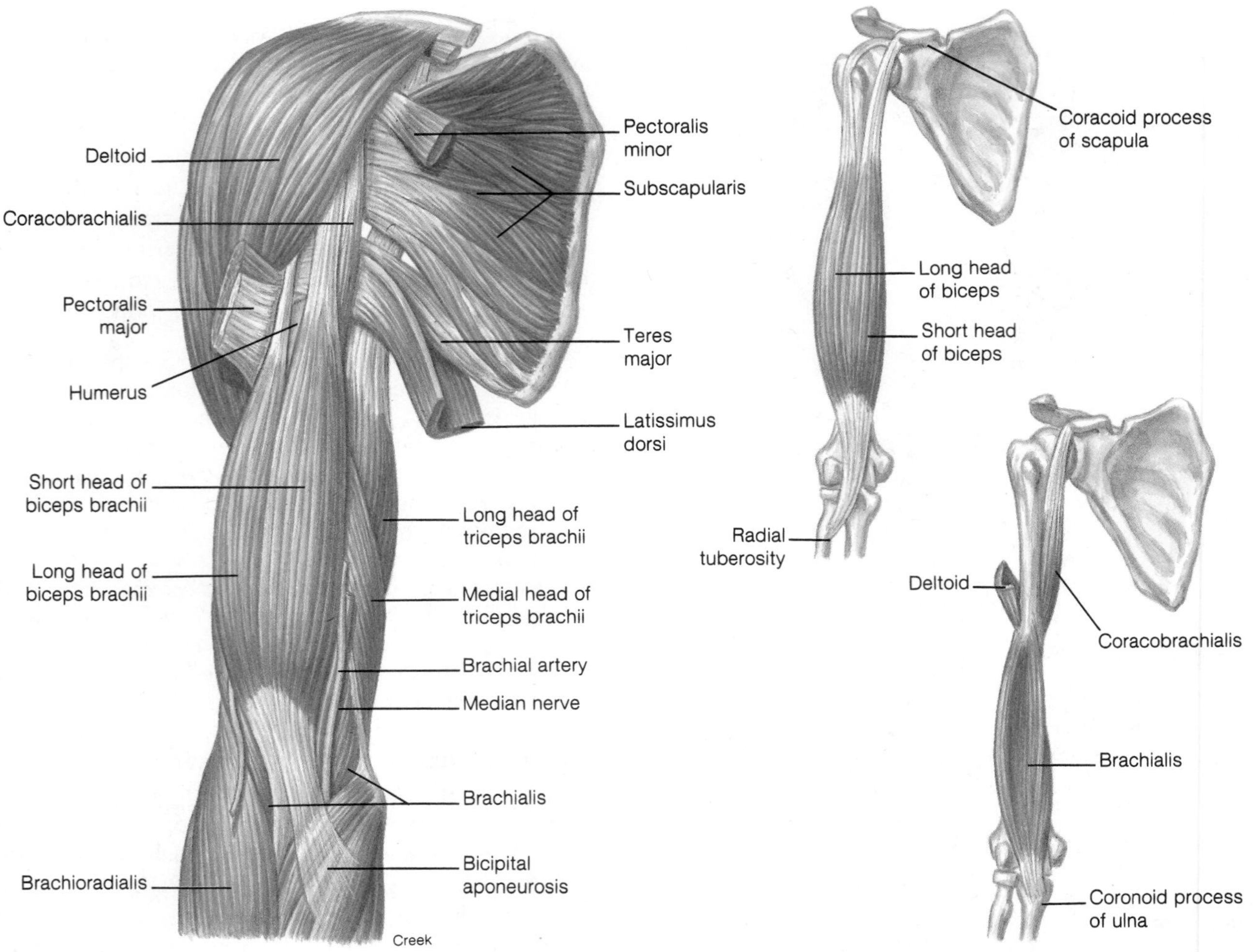

Figure 9.31 Muscles of the right anterior shoulder and brachium.

Refer to table 9.11 for a summary of the muscles that act on the forearm.

Muscles of the Forearm That Move the Wrist, Hand, and Fingers

The muscles that cause wrist, hand, and most finger movements are positioned along the forearm (figs. 9.34, 9.35). Several of these muscles act on two joints, the elbow and wrist. Others act on the joints of the wrist, hand, and digits. Still others produce rotational movement at the radioulnar joint. The precise actions of these muscles are complex, so only the basic movements will be described here. Most of these muscles perform four primary actions on the hand and digits: supination, pronation, flexion, and extension. Other actions of the hand include adduction and abduction.

Supination and Pronation of the Hand The **supinator** muscle is positioned around the upper posterior portion of the radius (fig. 9.35) where it works synergistically with the biceps brachii to supinate the hand. Two muscles are responsible for pronating the hand, the pronator teres and pronator quadratus. The **pronator teres** muscle is located on the upper medial side of the forearm, whereas the deep, anteriorly positioned **pronator quadratus** muscle extends between the ulna and radius on the distal fourth of the forearm. These two muscles work synergistically to rotate the palm of the hand posteriorly and position the thumb medially.

Flexion of the Wrist, Hand, and Fingers Six of the muscles that flex the wrist, hand, and fingers will be described from lateral to medial and from superficial to deep (see figs. 9.34, 9.35). Although four of the six arise from the medial epicondyle of the humerus (see table 9.12), their actions on the elbow joint are minimal. The brachioradialis, already described, is an obvious reference muscle for locating the muscles of the forearm that flex the wrist, hand, and fingers.

The **flexor carpi radialis** muscle extends diagonally across the anterior surface of the forearm, and its distal cordlike tendon crosses the wrist under the *flexor retinaculum.*

The narrow **palmaris longus** muscle is superficial in position on the anterior surface of the forearm. It has a long, slender tendon that attaches to the *palmar aponeurosis* where it assists in flexing the wrist.

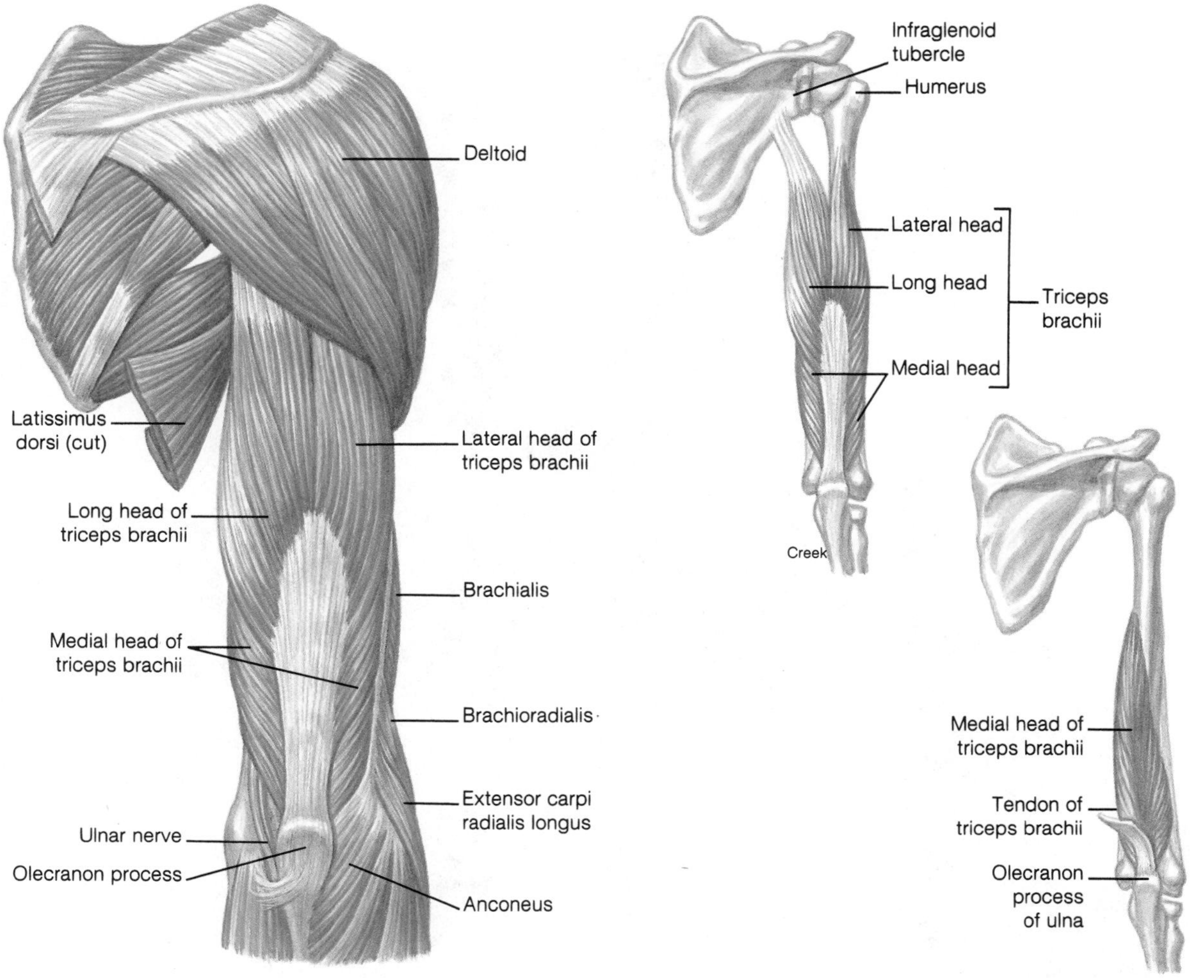

Figure 9.32 Muscles of the right posterior shoulder and brachium.

Table 9.10 Muscles that move the humerus

Muscle	Derivation	Origin	Insertion	Action	Innervation
Pectoralis major	*pectus* = chest; *major* = greater	Clavicle, sternum, costal cartilages of second to sixth rib; rectus sheath	Crest of greater tubercle of humerus	Flexes, adducts, and rotates arm medially	Medial and lateral pectoral
Latissimus dorsi	*latissimus* = widest; *dorsum* = back	Spines of sacral, lumbar, and lower thoracic vertebrae; iliac crest and lower four ribs	Intertubercular groove of humerus	Extends, adducts, and rotates humerus medially; retracts shoulder	Thoracodorsal
Deltoid	*delta* = triangular	Clavicle, acromion process; spine of scapula	Deltoid tuberosity of humerus	Abducts arm; extends or flexes humerus	Axillary
Supraspinatus	*supra* = above; *spina*=spine of scapula	Fossa—superior to spine of scapula	Greater tubercle of humerus	Abducts and laterally rotates humerus	Suprascapular
Infraspinatus	*infra* = below	Fossa—inferior to spine of scapula	Greater tubercle of humerus	Rotates arm laterally	Suprascapular
Teres major		Inferior angle and lateral border of scapula	Crest of lesser tubercle of humerus	Extends humerus, or adducts and rotates arm medially	Lower subscapular
Teres minor	*teres* = rounded; *minor* = lesser	Axillary border of scapula	Greater tubercle and groove of humerus	Rotates arm laterally	Axillary
Subscapularis	*sub* = below, under; *scapula* = scapula	Subscapular fossa	Lesser tubercle of humerus	Rotates arm medially	Subscapular
Coracobrachialis	*coraco* = referring to coracoid process; *brachium* = arm	Coracoid process of scapula	Shaft of humerus	Flexes and adducts shoulder joint	Musculocutaneous

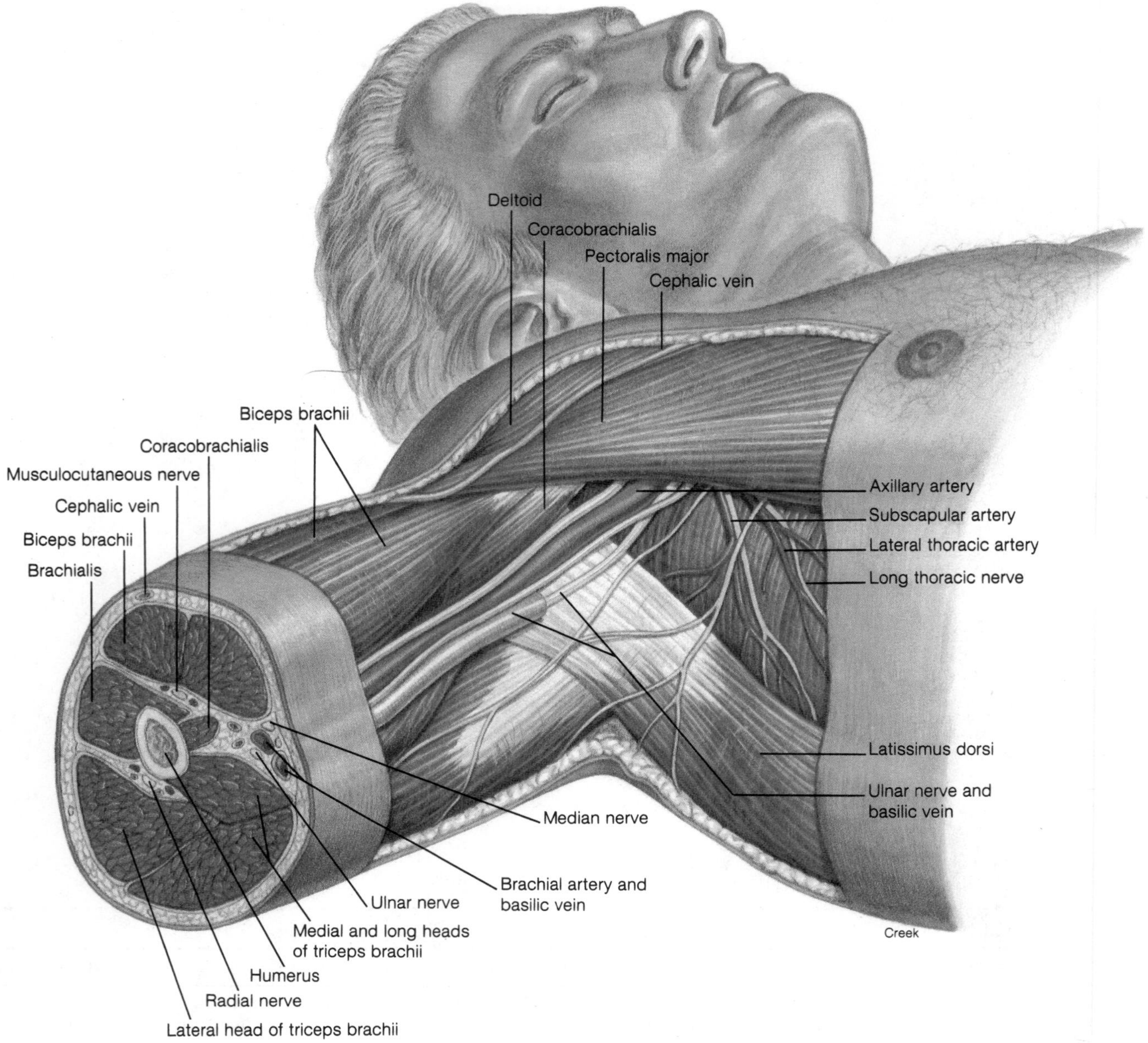

Figure 9.33 The axillary region and a transverse section through the brachium.

Table 9.11 Muscles that act on the forearm

Muscle	Derivation	Origin	Insertion	Action	Innervation
Biceps brachii	*biceps* = two heads;	Coracoid process; tuberosity above glenoid fossa of scapula	Radial tuberosity	Flexes and supinates forearm and hand	Musculocutaneous
Brachialis	*brachium* = upper arm	Anterior shaft of humerus	Coronoid process of ulna	Flexes forearm	Musculocutaneous, median, and radial
Brachioradialis	*radialis* = pertaining to radius	Lateral supracondylar ridge of humerus	Proximal to styloid process of radius	Flexes forearm	Radial
Triceps brachii	*triceps* = three heads	Tuberosity below glenoid fossa; lateral and medial surfaces of humerus	Olecranon process of ulna	Extends forearm	Radial
Anconeus	*ancon* = elbow	Lateral epicondyle of humerus	Olecranon process of ulna	Extends forearm	Radial

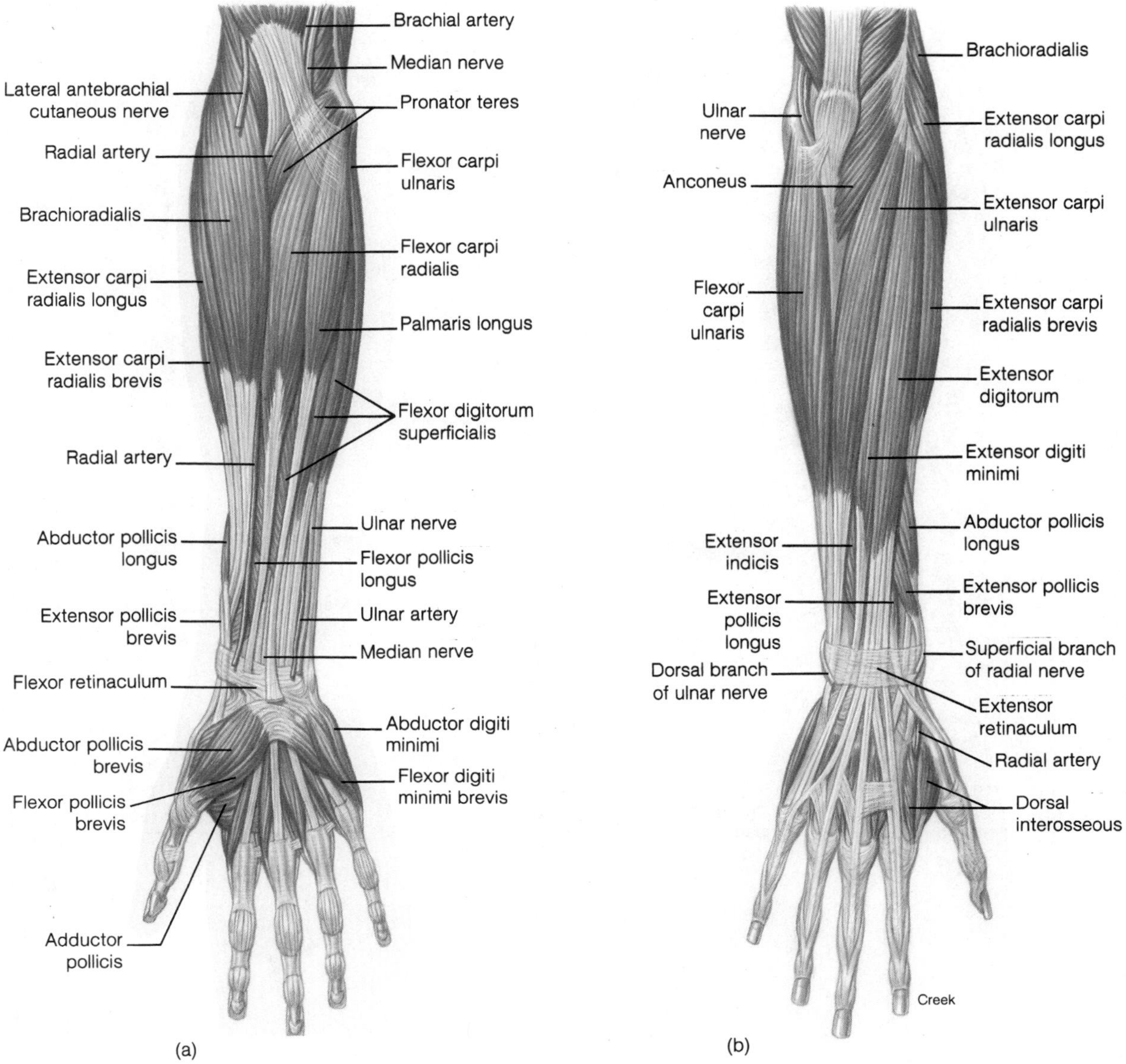

Figure 9.34 Muscles of the right forearm that move the wrist, hand, and digits. (*a*) An anterior view and (*b*) a posterior view.

The palmaris longus is the most variable muscle in the body. It is totally absent in approximately 8% of all persons. In 4% of all persons, it is absent in one or the other forearm. Furthermore, it is absent more often in females than males, and on the left side in both sexes. Because of the superficial position of the palmaris longus muscle, its presence or absence can be readily detected on oneself by clinching tightly the fist and examining for its tendon just proximal to the wrist (see fig. 10.26b).

The **flexor carpi ulnaris** muscle is positioned on the medial anterior side of the forearm where it assists in flexing the wrist and adducting the hand.

The broad **flexor digitorum superficialis (sublimis)** muscle lies directly beneath the three flexors just described (see figs. 9.34, 9.35). It has an extensive origin, involving all three of the long bones of the forearm (table 9.12). The tendon at the distal end of this muscle is united across the wrist joint but then splits to attach to the middle phalanx of digits two through five.

The **flexor digitorum profundus** muscle lies deep to the flexor digitorum superficialis muscle. These two muscles flex the wrist, hand and the second, third, fourth, and fifth fingers.

The **flexor pollicis longus** is a deep, lateral muscle of the forearm. It flexes the thumb, assisting the grasping mechanism of the hand.

The tendons of the muscles that flex the hand can be seen on the wrist as a fist is made. These tendons are securely positioned by the flexor retinaculum (fig. 9.34a), which crosses the wrist area transversely.

Extension of the Hand The muscles that extend the hand are located on the posterior side of the forearm. Most of the primary extensors of the hand can be seen superficially in figure 9.34b and will be discussed from lateral to medial.

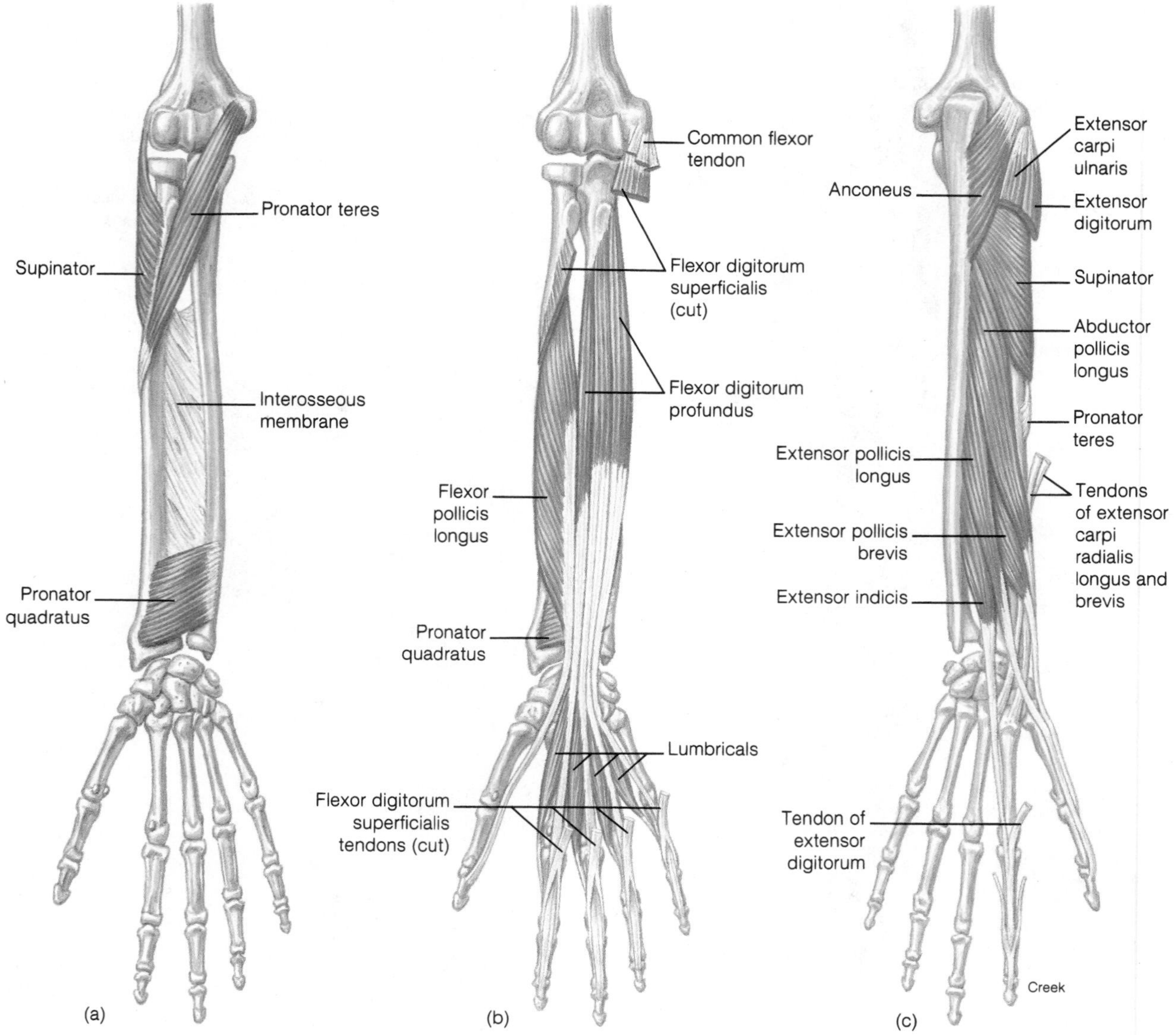

Figure 9.35 Deep muscles of the forearm. (*a*) Rotators, (*b*) flexors, and (*c*) extensors.

The long, tapered **extensor carpi radialis longus** muscle is located medial to the brachioradialis, where it extends and abducts the hand. The **extensor carpi radialis brevis** muscle is positioned immediately medial to the extensor carpi radialis longus muscle and performs approximately the same functions. Its origin and insertion, however, are different (see table 9.12).

The bipennate-fibered **extensor digitorum communis** muscle is positioned in the center of the forearm along the posterior surface. It originates on the lateral epicondyle of the humerus. Its tendon of insertion divides at the wrist, beneath the extensor retinaculum, into four tendons that attach to the distal tip of the medial phalanges of digits two through five.

The **extensor digiti minimi** is a long, narrow muscle located on the ulnar side of the extensor digitorum communis muscle. Its tendinous insertion fuses with the tendon of the extensor digitorum communis going to the fifth digit.

The **extensor carpi ulnaris** muscle is the most medial muscle on the posterior surface of the forearm. It inserts on the base of the fifth metacarpal bone where it functions to extend and adduct the hand.

The **extensor pollicis longus** muscle extends from the mid-ulnar region, across the lower two-thirds of the forearm, and inserts onto the base of the distal phalanx of the thumb (see fig. 9.35), where it extends the thumb and abducts the hand.

The **extensor pollicis brevis** muscle arises from the lower midportion of the radius and inserts on the base of the proximal phalanx of the thumb (fig. 9.35). The action of this muscle is similar to that of the extensor pollicis longus.

As its name implies, the **abductor pollicis longus** muscle abducts the thumb and hand. It originates on the interosseous ligament, between the ulna and radius, and inserts on the base of the first metacarpal bone.

The muscles that act on the wrist, hand, and digits are summarized in table 9.12.

Table 9.12 Muscles of the forearm that act on the wrist, hand, and digits

Muscle	Derivation	Origin	Insertion	Action	Innervation
Supinator	*supin* = bend back; palms up	Lateral epicondyle of humerus and crest of ulna	Lateral surface of radius	Supinates forearm	Radial
Pronator teres	*pron* = bend forward; palms down; *teres* = long, round	Medial epicondyle of humerus	Lateral surface of radius	Pronates forearm	Median
Pronator quadratus	*quad* = four (square)	Distal fourth of ulna	Distal fourth of radius	Pronates hand	Median
Flexor carpi radialis	*flexor* = decrease angle; *carpus* = wrist; *radial* = radius bone	Medial epicondyle of humerus	Base of second and third metacarpals	Flexes and abducts hand	Median
Palmaris longus	*palma* = flat of hand; *longus* = long	Medial epicondyle of humerus	Palmar aponeurosis	Flexes wrist	Median
Flexor carpi ulnaris		Medial epicondyle and olecranon process	Carpal and metacarpal bones	Flexes and adducts wrist	Ulnar
Flexor digitorum superficialis	*digitis* = finger or toe; *superficialis* = superficial	Medial epicondyle, coronoid process, and anterior border of radius	Middle phalanges of digits	Flexes wrist, hand, and digits	Median
Flexor digitorum profundus	*profundus* = deep	Proximal two-thirds of ulna and interosseous membrane	Distal phalanges	Flexes wrist, hand, and digits	Median and ulnar
Flexor pollicis longus	*pollex* = thumb	Shaft of radius, interosseous membrane, and coronoid process of ulna	Distal phalanx of thumb	Flexes thumb	Median
Extensor carpi radialis longus	*extensor* = increase joint angle	Lateral supracondylar ridge of humerus	Second metacarpal	Extends and abducts hand	Radial
Extensor carpi radialis brevis	*brevis* = short	Lateral epicondyle of humerus	Third metacarpal	Extends and abducts hand	Radial
Extensor digitorum communis	*communis* = common	Lateral epicondyle of humerus	Posterior surfaces of digits II-V	Extends wrist and phalanges	Radial
Extensor digiti minimi	*minimus* = very small	Lateral epicondyle of humerus	Extensor aponeurosis of fifth digit	Extends fifth digit and wrist	Radial
Extensor carpi ulnaris		Lateral epicondyle of humerus and olecranon process	Base of fifth metacarpal	Extends and adducts wrist	Radial
Extensor pollicis longus		Middle shaft of ulna, lateral side	Base of distal phalanx of thumb	Extends thumb; abducts hand	Radial
Extensor pollicis brevis		Distal shaft of radius and interosseous membrane	Base of first phalanx of thumb	Extends thumb; abducts hand	Radial
Abductor pollicis longus		Distal radius and ulna and interosseous membrane	Base of first metacarpal bone	Abducts thumb and hand	Radial

Notice that your hand is partially contracted even when relaxed. The muscles that extend the hand are not as strong as the muscles that flex it. This is why persons who receive strong electrical shocks through the arms will tightly flex their hands and hold on. All the muscles of the arm are stimulated to contract, but the flexors, being stronger, cause the hands to close tightly.

Muscles of the Hand

The hand is a marvelously complex structure adapted to permit an array of intricate movements. Flexion and extension movements of the hand and phalanges are accomplished by the muscles of the forearm just described. Precise finger movements that require coordinating abduction and adduction with flexion and extension are the function of the small intrinsic muscles of the hand. These muscles and associated structures of the hand are depicted in figure 9.36. The position and actions of the muscles of the hand are listed in table 9.13.

The muscles of the hand are divided into **thenar** *(the'nar)*, **hypothenar** *(hi-poth'ĕ-nar)*, and **intermediate** groups. The *thenar eminence* is the fleshy base of the thumb and is formed by three muscles: the **abductor pollicis brevis,** the **flexor pollicis brevis,** and the **opponens pollicis.** The most important of the thenar muscles is the opponens pollicis, which opposes the thumb to the palm of the hand.

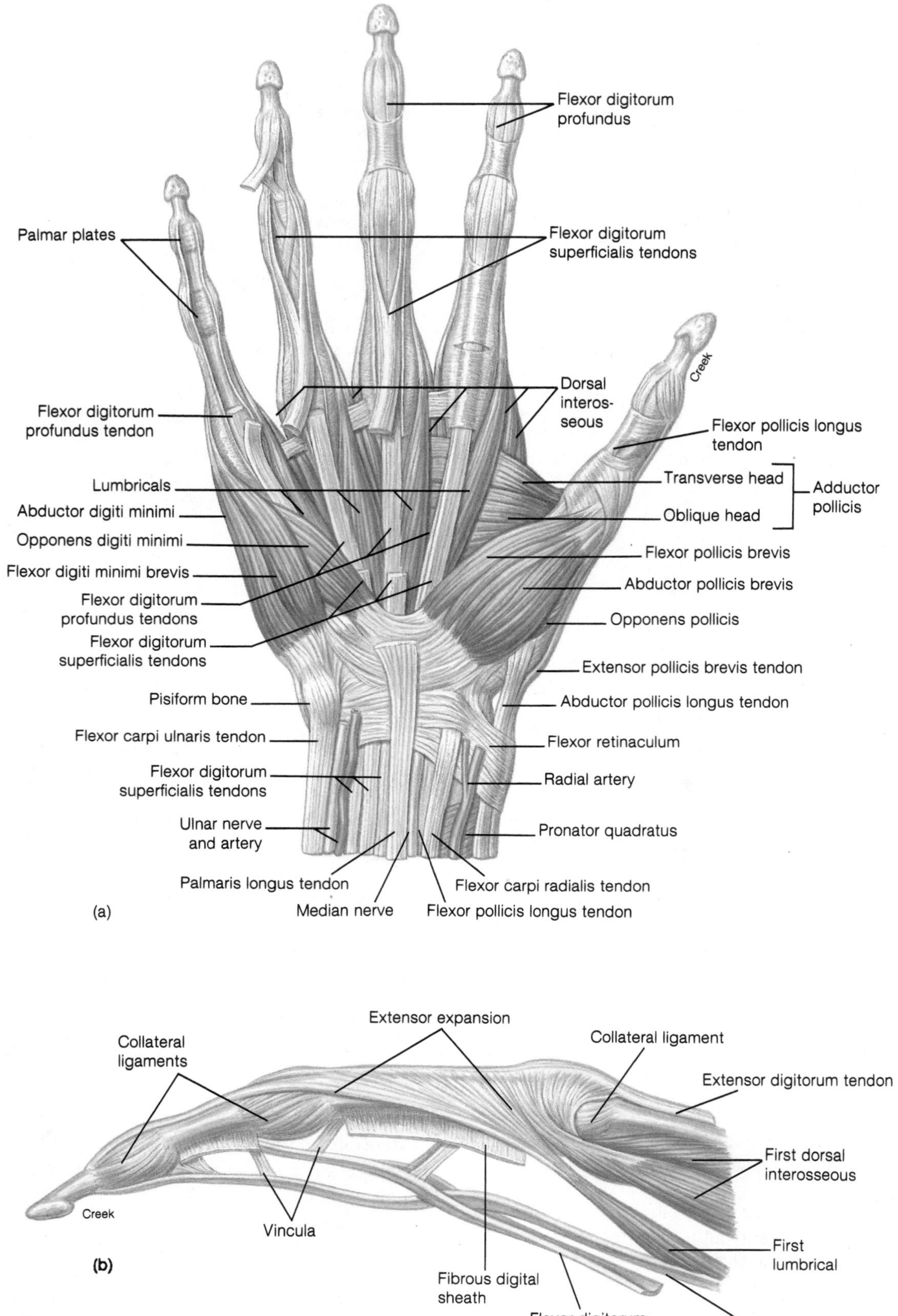

Figure 9.36 Muscles of the hand. (*a*) An anterior view and (*b*) a lateral view of the second (index) finger.

Table 9.13 Intrinsic muscles of the hand

Muscle	Derivation	Origin	Insertion	Action	Innervation
Thenar Muscles					
Abductor pollicis brevis	*pollex* = thumb; *brevis* = short	Flexor retinaculum, scaphoid, and trapezium	Proximal phalanx of thumb	Abducts thumb	Median
Flexor pollicis brevis		Flexor retinaculum and trapezium	Proximal phalanx of thumb	Flexes thumb	Median
Opponens pollicis	*opponens* = against	Trapezium and flexor retinaculum	First metacarpal	Opposes thumb	Median
Intermediate Muscles					
Adductor pollicis (oblique and transverse heads)		Oblique head, capitate; transverse head, second and third metacarpals	Proximal phalanx of thumb	Adducts thumb	Ulnar
Lumbricales (4)	*lumbricalis* = referring to digits	Tendons of flexor digitorum profundus	Extensor expansions of digits II-V	Flexes digits at metacarpophalangeal joints; extends digits at interphalangeal joints	Median and ulnar
Palmar interossei (3)	*palma* = flat of hand	Medial side of second metacarpal; lateral sides of fourth and fifth metacarpals	Proximal phalanges of index, ring, and little fingers and extensor digitorum communis	Adducts fingers toward middle finger at metacarpophalangeal joints	Ulnar
Dorsal interossei (4)		Adjacent sides of metacarpals	Proximal phalanges of index and middle fingers (lateral sides) plus proximal phalanges of middle and ring fingers (medial sides) and extensor digitorum communis	Abducts fingers away from middle finger at metacarpophalangeal joints	Ulnar
Hypothenar Muscles					
Abductor digiti minimi	*minimus* = very small	Pisiform and tendon of flexor carpi ulnaris	Proximal phalanx of digit V	Abducts digit V	Ulnar
Flexor digiti minimi		Flexor retinaculum and hook of hamate	Proximal phalanx of digit V	Flexes digit V	Ulnar
Opponens digiti minimi		Flexor retinaculum and hook of hamate	Fifth metacarpal	Opposes digit V	Ulnar

The *hypothenar eminence* is the elongated, fleshy bulge at the base of the little finger. It also is formed by three muscles: the **abductor digiti minimi,** the **flexor digiti minimi,** and the **opponens digiti minimi.**

Muscles of the intermediate group are positioned between the metacarpal bones in the region of the palm. This group includes the **adductor pollicis,** the **lumbricales** *(lum'bri-kalz),* and the **palmar** and **dorsal interossei.**

Muscles That Move the Thigh

The muscles that move the thigh originate from the pelvic girdle and insert on various places on the femur. These muscles stabilize a highly movable hip joint and provide support for the body during bipedal stance and locomotion. Found in this region are the most massive muscles of the body as well as some extremely small muscles. The muscles that move the thigh are divided into anterior, posterior, and medial groups.

Anterior Muscles The anterior muscles that move the thigh are the iliacus and psoas major (figs. 9.37, 9.38).

The triangular **iliacus** muscle arises from the iliac fossa and inserts on the lesser trochanter of the femur.

The long, thick **psoas major** muscle originates on the bodies and transverse processes of the lumbar vertebrae and inserts, along with the iliacus, on the lesser trochanter (fig. 9.38). The psoas major muscle and the iliacus work synergistically in flexing and rotating the thigh and flexing the vertebral column. They are frequently referred to as a single muscle, the **iliopsoas** *(il''e-o-so'as).*

Posterior and Lateral (Buttock) Muscles The posterior muscles that move the thigh include the gluteus maximus, gluteus medius, gluteus minimus, and tensor fasciae latae.

The large **gluteus maximus** muscle forms much of the prominence of the buttock (fig. 9.39). It is a powerful extensor muscle of the thigh and is very important for bipedal stance and locomotion. The gluteus maximus originates on the ilium, sacrum, coccyx, and aponeurosis of the lumbar region. It inserts on the gluteal tuberosity of the femur and the *iliotibial tract,* a tendinous band extending down the thigh (see fig. 9.41).

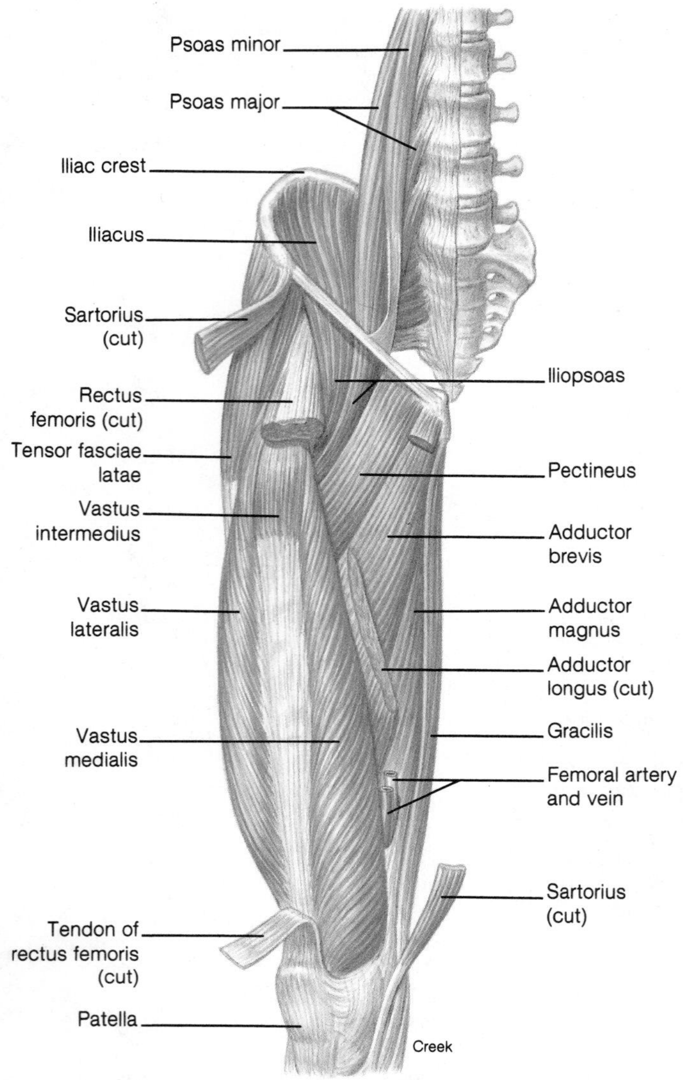

Figure 9.37 Muscles of the right anterior pelvic and thigh regions.

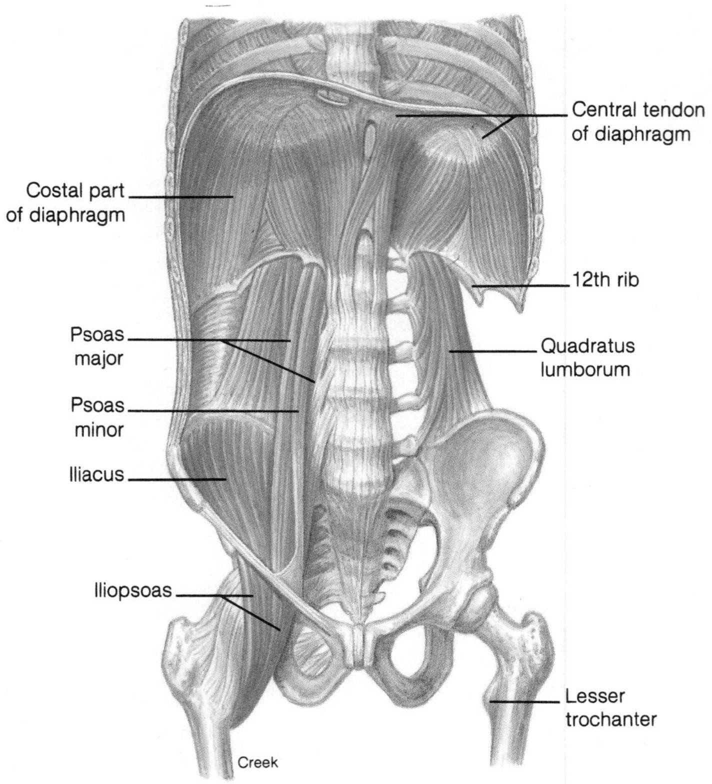

Figure 9.38 Anterior pelvic muscles that move the hip.

The **gluteus medius** muscle is located immediately deep to the gluteus maximus (fig. 9.39). It originates on the lateral surface of the ilium and inserts on the greater trochanter of the femur. The gluteus medius abducts and medially rotates the thigh. The mass of this muscle is of clinical significance as a site for intramuscular injections.

The **gluteus minimus** is the smallest and deepest of the gluteal muscles (fig. 9.39). It also arises from the lateral surface of the ilium and inserts on the lateral surface of the greater trochanter, where it acts synergistically with the gluteus medius to abduct and medially rotate the thigh.

The quadrangular **tensor fasciae latae** muscle is positioned superficially on the lateral surface of the hip (see fig. 9.41). It originates on the iliac crest and inserts on a broad lateral fascia of the thigh called the *fascia lata.* The fascia lata is continuous with the iliotibial tract. The tensor fasciae latae muscle is the principal abductor of the thigh.

A deep group of six lateral rotators of the thigh is positioned directly over the posterior aspect of the hip joint. These muscles are not discussed but are identified in figure 9.39 from superior to inferior as the **piriformis, superior gemellus, obturator internus, inferior gemellus, obturator externus,** and **quadratus femoris.**

The anterior and posterior group of muscles that move the thigh are summarized in table 9.14.

Medial, or Adductor, Muscles The medial muscles that move the thigh include the gracilis, pectineus, adductor longus, adductor brevis, and adductor magnus (figs. 9.40, 9.41, 9.42, 9.43).

The long, thin **gracilis** *(gras'il-is)* muscle is the most superficial of the medial thigh muscles. It is a two-joint muscle and can adduct the thigh or flex the leg.

The **pectineus** is the uppermost of the medial muscles that move the thigh. It is a flat, quadrangular muscle that flexes and adducts the thigh.

The **adductor longus** is immediately lateral to the gracilis on the upper third of the thigh, and is the most anterior of the adductor muscles. The **adductor brevis** is a triangular muscle that is behind and largely concealed by the adductor longus and pectineus muscles. The **adductor magnus** is a large, thick muscle, somewhat triangular in shape. It is located deep to the other two adductor muscles. The adductor longus, adductor brevis, and the adductor magnus are synergistic in adducting, flexing, and laterally rotating the thigh.

The muscles that adduct the thigh are summarized in table 9.15.

Muscles of the Thigh That Move the Leg

The muscles that move the leg originate on the pelvic girdle or thigh and are surrounded and compartmentalized by tough

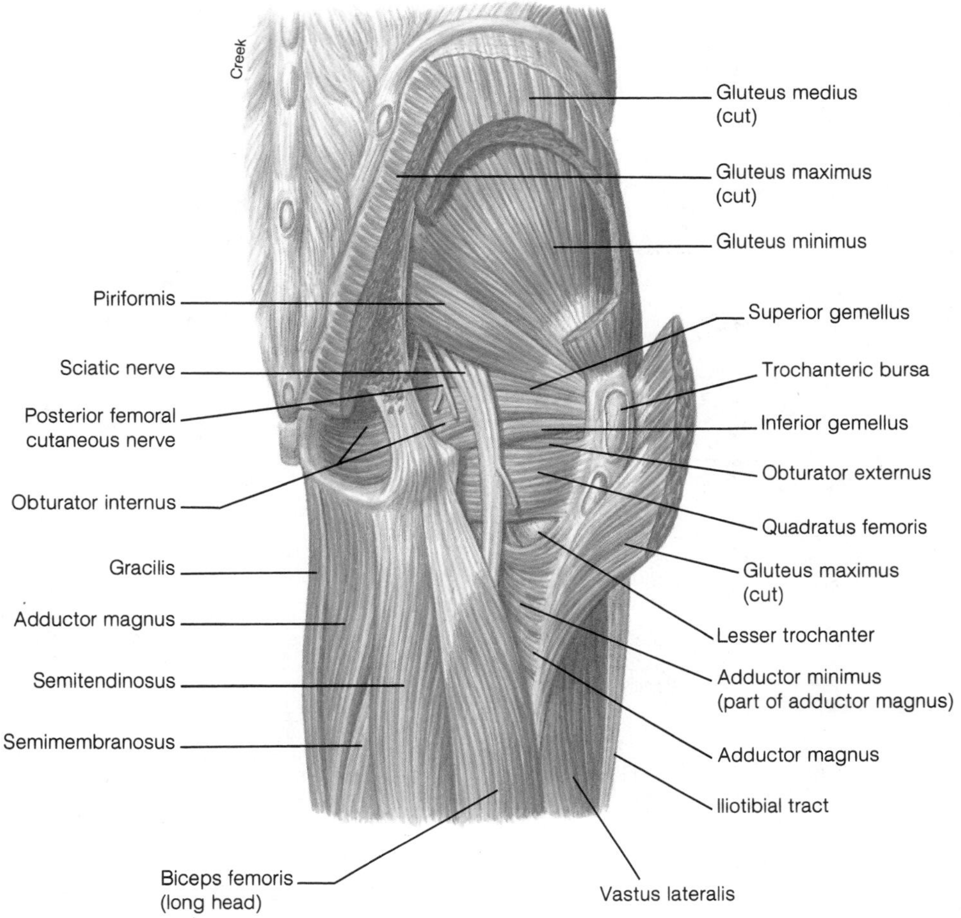

Figure 9.39 Deep gluteal muscles.

Table 9.14 Anterior and posterior muscles that move the thigh

Muscle	Derivation	Origin	Insertion	Action	Innervation
Iliacus	*iliacus* = pertaining to ilium (hip)	Iliac fossa	Lesser trochanter of femur	Flexes and rotates thigh laterally; flexes vertebral column	Femoral
Psoas major	*psoa* = muscle of loin	Transverse processes of all lumbar vertebrae	Lesser trochanter with iliacus	Flexes and rotates thigh laterally; flexes vertebral column	L2, L3
Gluteus maximus	*gloutos* = rump; *maximus* = greatest	Iliac crest, sacrum, coccyx, and aponeurosis of the lumbar region	Gluteal tuberosity and iliotibial tract	Extends and rotates thigh laterally	Inferior gluteal
Gluteus medius	*medius* = middle	Lateral surface of ilium	Greater trochanter	Abducts and rotates thigh medially	Superior gluteal
Gluteus minimus	*minimus* = very small	Lateral surface of lower half of ilium	Greater trochanter	Abducts and rotates thigh medially	Superior gluteal
Tensor fasciae latae	*tensor* = tightener; *fascia* = band; *latus* = wide	Anterior border of ilium and iliac crest	Iliotibial tract	Abducts thigh	Superior gluteal

fascial sheets, which are a continuation of the fascia lata and iliotibial tract. They are divided according to function and position into two groups: anterior extensors and posterior flexors.

Anterior, or Extensor, Muscles The anterior muscles that move the leg are the sartorius and quadriceps femoris (fig. 9.40).

The long, straplike **sartorius** muscle obliquely crosses the anterior aspect of the thigh. It can act on both the hip and knee joints to flex and rotate the thigh laterally and also to assist in flexing the leg and rotating it medially. The sartorius muscle is the longest muscle of the body and is frequently called the tailor's muscle because it enables one to assume a cross-legged sitting position.

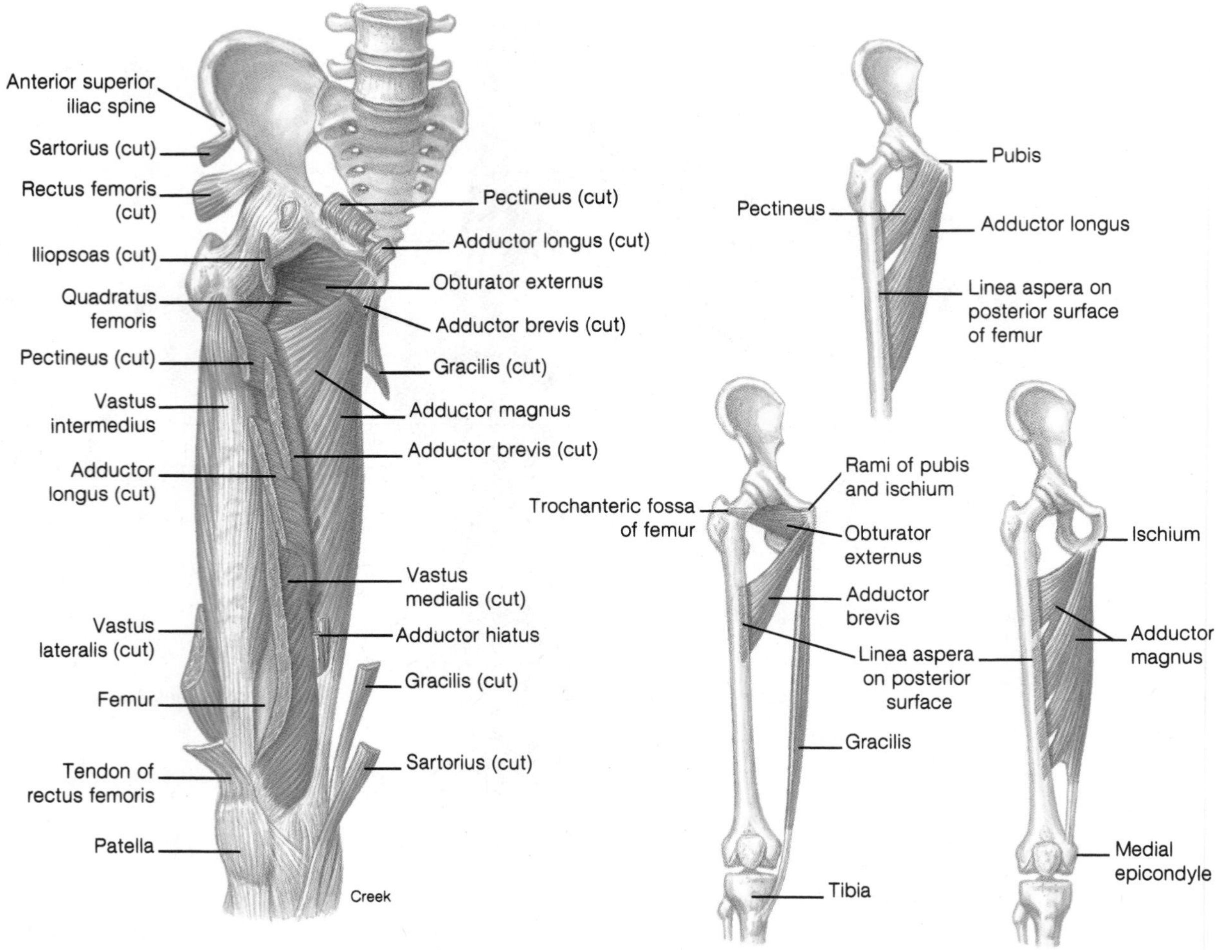

Figure 9.40 Adductor muscles of the right thigh.

Table 9.15 Medial muscles that move the thigh

Muscle	Derivation	Origin	Insertion	Action	Innervation
Gracilis	*gracilis* = slender	Inferior edge of symphysis pubis	Proximal medial surface of tibia	Adducts thigh; flexes and rotates leg at knee	Obturator
Pectineus	*pecten* = comb	Pectineal line of pubis	Distal to lesser trochanter of femur	Adducts and flexes thigh	Femoral
Adductor longus	*ad* = toward; *longus* = long	Pubis—below pubic crest	Linea aspera of femur	Adducts, flexes, and laterally rotates thigh	Obturator
Adductor brevis	*brevis* = short	Inferior ramus of pubis	Linea aspera of femur	Adducts, flexes, and laterally rotates thigh	Obturator
Adductor magnus	*magnus* = large	Inferior ramus of ischium and pubis	Linea aspera and medial epicondyle of femur	Adducts, flexes, and laterally rotates thigh	Obturator

The **quadriceps femoris** is actually a composite of four distinct muscles that have separate origins but a common insertion on the patella via the *patellar tendon.* The patellar tendon is continuous over the patella and becomes the *patellar ligament* as it attaches to the head of the tibial tuberosity (fig. 9.41). These muscles function synergistically to extend the leg, as in kicking a football. The four muscles of the quadriceps femoris muscle are the rectus femoris, vastus lateralis, vastus medialis, and vastus intermedius.

The **rectus femoris** occupies a superficial position and is the only one of the four quadriceps that functions in both the hip and knee joints. The laterally positioned **vastus lateralis** is the largest muscle of the quadriceps femoris. It is a common intramuscular injection site in infants who have small, underdeveloped buttock and shoulder muscles. The **vastus medialis** occupies a medial position along the thigh. The **vastus intermedius** lies deep to the rectus femoris.

The anterior thigh muscles that move the leg are summarized in table 9.16.

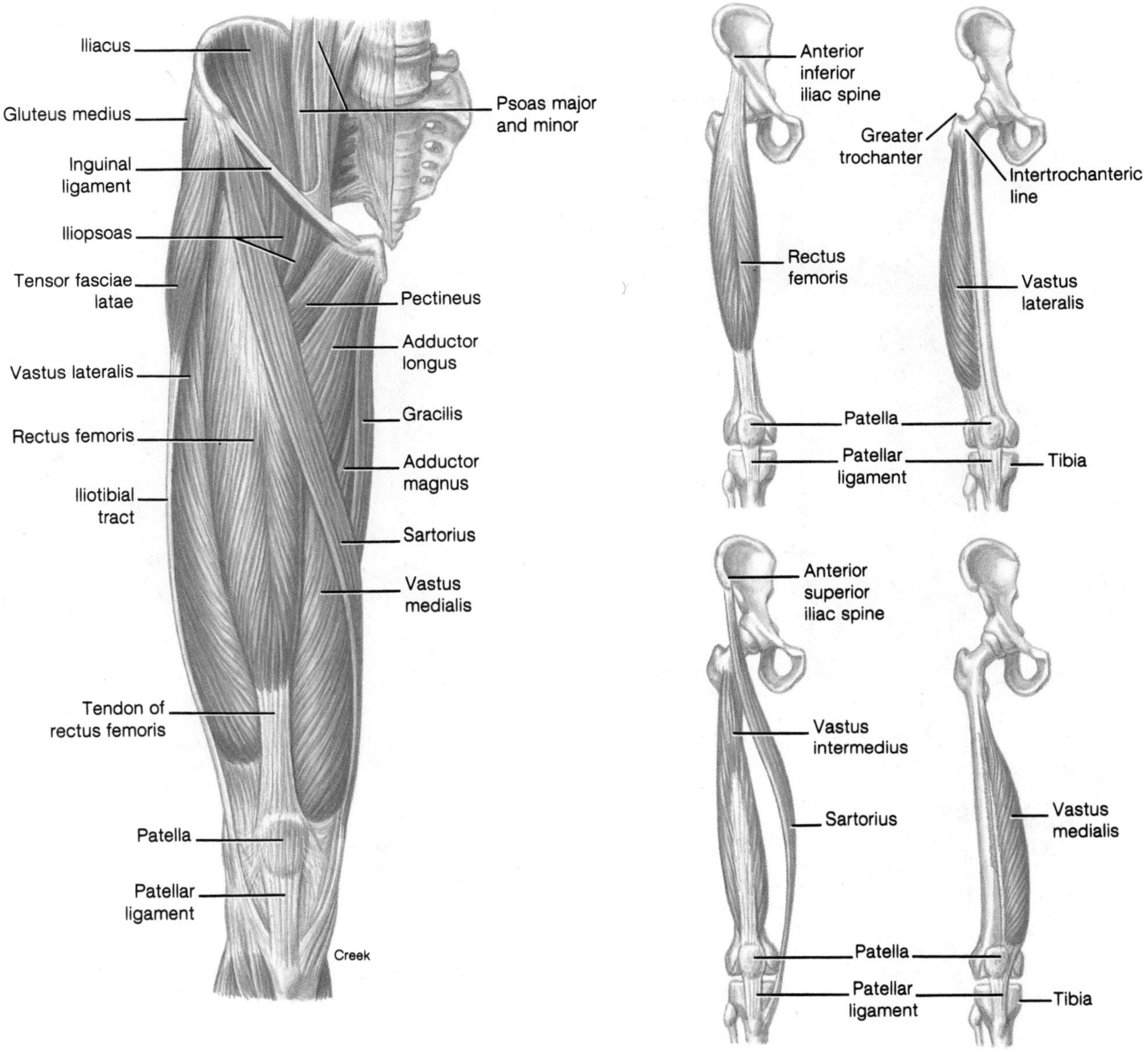

Figure 9.41 Muscles of the right anterior thigh.

Table 9.16 Anterior thigh muscles that move the leg

Muscle	Derivation	Origin	Insertion	Action	Innervation
Sartorius	*sartor* = tailor	Spine of ilium	Medial surface of tibia	Flexes leg and thigh, abducts thigh, rotates thigh laterally, and rotates leg medially	Femoral
Quadriceps femoris	*quad* = four; *cephalic* = head; *femoris* = femur		Patella by common tendon, which continues as patellar ligament to tibial tuberosity	Extends leg at knee	Femoral
Rectus femoris	*rectus* = straight fibers	Spine of ilium and lip of acetabulum			
Vastus lateralis	*vastus* = large; *lateralis* = lateral	Greater trochanter and linea aspera of femur			
Vastus medialis	*medialis* = medial	Medial surface and linea aspera of femur			
Vastus intermedius	*intermedius* = middle	Anterior and lateral surfaces of femur			

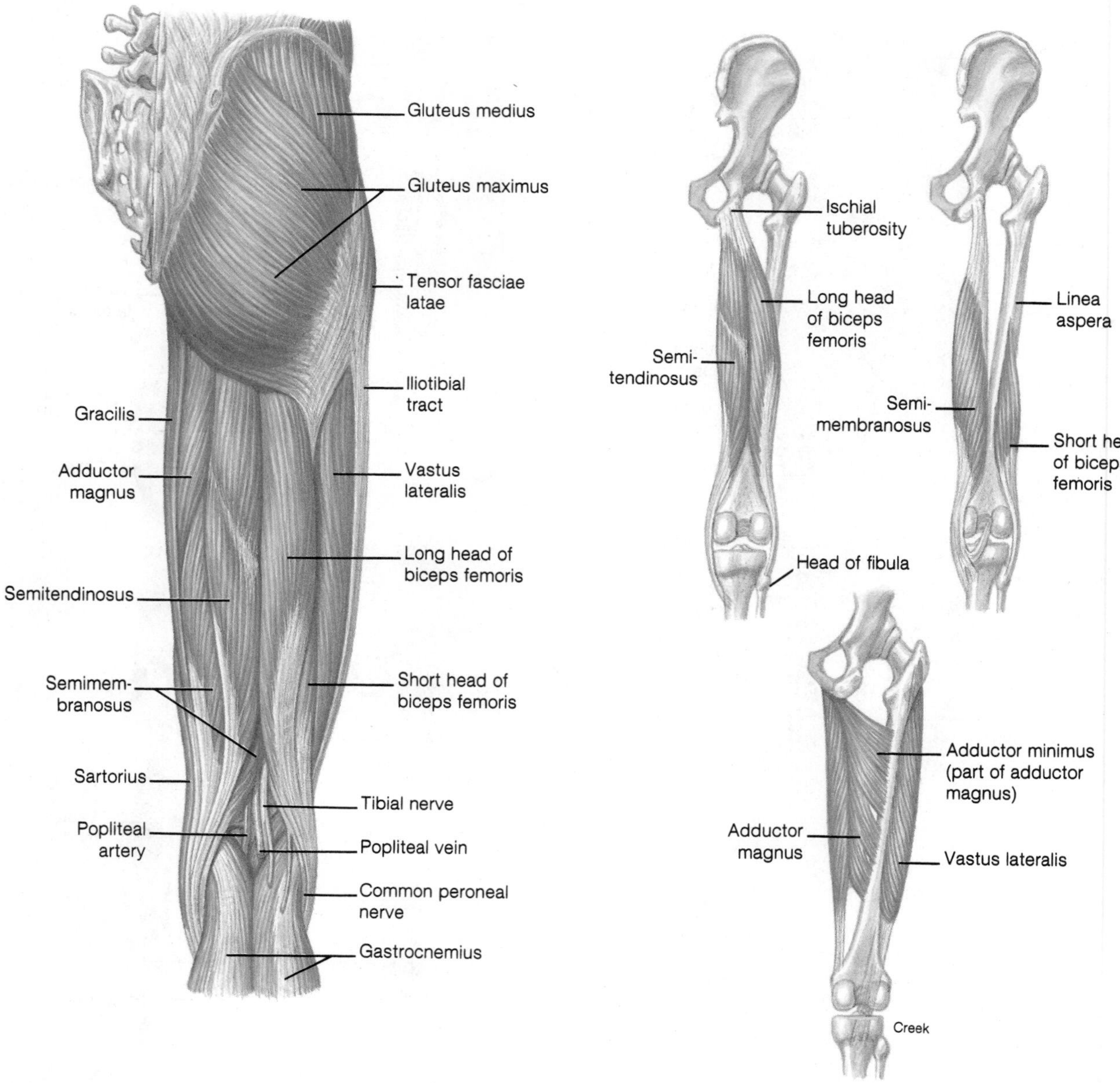

Figure 9.42 Muscles of the right posterior thigh.

Posterior, or Flexor, Muscles There are three posterior thigh muscles antagonistic to the quadriceps femoris muscles in flexing the knee. These muscles are known as the **hamstrings** (fig. 9.42). An interesting derivation for the name of this muscle group is from the old butchers' practice of using the tendons of these muscles to hang the hams of pork for smoke curing.

The **biceps femoris** muscle occupies the posterior lateral aspect of the thigh. It has a superficial long head and a deep short head, and causes movement at both the hip and knee joints. The superficial **semitendinosus** muscle is fusiform and is located on the posterior medial aspect of the thigh. It also works over two joints. The **semimembranosus** is a flat muscle that lies deep to the semitendinosus on the posterior medial aspect of the thigh.

The posterior thigh muscles that move the leg are summarized in table 9.17. The relationship of the muscles of the thigh is illustrated in figure 9.44.

Hamstring injuries are a common occurrence in some sports. The injury usually occurs when sudden lateral or medial stress to the knee joint tears the muscles or tendons. Because of its structure and the stress applied to it in competition, the knee joint is highly susceptible to injury. To reduce the number of knee injuries, the rules in contact sports should be altered or additional support and protection should be provided for this vulnerable joint.

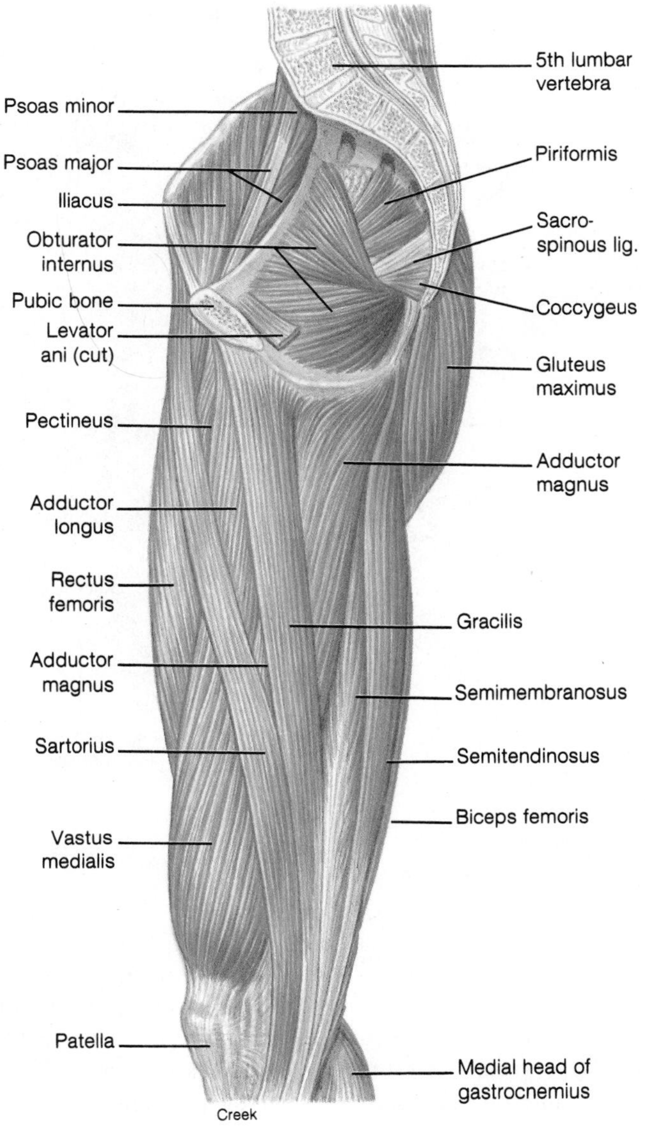

Figure 9.43 Muscles of the right medial thigh.

Table 9.17 Posterior thigh muscles that move the leg

Muscle	Derivation	Origin	Insertion	Action	Innervation
Biceps femoris	*bi* = two; *cephal* = head; *femoris* = femur	Long head—ischial tuberosity; short head—linea aspera of femur	Head of fibula and lateral condyle of tibia	Flexes leg at knee; extends and laterally rotates thigh at hip	Tibial
Semitendinosus	*semi* = half; *tendo* = tendon	Ischial tuberosity	Proximal portion of medial surface of body of tibia	Flexes leg at knee; extends and medially rotates thigh at hip	Tibial
Semimembranosus	*membran* = membrane	Ischial tuberosity	Medial condyle of tibia	Flexes leg at knee; extends and medially rotates thigh at hip	Tibial

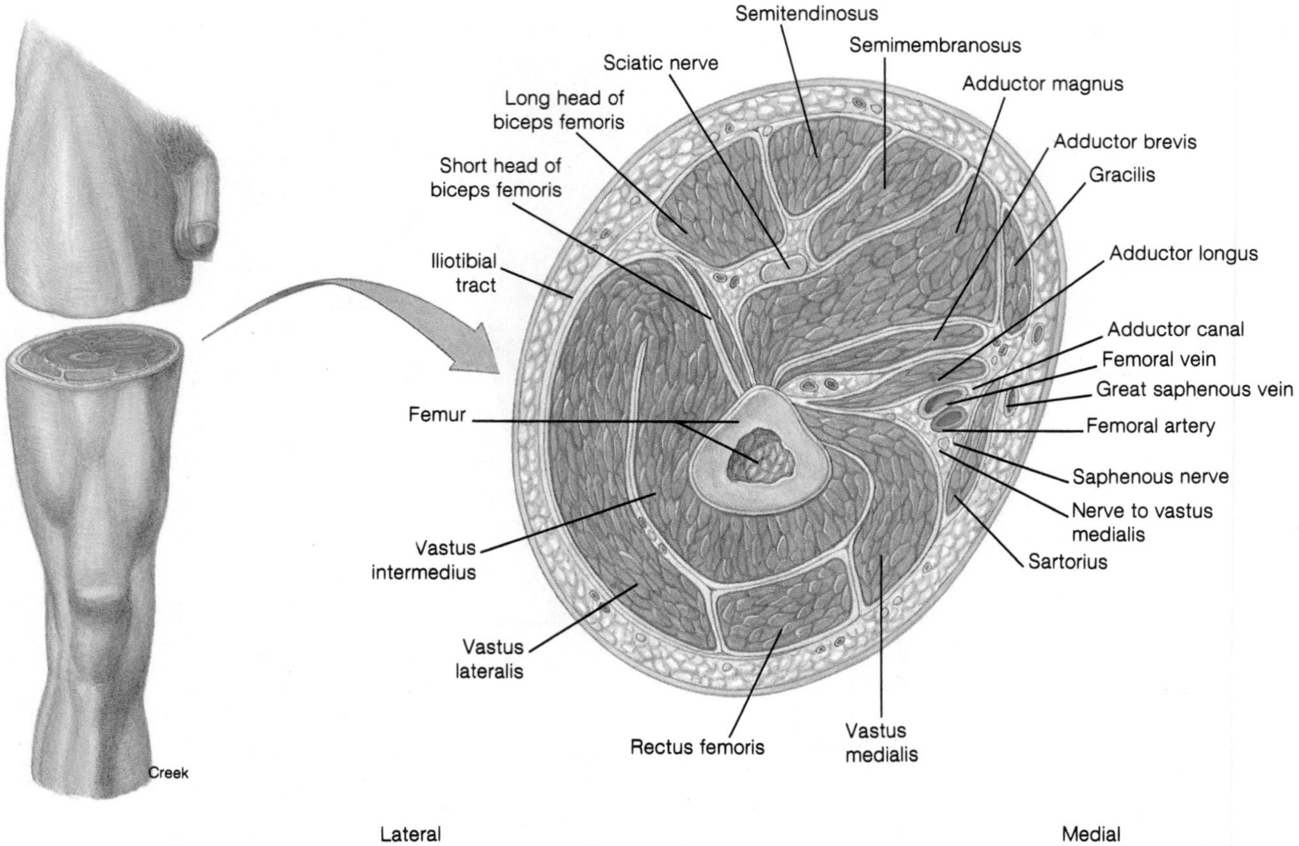

Figure 9.44 A transverse section of the right thigh as seen from above. (Note the position of the vessels and nerves.)

Muscles of the Leg That Move the Ankle, Foot, and Toes

The muscles of the leg, the **crural muscles,** are responsible for the movements of the foot. There are three groups of crural muscles: anterior, lateral, and posterior. The anteromedial aspect of the leg along the shaft of the tibia lacks muscle attachment.

Anterior Crural Muscles The anterior crural muscles include the tibialis anterior, extensor digitorum longus, extensor hallucis longus, and peroneus tertius (figs. 9.45, 9.46, 9.47).

The large, superficial **tibialis anterior** muscle can be easily palpated on the anterior lateral portion of the tibia (fig. 9.45). It parallels the prominent anterior crest of the tibia. The **extensor digitorum longus** muscle lies lateral to the tibialis anterior on the anterolateral surface of the leg. The **extensor hallucis longus** muscle is positioned deep between the tibialis anterior and the extensor digitorum longus. The **peroneus tertius** is a small muscle that is continuous with the distal portion of the extensor digitorum longus.

Lateral Crural Muscles The lateral crural muscles are the peroneus longus and peroneus brevis (figs. 9.45, 9.46). The long, flat **peroneus longus** muscle is a superficial lateral muscle that overlies the fibula. The **peroneus brevis** lies deep to the peroneus longus and is positioned closer to the foot. These two muscles are synergistic in flexion and eversion of the foot (see table 9.18).

Posterior Crural Muscles The seven posterior crural muscles can be grouped into a superficial and a deep group. The superficial group is composed of the gastrocnemius, soleus, and plantaris muscles (fig. 9.48). The four deep posterior crural muscles are the popliteus, flexor hallucis longus, flexor digitorum longus, and tibialis posterior (fig. 9.49).

The **gastrocnemius** *(gas''trok-ne'me-us)* is a large superficial muscle that forms the major portion of the calf of the leg. It consists of two distinct heads, which arise from the posterior surfaces of the medial and lateral condyles of the femur. This muscle and the deeper soleus muscle insert onto the calcaneus bone via the common *tendo calcaneus (tendon of Achilles).* The gastrocnemius acts over two joints to cause flexion of the knee and plantar flexion of the foot.

The **soleus** lies deep to the gastrocnemius. These two muscles are frequently referred to as a single muscle, the **triceps surae.** The soleus has a common insertion with the gastrocnemius but acts on only one joint, the ankle, in plantar flexing the foot.

The small **plantaris** muscle arises just above the origin of the lateral head of the gastrocnemius on the lateral condyle of the femur. It has a very long, slender tendon of insertion onto the calcaneus bone. The tendon of this muscle is frequently mistaken for a nerve by those dissecting it for the first time. The plantaris is a weak muscle with limited ability to flex the leg and plantar flex the foot.

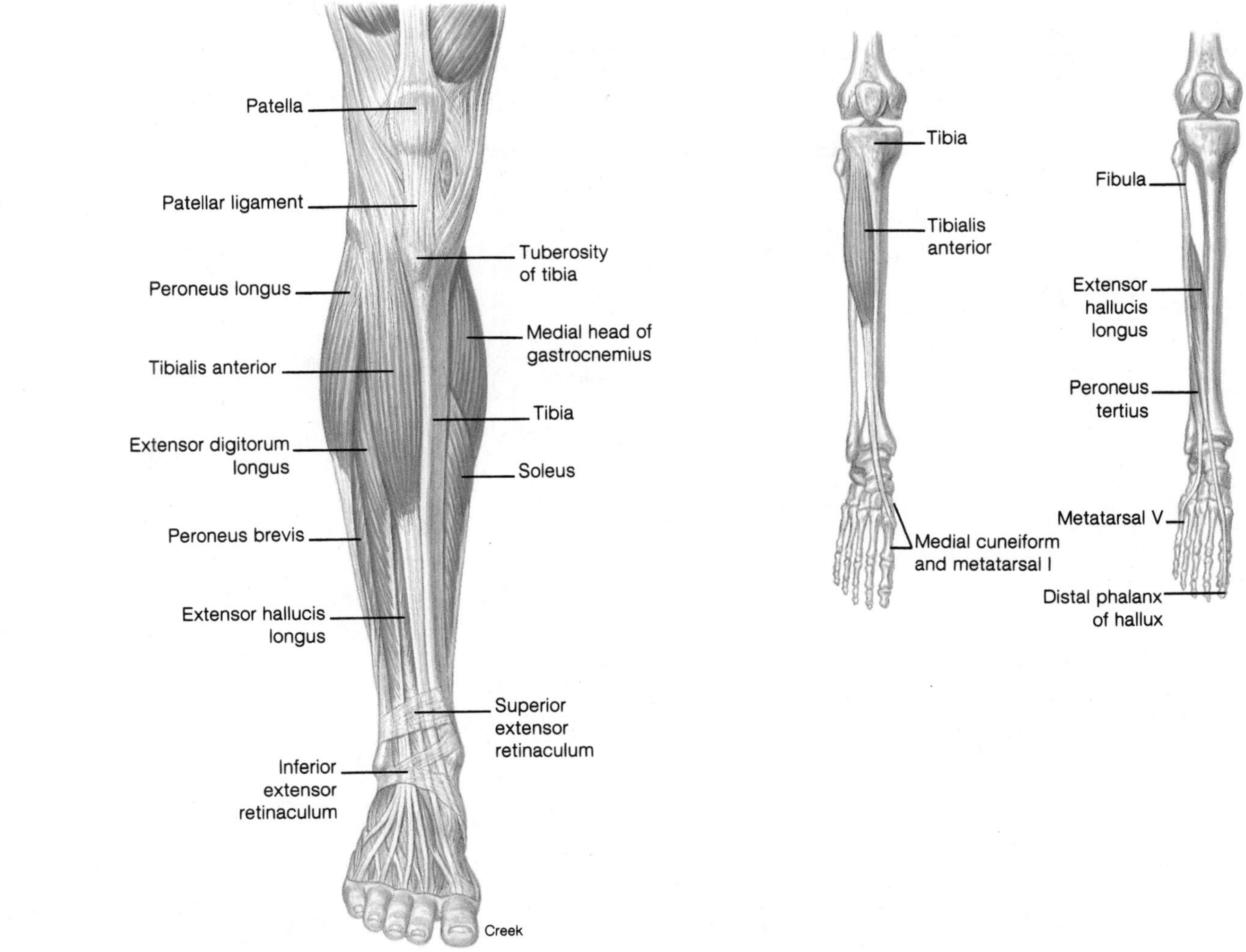

Figure 9.45 Anterior crural muscles.

The thin, triangular **popliteus** muscle is situated deep to the heads of the gastrocnemius, where it forms part of the floor of the popliteal fossa. The *popliteal fossa* is the depression on the back side of the knee joint (fig. 9.50). The **popliteus** muscle is a medial rotator of the tibia on the femur. The bipennate **flexor hallucis longus** muscle lies deep to the soleus on the posterolateral side of the leg. It flexes the big toe (hallux) and assists in plantar flexing and inverting the foot.

The **flexor digitorum longus** muscle also lies deep to the soleus and parallels the flexor hallucis longus on the medial side of the leg. Its distal tendon passes behind the medial malleolus and continues along the plantar surface of the foot where it branches into four tendinous slips that attach to the bases of the terminal phalanges of the second, third, fourth, and fifth toes (fig. 9.51). The flexor digitorum longus works over several joints, flexing four of the digits and assisting in plantar flexing and inverting the foot.

The **tibialis posterior** muscle is located deep to the soleus, directly on the interosseous membrane on the posterior surface of the leg. The distal tendon of the tibialis posterior passes behind the medial malleolus and inserts on the plantar surfaces of the navicular, the cuneiforms, the cuboid, and the second, third, and fourth metatarsals (fig. 9.51). The tibialis posterior plantar flexes and inverts the foot and gives support to the arches of the foot.

The crural muscles are summarized in table 9.18.

Muscles of the Foot

With the exception of one additional intrinsic muscle, the **extensor digitorum brevis,** the muscles of the foot are similar in number and nomenclature to those of the hand. The functions of the muscles of the foot are different, however, because the foot is adapted to provide support while bearing body weight rather than for grasping.

The muscles of the foot are topographically arranged into four layers (fig. 9.51) that are difficult to dissociate even in dissection. The muscles function either to move the toes or to support the arches of the foot through their contraction. Because of their complexity, the muscles of the foot will be presented only in illustrations (see figs. 9.51, 9.52).

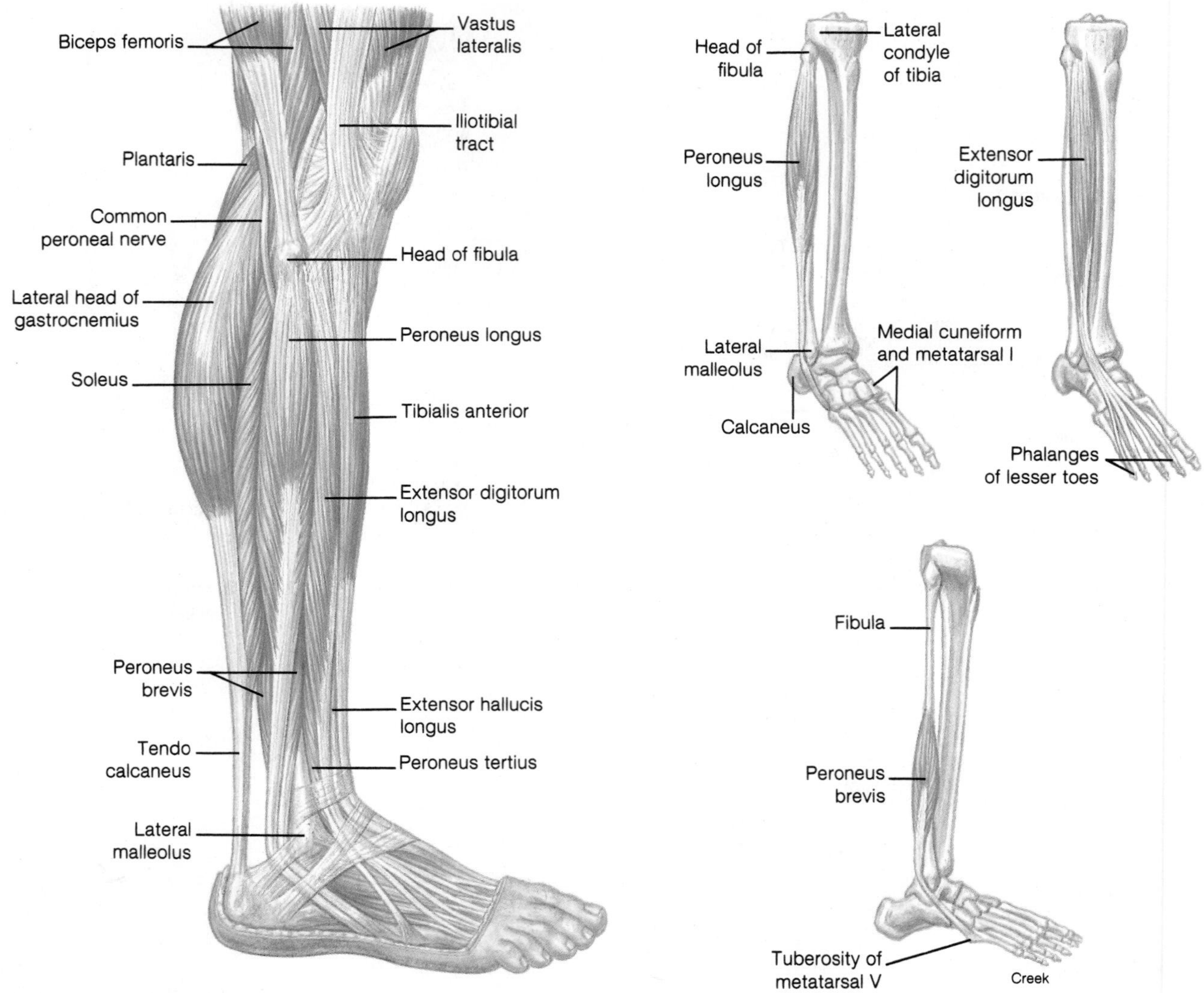

Figure 9.46 Lateral crural muscles.

Table 9.18 Muscles of the lower leg that move the ankle, foot, and toes

Muscle	Derivation	Origin	Insertion	Action	Innervation
Tibialis anterior	*tibialis* = tibia; *ante* = in front	Lateral condyle and body of tibia	First metatarsal and first cuneiform	Dorsiflexion of ankle and inversion of foot	Deep peroneal
Extensor digitorum longus	*extensor* = increase joint angle; *digit* = finger or toe; *longus* = long	Lateral condyle of tibia and anterior surface of fibula	Extensor expansions of digits II-V	Extends digits II-V and dorsiflexes foot	Deep peroneal
Extensor hallucis longus	*hallux* = great toe	Anterior surface of fibula and interosseous membrane	Distal phalanx of digit I	Extends big toe and assists dorsiflexion of foot	Deep peroneal
Peroneus tertius	*perone* = fibula; *tertius* = third	Anterior surface of fibula and interosseous membrane	Dorsal surface of fifth metatarsal	Dorsiflexes and everts foot	Deep peroneal
Peroneus longus	*longus* = long	Lateral condyle of tibia and head and shaft of fibula	First cuneiform and metatarsal I	Plantar flexes and everts foot	Superficial peroneal

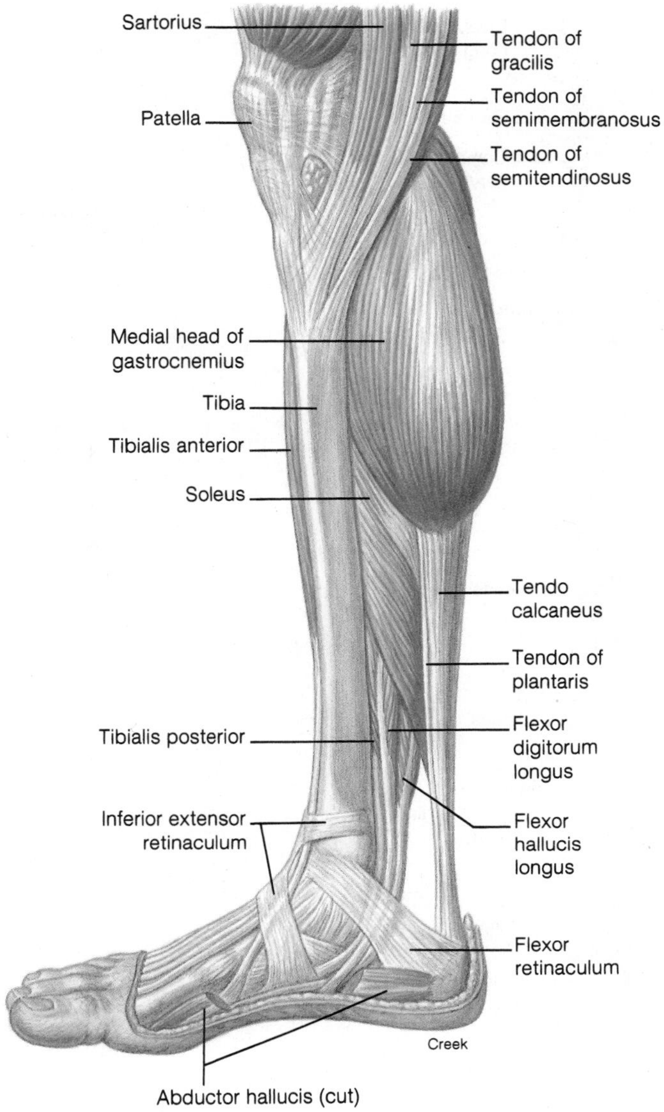

Figure 9.47 Medial view of crural muscles.

Muscle	Derivation	Origin	Insertion	Action	Innervation
Peroneus brevis	*brevis* = short	Lower aspect of fibula	Metatarsal V	Plantar flexes and everts foot	Superficial peroneal
Gastrocnemius	*gaster* = belly; *kneme* = leg	Lateral and medial condyle of femur	Posterior surface of calcaneus	Plantar flexes foot; flexes knee	Tibial
Soleus	*soleus* = sole of foot	Posterior aspect of fibula and tibia	Calcaneus	Plantar flexes foot	Tibial
Plantaris	*planta* = plantar surface of foot	Lateral supracondylar ridge of femur	Calcaneus	Plantar flexes foot	Tibial
Popliteus	*poples* = ham of the knee	Lateral condyle of femur	Upper posterior aspect of tibia	Flexes and medially rotates leg	Tibial
Flexor hallucis longus	*flexor* = decrease angle	Posterior aspect of fibula	Distal phalanx of big toe	Flexes distal phalanx of big toe	Tibial
Flexor digitorum longus		Posterior surface of tibia	Distal phalanges of digits II-V	Flexes distal phalanges of digits II-V	Tibial
Tibialis posterior	*posterior* = back	Tibia and fibula and interosseous membrane	Navicular, cuneiforms, cuboid, and metatarsals II-IV	Plantar flexes and inverts foot; supports arches	Tibial

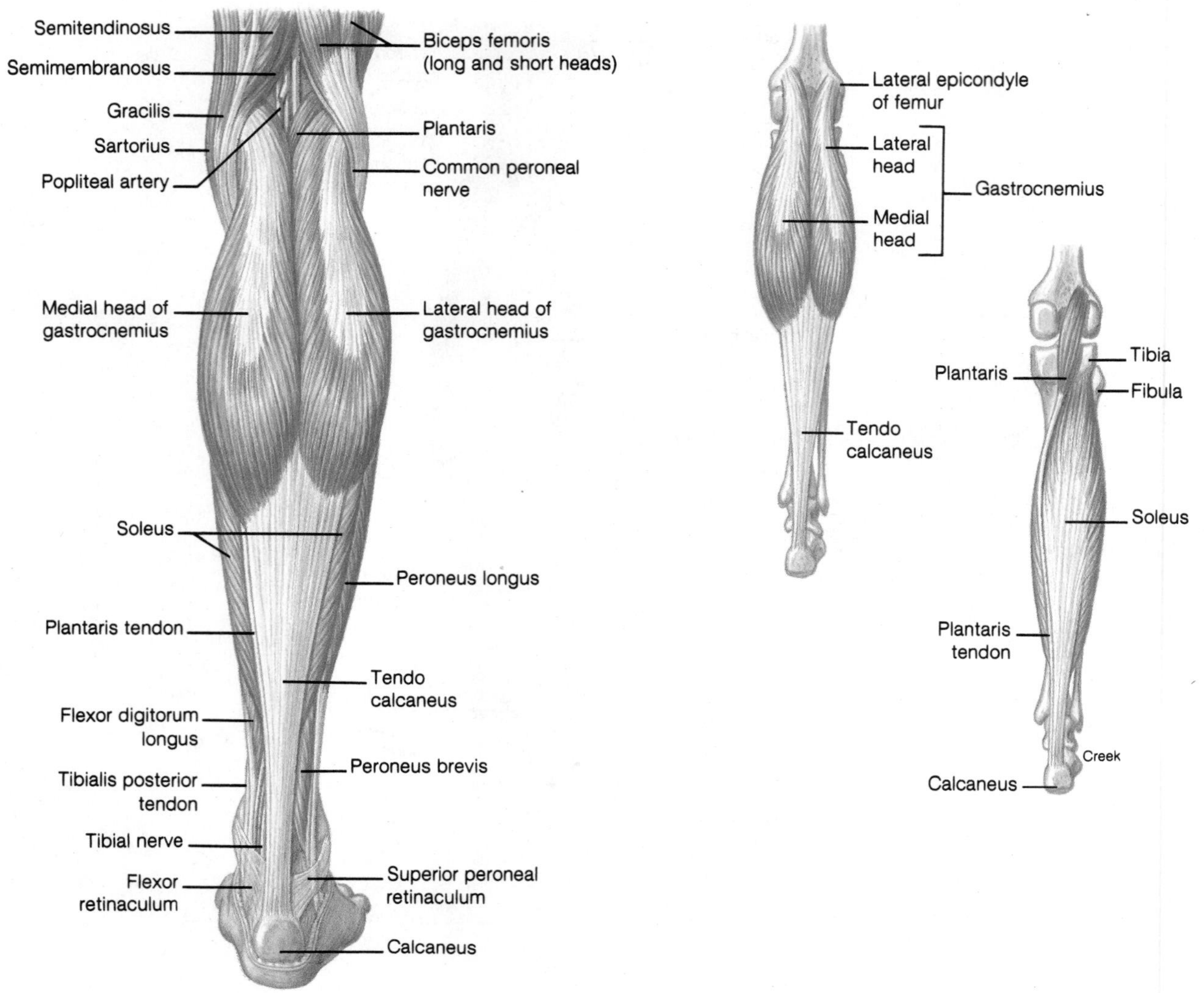

Figure 9.48 Posterior crural muscles and popliteal region.

---

1. List all the muscles that either originate from or insert on the scapula.
2. On the basis of function, list the muscles of the upper extremity into the categories of flexors, extensors, abductors, adductors, and rotators. (Each muscle may fit into two or more categories.)
3. Which muscles of the lower extremity span two joints and therefore have two different actions?

Clinical Considerations

Compared to the other systems of the body, the muscular system has few clinical afflictions. If properly conditioned, the muscles of the body adequately serve a person for a lifetime. Muscles are capable of doing incredible amounts of work and through exercise can become even stronger.

There are, however, several clinical considerations of muscles, including the diagnosis of muscle conditions, physiological conditions in muscles, and other diseases of muscles.

Diagnosis of Muscle Condition

The clinical symptoms of muscle diseases are usually weakness, loss of muscle mass (atrophy), and pain. The diagnosis and extent of the disease depends on the severity of the symptoms.

Several standard procedures are used to assess skeletal muscle dysfunction and disease. The most obvious is a clinical examination of the patient. Following this, it may be necessary to test the muscle function using **electromyography (EMG)** to measure conduction rates and motor unit activity within a muscle. Laboratory tests may be conducted, such as serum enzyme assays that reflect muscle destruction or muscle biopsy examinations under the microscope. Muscle biopsy is perhaps the definitive source of diagnostic information. Progressive atrophy, polymyositis, and metabolic diseases of muscles can be determined through a biopsy.

Functional Conditions in Muscles

Muscles depend on systematic, periodic contraction to maintain optimal health. Obviously overuse or disease will cause a change in muscle tissue. The immediate effect of overexertion of muscle

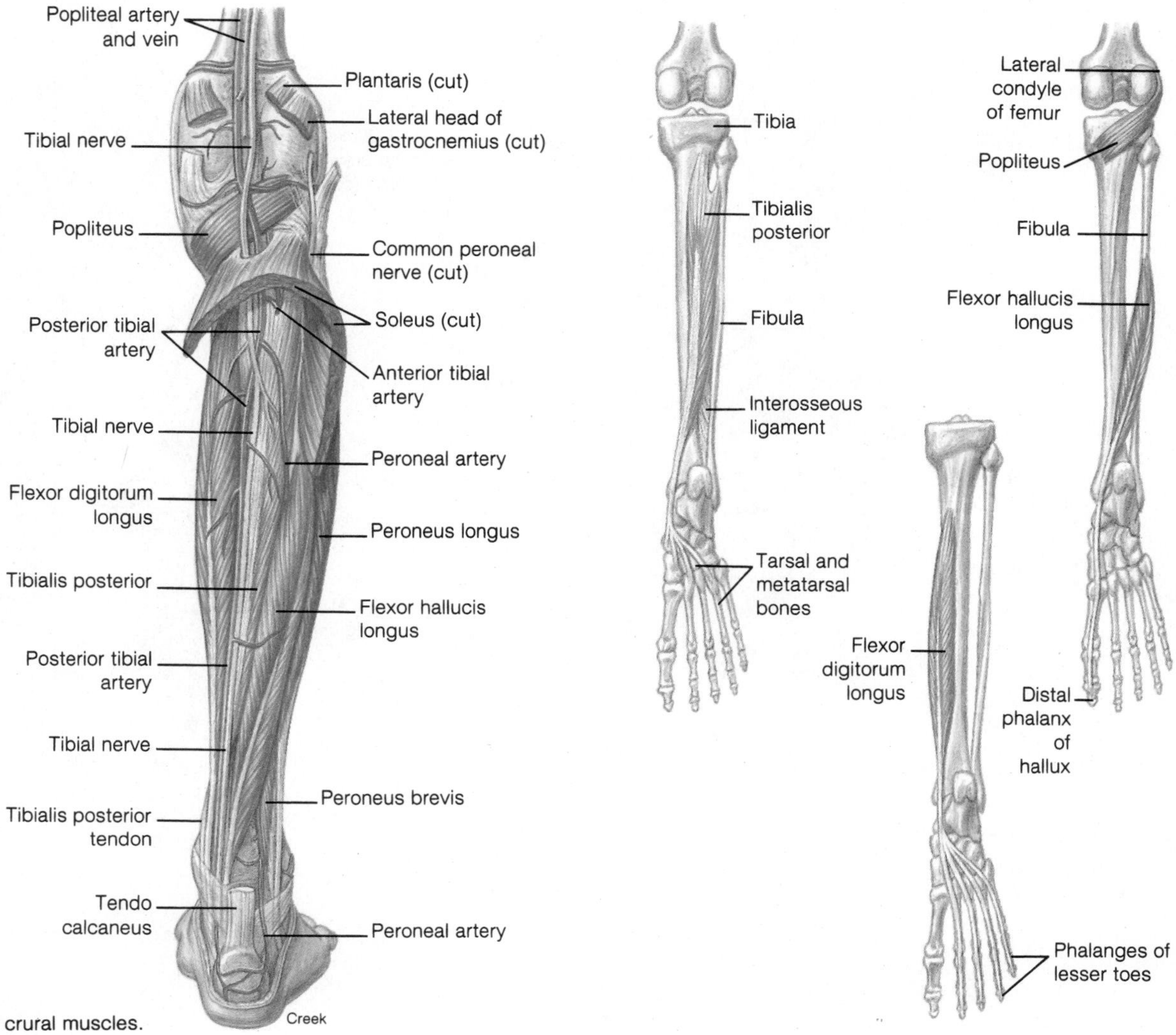
Popliteal artery and vein
Plantaris (cut)
Lateral head of gastrocnemius (cut)
Tibial nerve
Popliteus
Common peroneal nerve (cut)
Posterior tibial artery
Soleus (cut)
Anterior tibial artery
Tibial nerve
Peroneal artery
Flexor digitorum longus
Peroneus longus
Tibialis posterior
Flexor hallucis longus
Posterior tibial artery
Tibial nerve
Peroneus brevis
Tibialis posterior tendon
Tendo calcaneus
Peroneal artery
Creek
Tibia
Tibialis posterior
Fibula
Interosseous ligament
Tarsal and metatarsal bones
Lateral condyle of femur
Popliteus
Fibula
Flexor hallucis longus
Distal phalanx of hallux
Flexor digitorum longus
Phalanges of lesser toes

Figure 9.49 Deep posterior crural muscles.

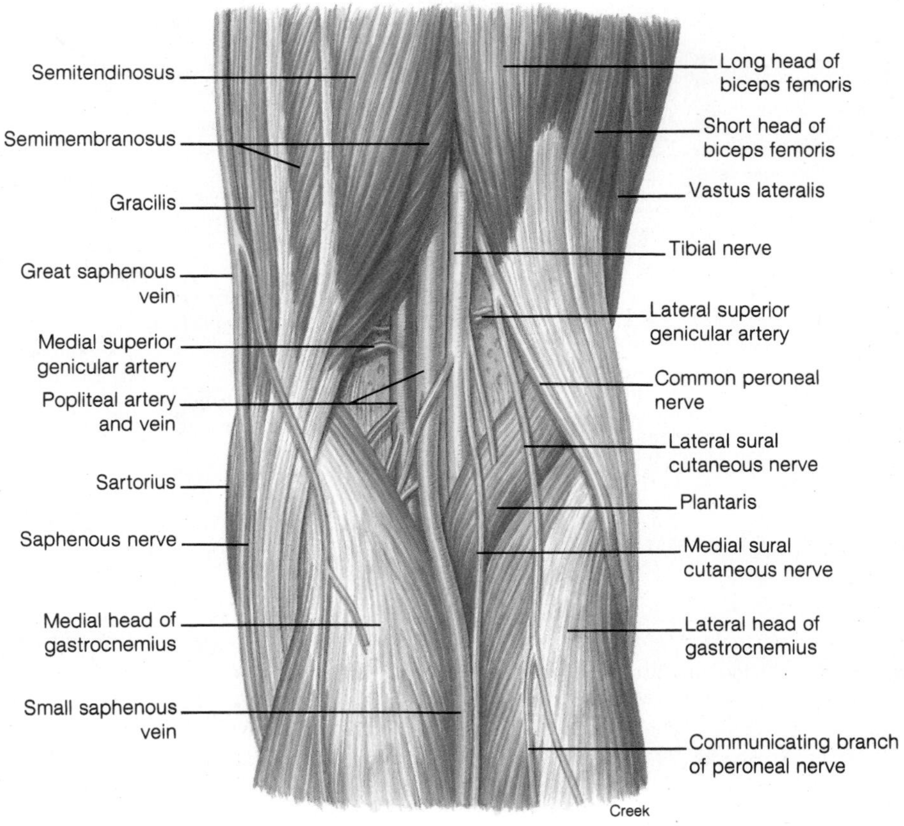
Semitendinosus
Long head of biceps femoris
Semimembranosus
Short head of biceps femoris
Gracilis
Vastus lateralis
Tibial nerve
Great saphenous vein
Lateral superior genicular artery
Medial superior genicular artery
Common peroneal nerve
Popliteal artery and vein
Lateral sural cutaneous nerve
Sartorius
Plantaris
Saphenous nerve
Medial sural cutaneous nerve
Medial head of gastrocnemius
Lateral head of gastrocnemius
Small saphenous vein
Communicating branch of peroneal nerve
Creek

Figure 9.50 Muscles that surround the popliteal fossa.

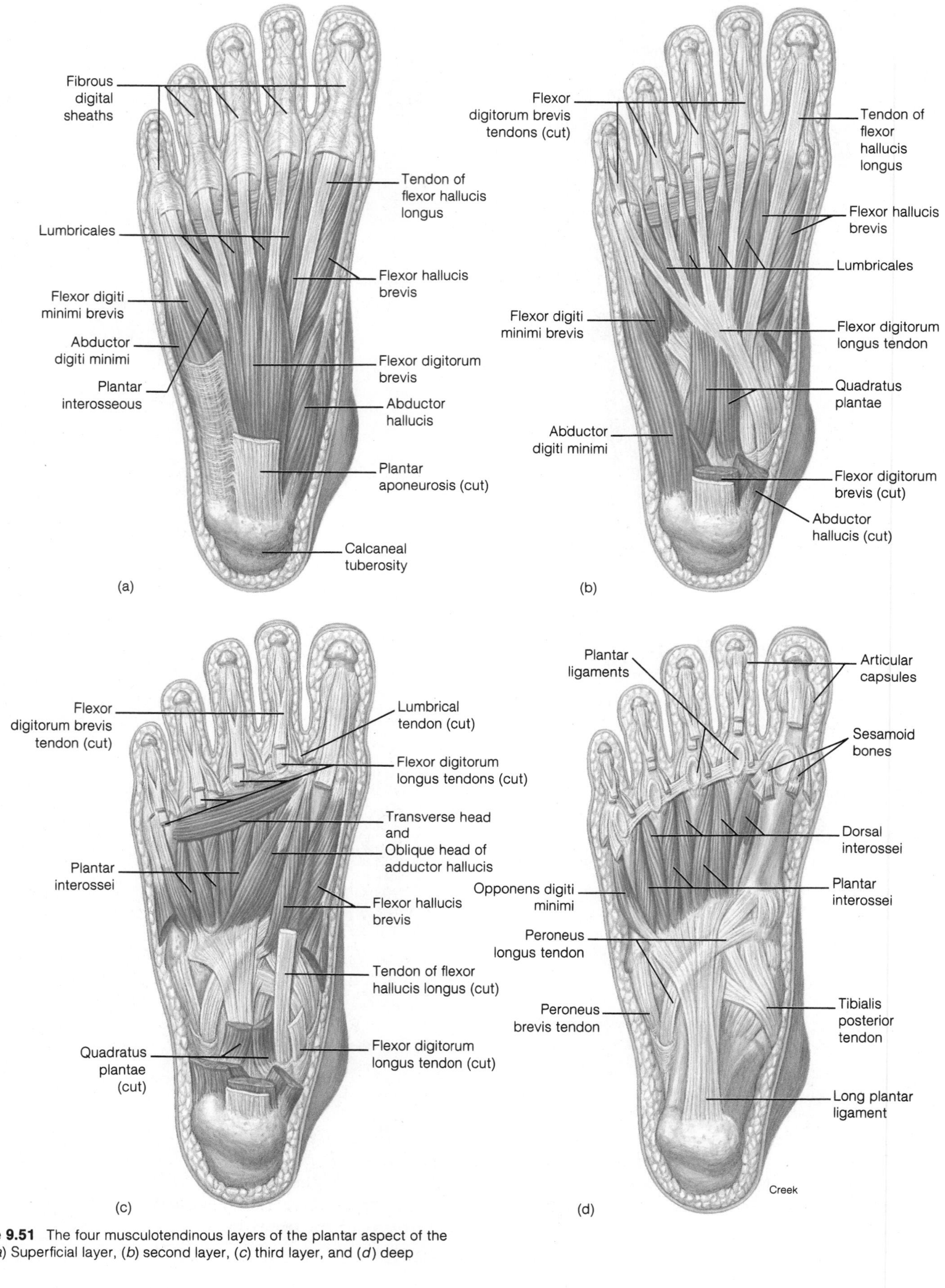

Figure 9.51 The four musculotendinous layers of the plantar aspect of the foot. (*a*) Superficial layer, (*b*) second layer, (*c*) third layer, and (*d*) deep layer.

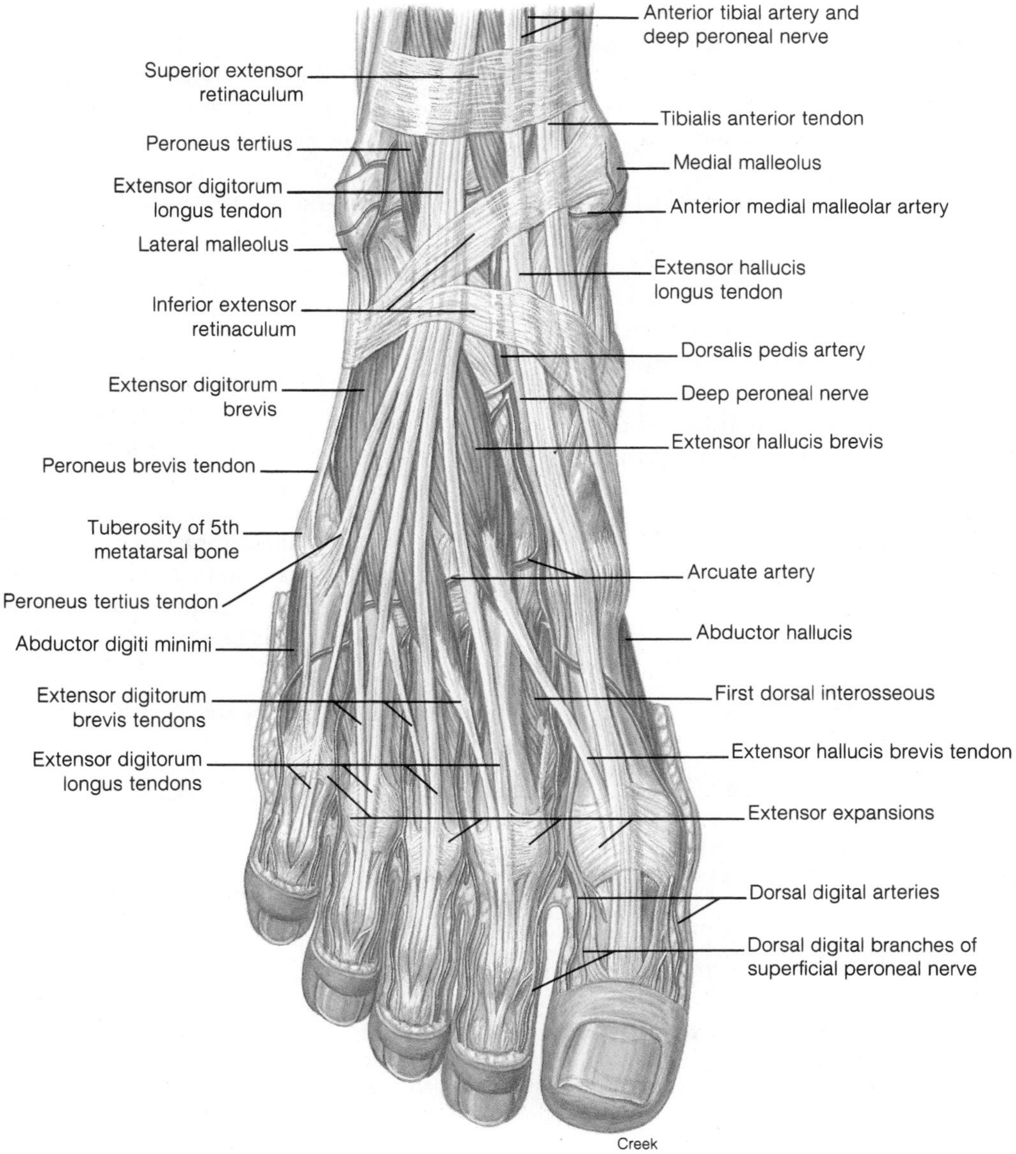

Figure 9.52 An anterior view of the dorsum of the foot.

tissue is the accumulation of lactic acid, resulting in fatigue and soreness. Excessive contraction of a muscle can also damage the fibers or associated connective tissue, resulting in a **strained muscle.**

When skeletal muscles are not contracted, either because the motor nerve supply is blocked or because the limb is immobilized (as when a broken bone is in a cast), the muscle fibers **atrophy,** or diminish in size. Atrophy is reversible if exercise is resumed, as after a healed fracture, but tissue death is inevitable if the nerves cannot be stimulated.

The fibers in healthy muscle tissue increase in size, or **hypertrophy,** if a muscle is systematically exercised. This increase in muscle size and strength is due not to an increase in the number of muscle cells but to the production of more myofibrils accompanied by a strengthening of the associated connective tissue.

A **cramp** within a muscle is an involuntary, painful, and prolonged contraction. Cramps can occur while muscles are in use or at rest. The precise cause of cramps is unknown, but evidence indicates that cramps may be related to conditions within the muscle (e.g., calcium or oxygen deficiencies) or to stimulation of the motor neurons.

A condition called **rigor mortis** (rigidity of death) affects skeletal muscle tissue several hours after death as depletion of ATP within the fibers causes a state of muscle contracture and stiffness of the joint. In a few days, however, as the muscle proteins decompose, the rigidity of the corpse disappears.

Diseases of Muscles

Fibromyositis is an inflammation of both skeletal muscular tissue and the associated connective tissue. Its causes are not fully understood. Fibromyositis frequently occurs in the extensor muscles of the lumbar region of the vertebral column where extensive aponeuroses exist. Fibromyositis of this region is called **lumbago,** or **rheumatism.**

lumbago: L. *lumbus,* loin

Muscular dystrophy is a genetic disease characterized by a gradual atrophy and weakening of muscle tissue. There are several kinds of muscular dystrophy, none of whose etiology is completely understood. The most frequent type affects children and is sex-linked to the male child. As muscular dystrophy progresses, the muscle fibers atrophy and are replaced by adipose tissue. Most children who have muscular dystrophy die before the age of twenty.

The disease **myasthenia gravis** is characterized by extreme muscle weakness and low endurance. There is a defective transmission of impulses at the neuromuscular junction. Myasthenia gravis is believed to be an autoimmune disease, and it typically affects women between the ages of twenty and forty.

Poliomyelitis (polio) is actually a viral disease of the nervous system that causes a paralysis of muscles. The viruses are usually localized in the ventral (anterior) horn of the spinal cord where they affect the motor nerve impulses to skeletal muscles.

myasthenia: Gk. *myos*, muscle; *astheneia*, weakness
poliomyelitis: Gk. *polios*, gray; *myolos*, marrow

Neoplasms of muscle are rare, but when they do occur they are usually malignant. **Rhabdomyosarcoma** *(rab''do-mi''o-sar-ko'mah)* is a malignant tumor of skeletal muscle. It can arise in any skeletal muscle, and most often afflicts young children and elderly persons.

CLINICAL CASE STUDY ANSWER

When cancer of the head or neck involves lymph nodes in the neck, a number of structures on the affected side are removed surgically. This procedure usually includes the sternocleidomastoid muscle. This muscle, which originates on the sternum and clavicle, inserts on the mastoid process. When it contracts, the mastoid, which is located posteriorly at the base of the skull, is pulled forward, causing the chin to rotate away from the contracting muscle. This explains why the patient who had his left sternocleidomastoid removed would have a disability in turning his head to the right. ■

rhabdomyosarcoma: Gk. *rhabdos*, rod; *myos*, muscle; *oma*, a growth

Important Clinical Terminology

convulsion An involuntary, spasmodic contraction of skeletal muscle.

fibrillation *(fi-bri-la'shun)* A series of rapid, uncoordinated, and spontaneous contractions of individual motor units within a muscle.

hernia The rupture or protrusion of a portion of the underlying viscera through muscle tissue. The most common hernias occur in the normally weak places of the abdominal wall. There are four basic types of hernias:
1. **femoral**—viscera passing through the femoral ring
2. **hiatal**—the superior portion of the stomach protruding through the esophageal opening of the diaphragm
3. **inguinal**—viscera protruding through the inguinal canal
4. **umbilical**—a hernia occurring at the navel

intramuscular injection A hypodermic injection at certain heavily muscled areas to avoid damaging nerves. The most common site is the buttock.

myalgia Pain within a muscle resulting from any muscular disorder or disease.

myokymia *(mi''o-kim'e-ah)* Continual quivering of a muscle.

myoma A tumor of muscle tissue.

myopathy Any muscular disease.

myotomy *(mi-ot'o-me)* Surgical cutting of muscle tissue.

myotonia A prolonged muscular spasm.

paralysis The loss of nervous control of a muscle.

shin splints Tenderness and pain on the anterior surface of the leg, caused by straining the tibialis anterior or extensor digitorium longus muscles.

torticollis (wryneck) A persistent contraction of a sternocleidomastoid muscle, drawing the head to one side and distorting the face. Torticollis may be acquired or congenital.

Chapter Summary

I. Organization and General Functions
   A. The contraction of skeletal muscle fibers results in body motion, heat production, and the maintenance of posture and body support.
   B. The four basic properties characteristic of all muscle tissue are irritability, contractility, extensibility, and elasticity.
   C. Axial muscles include facial muscles, neck muscles, and trunk muscles; appendicular muscles include those that act on the girdles and those that move the segments of the appendages.

II. Development of Skeletal Muscles
   A. Skeletal muscles begin development at four weeks after conception from blocks of mesoderm, called myotomes, in the trunk area and from loosely organized mesenchyme in the head area.
   B. The union of myoblasts forms a syncytial myotube, which becomes a skeletal muscle fiber as contractile filaments develop.
   C. The development of muscle fibers occurs within myotomes of the trunk region and loosely organized mesenchyme areas of the head region.

III. Structure of Skeletal Muscles
   A. The origin of a muscle is the more stationary attachment. The insertion is the more movable attachment.
   B. Individual muscle fibers are covered by endomysium. Fasciculi are covered by perimysium. The entire muscle is covered by epimysium.
   C. Synergistic muscles contract together. Antagonistic muscles perform in opposition to a particular group of muscles.
   D. Muscles may be classified according to fiber arrangement as parallel, convergent, pennate, or circular.
   E. Each muscle has motor and sensory innervation.

IV. Skeletal Muscle Fibers and Mode of Contraction
   A. Each skeletal muscle fiber is a multinucleated, striated cell, composed of myofibrils and enclosed by a sarcolemma.

1. Myofibrils are characterized by alternating A and I bands; each I band is bisected by a Z line, and the portion between two Z lines is the sarcomere.
2. Extending through the sarcoplasm is a network of membranous channels called the sarcoplasmic reticulum and a system of transverse tubules (T-tubules).

B. The activity of the cross-bridges causes sliding of the filaments.
1. The actin on each side of the A bands is pulled toward the center.
2. The H bands thus appear to be shorter as more actin overlaps the myosin.
3. The I bands also appear to be shorter as adjacent A bands are pulled closer together.
4. The A bands stay the same length because the filaments (both thick and thin) do not shorten during muscle contraction.

C. When a muscle exerts tension without shortening, the contraction is termed isometric; when shortening does occur, the contraction is isotonic.

D. The neuromuscular junction is the area consisting of the motor end plate and the sarcolemma of a muscle fiber. In response to a nerve impulse, the synaptic vesicles of the terminal end of an axon secrete a neurotransmitter, which diffuses across the neuromuscular cleft of the neuromuscular junction and stimulates the muscle fiber.

E. The motor neuron and the muscle fibers innervated by the motor neuron are called a motor unit.
1. When a muscle is composed of many motor units (such as in the hand), there is a fine control of muscle contraction.
2. The large muscles of the leg have relatively few motor units, which are correspondingly large.
3. Graded contractions are produced by the asynchronous stimulation of different motor units.

V. Naming of Muscles

A. Skeletal muscles are named on the basis of shape, location, attachment, orientation of fibers, relative position, and function.

B. Most of the muscles are paired; that is, the right side of the body is an image of the left.

VI. Muscles of the Axial Skeleton
The muscles of the axial skeleton include those of facial expression, mastication, eye movement, tongue movement, neck movement, respiration, the abdominal wall, the pelvic outlet, and the vertebral column.

VII. Muscles of the Appendicular Skeleton
The muscles of the appendicular skeleton include those of the pectoral girdle, humerus, forearm, wrist, hand, and fingers, and those of the pelvic girdle, thigh, leg, ankle, foot, and toes.

Review Activities

Objective Questions

1. Skeletal muscle fibers are formed from the fusion of specialized mesodermal cells called
(a) mesenchyme. (c) myotubes.
(b) myotomes. (d) myoblasts.
2. Muscles capable of highly dexterous movements contain
(a) one motor unit per muscle fiber.
(b) many muscle fibers per motor unit.
(c) few muscle fibers per motor unit.
(d) many motor units per muscle fiber.
3. Which of the following is *not* used as a means of naming muscles?
(a) location
(b) action
(c) shape
(d) attachment
(e) strength of contraction
4. Neurotransmitters are stored in synaptic vesicles within the
(a) sarcolemma.
(b) motor units.
(c) myofibrils.
(d) terminal ends of axons.
5. Which of the following muscles have motor units with the lowest innervation ratio?
(a) brachial muscles
(b) muscles of the forearm
(c) thigh muscles
(d) abdominal muscles
6. An eyebrow is drawn toward the midline of the face through contraction of which muscle? The
(a) corrugator. (c) nasalis.
(b) risorius. (d) frontalis.
7. A flexor of the shoulder joint is the
(a) pectoralis major.
(b) supraspinatus.
(c) teres major.
(d) trapezius.
(e) latissimus dorsi.
8. Which of the following muscles does *not* have either an origin or insertion upon the humerus? The
(a) teres minor.
(b) biceps brachii.
(c) supraspinatus.
(d) brachialis.
(e) pectoralis major.
9. Of the four quadriceps femoris muscles, which may act on the hip and knee joints? The
(a) vastus medialis.
(b) vastus intermedius.
(c) rectus femoris.
(d) vastus lateralis.
10. Which of the following muscles plantar flexes and inverts the foot as it supports the arches? The
(a) flexor digitorum longus.
(b) tibialis posterior.
(c) flexor hallucis longus.
(d) gastrocnemius.

Essay Questions

1. Describe how muscle fibers are formed, and explain why the fibers are multinucleated.
2. List the general functions of muscles, and discuss functions that are passive, involuntary events.
3. Distinguish between fascia, aponeurosis, and retinaculum.
4. Describe the structural arrangement of the muscle fibers and fasciculi within muscle.
5. What are the advantages and disadvantages of pennate-fibered muscles?
6. Why is it necessary to have dual (sensory and motor) innervation to a muscle?
7. List the major components of a skeletal muscle fiber, and describe the function of each part.
8. What is a motor unit, and what is its role in muscle contraction?
9. Give three examples of synergistic muscle groups within the upper extremity and identify the antagonistic muscle group for each.
10. Which muscles of the neck either originate or insert on the hyoid bone?
11. List all the muscles that either originate or insert on the scapula.
12. Discuss the position of flexor and extensor muscles relative to the shoulder, elbow, and wrist joints.
13. Based on function, describe exercises that would strengthen the following muscles: (*a*) the pectoralis major; (*b*) the deltoid; (*c*) the triceps; (*d*) the pronator teres; (*e*) the rhomboideus major; (*f*) the trapezius; (*g*) the serratus anterior; and (*h*) the latissimus dorsi.

14. Give three examples of synergistic muscle groups within the lower extremity, and identify the antagonistic muscle group for each.
15. Describe the flexor and extensor compartments of muscles within the forearm.
16. List the muscles that border the popliteal fossa. Describe the structures that are located within this region.
17. Describe the actions of the muscles of inspiration. Which muscles participate in forced expiration?
18. Which muscles of the pelvic outlet support the floor of the pelvic cavity, and which are associated with the genitalia?
19. Firmly press your fingers on the front, sides, and back of your ankle as you move your foot. The tendons of which muscles can be palpated anteriorly, laterally, and posteriorly?
20. Compare muscular dystrophy and myasthenia gravis as to causes, symptoms, and the effect they have on muscle tissue.

CHAPTER 10

Surface Anatomy

Outline and Concepts

Introduction to Surface Anatomy
Surface anatomy is a branch of gross anatomy concerned with identifying body structures through observation and palpation. The study of surface anatomy has tremendous application to physical fitness and medical diagnosis and treatment.

Surface Anatomy of the Newborn
The surface anatomy of a neonatal infant differs from that of an adult because it represents an early stage of human development. Certain aspects of the surface anatomy of a neonate are of clinical importance in ascertaining the degree of physical development, general health, and possible congenital abnormalities.
- General Appearance
- Palpable Structures

Head
The surface anatomy of the head is of concern because the head contains vital sense organs and provides openings into the respiratory and digestive systems. Of further concern are aesthetic appearances and the many medical problems that involve this part of the body including trauma, diseases, and dysfunctions.
- Cranium
- Face

Neck
Major organs are located within the flexible neck, and several structures (the spinal cord, vessels, trachea, and esophagus) that are essential for body sustenance pass through the neck.
- Surface Features
- Triangles of the Neck

Torso
The locations of vital visceral organs in the cavities of the torso make the surface anatomy of this body region especially important.
- Back
- Thorax
- Abdomen

Pelvis and Perineum
The surface features of the pelvic region are important primarily to identify reproductive organs and clinical problems of these organs.

Shoulder and Upper Extremity
The surface anatomy of the shoulder and upper extremity is of clinical importance because of frequent trauma to these body regions. Vessels of the upper extremity are also used as pressure points and for intravenous injections or blood withdrawal.
- Shoulder
- Axilla
- Brachium
- Forearm
- Hand

Buttock and Lower Extremity
The massive bones and muscles of the buttock and lower extremity are important as weight bearers and locomotors. Many of the surface features of these regions are important in relation to locomotion or locomotor dysfunction.
- Buttock
- Thigh
- Leg
- Foot

Chapter Summary

Review Activities

Introduction to Surface Anatomy

Surface anatomy is a branch of gross anatomy concerned with identifying body structures through observation and palpation. The study of surface anatomy has tremendous application to physical fitness and medical diagnosis and treatment.

Objective 1. Describe the value of surface anatomy in learning internal anatomical structures.

Objective 2. Discuss how surface anatomy is important in diagnosing and treating various diseases or conditions of the body.

Objective 3. Describe subcutaneous differences in adult males and females that may have a bearing on their surface anatomy.

It is amazing how much anatomical information can be learned by studying the surface anatomy of one's own body. Surface anatomy is the study of the structure and markings of the surface of the body that can be identified through observation or palpation. Surface features can be readily identified through visual *observation,* and anatomical features beneath the skin can be located by *palpation* (feeling with firm pressure, or perceiving by the sense of touch). Knowledge of surface anatomy is clinically important in locating precise sites for *percussion* (tapping sharply to detect resonating vibrations) and *auscultation* (listening to sounds emitted from organs).

With the exception of certain cranial bones, the bones of the entire skeleton can be palpated. Once the position, shape, and processes of these bones are identified, these skeletal features can serve as landmarks for other anatomical structures. One can observe the location of many skeletal muscles and their tendinous attachments as they are contracted and caused to move. The location and range of movement of each of the joints of the body can be determined as the articulating bones are moved by muscle contractions. On some persons, one can locate the positions of superficial veins and trace their courses. Even the location and function of the valves within the veins can be demonstrated on the surface of the skin (fig. 10.1). The locations of some of the arteries can be seen as they pulsate beneath the skin. The locations of these arterial pressure points are an important clinical aspect of surface anatomy. Other structures can be identified from the surface, including certain nerves, lymph nodes, glands, and other internal organs.

Surface anatomy is an essential part of the study of anatomy. Knowing the location of muscles and muscle groups as observed in surface anatomy can be extremely important in physical fitness and body conditioning. In many medical and paramedical professions the surface anatomy of a patient is of immeasurable value in diagnosis and treatment. Knowing where to record a pulse, insert needles and tubes, listen to the functioning of internal organs, take X rays, and perform physical therapy requires a knowledge of surface body landmarks.

The effectiveness of observation and palpation in studying a person's surface anatomy is dependent on the amount of subcutaneous adipose tissue present (fig. 10.2). In examining an obese person, it may be extremely difficult to observe or palpate certain internal structures that are readily discernible in a thin person. The hypodermis and subcutaneous tissue in a woman is normally thicker than in a man (fig. 10.3). This tends to smooth the surface contour of a woman and obscure the internal features, such as muscles, veins, and bony prominences, that are apparent in men.

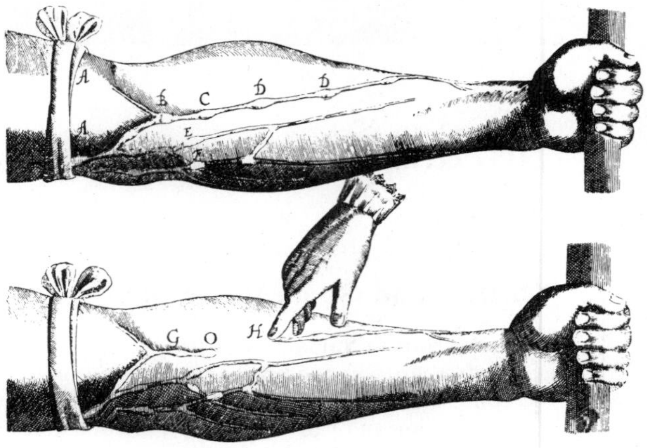

Figure 10.1 A demonstration of the presence and function of valves within the veins of the forearm, conducted by the great English anatomist William Harvey. (After William Harvey, *On the Motion of the Heart and Blood in Animals,* 1628.) Surface anatomy was extremely important to understanding the concept of a closed circulatory system (e.g., blood contained within vessels).

This chapter will be of great value in reviewing the bones of the skeleton, articulations, and muscles you have already studied. Refer back to this chapter to learn the anatomy of the remaining body systems. Reviewing this way will reinforce an external perspective of where the various organs and structures are located.

Make anatomy relevant and applicable by identifying your surface anatomy. Use yourself as a model from which to learn and review, and anatomy as a science will take on a new meaning. As you learn about a bone or a process on a bone, palpate that part of your body. Contract the muscles you are studying so that you better understand their locations, attachments, and actions. Do this and you will become better acquainted with your body, and anatomy will become more enjoyable and easier to learn. Your body is one crib sheet you can take with you to exams.

1. Define the terms *observation* and *palpation,* and explain the value of surface anatomy in learning the location of internal structures.
2. Discuss the medical importance of surface anatomy.
3. Describe the subcutaneous difference seen in the surface anatomy of men and women.

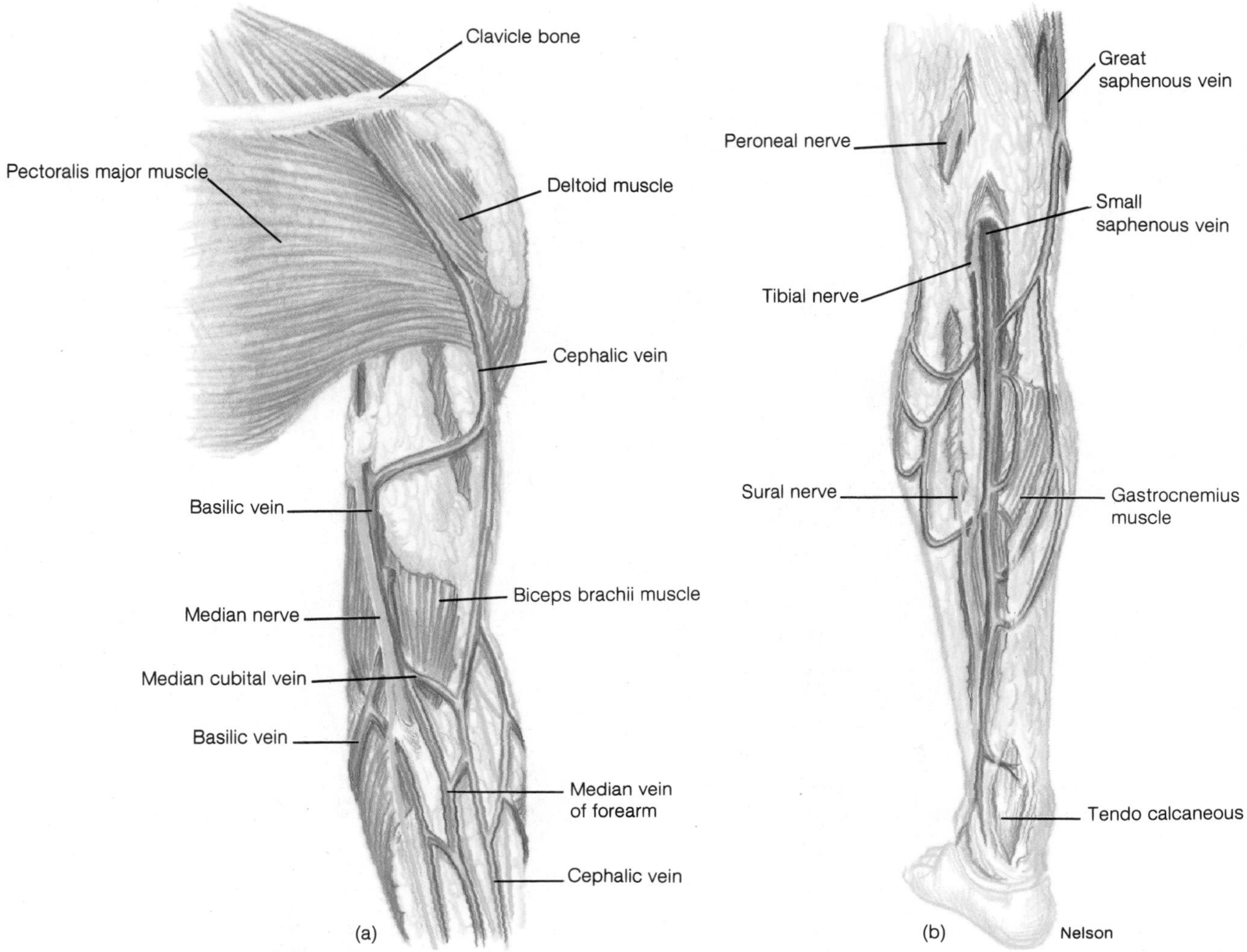

Figure 10.2 Subcutaneous adipose tissue. (*a*) An anterior view of the left brachial region and (*b*) a posterior view of the left leg.

Surface Anatomy of the Newborn

The surface anatomy of a neonatal infant differs from that of an adult because it represents an early stage of human development. Certain aspects of the surface anatomy of a neonate are of clinical importance in ascertaining the degree of physical development, general health, and possible congenital abnormalities.

Objective 4. Describe the surface anatomy of a normal, full-term neonate.

Objective 5. List some internal anatomical structures that can be palpated in a neonate.

The birth of a baby is a dramatic culmination of a nine-month gestation, during which the miraculous development of the fetus prepares it for extrauterine life. The normal, full-term neonate is physiologically prepared for life but is totally dependent on parental care. The physical assessment of the neonate is extremely important to insure its survival. Much of the assessment is performed through inspection and palpation of its surface anatomy. The surface anatomy of a neonate is obviously different than that of an adult because of the transitional stage of development from fetus to infant.

Although the surface anatomy of a neonate is discussed at this point in the book, prenatal development is discussed in chapter 22 and body growth with its accompanying physiological changes is discussed in chapter 23. A summary of the surface anatomy of the neonate is presented in table 10.1.

General Appearance

As a result of *in utero* position, the posture of the full-term neonate is one of flexion (fig. 10.4). The neonate born vertex (head first) keeps the neck and vertebral column flexed, with the chin resting on the upper chest. The hands are flexed (clenched), as are the arms, and held toward the chest. The legs are flexed at the knees, and the hips are flexed bringing the thighs toward the abdomen. The feet are dorsiflexed.

neonate: Gk. *neos,* new; L. *natalis,* birth

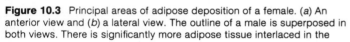

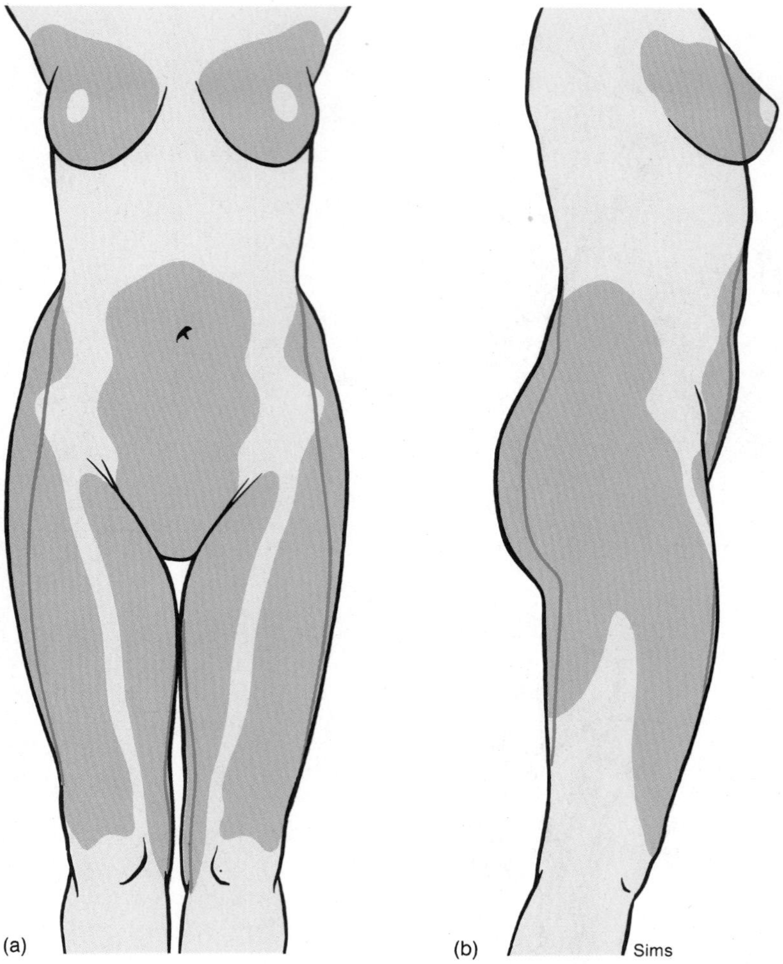

Figure 10.3 Principal areas of adipose deposition of a female. (*a*) An anterior view and (*b*) a lateral view. The outline of a male is superposed in both views. There is significantly more adipose tissue interlaced in the fascia covering the muscles, vessels, and nerves in a female than in a male. The hypodermis layer of the skin is also approximately 8 percent thicker in a female than in a male.

Table 10.1 Surface anatomy of the neonate

Body structure	Normal conditions	Common variations
General posture	Flexion of vertebral column and extremities	Extended legs and neck; abducted and rotated thighs (breech birth)
Skin	Red or pink skin with vernix caseosa and lanugo; edematous face, extremities, and genitalia	Neonatal jaundice; integumentary blisters; mongolian spots
Skull	Fontanels large, flat, and firm but soft to touch	Molded skull, bulging fontanels; cephalhematoma
Eyes	Lids edematous; color—gray, dark blue, or brown; absence of tears; corneal, pupillary, and blink reflexes	Conjunctivitis, subconjunctival hemorrhage
Ears	Auricle flexible, cartilage present; top of auricle positioned on horizontal line with outer canthus of eye	Auricle flat against head
Neck	Short and thick surrounded by neck folds	Torticollis
Chest	Equal anterioposterior and lateral dimensions; xiphoid process evident; breast enlargement	Funnel or pigeon chest; additional nipples (polythelia); secretions from breast (witch's milk)
Abdomen	Cylindric in shape; liver and kidneys palpable	Umbilical hernia
Genitalia	(♂ and ♀) Edematous and darkly pigmented; (♂) testes palpable in scrotum; periodic erection of penis	(♀) Blood-tinged discharge (pseudomenstruation); hymenal tag; (♂) testes palpable in inguinal canal; inability to retract prepuce; inguinal hernia
Extremities	Symmetry of extremities; 10 fingers and toes; soles flat with moderate to deep creases	Partial syndactyly; asymmetric length of toes

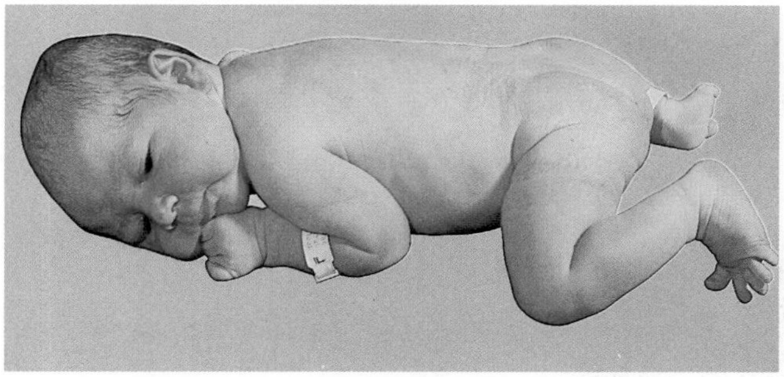

(a)

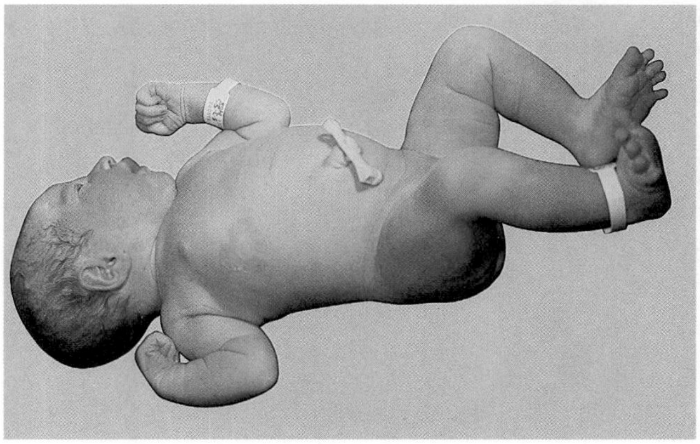

(b)

Figure 10.4 The flexion position of a neonate (*a* and *b*) is an indication of a healthy gestation and a normal delivery.

The skin is the one organ of the neonate that is completely visible and is therefore a source of considerable information concerning state of the development and the clinical condition of the newborn. At birth, the skin is covered with a grayish, cheeselike substance called **vernix caseosa.** If it is not washed away during bathing, it will dry and disappear within a couple of days. Fine, silklike hair, called **lanugo** *(lah-nu'go),* may be present on the forehead, cheeks, shoulders, and back. Distended sebaceous glands, called **milia** *(mil'e-ah),* may appear as tiny white papules on the nose, cheeks, and chin. Skin color depends on genetic background although certain areas, such as the genitalia, areola, and linea alba may appear darker because of a response to the maternal and placental hormones that enter the fetal circulation. **Mongolian spots** occur in about 90% of newborn blacks, Orientals, and American Indians. These blue-gray pigmented areas vary in size and are usually located in the lumbosacral region. Mongolian spots generally fade within the first year or two.

Abnormal skin color is clinically important in the physical assessment of the neonate. *Cyanosis* (bluish discoloration) is usually due to a pulmonary disease such as atelectasis or pneumonia or to congenital heart disease. Although *jaundice* (yellowish discoloration) is common in infants and is usually of no concern, it may indicate liver or bone marrow problems. *Pallor* (paleness) may indicate anemia, edema, or shock.

vernix caseosa: L. *vernix,* varnish; *caseus,* cheese
milia: L. *miliarius,* relating to millet

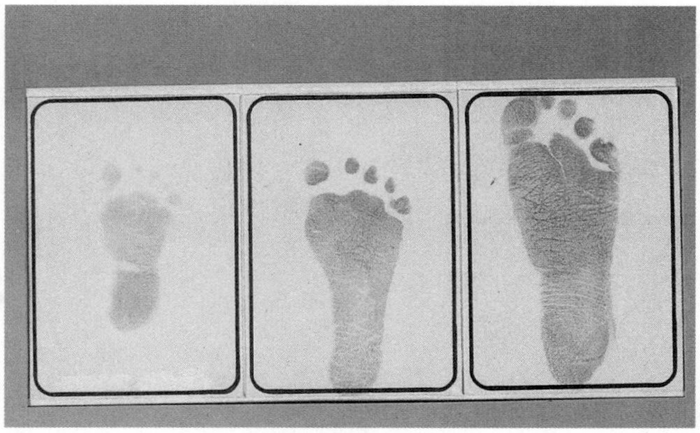

(a) (b) (c)

Figure 10.5 Sole creases at different ages of gestation as seen from footprints of two premature babies (*a* and *b*) and a full-term baby (*c*) (*a*) At twenty-six weeks, only an anterior transverse crease is present; (*b*) by thirty-three weeks, creases develop along the medial instep; and (*c*) the entire sole is creased by thirty-eight weeks.

The appearance of the nails and nail beds is especially valuable in determining body dysfunctions, certain genetical conditions, and even normal gestations. Cyanosis, pallor, and capillary pulsations are best observed at the nails. Yellow nails are found in postmature neonates.

Local edema (swelling) is not uncommon particularly in the skin of the face, legs, hands, feet, and genitalia. Creases on the palms of the hands and soles of the feet should be prominent; the absence of creases accompanies prematurity (fig. 10.5). The nose is usually flattened after birth, and bruises here or on other areas of the face are not uncommon. The auricle of the ear is flexible with the top edge positioned on a horizontal line with the outer canthus (corner) of the eye.

The neck of a neonate is short and thick and is surrounded by neck folds. The chest is rounded in cross section, and the abdomen is cylindrical. The abdomen may bulge in the upper right quadrant due to the large liver.

If the newborn is thin, peristaltic intestinal waves may be observed. At birth, the **umbilical** *(um-bil'i-kal)* **cord** appears bluish white and moist. After clamping it begins to dry and appears yellowish brown. It progressively shrivels and becomes greenish black prior to its falling off by the second week.

The genitalia of both sexes may appear darkly pigmented due to maternal hormonal influences. In a female neonate, a **hymenal** *(hi'men-al)* **tag** is frequently present and is visible from the posterior opening of the vagina. It is comprised of tissue from the hymen and labia minora. The hymenal tag usually disappears by the end of the first month.

Palpable Structures

The six **fontanels** can be lightly palpated as the "soft spots" on the infant's head. The liver is palpable 2–3 cm (1 in.) below the right costal margin. A physician doing a physical examination of a neonate will palpate both kidneys soon after delivery before the intestines fill with air. The suprapubic area is also palpated for an abnormal distended urinary bladder. The newborn should void urine within the first twenty-four hours after birth.

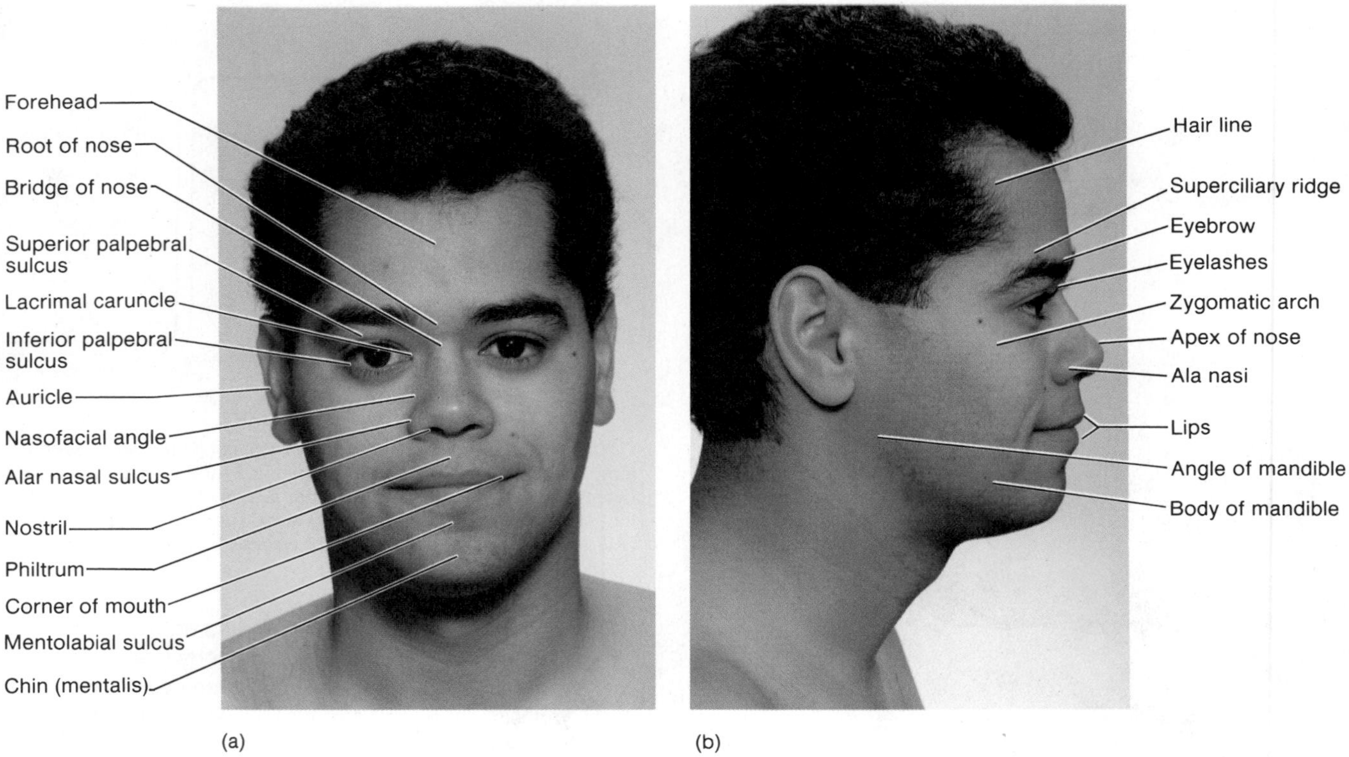

Figure 10.6 The surface anatomy of the facial region: (*a*) An anterior view and (*b*) a lateral view.

The testes of a male child should always be palpated in the scrotum. If the neonate is small or premature, the testes may be palpable in the inguinal canals. An examination for inguinal hernias is facilitated by the crying of an infant, which creates abdominal pressure.

---

1. Describe the appearance of each of the following in a normal neonate: skin, head, thorax, abdomen, genitalia, and extremities. What is meant by the normal flexion position of a neonate?
2. Which internal body organs are palpable in a neonate?

Head

The surface anatomy of the head is of concern because the head contains vital sense organs and provides openings into the respiratory and digestive systems. Of further concern are aesthetic appearances and the many medical problems that involve this part of the body including trauma, diseases, and dysfunctions.

Objective 6. Identify by observation or palpation various surface features of the cranial and facial regions.

The **head,** or **caput,** contains the brain and sense organs—the eyes, ears, and nose. It also provides openings into the respiratory and digestive systems. The head is structurally and developmentally divided into the cranium and the face.

Cranium

The **cranium** *(kra'ne-um),* also known as the **braincase,** is covered by the *scalp.* The scalp is attached anteriorly, at the level of the **eyebrows,** to the **supraorbital ridges.** The scalp continues posteriorly over the area commonly called the forehead and across the crown (vertex) of the top of the head to the **superior nuchal** *(nu'kal)* **line,** a ridge on the back of the skull. Both the supraorbital ridge above the **orbit,** or socket of the eye, and the superior nuchal line at the back of the skull can be easily palpated. Laterally, the scalp covers the **temporal region** and terminates at the fleshy portion of the ear called the **auricle,** or **pinna.** The temporal region is the attachment for the **temporalis muscle,** which can be palpated when repeatedly clenching the jaw. This region is clinically important since it is a point of entrance to the cranial cavity in many surgical procedures.

Only a portion of the scalp is covered with hair, and the variable **hairline** is genetically determined. The scalp is clinically important because of the dense connective tissue layer that supports nerves and vessels beneath the skin. When the scalp is cut, the wound is held together by the connective tissue, but at the same time the vessels are held open, resulting in profuse bleeding.

Face

The **face,** or **facies** *(fa'she-ēz)* (fig. 10.6), is composed of four regions: the **ocular region,** which includes the eye and associated structures; the **auricular region,** which includes the ear;

---

cranium: Gk. *kranion,* skull

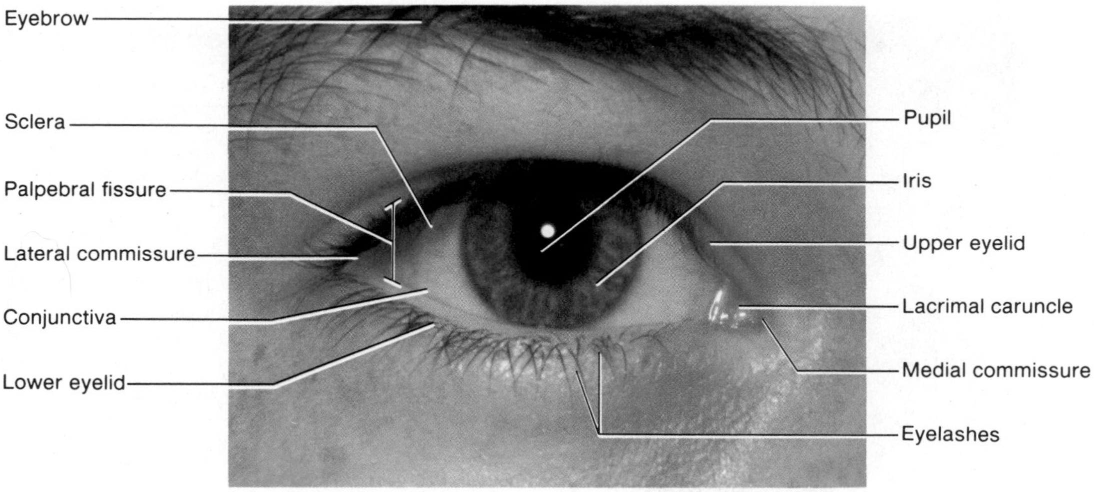

Figure 10.7 The surface anatomy of the ocular region.

Table 10.2 Surface anatomy of the ocular region

Structure	Comments	Structure	Comments
Eyebrow	Ridge of hair that superiorly arches the eye. It protects the eye against sunlight and is important in facial expression.	Iris	Circular, colored, muscular portion of the eyeball that surrounds the pupil. It reflexly regulates the amount of incoming light.
Eyelids	Movable folds of skin and muscle that cover the eyeball anteriorly. They assist in lubricating the anterior surface of the eyeball and reflexly close to protect the eyeball.	Pupil	Opening in the center of the iris through which light enters the eyeball
		Palpebral fissure	Space between the eyelids when they are open
Eyelashes	Row of hairs on the margin of eyelid. They prevent airborne substances from contacting eyeball.	Subtarsal sulcus	Groove beneath the eyelid that parallels the margin of the lid. It traps small foreign particles that contact the conjunctiva.
Conjunctiva	Thin mucous membrane that covers the anterior surface of the eyeball and lines the undersurface of the eyelids. It aids in reducing friction during blinking.	Medial commissure	Medial corner of the eye where the upper and lower eyelids come together
		Lateral commissure	Lateral corner of the eye where the upper and lower eyelids come together
Sclera	Outer fibrous layer of eyeball; the "white" of the eye that gives form to the eyeball	Lacrimal caruncle	Fleshy, pinkish elevation at the medial commissure. It contains sebaceous and sweat glands.
Cornea	Transparent anterior portion of eyeball. It is slightly convex to refract incoming light waves.		

the **nasal region,** which includes the external and internal structures of the nose; and the **oral region,** which includes the mouth and associated structures.

The skin of the face is relatively thin and contains many sensory receptors, particularly in the oral region. Certain facial regions also have numerous **sweat glands** and **sebaceous** *(se-ba'shus)* (oil-secreting) **glands.** Acne is a serious facial dermatological problem for many teenagers. Facial hair appears over most of the facial region in males after they go through puberty; unwanted facial hair may occur sparsely on some females and can be a social problem.

The muscles of facial expression are important for the way they affect surface features. Various emotions are conveyed when these muscles are contracted. These muscles originate on the different facial bones and insert into the dermis (second major layer) of the skin. Excessive contraction of these muscles may cause permanent crease lines in the skin.

Because the organs of the facial region are so complex and specialized, there are professional fields of speciality associated with the various regions. *Optometry* and *ophthalmology* are concerned with the structure and function of the eye. *Dentistry* is entirely devoted to the health and functional and cosmetic problems of the oral region, particularly the teeth. An *otorhinolaryngologist* is an ear, nose, and throat specialist.

The *ocular region* includes the eyeball and associated structures. Most of the surface features of the ocular region protect the eye. **Eyebrows** protect against potentially damaging sunlight and mechanical blows, **eyelids** reflexly close to protect against objects or visual stimuli, **eyelashes** prevent airborne objects from contacting the eyeball, and the **lacrimal** *(lak'rĭ-mal)* **secretions** (tears) wash away chemical or foreign materials and prevent the surface of the eyeball from drying. Refer to figure 10.7 for a depiction of many of the surface features of the ocular region, and table 10.2 for a description of these structures.

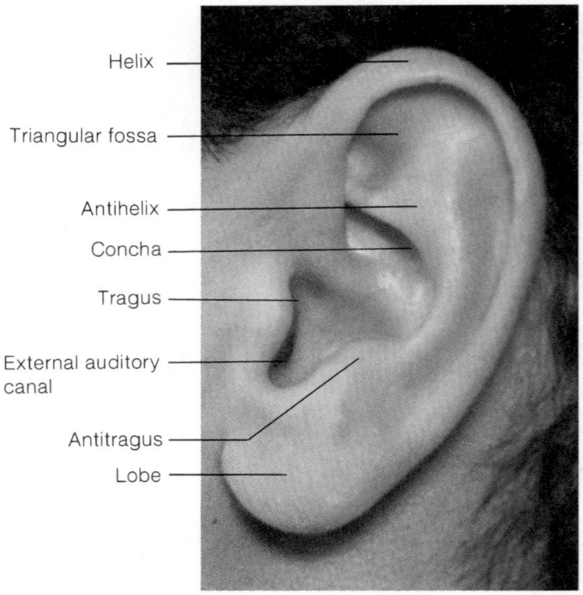

Figure 10.8 The surface anatomy of the auricular region.

The *auricular region* includes the visible surface structures as well as internal organs that function in hearing and in maintaining equilibrium. The fleshy **auricle** and the tubular opening into the middle ear, called the **external auditory canal** are the only observable surface features of the auricular region. The rim of the auricle is called the **helix,** and the inferior portion is referred to as the **lobe.** The lobule is composed primarily of connective and fatty tissue and therefore can be easily pierced. For this reason, it is sometimes used when obtaining blood for a blood count. The **tragus** *(tra'gus)* is a small, posteriorly directed projection partially covering and protecting the external auditory canal. Further protection is provided by the many fine hairs that surround the opening into this canal. The condyle of the mandible can be palpated at the opening of the external auditory canal by placing the little finger in the opening and then vigorously moving the jaw. Refer to figure 10.8 and table 10.3 for an illustration and discussion of other surface features of the auricular region.

The inspection of some of the internal structures of the ear is part of a routine physical examination and is performed using an otoscope. Earwax may accumulate in the canal, but this is a protective substance. It waterproofs the tympanic membrane (eardrum) and because of its bitter taste is thought to be an insect repellent.

A few structural features of the *nasal region* are apparent from its surface anatomy (fig. 10.9 and table 10.4). The principal function of the nose is associated with the respiratory system, and the need for a permanent body opening to permit gaseous ventilation accounts for its surface features. The **root** (nasion) is located at about the level of the eyebrows where the nose begins to protrude from the forehead. The firm, narrow portion between the eyes is the **bridge** of the nose and is formed by the union of the nasal bones. The nose below this level has a pliable cartilaginous framework that maintains an opening. The tip of the nose is called the **apex,** and the region between the bridge and the apex is the **dorsum nasi.** As the lateral portion of the nose blends with the face, it is called the **nasofacial angle.** The **nostrils,** or **external nares,** are the paired openings into the nose. The **nasal septum** forms a partition between the nostrils, and the **ala** *(a'lah)* forms the flaired outer margin of each nostril.

Table 10.3 Surface anatomy of the auricular region

Structure	Comments
Auricle (pinna)	Expanded, fleshy portion of the ear projecting from the side of the head that funnels sound waves into the external auditory canal
Helix	Outer rim of auricle. It gives form and shape to the pinna.
Lobe	Inferior, fleshy portion of auricle
Tragus	Posterior cartilaginous projection of auricle. It partially covers and protects the external auditory canal.
Antitragus	Small, cartilaginous anterior projection opposite the tragus
Antihelix	Semicircular ridge anterior to the greater portion of the helix
Concha	Depressed hollow of auricle. It funnels sound waves.
External auditory canal	Tube extending inward to the tympanic membrane. It is slightly S-shaped and contains glands that secrete earwax for protection.
Triangular fossa	Triangular depression in superior portion of antihelix

Structures of the *oral region* that are important in surface anatomy include the fleshy upper and lower **lips,** or **labia** *(la'be-ah),* and the structures of the **oral cavity** that can be observed when the mouth is open. The lips are shown in figure 10.9, and the structures of the oral cavity are seen in figure 10.10.

The color of the lips and other mucous membranes of the oral cavity are diagnostic of certain body dysfunctions. The lips may appear pale in patients with severe anemia and bluish or cyanotic in persons lacking in oxygen. Pernicious anemia may cause a lemon yellow tint to the lips, as does jaundice.

1. What are the boundaries of the cranial region, and why is this region clinically important?
2. Why do scalp wounds bleed so freely? How might this relate to infections?
3. What are the subdivisions of the facial region?

Neck

Major organs are located within the flexible neck, and several structures (the spinal cord, vessels, trachea, and esophagus) that are essential for body sustenance pass through the neck.

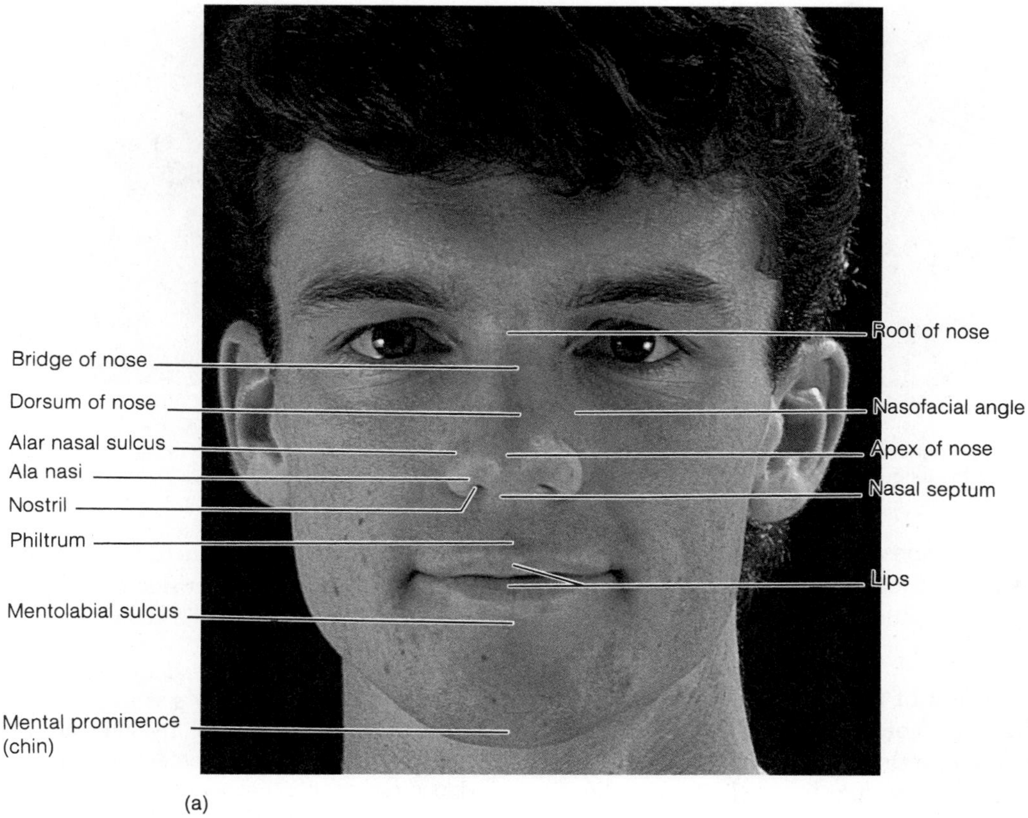

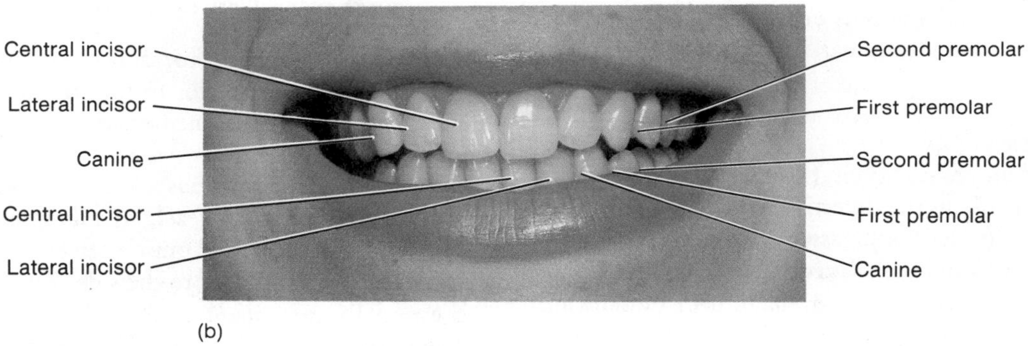

Figure 10.9 The surface anatomy of the nasal and oral regions: (*a*) The nose and lips and (*b*) the teeth.

Table 10.4 Surface anatomy of the nasal and oral regions

Structure	Comments	Structure	Comments
Root (nasion)	Superior attachment of nose to the cranium	Nostril	External opening into nasal cavity
Bridge	Anterior bony framework of nose formed by union of nasal bones	Nasal septum	Partition between external nares
		Ala	Laterally expanded border of nostril
Dorsum nasi	Anterior ridge of nose	Philtrum	Vertical depression in medial portion of upper lip
Nasofacial angle	Anterolateral boundary of nose as it blends with tissues of the face	Lip	Upper and lower anterior borders of the mouth
		Chin (mental)	Anterior portion of lower jaw
Apex	Terminal tip of nose		

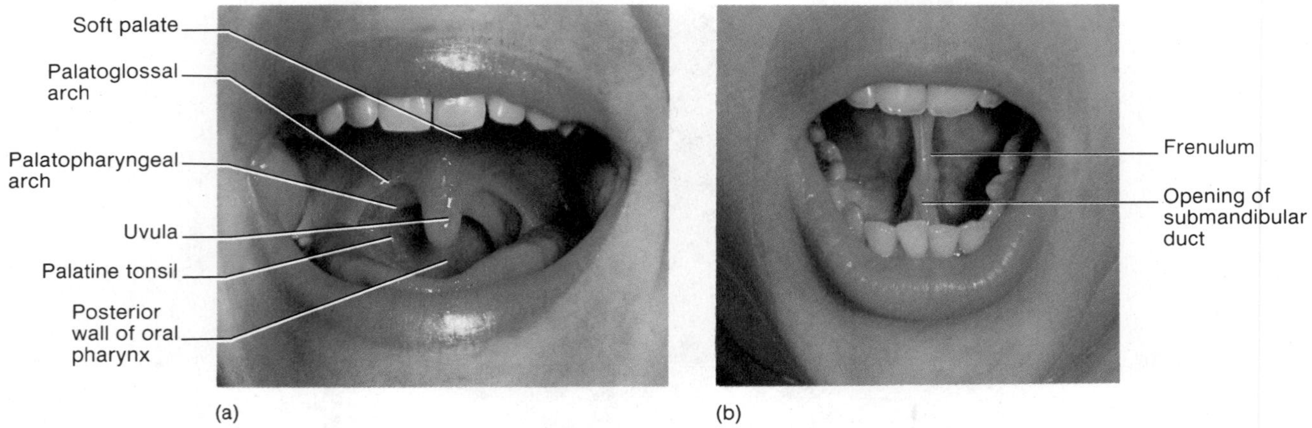

Figure 10.10 Surface structures of the oral cavity (*a*) with the mouth open, and (*b*) with the mouth open and the tongue elevated.

Objective 7. Discuss the functions of the neck.
Objective 8. List by name and location the triangles of the neck and the structures contained within these triangles.

The **neck** is a complex region of the body that connects the head to the thorax. The spinal cord, digestive and respiratory tracts, and major vessels traverse this highly flexible area. Several organs and glands are located here as well. Remarkable musculature in the neck produces an array of movements. Because of this complexity, the neck is a clinically important area. The surface features provide landmarks for determining the location of internal structures.

Surface Features

The neck is divided into four regions: (1) an *anterior region,* called the **cervix,** which contains the digestive and respiratory tracts, the **larynx** *(lar'inks)* (voice box), vessels to and from the head, nerves, and the **thyroid** and **parathyroid glands;** (2) right and (3) left *lateral regions,* each composed of major neck muscles and **cervical** *(ser'vĭ-kal)* **lymph nodes;** and (4) a *posterior region,* referred to as the **nucha** *(nu'kah),* which includes the spinal cord, cervical vertebrae, and associated structures.

The most prominent structure of the cervix of the neck is the **thyroid cartilage** (Adam's apple) of the larynx (fig. 10.11). The thyroid cartilage supports the vocal cords and is larger in sexually mature males than in females, which accounts for the male's deeper voice. The **hyoid bone** can be palpated just above the larynx. Both of these structures are elevated during swallowing, which is one of the actions that directs food and fluid into the esophagus. Note this action on yourself by gently cupping your fingers on the larynx, then swallowing. Below the thyroid cartilage is the **cricoid** *(kri'koid)* **cartilage** followed by the **trachea** *(tra'ke-ah)* (windpipe), both of which can be palpated. The trachea is clinically important in case a respiratory tube has to be inserted during a *tracheotomy.* The thyroid gland can be palpated on either side just below the level of the larynx. Pulsations of the **common carotid artery** can be observed and palpated on either side of the neck just lateral and a bit superior to the level of the larynx.

The arteries of the head and neck are rarely damaged because of their elasticity. In a severe lateral blow to the head, however, the internal carotid artery may rupture, resulting in the perception of a roaring sound as blood rushes into the cavernous sinuses of the temporal bone. Containment of carotid hemorrhage within the sinuses may actually be lifesaving.

The **suprasternal notch** is the depression in the midline of the cervix just superior to the sternum. The two **clavicles** are obvious in all persons because they are subcutaneous (just under the skin).

The **sternocleidomastoid** and **trapezius muscles** are the prominent structures of each lateral region (figs. 10.11, 10.12). The sternocleidomastoid muscle can be palpated its entire length when the head is turned to the side. The tendon of this muscle is especially prominent to the side of the suprasternal notch. The trapezius muscle can be felt if the shoulders are shrugged. An inflammation of the trapezius muscle causes a "stiff neck." If a person is angry or if a shirt collar is too tight, the **external jugular vein** can be seen as it courses obliquely across the sternocleidomastoid muscle. **Cervical lymph nodes** of the lateral neck region may become swollen and painful from infectious diseases of the oral or pharyngeal regions.

Most of the structures of the nucha are too deep to be of importance in surface anatomy. The spines of the lower cervical vertebrae (especially C7), however, can be observed and palpated when the neck is flexed. In this same position, a firm ridge is raised, called the **ligamentum nuchae** (not shown), that extends superiorly from C7 to the external occipital protuberance of the skull. Clinically the nucha is tremendously important because of the debilitating damage to it from whiplash injury or a broken neck.

Triangles of the Neck

The triangles of the neck, created by the arrangement of specific muscles and bones, are clinically important because of the specific structures included in each. The structures of the neck

cervix: L. *cervix,* neck
larynx: Gk. *larynx,* upper windpipe
hyoid: Gk. *hyoeides,* U-shaped

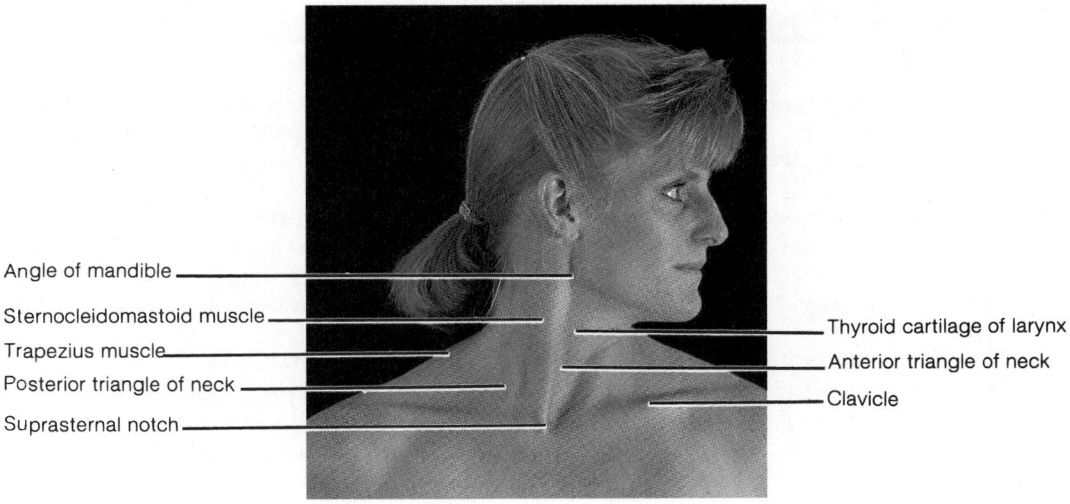

Figure 10.11 An anterolateral view of the neck.

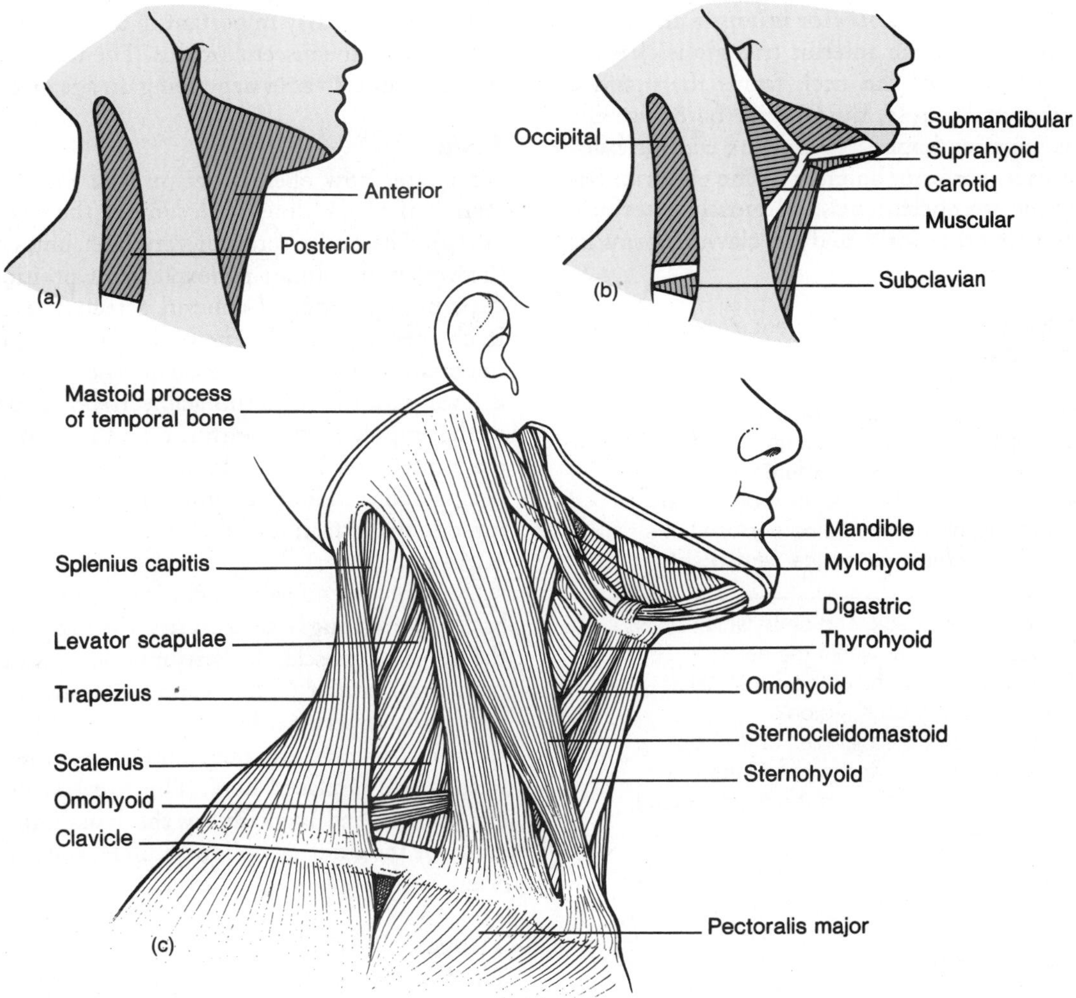

Figure 10.12 Triangles of the neck. (*a*) The two large triangular divisions, (*b*) the six lesser triangular subdivisions, and (*c*) the detailed muscular anatomy of the neck.

Table 10.5 Location and contents of the triangles of the neck

Triangle	Boundaries	Contents
Anterior	Sternocleidomastoid muscle, median line of neck, inferior border of mandible	Four lesser triangles, salivary glands, larynx, trachea, thyroid glands, various vessels and nerves
Carotid	Sternocleidomastoid, posterior digastric, and omohyoid muscles	Carotid arteries, internal jugular vein, vagus nerve
Submandibular	Digastric muscle (both heads), inferior border of mandible	Salivary glands
Suprahyoid	Digastric muscle, hyoid bone (This is the only unpaired triangle of the neck.)	Muscles of the floor of the mouth, salivary glands and ducts
Muscular	Sternocleidomastoid and omohyoid muscles, midline of neck	Larynx, trachea, thyroid gland, carotid sheath
Posterior	Sternocleidomastoid and trapezius muscles, clavicle	Nerves and vessels
Occipital	Sternocleidomastoid, trapezius, and omohyoid muscles	Cervical plexus, accessory nerve
Subclavian	Sternocleidomastoid and omohyoid muscles; clavicle	Brachial plexus, subclavian artery

that are important in surface anatomy have already been identified, however, so only a summary of the two major and six minor triangles will be depicted in figure 10.12 and presented in table 10.5. The sternocleidomastoid muscle obliquely transects the neck, dividing it into an **anterior triangle** and a **posterior triangle.** The apex of the anterior triangle is directed inferiorly. The median line of the neck forms the anterior boundary of the anterior triangle, the inferior border of the mandible forms its superior boundary, and the sternocleidomastoid muscle forms its posterior boundary. The posterior triangle is formed by the sternocleidomastoid muscle anteriorly and the trapezius muscle posteriorly, and the clavicle forms its base inferiorly.

Three structures traversing the neck are extremely important and potentially vulnerable. These structures are the common carotid artery, which carries blood to the head; the internal jugular vein, which drains blood from the head; and the vagus nerve, which conducts nerve impulses to visceral organs. These structures are protected in the neck by their deep position behind the sternocleidomastoid muscle and by being enclosed in a tough connective tissue sheath called the *carotid sheath.*

1. List four functions of the neck, and state which body systems are located, in part, within the neck.
2. What are the structural regions of the neck? What structures are included in each region?
3. Using your knowledge of the triangles of the neck, describe where to palpate to feel a pulse, the trachea, cervical lymph nodes, and the thyroid gland.

Torso

The locations of vital visceral organs in the cavities of the torso make the surface anatomy of this body region especially important.

Objective 9. Identify by observation or palpation various surface features of the torso.

Objective 10. List the auscultation sites of the thorax and abdomen.

The **torso,** or **trunk,** is divided into the **back** (dorsum), **thorax** (chest), **abdomen** (venter), and **pelvis.** A region called the **perineum** *(per''i-ne'um)* forms the floor of the pelvis and includes the external genitalia. The surface anatomy of these regions is particularly important in determining the location and condition of the visceral organs. The usability of some of these surface features varies depending on age, sex, and body weight.

Back

No matter how obese a person is, a **median furrow** can be seen on the back along with some of the **vertebral spines** (fig. 10.13). The entire series of vertebral spines can be observed if the vertebral column is flexed. This position is important in determining vertebral-column defects (see clinical comments in chapters 8 and 11). The back of the **scapula** presents other important surface landmarks. The base of the spine of the scapula is level with the third thoracic vertebra, and the inferior angle of the scapula is even with the seventh thoracic vertebra. Several muscles of the scapula can be observed on a lean, muscular person and are identified in figure 10.13. Many of the ribs and muscles that attach to the ribs can be seen in a lateral view (fig. 10.14).

There are two pairs of clinically important triangles on the back. The **triangle of auscultation** (fig. 10.13) is bound by the trapezius muscle, the latissimus dorsi muscle, and the medial (vertebral) border of the scapula. Because there is a space between the superficial back muscles in this area, heart and respiratory sounds are not muffled by the muscles when a stethoscope is placed here. The **lumbar triangle** (not illustrated) is bound medially by the latissimus dorsi muscle, laterally by the external oblique muscle, and inferiorly by the iliac crest.

Thorax

The leading causes of death in the United States are associated with disease or dysfunction of the thoracic organs. With the exception of the breasts and surrounding lymph nodes, the organs of the thorax are within the rib cage. Because of the location of the thoracic visceral organs and their clinical importance, the surface anatomy of the thorax is extremely important.

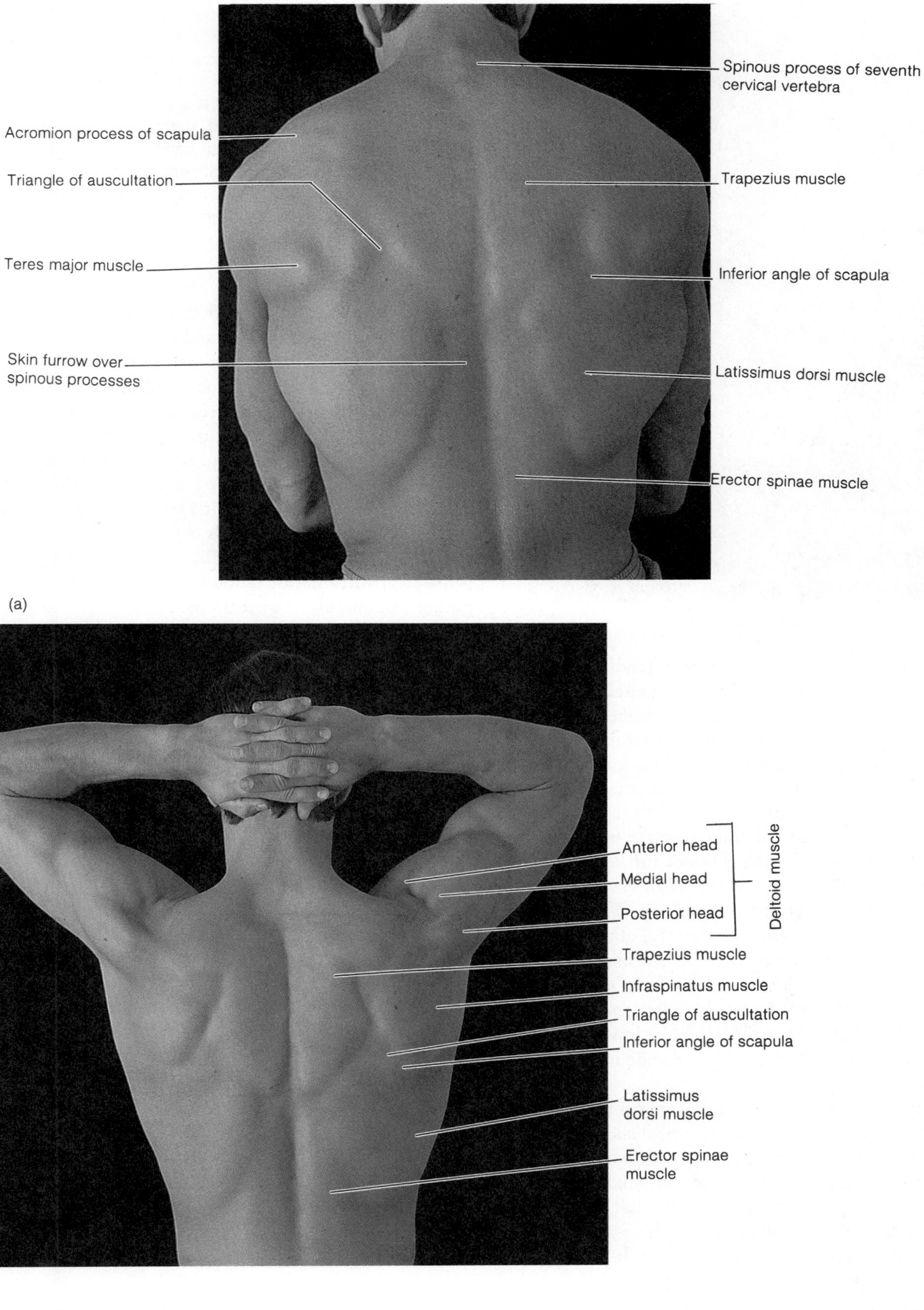

Figure 10.13 The surface anatomy of the back. (*a*) Flexion of the upper extremities and (*b*) abduction of the shoulders and adduction of the scapulae.

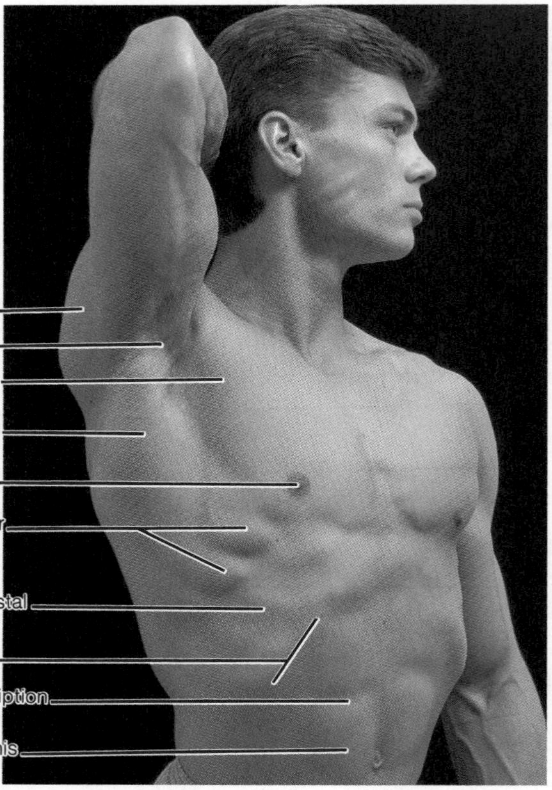

Figure 10.14 An anterolateral view of the torso and axilla.

The flexible rib cage presents several bony landmarks that can be observed or palpated (fig. 10.15). The paired clavicles and the suprasternal notch have already been identified as being important surface features of the neck. These structures are also important in the thoracic region as reference points for counting the ribs. Many of the ribs can be seen on a thin person. All of the ribs except the first, and at times the twelfth, can be palpated. The sternum is composed of three separate bones (manubrium, body, xiphoid process), each of which can be palpated. The **sternal angle** (angle of Louis) is felt as an elevation between the manubrium and body of the sternum. The sternal angle is important because it is located at the level of the second rib. The articulation between the body of the sternum and the xiphoid process, called the **xiphisternal** *(zif″i-ster′nal)* **joint,** is positioned over the lower border of the heart and the diaphragm. The **costal margin** of the rib cage is the lower oblique boundary and can be easily identified when a person inhales and holds his breath.

The nipples in the male are located at the fourth intercostal spaces (the area between the fourth and fifth ribs) and about 10 cm (4 in.) from the midline. They vary in position in sexually mature women according to age, size, and the pendulousness of the breasts (fig. 10.16). The position of the left nipple in males is an important landmark for knowing where to listen to various heart sounds and for determining if the heart is enlarged. For diagnostic purposes, an imaginary line, called the **midclavicular line,** can be extended vertically from the middle of the clavicle through the nipple. Several superficial chest muscles can be observed or palpated and are therefore important surface features. These muscles and the structures described above are depicted in figures 10.15 and 10.16.

In addition to helping one know where to listen with a stethoscope to heart sounds, surface features of the thorax are important for auscultations of the lungs, X rays, tissue biopsies, sternal taps for bone marrow studies, or thoracic surgery. Although the anatomical features of the rib cage are quite consistent, slight deformities and asymmetries do occur. These generally cause no disability and require no treatment. Most of the abnormalities are congenital and include conditions such as a projecting sternum (pigeon breast) or a receding sternum (funnel chest).

Abdomen

The **abdomen** is the portion of the body between the diaphragm and the pelvis. The abdomen does not have a bony framework as does the thorax, so the surface anatomical features are not as well defined. Bony landmarks of both the thorax and pelvis are used when referring to abdominal structures (fig. 10.17). The right costal margin of the rib cage is located over the liver on the right side, and the left costal margin is positioned over the stomach on the left. The xiphoid process is important because from this point a tendinous, midventral raphe, called the **linea alba,** extends the length of the abdomen to attach to the **symphysis pubis.** The symphysis pubis can be palpated at the anterior union of the two halves of the pelvic girdle. The **navel,** or **umbilicus,** is the former site of attachment of the fetal umbilical cord and is located along the linea alba. The linea alba separates the paired, straplike **rectus abdominis** muscles, which can be seen when a person flexes the abdomen (such as when doing sit-ups).

Clinically, the linea alba is a favored site for abdominal surgery because an incision made along this line severs no muscles and few vessels and nerves. The linea alba also heals readily (it has been said that only a zipper would provide a more convenient entry to the abdominal cavity).

The lateral margin of the rectus abdominis muscle can be observed on some persons, and the surface line produced is called the **linea semilunaris.** The **external oblique** muscle is the superficial layer of the muscular abdominal wall. The **iliac crest** is subcutaneous and can be palpated along its entire length. The highest point of the crest lies opposite the body of the fourth lumbar vertebra, an important level in spinal anesthesia. Another important landmark is **McBurney's point,** located one-third of the distance from the right anterior-superior iliac crest to the navel (see fig. 10.17). This point overlies the appendix of the intestinal tract.

The abdominal region is frequently divided into nine regions or four quadrants in order to describe the location of internal organs and to clinically identify the sites of various pains or conditions. These regions have been adequately described in chapter 2 (see figs. 2.13, 2.14).

linea alba: L. *linea,* line; *alba,* white
navel: O. E. *nafela,* umbilicus
umbilicus: L. *umbilicus,* navel
McBurney's point: from Charles McBurney, U.S. surgeon, 1845–1914

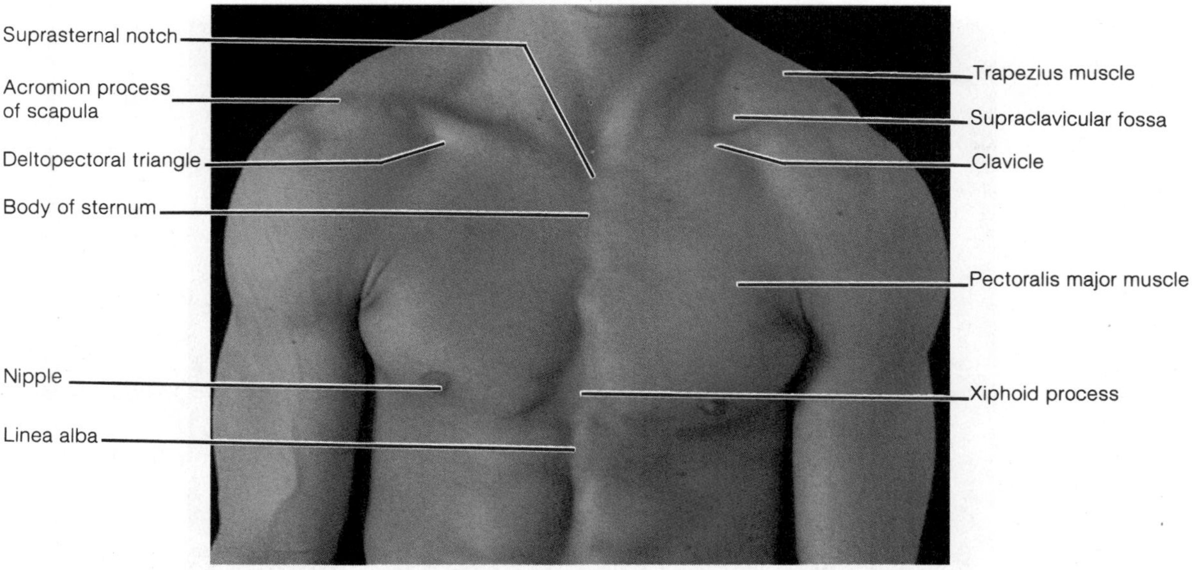

Figure 10.15 The surface anatomy of the anterior thoracic region of the male.

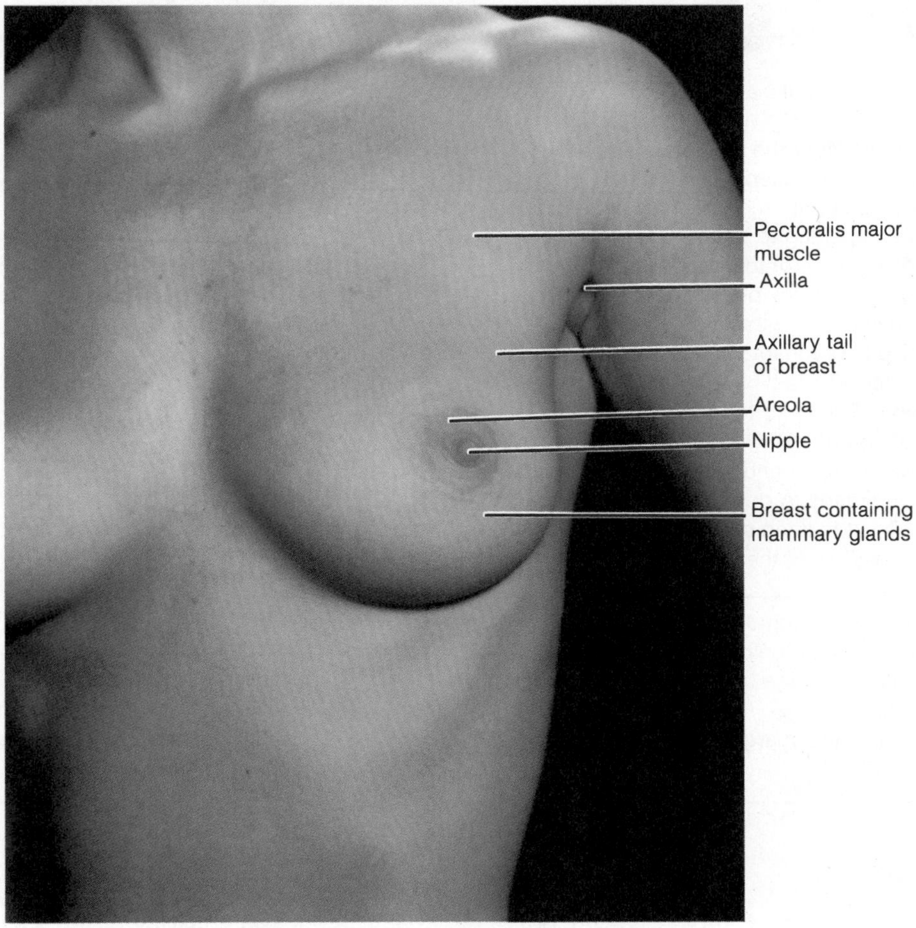

Figure 10.16 The surface anatomy of the female breast.

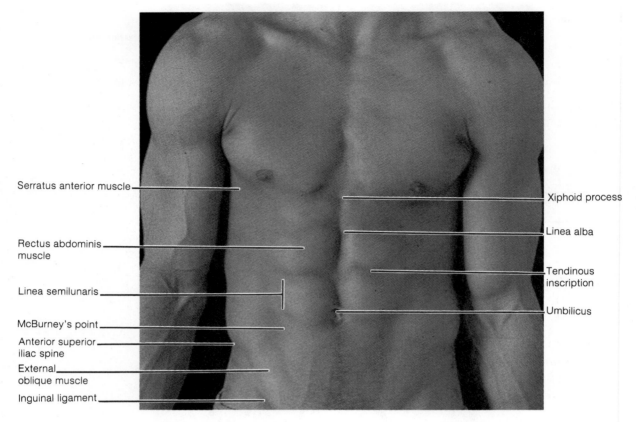

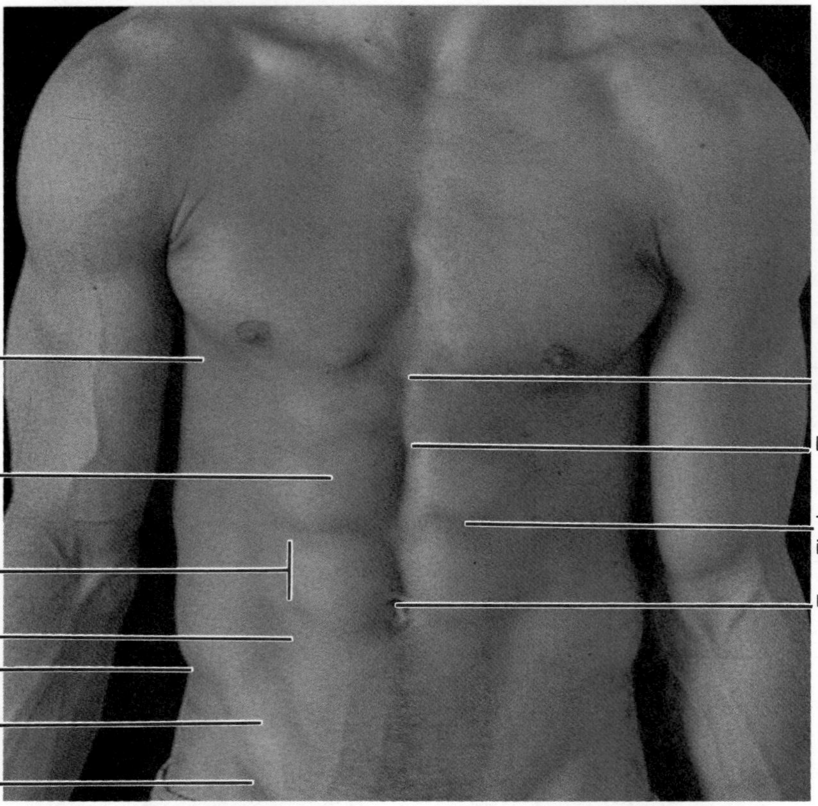

Figure 10.17 The surface anatomy of the anterior abdominal region.

Although the position of the umbilicus is relatively consistent in all persons, its shape and health is not. Embryological remains of the umbilicus may cause clinical problems such as an opening to the outside, called a *fistula,* or herniation of some of the abdominal contents. Acquired umbilical hernias may develop in children who have a weak abdominal wall in this area, or umbilical hernias may develop in pregnant women because of the extra pressure exerted at this time. The umbilicus is a common site for an incision into the abdominal cavity in a procedure called a *laparotomy.* A laparotomy is frequently performed in the examination and surgery of the internal female reproductive organs. A depressed umbilicus on an obese person is difficult to keep clean, and so various types of infections may occur there.

1. Which structures of the torso can be readily observed? Which can be palpated?
2. Describe the location of the common auscultation sites of the torso.
3. Define linea alba, costal margin, linea semilunaris, and McBurney's point.

Pelvis and Perineum

The surface features of the pelvic region are important primarily to identify reproductive organs and clinical problems of these organs.

Objective 11. Describe the location of the perineum, and list the organs of the pelvic and perineal regions.

The important bony structures of the **pelvis** include the crest of the ilium and symphysis pubis, located anteriorly, and the ischium and os coccyx, which are palpable posteriorly. An **inguinal** *(ing'gwĭ-nal)* **ligament** extends from the crest of the ilium to the symphysis pubis and is clinically important because hernias occur along it. Although the inguinal ligament cannot be seen, an oblique groove overlying the ligament is an apparent surface feature.

The **perineum** *(per''i-ne'um)* (see fig. 2.15) is the region that contains the external sex organs and the anal opening. The surface features of this region are further discussed in chapters 20 and 21. The surface anatomy of the perineum of a female becomes particularly important during parturition.

1. Define the term *perineum.* What structures are located within the perineum?
2. List three body systems that have openings within the pelvic region.

Shoulder and Upper Extremity

The surface anatomy of the shoulder and upper extremity is of clinical importance because of frequent trauma to these body regions. Vessels of the upper extremity are also used as

laparotomy: Gk. *lapara,* flank; *tome,* incision

inguinal: L. *inguinalis,* groin

pressure points and for intravenous injections or blood withdrawal.

Objective 12. Identify by observation or palpation various surface features of the shoulder and upper extremity.

Objective 13. Discuss the clinical importance of the axilla, cubital fossa, and wrist.

Shoulder

The scapula, clavicle, and proximal portion of the humerus form the **shoulder,** and portions of each of these bones are important surface landmarks in this region. Posteriorly, the spine of the scapula and acromion process are subcutaneous and easily located.

The acromion process and the clavicle, as well as several large shoulder muscles, can be seen anteriorly (fig. 10.18). The rounded curve of the shoulder is formed by the deltoid muscle covering the greater tuberosity of the humerus. The deltoid muscle is frequently a site for intramuscular injections. The large pectoralis major muscle is prominent as it crosses the shoulder joint and attaches to the humerus. A small depression, called the **deltopectoral triangle** (see fig. 10.15), is situated below the outer third of the clavicle and is bounded on either side by the deltoid and pectoralis major muscles.

Axilla

The **axilla** is commonly called the armpit. This depressed region of the shoulder supports axillary hair in sexually mature individuals. The axilla is clinically important because of the subcutaneous position of vessels, nerves, and lymph nodes here. Two muscles form the anterior and posterior borders of this region (fig. 10.19). The **anterior axillary fold** is formed by the pectoralis major muscle, and the **posterior axillary fold** consists primarily of the latissimus dorsi muscle extending from the lumbar vertebrae to the humerus. Axillary lymph nodes are palpable in some persons.

In sexually mature females, the axillary tail of the mammary gland, which is positioned on the pectoralis major muscle (see figs. 9.24 and 10.16), extends partially into the axilla. In doing a *breast self-examination* (see chapter 21), a woman should palpate the axillary area as well as the entire breast because the lymphatic drainage pathway is toward the axilla.

Brachium

Several muscles are clearly visible in the **brachium** (figs. 10.19, 10.20). The belly of the biceps brachii muscle becomes prominent when the elbow is flexed with the palm upward. While the arm is in this position, the deltoid muscle can be traced as it inserts upon the humerus. The triceps brachii muscle forms the bulk of the posterior surface of the brachium. A groove forms on the medial side of the brachium between the biceps and triceps muscles where pulsations of the brachial artery may be felt as it carries blood toward the forearm. This region is clinically important because it is where arterial blood pressure is taken with a sphygmomanometer. It is also the place to apply pressure in case of severe arterial hemorrhage in the forearm or hand.

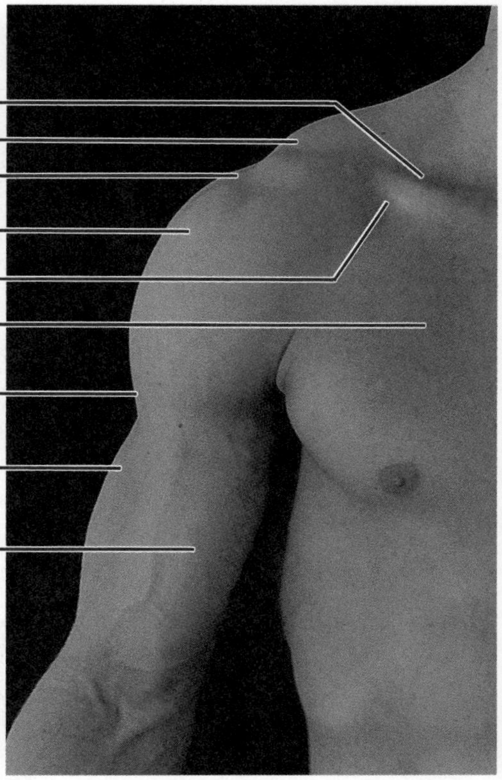

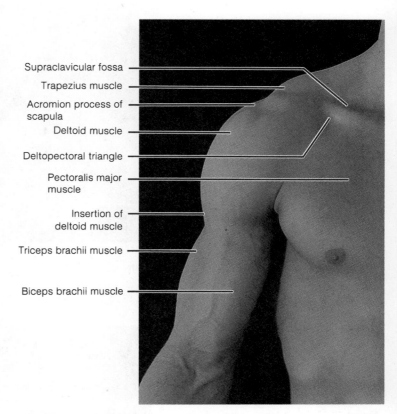

Figure 10.18 An anterior view of the shoulder region.

Three bony prominences can be located in the region of the elbow (fig. 10.21). The medial and lateral epicondyles are processes on the humerus, whereas the olecranon is a proximal process of the ulna. When the elbow is extended, these prominences lie on the same straight plane; when the elbow is flexed, these points form a triangle. The ulnar nerve can be palpated in the groove behind the medial epicondyle. This area is commonly known as the funny bone or crazy bone.

The **cubital** *(ku'bĭ-tal)* **fossa** is the triangular depression on the anterior surface of the elbow region where the median cubital vein links the cephalic and basilic veins. These veins are subcutaneous and become more conspicuous if a proximal compression is applied. For this reason, these veins, particularly the median cubital, are an important location for the removal of venous blood for analyses and transfusions or for intravenous therapy (fig. 10.22).

Forearm

Contained within the **forearm** are two parallel bones (the ulna and radius) and the muscles that control the movements of the hand. The forearm tapers distally toward the wrist, where the muscles give way to tendinous cords that attach to various bones of the hand. Several muscles of the forearm can be identified as surface features and are depicted in figures 10.22 and 10.23.

Because of the frequency of fractures involving the forearm, bony landmarks are clinically important when setting broken

acromion: Gk. *akros,* extreme, tip; *omion,* small shoulder

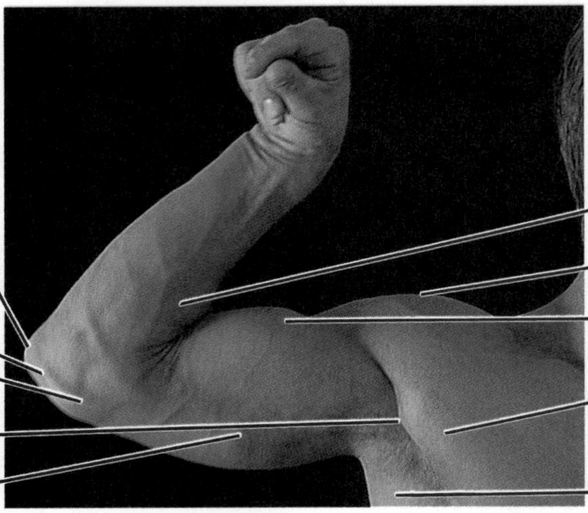

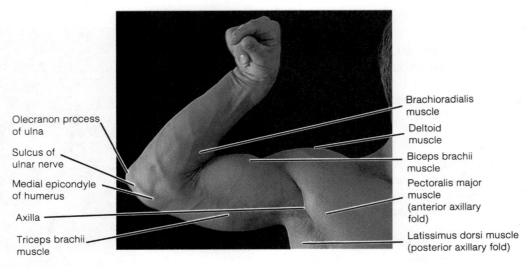

Figure 10.19 An anterior view of the upper extremity.

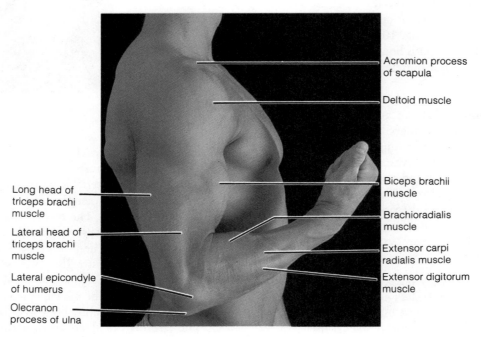

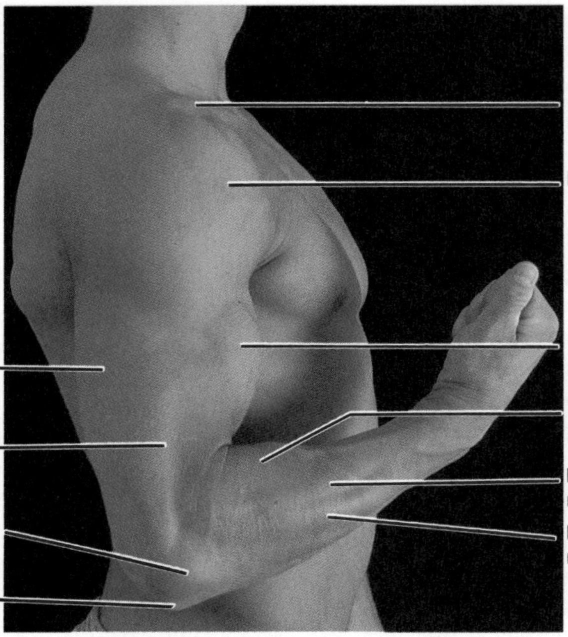

Figure 10.20 A lateral view of the upper extremity.

bones. The ulna can be palpated its entire length from the olecranon to the distal styloid process. The distal one-half of the radius is palpable as the forearm is rotated, and its styloid process can be located.

Nerves, tendons, and vessels are close to the surface at the wrist, making cuts to this area potentially dangerous. Tendons from four flexor muscles can be observed as surface features if the anterior forearm muscles are strongly contracted while making a fist. The tendons that can be observed along this surface, from lateral to medial, are from the following muscles: flexor carpi radialis, palmaris longus, flexor digitorum superficialis, and flexor carpi ulnaris. The median nerve going to the hand is located under the tendon of the palmaris longus muscle (see fig. 10.22), and the ulnar nerve is lateral to the tendon of the flexor carpi ulnaris. The radial artery lies along the surface of the radius immediately lateral to the tendon of the flexor carpi radialis. This is the artery commonly used when taking a pulse. By careful palpation, pulsations can also be detected in the ulnar artery lateral to the tendon of the flexor carpi ulnaris.

Two tendons that attach to the thumb can be seen on the posterior surface of the wrist as the thumb is extended backwards. The tendon of the extensor pollicis brevis is positioned anterolaterally along the thumb, and the tendon of the extensor pollicis longus lies posteromedially (fig. 10.24). The depression created between these two tendons as they are pulled taut is referred to as the anatomical snuffbox. Pulsations of the radial artery can be detected in this depression.

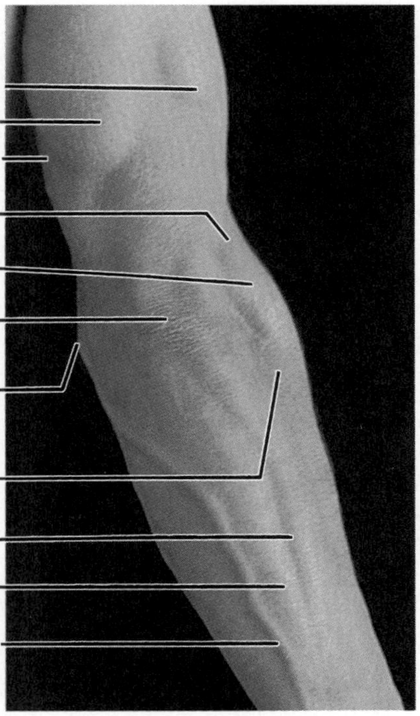

Figure 10.21 A posterior view of the elbow (joint slightly extended).

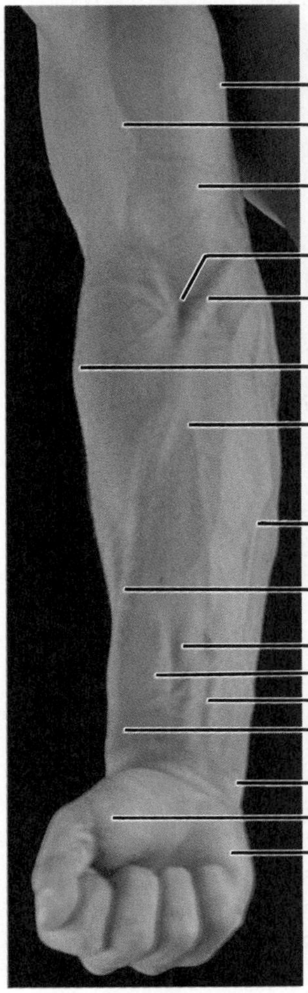

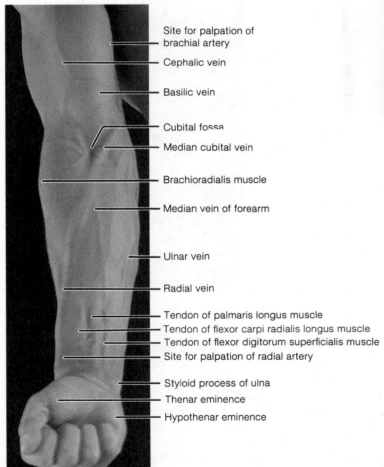

Figure 10.22 An anterior view of the forearm and hand.

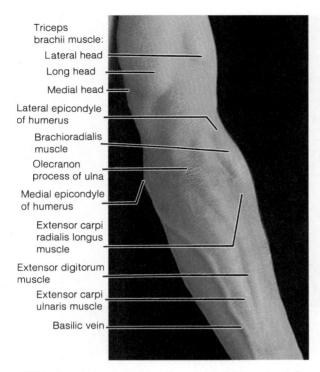

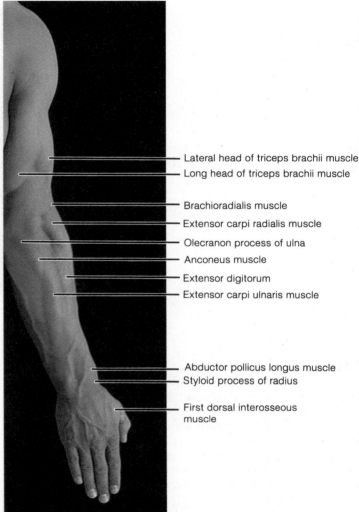

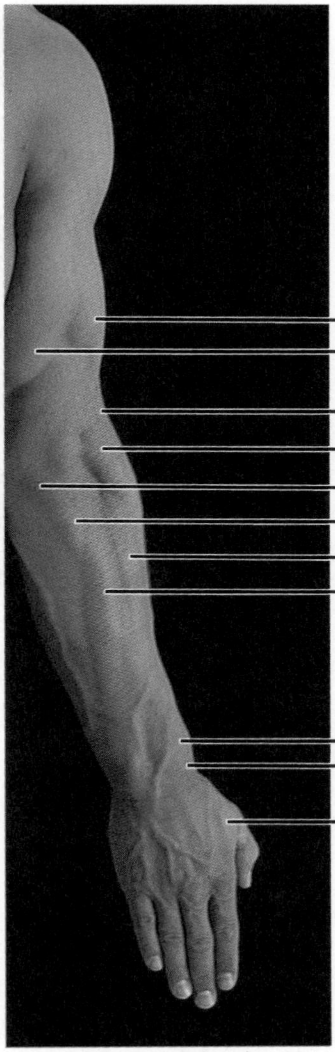

Figure 10.23 A posterior view of the forearm and hand.

The median nerve, which serves the thumb, is the nerve most commonly injured by stab wounds or the penetration of glass into the wrist or hand. If this nerve is severed, the muscles of the thumb are paralyzed and waste away, resulting in an inability to oppose the thumb in grasping.

Hand

Much of the surface anatomy of the **hand,** such as flexion creases, fingerprints, and fingernails, includes features of the skin that are discussed in chapter 5. Other surface features are the extensor tendons from the extensor digitorum muscle, which can be seen going to each of the fingers on the back side of the hand as the hand is extended (fig. 10.25). The knuckles of the hand are the distal ends of the second through the fifth metacarpal bones. Each of the joints of the fingers and the individual phalanges can be palpated. The **thenar** *(the'nar)* **eminence** is the thickened, muscular portion of the hand forming the base of the thumb (fig. 10.26).

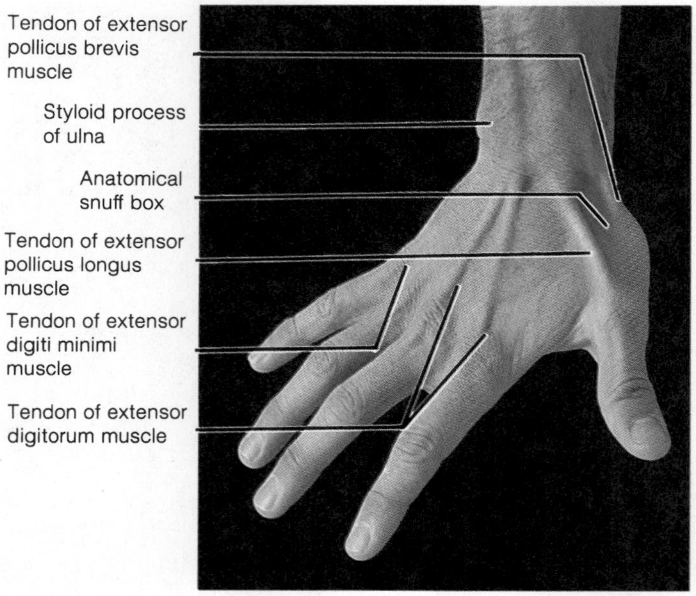

Figure 10.24 The anatomical snuffbox.

1. List the clinically important structures that can be observed or palpated in the shoulder and upper extremity.
2. Describe the locations of the axilla, brachium, cubital fossa, and wrist.
3. Bumping the ulnar nerve causes a tingling sensation along the medial portion of the forearm to the little finger of the hand. What does this tell you about its distribution?
4. Which of the two bones of the forearm is more stationary as the arm is rotated?

Buttock and Lower Extremity

The massive bones and muscles of the buttock and lower extremity are important as weight bearers and locomotors. Many of the surface features of these regions are important in relation to locomotion or locomotor dysfunction.

Objective 14. Identify by observation or palpation various surface features of the buttock and lower extremity.

Objective 15. Discuss the clinical importance of the buttock, femoral triangle, popliteal space, ankle, and arches of the foot.

Buttock

The superior borders of the **buttocks,** or **gluteal region,** are formed by the iliac crests (fig. 10.27). Each crest can be palpated medially to the level of the second sacral vertebra. From this point, the **natal cleft** extends vertically to separate the buttocks into two prominences, each formed by pads of fat as well as by the massive gluteal muscles. An ischial tuberosity can be palpated in the lower portion of each buttock. In a sitting position, the ischial tuberosities support the weight of the body; but when standing, these processes are covered by the gluteal muscles. The sciatic nerve, which is the major nerve to the lower extremity, lies under the gluteus maximus muscle. The inferior border of the gluteus maximus muscle forms the **fold of the buttock.**

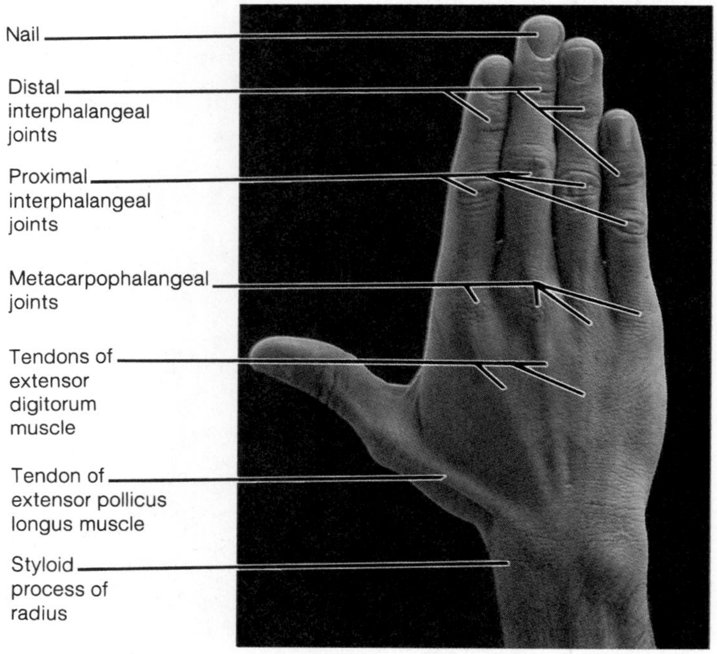

Figure 10.25 A posterior view of the hand.

buttock: O. E. *buttuc,* end or rump

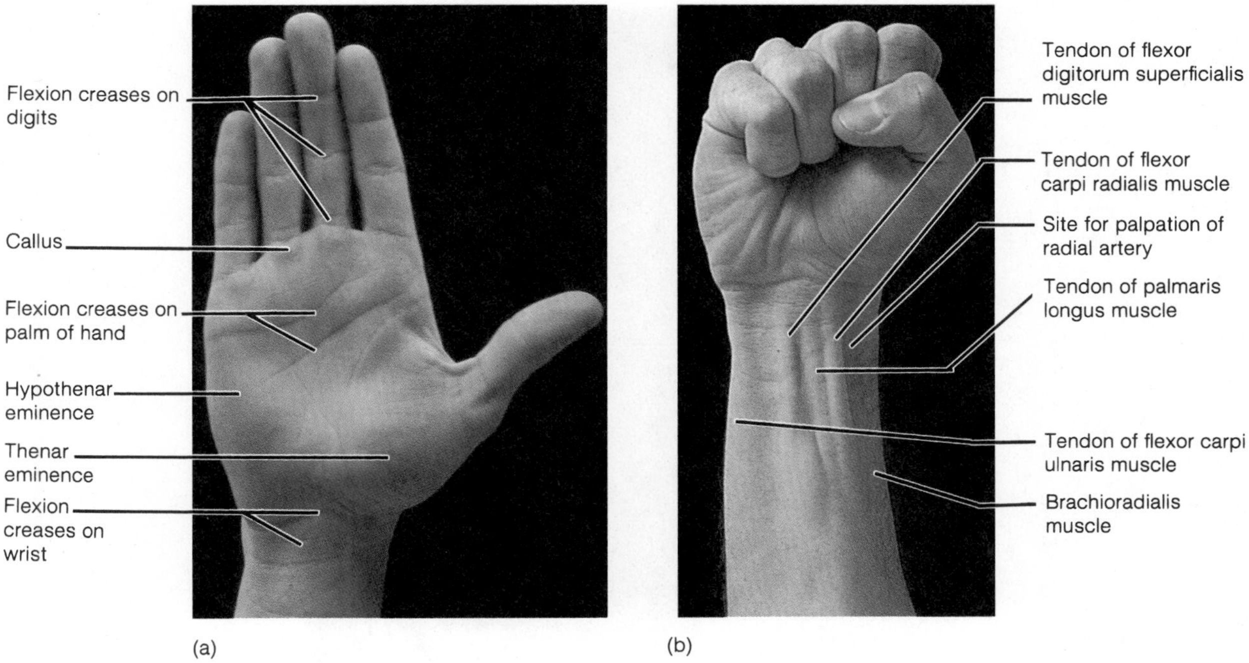

Figure 10.26 An anterior view of the wrist and hand; (*a*) with the hand open and (*b*) with a clenched fist.

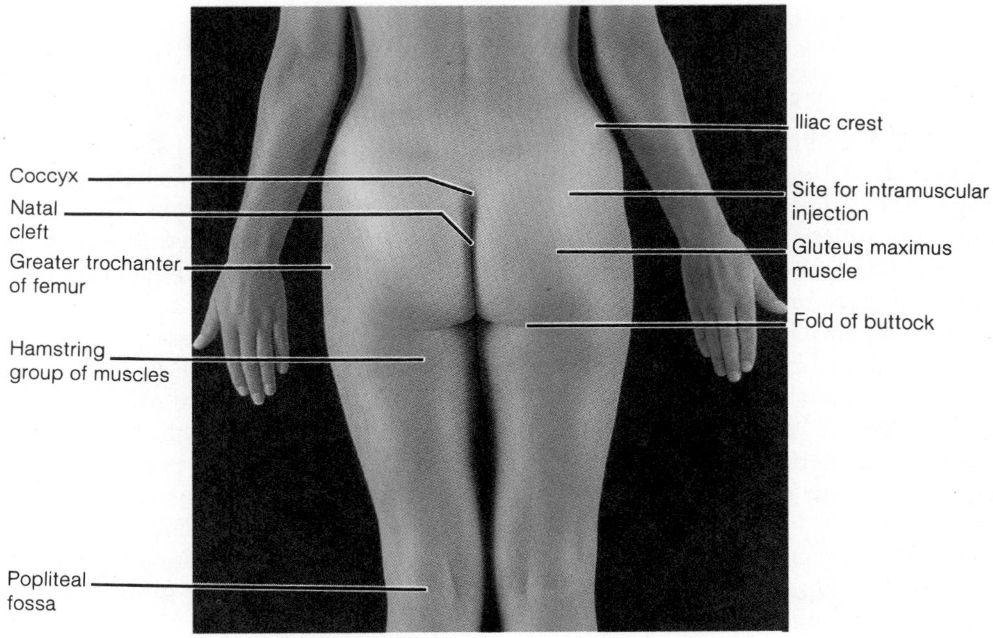

Figure 10.27 The buttocks and the posterior aspect of the thigh. (Note the relation of the angle of the elbow joint to the pelvic region, which is characteristic of females.)

Because of the thickness of the gluteal muscles and the rich blood supply, the buttock is a preferred site for intramuscular injections. Care must be taken, however, not to inject into the sciatic nerve. For this reason, the surface landmark of the iliac crest is important. The injection is usually administered 5–7 cm (2–3 in.) below the iliac crest in what is known as the upper lateral quadrant of the buttock.

Thigh

The femur is the only bone of the **thigh,** but there are three groups of thigh muscles. The anterior group of muscles, referred to as the **quadriceps,** extend the leg when they are contracted (fig. 10.28). The medial muscles are the **adductors,** and when contracted they draw the thigh medially. The **hamstrings** are positioned on the posterior aspect of the thigh (see fig. 10.26) and serve to extend the hip as well as flex the leg

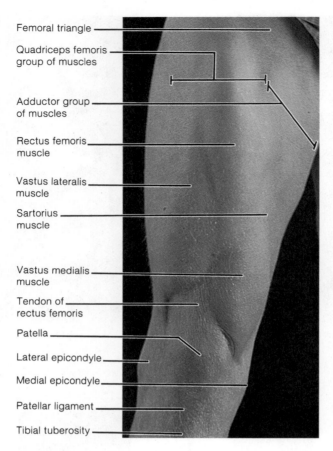

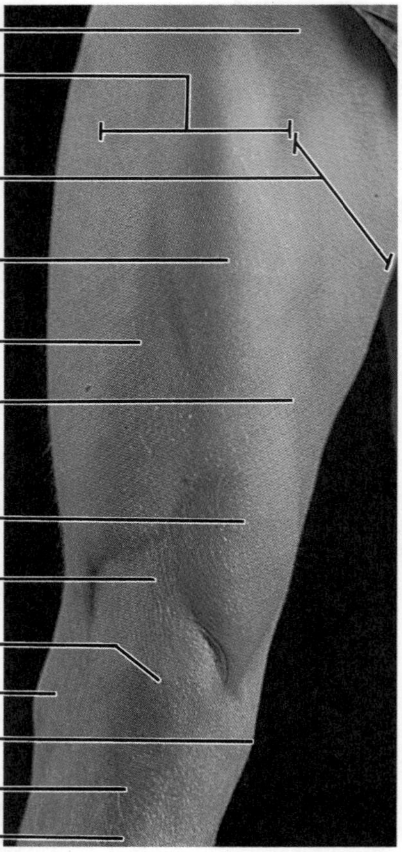

Figure 10.28 An anterior view of the right thigh and knee.

when they are contracted. The tendinous attachments of the hamstrings can be palpated along the posterior aspect of the knee joint when the leg is flexed. The hamstrings or their attachments are often injured in athletic events.

The **femoral triangle** is an extremely important part of the surface anatomy of the thigh. It can be seen as a depression inferior to the location of the inguinal ligament on the anterior surface in the upper part of the thigh (see fig. 16.34). The major vessels of the leg as well as the femoral nerve traverse through this region. Hernias are frequent in this area. More importantly, the femoral triangle is an arterial pressure point where it is vital to apply pressure in the case of uncontrolled hemorrhage of the lower extremity.

The greater trochanter of the femur can be palpated on the lateral, upper surface of the thigh (fig. 10.27). At the knee, the lateral and medial condyles of the femur and tibia can be identified (fig. 10.28). The patella (kneecap) can be easily located within the patellar tendon anterior to the knee joint. Stress or injury to this joint may cause swelling, commonly called water on the knee.

The depression on the posterior aspect of the knee joint is referred to as the **popliteal** *(pop''li-te'al)* **space** (fig. 10.29). This area becomes more clinically important in elderly persons who suffer degenerative conditions. Aneurysms of the popliteal artery are common, as are popliteal abscesses due to infected lymph nodes. The small saphenous vein as it traverses the popliteal space may become varicose in elderly persons.

Leg

Portions of the tibia and fibula, the bones of the **leg,** can be observed as surface features. The medial surface and anterior border (commonly called the shin) of the tibia are subcutaneous and are palpable throughout their length. At the ankle, the medial malleolus of the tibia and the lateral malleolus of the fibula are easy to observe as prominent eminences (fig. 10.30). Of clinical importance in setting fractures of the leg is knowing that the top of the medial malleolus lies about 1.3 cm (0.6 in.) proximal to the level of the tip of the lateral malleolus.

The heel is not part of the leg but the posterior portion of the calcaneus bone. It needs to be mentioned with the leg, however, because of its functional relationship to it. The **tendo calcaneus** (Achilles tendon) is the strong, cordlike tendon that attaches to the calcaneus from the calf of the leg. The muscle forming the belly of the calf is the gastrocnemius. Pulsations from the posterior tibial artery can be detected by palpating between the medial malleolus and the calcaneus.

The superficial veins of the leg can be observed on many persons (see fig. 10.2). The great saphenous vein can be seen subcutaneously along the medial aspect of the leg. The less conspicuous, small saphenous vein drains the lateral surface of the leg. If these veins become excessively enlarged, they are called varicose veins.

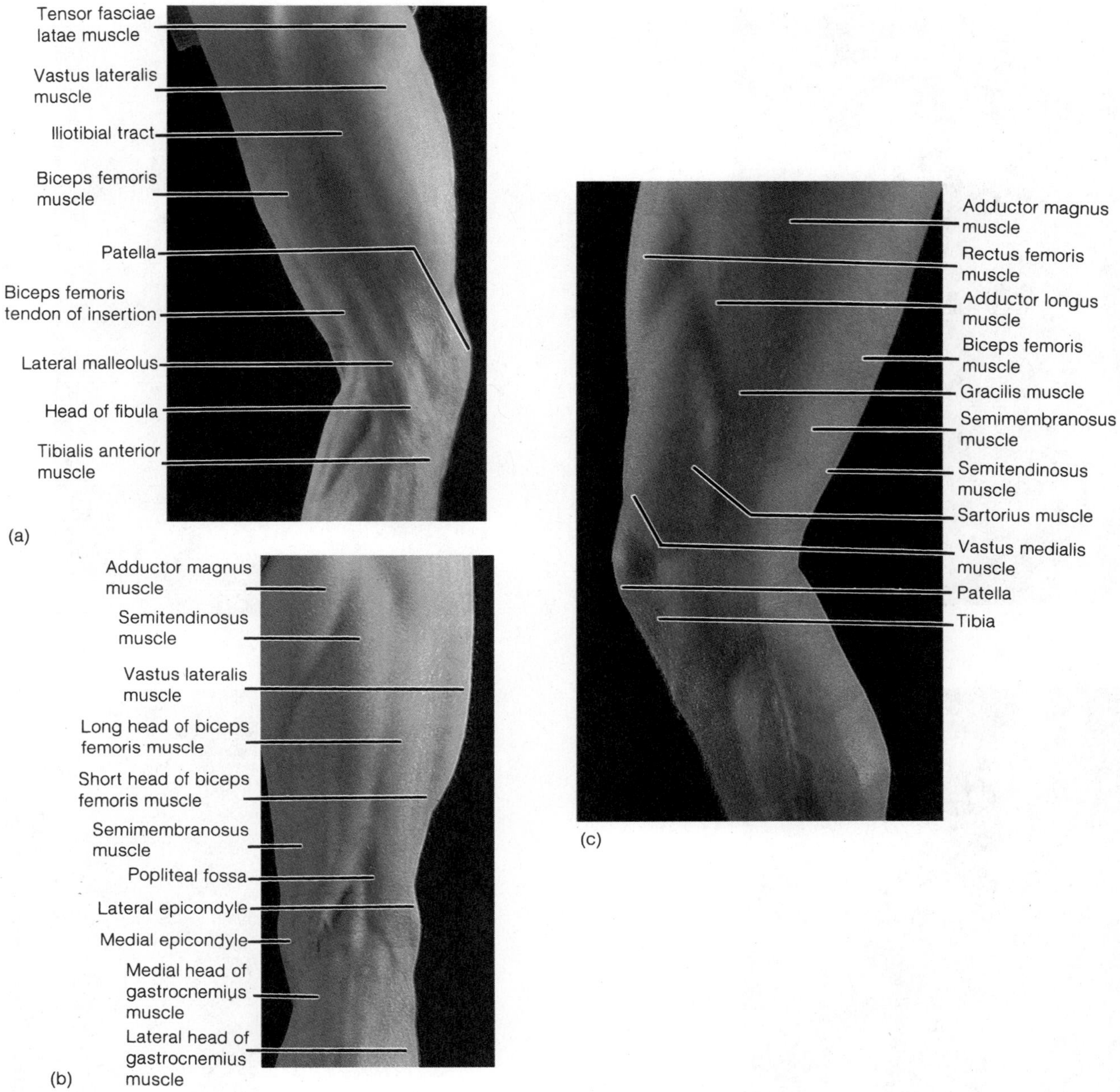

Figure 10.29 The (*a*) lateral, (*b*) posterior, and (*c*) medial surfaces of the leg.

Leg injuries are common among athletes. *Shin splints,* probably the result of a stress fracture or periosteum damage of the tibia, is a common condition in runners. A fracture of one or both malleoli is caused by a severe twisting of the ankle region. Skiing fractures are generally caused by strong torsion forces on the shafts of the tibia or fibula.

Foot

The feet are adapted to support the weight of the body, to maintain balance, and to function mechanically during locomotion. The structural features and surface anatomy of the **foot** are indicative of these functions. The **arch of the foot** is located on the medial portion of the plantar surface (fig. 10.30) and provides a spring effect when locomoting. The head of the first metatarsal bone forms the medial ball of the foot just proximal to the hallux (big toe).

The feet and toes are adapted to endure tremendous compression forces during locomotion. Appropriate shoes help to minimize trauma to the feet and toes, but still there is an array of common clinical conditions (fig. 10.31) that may impede walking or running. An *ingrown toenail* occurs as the sharp edge of a toenail embeds into and injures the skin fold resulting in inflammation and infection. *Hammer toe* is a condition resulting from a forceful hyperextension at the metatarsophalangeal joint with flexion at the proximal interphalangeal joint. A *corn* is a thickening of skin (callus) resulting from recurrent pressure on the skin over a bony prominence.

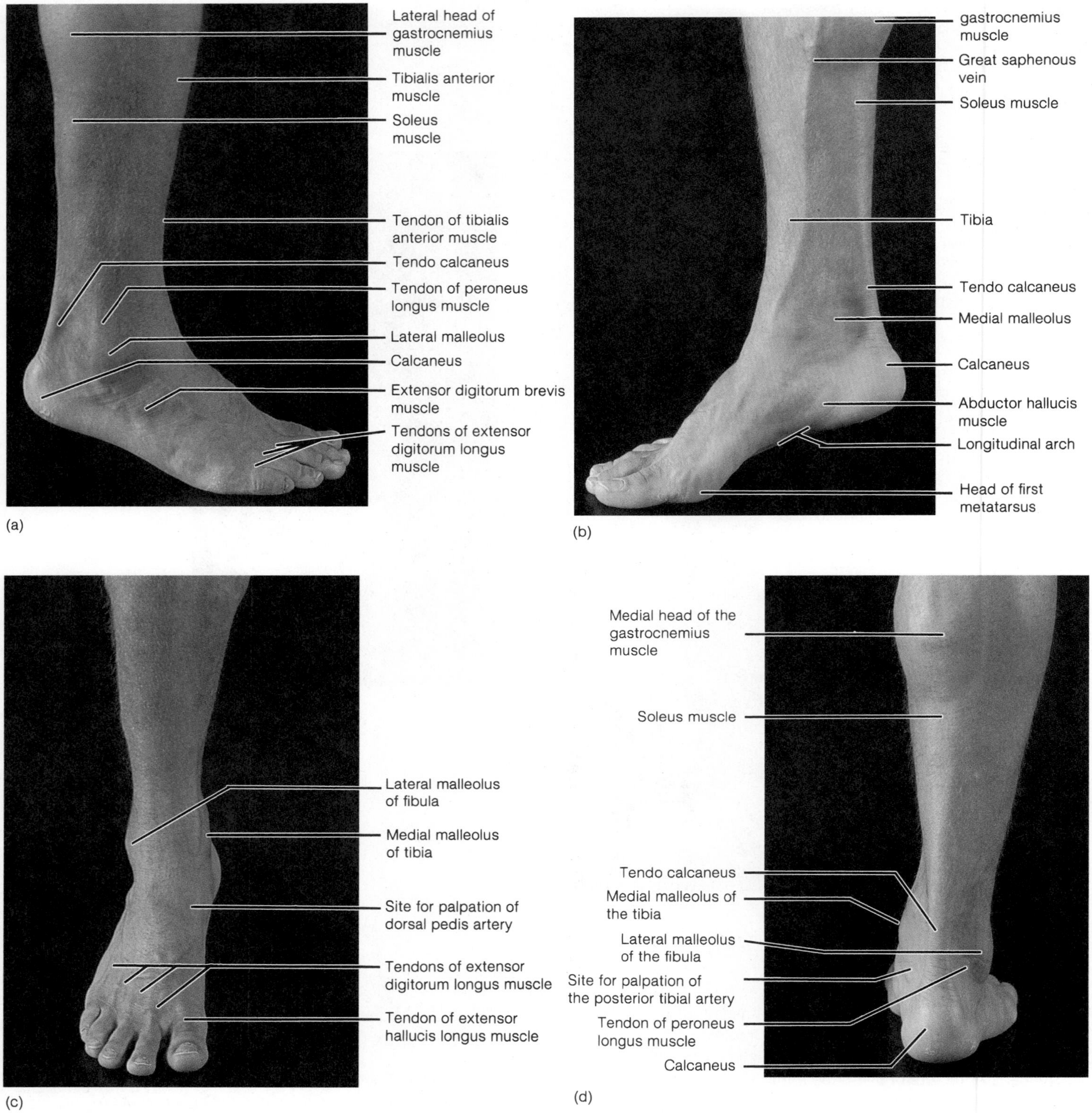
Lateral head of gastrocnemius muscle
Tibialis anterior muscle
Soleus muscle
Tendon of tibialis anterior muscle
Tendo calcaneus
Tendon of peroneus longus muscle
Lateral malleolus
Calcaneus
Extensor digitorum brevis muscle
Tendons of extensor digitorum longus muscle
(a)
gastrocnemius muscle
Great saphenous vein
Soleus muscle
Tibia
Tendo calcaneus
Medial malleolus
Calcaneus
Abductor hallucis muscle
Longitudinal arch
Head of first metatarsus
(b)
Lateral malleolus of fibula
Medial malleolus of tibia
Site for palpation of dorsal pedis artery
Tendons of extensor digitorum longus muscle
Tendon of extensor hallucis longus muscle
(c)
Medial head of the gastrocnemius muscle
Soleus muscle
Tendo calcaneus
Medial malleolus of the tibia
Lateral malleolus of the fibula
Site for palpation of the posterior tibial artery
Tendon of peroneus longus muscle
Calcaneus
(d)

Figure 10.30 The leg and foot: (*a*) a lateral view, (*b*) a medial view, (*c*) an anterior view, and (*d*) a posterior view.

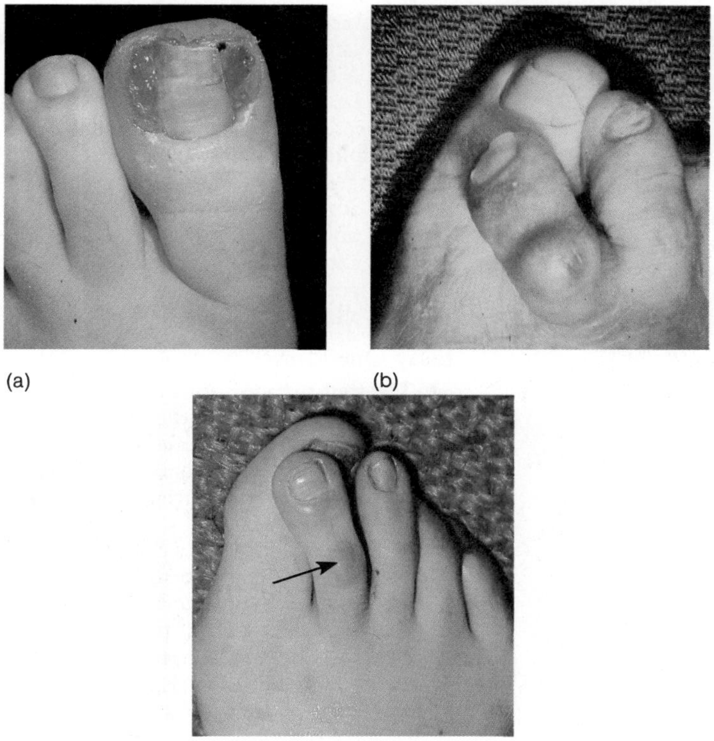

(a) (b) (c)

Figure 10.31 Common clinical conditions of the foot and toes: (*a*) ingrown toenail, (*b*) hammer toe, and (*c*) corn.

The fifth metatarsal bone forms much of the lateral border of the plantar surface of the foot. The tendons of the extensor digitorum longus muscle can be seen along the dorsal surface of the foot, especially if the toes are elevated. Pulsations of the dorsal pedis artery can be palpated on the dorsal surface of the foot between the first and second metatarsal bones. The individual phalanges of the toes, the joints between these bones, and the toenails are obvious surface landmarks.

1. What surface features form the boundaries of a buttock?
2. List the clinically important structures that can be observed or palpated in the buttock and lower extremity.
3. Describe the anatomical location where each of the following could be observed or palpated: the distal tendinous attachments of the hamstring muscles; the greater trochanter; the greater and lesser saphenous veins; the femoral, posterior tibial, and dorsal pedis arteries; and the medial malleolus.

Chapter Summary

I. Introduction to Surface Anatomy
   A. Surface anatomy is concerned with identifying body structures through observation and palpation. Surface anatomy has tremendous application to physical fitness and to medical diagnosis and treatment.
   B. Most of the bones of the skeleton are palpable and provide landmarks for other anatomical structures.
   C. The effectiveness of observation and palpation in studying a person's surface anatomy is dependent on the thickness of the hypodermis, which is due to the amount of subcutaneous adipose tissue present.

II. Surface Anatomy of the Newborn
   A. Certain aspects of the surface anatomy of a neonate are of clinical importance in ascertaining the degree of physical development, general health, and possible congenital abnormalities.
   B. The posture of a full-term, normal neonate is one of flexion.
   C. Portions of the skin and subcutaneous tissues are edematous. Vernix caseosa covers the body of a neonate, and lanugo may be present on the head, neck, and back.
   D. The fontanels, liver, kidneys, and testes of a male should be palpable.

III. Head
   A. Surface features of the cranium include the forehead, the crown, the temporalis muscles, and the hair and hairline.
   B. The face is composed of the ocular region that surrounds the eye, the auricular region of the ear, the nasal region serving the respiratory system, and the oral region of the mouth.

IV. Neck
   A. Major organs are located within the flexible neck, and structures that are essential for body sustenance pass through the neck to the torso.
   B. The neck consists of an anterior cervix, right and left lateral regions, and a posterior nucha.
   C. Two major and six minor triangles, which contain specific structures, are located on both sides of the neck.

V. Torso
   A. Vital visceral organs in the torso make the surface anatomy of this region especially important.
   B. The median furrow is observable and the vertebral spines and scapulae are palpable on the back.
   C. Palpable structures of the thorax include the sternum, the ribs, and the costal margins.
   D. The important surface anatomy of the abdomen includes the linea alba, umbilicus, costal margins, iliac crests, and pubis.

VI. Pelvis and Perineum
   A. The crest of the ilium, symphysis pubis, and inguinal ligament are important pelvic landmarks.
   B. The perineum is the region that contains the external genitalia and the anal opening.

VII. Shoulder and Upper Extremity
   A. The surface anatomy of the shoulder and upper extremity is important because of frequent trauma to these regions. Vessels of the upper extremity are also used as pressure points and for intravenous injections or blood withdrawal.

B. The scapula, clavicle, and humerus are palpable in the shoulder.
C. The axilla is clinically important because of the vessels, nerves, and lymph nodes located there.
D. The brachial artery is an important pressure point in the brachium, and the median cubital vein is important for the removal of blood or for intravenous therapy.
E. The ulna, radius, and their processes are palpable landmarks of the forearm.
F. The knuckles, fingernails, and tendons for the extensor muscles of the forearm can be observed on the posterior aspect of the hand.
G. Flexion creases and the thenar eminence are important surface features on the anterior surface of the hand.

VIII. Buttock and Lower Extremity
A. The massive bones and muscles in the buttock and lower extremity serve as weight bearers and locomotors. Many of the surface features of these regions are important in relation to locomotion or locomotor dysfunction.
B. The prominences of the buttocks are formed by the gluteal muscles and are separated by the natal cleft.
C. The thigh has three muscle groups: quadriceps, adductors, and hamstrings. The femoral triangle and popliteal spaces are clinically important surface landmarks.
D. The structures of the leg include the tibia and fibula, the muscles of the calf, and the saphenous veins.
E. The surface anatomy of the foot includes structures adapted to support the weight of the body, maintain balance, and function during locomotion.

Review Activities

Objective Questions

1. Eyebrows are located on the
(a) palpebral fissure.
(b) subtarsal sulcus.
(c) scalp.
(d) supraorbital ridges.
(e) c and d
2. Which of the following structures is *not* part of the auricle (pinna) of the ear? The
(a) tragus. (c) lobule.
(b) ala. (d) helix.
3. Which of the following clinical-structural word pairs is *incorrectly* matched?
(a) cleft lip—philtrum
(b) broken nose—nasion
(c) pierced ear—lobe
(d) black eye—concha
4. The conjunctiva
(a) covers the entire eyeball.
(b) is a thick nonmucous membrane.
(c) secretes tears.
(d) None of the above.
5. Which of the following could *not* be palpated within the cervix of the neck?
(a) larynx
(b) hyoid bone
(c) trachea
(d) cervical vertebrae
6. Palpation of a pulse to the head is best accomplished within the
(a) carotid triangle.
(b) occipital triangle.
(c) suprahyoid triangle.
(d) submandibular triangle.
(e) muscular triangle.
7. Which nerve traverses behind the medial epicondyle of the humerus?
(a) ulnar (d) brachial
(b) median (e) cephalic
(c) radial
8. Which of the following surface features could *not* be observed on obese people?
(a) suprasternal notch
(b) scapular muscles
(c) clavicles
(d) vertebral spines
(e) natal cleft
9. Which pairs of muscles form the anterior and posterior borders of the axilla?
(a) deltoid—pectoralis minor
(b) biceps brachii—triceps brachii
(c) latissimus dorsi—pectoralis major
(d) triceps brachii—pectoralis major
(e) latissimus dorsi—deltoid
10. Varicose veins occur when which of the following becomes excessively enlarged?
(a) saphenous veins
(b) tibial veins
(c) external iliac vein
(d) popliteal vein
(e) all of the above

Essay Questions

1. List four surface features of the cranium, and explain the relationship of the scalp to the cranium.
2. Identify the four regions of the face, and list at least two surface features of each region.
3. Which surface features can be observed on the torso on any person, regardless of how obese he or she is?
4. Identify the two major triangles of the neck, and list the associated six minor triangles. Discuss the importance of knowing the contents and boundaries of these triangles.
5. Name four structures that are palpable along the anterior midline of the neck.
6. What three bones are found in the shoulder region? List the surface features that can be either observed or palpated on each of these bones.
7. Identify the tendons or vessels that can be observed or palpated along the anterior surface of the wrist. Which nerves pass through this region?
8. Describe the locations of the arteries that can be palpated as they pulsate in the following regions: neck, brachium, forearm, thigh, and ankle. Which of these are considered clinical pressure points?
9. Describe the locations and clinical importance of the following regions: cubital fossa, femoral triangle, axilla, perineum, and popliteal space.
10. Explain why an understanding of surface anatomy is of immense importance in clinical application and in learning the basic gross anatomy of internal organs.

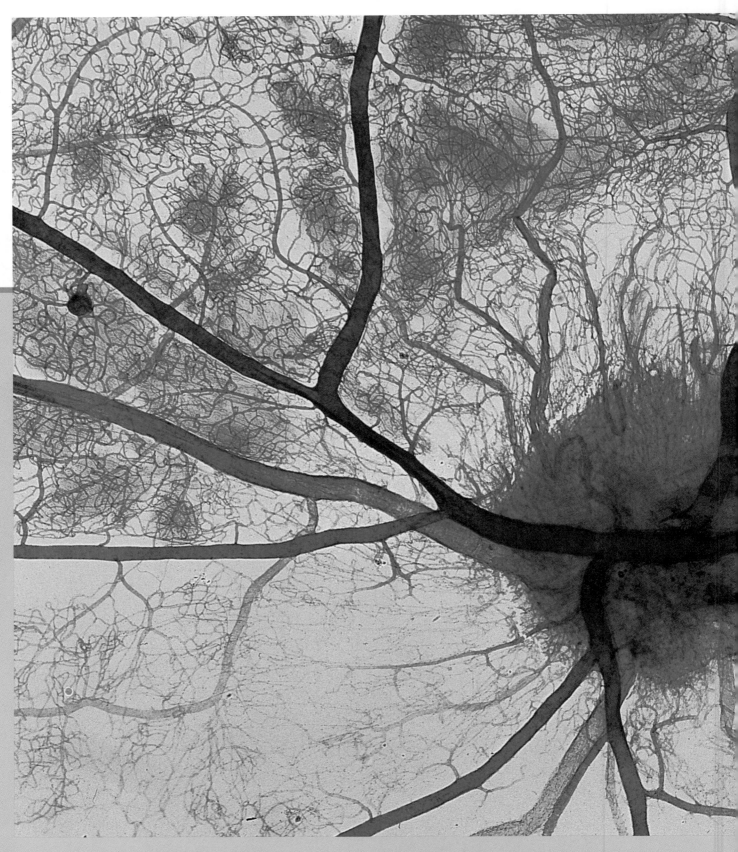

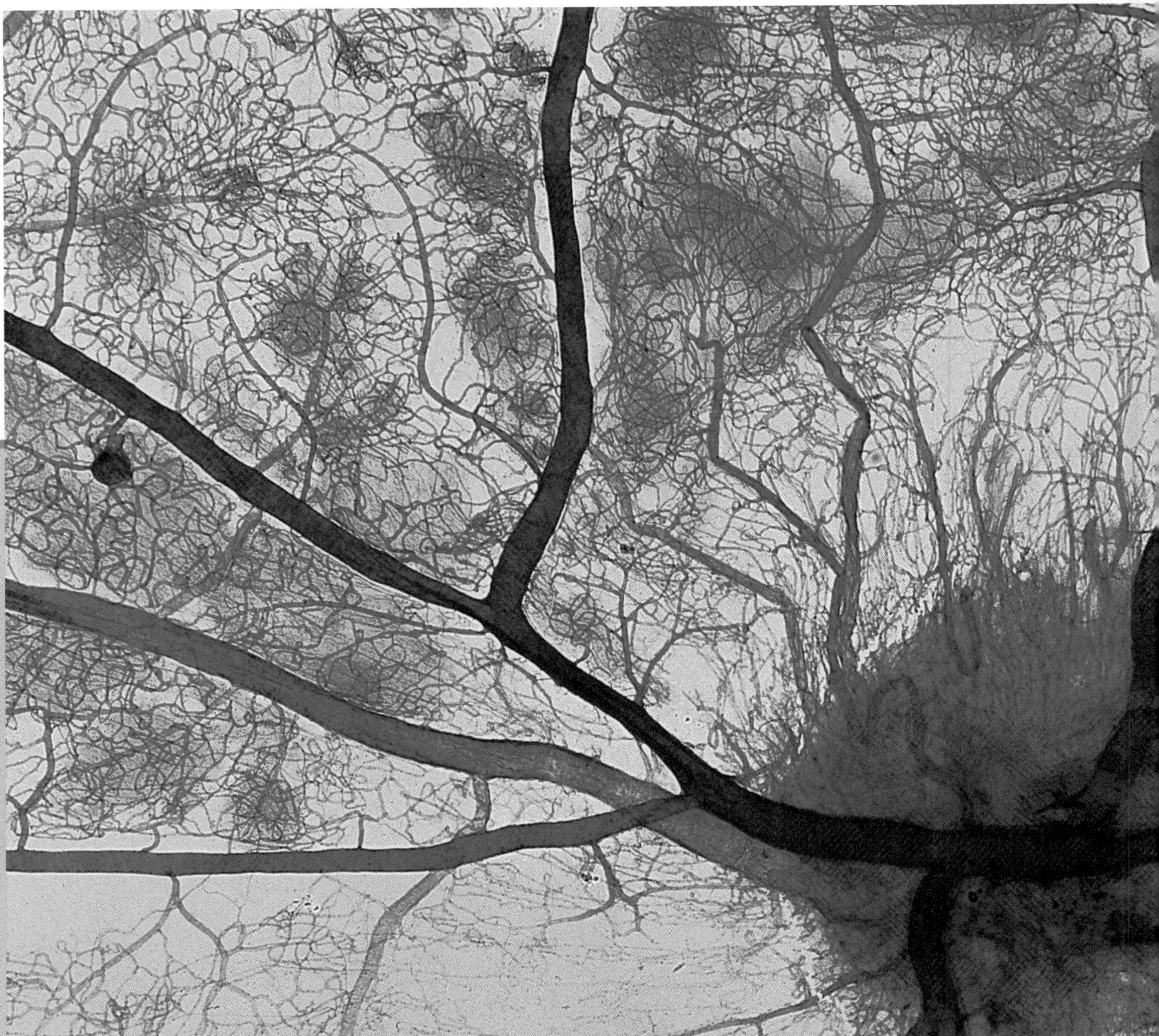

A photomicrograph of the optic disc (blind spot) at the back of the retina of the eye. The optic disc is the location at which the neurons from the rods and cones converge to form the optic nerve. The sensory organs, such as the eyes, provide sensations to the brain for perception and response.

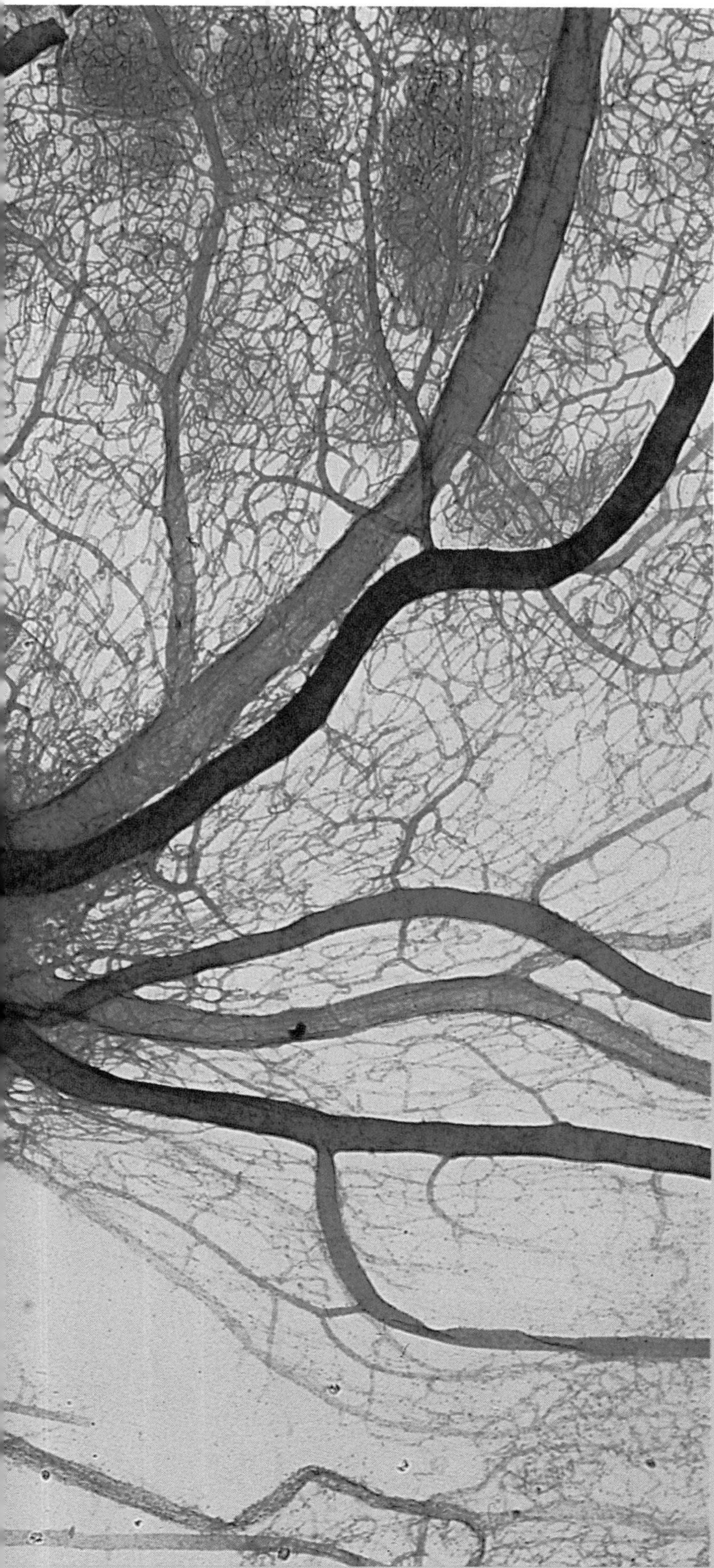

UNIT

5

Integration and Coordination

The body systems that regulate the activities of the other body systems are discussed in the chapters in unit 5. The nervous system and its sensory receptor organs are also discussed along with the endocrine system. These chapters include a description of how specific organs help a person to appropriately respond to external and internal stimuli and maintain homeostasis.

This unit includes:

CHAPTER 11

Nervous Tissue and the Central Nervous System

Outline and Concepts

CLINICAL CASE STUDY

A fifty-six-year-old woman visits her family doctor for evaluation of a headache that has been present for nearly a month. Upon questioning the patient, the doctor discovers that her left arm has been a bit unwieldy, "hard to control, and weak." Through examination, he determines that the entire left upper extremity is generally weak. He additionally finds weakness, although less significant, of the left lower extremity. Sensation in the limbs seems to be normal, although mild rigidity and hyperactive reflexes are present. With a concerned expression on his face, he tells the patient that she needs to have a CT scan of her head, explaining that there could be a problem within the brain, possibly a tumor or other lesion. The doctor then picks up the phone and contacts a radiologist, explaining the patient's case and parenthetically adding that he thinks he knows specifically where the problem is located.

Why does the doctor suggest to the patient that there may be a problem within her brain when the diagnosed symptoms are weakness to the extremities and only on one side of her body?

How does he know where within the brain the tumor resides? List the side and lobe. Explain the muscle weakness in terms of neuronal pathways from the brain to the periphery. ■

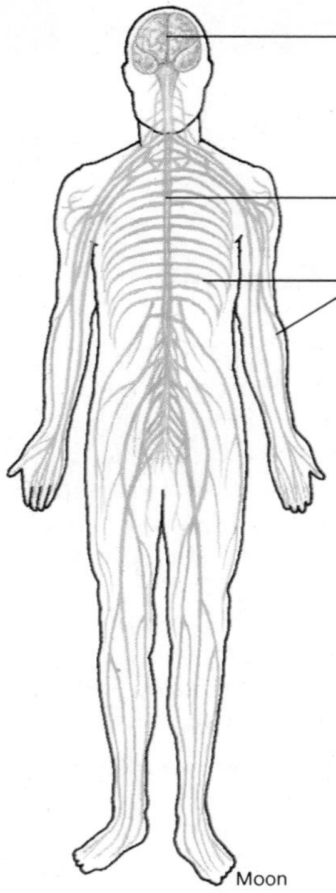

Figure 11.1 The nervous system consists of the central nervous system (brain and spinal cord) and the peripheral nervous system (cranial nerves and spinal nerves). The nerves within the extremities are also part of the peripheral nervous system.

Organization and Functions of the Nervous System

The central nervous system and the peripheral nervous system are structural components of the overall nervous system, whereas the autonomic nervous system is a functional component. Together they orient the body, coordinate body activities, permit the assimilation of experiences, and program instinctual behavior.

Objective 1. Describe the divisions of the nervous system.
Objective 2. Define *neurology;* define *neuron.*
Objective 3. List the functions of the nervous system.

The immensely complex brain and its myriad of connecting pathways constitute the nervous system. The nervous system, along with the endocrine system, regulates the functions of other body systems. The brain, however, does much more than that, and its potential is perhaps greatly underestimated. It is incomprehensible that one's personality, thoughts, and aspirations result from the functioning of a body organ. Plato referred to the brain as "the divinest part of us." The thought processes of this organ have devised the technology for launching rockets into space, curing diseases, and splitting atoms. But with all of these achievements, the brain still remains amazingly ignorant of its own workings.

Neurology, or the study of the nervous system, has been referred to as the last frontier of functional anatomy. Some basic questions concerning the functioning of the nervous system remain unanswered: How do nerve cells store and retrieve memory? What are the roles of the many chemical compounds within the brain? What causes mental illness or senility? Scientists are still developing the technology and skills necessary to understand the functional complexity of the nervous system. The next few decades should be very interesting as scientists come to a better understanding of this system.

Organization of the Nervous System

The nervous system is divided into the **central nervous system (CNS),** which includes the **brain** and **spinal cord,** and the **peripheral** *(pĕ-rif′er-al)* **nervous system (PNS),** which includes the **cranial nerves,** arising from the brain and the **spinal nerves,** arising from the spinal cord (fig. 11.1 and table 11.1).

The **autonomic nervous system (ANS)** is a functional subdivision of the entire nervous system. The controlling centers of the ANS are within the brain and are considered part of the CNS, and the nerve portions of the ANS are subdivided into the **sympathetic** and **parasympathetic divisions.**

Functions of the Nervous System

The nervous system is specialized for responding and perceiving to events in our internal and external environments. An awareness of one's environment is made possible by nerve cells called **neurons,** which are highly specialized in the properties of excitability and conductivity. The nervous system is present throughout the body as it functions with the endocrine system (see chap. 14) to closely coordinate the activities of the other

neurology: Gk. *neuron,* nerve; L. *logos,* study of

Table 11.1 Anatomical terms used in describing the nervous system

Term	Definition
Central nervous system (CNS)	Brain and spinal cord
Peripheral nervous system (PNS)	Nerves and ganglia
Sensory nerve fiber	Neuron that transmits impulses from a sensory receptor into the CNS (an afferent fiber)
Motor nerve fiber	Neuron that transmits impulses from the CNS to an effector organ, for example, muscle (efferent)
Nerve	Cablelike collection of nerve fibers; may be "mixed" (contain both sensory and motor fibers)
Plexus	Network of interlaced nerves
Somatic motor nerve	Nerve that stimulates contraction of skeletal muscles
Autonomic motor nerve	Nerve that stimulates contraction (or inhibits contraction) of smooth muscle and cardiac muscles, and stimulates secretion of glands
Ganglion	Collection of neuron cell bodies located outside the CNS
Nucleus	Grouping of neuron cell bodies within the CNS
Tract	Collection of nerve fibers that interconnect regions of the CNS

body systems. In addition to integrating body activities, the nervous system has the ability to store experiences *(memory)* and to establish patterns of response based on prior experiences *(learning)*.

The functions of the nervous system are the

1. orientation of the body to internal and external environments;
2. coordination and control of body activities;
3. assimilation of experiences requisite to memory, learning, and intelligence; and
4. programming of instinctual behavior (apparently more important in vertebrates other than humans).

An instinct may also be called a fixed action pattern and is typically genetically specified, little modified by the environment, and is triggered only by a specific stimulus. Some of the basic instincts in humans include survival, feeding, drinking, voiding, and specific vocalization. Some ethologists (scientists who study animal behavior) believe that reproduction becomes an instinctive behavior following puberty.

1. State the meaning of ANS, PNS, and CNS. Which of these is a functional component of the nervous system and is further subdivided into sympathetic and parasympathetic divisions?
2. Explain why neurology is considered a dynamic science.

Development of the Nervous System

The embryonic neural tube forms the brain and spinal cord, and the cranial and spinal nerves are formed predominately from the embryonic neural crest.

Objective 4. List the five developmental regions of the brain, and discuss the embryonic formation of these regions.

Objective 5. Describe the development of the spinal cord.

Objective 6. Define *dermatome,* and describe the clinical significance of the pattern of dermatome innervation in the body.

The first indication of nervous tissue development occurs about seventeen days following conception when a thickening appears along the entire dorsal length of the embryo. This thickening, called the **neural plate** (fig. 11.2), differentiates and eventually gives rise to all of the neurons and to most of the **neuroglial** *(nu-rog'le-al)* **cells,** which are the supporting cells of the nervous system.

As development progresses, the midline of the neural plate invaginates to become the **neural groove.** At the same time, there is a proliferation of cells along the lateral margins of the neural plate, which become the thickened **neural folds.** The neural groove continues to deepen as the neural folds elevate. By the twentieth day, the neural folds meet and fuse at the midline, and the neural groove becomes a **neural tube.** The neural tube, for a short time, is open both cranially and caudally. These openings, called **neuropores** (fig. 11.2), close during the fourth week. Once formed, the neural tube separates from the surface ectoderm and eventually develops into the CNS (brain and spinal cord).

The **neural crest** forms from the neural folds as they fuse longitudinally along the dorsal midline. The neural crest is positioned between the surface ectoderm and the neural tube. Most of the PNS (cranial and spinal nerves) forms from the neural crest. Some neural crest cells break away from the main tissue mass and migrate to other locations where they differentiate into motor nerve cells of the sympathetic nervous system or into neurolemmocytes, which are a type of neuroglia cell important in the PNS.

Development of the Brain

The brain begins its embryonic development as the cephalic end of the neural tube starts to grow rapidly and to differentiate (fig. 11.3). By the middle of the fourth week, three distinct swellings are evident: the **prosencephalon** *(pros''en-sef'ah-lon)* (forebrain), the **mesencephalon** (midbrain), and the **rhombencephalon** (hindbrain). Further development during the fifth week results in the formation of five specific regions: the **telencephalon** and the **diencephalon** *(di''en-sef'ah-lon)* derive from the forebrain; the mesencephalon remains unchanged; and the **metencephalon** and **myelencephalon** form from the hindbrain.

During the developmental transformation of the brain, hollow chambers known as **ventricles** form within each region. The **first** and **second ventricles** (lateral ventricles) form in the forebrain, a narrow **third ventricle** forms in the forebrain and

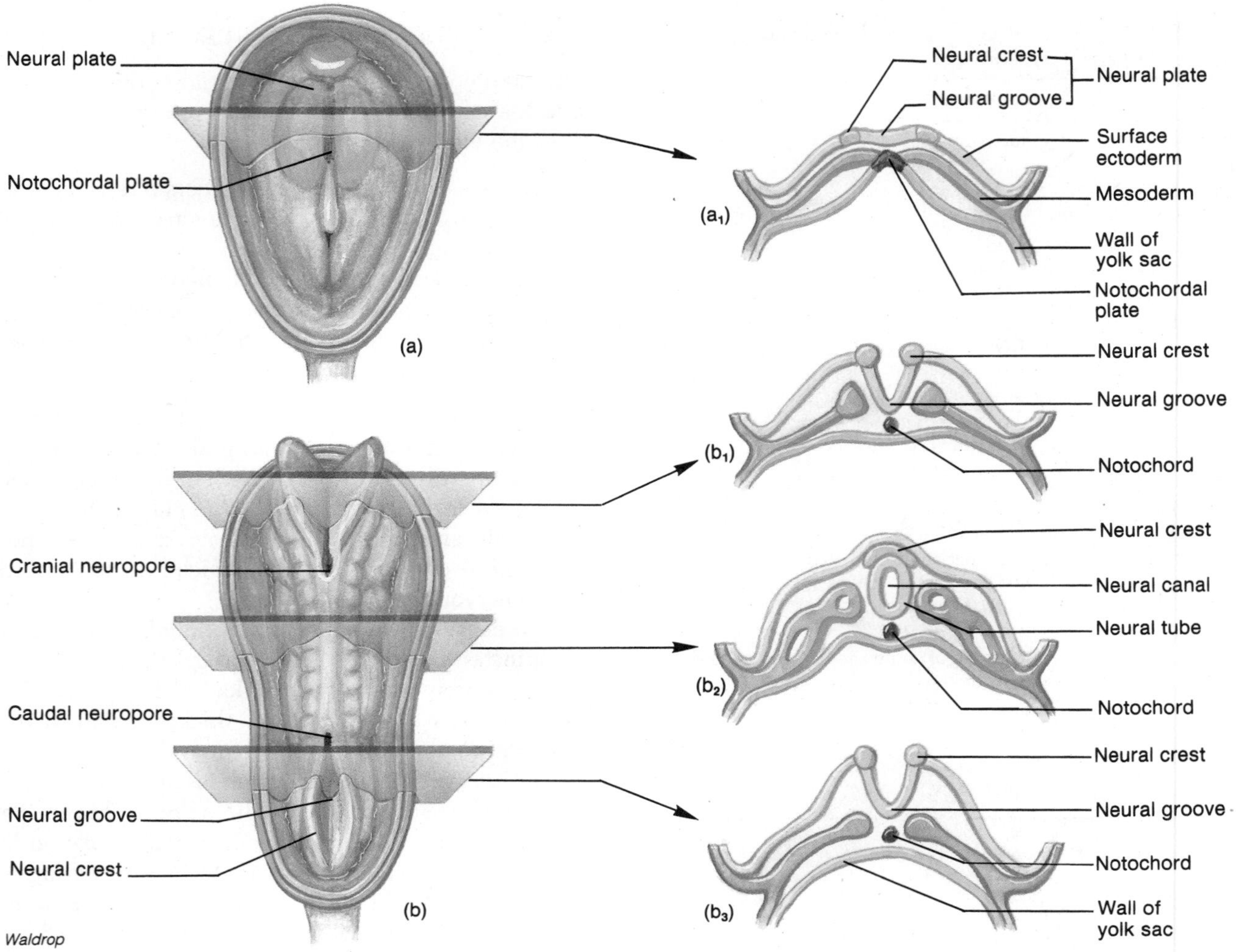

Figure 11.2 The early development of the nervous system from embryonic ectoderm. (*a*) A dorsal view of an eighteen-day-old embryo showing the formation of the neural plate and the position of a transverse cut indicated in (a_1). (*b*) A dorsal view of a twenty-two-day-old embryo showing cranial and caudal neuropores and the positions of three transverse cuts indicated in (b_1–b_3). Note the amount of fusion of the neural tube at the various levels of the twenty-two-day-old embryo. (Note also the relationship of the notochord to the neural tube.)

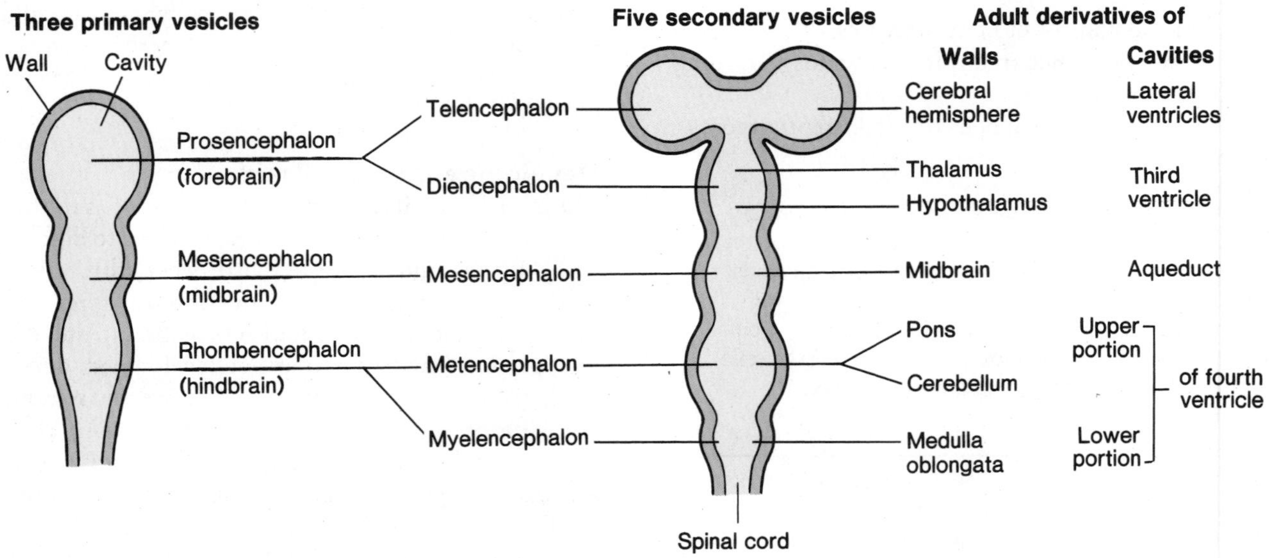

Figure 11.3 The developmental sequence of the brain. During the fourth week, the three principal regions of the brain are formed. During the fifth week, a five-regioned brain develops, and specific structures begin to form.

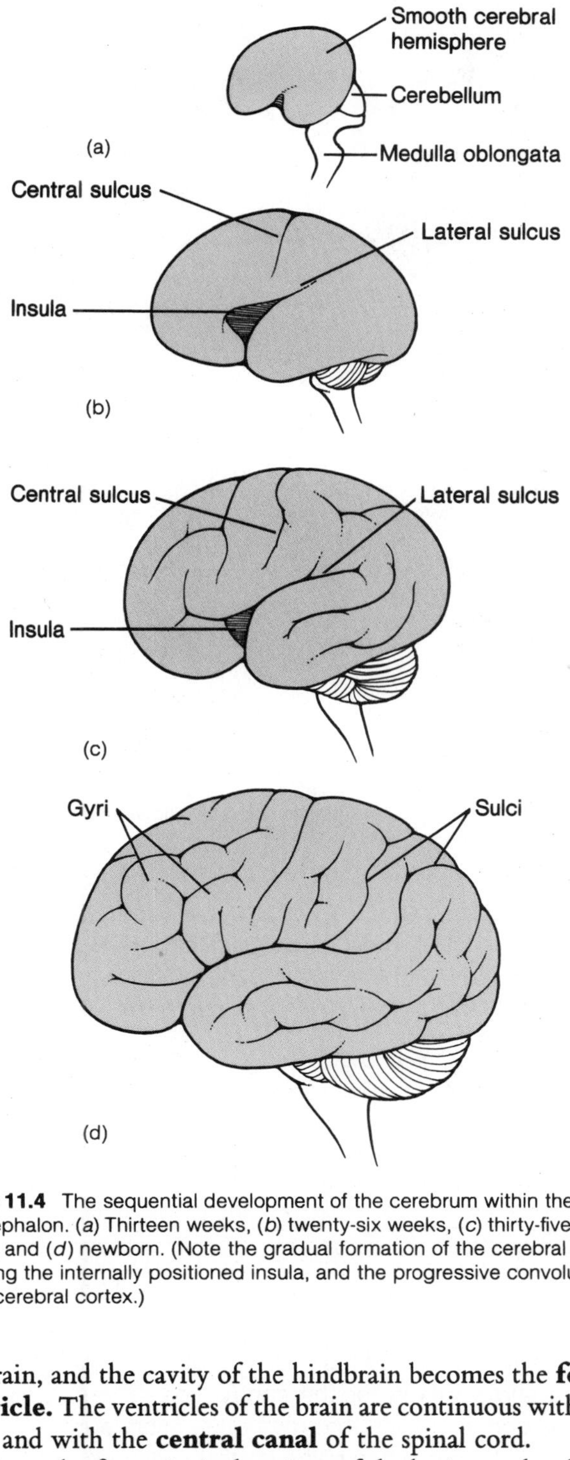

Figure 11.4 The sequential development of the cerebrum within the telencephalon. (*a*) Thirteen weeks, (*b*) twenty-six weeks, (*c*) thirty-five weeks, and (*d*) newborn. (Note the gradual formation of the cerebral lobes, including the internally positioned insula, and the progressive convolutions of the cerebral cortex.)

midbrain, and the cavity of the hindbrain becomes the **fourth ventricle.** The ventricles of the brain are continuous with each other and with the **central canal** of the spinal cord.

Once the five principal regions of the brain are developed, rapid differentiation occurs in each of the regions. Shortly after, all of the major structures of the brain are recognizable. The greatest amount of growth occurs within the telencephalon as its dorsal portions rapidly expand to form the two **hemispheres** of the **cerebrum** (fig. 11.4). The swelling of the cerebrum causes it to cover the diencephalon, the midbrain, and a portion of the hindbrain. By the thirty-fifth week, the lobes of the cerebrum are formed, and the surface of the cerebrum is distinctly convoluted. The outer, convoluted portion of the cerebrum consists of gray matter, and the inner portion consists of white matter.

Within the diencephalon three swellings form in the walls of the third ventricle and become the **epithalamus** *(ep''ĭ-thal'ah-mus)*, the **thalamus,** and the **hypothalamus** *(hi''po-thal'ah-mus)*. The **pineal** *(pin'e-al)* **gland** also forms in the diencephalon as a midline diverticulum (an outpouching) develops from the roof of the third ventricle. A portion of the **pituitary gland** develops from the ventral portion of the diencephalon.

Few developmental changes occur within the mesencephalon. The most obvious structures to develop are the four aggregates of neurons that become the paired **superior** and **inferior colliculi** (collectively, the corpora quadrigemina), which are concerned with visual and auditory functions respectively. Fibers from the cerebrum extend through the midbrain to form two tracts called the **cerebral peduncles** *(pe-dung'k'lz)*.

The walls of the metencephalon greatly expand to form the **cerebellum** dorsally and the **pons** ventrally. The cerebellum actually forms from the joining of a pair of dorsal swellings along the midline. The pons develops from numerous bands of nerve fibers. A vascular capillary network called the **choroid plexus** *(ko'roid plek'sus)* develops in the roof of the fourth ventricle. Similar plexuses form in the roof of the third ventricle and the medial walls of the lateral ventricles. These plexuses secrete **cerebrospinal** *(ser''ĕ-bro-spi'nal)* **fluid,** which circulates throughout the cavities of the CNS and within spaces outside the brain and spinal cord.

The caudal portion of the myelencephalon is continuous with and resembles the spinal cord. The entire adult myelencephalon is known as the **medulla oblongata** and forms from an aggregate of nerve fibers. The ventral **pyramids** of the myelencephalon form from the nerve fibers that extend from the developing cerebral cortex.

Development of the Spinal Cord

The spinal cord, like the brain, develops as the neural tube undergoes differentiation and specialization. Throughout the developmental process, the hollow central canal persists while the specialized white and gray matter forms (fig. 11.5). Changes in the neural tube become apparent during the sixth week as the lateral walls thicken to form a groove, called the **sulcus limitans,** along each lateral wall of the central canal. A pair of **alar plates** forms dorsal to the sulcus limitans, and a pair of **basal plates** forms ventrally. By the ninth week, the alar plates have specialized to become the **dorsal horns,** containing fibers of the afferent cell bodies, and the basal plates have specialized to form the **ventral** and **lateral horns,** containing efferent cell bodies. Afferent fibers of spinal nerves conduct impulses toward the spinal cord, whereas efferent fibers conduct impulses away from the spinal cord.

Mitotic activity within nervous tissue is completed during prenatal development. Thus, a person is born with all the neurons he is capable of producing. However, nervous tissue continues to grow and to specialize after a person is born, particularly in the first several years of postnatal life.

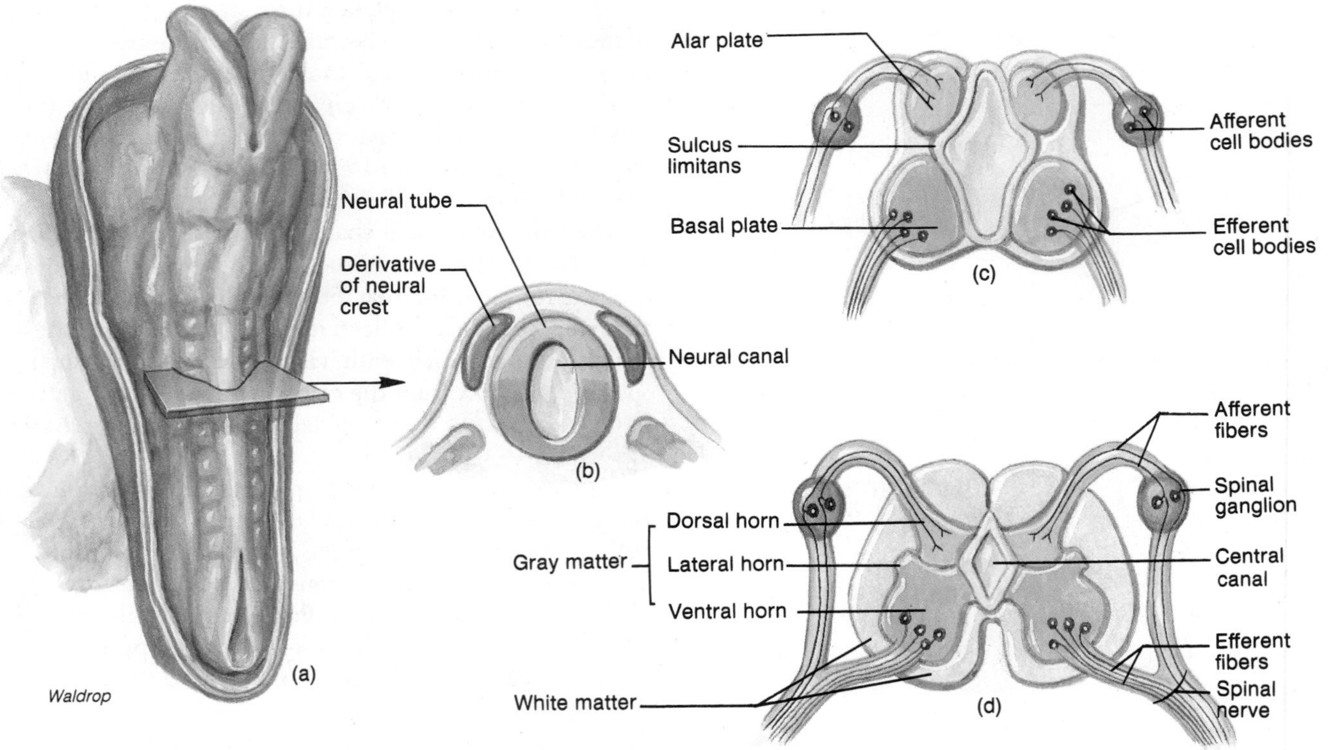

Figure 11.5 The development of the spinal cord. (*a*) A dorsal view of an embryo at twenty-three days and the position of a transverse cut indicated in (*b*). (*c*) The formation of the alar and basal plates is evident in a transverse section through the spinal cord at six weeks. (*d*) The central canal has reduced in size, and functional dorsal and ventral horns have formed at nine weeks.

Development of the Peripheral Nervous System

Development of the PNS produces the pattern of dermatomes within the body (fig. 11.6). A **dermatome** *(der'mah-tōm)* is an area of the skin innervated by all the cutaneous neurons of a certain spinal or cranial nerve. Most of the scalp and face is innervated by sensory fibers from the trigeminal nerve. With the exception of the first cervical nerve (C1), all of the spinal nerves are associated with specific dermatomes. Dermatomes are consecutive in the neck and torso regions. In the appendages, however, adjacent dermatome innervations overlap. The apparently uneven dermatome arrangement in the appendages is due to the uneven rate of nerve growth into the limb buds. Actually the limbs are segmented, and dermatomes overlap only slightly.

The pattern of dermatome innervation is of clinical importance when a physician desires to anesthetize a particular portion of the body. Because adjacent dermatomes overlap in the appendages, at least three spinal nerves must be blocked to produce complete anesthesia in these regions. Abnormally functioning dermatomes provide clues about injury to the spinal cord or specific spinal nerves. If a dermatome is stimulated but no sensation is perceived, the physician can infer that the injury involves the innervation to that dermatome.

dermatome: Gk. *derma,* skin; *tomia,* a cutting

1. List the five regions of the brain that develop during the fifth week, and discuss their formation.
2. Describe the locations of the ventricles within the brain.
3. Describe the embryonic origin of afferent and efferent cell bodies of spinal nerves.
4. Discuss the derivatives of the alar plates and the basal plates in the formation of the spinal cord.
5. Describe the dermatomes, and explain their clinical significance.

Neurons and Neuroglia

Neurons have many forms, but all contain dendrites for reception and an axon for the conduction of nerve impulses. Neurons contain axons that may be sensory and/or motor in function. Different types of neuroglial cells support the neurons, both structurally and functionally.

Objective 7. Describe the microscopic structure of a neuron.
Objective 8. Describe how a neurilemmal sheath and a myelin sheath are formed.
Objective 9. List the types of neuroglial cells, and describe their functions.
Objective 10. Describe the functions and locations of afferent and efferent nerve fibers.

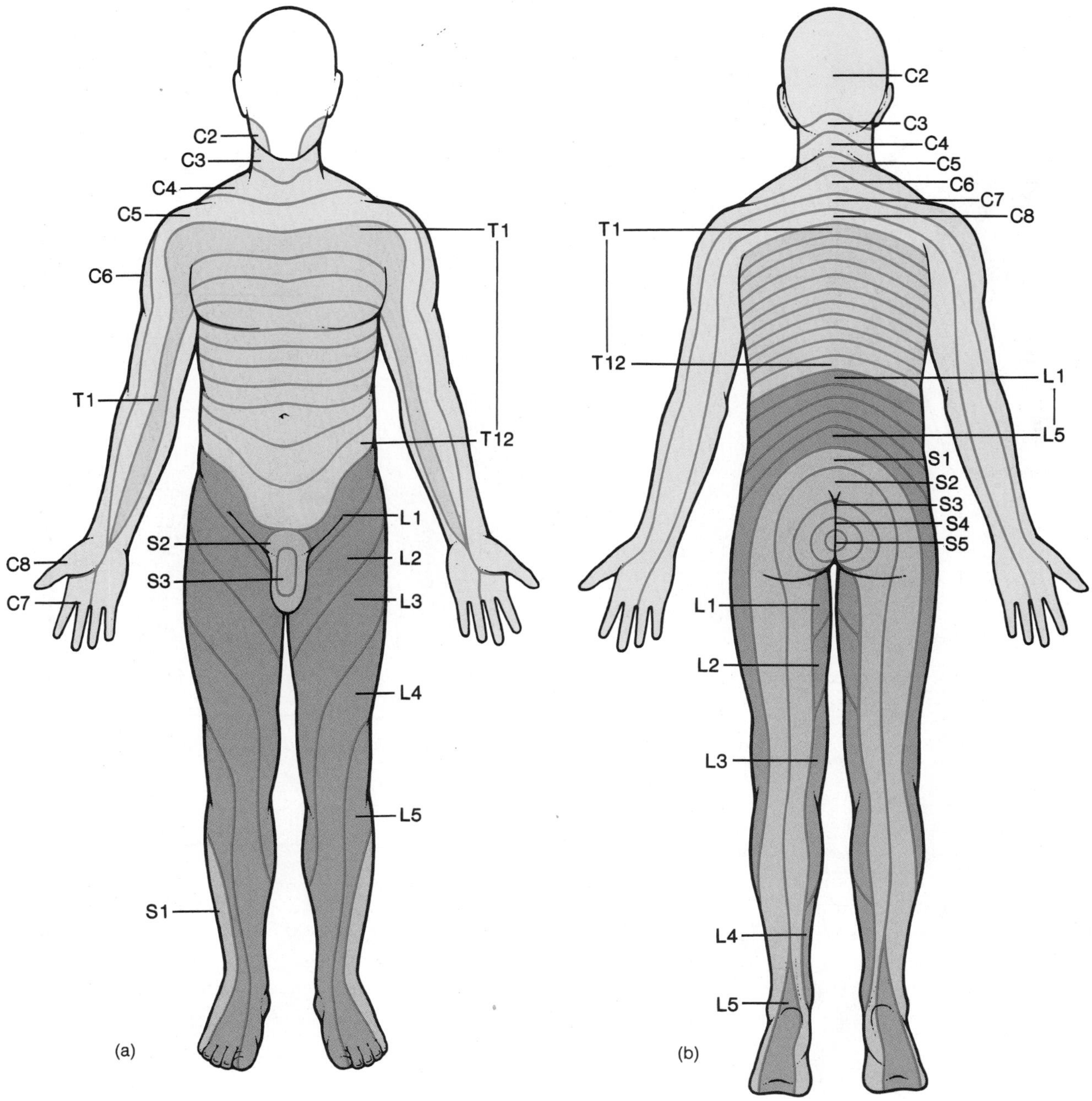

Figure 11.6 The pattern of dermatomes and the peripheral distribution of spinal nerves. (*a*) An anterior view and (*b*) a posterior view.

The highly specialized and complex nervous system is composed of only two principal categories of cells, neurons and neuroglia. **Neurons** are the basic structural and functional units of the nervous system. They are specialized to respond to physical and chemical stimuli, conduct impulses, and release specific chemical regulators. Through these activities, neurons perform functions such as storing memory, thinking, and regulating other organs and glands. Neurons cannot divide mitotically, although some neurons can regenerate a severed portion or sprout small new branches under some conditions.

Neuroglia, or **glial cells,** are supportive cells in the nervous system that aid the function of neurons. Neuroglia cells are about five times more abundant than neurons and have limited mitotic abilities.

Neurons

Although neurons vary considerably in size and shape, they generally have three principal components: (1) a cell body, (2) dendrites, and (3) an axon (figs. 11.7, 11.8).

The **cell body,** or **perikaryon** *(per″ĭ-kar′e-on),* is the enlarged portion of the neuron that contains the nucleus and nucleolus surrounded by cytoplasm. Besides containing organelles typically found in cells, the cytoplasm of neurons is characterized by the presence of **chromatophilic substances** (Nissl bodies) and filamentous strands of protein called **neurofibrils** *(nu″ro-fi′brilz).* Chromatophilic substances are specialized layers

Nissl body: from Franz Nissl, German neuroanatomist, 1860–1919

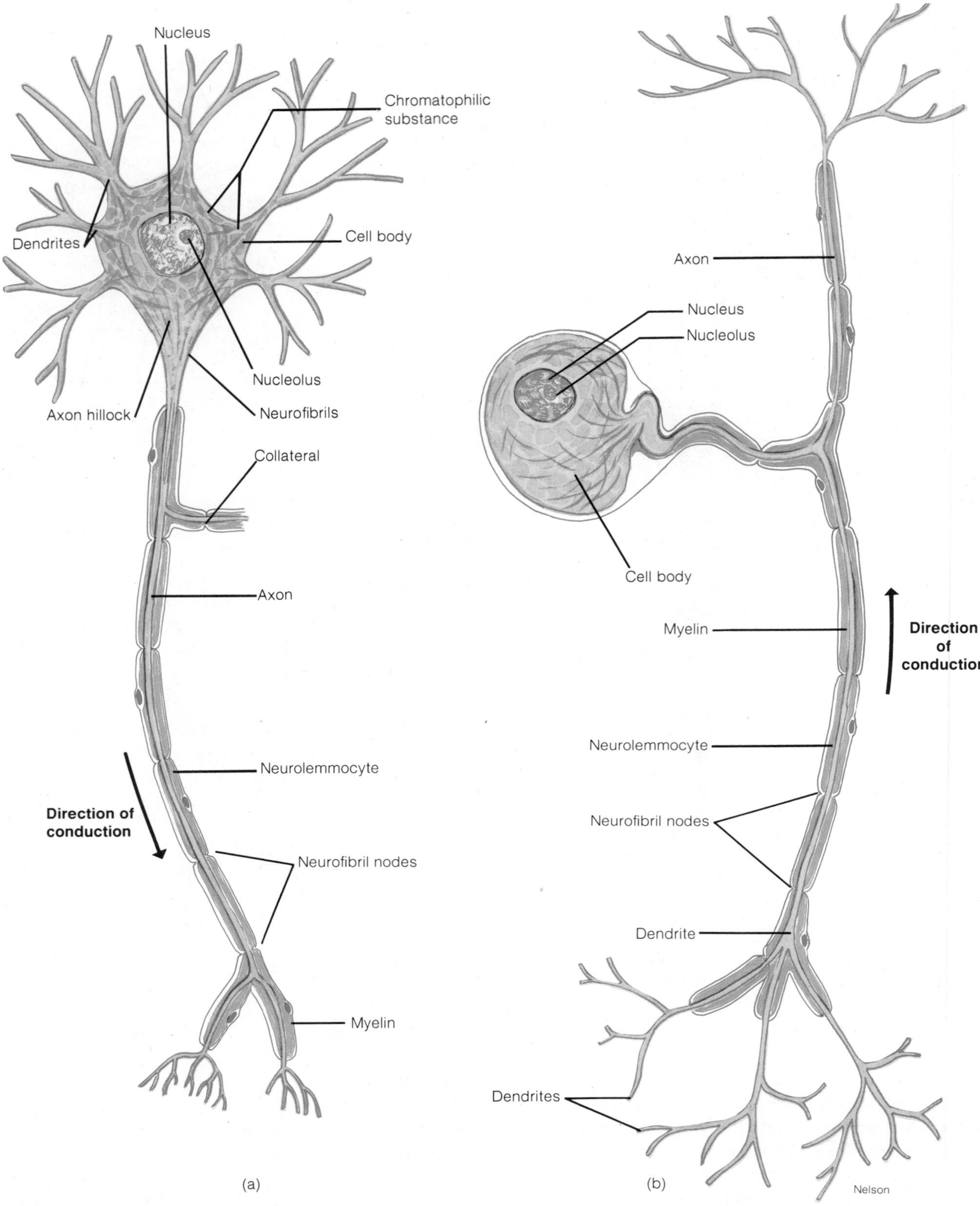

Figure 11.7 The structure of two kinds of neurons. (*a*) A motor neuron and (*b*) a sensory neuron.

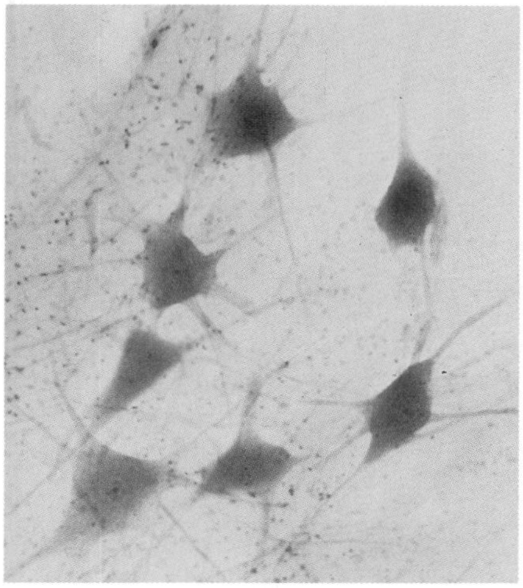

Figure 11.8 A photomicrograph of neurons from the anterior column of gray matter of the spinal cord (120×).

of granular (rough) endoplasmic reticulum that synthesize fibrils and minute **microtubules,** which appear to be involved in transporting material within the cell. The cell bodies within the CNS are frequently clustered into regions called **nuclei** (not to be confused with the nucleus of a cell). Cell bodies in the PNS generally occur in clusters called **ganglia.**

Dendrites are branched processes that extend from the cytoplasm of the cell body. Dendrites function to receive a stimulus and conduct impulses to the cell body. Some dendrites are covered with minute **dendritic spines,** which greatly increase their surface area and provide contact points for other neurons. The area occupied by dendrites is referred to as the **dendritic zone** of a neuron.

The **axon,** or **nerve fiber,** is the second type of cytoplasmic extension from the cell body. An axon is a relatively long, cylindrical process that conducts impulses away from the cell body. The conical tapering region of the axon, where it originates from the cell body, is referred to as the **axon hillock.** Axons vary in length from a few millimeters in the CNS to over a meter between the distal portions of the extremities and the spinal cord. Side branches called **axon collaterals** extend a short distance from the axon. The cytoplasm of an axon contains many mitochondria, microtubules, and neurofibrils.

Proteins and other molecules are transported rapidly through the axon by two different mechanisms: axoplasmic flow and axonal transport. *Axoplasmic flow,* the slower of the two, results from rhythmic waves of contraction that push cytoplasmic contents from the axon hillock to the nerve fiber endings. *Axonal transport,* which is more rapid and more selective, may occur in a retrograde as well as a forward direction. Indeed, such retrograde transport may be responsible for the movement of herpes virus, rabies virus, and tetanus toxin from nerve terminals into cell bodies.

ganglion: Gk. *ganglion,* swelling
dendrite: Gk. *dendron,* tree branch
axon: Gk. *axon,* axis

Neuroglia

Unlike other organs that are packaged in connective tissue derived from mesoderm, the supporting **neuroglial (glial) cells** of the nervous system (fig. 11.9) are derived from the same ectoderm that produces neurons. There are six categories of neuroglial cells: (1) **neurolemmocytes** (Schwann cells), which form myelin sheaths around axons in the PNS; (2) **oligodendrocytes** *(ol''ĭ-go-den'dro-sītz),* which form myelin sheaths around axons in the CNS; (3) **microglia** *(mi-krog'le-ah),* which are phagocytic cells that migrate through the CNS and remove foreign and degenerated material; (4) **astrocytes,** which help regulate the passage of molecules from the blood to the brain; (5) **ependyma** (ĕ-pen'dĭ-mah), which line the ventricles of the brain and the central canal of the spinal cord; and (6) **satellite cells,** which support neuron cell bodies within the ganglia of the PNS (table 11.2).

Myelination Neurons are either *myelinated* or *unmyelinated.* **Myelination** is the process in which a neuroglial cell surrounds a portion of the axon or dendrite to provide support and aid in the conduction of impulses (figs. 11.10, 11.11). The neuroglial cells that participate in the myelination process are composed of a lipid-protein substance called **myelin.** As several neuroglia are positioned in sequence, a **myelin sheath** forms that encloses the axon or dendrite. Myelinated neurons occur both in the CNS and the PNS. Because myelin appears white, it is responsible for the *white matter* of the brain and spinal cord and the white coloration of nerves.

Myelination in the PNS occurs as neurolemmocytes grow and wrap around an axon or dendrite (figs. 11.7, 11.11, 11.12). The outer surface of the myelin sheath is encased in a glycoprotein **neurolemmal sheath** that promotes neuron regeneration if the neuron becomes injured. Each neurolemmocyte only wraps about 1 mm of axon, leaving gaps of exposed axon between the adjacent neurolemmocyte. These gaps in the myelin sheath and neurolemmal sheath are known as the **neurolemmal nodes** (nodes of Ranvier). It is at the neurolemmal nodes that a nerve impulse is propagated along a neuron.

The myelin sheaths of the CNS are formed by **oligodendrocytes.** Unlike a neurolemmocyte, which forms a myelin sheath around only one axon, each oligodendrocyte has extensions that form myelin sheaths around several axons.

Regeneration of a Cut Axon When an axon in a peripheral nerve is cut, the distal region of the axon that was severed from the cell body degenerates and is phagocytosed by neurolemmocytes. The neurolemmocytes then form a *regeneration tube* (fig. 11.13), as the part of the axon that is connected to the cell body begins to grow and exhibit ameboid movement. The neurolemmocytes of the regeneration tube are believed to secrete

neuroglia: Gk. *neuron,* nerve; *glia,* glue
Schwann cell: from Theodor Schwann, German histologist, 1810–1882
oligodendrocyte: Gk. *oligos,* few; L. *dens,* tooth; Gk. *kytos,* hollow (cell)
microglia: Gk. *mikros,* small; *glia,* glue
astrocyte: Gk. *aster,* star; *kytos,* hollow (cell)
ependyma: Gk. *ependyma,* upper garment
satellite: L. *satelles,* attendant
myelin: Gk. *myelos,* marrow
nodes of Ranvier: from Louis A. Ranvier, French pathologist, 1835–1922

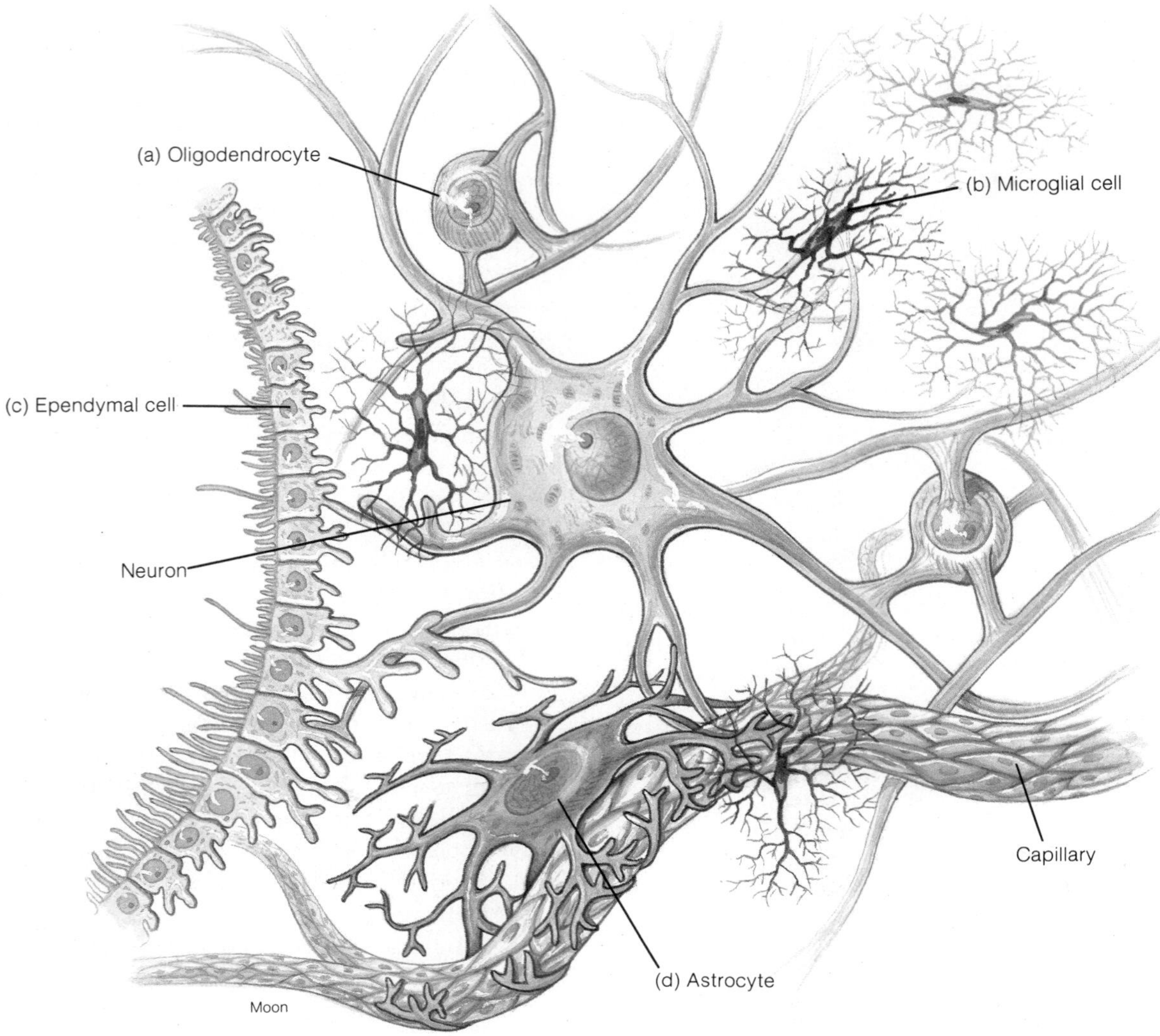

Figure 11.9 Types of neuroglia found within the CNS. (*a*) An oligodendrocyte, (*b*) a microglial cell, (*c*) an ependymal cell, and (*d*) an astrocyte.

Table 11.2 Structure and function of neuroglia

Type	Structure	Function	Type	Structure	Function
Astrocytes	Stellate with numerous processes	Form structural support between capillaries and neurons within the CNS; blood-brain barrier	Ependyma	Columnar cells that may have ciliated free surfaces	Line ventricles and the central canal within the CNS where cerebrospinal fluid is circulated by ciliary motion
Oligodendrocytes	Similar to astrocytes but with shorter and fewer processes	Form myelin in the CNS; guide development of neurons within the CNS	Satellite cells	Small, flattened cells	Support ganglia within the PNS
Microglia	Minute cells with few short processes	Phagocytize pathogens and cellular debris within the CNS	Neurolemmocytes (Schwann cells)	Flattened cells arranged in series around axons or dendrites	Form myelin within the PNS

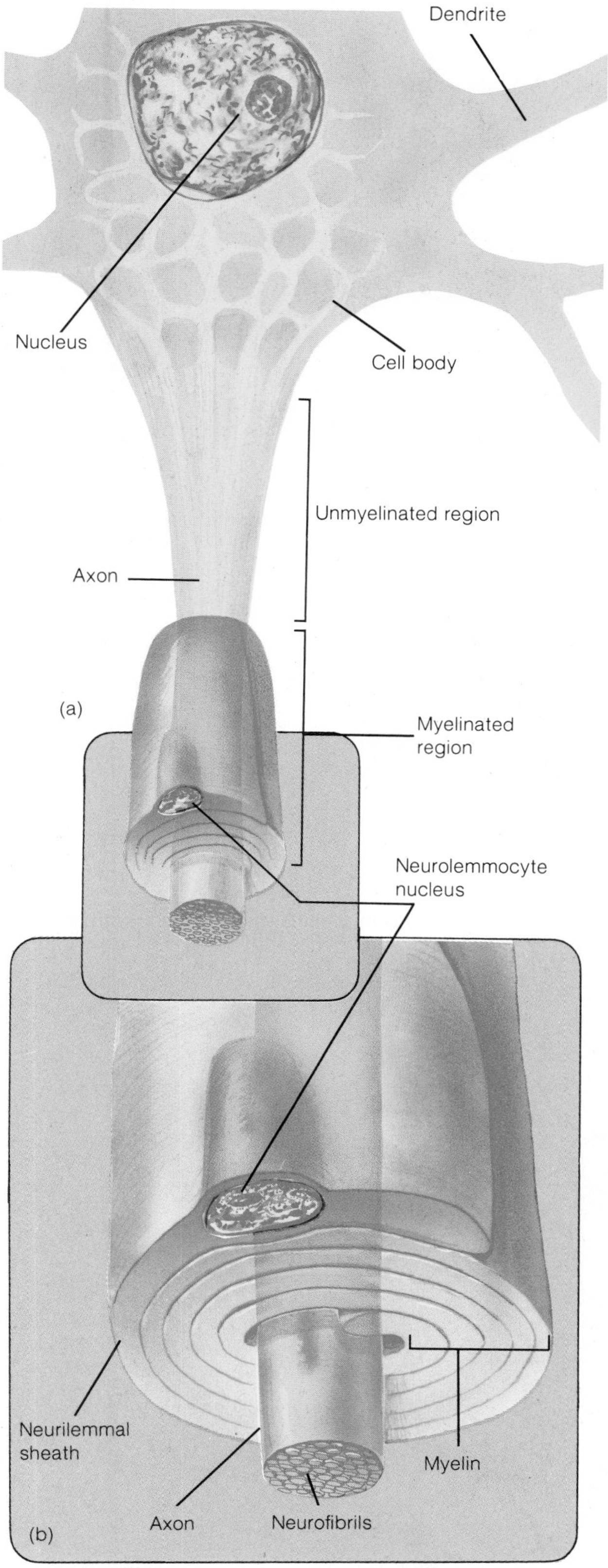

Figure 11.10 A myelinated neuron. (*a*) A myelin sheath forms to support and protect each peripheral neuron as neurolemmocytes tightly wrap around the axon. (*b*) Myelin is composed of a lipid-protein substance within the neurolemmocyte. The nucleus of the neurolemmocyte remains toward the outside of the myelin, embedded within the membranous neurilemma.

chemicals that attract the growing axon tip, and the tube helps to guide the regenerating axon to its proper destination. Even a severed major nerve may be surgically reconnected and the function of the nerve reestablished if the surgery is performed before tissue death. The CNS lacks neurolemmocytes, and central axons are generally believed to have a much more limited ability to regenerate than peripheral axons.

Astrocytes and the Blood-Brain Barrier Astrocytes are large, stellate cells with numerous cytoplasmic processes that radiate outward. These are the most abundant neuroglia in the CNS, and in some locations in the brain they constitute 90% of the nervous tissue.

Capillaries in the brain, unlike those of most other organs, do not have pores between adjacent endothelial cells. Molecules within these capillaries must thus be moved through the endothelial cells by active transport, endocytosis, and exocytosis. Astrocytes within the brain have numerous extensions, called **perivascular feet,** which surround most of the outer surface of the brain capillaries (fig. 11.14). Before molecules in the blood can enter neurons in the CNS, they may have to pass through both the endothelial cells and the astrocytes. Astrocytes, therefore, contribute to the **blood-brain barrier,** which is highly selective; some molecules are permitted to pass, whereas closely related molecules may not be allowed to cross the barrier.

The blood-brain barrier presents difficulties in the chemotherapy of brain diseases because drugs that could enter other organs may not be able to enter the brain. In the treatment of *Parkinson's disease,* for example, patients who need a chemical called dopamine in the brain must be given a precursor molecule called levodopa (L-dopa). This is because dopamine cannot cross the blood-brain barrier, whereas L-dopa can enter neurons and be changed to dopamine in the brain.

Classification of Neurons and Nerves

Neurons may be classified according to structure or function. The functional classification is based on the direction of conducted impulses. Sensory impulses originate in sensory receptors and are conducted by **sensory,** or **afferent, neurons** to the CNS. Motor impulses originate in the CNS and are conducted by **motor,** or **efferent, neurons** to a muscle or gland (fig. 11.15). Motor neurons may be *somatic* (nonvisceral) or *autonomic* (visceral). **Association neurons,** or **interneurons,** are located between sensory and motor neurons and are found within the spinal cord and brain.

The structural classification of neurons is based upon the number of processes that extend from the cell body of the neuron (fig. 11.16). The spindle-shaped **bipolar neuron** has a process at both ends; this type occurs in the retina of the eye. **Multipolar neurons** are the most common type and are characterized by several dendrites and one axon extending from the cell body. Motor neurons are a good example of this type. A **pseudounipolar neuron** has a single process that divides into

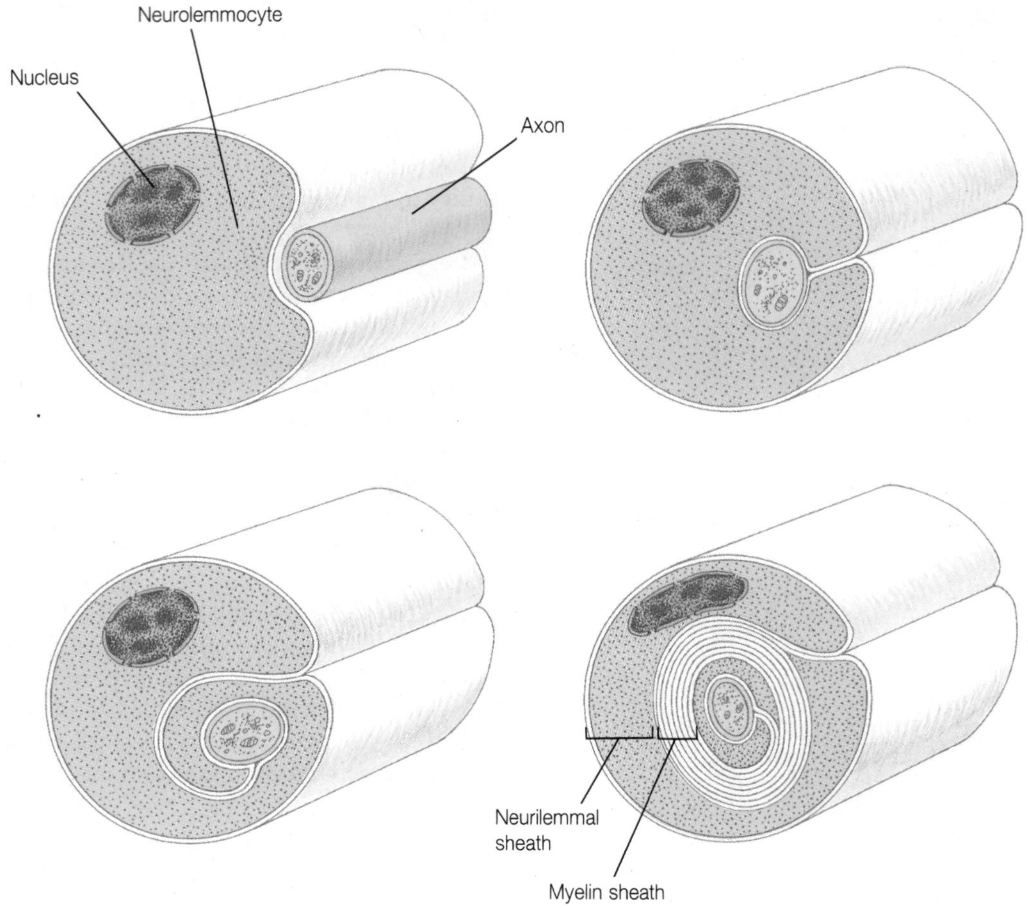

Figure 11.11 The formation of a myelin sheath in a peripheral axon. The myelin sheath is formed by successive wrappings of the neurolemmocyte membranes, leaving most of the neurolemmocyte cytoplasm outside the myelin. The neurilemmal sheath of neurolemmocyte is thus located outside the myelin sheath.

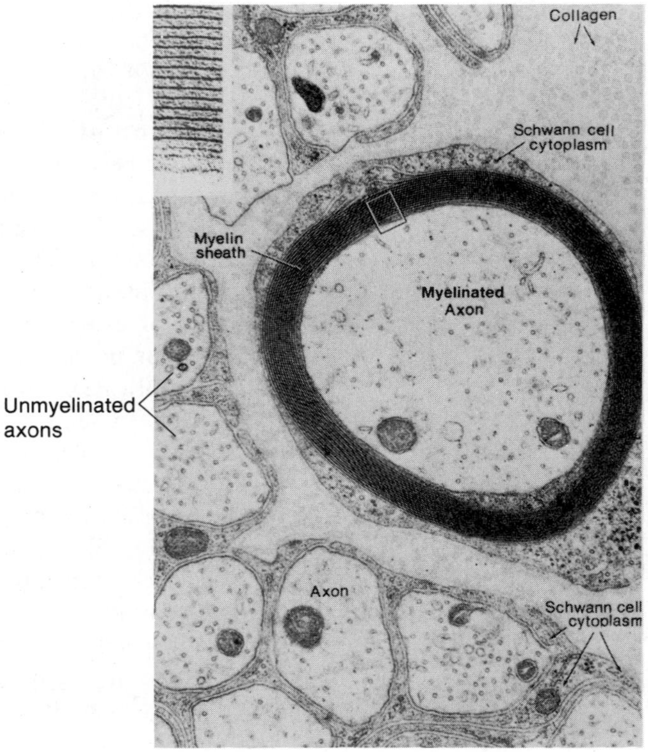

Figure 11.12 An electron micrograph of unmyelinated and myelinated axons.

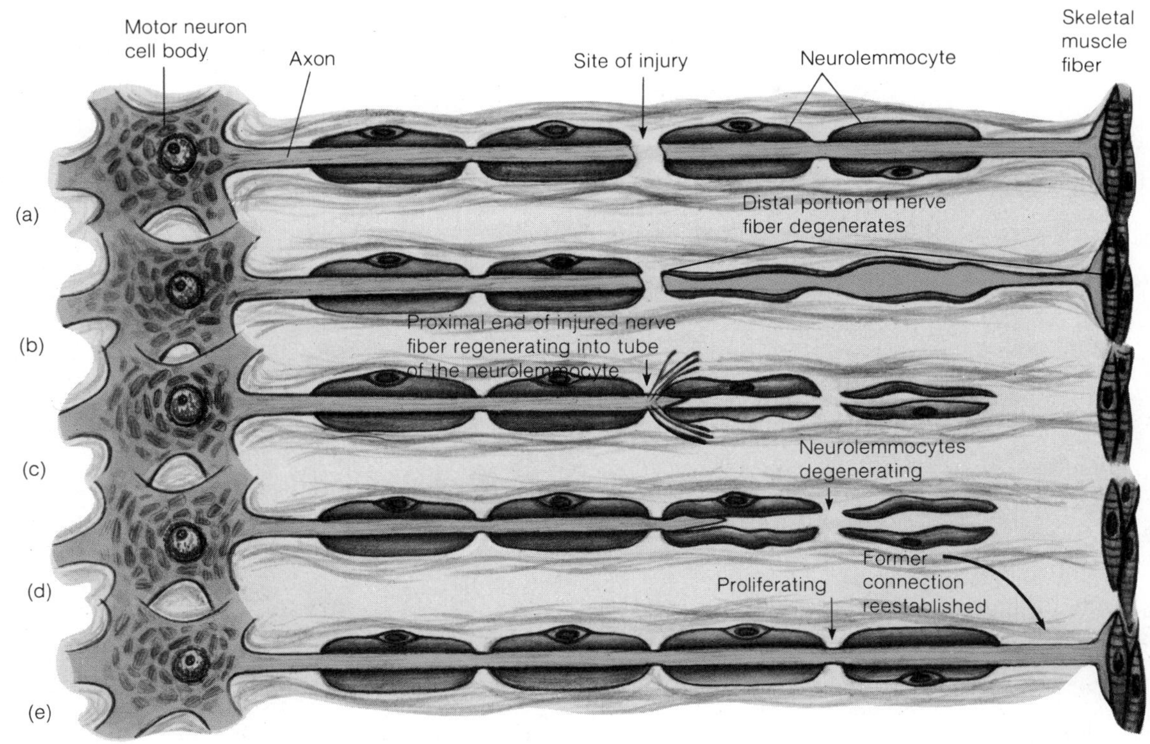

Figure 11.13 The process of neuron regeneration. (*a*) If a neuron is severed through a myelinated axon, the proximal portion may survive, but (*b*) the distal portion degenerates through phagocytosis. The myelin sheath provides a pathway for (*c* and *d*) the regeneration of an axon, and (*e*) innervation is restored.

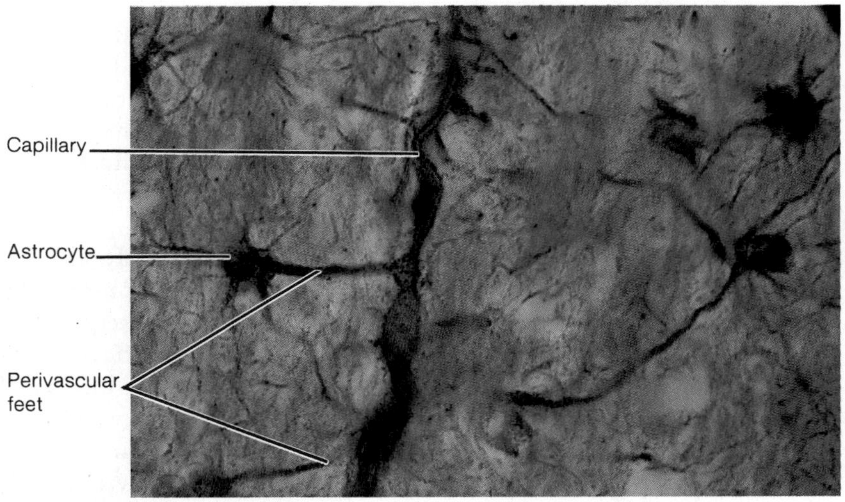

Figure 11.14 A photomicrograph showing the perivascular feet of astrocytes (a type of neuroglial cell), which cover most of the surface area of brain capillaries.

two. Sensory neurons are pseudounipolar and have their cell bodies located in dorsal root ganglia of spinal and cranial nerves.

A **nerve** is a collection of nerve fibers outside the CNS. Fibers within a nerve are held together and strengthened by loose fibrous connective tissue (fig. 11.17). Each individual nerve fiber is enclosed in a connective tissue sheath called the **endoneurium** *(en''do-nu're-um)*. A group of fibers, called a **fasciculus** *(fah-sik'u-lus)*, is surrounded by a connective tissue sheath called a **perineurium.** The entire nerve is surrounded and supported by connective tissue called the **epineurium,** which contains tiny blood vessels and often adipose cells. Perhaps less than a quarter of the bulk of a nerve consists of nerve fibers: more than half is associated connective tissue, and approximately a quarter is the myelin that surrounds the nerve fibers.

Most nerves are composed of both motor and sensory fibers and thus are called **mixed nerves.** Some of the cranial nerves, however, are composed either of sensory neurons only **(sensory nerves)** or of motor neurons only **(motor nerves).** Sensory nerves serve the special senses, such as taste, smell, sight,

fasciculus: L. diminutive of *fascis,* bundle

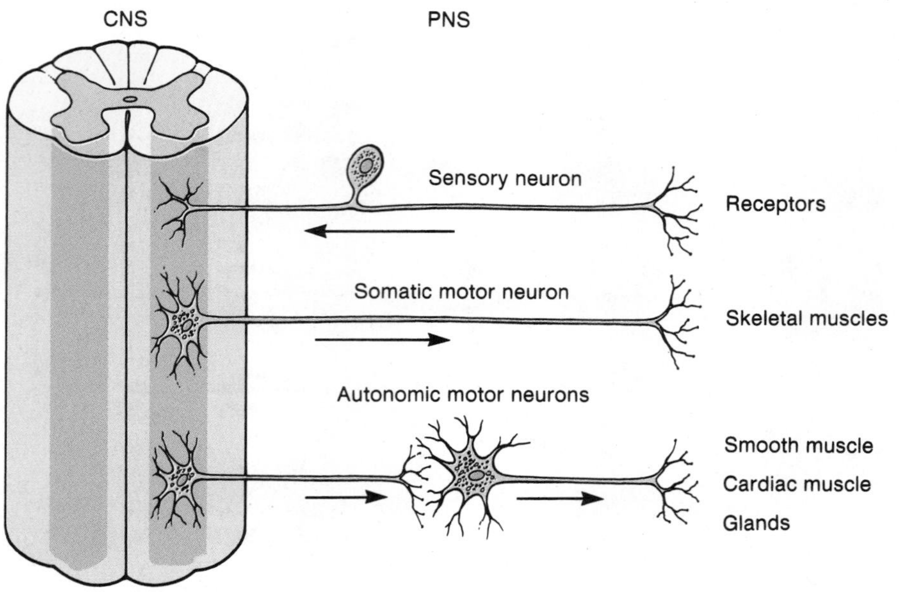

Figure 11.15 The relationship between sensory and motor fibers of the peripheral nervous system (*PNS*) and the central nervous system (*CNS*).

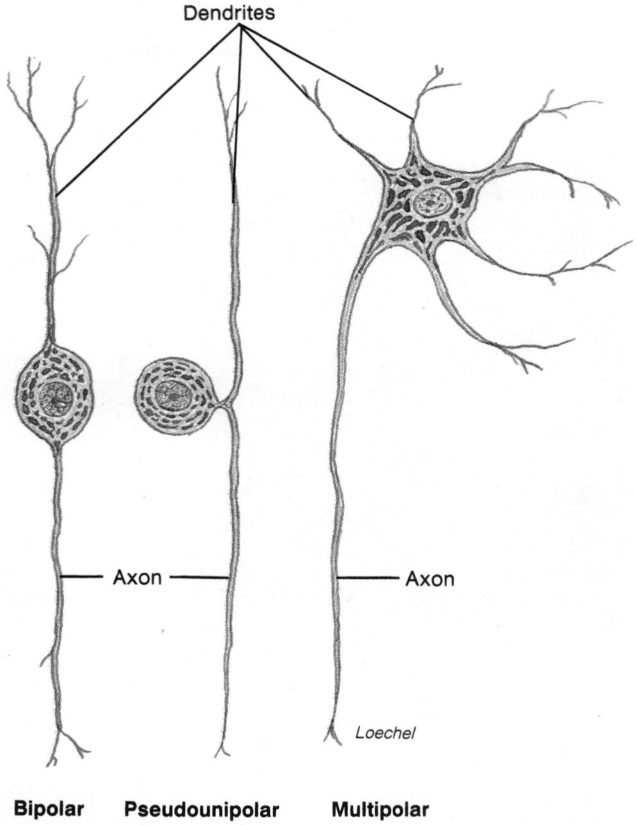

Figure 11.16 Three different types of neurons.

and hearing. Motor nerves conduct impulses to muscles, causing them to contract, or to glands, causing them to secrete.

Neurons and their fibers within nerves may be classified according to the area of innervation into the following scheme (fig. 11.18):

1. **Somatic afferent.** Sensory receptors within the skin, muscles, and joints receive stimuli and convey nerve impulses through somatic afferent fibers to the CNS for interpretation.
2. **Somatic efferent.** Impulses from the CNS through somatic efferent fibers cause the contraction of skeletal muscles.
3. **Visceral afferent.** Visceral afferent fibers convey impulses from visceral organs and blood vessels to the CNS for interpretation.
4. **Visceral efferent.** Visceral efferent fibers, also called **autonomic efferent fibers,** are part of the autonomic nervous system. They originate in the CNS and innervate cardiac muscle, glands, and smooth muscle within the visceral organs.

---

1. Briefly describe the functions of dendrites, cell bodies, and axons, and describe the differences between bipolar, pseudounipolar, and multipolar neurons.
2. Distinguish between myelinated and unmyelinated axons, and describe how a myelin sheath is formed.
3. Discuss specific ways that neuroglia aid neurons.
4. Explain the nature of the blood-brain barrier, and describe its structure.
5. Explain the differences between a neuron, nerve fiber, and nerve. Describe how nerves are classified.

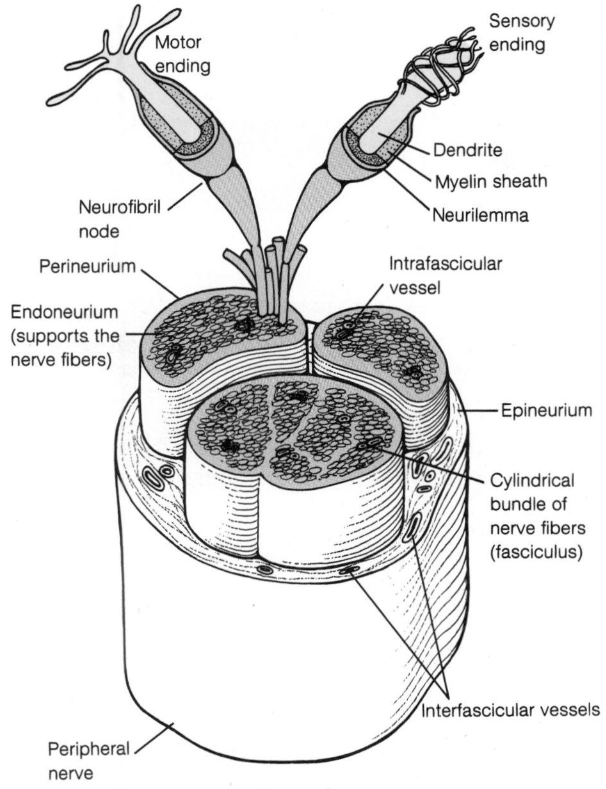

Figure 11.17 The structure of a nerve.

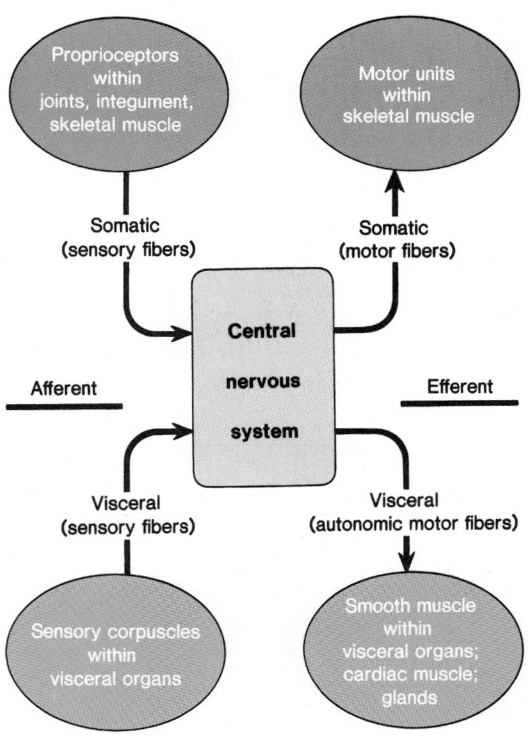

Figure 11.18 The classification of nerve fibers by origin and function.

Transmission of Impulses

Movements of sodium and potassium ions result in an impulse traveling through a neuron toward a synapse. Synaptic transmission is facilitated by the secretion of a neurotransmitter chemical.

Objective 11. Explain how a nerve fiber first becomes depolarized and then repolarized.

Objective 12. Describe the structure of a presynaptic nerve fiber ending, and explain how neurotransmitters are released.

Nerve Impulse

Two properties for which neurons are specialized are irritability and conductivity, both of which are involved in the transmission of a nerve impulse. *Irritability* is the ability of dendrites and cell bodies to respond to a stimulus and convert it into an impulse. *Conductivity* is the transmission of an impulse along the axon or a dendrite of a neuron. A *nerve impulse* is the actual movement, or exchange, of sodium (Na^+) and potassium (K^+) ions progressively along a nerve fiber, resulting in the creation of a stimulus to another neuron or another tissue.

Before a nerve fiber can respond to a stimulus, it must be *polarized.* A polarized nerve fiber has an abundance of sodium ions on the outside of the axon membrane, which produces an electrical charge called the *resting potential.* When a stimulus of sufficient strength arrives at the receptor portion of the neuron, the polarized nerve fiber becomes *depolarized* and a nerve impulse is initiated (fig. 11.19). Once depolarization has started, a sequence of ionic exchange occurs along the axon, and the nerve impulse is transmitted (fig. 11.20). *Repolarization* occurs after an impulse has passed through a nerve fiber and the original concentrations of sodium and potassium are reestablished. As a result, the resting potential returns (fig. 11.21), and the nerve fiber is again ready to send another impulse.

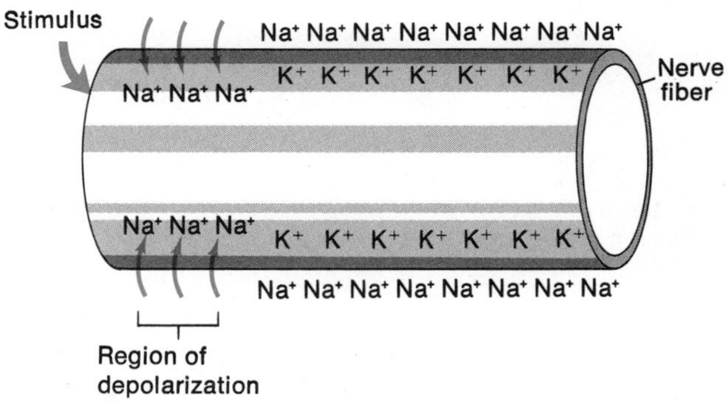

Figure 11.19 A nerve impulse is initiated if a sufficient stimulus occurs at the receptor site, causing movement of ions across the membrane of the nerve fiber.

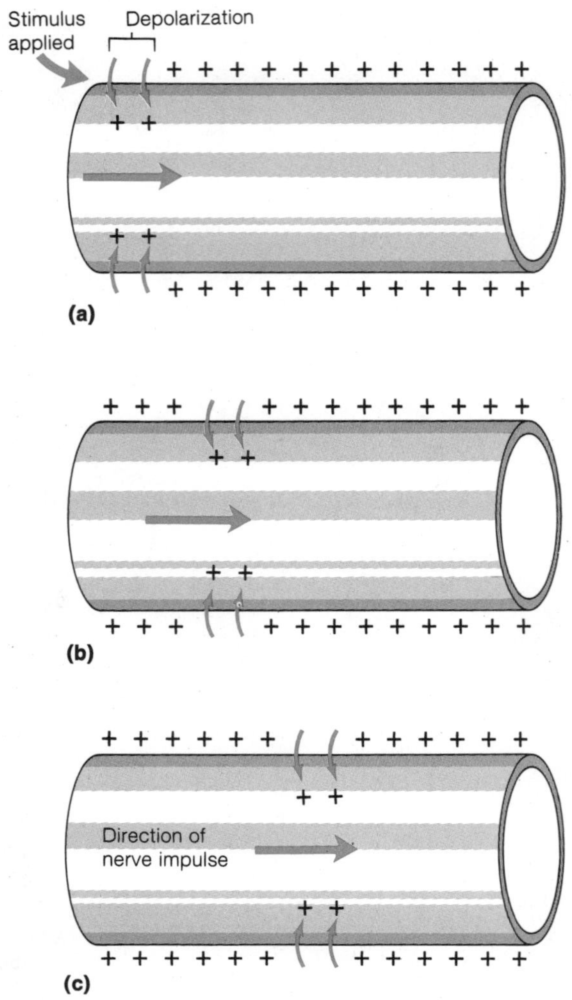

Figure 11.20 Depolarization occurs in a sequential pattern along a nerve fiber as sodium ions diffuse inward and a wave of action potentials, or a nerve impulse, moves in one direction along the fiber.

A nerve impulse travels in only one direction and is an *all-or-none* response. This means that if an impulse is initiated, it will travel the length of the nerve fiber and proceed without a loss in voltage. The speed of a nerve impulse is determined by the diameter of a nerve fiber, the presence and thickness of myelin, and the general physiological condition of the neuron. For example, unmyelinated nerve fibers with small diameters conduct impulses at the rate of about 0.5 m/second; myelinated neurons conduct impulses at speeds up to 130 m/second.

Synapse

A **synapse** is a minute space of 200 to 300 Å between the synaptic knob of a **presynaptic neuron** and a dendrite of a **postsynaptic neuron** (fig. 11.22). The **synaptic knob,** or **terminal button,** is the distal portion of the presynaptic neuron at the end of the axon; it is characterized by the presence of numerous mitochondria and **synaptic vesicles.** Synaptic vesicles contain a neurotransmitter chemical, the most common of which is *acetylcholine.* When a nerve impulse reaches the synaptic knob,

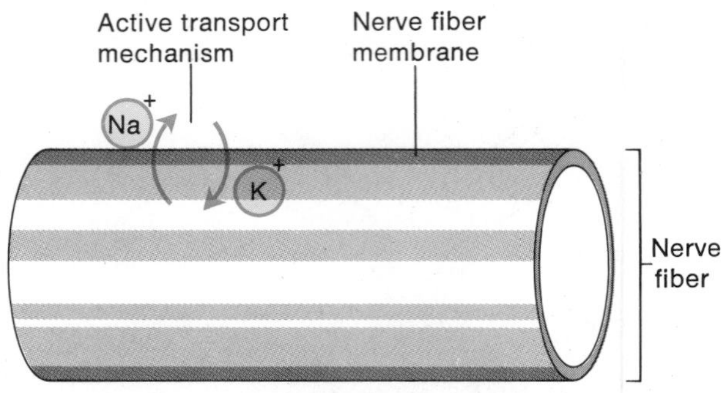

Figure 11.21 A nerve fiber is repolarized following ionic exchange. Ions move across the cell membrane primarily by active transport.

some of the vesicles respond by releasing their neurotransmitter. If a sufficient number of nerve impulses occur in a short interval of time, enough neurotransmitter accumulates within the synaptic cleft to stimulate the postsynaptic neuron to depolarize.

Neurotransmitters are decomposed by enzymes present in the synaptic cleft. *Cholinesterase* is the enzyme that decomposes acetylcholine and prepares the synapse to receive additional neurotransmitter with the next impulse.

Synaptic transmission may be affected by various drugs or diseases. Caffeine is a stimulant that increases the rate of transmission across the synapse. Aspirin causes a moderate decrease in the transmission rate. The drug strychnine profoundly reduces synaptic transmission. It affects breathing and paralyzes respiratory structures, causing death. *Parkinson's disease* is caused by a deterioration of neurons within the brain that synthesize a neurotransmitter called *dopamine.*

1. Define the meaning of the terms *depolarization* and *repolarization,* and illustrate these processes with a figure.
2. How does a neurotransmitter chemical affect impulse conduction across a synapse?

General Features of the Brain

The brain is enclosed by the skull and meninges and is bathed in cerebrospinal fluid. The brain has a tremendous metabolic rate and is susceptible to oxygen deprivation and certain toxins.

Objective 13. Describe the general features of the brain.
Objective 14. Explain the importance of neuropeptides.
Objective 15. List the metabolic needs of the brain.

The entire delicate CNS is protected by a bony encasement—the cranium surrounding the brain and the vertebral column surrounding the spinal cord. The **meninges** *(mĕ-nin'jēz)* form a protective membrane between the bone and the soft tissue of the CNS. The CNS is bathed in **cerebrospinal fluid,** which circulates within the hollow **ventricles** of the brain, the **central canal** of the spinal cord (fig. 11.23), and the **subarachnoid** *(sub"ah-rak'noid)* **space** surrounding the entire CNS (see figs. 11.37, 11.39).

synapse: Gk. *syn,* together; *haptein,* to fasten

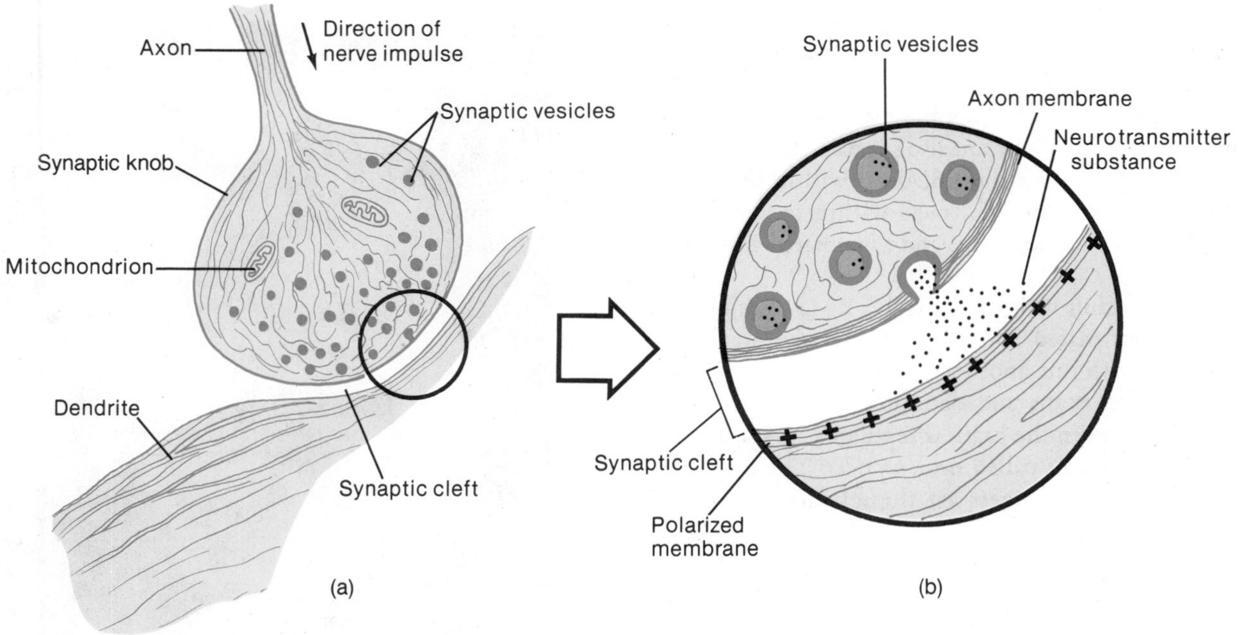

Figure 11.22 Synaptic transmission occurs if sufficient amounts of neurotransmitter chemicals are released from synaptic vesicles. (*a*) A nerve impulse reaches the synaptic knob and (*b*) causes the neurotransmitter chemical to be released into the synaptic cleft.

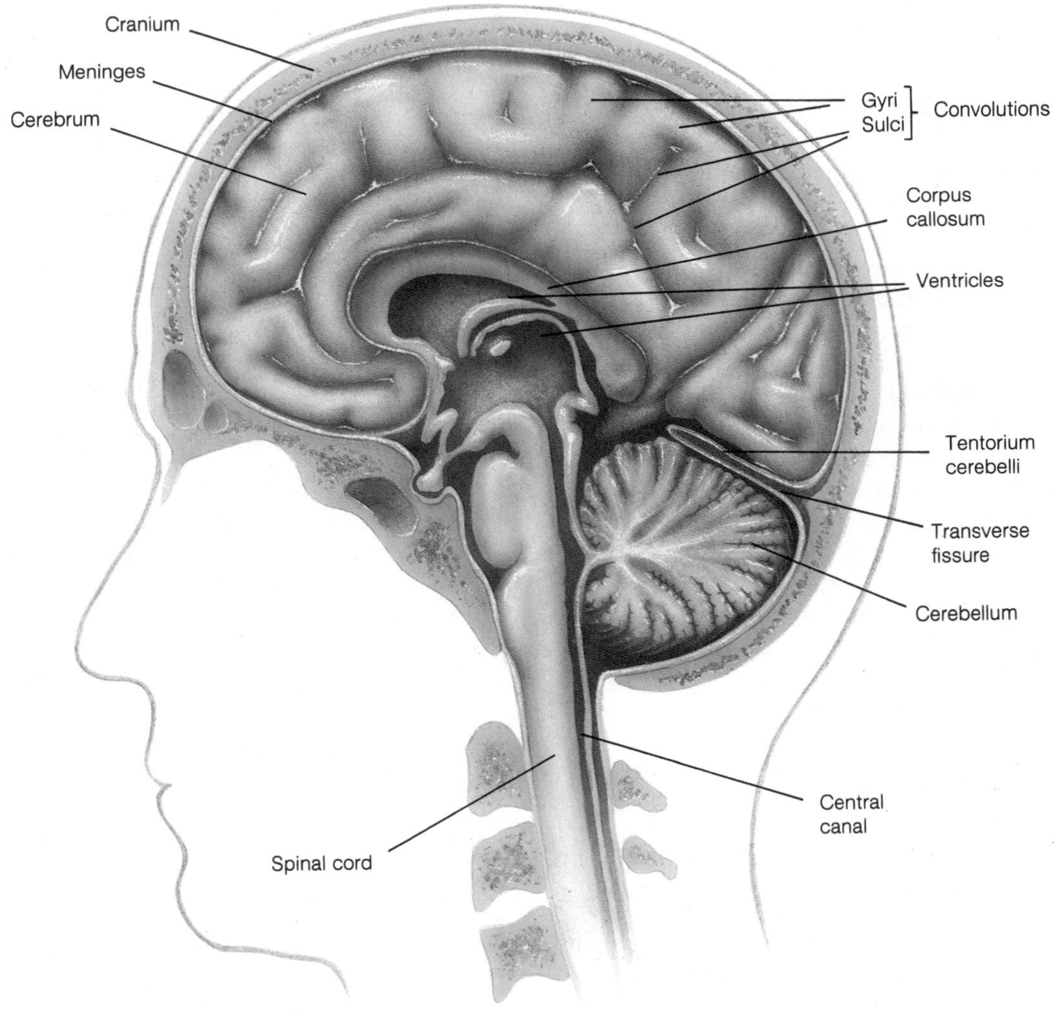

Figure 11.23 The CNS consists of the brain and the spinal cord, both of which are covered with meninges and bathed in cerebrospinal fluid.

The CNS is composed of gray and white matter. **Gray matter** consists of either nerve cell bodies and dendrites or bundles of unmyelinated axons and neuroglia. The gray matter of the brain exists as the outer convoluted **cortex layer** of the cerebrum and cerebellum. There are also specialized gray matter clusters of nerve cells, called **nuclei,** deep within the white matter. **White matter** forms the tracts within the CNS and consists of aggregations of dendrites and myelinated axons and associated neuroglia.

The brain of an adult weighs nearly 1.5 kg (3–3.5 lbs) and is composed of an estimated 100 billion (10^{11}) neurons. Neurons communicate with one another by means of innumerable synapses between the axons and dendrites within the brain. Neurotransmission within the brain is regulated by specialized neurotransmitter chemicals called *neuropeptides.* These specialized chemical protein messengers are thought to account for specific mental functions.

Over 200 neuropeptides have been identified within the brain, yet most of their functions are little understood. Two groups of neuropeptides that have received considerable attention are *enkephalins* and *endorphins.* These classes of substances numb the brain to pain, functioning in a manner similar to morphine. They are released in response to stress or pain in a traumatized person.

The brain has a tremendous metabolic rate and has to have a continual supply of oxygen and nutrients. The brain accounts for only 2% of a person's body weight, and yet it receives approximately 20% of the total resting cardiac output. This amounts to a flow of about 750 ml of blood per minute. The volume remains relatively constant even during physical or mental activity. This continuous flow is so crucial that a failure of cerebral circulation for as short as ten seconds causes unconsciousness. The brain is composed of perhaps the most sensitive tissue of the body. Due to its high metabolic rate, not only does the brain have to have continuous oxygen, but it also has to have a continuous nutrient supply and rapid removal of wastes. The brain is also very sensitive to certain toxins and drugs. The cerebrospinal fluid aids the metabolic needs of the brain through the distribution of nutrients and the removal of wastes. Cerebrospinal fluid also maintains a protective homeostatic environment within the brain. The blood-brain barrier and the secretory activities of neural tissue also help to maintain homeostasis. The brain has an extensive vascular supply through the paired internal carotid and vertebral arteries that unite at the arterial circle (circle of Willis) (see chap. 16 and fig. 16.29).

The brain of a newborn is especially sensitive to oxygen deprivation or to excessive oxygen. If complications arise during childbirth and the oxygen supply from the mother's blood to the baby is interrupted while it is still in the birth canal, the infant may be stillborn or suffer brain damage that can result in cerebral palsy, epilepsy, paralysis, or mental retardation. Excessive oxygen administered to a newborn may cause blindness.

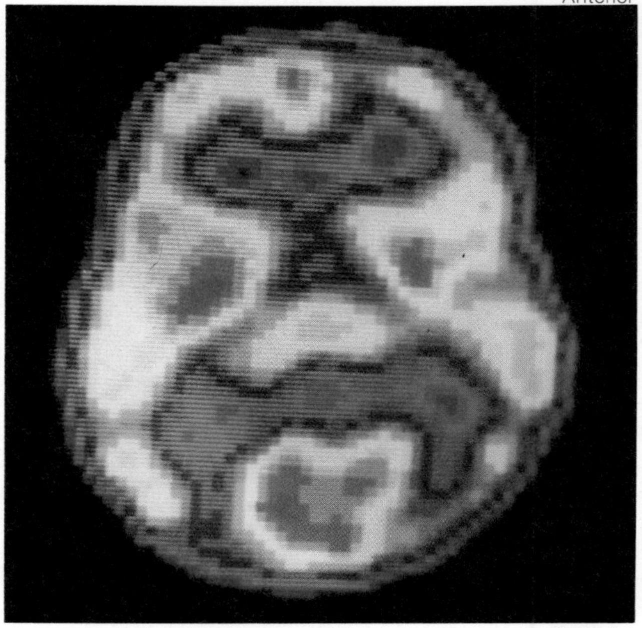

Figure 11.24 Positron emission tomographic (PET) brain scan (in transverse section) of a schizophrenic patient receiving shocks to the right forearm while the eyes are closed. Patient was not medicated. Red areas indicate high glucose use (uptake of 18-F-deoxyglucose). The scan shows highest glucose uptake in the posterior region, where the brain's visual center is located.

There are measurable increases within the brain in regional blood flow and in glucose and oxygen metabolism that accompany mental functions, including perception and emotion. These metabolic changes can be assessed through the use of *positron emission tomography (PET).* The technique of a PET scan (fig. 11.24) is based on injecting radioactive tracer molecules labeled with carbon-11, fluorine-18, and oxygen-15 into the bloodstream and photographing the gamma rays that are emitted from the patient's brain through the skull. PET scans are of value in studying neurotransmitters and neuroreceptors as well as the substrate metabolism of the brain.

From the five basic regions of the brain, the development of which was described earlier in this chapter, distinct functional structures are formed (table 11.3). These structures will be discussed in greater detail in the following sections.

1. What characteristics do the brain and spinal cord have in common? Describe the general features of each.
2. Discuss how the study of neuropeptides will aid in an understanding of brain function.
3. Using specific examples, describe the metabolic requirements of the brain.

Table 11.3 Derivation and functions of the major brain structures

	Region	Structure	Function
Prosencephalon (forebrain)	Telencephalon	Cerebrum	Controls most sensory and motor activities; reasoning, memory, intelligence, etc.; instinctual and limbic functions
	Diencephalon	Thalamus	Relay center: all impulses (except olfactory) going into cerebrum synapse here; some sensory interpretation; initial autonomic response to pain
		Hypothalamus	Regulation of renal water flow, body temperature, hunger, heartbeat, etc.; control of secretory activity in anterior pituitary gland; instinctual and limbic functions
		Pituitary gland	Regulation of other endocrine glands
Mesencephalon (midbrain)	Mesencephalon	Superior colliculi	Visual reflexes
		Inferior colliculi	Auditory reflexes
		Cerebral peduncles	Coordinating reflexes; contain many motor fibers
Rhombencephalon (hindbrain)	Metencephalon	Cerebellum	Balance and motor coordination
		Pons	Relay center; contains nuclei (pontine nuclei)
	Myelencephalon	Medulla oblongata	Relay center; contains many nuclei; visceral autonomic center (e.g., respiration, heart rate, vasoconstriction)

Cerebrum

The cerebrum, consisting of five paired lobes within two convoluted hemispheres, is concerned with higher brain functions, such as the perception of sensory impulses, the instigation of voluntary movement, the storage of memory, thought processes, and reasoning ability. The cerebrum is also concerned with instinctual and limbic (emotional) functions.

Objective 16. Describe the structure of the cerebrum, and list the functions of the cerebral lobes.

Objective 17. Define the term *electroencephalogram,* and discuss its clinical importance.

Objective 18. Describe the fiber tracts within the cerebrum.

Structure of the Cerebrum

The **cerebrum,** located in the region of the telencephalon, is the largest and most obvious portion of the brain. It accounts for about 80% of the mass of the brain and is responsible for the higher mental functions, including memory and reason. The cerebrum consists of the **right** and **left hemispheres,** which are incompletely separated by a **longitudinal fissure** (fig. 11.25). Portions of the two hemispheres are connected internally by the **corpus callosum** *(kah-lo'sum),* a large tract of white matter (see fig. 11.23). A portion of the meninges, called the **falx** *(falks)* **cerebri,** extends into the longitudinal fissure. Each cerebral hemisphere contains a central cavity called the **lateral ventricle** (fig. 11.26), which is lined with ependymal cells and filled with cerebrospinal fluid.

The two cerebral hemispheres carry out different functions. In most people, the left hemisphere controls analytical and verbal skills such as reading, writing, and mathematics. The right hemisphere is the source of spatial and artistic kinds of intelligence. The corpus callosum unifies attention and awareness between the two hemispheres and permits a sharing of learning and memory. Severing the corpus callosum is a radical treatment to control severe epileptic seizures. Although this surgery is successful, it results in the cerebral hemispheres functioning as separate structures, each with its own memories and thoughts, competing for control. A more recent and effective technique of controlling epileptic seizures is a precise laser treatment of the corpus callosum.

The cerebrum consists of two layers. The surface layer, referred to as the **cerebral cortex,** is composed of gray matter that is 2–4 mm (0.08–0.16 in.) thick (fig. 11.26). Beneath the cerebral cortex is the thick **white matter** of the cerebrum, which constitutes the second layer. The cerebral cortex is characterized by numerous folds and grooves called **convolutions.** Convolutions form during early fetal development, when brain size increases rapidly and the cortex enlarges out of proportion to the underlying white matter. The elevated folds of the convolutions are the **gyri** (singular, *gyrus*), and the depressed grooves are the **sulci** *(sul'si)* (singular, *sulcus*). The convolutions greatly increase the area of the gray matter, which is comprised of nerve cell bodies.

Recent studies indicate that with increased learning, there is an increase in the number of synapses between neurons within the cerebrum. Although the number of neurons is established during prenatal development, the number of synapses is variable depending upon the learning process. The number of cytoplasmic extensions from the cell body of a neuron determines the extent of nerve impulse conduction and the associations that can be made to cerebral areas already containing stored information.

cerebrum: L. *cerebrum,* brain

gyrus: Gk. *gyros,* circle
sulcus: L. *sulcus,* a furrow or ditch

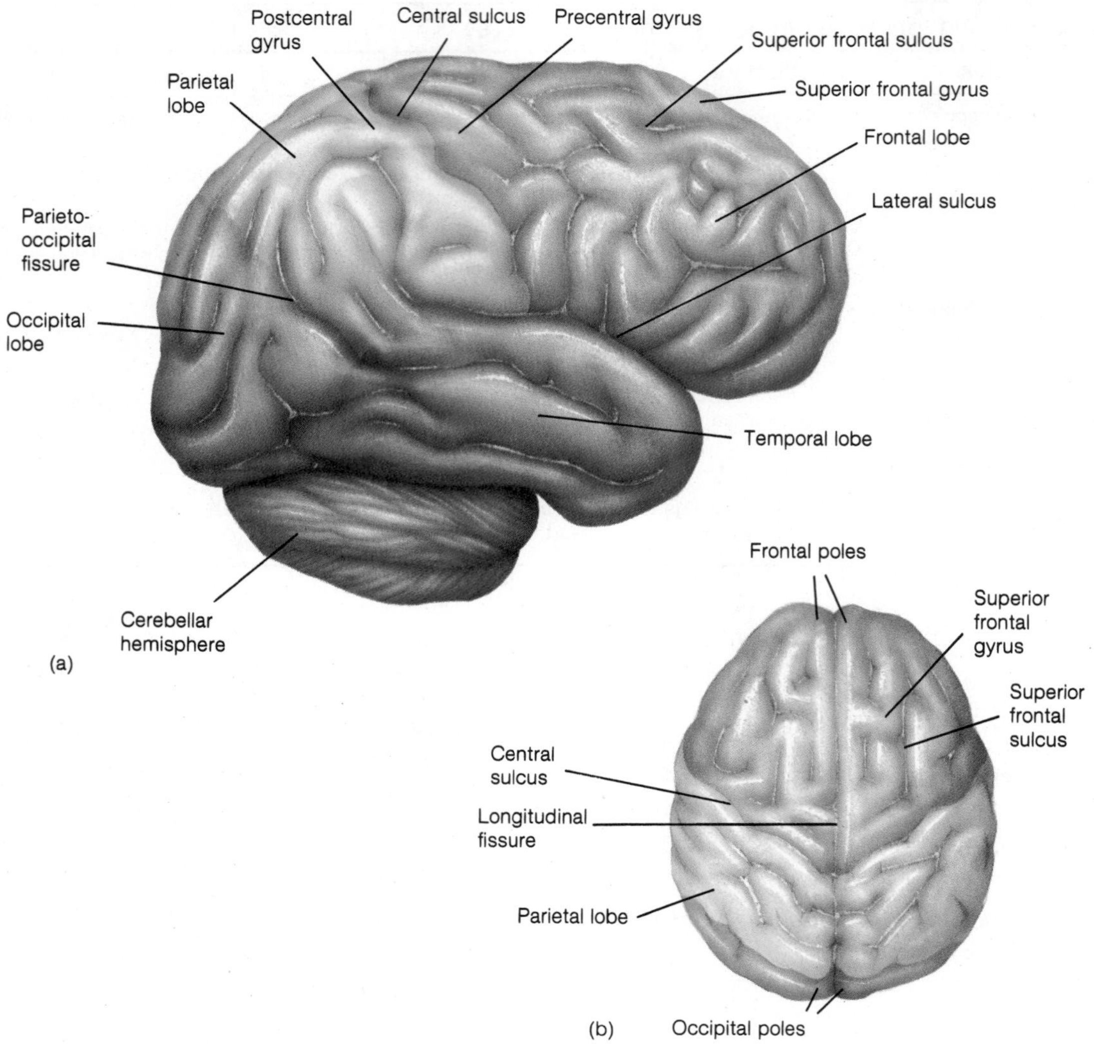

Figure 11.25 The cerebrum. (*a*) A lateral view and (*b*) a superior view.

Lobes of the Cerebrum

Each cerebral hemisphere is subdivided by deep sulci, or fissures, into the five lobes, four of which appear on the surface of the cerebrum and are named according to the overlying cranial bones (fig. 11.27). The reasons for the separate cerebral lobes, as well as two cerebral hemispheres, have to do with specificity of function (table 11.4).

Frontal Lobe The frontal lobe forms the anterior portion of each cerebral hemisphere (fig. 11.27). A prominent deep furrow, called the **central sulcus** (fissure of Rolando), separates the frontal lobe from the parietal lobe. The central sulcus extends at right angles from the longitudinal fissure to the lateral sulcus. The **lateral sulcus** (fissure of Sylvius) extends laterally from the inferior surface of the cerebrum to separate the frontal and temporal lobes. The **precentral gyrus** (see figs. 11.25, 11.27), an important motor area, is positioned immediately in front of the central sulcus. The frontal lobe's functions include initiating voluntary motor impulses for the movement of skeletal muscles, analyzing sensory experiences, and providing responses relating to personality. The frontal lobes also involve responses related to memory, emotions, reasoning, judgment, planning, and verbal communication.

Parietal Lobe The parietal lobe is posterior to the frontal lobe. Except for the central sulcus along its anterior border, the parietal lobe lacks distinct boundaries. An important sensory area called the **postcentral gyrus** (see figs. 11.25, 11.27) is positioned immediately behind the central sulcus. The postcentral gyrus is designated as a somatesthetic area because it responds to stimuli from cutaneous and muscular receptors throughout the body.

The size of the portions of the precentral gyrus responsible for motor movement and the size of the portions of the postcentral gyrus that respond to sensory stimuli do not correspond to the size of the part of the body being served but to the number of motor units activated or the density of receptors (fig. 11.28). For example, because the hand has many motor units and sensory receptors, larger portions of the precentral and postcentral gyri serve it than serve the thorax, which is much larger in size.

fissure of Rolando: from Luigi Rolando, Italian anatomist, 1773–1831
fissure of Sylvius: from Franciscus Sylvius de la Boe, Dutch anatomist, 1614–72

Table 11.4 Functions of the cerebral lobes

Lobe	Functions	Lobe	Functions
Frontal	Voluntary motor control of skeletal muscles; personality; higher intellectual processes (e.g., concentration, planning, and decision making); verbal communication	Temporal	Interpretation of auditory sensations; storage (memory) of auditory and visual experiences
Parietal	Somatesthetic interpretation (e.g., cutaneous and muscular sensations); understanding speech and formulating words to express thoughts and emotions; interpretation of textures and shapes	Occipital	Integrates movements in focusing the eye; correlating visual images with previous visual experiences and other sensory stimuli; conscious perception of vision
		Insula	Memory; integrates other cerebral activities

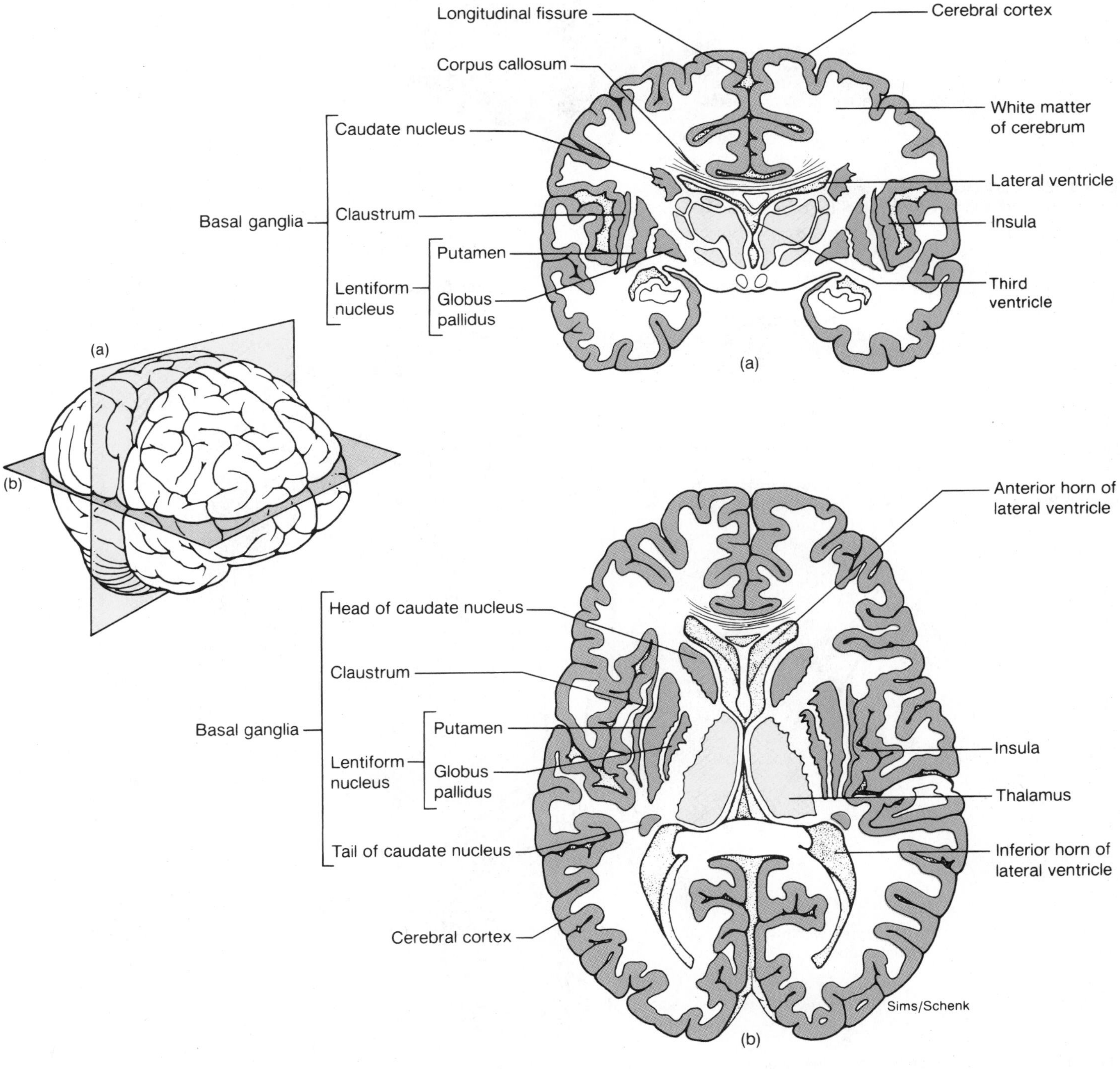

Figure 11.26 Sections through the cerebrum and diencephalon. (*a*) A coronal section and (*b*) a transverse section.

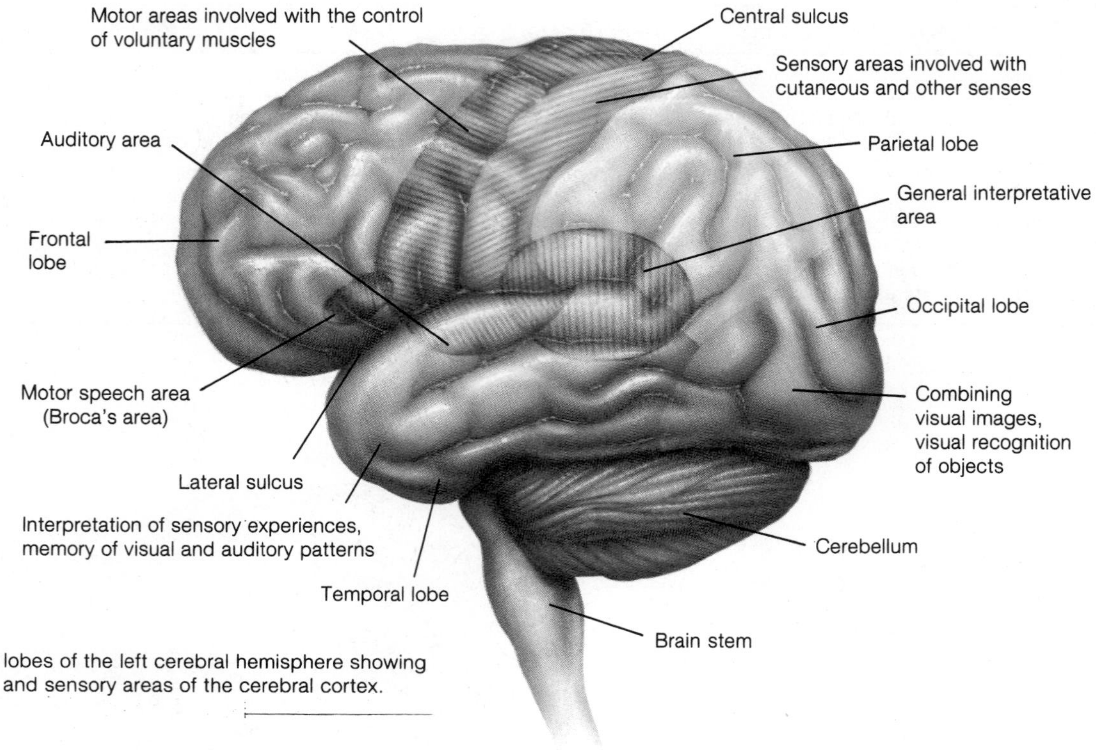

Figure 11.27 The lobes of the left cerebral hemisphere showing the principal motor and sensory areas of the cerebral cortex.

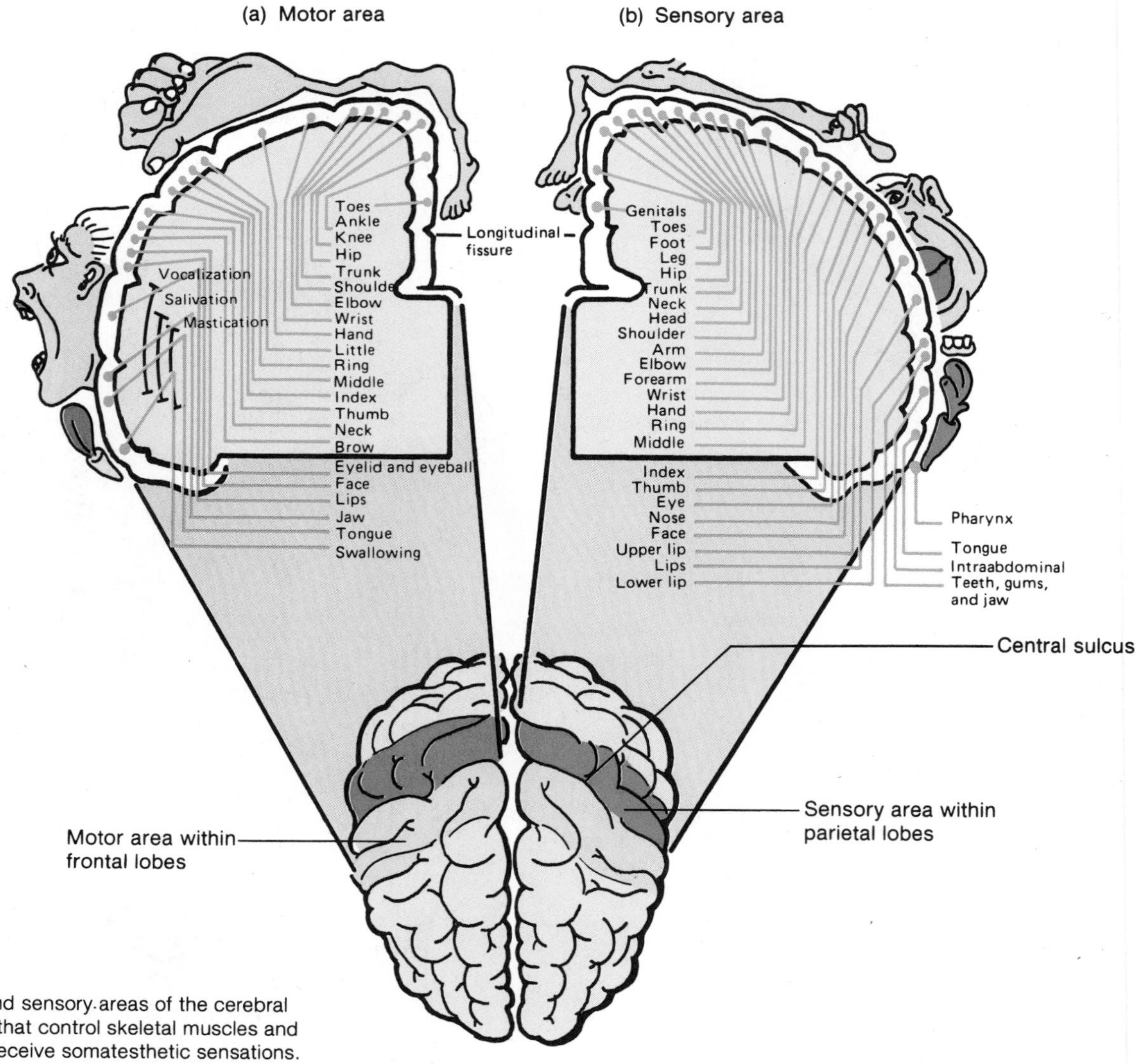

Figure 11.28 Motor and sensory areas of the cerebral cortex. (*a*) Motor areas that control skeletal muscles and (*b*) sensory areas that receive somatesthetic sensations.

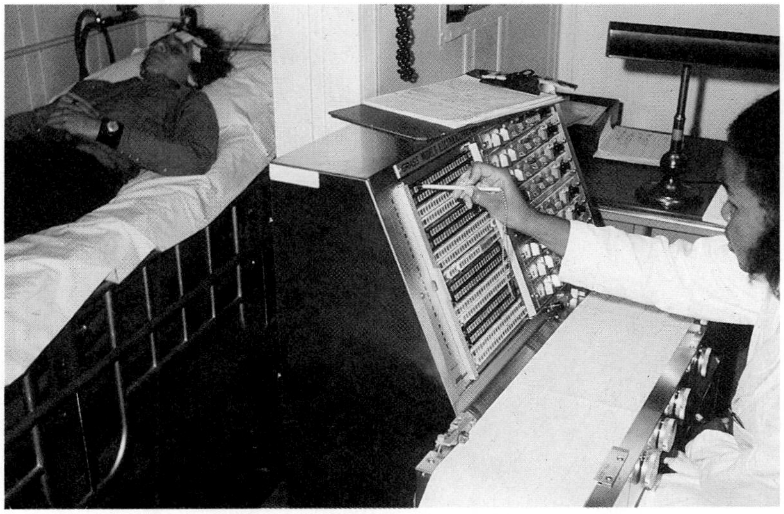

(a)

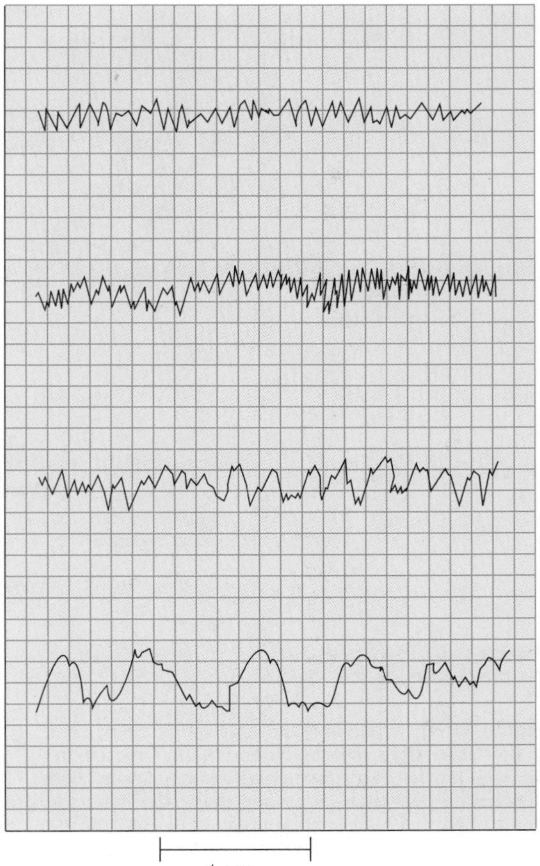

(b)

Figure 11.29 Brain waves. (*a*) A technician using an *electroencephalograph* to take the EEG of a patient. (*b*) Types of electroencephalograms.

In addition to responding to somatesthetic stimuli, the parietal lobe functions in the understanding of speech and in verbal articulation of thoughts and emotions. The parietal lobe also interprets the textures and shapes of objects as they are handled.

Temporal Lobe The temporal lobe is located below the parietal lobe and the posterior portion of the frontal lobe. It is separated from both by the lateral sulcus (see fig. 11.27). The temporal lobe contains auditory centers that receive sensory fibers from the cochlea of the ear. This lobe also interprets some sensory experiences and stores memories of both auditory and visual experiences.

Occipital Lobe The occipital lobe forms the posterior portion of the cerebrum and has no distinct separation from the temporal and parietal lobes (see fig. 11.27). The occipital lobe is superior to the cerebellum and is separated from it by an infolding of the meningeal layer called the **tentorium cerebelli** (see fig. 11.23). The principal functions of the occipital lobe concern vision. The occipital lobe integrates eye movements by directing and focusing the eye. It is also responsible for visual association, correlating visual images with previous visual experiences and other sensory stimuli.

Insula The insula is a deep lobe of the cerebrum that cannot be viewed on the surface (see fig. 11.26). It is deep to the lateral sulcus and is covered by portions of the frontal, parietal, and temporal lobes. Little is known of the function of the insula, though it primarily integrates other cerebral activities and may have some function in memory.

Because of its size and position, portions of the cerebrum frequently suffer brain trauma. A concussion to the brain may cause a temporary or permanent impairment of cerebral functions; a stroke usually affects cerebral function. Much of what is known about cerebral function comes from observing body dysfunctions when specific regions of the cerebrum are traumatized.

Brain Waves Neurons within the cerebral cortex continuously generate electrical activity, which can be recorded by electrodes attached to precise locations on the scalp, producing an **electroencephalogram (EEG).** An EEG pattern, commonly called *brain waves,* is the collective expression of millions of action potentials from neurons.

Brain waves are first emitted from a developing brain during early fetal development and continue throughout a person's life. The cessation of brain-wave patterns may be a decisive factor in the legal determination of death.

Certain distinct EEG patterns signify healthy mental functions. Deviations from these patterns are of clinical significance in diagnosing trauma, mental depression, hematomas, and various diseases, such as tumors, infections, and epilepsy. Normally, there are four kinds of EEG patterns (fig. 11.29).

1. **Alpha waves** are best recorded from the parietal and occipital regions while a person is awake and relaxed but with the eyes closed. These waves are rhythmic oscillations of about 10–12 cycles/second. The alpha rhythm of a child younger than eight years old occurs at a slightly lower frequency of 4–7 cycles/second.
2. **Beta waves** are strongest from the frontal lobes, especially the area near the precentral gyrus. These waves are sensory evoked and respond to visual and mental activity. Because they respond to stimuli from receptors and are superimposed on the continuous activity patterns of the alpha waves, they constitute *evoked activity.* The frequency of beta waves is 13–25 cycles/second.

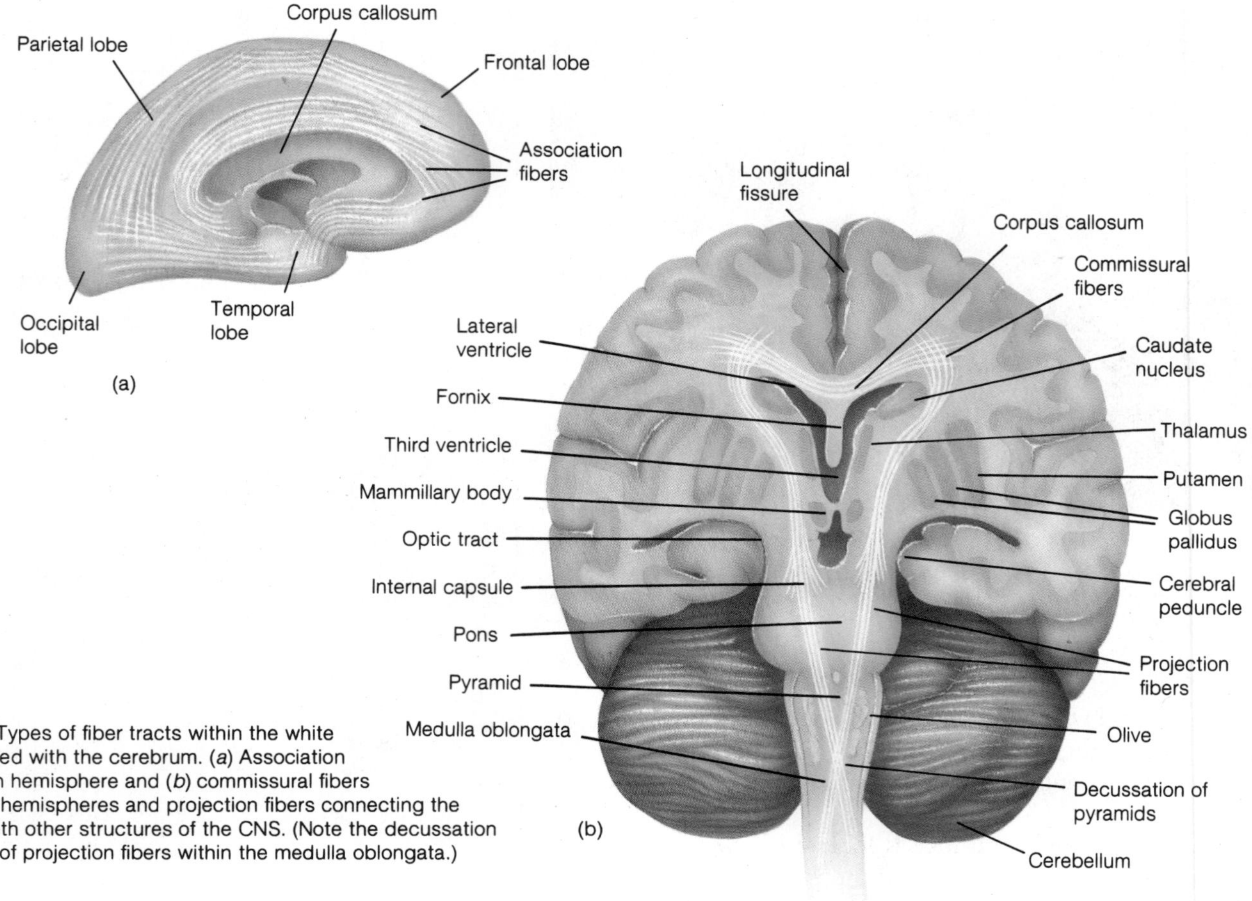

Figure 11.30 Types of fiber tracts within the white matter associated with the cerebrum. (*a*) Association fibers of a given hemisphere and (*b*) commissural fibers connecting the hemispheres and projection fibers connecting the hemispheres with other structures of the CNS. (Note the decussation (crossing over) of projection fibers within the medulla oblongata.)

3. **Theta waves** are emitted from the temporal and occipital lobes. They have a frequency of 5–8 cycles/second and are common in newborn infants. The recording of theta waves in adults generally indicates severe emotional stress and can be a forewarning of a nervous breakdown.
4. **Delta waves** are seemingly emitted in a general pattern from the cerebral cortex. These waves have a frequency of 1–5 cycles/second and are common during sleep and in an awake infant. The presence of delta waves in an awake adult indicates brain damage.

White Matter of the Cerebrum

The thick white matter of the cerebrum is deep to the cortex (see fig. 11.26) and consists of dendrites, myelinated axons, and associated neuroglia. These fibers form the billions of connections within the brain by which information in the form of electrical impulses is transmitted to the appropriate places. There are three types of fiber tracts within the white matter, which are named according to location and the direction in which they conduct impulses (fig. 11.30).

1. **Association fibers** are confined to a given cerebral hemisphere and conduct impulses between neurons within that hemisphere.
2. **Commissural** *(kom-mis'u-ral)* **fibers** connect the neurons and gyri of one hemisphere with those of the other. The **corpus callosum** and **anterior commissure** (fig. 11.31) are composed of commissural fibers.
3. **Projection fibers** form the ascending and descending tracts that transmit impulses from the cerebrum to other parts of the brain and spinal cord and from the spinal cord and other parts of the brain to the cerebrum.

Basal Ganglia

The **basal ganglia** are specialized paired masses of gray matter located deep within the white matter of the cerebrum (see fig. 11.26). The most prominent of the basal ganglia is the **corpus striatum,** so named because of its striped appearance. The corpus striatum is composed of several masses of nuclei. The **caudate nucleus** is the upper mass. A thick band of white matter lies between the caudate nucleus and the next lower two masses, collectively called the **lentiform nucleus.** The lentiform nucleus consists of a lateral portion, called the **putamen** *(pu-ta'men),* and a medial portion, called the **globus pallidus** (fig. 11.26). The claustrum is another portion of the basal ganglia. It is a thin layer of gray matter just deep to the cerebral cortex of the insula.

The basal ganglia are associated with other structures of the brain, particularly within the mesencephalon. The caudate nucleus and putamen of the basal ganglia control unconscious contractions of certain skeletal muscles, such as those of the upper extremities involved in involuntary arm movements

---

corpus striatum: L. *corpus*, body; *striare*, striped
lentiform: L. *lentis*, elongated
putamen: L. *putare*, to cut, prune
globus pallidus: L. *globus*, sphere; *pallidus*, pale

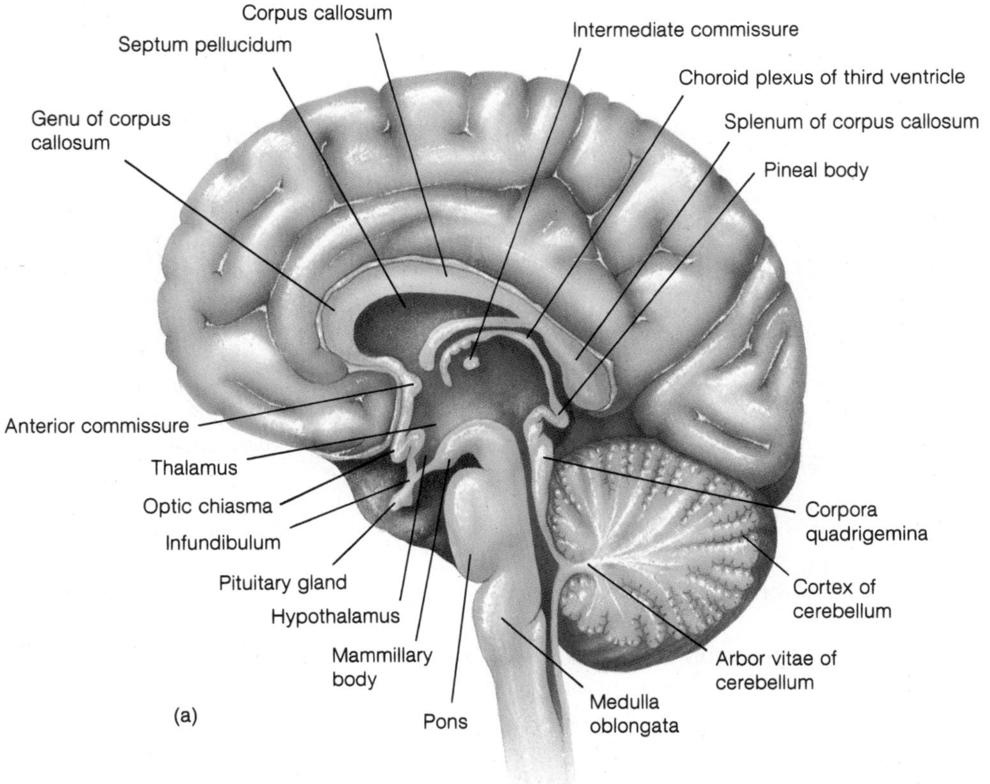

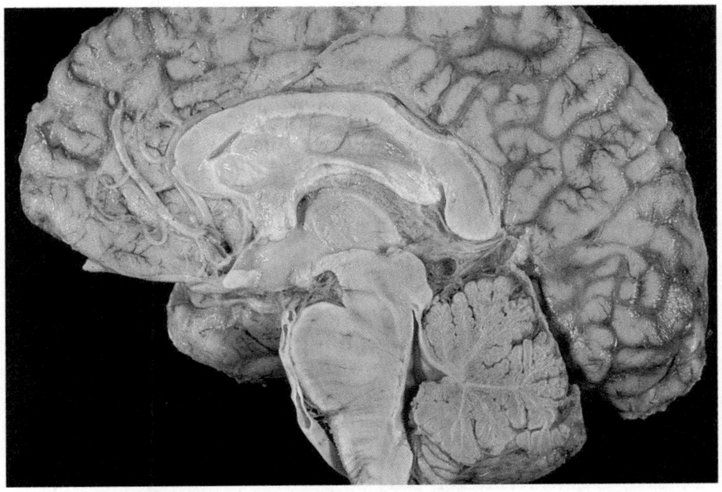

(b)

Figure 11.31 A midsagittal section through the brain. (*a*) A diagram and (*b*) a photograph.

during walking. The globus pallidus regulates the muscle tone necessary for specific, intentional body movements. Neural diseases or physical trauma to the basal ganglia generally cause a variety of motor movement dysfunctions, including rigidity, tremor, and rapid and aimless movements.

Language

Knowledge of the brain regions involved in language has been gained primarily by the study of *aphasias*—speech and language disorders caused by damage to specific language areas of the brain. These areas (fig. 11.32) are generally located in the cerebral cortex of the left hemisphere in both right-handed and left-handed people.

The **motor speech area** (Broca's area) is located in the left inferior gyrus of the frontal lobe. Neural activity in the motor speech area causes selective stimulation of motor impulses in motor centers elsewhere in the frontal lobe, which in turn causes coordinated skeletal muscle movement in the pharynx and larynx. At the same time, motor impulses are sent to the respiratory muscles to regulate air movement across the vocal cords. The combined muscular stimulation translates thought patterns into speech.

Wernicke's area is located in the superior gyrus of the temporal lobe and is directly connected to the motor speech area by a fiber tract called the **arcuate fasciculus.** People with *Wernicke's aphasia* produce speech that has been described as a "word salad." The words used may be real words that are chaotically mixed together, or they may be made-up words. Language comprehension has been destroyed in people with Wernicke's aphasia; they cannot understand either spoken or written language.

It appears that the concept of words to be spoken originates in Wernicke's area and is then communicated to the motor speech area through the arcuate fasciculus. Damage to the arcuate fasciculus produces *conduction aphasia,* which is fluent but nonsensical speech as in Wernicke's aphasia, even though both the motor speech area and Wernicke's areas are intact.

aphasia: L. *a,* without; Gk. *phasis,* speech

Broca's area: from Pierre P. Broca, French neurologist, 1824–1880
Wernicke's area: from Karl Wernicke, German neurologist, 1848–1905

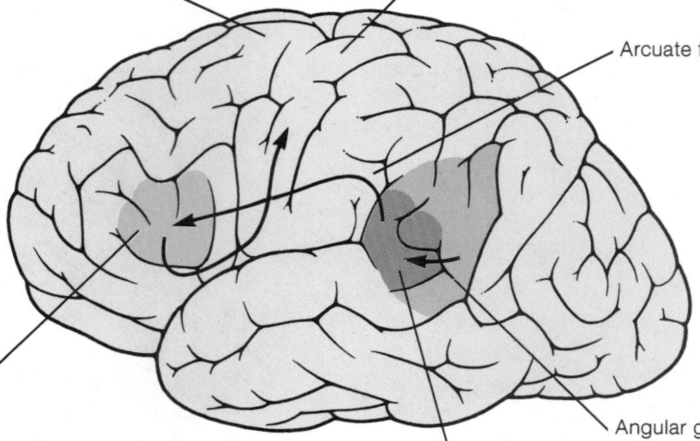

Figure 11.32 Brain areas involved in the control of speech. Arrows indicate the direct communication between these areas.

The **angular gyrus,** located at the junction of the parietal, temporal, and occipital lobes, is believed to be a center for the integration of auditory, visual, and somatesthetic information. Damage to the angular gyrus produces aphasias, which suggests that this area projects to Wernicke's area. Some patients with damage to the left angular gyrus can speak and understand spoken language but cannot read or write. Other patients can write a sentence but cannot read it, presumably due to damage to the projections from the occipital lobe (involved in vision) to the angular gyrus.

Recovery of language ability, by transfer to the right hemisphere after damage to the left hemisphere, is very good in children but decreases after adolescence. Recovery is reported to be faster in left-handed people, possibly because language ability is more evenly divided between the two hemispheres in left-handed people. Some recovery usually occurs after damage to the motor speech area, but damage to Wernicke's area produces more severe and permanent aphasias.

1. Diagram a lateral view of the cerebrum, and label the four superficial lobes and the fissures that separate them.
2. List the functions of each of the paired cerebral lobes.
3. What is a brain-wave pattern? How are these patterns monitored clinically?
4. Describe the arrangement of the fiber tracts within the cerebrum.
5. Define *basal ganglia,* and list their functions.
6. Describe the aphasias produced by damage to the motor speech and Wernicke's areas, by damage to the arcuate fasciculus, and by damage to the angular gyrus. Explain how these areas may interact in the production of speech.

Diencephalon

The diencephalon is a major autonomic region of the brain that consists of vital structures such as the thalamus, hypothalamus, epithalamus, and pituitary gland.

Objective 19. List the autonomic functions of the thalamus and the hypothalamus.

Objective 20. Describe the location and structure of the pituitary gland.

The **diencephalon** *(di''en-sef'ah-lon)* is the second subdivision of the forebrain and is almost completely surrounded by the cerebral hemispheres of the telencephalon. The third ventricle (see fig. 11.43) forms a midplane cavity within the diencephalon. The most important structures of the diencephalon are the thalamus, hypothalamus, epithalamus, and pituitary gland.

Thalamus

The **thalamus** *(thal'ah-mus)* is a large ovoid mass of gray matter, constituting nearly four-fifths of the diencephalon. It is actually a paired organ, with each portion positioned immediately below the lateral ventricle of its respective cerebral hemisphere (see figs. 11.26, 11.31). The principal function of the thalamus is to act as a relay center for all sensory impulses, except smell, to the cerebral cortex. Specialized masses of nuclei relay the incoming impulses to precise locations within the cerebral lobes for interpretation.

The thalamus also performs some sensory interpretation. The cerebral cortex discriminates pain and other tactile stimuli, but the thalamus responds to general sensory stimuli and provides crude awareness. The thalamus probably plays a role in the initial autonomic response of the body to intense pain and is, therefore, partially responsible for the physiological shock that frequently follows serious trauma.

thalamus: L. *thalamus,* inner room

Hypothalamus

The **hypothalamus** *(hi''po-thal'ah-mus)* is a small portion of the diencephalon located below the thalamus, where it forms the floor and part of the lateral walls of the third ventricle (fig. 11.31). The hypothalamus consists of several masses of nuclei interconnected with other parts of the nervous system. Despite its small size, the hypothalamus performs numerous vital functions, most of which relate directly or indirectly to the regulation of visceral activities. It also performs emotional and instinctual functions (see also fig. 14.15).

The hypothalamus has been described as an autonomic nervous center because of its role in accelerating or decreasing certain body functions. Another major function of the hypothalamus is to regulate the release of hormones from the pituitary gland. The principal autonomic and limbic (emotional) functions of the hypothalamus are described below:

1. **Cardiovascular regulation.** Although the heart has an innate pattern of contraction, impulses from the hypothalamus cause autonomic acceleration or deceleration of the heart. Impulses from the posterior hypothalamus produce a rise in arterial blood pressure and an increase of the heart rate. Impulses from the anterior portion have an opposite effect. The impulses from these regions do not travel directly to the heart but pass first to the cardiovascular centers of the medulla oblongata.
2. **Body-temperature regulation.** Specialized nuclei within the anterior portion of the hypothalamus are sensitive to changes in body temperature. If the arterial blood flowing through this portion of the hypothalamus is above normal temperature, the hypothalamus initiates impulses that cause heat loss through sweating and vasodilation of cutaneous vessels of the skin. A blood temperature below normal causes the hypothalamus to relay impulses that result in heat production and retention through the initiation of shivering, the contraction of cutaneous blood vessels, and the cessation of sweating.
3. **Regulation of water and electrolyte balance.** Specialized *osmoreceptors* in the hypothalamus continuously monitor the osmotic concentration of the blood. An increased osmotic concentration due to lack of water causes antidiuretic hormone (ADH) to be produced by the hypothalamus and released from the posterior pituitary gland. At the same time, a *thirst center* within the hypothalamus produces feelings of thirst.
4. **Regulation of hunger and control of gastrointestinal activity.** The *feeding center* is a specialized portion of the lateral hypothalamus that monitors the blood glucose, fatty-acid, and amino-acid levels. Low levels of these substances in the blood are partially responsible for a sensation of hunger elicited from the hypothalamus. When sufficient amounts of food have been ingested, the *satiety (sah-ti'e-te) center* in the midportion of the hypothalamus inhibits the feeding center. The hypothalamus also receives sensory impulses from the abdominal viscera and regulates glandular secretions and the peristaltic movements of the digestive tract.
5. **Regulation of sleeping and wakefulness.** The hypothalamus has both a *sleep center* and a *wakefulness center* that function with other parts of the brain to determine the level of conscious alertness.
6. **Sexual response.** Specialized *sexual center* nuclei within the dorsal portion of the hypothalamus respond to sexual stimulation of the tactile receptors within the genital organs. The experience of orgasm involves neural activity within the sexual center of the hypothalamus.
7. **Emotions.** Found within the hypothalamus are a number of nuclei associated with specific emotional responses such as anger, fear, pain, and pleasure (see fig. 14.15).
8. **Control of endocrine functions.** The hypothalamus produces neurosecretory chemicals that stimulate the anterior and posterior pituitary to release various hormones.

Obviously the hypothalamus is one of the most vital structures of the body. Dysfunction of the hypothalamus may seriously affect the autonomic, somatic, or psychic body functions. Unsurprisingly, the hypothalamus is implicated as a principal factor in *psychosomatic illness.* Insomnia, peptic ulcers, palpitation of the heart, diarrhea, and constipation are a few symptoms of psychosomatic problems.

Epithalamus

The **epithalamus** is the dorsal portion of the diencephalon that includes a thin roof over the third ventricle. The inside lining of the roof consists of a vascular **choroid plexus** where cerebrospinal fluid is produced (see fig. 11.31). A small cone-shaped mass, called the **pineal gland,** (epiphysis), which has a neuroendocrine function, extends outward from the posterior end of the epithalamus (see fig. 11.31). The posterior commissure, located ventral to the pineal gland, is a tract of commissural fibers that connects the superior colliculi (see fig. 11.36).

Pituitary Gland

The **pituitary gland,** or **hypophysis** *(hi-pof'i-sis),* is positioned on the inferior aspect of the diencephalon and is attached to the hypothalamus by a stalklike structure called the **infundibulum** *(in''fun-dib'u-lum)* (see fig. 11.31). The pituitary is a rounded, pea-shaped gland measuring about 1.3 cm (0.5 in.) in diameter. It is covered by the dura mater and is supported by the sella turcica of the sphenoid bone (fig. 11.33). The arterial circle surrounds the highly vascular pituitary gland, providing it with a rich blood exchange (see fig. 16.29). The pituitary, which has an endocrine function, is structurally and

pineal: L. *pinea,* pine cone
pituitary: L. *pituita,* phlegm (this gland was originally thought to secrete mucus into nasal cavity)
infundibulum: L. *infundibulum,* funnel

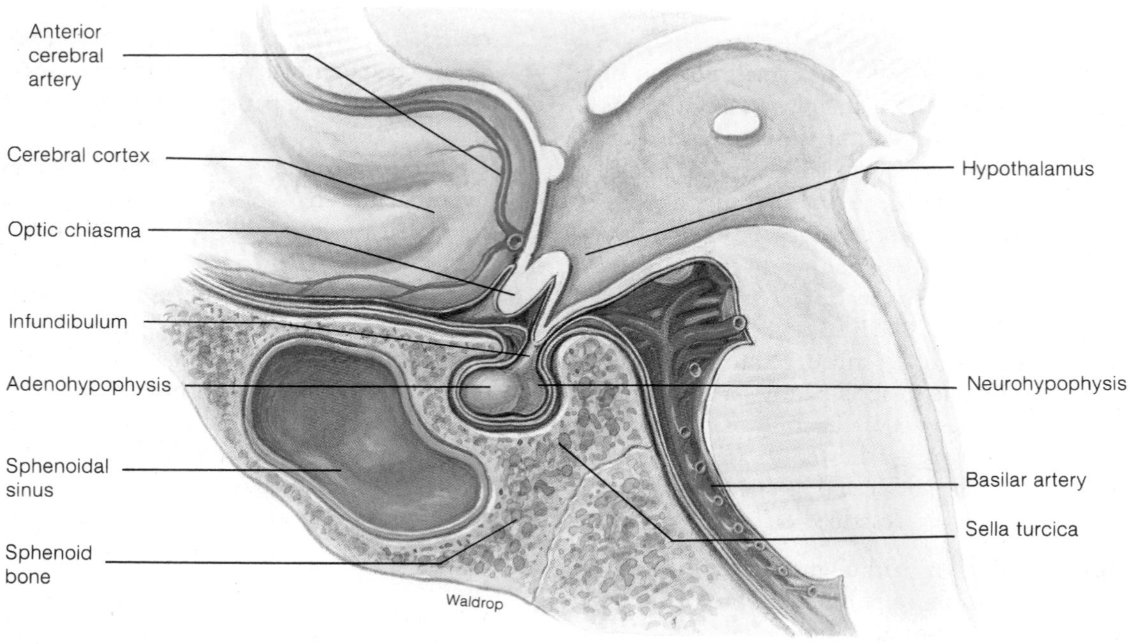

Figure 11.33 The pituitary gland is positioned within the sella turcica of the sphenoid bone and is attached to the brain by the infundibulum.

functionally divided into an anterior portion, called the **adenohypophysis** *(ad''ĕ-no-hi-pof'ĭ-sis),* and a posterior portion, called the **neurohypophysis** (see chap. 14).

1. Discuss what is meant by the statement that the thalamus is the pain center of the brain.
2. List the body systems and the functions over which the hypothalamus has some control.
3. Describe the location of the pituitary gland relative to the rest of the diencephalon and the rest of the brain. What are the two portions of the pituitary gland, and how are they supported?

Mesencephalon

The mesencephalon contains the corpora quadrigemina, concerned with visual and auditory reflexes, and the cerebral peduncles, composed of fiber tracts. It also contains specialized nuclei that help to control posture and movement.

Objective 21. List the structures of the mesencephalon, and explain their functions.

The **mesencephalon** *(mes''en-sef'ah-lon),* or midbrain, is a short section of the brain stem between the diencephalon and the pons (see fig. 11.36). Within the midbrain is the **cerebral aqueduct** (aqueduct of Sylvius) (see fig. 11.43), which connects the third and fourth ventricle. The midbrain also contains the corpora quadrigemina (fig. 11.31), cerebral peduncles (see figs. 11.30, 11.36), red nucleus, and substantia nigra.

The **corpora quadrigemina** are the four rounded elevations on the dorsal portion of the midbrain. The two upper eminences, the **superior colliculi,** are concerned with visual reflexes. The two posterior eminences, the **inferior colliculi,** are responsible for auditory reflexes. The **cerebral peduncles** are a pair of cylindrical structures composed of ascending and descending projection fiber tracts that support and connect the cerebrum to the other regions of the brain.

The **red nucleus** is deep within the midbrain between the cerebral peduncle and the cerebral aqueduct. The red nucleus is gray matter that connects the cerebral hemispheres and the cerebellum and functions in reflexes concerned with motor coordination and maintaining posture. Another nucleus called the **substantia nigra** is ventral to the red nucleus. The substantia nigra is thought to inhibit forced involuntary movements.

1. Discuss what possible symptoms might be apparent as a result of a tumor in the midbrain.
2. Which structures of the midbrain function with the cerebellum in controlling posture and movement?

Metencephalon

The metencephalon contains the pons, which relays impulses, and the cerebellum, which coordinates skeletal muscle contractions.

Objective 22. Describe the location and structure of the pons and cerebellum, and list their functions.

The **metencephalon** *(met''en-sef'ah-lon)* is the most superior portion of the hindbrain. Two vital structures of the metencephalon are the pons and cerebellum. The cerebral aqueduct of the mesencephalon enlarges to become the **fourth ventricle** (see fig. 11.43) within the metencephalon and myelencephalon.

aqueduct of Sylvius: from Jacobus Sylvius, French anatomist, 1478–1555
corpora quadrigemina: L. *corpus,* body; *quadri,* four; *geminus,* twin

colliculus: L. *colliculus,* small mound

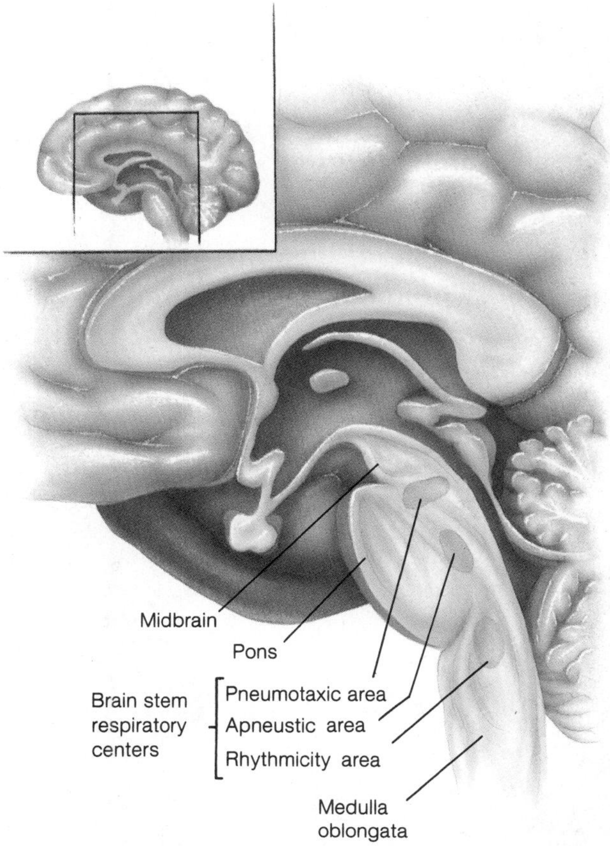

Figure 11.34 Nuclei within the pons and medulla oblongata that constitute the respiratory center.

Pons

The **pons** can be observed as a rounded bulge on the underside of the brain, between the midbrain and the medulla oblongata (fig. 11.34). The pons consists of white fiber tracts that course in two principal directions. The surface fibers extend transversely to connect with the cerebellum through the middle cerebellar peduncles. The deeper longitudinal fibers are part of the motor and sensory tracts that connect the medulla with the tracts of the midbrain.

Scattered throughout the pons are several nuclei associated with specific cranial nerves. The cranial nerves that have nuclei within the pons include the trigeminal (V), which transmits impulses for chewing and sensory sensations from the head; the abducens (VI), which controls certain movements of the eyeball; the facial (VII), which transmits impulses for facial movements and sensory sensations from the taste buds; and the vestibular branches of the vestibulocochlear (VIII), which maintain equilibrium.

Other nuclei of the pons function with nuclei of the medulla oblongata to regulate the rate and depth of breathing. The two respiratory centers of the pons are called the **apneustic** and the **pneumotaxic areas** (fig. 11.34).

pons: L. *pons,* bridge

Cerebellum

The **cerebellum** is the second largest structure of the brain. It is located in the metencephalon and occupies the inferior and posterior aspect of the cranial cavity. The cerebellum is separated from the overlying cerebrum by a **transverse fissure** (see fig. 11.23). A portion of the meninges called the **tentorium cerebelli** extends into the transverse fissure. The cerebellum consists of two **hemispheres** and a central constricted area called the **vermis** (fig. 11.35). The **falx cerebelli** is the portion of the meninges that partially extends between the hemispheres.

Like the cerebrum, the cerebellum has a thin, outer layer of gray matter, called the **cerebellar cortex,** and a thick, deeper layer of white matter. The cerebellum is convoluted into a series of slender, parallel **gyri.** The tracts of white matter within the cerebellum have a distinctive branching pattern called the **arbor vitae,** which can be seen in a sagittal view (fig. 11.35).

Three paired bundles of nerve fibers, called **cerebellar peduncles,** support the cerebellum and provide it with tracts for communicating with the rest of the brain (fig. 11.36). The cerebellar peduncles are as follows:

1. **Superior cerebellar peduncles** connect the cerebellum with the midbrain. The fibers within these peduncles originate primarily from specialized **dentate nuclei** within the cerebellum and pass through the red nucleus to the thalamus and then to the motor areas of the cerebral cortex. Impulses through the fibers of these peduncles provide feedback to the cerebrum.
2. **Middle cerebellar peduncles** convey impulses of voluntary movement from the cerebrum through the pons and to the cerebellum.
3. **Inferior cerebellar peduncles** connect the cerebellum with the medulla oblongata and the spinal cord. They contain both incoming vestibular and proprioceptive fibers and outgoing motor fibers.

The principal function of the cerebellum is coordinating skeletal muscle contractions by recruiting precise motor units within the muscles. Impulses for voluntary muscular movement originate in the cerebral cortex and are coordinated by the cerebellum. The cerebellum constantly initiates impulses to selective motor units for maintaining posture and muscle tone. The cerebellum also adjusts to incoming impulses from **proprioceptors** *(pro''pre-o-sep'tor)* within muscles, tendons, joints, and special sense organs. A proprioceptor is a sensory nerve ending that is sensitive to changes in the tension of a muscle or tendon.

Trauma or diseases of the cerebellum, such as a *stroke* or *cerebral palsy,* frequently cause an impairment of skeletal muscle function. Movements become jerky and uncoordinated in a condition known as *ataxia.* There is also a loss of equilibrium, resulting in a disturbance of gait. *Alcohol intoxication* causes similar uncoordinated body movements.

cerebellum: L. *cerebellum,* diminutive of *cerebrum,* brain
vermis: L. *vermis,* worm
arbor vitae: L. *arbor,* tree; *vitae,* life
peduncle: L. *peduncle,* diminutive of *pes,* foot

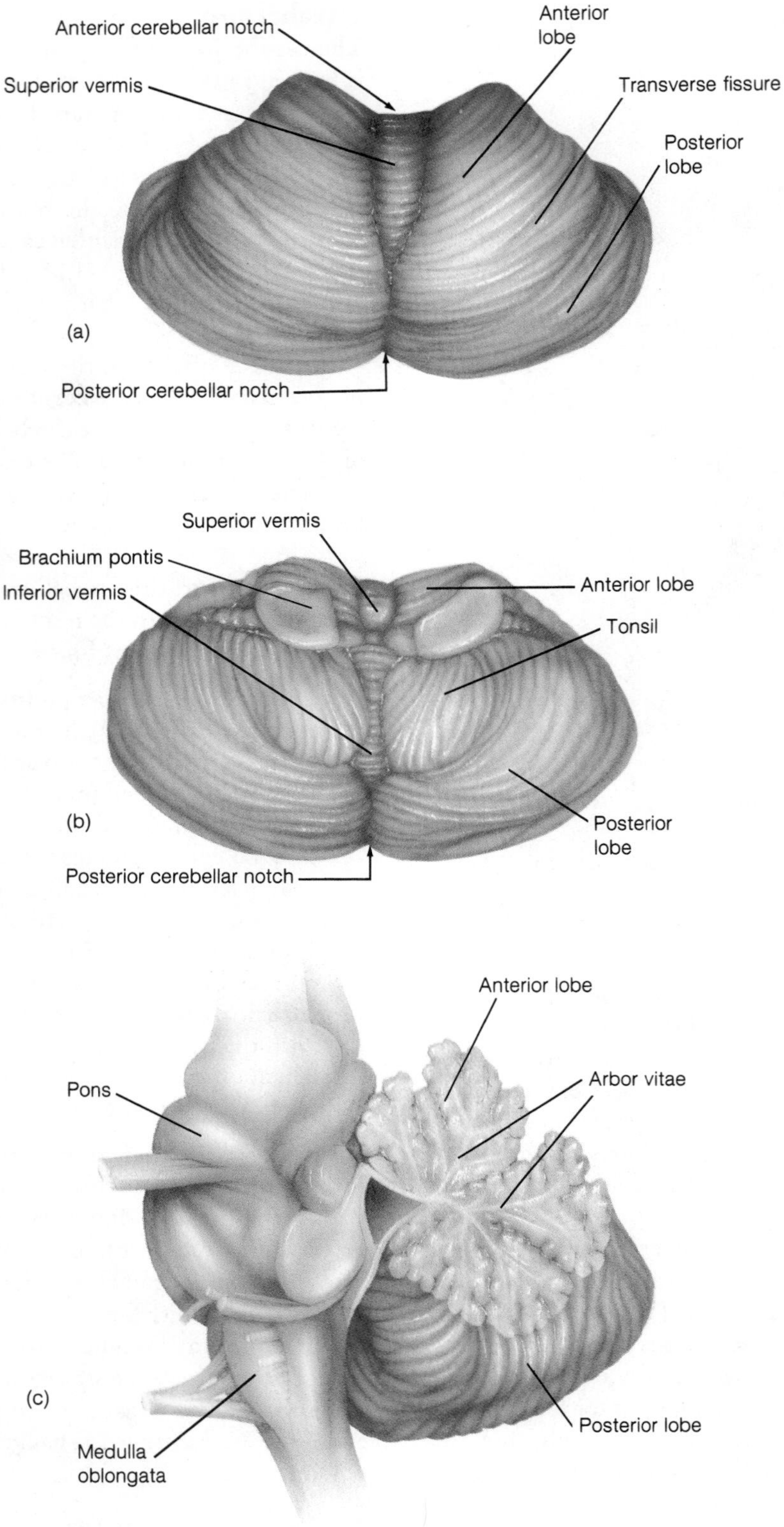

Figure 11.35 The structure of the cerebellum. (*a*) A superior view, (*b*) an inferior view, and (c) a sagittal view.

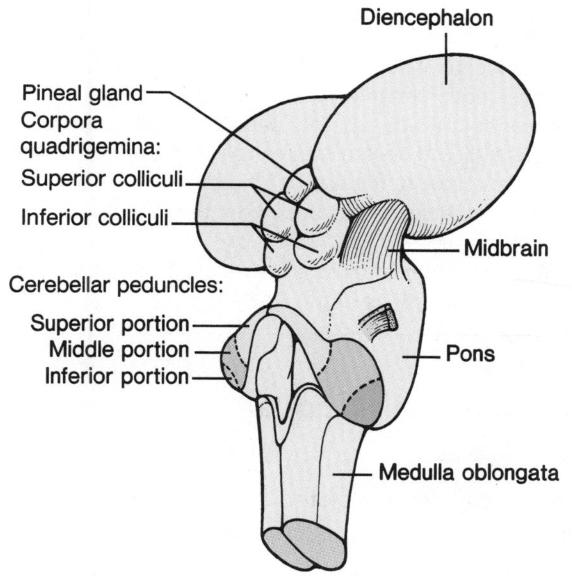

Figure 11.36 The cerebellar peduncles can be seen when the cerebellum has been removed from its attachment to the brain stem.

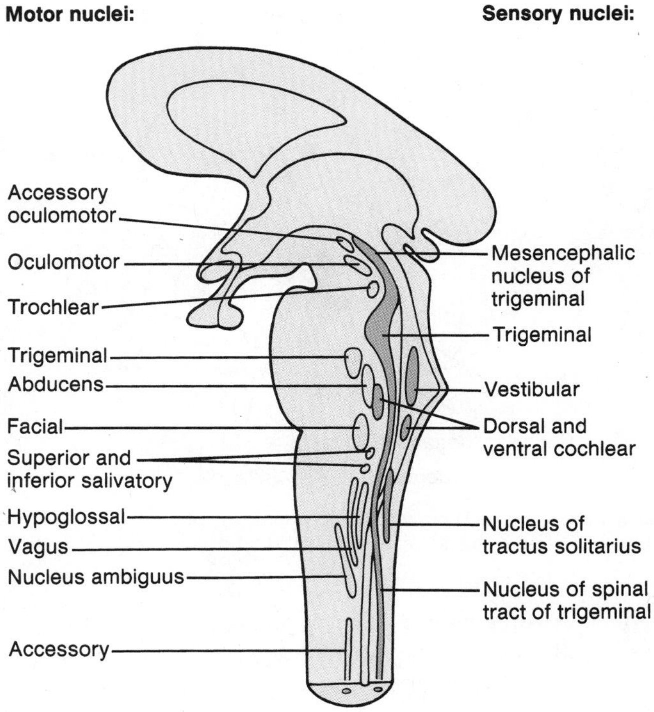

Figure 11.37 A sagittal section of the medulla oblongata and pons showing the cranial nerve nuclei of gray matter.

1. Describe the locations and relative sizes of the pons and the cerebellum.
2. Which cranial nerves have nuclei located within the pons?
3. Define the terms *tentorium cerebelli, vermis, arbor vitae,* and *cerebellar peduncles.*
4. List the functions of the pons and the cerebellum.

Myelencephalon

The medulla oblongata, contained within the myelencephalon, connects to the spinal cord and contains nuclei for the cranial nerves and vital autonomic functions.

Objective 23. Describe the location and structure of the medulla oblongata, and list its functions.

Objective 24. Define the term *reticular formation,* and explain its function.

Medulla Oblongata

The **medulla oblongata,** or simply **medulla,** is a bulbous structure about 3 cm (1 in.) long that is continuous with the pons anteriorly and the spinal cord posteriorly at the level of the foramen magnum (see figs. 11.30, 11.31). Externally, the medulla resembles the spinal cord, except for the two triangular, elevated structures, called **pyramids,** on the ventral side and an oval enlargement, called the **olive** (see fig. 11.30), on each lateral surface. The **fourth ventricle,** the space within the medulla, is continuous posteriorly with the central canal of the spinal cord and anteriorly with the cerebral aqueduct (see fig. 11.43).

The medulla is composed of vital nuclei and white matter that form all of the descending and ascending tracts communicating between the spinal cord and various parts of the brain. Most of the fibers within these tracts cross over to the opposite side through the pyramidal region of the medulla, permitting one side of the brain to receive information from and send information to the opposite side of the body (see fig. 11.30).

The gray matter of the medulla consists of several important nuclei for the cranial nerves, sensory relay, and for autonomic functions (fig. 11.37). The **nucleus ambiguus** and the **hypoglossal nucleus** are the centers from which arise the vestibulocochlear (VIII), glossopharyngeal (IX), accessory (XI), and hypoglossal (XII) nerves. The vagus nerves (X) arise from **vagus nuclei,** one on each lateral side of the medulla adjacent to the fourth ventricle. The **nucleus gracilis** and the **nucleus cuneatus** relay sensory information to the thalamus, and then the impulses are relayed to the cerebral cortex via the thalamic nuclei (not illustrated). The **inferior olivary nuclei** and the **accessory olivary nuclei** of the olive mediate impulses passing from the forebrain and midbrain through the inferior cerebellar peduncles to the cerebellum.

Three other nuclei within the medulla function as autonomic centers for controlling vital visceral functions.

1. **Cardiac center.** Both *inhibitory* and *accelerator fibers* arise from nuclei of the cardiac center. Inhibitory impulses constantly travel through the vagus nerves to slow the heartbeat. Accelerator impulses travel through the spinal cord and eventually innervate the heart through fibers within spinal nerves T1–T5.
2. **Vasomotor center.** Nuclei of the vasomotor center send impulses via the spinal cord and spinal nerves to the smooth muscles of arteriole walls, causing them to constrict and elevate arterial blood pressure.

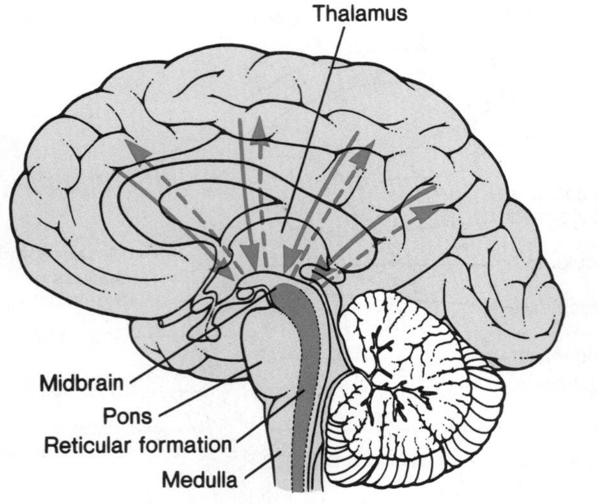

Figure 11.38 The reticular activating system. The arrows indicate the direction of impulses along nerve pathways that connect with the RAS.

3. **Respiratory center.** The respiratory center of the medulla controls the rate and depth of breathing and functions with nuclei of the pons (see fig. 11.34) to produce rhythmic breathing.

Other nuclei of the medulla function as centers for nonvital respiratory movements such as sneezing, coughing, swallowing, and vomiting. Some of these activities, such as swallowing, may be initiated voluntarily, but once they progress to a certain point they become involuntary and cannot be stopped.

Reticular Formation

The **reticular formation** is a complex network of nuclei and nerve fibers within the brain stem that functions as the *reticular activating system (RAS),* which arouses the cerebrum. Portions of the reticular formation are located in the spinal cord, pons, midbrain, and parts of the thalamus and hypothalamus (fig. 11.38). The reticular formation contains ascending and descending fibers from most of the structures within the brain.

Nuclei within the reticular formation generate a continuous flow of impulses unless they are inhibited by other parts of the brain. The principal functions of the RAS are to keep the cerebrum in a state of alert consciousness and to monitor selectively the afferent impulses perceived by the cerebrum. The RAS also helps the cerebellum activate selected motor units to maintain muscle tonus and produce smooth, coordinated contractions of skeletal muscles.

The RAS is sensitive to changes in and trauma to the brain. The sleep response is thought to occur because of a decrease in activity within the RAS, perhaps due to the secretion of specific neurotransmitters. A blow to the head or certain drugs and diseases may damage the RAS, causing unconsciousness. A *coma* is a state of unconsciousness and inactivity of the RAS that even the most powerful external stimuli cannot disturb.

---

1. When a nurse or physician prepares a case study, the location, structure, function, and possible dysfunction of an afflicted structure or organ is reported. Prepare a brief case study of a patient who has suffered severe trauma to the medulla oblongata from a blow to the back of the skull.
2. Which cranial nerves arise from the medulla oblongata?
3. Describe the location of the reticular formation. What are two functions of the RAS?

Meninges of the Central Nervous System

The CNS is covered by protective meninges, consisting of a dura mater, an arachnoid membrane, and a pia mater.

Objective 25. Describe the position of the meninges as they protect the CNS.

The entire delicate CNS is protected by a bony encasement: the cranium, surrounding the brain, and the vertebral column, surrounding the spinal cord. It is also protected by three connective tissue membranous coverings called the **meninges.** See figures 11.39, 11.40, 11.41, 11.42. Individually, from the outside in, they are known as the dura mater, the arachnoid membrane, and the pia mater.

Dura Mater

The **dura mater** is in contact with the bone and is composed primarily of dense fibrous connective tissue. The **cranial dura mater** is a double-layered structure. The thicker, outer **periosteal** *(per''e-os'te-al)* **layer** adheres lightly to the cranium where it is the periosteum (fig. 11.40). The thinner, inner **meningeal layer** follows the general contour of the brain. The **spinal dura mater** is not double layered but is similar to the meningeal layer of the cranial dura mater (fig. 11.41).

The two layers of the cranial dura mater are fused and cover most of the brain. In certain regions, however, the layers are separated, enclosing **dural sinuses** (see fig. 11.40), which collect venous blood and drain it to the internal jugular veins of the neck.

In four locations, the meningeal layer of the cranial dura forms distinct septa to partition major structures on the surface of the brain and anchor the brain to the inside of the cranial case. These septa have been previously identified in this chapter and are reviewed in table 11.5.

The spinal dura mater forms a tough, tubular **dural sheath** that continues into the vertebral canal and surrounds the spinal cord. There is no connection between the dural sheath and the vertebrae forming the vertebral canal, but instead there is a potential cavity called the **epidural space** (see fig. 11.41). The epidural space is highly vascular and contains areolar and adipose connective tissue, which form a protective pad around the spinal cord.

---

meninges: L. plural form of *meninx,* membrane
dura mater: L. *dura,* hard; *mater,* mother

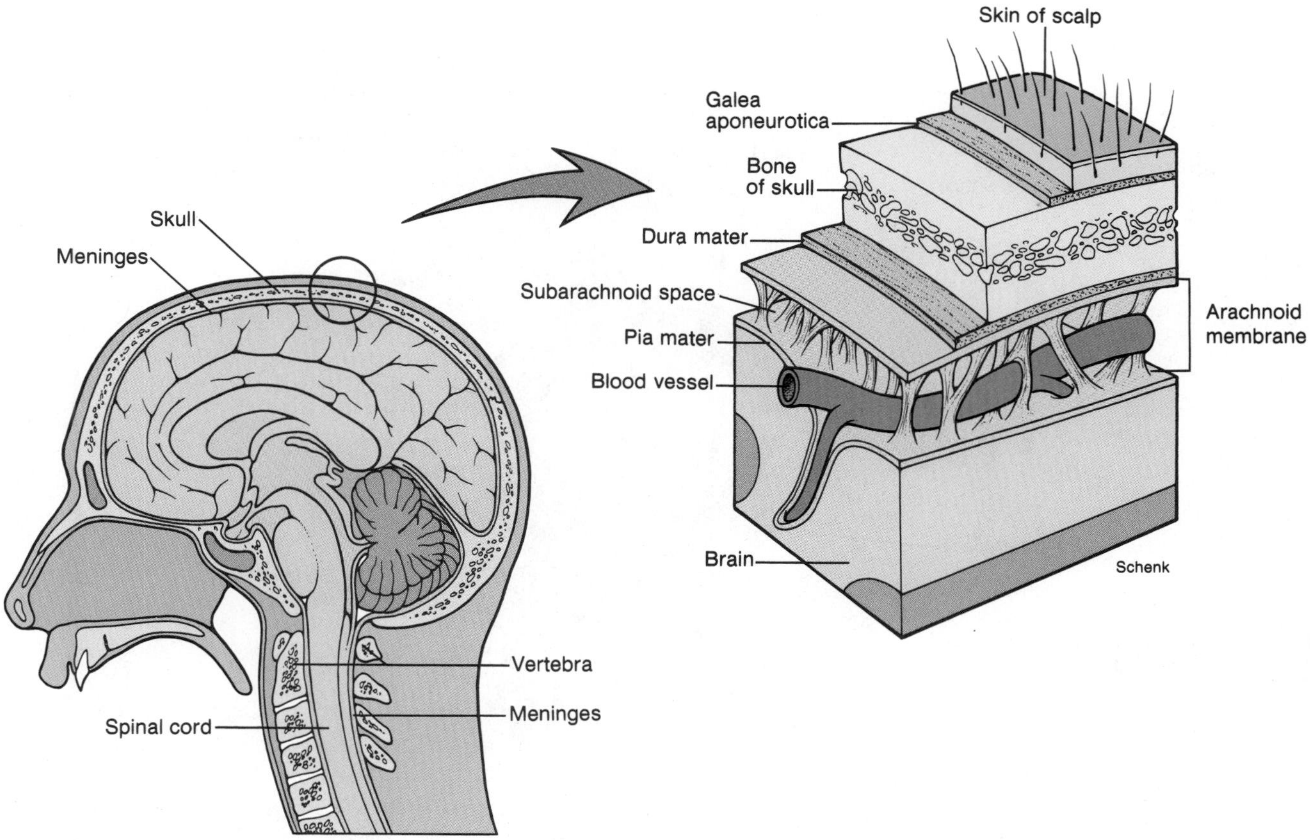

Figure 11.39 The meninges of the brain.

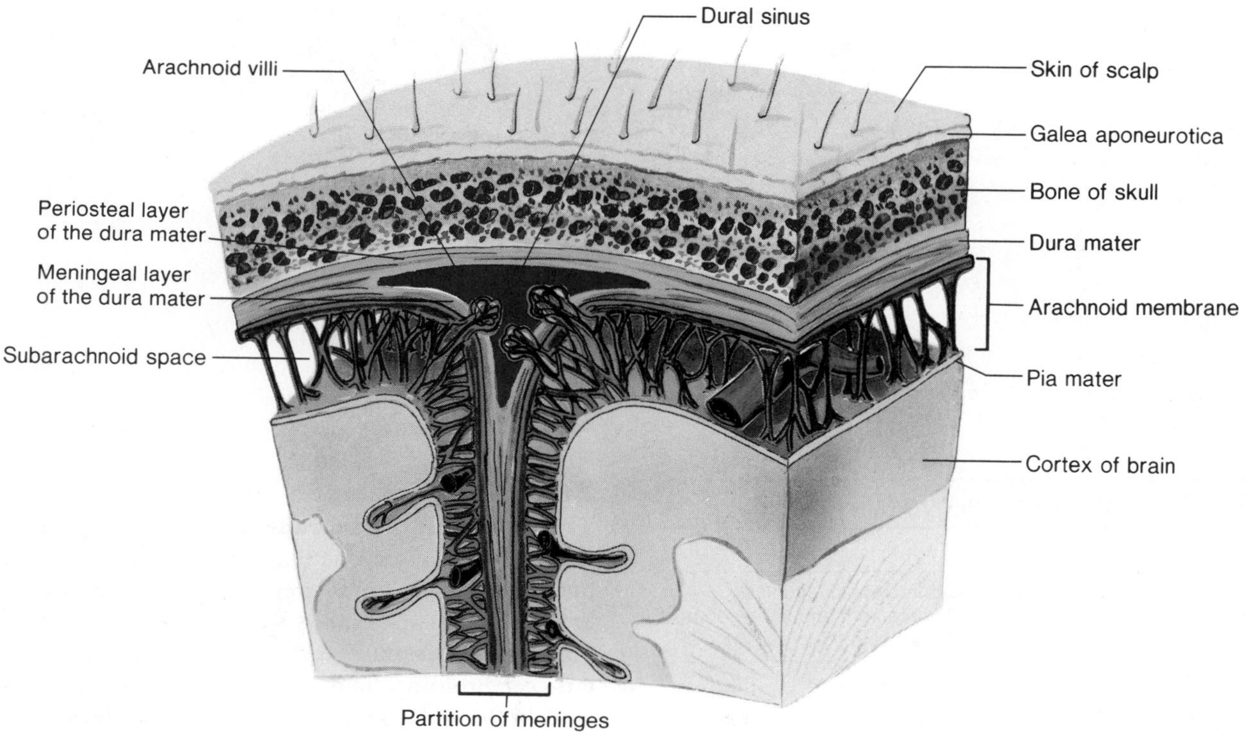

Figure 11.40 The dural sinuses.

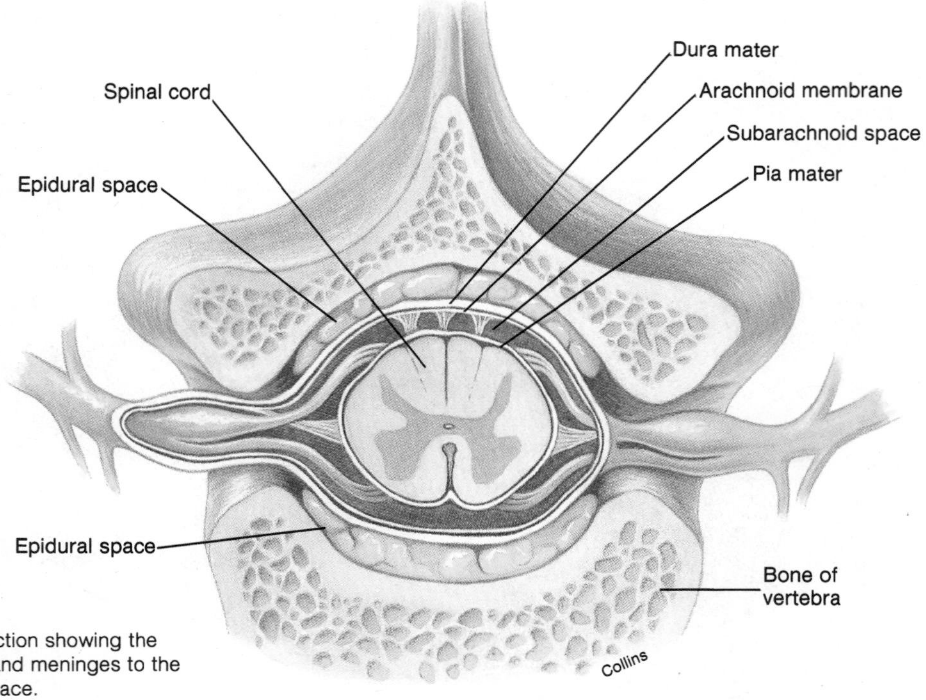

Figure 11.41 A transverse section showing the relationship of the spinal cord and meninges to the protective fat in the epidural space.

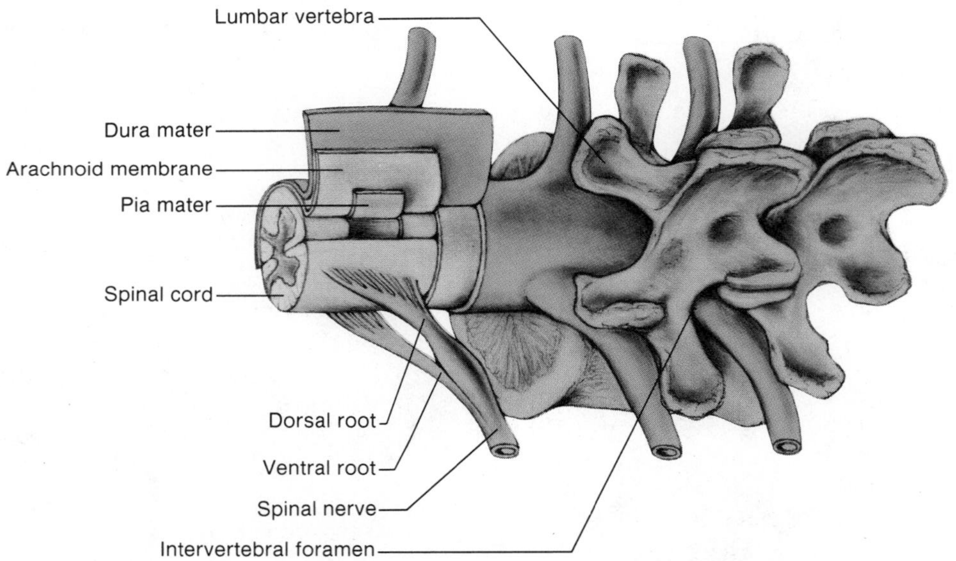

Figure 11.42 The spinal cord and the meninges.

Table 11.5 Septa of the cranial dura mater

Septa	Location
Falx cerebri	Extends downward into the longitudinal fissure to partition the right and left cerebral hemispheres; anchored anteriorly to the crista galli of the ethmoid bone and posteriorly to the tentorium
Tentorium cerebelli	Separates the occipital lobes of the cerebrum from the cerebellum; anchored to the tentorium, petrous bones, and occipital bone
Falx cerebelli	Partitions the right and left cerebellar hemispheres; anchored to occipital crest
Diaphragma sellae	Forms the roof of the sella turcica

Arachnoid Membrane

The **arachnoid membrane** is the middle of the three meninges. This delicate, netlike membrane spreads over the CNS but generally does not extend into the sulci or fissures of the brain. The **subarachnoid space,** located between the arachnoid membrane and the deepest meninx, the pia mater, contains cerebrospinal fluid. The subarachnoid space is maintained by delicate, web-like strands that connect the arachnoid membrane and pia mater (see fig. 11.40).

arachnoid: L. *arachnoides,* like a cobweb

Pia Mater

The thin **pia mater** is attached to the surfaces of the CNS and follows the irregular contours of the brain and spinal cord. The pia mater is composed of modified loose fibrous connective tissue. It is highly vascular and functions to support the vessels that nourish the underlying cells of the brain and spinal cord. The pia mater is specialized over the roofs of the ventricles, where along with the arachnoid membrane, it contributes to the formation of the choroid plexuses. Lateral extensions of the pia mater along the spinal cord form the **ligamentum denticulatum,** which attaches the cord to the dura mater.

Meningitis is an inflammation of the meninges, usually caused by certain bacteria or viruses. The arachnoid membrane and the pia mater are the two meninges most frequently affected. Meningitis is accompanied by high fever and severe headache. Complications may cause sensory impairment, paralysis, or mental retardation. Untreated meningitis generally results in coma and death.

---

1. Contrast cranial and spinal dura mater.
2. Explain how the meninges support and protect the CNS.
3. Describe the location of the dural sinuses and the epidural space.

Ventricles and Cerebrospinal Fluid

The ventricles, central canal, and subarachnoid space contain cerebrospinal fluid, formed by the active transport of substances from blood plasma in the choroid plexuses.

Objective 26. Discuss the formation, function, and flow of cerebrospinal fluid.

Objective 27. Explain the importance of the blood-brain barrier in maintaining homeostasis within the brain.

Cerebrospinal fluid, *CSF,* is a clear, lymphlike fluid that forms a protective cushion around and within the CNS. The fluid also buoys the brain. CSF circulates through the various **ventricles** of the brain, the **central canal** of the spinal cord, and the **subarachnoid space** around the entire CNS. The cerebrospinal fluid returns to the circulatory system by draining through the walls of the **arachnoid villi,** which are venous capillaries.

Ventricles of the Brain

The ventricles of the brain are connected to one another and to the central canal of the spinal cord (figs. 11.43, 11.44). Each of the two **lateral ventricles** (first and second ventricles) is located in each hemisphere of the cerebrum, inferior to the corpus callosum. The **third ventricle** is located in the diencephalon between the thalami. Each lateral ventricle is connected to the third ventricle by a narrow, oval opening called the **interventricular foramen** (foramen of Monro). The **fourth ventricle** is located in the brain stem, within the pons, cerebellum, and medulla oblongata. The **cerebral aqueduct** (aqueduct of Sylvius) passes through the midbrain to link the third and fourth ventricles. The fourth ventricle also communicates posteriorly with the central canal. Cerebrospinal fluid exits from the fourth ventricle into the subarachnoid space (fig. 11.45) through three foramina: the **foramen of Magendie,** a medial opening, and two lateral **foramina of Luschka** (not illustrated). Cerebrospinal fluid returns to the venous blood through the arachnoid villi.

Internal hydrocephalus is a condition in which cerebrospinal fluid builds up within the ventricles of the brain (fig. 11.44b). It is more common in infants whose cranial sutures have not yet strengthened or ossified. If the pressure is excessive, the condition may have to be treated surgically.

External hydrocephalus, an accumulation of fluid within the subarachnoid space, usually results from an obstruction of drainage at the arachnoid villi.

Cerebrospinal Fluid

CSF buoys the CNS and protects it from mechanical injury. CSF has a specific gravity of 1.007, which is a density close to that of brain tissue. The brain weight is about 1,500 g, but suspended in CSF its buoyed weight is about 50 g. This means that the brain has a near neutral buoyancy and can therefore function effectively as a relatively heavy organ. At a true neutral buoyancy, an object does not float or sink but is suspended in its fluid environment.

In addition to buoying the CNS, CSF reduces the damaging effect of an impact to the head by spreading the force over a larger area. CSF also helps to remove metabolic wastes from nervous tissue. Since the CNS lacks lymphatic circulation, the CSF moves cellular wastes into the venous return at its places of drainage.

The clear, watery CSF is continuously produced from materials within the blood by masses of specialized capillaries called **choroid plexuses** and, to a lesser extent, by secretions of the ependymal cells. The ciliated ependymal cells cover the choroid plexuses as well as line the central canal and presumably aid the movement of the CSF.

CSF is formed mainly by the active transport and ultrafiltration of substances within the blood plasma. CSF has more sodium, chloride, magnesium, and hydrogen ions than blood plasma and less calcium, potassium, and glucose. In addition, CSF contains some proteins, urea, and white blood cells.

As much as 800 ml of cerebrospinal fluid are produced each day, although only 140–200 ml are bathing the CNS at any moment. A person lying in a horizontal position has a slow but continuous circulation of cerebrospinal fluid, with a fluid pressure of about 10 mm Hg.

---

pia mater: L. *pia,* soft or tender; *mater,* mother
foramen of Monro: from Alexander Monro, Jr., Scottish anatomist, 1733–1817

foramen of Magendie: from Francois Magendie, French physiologist, 1783–1855
foramen of Luschka: from Hubert Luschka, German anatomist, 1820–1875

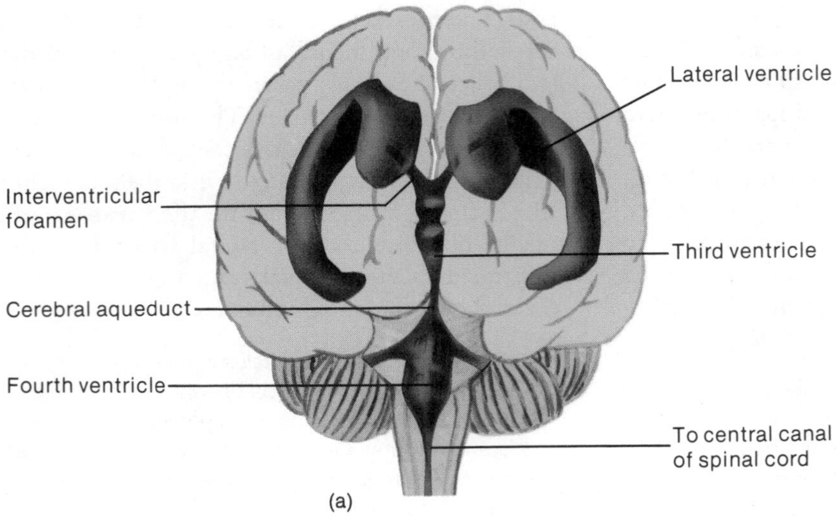

(a)

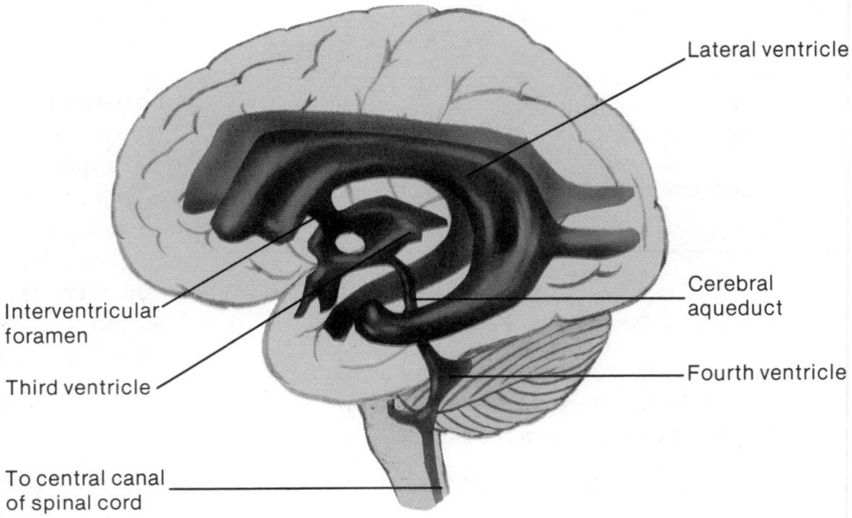

(b)

Figure 11.43 The ventricles of the brain. (*a*) An anterior view and (*b*) a lateral view.

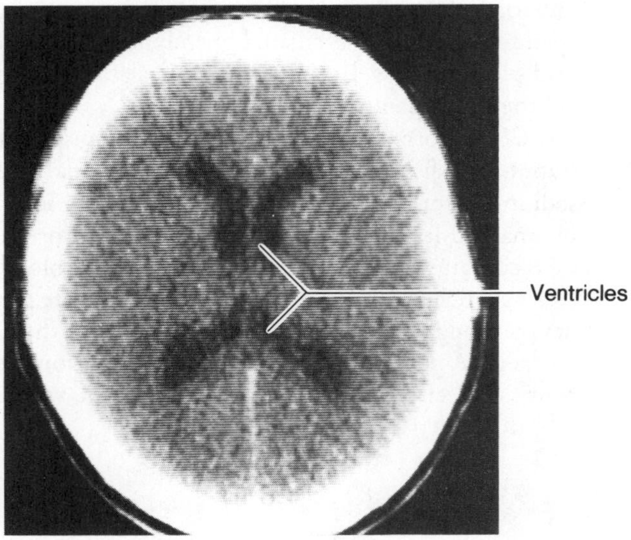

(a)

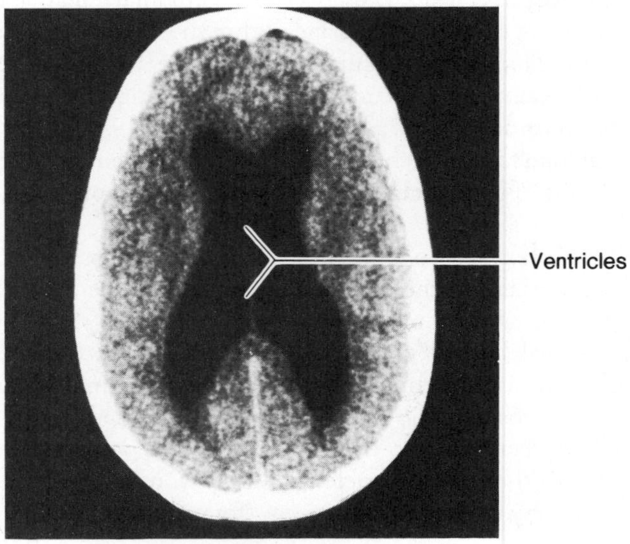

(b)

Figure 11.44 CT scans of the brain showing (*a*) a normal configuration of the ventricles and (*b*) an abnormal configuration of the ventricles due to hydrocephalism.

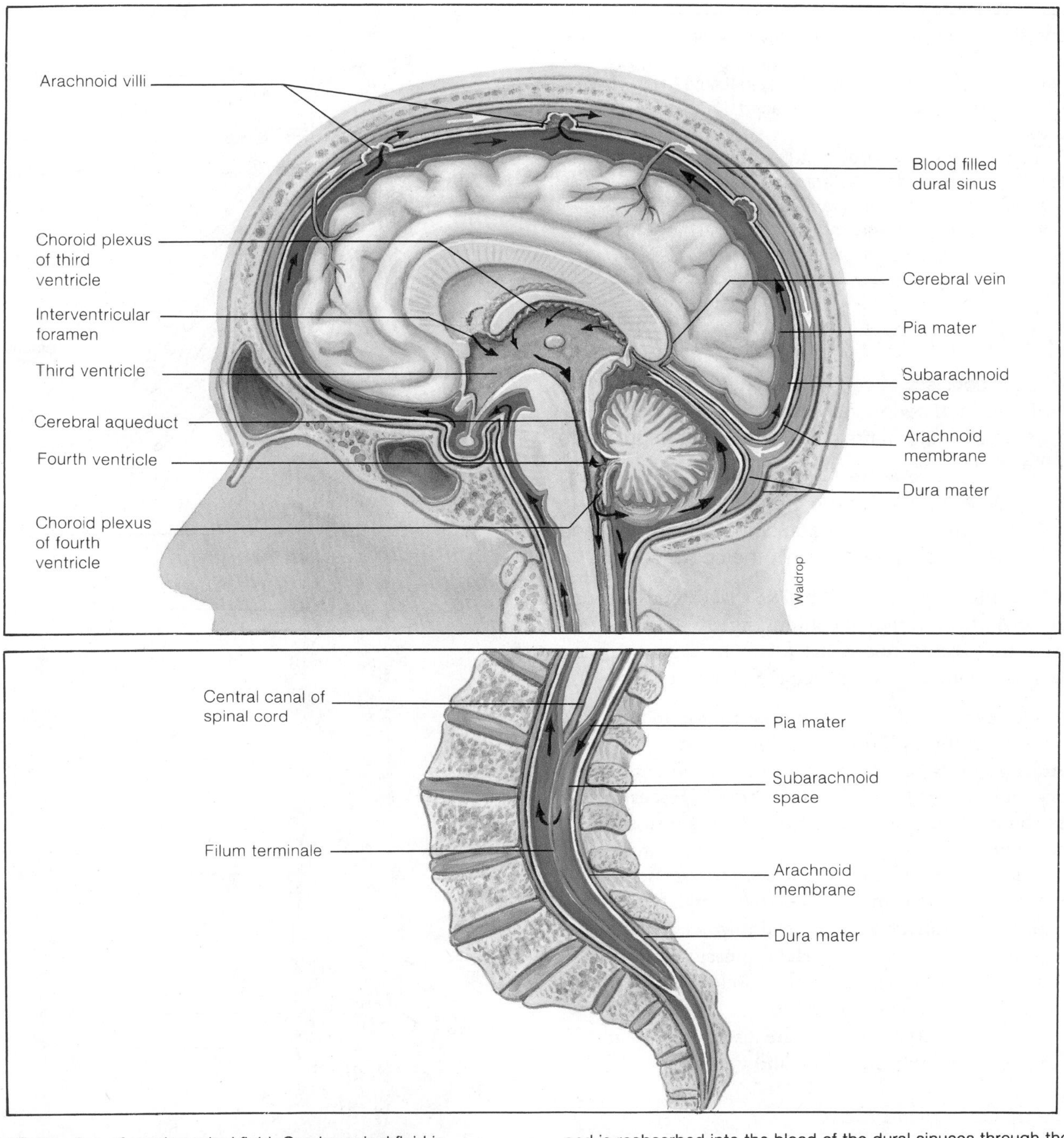

Figure 11.45 The flow of cerebrospinal fluid. Cerebrospinal fluid is secreted by choroid plexuses in the ventricular walls. The fluid circulates through the ventricles and central canal, enters the subarachnoid space, and is reabsorbed into the blood of the dural sinuses through the arachnoid villi.

The homeostatic consistency of the CSF composition is critical and a chemical unbalance may have marked effects on CNS functions. An increase in amino acid glycine concentration, for example, produces hypothermia and hypotension as temperature and blood pressure regulatory mechanisms are disrupted. A slight change in pH may affect the respiratory rate and depth.

Blood-Brain Barrier

The **blood-brain barrier (BBB)** is a structural arrangement of capillaries, surrounding connective tissue, and specialized neuroglial cells, called astrocytes (see fig. 11.9), that selectively determine which substances can move from the plasma of the blood to the extracellular fluid of the brain. Certain substances such as H_2O, O_2, CO_2, glucose, and lipid-soluble compounds (alcohol, for example) pass readily through the BBB. Certain inorganic ions such as Ca^+ and K^+ pass more slowly so that the concentrations of these ions are different in the brain than in the plasma. Other substances, such as proteins, lipids, creatine, urea, inulin, certain toxins, and most antibiotics, are restricted in passage. The BBB is an important factor to consider when planning drug therapy for neurological disorders.

1. Describe the location of the ventricles within the brain.
2. What are the functions of cerebrospinal fluid?
3. Where is cerebrospinal fluid produced, and where does it drain?
4. Construct a flow chart depicting a substance that can flow freely across the BBB from the plasma to the nervous tissue. Indicate also a substance that passes more slowly and one that is restricted in its passage. Why is the BBB functionally important?

Spinal Cord

The spinal cord consists of centrally located gray matter, involved in reflexes, and peripherally located ascending and descending tracts of white matter, which conduct impulses to and from the brain.

Objective 28. Describe the structure of the spinal cord.
Objective 29. Describe the arrangement of ascending and descending tracts within the spinal cord.

The **spinal cord** is the portion of the CNS that extends through the neural canal of the vertebral column (fig. 11.46). It is continuous with the brain through the foramen magnum of the skull. The spinal cord has two principal functions:

1. It provides a means of neural communication to and from the brain through tracts of white matter. **Ascending tracts** conduct impulses from the peripheral sensory receptors of the body to the brain. **Descending tracts** conduct motor impulses from the brain to the muscles and glands.
2. It serves as a center for spinal reflexes. Specific nerve pathways enable some movements to be reflexive rather than initiated voluntarily by the brain. Movements of this type are not confined to skeletal muscles; reflexive movements of cardiac and smooth muscles control heart rate, breathing rate, blood pressure, and digestive activities. Spinal nerve pathways are also involved in swallowing, coughing, sneezing, and vomiting.

Structure of the Spinal Cord

The spinal cord extends inferiorly from the position of the foramen magnum of the occipital bone to the level of the first lumbar vertebra (L1). The spinal cord is somewhat flattened dorsoventrally, making it oval in cross section. Two prominent enlargements can be seen in a dorsal view (see fig. 11.46). The **cervical enlargement** is located between the third cervical and the second thoracic vertebrae. Nerves emerging from this region serve the upper extremities. The **lumbar enlargement** lies between the ninth and twelfth thoracic vertebrae. Nerves from the lumbar enlargement supply the lower extremities.

The embryonic spinal cord develops more slowly than the associated vertebral column; thus, in the adult, the cord does not extend beyond L1. The tapering, terminal portion of the spinal cord is called the **conus medullaris.** The **filum terminale,** a fibrous strand composed mostly of pia mater, extends

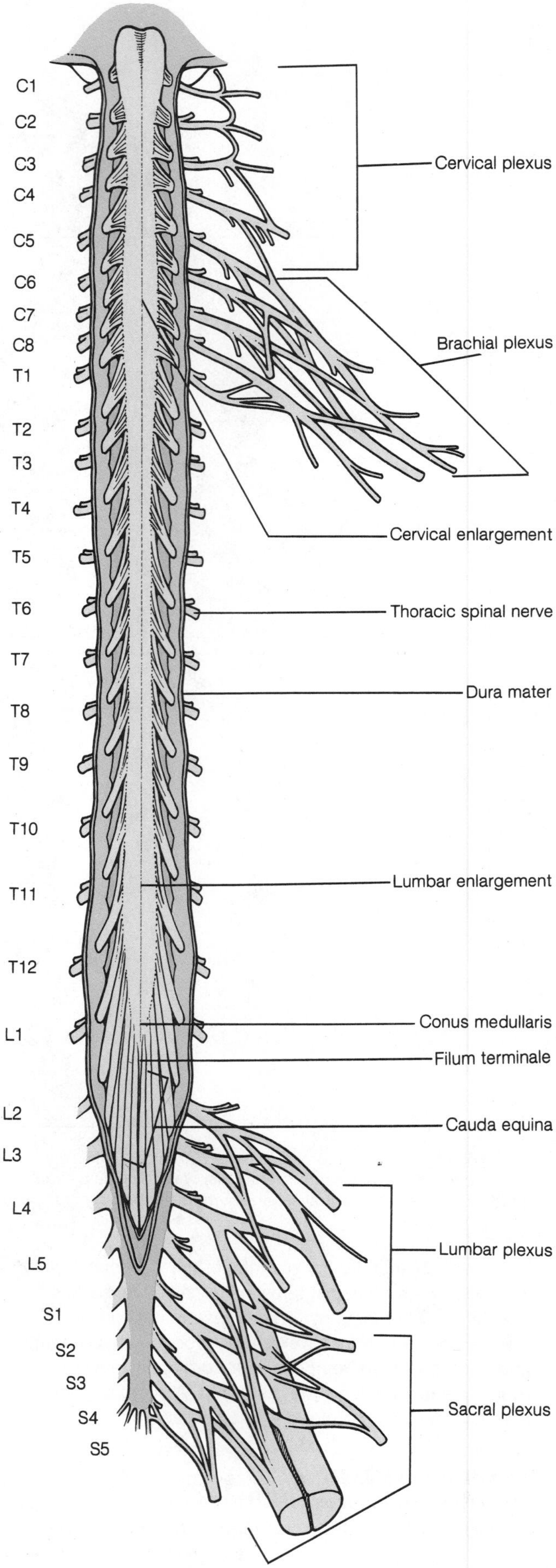

Figure 11.46 The spinal cord and plexuses.

filum terminalis: L. *filum,* filament; *terminus,* end

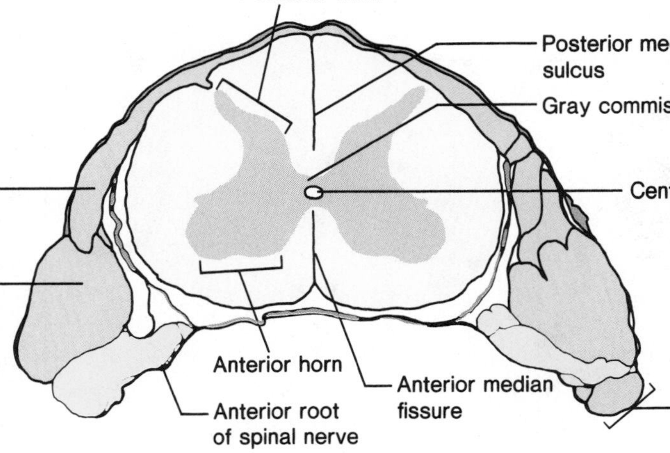

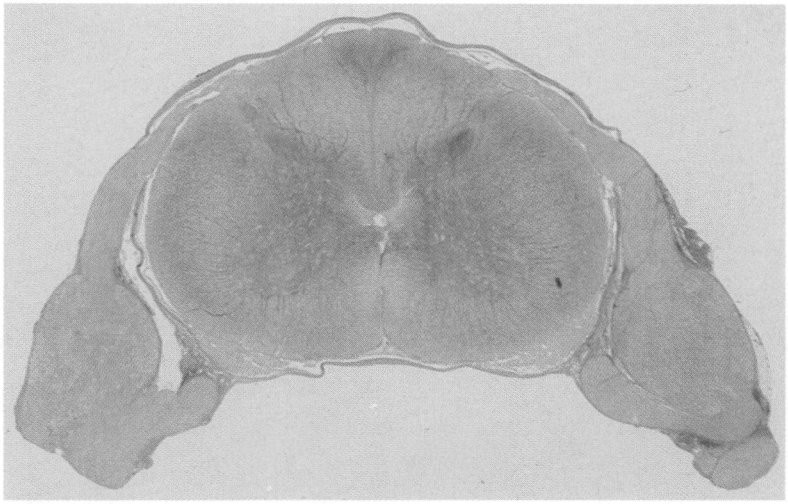

Figure 11.47 The (*a*) spinal cord in a transverse section and (*b*) a photograph of the spinal cord and nerve.

inferiorly from the conus medullaris at the level of L1 to the coccyx. Nerve roots also radiate inferiorly from the conus medullaris through the vertebral canal. These nerves are collectively referred to as the **cauda equina** because they resemble a horse's tail.

The spinal cord develops as thirty-one segments, each of which gives rise to a pair of **spinal nerves** that emerge from the cord through the intervertebral foramina. Two grooves, an **anterior median fissure** and a **posterior median sulcus,** extend the length of the spinal cord and partially divide the cord into right and left portions. The spinal cord, like the brain, is protected by three distinct meninges and is cushioned by cerebrospinal fluid. The pia mater contains an extensive vascular network.

The **gray matter** of the spinal cord is centrally located and surrounded by white matter. It is composed of nerve cell bodies, neuroglia, and unmyelinated internuncial (association) neurons. The **white matter** consists of bundles, or tracts, of myelinated fibers of sensory and motor neurons.

The relative size and shape of the gray and white matter varies throughout the spinal cord. The amount of white matter increases toward the brain as the nerve tracts become thicker. More gray matter exists in the cervical and lumbar enlargements where innervations from the upper and lower extremities respectively make connections.

The core of gray matter roughly resembles the letter H (fig. 11.47). Projections of the gray matter within the spinal cord are called horns and are named according to the direction in which they project. The paired **posterior (dorsal) horns** extend posteriorly, and the paired **anterior (ventral) horns** project anteriorly. A pair of short **lateral horns** extend to the sides and are located between the other two pairs. Lateral horns are prominent only in the thoracic and upper lumbar regions. The transverse bar of gray matter that connects the paired horns across the center of the spinal cord is called the **gray commissure.** Contained within the gray commissure is the **central canal,** which is continuous with the ventricles of the brain and filled with cerebrospinal fluid.

Spinal Cord Tracts

Impulses are conducted through the ascending and descending tracts of the spinal cord within columns of white matter. The

cauda equina: L. *cauda,* tail; *equus,* horse

commissure: L. *commissura,* a joining

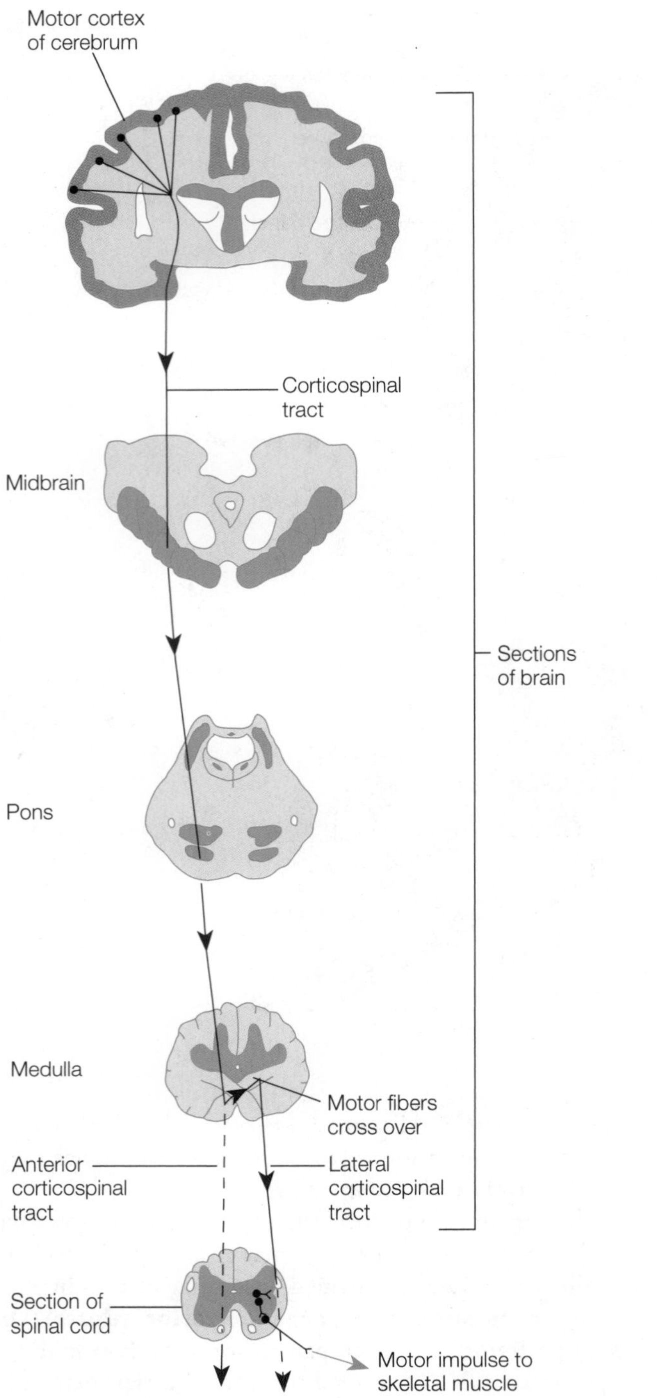

Figure 11.48 Descending tracts composed of motor fibers that cross over within the medulla oblongata.

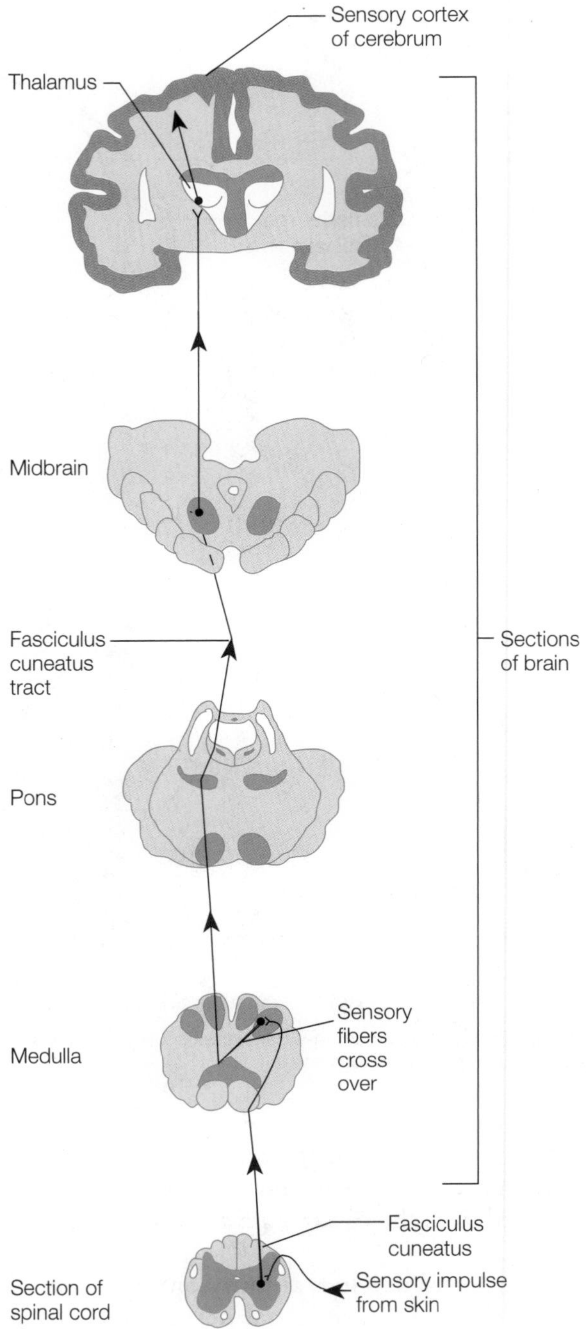

Figure 11.49 Ascending tracts composed of sensory fibers that cross over within the medulla.

spinal cord has six columns of white matter called **funiculi** *(fu-nik'u-li)*, which are named according to their relative position within the cord. The two **anterior (ventral) funiculi** are located between the two anterior horns of gray matter to either side of the anterior median fissure (fig. 11.47). The two **posterior (dorsal) funiculi** are located between the two posterior horns of gray matter to either side of the posterior median sulcus. Two **lateral funiculi** are located between the anterior and posterior horns of gray matter.

funiculus: L. diminutive of *funis*, cord, rope

Each funiculus consists of both ascending and descending tracts. The nerve fibers within the tracts are generally myelinated and have specific sites of origin and termination. In fact, the names of the various tracts reflect their origin and termination. The fibers of the tracts either remain on the same side of the brain and spinal cord or cross over within the medulla or the spinal cord. The crossing over of nerve tracts is referred to as *decussation (de''kus-sa'shun)*. Illustrated in figure 11.48 is a descending tract that decussates within the medulla, and illustrated in figure 11.49 is an ascending tract that decussates within the medulla.

Table 11.6 Principal ascending and descending tracts of spinal cord

Tract	Funiculus	Origin	Termination	Function
Ascending tracts				
Anterior spinothalamic	Anterior	Posterior horn on one side of cord but crosses to opposite side	Thalamus, then cerebral cortex	Conducts sensory impulses for crude touch and pressure
Lateral spinothalamic	Lateral	Posterior horn on one side of cord but crosses to opposite side	Thalamus, then cerebral cortex	Conducts pain and temperature impulses that are interpreted within cerebral cortex
Fasciculus gracilis and fasciculus cuneatus	Posterior	Peripheral afferent neurons; does not cross over	Nucleus gracilis and nucleus cuneatus of medulla; eventually thalamus, then cerebral cortex	Conducts sensory impulses from skin, muscles, tendons, and joints, which are interpreted as sensations of fine touch, precise pressures, and body movements.
Posterior spinocerebellar	Lateral	Posterior horn; does not cross over	Cerebellum	Conducts sensory impulses from one side of body to same side of cerebellum for subconscious proprioception necessary for coordinated muscular contractions
Anterior spinocerebellar	Lateral	Posterior horn; some fibers cross, others do not	Cerebellum	Conducts sensory impulses from both sides of body to cerebellum for subconscious proprioception necessary for coordinated muscular contractions
Descending tracts				
Anterior corticospinal	Anterior	Cerebral cortex on one side of brain; crosses to opposite side of cord	Anterior horn	Conducts motor impulses from cerebrum to spinal nerves and outward to cells of anterior horns for coordinated, precise voluntary movements of skeletal muscle
Lateral corticospinal	Lateral	Cerebral cortex on one side of brain; crosses in base of medulla to opposite side of cord	Anterior horn	Conducts motor impulses from cerebrum to spinal nerves and outward to cells of anterior horns for coordinated, precise voluntary movements
Tectospinal	Anterior	Midbrain; crosses to opposite side of cord	Anterior horn	Conducts motor impulses to cells of anterior horns and eventually to muscles that move the head in response to visual, auditory, or cutaneous stimuli
Rubrospinal	Lateral	Midbrain (red nucleus); crosses to opposite side of cord	Anterior horn	Conducts motor impulses concerned with muscle tone and posture
Vestibulospinal	Anterior	Medulla; does not cross over	Anterior horn	Conducts motor impulses that regulate body tone and posture (equilibrium) in response to movements of head
Anterior and medial reticulospinal	Anterior	Reticular formation of brain stem; does not cross over	Anterior horn	Conducts motor impulses that control muscle tone and sweat gland activity
Lateral reticulospinal	Lateral	Reticular formation of brain stem; does not cross over	Anterior horn	Conducts motor impulses that control muscle tone and sweat gland activity

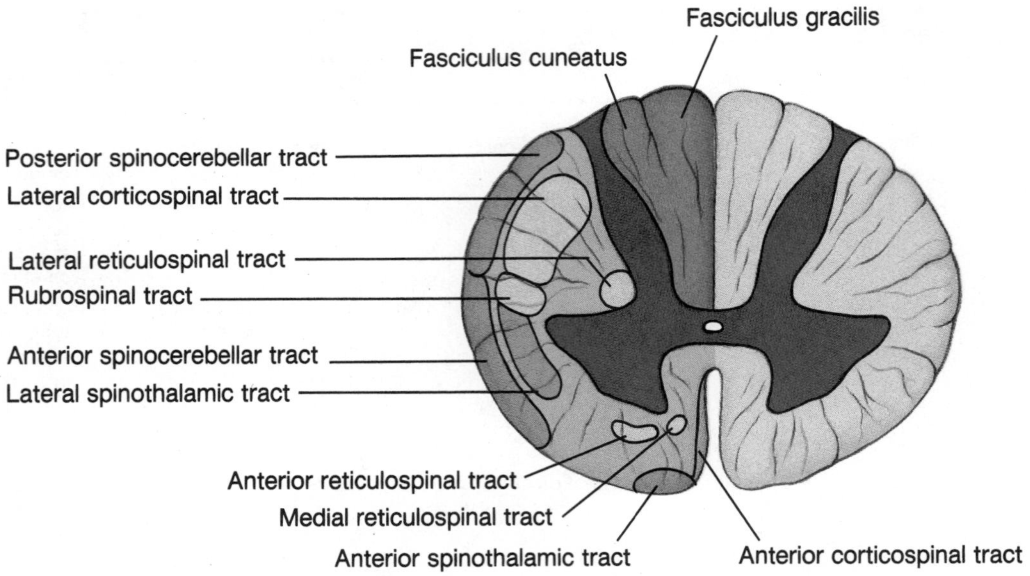

Figure 11.50 A transverse section showing the principal ascending and descending tracts within the spinal cord.

The principal ascending and descending tracts within the funiculi are presented with their functions in table 11.6 and are illustrated in figure 11.50.

Descending tracts are grouped according to place of origin as either corticospinal or extrapyramidal. **Corticospinal (pyramidal) tracts** descend directly, without synaptic interruption, from the cerebral cortex to the lower motor neurons. The cell bodies of the neurons that contribute fibers to these tracts are located primarily in the precentral gyrus of the frontal lobe. Most (about 85%) of the corticospinal fibers decussate in the pyramids of the medulla oblongata (see fig. 11.31). The remaining 15% do not cross from one side to the other. The fibers that cross comprise the **lateral corticospinal tracts,** and the remaining uncrossed fibers comprise the **anterior corticospinal tracts.** Because of the crossing of fibers from higher motor neurons in the pyramids, the right hemisphere primarily controls the musculature on the left side of the body, whereas the left hemisphere controls the right musculature.

The corticospinal tracts appear to be particularly important in voluntary movements that require complex interactions between the motor cortex and sensory input. Speech, for example, is impaired when the corticospinal tracts are damaged in the thoracic region of the spinal cord, whereas involuntary breathing continues. Damage to the pyramidal motor system can be detected clinically by a positive *Babinski's reflex,* in which stimulation of the sole of the foot causes extension (upward movement) of the toes. This positive Babinski reflex is normal in infants because neural control is not yet fully developed.

The remaining descending tracts are **extrapyramidal tracts,** which originate in the midbrain and brain stem regions (fig. 11.51). Electrical stimulation of the cerebral cortex, the cerebellum, and the basal ganglia indirectly evokes movements because of synaptic connections within extrapyramidal tracts.

The **reticulospinal** *(rĕ-tik″u-lo-spi′nal)* **tracts** are the major descending pathways of the extrapyramidal system. These tracts originate in the reticular formation of the brain stem. Neurostimulation of the reticular formation by the cerebrum or cerebellum either facilitates or inhibits (depending on the area stimulated) the activity of lower motor neurons (fig. 11.52).

There are no descending tracts from the cerebellum. The cerebellum can only influence motor activity indirectly, through the vestibular nuclei, red nucleus, and basal ganglia. These structures, in turn, affect lower motor neurons via the **vestibulospinal tracts, rubrospinal tracts,** and reticulospinal tracts. Damage to the cerebellum interferes with the coordination of movements with spatial judgment. Underreaching or overreaching for an object may occur, followed by *intention tremor,* in which the limb moves back and forth in a pendulumlike motion.

The basal ganglia, acting through synapses in the reticular formation particularly, appear normally to exert an inhibitory influence on the activity of lower motor neurons. Damage to the basal ganglia thus results in increased muscle tone. People with such damage display *akinesia (ah″ki-ne′ze-ah)* (lack of desire to use the affected limb) and *chorea* (sudden and uncontrolled random movements).

Paralysis agitans, better known as *Parkinson's disease,* is a disorder of the basal ganglia involving the degeneration of fibers from the substantia nigra. These fibers, which use dopamine as a neurotransmitter, are required to antagonize the effects of other fibers that use acetylcholine (ACh) as a transmitter. The relative deficiency of dopamine compared to ACh is believed to produce the symptoms of Parkinson's disease, including *resting tremor.* This shaking of the limbs tends to disappear during voluntary movements and then reappear when the limb is again at rest.

Parkinson's disease is treated with drugs that block the effects of ACh and by the administration of L-dopa, which can be converted to dopamine in the brain (dopamine cannot be given directly because it does not cross the blood-brain barrier).

akinesia: Gk. *a,* without; *kinesis,* movement
chorea: Fr. *choros,* a dance

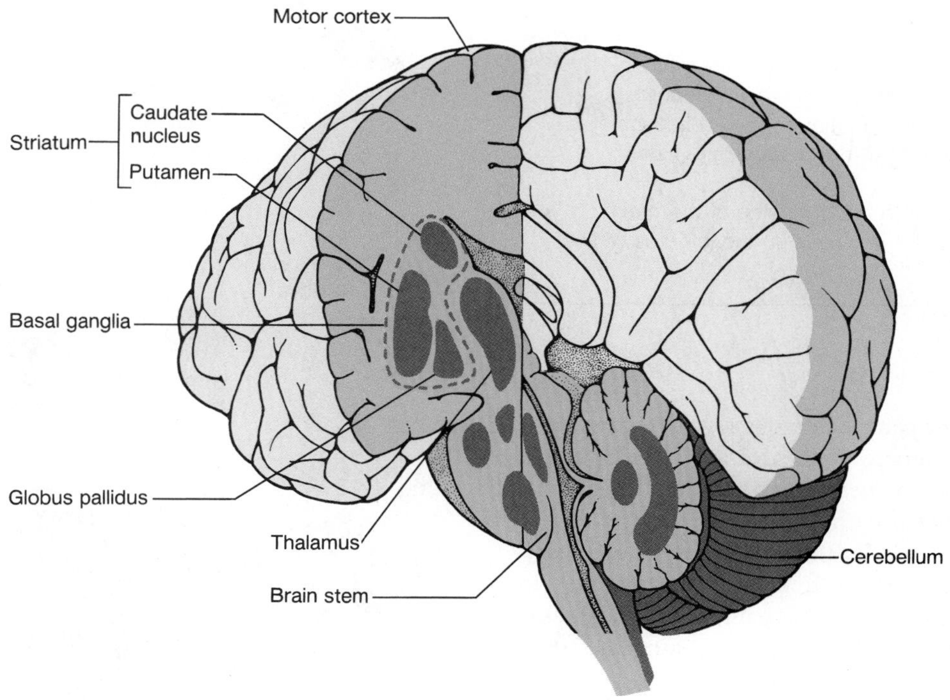

Figure 11.51 Areas of the brain containing neurons involved in the control of skeletal muscles (higher motor neurons). The thalamus is a relay center between the motor cortex and other brain areas.

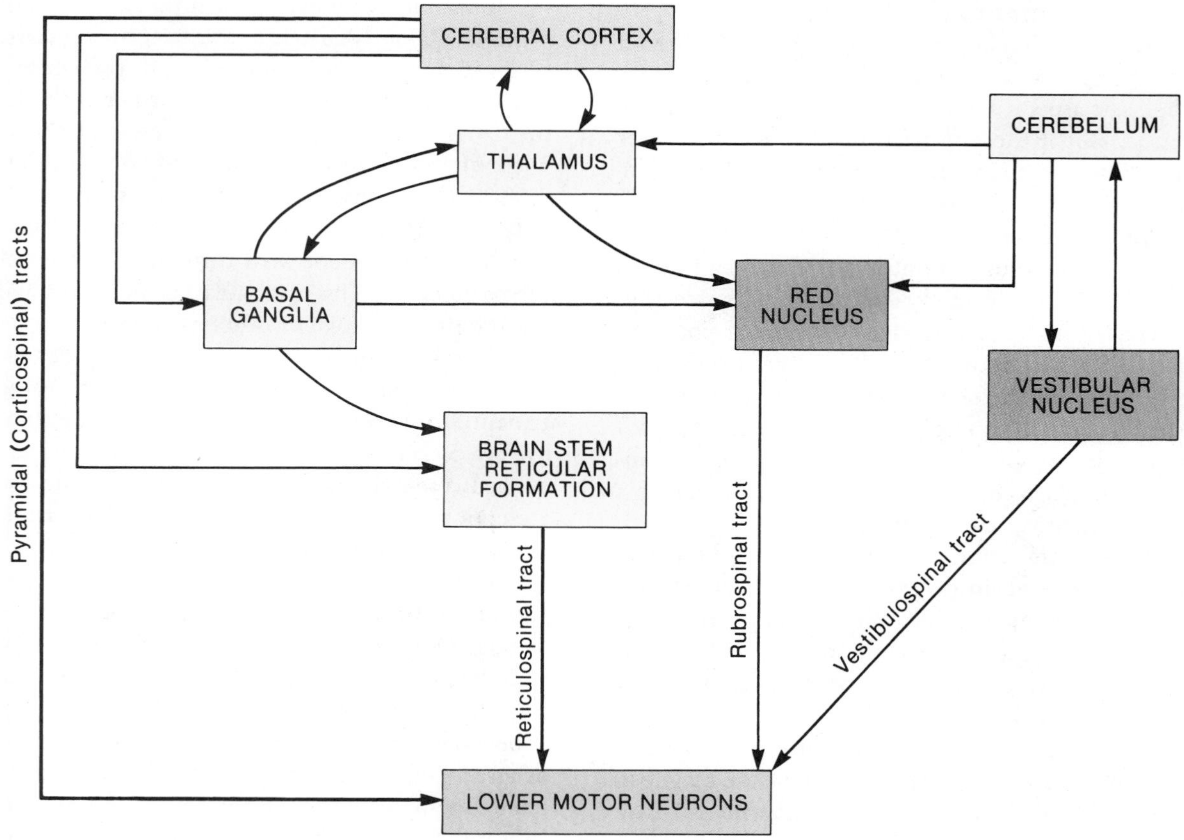

Figure 11.52 Pathways involved in the higher motor neuron control of skeletal muscles.

1. Diagram a cross section of the spinal cord, and label the structures of the gray matter and the white matter. Describe the location of the spinal cord.
2. List the structures and function of the corticospinal tracts. Make a similar list for the extrapyramidal tracts.
3. Explain why damage to the right side of the brain primarily affects motor activities on the left side of the body.

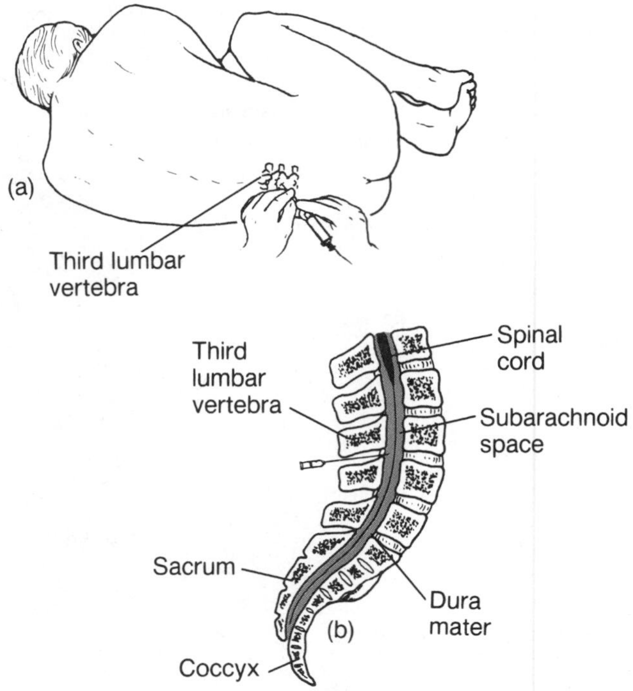

Figure 11.53 (*a*) A lumbar puncture is performed by inserting a needle between the third and fourth lumbar vertebrae (L3–L4) and (*b*) withdrawing cerebrospinal fluid from the subarachnoid space.

Clinical Considerations

The clinical aspects of the central nervous system are extensive and usually complex. Numerous diseases and developmental problems directly involve the nervous system, and the nervous system is indirectly involved with most of the diseases that afflict the body because of the location and activity of sensory pain receptors. Pain receptors are free nerve endings that are present throughout living tissue. The pain sensations elicited by disease or trauma are important for localizing and diagnosing specific diseases or dysfunctions.

Only a few of the many clinical considerations of the central nervous system will be discussed, including neurological assessment and drugs, developmental problems, injuries, infections and diseases, and degenerative disorders.

Neurological Assessment and Drugs

Neurological assessment has become exceedingly sophisticated and accurate in the past few years. In most physical examinations, only the basic aspects such as reflexes and sensory functions are assessed. But if the physician suspects abnormalities involving the nervous system, further neurological tests are done, employing the following techniques.

A **lumbar puncture** is performed by inserting a fine needle between the third and fourth lumbar vertebrae and withdrawing a sample of CFS from the subarachnoid space (fig. 11.53). A **cisternal puncture** is similar to a lumbar puncture except that the CFS is withdrawn from the cisterna magna near the foramen magnum of the skull. The pressure of the CFS, which is normally about 10 mm of mercury, is measured with a *manometer.* Samples of CFS may also be examined for abnormal constituents. Also, excessive fluids, accumulated as a result of disease or trauma, may be drained.

The condition of the arteries of the brain can be determined through a **cerebral angiogram.** In this technique, a radiopaque substance is injected into the common carotid arteries and allowed to disperse through the cerebral vessels. Aneurysms and vascular constrictions or displacements by tumors may then be revealed on X rays.

The development of the **CT scanner,** or **computerized axial tomographic scanner,** has revolutionized the diagnosis of brain disorders. The CT scanner projects a sharply focused, detailed tomogram, or cross section, of a patient's brain onto a television screen. The versatile CT scanner allows quick and accurate diagnoses of tumors, aneurysms, blood clots, and hemorrhage. The CT scanner may also be used to detect certain types of birth defects, brain damage, scar tissue, and old or recent strokes.

A machine with even greater potential than the CT scanner is the **DSR,** or **dynamic spatial reconstructor.** Like the CT scanner, the DSR is computerized to transform x-ray pictures into composite video images. The advantage of the DSR over the CT is the three-dimensional view it presents and the speed with which the image is portrayed. DSR can produce 75,000 cross-sectional images in five seconds, whereas CT can produce only one. At that speed, body functions as well as structures may be studied. Blood flow through blood vessels of the brain can be observed. This type of data is important in detecting early symptoms of a stroke or other disorders.

Certain disorders of the brain may be diagnosed more simply by examining brain-wave patterns using an **electroencephalogram** (see fig. 11.29). Sensitive electrodes placed on the scalp record particular EEG patterns being emitted from evoked cerebral activity. EEG recordings are used to monitor epileptic patients to predict seizures and determine proper drug therapy and also to monitor comatose patients.

The fact that the nervous system is extremely sensitive to various drugs is both fortunate and potentially disastrous to a person. *Drug abuse* is a major clinical concern because of the addictive and devastating effect that certain drugs have on the nervous system. Much has been written on drug abuse, and it is beyond the scope of this text to elaborate on the effects of drugs. A positive aspect of drugs is their administration in medicine to temporarily interrupt the passage or perception of sensory impulses. Injecting an anesthetic drug near a nerve, as in dentistry, desensitizes a specific area and causes a *nerve block.* Nerve blocks of a limited extent occur if an appendage is cooled or if a nerve is compressed for a period of time. Before the discovery of pharmacological drugs, physicians would frequently cool an affected appendage with ice or snow before performing surgery. **General anesthetics** affect the brain and render a person unconscious. A **local anesthetic** causes a nerve block by desensitizing a specific area.

Developmental Problems

Congenital malformations of the CNS are common and frequently involve overlying bone, muscle, and connective tissue. The more severe abnormalities make life impossible, and the less severe malformations frequently result in functional disability.

Most congenital malformations of the nervous system occur during the sensitive embryonic period. Neurological malformations are generally caused by genetic abnormalities but may result from environmental factors such as anoxia, infectious agents, drugs, and ionizing radiation.

Spina bifida is a defective fusion of the vertebral elements and may or may not involve the spinal cord. **Spina bifida occulta** is the most common and least serious type of spina bifida. This defect usually involves few vertebrae, is not externally apparent except for perhaps a pigmented spot with a tuft of hair, and usually does not cause neurological disturbances. **Spina bifida cystica,** a severe type of spina bifida, is a saclike protrusion of skin and underlying meninges that may contain portions of the spinal cord and nerve roots. Spina bifida cystica is most common in the lower thoracic, lumbar, and sacral regions (see fig. 6.40). The position and extent of the defect determines the degree of neurological impairment.

Anencephalia *(an″en-sĕ-fa′le-ah)* is a markedly defective development of the brain and the surrounding cranial bones. Anencephalia occurs once per thousand births and makes sustained extrauterine life impossible. This congenital defect apparently results from the failure of the neural folds at the cranial portion of the neural plate to fuse and form the prosencephalon.

Microcephaly is an uncommon condition in which brain development is not completed. If enough neurological tissue is present, the infant will survive but will be severely mentally retarded.

Defective skull development frequently causes **cranial encephalocele** *(en-sef′ah-lo-sēl)*. This condition occurs approximately once per two thousand births. Cranial encephalocele is characterized by a herniated portion of the brain and meninges through a defect in the skull, usually in the occipital region. Occasionally the herniation contains fluid with the meninges and not the brain tissue. In this case, it is referred to as a **cranial meningocele** *(mĕ-ning′go-sēl)*.

Hydrocephalus is the abnormal accumulation of cerebrospinal fluid in the ventricles and subarachnoid or subdural space. Hydrocephalus may be caused by the excessive production of or blocked flow of cerebrospinal fluid. It may also be associated with other congenital problems such as spina bifida cystica or encephalocele. Hydrocephalus frequently causes the cranial bones to thin and the cerebral cortex to atrophy.

Many congenital disorders cause an impairment of intelligence, known as **mental retardation.** Chromosomal abnormalities, maternal and fetal infections such as syphilis and German measles, and excessive irradiation of the fetus are all commonly associated with mental retardation.

Injuries

Although the brain and spinal cord seem to be well protected within a bony encasement, they are sensitive organs highly susceptible to injury.

Certain symptomatic terms are used when determining possible trauma within the CNS. **Headaches** are the most common ailment of the CNS. Most headaches are due to dilated blood vessels within the meninges of the brain. Headaches are generally asymptomatic of brain disorders, associated rather with physiological stress, eyestrain, or fatigue. Persistent and intense headaches may indicate a more serious problem such as a brain tumor. A **migraine** is a specific type of headache commonly preceded or accompanied by visual impairments and gastrointestinal unrest. It is not known why only 5%–10% of the population periodically suffer from migraines or why they are more common in women. Fatigue, allergy, or emotional stress tends to trigger migraines.

Fainting is a brief loss of consciousness that may result from a rapid pooling of blood in the lower extremities. It may occur when a person rapidly arises from a reclined position, receives a blow to the head, or experiences an intense psychologic stimulus, such as viewing a cadaver for the first time. Fainting is of more concern when it is symptomatic of a particular disease.

A **concussion** is a sudden movement of the brain caused by a violent blow to the head, which may or may not fracture bones of the skull. A concussion usually results in a brief period of unconsciousness followed by mild **delirium,** in which the patient is in a state of confusion. **Amnesia** is a more intense disorientation in which the patient suffers various degrees of memory loss.

A person who survives a severe head injury may be **comatose** for a short or an extended period of time. A coma is a state of unconsciousness from which the patient cannot be aroused by even the most intense external stimuli. The area of the brain most likely to cause a coma from trauma is the reticular activating system. Although a head injury is the most common cause of coma, certain diseases, drugs, poisons, infections, and some chemical imbalances (e.g., diabetes) may also be responsible.

The flexibility of the vertebral column is essential for body movements, but because of this flexibility the spinal cord and spinal nerves are somewhat vulnerable to trauma. Falls or severe blows to the back are a common cause of injury. A skeletal injury, such as a fracture, dislocation, or compression of the vertebrae, usually traumatizes nervous tissue as well. Other frequent causes of trauma to the spinal cord include gunshot wounds, stabbings, herniated discs, and birth injuries. The consequences of the trauma depend upon the severity and location of the injury and the medical treatment the patient receives. If nerve fibers of the spinal cord are severed, motor or sensory functions will be permanently lost.

Paralysis is a permanent loss of motor control, usually resulting from disease or a lesion of the spinal cord or specific

---

amnesia: L. *amnesia,* forgetfulness
comatose: Gk. *koma,* deep sleep
paralysis: Gk. *paralysis,* loosening

nerves. Paralysis of both lower extremities is called **paraplegia.** Paralysis of both the upper and lower extremity on the same side is called **hemiplegia,** and paralysis of all four extremities is **quadriplegia.** Paralysis may be flaccid or spastic. **Flaccid** *(flak'sid)* **paralysis** generally results from a lesion of the anterior horn cells and is characterized by noncontractible muscles that atrophy. **Spastic paralysis** results from lesions of the corticospinal tracts of the spinal cord and is characterized by hypertonicity of the skeletal muscles.

Whiplash is a sudden hyperextension and flexion of the cervical vertebrae (fig. 11.54) such as may occur during a rear-end automobile collision. Recovery of a minor whiplash (muscle and ligament strains) is generally complete but requires a long time. Severe whiplash (spinal cord compression) may cause permanent paralysis to the structures below the level of injury.

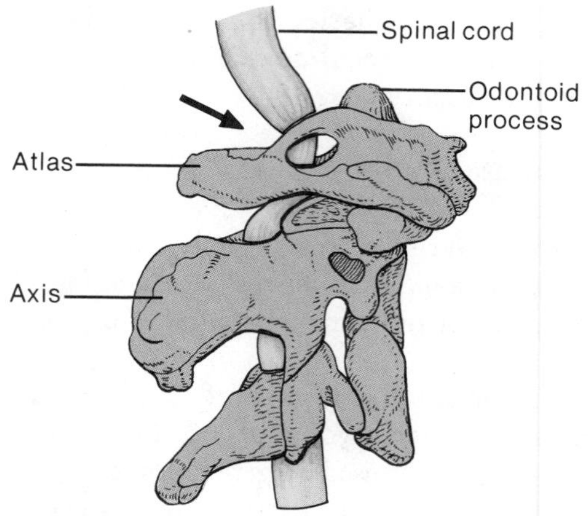

Figure 11.54 Whiplash varies in severity from muscle and ligament strains to dislocation of the vertebrae and compression of the spinal cord. Injuries such as this may cause permanent loss of some or all of the spinal cord functions.

Diseases and Infections

Mental illness is a major clinical consideration of the nervous system and is perhaps the least understood. The two principal categories of mental disorders are neurosis and psychosis. In *neurosis,* a maladjustment to certain aspects of life interferes with normal functioning, but contact with reality is maintained. An irrational fear (*phobia*) is an example of neurosis. Neurosis frequently causes intense anxiety or abnormal distress that brings about increased sympathetic stimulation. *Psychosis,* a more serious mental condition, is typified by a withdrawal from reality and is usually socially unacceptable. The more common forms of psychosis include *schizophrenia,* in which a person withdraws into a world of fantasy; *paranoia,* in which a person has systematized delusions often of a persecutory nature; and *manic-depressive psychosis,* in which a person's moods swing widely from intense elation to deepest despair.

Epilepsy is a relatively common brain disorder with a strong hereditary basis but is also caused by head injuries, tumors, childhood infectious diseases, or it can be idiopathic (without demonstrable cause). A person with epilepsy may periodically suffer from an *epileptic seizure,* which has various symptoms depending on the type of epilepsy.

The most common kinds of epilepsy are petit mal, psychomotor epilepsy, and grand mal. **Petit mal** *(pē-te'mahl')* occurs almost exclusively in children between the ages of three and twelve. A child experiencing a petit mal seizure loses contact with reality for 5–30 seconds but does not lose consciousness or display convulsions. There may, however, be slight uncontrollable facial gestures or eye movements, and the child stares as if in a daydream. During a petit mal seizure, the thalamus and hypothalamus produce an extremely slow EEG pattern of 3 waves per second. Children with petit mal usually outgrow the condition by age nine or ten and generally require no medication.

Psychomotor epilepsy is often confused with mental illness because of the symptoms characteristic of the seizure. During such a seizure, EEG activity accelerates in the temporal lobes, causing a person to become disoriented and lose contact with reality. Occasionally during a seizure, specific cerebral motor areas cause a person to smack his lips or clap his hands involuntarily. If motor areas in the brain are not stimulated, a person having a psychomotor epileptic seizure may wander aimlessly until the seizure subsides.

Grand mal is a more serious form of epilepsy characterized by periodic convulsive seizures that generally render a person unconscious. Grand mal epileptic seizures are accompanied by rapid EEG patterns of 25–30 waves per second. This sudden increase of EEG patterns from the normal of about 10 waves per second may cause an extensive stimulation of motor units and, therefore, uncontrollable skeletal-muscle activity. During a grand mal seizure, a person loses consciousness, convulses, and may lose urinary and bowel control. The unconsciousness and convulsions usually last a few minutes, after which the muscles relax and the person awakes but remains disoriented for a short time.

Epilepsy almost never affects intelligence and can be effectively treated with drugs in about 85% of the patients.

Cerebral Palsy Cerebral palsy is a condition of motor disorders characterized by paresis (partial paralysis) and lack of muscular contraction. It is caused by damage to the motor areas of the brain during prenatal development, birth, or infancy. During neural development within an embryo, radiation or bacterial toxins (such as from German measles), transferred through the placenta of a pregnant mother, may cause cerebral palsy. Oxygen deprivation due to complications at birth and hydrocephalus in a newborn may also cause cerebral palsy. The three areas of the brain most severely affected by this disease are the cerebral cortex, the basal ganglia, and the cerebellum. The type of cerebral palsy is determined by which region of the brain is affected.

Some degree of mental retardation occurs in 60%–70% of cerebral palsy victims. Partial blindness, deafness, and speech problems frequently accompany this disease. Cerebral palsy is nonprogressive (i.e., these impairments do not worsen as a person ages), but neither are there organic improvements.

---

paraplegia: Gk. *para,* beside; *plessein,* to strike
epilepsy: Gk. *epi,* upon; *lepsis,* seize
petit mal: L. *pitinnus,* small child; *malus,* bad

Neoplasms of the CNS Neoplasms of the CNS are either **intracranial tumors,** which affect cells within or associated with the brain, or they are **intravertebral (intraspinal) tumors,** which affect cells within or near the spinal cord. **Primary neoplasms** develop within the CNS. Approximately one-half are benign, but they may become lethal because of the pressure they apply upon vital centers as they grow. Patients with **secondary,** or **metastatic, neoplasms** within the brain have a poor prognosis because the cancer has already established itself in another body organ, frequently the liver, lung, or breast, and has only secondarily spread to the brain. The symptoms of a brain tumor include headache, convulsions, pain, paralysis, or a change in behavior.

Neoplasms of the CNS are classified according to the tissues in which the cancer occurs. Tumors arising in neuroglia cells are called **gliomas** *(gli-o'mahz)* and account for about one-half of all primary neoplasms within the brain. Gliomas are frequently spread throughout cerebral tissue, develop rapidly, and usually cause death within a year after diagnosis. **Astrocytoma** *(as''tro-si-to'mah),* **oligodendroglioma** *(ol''i-go-den''dro-gli-o'mah),* and **ependymoma** *(ĕ-pen''dĭ-mo'mah)* are common types of gliomas.

Meningiomas arise from meningeal coverings of the brain and account for about 15% of primary intracranial tumors. Meningiomas are usually benign if they can be treated readily.

Intravertebral tumors are classified as **extramedullary** when they develop on the outside of the spinal cord and as **intramedullary** when they develop within the substance of the spinal cord. Extramedullary neoplasms may cause pain and numbness in body structures away from the tumor as the growing tumor compresses the spinal cord. An intramedullary neoplasm causes a gradual loss of sensory, temperature, and motor function below the spinal-segmental level of the affliction.

The methods of detecting and treating cancers within the CNS have greatly improved in the last few years. Early detection and competent treatment have lessened the likelihood of death from this disease and reduced the probability of physical impairment.

Dyslexia Dyslexia is a defect in the language center within the brain. In dyslexia, otherwise intelligent people reverse the order of letters in syllables, of syllables in words, and of words in sentences. The sentence: "The man saw a red dog," for example might be read by the dyslexic as "A red god was the man." Dyslexia is believed to result from the failure of one cerebral hemisphere to respond to written language, perhaps due to structural defects. Dyslexia can usually be overcome by intense remedial instruction in reading and writing.

Meningitis The nervous system is vulnerable to a variety of organisms and viruses that may cause abscesses or infections. Meningitis is an infection of the meninges, which may be confined to the spinal cord, in which case it is referred to as **spinal meningitis,** or it may involve the brain and associated meninges, in which case it is known as **encephalitis** or **encephalomyelitis** *(en-sef''ah-lo-mi''ĕ-li'tis),* respectively. The microorganisms that most commonly cause meningitis are meningococci, streptococci, pneumococci, and tubercle bacilli. Viral meningitis is more serious than bacterial meningitis. Nearly 20% of viral encephalitides are fatal. The organisms that cause meningitis probably enter the body through respiratory passageways.

Poliomyelitis Poliomyelitis, or infantile paralysis, is primarily a childhood disease caused by a virus that destroys nerve cell bodies within the anterior horn of the spinal cord, especially those within the cervical and lumbar enlargements. This degenerative disease is characterized by fever, severe headache, stiffness and pain, and the loss of certain somatic reflexes. Muscle paralysis follows within several weeks and eventually the muscles atrophy. Death results if the virus invades the vasomotor and respiratory nuclei within the medulla oblongata or anterior horn cells controlling respiratory muscles. Poliomyelitis has been effectively controlled with immunization.

Syphilis Syphilis is a sexually transmitted disease that, if untreated, progressively destroys body organs. When syphilis causes organ degeneration, it is said to be in the *tertiary stage* (ten to twenty years after the primary infection). The organs of the nervous system are frequently infected, causing a condition called **neurosyphilis.** Neurosyphilis is classified according to the tissue involved, and the symptoms vary correspondingly. If the meninges are infected, the condition is termed **chronic meningitis. Tabes dorsalis** is a form of neurosyphilis in which there is a progressive degeneration of the posterior funiculi of the spinal cord and dorsal roots of spinal nerves. Motor control is gradually lost and patients eventually become bedridden, unable even to feed themselves.

Degenerative Disorders of the Nervous System

Degenerative diseases of the CNS are characterized by a progressive, symmetrical deterioration of vital structures of the brain or spinal cord. The etiologies of these diseases are poorly understood, but it is thought that most of them are genetic.

Cerebrovascular Accident (CVA) Cerebrovascular accident is the most common disease of the nervous system. It is the third highest cause of death in the United States and perhaps the major cause of disability. The term **stroke** is frequently used synonymously with CVA, but actually a stroke refers to the sudden and dramatic appearance of a neurological defect. Cerebral thrombosis, in which a thrombus, or clot, forms in an artery of the brain, is the most common cause of CVA. Other causes of CVA include intracerebral hemorrhages, aneurysms, atherosclerosis, and arteriosclerosis of the cerebral arteries.

Patients who recover from CVA frequently suffer partial paralysis and mental disorders, such as loss of language skills. The dysfunction depends upon the severity of the CVA and the regions of the brain that were injured. Patients surviving a CVA can often be rehabilitated, but approximately two-thirds die within three years of the initial damage.

Multiple Sclerosis Multiple sclerosis (MS) is a relatively common neurological disease in persons between the ages of twenty and forty. MS is a chronic, degenerating, remitting, and relapsing disease that progressively destroys the myelin sheaths of neurons in multiple areas of the CNS. Initially, lesions form on the myelin sheaths and soon develop into hardened *scleroses* or scars (hence the name). The destruction of myelin sheaths prohibits the normal conduction of impulses, resulting in a progressive loss of functions. Because myelin degeneration is widely distributed, MS has a greater variety of symptoms than any other neurologic disease. This characteristic, coupled with remission, frequently causes misdiagnosis of this disease.

During the early stages of MS, many patients are considered neurotic because of the wide variety and temporary nature of their symptoms. As the disease progresses, the symptoms may include double vision (diplopia), spots in the visual field, blindness, tremor, numbness of appendages, and locomotor difficulty. Eventually the patient is bedridden, and death may occur anytime from seven to thirty years after the first symptoms appear.

Syringomyelia Syringomyelia *(sĭ-ring''go-mi-e'le-ah)* is a relatively uncommon condition characterized by the appearance of cystlike cavities, called *syringes,* within the gray matter of the spinal cord. These syringes progressively destroy the cord from the inside out. Syringomyelia is a chronic, slow progressing disease of unknown cause. As the spinal cord deteriorates, the patient experiences muscular weakness and atrophy and sensory loss, particularly the senses of pain and temperature.

Tay-Sachs Disease In Tay-Sachs disease, the myelin sheaths are destroyed by the excessive accumulation of one of the lipid components of the myelin. This results from an enzyme defect due to the inheritance of genes that are carried by the parents in a recessive state. Tay-Sachs disease, which is inherited primarily in Jewish families, appears when the infant is under one year of age. It causes blindness, loss of mental and motor ability, and ultimately death by the age of three. Potential parents can tell if they are carriers for this condition by the use of a special blood test for the defective enzyme.

Diseases Involving Neurotransmitters **Parkinson's disease,** or **paralysis agitans,** is a major cause of neurological disability in people over sixty years of age. It is a progressively degenerative disease of unknown cause. Nerve cells within the substantia nigra, an area within the basal ganglia of the brain, are destroyed. This causes muscle tremors, muscular rigidity, speech defects, and other severe problems. The symptoms of this disease can be partially treated by altering the neurotransmitter status of the brain. These patients are given L-dopa, to increase the production of dopamine in the brain, and may also be given anticholinergic drugs to decrease the production of acetylcholine.

Alzheimer's disease is the most common cause of dementia, often beginning in middle age and producing progressive mental deterioration. The cause of Alzheimer's disease is unknown, but evidence suggests that it is associated with the decreased ability of the brain to produce acetylcholine. Attempts to increase ACh production by increased ingestion of precursor molecules (choline or lecithin) have thus far not been successful. Drugs that block AChE and thus enhance the action of endogenously produced ACh offer promise but are so far still in the experimental stage.

At least some psychiatric disorders may be produced by dysfunction of neurotransmitters. There is evidence that **schizophrenia** *(skiz''o-fre'ne-ah)* is associated with hyperactivity of the neurons that use dopamine as a neurotransmitter; drugs that are effective in the treatment of schizophrenia (e.g., chlorpromazine) act by blocking dopamine receptor proteins. **Depression** is associated with decreased activity of the neurons that use monoamines—norepinephrine and serotonin—as neurotransmitters; antidepressant drugs enhance the action of these neurotransmitters. In a similar way, barbiturates and benzodiazepine (e.g., Valium), which decrease **anxiety,** act by enhancing the action of the neurotransmitter GABA in the CNS.

CLINICAL CASE STUDY ANSWER

The affected upper motor neurons have cell bodies that reside in the right cerebral hemisphere. They give rise to fibers that, as they course downward, cross in the medulla to the left side of the brain stem and spinal cord. They continue on the left side, eventually synapsing at the appropriate level with lower motor neurons in the anterior horn of the spinal cord. The lower motor neurons then give rise to fibers that travel to the periphery where they innervate end organs, in other words, muscle cells in the left side of the body. Since the patient's neurological deficits are all motor as opposed to sensory, the tumor is most likely located in the right frontal lobe. The parietal lobe contains sensory neurons. The persistent headache is due to the pressure that the tumorous mass exerts on the meninges, which are heavily sensory innervated. ■

multiple sclerosis: L. *multiplus,* many parts; Gk. *skleros,* hardened
Tay-Sachs disease: from Warren Tay, English physician, 1843–1927, and Bernard Sachs, U.S. neurologist, 1858–1944

Alzheimer's disease: from Alois Alzheimer, German neurologist, 1864–1915

Chapter Summary

I. Organization and Functions of the Nervous System
   A. The central nervous system (CNS), consisting of the brain and spinal cord, is covered with meninges, is bathed in cerebrospinal fluid, and contains gray and white matter.
   B. The functions of the nervous system are orientation, coordination, assimilation, and instinctual behavior.

II. Development of the Nervous System
   A. The prosencephalon, mesencephalon, and rhombencephalon develop from the embryonic neural tube.
      1. The telencephalon and diencephalon develop from the forebrain (prosencephalon).
      2. The metencephalon and myelencephalon develop from the hindbrain (rhombencephalon).
   B. The spinal cord develops from the neural tube as afferent and efferent neurons are formed that conduct into and out of the spinal cord, respectively.
   C. The development of the peripheral nervous system from neural crest tissue produces a pattern of dermatomes in the skin.

III. Histology of the Nervous System
   A. The dendrites of a neuron receive stimulation, and the axon conducts nerve impulses. Neuroglial cells provide structural and functional support for the activities of neurons.
   B. A neuron contains dendrites, a cell body, and an axon.
      1. The cell body contains the nucleus, chromatophilic substances, neurofibrils, and other organelles.
      2. Dendrites receive stimuli, and the axon conducts nerve impulses away from the cell body.
   C. There are six different categories of neuroglial cells.
      1. The glial cells that surround an axon form an outer covering called a neurilemma; this is produced by neurolemmocytes in the PNS.
      2. Some neurons are surrounded by successive wrappings of glial cell membranes called a myelin sheath; this is produced by neurolemmocytes in the PNS and oligodendrocytes in the CNS.
   D. A nerve is a collection of dendrites and axons in the PNS.
      1. Sensory neurons are afferent and pseudounipolar.
      2. Motor neurons are efferent and multipolar.
      3. Interneurons, or association neurons, are located entirely within the CNS.
      4. Somatic motor nerves innervate skeletal muscle; visceral motor (autonomic) nerves innervate smooth muscle, cardiac muscle, and glands.

IV. Transmission of Impulses
   A. Irritability and conductivity are specialized properties of neurons that permit nerve impulse transmission.
   B. Neurotransmitters facilitate synaptic impulse transmission.

V. General Features of the Brain
   A. The brain, composed of gray matter and white matter, is protected by meninges and is bathed in cerebrospinal fluid.
   B. About 750 ml of blood flows to the brain per minute.

VI. Cerebrum
   A. The cerebrum, consisting of two convoluted hemispheres, is concerned with higher brain functions, such as the perception of sensory impulses, the instigation of voluntary movement, the storage of memory, thought processes, and reasoning ability.
   B. The cerebral cortex of the cerebral hemispheres is convoluted with gyri and sulci.
   C. Each cerebral hemisphere contains frontal, parietal, temporal, occipital, and insula lobes.
   D. Brain waves generated by the cerebral cortex are recorded as an electroencephalogram and may provide valuable diagnostic information.
   E. The white matter of the cerebrum consists of association, commissural, and projection fibers.
   F. The basal ganglia are specialized masses of gray matter located within the white matter of the cerebrum.

VII. Diencephalon
   A. The diencephalon is a major autonomic region of the brain.
   B. The thalamus is an ovoid mass of gray matter that functions as a relay center for sensory impulses and responds to pain.
   C. The hypothalamus is an aggregation of specialized nuclei that regulate many visceral activities. It also performs emotional and instinctual functions.
   D. The epithalamus contains the pineal gland and the vascular choroid plexus over the roof of the third ventricle.

VIII. Mesencephalon
   A. The mesencephalon contains the corpora quadrigemina, the cerebral peduncles, and specialized nuclei that help to control posture and movement.
   B. The superior colliculi of the corpora quadrigemina are concerned with visual reflexes and the inferior colliculi are concerned with auditory reflexes.
   C. The red nucleus and the substantia nigra are concerned with motor activities.

IX. Metencephalon
   A. The pons consists of fiber tracts connecting the cerebellum and medulla to other structures of the brain. The pons also contains nuclei for certain cranial nerves and the regulation of respiration.
   B. The cerebellum consists of two hemispheres connected by the vermis and supported by three paired peduncles.
      1. It is composed of white matter surrounded by a thin convoluted cortex of gray matter.
      2. The cerebellum is concerned with coordinated contractions of skeletal muscle.

X. Myelencephalon
   A. The medulla oblongata is composed of the ascending and descending tracts of the spinal cord and contains nuclei for several autonomic functions.
   B. The reticular formation functions as the reticular activating system in arousing the cerebrum.

XI. Meninges of the Central Nervous System
  A. The cranial dura mater consists of an outer periosteal layer and an inner meningeal layer. The spinal dura mater is a single layer that is surrounded by the vascular epidural space.
  B. The arachnoid membrane is a netlike meninx surrounding the subarachnoid space, which contains cerebrospinal fluid.
  C. The thin pia mater adheres to the contour of the CNS.

XII. Ventricles and Cerebrospinal Fluid
  A. The lateral (first and second), third, and fourth ventricles are interconnected chambers within the brain that are continuous with the central canal of the spinal cord.
  B. These chambers are filled with cerebrospinal fluid, which also flows throughout the subarachnoid space.
  C. Cerebrospinal fluid is continuously secreted by the choroid plexuses and is absorbed into the blood at the arachnoid villi.
  D. The blood-brain barrier determines which substances within blood plasma can enter the extracellular fluid of the brain.

XIII. Spinal Cord
  A. The spinal cord is composed of thirty-one segments, each of which gives rise to a pair of spinal nerves.
    1. It is characterized by a cervical enlargement, a lumbar enlargement, and two longitudinal grooves, which partially divide it into right and left halves.
    2. The conus medullaris is the terminal portion of the spinal cord, and the cauda equina are nerve roots that radiate inferiorly from that point.
  B. Ascending and descending spinal cord tracts are referred to as funiculi.
    1. Descending tracts are grouped as either corticospinal (pyramidal) or extrapyramidal.
    2. Many of the fibers in the funiculi decussate (cross over) in the spinal cord or in the medulla oblongata.

Review Activities

Objective Questions

Match the following structures of the brain to the region in which they are located:

1. Cerebellum (a) telencephalon
2. Cerebral cortex (b) diencephalon
3. Medulla oblongata (c) mesencephalon
(d) metencephalon
(e) myelencephalon

4. The neuroglial cells that form myelin sheaths in the peripheral nervous system are
   (a) oligodendrocytes.
   (b) satellite cells.
   (c) neurolemmocytes.
   (d) astrocytes.
   (e) microglia.
5. A collection of neuron cell bodies located outside the CNS is called a
   (a) tract. (c) nucleus.
   (b) nerve. (d) ganglion.
6. Which of the following neurons is pseudounipolar?
   (a) sensory neurons
   (b) somatic motor neurons
   (c) neurons in the retina
   (d) autonomic motor neurons
7. Depolarization of an axon is produced by the
   (a) inward diffusion of Na^+.
   (b) active extrusion of K^+.
   (c) outward diffusion of K^+.
   (d) inward active transport of Na^+.
8. The principal connection between the cerebral hemispheres is the
   (a) corpus callosum.
   (b) pons.
   (c) intermediate mass.
   (d) vermis.
   (e) precentral gyrus.
9. The structure of the brain that is most directly involved in the autonomic response to pain is the
   (a) pons.
   (b) hypothalamus.
   (c) medulla oblongata.
   (d) thalamus.
10. Which statement is *false* concerning the basal ganglia?
   (a) They are located within the cerebrum.
   (b) They regulate the basal metabolic rate.
   (c) They consist of the caudate nucleus, lentiform nucleus, putamen and globus pallidus.
   (d) They indirectly exert an inhibitory influence on lower motor neurons.
11. The corpora quadrigemina, red nucleus, and substantia nigra are structures of the
   (a) diencephalon.
   (b) metencephalon.
   (c) mesencephalon.
   (d) myelencephalon.
12. The fourth ventricle is contained within the
   (a) cerebrum. (c) midbrain.
   (b) cerebellum. (d) metencephalon.

Essay Questions

1. Describe the formation of the neural crest, and explain how derivatives of neural crest tissue can be located in different parts of the body.
2. Diagram and label a neuron. Beside each label, list the function of the identified structure. Why are neurons considered the basic functional units of the nervous system?
3. What is meant by a myelinated neuron? Describe how an injured nerve fiber may regenerate.
4. List the six principal types of neuroglia and discuss the location, structure, and function of each.
5. Define a nerve impulse and relate this to polarization, depolarization, and action potential.
6. What is a synapse and what is the role of a neurotransmitter in relaying a nerve impulse?
7. List the types of brain waves recorded on an electroencephalogram, and explain the diagnostic value of each.
8. List the functions of the hypothalamus. Why is the hypothalamus considered a major part of the autonomic nervous system?
9. What structures are within the midbrain? List the nuclei located in the midbrain, and give the function of each.
10. Describe the location and structure of the medulla oblongata. List the nuclei found within this structure. What are the functions of the medulla oblongata?
11. What is cerebrospinal fluid? Where is it produced, and what is its pathway of circulation?
12. Define the following abbreviations: EEG, ANS, CSF, PNS, RAS, CT scan, MS, DSR, and CVA.
13. List and define the various psychological terms used to describe mental illness.
14. What is epilepsy? What causes it, and how is it controlled?
15. What do meningitis, poliomyelitis and neurosyphilis have in common, and how do these conditions differ?

CHAPTER 12

Peripheral Nervous System

Outline and Concepts

Clinical Case Study

Introduction to the Peripheral Nervous System

The peripheral nervous system consists of nerves that convey impulses to and from the central nervous system.

Cranial Nerves

Twelve pairs of cranial nerves emerge from the inferior surface of the brain and pass through the foramina of the skull to innervate structures in the head, neck, and visceral organs of the torso.

Structure and Function of the Cranial Nerves

Neurological Assessment of Cranial Nerves

Spinal Nerves

Each of the thirty-one pairs of spinal nerves is formed by the union of a dorsal and a ventral spinal root, which emerges from the spinal cord through an intervertebral foramen to innervate a body dermatome.

Nerve Plexuses

Except in the thoracic nerves T2–T12, the ventral rami of the spinal nerves combine and then split again as networks of nerves referred to as plexuses. There are four plexuses of spinal nerves: the cervical, the brachial, the lumbar, and the sacral. Nerves emerge from the plexuses and are named according to the structures they innervate or the general course they take.

Reflex Arcs and Reflexes

The conduction pathway of a reflex arc consists of a receptor, a sensory neuron, a motor neuron and its innervation in the PNS, and an association neuron in the CNS. The reflex arc provides the mechanism for a rapid, automatic response to a potentially threatening stimulus.

Components of the Reflex Arc

Kinds of Reflexes

Clinical Case Study Answer

Chapter Summary

Review Activities

CLINICAL CASE STUDY

A twenty-three-year-old male involved in an auto accident is brought to the emergency room for treatment of a fracture of the right humerus. Although the skin is not broken, there is obvious deformity caused by an angulated fracture at the midshaft. Before initiating the appropriate treatment, the orthopedist notices that the patient is unable to extend his hand and fingers.

What structure has been injured? Describe other movements of the upper extremity which might be diminished and mention the associated muscles. Would you suspect any other neurological deficits? Explain.

Introduction to the Peripheral Nervous System

The peripheral nervous system consists of nerves that convey impulses to and from the central nervous system.

Objective 1. Define *peripheral nervous system,* and distinguish between sensory, motor, and mixed nerves.

The **peripheral nervous system (PNS)** is that portion of the nervous system outside the central nervous system. The PNS

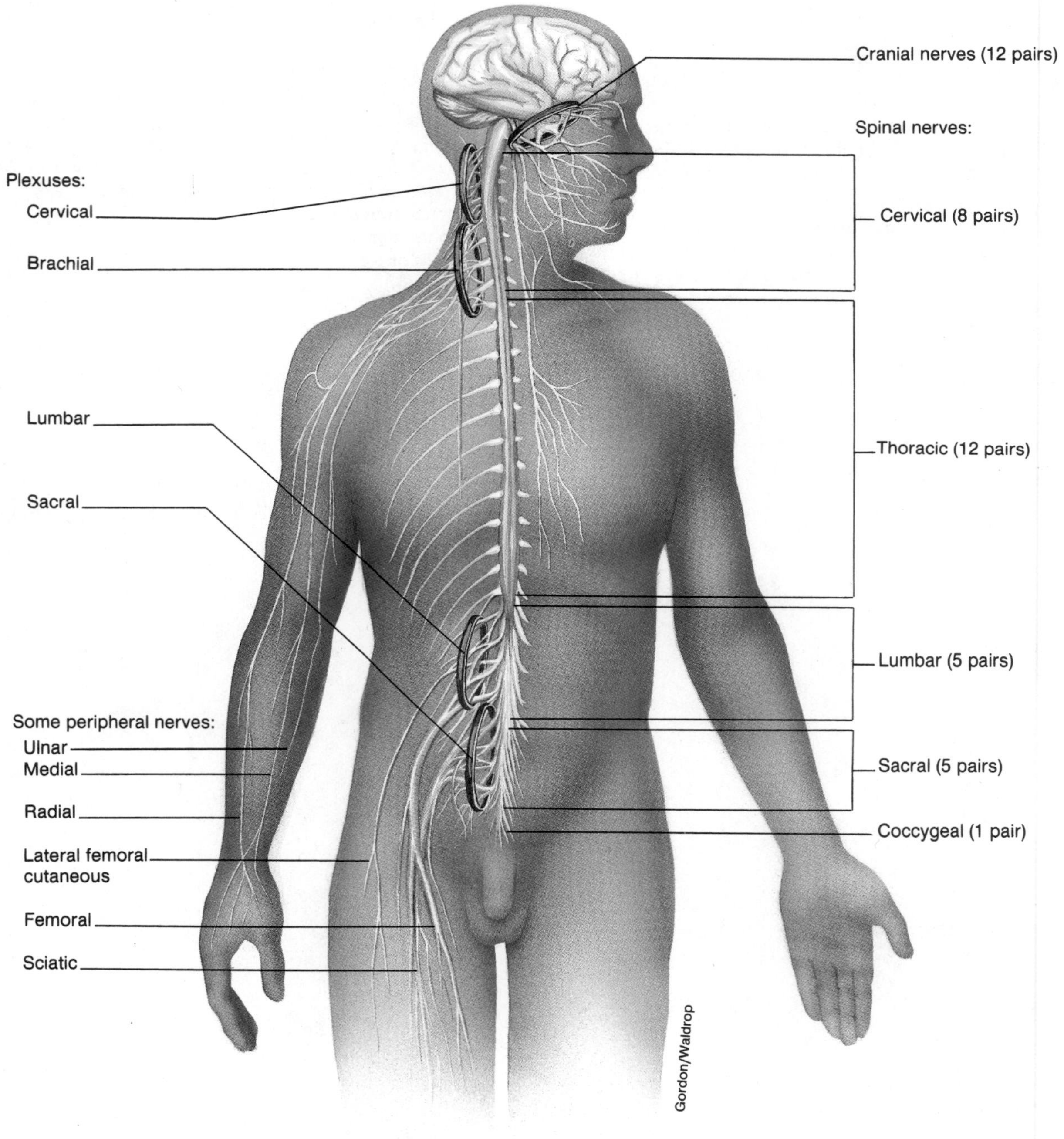

Figure 12.1 The peripheral nervous system consists of cranial nerves, spinal nerves, plexuses, ganglia (not shown), and the peripheral nerves.

functions to convey impulses to and from the brain or spinal cord. Sensory receptors within the sensory organs, neurons, nerves, ganglia, and plexuses are all part of the PNS, which serves virtually every part of the body (fig. 12.1). The sensory receptors are discussed in chapter 15.

The nerves of the PNS are classified as cranial nerves or spinal nerves, depending on whether they arise from the brain or the spinal cord. The terms *sensory nerve, motor nerve,* and *mixed nerve* relate to the direction in which the nerve impulses are being conducted. **Sensory nerves** consist of sensory (afferent) fibers that convey impulses toward the CNS. **Motor nerves** consist of motor (efferent) fibers that convey impulses away from the CNS. **Mixed nerves** are composed of both sensory and motor fibers and therefore convey impulses in both directions. Reflexes that are considered in this chapter involve sensory and motor nerves of the PNS and specific portions of the CNS. A transverse section of a spinal nerve is depicted in figure 12.2.

1. The tongue responds to tastes and pain and moves to manipulate food. Make a quick sketch of the brain and the tongue to depict the relationship between the CNS and the PNS. Indicate diagrammatically with lines and arrows the sensory and motor innervation of the tongue. Define *mixed nerve.* What kinds of sensory stimulation arise from the tongue? What type of response is caused by motor stimulation to the tongue?
2. List the structures of the nervous system that are considered a part of the PNS.

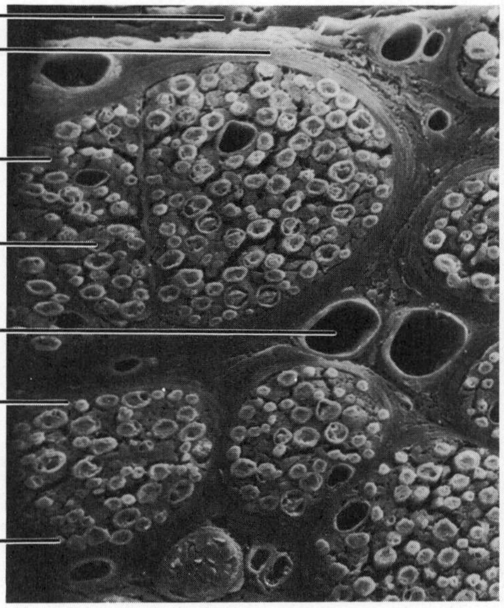

Figure 12.2 Scanning electron micrograph of a spinal nerve seen in transverse section (about 1000×). (From: *Tissues and Organs: A Text-Atlas of Scanning Electron Microscopy,* by R. G. Kressel and R. H. Kardon. © 1979 W. H. Freeman and Company.)

Cranial Nerves

Twelve pairs of cranial nerves emerge from the inferior surface of the brain and pass through the foramina of the skull to innervate structures in the head, neck, and visceral organs of the torso.

Objective 2. List the twelve pairs of cranial nerves, and describe the location and function of each.

Objective 3. Describe the clinical methods for determining cranial nerve dysfunction.

Structure and Function of the Cranial Nerves

Of the twelve pairs of cranial nerves, two pairs arise from the forebrain and ten pairs arise from the midbrain and brain stem (fig. 12.3). The cranial nerves are designated by Roman nu-

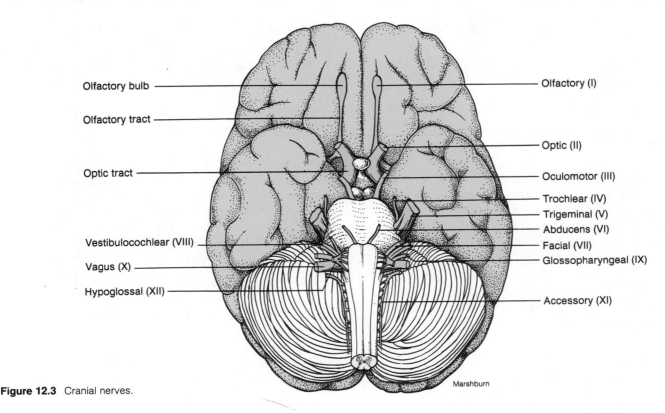

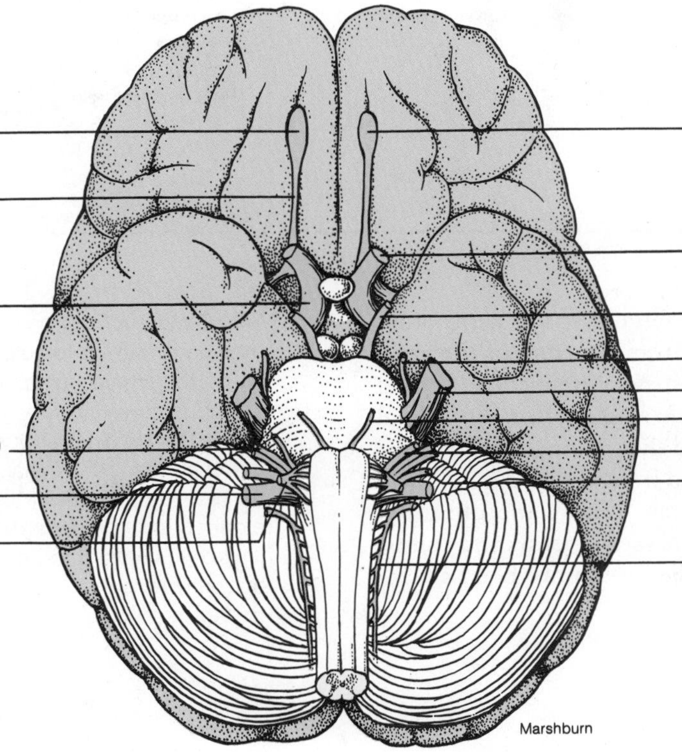

Figure 12.3 Cranial nerves.

Table 12.1 Summary of cranial nerves

Number and name	Foramen transmitting	Composition	Location of cell bodies	Function
I Olfactory	Foramina in ethmoidal cribriform plate	Sensory	Bipolar cells in nasal mucosa	Olfaction
II Optic	Optic canal	Sensory	Ganglion cells of retina	Vision
III Oculomotor	Superior orbital fissure	Motor	Oculomotor nucleus	Motor impulses to levator palpebrae superioris and extrinsic eye muscles except superior oblique and lateral rectus. Innervation to
		Motor: parasympathetic		muscles that regulate amount of light entering eye and that focus the lens
		Sensory: proprioception		Proprioception from muscles innervated with motor fibers
IV Trochlear	Superior orbital fissure	Motor	Trochlear nucleus	Motor impulses to superior oblique muscle of eyeball
		Sensory: proprioception		Proprioception from superior oblique muscle of eyeball
V Trigeminal				
Ophthalmic division	Superior orbital fissure	Sensory	Semilunar ganglion	Sensory impulses from cornea, skin of nose, forehead, and scalp
Maxillary division	Foramen rotundum	Sensory	Semilunar ganglion	Sensory impulses from nasal mucosa, upper teeth and gums, palate, upper lip, and skin of cheek
Mandibular division	Foramen ovale	Sensory	Semilunar ganglion	Sensory impulses from temporal region, tongue, lower teeth and gum, and skin of chin and lower jaw
		Sensory: proprioception		Proprioception from muscles of mastication
		Motor	Motor trigeminal nucleus	Motor impulses to muscles of mastication and muscle that tenses tympanum
VI Abducens	Superior orbital fissure	Motor	Abducens nucleus	Motor impulses to lateral rectus muscle of eyeball
		Sensory: proprioception		Proprioception from lateral rectus muscle of eyeball
VII Facial	Stylomastoid foramen	Motor	Motor facial nucleus	Motor impulses to muscles of facial expression and muscle that tenses the stapes
		Motor: parasympathetic	Superior salivatory nucleus	Secretion of tears from lacrimal gland and salivation from sublingual and submandibular salivary glands
		Sensory	Geniculate ganglion	Sensory impulses from taste buds on anterior two-thirds of tongue; nasal and palatal sensation
		Sensory: proprioception		Proprioception from muscles of facial expression

merals and with names. The Roman numerals refer to the order in which the nerves are positioned from the front of the brain to the back. The names indicate the structures innervated or the principal functions of the nerves. A summary of the cranial nerves is presented in table 12.1, and the locations of the nuclei from which they arise are illustrated in figure 11.37.

Although most cranial nerves are mixed, the olfactory, optic, and vestibulocochlear serve the special senses and consist of sensory fibers only. The cell bodies of sensory fibers are located in ganglia outside the brain.

Generations of anatomy students have used a mnemonic device to help them remember the order in which the cranial nerves emerge from the brain: "On old Olympus' towering top, a Finn and German viewed a hop." The initial letter of each word in this jingle corresponds to the initial letter of each pair of cranial nerves. A problem with this classic verse is that the eighth cranial nerve represented by the word *and* in the jingle, which used to be referred to as auditory, is currently recognized as the vestibulocochlear cranial nerve.

Number and name	Foramen transmitting	Composition	Location of cell bodies	Function
VIII Vestibulocochlear	Internal acoustic meatus	Sensory	Vestibular ganglion	Sensory impulses associated with equilibrium
			Spiral ganglion	Sensory impulses associated with hearing
IX Glossopharyngeal	Jugular foramen	Motor	Nucleus ambiguus	Motor impulses to muscles of pharynx used in swallowing
		Parasympathetic	Inferior salivatory nucleus	Salivation from parotid salivary gland
X Vagus	Jugular foramen	Motor	Nucleus ambiguus	Contraction of muscles of pharynx (swallowing) and larynx (phonation)
		Sensory: proprioception		Proprioception from visceral muscles
		Motor: parasympathetic	Dorsal motor nucleus	Regulate visceral motility
XI Accessory	Jugular foramen	Motor	Nucleus ambiguus	Laryngeal movement; soft palate
			Spinal accessory nucleus	Motor impulses to trapezius and sternocleidomastoid muscles for movement of head, neck, and shoulders
		Sensory: proprioception		Proprioception from muscles that move head, neck, and shoulders
XII Hypoglossal	Hypoglossal canal	Motor	Hypoglossal nucleus	Motor impulses to intrinsic and extrinsic muscles of tongue and infrahyoid muscles
		Sensory: proprioception		Proprioception from muscles of tongue

I Olfactory Nerve Actually, numerous olfactory nerves relay sensory impulses of smell from the mucous membranes of the nasal cavity (fig. 12.4). Olfactory nerves are composed of bipolar neurons that function as *chemoreceptors,* responding to volatile chemical particles breathed into the nasal cavity. The dendrites and cell bodies of olfactory neurons are positioned within the mucosa, primarily that which covers the superior nasal conchae and adjacent nasal septum. The axons of these neurons pass through the cribriform plate of the ethmoid to the **olfactory bulb** where synapses are made, and the sensory impulses are passed through the **olfactory tract** to the primary olfactory area in the cerebral cortex.

II Optic Nerve The optic nerve, another sensory nerve, conducts impulses from the *photoreceptors* (rods and cones) in the retina of the eye. Each optic nerve is composed of an estimated 1.25 million nerve fibers that converge at the back of the eyeball and enter the cranial cavity through the optic foramen. The

olfactory: L. *olfacere,* smell out

optic: L. *optica,* see

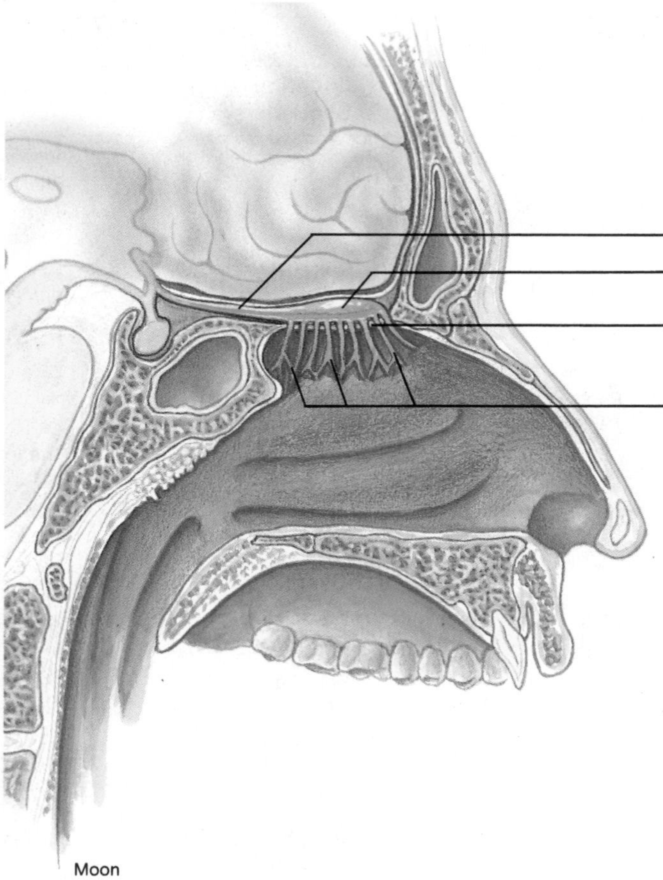

Figure 12.4 The olfactory nerve.

two optic nerves unite on the floor of the diencephalon to form the **optic chiasma** *(ki-as'mah)* (fig. 12.5). Nerve fibers that arise from the medial half of each retina cross at the chiasma to the opposite side of the brain, whereas fibers arising from the lateral half remain on the same side of the brain. The optic nerve fibers pass posteriorly from the chiasma in the **optic tracts.** The optic tracts lead to the thalami, where a majority of the fibers terminate within certain thalamic nuclei. A few of the ganglion-cell axons that reach the thalamic nuclei have collaterals that convey impulses to the superior colliculi. Synapses within the thalamic nuclei, however, permit impulses to pass through neurons to the **visual cortex** within the occipital lobes. Other synapses permit impulses to reach the nuclei for the oculomotor, trochlear, and abducens nerves, which regulate intrinsic (internal) and extrinsic (from orbit to eyeball) eye muscles. The visual pathway into the eyeball functions reflexively to produce motor responses to light stimuli. If an optic nerve is damaged, the eyeball served by that nerve is blinded.

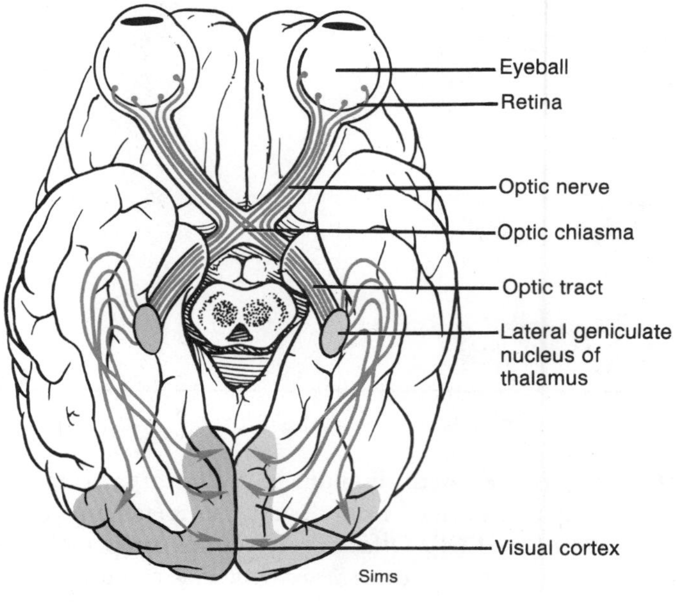

Figure 12.5 The optic nerve.

III Oculomotor The oculomotor nerve produces certain extrinsic and intrinsic movements of the eyeball. The oculomotor is primarily a motor nerve that arises from nuclei within the midbrain. The oculomotor divides into superior and inferior branches as it passes through the superior orbital fissure in the orbit (fig. 12.6). The superior branch innervates the **superior rectus** muscle, which moves the eyeball superiorly, and the **levator palpebrae** *(le-va'tor pal'pĕ-bre)* **superioris** muscle, which raises the upper eyelid (see fig. 9.19b). The inferior branch innervates the **medial rectus, inferior rectus,** and **inferior oblique** eye muscles for medial, inferior, and superior and lateral movement of the eyeball, respectively. In addition,

chiasma: Gk. *chiasma,* an X-shaped arrangement

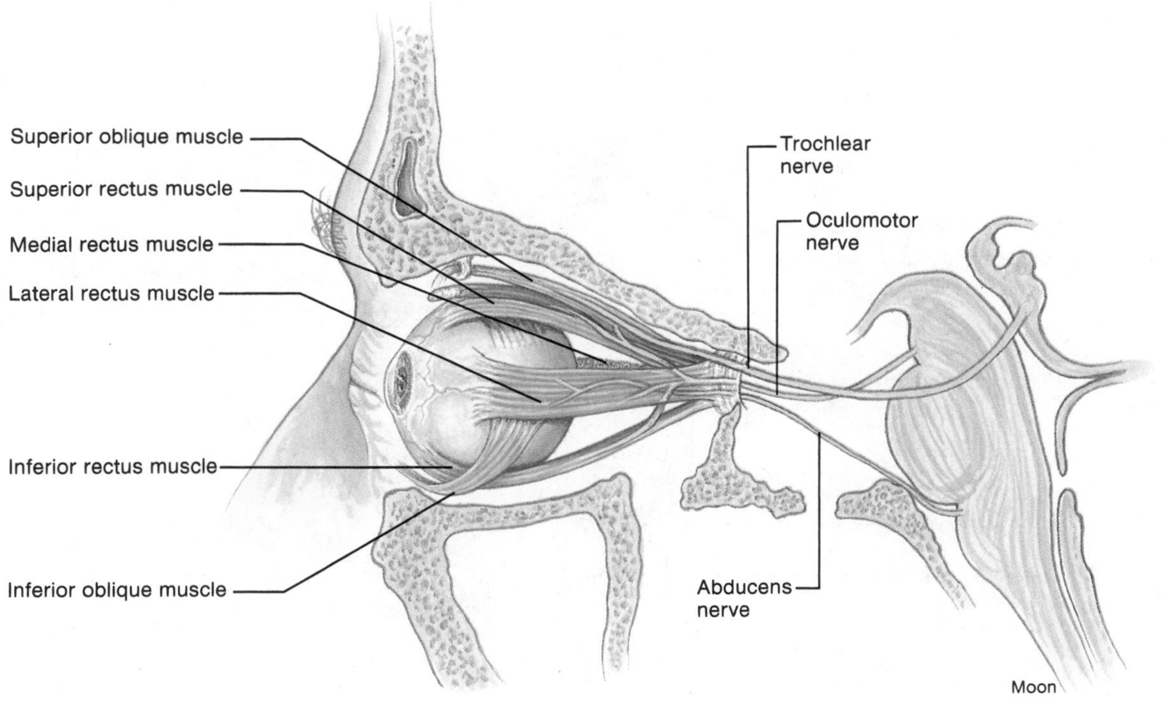

Figure 12.6 The oculomotor, trochlear, and abducens cranial nerves.

fibers from the inferior branch of the oculomotor enter the eyeball to supply motor innervation to the intrinsic smooth muscles of the iris for pupil dilation and to the muscles within the ciliary body for lens accommodation.

A few sensory fibers of the oculomotor originate from proprioceptors within the intrinsic muscles of the eyeball. These fibers convey impulses that affect the position and activity of the muscles they serve. A person whose oculomotor nerve is damaged may have a drooping upper eyelid or a dilated pupil or be unable to move the eyeball in the directions permitted by the four extrinsic muscles innervated by this nerve.

IV Trochlear The trochlear is a very small mixed nerve that emerges from a nucleus within the midbrain and passes from the cranium through the superior orbital fissure of the orbit. The trochlear innervates the **superior oblique** muscle of the eyeball with both motor and sensory fibers (fig. 12.6). Motor impulses to the superior oblique cause the eyeball to rotate downward and away from the midline. Sensory impulses originate in proprioceptors of the superior oblique muscle and provide information about its position and activity. Damage to the trochlear nerve impairs movement in the direction permitted by the superior oblique eye muscle.

V Trigeminal The large trigeminal nerve is a mixed nerve with motor functions originating from the nuclei within the pons and sensory functions terminating in nuclei within the midbrain, pons, and medulla (fig. 12.3). Two roots of the trigeminal are apparent as they emerge from the ventrolateral side of the pons. The larger, **sensory root,** immediately enlarges into a swelling called the **semilunar** (gasserian) **ganglion,** located in a bony depression on the inner surface of the petrous portion of the temporal bone. Three large branches arise from the semilunar ganglion (fig. 12.7): the **ophthalmic branch** enters the orbit through the superior orbital fissure; the **maxillary branch** extends through the foramen rotundum; and the **mandibular branch** passes through the foramen ovale. The smaller, **motor root,** consists of motor fibers of the trigeminal nerve which accompany the mandibular branch through the foramen ovale and innervate the muscles of mastication and certain muscles in the floor of the mouth. Impulses through the motor branch of the mandibular portion of the trigeminal nerve stimulate contraction of the muscles involved in chewing, including the medial and lateral pterygoids, masseter, temporalis, mylohyoid, and the anterior belly of the digastric muscle.

Although the trigeminal is a mixed nerve, its sensory functions are much more extensive than its motor functions. The three sensory branches of the trigeminal respond to touch, temperature, and pain sensations from the face. More specifically, the ophthalmic branch consists of sensory fibers from the anterior half of the scalp, skin of the forehead, upper eyelid, surface of the eyeball, lacrimal (tear) gland, side of the nose, and upper mucosa of the nasal cavity. The maxillary branch is composed of sensory fibers from the lower eyelid, lateral and inferior mucosa of the nasal cavity, palate and portions of the pharynx, teeth and gums of the upper jaw, upper lip, and skin of the cheek. Sensory fibers of the mandibular branch transmit impulses from the teeth and gums of the lower jaw, anterior

trochlear: Gk. *trochos,* a wheel
trigeminal: L. *trigeminus,* three born together

gasserian ganglion: from Johann L. Gasser, Vienna anatomist, 18th century
ophthalmic: L. *ophthalmia,* region of the eye

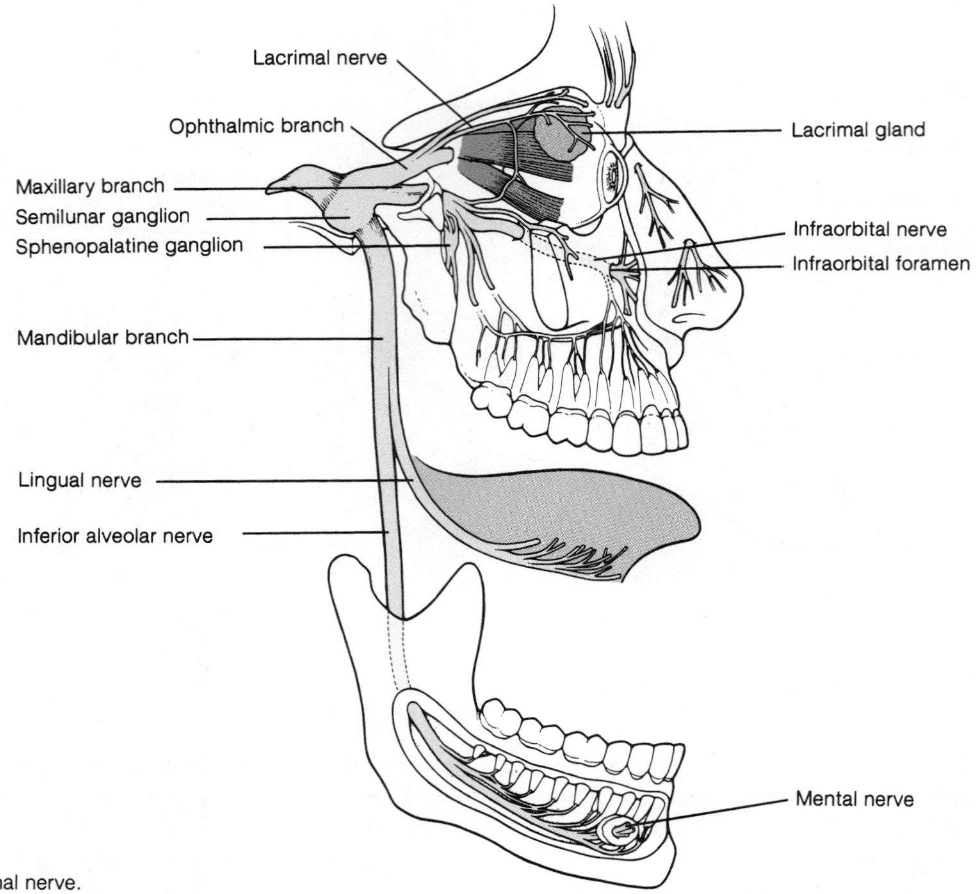

Figure 12.7 The trigeminal nerve.

two-thirds of the tongue (not taste), mucosa of the mouth, auricle of the ear, and lower part of the face. Trauma to the trigeminal nerve causes a lack of sensation from specific facial structures, whereas damage to the mandibular branch impairs chewing.

> The trigeminal nerve is the principal nerve relating to the practice of dentistry. Before teeth are filled or extracted, anesthetic is injected into the appropriate nerve to block sensation. A *maxillary,* or *second-division, nerve block,* performed by injecting near the sphenopalatine (pterygopalatine) ganglion (see fig. 12.7), desensitizes the teeth in the upper jaw. A *mandibular,* or *third-division, nerve block* desensitizes the lower teeth. This is performed by injecting anesthetic into the inferior alveolar branch of the mandibular nerve where it enters the mandible through the mandibular foramen.

VI Abducens The small abducens nerve originates from a nucleus within the pons and emerges from the lower portion of the pons and the anterior border of the medulla oblongata. It is a mixed nerve that traverses the superior orbital fissure of the orbit to innervate the **lateral rectus** eye muscle (see figs. 12.6, 19.9b). Impulses through the motor fibers of the abducens nerve cause the lateral rectus eye muscle to contract and the eyeball to move laterally away from the midline. Sensory impulses through the abducens nerve originate in proprioceptors in the lateral rectus muscle and are conveyed to the pons where muscle contraction is mediated. If the abducens nerve were damaged, not only would the patient be unable to turn the eyeball laterally, but because of the lack of muscle tonus to the lateral rectus, the eyeball would be pulled medially.

VII Facial The facial nerve arises from nuclei within the lower portion of the pons, traverses the petrous portion of the temporal bone, and emerges on the side of the face near the parotid salivary gland. The facial nerve is mixed. Impulses through the motor fibers cause contraction of the posterior belly of the digastric muscle and the muscles of facial expression, including the scalp and platysma muscles (fig. 12.8). The submandibular and sublingual salivary glands also receive some autonomic motor innervation from the facial nerve.

Sensory fibers of the facial nerve arise from taste buds on the anterior two-thirds of the tongue. Taste buds function as *chemoreceptors* because they respond to specific chemical stimuli. The **geniculate ganglion** is the enlargement of the facial nerve prior to the entry of the sensory portion into the pons. Sensory sensations of taste are conveyed into nuclei within the medulla, through the thalamus, and finally to the gustatory (taste) area of the cerebral cortex of the insula.

Trauma to the facial nerve destroys the ability to contract facial muscles on the affected side of the face and distorts taste perception, particularly of sweets. The affected side of the face tends to sag since muscle tonus is lost. *Bell's palsy* is a functional disorder (probably of viral origin) of the facial nerve.

Bell's palsy: from Sir Charles Bell, Scottish physician, 1774–1842

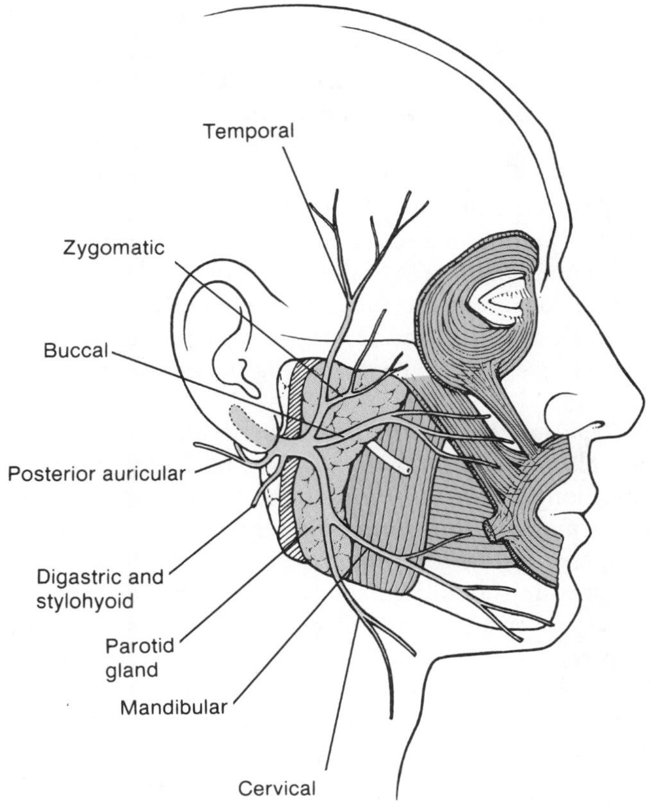

Figure 12.8 Facial nerve branches.

VIII Vestibulocochlear The vestibulocochlear nerve is also referred to as the **auditory, acoustic,** or **statoacoustic** nerve. It is purely sensory and is composed of two branches that arise within the inner ear (fig. 12.9). The **vestibular branch** arises from the **vestibular organs** of equilibrium and balance. Bipolar neurons from the vestibular organs (saccule, utricle, and semicircular canals) extend to the **vestibular ganglion,** where cell bodies are contained. From there, fibers convey impulses to the **vestibular nuclei** within the pons and medulla oblongata. Fibers from there extend to the thalamus and the cerebellum.

The **cochlear branch** arises from the **organ of Corti** within the cochlea and is associated with hearing. The cochlear branch is composed of bipolar neurons that convey impulses through the **spiral ganglion** to the **cochlear nuclei** within the medulla oblongata. From there, fibers extend to the thalamus and synapse with neurons that convey the impulses to the auditory areas of the cerebral cortex.

Injury to the cochlear portion of the vestibulocochlear nerve results in perception deafness, whereas damage to the vestibular portion causes dizziness and an inability to maintain balance.

IX Glossopharyngeal The glossopharyngeal nerve is a mixed nerve that innervates part of the tongue and pharynx (fig. 12.10). The motor fibers of this nerve originate in a nucleus within the medulla oblongata and pass through the jugular foramen. The motor fibers innervate the muscles of the pharynx and the parotid salivary gland to stimulate the swallowing reflex and the secretion of saliva.

The sensory fibers of the glossopharyngeal nerve arise from the pharyngeal region, the parotid salivary gland, the middle ear cavity, and the taste buds on the posterior one-third of the tongue. These taste buds, like those innervated by the facial nerve, are *chemoreceptors.* Some sensory fibers also arise from sensory receptors within the carotid sinus of the neck and help regulate blood pressure. Impulses from the glossopharyngeal nerve travel through the medulla and into the thalamus where they synapse with fibers that convey the impulses to the gustatory area of the cerebral cortex.

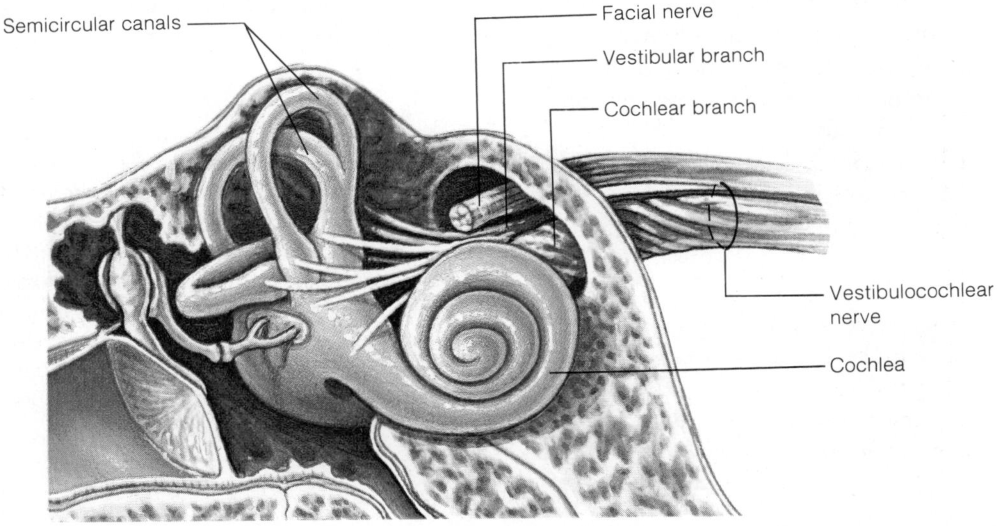

Figure 12.9 The vestibulocochlear nerve.

vestibulocochlear: L. *vestibulum,* chamber; *cochlea,* snail shell

organ of Corti: from Alfonso Corti, Italian anatomist, 1822–1888
glossopharyngeal: L. *glossa,* tongue; Gk. *pharynx,* throat

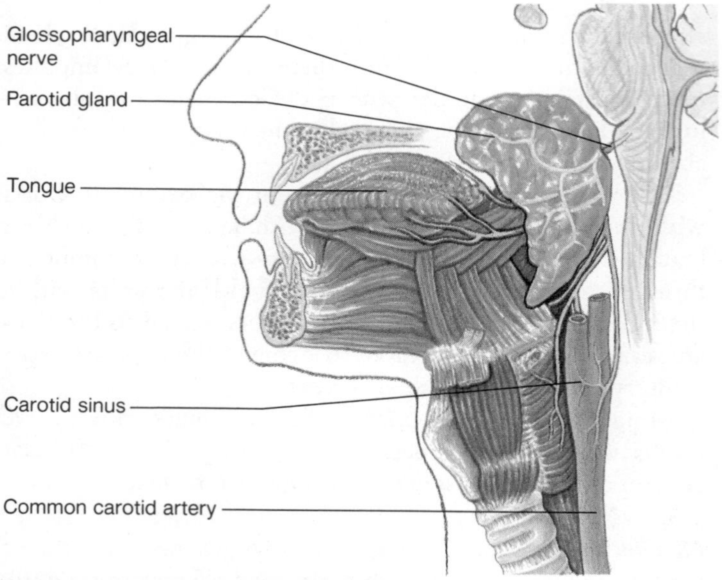

Figure 12.10 The glossopharyngeal nerve.

Damage to the glossopharyngeal nerve results in the loss of perception of bitter and sour taste from taste buds on the posterior portion of the tongue. If the motor portion of this nerve is damaged, swallowing becomes difficult.

X Vagus The vagus nerve has motor and sensory fibers that innervate visceral organs of the thoracic and abdominal cavities (fig. 12.11). The vagus nerve passes through the jugular foramen. The motor portion of the vagus arises from the **nucleus ambiguus** and **dorsal motor nucleus** of the vagus within the medulla. Through various branches, it innervates the muscles of the pharynx, larynx, respiratory tract, lungs, heart, esophagus, and abdominal viscera with the exception of the lower portion of the large intestine. One motor branch of the vagus nerve, the **recurrent laryngeal nerve,** innervates the larynx, enabling one to speak.

Sensory fibers of the vagus convey impulses from essentially the same organs served by motor fibers. Impulses through the sensory fibers relay specific sensations such as hunger pangs, distension, intestinal discomfort, or laryngeal movements. Sensory fibers also arise from proprioceptors in the muscles innervated by the motor fibers of this nerve.

If both vagus nerves are damaged, death ensues rapidly because vital autonomic functions stop, but the injury of one nerve causes vocal impairment, difficulty in swallowing, or other visceral disturbances.

XI Accessory The accessory nerve is principally a motor nerve, but it does contain some sensory fibers from proprioceptors within the muscles it innervates. The accessory nerve is unique in that it arises from both the brain and the spinal cord (fig. 12.12). The **cranial motor component,** also called the **bulbar (medullary) portion,** arises from nuclei within the medulla (ambiguus and spinal accessory), passes through the jugular foramen with the vagus nerve, and innervates the skeletal muscles of the soft palate, pharynx, and larynx, which contract reflexively during swallowing. The **spinal motor component** arises from the first five segments of the cervical portion of the spinal cord, passes cranially through the foramen magnum to join with the bulbar portion, and passes through the jugular foramen. The spinal motor component of the accessory nerve innervates the sternocleidomastoid and the trapezius muscles that move the head, neck, and shoulders. Damage to an accessory nerve makes it difficult to move the head or shrug the shoulders.

XII Hypoglossal The hypoglossal nerve is a mixed nerve. The motor fibers arise from the hypoglossal nucleus within the medulla oblongata and pass through the hypoglossal canal of the skull to innervate both the extrinsic and intrinsic muscles of the tongue (see fig. 12.12). Motor impulses along these fibers account for the coordinated contraction of the tongue muscles necessary for such activities as food manipulation, swallowing, and speech.

The sensory portion of the hypoglossal nerve arises from proprioceptors within the same tongue muscles and conveys impulses to the medulla oblongata regarding the position and function of the muscles.

If a hypoglossal nerve were damaged, a person would have difficulty in speaking, swallowing, and protruding the tongue.

Neurological Assessment of Cranial Nerves

Head injuries and brain concussions are common occurrences in automobile accidents. The cranial nerves would seem to be well protected on the inferior side of the brain. But the brain, immersed in and filled with cerebrospinal fluid, is like a water-sodden log; a blow to the top of the head can cause a serious rebound of the brain from the floor of the cranium. Routine neurological examinations involve testing for cranial-nerve dysfunction.

Commonly used clinical methods for determining cranial-nerve dysfunction are presented in table 12.2.

1. Which cranial nerves consist of sensory fibers only?
2. Which cranial nerves pass through the superior orbital fissure? through the jugular foramen?
3. Which cranial nerves have to do with tasting, chewing and manipulating food, and swallowing?
4. Which cranial nerves have to do with the structure, function, or movement of the eyeball?
5. List the cranial nerves, and indicate how each would be tested (both motor and sensory) for possible dysfunction.

vagus: L. *vagus*, wandering

hypoglossal: Gk. *hypo*, under; L. *glossa*, tongue

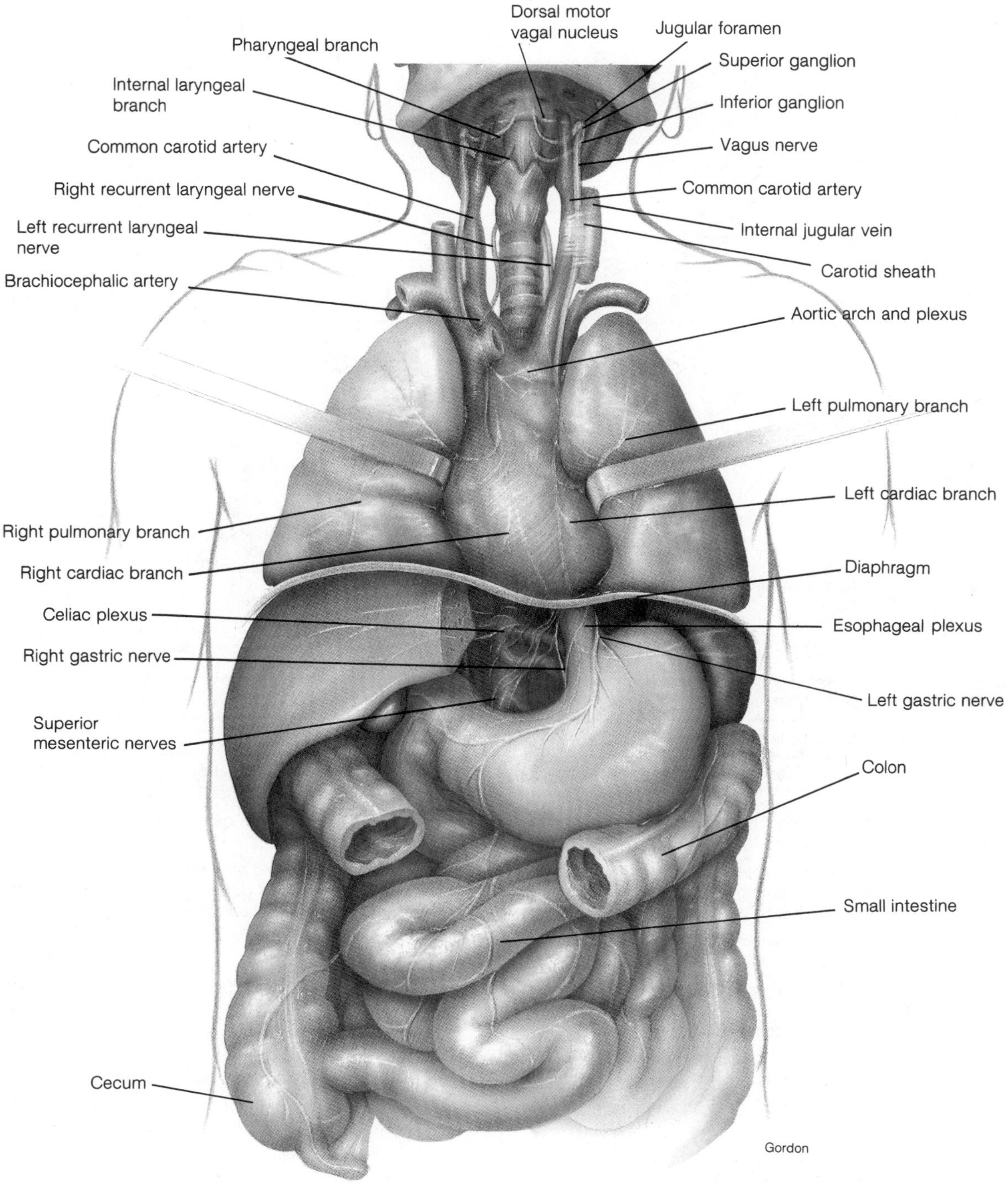

Figure 12.11 The visceral motor components of the vagus nerve.

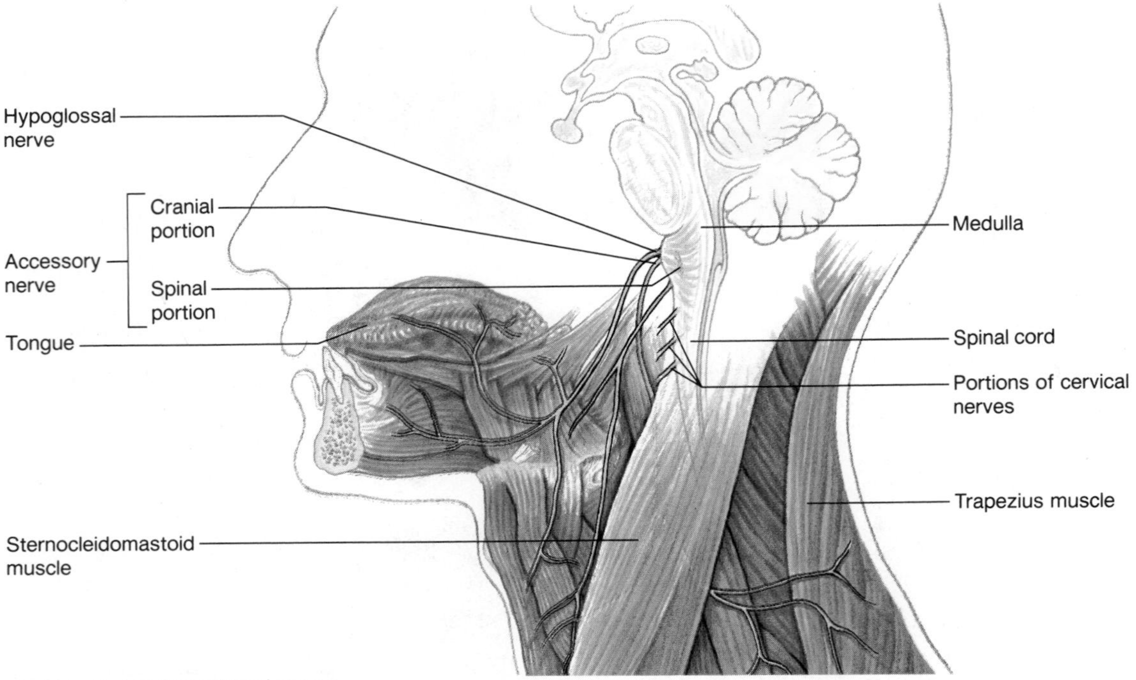

Figure 12.12 The accessory and hypoglossal nerves.

Table 12.2 Methods of determining cranial-nerve dysfunction

Nerve	Techniques of examination	Comments
Olfactory	With eyes closed, patient differentiates odors (tobacco, coffee, soap, etc.).	Nasal passages must be patent and tested separately by occluding the opposite side.
Optic	Examine the optic fundi with ophthalmoscope; test visual acuity by reading eye charts.	Visual acuity must be determined with lenses on, if patient wears them.
Oculomotor	Have patient follow examiner's finger movement with eyes—especially cross-eyed movement; observe pupillary change by shining light into each eye separately.	Examiner should note rate of pupillary change and coordinated constriction of pupils. Light in one eye should cause a similar pupillary change in other eye but to a lesser degree.
Trochlear	Have patient follow examiner's finger movement with eyes—especially lateral and downward movement.	
Trigeminal	Motor portion: examiner palpates temporalis and masseter muscles as patient clenches teeth; patient is asked to open mouth against resistance applied by examiner.	Muscles of both sides of the jaw should show equal contractile strength.
	Sensory portion: tactile and pain receptors are tested by lightly touching entire face with cotton and then with pin stimulus.	Patient's eyes should be closed and innervation areas for all three trigeminal branches should be tested.
Abducens	Have patient follow examiner's finger movement—especially lateral movement.	Motor functioning of cranial nerves III, IV, and VI may be tested simultaneously through selective movements of eyeball.
Facial	Motor portion: have patient raise eyebrows, frown, tightly constrict eyelids, smile, puff out cheeks, and whistle.	Examiner should note lack of tonus expressed by sagging regions of face.
	Sensory portion: examiner should place sugar on each side of tip of tongue.	Not reliable test for specific facial-nerve dysfunction because of tendency to stimulate taste buds on both sides of tip of tongue.
Vestibulocochlear	Vestibular portion: patient asked to walk a straight line.	Not usually tested unless patient complains of dizziness or balance problems.
	Cochlear portion: test with tuning fork.	Note ability to discriminate sounds.
Glossopharyngeal and vagus	Motor: examiner should note disturbances in swallowing, talking, and movement of soft palate; stimulate back of throat and note gag reflex.	Visceral innervation of vagus cannot be examined except for innervation to larynx, which is also served by glossopharyngeal.
Accessory	Patient is asked to shrug shoulders against resistance of examiner's hand and to rotate head against resistance.	Sides should show uniformity of strength.
Hypoglossal	Patient is requested to protrude tongue; tongue thrust may be resisted with tongue blade.	Tongue should protrude straight; deviation to side indicates ipsilateral-nerve dysfunction; asymmetry, atrophy, or lack of strength should be noted.

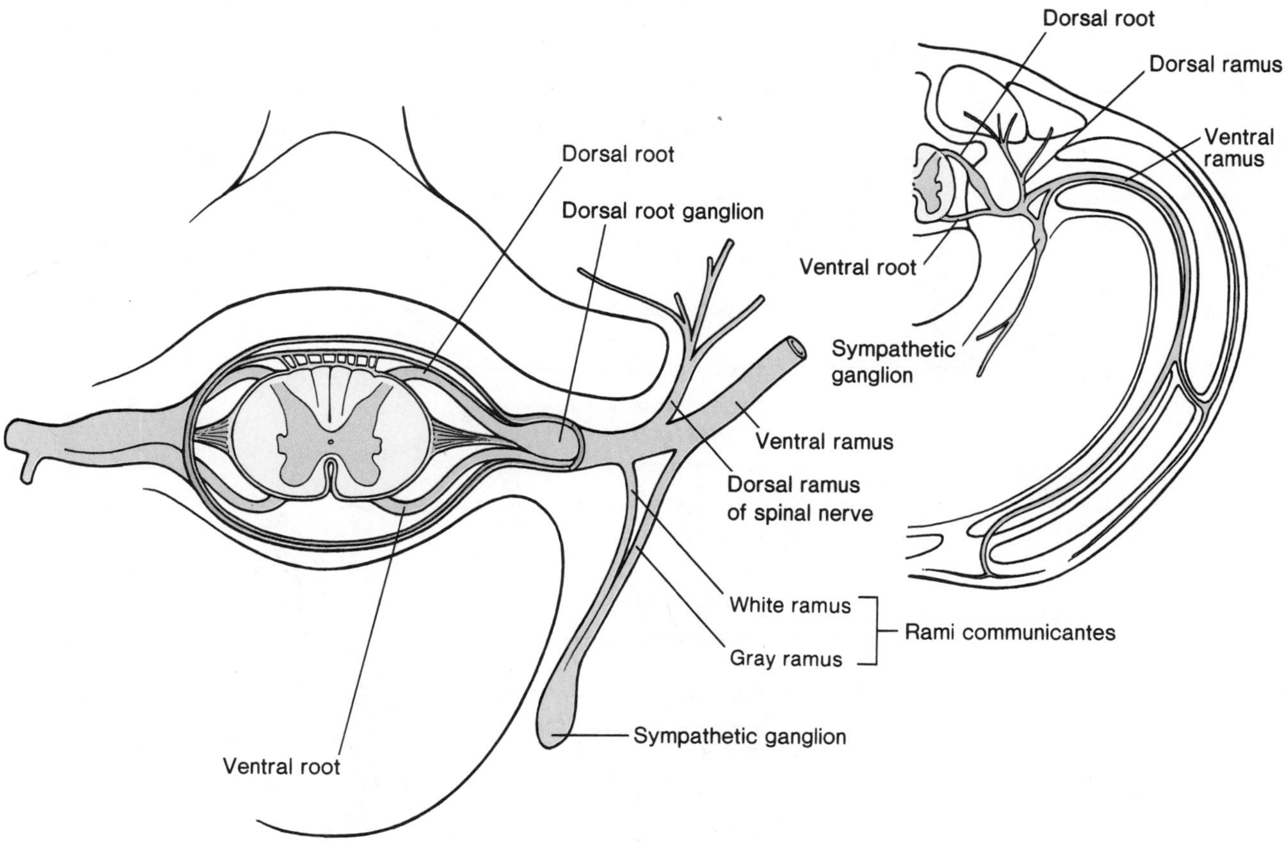

Figure 12.13 A transverse section of the spinal cord and the distribution of a spinal nerve.

Spinal Nerves

Each of the thirty-one pairs of spinal nerves is formed by the union of a dorsal and a ventral spinal root, which emerges from the spinal cord through an intervertebral foramen to innervate a body dermatome.

Objective 4. Discuss how the spinal nerves are grouped.
Objective 5. Describe the general distribution of a spinal nerve.

The thirty-one pairs of **spinal nerves** (see fig. 12.1) are grouped as follows: eight cervical, twelve thoracic, five lumbar, five sacral, and one coccygeal. With the exception of the first cervical nerve, the spinal nerves leave the spinal cord and vertebral canal through intervertebral foramina. The first pair of cervical nerves emerges between the occipital bone of the skull and the atlas vertebra. The second through the seventh pairs of cervical nerves emerge above the vertebrae for which they are named, whereas the eighth pair of cervical nerves passes between the seventh cervical and first thoracic vertebrae. The remaining pairs of spinal nerves emerge below the vertebrae for which they are named.

A spinal nerve is a mixed nerve attached to the spinal cord by a **dorsal (posterior) root** composed of sensory fibers and a **ventral (anterior) root** composed of motor fibers (fig. 12.13). The dorsal root contains an enlargement called the **dorsal-root ganglion,** where the cell bodies of sensory neurons are located. The axons of sensory neurons convey sensory impulses through the dorsal root and into the spinal cord where synapses occur with dendrites of other neurons. The ventral root consists of axons of motor neurons that convey motor impulses away from the CNS. A spinal nerve is formed as the fibers from the dorsal and ventral roots converge and emerge through an intervertebral foramen.

The disease *herpes zoster,* also known as *shingles,* is a viral infection of the dorsal-root ganglia. Herpes zoster causes painful, often unilateral, clusters of fluid-filled vesicles in the skin along the paths of the affected peripheral sensory neurons. This disease requires no special treatment, as the lesions gradually heal, and is usually not serious except in elderly debilitated patients, who may die from exhaustion.

A spinal nerve divides into several branches immediately after it emerges through the intervertebral foramen. The small **meningeal branch** reenters the vertebral canal to innervate the meninges, vertebrae, and vertebral ligaments. A larger branch, called the **dorsal ramus,** innervates the muscles, joints, and skin of the back along the vertebral column (fig. 12.13). A **ventral ramus** of a spinal nerve innervates the muscles and skin on the lateral and ventral (anterior) side of the torso. Combinations of ventral rami innervate the limbs.

The **rami communicantes** are two branches off the spinal nerve that connect to a **sympathetic ganglion,** which is part of the autonomic nervous system. The rami communicantes are composed of a **gray ramus,** containing unmyelinated fibers, and a **white ramus,** containing myelinated fibers. All spinal nerves have a gray ramus, but only the spinal nerves T1 through T3 have both gray and white rami.

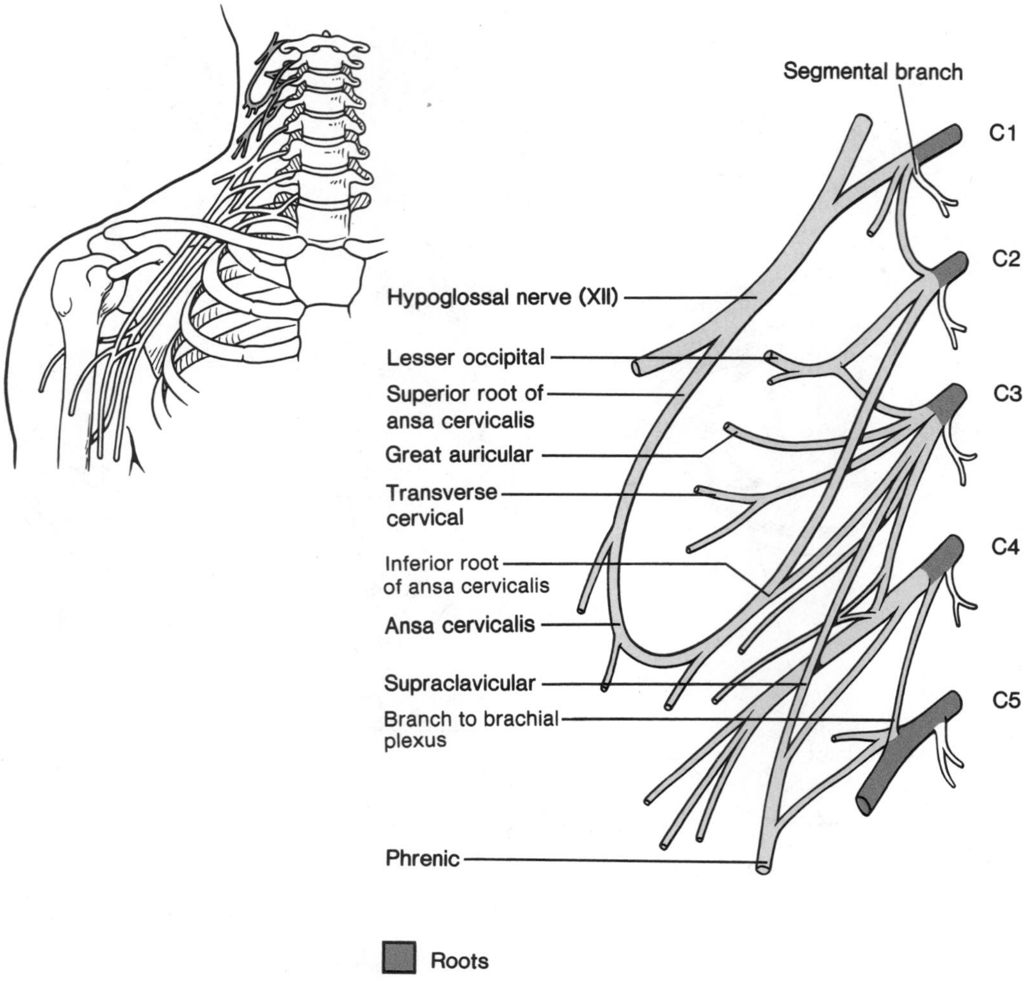

Figure 12.14 The cervical plexus.

---

1. List the number of nerves in each of the five regions of the vertebral column.
2. What are the four principal branches, or rami, from a spinal nerve, and what does each innervate?

Nerve Plexuses

Except in the thoracic nerves T2–T12, the ventral rami of the spinal nerves combine and then split again as networks of nerves referred to as plexuses. There are four plexuses of spinal nerves: the cervical, the brachial, the lumbar, and the sacral. Nerves emerge from the plexuses and are named according to the structures they innervate or the general course they take.

Objective 6. List the spinal nerve composition of each of the plexuses arising from the spinal cord.

Objective 7. List the principal nerves that emerge from the plexuses, and describe their general innervation.

Cervical Plexus The cervical plexus *(plek'sus)* is positioned deep on the side of the neck, lateral to the first four cervical vertebrae (fig. 12.14). It is formed by the ventral rami of the first four cervical nerves (C1–C4) and a portion of C5. Branches from the cervical plexus innervate the skin and muscles of the neck and portions of the head and shoulders. Branches of the cervical plexus also combine with the accessory and hypoglossal cranial nerves to supply dual innervation to some specific neck and pharyngeal muscles. Fibers from the third, fourth, and fifth cervical nerves unite to become the **phrenic** *(fren'ik)* **nerve,** which innervates the diaphragm. Motor impulses through the paired phrenic nerves cause the diaphragm to contract, inspiring air into the lungs.

The nerves of the cervical plexus are summarized in table 12.3.

Brachial Plexus The brachial plexus is positioned to the side of the last four cervical and first thoracic vertebrae. It is formed by the ventral rami of C5 through T1 and some fibers from C4 and T2. From its emergence, the brachial plexus extends downward and laterally, passes over the first rib behind the clavicle, and enters the axilla. Each brachial plexus innervates the entire upper extremity of one side plus a number of shoulder and neck muscles.

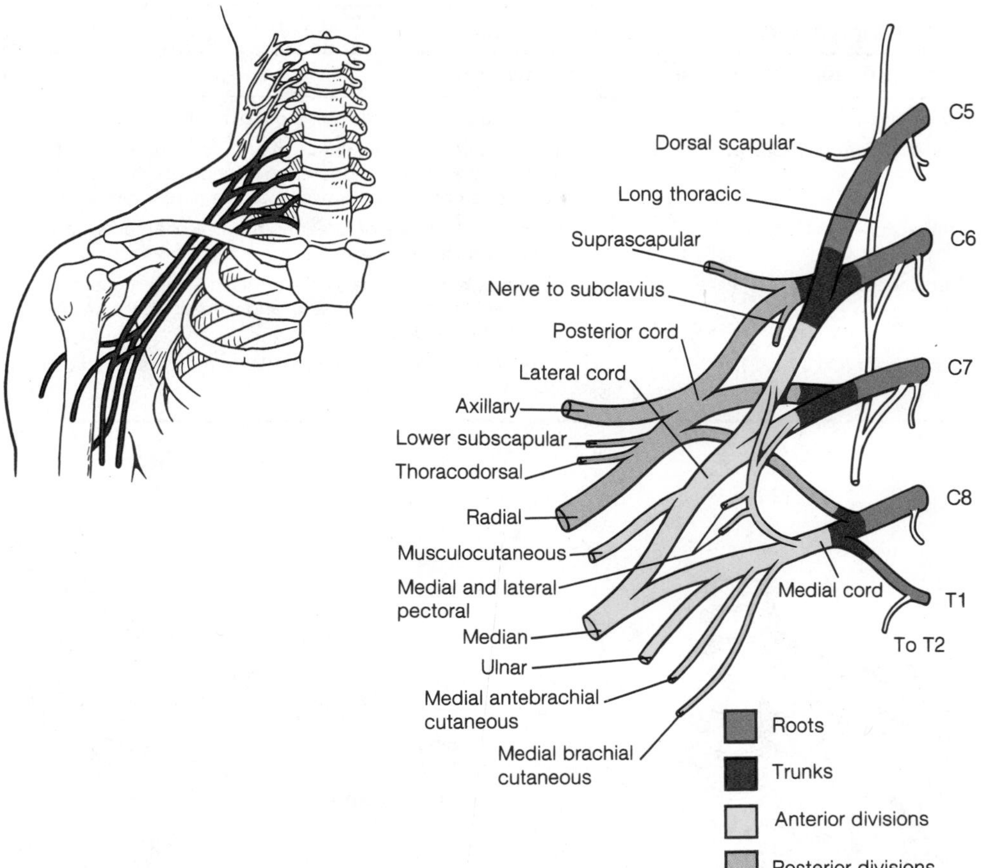

Figure 12.15 The brachial plexus.

Table 12.3 Branches of cervical plexus

Nerve	Spinal component	Innervation
Superficial Cutaneous Branches		
Lesser occipital	C2,C3	Skin of scalp above and behind ear
Greater auricular	C2,C3	Skin in front of, above, and below ear
Transverse cervical	C2,C3	Skin of anterior aspect of neck
Supraclavicular	C3,C4	Skin of upper portion of chest and shoulder
Deep Motor Branches		
Ansa cervicalis		
Superior root	C1,C2	Geniohyoid, thyrohyoid, and infrahyoid muscles of neck
Inferior root	C3,C4	Omohyoid, sternohyoid, and sternothyroid muscles of neck
Phrenic	C3,C4,C5	Diaphragm
Segmental branches	C1–C5	Deep muscles of neck (levator scapulae ventralis, trapezius, scalenus, and sternocleidomastoid)

Structurally, the brachial plexus is divided into *roots, trunks, divisions,* and *cords* (fig. 12.15). The roots of the brachial plexus are simply continuations of the ventral rami of the cervical nerves. The ventral rami of C5 and C6 converge to become the **upper trunk,** the C7 ramus becomes the **middle trunk,** and the ventral rami of C8 and T1 converge to become the **lower trunk.** Each of the three trunks immediately divides into an **anterior division** and a **posterior division.** The divisions then converge to form three cords. The **posterior cord** is formed by the convergence of the posterior divisions of the upper, middle, and lower trunks and hence contains fibers from C5 through C8. The **medial cord** is a continuation of the anterior division of the lower trunk and primarily contains fibers from C8 and T1. The **lateral cord** is formed by the convergence of the anterior division of the upper and middle trunk and consists of fibers from C5 through C7.

In summary, the brachial plexus is composed of nerve fibers from spinal nerves C5 through T1 and a few fibers from C4 and T2. Roots are continuations of the ventral rami. The roots converge to form trunks, and the trunks branch into divisions. The divisions in turn form cords, and the nerves of the upper extremity arise from the cords.

Five major nerves and several smaller ones arise from the three cords of the brachial plexus. The principal nerves of the brachial plexus are summarized in table 12.4.

Table 12.4 Branches of brachial plexus

Nerve	Cord and spinal components	Innervation
Axillary	Posterior cord (C5,C6)	Skin of shoulder; shoulder joint, deltoid and teres minor muscles
Radial	Posterior cord (C5–C8,T1)	Skin of posterior lateral surface of arm, forearm, and hand; extensor muscles of posterior upper arm and forearm (triceps brachii, supinator, anconeus, brachioradialis, extensor carpi radialis brevis, extensor carpi radialis longus, extensor carpi ulnaris)
Musculocutaneous	Lateral cord (C5,C6,C7)	Skin of lateral surface of forearm; muscles of anterior upper arm (coracobrachialis, biceps brachii, brachialis)
Ulnar	Medial cord (C8,T1)	Skin of medial third of hand; flexor muscles of anterior forearm (flexor carpi ulnaris, flexor digitorum), medial palm, and intrinsic flexor muscles of hand (profundus, third and fourth lumbricales)
Median	Medial cord (C6,C7,C8,T1)	Skin of lateral two-thirds of hand; flexor muscles of anterior forearm, lateral palm, and first and second lumbricales

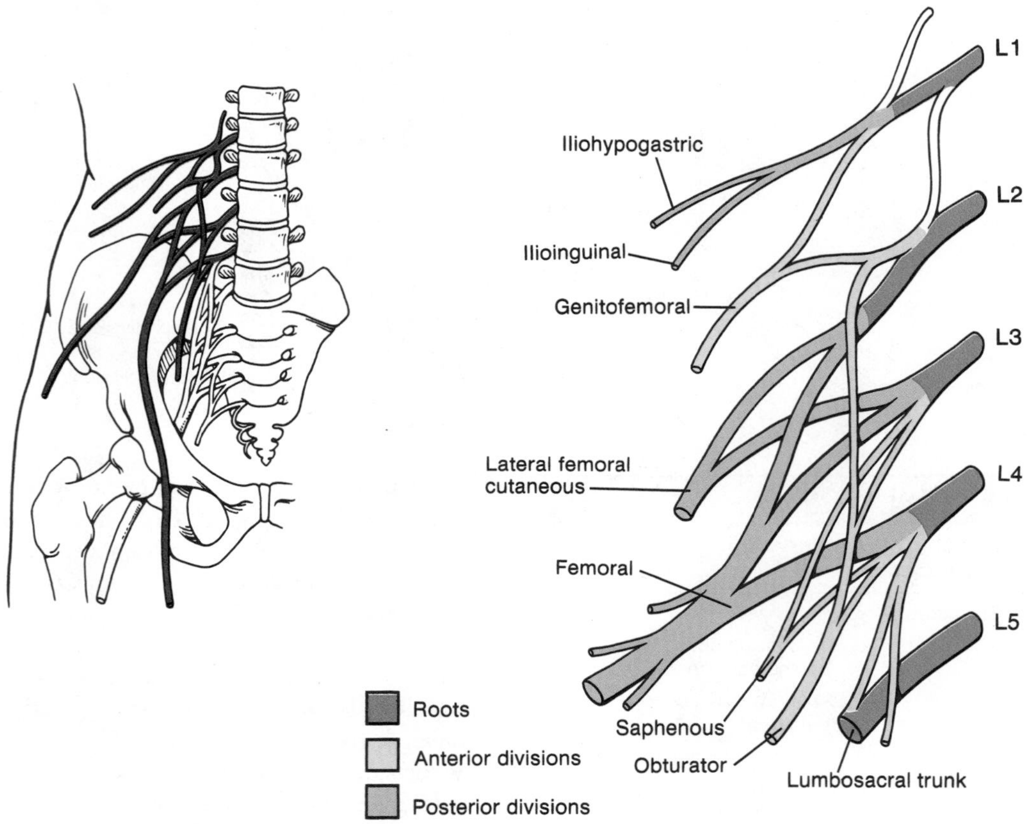

Figure 12.16 The lumbar plexus.

Trauma to the brachial plexus is not uncommon, especially if the clavicle, upper ribs, or lower cervical vertebrae are seriously fractured. Occasionally, the brachial plexus of a newborn is severely strained during a difficult delivery when pulling the baby through the birth canal. In such cases, the arm of the injured side is paralyzed and withers as the muscles atrophy in relation to the extent of the injury.

Lumbar Plexus The lumbar plexus is positioned to the side of the first four lumbar vertebrae. It is formed by the ventral rami of spinal nerves L1–L4 and some fibers from T12 (fig. 12.16). The nerves that arise from the lumbar plexus innervate structures of the lower abdomen and anterior and medial portions of the lower extremity. The lumbar plexus is not as complex as the brachial plexus, having only roots and divisions rather than the roots, trunks, divisions, and cords as in the brachial plexus.

Structurally, the **posterior division** of the lumbar plexus passes obliquely outward, deep to the psoas major muscle, whereas the **anterior division** is superficial to the quadratus lumborum muscle. From these two divisions arise the large **femoral nerve,** which innervates the anterior muscles of the thigh and the **obturator nerve,** which innervates the adductor muscles of the leg. Several smaller nerves also arise from the lumbar plexus and are summarized in table 12.5.

Table 12.5 Branches of lumbar plexus

Nerve	Spinal components	Innervation
Iliohypogastric	T12–L1	Skin of lower abdomen and buttock; muscles of anterolateral abdominal wall (external oblique, internal oblique, transversus abdominis)
Ilioinguinal	L1	Skin of upper median thigh, scrotum and root of penis in male, and labia majora in female; muscles of anterolateral abdominal wall with iliohypogastric nerve
Genitofemoral	L1,L2	Skin of middle anterior surface of thigh, scrotum in male, and labia majora in female; cremaster muscle in male
Lateral femoral cutaneous	L2,L3	Skin of anterior, lateral, and posterior aspects of thigh
Femoral	L2–L4	Skin of anterior and medial aspect of thigh and medial aspect of leg and foot; anterior muscles of thigh (iliacus, psoas major, pectineus, rectus femoris, sartorius) and extensor muscles of leg (rectus femoris, vastus lateralis, vastus medialis, vastus intermedius)
Obturator	L2–L4	Skin of medial aspect of thigh; adductor muscles of leg (external obturator, pectineus, adductor longus, adductor brevis, adductor magnus, gracilis)
Saphenous	L2–L4	Skin of medial aspect of leg

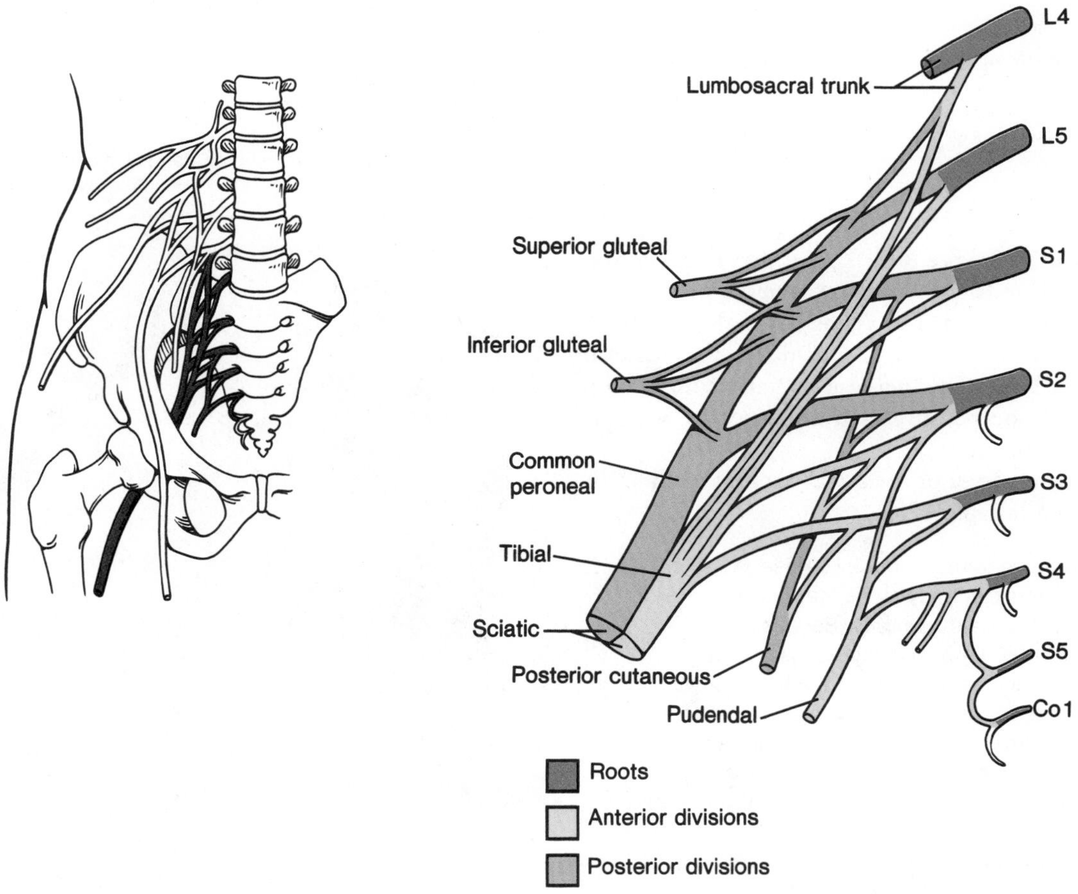

Figure 12.17 The sacral plexus.

A herniated intervertebral disc in the lumbar region may cause a condition called *sciatica (si-at'ĭ-kah).* This condition is characterized by a sharp pain in the gluteal region and by pains that extend down the posterior side of the thigh, caused by the disc compressing the spinal roots that form the sciatic nerve.

Sacral Plexus The sacral plexus is positioned immediately caudal to the lumbar plexus. It is formed by the ventral rami of spinal nerves L4, L5, and S1–S4 (fig. 12.17). The nerves arising from the lumbar plexus innervate the lower back, pelvis, perineum, posterior surface of the thigh and leg, and the dorsal and ventral surfaces of the foot. Like the lumbar plexus, the sacral plexus consists of **roots** and **anterior** and **posterior divisions** from which nerves arise. Because some of the nerves of the sacral plexus also contain fibers from the nerves of the lumbar plexus through the **lumbosacral trunk,** these two plexuses are frequently described collectively as the **lumbosacral plexus.**

sacral: L. *sacris,* sacred

Table 12.6 Branches of sacral plexus

Nerve	Spinal components	Innervation
Superior gluteal	L4,L5,S1	Abductor muscles of thigh (gluteus minimus, gluteus medius, tensor fasciae latae)
Inferior gluteal	L5–S2	Extensor muscle of thigh (gluteus maximus)
Nerve to piriformis	S1,S2	Abductor and rotator of thigh (piriformis)
Nerve to quadratus femoris	L4,L5,S1	Rotators of thigh (gemellus inferior, quadratus femoris)
Nerve to internal obturator	L5–S2	Rotators of thigh (gemellus superior, internal obturator)
Perforating cutaneous	S2,S3	Skin over lower medial surface of buttock
Posterior cutaneous	S1–S3	Skin over lower lateral surface of buttock, anal region, upper posterior surface of thigh, upper aspect of calf, scrotum in male, and labia majora in female
Sciatic	L4–S3	Composed of two nerves (tibial and common peroneal) within sciatic sheath; splits into two portions at popliteal fossa; branches from sciatic in thigh region to hamstring muscles (biceps femoris, semitendinosus, semimembranosus) and adductor magnus
Tibial (sural, medial and lateral plantar)	L4–S3	Skin of posterior surface of leg and sole of foot; muscle innervation includes gastrocnemius, soleus, flexor digitorum longus, flexor hallucis longus, tibialis posterior, popliteus, and intrinsic muscles of the foot
Common peroneal (superficial and deep peroneal)	L4–S2	Skin of anterior surface of the leg and dorsum of foot; muscle innervation includes peroneus tertius, peroneus brevis, peroneus longus, tibialis anterior, extensor hallucis longus, extensor digitorum longus, extensor digitorum brevis
Pudendal	S2–S4	Skin of penis and scrotum in male and skin of clitoris, labia majora, labia minora, and lower vagina in female; muscles of perineum

The **sciatic** *(si-at'ik)* **nerve** is the largest nerve arising from the sacral plexus and, in fact, is the largest nerve in the body. The sciatic nerve passes from the pelvis through the greater sciatic notch of the os coxa and extends down the posterior aspect of the thigh. The sciatic nerve is actually composed of two nerves, the common peroneal and tibial nerves, wrapped in a common sciatic sheath.

A summary of the nerves of the sacral plexus and their distribution is presented in table 12.6.

The sciatic nerve in the buttock lies deep to the gluteus maximus muscle, midway between the greater trochanter and the ischial tuberosity. Because of its position, the sciatic nerve is of tremendous clinical importance. A posterior dislocation of the hip joint will generally injure the sciatic nerve. A herniated disc (fig. 12.18) or pressure from the uterus during pregnancy may damage the roots of the nerves forming the sciatic nerve. An improperly administered injection into the buttock may injure the sciatic nerve itself.

1. Define *nerve plexus.* What are the four spinal nerve plexuses, and which spinal nerves contribute to each?
2. Which of the spinal nerves are not involved in a plexus?
3. Distinguish between a dorsal ramus and a ventral ramus. Which is involved in the formation of plexuses?
4. Construct a table that lists the plexus of origin and the general region of innervation for the following nerves: pudendal, phrenic, femoral, ulnar, median, sciatic, saphenous, axillary, radial, ansa cervicalis, tibial, and common peroneal.

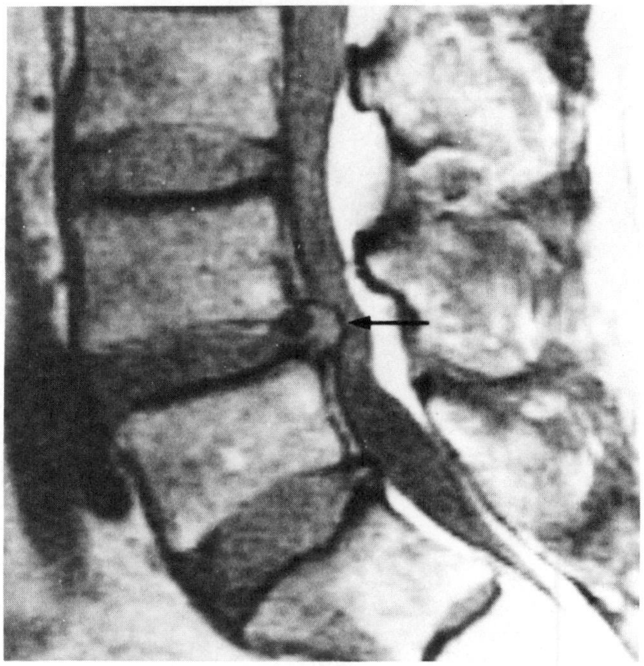

Figure 12.18 An MR scan of a herniated disc in the lumbar region.

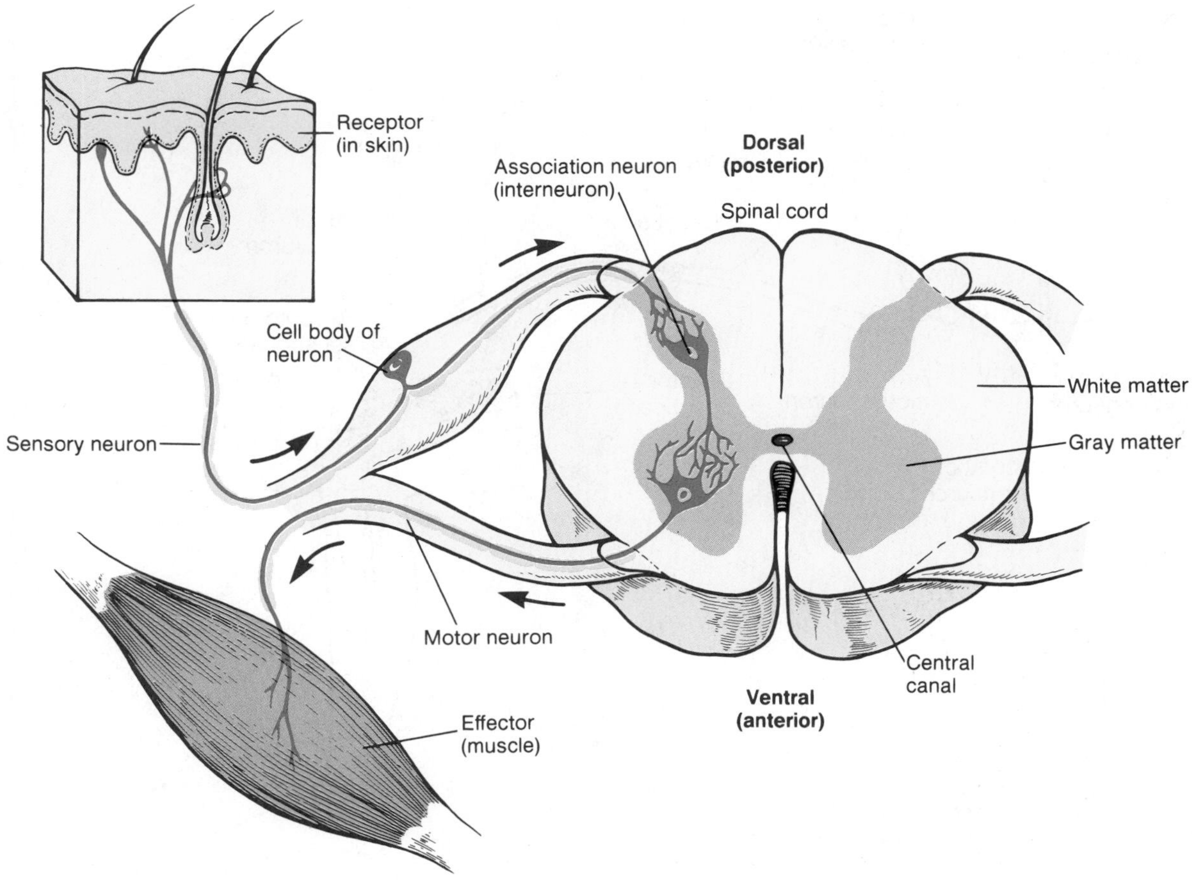

Figure 12.19 The reflex arc.

Reflex Arcs and Reflexes

The conduction pathway of a reflex arc consists of a receptor, a sensory neuron, a motor neuron and its innervation in the PNS, and an association neuron in the CNS. The reflex arc provides the mechanism for a rapid, automatic response to a potentially threatening stimulus.

Objective 8. Define *reflex arc,* and list its five components.
Objective 9. Distinguish between the various kinds of reflexes.

Specific **nerve pathways** provide routes by which impulses travel through the nervous system. Frequently, a nerve pathway begins with impulses being conducted to the CNS through sensory receptors and sensory neurons of the PNS. Once within the CNS, impulses may immediately travel back through motor portions of the PNS to activate specific skeletal muscles, glands, or smooth muscles, and also be sent simultaneously to other parts of the CNS through ascending tracts within the spinal cord.

Components of the Reflex Arc

The simplest type of nerve pathway is a **reflex arc** (fig. 12.19). A reflex arc implies an automatic, unconscious, protective response to a situation in an attempt to maintain body homeostasis. A reflex arc leads by a short route from sensory to motor neurons and includes only two or three neurons. The five components of a reflex arc are the receptor, sensory neuron, center, motor neuron, and effector. The **receptor** includes the dendrite of a sensory neuron and the place where the electrical impulse is initiated. The sensory neuron relays the impulse through the posterior root to the CNS. The **center** is within the CNS and usually involves an association neuron (interneuron). It is here that the arc is made and other impulses are sent through synapses to other parts of the body. The **motor neuron** conducts the impulse to an effector organ (generally a skeletal muscle) that responds. The result of an impulse through a reflex arc is called a reflex action or, simply, a *reflex.*

Kinds of Reflexes

Visceral Reflexes Reflexes that cause smooth or cardiac muscle to contract or glands to secrete are **visceral (autonomic) reflexes.** Visceral reflexes help control the body's many involuntary processes such as heart rate, respiratory rate, blood pressure, and digestion. Swallowing, sneezing, coughing, and vomiting may also be reflexive, although they involve skeletal muscles that function involuntarily.

Somatic Reflexes Somatic reflexes are those that result in the contraction of skeletal muscles. There are three principal kinds of somatic reflexes, named according to the response they produce.

The **stretch reflex** involves only two neurons and one synapse in the pathway and is therefore called a *monosynaptic reflex arc.* Slight stretching of neuromuscular spindle receptors

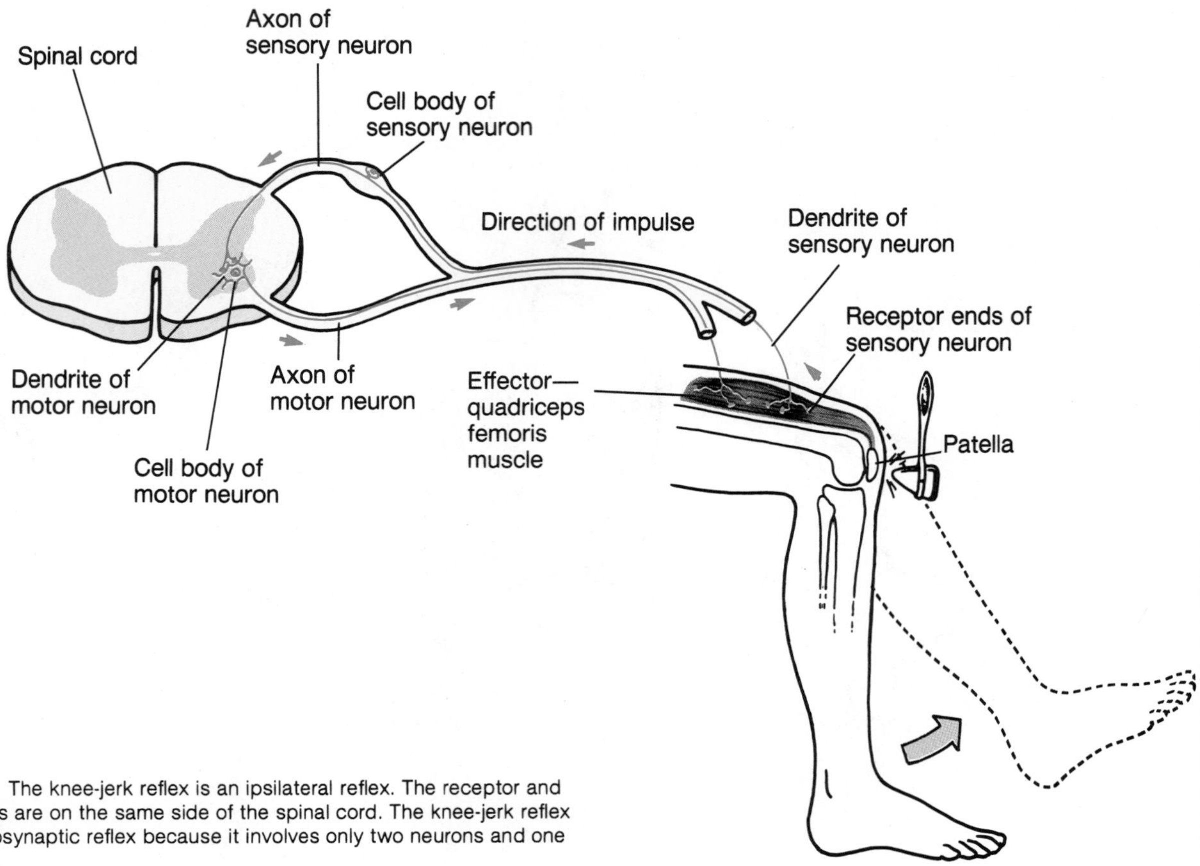

Figure 12.20 The knee-jerk reflex is an ipsilateral reflex. The receptor and effector organs are on the same side of the spinal cord. The knee-jerk reflex is also a monosynaptic reflex because it involves only two neurons and one synapse.

within a muscle initiates an impulse along an afferent neuron to the spinal cord. A synapse with a motor neuron occurs in the anterior gray horn and a motor unit is activated, causing specific muscle fibers to contract. Since the receptor and effector organs of the stretch reflex involve structures on the same side of the spinal cord, the reflex arc is an *ipsilateral reflex arc.* The knee-jerk reflex is an ipsilateral reflex (fig. 12.20), as are all monosynaptic reflex arcs.

A **flexor reflex,** or **withdrawal reflex,** consists of a *polysynaptic reflex arc* (fig. 12.21). Flexor reflexes involve association neurons in addition to the sensory and motor neurons. A flexor reflex is initiated as a person encounters a painful stimulus such as a hot or sharp object. As a receptor organ is stimulated, afferent neurons transmit the impulse to the spinal cord where association neurons are activated. There the impulses are directed through motor neurons to flexor muscles that contract in response. Simultaneously, antagonistic muscles are inhibited (relaxed) so that the traumatized extremity can be quickly withdrawn from the harmful source of stimulation.

Several additional reflexes may be activated while a flexor reflex is in progress. In an *intersegmental reflex arc,* motor units from several segments of the spinal cord are activated by impulses coming in from the receptor organ. An intersegmental reflex arc produces stimulation of more than one effector organ. Frequently, afferent impulses from a receptor organ cross over through the spinal cord to activate effector organs within the opposite (*contralateral*) limb. This type of reflex is called a **crossed extensor reflex** (fig. 12.22) and is important for maintaining body balance while a flexor reflex is in progress. The reflexive inhibition of certain muscles to contract, called *reciprocal inhibition,* also helps maintain balance while either flexor or crossed extensor reflexes are in progress.

Certain reflexes are important for physiological functions, while others are important for avoiding injury. Some common reflexes are presented in table 12.7 and illustrated in figure 12.23.

Part of a routine physical examination involves testing a person's reflexes. The condition of the nervous system, particularly the functioning of the synapses, may be determined by examining reflexes. In case of injury to some portion of the nervous system, testing certain reflexes may indicate the location and extent of the injury. Also, an anesthesiologist may try to initiate a reflex to ascertain the effect of an anesthetic.

1. How are reflexes important in maintaining body homeostasis?
2. List the five components of a reflex arc.
3. Define or describe the following: visceral reflex, somatic reflex, stretch reflex, flexor reflex, crossed extensor reflex, ipsilateral reflex, and contralateral reflex.

anesthesia: Gk. *an,* without; *aisthesis,* sensation

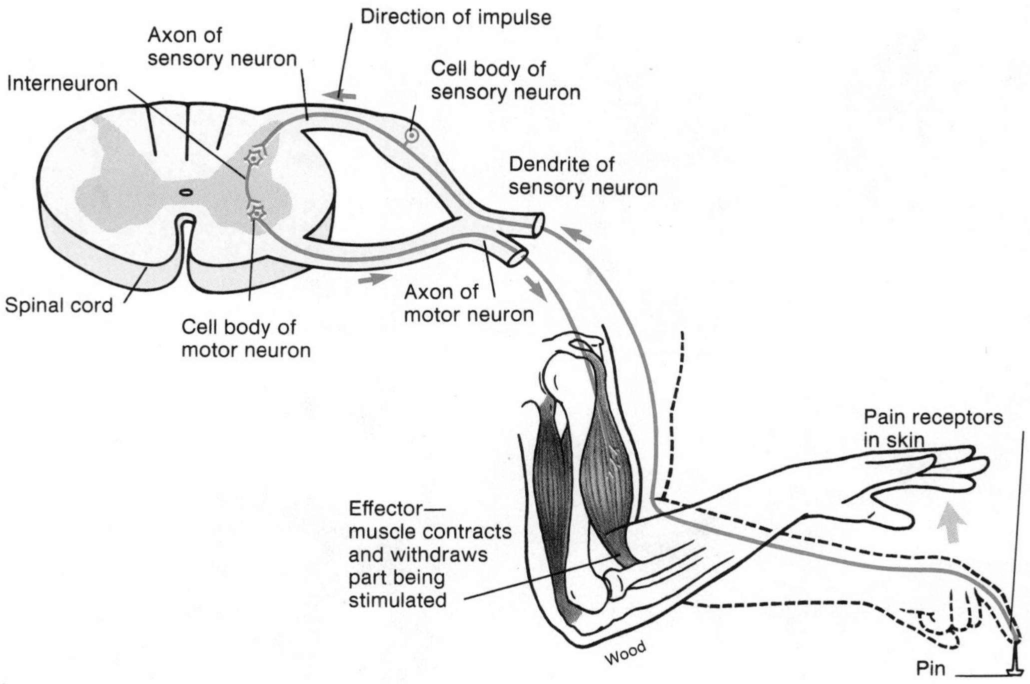

Figure 12.21 The flexor, or withdrawal, reflex is a polysynaptic reflex because it involves association neurons (interneurons) in addition to sensory and motor neurons.

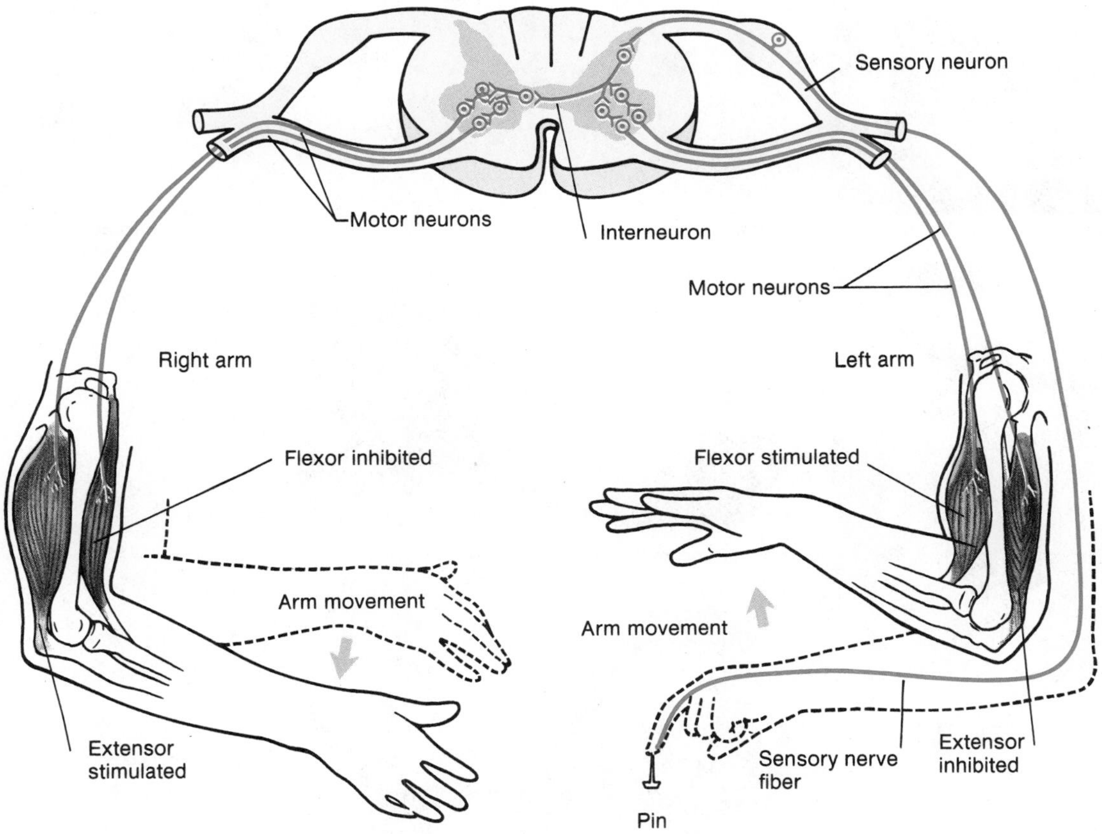

Figure 12.22 A crossed extensor reflex causes a reciprocal inhibition of muscles of the opposite appendage. This type of reflex inhibition is important in maintaining balance.

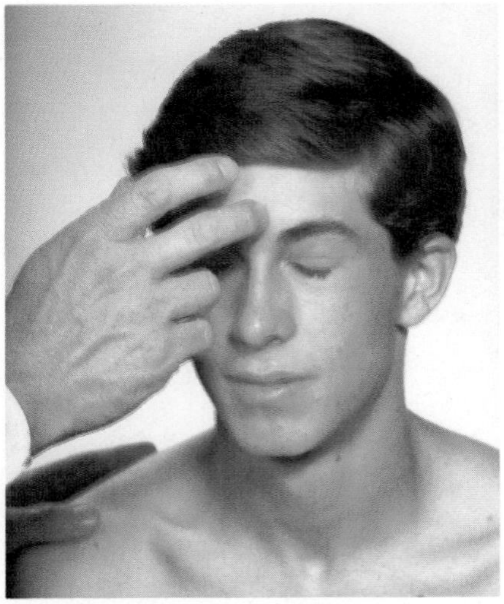

Glabellar reflex

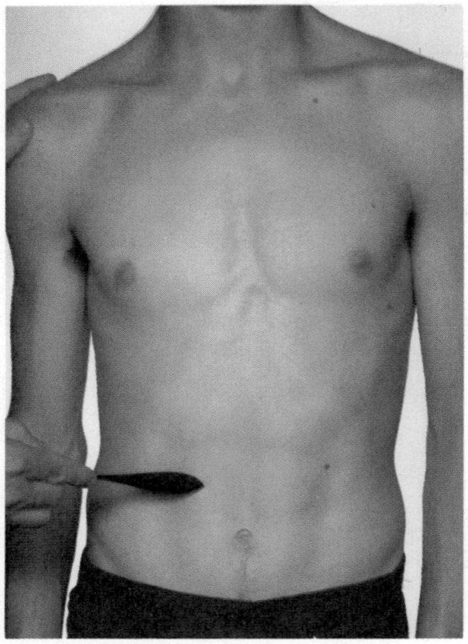

Abdominal reflex

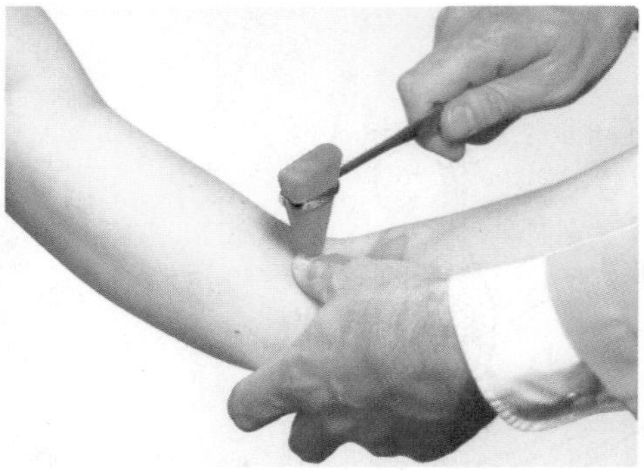

Biceps reflex

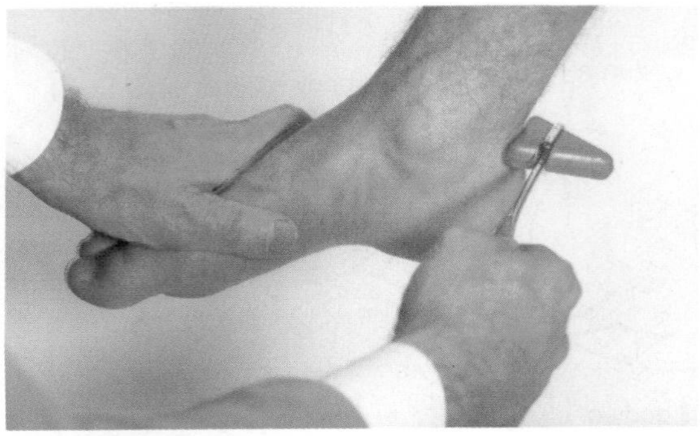

Ankle (Achilles) reflex

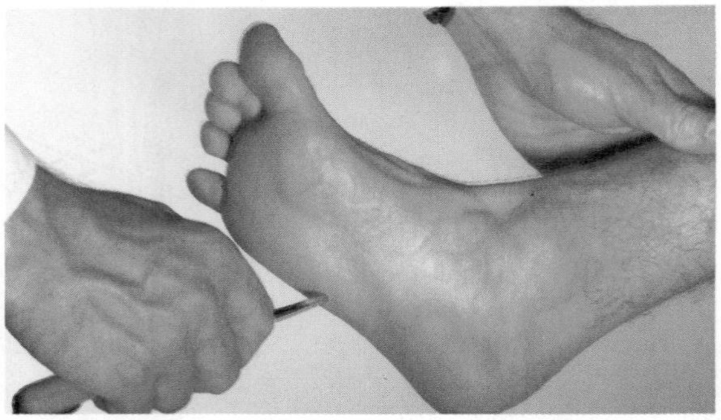

Babinski reflex

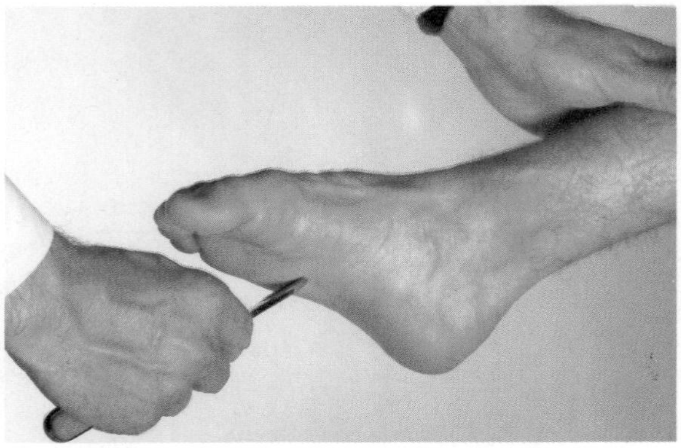

Plantar reflex

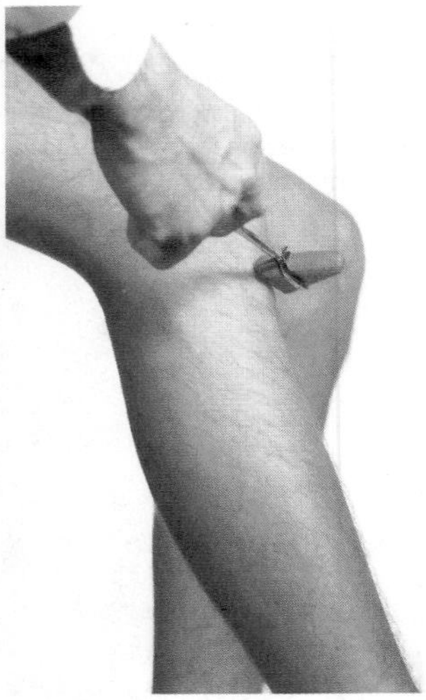

Knee (patellar) reflex

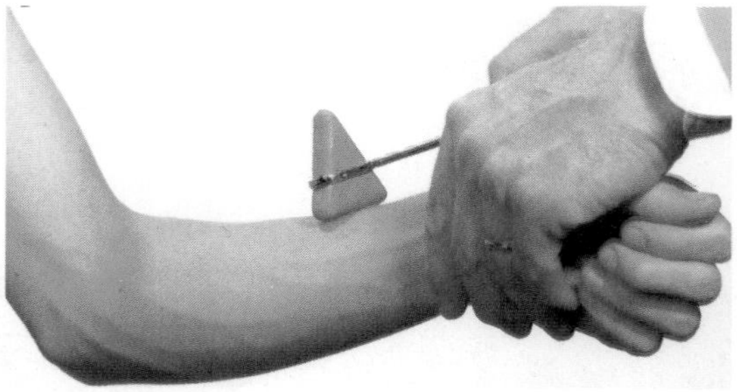

Supinator (brachioradialis) reflex

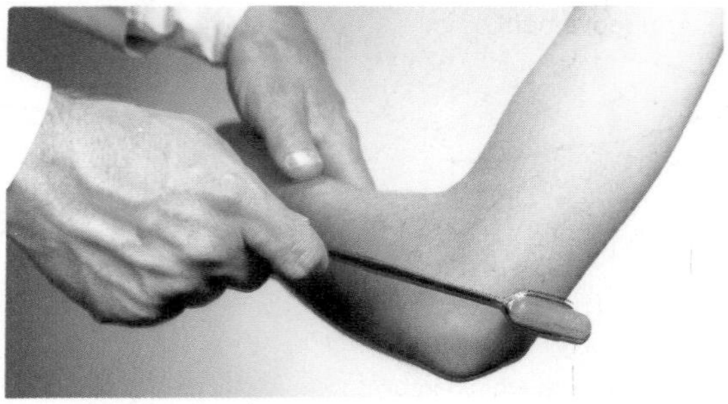

Triceps reflex

Figure 12.23 Some reflexes of clinical importance.

Table 12.7 Some clinically important reflexes

Reflex	Spinal segment	Site of receptor stimulation	Effector action
Biceps reflex	C5, C6	Biceps tendon near attachment on radial tuberosity	Contracts biceps brachii to flex elbow
Triceps reflex	C7, C8	Triceps tendon near attachment on olecranon process	Contracts triceps brachii to extend elbow
Supinator or brachioradialis reflex	C5, C6	Radial attachment of supinator and brachioradialis muscles	Supinates forearm and hand
Knee reflex	L2, L3, L4	Patellar tendon just below patella	Contracts quadriceps to extend the knee
Ankle reflex	S1, S2	Tendo calcaneus near attachment on calcaneus	Plantar flexes ankle
Plantar reflex	L4, L5, S1, S2	Lateral aspect of sole from heel to ball of foot	Plantar flexes foot and flexes toes
Babinski reflex*	L4, L5, S1, S2	Lateral aspect of sole from heel to ball of foot	Dorsiflexes great toe and fans other toes
Abdominal reflexes	T8, T9, T10 above umbilicus and T10, T11, T12 below umbilicus	Sides of abdomen above and below level of umbilicus	Contracts abdominal muscles and deviates umbilicus toward stimulus
Cremasteric reflex	L1, L2	Stroking upper inside of thigh in males	Contracts cremasteric muscle and elevates testis on same side of stimulation

*If the Babinski reflex rather than the plantar reflex occurs as the sole of the foot is stimulated, it may indicate damage to the corticospinal tract within the spinal cord. However, the Babinski reflex is present in infants up to 12 months of age because of the immaturity of their corticospinal tracts.

CLINICAL CASE STUDY ANSWER

The radial nerve is susceptible to injury in cases of midshaft humerus fractures, owing to its close association to the bone as it lies in the radial groove. In addition to wrist extension (extensor carpi radialis longus and brevis, extensor carpi ulnaris), wrist adduction (extensor carpi ulnaris) and abduction (extensor carpi radialis brevis and longus), supination (supinator), and finger extension (finger extensors). A detectable weakness in arm flexion would also be possible, due to impairment of the brachioradialis. The triceps would be spared because of the more proximal branching point of its motor nerve supply. The radial nerve is a mixed nerve, carrying both motor and sensory fibers and, therefore, a sensory deficit would be present. Decreased sensation would be detectable on the dorsal radial aspect of the thumb and hand (excluding the fingers).

Babinski reflex: from Joseph F. Babinski, French neurologist, 1857–1932

Chapter Summary

I. Introduction to the Peripheral Nervous System
   A. The peripheral nervous system consists of nerves that convey impulses to and from the central nervous system.
   B. The cranial nerves arise from the brain, and the spinal nerves arise from the spinal cord.
   C. Sensory (afferent) nerves convey impulses toward the CNS, motor (efferent) nerves convey impulses away from the CNS, and mixed nerves are composed of both sensory and motor fibers.

II. Cranial Nerves
   A. Twelve pairs of cranial nerves emerge from the inferior surface of the brain and pass through foramina of the skull to innervate structures in the head, neck, and visceral organs of the torso.
   B. The names of the cranial nerves indicate their primary function or the general distribution of their fibers.
   C. Most of the cranial nerves are mixed, some consist of sensory fibers only, and others are primarily motor.
   D. Some of the cranial nerve fibers are somatic, and others are visceral.
   E. A test for cranial-nerve dysfunction is clinically important in a neurological examination.

III. Spinal Nerves
   A. Each of the thirty-one pairs of spinal nerves is formed by the union of a dorsal and ventral spinal root, which emerges from the spinal cord through an intervertebral foramen to innervate a body dermatome.
   B. The spinal nerves are grouped according to the levels of the spinal column from which they arise, and they are numbered in sequence.
   C. Each spinal nerve is a mixed nerve consisting of a dorsal root of sensory fibers and a ventral root of motor fibers.
   D. Just beyond its intervertebral foramen, each spinal nerve divides into several branches.

IV. Nerve Plexuses
   A. Except in the thoracic nerves T2–T12, the ventral rami of the spinal nerves combine and then split again as networks of nerves referred to as plexuses. There are

four plexuses of spinal nerves: the cervical, the brachial, the lumbar, and the sacral. Nerves emerge from the plexuses and are named according to the structures they innervate or the general course they take.

B. The cervical plexus is formed by the ventral rami of C1–C4 and by a portion of C5.

C. The brachial plexus is formed by the ventral rami of C5–T1 and by some fibers from C4 and T2.
   1. The brachial plexus is divided into roots, trunks, divisions, and cords.
   2. The axillary, radial, musculocutaneous, ulnar, and median are the principal nerves arising from the brachial plexus.

D. The lumbar plexus is formed by the ventral rami of L1–L4 and by some fibers from T12.
   1. The lumbar plexus is divided into roots and divisions.
   2. The femoral and obturator are the principal nerves arising from the lumbar plexus.

E. The sacral plexus is formed by the ventral rami of L4, L5, and S1–S4.
   1. The sacral plexus is divided into roots and divisions.
   2. The sciatic nerve, composed of the common peroneal and tibial nerves, arises from the sacral plexus.

V. Reflex Arcs and Reflexes

A. The conduction pathway of a reflex arc consists of a receptor, a sensory neuron, a motor neuron and its innervation in the PNS, and an association neuron in the CNS. The reflex arc provides mechanisms for a rapid, automatic response to a potentially threatening stimulus.

B. A reflex arc is the simplest type of nerve pathway.

C. Visceral reflexes cause smooth or cardiac muscle to contract or glands to secrete.

D. Somatic reflexes cause skeletal muscles to contract.
   1. The stretch reflex is a monosynaptic reflex arc.
   2. The flexor reflex is a polysynaptic reflex arc.

Review Activities

Objective Questions

1. Which of the following is a *false* statement concerning the peripheral nervous system?
   (a) It consists of cranial and spinal nerves only.
   (b) It contains components of the autonomic nervous system.
   (c) Sensory receptors, neurons, nerves, ganglia, and plexuses are all part of the PNS.
2. An inability to look cross-eyed would most likely indicate a problem with which cranial nerve? The
   (a) optic. (c) abducens.
   (b) oculomotor. (d) facial.
3. Which cranial nerve innervates the muscle that raises the upper eyelid? The
   (a) trochlear. (c) abducens.
   (b) oculomotor. (d) facial.
4. The inability to walk a straight line may indicate damage to which cranial nerve? The
   (a) trigeminal.
   (b) facial.
   (c) vestibulocochlear.
   (d) vagus.
5. Which cranial nerve passes through the stylomastoid foramen? The
   (a) facial.
   (b) glossopharyngeal.
   (c) vagus.
   (d) hypoglossal.
6. Which of the following cranial nerves does not contain parasympathetic fibers? The
   (a) oculomotor. (c) vagus.
   (b) accessory. (d) facial.
7. Which of the following is not a spinal nerve plexus? The
   (a) cervical. (d) lumbar.
   (b) brachial. (e) sacral.
   (c) thoracic.
8. Roots, trunks, divisions, and cords are characteristic of the
   (a) sacral plexus.
   (b) thoracic plexus.
   (c) lumbar plexus.
   (d) brachial plexus.
9. Which of the following nerve-plexus association is incorrect?
   (a) median—sacral
   (b) phrenic—cervical
   (c) axillary—brachial
   (d) femoral—lumbar
10. Extending the leg when the patellar tendon is tapped is an example of a(n)
   (a) visceral reflex.
   (b) flexor reflex.
   (c) ipsilateral reflex.
   (d) crossed extensor reflex.

Essay Questions

1. Explain the structural and functional relationship between the central nervous system, the autonomic nervous system, and the peripheral nervous system.
2. List the cranial nerves, and describe the major function(s) of each. How is each cranial nerve tested for dysfunction?
3. Describe the structure of a spinal nerve.
4. List the roots of each of the spinal plexuses, and describe where each is located and the nerves which originate from them.
5. What is a reflex arc? Explain how reflexes are important in maintaining body homeostasis.
6. Distinguish between monosynaptic, polysynaptic, ipsilateral, stretch, and flexor reflexes.

CHAPTER 13

Autonomic Nervous System

Outline and Concepts

Clinical Case Study

Introduction to the Autonomic Nervous System

The action of effectors, muscles, and glands is controlled to a large extent by motor neuron impulses. Skeletal muscles, which are the voluntary effectors, are regulated by somatic motor impulses. The involuntary effectors, including smooth muscle, cardiac muscle, and glands, are regulated by autonomic motor impulses through the autonomic nervous system.

Organization of the Autonomic Nervous System
Visceral Effector Organs

Structure of the Autonomic Nervous System

The sympathetic and parasympathetic divisions of the autonomic nervous system consist of preganglionic neurons that originate in the CNS and postganglionic neurons that originate outside of the CNS in ganglia. The specific origin of the preganglionic neurons and the location of the ganglia, however, are different in the two subdivisions of the autonomic nervous system.

Sympathetic (Thoracolumbar) Division
Parasympathetic (Craniosacral) Division

Functions of the Autonomic Nervous System

The actions of the autonomic nervous system, together with the effects of hormones, help to maintain a state of dynamic constancy in the internal environment. The sympathetic division activates the body to "fight or flight" through adrenergic effects; the parasympathetic division often manifests antagonistic actions through cholinergic effects. Homeostasis thus depends, in large part, on the complementary and often antagonistic effects of sympathetic and parasympathetic innervation.

Neurotransmitters of the Autonomic System
Responses to Adrenergic Stimulation
Responses to Cholinergic Stimulation
Organs with Dual Innervation
Organs without Dual Innervation

Control of the Autonomic Nervous System by Higher Brain Centers

Visceral functions are mainly regulated by autonomic reflexes. In most autonomic reflexes, sensory input is directed to brain centers, which in turn regulate the activity of descending pathways to preganglionic autonomic neurons. The neural centers that directly control the activity of autonomic nerves are influenced by higher brain areas as well as by sensory input.

Medulla Oblongata
Hypothalamus
Limbic System, Cerebellum, and Cerebrum

Clinical Considerations

Autonomic Dysreflexia

Clinical Case Study Answer

Chapter Summary

Review Activities

A sixty-seven-year-old female with unresectable lung cancer is seen for a routine checkup at a tumor clinic. Her only trouble since her diagnosis six months prior is occasional coughing, slight shortness of breath, and a mild but nagging pain in the right side of her chest. She tells her doctor that recently the right side of her face has felt funny. These symptoms, upon closer questioning, include lack of perspiration on the right side of her face during her regular morning walk. Examination of the patient reveals a right pupil that is much smaller than the left, along with a drooping of the right eyelid. Recognizing the patient's peculiar face and eye findings as Horner's Syndrome, the doctor turns to a previous chest CT scan to look for the cause. He knowlingly nods his head as he examines the copy of the scan, noting that the lung tumor, located in the medial aspect of the right upper lobe, invades the mediastinum and tracts upward almost into the base of the neck.

What area or general structure(s) within the autonomic nervous system has the tumor invaded and compromised? Explain, using autonomic nervous system terminology, the reason for each of the three findings (lack of right facial sweating, pupillary constriction, and eyelid droop) including a description of neural pathways and end organs. ▪

Introduction to the Autonomic Nervous System

The action of effectors, muscles, and glands is controlled to a large extent by motor neuron impulses. Skeletal muscles, which are the voluntary effectors, are regulated by somatic motor impulses. The involuntary effectors, including smooth muscle, cardiac muscle, and glands, are regulated by autonomic motor impulses through the autonomic nervous system.

Objective 1. Define the terms *preganglionic neuron* and *postganglionic neuron,* and explain the difference between the efferent pathways of the somatic motor and autonomic motor systems.

Objective 2. Describe how the neural regulation of visceral effector organs can differ from the neural regulation of skeletal muscles.

Objective 3. Compare the structure and regulation of single-unit smooth muscle with that of multi-unit smooth muscle.

Organization of the Autonomic Nervous System

The autonomic portion of the nervous system is concerned with maintaining homeostasis within the body by increasing or decreasing the activity of various organs in response to changing physiological conditions. Although the autonomic nervous system (ANS) is composed of portions of both the central nervous system and peripheral nervous system, it functions independently and without the conscious control of the person.

Autonomic motor nerves innervate organs whose functions are not usually under voluntary control. The effectors that respond to autonomic regulation include **cardiac muscle** (the

Table 13.1 Comparisons of the somatic motor system with the autonomic motor system

Feature	Somatic motor	Autonomic motor
Effector organs	Skeletal muscles	Cardiac muscle, smooth muscle, and glands
Presence of ganglia	No ganglia outside CNS	Cell bodies of postganglionic autonomic fibers located in paravertebral, prevertebral (collateral), and terminal ganglia
Number of neurons from CNS to effector	One	Two
Type of neuromuscular junction	Specialized motor end plate	No specialization of postsynaptic membrane; all areas of smooth muscle cells contain receptor proteins for neurotransmitters
Effect of nerve impulse on muscle	Excitatory only	Either excitatory or inhibitory
Type of nerve fibers	Fast-conducting, thick (9–13 μm), and myelinated	Slow conducting; preganglionic fibers, lightly myelinated but thin (3 μm); postganglionic fibers, unmyelinated and very thin (about 1.0 μm)
Effect of denervation	Flaccid paralysis and atrophy	Much muscle tone and function remains; target cells show denervation hypersensitivity

From Stuart Ira Fox, *Human Physiology,* 3d ed. Copyright © 1990 Wm. C. Brown Publishers, Dubuque, Iowa. All Rights Reserved. Reprinted by permission.

heart), **smooth** (visceral) **muscles,** and glands. These effectors are part of the organs of the *viscera,* of blood vessels, and of specialized structures within other organs. The involuntary effects of autonomic innervation contrast with the voluntary control of skeletal muscles by way of somatic motor innervation.

Unlike somatic motor neurons, which conduct impulses along a single axon from the spinal cord to the neuromuscular junction, the autonomic motor pathway involves two neurons in the efferent flow of impulses (table 13.1).

The first of these autonomic motor neurons has its cell body in the gray matter of the brain or spinal cord. The axon of this neuron does not directly innervate the effector organ but instead synapses with a second neuron within an *autonomic ganglion* (a ganglion is a collection of neuron cell bodies outside the CNS). The first neuron is thus called a **preganglionic,** or **presynaptic, neuron.** The second neuron in this pathway,

viscus, pl. viscera: L. *viscera,* internal organs
autonomic: Gk. *auto,* self; *nomos,* law
ganglion: Gk. *ganglion,* a swelling or knot

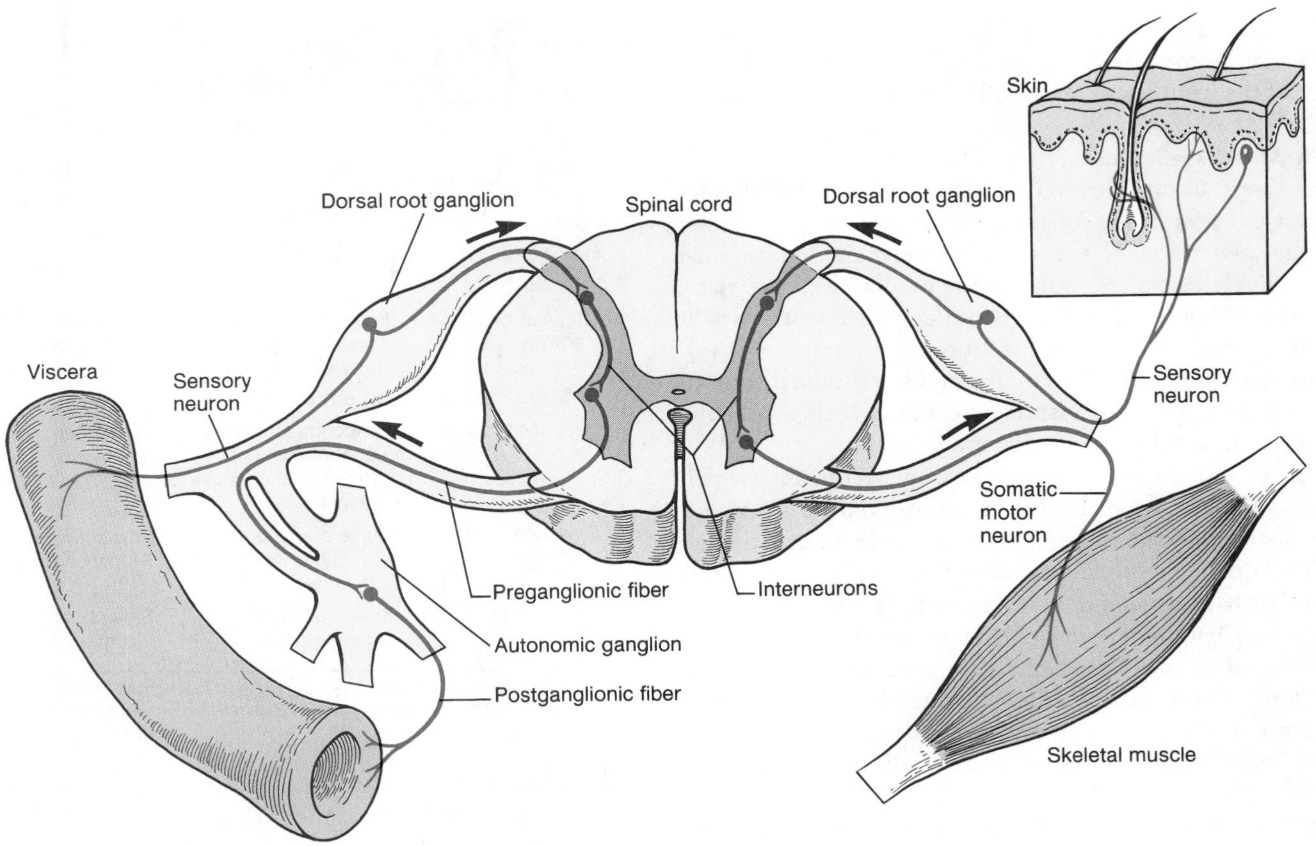

Figure 13.1 A comparison of a somatic motor reflex with an autonomic motor reflex. Although, for the sake of clarity, each is shown on different sides of the spinal cord, both visceral and somatic sensory neurons are found bilaterally in the dorsal roots of spinal nerves, and somatic and autonomic motor neurons are found bilaterally in the ventral roots.

called a **postganglionic,** or **postsynaptic, neuron,** has an axon that leaves the autonomic ganglion and synapses with the cells of an effector organ (fig. 13.1).

Preganglionic autonomic neurons originate in the midbrain and hindbrain and from the upper thoracic to the fourth sacral portions of the spinal cord with the exceptions of the area between L3 and S1. Autonomic ganglia are located in the head, neck, and abdomen; chains of autonomic ganglia also parallel the right and left sides of the spinal cord. The origin of the preganglionic neurons and the location of the autonomic ganglia help to differentiate the **sympathetic** and **parasympathetic** subdivisions of the autonomic system, which will be discussed in later sections of this chapter.

Visceral Effector Organs

Unlike skeletal muscles, which enter a state of flaccid paralysis when their motor nerves are severed, the involuntary effectors are somewhat independent of their innervation. Smooth muscles maintain a resting tone (tension) in the absence of nerve stimulation. Damage to an autonomic nerve, in fact, makes its target muscle more sensitive than normal to stimulating agents.

In addition to their intrinsic (built-in) muscle tone, many smooth muscles and the cardiac muscle contract rhythmically, even in the absence of nerve stimulation, in response to electrical waves of depolarization initiated by the muscles themselves. Autonomic nerves also maintain a resting tone, in the sense that they maintain a baseline firing rate that can be either increased or decreased. Changes in tonic neural activity produce changes in the intrinsic activity of the effector organ. A decrease in the excitatory input to the heart, for example, will slow its rate of beat.

Cardiac Muscle Like skeletal muscle fibers, cardiac muscle fibers are striated. The long, fibrous skeletal muscle fibers, however, are structurally and functionally separated from each other, whereas the cardiac fibers are short, branched, and interconnected by **intercalated** *(in-ter'kah-lat-ed)* **discs.**

Electrical impulses that originate at any point in the mass of cardiac fibers, called the **myocardium** *(mi''o-kar'de-um),* can spread to all cells in the mass that are joined by intercalated discs. Because all of the cells in the myocardium are electrically joined, the myocardium behaves as a single functional unit, or a *functional syncytium (sin-sish'e-um).* Unlike skeletal muscles, which can produce graded contractions with a strength that depends on the number of cells stimulated, the heart contracts with an *all-or-none contraction.*

Unlike skeletal muscles, which require external stimulation by somatic motor nerves before they can produce action potentials and contract, cardiac muscle is able to produce action potentials automatically. Cardiac action potentials normally originate in a specialized group of cells called the *pacemaker.* However, the rate of this spontaneous depolarization and, thus, the rate of the heartbeat is regulated by autonomic innervation.

Smooth Muscles Smooth (visceral) muscles are arranged in circular layers around the walls of blood vessels, bronchioles (small air passages in the lungs), and in the sphincter muscles of the digestive tract. However, both circular and longitudinal smooth muscle layers are found in the tubular digestive tract, the ureters (which transport urine), the ductus deferentia (which transports sperm), and the uterine tubes (which transport ova). The alternate contraction of circular and longitudinal smooth muscle layers produces **peristaltic waves,** which propel the contents of these tubes in one direction.

Smooth muscle fibers do not contain sarcomeres (which produce striations in skeletal and cardiac muscle). Smooth muscle fibers do, however, contain a great amount of actin and some myosin, which produces a ratio of thin-to-thick filaments of about 16:1 (in striated muscles the ratio is 2:1).

The long length of myosin filaments and the fact that they are not organized into sarcomeres may be advantageous for the function of smooth muscles. Smooth muscles must be able to exert tension even when greatly stretched—in the urinary bladder, for example, the smooth muscle cells may be stretched up to two and a half times their resting length. Skeletal muscles, in contrast, lose their ability to contract when the sarcomeres are stretched to the point where actin and myosin no longer overlap.

Single-Unit and Multi-Unit Smooth Muscles Smooth muscles are often grouped into two functional categories: **single-unit** and **multi-unit.** Single-unit smooth muscles have numerous gap junctions (electrical synapses) between adjacent cells that weld them together electrically; they thus behave as a single unit. Multi-unit smooth muscles have few, if any, gap junctions; the individual cells must thus be stimulated separately by impulses through nerve endings. This is similar to the numerous motor units required for the control of skeletal muscles.

Single-unit smooth muscles display *pacemaker* activity in which certain cells stimulate others in the mass. Single-unit smooth muscles also display intrinsic, or *myogenic (mi''o-jen'ik),* electrical activity and contraction in response to stretch. For example, the stretch induced by an increase in the luminal contents of a small artery or a section of the gastrointestinal tract can stimulate myogenic contraction. Such contraction does not require stimulation by autonomic nerves. Contraction of multi-unit smooth muscles, in contrast, requires nerve stimulation. Comparisons of single-unit and multi-unit smooth muscles are given in table 13.2.

Autonomic Innervation of Smooth Muscles There are significant differences between the neural control of skeletal muscles and that of smooth muscles. A skeletal muscle fiber has only one junction with a somatic nerve fiber, and the receptors for the neurotransmitter are localized at the neuromuscular junction in the membrane of the skeletal muscle fiber. In contrast, the entire surface of smooth muscle fibers contains transmitter receptor proteins. Neurotransmitter molecules are released along a stretch of an autonomic nerve fiber that is located some distance from the smooth muscle fibers. The regions of the autonomic fiber that release transmitters appear as bulges, or *varicosities,* and the neurotransmitters released from these varicosities stimulate a number of smooth muscle fibers.

Table 13.2 Some comparisons between single-unit and multi-unit smooth muscles

	Single-unit muscle	Multi-unit muscle
Location	Gastrointestinal tract; uterus; ureter; small arteries (arterioles)	Arrector pili muscles of hair follicles; ciliary muscle (attached to lens); iris; ductus deferens; large arteries
Origin of electrical activity	Spontaneous activity by pacemakers (myogenic)	Not spontaneously active; potentials are neurogenic
Type of potentials	Action potentials	Graded depolarizations
Response to stretch	Responds to stretch by contraction; not dependent on nerve stimulation	No inherent response to stretch
Presence of gap junctions	Numerous gap junctions join all cells together electrically	Few (if any) gap junctions
Type of contraction	Slow, sustained contractions	Slow, sustained contractions

1. Describe how the regulation of the contraction of cardiac and smooth muscle fibers differs from that of skeletal muscle fibers and how these muscles are affected by the experimental removal of their innervation.
2. Define the terms *preganglionic* and *postganglionic neurons* in the ANS, and use a diagram to illustrate the difference in efferent outflow between somatic and autonomic nerves.
3. Compare the structure of single-unit and multi-unit smooth muscles, and explain how these two types differ in the way they are regulated by autonomic nerves.

Structure of the Autonomic Nervous System

The sympathetic and parasympathetic divisions of the autonomic nervous system consist of preganglionic neurons that originate in the CNS and postganglionic neurons that originate outside of the CNS in ganglia. The specific origin of the preganglionic neurons and the location of the ganglia, however, are different in the two subdivisions of the autonomic nervous system.

myogenic: Gk. *mys,* muscle; *genesis,* origin

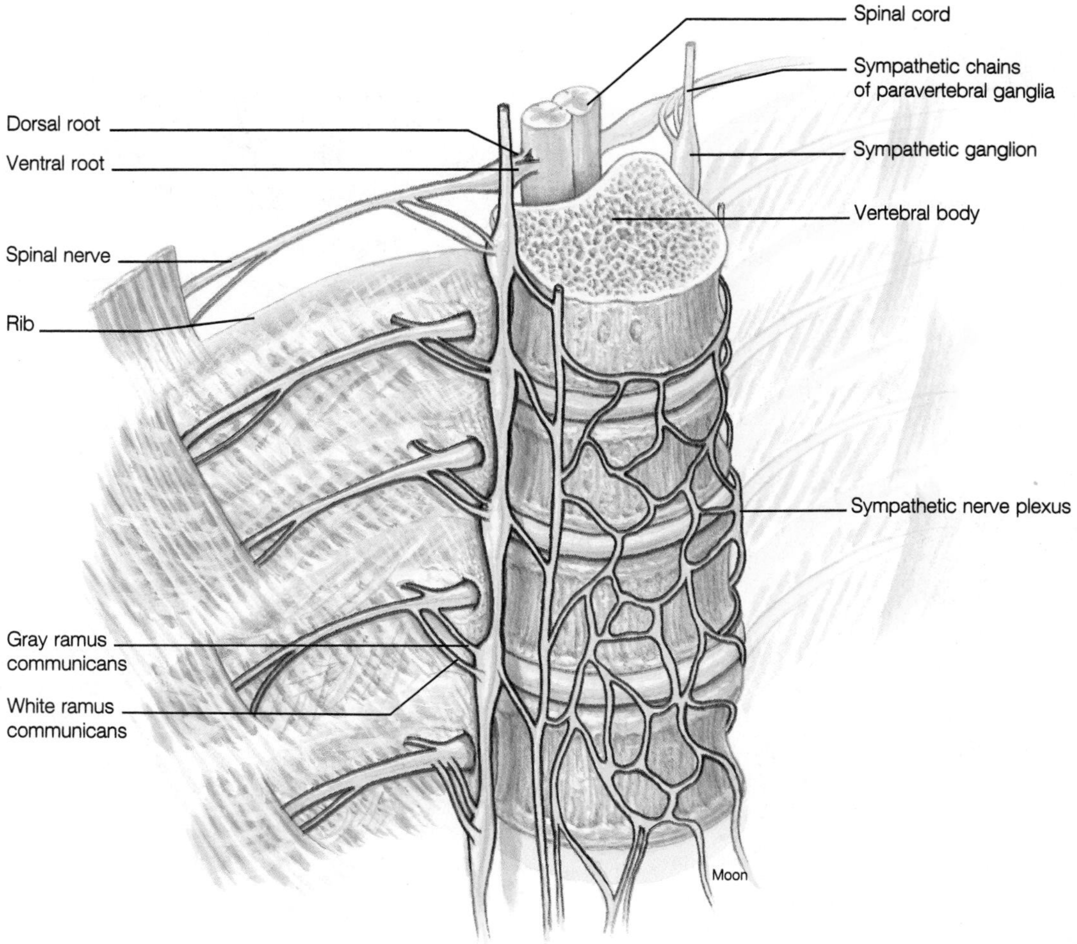

Figure 13.2 The sympathetic chain of paravertebral ganglia, showing its relationship to the vertebral column and the spinal cord.

Objective 4. Describe the origin of preganglionic sympathetic neurons and the location of sympathetic ganglia.
Objective 5. Explain the relationship between the sympathetic system and the adrenal medulla.
Objective 6. Describe the origin of the preganglionic parasympathetic neurons and the location of the parasympathetic ganglia.
Objective 7. Describe the distribution of the vagus nerve, and explain its significance within the parasympathetic division.

Sympathetic (Thoracolumbar) Division

The **sympathetic division** is also called the *thoracolumbar division* of the ANS because its preganglionic neurons exit the vertebral column from the first thoracic (T1) to the second lumbar (L2) levels. Most sympathetic neurons, however, separate from the somatic motor neurons and synapse with postganglionic neurons within a chain of sympathetic ganglia, or **paravertebral ganglia,** located on both sides of the vertebral column (fig. 13.2).

Since the preganglionic sympathetic neurons are myelinated and thus appear white, they are called **white rami communicantes** (fig. 13.3). Some of these preganglionic sympathetic neurons synapse with postganglionic neurons located at their same level in the chain of sympathetic ganglia. Other preganglionic neurons travel up or down within the sympathetic chain before synapsing with postganglionic neurons. Since the postganglionic sympathetic neurons are unmyelinated and thus appear gray, they form the **gray rami communicantes.** Postganglionic axons in the gray rami communicantes extend directly back to the ventral roots of the spinal nerves and travel distally within the spinal nerves to innervate their effector organs.

Within the chain of paravertebral ganglia, *divergence* is seen when preganglionic neurons branch to synapse with many postganglionic neurons located at different levels in the chain. *Convergence* is seen when a given postganglionic neuron receives synaptic input from a large number of preganglionic neurons. The divergence of impulses from the spinal cord to the ganglia and the convergence of impulses within the ganglia usually results in the *mass activation* of almost all of the postganglionic neurons. This explains why the sympathetic system is usually activated as a unit and affects all of its effector organs at the same time.

Many preganglionic neurons that exit the spinal cord in the upper thoracic level travel through the sympathetic chain into the neck, where they synapse in cervical sympathetic ganglia (fig. 13.4). Postganglionic neurons from here innervate the smooth muscles and glands of the head and neck.

ramus: L. *ramus,* a branch

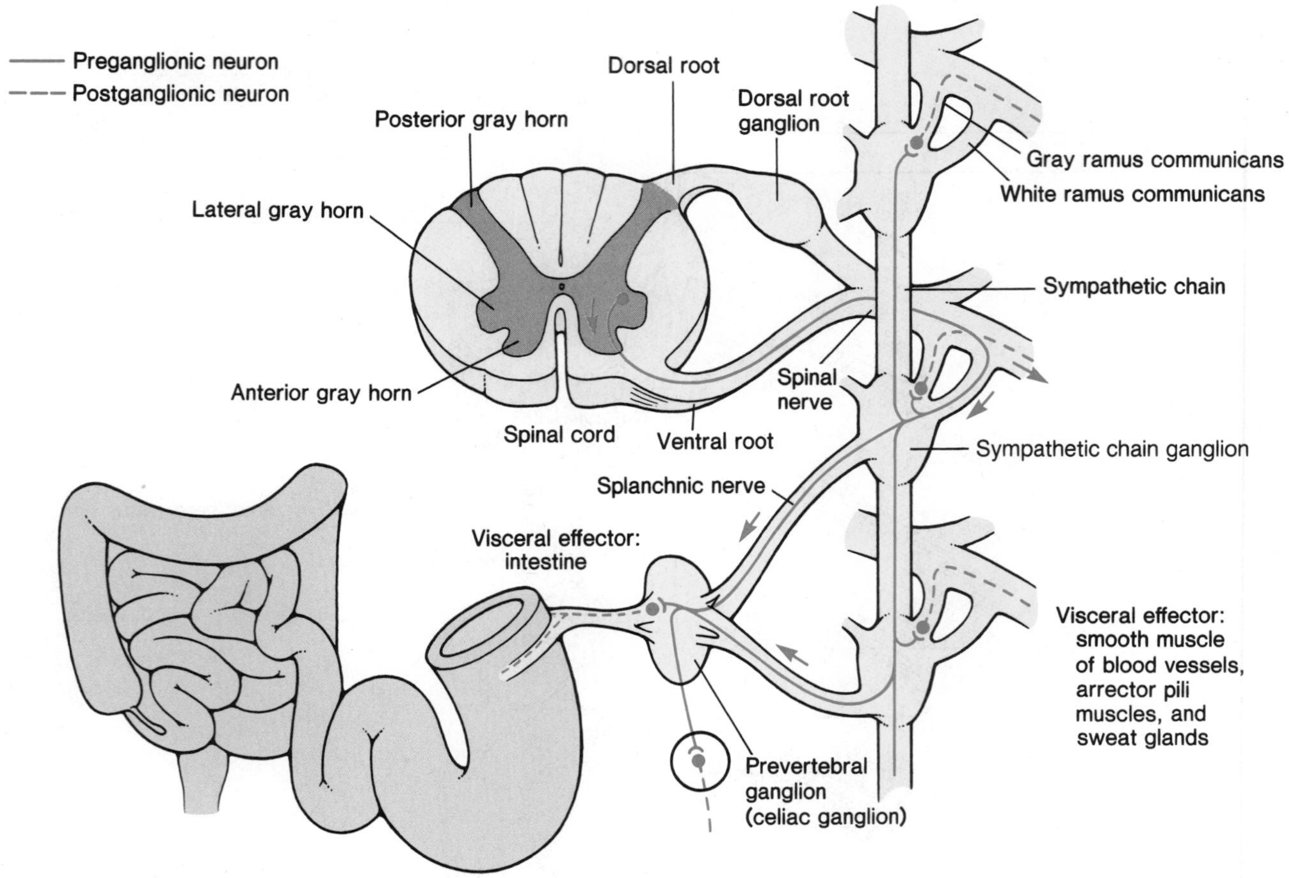

Figure 13.3 Sympathetic chain ganglia, the sympathetic chain, and rami communicantes of the sympathetic division of the ANS. (Solid lines = preganglionic fibers; dashed lines = postganglionic fibers.)

Collateral Ganglia Many preganglionic neurons that exit the spinal cord below the level of the diaphragm pass through the sympathetic chain of ganglia without synapsing. Beyond the sympathetic chain, these preganglionic neurons form **splanchnic** *(splank'nik)* **nerves** (fig. 13.3). Preganglionic neurons in the splanchnic nerves synapse in **collateral ganglia,** which are also called **prevertebral ganglia.** These include the **celiac** *(se'le-ak),* **superior mesenteric** *(mes''en-ter'ik),* and **inferior mesenteric ganglia** (figs. 13.5, 13.6).

The *greater splanchnic nerve* arises from preganglionic sympathetic neurons T4–T9 and synapses in the celiac ganglion. These neurons contribute to the *celiac (solar) plexus.* Postganglionic neurons from the celiac ganglion innervate the stomach, spleen, liver, small intestine, and kidneys. The *lesser splanchnic nerve* terminates in the superior mesenteric ganglion. Postganglionic neurons from here innervate the small intestine and colon. The *lumbar splanchnic nerve* synapses in the inferior mesenteric ganglion, and the postganglionic neurons innervate the distal colon and rectum, urinary bladder, and genital organs.

Adrenal Glands The paired adrenal glands are located above each kidney. Each adrenal is composed of two parts: an outer **cortex** and an inner **medulla.** These two parts are really two functionally different glands with different embryonic origins, different hormones, and different regulatory mechanisms. The adrenal cortex secretes steroid hormones; the adrenal medulla secretes the hormone **epinephrine** (adrenalin) and, to a lesser degree, **norepinephrine,** when it is stimulated by the sympathetic system.

The adrenal medulla is a modified sympathetic ganglion whose cells are derived from postganglionic sympathetic neurons. The cells of the adrenal medulla are innervated by preganglionic sympathetic neurons originating in the thoracic level of the spinal cord; they secrete epinephrine into the blood in response to sympathetic stimulation. The effects of epinephrine are complementary to those of the neurotransmitter norepinephrine, which is released from postganglionic sympathetic nerve endings. For this reason and because the adrenal medulla is stimulated as part of the mass activation of the sympathetic system, the two are often grouped together as the **sympathoadrenal system.**

Parasympathetic (Craniosacral) Division

The **parasympathetic division** is also known as the *craniosacral division* of the autonomic system. This is because its preganglionic neurons originate in the brain (specifically, the midbrain and the medulla oblongata of the brain stem) and in the second through fourth sacral levels of the spinal cord. These preganglionic parasympathetic neurons synapse in ganglia that are located next to—or actually within—the organs inner-

splanchnic: Gk. *splanchno,* relating to viscera
adrenal: L. *ad,* to; *renes,* kidney
cortex: L. *cortex,* bark
medulla: L. *medulla,* marrow

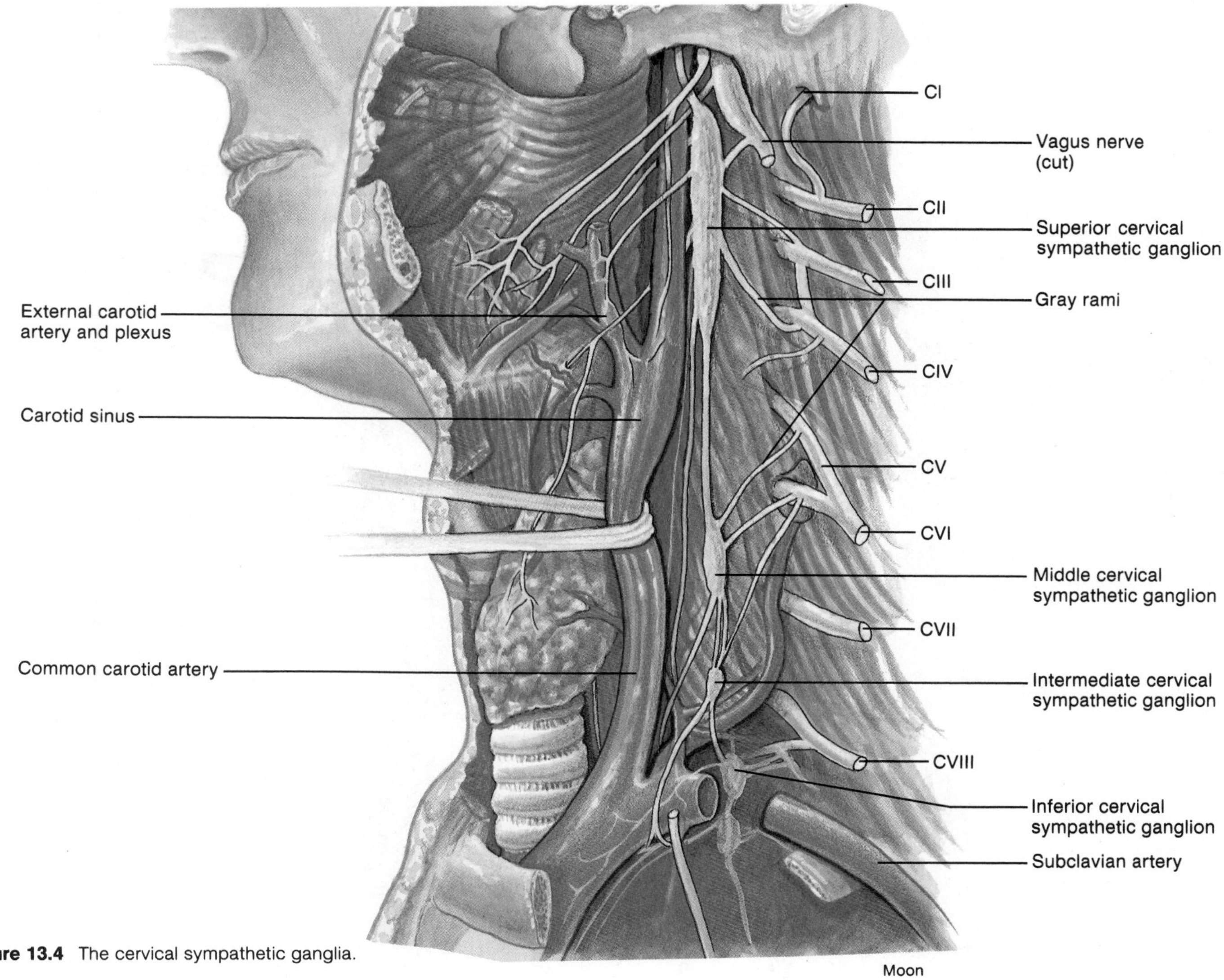

Figure 13.4 The cervical sympathetic ganglia.

vated. These parasympathetic ganglia, which are called **terminal ganglia,** supply the postganglionic neurons that synapse with the effector cells. Tables 13.3 and 13.4 show the comparative structures of the sympathetic and parasympathetic divisions. It should be noted that, unlike sympathetic neurons, most parasympathetic neurons do not travel within spinal nerves. Cutaneous effectors (blood vessels, sweat glands, and arrector pili muscles) and blood vessels in skeletal muscles thus receive sympathetic but not parasympathetic innervation.

Four of the twelve pairs of cranial nerves contain preganglionic parasympathetic neurons. These are the oculomotor (III), facial (VII), glossopharyngeal (IX), and vagus (X) nerves. Parasympathetic neurons within the first three of these cranial nerves synapse in ganglia located in the head; neurons in the vagus nerve synapse in terminal ganglia located in many regions of the body.

The oculomotor cranial nerve contains somatic motor and parasympathetic neurons that originate in the oculomotor nuclei of the midbrain. These parasympathetic neurons synapse in the **ciliary ganglion,** whose postganglionic neurons innervate the ciliary muscle and constrictor neurons in the iris of the eye. Preganglionic neurons that originate in the medulla oblongata travel in the facial nerve to the **pterygopalatine** *(ter''ĭ-go-pal'ah-tēn)* **ganglion,** also called the sphenopalatine ganglion, which sends postganglionic neurons to the nasal mucosa, pharynx, palate, and lacrimal glands. Another group of neurons in the facial nerve terminate in the **submandibular ganglion,** which sends postganglionic neurons to the submandibular and sublingual salivary glands. Preganglionic neurons of the glossopharyngeal nerve synapse in the **otic ganglion,** which sends postganglionic neurons to innervate the parotid salivary gland.

Other nuclei in the medulla oblongata contribute preganglionic neurons to the very long vagus cranial nerves, which provide the most extensive parasympathetic innervation in the body. As the paired vagus nerves pass through the thorax, they contribute to the *pulmonary plexuses* within the mediastinum. Branches of the pulmonary plexuses accompany blood vessels and bronchi into the lungs. Below the pulmonary plexuses, the vagus nerves merge to form the *esophageal (ĕ-sof''ah-je'al) plexuses.*

At the lower end of the esophagus, vagal neurons collect to form an **anterior** and **posterior vagal** *(va'gal)* **trunk,** each composed of neurons from both vagus nerves. The vagal trunks enter the abdominal cavity through the esophageal opening in the diaphragm. Neurons from the vagal trunks innervate the stomach on the anterior and posterior sides. Branches of the

vagus: L. *vagus,* wandering

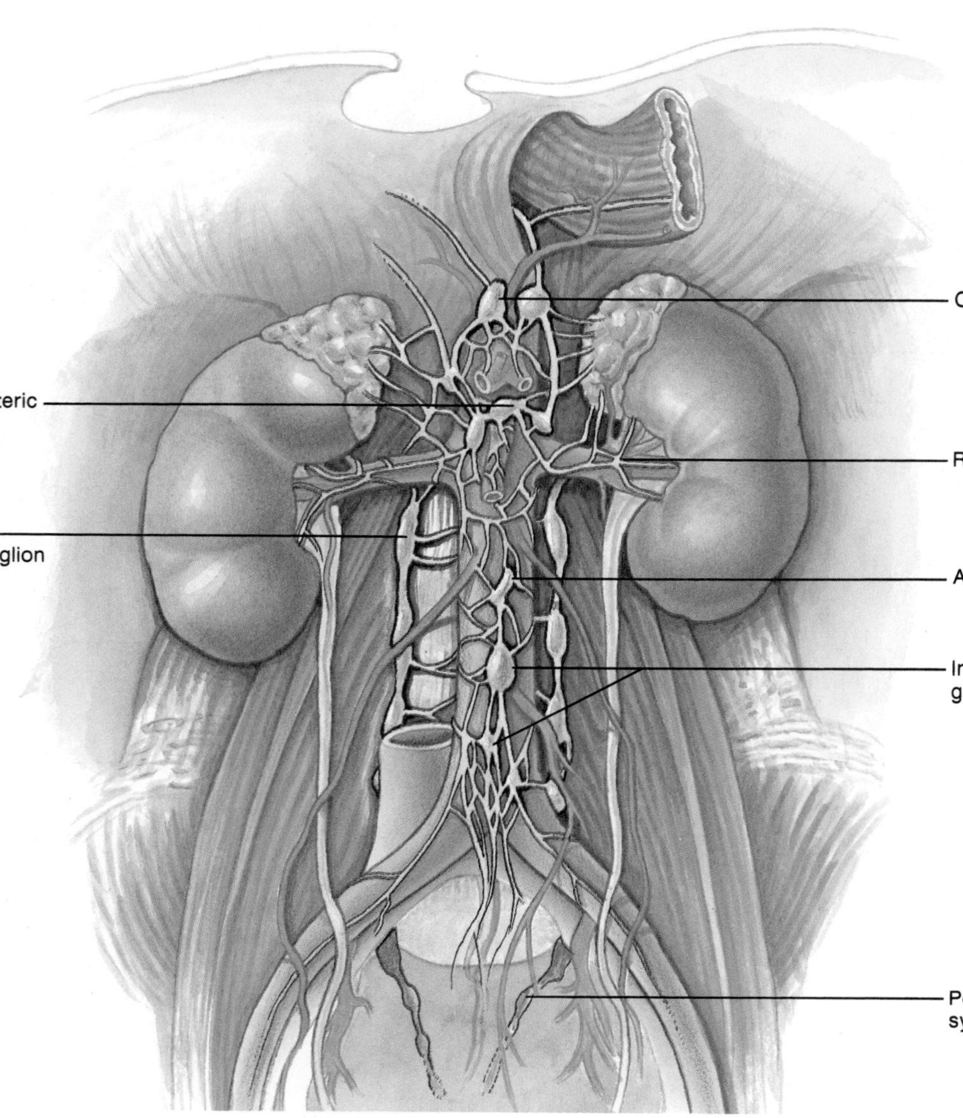

Figure 13.5 The collateral sympathetic ganglia: the celiac and the superior and inferior mesenteric ganglia.

vagus nerves within the abdominal cavity also contribute to the *celiac plexus* and *plexuses of the abdominal aorta.*

The preganglionic neurons in the vagus synapse with postganglionic neurons that are actually located *within* the innervated organs. These preganglionic neurons are thus quite long, and provide parasympathetic innervation to the heart, lungs, esophagus, stomach, pancreas, liver, small intestine, and upper half of the large intestine. Postganglionic parasympathetic neurons arise from terminal ganglia within these organs and synapse with effector cells.

Preganglionic neurons from the sacral levels of the spinal cord provide parasympathetic innervation to the lower half of the large intestine, rectum, and to the urinary and reproductive systems. These neurons, like those of the vagus, synapse with terminal ganglia located within the effector organs.

Parasympathetic nerves to the visceral organs thus consist of preganglionic neurons, whereas sympathetic nerves to these organs contain postganglionic neurons. A composite view of the sympathetic and parasympathetic divisions is provided in figure 13.6, and these comparisons are summarized in table 13.5.

1. Compare the origin of preganglionic sympathetic and parasympathetic neurons and the location of sympathetic and parasympathetic ganglia.
2. Using a simple line drawing, illustrate the sympathetic pathway from the spinal cord to the heart. Label the preganglionic neuron, postganglionic neuron, and the ganglion.
3. Use a simple line diagram to show the parasympathetic innervation of the heart. Label the preganglionic and postganglionic neurons, the nerve involved, and the terminal ganglion.
4. Describe the distribution of the vagus nerve and its functional significance.
5. Define the terms *white rami communicantes* and *gray rami communicantes,* and explain why blood vessels in the skin and skeletal muscles receive sympathetic but not parasympathetic innervation.
6. Describe the structure of the adrenal gland, and explain its relationship to the sympathetic division of the ANS.

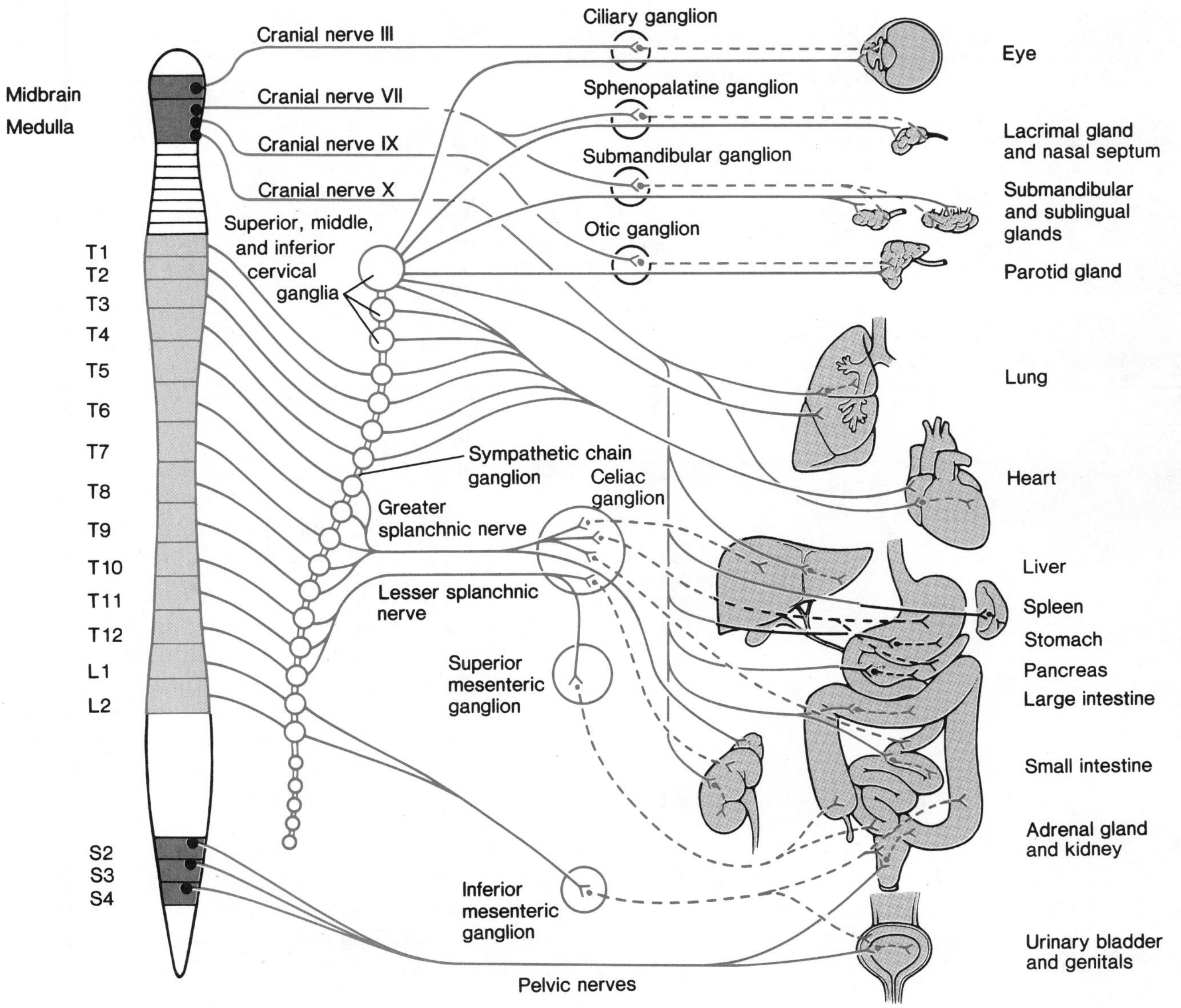

Figure 13.6 A simplified version of the autonomic nervous system. The preganglionic fibers of the sympathetic division arise from the thoracic and lumbar regions of the spinal cord, whereas those of the parasympathetic division arise from the brain and sacral region of the cord.

Table 13.3 The sympathetic (thoracolumbar) division

Parts of body innervated	Spinal origin of preganglionic neurons	Origin of postganglionic neurons
Eye	C-8 and T-1	Cervical ganglia
Head and neck	T-1 to T-4	Cervical ganglia
Heart and lungs	T-1 to T-5	Upper thoracic (paravertebral) ganglia
Upper extremity	T-2 to T-9	Lower cervical and upper thoracic (paravertebral) ganglia
Upper abdominal viscera	T-4 to T-9	Celiac and superior mesenteric (collateral) ganglia
Adrenal	T-10 and T-11	Adrenal medulla
Urinary and reproductive systems	T-12 to L-2	Celiac and inferior mesenteric (collateral) ganglia
Lower extremities	T-9 to L-2	Lumbar and upper sacral (paravertebral) ganglia

Table 13.4 The parasympathetic (craniosacral) division

Effector organs	Origin of preganglionic neurons	Nerve	Origin of postganglionic neurons
Eye (ciliary and iris muscles)	Midbrain (cranial)	Oculomotor (third cranial) nerve	Ciliary ganglion
Lacrimal, mucus, and salivary glands in head	Medulla oblongata (cranial)	Facial (seventh cranial) nerve	Sphenopalatine and submandibular ganglia
Parotid (salivary) gland	Medulla oblongata (cranial)	Glossopharyngeal (ninth cranial) nerve	Otic ganglion
Heart, lungs, gastrointestinal tract, liver, pancreas	Medulla oblongata (cranial)	Vagus (tenth cranial) nerve	Terminal ganglia in or near organ
Lower half of large intestine, rectum, urinary bladder, and reproductive organs	S-2 to S-4 (sacral)	Through pelvic spinal nerves	Terminal ganglia near organs

Table 13.5 Some comparisons between the structure of the sympathetic and parasympathetic divisions

	Sympathetic	Parasympathetic
Origin of preganglionic outflow	Thoracolumbar levels of spinal cord	Midbrain, hindbrain, and sacral levels of spinal cord
Location of ganglia	Chain of paravertebral ganglia and prevertebral (collateral) ganglia	Terminal ganglia in or near effector organs
Distribution of postganglionic neurons	Throughout the body	Mainly limited to the head and the viscera of the chest, abdomen, and pelvis
Divergence of impulses from pre- to postganglionic neurons	Great divergence (one preganglionic may activate twenty postganglionic neurons)	Little divergence (one preganglionic only activates a few postganglionic neurons)
Mass discharge of system as a whole	Yes	Not normally

Functions of the Autonomic Nervous System

The actions of the autonomic nervous system, together with the effects of hormones, help to maintain a state of dynamic constancy in the internal environment. The sympathetic division activates the body to "fight or flight" through adrenergic effects; the parasympathetic division often manifests antagonistic actions through cholinergic effects. Homeostasis thus depends, in large part, on the complementary and often antagonistic effects of sympathetic and parasympathetic innervation.

Objective 8. List the neurotransmitters of the preganglionic and postganglionic neurons of the sympathetic and parasympathetic divisions.

Objective 9. Describe the effects of acetylcholine released by postganglionic parasympathetic neurons.

Objective 10. Explain the antagonistic, complementary, and cooperative effects of sympathetic and parasympathetic innervation.

The sympathetic and parasympathetic divisions of the ANS (fig. 13.6) affect the visceral organs in different ways. Mass activation of the *sympathetic division* prepares the body for intense physical activity in emergencies; the heart rate increases, blood glucose rises, and blood is diverted to the skeletal muscles (away from the visceral organs and skin). These and other effects are listed in table 13.6. The theme of the sympathetic division has been aptly summarized in a phrase: **fight or flight.**

The effects of parasympathetic nerve stimulation are in many ways opposite to the effects of sympathetic stimulation. The parasympathetic division, however, is not normally activated as a whole. Stimulation of separate parasympathetic nerves can result in slowing of the heart, dilation of visceral blood vessels, and an increased activity of the gastrointestinal (GI) tract (table 13.6). The different responses of visceral organs to sympathetic and parasympathetic nerve activity is due to the fact that the postganglionic neurons of these two divisions release different neurotransmitters.

Neurotransmitters of the Autonomic System

The neurotransmitter released by most postganglionic sympathetic neurons is **norepinephrine** *(noradrenalin)*. Transmission at these synapses is thus said to be **adrenergic** *(ad''ren-er'jik)*. There are a few exceptions to this rule: some sympathetic neurons that innervate blood vessels in skeletal muscles, as well as sympathetic neurons to sweat glands, release ACh (are cholinergic).

Acetylcholine *(as''ĕ-til-kō'lēn) (ACh)* is the neurotransmitter of all preganglionic neurons (both sympathetic and parasympathetic). Acetylcholine is also the transmitter released by all parasympathetic postganglionic neurons at their synapses with effector cells (fig. 13.7). Transmission at the autonomic ganglia and at synapses of postganglionic neurons is thus said to be **cholinergic** *(ko''lin-er'jik)*. In other words, a *cholinergic fiber* is a neuron that secretes ACh at the terminal end of its axon.

cholinergic: Gk. *chole,* bile; *ergon,* work

Table 13.6 Effects of autonomic nerve stimulation on various visceral effector organs

Effector organ	Sympathetic effect	Parasympathetic effect
Eye		
Iris (radial muscle)	Dilates pupil	—
Iris (sphincter muscle)	—	Constricts pupil
Ciliary muscle	Relaxes (for far vision)	Contracts (for near vision)
Glands		
Lacrimal (tear)	—	Stimulates secretion
Sweat	Stimulates secretion	—
Salivary	Decreases secretion; saliva becomes thick	Increases secretion; saliva becomes thin Stimulates secretion
Stomach	—	Stimulates secretion
Intestine	—	—
Adrenal medulla	Stimulates secretion of hormones	
Heart		
Rate	Increases	Decreases
Conduction	Increases rate	Decreases rate
Strength	Increases	—
Blood vessels	Mostly constricts; affects all organs	Dilates in a few organs (e.g., penis)
Lungs		
Bronchioles (tubes)	Dilates	Constricts
Mucous glands	Inhibits secretion	Stimulates secretion
Gastrointestinal tract		
Motility	Inhibits movement	Stimulates movement
Sphincters	Stimulates closing	Inhibits closing
Liver	Stimulates hydrolysis of glycogen	—
Adipose (fat) cells	Stimulates hydrolysis of fat	—
Pancreas	Inhibits exocrine secretions	Stimulates exocrine secretions
Spleen	Stimulates contraction	—
Urinary bladder	Helps set muscle tone	Stimulates contraction
Arrector pili muscles	Stimulates erection of hair and goosebumps	—
Uterus	If pregnant: contraction if not pregnant: relaxation	—
Penis	Ejaculation	Erection (due to vasodilation)

Responses to Adrenergic Stimulation

Adrenergic stimulation—by epinephrine in the blood and by norepinephrine released from sympathetic nerve endings—has both excitatory and inhibitory effects. The heart, dilatory muscles of the iris, and the smooth muscles of many blood vessels are stimulated to contract. The smooth muscles of the bronchioles and of some blood vessels, however, are inhibited from contracting; adrenergic chemicals, therefore, cause these structures to dilate.

Responses to Cholinergic Stimulation

Somatic motor neurons, all preganglionic autonomic neurons, and all postganglionic parasympathetic neurons are cholinergic—they use acetylcholine as a neurotransmitter. The cholinergic effects of somatic motor neurons and preganglionic autonomic neurons are always excitatory. The cholinergic effects of postganglionic parasympathetic neurons are usually excitatory, with some notable exceptions; the parasympathetic neurons innervating the heart, for example, cause slowing of the heart rate.

The drug *muscarine (mus'kar-in),* derived from some poisonous mushrooms, blocks the cholinergic effects of parasympathetic nerves in the heart, smooth muscles, and glands. This drug, however, does not block the effects of somatic motor nerves or the cholinergic transmission in autonomic ganglia. The acetylcholine receptors in visceral organs are, therefore, said to promote the *muscarinic effects* of ACh. These effects are identical to those produced by parasympathetic nerve stimulation and are believed to be mediated by different receptors than those that mediate other cholinergic effects.

The muscarinic effects of ACh are specifically inhibited by the drug *atropine,* derived from the deadly nightshade plant *(Atropa belladonna).* Indeed, extracts of this plant were used by women during the middle ages to dilate their pupils (atropine inhibits parasympathetic stimulation of the iris). This was done to enhance their beauty (belladonna—beautiful woman). Atropine is used clinically today to dilate pupils during eye examinations, to dry mucous membranes of the respiratory tract prior to general anesthesia, and to inhibit spasmodic contractions of the lower digestive tract.

Organs with Dual Innervation

Many organs receive a dual innervation—they are innervated by both sympathetic and parasympathetic neurons. When this occurs, the effects of these two divisions may be antagonistic, complementary, or cooperative.

Antagonistic Effects The effects of sympathetic and parasympathetic innervation on the pacemaker region of the heart is the best example of the antagonism of these two systems. In this case sympathetic and parasympathetic neurons innervate the same cells. Adrenergic stimulation from sympathetic neurons increases the heart rate, and cholinergic stimulation from parasympathetic neurons inhibits the pacemaker cells and, thus, decreases the heart rate. Antagonism is also seen in the GI tract, where impulses through sympathetic nerves inhibit and impulses through parasympathetic nerves stimulate intestinal movements and secretions.

The effects of sympathetic and parasympathetic stimulation on the diameter of the pupil of the eye is analogous to the reciprocal innervation of flexor and extensor skeletal muscles by somatic motor neurons. This is because the iris contains antagonistic muscle layers. Contraction of the radial muscles, which is stimulated by impulses through sympathetic nerves, causes dilation; contraction of the circular muscles, which are innervated by parasympathetic nerve endings, causes constriction of the pupils (fig. 13.8).

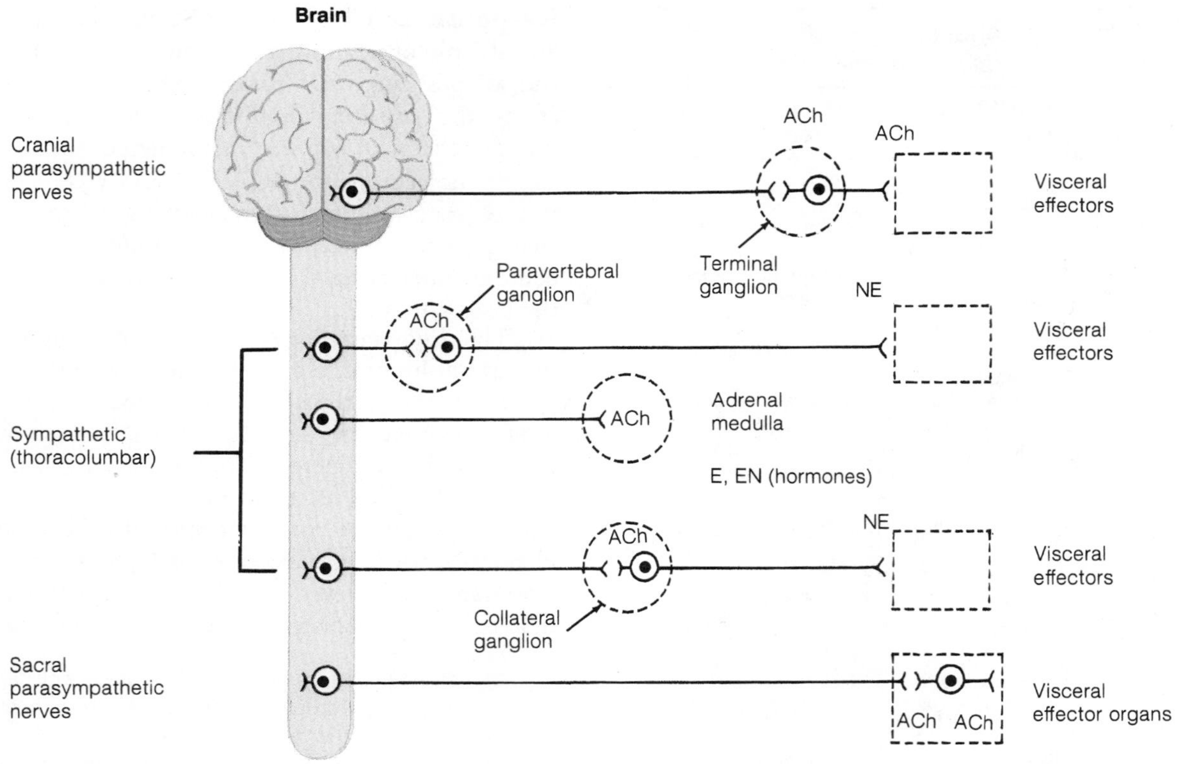

Figure 13.7 Neurotransmitters of the autonomic motor system. ACh = acetylcholine; NE = norepinephrine; E = epinephrine. Those nerves that release ACh are called cholinergic; those nerves that release NE are called adrenergic. The adrenal medulla secretes both epinephrine (85%) and norepinephrine (15%) as hormones into the blood.

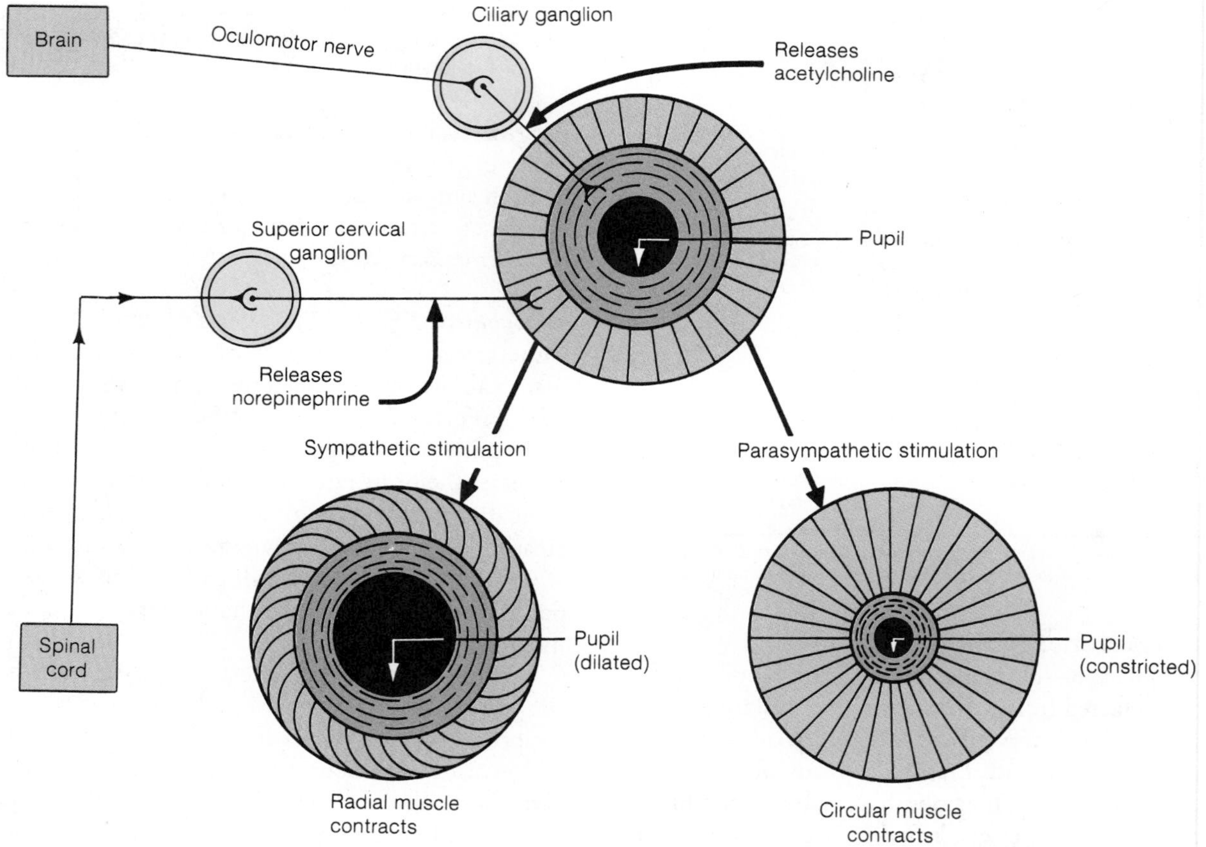

Figure 13.8 Reciprocal innervation of the iris muscles by the sympathetic and parasympathetic divisions. Stimulation of sympathetic nerves produces contraction of the dilator (radial) muscles, which enlarges the pupil. Stimulation of the parasympathetic nerves produces contraction of the constrictor (circular) muscle layer, which makes the pupil smaller.

Complementary Effects The effects of sympathetic and parasympathetic stimulation on salivary gland secretion are complementary. The secretion of watery saliva is stimulated through parasympathetic nerves, which also stimulate the secretion of other exocrine glands in the GI tract. Impulses through sympathetic nerves stimulate the constriction of blood vessels throughout the GI tract. The resultant decrease in blood flow to the salivary glands causes the production of a thicker, more viscous saliva.

Cooperative Effects The effects of sympathetic and parasympathetic stimulation on the urinary and reproductive systems are cooperative. Erection of the penis, for example, is due to vasodilation resulting from parasympathetic nerve stimulation; ejaculation is due to stimulation through sympathetic nerves. Although the contraction of the urinary bladder is myogenic (independent of nerve stimulation), it is promoted in part by the action of parasympathetic nerves. This *micturition (mik''tu-rish'un),* or urination, urge and reflex is also enhanced by sympathetic nerve activity, which increases the tone of the bladder muscles. Emotional states that are accompanied by high sympathetic nerve activity may thus result in reflex urination at bladder volumes that are normally too low to trigger this reflex.

Organs without Dual Innervation

Although most organs are innervated by both sympathetic and parasympathetic nerves, some—including the adrenal medulla, arrector pili muscles, sweat glands, and most blood vessels—receive only sympathetic innervation. In these cases regulation is achieved by increases or decreases in the "tone" (firing rate) of the sympathetic neurons. Constriction of blood vessels, for example, is produced by increased sympathetic activity, which stimulates adrenergic receptors, and vasodilation results from decreased sympathetic nerve stimulation.

The sympathetic division is required for proper thermoregulatory responses to heat. In a hot room, for example, decreased sympathetic stimulation produces dilation of the blood vessels in the surface of the skin, which increases cutaneous blood flow and provides better heat radiation. During exercise, in contrast, there is increased sympathetic activity, which causes constriction of the blood vessels in the skin of the limbs and stimulation of sweat glands in the trunk.

The eccrine sweat glands in the trunk secrete a watery fluid in response to sympathetic stimulation. Evaporation of this dilute sweat helps to cool the body. The eccrine sweat glands also secrete a chemical called **bradykinin** *(brad''e-ki'nin)* in response to sympathetic stimulation. Bradykinin stimulates dilation of the surface blood vessels near the sweat glands, helping to radiate heat. At the conclusion of exercise, sympathetic stimulation is reduced and blood flow to the surface of the limbs is increased, which aids in the elimination of metabolic heat. Notice that all of these thermoregulatory responses are achieved without the direct involvement of the parasympathetic division.

1. Define the meaning of the terms *adrenergic* and *cholinergic,* and use these terms to describe the neurotransmitters of different autonomic neurons.
2. Describe the effects of the drug *atropine,* and explain these effects in terms of the actions of the parasympathetic division.
3. Explain and give examples of how the sympathetic and parasympathetic divisions can have antagonistic, cooperative, and complementary effects.

Control of the Autonomic Nervous System by Higher Brain Centers

Visceral functions are mainly regulated by autonomic reflexes. In most autonomic reflexes, sensory input is directed to brain centers, which in turn regulate the activity of descending pathways to preganglionic autonomic neurons. The neural centers that directly control the activity of autonomic nerves are influenced by higher brain areas as well as by sensory input.

Objective 11. Describe the area of the brain that most directly controls the activity of autonomic nerves, and describe the higher brain areas that influence autonomic activity.

Objective 12. Explain, in terms of the structures involved, how the activity of the autonomic system and the activity of the endocrine system can be coordinated.

Objective 13. Explain, in terms of the structures involved, how autonomic functions can be affected by emotions.

Medulla Oblongata

The **medulla oblongata** of the brain stem is the structure that most directly controls the activity of the ANS. Almost all autonomic responses can be elicited by experimental stimulation of the medulla, which contains centers for the control of the cardiovascular, pulmonary, urinary, reproductive, and digestive systems. Much of the sensory input to these centers travels through the sensory neurons of the vagus nerve. These reflexes are listed in table 13.7.

Hypothalamus

The **hypothalamus** (fig. 13.9) is an extremely important autonomic structure within the brain located just above (superior to) the pituitary gland. By means of motor fibers to the brain stem and to the posterior pituitary and also by means of hormones that regulate the anterior pituitary, the hypothalamus serves to orchestrate somatic, autonomic, and endocrine responses during various behavioral states.

Table 13.7 Some reflexes stimulated by input from sensory neurons in the vagus nerve that are transmitted to centers in the medulla oblongata

Organs	Type of receptors	Reflex effects
Lungs	Stretch receptors	Inhibits further inhalation; stimulates an increase in cardiac rate and vasodilation
	Type J receptors	Stimulated by pulmonary congestion—produces feelings of breathlessness, and causes a reflex fall in cardiac rate and blood pressure
Aorta	Chemoreceptors	Stimulated by rise in CO_2 and fall in O_2—produces increased rate of breathing, fall in heart rate, and vasoconstriction
Aorta (cont.)	Baroreceptors	Stimulated by increased blood pressure—produces a reflex decrease in heart rate
Heart	Atrial stretch receptors	Inhibits antidiuretic hormone secretion, thus increasing the volume of urine excreted
	Stretch receptors in ventricles	Produces a reflex decrease in heart rate and vasodilation
Gastrointestinal tract	Stretch receptors	Feelings of satiety, discomfort, and pain

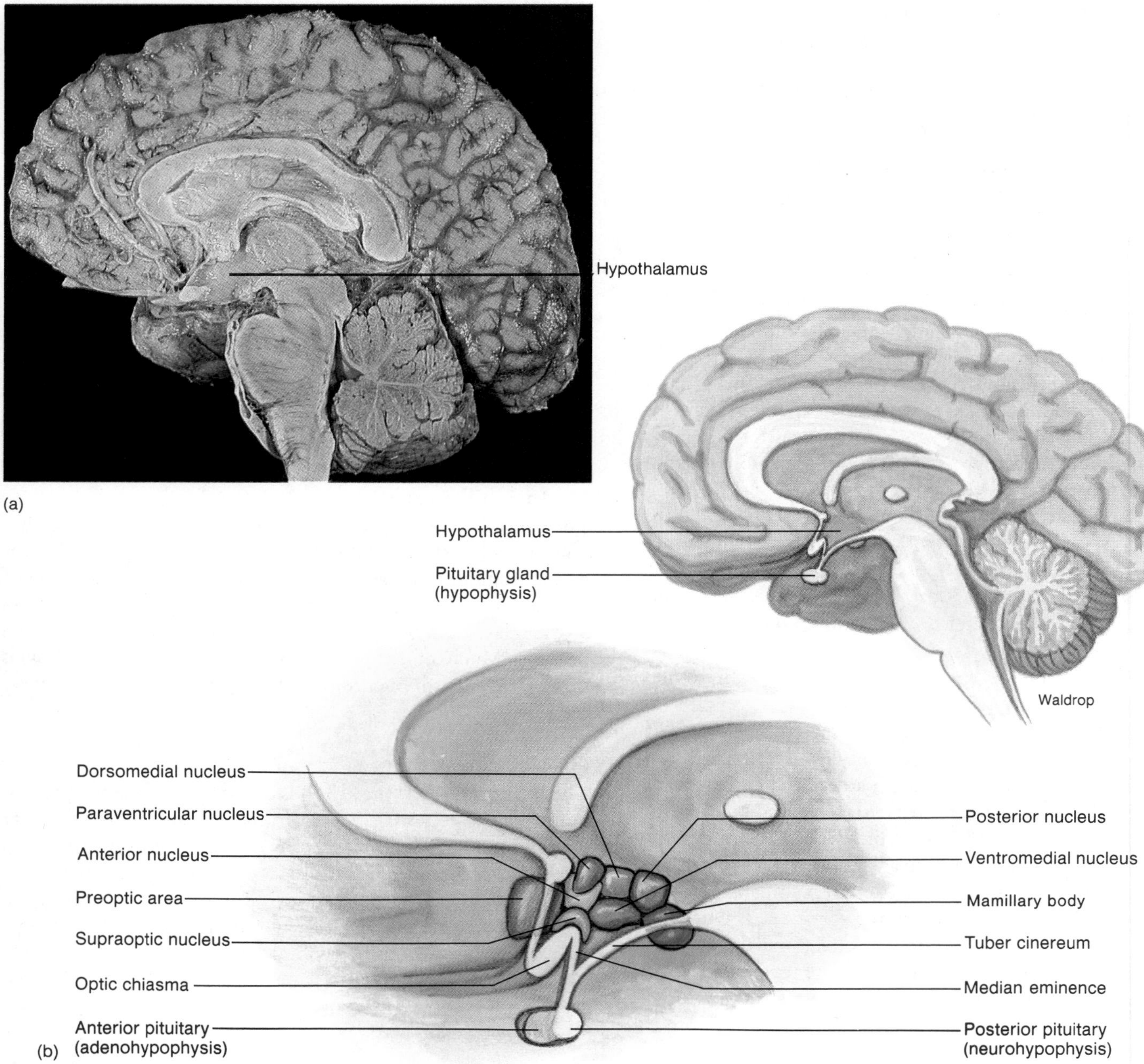

Figure 13.9 (*a*) A photograph of a sagittal section of the brain and (*b*) an illustration of the area of the hypothalamus.

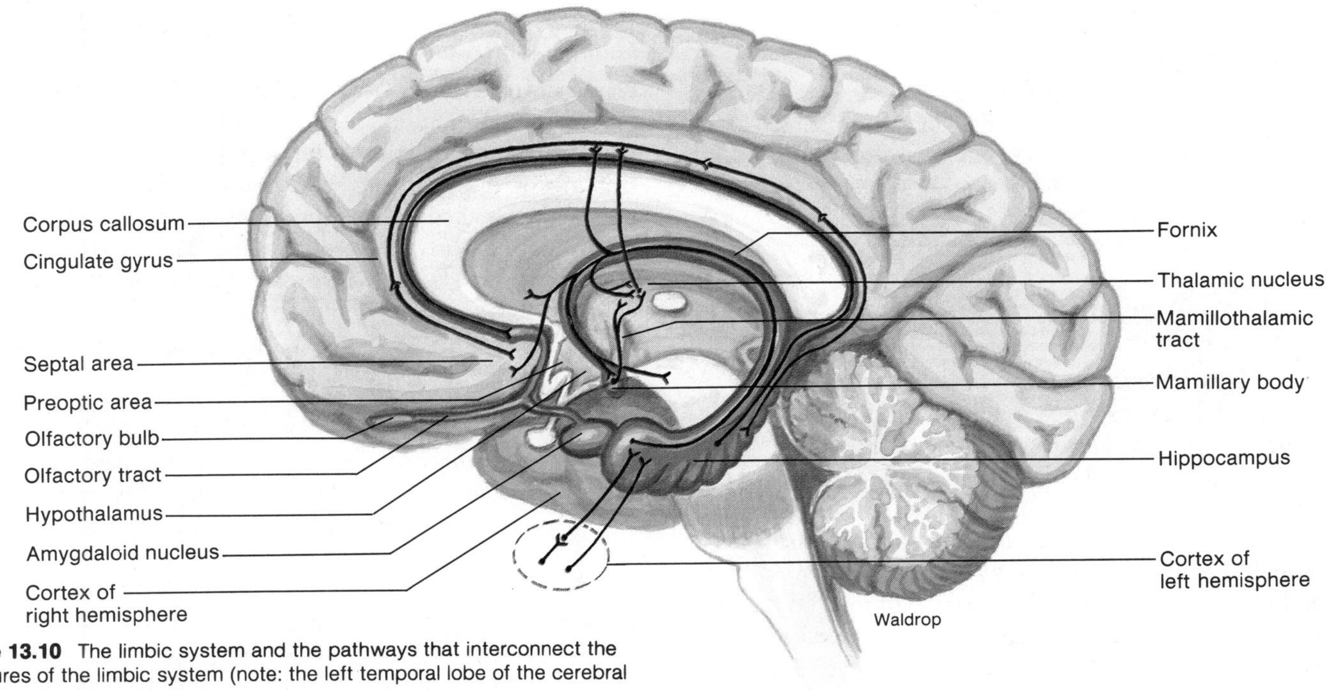

Figure 13.10 The limbic system and the pathways that interconnect the structures of the limbic system (note: the left temporal lobe of the cerebral cortex has been removed).

Experimental stimulation of different areas of the hypothalamus can evoke the autonomic responses characteristic of aggression, sexual behavior, eating, or satiety. Chronic stimulation of the lateral hypothalamus, for example, can make an animal eat and become obese, whereas stimulation of the medial hypothalamus inhibits eating. Other areas contain osmoreceptors that stimulate thirst and the secretion of antidiuretic hormone (ADH) from the posterior pituitary.

The hypothalamus is also where the body's thermostat is located. Experimental cooling of the preoptic-anterior hypothalamus causes shivering (a somatic response) and nonshivering thermogenesis (sympathetic responses). Experimental heating of this hypothalamic area results in hyperventilation (stimulated by somatic motor nerves), vasodilation, salivation, and sweat gland secretion (stimulated by autonomic nerves).

The coordination of sympathetic and parasympathetic reflexes by the medulla oblongata is thus integrated with the control of somatic and endocrine responses by the hypothalamus. The activities of the hypothalamus are in turn influenced by higher brain centers.

Limbic System, Cerebellum, and Cerebrum

The **limbic system** is a group of fiber tracts and nuclei that form a ring (a limbus) around the brain stem. It includes the cingulate gyrus of the cerebral cortex, the hypothalamus, the fornix (a fiber tract), the hippocampus, and the amygdaloid nucleus (fig. 13.10). These structures, which were derived early in the course of vertebrate evolution, were once called the *rhinencephalon (ri''nen-sef'ah-lon),* or "smell brain," because of their importance in the central processing of olfactory information.

In higher vertebrates, these structures are now recognized as centers involved in basic emotional drives—such as anger, fear, sex, and hunger—and in short-term memory. Complex circuits between the hypothalamus and other parts of the limbic system (illustrated in figure 13.10) contribute visceral responses to emotions.

The autonomic correlates of motion sickness—nausea, sweating, and cardiovascular changes—are abolished by cutting the motor tracts of the cerebellum. This demonstrates that impulses from the cerebellum to the medulla influence activity of the ANS. Experimental and clinical observations have also demonstrated that the frontal and temporal lobes of the cerebral cortex influence lower brain areas as part of their involvement in emotion and personality.

One of the most dramatic examples of the role of higher brain areas in personality and emotion is the famous crowbar accident, which occurred in 1848. A twenty-five-year-old railroad foreman, Phineas P. Gage, was tamping gunpowder into a hole in a rock when it exploded. The rod—three feet, seven inches long and one and one-fourth inches thick—was driven through his left eye, passed through his brain, and emerged through the back of his skull.

After a few minutes of convulsions, Gage got up, rode a horse three-quarters of a mile into town, and walked up a long flight of stairs to see a doctor. He recovered well, with no noticeable sensory or motor deficits. His associates, however, noted striking personality changes. Before the accident Gage was a responsible, capable, and financially prudent man. Afterwards, he appeared to lose social inhibitions, engaged in gross profanity (which he did not do before the accident), and seemed to be tossed about by chance whims. He was eventually fired from his job, and his previous friends remarked that he was "no longer Gage."

limbic: L. *limbus,* edge or border

1. Describe the role of the medulla oblongata in the regulation of the ANS.
2. Describe the role of the hypothalamus in the regulation of the autonomic and endocrine systems.
3. Explain, in terms of the neural pathways involved, how emotional states can be associated with particular autonomic responses (such as blushing).

Clinical Considerations

Autonomic Dysreflexia

Autonomic dysreflexia, a serious condition producing rapid elevations in blood pressure that can lead to stroke (cerebrovascular accident), occurs in 85% of people with quadriplegia and others with spinal cord lesions above the sixth thoracic level. Lesions to the spinal cord first produce the symptoms of spinal shock, characterized by the loss of both skeletal muscle and autonomic reflexes. After a period of time both types of reflexes return in an exaggerated state; the skeletal muscles may become spastic, due to the absence of higher inhibitory influences, and the visceral organs experience denervation hypersensitivity. Patients in this state have difficulty emptying their urinary bladders and must often be catheterized.

Noxious stimuli, such as overdistension of the urinary bladder, can result in reflex activation of the sympathetic nerves below the spinal cord lesion. This produces goose bumps, cold skin, and vasoconstriction in the regions served by the spinal cord below the level of the lesion. The rise in blood pressure resulting from this vasoconstriction activates pressure receptors that transmit impulses along sensory neurons to the medulla. In response to this sensory input, the medulla directs a reflex slowing of the heart and vasodilation. Since descending impulses are blocked by the spinal lesion, however, the skin is warm and moist (due to vasodilation and sweat gland secretion) above the lesion, but cold below the level of spinal cord damage.

CLINICAL CASE STUDY ANSWER

The tumor has invaded the sympathetic ganglia and connecting pathways within the base of the neck, known as the cervical ganglia and sympathetic tracts (chain). Preganglionic sympathetic neurons originating in the upper thoracic spinal cord travel upward through the sympathetic chain where they synapse on a postganglionic neuron in one of the cervical ganglia. From that point, postganglionic neurons travel through various routes (mostly paralleling arteries) to innervate the end organs involved. In the case of our patient, these include the radial muscle of the iris, the smooth muscle portion of the levator palpebrae superioris, and sweat glands on the right side of the face. ■

Chapter Summary

I. Introduction to the Autonomic Nervous System
   A. Preganglionic autonomic neurons originate from the brain or spinal cord; postganglionic neurons originate from ganglia located outside the CNS.
   B. Smooth muscle, cardiac muscle, and glands receive autonomic innervation.
      1. The involuntary effectors are somewhat independent of their innervation and become hypersensitive when their innervation is removed.
      2. Myocardial cells are interconnected by electrical synapses, or gap junctions, to form a functional syncytium with independent pacemaker activity.
      3. Single-unit smooth muscles have gap junctions and pacemaker activity; multi-unit smooth muscles lack gap junctions and pacemaker activity.

II. Structure of the Autonomic Nervous System
   A. Preganglionic neurons of the sympathetic (thoracolumbar) division originate in the spinal cord between the thoracic and lumbar levels.
      1. Many of these neurons synapse with postganglionic neurons whose cell bodies are located in a chain of sympathetic (paravertebral) ganglia outside the spinal cord.
      2. Some preganglionic neurons synapse in collateral ganglia; these are the celiac, superior mesenteric, and the inferior mesenteric ganglia.
      3. Some preganglionic neurons innervate the adrenal medulla, which secretes epinephrine (and some norepinephrine) into the blood in response to this stimulation.
   B. Preganglionic parasympathetic neurons originate in the brain and in the sacral levels of the spinal cord.
      1. Preganglionic parasympathetic neurons contribute to cranial nerves III, VII, IX, and X.
      2. Preganglionic neurons of the vagus (X) nerve are very long and synapse in terminal ganglia located next to or within the innervated organ; short postganglionic neurons then innervate the effector cells.
      3. The vagus provides parasympathetic innervation to the heart, lungs, esophagus, stomach, liver, small intestine, and upper half of the large intestine.
      4. Parasympathetic outflow from the sacral levels of the spinal cord innervates terminal ganglia in the lower half of the large intestine, the rectum, and the urinary and reproductive systems.

III. Functions of the Autonomic Nervous System
   A. The sympathetic and parasympathetic divisions of the ANS may act in a complementary or antagonistic way to help maintain homeostasis. The sympathetic division activates the body to "fight or flight" through

adrenergic effects; the parasympathetic division often manifests antagonistic actions through cholinergic effects.

B. All preganglionic autonomic neurons are cholinergic (use ACh as a neurotransmitter).
   1. All postganglionic parasympathetic neurons are cholinergic.
   2. Most postganglionic sympathetic neurons are adrenergic (use norepinephrine at their synapses).
   3. Sympathetic neurons that innervate sweat glands and those that innervate blood vessels in skeletal muscles are cholinergic.

C. Adrenergic effects include stimulation of the heart, vasoconstriction in the viscera and skin, bronchodilation, and glycogenolysis in the liver.

D. Cholinergic effects of parasympathetic nerves are promoted by the drug muscarine and inhibited by atropine.

E. In organs with dual innervation, the actions of the sympathetic and parasympathetic divisions can be antagonistic, complementary, or cooperative.
   1. The effects are antagonistic in the heart and pupils.
   2. The actions are complementary in the regulation of salivary gland secretion and are cooperative in the regulation of the reproductive and urinary systems.

F. In organs without dual innervation (such as most blood vessels), regulation is achieved by variations in sympathetic nerve activity.

IV. Control of the Autonomic System by Higher Brain Centers

A. Visceral sensory input to the brain may result in the activity of the descending pathways to the preganglionic autonomic neurons. The centers in the brain that control autonomic activity are influenced by higher brain areas as well as by sensory input.

B. The medulla oblongata of the brain stem is the structure that most directly controls the activity of the ANS.
   1. The medulla oblongata is in turn influenced by sensory input and by input from the hypothalamus.
   2. The hypothalamus orchestrates somatic, autonomic, and endocrine responses during various behavioral states.

C. The activity of the hypothalamus is influenced by input from the limbic system, cerebellum, and cerebrum; these interconnections provide an autonomic component to changes in body position, emotion, and various expressions of personality.

Review Activities

Objective Questions

1. Which of the following statements about the superior mesenteric ganglion is *true*?
   (a) It is a parasympathetic ganglion.
   (b) It is a paravertebral sympathetic ganglion.
   (c) It is located in the head.
   (d) It contains postganglionic sympathetic neurons.
2. The pterygopalatine, ciliary, submandibular, and otic ganglia are
   (a) collateral sympathetic ganglia.
   (b) cervical sympathetic ganglia.
   (c) parasympathetic ganglia that receive neurons from the vagus.
   (d) parasympathetic ganglia that receive neurons from the third, seventh, and ninth cranial nerves.
3. Parasympathetic ganglia are located
   (a) in a chain parallel to the spinal cord.
   (b) in the dorsal roots of spinal nerves.
   (c) next to or within the organs innervated.
   (d) in the brain.
4. The neurotransmitter of preganglionic sympathetic neurons is
   (a) norepinephrine.
   (b) epinephrine.
   (c) acetylcholine.
   (d) dopamine.
5. In which location would autonomic ganglia *not* be found? In
   (a) the head.
   (b) the axilla.
   (c) the neck.
   (d) the abdomen.
   (e) parallel to the spinal cord.
6. Which of the following neurons release norepinephrine?
   (a) preganglionic parasympathetic neurons.
   (b) postganglionic parasympathetic neurons.
   (c) postganglionic sympathetic neurons in the heart.
   (d) postganglionic parasympathetic neurons in sweat glands.
   (e) All of these.
7. The actions of sympathetic and parasympathetic neurons are cooperative in the
   (a) heart.
   (b) reproductive system.
   (c) digestive system.
   (d) eyes.
8. Which of the following is *not* a result of parasympathetic nerve stimulation?
   (a) increased movement of the digestive tract
   (b) increased mucus secretion
   (c) constriction of the pupils
   (d) constriction of visceral blood vessels
9. Atropine blocks parasympathetic nerve effects. It would therefore cause
   (a) dilation of the pupils.
   (b) decreased mucus secretion.
   (c) decreased movements of the GI tract.
   (d) increased heart rate.
   (e) All of these.
10. The area of the brain that is most directly involved in the reflex control of the autonomic system is the
   (a) hypothalamus.
   (b) cerebral cortex.
   (c) medulla oblongata.
   (d) cerebellum.

Essay Questions

1. Compare the sympathetic and parasympathetic divisions in terms of the location of their ganglia and the distribution of their nerves.
2. Explain the structural and functional relationship between the sympathetic nervous system and the adrenal glands.
3. Compare the effects of adrenergic and cholinergic stimulation on the cardiovascular and digestive systems.
4. Explain how effectors that receive only sympathetic innervation are regulated by the ANS.
5. Explain why a person may sweat more profusely immediately after exercise than during exercise.

CHAPTER 14

Endocrine System

Outline and Concepts

CLINICAL CASE STUDY

A thirty-eight-year-old female visits her family doctor complaining of chronic fatigue and weakness, especially in her legs. Upon greeting the patient, the doctor notes that although she is mildly obese, there is an unusually round contour to her face. During questioning, he learns that at her recent twenty-year high school reunion, nobody recognized her because her face looked so different. He also learns that she has a chronic unrelenting headache. Physical examination yields, in addition to the facial findings, an unusual fat distribution consisting of a hump on the upper back and marked truncal obesity. Additional findings include hypertension and blindness in the lateral visual fields of both eyes (bitemporal hemianopsia). Laboratory values are remarkable for elevated blood glucose and other evidence that cortisol levels are high.

Can a correlation be drawn between the patient's visual findings and the rest of her clinical picture? Explain. In which endocrine organ does the primary disease process reside? Does that same organ produce the hormones that directly cause the patient's signs and symptoms? Explain. Do you suppose that similar findings could occur due to a tumor elsewhere in the endocrine system? ■

Introduction to the Endocrine System

Hormones are regulatory chemicals secreted by the endocrine glands into the blood, which transports them to their target tissues, where they are controlled by feedback mechanisms.

Objective 1. Distinguish between endocrine and exocrine glands.

Objective 2. Discuss the similarities and differences between the nervous system and the endocrine system in regulating body functions and maintaining homeostasis.

Objective 3. Define *mixed gland,* and identify which endocrine glands are mixed.

Objective 4. Describe the action of a hormone on its target tissue, and explain how negative feedback regulates hormonal secretion.

Objective 5. Differentiate between the three principal kinds of hormones: steroids, proteins, and amines.

The numerous glands of the body can be classified into two types based on structure and function, *exocrine* and *endocrine.* Exocrine glands, such as sweat, salivary, or mucous glands, produce secretions that are transported through ducts to their respective destinations. Each of the exocrine glands functions within a particular system of the body. The endocrine glands constitute their own system, the endocrine system, and have several distinctions. Endocrine glands are ductless glands that secrete specific chemicals called *hormones* directly into the blood (fig. 14.1). The blood then transports these hormones to specific sites called *target tissues,* where they perform precise functions.

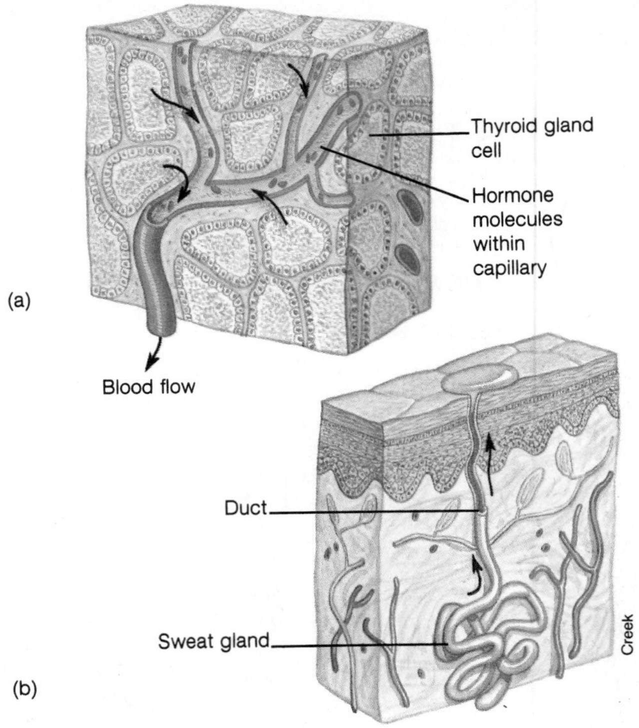

Figure 14.1 Comparison of (*a*) an endocrine gland such as the thyroid with (*b*) an exocrine gland such as a sweat gland in the skin. An endocrine gland is a ductless gland that releases hormones into the blood, whereas exocrine glands secrete their products into ducts that lead to body surfaces.

The endocrine system functions closely with the nervous system in regulating and integrating body processes and maintaining homeostasis. The nervous system regulates body activities through the action of electrochemical impulses that are transmitted by means of neurons, resulting in rapid, but usually brief, responses. In contrast, the glands of the endocrine system secrete chemical regulators that travel through the bloodstream to their intended sites where they have a slow but relatively long-lasting effect. Neurological responses are measured in milliseconds, but hormonal action requires seconds or days to elicit a response. Some hormones may have an effect that lasts for minutes and others, for weeks or months.

The nervous and endocrine systems are closely coordinated in autonomically controlling the functions of the body. Three endocrine glands are located within the cranial cavity, where certain structures of the brain routinely stimulate or inhibit the release of hormones. Likewise, certain hormones may stimulate or inhibit the activities of the nervous system. The functions of the two body control systems are compared in table 14.1.

Glands of the Endocrine System

The endocrine glands are located throughout the body (fig. 14.2). The **pituitary (hypophysis),** the **hypothalamus,** and the **pineal gland** are associated with the brain within the cranial cavity. The **thyroid gland** and **parathyroid glands** are located in the neck. The **adrenal glands** and **pancreas** are

endocrine: Gk. *endon,* within; *krinein,* to separate
exocrine: Gk. *exo,* outside, *krinein,* to separate
hormone: Gk. *hormon,* to set in motion

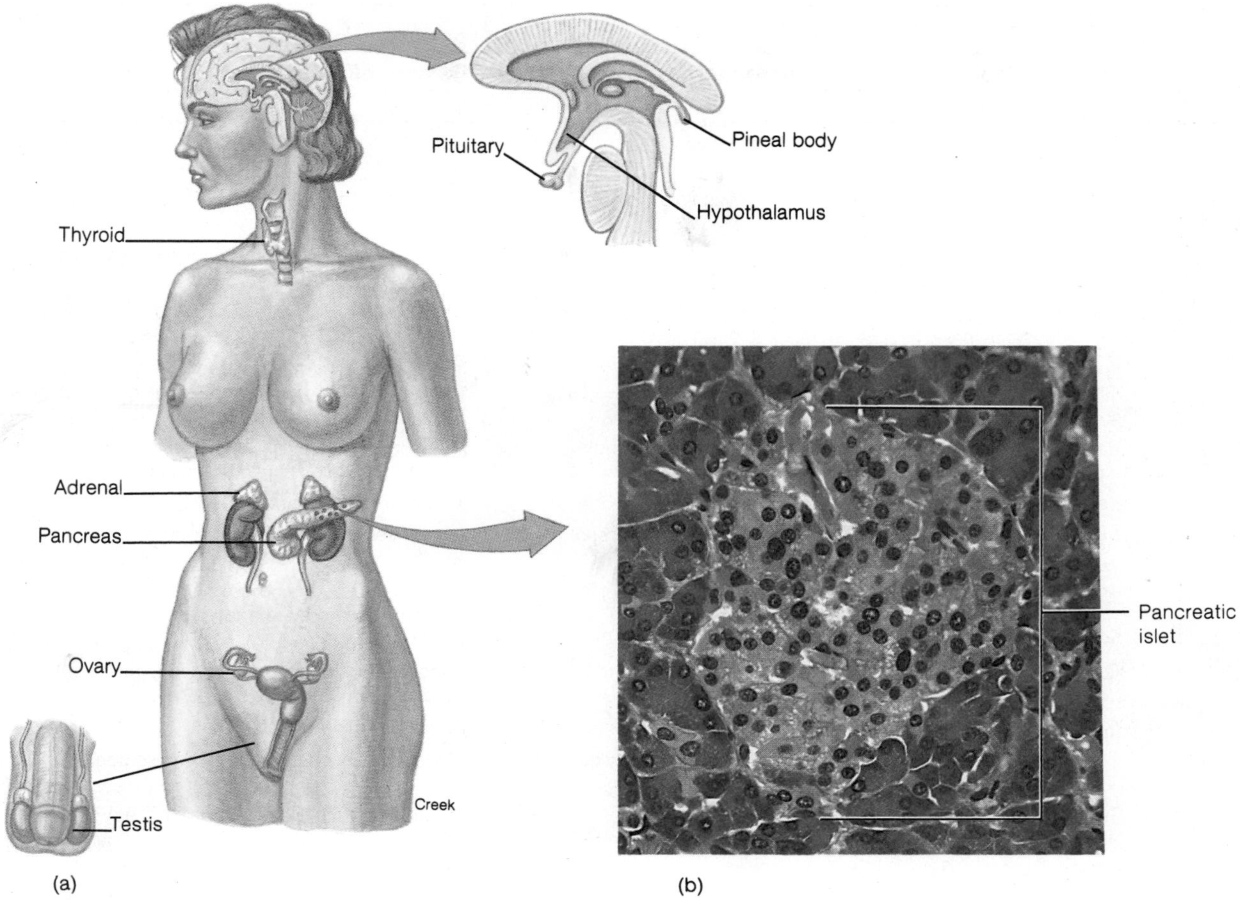

Figure 14.2 (*a*) The location of some of the endocrine glands. (*b*) The pancreatic islets (islets of Langerhans) within the pancreas.

Table 14.1 Comparison of the basic functions of the endocrine and nervous systems

Endocrine system	Nervous system
Secretes hormones that are transported to target tissues via the blood	Transmits neurochemical impulses via nerve fibers
Causes changes in the metabolic activities in specific cells	Causes muscles to contract or glands to secrete
Exerts effects relatively slowly (seconds or even days)	Exerts effects relatively rapidly (milliseconds)
Has generally prolonged effects	Has generally brief effects

within the abdominal region. The **gonads** (ovaries) of the female are within the pelvic cavity, whereas the **gonads** (testes) of the male are in the scrotum. The pancreas and gonads are frequently classified as *mixed glands* because they have exocrine as well as endocrine functions.

In addition to the aforementioned glands, the thymus in the lower median neck region, the stomach, the mucosal cells of the duodenum, and the placenta, associated with the fetus, may be considered part of the endocrine system because each has endocrine functions. The principal endocrine glands of the body, their hormones, and the effects of the hormones are listed in table 14.2.

Hormones and Their Actions

Hormones are specific organic substances that are the chemical messengers of the endocrine system. There are three basic kinds of hormones (steroids, proteins, and amines), derived either from cholesterol or from amino acids (fig. 14.3).

A *steroid* is a lipid synthesized from cholesterol. Steroids exist as complex rings of carbon and hydrogen atoms (fig. 14.3b). The types of atoms attached to the rings determine the specific kind of steroid. There are more than twenty steroid hormones in the body, including such common ones as *cortisol, cortisone, estrogen, progesterone,* and *testosterone.* The sex hormones produced by the gonads and the hormones produced by the adrenal cortex are all steroids. Steroids can be taken orally or intravenously to regulate body activity if glandular dysfunction prevents normal amounts from being produced naturally.

Proteins are composed of amino acids bonded together in peptide chains (fig. 14.3a). Most of the hormones of the body are proteins, including calcitonin from the thyroid gland and hormones secreted by the pituitary gland, the pancreas, and the parathyroid glands. Protein hormones cannot be administered orally because the peptide bonds would be split during the hydrolytic reaction of digestion; thus, they must be injected intravenously, intramuscularly, or subcutaneously.

Amines are produced from amino acids but do not contain peptide bonds. Amine molecules contain atoms of carbon, hydrogen, and nitrogen and always have an amine group ($-NH_2$)

Table 14.2 A partial list of the endocrine glands

Endocrine gland	Major hormones	Primary target organs	Primary effects
Adrenal cortex	Cortisol Aldosterone	Liver, muscles Kidneys	Glucose metabolism; Na^+ retention, K^+ excretion
Adrenal medulla	Epinephrine	Heart, bronchioles, blood vessels	Adrenergic stimulation
Hypothalamus	Releasing and inhibiting hormones	Anterior pituitary	Regulates secretion of anterior pituitary hormones
Intestine	Secretin and cholecystokinin	Stomach, liver, and pancreas	Inhibits gastric motility; stimulates bile and pancreatic juice secretion
Pancreatic islets	Insulin Glucagon	Many organs Liver and adipose tissue	Insulin promotes cellular uptake of glucose and formation of glycogen and fat; glucagon stimulates hydrolysis of glycogen and fat
Ovaries	Estradiol-17β and progesterone	Female genital tract and mammary glands	Maintains structure of genital tract; promotes secondary sexual characteristics
Parathyroids	Parathyroid hormone	Bone, intestine, and kidneys	Increases Ca^{++} concentration in blood
Pineal	Melatonin	Hypothalamus and anterior pituitary	Affects secretion of gonadotrophic hormones
Pituitary, anterior	Trophic hormones	Endocrine glands and other organs	Stimulates growth and development of target organs; stimulates secretion of other hormones
Pituitary, posterior	Antidiuretic hormone Oxytocin	Kidneys, blood vessels Uterus, mammary glands	Antidiuretic hormone promotes water retention and vasoconstriction. Oxytocin stimulates contraction of uterus and mammary secretory units
Stomach	Gastrin	Stomach	Stimulates acid secretion
Testes	Testosterone	Prostate, seminal vesicles, other organs	Stimulates secondary sexual development
Thymus	Thymosin	Lymph nodes	Stimulates white blood cell production
Thyroid	Thyroxine (T_4) and triiodothyronine (T_3); Calcitonin	Most organs; blood	Growth and development; stimulates basal rate of cell respiration (basal metabolic rate or BMR); regulates Ca^{++} levels within blood by inhibiting bone decalcification

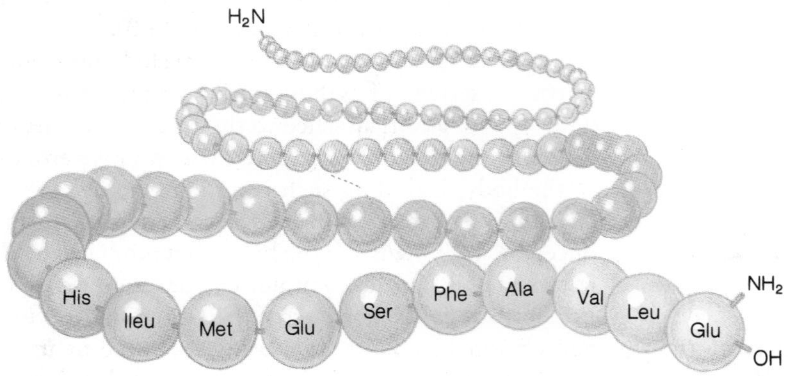

(a) Parathyroid hormone (PTH)

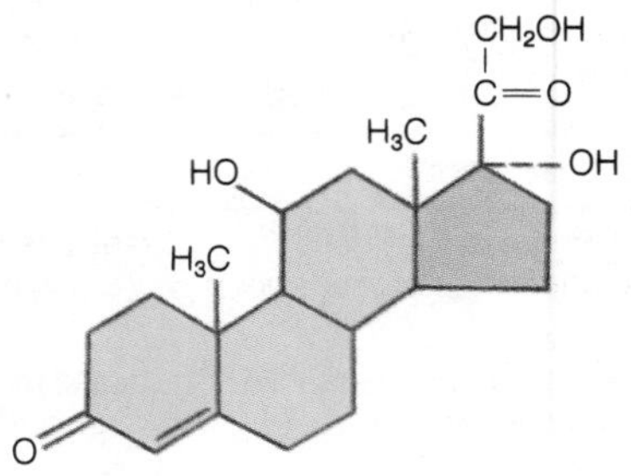

(b) Cortisol

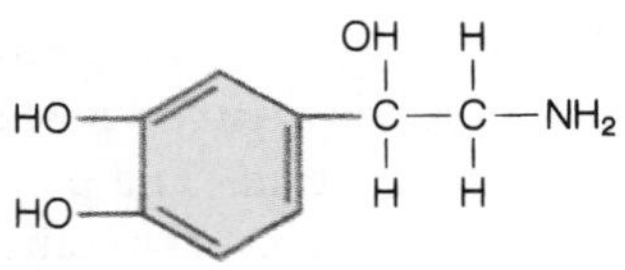

(c) Norepinephrine

Figure 14.3 Structural formulas of the three basic kinds of hormones: (*a*) protein hormone, (*b*) steroid hormone, and (*c*) amine hormone.

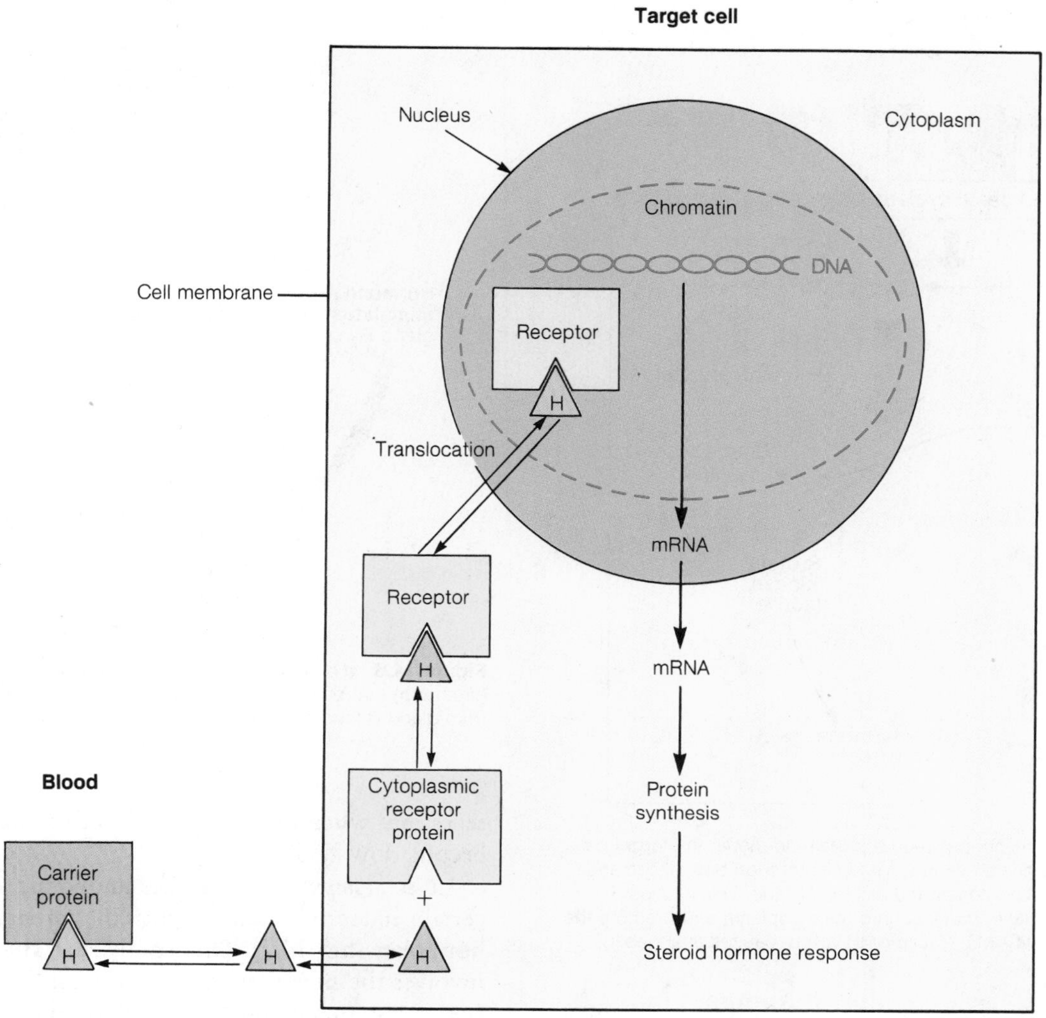

Figure 14.4 The mechanism of the action of a steroid hormone (*H*) on the target cells. A steroid hormone passes through a cell membrane and unites with a receptor protein in the cytoplasm. The steroid-protein complex then enters the cell nucleus and activates the synthesis of messenger RNA. The messenger RNA then disperses into the cytoplasm where it activates the synthesis of proteins by means of ribosomes.

Table 14.3 Summary of the classes of hormones

Type of hormone	Composition	Examples	Method of administration
Steroid	Lipid with a cholesterol-type nucleus	Sex hormones and hormones from adrenal cortex	Orally, intravenously, or intramuscularly
Protein	Amino acids bonded by peptide chains	Pituitary hormones, pancreatic hormones, parathyroid hormones, and calcitonin from thyroid gland	Intravenously, intramuscularly, or subcutaneously
Amine	Amino acids with no peptide bonds; molecule contains $-NH_2$ group	Thyroxine from thyroid gland, adrenaline from adrenal gland, and melatonin from the pineal gland	Orally, subcutaneously, intravenously, or as an inhalant

associated with the molecule (fig. 14.3c). Examples of amines include thyroxine produced in the thyroid gland, epinephrine (adrenaline) and norepinephrine produced in the adrenal gland, and melatonin produced in the pineal gland.

Thyroxine is usually administered orally, whereas epinephrine is administered intravenously to produce a fast response. Epinephrine can also be administered as an inhalant when needed to enlarge the ductules of the lungs.

A summary of the basic types of hormones is presented in table 14.3.

The usual action of hormones is to speed up or slow down cellular processes in the target tissue. Hormones are extremely specific as to which cells they affect and the cellular changes they elicit. The ability of a hormone to affect a particular cell depends on the presence of receptor molecules in the cell or specific receptor sites on its cell membrane.

Steroid hormones are soluble in lipids and can readily pass through a cell membrane. Once in a cell, the steroid combines with a protein to form a steroid-protein complex (fig. 14.4), which is necessary for the synthesis of specific kinds of messenger RNA molecules.

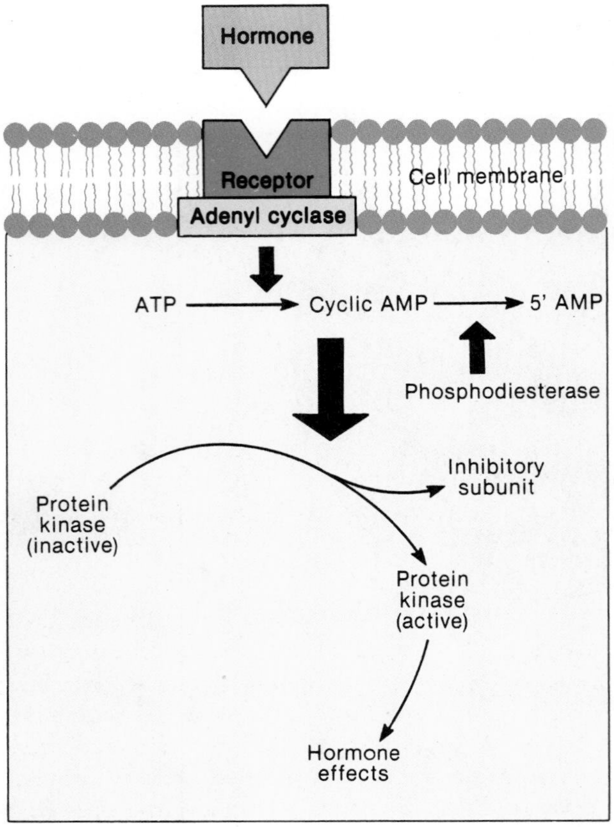

Figure 14.5 The relationship between a protein hormone, the target cells, and cyclic AMP. Hormones arrive at target cells through body fluids and attach to specific receptor sites on the cell membrane. This causes molecules of adenyl cyclase to diffuse into the cytoplasm and promote the change of ATP into cyclic AMP. Cyclic AMP in turn causes changes in cellular processes.

Protein and amine hormones are insoluble in lipids and therefore must attach to specific receptor sites on the cell membrane. These hormones do not enter the cell, but their attachment increases the activity of an enzyme in the cell membrane called *adenyl cyclase.* In the presence of adenyl cyclase, ATP molecules in the cell are converted to *cyclic AMP* (adenosine monophosphate), which in turn disperses throughout the cell to cause changes in cellular processes (fig. 14.5). These changes may include increasing protein synthesis, altering membrane permeability, or activating certain cellular enzymes.

Control of Hormone Secretion

A close balance exists between the rate of secretion of a particular hormone and the rate of usage by the target tissue. This stable concentration is maintained by a negative feedback system and autonomic neural impulses.

Negative feedback (fig. 14.6) is a homeostatic mechanism that maintains the status quo of supply and demand between the normal level and the needs of the target tissue. An endocrine gland will continue to secrete hormones that affect a target tissue until messages come back from the tissue reporting that sufficient hormones are present. These messages are generally in the form of hormones secreted by the target tissue. This chemical feedback information signals the endocrine gland to inhibit secretion (fig. 14.6). The primary endocrine gland will resume secretions when the levels of inhibiting chemicals in the blood become low again.

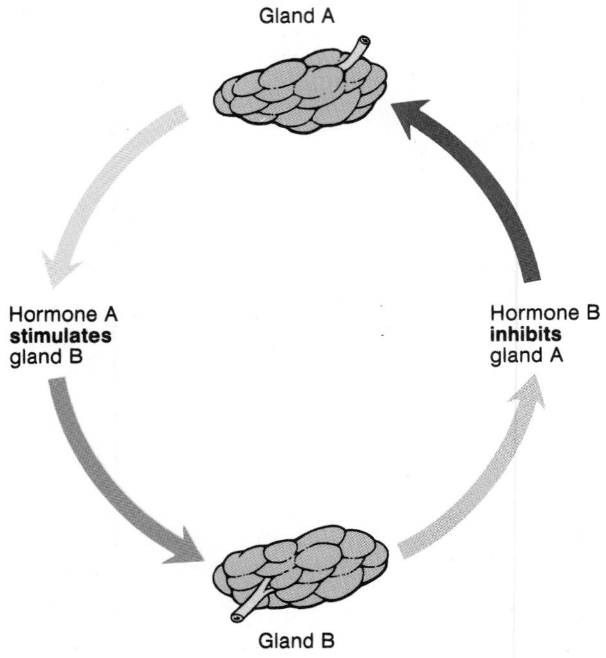

Figure 14.6 In a negative feedback system, gland A secretes hormone A (first step) that stimulates gland B to release its hormone; the hormone from gland B (second step) inhibits the action of gland A.

Neural impulses through the autonomic nervous system cause certain endocrine glands, such as the adrenal medulla, to secrete hormones (fig. 14.7). One specialized kind of neural impulse involves the hypothalamus and the pituitary gland (fig. 14.7). In this system, neurosecretory cells in the hypothalamus secrete *releasing-factor* molecules that influence specific target cells in the pituitary gland. Stimulation of the pituitary by the releasing factor causes the secretion of specific hormones. The hypothalamus continues to secrete the releasing factor until a given level of hormones is present in the body fluids and is detected by the hypothalamus through a negative feedback mechanism.

1. What are some structural differences between endocrine and exocrine glands?
2. How are the nervous and endocrine systems functionally related?
3. Make a list of the body organs that are exclusively endocrine in function; and also a list of the organs that secrete hormones and serve additional body functions.
4. Using diagrams, describe the mechanism of steroid hormone action and the mechanism of protein hormone action within cells.
5. List the three kinds of hormones, give examples of each, and describe their chemical compositions.

Development of the Endocrine System

The embryonic development of the endocrine glands establishes the anatomical relationships between the glands and

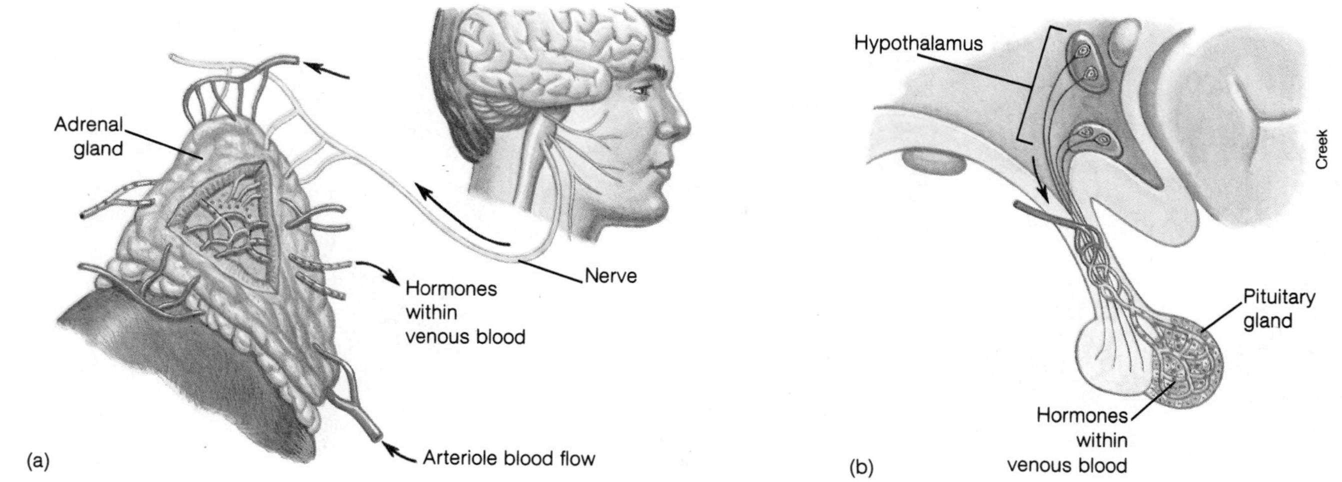

Figure 14.7 Neural control of endocrine secretion. (*a*) The adrenal medulla secretes hormones in response to autonomic stimulation. (*b*) The hypothalamus has a neurosecretory role in stimulating target cells in the pituitary gland to secrete hormones.

associated structures. A knowledge of the origin and nature of these structural relationships aids in the understanding of endocrine function.

Objective 6. Describe the structure and embryonic development of the pituitary gland.
Objective 7. Describe the embryonic development of the thyroid gland and pancreas.
Objective 8. Describe the embryonic development of the adrenal cortex and adrenal medulla.

The endocrine system is the only anatomical body system whose organs are not structurally connected. Because these organs are isolated from each other and are distributed throughout the body, each endocrine organ has a separate and independent development. All three embryonic germ layers (endoderm, mesoderm, and ectoderm) contribute to the development of the endocrine system.

The development of the pituitary, thyroid, pancreas, and adrenal glands will be discussed separately. The development of the testes and ovaries is discussed in chapters 20 and 21, respectively.

Pituitary Gland

Although the **pituitary gland** is a single organ, it is actually composed of two distinct types of tissues that have different embryonic origins, that secrete different hormones, and that are under different control systems. The pituitary gland, or **hypophysis** *(hi-pof'ĭ-sis),* has two parts. The anterior portion, called the **adenohypophysis** *(ad"ĕ-no-hi-pof'ĭ-sis),* develops from ectoderm that lines the primitive oral cavity. The posterior portion, called the **neurohypophysis,** develops from neuroectoderm of the developing brain (fig. 14.8).

The adenohypophysis begins to develop during the third week as a diverticulum, a pouchlike extension, called the **hypophyseal** (Rathke's) **pouch.** It arises from the roof of the primitive oral cavity and grows toward the brain. At the same time, another diverticulum called the **infundibulum** *(in"fun-dib'u-lum)* forms from the diencephalon on the inferior aspect of the brain. As the two diverticula come in contact, the hypophyseal pouch loses its connection with the oral cavity and the primordial tissue of the adenohypophysis is formed. The fully developed adenohypophysis includes the **pars distalis, pars tuberalis,** and **pars intermedia.**

The neurohypophysis develops as the infundibulum extends inferiorly from the diencephalon to come in contact with the developing adenohypophysis. The fully formed neurohypophysis consists of the infundibulum and the **pars nervosa.** Specialized nerve fibers that connect the hypothalamus with the pars nervosa develop within the infundibulum.

Notice that the neurohypophysis is essentially an extension of the brain, and indeed the entire pituitary gland, like the brain, is surrounded by the meninges. The adenohypophysis, in contrast, is derived from nonneural tissue (the same embryonic tissue that will form the epithelium over the roof of the mouth). These different embryologies have important consequences with regard to the functions of the two parts of the pituitary gland.

Thyroid Gland

The **thyroid gland** is derived from endoderm and begins its development during the third week as a thickening in the floor of the primitive pharynx. The thickening soon out-pouches downward as the **thyroid diverticulum** (fig. 14.9). A narrow **thyroglossal duct** connects the descending primordial thyroid tissue to the pharynx. As the descent continues, the tongue starts to develop, and the opening into the thyroglossal duct, called the **foramen cecum** *(se'kum),* pierces through the base of the tongue. By the seventh week, the thyroid gland occupies a position immediately inferior to the larynx and around the front and lateral sides of the trachea. Once in position, the thyroglossal duct disappears and the foramen cecum regresses in size to a vestigial pit that persists throughout life.

pituitary: L. *pituita,* phlegm (this gland was originally thought to secrete mucus into the nasal cavity)
adenohypophysis: Gk. *adeno,* gland; *hypo,* under; *physis,* a growing
Rathke's pouch: from Martin H. Rathke, German anatomist, 1793–1860

infundibulum: L. *infundibulum,* a funnel
foramen cecum: L. *foramen,* open passage; *caecum,* blind tube

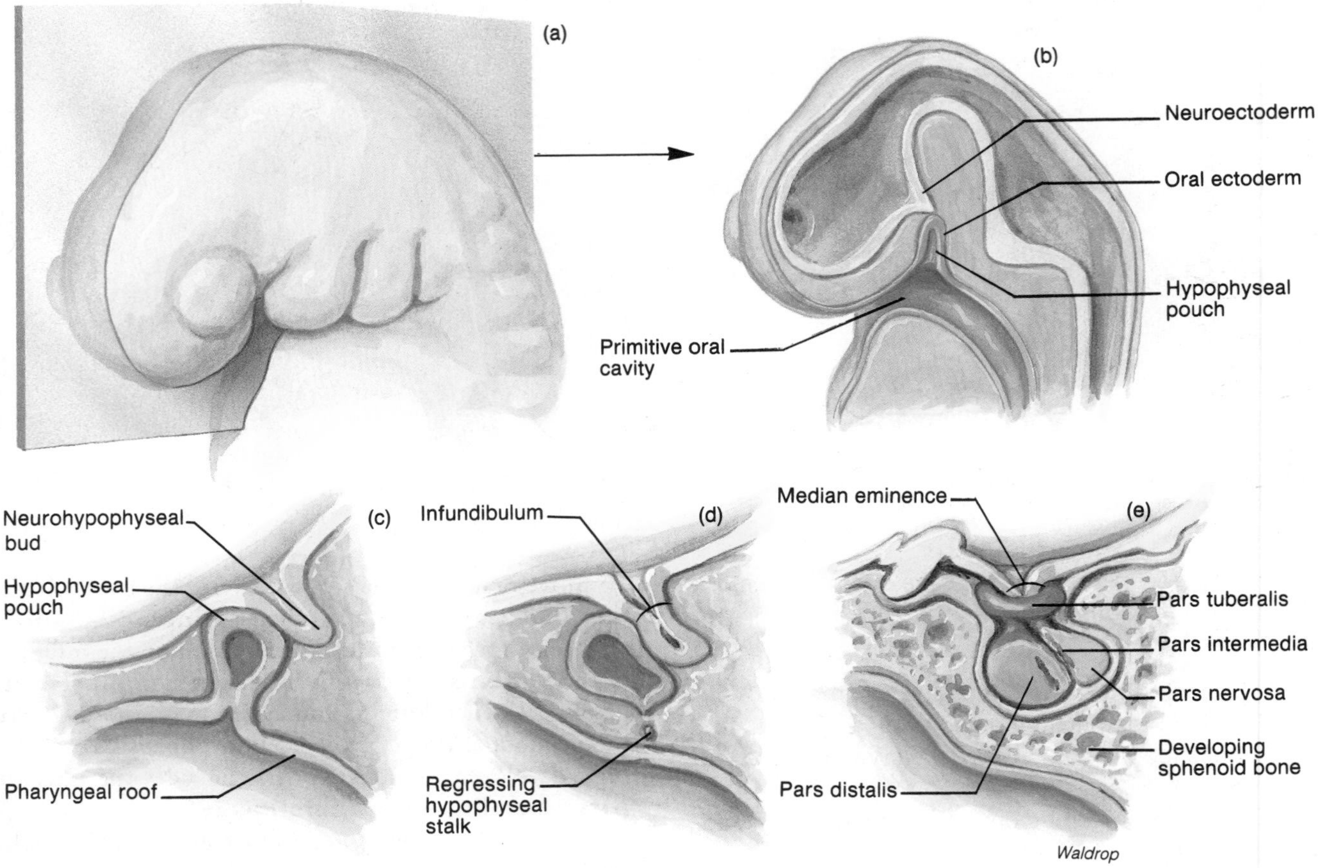

Figure 14.8 The development of the pituitary gland. (*a*) The head end of an embryo at four weeks showing the position of a midsagittal cut seen in the developmental sequence (*b–e*). The pituitary gland arises from a specific portion of the neuroectoderm, called the neurohypophyseal bud, which evaginates downward during the fourth and fifth weeks respectively in (*b*) and (*c*) and a specific portion of the oral ectoderm, called the hypophyseal (Rathke's) pouch, which evaginates upward from a specific portion of the primitive oral cavity. By eight weeks (*d*), the hypophyseal pouch is no longer connected to the pharyngeal roof of the oral cavity. During the fetal stage (*e*), the development of the pituitary gland is completed.

Pancreas

The **pancreas** begins development during the fifth week, as dorsal and ventral pancreatic buds of endoderm arise from the caudal portion of the foregut. These primordial pancreatic tissues continue to grow independently until the ventral bud is carried dorsally and fuses with the dorsal bud as the duodenum rotates to the right. The fusion of the two portions of the pancreas occurs during the seventh week.

The pancreas develops from and maintains connections with the intestine (via the pancreatic duct). The exocrine products of the pancreas, contained in *pancreatic juice,* are channeled through the pancreatic duct to the intestine. The endocrine parts of the pancreas—the **pancreatic islets** (islets of Langerhans)—however, secrete their products (insulin and glucagon) into the blood. The precise origin of the endocrine cells of the pancreas remains uncertain. It is believed that some of the endocrine cells develop as buds from pancreatic ductules and that others arise from neuroectodermal cells that migrate to the pancreas to form these endocrine cells and the autonomic innervation of the pancreas.

Adrenal Glands

The **adrenal glands** begin development during the fifth week from two different germ layers. Each adrenal gland has an outer part, or **cortex,** which develops from mesoderm, and an inner part, or **medulla,** which develops from neuroectoderm. The mesodermal ridge that forms the adrenal cortex is in the same region from which the gonads develop.

The neuroectodermal cells that form the adrenal medulla are derived from the neural crest of the neural tube. The developing adrenal medulla is gradually encapsulated by the adrenal cortex, a process that continues into the fetal stage. The formation of the adrenal gland is not completed until the end of the third year of age.

Notice that the adrenal gland, like the pituitary, has a dual origin; part is neural and part is not. Like the pituitary, the adrenal cortex and medulla are really two different endocrine tissues. They are located in the same organ but secrete different hormones and are regulated by different control systems.

---

pancreas: Gk. *pan,* all; *kreas,* flesh

adrenal: L. *ad,* to; *renes,* kidney

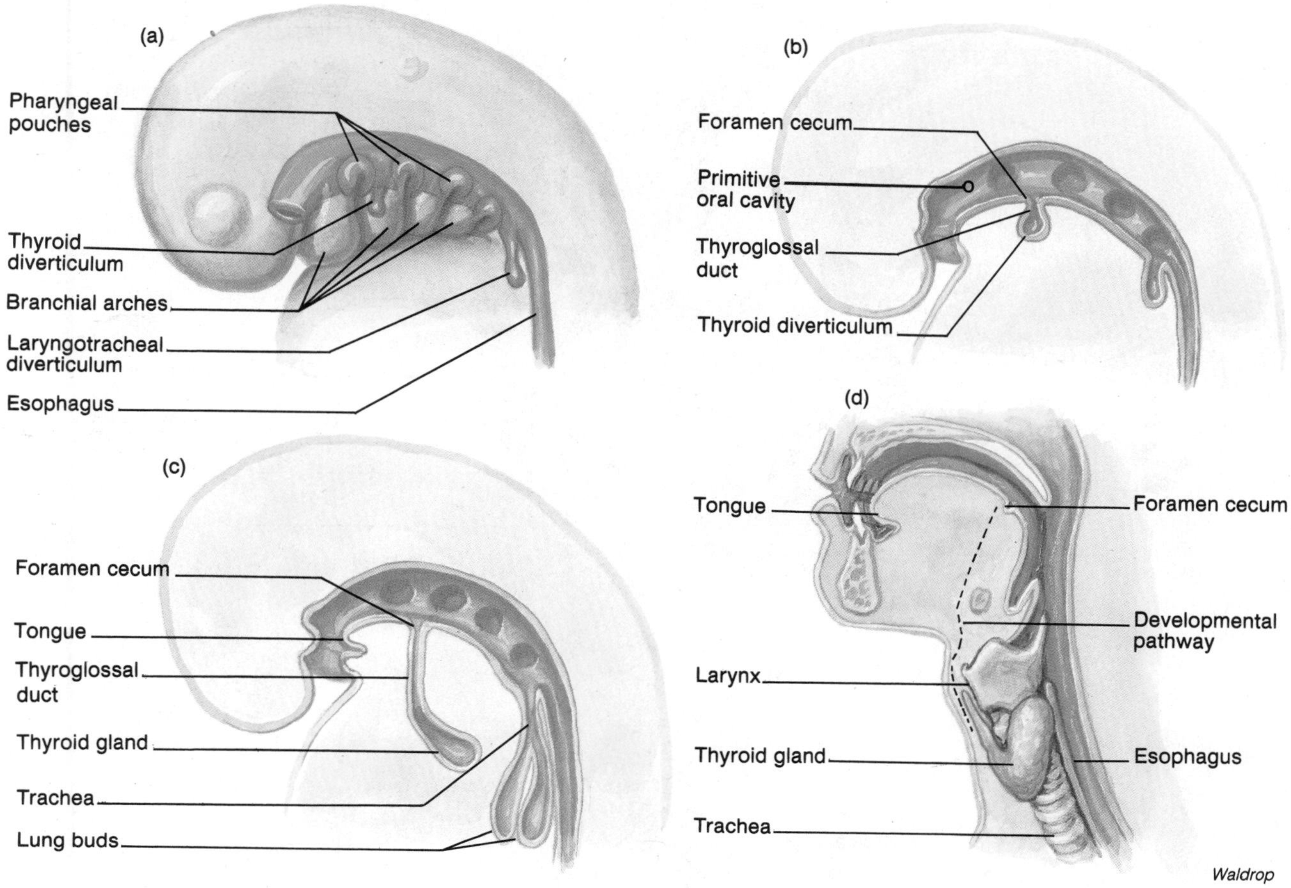

Figure 14.9 The embryonic development of the thyroid gland. (*a*) At four weeks, a thyroid diverticulum begins to form in the floor of the pharynx at the level of the second brachial arch. (*b,c*) The thyroid diverticulum extends downward during the fifth and sixth weeks. (*d*) A sagittal section through the head and neck of an adult shows the path of development and final position of the thyroid gland.

1. Describe the structure and embryonic origins of the adenohypophysis and neurohypophysis.
2. Name two endocrine glands that develop from endoderm, and describe the developmental process shared in common by these two glands.
3. Describe the embryonic formation of the adrenal glands, and explain how this is similar to the formation of the pituitary gland.

Pituitary Gland

The neurohypophysis of the pituitary gland secretes hormones that are produced by the hypothalamus, whereas the adenohypophysis of the pituitary gland secretes its own hormones in response to regulation from hypothalamic hormones. The secretions of the pituitary gland are thus controlled by the hypothalamus, as well as by negative feedback influences from the target glands.

Objective 9. List the hormones secreted by the adenohypophysis and neurohypophysis.

Objective 10. Describe, in a general way, the actions of anterior pituitary hormones.

Objective 11. Describe how the secretions of anterior and posterior pituitary hormones are controlled by the hypothalamus.

Objective 12. Explain how the secretion of anterior pituitary hormones is regulated by negative feedback.

Description of the Pituitary Gland

The **pituitary gland,** or **hypophysis,** is located on the inferior aspect of the brain in the region of the diencephalon and is attached to the brain by a stalklike structure called the *pituitary stalk* (fig. 14.10). The *infundibulum* is the portion of the pituitary stalk that connects the hypothalamus to the posterior lobe. The pituitary is a rounded, pea-shaped gland measuring about 1.3 cm (0.5 in.) in diameter. It is covered by the dura mater and is supported by the sella turcica of the sphenoid bone. The arterial circle surrounds the highly vascular pituitary gland, providing it with a rich blood supply (see fig. 16.29).

The pituitary gland is structurally and functionally divided into an anterior lobe, or **adenohypophysis,** and a posterior lobe, called the **neurohypophysis.** The adenohypophysis consists of three parts: (1) the **pars distalis** is the bulbar portion; (2) the **pars tuberalis** is the thin extension in contact with the infundibulum; and (3) the **pars intermedia** is between the anterior

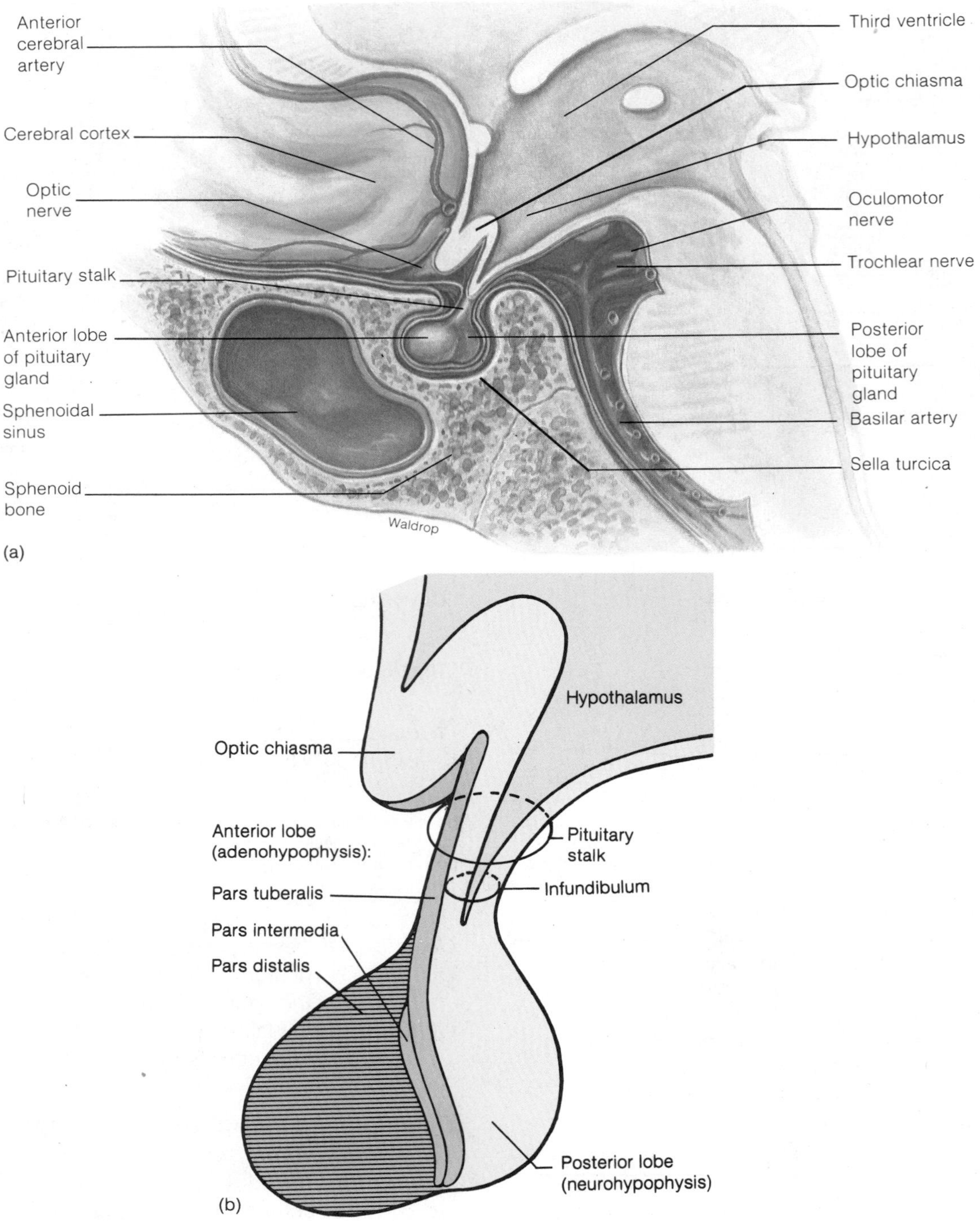

Figure 14.10 The pituitary gland. (*a*) The pituitary gland is attached to the hypothalamus and lies in the sella turcica of the sphenoid bone. (*b*) A diagram of the pituitary gland shows the various portions.

and posterior parts of the pituitary. These parts are illustrated in figure 14.10. Refer to figure 14.11 for the histological structure of the adenohypophysis.

The neurohypophysis is the neural part of the pituitary gland. It consists of the bulbar **pars nervosa,** which is in contact with the pars intermedia and the pars distalis of the adenohypophysis, and the **infundibulum,** which is the connecting stalk to the hypothalamus. Nerve fibers extend through the infundibulum along with minute neuroglia-like cells, called **pituicites** (*pĭ-tu'ĭ-sītz*).

The pituitary is the structure of the brain perhaps most subject to neoplasms. A tumor of the pituitary is generally easily detected as it begins to grow and interfere with hormonal activity. If a tumor of the pituitary grows superiorly, it may exert sufficient pressure on the optic chiasma to cause *bitemporal hemianopia,* which is blindness in the temporal field of vision of both eyes. Surgical removal of a neoplasm of the pituitary gland *(hypophysectomy)* may be performed transcranially through the frontal bone or through the nasal cavity and sphenoidal air sinus.

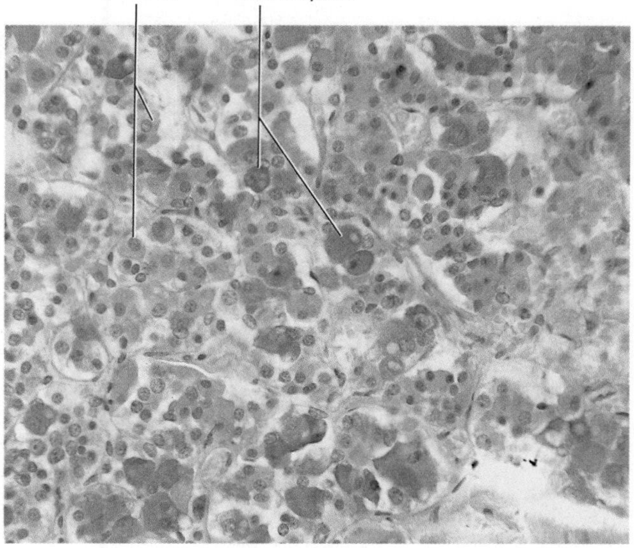

Figure 14.11 The histology of the adenohypophysis.

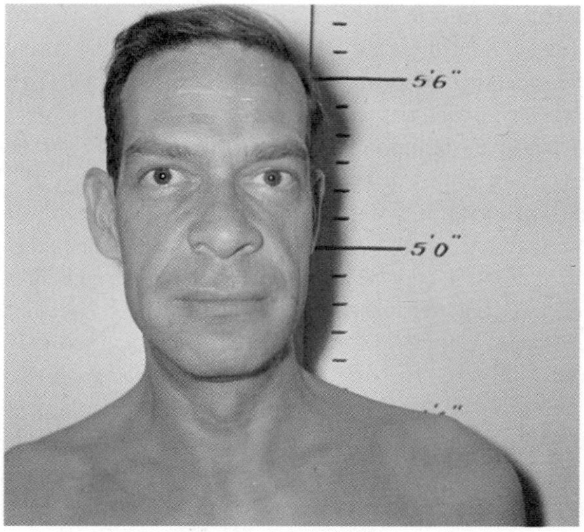

Figure 14.12 Acromegaly is caused by a hypersecretion of growth hormone during adulthood.

Pituitary Hormones

There are nine important hormones secreted by the pituitary gland. The first six in the following list are secreted by the anterior pituitary (the pars distalis of the adenohypophysis). The seventh hormone in the list is secreted by the intermediate lobe (pars intermedia) of the pituitary gland. The last two hormones in the list are produced in the hypothalamus, transported through axons in the infundibulum to the posterior pituitary, and secreted by the posterior pituitary (the pars nervosa of the neurohypophysis) into the blood.

The hormones secreted by the pars distalis are called **trophic** *(trof'ik)* **hormones.** The term *trophic* means food. This term is used because high amounts of the anterior pituitary hormones make their target glands hypertrophy.

1. *Growth hormone (GH).* Growth hormone, or *somatotropin,* regulates the rate of growth of all body cells and promotes mitotic activity. The secretion of GH is regulated by growth hormone-releasing factor (GRF) and growth hormone release-inhibiting factor (GR-IH) from the hypothalamus. The precise mechanism of GH is not understood, but it seems to promote the movement of amino acids through cell membranes and the utilization of these substances in protein synthesis. Pathological hyposecretion of GH during adolescence limits body growth, causing a type of *dwarfism.* Hypersecretion of GH during adolescence may result in *gigantism.* In both of these conditions, body proportions are greatly distorted. Hypersecretion of GH within an adult, after the epiphyses (see chap. 6) are fused, causes *acromegaly.* In this condition, the soft tissues rapidly proliferate and certain body features, such as the hands, feet, nose, jaw, and tongue, are greatly distorted (fig. 14.12).
2. *Thyroid-stimulating hormone (TSH).* TSH, frequently called *thyrotropin,* regulates the hormonal activity of the thyroid gland. The secretion of TSH, however, is partly regulated by the hypothalamus through the secretion of *thyrotropin-releasing factor (TRF).* External factors may influence the release of TSH as well. Exposure to cold, certain illnesses, and emotional stress may trigger an increased output of TSH.
3. *Adrenocorticotropic hormone (ACTH).* ACTH promotes normal functioning of the adrenal cortex. It also acts on all body cells by assisting in the breakdown of fats. The release of ACTH is controlled by a *corticotropin-releasing factor (CRF)* produced in the hypothalamus. As with TSH, stress further influences the release of ACTH.
4. *Follicle-stimulating hormone (FSH).* In the male, FSH stimulates the testes to produce sperm. In the female, FSH regulates the monthly development of the follicle and egg. It also stimulates the secretion of the female sex hormone *estrogen.*
5. *Luteinizing hormone (LH).* LH works with FSH, and together they are referred to as *gonadotrophins,* which means their target tissues are within the gonads or reproductive organs. In the female, LH works with FSH in bringing about ovulation. It also stimulates the formation of the corpus luteum and the production of another female sex hormone, *progesterone* (see chap. 21).

   In the male, the luteinizing hormone is called *interstitial cell stimulating hormone (ICSH)* and stimulates the interstitial cells of the testes to develop and secrete the male sex hormone testosterone (see chap. 20).

   The mechanism that controls the production and release of gonadotrophic hormones is not well understood. It is known, however, that following puberty (see chaps. 20 and 21) the hypothalamus releases a hormone called *gonadotrophin releasing hormone (GnRH)* that regulates the secretion of both LH and FSH.
6. *Prolactin.* Prolactin is secreted in both males and females, but its major functions are only in females after parturition. Prolactin assists other hormones in initiating and sustaining milk production by the mammary glands. The hypothalamus plays an important role in the release

of this hormone through the production of *prolactin release-inhibiting factor (PR-IF)*. When PR-IF is secreted, the secretion of prolactin is inhibited; when PR-IF is not secreted, prolactin is released.

7. *Melanocyte-stimulating hormone (MSH)*. MSH increases skin pigmentation by stimulating the dispersal of melanin granules within melanocytes. Secretion of MSH is regulated by *melanocyte-stimulating hormone releasing factor (MRF)* and *melanocyte-stimulating hormone inhibiting factor (MIF)*, which both come from the hypothalamus.
8. *Oxytocin.* Oxytocin is produced by specialized cells in the hypothalamus. It then travels through axons in the infundibulum to the pars nervosa where it is stored and secreted in response to neural impulses from the hypothalamus. Oxytocin influences physiological activity in the female reproductive system but has no known function in males. Oxytocin is released near the end of gestation and causes uterine contractions during labor. It also has an effect on the mammary glands, causing contractions in the tissue surrounding the alveoli and lactiferous ducts (see chap. 21). This results in the let-down of milk in the mammary glands and is necessary before milk can be removed by an infant's sucking.

Injections of oxytocin may be given to a woman during labor if she is having difficulties in parturition. Increased amounts of oxytocin assist uterine contractions and generally speed up delivery. Oxytocin administration after parturition causes the uterus to regress in size and squeezes the blood vessels, thus minimizing the danger of hemorrhage.

9. *Antidiuretic hormone.* Antidiuretic hormone (ADH) is similar to oxytocin in its site of production and release and in being a polypeptide. The major function of ADH is to act on the kidneys to inhibit the formation of urine, or more specifically, to reduce the amount of water excreted from the kidneys. This hormone also acts as a *vasopressin* to constrict blood vessels, which increases blood pressure.

Diabetes insipidus results from a marked decrease in ADH output caused by trauma or disease to the hypothalamus or neurohypophysis. The symptoms of this disease are polyuria (voiding excessive dilute urine), concentrated body fluids with dehydration, and an immense sensation of thirst.

Oxytocin and antidiuretic hormone are released by the posterior pituitary (pars nervosa of the neurohypophysis). These two hormones, however, are actually produced in neuron cell bodies of the **supraoptic nuclei** and **paraventricular nuclei** of the hypothalamus. These nuclei within the hypothalamus are thus endocrine glands; the hormones they produce are transported along axons of the **hypothalamo-hypophyseal tract** (fig. 14.13) to the posterior pituitary, which stores and later secretes these hormones. The posterior pituitary is thus more a storage organ than a true gland.

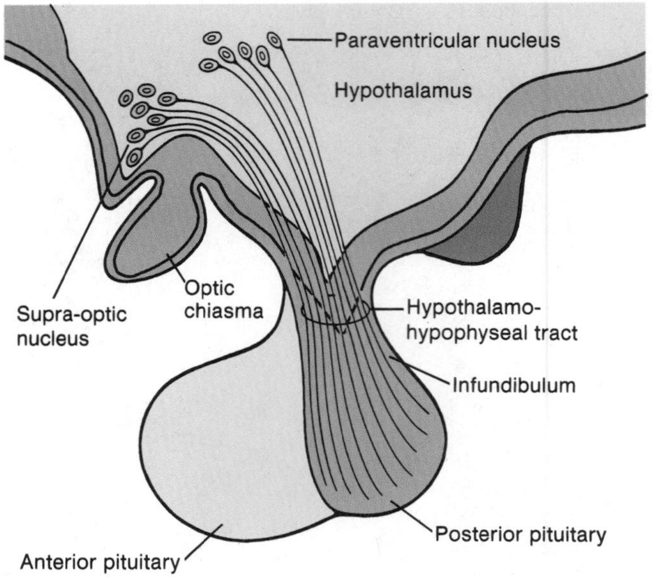

Figure 14.13 The posterior pituitary, or neurohypophysis, stores and secretes oxytocin and antidiuretic hormone produced in neuron cell bodies within the supraoptic and paraventricular nuclei of the hypothalamus. These hormones are transported to the posterior pituitary by nerve fibers of the hypothalamo-hypophyseal tract.

The secretion of oxytocin and ADH from the posterior pituitary is controlled by **neuroendocrine reflexes.** In nursing mothers, for example, the stimulus of sucking acts via sensory nerve impulses to the hypothalamus to stimulate the reflex secretion of oxytocin. The secretion of ADH is stimulated by osmoreceptor neurons in the hypothalamus in response to a rise in blood osmotic pressure; its secretion is inhibited by sensory impulses from stretch receptors in the left atrium of the heart in response to a rise in blood volume.

At one time the anterior pituitary was called the master gland because it secretes hormones that regulate some other endocrine glands (see table 14.4 and fig. 14.14). Adrenocorticotrophic hormone (ACTH), thyroid-stimulating hormone (TSH), and the gonadotrophic hormones (FSH and LH) stimulate the adrenal cortex, thyroid, and gonads, respectively, to secrete their hormones. The anterior pituitary hormones also have a trophic effect on their target glands, in that the structure and health of the target glands depend on adequate stimulation by anterior pituitary hormones. The anterior pituitary, however, is not really the master gland, because secretion of its hormones is in turn controlled by hormones secreted by the hypothalamus.

Releasing and inhibiting hormones from the hypothalamus travel through the **hypothalamo-hypophyseal portal system** to control the secretion of hormones from the anterior pituitary. Neurons in the hypothalamus secrete hormones into the region of the **median eminence** where they enter a network of primary capillaries (fig. 14.15). Venous drainage through the pituitary stalk transports the hypothalamic hormones to a network of secondary capillaries within the anterior pituitary. This system is considered a *portal system* because the

Table 14.4 Hormones released by the pituitary gland

Hormone	Action	Source of regulation
Adenohypophysis		
Growth hormone (GH), or somatotropin	Regulates mitotic activity and growth of body cells; promotes movement of amino acids through plasma membranes	Growth hormone-releasing factor (GRF) and growth hormone release-inhibiting factor (GR-IH) from the hypothalamus
Thyroid-stimulating hormone (TSH), or thyrotropin	Regulates hormonal activity of thyroid gland	Thyrotropin-releasing factor (TRF) from the hypothalamus
Adrenocorticotropic hormone (ACTH)	Controls secretion of certain hormones from the adrenal cortex; assists in breakdown of fats	Corticotropin-releasing factor (CRF) from the hypothalamus
Follicle-stimulating hormone (FSH)	In male: stimulates production of sperm cells; in female: regulates follicle development in ovary and stimulates secretion of estrogen	Gonadotrophin-releasing factor from the hypothalamus
Luteinizing hormone (LH, or ICSH in males)	Promotes secretion of sex hormones; in females, plays role in release of ovum; stimulates formation of corpus luteum and production of progesterone; in males, stimulates testosterone secretion	Gonadotrophin-releasing factor from the hypothalamus
Prolactin	Promotes secretion of milk from mammary glands (lactation)	Hypothalamus through production of prolactin release-inhibiting factor (PR-IF)
Melanocyte-stimulating hormone (MSH)	Stimulates pigmentation within the melanocytes of the skin	Melanocyte-stimulating hormone releasing factor (MRF) and melanocyte-stimulating hormone inhibiting factor (MIF), both from the hypothalamus
Neurohypophysis		
Oxytocin	Stimulates contractions of muscles in uterine wall; causes contraction of muscles in mammary glands	Hypothalamus in response to stretch in uterine walls and stimulation of breasts
Antidiuretic hormone	Reduces water loss from kidneys; elevates blood pressure	Hypothalamus in response to changes in blood-water concentration

network of secondary capillaries in the anterior pituitary is downstream from the network of primary capillaries in the median eminence. (This is analogous to the hepatic portal system that delivers venous blood from the intestine to the liver—chapter 16.)

1. List the hormones secreted by the anterior pituitary, and describe how the hypothalamus controls the secretion of each hormone.
2. Which hormone secreted by the anterior pituitary does not affect another endocrine gland?
3. List the hormones secreted by the posterior pituitary. Describe the site of origin of these hormones and the mechanisms by which their secretions are regulated.
4. Which of the hormones secreted by the pituitary gland affect only females?
5. How does the hypothalamus regulate hormonal secretion from the pituitary?

Thyroid and Parathyroid Glands

The thyroid gland secretes thyroxine and triiodothyronine, which participate in the regulation of energy metabolism and are critically important for proper growth and development. The thyroid also secretes calcitonin, which may antagonize the action of parathyroid hormone in the regulation of calcium and phosphate balance.

Objective 13. Describe the location and structure of the thyroid gland, and list the actions of the thyroid hormones.

Objective 14. Describe the location and structure of the parathyroid glands, and list the actions of parathyroid hormone.

Description of the Thyroid Gland

The **thyroid gland** is positioned just below the larynx (fig. 14.16). This gland consists of two lobes that lie on both sides of the trachea and are connected anteriorly by a broad **isthmus.** The thyroid is the largest of the endocrine glands, weighing between 20 and 25 g. It receives an abundant blood supply (80–120 ml/minute) through the paired superior thyroid branches of the external carotid arteries and the paired inferior thyroid branches of the subclavian arteries. The venous return is through the paired superior and middle thyroid veins that pass into the internal jugular veins and through the inferior thyroid veins that empty into the brachiocephalic veins.

On a microscopic level, the thyroid gland consists of many spherical hollow sacs called **thyroid follicles** (fig. 14.17). These follicles are lined with a simple cuboidal epithelium

thyroid: Gk. *thyreos,* oblong shield
isthmus: L. *isthmus,* narrow portion

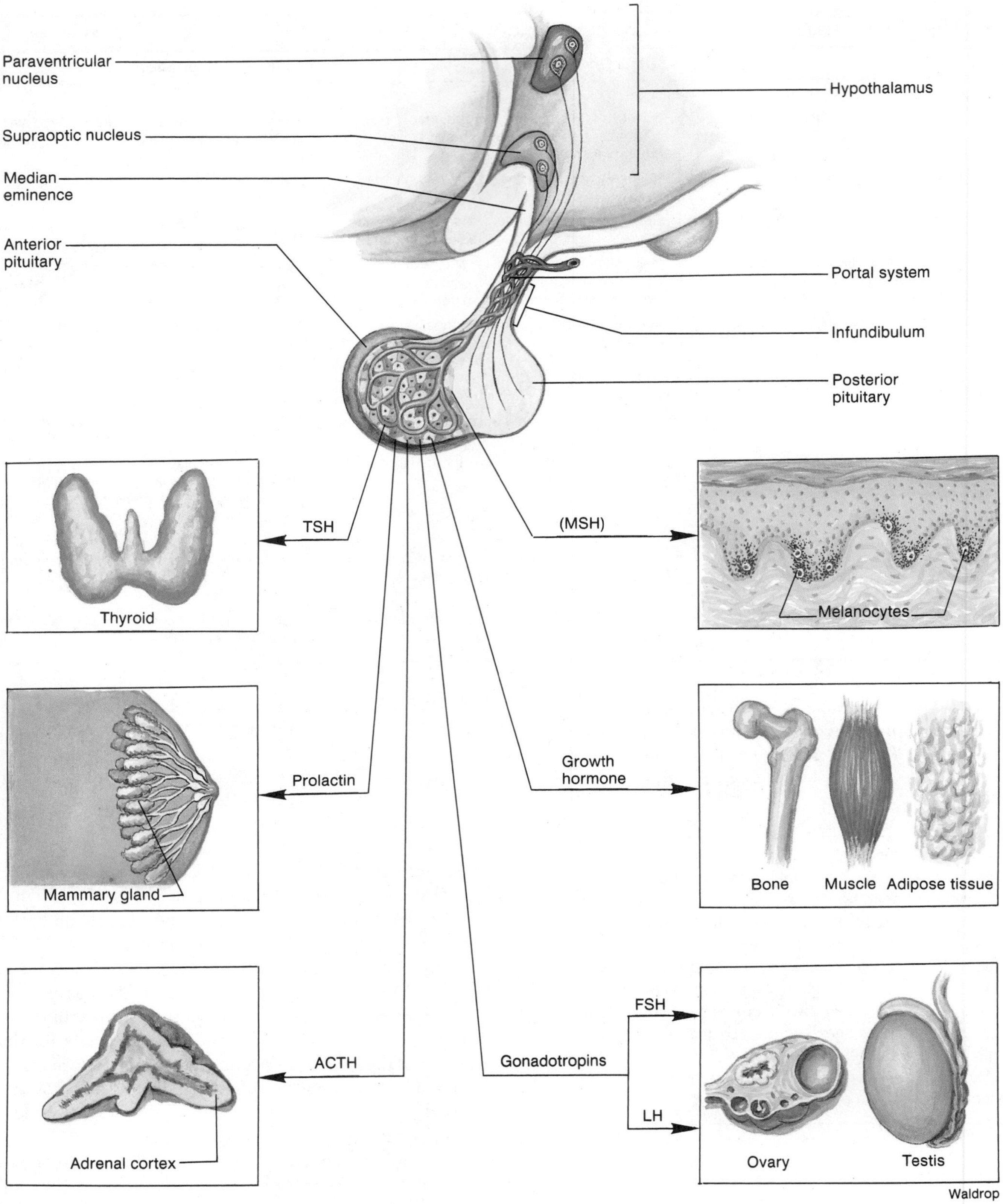

Figure 14.14 The hormones secreted by the anterior pituitary and the target organs for those hormones.

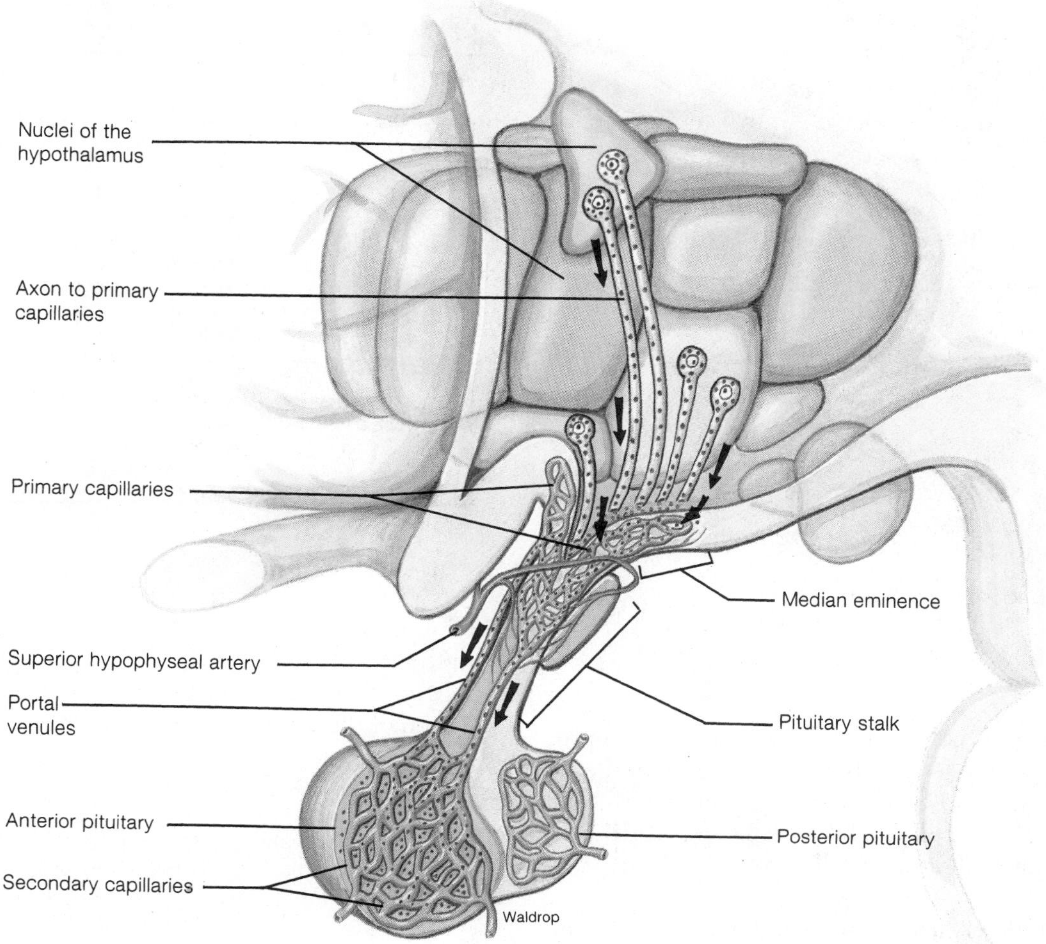

Figure 14.15 The hypothalmo-hypophyseal portal system. Hypothalmic hormones (shown as dots) enter this system in the primary capillaries of the median eminence and are transported through the portal veins of the pituitary stalk to the secondary capillaries of the anterior pituitary.

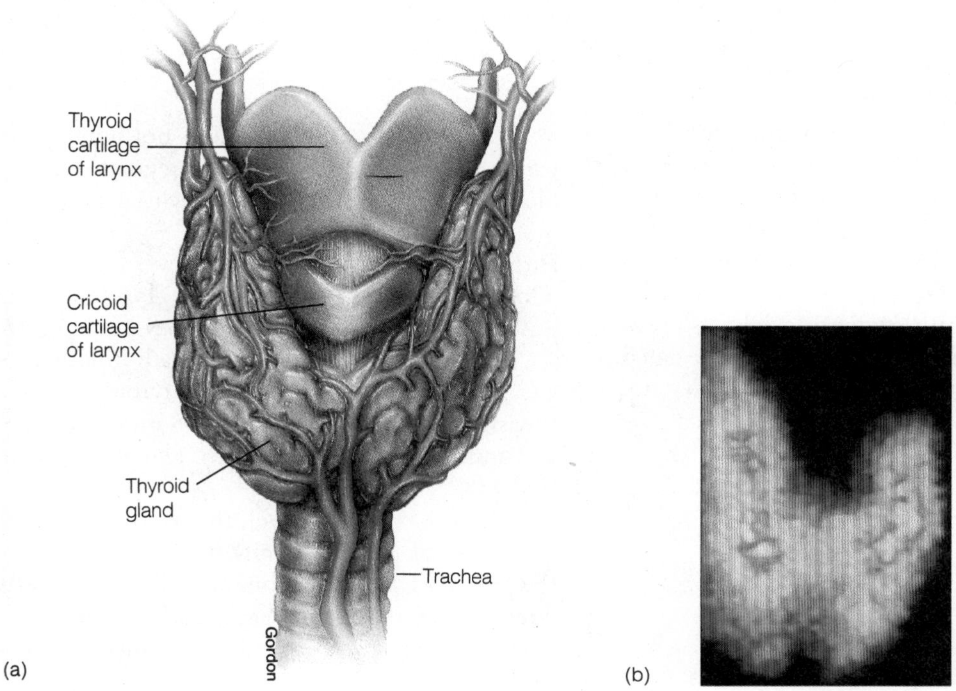

Figure 14.16 The thyroid gland. (*a*) Its relationship to the larynx and trachea and (*b*) a scan of the thyroid gland twenty-four hours after the intake of radioactive iodine.

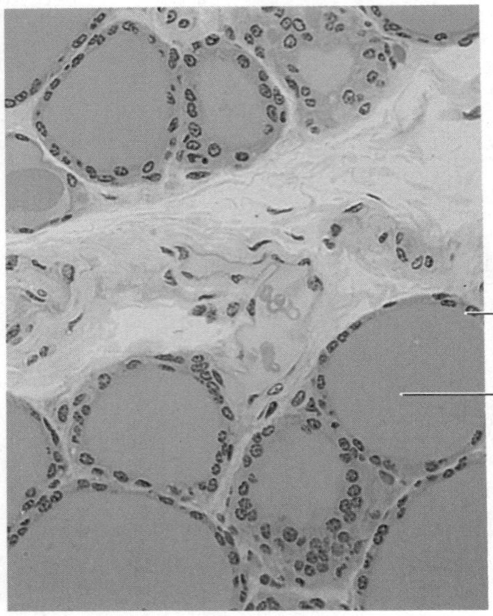

Figure 14.17 The histology of the thyroid gland, showing numerous thyroid follicles. Each follicle consists of principal cells surrounding the fluid known as colloid.

Table 14.5 Hormones of the thyroid gland

Hormone	Action	Source of regulation
Thyroxine (T_4)	Increases rate of protein synthesis and rate of energy release from carbohydrates; regulates rate of growth; stimulates maturity of nervous system	Hypothalamus and release of TSH from adenohypophysis of the pituitary gland
Triiodothyronine (T_3)	Same as above	Same as above
Calcitonin (thyrocalcitonin)	Lowers blood calcium by inhibiting the release of calcium from bone tissue	Calcium levels in blood

composed of **principal cells,** which synthesize the two principal thyroid hormones (see table 14.5). The interior of the follicles contains **colloid** *(kol'oid),* which is a protein-rich fluid. Between the follicles are epithelial cells called **parafollicular cells,** which produce a hormone called *calcitonin* (or *thyrocalcitonin*).

The thyroid is innervated by postganglionic neurons from the superior and middle cervical sympathetic ganglia and preganglionic neurons from ganglia derived from the second through the seventh thoracic segment of the spinal cord.

Functions of the Thyroid Gland

The thyroid gland produces two major hormones, *thyroxine* (T_4) and *triiodothyronine* (T_3), and a minor hormone, *calcitonin (thyrocalcitonin).* The release of thyroxine and triiodothyronine is controlled by the hypothalamus and by the TSH secreted from the adenohypophysis of the pituitary gland. Thyroxine and triiodothyronine are stored in the thyroid follicles and released as needed to control the metabolic rate of the body. More specifically, they act to increase the rate of protein synthesis and the rate of energy release from the carbohydrates. They also regulate the rate of growth in young persons and are associated with sexual maturity and early maturation of the nervous system.

Iodine is the most common component of thyroxine and triiodothyronine, and a continual intake of iodine is essential for normal thyroid function. Seafood contains adequate amounts of iodine, and commercial salt generally has iodine as an additive. Absorbed iodine is transported through the blood to the thyroid gland, where an active transport mechanism called an iodine pump moves the iodides into the follicle cells. Here they combine with amino acids in the synthesis of thyroid hormones.

colloid: Gk. *kolla,* glue

Most of the thyroxine in blood is attached to carrier proteins. Only the very small percentage of thyroxine that is free in the plasma can enter the target cells. There it is converted to triiodothyronine and attached to nuclear receptor proteins. Through the activation of genes, thyroid hormones stimulate protein synthesis, promote maturation of the nervous system, and increase the rate of energy utilization by the body.

Thyroid-stimulating hormone (TSH) from the anterior pituitary stimulates the thyroid to secrete thyroxine and exerts a trophic effect on the thyroid gland. This trophic effect is dramatically revealed in people who develop an *iodine-deficiency (endemic) goiter.* In the absence of sufficient dietary iodine the thyroid cannot produce adequate amounts of T_4 and T_3. The resulting lack of negative feedback inhibition causes abnormally high levels of TSH secretion, which in turn stimulate the abnormal growth of the thyroid (a goiter). These events are summarized in figure 14.18.

Calcitonin is a polypeptide hormone produced by the parafollicular cells. It functions to regulate calcium levels in the blood by inhibiting the rate at which calcium leaves bone tissue.

Parathyroid Glands

The small, flattened **parathyroid glands** are embedded in the posterior surfaces of the lateral lobes of the thyroid gland (fig. 14.19). There are usually four parathyroid glands: a **superior** and an **inferior** pair. Each parathyroid gland is a small yellowish brown body measuring 3–8 mm (0.1–0.3 in.) in length, 2–5 mm (.07–0.2 in.) in width, and about 1.5 mm (0.05 in.) in depth.

On a microscopic level, the parathyroids are composed of two types of epithelial cells (fig. 14.20). The cells that synthesize parathyroid hormone are called **chief cells** and are scattered among **oxyphil** *(ok'se-fil)* **cells.** Oxyphil cells support the chief cells and are believed to produce reserve quantities of parathyroid hormone.

The blood supply and drainage of the parathyroid glands is similar to that of the thyroid gland. The innervation of the parathyroids, however, is a bit different. The parathyroids re-

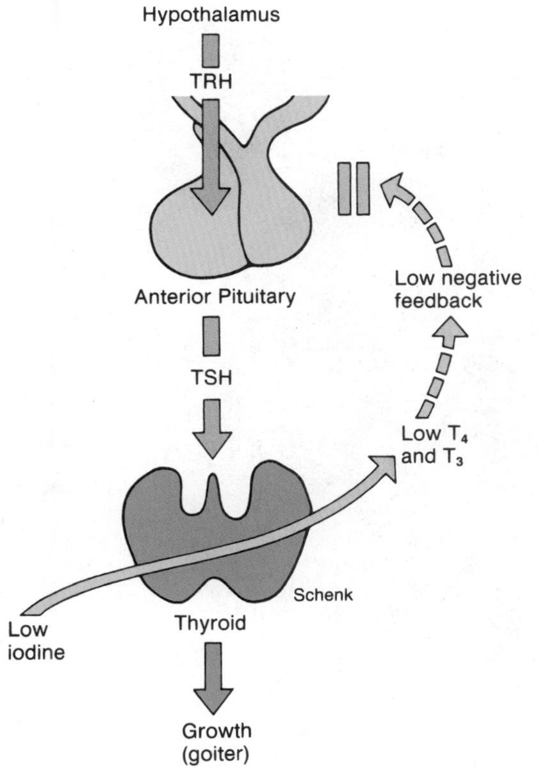

Figure 14.18 The mechanism of goiter formation in iodine deficiency.

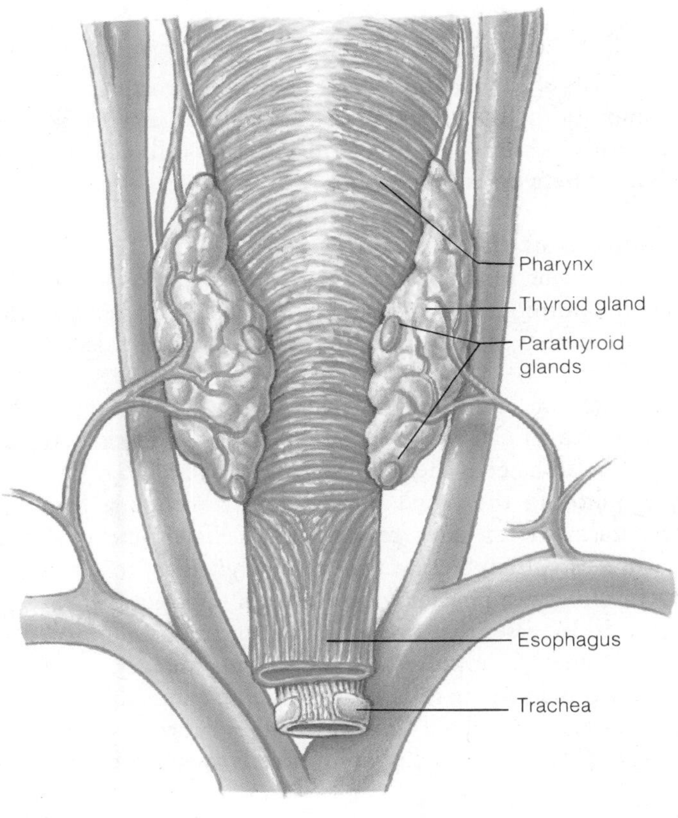

Figure 14.19 A posterior view of the parathyroid glands.

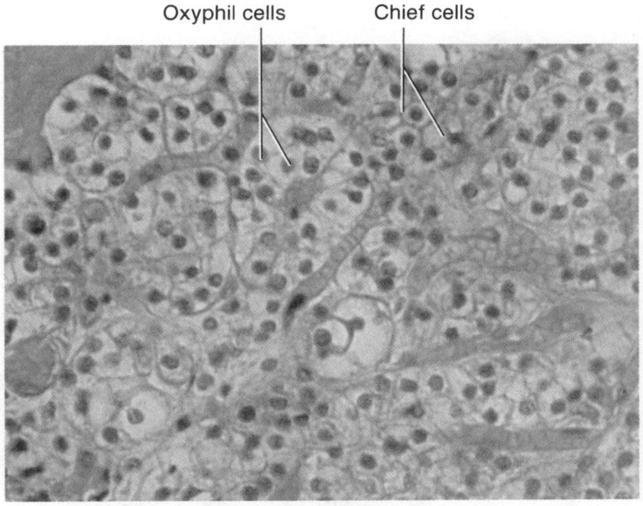

Figure 14.20 The histology of the parathyroid gland.

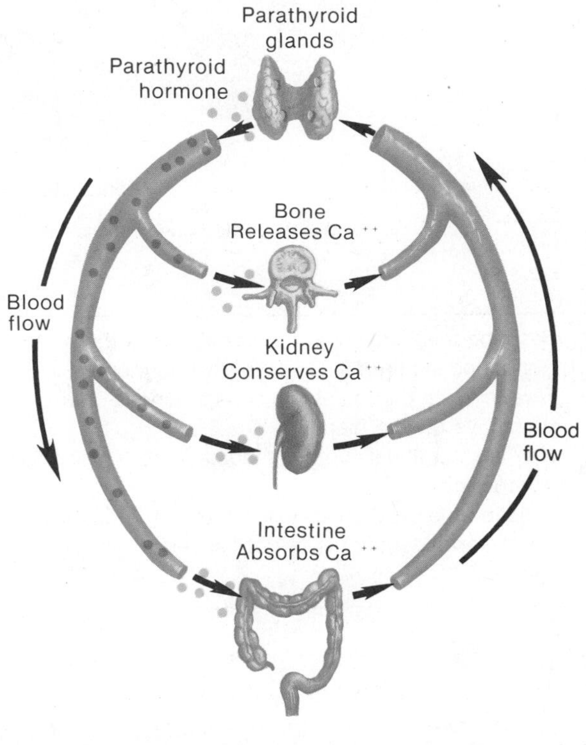

Figure 14.21 Actions of parathyroid hormone. An increased level of parathyroid hormone causes the bones to release calcium, the kidneys to conserve calcium loss through the urine, and the absorption of calcium through the intestinal wall. Negative feedback of increased calcium levels in the blood inhibits the secretion of this hormone.

ceive neurons from the pharyngeal branches of the vagus nerves as well as neurons arising from the cervical sympathetic ganglia.

The parathyroid glands secrete one hormone called **parathyroid hormone (PTH).** This hormone promotes a rise in blood calcium levels by acting on the bones, kidneys, and intestine (fig. 14.21).

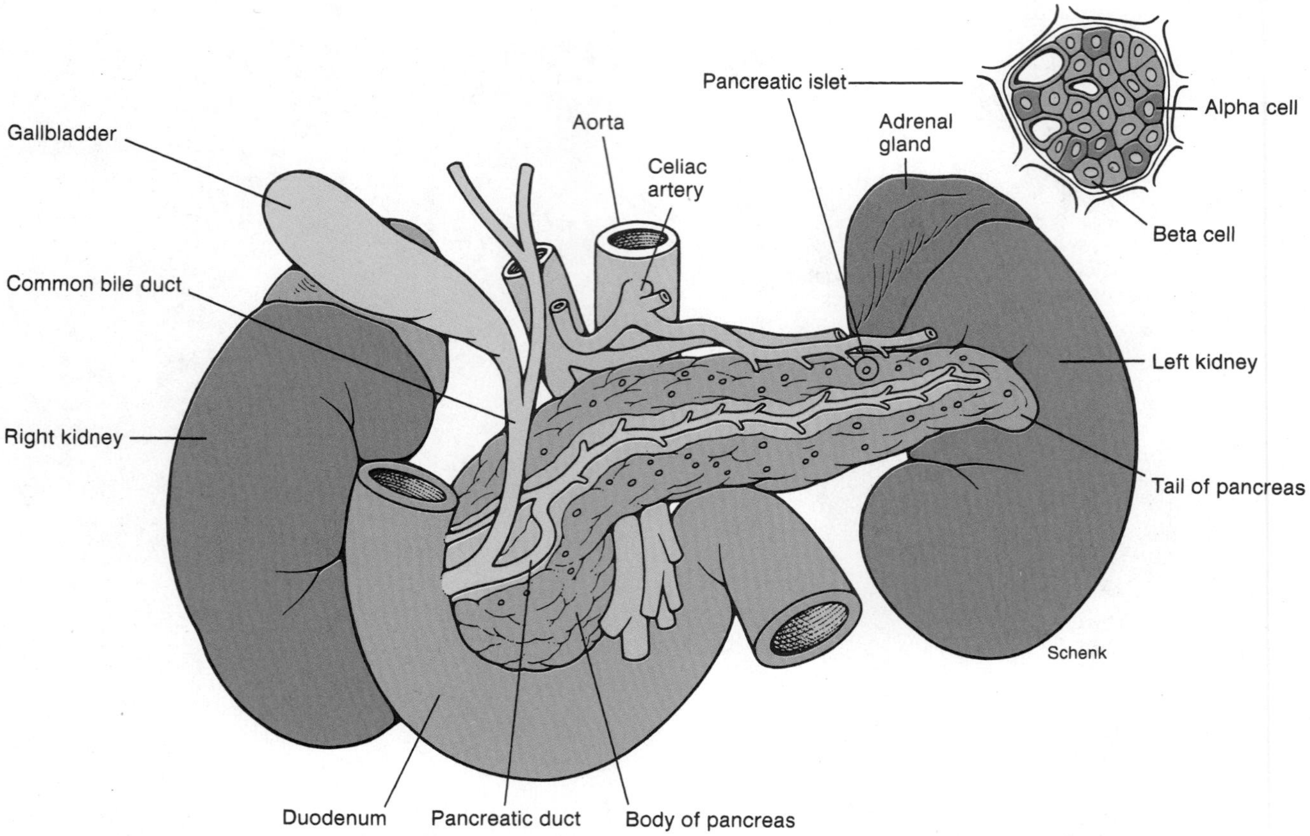

Figure 14.22 The pancreas and the associated pancreatic islets.

1. Describe the location and structure of the thyroid gland, and list the effects of thyroid hormones.
2. What may be the metabolic consequences of an overactive thyroid gland?
3. Why is a continual supply of iodine important for body metabolism?
4. Describe the location and structure of the parathyroid glands, and identify the target tissues of parathyroid hormone.

Pancreas

The pancreatic islets in the pancreas secrete two hormones, insulin and glucagon, which are critically involved in the regulation of metabolic balance in the body.

Objective 15. Describe the structure of the endocrine portion of the pancreas and the origin of insulin and glucagon.

Objective 16. Describe the actions of insulin and glucagon.

Description of the Pancreas

The **pancreas** is both an endocrine and an exocrine gland. The gross structure of this gland and its exocrine functions in digestion are described in chapter 18. The endocrine portion of the pancreas consists of scattered clusters of cells called **pancreatic islets** (islets of Langerhans). These endocrine structures are most common in the body and tail of the pancreas (fig. 14.22) and principally consist of two types of secretory cells called **alpha cells** and **beta cells** (fig. 14.23).

Functions of the Pancreas

The endocrine function of the pancreas is to produce and secrete the hormones *glucagon* and *insulin.* The alpha cells of the pancreatic islets secrete glucagon, and the beta cells secrete insulin.

Glucagon stimulates the liver to convert glycogen into glucose, which causes the blood glucose level to rise. Apparently, the alpha cells themselves regulate the glucagon output by monitoring the blood glucose level through a feedback mechanism. If for some reason alpha cells secrete glucagon continuously, high blood sugar, or *hyperglycemia,* may result.

Insulin has a physiological function opposite that of glucagon: it decreases the level of blood sugar. Insulin promotes the movement of glucose through cell membranes, especially in muscle and adipose cells. As the glucose enters the cells, the sugar level of the blood decreases. Other functions of insulin include stimulating muscle and liver cells to convert glucose to

islets of Langerhans: from Paul Langerhans, German anatomist, 1847–1888
insulin: L. *insula,* island

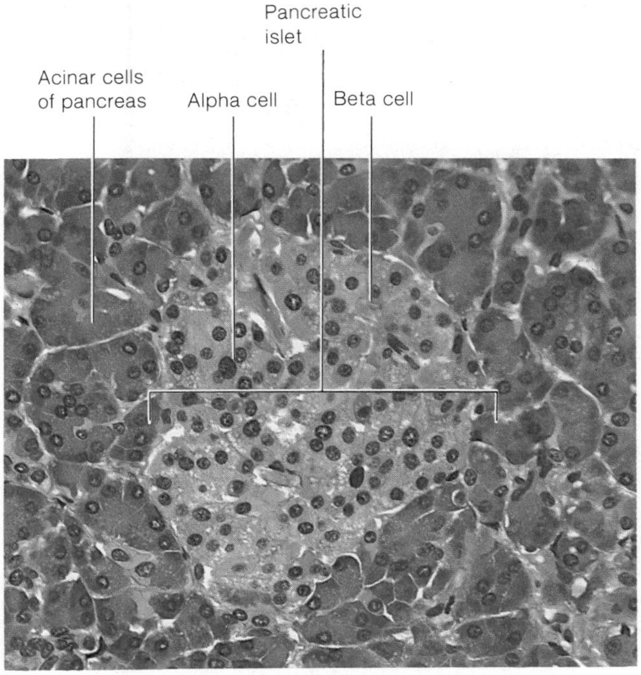

Figure 14.23 The histology of the pancreatic islets.

Table 14.6 Hormones of the pancreas

Hormone	Action	Source of regulation
Glucagon	Stimulates the liver to convert glycogen into glucose, causing the blood glucose level to rise	Blood glucose level through negative feedback in pancreas
Insulin	Promotes movement of glucose through cell membranes; stimulates liver to convert glucose into glycogen; promotes the transport of amino acids into cells; assists in synthesis of proteins and fats	Blood glucose level through negative feedback in pancreas

glycogen, helping amino acids to enter cells, and assisting the synthesis of proteins and fats. Failure of beta cells to produce insulin causes the common hereditary disease *diabetes mellitus.*

The action of the hormones from the pancreatic islets is summarized in table 14.6.

1. Where are the specific sites of glucagon and insulin production?
2. What is the function of glucagon? Of insulin?
3. What disease is caused by the insufficient production of insulin?

Adrenal Glands

The adrenal cortex and adrenal medulla are structurally and functionally different. The adrenal medulla secretes catecholamine hormones, which complement the action of the sympathetic division of the ANS. The adrenal cortex secretes corticosteroids, which participate in the regulation of mineral balance, energy balance, and reproductive function.

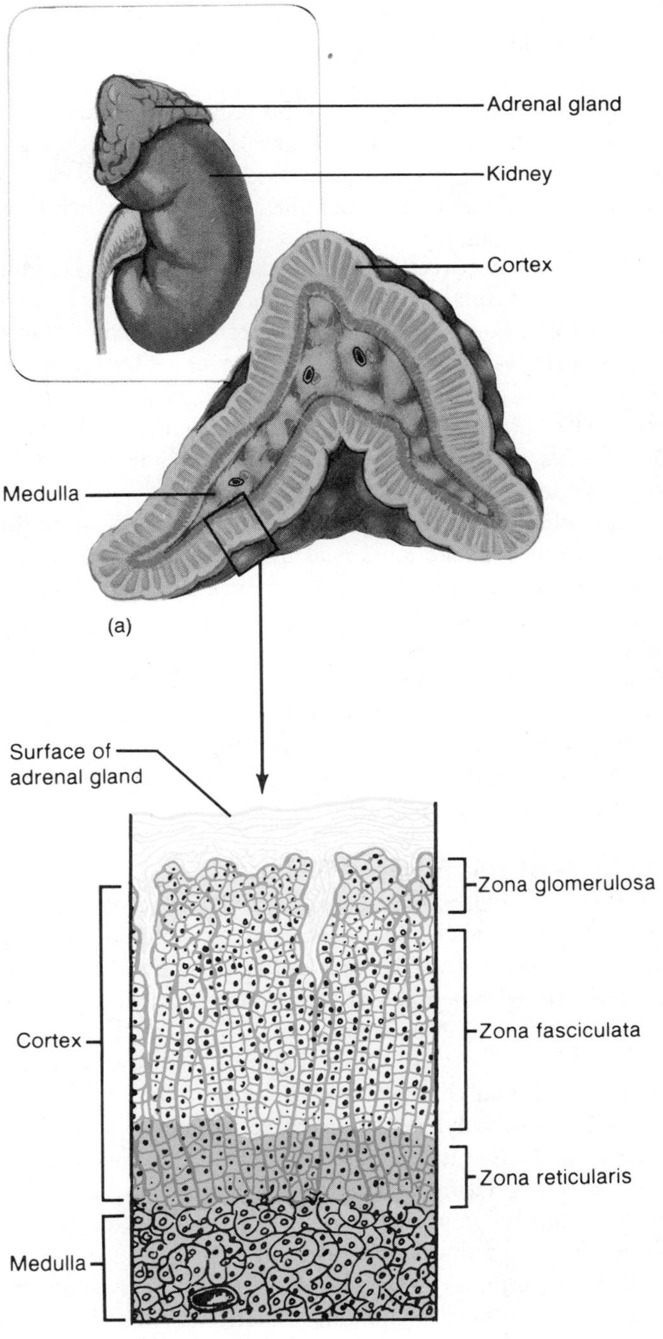

Figure 14.24 The structure of the adrenal gland. (*a*) The gross structure and (*b*) the histological structure showing the three zones of the adrenal cortex.

Objective 17. Describe the location of the adrenal glands, and distinguish between the adrenal cortex and the adrenal medulla.

Objective 18. List the hormones secreted by the adrenal glands, and discuss their effects in the body.

Description of the Adrenal Glands

The **adrenal glands** (also called **suprarenal glands**) are paired organs that cap the superior borders of the kidneys (fig. 14.24).

The adrenal glands, along with the kidneys, are retroperitoneal and are embedded against the muscles of the back in a protective pad of fat.

Each adrenal gland is about 50 mm (2 in.) in length, 30 mm (1.1 in.) in width, and 10 mm (0.4 in.) in depth, and is generally pyramidal in appearance. Each adrenal gland consists of an outer cortex and inner medulla (fig. 14.24), which function as separate glands.

The **adrenal cortex** makes up the bulk of the gland and is histologically subdivided into three zones: an outer **zona glomerulosa** *(glo-mer''u-lo'sah),* an intermediate **zona fasciculata** *(fah-sik''u-la'tah),* and an inner **zona reticularis.** The **adrenal medulla** is composed of tightly packed clusters of **chromaffin** *(kro-maf'in)* **cells,** which are arranged around blood vessels. Each cluster of chromaffin cells receives direct autonomic innervation.

Like other endocrine glands, the adrenal glands are highly vascular. Three separate suprarenal arteries supply blood to each adrenal gland. One arises from the inferior phrenic artery, another from the aorta, and a third is a branch of the renal artery. The venous drainage goes through a suprarenal vein into the inferior vena cava for the right adrenal gland and through the suprarenal vein into the left renal vein for the left adrenal gland.

The adrenal glands are innervated by preganglionic neurons of the splanchnic nerves and by fibers of the celiac and associated sympathetic plexuses.

Functions of the Adrenal Glands

Over thirty hormones have been identified as being produced by the adrenal cortex. These hormones are called **corticosteroids** *(kor''ti-ko-ste'roids),* or **corticoids,** for short. The adrenal corticoids are grouped into three types: mineralocorticoids, glucocorticoids, and gonadocorticoids.

The *mineralocorticoids* are produced by the zona glomerulosa of the adrenal cortex and regulate the concentrations of extracellular electrolytes. Of the three hormones secreted by this layer, *aldosterone* is the most important. Aldosterone affects the kidneys, regulating the amounts of sodium and potassium that are eliminated in urine. At the same time, it promotes water reabsorption and reduces urine output.

The *glucocorticoids* are produced primarily by the zona fasciculata of the adrenal gland and influence the metabolism of carbohydrates, proteins, and fats. The glucocorticoids also help the body resist stress, promote vasoconstriction, and act as anti-inflammatory compounds. The most abundant and physiologically important glucocorticoid is *cortisol (hydrocortisone).*

Cortisol and related anti-inflammatory compounds are commonly used to treat patients suffering from arthritis and various allergies. They are also frequently used in sports medicine to treat traumatized joints and in organ transplants and tissue grafts to help keep the host from rejecting donor tissues. Unfortunately, cortisol inhibits the regeneration of connective tissues and therefore should be used sparingly and only when necessary.

Table 14.7 Hormones of the adrenal cortex

Hormone	Action	Source of regulation
Mineralocorticoids	Regulate the concentration of extracellular electrolytes, especially sodium and potassium	Electrolyte concentration in blood
Glucocorticoids	Influence the metabolism of carbohydrates, proteins, and fats; promote vasoconstriction; anti-inflammatory compounds	ACTH from the adenohypophysis of the pituitary gland in response to stress
Gonadocorticoids	Supplement the sex hormones from the gonads	

Gonadocorticoids are the sex hormones that are secreted by the zona reticularis of the adrenal cortex. The majority of these hormones are adrenal *androgens,* but small quantities of adrenal *estrogens* and *progesterones* are present as well. It is thought that these hormones supplement the sex hormones produced in the gonads. There is also evidence that the concentration of androgens in both the male and female plays a role in determining the sex drive.

A summary of the hormones produced by the adrenal cortex is presented in table 14.7.

The chromaffin cells of the adrenal medulla produce two closely related hormones: *epinephrine* and *norepinephrine.* Both of these hormones are classified as amines (more specifically, as catecholamines), because they consist of amino acids without peptide bonds.

The effects of these hormones are similar to those caused by stimulation of the sympathetic division of the ANS, except that the hormonal effect lasts about ten times longer. The hormones from the adrenal medulla increase cardiac output and heart rate, dilate coronary blood vessels, increase mental alertness, increase the respiratory rate, and elevate metabolic rate. A comparison of the effects of epinephrine and norepinephrine are presented in table 14.8.

The adrenal medulla is innervated by sympathetic neurons. The impulses are initiated from the hypothalamus via the spinal cord in response to stress. Stress therefore activates the adrenal medulla as well as the adrenal cortex. Activation of the adrenal medulla prepares the body for greater physical performance—the *fight or flight* response (fig. 14.25).

Excessive stimulation of the adrenal medulla can result in depletion of the body's energy reserves, and high levels of corticosteroid secretion from the adrenal cortex can significantly impair the immune system. It is reasonable to expect, therefore, that prolonged stress can result in increased susceptibility to disease. Indeed, many studies show that prolonged stress results in an increased incidence of cancer and other diseases.

Table 14.8 Comparison of the hormones from the adrenal medulla

Epinephrine	Norepinephrine
Elevates blood pressure because of increased cardiac output and peripheral vasoconstriction	Elevates blood pressure because of generalized vasoconstriction
Accelerates respiratory rate and dilates respiratory passageways	Similar effect but to a lesser degree
Increases efficiency of muscular contraction	Similar effect but to a lesser degree
Increases rate of glycogen breakdown into glucose, so level of blood glucose rises	Similar effect but to a lesser degree
Increases rate of fatty acid released from fat, so level of blood fatty acids rises	Similar effect but to a lesser degree
Increases release of ACTH and TSH from the adenohypophysis of the pituitary gland	No effect

1. Describe the location and appearance of the adrenal gland. Diagram and label the layers of the adrenal gland that can be seen in a sagittal section as, for example, in a histological slide.
2. List the hormones of the adrenal medulla, and describe their effects.
3. List the categories of corticosteroids and the zones of the adrenal cortex that secrete these hormones.

Gonads and Other Endocrine Glands

The gonads produce sex hormones, which control the development and function of the male and female reproductive systems. Additionally, many other organs secrete hormones that help regulate digestion, metabolism, growth, and immunity.

Objective 19. Discuss the endocrine functions of the gonads.

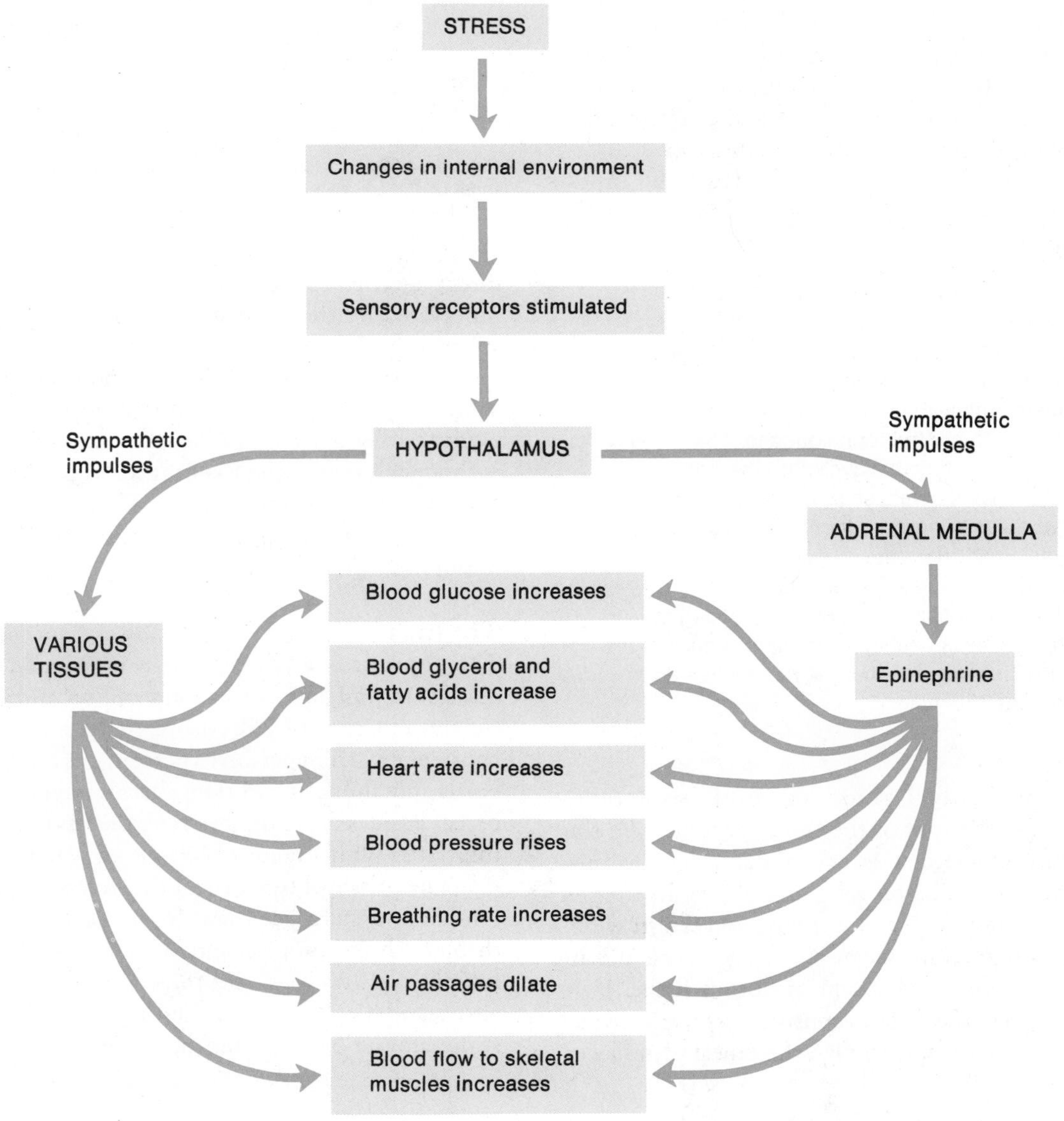

Figure 14.25 Stress is the physiological change in the body that prepares it for fight or flight. These changes are elicited by neural impulses from the hypothalamus that directly influence body tissue and by impulses to the adrenal medulla that secondarily influence the same body tissues through the release of epinephrine.

Objective 20. Describe the structure and anatomical location of the pineal and thymus glands, and indicate their endocrine functions.

Gonads

The **gonads** are the male and female primary sex organs. The male gonads are called **testes,** and the female gonads are called **ovaries.** They are mixed glands in that they produce both sex hormones and sex cells, or **gametes** (see chaps. 20 and 21).

Testes The **interstitial cells** of the testes produce and secrete the male sex hormone *testosterone.* Testosterone controls the development and function of the male secondary sexual organs such as the penis and accessory glands. It also promotes the male secondary sexual characteristics (see chap. 20) and somewhat determines the sexual drive.

Ovaries The endocrine function of the ovaries is the production of the female sex hormones, *estrogen* and *progesterone.* Estrogen is produced in the **ovarian (graafian) follicles** and **corpus luteum** of the ovaries. It is also produced in the placenta, adrenal cortex, and even the testes of the male. Estrogen is responsible for (1) development and function of the secondary sexual organs, (2) menstrual changes of the uterus, (3) development of the female secondary sexual characteristics (see chap. 21), and (4) regulation of the sexual drive.

Progesterone is produced by the corpus luteum and is primarily associated with pregnancy in preparing the uterus for implantation and preventing abortion of the fetus.

The desire to avoid conception and prevent pregnancy in humans is almost as ancient and widespread as the behavior for promoting it. The age of hormonal biochemistry has produced a science of birth control. The female, rather than the male, has been the target of hormonal birth control techniques for the following reasons: (1) ovulation is cyclic; (2) the genetic structure of each ovum is established by the time of the female's birth, whereas sperm production is a continuous process in the male and therefore is more vulnerable to genetic damage; (3) the female system has more potential sites for hormonal interference than does the male; and (4) the female is usually more conscientious about birth control because she has far more invested in pregnancy than does the male.

Pineal Gland

The small, cone-shaped **pineal gland,** or **epiphysis cerebri** (fig. 14.2), is located in the roof of the third ventricle near the corpora quadrigemina, where it is encapsulated by the meninges covering the brain. The pineal gland of a child weighs about 0.2 g and is 5–8 mm (0.2–0.3 in.) long and 9 mm wide. The gland begins to regress in size at about age seven and in the adult appears as a thickened strand of fibrous tissue. Histologically speaking, the pineal gland consists of specialized parenchymal and neuroglial cells. Although the pineal gland lacks direct nervous connection to the rest of the brain, it is highly innervated by the sympathetic division of the ANS from the superior cervical ganglion.

The function of the pineal gland in some other vertebrates is well known but is not well understood in humans. There is evidence that it secretes a hormone called *melatonin.* Melatonin is thought to affect the hypothalamus by stimulating the secretion of certain releasing factors. These factors in turn affect the secretion of gonadotrophin and the ACTH from the adenohypophysis of the pituitary gland.

pineal: L. *pinea,* pine cone

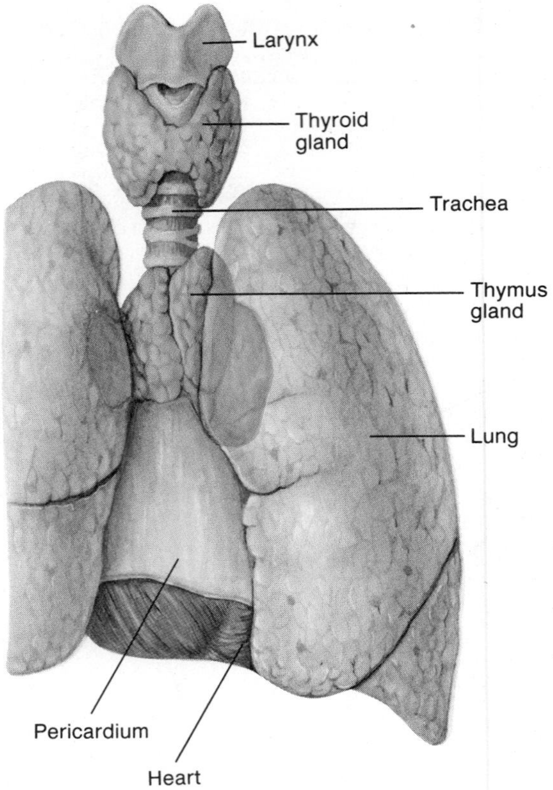

Figure 14.26 The thymus is a bilobed organ within the mediastinum of the thorax.

Thymus

The **thymus** is a bilobed organ positioned in the upper mediastinum, in front of the aorta and behind the manubrium of the sternum (fig. 14.26). Although the size of the thymus varies considerably from person to person, it is relatively large in newborns and children and then sharply regresses in size after puberty. Besides decreasing in size, the thymus of adults becomes infiltrated with strands of fibrous and fatty connective tissue.

The principal function of the thymus is associated with the lymphatic system (see chap. 16) in maintaining body immunity through the maturation and discharge of a specialized group of lymphocytes called *T-cells (thymus-dependent cells).* The thymus also secretes a hormone called *thymosin,* which is believed to stimulate the T-cells after they leave the thymus.

thymus: Gk. *thymos,* a warty excrescence

Table 14.9 Summary of the physiological effects of gastrointestinal hormones

Hormone	Secreted by	Effects
Gastrin	Stomach	Stimulates parietal cells to secrete HCl Stimulates chief cells to secrete pepsinogen Maintains structure of gastric mucosa
Secretin	Small intestine	Stimulates water and bicarbonate secretion in pancreatic juice Potentiates actions of cholecystokinin on pancreas
Cholecystokinin (CCK)	Small intestine	Stimulates contraction of the gallbladder Stimulates secretion of pancreatic juice enzymes Potentiates action of secretin on pancreas Maintains structure of exocrine pancreas (acini)
Gastric inhibitory peptide (GIP)	Small intestine	Inhibits gastric emptying Inhibits gastric acid secretion Stimulates secretion of insulin from pancreatic islets

From Stuart Ira Fox, *A Laboratory Guide to Human Physiology: Concepts and Clinical Applications,* 4th ed.

Stomach and Small Intestine

Certain cells of the mucosal linings of the stomach and small intestine secrete hormones that promote digestive activities. These hormones are summarized in table 14.9.

Placenta

The **placenta** is the organ responsible for nutrient and waste exchange between the fetus and the mother (see chap. 22). The placenta is also an endocrine gland; it secretes large amounts of estrogens and progesterone, as well as a number of polypeptide and protein hormones that are similar to some hormones secreted by the anterior pituitary gland. These latter hormones include *human chorionic gonadotrophin (hCG),* which is similar to LH, and *somatomammotrophin,* which is similar in action to growth hormone and prolactin. Detection of hCG in urine is an indication of pregnancy and is the basis of a home pregnancy test.

---

1. Where is testosterone produced? What are its functions?
2. What are the functions of estrogen and progesterone? Where are they produced?
3. Describe the location of the pineal gland, and discuss the possible function of melatonin.
4. Describe the location and function of the thymus gland.

Clinical Considerations

Endocrinology is the science concerned with the structure and function of the glands of the endocrine system as well as the consequences of glandular dysfunctions. The most frequent endocrine disorders result from an unusual increase (hypersecretion) or decrease (hyposecretion) of hormones. Hypersecretion of an endocrine gland is generally caused by hyperplasia (increase in size) of a gland, whereas hyposecretion of hormones is caused by destruction or atrophy of a gland.

The diagnosis and treatment of endocrine problems can be difficult because of three complex physiological effects of hormones. (1) Hormones are extremely specific, and their target tissues are sensitive to minute changes in hormonal levels. (2) The responses of hormones are usually a type of chain reaction; that is, the hormones secreted from one gland cause the secretion of hormones by another gland. Thus, the clinical symptoms obscure the source of the problem. (3) Endocrine problems frequently manifest as metabolic problems because the primary action of hormones is to regulate metabolism.

Common diagnostic methods will be discussed in this section as well as the more important endocrine disorders that affect the major endocrine glands.

Diagnostic Tests for Endocrine Disorders

Certain endocrine disorders affect the physical appearance and behavior of the patient, so observation is very important in diagnosis. The clinical history is also important in evaluating the speed of progression and the stage of development of an endocrine disorder.

The confirmation of endocrine disorders requires laboratory tests, particularly of blood and urine samples. These samples are important because hormones are distributed via the blood, and urine is produced from the metabolic wastes filtered from the blood. A **radioimmunoassay (RIA)** is a laboratory test to determine the concentration of hormones in blood and urine. Other blood tests include the **protein-bound iodine (PBI)** test to determine the iodine level within the blood and hence thyroid problems; measuring blood cholesterol content for thyroid problems; measuring sodium-potassium ratios to detect Addison's disease; and blood sugar tests, including testing glucose tolerance in a fasting patient to examine for diabetes mellitus.

A urinalysis can be important in the diagnosis of several endocrine disorders. A high level of glucose in a fasting patient indicates diabetes mellitus. A patient who has diabetes insipidus will have frequent urinations that are dilute and of low specific gravity. Certain diseases of the adrenal glands can also be detected by examining for changes in urine samples collected over a twenty-four-hour period.

---

placenta: L. *placenta,* flat cake

Basal metabolism rate (BMR) and **thyroid scans** are tests for thyroid disorders. A visual-field examination may be an important test in detecting a tumor of the pituitary gland. Other laboratory studies such as X rays or electrocardiograms assist in diagnosing endocrine disorders.

Disorders of the Pituitary Gland

The pituitary is a remarkable gland. It simultaneously carries out several functions, yet more than 90% of the gland must be destroyed before pituitary function becomes severely impaired.

Panhypopituitarism A reduction in the activity of the pituitary gland is called **hypopituitarism** *(hi''po-pĭ-tu'ĭ-tah-rizm'')* and can result from intracranial hemorrhage, a blood clot, prolonged steroid treatments, or a tumor. Total pituitary impairment, termed **panhypopituitarism,** brings about a progressive and general loss of hormonal activity. For example, the gonads stop functioning and the person suffers from amenorrhea (lack of menstruation) or aspermia (no sperm production) and loss of pubic and axillary hair. The thyroid and adrenals also eventually stop functioning. People with this condition, and those who have had their pituitary surgically removed (a procedure called **hypophysectomy**) receive thyroxine, cortisone, growth hormone, and gonadal hormones throughout life to maintain normal body function.

Abnormal Growth Hormone Secretion Inadequate growth hormone secretion during childhood causes **pituitary dwarfism.** Hyposecretion of growth hormone in an adult produces a rare condition called **pituitary cachexia** *(kah-kek'se-ah)* **(Simmonds' disease).** One of the symptoms of this disease is premature aging caused by tissue atrophy. Oversecretion of growth hormone during childhood, in contrast, causes **gigantism.** Excessive growth hormone secretion in an adult does not cause further growth in length because the epiphyseal discs have ossified. Hypersecretion of growth hormone in an adult causes **acromegaly** *(ak''ro-meg'ah-le),* in which the person's appearance gradually changes as a result of thickening of bones and growth of soft tissues, particularly in the face, hands, and feet.

Inadequate ADH Secretion A dysfunction of the neurohypophysis results in a deficiency in ADH secretion, causing a condition called **diabetes insipidus.** Symptoms of this disease include polyuria (excessive urination), polydipsia (drinking large amounts of water), and severe ionic imbalances. Diabetes insipidus is treated by injections of ADH.

Disorders of the Thyroid and Parathyroids

Hypothyroidism The infantile form of hypothyroidism is known as **cretinism** *(kre'tin-izm).* An affected child usually appears normal at birth because thyroxine is received from the mother through the placenta. The clinical symptoms of cretinism are stunted growth, thickened facial features, abnormal bone development, mental retardation, low body temperature, and general lethargy. If cretinism is diagnosed early, it can be successfully treated by administering thyroxine.

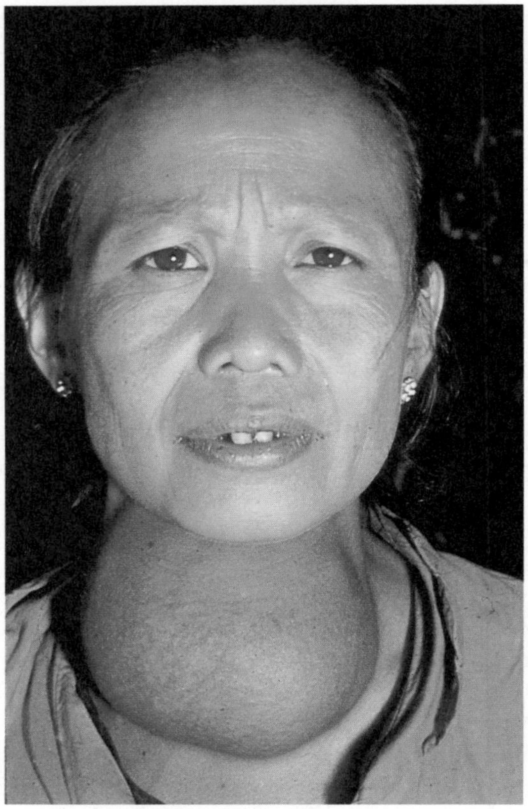

Figure 14.27 A simple or endemic goiter is caused by insufficient iodine in the diet.

Myxedema Hypothyroidism in an adult causes myxedema *(mik''sĕ-de'mah).* This disorder affects body fluids, causing edema and increasing blood volume, hence increasing blood pressure. A person with myxedema has a low metabolic rate, lethargy, and a tendency to gain weight. This condition is treated with thyroxine or with triiodothyronine, which is taken orally (as pills).

Endemic Goiter A goiter is an abnormal growth of the thyroid gland. When this is a result of inadequate dietary intake of iodine, the condition is called endemic goiter (fig. 14.27). In this case, growth of the thyroid is due to excessive TSH secretion, which results from low levels of thyroxine secretion. Endemic goiter is thus associated with hypothyroidism.

Graves' Disease Graves' disease, also called **toxic goiter,** involves growth of the thyroid associated with hypersecretion of thyroxine. This hyperthyroidism is produced by antibodies that act like TSH and stimulate the thyroid; it is an autoimmune disease. As a consequence of high levels of thyroxine secretion, the metabolic rate and heart rate increase, there is loss of weight,

---

Simmonds' disease: from Morris Simmonds, German physician, 1855–1925
acromegaly: Gk. *akron,* extremity; *megas,* large
diabetes: Gk. *diabetes,* to pass through a siphon

myxedema: Gk. *myxa,* mucus; *oidema,* swelling
Graves' disease: from Robert James Graves, Irish physician, 1796–1853

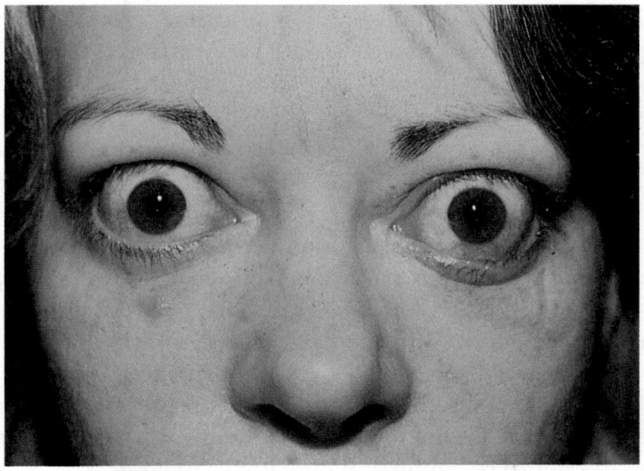

Figure 14.28 Hyperthyroidism is characterized by an increased metabolic rate, weight loss, muscular weakness, and nervousness. The eyes may also protrude.

and the autonomic nervous system induces excessive sweating. In about half of the cases, **exophthalmos** *(ek″sof-thal′mos)* (bulging of the eyes) also develops (fig. 14.28) because of edema in the tissues of the eye sockets and swelling of the extrinsic eye muscles.

Disorders of the Parathyroid Glands Surgical removal of the parathyroids sometimes unintentionally occurs when the thyroid is removed because of a tumor or the presence of Graves' disease. The resulting fall in parathyroid hormone (PTH) causes a decrease in plasma calcium levels, which can lead to severe muscle tetany. Hyperparathyroidism is usually caused by a tumor that secretes excessive amounts of PTH. This stimulates demineralization of bone, which makes the bones soft and raises the blood levels of calcium and phosphate. As a result of these changes, bones are subject to deformity and fracture, and stones composed of calcium phosphate are likely to develop in the urinary tract.

Disorders of the Pancreatic Islets

Diabetes Mellitus Diabetes mellitus is characterized by fasting hyperglycemia and the presence of glucose in the urine. There are two forms of this disease. **Type I,** or **juvenile-onset** diabetes mellitus, is caused by destruction of the beta cells and the resulting lack of insulin secretion. **Type II,** or **maturity-onset** diabetes mellitus (which is the more common form), is caused by decreased tissue sensitivity to the effects of insulin, so that larger amounts of insulin are required to produce a normal effect. Both types of diabetes mellitus are also associated with abnormally high levels of glucagon secretion.

Treatment of diabetes mellitus requires administering the necessary amounts of insulin to maintain a balanced carbohydrate metabolism. If excessive amounts of insulin are given, the person will experience extreme nervousness and tremors, perhaps followed by convulsion and loss of consciousness. This condition is commonly called insulin shock and can be treated by administering glucose intravenously.

Reactive Hypoglycemia People who have a genetic predisposition for type II diabetes mellitus often first develop reactive hypoglycemia. In this condition, the rise in blood glucose that follows the ingestion of carbohydrates stimulates an excessive secretion of insulin, which in turn causes the blood glucose levels to fall below the normal range. This can result in weakness, changes in personality, and mental disorientation.

Disorders of the Adrenal Glands

Tumors of the Adrenal Medulla Tumors of the chromaffin cells of the adrenal medulla are referred to as **pheochromocytomas** *(fe-o-kro″mo-si-to′mahz).* These tumors cause hypersecretion of epinephrine and norepinephrine, which produce an effect similar to continuous sympathetic nervous stimulation. The symptoms of this condition are hypertension, elevated metabolism, hyperglycemia and sugar in the urine, nervousness, digestive problems, and sweating. It does not take long for the body to become totally fatigued under these conditions, making the patient susceptible to other diseases.

Addison's Disease This disease is caused by an inadequate secretion of both glucocorticoids and mineralocorticoids, which results in hypoglycemia, sodium and potassium imbalance, dehydration, hypotension, rapid weight loss, and generalized weakness. A person with this condition who is not treated with corticosteroids will die within a few days because of the severe electrolyte imbalance and dehydration. Another symptom of this disease is darkening of the skin. This is caused by high secretion of ACTH and possibly MSH (because MSH is derived from the same parent molecule as ACTH), which is a result of lack of negative feedback inhibition of the pituitary by corticosteroids.

Cushing's Syndrome Hypersecretion of corticosteroids results in Cushing's syndrome. This is generally caused by a tumor of the adrenal cortex or by oversecretion of ACTH from the adenohypophysis. Cushing's syndrome is characterized by changes in carbohydrate and protein metabolism, hyperglycemia, hypertension, and muscular weakness. Metabolic problems give the body a puffy appearance and can cause structural changes characterized as buffalo hump and moon face. Similar effects are also seen when people with chronic inflammatory diseases receive prolonged treatment with corticosteroids, which are given to reduce inflammation and inhibit the immune response.

Adrenogenital Syndrome Usually associated with Cushing's syndrome, this condition is caused by hypersecretion of adrenal sex hormones, particularly the androgens. Adrenogenital syndrome in young children causes premature puberty and enlarged genitals, especially the penis in a male and the clitoris in a female. An increase in body hair and a deeper voice are other characteristics. This condition in a mature woman can cause growth of a beard.

exophthalmos: Gk. *ex,* out; *opthalmos,* eyeball

pheochromocytomas: Gk. *phaios,* dusky; *chroma,* color; *oma,* tumor
Addison's disease: from Thomas Addison, English physician, 1793–1860
Cushing's syndrome: from Harvey Cushing, U.S. physician, 1869–1939

As with most other endocrine disorders, problems of the adrenal cortex are treated either through surgery, if there is a tumor causing hypersecretion, or by the administration of the necessary levels of appropriate hormones, if there is tissue dysfunction with hyposecretion.

CLINICAL CASE STUDY ANSWER

The visual findings are likely caused by a pituitary tumor (adenoma) eroding and pushing upward on the optic chiasma. The chiasma contains nerve tracts that eminate from the medial portions of each retina, which receive light from the lateral (temporal) visual fields. The pituitary tumor is secreting either increased or unregulated amounts of ACTH (adrenocorticotropic hormone), which in turn cause overproduction of adrenal cortical hormones. These include glucocorticoids, mineralocorticoids, and various androgens, all of which combine to produce the patient's constellation of findings.

This specific condition, hypersecretion of adrenal cortical hormones secondary to an ACTH-producing pituitary adenoma, is known as *Cushing's Disease.* This condition is usually treated by transsphenoidal excision of the tumor. In some cases, adrenal-suppressing drugs are used. Elevated adrenal corticoid levels can be a result of endocrine tumors other than pituitary. An example would be an adrenal tumor(s), either cancerous or benign, that produces adrenal cortical hormones. The physical effects of this condition and all others that involve high adrenal corticoid levels are known collectively as *Cushing's Syndrome,* which is not to be confused with *Cushing's Disease,* a term specifically reserved for hyperadrenalism caused by a pituitary adenoma. ■

Chapter Summary

I. Introduction to the Endocrine System
   A. Hormones are regulatory molecules secreted into the blood by endocrine glands. The action of a hormone on its target tissue is dependent on its concentration and the specific receptor sites on cell membranes.
   B. Hormones are classified chemically as steroids, proteins, and amines.
   C. A negative feedback occurs when there is an imbalance in hormone concentration, and that information is fed back to an organ, which acts to correct the imbalance.

II. Development of the Endocrine System
   A. The pituitary consists of an adenohypophysis, derived from an oral diverticulum, and a neurohypophysis, derived as a downgrowth of the brain. Both parts are derived from ectoderm.
   B. The thyroid is derived from endoderm in the primitive pharynx.
   C. The pancreas is derived from endoderm in the foregut.
   D. The adrenal cortex develops from mesoderm; the adrenal medulla develops from neural crest cells of the neural tube.

III. Pituitary Gland
   A. The pituitary gland, or hypophysis, is divided into an anterior adenohypophysis and a posterior neurohypophysis.
      1. The adenohypophysis has a glandular pars distalis, a thin proximal extension called the pars tuberalis, and a narrow pars intermedia.
      2. The neurohypophysis consists of the pars nervosa and the infundibulum.
   B. The anterior pituitary produces and secretes growth hormone, thyroid-stimulating hormone, adrenocorticotropic hormone, follicle-stimulating hormone, luteinizing hormone, prolactin, and melanocyte-stimulating hormone.
   C. The posterior pituitary releases antidiuretic hormone and oxytocin.
   D. Secretions of the anterior pituitary are controlled by hypothalamic hormones, which stimulate or inhibit secretions of the anterior pituitary, and regulated by the feedback of hormones from the target glands; secretions of the posterior pituitary are controlled by the hypothalamo-hypophyseal nerve tract.

IV. Thyroid and Parathyroid Glands
   A. The bilobed thyroid gland is located in the neck just below the larynx, and there are usually four small parathyroid glands embedded on its posterior surface.
   B. Thyroid follicles secrete thyroxine and triiodothyronine, which increase the rate of protein synthesis and the rate of energy release from carbohydrates. They also regulate the rate of growth and the rate of maturity of the nervous system.
   C. Parafollicular cells of the thyroid secrete the hormone calcitonin, which lowers the blood calcium by inhibiting the release of calcium from bone tissue.
   D. Parathyroid hormone causes an increase in blood calcium and a decrease in blood phosphate levels. It acts on the intestines, kidneys, and bones.

V. Pancreas
   A. The pancreas is a mixed endocrine and exocrine gland located in the abdominal cavity.
   B. The pancreatic islets within the pancreas contain beta cells, which secrete insulin, and alpha cells, which secrete glucagon.
   C. Insulin lowers blood glucose and stimulates the production of glycogen, fat, and protein.
   D. Glucagon raises blood glucose by stimulating the breakdown of liver glycogen.

VI. Adrenal Glands
   A. An adrenal gland consists of an adrenal cortex and an adrenal medulla and is positioned along the superior border of the kidney.
   B. Hormones of the adrenal cortex include mineralocorticoids, which regulate sodium reabsorption and potassium excretion; glucocorticoids, which influence metabolism by promoting vasoconstriction and resistance to stress; and gonadocorticoids, which supplement gonadal hormones.
   C. Epinephrine and norepinephrine, secreted from the adrenal medulla, produce effects similar to those of the sympathetic division of the ANS.

VII. Gonads and Other Endocrine Glands
   A. The testes are the male gonads, which produce the male sex hormone testosterone within the interstitial cells.
   B. The ovaries are the female gonads, which produce estrogen within the ovarian follicles and corpus luteum.
   C. The pineal gland, located in the roof of the third ventricle of the brain, secretes melatonin, which seems to have an effect upon the hypothalamus in the secretion of gonadotrophin.
   D. The thymus lies behind the sternum within the mediastinum, where it produces T-cells, which are important in maintaining body immunity.
   E. Certain gastrointestinal cells secrete hormones that aid digestion.
   F. The maternal placenta secretes hormones that maintain pregnancy.

Review Activities

Objective Questions

Match the gland to its embryonic origin.

1. Adenohypophysis	(a) endoderm of pharynx
2. Neurohypophysis	(b) diverticulum from brain
3. Adrenal medulla	(c) endoderm of foregut
4. Pancreas	(d) neural crest ectoderm
5. Thyroid gland	(e) hypophyseal pouch

6. The sella turcica that supports the pituitary gland is located in which bone? The
   (a) ethmoid. (b) frontal. (c) sphenoid. (d) occipital.
7. The hormone primarily responsible for setting the basal metabolic rate and for promoting the maturation of the brain is
   (a) cortisol. (b) ACTH. (c) TSH. (d) thyroxine.
8. Which of the following statements about the adrenal cortex is *true*?
   (a) It is developed from mesoderm.
   (b) It secretes some androgens.
   (c) Its secretion prepares the body for the fight or flight response.
   (d) The zona fasciculata is stimulated by ACTH.
   (e) All of the above are true.
9. The hormone insulin
   (a) is secreted by alpha cells in the pancreatic islets.
   (b) is secreted in response to a rise in blood glucose.
   (c) stimulates the production of glycogen and fat.
   (d) Both a and b.
   (e) Both b and c.

Match the hormone with the primary agent that stimulates its secretion.

10. Epinephrine	(a) TSH
11. Thyroxine	(b) ACTH
12. Corticosteroids	(c) Growth hormone
13. ACTH	(d) Sympathetic nerve impulse
	(e) CRF

14. Steroid hormones are secreted by
    (a) the adrenal cortex.
    (b) the gonads.
    (c) the thyroid.
    (d) Both a and b.
    (e) Both b and c.
15. The secretion of which of the following hormones would be *increased* in a person with endemic goiter?
    (a) TSH
    (b) thyroxine
    (c) triiodothyronine
    (d) All of the above.

Essay Questions

1. Compare and contrast the roles of the nervous system and the endocrine system in maintaining body homeostasis.
2. List the glands of the endocrine system and describe their general locations.
3. Give examples of steroids, proteins, and amines, and describe the various ways they are usually administered.
4. Define negative feedback and explain why this is a reliable mechanism for controlling hormonal secretion.
5. Which endocrine glands, or portions of endocrine glands, develop as a process of germ layer invagination, and which develop as an outgrowth or budding of germ layer tissue?
6. Why is the pituitary frequently considered two separate glands?
7. List the hormones secreted by the adenohypophysis, and describe the general functions of each.
8. Describe the gross structure and position of the neurohypophysis. What hormones are released from this portion of the pituitary gland, and how is their release regulated?
9. List the hormones secreted by the thyroid gland, and discuss the general function of each.
10. Describe the location of the parathyroid glands. What specialized cells constitute these glands and what hormone do they secrete?
11. Describe the structure and position of the pancreatic islets, list the hormones they produce, and discuss the general function of each.
12. Distinguish between the adrenal cortex and adrenal medulla in structure, function, and value to the body.
13. Define *chromaffin cells, interstitial cells, alpha* and *beta cells, oxyphil cells, principal cells,* and *pituicytes.*
14. Explain the causes of each of the following endocrine diseases: *gigantism, acromegaly, pituitary cachexia,* and *pituitary dwarfism.*
15. Distinguish between *cretinism, myxedema,* and *Graves' disease.*

CHAPTER 15

Sensory Organs

Outline and Concepts

Clinical Case Study

Overview of Sensory Perception

Sensory organs are highly specialized extensions of the nervous system in that they contain sensory neurons adapted to respond to specific stimuli and conduct nerve impulses to the brain for interpretation.

Classification of the Senses

The senses are classified as general or special according to the simplicity or complexity of the receptors and neural pathways. They are also classified as somatic or visceral according to the location of the receptors.

Somatic Senses

Somatic senses include cutaneous receptors and proprioceptors. The perception of somatic sensations is determined by the density of the receptors in the stimulated receptive field and the intensity of the sensation.

Tactile and Pressure Receptors
Thermoreceptors
Pain Receptors
Proprioceptors
Neural Pathways for Somatic Sensation

Olfactory Sense

Olfactory receptors are the dendritic endings of the olfactory (first cranial) nerve that respond to chemical stimuli and transmit the sensation of olfaction directly to the olfactory portion of the cerebral cortex.

Gustatory Sense

Taste receptors are specialized epithelial cells clustered together into taste buds that respond to chemical stimuli and transmit the sense of taste through the glossopharyngeal (ninth cranial) nerve or the facial (seventh cranial) nerve to the cortex of the parietal cerebral lobe for interpretation.

Visual Sense

Rods and cones are the photoreceptors within the eyeball that are sensitive to light energy and are stimulated to transmit nerve impulses through the optic nerve and optic tract to the visual cortex of the occipital lobes, where the interpretation of vision occurs. The sensory components of the eye are formed by twenty weeks, and the accessory structures are formed by thirty-two weeks.

Development of the Eye
Associated Structures of the Eye
Structure of the Eyeball
Function of the Eyeball
Neural Pathways of Vision, Eye Movements, and Processing Visual Information

Senses of Hearing and Balance

Structures of the outer, middle, and inner ear are involved in the sense of hearing. The inner ear also contains structures that provide a sense of balance, or equilibrium. The development of the ear begins during the fourth week and is completed by the thirty-second week.

Development of the Ear
Structure of the Ear
Sound Waves and Neural Pathways for Hearing
Mechanics of Equilibrium

Clinical Considerations

Diagnosis of Sensory Organs
Developmental Problems of the Eyes and Ears
Functional Impairments of the Eye
Infections and Diseases of the Eye
Infections, Diseases, and Functional Impairments of the Ear

Clinical Case Study Answer

Chapter Summary

Review Activities

CLINICAL CASE STUDY

A fifty-year-old man visits his family doctor and complains of progressive hearing loss in his right ear. In order to rule out visible abnormalities or wax build-up, the physician performs an otoscopic examination that reveals no abnormalities. He then strikes a tuning fork and places it on the man's right mastoid process, which causes the patient to remark: "I can really hear that well in my bad ear." After a few moments the patient notes that the tone has nearly died out. The doctor then holds the barely-vibrating instrument two centimeters away from the same ear; the patient is unable to hear. The doctor explains that a patient with normal hearing can hear the vibrating fork when it is held onto the mastoid bone but can hear better when it is held just outside the external auditory canal.

What components of the hearing mechanism are being bypassed when the handle of the tuning fork is placed on the mastoid process? (Hint: The hearing organs of the inner ear can receive and effectively process sound waves directly from the bone they are encased in.) Name the structures being bypassed beginning with the first to receive sound waves. What kind of hearing problem exists in our patient—one that involves the conductive structures, or one that involves the inner ear organs and/or neural pathways? ■

Overview of Sensory Perception

Sensory organs are highly specialized extensions of the nervous system in that they contain sensory neurons adapted to respond to specific stimuli and conduct nerve impulses to the brain for interpretation.

Objective 1. Explain what is necessary for perceiving a sensation.

Objective 2. Discuss the selectivity of sensory receptors for specific stimuli.

The sense organs are actually extensions of the nervous system that allow humans to perceive their internal and external environments. These sense organs have been described as windows for the brain because it is through them that awareness of the environment is possible. A stimulus must first be received before the sensation can be interpreted and the necessary body adjustments dictated by the central nervous system can be made. The sense organs permit daily experiences of pleasure and are vital to the daily survival of humans by allowing them to hear sounds, see dangers, taste undesirable foods, and perceive pain, hunger pangs, or intestinal aches.

A **sensation** is the arrival of a sensory impulse to the brain. The interpretation of a sensation is referred to as **perception.** In other words, one feels, sees, hears, tastes, and smells with the brain. In order to perceive a sensation, the following four conditions are necessary.

1. A **stimulus** sufficient to initiate a response in the nervous system must be present.
2. A **receptor** must convert the stimulus to a nerve impulse. A receptor is a specialized, peripheral dendritic ending of a sensory nerve fiber or the specialized receptor cell associated with it.
3. The **conduction of the nerve impulse** must occur from the receptor to the brain along a nervous pathway.
4. The **interpretation of the impulse** in the form of a perception must occur within a specific portion of the brain.

Only impulses reaching the cerebral cortex of the brain are consciously interpreted as sensations. If impulses terminate in the spinal cord or brain stem, they initiate a reflexive motor activity response rather than a conscious sensation. Impulses reaching the higher brain centers travel through nerve fibers composing *sensory,* or *ascending, tracts.* Clusters of neuron cell bodies, called **nuclei,** are synaptic sites along sensory tracts within the CNS. The nuclei that sensory impulses pass through before reaching the cerebral cortex are located in the spinal cord, medulla oblongata, pons, and thalamus.

Through the use of scientific instruments, it is known that the senses act as energy filters that allow perception of only a narrow range of energy. Vision, for example, is limited to light waves in the visible spectrum. Other types of waves of the same type of energy as visible light, such as X rays, radio waves, and ultraviolet and infrared light, cannot normally excite the sensory receptors in the eyes. Although filtered and distorted by the limitations of sensory functions, perceptions allow humans effective interactions with the environment and are of obvious survival value.

---

1. Distinguish between sensation and perception.
2. List the four conditions necessary for perception, and identify which of the four must always involve consciousness in order for perception to occur.
3. Using examples, explain the statement that each of the senses acts as a filter, allowing the perception of only a narrow range of energy.

Classification of the Senses

The senses are classified as general or special according to the simplicity or complexity of the receptors and neural pathways. They are also classified as somatic or visceral according to the location of the receptors.

Objective 3. Compare and contrast somatic, visceral, and special senses.

Objective 4. Define and give examples of *exteroceptors, visceroceptors,* and *proprioceptors.*

Structurally, the sensory receptor can be the dendrites of sensory neurons, which are either free (such as those in the skin that mediate pain and temperature) or are encapsuled within

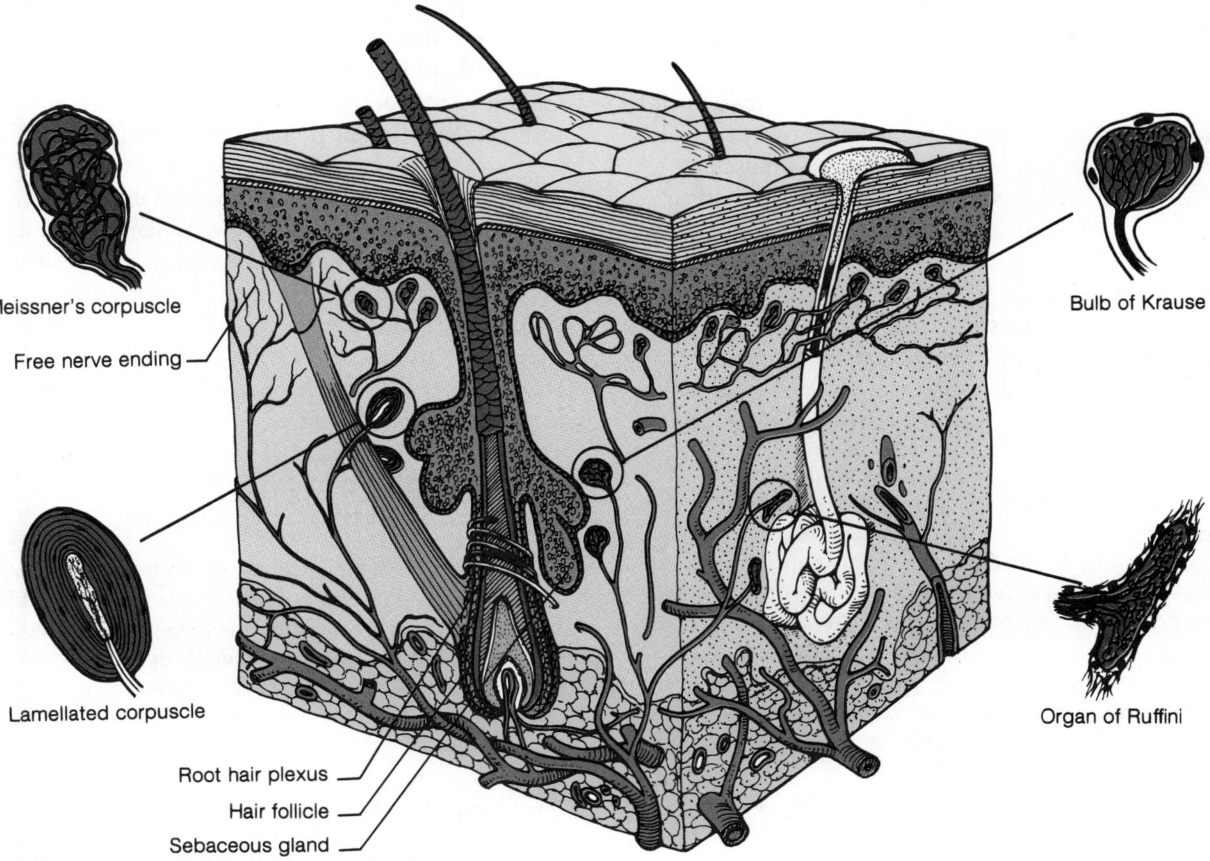

Figure 15.1 A diagrammatic section of the skin showing the general location and magnified structure of cutaneous receptors.

non-neural structures, such as lamellated corpuscles, or pressure receptors in the skin (fig. 15.1). Other receptors, including taste buds, photoreceptors in the eyes, and hair cells in the inner ears, derive from epithelial cells that synapse with sensory dendrites.

The senses of the body can be classified as general or special according to the simplicity or complexity of the receptors and neural pathways involved. **General senses** are widespread through the body and are simple in structure. Examples are touch, pressure, cold-heat, and pain. **Special senses** are localized in complex receptor organs and have extensive neural pathways. Among the special senses are taste, smell, sight, hearing, and balance.

The senses can also be classified as somatic or visceral according to the location of the receptors. **Somatic senses** are those in which the receptors are localized within the body wall. These include the cutaneous receptors and those within muscles, tendons, and joints. **Visceral senses** are those in which the receptors are located within visceral organs. These classes of senses intersect in certain organs. Hearing, for example, is a special somatic sense, whereas pain from the digestive tract is a general visceral sense.

Senses are also classified according to the location of the receptors and the types of stimuli to which they respond. There are three basic kinds of receptors: exteroceptors, visceroceptors, and proprioceptors.

Exteroceptors Exteroceptors *(eks''ter-o-sep'torz)* are located near the surface of the body where they respond to stimuli from the external environment. They include the following:

1. Cones and rods in the retina of the eye—*photoreceptors.*
2. Hair cells in the organ of Corti within the inner ear—*pressure receptors.*
3. Olfactory receptors in the nasal epithelium of the nasal cavity—*chemoreceptors.*
4. Taste receptors on the tongue—*chemoreceptors.*
5. Skin receptors within the dermis—*tactile receptors* for touch; *pressure receptors,* or *mechanoreceptors,* for pressure; *thermoreceptors* for temperature; and *nocireceptors (no-ci're-sep'torz),* for pain.

Pain receptors are stimulated by chemicals released from damaged tissue cells and thus are a type of chemoreceptor. Although there are specific pain receptors, nearly all types of receptors when intensely stimulated transmit impulses that are perceived as pain. Pain receptors exist throughout the body, but only those located within the skin are classified as exteroceptors.

Visceroceptors (Enteroceptors) As the name implies, visceroceptors *(vis''er-o-sep'torz)* produce sensations arising from the viscera, such as internal pain, hunger, thirst, fatigue, or nausea.

somatic: Gk. *somatikos,* body
visceral: L. *viscera,* body organs

nocirecepor: L. *nocco,* to injure; *ceptus,* taken

Specialized visceroceptors located within the circulatory system are sensitive to changes in blood pressure (baroreceptors).

Proprioceptors Proprioceptors *(pro''pre-o-sep'torz)* relay information about body position, equilibrium, and movement. They are located in the inner ear, joints, tendons, and muscles.

The three types of receptors may be further classified on the basis of sensory adaptation (accommodation). Some receptors respond with a burst of activity when a stimulus is first applied but quickly decrease their firing rate as they adapt to the stimulus even though the stimulus is maintained. Receptors with this response pattern are called **phasic receptors.** Receptors that produce a relatively constant rate of firing as long as the stimulus is maintained are known as **tonic receptors.**

Phasic receptors alert us to changes in sensory stimuli and are in part responsible for the fact that we can cease paying attention to constant stimuli. This ability is called *sensory adaptation.* Odor and touch, for example, adapt rapidly; bath water feels hotter when we first enter it. Sensations of pain, in contrast, adapt little if at all.

1. Using examples, explain how sensory receptors can be classified according to complexity, location, structure, and the type of stimuli to which they respond.
2. Describe sensory adaptation in olfactory and pain receptors.

Somatic Senses

Somatic senses include cutaneous receptors and proprioceptors. The perception of somatic sensations is determined by the density of the receptors in the stimulated receptive field and the intensity of the sensation.

Objective 5. Describe the structure, function, and location of the various tactile receptors and the neural pathway for somatic sensation.

Objective 6. Explain the purpose of pain, the receptors that respond to pain, and the impulse pathway to the brain.

Objective 7. Define and give examples of *referred pain* and *phantom pain.*

The **somatic senses,** or **somesthetic senses,** include the receptors that are located in the skin and cause cutaneous sensations such as touch, tickling, pressure, cold, heat, and pain. The somatic senses also include proprioceptors that are located in the inner ear, joints, tendons, and muscles, and relay information about body position, equilibrium, and movement.

Tactile and Pressure Receptors

Both tactile receptors and pressure receptors are sensitive to mechanical forces that distort or displace the tissue in which they are located. **Tactile receptors** respond to *fine,* or *light, touch* and are located primarily in the dermis and hypodermis of the skin. **Pressure receptors** respond to *deep pressure* and are commonly found in the hypodermis of the skin and in the tendons and ligaments of joints. There are three kinds of tactile receptors and one kind of pressure receptor (see fig. 15.1).

Meissner's Corpuscles A Meissner's corpuscle *(kor'pus'l)* is an oval receptor composed of a mass of dendritic endings from two or three nerve fibers enclosed by connective tissue sheaths. These corpuscles are numerous in the hairless portions of the body, such as the eyelids, lips, tip of the tongue, fingertips, palms of the hands, nipples, and external genital organs. Meissner's corpuscles lie within the papillary layer of the dermis where they are especially sensitive to the motion of objects that barely contact the skin. Sensations of fine or light touch are perceived as these receptors are stimulated. They also function when a person touches an object to determine its texture.

The high tactile acuity of the fingertips is exploited in the reading of Braille. Braille symbols consist of dots that are raised 1 mm up from the page and separated from each other by 2.5 mm. Experienced Braille readers can scan words at about the same speed that a sighted person can read aloud—a rate of about 100 words per minute.

Free Nerve Endings Free nerve endings are the least modified and the most superficial of the tactile receptors. These receptors extend into the lower layers of the epidermis where they end as knobs between the epithelial cells. Free nerve endings are most important as pain receptors, although they also respond to objects that are in continuous contact with the skin, such as clothing.

Hair Root Plexuses Hair root plexuses are a specialized type of free nerve ending. They are coiled around hair follicles where they respond to movement of the hair.

Lamellated (Pacinian) Corpuscles Lamellated corpuscles are large, onion-shaped receptors composed of the dendritic endings of several sensory nerve fibers enclosed by connective tissue layers. They are commonly found within the synovial membranes of diarthrotic joints, in the perimysium of muscle tissue, and in certain visceral organs. Lamellated corpuscles are also abundant in the skin of the palms and fingers of the hand, soles of the feet, external genital organs, and breasts.

Lamellated corpuscles respond to heavy pressures, generally those that are constantly applied. They can also detect vibrations in tissues and organs.

proprioceptor: L. *proprius,* one's own; *ceptus,* taken

Meissner's corpuscle: from George Meissner, German histologist, 1829–1905
corpuscle: L. *corpusculum,* diminutive of *corpus,* body
Braille: from Louis Braille, French teacher of the blind, 1809–1852
Pacinian corpuscle: from Filippo Pacini, Italian anatomist, 1812–1883

Table 15.1 Cutaneous receptors

Type	Location	Function	Sensation
Meissner's corpuscles (mechanoreceptors)	Papillae of dermis; numerous in hairless portions of body (eyelids, fingertips, lips, nipples, external genitalia)	Detect light motion against surface of skin	Fine touch; texture
Free nerve endings (pressure receptors; pain receptors)	Lower layers of epidermis	Detect changes in pressure; detect tissue damage	Touch, pressure; pain
Hair root plexuses (tactile receptors)	Around hair follicles	Detect movement of hair	Touch
Lamellated (Pacinian) corpuscles (mechanoreceptors)	Hypodermis; synovial membranes; perimysium; certain visceral organs	Detect changes in pressure	Deep pressure; vibrations
Organs of Ruffini (thermoreceptors)	Lower layers of dermis	Detect changes in temperature	Heat
Bulbs of Krause (thermoreceptors)	Dermis	Detect changes in temperature	Cold

Thermoreceptors

Thermoreceptors are widely distributed throughout the dermis of the skin but are especially abundant in the lips and the mucous membranes of the mouth and anal regions. There are two kinds of thermoreceptors—one that responds to heat and the other to cooler temperatures (see fig. 15.1).

Organs of Ruffini The organs of Ruffini are heat receptors located deep within the dermis. Heat receptors are elongated, oval structures that are most sensitive to temperatures above 25° C (77° F). Temperatures above 45° C (113° F) elicit impulses through the organs of Ruffini that are perceived as painful, burning sensations.

Bulbs of Krause The bulbs of Krause are receptors for the sensation of cold. They are more abundant than heat receptors and are closer to the surface of the skin. The bulbs of Krause are most sensitive to temperatures below 20° C (68° F). Temperatures below 10° C elicit responses through the bulbs of Krause that are perceived as painful, freezing sensations.

Pain Receptors

The principal receptors for pain are the **free nerve endings.** Several million free nerve endings are distributed throughout the skin and internal tissues. Pain receptors are sparse in certain visceral organs and absent within the nervous tissue of the brain. Although the free nerve endings are specialized to respond to tissue damage, all of the cutaneous receptors will relay impulses that are interpreted as pain if stimulated excessively. The cutaneous receptors are summarized in table 15.1.

The protective value of pain receptors is obvious. Unlike other cutaneous receptors, free nerve endings have little accommodation (few phasic receptors), so impulses are relayed continuously to the CNS as long as the irritating stimulus is present. Although pain receptors can be activated by all types of stimuli, they are particularly sensitive to chemical stimulation. Muscle spasms, muscle fatigue, or an inadequate supply of blood to an organ may also cause pain.

Impulses for pain are conducted to the spinal cord through sensory neurons. The pain sensations are then conducted to the thalamus along the *lateral spinothalamic tract* of the spinal cord and from there to the somatesthetic area of the cerebral cortex. Although an awareness of pain occurs in the thalamus, the type and intensity of pain is interpreted in the specialized areas of the cerebral cortex.

The sensation of pain can be clinically classified as **somatic pain** and **visceral pain.** Stimulation of the cutaneous pain receptors results in the perception of superficial somatic pain. Deep somatic pain comes from stimulation of receptors in skeletal muscles, joints, and tendons.

Stimulation of the receptors within the viscera causes the perception of visceral pain. Through precise neural pathways, the brain is able to perceive the area of stimulation and project the pain sensation back to that area. The sensation of pain from certain visceral organs, however, may not be perceived as arising from those organs but from other somatic locations. This phenomenon is known as **referred pain** (fig. 15.2). The sensation of referred pain is consistent from one person to another and is clinically important in diagnosing organ dysfunctions. The pain of a heart attack, for example, may be perceived subcutaneously over the heart and down the medial side of the left arm. Ulcers of the stomach may cause pain that is perceived as coming from the upper central (epigastric) region of the torso. Pain from problems of the liver or gallbladder may be perceived as localized visceral pain or perceived as referred pain arising from the right neck and shoulder regions.

Referred pain is not totally understood but seems to be related to the development of the tracts within the spinal cord. It is thought that there are some *common nerve pathways* used by sensory impulses coming from both the cutaneous areas and from visceral organs (fig. 15.3). Consequently, impulses along these pathways may be incorrectly interpreted as arising cutaneously rather than from within a visceral organ.

organs of Ruffini: from Angelo Ruffini, Italian anatomist, 1864–1929
bulbs of Krause: from Wilhelm J. F. Krause, German anatomist, 1833–1910

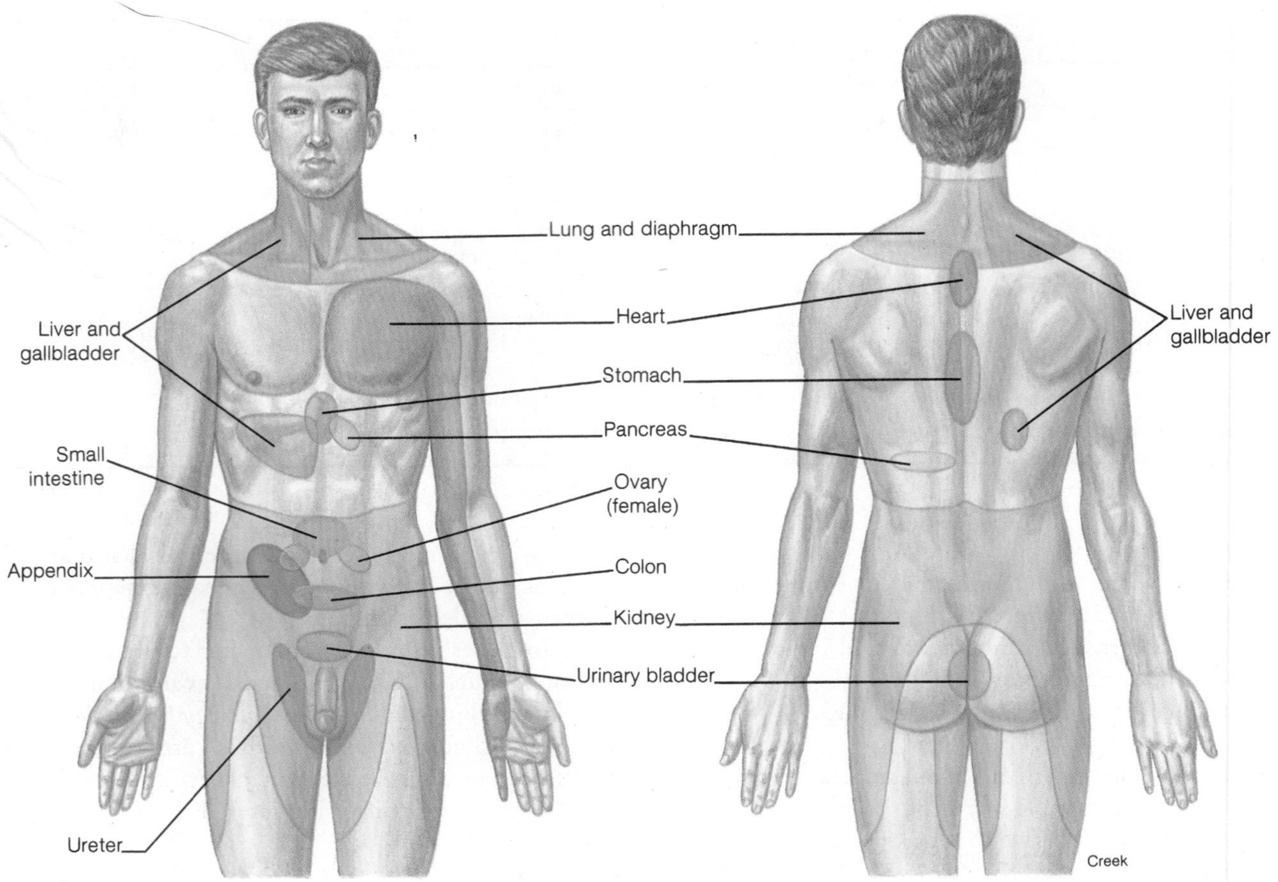

Figure 15.2 Sites of referred pain are perceived cutaneously but actually originate from specific visceral organs.

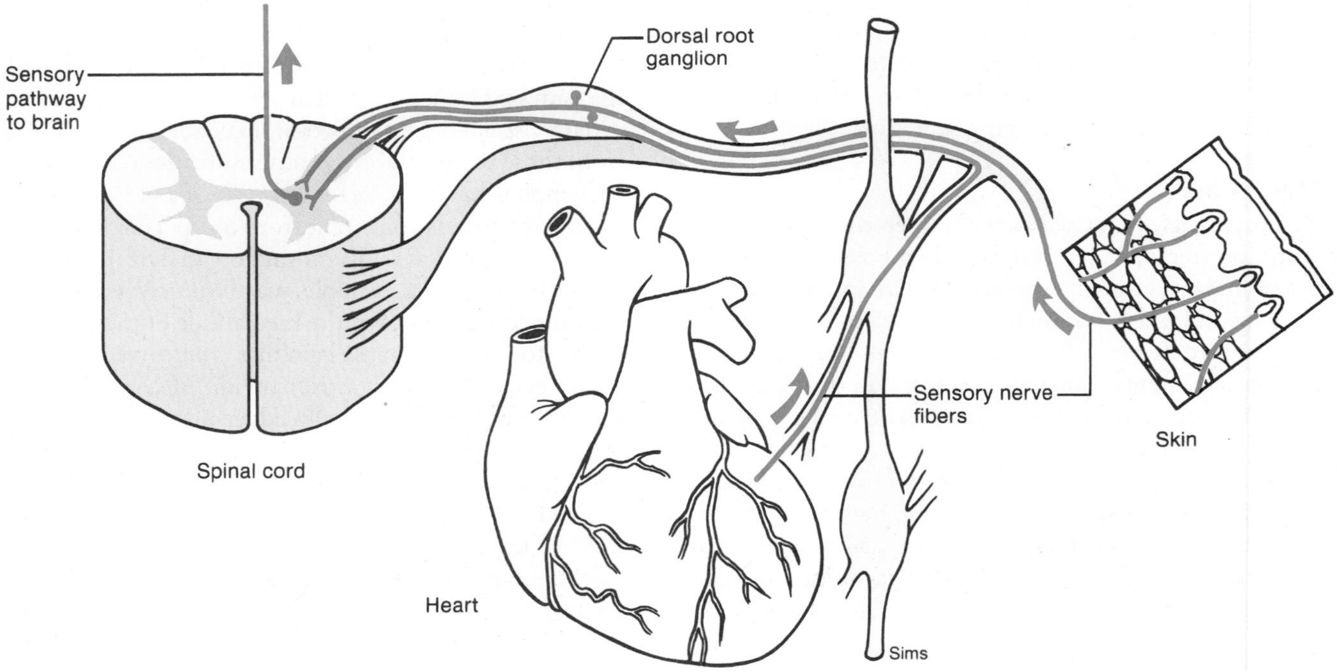

Figure 15.3 An explanation of referred pain. Pain originating from the myocardium of the heart may be perceived as coming from the skin of the left arm because sensory impulses from these two organs are conducted through common nerve pathways to the brain.

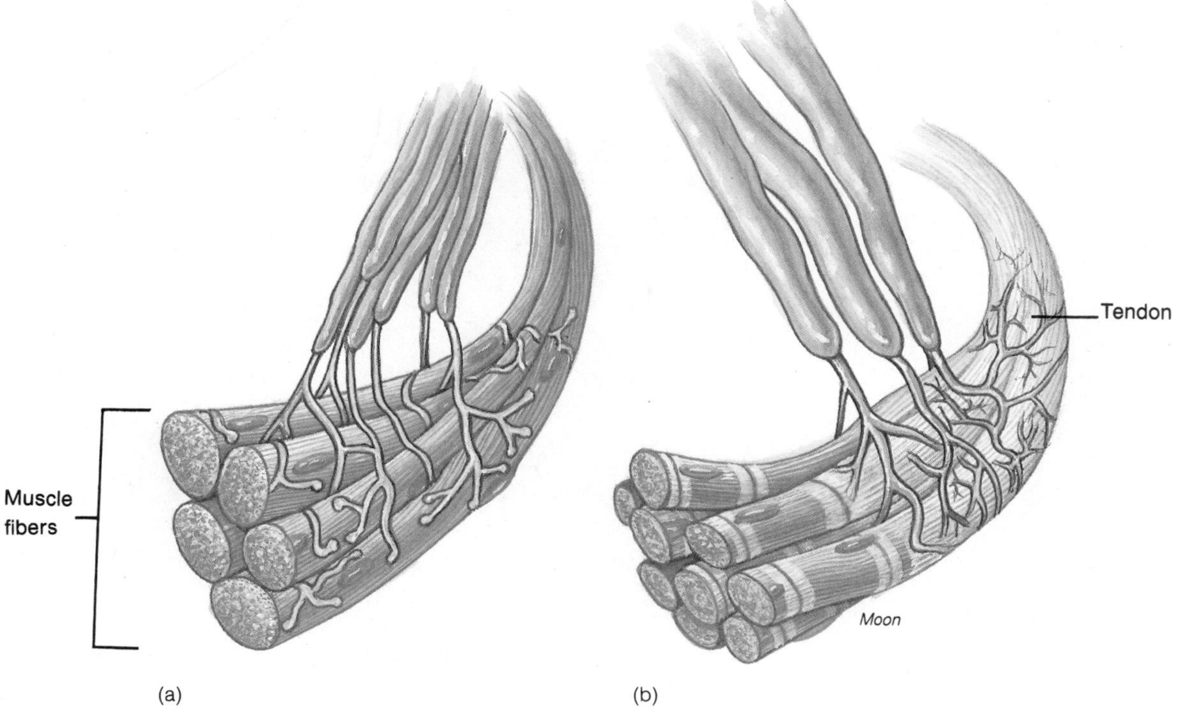

Figure 15.4 Proprioreceptors. (*a*) A neuromuscular spindle and (*b*) a neurotendinous receptor (Golgi tendon organ).

The perception of pain is of survival value because it alerts the body to an injury, disease, or organ dysfunction. *Acute pain* is sudden, usually short-term, and can generally be endured and attributed to a known cause. *Chronic pain,* however, is long-term and tends to weaken a person as it interferes with the ability to function effectively. Certain diseases, such as arthritis, are characterized by chronic pain. In these patients, relief of pain is of paramount concern. Treatment of chronic pain often requires the use of moderate pain-reducing drugs (analgesics) or intense narcotic drugs. Treatment in severely tormented chronic pain patients may include severing sensory nerves or implanting stimulating electrodes in appropriate nerve tracts.

Phantom pain is frequently experienced by an amputee who continues to feel pain from the body part that was amputated. After amputation, the severed sensory neurons heal and function in the remaining portion of the appendage. Although it is not known why impulses that are interpreted as pain are sent periodically through these neurons, the sensations evoked in the brain are projected to the amputated region, resulting in phantom pain.

Proprioceptors

Proprioceptors monitor internal conditions, frequently those under voluntary control. They are located within skeletal muscle tissue, tendons, and the synovial membranes and connective tissue surrounding joints. Proprioceptors are sensitive to changes in stretch and tension. Some of the sensory impulses from proprioceptors reach the level of consciousness as the **kinesthetic sense,** by which the position of the body parts is perceived. Other proprioceptor information is not consciously interpreted and is used to adjust the intensity and timing of muscle contractions to provide coordinated movements. With the kinesthetic sense, the position and movement of the limbs can be determined without visual sensations, such as when dressing or walking in the dark. The kinesthetic sense, along with hearing, becomes keenly developed in a blind person.

High-speed transmission is a vital characteristic of the kinesthetic sense, because rapid feedback to various body parts is essential for quick, smooth, coordinated body movements.

There are three types of proprioceptors: joint kinesthetic receptors, neuromuscular spindles, and neurotendinous receptors.

Joint Kinesthetic Receptors Joint kinesthetic receptors are located in the connective tissue capsule in diarthrotic joints where they are stimulated by changes in position caused by movement at the joints.

Neuromuscular Spindles Neuromuscular spindles are located in skeletal muscle, particularly in the muscles of the extremities. They consist of the endings of sensory neurons spiraled around specialized individual muscle fibers (fig. 15.4). Neuromuscular spindles are stimulated by an increase in muscle tension due to the lengthening or stretching of the individual spindles and thus provide information about the length of the muscle and the rate of change of length (i.e., velocity of contraction).

Neurotendinous Receptors Neurotendinous receptors, also called *Golgi (gol'je) tendon organs,* are located where a muscle attaches to a tendon (fig. 15.4). They are stimulated by the tension produced in a tendon when the attached muscle is either stretched or contracted.

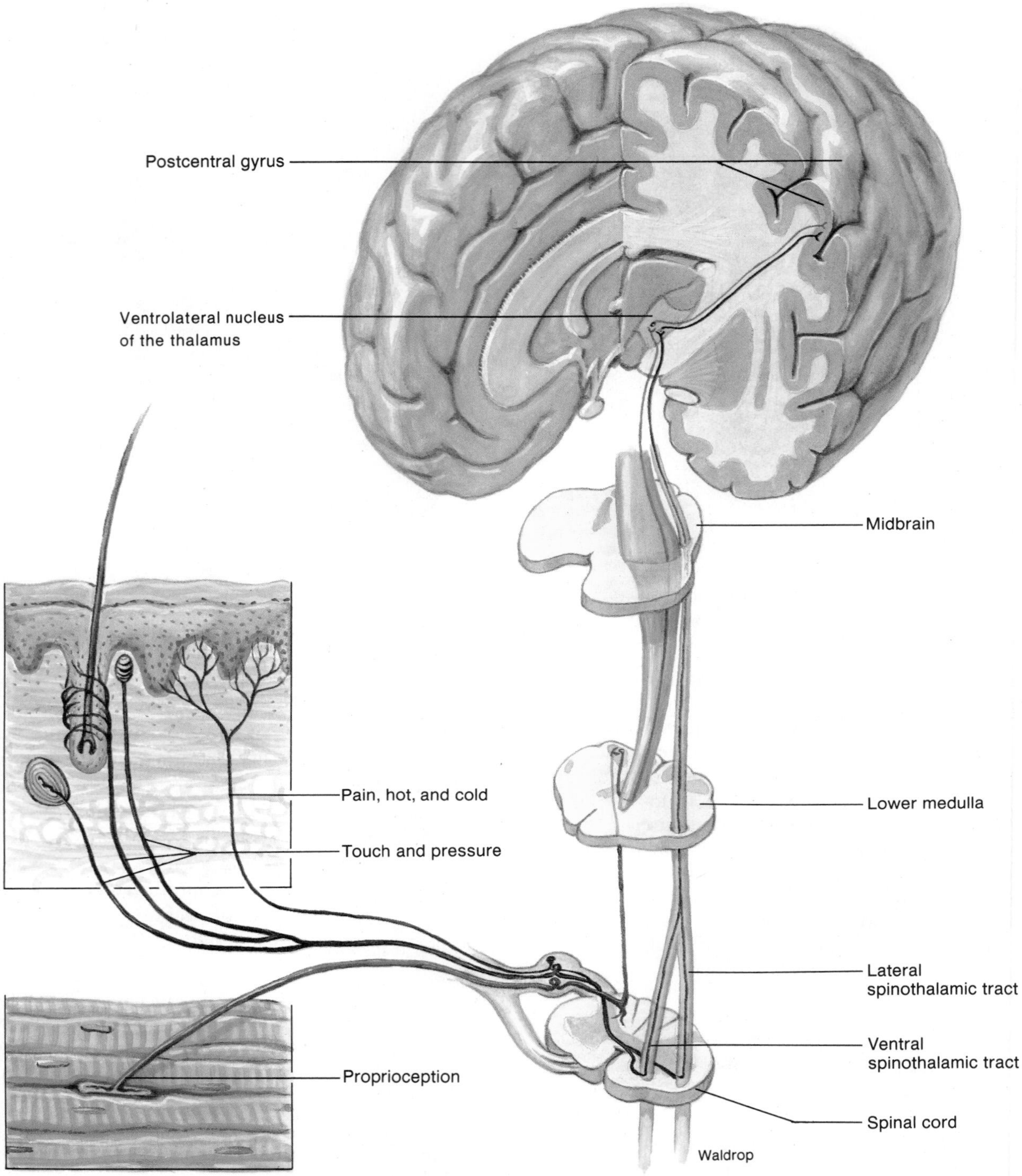

Figure 15.5 Pathways that lead from the cutaneous receptors and proprioreceptors into the postcentral gyrus in the cerebral cortex.

Neural Pathways for Somatic Sensation

The conduction pathways for the somatic senses are shown in figure 15.5. Sensations of proprioception and touch and pressure are carried by large, myelinated nerve fibers that ascend in the dorsal columns of the spinal cord on the ipsilateral (same) side. These fibers do not synapse until they reach the medulla oblongata of the brain stem; fibers that carry these sensations from the feet are thus incredibly long. After synapsing in the medulla with other, second-order sensory neurons, information in the latter neurons crosses over to the contralateral (opposite) side as it ascends via a fiber tract, called the **medial lemniscus,** to the thalamus. Third-order sensory neurons in the thalamus that receive this input in turn project to the **postcentral gyrus.**

Sensations of hot, cold, and pain are carried by thin, unmyelinated sensory neurons into the spinal cord. These synapse within the spinal cord with second-order interneurons, which cross over to the contralateral side and ascend to the brain in

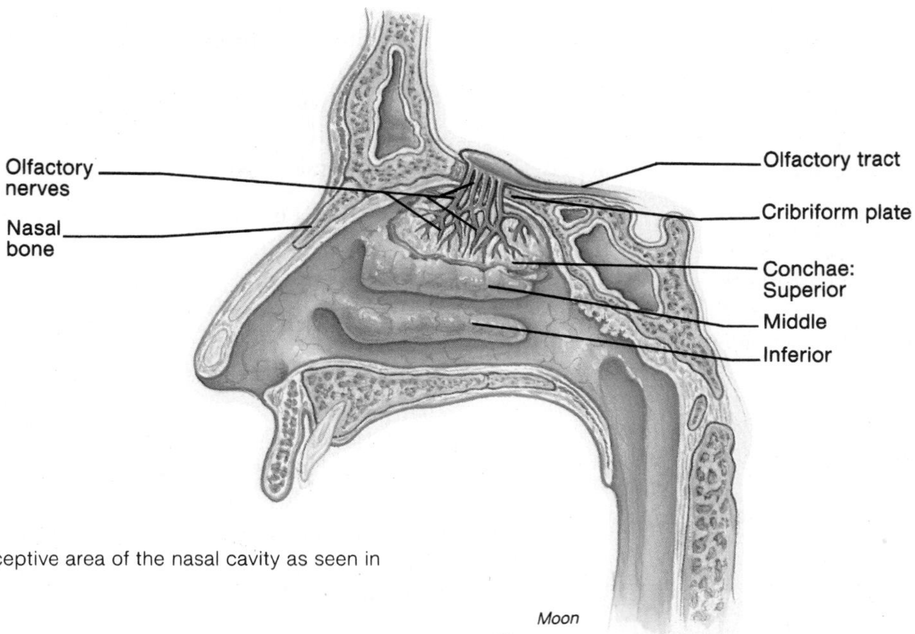

Figure 15.6 The olfactory receptive area of the nasal cavity as seen in sagittal view.

the **lateral spinothalamic tract.** Fibers that mediate touch and pressure ascend in the **ventral spinothalamic tract.** Fibers of both spinothalamic tracts synapse in the thalamus with third-order neurons, which in turn project to the postcentral gyrus of the parietal cerebral lobe. Note that, in all cases, somatic information is carried to the postcentral gyrus in third-order neurons. Also, because of crossing over, somatic information from each side of the body is projected to the postcentral gyrus of the contralateral cerebral hemisphere.

All somatic information from the same area of the body projects to the same area of the postcentral gyrus, so that a *sensory homunculus* map of the body can be drawn on the postcentral gyrus to represent sensory projection points (see fig. 11.29). This map is greatly distorted, however, because it shows larger areas of cerebral cortex devoted to sensation in the face and hands than in other areas of the body. This disproportionately larger area of the cortex devoted to the face and hands reflects the fact that there is a higher density of sensory receptors in these regions.

1. List the types of tactile receptors, and state where they are located. What portion of the brain interprets tactile sensations?
2. Discuss the importance of pain. List the receptors that respond to pain and the structures of the brain that are particularly important in the perception of pain sensation.
3. Using examples, distinguish between referred pain and phantom pain. Discuss why it is important for a physician to know the referred pain sites.
4. Using a flow diagram, describe the neural pathways leading from cutaneous pain and pressure receptors to the postcentral gyrus. Indicate where crossing over occurs.

Olfactory Sense

Olfactory receptors are the dendritic endings of the olfactory (first cranial) nerve that respond to chemical stimuli and transmit the sensation of olfaction directly to the olfactory portion of the cerebral cortex.

Objective 8. Describe the sensory pathway of olfaction.

Olfactory reception in humans is not highly developed compared to that of certain other vertebrates. Because humans do not rely on smell for communicating or for finding food, the olfactory sense is probably the least important of the senses; it is more important in detecting the presence of an odor rather than its intensity. Accommodation occurs relatively rapidly with this sense. Olfaction functions closely with gustation (taste) in that the receptors of both are chemoreceptors and require dissolved substances for stimuli.

Olfactory receptor cells are located in the nasal epithelium within the roof of the nasal cavity on both sides of the nasal septum (fig. 15.6). Olfactory cells are moistened and stabilized by the surrounding glandular goblet cells and the supporting columnar epithelial cells. The olfactory cells are actually bipolar neurons whose cell bodies lie between the supporting columnar cells. The free end of each olfactory cell contains several dendritic endings called **olfactory hairs,** which are the sensitive portion of the receptor cell.

The sensory pathway for olfaction consists of several neural segments. The unmyelinated axons of the olfactory cells unite to form the **olfactory nerves,** which traverse the foramina of the cribriform plate and terminate in the paired masses of gray and white matter called the **olfactory bulbs.** The olfactory bulbs lie on both sides of the crista galli of the ethmoid bone beneath the frontal lobes of the cerebrum. The neurons of the olfactory nerves synapse within the olfactory bulb with dendrites of the **olfactory tract.** Sensory impulses are conveyed along the olfactory tract and into the olfactory portion of the cerebral cortex where they are interpreted as odor and cause the sensation of smell.

Unlike taste, which is divisible into only four modalities, thousands of distinct odors can be distinguished by people who are trained in this capacity (as in the perfume industry). The molecular basis of olfaction is not understood, but it is known that a single odorant molecule is sufficient to excite an olfactory receptor.

Only about 2% of inhaled air comes in contact with the olfactory receptors, which are positioned above the main stream of air flow. Olfactory sensitivity can be increased by forceful sniffing, which draws the air into contact with the receptors.

Certain chemicals activate the trigeminal (fifth) as well as the olfactory (first) cranial nerves and cause reactions. Pepper, for example, may cause sneezing; onions cause the eyes to water; and smelling salts (ammonium salts) initiate respiratory reflexes and are used to revive unconscious persons.

---

1. What are olfactory hairs? Where are they located?
2. Trace the pathway of an olfactory stimulation from the olfactory hairs to the cerebrum, where interpretation occurs.

Gustatory Sense

Taste receptors are specialized epithelial cells, clustered together into taste buds, that respond to chemical stimuli and transmit the sense of taste through the glossopharyngeal (ninth cranial) nerve or the facial (seventh cranial) nerve to the cortex of the parietal cerebral lobe for interpretation.

Objective 9. List the three types of papillae, and explain how they function in the perception of taste.

Objective 10. Identify the cranial nerves and the sensory pathways of gustation.

The receptors of taste are located in the **taste buds.** Taste buds are specialized sensory organs that are most numerous on the dorsum of the tongue but are also present on the soft palate and on the walls of the oropharynx. The cylindrical taste bud is composed of many sensory **gustatory cells** that are encapsulated by **supporting cells** (fig. 15.7). Each gustatory cell contains a dendritic ending called a **gustatory hair** that projects to the surface through an opening in the taste bud called the **taste pore.** The gustatory hairs are the sensitive portion of the receptor cells. Saliva provides a moistened environment necessary for a chemical stimulus to activate the gustatory cells.

Taste buds are elevated by surrounding connective tissue and epithelium to form *papillae (pah-pil'e)* (fig. 15.7). Three types of papillae can be identified:

1. **Vallate papillae.** The largest but least numerous are the vallate papillae, which are arranged in an inverted V-shape pattern on the back of the tongue.
2. **Fungiform papillae.** Knoblike fungiform papillae are present on the tip and sides of the tongue.
3. **Filiform papillae.** Short, thickened, threadlike filiform papillae are located on the anterior two-thirds of the tongue.

Taste buds are found only in the circumvallate and fungiform papillae. Filiform papillae contain gustatory cells, which are not clustered into taste buds.

There are only four basic modalities of taste, known as *gustation (gus-ta'shun),* which are sensed most acutely on particular regions of the tongue (fig. 15.8). These are *sweet* (tip of tongue), *sour* (sides of tongue), *bitter* (back of tongue), and *salty* (over most of the tongue but concentrated on the sides).

Sour taste is produced by hydrogen ions (H^+); all acids therefore taste sour. Most organic molecules, particularly sugars, taste sweet to varying degrees. Only pure table salt (NaC1) has a pure salty taste. Other salts, such as KC1 (commonly used in place of NaC1 by people with hypertension), taste salty but have bitter overtones. Bitter taste is evoked by quinine and seemingly unrelated molecules.

The sensory pathway of taste receptors to the brain involves mainly two cranial nerves (fig. 15.9). Taste buds on the posterior one-third of the tongue have a sensory pathway through the glossopharyngeal nerve, whereas the anterior two-thirds of the tongue is served by the chorda tympani branch of the facial nerve.

Taste sensations passing through the nerves just mentioned are conveyed through the medulla and thalamus to the parietal lobe of the cerebral cortex, where they are interpreted.

Because taste and smell are both chemoreceptors, they complement each other. We often confuse a substance's smell with its taste; and if we have a head cold or hold our nose while eating, food seems to lose its flavor.

---

1. Distinguish between papillae, taste buds, and gustatory cells. Discuss the function of each as related to taste.
2. List the three types of papillae.
3. Which cranial nerves have sensory innervation associated with taste? What are the afferent pathways to the brain where the perception of taste occurs?

Visual Sense

Rods and cones are the photoreceptors within the eyeball that are sensitive to light energy and are stimulated to transmit nerve impulses through the optic nerve and optic tract to the visual cortex of the occipital lobes, where the interpretation of vision occurs. The sensory components of the eye are formed by twenty weeks, and the accessory structures are formed by thirty-two weeks.

---

gustatory: L. *gustare,* to taste
papilla: L. *papilla,* nipple

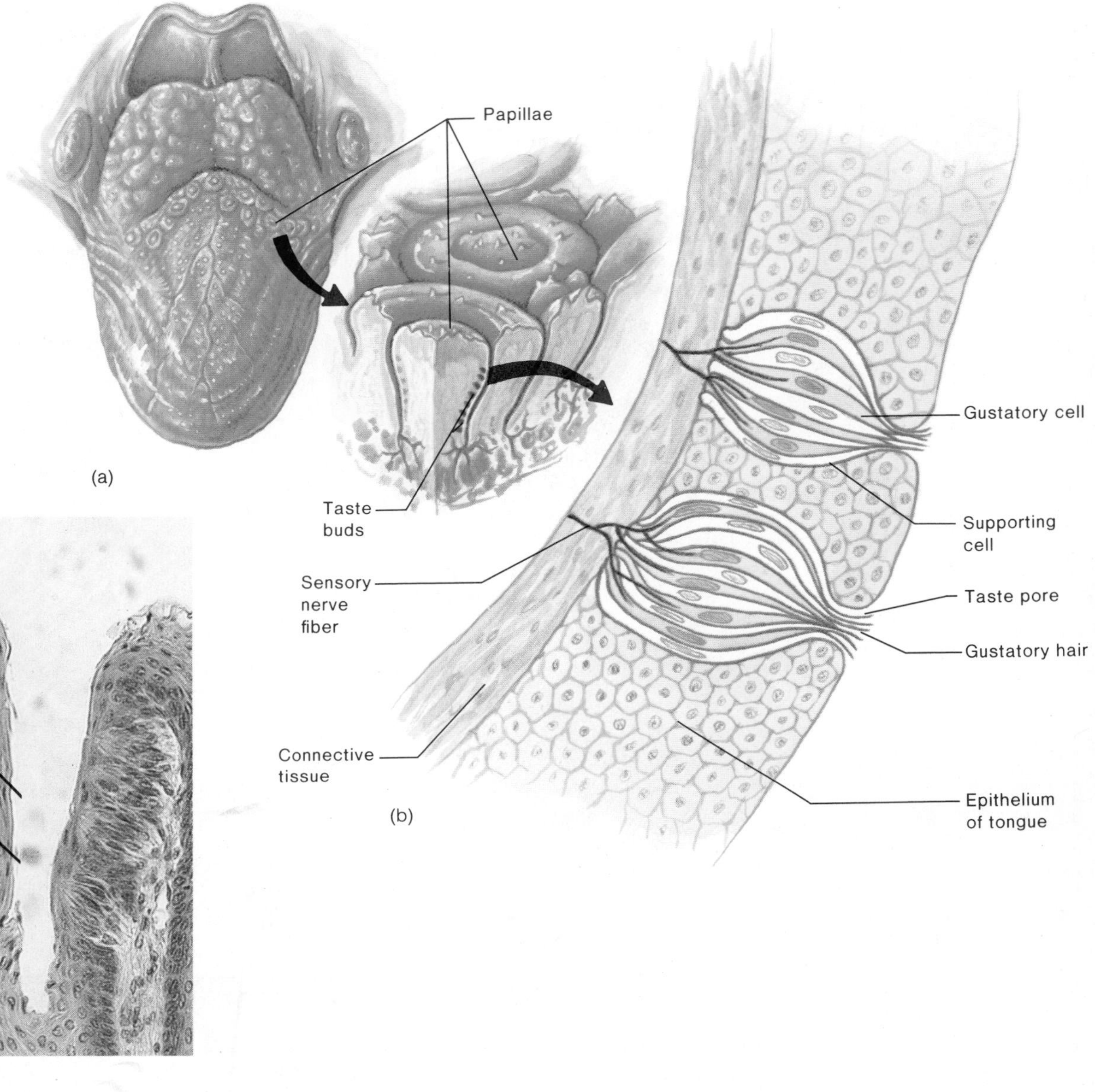

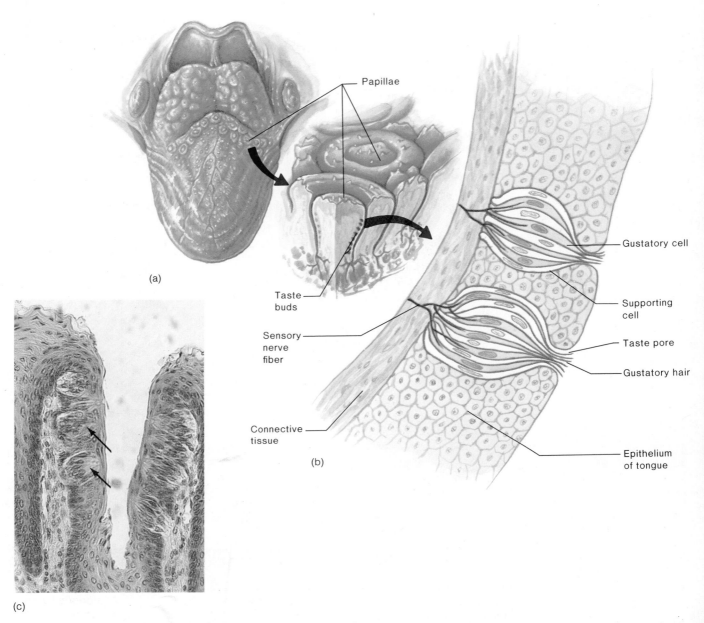

Figure 15.7 Papillae of the tongue and associated taste buds. (*a*) Numerous taste buds are positioned within each papilla. (*b*) Each gustatory cell and its associated gustatory hair is encapsuled by supporting cells. (*c*) A photomicrograph of some taste buds (identified with arrows).

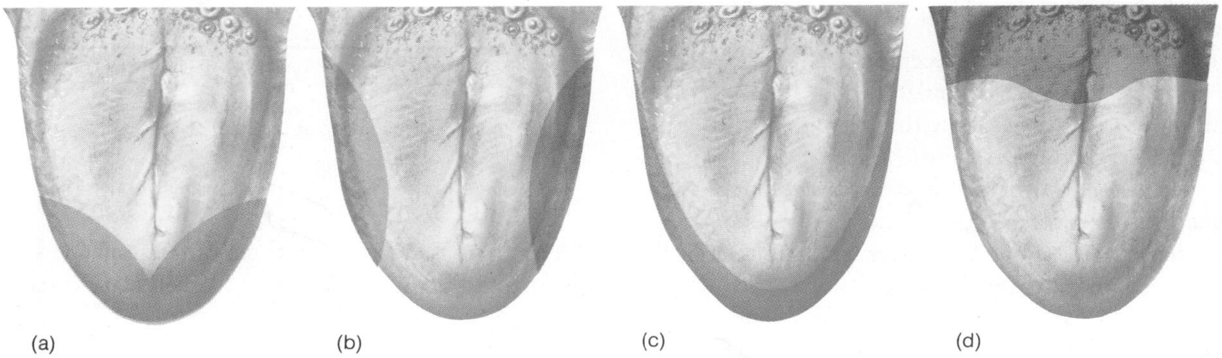

Figure 15.8 Patterns of taste receptor distribution on the dorsum of the tongue. (*a*) Sweet receptors, (*b*) sour receptors, (*c*) salt receptors, and (*d*) bitter receptors.

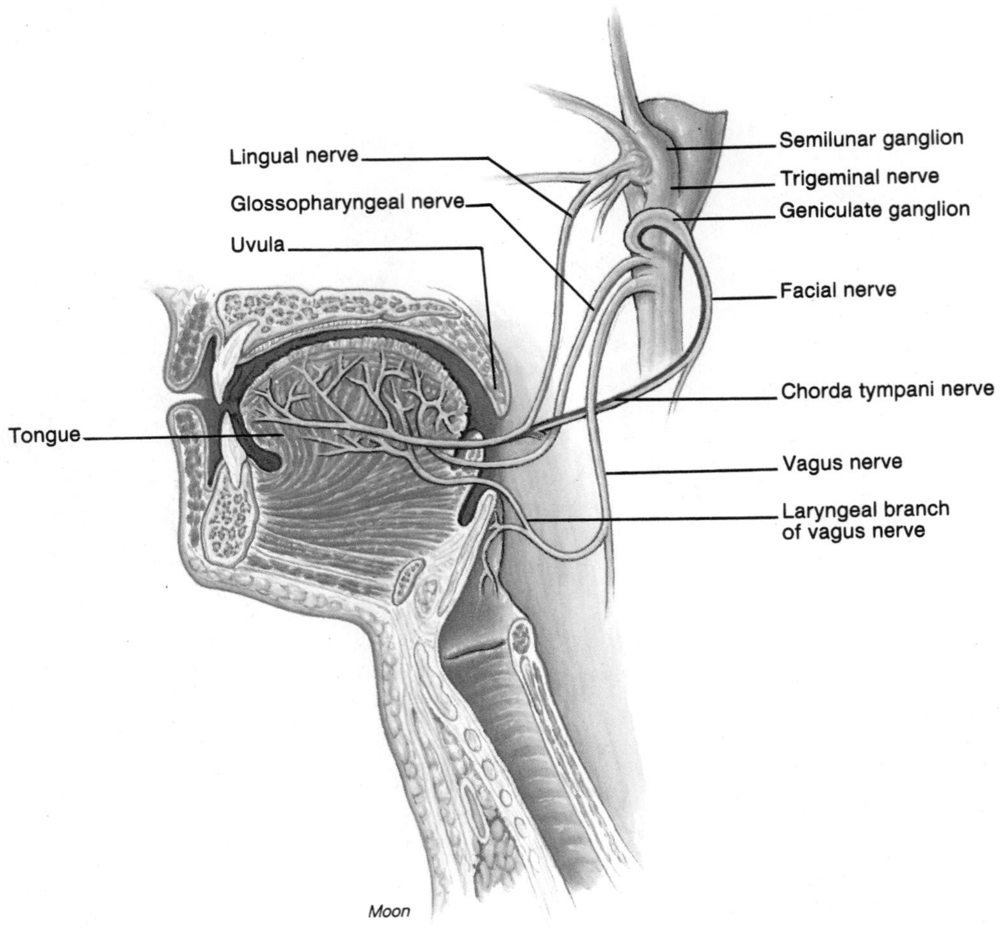

Figure 15.9 The principal viscerosensory innervation (taste) is provided by the facial (VII) and the glossopharyngeal (IX) cranial nerves. The chorda tympani nerve is the branch of the facial nerve innervating the tongue. Branches from the vagus and the trigeminal nerves also provide some sensory innervation. The hypoglossal (XII) cranial nerve (not shown) provides motor innervation to the tongue.

Objective 11. Describe the development of the eye in the human embryo.

Objective 12. Describe the associated structures of the eye and the structure of the eyeball.

Objective 13. Trace the path of light waves through the eye, and explain how they are focused on distant and near objects.

Objective 14. Describe the neural pathway of a visual impulse, and discuss the neural processing of visual information.

The eyes are organs that refract (bend) and focus the incoming light waves onto the sensitive photoreceptors at the back of each eye. Nerve impulses from the stimulated photoreceptors are conveyed through visual pathways within the brain to the occipital lobes of the cerebrum where the sense of vision is perceived. The specialized photoreceptor cells can respond to an incredible one billion different stimuli each second. Further, these cells are sensitive to about 10 million gradations of light intensity and 7 million different shades of color.

The eyes are anteriorly positioned on the skull and set just far enough apart to achieve *binocular (stereoscopic) vision* when focusing on an object. This three-dimensional perspective allows a person to assess depth. Often likened to a camera (table 15.2) the eyes are responsible for approximately 80% of all knowledge that is assimilated.

Table 15.2 Comparisons between eye structures and analogous structures in a camera

Eye structure	Analogous camera structure
Cornea and lens	Lens system
Iris and pupil	Variable aperture system
Eyelid	Lens cap
Sclera	Camera frame
Pigment epithelium and choroid	Black interior of camera
Retina*	Film

*Since neural processing begins in the retina, this eye structure may also be considered analogous (in part) to the photographer.

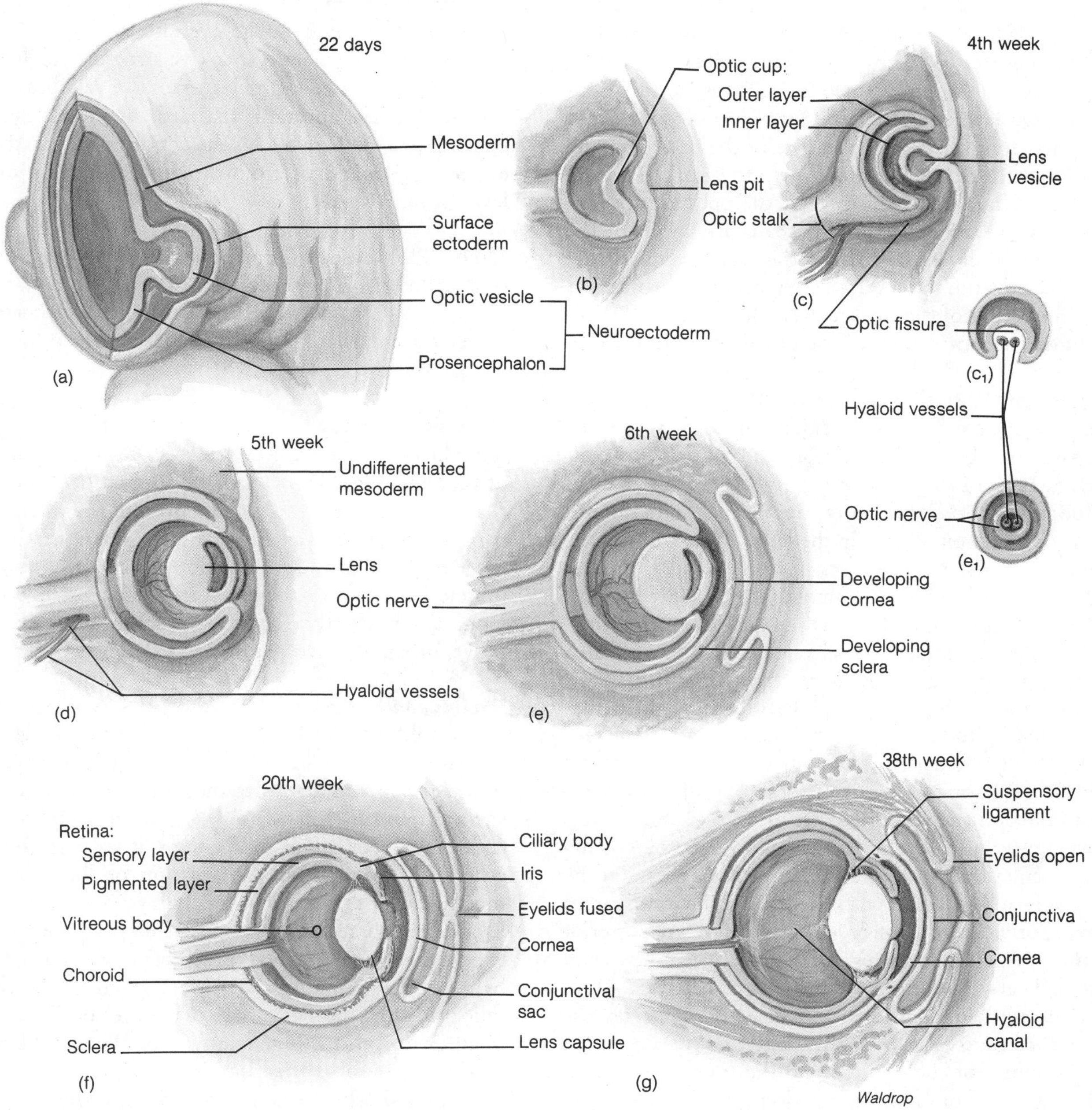

Figure 15.10 The development of the eye. (*a*) An anterior view of the developing head of a twenty-two-day-old embryo and the formation of the optic vesicle from the neuroectoderm of the prosencephalon (forebrain). (*b*) The development of the optic cup. The lens vesicle is formed (*c*) as the ectodermal lens placode invaginates during the fourth week. The hyaloid vessels become enclosed (c_1) and (e_1) within the optic nerve as there is fusion of the optic fissure. (*d*) The basic shape of the eyeball and the position of its internal structures are established during the fifth week. The successive development of the eye is shown at (*e*) six weeks and at (*f*) twenty weeks, respectively. (*g*) The eye of the newborn.

The eyes of other vertebrates are basically similar to yours. Certain species, however, have adaptive modifications. Consider, for example, the extremely keen eyesight of a hawk, which soars high in the sky searching for food, or the eyesight of a nocturnal animal such as an owl, which feeds only at night. Note how the location of the eyes on the head corresponds to behavior. Predatory species, like cats, have eyes that are directed forward, allowing depth perception. Prey species, like deer, have eyes positioned high on the sides of their heads, allowing panoramic vision to detect distant threatening movements, even while grazing.

Development of the Eye

The development of the eye is a complex, rapid process that involves the precise interaction of neuroectoderm, surface ectoderm, and mesoderm. Differentiation of these germ layers is evident early in the fourth week. The initial differentiation involves neuroectoderm forming a lateral diverticulum on each side of the prosencephalon (forebrain). As the diverticulum increases in size, the distal portion dilates to become the **optic vesicle,** and the proximal portion constricts to become the **optic stalk** (fig. 15.10). Once the optic vesicle is formed, the overlying surface ectoderm thickens and invaginates. The thickened portion is the **lens placode** *(plak'ōd),* and the invagination is the **lens pit.**

During the fifth week, the lens placode is depressed and eventually cut off from the surface ectoderm, causing the formation of the **lens vesicle.** Simultaneously with the formation of the lens vesicle, the optic vesicle invaginates and differentiates into the two-layered **optic cup.** A groove called the **optic fissure** appears along the inferior surface of the optic cup and is continuous with a depression along the optic stalk. The **hyaloid artery** and **hyaloid vein** traverse the optic fissure longitudinally to serve the developing eyeball. The walls of the optic fissure eventually close so that the hyaloid vessels are within the tissue of the optic stalk. They become the **central vessels of the retina** of the mature eye. The optic stalk eventually becomes the **optic nerve,** composed of sensory axons from the retina.

By the end of the sixth week and the early part of the seventh week, the optic cup has differentiated into two sheets of epithelial tissue that become the sensory and pigmented layers of the **retina.** Both of these layers also line the entire vascular coat, including the **ciliary body, iris,** and the **choroid** *(ko'roid)*. A proliferation of cells in the lens vesicle leads to the formation of the **lens.**

To this point of eye development, the tissue mass surrounding the differentiating ectoderm is undifferentiated mesoderm. Once the primordial ectodermal structures are formed, however, the mesodermal germ tissue begins to differentiate. The **lens capsule** (see figs. 15.10, 15.18) forms from the mesoderm surrounding the lens. The mesoderm between the lens and retina, through which the hyaloid vessels pass, is transformed into the **vitreous humor** (see fig. 15.26). Mesoderm surrounding the optic cup differentiates into two distinct layers of the developing eyeball. The inner layer of mesoderm becomes the vascular **choroid,** and the outer layer becomes the toughened **sclera** *(skle'rah)* posteriorly and the transparent **cornea** anteriorly. The anterior chamber and the posterior chamber are the fluid-filled spaces that form between the cornea and the iris and between the iris and the lens, respectively. Once the cornea has formed, additional surface ectoderm gives rise to the thin **conjunctiva** covering the anterior surface of the eyeball. Epithelium of the **eyelids** and the **lacrimal glands** and **ducts** develop from surface ectoderm, whereas the **extrinsic eye muscles** and all connective tissue associated with the eye develop from mesoderm. These accessory structures of the eye gradually develop during the embryonic period and into the fetal period as late as the fifth month.

The developing eye is extremely sensitive to and may be impaired by teratogenic agents, particularly *rubella (German measles).* If a pregnant woman contracts rubella, there is a 90% chance the embryo or fetus will contract it also. An embryo afflicted with rubella has more than a 30% chance of being aborted, stillborn, or congenitally deformed. Rubella interferes with the mitotic process and thus causes underdeveloped organs. An embryo with rubella may suffer from a number of physical deformities, *cataracts* and *glaucoma* being common deformities of the eye.

hyaloid: Gk. *hyalos,* glass; *eiodos,* form

Associated Structures of the Eye

Associated structures of the eye either protect the eyeball or provide eye movement. Protective structures include the bony orbit, eyebrow, facial muscles, eyelids, eyelashes, conjunctiva, and the lacrimal apparatus that produces tears. Eyeball movements are made possible by the actions of the extrinsic ocular eye muscles that arise from the orbit and insert on the outer layer of the eyeball.

Orbit Each eyeball is positioned in a bony depression in the skull called the orbit (see fig. 6.19 and table 6.4). Seven bones of the skull (frontal, lacrimal, ethmoid, zygomatic, maxilla, sphenoid, and palatine) form the walls of the orbit that support and protect the eye.

Eyebrow Eyebrows consist of short, thick hair positioned transversely above both eyes along the superior orbital ridges of the skull (figs. 15.11, 15.12). Located here they effectively shade the eyes from the sun and prevent perspiration or falling particles from getting into the eyes. Underneath the skin of each eyebrow is the orbital portion of the orbicularis oculi muscle and a portion of the corrugator muscle (see fig. 9.15). Contraction of either of these muscles causes the eyebrow to move, often reflexively, to protect the eye.

Eyelids and Eyelashes **Eyelids (palpebrae)** develop as reinforced folds of skin with attached skeletal muscle so that they are movable. In addition to the orbicularis oculi muscle attached to the skin that surrounds the front of the eye, the **levator palpebrae** *(le-va'tor pal'pĕ-bre)* **superioris** muscle attaches along the upper eyelid and provides it with greater movability than the lower eyelid. Contraction of the orbicularis oculi muscle closes the eyelids over the eye, and contraction of the levator palpebrae superioris muscle elevates the upper eyelid to expose the eye. The eyelids protect the eyeball from desiccation by reflexively blinking about every seven seconds and moving fluid across the anterior surface of the eyeball. To avoid a blurred image, the eyelid will generally blink when the eyeball moves to a new position of fixation.

The **palpebral fissure** (fig. 15.12) is the interval between the upper and lower eyelids. The **commissures (canthi)** of the eye are the medial and lateral angles where the eyelids come together. The **medial commissure** is broader than the **lateral commissure** and is characterized by a small, reddish, fleshy elevation called the **lacrimal caruncle** *(kar'ung-kl)* (fig. 15.13), which contains sebaceous and sudoriferous glands. The shape of the palpebral fissure is elliptical when the eyes are open.

Each eyelid supports a row of numerous **eyelashes,** which protect the eye from airborne particles. The shaft of each eyelash is surrounded by a root hair plexus that provides the hair with the sensitivity necessary to elicit a reflexive closure of the lids. Eyelashes of the upper lid are long and turn upward, whereas those of the lower lid are short and turn downward.

palpebra: L. *palpebra,* eyelid (related to *palpare,* to pat gently)
commissure: L. *commissura,* a joining
caruncle: L. *caruncula,* diminutive of caro, flesh

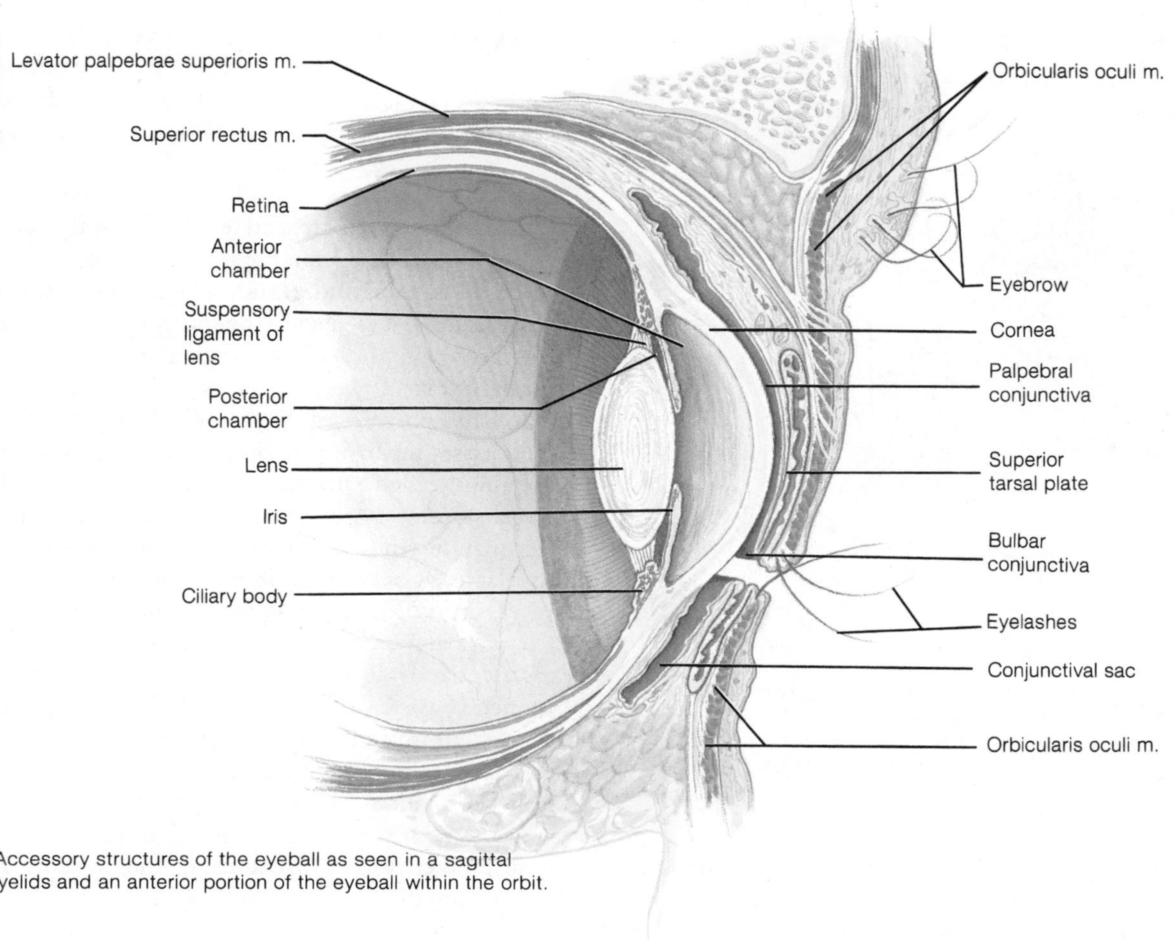

Figure 15.11 Accessory structures of the eyeball as seen in a sagittal section of the eyelids and an anterior portion of the eyeball within the orbit.

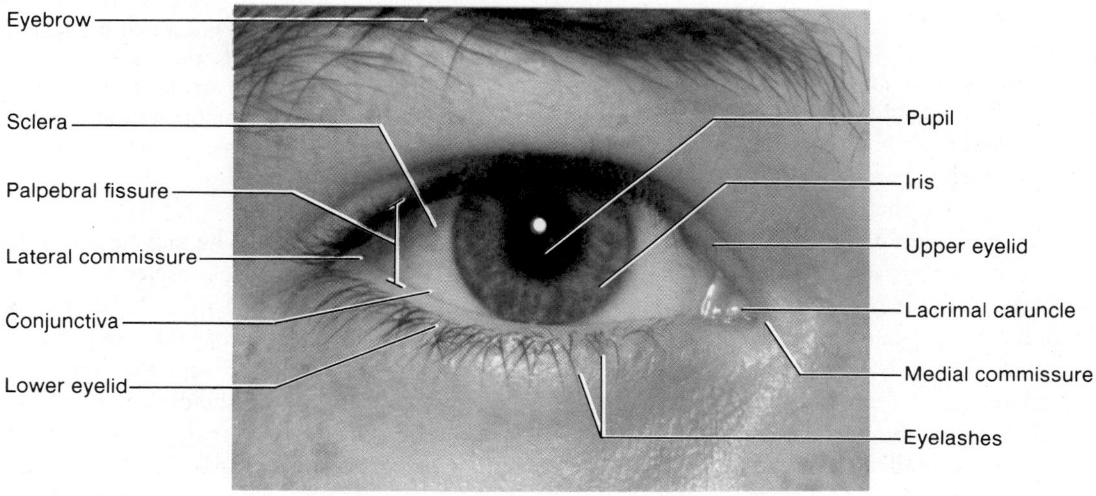

Figure 15.12 The surface anatomy of the eye.

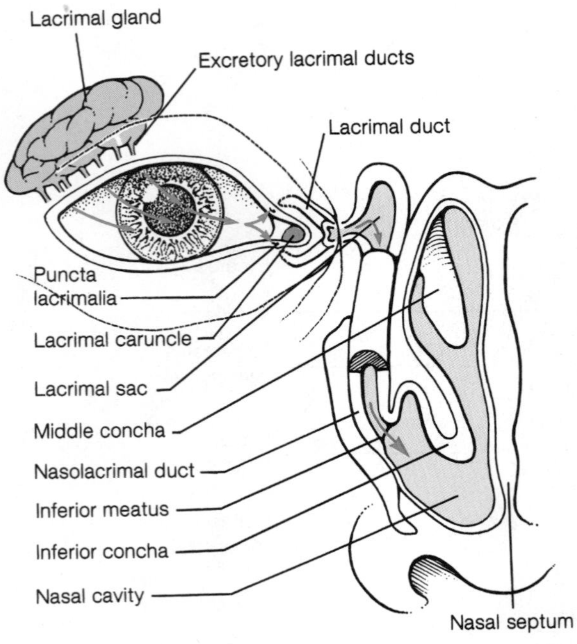

Figure 15.13 The lacrimal apparatus consists of the lacrimal gland and the ducts that drain lacrimal fluid into the nasal cavity. The lacrimal gland produces lacrimal fluid, which moistens and cleanses the conjunctiva that lines the underside of the eyelids and covers the anterior surface of the eyeball.

In addition to the layers of the skin and the underlying connective tissue and orbicularis oculi muscle fibers, each eyelid contains a tarsal plate, tarsal glands, and conjunctiva. The **tarsal plates,** composed of dense fibrous connective tissue, are important in maintaining the shape of the eyelids (fig. 15.11). Specialized sebaceous glands called **tarsal** (meibomian) **glands** are embedded within the tarsal plates along the exposed inner surfaces of the eyelids. The ducts of the tarsal glands open onto the edges of the eyelids, and their oily secretions help keep the eyelids from adhering to each other. A *chalazion (kah-la'ze-on)* is a tumor or cyst on the eyelid that results from an infection of the tarsal glands. Modified sweat glands called **ciliary glands** are also located within the eyelids along with additional sebaceous glands at the bases of the hair follicles of the eyelashes. An infection of these sebaceous glands is referred to as a *sty.*

Conjunctiva The conjunctiva *(kon''junk-ti'vah)* is a thin mucus-secreting epithelial membrane that lines the interior surface of each eyelid and exposed anterior surface of the eyeball (fig. 15.11). It consists of stratified squamous epithelium that varies in thickness in different regions. The **palpebral conjunctiva** is thick and adheres to the tarsal plates of the eyelids. As the conjunctiva reflects onto the anterior surface of the eyeball, it is known as the **bulbar conjunctiva.** This portion is transparent and especially thin where it covers the cornea. Because the conjunctiva is continuous and reflects from the eyelids to the anterior surface of the eyeball, a space called the **conjunctival sac** exists when the eyelids are closed. The conjunctival sac protects the eyeball by preventing objects (including a contact lens) from passing beyond the confines of the sac. The conjunctiva can rapidly repair itself if it is scratched.

tarsal: Gk. *tarsos,* flat basket
meibomian glands: from Heinrich Meibom, German anatomist, 1638–1700
chalazion: Gk. *chalazion,* hail; a small tubercle

Lacrimal Apparatus The lacrimal apparatus consists of the lacrimal gland, which secretes the *lacrimal fluid* (tears), and a series of ducts that drain the secretion into the nasal cavity (fig. 15.13). The **lacrimal gland,** which is about the size and shape of an almond, is located in the superolateral portion of the orbit. It is a compound tubuloacinar gland that secretes lacrimal fluid through several excretory lacrimal ducts into the conjunctival sac of the upper eyelid. With each blink of the eyelids, lacrimal fluid passes medially and downward and drains into two small openings, called **puncta lacrimalia,** on both sides of the lacrimal caruncle. From here the lacrimal fluid drains through the lacrimal duct into the lacrimal sac and continues through the nasolacrimal duct to the inferior meatus of the nasal cavity.

Lacrimal fluid is an aqueous, mucus secretion that contains a bactericidal enzyme called *lysozyme.* Lacrimal fluid not only moistens and lubricates the conjunctival sac but also reduces the chance of eye infections. Normally about one milliliter of lacrimal fluid is produced each day by the lacrimal gland of each eye. If irritating substances, such as particles of sand or chemicals from onions, make contact with the conjunctiva, the lacrimal glands are stimulated to oversecrete. The extra lacrimal fluid protects the eye by diluting and washing away the irritating substance.

Humans are the only animals known to weep in response to emotional stress. While crying, the volume of lacrimal secretion is so great that the tears may spill over the edges of the eyelids and the nasal cavity fill with fluid. The crying response is an effective means of communicating one's emotions and results from stimulation of the lacrimal glands by parasympathetic motor neurons.

Extrinsic Eye Muscles The movements of the eyeball are controlled by six extrinsic eye muscles called the **extrinsic ocular muscles** (figs. 15.14, 15.15). Each extrinsic ocular muscle originates from the bony orbit and inserts by a tendinous attachment to the tough outer tunic of the eyeball. Four **recti muscles** maneuver the eyeball in the direction indicated by their names **(superior, inferior, lateral,** and **medial),** and two **oblique muscles (superior** and **inferior)** rotate the eyeball on its axis (see also fig. 9.19). One of the extrinsic ocular muscles, the superior oblique, passes through a pulleylike cartilaginous loop, called the **trochlea** *(trok'le-ah),* before attaching to the eyeball. Although stimulation of each muscle causes a precise movement of the eyeball, most of the movements involve the combined contraction of usually two muscles.

The motor units of the extrinsic ocular muscles are the smallest in the body. This means that a single motor neuron serves about ten muscle fibers, resulting in precision movements. The eyes move in synchrony by contracting synergistic muscles while relaxing antagonistic muscles.

trochlea: Gk. *trochos,* a wheel

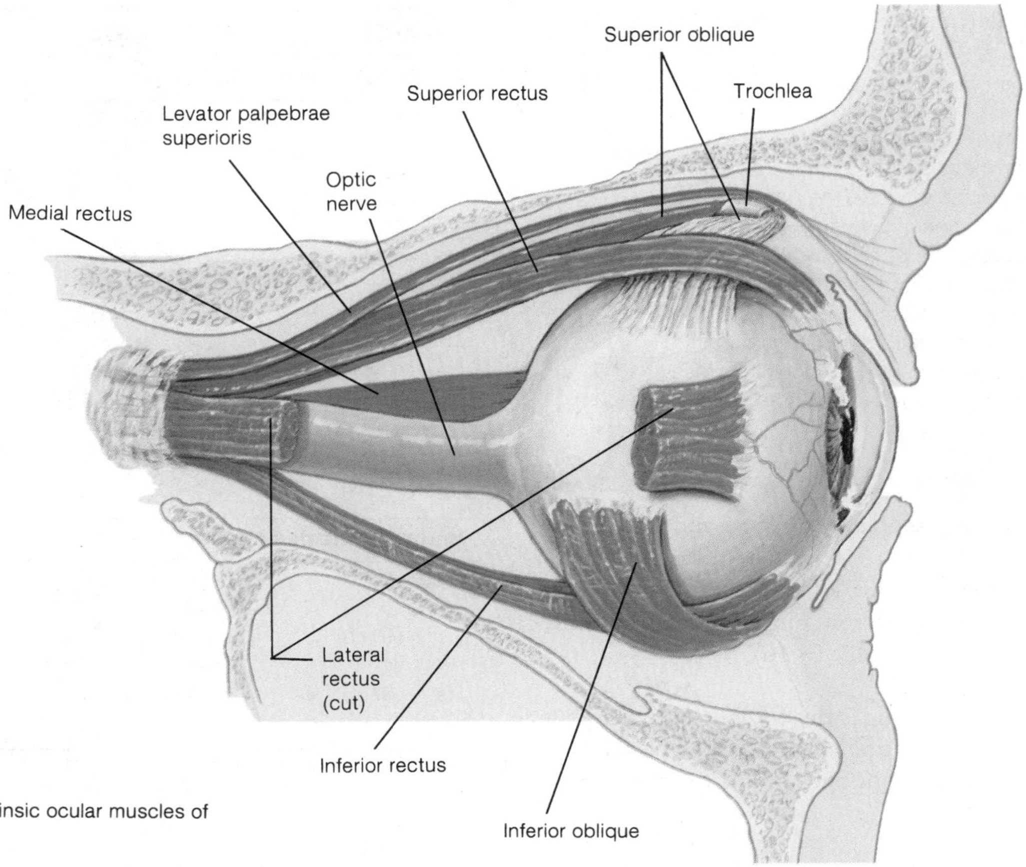

Figure 15.14 The extrinsic ocular muscles of the right eyeball.

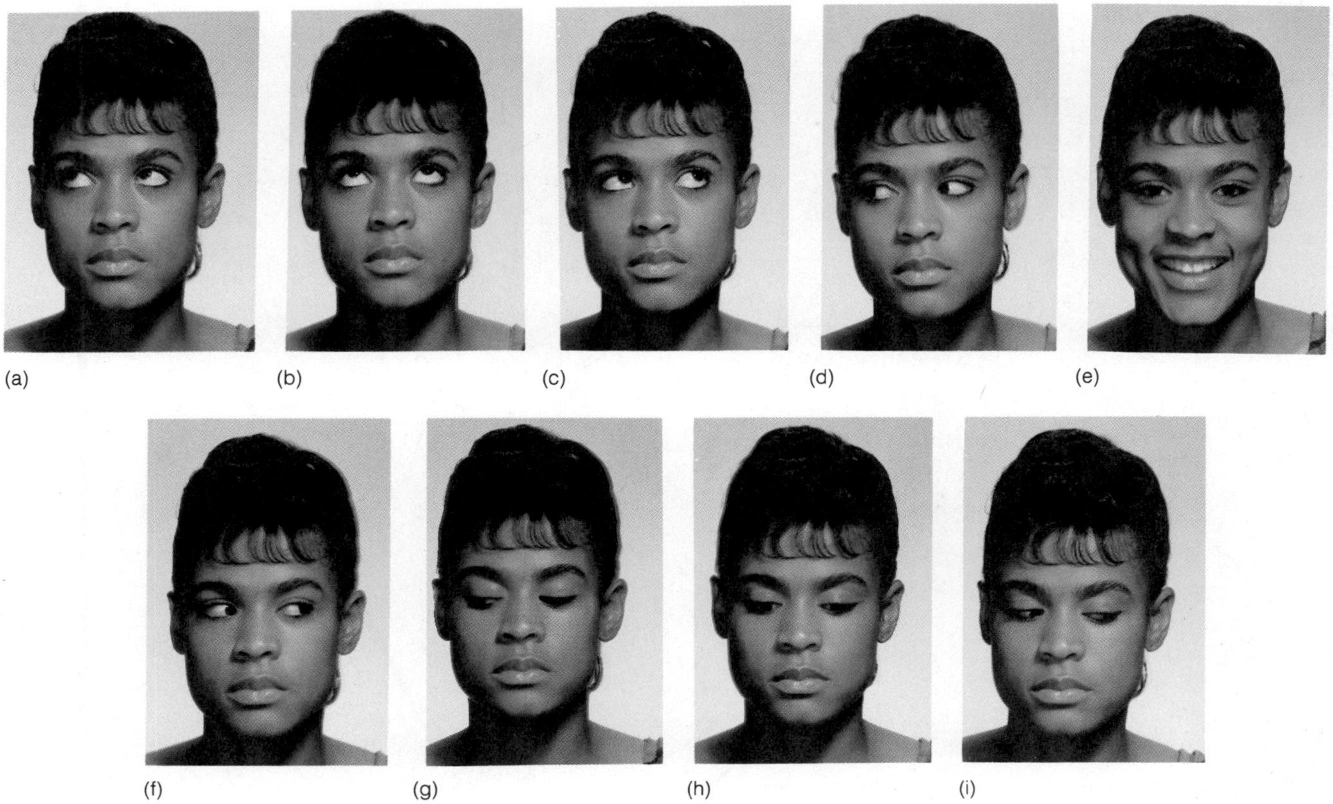

Figure 15.15 The positions of the eyes as the ocular muscles are contracted. (*a*) Right eye, inferior oblique muscle; left eye, superior and medial recti muscles. (*b*) Both eyes, superior recti and inferior oblique muscles. (*c*) Right eye, superior and medial recti muscles; left eye, inferior oblique muscle. (*d*) Right eye, lateral rectus muscle; left eye, medial rectus muscle. (*e*) Primary position with the eyes fixed on a distant fixation point. (*f*) Right eye, medial rectus muscle; left eye, lateral rectus muscle. (*g*) Right eye, superior oblique muscle; left eye; inferior and medial recti muscles. (*h*) Both eyes, inferior recti and superior oblique muscles. (*i*) Right eye, inferior and medial recti muscles; left eye, superior oblique muscle.

The extrinsic ocular muscles are innervated by three cranial nerves (table 15.3). Innervation of the other skeletal and smooth muscles that serve the eye is also indicated in table 15.3.

A physical examination may include an eye movement test. As the patient's eyes follow the movement of a physician's finger, the physician can assess weaknesses in specific muscles or dysfunctions of specific cranial nerves. The patient experiencing *double vision (diplopia)* when moving his eyes may be suffering from muscle weakness. Looking laterally tests the abducens cranial nerve, looking inferiorly and laterally tests the trochlear cranial nerve, and looking cross-eyed tests the oculomotor and trochlear cranial nerves of both eyes.

Structure of the Eyeball

The eyeball of an adult is essentially spherical and is approximately 25 mm (1 in.) in diameter. About four-fifths of the eyeball is positioned within the orbit of the skull. The eyeball consists of three basic layers: the fibrous tunic, the vascular tunic, and the internal tunic, or retina (fig. 15.16).

Fibrous Tunic The fibrous tunic is the outer layer of the eyeball. It is divided into two regions: the posterior five-sixths is the opaque sclera, and the anterior one-sixth is the transparent cornea (fig. 15.17).

Table 15.3 Muscles of the eye

Extrinsic muscles (skeletal)		
Superior rectus	Oculomotor nerve (III)	Rotates eye upward and toward midline
Inferior rectus	Oculomotor nerve (III)	Rotates eye downward and toward midline
Medial rectus	Oculomotor nerve (III)	Rotates eye toward midline
Lateral rectus	Abducens nerve (VI)	Rotates eye away from midline
Superior oblique	Trochlear nerve (IV)	Rotates eye downward and away from midline
Inferior oblique	Oculomotor nerve (III)	Rotates eye upward and away from midline
Intrinsic muscles (smooth)		
Ciliary muscles	Oculomotor nerve (III) parasympathetic fibers	Causes suspensory ligament to relax
Iris, circular muscles	Oculomotor nerve (III) parasympathetic fibers	Causes size of pupil to decrease
Iris, radial muscles	Sympathetic fibers	Causes size of pupil to increase

From Stuart Ira Fox, *Human Physiology*, 3d ed. Copyright © 1990 Wm. C. Brown Publishers, Dubuque, Iowa. All Rights Reserved. Reprinted by permission.

Tunicas:
- Internal
- Vascular
- Fibrous

Optic disc

Optic nerve

(a)

Cornea

Lens

(b)

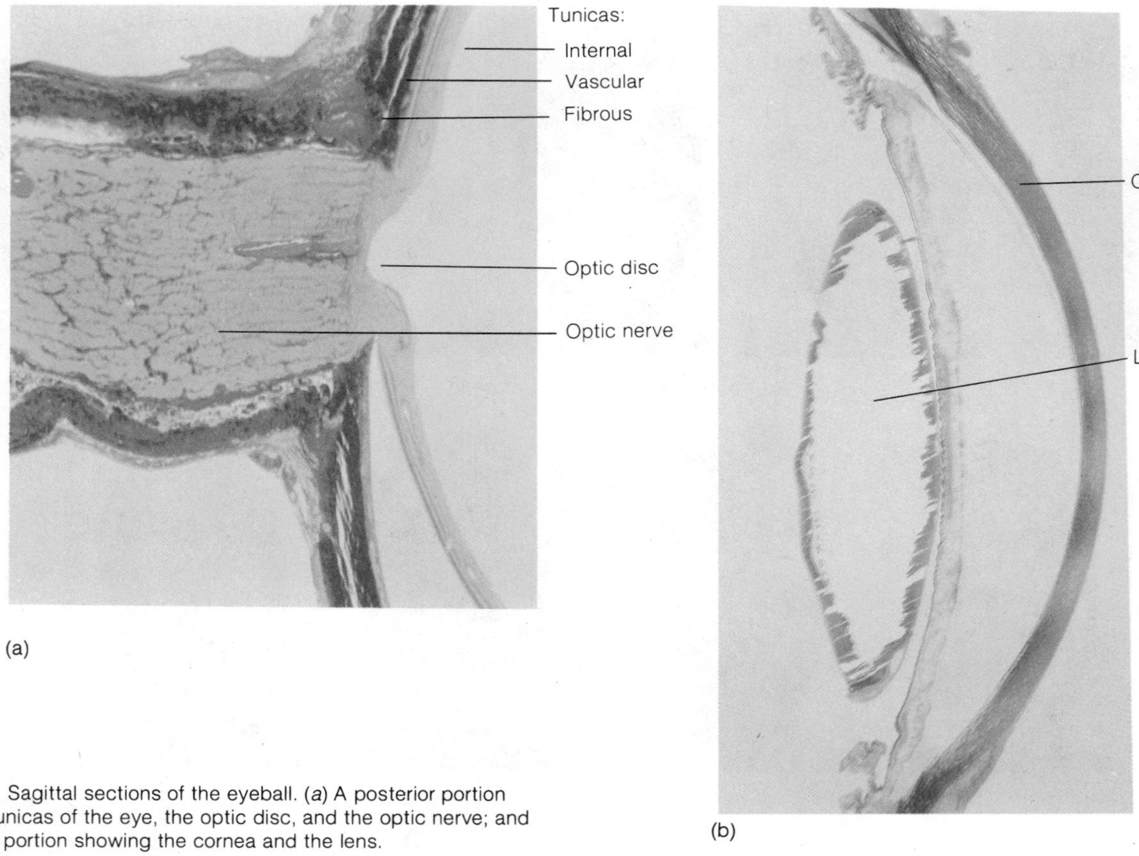

Figure 15.16 Sagittal sections of the eyeball. (*a*) A posterior portion showing the tunicas of the eye, the optic disc, and the optic nerve; and (*b*) an anterior portion showing the cornea and the lens.

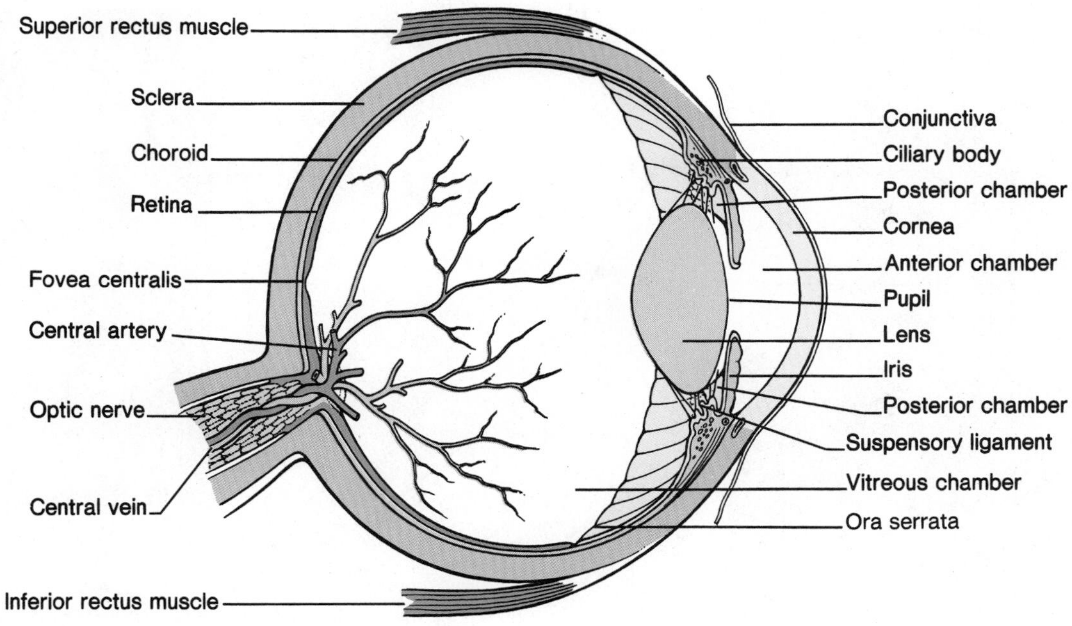

Figure 15.17 The internal anatomy of the eyeball.

The toughened **sclera** is the white of the eye. It is composed of tightly bound elastic and collagenous fibers, which give shape to the eyeball and protect its inner structures. The sclera is avascular but does contain sensory receptors for pain. The large **optic nerve** exits through the sclera at the posterior portion of the eyeball.

The **cornea** is transparent and convex to permit the passage and cause the refraction (bending) of incoming light waves. The transparency of the cornea is due to tightly packed, avascular, dense connective tissue. Also, the relatively few cells that are present in the cornea are arranged in unusually regular patterns. The circumferential edge of the cornea is continuous structurally with the sclera. The outer surface of the cornea is covered with a thin, nonkeratinized, stratified squamous epithelial layer called the **bulbar (corneal) epithelium,** which is actually a continuation of the conjunctiva of the sclera (fig. 15.11).

A defective cornea can be replaced with a donor cornea in a surgical procedure called a *corneal transplant (keratoplasty).* A defective cornea is one that does not transmit or refract light effectively due to its shape, scars, or disease. During a corneal transplant, the defective cornea is excised and replaced with a transplanted cornea that is sutured into place. It is considered to be the most successful type of homotransplant (between individuals of the same species).

Vascular Tunic The vascular tunic, or **uvea** *(u've-ah)* of the eyeball consists of the choroid, the ciliary body, and the iris (fig. 15.17).

The **choroid** is a thin, highly vascular layer that lines most of the internal surface of the sclera. The choroid contains numerous pigment-producing melanocytes, which give it a dark brownish color that prevents light waves from being reflected out of the eyeball. There is an opening in the choroid at the back of the eyeball where the optic nerve is located.

The **ciliary body** is the thickened, anterior portion of the vascular tunic that forms an internal muscular ring toward the front of the eyeball (fig. 15.18). Three distinct planes of smooth muscle fibers called **ciliary muscles** are found within the ciliary body.

Numerous extensions of the ciliary body called **ciliary processes** attach to the **zonular fibers,** which in turn attach to the lens. Collectively, the zonular fibers constitute the **suspensory ligament.** The transparent **lens** consists of tight layers of protein fibers arranged like the layers of an onion. A thin, clear **lens capsule** encloses the lens and provides attachment for the suspensory ligament (see figs. 15.18, 15.26).

The shape of the lens determines how much the light waves that pass through will be refracted. Constant tension of the suspensory ligament, when the ciliary muscles are relaxed, flattens the lens somewhat (fig. 15.19). Contraction of the ciliary muscles relaxes the suspensory ligament and makes the lens more spherical. The constant tension within the lens capsule results in the surface of the lens becoming more convex when the suspensory ligament is not taut. A flattened lens facilitates viewing a distant object, whereas a rounded lens permits viewing a close object.

The **iris** is the anterior portion of the vascular tunic and is continuous with the choroid. The iris is viewed from the outside as the colored portion of the eyeball (figs. 15.17, 15.18).

sclera: Gk. *skleros,* hard
cornea: L. *cornu,* horn
uvea: L. *uva,* grape

choroid: Gk. *chorion,* membrane
zonular fibers: L. *zona,* a girdle
iris: Gk. *irid,* rainbow

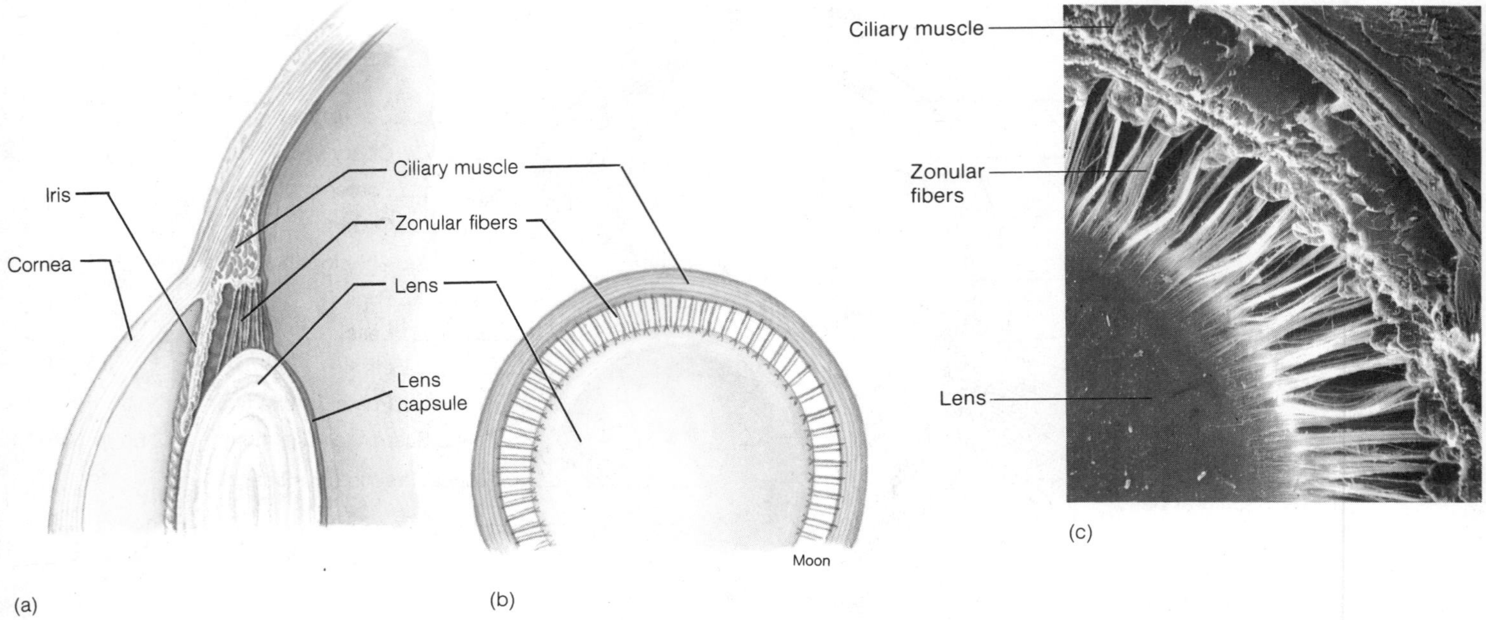

Figure 15.18 Structure of the anterior portion of the eyeball. (*a*) A sagittal view; (*b*) an anterior view of the lens and supporting structures; and (*c*) a scanning electron micrograph in anterior view showing the relationship between the lens, zonular fibers, and ciliary muscles of the eye.

(*a*) and (*b*): Adapted from "How the Eye Focuses," by James F. Koretz and George Handleman.

Table 15.4 Location and functions of the structures of the eyeball

Tunic and structure	Location	Composition	Function
Fibrous tunic	Outer layer of eyeball	Avascular connective tissue	Provides shape of eyeball
Sclera	Posterior, outer layer; white of the eye	Tightly bound elastic and collagen fibers	Supports and protects eyeball
Cornea	Anterior surface of eyeball	Tightly packed dense connective tissue—transparent and convex	Transmits and refracts light
Vascular tunic (uvea)	Middle layer of eyeball	Highly vascular pigmented tissue	Supplies blood; prevents reflection
Choroid	Middle layer in posterior portion of eyeball	Vascular layer	Supplies blood to eyeball
Ciliary body	Anterior portion of vascular tunic	Smooth muscle fibers and glandular epithelium	Supports the lens through suspensory ligament and determines its shape; secretes aqueous humor
Iris	Anterior portion of vascular tunic continuous with ciliary body	Pigment cells and smooth muscle fibers	Regulates the diameter of the pupil and hence the amount of light entering the vitreous chamber
Retina	Inner layer in posterior portion of eyeball	Photoreceptor neurons (rods and cones), bipolar neurons, and ganglion neurons	Photoreception; transmits impulses
Lens (not part of any tunic)	Between posterior and vitreous chambers; supported by suspensory ligament of ciliary body	Tightly arranged protein fibers; transparent	Refracts light and focuses onto fovea centralis

It consists of smooth muscle fibers arranged in a circular and a radial pattern. The contraction of the smooth muscle fibers regulates the diameter of the **pupil** (table 15.4 and fig. 15.20), an opening in the center of the iris. Contraction of the circularly arranged fibers of the iris, stimulated by bright light, constricts the pupil and diminishes the amount of light entering the eyeball (see fig. 13.8). Contraction of the radially arranged fibers, in response to dim light, dilates the pupil and permits more light to enter.

Retina The retina covers the choroid as the innermost layer of the eyeball. It consists of a thin, outer **pigmented layer** in contact with the choroid, and a thick, inner **nervous layer,** or visual portion. The retina is principally in the posterior portion of the eyeball (fig. 15.17). The visual layer of the retina terminates in a jagged margin near the ciliary body called the **ora serrata.** The pigmented layer extends anteriorly over the back of the ciliary body and iris.

ora serrata: L. *ora,* margin; *serra,* saw

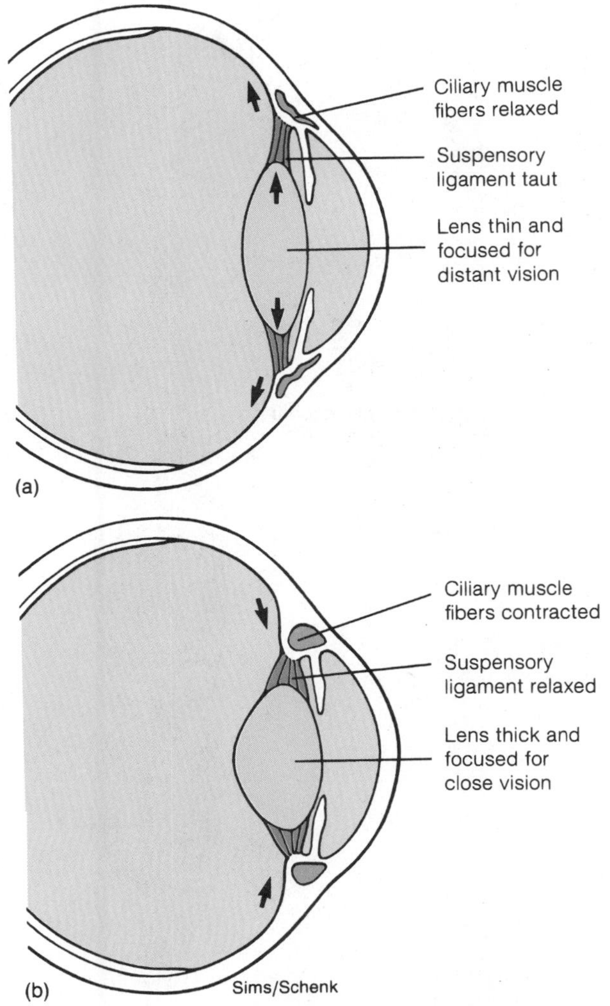

Figure 15.19 Changes in the shape of the lens during accommodation. (*a*) The lens is flattened for distant vision when the ciliary muscle fibers are relaxed and the suspensory ligament is taut. (*b*) The lens is more spherical for close-up vision when the ciliary muscle fibers are contracted and the suspensory ligament is relaxed.

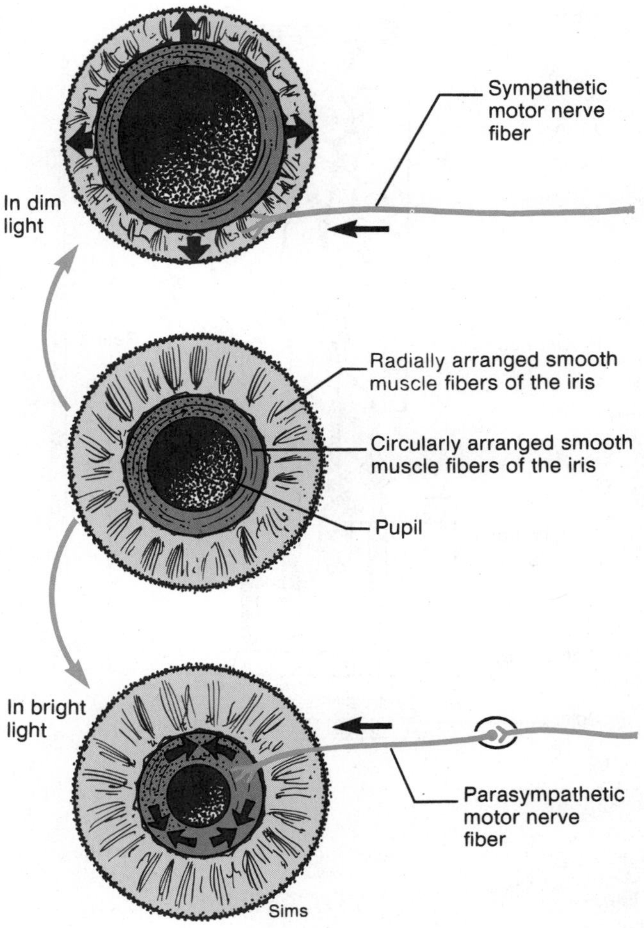

Figure 15.20 Dilation and constriction of the pupil. In dim light, the radially arranged smooth muscle fibers are stimulated to contract by sympathetic stimulation, dilating the pupil. In bright light, the circularly arranged smooth muscle fibers are stimulated to contract by parasympathetic stimulation, constricting the pupil.

The pigmented layer and nervous layer of the retina are not attached to each other except surrounding the optic nerve and at the ora serrata. Because of this loose connection, the two layers may become separated as a *detached retina.* Such a separation can be corrected by fusing the layers with a laser.

The nervous layer of the retina is composed of three principal layers of neurons. Listing them in the order in which they conduct impulses, they are rod and cone cells, bipolar neurons, and ganglion neurons (fig. 15.21). In terms of the passage of light, however, the order is reversed. Light must first pass through the layer of ganglion cells and then bipolar cells before reaching the photoreceptors.

Rods and **cones** are photoreceptors. Rods number over 100 million per eye and are more slender and elongated than cones (fig. 15.22). Rods are positioned on the peripheral parts of the retina, where they respond to dim light for black-and-white vision. They also respond to form and movement but provide poor visual acuity. Cones number about 7 million per eye. They provide daylight color vision and are responsible for visual acuity. The photoreceptors synapse with **bipolar neurons,** which in turn synapse with the **ganglionic neurons.** The axons of ganglionic neurons leave the eye as the optic nerve.

Cones are concentrated in a depression near the center of the retina called the **fovea centralis,** which is the area of keenest vision (figs. 15.17, 15.23). Surrounding the fovea centralis is the yellowish **macula lutea,** which also has an abundance of cones (fig. 15.23). There are no photoreceptors at the point where the optic nerve is attached. This area is a blind spot and is referred to as the **optic disc** (figs. 15.23, 15.24). Normally a person is unaware of the blind spot because the eyes continually move about and because an object is viewed from a different angle with each eye. The blind spot can easily be demonstrated as described in figure 15.25.

fovea: L. *fovea,* small pit
macula lutea: L. *macula,* spot; *luteus,* yellow

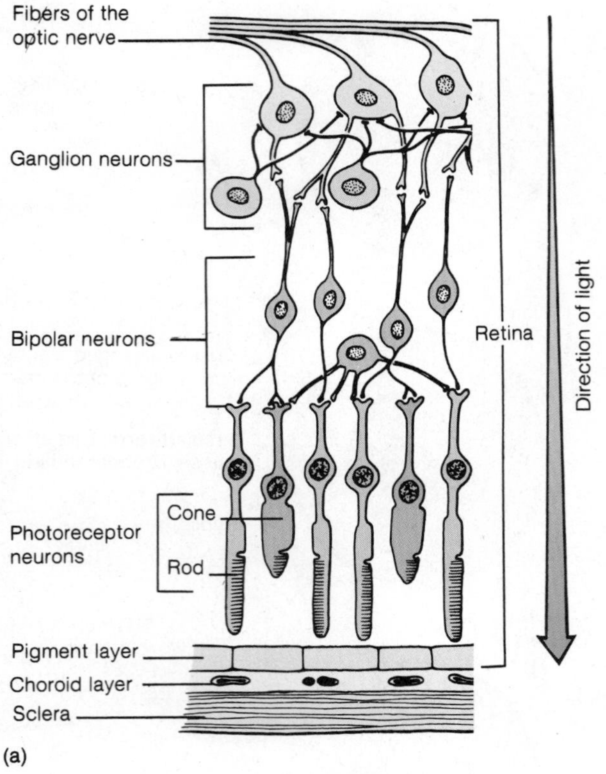

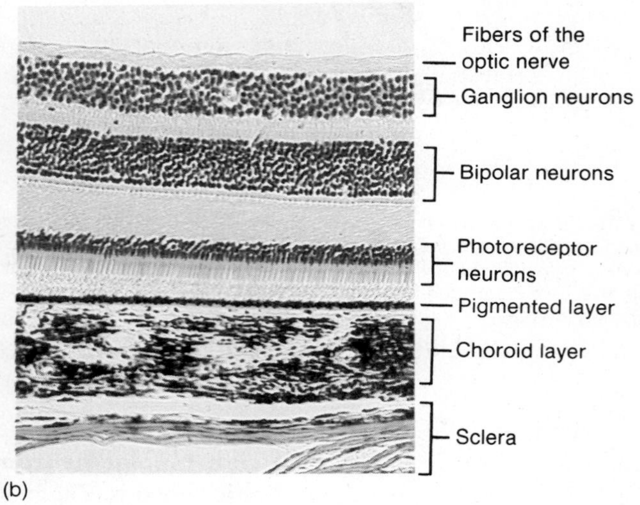

Figure 15.21 The layers of the retina. The retina is inverted, so that light must pass through various layers of nerve cells before reaching the photoreceptors (rods and cones): (*a*) a schematic diagram and (*b*) a light micrograph.

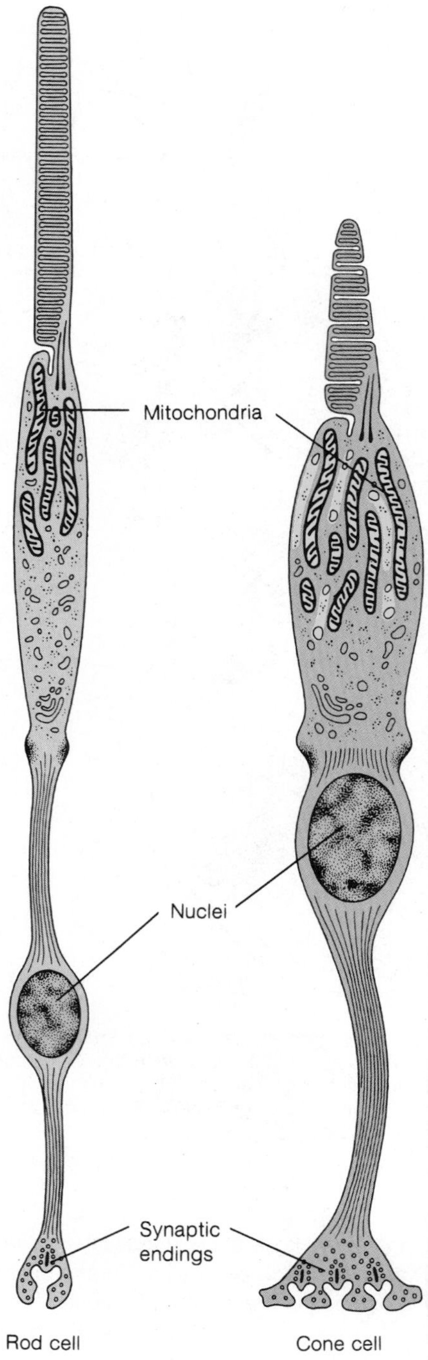

Figure 15.22 Photoreceptor cells of the eyeball.

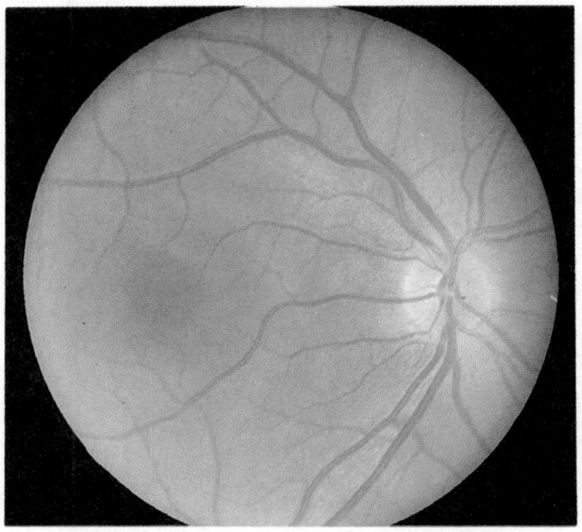

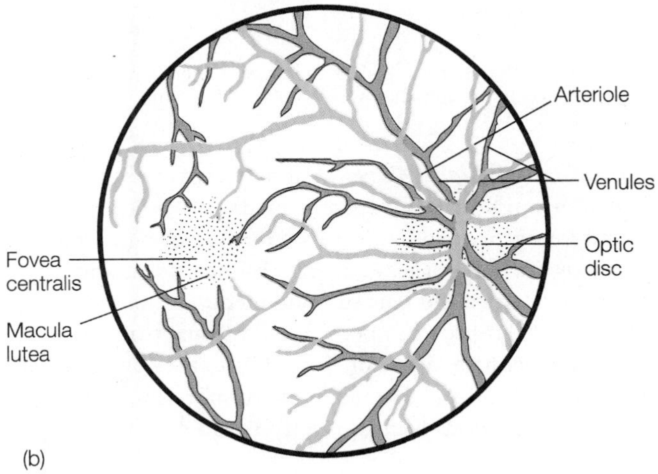

Figure 15.23 A view of the retina (*a*) as seen with an *ophthalmoscope and* (*b*) a diagrammatic view. Optic nerve fibers leave the eyeball at the optic disc to form the optic nerve. (Note the blood vessels that can be seen entering the eyeball at the optic disc.)

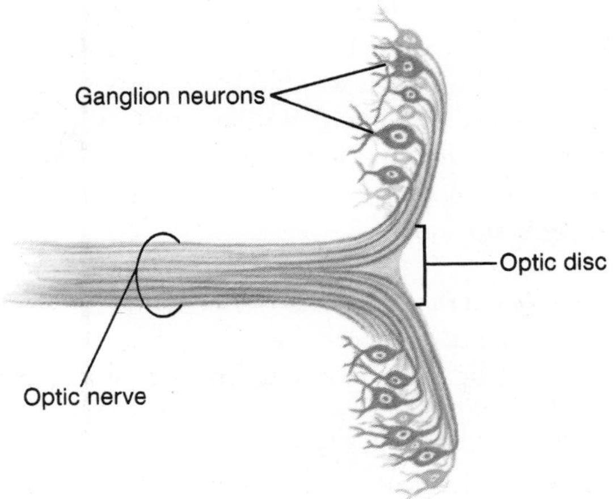

Figure 15.24 The optic disc is a small area of the retina where the fibers of the ganglion neurons emerge forming the optic nerve. The optic disc is frequently called the blind spot because it is devoid of rods and cones.

Figure 15.25 The blind spot. Hold the drawing about *twenty inches* from your face, with your left eye closed and your right eye focused on the circle. Slowly move the drawing closer to your face until the cross disappears. This occurs because the image of the cross is focused on the optic disc, where photoreceptors are absent.

Blood Supply to the Eyeball Both the choroid and the retina are richly supplied with blood. Two **ciliary arteries** pierce the sclera at the posterior aspect of the eyeball and traverse the choroid to the ciliary body and base of the iris. Although the ciliary arteries enter the eyeball independently, they anastomose (connect) extensively throughout the choroid.

The **central artery** (central retinal artery) branches from the ophthalmic artery and enters the eyeball in contact with the optic nerve. As the central artery passes through the optic disc, it divides into superior and inferior branches, each of which then divides into temporal and nasal branches to serve the inner layers of the retina (see fig. 15.17). The **central vein** drains blood from the eyeball through the optic disc. The branches of the central artery can be observed within the eyeball through an ophthalmoscope (fig. 15.23).

An examination of the internal eyeball with an ophthalmoscope is frequently part of a routine physical examination. Capillary vessels can be seen within the eyeball. If they appear abnormal, such as constricted, dilated, or hemorrhaged, they may be symptomatic of certain diseases or body dysfunctions. Diseases such as arteriosclerosis, diabetes, cataracts, or glaucoma can be detected during an examination of the internal eyeball.

Chambers of the Eyeball The interior of the eyeball is separated by the lens into two main cavities (see fig. 15.17). The cavity anterior to the lens is subdivided by the iris into an **anterior chamber** and a **posterior chamber.** The anterior chamber is located in front of the iris and behind the cornea. The posterior chamber is located between the iris and the suspensory ligament and the lens. The anterior and posterior chambers connect through the pupil and are filled with a watery fluid called **aqueous humor.** The constant production of aqueous humor maintains an *intraocular pressure* of about 12 mm of mercury within the anterior and posterior chambers. Aqueous humor also provides nutrients and oxygen to the avascular lens and cornea. An estimated 5–6 ml of aqueous humor is secreted each day from the vascular epithelium of the ciliary body (fig. 15.26). From its site of secretion within the posterior chamber, the aqueous humor passes through the pupil into the anterior chamber. From here, it drains from the eyeball through the **venous sinus** (canal of Schlemm) into the bloodstream. The venous sinus is located at the junction of the cornea and iris.

canal of Schlemm: from Friedrich S. Schlemm, German anatomist, 1795–1858

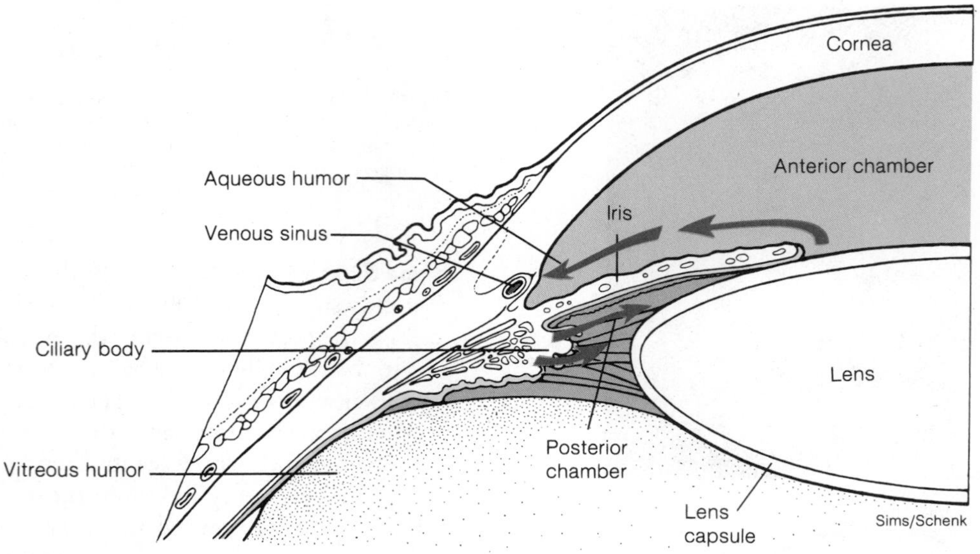

Figure 15.26 Aqueous humor maintains the intraocular pressure within the anterior and posterior chambers. It is secreted into the posterior chamber, flows through the pupil into the anterior chamber, and drains from the eyeball through the venous sinus (canal of Schlemm).

Between the lens and retina is the large **vitreous chamber,** which is filled with a transparent jellylike **vitreous humor.** Vitreous humor also contributes to the intraocular pressure to maintain the shape of the eyeball and to hold the retina against the choroid. Unlike aqueous humor, vitreous humor is not continuously produced but formed prenatally.

The location and functions of the structures within the eyeball are summarized in table 15.4.

Puncture wounds to the eyeball are especially dangerous and frequently cause blindness. Protective equipment such as goggles, shields, and shatterproof lenses should be used in hazardous occupations and certain sports. If the eye is punctured, the principal thing to remember is to *not remove the object* if it is still impaling the eyeball. Removal may allow the fluids to drain from the eyeball and cause the loss of intraocular pressure. This is particularly serious if the nonreplaceable vitreous humor drains from the eyeball.

Function of the Eyeball

The focusing of light waves and stimulation of photoreceptors of the retina require five basic processes:

1. The transmission of light waves through transparent media of the eyeball.
2. The refraction of light waves through media of different densities.
3. Accommodation of the lens to focus the light waves.
4. Constriction of the pupil by the iris to regulate the amount of light entering the vitreous chamber.
5. Convergence of the eyeballs so that visual acuity is maintained.

Visual impairment may exist if one or more of these processes does not function properly (see Clinical Considerations).

vitreous: L. *vitreus,* glassy

Transmission of Light Waves Light waves entering the eyeball pass through four transparent media before they stimulate the photoreceptors. In sequence, the media through which light waves pass are the cornea, aqueous humor, lens, and vitreous humor. The cornea and lens are solid media composed of tightly packed, avascular protein fibers. An additional thin, transparent membranous continuation of the conjunctiva covers the outer surface of the cornea. The aqueous humor is a low viscosity fluid and the vitreous humor is jellylike in consistency.

Refraction of Light Waves Refraction is the bending of light waves. Refraction occurs as light waves pass at an oblique angle from a medium of one optical density to a medium of another optical density (fig. 15.33). The convex cornea is the principal refractive medium; the fluids within the various chambers produce minimal refraction. The lens is particularly important for refining and altering refraction. Of the refractive media, only the lens can be altered in shape to achieve precise refraction.

The refraction of light waves is so extensive that the visual image is formed upside down on the retina (fig. 15.27). Nerve impulses of the image in this position are relayed to the visual cortex of the occipital lobe where the inverted image is interpreted as right side up.

Accommodation of the Lens Accommodation is the automatic adjustment of the curvature of the lens by contraction of ciliary muscles to bring light waves into sharp focus on the retina. The lens of the eyeball is biconvex. When an object is viewed closer than about twenty feet, the lens must make an adjustment, or accommodation, if a clear focus is to appear on the retina. Contraction of the smooth muscle fibers of the ciliary body causes the suspensory ligament to relax and the lens to become thicker (see fig. 15.19). A thicker, more convex lens causes a greater convergence of the refracted light waves necessary for viewing close objects.

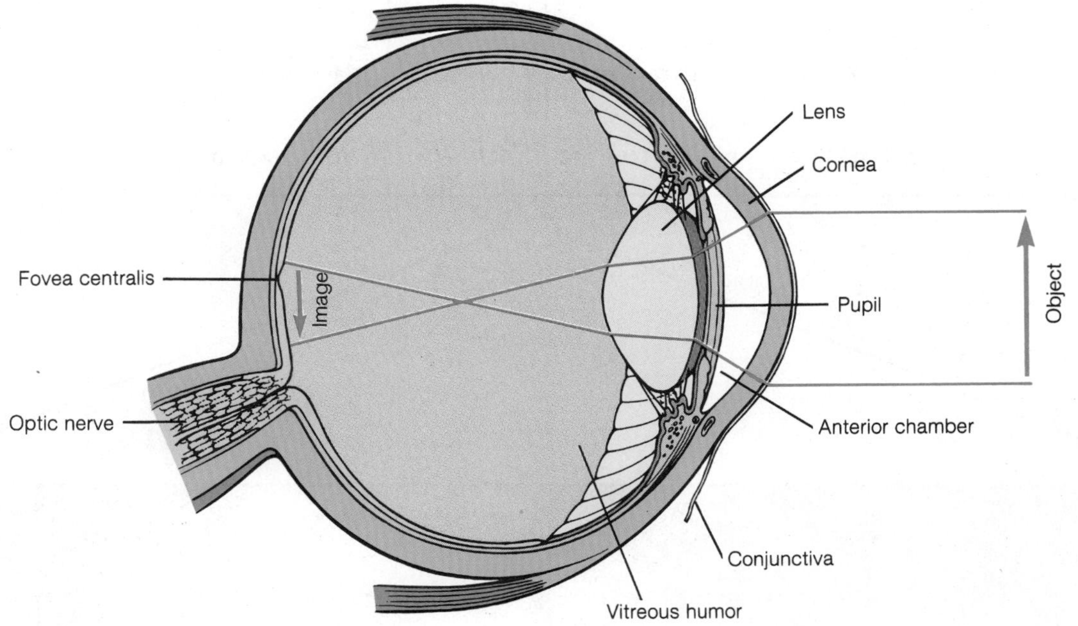

Figure 15.27 The refraction of light waves within the eyeball causes the image of an object to be inverted on the retina.

Constriction of the Pupil Constriction of the pupil through parasympathetic stimulation that causes contraction of the circular muscle fibers of the iris (see fig. 15.20) is important for two reasons. One is the reduction in the amount of light that enters the vitreous chamber. A reflexive pupil constriction protects the retina from sudden or intense bright light. Another more important function is that a reduced pupil diameter prevents light waves from entering the vitreous chamber through the periphery of the lens. Light waves refracted from the periphery would not be brought into focus on the retina and would cause blurred vision. Autonomic constriction of the pupil and accommodation of the lens occur simultaneously.

Convergence of the Eyeballs Convergence of the eyeballs means that the eyes rotate more medially when viewing closer objects. In fact, focusing on an object close to the tip of the nose causes a person to appear cross-eyed. The eyeballs must converge when viewing close objects because only then can the light waves focus on the same portions in both retinas.

Amblyopia exanopsia, commonly called lazy eye, is a condition of ocular muscle weakness causing a deviation of one eye so that there is not a concurrent convergence of both eyeballs. Because of disconjugate fixation, two images are received by the optic cortex—one of which is suppressed to avoid *diplopia (double vision)* or images of unequal clarity. A person who has amblyopia will experience dimness of vision and partial loss of sight. Amblyopia is frequently tested for in young children because if untreated before age six, there is little that can be done to strengthen the afflicted muscle.

amblyopia: Gk. *amblys,* dull; *ops,* vision

Visual Spectrum The eyes transduce energy in the *electromagnetic spectrum* (fig. 15.28) into nerve impulses. Only a limited part of this spectrum can excite the photoreceptors. Electromagnetic energy with wavelengths between 400 and 700 nanometers (nm) comprise *visible light.* Light of longer wavelengths, which are in the infrared regions of the spectrum, do not have sufficient energy to excite photoreceptors but are perceived as heat. Ultraviolet light, which has shorter wavelengths and more energy than visible light, is filtered out by the yellow color of the eye's lens. Certain insects, such as honeybees, and people who have had their lenses removed, can see light in the ultraviolet range.

Three distinct and specialized types of cones within the retina permit color vision. Different photosensitive pigments enable each type to absorb light waves primarily in the red, green, or blue portion of the color spectrum. Red cones are stimulated by wavelengths between 475 and 700 nanometers; green cones are stimulated by wavelengths between 400 and 550 nanometers. *Color blindness* is the inability to distinguish colors, particularly reds and greens. It affects about 5% of Americans. Color blindness for the majority of these 5% is a misnomer, however, since in nearly all cases there is a deficiency in the number of specific cones, not a total lack of them. Hence, the persons have some ability to distinguish color.

Neural Pathways of Vision, Eye Movements, and Processing Visual Information

The photoreceptor neurons, rods and cones, are the functional units of sight in that they respond to light waves and produce nerve impulses. Nerve impulses from the rods and cones pass through bipolar neurons to ganglion neurons (see fig. 15.21).

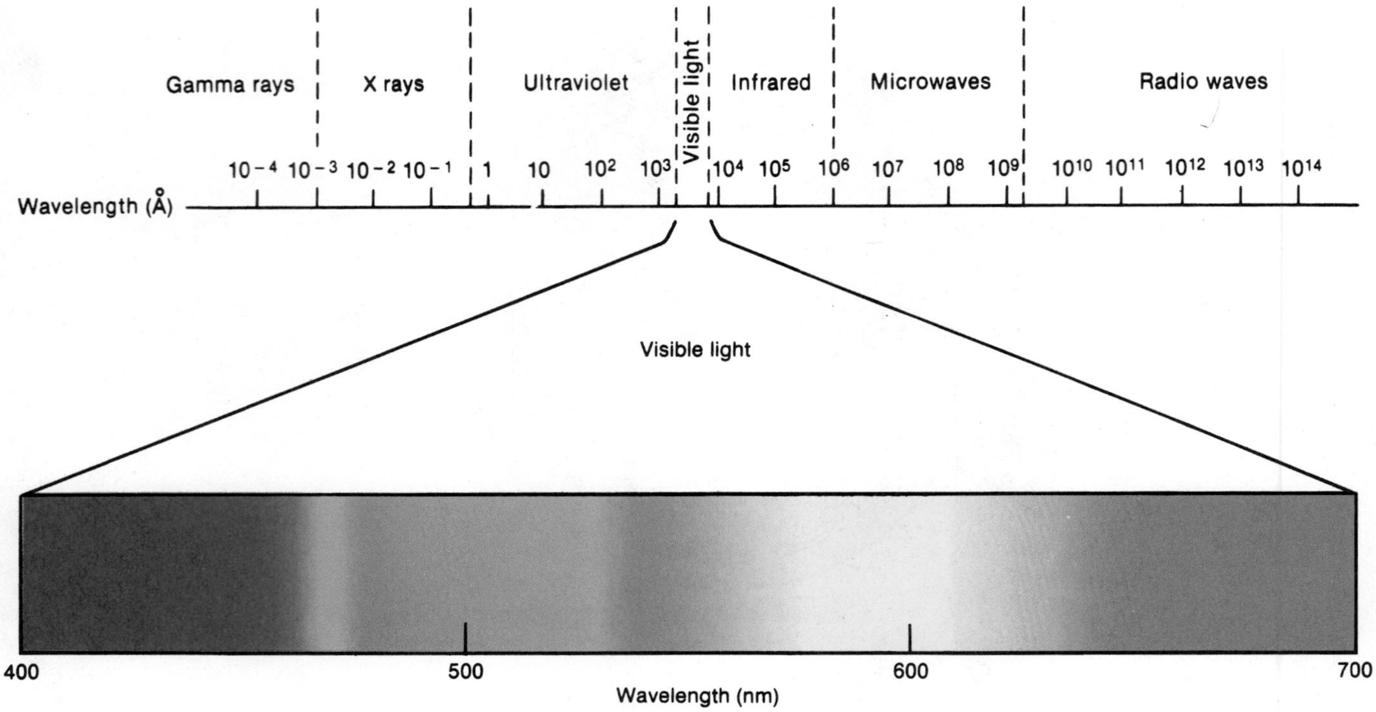

Figure 15.28 The electromagnetic spectrum (*top*) is shown in Angstrom units ($1Å = 10^{-10}$ meter). The visible spectrum comprises only a small range of this spectrum (*bottom*), shown in nanometer units ($1 \text{ nm} = 10^{-9}$ meter).

The optic nerve consists of axons of aggregated ganglion neurons that emerge through the posterior aspect of the eyeball. The two optic nerves (one from each eyeball) converge at the **optic chiasma** *(ki-as'mah)* (fig. 15.29). The fibers partially cross at the optic chiasma so that all the fibers arising from the medial (nasal) half of each retina cross to the opposite side. The fibers of the optic nerve that arise from the lateral (temporal) half of the retina do not cross. The **optic tract** is a continuation of optic nerve fibers from the optic chiasma and is composed of fibers arising from the retinas of both eyeballs.

As the optic tracts enter the brain, some of the fibers in the tracts terminate in the **superior colliculi** *(ko-lik'u-li)*. These fibers and the motor pathways they activate constitute the **tectal system,** which is responsible for body-eye coordination.

Approximately 70%–80% of the fibers in the optic tract pass to the **lateral geniculate** *(jĕ-nik'u-lat)* **body** of the thalamus (fig. 15.29). Here the fibers synapse with neurons whose axons constitute a pathway called the **optic radiation.** The optic radiation transmits impulses to the **striate** *(stri'āt)* **cortex** area of the occipital cerebral lobe. This entire arrangement of visual fibers is known as the **geniculostriate system,** which is responsible for perception of the visual field.

The nerve fibers that cross at the optic chiasma arise from the retinas in the medial portions of the eyeballs. The photoreceptors of these fibers are stimulated by light entering the eyeball from the periphery. If the optic chiasma were severed longitudinally, peripheral vision would be lost and leave only "tunnel vision." If an optic tract were severed, both eyes would be partially blind, the lateral field of vision lost for one eye and the medial field of vision lost for the other.

Superior Colliculi and Eye Movements Neural pathways from the superior colliculi to motor neurons in the spinal cord help mediate the startle response to the sight of an unexpected intruder. Other nerve fibers from the superior colliculi stimulate the **extrinsic eye muscles** (table 15.3), which are the skeletal muscles that move the eyes.

There are two types of eye movements coordinated by the superior colliculi. *Smooth pursuit movements* track moving objects and keep the image focused on the fovea centralis. *Saccadic (sah-kad'ik) eye movements* are short (lasting 20–50 msec), jerky movements that occur while the eyes appear to be still. These saccadic movements are believed to be important in maintaining visual acuity.

The tectal system is also involved in the control of the **intrinsic eye muscles**—the iris and the muscles of the ciliary body. Shining a light into one eye stimulates the *pupillary reflex* in which both pupils constrict. This is caused by activation of parasympathetic neurons by fibers from the superior colliculi. Postganglionic neurons in the ciliary ganglia behind the eyes, in turn, stimulate constrictor fibers in the iris. Contraction of the ciliary body during *accommodation* also involves stimulation of the superior colliculi.

Processing of Visual Information In order for visual information to have meaning it must be associated with past experience and integrated with information from other senses. Some of this higher processing occurs in the inferior temporal lobes of the cortex. Experimental removal of these areas from monkeys impairs their ability to remember visual tasks that they previously learned and hinders their ability to associate visual images with the significance of the object. Monkeys with their

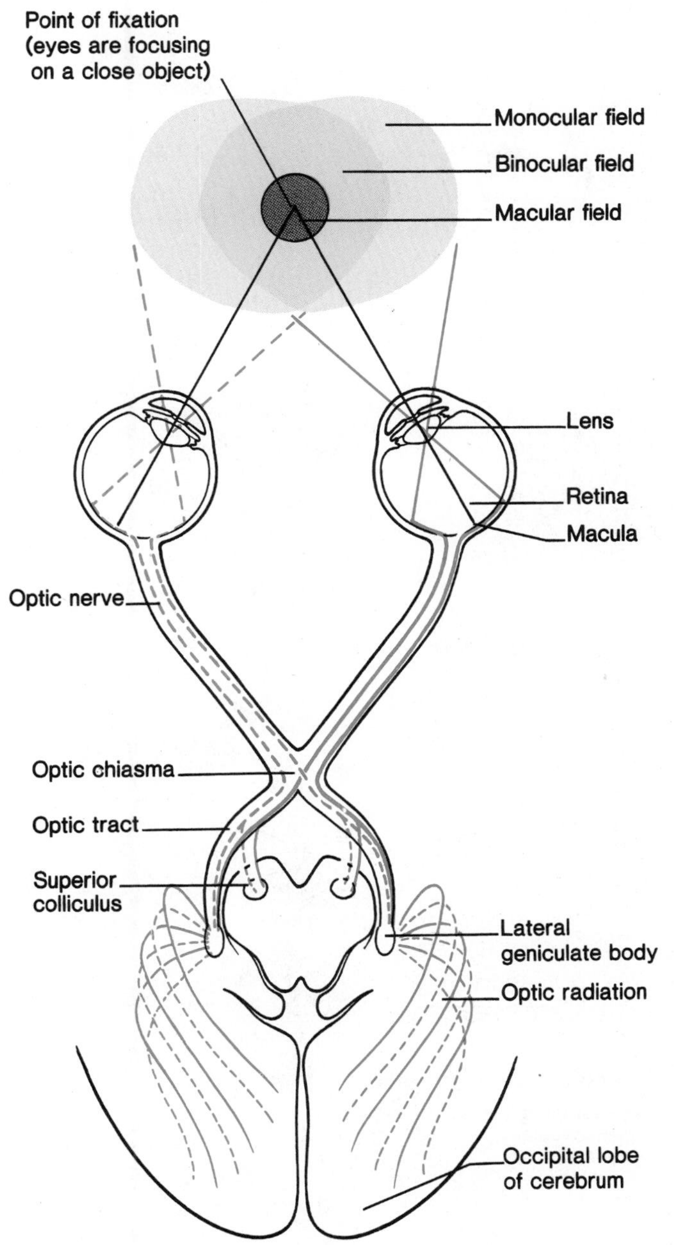

Figure 15.29 Visual fields of the eyes and neural pathways for vision. An overlapping of the visual field of each eye provides binocular vision, which is the ability to perceive depth.

inferior temporal lobes removed, for example, will handle a snake without fear. The symptoms produced by loss of the inferior temporal lobes are known as the *Kluver-Bucy syndrome.*

In an attempt to reduce the symptoms of severe epilepsy, surgeons at one time cut the corpus callosum in some patients. This fiber tract, as previously described, transmits impulses between the right and left cerebral hemispheres. The right cerebral hemisphere of patients with such *split brains,* therefore, receives sensory information from only the left half of the external world. The left hemisphere, similarly cut off from communication with the right hemisphere, receives sensory information from only the right half of the external world.

Experiments with split-brain patients have revealed that the functions of the two hemispheres are different. This is true even though each hemisphere would normally receive input from both halves of the external world through the corpus callosum. If the sensory image of an object, such as a key, is delivered to only the left hemisphere (by showing it to only the right visual field), the object can be named. If the object is presented to the right cerebral cortex, the person knows what the object is but cannot name it. Experiments such as this suggest that (in right-handed people) the left hemisphere is needed for language and the right hemisphere is responsible for pattern recognition.

1. List the associated structures of the eye that either cause the eye to move or protect it within the orbit.
2. Diagram the structure of the eye, and label the following: sclera, cornea, choroid, retina, fovea centralis, iris, pupil, lens, and ciliary body. What are the principal cells or tissues in each of the three layers of the eye?
3. Trace the path of light through the three chambers of the eye, and explain the mechanism of light refraction and how the eye is focused for viewing distant and near objects.
4. List the different layers of the retina, and describe the path of light and of nerve activity through these layers. Continue tracing the path of a visual impulse to the cerebral cortex, and list in order the structures traversed.

Senses of Hearing and Balance

Structures of the outer, middle, and inner ear are involved in the sense of hearing. The inner ear also contains structures that provide a sense of balance, or equilibrium. The development of the ear begins during the fourth week and is completed by the thirty-second week.

Objective 15. Describe the development of the ear.
Objective 16. Describe the structures of the ear that relate to hearing, and list their locations and functions.
Objective 17. Trace the path of sound waves through the ear, and describe how they are transmitted and converted to nerve impulses.
Objective 18. Explain the mechanisms by which equilibrium is maintained.

Development of the Ear

The ear begins to develop at the same time as the eye, early during the fourth week. All three embryonic germ layers—ectoderm, mesoderm, and endoderm—are involved in the formation of the ear. Both types of ectoderm (neuroectoderm and surface ectoderm) play a role.

The ear of an adult is structurally and functionally divided into an **outer ear,** a **middle ear,** and an **inner ear,** each of which has a separate embryonic origin. The inner ear does not develop from deep embryonic tissue as one might expect but

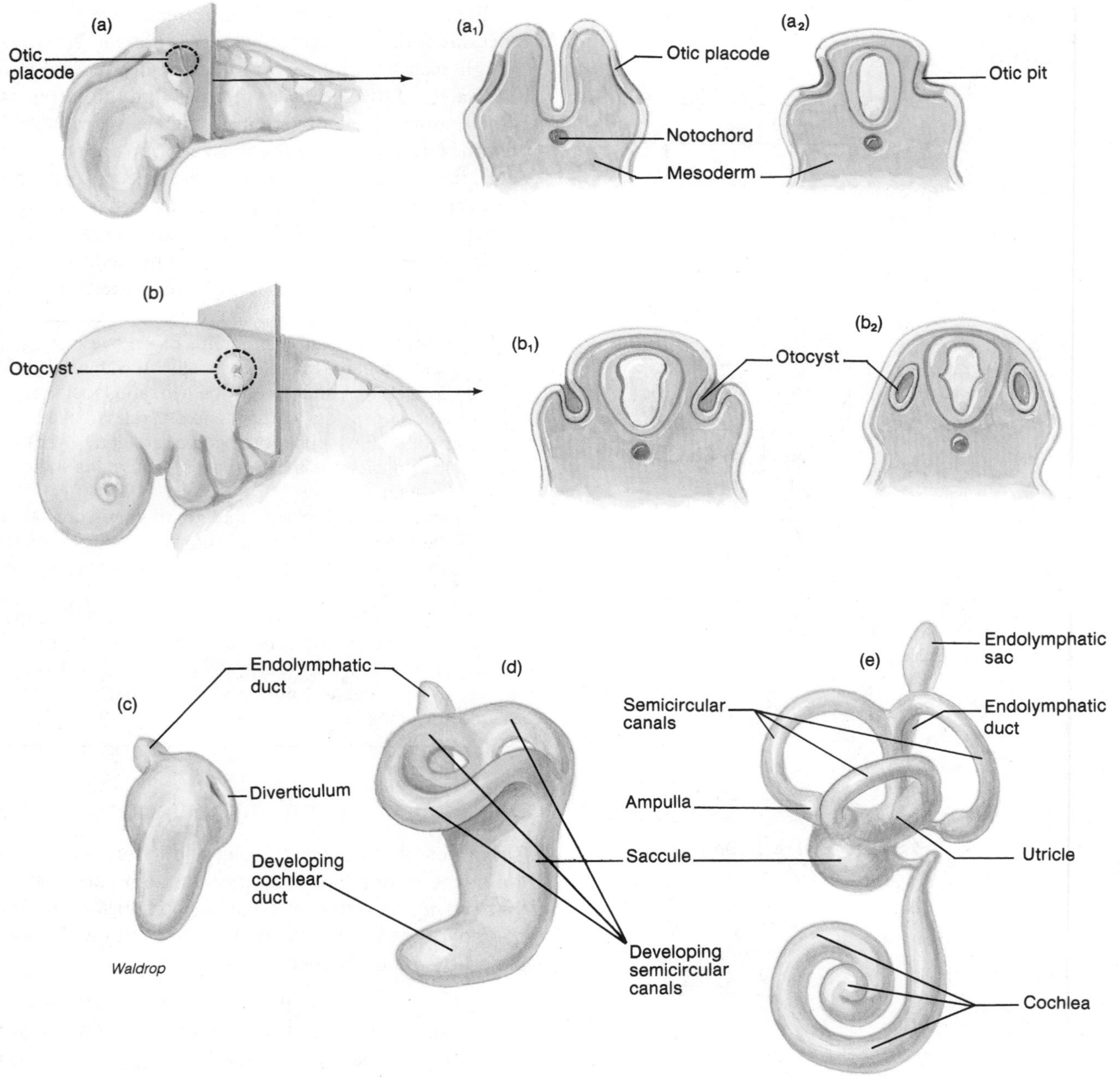

Figure 15.30 Development of the inner ear. (*a*) A lateral view of a twenty-two-day-old embryo showing the position of a transverse cut through the otic placode. (a_1) The otic placode of surface ectoderm begins to invaginate at twenty-two days. (a_2) By twenty-four days, a distinct otic pit is formed and the neural ectoderm is positioned to give rise to the brain. (*b*) A lateral view of a twenty-eight-day-old embryo showing the position of a transverse cut through the otocyst. (b_1) By twenty-eight days, the otic pit has become a distinct otocyst. (b_2) The otocyst is in position in the thirty-day-old embryo where it gives rise to the structures of the inner ear. (*c–e*) Lateral views of the differentiating otocyst into the cochlea and semicircular canals from the fifth to the eighth week.

rather begins to form early in the fourth week when a plate of surface ectoderm called the **otic** *(o'tik)* **placode** appears lateral to the developing hindbrain (fig. 15.30). The otic placode soon invaginates and forms an **otic pit.** Toward the end of the fourth week, the outer edges of the invaginated otic pit come together and fuse to form an **otocyst (otic vesicle).** The otocyst soon pinches off and separates from the surface ectoderm. The otocyst further differentiates to form a dorsal **utricular portion** and a ventral **saccular portion.** Three separate diverticula extend outward from the utricular portion and develop into the **semicircular canals,** which later function in balance and equilibrium. A tubular diverticulum, called the **cochlear duct,** extends in a coiled fashion from the saccular portion and forms the membranous portion of the **cochlea** of the ear (fig. 15.30). The **organ of Corti,** which is the functional portion of the cochlea, differentiates from cells along the wall of the cochlear duct (fig. 15.31). The sensory nerves that innervate the inner ear are derived from neuroectoderm from the developing brain.

The differentiating otocyst is surrounded by mesodermal tissue that soon forms a cartilaginous **otic capsule** (fig. 15.32). As the otocyst and surrounding otic capsule grow in size, vacuoles containing the fluid **perilymph** form within the otic

otic: Gk. *otikos,* ear

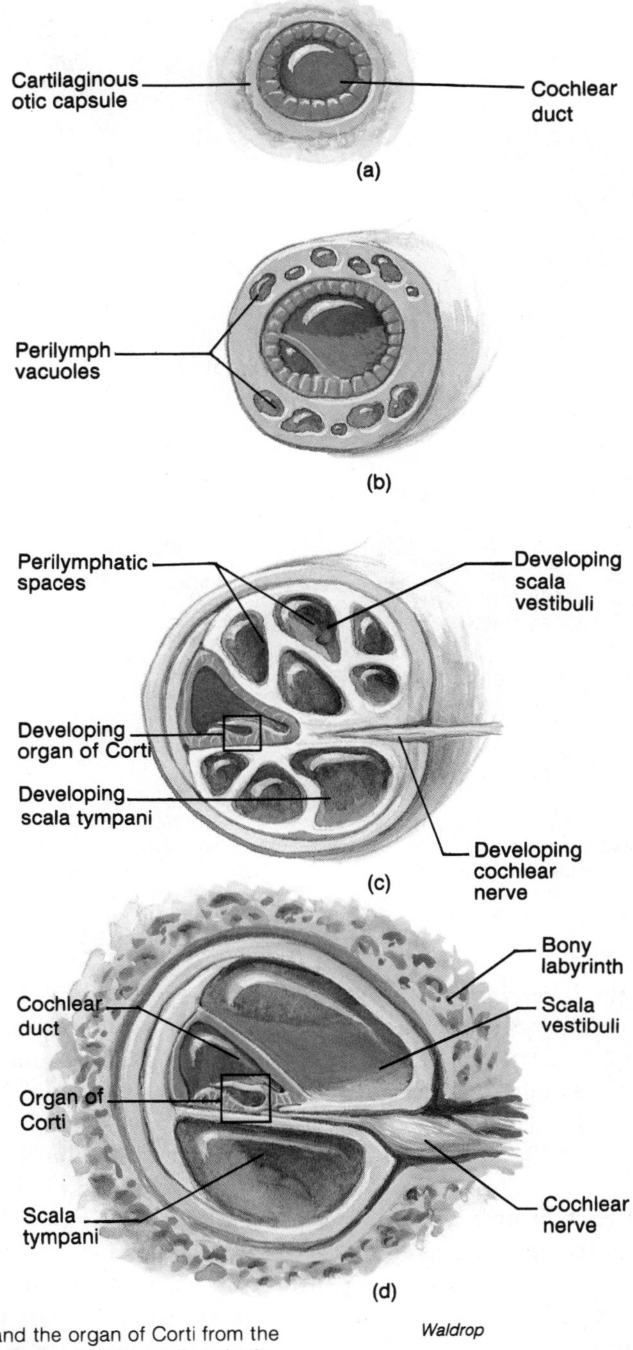

Figure 15.31 The formation of the cochlea and the organ of Corti from the otic capsule. (*a–d*) Successive stages of development of the perilymphatic space and the organ of Corti from the eighth to the twentieth week.

capsule. The vacuoles soon enlarge and coalesce to form the **perilymphatic space,** which divides into the **scala tympani** and the **scala vestibuli.** Eventually, the cartilaginous otic capsule ossifies to form the **bony (osseous) labyrinth** of the inner ear. The middle ear chamber is referred to as the **tympanic cavity** and derives from the first pharyngeal pouch (see fig. 15.32). The **ossicles,** which amplify incoming sound waves, derive from the first and second pharyngeal arch cartilages. As the tympanic cavity enlarges, it surrounds and encloses the developing ossicles (fig. 15.32). The connection of the tympanic cavity to the pharynx gradually elongates to develop into the **auditory** (Eustachian) **tube,** which remains patent throughout life and is important in maintaining an equilibrium of air pressure between the pharyngeal and tympanic cavities.

The outer ear includes the fleshy **auricle** attached to the side of the head and the tubular **external auditory meatus** that extends into the temporal bone of the skull. The external auditory meatus forms from the surface ectoderm that covers the dorsal end of the first branchial groove (see fig. 15.32). A solid epithelial plate called the **meatal plug** soon develops at the bottom of the funnel-shaped branchial groove. The meatal

Eustachian tube: from Bartolommeo E. Eustachio, Italian anatomist, 1520–1574

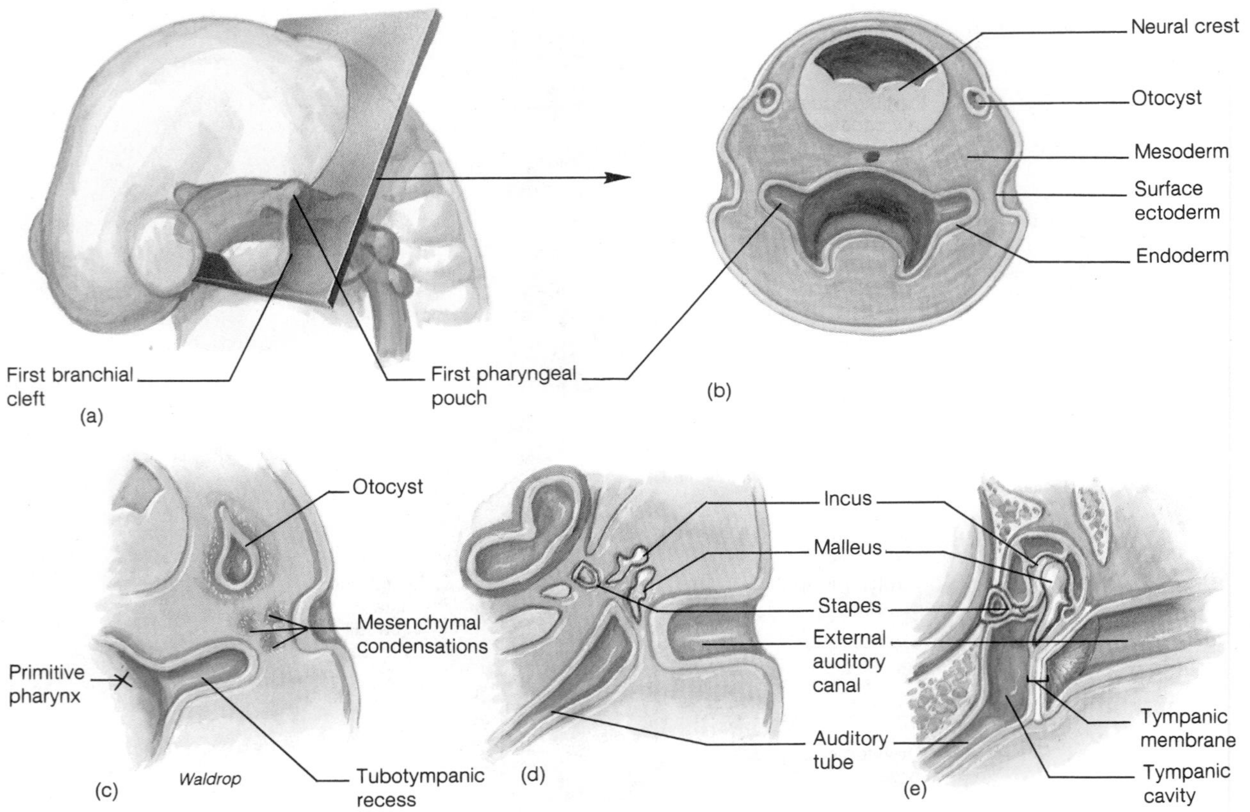

Figure 15.32 Development of the outer and middle ear regions and the ear ossicles. (*a*) A lateral view of a four-week-old embryo showing the position of the cut depicted in the sequential development (*b–e*). (*b*) The embryo at four weeks illustrating the invagination of the surface ectoderm and the evagination of the endoderm at the level of the first pharyngeal pouch. (*c*) During the fifth week, mesenchymal condensations are apparent, which will derive the ear ossicles. (*d*) Further invagination and evagination at six weeks correctly positions the structures of the outer and middle ear regions. (*e*) By the end of the eighth week, the ear ossicles, tympanic membrane, auditory tube, and external auditory canal are formed.

plug is involved in the formation of the inner wall of the external auditory meatus and contributes to the **tympanic membrane** (eardrum). The tympanic membrane has a dual origin from surface ectoderm and the endoderm lining the first pharyngeal pouch (see fig. 15.32).

Structure of the Ear

The ear is the organ of hearing and equilibrium. It contains receptors that respond to movements of the head and receptors that convert sound waves into nerve impulses. Impulses from both receptor types are transmitted through the vestibulocochlear (VIII) cranial nerve to the brain for interpretation. The ear consists of three principal regions: the outer ear, the middle ear, and the inner ear (fig. 15.33).

Outer Ear The outer ear consists of the auricle, or pinna, and the external auditory canal. The external auditory canal is the fleshy tube that is fitted into the bony tube called the *external auditory meatus.* The **auricle** *(aw'rĕ-kl)* is the visible fleshy appendage attached to the side of the head. It consists of a cartilaginous framework of elastic connective tissue covered with skin. The rim of the auricle is called the **helix,** and the inferior fleshy portion is the **lobe** (fig. 15.34). The lobe is the only portion of the auricle that is not supported with cartilage. The auricle has a ligamentous attachment to the skull and poorly developed auricular muscles inserting anteriorly, superiorly, and posteriorly within it. The blood supply to the auricle is from the posterior auricular and occipital arteries, which branch from the external carotid and superficial temporal arteries, respectively. The structure of the auricle directs sound waves into the external auditory canal.

The **external auditory canal** is a slightly S-shaped tube about 2.5 cm (1 in.) in length, extending slightly upward from the auricle to the tympanic membrane (fig. 15.33). The skin that lines the canal contains fine hairs and sebaceous glands near the entrance. Specialized wax-secreting glands, called **ceruminous** *(sĕ-roo'mĭ-nus)* **glands,** are located in the skin, deep within the canal. *Cerumen* (earwax) secreted from ceruminous glands keeps the tympanic membrane soft and waterproof. Cerumen and the hairs also help to prevent small foreign objects from reaching the tympanic membrane. The bitter cerumen is probably an insect repellent as well.

The **tympanic** *(tim-pan'ik)* **membrane** (eardrum) is a thin, double-layered, epithelial partition between the external auditory canal and the middle ear. It is approximately 1 cm in diameter and is composed of an outer concave layer of stratified squamous epithelium and an inner convex layer of low columnar epithelium. The tympanic membrane is extremely sensitive to pain and is innervated by the auriculotemporal nerve (a branch of the mandibular portion of the trigeminal [V] cranial nerve) and the auricular nerve (a branch of the vagus [X] cranial nerve).

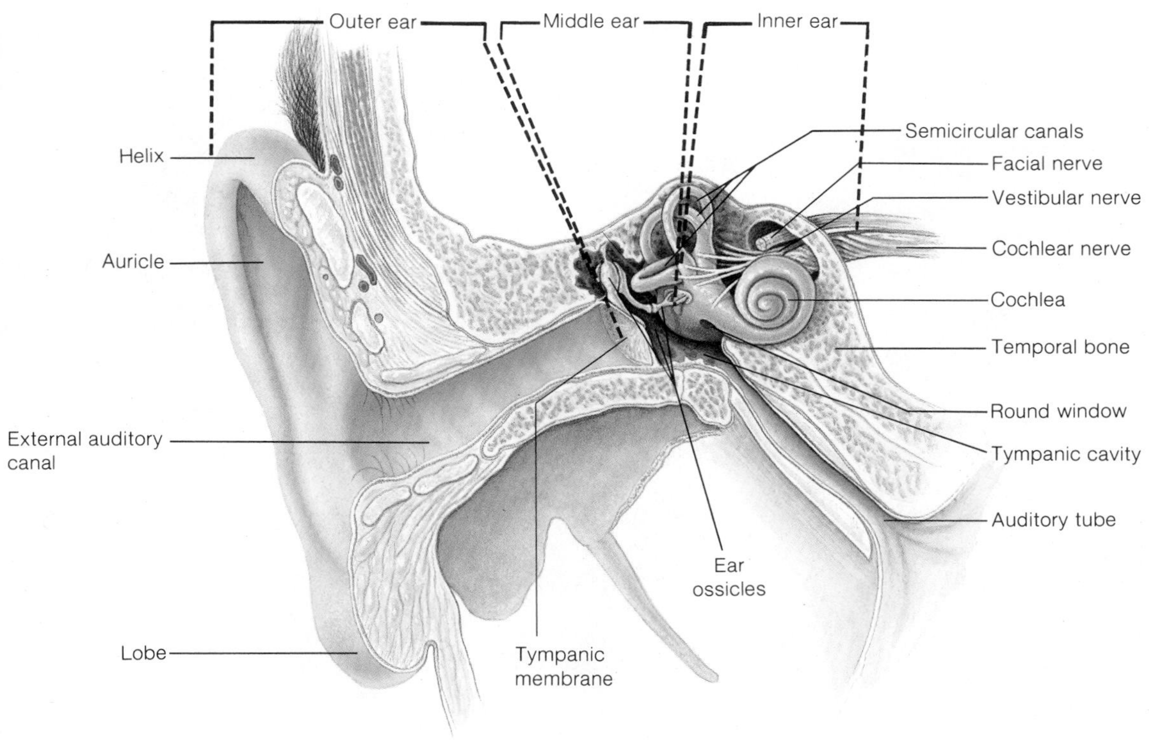

Figure 15.33 The ear. Note the outer, middle, and inner regions of the ear.

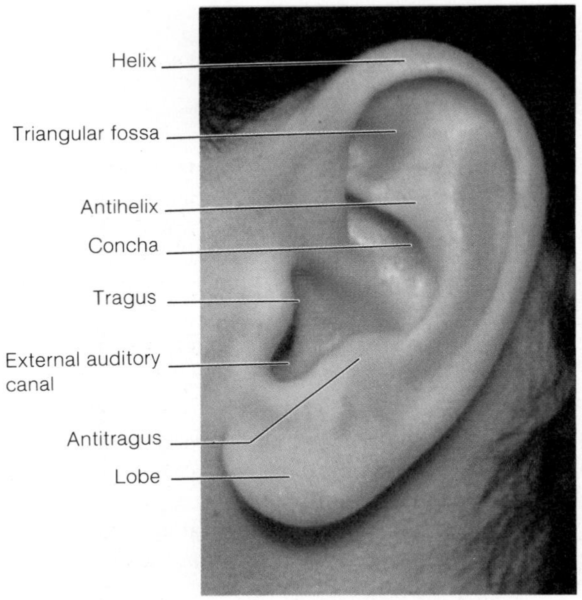

Figure 15.34 The surface anatomy of the auricle of the ear.

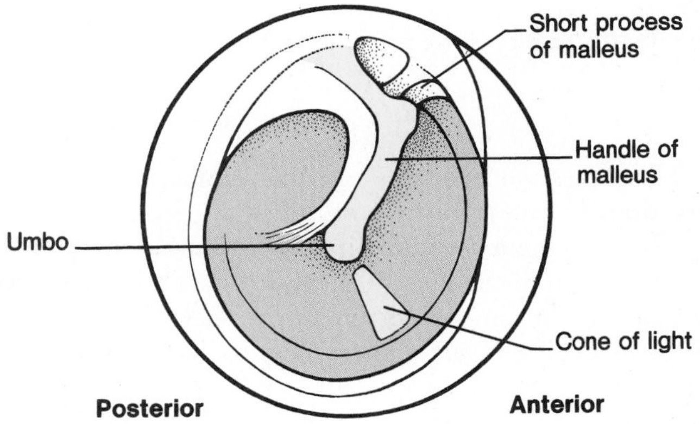

Figure 15.35 The right tympanic membrane as seen with an *auriscope* (a form of otoscope). A small depression, the *umbo,* occurs at the tip of the handle of the malleus. When the tympanic membrane is illuminated with an auriscope, the concavity of the umbo produces a cone of light that radiates anteriorly and inferiorly.

Inspecting the tympanic membrane with an otoscope during a physical examination provides significant information about the condition of the middle ear. A diagram of the normal tympanic membrane is presented in figure 15.35. The color, curvature, presence of lesions, or position of the malleus of the middle ear are features of particular importance. If ruptured, the tympanic membrane can generally regenerate and readily heal itself.

otoscope: Gk. *otikos,* ear; *skopein,* to examine

Middle Ear The laterally compressed middle ear is an air-filled chamber called the **tympanic cavity** in the petrous portion of the temporal bone (see figs. 15.33, 15.36). The tympanic membrane separates the middle ear from the external auditory canal of the outer ear. A bony partition containing the **oval window (fenestra vestibuli)** and the **round window (fenestra cochlea)** separates the middle ear from the inner ear.

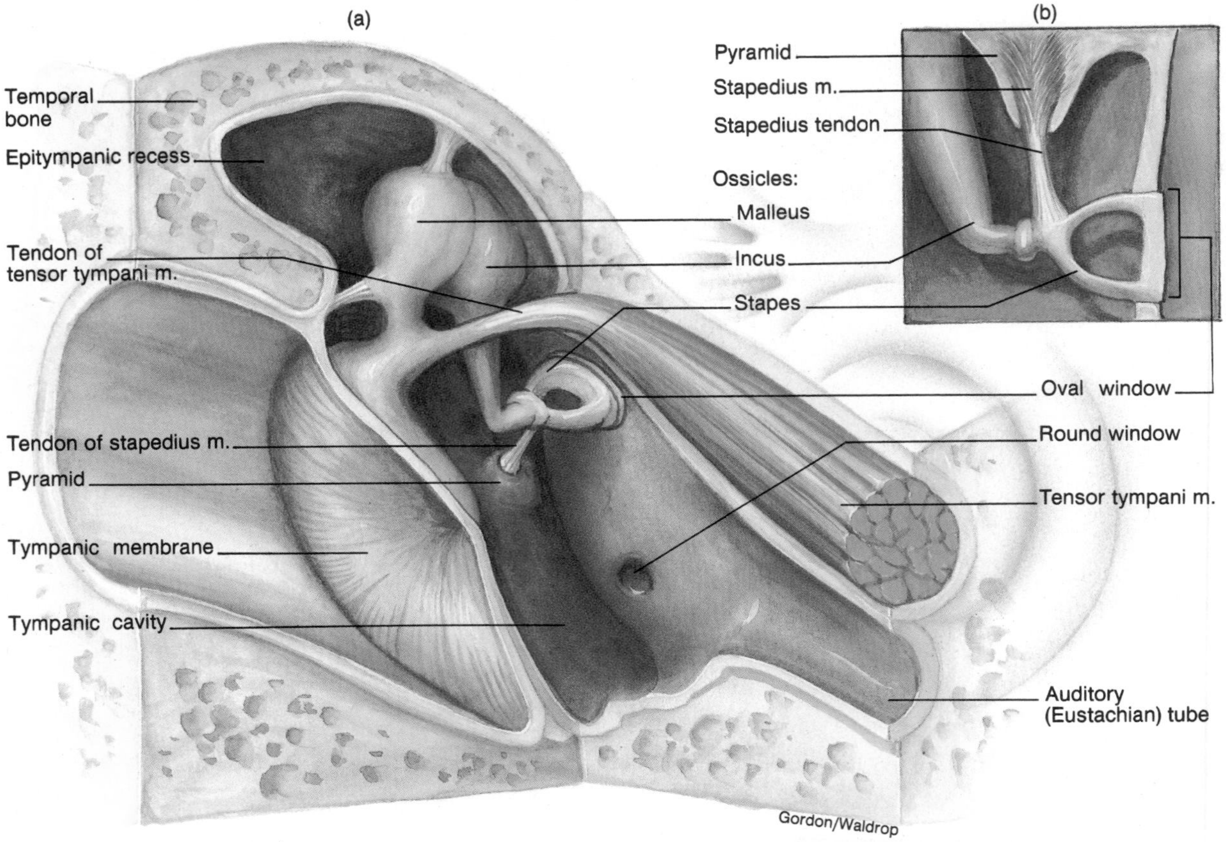

Figure 15.36 The ear ossicles and associated structures within the (*a*) tympanic cavity. (*b*) The stapedious muscle arises from a bony protrusion called the pyramid.

There are two openings into the tympanic cavity. The **epitympanic recess** in the posterior wall connects the tympanic cavity to the **mastoidal air cells** within the mastoid process of the temporal bone. The **auditory** Eustachian, **tube** connects the tympanic cavity anteriorly with the nasopharynx and equalizes air pressure on both sides of the tympanic membrane.

A series of three **ear ossicles** crosses the tympanic cavity from the tympanic membrane to the oval window (fig. 15.36). These tiny bones, from outer to inner, are the **malleus** *(mal'e-us)* (hammer), **incus** (anvil), and **stapes** (stirrup). The ear ossicles are attached to the wall of the tympanic cavity by ligaments. Vibrations of the tympanic membrane cause the ear ossicles to move and transmit sound waves across the tympanic cavity to the oval window. Vibration of the oval window moves a fluid within the inner ear and stimulates the receptors of hearing.

As the ear ossicles transmit vibrations, they act as a lever system to increase the force of vibration. In addition, the force of vibration is intensified as it is transmitted from the relatively large surface of the tympanic membrane to the smaller surface area of the oval window. The combined effect increases the force of the vibrations by about twenty times.

Two small skeletal muscles, the **tensor tympani** and the **stapedius** *(sta-pe'de-us)* (fig. 15.36), attach to the malleus and stapes, respectively, and contract reflexively to protect the inner ear against loud noises. When contracted, the tensor tympani pulls the malleus inward, and the stapedius pulls the stapes outward. This combined action reduces the force of vibration of the ossicles.

The mucous membranes that line the tympanic cavity, the mastoidal air cells, and the auditory tube are continuous with those of the nasopharynx. For this reason, infections of the nose or throat may spread to the tympanic cavity and cause a middle ear infection or even to the mastoidal air cells and cause *mastoiditis.* Forcefully blowing the nose advances the spread of the infection.

An equalization of air pressure on both sides of the tympanic membrane is of functional importance to hearing. When atmospheric pressure is reduced, as occurs at higher altitudes, the tympanic membrane bulges outward in response to the greater air pressure within the tympanic cavity. The bulging is painful and may impair hearing by reducing flexibility. The auditory tube, which is collapsed most of the time in adults, opens during swallowing or yawning and allows the air pressures to equalize.

malleus: L. *malleus,* hammer
incus: L. *incus,* anvil
stapes: L. *stapes,* stirrup

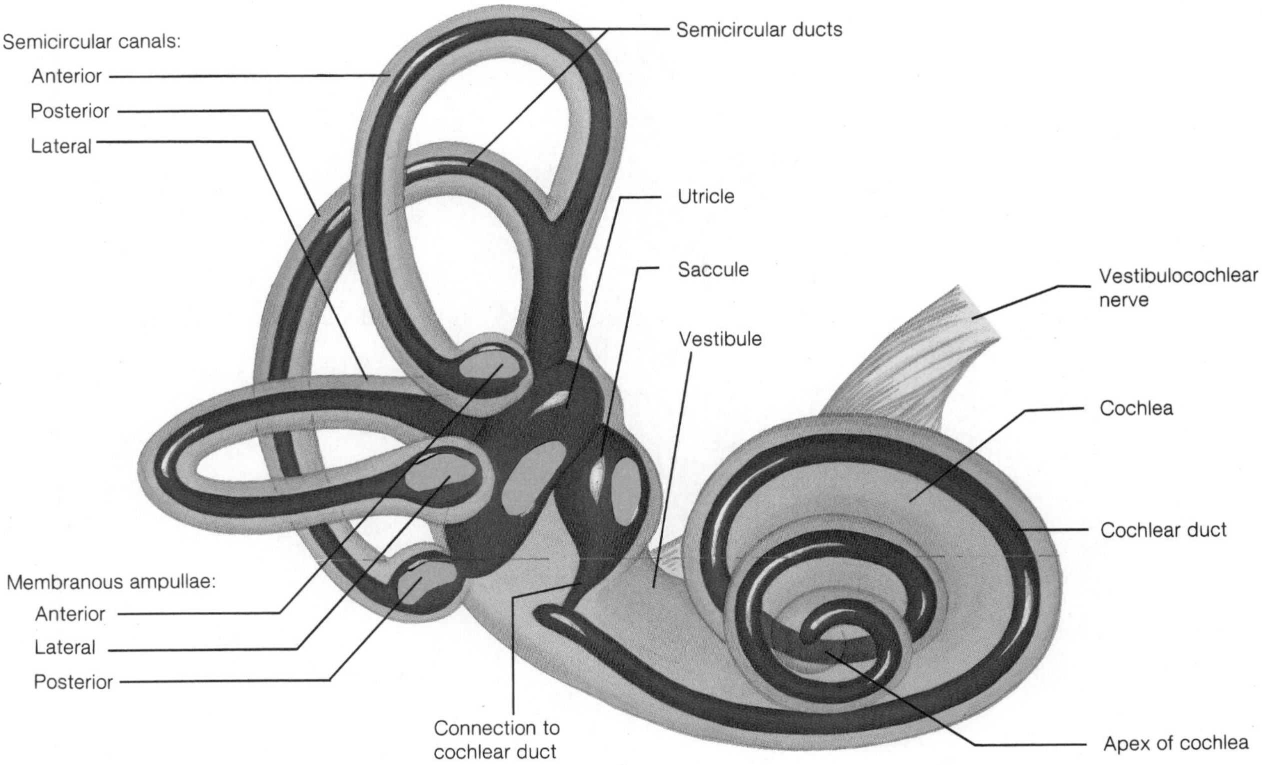

Figure 15.37 The labyrinths of the inner ear. The membranous labyrinth is contained within the bony labyrinth.

Inner Ear The inner ear, or **labyrinth** *(lab'ĭ-rinth),* contains the functional organs for hearing and equilibrium. The labyrinth consists of two parts: an outer **bony labyrinth** and a **membranous labyrinth** contained within the bony labyrinth (fig. 15.37). The space between the bony labyrinth and the membranous labyrinth is filled with a fluid, called *perilymph,* secreted by cells lining the bony canals. Within the tubular chambers of the membranous labyrinth is another fluid called *endolymph.* These two fluids provide a liquid-conducting medium for the vibrations involved in hearing and the maintenance of equilibrium.

The bony labyrinth is structurally and functionally divided into three areas: vestibule, semicircular canals, and cochlea.

Vestibule The vestibule is the central portion of the bony labyrinth that contains the oval window, into which the stapes fits, and the round window on the opposite end (fig. 15.37).

The membranous labyrinth within the vestibule consists of two connected sacs called the **utricle** *(u'tre-k'l)* and the **saccule** *(sak'ūl.)* The utricle is larger than the saccule and lies in the upper back portion of the vestibule. Both the utricle and saccule contain receptors that are sensitive to gravity and linear movement of the head.

Semicircular Canals The three bony semicircular canals of each ear are at right angles to each other and are positioned posterior to the vestibule. The thinner **semicircular ducts** form the membranous labyrinth within the semicircular canals (fig. 15.37). Each of the three semicircular ducts has a **membranous ampulla** *(am-pul'lah)* at one end and connects with the upper back part of the utricle. Receptors within the semicircular ducts are sensitive to angular acceleration and deceleration of the head, as in rotational movement.

Cochlea The snail-shaped cochlea *(kok'le-ah)* is coiled two and a half times around a central axis of bone (fig. 15.38). There are three chambers in the cochlea (fig. 15.39). The upper chamber, the **scala** *(ska'lah)* **vestibuli,** begins at the oval window and is continuous with the vestibule. The lower chamber, the **scala tympani,** terminates at the round window. The **cochlear duct** is the triangular middle chamber. The roof of the cochlear duct is called the **vestibular membrane,** and the floor is called the **basilar** *(bas'ĭ-lar)* **membrane.** The scala vestibuli and the scala tympani contain perilymph and are completely separated, except at the narrow apex of the cochlea, called the **helicotrema** *(hel''ĭ-ko-tre'mah),* where they are continuous (see fig. 15.41). The cochlear duct is filled with endolymph and ends at the helicotrema.

---

cochlea: L. *cochlea,* snail shell
scala: Gk. *scala,* staircase
helicotrema: Gk. *helix,* a spiral; *trema,* a hole

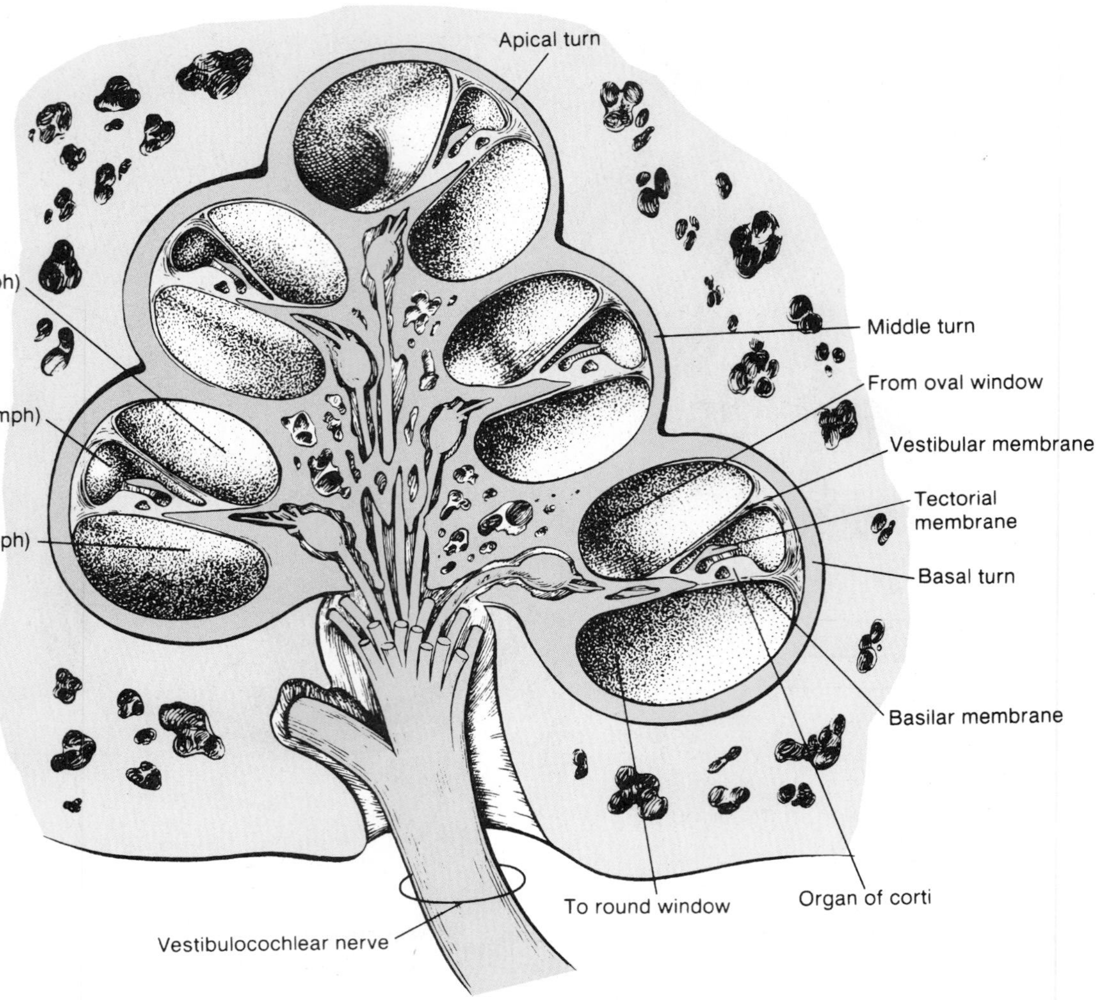

Figure 15.38 A cross section of the cochlea showing its three turns and its three compartments—scala vestibuli, cochlear duct (scala media), and scala tympani.

The **organ of Corti** is contained within the cochlear duct of the cochlea. The sound receptors that transform mechanical vibrations into nerve impulses are located along the basilar membrane of this structure, making it the functional unit of hearing. The epithelium of the organ of Corti consists of supporting cells and hair cells (figs. 15.39, 15.40). The bases of the hair cells are anchored in the basilar membrane, and their tips are embedded in the tectorial membrane that forms a gelatinous canopy over them.

Sound Waves and Neural Pathways for Hearing

Sound Waves Sound waves travel in all directions from their source, like ripples in a pond after a stone is dropped. These waves are characterized by their frequency and their intensity. The **frequency,** or distances between crests of the sound waves, is measured in *hertz* (*Hz*), which is the modern designation for *cycles per second* (*cps*). The *pitch* of a sound is directly related to its frequency—the higher the frequency of a sound, the higher its pitch.

The **intensity,** or loudness of a sound, is directly related to the amplitude of the sound waves. This is measured in units known as *decibels* (*db*). A sound that is barely audible—at the threshold of hearing—has an intensity of zero decibels. Every 10 decibels indicates a tenfold increase in sound intensity: a sound is ten times higher than threshold at 10 db, 100 times higher at 20 db, a million times higher at 60 db and ten billion times higher at 100 db.

The ear of a trained, young individual can hear sound over a frequency range of 20,000–30,000 Hz, yet can distinguish between two pitches that have only a 0.3% difference in frequency. The human ear can detect differences in sound intensities of only 0.1 to 0.5 db, whereas the range of audible intensities covers twelve orders of magnitude (10^{12}), from the barely audible to the limits of painful loudness.

Sound waves funneled through the external auditory canal produce extremely small vibrations of the tympanic membrane. Movements of the tympanum during speech (average intensity of 60 db) are estimated to be equal to the diameter of a molecule of hydrogen.

organ of Corti: from Alfonso Corti, Italian anatomist, 1822–1888

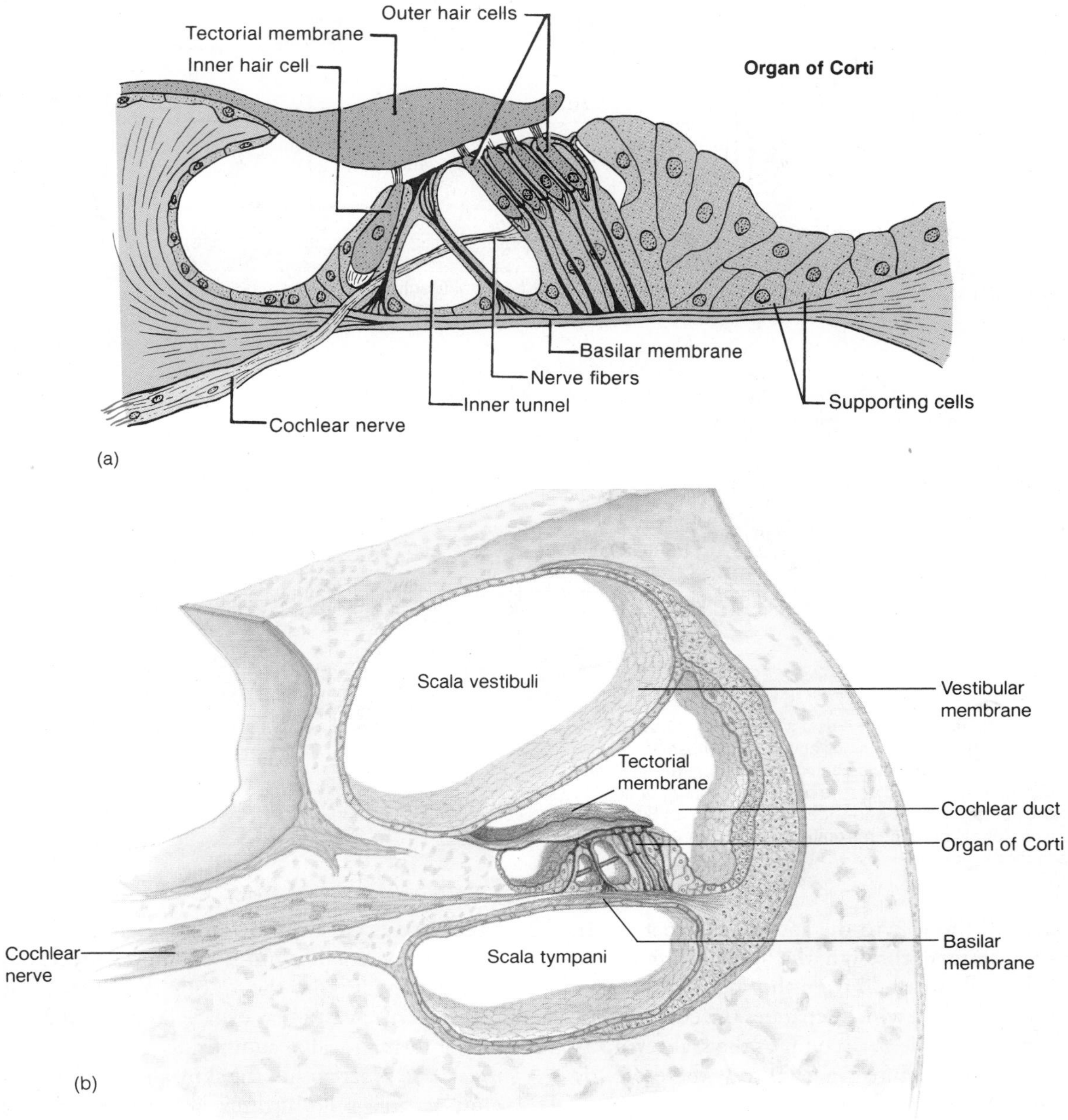

Figure 15.39 The organ of Corti (*a*) is located within the cochlear duct of the cochlea (*b*).

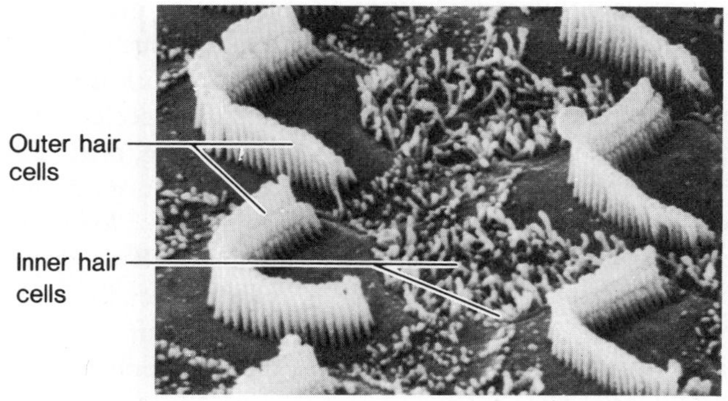

Figure 15.40 A scanning electron micrograph of the hair cells of the organ of Corti.

Sound waves passing through the solid medium of the ear ossicles are amplified by about twenty times as they reach the footplate of the stapes against the oval window. As the oval window is displaced, pressure waves pass through the fluid medium of the scala vestibuli (fig. 15.41), which pass around the helicotrema to the scala tympani. Movements of perilymph within the scala tympani, in turn, displace the round window into the tympanic cavity.

When the sound frequency (pitch) is sufficiently low, there is adequate time for the pressure waves of perilymph within the upper scala vestibuli to travel through the helicotrema to the scala tympani. As the sound frequency increases, however, pressure waves of perilymph within the scala vestibuli do not

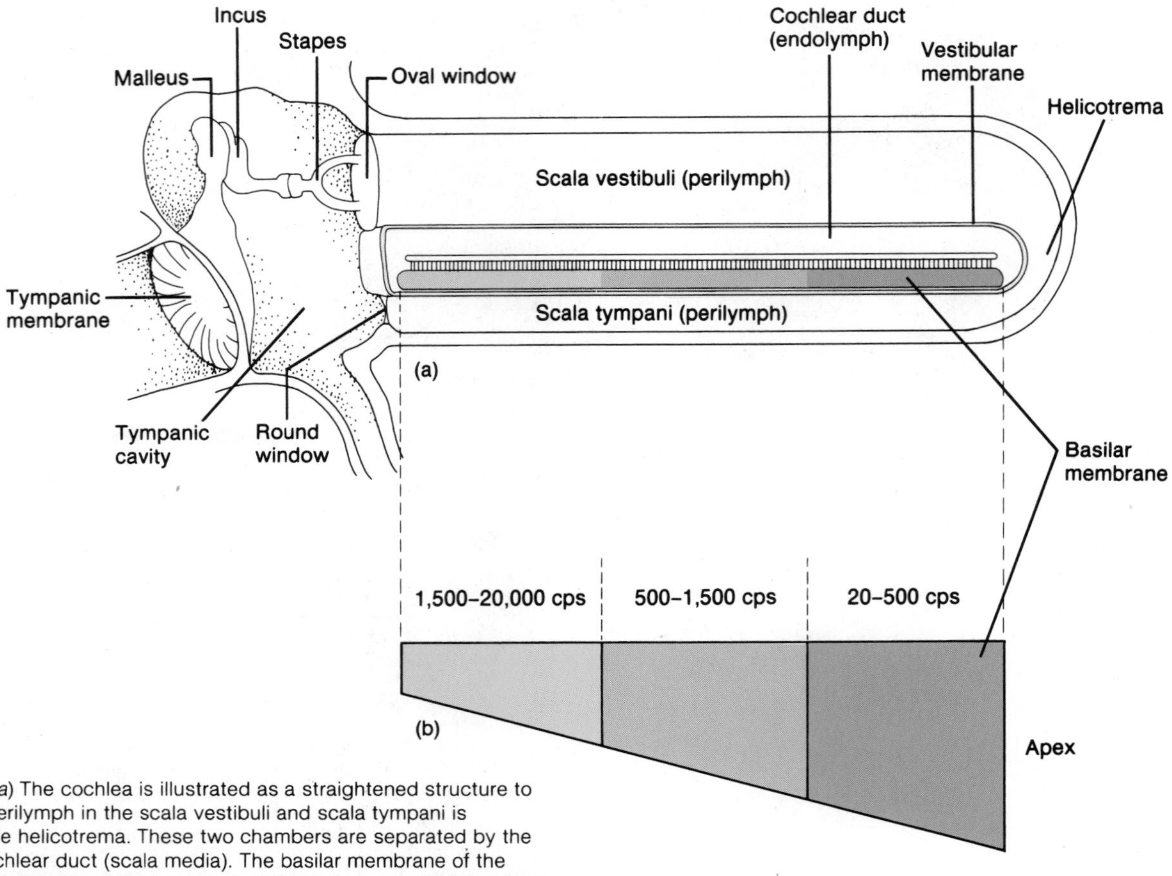

Figure 15.41 (*a*) The cochlea is illustrated as a straightened structure to show that the perilymph in the scala vestibuli and scala tympani is continuous at the helicotrema. These two chambers are separated by the blind-ending cochlear duct (scala media). The basilar membrane of the cochlear duct is narrower at the base and wider at the apex. (*b*) Sounds of different frequencies (pitches) produce maximum displacement of the basilar membrane in these regions.

have time to travel all the way to the apex of the cochlea. Instead, they are transmitted through the *vestibular membrane,* which separates the scala vestibuli from the cochlear duct, and through the *basilar membrane,* which separates the cochlear duct from the scala tympani, to the perilymph of the scala tympani. The distance that these pressure waves travel, therefore, decreases as the sound frequency increases.

Sounds of low pitch (with frequencies below about 50 Hz) cause movements of the entire length of the basilar membrane—from the base to the apex. Higher sound frequencies result in maximum displacement of the basilar membrane closer to its base, as illustrated in figure 15.41.

Displacement of the basilar membrane and hair cells by movements of perilymph causes the microvilli that are embedded in the tectorial membrane to bend. This stimulation excites the sensory cells, which causes the release of an unknown neurotransmitter that excites sensory endings of the cochlear nerve.

Neural Pathways for Hearing Cochlear sensory neurons in the eighth cranial nerve synapse with neurons in the medulla (fig. 15.42), which project to the inferior colliculi of the midbrain. Neurons in this area in turn project to the thalamus, which sends axons to the auditory cortex of the temporal lobe.

Mechanics of Equilibrium

Maintaining equilibrium is a complex process that depends on constant input from the vestibular organs of both inner ears. Although the vestibular organs are the principal source of sensory information for equilibrium, visual input from the eyes, tactile receptors within the skin, and proprioceptors of tendons, muscles, and joints also provide sensory input that is vital for maintaining equilibrium (fig. 15.43).

The vestibular organs provide the CNS with two kinds of receptor information. One kind is provided by receptors within the saccule and utricle, which are sensitive to gravity and to linear acceleration and deceleration of the head, such as in riding in a car. The other is provided by receptors within the semicircular canals, which are sensitive to angular acceleration and deceleration, such as turning the head, spinning, or tumbling.

The receptor hair cells of the vestibular organs contain twenty to fifty microvilli and one cilium, called a **kinocilium** *(ki''no-sil'e-um)* (fig. 15.44). When the hair cells are displaced in the direction of the kinocilium, the cell membrane is depressed and becomes depolarized. When the hair cells are displaced in the opposite direction, the membrane becomes hyperpolarized.

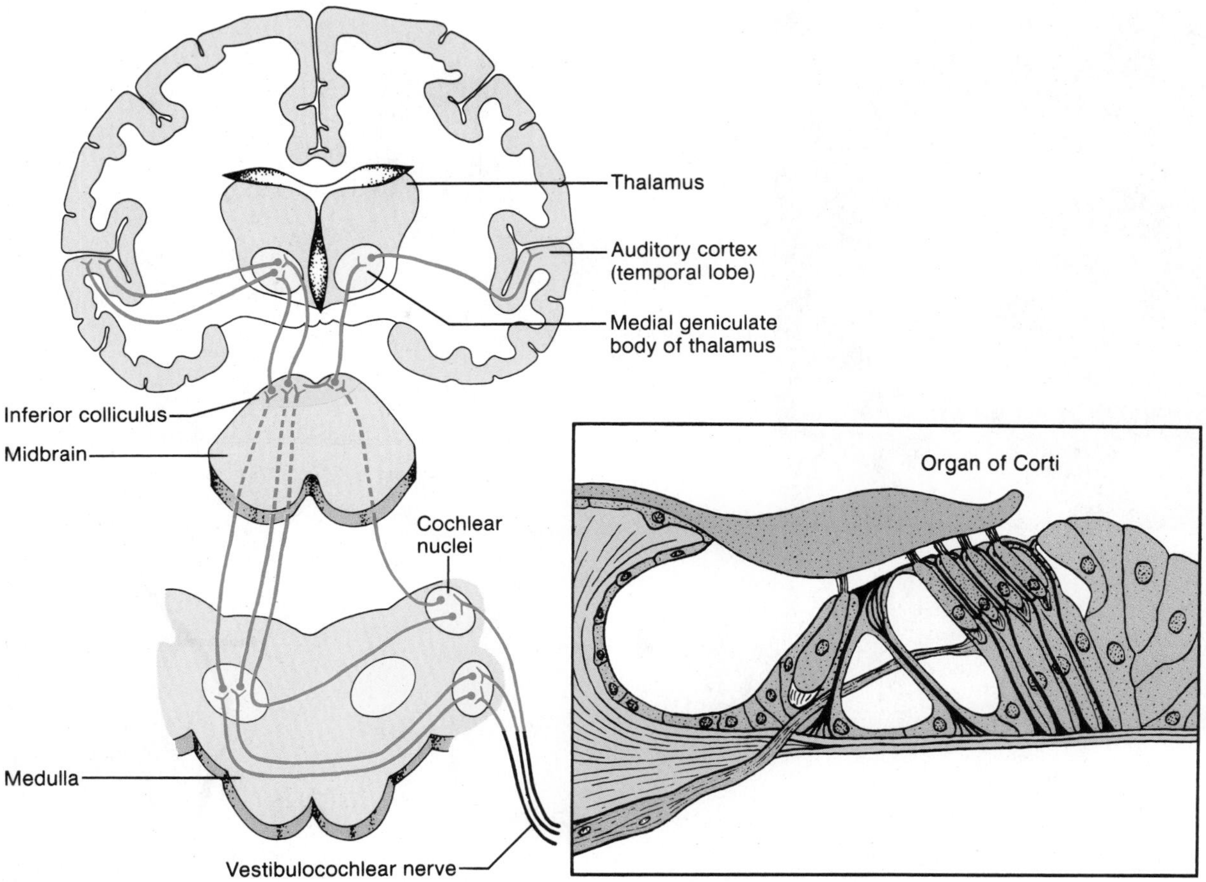

Figure 15.42 Neural pathways from the spiral ganglia of the cochlea of the auditory cortex.

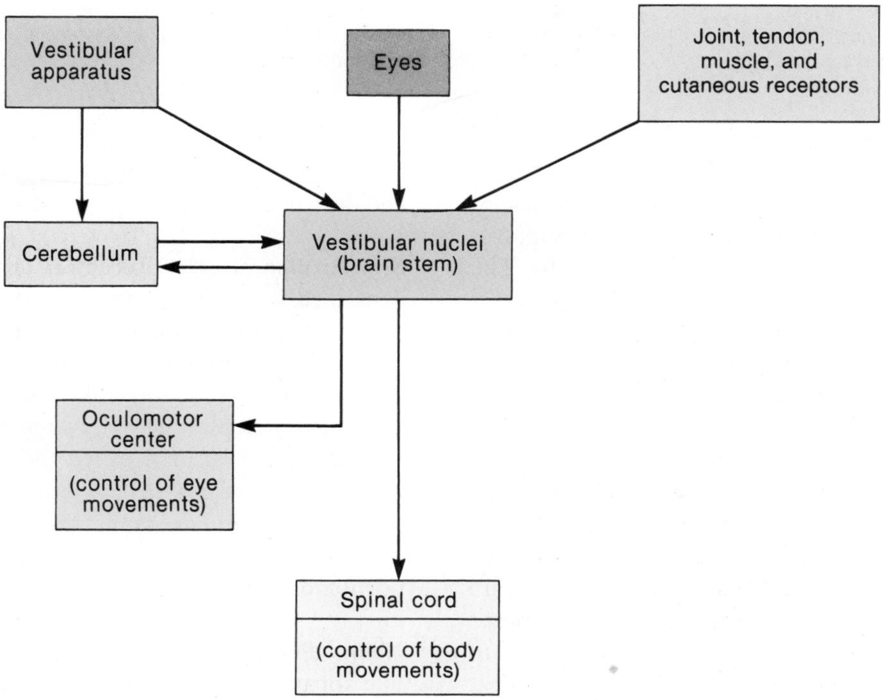

Figure 15.43 Neural processing involved in maintenance of equilibrium and balance.

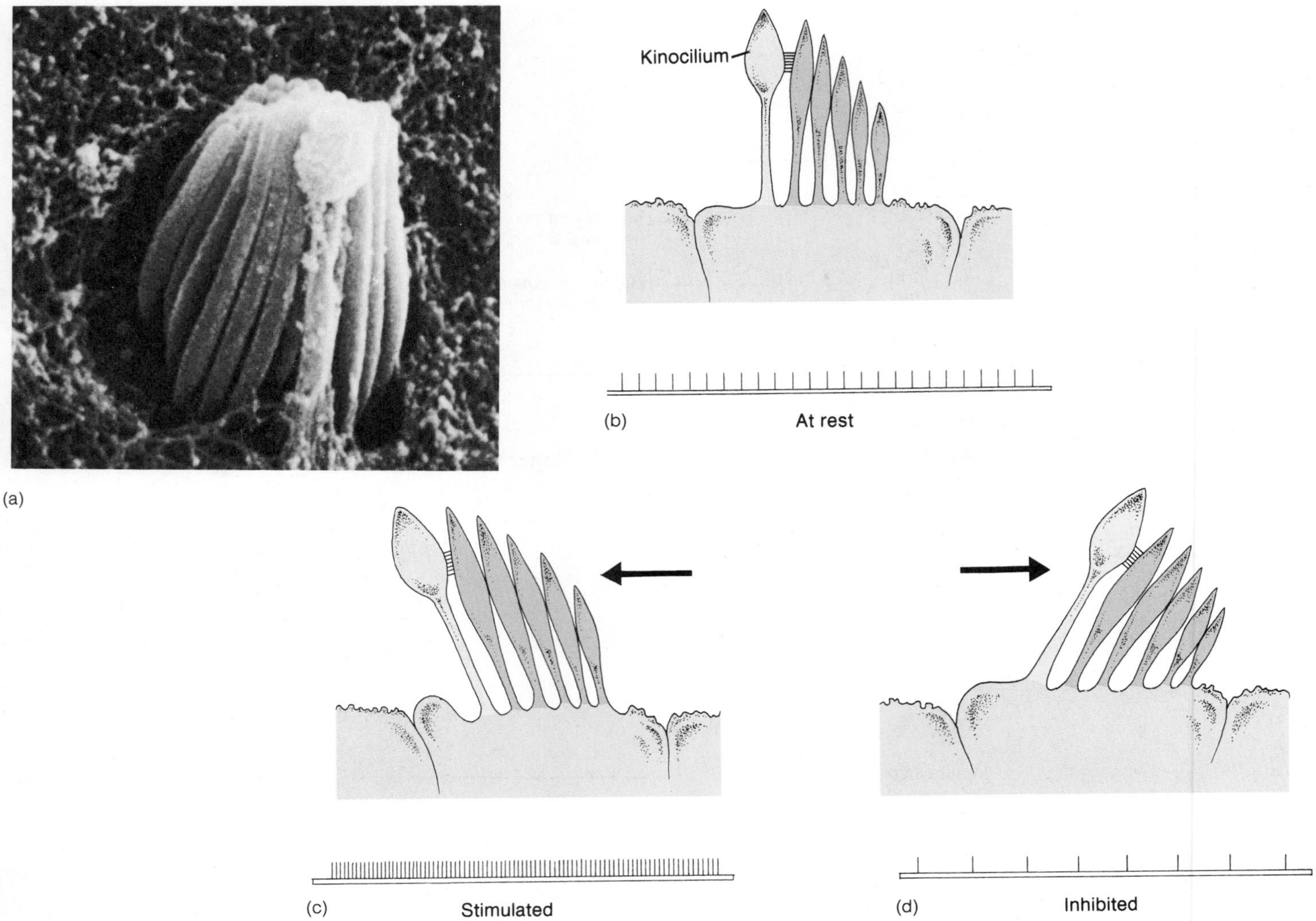

Figure 15.44 Sensory hair cells of a vestibular organ. (*a*) A scanning electron photomicrograph of a kinocilium. (*b*) Sensory hairs (microvilli) and one kinocilium. (*c*) When hair cells are bent in the direction of the kinocilium, the cell membrane is depressed (see arrow) and the sensory neuron innervating the hair cell is stimulated. (*d*) When the hairs are bent in a direction opposite to the kinocilium, the sensory neuron is inhibited.

Saccule and Utricle Receptor hair cells of the saccule and utricle are located in a small, thickened area of the walls of these organs called the **macula** (fig. 15.45). Cytoplasmic extensions of the hair cells project into a gelatinous material that supports microscopic crystals of calcium carbonate, called **otoliths.** The otoliths increase the weight of the gelatinous mass, which results in a higher *inertia* (resistance to change in movement).

When a person is upright, the hairs of the utricle are oriented vertically, whereas those of the saccule are oriented horizontally into the otolith membrane. During forward acceleration, the otolith membrane lags behind the hair cells, so the hairs are pushed backward. This is similar to the backward thrust of the body when a car accelerates rapidly forward. The inertia of the otolith membrane similarly causes the hairs of the utricle to be pushed upward when a person jumps from a raised platform. These effects, and the opposite ones that occur when a person accelerates backward or upward, allow us to maintain our equilibrium with respect to gravity during linear acceleration.

macula: L. *macula,* spot
otoliths: Gk. *otos,* ear; *lithos,* stone

Sensory impulses from the vestibular organs are conveyed to the brain by way of the vestibular portion of the vestibulocochlear nerve.

Semicircular Canals Receptors of the semicircular canals are contained within the ampulla at the base of each semicircular duct. An elevated area of the ampulla called the **crista ampullaris** *(am''pu-lar'is)* contains numerous hair cells and supporting cells (fig. 15.46). Like the saccule and utricle, the hair cells have cytoplasmic extensions that project into a dome-shaped gelatinous mass called the **cupula** *(ku'pu-lah)*. When the hair cells within the cupula are bent by rapid displacement of the fluid within the semicircular ducts, as in spinning around, sensory impulses travel to the brain by way of the vestibular portion of the vestibulocochlear nerve.

Neural Pathways Stimulation of the hair cells in the vestibular apparatus activates the sensory neurons of the *vestibulocochlear* (*eighth cranial*) *nerve.* These fibers transmit impulses to the

cupula: L. *cupula,* cup-shaped

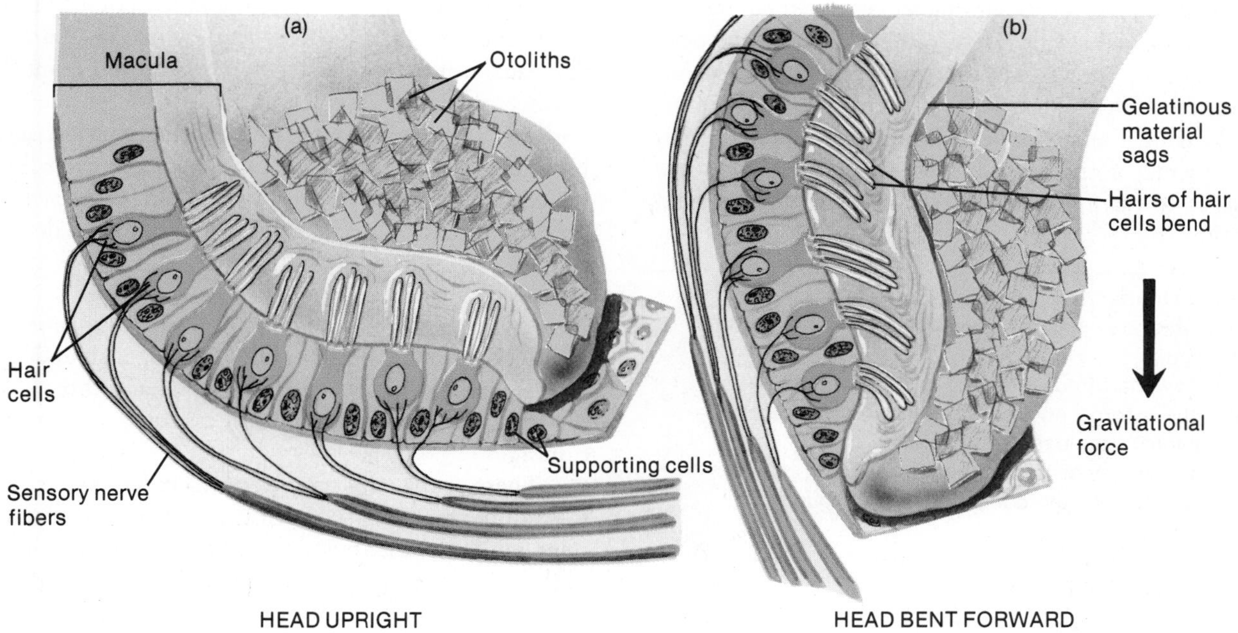

Figure 15.45 The macula of the utricle. (*a*) When the head is in an upright position, the weight of the otoliths applies direct pressure to the sensitive cytoplasmic extensions of the hair cells. (*b*) As the head is tilted forward, the extensions of the hair cells bend in response to gravitational force and cause the hair cells to be stimulated.

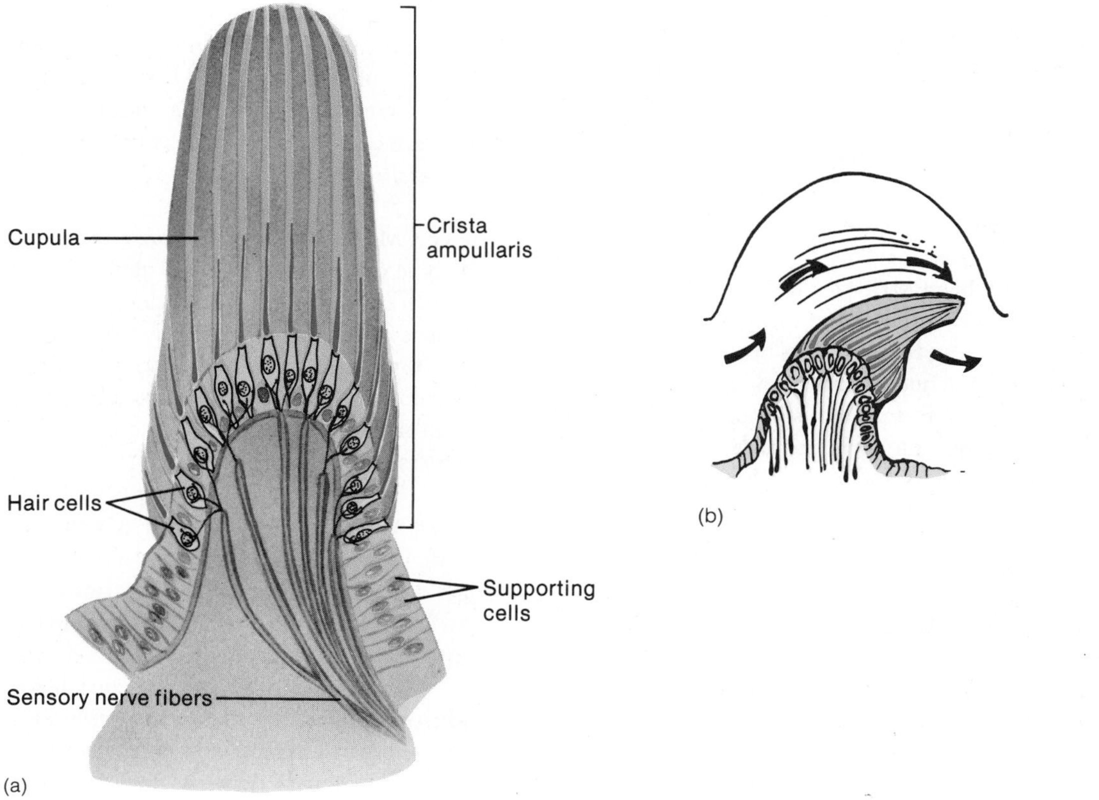

Figure 15.46 (*a*) A crista ampullaris within an ampulla. (*b*) Movement of the endolymph during rotation causes the cupula to displace, thus stimulating the hair cells.

cerebellum and to the vestibular nuclei of the medulla oblongata. The vestibular nuclei, in turn, send fibers to the oculomotor center of the brain stem and to the spinal cord. Neurons in the oculomotor center control eye movements, and neurons in the spinal cord stimulate movements of the head, neck, and limbs. Movements of the eyes and body produced by these pathways serve to maintain balance and track the visual field during rotation.

---

1. List the structures of the outer ear, middle ear, and inner ear, and explain the function of each as related to hearing.
2. Use a flowchart to describe how sound waves in air within the external auditory canal are transduced into movements of the basilar membrane.
3. Explain how movements of the basilar membrane of the organ of Corti can code for different sound frequencies (pitches).
4. Explain how the vestibular organs maintain a sense of balance and equilibrium.

Clinical Considerations

Numerous disorders and diseases afflict the sensory organs. Some of these occur during the sensitive period of prenatal development and have a congenital effect. Other sensory impairments, some of which are avoidable, can occur at any time of life. Still others are the result of changes during the natural aging process. The loss of a sense is frequently a traumatic adjustment. Fortunately, however, when a sensory function is impaired or lost, the other senses seem to become more keen to lessen the extent of the handicap. A blind person, for example, compensates somewhat for the loss of sight by developing a remarkable hearing ability.

Many books have been written on diseases and dysfunctions of the sensory organs, and entire specialities within medicine are devoted to treating the disorders of the specific sensory organs. It is beyond the scope of this book to attempt a comprehensive discussion of the numerous aspects of the sensory organs. Some general comments will be made, however, on the diagnosis of sensory disorders and on developmental problems that can affect the eyes or ears. In addition, the more common diseases and dysfunctions of these organs will be discussed.

Diagnosis of Sensory Organs

Eye There are two distinct professional specialities concerned with the structure and function of the eye. *Optometry* is the paramedical profession concerned with assessing vision and treating visual problems. An *optometrist* is not a physician and does not treat eye diseases but prescribes lenses or visual training. *Ophthalmology (of''thal-mol'o-je)* is the speciality of medicine concerned with diagnosing and treating eye diseases.

Although the eyeball is an extremely complex organ, it is quite accessible to examination. Various techniques and instruments are used during an eye examination, but the following are used most frequently: (1) a *cycloplegic drug,* which is instilled into the eyes to dilate the pupils and temporarily inactivate the ciliary muscles; (2) a *Snellen's chart,* which is used to determine the visual acuity of a person standing 20 feet from the chart (a reading of 20/20 is considered normal for the test); (3) an *ophthalmoscope,* which contains a light, mirrors, and lenses to illuminate and magnify the interior of the eyeball so that the structures within may be examined; and (4) a *tonometer,* used to measure ocular tension, which is important in detecting glaucoma.

Ear *Otorhinolaryngology (o''to-ri''no-lar''in-gol'o-je)* is the speciality of medicine dealing with the diagnosis and treatment of diseases or conditions of the ear, nose, and throat. *Audiology* is the study of hearing, particularly assessing the ear and its functioning.

There are three common instruments or techniques used to examine the ears to determine auditory function: (1) an *otoscope* is an instrument used to examine the tympanic membrane of the ear; abnormalities of this membrane are informative when diagnosing specific auditory problems, including middle ear infections; (2) *tuning fork tests* are useful in determining hearing acuity and especially for discriminating the various kinds of hearing loss; (3) *audiometry* is a functional examination for hearing sensitivity and speech discrimination.

Developmental Problems of the Eyes and Ears

Although there are many congenital abnormalities of the eyes and ears, most of them are rare. The sensitive period of development of these organs is from about twenty-four to forty-five days after conception. Most congenital disorders of the eyes and ears are caused by genetic factors or intrauterine infections such as *rubella virus.*

Eye Most **congenital cataracts** are hereditary, but they may also be caused by maternal rubella infection during the critical fourth to sixth week of eye development. In this condition, the lens is opaque and frequently appears grayish white.

Cyclopia is a rare condition in which the eyes are partially fused into a median eye enclosed by a single orbit. Other severe malformations, which are incompatible with life, are generally expressed with this condition.

Ear **Congenital deafness** is generally caused by an autosomal recessive gene but may also be caused by a maternal rubella infection. The actual functional impairment is generally either a defective set of ossicles or improper development of the neurosensory structures of the inner ear.

Although the shape of the auricle varies widely, **auricular abnormalities** are not uncommon, especially in infants with chromosomal syndromes causing mental deficiencies. In addition, the external auditory canal frequently does not develop in these children, producing a condition called **atresia** *(ah-tre'ze-ah)* of the external auditory canal.

Functional Impairments of the Eye

Few people have perfect vision. Slight variations in the shape of the eyeball or curvature of the cornea or lens cause an imperfect focal point of light waves onto the retina. Most variations are slight, however, and the error of refraction goes unnoticed. Severe deviations that are not corrected may cause blurred vision, fatigue, chronic headaches, and depression.

There are four principal clinical considerations caused by defects in the refractory structures or general shapes of the eyeball. **Myopia** (nearsightedness) is an elongation of the eyeball that causes light waves to focus at a point in the vitreous humor in front of the retina (fig. 15.47). Only light waves from close objects can be focused clearly on the retina; distant objects appear blurred. **Hyperopia** (farsightedness) is a condition in which the eyeball is too short, which causes light waves to have a focal point behind the retina. Although visual accommodation aids a hyperopic person, it generally does not help enough for the person to clearly view very close or distant objects. **Presbyopia** is a condition in which the lens tends to lose its elasticity and ability to accommodate. It is relatively common in elderly people. In order to read print on a page, a person with presbyopia must hold the page farther from the eyes. **Astigmatism** is a condition in which an irregular curvature of the cornea or lens of the eye distorts the refraction of light waves. If a person with astigmatism views a circle, the image will not appear clear in all 360 degrees; the parts of the circle that appear blurred can thus be used to map the astigmatism.

Various glass or plastic lenses can generally aid persons with the above visual impairments. Myopia may be corrected with a biconcave lens; hyperopia with a biconvex lens; and presbyopia with bifocals, or a combination of two lenses adjusted for near and distant vision. Correction for astigmatism requires a careful assessment of the irregularities and a prescription of a specially ground corrective lens.

The condition of myopia may also be treated by a surgical procedure called **radial keratotomy.** In this technique, eight to sixteen microscopic slashes, like the spokes of a wheel, are made in the cornea from the center to the edge. The ocular pressure inside the eyeball bulges the weakened cornea and flattens its center, changing the focal length of the eyeball. Side effects may include vision fluctuations, sensitivity to glare, and incorrect corneal alteration.

Cataracts A cataract is a clouding of the lens that leads to a gradual blurring of vision and the eventual loss of sight. A cataract is not a growth within or upon the eye but a chemical change in the protein of the lens caused by injury, poisons, infections, or age degeneration.

Cataracts are the leading cause of blindness. A cataract can be removed surgically, however, and vision restored by implanting a tiny intraocular lens that either clips to the iris or is secured into the vacant lens capsule. Special contact lenses or thick lenses for glasses are other options.

Detachment of the Retina Retinal detachment is a separation of the nervous or visual layer of the retina from the underlying pigment epithelium. It generally begins as a minute tear in the retina that gradually extends as vitreous fluid accumulates between the layers. Retinal detachment may result from hemorrhage, a tumor, degeneration, or trauma from a violent blow to the eye. A detached retina may be repaired by using laser beams, cryoprobes, or intense heat to destroy the tissue beneath the tear and rejoin the layers.

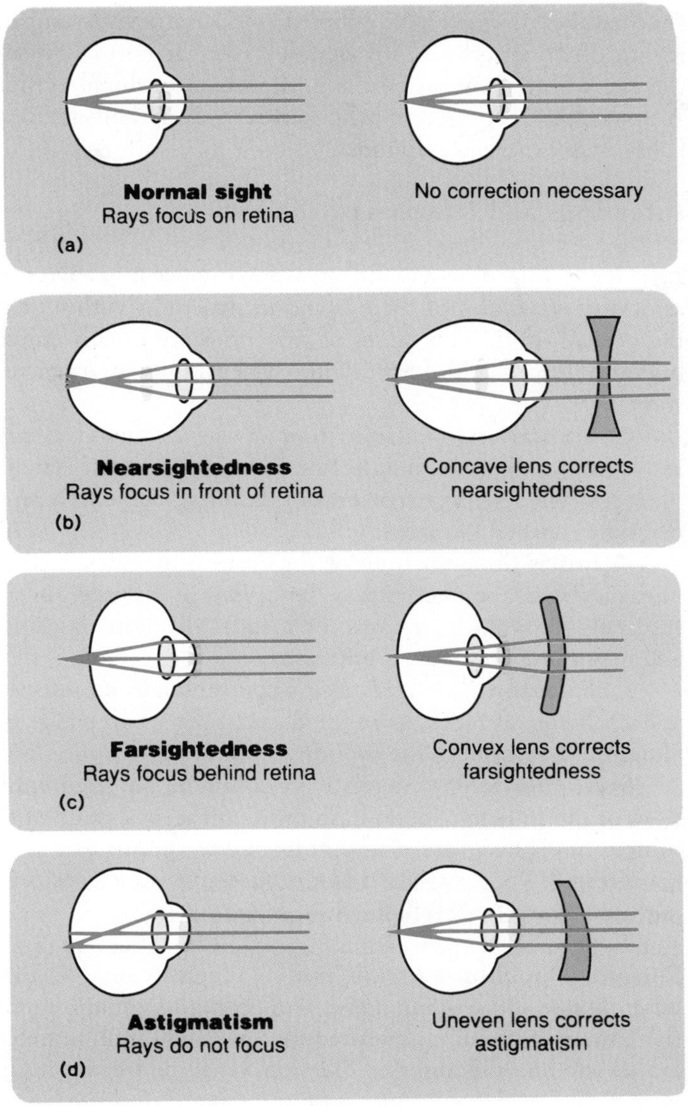

Figure 15.47 In a normal eye (*a*), parallel rays of light are brought to a focus on the retina by refraction in the cornea and lens. If the eye is too long, as in myopia (*b*), the focus is in front of the retina. This can be corrected by a concave lens. If the eye is too short (*c*), as in hyperopia, the focus is behind the retina. This is corrected by a convex lens. In astigmatism (*d*), light refraction is uneven due to an abnormal shape of the cornea or lens.

Strabismus Strabismus is a condition in which both eyes do not focus upon the same axis of vision. This prevents stereoscopic vision, and the persons afflicted will have varied visual impairments. Strabismus is usually caused by a weakened extrinsic eye muscle.

Strabismus is assessed while the patient attempts to look straight ahead. If the afflicted eye is turned toward the nose, the condition is called convergent strabismus **(esotropia).** If the eye is turned outward, it is called divergent strabismus

myopia: Gk. *myein,* to shut; *ops,* eye
hyperopia: Gk. *hyper,* over; *ops,* eye
presbyopia: Gk. *presbys,* old man; *ops,* eye
astigmatism: Gk. *a,* without; *stigma,* point
cataract: Gk. *katarrhegnynai,* to break down

(exotropia). Disuse of the afflicted eye causes a visual impairment called **amblyopia.** Visual input from the normal eye and the eye with strabismus results in **diplopia,** or double vision. A normal, healthy person who has excessively consumed alcohol may experience diplopia.

Infections and Diseases of the Eye

Infections Infections and inflammation can occur in any of the associated structures of the eye and in structures within or on the eyeball itself. The causes of infections are usually microorganisms, mechanical irritation, or sensitivity to particular substances.

Conjunctivitis (inflammation of the conjunctiva) may result from sunburn or an infection by organisms such as staphylococci, viruses, or streptococci. Bacterial conjunctivitis is commonly called "pinkeye."

Keratitis (inflammation of the cornea) may develop secondarily from conjunctivitis or be caused by diseases such as tuberculosis, syphilis, mumps, or measles. Keratitis is painful and may cause blindness if untreated.

A **chalazion** *(kah-la'ze-on)* is a cyst caused by an infection and a subsequent blockage in the ducts of the sebaceous glands along the free edge of the eyelids.

Styes (hordeola) are relatively common, but mild, infections of the follicle of an eyelash or the sebaceous gland of the follicle. Styes may spread readily from one eyelash to another if untreated. Poor hygiene and the excessive use of cosmetics may contribute to development of styes.

Diseases **Trachoma** *(trah-ko'mah)* is a highly contagious bacterial disease of the conjunctiva and cornea. Although rare in the United States, it is estimated that over 500 million people are afflicted by this disease. Trachoma may be treated readily with sulfonamides and some antibiotics, but if untreated it will spread progressively until it covers the cornea. At this stage, vision is lost and the eye undergoes degenerative changes.

Glaucoma is the second leading cause of blindness and is particularly common in underdeveloped countries. Although it can afflict persons of any age, 95% of the cases involve persons over the age of forty. Glaucoma is an abnormal increase in the intraocular pressure of the eyeball. Aqueous humor does not drain through the venous sinus as quickly as it is produced. Glaucoma can be caused by several things, including malnutrition. Glaucoma causes compression of the blood vessels in the eyeball and compression of the optic nerve. Retinal cells die and the optic nerve may atrophy, producing blindness.

Infections, Diseases, and Functional Impairments of the Ear

Disorders of the ear are common and may affect both hearing and the vestibular functions. The ear is afflicted by numerous diseases as well as environmental factors—some of which can be prevented.

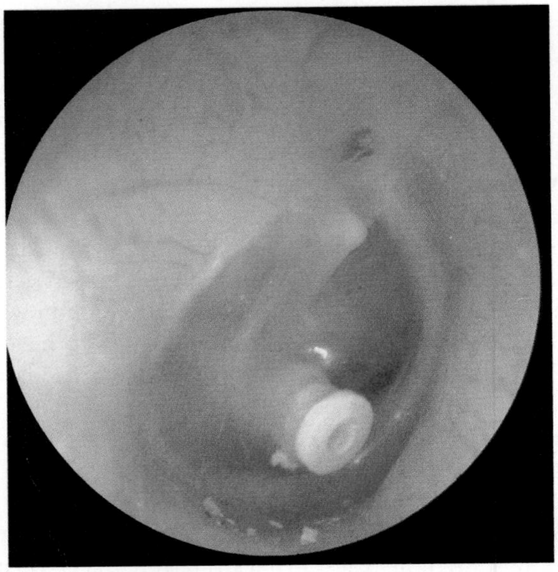

Figure 15.48 An implanted ventilation tube in the tympanic membrane following a *myringotomy.*

Infections and Diseases **External otitis** is a general term for infections of the outer ear. The causes of external otitis range from dermatitis to fungal and bacterial infections.

Acute purulent otitis media is a middle ear infection. Pathogens of this disease usually enter through the auditory tube, most often following a cold or tonsillitis. Children frequently have middle ear infections because of their susceptibility to infections and their short and straight auditory tubes. As a middle ear infection progresses to the inflammatory stage, the auditory tube is closed and drainage prohibited. An intense earache is a common symptom of a middle ear infection. The pressure from the inflammation may eventually rupture the tympanic membrane to permit drainage.

Repeated middle ear infections, particularly in children, usually necessitate an incision of the tympanic membrane, called a **myringotomy** *(mir''in-got'o-me),* and the implantation of a tiny tube within the tympanic membrane (fig. 15.48) to assist the patency of the auditory tube. The tubes, which are eventually sloughed out of the ear, permit the infection to heal and help prohibit further infections by keeping the auditory tube open.

Perforation of the tympanic membrane may occur as the result of infections or trauma. Sudden, intense noise might rupture the membrane. Spontaneous perforation of the membrane usually heals rapidly, but scar tissue may form and lessen the sensitivity to sound vibrations.

Otosclerosis is a progressive deterioration of the normal bone of the bony labyrinth and its replacement with vascular spongy bone. This frequently causes hearing loss as the ear ossicles are immobilized. Surgical scraping of the bone growth and replacing the stapes with a prosthesis is the most frequent treatment of otosclerosis.

Meniere's disease afflicts the inner ear and may cause hearing loss as well as equilibrium disturbance. The causes of Meniere's disease vary and are not completely understood, but

glaucoma: Gk. *glaukos,* gray

Meniere's disease: from Prosper Meniere, French physician, 1799–1862 they are thought to be related to a dysfunction of the autonomic nervous system that causes a vasoconstriction within the inner ear. The disease is characterized by recurrent periods of **vertigo** (dizziness and a sensation of rotation), **tinnitus** *(ti-ni'tus)* (ringing in the ear), and progressive deafness in the affected ear. Meniere's disease is chronic and affects both sexes equally. It is more common in middle-aged and elderly persons.

Nystagmus and Vertigo When a person first begins to spin, the inertia of the endolymph within the semicircular canals causes the cupula to bend in the opposite direction. As the spin continues, however, the inertia of the endolymph is overcome and the cupula straightens. At this time the endolymph and the cupula are moving in the same direction and at the same speed. If movement is suddenly stopped, the greater inertia of the endolymph causes it to continue moving in the previous direction of spin and to bend the cupula in that direction.

Bending of the cupula after movement has stopped affects muscular control of the eyes and body. The eyes slowly drift in the direction of the previous spin and are then rapidly jerked back to the midline position, producing involuntary oscillations. These movements are called post-rotatory **vestibular nystagmus** *(nis-tag'mus)*. People experiencing this may feel that they, or the room, are spinning. The loss of equilibrium that results is called **vertigo.** If the vertigo is sufficiently severe or the person particularly susceptible, the autonomic system may become involved. This can produce dizziness, pallor, sweating, and nausea.

Auditory Impairment Loss of hearing results from disease, trauma, or developmental problems involving any portion of the auditory apparatus, cochlear nerve and auditory pathway, or areas of auditory perception within the brain. Hearing impairment varies from slight disablement, which may or may not worsen, to total deafness. Some types of hearing impairment, including deafness, can be mitigated through hearing aids or surgery.

There are two principal types of auditory impairments. **Conduction deafness** is caused by an interference with the sound waves through the outer or middle ear. Conduction problems include impacted cerumen (wax) ruptured tympanic membrane, a severe middle ear infection, or adhesions of one or more ear ossicles *(otosclerosis)*. Conductive deafness can be successfully treated.

Perceptive (sensorineural) deafness results from disorders that affect the inner ear, the cochlear nerve or nerve pathway, or auditory centers within the brain. Perceptive impairment ranges in severity from the inability to hear certain frequencies to deafness. Such deafness may be caused by a number of factors, including diseases, trauma, and genetic or developmental problems. Hearing impairments of this type, which are common in elderly persons, are related to age. The ability to perceive high-frequency sounds is generally affected first. Hearing aids may help patients with perceptive deafness. This type of deafness is permanent, however, because it involves sensory structures that cannot heal themselves through regeneration or replication.

vertigo: L. *vertigo,* dizziness
tinnitus: L. *tinnitus,* ring or tingle

CLINICAL CASE STUDY ANSWER

When the handle of the vibrating tuning fork is placed on the mastoid process, the temporal bone transmits sound waves directly to the inner ear. This bypasses the conductive components of the middle ear, which are (beginning with the first to receive sound waves) the tympanic membrane, malleus, incus, and stapes. Since the patient can hear well when the sound waves are transmitted through the temporal bone, the inner ear organs, as well as the neural pathways, can be assumed to be functioning. It can, therefore, be concluded that the problem lies in the conductive components. This is often the result of otosclerosis (a spongy proliferation of bone) or other conditions affecting the ossicles and/or tympanic membrane, many of which are surgically treatable.

Chapter Summary

I. Overview of Sensory Perception
   A. Sensory organs are specialized extensions of the nervous system that respond to specific stimuli and conduct nerve impulses.
   B. A stimulus to a receptor that conducts an impulse to the brain is necessary for perception.
   C. Senses act as energy filters that permit perception of only a narrow range of energy.

II. Classification of the Senses
   A. The senses are classified according to receptors or on the basis of the stimulus energy they transduce.
   B. General senses are widespread throughout the body and are simple in structure. Special senses are localized in complex receptor organs and have extensive neural pathways.
   C. Somatic senses are localized within the body wall, and visceral senses are located within the visceral organs.
   D. Phasic receptors respond quickly to a stimulus but then adapt and decrease firing rate. Tonic receptors produce a constant rate of firing.

III. Somatic Senses
   A. Meissner's corpuscles, free nerve endings, and hair root plexuses are tactile receptors, responding to light touch.
   B. Lamellated corpuscles are pressure receptors in the dermis.
   C. Thermoreceptors include the organs of Ruffini (heat) and bulbs of Krause (cold).
   D. Free nerve endings are the principal pain receptors, serving to protect the body from injury.
   E. Joint kinesthetic receptors, neuromuscular spindles, and neurotendinous receptors are proprioceptors that are sensitive to changes in stretch and tension.
   F. The acuity of sensation depends on receptor density and lateral inhibition.

IV. Olfactory Sense
   A. Olfactory receptors of the olfactory nerve respond to chemical stimuli and transmit the sensation of olfaction to the cerebral cortex.

B. Olfaction functions closely with gustation in that the receptors of both are chemoreceptors, requiring dissolved substances for stimuli.

V. Gustatory Sense
  A. Taste receptors in taste buds respond to chemical stimuli and transmit the sensation of gustation to the cerebral cortex.
  B. Gustatory cells compose the taste buds located in circumvallate, fungiform, and filiform papillae.
  C. The four modalities of taste sensation are sweet, salty, sour, and bitter.

VI. Visual Sense
  A. The sensory components of the eye are formed by twenty weeks, and the accessory structures are formed by thirty-two weeks.
  B. Accessory structures of the eye (the orbit, eyebrow, eyelids, eyelashes, conjunctiva, lacrimal gland, and ocular muscles) either protect the eyeball or provide eye movement.
  C. The eyeball consists of the fibrous tunic, which is divided into the sclera and cornea; the vascular tunic, which consists of the choroid, the ciliary body, and the iris; and the internal tunic, or retina, which consists of an outer pigmented layer and an inner nervous layer.
  D. Rods and cones, which are the photoreceptors in the nervous layer of the retina, respond to dim and colored light, respectively. Cones are concentrated in the fovea centralis, the area of keenest vision.
  E. Rods and cones contain specific pigments that provide sensitivity to different light waves.
  F. The anterior and posterior chambers contain aqueous humor, and the vitreous chamber contains vitreous humor.
  G. The visual process includes the transmission and refraction of light waves, accommodation of the lens, constriction of the pupil, and convergence of the eyes.
  H. Neural pathways from the retina to the superior colliculus help regulate eye and body movements. Most fibers from the retina project to the lateral geniculate body and from there to the striate cortex.

VII. Senses of Hearing and Balance
  A. The development of the ear begins during the fourth week and is completed by the thirty-second week.
  B. The outer ear consists of the auricle and the external auditory canal.
  C. The middle ear, bounded by the tympanic membrane and the oval and round windows, contains the ear ossicles (malleus, incus, and stapes) and the auditory muscles (tensor tympani and stapedius).
  D. The vestibule of the middle ear connects to the pharynx through the auditory tube.
  E. The inner ear, or labyrinth, contains the functional organs for hearing and equilibrium: the organ of Corti in the cochlea, and the semicircular canals, saccule, and utricle.

Review Activities

Objective Questions

1. Which of the following must occur for the perception of a sensation to take place?
   (a) stimulus present
   (b) nerve impulse conduction
   (c) receptor activated
   (d) all of the above
2. The cutaneous receptor sensitized for deep pressure detection is the
   (a) hair root plexus.
   (b) lamellated corpuscle.
   (c) organ of Ruffini.
   (d) free nerve endings.
3. Proprioceptors which are located within the connective tissue capsule in diarthrotic joints are called
   (a) neuromuscular spindles.
   (b) Golgi tendon organs.
   (c) neurotendinous receptors.
   (d) joint kinesthetic receptors.
4. The sensation of visceral pain perceived as arising from another somatic location is known as
   (a) related pain.
   (b) phantom pain.
   (c) referred pain.
   (d) parietal pain.
5. When a person with normal vision views an object from a distance of at least 20 feet,
   (a) the ciliary muscles are relaxed.
   (b) the suspensory ligament is taut.
   (c) the lens is flat, having the least convex shape.
   (d) All of the above.
6. Which of the following is an avascular ocular tissue?
   (a) sclera (c) ciliary body
   (b) choroid layer (d) iris
7. Pupils are dilated in response to the contraction of the
   (a) ciliary muscles.
   (b) radial iris muscles.
   (c) circular iris muscles.
   (d) orbicularis oculi muscles.
8. The stimulation of hair cells in the semicircular canals results from the movement of
   (a) endolymph.
   (b) perilymph.
   (c) the otolith membrane.
9. The middle ear is separated from the inner ear by the
   (a) round window.
   (b) tympanic membrane.
   (c) oval window.
   (d) Both a and c.
10. Glasses with concave lenses help to correct
   (a) presbyopia. (c) hyperopia.
   (b) myopia. (d) astigmatism.

Essay Questions

1. What four events are necessary for perception of a sensation? What is considered a sensation and how does it differ from a reflex?
2. Outline the major events in the development of the eye and ear. When are congenital deformities most likely to occur?
3. Name the senses of the body. Differentiate between the special and somatic senses. What are the similarities between these two senses?
4. List the functions of proprioceptors, and differentiate between the three types. What role do proprioceptors play in the kinesthetic sense?
5. Compare and contrast olfaction and gustation. Identify the cranial nerves that serve both of these senses and trace the afferent pathways of each.
6. Identify and describe the accessory structures of the eye and list their functions.
7. Diagram the structure of the eyeball, and label the sclera, cornea, choroid, macula lutea, ciliary body, suspensory ligament, lens, iris, pupil, retina, optic disc, and fovea centralis.
8. Outline and explain the process of focusing light waves onto the fovea centralis.
9. Diagram the ear and label the structures of the outer, middle, and inner ear.
10. Trace a sound wave through the structures of the ear, and explain how the mechanism of hearing functions.
11. Explain the mechanism by which equilibrium is maintained and the role played in this by the two kinds of receptor information.
12. List some techniques used to examine the eye and the ear. Give examples of congenital abnormalities of the eye and ear.

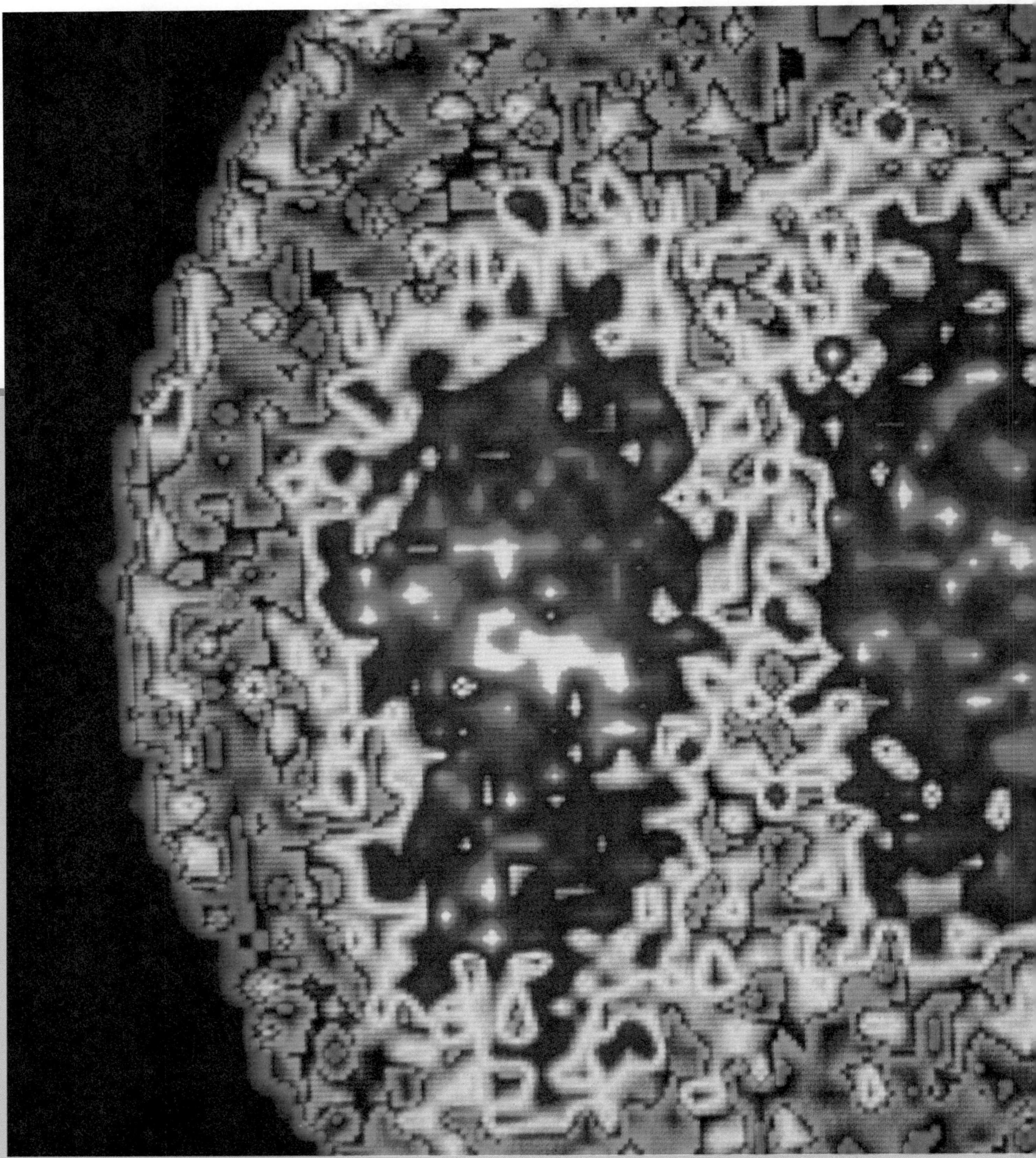

A color-enhanced scintigram (gamma camera scan) of the lungs of an asthmatic patient, obtained following the inhalation of a radioactive gas.

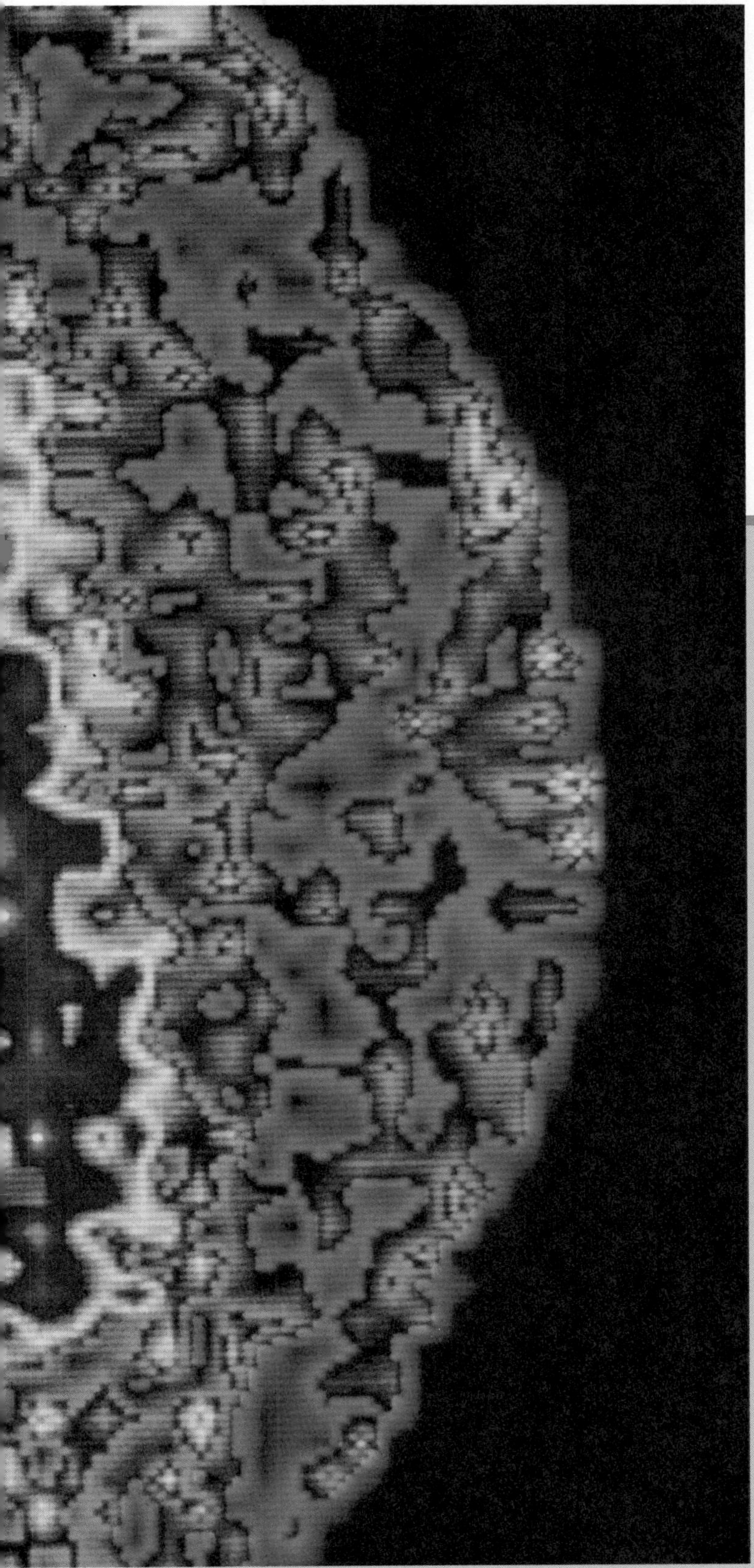

UNIT

6

Maintenance of the Body

The discussions in the chapters in unit 6 revolve around the anatomy of those body systems whose principal function is to maintain homeostasis. The blood of the circulatory system transports materials for each of these body systems. Oxygen and nutrients are made available to the living cells of the body through the blood. The blood also removes metabolic wastes from the cells and distributes hormones to the tissues within the appropriate organs to regulate metabolic activities.

This unit includes:

CHAPTER 16

Circulatory System

Outline and Concepts

CLINICAL CASE STUDY

A sixty-five-year-old woman recently discharged from the hospital following a myocardial infarction (heart attack) returns to the hospital emergency room complaining of the sudden onset of pain in her right lower extremity. The E.R. doctor notices that the patient's leg is also pale and cool from the knee distally. Furthermore, he is unable to detect a pulse over the posterior tibial, dorsalis pedis, and popliteal arteries. A good femoral pulse, however, is palpable in the inguinal region. Upon questioning, the patient states that she was discharged from the hospital with blood thinning medication because tests had revealed that the heart attack had caused a blood clot to form within her heart. She added that the doctors were worried that "a piece of the clot could break off and go to other parts of the body."

Explain how the patient's heart attack could have, several weeks later, led to her leg complaints. In which side of her heart (right or left) did the previously diagnosed blood clot most likely reside? Explain anatomically where the cause of her new problem is located and describe how it arrived there. ■

Functions and Major Components of the Circulatory System

The circulatory system consists of components that provide functions essential to the life of complex, multicellular organisms.

Objective 1. Describe the functions of the circulatory system.
Objective 2. Describe the major components of the circulatory system.

A unicellular organism can provide for its own maintenance and continuity by performing the wide variety of functions needed for life. By contrast, the complex human body is composed of trillions of specialized cells that demonstrate a division of labor. Cells of a multicellular organism depend on one another for the very basics of their existence. The majority of the cells of the body are firmly implanted in tissues and are incapable of procuring food and oxygen or even moving from their own wastes. Therefore, a highly specialized and effective means of transporting materials within the body is needed.

The blood serves this transportation function. An estimated 60,000 miles of vessels throughout the body of an adult ensure that continued sustenance reaches each of the trillions of living cells. The blood can also, however, serve to transport disease-causing viruses, bacteria, and their toxins. To ensure against this, the circulatory system has protective mechanisms: the white blood cells and lymphatic system. In order to perform its various functions, the circulatory system works together with the respiratory, urinary, digestive, endocrine, and integumentary systems in maintaining homeostasis (fig. 16.1).

Functions of the Circulatory System

The functions of the circulatory system can be divided into three areas: transportation, regulation, and protection.

1. **Transportation.** All of the substances involved in cellular metabolism are transported by the circulatory system. These substances can be categorized as follows:
   a. Respiratory. Red blood cells, called **erythrocytes** *(ĕ-rith'ro-sīts)*, transport oxygen to the tissue cells. In the lungs, oxygen from the inhaled air attaches to hemoglobin molecules within the erythrocytes and is transported to the cells for aerobic respiration. Carbon dioxide produced by cellular respiration is carried by the blood to the lungs for elimination in the exhaled air (fig. 16.2).
   b. Nutritive. The digestive system is responsible for the mechanical and chemical breakdown of food so that it can be absorbed through the intestinal wall and into the blood vessels of the circulatory system. The blood then carries these absorbed products of digestion through the liver and to the cells of the body.
   c. Excretory. Metabolic wastes, excessive water and ions, as well as other molecules in plasma (the fluid portion of blood) are filtered through the capillaries of the kidneys into kidney tubules and excreted in urine.
2. **Regulation.** The blood carries hormones and other regulatory molecules from their site of origin to distant target tissues.
3. **Protection.** The circulatory system protects against injury and foreign microbes or toxins introduced into the body. The clotting mechanism protects against blood loss when vessels are damaged, and white blood cells, called **leukocytes** *(lu'ko-sīts)*, provide immunity to many disease-causing agents.

Major Components of the Circulatory System

The circulatory system is frequently divided into the **cardiovascular system,** which consists of the heart and blood vessels, and the **lymphatic system,** which consists of lymph vessels and lymph nodes.

The **heart** is a four-chambered, double pump. Its pumping action creates the pressure needed to push blood in the vessels to the lungs and body cells. At rest, the heart of an adult pumps about 5.0 L of blood/min. It takes about one minute for blood to be circulated to the most distal extremity and back to the heart.

Blood vessels form a tubular network that permits blood to flow from the heart to all living cells of the body and then back to the heart. *Arteries* carry blood away from the heart, while *veins* return blood to the heart. Arteries and veins are continuous with each other through smaller blood vessels.

Arteries branch extensively to form a tree of progressively smaller vessels. Those that are microscopic in diameter are called *arterioles.* Conversely, microscopic-sized veins, called *venules (ven'ūls),* deliver blood into progressively larger vessels that empty into the large veins. Blood passes from the arterial to the venous system in *capillaries (kap'ĭ-lar''es),* which are the thinnest and most numerous blood vessels. All exchanges of fluid, nutrients, and wastes between the blood and tissue cells occur across the walls of capillaries.

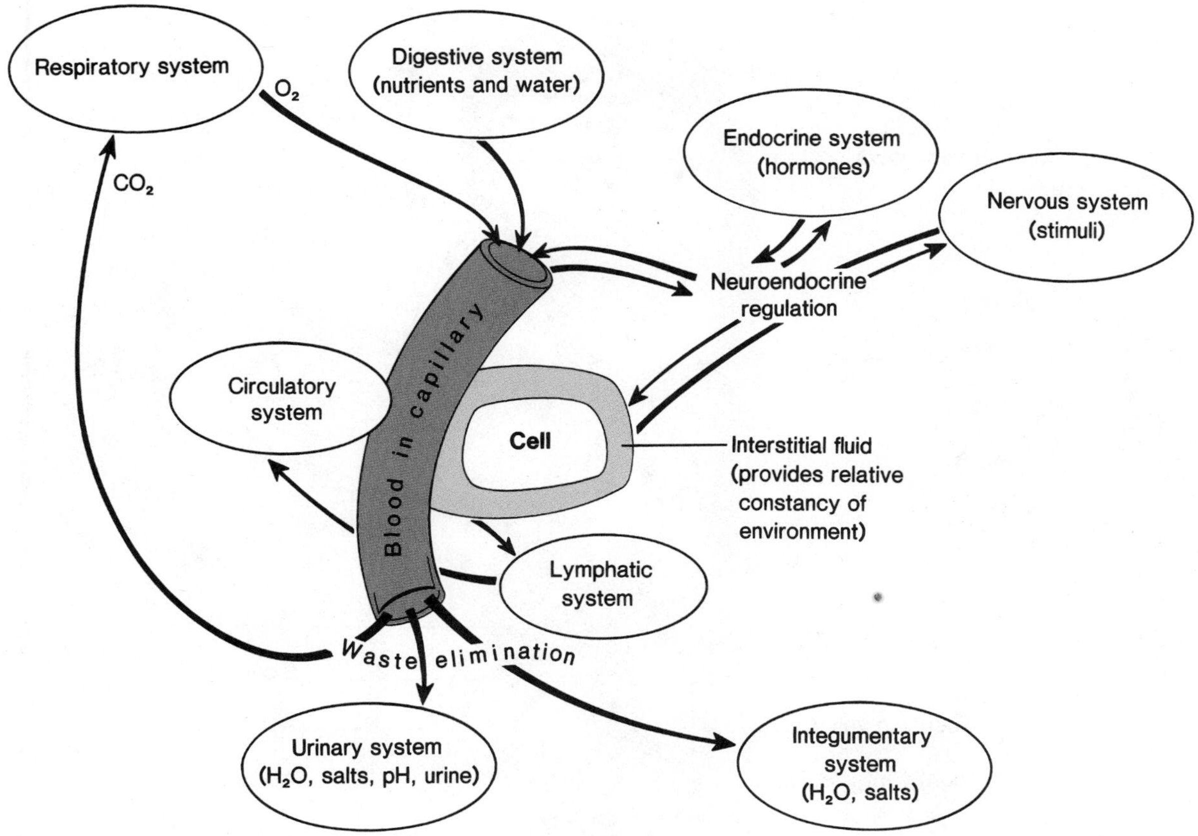

Figure 16.1 The relationship of the circulatory system to the other body systems in maintaining homeostasis.

Fluid derived from plasma that passes out of capillary walls into the surrounding tissues is called *tissue fluid,* or *interstitial (in''ter-stish'al) fluid.* Some of this fluid returns directly to capillaries, and some enters into **lymph vessels** located in the connective tissues around the blood vessels. Fluid in lymph vessels is called *lymph (limf)*. This fluid is returned to the venous blood at particular sites. **Lymph nodes,** positioned along the way, cleanse the lymph prior to its return to the venous blood.

Endothermic (warm-blooded) animals, including humans, need an efficient circulatory system to transport oxygen-rich blood rapidly to all parts of the body. Of the vertebrates, only birds and mammals with their consistently warm body temperatures are considered endothermic, and only birds, mammals, and a few reptiles (crocodiles and alligators) have a four-chambered heart.

1. Name the components of the circulatory system that function in oxygen transport, in the transport of nutrients from the digestive system, and in protection.
2. Define the terms *artery, vein,* and *capillary,* and describe the function of each.
3. Define the terms *interstitial fluid* and *lymph.* Describe their relationships to blood plasma.

Development of the Cardiovascular System

The heart and blood vessels develop from mesoderm. The pattern of this developmental process reflects evolutionary relationships among different vertebrate groups, and a knowledge of this developmental process helps to explain certain congenital defects sometimes found in the human cardiovascular system.

Objective 3. Describe the events in the development of the embryonic heart from the eighteenth through the twenty-fifth day.

Objective 4. Trace the formation and fate of the six aortic arches during embryonic development.

The cardiovascular system is one of the first systems to form in the embryo, delivering nutrients to the mitotically active cells and disposing of waste products through its association with the maternal blood vessels in the placenta. Blood is formed and is circulated through the vessels by the pumping action of the heart on about the twenty-fifth day after conception.

Throughout pregnancy, the fetus is dependent on the mother's circulatory system for nutrient, gaseous, and waste exchange. Therefore, some unique structures of the fetal circulatory system develop to accommodate the placental arrangement. This section of the chapter is concerned with the

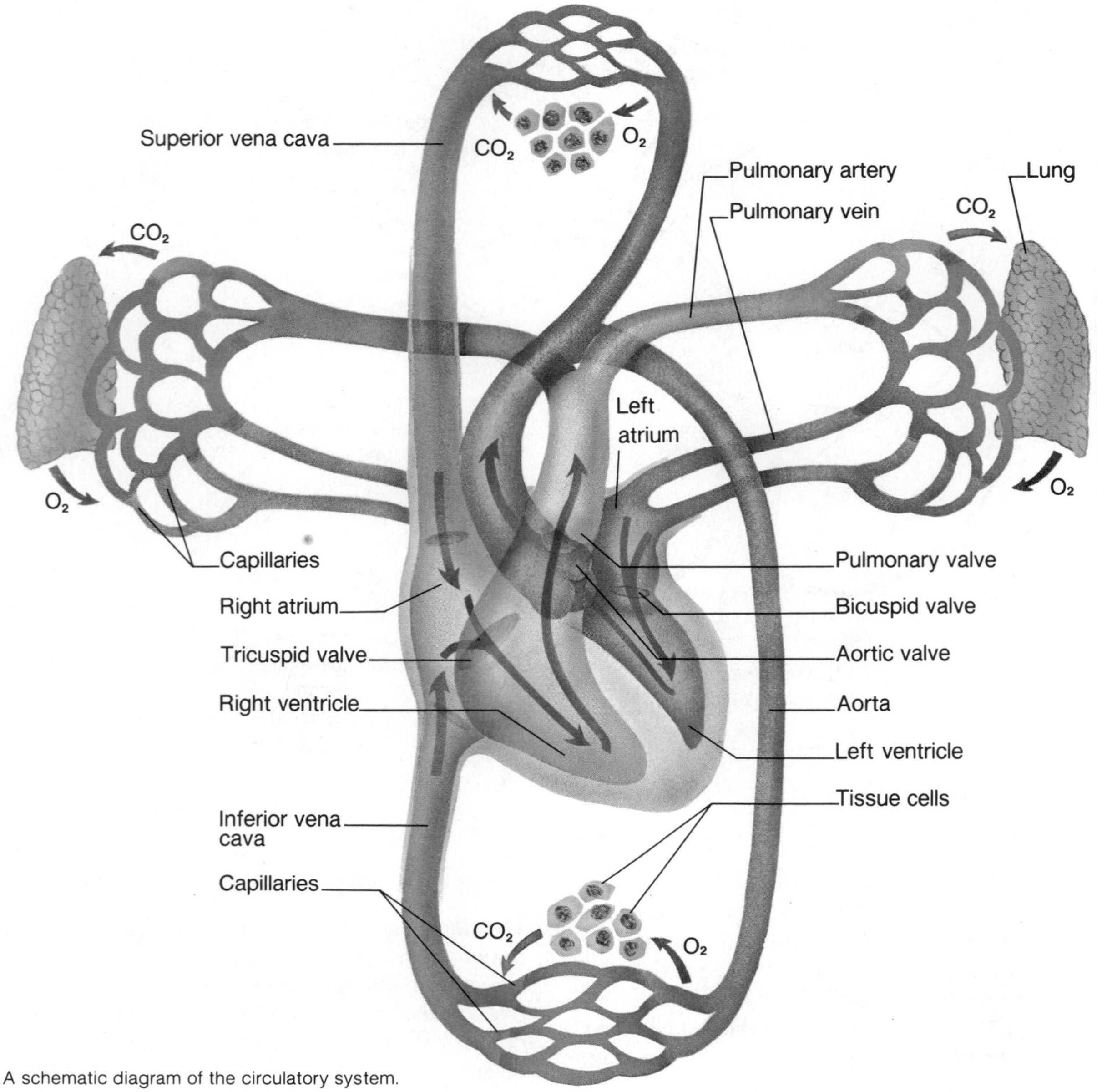

Figure 16.2 A schematic diagram of the circulatory system.

development of the fetal heart and associated major vessels. The circulation of blood through the fetal cardiovascular system is discussed at the end of the chapter after some basic concepts of the circulatory system are introduced.

Development of the Heart

The development of the heart requires only six to seven days. Heart development is first apparent at the eighteenth or nineteenth day in the *cardiogenic (kar''de-o-jen'ik) area* of the mesoderm layer (fig. 16.3). A small paired mass of specialized cells called *heart cords* form here. Shortly after, a hollow center develops in each heart cord, and each structure is then referred to as a *heart tube.* The heart tubes begin to migrate toward each other during the twenty-first day and soon fuse to form a single median *endocardial heart tube.* During this time the endocardial heart tube undergoes dilations and constrictions so that when fusion is completed during the fourth week, five distinct regions of the heart can be identified. These are the **truncus arteriosus, bulbus cordis, ventricle, atrium,** and **sinus venosus** (fig. 16.4).

After the fusion of the heart tubes and the formation of distinct dilations, the heart begins to pump blood. Partitioning of the heart chambers begins during the middle of the fourth week and is completed by the end of the fifth week. During this crucial time many congenital heart problems develop.

Major changes occur in each of the five primitive dilations of the developing heart during the week-and-a-half embryonic period beginning in the middle of the fourth week. The truncus arteriosus differentiates to form a partition between the aorta and the pulmonary trunk. The bulbus cordis is incorporated in the formation of the walls of the ventricles. The sinus venosus forms the **coronary sinus** and a portion of the wall of the right atrium. The ventricle is divided into the right and left chambers by the growth of the **interventricular septum.** The atrium is partially partitioned into right and left chambers by the **septum secundum.** An opening between the two atria

cardiogenic: Gk. *kardia,* heart; *genesis,* be born (origin)

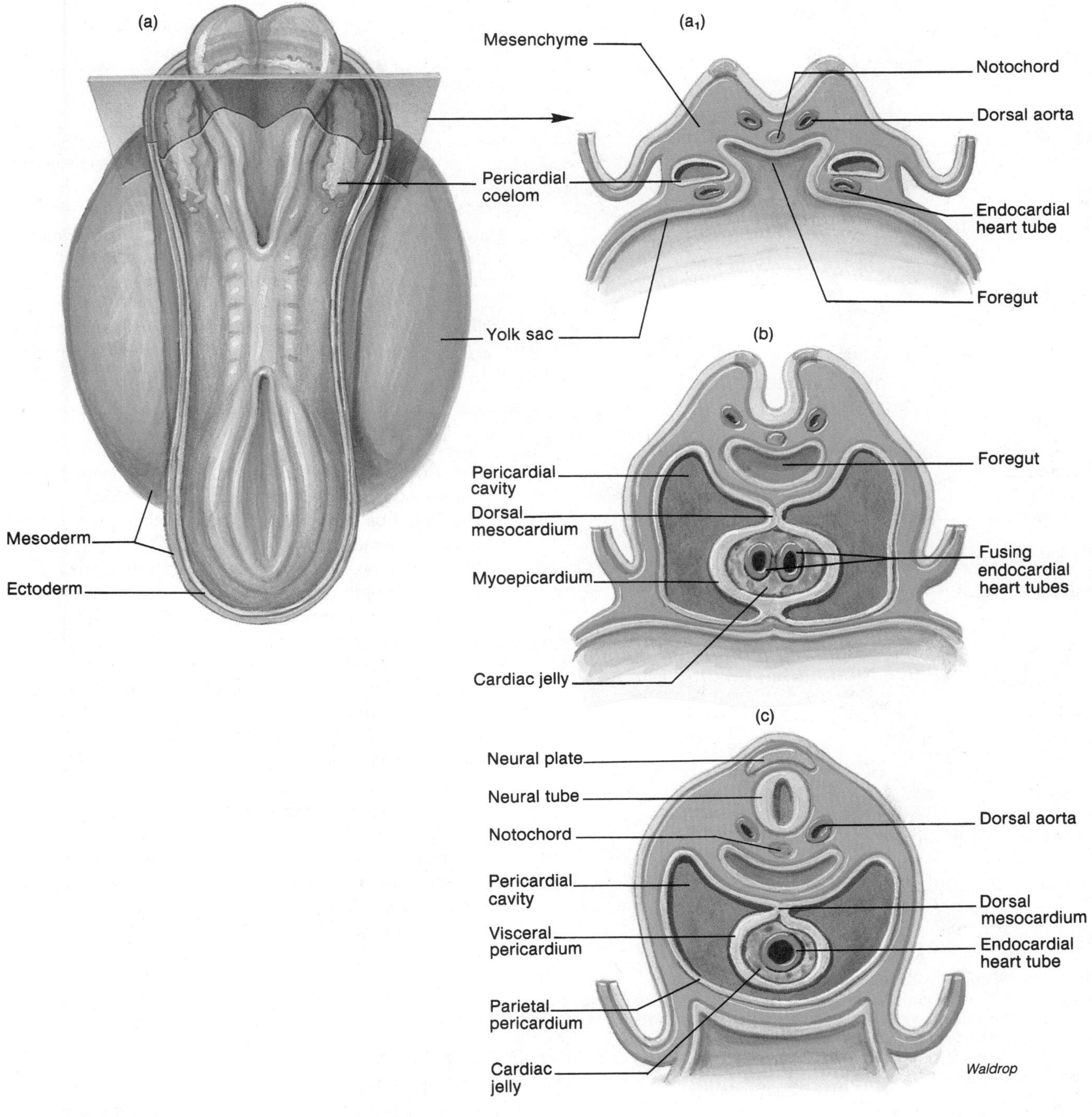

Figure 16.3 The early development of the heart from embryonic mesoderm. (*a*) A dorsal view of an embryo at day 20 showing the position of a transverse cut depicted in (a_1), (*b*), and (*c*). (a_1) At day 20, the endocardial heart tubes are formed from the heart cords. (*b*) By day 21, the medial migration of the endocardial heart tubes brings them together within the forming pericardial cavity. (*c*) A single endocardial heart tube is completed at day 22 and is suspended in the pericardial cavity by the dorsal mesocardium.

called the **foramen ovale** persists throughout fetal development. This opening is covered by a flexible valve, which permits blood to pass from the right to the left side of the heart.

The development of the heart is summarized in table 16.1.

Congenital defects of the cardiac septa are a relatively common form of birth abnormality. Approximately 0.7% of live births and 2.7% of stillbirths show cardiac abnormalities. Ventricular septal defects are the most common of the cardiac defects. An infant with a congenital cardiac defect may suffer from inadequate oxygenation of blood and is termed a "blue baby."

Development of the Major Arteries

The formation of the major arteries occurs simultaneously with the development of the heart. The most complex and fascinating vascular formation is the development of the **aortic arches** associated with the pharyngeal pouches and branchial arches in the neck region (fig. 16.5). These aortic arches arise from the truncus arteriosus. Although six pairs of aortic arches develop, they are not all present at the same time, and none of them persists in entirety through fetal development.

The transformation of the six aortic arches into the basic adult arterial arrangement occurs during the sixth to the eighth

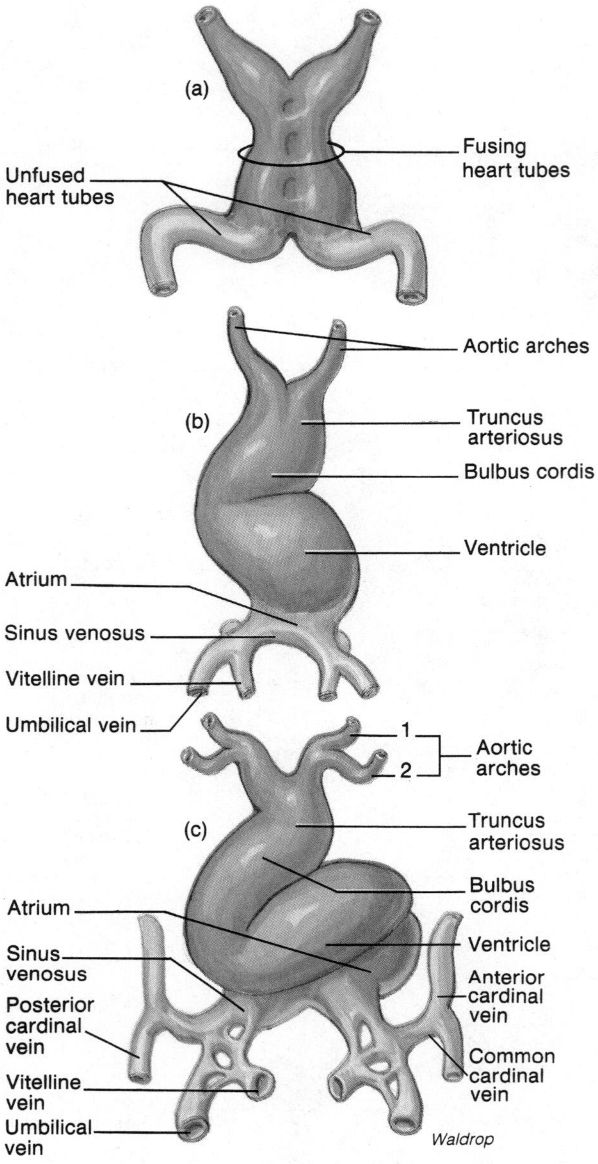

Figure 16.4 Formation of the heart chambers. (*a*) The heart tubes fuse during days 21 and 22. (*b*) The developmental chambers are formed during day 23. (*c*) Differential growth causes folding between the chambers during day 24, and vessels are developed to transport blood to and from the heart. The embryonic heart generally begins rhythmic contractions and pumping blood by day 25.

week of embryonic development. The first and second pairs of aortic arches disappear before the formation of the sixth. The third pair of aortic arches form the common carotid arteries and the internal carotid arteries. The right fourth aortic arch forms the base of the right subclavian artery, and the left fourth arch contributes to part of the arch of the aorta. The first, second, and fifth pairs of aortic arches have no derivatives and soon atrophy. The right sixth aortic arch forms the proximal portion of the right pulmonary artery. The left sixth arch forms the proximal portion of the left pulmonary artery, and the distal portion of this arch persists as an embryonic shunt between the pulmonary trunk and the aorta called the **ductus arteriosus.**

The derivatives of the aortic arches are summarized in table 16.2.

Table 16.1 Structural changes in the development of the heart

Primitive cardiac dilation	Developmental fate
Truncus arteriosus	Aorticopulmonary septum, which forms the partition between the ascending aorta and the pulmonary trunk
Bulbus cordis	Incorporated into the walls of the ventricles
Sinus venosus	Differentiates to form the coronary sinus and a portion of the wall of the right atrium
Ventricle	Development of the interventricular septum to form the right and left ventricles
Atrium	Development of the septum secundum to partially partition the right and left atria, and formation of the foramen ovale with the valve of the foramen ovale

Although the embryonic vessels that develop from the truncus arteriosus are called aortic arches, they should not be confused with the aortic arch that is the major systemic artery leaving the heart in the adult. Only the left fourth arch participates in the formation of the adult aortic arch.

An examination of the persisting aortic arches in the adults of different classes of vertebrates reveals interesting evolutionary relationships. Fish have six aortic arches persisting in the gill region. Reptiles have only one pair, a branch to the left and one to the right. All birds have a single right aortic arch, and all mammals have a single left aortic arch.

1. When does the heart begin development, and when does it begin to pump blood?
2. List the five regions of the embryonic heart tube, and indicate which regions of the heart they help form.
3. Describe the fate of the six embryonic aortic arches in humans.

Blood

Blood, a highly specialized connective tissue, consists of formed elements (erythrocytes, leukocytes, and thrombocytes) that are suspended and carried in the plasma. The constituents of blood function in transport, immunity, and blood-clotting mechanisms.

Objective 5. List the different types of formed elements of blood, and describe their appearance.

Objective 6. Describe the origin of erythrocytes, leukocytes, and thrombocytes.

Objective 7. List the different types of substances present in plasma.

Objective 8. Describe the origin and function of the different categories of plasma proteins.

The average-sized adult has about 5 liters of blood, which constitute about 7.7% of the total body weight. Blood leaving the heart is referred to as *arterial blood.* Arterial blood, with the ex-

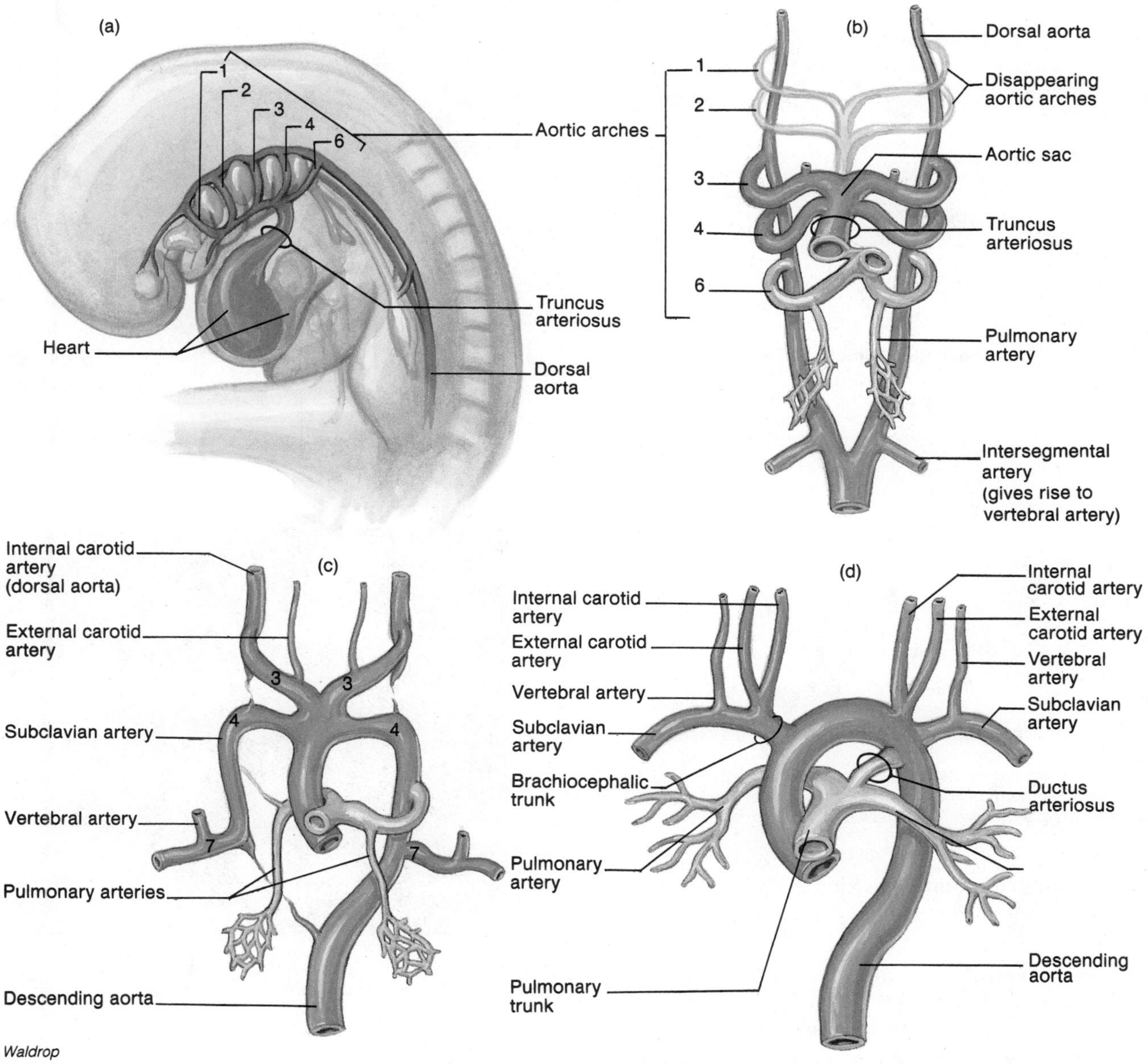

Figure 16.5 Formation of the aortic arch and major arteries of the thoracic region. (*a*) Schematic lateral view of the pharyngeal pouches and aortic arches within the neck region of a 28-day embryo. (*b*) Aortic arches at six weeks. (Note that by this time arches one, two, and five have largely disappeared.) (*c*) Aortic arches at seven weeks. (*d*) Arterial configuration at eight weeks.

ception of that going to the lungs, is bright red in color due to a high concentration of oxyhemoglobin (the combination of oxygen with hemoglobin) in the erythrocytes. *Venous blood* is blood returning to the heart and, except for the venous blood from the lungs, has a darker color (due to increased amounts of hemoglobin that are no longer combined with oxygen). Blood has a viscosity that ranges between 4.5 and 5.5. This means that it is thicker than water, which has a viscosity of 1. Blood has a pH range of 7.35 to 7.45 and a temperature within the thorax of the body of about 38° C (100.4° F).

Blood is composed of a cellular portion, called **formed elements,** and a fluid portion, called **plasma.** When a blood sample is centrifuged, the heavier formed elements are packed into the bottom of the tube, leaving plasma at the top (fig. 16.6).

plasma: Gk. *plasma,* to form or mold

Table 16.2 Derivatives of the aortic arches

Aortic arch	Derivative
First pair	No derivative
Second pair	No derivative
Third pair	Proximal portions form the common carotid arteries. Distal portions form the internal carotid arteries.
Fourth pair	Right arch forms the base of the right subclavian artery. Left arch forms a portion of the arch of the aorta.
Fifth pair	No derivative
Sixth pair	Right arch: proximal portion persists as proximal part of the right pulmonary artery; distal portion degenerates. Left arch: proximal portion forms the proximal part of the left pulmonary artery; distal part persists as the ductus arteriosus.

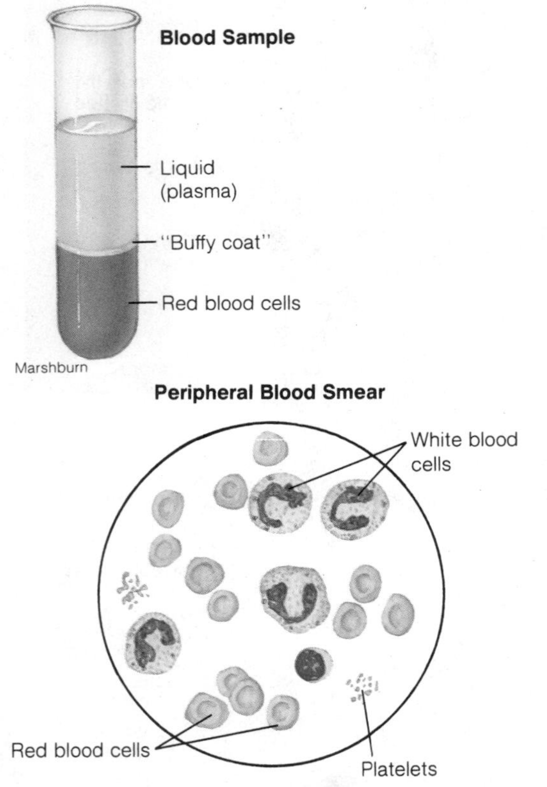

Figure 16.6 Blood cells become packed at the bottom of the test tube when whole blood is centrifuged, leaving the fluid plasma at the top of the tube. Red blood cells are the most abundant of the blood cells—white blood cells and platelets form only a thin, light-colored buffy coat at the interface between the packed red blood cells and the plasma.

The formed elements constitute approximately 45% of the total blood volume (the *hematocrit*), and the plasma accounts for the remaining 55%.

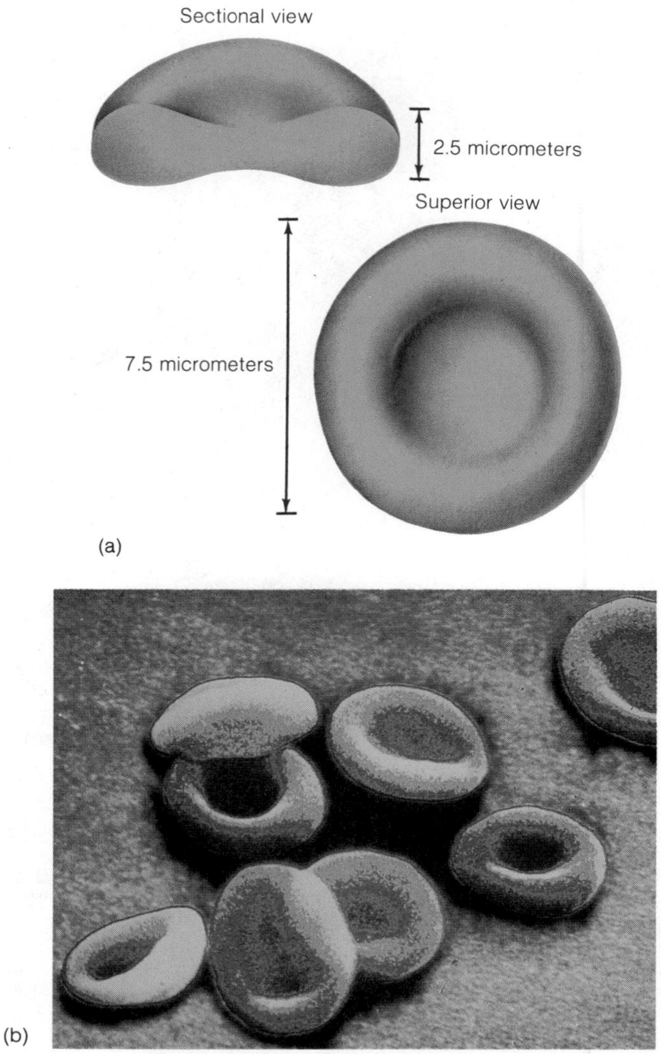

Figure 16.7 Erythrocytes. (a) A diagrammatic appearance and (b) scanning electron micrograph of erythrocytes.

Formed Elements of Blood

The formed elements of blood include **erythrocytes,** or red blood cells; **leukocytes,** or white blood cells; and **thrombocytes,** or platelets. Erythrocytes are by far the most numerous of these three types: a cubic millimeter of blood contains 5.1 million to 5.8 million erythrocytes in males and 4.3 million to 5.2 million erythrocytes in females. The same volume of blood, in contrast, contains only 5,000 to 9,000 leukocytes and 250,000 to 500,000 thrombocytes.

Erythrocytes Erythrocytes are flattened, biconcave discs, about 7.5 micrometers in diameter and 2.5 micrometers thick. Their unique shape relates to their function of transporting oxygen and provides an increased surface area through which gas can diffuse (fig. 16.7). Mature erythrocytes lack a nucleus and mitochondria (they get energy from anaerobic respiration). Because of these deficiencies, erythrocytes have a circulating life span of only about 120 days before they are destroyed by phagocytic cells in the liver, spleen, and bone marrow.

Each erythrocyte contains approximately 280 million *hemoglobin* molecules, which give blood its red color. Each hemoglobin molecule consists of a protein, called *globin,* and an iron-containing pigment, called *heme.* The iron group of heme is able to combine with oxygen in the lungs and release oxygen in the tissues.

Leukocytes Leukocytes differ from erythrocytes in several ways. Leukocytes contain nuclei and mitochondria and can move in an amoeboid fashion (erythrocytes are not able to move independently). Because of their amoeboid ability, leukocytes can squeeze through pores in capillary walls and get to a site of infection, whereas erythrocytes usually remain confined within blood vessels. The movement of leukocytes through capillary walls is called *diapedesis.*

Leukocytes, which are almost invisible under the microscope unless they are stained, are classified according to their stained appearance. Those leukocytes that have granules in their

hematocrit: Gk. *haima,* blood; *krino,* to separate
erythrocytes: Gk. *erythros,* red; *kytos,* hollow (cell)
leukocytes: Gk. *leukos,* white; *kytos,* hollow (cell)
hemoglobin: Gk. *haima,* blood; *globus,* globe
diapedesis: Gk. *dia,* through; *pedester,* on foot

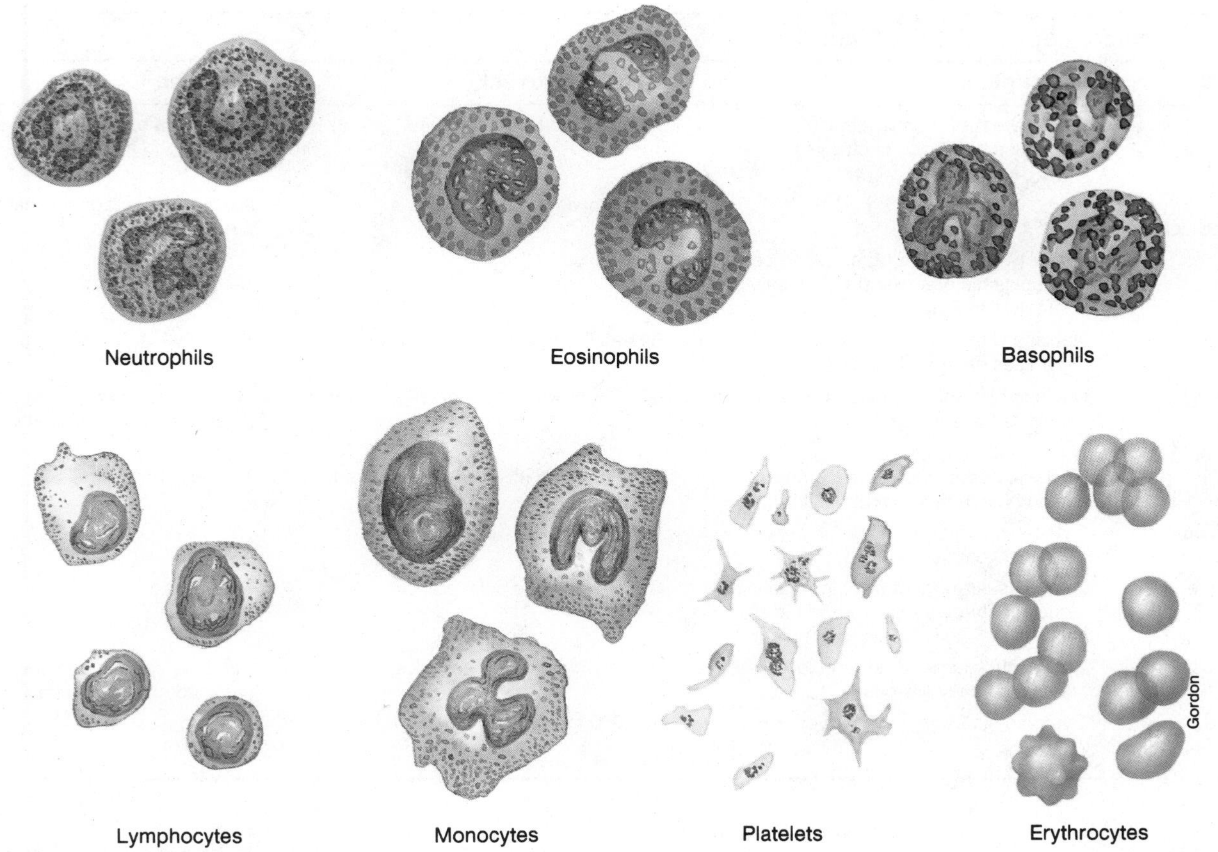

Figure 16.8 Types of formed elements in blood.

cytoplasm are called **granular leukocytes,** and those that do not are **agranular** (or nongranular) **leukocytes.** The granular leukocytes are also identified by their oddly shaped nuclei, which in some cases are contorted into lobes separated by thin strands. The granular leukocytes are therefore known as **polymorphonuclear** *(pol''e-mor''fo-nu'kle-ar)* **leukocytes (PMN).**

The stain used to identify leukocytes is usually a mixture of a pink-to-red stain called eosin and a blue-to-purple (hematoxylin) stain called a basic stain. Granular leukocytes with pink-staining granules are therefore called **eosinophils** *(e''o-sin'o-fils)* (fig. 16.8), and those with blue-staining granules are called **basophils.** Those with granules that have little affinity for either stain are **neutrophils.** Neutrophils are the most abundant type of leukocyte, comprising 54%–62% of the leukocytes in the blood.

The leukocytes of the agranular group include **monocytes** and **lymphocytes.** Monocytes are the largest cells found in the blood, and their large nuclei may vary considerably in shape. Lymphocytes have large nuclei surrounded by a relatively thin layer of cytoplasm.

Thrombocytes Thrombocytes are the smallest of the formed elements and are actually fragments of large cells called **megakaryocytes** *(meg''ah-kar'e-o-sīts),* found in bone marrow. (This is why the term *formed elements* is used rather than *blood cells* to describe erythrocytes, leukocytes, and thrombocytes.) The fragments that enter the circulation as platelets lack nuclei but, like leukocytes, are capable of amoeboid movement. The platelet count per cubic millimeter of blood is 130,000 to 360,000. Platelets survive about five to nine days and then are destroyed by the spleen and liver.

Platelets play an important role in blood clotting. They constitute the major portion of the mass of the clot, and phospholipids in their cell membranes serve to activate the clotting factors in plasma that result in threads of fibrin, which reinforce the platelet plug. Platelets that attach together in a blood clot also release a chemical called *serotonin,* which stimulates constriction of blood vessels and thus reduces the flow of blood to the injured area.

The appearance of the formed elements of the blood is shown in figure 16.8, and a summary of the characteristics of these formed elements is presented in table 16.3.

Blood cell counts are an important source of information in determining the health of a person. An abnormal increase in erythrocytes, for example, is termed *polycythemia (pol''e-si-the'me-ah)* and is indicative of several dysfunctions. An abnormally low red blood cell count is *anemia.* An elevated leukocyte count, called *leukocytosis,* is often associated with localized infection. A large number of immature leukocytes within a blood sample is diagnostic of the disease *leukemia.*

Table 16.3 Formed elements of the blood

Component	Description	Number present	Function
Erythrocyte (red blood cell)	Biconcave disc without nucleus; contains hemoglobin; survives 100–120 days	4,000,000 to 6,000,000/mm³	Transports oxygen and carbon dioxide
Leukocytes (white blood cells)		5,000 to 10,000/mm³	Aid in defense against infections by microorganisms
Granulocytes	About twice the size of red blood cells; cytoplasmic granules present; survive 12 hours to 3 days		
1. Neutrophil	Nucleus with 2–5 lobes; cytoplasmic granules stain slightly pink	54%–62% of white cells present	Phagocytic
2. Eosinophil	Nucleus bilobed; cytoplasmic granules stain red in eosin stain	1%–3% of white cells present	Helps to detoxify foreign substances; secretes enzymes that break down clots
3. Basophil	Nucleus lobed; cytoplasmic granules stain blue in hematoxylin stain	Less than 1% of white cells present	Releases anticoagulant heparin
Agranulocytes	Cytoplasmic granules absent; survive 100–300 days		
1. Monocyte	2–3 times larger than red blood cell; nuclear shape varies from round to lobed	3%–9% of white cells present	Phagocytic
2. Lymphocyte	Only slightly larger than red blood cell; nucleus nearly fills cell	25%–33% of white cells present	Provides specific immune response (including antibodies)
Thrombocyte (platelet)	Cytoplasmic fragment; survives 5–9 days	130,000 to 360,000/mm³	Clotting

Hemopoiesis

Blood cells are constantly formed through a process called *hemopoiesis* (fig. 16.9). The term **erythropoiesis** *(ĕ-rith''ro-poi-e'sis)* refers to the formation of erythrocytes, and **leukopoiesis** to the formation of leukocytes. These processes occur in two classes of tissues. **Myeloid tissue** is the red bone marrow of the humeri, femora, ribs, sternum, pelvis, and portions of the skull that produces erythrocytes, granular leukocytes, and platelets.

Lymphoid tissue—including the lymph nodes, tonsils, spleen, and thymus—produces the agranular leukocytes (monocytes and lymphocytes). During embryonic and fetal development the hemopoietic centers are located in the yolk sac, liver, and spleen. After birth, the liver and spleen serve as sites of blood cell destruction.

Erythropoiesis is an extremely active process. It is estimated that about 2.5 million erythrocytes are produced every second in order to replace the number that are continuously destroyed by the spleen and liver. During the destruction of erythrocytes, iron is salvaged and returned to the red bone marrow where it is used again in the formation of erythrocytes. The life span of an erythrocyte is approximately 120 days. Agranular leukocytes remain functional for 100–300 days under normal body conditions. Granular leukocytes, in contrast, have an extremely short life span of 12 hours to 3 days.

Hemopoiesis begins the same way in both myeloid and lymphoid tissues (fig. 16.9). Undifferentiated mesenchymal-like cells develop into stem cells called **hemocytoblasts.** These stem cells are able to divide rapidly; some of the daughter cells become new stem cells (so the stem cell population is never depleted), whereas other daughter cells become specialized along different paths of blood cell formation. Hemocytoblasts, for example, may develop into **proerythroblasts** *(pro''ĕ-rith'ro-blasts),* which form erythrocytes; **myeloblasts,** which form granular leukocytes (neutrophils, eosinophils, and basophils); **lymphoblasts,** which form lymphocytes; **monoblasts,** which form monocytes; or **megakaryoblasts** *(meg''ah-kar'e-o-blasts),* which form thrombocytes.

Plasma

Plasma is the fluid portion of the blood, or blood with the formed elements removed. It constitutes approximately 55% of a given volume of blood. Plasma is a straw-colored liquid that is about 90% water. The remainder of the plasma consists of proteins, inorganic salts, carbohydrates, lipids, amino acids, vitamins, and hormones. The functions of plasma include transporting nutrients, gases, and vitamins, regulating electrolyte and fluid balances, and maintaining a consistent blood pH.

Plasma proteins constitute 7%–9% of the plasma. These proteins remain within the blood and interstitial fluid and assist in maintaining body homeostasis. The three types of plasma proteins are albumins, globulins, and fibrinogen. **Albumins** *(albu'mins)* account for about 60% of the plasma proteins and are the smallest of the three types. They are produced by the liver and provide the blood with the necessary viscosity that is directly related to maintaining and regulating blood pressure. **Globulins** *(glob'u-lins)* make up about 36% of the plasma proteins. The three types of globulins are **alpha globulins, beta**

hemopoiesis: Gk. *haima,* blood; *poiesis,* production
erythropoiesis: Gk. *erythros,* red; *poiesis,* production
leukopoiesis: Gk. *leukos,* white; *poiesis,* production

megakaryoblasts: Gk. *megas,* great; *karyon,* nut; *blastos,* germ
albumin: L. *albumen,* white
globulin: L. *globulus,* small globe

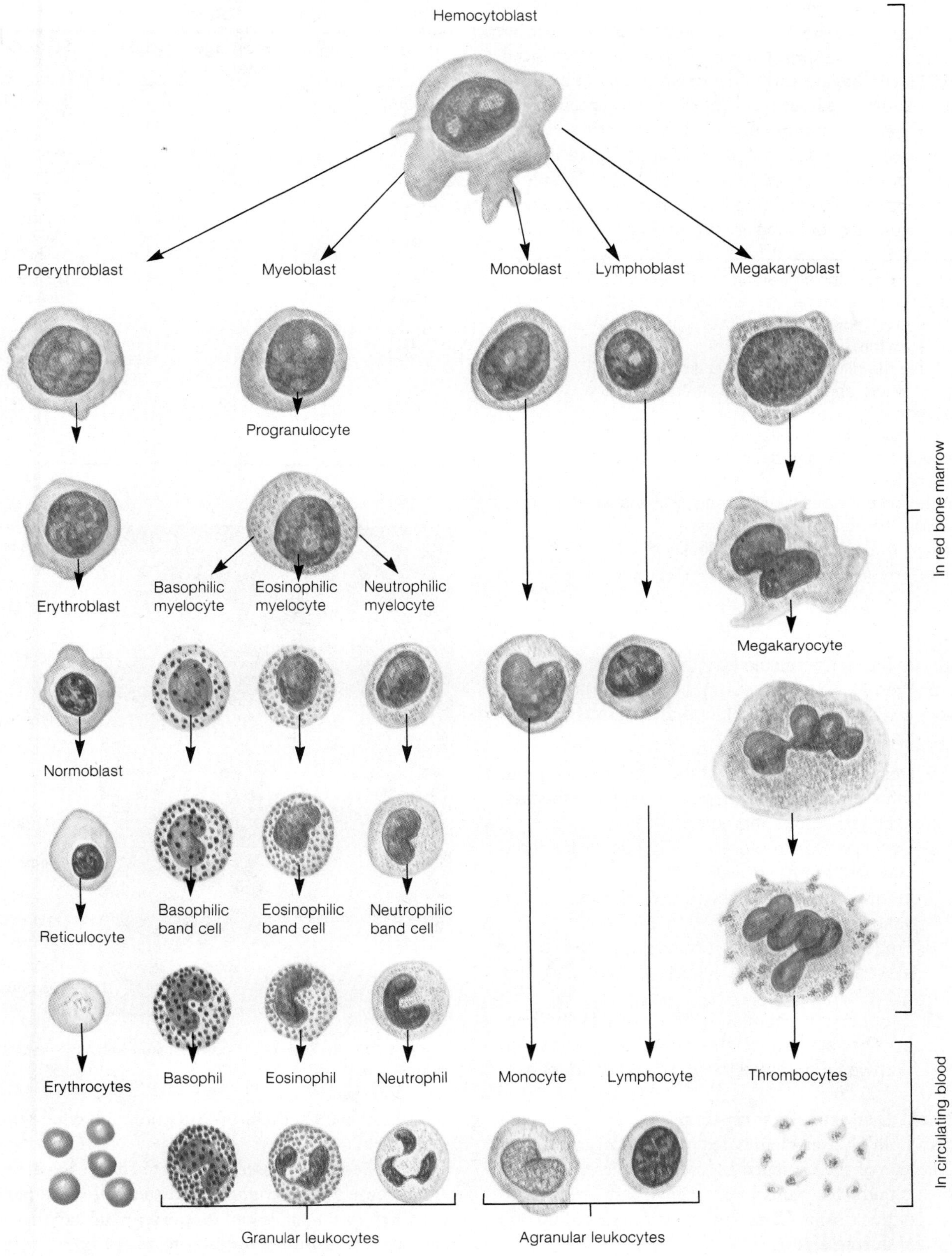

Figure 16.9 The processes of hemopoiesis. Formed elements begin as hemocytoblasts and differentiate into the various kinds of blood cells depending on the needs of the body.

globulins, and **gamma globulins.** The alpha and beta globulins are synthesized in the liver and function in transporting lipids and fat-soluble vitamins. Gamma globulins are produced by lymphoid tissues and are antibodies of immunity. The third type of plasma protein, **fibrinogen** *(fi-brin'o-jen),* accounts for only about 4% of the protein content. Fibrinogens are large molecules that are synthesized in the liver and play an important role with thrombocytes in clotting the blood. When fibrinogen is removed, the remaining fluid is referred to as *serum.*

The plasma proteins of blood are summarized in table 16.4.

1. Describe erythropoiesis in terms of its location, rate, and the specific steps that lead to the production of mature erythrocytes. What is the appearance of a mature erythrocyte?
2. Describe the different types of leukocytes, and explain where and how these different types are produced.
3. Explain how platelets differ from other formed elements of blood, and describe how they are produced.
4. List the different classes of plasma proteins, and describe their origin and functions.
5. Explain how leukocytes get to the site of an infection.

Table 16.4 Plasma proteins

Protein	Percentage of total	Origin	Function
Albumin	60	Liver	Gives viscosity to blood and assists in maintaining blood osmotic pressure
Globulin	36		
Alpha globulins		Liver	Transport lipids and fat-soluble vitamins
Beta globulins		Liver	Transport lipids and fat-soluble vitamins
Gamma globulins		Lymphoid tissue	Constitute antibodies of immunity
Fibrinogen	4	Liver	Assist thrombocytes in the formation of clots

Heart

The structure of the heart muscle and the action of its valves allow it to function as a double pump. In this way, the heart pumps blood low in oxygen to the lungs and pumps oxygen-rich blood to the rest of the body.

Objective 9. Describe the location of the heart in relation to other organs of the thoracic cavity and their associated serous membranes.
Objective 10. Describe the structure and functions of the three layers of the heart wall.
Objective 11. Identify the chambers, valves, and associated vessels of the heart.
Objective 12. Trace the flow of blood through the heart, and distinguish between the pulmonary and systemic circulations.
Objective 13. Describe the location of the parts of the conduction system of the heart, and trace the pathway of impulse initiation and conduction.

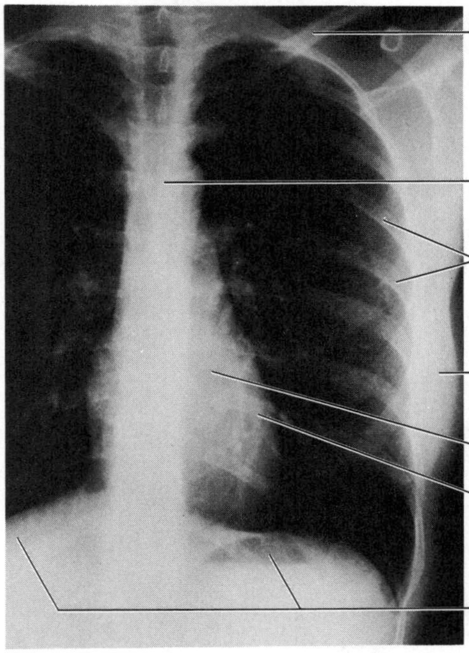

Figure 16.10 An X ray of the heart situated within the mediastinum between the lungs.

Location and General Description

The **heart** is a hollow, four-chambered muscular organ, which is roughly the size of the clenched fist and averages 255 grams in adult females and 310 grams in adult males. It is estimated that the heart contracts some 42 million times and ejects 700,000 gallons of blood during a year.

The heart is located within the thoracic cavity between the lungs in the mediastinum (fig. 16.10). About two-thirds of the heart is located left of the midline, with its *apex,* or cone-shaped end, pointing downward in contact with the diaphragm. The *base* of the heart is the broad superior end, where the large vessels attach.

The heart is enclosed and protected by a loose-fitting, serous sac of dense fibrous connective tissue, called the **parietal pericardium** *(per"ĭ-kar'de-um),* or **pericardial sac** (fig. 16.11). The parietal pericardium separates the heart from the other thoracic organs and forms the wall of the *pericardial cavity,* which contains a watery, lubricating *pericardial fluid.* The parietal pericardium is actually composed of an outer *fibrous layer* and an inner *serous layer.* The serous layer secretes the pericardial fluid.

fibrinogen: L. *fibra,* fibrous

pericardium: Gk. *peri,* around; *kardia,* heart

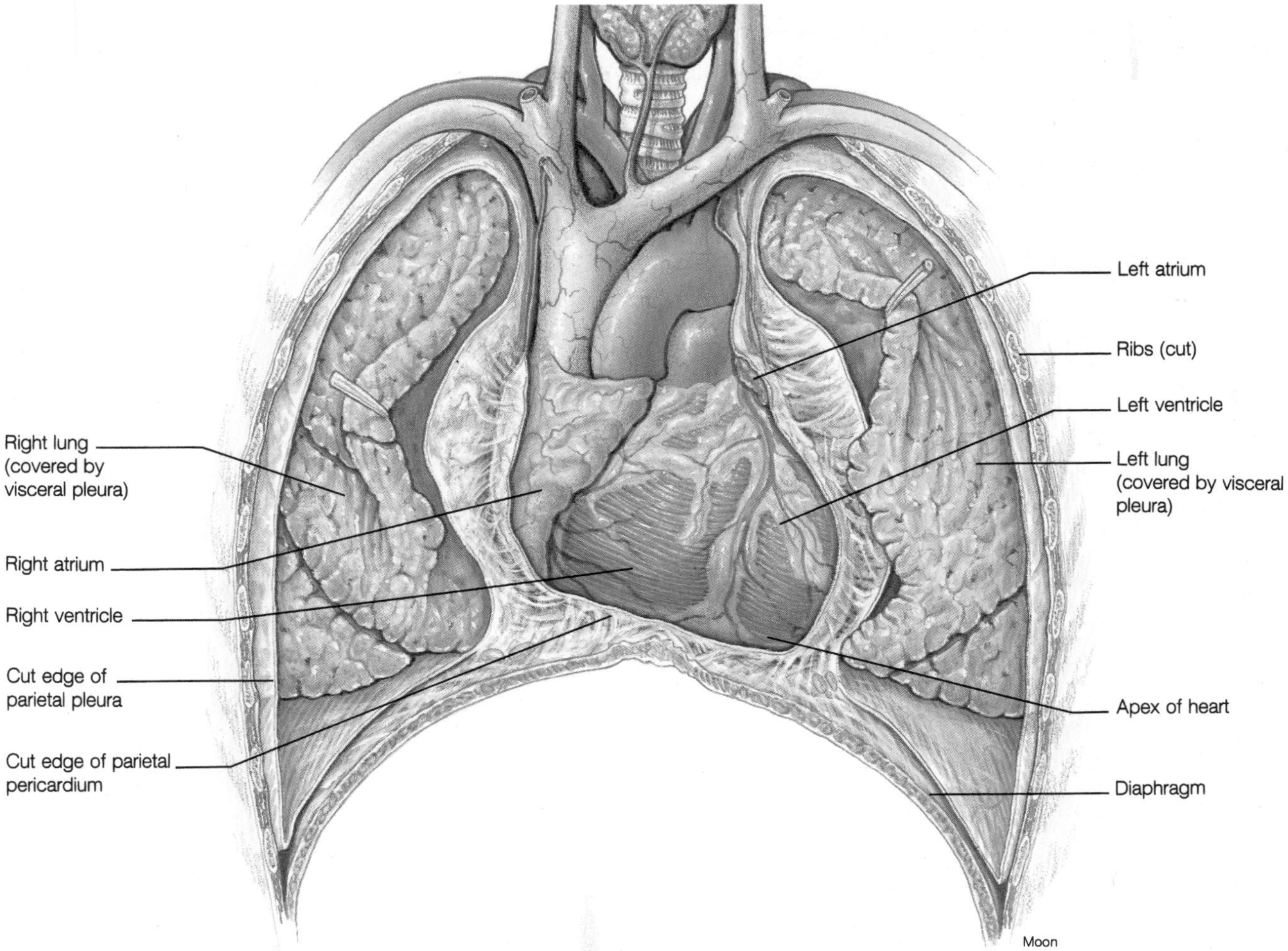

Figure 16.11 The position of the heart and associated serous membranes within the thoracic cavity.

Pericarditis is an inflammation of the parietal pericardium, which results in an increased secretion of fluid into the pericardial cavity. Because the tough, fibrous portion of the parietal pericardium is inelastic, an increase in fluid pressure impairs the movement of blood into and out of the chambers of the heart. Some of the pericardial fluid may be withdrawn for analysis by injecting a needle to the left of the xiphoid process to pierce the parietal pericardium.

Heart Wall

The wall of the heart is composed of three distinct layers (fig. 16.12). The outer layer is the **epicardium,** also called the *visceral pericardium.* The space between this layer and the parietal pericardium is the pericardial cavity, just described. The thick middle layer of the heart is called the **myocardium.** It is composed of cardiac muscle tissue (see chap. 4) and arranged in such a way that the contraction of the muscle bundles results in squeezing or wringing of the heart chambers. The thickness of the myocardium varies depending on the force needed to eject blood from the particular chambers. The thickest portion of the myocardium, therefore, surrounds the left ventricle. The atrial walls are relatively thin. The inner layer, called the **endocardium,** is continuous with the endothelium of blood vessels. The endocardium also forms part of the valves of the heart. Inflammation of the endocardium is called *endocarditis.*

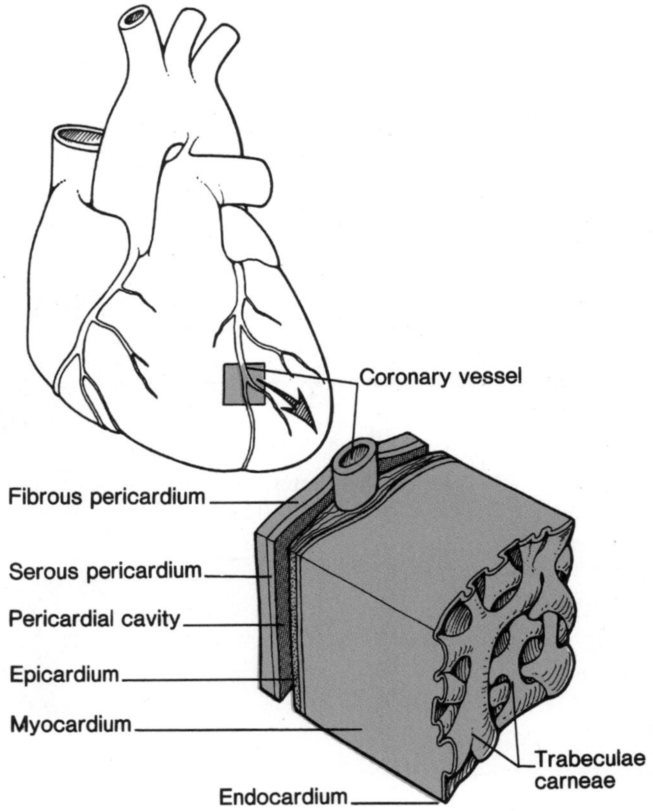

Figure 16.12 The structure of the heart wall and the pericardium. A section of the heart has been cut out and rotated to show the layers of the heart and the pericardium.

Table 16.5 Layers of the heart wall

Layer	Characteristics	Function
Epicardium (visceral pericardium)	Serous membrane of connective tissue covered with epithelium and including blood capillaries, lymph capillaries, and nerve fibers	Lubricative outer covering
Myocardium	Cardiac muscle tissue separated by connective tissues and including blood capillaries, lymph capillaries, and nerve fibers	Muscular contractions that eject blood from the heart chambers
Endocardium	Serous membrane of epithelium and connective tissues, including elastic and collagenous fibers, blood vessels, and specialized muscle fibers	Lubricative inner lining of the chambers and valves

The characteristics of the three layers of the heart wall are summarized in table 16.5.

Chambers and Valves

The heart is a four-chambered, double pump, consisting of upper right and left **atria** and lower right and left **ventricles;** the atria contract and empty simultaneously into the ventricles (fig. 16.13), which also contract in unison. Each atrium has an ear-shaped appendage called an **auricle.** The atria are separated by the thin muscular *interatrial septum,* and the ventricles are separated by the thick muscular *interventricular septum.* **Atrioventricular valves** *(bicuspid* and *tricuspid)* are located between the atria and ventricles, and **semilunar valves** are present at the bases of the two large vessels leaving the heart. These valves maintain the flow of blood in one direction.

atrium: L. *atrium,* chamber
ventricle: L. *ventriculus,* diminutive of *venter,* belly
auricle: L. *auricula,* a little ear

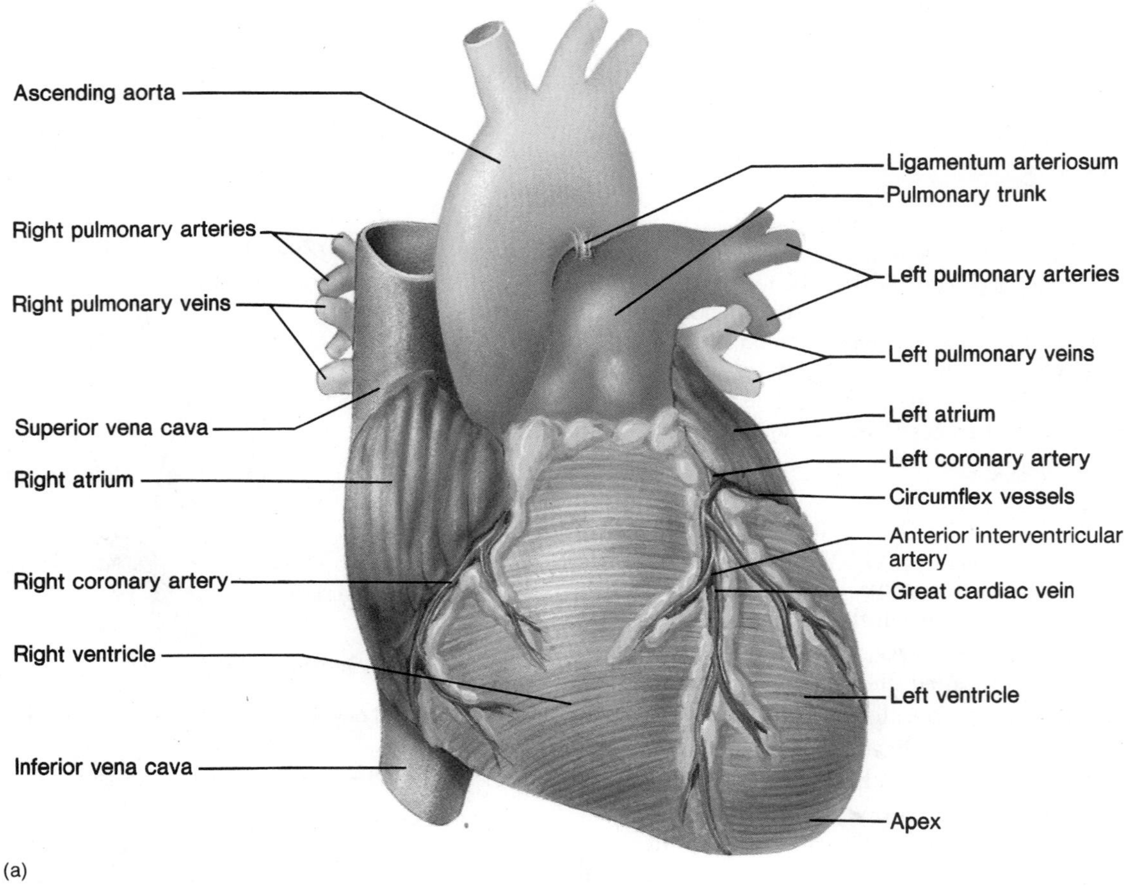

Figure 16.13 The structure of the heart. (*a*) An anterior view, (*b*) a posterior view, and (*c*) an internal view.

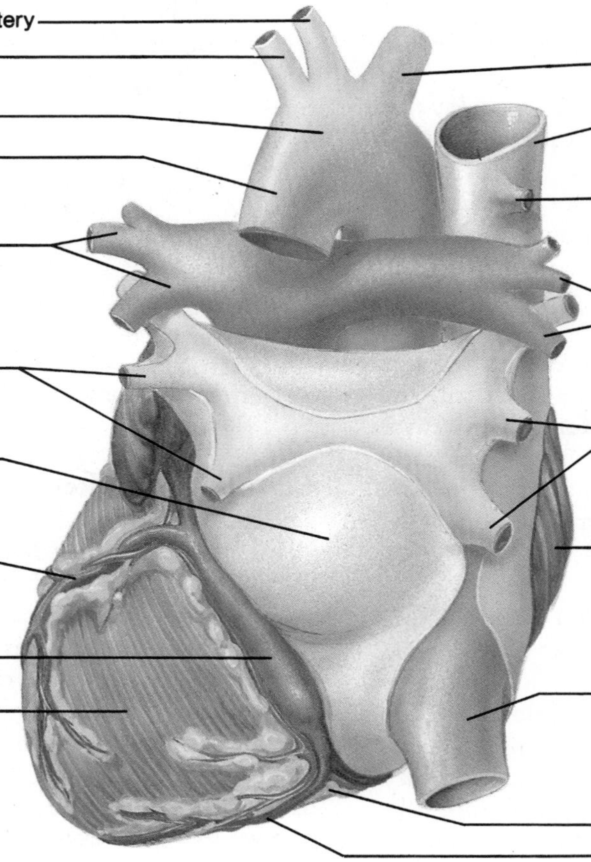
Left common carotid artery
Left subclavian artery
Arch of the aorta
Descending aorta
Left pulmonary arteries
Left pulmonary veins
Left atrium
Posterior cardiac vein
Coronary sinus
Left ventricle
Brachiocephalic artery (innominate)
Superior vena cava
Azygos vein
Right pulmonary arteries
Right pulmonary veins
Right atrium
Inferior vena cava
Right ventricle
Middle cardiac vein

(b)

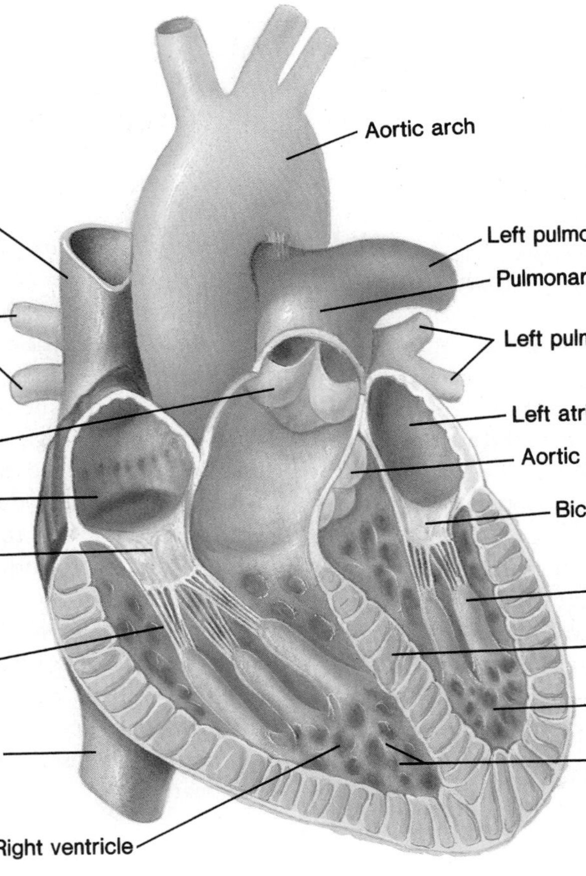
Aortic arch
Superior vena cava
Right pulmonary veins
Pulmonary semilunar valve
Right atrium
Tricuspid valve
Chordae tendineae
Inferior vena cava
Right ventricle
Left pulmonary artery
Pulmonary trunk
Left pulmonary veins
Left atrium
Aortic semilunar valve
Bicuspid valve
Papillary muscle
Interventricular septum
Left ventricle
Trabeculae carneae

(c)

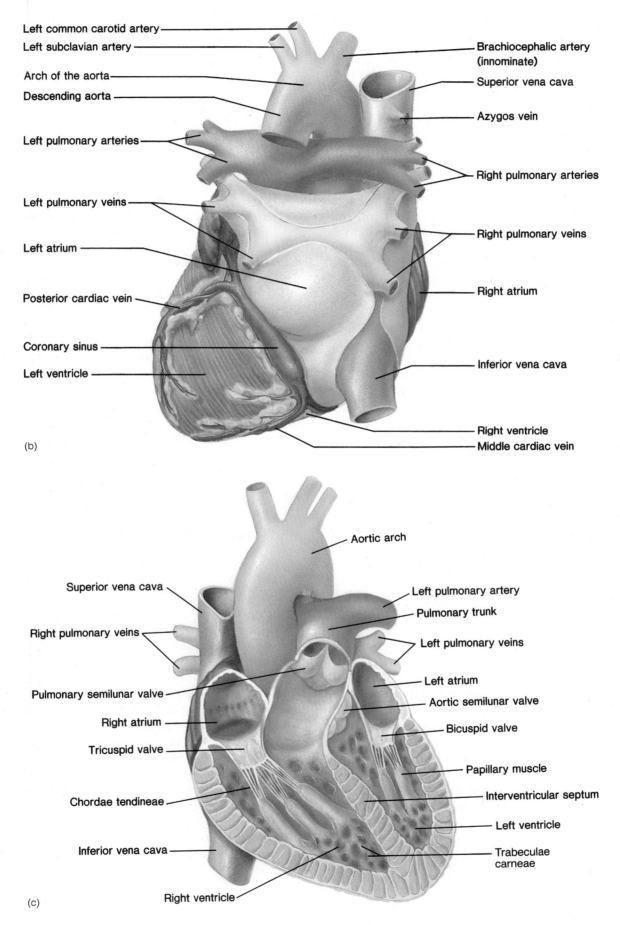

Figure 16.13—*Continued*

Grooved depressions on the surface of the heart indicate the partitions between the chambers and also contain **cardiac vessels** that supply blood to the muscular wall of the heart. The most prominent groove is the *atrioventricular* (*coronary*) *sulcus* that encircles the heart and marks the division between the atria and ventricles. The partition between right and left ventricles is denoted by two (anterior and posterior) *interventricular sulci.*

The following discussion follows the sequence in which blood flows through the structures of the atria, ventricles, and valves.

Right Atrium The right atrium *(a'tre-um)* receives systemic venous blood from the **superior vena cava,** which drains the upper portion of the body, and from the **inferior vena cava,** which drains the lower portion (fig. 16.13). There is an additional venous opening into the right atrium from the *coronary sinus,* which returns venous blood from the myocardium of the heart itself.

Right Ventricle Blood from the right atrium passes through the **tricuspid valve** to fill the right ventricle. The tricuspid valve is an atrioventricular (AV) valve that gets its name from its three valve leaflets, or cusps. Each cusp is held in position by strong tendinous cords called **chordae tendineae,** which are secured to the ventricular wall by cone-shaped **papillary muscles.** These structures prevent the valves from everting, like an umbrella in a strong wind, when the ventricles contract.

Ventricular contraction causes the tricuspid valve to close and the blood to leave the right ventricle through the **pulmonary trunk** and to enter the capillaries of the lungs via the right and left **pulmonary arteries.** The **pulmonary semilunar** (half-moon-shaped) **valve** lies at the base of the pulmonary trunk, where it prevents the backflow of ejected blood into the right ventricle.

Left Atrium After gas exchange has occurred within the capillaries of the lungs, oxygenated blood is transported to the left atrium through four **pulmonary veins,** two from each lung.

Left Ventricle The left ventricle receives blood from the left atrium. These two chambers are separated by the **bicuspid,** or **mitral** *(mi'tral),* **valve.** When the left ventricle is relaxed, the valve is open and allows blood to flow from the atrium to the ventricle; when the left ventricle contracts, the valve closes. Closing of the valve during ventricular contraction prevents the backflow of blood into the atrium.

The walls of the left ventricle are thicker than those of the right ventricle. The endocardium of both chambers has distinct ridges called **trabeculae** *(trah-bek'u-le)* **carneae** (fig. 16.13c). Oxygenated blood leaves the left ventricle through the **ascending aorta** *(a-or'tah).* The **aortic semilunar valve,** located at the base of the ascending aorta, closes as a result of the pressure of the blood when the left ventricle relaxes and thus prevents the backflow of blood into the relaxed ventricle.

The valves are shown in figure 16.14, and their actions are summarized in table 16.6.

Cardiac catheterization can aid in the diagnosis of certain heart disorders. In this procedure, a tube, or catheter, is inserted into a blood vessel in the leg or the arm and threaded through the lumen of the vessel until it enters the heart. The location of the catheter can be monitored by means of a fluoroscope. When the catheter reaches the desired location within the heart, a radiopaque dye is released and x-rayed, blood samples are removed for analyses, or blood pressures are determined.

Circulatory Routes

The circulatory routes of the blood are illustrated in figure 16.2. There are two principal divisions of the circulatory blood flow: the pulmonary and systemic circulations.

The **pulmonary circulation** consists of blood vessels that transport blood to the lungs for gas exchange and then back to the heart. It consists of the right ventricle that ejects the blood, the pulmonary trunk with its semilunar valve, the pulmonary arteries that transport blood to the lungs, the pulmonary capillaries within each lung, four pulmonary veins that transport oxygenated blood back to the heart, and the left atrium that receives the blood from the pulmonary veins.

The **systemic circulation** is composed of all of the remaining vessels of the body that are not part of the pulmonary circulation. This includes the right atrium, the left ventricle, the aorta with its semilunar valve, all of the branches of the aorta, all capillaries other than those in the lungs, and all veins other than the pulmonary veins. The right atrium receives all of the venous return of oxygen-depleted blood from the systemic veins.

Coronary Circulation The myocardium is supplied with blood by the right and left **coronary arteries** (fig. 16.15). These two vessels arise from the aorta immediately beyond the aortic semilunar valve. The coronaries encircle the heart within the atrioventricular sulcus, the depression between the atria and ventricles. Two branches arise from both the right and left coronaries to serve the atrial and ventricular walls. The left coronary artery gives rise to the **anterior interventricular artery,** which courses within the anterior interventricular sulcus to serve both ventricles, and the **circumflex artery,** which supplies oxygenated blood to the walls of the left atrium and left ventricle. The right coronary artery gives rise to the **marginal artery,** which serves the walls of the right atrium and right ventricle and the **posterior interventricular artery,** which courses through the posterior interventricular sulcus to serve the two ventricles. The main trunks of the right and left coronaries anastomose (join together) on the posterior surface of the heart.

---

vena cava: L. *vena,* vein; *cava,* empty
chordae tendineae: L. *chorda,* string; *tendere,* to stretch
bicuspid: L. *bi,* two; *cuspis,* tooth point or spike
mitral: L. *mitra,* like a bishop's mitre
trabeculae carneae: L. *trabecula,* small beams; *carneus,* flesh

circumflex: L. *circum,* around; *flectere,* to bend

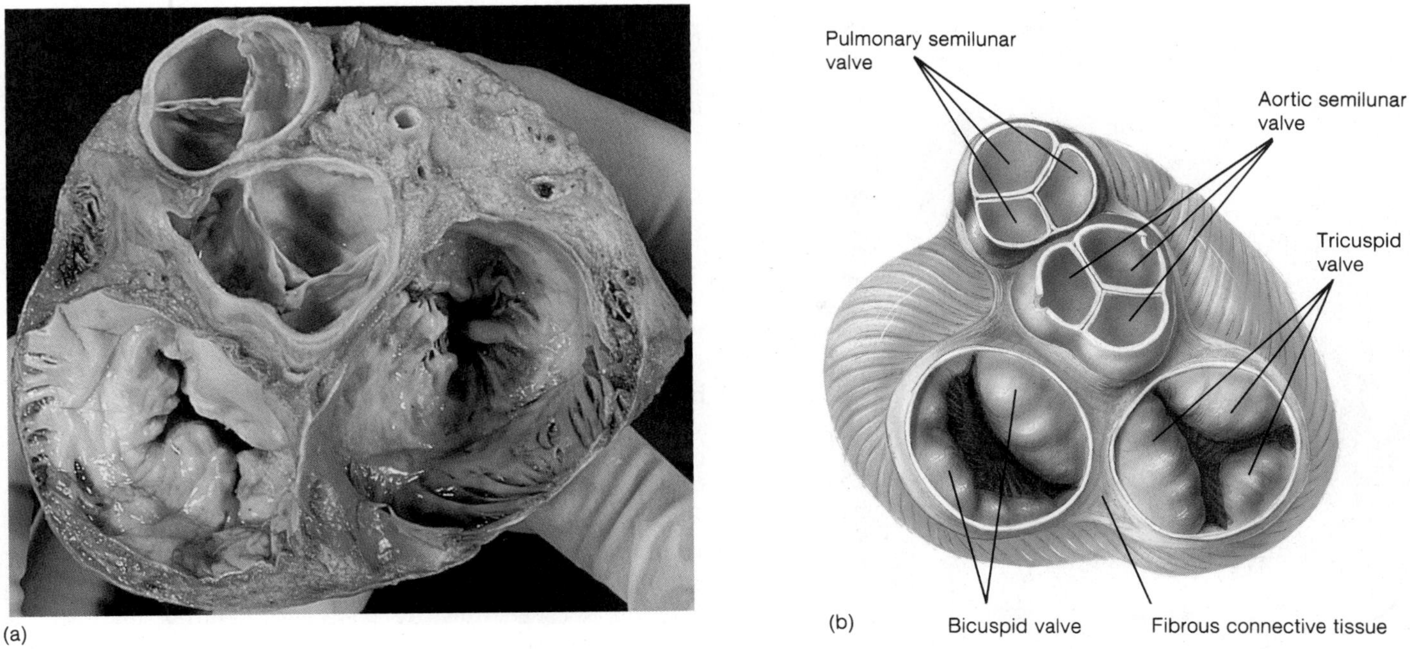

Figure 16.14 A superior view of the heart valves. (*a*) A photograph of the semilunar valves. (*b*) A diagram of the relative position of the valves and the supporting fibrous connective tissue.

Table 16.6 Valves of the heart

Valve	Location	Comments
Tricuspid valve	Between right atrium and right ventricle	Composed of three cusps that prevent backflow of blood from right ventricle into right atrium during ventricular contraction
Pulmonary semilunar valve	Entrance to pulmonary trunk	Composed of three half-moon-shaped flaps that prevent backflow of blood from pulmonary trunk into right ventricle during ventricular relaxation
Bicuspid (mitral) valve	Between left atrium and left ventricle	Composed of two cusps that prevent backflow of blood from left ventricle into left atrium during ventricular contraction
Aortic semilunar valve	Entrance to ascending aorta	Composed of three half-moon-shaped flaps that prevent backflow of blood from aorta into left ventricle during ventricular relaxation

From the capillaries in the myocardium, the blood enters the **cardiac veins.** The course of these vessels parallels that of the coronary arteries. The cardiac veins, however, have thinner walls and are more superficial than the arteries. The two principal cardiac veins are the **great cardiac vein,** which returns blood from the anterior aspect of the heart, and the **middle cardiac vein,** which drains the posterior aspect of the heart. These cardiac veins converge into a large venous channel on the posterior surface of the heart called the **coronary sinus** (figs. 16.13, 16.15). The coronary venous blood then enters the heart through an opening into the right atrium.

Heart attacks are the most common cause of death in the United States. The most common type of heart attack involves an occlusion of a coronary artery, which reduces the delivery of oxygen to the myocardium. There are several things that may be done to minimize the risk of heart attack: (1) reduce excess weight; (2) avoid hypertension through proper diet, stress reduction, or medication if necessary; (3) maintain a low cholesterol level through proper diet; (4) avoid smoking; and (5) exercise regularly.

Conduction System of the Heart

Cardiac muscle possesses an intrinsic rhythmicity that allows the heartbeat to originate in and be conducted through the heart without extrinsic stimulation. Specialized strands of cardiac muscle tissue that coordinate cardiac contraction constitute the **conduction system.** The conduction system makes possible the **cardiac cycle,** which is defined as the events surrounding the filling and emptying of the chambers of the heart. The conduction system consists of specialized tissues that generate and distribute electrical impulses through the heart. The components of the conduction system are the **sinoatrial node (SA node), atrioventricular node (AV node), atrioventricular bundle (bundle of His),** and **Purkinje** *(pur-kin'je)* **fibers.**

bundle of His: from Wilhelm His, Jr., Swiss physician, 1863–1934

Purkinje fibers: from Johannes E. von Purkinje, Bohemian anatomist, 1787–1869

None of these tissues can be seen grossly, but their locations can be noted (fig. 16.16). The SA node, or pacemaker, is located in the posterior wall of the right atrium. The SA node initiates the cardiac cycle by producing an electrical impulse that spreads over both atria, causing them to contract simultaneously and force blood into the ventricles. The basic depolarization rate of the SA node is 70–80 times per minute. The impulse then passes to the AV node located in the inferior portion of the interatrial septum. From here, the impulse continues through the atrioventricular bundle, located at the top of the interventricular septum, where it divides into right and left branches. These branches are continuous with the Purkinje fibers within the ventricular walls. Stimulation of these fibers causes the ventricles to contract simultaneously.

Contraction of the ventricles is referred to as **systole** *(sis'to-le)*. Systole, together with the tension of the elastic fibers and contraction of smooth muscles within the systemic arteries, accounts for the systolic pressure within arteries. Ventricular relaxation is called **diastole** *(di-as'to-le)*. During diastole, the diastolic pressure within arteries can be recorded.

Variation in functioning within vital organs should be regarded as normal for a healthy adult. The heartbeat can vary by as much as 20 or 30 beats per minute in 24 hours, but the heart maintains an average of about 70 beats per minute during a day. Blood pressure recorded at 120 over 80 in the morning can rise to 140 over 100 by evening. Normal body temperature does not stay at 98.6° F, but varies by one and a half to two degrees during the day—from 97 to 99 degrees.

Although the heart does have an innate contraction pattern, it is also innervated from the autonomic nervous system in order to respond to the ever-changing physiological needs of the body. The SA and AV nodes have both sympathetic and parasympathetic innervations. Sympathetic stimulation accelerates the heart rate and dilates the coronary arteries enabling the heart to meet its own increased metabolic demands as well as those of the rest of the body. Parasympathetic stimulation has the opposite effect. Sympathetic innervation is through fibers from the cervical and upper thoracic ganglia. Parasympathetic innervation is through branches of the vagus nerves. Branches from the right vagus innervate the SA node, and branches from the left vagus innervate the AV node.

Cardiac output is the volume of blood ejected from the heart during one minute. It is determined by the stroke rate, or heart rate, times the stroke volume of blood. The stroke volume is the amount of blood pumped from the heart with each ventricular contraction, which amounts to about 70 ml of blood. A normal resting cardiac output ranges between 4.2 and 5.6 liters per minute. Exercise increases the heart rate, as do the hormones epinephrine and thyroxine. Atropine, caffeine, and camphor are drugs that have a stimulatory effect. Blood pressure and body temperature also have a profound effect on the heart rate.

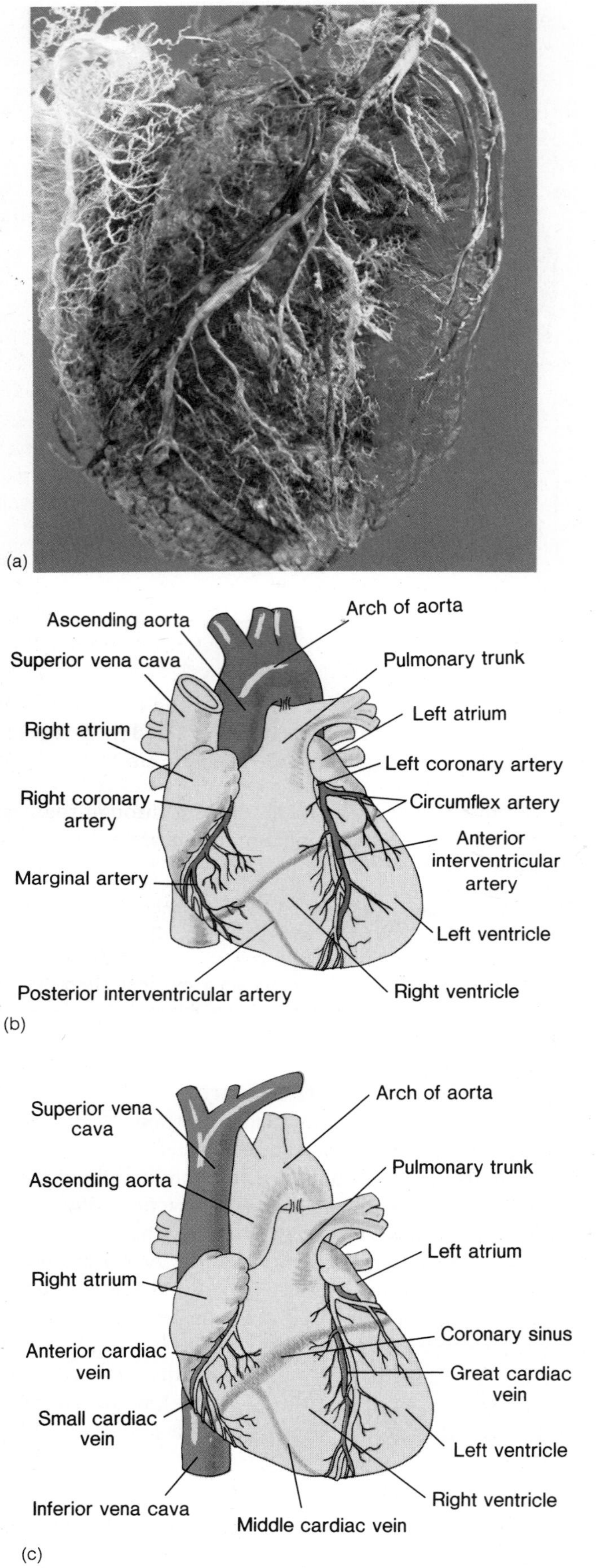

Figure 16.15 Coronary circulation. (*a*) A plastic cast of the coronary vessels and their major branches. (*b*) An anterior view of the arterial supply to the heart. (*c*) An anterior view of the venous drainage.

systole: Gk. *systole,* contraction
diastole: Gk. *diastole,* prolonged or expansion

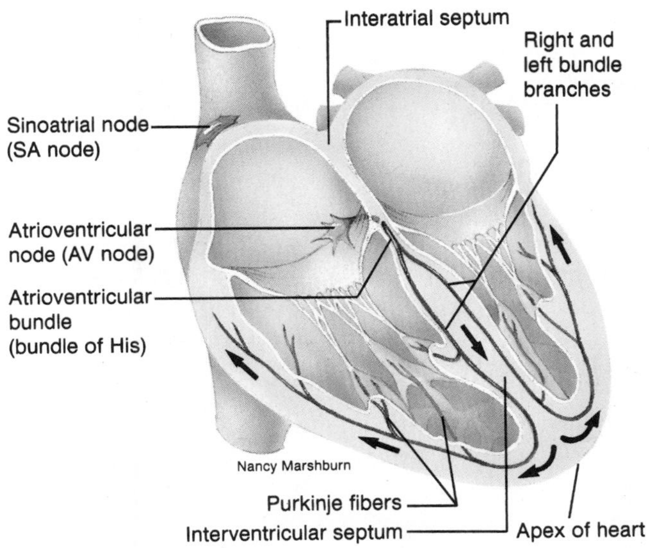

Figure 16.16 The conduction system of the heart.

Electrocardiogram

The electrical impulses that pass through the conduction system of the heart during the cardiac cycle can be recorded as an *electrocardiogram* (ECG or EKG). The electrical changes result from depolarization and repolarization of cardiac muscle fibers and can be detected on the surface of the skin using an instrument called the *electrocardiograph.*

An ECG of a normal cardiac cycle is illustrated in figure 16.17. The several waves, or deflections, are each produced as specific events of the cardiac cycle occur. The wave deflections have been designated as P, Q, R, S, and T. Any pathologic condition that disturbs the electrical activity of the heart will produce characteristic changes in one or more of the waves, so understanding the normal wave deflection patterns is clinically important.

P Wave The P wave is a small upward deflection that accompanies depolarization of the atrial fibers through stimulation of the SA node. The actual contraction of the atria follows the P wave by a fraction of a second. The ventricles of the heart are in diastole during the expression of the P wave. A missing or otherwise abnormal P wave may indicate a dysfunction of the SA node.

P-R Interval The period of time from the start of the P wave to the beginning of the QRS complex is known as the P-R interval. This period indicates the time required for the SA depolarization to reach the ventricles. A prolonged P-R interval suggests a conduction delay in the AV node.

QRS Complex The QRS wave begins as a short downward deflection, followed by a sharp upward spike, and ends as a downward deflection at the base. The QRS complex represents the depolarization of the ventricles. During this interval, the ventricles of the heart are in systole and blood is being ejected from the heart. An abnormal QRS complex generally indicates cardiac problems involving the ventricles. An enlarged R spike, for example, generally indicates enlarged ventricles.

S-T Segment The time duration known as the S-T segment represents the period between the completion of ventricular depolarization and initiation of repolarization. The S-T segment is depressed when the heart receives insufficient oxygen and is elevated in acute myocardial infarction.

T Wave The T wave is produced by ventricular repolarization. Various heart diseases, such as an arteriosclerotic heart, will produce altered T waves.

Heart Sounds

Closing of the AV and semilunar valves produces sounds that can be heard at the surface of the chest with a stethoscope. These sounds are often described phonetically as *lub-dub.* The "lub," or **first sound,** is produced by closing of the AV valves. The "dub," or **second sound,** is produced by closing of the semilunar valves. The first sound is thus heard when the ventricles contract at systole, and the second sound is heard when the ventricles relax at the beginning of diastole.

Heart sounds are of clinical importance because they provide information about the condition of the heart valves and other heart problems. Abnormal sounds are referred to as *heart murmurs* and are caused by valvular leakage or turbulence of the blood as it passes through the heart. In general, three basic conditions cause murmurs: (1) *valvular insufficiency,* in which the cusps of the valves do not form a tight seal; (2) *stenosis,* in which the walls surrounding a valve are roughened or constricted; and (3) a *functional murmur,* which is frequent in children and is caused by turbulence of the blood moving through the heart during heavy exercise. Functional murmurs are not pathological and are considered normal.

The valves of the heart are positioned directly deep to the sternum, which tends to obscure and dissipate valvular sounds. For this reason, a physician will listen with a stethoscope for the heart sounds at locations designated as **valvular auscultatory areas,** which are named according to the valve that can be detected (fig. 16.18). The **aortic area** is at the right second intercostal space near the sternum. The **pulmonic area** is directly across from the aortic area at the left second intercostal space near the sternum. The **tricuspid** and **bicuspid (mitral) areas** are both at the fifth intercostal space, with the bicuspid area further out. Surface landmarks are extremely important in identifying auscultatory areas.

1. Distinguish between the pericardial and the thoracic cavities. Describe the cardiac serous membranes.
2. List the three layers of the heart wall, and describe the structures associated with each layer.
3. List the valves that control blood flow through the heart, and describe their locations and functions.
4. Trace the flow of electrical impulses through the conduction system of the heart.

stenosis: Gk. *stenosis,* a narrowing

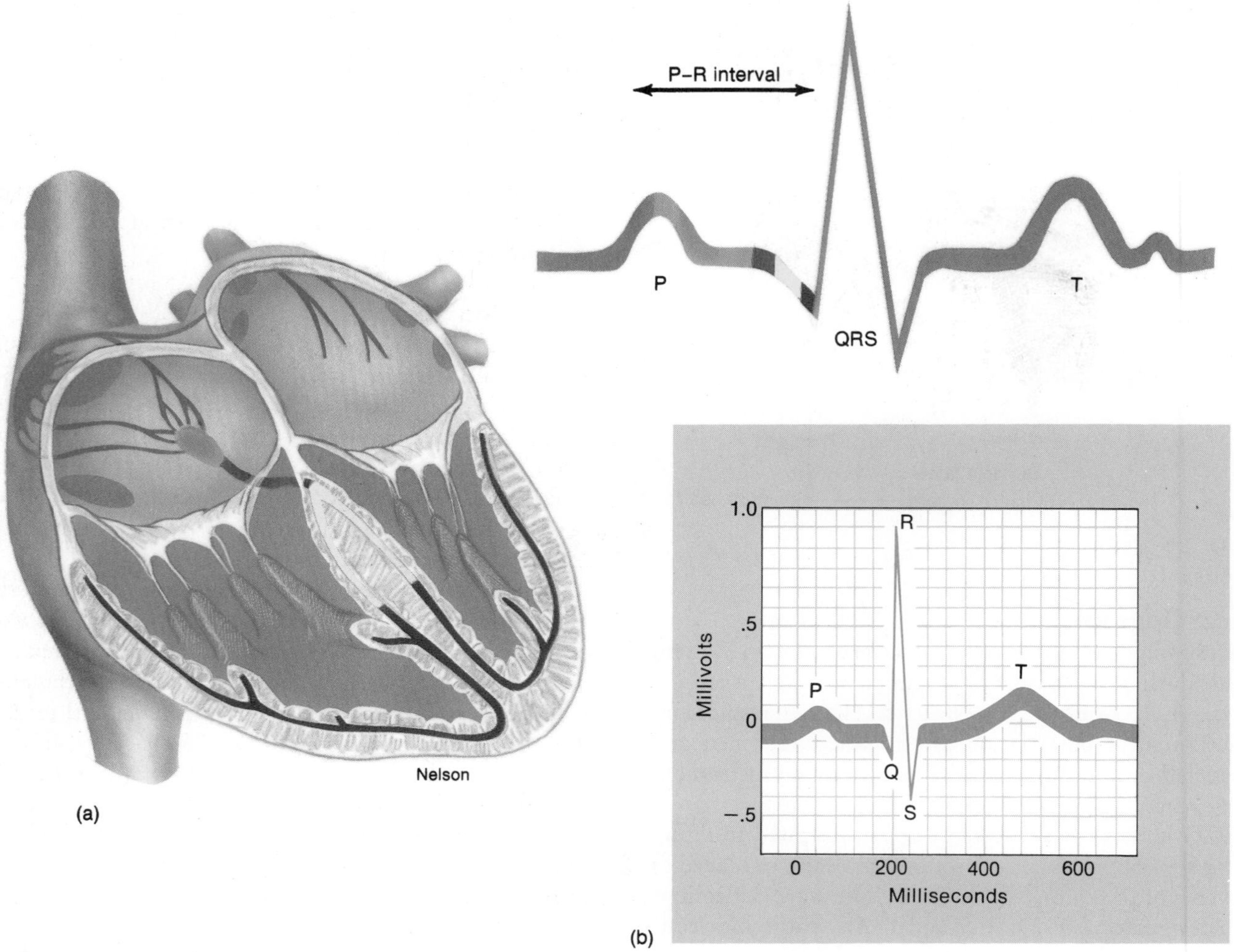

Figure 16.17 The electrocardiogram indicates the conduction of electrical impulses through the heart (*a*) and measures and records both the intensity of this electrical activity (in millivolts) and the time intervals involved (*b*).

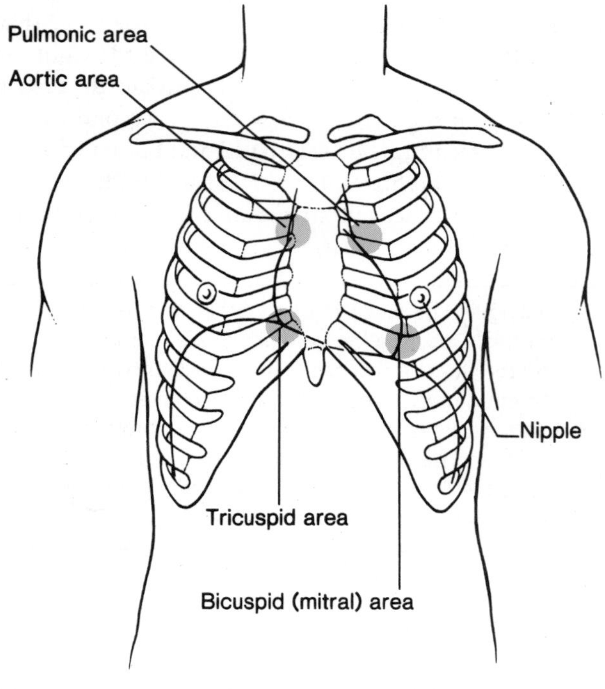

Figure 16.18 Routine stethoscope positions for listening to the heart sounds.

Blood Vessels

The structure of arteries and veins allows them to function in the transport of blood from the heart to the capillaries and from the capillaries back to the heart. The structure of capillaries provides the mechanism for the exchange of plasma fluid and dissolved molecules between the blood and the surrounding tissues.

Objective 14. Describe the structure, size, and function of arteries, capillaries, and veins.

Objective 15. Explain why capillaries are considered the functional units of the circulatory system.

Blood vessels form a tubular network throughout the body that permits blood to flow from the heart to all the living cells of the body and then back to the heart. Blood leaving the heart passes through vessels of progressively smaller diameters referred to as **arteries, arterioles,** and **capillaries.** Capillaries are microscopic vessels that join the arterial flow to the venous flow. Blood returning to the heart from the capillaries passes through vessels of progressively larger diameters called **venules** and **veins.**

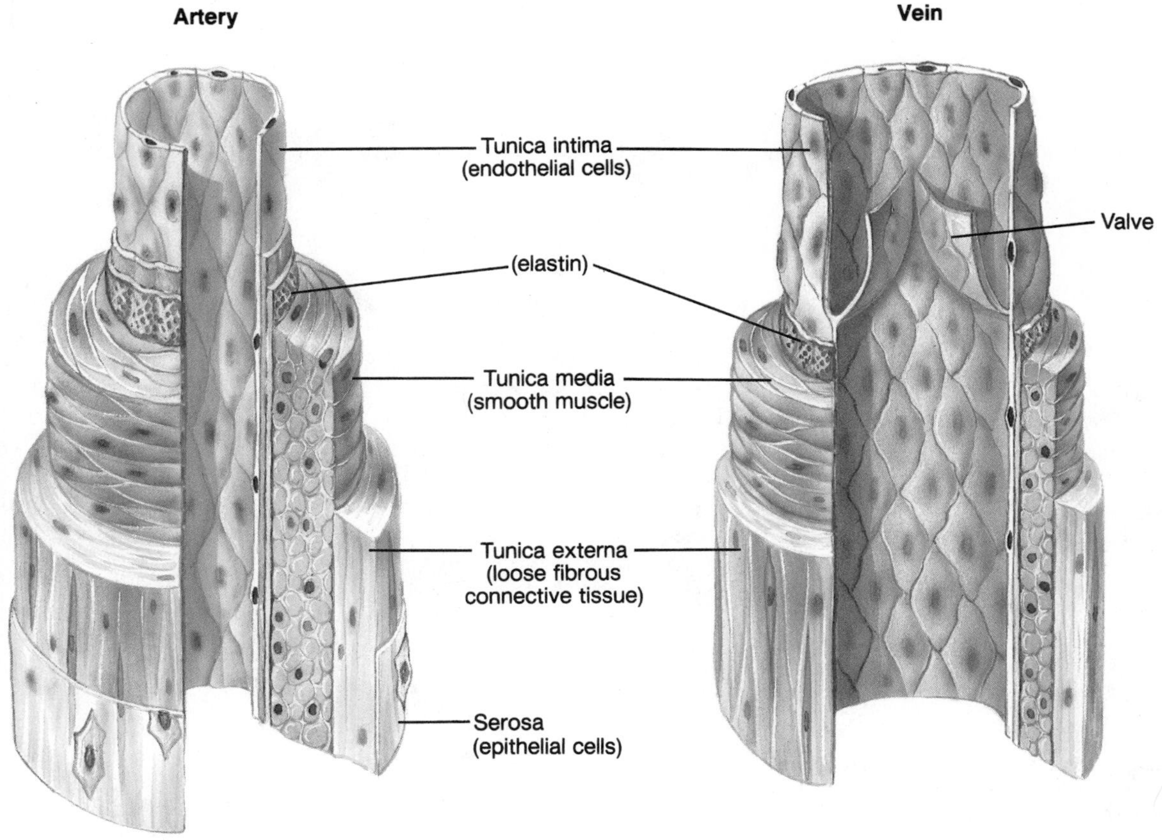

Figure 16.19 The structure of a medium-sized artery and vein showing the relative thickness and composition of the tunics.

The walls of arteries and veins are composed of three coats, or tunics (fig. 16.19). The outermost layer is the **tunica** *(tu'ni-kah)* **externa,** the middle layer is the **tunica media,** and the inner layer is the **tunica intima.** The tunica externa is composed of connective tissue. The tunica media is composed primarily of smooth muscle. The tunica intima consists of two parts: (1) an innermost simple squamous epithelium, the *endothelium (en''do-the'le-um),* which lines the lumina of all blood vessels and (2) a layer of elastic fibers, or *elastin.*

Although both arteries and veins have the same basic structure, there are some important differences between the two types. Arteries have more muscle for their diameters than do comparably sized veins. Also, arteries appear rounder than veins in cross section, whereas veins are usually partially collapsed. This is due to the fact that veins are not usually filled to their capacity; they can stretch when they receive more blood and thus function as reservoirs or capacitance vessels. In addition, many veins have valves, which are absent in arteries.

Arteries

The aorta and other large arteries contain numerous layers of elastin fibers between smooth muscle cells in the tunica media. These large arteries expand when the pressure of the blood rises as a result of the heart's contraction; they recoil, like a stretched rubber band, when the blood pressure falls during relaxation of the heart. This elastic recoil helps to produce a smoother, less pulsatile flow of blood through the smaller arteries and arterioles.

The small arteries and arterioles are less elastic than the larger arteries and have a thicker layer of smooth muscle for their diameters. Unlike the larger *elastic arteries,* therefore, the smaller *muscular arteries* retain almost the same diameter as the pressure of the blood rises and falls during the heart's pumping activity. Since small muscular arteries and arterioles have narrow lumina, they provide the greatest resistance to blood flow through the arterial system.

Small muscular arteries that are 100 micrometers or less in diameter branch to form smaller arterioles (20–30 micrometers in diameter). In some tissues, blood from the arterioles can enter the venules directly through **arteriovenous anastomoses,** called **metarterioles** (fig. 16.20). In most cases, however, blood from arterioles passes into capillaries. Capillaries are the narrowest of blood vessels (7–10 micrometers in diameter), and serve as the functional unit of the circulatory system, in which exchanges of gases and nutrients between the blood and the tissues occur.

tunic: L. *tunica,* covering or coat
lumina: L. *lumen,* opening

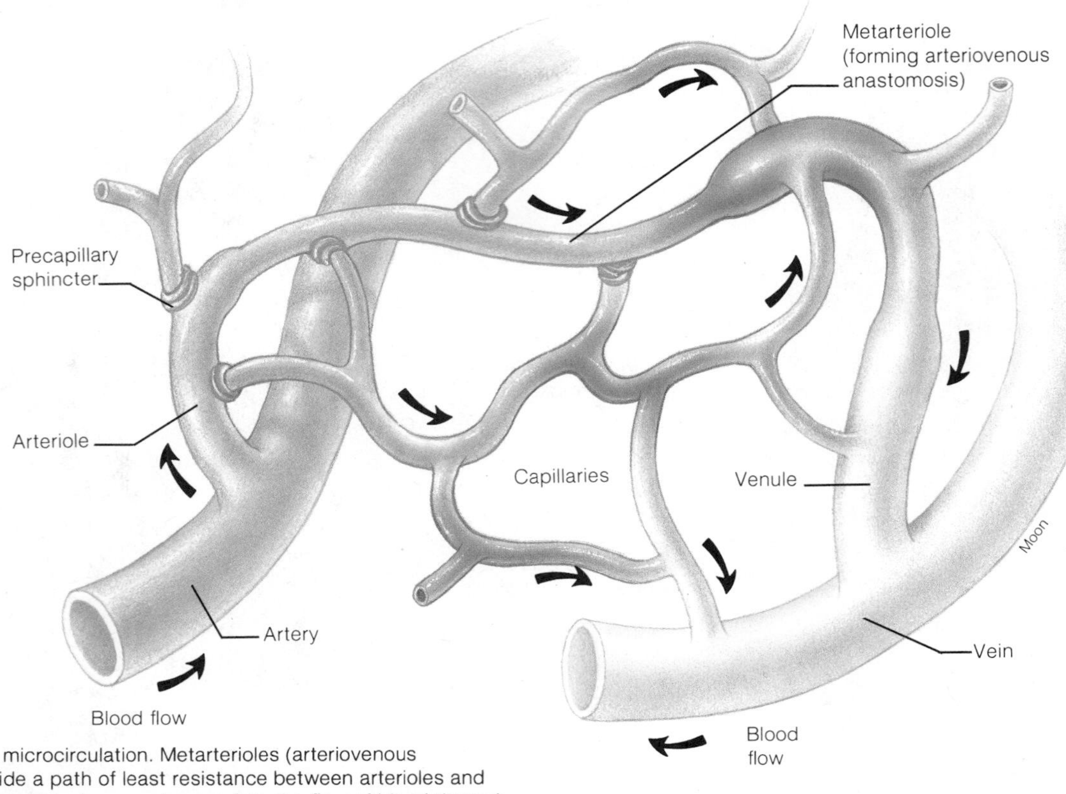

Figure 16.20 The microcirculation. Metarterioles (arteriovenous anastomoses) provide a path of least resistance between arterioles and venules. Precapillary sphincter muscles regulate the flow of blood through the capillaries.

Capillaries

The arterial system branches extensively to deliver blood to over forty billion capillaries in the body. The extensiveness of these branchings is indicated by the fact that all tissue cells are located within a distance of only 60–80 micrometers of a capillary and by the fact that capillaries provide a total surface area of 1,000 square miles for exchanges between blood and tissue fluid.

Despite their large number, capillaries contain only about 250 ml of blood at any time, out of a total blood volume of about 5,000 ml (most is contained in the venous system). The amount of blood flowing through a particular capillary bed is determined in part by the action of the **precapillary sphincter muscles** (fig. 16.20). These muscles allow only 5%–10% of the capillary beds in skeletal muscles, for example, to be open at rest. Blood flow to an organ is regulated by the action of these precapillary sphincters and by the degree of resistance to blood flow (due to constriction or dilation) provided by the small arteries and arterioles in the organ.

Unlike the vessels of the arterial and venous systems, the walls of capillaries are composed of only one cell layer—a simple squamous epithelium, or endothelium. The absence of smooth muscle and connective tissue layers permits a more rapid rate of transport of materials between the blood and the tissues.

Types of Capillaries Different organs have different types of capillaries, which are distinguished by significant differences in structure. In terms of their endothelial lining, these capillary types include those that are *continuous,* those that are *discontinuous,* and those that are *fenestrated (fen'es-trāt''ed)* (fig. 16.21).

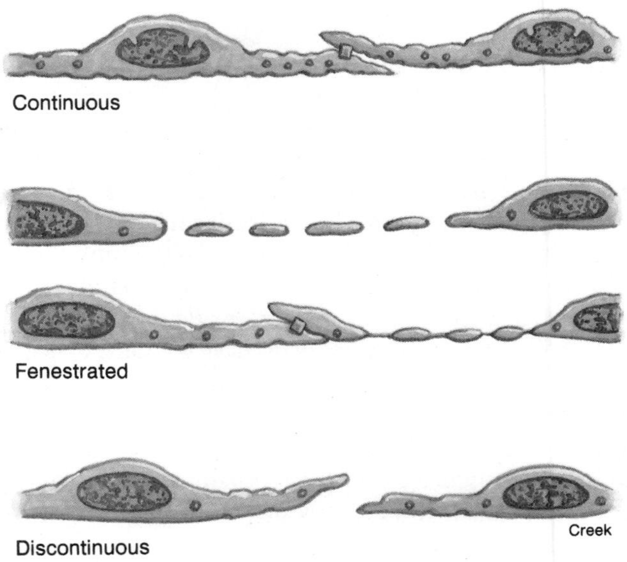

Figure 16.21 Diagrams of continuous, fenestrated, and discontinuous capillaries as they appear in the electron microscope. This classification is derived from the continuity of the endothelial layer. (Dark circles in the cytoplasm indicate pinocytotic vesicles.)

Continuous capillaries are those in which adjacent endothelial cells are closely joined together. These are found in muscles, lungs, adipose tissue, and in the central nervous system. Continuous capillaries in the CNS lack intercellular channels; this fact contributes to the blood-brain barrier. Continuous capillaries in other organs have narrow intercellular channels (about

40–45 Å in width) that allow the passage of molecules other than protein between the capillary blood and tissue fluid.

The examination of endothelial cells with an electron microscope has revealed the presence of pinocytotic vesicles, which suggests that the intracellular transport of material may occur across the capillary walls. This type of transport appears to be the only mechanism of capillary exchange available within the central nervous system and may account, in part, for the selective nature of the blood-brain barrier.

The kidneys, endocrine glands, and intestines have **fenestrated capillaries,** characterized by wide intercellular pores (800–1,000 Å) that are covered by a layer of mucoprotein, which may serve as a diaphragm. In the bone marrow, liver, and spleen, the distance between endothelial cells is so great that these **discontinuous** capillaries appear as little cavities (*sinusoids*) in the organ.

Veins

The average pressure in the veins is only 2 mm Hg (millimeters of mercury), compared to a much higher average arterial pressure of about 100 mm Hg. These pressures represent the hydrostatic pressure that the blood exerts on the walls of the vessels, and the numbers indicate the differences from atmospheric pressure.

The low venous pressure is insufficient to return blood to the heart, particularly from the lower limbs. Veins, however, pass between skeletal muscle groups that produce a massaging action as they contract (fig. 16.22). As the veins are squeezed by contracting skeletal muscles, a one-way flow of blood to the heart is insured by the presence of **venous valves.**

The effect of the massaging action of skeletal muscles on venous blood flow is often described as the *skeletal muscle pump.* The rate of venous return to the heart is dependent, in large part, on the action of skeletal muscle pumps. When these pumps are less active, as when a person stands still or is bedridden, blood accumulates in the veins and causes them to bulge. When a person is more active, blood returns to the heart at a faster rate and less is left in the venous system.

The accumulation of blood in the veins of the legs over a long period of time, as may occur in people with occupations that require standing still all day, can cause the veins to stretch to the point where the venous valves are no longer efficient. This can produce *varicose veins.* During walking the movements of the foot activate the soleus muscle pump. This effect can be produced in bedridden people by upward and downward manipulations of the feet.

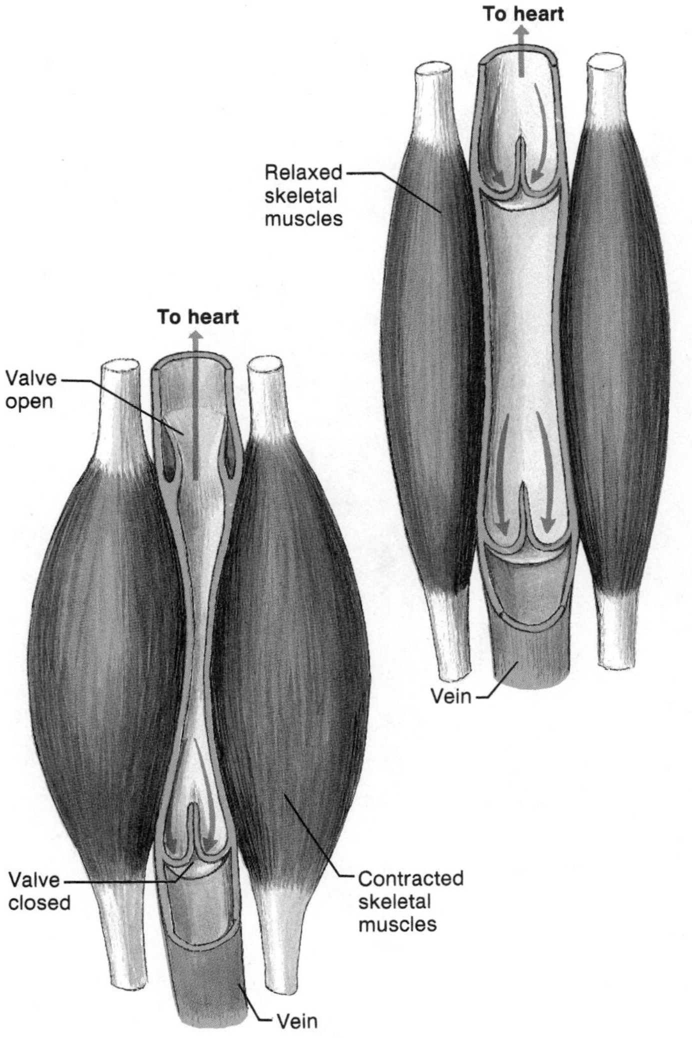

Figure 16.22 The action of the one-way venous valves. The contraction of skeletal muscles helps to pump blood toward the heart, but it is prevented from pushing blood away from the heart by closure of the venous valves.

Blood Pressure

Blood pressure exists throughout the vascular system but is greater within arteries where it is commonly measured and used as an indication of health. Blood pressure is created primarily by the action of the heart and is the force exerted by the blood against the vascular walls.

Arterial blood pressure can be measured by using a device called a *sphygmomanometer* (fig. 16.23). This instrument has an inflatable cuff, which is used to constrict an artery at a pressure point (fig. 16.24). The most common site of application is the brachial artery compressed against the humerus. As the blood pulsates through the constricted brachial artery, the pressure produced is indicated by the sphygmomanometer.

The normal blood pressure of an adult is about 120/80. This is an expression of the **systolic pressure** (120 mm Hg) over the **diastolic pressure** (80 mm Hg). The systolic pressure is produced as blood is ejected from the heart during ventricular systole. When the ventricles relax during ventricular diastole, the arterial pressure drops and the diastolic pressure is recorded. For the most part, diastolic pressure results from contraction of the smooth muscles within the arterial walls. The difference between the systolic and diastolic pressure is called the **pulse pressure** and is generally about 40 mm Hg.

Several factors can influence the arterial blood pressure. Blood pressure decreases as the distance from the heart increases (fig. 16.25). An increased cardiac output (stroke volume $\times$ heart rate) increases blood pressure. Blood volume and blood viscosity both influence blood pressure, as do various drugs. Changes in the diameters of the vascular lumina through autonomic vasoconstriction or vasodilation have a direct bearing on

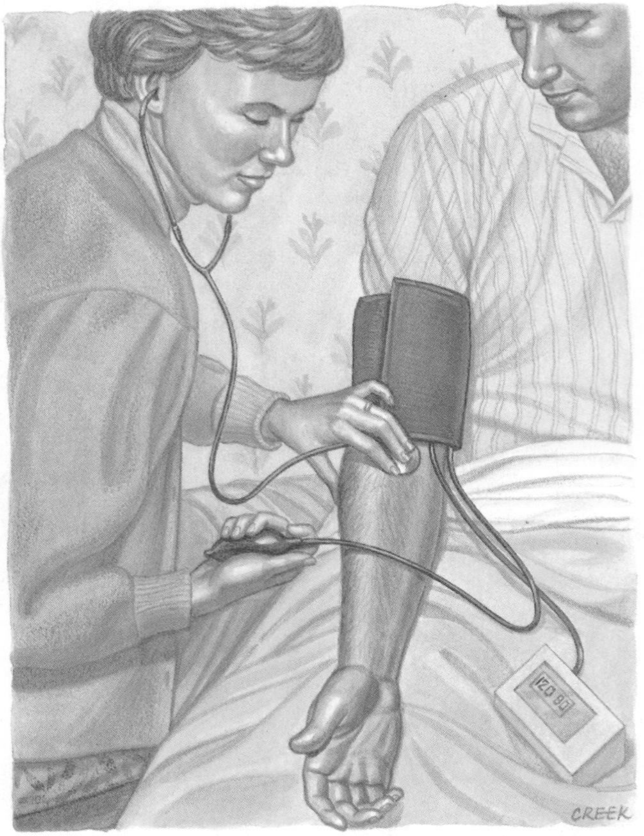

Figure 16.23 The use of a pressure cuff and a sphygmomanometer to measure blood pressure.

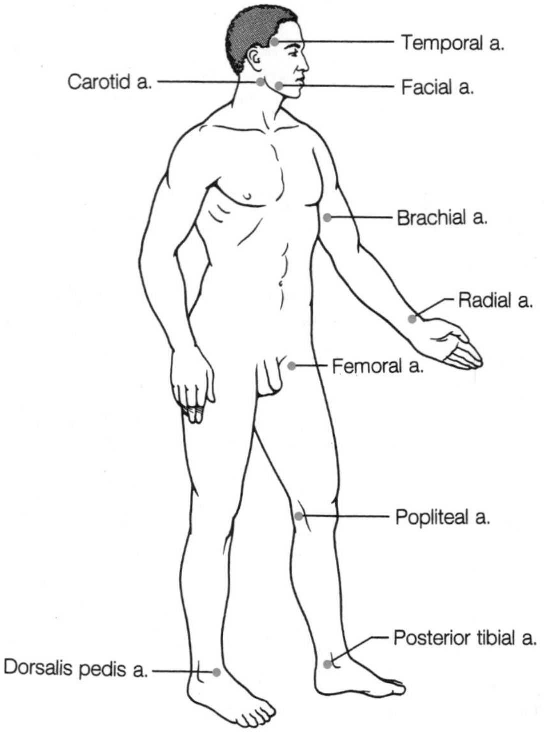

Figure 16.24 Important arterial pressure points and locations where arterial pulsations can best be detected (a. = artery).

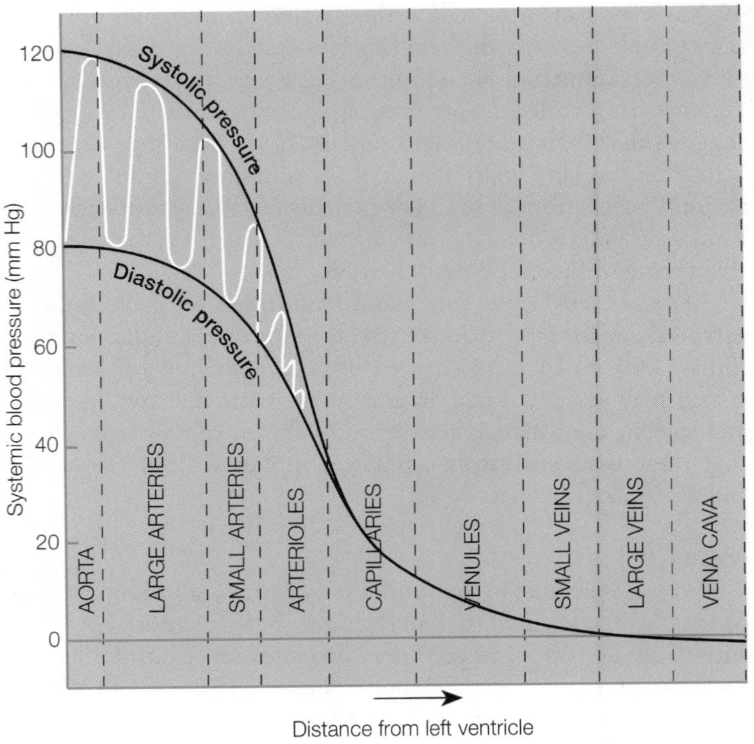

Figure 16.25 Blood pressure as recorded at various vascular sites.

blood pressure. Blood pressures are also greatly influenced by the general health of the cardiovascular system. Elevated blood pressure is referred to as *hypertension* and is potentially dangerous because of the extra strain it places on the heart.

One of the harmful effects of nicotine inhaled from cigarette smoking is that it causes vasoconstriction of arterioles. It also is a stimulant to the heart and increases cardiac output. Both of these responses raise the blood pressure and increase the strain on the heart.

---

1. Describe the basic structural pattern of arteries and veins. Describe how arteries and veins differ in structure and how these differences contribute to the resistance function of arteries and the capacitance function of veins.
2. Describe the functional significance of the skeletal muscle pump, and explain the action of venous valves.
3. Discuss the functions of capillaries, and describe the structural differences between capillaries in different organs.
4. Describe the cardiovascular events that determine systolic and diastolic blood pressures.

Principal Arteries of the Body

The aorta ascends from the left ventricle to a position just above the heart where it arches to the left and then descends through the thorax and abdomen. Branches of the aorta carry oxygenated blood to all of the cells of the body.

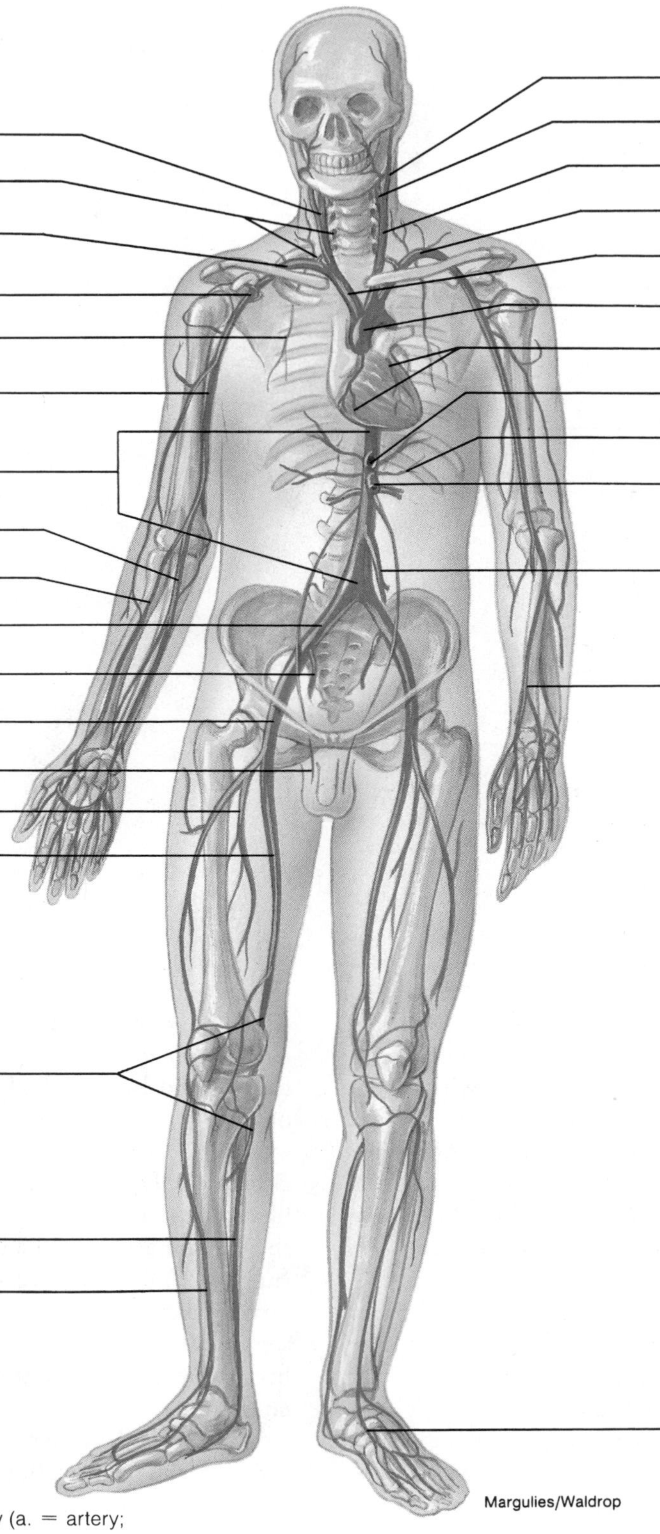

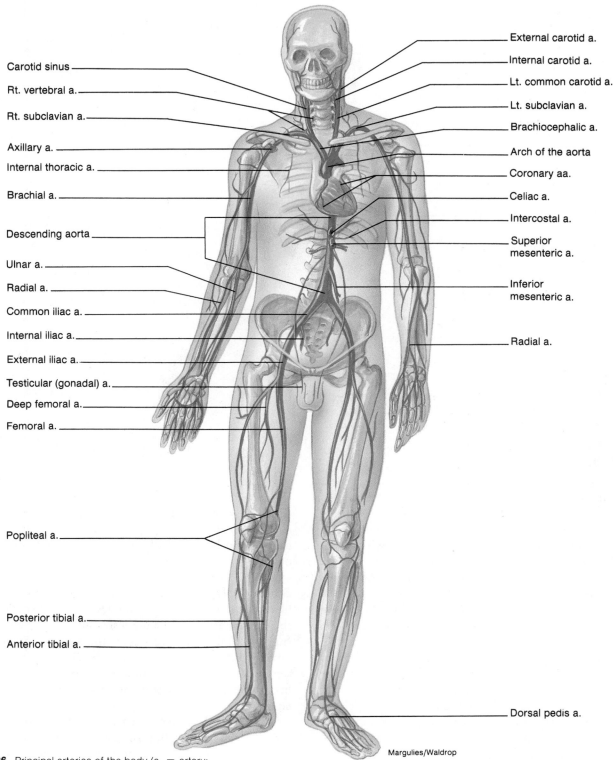

Figure 16.26 Principal arteries of the body (a. = artery; aa. = arteries).

Objective 16. In the form of a flow chart, list the arterial branches of the ascending aorta and aortic arch.
Objective 17. Describe the arterial supply to the brain.
Objective 18. Describe the arterial pathways that supply the upper extremity.
Objective 19. Describe the major arteries to the thorax, abdomen, and lower extremity.

Contraction of the left ventricle forces oxygenated blood into the arteries of the systemic circulation. The principal arteries of the body are shown in figure 16.26. They will be described by region and identified in order from largest to smallest, or as the blood flows through the system. The major systemic artery is the **aorta,** from which all the primary arteries arise.

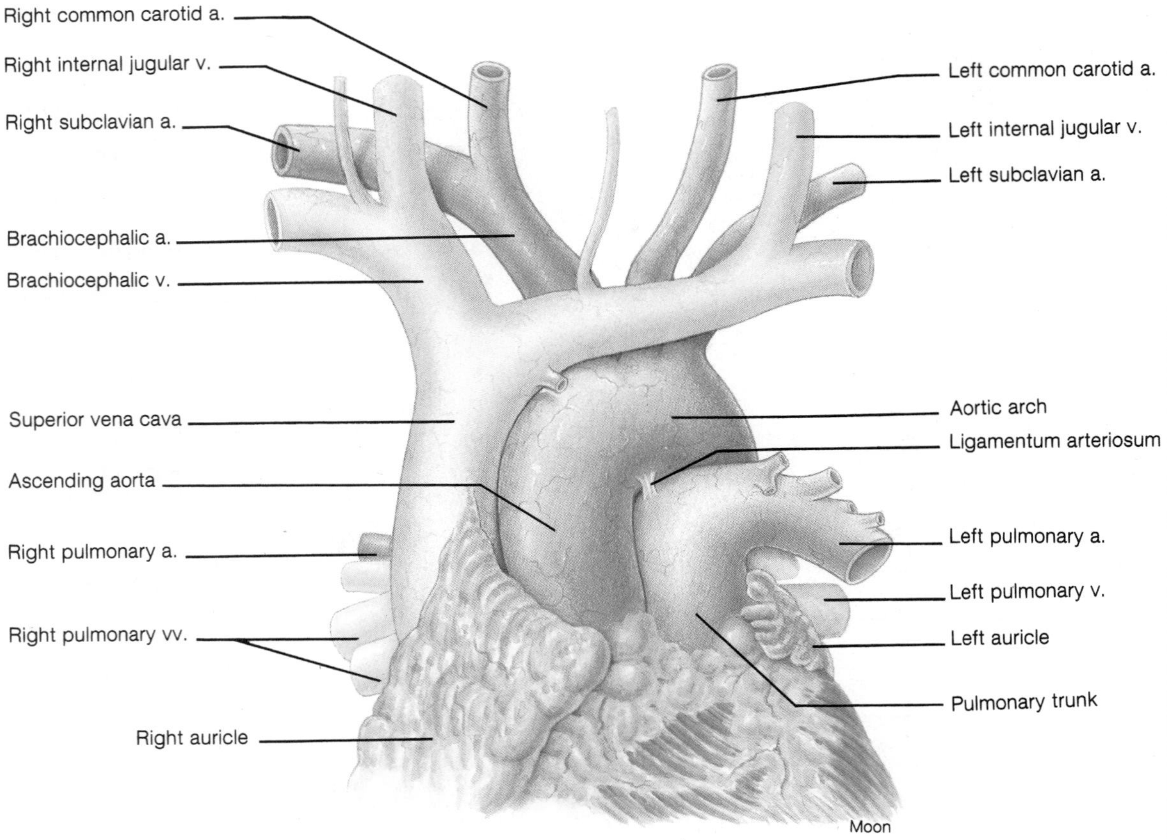

Figure 16.27 The structural relationship between the major arteries and veins to and from the heart (v. = vein; vv. = veins).

Arch of the Aorta

The systemic vessel that ascends from the left ventricle of the heart is called the **ascending aorta.** The right and left **coronary arteries** serving the myocardium of the heart with blood are the only branches that arise off the ascending aorta. The aorta arches to the left and posteriorly over the pulmonary arteries as the **aortic arch** (fig. 16.27). Three vessels arise from the aortic arch: the **brachiocephalic** *(brak''e-o-se-fal'ik),* the **left common carotid** *(kah-rot'id),* and the **left subclavian.**

The brachiocephalic is the first vessel to branch from the aortic arch and, as its name suggests, supplies blood to the tissues of the arm and head on the right side of the body. It is a short vessel rising superiorly through the mediastinum to a point near the junction of the sternum and the right clavicle. There it bifurcates into the **right common carotid,** which extends to the right side of the neck and head, and the **right subclavian,** which carries blood to the right upper extremity.

The remaining two branches from the arch of the aorta are the left common carotid and the left subclavian arteries. The left common carotid transports blood to the left side of the neck and head, and the left subclavian supplies the left upper extremity.

Arteries of the Neck and Head

The common carotids course upward in the neck along the lateral sides of the trachea (fig. 16.28). The common carotid bifurcates into the **internal** and **external carotid arteries** slightly below the angle of the mandible. By pressing gently in this area, a pulse can be detected (see fig. 16.24). At the base of the internal carotid, near the bifurcation, is a slight dilation called the **carotid sinus.** The carotid sinus contains *baroreceptors,* which monitor blood pressure, and *chemoreceptors* within the **carotid body,** which respond to chemical changes in the blood.

Blood Supply to the Brain The brain is supplied with arterial blood that arrives through four vessels that eventually unite on the inferior aspect of the brain surrounding the pituitary gland (fig. 16.29). The four vessels are the paired internal carotid arteries and the paired vertebral arteries. The value of having four separate vessels that anastomose at one location is that if one becomes occluded, the three alternate routes may insure an adequate blood supply to the brain.

The **vertebral arteries** arise from the subclavian arteries at the base of the neck (fig. 16.28). They pass superiorly through the transverse foramina of the cervical vertebrae and enter the skull through the foramen magnum. Within the braincase, the two vertebral arteries unite to form the **basilar artery** at the level of the pons. The basilar ascends along the inferior surface of the brain stem and inner ear; it terminates by forming two **posterior cerebral arteries,** which supply the posterior portion of the cerebrum (fig. 16.29). The **posterior communicating arteries** are branches that arise from the posterior cerebral arteries and participate in forming the **arterial circle** around the pituitary gland, also called the *circle of Willis.*

---

circle of Willis: from Thomas Willis, English physician, 1621–75

Figure 16.28 Arteries of the neck and head. (*a*) Major branches of the right common carotid and right subclavian arteries. (*b*) A radiograph of the head following a radiopaque injection of the arteries.

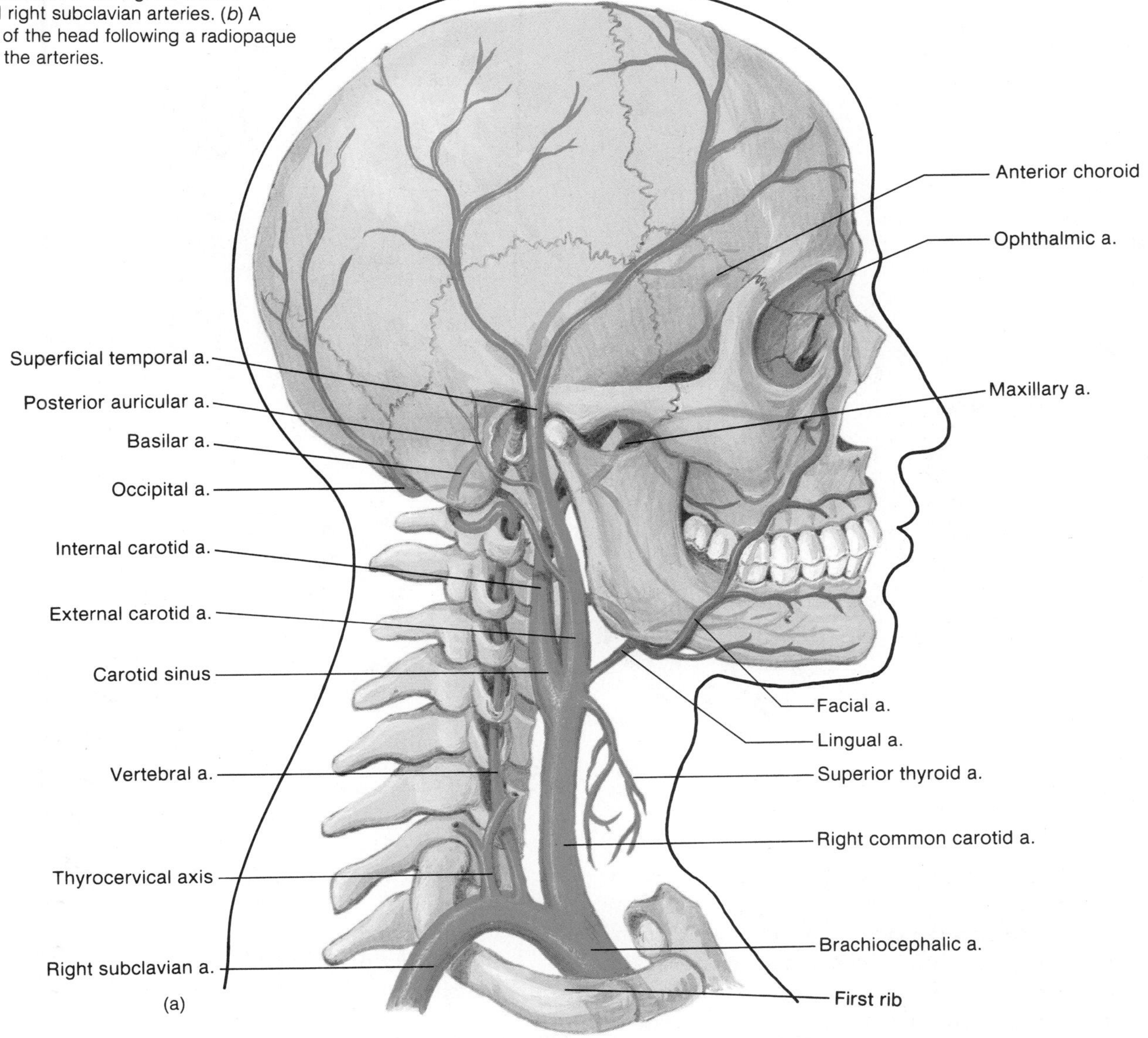

(a)

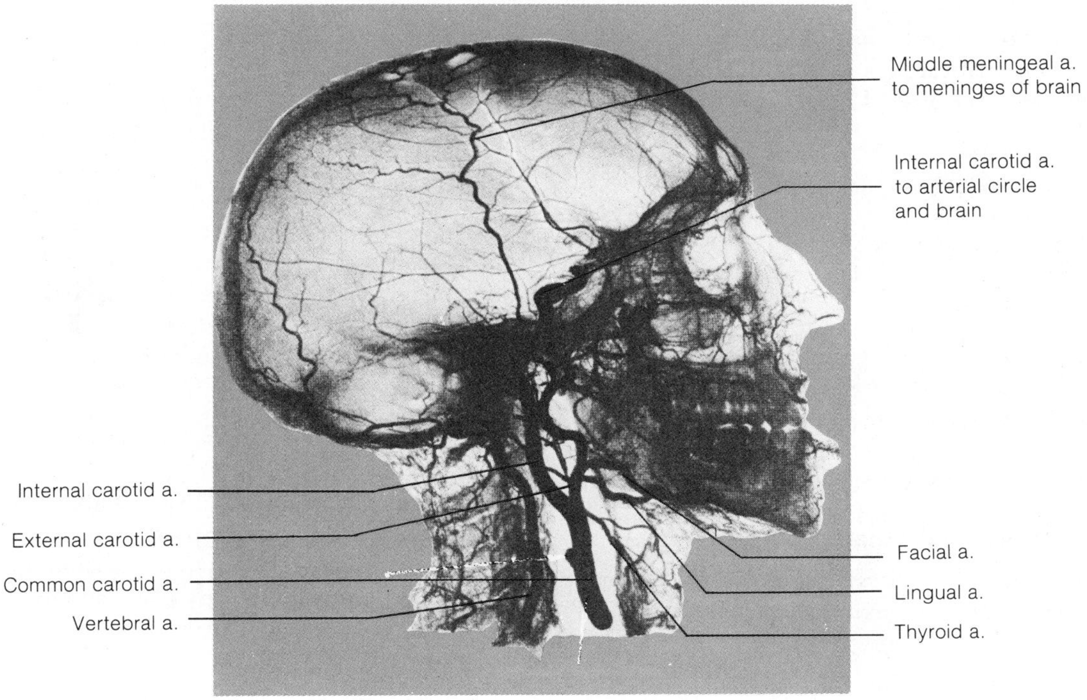

(b)

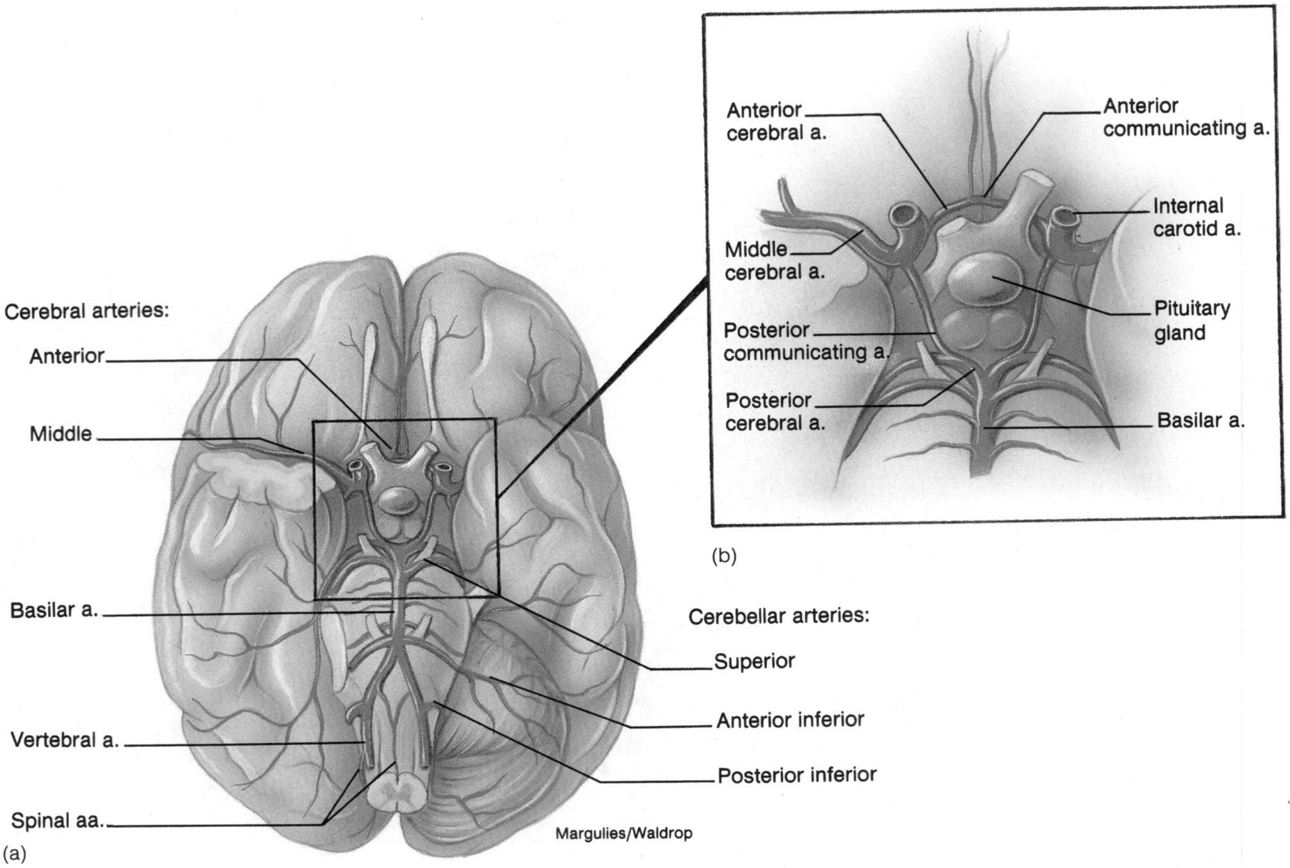

Figure 16.29 Arteries that supply blood to the brain. (*a*) Inferior view of the brain and (*b*) a closeup view of the region of the pituitary gland.

The **internal carotid artery** branches from the common carotid artery and ascends in the neck until it reaches the base of the skull, where it enters the carotid canal of the temporal bone. Several branches arise from the internal carotid once it is on the inferior surface of the brain. Three of the more important ones are the **ophthalmic** *(of'thal'mik)* **artery** (fig. 16.28a), which supplies the eye and associated structures, and the **anterior** and **middle cerebral arteries,** which provide blood to the cerebrum. The internal carotids are connected to the posterior cerebral arteries at the arterial circle.

Capillaries within the pituitary gland receive both arterial and venous blood. The venous blood arrives from venules immediately superior to the pituitary, which drain capillaries in the hypothalamus of the brain. This arrangement of two capillary beds in series—where the second capillary bed receives venous blood from the first—is called a *portal system.* The venous blood that travels from the hypothalamus to the pituitary contains hormones from the hypothalamus that help regulate pituitary gland hormone secretion.

External Carotid Artery The external carotid artery gives off several branches as it extends upward along the side of the neck and head (see fig. 16.28). The names of these branches are determined by the areas or structures that they serve. The principal vessels that arise from the external carotid are the following:

1. **Superior thyroid artery,** which serves the muscles of the hyoid region, the larynx and vocal cords, and the thyroid gland.
2. **Ascending pharyngeal artery** (not seen in illustration), which serves the pharyngeal area and various lymph nodes.
3. **Lingual artery,** which provides extensive vascularization of the tongue and sublingual salivary gland.
4. **Facial artery,** which traverses a notch on the inferior margin of the mandible to serve the pharyngeal area, palate, chin, lips, and nasal region. The facial artery is an important vessel to apply pressure when controlling bleeding from the face.
5. **Occipital artery,** which serves the posterior portion of the scalp, the meninges over the brain, the mastoid process, and certain posterior neck muscles.
6. **Posterior auricular artery,** which serves the ear and scalp over the ear.

The external carotid artery terminates at a level near the mandibular condyle by dividing into **maxillary** and **superficial temporal arteries.** The maxillary artery gives off branches to the teeth and gums, the muscles of mastication, nasal cavity, eyelids, and meninges. The superficial temporal artery supplies blood to the parotid salivary gland and to the superficial structures on the side of the head. Pulsations through the temporal

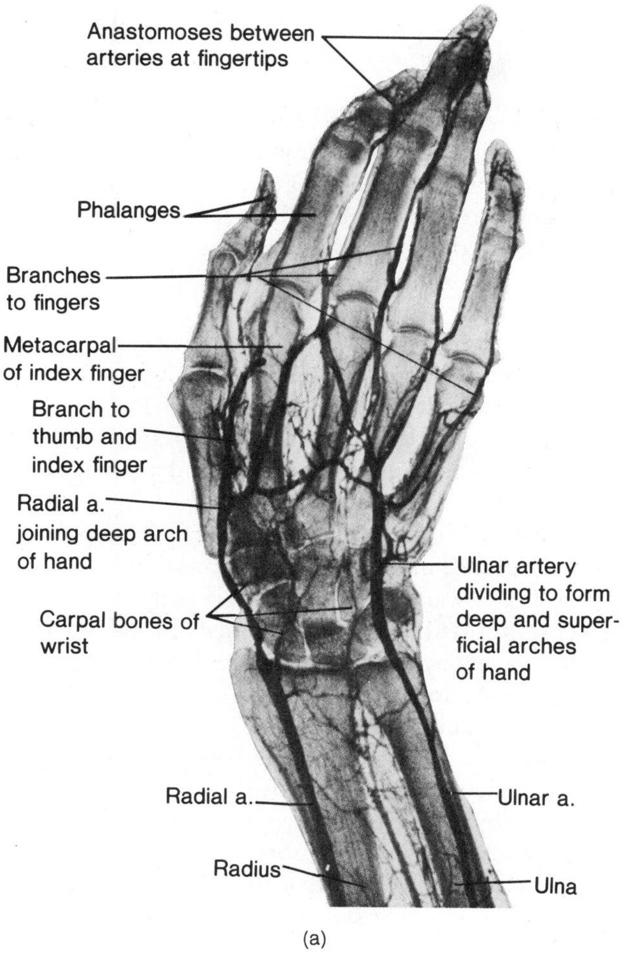

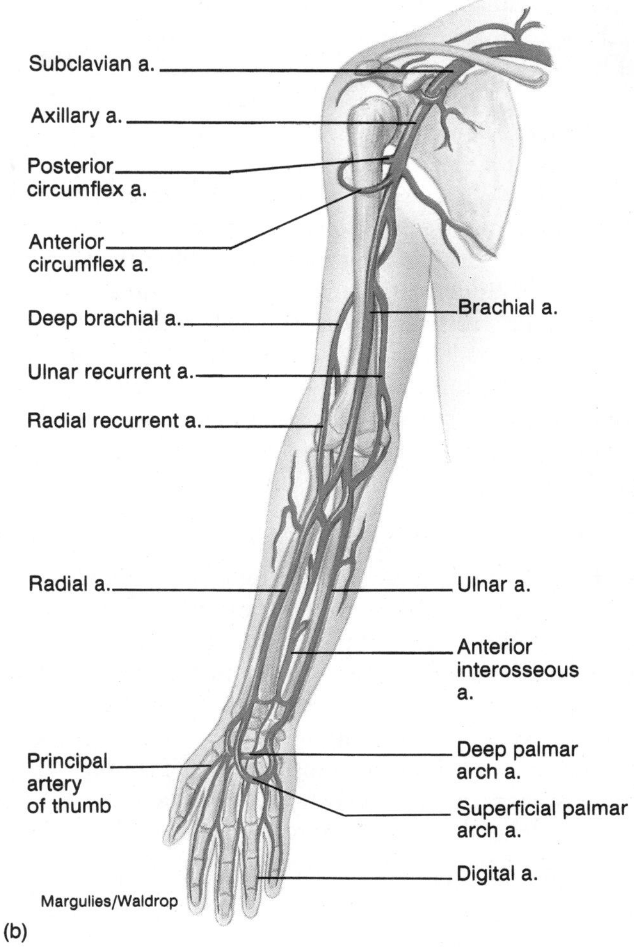

Figure 16.30 Arteries of the upper extremity. (*a*) A radiograph of the forearm and hand following a radiopaque injection of the arteries. (*b*) An anterior view of major arteries.

artery can be easily detected by placing the fingertips immediately in front of the ear at the level of the eye. This vessel is frequently used by anesthesiologists to check a patient's pulse rate during surgery.

Headaches are usually caused by vascular pressure on the sensitive meninges covering the brain. The two principal vessels serving the meninges are the occipital and maxillary arteries. Vasodilation of these vessels creates excessive pressure on the sensory receptors within the meninges, resulting in a headache.

Arteries of the Upper Extremity

The right subclavian artery branches off the brachiocephalic, and the left subclavian artery arises directly from the arch of the aorta (see fig. 16.27). The **subclavian artery** passes laterally deep to the clavicle, carrying blood toward the arm (fig. 16.30). The pulsations of the subclavian artery can be detected by pressing firmly on the tissue just above the medial portion of the clavicle. From each subclavian artery arises a **vertebral artery** that carries blood to the brain (already described), a short **thyrocervical trunk** that serves the thyroid gland, trachea, and larynx, and an **internal thoracic (mammary) artery** that descends into the thorax to serve the thoracic wall, thymus, and pericardium. A branch of the internal thoracic artery supplies blood to the muscles and tissues (mammary glands) of the anterior thorax.

The subclavian artery becomes the **axillary** *(ak'sĭ-lar"e)* **artery** as it passes into the axillary region. The axillary artery is that portion of the major artery of the upper extremity between the outer border of the first rib and the lower border of the teres major muscle. Several small branches arise from the axillary artery and supply blood to the tissues of the upper thorax and shoulder region.

The **brachial** *(bra'ke-al)* **artery** is the continuation of the axillary artery through the brachial region. The brachial artery courses on the medial side of the humerus, where it is a major pressure point and the most common site for determining blood pressure. A **deep brachial artery** branches from the brachial artery and curves posteriorly near the radial nerve to supply the triceps muscle. Two additional branches from the brachial, the **anterior** and **posterior circumflex arteries,** form a continuous ring of vessels around the proximal portion of the humerus.

Just proximal to the cubital fossa, the brachial bifurcates into the **radial** and **ulnar arteries,** which supply blood to the forearm and a portion of the hand and digits. The radial artery courses down the lateral, or radial, side of the arm, where it sends numerous small branches to the muscles of the forearm. The **radial recurrent artery** serves the region of the elbow and is the first and largest off of the radial artery. The radial artery is important as a site to record the pulse near the wrist.

The ulnar artery is a bit larger in caliber than the radial artery. It extends down the ulnar side of the forearm and gives off many small branches to the muscles on the medial side of the forearm. It too has an initial large branch, which arises from the proximal portion near the elbow and is called the **ulnar recurrent artery.** At the wrist, the ulnar and radial arteries anastomose to form the **palmar arches** in the hand, from which **digital arteries** extend into the fingers.

Branches of the Thoracic Aorta

The **thoracic aorta** is a continuation of the arch of the aorta as it descends through the thoracic cavity to the diaphragm. This large vessel gives off branches to the organs and muscles of the thoracic region. These branches include **pericardial arteries,** going to the pericardium of the heart; **bronchial arteries** for systemic circulation to the lungs; **esophageal arteries,** going to the esophagus as it passes through the mediastinum; segmental **intercostal arteries,** serving the intercostal muscles and structures of the wall of the thorax (see fig. 16.27); and **phrenic** *(fren'ik)* **arteries,** supplying blood to the diaphragm.

These vessels are summarized according to their location and function in table 16.7.

Branches of the Abdominal Aorta

The **abdominal aorta** is the segment of the aorta between the diaphragm and the level of the fourth lumbar vertebra, where it divides into the right and left **common iliac arteries.** The first branch of the abdominal aorta is called the **celiac** *(se'le-ak)* **artery.** It is a short, thick trunk that arises anteriorly just below the diaphragm. The celiac divides immediately into three arteries: the **splenic,** going to the spleen; the **left gastric,** going to the stomach; and the **hepatic,** going to the liver (fig. 16.31).

The **superior mesenteric artery** is another unpaired vessel and arises anteriorly from the abdominal aorta just below the celiac. The superior mesenteric supplies blood to the small intestine (except for a portion of the duodenum), the cecum, the appendix, the ascending colon, and the transverse colon.

The next major vessels to arise from the abdominal aorta are the paired **renal arteries,** which carry blood to the kidneys. Smaller **suprarenal arteries** are located just above the renal arteries and serve the suprarenal (adrenal) glands. The **testicular** (internal spermatic) **arteries** in the male and the **ovarian arteries** in the female are small, paired vessels that arise from the abdominal aorta just below the renals and serve the gonads.

The **inferior mesenteric artery** is the last major branch of the abdominal aorta. It is an unpaired anterior vessel that arises just before the iliac bifurcation. The inferior mesenteric supplies blood to the descending colon, sigmoid colon, and rectum.

Several **lumbar arteries** branch posteriorly from the abdominal aorta throughout its length and serve the muscles and the spinal cord in the lumbar region. In addition, an unpaired **middle sacral artery** (fig. 16.32) arises from the posterior terminal portion of the abdominal aorta to supply the sacrum and coccyx.

Arteries of the Pelvis and Lower Extremity

The aorta terminates in the posterior pelvic area as it bifurcates into the right and left common iliac arteries. These vessels pass downward approximately 5 cm on their respective sides and terminate by dividing into the **internal** and **external iliac arteries.**

The internal iliac has extensive branches to supply arterial blood to the gluteal muscles and the organs of the pelvic region (fig. 16.32). The wall of the pelvis is served by the **iliolumbar** and **lateral sacral arteries.** The internal visceral organs of the pelvis are served by the **middle rectal** and the **superior, middle,** and **inferior vesicular arteries** to the urinary bladder. In addition, in females **uterine** and **vaginal arteries** branch from the internal iliacs to serve the reproductive organs. The muscles of the buttock are served by the **superior** and **inferior gluteal arteries.** Some of the upper medial thigh muscles are supplied with blood from the **obturator artery.** The **internal pudendal artery** of the internal iliac serves the musculature of the perineum and the external genitalia of male and female. The internal pudendal artery is an important vessel in sexual activity. Erection of the penis in the male and corresponding vascular engorgement of the genitalia of the female are vascular phenomena controlled by the autonomic nervous system (see chap. 13).

The external iliac artery passes out of the pelvic cavity beneath the inguinal ligament (fig. 16.33) and becomes the **femoral artery.** Two branches arise from the external iliac, however, before it passes beneath the inguinal ligament. An **inferior epigastric artery** branches from the external iliac and passes superiorly to supply the skin and muscles of the abdominal wall. The **deep circumflex iliac artery** is a small branch that extends laterally to supply the muscles attached to the iliac fossa.

The femoral artery passes through an area called the **femoral triangle** on the upper medial portion of the thigh (figs. 16.33, 16.34). At this point, the femoral artery is close to the surface and not only can be palpated but is an important pressure point. Several vessels arise from the femoral to serve the thigh region. The largest of these, the **deep femoral artery,** passes posteriorly to serve the hamstring muscles. The **lateral** and **medial femoral circumflex arteries** encircle the proximal end of the femur and serve muscles in this region. The femoral artery becomes the **popliteal** *(pop-lit'e-al)* **artery** as it passes across the posterior aspect of the knee.

Table 16.7 Segments and branches of the aorta

Segment of aorta	Arterial branch	General region or organ served
Ascending aorta	Right and left coronary	Heart
Arch of aorta	Brachiocephalic	
	Right common carotid	Right side of head and neck
	Right subclavian	Right shoulder and right upper extremity
	Left common carotid	Left side of head and neck
	Left subclavian	Left shoulder and left upper extremity
Thoracic aorta	Pericardials	Pericardium of heart
	Intercostals	Intercostal and thoracic muscles, pleurae
	Superior phrenics	Diaphragm
	Bronchials	Bronchi of lungs
	Esophageals	Esophagus
	Inferior phrenics	Inferior surface of diaphragm
Abdominal aorta	Celiac	
	Common hepatic	Liver, upper pancreas, duodenum
	Left gastric	Stomach, esophagus
	Splenic	Spleen, pancreas, stomach
	Superior mesenteric	Small intestine, pancreas, cecum, appendix, ascending and transverse colons
	Suprarenals	Suprarenal (adrenal) glands
	Renals	Kidneys
	Gonadals	
	Testiculars	Testes
	Ovarians	Ovaries
	Inferior mesenteric	Transverse, descending, and sigmoid colons; rectum
	Common iliacs	
	External iliacs	Lower extremities
	Internal iliacs (hypogastrics)	Genital organs, gluteal muscles

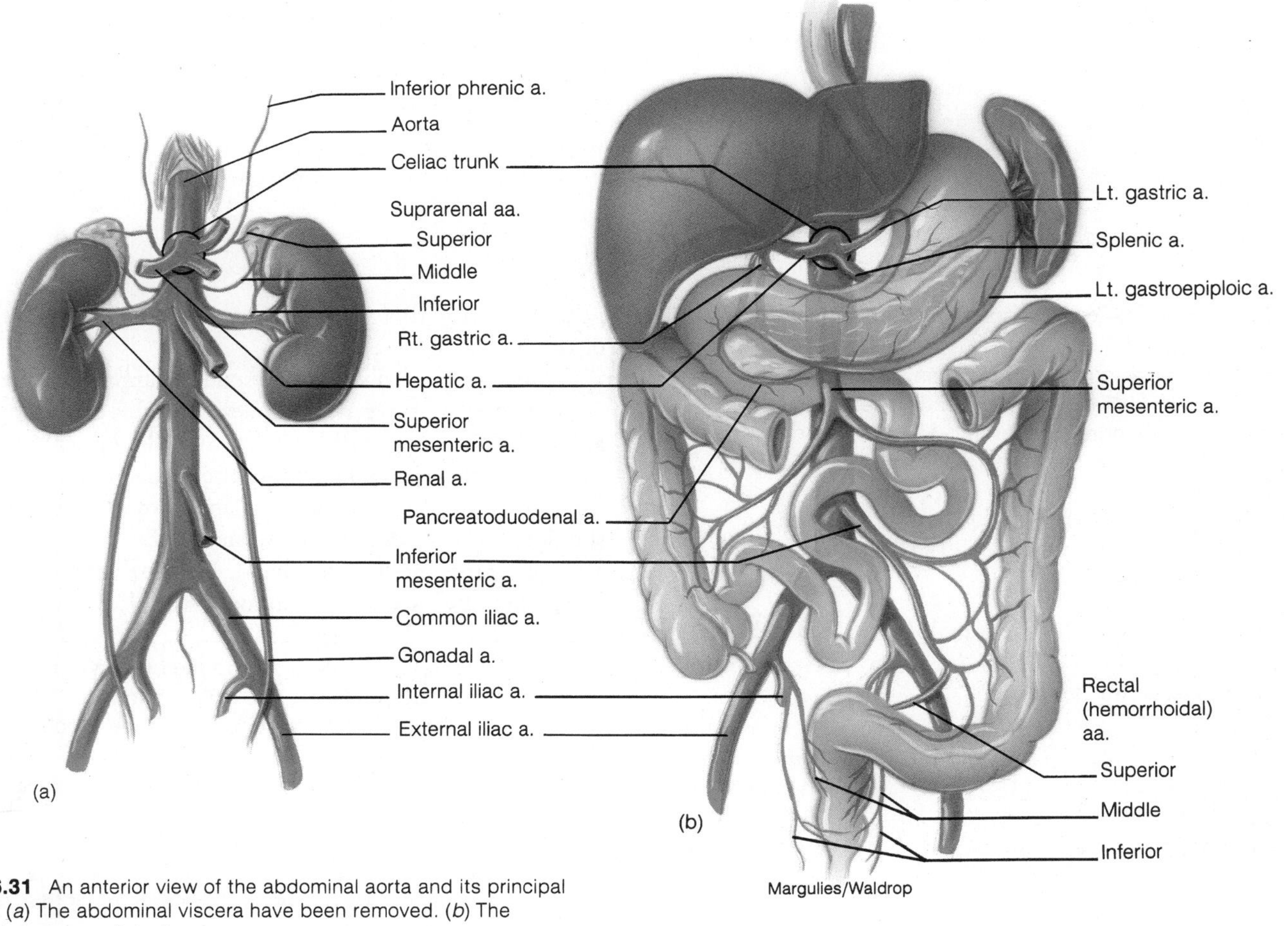

Figure 16.31 An anterior view of the abdominal aorta and its principal branches. (*a*) The abdominal viscera have been removed. (*b*) The abdominal viscera are intact.

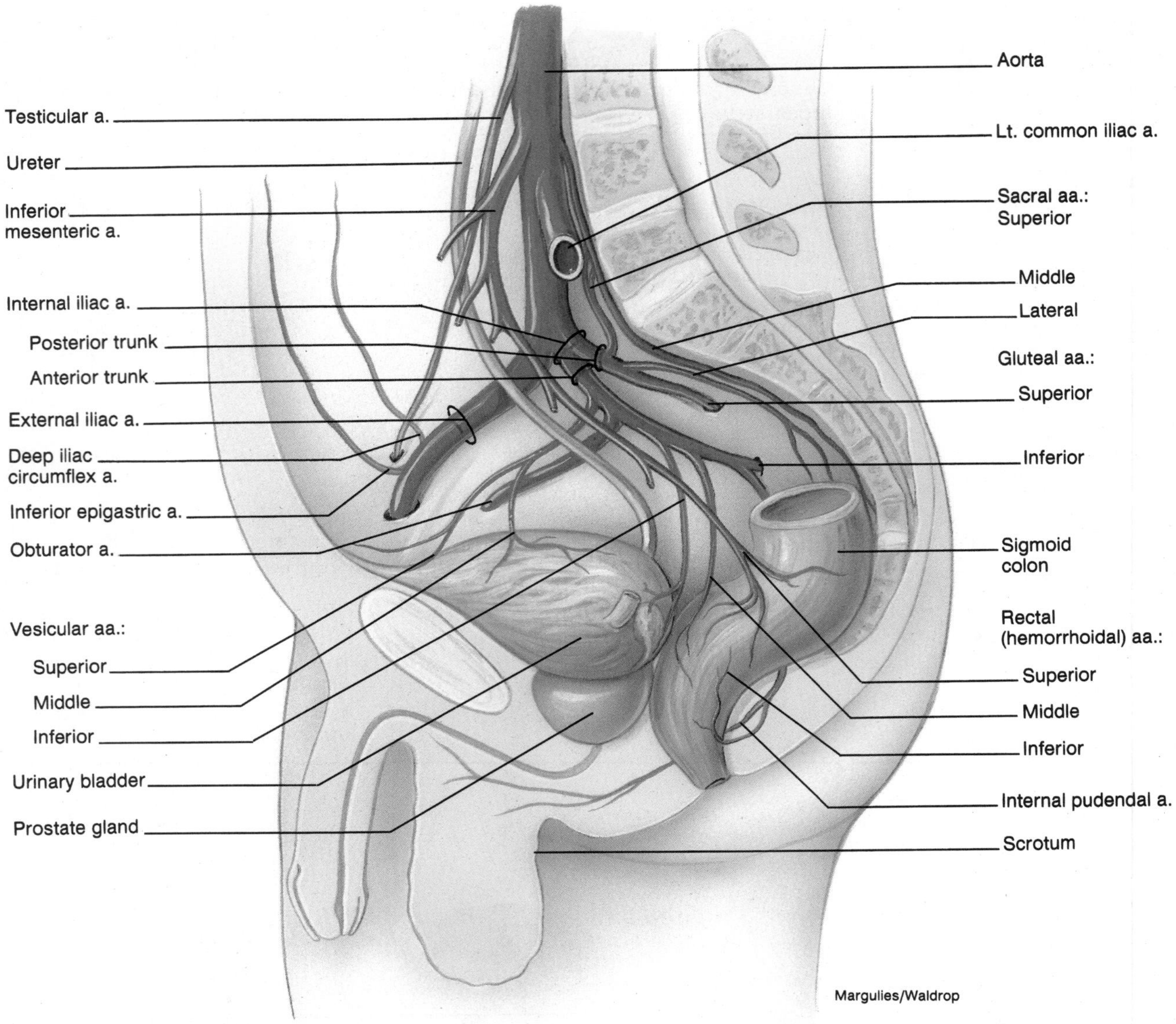

Figure 16.32 Arteries of the pelvic region.

Hemorrhage can be a serious problem in many accidents. Therefore, one should know the pressure points where the arterial blood flow can be curtailed to prevent a victim from bleeding to death. The pressure points for the appendages are the brachial artery on the medial side of the arm and the femoral artery in the groin. Firmly applied pressure to these regions greatly diminishes the flow of blood to traumatized areas below. A tourniquet may have to be applied if bleeding is severe enough to endanger life.

The popliteal artery supplies small branches to the knee joint and then divides into an **anterior tibial artery** and **posterior tibial artery** (fig. 16.33). These vessels traverse the anterior and posterior aspects of the leg, respectively, providing blood to the muscles of these regions and to the foot.

At the ankle, the anterior tibial artery becomes the **dorsalis pedis artery,** which serves the ankle and dorsum of the foot, after which it contributes to the formation of the **plantar arch** of the foot. Clinically, palpation of the dorsalis pedis artery can provide information about circulation to the foot; but more important, it can provide information about the circulation in general because these pulses are taken at the most distal portion of the body.

The posterior tibial artery sends a large **peroneal** *(per''o-ne'al)* **artery** to serve the peroneal muscles of the leg. At the ankle, the posterior tibial bifurcates into the **lateral** and **medial plantar arteries,** which supply the muscles and structures on the sole of the foot. The lateral plantar artery anastomoses with the dorsal pedis artery to form the plantar arch, similar to the arterial arrangement in the hand. **Digital arteries** arise from the plantar arch to supply the toes with blood.

1. Describe the blood supply to the brain. Where is the arterial circle located, and how is it formed?
2. Describe the clinical significance of the brachial and radial arteries.
3. Describe the arterial pathway from the subclavian artery to the digital arteries.
4. List the arteries that supply blood to the lower abdominal wall, the external genitalia, the hamstring muscles, the knee joint, and the dorsum of the foot.

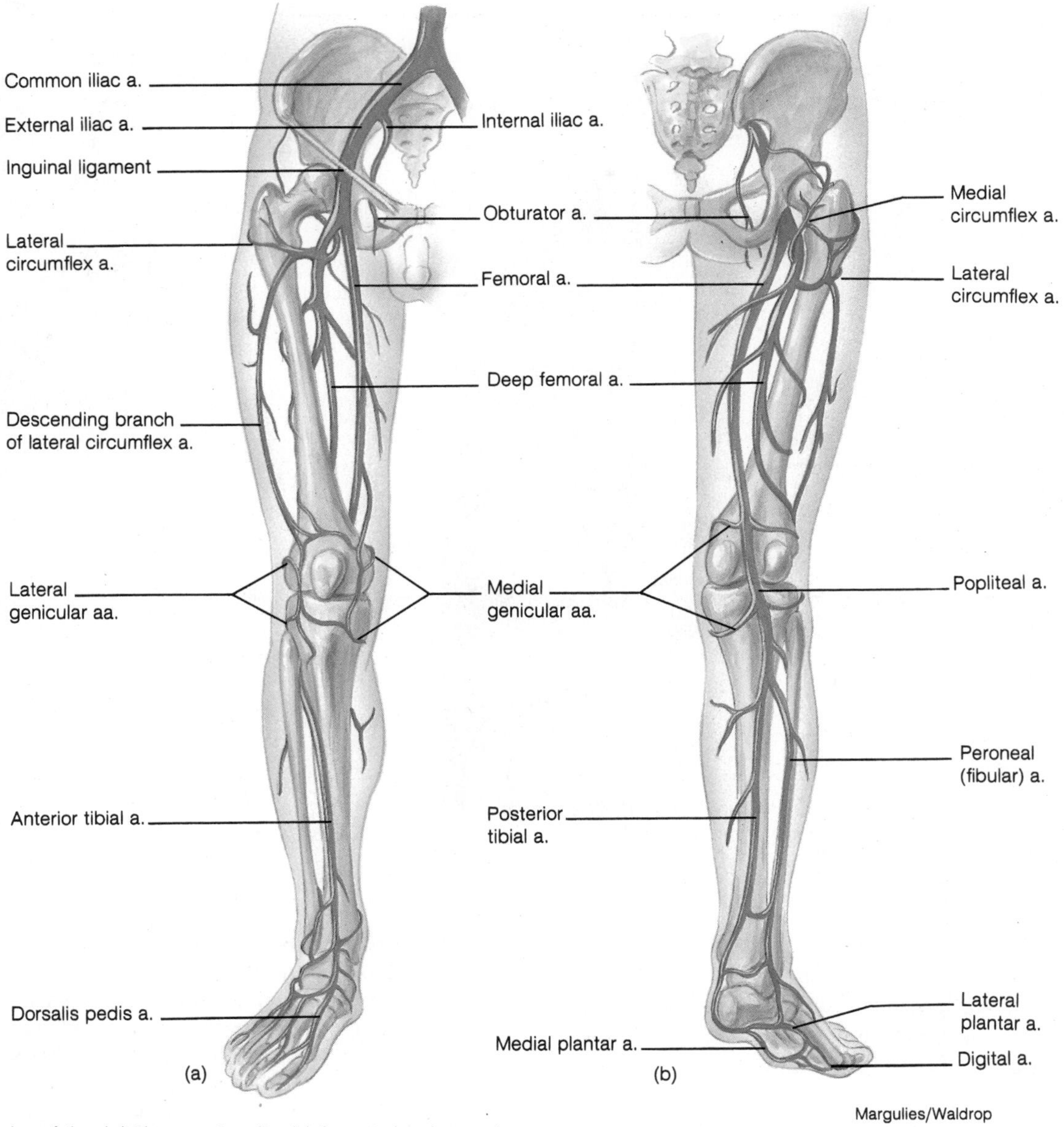

Figure 16.33 Arteries of the right lower extremity. (*a*) An anterior view and (*b*) a posterior view.

Principal Veins of the Body

After systemic blood has passed through the tissues and become depleted of oxygen, it returns through veins of progressively larger diameters to the right atrium of the heart.

Objective 20. Describe the venous drainage of the head, neck, and upper extremity.

Objective 21. Describe venous drainage of the thorax, lower extremities, and abdominal region.

Objective 22. Describe the vessels involved in the hepatic portal system.

In the venous portion of the systemic circulation, blood flows from smaller vessels into larger ones, so that a vein receives smaller tributaries instead of giving off branches as an artery does. The veins from all parts of the body converge into two major vessels that empty into the right atrium: the **superior** and **inferior vena cavae.** Veins are more numerous than arteries and are both superficial and deep. Superficial veins can generally be seen just beneath the skin and are clinically important in drawing blood and giving injections. Deep veins are close to the principal arteries and are usually similarly named. As with arteries, veins are named according to the region in which they are found or the organ that they serve (when a vein serves an organ, it drains blood away from it).

The principal systemic veins of the body are illustrated in figure 16.35.

Veins Draining the Head and Neck

Blood from the scalp, portions of the face, and the superficial neck regions is drained by the **external jugular veins** (fig. 16.36). These vessels descend on either side of the neck, superficial to the sternocleidomastoid muscle and deep to the platysma. They drain into the right and left **subclavian veins** located just behind the clavicles.

jugular: L. *jugulum,* throat or neck

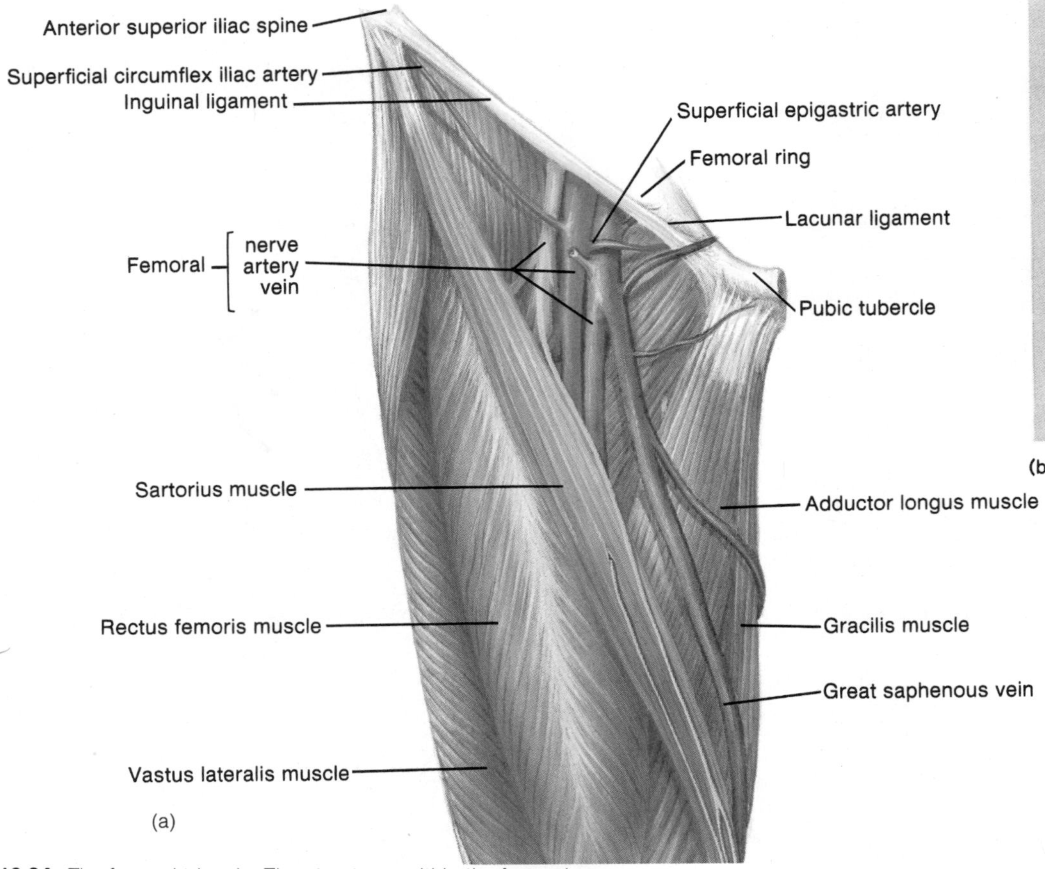

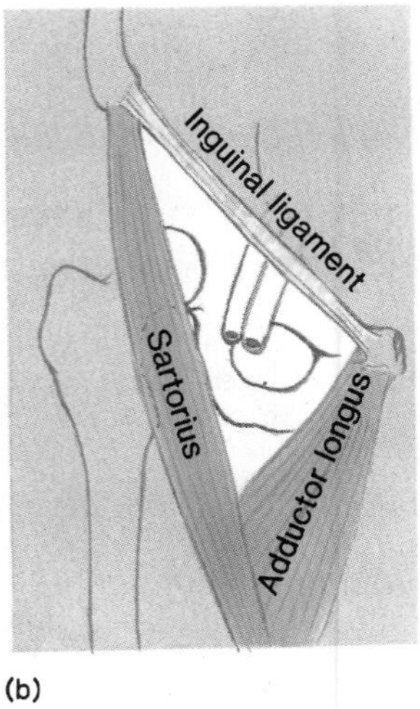

Figure 16.34 The femoral triangle. The structures within the femoral triangle are shown in (*a*) and the boundries of the triangle are shown in (*b*).

The paired **internal jugular veins** drain blood from the brain, meninges, and deep regions of the face and neck. The internal jugular veins are larger and deeper than the external jugular veins. They arise from numerous cranial **venous sinuses,** which are a series of both paired and unpaired channels positioned between the two layers of dura mater. The venous sinuses, in turn, receive venous blood from the **cerebral,** the **cerebellar,** the **ophthalmic** and the **meningeal veins.** The principal cranial venous sinuses and the formation of the internal jugular vein are illustrated in figure 16.37.

The internal jugular vein passes inferiorly down the neck adjacent to the common carotid artery and the vagus nerve. All three are surrounded by the protective *carotid sheath* and are positioned beneath the sternocleidomastoid muscle. The internal jugular empties into the subclavian vein, and the union of these two vessels forms the large **brachiocephalic vein** on each side. The two brachiocephalic veins then merge to form the **superior vena cava,** which drains into the right atrium of the heart (see fig. 16.35).

The external jugular vein is the vessel that is seen on the side of the neck when a person is angry or wearing a tight collar. You can voluntarily distend this vein by performing the *Valsalva maneuver.* To do this, take a deep breath and hold it while you forcibly contract your abdominal muscles as in a forced exhalation. This procedure is automatically performed when lifting a heavy object or defecating. The increased thoracic pressure that results compresses the vena cavae and interferes with the return of blood to the right atrium. The Valsalva maneuver, as a result, can be dangerous if performed by people with cardiovascular disease.

Valsalva maneuver: from Antonio Valsalva, Italian anatomist, 1666–1723

Veins of the Upper Extremity

The upper extremity has both superficial and deep venous drainage (fig. 16.38). The superficial veins are highly variable and form an extensive network just below the skin. The deep veins accompany the arteries of the same region and are given similar names. The deep veins of the upper extremity will be described first. Both the **radial vein** on the lateral side of the forearm and the **ulnar vein** on the medial side drain blood from the **deep** and **superficial palmar** (volar) **arches** of the hand. The radial and ulnar veins join in the cubital fossa to form the **brachial vein,** which continues up the medial side of the brachium.

The main superficial vessels of the upper extremity are the **basilic vein** and the **cephalic vein.** The basilic vein passes on the ulnar side of the forearm and the medial side of the arm. Near the head of the humerus, the basilic vein merges with the brachial vein and forms the **axillary vein.**

The cephalic vein drains the superficial portion of the hand and forearm on the radial side and then continues up the lateral side of the arm. In the shoulder region, the cephalic vein pierces

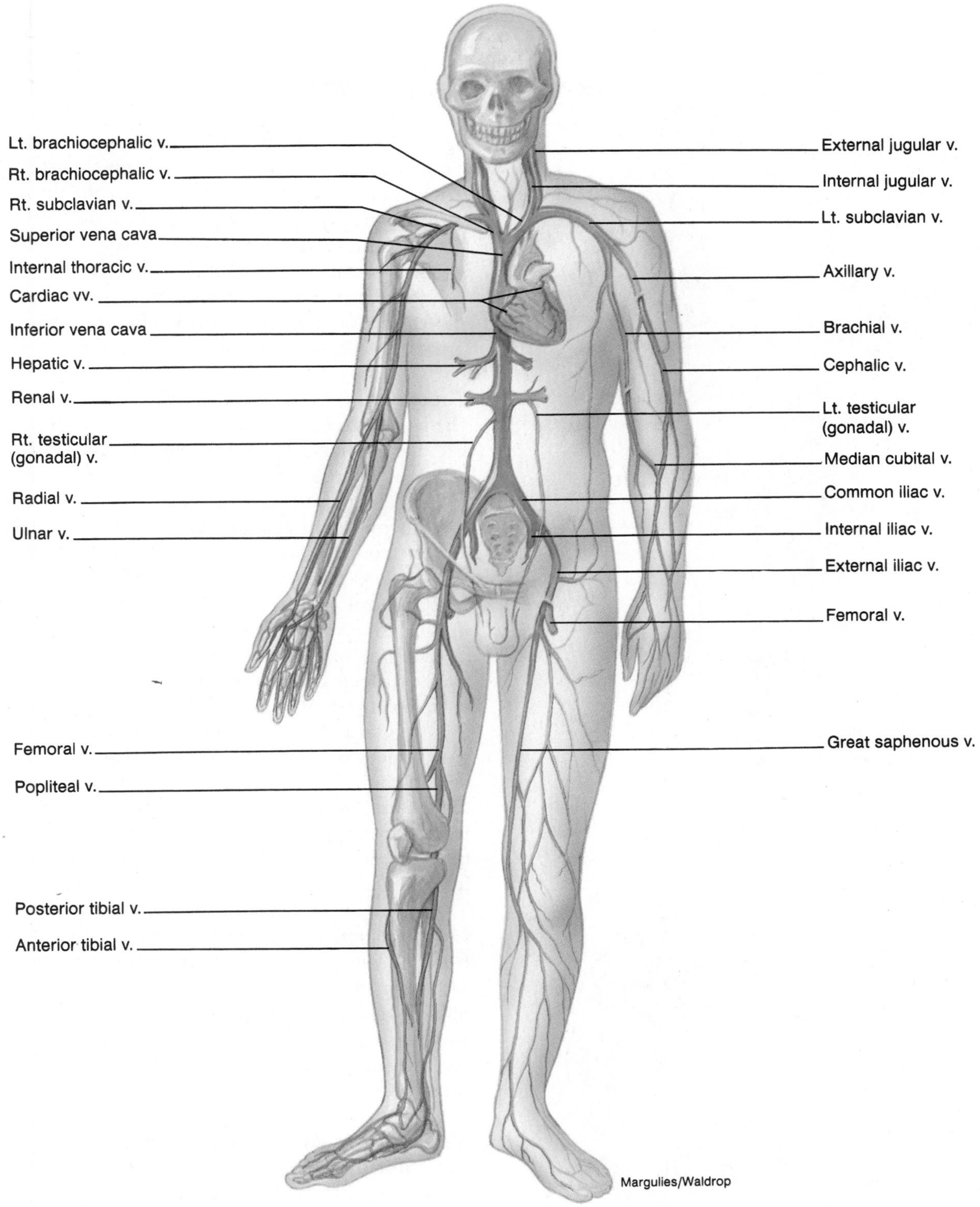

Figure 16.35 Principal veins of the body. Superficial veins are depicted in the left extremities and deep veins in the right extremities (v. = vein; vv. = veins).

the fascia and joins the axillary vein. The axillary vein then passes the first rib to form the subclavian vein, which unites with the internal jugular to form the brachiocephalic vein of that side.

Superficially, in the cubital fossa of the elbow, the **median cubital vein** ascends from the cephalic vein on the lateral side to connect with the basilic vein on the medial side. The median cubital vein is a frequent site for venipuncture in order to remove a sample of blood or add fluids to the blood.

Veins of the Thorax

The **superior vena cava,** formed by the union of the two brachiocephalic veins, empties venous blood from the head, neck, and upper extremities directly into the right atrium of the heart.

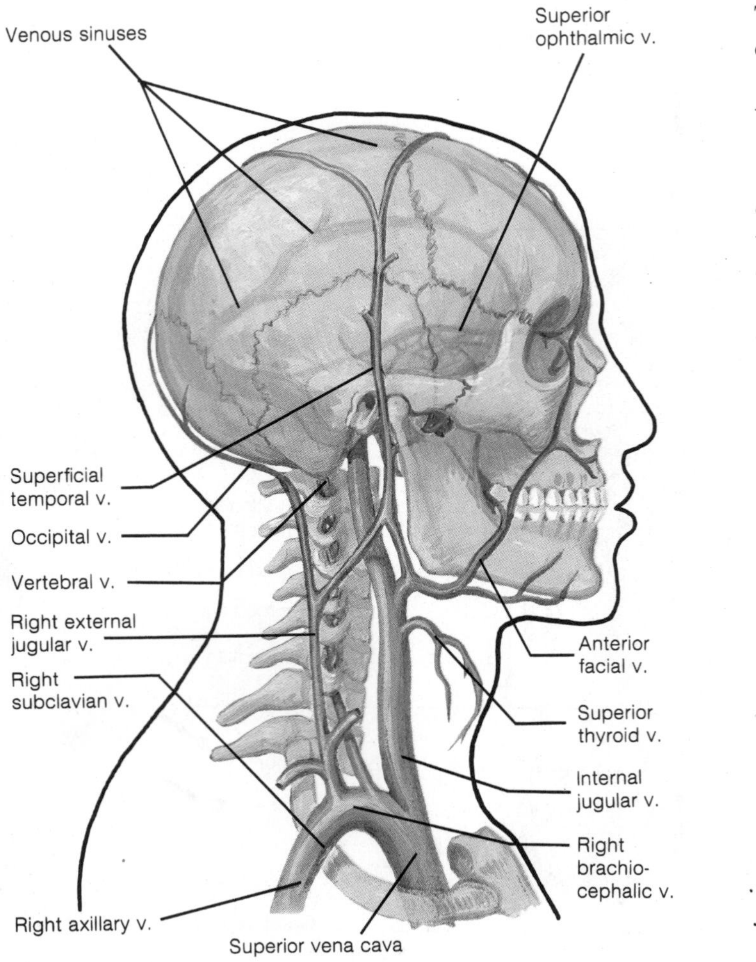

Figure 16.36 Veins that drain the head and neck.

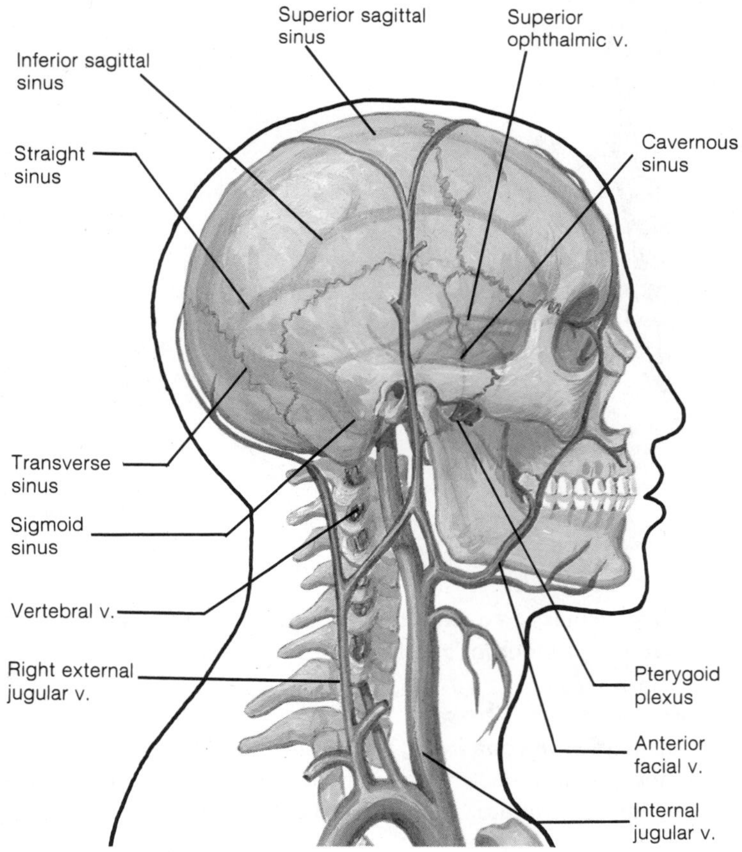

Figure 16.37 Cranial venous sinuses and the internal jugular vein.

These large vessels lack the valves that are characteristic of most other veins in the body.

In addition to receiving blood from the brachiocephalic veins, the superior vena cava collects blood from the azygos *(az'i-gos)* system of veins arising from the posterior thoracic wall (fig. 16.39). The **azygos vein** extends superiorly along the dorsal abdominal and thoracic walls on the right side of the vertebral column. The azygos vein ascends through the mediastinum to join the superior vena cava at the level of the fourth thoracic vertebra. Tributaries of the azygos vein include the **ascending lumbar veins** that drain the lumbar and sacral regions, **intercostal veins** draining from the intercostal muscles, and the **accessory hemiazygos** and **hemiazygos veins,** which form the major tributaries to the left of the vertebral column.

Veins of the Lower Extremity

The lower extremities, like the upper extremities, have both a deep and a superficial group of veins (fig. 16.40). The deep veins accompany corresponding arteries and have more valves than do the superficial veins. The deep veins will be described first.

The **posterior** and **anterior tibial veins** originate in the foot and course upward behind and in front of the tibia to the back of the knee, where they merge to form the **popliteal vein.** The popliteal vein receives blood from the knee region. Just above the knee, this vessel becomes the **femoral vein.** The femoral vein in turn continues up the thigh and receives blood from the **deep femoral vein** near the groin. Just above this, the femoral vein receives blood from the **great saphenous** *(sah-fe'nus)* **vein** and then becomes the **external iliac vein** as it passes under the inguinal ligament. The external iliac curves upward to the level of the sacroiliac joint. There it merges with the **internal iliac** (hypogastric) **vein** at the pelvic and genital regions to form the **common iliac vein.** At the level of the fifth lumbar vertebra, the right and left common iliacs unite to form the large **inferior vena cava** (fig. 16.40).

The superficial veins of the lower extremity are the **small** and **great saphenous veins.** The small saphenous vein arises from the lateral side of the foot, courses posteriorly along the surface of the calf of the leg, and descends deep to enter the popliteal vein behind the knee. The great saphenous vein is the longest vessel in the body. It originates at the arch of the foot and ascends superiorly along the medial aspect of the leg and thigh before draining into the femoral vein.

Veins of the Abdominal Region

The **inferior vena cava** parallels the abdominal aorta on the right side as it ascends through the abdominal cavity to penetrate the diaphragm and enter the right atrium (see fig. 16.35). It is the largest vessel in diameter in the body and is formed by the union of the two common iliac veins draining the lower extremities. As the inferior vena cava ascends through the abdominal cavity, it receives tributaries from veins that correspond in name and position to arteries previously described.

Four paired **lumbar veins** (not shown) drain the posterior abdominal wall, the vertebral column, and the spinal cord. The

azygos: Gk. *a,* without; *zygon,* yoke
saphenous: L. *saphena,* the hidden one

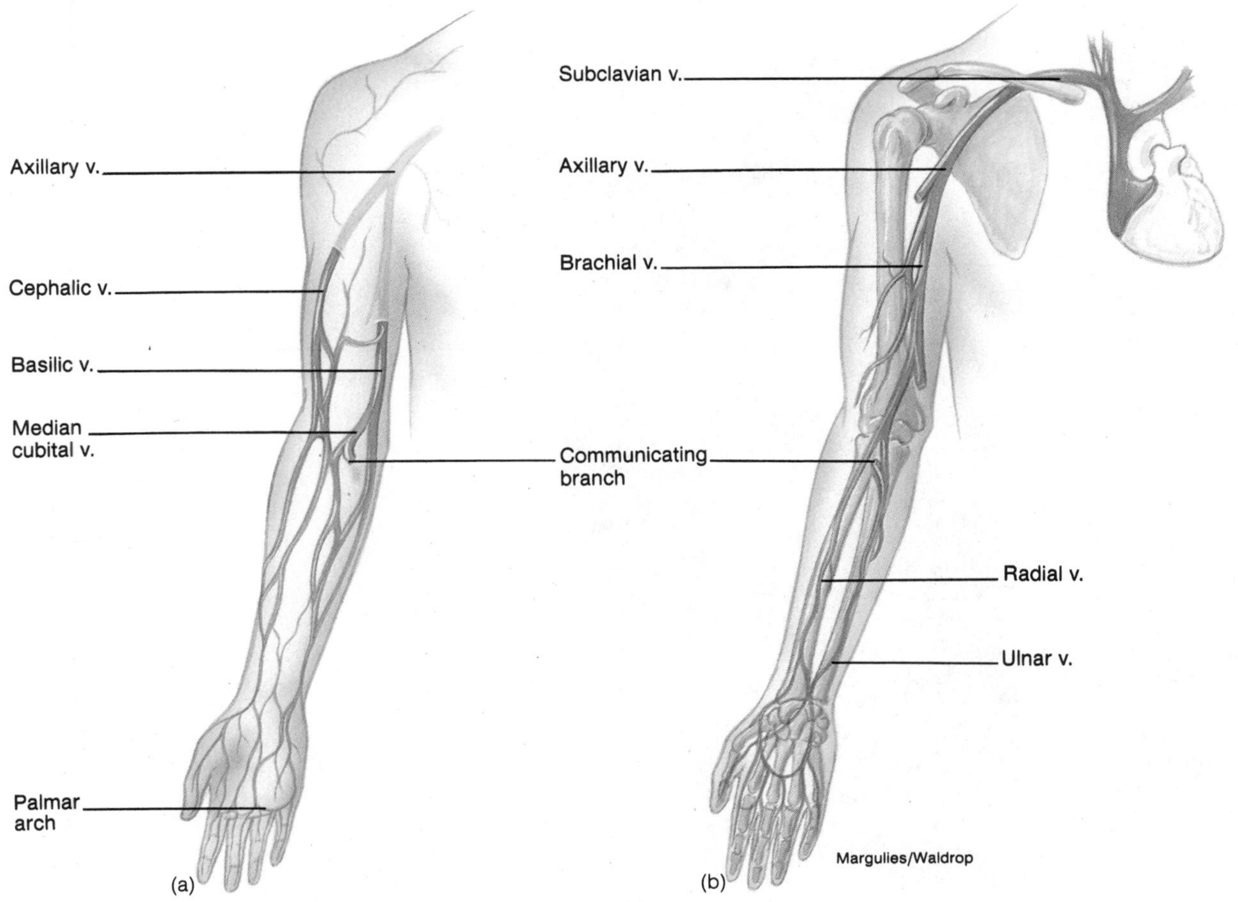

Figure 16.38 An anterior view of the veins that drain the upper right extremity. (*a*) Superficial veins and (*b*) deep veins.

renal veins drain blood from the kidneys and ureters into the inferior vena cava. The **right testicular vein** or the **right ovarian vein,** draining the corresponding gonads, and the **right suprarenal vein,** draining the right adrenal gland, each empty into the inferior vena cava. The **left testicular vein** or **left ovarian vein,** and the **left suprarenal vein,** in contrast, each drain into the left renal vein. The **inferior phrenic veins** receive blood from the inferior side of the diaphragm and drain into the inferior vena cava. **Right** and **left hepatic veins** originate from the capillary sinusoids of the liver and empty into the inferior vena cava immediately below the diaphragm.

Note that the inferior vena cava does not receive blood directly from the GI tract, pancreas, or spleen. Instead, the venous outflow from these organs first passes through capillaries in the liver.

Hepatic Portal System

A *portal system* is one in which the veins that drain one group of capillaries deliver blood to another group of capillaries, which in turn are drained by more usual systemic veins that carry blood to the vena cavae and the right atrium of the heart. There are thus two capillary beds in series. The hepatic portal system is composed of veins that drain blood from capillaries in the intestines, pancreas, spleen, stomach, and gallbladder into capillaries in the liver (called *sinusoids*) and of the right and left **hepatic veins** that drain the liver and empty into the inferior vena cava (fig. 16.41). As a result of the hepatic portal system, the absorbed products of digestion must first pass through the liver before entering the general circulation.

The **hepatic portal vein** is the large vessel that receives blood from the digestive organs. It is formed by a union of the **superior mesenteric vein,** which drains nutrient-rich blood from the small intestine, and the **splenic vein.** The splenic vein drains the spleen but is enlarged because of a convergence of the following four tributaries: (1) the **inferior mesenteric vein** from the large intestine; (2) the **pancreatic vein** from the pancreas; (3) the **left gastroepiploic vein** from the stomach; and (4) the **right gastroepiploic vein,** also from the stomach.

Three additional veins empty into the portal vein. The **right** and **left gastric veins** drain the lesser curvature of the stomach, and the **cystic vein** drains blood from the gallbladder.

gastroepiploic: Gk. *gastros,* stomach; *epiplein,* to float on (referring to greater omentum)

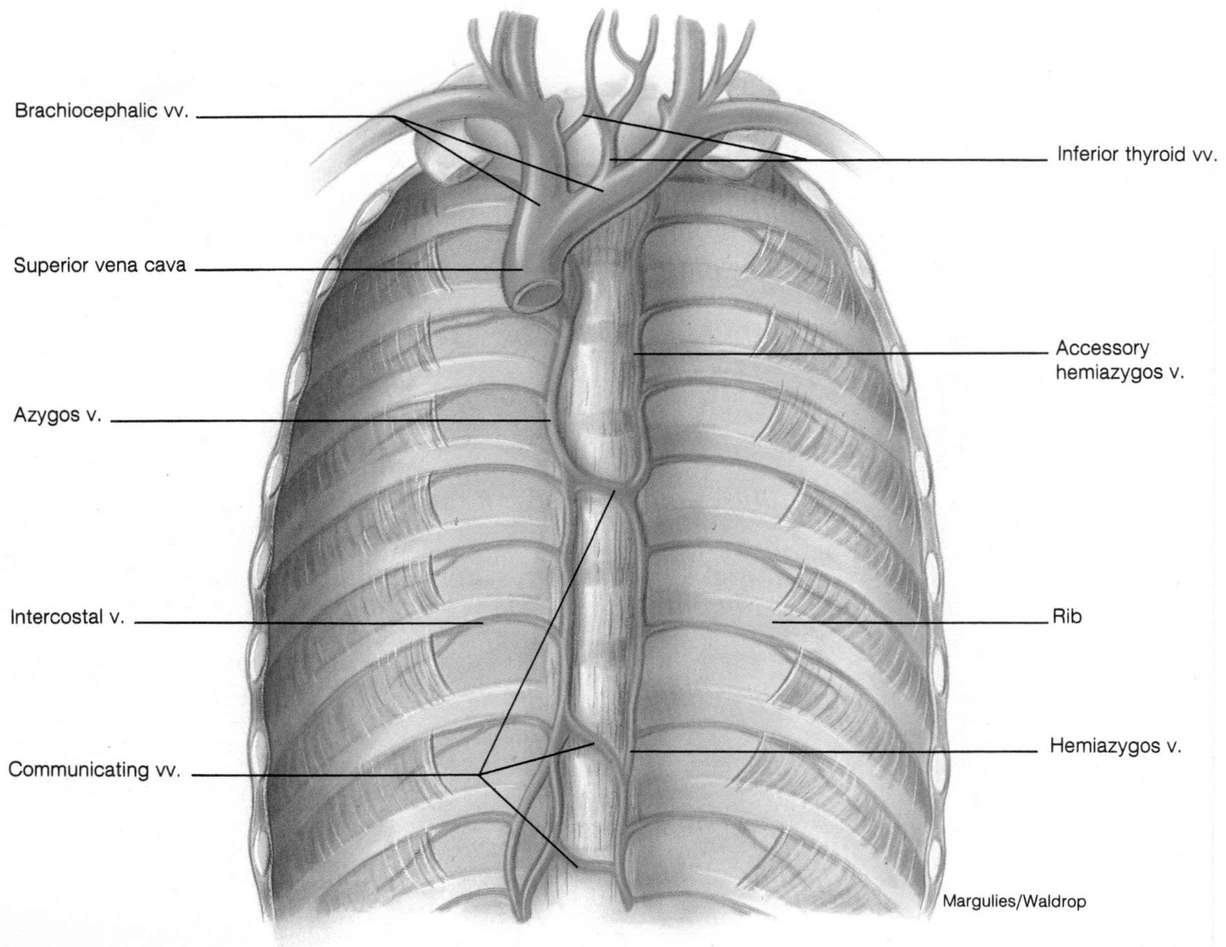

Figure 16.39 Veins of the thoracic region. The lungs and heart have been removed.

One of the functions of the liver is to detoxify harmful substances, such as alcohol, that are absorbed into the blood from the small intestine. However, excessive quantities of alcohol cannot be processed during a single pass through the liver, and so a person becomes intoxicated. Eventually, the liver is able to process the alcohol as the circulating blood is repeatedly exposed to the liver sinusoids via the hepatic artery. Alcoholics may eventually suffer from *cirrhosis* of the liver as the normal liver tissue is destroyed.

In summary, it is important to note that the sinusoids of the liver receive blood from two sources. The hepatic artery supplies oxygen-rich blood to the liver, whereas the hepatic portal vein transports nutrient-rich blood from the intestine for processing. These two blood sources become mixed in the liver sinusoids. Liver cells exposed to this blood obtain nourishment from it and are uniquely qualified (because of their anatomical position and enzymatic ability) to modify the chemical nature of the venous blood that enters the general circulation from the digestive tract.

1. Using a flowchart, list the venous drainage from the head and neck to the superior vena cava. Indicate which vein may bulge in the side of the neck when a person performs the Valsalva maneuver and which vein is commonly used as a site for venipuncture.
2. Describe the positions, sources, and drainages of the small and great saphenous veins.
3. Describe the hepatic portal system, and explain the functional importance of this system.

Fetal Circulation

The fetal circulation is adapted to the fact that the fetal lungs are nonfunctional and that oxygen and nutrients are obtained from the placenta.

Objective 23. Describe the fetal circulation to and from the placenta.

Objective 24. Describe the structure and function of the foramen ovale, ductus venosus, and ductus arteriosus.

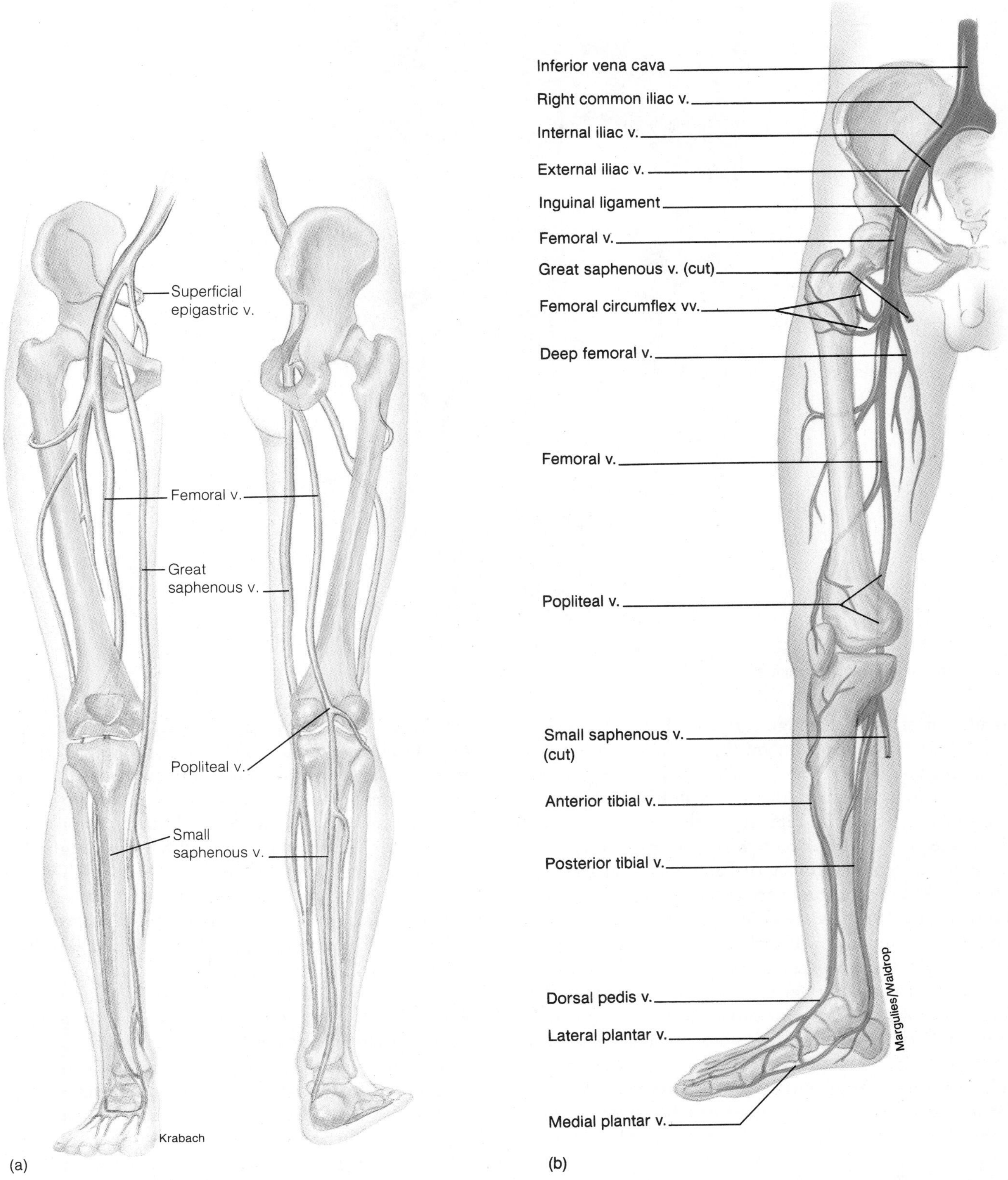

Figure 16.40 Veins of the lower extremity. (*a*) Superficial veins, medial and posterior aspects and (*b*) deep veins, medial view.

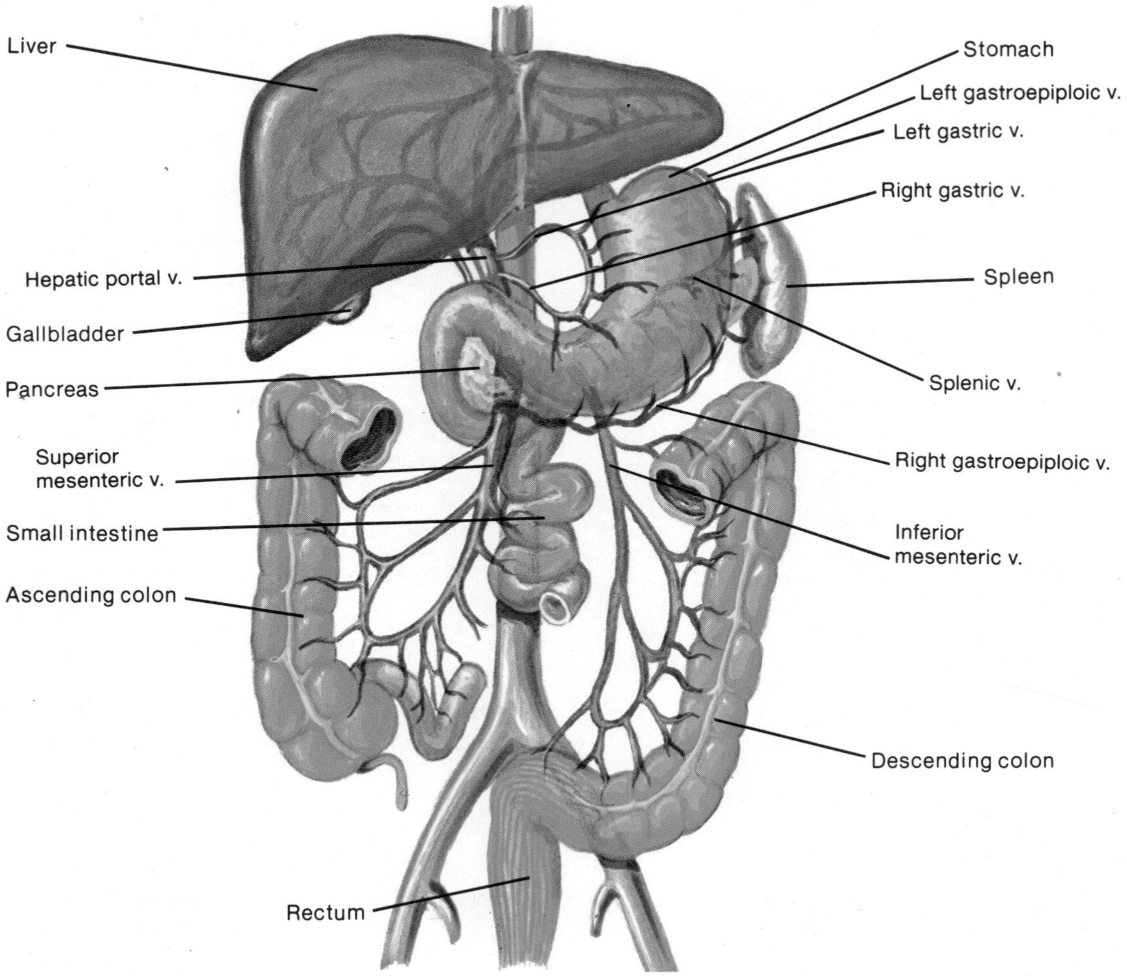

Figure 16.41 The hepatic portal system.

The circulation of blood through a fetus is by necessity different from that of a newborn infant (fig. 16.42). Respiration, the procurement of nutrients, and the elimination of metabolic wastes occur through the maternal blood instead of through the organs of the fetus. The capillary exchange between the maternal and fetal circulation occurs within the **placenta.** This remarkable structure includes maternal and fetal capillary beds and is discharged following delivery as the afterbirth.

The **umbilical cord** is the connection between the placenta and the fetal umbilicus. It includes one **umbilical vein** and two **umbilical arteries** surrounded by a gelatinous substance. Oxygenated and nutrient-rich blood flows through the umbilical vein toward the inferior surface of the liver. At this point, the umbilical vein bifurcates into one branch that joins with the portal vein, while the other branch, called the **ductus venosus,** enters the inferior vena cava. Thus, oxygenated blood is mixed with venous blood returning from the lower extremities of the fetus before it enters the heart. The umbilical vein is the only vessel of the fetus that carries fully oxygenated blood.

The inferior vena cava empties into the right atrium of the fetal heart. From here, most of the blood is directed through the **foramen ovale,** an opening between the two atria. Here it mixes with a small quantity of blood returning through the pulmonary circulation. The blood then passes into the left ventricle, from which it is pumped into the aorta and through the body of the fetus. Some blood entering the right atrium passes into the right ventricle and out of the heart via the pulmonary trunk. Since the lungs of the fetus are not functional, only a small portion of blood continues through the pulmonary circulation (the resistance to blood flow is very high in the collapsed fetal lungs). Most of the blood in the pulmonary trunk passes through the **ductus arteriosus** into the aortic arch, where it mixes with blood coming from the left ventricle. Blood is returned to the placenta by the two **umbilical arteries** that arise from the internal iliac arteries.

Notice that in the fetus oxygen-rich blood is transported by the inferior vena cava to the heart and via the foramen ovale and ductus arteriosus to the systemic circulation.

Important changes occur in the cardiovascular system at birth. The foramen ovale, ductus arteriosus, ductus venosus, and the umbilical vessels are no longer necessary. The foramen ovale abruptly closes with the first breath of air because the reduced pressure in the right side of the heart causes a flap to cover the opening. This reduced pressure is because (1) when the lungs fill with air, the vascular resistance to blood flow in the pulmonary circulation falls far below that of the systemic circulation; and (2) the pressure in the inferior vena cava and right atrium falls as a result of the loss of the placental circulation.

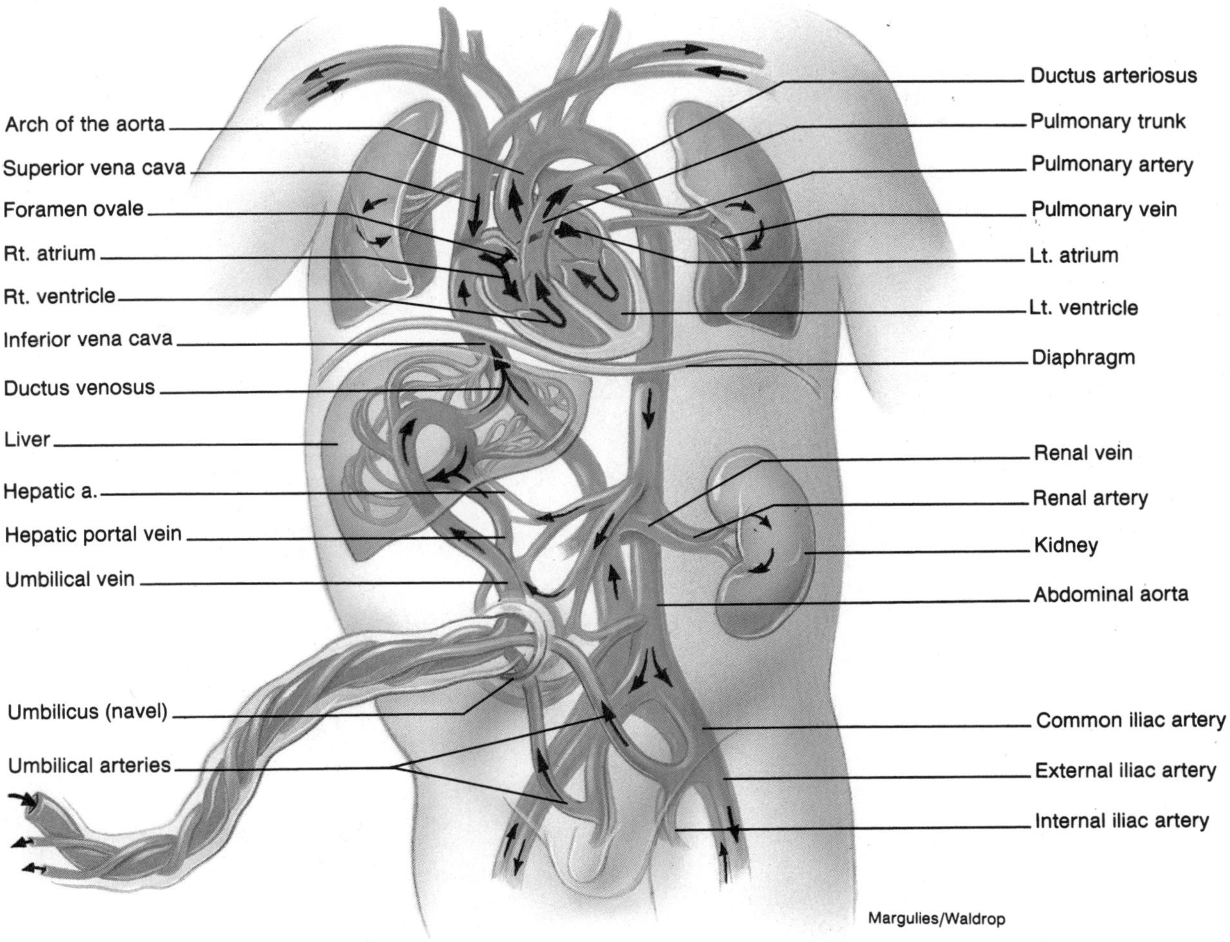

Figure 16.42 Fetal circulation. Arrows indicate the direction of blood flow.

Table 16.8 Cardiovascular structures of the fetus and changes in the neonatal infant

Structure	Location	Function	Fate in neonatal infant
Umbilical vein	Connects the placenta to the liver; forms a major portion of umbilical cord	Transports nutrient-rich, oxygenated blood from the placenta to the fetus	Forms the round ligament of the liver
Ductus venosus	Venous shunt within the liver to connect with the inferior vena cava	Transports oxygenated blood directly into the inferior vena cava	Forms the ligamentum venosum, a fibrous cord in the liver
Foramen ovale	Opening between the right and left atria	A shunt to bypass the pulmonary circulation	Closes at birth and becomes the fossa ovalis, a depression in the interatrial septum
Ductus arteriosus	Between pulmonary trunk and arch of the aorta	A shunt to bypass the pulmonary circulation	Closes shortly after birth, atrophies, and becomes the ligamentum arteriosum
Umbilical arteries	Arise from internal iliac arteries and associated with umbilical cord	Transport blood from the fetus to the placenta	Atrophies to become the lateral umbilical ligaments

The constriction of the ductus arteriosus occurs gradually over a period of about six weeks after birth as the vascular muscle cells constrict in response to the higher oxygen concentration in the postnatal blood. The remaining structure of the ductus gradually atrophies and becomes nonfunctional as a blood vessel. The fate of the unique fetal cardiovascular structures is summarized in table 16.8.

1. Describe the path of blood from the fetal heart, through the placenta, and back to the fetal heart.
2. Trace the pathway of oxygenated blood through the fetal circulation.
3. Describe the foramen ovale and ductus arteriosus, and explain why blood flows the way it does through these structures in the fetal circulation.

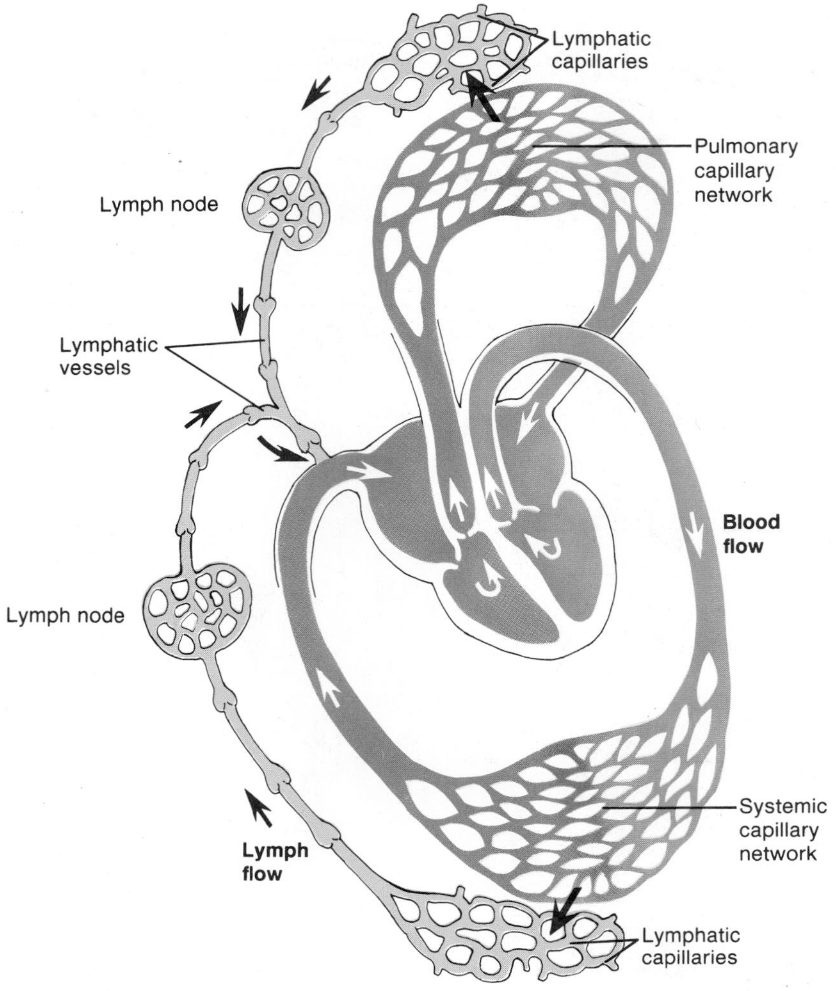

Figure 16.43 The schematic relationship of the circulatory and lymphatic systems. Lymphatic vessels transport lymph fluid from interstitial spaces to the venous bloodstream.

Lymphatic System

The lymphatic system consists of lymphatic vessels and lymph nodes. It functions to return tissue fluid to the venous system and to help protect the body from diseases.

Objective 25. Describe the pattern of lymph flow from the lymphatic capillaries to the venous system.

Objective 26. Describe the structure and location of the lymph nodes, and name two lymphatic organs.

The lymphatic system is closely interrelated anatomically and physiologically to the circulatory system (fig. 16.43). The functions of the lymphatic system are basically threefold: (1) it transports excess interstitial (tissue) fluid, which was initially formed as a blood filtrate, back to the bloodstream; (2) it serves as the route by which absorbed fat from the intestine is transported to the blood; and (3) it helps provide immunological defenses against disease-causing agents.

Lymph and Lymph Capillaries

The smallest vessels of the lymphatic system are the **lymph capillaries** (fig. 16.44). Lymph capillaries are microscopic, closed-ended tubes that form vast networks in the intercellular spaces within most tissues. Within the villi of the small intestine, for example, lymph capillaries, called *lacteals (lak'te-al),* transport the products of fat absorption away from the digestive tract.

Because the walls of lymph capillaries are composed of a highly permeable simple squamous epithelium, tissue fluid can readily enter. This fluid is formed as a filtrate of plasma through blood capillaries and is identical in composition to plasma except for a lower protein concentration (proteins are largely held back by the filtration process). Adequate lymphatic drainage is needed to prevent the accumulation of tissue fluid, a condition called *edema (ĕ-de'mah).* Once fluid enters the lymphatic capillaries, it is referred to as **lymph.**

lymph: L. *lympha,* clear water
lacteal: L. *lacteus,* milk
edema: Gk. *edema,* a swelling

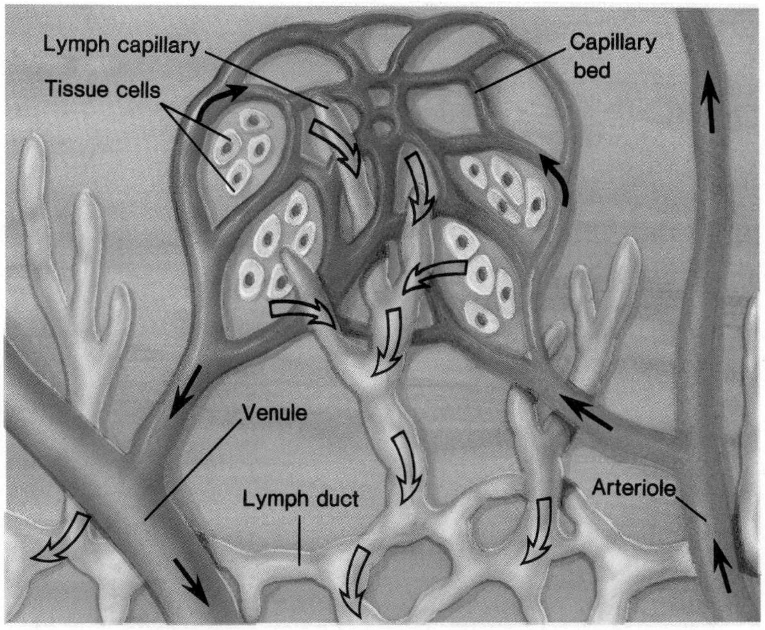

Figure 16.44 A schematic diagram showing the structural relationship of a capillary bed and a lymph capillary.

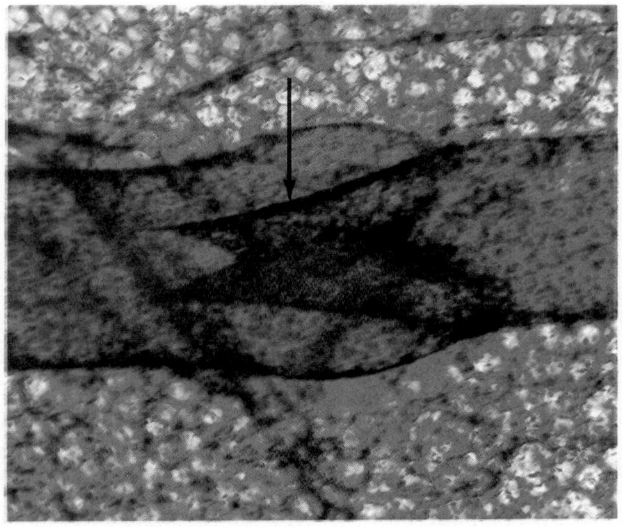

Figure 16.45 A photomicrograph of a valve (*arrow*) within a lymphatic vessel.

Lymph Ducts

From merging lymph capillaries, the lymph is carried into larger **lymph ducts.** The walls of lymph ducts are similar to those of veins in that they have the same three layers and contain valves to prevent the backflow of lymph (fig. 16.45). Interconnecting lymph ducts eventually empty into one of the two principal vessels: the **thoracic duct** and the **right lymphatic duct** (fig. 16.46).

The larger thoracic duct drains lymph from the lower extremities, abdomen, left thoracic region, left upper extremity, and left side of the head and neck. The main trunk of this vessel ascends along the spinal column and drains into the left subclavian vein. In the abdominal area there is a saclike enlargement of the thoracic duct called the **cisterna chyli** *(sis-ter'nah ki'le)*. The shorter right lymphatic duct drains lymph vessels from the right upper extremity, right thoracic region, and right side of the head and neck. The right lymphatic duct empties into the right subclavian vein near the internal jugular vein.

The pressure that pushes lymph through the lymph ducts comes from the massaging actions produced by skeletal muscle contractions, intestinal movements, and other body movements. Valves in the walls of lymph vessels keep lymph moving in one direction.

---

cisterna chyli: L. *cisterna*, box, Gk. *chylos*, juice

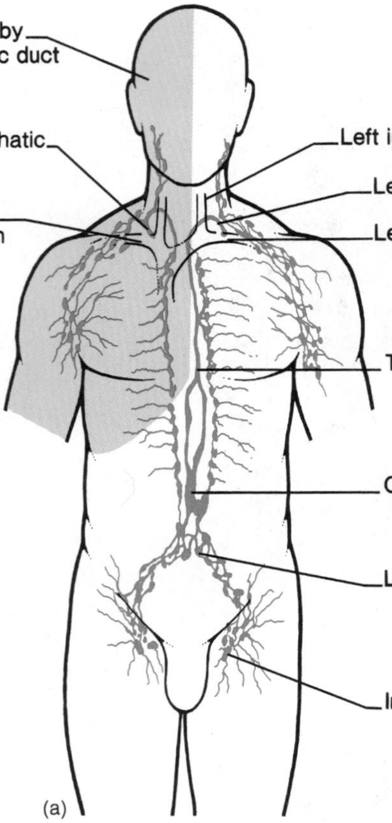

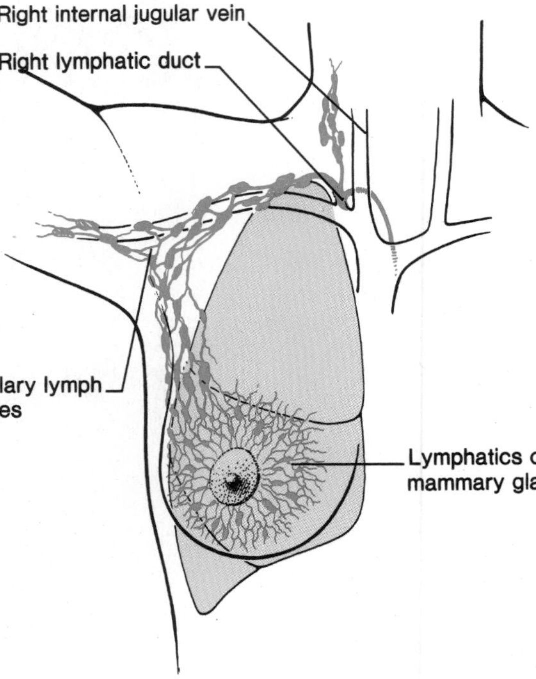

Figure 16.46 Lymphatic vessels. (*a*) The major lymph drainage of the body and (*b*) a magnified view of the upper right quadrant showing the lymph drainage of the right breast and axilla.

Lymph Nodes

Lymph filters through the reticular tissue of **lymph nodes** (fig. 16.47), which contain phagocytic cells that help purify the fluid. Lymph nodes are small, oval bodies enclosed within fibrous connective tissue *capsules.* Specialized connective tissue bands called *trabeculae* divide the node. *Afferent lymphatic vessels* carry lymph into the node, where it is circulated through sinuses in the *cortical tissue.* Lymph leaves the node through the *efferent lymphatic vessel,* which emerges from the *hilum,* the depression on the concave side. **Germinal centers** within the node are the sites of lymphocyte production.

Lymph nodes usually occur in clusters in specific regions of the body (fig. 16.48). Some of the principal groups of lymph nodes are the **popliteal** (not illustrated) and **inguinal nodes** of the lower extremity; the **lumbar nodes** of the pelvic region; the **cubital** and **axillary nodes** of the upper extremity; the **thoracic nodes** of the chest; and the **cervical nodes** of the neck. The submucosa of the small intestine contains numerous scattered lymphocytes and lymphatic nodules and larger clusters of lymphatic tissue called **mesenteric** (Peyer's) **patches.**

Migrating cancer cells (metastases) are especially dangerous if they enter the lymphatic system, which can disperse them widely. On entering the lymph nodes, the cancer cells can multiply and establish secondary tumors in organs that are distal to the lymphatic drainage of the site of the primary tumor.

Peyer's patches: from Johann K. Peyer, Swiss anatomist, 1653–1712

Lymphoid Organs

The **spleen** and the **thymus** are lymphoid organs. The spleen is located posterior and lateral to the stomach, from which it is suspended (fig. 16.49). The spleen is not a vital organ in an adult, but it does assist other body organs in producing lymphocytes, filtering the blood, and destroying old erythrocytes. In an infant, it is an important site for erythrocyte production.

The thymus is located in the anterior thorax, deep to the manubrium of the sternum (fig. 16.50). It is much larger in a fetus and child than in an adult, because it regresses in size during puberty. The thymus plays a key role in the immune system.

The lymphatic organs are summarized in table 16.9.

The tonsils, of which there are three pairs, are actually lymphatic organs of the pharyngeal region (see chap. 18). The function of the tonsils is to combat infection of the ear, nose, and throat regions. Because of the persistent infections that some children suffer, the tonsils may become so overrun with infections that they actually become the source of infections. A *tonsillectomy* may then have to be performed. This operation is not as common as in the past because of the availability of powerful antibiotics and because the functional value of the tonsils is more greatly appreciated.

spleen: L. *splen,* low spirits (thought to cause melancholy)
thymus: Gk. *thymos,* thyme (compared to the flower of this plant by Galen)

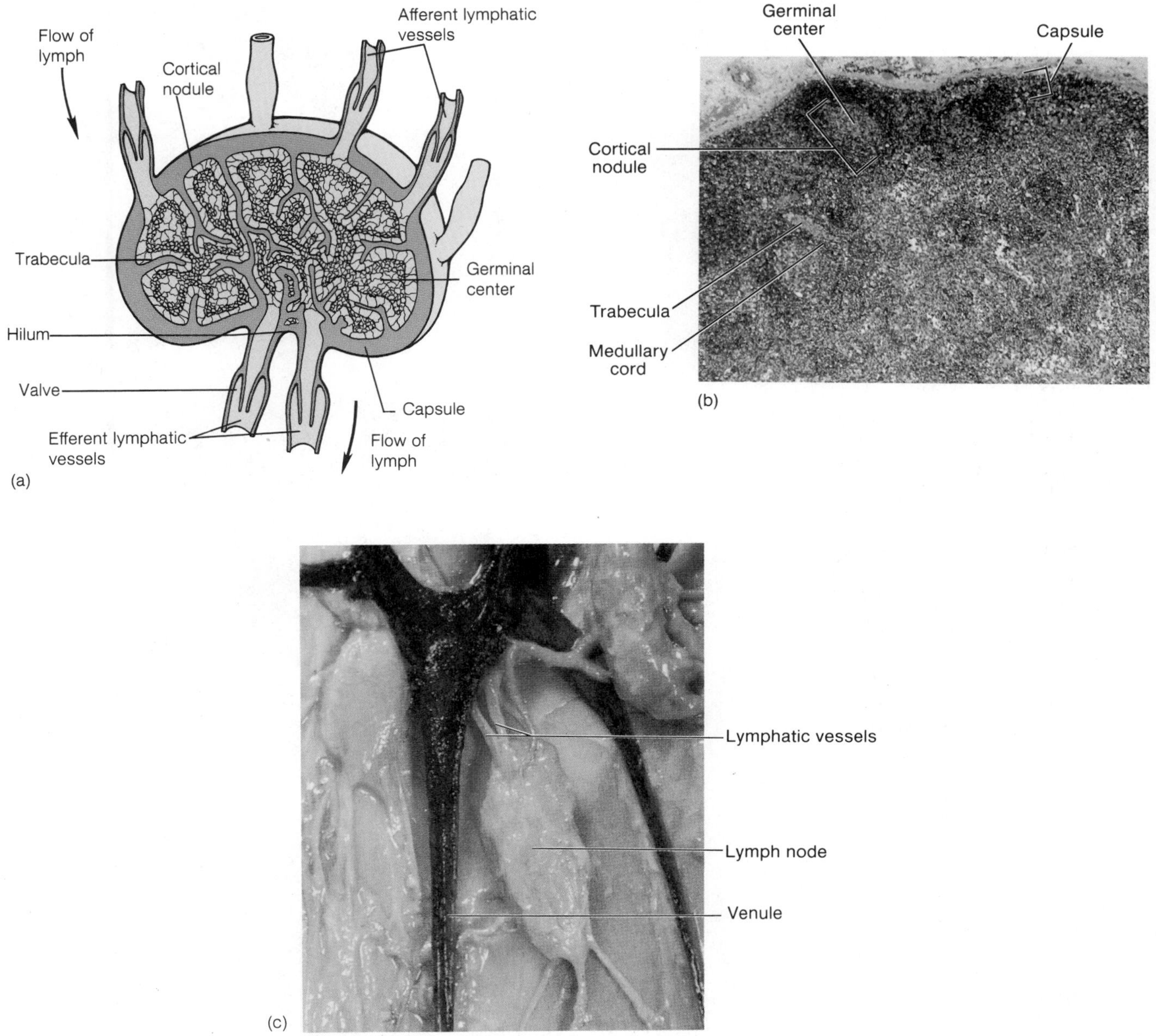

Figure 16.47 The structure of a lymph node. (*a*) A schematic diagram of a sectioned lymph node and associated vessels. (*b*) A photomicrograph of lymphatic tissue. (*c*) A photograph of a lymph node positioned near a blood vessel.

---

1. Describe the relationship between lymph and blood in terms of their origin, composition, and fate.
2. List the two major lymph vessels of the body, and describe their relationships to the vascular system.
3. Describe the structure, location, and function of the lymph nodes, and list the lymphoid organs.

Clinical Considerations

It has been said that as the circulatory system goes, so goes the rest of the body. The circulatory system is the lifeline to the other organs of the body, and if the vital organs are deprived of the essentials for life, they will falter in their functions and bring about disease or death. Some gerontologists believe that the first system of the body to show signs of aging (see chap. 23) is the circulatory system and that this accelerates the aging process.

Clinically speaking, the circulatory system is extremely important because of the frequent irregularities and diseases that afflict it and the effect of its dysfunction on maintaining homeostasis. Lengthy books have been written on diseases and dysfunctions of the cardiovascular system, and entire specialities within medicine are devoted to its treatment. It is not practical in this book to attempt a comprehensive discussion of the numerous clinical aspects of the cardiovascular system. However, some general comments will be made on cardiovascular diagnosis, on dysfunctions of the blood, lymphatics, vessels, and heart, and on the prevention and treatment of heart problems.

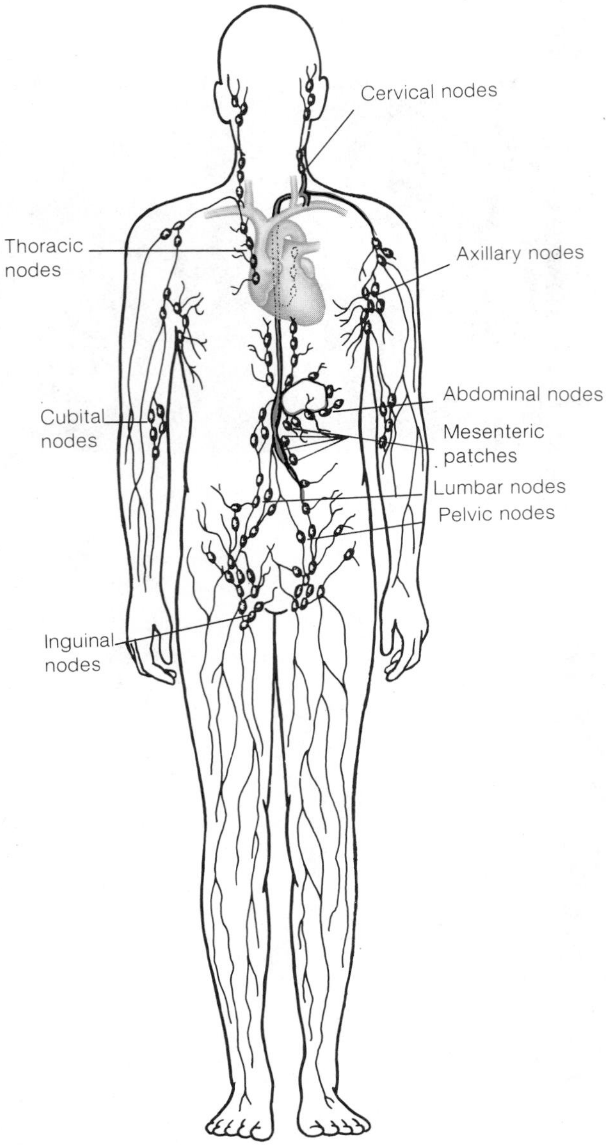

Figure 16.48 Major locations of lymph nodes.

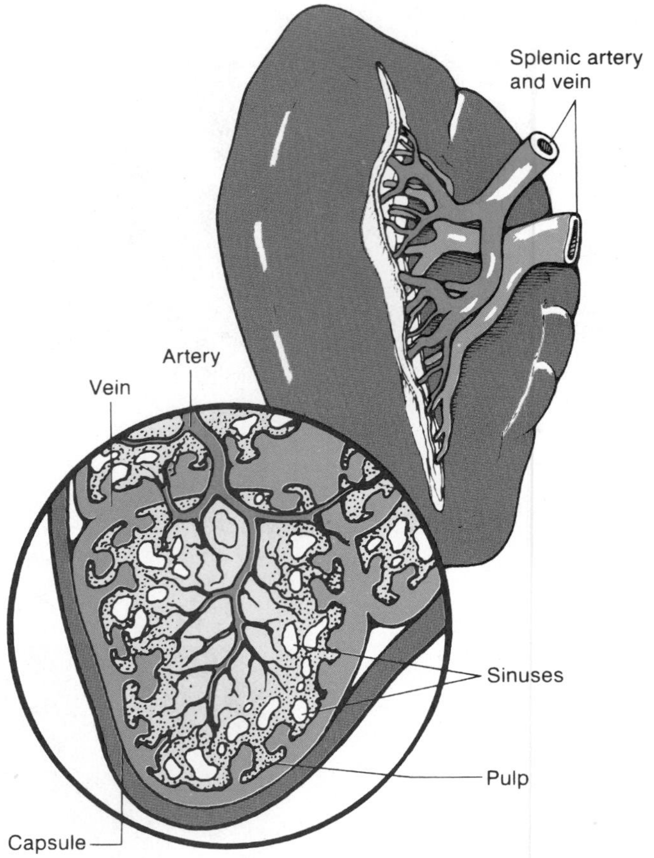

Figure 16.49 The structure of the spleen.

Table 16.9 Lymphatic organs

Organ	Location	Function
Lymph nodes	In clusters or chains along the paths of larger lymphatic vessels	Lymphocyte production; house T-lymphocytes and B-lymphocytes that are responsible for immunity; phagocytes filter foreign particles and cellular debris from lymph
Spleen	In upper left portion of abdominal cavity beneath the diaphragm and suspended from the stomach	Serves as blood reservoir; phagocytes filter foreign particles, cellular debris, and worn erythrocytes from the blood; houses lymphocytes
Thymus	Within the mediastinum behind the manubrium	Important site of immunity in a child; houses lymphocytes; changes undifferentiated lymphocytes into T-lymphocytes

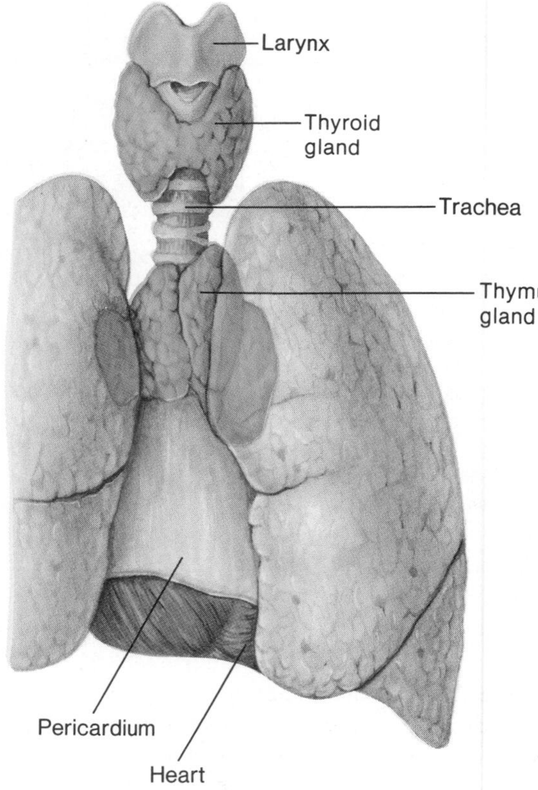

Figure 16.50 The location of the thymus within the mediastinum between the lungs.

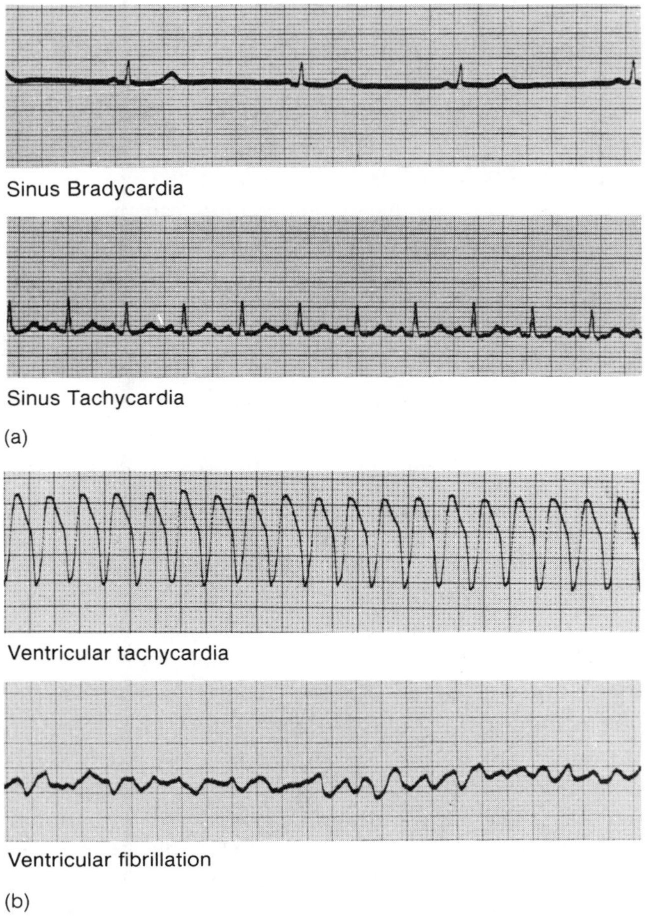

Figure 16.51 In (*a*) the heartbeat is paced by the normal pacemaker—the SA node (hence the name sinus rhythm). This can be abnormally slow (bradycardia—46 beats per minute in this example) or fast (tachycardia—136 beats per minute in this example). Compare the pattern of tachycardia in (*a*) with the tachycardia in (*b*), which is produced by an ectopic pacemaker in the ventricles. This dangerous condition can quickly lead to ventricular fibrillation, also shown in (*b*).

Cardiovascular Assessment

Cardiovascular assessment is an extremely important aspect of a physical examination for identifying existing and potential health problems. Several techniques are routinely used in gathering information. Auscultation of heart sounds with a **stethoscope** is not only important for detecting murmurs, but may help one evaluate general cardiac functioning or the progress of a patient with a cardiovascular problem.

An **electrocardiograph** (ECG) is an important way of assessing electrical activity within the heart. The most important diagnostic use of the resting ECG is identifying abnormal cardiac rhythms. Injury to cardiac muscle, as in myocardial infarction, elevates or depresses the ST segment while inverting and shortening the T wave.

There are several diagnostic techniques for determining the size and position of the heart, the condition of the pericardial sac, and even the internal structure and condition of the vessels, chambers, and valves. Photographs from X rays, CT scans, and ultrasound are important for gaining a perspective of how the heart functions and relates to other thoracic structures. The accumulation of fluids is easy to detect using these techniques. **Catheterization** of the heart in conjunction with a **fluoroscopic analysis** permits observation of the vessels, chambers, and valves of the heart. In addition, internal blood pressures can be recorded and samples of blood taken. Selective **arteriography** enables a physician to study the wall of a particular vessel. An **angiocardiogram** is a radiograph of the chambers of the heart and its vessels made after an intravenous injection of a radiopaque substance.

Arterial blood pressure can be measured using a **sphygmomanometer.** Blood pressure is indicative of the general health of the cardiovascular system. **Hypertension** is sustained high blood pressure and can result in mechanical damage to the cardiovascular system. Hypertension can be caused by resistance in the vessels from physiological actions (vasoconstriction) or by various vascular diseases (arteriosclerosis). Certain characteristics of the blood itself may cause hypertension.

Cardiac Arrhythmia Arrhythmias, or abnormal heart rhythms, can be detected and described by the abnormal ECG patterns they produce. Since a heartbeat occurs whenever a normal QRS complex (see fig. 16.17) is seen and since the ECG chart paper moves at a known speed, its x-axis indicates time, and the cardiac rate (beats per minute) can easily be obtained from the ECG recording. A cardiac rate slower than 60 beats per minute indicates **bradycardia** *(brad''e-kar'de-ah)*; a rate faster than 100 beats per minute is described as **tachycardia** *(tak''e-kar'de-ah)*.

Both bradycardia and tachycardia can occur normally (fig. 16.51a). Endurance-trained athletes, for example, commonly have a slower heart rate than the general population. This *athlete's bradycardia* occurs as a result of higher levels of parasympathetic inhibition of the SA node and is a beneficial adaptation. Activation of the sympathetic division of the ANS during exercise or emergencies causes a normal tachycardia to occur.

An abnormal bradycardia may be caused by a heart block, various drugs, shock, or increased intracranial pressure. An abnormal tachycardia occurs if the rate increases when a person is at rest. This may result from abnormally fast pacing by the atria, due to drugs or to the development of abnormally fast *ectopic pacemakers*—cells located outside the SA node that assume a pacemaker function. This abnormal atrial tachycardia thus differs from normal "sinus" (SA node) tachycardia. *Ventricular tachycardia* results when abnormally fast ectopic pacemakers in the ventricles cause them to beat rapidly and independently of the atria (fig. 16.51b). This is very dangerous because it can quickly degenerate into a lethal condition known as ventricular fibrillation.

Ventricular Fibrillation Fibrillation is caused by a continuous recycling of electrical waves through the myocardium. This recycling is normally prevented by the fact that the myocardium enters a refractory period simultaneously at all regions. If some cells emerge from their refractory periods before others, however, electrical waves can be continuously regenerated and conducted. The recycling of electrical waves along continuously

stethoscope: Gk. *stethos*, breast; *skopos*, watch, look at

sphygmomanometer: Gk. *sphygmos*, throbbing; *manos*, at intervals; L. *metrum*, measure
bradycardia: Gk. *bradys*, slow; *kardia*, heart
tachycardia: Gk. *tachys*, rapid; *kardia*, heart

changing pathways produces uncoordinated contraction and an impotent pumping action. These effects can be produced by damage to the myocardium.

Fibrillation can sometimes be stopped by a strong electric shock delivered to the chest, a procedure called **electrical defibrillation.** The electric shock depolarizes all the myocardial cells at the same time, causing them to enter a refractory state. The conduction of random, recirculating impulses thus stops, and—within a short time—the SA node can begin to stimulate contraction in a normal fashion. This does not correct the initial problem that caused the abnormal electrical patterns, but it does keep the person alive long enough to take other corrective measures.

Blood Disorders

Since blood is the functional component of the circulatory system and the circulatory system works in such close association with other body systems, blood analysis is perhaps the most informative part of a physical exam. Peripheral arterial pulsations, usually obtained at the radial artery, are informative of blood flow. Capillary filling, following **blanching,** is an indicator of peripheral arterial circulation and is generally tested at the nail bed. This is done by firmly pressing the thumbnail against the patient's toenail or fingernail and then quickly releasing the pressure. If the pinkish color returns quickly to the whitened, or blanched, area, circulation is considered normal. Lack of peripheral coloration indicates vascular insufficiency.

Certain cardiovascular and blood abnormalities are expressed through the skin. **Cyanosis** is a condition characterized by a bluish coloration of the skin due to a decreased oxygen concentration. **Anemia** is a condition characterized by a pallor of the skin due to a deficiency of erythrocytes or hemoglobin. **Jaundice** is a condition in which the skin is yellowed because of excessive bile pigment (bilirubin) in the blood. **Edema** is an excessive accumulation of fluid within the tissue spaces, causing a swelling of a portion of the body. **Erythema** is a redness of the skin usually due to an infection, inflammation, toxic reaction, sunburn, or a lesion. None of the aforementioned conditions are themselves diseases, but each is symptomatic of problems that may involve the cardiovascular system.

Actual examination of the blood is very informative and is a part of any thorough physical examination. Blood cell counts are used to determine the percentage of formed elements in the blood. An excess of red blood cells, called **polycythemia** (more than 6 million/mm^3), may indicate certain bone diseases. Excessive leukocyte production, called **leukocytosis,** is generally diagnostic of infections or diseases within the body. **Leukopenia** is a decrease in leukocyte count. The disease **leukemia** causes the unrestricted reproduction of immature leukocytes, which depresses erythrocyte and platelet formation and causes anemia and a tendency to bleed. Coagulation time, blood sedimentation rates, prothrombin time, and various serum analyses are other blood tests that provide specific information about body function or dysfunction.

Several blood diseases are distinguished by their rate of occurrence. **Sickle-cell disease** is an autosomal recessive disease that is more prevalent in blacks. Although about 10% of American blacks have the sickle cell trait, fortunately less than 1% have sickle-cell disease. Because of the disfigured shape of the diseased cells, there is an abnormally high destruction of erythrocytes. With the decrease in erythrocytes, the patient becomes anemic. **Mononucleosis** is an infectious disease that is transmitted by a virus in saliva and is therefore commonly called the kissing disease. Mononucleosis is characterized by atypical lymphocytes. It affects primarily adolescents, causing fever, sore throat, enlarged lymph glands, and fatigue.

Heart Diseases

Heart diseases can be classified as congenital or acquired. **Congenital heart problems** result from abnormalities in embryonic development and may be attributed to heredity, nutritional problems (poor diet) of the pregnant mother, or viral infections such as rubella. Congenital heart diseases occur in approximately 3 of every 100 births and account for about 50% of early childhood deaths. Many congenital heart defects can be corrected surgically, however, and others are not of a serious nature.

Heart murmurs are both congenital and acquired but are generally of no clinical significance. Nearly 10% of the population have heart murmurs, ranging from slight to severe. A **septal defect** is the most common type of congenital heart problem (fig. 16.52). An **atrial septal defect,** or **patent foramen ovale,** is a failure of the fetal foramen ovale to close at the time of birth. A **ventricular septal defect** is caused by an abnormal development of the interventricular septum. This condition may interfere with closure of the atrioventricular valves and may be indicated by cyanosis and abnormal heart sounds. **Pulmonary stenosis** is a narrowing of the opening into the pulmonary trunk from the right ventricle. It may lead to a pulmonary embolism and is usually recognized by extreme lung congestion.

The **tetralogy of Fallot** is a combination of four defects within a newborn and immediately causes a cyanotic condition, leading to the newborn being termed a blue baby. The four characteristics of this anomaly are (1) a ventricular septal defect, (2) an overriding aorta, (3) pulmonary stenosis, and (4) right ventricular hypertrophy (fig. 16.53). The pulmonary stenosis obstructs blood flow to the lungs and causes hypertrophy of the right ventricle. In an overriding aorta, the ascending portion arises midway between the right and left ventricles. Open-heart surgery is necessary to correct tetralogy of Fallot, and the overall mortality rate is about 5%.

Acquired heart disease may develop suddenly or gradually. Heart attacks are included in this category and are of immense clinical importance because they are the leading cause of death in the United States. It is estimated that one in five persons over the age sixty will succumb to a heart attack. The immediate cause of heart attacks is generally one of the following: inadequate coronary blood supply, anatomical disorders, or conduction disturbances.

Other types of acquired heart diseases affect the layers of the heart. **Bacterial endocarditis** is a disease of the lining of the heart, especially the valve cusps. It is caused by infectious

tetralogy of Fallot: from Etienne-Louis A. Fallot, French physician, 1850–1922

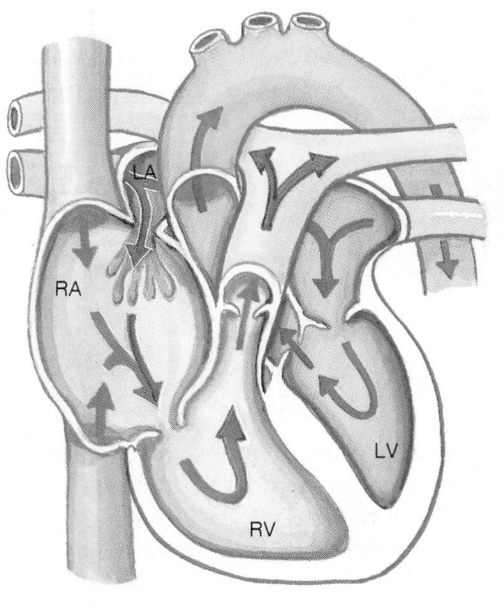

Septal defect
in atria

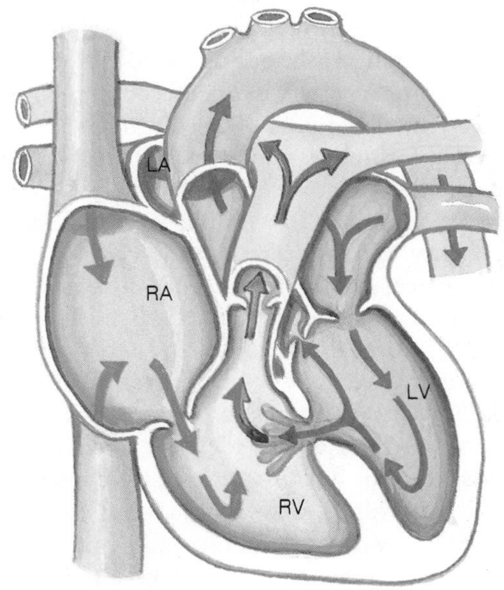

Septal defect
in ventricles

BECK

Figure 16.52 Abnormal patterns of blood flow due to septal defects. Left-to-right shunting of blood is shown *(shaded circles)* because the left pump is at a higher pressure than the right pump. Under other abnormal conditions (not illustrated here), however, the pressure in the right atrium may exceed that of the left, causing right-to-left shunting of blood through a septal defect in the atria.

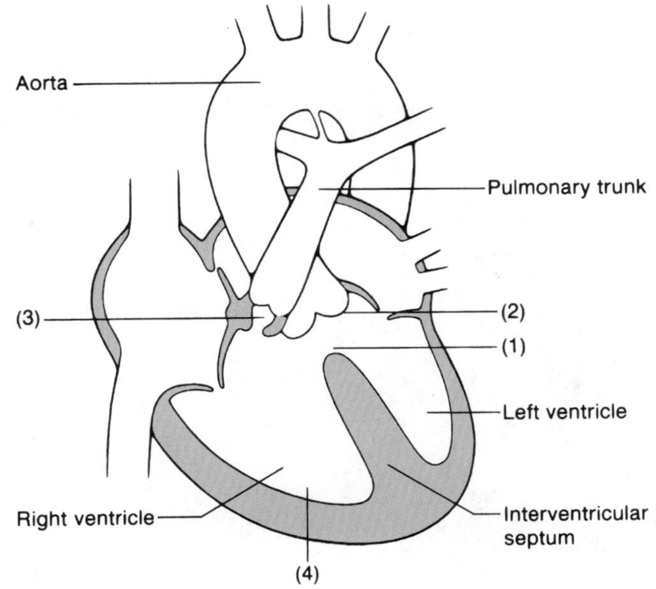

Figure 16.53 The tetralogy of Fallot. The four defects of this anomaly are (*1*) a ventricular septal defect, (*2*) an overriding aorta, (*3*) pulmonary stenosis, and (*4*) right ventricular hypertrophy.

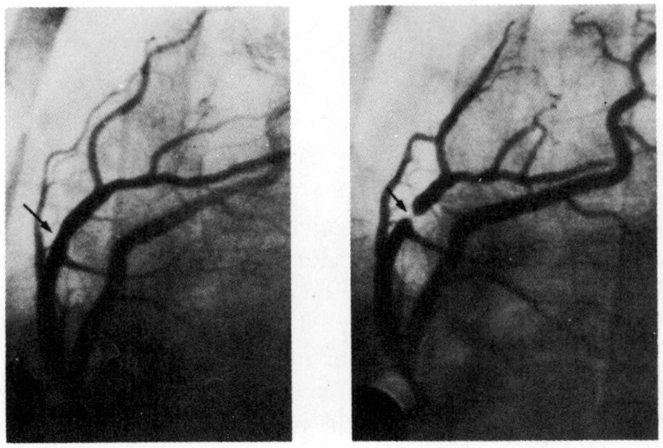

(a) (b)

Figure 16.54 An arteriogram of the left coronary artery in a patient (*a*) when the ECG was normal and (*b*) when the ECG showed evidence of myocardial ischemia. Notice that a coronary artery spasm—see arrow in (*b*)—appears to accompany the ischemia.

organisms that enter the bloodstream. **Myocardial disease** is an inflammation of the heart muscle followed by cardiac enlargement and congestive heart failure. **Pericarditis** causes an inflammation of the pericardium, the covering membrane of the heart. Its distinctive feature is pericardial friction rub, a transitory scratchy sound heard during auscultation.

A tissue is said to be **ischemic** *(is-kem'ik)* when it receives an inadequate supply of oxygen because of an inadequate blood flow. The most common cause of myocardial ischemia is atherosclerosis of the coronary arteries. The adequacy of blood flow is relative—it depends on the metabolic requirements of the tissue for oxygen. An obstruction in a coronary artery, for example, may allow sufficient blood flow at rest but may produce ischemia when the heart is stressed by exercise or emotional conditions (fig. 16.54). In patients with this condition, coronary artery bypass surgery may be performed (fig. 16.55).

Myocardial ischemia is associated with increased concentrations of blood lactic acid produced by anaerobic respiration of the ischemic tissue. This condition often causes substernal pain, which may also be referred to the left shoulder and arm, as well as other areas. This referred pain is called **angina pectoris.** People with angina frequently take nitroglycerin or related drugs that help relieve the ischemia and pain. These drugs

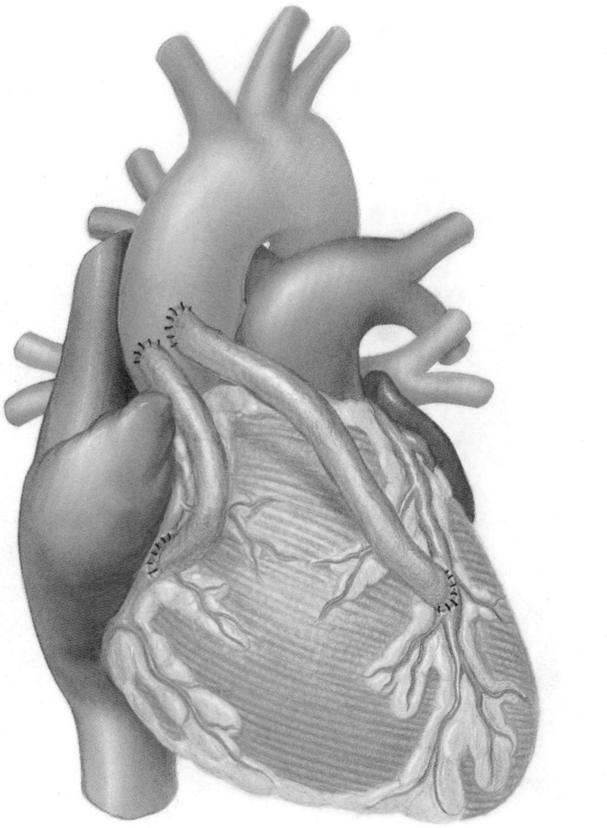

Figure 16.55 An illustration of double coronary artery bypass surgery. Several vessels may be used in the autotransplant including the internal thoracic artery or the great saphenous vein.

are effective because they stimulate vasodilation, which improves circulation to the heart and decreases the work that the heart must perform to eject blood into the arteries.

Myocardial cells are adapted to respire aerobically and cannot respire anaerobically for more than a few minutes. If ischemia and anaerobic respiration continue for more than a few minutes, *necrosis* (cellular death) may occur in the areas most deprived of oxygen. A sudden, irreversible injury of this kind is called a **myocardial infarction,** or **MI.**

Myocardial ischemia may be detected by characteristic changes in the ECG (fig. 16.56). The diagnosis of myocardial infarction is aided by the blood concentration of various enzymes that are released from the damaged cells.

Vascular Disorders

Hypertension, or high blood pressure, is the most common type of vascular disorder. In hypertension, the resting systolic blood pressure exceeds 140 mm Hg. An estimated 22 million adult Americans are afflicted by hypertension. About 15% of the cases are the result of other body problems, such as kidney diseases, adrenal hypersecretion, or arteriosclerosis and are diagnosed as secondary hypertension. Primary hypertension is more common and cannot be attributed to any particular body dysfunction. If hypertension is not controlled by diet, exercise, or drugs that reduce the blood pressure, it can damage various vital body organs such as the heart or kidneys.

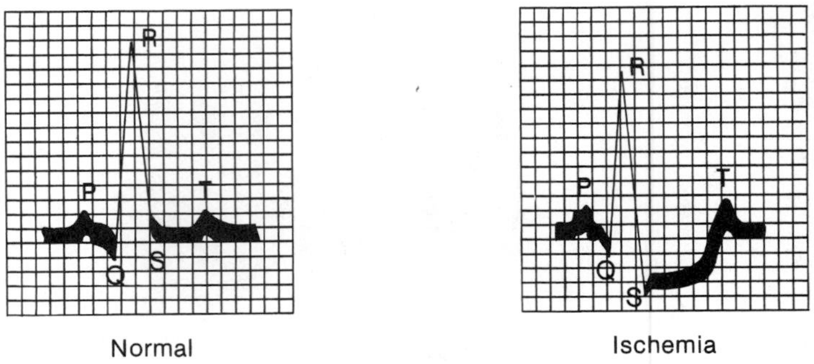

Figure 16.56 Depression of the S–T segment of the electrocardiogram as a result of myocardial ischemia.

Arteriosclerosis, or hardening of the arteries, is a generalized degenerative vascular disorder that results in loss of elasticity and thickening of the arteries. **Atherosclerosis** is a type of arteriosclerosis in which plaque material called **atheroma** forms on the tunica intima, narrowing the lumina of the arteries and prohibiting the normal flow of blood (fig. 16.57). Furthermore, an atheroma often creates a rough surface that can initiate the formation of a blood clot called a **thrombus.** An **embolism** is a thrombus that has dislodged from the wall of a vessel and is moving through the bloodstream. Both a thrombus and an embolism can occlude blood flow. An embolism lodged in a coronary artery is called a coronary thrombus; in a vessel of the lung it is a pulmonary thrombus; and in the brain it is a cerebral thrombus, which could cause a **stroke.**

The causes of atherosclerosis are not well understood, but the disease seems to be associated with improper diet, smoking, hypertension, obesity, lack of exercise, and heredity.

Aneurysms, coarctations, and varicose veins are all types of vascular disfigurations. An **aneurysm** *(an'u-rizm)* is an expansion or bulging of the heart, aorta, or any other artery (fig. 16.58). Aneurysms are caused by weakening of the tunicas and may rupture or lead to embolisms. A **coarctation** is a constriction of a segment of a vessel, usually the aorta, and is frequently caused by a remnant of the ductus arteriosus tightening around the vessel. **Varicose veins** are weakened veins that become stretched and swollen. Varicose veins are most common in the legs because the force of gravity tends to weaken the valves and overload the veins. Varicose veins can also occur in the rectum, in which case they are called **hemorrhoids.** *Vein stripping* is the surgical removal of superficial weakened veins. **Phlebitis** *(flĕ-bi'tis)* is inflammation of a vein. It may develop as a result of trauma or an aftermath of surgery. Frequently, however, it appears for no apparent reason. Phlebitis interferes with normal venous circulation.

Disorders of the Lymphatic System

Infections of the body are generally accompanied by a swelling and tenderness of lymph nodes near the infection. An inflammation of lymph nodes is referred to as **lymphadenitis.** In prolonged lymphadenitis, an **abscess** usually forms in the nodal tissue. An abscess is a localized pocket of pus formed by tissue

arteriosclerosis: Gk. *arterio,* artery; *skleros,* hard
atheroma: Gk. *athere,* mush; *oma,* tumor
thrombus: Gk. *thrombos,* a clot
embolism: Gk. *embolos,* a plug

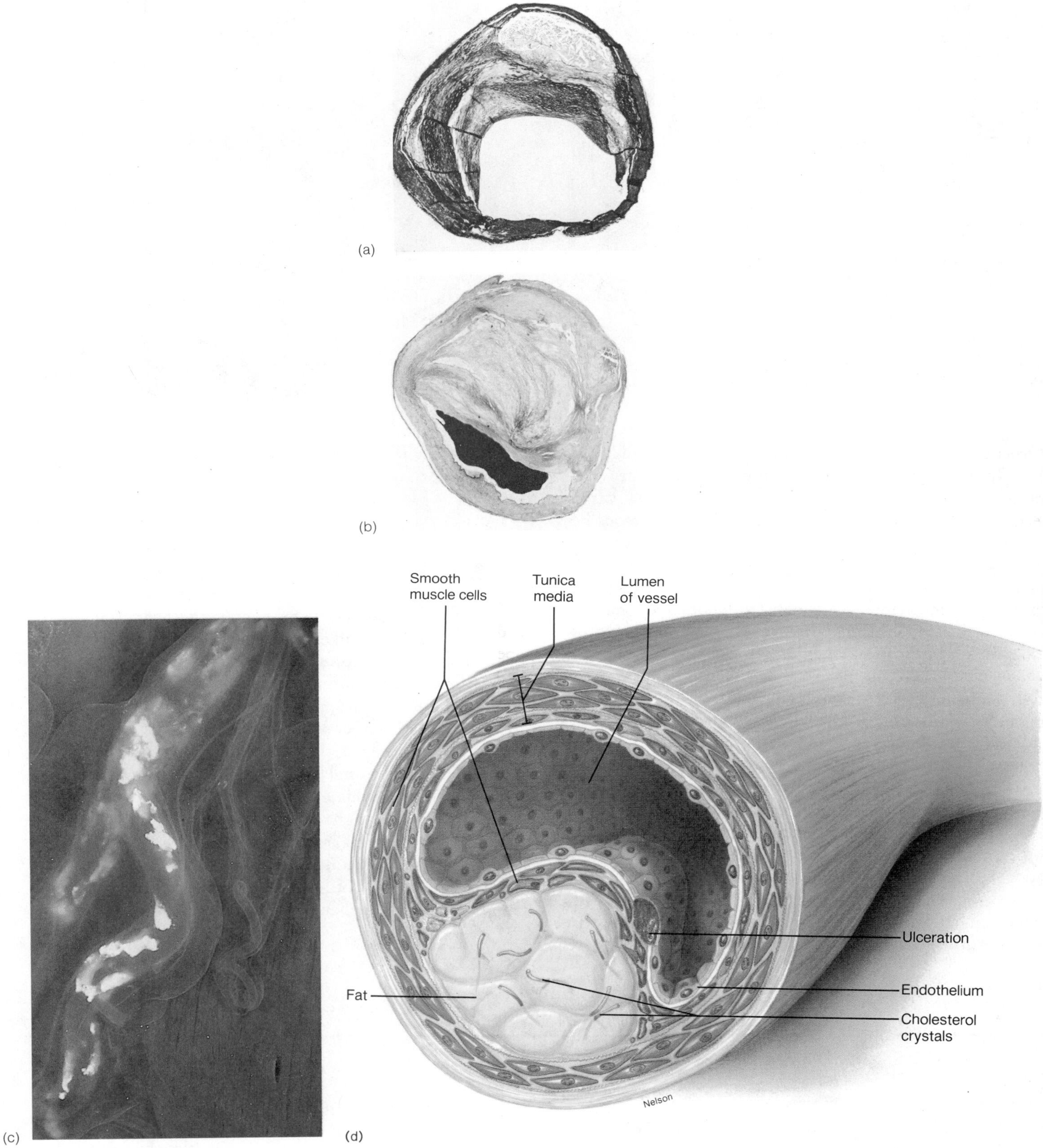

Figure 16.57 Atherosclerosis. (*a*) The lumen of a human coronary artery is partially occluded by an atheroma and (*b*) almost completely occluded by a thrombus. (*c*) A close-up view of the cleared left anterior descending coronary artery containing calcified atherosclerotic plaques. The heart is of an 85-year-old female. (*d*) The structure of an atheroma is diagrammed.

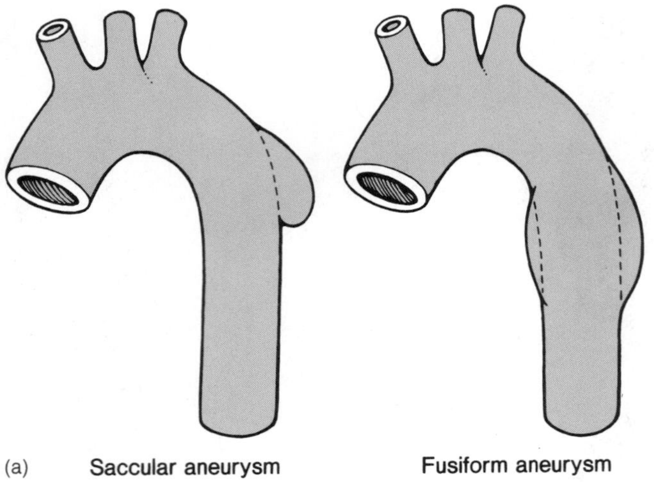

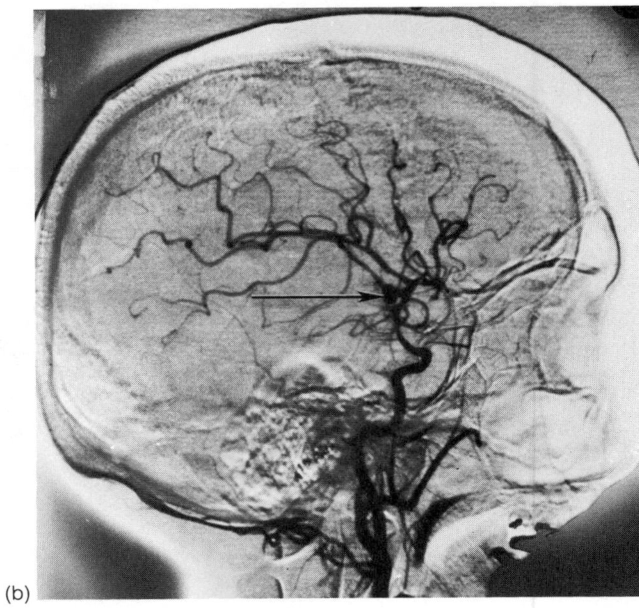

Figure 16.58 An aneurysm is an outpouching of the vascular wall. (*a*) A diagram of two types of aortic aneurysms and (*b*) an X ray of a saccular aneurysm of the middle cerebral artery.

destruction. If an infection is not contained by localized lymph nodes, **lymphangitis** may ensue. In this condition, narrow red streaks can be seen through the skin extending proximally from the infected area. Lymphangitis is potentially dangerous because it means that the infection is not contained and may spread, infecting the blood and causing **septicemia** (blood poisoning).

The term **lymphoma** is used to describe primary malignancies within lymphoid tissues. Lymphomas are generally classified as **Hodgkin's disease lymphomas** or **non-Hodgkin's lymphomas.** Hodgkin's disease manifests itself as swollen lymph nodes in the neck and then progresses to involve the spleen, liver, and bone marrow. The prognosis for Hodgkin's disease is good if it is detected early. Non-Hodgkin's lymphomas include an array of specific and more obscure lymphatic cancers.

The lymphatic system is also frequently infected by metastasizing carcinomas. Fragmented cells from the original tumor may enter the lymphatic ducts with the lymph and travel to the lymph nodes where they cause secondary cancerous growths. Breast cancer will characteristically do this. The surgical treatment involves the removal of the infected nodes as well as some of the healthy nodes downstream to ensure that the cancer is eradicated.

septicemia: Gk. *septikos,* septic; *haima,* blood
Hodgkin's disease: from Thomas Hodgkin, English physician, 1798–1866

CLINICAL CASE STUDY ANSWER

Mural thrombus (a blood clot adherent to the inner surface of one of the heart's chambers) is a fairly common complication of myocardial infarction. Once a thrombus forms within the heart, a piece may break off (embolus), which can travel throughout the body. This is the most likely cause of the symptoms in the right leg of our patient. Because the embolus traveled and lodged in the systemic circulation (as opposed to the pulmonary circulation), the mural thrombus was probably located in the left side of the heart. The embolus traveled to a point in the femoral artery where it lodged, thus occluding blood flow to the popliteal artery and its distal branches. The route of travel was: left side of heart → ascending aorta → descending thoracic aorta → abdominal aorta → right common iliac artery → right external iliac artery → femoral artery.

The standard treatment of this problem is emergency surgery to extract the clot from the leg. Anticoagulation (blood-thinning) therapy is then continued or instituted. ■

Chapter Summary

I. Functions and Major Components of the Circulatory System
  A. Blood transports oxygen and nutritive molecules to the tissue cells and carbon dioxide and other wastes away from tissue cells; it also carries hormones and other regulatory molecules to their target organs.
  B. Leukocytes and their products serve to protect the body from infection.

II. Development of the Cardiovascular System
  A. The heart develops from a heart tube that becomes subdivided into regions called the truncus arteriosus, bulbus cordis, ventricle, atrium, and sinus venosus.
    1. The ventricle is divided into right and left chambers by the growth of an interventricular septum.
    2. A septum also forms in the atrium, but in the fetus it contains an opening called the foramen ovale, which allows communication between the right and left atria.

B. The major arteries are formed from six pairs of aortic arches.
   1. Different members of these aortic arches form or contribute to the common carotid, internal carotid, right subclavian, and pulmonary arteries.
   2. Only the left fourth arch contributes to the aortic arch of the adult.
   3. A part of the sixth left arch persists as a shunt between the pulmonary trunk and the aorta called the ductus arteriosus.

III. Blood
   A. Blood, a highly specialized connective tissue, consists of formed elements (erythrocytes, leukocytes, and thrombocytes) that are suspended and carried in the plasma. The constituents of blood function in transport, immunity, and blood-clotting mechanisms.
   B. Erythrocytes are disc-shaped cells that lack nuclei but do contain hemoglobin. Erythrocytes live about 120 days, and there are between 4.3 and 5.8 million per cubic millimeter of blood.
   C. Leukocytes have nuclei and are classified as granular (eosinophils, basophils, and neutrophils) or agranular (monocytes and lymphocytes). Leukocytes function to defend the body against infections by microorganisms.
   D. Thrombocytes, or platelets, are cytoplasmic fragments that assist in the formation of clots to prevent blood loss.
   E. Erythrocytes are formed through a process called erythropoiesis, and leukocytes are formed through leukopoiesis.
   F. Prenatal hemopoietic centers are the yolk sac, liver, and spleen, whereas bone marrow and lymphoid tissues perform this function in the adult.

IV. Heart
   A. The heart is enclosed within a pericardial sac; the wall of the heart consists of the epicardium, myocardium, and endocardium.
      1. The right atrium receives blood from the superior and inferior venae cavae, and the right ventricle pumps blood into the pulmonary trunk to the pulmonary arteries.
      2. The left atrium receives blood from the pulmonary veins and pumps blood into the aorta.
      3. Atrioventricular valves (tricuspid and bicuspid) are located between the atria and ventricles; semilunar valves are located at the base of the pulmonary trunk and aorta.
   B. There are two principal circulatory divisions, the pulmonary and the systemic, as well as the coronary system, which serves the heart.
      1. The pulmonary circulation includes the vessels that carry blood from the right ventricle, through the lungs, and from there to the left atrium.
      2. The systemic circulation includes all other arteries, capillaries, and veins in the body, which carry blood from the left ventricle, through the body, and return blood to the right atrium.
      3. The coronary arteries arise from the aorta immediately beyond the semilunar valve and provide arterial blood to the myocardium via several branches; venous blood ultimately enters a coronary sinus that drains into the right atrium.
   C. Contraction of the atria and ventricles is produced by action potentials that originate in the sinoatrial (SA) node.
      1. These electrical waves spread over the atria and then enter the atrioventricular (AV) node.
      2. From here, the impulses are conducted by the atrioventricular bundle and Purkinje fibers into the ventricular walls.
   D. During contraction of the ventricles, the intraventricular pressure rises and causes the AV valves to close; during relaxation, the semilunar valves close because the pressure is greater in the arteries than in the ventricles.
   E. Closing of the AV valves causes the first sound (lub), and closing of the semilunar valves causes the second sound (dub); heart murmurs are commonly caused by abnormal valves or by septal defects.
   F. The pattern of electrical conduction is observed by an electrocardiogram (ECG or EKG).

V. Blood Vessels
   A. Arteries and veins have a tunica externa, tunica media, and tunica intima.
      1. Arteries have thicker muscle layers for their diameters than do veins because they must withstand a higher blood pressure.
      2. Veins have venous valves that direct blood to the heart when the veins are compressed by the skeletal muscle pumps.
   B. Capillaries are composed only of endothelial cells and may be continuous or discontinuous.

VI. Principal Arteries of the Body
   A. Three arteries arise from the aortic arch: the brachiocephalic, the left common carotid, and the left subclavian; the brachiocephalic divides into the right common carotid and the right subclavian.
   B. The head and neck receive an arterial supply from branches of the internal and external carotid arteries and the vertebral arteries.
      1. The brain receives blood from the paired internal carotid arteries and the paired vertebral arteries, which form the arterial circle around the pituitary gland.
      2. The external carotid artery gives off many branches that supply the head and neck.
   C. The upper extremity is served by the subclavian artery and its derivatives.
      1. The subclavian artery becomes first the axillary and then the brachial artery as it enters the arm.
      2. The brachial artery bifurcates to form the radial and ulnar arteries, which supply blood to the forearm and a portion of the hand and digits.
   D. The abdominal aorta produces the following branches: the celiac, superior mesenteric, renal, suprarenal, testicular (or ovarian), and inferior mesenteric arteries; the abdominal aorta divides into the right and left common iliac arteries.
   E. The common iliac arteries divide into the internal and external iliac arteries, which supply arteries to the pelvis and lower extremities.

VII. Principal Veins of the Body
   A. Blood from the head and neck is drained by the external and internal jugular veins; blood from the brain is drained by the internal jugular veins.
   B. The upper extremity is drained by superficial and deep veins.
   C. In the thorax, the superior vena cava is formed by the union of the two brachiocephalic veins and also receives blood from the azygos vein.
   D. The lower extremity is drained by both superficial and deep veins; at the level of the fifth lumbar vertebra, the right and left common iliac veins unite to form the inferior vena cava.
   E. Blood from capillaries in the digestive tract is drained via the hepatic portal vein to the liver.
      1. This venous blood then passes through hepatic sinusoids and is drained from the liver in the hepatic veins.
      2. The pattern of circulation in which two capillary beds are in a series is called a portal system.

VIII. Fetal Circulation
   A. There are structural adaptations in the fetal cardiovascular system due to the fact that oxygen and nutrients are obtained from the placenta rather than from the fetal lungs and digestive tract.
   B. Fully oxygenated blood is carried only in the umbilical vein, which drains the placenta; this blood is carried via the ductus venosus to the inferior vena cava of the fetus.
   C. Partially oxygenated blood is shunted from the right to the left atrium via the foramen ovale and from the pulmonary trunk to the aorta via the ductus arteriosus.

IX. Lymphatic System
   A. The lymphatic system functions to return tissue fluid to the venous system and to help protect the body from disease.
   B. Lymph capillaries drain tissue fluid, which is formed from plasma; when this fluid enters lymph capillaries, it is called lymph.
   C. Lymph is returned to the venous system via two large lymph ducts—the thoracic duct and the right lymphatic duct.
   D. Lymph filters through lymph nodes, which contain phagocytic cells and germinal centers that produce lymphocytes.
   E. The spleen and thymus are lymphoid organs because they produce lymphocytes.

Review Activities

Objective Questions

1. The heart derives from mesoderm and is completed in embryonic form by the
   (a) third week.
   (b) fifth week.
   (c) tenth week.
   (d) seventeenth week.
2. Which of the following is *not* a type of blood cell?
   (a) leukocyte (c) fibrinogen
   (b) eosinophil (d) thrombocyte
3. An elevated white blood cell count is referred to as
   (a) leukocytosis. (c) leukopoiesis.
   (b) polycythemia. (d) leukemia.
4. All arteries in the body contain oxygen-rich blood with the exception of the
   (a) aorta.
   (b) pulmonary artery.
   (c) renal artery.
   (d) coronary arteries.
5. Blood from the coronary circulation directly enters the
   (a) inferior vena cava.
   (b) superior vena cava.
   (c) right atrium.
   (d) left atrium.
6. The "lub," or first heart sound, is produced by the closing of the
   (a) aortic semilunar valve.
   (b) pulmonary semilunar valve.
   (c) tricuspid valve.
   (d) bicuspid valve.
   (e) Both (a) and (b).
   (f) Both (c) and (d).
7. The QRS wave of an ECG is produced by
   (a) depolarization of the atria.
   (b) repolarization of the atria.
   (c) depolarization of the ventricles.
   (d) repolarization of the ventricles.
8. Which of the following arteries do *not* arise from the aortic arch? The
   (a) brachiocephalic.
   (b) coronary.
   (c) left common carotid.
   (d) left subclavian.
9. Which of the following arteries do *not* supply blood to the brain? The
   (a) external carotid.
   (b) internal carotid.
   (c) vertebral.
   (d) basilar.
10. The maxillary and superficial temporal arteries are derived from the
   (a) external carotid.
   (b) internal carotid.
   (c) vertebral.
   (d) facial.
11. Which of the following organs is/are served by a portal vein?
   (a) the liver
   (b) the pituitary
   (c) both (a) and (b)
   (d) neither (a) nor (b)
12. Which of the following fetal blood vessels carries fully oxygenated blood? The
   (a) ductus arteriosus.
   (b) ductus venosus.
   (c) umbilical vein.
   (d) aorta.

Essay Questions

1. What are the functions of the circulatory system? Identify the body systems with which the circulatory system closely functions in maintaining homeostasis.
2. Briefly describe the development of the heart.
3. Explain how the development of the aortic arches contributes to the formation of the major vessels associated with the heart.
4. Distinguish between granulocytes and agranulocytes. What are the functions of leukocytes? How do they differ from erythrocytes and thrombocytes?
5. Describe how the heart is compartmentalized within the thoracic cavity. What is the function of the pericardium?
6. Describe the cardiac cycle. Why is the sinoatrial node called the pacemaker of the heart?
7. Diagram and label a normal electrocardiogram. What is the significance of the deflection waves of an ECG?
8. Compare the structure, size, and function of arteries, capillaries, and veins.
9. What is the difference between blood pressure and pulse pressure? What cardiac event determines the systolic blood pressure? What accounts for the diastolic blood pressure?
10. Name the vessels of the thoracic and shoulder regions that are not symmetrical, that is, that do not have a counterpart on the opposite side of the body.
11. Trace the path to the brain of a glucose injection into the median cubital vein of the arm, and name in proper sequence all the blood vessels and chambers of the heart through which it passes.
12. What is significant about the hepatic portal system? What is meant by saying the liver has two blood supplies?
13. List the functions of the lymphatic system and describe the relationship between the lymphatic system and the circulatory system.
14. What are five fetal circulatory structures that cease to function in a newborn?
15. Distinguish between congenital and acquired heart diseases and identify several kinds of each.

CHAPTER 17

Respiratory System

Outline and Concepts

CLINICAL CASE STUDY

A thirty-seven-year-old male comes to the emergency room after having been stabbed above the left clavicle with an ice pick. The patient's main complaint is pain in the left side of the chest. Initial evaluation reveals a minute puncture wound superior to the left clavicle, just lateral to the sternocleidomastoid muscle. The vital signs are normal except for a moderately high respiratory rate. The chest X ray shows the left lung to be collapsed to half its normal size and surrounded by air and blood. (This condition is known as *hemopneumothorax,* meaning: blood and air in the thoracic cavity.)

Explain how the air arrived in the thoracic cavity (assuming it did not enter through the skin puncture wound). What is the name of the space where the air and blood are located, and what membranes contain the space?

Functions and Development of the Respiratory System

The respiratory system can be divided into upper and lower divisions on the basis of function and embryological development.

Objective 1. Describe the functions included in the term *respiration.*
Objective 2. Identify the organs of the respiratory system, and describe their locations.
Objective 3. List the functions of the respiratory system.
Objective 4. Compare the development of the upper respiratory tract with that of the lower.

The term *respiration* includes three separate but related functions: (1) **ventilation** (breathing); (2) **gas exchange** between the air and blood in the lungs and between the blood and tissues; and (3) **oxygen utilization** by the tissues in the energy-liberating reactions of cell respiration. The exchange of gases (oxygen and carbon dioxide) between the air and blood is called *external respiration.* Gas exchange between the blood and tissues is known as *internal respiration.*

It is estimated that a relaxed adult breathes about nine to twenty times a minute, ventilating about 5 to 6 liters of air during this period. Strenuous exercise increases the demand for oxygen and increases the respiratory rate fifteenfold to twentyfold, so that about 100 liters of air are breathed each minute. If breathing stops a person will lose consciousness after four or five minutes, may suffer brain damage after seven to eight minutes, and will die after ten minutes. A knowledge of the structure and function of the respiratory system, therefore, is of the greatest clinical importance.

Physical Requirements of the Respiratory System

The respiratory system includes those organs and structures that function together to bring gases in contact with the blood of the circulatory system. In order to be effective, the respiratory system must comply with certain physical requirements:

1. The structures through which gas is exchanged between the circulatory and respiratory system must be thin-walled and selectively permeable so that diffusion can occur easily.
2. This membrane must be kept moist so that oxygen and carbon dioxide can be dissolved in water to facilitate diffusion.
3. A rich blood supply must be present.
4. The surface for gas exchange must be located deep in the body so that the incoming air can be sufficiently warmed, moistened, and cleansed of airborne particles in order not to damage this delicate membrane.
5. There must be an effective ventilation mechanism to constantly replenish the air.

The respiratory system adequately meets all of these requirements to ensure that all of the trillions of cells of the body can carry on energy metabolism.

Functions of the Respiratory System

The respiratory system has four basic functions, not all of which are associated with breathing. They are:

1. Providing oxygen to the bloodstream and removing carbon dioxide.
2. Sound production or vocalization as expired air passes over the vocal cords.
3. Assisting in abdominal compression during micturition (urination), defecation (passing of feces), and parturition (childbirth). The abdominal muscles become more effective during a deep breath when the air is held in the lungs by closing the glottis and fixing the diaphragm. This same technique is used when lifting a heavy object, in which case the diaphragm indirectly assists the back muscles.
4. Providing protective and reflexive nonbreathing air movements, such as coughing and sneezing, to keep the air passageways clean.

Basic Structure of the Respiratory System

The major passages and structures of the respiratory system are the **nasal cavity, pharynx, larynx** and **trachea,** and the **bronchi, bronchioles,** and **alveoli** within the **lungs** (fig. 17.1). The respiratory system is frequently divided into the *conducting division* and the *respiratory division.* The conducting portion includes all of the cavities and structures that transport gases to and from alveoli. The alveoli, the functional units of the system, constitute the respiratory portion.

Early Greek and Roman scientists placed great emphasis on the invisible material that was breathed in. They knew nothing about oxygen or the role of the blood in transporting this vital substance to cells. For that matter, they knew nothing about microscopic structures such as cells because the microscope had not yet been invented. They did know, however, that respiration was essential for life. Early Greeks referred to air as an intangible, divine spirit called *pneuma.* In Latin, the term for breath, *spiritus,* meant life-force.

respiration: L. *re,* back; *spirare,* to breathe

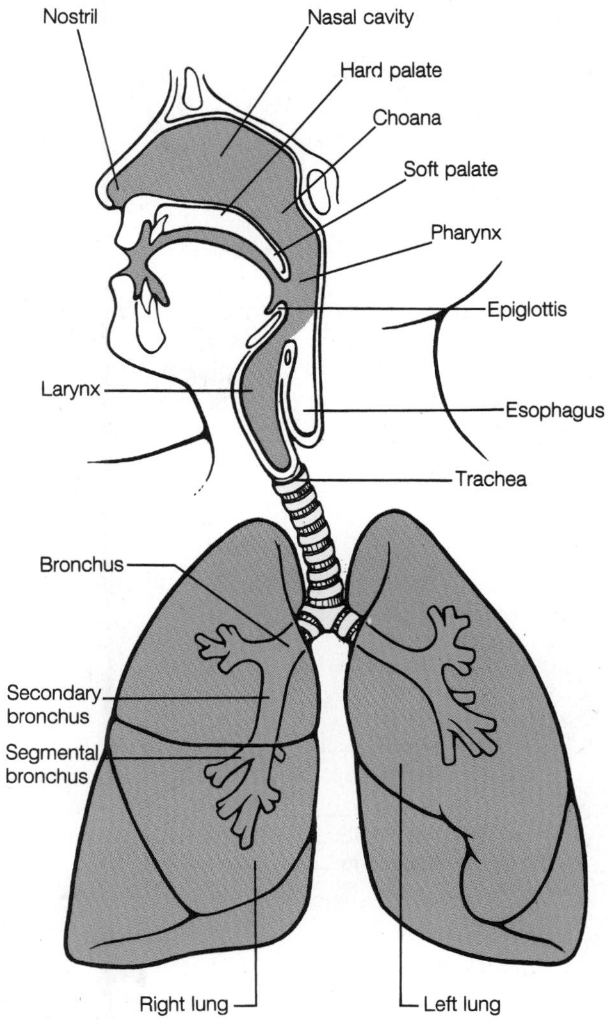

Figure 17.1 The basic anatomy of the respiratory system.

Embryological Development of the Respiratory System

The development of the respiratory system is initiated early in embryonic development and involves both ectoderm and endoderm. Although the structures of the upper respiratory system (nose and pharynx) develop simultaneously with those of the lower (trachea and lungs), it is best to discuss the development separately because of the different germ layers involved.

Development of the Upper Respiratory System *Cephalization* means that the cephalic, or head, end of an organism structurally and functionally differentiates from the rest of the body. In humans, cephalization is apparent early in development. One of the initial events is the formation of the nasal cavity at three and one-half to four weeks of embryonic life. A region of thickened ectoderm called the **olfactory** (nasal) **placode** *(plak'ōd)* appears on the front and inferior part of the head (fig. 17.2). The placode invaginates to form the **olfactory pit,** which extends posteriorly to connect with the **foregut.** The foregut, derived of endoderm, later develops into the pharynx.

cephalization: Gk. *kephale,* head

The mouth, or oral cavity, develops at the same time as the nasal cavity, and for a short time there is a thin **oronasal** *(or''o-na'zal)* **membrane** separating the two cavities. This membrane ruptures during the seventh week, and a single, large **oronasal cavity** forms. Shortly after, tissue plates of mesoderm begin to grow horizontally across the cavity. At approximately the same time, a vertical plate develops inferiorly from the roof of the nasal cavity. These plates complete their formation by three months of development. The vertical plate forms the nasal septum, and the horizontal plates form the hard palate.

A *cleft palate* forms when the horizontal plates fail to meet in the midline. This condition can be surgically and cosmetically treated. The more immediate and serious problem facing an infant with a cleft palate is that it may be unable to create enough suction to nurse properly.

Development of the Lower Respiratory System The lower respiratory system begins as a diverticulum *(di''ver-tik'u-lum),* or outpouching, from the ventral surface of endoderm along the lower pharyngeal region (fig. 17.3). This diverticulum forms during the fourth week of development and is referred to as the **laryngotracheal** *(lah-ring''go-tra'ke-al)* **bud.** As the bud grows, the proximal portion forms the trachea and the distal portion bifurcates (splits) into a right and left bronchus.

The buds continue to elongate and split until all the tubular network within the lower respiratory tract is formed (fig. 17.3). As the terminal portion forms air sacs, called **alveoli,** at about eight weeks of development, the supporting lung tissue begins to form. The complete structure of the lungs, however, is not fully developed until about twenty-six weeks of age, so premature infants born prior to this time require special artificial respiratory equipment to live.

1. Define the terms *external respiration* and *internal respiration.*
2. Differentiate between the physical requirements and the functions of the respiratory system.
3. List in order the major passages and structures through which inspired air would pass from the nostrils to the alveoli of the lungs.
4. Explain the role of ectoderm and endoderm in the embryological development of the respiratory pathway.
5. At what age might a cleft palate develop? At what age are the lungs sufficiently developed to sustain independent life?
6. Explain the difference between an invagination and a diverticulum in relation to the development of the respiratory system.

Conducting Passages

Air is conducted through the oral and nasal cavities to the pharynx and then through the larynx to the trachea and bronchial tree. These structures deliver warmed and humidified air to the respiratory division in the lungs.

alveolus: L. diminutive of *alveus,* cavity

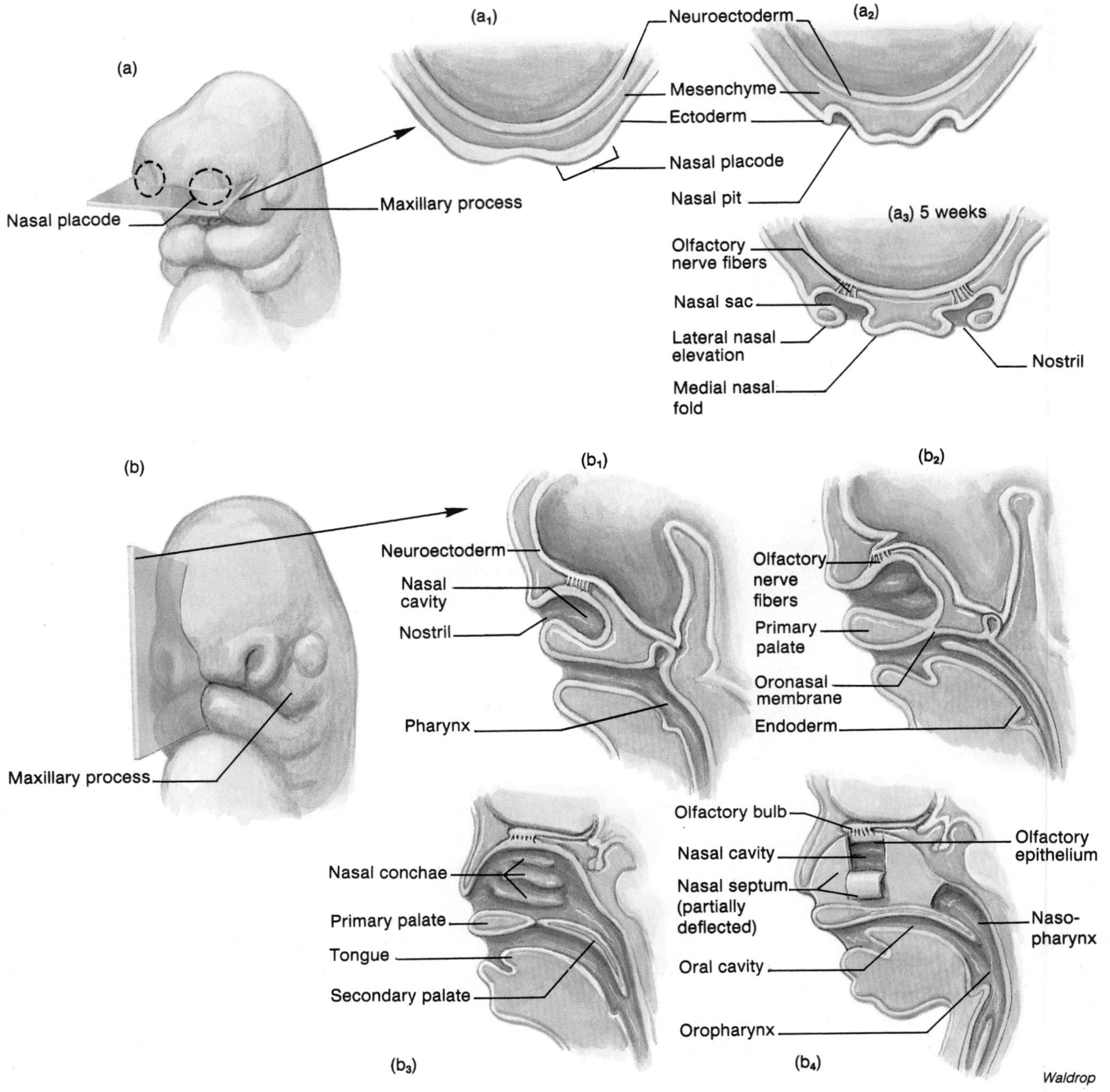

Figure 17.2 The development of the upper respiratory system. (*a*) An anterior view of the developing head of an embryo at four weeks showing the position of a transverse cut depicted in (a_1, a_2, and a_3). (a_2) Development at five weeks and (a_3) at five and one-half weeks. (*b*) An anterior view of the developing head of an embryo at six weeks showing the position of a sagittal cut depicted in (b_1, b_2, b_3 and b_4) at fourteen weeks.

Objective 5. List the types of epithelial tissue present in each region of the respiratory tract, and discuss the significance of these differences.

Objective 6. Identify the boundaries of the nasal cavity, and discuss the relationship of the paranasal sinuses to the rest of the respiratory system.

Objective 7. Describe the three regions of the pharynx, and identify the structures located in each.

Objective 8. Discuss the role of the laryngeal region in digestion and respiration.

Objective 9. Identify the anatomical features of the larynx associated with sound production and respiration.

The conducting passages serve to transport air to the respiratory structures of the lungs. The passageways are lined with various types of epithelia to prepare the air properly for utilization. The majority of the conducting passages are held permanently open by muscle or a bony or cartilaginous framework.

Nose

The **nose** includes an external portion that juts out from the face and an internal **nasal cavity** for the passage of air. The external portion of the nose is covered with skin and supported

nose: O.E. *nosu,* nose

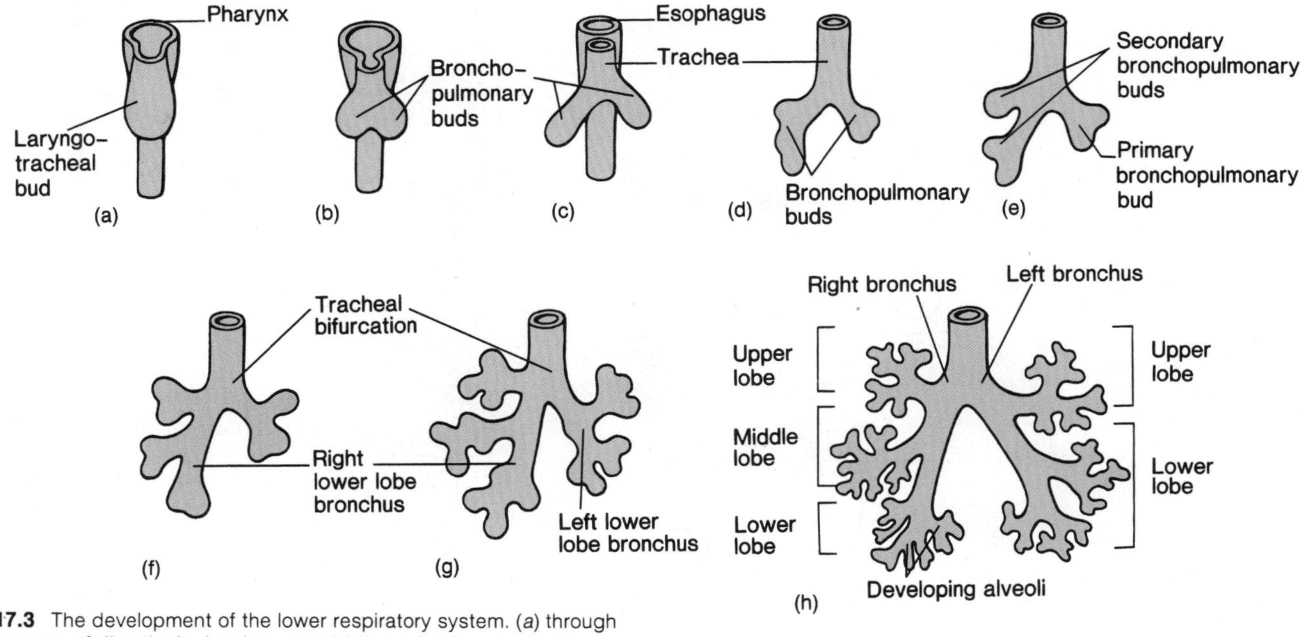

Figure 17.3 The development of the lower respiratory system. (*a*) through (*h*): a sequence of diverticula development. (*a*) through (*d*) occur by the fourth week; (*e*) through (*h*) occur by the eighth week.

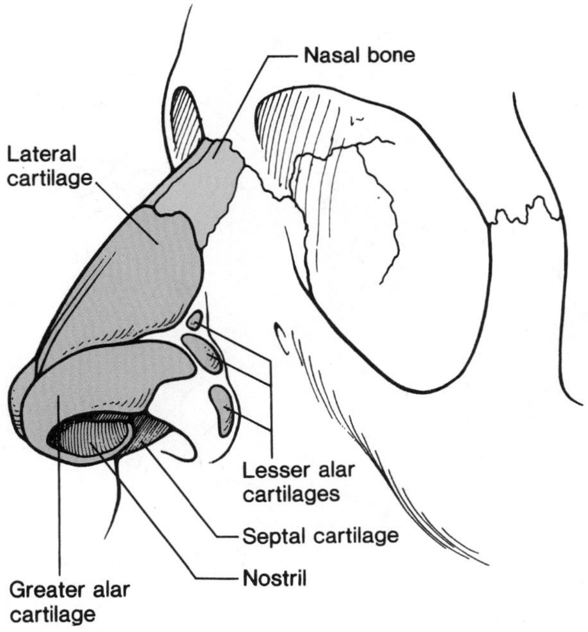

Figure 17.4 The supporting framework of the external nose.

by paired **nasal bones,** forming the bridge, and pliable cartilage, forming the distal portions (figs. 17.4 and 17.5). The **septal cartilage** forms the anterior portion of the **nasal septum,** and the paired **lateral cartilages** and **alar cartilages** form the framework around the **nostrils.**

The perpendicular plate of the **ethmoid** bone and the **vomer** bone together with the septal cartilage constitute the supporting framework of the **nasal septum,** which divides the nasal cavity into two lateral halves, each of which is referred to as a **nasal fossa.** The vestibule is the anterior expanded portion of the nasal fossa (fig. 17.6). Each nasal fossa opens anteriorly through the **nostril,** or **external naris** and communicates posteriorly with the **nasopharynx** through the **choanae** *(ko-a'ne),* or **internal naris.** The roof of the nasal cavity is formed anteriorly by the frontal bone and paired nasal bones, medially by the cribriform plate of the ethmoid, and posteriorly by the sphenoid bone (fig. 17.5). The palatine and maxillary bones form the floor of the cavity. On the lateral walls are the **superior, middle,** and **inferior conchae** *(kong'ke)* (turbinates).

nostril: O.E. *nosu,* nose; *thyrel,* hole
choana: Gk. *choane,* funnel
concha: L. a shell

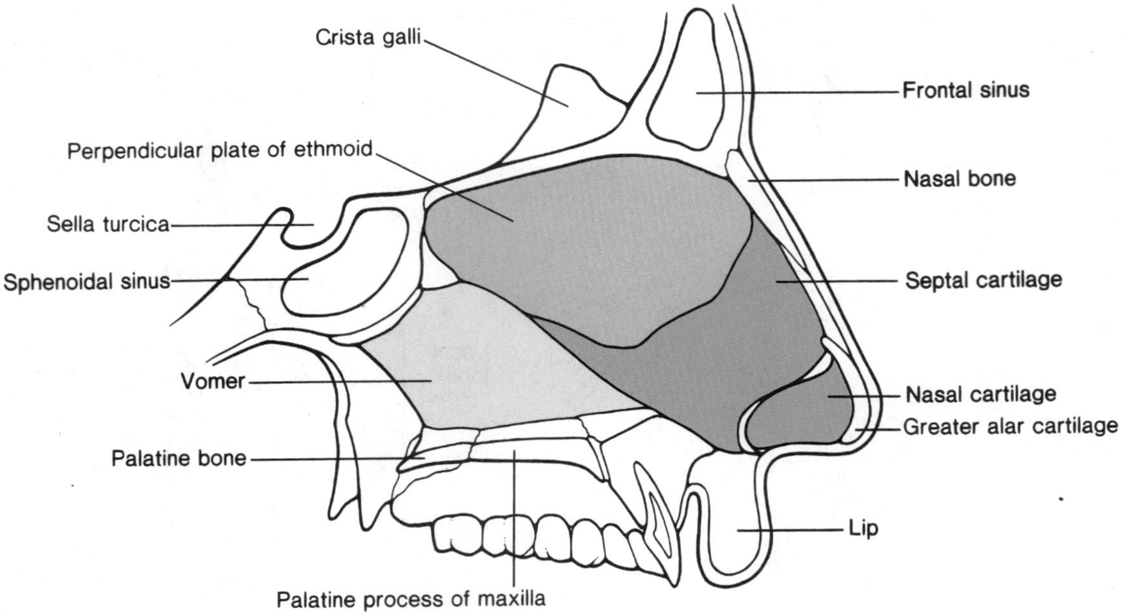

Figure 17.5 A sagittal section through the nose showing the parts of the nasal septum.

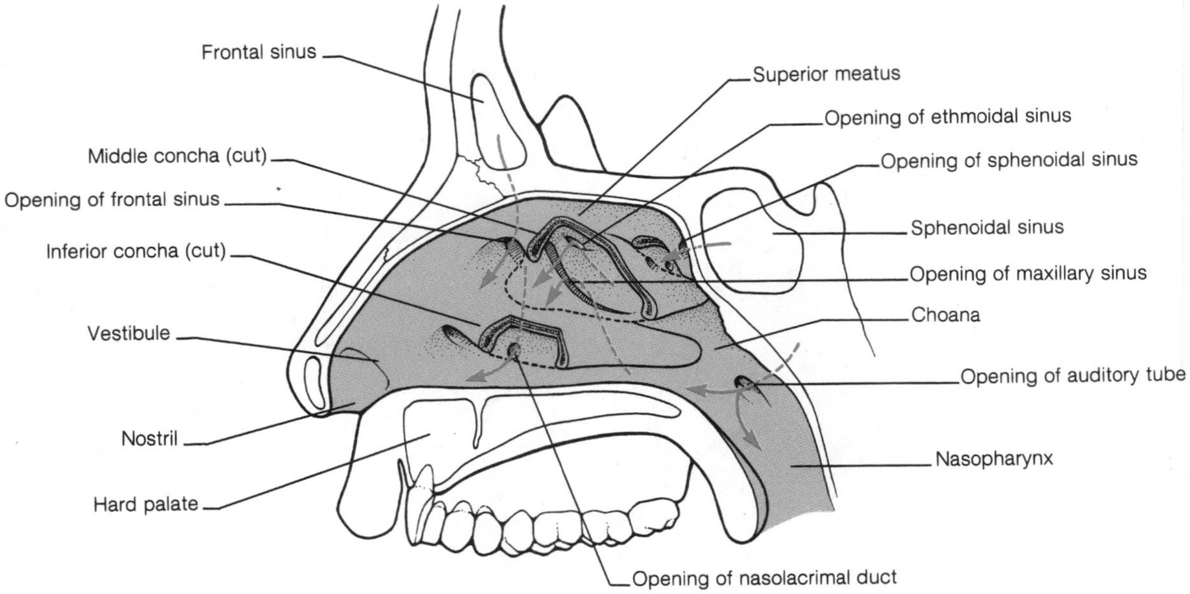

Figure 17.6 The lateral wall of the nasal cavity. There are several openings into the nasal cavity, including the openings of the various paranasal sinuses and the nasolacrimal ducts that drain from the eyes.

Air passages between the conchae are referred to as **meatuses** (fig. 17.6). The anterior openings of the nasal cavity are lined with stratified squamous epithelium, whereas the conchae are lined with pseudostratified ciliated columnar epithelium (figs. 17.7, 17.8). Mucus-secreting goblet cells are present in great abundance throughout both regions.

There are three functions of the nasal cavity and its contents:

1. The nasal epithelium covering the conchae serves to warm, moisten, and cleanse the air. The nasal epithelium is highly vascular and covers an extensive surface area. This is important for warming the air but unfortunately also makes humans susceptible to nosebleeds. Nasal hairs called **vibrissae** *(vi-bris'e)*, which often extend from the nostrils, filter macroparticles that might otherwise be inhaled. Fine particles such as dust, pollen, or smoke are trapped along the moist mucous membrane lining the nasal cavity.
2. Olfactory epithelium in the upper medial portion of the nasal cavity is concerned with the sense of smell.

vibrissa: L. *vibrare,* to vibrate

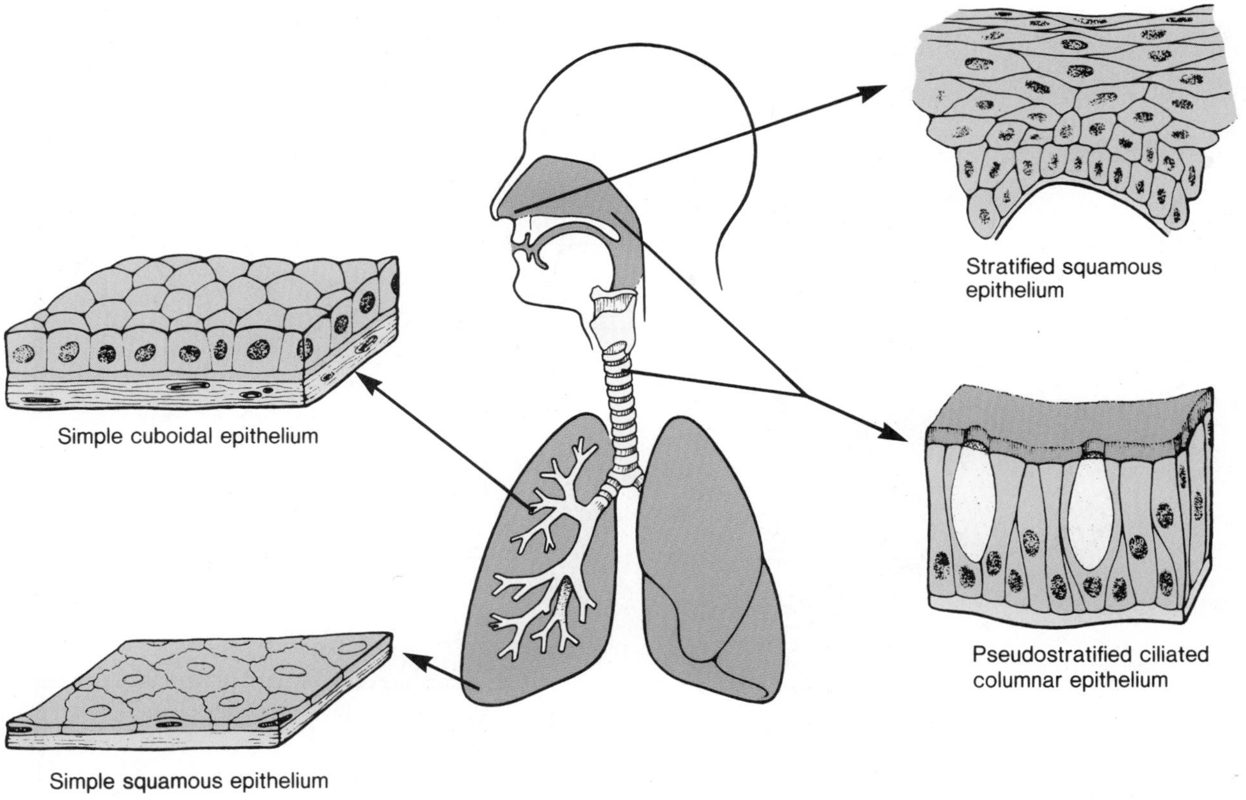

Figure 17.7 The various types of epithelial tissues present throughout the respiratory system.

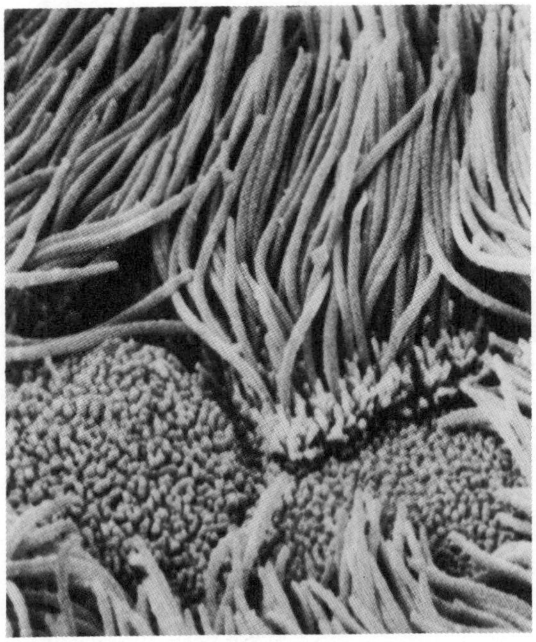

Figure 17.8 A scanning electron micrograph of a bronchial wall showing cilia. In the trachea and bronchi, there are about three hundred cilia per cell that move in a coordinated fashion, moving mucus and trapped particles toward the pharynx, where it can either be swallowed or expectorated.

3. The nasal cavity is associated with voice phonetics by functioning as a resonating chamber.

There are several drainage openings into the nasal cavity (see fig. 17.6). The paranasal ducts drain mucus from the paranasal (the frontal, ethmoidal, sphenoidal, and maxillary) sinuses, and the nasolacrimal ducts drain tears from the eyes. An excessive secretion of tears causes the nose to run as the tears drain into the nasal cavity. The auditory tube from the middle ear enters the upper respiratory tract posterior to the nasal cavity in the nasopharynx. With all these accessory connections, it is no wonder that infections can spread so easily from one chamber to another throughout the facial area. To avoid causing damage or spreading infections to other areas, one must be careful not to blow the nose too forcefully.

Paranasal Sinuses

Paired air spaces in certain bones of the skull are called **paranasal sinuses.** These sinuses are named according to the bones in which they are found: thus there are the **maxillary, frontal, sphenoidal,** and **ethmoidal sinuses** (fig. 17.9). Each sinus communicates via drainage ducts within the nasal cavity on its own side (see fig. 17.6). Paranasal sinuses may help to warm and moisten the inspired air. These sinuses are responsible for some sound resonance, but most importantly, they function to decrease the weight of the skull while giving structural strength.

sinus: L. *sinus,* bend or curve

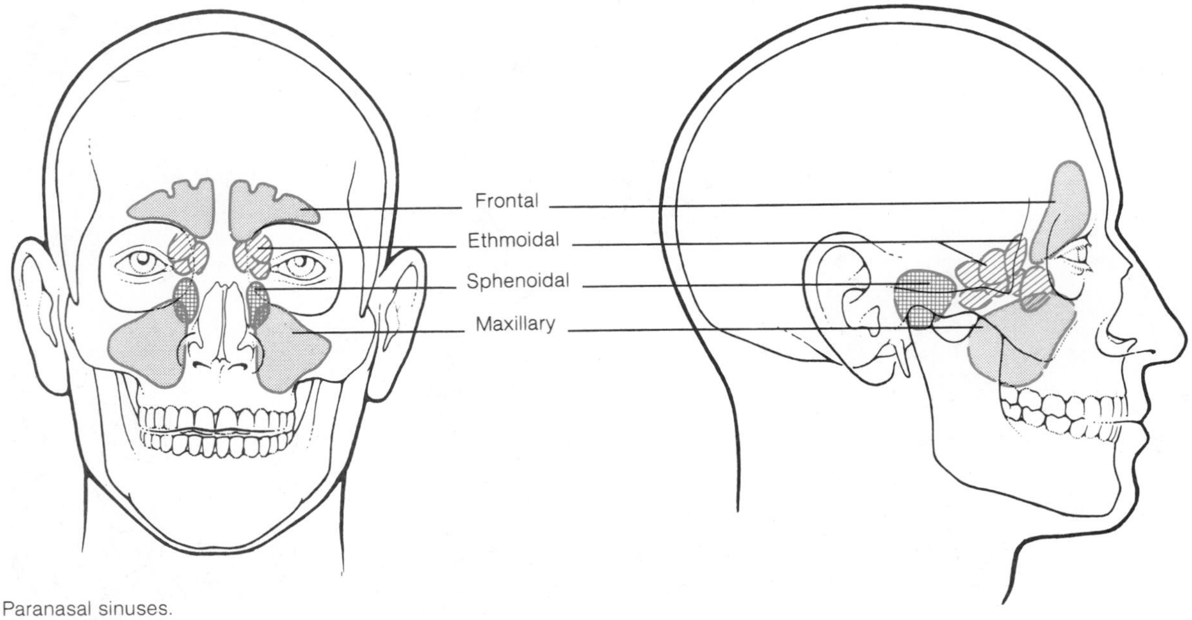

Figure 17.9 Paranasal sinuses.

You can observe your own paranasal sinuses. Face a mirror in a darkened room and shine a small flashlight into your face. The frontal sinuses will be illuminated by directing the light just below the eyebrow. The maxillary sinuses are illuminated by shining the light into the oral cavity and closing your mouth around the flashlight.

Pharynx

The **pharynx** *(far'inks)* is a funnel-shaped passageway approximately 13 cm (5 in.) in length that connects the nasal and oral cavities to the larynx at the base of the skull. The supporting walls of the pharynx are composed of skeletal muscle, and the lumen is lined with a mucous membrane. There are several paired lymphoid organs called **tonsils** in the pharynx (fig. 17.10). The pharynx, commonly referred to as the throat or gullet, has both respiratory and digestive functions. It also provides a resonating chamber for certain speech sounds. The pharynx is divided on the basis of location and function into three regions: nasal, oral, and laryngeal.

The **nasopharynx** *(na''zo-far'inks)* has only a respiratory function. It is the uppermost portion of the pharynx, directly posterior to the nasal cavity and above the soft palate. A pendulous **uvula** hangs from the middle, lower border of the soft palate. The paired **auditory,** or eustachian, **tubes** connect the nasopharynx with the middle ear cavities. **Pharyngeal tonsils,** or **adenoids,** are situated in the posterior wall of this cavity. During the act of swallowing, the soft palate and uvula are elevated to block the nasal cavity and prevent food from entering. Occasionally a person may suddenly exhale air (as with a laugh) while in the process of swallowing fluid. If this occurs before the uvula effectively blocks the nasopharynx, fluid will be discharged through the nasal cavity.

The **oropharynx** *(o''ro-far'inks)* is the middle portion of the pharynx between the soft palate and the level of the hyoid bone. The base of the tongue forms the anterior wall of the oropharynx. Paired **palatine tonsils** are located along the posterior lateral wall of the oropharynx, and the **lingual tonsils** are found on the base of the tongue. This portion of the pharynx has both a respiratory and a digestive function.

The **laryngopharynx** *(lah-ring''go-far'inks)* is the lowermost portion of the pharynx. It extends posteriorly from the level of the hyoid bone to the larynx and opens into the esophagus and larynx. It is at the lower laryngopharynx that the respiratory and digestive systems become distinct. Food is directed into the esophagus, whereas air is moved anteriorly into the larynx.

During a physical examination, a physician commonly depresses the patient's tongue and examines the condition of the palatine tonsils. Tonsils are pharyngeal lymphoid organs and tend to become swollen and inflamed after persistent infections. Tonsils have to be surgically removed when they become so overrun with pathogens after repeated infections that they themselves become the source of the infection. The removal of the palatine tonsils is called a *tonsillectomy,* whereas the removal of the pharyngeal tonsils is called an *adenoidectomy.*

Larynx

The **larynx** *(lar'inks),* or voice box, forms the entrance into the lower respiratory system as it connects the laryngopharynx with the trachea. It is positioned in the anterior midline of the neck at the level of the fourth through sixth cervical vertebrae. The larynx has two functions. Its primary function is to prevent food

pharynx: L. *pharynx,* throat
tonsil: L. *toles,* goiter or swelling
uvula: L. *uvula,* small grape
adenoid: Gk. *adenoeides,* glandlike

larynx: Gk. *larynx,* upper windpipe

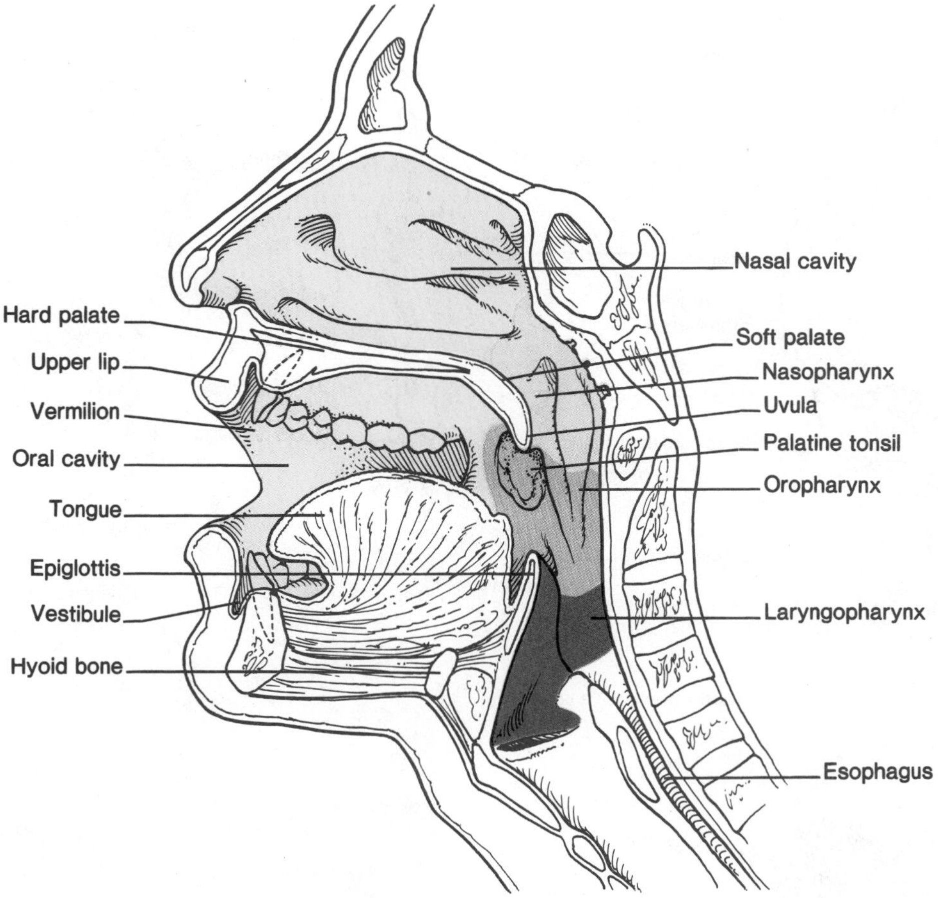

Figure 17.10 A sagittal section of the head showing the structures of the upper respiratory tract.

or fluid from entering the trachea and lungs during swallowing and to permit passage of air while breathing. A secondary function is to produce sounds.

> *Laryngitis* is the inflammation of the mucosal epithelium of the larynx and vocal cords, which causes a hoarseness of a person's voice or an inability to speak above a whisper. Laryngitis may result from overuse of the voice, inhalation of an irritating chemical, or a bacterial or viral infection. Mild cases are temporary and seldom serious.

In structure, the larynx is shaped like a triangular box (fig. 17.11). It is composed of a framework involving nine cartilages: three are large, single structures and six are smaller, paired structures. The largest of the unpaired cartilages is the anterior **thyroid cartilage.** The **laryngeal prominence** of the thyroid cartilage, commonly called the Adam's apple, forms an anterior vertical ridge along the larynx that can be palpated on the midline of the neck. The thyroid cartilage is typically larger and more prominent in males than females because of the effect of male sex hormones on the development of the larynx during puberty.

The spoon-shaped **epiglottis** has a cartilaginous framework. It is behind the root of the tongue and aids in closing the **glottis,** or laryngeal opening, during swallowing.

The lower end of the larynx is formed by the ring-shaped **cricoid** *(kri'koid)* **cartilage.** This third unpaired cartilage connects the thyroid cartilage above and the trachea below. The paired **arytenoid** *(ar''ĕ-te'noid)* **cartilages,** located above the cricoid and behind the thyroid, furnish the posterior attachment of the **vocal cords.** The other paired **cuneiform cartilages** and **corniculate** *(kor-nik'u-lat)* **cartilages** are small accessory cartilages that are closely associated with the arytenoid cartilages (fig. 17.12).

Two pairs of strong connective tissue bands are stretched across the upper opening of the larynx from the thyroid cartilage anteriorly to the paired arytenoid cartilages posteriorly. These are the **true vocal cords** and the **false vocal cords** (figs. 17.12, 17.13). The false vocal cords, or **ventricular folds,** support the true vocal cords and, as their name implies, are not

thyroid: Gk. *thyreos,* shieldlike

cricoid: Gk. *krikos,* ring; *eidos,* form
arytenoid: Gk. *arytaina,* ladle- or cup-shaped
cuneiform: L. *cuneus,* wedge-shaped
corniculate: L. *corniculum,* diminutive of *cornu,* horn

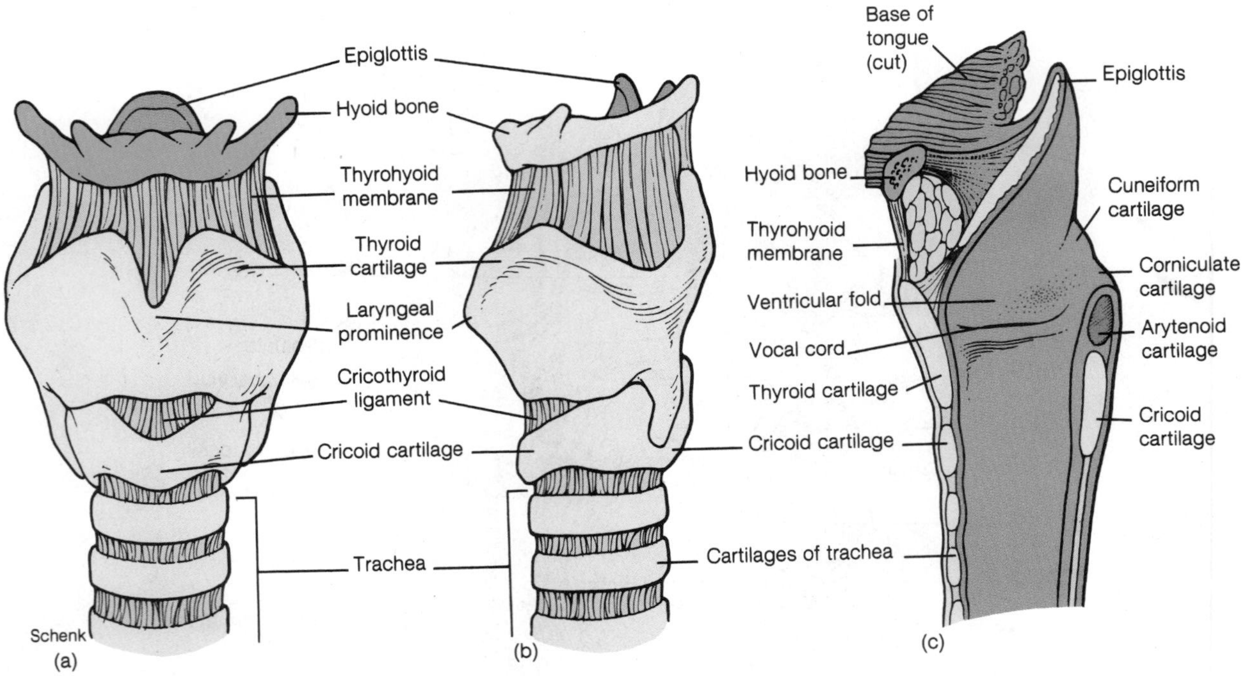

Figure 17.11 The structure of the larynx: (*a*) an anterior view, (*b*) a lateral view, and (*c*) a sagittal view.

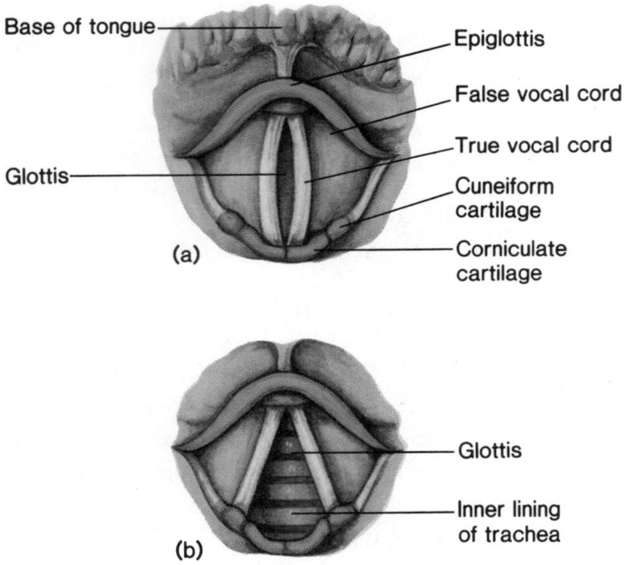

Figure 17.12 A superior view of the vocal cords as seen with a laryngoscope: (*a*) vocal cords are taut and (*b*) vocal cords are relaxed and the glottis is opened.

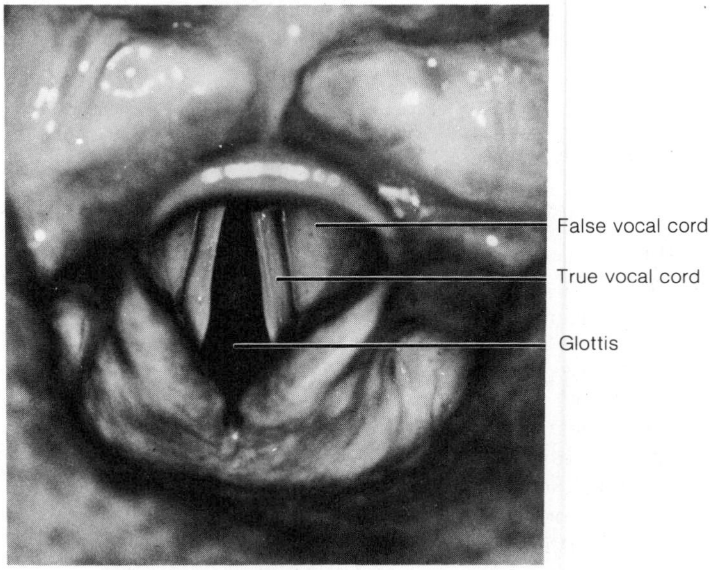

Figure 17.13 A photograph of the larynx, showing the true and false vocal cords and the glottis.

used in sound production. Stratified squamous epithelium lines the true vocal cords whereas the rest of the larynx is lined with pseudostratified ciliated columnar epithelium. This is an important anatomical modification considering the tremendous vibratory action of the true vocal cords in the production of sound.

The **laryngeal muscles** are extremely important in closing the glottis during swallowing and in speech. There are two groups of laryngeal muscles: **extrinsic muscles,** responsible for elevating the larynx during swallowing, and **intrinsic muscles** that, when contracted, change the length, position, and tension of the vocal cords. Various pitches are produced as air passes over the altered vocal cords. If the vocal cords are taut, vibration is more rapid and causes a higher pitch. Less tension on the cords produces lower sounds. Mature males generally have thicker and longer vocal cords than females; therefore, the vocal cords of males vibrate more slowly and produce lower pitches. The loudness of vocal sound is determined by the force of the air passed over the vocal cords and the amount of vibration. The vocal cords do not vibrate when whispering.

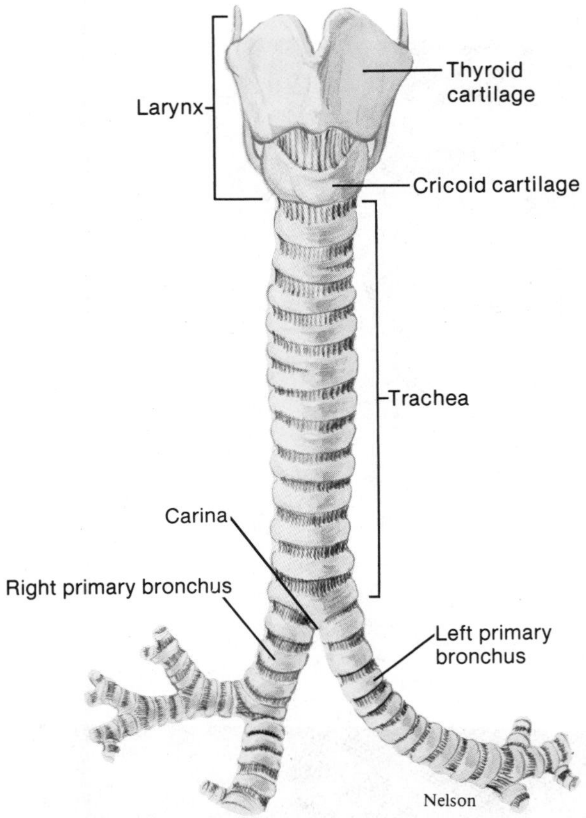

Figure 17.14 An anterior view of the larynx, trachea, and bronchi.

Sounds originate in the larynx, but other structures are necessary to convert sound into recognizable speech. Vowel sounds, for example, are produced by constriction of the walls of the pharynx. The pharynx, paranasal sinuses, and oral and nasal cavities act as resonating chambers. The final enunciation of words is accomplished through movements of the lips and tongue.

Swallowing is a complex action involving several neurological pathways and various groups of muscles. One of the swallowing responses is the elevation of the larynx to close the glottis against the epiglottis. This movement can be noted by cupping the fingers lightly over the larynx and then swallowing. Food may become lodged within the glottis if it is not closed as it should be. In this case, the *abdominal thrust (Heimlich) maneuver* can be used to prevent suffocation.

Trachea

The **trachea** *(tra'ke-ah),* commonly called the windpipe, is a rigid tube, approximately 12 cm (4½ in.) long and 2.5 cm (1 in.) in diameter, connecting the larynx to the primary bronchi (fig. 17.14). It is positioned anterior to the esophagus as it extends into the thoracic cavity. A series of sixteen to twenty C-shaped rings of hyaline cartilage form the walls of the trachea (figs. 17.15, 17.16). The open part of the C is positioned posteriorly and is covered by fibrous connective tissue and smooth muscle. These cartilages provide a rigid but flexible tube in which the lumen, or airway, is permanently open. The mucosa (surface lining of the lumen) consists of pseudostratified ciliated columnar epithelium containing many mucus-secreting **goblet cells** (see figs. 17.7, 17.8). Inhaled dust particles stick to the mucus, and the cilia sweep it upward to the pharynx, where it is removed through a cough reflex. The tracheal bifurcation forming the right and left **bronchi** is reinforced by the **carina** *(kah-ri'nah),* a keel-like cartilage plate (fig. 17.14).

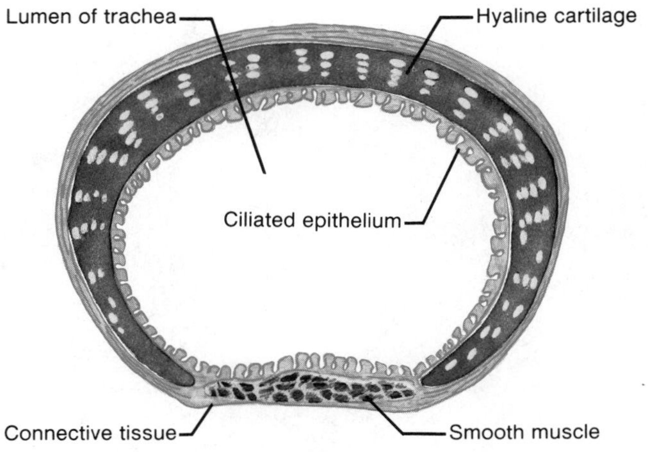

Figure 17.15 A transverse section of the trachea. The connective tissue portion faces posteriorly in contact with the esophagus.

If the trachea becomes occluded through inflammation, excessive secretion, trauma, or aspiration of a foreign object, it may be necessary to create an emergency opening into this tube so that ventilation can still occur. A *tracheotomy* is the process of surgically opening the trachea, and a *tracheostomy* is the procedure of inserting a tube into the trachea to permit breathing and to keep the passageway open (fig. 17.17). A tracheotomy should be performed only by a competent physician as there is a great risk of cutting a recurrent laryngeal nerve or carotid artery, which is also in this area.

Bronchial Tree

The **bronchial tree** is so named because it is composed of a series of respiratory tubes that branch into progressively narrower tubes as they extend into the lung (fig. 17.18). The trachea bifurcates into a right and left **primary bronchus** *(brong'kus)* at the level of the sternal angle behind the manubrium. Each bronchus has hyaline cartilage wedges within its wall surrounding the lumen to keep it open as it extends into the lung. Because of the more vertical position of the right bronchus, foreign particles are more likely to lodge here than in the left bronchus.

The bronchus divides deeper in the lungs to form **secondary bronchi** and **segmental (tertiary) bronchi** (fig. 17.19). The bronchial tree continues to branch into yet smaller

trachea: L. *trachia,* rough air vessel

bronchus: L. *bronchus,* windpipe
carina: L. *carina,* keel

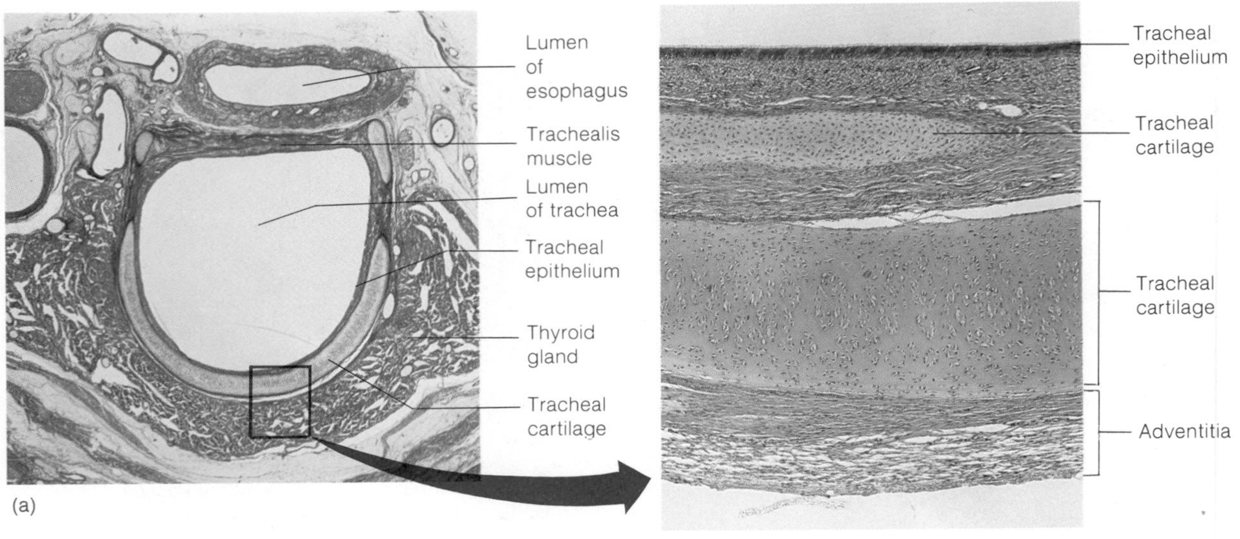

Figure 17.16 Histology of the trachea. (*a*) Photomicrograph showing the relationship of the trachea to the esophagus (3×) and (*b*) photomicrograph of tracheal cartilage (63×).

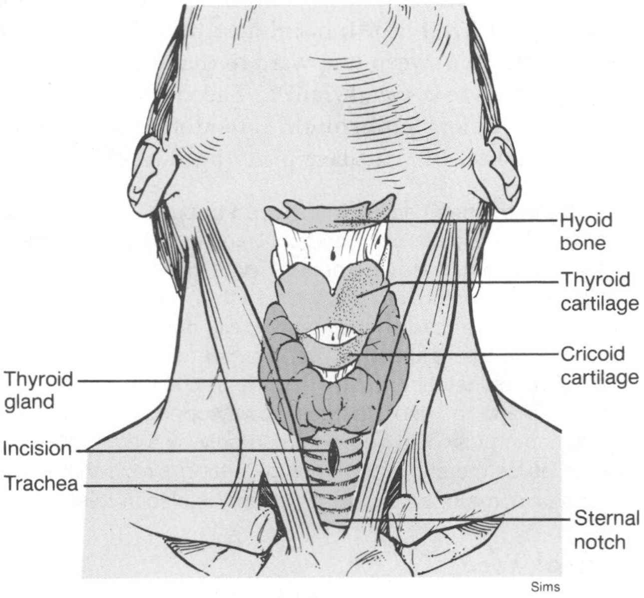

Figure 17.17 The site for a surgical tracheostomy.

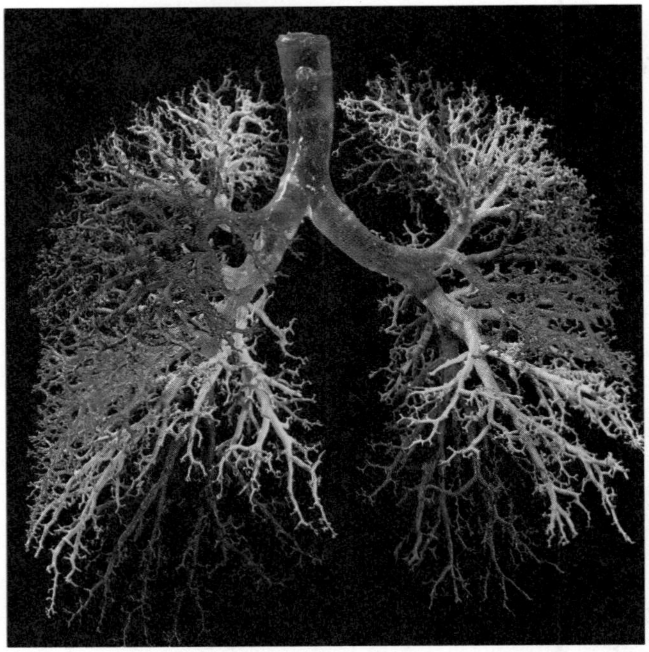

Figure 17.18 A photograph of a plastic cast of the conducting airways from the trachea to the terminal bronchioles.

tubules called **bronchioles** (fig. 17.20). There is little cartilage in the bronchioles, which contain thick smooth muscle that can constrict or dilate these airways. Bronchioles provide the greatest resistance to air flow in the conducting passages and thus are analogous to the function of arterioles in the vascular system. A simple cuboidal epithelium lines the bronchioles rather than a pseudostratified columnar epithelium as in the bronchi (fig. 17.7). Numerous **terminal bronchioles** connect to **respiratory bronchioles** that lead into the **alveolar sacs** (see fig. 17.22). The conduction portion of the respiratory system ends at the terminal bronchioles, and the respiratory portion begins at the respiratory bronchioles.

Asthma is an infectious or allergenic condition that involves the bronchi. During an asthmatic attack, there is a spasm of the smooth muscles in the lower bronchioles. Because of an absence of cartilage at this level, the air passageways constrict.

A fluoroscopic examination of the bronchi with a radiopaque medium is termed *bronchography.* This technique enables the physician to visualize the bronchial tree by x-ray film. A bronchogram of the lungs is shown in figure 17.20, and the change in the bronchial tree during respiration is demonstrated in figure 17.21.

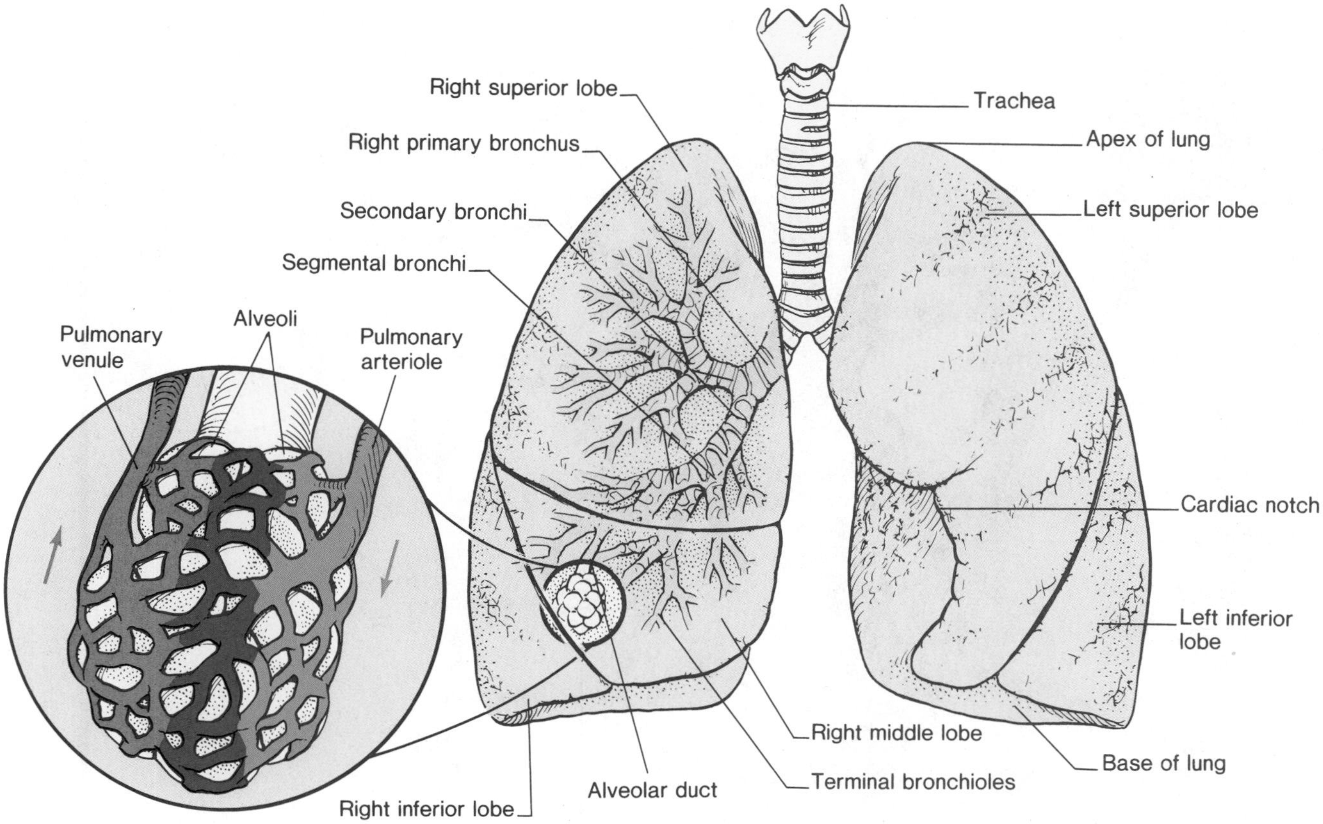

Figure 17.19 An anterior view of the lower respiratory system showing the bronchial tree, alveoli, and lungs.

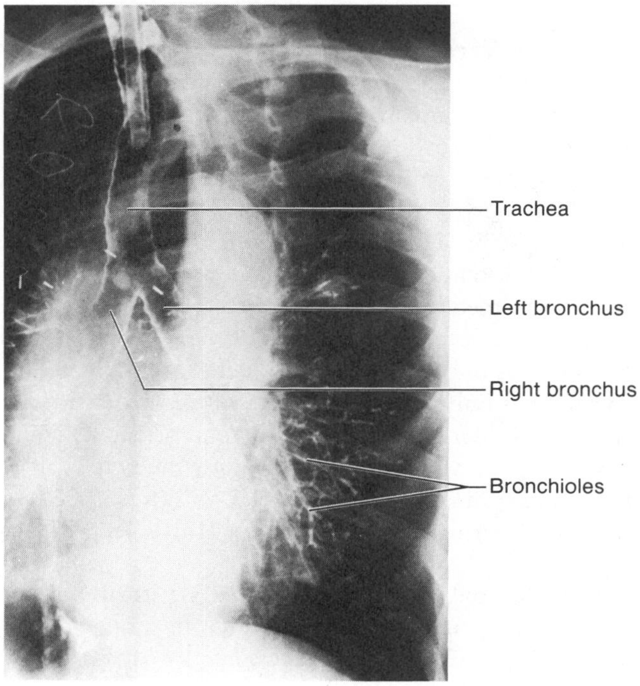

Figure 17.20 An anteroposterior bronchogram of the lungs.

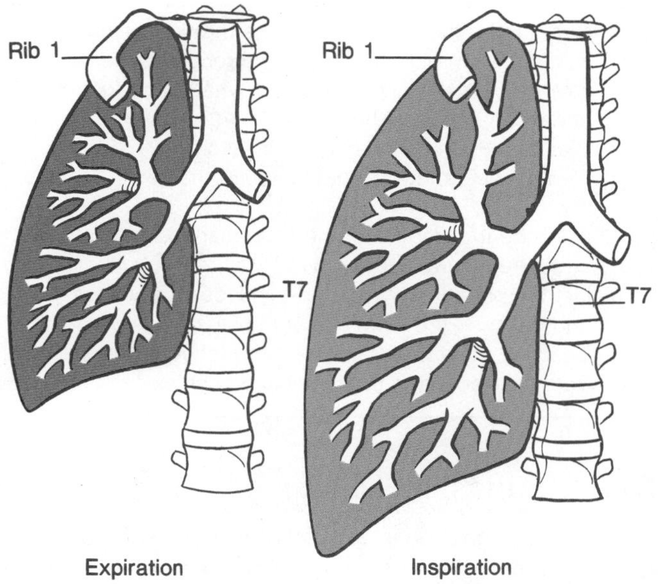

Figure 17.21 A diagram of the change in the bronchial tree during respiration. During inspiration, the bronchi lengthen and increase in diameter; whereas during expiration, they return to their original length and diameter.

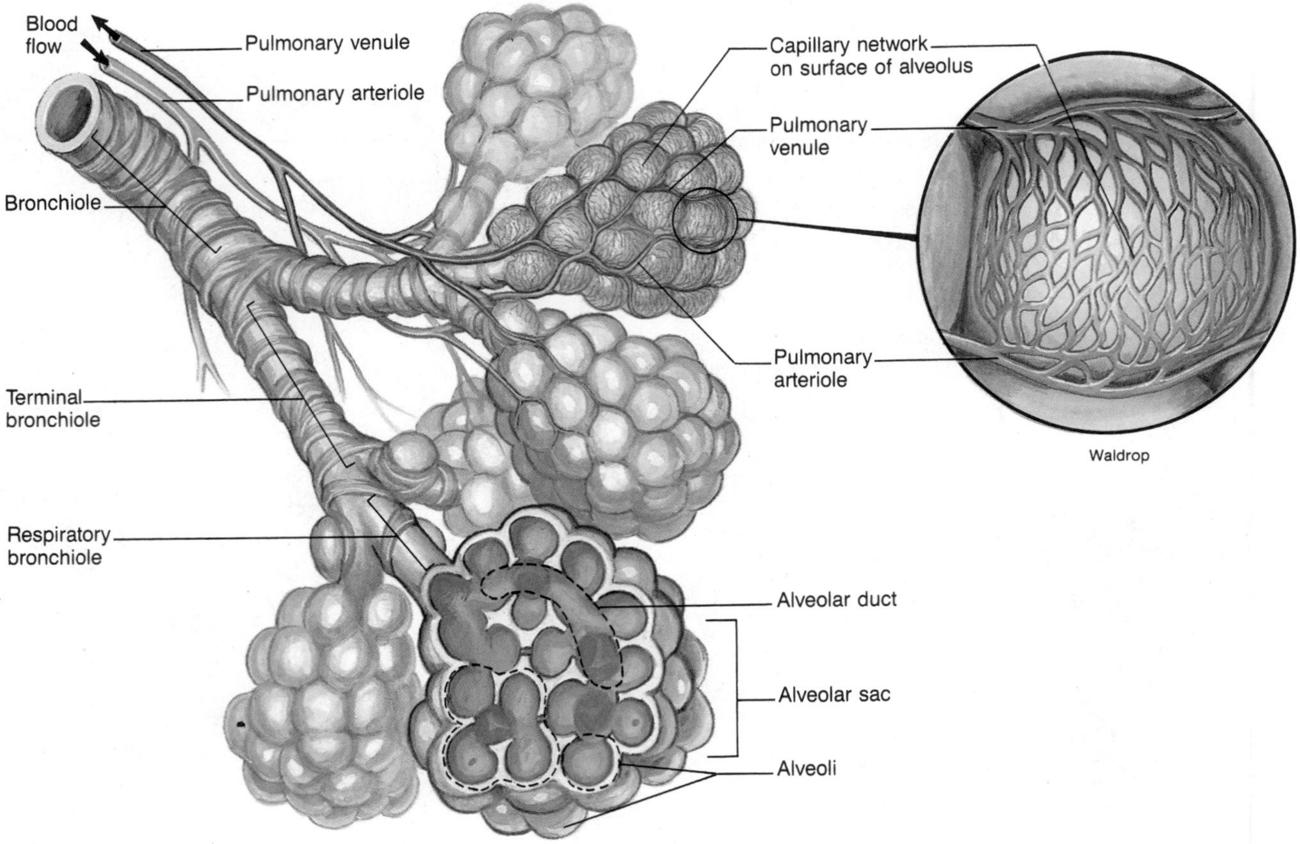

Figure 17.22 The respiratory division of the respiratory system. The respiratory tubes end in minute alveoli, each of which is surrounded by an extensive capillary network.

1. List in order the types of epithelia through which inspired air would pass in going through the nasal cavity to the alveoli of the lungs. What is the function of each of these epithelia?
2. What are the functions of the nasal cavity?
3. Identify the structures that make up the nasal septum.
4. Describe the location of the nasopharynx, and list the structures found in this cavity.
5. List the paired and unpaired cartilages of the larynx, and describe the functions of the larynx.
6. Describe the structure of the conducting airways from the trachea to the terminal bronchioles.

Alveoli, Lungs, and Pleurae

Alveoli are the functional units of the lungs where gas exchange occurs. Right and left lungs are separately contained in pleural membranes.

Objective 10. Describe the structure and function of the alveoli.

Objective 11. Describe the surface anatomy of the lungs in relation to the thorax.

Objective 12. Discuss the structural arrangement of the thoracic serous membranes, and explain their functions.

Respiratory Sacs

The respiratory bronchioles branch into many **alveolar ducts,** which lead directly into **alveolar sacs** and **alveoli** (fig. 17.22). The latter three structures comprise the *respiratory division* of the lungs. Gas exchange occurs across the walls of the tiny (0.25–0.50 mm in diameter) alveoli, the functional units of the respiratory system. The enormous number of these structures (about 350 million per lung) provides a very high surface area (60–80 square meters, or 760 square feet) for the diffusion of gases. The diffusion rate is further increased by the fact that each alveolus is only one cell layer thick, so that the total air-blood barrier is only one alveolar cell with its basement membrane and one blood capillary cell across, or about 2 micrometers. This is an average distance because the type II alveolar cells are thicker than the type I alveolar cells (fig. 17.23). Type I alveolar cells permit diffusion, and type II alveolar cells secrete *surfactant.*

Alveoli are polyhedral in shape and are usually clustered together, like the units of a honeycomb, in groups called alveolar sacs (fig. 17.24). These clusters usually occur at the ends of alveolar ducts. Individual alveoli also occur as separate outpouchings along the length of the alveolar ducts. Although the distance between each alveolar duct and its terminal alveoli is only about 0.5 mm, these units together comprise most of the mass of the lungs.

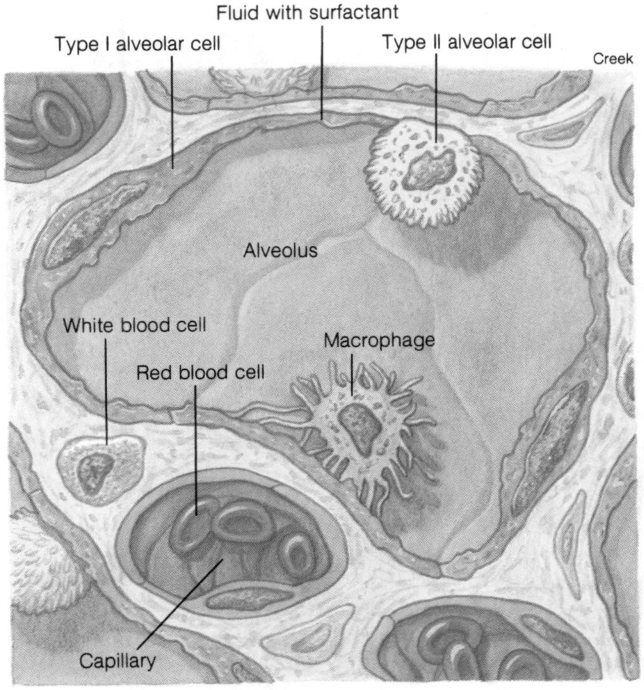

Figure 17.23 A diagram showing the relationship between lung alveoli and pulmonary capillaries.

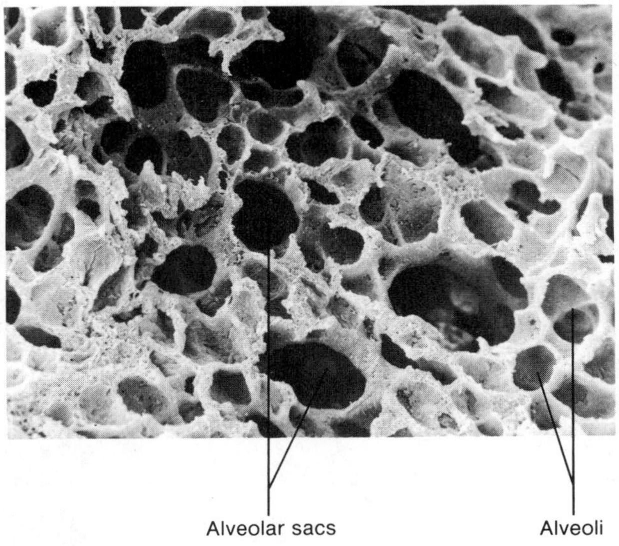

Figure 17.24 A scanning electron micrograph of lung tissue showing alveolar sacs and alveoli.

Lungs

The **lungs** are large, spongy, paired organs within the thoracic cavity. They lie against the rib cage anteriorly and posteriorly and extend from the diaphragm to a point just above the clavicles. The lungs are separated from one another by the heart and other structures of the **mediastinum** *(me''de-as-ti'num),* as shown in figure 17.25. The mediastinum is the area between the lungs. All structures of the respiratory system beyond the primary bronchi, including the bronchial tree and alveoli, are contained in the lungs.

There are some basic similarities and distinct differences between the right and left lung. Each lung presents four borders or surfaces that match the contour of the thoracic cavity. The **mediastinal (medial) surface** of each lung is slightly concave and contains a vertical slit, the **hilum** *(hi'lum),* through which pulmonary vessels, nerves, and bronchi traverse. The inferior border of the lung, called the **base,** is concave as it fits over the convex dome of the diaphragm. Anteriorly, the portion of the lungs that extends above the level of the clavicle is called the **apex,** or **cupola.** Finally, the broad, rounded surface in contact with the membranes covering the ribs is called the **costal surface.**

The left lung is somewhat smaller than the right and has a **cardiac notch** on its medial surface to accommodate the heart. The left lung is subdivided into a **superior lobe** and an **inferior lobe** by a single fissure. The right lung is subdivided by two fissures into three lobes: **superior, middle,** and **inferior** (see fig. 17.19). Each lobe of the lung is divided into many small lobules, which in turn contain the alveoli. Lobular divisions of the lungs comprise specific bronchial segments. The right lung contains ten bronchial segments, and the left lung contains eight (fig. 17.26).

The lungs of a newborn are pink in color but may become discolored in an adult as a result of smoking or air pollution. Smoking not only discolors the lungs, it may also cause deterioration of the alveoli. *Emphysema (em''fi-se'mah)* and *lung cancer* are diseases that have been linked to smoking. If a person moves to a less polluted environment or gives up smoking, the lungs will become more pinkish and healthy if they have not been permanently damaged by disease.

Pleurae

Pleurae are serous membranes investing the lungs (figs. 17.25, 17.27). The pleura of each lung is composed of both a visceral and parietal portion. The **visceral pleura** *(ploo'rah)* adheres to the outer surface of the lung and extends into each of the interlobar fissures. The **parietal pleura** lines the thoracic walls and the thoracic surface of the diaphragm. A continuation of the parietal pleura around the heart and between the lungs forms the boundary of the mediastinum. Between the visceral and parietal pleurae is a moistened space called the **pleural cavity.** An inferiorly extending reflection of the pleural layers around the roots of each lung is called the **pulmonary ligament.** The pulmonary ligaments help to support the lungs.

---

mediastinum: L. *mediastinus,* intermediate
hilum: L. *hilum,* a trifle (little significance)
cupola: L. *cupula,* diminutive of *cupa,* dome, tub

pleura: Gk. *pleura,* side or rib
pulmonary: Gk. *pleumon,* lung

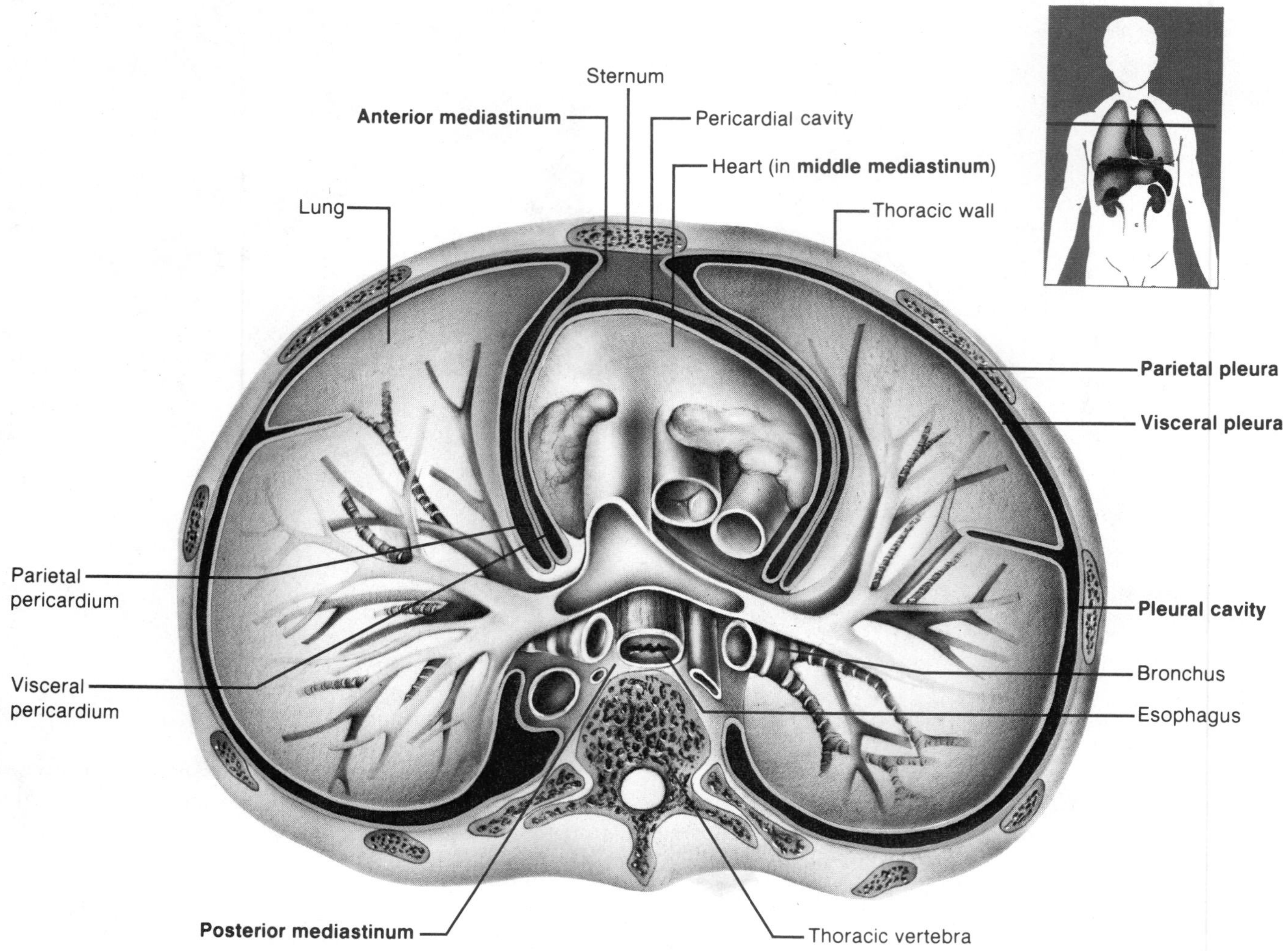

Figure 17.25 A transverse section of the thoracic cavity, viewed from a superior aspect, showing the mediastinum and pleural membranes.

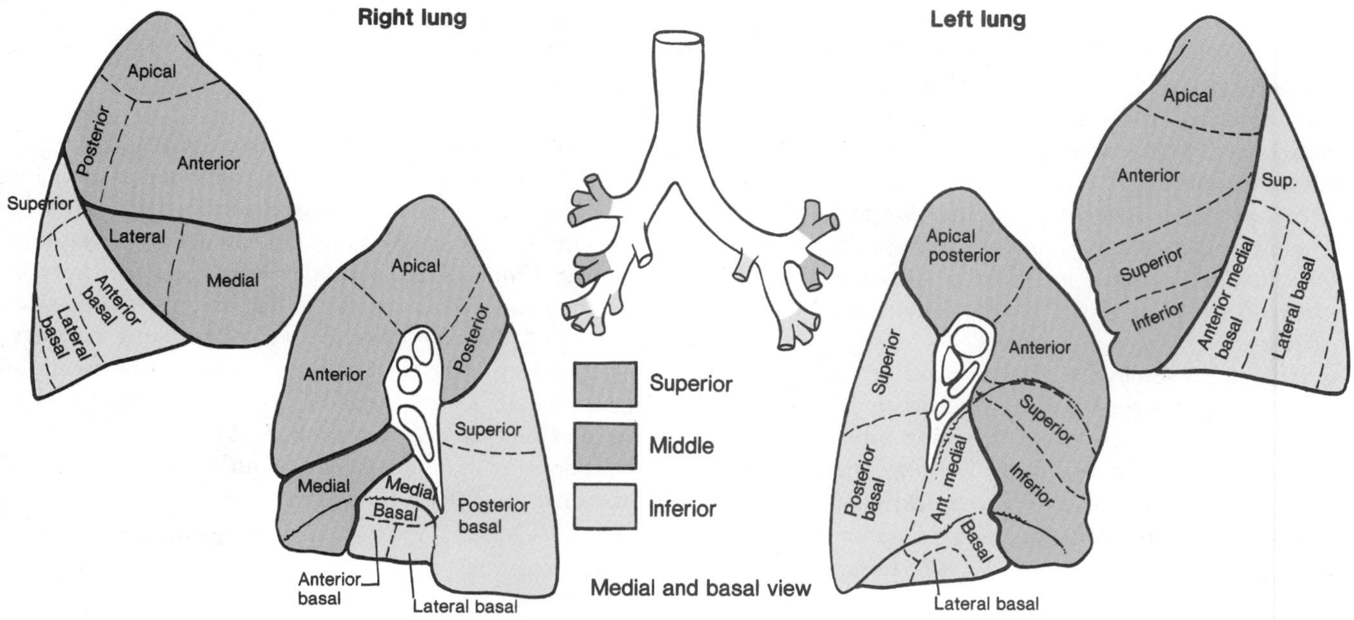

Figure 17.26 Lobes, lobules, and bronchopulmonary segments of the lungs.

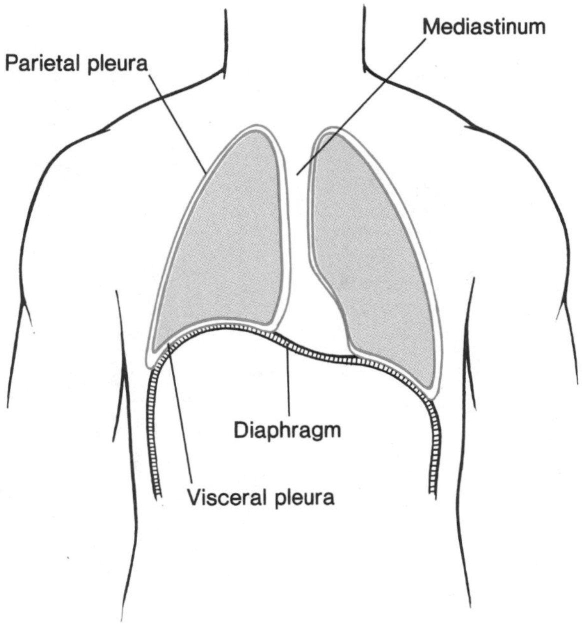

Figure 17.27 The position of the lungs and associated pleurae.

Table 17.1 Major structures of the respiratory system

Structure	Description	Function
Nose	Jutting external portion that is part of the face, plus internal nasal cavity	Warms, moistens, and filters inhaled air as it is conducted to the pharynx
Paranasal sinuses	Air spaces in certain facial bones	Produce mucus, provide sound resonance, lighten the skull
Pharynx	Chamber connecting oral and nasal cavities to the larynx	Passageway for air into the larynx and for food into the esophagus
Larynx	Voice box; short passageway that connects the pharynx to the trachea	Passageway for air; produces sound; prevents foreign materials from entering the trachea
Trachea	Flexible, tubular connection between the larynx and bronchial tree	Passageway for air; pseudostratified ciliated columnar epithelium cleanses the air
Bronchial tree	Bronchi and branching bronchioles in the lung; tubular connection between the trachea and alveoli	Passageway for air; continued cleansing of air
Alveoli	Microscopic, membranous air sacs within the lungs	Functional units of respiration: site of gaseous exchange between the respiratory and circulatory systems
Lungs	Major organs of the respiratory system; located in pleural cavities in the thorax	Contain bronchial trees, alveoli, and associated pulmonary vessels
Pleurae	Serous membranes covering the lungs and lining the thoracic cavity	Compartmentalize, protect, and lubricate the lungs

The arrangement of the pleurae serve three functions. First, a small amount of fluid within the pleural cavities acts as a lubricant to allow the lungs to slide along the chest wall. Second, the pressure in the pleural cavity is lower than the pressure in the lungs, which is needed for ventilation. The third function of the pleurae is the effective separation of the thoracic organs. Although the serous membrane is composed of simple squamous epithelium and fibrous connective tissue, it provides tight compartments for each major thoracic organ.

There are four distinct compartments in the thoracic cavity. A pleural cavity surrounds each lung, the heart is situated in the pericardial cavity, and the esophagus, thoracic duct, major vessels, various nerves, and portions of the respiratory tract are located in the mediastinum. This *compartmentalization* has a protective value because infections are usually confined to one compartment and damage to one organ usually will not involve another. *Pleurisy,* for example, which is an inflamed pleura, is generally confined to one side. A penetrating injury to one side, like a knife wound, might cause one lung to collapse but not the other.

The moistened serous membranes of the visceral and parietal pleurae are normally against each other, so that the lungs are stuck to the thoracic wall in the same manner that two wet pieces of glass stick to each other. The pleural cavity (intrapleural space) between the two moistened membranes contains only a thin layer of fluid secreted by the pleural membranes. The pleural cavity in a healthy, living organism is thus potential rather than real; it can become real only in abnormal situations when air enters the intrapleural space. Since the lungs normally remain against the thoracic wall, they get larger and smaller together with the thoracic cavity during respiratory movements.

The functions of the major organs of the respiratory system along with a description of each is presented in table 17.1.

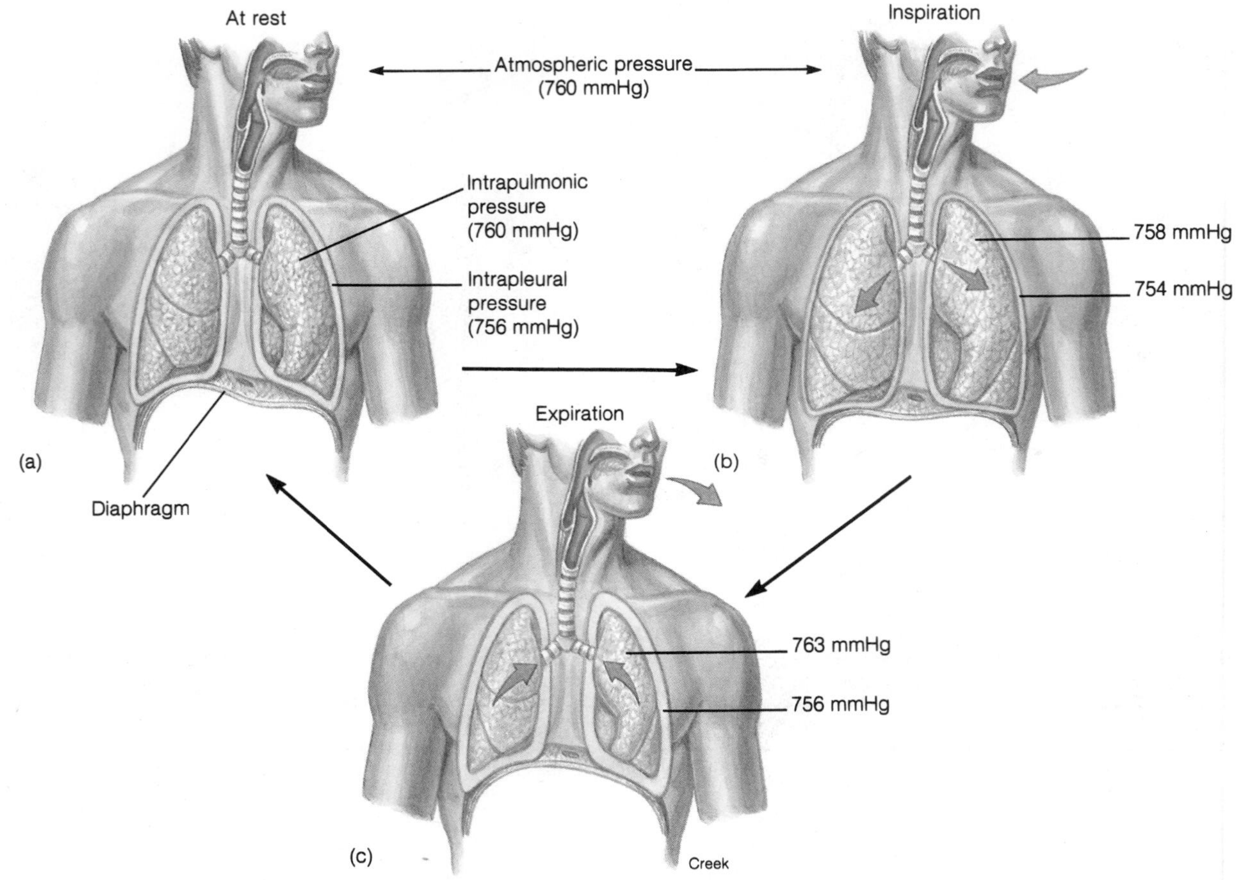

Figure 17.28 Mechanics of pulmonary ventilation.

1. Describe the structure of the respiratory division of the lungs, and explain how this structure aids in gas exchange.
2. Compare the structure of the right and left lung.
3. Describe the structure and location of the mediastinum, and list the organs it contains.
4. Describe the location of the visceral pleura, parietal pleura, and pleural cavity. Explain the functional significance of the pleural membranes.

Mechanics of Breathing

Normal, quiet inspiration results from muscle contraction, normal expiration from muscle relaxation and elastic recoil. These actions can be forced by contractions of the accessory respiratory muscles. The amount of air inspired and expired can be measured by different procedures and used as tests of pulmonary function.

Objective 13. Identify and describe the actions of the muscles involved in both quiet and forced inspiration.

Objective 14. Describe how quiet expiration is produced, and identify and describe the actions of the muscles involved in forced expiration.

Objective 15. Identify different lung volumes and capacities, and describe how pulmonary function tests help diagnose lung disorders.

Breathing, or **pulmonary ventilation**, entails the movement of air through the respiratory passageways into the alveoli of the lungs, followed by a reversal of air movement out of the lungs. It consists of two phases, inspiration and expiration. Inspiration (inhaling) and expiration (exhaling) are accomplished by alternately increasing and decreasing the volume of the thoracic cavity (fig. 17.28). Pulmonary ventilation is accomplished because pressure gradients are created. Breathing in takes place when the air pressure within the lungs is less than the atmospheric pressure, and breathing out takes place when the air pressure within the lungs is greater than the atmospheric pressure.

Pressure gradients change as the size of the thoracic cavity changes. The thorax, therefore, must be constructed so that it has sufficient rigidity to protect the vital organs it contains, provide extensive attachment surface for many short, powerful muscles, and flexible enough to function as a bellows during the ventilatory cycle. The rigidity and the extensive surfaces for muscle attachment are provided by the bony composition of the rib cage. The rib cage has pliability because the ribs are separated from one another and because most ribs (upper ten of

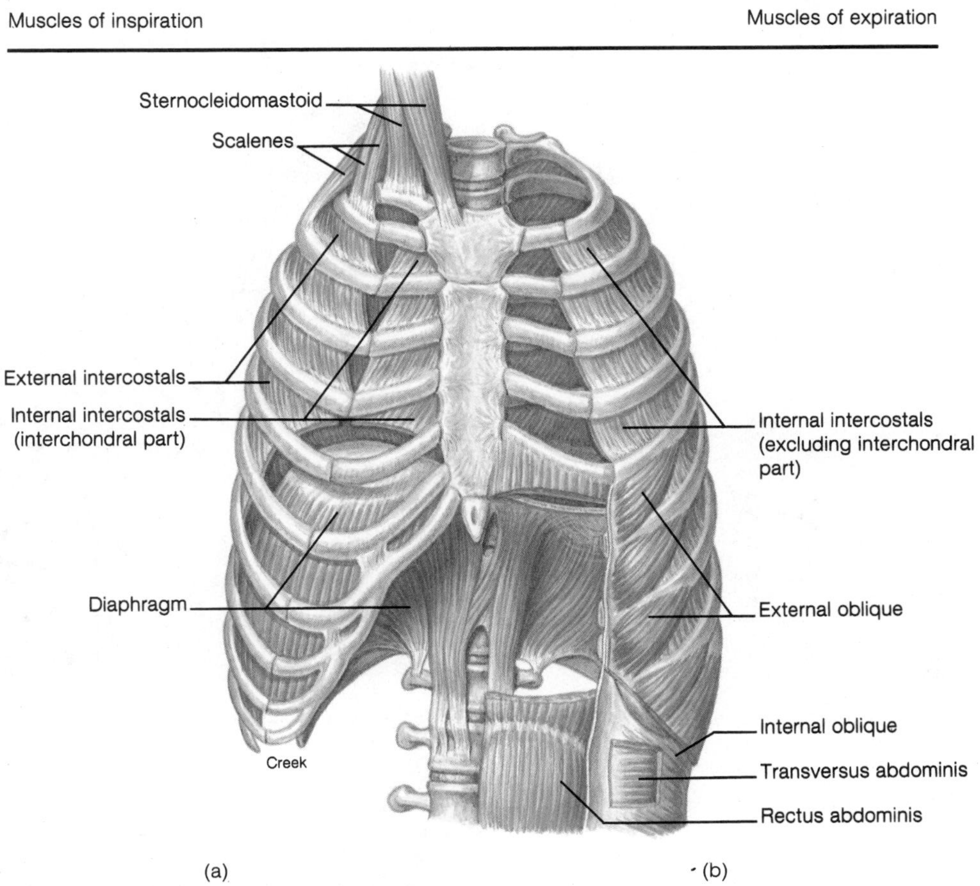

Figure 17.29 The muscles of respiration: (*a*) the principal muscles of inspiration and (*b*) the principal muscles of forced expiration. For the most part, expiration is passive.

the twelve pairs) are attached to the sternum by resilient costal cartilage. The vertebral attachment likewise provides considerable mobility. The structure of the rib cage and associated cartilage provides continuous elastic tension so that when the thorax expands it will return passively to its resting position when relaxed.

Inspiration

The thoracic cavity increases in size with inspiration. This increase in dimension normally takes place in three directions: anteroposteriorly, laterally, and vertically.

During normal inspiration, the muscles of importance are the *diaphragm,* the *external intercostals,* and the interchondral portion of the *internal intercostals* (fig. 17.29). Downward contraction of the dome-shaped diaphragm vertically increases thoracic volume. A simultaneous contraction of the external intercostals and interchondral portion of the internal intercostals increases the lateral and anteroposterior dimensions of the thorax.

The *scalenes* and *sternocleidomastoid* muscles are involved in deep inspiration or forced breathing. When these muscles are contracted, the ribs are elevated in an anteroposterior direction, while at the same time the upper rib cage is stabilized so that the intercostals become more effective. The expanded thoracic cavity decreases the air pressure within the pleural spaces to below that of the atmosphere. This pressure difference causes the lungs to become inflated.

Expiration

For the most part, expiration is a passive process that occurs as the muscles of inspiration are relaxed and the rib cage rebounds back to its original position. Even the lungs recoil during expiration as the alveoli draw together to eliminate gaseous wastes. Lowering of the surface tension in the alveoli, which brings on elastic recoiling, is due to a lipoprotein substance called *surfactant* that lines the walls of the alveoli (fig. 17.30). Surfactant is extremely important not only in causing the alveoli to recoil during expiration, but also in equalizing the overall surface tension as the alveoli expand and contract. A deficiency in surfactant in premature infants can cause *respiratory distress syndrome* (RDS), or as it is commonly called, *hyaline membrane disease.*

diaphragm: Gk. *dia,* across; *phragma,* fence

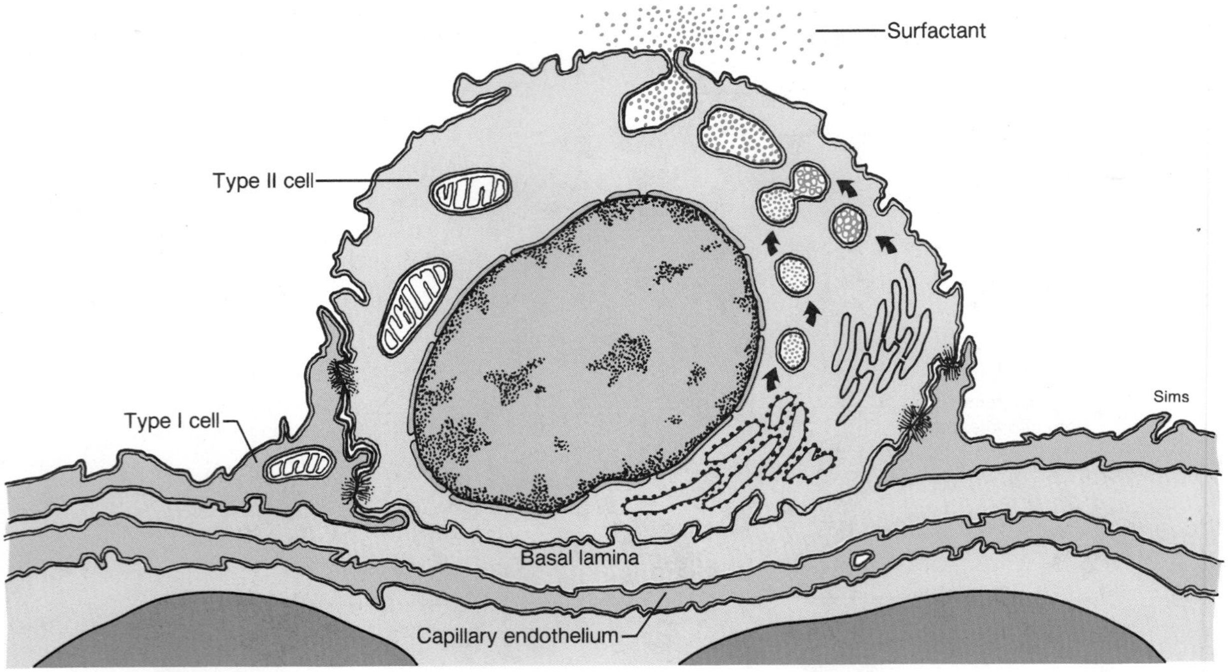

Figure 17.30 The production of surfactant by type II alveolar cells. Since surfactant does not start to be produced until about the eighth month, premature babies are sometimes born with lungs that lack sufficient surfactant, and their alveoli are collapsed as a result in a condition known as respiratory distress syndrome.

Even under normal conditions, the first breath of life is a difficult one because the newborn must overcome large surface tension forces in order to inflate its partially collapsed alveoli. The transpulmonary pressure required for the first breath is fifteen to twenty times that required for subsequent breaths, and an infant with *respiratory distress syndrome* must duplicate this effort with every breath. Fortunately, many babies with this condition can be saved by mechanical ventilators that keep them alive long enough for their lungs to mature and manufacture sufficient surfactant.

During forced expiration, such as coughing or sneezing, the interosseous portion of the *internal intercostal* muscles contract, causing the rib cage to be depressed. The *abdominal* muscles may also aid expiration, because when contracted, they force abdominal organs up against the diaphragm and further decrease the volume of the thorax. Thus intrapulmonary pressure can rise 20 or 30 mm Hg above the atmospheric pressure. The events that occur during inspiration and expiration are summarized in table 17.2.

Respiratory Air Volumes

The respiratory system is somewhat inefficient because the air enters and exits at the same place, through either the nose or the mouth. Consequently, there is an incomplete exchange of gas during each ventilatory cycle, and approximately five-sixths of the air present in the lungs still remains when the next inspiration begins.

Clinically speaking, it is important to know the amount of air that is breathed in a given time as well as the degree of difficulty in breathing. The amount of air exchanged during pulmonary ventilation varies from person to person according to

Table 17.2 Events of inspiration and expiration during a pulmonary ventilatory cycle*

Nerve stimulus	Event
Inspiration	
Phrenic nerves	The diaphragm contracts inferiorly, increasing the volume of the thorax. The diaphragm is the principal muscle involved in quiet inspiration.
Intercostal nerves	Contraction of the external intercostal, and the interchondral portion of the internal intercostal muscles elevates the ribs, thus increasing the capacity of the thoracic cavity.
Accessory, cervical, and thoracic nerves	Forced inspiration is accomplished through contraction of the sternocleidomastoid, and scalenus muscles, which increases the dimension of the thoracic cavity anteroposteriorly. Pulmonary ventilation during forced inspiration usually occurs through the mouth rather than the nose.
	As the dimension of the thoracic cavity is increased, the pressure within the pleural cavities decreases and the lungs inflate because of the greater atmospheric pressure.
Expiration	
	Nerve stimuli to the inspiratory muscles cease and the muscles relax.
	The rib cage and lungs recoil as air is forced out of the lungs due to the increased pressure.
Intercostal and lower spinal nerves	Forced expiration occurs when the interosseus portion of the internal intercostal and abdominal muscles are contracted.

*Some of the events during inspiration and expiration may occur simultaneously.

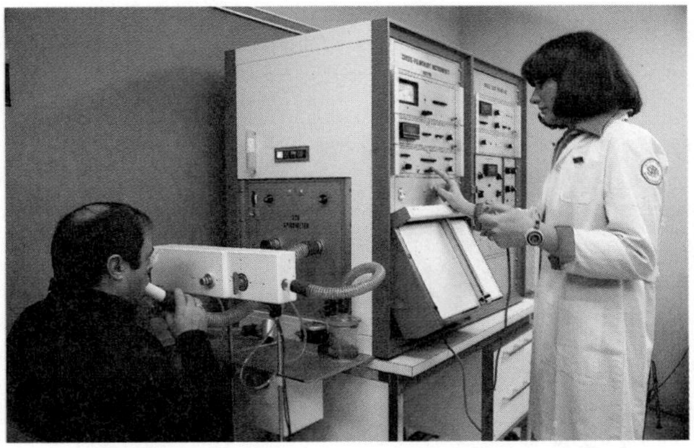

Figure 17.31 A spirometer. With the exception of the residual volume, which is measured using special techniques, this instrument can determine respiratory air volumes.

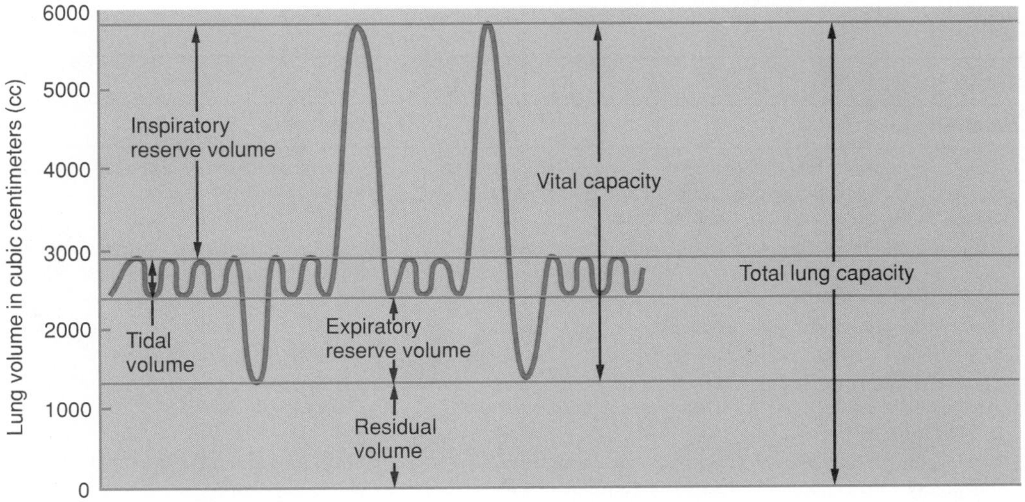

Figure 17.32 Respiratory air volumes.

age, sex, activity, and general health. Respiratory air volumes are measured with a *spirometer* (fig. 17.31). Any ventilatory abnormalities can then be compared to what is accepted as normal. The normal adult respiratory air volumes are presented in table 17.3 and figure 17.32.

People with pulmonary disorders frequently complain of *dyspnea (disp′ne-ah),* which is a subjective feeling of shortness of breath. Dyspnea may occur even when ventilation is normal, however, and may not occur even when the total volume of air movement is very high as in exercise. Some of the terms used to define ventilation are defined in table 17.4.

Nonrespiratory Air Movements

Air movements through the respiratory system other than for pulmonary ventilation are termed nonrespiratory movements. Such movements are part of emotional displays such as laughing, sighing, crying, or yawning, or they may function to expel foreign matter from the respiratory tract, as in coughing and sneezing. Generally, nonrespiratory movements are reflexive.

dyspnea: Gk. *dys,* bad; *pnoe,* breathing

Table 17.3 Respiratory air volumes for a healthy adult male

Volume	Quantity of air	Description
Tidal volume (TV)	500 cc	Volume moved in or out of the lungs during quiet breathing
Inspiratory reserve volume (IRV)	3000 cc	Volume that can be inhaled during forced breathing in addition to tidal volume
Expiratory reserve volume (ERV)	1000 cc	Volume that can be exhaled during forced breathing in addition to tidal volume
Vital capacity (VC)	4500 cc	Maximum volume of air that can be exhaled after taking the deepest breath possible: VC = TV + IRV + ERV
Residual volume (RV)	1500 cc	Volume that remains in the lungs at all times
Total lung capacity (TLC)	6000 cc	Total volume of air that the lungs can hold: TLC = VC + RV

Table 17.4 Definitions of some terms used to describe ventilation

Term	Definition	Term	Definition
Air spaces	Alveolar ducts, alveolar sacs, and alveoli	Hyperventilation	An alveolar ventilation that is excessive in relation to metabolic rate; results in abnormally low alveolar CO_2
Airways	Structures that conduct air from the mouth and nose to the respiratory bronchioles	Hypoventilation	An alveolar ventilation that is low in relation to metabolic rate; results in abnormally high alveolar CO_2
Alveolar ventilation	Removal and replacement of gas in alveoli; equal to the tidal volume minus the volume of dead space	Physiological dead space	Combination of anatomical dead space and underventilated alveoli that do not contribute normally to blood-gas exchange
Anatomical dead space	Volume of the conducting airways to the zone where gas exchange occurs	Pneumothorax	Presence of gas in the pleural space (the space between the visceral and parietal pleural membranes) causing lung collapse
Apnea	Cessation of breathing	Torr	Synonymous with millimeters of mercury (760 mm Hg = 760 torr)
Dyspnea	Unpleasant subjective feeling of difficult or labored breathing		
Eupnea	Normal, comfortable breathing at rest		

Table 17.5 Nonrespiratory air movements

Air movement	Mechanism	Comments
Coughing	A deep inspiration is taken, followed by a closure of the glottis. Suddenly the glottis is opened, and a forceful expiration of air is sent through the upper respiratory tract.	Reflexive or voluntary. Stimulus may be foreign material irritating larynx or trachea.
Sneezing	Similar to a cough except the forceful expired air is directed primarily through the nasal cavity. The eyelids are reflexively closed during a sneeze.	Reflexive response to irritating stimulus of the nasal mucosa. Sneezing clears the upper respiratory passages.
Sighing	A deep, prolonged inspiration followed by a rapid, forceful expiration.	Reflexive or voluntary, usually in response to boredom or sadness.
Yawning	A deep inspiration through a widely opened mouth. The inspired air is usually held for a short period before sudden expiration.	Usually reflexive or psychological. May display boredom or tiredness, but exact stimulus-receptor cause is unknown.
Laughing	Deep inspiration followed by a rapid convulsive expiration. Air movements are accompanied by expressive facial distortions.	Reflexive or voluntary to express emotional feelings.
Crying	Similar to laughing, but glottis remains open during entire expiration and there is an expression of different facial muscles.	Somewhat reflexive but under voluntary control.
Hiccuping	Spasmodic contraction of the diaphragm while the glottis is closed, producing a sharp inspiratory sound.	Reflexive and serves no known function.

Some of them, however, can be voluntarily initiated. These types of air movements and the reflexive mechanisms involved are summarized in table 17.5.

1. Describe the actions of the diaphragm and intercostal muscles during relaxed inspiration.
2. Describe how forced inspiration and forced expiration are produced.
3. Define the terms *tidal volume* and *vital capacity.*
4. Describe which respiratory volumes are being used during the following: a sneeze; a deep inspiration prior to jumping into a swimming pool; maximum ventilation during running; sleep.

Regulation of Breathing

The rhythm of breathing is controlled by centers in the brain stem. These centers are influenced by higher brain areas and regulated by sensory information that makes breathing responsive to the changing respiratory needs of the body.

Objective 16. Describe the functions of the pneumotaxic, apneustic, and rhythmicity centers in the brain stem.

Objective 17. Identify the chemoreceptors, and describe their pathway of innervation.

Pulmonary ventilation is primarily an involuntary, rhythmic action so effective that it continues to function even when a person is unconscious. In order for the neural control center of

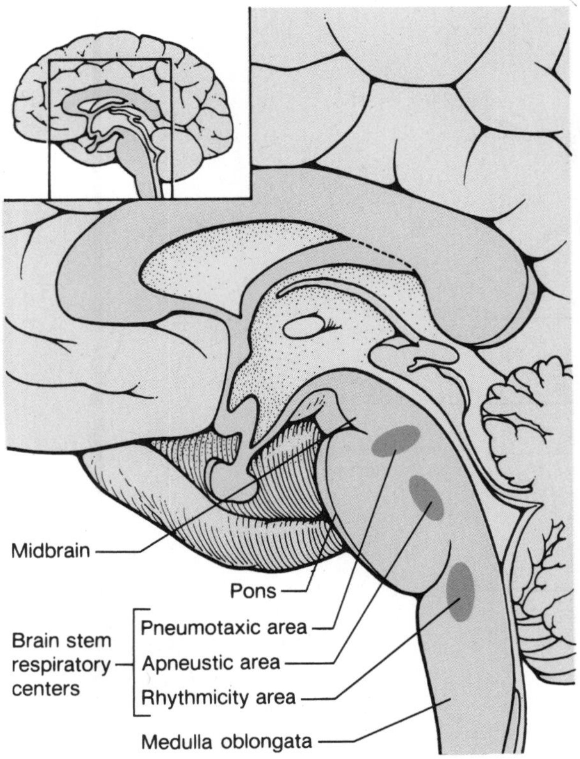

Figure 17.33 Approximate locations of the brain stem respiratory centers.

respiration to function effectively, it must possess monitoring, stimulating, and inhibiting properties so that the body can respond appropriately to increased or decreased metabolic needs. In addition, the center must be connected to the cerebrum to receive the voluntary impulses from a person who wants to change the rate of respiration.

Three portions of the brain compose the **respiratory center** (fig. 17.33): the rhythmicity, the apneustic, and the pneumotaxic areas. The **rhythmicity area** is located in the medulla oblongata of the brain and contains two aggregations of nerve cell bodies that form the **inspiratory** and **expiratory portions.** Nerve impulses from the inspiratory portion travel through the phrenic and intercostal nerves to stimulate the diaphragm and intercostal muscles. Impulses from the expiratory portion stimulate the muscles of expiration. These two portions act reciprocally; that is, when one is stimulated, the other is inhibited. Stretch receptors in the visceral pleura of the lungs provide feedback through the vagus nerves to stimulate the expiratory portion.

The **apneustic** and **pneumotaxic areas** are located in the pons of the brain. These areas function to modify and control the activity of the rhythmicity area. The apneustic center promotes inspiration, and the pneumotaxic center inhibits the activity of inspiratory neurons.

The brain stem respiratory centers produce rhythmic breathing even in the absence of other neural input. This intrinsic respiratory pattern, however, is modified by input from higher brain centers and by input from receptors sensitive to the chemical composition of the blood. The influence of higher brain centers is evidenced by the fact that you can voluntarily hyperventilate and hypoventilate (as in breath-holding). The inability to hold your breath longer than a short period is due to reflex breathing in response to input from the chemoreceptors *(ke''mo-re-cep'tors)*.

There are two groups of chemoreceptors that respond to changes in blood chemistry. These are the *central chemoreceptors* in the medulla oblongata and the *peripheral chemoreceptors.* The peripheral chemoreceptors include the **aortic bodies,** located in the aortic arch, and the **carotid bodies,** located at the junctions of the internal and external carotid arteries (fig. 17.34). These peripheral chemoreceptors control breathing indirectly via sensory neurons to the medulla. The aortic bodies send sensory information to the medulla in the vagus (tenth) cranial nerve; the carotid bodies stimulate sensory fibers in the glossopharyngeal (ninth) cranial nerve.

1. Identify the two structures within the brain where the three respiratory areas are located. Which of these three areas is responsible for autonomic rhythmic breathing?
2. Identify the locations of the peripheral chemoreceptors, and the two cranial nerves which carry sensory impulses from these sites to the respiratory centers within the brain stem.

Clinical Considerations

The respiratory system is particularly vulnerable to inflammatory and infectious diseases simply because many pathogens are airborne, humans are highly social, and the warm, moistened environment along the respiratory tract is very susceptible. Injury and trauma are likewise frequent problems. Protruding noses may become broken, the large, spongy lungs are easily penetrated by broken ribs, and portions of the respiratory tract are liable to become occluded because they also have a digestive function.

Developmental Problems

Birth defects, inherited disorders, and premature births commonly cause problems in the respiratory system of infants. A **cleft palate** is a developmental deformity of the hard palate of the mouth in which an opening persists between the oral and nasal cavities, making it difficult, if not impossible, for an infant to nurse. A cleft palate may be hereditary or a complication of some disease (e.g., German measles) contracted by the mother during pregnancy. A **cleft lip** is a genetically based developmental disorder in which the two sides of the upper lip fail to fuse. Cleft palates and cleft lips can be treated very effectively with cosmetic surgery.

As mentioned earlier in this chapter, **hyaline membrane disease** is a fairly common respiratory disorder that accounts

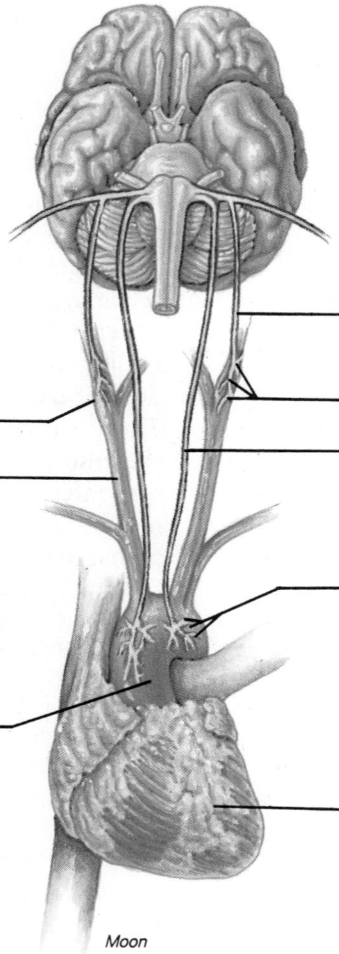

Figure 17.34 The peripheral chemoreceptors (aortic and carotid bodies) regulate the brain stem respiratory centers by means of sensory nerve stimulation.

for about one-third of neonatal deaths. This condition results from the deficient production of surfactant. **Cystic fibrosis** *(sis'tik fi-bro'sis)* is a genetic disorder that affects the respiratory system, as well as other systems of the body, and accounts for approximately one childhood death in twenty. The effect of this disease on the respiratory system is usually a persistent inflammation and infection of the respiratory tract.

Alveoli of the lungs are not developed sufficiently to sustain life until after the twentieth week of gestation. Thus, extrauterine life prior to that time is extremely difficult even with life-supporting devices.

Trauma or Injury

Humans are especially susceptible to **epistaxis** *(ep"ĭ-stak'sis)* (nosebleeds) because the nose is located where it can be easily bumped and because the nasal mucosa has extensive vascularity for warming the inspired air. Certain conditions or diseases such as high blood pressure or leukemia may cause epistaxis.

When air enters the pleural cavity surrounding either lung and causes the lung to collapse, it is referred to as a **pneumothorax** (fig. 17.35). A pneumothorax can result from an ex-

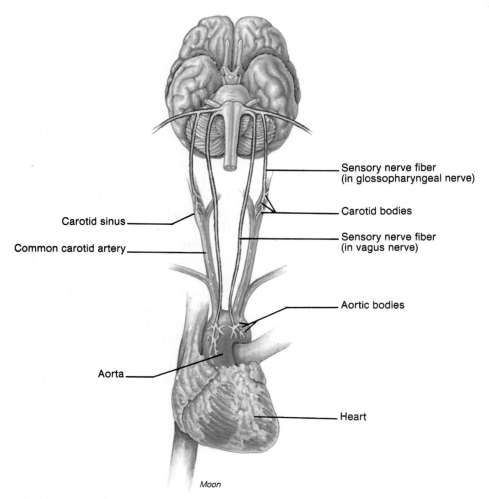

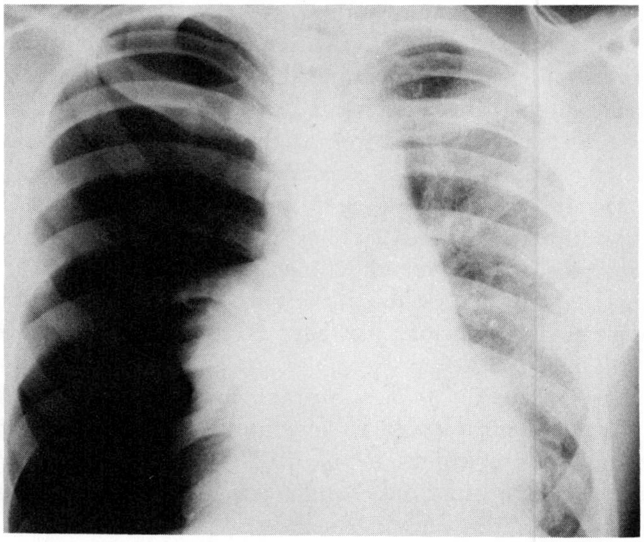

Figure 17.35 A pneumothorax of the right lung. The right side of the thorax appears uniformly dark because it is filled with air; the spaces between the ribs are also greater than on the left due to release from the elastic tension of the lungs. The left lung appears denser (less dark) because of shunting of blood from the right to the left lung.

ternal injury, such as a stabbing, bullet wound, or penetrating fractured rib, or it can occur internally. A severely diseased lung, as in emphysema, can create a pneumothorax as the wall of the lung deteriorates along with the visceral pleura and permits air to enter the pleural cavity.

Choking on a foreign object such as aspirated food is a common serious trauma to the respiratory system. More than eight Americans choke to death each day on food lodged in their trachea. A simple process termed the **abdominal thrust** (Heimlich) **maneuver** can save the life of a person who is choking. The abdominal thrust maneuver is performed as follows.

If the victim is standing or sitting:

1. Stand behind the victim or the victim's chair, and wrap your arms around his or her waist.
2. Grasp your fist with your other hand, and place the fist against the victim's abdomen, slightly above the navel and below the rib cage.
3. Press your fist into the victim's abdomen with a quick upward thrust.
4. Repeat several times if necessary.

If the victim is lying down:

1. Position the victim on his or her back.
2. Face the victim, and kneel on his or her hips.
3. With one of your hands on top of the other, place the heel of your bottom hand on the abdomen, slightly above the navel and below the rib cage.
4. Press into the victim's abdomen with a quick upward thrust.
5. Repeat several times if necessary.

If you are alone and choking: Use anything that applies force just below your diaphragm. Press into a table or a sink, or use your own fist.

Persons saved from drowning and shock victims frequently experience apnea (cessation of breathing) and will soon die if not revived by someone performing artificial respiration. The accepted treatment for reviving a person who has stopped breathing is illustrated in figure 17.36.

Common Respiratory Disorders

A cough is the most common symptom of respiratory disorders. Acute problems may be accompanied by dyspnea or wheezing. Respiratory or circulatory problems may cause **cyanosis,** which is a blue discoloration of the skin caused by blood with a low oxygen content.

The **common cold** is the most widespread of all respiratory diseases, and yet no cure, but only symptomatic relief, is available thus far. Colds occur repeatedly because acquired immunity for one virus does not protect against other viruses. Cold viruses generally incite acute inflammation in the respiratory mucosa, causing flow of mucus, a fever, and often headache.

cyanosis: Gk. *kyanosis,* dark-blue color

Nearly all of the structures and regions of the respiratory passageways can become infected and inflamed. **Influenza** is a viral disease that causes an inflammatory condition of the upper respiratory tract. Influenza can be epidemic, but fortunately vaccines are available. **Sinusitis** is an inflammation of the paranasal sinuses. Sinusitis can be quite painful if the drainage ducts from the sinuses into the nasal cavity become blocked. **Tonsillitis** may involve any or all of the tonsils and frequently follows other lingering diseases of the oral or pharyngeal regions. Chronic tonsillitis generally requires a tonsillectomy. **Laryngitis** is inflammation of the larynx, which often produces a hoarse voice and limits the ability to talk. **Tracheobronchitis** and **bronchitis** are infections of the regions that give them their names. Severe inflammation in these areas can cause smaller respiratory tubules to collapse, blocking the passage of air.

Diseases of the lungs are likewise common and usually serious. **Pneumonia** is an acute infection and inflammation of lung tissue accompanied by exudation (accumulation of fluid). It can have many causes but the most common is a bacterial pneumonia. **Tuberculosis** (TB) is an inflammatory disease of the lungs contracted by inhaling air sneezed or coughed by someone who has active tuberculosis bacteria. Tuberculosis softens and leads to ulceration of lung tissue. **Asthma** is a disease that affects persons who are allergic to certain inhaled antigens such as pollen. Asthma causes a swelling and blocking of lower respiratory tubes, often accompanied by the formation of mucous plugs. **Pleurisy** is an inflammation of the pleura and is usually secondary to some other respiratory disease. Inspiration may become painful, and fluid may collect within the pleural space. **Emphysema** is a disease that causes the breakdown of the alveoli, thus increasing the size of air spaces and decreasing the surface area (fig. 17.37). It is a frequent cause of death among heavy cigarette smokers.

Cancer in the respiratory system is known to be caused by the repeated inhalation of irritating substances such as cigarette smoke. Cancers of the lip, larynx, and lungs are especially common in smokers over the age of fifty.

Disorders of Respiratory Control

There are a variety of disease processes that can produce cessation of breathing during sleep, or *sleep apnea.* **Sudden infant death syndrome** (SIDS) is an especially tragic form of sleep apnea that claims the lives of about ten thousand babies annually in the United States. Victims of this condition are apparently healthy two-to-five-month-old babies who die in their sleep without apparent reason—hence, the layman's term of *crib death.* These deaths seem to be caused by failure of the respiratory control mechanisms in the brain stem and/or by failure of the carotid bodies to be stimulated by reduced arterial oxygen.

influenza: L. *influentia,* a flowing in
tuberculosis: L. *tuberculum,* diminutive of tuber, swelling
asthma: Gk. *asthma,* panting
emphysema: Gk. *emphysan,* blow up, inflate

Mouth-to-Mouth Method

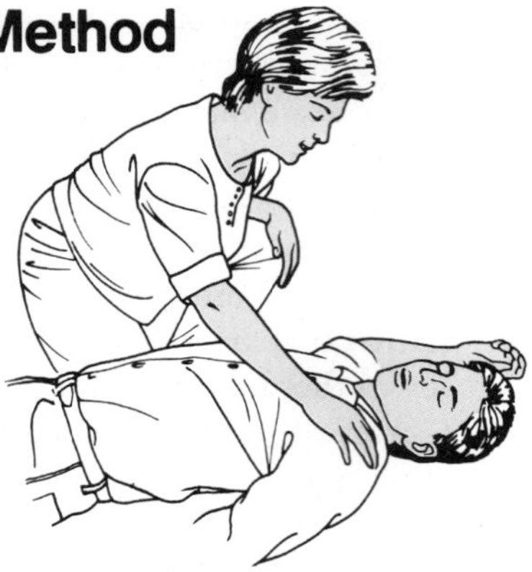

1. **Check the victim for unresponsiveness.** Gently shake him and shout, "Are you okay?" If no response, get the attention of someone who can phone for help. Make sure that the victim is on his back.

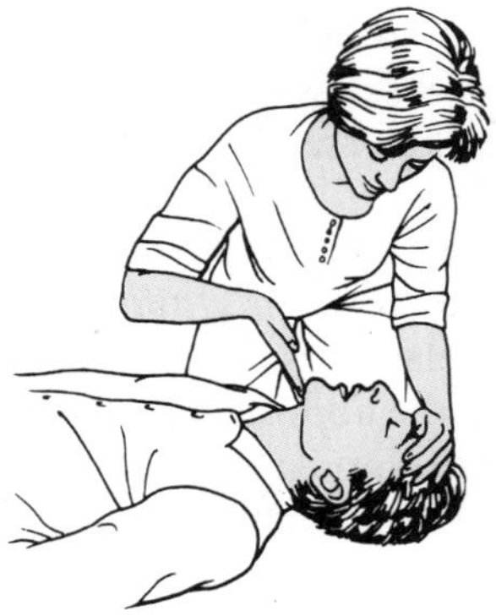

2. **Open the airway.** Tilt the victim's head back by pushing on his forehead with your hand and lifting his chin with your fingers under his jaw. This will open his airway by moving his tongue away from the back of his throat.

3. **Check for breathing.** Put your ear close to the victim's face to listen and feel for any return of air. At the same time, look to see if there is chest movement. Check for breathing for about five seconds.

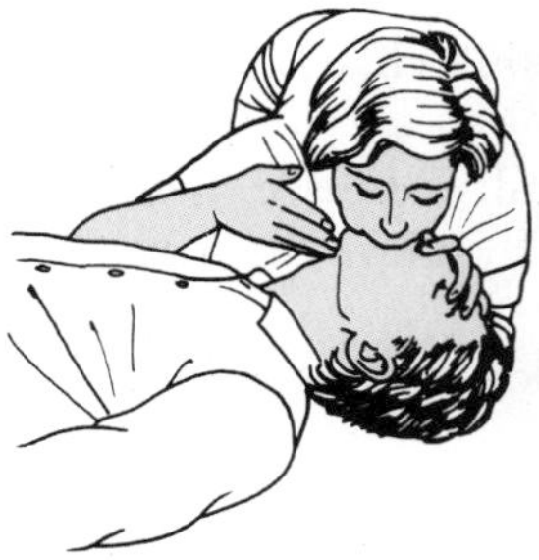

4. **If no breathing, give two full breaths.** While maintaining the victim in the head-tilt position, pinch his nose to close off the nasal passageway. Take a deep breath, then seal your mouth around the victim's mouth and give two full breaths. (After the first breath, raise your head slightly to inhale quickly and then give the second breath.)

5. **Check for pulse.** While maintaining head tilt, feel for a carotid pulse for five to ten seconds on the side of the victim's neck.

6. **Continue rescue breathing.** With the victim in the head-tilt position and his nostrils pinched, give one breath every five seconds. Observe for signs of breathing between breaths. For an infant, give one gentle puff every three seconds.

7. **Recheck for pulse.** Feel for a carotid pulse at one-minute intervals. If the victim has a pulse but is not breathing, continue rescue breathing.

Mouth-to-Nose Method

1. **Open the airway.** Place the victim in the head-tilt position as described above.

2. **Blow into the victim's nose.** Using the same sequence described above, blow into the victim's nose while holding his mouth closed.

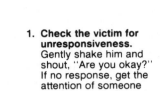

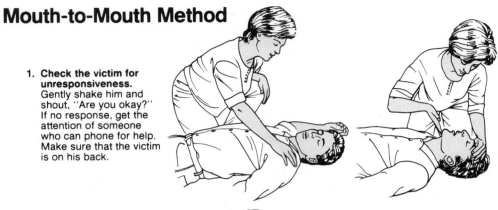

3. **Feel and observe for breathing.** With the victim's mouth held open, detect for breathing between giving forced breaths.

Figure 17.36 Artificial respiration.

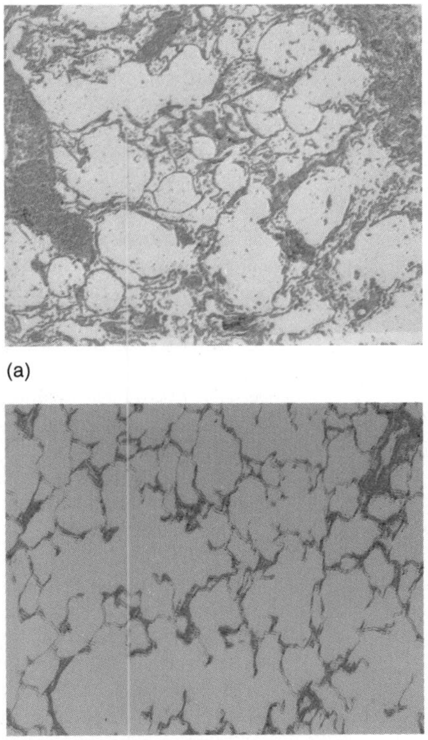

(a)

(b)

Figure 17.37 Photomicrographs of tissue from a normal lung (*a*) and from the lung of a person with emphysema (*b*). In emphysema, lung tissue is destroyed, resulting in the presence of fewer and larger alveoli.

Abnormal breathing patterns often appear prior to death by brain damage or heart disease. The most common of these abnormal patterns is **Cheyne-Stokes breathing,** in which the depth of breathing progressively increases and then progressively decreases. These cycles of increasing and decreasing tidal volumes may be followed by periods of apnea of varying durations. Cheyne-Stokes breathing may be caused by neurological damage or by insufficient oxygen delivery to the brain. The latter may result from heart disease or from a brain tumor that diverts a large part of the vascular supply from the respiratory centers.

CLINICAL CASE STUDY ANSWER

The ice pick traversed the parietal pleural, the visceral pleura, and then entered the airway, at least at the alveolar-terminal bronchiole level, but possibly at larger airways. This allowed inspired air to escape from the airway into the pleural space. The route of the air is: pharynx → larynx → trachea → left primary bronchus → secondary bronchus (of left upper lobe) → apical segmental (teriary) bronchus → bronchioles → air escapes through laceration into pleural space. This condition is treated with insertion of a tube (tube thoracostomy) into the pleural space to allow suction evacuation of air and blood resulting in re-expansion of the lung. The laceration usually seals over within one to two days. Persistent bleeding may necessitate thoracotomy for repair. ■

Chapter Summary

I. Functions and Development of the Respiratory System
   A. Respiration refers not only to simply breathing, but also to the exchange of gases between the atmosphere, the blood, and individual cells.
   B. In order for the respiratory system to be functional, the respiratory membranes must be moist, thin-walled, highly vascular, and differentially permeable.
   C. The functions of the respiratory system are gaseous exchange, sound production, assistance in abdominal compression, and coughing and sneezing.
   D. The upper respiratory pathway develops from ectoderm lining the oronasal cavity, whereas the lower respiratory system develops as an endodermal outpouching from the foregut.

II. Conducting Passages
   A. The nose is supported by nasal bones and cartilages.
      1. The nasal epithelium warms, moistens, and cleanses the inspired air.
      2. Olfactory epithelium provides a sense of smell, and the nasal cavity acts as a resonating chamber for the voice.
   B. The paranasal sinuses are found in the maxillary, frontal, sphenoid, and ethmoid bones.
   C. The pharynx is a funnel-shaped passageway that connects the oral and nasal cavities with the larynx.
      1. The nasopharynx is connected by the auditory tubes to the middle ear cavities and contains the pharyngeal tonsils, or adenoids.
      2. The oropharynx is the middle portion, extending from the soft palate to the level of the hyoid bone, which contains the palatine and lingual tonsils.
      3. The laryngopharynx extends from the hyoid bone to the larynx.
   D. The larynx is composed of a number of cartilages that keep the passageway to the trachea open during breathing and close the respiratory passageway during swallowing.
      1. The epiglottis is a spoon-shaped structure that aids in closing the laryngeal opening, or glottis, during swallowing.
      2. The larynx contains vocal cords that are controlled by intrinsic muscles and used in sound production.
   E. The trachea is a rigid tube, supported by rings of cartilage, that leads from the larynx to the bronchial tree.
   F. The bronchial tree includes a right and left primary bronchus, which divides to produce secondary bronchi, tertiary bronchi, and bronchioles; the conducting division ends with the respiratory bronchioles, which connect to the alveoli.

III. Alveoli, Lungs, and Pleura
   A. Alveoli are the functional units of the lungs where gas exchange occurs; they are small, numerous, thin-walled air sacs.
   B. The right and left lungs are separated by the mediastinum; each lung is divided into lobes and lobules.

C. The lungs are covered by visceral pleura, and the thoracic cavity is lined by a parietal pleura.
   1. There is a potential space between these two pleural membranes called the pleural cavity.
   2. The pleural membranes package each lung separately and exclude the structures located in the mediastinum.

IV. Mechanics of Breathing
   A. Quiet (unforced) inspiration is due to contraction of the diaphragm and certain intercostal muscles. Forced inspiration is aided by the scalenus, pectoralis minor, and sternocleidomastoid muscles.
   B. Quiet expiration is produced by relaxation of the respiratory muscles and elastic recoil of the lungs and thorax; forced expiration is aided by certain intercostal muscles and the abdominal muscles.
   C. Among the air volumes exchanged in ventilation are tidal, inspiratory reserve, and expiratory reserve volumes.
   D. Any air movement for purposes other than respiration is nonrespiratory and includes coughing, sneezing, sighing, yawning, laughing, crying, and hiccuping.

V. Regulation of Breathing
   A. Ventilation is directly controlled by the rhythmicity center in the medulla oblongata, which in turn is influenced by the pneumotaxic and apneustic centers in the pons.
   B. These brain stem areas are affected by higher brain function and by sensory input from chemoreceptors.
   C. There are central chemoreceptors in the medulla oblongata and peripheral chemoreceptors in the aortic and carotid bodies.

Review Activities

Objective Questions

1. The laryngotracheal bud forms
   (a) the alveoli.
   (b) from endoderm.
   (c) the trachea.
   (d) during the fourth week of development.
   (e) All of the above.
2. Which is *not* a component of the nasal septum?
   (a) palatine (c) ethmoid
   (b) vomer (d) septal cartilage
3. An adenoidectomy is the removal of the
   (a) uvula.
   (b) pharyngeal tonsils.
   (c) palatine tonsils.
   (d) lingual tonsils.
4. Which is *not* a paranasal sinus?
   (a) palatine sinus
   (b) ethmoidal sinus
   (c) sphenoidal sinus
   (d) frontal sinus
   (e) maxillary sinus
5. Which of the following is *not* a structure of the left lung?
   (a) cardiac notch
   (b) superior lobe
   (c) a single fissure
   (d) inferior lobe
   (e) middle lobe
6. The epithelial lining of the wall of the thorax is called the
   (a) parietal pleura.
   (b) pleural peritoneum.
   (c) mediastinal pleura.
   (d) visceral pleura.
   (e) costal pleura.
7. The muscle group combination that permits inspiration is the
   (a) diaphragm, abdominal complex.
   (b) internal intercostals, diaphragm.
   (c) external intercostals, internal intercostals.
   (d) external intercostals, diaphragm, internal intercostals (interchondral part).
8. The vocal cords are attached upon the
   (a) cricoid and thyroid cartilages.
   (b) cuneiform and cricoid cartilages.
   (c) corniculate and thyroid cartilages.
   (d) arytenoid and thyroid cartilages.
9. The maximum amount of air that can be expired after a maximum inspiration is the
   (a) tidal volume.
   (b) forced expiratory volume.
   (c) vital capacity.
   (d) maximum expiratory flow rate.
   (e) residual volume.
10. The rhythmic control of breathing is produced by the activity of inspiratory and expiratory neurons in the
    (a) medulla oblongata.
    (b) apneustic center of the pons.
    (c) pneumotaxic center of the pons.
    (d) cerebral cortex.
    (e) hypothalamus.

Essay Questions

1. Why is warming, moistening, and filtering the air so important for a healthy, functioning respiratory system?
2. Explain the structural and functional differences between the three regions of the pharynx.
3. What is a bronchial tree? Why are alveoli rather than bronchial trees considered the functional units of the respiratory system?
4. Define each of the following structures of a lung: *base, hilum, apex, costal surface, fissure, lobe, lobule, pulmonary ligament,* and *bronchial segment.*
5. Diagram the location of the lungs in the thoracic cavity with respect to the heart. Identify the various thoracic serous membranes.
6. List the kinds of epithelial tissues found within the respiratory system and describe the location of each.
7. Identify two places in the respiratory system where there are large surface areas of capillary networks. What is the function of each of these areas?
8. What protective devices of the respiratory system guard against pollutants, the spread of infections, or dual collapse of the lungs?
9. What are the functions of the larynx? List some possible clinical conditions of this organ.
10. What are the advantages of compartmentalization of the thoracic organs?
11. Explain the sequence of pulmonary ventilation. Discuss the mechanisms of inspiration and expiration. How are air pressures related to ventilation?
12. Distinguish between tidal volume, vital capacity, and total lung capacity.
13. What is meant by a rhythmicity respiratory area? How are the apneustic and pneumotaxic areas related to the rhythmicity area?
14. Describe how the abdominal thrust maneuver is performed. When should it be used? What is the principle behind the success of the abdominal thrust maneuver?
15. When would a person use mouth-to-mouth respiration rather than the abdominal-thrust maneuver to revive a person?

CHAPTER 18

Digestive System

Outline and Concepts

CLINICAL CASE STUDY

A twenty-five-year-old male is admitted to the emergency room after suffering a blow to the upper abdomen by a swinging beam while working heavy construction. Initial assessment is significant for marked tenderness in the right upper quadrant of the abdomen as well as vital signs and examination findings consistent with mild hemorrhagic shock. Intravenous fluids are administered, causing stabilization of vital signs. Chest X ray reveals no abnormalities. Peritoneal lavage (see chap. 2 "Case History") is likewise negative for blood. There are no externally detectable signs of hemorrhage. A CT scan demonstrates a significant hematoma (collection of blood) deep within the substance of the liver as well as a notable amount of blood in the small intestine. The decision is made to operate. Initial exploration reveals no trauma to the stomach or small intestine.

What is the likely source of bleeding? Explain anatomically how the blood found its way into the small intestine. Use anatomical terms to describe each step of the blood's travel, from the source of bleeding to the small intestine. Given that the hepatic arterial system and hepatic venous system are possible sources of the bleeding, is there another system of blood vessels relative to the liver that could also be a source of hemorrhage? ■

Introduction to the Digestive System

The organs of the digestive system are specialized for the digestion and absorption of food. The digestive system consists of a tubular gastrointestinal tract and accessory digestive organs.

Objective 1. List the activities of the digestive system, and distinguish between digestion and absorption.

Objective 2. Identify the major structures and regions of the digestive system.

Objective 3. Define the terms *viscera* and *gut.*

It is apparent that food is necessary to sustain life and that many of our daily endeavors involve procuring and eating food. The food is utilized at the cellular level, where nutrients are required for chemical reactions involving synthesis of enzymes, cellular division and growth, repair, and the production of heat energy. Most of the food eaten, however, is not suitable for cellular utilization until it is mechanically and chemically reduced to forms that can be absorbed through the intestinal wall and transported to the cells by the blood. Ingested food is not technically in the body until it is absorbed; and, in fact, a large portion of consumed food is not digested and passes through as waste material.

The principal function of the digestive system is to prepare food for cellular utilization. This involves several functional activities listed as follows:

1. **Ingestion,** the taking of food into the digestive system by way of the mouth
2. **Mastication,** chewing movements to pulverize food and mix it with saliva
3. **Deglutition,** swallowing of food to move it from the mouth to the stomach
4. **Digestion,** mechanical and chemical breakdown of food material
5. **Absorption,** passage of molecules of food through the mucous membrane of the small intestine and into the circulatory or lymphatic systems for distribution to cells
6. **Peristalsis,** rhythmic, wavelike intestinal contractions that move food through the digestive tract
7. **Defecation,** the discharge of indigestible wastes, called feces, from the body

Anatomically and functionally, the digestive system can be divided into a tubular **alimentary canal,** or **gastrointestinal tract** (GI tract), and **accessory organs.** The GI tract is approximately 9 m (30 ft) long and extends from the mouth to the anus. It traverses the thoracic cavity and enters the abdominal cavity at the level of the diaphragm. The anus is located at the inferior portion of the pelvic cavity. The organs of the GI tract include the **oral (buccal) cavity, pharynx, esophagus, stomach, small intestine,** and **large intestine** (fig. 18.1). The accessory digestive organs include the **teeth, tongue, salivary glands, liver, gallbladder,** and **pancreas.** The term **viscera** is frequently used to refer to the abdominal organs of digestion, but actually viscera can be any of the organs (lungs, stomach, spleen, etc.) of the thoracic and abdominal cavities. **Gut** is an anatomical term that generally refers to the developing stomach and intestines in the embryo.

Food usually requires about twenty-four hours to travel the length of the GI tract. Food ingested through the mouth passes in assembly-line fashion through the GI tract where complex molecules are progressively broken down. Each region of the GI tract has specific functions in preparing food for utilization (table 18.1).

Although there is an abundance of food in the United States, too many people eat the wrong kinds of foods, eat irregularly, or overeat, to the point that eating patterns have become a critical public health concern. Obesity is a major health problem; obese persons are at greater risk for cardiovascular disease, hypertension, osteoarthritis, and diabetes mellitus. People with good nutritional habits are better able to withstand trauma, are less likely to get sick, and are usually less seriously ill when they do become sick.

---

ingestion: L. *ingerere,* carry in
mastication: Gk. *mastichan,* gnash the teeth
deglutition: L. *deglutire,* swallow down
peristalsis: Gk. *peri,* around; *stellein,* compress
defecation: L. *de,* from, away; *faecare,* cleanse

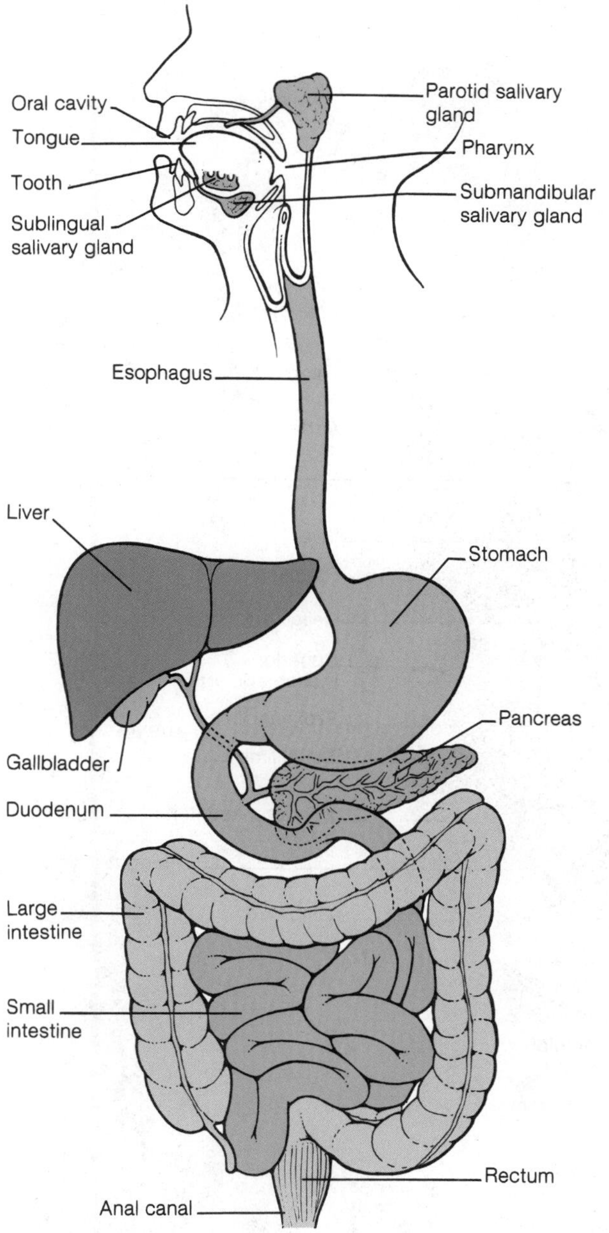

Figure 18.1 The digestive system, including the gastrointestinal tract and the accessory digestive organs.

---

1. Which functional activities of the digestive system break down food? Define *absorption.* Which functional activities move the food through the GI tract?
2. List in order the regions of the GI tract through which ingested food would pass from the mouth to the anus.
3. List the abdominal visceral organs of the digestive system.
4. Write a sentence correctly using the term *gut.*

Table 18.1 Regions of the GI tract and basic functions

Region	Function
Oral cavity	Ingests food; receives saliva; mastication; initiates digestion of carbohydrates; forms bolus; deglutition
Pharynx	Receives bolus from oral cavity; autonomically continues deglutition of bolus to esophagus
Esophagus	Transports bolus to stomach by peristalsis; esophageal sphincter restricts backflow of food
Stomach	Receives bolus from esophagus; churns bolus with gastric juice; initiates digestion of proteins; limited absorption; moves chyme into duodenum and prohibits backflow of chyme; regurgitates when necessary
Small intestine	Receives chyme from stomach and secretions from liver and pancreas; chemically and mechanically breaks down chyme; absorbs nutrients; transports wastes through peristalsis to large intestine; prohibits backflow of intestinal wastes from large intestine
Large intestine	Receives undigested wastes from small intestine; absorbs water and electrolytes; forms, stores, and expels feces through defecation reflex

Embryological Development of the Digestive System

The primitive gut is derived primarily from endoderm and differentiates during the fourth week to give rise to the specific regions of the GI tract and the accessory digestive organs.

Objective 4. Describe the embryological development of the GI tract and the accessory digestive organs.

The entire digestive system develops from modifications of an elongated tubular structure called the **primitive gut.** These modifications are initiated during the fourth week of embryonic development. The primitive gut is composed solely of endoderm and for descriptive purposes can be divided into three regions: foregut, midgut, and hindgut (fig. 18.2).

Foregut The **stomodeum** *(sto''moh-de'um),* or **oral pit,** is not part of the foregut but an invagination of ectoderm that breaks through a thin **oral membrane** to become continuous with the foregut and form part of the oral cavity or mouth. Structures in the mouth, therefore, are ectodermal in origin. The esophagus, pharynx, stomach, a portion of the duodenum, the pancreas, liver, and gallbladder are the organs that develop from the foregut (fig. 18.3). Along the alimentary canal, only the inside epithelial lining of the lumen is derived from the endoderm of the primitive gut. The vascular portion and smooth muscle layers are formed from mesoderm that develops from the surrounding splanchnic mesenchyme. The stomach first appears as an elongated dilation of the foregut. The dorsal border of the stomach undergoes more rapid growth than the ventral border, forming a distinct curvature. The caudal portion of the

---

stomodeum: Gk. *stoma,* mouth; *hodaios,* on the way to

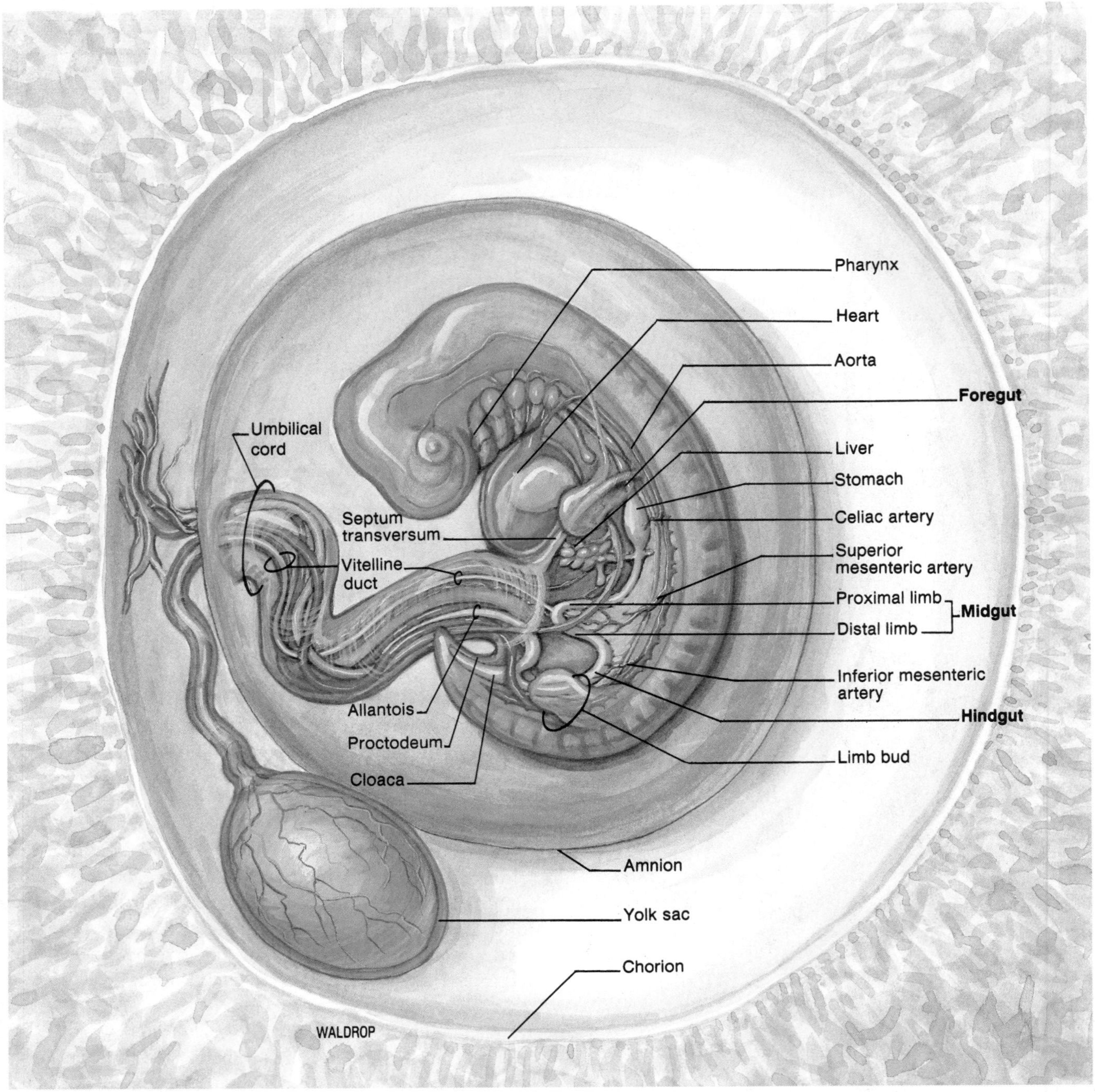

Figure 18.2 A sagittal section of a five-week-old embryo showing the development of the digestive system and its association with the extraembryonic membranes and organs.

foregut and the cranial portion of the midgut form the duodenum. The liver and pancreas arise as small **hepatic** and **pancreatic buds,** respectively, from the wall of the duodenum. The hepatic bud experiences incredible growth to form the gallbladder, associated ducts, and the various lobes of the liver (fig. 18.3). By the sixth week, the liver is carrying out hemopoiesis (the formation of blood cells); and by the ninth week, the liver represents 10% of the total weight of the fetus.

The pancreas develops from dorsal and ventral pancreatic buds of endodermal cells. As the duodenum grows, it rotates clockwise and the two pancreatic buds fuse (fig. 18.3).

Midgut During the fourth week of the embryonic stage (fig. 18.2), the midgut is continuous with the yolk sac. By the fifth week, the midgut forms a ventral U-shaped **midgut loop,** which projects into the umbilical cord (fig. 18.4). As development continues, the anterior limb of the midgut loop coils to form most of the small intestine. The posterior limb of the midgut loop expands to form a portion of the small and large intestines. A **cecal diverticulum** *(di''ver-tik'u-lum)* appears during the fifth week.

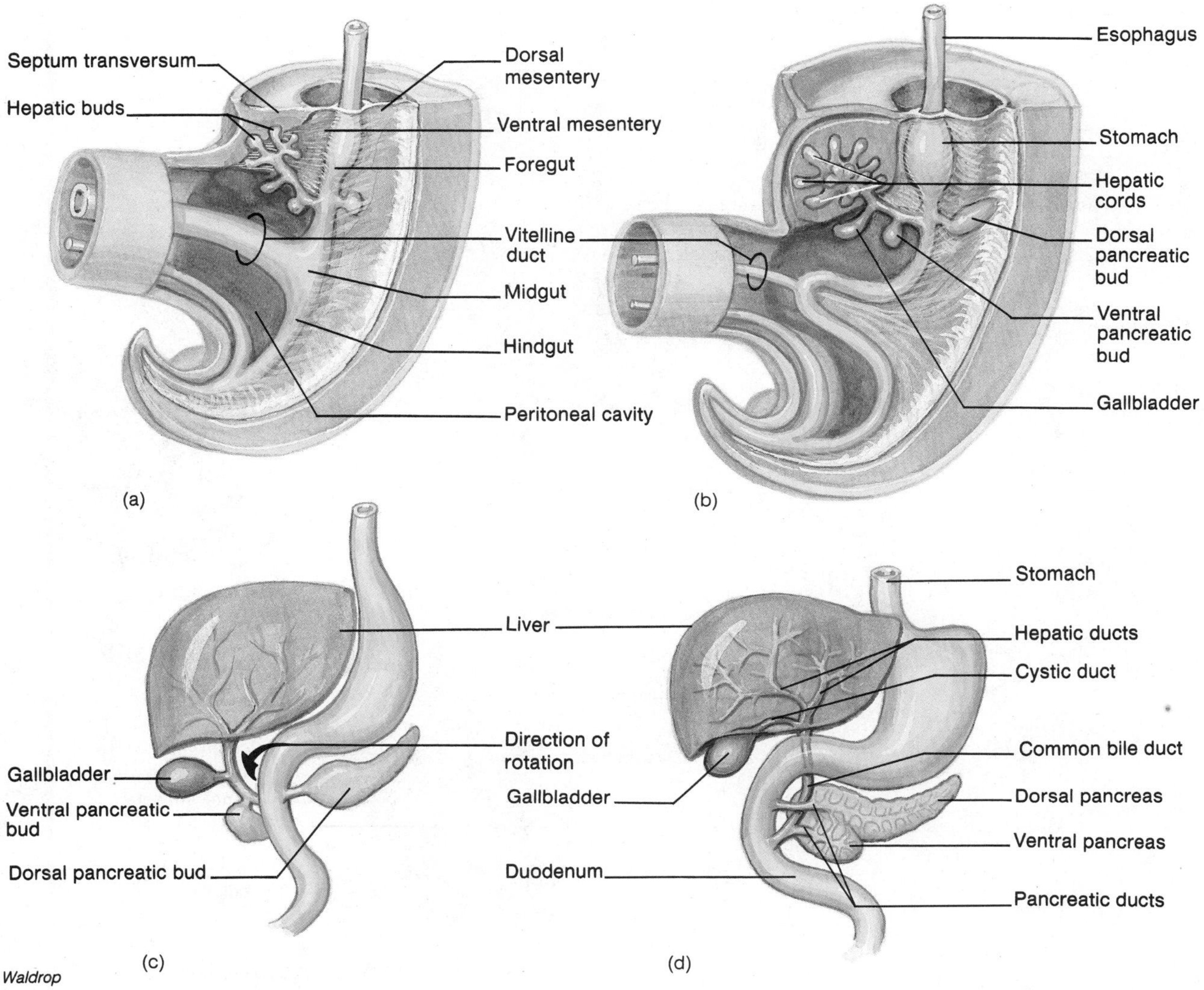

Figure 18.3 Progressive stages of development of the foregut to form the stomach, duodenum, liver, gallbladder, and pancreas. (*a*) Four weeks, (*b*) five weeks, (*c*) six weeks, and (*d*) seven weeks.

During the tenth week, the intestines are withdrawn to the abdominal cavity and further differentiation and rotation occur. The cecal diverticulum continues to develop, forming the cecum and appendix. The remainder of the midgut gives rise to the ascending colon and hepatic flexure (fig. 18.4).

Hindgut The hindgut extends from the midgut to the **cloacal membrane** (fig. 18.5). The **proctodeum** *(prok''to-de'um),* or **anal pit,** is a depression in the anal region formed from an invagination of ectoderm that contributes to the cloacal membrane. The *allantois (ah-lan'to-is),* which receives urinary wastes from the fetus, connects to the hindgut at a region called the **cloaca** *(klo-a'kah),* as seen in figure 18.5. A band of mesenchymal cells called the **urorectal septum** grows caudally during the fourth through seventh week until a complete partition separates the cloaca into a dorsal **anal canal** and a ventral **urogenital sinus.** With the completion of the urorectal septum, the cloacal membrane is divided into an anterior **urogenital membrane** and a posterior **anal membrane.** Toward the end of the seventh week, the anal membrane perforates and forms the anal opening, which is lined with ectodermal cells. About this time, the urogenital membrane ruptures to provide further development of the urinary and reproductive systems.

proctodeum: Gk. *proktos,* anus; *hodaios,* on the way to
cloaca: L. *cloaca,* sewer

1. Describe the embryonic derivation of the mouth and anus.
2. Identify the portion of the gut that gives rise to the pancreas and liver.
3. Define the terms *stomodeum, proctodeum, allantois,* and *cloaca.*

Serous Membranes and Tunics of the Gastrointestinal Tract

Serous membranes line the abdominal cavity and cover the visceral organs. The wall of the GI tract is composed of four tunics.

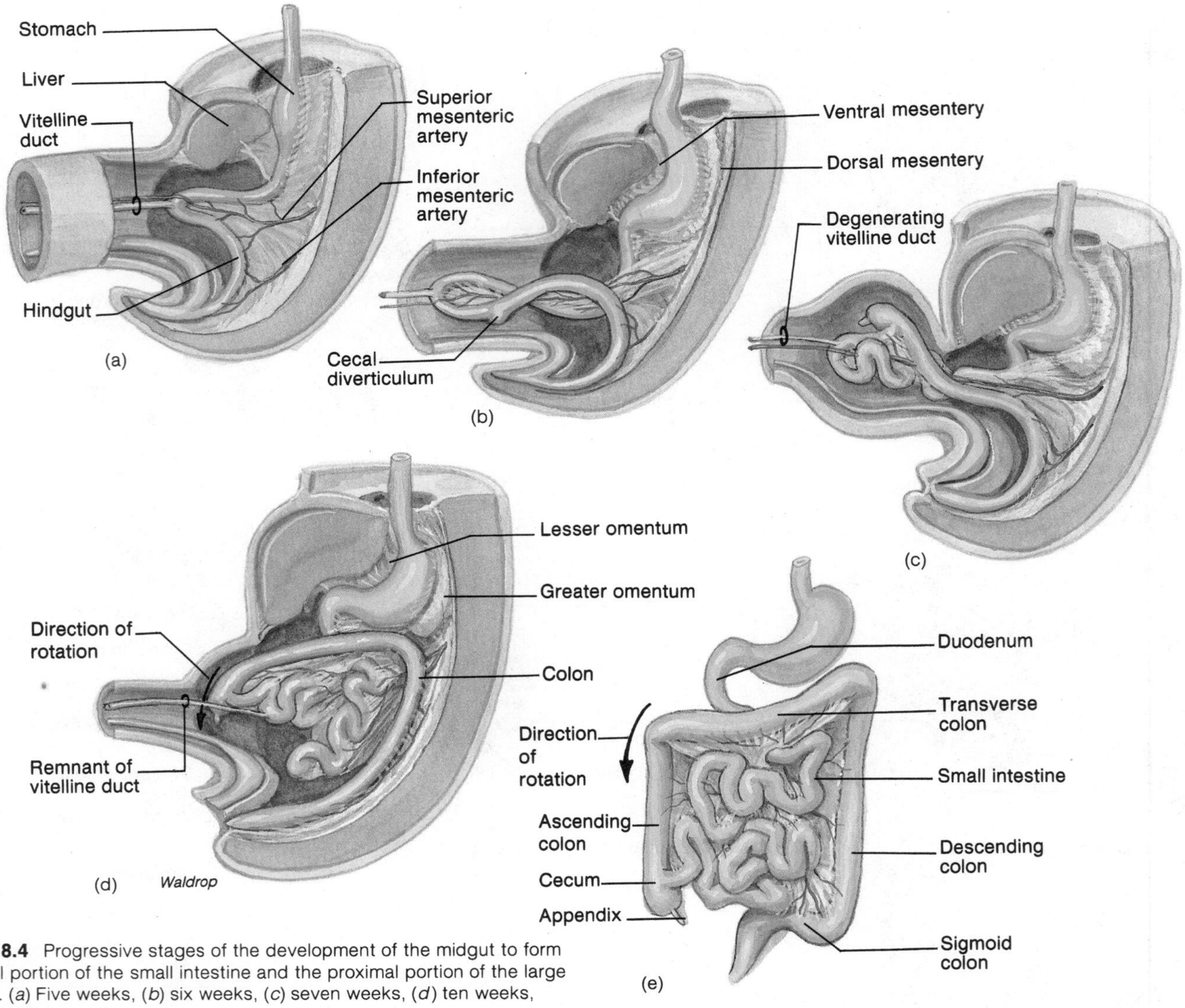

Figure 18.4 Progressive stages of the development of the midgut to form the distal portion of the small intestine and the proximal portion of the large intestine. (*a*) Five weeks, (*b*) six weeks, (*c*) seven weeks, (*d*) ten weeks, and (*e*) eighteen weeks.

Objective 5. Describe the arrangement of the serous membranes within the abdominal cavity.

Objective 6. Describe the generalized structure of the four tunics composing the wall of the GI tract.

Serous Membranes

Most of the GI tract and abdominal accessory digestive organs are positioned within the **abdominal cavity.** These organs are not firmly embedded in solid tissue but are supported and covered by **serous membranes.** A serous membrane is an epithelial membrane that lines the thoracic and abdominal cavities and covers the organs that lie within these cavities. A serous membrane has a parietal portion lining the body wall and a visceral portion covering the internal organs. The serous membranes associated with the lungs are called pleurae. The serous membranes of the abdominal cavity are called **peritoneal membranes,** or **peritoneum** *(per''i-to-ne'um)*. The peritoneum is the largest serous membrane of the body. It is composed of simple squamous epithelium with portions reinforced with connective tissue.

The **parietal peritoneum** lines the wall of the abdominal cavity (fig. 18.6). Along the dorsal, or posterior, aspect of the abdominal cavity the parietal peritoneum comes together to form a double-layered peritoneal fold called the **mesentery,** which supports the GI tract. The **dorsal mesentery** gives the pendulous small intestine freedom for peristaltic movement and provides a structure through which intestinal nerves and vessels traverse. The **mesocolon** is a specific portion of the mesentery that supports the large intestine (fig. 18.7c and d). The peritoneal covering continues around the intestinal viscera as the **visceral peritoneum.** The **peritoneal cavity** is the space between the parietal and visceral portions of the peritoneum.

Extensions of the parietal peritoneum, located in the peritoneal cavity, serve specific functions (fig. 18.7). The **falciform ligament,** a serous membrane reinforced with connective

peritoneum: Gk. *peritonaion,* stretched over

mesentery: Gk. *mesos,* middle; *enteron,* intestine

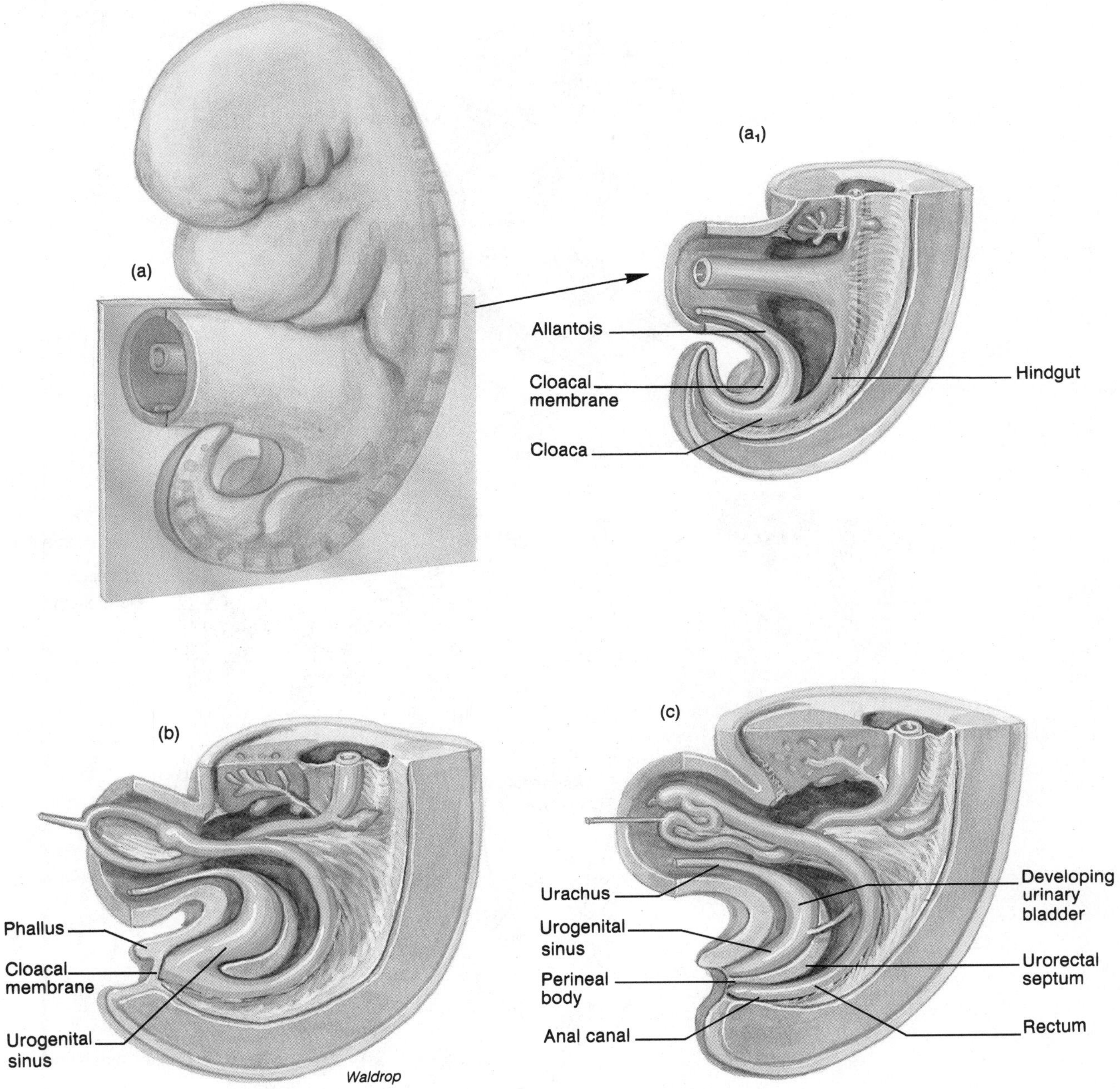

Figure 18.5 The progressive development of the hindgut, illustrating the developmental separation of the digestive system from the urogenital system. (*a*) An anterolateral view of an embryo at four weeks showing the position of a sagittal cut depicted in (a_1, *b*, and *c*). (a_1) At four weeks, the hindgut, cloaca, and allantois are connected. (*b*) At six weeks, the connections between the gut and extraembryonic structures are greatly diminished. (*c*) By seven weeks, structural and functional separation between the digestive system and the urogenital system is nearly completed.

tissue, attaches the liver to the diaphragm and anterior abdominal wall. The **lesser omentum** passes from the lesser curvature of the stomach and the upper duodenum to the inferior surface of the liver. The **greater omentum** extends from the greater curvature of the stomach to the transverse colon, forming an apronlike structure over most of the small intestine. Functions of the greater omentum include storing fat, cushioning visceral organs, supporting lymph nodes, and protecting against the spread of infections. In cases of localized inflammation, such as appendicitis, the greater omentum may compartmentalize the inflamed area, sealing it off from the rest of the peritoneal cavity.

omentum: L. *omentum,* apron

Certain abdominal organs are not within the peritoneal cavity and are therefore covered by the parietal peritoneum. These organs are said to be *retroperitoneal* and include most of the pancreas, kidneys, a portion of the duodenum, and the abdominal aorta.

Peritonitis is an inflammation of the peritoneum. Peritonitis may be caused by trauma, rupture of a visceral organ, ectopic pregnancy, or postoperative complications. Peritonitis is usually extremely painful and serious. Treatment usually involves the injection of massive doses of antibiotics and perhaps peritoneal intubation to permit drainage.

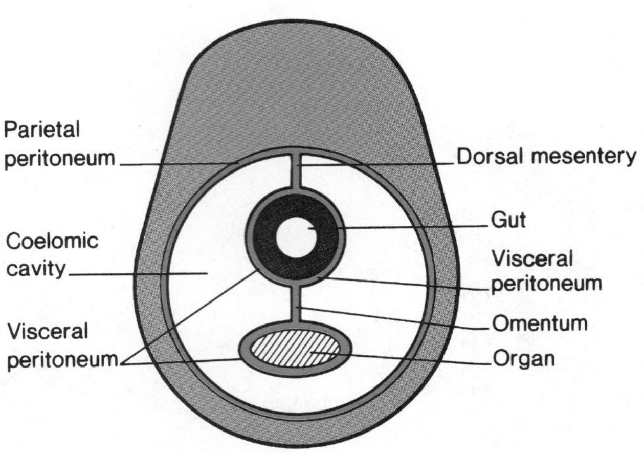

Figure 18.6 A diagrammatic representation of the serous membranes.

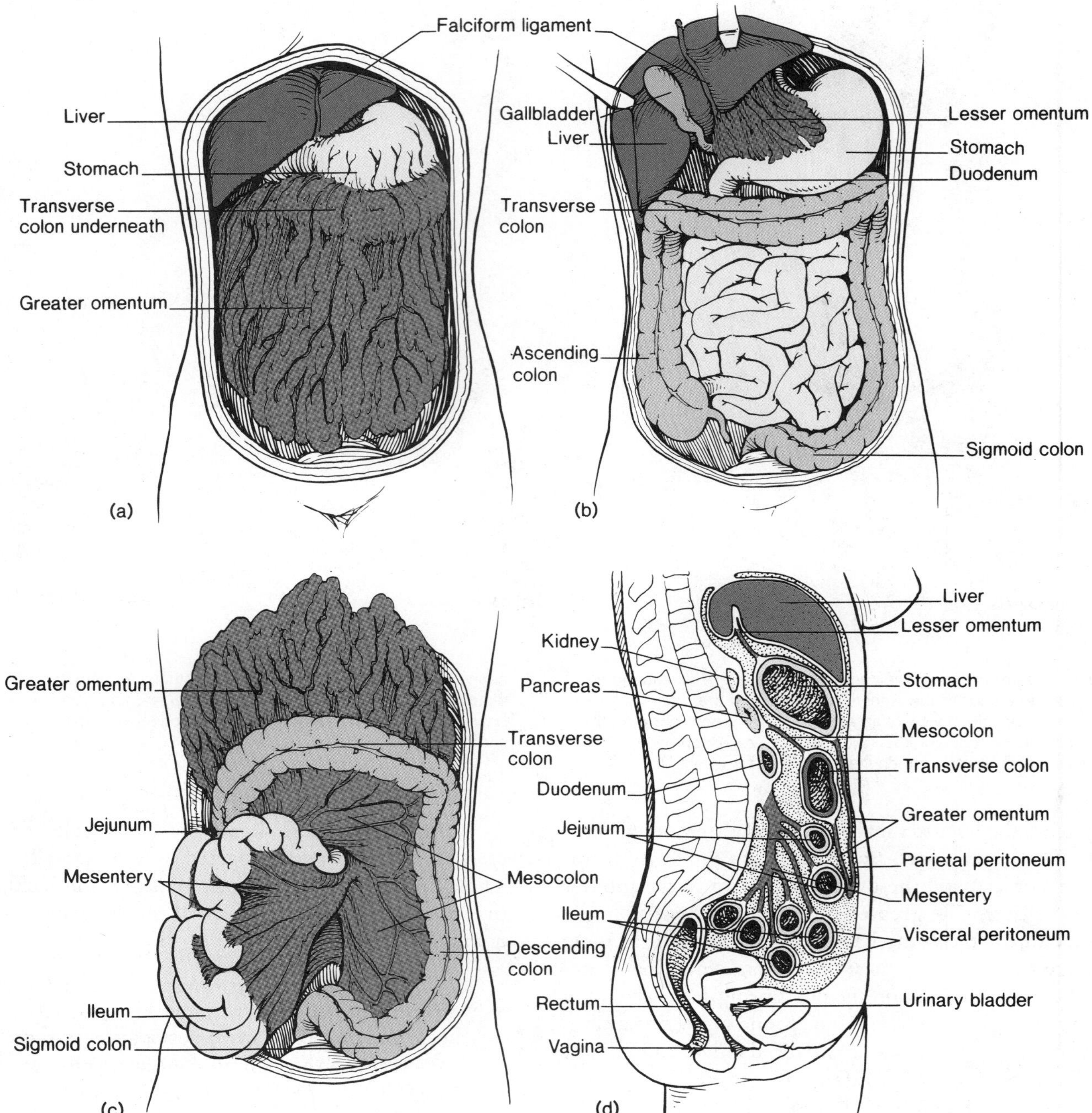

Figure 18.7 The structural arrangement of the abdominal organs and peritoneal membranes within the peritoneal cavity. (*a*) The greater omentum, (*b*) the lesser omentum with the liver lifted, (*c*) the mesentery with the greater omentum lifted, and (*d*) the relationship of the peritoneal membranes to the visceral organs as shown in a sagittal view.

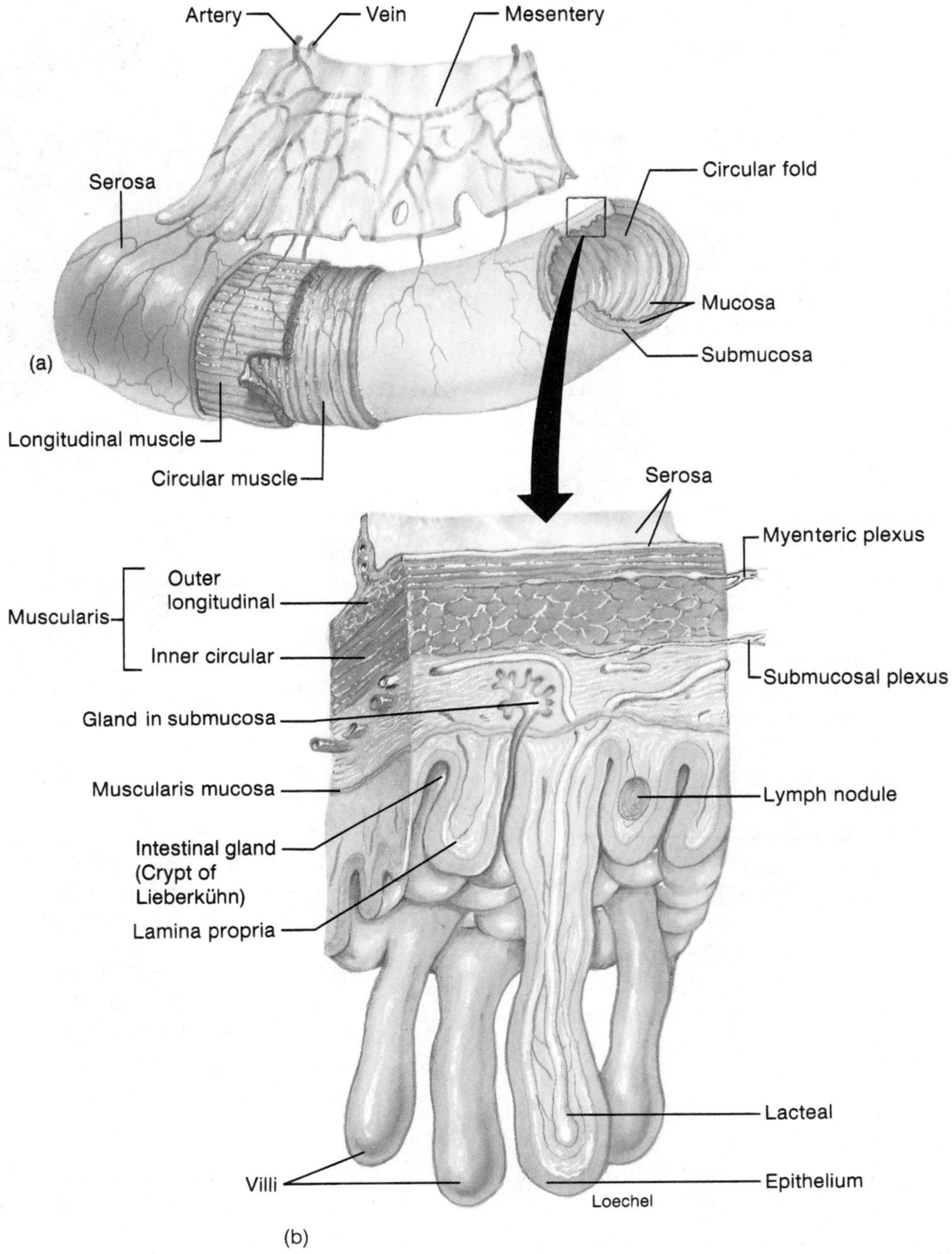

Figure 18.8 The tunics (layers) of the GI tract. (*a*) A section of the small intestine with each of the four tunics exposed. (*b*) A section showing the detailed structure and relative thickness of each tunic. (Note the location of the exocrine gland and the innervation of the small intestine.)

Layers of the Gastrointestinal Tract

The GI tract from the esophagus to the anal canal is composed of four layers, or **tunics.** Each tunic contains a dominant tissue type that performs specific functions in the digestive process. The four tunics of the GI tract, from the inside out, are the **mucosa, submucosa, muscularis,** and **serosa** (fig. 18.8a).

Mucosa The mucosa surrounds the lumen of the GI tract and is the absorptive and major secretory layer. It consists of a simple columnar epithelium supported by the **lamina propria,** which is a thin layer of connective tissue. The lamina propria contains numerous lymph nodules, which are important in protecting against disease (fig. 18.8b). External to the lamina propria are thin layers of smooth muscle called the **muscularis mucosa.** This is the muscle layer that causes the small intestine portion of the GI tract to have numerous small folds, called plicae circulares (see fig. 18.23), which greatly increase the absorptive surface area. Specialized **goblet cells** in the mucosa secrete mucus throughout most of the GI tract.

Submucosa The relatively thick submucosa is a highly vascular layer of connective tissue serving the mucosa. Absorbed molecules that pass through the columnar epithelial cells of the mucosa enter into blood vessels or lymph ductules of the submucosa. In addition to blood vessels, the submucosa contains

tunica: L. *tunica,* covering or coat

glands and nerve plexuses. The *submucosal plexus (Meissner's plexus)* (fig. 18.8b) provides autonomic innervation to the muscularis mucosa.

Muscularis The muscularis (also called the muscularis externa) is responsible for segmental contractions and peristaltic movement through the GI tract. The muscularis has an inner circular and an outer longitudinal layer of smooth muscle. Contractions of these layers move the food peristaltically through the tract and physically pulverize and churn the food with digestive enzymes. The *myenteric plexus (Auerbach's plexus)* located between the two muscle layers provides the major nerve supply to the GI tract and includes neurons and ganglia from both the sympathetic and parasympathetic divisions of the ANS.

Serosa The outer serosa layer completes the wall of the GI tract. It is a binding and protective layer consisting of loose fibrous connective tissue covered with a layer of simple squamous epithelium and subjacent connective tissue. The simple squamous epithelium is actually the visceral peritoneum.

The body has several defense mechanisms to protect against ingested material that may be harmful if absorbed. The acidic environment of the stomach and the lymphatic system kill many harmful bacteria. A mucous lining throughout the GI tract serves as a protective layer. Vomiting and in certain cases diarrhea are reactions to substances that irritate the GI tract. Vomiting is a reflexive response to many toxic chemicals and as such can be beneficial even though unpleasant.

Innervation The GI tract is innervated by the sympathetic and parasympathetic divisions of the ANS. The vagus nerve is the source of parasympathetic activity in the esophagus, stomach, pancreas, gallbladder, small intestine, and upper portion of the large intestine. The lower portion of the large intestine receives parasympathetic innervation from spinal nerves in the sacral region. The submucosal plexus and myenteric plexus are the sites where preganglionic neurons synapse with postganglionic neurons that innervate the smooth muscle of the GI tract. Stimulation of the parasympathetic neurons increases peristalsis and the secretions of the GI tract.

Postganglionic sympathetic fibers pass through the submucosal and myenteric plexuses and innervate the GI tract. Sympathetic nerve stimulation acts antagonistically to the effects of parasympathetic nerves by reducing peristalsis and secretions and stimulating the contraction of sphincter muscles along the GI tract.

1. Describe the functions of the dorsal mesentery and greater omentum. Identify the location of the peritoneum, and list the organs that are retroperitoneal.
2. List the four tunics of the GI tract, and identify their major tissue types. Explain the functions of these four tunics.

Meissner's plexus: from Georg Meissner, German histologist, 1829–1905
Auerbach's plexus: from Leopold Auerbach, German anatomist, 1828–97

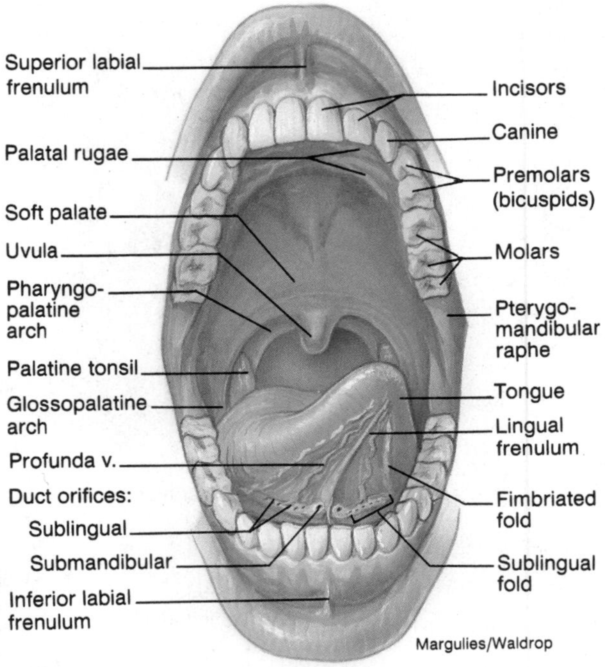

Figure 18.9 The superficial structures of the oral cavity.

Mouth, Pharynx, and Associated Structures

Ingested food is changed by the mechanical action of teeth and by the chemical activity of saliva into a bolus, which is swallowed in the process of deglutition.

Objective 7. Describe the anatomy of the oral cavity.
Objective 8. Contrast deciduous and permanent dentitions, and describe the structure of a typical tooth.
Objective 9. Describe the location and histological structures of the salivary glands, and list the functions of saliva.

The functions of the **mouth** and associated structures are to form a receptacle for food, to initiate digestion through mastication, to swallow food, and to form words in speech. The mouth can also assist the respiratory system in the passage of air. The **pharynx,** which is posterior to the mouth, serves as a common passageway for both the respiratory and digestive systems. Both the mouth and pharynx are lined with noncornified, stratified squamous epithelium, which is constantly moistened by the secretion of saliva. The mouth is also known as the **oral,** or **buccal** *(buk'al),* cavity (fig. 18.9). It is formed by the **cheeks, lips, hard** and **soft palates,** and **tongue.** The **vestibule** of the oral cavity is the depression between the cheeks and lips externally and the gums and teeth internally (fig. 18.10). The opening of the oral cavity is referred to as the **oral orifice,** and the opening between the oral cavity and the pharynx is called the **fauces.**

pharynx: L. *pharynx,* throat
buccal: L. *bucca,* cheek
fauces: L. *fauces,* throat

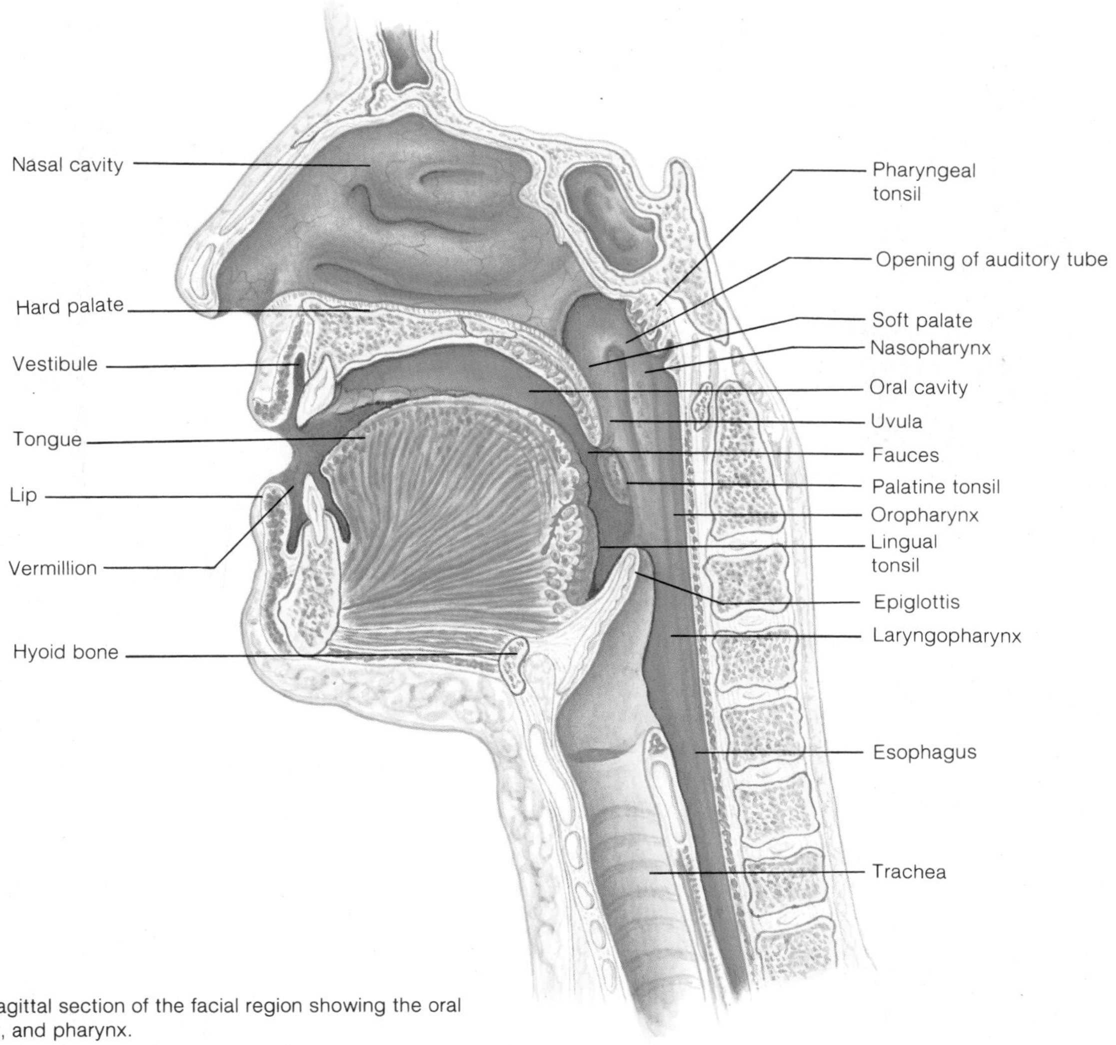

Figure 18.10 A sagittal section of the facial region showing the oral cavity, nasal cavity, and pharynx.

Cheeks and Lips

The **cheeks** form the lateral walls of the oral cavity and consist of outer layers of skin, subcutaneous fat, facial muscles that assist in manipulating food in the oral cavity, and inner linings of moistened, stratified squamous epithelium. The anterior portion of the cheeks terminates in the superior and inferior lips that surround the oral orifice.

The **lips** are fleshy, highly mobile organs whose principal function in humans is associated with speech. Lips also serve for suckling, manipulating food, and keeping food between the upper and lower teeth. Each lip is attached from its inner surface to the gum by a midline fold of mucous membrane called the **labial frenulum** *(fren'u-lum)* (fig. 18.9). The lips are formed from the orbicularis oris muscle and associated connective tissue and covered with soft, pliable skin. Between the outer skin and the mucous membrane of the oral cavity is a transition zone called the **vermilion.** Lips are red to reddish-brown in color because of blood vessels close to the surface. The many sensory receptors in the lips aid in determining the temperature and texture of food.

vermilion: O.E. *vermeylion*, red colored

Suckling is an innate characteristic of newborn infants. The lips are well formed for this activity and even contain blisterlike milk pads that aid in suckling. The wide nostrils and receding lower jaw of infants also facilitate suckling.

Tongue

As a digestive organ, the **tongue** functions to move food around in the mouth during mastication and to assist in swallowing food. It contains **taste buds,** that sense various food tastes. The tongue is also essential in producing articulated speech. The tongue is a mass of skeletal muscle covered with a mucous membrane. Extrinsic tongue muscles (those that insert upon the tongue) move the tongue from side to side and in and out. Only the anterior two-thirds of the tongue lies in the oral cavity; the remaining one-third lies in the pharynx (fig. 18.10) and is attached to the hyoid bone. Rounded masses of **lingual tonsils** are located on the dorsal surface of the base of the tongue (fig. 18.11). The undersurface of the tongue is connected along the midline anteriorly to the floor of the mouth by the vertically positioned **lingual frenulum** (fig. 18.9).

tonsil: L. *toles*, swelling
frenulum: L. diminutive of *frenum*, bridle

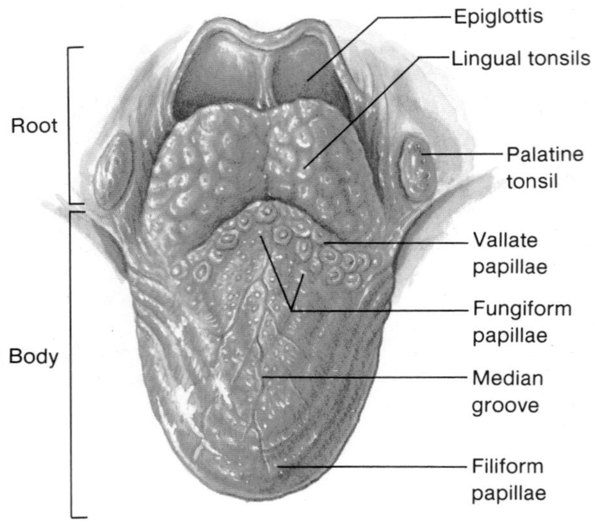

Figure 18.11 The dorsum of the tongue.

When a short lingual frenulum restricts tongue movements, the person is said to be *tongue-tied.* If this developmental problem is severe, the infant may have difficulty suckling. Older children with this problem may have faulty speech. These functional problems can be easily corrected through surgery.

The dorsal surface of the tongue has numerous small elevations, called **papillae,** that give the tongue a distinct roughened surface, which aids the handling of food. These papillae also contain taste buds that can distinguish sweet, salty, sour, and bitter sensations. Three types of papillae are present on the dorsum of the tongue: **filiform, fungiform,** and **vallate (circumvallate)** (fig. 18.11). Filiform papillae are sensitive to touch, have tapered tips, and are by far the most numerous. The larger and rounded fungiform papillae are scattered among the filiform type. The few vallate papillae are arranged in a V-shape on the posterior surface of the tongue.

Palate

The **palate** is the roof of the oral cavity and consists of the bony hard palate anteriorly and the soft palate posteriorly (figs. 18.9, 18.10). The hard palate is formed by the palatine processes of the maxillae and the horizontal plates of the palatine bones and is covered with a mucous membrane. Transverse ridges called **palatal rugae** *(roo'je)* are located along the mucous membrane of the hard palate. These structures serve as friction ridges against which the tongue is placed during swallowing. The soft palate is a muscular arch covered with mucous membrane and is anteriorly continuous with the hard palate. Suspended from the middle lower border of the soft palate is a cone-shaped projection called the **uvula.** During swallowing, the soft palate and uvula are drawn upward, closing the nasopharynx and preventing food and fluid from entering the nasal cavity.

papilla: L. *papula,* little nipple
filiform: L. *filum,* thread; *forma,* form
fungiform: L. *fungus,* fungus; *forma,* form
vallate: L. *vallatus,* surrounded with a rampart
uvula: L. *uvula,* small grapes

Two muscular folds extend downward from both sides of the base of the uvula (fig. 18.9). The anterior fold is called the **glossopalatine arch,** and the posterior fold is the **pharyngopalatine** *(fah-ring''go-pal'ah-tīn)* **arch.** Between these two arches, toward the posterior lateral portion of the oral cavity, is the **palatine tonsil.**

Teeth

Humans, being mammals, have *heterodont dentition,* which means **teeth,** or **dentes** *(den'tēz),* that differ in structure and are adapted to handle food in different ways (fig. 18.12). The four pairs (upper and lower jaws) of anteriormost teeth are the **incisors** *(in-si'zerz).* The chisel-shaped incisors are adapted for cutting and shearing food. The two pairs of cone-shaped **canines,** or **cuspids,** are located at the anterior corners of the mouth and are adapted for holding and tearing. Incisors and canines are further characterized by a single root on each tooth. Located behind the canines are the **premolars,** or **bicuspids,** and **molars.** These teeth have two or three roots and have somewhat rounded, irregular surfaces called **cusps** for crushing and grinding food.

Humans are *diphyodont;* that is, normally two sets of teeth develop in a person's lifetime. There are twenty **deciduous (milk) teeth,** which begin to erupt at about six months of age (fig. 18.13 and tables 18.2, 18.3), beginning with the incisors. All of the deciduous teeth have erupted by the age of two and a half. There are thirty-two **permanent teeth,** which replace the deciduous teeth in a predictable sequence. This process begins at about age six and continues until about age seventeen. The **third molars,** or **wisdom teeth,** are the last to erupt. Wisdom teeth are less predictable, and if they do erupt, it is between the ages of seventeen and twenty-five. Because the jaws are formed by this time and other teeth are in place, the eruption of wisdom teeth may cause serious problems of crowding or impaction.

A **dental formula** is a graphic representation of the types, number, and position of teeth in the oral cavity. Most mammals are *heterodont* and have a constant tooth count, so a dental formula can be written for each species of mammal that has heterodontia. Following are the deciduous and permanent dental formulae for humans:

Formula for deciduous dentition:
I 2/2, **C** 1/1, **DM** 2/2 = 10 × 2 = 20 teeth

Formula for permanent dentition:
I 2/2, **C** 1/1, **P** 2/2, **M** 3/3 = 16 × 2 = 32 teeth

(**I** = incisor; **C** = canine; **P** = premolar;
DM = deciduous molar; **M** = molar)

The cusps of the upper and lower premolar and molar teeth occlude for chewing food in a process called *mastication,* whereas the upper incisors normally form an overbite with the incisors of the lower jaw. An overbite of the upper incisors creates a shearing action as these teeth slide past one another. Masticated

incisor: L. *incidere,* to cut
canine: L. *canis,* dog
molar: L. *mola,* millstone
deciduous: L. *deciduus,* to fall away
heterodont: Gk. *heteros,* other; *odous,* tooth

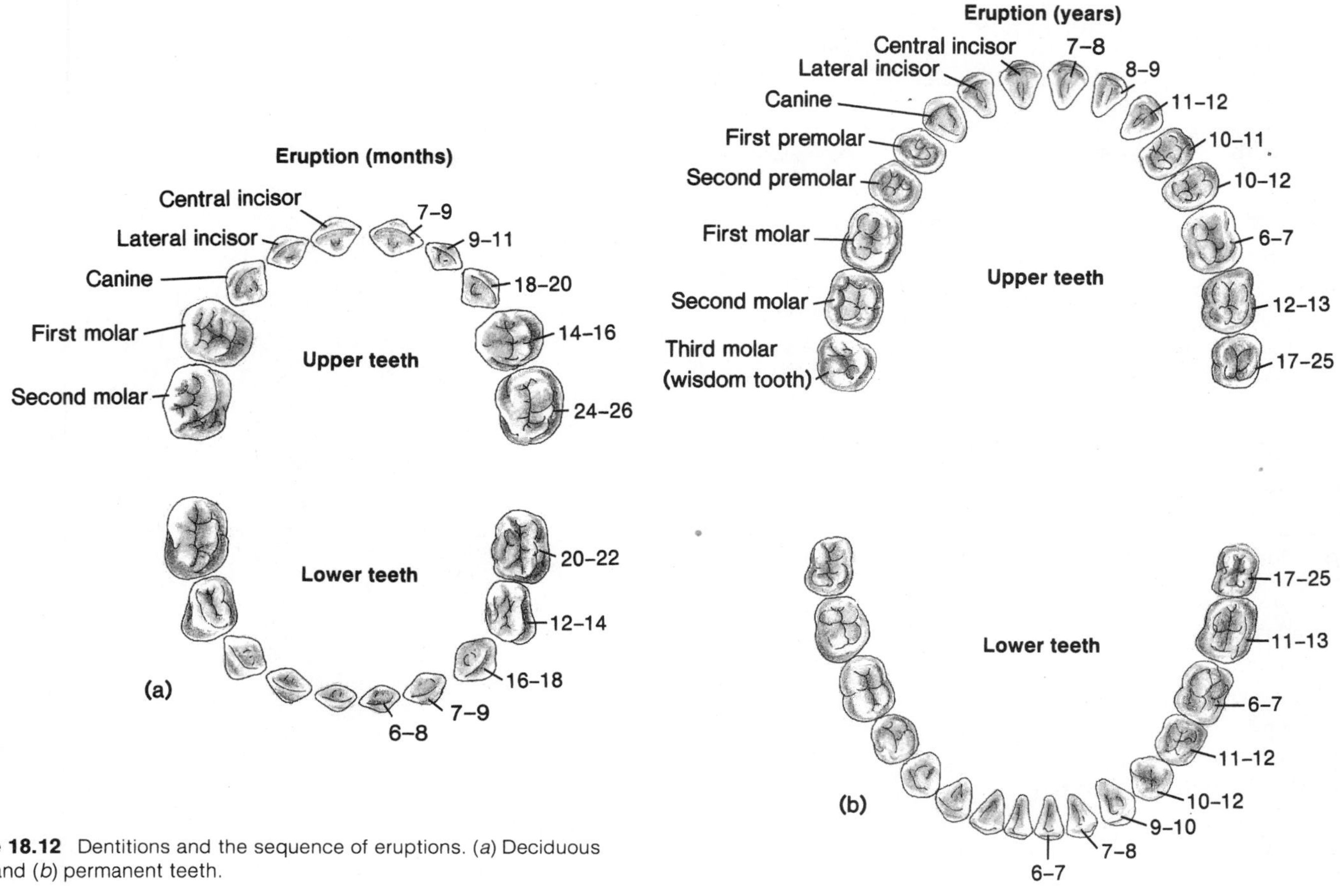

Figure 18.12 Dentitions and the sequence of eruptions. (*a*) Deciduous teeth and (*b*) permanent teeth.

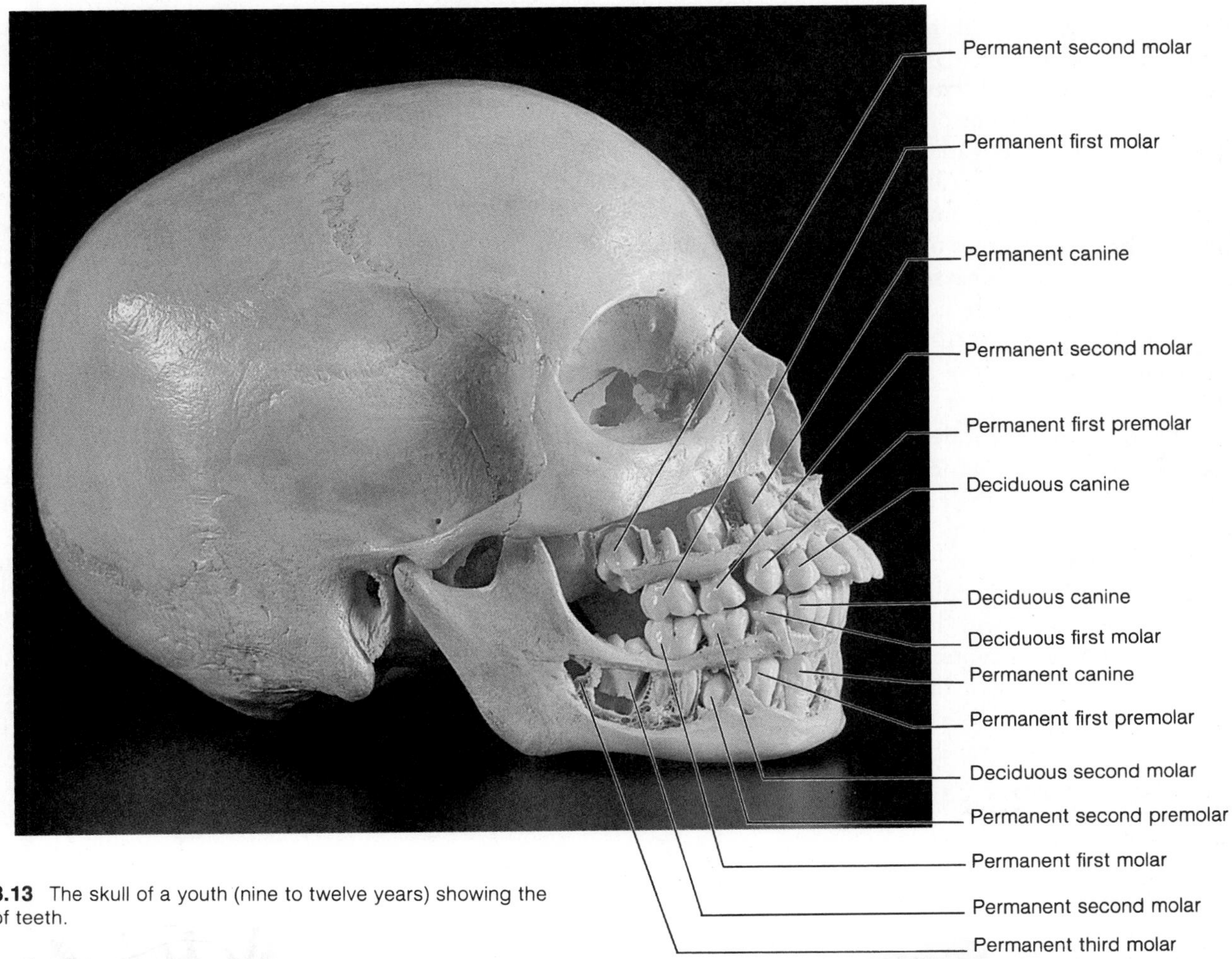

Figure 18.13 The skull of a youth (nine to twelve years) showing the eruption of teeth.

Table 18.2 Eruption sequence and loss of deciduous teeth

	Average age of eruption		
Type of tooth	**Lower**	**Upper**	**Average age of loss**
Central incisors	6–8 mos	7–9 mos	7 yrs
Lateral incisors	7–9 mos	9–11 mos	8 yrs
First molars	12–14 mos	14–16 mos	10 yrs
Canines (cuspids)	16–18 mos	18–20 mos	10 yrs
Second molars	20–22 mos	24–26 mos	11–12 yrs

Table 18.3 Eruption sequence of permanent teeth

	Average age of eruption	
Type of tooth	**Lower**	**Upper**
First molars	6–7 yrs	6–7 yrs
Central incisors	6–7 yrs	7–8 yrs
Lateral incisors	7–8 yrs	8–9 yrs
Canines (cuspids)	9–10 yrs	11–12 yrs
First premolars (bicuspids)	10–12 yrs	10–11 yrs
Second premolars (bicuspids)	11–12 yrs	10–12 yrs
Second molars	11–13 yrs	12–13 yrs
Third molars (wisdom)	17–25 yrs	17–25 yrs

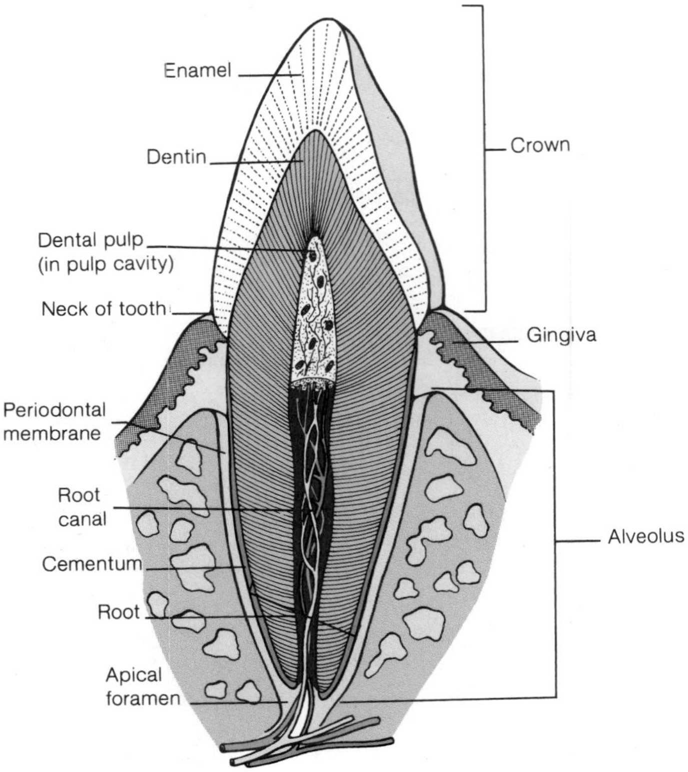

Figure 18.14 The structure of a tooth shown in a vertical section through a canine tooth.

food is mixed with saliva, which initiates chemical digestion and facilitates swallowing. The soft, flexible mass of food that is swallowed is called a *bolus.*

A tooth consists of an exposed **crown,** supported by a **neck,** anchored firmly into the jaw by one or more **roots** (fig. 18.14). The roots of teeth fit into sockets, called **alveoli,** in the alveolar processes of the mandible and maxillae. Each socket is lined with a connective tissue periosteum, specifically called the **periodontal membrane.** The root of a tooth is covered with a bonelike material called the **cementum;** fibers in the periodontal membrane insert into the cementum and fasten the tooth in its socket. The **gingiva** *(jin-ji'vah),* or **gum,** is the mucous membrane surrounding the alveolar processes in the oral cavity.

The bulk of a tooth consists of **dentin,** a substance similar to bone but harder. Covering the dentin on the outside and forming the crown is a tough, durable layer of **enamel.** Enamel is composed primarily of calcium phosphate and is the hardest substance in the body. The central region of the tooth contains the **pulp cavity.** The pulp cavity contains the **pulp,** which is composed of connective tissue with blood vessels, lymph vessels, and nerves. A **root canal** is continuous with the pulp cavity through the root to an opening at the base called the **apical foramen.** The tooth receives nourishment through vessels traversing the apical foramen. Proper nourishment is particularly important during embryonic development. The diet of the mother should contain an abundance of calcium and vitamin D during pregnancy to insure the proper development of the baby's teeth.

Enamel is the hardest substance in the body, but it can be dissolved by bacterial activity that results in *dental caries (cavities).* These caries must be artificially filled because new enamel is not produced after a tooth erupts. The rate of tooth decay decreases after age thirty-five, but then periodontal diseases may develop. *Periodontal diseases* result from plaque or tartar buildup at the gum line, which wedges the gum away from the teeth, allowing bacterial infections to develop.

Salivary Glands

Salivary glands are accessory digestive glands that produce a fluid secretion called *saliva.* Saliva functions as a solvent in cleansing the teeth and dissolving food chemicals so they can be tasted. Saliva also contains enzymes, which digest starch, and mucus, which lubricates the pharynx to facilitate swallowing. Saliva is secreted continuously but usually only in sufficient amounts to keep the mucous membranes of the oral cavity moist. Numerous minor salivary glands, called **buccal glands,** are located in the mucous membranes of the palatal region of the oral cavity. But most of the saliva is produced by three pairs of sal-

bolus: Gk. *bolos,* lump
gingiva: L. *gingiva,* gum
dentin: L. *dens,* tooth

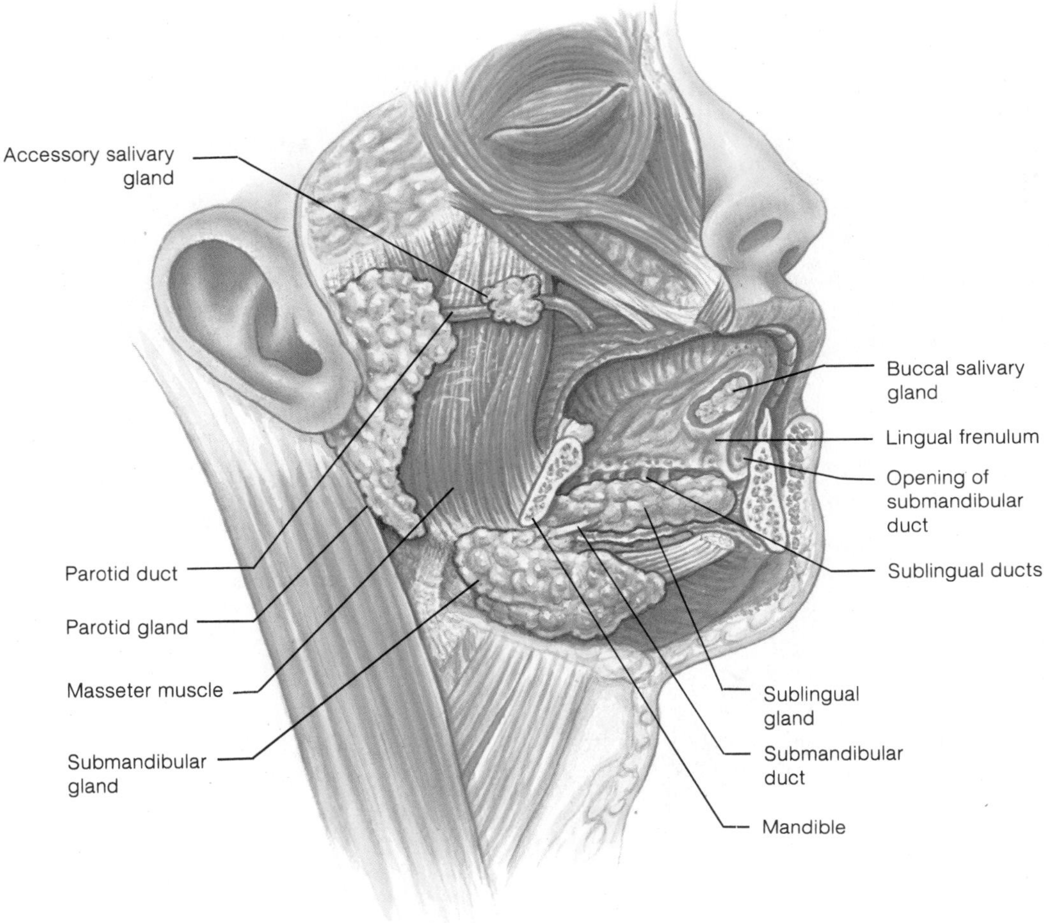

Figure 18.15 The salivary glands.

ivary glands outside of the oral cavity and is transported to the mouth via **salivary ducts.** The three major pairs of salivary glands are the parotid, submandibular, and sublingual (fig. 18.15).

The **parotid** *(pah-rot'id)* **gland** is the largest and is positioned below and in front of the ear, between the skin and the masseter muscle. The **parotid** (Stensen's) **duct** parallels the zygomatic arch across the masseter muscle, pierces the buccinator muscle, and drains into the oral cavity opposite the second upper molar. It is the parotid gland that becomes infected and swollen with the mumps.

The **submandibular** (submaxillary) **gland** is inferior to the body of the mandible about midway along the inner side of the jaw. This gland is covered by the more superficial mylohyoid muscle. The **submandibular** (Wharton's) **duct** empties into the floor of the mouth on both sides of the lingual frenulum.

The **sublingual gland** lies under the mucosa in the floor of the mouth on the side of the tongue. Each sublingual gland possesses several small ducts that empty into the floor of the mouth in an area posterior to the papilla of the submandibular duct.

Two types of secretory cells, called **serous** and **mucous cells,** are found in all salivary glands in various proportions (fig. 18.16). Serous cells produce a watery fluid containing digestive enzymes; mucous cells secrete a thick, stringy mucus. Cuboidal epithelial cells line the lumina of the salivary ducts.

The salivary glands are innervated by both divisions of the autonomic nervous system. Sympathetic impulses stimulate the secretion of small amounts of viscous saliva. Parasympathetic stimulation causes the secretion of large volumes of watery saliva. Physiological responses of this type occur whenever a person sees, smells, tastes, or even thinks about desirable food. The amount of saliva secreted daily ranges from 1,000 to 1,500 ml. Information about the salivary glands is summarized in table 18.4.

Pharynx

The funnel-shaped **pharynx** *(far'inks)* is a passageway approximately 13 cm (5 in.) in length connecting the oral and nasal cavities to the esophagus and trachea. The pharynx has both digestive and respiratory functions. The supporting walls of the pharynx are composed of skeletal muscle, and the lumen is lined with a mucous membrane of stratified squamous epithelium.

parotid: Gk. *para,* beside; *otos,* ear
Stensen's duct: from Nicholaus Stensen, Danish anatomist, 1638–86

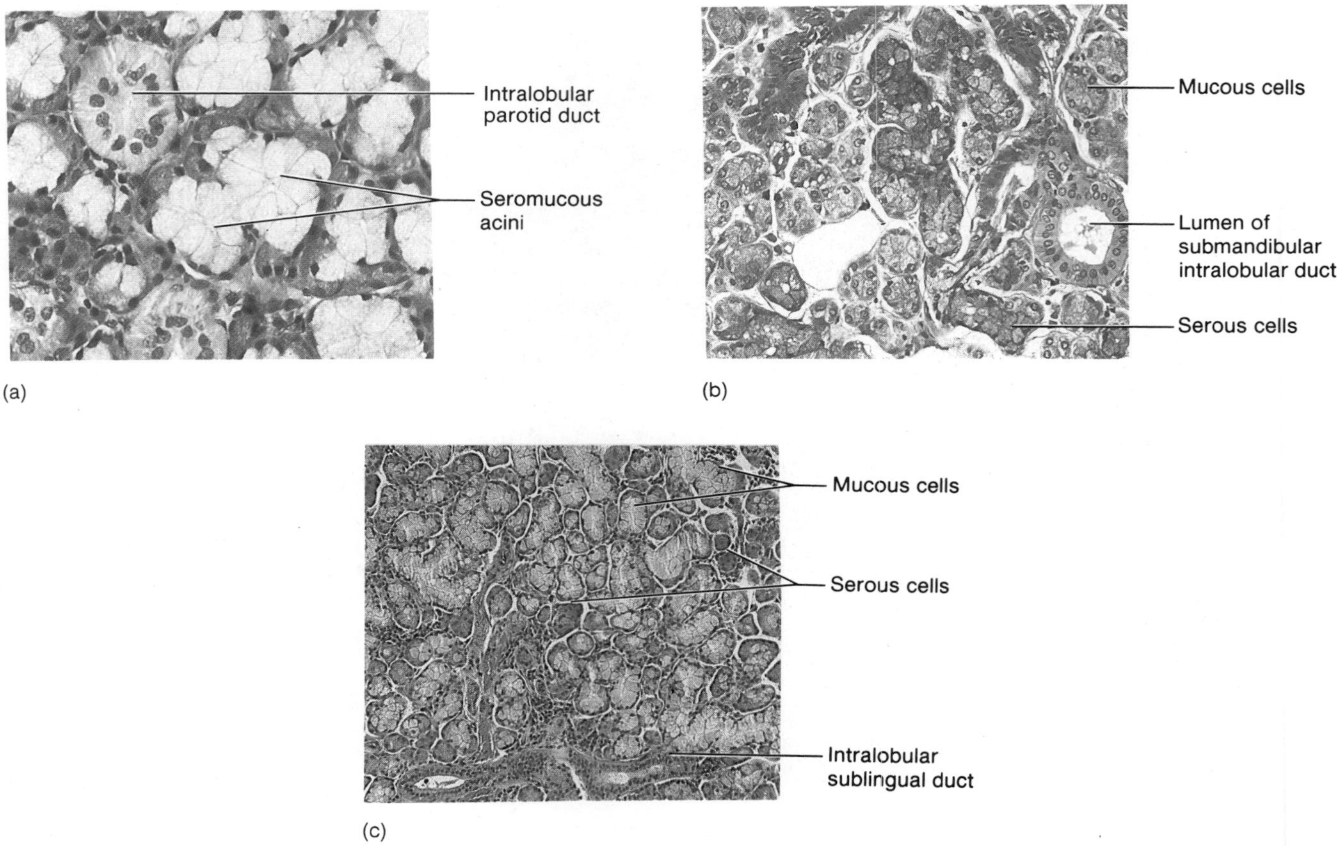

Figure 18.16 The histology of the salivary glands. (*a*) The parotid gland, (*b*) the submandibular gland, and (*c*) the sublingual gland.

Table 18.4 Major salivary glands

Gland	Location	Duct	Entry into oral cavity	Type of secretion
Parotid	Anterior and inferior to auricle; subcutaneous over masseter muscle	Parotid (Stensen's duct)	Lateral to upper second molar	Watery serous fluid, salts, and enzyme
Submandibular	Inferior to the base of the tongue	Submandibular (Wharton's duct)	Papilla lateral to lingual frenulum	Watery serous fluid with some mucus
Sublingual	Anterior to submandibular; under tongue	Several small ducts (Rivinus's ducts)	Ducts along the base of the tongue	Mostly thick, stringy mucus, salts, and enzyme

The external, *circular layer* of pharyngeal muscles, called **constrictors** (fig. 18.17), serves to compress the lumen of the pharynx involuntarily during swallowing. The **superior constrictor muscle** attaches to bony processes of the skull and mandible and encircles the upper portion of the pharynx. The **middle constrictor muscle** arises from the hyoid bone and stylohyoid ligament and encircles the middle portion of the pharynx. The **inferior constrictor muscle** arises from the cartilages of the larynx and encircles the lower portion of the pharynx. During breathing, the lower portion of the inferior constrictor muscle is contracted, preventing air from entering the esophagus.

The motor and most of the sensory innervation to the pharynx is via the **pharyngeal plexus,** situated chiefly on the middle constrictor muscle. It is formed by the pharyngeal branches of the glossopharyngeal (ninth) and the vagus (tenth) cranial nerves, together with a deep sympathetic branch from the superior cervical ganglion.

The pharynx is served by the ascending pharyngeal and inferior thyroid arteries, both of which are branches of the external carotid arteries. Venous return is via the internal jugular veins.

1. Define the terms *diphyodont* and *heterodont.* Which of the four kinds of teeth is absent in deciduous dentition?
2. Identify the locations of the enamel, dentin, cementum, and pulp of a tooth. Explain how a tooth is anchored into its socket.
3. Identify the location of the parotid and the submandibular ducts, and describe where they empty in the oral cavity.
4. Describe the effects of autonomic stimulation on salivary secretion.

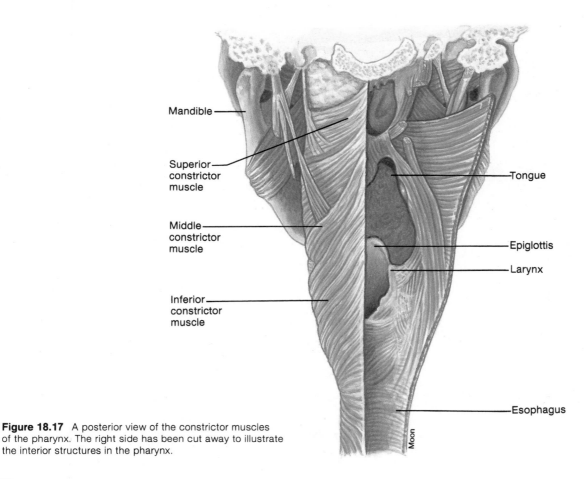

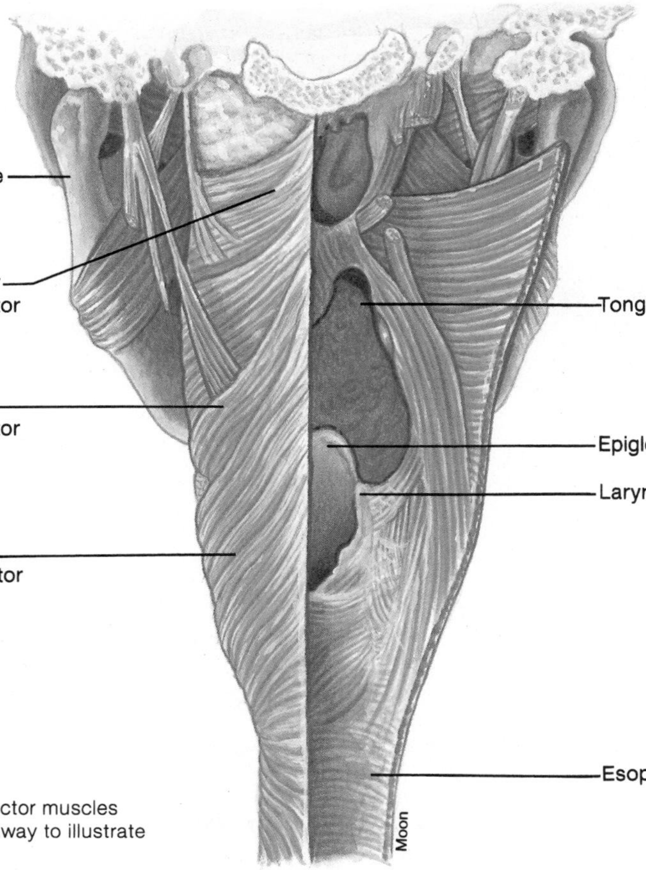

Figure 18.17 A posterior view of the constrictor muscles of the pharynx. The right side has been cut away to illustrate the interior structures in the pharynx.

Esophagus and Stomach

A bolus of food is passed from the esophagus to the stomach, where it is churned and mixed with gastric secretions. The chyme thus produced is sent past the pyloric sphincter of the stomach to the duodenum.

Objective 10. Describe the steps in deglutition.
Objective 11. Describe the location, gross structure, and functions of the stomach.
Objective 12. Describe the histological structure of the esophagus and stomach. List the cell types in the gastric mucosa and their secretory products.

Esophagus

The **esophagus** *(ĕ-sof'ah-gus)* is that portion of the GI tract that connects the pharynx to the stomach. It is a collapsible muscular tube approximately 25 cm (10 in.) long, originating at the larynx and located posterior to the trachea.

The esophagus is located within the mediastinum of the thorax and passes through the diaphragm just above the opening into the stomach. The opening through the diaphragm is called the **esophageal hiatus** *(ĕ-sof''ah-je'al hi-a'tus)*. The esophagus is lined with a nonkeratinized stratified squamous epithelium (fig. 18.18); its walls contain either skeletal or smooth muscle, depending on the location. The upper third of the esophagus contains skeletal muscle; the middle third contains a mixture of skeletal and smooth muscle, and the terminal portion contains only smooth muscle.

The lumen of the terminal portion of the esophagus is slightly narrowed because of a thickening of the circular muscle fibers in its wall. This portion is referred to as the **lower esophageal (gastroesophageal) sphincter.** The muscle fibers of this region constrict after food passes into the stomach to help prevent the stomach contents from regurgitating into the esophagus. Regurgitation would occur because the pressure in the abdominal cavity is greater than the pressure in the thoracic cavity as a result of respiratory movements. The lower esophageal sphincter must remain closed, therefore, until food is pushed through it by peristalsis into the stomach.

The lower esophageal sphincter is not a true sphincter muscle that can be identified histologically, and it does at times permit the acidic contents of the stomach to enter the esophagus. This can create a burning sensation commonly called *heartburn,* although the heart is not involved. Infants under one year old have difficulty controlling their lower esophageal sphincter and thus often "spit up" following meals. Certain mammals, such as rodents, have a true lower esophageal sphincter and cannot regurgitate, which is why poison grains can kill mice and rats effectively.

esophagus: Gk. *oisein,* to carry; *phagema,* food

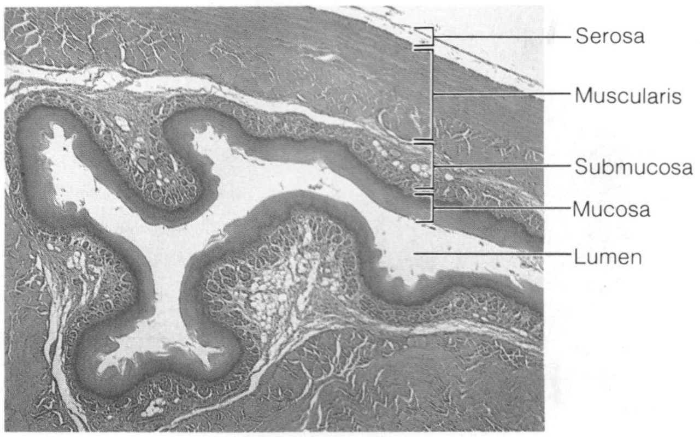

Figure 18.18 The histology of the esophagus.

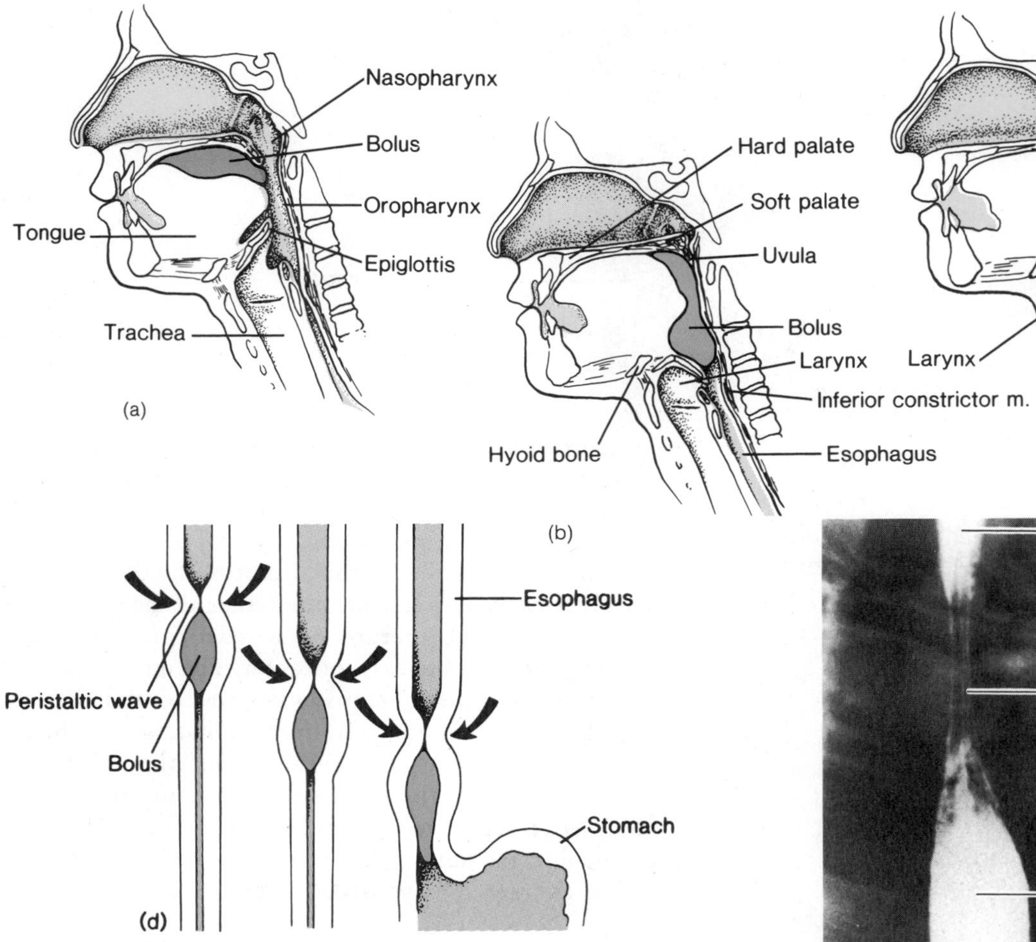

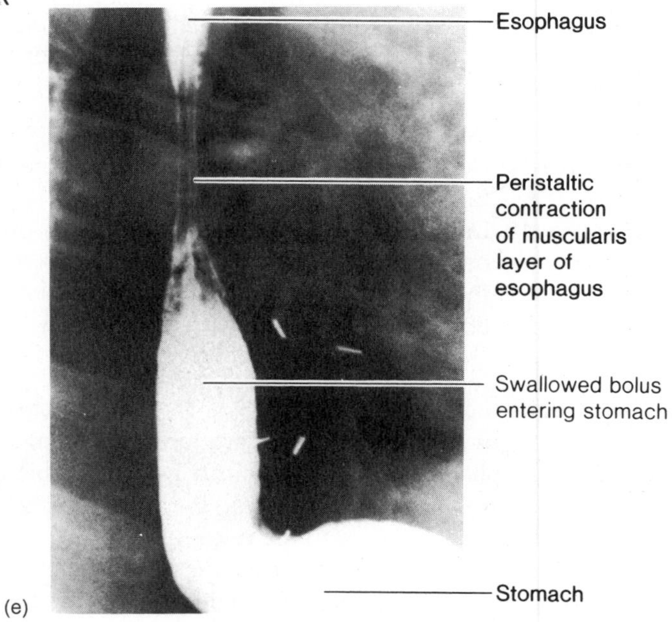

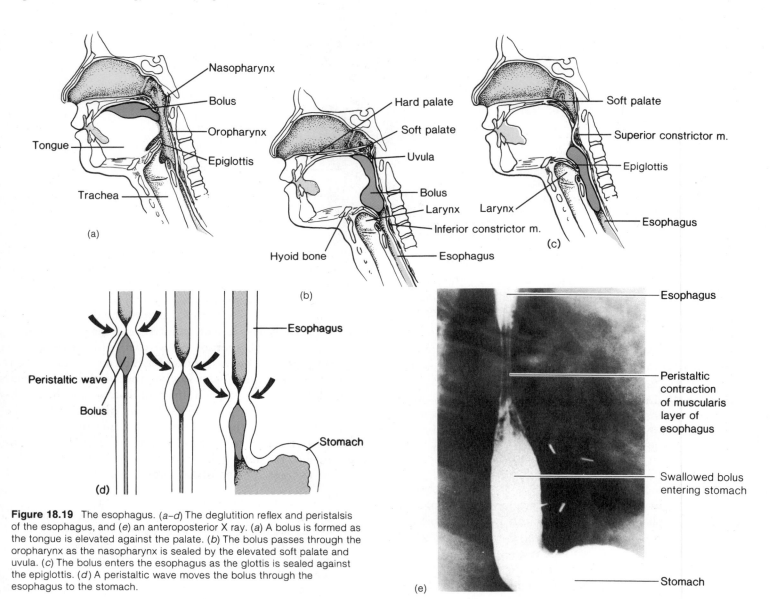

Figure 18.19 The esophagus. (*a–d*) The deglutition reflex and peristalsis of the esophagus, and (*e*) an anteroposterior X ray. (*a*) A bolus is formed as the tongue is elevated against the palate. (*b*) The bolus passes through the oropharynx as the nasopharynx is sealed by the elevated soft palate and uvula. (*c*) The bolus enters the esophagus as the glottis is sealed against the epiglottis. (*d*) A peristaltic wave moves the bolus through the esophagus to the stomach.

Mechanism of Swallowing

Swallowing, or **deglutition** *(deg''loo-tish'un),* is the complex mechanical and physiological act of moving food or fluid from the oral cavity to the stomach. For descriptive purposes, deglutition is divided into three stages (fig. 18.19).

The first stage is voluntary and follows mastication, if food is involved. During this stage, the mouth is closed and breathing is temporarily interrupted. A bolus is formed as the tongue is elevated against the palate through contraction of the mylohyoid and styloglossus muscles and the intrinsic muscles of the tongue.

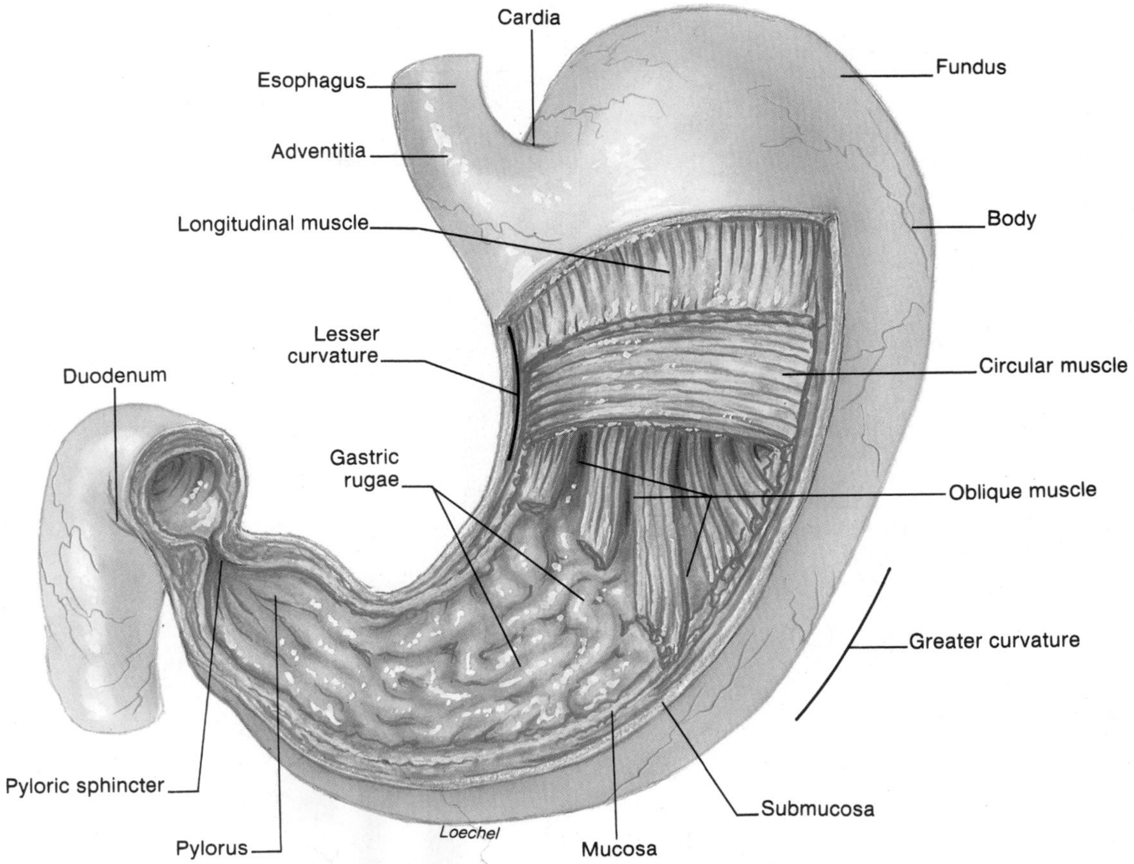

Figure 18.20 The major regions and structures of the stomach.

The second stage of deglutition is the passage of the bolus through the pharynx. The events of this stage are involuntary and are elicited by stimulation of sensory receptors located at the opening of the oropharynx. Pressure of the tongue against the rugae of the hard palate seals the nasopharynx from the oral cavity, creates a pressure, and forces the bolus into the oropharynx. The soft palate and pendulant uvula are elevated to close the nasopharynx as the bolus passes. The hyoid bone and the larynx are also elevated. Elevation of the larynx against the epiglottis seals the glottis so that food or fluid is less likely to enter the trachea. Sequential contraction of the constrictor muscles of the pharynx moves the bolus through the pharynx to the esophagus. This stage is completed in one second or less.

The third stage, the entry and passage of food through the esophagus, is also involuntary. The bolus is moved through the esophagus by peristalsis. The entire time for deglutition varies, but it is slightly more than one second in the case of fluids and five to eight seconds with solid food material.

Stomach

The **stomach** is the most distensible part of the GI tract and is positioned in the upper left quadrant of the peritoneal cavity immediately below the diaphragm. It is a J-shaped pouch that is continuous with the esophagus superiorly and empties into the duodenal portion of the small intestine inferiorly. The functions of the stomach are to store food as it is mechanically churned with gastric secretions; to initiate the digestion of proteins; to carry on limited absorption; and to move food into the small intestine as a pasty material called *chyme (kīm)*.

The stomach is divided into four regions: the cardia, fundus, body, and pylorus (fig. 18.20). The **cardia** is the upper, narrow region immediately below the lower esophageal sphincter. The **fundus** is the dome-shaped portion to the left and in direct contact with the diaphragm. The **body** is the large central portion, and the **pylorus** is the funnel-shaped terminal portion. The pylorus communicates with the duodenal portion of the small intestine through a **pyloric sphincter.** *Pylorus* is a Greek word meaning gatekeeper, and this junction is just that, regulating the movement of chyme into the small intestine and prohibiting backflow.

The stomach has two surfaces and two borders. The broadly rounded surfaces are referred to as the **anterior** and **posterior surfaces.** The medial concave border is the **lesser curvature** (fig. 18.20), and the lateral convex border is the **greater curvature.** The lesser omentum extends between the lesser curvature and the liver, and the greater omentum is attached to the greater curvature.

The wall of the stomach consists of the same four layers found in other regions of the GI tract with certain modifications. The muscularis is composed of three layers of smooth

---

chyme: L. *chymus*, juice
cardia: Gk. *kardia*, heart (upper portion, nearer the heart)
fundus: L. *fundus*, bottom
pylorus: Gk. *pyloros*, gatekeeper

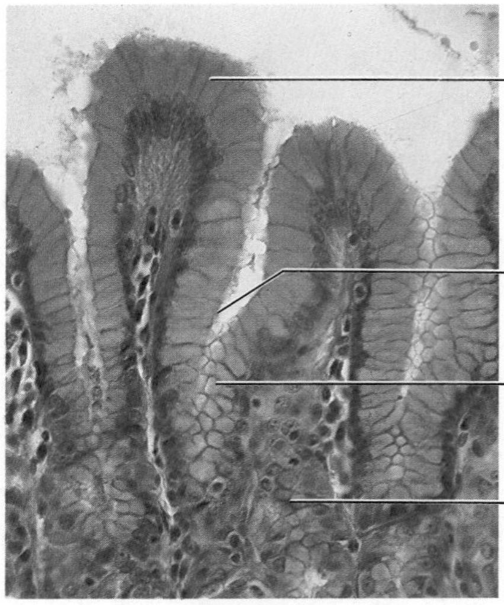

Figure 18.21 Microscopic structures of the mucosa of the stomach.

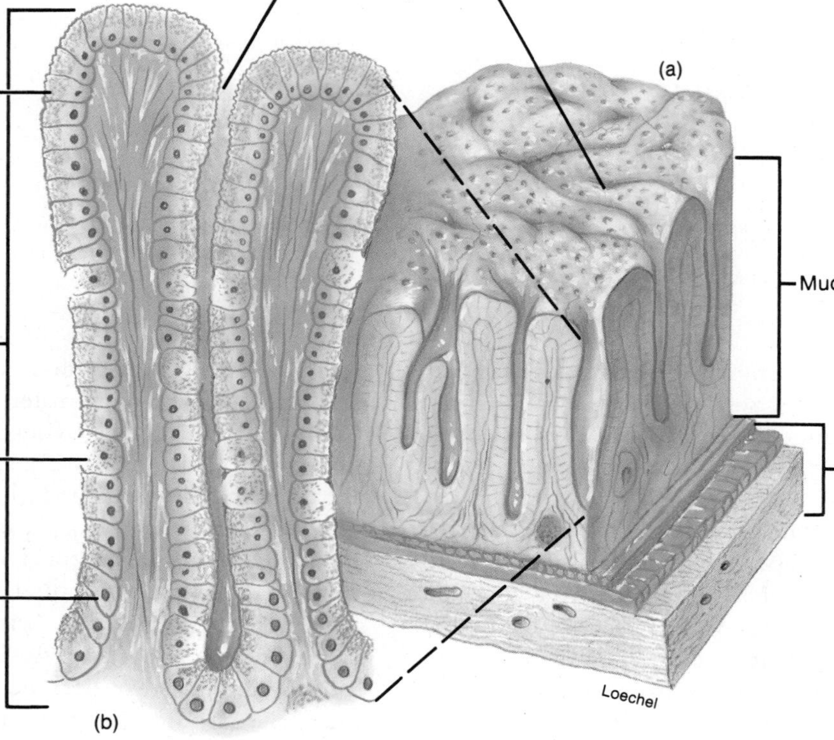

Figure 18.22 Gastric pits and gastric glands of the mucosa. (*a*) Gastric pits are the openings of the gastric glands. (*b*) Gastric glands consist of mucous cells, chief cells, and parietal cells, each of which produces a specific secretion.

muscle named according to the direction of fiber arrangement: an outer **longitudinal layer,** a middle **circular layer,** and an inner **oblique layer** (fig. 18.20). The circular muscle layer is further thickened at the gastroduodenal junction to form the pyloric sphincter.

The mucosa is shaped into numerous longitudinal folds, called **gastric rugae,** which permit stomach distension. The gastric rugae gradually smooth out as the stomach fills. The mucosal layer contains a simple columnar epithelium. The entire mucosal layer is folded to form **gastric pits** and **gastric glands** (figs. 18.21, 18.22).

Gastric glands have several kinds of secreting cells: *chief,* or *zymogenic, cells, parietal cells, mucous cells, argentaffin cells,* and *G cells.* The chief cells secrete various digestive enzymes, and the parietal cells produce hydrochloric acid. The product of these two cell types is called *gastric juice.* Mucous cells secrete large quantities of protective mucus. Argentaffin cells secrete serotonin and histamine, and G cells secrete the hormone gastrin. In addition to these products, the gastric mucosa (probably the parietal cells) secretes a polypeptide called *intrinsic factor,* which is required for absorption of vitamin B_{12} in the intestine.

argentaffin cells: L. *argentum,* silver; *affinis,* attraction (become colored with silver stain)

Table 18.5 Phases of gastric secretion

Phase	Response
Cephalic phase	Parasympathetic response via vagus nerves; reflexes respond to sight, taste, smell, or mental stimuli; 50–150 ml of gastric juice is secreted
Gastric phase	Food in stomach stretches the mucosa, and chemical breakdown of protein stimulates release of gastrin; gastrin causes production of 600–750 ml of gastric juice
Intestinal phase	Chyme entering duodenum stimulates intestinal cells to release intestinal gastrin; intestinal gastrin promotes production of additional small quantities of gastric juice

Mucus, secreted by mucous cells of the stomach, is important in preventing hydrochloric acid and the digestive enzyme pepsin from eroding the stomach wall. The stomach is sensitive to emotional stress. *Peptic ulcers* may be caused by an increase in cellular secretion or by not enough protective mucus being secreted. Another protective feature is the rapid mitotic activity of the columnar epithelium of the stomach. The entire lining of the stomach is usually replaced every few days. Nevertheless, approximately 10% of American men and 4% of American women develop peptic ulcers.

Regulation of gastric activity is autonomic. The sympathetic neurons arise from the celiac plexus, the parasympathetic neurons from the vagus nerves. Parasympathetic neurons synapse in the myenteric plexus between the muscular layers and in the submucosal plexus in the submucosa. Parasympathetic impulses promote gastric activity, the phases of which are presented in table 18.5.

Vomiting is a reflexive response of emptying the stomach through the esophagus, pharynx, and oral cavity. This action is controlled by the **vomiting center** of the medulla oblongata. Stimuli within the GI tract, especially the duodenum, may activate the vomiting center, as may nauseating odors or sights, motion sickness, and body stress. Various drugs called *emetics* can also stimulate a vomiting reflex. The mechanics of vomiting are as follows: (1) strong, sustained contractions of the upper small intestine, followed by a contraction of the pyloric sphincter; (2) relaxation of the lower esophageal sphincter and contraction of the pyloric portion of the stomach; (3) a shallow inspiration and closure of the glottis; (4) compression of the stomach against the liver by contraction of the diaphragm and the abdominal muscles. This reflexive sequence causes a forceful ejection of vomit.

The feeling of *nausea* is caused by stimuli in the vomiting center and may or may not cause vomiting.

The only function of the stomach that appears to be essential for life is the secretion of *intrinsic factor.* This polypeptide is needed for the intestinal absorption of vitamin B_{12}, required for maturation of red blood cells in the bone marrow. A patient with a gastrectomy has to receive B_{12} orally (together with intrinsic factor) or through injections so that he or she will not develop *pernicious anemia.*

1. Describe the three stages in deglutition, including the structures involved.
2. Describe the structure and function of the lower esophageal sphincter.
3. List the functions of the stomach. What is the function of the gastric rugae?
4. Identify the modifications of the stomach that aid in mechanical and chemical digestion.

Small Intestine

The small intestine, consisting of the duodenum, jejunum, and ileum, is the site where digestion is completed and nutrients are absorbed. The surface area of the intestinal wall is increased by plicae circulares, villi, and microvilli.

Objective 13. Describe the location, support, and regions of the small intestine.

Objective 14. List the functions of the small intestine, and describe the anatomical specializations that accomplish these functions.

Objective 15. Describe the movements that occur within the small intestine.

The **small intestine** is that portion of the GI tract between the pyloric sphincter of the stomach and the ileocecal valve opening into the large intestine. It is positioned in the central and lower portions of the abdominal cavity and is supported, except for the first portion, by **mesentery** (fig. 18.23). The attachment of the fan-shaped mesentery to the small intestine maintains mobility but allows little chance of the intestine becoming twisted or kinked. Enclosed within the mesentery are blood vessels, nerves, and lymphatic vessels that supply the intestinal wall.

The small intestine is approximately 3 m (10 ft) long and 2.5 cm (1 in.) wide in a living person, but it will measure nearly twice this length in a cadaver when the muscular wall is relaxed. It is called the small intestine because of its relatively small diameter compared to that of the large intestine. The functions of the small intestine are the reception of the secretions from the liver and pancreas, chemical and mechanical breakdown of chyme, absorption of nutrients, and transportation of the remaining undigested material to the large intestine. A summary of the digestive enzymes of the GI tract is presented in table 18.6.

The small intestine is innervated by the **superior mesenteric** *(mes''en-ter'ik)* **plexus.** The branches of the plexus contain sensory neurons, postganglionic sympathetic neurons, and preganglionic parasympathetic neurons. The small intestine's arterial blood supply comes through the superior mesenteric artery and small branches from the celiac and inferior mesenteric arteries. Venous drainage is through the superior mesenteric vein, which unites with the splenic vein to form the hepatic portal vein carrying nutrient-rich blood to the liver.

Regions of the Small Intestine

The small intestine is divided into three regions on the basis of function and histological structure. These three regions are the duodenum, jejunum, and ileum.

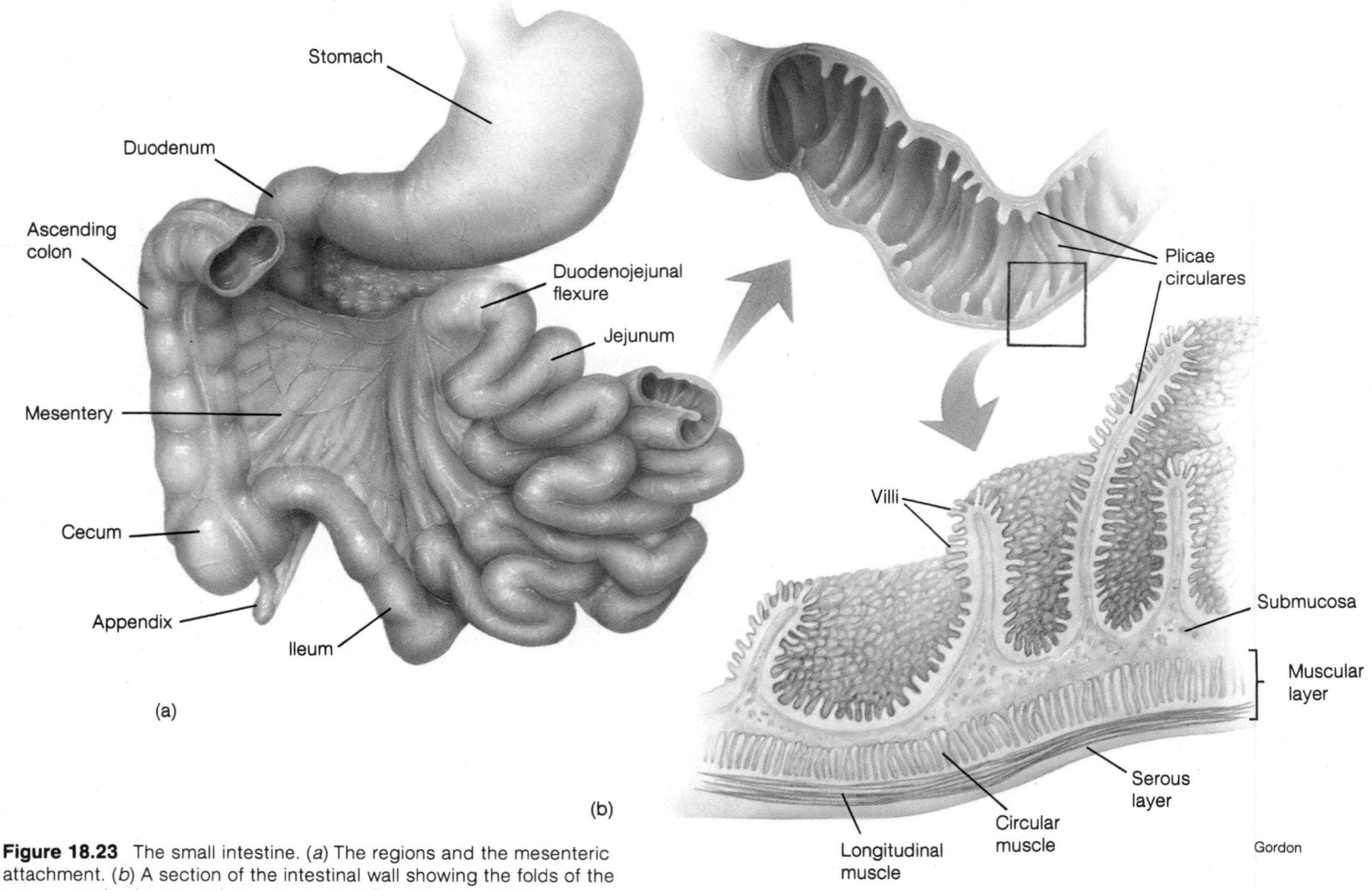

Figure 18.23 The small intestine. (*a*) The regions and the mesenteric attachment. (*b*) A section of the intestinal wall showing the folds of the mucosa and submucosa into structures called plicae circulares.

Table 18.6 Summary of the digestive enzymes

Enzyme	Source	Digestive action
Salivary enzyme		
Amylase	Salivary glands	Begins carbohydrate digestion by converting starch and glycogen to disaccharides
Gastric juice enzyme		
Pepsin	Gastric glands	Begins the digestion of nearly all types of proteins
Intestinal juice enzymes		
Peptidase	Intestinal glands	Converts proteins into amino acids
Sucrase	Intestinal glands	Converts disaccharides into monosaccharides
Maltase		
Lactase		
Lipase	Intestinal glands	Converts fats into fatty acids and glycerol
Amylase	Intestinal glands	Converts starch and glycogen into disaccharides
Nuclease	Intestinal glands	Converts nucleic acids into nucleotides
Enterokinase	Intestinal glands	Activates trypsin
Pancreatic juice enzymes		
Amylase	Pancreas	Converts starch and glycogen into disaccharides
Lipase	Pancreas	Converts fats into fatty acids and glycerol
Peptidases	Pancreas	Converts proteins or partially digested proteins into amino acids
Trypsin		
Chymotrypsin		
Carboxypeptidase		
Nuclease	Pancreas	Converts nucleic acids into nucleotides

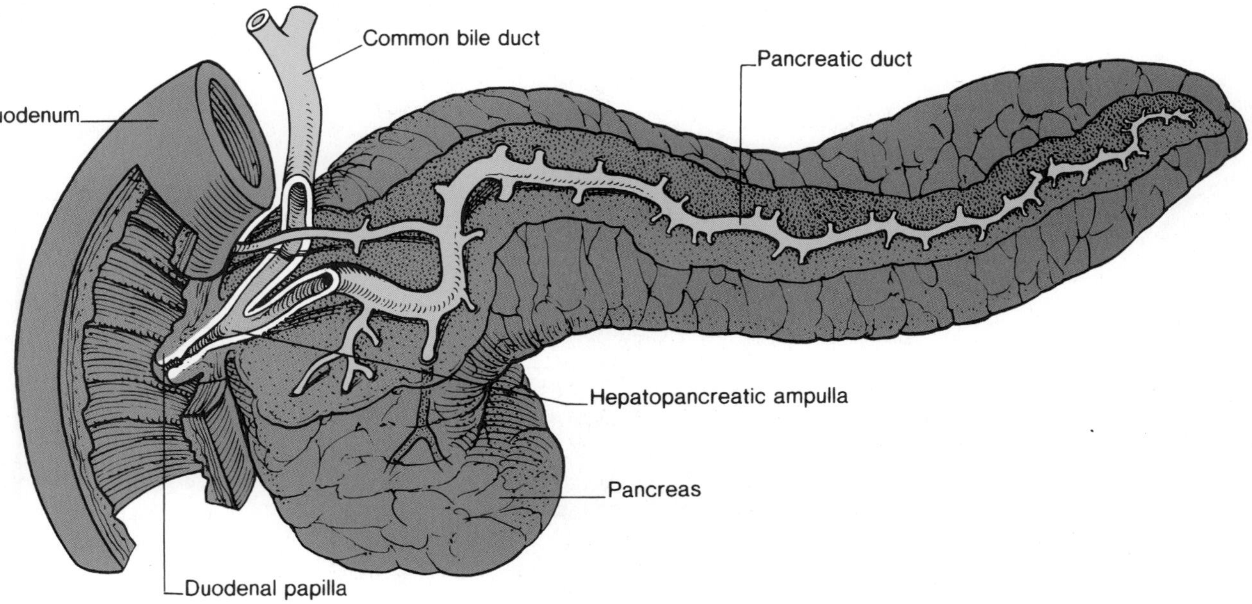

Figure 18.24 The duodenum and associated structures.

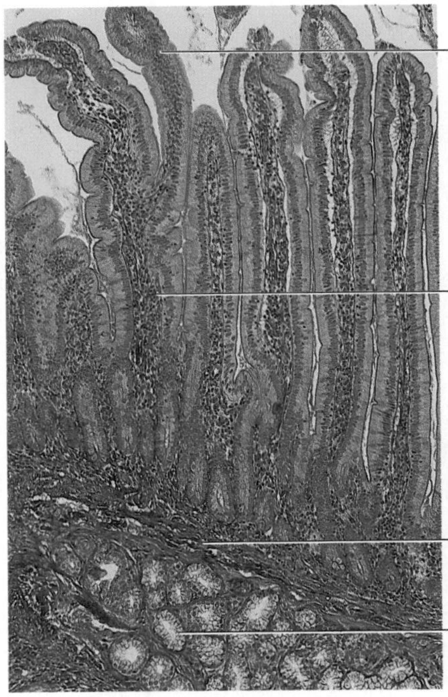

Figure 18.25 The histology of the duodenum.

The **duodenum** *(du''o-de'num* or *du-od'ĕ-num)* is a relatively fixed, C-shaped tube, measuring approximately 25 cm (10 in.) from the pyloric sphincter of the stomach to the **duodenojejunal** *(du-od''ĕ-no''jĕ-joo'nal)* **flexure.** The concave surface of the duodenum faces to the left, where it receives bile secretions through the **common bile duct** from the liver and gallbladder and pancreatic secretions through the **pancreatic duct** of the pancreas (fig. 18.24). These two ducts unite to form a common entry into the duodenum called the **hepatopancreatic ampulla** *(hep''ah-to-pan''kre-at'ik am-pul'lah)* (or ampulla of Vater), which pierces the duodenal wall and drains into the duodenum from an elevation called the **duodenal papilla.** The duodenum is retroperitoneal except for a short portion near the stomach. The duodenum histologically differs from the rest of the small intestine by the presence of **duodenal** (Brunner's) **glands** in the submucosa (fig. 18.25). These compound tubuloalveolar glands secrete mucus and are most numerous near the superior end of the duodenum.

The **jejunum** *(je-joo'num)* is approximately 1 m (3 ft) long and extends from the duodenum to the ileum. The jejunum has a slightly larger lumen and more internal folds than does the ileum but lacks unique histological structures.

The **ileum** *(il'e-um)* makes up the remaining 2 m (6–7 ft) of the small intestine. The terminal portion of the ileum empties into the medial side of the cecum through the **ileocecal** *(il''e-o-se'kal)* **valve.** The walls of the ileum have an abundance of lymphatic tissue. Aggregates of lymph nodules, called **mesenteric** (Peyer's) **patches,** are characteristic of the ileum.

Structural Modifications of the Small Intestinal Wall

The products of digestion are absorbed across the epithelial lining of the intestinal mucosa. Absorption occurs primarily in the jejunum, although some also occurs in the duodenum and ileum. Absorption occurs at a rapid rate as a result of the high mucosal surface area in the small intestine, provided by folds. The mucosa and submucosa form large folds called the **plicae** *(pli'se)* **circulares,** which can be observed with the unaided eye.

duodenum: L. *duodeni,* twelve each (length of twelve fingers' breadth)

ampulla of Vater: from Abraham Vater, German anatomist, 1684–1751
Brunner's glands: from Johann C. Brunner, Swiss anatomist, 1653–1727
Peyer's patches: from Johann K. Peyer, Swiss anatomist, 1653–1712
plica: L. *plicatus,* folded

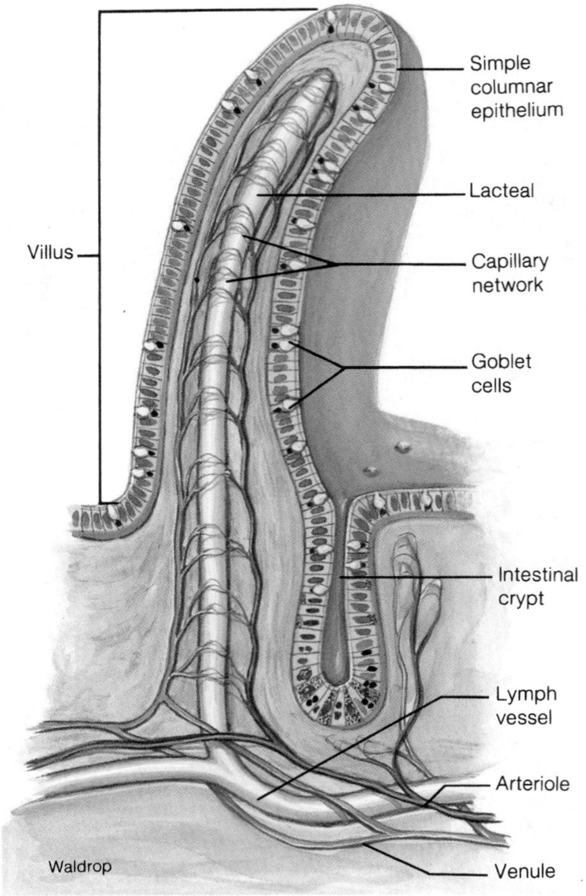

Figure 18.26 A diagram of the structure of an intestinal villus.

The surface area is further increased by the microscopic folds of mucosa called villi and by the foldings of the apical cell membrane of epithelial cells (which can only be seen with an electron microscope), called microvilli.

Each **villus** is a fingerlike fold of mucosa that projects into the intestinal lumen. The villi are covered with columnar epithelial cells and interspersed among these are the mucus-secreting goblet cells. The lamina propria forms a connective tissue core of each villus and contains numerous lymphocytes, blood capillaries, and a lymphatic vessel called the **lacteal** (fig. 18.26). Absorbed monosaccharides and amino acids enter the blood capillaries; absorbed fatty acids enter the lacteals.

Exposed epithelial cells of the villi are continuously exfoliated and are replaced by deeper cells that are pushed up from the germinal layer. The epithelium at the base of the villi invaginates downward at various points to form tubular **intestinal glands** (crypts of Lieberkühn), which extend downward into the mucous membrane (fig. 18.27). The intestinal glands secrete digestive enzymes and are also the site for the production of new epithelial cells. In a light microscope, the **microvilli** produce a somewhat vague **brush border** on the edges of the columnar epithelial cells. The term *brush border* is thus often used synonymously with microvilli in descriptions of the intestine (fig. 18.28).

Mechanical Activities of the Small Intestine

Contractions of the longitudinal and circular muscles of the small intestine produce three distinct types of movement: rhythmic segmentation, pendular movements, and peristalsis.

Rhythmic segmentations are local, ringlike contractions of the circular muscular layer that occur at the rate of about twelve to sixteen per minute in regions containing chyme. This action churns chyme with digestive juices and brings it in contact with the mucosa. During these contractions, the vigorous motion of the villi stirs the chyme and facilitates absorption.

Pendular movements primarily occur in the longitudinal muscle layer. In this motion, a constrictive wave moves along a segment of the intestine and then reverses and moves in the opposite direction, moving the chyme back and forth. Pendular movements also mix the chyme but do not seem to have a particular frequency.

---

villus: L. *villosus*, shaggy
lacteal: L. *lacteus*, milk
crypts of Lieberkühn: from Johann N. Lieberkühn, German anatomist, 1711–56

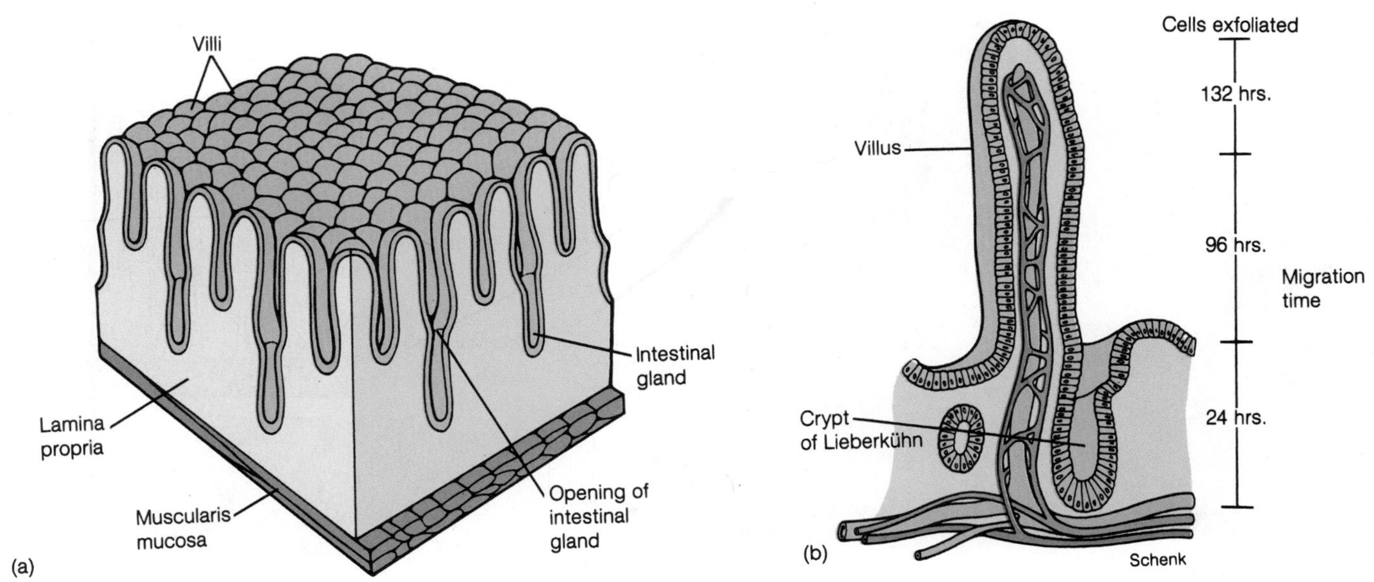

Figure 18.27 Intestinal villi and intestinal glands, or crypts of Lieberkühn, are shown in (*a*). The intestinal glands serve as sites for production of new epithelial cells. The time required for migration of these new cells to the tip of the villus is shown in (*b*). Epithelial cells are exfoliated from the tips of the villi.

Peristalsis is responsible for the propulsive movement of the chyme through the small intestine. These wavelike contractions are usually weak and relatively short and occur at a frequency of about fifteen to eighteen per minute. Chyme requires three to ten hours to travel the length of the small intestine. Both muscle layers are involved in peristalsis.

The sounds of digestive peristalsis can be easily listened to by placing a stethoscope at various abdominal locations. These sounds can be detected even through clothing. The sounds, mostly clicks and gurgles, occur at a frequency of five to thirty per minute.

1. Describe the small intestine. Where is it located? How is it subdivided? How is it supported?
2. What are the functions of the small intestine?
3. List three structural modifications of the small intestine that increase the absorptive surface area.
4. Describe the movements of the small intestine. Which movements are produced by the circular layer of the tunica muscularis?
5. Which region of the small intestine is the longest? Which is the shortest? How long does it take a portion of chyme to move through the small intestine?

Large Intestine

The large intestine absorbs water and electrolytes from the chyme it receives from the small intestine and, through the action of sphincter muscles, passes waste products out of the body through the rectum and anal canal.

Objective 16. Identify the regions of the large intestine, and describe its gross and histological structure.

Objective 17. Describe the functions of the large intestine, and explain how defecation is accomplished.

The **large intestine** is about 1.5 m (5 ft) long and 6.5 cm (2.5 in.) in diameter. It is named the large intestine because its diameter is larger than that of the small intestine. The large intestine begins at the terminal end of the ileum in the lower right quadrant of the abdominal cavity. From there, the large intestine leads superiorly on the right side to just below the liver where it crosses to the left and descends into the pelvis and terminates at the anus. A specialized portion of the mesentery, called the **mesocolon,** supports the transverse portion of the large intestine along the posterior abdominal wall.

The large intestine has little or no digestive function, but it does absorb water and electrolytes from the remaining chyme. In addition, the large intestine functions to form, store, and expel *feces* from the body.

Regions and Structures of the Large Intestine

The large intestine is structurally divided into the cecum, colon, rectum, and anal canal (figs. 18.29, 18.30).

The **cecum** *(se'kum)* is a dilated pouch that hangs inferiorly, slightly below the ileocecal valve. The ileocecal valve is a fold of mucous membrane at the junction of the small and large intestine, which prohibits the backflow of chyme. A fingerlike projection called the **appendix** is attached to the inferior medial margin of the cecum. The 8 cm (3 in.) appendix has an abundance of lymphatic tissue (fig. 18.31), which may serve to resist infection.

cecum: L. *caecum,* blind pouch
appendix: L. *appendix,* attachment

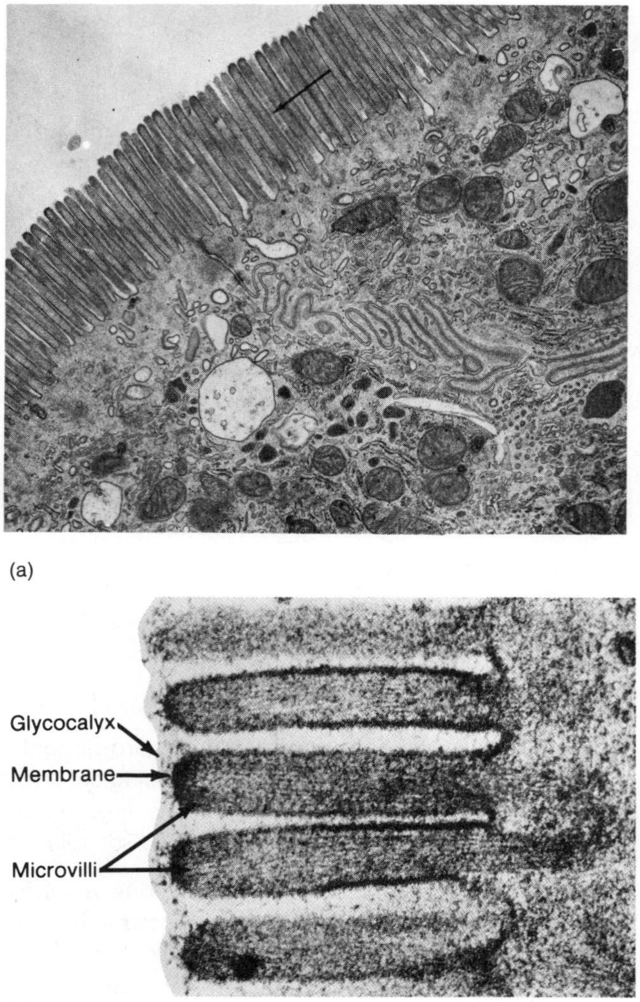

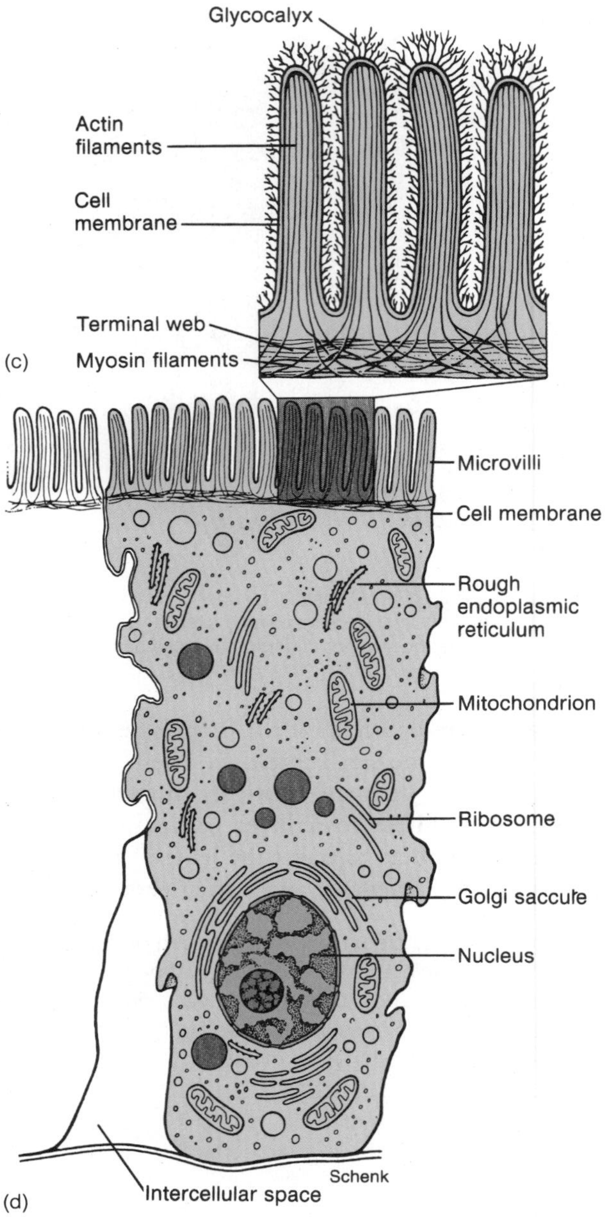

Figure 18.28 *(a)* The cell membrane of epithelial cells that line the small intestine is folded into microvilli. *(b)* Glycoproteins within the membranes of the microvilli contain polysaccharides that extend out into the lumen and form the glycocalyx, which covers the brush border epithelium. *(c)* A diagram of an intestinal epithelial cell. *(d)* The microvilli contain actin filaments attached to a terminal web of myosin filaments. This allows the microvilli to shorten.

A common disorder of the large intestine is inflammation of the appendix, or *appendicitis.* Wastes that accumulate in the appendix cannot be moved easily by peristalsis since the appendix has only one opening. The symptoms of appendicitis include muscular rigidity, localized pain in the lower right quadrant, loss of appetite, and occasionally vomiting. The chief danger of appendicitis is that the appendix might rupture and produce *peritonitis.*

The open, superior portion of the cecum is continuous with the colon. The colon consists of ascending, transverse, descending, and sigmoid portions (fig. 18.29). The **ascending colon** extends superiorly from the cecum along the right abdominal wall to the inferior surface of the liver. Here the colon bends sharply to the left at the **hepatic flexure** and transversely crosses the upper abdominal cavity as the **transverse colon.** At the left abdominal wall, another right angle bend, called the **splenic flexure,** denotes the beginning of the **descending colon.** The descending colon traverses inferiorly along the left abdominal wall to the pelvic region. The colon then angles medially from the brim of the pelvis to form an S-shaped bend, known as the **sigmoid colon.**

The terminal 20 cm (7.5 in.) of the GI tract is the **rectum,** and the last 2 to 3 cm of the rectum is referred to as the **anal canal** (fig. 18.32). The rectum lies anterior to the sacrum where it is firmly attached by peritoneum. The **anus** is the external opening of the anal canal. Two sphincter muscles guard the anal opening: the **internal anal sphincter,** which is composed

colon: Gk. *kolon,* member of the whole

sigmoid: Gk. *sigmoeides,* shaped like the sigma symbol, Σ
rectum: L. *rectum,* straight tube
anus: L. *anus,* ring

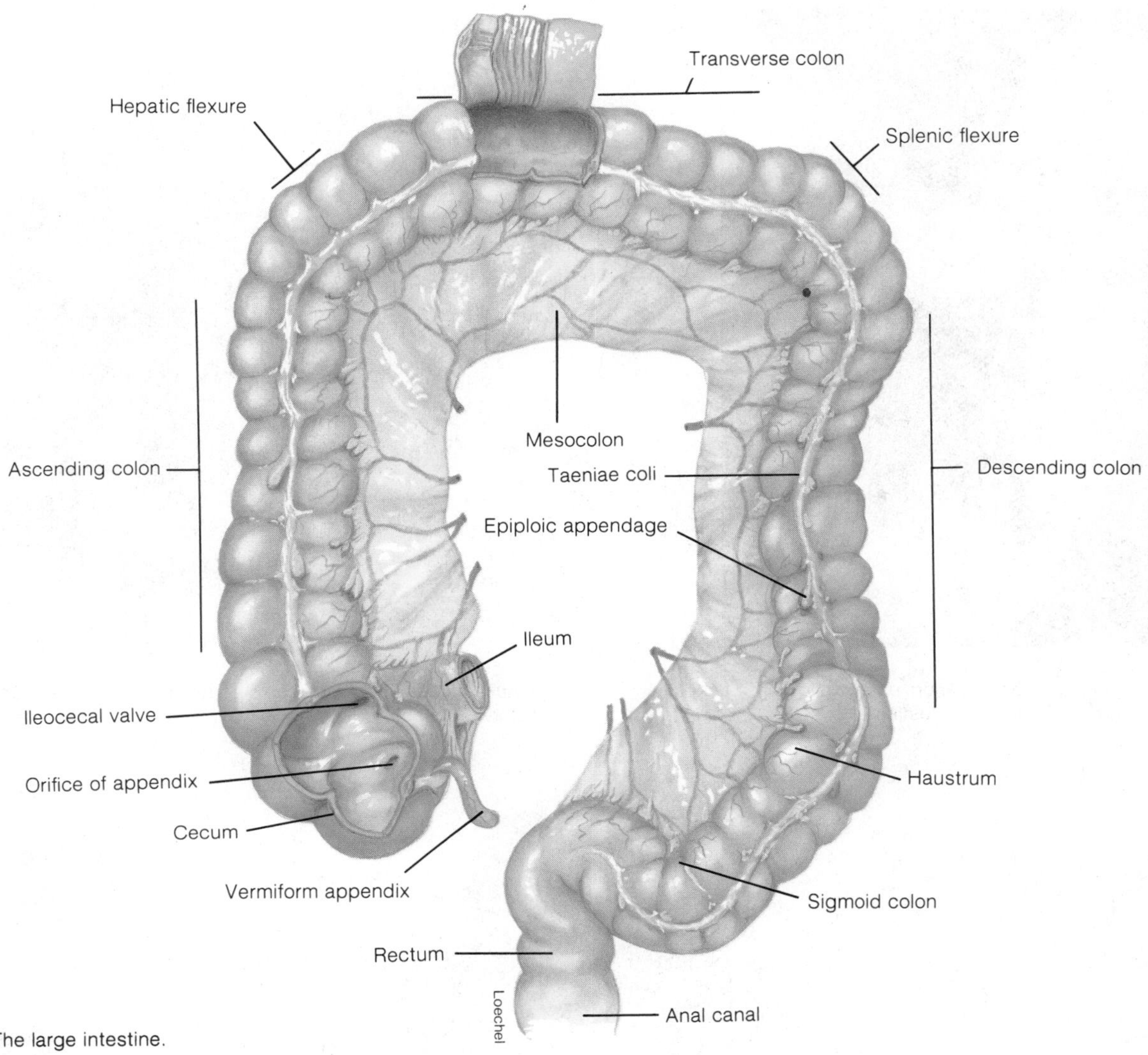

Figure 18.29 The large intestine.

of smooth muscle fibers, and the **external anal sphincter,** consisting of skeletal muscle. The mucous membrane of the anal canal is arranged into highly vascular, longitudinal folds called **anal columns.**

A *hemorrhoid (hem'o-roid),* or *pile,* is a varicose vein in the anal area caused, in part, by difficulty in defecating. A hemorrhoid is classified according to the severity of vessel distension and exposure. A first-degree hemorrhoid is internally contained within the anal canal. A second-degree hemorrhoid prolapses, or extends outward during defecation. A third-degree hemorrhoid remains prolapsed through the anal orifice. Rubber band constriction is a common medical treatment for a prolapsed hemorrhoid. In this technique, a rubber band is tied around the hemorrhoid, constricting its blood supply causing the tissue to dry and fall off. In a new treatment, called infrared photocoagulation, a high-energy light beam coagulates the hemorrhoid.

Although the large intestine consists of the same tunicas as the small intestine, there are some structural differences. The large intestine lacks villi but does have numerous goblet cells in the mucosal layer (fig. 18.33). The longitudinal muscle layer of the muscularis forms three distinct muscle bands, called **taeniae coli** *(te'ne-ah co'li),* which run the length of the large intestine. A series of bulges in the walls of the large intestine form sacculations, or **haustra** *(hows'tra),* along its entire length (see figs. 18.29, 18.30). Finally, the large intestine has small but numerous fat-filled pouches called **epiploic appendages** (see fig. 18.29), which are attached superficially to the taeniae coli in the serous layer.

The large intestine has both types of autonomic innervation. The sympathetic innervation arises from superior and inferior mesenteric plexuses as well as from the celiac plexus. The parasympathetic innervation arises from the paired pelvic splanchnic and vagus nerves. Sensory fibers from the large intestine respond to bowel pressure and signal the need to defecate.

Branches from the superior mesenteric and inferior mesenteric arteries supply blood to the large intestine. Venous blood is returned through the superior and inferior mesenteric veins, which are tributaries to the hepatic portal vein that drains the liver.

---

taenia: L. *tainia,* a ribbon
haustrum: L. *haustrum,* bucket or scoop
epiploic: Gk. *epiplein,* to float on

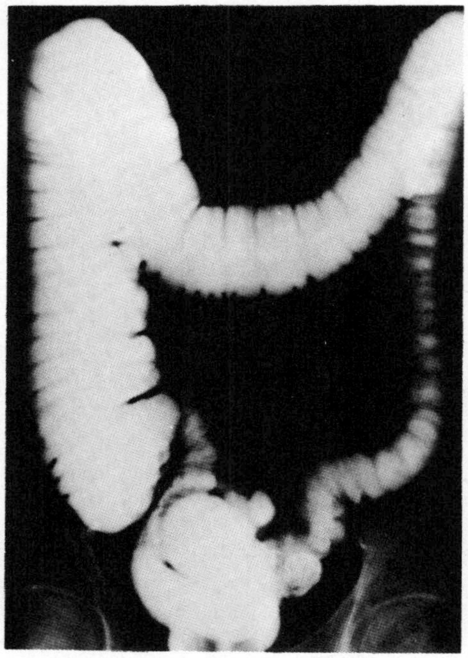

Figure 18.30 A radiograph after a barium enema showing the regions, flexures, and the haustra of the large intestine.

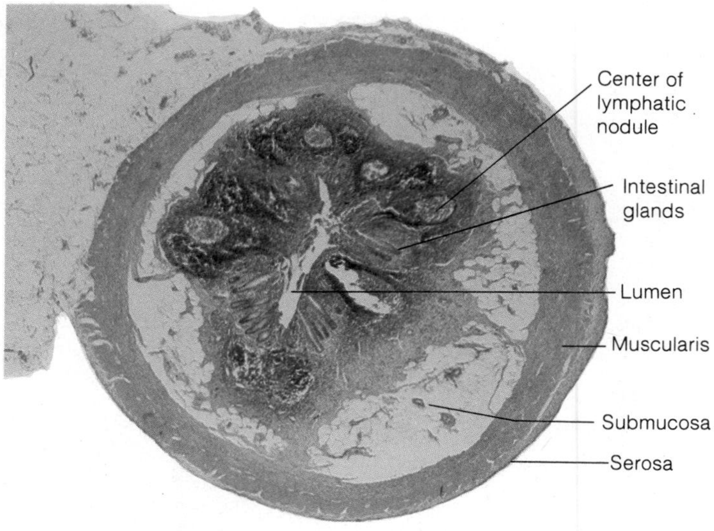

Figure 18.31 The microscopic appearance of a cross section of the human appendix.

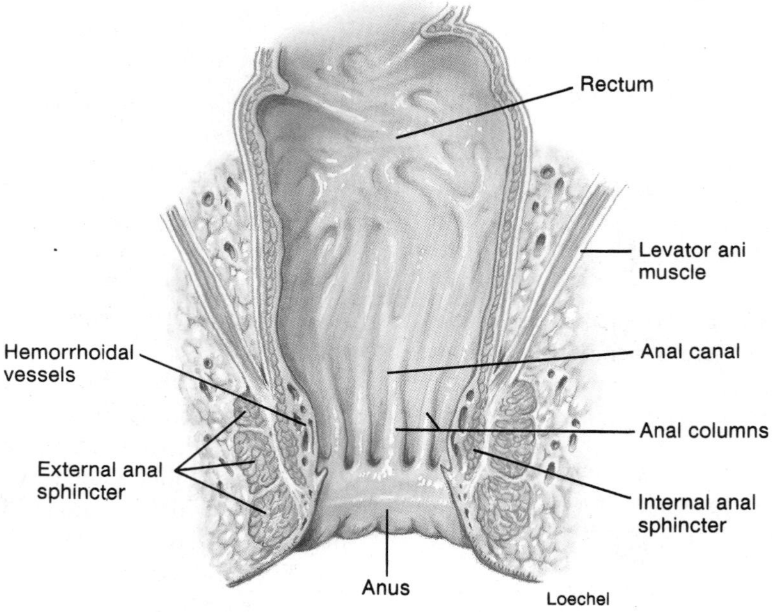

Figure 18.32 The anal canal.

Mechanical Activities of the Large Intestine

Chyme enters the large intestine through the ileocecal valve. About 15 ml of pasty material enters the cecum with each rhythmic opening of the valve. The ingestion of food intensifies peristalsis of the ileum and increases the frequency with which the ileocecal valve opens; this is called the **gastroileal reflex.** Material entering the large intestine accumulates in the cecum and ascending colon.

Three types of movements occur throughout the large intestine: peristalsis, haustral churning, and mass movement. **Peristaltic movements** of the colon are similar to those of the small intestine, though they are usually more sluggish. In **haustral churning,** a relaxed haustrum is filled until it reaches a certain point of distension, and then the muscularis layer is stimulated to contract. This contraction not only moves the material to the next haustrum but churns the contents and exposes it to the mucosa where water and electrolytes are absorbed. **Mass movement** is a very strong peristaltic wave, involving the action of the taeniae coli, which moves the colonic contents toward the rectum. Mass movements generally occur only two or three times a day, usually during or shortly after a meal. This response to eating is called the **gastrocolic**

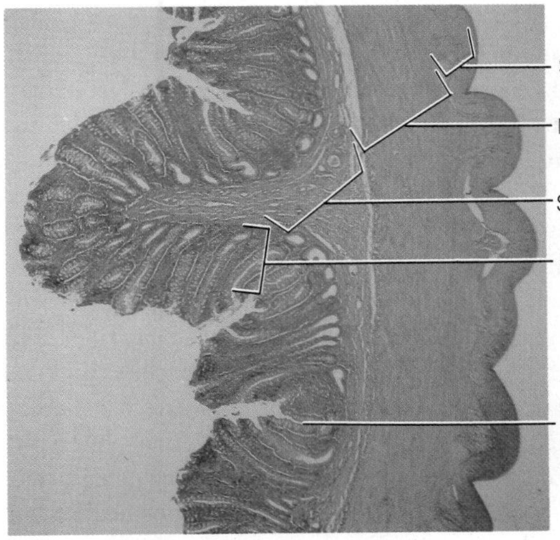

(a)

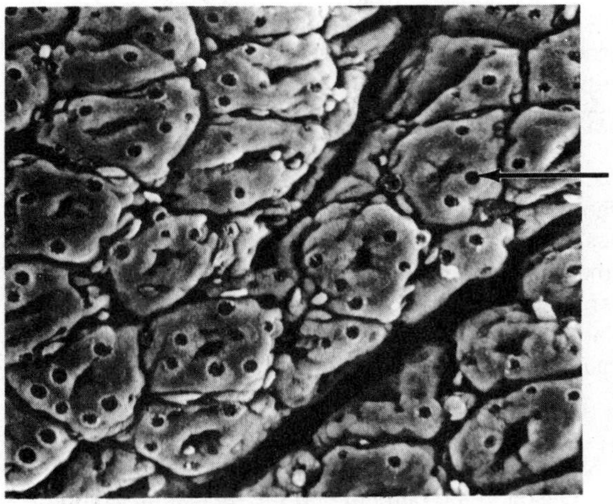
(b)

Figure 18.33 The histology of the large intestine. (*a*) A photomicrograph of the tunics. (*b*) A scanning electron micrograph of the mucosa. The arrow indicates the opening of a goblet cell into the intestinal lumen. (From: *Tissues and Organs: A Text-Atlas of Scanning Electron Microscopy* by R. G. Kessel and R. H. Kardon. W. H. Freeman and Company. © 1979.)

Table 18.7 Summary of the mechanical activity in the GI tract

Region	Type of motility	Frequency	Stimulus	Result
Oral cavity	Mastication	Variable	Initiated voluntarily, proceeds reflexly	Subdivision, mixing with saliva
Oral cavity and pharynx	Deglutition	Maximum of 20 per min	Initiated voluntarily, reflexly controlled by swallowing center	Clears oral cavity of food
Esophagus	Peristalsis	Depends on frequency of swallowing	Initiated by swallowing	Movement through the esophagus
Stomach	Receptive relaxation	Matches frequency of swallowing	Unknown	Permits filling of stomach
	Tonic contraction	15–20 per min	Autonomic plexuses	Mixing and churning
	Peristalsis	1–2 per min	Autonomic plexuses	Evacuation of stomach
	Hunger contractions	3 per min	Low blood sugar level	Feeding
Small intestine	Peristalsis	15–18 per min	Autonomic plexuses	Transfer through intestine
	Rhythmic segmentation	12–16 per min	Autonomic plexuses	Mixing
	Pendular movements	Variable	Autonomic plexuses	Mixing
Large intestine	Peristalsis	3–12 per min	Autonomic plexuses	Transport
	Mass movement	2–3 per day	Stretch	Fills sigmoid colon
	Haustral churning	3–12 per min	Autonomic plexuses	Mixing
	Defecation	Variable: 1 per day to 3 per week	Reflex triggered by rectal distension	Defecation

reflex and can best be observed in infants who have a bowel movement during or shortly after feeding.

As material passes through the large intestine, Na^+, K^+, and water are absorbed. It has been estimated that an average of 850 ml of water per day is absorbed across the mucosa of the colon. The waste material that is left then passes to the rectum, leading to an increase in rectal pressure and the urge to defecate. If the urge to defecate is denied, feces are prevented from entering the anal canal by the internal anal sphincter. In this case the feces remain in the rectum and may even back up into the sigmoid colon. The **defecation reflex** normally occurs when the rectal pressure rises to a particular level that is determined, to a large degree, by habit. At this point the internal anal sphincter relaxes to admit feces into the anal canal.

During the act of defecation the longitudinal rectal muscles contract to increase rectal pressure and the internal and external anal sphincter muscles relax. Excretion is aided by contractions of abdominal and pelvic skeletal muscles, which raise the intra-abdominal pressure and help push the feces from the rectum through the anal canal and out the anus.

The various mechanical activities of the GI tract are summarized in table 18.7.

Constipation occurs when fecal material accumulates because of longer-than-normal periods between defecations. The slower rate of elimination allows more time for water absorption, so that the waste products become harder. Although uncomfortable and sometimes painful, this condition is usually not dangerous. *Diarrhea* is produced when waste material passes too quickly through the colon, so that insufficient time is allowed for water absorption. Excessive diarrhea can result in dangerous levels of dehydration and electrolyte imbalance, particularly in infants because of their small body size.

1. Identify the four principal regions of the large intestine, and describe the functions of the colon.
2. Describe the haustra and the taeniae coli, and explain how they participate in the movements of the large intestine.
3. Describe the location of the rectum, anal canal, and anal sphincter muscles, and explain how defecation is accomplished.

Liver, Gallbladder, and Pancreas

The liver, consisting of four lobes, processes nutrients and secretes bile, which is stored and concentrated in the gallbladder prior to discharge into the duodenum. The pancreas, consisting of endocrine (islet) cells and exocrine (acini) cells, secretes important hormones into the blood and essential digestive enzymes into the duodenum.

Objective 18. Describe the location, structure, and functions of the liver.
Objective 19. Describe the location of the gallbladder, and discuss the flow of bile through the systems of ducts into the duodenum.
Objective 20. Describe the location, structure, and functions of the pancreas.

Three accessory digestive organs in the abdominal cavity aid in the chemical breakdown of food. These are the liver, gallbladder, and pancreas. The liver and pancreas function as exocrine glands in this process because their secretions are transported to the lumen of the GI tract via ducts.

Liver

The **liver** is the largest internal organ of the body, weighing about 1.3 kg (3.5–4.0 lbs) in an adult. It is positioned immediately beneath the diaphragm in the epigastric and right hypochondriac regions (see fig. 2.14 and table 2.3) of the abdominal cavity. The liver is reddish-brown in color because of its great vascularity.

The liver has two major lobes and two minor lobes. Anteriorly, the **right lobe** is separated from the smaller **left lobe** by the **falciform ligament** (fig. 18.34). Inferiorly, the **caudate** *(kaw'dāt)* **lobe** is near the inferior vena cava, and the **quadrate lobe** is adjacent to the gallbladder. The falciform ligament attaches the liver to the anterior abdominal wall and the diaphragm. A **ligamentum teres** (round ligament) is continuous along the free border of the falciform ligament to the umbilicus. The ligamentum teres is the remnant of the umbilical vein of the fetus. The **porta** of the liver is where the hepatic artery, portal vein, lymphatics, and nerves enter the liver and where the **hepatic ducts** exit.

Although the liver is the largest internal organ, it is, in a sense, only one to two cells thick. This is because the liver cells, or **hepatocytes,** form **plates** that are one to two cells thick and separated from each other by large capillary spaces called **sinusoids** (fig. 18.35). The sinusoids are lined with phagocytic **Kupffer cells,** but the large intercellular gaps between adjacent Kupffer cells make these sinusoids more highly permeable than other capillaries. The plate structure of the liver and the high permeability of the sinusoids allow each hepatocyte to have direct contact with the blood.

The hepatic plates are arranged into functional units called **liver lobules** (fig. 18.36). In the middle of each lobule is a **central vein,** and at the periphery of each lobule are branches of the hepatic portal vein and of the hepatic artery, which open into the spaces *between* hepatic plates. Arterial blood and portal venous blood, containing molecules absorbed in the GI tract, thus mix as the blood flows within the sinusoids from the periphery of the lobule to the central vein (fig. 18.37). The central veins of different liver lobules converge to form the hepatic vein, which carries blood from the liver to the inferior vena cava.

The liver lobules carry out numerous functions, including synthesis, storage, and release of vitamins; synthesis, storage, and release of glycogen; synthesis of blood proteins; phagocytosis of old red blood cells and certain bacteria; removal of toxic substances; and production of bile. Bile is stored in the gallbladder and is eventually secreted into the duodenum for the emulsification (breaking into smaller particles) and absorption of fats.

Bile is produced by the hepatocytes and secreted into thin channels called **bile canaliculi** *(kan"ah-lik'u-li)* located *within* each hepatic plate (fig. 18.37). These bile canaliculi are drained at the periphery of each lobule by **bile ducts,** which in turn drain into **hepatic ducts** that carry bile away from the liver. Since blood travels in the sinusoids and bile travels in the opposite direction within the hepatic plates, blood and bile do not mix in the liver lobules.

The liver receives parasympathetic innervation from the vagus nerves and sympathetic innervation from thoracolumbar nerves through the celiac ganglia.

Gallbladder

The **gallbladder** is a saclike organ attached to the inferior surface of the liver. This organ stores and concentrates bile, which drains to it from the liver by way of the bile ducts, hepatic duct, and **cystic duct,** respectively. A sphincter valve at the neck of

falciform: L. *falcis,* sickle; *forma,* form

porta: L. *porta,* gate
hepatic: Gk. *hepatos,* liver
cystic: Gk. *kystis,* pouch

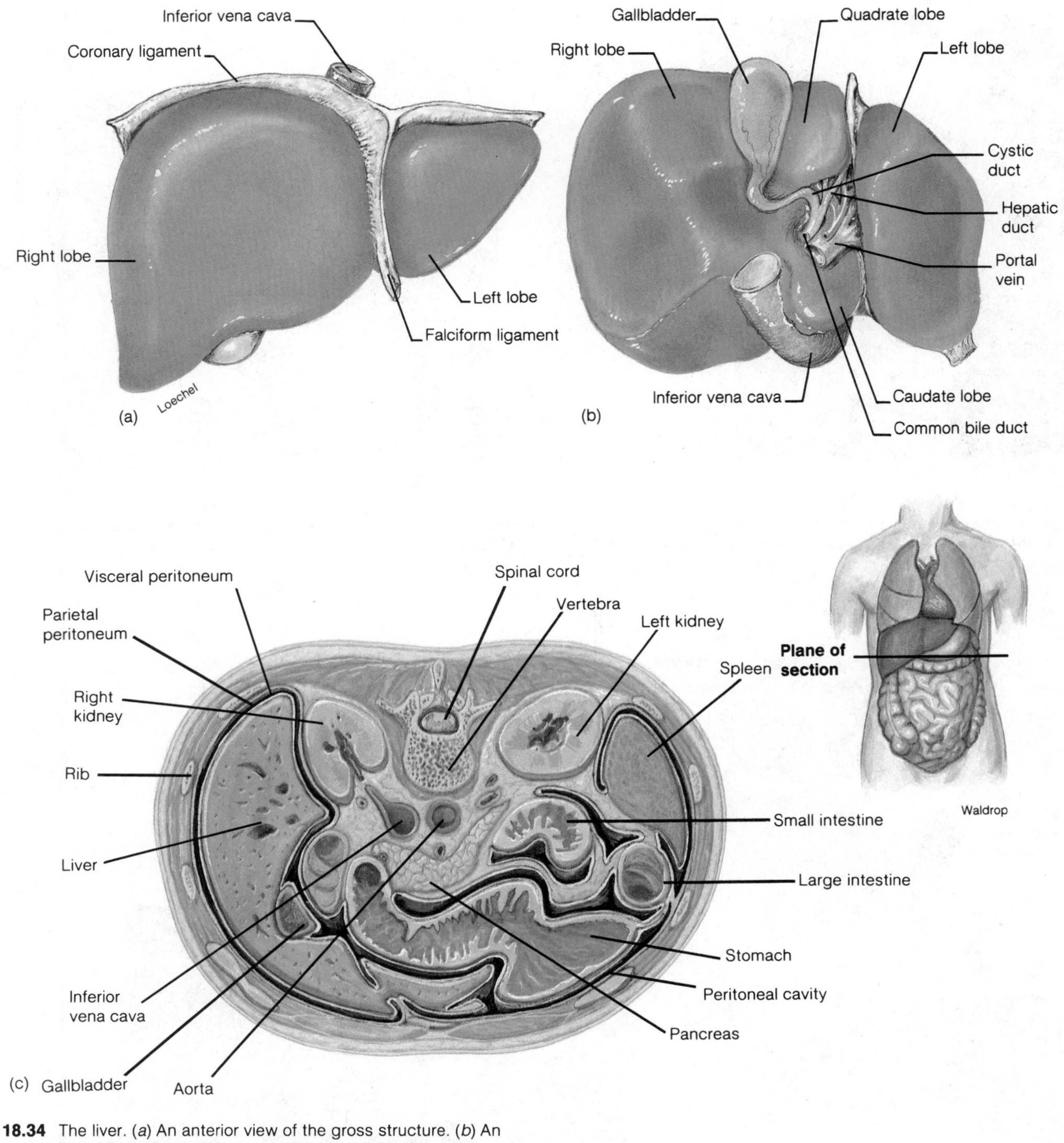

Figure 18.34 The liver. (*a*) An anterior view of the gross structure. (*b*) An inferior view showing the lobes of the liver, gallbladder, hepatic vessels, and ducts. (*c*) A transverse section of the abdomen showing the relative position of the liver to other abdominal organs.

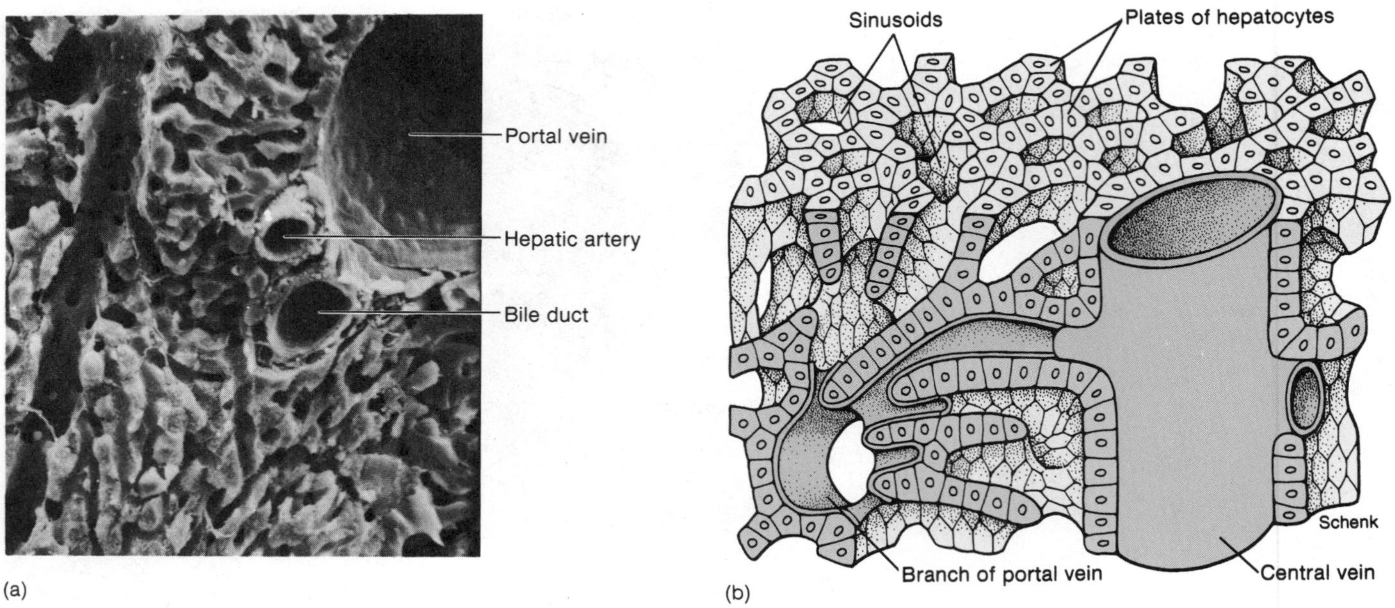

Figure 18.35 The structure of the liver. (*a*) A scanning electron micrograph of the liver. Hepatocytes are arranged in plates so that blood that passes through sinusoids (*b*) will be in contact with each liver cell.

(Photo from: *Tissues and Organs: A Text-Atlas of Scanning Electron Microscopy* by R. G. Kessel and R. H. Kardon. W. H. Freeman and Company. © 1979.)

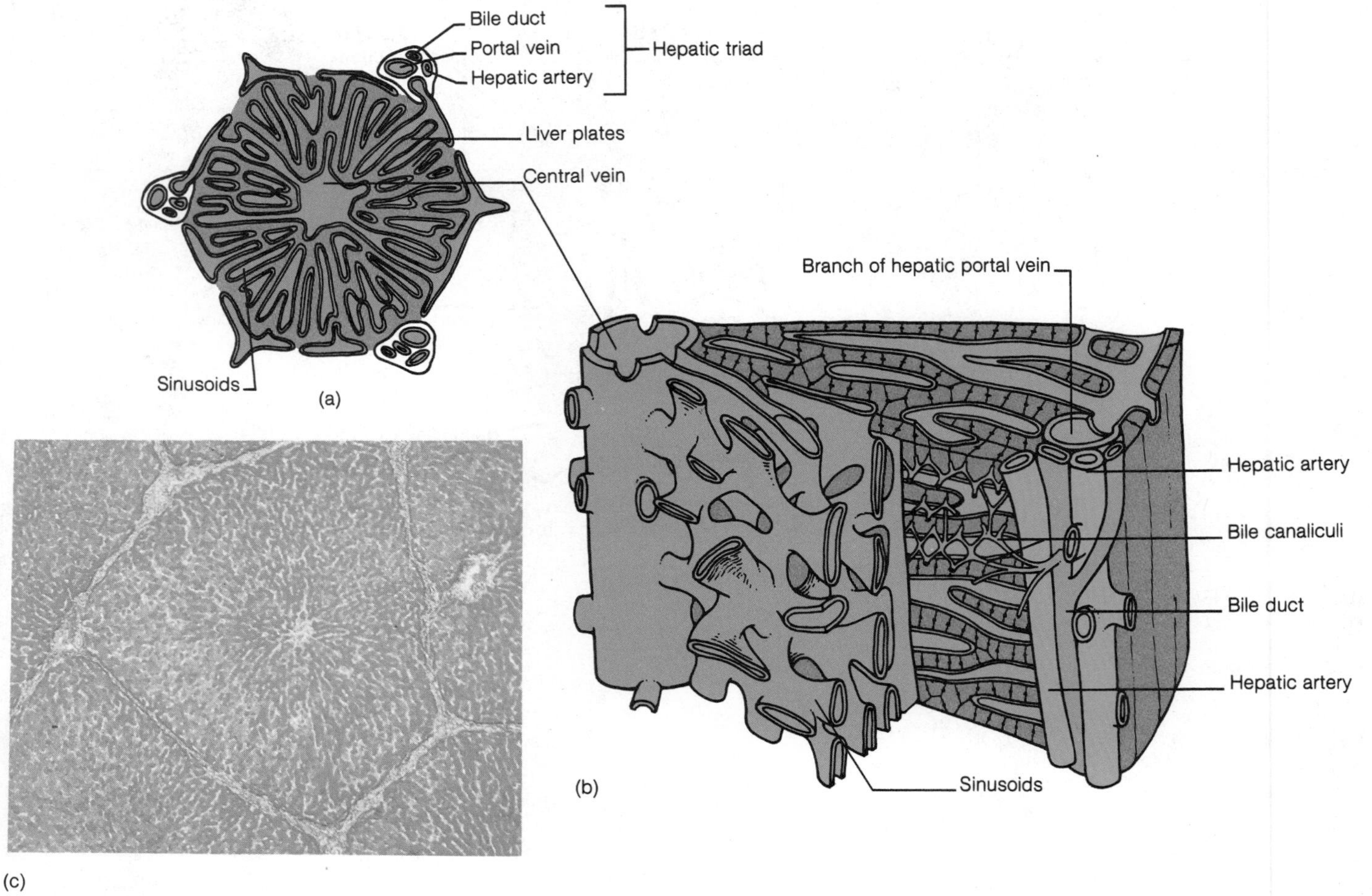

Figure 18.36 A liver lobule and the histology of the liver. (*a*) A liver lobule seen in cross section and (*b*) longitudinal section. Blood enters a liver lobule through the vessels in a hepatic triad, passes through hepatic sinusoids, and leaves the lobule through a central vein. The central veins converge to form hepatic veins that transport venous blood from the liver. (*c*) A photomicrograph of a liver lobule in transverse section.

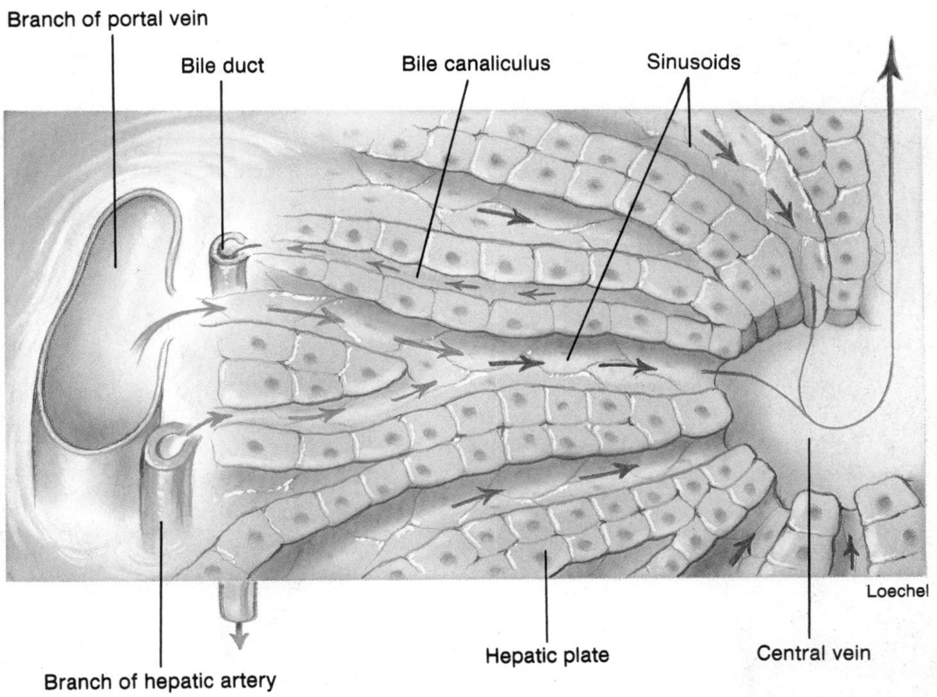

Figure 18.37 The flow of blood and bile in a liver lobule. Blood flows within sinusoids from a portal vein to the central vein (from the periphery to the center of a lobule). Bile flows within hepatic plates from the center to bile ducts at the periphery of a lobule.

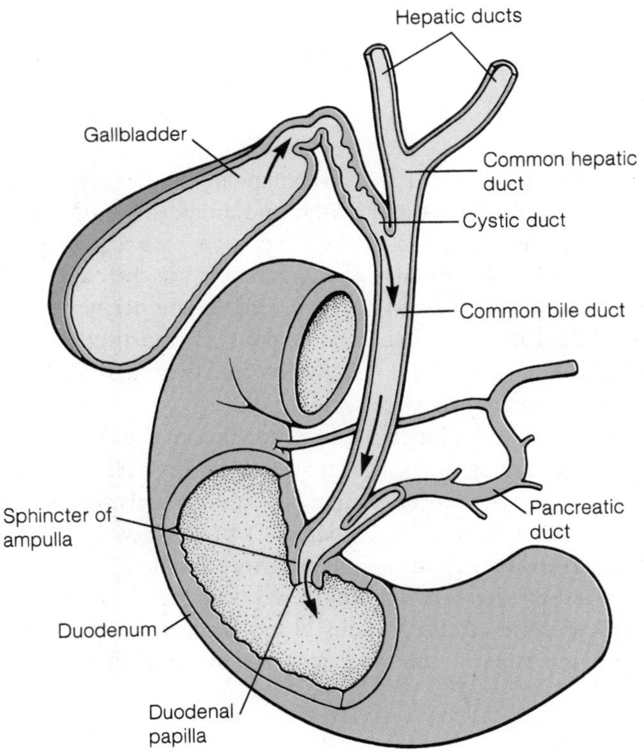

Figure 18.38 The pancreatic duct joins the common bile duct to empty its secretions through the duodenal papilla into the duodenum. The release of bile and pancreatic juice into the duodenum is controlled by the sphincter of ampulla (Oddi).

the gallbladder allows a storage capacity of about 35 to 50 ml. The inner mucosal layer of the gallbladder is arranged in rugae similar to those of the stomach. When the gallbladder fills with bile, it expands to the size and shape of a small pear. Bile is a yellowish-green fluid containing bile salts, bilirubin, (a product resulting from a breakdown of blood), cholesterol, and other compounds. Contraction of the muscularis ejects bile from the cystic duct into the **common bile duct,** which conveys bile into the duodenum (fig. 18.38).

Bile is continuously produced by the liver and drains through the hepatic and common bile ducts to the duodenum. When the small intestine is empty of food, the **sphincter of ampulla** (Oddi) constricts, and bile is forced up the cystic duct to the gallbladder for storage.

The gallbladder is supplied with blood from the cystic artery, which branches from the right hepatic artery. Venous blood is returned through the cystic vein, which empties into the hepatic portal vein. Autonomic innervation of the gallbladder is similar to the liver; both receive parasympathetic innervation from the vagus nerves and sympathetic innervation from thoracolumbar nerves through the celiac ganglia.

A common clinical problem of the gallbladder is the development of *gallstones.* Bile is composed of various salts, pigments, and cholesterols that become concentrated as water is removed. Normally cholesterols remain in solution, but under certain conditions they precipitate to form solid crystals. Large crystals may block the bile duct and have to be surgically removed. The X ray in figure 18.39 shows gallstones in position, and the photograph shows removed gallstones.

---

sphincter of Oddi: from Ruggero Oddi, Italian physician, 19th century

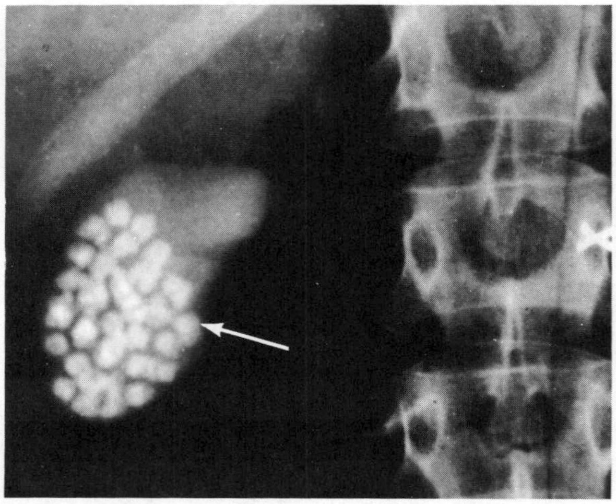
(a)

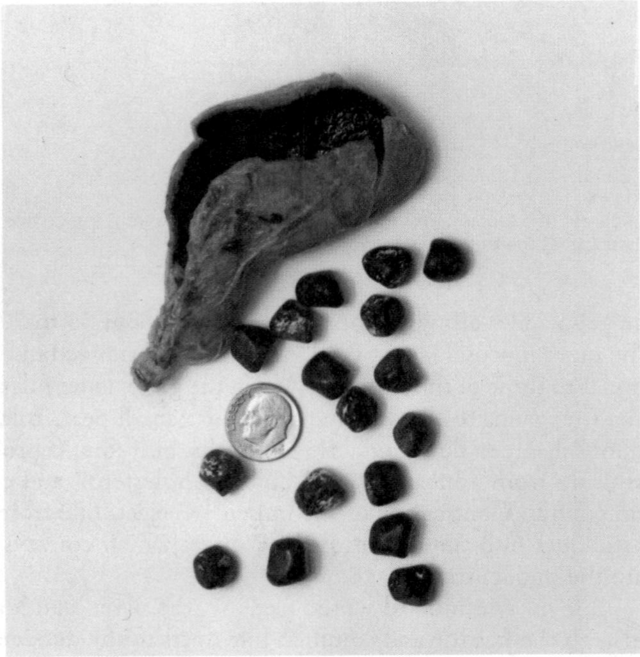
(b)

Figure 18.39 (*a*) An X ray of a gallbladder that contains gallstones. (*b*) A posterior view of a gallbladder that has been removed (cholecystectomy) and cut open to reveal its gallstones (biliary calculi). A dime is placed in the photo to show relative size.

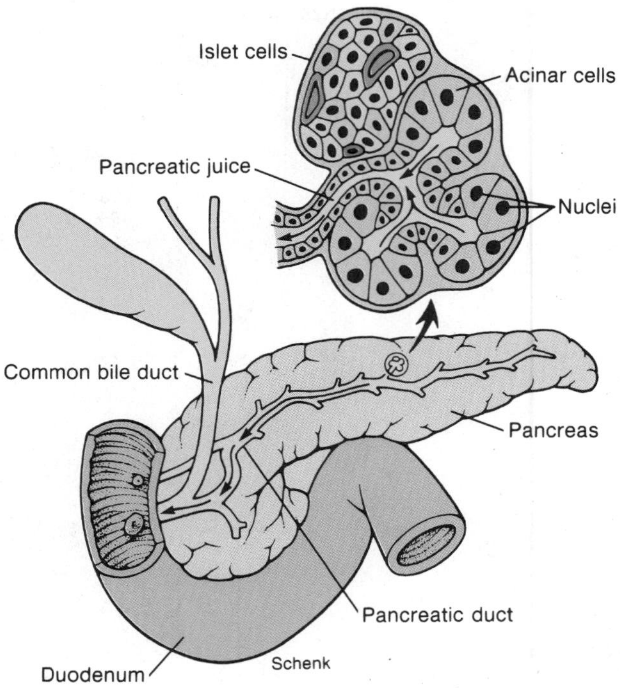

Figure 18.40 The pancreas is both an exocrine and an endocrine gland. Pancreatic juice—the exocrine product—is secreted by acinar cells into the pancreatic duct. Scattered islands of cells, called pancreatic islets (islets of Langerhans), secrete the hormones insulin and glucagon into the blood.

Pancreas

The **pancreas** is a soft, lobulated, glandular organ that has both exocrine and endocrine functions. The endocrine function is performed by clusters of cells, called the **pancreatic islets** (islets of Langerhans) that secrete the hormones insulin and glucagon into the blood. As an exocrine gland, the pancreas secretes pancreatic juice through the pancreatic duct (fig. 18.40) into the duodenum. The pancreas is horizontally positioned along the posterior abdominal wall, adjacent to the greater curvature of the stomach. The pancreas is about 12.5 cm (6 in.) long and 2.5 cm (1 in.) thick. It has an expanded **head** near the duodenum, a centrally located **body,** and a tapering **tail** near the spleen. All but a portion of the head is positioned retroperitoneally. Within the lobules of the pancreas are the exocrine secretory units, called **acini** *(as'ĭ-ni),* and the endocrine secretory units, called *islet cells.* Each acinus consists of a single layer of epithelial cells surrounding a lumen into which the constituents of pancreatic juice are secreted.

The pancreas is innervated by branches of the celiac plexus. The glandular portion of the pancreas receives parasympathetic innervation, whereas the pancreatic blood vessels receive sympathetic innervation. The pancreas is supplied with blood by the pancreatic branch of the splenic artery arising from the celiac artery and by the pancreatoduodenal branches from the superior mesenteric artery. Venous blood is returned through the splenic and superior mesenteric veins into the hepatic portal vein.

Pancreatic cancer has the worst prognosis of all types of cancer. This is probably because of the spongy, vascular nature of this organ and its vital exocrine and endocrine functions. Pancreatic surgery is a problem because the soft, spongy tissue is difficult to suture.

---

pancreas: Gk. *pan,* all; *kreas,* flesh
islets of Langerhans: from Paul Langerhans, German anatomist, 1847–88

acinus: L. *acinus,* grape

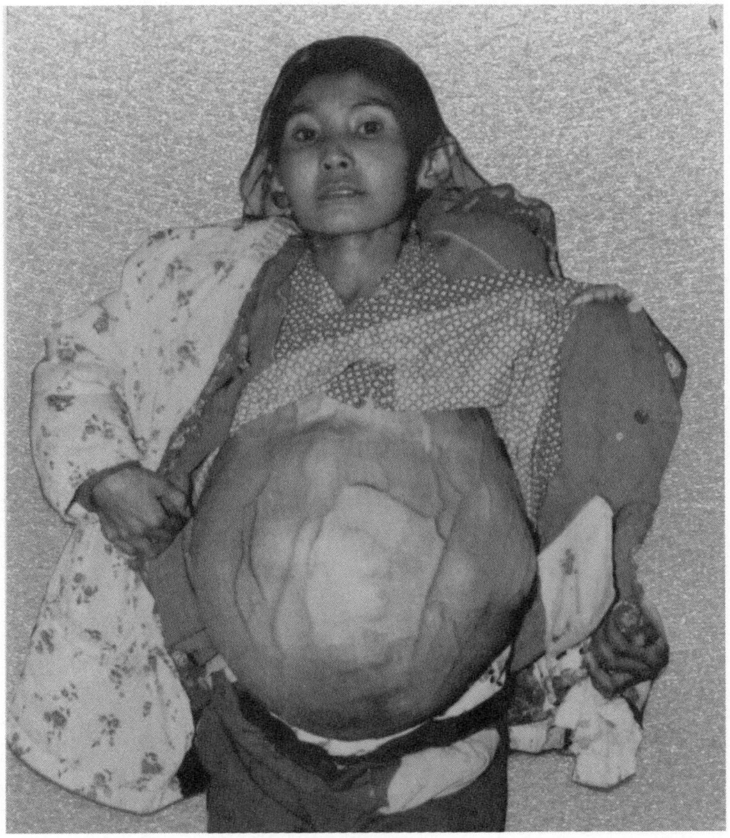

Figure 18.41 The gastrointestinal tract and the accessory digestive organs are susceptible to an array of pathogenic (disease causing) agents. This woman is afflicted with hyatid cysts in her liver caused from ingesting an egg from the tapeworm *Echinococcus multiocularis*. If untreated, conditions such as this are eventually fatal.

1. Describe the liver. Where is it located? List the lobes of the liver and the supporting ligaments.
2. List the principal functions of the liver.
3. Describe the structure of liver lobules, and trace the pathways of the flow of blood and bile in the lobules.
4. Explain the double blood supply to the liver.
5. Discuss how the gallbladder is filled with bile secretions, and the mechanism permitting bile and pancreatic secretion into the duodenum.
6. Identify the exocrine and endocrine functions of the pancreas. What are the various cellular secretory units of the pancreas?

Clinical Considerations

Developmental Disorders

Most of the congenital disorders of the digestive system have unknown causes and occur during the fourth or fifth week of embryonic development. A **cleft palate** is a congenital opening between the oral and nasal cavities and therefore involves both the digestive and respiratory systems. It occurs in about one in 1,000 live births and has immediate importance because it interferes with suckling and swallowing mechanisms. **Esophageal atresia,** or failure to develop the normal structure of the esophageal-stomach area, is another disorder of the upper digestive tract that requires surgery to correct. **Pyloric stenosis** is a common abnormality in which the pyloric sphincter muscle is hypertrophied, reducing the size of the lumen. This condition affects one in about 200 newborn males and one in 1,000 newborn females. Stenoses, atresias, and malrotations of various portions of the GI tract may occur as the gut develops. Umbilical problems involving the GI tract are fairly common, as is some form of *imperforate anus,* which occurs in about one in 5,000 births.

Pathogens and Poisons

The GI tract is a suitable environment for an array of parasitic helminths and microorganisms (fig. 18.41). Many of the microorganisms are beneficial, but some bacteria and protozoa can cause diseases. Only a few examples of the pathogenic microorganisms will be discussed.

Dysentery *(dis'en-ter''e)* is an inflammation of the intestinal mucosa, characterized by the discharge of loose stools that contain mucus, pus, and blood. The most common dysentery is **amoebic dysentery,** which is caused by the protozoan *Entamoeba histolytica.* Cysts from this organism are ingested in contaminated food, and after the protective coat is removed by HCl in the stomach, the vegetative form invades the mucosal walls of the ileum and colon.

Food poisoning comes from ingesting pathogenic bacteria or their toxins. *Salmonella* is a bacterium that commonly causes food poisoning. **Botulism** is the most serious type of food poisoning and is caused by ingesting food contaminated with the bacterium *Clostridium botulinum.* This organism is widely distributed in nature, and the spores from it are frequently on food being processed by canning. For this reason food must be heated to 120°C (248°F) before it is canned. It is the toxins produced by the bacterium growing in the food that are pathogenic rather than the organisms themselves. The poison is a neurotoxin that is readily absorbed in the blood where it affects the nervous system.

Clinical Problems of the Teeth and Salivary Glands

Dental caries, or tooth decay, is the gradual decalcification of tooth enamel (fig. 18.42) and underlying dentin, produced by the acid products of bacteria. These bacteria thrive between teeth where food particles accumulate and form part of the thin layer of bacteria, proteins, and other debris called *plaque* that covers teeth. The development of dental caries can thus be reduced by brushing at least once a day and by flossing between teeth at regular intervals.

People over the age of thirty-five are particularly susceptible to **periodontal diseases.** There are many types of periodontal disease, but basically they all cause inflammation and deterioration of the gingiva, alveolar sockets, periodontal membranes, and teeth cementum. Some of the symptoms are loosening of the teeth, bad breath, bleeding gums when

dysentery: Gk. *dys,* bad; *entera,* intestine

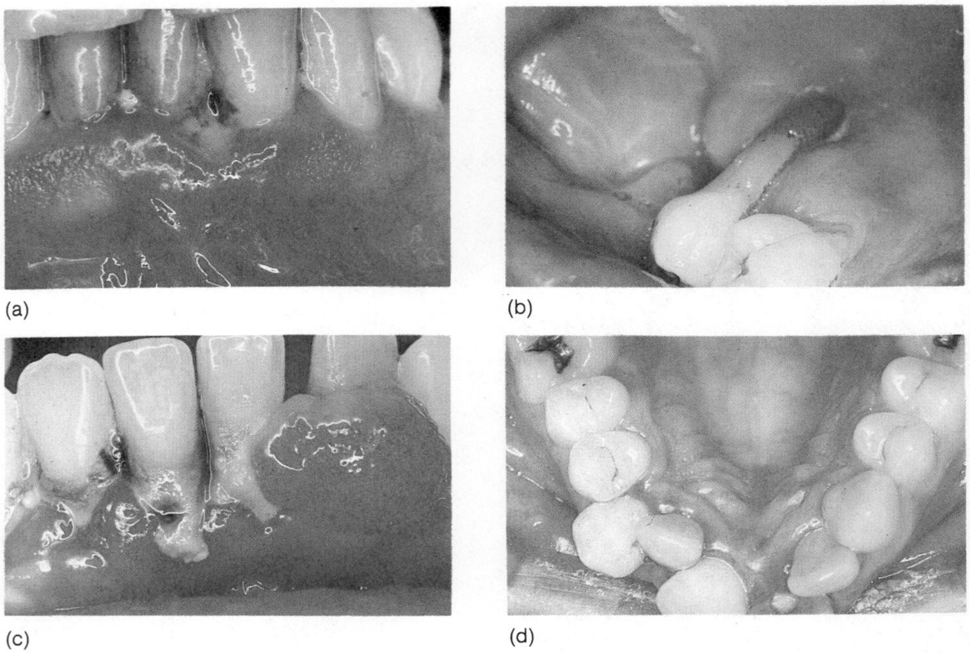

Figure 18.42 Clinical problems of the teeth. (*a*) Trench mouth and dental caries, (*b*) severe alveolar bone destruction from periodontitis, (*c*) pyogenic granuloma and dental caries, and (*d*) malposition of teeth.

brushing, and some edema. Periodontal diseases are caused by many things, including impacted plaque, cigarette smoking, malocclusion, and poor diet.

Mumps is a viral disease of the parotid salivary glands and in advanced stages may involve the pancreas and testes. It is generally not serious in children but can be very serious in adults, in whom it may cause deafness and destroy the pancreatic islet tissue or testicular cells.

Disorders of the Liver

The liver is a remarkable organ that has the ability to regenerate itself even if up to 80% has been removed. The most serious diseases of the liver (hepatitis, cirrhosis, and hepatomas) affect the liver throughout so that it cannot repair itself. **Hepatitis** is inflammation of the liver. Certain chemicals may cause hepatitis, but generally it is caused by infectious viral agents. **Infectious hepatitis** is a viral disease transmitted through contaminated foods and liquids. **Serum hepatitis** is also caused by a virus and is transmitted in serum or plasma during transfusions or by improperly sterilized needles and syringes.

In **cirrhosis,** large numbers of liver lobules are destroyed and replaced with permanent connective tissue and regenerative nodules of hepatocytes. These nodules do not have the platelike structure of normal liver tissue and are therefore less functional. Cirrhosis may be caused by chronic alcohol abuse, viral hepatitis, and other agents that attack liver cells.

Jaundice is a yellow staining of the tissue produced by high blood concentrations of either free or conjugated bilirubin. Since free bilirubin is derived from heme, abnormally high concentrations of this pigment may result from an unusually high rate of red blood cell destruction. This can occur, for example, as a result of Rh disease (erythroblastosis fetalis) in an Rh positive baby born to a sensitized Rh negative mother. Jaundice may also occur in otherwise healthy infants because of the fact that red blood cells are normally destroyed at about the time of birth (hemoglobin concentrations decrease from 19 g per 100 ml to 14 g per 100 ml near the time of birth). This condition is called *physiological jaundice of the newborn* and is not indicative of disease. Premature infants may also develop jaundice due to inadequate amounts of hepatic enzymes necessary to conjugate bilirubin and excrete it in the bile. Jaundice due to high levels of conjugated bilirubin in the blood is commonly produced in adults when the excretion of bile is blocked by gallstones.

Hepatomas *(hep''ah-to'mahz)* are malignant tumors that originate in or secondarily invade the liver. Those that originate in the liver (primary hepatomas) are relatively rare, but those that secondarily metastasize to the liver from other organs (secondary hepatomas) are common. Carcinoma of the liver is usually fatal.

Disorders of the GI Tract

Peptic ulcers are erosions of the mucous membranes of the stomach or duodenum produced by the action of HCl. Agents that weaken the mucosal lining of the stomach, such as alcohol and aspirin, and abnormally high secretions of HCl thus increase the likelihood of developing peptic ulcers. Many people subject to chronic stress produce too much gastric acid and develop a peptic ulcer as a result (fig. 18.43).

---

cirrhosis: Gk. *kirrhos,* yellow-orange
jaundice: L. *galbus,* yellow

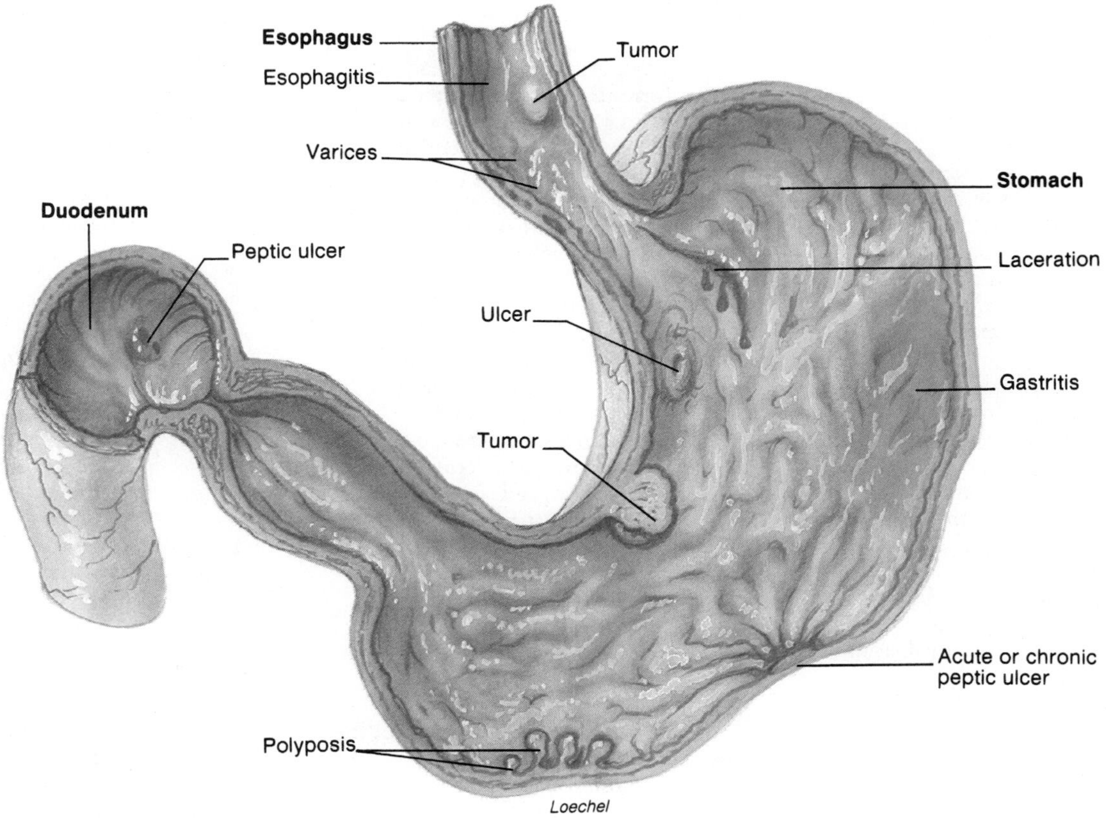

Figure 18.43 Common sites of upper GI disorders.

Enteritis is inflammation of the intestinal mucosa and is frequently referred to as intestinal flu. Causes of enteritis include bacterial or viral infections, irritating foods or fluids (including alcohol), and emotional stress. The symptoms are abdominal pain, nausea, and diarrhea. **Diarrhea** is the passage of watery, unformed stools. This condition is symptomatic of inflammation, stress, and other body dysfunctions.

A **hernia** is a protrusion of a portion of a visceral organ, usually the small intestine, through a weakened portion of the abdominal wall. Inguinal, femoral, umbilical, and hiatal hernias are the most common. A **hiatal hernia** is when a portion of the stomach pushes superiorly through the esophageal hiatus in the diaphragm and into the thorax. The potential dangers of a hernia are strangulation of the blood supply followed by gangrene, blockage of chyme, or rupture—each of which can threaten life.

Diverticulosis is a condition where the intestinal wall weakens and an outpouching occurs. Studies suggest that suppressing the passage of flatus (intestinal gas) may contribute to diverticulosis, especially in the sigmoid colon. **Diverticulitis,** or inflammation of a diverticulum, can develop if fecal material becomes impacted in these pockets.

Peritonitis is inflammation of the peritoneum lining the abdominal cavity and covering the viscera. The causes of peritonitis include bacterial contamination of the abdominal cavity through accidental or surgical wounds in the abdominal wall or perforation of the intestinal wall (as with ulcers or a ruptured appendix).

CLINICAL CASE STUDY ANSWER

The likely source of bleeding in such a case is a rupture of the internal portion of the liver involving significant blood vessels, either of the hepatic arterial or venous circulation, but also possibly the portal venous circulation. Nearby lacerated bile vessels then receive blood into the biliary tree. From there, the course of the blood is as follows: large hepatic ducts to common hepatic duct, to common bile duct through duodenal papilla and into the duodenum of the small intestine.

This type of bleeding in some instances may be stopped with the use of x-ray-guided catheters placed into the arterial system (angiographic embolization). However, sometimes surgery is required.

Important Clinical Terminology

chilitis Inflammation of the lips.

colitis Inflammation of the colon.

colostomy The formation of an abdominal exit from the GI tract by bringing a loop of the colon to the surface of the abdomen. If the rectum is removed because of cancer, the colostomy provides a permanent outlet for the feces.

cystic fibrosis An inherited disease of the exocrine glands, particularly the pancreas. Pancreatic secretions are too thick to drain easily, causing the ducts to become inflamed and promoting connective tissue formation that occludes drainage from the ducts still further.

gingivitis Inflammation of the gums. It may result from improper hygiene, poorly fitted dentures, improper diet, or certain infections.

halitosis Offensive breath odor. It may result from dental caries, certain diseases, or eating particular foods.

heartburn A burning sensation of the esophagus and stomach. It may result from the regurgitation of gastric juice into the lower portion of the esophagus.

hemorrhoids Varicose veins of the rectum and anus.

nausea Gastric discomfort and sensations of illness with a tendency to vomit. This feeling is symptomatic of many conditions (e.g., motion sickness, foul odors or sights, pregnancy, diseases, etc.).

pyorrhea The discharge of pus at the base of the teeth at the gum line.

regurgitation (vomiting) The forceful expulsion of gastric contents into the mouth. Nausea and vomiting are common symptoms of almost any dysfunction of the digestive system.

trench mouth A contagious bacterial infection that causes inflammation, ulceration, and painful swelling of the floor of the mouth (fig. 18.42). Generally it is contracted through direct contact by kissing an infected person. Trench mouth can be treated by penicillin and other medications.

vagotomy The surgical removal of a section of the vagus nerve where it enters the stomach to eliminate nerve impulses that stimulate gastric secretion. This procedure is used to treat ulcers.

Chapter Summary

I. Introduction to the Digestive System
   A. The digestive system prepares ingested food for cellular utilization through a series of mechanical and chemical processes that reduce food to forms that can be absorbed through the intestinal wall and transported by the blood and lymph.
   B. The digestive system consists of a gastrointestinal tract and accessory digestive organs.

II. Embryological Development of the Digestive System
   A. The digestive tract is embryologically derived from the primitive gut, which is composed of endoderm.
   B. The structures of the mouth are derived from ectoderm of the stomodeum, and the epithelium of the anal canal is derived from ectoderm of the proctodeum; these structures become continuous with the gut endoderm.
   C. The liver and pancreas develop from buds that arise from the wall of the small intestine.

III. Serous Membranes and Tunics of the Gastrointestinal Tract
   A. Peritoneal membranes line the abdominal wall and cover the visceral organs; the GI tract is supported by a double layer of peritoneum called the mesentery.
      1. The lesser omentum and greater omentum are folds of peritoneum that extend from the stomach.
      2. Visceral organs that are not in the peritoneal cavity and not covered with peritoneum are retroperitoneal.
   B. The layers (tunics) of the abdominal GI tract are, from the inside outward, mucosa, submucosa, muscularis, and serosa.
      1. The mucosa consists of a simple columnar epithelium, a thin layer of connective tissue called the lamina propria, and a thin layer of smooth muscle called the muscularis mucosa.
      2. The submucosa is composed of connective tissue, the muscularis consists of layers of smooth muscles, and the serosa is connective tissue covered with the visceral peritoneum.
      3. The submucosa contains the submucosal plexus, and the muscularis contains the myenteric plexus of autonomic nerves.

IV. Mouth, Pharynx, and Associated Structures
   A. The oral cavity is formed by the cheeks, lips, hard and soft palates, and tongue.
      1. The tongue contains lingual tonsils and papillae with taste buds.
      2. The palate contains rugae, a cone-shaped projection called the uvula, and palatine tonsils.
   B. The four most anterior pairs of teeth are the incisors and canines, which have one root each; the bicuspids and molars have two or three roots.
      1. Humans are diphyodont; they have deciduous and permanent sets of teeth.
      2. The tooth is anchored to its bony alveolar socket by a periodontal membrane, which contains fibers that insert into the cementum of the tooth root.
      3. Enamel forms the outer layer of the tooth crown; beneath the enamel is dentin.
      4. The interior of a tooth contains a pulp cavity, which is continuous through the apical foramen of the root with the connective tissue around the tooth.
   C. The major salivary glands are the parotid glands and the submandibular glands.
   D. The muscular pharynx is a passageway connecting the oral and nasal cavities to the esophagus and trachea.

V. Esophagus and Stomach
   A. Peristaltic waves of contraction push food through the lower esophageal sphincter into the stomach.
   B. Swallowing, or deglutition, occurs in three phases, and involves structures of the buccal cavity, pharynx, and esophagus.
   C. The stomach consists of a cardia, fundus, body, and pylorus, which ends with the pyloric sphincter.
      1. The lining of the stomach is characterized by distensible folds, or rugae, and the mucosa is formed into gastric pits and gastric glands.
      2. The parietal cells of the gastric glands secrete HCl, and the chief cells secrete pepsinogen.

VI. Small Intestine
   A. Regions of the small intestine include the duodenum, jejunum, and ileum; the common bile duct and pancreatic duct empty into the duodenum.
   B. Fingerlike extensions of mucosa, called villi, project into the lumen, and at the bases of the villi the mucosa forms intestinal glands.
      1. New epithelial cells are formed in the crypts.
      2. The membrane of intestinal epithelial cells is folded to form microvilli called the brush border of the mucosa, which increases the absorptive surface area.
   C. Movements of the small intestine include rhythmic segmentation, pendular movement, and peristalsis.

VII. Large Intestine
   A. The large intestine absorbs water and electrolytes from the chyme and passes waste products out of the body through the rectum and anal canal.
   B. The large intestine is divided into the cecum, colon, rectum, and anal canal.
      1. The appendix is attached to the inferior medial margin of the cecum.
      2. The colon consists of ascending, transverse, descending, and sigmoid portions.
      3. Haustra are bulges in the walls of the large intestine.
   C. Movements of the large intestine include peristalsis, haustral churning, and mass movement.

VIII. Liver, Gallbladder, and Pancreas
   A. The liver is divided into four lobes, and each lobe contains hepatic lobules, the functional units of the liver.
      1. Liver lobules consist of plates of hepatic cells separated by capillary sinusoids.
      2. Blood flows from the periphery of each lobule, where the hepatic artery and portal vein empty, through the sinusoids, and out the central vein.
      3. Bile flows within the hepatocyte plates, in canaliculi, to the bile ducts at the periphery of each lobule.
   B. The gallbladder serves to store and concentrate the bile, and it releases bile through the cystic duct and common bile duct to the duodenum.
   C. The pancreas is both an exocrine and an endocrine gland.
      1. The endocrine portion is known as the pancreatic islets, and secretes the hormones insulin and glucagon.
      2. The exocrine acini of the pancreas produce pancreatic juice, which contains various digestive enzymes.

Review Activities

Objective Questions

1. The epithelium of the mouth is derived from the
   (a) ectoderm of the stomodeum.
   (b) ectoderm of the proctodeum.
   (c) endoderm of the foregut.
   (d) All of the above.
2. Which of the following types of teeth are found in the permanent but not in the deciduous dentition?
   (a) incisors (c) premolars
   (b) canines (d) molars
3. A double layer of peritoneum that supports the GI tract is called the
   (a) visceral peritoneum.
   (b) mesentery.
   (c) greater omentum.
   (d) lesser omentum.
4. Which of the following tissue layers in the small intestine contains the lacteals?
   (a) submucosa
   (b) muscularis mucosa
   (c) lamina propria
   (d) muscularis
5. Which of the following organs is *not* considered a part of the digestive system?
   (a) pancreas (c) tongue
   (b) spleen (d) gallbladder
6. The numerous small elevations on the dorsal surface of the tongue that support taste buds and aid in handling food are called
   (a) cilia. (c) villi.
   (b) rugae. (d) papillae.
7. Most digestion occurs in the
   (a) mouth.
   (b) stomach.
   (c) small intestine.
   (d) large intestine.
8. Stenosis (constriction) of the sphincter of ampulla (Oddi) would interfere with
   (a) transport of bile and pancreatic juice.
   (b) secretion of mucus.
   (c) passage of chyme into the small intestine.
   (d) peristalsis.
9. The first organ to receive the blood-borne products of digestion is the
   (a) liver. (c) heart.
   (b) pancreas. (d) brain.
10. Which of the following statements about hepatic portal blood is true?
   (a) It contains absorbed fat.
   (b) It contains ingested proteins.
   (c) It is mixed with bile in the liver.
   (d) It is mixed with blood from the hepatic artery in the liver.

Essay Questions

1. Define *digestion*. Differentiate between the mechanical and chemical aspects of digestion.
2. Differentiate between the following terms: *gastrointestinal tract, viscera, accessory digestive organs, alimentary canal,* and *gut.*
3. List the specific portions or structures of the digestive system formed by each of the three embryonic germ layers.
4. Define *serous membrane.* How are the serous membranes of the abdominal cavity classified and what are their functions?
5. Describe the structures of the four tunics in the wall of the GI tract.
6. Why are there two autonomic innervations to the GI tract? Identify the specific sites of autonomic stimulation in the tunic layers.
7. Define the following: *dental formula, diphyodont, deciduous teeth, permanent teeth, and wisdom teeth.*
8. List the stages of deglutition. What biomechanical roles do the tongue, hard and soft palates, pharynx, and hyoid bone perform in deglutition?
9. What are the specializations of the mucosa of the stomach that permit distension and secretion?
10. Describe the structural modifications of the small intestine that increase the surface area for absorption.
11. Diagram a villus and explain why villi are considered the functional units of the digestive system.
12. What are the regions of the large intestine? In what portion of the abdominal cavity and pelvic cavity is each region located?
13. Describe the location and the gross structure of the liver. What structures are associated with the porta of the liver?
14. Discuss how the gallbladder is filled with and emptied of bile fluid. What is the function of bile?
15. List the functions of the large intestine. What are the biomechanical movements of the large intestine that make these functions possible?
16. Define *cirrhosis* and explain why it is so devastating to the liver. What are some of the causes of cirrhosis?

CHAPTER 19

Urinary System

Outline and Concepts

CLINICAL CASE STUDY

A seventeen-year-old male is involved in a knife fight, in which he sustains two stab wounds to the anterior abdomen. He is brought to the emergency room complaining of mild abdominal pain and an urgent need to urinate. Although neither of the stab wounds is externally bleeding, examination by the surgeon reveals signs of moderate hemorrhagic shock. One wound is 3 cm below the right costal margin at the midclavicular line, and the other is just medial to the right anterior superior iliac spine. The surgeon, without delay, orders preparations for emergency exploratory laparotomy. She parenthetically adds that the patient's urinary bladder is quite full and chides the intern for not having placed a urinary catheter. Placement of the catheter yields a brisk flow of bright red blood.

How would you explain the phenomenon of hematuria (blood in the urine) in this case? Which of the two stab wounds would most likely be associated with the hematuria? Regarding the blood draining into the catheter, trace and explain its course of flow. Begin at a point in the abdominal aorta, ending with its arrival into the catheter. Supposing the surgeon is prompted to remove the kidney in order to quickly control hemorrhage, what possible anatomical variant should she keep in mind? ■

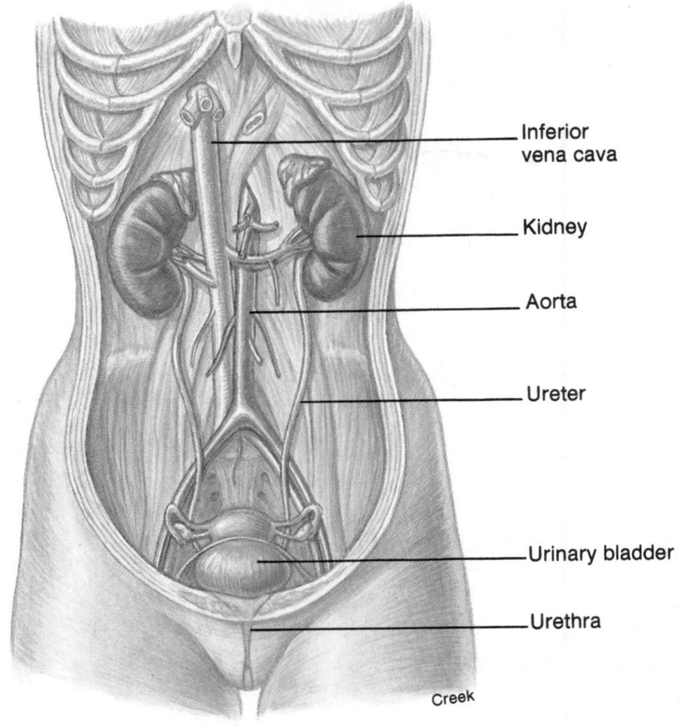

Figure 19.1 The anatomical locations of the kidneys, ureters, and urinary bladder.

Introduction to the Urinary System

The urinary system maintains the composition and properties of the body fluid, which forms the internal environment of the body cells. The end product of the urinary system is urine, which is voided from the body during micturition.

Objective 1. List the functions of the urinary system.
Objective 2. Identify the arteries that transport blood to the urinary system for filtration.

The urinary system, along with the respiratory, digestive, and integumentary systems, excretes substances from the body. For this reason, these systems are occasionally referred to as excretory systems. In the process of cellular metabolism, nutrients taken in by the digestive system and oxygen from inhaled air are utilized to synthesize a variety of substances and provide the energy needed for body maintenance. Metabolic processes, however, produce cellular wastes that must be eliminated if homeostasis is to be maintained. Just as the essential elements are transported to the cells by the blood, the cellular wastes are removed through the circulatory system to the appropriate excretory system. Carbon dioxide is eliminated through the respiratory system; excessive water, salts, nitrogenous wastes, and even excessive metabolic heat are removed through the integumentary system; and various bile digestive wastes are eliminated through the digestive system.

The urinary system is the principal system responsible for water-and-electrolyte balance. Electrolytes are compounds that separate into ions when dissolved in water, and a balance is achieved when the number of electrolytes entering the body equals the number leaving. Hydrogen ions are one type that is maintained in precise concentration so that an acid-base, or pH, balance exists in the body.

A second major function of the urinary system is the excretion of toxic metabolic products that are nitrogenous compounds—specifically urea and creatinine. Other functions of the urinary system include the elimination of toxic wastes that may result from bacterial action and the removal of various drugs that have been taken into the body. The end product of all of these processes is **urine.**

The urinary system consists of two kidneys, two ureters, the urinary bladder, and the urethra (fig. 19.1). Tubules in the kidneys are intertwined with vascular networks of the circulatory system to make possible the production of urine. After the urine is formed, it is moved through the ureters to the urinary bladder for storage. **Micturition** *(mik''tu-rish'un),* or voiding of urine from the urinary bladder, occurs through the urethra.

Blood that is to be processed by a kidney enters through the large **renal artery** and after the filtration process (review filtration in chap. 3) exits through the **renal vein.** The importance of filtration of blood is shown by the fact that during normal resting conditions the kidneys receive approximately 20%–25% of the entire cardiac output. Every minute the kidneys process approximately 1,200 ml of blood.

1. Knowing the functions of the urinary system, list the basic substances that compose normal urine.
2. Explain the role of the renal vessels in maintaining homeostasis. What percent of the body blood is filtered in the kidneys each minute during normal resting conditions?

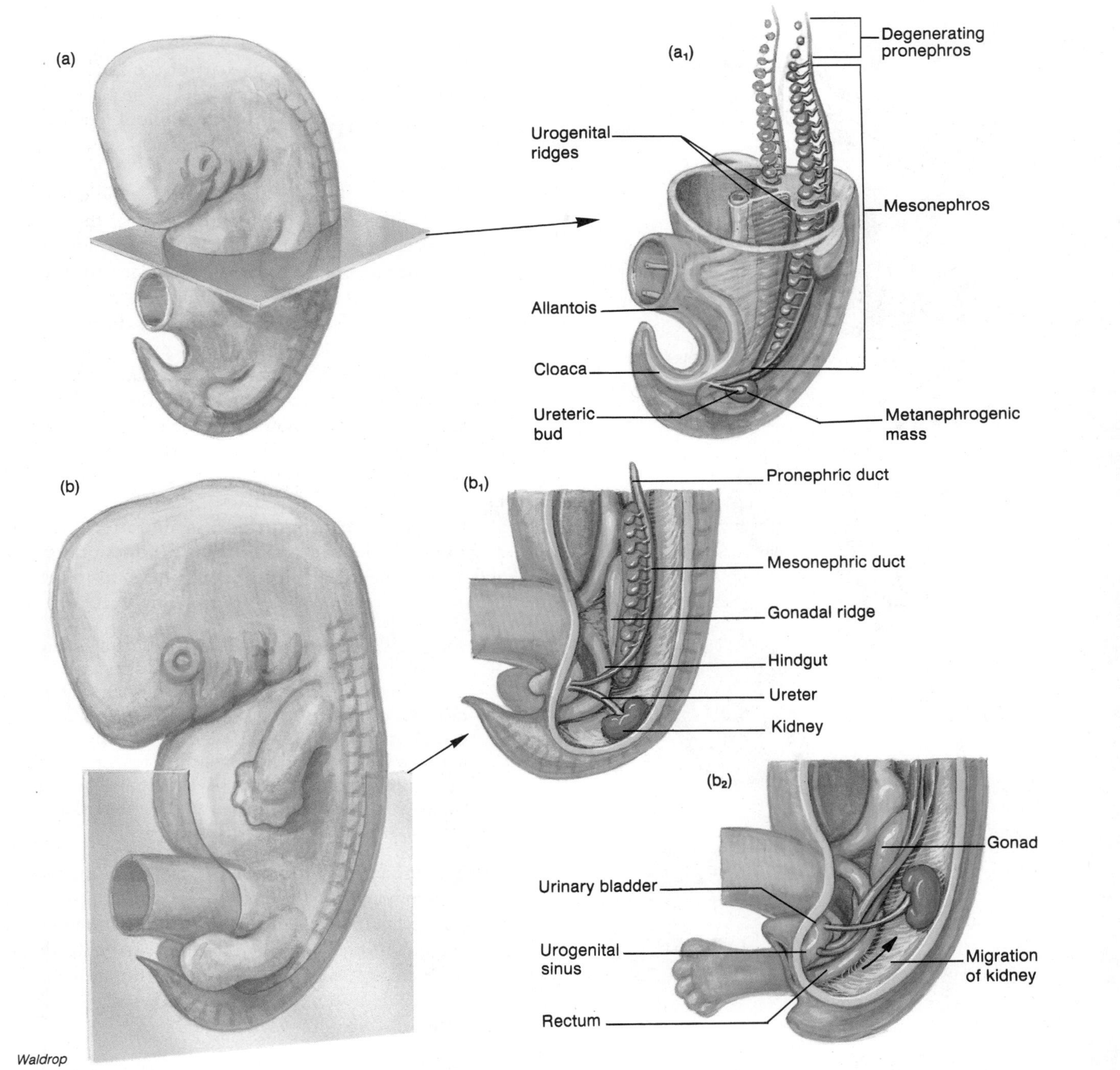

Figure 19.2 The embryonic development of the kidney. (*a*) An anterolateral view of an embryo at five weeks showing the position of a transverse cut depicted in (a_1). During the sixth week (*b* and b_1), the kidney is forming in the pelvic region and the mesonephric duct and gonadal ridge are prominent. The kidney begins migrating during the seventh week (b_2), and the urinary bladder and gonad are formed.

Development of the Urinary System

Humans have metanephric kidneys, which become functional after pronephric and mesonephric kidneys degenerate during embryonic development.

Objective 3. Describe the mesodermal site of origin of the urinary system.

Objective 4. Describe development of the pronephric, mesonephric, and metanephric kidneys.

The urinary and reproductive systems originate from a specialized elevation of mesodermal tissue called the **urogenital ridge.** The two systems share common structures during a portion of development, but by the time of birth two separate systems have formed. The separation in the male is not totally complete, however, since the urethra serves to transport both urine and semen. The development of both systems is initiated during the embryonic stage, but the development of the urinary system starts and finishes sooner than the reproductive system.

Three successive types of kidneys develop in the human embryo: the pronephros, mesonephros, and metanephros (fig. 19.2). The third type, or metanephric kidney, remains as the permanent kidney.

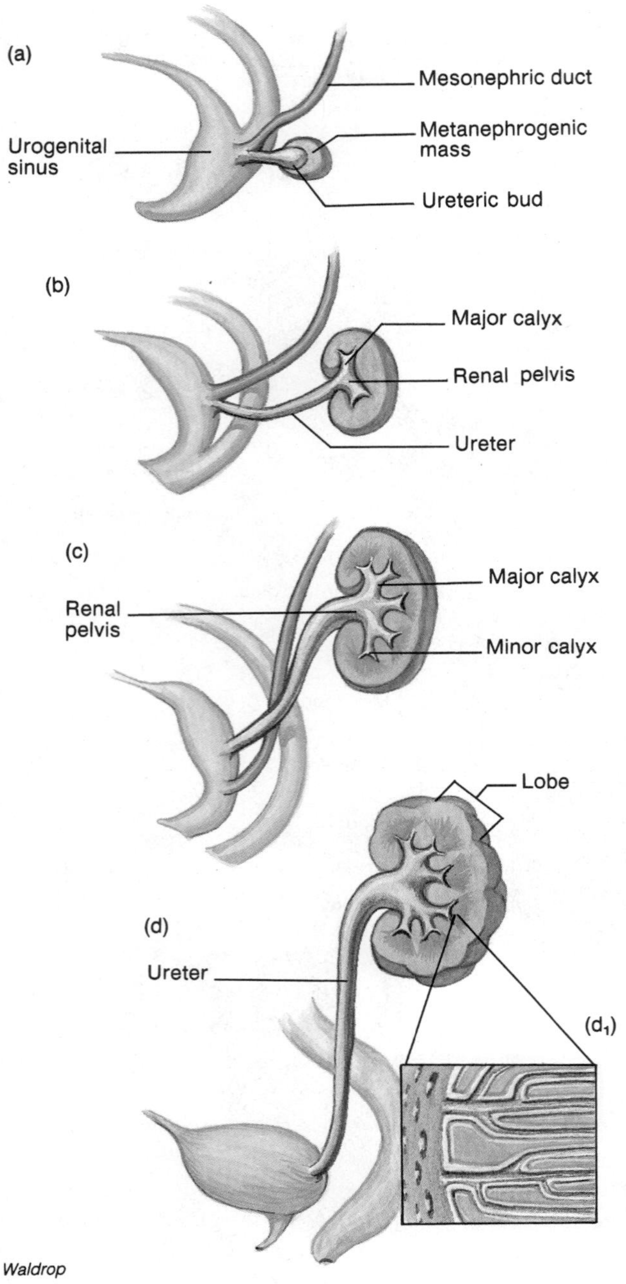

Figure 19.3 The development of the metanephric kidney at *(a)* four weeks, *(b)* five weeks, *(c)* seven weeks, *(d)* birth. *(d_1)* Magnified view of the arrangement of collecting ducts within a papilla.

The **pronephros** *(pro-nef'ros)* develops during the fourth week after conception and persists only through the sixth week. It is the most superior in position on the urogenital ridge of the three kidneys and is connected to the embryonic **cloaca** by the **pronephric duct.** Although the pronephros is nonfunctional and degenerates in humans, most of its duct is used by the mesonephric kidney (fig. 19.2), and a portion of it is important in the formation of the metanephros.

The **mesonephros** *(mes''o-nef'ros)* develops toward the end of the fourth week as the pronephros becomes vestigial. The mesonephros forms from an intermediate portion of the urogenital ridge and functions throughout the embryonic period of development.

Although the **metanephros** *(met''ah-nef'ros)* begins its formation during the fifth week, it does not become functional until immediately before the start of the fetal stage of development at the end of the eighth week. The paired metanephric kidneys continue to form urine throughout fetal development. The urine is expelled through the urinary system into the amniotic fluid.

The metanephros develops from two mesodermal sources (fig. 19.3). The meaty (glomerular) part of the kidney forms from a specialized caudal portion of the urogenital ridge called the **metanephrogenic mass.** The tubular drainage portion of the kidneys forms as a diverticulum emerging from the wall of the mesonephric duct near the cloaca. This diverticulum, known as the **ureteric** *(u''rĕ-ter'ik)* **bud,** expands into the metanephrogenic mass to form the drainage pathway for urine. The stalk of the ureteric bud develops into the ureter, whereas the expanded terminal portion forms the renal pelvis, calyces, and collecting tubules. A combination of the ureteric bud and metanephrogenic mass forms the other tubular channels within the kidney.

Once the metanephric kidneys are formed, they begin to migrate from the pelvis to the upper, posterior portion of the abdomen. The renal blood supply develops as the kidneys become positioned in the posterior body wall.

The development of the kidneys illustrates the concept that ontogeny (embryonic development) recapitulates phylogeny (evolution). This means that the development of the three kidney types follows an evolutionary pattern. The larvae of a few of the more primitive vertebrates have functional pronephric kidneys. Adult fishes and amphibians have mesonephric kidneys, whereas adult reptiles, birds, and mammals have metanephric kidneys.

The urinary bladder develops from the urogenital sinus, which is connected to the embryonic umbilical cord by the fetal membrane called the allantois (fig. 19.2). By the twelfth week, the two ureters are emptying into the urinary bladder, the urethra is draining, and the connection of the urinary bladder to the allantois has been reduced to a supporting structure called the **urachus.**

Occasionally a *patent urachus* is present in a newborn and is discovered when urine is passed through the umbilicus, especially if there is a urethral obstruction. Usually, however, it remains undiscovered until old age, when an enlarged prostate obstructs the urethra and forces urine through the patent urachus and out the umbilicus (navel).

1. Discuss the basic developmental similarities between the urinary and the reproductive systems.
2. Which type of kidney is functional during the embryonic stage of development? When do the metanephric kidneys form and become functional?
3. Explain how kidney development illustrates the concept that ontogeny recapitulates phylogeny.

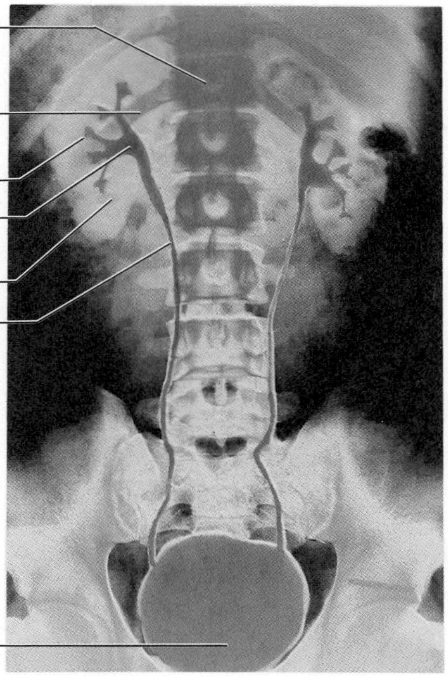

Figure 19.4 Color-enhanced X ray of the calyces and renal pelvises of the kidneys, the ureters, and the urinary bladder. Notice the position of the kidneys relative to the vertebral column and ribs.

Kidneys

The kidney consists of an outer, vascular cortex, and an inner medulla containing the renal pyramids. Urine is formed as a filtrate from the blood at the nephrons and collects in the calyces and renal pelvis before flowing from the kidney via the ureter.

Objective 5. Describe the gross structure of the kidney.
Objective 6. Describe the tubular and vascular components of a nephron.
Objective 7. Locate the position of cortical and juxtamedullary nephrons in the gross structure of the kidney.

Position and Appearance

The reddish-brown **kidneys** lie on each side of the vertebral column, high in the abdominal cavity, from the level of the twelfth thoracic to the third lumbar vertebrae (fig. 19.4). The right kidney is usually 1.5–2.0 cm lower than the left because of the large area occupied by the liver.

The kidneys are retroperitoneal, which means that they are positioned behind the peritoneum (fig. 19.5) and are thus not, in a strict sense, within the peritoneal cavity. Each adult kidney is an anteroposteriorly compressed, lima-bean-shaped organ, which measures about 11.25 cm (4 in.) long, 5.5–7.7 cm (2–3 in.) wide, and 2.5 cm (1 in.) thick. The lateral surface of each kidney is convex, whereas the medial surface is strongly concave (fig. 19.6). The **hilum** of the kidney is the depression along the medial surface through which the **renal artery** enters and the **renal vein** and **ureter** exit. Innervation to the kidney is also at the hilum.

hilum: L. *hilum,* a trifle

Each kidney is embedded in a fatty, fibrous pouch consisting of three layers. The **renal capsule** is the innermost layer that forms a strong, transparent fibrous attachment to the kidney. The renal capsule protects the kidney from trauma and the spread of infections. The second layer is formed by a firm, protective layer of adipose tissue called the **adipose capsule** (see fig. 19.5). The outermost layer, called the **renal fascia,** is a supportive layer that anchors the kidney to the peritoneum and the abdominal wall.

Although the kidneys are firmly supported by the adipose capsule, renal fascia, and even the renal vessels, under certain conditions these structures may give in to the force of gravity and the kidneys may drop a bit in position. This condition is called *renal ptosis (to'sis)* and generally occurs in elderly persons who are extremely thin and have insufficient amounts of supportive fat in the adipose capsular layer. It may also affect victims of *anorexia nervosa* who suffer from extreme weight loss. The potential danger of renal ptosis is that the ureter may kink, blocking the flow of urine from the affected kidney.

Gross Structure

A coronal section of the kidney shows two distinct regions and a major cavity (fig. 19.6). The outer **cortex,** in contact with the capsule, is reddish brown and appears granular because of its many capillaries. The deeper region, or **medulla,** is darker and striped because of the presence of microscopic tubules and blood vessels. The medulla is composed of eight to fifteen conical **renal pyramids** separated by **renal columns.** The apexes of the renal pyramids are all directed into the renal sinus and are known as the renal papillae.

The cavity of the kidney collects and transports urine from the kidney to the ureter. It is divided into several portions. Each papilla of a pyramid projects into a small depression called the **minor calyx** *(ka'liks)*—the plural form is *calyces.* Several minor calyces unite to form a **major calyx.** In turn, the major calyces join to form the funnel-shaped **renal pelvis.** The renal pelvis is actually an expanded portion of the ureter in the kidney and serves to collect urine from the calyces and transport it to the ureter.

Microscopic Structure

The **nephron** *(nef'ron)* is the functional unit of the kidney that is responsible for the formation of urine. Each kidney contains more than a million nephrons. A nephron consists of **urinary tubules** and associated small blood vessels. Fluid formed by capillary filtration enters the tubules and is subsequently modified by transport processes; the resulting fluid that leaves the tubules is urine.

Renal Blood Vessels Arterial blood enters the kidney at the hilum through the **renal artery,** which divides into **interlobar** *(in"ter-lo'bar)* **arteries** (fig. 19.7), which pass between the pyramids through the renal columns. **Arcuate** *(ar'ku-āt)* **arteries** branch from the interlobar arteries at the boundary of

ptosis: Gk. *ptosis,* a falling
arcuate: L. *arcuare,* to bend

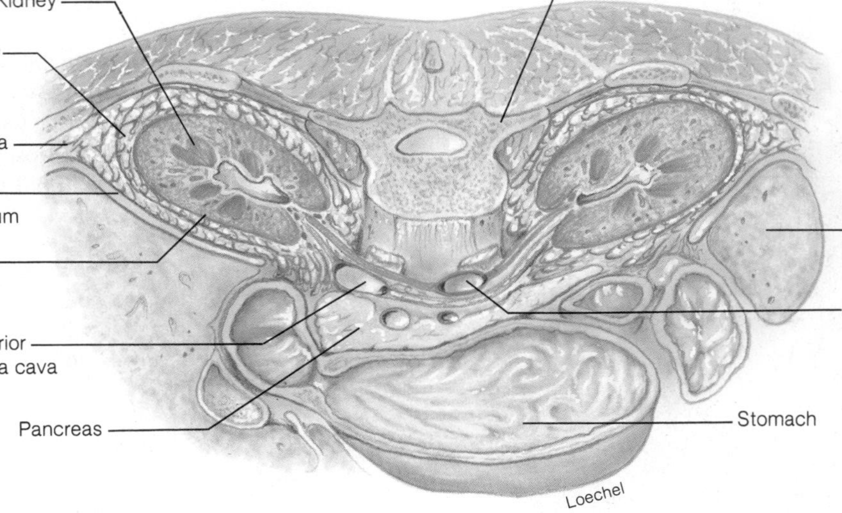

Figure 19.5 The positions of the kidneys as seen in transverse section through the upper abdominal cavity. The kidneys are embedded in adipose tissue behind the parietal peritoneum.

the cortex and medulla. Many **interlobular arteries** radiate from the arcuate arteries and subdivide into numerous **afferent arterioles,** which are microscopic in size. The afferent arterioles deliver blood into capillary networks, called **glomeruli,** which produce a blood filtrate that enters the urinary tubules. The blood remaining in the glomerulus leaves through an **efferent arteriole,** which delivers the blood into either the **peritubular capillaries** surrounding the convoluted tubules or into the **vasa recta** surrounding the ascending and descending tubules (fig. 19.8). From these capillary networks, the blood is drained into veins that parallel the course of the arteries in the kidney and are named the **interlobular veins, arcuate veins,** and **interlobar veins.** The interlobar veins descend between the pyramids, converge, and leave the kidney as a single **renal vein** that empties into the inferior vena cava.

In summary, vessels are at a capillary level between the arterioles in each glomerulus of the kidney. The blood pressure at this level is strong enough to force considerable water, including the wastes, from the blood into the urinary tubular portion of the nephron. There is a secondary capillary network of peritubular capillaries that surround various tubular portions of the nephron. This capillary plexus, however, is between arterioles and venules and functions in the reabsorption of water and other substances that should not be excreted with the urine.

Although the kidneys are generally well protected by being encapsulated retroperitoneally, they may be injured by a hard blow to the lumbar region. The immense vascularity of the kidney makes it highly susceptible to hemorrhage, and as a result such kidney damage can produce blood in the urine.

Nephron Tubules The tubular portion of a nephron consists of a glomerular capsule, proximal convoluted tubule, descending limb of the nephron loop (loop of Henle), ascending limb of the nephron loop, and distal convoluted tubule (fig. 19.8).

The **glomerular** (Bowman's) **capsule** surrounds the glomerulus. The glomerular capsule and its associated glomerulus are located in the cortex of the kidney and together constitute the **renal corpuscle** (fig. 19.9). The glomerular capsule contains an inner visceral layer of epithelium around the glomerular capillaries and an outer parietal layer. The space between these two layers, called the **capsular space,** receives the glomerular filtrate.

The glomerular epithelium contains pores called **fenestrae** *(feh'neh-stray)* that permit the filtrate to pass from the blood into the glomerular capsule (fig. 19.9). Though the fenestrae are large, they are still small enough to prevent the passage of blood cells, platelets, and most plasma proteins into the filtrate. The inner layer of the glomerular capsule is composed of unique cells called **podocytes** *(pod'o-sits)* with numerous cytoplasmic extensions known as **pedicels** *(ped'i-sils)*. Pedicels inderdigitate, like fingers wrapped around the glomerular capillaries. The narrow slits between adjacent pedicels provide the passageways through which filtered molecules must pass to enter the interior of the glomerular capsule.

Filtrate in the glomerular capsule passes into the lumen of the **proximal convoluted tubule.** The wall of the proximal convoluted tubule consists of a single layer of cuboidal cells containing millions of microvilli; these serve to increase the surface area for reabsorption. In the process of reabsorption, salt,

glomerulus: L. diminutive of *glomus,* ball

Bowman's capsule: from Sir William Bowman, English anatomist, 1816–92

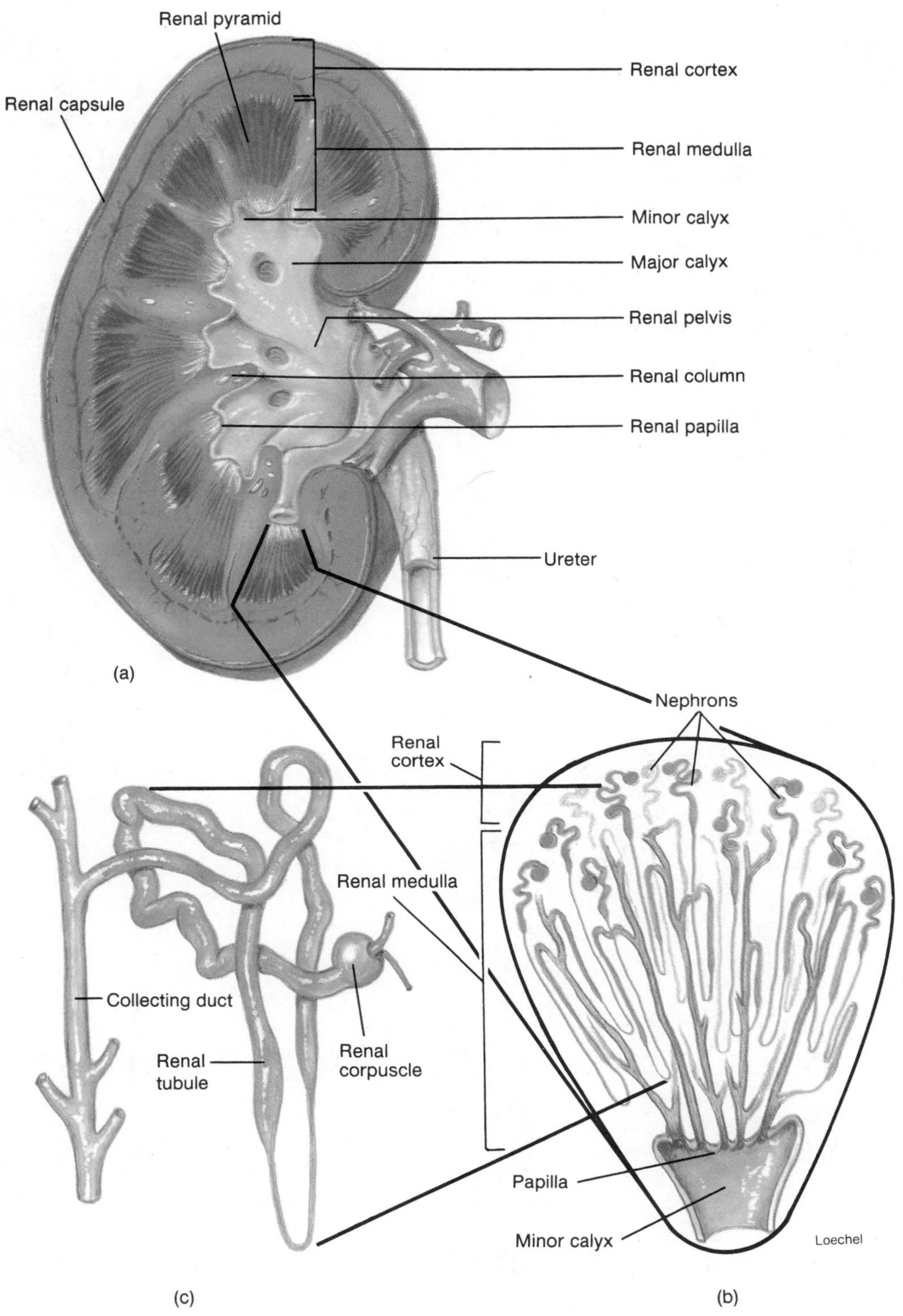

Figure 19.6 The internal structures of a kidney. (*a*) A coronal section showing the structure of the cortex, medulla, and renal pelvis. (*b*) A diagrammatic magnification of a renal pyramid and cortex to depict the tubules. (*c*) A diagrammatic view of a single nephron and a collecting duct.

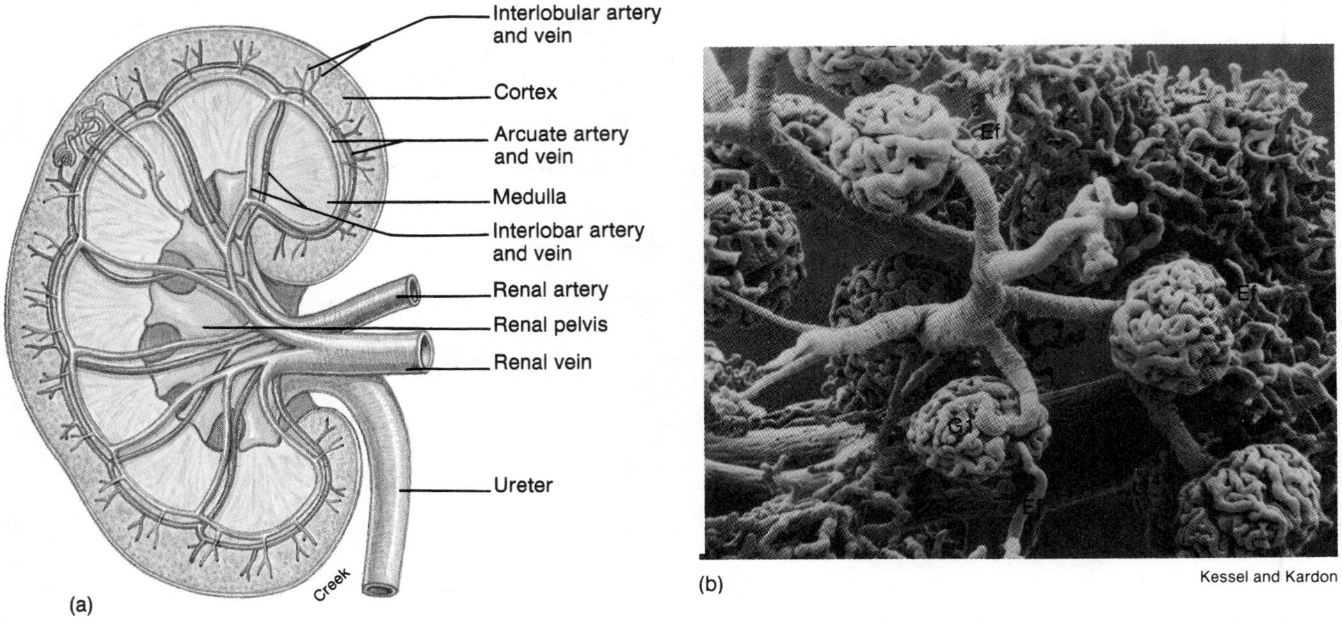

Figure 19.7 The vascular structure of the kidneys. (*a*) An illustration of the major arterial supply and (*b*) a scanning electron micrograph of the glomeruli.

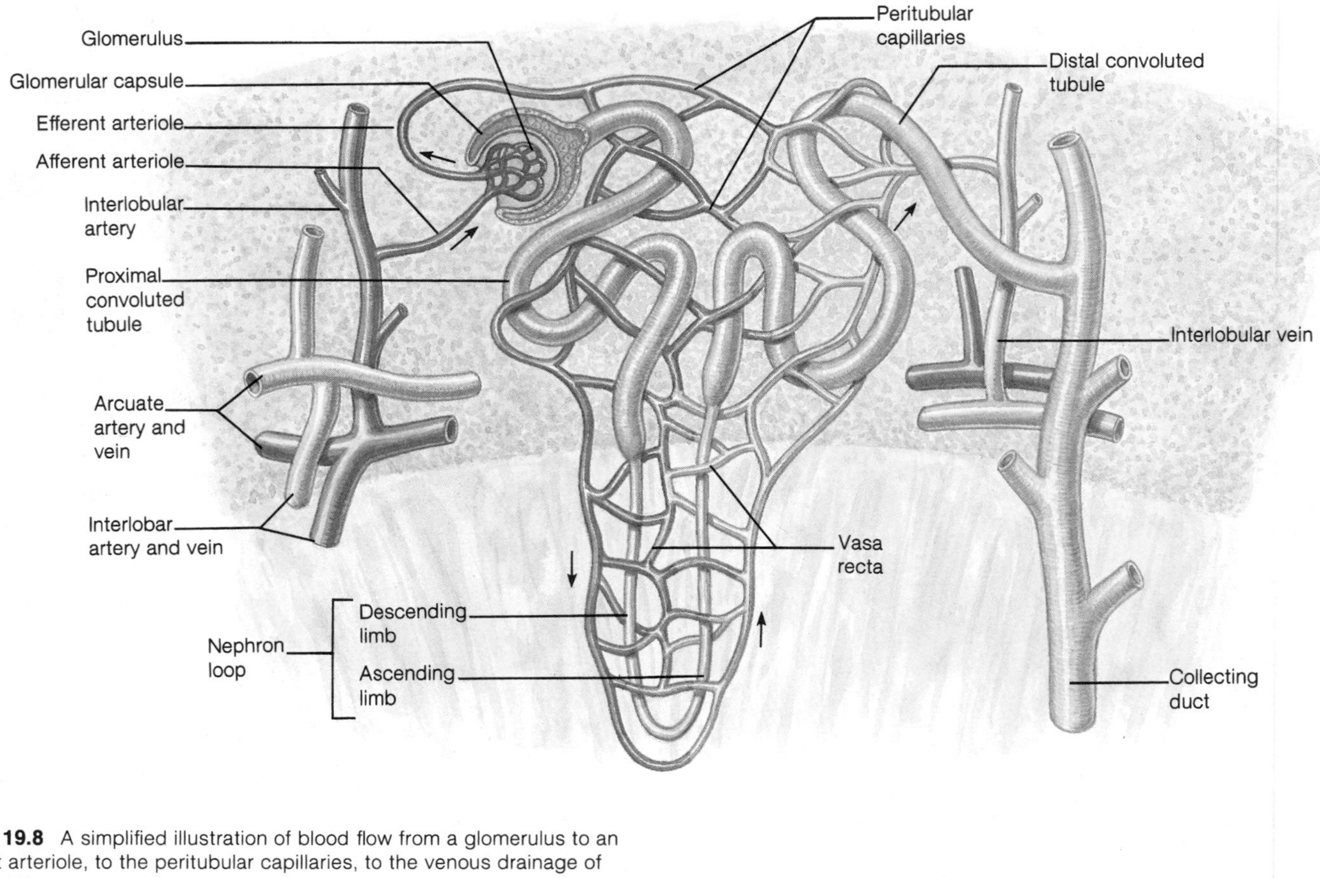

Figure 19.8 A simplified illustration of blood flow from a glomerulus to an efferent arteriole, to the peritubular capillaries, to the venous drainage of the kidneys.

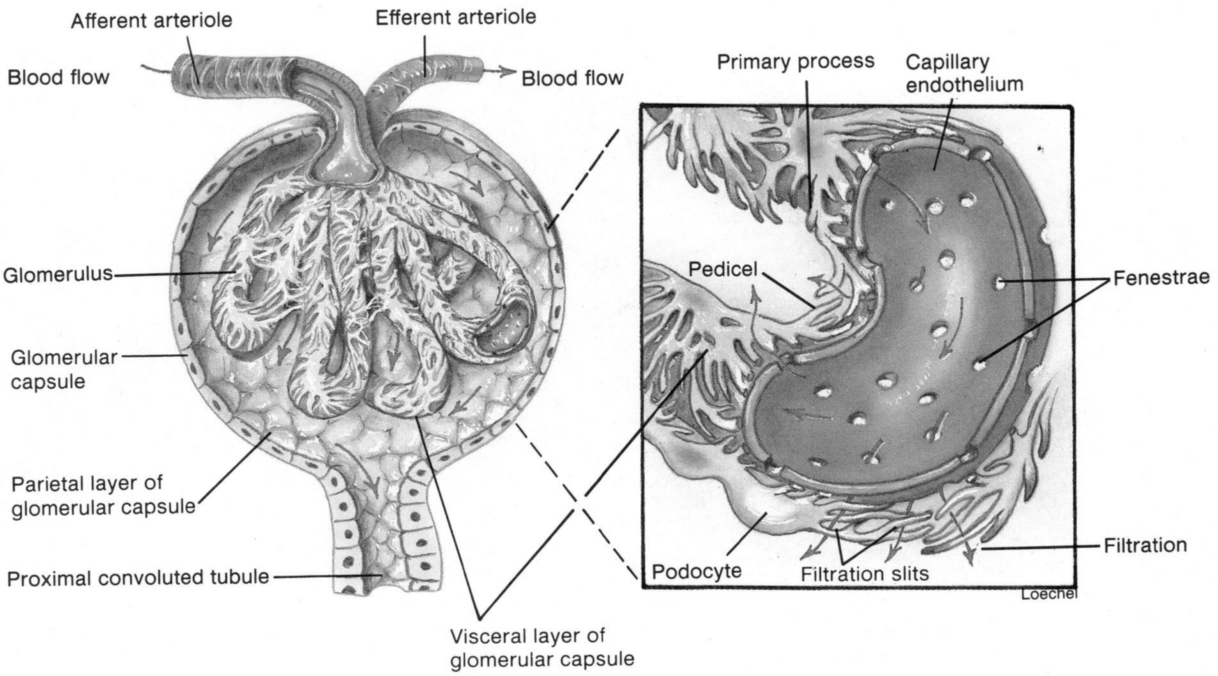

Figure 19.9 A renal corpuscle. Note that the diameter of the efferent arteriole is smaller than that of the afferent arteriole. This is one factor that maintains a high blood pressure within the glomerulus. The first step of urine formation is the filtration through the glomerular membrane into the glomerular (Bowman's) capsule.

water, and other molecules needed by the body are transported from the lumen, through the tubular cells, and into the surrounding peritubular capillaries.

The glomerulus, glomerular capsule, and proximal convoluted tubule are located in the renal cortex. Fluid passes from the proximal convoluted tubule to the **nephron loop** (loop of Henle). This fluid is carried into the medulla in the **descending limb** of the loop and returns to the cortex in the **ascending limb** of the loop. Back in the cortex, the tubule becomes coiled again and is called the **distal convoluted tubule.** In contrast to the proximal tubule, the distal convoluted tubule is shorter and has relatively few microvilli. The distal convoluted tubule is the last segment of the nephron and terminates as it empties into a collecting duct.

There are two types of nephrons, which are classified according to their position in the kidney and the lengths of their nephron loops. Nephrons that originate in the inner one-third of the cortex—called **juxtamedullary nephrons**—have longer loops than the **cortical nephrons** that originate in the outer two-thirds of the cortex (fig. 19.10).

The distal convoluted tubules of several nephrons drain into a **collecting duct.** Fluid is then drained by the collecting duct from the cortex into the medulla as the collecting duct passes through a renal pyramid. This fluid, now called urine, passes out of a renal papilla into a minor calyx. Urine is then funneled through the renal pelvis and out of the kidney into the ureter.

The kidneys have an autonomic nerve supply derived from the **renal plexus** of the tenth, eleventh, and twelfth thoracic nerves. Sympathetic stimulation of the renal plexus produces a vasomotor vascular network response in the kidney. This response determines the circulation of the blood by regulating the diameters of arterioles.

Table 19.1 Physiological processes of the nephron in the formation of urine

Structure	Function
Renal corpuscle	
Glomerulus	Filtration of water and dissolved substances from blood plasma
Glomerular capsule	Receives glomerular filtrate
Nephron tubules	
Proximal convoluted tubule	Reabsorption of water by osmosis
	Reabsorption of glucose, amino acids, creatine, lactic acid, citric acid, uric acid, ascorbic acid, phosphate ions, sulfate ions, calcium ions, potassium ions, and sodium ions by active transport
	Reabsorption of proteins by pinocytosis
	Reabsorption of chloride ions and other negatively charged ions by electrochemical attraction
	Active secretion of substances such as penicillin, histamine, and hydrogen ions
Descending limb of the nephron loop	Reabsorption of water by osmosis
Ascending limb of the nephron loop	Reabsorption of chloride ions by active transport and passive reabsorption of sodium ions
Distal convoluted tubule	Reabsorption of water by osmosis
	Reabsorption of sodium ions by active transport
	Active secretion of hydrogen ions
	Passive secretion of potassium ions by electrochemical attraction

The physiological function of the nephron is summarized in table 19.1.

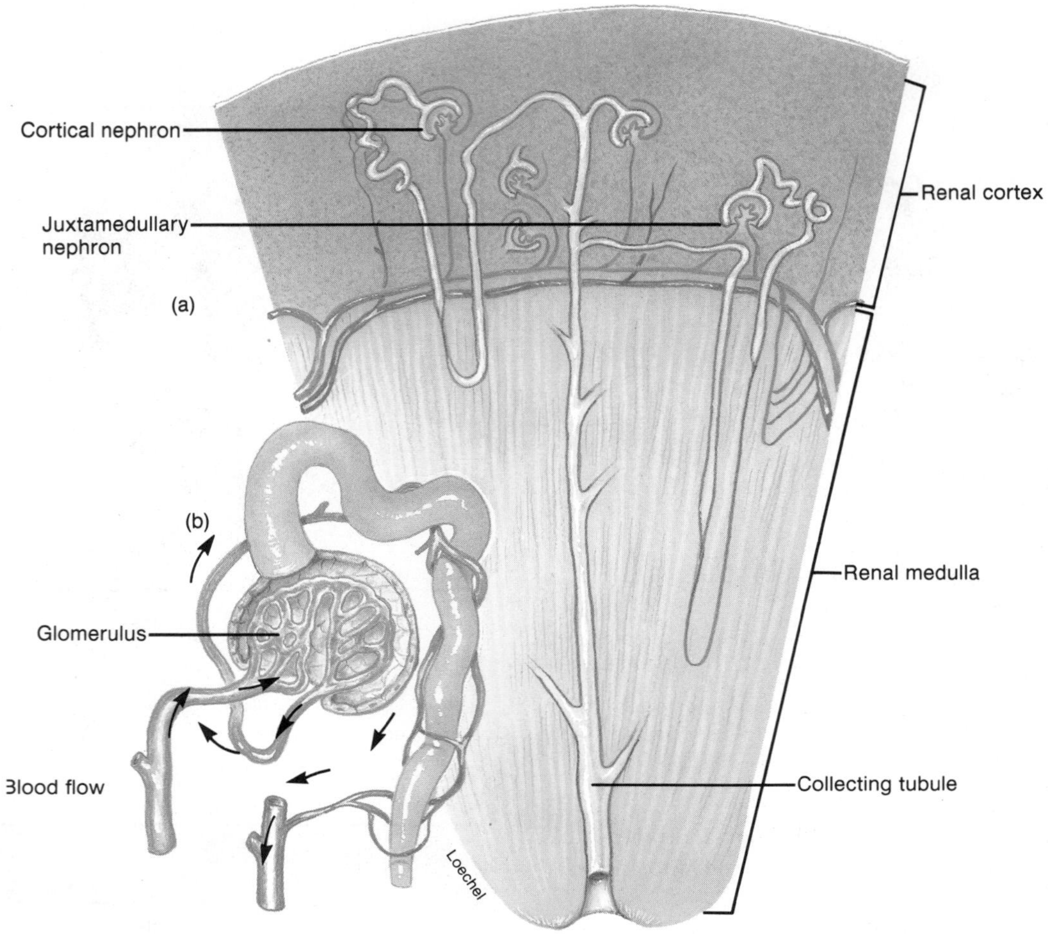

Figure 19.10 Cortical nephrons are located almost exclusively within the renal cortex, and juxtamedullary nephrons are located, for the most part, within the outer portion of the renal medulla.

Urine from a healthy individual is virtually bacteria free but easily becomes contaminated after voiding because of its organic components. The breakdown of urine by bacterial action produces ammonia. A urinalysis is a common procedure in any routine physical examination. Such things as appearance, pH, or the presence of albumin, blood, glucose, and acetone are studied during this test. Abnormal urine is symptomatic of several kinds of diseases or problems in the urinary system.

1. Describe the gross appearance of the renal cortex and medulla, and indicate the microscopic structures found in each layer.
2. Trace the course of blood flow through the kidney from the renal artery to the renal vein.
3. Trace the course of tubular fluid from the glomerular capsules to the ureter.
4. Draw a diagram of the tubular component of a nephron. Label the segments, and indicate which parts are in the cortex and which are in the medulla.

Ureters, Urinary Bladder, and Urethra

Urine is channeled from the kidneys to the urinary bladder by the ureters and expelled from the body through the urethra. The mucosa of the urinary bladder permits distension, and the muscles of the urinary bladder and urethra are used in the control of micturition.

Objective 8. Describe the location, structure, and function of the ureters.
Objective 9. Discuss the gross and histological structure of the urinary bladder and its innervation.
Objective 10. Explain the process of micturition.
Objective 11. Compare and contrast the structure of the male and female urethra.

Ureters

The **ureters,** like the kidneys, are retroperitoneal in location. Each ureter is a tubular organ about 25 cm (10 in.) long, which begins at the renal pelvis and courses inferiorly to enter the urinary bladder at the posteriolateral angle of its base. The thickest portion of the ureter is near where it enters the urinary bladder and is approximately 1.7 cm (0.5 in.) in diameter.

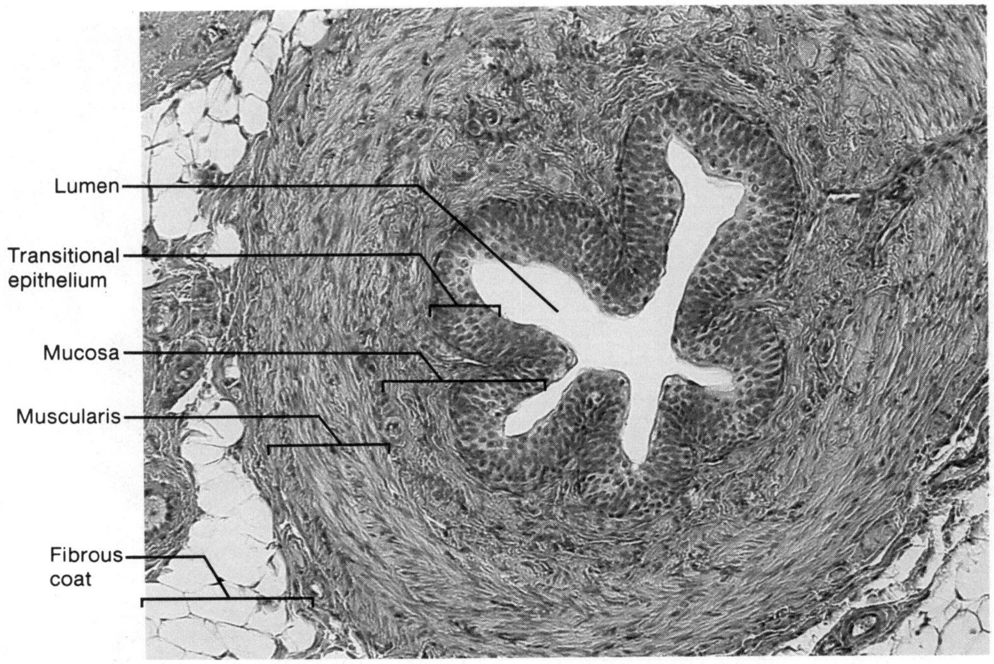

Figure 19.11 A photomicrograph of a ureter in transverse section. Note the transitional epithelium of the mucosa layer.

The wall of the ureter consists of three layers, or tunicas. The inner **mucosa** is continuous with the linings of the renal tubules and the urinary bladder. The mucosa consists of transitional epithelium (fig. 19.11). The cells of this layer secrete a mucus that lubricates the walls of the ureter with a protective film. The middle layer of the ureter is called the **muscularis.** It consists of an inner longitudinal and an outer circular layer of smooth muscle. In addition, the proximal one-third of the ureter contains another longitudinal layer outside of the circular layer. Muscular peristaltic waves move the urine through the ureter. The peristaltic waves are initiated by the presence of urine in the renal pelvis, and their frequency is determined by the volume of urine. The waves, which occur from every few seconds to every few minutes, force urine through the ureter and cause it to spurt into the urinary bladder. The outer layer of the ureter is called the **fibrous coat.** The fibrous coat is composed of loose fibrous connective tissue that not only covers the ureter but has extensions that anchor the ureter in place.

The arterial supply of the ureter comes from several sources. Branches from the renal artery serve the superior portion. The testicular (or ovarian) artery supplies the middle portion, and the superior vesicular artery serves the pelvic region. The venous return is through corresponding veins.

A *calculus*, or *kidney stone*, may obstruct the ureter and produce a tremendous amount of peristaltic waves in an attempt to pass the stone. The pain from a lodged calculus is extreme and extends throughout the pelvic area. A lodged calculus also causes a sympathetic ureterorenal reflex that results in constriction of renal arterioles, which reduces the production of urine in the kidney on the affected side.

Urinary Bladder

The **urinary bladder** is a storage sac for urine. It is located posterior to the symphysis pubis and anterior to the rectum (fig. 19.12). In females, the urinary bladder is in contact with the uterus and vagina. In males, the prostate gland is positioned below the urinary bladder (fig. 19.13).

The shape of the urinary bladder is determined by the volume of urine it contains. An empty urinary bladder is pyramidal in shape, having an anteroinferior apex, a superior surface, two inferolateral surfaces, a base (posterior surface), and a neck. The apex of the urinary bladder is superior to the symphysis pubis and is secured to the **median umbilical ligament,** a fibrous remnant of the embryonic urachus. The base of the urinary bladder receives the ureters along the superolateral angles, and the urethra exits at the inferior angle. The urethra is a tubular continuation of the neck of the urinary bladder.

The wall of the urinary bladder consists of four layers: the mucosa, submucosa, muscularis, and serosa (adventitia). The **mucosa** is the innermost layer. It is composed of transitional epithelium, which decreases in thickness as the urinary bladder distends and the cells are stretched. Further distension is permitted by folds of the mucosa, called **rugae,** which can be seen when the urinary bladder is empty (fig. 19.13). Fleshy flaps of mucosa located where the ureters pierce into the urinary bladder act as valves over the openings of the ureters to prevent a reverse flow of urine toward the kidneys as the bladder fills. A triangular area, known as the **trigone** *(tri'gōn)*, is formed on the mucosa between the two ureter openings and the single

calculus: L. *calculus*, small stone

trigone: L. *trigonum*, triangle

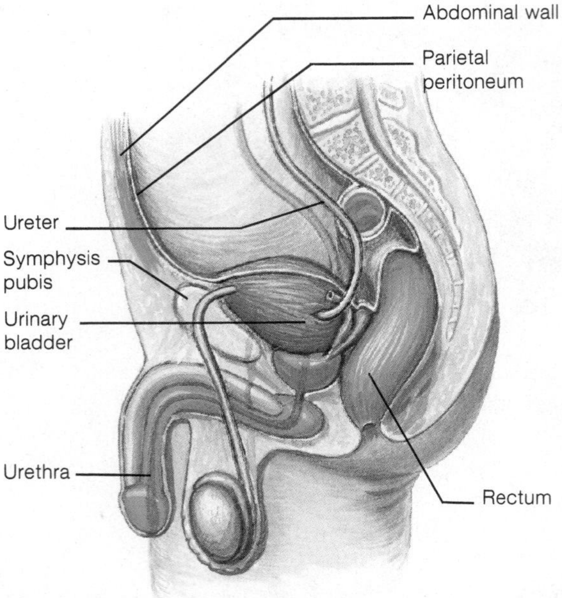

Figure 19.12 The position of the urinary bladder. In the male, the urinary bladder is located within the pelvic cavity and behind the symphysis pubis.

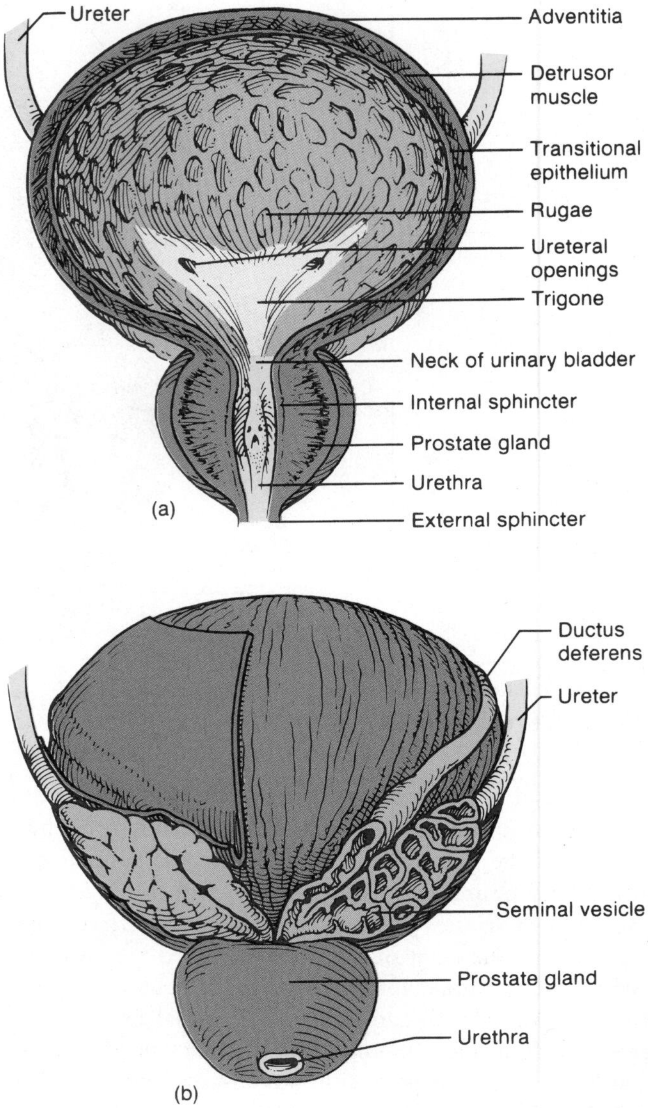

Figure 19.13 The male urinary bladder and urethra. (*a*) A coronal view, and (*b*) a posterior view.

urethral opening. The internal trigone lacks rugae and is therefore smooth in appearance and remains relatively fixed in position as the urinary bladder changes shape during distension and contraction.

The second layer of the urinary bladder is the **submucosa,** which functions to support the mucosa. The **muscularis** consists of three interlaced smooth muscle layers and is referred to as the **detrusor muscle.** At the neck of the urinary bladder, the detrusor muscle is modified to form the upper (the internal) of two muscular sphincters surrounding the urethra. The outer covering of the urinary bladder is the **serosa.** It appears only on the superior surface of the bladder and is actually a continuation of the peritoneum.

The arterial supply to the urinary bladder comes from the **superior** and **inferior vesicular arteries,** which arise from the internal iliac arteries. Blood draining the urinary bladder enters from a **vesicular venous plexus** and empties into the internal iliac veins.

The autonomic nerves serving the urinary bladder are derived from pelvic plexuses. Sympathetic innervation arises from the last thoracic and first and second lumbar spinal nerves to serve the trigone, ureteral openings, and blood vessels of the urinary bladder. Parasympathetic innervation arises from the second, third, and fourth sacral nerves to serve the detrusor muscle. The sensory receptors of the urinary bladder respond to distension and relay impulses to the central nervous system via the pelvic spinal nerves.

The urinary bladder becomes infected easily, particularly in women because of their short urethra, which increases the possibility of contamination. A urinary bladder infection, called *cystitis,* may easily ascend from the bladder to the ureters since the mucous linings are continuous. An infection that involves the renal pelvis is called *pyelitis;* if it continues into the nephrons, it is known as *pyelonephritis.*

Urethra

The tubular **urethra** conveys urine from the urinary bladder to the outside of the body. The urethral wall has an inside lining of mucous membrane, surrounded by a relatively thick layer of smooth muscle, whose fibers are directed longitudinally. Specialized **urethral glands,** embedded in the urethral wall, secrete mucus into the urethral canal.

Two muscular sphincters surround the urethra. The upper, involuntary smooth muscle sphincter is the **internal urethral sphincter** (sphincter vesicae), which is formed from the detrusor muscle of the urinary bladder. The lower sphincter is composed of voluntary, skeletal muscle fibers and is called the **external urethral sphincter** (sphincter urethrae) (fig. 19.13).

---

detrusor: L. *detrudere,* thrust or forced down

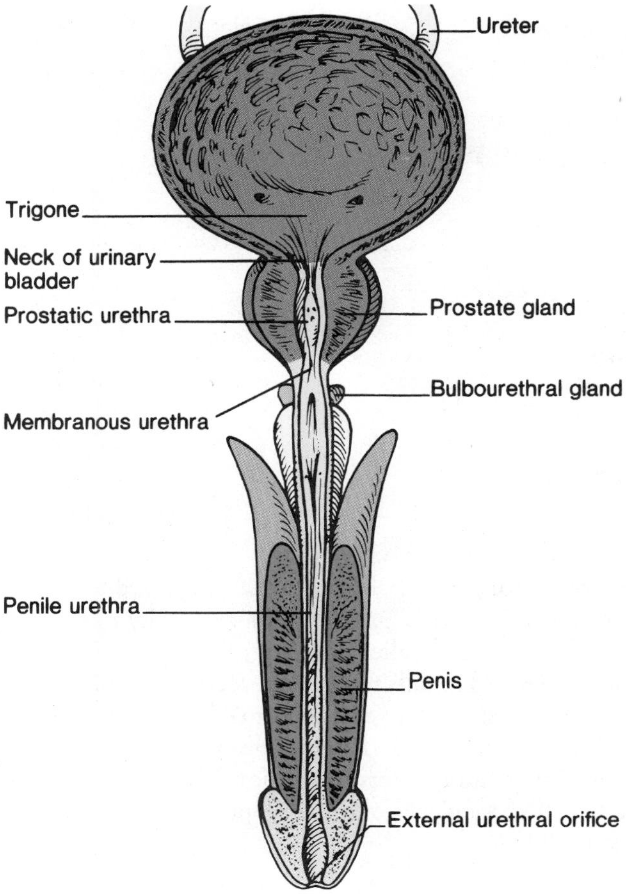

Figure 19.14 A longitudinal section of a male urethra.

The urethra of the female is a simple tube about 4 cm (1.5 in.) long, which empties urine through the **urethral orifice** *(or'ĭ-fis)* into the vestibule between the labia minora. The urethral orifice is positioned anterior to the vaginal orifice and about 2.5 cm posterior to the clitoris.

The urethra of the male serves both the urinary and reproductive systems. It is about 20 cm (8 in.) long and S-shaped because of the shape of the penis. Three regions can be identified in the male urethra: the prostatic urethra, membranous urethra, and penile urethra (fig. 19.14).

The **prostatic urethra** is the proximal portion, about 2.5 cm long, that passes through the **prostate gland** located near the neck of the urinary bladder. The prostatic urethra receives drainage from small ducts of the prostate gland and two **ejaculatory ducts** of the reproductive system.

The **membranous urethra** is the short (0.5 cm) portion of the urethra that passes through the urogenital diaphragm. The external urethral muscle is located in this region.

The **penile urethra** (cavernous urethra) is the longest portion (15 cm), extending from the outer edge of the urogenital diaphragm to the external urethral orifice on the glans penis. This portion is surrounded by erectile tissue as it passes through the corpus spongiosum of the penis. The paired ducts of the **bulbourethral glands** (Cowper's glands) of the reproductive system attach to the penile urethra near the urogenital diaphragm.

Micturition

Micturition *(mik''tu-rish'un)*, commonly called urination or voiding, is a reflex action that expels urine from the urinary bladder. It is a complex function that requires a stimulus from the urinary bladder and a combination of involuntary and voluntary nerve impulses to the appropriate muscular structures of the bladder and urethra.

In young children, micturition is a simple reflex action that occurs when the urinary bladder becomes sufficiently distended. Voluntary control of micturition is normally developed at the time a child is two or three years old. Voluntary control requires the development of an inhibitory ability by the cerebral cortex and a maturing of various portions of the spinal cord.

The volume of urine produced by an adult averages about 1,200 ml per day, but it can vary from 600–2,500 ml. The average capacity of the urinary bladder is 700–800 ml. A volume of 200–300 ml will distend the bladder enough to stimulate stretch receptors and trigger the micturition reflex, creating a desire to urinate.

The micturition reflex center is located in the second, third, and fourth sacral segments of the spinal cord. Following stimulation of this center by impulses arising from stretch receptors in the urinary bladder, parasympathetic nerves that stimulate the detrusor muscle and the internal urethral sphincter are activated. Stimulation of these muscles causes a rhythmic contraction of the bladder wall and a relaxation of the internal sphincter. At this point, a sensation of urgency is perceived in the brain, but there is still voluntary control over the external urethral sphincter. At the appropriate time, the conscious activity of the brain activates the motor nerve fibers (S4) to the external urethral sphincter via the pudendal nerve (S2, S3, and S4), causing the sphincter to relax and urination to occur.

The innervation of the ureter, urinary bladder, and urethra is shown in figure 19.15, and a summary of the process of micturition is presented in table 19.2.

Urinary retention, or the inability to void, may occur postoperatively, especially following surgery of the rectum, colon, or internal reproductive organs. The difficulty may be due to nervous tension, the effects of anesthetics, or pain and edema at the site of the operation. If urine is retained beyond six to eight hours, *catheterization* may become necessary. In this procedure, a tube or catheter is passed through the urethra into the urinary bladder so that urine can flow freely.

---

1. Describe the location and the structure of a ureter, and indicate the function of its muscularis layer.
2. Describe the structure of the urinary bladder, and indicate the anatomical structures that permit the organ to be distended.
3. Compare the male urinary system with that of the female.
4. Explain the structures and processes involved in the control of micturition.

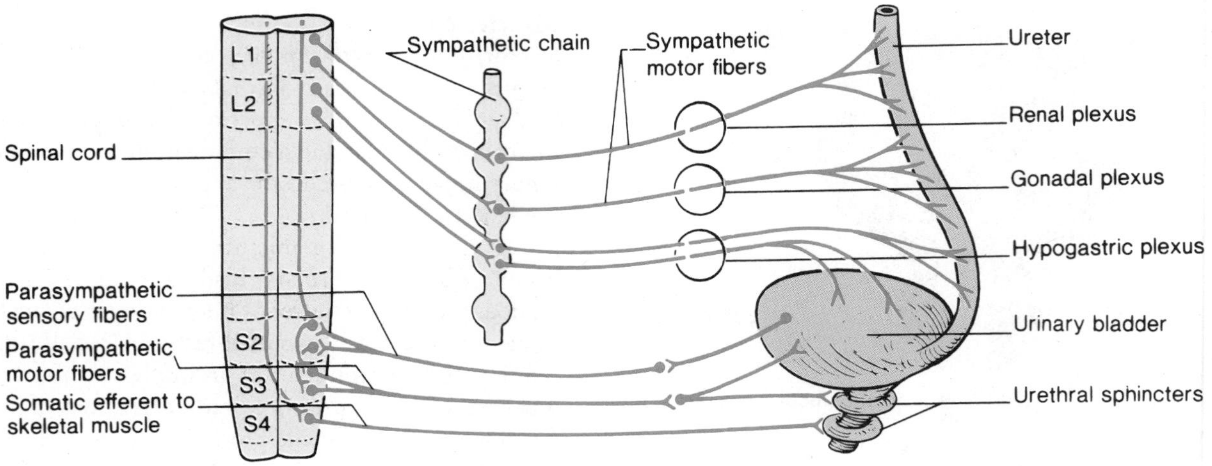

Figure 19.15 Innervation of the ureter, urinary bladder, and urethra.

Table 19.2 Events of micturition

1. The urinary bladder becomes distended as it fills with urine.
2. Stretch receptors in the bladder wall are stimulated, and impulses are sent to the micturition center in the spinal cord.
3. Parasympathetic nerve impulses travel to the detrusor muscle and the internal urethral sphincter.
4. The detrusor muscle contracts rhythmically, and the internal urethral sphincter relaxes.
5. The need to urinate is sensed as urgent.
6. Urination is prevented by voluntary contraction of the external urethral sphincter and by inhibition of the micturition reflex by impulses from the midbrain and cerebral cortex.
7. Following the decision to urinate, the external urethral sphincter is relaxed, and the micturition reflex is facilitated by impulses from the pons and the hypothalamus.
8. The detrusor muscle contracts, and urine is expelled through the urethra.
9. Neurons of the micturition reflex center are inactivated, the detrusor muscle relaxes, and the bladder begins to fill with urine.

Clinical Considerations

The importance of kidney function in maintaining homeostasis and the ease with which urine can be collected and used as a mirror of the plasma's chemical composition make the clinical study of renal function and urine composition particularly significant. **Urology** is the medical specialty concerned with dysfunctions of the urinary system. Urinary dysfunctions can be congenital, acquired, due to physical trauma, or the result of conditions that secondarily involve the urinary organs.

Congenital Malformations of the Urinary Organs

Abnormalities of the organs of the urinary system occur in about 12% of newborn babies. These deviations range from insignificant anomalies to those that are incompatible with life.

Kidneys Common malformations of the kidneys are illustrated in figure 19.16. One common deformity is the unilateral absence of a kidney, called **renal agenesis** *(ah-jen'ĕ-sis),* caused when a uretic bulb fails to develop. **Renal ectopia** means that one or both kidneys are in an abnormal position. Generally, this condition occurs when a kidney remains in the pelvic area. **Horseshoe kidney** refers to a fusion of the kidneys across the midline. The incidence of this asymptomatic condition is about one in 600.

Ureters **Duplication of the ureters** and the associated renal pelvis is a frequent anomaly of the urinary tract. Unilateral duplication occurs in one in 200 births, whereas bilateral duplication occurs in one in 1,200 births. Occasionally there is partial duplication, which may become very important because of the propensity for urinary infections. A completely duplicated ureter frequently opens into areas other than the urinary bladder and requires surgical correction.

Urinary Bladder Protrusion of the posterior wall of the urinary bladder is called **exstrophy of the bladder** and occurs when the wall of the perineum fails to close. Associated with this condition, which occurs in about one in 40,000 births, are defective urethral sphincter muscles.

Urethra The most common anomaly of the urethra is a condition called **hypospadias** *(hi''po-spa'de-ahs),* where the urethra of the male opens on the ventral surface of the penis instead of at the tip of the glans (fig. 19.17); there is a similar defect in the female in which the urethra opens into the vagina. **Epispadias** is a failure of closure on the dorsal surface of the penis. Hypospadias and epispadias can be corrected surgically.

Symptoms and Diagnosis of Urinary Disorders

Normal micturition is painless. **Dysuria** *(dis-u're-ah),* or painful urination, is a sign of a urinary tract infection or obstruction of the urethra—as in an enlarged prostate gland in a male. **He-**

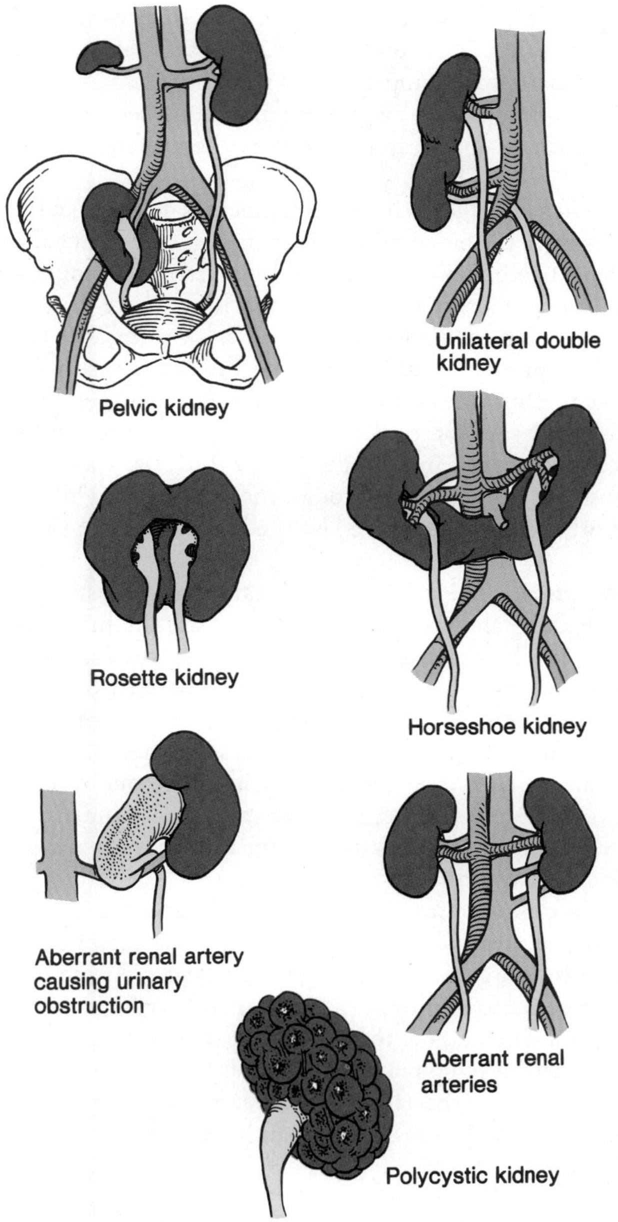

Figure 19.16 Various congenital anomalies involving the kidneys.

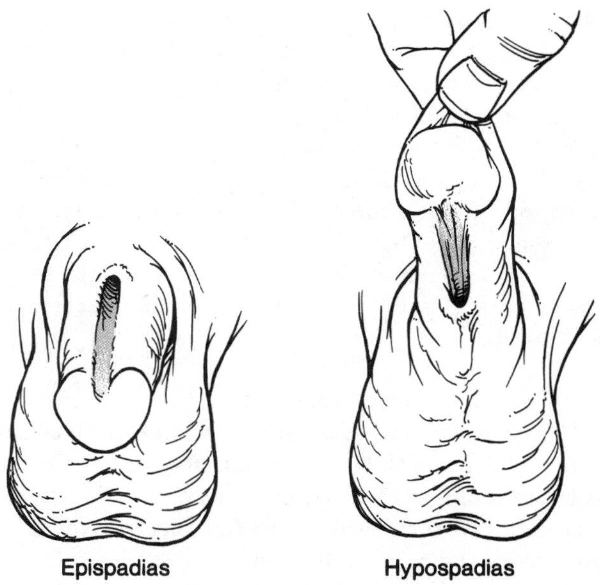

Figure 19.17 Epispadias and hypospadias of the male urethra.

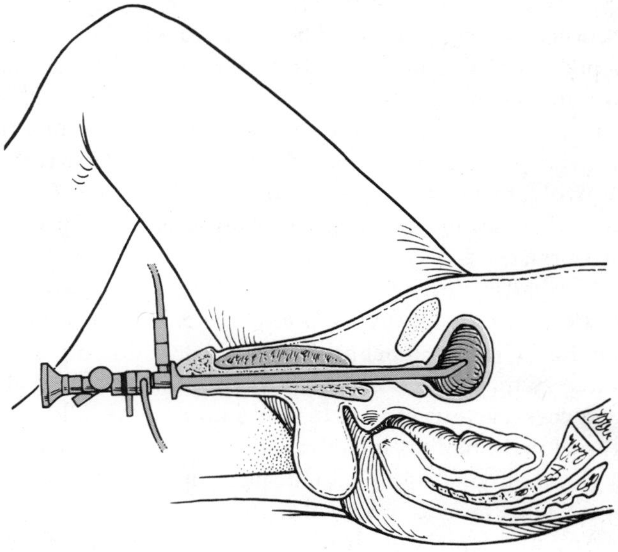

Figure 19.18 Cystoscopic examination of a male.

maturia means blood in the urine and is usually associated with trauma. **Bacteriuria** means bacteria in the urine, and **pyuria** is the term for pus in the urine, which may result from a prolonged infection. **Oliguria** is a scanty output of urine, whereas **polyuria** is an excessive output. Low blood pressure and kidney failure are two causes of oliguria. **Uremia** is a condition in which substances ordinarily excreted in the urine accumulate in the blood. **Enuresis** *(en''u-re'sis)*, or **incontinence,** is the inability to control micturition. Its causes range from psychosomatic sources to actual physical impairment.

The palpation and inspection of urinary organs is an important aspect of physical assessment. The right kidney is palpable in the supine position; the left kidney usually is not. The distended urinary bladder is palpable along the superior pelvic rim.

The urinary system may be examined using x-ray techniques. An **intravenous pyelogram** (IVP) permits x-ray examination of the kidneys following the injection of radiopaque dye. In this procedure, the dye that has been injected intravenously is excreted by the kidneys so that the renal pelvises and the outlines of the ureters and urinary bladder can be observed in an X ray.

Cystoscopy *(sis-tos'ko-pe)* is the inspection of the inside of the urinary bladder by means of an instrument called a cystoscope (fig. 19.18). By using this technique, tissue samples can be obtained as well as urine samples from each kidney prior to mixing in the bladder. Once the cystoscope is in the bladder, the ureters and pelvis can be viewed through urethral catheterization and inspection for obstructions.

A **renal biopsy** is a diagnostic test for evaluating certain types and stages of kidney diseases. The biopsy is performed either through a skin puncture (closed biopsy) or through a surgical incision (open biopsy).

Urinalysis is a simple and important laboratory aspect of a physical examination. The voided urine specimen is tested for color, specific gravity, chemical composition, and for the presence of microscopic bacteria, crystals, and *casts.* Casts are accumulations of proteins that have leaked through the glomeruli and have been pushed through the tubules like toothpaste through a tube.

Infections of Urinary Organs

Urinary tract infections are a significant cause of illness and are also a major factor in the development of chronic renal failure. Females are more predisposed to urinary tract infections than are males, and the incidence of infection increases directly with sexual activity and aging. The higher infection rate in females has been attributed to their shorter urethra, which has a close proximity to the rectum, and to the lack of protection offered by prostatic secretions in males. To reduce the risk of urinary infections, a young girl should be taught to wipe her anal region in a posterior direction, away from the urethral orifice, after a bowel movement.

Infections of the urinary tract are named according to the infected organ. An infection of the urethra is called **urethritis,** and involvement of the urinary bladder is **cystitis.** Cystitis is frequently a secondary infection from some other part of the urinary tract.

Nephritis means inflammation of the kidney tissue. **Glomerulonephritis** *(glo-mer''u-lo-nĕ-fri'tis)* is inflammation of the glomeruli. Glomerulonephritis may occur following an upper respiratory tract infection, because antibodies produced against streptococci bacteria can produce an autoimmune inflammation in the glomeruli. This inflammation may permanently change the glomeruli and figure significantly in the development of chronic renal disease and renal failure.

Any interference with the normal flow of urine, such as from a kidney stone or an enlarged prostate gland in a male, causes stagnation of urine in the renal pelvis and the development of pyelitis. **Pyelitis** is an inflammation of the renal pelvis and its calyces. **Pyelonephritis** is inflammation involving the renal pelvis, the calyces, and the tubules of the nephron within one or both kidneys. Bacterial invasion from the blood or from the lower urinary tract is another cause of both pyelitis and pyelonephritis.

Trauma to Urinary Organs

A sharp blow to a lumbar region of the back may cause a contusion or rupture of a kidney. Symptoms of kidney trauma include hematuria and pain of the upper abdominal quadrant and flank on the injured side.

Pelvic fractures from accidents may result in perforation of the urinary bladder and urethral tearing. When driving an automobile, it is advisable to stop periodically to urinate because an attached seat belt over the region of a full urinary bladder can cause it to rupture in even a relatively minor accident. Urethral injuries are more common in men than women because of the position of the urethra in the penis. Straddle injuries are those where, for example, a man walking along a raised beam slips and compresses his urethra and penis between the hard surface and his pubic arch, rupturing the urethra.

Hemodialysis Hemodialysis equipment is designed to filter the wastes from the blood of a patient who has chronic renal failure. During hemodialysis, the blood of a patient is pumped via a tube from an artery through the machine where it is cleansed and then returned to the body via a vein (fig. 19.19). The cleaning process involves pumping the blood past a semipermeable cellophane membrane, which separates the blood from an isotonic solution containing molecules needed by the body (such as glucose). In a process called **dialysis,** waste products diffuse out of the blood through the membrane while glucose and other molecules needed by the body remain in the blood.

Obstruction The urinary system can become obstructed anywhere along the tract. Calculi (stones) are the most common cause, but blockage can also come from trauma, strictures, tumors or cysts, spasms or kinks of the ureters, or congenital anomaly. If not corrected, an obstruction causes urine to collect behind the blockage and generate pressure that may cause permanent functional and anatomic damage to one or both kidneys. Pressure buildup in a ureter causes a distended ureter to develop, called a **hydroureter.** Dilation in the renal pelvis is called **hydronephrosis.**

Calculi, or kidney stones, are generally the result of infections or metabolic disorders that cause the excretion of large amounts of organic and inorganic substances. As the urine becomes concentrated these substances may crystalize and form granules. The granule then serves as a core for further precipitation and development into a larger calculus. This becomes dangerous when the calculus is sufficiently large to cause an obstruction. It also causes intense pain when it passes through the urinary tract.

Renal Failure An output of 50–60 cc of urine per hour is considered normal. If the output drops to less than 30 cc per hour, it may indicate renal failure. Renal failure is the loss of the kidney's ability to maintain fluid and electrolyte balance and to excrete waste products.

Renal failure can be either acute or chronic. **Acute renal failure** is the sudden loss of kidney function caused by shock and hemorrhage, thrombosis, or other physical trauma to the kidneys. The kidneys may sustain a 90% loss of their nephrons through tissue death and still not have an obvious loss of function. If a patient suffering acute renal failure is stabilized, the nephrons have an excellent capacity to regenerate.

A person with **chronic renal failure** cannot sustain life independently. Chronic renal failure is the end result of kidney disease in which the kidney tissue is progressively destroyed. As renal tissue continues to deteriorate, the options for sustaining life are hemodialysis or kidney transplantation.

Urinary Incontinence The inability to voluntarily retain urine in the urinary bladder is known as urinary incontinence. It has

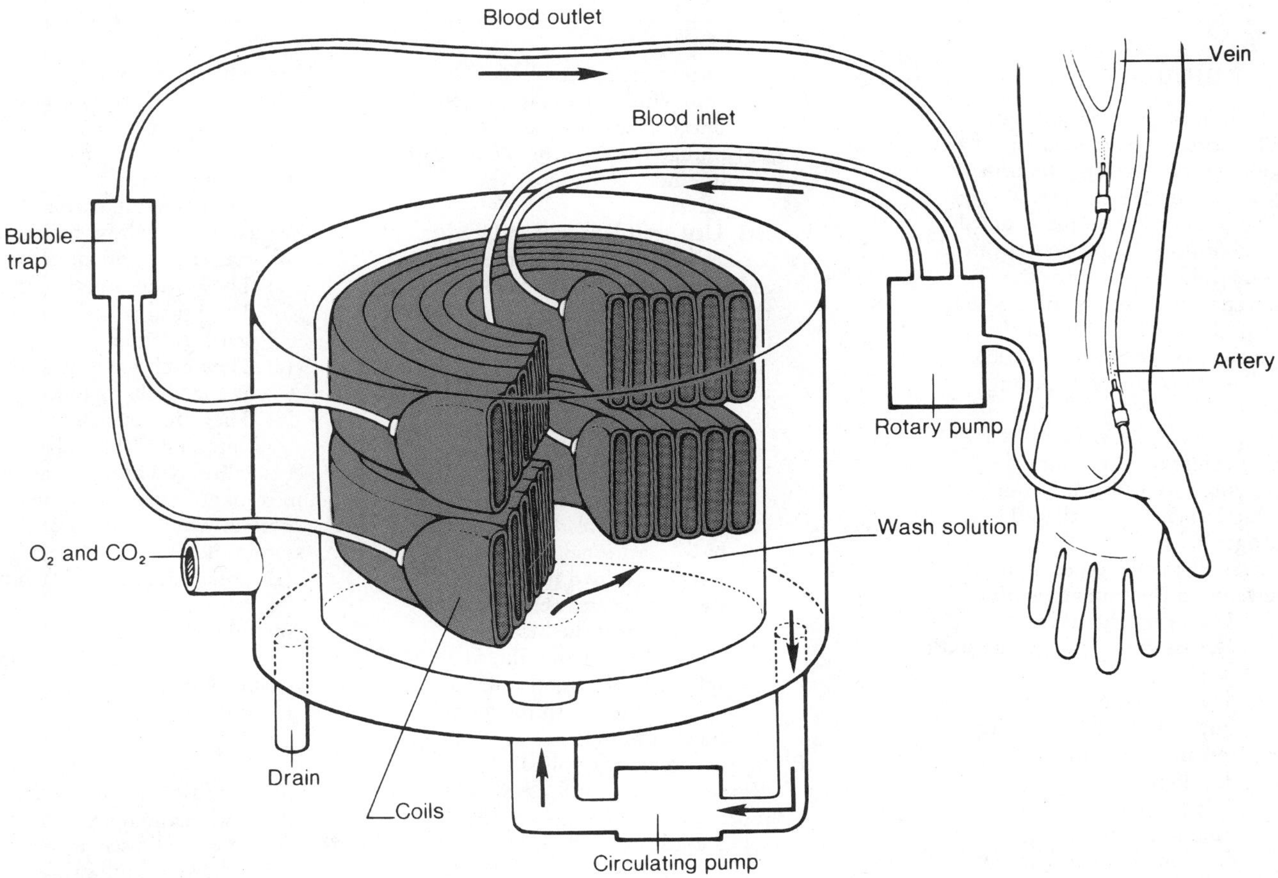

Figure 19.19 A diagram of a hemodialysis machine.

a number of temporary or permanent causes. Emotional stress is a cause of temporary incontinence in adults. Causes of permanent incontinence include neurological trauma, various urinary diseases, and tissue damage within the urinary bladder or urethra. Remarkable advances have been made in treating permanent urinary incontinence through the implantation of an artificial urethral sphincter (fig. 19.20).

CLINICAL CASE STUDY ANSWER

The hematuria is likely the result of a right renal laceration caused by the upper abdominal knife wound. The right lower quadrant stab most likely did not violate the urinary tract. The course of blood seen in the catheter begins and proceeds as follows: Abdominal aorta → right renal artery → smaller parenchymal artery → through the lacerated vessel(s) → into the lacerated urinary collecting system either at the calyx or the renal pelvis or proximal right ureter → urinary bladder → into catheter. During the operation, the surgeon should keep in mind that the presence of only one kidney is a relatively common occurrence (2–4% of the population). If, therefore, she is prompted to remove the damaged kidney, she should first confirm the presence of a second functioning kidney. If a second kidney is not present, every effort should be made to correct the problem without performing a nephrectomy, which would consign the patient to chronic hemodialysis or kidney transplant. ■

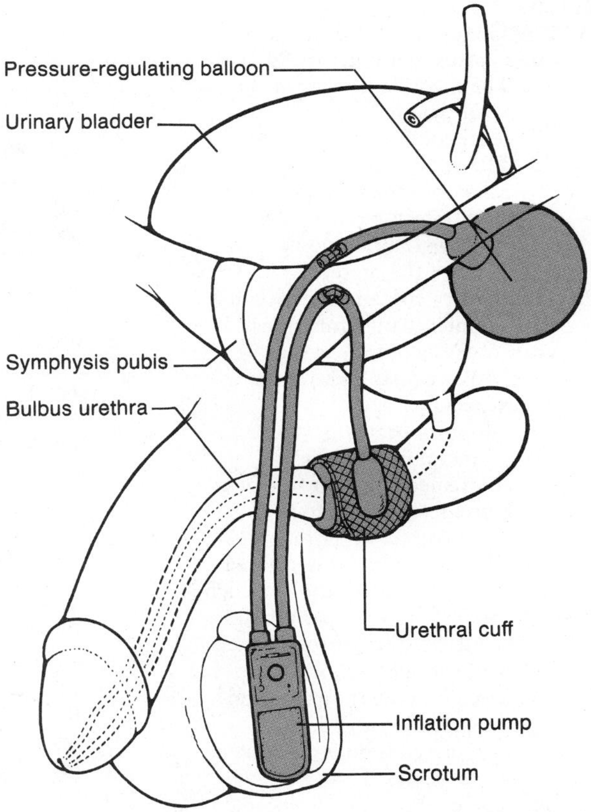

Figure 19.20 An artificial urethral sphincter. The entire device is implanted internally and consists of an inflation bulb that is manually squeezed to inflate the urethral cuff from the fluid stored in a reservoir (pressure-regulating balloon). Manual pumping of the deflation bulb (implanted in the opposite side of the scrotum) releases the pressure of the urethral cuff and allows urine to flow from the bladder. The device for a female is similar except that the inflation and deflation bulbs are implanted in the tissue of the labia minora.

Chapter Summary

I. Introduction to the Urinary System
   A. The urinary system consists of two kidneys, two ureters, the urinary bladder, and the urethra.
   B. The urinary system maintains the composition and properties of the body fluid, which form the internal environment of the body cells. The end product of the urinary system is urine, which is voided from the body during micturition.

II. Development of the Urinary System
   A. The urinary and reproductive systems both originate from mesodermal tissue called the urogenital ridge.
   B. Three successive types of kidneys develop in the human embryo.
      1. The pronephric kidney persists only through the sixth week.
      2. The mesonephric kidney functions throughout the remainder of embryonic development.
      3. The metanephric kidney functions during fetal development and after birth.

III. Kidneys
   A. The kidney is divided into an outer cortex and inner medulla.
      1. The medulla is composed of renal pyramids separated by renal columns.
      2. The renal papillae empty urine into the calyces, which drain into the renal pelvis; from there urine flows through the ureter.
   B. Each kidney contains more than a million microscopic functional units called nephrons, which consist of vascular and tubular components.
      1. Filtration occurs in the glomerulus (vascular component), which receives blood from an afferent arteriole.
      2. Glomerular blood is drained by an efferent arteriole, which delivers blood to peritubular capillaries that surround the nephron tubules.
      3. The glomerular capsule and the proximal and distal convoluted tubules are located in the cortex.
      4. The nephron loop is located in the medulla.
      5. Filtrate from the distal convoluted tubule is drained into collecting ducts, which plunge through the medulla to empty urine into the calyces.

IV. Ureters, Urinary Bladder, and Urethra
   A. Urine is channeled from the kidneys to the urinary bladder by the ureters and expelled from the bladder through the urethra. Muscles of the urinary bladder and urethra are used in the control of micturition.
      1. The ureters contain three layers: the mucosa, muscularis, and fibrous coats.
      2. The urinary bladder is lined by a transitional epithelium, which is folded into rugae; these structures permit great distension of the bladder.
      3. The urethra has an internal sphincter of smooth muscle and an external sphincter of skeletal muscle.
   B. Micturition is controlled by reflex centers in the second, third, and fourth sacral segments of the spinal cord.

Review Activities

Objective Questions

1. Which of the following statements about metanephric kidneys is *true?*
   (a) They become functional at the end of the eighth week.
   (b) They are active throughout fetal development.
   (c) They are the third pair of kidneys to develop.
   (d) All of the above are true.
2. Which of the following statements about the renal pyramids is *false?*
   (a) They are located in the medulla.
   (b) They contain glomeruli.
   (c) They contain collecting ducts.
   (d) They open through the renal papillae into the renal calyces.
3. Renal vessels and the ureter attach at the concave surface of the kidney called the
   (a) renal pelvis. (c) calyx.
   (b) urachus. (d) hilum.
4. The medulla of the kidney contains
   (a) glomerular capsules.
   (b) glomeruli.
   (c) renal pyramids.
   (d) adipose capsules.
5. Urine flowing from the collecting tubules enters directly into the
   (a) renal calyces.
   (b) ureter.
   (c) renal pelvis.
   (d) distal convoluted tubules.
6. Which of the following statements is *false* concerning the kidneys?
   (a) They are retroperitoneal.
   (b) They each contain eight to fifteen renal pyramids.
   (c) They each have a distinct cortex and medulla region.
   (d) They are positioned between the third and fifth lumbar vertebrae.
7. A calculus, or kidney stone, would most likely cause stagnation of the urine in which portion of the urinary system? The
   (a) urinary bladder. (d) renal pelvis.
   (b) renal column.
   (c) ureter. (e) urethra.
8. Distention of the urinary bladder is possible because of the presence of the
   (a) rugae.
   (b) trigone.
   (c) fibrous coat.
   (d) transitional epithelium.
   (e) Both a and d are correct.
9. The detrusor muscle is located in the
   (a) kidneys. (c) urinary bladder.
   (b) ureters. (d) urethra.
10. The internal urethral sphincter is innervated by
   (a) sympathetic neurons.
   (b) parasympathetic neurons.
   (c) somatic motor neurons.
   (d) All of the above.

Essay Questions

1. What is a metanephrogenic mass and a ureteric bud? Discuss the sequential development of these embryonic structures to form the urinary system. How can a greater knowledge of this development process lead to a better understanding of congenital abnormalities?
2. Describe the location of the kidneys in relation to the abdominal cavity and the peritoneal membranes.
3. Diagram the structures of a kidney that can be identified in a coronal section.
4. Describe how the kidney is supported against the posterior abdominal wall. How is this support related to the condition called *renal ptosis?*
5. Trace a drop of blood from an interlobular artery through a glomerulus and to an interlobular vein.

List in order all the vessels through which it would pass. How does the structural difference between an afferent and efferent arteriole ensure the necessary high blood pressure for filtrate formation?

6. In a male, trace the pathway of urine from the site of filtration at the renal corpuscle, through the urinary tubules of the nephron, and to the outside of the body. List in order all the structures through which the urine passes.
7. What is a nephron? Describe the two types of nephrons found in a kidney. Why are nephrons considered the functional units of the urinary system?
8. Describe the mechanism involved in the passage of urine from the renal pelvis to the urinary bladder.
9. Describe the urinary bladder with regard to position, histological structure, blood supply, and innervation.
10. Discuss the similarities and differences between the male and female urethra.
11. What is the micturition reflex? Discuss the physiological and functional events of a voluntary micturition response.
12. Define *urology, cystoscopy, renal calculi,* and *urinalysis.*
13. List four common congenital malformations of the urinary system. Which of these require surgical correction?
14. Define each of the following: *dysuria, hematuria, bacteriuria, pyuria, oliguria, polyuria, uremia,* and *enuresis.*

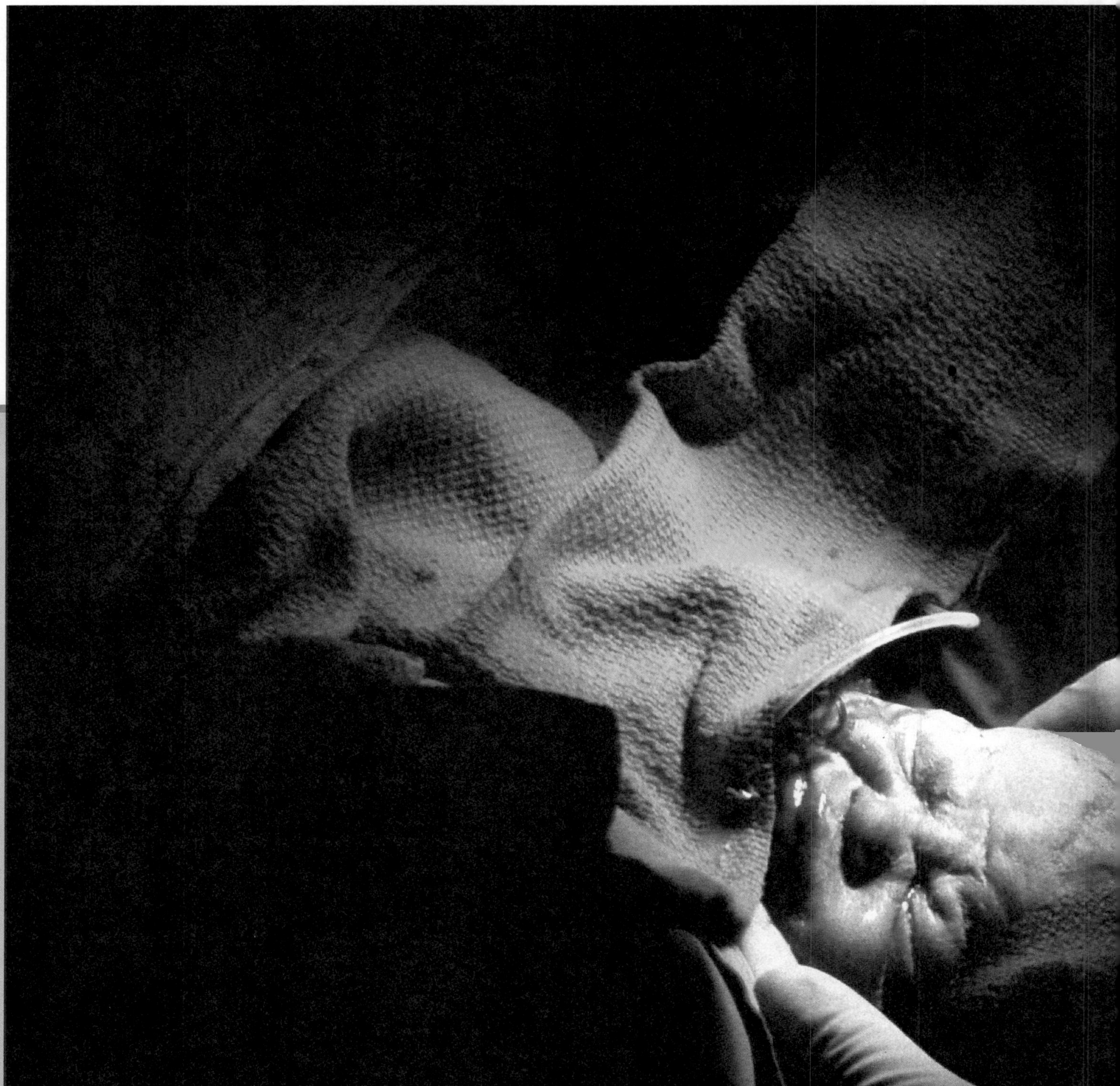

Parturition (childbirth) is a transitional event between prenatal and postnatal development.

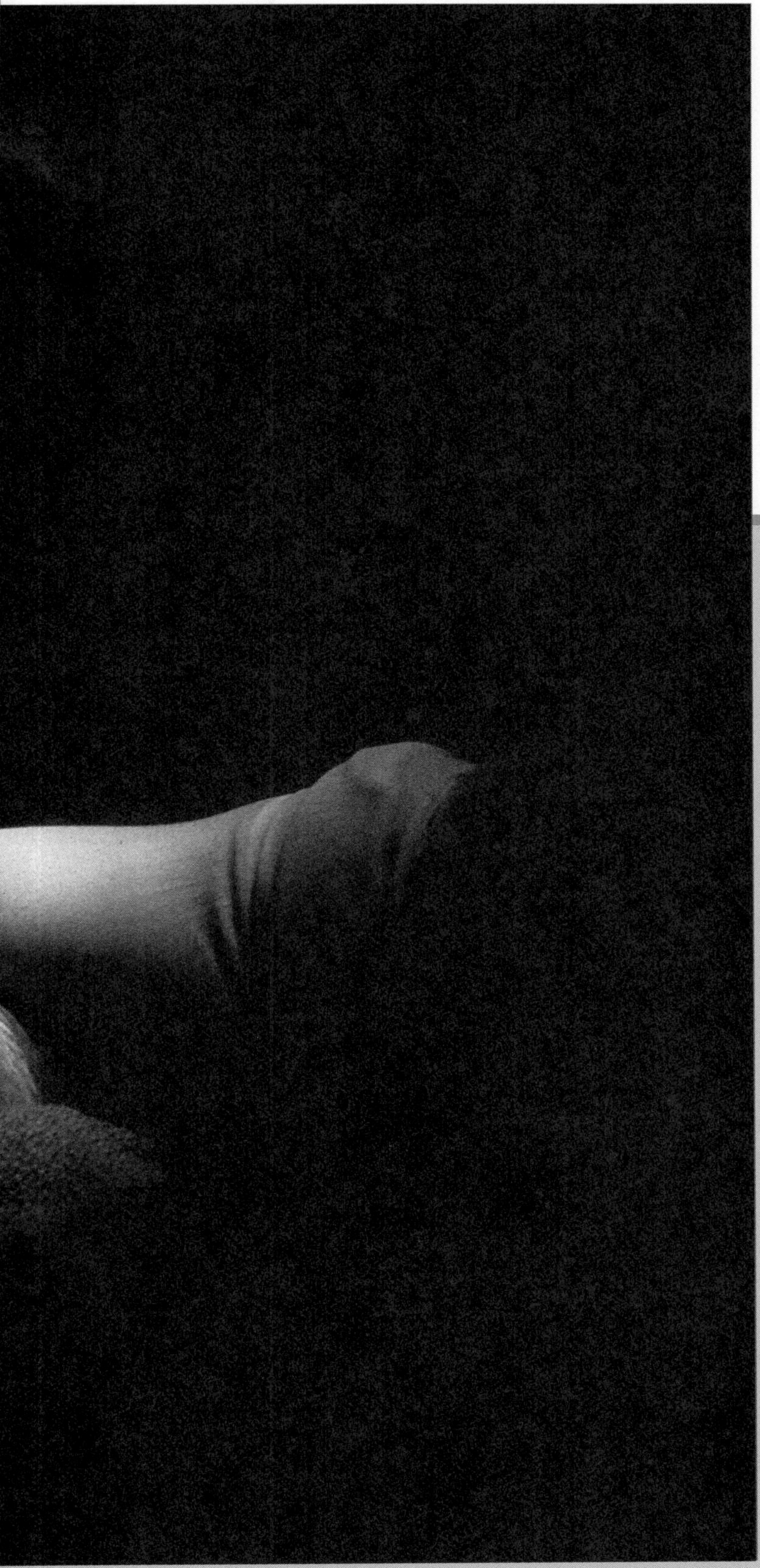

UNIT

7

Continuance of the Species

Reproduction, genetics, development, growth, and aging of humans are discussed in the chapters in unit 7. Described in these chapters are the anatomy of the male and female reproductive systems and how these systems produce a new and genetically unique individual. The fascinating process of prenatal development is also described. The final chapter of unit 7 discusses the structural and functional changes that occur in each body system as a person ages.

This unit includes:

CHAPTER 20

Male Reproductive System

Outline and Concepts

CLINICAL CASE STUDY

A twenty-seven-year-old man visits his family doctor, stating that he and his wife have been unable to conceive a child after nearly two years of trying. Furthermore, he states that his wife has undergone a thorough gynecological evaluation revealing no female cause for the infertility. Upon palpating the patient's testes, the doctor states that there seems to be nothing abnormal. Further examination of the scrotal sac above the testes, however, causes a look of perplexion on the doctor's face. He states that there is a tubular structure, one for each testis, for which he is palpating but cannot find. He adds that these structures have probably been missing from birth. During a follow-up visit, the doctor informs the patient that examination of his ejaculate revealed azoospermia (no viable sperm).

Explain the result of the semenalysis and how it relates to the patient's physical exam findings. List the structures involved. Does it seem unusual that the patient is capable of producing an ejaculate? Explain. ■

Introduction to the Male Reproductive System

The organs of the male and female reproductive systems are adapted to produce and allow the union of gametes that contain specific genes. A random combination of the genes during sexual reproduction permits the propagation of individuals with genetic differences.

Objective 1. Explain why sexual reproduction is biologically advantageous.

Objective 2. List the functions of the male reproductive system, and compare them with those of the female.

Objective 3. Distinguish between primary and secondary sex organs.

Sexual reproduction is common in organisms. Through this process, individuals of a species are propagated and have a genetic diversity inherited from both parents. Sexual reproduction provides the important advantage of genetic recombination, whereby a genetic change becomes a part of the gene pool. Genetic differences in the offspring help some of them adapt to environmental changes that might otherwise cause the extinction of a population.

Because sexual reproduction requires the production of two types of **gametes** *(gam'ēts)*, or sex cells, the species has a male and female form, each with its own unique reproductive system. The functions of the female reproductive system are more complex than those of the male. The functions of the male reproductive system are to produce the male gametes, **spermatozoa** (sperm), and to transfer them to the female through the process of *coitus (sexual intercourse)*, or *copulation*. The female not only produces her own gametes, called **ova,** and receives the sperm from the male, but her reproductive organs are specialized to provide a site for fertilization, implantation of the blastocyst, pregnancy, and delivery of a baby. The reproductive system of the female also provides a means of nourishing the baby through the secretions of the mammary glands in the breast.

The reproductive system of the male and female is a unique body system in three respects. First, whereas all of the other body systems function to sustain the individual, the reproductive system is specialized to perpetuate the species and pass genetic material from generation to generation. Second, the anatomy and physiology of the reproductive organs are the major differences between the male and female. Other systems, such as the integumentary, skeletal, and urinary, have minor sexual differences, but none to the extent of the reproductive system. The third uniqueness of the reproductive system is its latent development under hormonal control. The other body systems are totally functional at birth or shortly thereafter, whereas the reproductive system does not become functional until it is acted on by hormones during puberty. *Puberty* is the period of human development during which the reproductive organs become functional.

The structures of the male reproductive system can be divided into three categories based on function:

1. **Primary sex organs.** The primary sex organs are called **gonads,** specifically the **testes** in the male. Gonads produce the gametes, or sperm, and produce and secrete sex hormones. The appropriate amounts and timing of the production and secretion of male sex hormones cause the development of secondary sex organs and the expression of secondary sex characteristics.
2. **Secondary sex organs.** Secondary sex organs are those structures that are essential in caring for and transporting sperm. The three categories of secondary sex organs are the sperm-transporting ducts, the accessory glands, and the copulatory organ. The ducts that transport sperm include the **epididymides, ductus deferentia, ejaculatory ducts,** and **urethra.** The accessory glands are the **seminal vesicles,** the **prostate gland,** and the **bulbourethral** *(bul''bo-u-re'thral)* **glands.** The **penis,** which contains erectile tissue, is the copulatory organ. The **scrotum** is a pouch of skin that encloses the testes.
3. **Secondary sex characteristics.** Secondary sex characteristics are features that are not essential for the reproductive process but are generally considered sexually attractant features. Body physique, body hair, and voice pitch are examples.

The organs of the male reproductive system are depicted in figure 20.1, and their functions are summarized in table 20.1.

gamete: Gk. *gameta,* husband or wife
spermatozoa: Gk. *sperma,* seed; *zoon,* animal

puberty: L. *puberty,* grown up

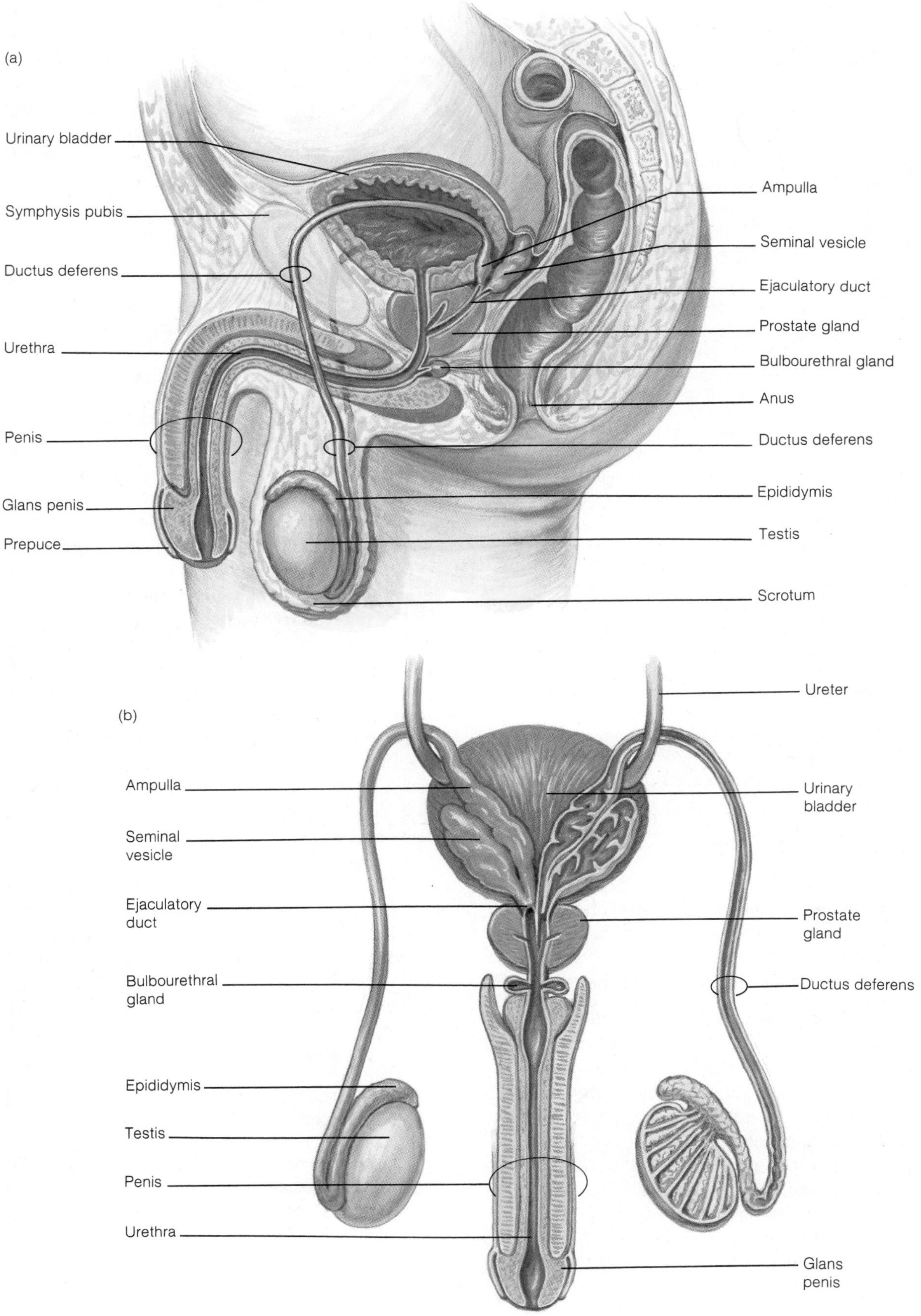

Figure 20.1 Organs of the male reproductive system. (*a*) A sagittal view and (*b*) a posterior view.

Table 20.1 Functions of the organs of the male reproductive system

Organ	Function
Testes	
Seminiferous tubules	Produce spermatozoa
Interstitial cells	Produce and secrete male sex hormones
Epididymides	Storage and maturation of spermatozoa; convey spermatozoa to ductus deferentia
Ductus deferentia	Store spermatozoa; convey spermatozoa to ejaculatory ducts
Ejaculatory ducts	Receive spermatozoa and additives to produce seminal fluid
Seminal vesicles	Secrete alkaline fluid containing nutrients and prostaglandins
Prostate gland	Secretes alkaline fluid that helps neutralize acidic seminal fluid and enhances motility of spermatozoa
Bulbourethral glands	Secrete fluid that lubricates urethra and end of penis
Scrotum	Encloses and protects testes
Penis	Conveys urine and seminal fluid to outside of body; organ of coitus

1. What is the principal value of sexual reproduction?
2. What are the functions of the male reproductive system?
3. List the organs of the male reproductive system, and indicate whether they are primary or secondary sex organs.
4. Define latent development with respect to the reproductive system.

Development of the Male Reproductive System

Sexual distinction is determined upon fertilization of an ovum by a sperm containing either an X or a Y chromosome. Sexual differentiation under hormonal control is a progressive process through embryonic and early fetal development.

Objective 4. Discuss how the increased production of testosterone affects the development of the secondary sex organs and secondary sex characteristics.

Objective 5. Explain how the genetic sex of a person is determined by the action of an XX or XY chromosome combination.

Objective 6. Trace the development of the reproductive organs to the indifferent stage.

Objective 7. Describe the sequence of events in the development of the male reproductive organs during the fetal period—including the descent of the testes into the scrotum.

Sex Determination

Sexual identity is initiated at the moment of conception when the *genetic sex* of the zygote (fertilized egg) is determined. The ovum is fertilized by a sperm containing either an X or a Y chromosome. If the sperm contains an X chromosome, it will pair with the X chromosome of the ovum and a female child will develop. A sperm carrying a Y chromosome results in an XY combination, and a male child will develop.

Genetic sex determines whether the gonads will be testes or ovaries. If testes develop, they will produce and secrete male sex hormones during late embryonic and early fetal development and cause the secondary sex organs of the male to develop.

Embryonic Development

The first sign of development of either the male or the female reproductive organs occurs during the fifth week as the medial aspect of each mesonephros enlarges to form the **gonadal ridge.** The gonadal ridge continues to grow behind the developing peritoneal membrane. By the sixth week, stringlike masses called **primary sex cords** form within the enlarging gonadal ridge (fig. 20.2). The primary sex cords in the male will eventually mature to become the sperm-nurturing seminiferous tubules. Externally, a swelling called the **genital tubercle** appears cephalad to the cloacal membrane (see fig. 20.6).

Reproductive development is well progressed by the eighth week, but it is still in what is known as the *indifferent stage,* because although the sex organs are apparent, external distinction is not (see fig. 20.6). The gonads at six weeks are relatively large and have a distinct outer cortex composed of primary sex cords and an inner medulla (fig. 20.3). During this time, specialized **primordial germ cells** are forming and migrating from the yolk sac to the embryonic gonads. These primordial cells are more specifically called *spermatogonia* in the developing male and *oogonia* in the developing female. Prior to approximately seven weeks of development, the gonads have the potential to become either testes or ovaries (fig. 20.4). The significance of embryonic development is that the sexual organs for both male and female are derived from the same developmental tissues and are considered *homologous* structures.

The Y chromosome in XY embryos masculinizes the gonads (causes them to develop into testes) through the production of cell-surface proteins coded by genes in the Y chromosome. These male markers are a type of histocompatibility antigen and are therefore known as **H-Y antigens.** Embryos that are chromosomal females (XX) lack Y chromosomes and therefore lack H-Y antigens. In the absence of these antigens the gonads develop into ovaries.

Notice that it is the presence or absence of the Y chromosome that determines whether the embryo will have testes or ovaries. This point is well illustrated by two genetic abnormalities. In *Klinefelter's syndrome* the affected person has forty-seven instead of forty-six chromosomes because of the presence of an extra X chromosome. These people, with XXY genotypes, develop testes despite the presence of two X chromosomes. Patients with *Turner's syndrome,* who have the genotype XO (and therefore have only forty-five chromosomes) develop ovaries.

spermatogonia: Gk. *sperma,* seed; *gone,* generation
oogonia: Gk. *oon,* egg; *gone,* generation
homologous: Gk. *homos,* the same

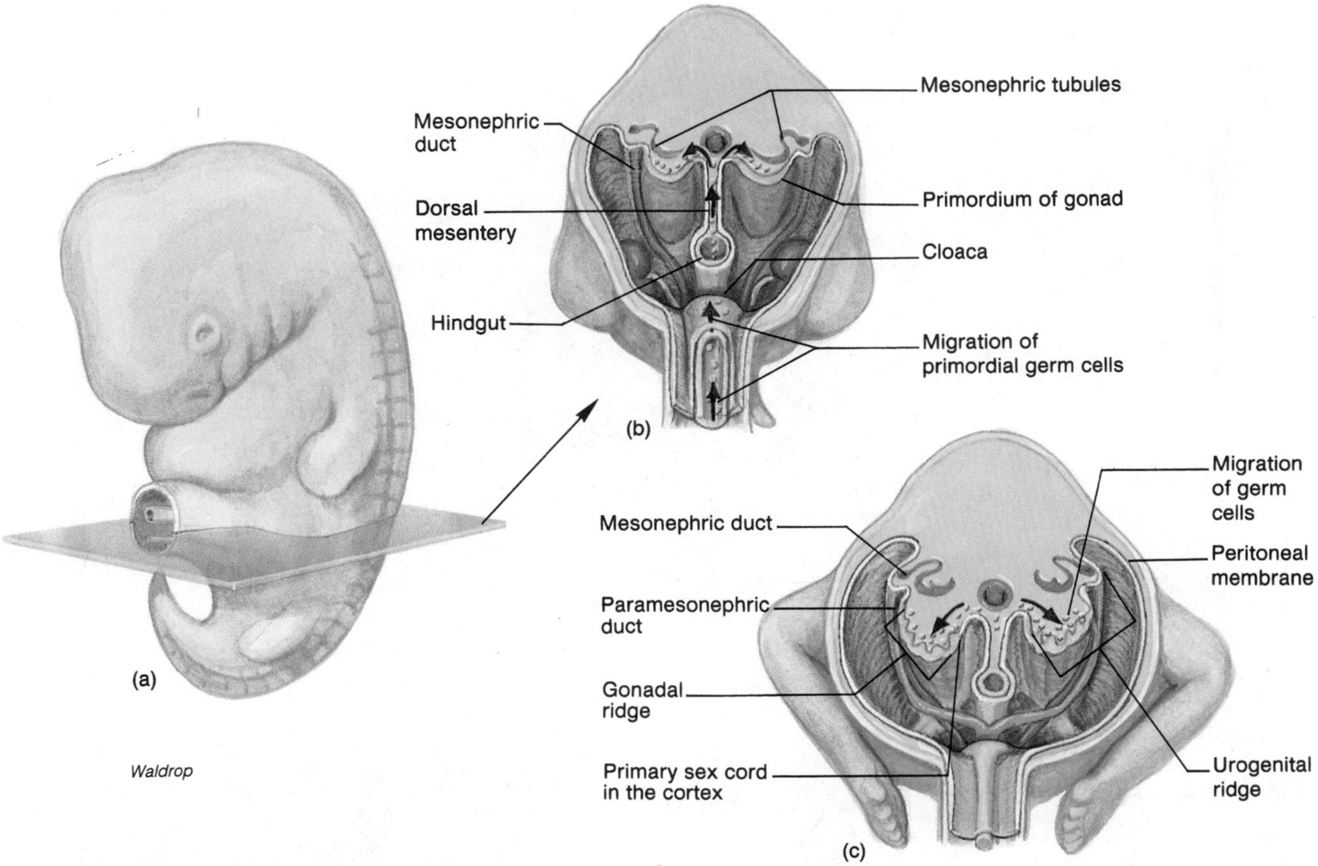

Figure 20.2 The development of the gonadal ridge and the primary sex cords. (*a*) An embryo at five weeks showing the position of a transverse cut depicted in (*b*) and (*c*). (*b*) At five weeks, the primordia of the gonads are forming along the gonadal ridges. (*c*) At seven weeks, the germ cells are migrating into the developing gonads.

Once the testes have differentiated, large amounts of male sex hormones, called *androgens (an'dro-jens),* are secreted from **interstitial cells** (cells of Leydig). The major androgen secreted by these cells is *testosterone (tes-tos'tĕ-rōn).* Testosterone secretion begins as early as eight to ten weeks after conception, reaches a peak at twelve to fourteen weeks, and thereafter declines to very low levels by the end of the second trimester (about twenty-one weeks). High levels of testosterone will not appear again in the life of the individual until the time of puberty.

The **seminiferous** *(se''mĭ-nif'er-us)* **tubules,** which will eventually produce sperm within the testes, appear about forty-five to fifty days following conception. Although spermatogenesis begins during embryonic life, it is arrested until the onset of puberty.

In addition to gonads, various internal accessory sexual organs are needed for reproductive functioning. Most of these sex accessory organs are derived from two systems of embryonic ducts. Male accessory organs are derived from mesonephric (wolffian) ducts, and female accessory organs are derived from paramesonephric (müllerian) ducts (figs. 20.3 and 20.5). Interestingly, both male and female embryos between day twenty-five and day fifty have both duct systems and, therefore, have the potential to form the accessory organs characteristic of either sex.

The experimental removal of the testes (castration) from male embyronic animals results in the regression of the mesonephric ducts and the development of the paramesonephric ducts into female accessory organs: the uterus and uterine tubes. Female sex accessory organs, therefore, develop as a result of the absence of testes rather than as a result of the presence of ovaries.

The developing seminiferous tubules within the testes secrete a polypeptide called *müllerian inhibition factor* (MIF), which causes the regression of the paramesonephric ducts beginning about day sixty. The secretion of testosterone by the interstitial cells of the testes subsequently causes the growth and development of the mesonephric ducts into male secondary sex organs.

Other embryonic reproductive structures **(urogenital sinus, genital tubercle, urogenital folds,** and **labioscrotal folds)** are also masculinized by secretions of the testes. The prostate gland derives from the urogenital sinus, and the other embryonic structures differentiate into the external genitalia (fig. 20.6). In the absence of testicular secretions, the female genitalia are formed.

In summary, the genetic sex is determined by whether a Y-bearing or an X-bearing sperm fertilizes the ovum; the presence or absence of a Y chromosome in turn determines whether

androgen: Gk. *andros,* male producing

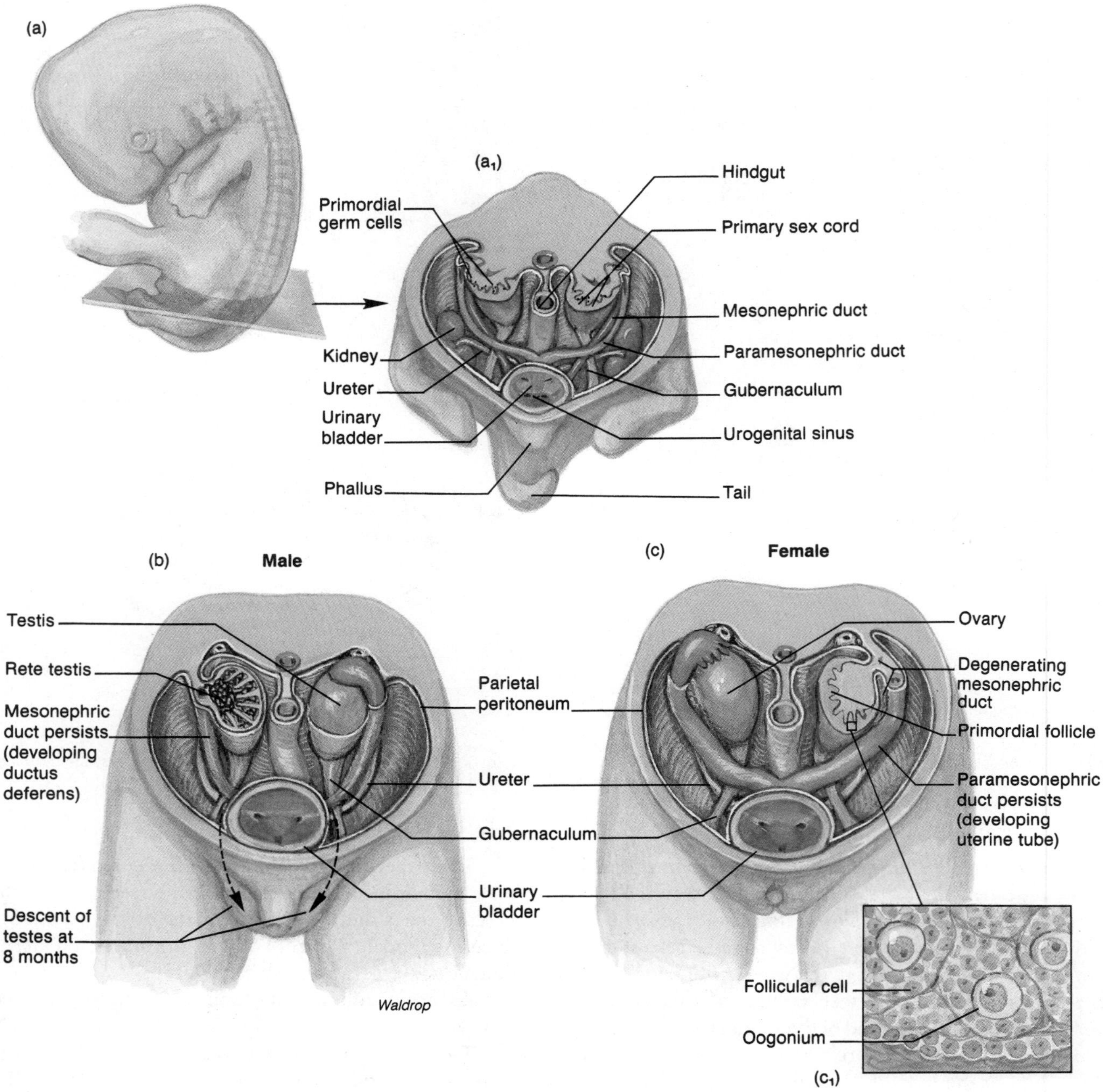

Figure 20.3 Differentiation of the male and female gonads. (*a*) An embryo at six weeks showing the positions of a transverse cut depicted in (a_1, *b*, and *c*). (a_1) At six weeks, the developing gonads (primary sex cords) are still indifferent. By four months, the gonads have differentiated into male (*b*) or female (*c*). The oogonia are formed within the ovaries (c_1) by six months.

the gonads of the embryo will be testes or ovaries; and the presence or absence of testes, (which secrete androgens), finally, determines whether the sex accessory organs and external genitalia will be male or female. This regulatory pattern of sex determination makes sense in light of the fact that both male and female embryos develop within an environment high in estrogen, which is secreted by the mother's ovaries and the placenta. If estrogen determined the sex, all embryos would become feminized.

Fetal Development and Descent of the Testes

By the beginning of the fetal period at nine weeks, male differentiation of the gonads into testes is well under way. Internal changes include the formation of the tubular **seminiferous tubules** and **rete** *(re'te)* **testis** from the primary sex cord (see fig. 20.3). Developing on the outside surface of each testis is a fibromuscular cord called the **gubernaculum** *(gu''ber-nak'u-lum)* (fig. 20.7), which attaches to the inferior portion of the testis and extends to the labioscrotal fold of the same side. At the

rete: L. *rete,* a net
gubernaculum: L. *gubernaculum,* helm

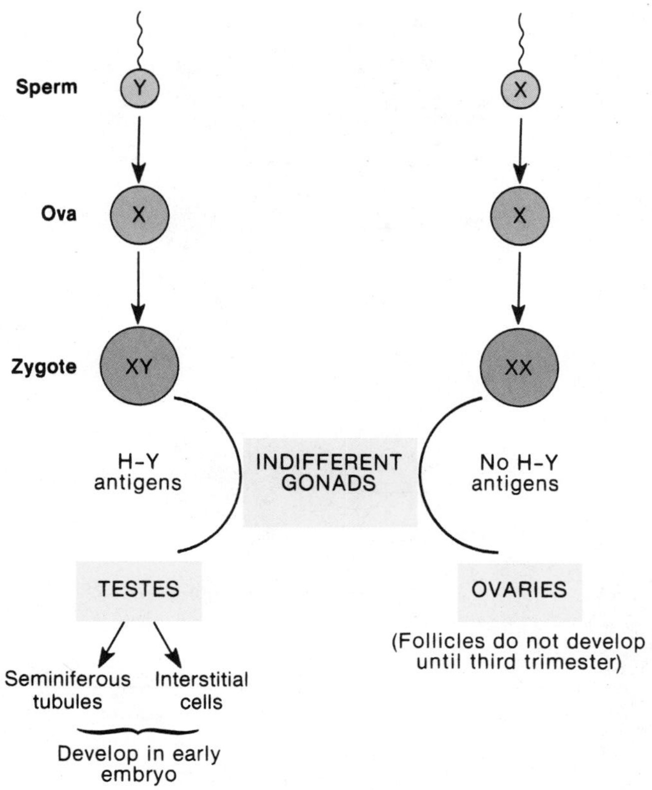

Figure 20.4 The formation of the chromosomal sex of the embryo and the development of the gonads. The very early embryo has indifferent gonads that can develop into either testes or ovaries. If the embryonic cells have Y chromosomes, which code for H-Y antigens, the gonads become testes. If no Y chromosomes and therefore no H-Y antigens are present, the gonads become ovaries. Embryonic testes develop quickly, forming seminiferous tubules (which will produce sperm later in life) and interstitial cells (which secrete testosterone during embryonic development).

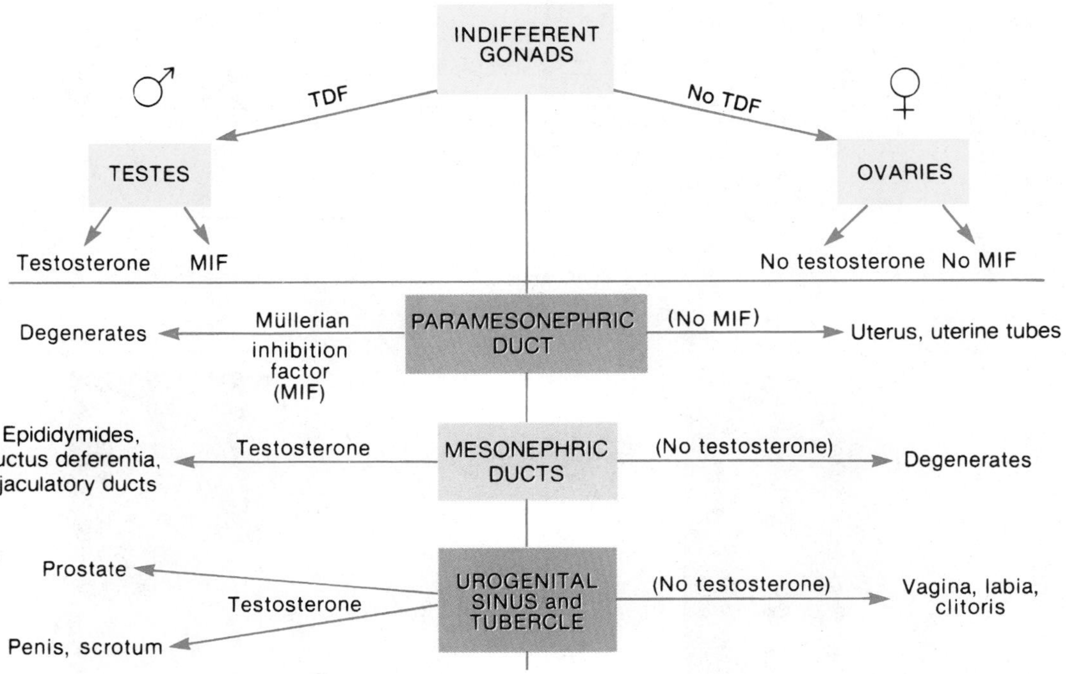

Figure 20.5 The embryonic development of male and female accessory organs and external genitalia. In the presence of testosterone and müllerian inhibition factor (MIF) secreted by the testes, male structures develop. In the absence of these secretions, female structures develop.

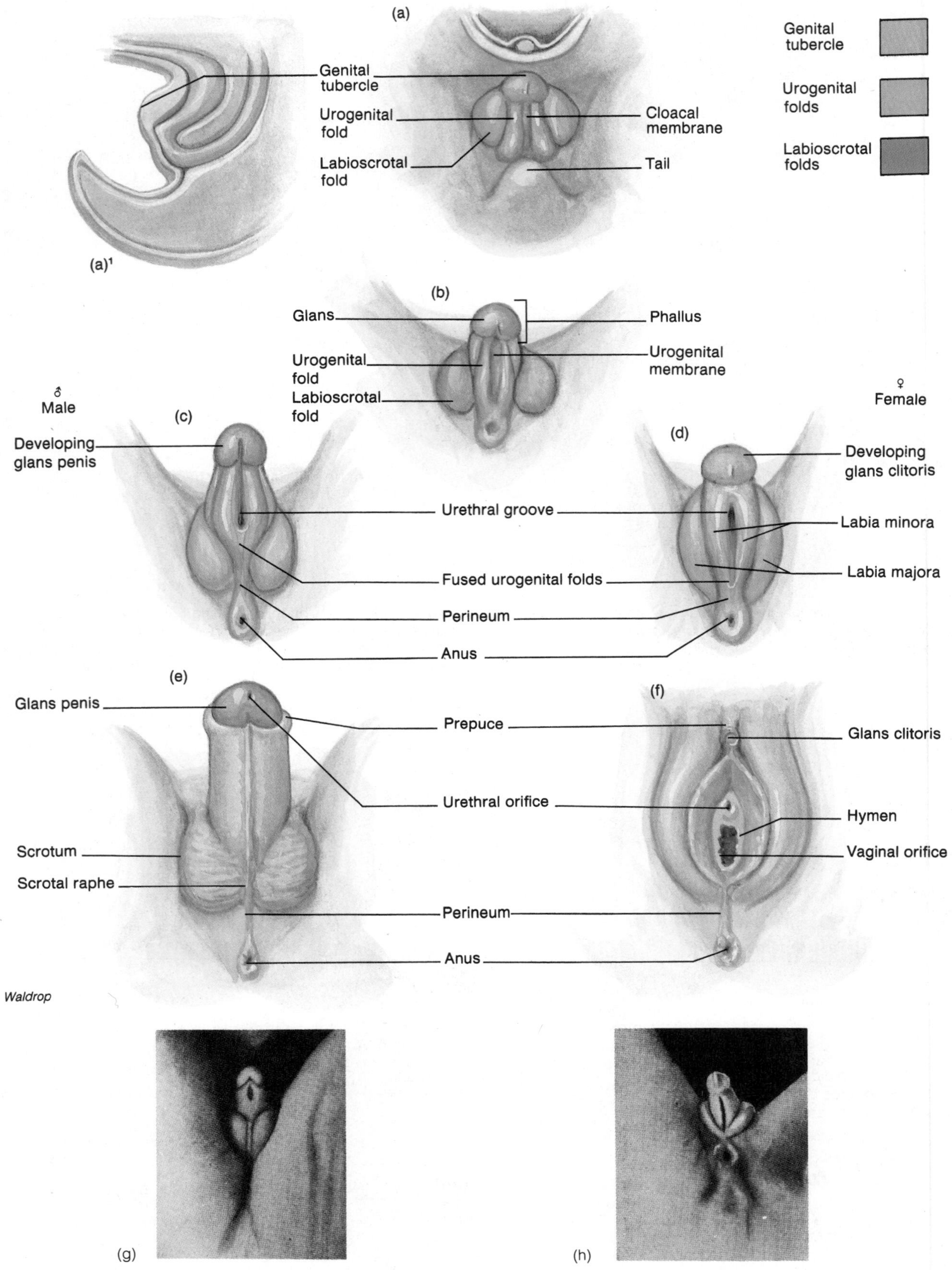

Figure 20.6 Differentiation of the external genitalia in the male and female. (*a*) and (a_1, sagittal view) At six weeks, the genital tubercle, urogenital fold, and labioscrotal swelling have differentiated from the genital tubercle. (*b*) At eight weeks, a distinct phallus is present during the indifferent stage. By the twelfth week, the genitalia are distinctly male (*c*) or female (*d*), being derived from homologous structures. (*e* and *f*) At sixteen weeks, the genitalia are formed. (*g* and *h*) Photographs at week ten of male and female genitalia, respectively.

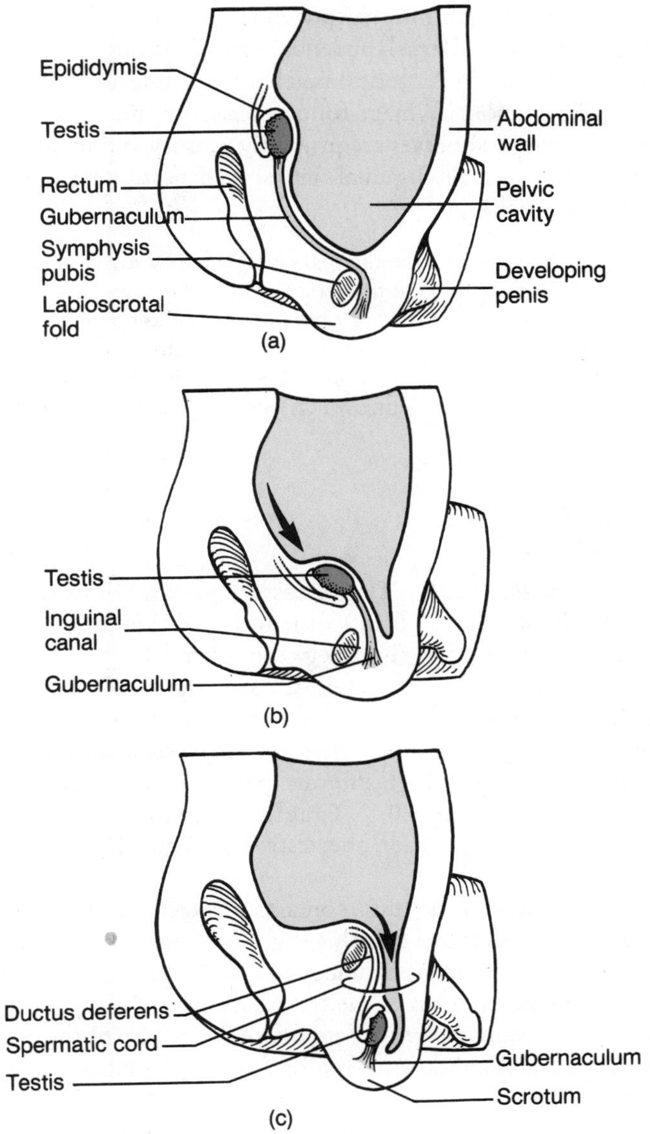

Figure 20.7 The descent of the testes. (*a*) At ten weeks, (*b*) at eighteen weeks, and (*c*) at twenty-eight weeks. During development, each testis descends through an inguinal canal in front of the symphysis pubis and enters the scrotum.

same time, a portion of the embryonic mesonephric duct adjacent to the testis attaches itself to the testis, becomes convoluted, and forms the **epididymis.** Another portion of the mesonephric duct becomes the **ductus deferens.**

The accessory reproductive glands in a male develop from two different sources. The **seminal vesicles** form from lateral outgrowths of the caudal ends of each mesonephric duct. The **prostate gland** arises from an endodermal outgrowth of the urogenital sinus. The **bulbourethral glands** develop from outgrowths in the membranous portion of the urethra.

The external genitalia of the male become distinct from those of a female by the end of the ninth week (see fig. 20.6). Prior to that, the genital tubercle in both sexes elongates and is called a **phallus** *(fal'us)*. A **urethral groove** forms on the ventral surface of the phallus.

phallus: Gk. *phallus,* penis

The differentiation of the external genitalia of a male from the indifferent stage to discernible organs is caused by the androgens produced and secreted by the testes. These changes include the elongation and differentiation of the phallus into a **penis,** a fusion of the **urogenital folds** surrounding the urethral groove along the ventral surface of the penis, and a midventral fusion of the labioscrotal folds to form the wall of the **scrotum.** The external genitalia of a male are completely formed by the end of the twelfth week.

The descent of the testes from the site of development begins between the sixth and tenth week. Descent into the scrotal sac, however, does not occur until about the twenty-eighth week, when paired inguinal canals form in the abdominal wall to provide openings from the pelvic cavity to the scrotal sac. The process by which a testis descends is not well understood, but it seems to be associated with the shortening and differential growth of the gubernaculum, which is attached to the testis and extends through the inguinal canal to the wall of the scrotum (fig. 20.7). As the testis descends, it passes to the side of the urinary bladder and anterior to the symphysis pubis. It carries with it the ductus deferens, the testicular vessels and nerve, a portion of the abdominal muscle, and lymph vessels. All of these structures remain attached to the testis and form what is known as the **spermatic cord** (see fig. 20.19). By the time the testis is in the scrotal sac, the gubernaculum is no more than a remnant of scarlike tissue.

During the physical examination of a neonatal male child, a physician will palpate the scrotum to determine if the testes are in position. If one or both are not in the scrotal sac, it may be possible to induce descent by administering certain hormones. If this procedure does not work, surgery is necessary. The surgery is generally performed before the age of five. Failure to correct the situation may result in sterility and possibly the development of a tumorous testicle.

1. Describe the hormonal process of genital masculinization.
2. Explain how the chromosomal sex determines whether testes or ovaries will be formed.
3. What early embryonic structures give rise to the gonads, and what structures give rise to the external genitalia?
4. List the homologous structures of the male and female reproductive systems.
5. Describe the reproductive changes that occur during the fetal stage of development. How and why do the testes descend into the scrotum?

Structure and Function of the Testes

The testes within the scrotum produce spermatozoa and androgens. Androgens regulate spermatogenesis and the development and functioning of the secondary sex organs.

Objective 8. Describe the location, structure, and functions of the testes.
Objective 9. Explain how hormones control the activities of the male reproductive organs and how they are related to the development of male secondary sex characteristics.
Objective 10. List the events of spermatogenesis, and distinguish between spermatogenesis and spermiogenesis.
Objective 11. Diagram the structure of a sperm, and explain the function of each of its parts.

Scrotum

The saclike **scrotum** is suspended immediately behind the base of the penis, anterior to the anal opening, in a region known as the **perineum** *(per''i-ne'um)* (see fig. 2.15). The functions of the scrotum are to support and protect the testes and to regulate their position relative to the pelvic region of the body. The soft textured skin of the scrotum is covered with sparse hair in mature males and is darker in color than most of the other skin of the body. It also contains numerous sebaceous glands. The external appearance of the scrotum varies at different times in the same individual, depending on environmental conditions and the contraction of the dartos and cremaster muscles. The **dartos** *(dar'tos)* is a layer of smooth muscle fibers in the subcutaneous tissue of the scrotum, and the **cremaster** *(kre-mas'ter)* is a thin strand of skeletal muscle associated with the testis and spermatic cord. Both muscles involuntarily contract in response to cold temperatures to move the testes closer to the heat of the body in the pelvic region. The cremaster muscle is a continuation of the internal oblique muscle of the abdominal wall, from which it is derived as the testes descend into the scrotum. Because it is a skeletal muscle, it can contract voluntarily as well. When these muscles are contracted, the scrotum appears tightly wrinkled. Warm temperatures cause the dartos and cremaster muscles to relax and the testes to become pendent in the flaccid scrotum. The temperature of the testes is maintained at about 35°C (95°F, or about 3.6°F below normal body temperature) by the contraction or relaxation of the scrotal muscles. This temperature has been determined to be an optimum temperature for the production and storage of sperm.

The scrotum is subdivided into two longitudinal compartments by a fibrous **median septum.** The purpose of the median septum is to compartmentalize each testis so that the trauma or infections of one will generally not affect the other. Another protective feature is that the left testis is generally suspended lower in the scrotum than the right so that the two will not as likely be compressed forcefully together. The site of the median septum is apparent on the surface of the scrotum along a median longitudinal ridge called the **perineal raphe** *(ra'fe).* The perineal raphe extends forward to the undersurface of the penis and backward to the anal opening. The blood supply and innervation of the scrotum are extensive. The arteries that serve the scrotum are the internal pudendal branch of the internal iliac artery, the external pudendal branch from the femoral artery, and the cremasteric branch of the inferior epigastric artery. The venous drainage follows a pattern similar to the arteries. The scrotal nerves are primarily sensory and include the pudendal nerves, ilioinguinal nerves, and posterior cutaneous nerves of the thigh.

Although uncommon, *male infertility* may result from an excessively high temperature of the testes over an extended period of time. Tight clothing that keeps the testes close to the body or frequent hot baths or saunas may destroy sperm to the extent that the sperm count of discharged semen is below that necessary to cause fertilization.

Testes

Structure of the Testis The **testes** are paired, whitish, ovoid organs, about 4 cm (1.5 in.) in length and 2.5 cm (1 in.) in diameter. Each testis weighs between 10 and 14 g.

Two tissue layers, or tunicas, cover the testis. The outer **tunica vaginalis** is a thin, serous sac derived from the peritoneum during the descent of the testis. The **tunica albuginea** *(al''bu-jin'e-ah)* is a tough, fibrous membrane that directly encapsules the testis (fig. 20.8). Fibrous, inward extensions of the tunica albuginea partition the testis into 250 to 300 wedge-shaped **lobules.**

Each lobule of the testis contains one to three tightly convoluted **seminiferous tubules** (fig. 20.9), which may exceed 70 cm (28 in.) in length if uncoiled. The seminiferous tubules are the functional units of the testis because it is here that *spermatogenesis* occurs. Sperm are produced at the rate of thousands per second throughout the life of a healthy, sexually mature male.

Various stages of meiosis can be observed in a section of seminiferous tubules (see fig. 20.12). The germinal cells, called **spermatogonia** *(sper''mah-to-go'ne-ah),* are in contact with the basement membrane. Spermatogonia are in a constant state of mitotic division to produce, in order of advancing maturity, the primary spermatocytes, secondary spermatocytes, and spermatids (see fig. 20.11). Forming the walls of the seminiferous tubules are **nurse cells** (Sertoli cells) which produce and secrete nutrients to the developing spermatozoa embedded between them (see fig. 20.12). The spermatozoa are formed, but not fully matured, by the time they reach the lumina of the seminiferous tubules.

Between the seminiferous tubules are specialized endocrine cells called **interstitial cells** (cells of Leydig). The function of these cells is to produce and secrete the male sex hormones. The testes are considered mixed exocrine and endocrine glands because they produce both sperm and androgens.

dartos: Gk. *dartos,* skinned or flayed
cremaster: Gk. *cremaster,* a suspender, to hang
septum: L. *septum,* a partition
raphe: Gk. *raphe,* a seam

pudendal: L. *pudeo,* to feel ashamed
tunica: L. *tunica,* a coat
vaginalis: L. *vagina,* a sheath
albuginea: L. *albus,* white
Sertoli cells: from Enrico Sertoli, Italian histologist, 1842–1910

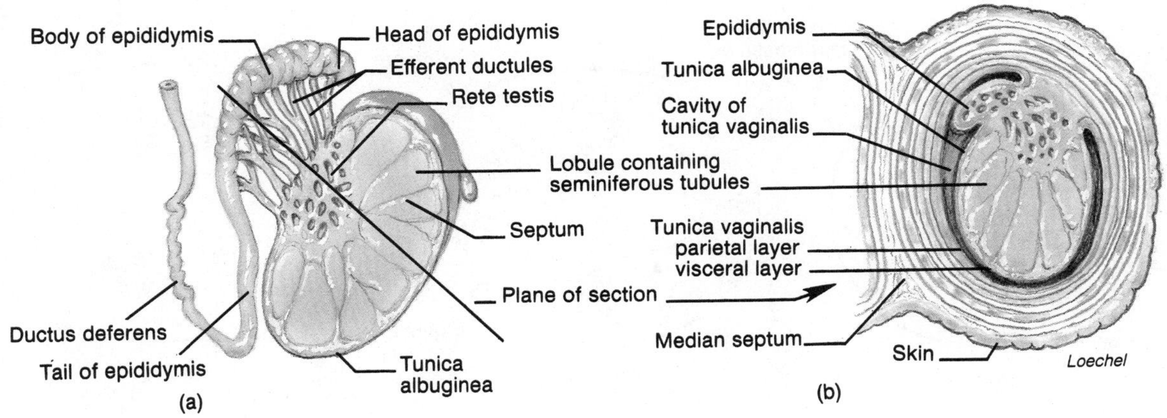

Figure 20.8 Structural features of the testis and epididymis: (*a*) a longitudinal view and (*b*) a transverse view.

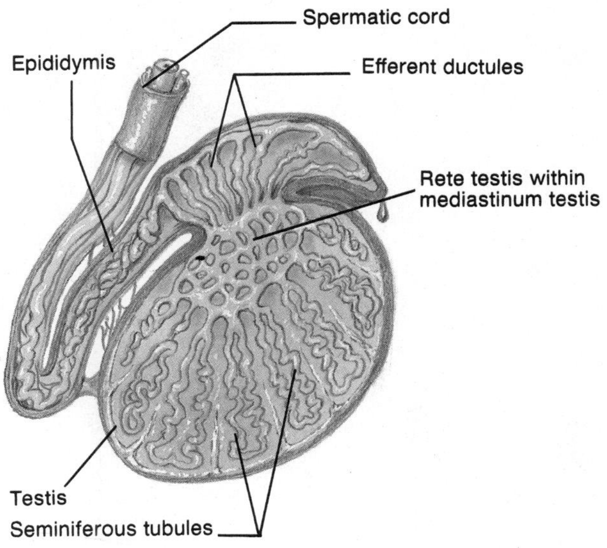

Figure 20.9 The testis and epididymis seen in sagittal section.

Table 20.2 Summary of some of the actions of androgens in the male

Category	Action
Sex determination	Growth and development of mesonephric ducts into epididymides, ductus deferentia, seminal vesicles, and ejaculatory ducts Development of the urogenital sinus and tubercle into prostate gland Development of male external genitalia
Spermatogenesis	At puberty: meiotic division and early maturation of spermatids After puberty: maintenance of spermatogenesis
Secondary sex characteristics	Growth and maintenance of accessory sex organs Growth of the penis Growth of facial and axillary hair Body growth
Anabolic effects	Protein synthesis and muscle growth Growth of bones Growth of other organs (including the larynx) Erythropoiesis (red blood cell formation)

Once the sperm are produced, they move through the seminiferous tubules and enter the **rete testis** for further maturation (fig. 20.9). Cilia are located on some of the cells of the rete testis, presumably for moving the sperm. The sperm are transported out of the testis and into the epididymis through a series of **efferent ductules.**

The testes receive blood through the **testicular arteries,** which arise from the abdominal aorta immediately below the origin of the renal arteries. The **testicular veins** drain the testes. The testicular vein of the right side enters directly into the inferior vena cava, whereas the testicular vein of the left side drains into the left renal vein (see fig. 16.35).

Testicular nerves innervate the testes with both motor and sensory neurons arising from the tenth thoracic segment of the spinal cord. Innervation is primarily through sympathetic neurons, but there is limited parasympathetic stimulation as well.

The primary cause of male infertility is a condition called *varicocele* (var'ĭ-ko-sēl''). Varicocele occurs when one or both of the testicular veins draining from the testes becomes swollen, resulting in poor vascular circulation in the testes. A varicocele generally occurs on the left side, because the left testicular vein drains into the renal vein where the blood pressure is higher than it is in the inferior vena cava, into which the right testicular vein empties.

Endocrine Functions of the Testis Testosterone is by far the major androgen secreted by the adult testis. This hormone is responsible for the initiation and maintenance of the body changes of puberty in males. Androgens are sometimes called *anabolic steroids* because they stimulate the growth of muscles and other structures (table 20.2). Increased testosterone secretion during puberty is also required for the growth of the sex

efferent ductules: L. *efferre,* to bring out; *ducere,* to lead

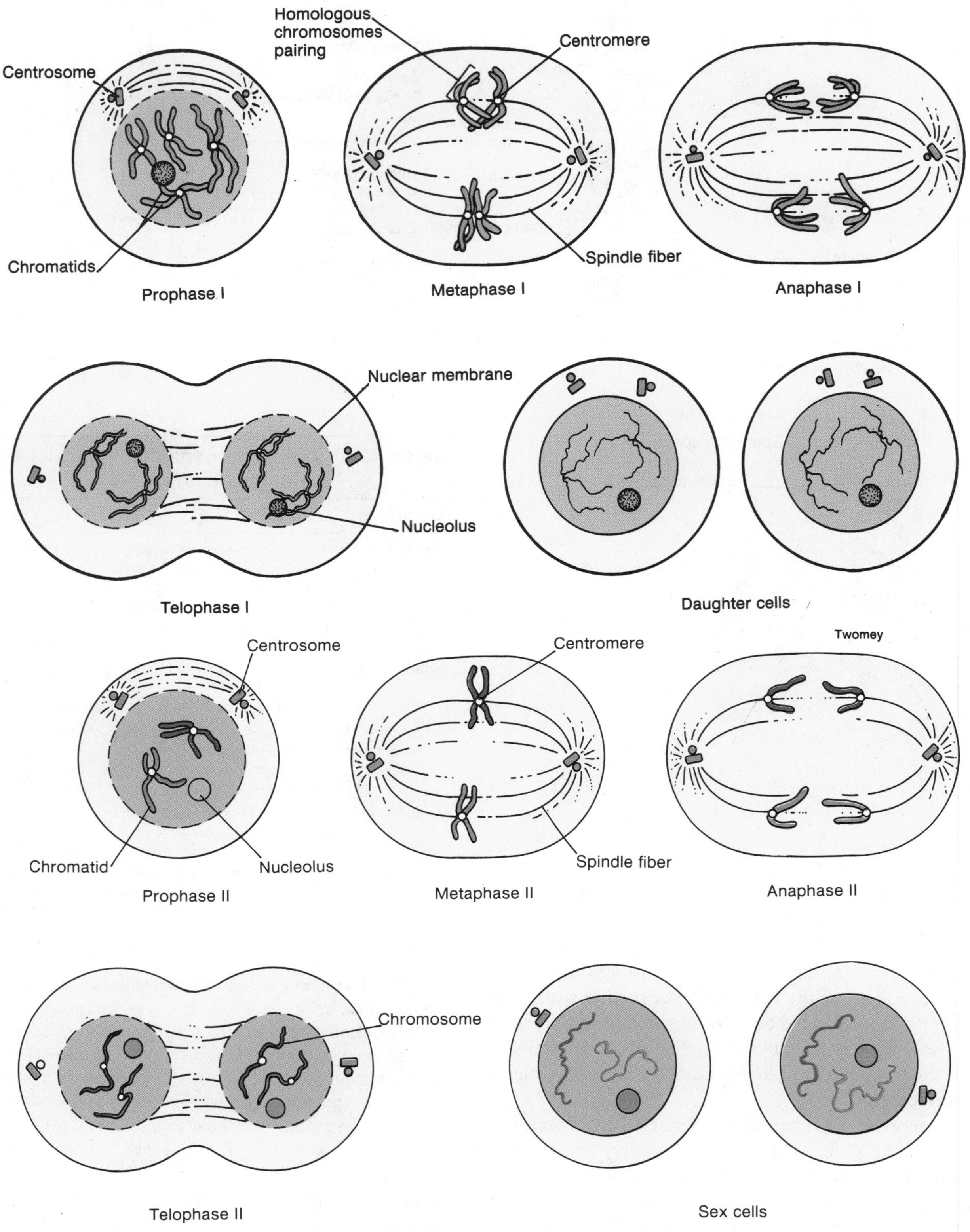

Figure 20.10 Meiosis, or reduction division. In the first meiotic division the homologous chromosomes of a diploid parent cell are separated into two haploid daughter cells. Each of these chromosomes contains duplicate strands, or chromatids. In the second meiotic division these chromatids are distributed to two new haploid daughter cells.

Table 20.3 Stages of meiosis

Stage	Events	Stage	Events
First meiotic division		**Second meiotic division**	
Prophase I	Chromosomes appear double-stranded. Each strand, called a chromatid, contains duplicate DNA joined together by a structure known as a centromere. Homologous chromosomes pair up side by side.	Prophase II	Chromosomes appear, each containing two chromatids.
Metaphase I	Homologous chromosome pairs line up at equator. Spindle apparatus is completed.	Metaphase II	Chromosomes line up single file along equator as spindle formation is completed.
Anaphase I	Homologous chromosomes are separated; each member of a homologous pair moves to opposite poles.	Anaphase II	Centromeres split and chromatids move to opposite poles.
Telophase I	Cytoplasm divides to produce two haploid cells.	Telophase II	Cytoplasm divides to produce two haploid cells from each of the haploid cells formed at telophase I.

accessory organs, primarily the seminal vesicles and prostate gland. The removal of androgens by castration results in atrophy of these organs.

Androgens stimulate the growth of the larynx (causing lowering of the voice), increased hemoglobin synthesis (males have higher hemoglobin levels than females), and bone growth. The effect of androgens on bone growth is self-limiting, however, because androgens ultimately cause the conversion of cartilage to bone in the epiphyseal plates, thus sealing the plates and preventing further lengthening of the bones.

Spermatogenesis The germ cells that migrate from the yolk sac to the testes during early embryonic development become stem cells called **spermatogonia** within the outer region of the seminiferous tubules. Spermatogonia are diploid cells (with forty-six chromosomes) that ultimately give rise to mature haploid gametes by a process of cell division called *meiosis (mi-o'sis)*.

Meiosis occurs within the testes of males who have gone through puberty. Meiosis progresses in two parts as summarized in table 20.3. During the first part of this process, the DNA duplicates (prophase I), and homologous chromosomes are separated (during anaphase I) into two daughter cells (telophase I). Since each daughter cell contains only one of each homologous pair of chromosomes, the cells formed at the end of this *first meiotic division* contain twenty-three chromosomes each and are haploid (fig. 20.10). Each of the twenty-three chromosomes at this stage, however, consists of two strands (called *chromatids*) of identical DNA. During the *second meiotic division,* these duplicate chromatids are separated (at anaphase II) into daughter cells at telophase II. Meiosis of one diploid spermatogonia cell therefore produces four haploid daughter cells.

Actually, only about 1,000–2,000 stem cells migrate from the yolk sac into the embryonic testes. In order to produce many millions of sperm throughout adult life, these spermatogonia cells duplicate themselves by mitotic division, and only one of the two cells—now called a **primary spermatocyte**—undergoes meiotic division (fig. 20.11). In this way, spermatogenesis can occur continuously without exhausting the number of spermatogonia.

When a diploid primary spermatocyte completes the first meiotic division (at telophase I), the two haploid daughter cells thus produced are called **secondary spermatocytes.** At the end

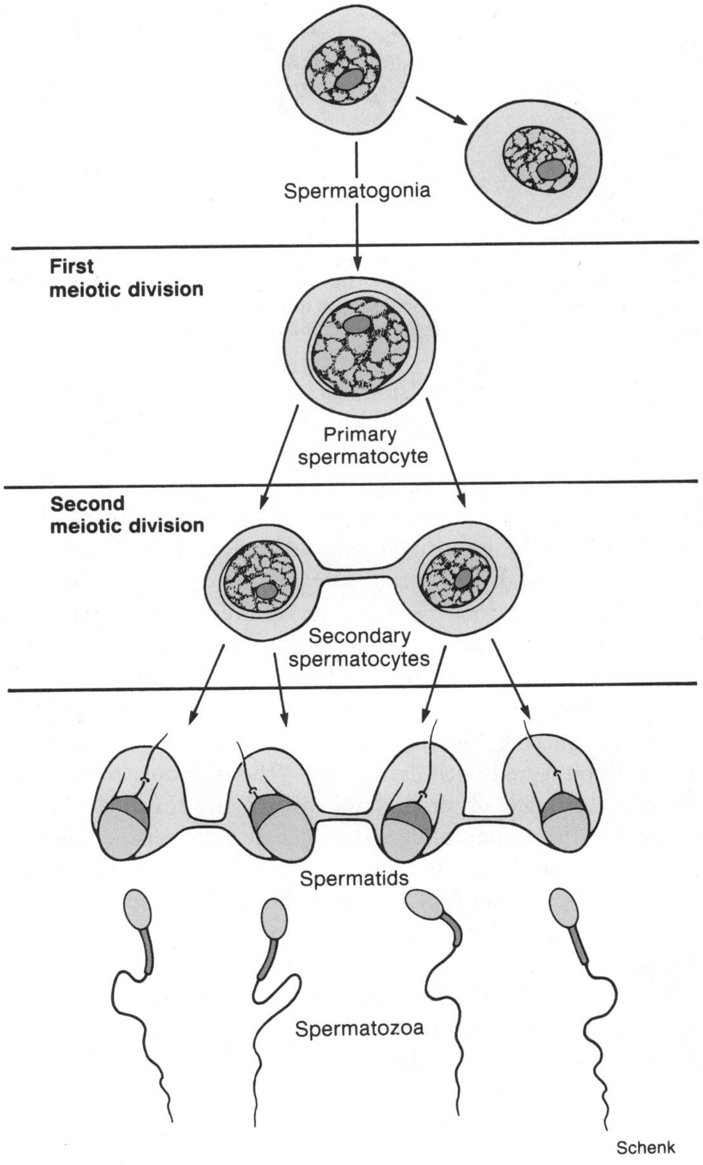

Figure 20.11 Spermatogonia undergo mitotic division to replace themselves and produce a daughter cell that will undergo meiotic division. This cell is called a primary spermatocyte. Upon completion of the first meiotic division, the daughter cells are called secondary spermatocytes. Each of these completes a second meiotic division to form spermatids. Notice that the four spermatids produced by the meiosis of a primary spermatocyte are interconnected. Each spermatid forms a mature spermatozoon.

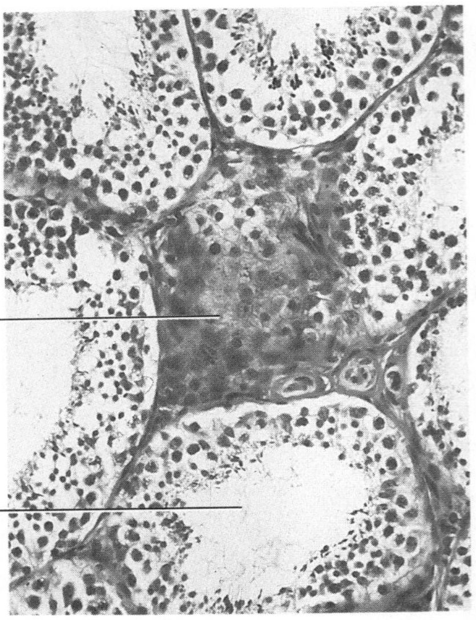

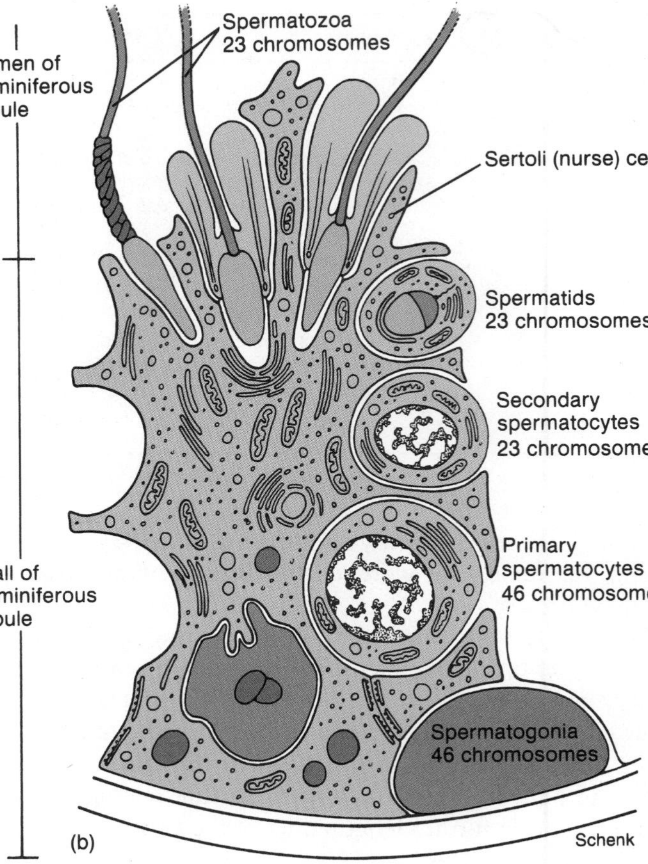

Figure 20.12 Seminiferous tubules. (*a*) A transverse section with surrounding interstitial tissue and (*b*) the stages of spermatogenesis within the germinal epithelium of a seminiferous tubule, showing the relationship between nurse cells and developing spermatozoa.

of the second meiotic division, each of the two secondary spermatocytes produce two haploid **spermatids.** One primary spermatocyte therefore produces four spermatids.

The stages of spermatogenesis are arranged sequentially in the wall of the seminiferous tubule. The epithelial wall of the tubule, called the **germinal epithelium,** is indeed composed of germ cells in different stages of spermatogenesis. The spermatogonia and primary spermatocytes are located toward the outer side of the tubule, whereas the spermatids and mature spermatozoa are located toward the lumen (fig. 20.12).

At the end of the second meiotic division, the four spermatids produced by the meiosis of two secondary spermatocytes are interconnected with each other—their cytoplasm does not completely pinch off at the end of each division. The development of these interconnected spermatids into separate, mature **spermatozoa** (a process called *spermiogenesis*) requires the participation of another type of cell in the tubules, the nurse cells.

Nurse Cells The nurse (Sertoli) cells are the only nongerminal cell type in the tubules. They form a continuous layer, connected by tight junctions, around the circumference of each tubule. In this way the nurse cells comprise a *blood-testis barrier,* because molecules from the blood must pass through the cytoplasm of nurse cells before entering germinal cells. The cytoplasm of the nurse cells extends through the width of the tubule and envelops the developing germ cells, so that it is often difficult to tell where the cytoplasm of nurse cells and germ cells are separated.

In the process of *spermiogenesis* (the conversion of spermatids to spermatozoa), most of the spermatid cytoplasm is eliminated (fig. 20.12). This occurs through phagocytosis by the nurse cells of the residual bodies of cytoplasm from the spermatids. It is believed that phagocytosis of the residual bodies may transmit informational molecules from the germ cells to the nurse cells. The nurse cells, in turn, may provide many molecules needed by the germ cells. It is known, for example, that the X chromosome of the germ cells is inactive during meiosis. Since this chromosome contains the genes needed to produce many essential molecules, it is believed that these molecules are provided by the nurse cells during this time.

Structure of Spermatozoa

A mature sperm cell, or **spermatozoon,** is a microscopic, tadpole-shaped structure about 0.06 mm long (fig. 20.13). It consists of an oval-shaped **head,** a cylindrical **body,** and an elongated **tail.** The head of a sperm contains a nucleus with twenty-three chromosomes. The tip of the head, called the *acrosome,* contains enzymes that help the sperm penetrate into the ovum. The body of the sperm contains numerous mitochondria spiraled around a filamentous core. The mitochondria provide the energy necessary for locomotion. The tail of the sperm is a flagellum that propels the sperm with a lashing movement. The maximum unassisted rate of sperm movement is about 3 mm per hour.

The life expectancy of ejaculated sperm is between forty-eight and seventy-two hours at body temperature. Many of the ejaculated sperm, however, are defective and are of no value. It is not uncommon for sperm to have enlarged heads, dwarfed and misshapen heads, two flagella, or a flagellum that is bent. Sperm such as these are unable to propel themselves adequately.

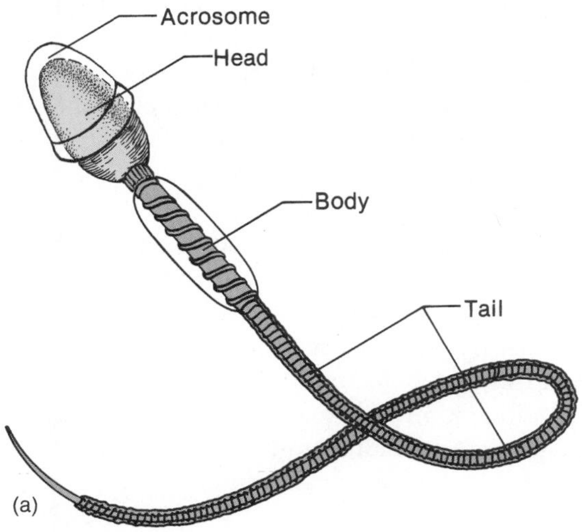

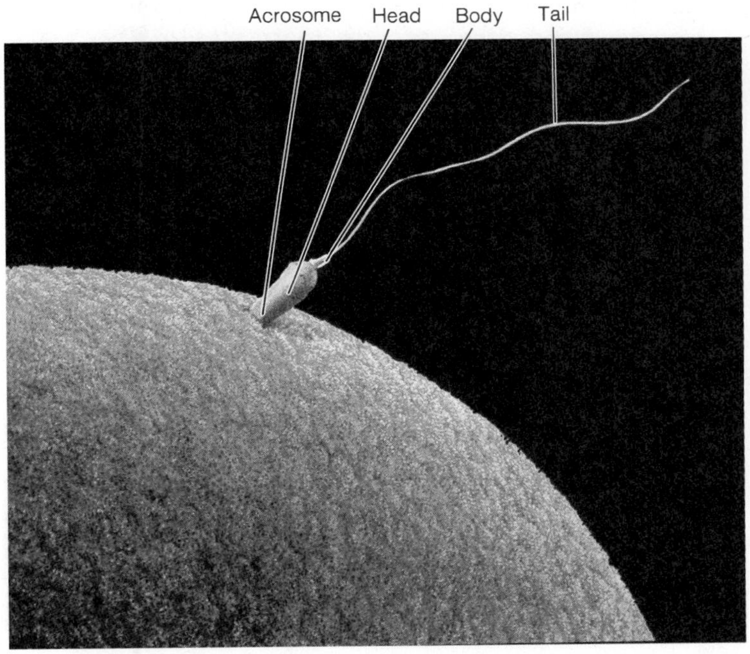

(b)

Figure 20.13 A human spermatozoon: (*a*) A diagrammatic representation and (*b*) a scanning electron micrograph of a spermatozoon in contact with an egg.

1. Describe the location and structure of the testes.
2. What is the function of the seminiferous tubules, the germinal epithelial cells, and the interstitial cells?
3. Discuss the effect of testosterone on the production of sperm and the development of the secondary sex characteristics.
4. Using a diagram, describe the stages of spermatogenesis. Explain why spermatogenesis can continue throughout life without using up all of the spermatogonia.
5. Diagram a sperm and its adjacent nurse cell, and explain the functions of nurse cells in the seminiferous tubules.

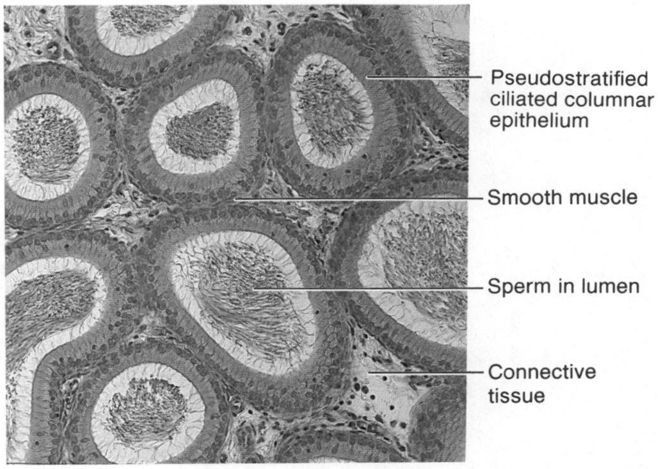

Figure 20.14 The histology of the epididymis showing sperm in the lumen (50×).

Spermatic Ducts, Accessory Glands, and the Penis

The spermatic ducts store spermatozoa and transport them from the testes to the urethra. The accessory reproductive glands provide additives to the spermatozoa in the formation of semen, which is discharged from the erect penis during ejaculation.

Objective 12. List the various spermatic ducts, and describe the location and structure of each segment.
Objective 13. Describe the structure and contents of the spermatic cord.
Objective 14. Describe the location, structure, and function of the ejaculatory ducts, seminal vesicles, prostate gland, and bulbourethral glands.
Objective 15. Describe the structure and function of the penis.

Spermatic Ducts

The duct system, which stores and transports spermatozoa from the testes to the urethra, includes the epididymides, the ductus deferentia, and the ejaculatory ducts.

Epididymis The **epididymis** *(ep''ĭ-did'ĭ-mis)*—pl., *epididymides*—is a long, flattened organ, attached to the posterior surface of the testis (see figs. 20.1, 20.8). The tubular portion of the epididymis is highly coiled and contains millions of sperm in their final stages of maturation. It is estimated that if the epididymis were uncoiled, it would measure 5–6 m (about 17 ft). The upper, expanded portion of the epididymis is the **head,** the tapering middle section is the **body,** and the lower, tubular portion is the **tail.** The tail of the epididymis is continuous with the ductus deferens. Both the tail of the epididymis and the ductus deferens store the sperm that is to be discharged during ejaculation (fig. 20.14). The time required to produce mature sperm—from meiosis in the seminiferous tubules to storage in the ductus deferens—is approximately two months.

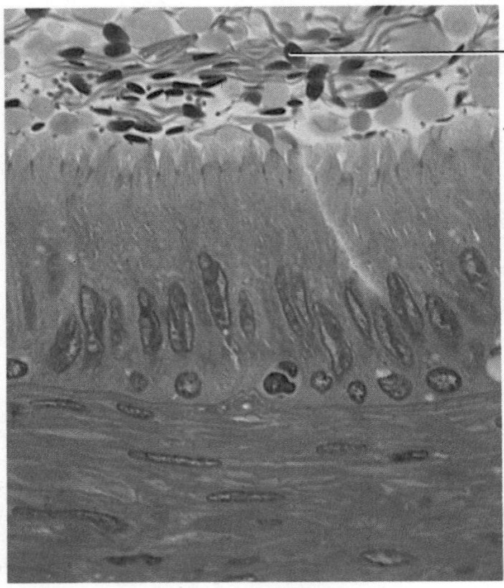

Figure 20.15 A photomicrograph of the ductus deferens (250×).

Ductus Deferens The **ductus deferens** *(duk'tus def'er-enz)*—pl., *ductus deferentia*—also called *vas deferens,* is a fibromuscular tube about 45 cm (18 in.) long and 2.5 mm thick (see fig. 20.1), which conveys sperm from the epididymis to the ejaculatory duct. The ductus deferens originates where the tail of the epididymis becomes less convoluted and is no longer attached to the testis. This first portion of the ductus deferens is important for the storage of sperm. The ductus deferens exits from the scrotum as it ascends along the posterior border of the testis. From here, it penetrates the inguinal canal, enters the pelvic cavity, and passes to the side of the urinary bladder on the medial side of the ureter. The **ampulla** of the ductus deferens is the terminal portion that joins the ejaculatory duct.

The histological structure of the ductus deferens includes a layer of pseudostratified ciliated columnar epithelium in contact with the tubular lumen and surrounded by three layers of tightly packed smooth muscle (fig. 20.15). Sympathetic nerves from the pelvic plexus serve the ductus deferens. Stimulation through these nerves causes peristaltic contractions of the muscular layer, which forcefully ejects the stored sperm toward the ejaculatory duct.

Much of the ductus deferens is located within a structure known as the **spermatic cord** (see figs. 20.9, 20.19). The spermatic cord extends from the testis to the inguinal ring and consists of the ductus deferens, spermatic vessels, nerves, cremaster muscle, lymph vessels, and connective tissue. The portion of the spermatic cord that is anterior to the pubic bone can be palpated on a male as it is compressed between the skin and the bone. The **inguinal canal** is an opening for the spermatic cord to traverse the inguinal ligament and is a potentially weak area and a common site for a hernia to develop.

Ejaculatory Duct The ejaculatory *(e-jak'u-lah-to''re)* duct is 2 cm (1 in.) long and is formed by the union of the ampulla of the ductus deferens and the duct of the seminal vesicle. The ejaculatory duct then pierces the capsule of the prostate gland on its posterior surface and continues through this gland (fig. 20.1). Both ejaculatory ducts receive secretions from the seminal vesicles and prostate gland and then eject the sperm with its additives into the prostatic urethra.

Accessory Glands

Accessory reproductive glands include the seminal vesicles, the prostate gland, and the bulbourethral glands (see fig. 20.1). The contents of the seminal vesicles and the prostate gland are mixed with the sperm during ejaculation to form *semen (seminal fluid).* The fluid from the bulbourethral glands is released in response to sexual stimulation prior to ejaculation. Both the seminal vesicles and prostate are androgen-dependent organs. They atrophy if androgen is deprived by castration.

Seminal Vesicles The seminal vesicles, which are convoluted, club-shaped glands about 5 cm (2 in.) long, are immediately posterior to and at the base of the urinary bladder. They secrete a sticky, slightly alkaline, yellowish substance, which serves as a fluid medium to enhance sperm movement and longevity. The secretion from the seminal vesicles contains a variety of nutrients, including fructose, a monosaccharide that provides sperm with an energy source. It also contains citric acid, coagulation proteins, and prostaglandins. The discharge from the seminal vesicles makes up about 60% of the volume of semen.

Histologically, the seminal vesicle appears as a mass of cuts embedded in connective tissue (fig. 20.16). The extensively coiled mucosal layer breaks the lumen into numerous intercommunicating spaces that are lined by pseudostratified columnar and cuboidal secretory epithelia (referred to as glandular epithelium).

Blood is supplied to the seminal vesicles by branches from the middle rectal arteries. The seminal vesicles are innervated by both sympathetic and parasympathetic neurons. Sympathetic stimulation causes the contents of the seminal vesicles to empty into the ejaculatory ducts of their respective sides.

Prostate Gland The firm prostate *(pros'tāt)* gland is the size and shape of a horse chestnut. It is about 4 cm (1.6 in.) across and 3 cm (1.2 in.) thick and positioned immediately below the urinary bladder, surrounding the beginning of the urethra (see fig. 20.1). It is enclosed by a fibrous capsule and divided into five distinct lobules. The lobules are formed by the urethra and the ejaculatory ducts that extend through the gland. The ducts from the lobules open into the urethra. Extensive bands of smooth muscular tissue course throughout the prostate to form a meshwork that supports the glandular tissue (fig. 20.17). Contraction of the smooth muscle within the prostate empties the contents from the gland and provides part of the propulsive force needed to ejaculate the semen. The thin, milky-colored, prostatic secretion assists sperm motility as a liquefying agent, and its alkalinity protects sperm in their passage through the acidic environment of the female vagina. The prostate also secretes the enzyme *acid phosphatase,* which is often measured clinically to assess prostate function. The discharge from the prostate gland makes up about 40% of the volume of the semen.

deferens: L. *deferens,* conducting away
ampulla: L. *ampulla,* a two-handled bottle

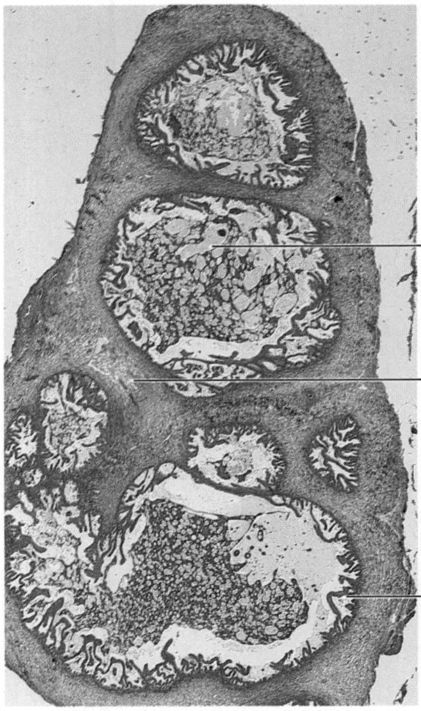

Figure 20.16 The histology of the seminal vesicle.

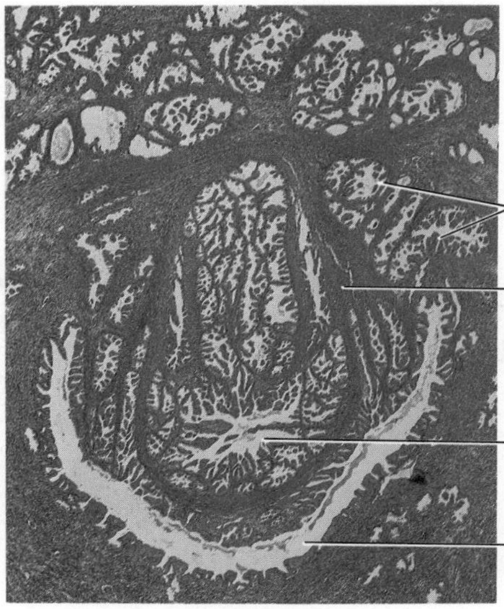

Figure 20.17 The histology of the prostate gland.

Blood is supplied to the prostate from branches of the middle rectal and inferior vesical arteries. The venous return forms the prostatic venous plexus along with blood draining from the penis. The prostatic venous plexus drains into the internal iliac veins. The prostate gland has both sympathetic and parasympathetic innervation arising from the pelvic plexuses.

A routine physical examination of the male includes rectal palpation of the prostate gland. Enlargement or overgrowth of the glandular substance of the prostate, called *benign prostatic hypertrophy,* is relatively common in older men. This may constrict the urethra and cause difficult micturition. An enlarged prostate gland usually requires surgery. If the obstruction is slight, the surgery may be accomplished through the urethral canal using a technique called a *transurethral prostatic resection,* in which excessive tissue is cut and cauterized.

Bulbourethral Glands The paired, pea-sized bulbourethral (Cowper's) glands are inferior to the prostate gland. Each bulbourethral gland is brownish in color and about 1 cm in diameter, and it drains by a 2.5 cm (1 in.) duct into the urethra (see fig. 20.1). Upon sexual excitement and prior to ejaculation, the bulbourethral glands are stimulated to secrete a mucoid substance, which coats the lining of the urethra to neutralize the pH of the urine residue and lubricates the tip of the penis in preparation for sexual intercourse.

Urethra

The **urethra** of the male serves as a common tube for both the urinary and reproductive systems. However, urine and semen cannot simultaneously pass through the urethra because the nervous reflex during ejaculation automatically inhibits micturition. The urethra of the male is about 20 cm (8 in.) long and S-shaped due to the shape of the penis. Three regions can be identified—the prostatic urethra, the membranous urethra, and the penile urethra (see fig. 19.14).

The **prostatic urethra** is the proximal (2.5 cm) portion of the urethra that passes through the prostate gland. The prostatic urethra receives drainage from the small ducts of the prostate gland and the two ejaculatory ducts.

The **membranous urethra** is the short (0.5 cm) portion of the urethra that passes through the urogenital diaphragm. The external urethral sphincter muscle is located in this region.

The **penile urethra** (cavernous urethra) is the longest portion (15 cm), extending from the outer edge of the urogenital diaphragm to the external urethral orifice on the glans penis. This portion is surrounded by erectile tissue as it passes through the corpus spongiosum of the penis. The paired ducts from the bulbourethral glands attach to the penile urethra near the urogenital diaphragm.

The wall of the urethra has an inside lining of mucous membrane, composed of transitional epithelium (fig. 20.18) and surrounded by a relatively thick layer of smooth muscle tissue, called *tunica muscularis.* Specialized **urethral glands** are embedded in the urethral wall and function to secrete mucus into the urethral canal.

Penis

The **penis** is composed mainly of erectile tissue and, when distended, serves as the copulatory organ of the male reproductive system. **Erectile tissue** contains numerous vascular spaces that become engorged with blood. The penis is a pendent structure, which is anterior to the scrotum and attached to the pubic arch. It is divided into a proximal attached root, an elongated tubular shaft, and a distal cone-shaped glans penis (fig. 20.19).

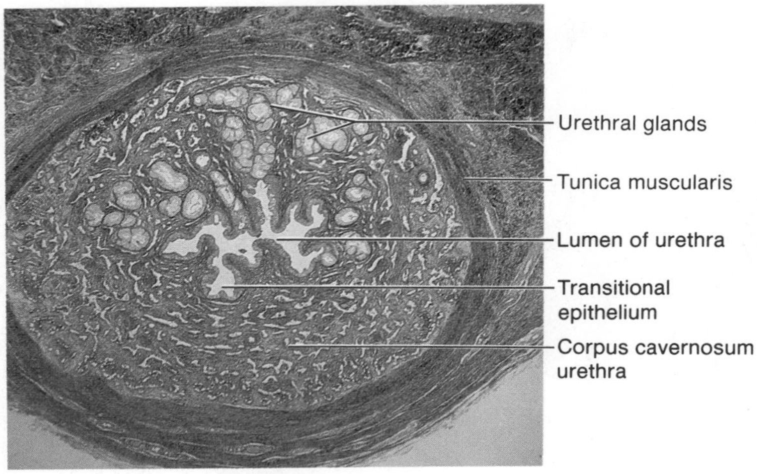

Figure 20.18 The histology of the urethra (10×).

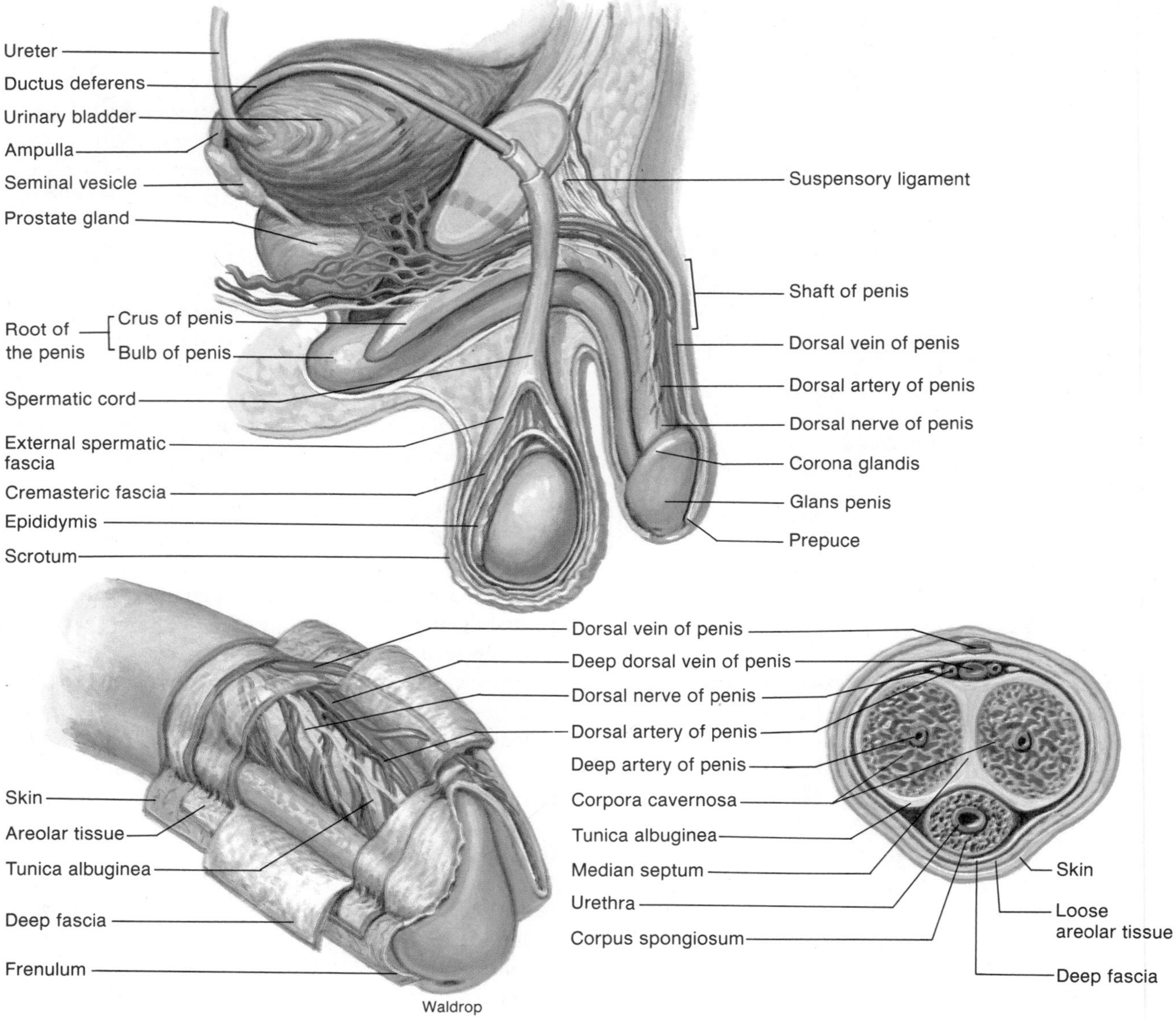

Figure 20.19 The structure of the penis showing the attachment, blood and nerve supply, and the arrangement of the erectile tissue.

The **root of the penis** expands posteriorly to form the **bulb of the penis** and the **crus** *(krus)* **of the penis.** The bulb is positioned in the urogenital triangle of the perineum, where it is attached to the undersurface of the urogenital diaphragm and enveloped by the bulbocavernosus muscle (see fig. 9.26). The crus, in turn, attaches the root of the penis to the pubic arch (ischiopubic ramus) and to the perineal membrane. The crus, which is superior to the bulb, is enveloped by the ischiocavernosus muscle (see fig. 9.25).

The **shaft,** or body, **of the penis** is composed of three cylindrical columns of erectile tissue that are bound together by fibrous tissue and covered with skin (fig. 20.19). The paired dorsally positioned masses are named the **corpora cavernosa penis.** The fibrous tissue between the two corpora forms a **median septum.** The **corpus spongiosum penis (corpus cavernosum urethrae)** is ventral to the other two and surrounds the penile urethra. The penis is flaccid and relaxed when the spongelike tissue is not engorged with blood but becomes firm and erect when the spaces are filled.

The **glans penis** is the cone-shaped terminal portion of the penis, which is formed from the expanded corpus spongiosum. The opening of the urethra at the tip of the glans is called the **external urinary meatus.** The **corona glandis** is the prominent posterior ridge of the glans. On the undersurface of the glans, a vertical fold of tissue, called the **frenulum** *(fren'u-lum),* attaches the skin covering the penis to the glans.

The skin covering the penis is hairless, contains no fat cells, and generally has a darker pigment than the other body skin. The skin of the shaft is loosely attached and is continuous over the glans as a retractable sheath called the **prepuce** *(pre'pūs),* or **foreskin.** The prepuce is commonly removed in a newborn by a surgical procedure called *circumcision.*

The penis is supplied with blood through the superficial external pudendal branch of the femoral artery and the internal pudendal branch of the internal iliac artery. The venous return is through a superficial median dorsal vein that drains into the great saphenous vein in the thigh and through the deep median vein that drains into the prostatic plexus.

A *circumcision* is generally performed for hygienic purposes because the glans is easier to clean if exposed. A sebaceous secretion from the glans, called *smegma,* will accumulate along the border of the corona glandis if good hygiene is not practiced. Smegma can foster bacteria that may cause infections and therefore should be removed through washing. Cleaning the glans of an uncircumcised male requires retraction of the prepuce. Occasionally, a child is born with a prepuce that is too tight to permit retraction. This condition is called *phimosis* and necessitates circumcision.

---

crus: L. *crus,* leg, resembling a leg
cavernosa: L. *cavus,* hollow
glans: L. *glans,* acorn
corona: L. *corona,* garland, crown
frenulum: L. diminutive of *frenum,* a bridle
prepuce: L. *prae,* before; *putium,* penis
phimosis: Gk. *phimosis,* a muzzling

The penis has many sensory tactile receptors, especially in the glans, that make it a highly sensitive organ. In addition, the penis has extensive sympathetic and parasympathetic motor innervation.

---

1. What are the functions of each of the regions of the spermatic duct?
2. Differentiate between the ductus deferens and the spermatic cord.
3. Describe the structure and location of each of the accessory glands.
4. Why is the position of the prostate gland of clinical importance?
5. What do each of the accessory glands secrete?
6. Describe the external structure of the penis and the internal arrangement of the erectile tissue within the penis.

Mechanisms of Erection, Emission, and Ejaculation

Erection of the penis results from parasympathetic induced vasodilation of arteries within the penis. Emission and ejaculation are stimulated by sympathetic impulses, resulting in the forceful expulsion of semen from the penis.

Objective 16. Distinguish between erection, emission, and ejaculation.
Objective 17. Describe the events that result in erection of the penis.
Objective 18. Explain the physiological process of ejaculation.
Objective 19. Describe the characteristic properties of semen.

Erection, emission, and ejaculation are interrelated events that are necessary for the deposition of semen into the vagina by the natural means of coitus (sexual intercourse). *Erection* usually occurs as a male becomes sexually aroused; the erectile tissue of the penis becomes engorged with blood, causing the penis to become wider, longer, and firmer. *Emission* is the movement of spermatozoa from the epididymides to the ejaculatory ducts. *Ejaculation* is the forceful expulsion of the ejaculate, or *semen* (seminal fluid), from the ejaculatory ducts and urethra of the penis. Emission and ejaculation do not have to follow erection of the penis and only occur if there is sufficient stimulation of the sensory receptors in the penis to elicit the ejaculatory response.

Erection of the Penis

Erection of the penis depends on the volume of blood that enters the arteries of the penis as compared to the volume that exits through venous drainage. Normally, constant sympathetic stimuli to the arterioles of the penis maintain a partial constriction of smooth muscles within the arteriole walls so that there is an even flow of blood throughout the penis. During sexual excitement, however, parasympathetic impulses cause marked vasodilation within the arterioles of the penis, resulting in more

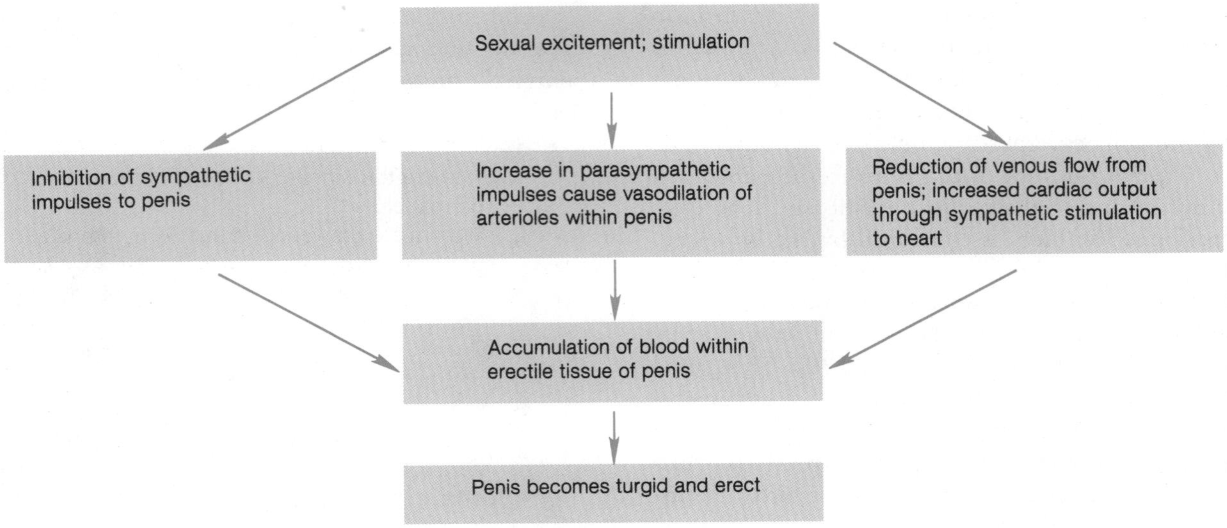

Figure 20.20 The mechanism of erection of the penis.

blood entering than venous blood draining. And during parasympathetic stimulation, there is inhibition of sympathetic impulses to arterioles of the penis. At the same time, there may be slight vasoconstriction of the dorsal vein of the penis and an increase in cardiac output. These combined events cause the spongy tissue of the corpora cavernosa and the corpus spongiosum to become distended with blood and the penis to become turgid. In this condition, the penis can be inserted into the vagina of the female and function as a copulatory organ to discharge semen.

Erection is controlled by two portions of the central nervous system—the hypothalamus in the brain and the sacral portion of the spinal cord. The hypothalamus controls conscious sexual thoughts that originate in the cerebral cortex. Nerve impulses from the hypothalamus elicit parasympathetic responses from the sacral region, which cause vasodilation of the arterioles within the penis. Conscious thought is not required for an erection, however, and stimulation of the penis can cause an erection because of a reflex response in the spinal cord. This reflexive action makes possible an erection in a sleeping male or in an infant—perhaps from the stimulus of a diaper.

The mechanism of erection of the penis is summarized in figure 20.20.

Ejaculation is the expulsion of semen through the urethra of the penis. In contrast to erection, ejaculation is a response involving the sympathetic innervation of the accessory reproductive organs. Ejaculation is preceded by continued sexual stimulation, usually through activated tactile receptors in the glans penis and the skin of the shaft. Rhythmic friction of these structures during coitus causes sensory impulses to be transmitted to the thalamus and cerebral cortex. The first sympathetic response, which occurs prior to ejaculation, is the discharge from the bulbourethral glands. The fluids from these glands are usually discharged before penetration into the vagina and serve to lubricate the urethra and the glans penis.

Emission and Ejaculation of Semen

Emission Continued sexual stimulation following erection of the penis causes emission. Emission is the movement of sperm from the epididymides to the ejaculatory ducts and the secretions of the accessory glands into the ejaculatory ducts and urethra in the formation of semen. Emission occurs as sympathetic impulses from the pelvic plexus cause a rhythmic contraction of the smooth muscle of the epididymides, ductus deferentia, ejaculatory ducts, seminal vesicles, and prostate gland.

Ejaculation Ejaculation immediately follows emission and is accompanied by *orgasm,* which is considered the climax of the sex act. Ejaculation occurs in a series of spurts of semen from the urethra. This takes place as parasympathetic impulses traveling through the pudendal nerves stimulate the bulbocavernous muscles (see fig. 9.25) at the base of the penis and cause them to contract rhythmically. There is also sympathetic stimulation of the smooth muscles in the urethral wall that peristaltically contract to help eject the semen.

Sexual function in the male thus requires the synergistic action (rather than antagonistic action) of the parasympathetic and sympathetic divisions of the ANS. The mechanism of emission and ejaculation is summarized in figure 20.21.

Immediately following ejaculation or a cessation of sexual stimulus, sympathetic impulses cause vasoconstriction of the arterioles within the penis, reducing the inflow of blood. At the same time, cardiac output returns to normal as does venous return of blood from the penis. With the normal flow of blood through the penis, it returns to its flaccid condition. If an ejaculation of semen occurred while the penis was erect, another erection and ejaculation cannot be triggered for ten to thirty minutes or longer.

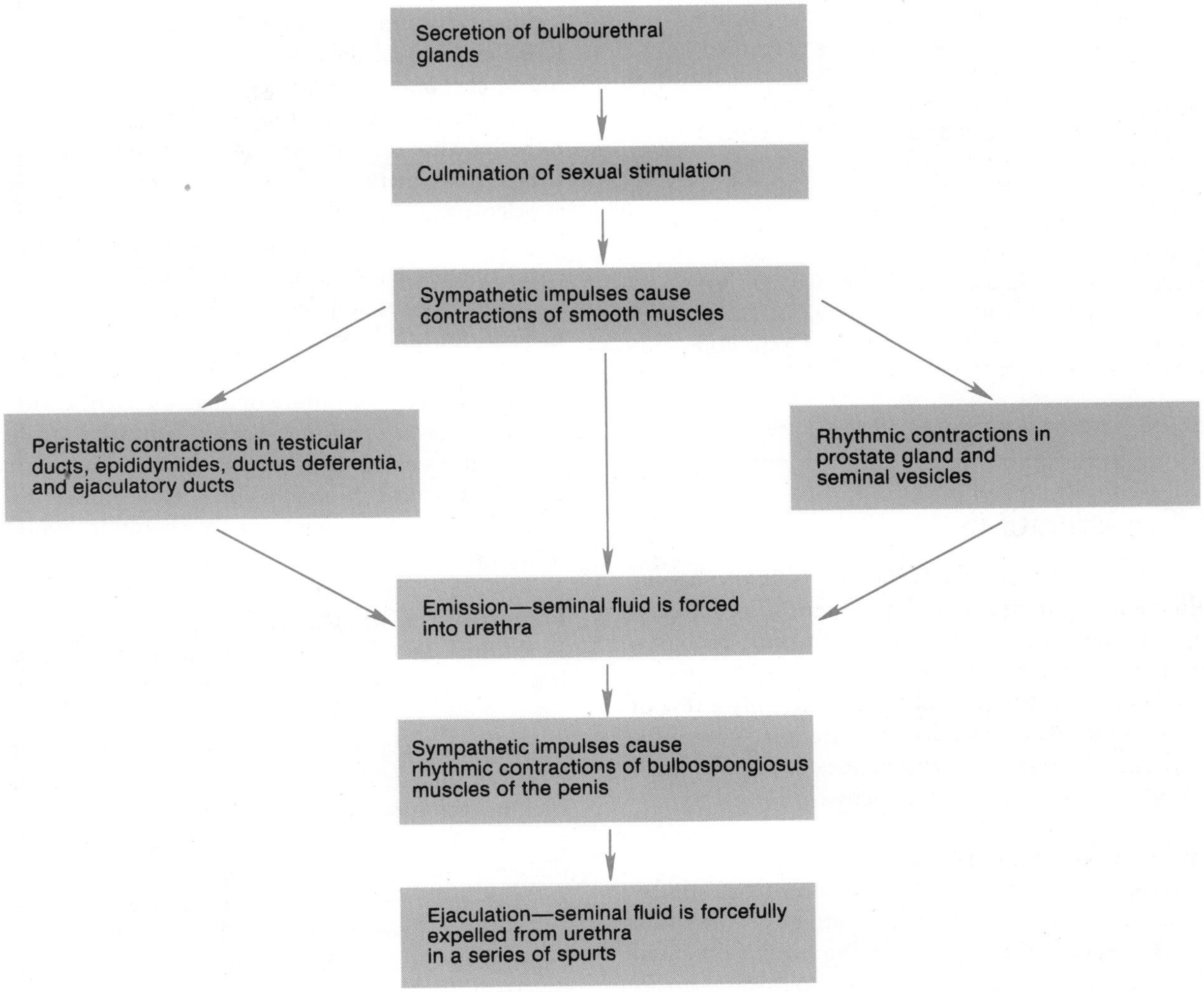

Figure 20.21 The mechanism of emission and ejaculation.

Adolescent males may experience erection of the penis and spontaneous emission and ejaculation of semen during sleep. These *nocturnal emissions* are thought to be caused by changes in hormonal concentrations that accompany adolescent development.

Semen Semen, also called seminal *(sem'ĭ-nal)* fluid, which consists of spermatozoa plus the additives from the accessory reproductive glands, is the substance discharged during ejaculation (table 20.4). Generally, between 1.5 and 5.0 ml of semen are ejected during ejaculation. The bulk of the fluid (about 60%) is produced by the seminal vesicles, and the remaining (about 40%) is contributed by the prostate gland. Spermatozoa constitute less than 1% of the volume of an ejaculate. There are usually between 60 and 150 million sperm per milliliter of ejaculate. In the condition of *oligospermia,* the male ejaculates fewer than 10 million sperm per milliliter and is likely to have fertility problems.

Table 20.4 Some characteristics and reference values used in clinical examination of semen

Characteristic	Reference value
Volume of ejaculate	1.5–5.0 ml
Sperm count	40–250 million/ml
Sperm motility	
Percent of motile forms:	
1 hour after ejaculation	70% or more
3 hours after ejaculation	60% or more
Leukocyte count	0–2000/ml
pH	7.2–7.8
Fructose concentration	150–600 mg/100 ml

Modified from Glasser, L. "Seminal Fluid and Subfertility." *Diagnostic Medicine* July/August 1981, p. 28. By permission.

Human semen can be frozen and stored in sperm banks for future artificial insemination. In this procedure, the semen is diluted with 10% glycerol, monosaccharide, and distilled water buffer and frozen in liquid nitrogen. The freezing process destroys defective and abnormal sperm. For some unknown reason, however, not all human sperm is suitable for freezing.

1. Define the terms *erection, emission* and *ejaculation.*
2. Explain the statement that male sexual function is an autonomic synergistic action.
3. Compose a flow chart to explain the physiological and physical events of erection, emission, and ejaculation.
4. Describe the components of a normal ejaculate.

Clinical Considerations

Sexual dysfunction is a broad area of medical concern that includes developmental and psychogenic problems as well as conditions resulting from various diseases. Psychogenic problems of the reproductive system are extremely complex, poorly understood, and beyond the scope of this book. Only a few of the principal developmental conditions, functional disorders, and diseases that affect the physical structure and function of the male reproductive system will be discussed.

Developmental Abnormalities

The reproductive organs of both sexes develop from similar embryonic tissue that follows a consistent pattern of formation well into the fetal period. Because an embryo has the potential to differentiate into a male or a female, developmental errors can result in various degrees of intermediate sex, or **hermaphroditism** *(her-maf'ro-di-tizm'')*. A person with undifferentiated or ambiguous external genitalia is called a **hermaphrodite.**

True hermaphroditism—in which both male and female gonadal tissues are present, in either the same or opposite gonads—is a rare anomaly. True hermaphrodites usually have a forty-six, XX chromosome constitution. **Male pseudohermaphroditism** occurs more commonly and generally results from hormonal influences during early fetal development. This condition is caused either by inadequate amounts of androgenic hormones being secreted or by the delayed development of the reproductive organs after the period of tissue sensitivity has passed. These persons have a forty-six, XY chromosome constitution and male gonads, but the genitalia are intersexual and variable. The treatment of hermaphroditism varies, depending on the extent of ambiguity of the reproductive organs. These persons are sterile but may marry and live a normal life following hormonal therapy and plastic surgery.

Chromosomal anomalies result from the improper separation of the chromosomes during meiosis and are usually expressed in deviations of the reproductive organs. The two most frequent chromosomal anomalies cause Turner's syndrome and Klinefelter's syndrome. **Turner's syndrome** occurs when only one X chromosome is present. About 97% of embryos lacking an X chromosome die; the remaining 3% survive and appear to be females, but their gonads are rudimentary or absent, and they do not mature at puberty. A person with **Klinefelter's syndrome** has an XXY chromosome constitution, develops breasts and male genitalia, but has underdeveloped seminiferous tubules and is generally retarded.

A more common developmental problem than genetic abnormalities, and fortunately less serious, is cryptorchidism. **Cryptorchidism** *(krip-tor'kĭ-dizm)* means hidden testis and is characterized by the failure of one or both testes to descend into the scrotum. A cryptorchid testis is usually located along the path of descent but can be anywhere in the pelvic cavity (fig. 20.22). It occurs in about 3% of male infants and should be treated before the infant is five years old to reduce the chance of infertility or other complications.

Functional Considerations

Functional disorders of the male reproductive system include impotence, infertility, and sterility. **Impotence** *(im'po-tens)* is the inability of a sexually mature male to achieve penile erection or to achieve ejaculation. The causes of impotence may be physical, such as abnormalities of the penis, vascular irregularities, neurological disorders, or the result of diseases. Generally, however, the cause of impotence is psychological, and the patient requires skilled counseling by a sex therapist.

Infertility is the inability of the sperm to fertilize the ovum and may be the fault of the male or female, or both. The term *impotence* should not be used when referring to infertility. Infertility in males may be caused by a number of things, the most common of which is the inadequate production of viable sperm. Some of the causes of infertility in males are alcoholism, dietary deficiencies, local injury, varicocele, excessive heat, or exposure to X rays. A hormonal imbalance may also contribute to infertility. Many of the causes of infertility can be treated through proper nutrition, gonadotrophic hormone treatment, or microsurgery. If corrective treatment is not possible, however, it may be possible to concentrate the sperm obtained through *masturbation* (in males, self-stimulation to the point of ejaculation) and use this concentrate to artificially inseminate the woman.

Sterility is similar to infertility except that it is a permanent condition. Sterility may be genetically caused, or it may be the result of degenerative changes in the seminiferous tubules (for example, mumps in a mature male may secondarily infect the testes and cause irreversible tissue damage).

Voluntary sterilization of the male in a procedure called a **vasectomy** is a common technique of birth control. In this procedure, a small section of each ductus deferens near the ep-

hermaphrodite: Gk. (mythology) *Hermaphroditos,* son of Hermes (Mercury)

Turner's syndrome: from Henry H. Turner, American endocrinologist, 1892–1970

Klinefelter's syndrome: from Harry F. Klinefelter, Jr., American physician, b. 1912

cryptorchidism: Gk. *crypto,* hidden; *orchis,* testis

impotence: L. *im,* not; *potens,* potent

sterility: L. *sterilis,* barren

vasectomy: L. *vas,* vessel; Gk. *ektome,* excision

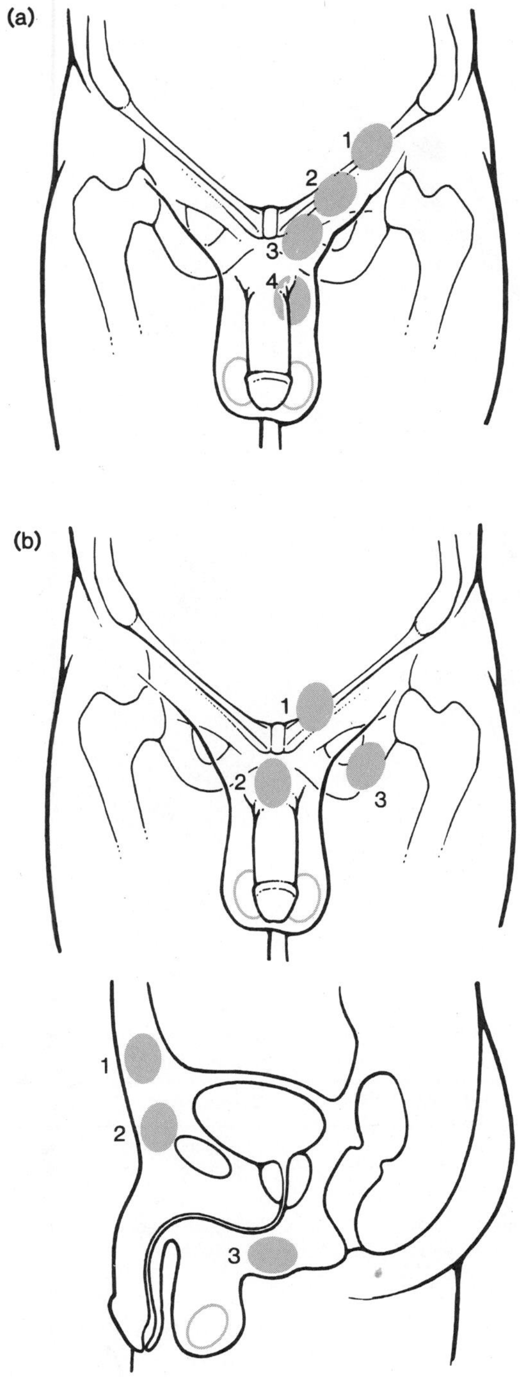

Figure 20.22 Cryptorchidism. (*a*) Incomplete descent of a testis may involve four separate regions: (*1*) in the pelvic cavity, (*2*) in the inguinal canal, (*3*) at the superficial inguinal ring, (*4*) in the upper scrotum. (*b*) An ectopic testis may be (*1*) in the superficial fascia of the anterior pelvic wall, (*2*) at the root of the penis, (*3*) in the perineum, in the thigh alongside the femoral vessels.

ididymis is surgically removed, and the cut ends of the ducts are tied (fig. 20.23). A vasectomy interferes with sperm transport but does not directly affect the secretion of androgens from interstitial cells in the interstitial tissue. Since spermatogenesis continues, the sperm cannot be drained from the testes and instead accumulate in the crypts that form in the seminiferous tubules and ductus deferens. These crypts present sites of inflammatory reactions in which spermatozoa are phagocytosed and destroyed by the immune system.

Diseases of the Male Reproductive System

Sexually Transmitted Diseases Sexually transmitted diseases, frequently called venereal diseases (VD), are contagious diseases that affect the reproductive systems of both the male and the female (table 20.5) and are transmitted during sexual activity. The frequency of sexually transmitted diseases in the United States is regarded by health authorities as epidemic. These diseases have not been eradicated, mainly because humans cannot develop immunity to them and increased sexual promiscuity increases the chances of reinfection.

Gonorrhea *(gon''o-re'ah),* commonly called clap, is caused by the bacterium gonococcus, or *Neisseria gonorrhoeae.* Males with this disease suffer inflammation of the urethra, accompanied by painful urination and frequently the discharge of pus. In females, the condition is usually asymptomatic, and therefore many women may be unsuspecting carriers of the disease. Advanced stages of gonorrhea in females may infect the uterus and the uterine tubes. A pregnant woman with gonorrhea that is not treated may transmit the disease to the eyes of her newborn, causing blindness.

Syphilis *(sif'ĭ-lis)* is caused by the bacterium *Treponema pallidum.* Syphilis is less common than gonorrhea but is the more serious of the two diseases. During the *primary stage* of syphilis, a lesion called a *chancre* develops at the point where contact was made with a similar sore from an infected person. The chancre is an ulcerated sore that has hard edges and endures for ten days to three months. It is only during the primary stage that syphilis can be spread to another sexual partner. The chancre will heal with time, but if not treated, it will be followed by secondary and tertiary stages of syphilis. During the initial contact, the bacteria enter the bloodstream and spread throughout the body. The *secondary stage* of syphilis is expressed by lesions or a rash of the skin and mucous membranes, accompanied by fever (fig. 20.24). This stage lasts from two weeks to six months, and the symptoms disappear of their own accord. The *tertiary stage* occurs ten to twenty years following the primary infection. The circulatory, integumentary, skeletal, and nervous systems are particularly vulnerable to the degenerative changes caused by this disease. The end result of untreated syphilis is blindness, insanity, and eventual death.

AIDS, or **acquired immune deficiency syndrome,** is a viral disease that is transmitted primarily through coitus and drug abuse (by sharing contaminated syringe needles). Additional information about this fatal disease, for which there is no known cure, is presented in table 20.5.

Disorders of the Prostate Gland The prostate gland is subject to several disorders, most of which are common in older men. The four most frequent prostatic problems are acute prostatitis, chronic prostatitis, benign prostatic hyperplasia, and carcinoma of the prostate.

---

venereal: L. (mythology) from *Venus,* the goddess of love
gonorrhea: L. *gonos,* seed; *rhoia,* a flow
chancre: Fr. *chancre,* indirectly from L. *cancer,* a crab

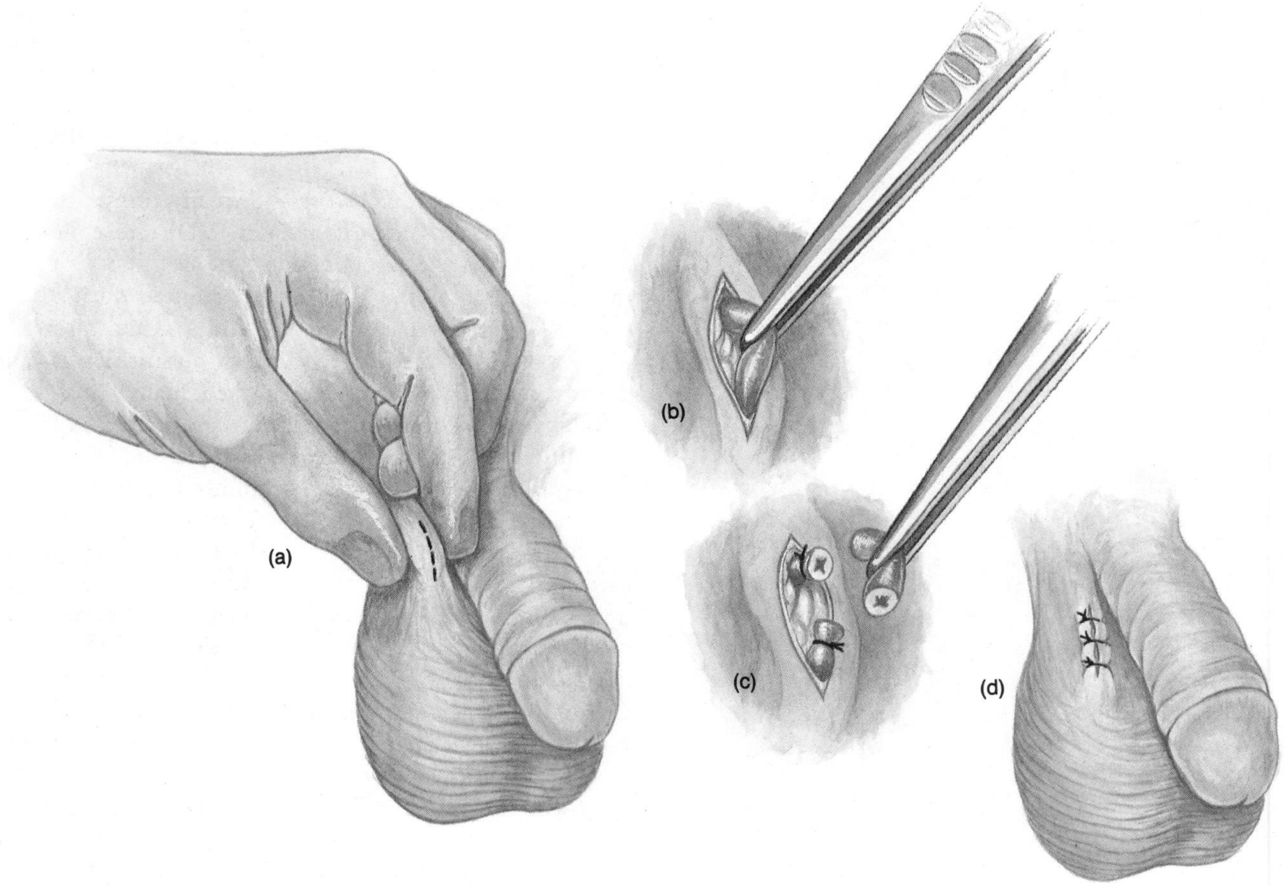

Figure 20.23 A simplified illustration of a vasectomy, in which a segment of the ductus deferens is removed through an incision in the scrotum. The procedure is then repeated on the opposite side.

Table 20.5 Kinds of sexually transmitted diseases

Name	Organism	Resulting condition	Treatment
Gonorrhea	*Gonococcus* (bacterium)	Adult: sterility due to scarring of epididymides and tubes; rarely: septicemia; newborn: blindness	Penicillin injections; tetracycline tablets; eye drops (silver nitrate or penicillin)
Syphilis	*Treponema pallidum* (bacterium)	Adult: gummas, cardiovascular neurosyphilis; newborn: congenital syphilis (abnormalities, blindness)	Penicillin injections; tetracycline tablets
Chancroid (soft chancre)	*Hemophilus ducreyi* (bacterium)	Chancres, buboes	Tetracycline; sulfa drugs
Urethritis in men	Various microorganisms	Clear discharge	Tetracycline
Vaginitis	*Trichomonas* (protozoan)	Frothy white or yellow discharge	Metronidazole
	Candida albicans (yeast)	Thick, white, curdy discharge (moniliasis)	Nystatin
Acquired immunity deficiency syndrome (AIDS)	Human immunodeficiency virus (HIV)	Early symptoms include extreme fatigue, weight loss, fever, diarrhea; severe susceptibility to pneumonia, rare infections, and cancer	Azidothymidine (AZT, or Retrovir); no cure available
Chlamydia	*Chlamydia trachomatis* (bacterium)	Whitish discharge from penis or vagina; pain during urination	Tetracycline and sulfonamides
Lymphogranuloma venereum (LGV)	Microorganism	Ulcerating buboes; rectal stricture	Tetracycline; sulfa drugs
Granuloma venereum (inguinale)	*Donovania granulomatis*	Raw, open, extended sore	Tetracycline
Venereal warts	Virus	Warts	Podophyllin
Genital herpes	Herpes simplex virus	Sores	Palliative treatment
Crabs	Arthropod	Itching	Gamma benzene hexachloride

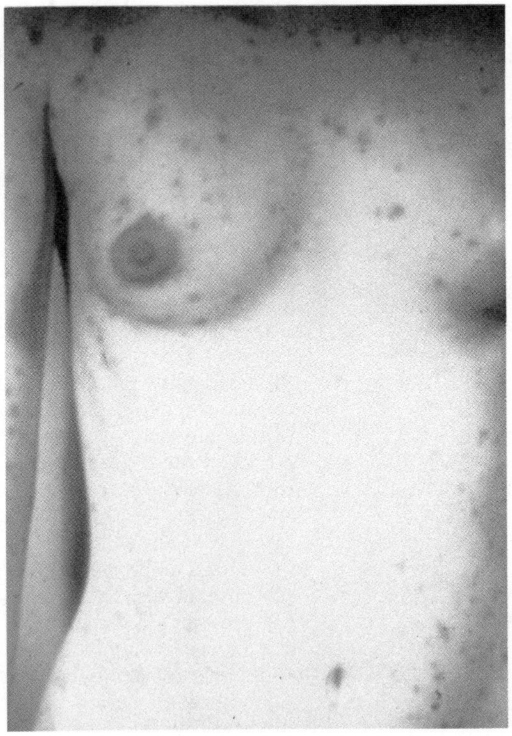

Figure 20.24 The secondary stages of syphilis as expressed by lesions of the skin of this young woman.

Acute prostatitis is common in sexually active young men through infections acquired from a gonococcus bacterium. The symptoms of acute prostatitis are a swollen and tender prostate gland, painful urination, and in extreme conditions, pus dripping from the penis. It is treated with penicillin, bed rest, and increased fluid intake.

Chronic prostatitis is one of the most common afflictions of middle-aged and elderly men. The symptoms of this condition vary considerably from irritation and slight difficulty in urination to extreme pain and urine blockage, which commonly causes secondary renal infections. In this disease, several kinds of infectious microorganisms are believed to be harbored in the prostate gland and are responsible for inflammations elsewhere in the body, such as in the nerves (neuritis), the joints (arthritis), the muscles (myositis), and the iris (iritis).

Prostatic hyperplasia, or an enlarged prostate gland, occurs in approximately one-third of all males over the age of sixty. In this condition, an overgrowth of granular material compresses the prostatic urethra. The cause of prostatic hyperplasia is not known. As the prostate enlarges, urination becomes painful and difficult. If the urinary bladder is not emptied completely, cystitis eventually occurs. Persons with cystitis may become incontinent and dribble urine continuously. Prostatic hypertrophy is usually treated by the surgical removal of portions of the gland through transurethral curetting (cutting and removal of a small section) or the removal of the entire prostate gland, called **prostatectomy.**

Prostatic carcinoma, or cancer of the prostate gland, is the second leading cause of death from cancer in males in the United States. It is common in males over sixty and accounts for 19,000 deaths annually. When prostatic cancer is confined to the prostate gland, it is generally small and asymptomatic. But as the cancer grows and invades surrounding nerve plexuses, it becomes extremely painful and easily detected. The metastases of this cancer to the spinal column and brain are generally what kills the patient.

As prostatic carcinoma develops, it has symptoms nearly identical to prostatic hyperplasia—painful urination and cystitis. When examined by rectal palpation with a gloved finger, however, a hard cancerous mass can be detected in contrast to the enlarged, soft, and tender prostate diagnostic of prostatic hypertrophy. Prostatic carcinoma is treated by prostatectomy and frequently by the removal of the testes (called **orchiectomy**) as well. An orchiectomy inhibits metastases by eliminating testosterone secretion.

Disorders of the Testes and Scrotum A **hydrocele** *(hi'dro-sēl)* is a benign fluid mass within the tunica vaginalis that causes swelling of the scrotum. It is a frequent, minor disorder in infant boys as well as in adults. The cause is unknown.

An infection in the testes is called **orchitis.** Orchitis may develop from a primary infection from a tubercle bacterium or as a secondary complication of mumps contracted after puberty. If orchitis from mumps involves both testes, it usually causes sterility.

Trauma to the testes and scrotum is common because of their pendent position. The testes are extremely sensitive to pain, and a male responds reflexively to protect the groin area.

CLINICAL CASE STUDY ANSWER

The tubular structures that are apparently absent in our patient are the ductus deferentia. This condition, known as *congenital bilateral absence of the ductus deferentia,* prevents spermatozoa from being transported from the testes to the ejaculatory ducts. This explains the absence of spermatozoa in the patient's ejaculate since the accessory reproductive glands that contribute to the production of seminal fluid add their secretions at a point in the reproductive tract distal to the absent duct. The accessory reproductive glands include the seminal vesicles (which in many cases are also absent or nonfunctional in this deformity) and the prostate gland. Until recently, this condition would have categorically prevented our patient from becoming a father. Microsurgical extraction of spermatozoa from the epididymes is now possible, however, and has allowed many afflicted men to father children. ■

Chapter Summary

I. Introduction to the Male Reproductive System
   A. The functions of the male reproductive system are to produce sperm, secrete testosterone, and transfer sperm to the reproductive system of the female.
   B. The male reproductive system is divided into primary sex organs (the testes), secondary sex organs (those that are essential for sexual reproduction), and secondary sex characteristics (features that are sexual attractants, which are expressed after puberty).

II. Development of the Male Reproductive System
   A. An XY chromosome combination produces a male, and an XX chromosome combination produces a female.
   B. The appearance of the gonadal ridge during the fifth week is the first indication of sex organ formation.
      1. Primary sex cords and a genital tubercle develop during the sixth week.
      2. Reproductive development is well progressed by the eighth week but is in the indifferent stage because external sexual distinction is not apparent.
   C. The testes secrete testosterone, which stimulates the development of male accessory sex organs and male genitalia.
   D. The penis and scrotum are formed by the end of the twelfth week, and the descent of the testes occurs during the twenty-eighth week.

III. Structure and Function of the Testes
   A. The saclike scrotum supports and protects the testes and regulates their position relative to the pelvic region of the body.
   B. The testes are separated into lobules, composed of seminiferous tubules, which produce sperm and interstitial tissue that produces androgens.
   C. Spermatogenesis occurs by meiotic division of the cells that line the seminiferous tubules.
      1. At the end of the first meiotic division, two secondary spermatocytes are produced.
      2. At the end of the second meiotic division, four haploid spermatids are produced.
   D. The conversion of spermatids to spermatozoa is called spermiogenesis.
   E. A sperm consists of a head, body, and tail and matures in the epididymides prior to ejaculation.

IV. Spermatic Ducts, Accessory Glands, and the Penis
   A. The epididymides, ductus (vas) deferentia, and ejaculatory ducts are the components of the spermatic ducts.
      1. The highly coiled epididymides are the tubular structures on the testes where sperm mature and are stored.
      2. The ductus deferentia convey sperm from the epididymides to the ejaculatory ducts during emission. Each ductus deferens forms a component of a spermatic cord.
   B. The seminal vesicles and prostate gland provide additives to the sperm in the formation of semen.
      1. The seminal vesicles are posterior to the base of the urinary bladder and secrete about 60% of the additive fluid of semen.
      2. The prostate gland surrounds the urethra just below the urinary bladder and secretes about 40% of the additive fluid of semen.
      3. The small bulbourethral glands secrete fluid that serves as a lubricant for the erect penis in preparation for coitus.
   C. The male urethra, which serves both the urinary and reproductive systems, is divided into the prostatic, membranous, and penile portions.
   D. The penis is specialized to become erect for insertion into the vagina during coitus.
      1. The body of the penis is composed of three columns of erectile tissue, the penile urethra, and associated vessels and nerves.
      2. The root of the penis is attached to the pubic arch and urogenital diaphragm.
      3. The glans penis is the terminal end, which is covered with the prepuce in an uncircumcised male.

V. Mechanisms of Erection, Emission, and Ejaculation
   A. Erection of the penis occurs as the erectile tissue becomes engorged with blood; emission is the movement of the spermatozoa from the epididymides to the ejaculatory ducts; and ejaculation is the forceful expulsion of semen from the ejaculatory ducts and urethra of the penis.
   B. Parasympathetic stimuli to arteries in the penis cause the erectile tissue to engorge with blood as arteriole flow increases and venous drainage decreases.
   C. Ejaculation is the result of sympathetic reflexes in the smooth muscles of the reproductive organs.

Review Activities

Objective Questions

1. An embryo with the genotype XY develops male accessory sex organs because of
   (a) androgens.
   (b) estrogens.
   (c) the absence of androgens.
   (d) the absence of estrogens.
2. Which of the following does *not* arise from the embryonic mesonephric duct? The
   (a) epididymis.
   (b) ductus deferens.
   (c) seminal vesicle.
   (d) prostate gland.
3. The external genitalia of a male are completely formed by the end of the
   (a) embryonic period.
   (b) ninth week.
   (c) tenth week.
   (d) twelfth week.
4. The interstitial cells (cells of Leydig)
   (a) nourish spermatids.
   (b) produce testosterone.
   (c) produce spermatozoa.
   (d) secrete alkaline fluid.
   (e) Both b and d.
5. The bulb and crus of the penis are located at the
   (a) glans.
   (b) corona glandis.
   (c) shaft.
   (d) prepuce.
   (e) root.
6. Which of the following is *not* a spermatic duct? The
   (a) epididymis.
   (b) spermatic cord.
   (c) ejaculatory duct.
   (d) ductus deferens.
7. Spermatozoa are stored prior to emission and ejaculation in the
   (a) epididymides.
   (b) seminal vesicles.
   (c) penile urethra.
   (d) prostate gland.
8. Urethral glands function to
   (a) secrete mucus.
   (b) produce nutrients.
   (c) secrete hormones.
   (d) regulate sperm production.
9. Which statement is *false* regarding erection of the penis?
   (a) It is a parasympathetic response.
   (b) It may be both a voluntary and involuntary response.
   (c) It has to be followed by emission and ejaculation.
   (d) It is controlled by the hypothalamus of the brain and sacral portion of the spinal cord.

10. The condition where one or both testes fail to descend into the scrotum is
    (a) cryptorchidism.
    (b) Turner's syndrome.
    (c) hermaphroditism.
    (d) Klinefelter's syndrome.

Essay Questions

1. What are the functions of the male reproductive system? How are they similar to and how do they differ from those of the female reproductive system?
2. Discuss how and when the genetic sex of a zygote is determined.
3. Discuss the significance of the indifferent stage in reproductive development. When is the indifferent stage, and what are its implications in the potential development of abnormalities of the reproductive organs?
4. Explain what is meant by latent development of the reproductive organs during puberty and identify the hormone that causes puberty in males.
5. Define what is meant by homologous structures. List the structures of the male reproductive system that form from the phallus, urogenital folds, labioscrotal folds, and the gonadal ridge.
6. Describe the location and structure of the scrotum. Explain how the scrotal muscles regulate the position of the testes in the scrotum. Why is this important?
7. Describe the internal structure of a testis. Discuss the function of the nurse cells, interstitial cells, seminiferous tubules, rete testis, and efferent ductules.
8. List the structures that constitute the spermatic cord. Where is the inguinal canal? Why is the inguinal canal clinically important?
9. Diagram a spermatozoon and list the function of each of its principal parts. Define *semen.* What amount of semen is ejected during ejaculation and what are the properties and substances of semen?
10. Describe the position, function, and histological structure of the epididymis, ductus deferens, and ejaculatory duct.
11. Compare the seminal vesicles and the prostate gland in terms of location, structure, and function.
12. Describe the structure of the penis, and explain the mechanisms that result in erection, emission, and ejaculation.
13. Distinguish between impotence, infertility, and sterility.
14. Distinguish between gonorrhea and syphilis. Describe the stages through which syphilis will progress if untreated.
15. Define *hydrocele, orchitis,* and *orchiectomy.* What conditions would warrant an orchiectomy?

CHAPTER 21

Female Reproductive System

Outline and Concepts

Clinical Case Study

Introduction to the Female Reproductive System

The female reproductive system produces ova, secretes sex hormones, receives spermatozoa from the male, and provides sites for fertilization and implantation. Parturition follows gestation, and secretion from the mammary glands provides nourishment for the baby.

Development of the Female Reproductive System

In the course of embryonic development, follicles containing oogonia appear in the ovaries, and the sex accessory organs are formed. The female external genitalia develop from the same indifferent structures that give rise to the external genitalia of a male.

Structure and Function of the Ovaries

The ovary contains numerous primary follicles with primary oocytes. During an ovarian cycle some of these become secondary follicles, and one becomes a mature vesicular ovarian follicle, which ruptures and releases its oocyte in ovulation. The ruptured follicle becomes a corpus luteum and regresses to become a corpus albicans. These cyclic changes in structure are associated with cyclic changes in hormone secretion.

Position and Structure of the Ovaries
Ovarian Cycle

Secondary Sex Organs

The uterine tube conduct the zygote to the uterus, where implantation into the endometrial layer may occur. The muscular layer of the uterus, or myometrium, is functional in labor and delivery. Sperm enter the female genital duct through the vagina, which also serves as the birth canal during parturition.

Uterine Tubes
Uterus
Vagina
Vulva
Mechanism of Erection and Orgasm

Mammary Glands

Mammary glands are modified sweat glands composed of secretory alveoli and ducts. The glands develop in the female breasts at puberty and function in lactation.

Structure of the Breast and the Mammary Glands

Ovulation and Menstruation

Ovulation and menstruation are reproductive cyclic events that are regulated by follicle-stimulating hormone (FSH) and luteinizing hormone (LH) secreted by the anterior pituitary gland, and by estrogen and progesterone, secreted by structures of the ovaries.

Clinical Considerations

Diagnostic Procedures
Developmental Problems
Problems Involving the Ovaries and Uterine Tubes
Problems Involving the Uterus
Diseases of the Vagina and Vulva
Diseases of the Breasts and Mammary Glands
Methods of Contraception

Clinical Case Study Answer

Chapter Summary

Review Activities

CLINICAL CASE STUDY

A twenty-eight-year-old female is brought to the emergency room with a four-day history of moderate right-sided pelvic pain. On the day of admission the pain increased in intensity, which caused the patient to seek medical attention. She complains of weakness and light-headedness and states that she missed her last period, which was due four weeks prior. A urine pregnancy test is positive. The consulting gynecologist says that a ruptured ectopic pregnancy is likely. He orders a blood test that suggests that the patient has suffered a slight amount of hemorrhage. A *culdocentesis* (needle sampling of the peritoneal cavity via the posterior vaginal wall to detect pooled blood) is performed, which is positive. The patient then undergoes preparation for surgery.

What is an ectopic pregnancy? Where is it most likely to occur? Briefly explain the sequence of events leading up to the rupture of the ectopic pregnancy beginning with ovulation. Differentiate normal from abnormal events. Explain how blood from a ruptured ectopic pregnancy can be aspirated through the vagina. ■

Introduction to the Female Reproductive System

The female reproductive system produces ova, secretes sex hormones, receives spermatozoa from the male, and provides sites for fertilization and implantation. Parturition follows gestation, and secretion from the mammary glands provides nourishment for the baby.

Objective 1. Explain the functional differences between the male and female reproductive systems.

Objective 2. Define *puberty, menstruation, ovulation,* and *menopause.*

Objective 3. Identify the primary and secondary sex organs in a female, and describe the secondary sex characteristics.

The reproductive systems of the male and female have some basic similarities and some specialized differences. The similarities are that (1) most of the reproductive organs of both sexes develop from similar embryonic tissues and are therefore *homologous;* (2) both systems have gonads that produce gametes and sex hormones; and (3) both systems experience latent development of the reproductive organs as they mature and become functional during puberty because of the effects of sex hormones secreted by the gonads.

The differences between the reproductive systems of the female and male are based on the specific functions of each in sexual reproduction and on the cyclic events that are characteristic of the female. The reproductive organs of the sexually mature, healthy male continuously produce male gametes, or sperm, and transfer them to the female during *coitus (sexual intercourse).* A male does not produce any sperm until puberty (about age thirteen), but is then capable of producing viable sperm throughout his life if he remains healthy. The gametes, or ova, of a female are completely formed, but not totally matured, during fetal development of the ovaries. The ova are generally discharged, or *ovulated,* one at a time in a cyclic pattern throughout the reproductive period of the female, which extends from puberty to menopause. *Menstruation* is the discharge of *menses* (blood and solid tissue) from the uterus at the end of each ovulation cycle. *Menopause* is the termination of ovulation and menstruation. The reproductive period in females generally extends from about age twelve to age forty-seven. The cyclic reproductive pattern of ovulation and the age span of fertility are determined by hormonal action.

The functions of the female reproductive system are (1) to produce ova; (2) to secrete sex hormones; (3) to receive the sperm from the male during coitus; (4) to provide sites for fertilization, implantation of the blastocyst (see chap. 22), and development of pregnancy; (5) to facilitate *parturition,* or delivery of the baby; and (6) to provide nourishment for the baby through the secretion of milk from the mammary glands in the breasts.

The organs of the female reproductive system are divided, like those of the male, into three functional categories:

1. **Primary sex organs.** The primary sex organs are called **gonads,** and in the female are known more specifically as the **ovaries.** Ovaries produce the gametes (ova), or eggs, and produce and secrete sex steroid hormones. Secretion of the female sex hormones at puberty contributes to the development of secondary sexual characteristics and causes cyclic changes in the secondary sex organs that are required for reproductive function.
2. **Secondary sex organs.** Secondary sex organs (fig. 21.1 and table 21.1) are those structures that are essential for successful fertilization of the ovum, implantation of the blastocyst, development of the embryo and fetus, and parturition. The secondary sex organs include the **vagina,** which receives the penis and ejaculated semen during coitus and through which the baby passes during delivery; the external genitalia, which protect the vaginal orifice (opening); the **uterine** (fallopian) **tubes,** through one of which an egg is transported toward the uterus after ovulation and is the site of fertilization; and the **uterus** (womb), where implantation and development occur. The muscular walls of the uterus play an active role in parturition. **Mammary glands** are also considered secondary sex organs because the milk secreted by them after parturition provides nourishment to the child.
3. **Secondary sex characteristics.** Secondary sex characteristics are features that are not essential for the reproductive process but are generally considered sexual attractants. Body physique, the pattern of body hair, and the development of breasts are examples. Although breasts contain the mammary glands, large breasts are not

coitus: L. *coitio,* a coming together

menopause: Gk. *men,* month; *pausis,* cessation
vagina: L. *vagina,* sheath or scabbard

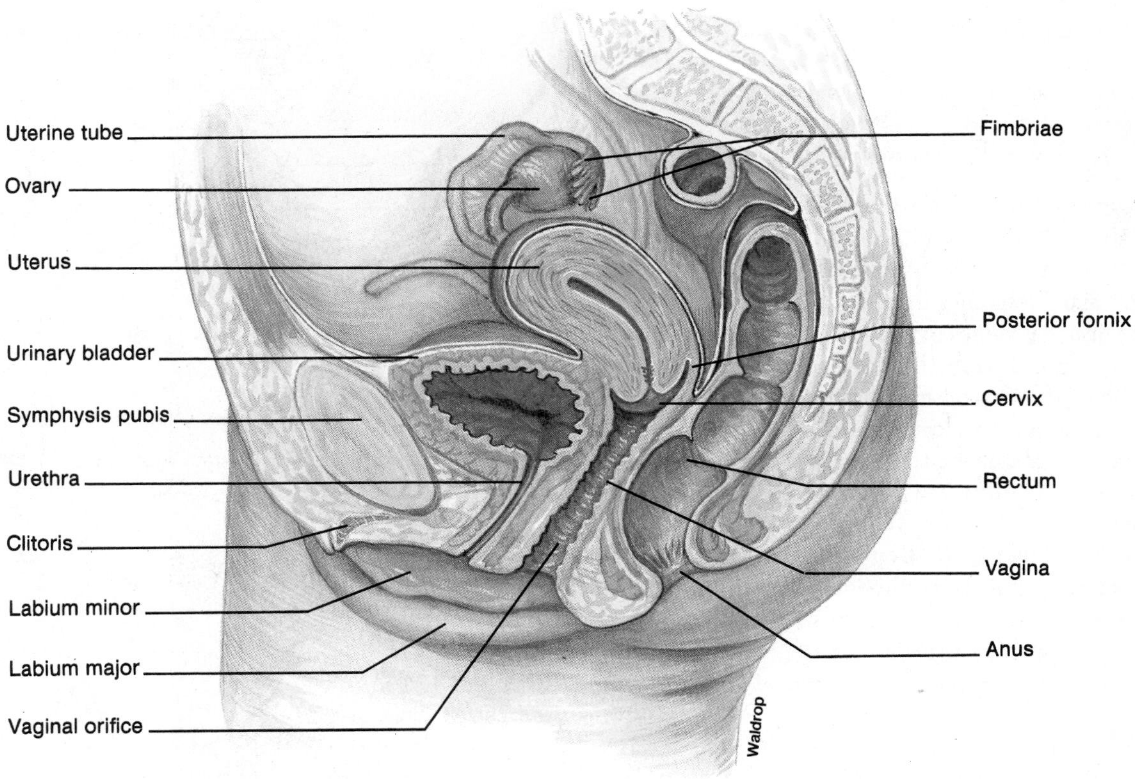

Figure 21.1 Organs of the female reproductive system seen in sagittal section.

Table 21.1 Functions of the organs of the female reproductive system

Organ(s)	Function(s)
Ovaries	Produce ova and female sex hormones
Uterine tubes	Convey ova toward uterus; site of fertilization; convey developing blastocyst to uterus
Uterus	Site of implantation; protects and sustains life of embryo and fetus during pregnancy; active role in parturition
Vagina	Conveys uterine secretions to outside of body; receives erect penis and semen during coitus and ejaculation; passage for fetus during parturition
Labia majora	Form margins of pudendal cleft; enclose and protect other external reproductive organs
Labia minora	Form margins of vestibule; protect openings of vagina and urethra
Clitoris	Glans of the clitoris is richly supplied with sensory nerve endings associated with feeling of pleasure during sexual stimulation
Vestibule	Cleft between labia minora that includes vaginal and urethral openings
Vestibular glands	Secrete fluid that moistens and lubricates the vestibule and vaginal opening during coitus
Mammary glands	Produce and secrete milk for nourishment of an infant

essential for nursing the young. In fact, all female mammals have mammary glands, but only human females have protruding breasts that function as a sexual attractant.

Puberty occurs at age twelve to fourteen, varying with the nutritional condition, genetic background, and even sexual exposure of the individual. Generally, girls attain puberty six months to one year earlier than boys, and the transition is more abrupt because of the onset of menstruation, or *menarche (mĕ-nar'ke).* Puberty results from the increased secretion of gonadotrophins from the anterior pituitary, which stimulates the ovaries to begin their cycles of ova development and sex steroid secretion.

1. List the functions of the female reproductive system.
2. Define *puberty,* and explain the process by which it occurs. Define *menstruation* and *ovulation.* What is the usual age span of female fertility? Define *menses* and *menopause.*
3. Distinguish between the primary sex organs, secondary sex organs, and secondary sexual characteristics.

Development of the Female Reproductive System

In the course of embryonic development, follicles containing oogonia appear in the ovaries, and the sex accessory organs are formed. The female external genitalia develop from the same indifferent structures that give rise to the external genitalia of a male.

Objective 4. Describe the embryonic origin of the female genital ducts.

Objective 5. Explain why the glans penis and clitoris are considered homologous structures, and list other structures that are homologous in the two sexes.

Although the genetic sex is determined at fertilization, both sexes develop similarly through the indifferent stage of the eighth week. The gonads of both sexes develop from the **gonadal ridges** medial to the mesonephros. **Primary sex cords** form within the gonadal ridges by the sixth week. The **genital tubercle** also develops during the sixth week as an external swelling cephalic to the cloacal membranes.

The ovaries develop more slowly than do the testes. Ovarian development begins at about the tenth week when **primordial follicles** begin to form within the medulla of the gonads. Each of the primordial follicles consists of an **oogonium** *(o''o-go'ne-um)* surrounded by a layer of **follicular cells.** Mitosis of the oogonia occurs during fetal development, so that thousands of germ cells are formed. Unlike the male, in which spermatogonia are formed by mitosis throughout life, all oogonia are formed prenatally and their number continuously decreases after birth.

The **genital ducts** include the uterus and the uterine tubes. These organs develop from a pair of embryonic tubes called the **paramesonephric** (müllerian) **ducts.** The paramesonephric ducts form to the sides of the mesonephric ducts that develop into the kidneys. As the mesonephric ducts regress, the paramesonephric ducts develop into the female genital tract (fig. 21.2). The lower portion of both ducts fuse to form the uterus. The upper portions remain unfused and give rise to the uterine tubes.

The epithelial lining of the vagina develops from the endoderm of the urogenital sinus. A thin membrane called the **hymen** forms to separate the lumen of the vagina from the urethral sinus. The hymen usually is perforated during later fetal development.

The external genitalia of both sexes appear the same during the indifferent stage of the eighth week (see fig. 20.6). A prominent **phallus** *(fal'us)* forms from the genital tubercle, and a **urethral groove** forms on the ventral side of the phallus. Paired **urogenital folds** surround the urethral groove on the lateral sides. In the male embryo these indifferent structures become masculinized by testosterone secreted by the testes. In the female embryo, in the absence of testosterone, feminization occurs.

These feminizing changes include an inhibition in the size of the phallus to form the relatively small **clitoris** *(kli'tor-is);* the urogenital folds remain unfused to form an inner **labia minora;** and the labioscrotal folds remain unfused to form the prominent **labia majora** (table 21.2). The external genitalia of a female are completely formed by the end of the twelfth week.

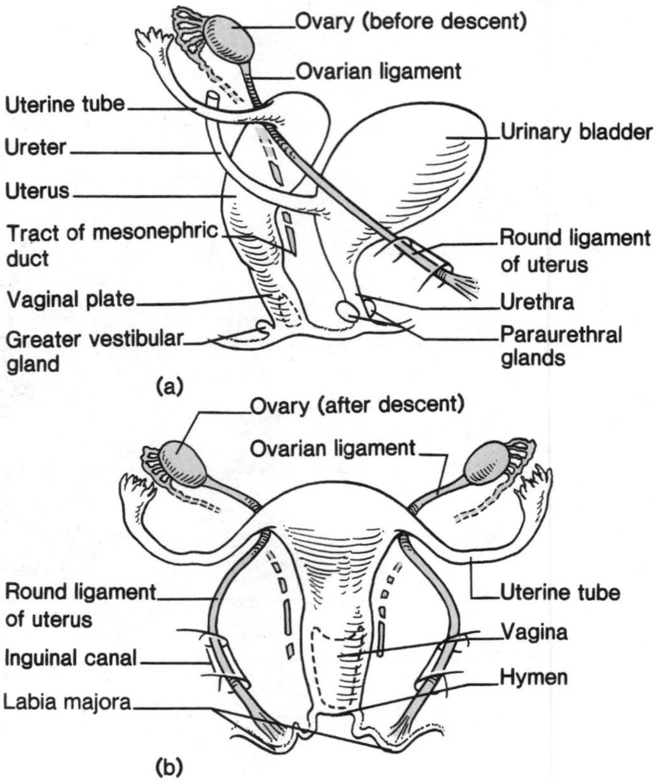

Figure 21.2 The development of the female genital tract: (*a*) a lateral view and (*b*) an anterior view.

Table 21.2 Homologous reproductive organs in the male and female and the undifferentiated structures from which they develop

Indifferent stage	Male	Female
Gonads	Testes	Ovaries
Urogenital groove	Membranous urethra	Vestibule
Genital tubercle	Glans penis	Clitoris
Urogenital folds	Penile urethra	Labia minora
Genital swelling	Scrotum Bulbourethral glands	Labia majora Vestibular glands

There is tremendous individual variation in the structure of the hymen. The hymen of a baby girl may be absent, or it may partially, or occasionally completely, cover the vaginal orifice, in which case it is called an *imperforate hymen.* An imperforate hymen is usually not detected until the first menstruation (menarche) when the discharge cannot be expelled. If the hymen is present, it may be ruptured during childhood in the course of normal exercise. On the other hand, a hymen may be so elastic that it persists even after coitus. The presence of a hymen is not, therefore, a reliable sign of virginity.

primordial: L. *prima,* first; *ordior,* to begin
follicle: L. diminutive of *follis,* bag
hymen: Gk. (mythology) Hymen was god of marriage; *hymen,* thin skin or membrane

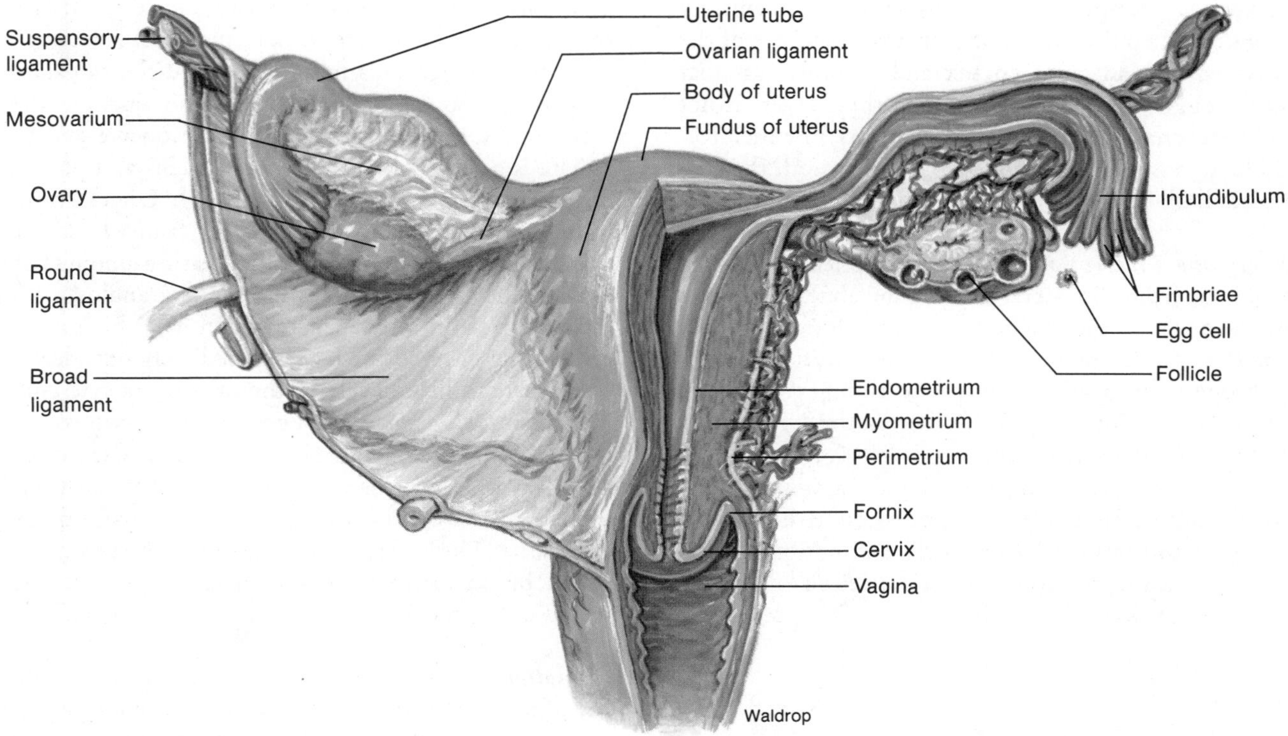

Figure 21.3 An anterior view of the female reproductive organs showing the relationship of the ovaries, uterine tubes, uterus, cervix, and vagina.

---

1. Describe the location and fate of the paramesonephric ducts in females.
2. Explain why the external genitalia of males and females are considered to be homologous, and explain what may happen if a female embryo is exposed to androgens at a critical stage of development.

Structure and Function of the Ovaries

The ovary contains numerous primary follicles with primary oocytes. During an ovarian cycle some of these become secondary follicles, and one becomes a mature vesicular ovarian follicle, which ruptures and releases its oocyte in ovulation. The ruptured follicle becomes a corpus luteum and regresses to become a corpus albicans. These cyclic changes in structure are supporting cyclic changes in hormone secretion.

Objective 6. Describe the position of the ovaries and the ligaments supporting the ovaries and genital ducts.

Objective 7. Describe the structural changes that occur in the ovaries leading to and following ovulation.

Objective 8. Describe oogenesis, and explain why meiosis of one primary oocyte results in the formation of only one mature ovum.

Objective 9. Discuss the hormonal secretions of the ovaries during an ovarian cycle.

Position and Structure of the Ovaries

Ovaries are the paired primary sex organs of the female that produce *gametes,* or *ova,* and the sex hormones *estrogen* and *progesterone.* The ovaries of a sexually mature female are solid, ovoid structures about 3.5 cm (1.4 in.) long, 2 cm (0.8 in.) wide, and 1 cm (0.4 in.) thick. The color and texture of the ovaries vary according to the age and reproductive stage of the female. The ovaries of a young girl are smooth and pinkish in color. Following puberty, the ovaries are pinkish-gray and have an irregular surface because of the scarring caused by ovulation. On the medial portion of each ovary is a **hilum,** which is the point of entrance for ovarian vessels and nerves. The lateral portion of the ovary is in the open end of the uterine tube (fig. 21.3).

The ovaries are positioned in the upper pelvic cavity on both sides of the uterus. Each ovary is situated in a shallow depression of the posterior body wall, called the **ovarian fossa,** and secured by several membranous attachments. The principal supporting membrane of the female reproductive tract is the **broad ligament.** The broad ligament is the parietal peritoneum that supports the uterine tubes and uterus. The **mesovarium** *(mes''o-va're-um)* is a specialized posterior extension of the broad ligament that attaches to an ovary. Each ovary is additionally supported by an **ovarian ligament,** anchored to the uterus, and a **suspensory ligament,** attached to the pelvic wall (fig. 21.3).

Each ovary consists of four layers. The **germinal epithelium** is the thin, outermost layer composed of cuboidal epithelial cells (see fig. 21.6). A collagenous connective tissue layer

---

ovaries: L. *ovum,* egg

called the **tunica albuginea** *(al''bu-jin'e-ah)* is immediately below the germinal epithelium. The principal substance of the ovary is divided into an outer **cortex** and an inner, vascular **medulla,** although the boundary between these layers is not distinct. The **stroma** is the material of the ovary in which follicles and blood vessels are embedded and lies in both the cortical and medullary layers.

Blood Supply and Innervation Blood is supplied by ovarian arteries that arise from the lateral sides of the abdominal aorta just below the origin of the renal arteries. An additional supply comes from the ovarian branches of the uterine arteries. Venous return is through the ovarian veins. The right ovarian vein empties into the inferior vena cava, whereas the left ovarian vein drains into the left renal vein.

The ovaries have both sympathetic and parasympathetic innervation from the ovarian plexus. Innervation to the ovaries, however, is only to the vascular networks and not to the follicular substance within the stroma. All of the vessels and nerves to an ovary enter by way of the hilum, which is supported by the ovarian ligament.

Normal, healthy ovaries usually cannot be palpated either by vaginal or abdominal examination. If the ovaries become swollen or displaced, however, they are palpable through the vagina. There is a great variety of nonmalignant tumors of the ovaries, most of which cause swelling and some localized tenderness. The ovaries atrophy during menopause, and ovarian enlargement in postmenopausal women is usually cause for concern.

Ovarian Cycle

The germ cells that migrate into the ovaries during early embryonic development multiply, so that by about five months of gestation the ovaries contain approximately six to seven million oogonia. Production of new oogonia stops at this point and never resumes again. Toward the end of gestation the oogonia begin meiosis, at which time they are called **primary oocytes** *(o'o-sīts)*. Oogenesis progresses to prophase I of the first meiotic division at which point it is arrested. The number of primary oocytes decreases throughout a woman's reproductive years. The ovaries of a newborn girl contain about two million oocytes, but this number is reduced to about 300,000–400,000 by the time the girl enters puberty. Oogenesis ceases entirely at menopause (when menstruation stops).

Primary oocytes that are not stimulated to complete the first meiotic division are contained within tiny follicles called **primordial follicles.** In response to gonadotrophin stimulation, some of these oocytes and follicles get larger, and the follicular cells divide to produce numerous small **granulosa cells** that surround the oocyte and fill the follicle. A follicle at this stage in development is called a **primary follicle.**

Some primary follicles will be stimulated to grow still bigger and develop a fluid-filled cavity, called an *antrum,* at which time they are called **secondary follicles** (fig. 21.4). The granulosa (follicle) cells of secondary follicles form a ring around the circumference of the follicle and form a mound that supports the oocyte. This mound is called the *cumulus oophorous (o-of'o-rus)*. Some granulosa cells also encircle the oocyte, forming a *corona radiata.* Between the oocyte and the corona radiata is a thin gel-like layer of proteins and polysaccharides called the *zona pellucida (pel-lu'sĭ-dah)*. Under stimulation of follicle-stimulating hormone (FSH) from the anterior pituitary, the granulosa cells secrete increasing amounts of estrogen as the follicles grow. Interestingly, the granulosa cells produce estrogen from its precursor testosterone, which is supplied by cells of the *theca interna* layer immediately outside the follicle.

As the follicle develops, the primary oocyte completes its first meiotic division. This does not form two complete cells, however, because only one cell—the **secondary oocyte**—gets all the cytoplasm. The other cell formed at this time becomes a small *polar body* (fig. 21.5), which eventually fragments and disappears. The secondary oocyte enters the second meiotic division, but meiosis is arrested at metaphase II and is never completed unless fertilization occurs.

Ovulation Usually, by about ten to fourteen days after the first day of menstruation, only one follicle has continued its growth to become a mature **vesicular ovarian** (graafian) **follicle** (fig. 21.6); other secondary follicles during that cycle regress and become *atretic (ah-tret'ik)*. The vesicular ovarian follicle is so large that it forms a bulge on the surface of the ovary. Under proper hormonal stimulation this follicle will rupture—much like the popping of a blister—and extrude its secondary oocyte near the opening of the uterine tube in the process of *ovulation* (fig. 21.7).

The released secondary oocyte is surrounded by the zona pellucida and corona radiata. If it is not fertilized it disintegrates in a couple of days. If a sperm passes through the corona radiata and zona pellucida and enters the cytoplasm of the secondary oocyte, the oocyte completes the second meiotic division becoming a mature ovum. In this process the cytoplasm is again not divided equally; most of the cytoplasm remains in the zygote (fertilized egg), leaving another polar body, which like the first, disintegrates (fig. 21.8).

Changes continue in the ovary following ovulation. The empty follicle, under the influence of luteinizing hormone from the anterior pituitary, undergoes structural and biochemical changes to become a **corpus luteum.** Unlike the ovarian follicles, which secrete only estrogen, the corpus luteum secretes two sex steroid hormones: estrogen and progesterone. Toward the end of a nonfertile cycle the corpus luteum regresses and is changed into a nonfunctional **corpus albicans.** These cyclic changes in the ovary are summarized in figure 21.9.

stroma: Gk. *stroma,* a couch or bed
oogonium: Gk. *oion,* egg; *gonos,* generation

cumulus oophorous: L. *cumulus,* a mound; G. *oophoros,* egg-bearing
corona radiata: Gk. *korone,* crown; L. *radiata,* radiate
zona pellucida: L. *zone,* girdle; L. *pellis,* skin
theca interna: Gk. *theke,* a box; L. *internus,* interior
vesicular ovarian (graafian) follicle: from Regnier de Graaf, Dutch anatomist and physician, 1641–73
atretic: Gk. *atretos,* not perforated
corpus luteum: L. *corpus,* body; *luteum,* yellow
albicans: L. *albicare,* to whiten

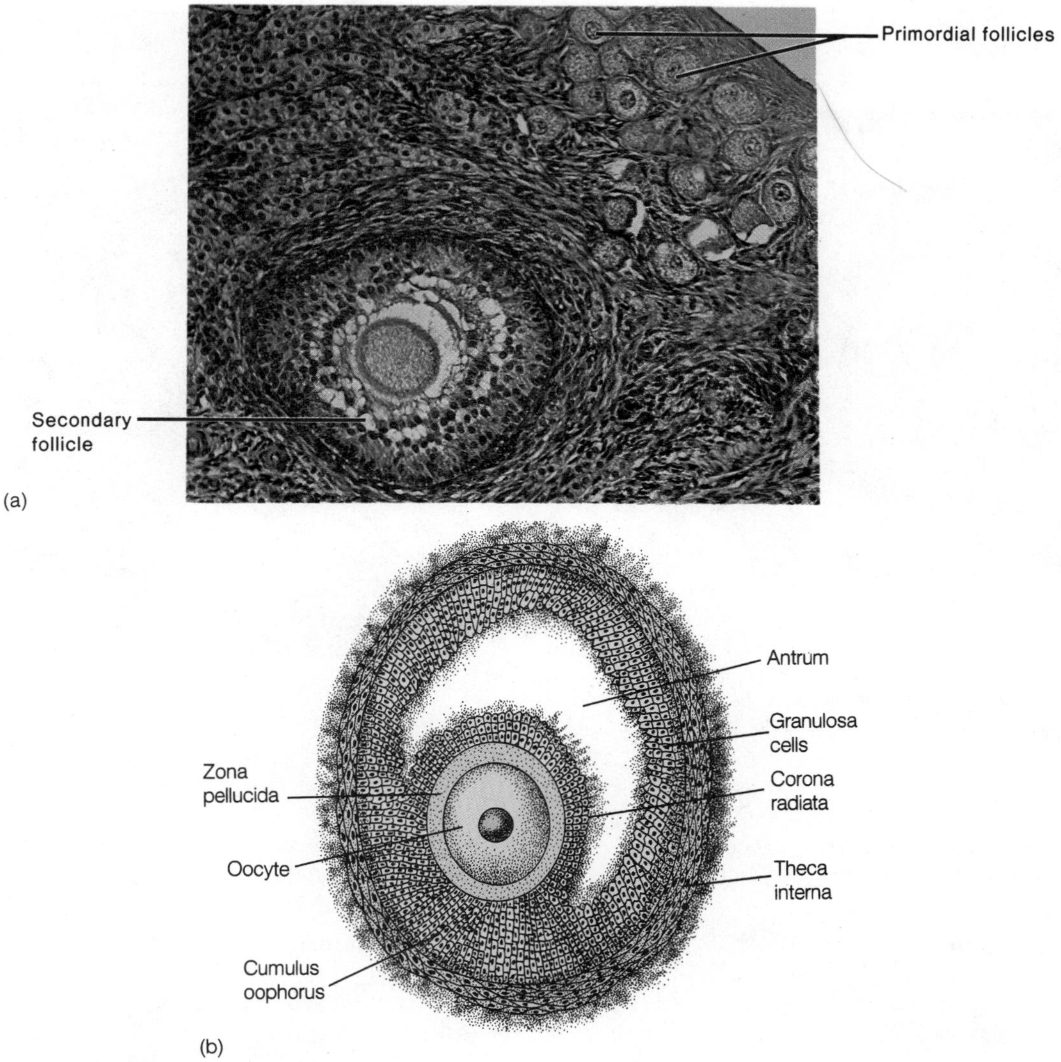

Figure 21.4 (*a*) A photomicrograph of primordial and secondary follicles. (*b*) A diagram of the parts of a secondary follicle.

1. Describe the position of the ovaries relative to the uterine tubes, and describe the position and functions of the broad ligament and mesovarium.
2. Compare the structure and contents of a primordial follicle, primary follicle, secondary follicle, and vesicular ovarian follicle.
3. Define *ovulation,* and describe the changes that occur in the ovary following ovulation in a nonfertile cycle.
4. Describe oogenesis, and explain why only one mature ovum is produced by this process.
5. Compare the hormonal secretions of the vesicular ovarian follicle with those of a corpus luteum.

Secondary Sex Organs

The uterine tube conduct the zygote to the uterus, where implantation into the endometrial layer may occur. The muscular layer of the uterus, or myometrium, is functional in labor and delivery. Sperm enter the female genital duct through the vagina, which also serves as the birth canal during parturition.

Objective 10. Describe how an ovum is moved through a uterine tube to the uterus.

Objective 11. Describe the structure and function of each of the three layers of the uterus.

Objective 12. Describe the structure and functions of the vulva and vagina.

Objective 13. Discuss the changes that occur in the female reproductive system during sexual excitement and coitus.

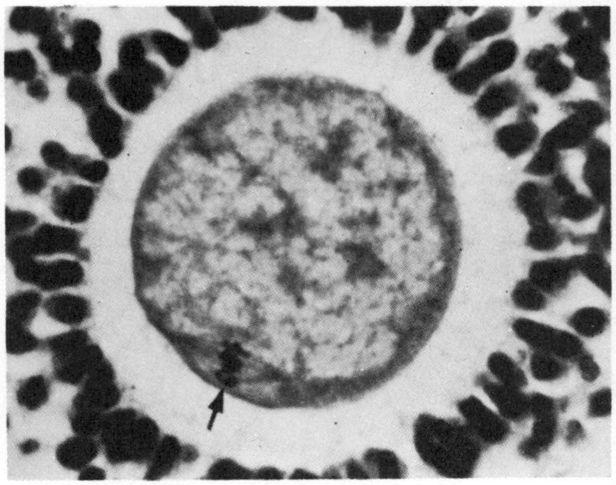

(a)

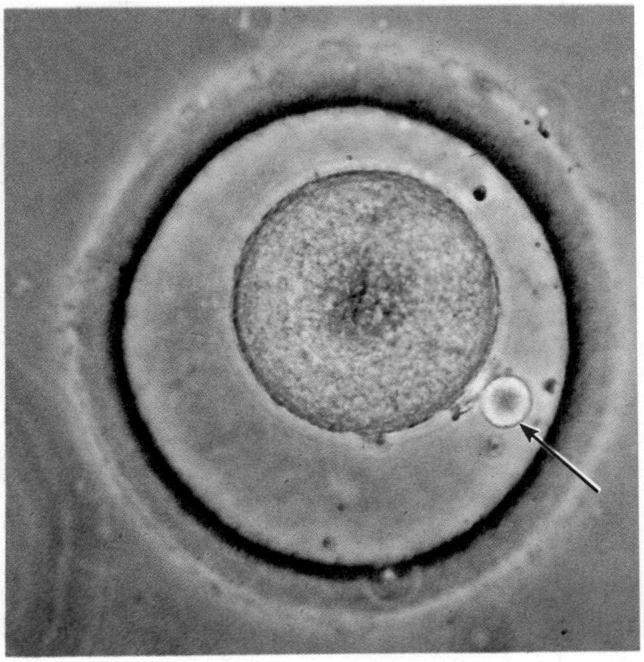

(b)

Figure 21.5 (*a*) A primary oocyte at metaphase I of meiosis. Note the alignment of chromosomes (*arrow*). (*b*) A human secondary ooctye formed at the end of the first meiotic division and the first polar body (*arrow*).

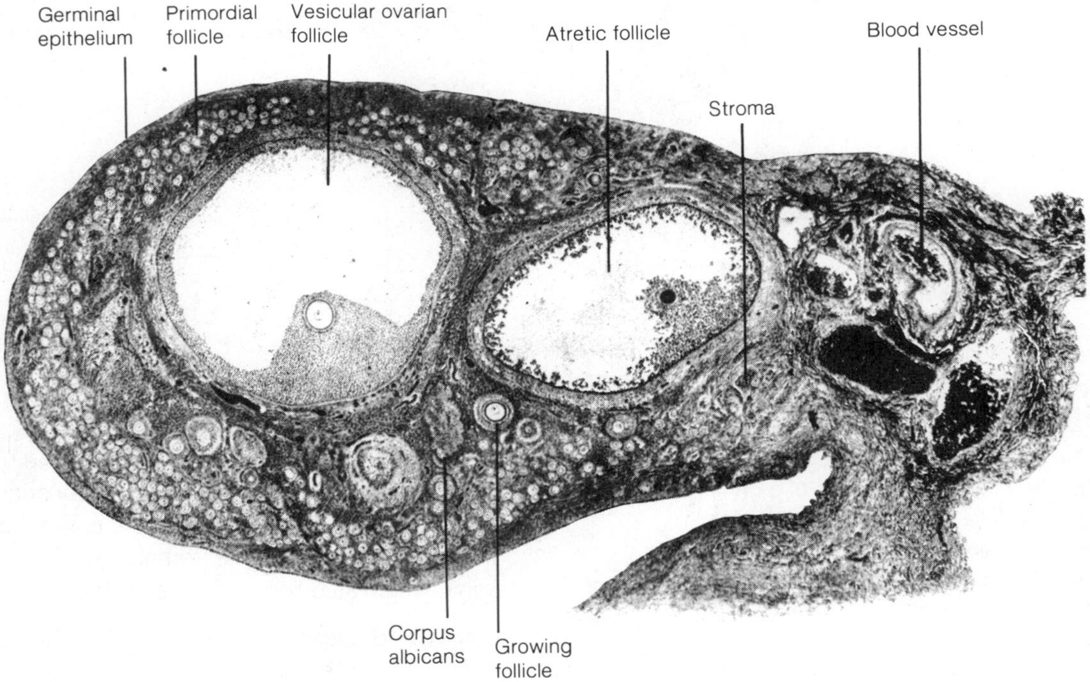

Figure 21.6 A vesicular ovarian (graafian) follicle within the ovary of a monkey.

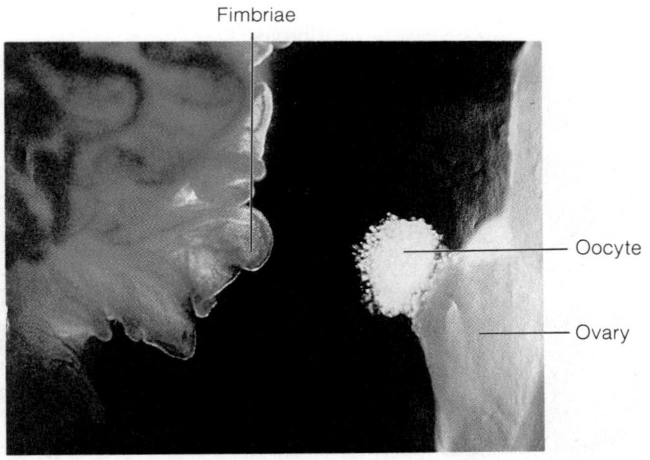

Figure 21.7 Ovulation from a human ovary.

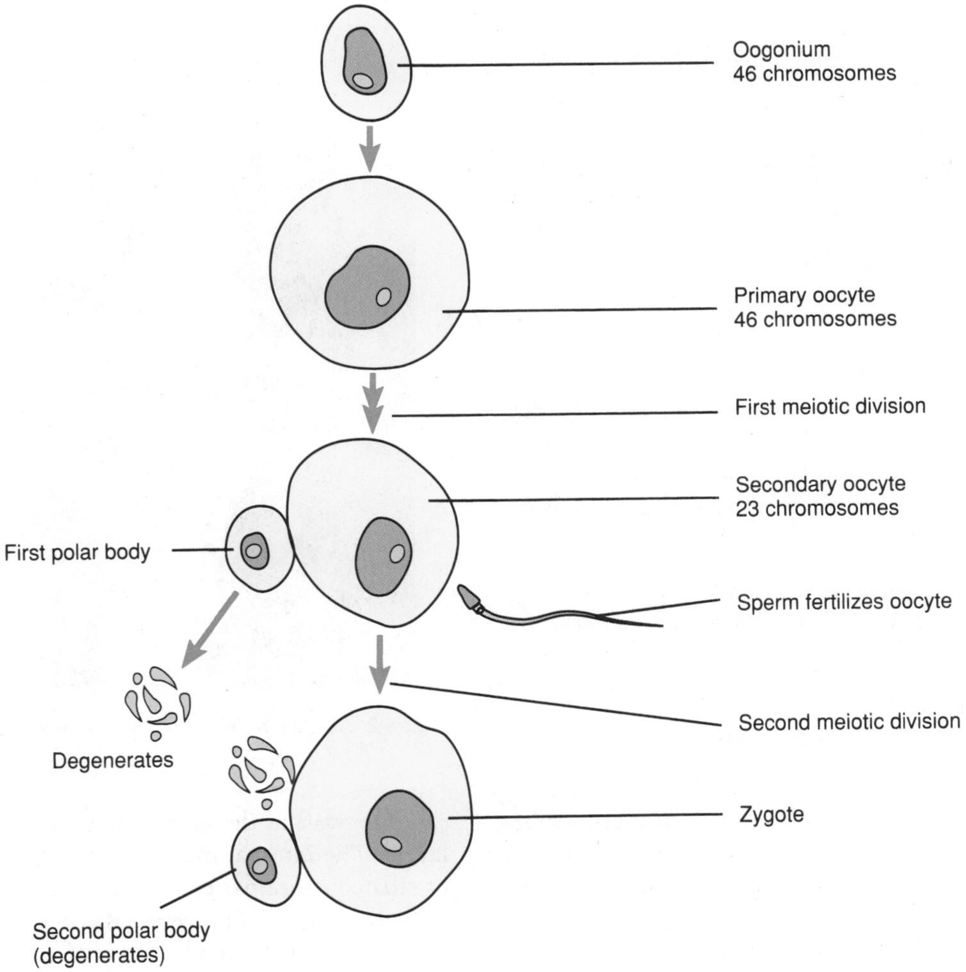

Figure 21.8 A schematic diagram of the process of oogenesis. During meiosis, each primary oocyte produces a single haploid gamete. If the secondary oocyte is fertilized, it forms a secondary polar body and becomes a zygote.

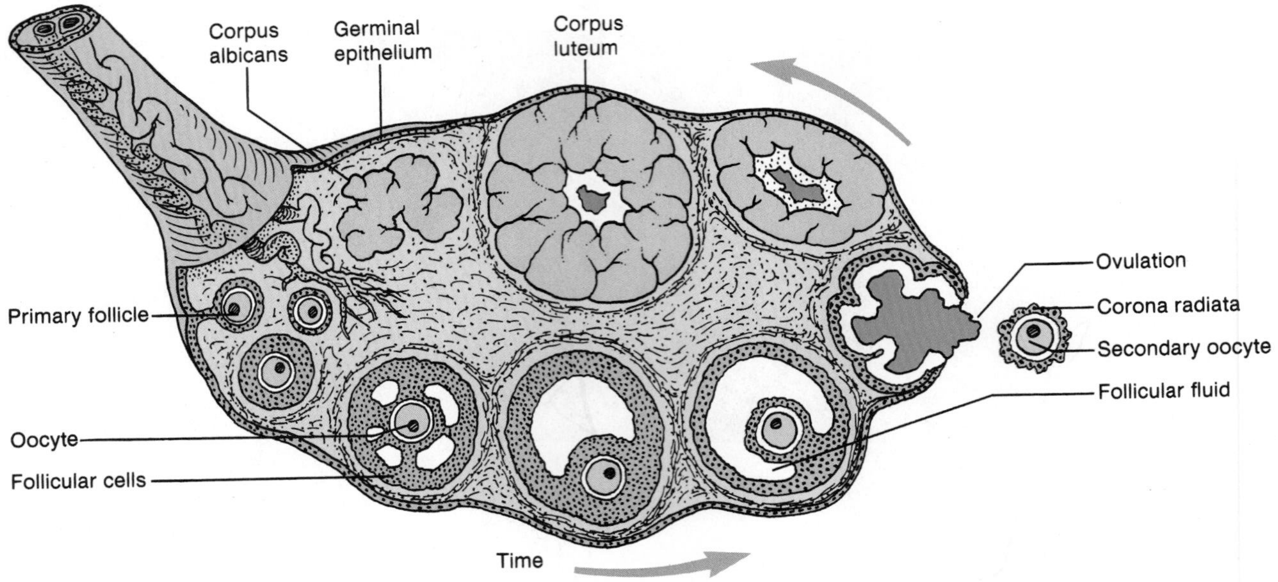

Figure 21.9 A schematic diagram of an ovary showing the various stages of ovum and follicle development.

The secondary sex organs described in this section include the internal organs of the uterine tubes, uterus, and vagina, as well as the external genitalia (vulva). The mammary glands, which are also considered to be secondary sex organs, are described in a separate section.

Uterine Tubes

The paired **uterine tubes,** also known as the fallopian *(fal-lo'pe-an)* tubes, or **oviducts,** transport ova from the ovaries to the uterus. Fertilization normally occurs within the uterine tube. Each uterine tube is approximately 10 cm (4 in.) long and 0.7 cm (0.3 in.) in diameter and is positioned between the folds of the broad ligament of the uterus (see fig. 21.3).

The term *salpinx* is occasionally used to refer to the uterine tubes. It is a Greek word meaning trumpet or tube and is the root of such clinical terms as *salpingitis (sal'' pin-ji' tis),* or inflammation of the uterine tubes, *salpingography* (radiography of the uterine tubes), and *salpingolysis* (the breaking up of adhesions of the uterine tube to correct female infertility).

The funnel-shaped, open-ended portion of the uterine tube is called the **infundibulum** (see fig. 21.3). Although the infundibulum is close to the ovary, it is not attached. A number of fringed, fingerlike processes, called **fimbriae,** project from the margins of the infundibulum over the lateral surface of the ovary. Movement of the fimbriae directs oocytes into the lumen of the uterine tube. From the infundibulum, the uterine tube extends medially and inferiorly to open into the superior-lateral cavity of the uterus at the **uterine opening.** The **ampulla** *(am-pul'ah)* is the longest and widest portion of the uterine tube (see fig. 21.3).

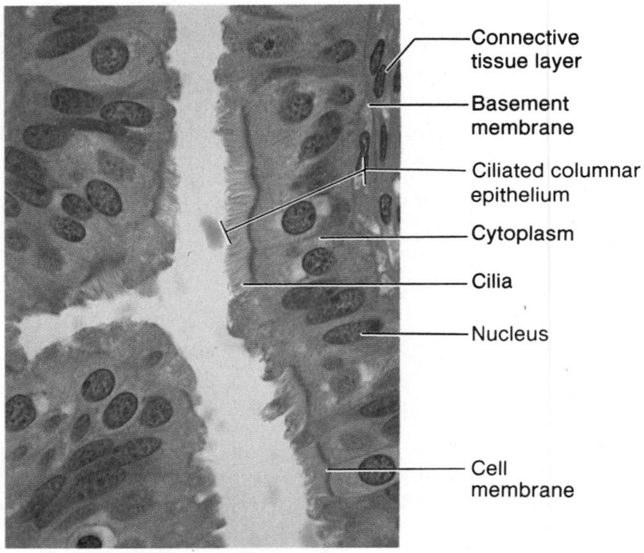

Figure 21.10 The histology of the uterine tube.

The wall of the uterine tube consists of three histological layers. The internal **mucosa** lines the lumen and is composed of ciliated columnar epithelium (fig. 21.10) that is drawn into numerous folds. The **muscularis** is the middle layer, composed of a thick, circular layer of smooth muscle and a thin, outer layer of smooth muscle. Peristaltic contractions of the muscularis and ciliary action of the mucosa move the oocyte through the lumen of the uterine tube. The outer **serous layer** of the uterine tube is part of the visceral peritoneum.

The oocyte takes four to five days to move through the uterine tube. If enough viable sperm are ejaculated into the

fallopian tubes: from Gabriele Fallopius, Italian anatomist, 1523–62
oviduct: L. *ovum,* egg; *ductus,* a leading

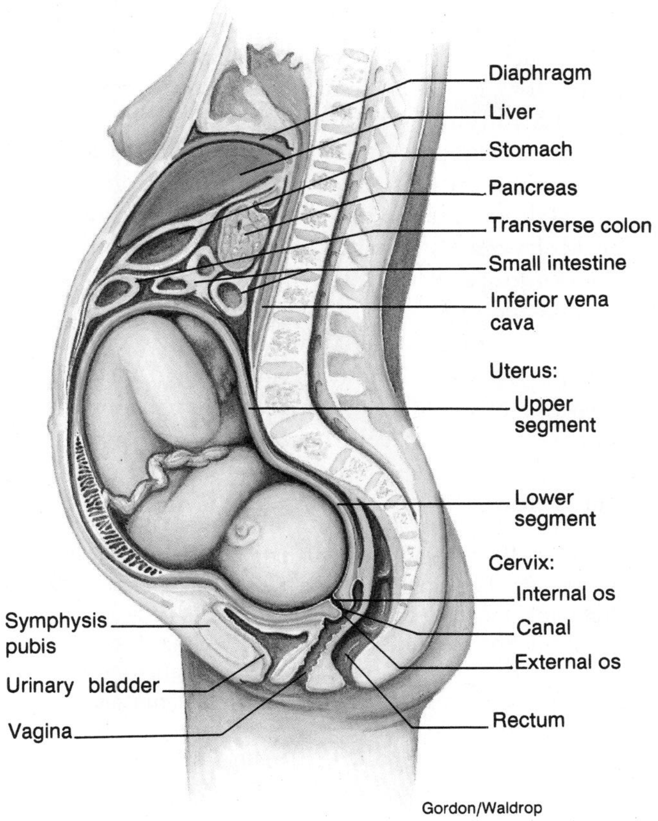

Figure 21.11 The size and position of the uterus in a full-term pregnant woman in sagittal section.

vagina during coitus and an oocyte is in the uterine tube, fertilization will occur within hours after discharge of the semen. The zygote will move toward the uterus where implantation occurs. If the embryo (called a *blastocyst*) implants into the uterine tube instead, an *ectopic pregnancy* will be produced. An ectopic pregnancy is an implantation of a blastocyst in a site other than the uterus.

Since the infundibulum of the uterine tube is unattached, it provides a pathway for pathogens to enter the peritoneal cavity. The mucosa of the uterine tube is continuous with that of the uterus and vagina, and it is possible for infectious agents to enter the vagina and cause infections that may ultimately spread to the peritoneal linings resulting in *pelvic inflammatory disease* (PID). There is no opening into the peritoneal cavity other than through the uterine tubes. The abdominopelvic cavity of a male is totally sealed from external contamination.

The uterine tubes are supplied with blood through the ovarian and uterine arteries. Venous drainage is through uterine veins that parallel the arteries. Both the uterine artery and vein can be observed in the broad ligament that supports the uterine tube (see fig. 21.3).

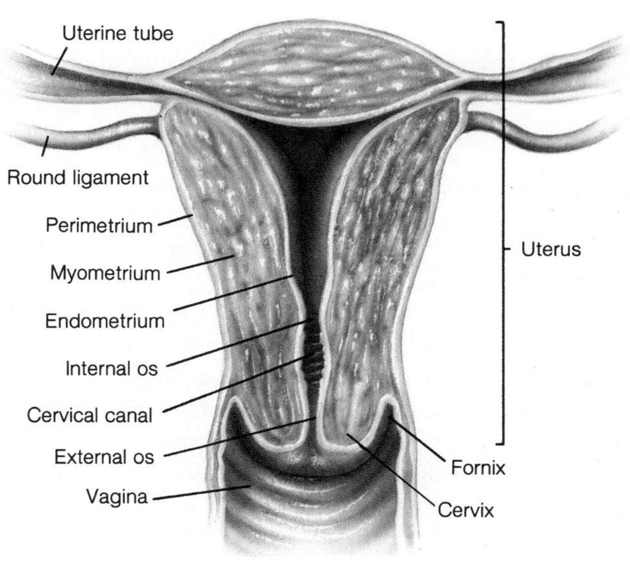

Figure 21.12 Layers of the uterine wall.

The uterine tubes have both sympathetic and parasympathetic innervation from the hypogastric plexus and pelvic splanchnic nerves. The nerve supply to the uterine tubes regulates the activity of the smooth muscles and blood vessels.

Uterus

The **uterus** receives the blastocyst that develops from a fertilized oocyte and provides a site for implantation. Prenatal development continues within the uterus until gestation is completed, at which time the uterus plays an active role in the delivery of the baby.

Structure The uterus is a hollow, thick-walled, muscular organ that is shaped like an inverted pear. Although the shape and position of the uterus changes immensely during pregnancy (fig. 21.11), in its nonpregnant state it is about 7 cm (2.8 in.) long, 5 cm (2 in.) wide (through its broadest region), and 2.5 cm (1 in.) in diameter. The anatomical regions of the uterus include the uppermost, dome-shaped portion above the entrance of the uterine tubes called the **fundus,** the enlarged main portion called the **body,** and the inferior constricted portion opening into the vagina called the **cervix** (fig. 21.12). The uterus is located between the urinary bladder anteriorly and the rectum and sigmoid colon posteriorly. The fundus projects anteriorly and slightly superiorly over the urinary bladder. The cervix projects posteriorly and inferiorly, joining the vagina at nearly a right angle.

The **uterine cavity** is the space within the regions of the fundus and body. The lumina of the uterine tubes open into the uterine cavity on the superolateral portions. The uterine cavity is continuous inferiorly with the **cervical canal,** which extends through the cervix and opens into the lumen of the

ectopic: Gk. *ex,* out; *topos,* place

fundus: L. *fundus,* bottom
cervix: L. *cervix,* neck

vagina. The junction of the uterine cavity with the cervical canal is called the **internal os,** whereas the opening of the cervical canal into the cavity of the vagina is called the **external os** (fig. 21.12).

Support of the Uterus The uterus is maintained in position by muscular support and ligaments that extend from the pelvic girdle or body wall to the uterus. Muscles of the perineum, especially the levator ani muscle (see fig. 9.25), provide the principal muscular support. The ligaments that support the uterus undergo marked hypertrophy during pregnancy, regress in size after parturition, and atrophy after menopause.

Four paired ligaments support the uterus in position within the pelvic cavity. The paired **broad ligaments** are folds of the peritoneum that extend from the pelvic walls and floor to the lateral walls of the uterus (see fig. 21.3). The ovaries and uterine tubes are also supported by the broad ligaments. The paired **uterosacral ligaments** (not illustrated) are also folds of peritoneum that curve along the lateral pelvic wall on both sides of the rectum to connect the uterus to the sacrum. The **cardinal (lateral cervical) ligaments** (not illustrated) are fibrous sheets of peritoneum that extend laterally from the cervix and vagina across the pelvic floor where they attach to the wall of the pelvis. The cardinal ligaments contain some smooth muscle as well as vessels and nerves to the cervix and vagina. The fourth paired ligaments are the **round ligaments** (see fig. 21.3). The round ligaments are actually continuations of the ovarian ligaments, which support the ovaries. Each round ligament extends from the lateral border of the uterus just below the position where the uterine tube attaches to the lateral pelvic wall. Similar to the course taken by the ductus deferens in the male, the round ligaments continue through the inguinal canal of the abdominal wall where they attach to the deep tissues of the labium majus.

Although the uterus has extensive support, considerable movement is possible. The uterus tilts slightly posteriorly as the urinary bladder fills and moves anteriorly during defecation. In some women, the uterus may go out of position and interfere with the normal progress of pregnancy. A posterior tilting of the uterus is called *retroflexion,* whereas an anterior tilting is called *anteflexion.*

Uterine Wall The wall of the uterus is composed of three layers: the perimetrium, myometrium, and endometrium (fig. 21.12).

The **perimetrium** is the thin, outer serosal covering and a part of the peritoneum. The lateral portion of the perimetrium merges with the broad ligament. A shallow pouch called the **vesicouterine** *(ves''ĭ-ko-u'-ter-in)* **pouch** is formed as the peritoneum is reflected over the urinary bladder. The **rectouterine** *(rek''to-u'ter-in)* **pouch** (pouch of Douglas) is formed as the peritoneum is reflected onto the rectum. The rectouterine pouch is the lowest point in the pelvic cavity and provides a site for surgical entry into the peritoneal cavity.

The thick **myometrium** is composed of three poorly defined layers of smooth muscle arranged in longitudinal, circular, and spiral patterns. The myometrium is thickest in the fundus and thinnest in the cervix. During parturition, the muscles of this layer are stimulated to contract forcefully.

The **endometrium** is the inner mucosal lining of the uterus. The endometrium has two distinct layers. The superficial **stratum functionale** layer, composed of columnar epithelium and containing secretory glands, is shed as *menses* during menstruation and built up again under the stimulation of ovarian steroid hormones. The deeper **stratum basale** layer is highly vascular and serves to regenerate the stratum functionale after each menstruation.

The size of the uterus changes tremendously during pregnancy. Its weight increases more than sixteen times (from about 60 g to about 1,000 g), and its capacity increases from about 2.5 ml to over 5,000 ml. The principal change in the myometrium is a marked hypertrophy, or elongation, of the individual muscle cells to as much as ten times their original length. There is some atrophy of the muscle cells after parturition, but the uterus never returns to its original size.

Uterine Blood Supply and Innervation The uterus is supplied with blood through the **uterine arteries** that arise from the internal iliac arteries and by the **uterine branches** of the **ovarian arteries** (fig. 21.13). Each pair of these two vessels anastomose on the upper lateral margin of the uterus. Branches from the uterine arteries penetrate the perimetrium to form a vascular arch, called the **arcuate artery,** along the outside of the myometrium. Numerous **radial arteries** arise from the arcuate artery to penetrate the myometrium and supply it with blood. **Spiral arterioles** arise from the radial arteries to serve the endometrium. The blood from the uterus returns through uterine veins that parallel the pattern of the arteries.

The uterus has both sympathetic and parasympathetic innervation from the pelvic and hypogastric plexuses. Both autonomic innervations serve the arteries of the uterus, whereas the smooth muscle of the myometrium receives only sympathetic innervation.

Vagina

The **vagina** is the organ that receives sperm through the urethra of the erect penis during coitus. It also serves as the birth canal during parturition and provides for the passage of menses to the outside. The musculomembranous vagina is a tubular organ about 9 cm (3.6 in.) in length, passing from the cervix of the uterus to the vestibule. The vagina is positioned between the urinary bladder and urethra anteriorly and the rectum pos-

os: L. *os,* mouth
perimetrium: Gk. *peri,* around; *metra,* uterus
vesicouterine: L. *vesico,* bladder; *uterus,* womb
pouch of Douglas: from James Douglas, English anatomist and physician, 1675–1742

myometrium: Gk. *mys,* muscle; *metra,* uterus
endometrium: Gk. *endon,* within; *metra,* uterus
menses: L. *menses,* plural of *mensis,* monthly

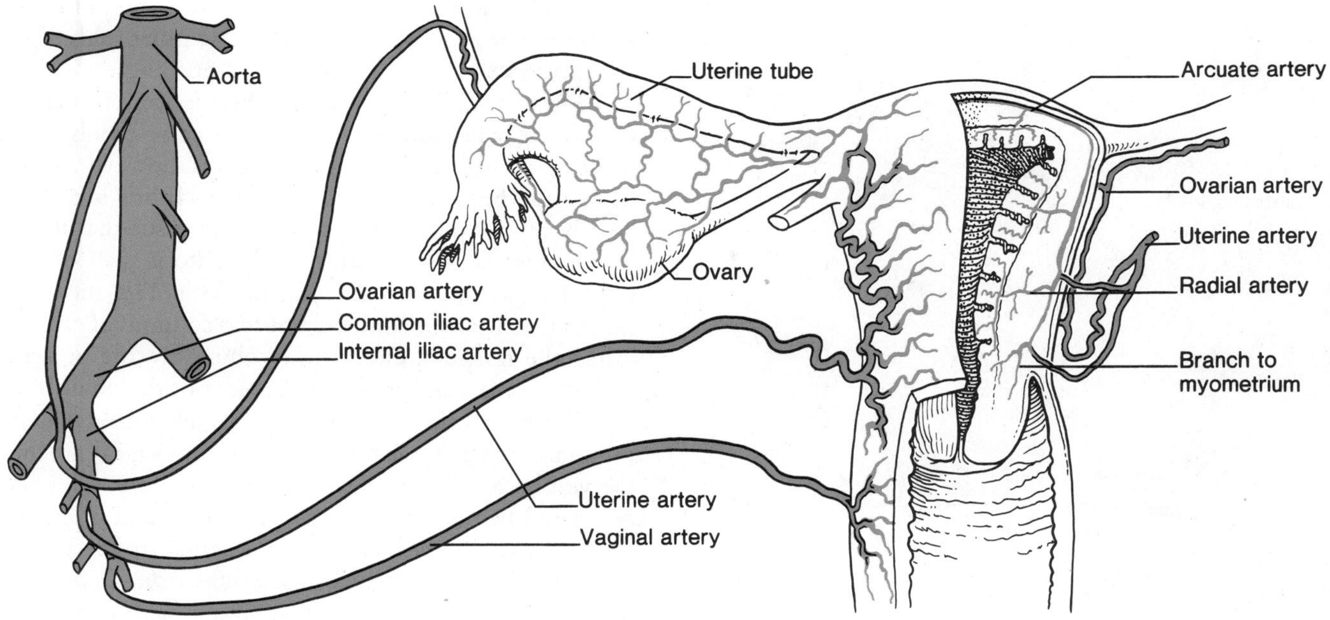

Figure 21.13 Vascular supply to the uterus.

teriorly, where it is continuous with the cervical canal of the uterus. The cervix attaches to the vagina at a nearly 90-degree angle. The deep recess behind the protrusion of the cervix into the vagina is called the **posterior fornix.** The smaller recesses in front and to the sides are called the **anterior** and **lateral fornices** (see fig. 21.3).

The fornices are of clinical importance since they permit the cervix to be palpated during a gynecological examination. Occasionally, the deep posterior fornix provides surgical access to the pelvic cavity through the vagina. In addition, the fornices are important in the placement of two forms of *birth-control devices*—the cervical cap and the diaphragm.

The exterior opening of the vagina, at its lower end, is called the **vaginal orifice.** A thin fold of mucous membrane, called the **hymen,** may partially cover the vaginal orifice.

The vaginal wall is composed of three layers: an inner mucosa, a middle muscularis, and an outer fibrous layer. The **mucosal layer** consists of nonkeratinized, stratified squamous epithelium, which forms a series of transverse folds called **vaginal rugae** (fig. 21.14). The vaginal rugae permit considerable distension of the vagina for penetration of the erect penis. They also provide friction ridges for stimulation of the erect penis during coitus. The mucosal layer contains few glands; the acidic mucus that is present in the vagina comes primarily from glands within the uterus. The acidic environment of the vagina retards microbial growth. The additives within semen, however, temporarily neutralize the acidity of the vagina to insure the survival of the ejaculated sperm deposited within the vagina.

The **muscularis layer** consists of longitudinal and circular bands of smooth muscle interlaced with distensible connective tissue. The distension of this layer is especially important during parturition. Skeletal muscle strands, including the levator ani muscle, near the vaginal orifice partially constrict this opening.

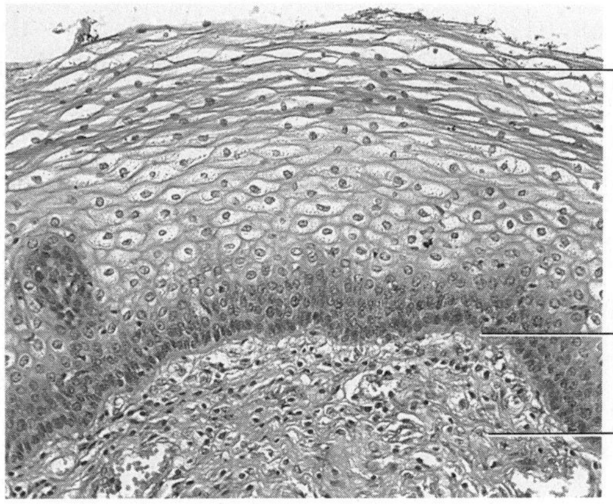

Figure 21.14 The histology of a vaginal ruga.

The **fibrous layer** covers the vagina and attaches it to surrounding pelvic organs. This layer consists of dense fibrous connective tissue interlaced with strands of elastic fibers.

The blood supply to the highly vascular vagina is primarily from the vaginal branches of the internal iliac artery. Blood is also supplied to the vagina through branches of the uterine, middle rectal, and internal pudendal arteries. All of these vessels form an extensive plexus surrounding the vagina called the **vaginal azygos arteries.** Blood draining from the vagina returns through vaginal veins that parallel the course of the arteries.

The vagina has sympathetic innervation from the hypogastric plexus and parasympathetic innervation from the second and third sacral nerves. Sensory innervation is through the pudendal plexus and is especially well developed near the vaginal orifice.

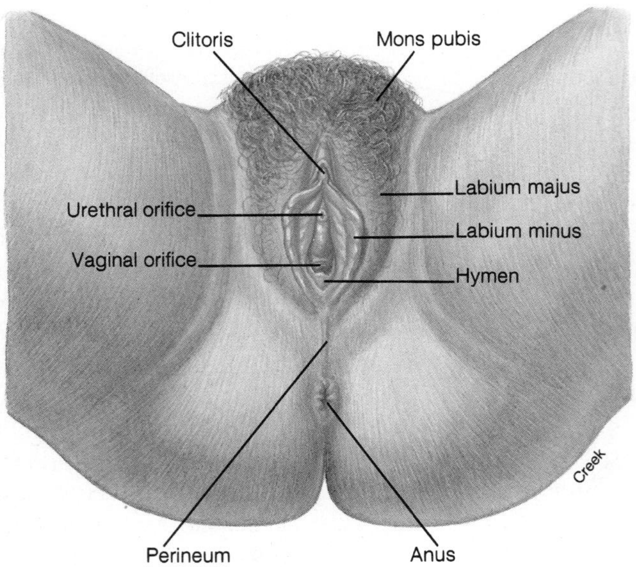

Figure 21.15 The external female genitalia.

Vulva

The external genitalia of the female are referred to as the **vulva,** or **pudendum** *(pu-den'dum),* which is shown in figure 21.15. The structures of the vulva surround the vaginal orifice and include the mons pubis, labia majora, labia minora, clitoris, vestibule, and vestibular glands.

The **mons pubis** is the subcutaneous pad of adipose connective tissue covering the symphysis pubis. At puberty, the mons pubis becomes covered with a pattern of coarse pubic hair that is somewhat triangular and usually has a horizontal upper border. The elevated and padded mons pubis cushions the symphysis pubis and vulva during coitus.

The **labia majora** (labium majus, singular) are two thickened longitudinal folds of skin that contain loose fibrous connective tissue and adipose tissue as well as some smooth muscle. The labia majora are continuous anteriorly with the mons pubis, are separated longitudinally by the **pudendal cleft,** and converge again posteriorly on the **perineum** *(per''ĭ-ne'um).* They are also covered with hair and contain numerous sebaceous and sweat glands. The labia majora are homologous to the scrotum of the male and function to enclose and protect the other organs of the vulva.

An *episiotomy (e-piz''e-ot'o-me)* is a surgical incision, for obstetrical purposes, of the vaginal orifice that extends into the perineum. An episiotomy may be done during parturition to facilitate delivery and accommodate the head of an emerging fetus when laceration seems imminent. After delivery the cut is sutured.

The **labia minora** (labium minus, singular) are two smaller longitudinal folds positioned close together between the labia majora. The labia minora are hairless but do contain sebaceous glands. On the anterior side, the labia minora split to form the **prepuce** *(pre'pūs)* of the clitoris. These inner folds of skin further protect the vaginal and urethral openings.

The **clitoris** is a small rounded projection at the upper portion of the pudendal cleft. The clitoris corresponds in structure and origin to the penis in the male; it is, however, smaller and has no urethra. Although most of the clitoris is embedded, it does have an exposed **glans** of erectile tissue that is richly innervated with sensory endings. The clitoris is about 2 cm (0.8 in.) long and 0.5 cm (0.2 in.) in diameter. The unexposed portion of the clitoris is composed of two columns of erectile tissue called the **corpora cavernosa,** which diverge posteriorly to form the **crura** and attach to the sides of the pubic arch.

The **vestibule** is the longitudinal cleft enclosed by the labia minora. The openings for the urethra and vagina are located in the vestibule. The external opening of the urethra is about 2.5 cm (1 in.) behind the glans of the clitoris and immediately in front of the vaginal orifice. The vaginal orifice is lubricated during sexual excitement by secretions from a pair of **vestibular glands** (Bartholin's glands) located within the wall of the region immediately inside the vaginal orifice. The ducts from these glands open into the vestibule near the lateral margins of the vaginal orifice. Bodies of vascular erectile tissue, called **vestibular bulbs,** are located immediately below the skin forming the lateral walls of the vestibule. The vestibular bulbs are separated from each other by the vagina and urethra and extend from the level of the vaginal orifice to the clitoris.

The vulva is highly vascular and is supplied with arterial blood from internal pudendal branches of the internal iliac arteries and external pudendal branches from the femoral arteries. Extensive vascular networks exist within most of the organs of the vulva. The venous return is through vessels that correspond in name and position to the arteries.

During pregnancy the vulva becomes swollen and bluish, especially the labia minora, due to increased vascularity and venous congestion. This discoloration is an important *diagnosis of pregnancy,* appears at about the eighth to the twelfth week, and becomes more apparent as pregnancy progresses.

The vulva has both sympathetic and parasympathetic innervation, as well as extensive somatic neurons that respond to sensory stimulation. Parasympathetic stimulation causes a response similar to that of the male: dilation of the arterioles of the genital erectile tissue and compression of the venous return.

Mechanism of Erection and Orgasm

The homologous structures of the male and female reproductive systems respond to sexual stimulation in a similar fashion. The erectile tissues of a female, like those of a male, become engorged with blood and swollen during sexual arousal. During sexual excitement, the hypothalamus of the brain sends parasympathetic nerve impulses through the sacral segments of the spinal cord, which cause dilation of the arteries serving the cli-

---

vulva: L. *volvere,* to roll, wrapper
pudendum: L. *pudere,* to be ashamed
mons pubis: L. *mons,* mountain; *pubis,* genital area

vestibule: L. *vestibule,* an entrance, court
Bartholin's glands: from Casper Bartholin, Jr., Danish anatomist, 1655–1738

toris and vestibular bulbs. This increased blood flow causes the erectile tissues to swell. In addition, the erectile tissues in the areola of the breasts become engorged.

Simultaneous with the erection of the clitoris and vestibular bulbs, the vagina expands and elongates to accommodate the erect penis of the male, and parasympathetic impulses cause the vestibular glands to secrete mucus near the vaginal orifice. The vestibular secretion moistens and lubricates the tissues of the vestibule, thus facilitating the penetration of the erect penis into the vagina during coitus. Mucus continues to be secreted during coitus so that the male and female genitalia do not become irritated as they would if the vagina became dry.

The position of the sensitive clitoris usually permits its being stimulated during coitus. If stimulation of the clitoris is of sufficient intensity and duration, a woman will experience a culmination of pleasurable psychological and physiological release called *orgasm.*

If orgasm occurs, there is an associated rhythmic contraction of the muscles of the perineum and the muscular walls of the uterus and uterine tubes. These reflexive muscular actions are thought to aid the movement of sperm through the female reproductive tract toward the upper end of a uterine tube where an ovum might be located.

Following an orgasm or completion of the sexual act, sympathetic impulses cause a reduction in arterial flow to the erectile tissues, and their size diminishes to that prior to sexual stimulation.

---

1. Describe the structure and position of the uterine tubes, and explain how an ovum is moved through these tubes to the uterus.
2. Describe the structure and function of the three layers of the uterus, and explain why the endometrium is subdivided into a stratum functionale and a stratum basale.
3. Describe the structures of the vagina and the vulva, and explain how these structures change during sexual excitement and coitus.

Mammary Glands

Mammary glands are modified sweat glands composed of secretory alveoli and ducts. The glands develop in the female breasts at puberty and function in lactation.

Objective 14. Distinguish between the mammary glands and the breast, and describe the structure of the mammary glands.

In structure, the **mammary glands,** located in the **breasts,** are modified sweat glands and are a part of the integumentary system. In function, however, these glands are associated with the reproductive system since they secrete milk for the nourishment of the young. The size and shape of the breasts vary considerably from person to person because of genetic differences, age, percentage of body fat, or pregnancy. At puberty, estrogen from the ovaries stimulates growth of the mammary glands and the deposition of adipose tissue within the breasts. Mammary glands hypertrophy in pregnant and lactating women and usually atrophy somewhat after menopause.

Structure of the Breast and the Mammary Glands

Each breast is positioned over the second to the sixth rib and overlies the pectoralis major muscle, the pectoralis minor muscle, and portions of the serratus anterior and external oblique muscles (see figs. 13.12 and 21.16). The medial boundary of the breast is over the lateral margin of the sternum, and the lateral margin of the breast is along the anterior border of the axilla. The **axillary tail** of the breast extends upward and laterally toward the axilla where it comes into close relationship with the axillary vessels. This region of the breast is clinically significant because of the high incidence of breast cancer within the lymphatic drainage of the axillary tail.

Each mammary gland is composed of fifteen to twenty **lobes,** divided by adipose tissue and each with its own drainage pathway to the outside. The amount of adipose tissue determines the size and shape of the breast but has nothing to do with the ability of a woman to nurse. Each lobe is subdivided into **lobules,** which contain the glandular **alveoli** (fig. 21.17) that secrete the milk of a lactating female. **Suspensory ligaments** between the lobules extend from the skin to the deep fascia overlying the pectoralis muscle and support the breasts. The clustered alveoli secrete milk into a series of **secondary tubules.** The secondary tubules in each lobe converge to form mammary ducts, which in turn converge to form **lactiferous** *(lak-tif'er-us)* **ducts** (fig. 21.16). The lumen of each lactiferous duct expands near the nipple to form an **ampulla,** where milk may be stored before it drains at the tip of the nipple.

The **nipple** is a cylindrical projection containing some erectile tissue. A circular pigmented **areola** *(ah-re'o-lah)* surrounds the nipple. The surface of the areola may appear rough because of the sebaceous **areolar glands** close to the surface. The secretions of the areolar glands keep the nipple pliable. The color of the areola and nipple varies with the complexion of the woman and whether or not she is pregnant. During pregnancy, the areola becomes darker and enlarges somewhat, presumably to become more conspicuous to a nursing infant.

Blood is supplied to the mammary gland through the perforating branches of the internal thoracic artery, which enter the breast through the second, third, and fourth intercostal spaces just lateral to the sternum, and through the more superficial mammary artery, branching from the lateral thoracic artery. Venous return is through a series of veins that parallel the pattern of the arteries. A superficial venous plexus may be apparent through the skin of the breast, especially during pregnancy and lactation.

The breast is innervated primarily through sensory somatic neurons that are derived from the anterior and lateral cutaneous branches of the fourth, fifth, and sixth thoracic nerves. Sensory nerve endings in the nipple and areola are especially important in stimulating the release of milk from the mammary glands to a suckling infant (see chap. 14).

---

orgasm: Gk. *orgasmos,* to swell; to become excited

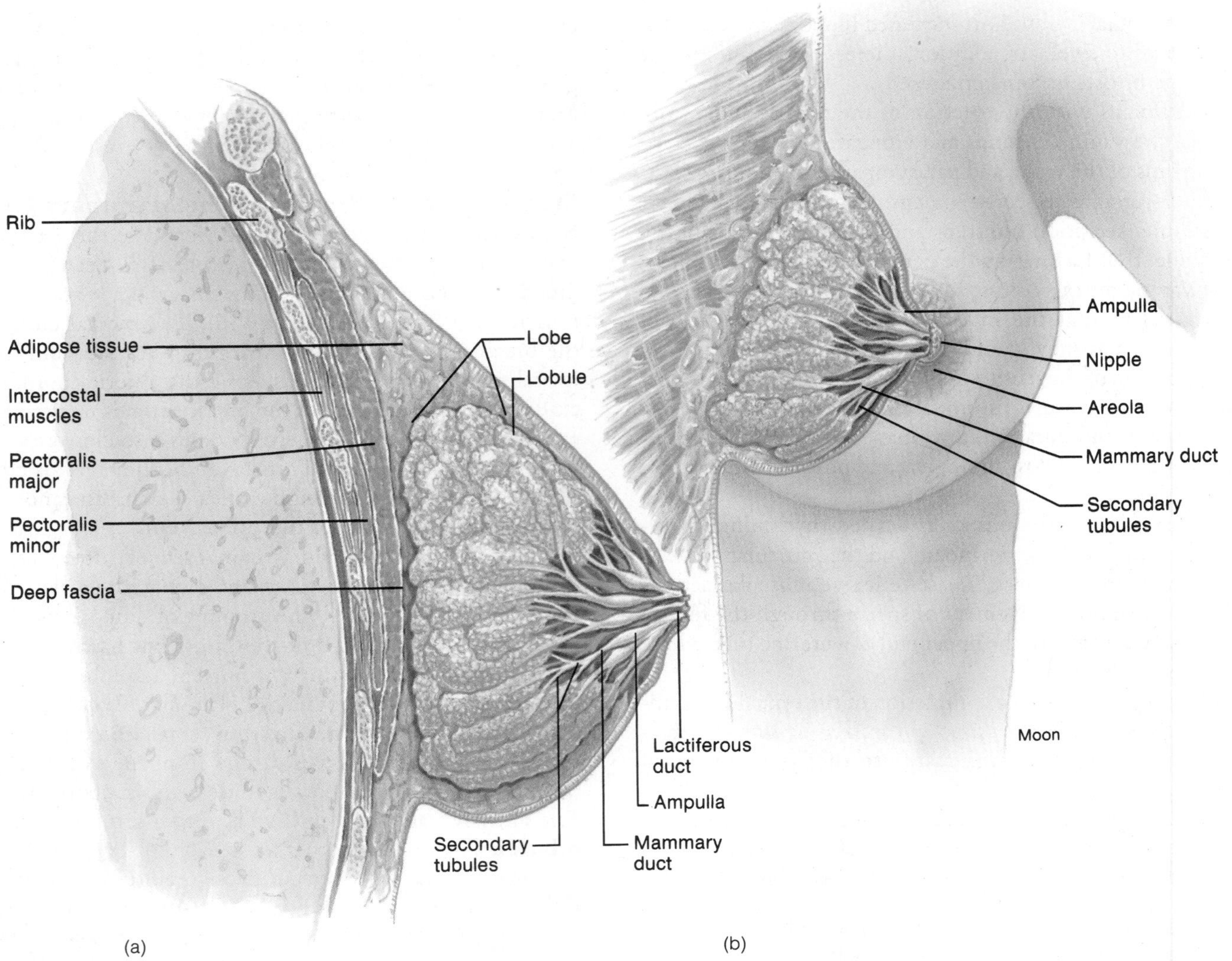

Figure 21.16 The structure of the breast and mammary glands: (*a*) a sagittal section and (*b*) an anterior view partially sectioned.

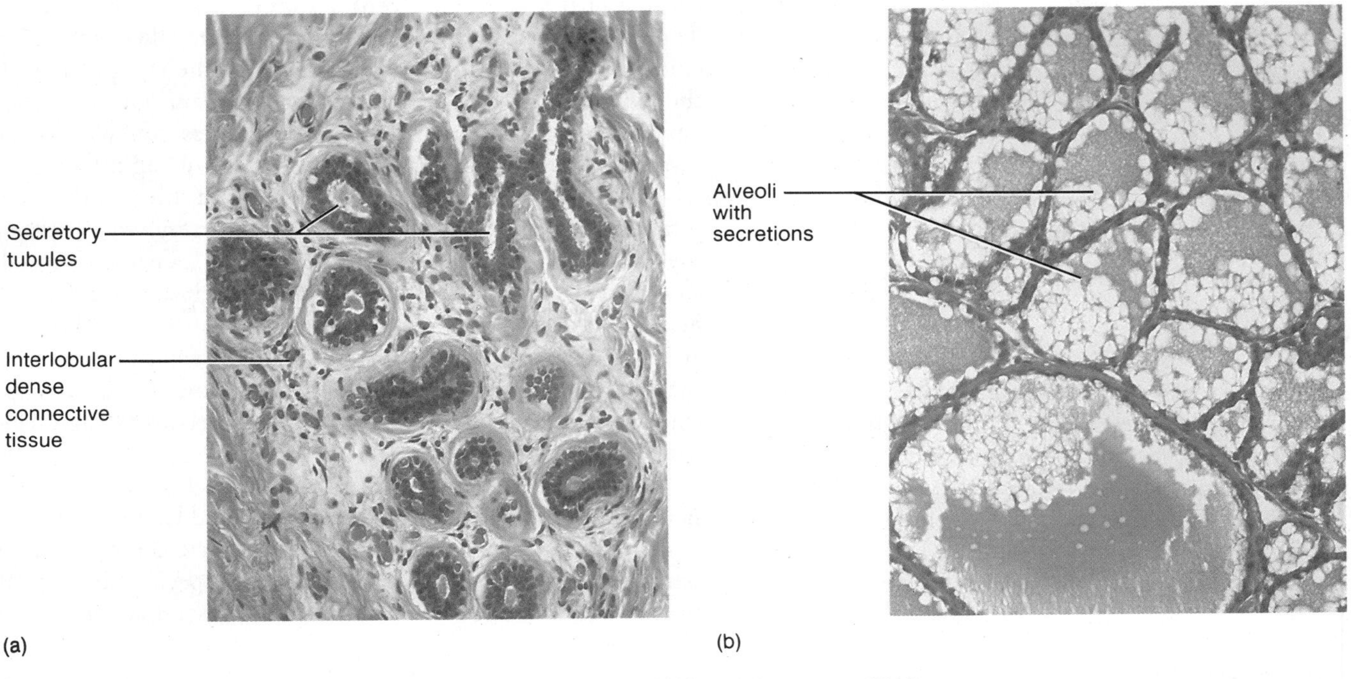

Figure 21.17 The histology of the mammary gland: (*a*) nonlactating (63×) and (*b*) lactating (63×).

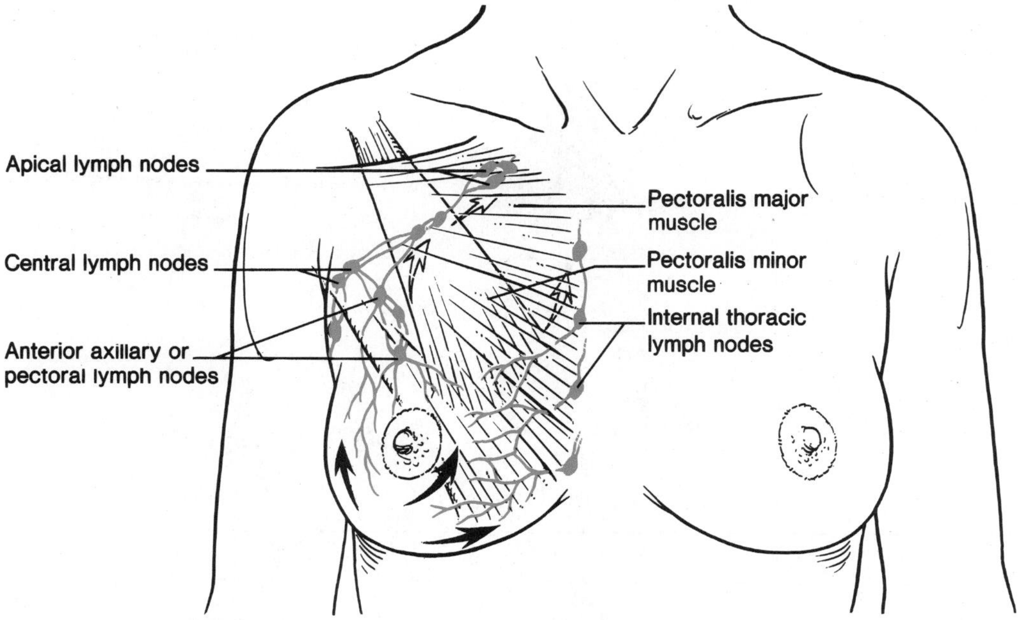

Figure 21.18 Lymphatic drainage of the mammary gland.

Lymphatic drainage and the location of lymph nodes within the breast are of considerable clinical importance because of the frequency of *breast cancer* and the high incidence of metastases. About 75% of the lymph drains through the axillary tail of the breast into the axillary lymph nodes (fig. 21.18). Some 20% of the lymph passes toward the sternum to the internal thoracic lymph nodes. The remaining 5% of the lymph is subcutaneous and follows the lymph drainage pathway in the skin toward the back where it reaches the intercostal nodes near the neck of the ribs.

---

1. Describe the structure of the breasts and mammary glands, and explain why variations in breast size do not affect the ability to lactate.
2. List in order the parts of the mammary glands through which milk passes during lactation.

Ovulation and Menstruation

Ovulation and menstruation are reproductive cyclic events that are regulated by follicle-stimulating hormone (FSH) and luteinizing hormone (LH) secreted by the anterior pituitary gland, and by estrogen and progesterone, secreted by structures of the ovaries.

Objective 15. Describe the hormonal changes that occur during the cycles of ovulation and menstruation.

Objective 16. Describe the structural changes that occur in the endometrium during the cycles of ovulation and menstruation, and explain the hormonal control of these changes.

Both ovulation and menstruation are reproductive functions of sexually mature females and are largely regulated by hormones secreted by the anterior pituitary gland and the ovaries. Both occur approximately every twenty-eight days, as menstruation follows ovulation and is regulated by the hormonal activity of the ovaries.

In ovulation, the follicular wall of the ovary ruptures, and a secondary oocyte is released and passes into the uterine tube. Ovulation generally occurs from alternate ovaries. If fertilization occurs in the uterine tube, mitotic divisions are initiated and the blastocyst implants on the uterine wall (see chap. 22).

If the egg is not fertilized, the menstrual cycle is initiated usually fourteen days after ovulation. The cycle is divided into three phases (fig. 21.19):

1. *Menstrual phase (menstruation).* The menstrual phase is characterized by a bloody discharge of endometrial tissue from the uterus during the first three to five days of the cycle.
2. *Proliferative phase.* During the proliferative phase, days five to fourteen of the cycle, endometrial tissue regrows.
3. *Secretory phase.* The secretory phase, lasting from day fourteen to day twenty-eight, is characterized by an increase in glandular secretions and blood to the endometrium in preparation for nourishing a blastocyst. The last two or three days of the secretory phase may be characterized by cramping and external spotting of blood. This period of the cycle is sometimes referred to as the *premenstrual phase* (not shown in figure 21.19) and involves the initial breakdown of the endometrial lining.

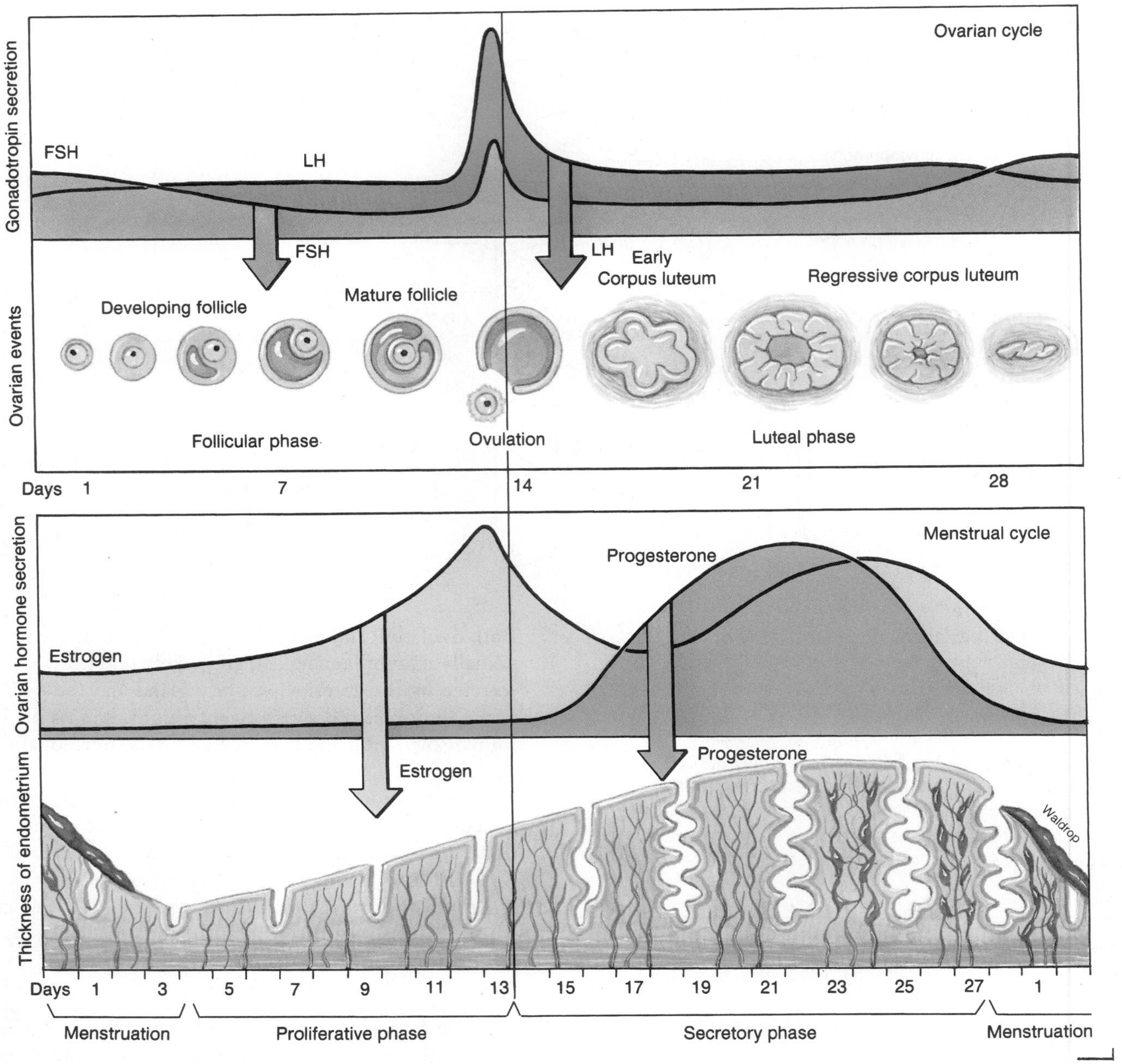

Figure 21.19 The cycle of ovulation and menstruation.

The controlling center for ovulation and menstruation is the hypothalamus. On a regular cycle, the hypothalamus releases a *gonadotropin-releasing factor* (GRF), which in turn stimulates the anterior pituitary gland to secrete *follicle-stimulating hormone* (FSH) and *luteinizing hormone* (LH) at the appropriate time. The FSH stimulates the maturation of a follicle within an ovary. During the middle of the menstrual cycle, the anterior pituitary releases a relatively large quantity of LH and an increased amount of FSH. This increased concentration of hormones causes the mature follicle to swell rapidly and rupture. Ovulation is complete as the oocyte is discharged, along with its follicular fluid, toward the uterine tube.

Although there are several different female sex hormones, they all belong to two major groups that are referred to as *estrogen* and *progesterone*. The principal source of estrogen (in a nonpregnant female) is the ovaries. Estrogen, as associated with the menstrual cycle, causes the endometrium to thicken. Estrogen also plays an important role in the development and maintenance of the secondary sexual organs and secondary sexual characteristics. Progesterone is secreted by the ovaries also (in a nonpregnant female) and helps estrogen maintain the endometrium.

The principal events of ovulation and menstruation are outlined in table 21.3.

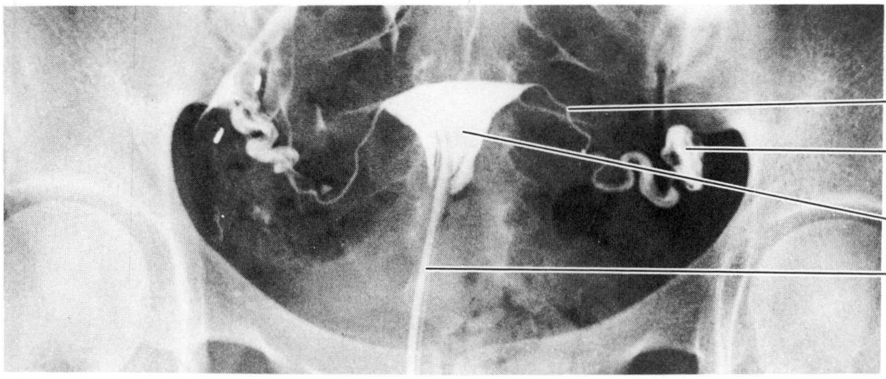

Figure 21.20 A hysterosalpingogram showing the cavities of the uterus and uterine tubes.

Table 21.3 Principal events in an ovulation and menstruation cycle

1. The hypothalamus releases GRF, which stimulates the anterior pituitary.
2. The anterior pituitary gland secretes FSH.
3. FSH stimulates the maturation of a follicle.
4. Follicular cells produce and secrete estrogen.
   a. Estrogen maintains secondary sexual traits.
   b. Estrogen causes the uterine lining to thicken.
5. The anterior pituitary gland secretes relatively large amounts of LH and FSH, which stimulate ovulation.
6. Follicular cells become corpus luteum cells, which secrete estrogen and progesterone.
   a. Estrogen continues to stimulate the uterine wall development.
   b. Progesterone stimulates the uterine lining to become more glandular and vascular.
   c. Estrogen and progesterone inhibit the secretion of FSH and LH from the anterior pituitary gland.
7. If the oocyte is not fertilized, the corpus luteum degenerates.
8. As concentrations of estrogen and progesterone decline, blood vessels in the uterine lining constrict.
9. The uterine lining disintegrates and sloughs away, producing menstrual flow.
10. The anterior pituitary gland, which is no longer inhibited, again secretes FSH.
11. The cycle is repeated.

1. Define *ovulation.* Define *menstruation.*
2. Explain the role of *GRF* in regulating female reproductive functions.
3. Diagram the relative thickness of the endometrium during the three phases of the menstrual cycle.
4. Summarize the major events in a menstrual cycle.

Clinical Considerations

Females are more prone to dysfunctions and diseases of the reproductive organs than are males because of cyclic changes in reproductive events, problems associated with pregnancy, and the susceptibility of the breasts to infections and neoplasms. The termination of reproductive capabilities at menopause can also cause complications due to hormonal alterations. *Gynecology* is the specialty of medicine concerned with dysfunction and diseases of the female reproductive system, whereas *obstetrics* is the specialty dealing with pregnancy and childbirth. Frequently a physician will specialize in both obstetrics and gynecology (OBGYN).

There are numerous clinical aspects of the female reproductive system, but only the most important conditions will be discussed. This portion of the chapter is divided into diagnostic procedures, developmental abnormalities, problems involving the ovaries and uterine tubes, problems involving the uterus, diseases of the vagina and vulva, and diseases of the breasts and mammary glands.

Diagnostic Procedures

A gynecological, or pelvic, examination is generally given in a thorough physical examination, especially prior to marriage, during pregnancy, or if problems involving the reproductive organs are suspected. In a gynecological examination, the physician inspects the vulva for irritations, lesions, or abnormal vaginal discharge, and palpates the vulva and internal organs. Most of the internal organs can be palpated through the vagina, especially if they are enlarged or tender. Inserting a lubricated *speculum* into the vagina allows visual examination of the cervix and vaginal walls. A speculum is an instrument for opening or distending a body opening to permit visual inspection.

In special cases, it may be necessary to examine the cavities of the uterus and uterine tubes by *hysterosalpingography (his''ter-o-sal''ping-gog'rah-fe)* (fig. 21.20). This technique involves injecting a radiopaque dye into the reproductive tract. The patency of the uterine tubes, irregular pregnancies, and various types of tumors may be detected using this technique. A *laparoscopy (lap''ah-ros'ko-pe)* permits *in vivo* visualization of the internal reproductive organs. The entrance for the laparoscope may be the umbilicus, a small incision in the lower abdominal wall, or through the posterior fornix of the vagina into the rectouterine pouch. Although a laparoscope is used primarily in diagnosis, it can be used when performing a **tubal ligation** (fig. 21.21), which is a method of sterilizing a female.

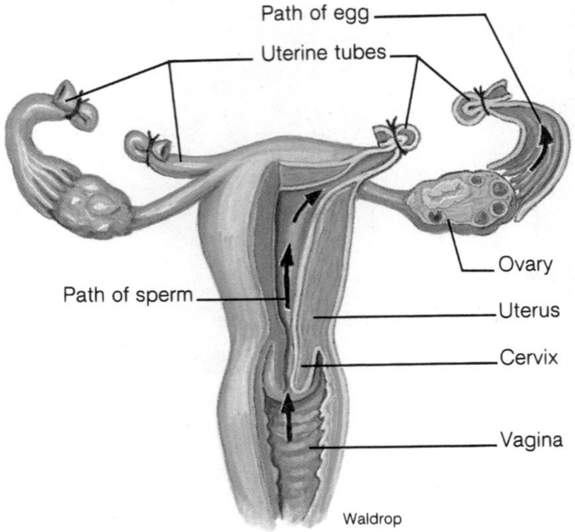

Figure 21.21 Tubal ligation involves removal of a portion of each uterine tube.

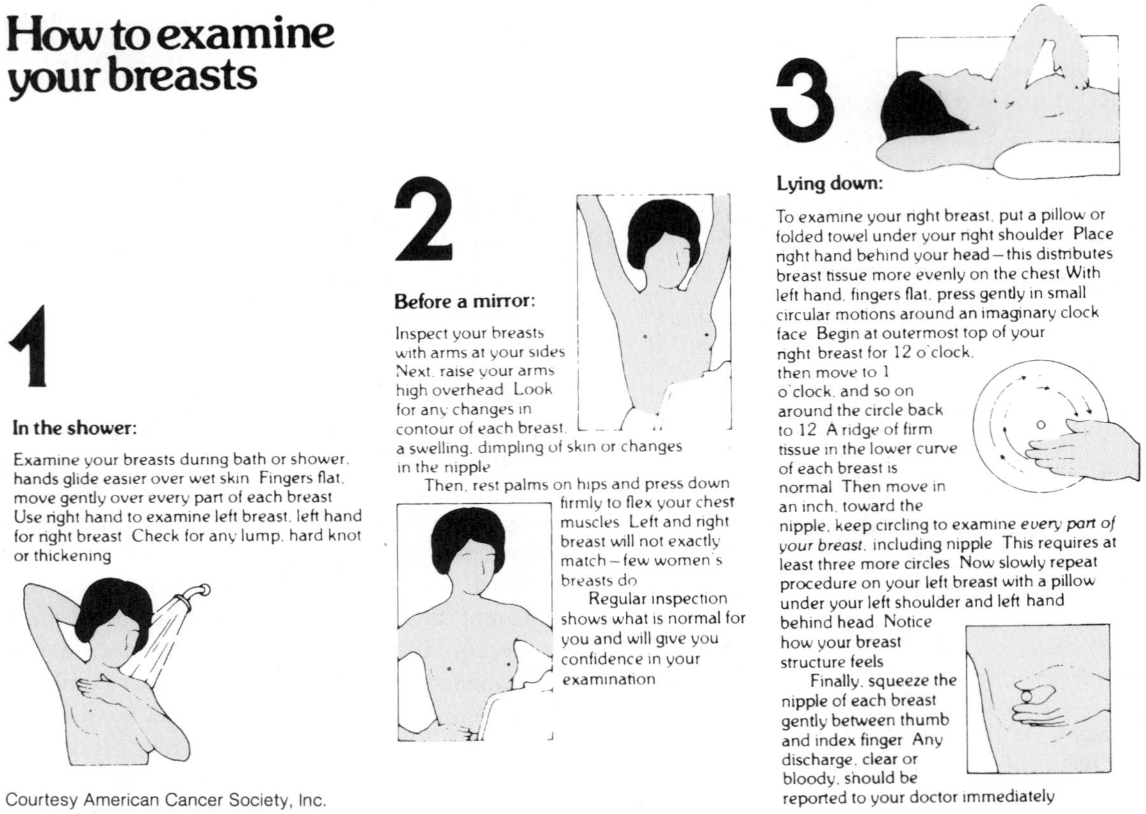

How to examine your breasts
1
In the shower:
Examine your breasts during bath or shower, hands glide easier over wet skin. Fingers flat, move gently over every part of each breast. Use right hand to examine left breast, left hand for right breast. Check for any lump, hard knot or thickening.
2
Before a mirror:
Inspect your breasts with arms at your sides. Next, raise your arms high overhead. Look for any changes in contour of each breast, a swelling, dimpling of skin or changes in the nipple.
Then, rest palms on hips and press down firmly to flex your chest muscles. Left and right breast will not exactly match—few women's breasts do.
Regular inspection shows what is normal for you and will give you confidence in your examination.
3
Lying down:
To examine your right breast, put a pillow or folded towel under your right shoulder. Place right hand behind your head—this distributes breast tissue more evenly on the chest. With left hand, fingers flat, press gently in small circular motions around an imaginary clock face. Begin at outermost top of your right breast for 12 o'clock, then move to 1 o'clock, and so on around the circle back to 12. A ridge of firm tissue in the lower curve of each breast is normal. Then move in an inch, toward the nipple, keep circling to examine *every part of your breast*, including nipple. This requires at least three more circles. Now slowly repeat procedure on your left breast with a pillow under your left shoulder and left hand behind head. Notice how your breast structure feels.
Finally, squeeze the nipple of each breast gently between thumb and index finger. Any discharge, clear or bloody, should be reported to your doctor immediately.
Courtesy American Cancer Society, Inc.

Figure 21.22 The procedures for a breast self-examination (BSE).

One diagnostic procedure that should be routinely performed by a woman is a *breast self-examination* (BSE). The importance of a BSE is not to prevent diseases of the breast but to detect any problems before they become serious. A BSE should be performed monthly one week after the cessation of menstruation so that the breast will not be swollen or especially tender. The procedure for a BSE is presented in figure 21.22.

Another important diagnostic procedure is a *Papanicolaou (Pap) smear.* The Pap smear permits a microscopic examination of cells obtained near the external os of the cervix. Samples of cells are obtained by using a vaginal pipette or a specially designed wooded spatula. Women should have routine Pap smears for the early detection of cervical cancer.

Pap smear: from George N. Papanicolaou, American anatomist and physician, 1883–1962

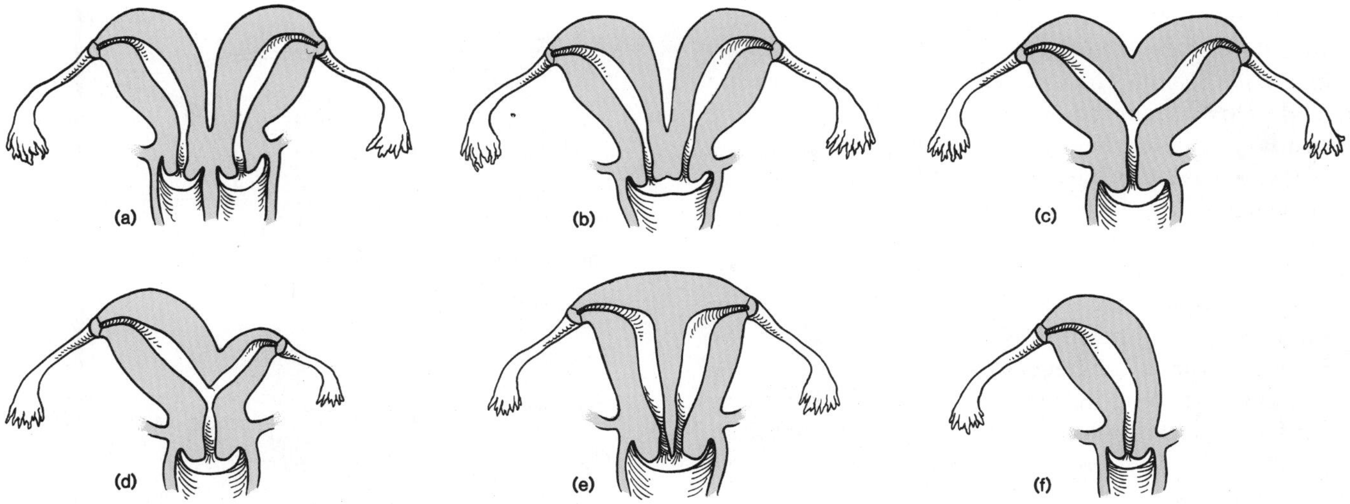

Figure 21.23 Congenital uterine abnormalities. (*a*) A double uterus and double vagina, (*b*) a double uterus with single vagina, (*c*) a bicornuate uterus, (*d*) a bicornuate uterus with a rudimentary left horn, (*e*) a septate uterus, and (*f*) a unicornuate uterus.

Developmental Problems

Many of the developmental abnormalities of the female reproductive system also occur in the reproductive system of the male and are discussed in the previous chapter. Hermaphroditism and irregularities of the sex chromosomes, for example, are developmental conditions that cause a person to develop both male and female characteristics.

Other developmental abnormalities of the female reproductive system may occur during the formation of the uterus and vagina. Failure of the paramesonephric ducts to fuse normally can result in a **double uterus,** a **bicornuate** *(bi-kor'nu-āt)* **uterus,** or a **unicornuate uterus.** These types of uterine abnormalities and others are diagrammed in figure 21.23. In about one in 4,000 females, the vagina is absent. This is usually accompanied by an absence of the uterus as well.

Problems Involving the Ovaries and Uterine Tubes

Most **ovarian neoplasms** are nonmalignant ovarian cysts lined by cuboidal epithelium and filled with a serous albuminous fluid. These tumors may frequently be palpated during a gynecological examination and may require surgical removal if they exceed about 4 cm in diameter. They are generally removed as a precaution because it is impossible to determine by palpation whether the mass is malignant or benign.

Malignant ovarian tumors, which are most likely to occur in women over the age of sixty, may reach massive size. Ovarian tumors of 5 kg (14 lbs) are not uncommon and ovarian tumors of 110 kg (300 lbs) have been reported. Some ovarian tumors produce estrogen and thus cause feminization in elderly women, including recommencement of menstrual periods. The prognosis for women with ovarian tumors varies depending on the type of tumor and whether or not metastasis has occurred.

Two frequent problems involving the uterine tubes are salpingitis and ectopic pregnancies. **Salpingitis** is an inflammation of one or both uterine tubes. Infection of the uterine tubes is generally caused by a sexually transmitted disease, although secondary bacterial infections from the vagina may also cause salpingitis. Salpingitis may cause sterility if the uterine tubes become occluded.

Ectopic pregnancy results from implantation of the blastocyst in a location other than the body or fundus of the uterus. The most frequent ectopic site is in the uterine tube, where an implanted blastocyst causes what is commonly called a **tubular pregnancy.** One danger of a tubular pregnancy is the enlargement, rupture, and subsequent hemorrhage of the uterine tube where implantation has occurred. A tubular pregnancy is frequently treated by removing the affected tube.

Infertility, or the inability to conceive, is a clinical problem that involves the male or female reproductive system and affects about 10% of couples. Generally, when a male is infertile, it is because of inadequate sperm counts. Female infertility is frequently caused by an obstruction of the uterine tubes or abnormal ovulation.

Problems Involving the Uterus

Abnormal menstruations are among the most common disorders of the female reproductive system. Abnormal menstruations may be directly related to problems of the reproductive organs and pituitary gland or associated with emotional and psychological stress.

Amenorrhea *(ah-men''o-re'ah)* is the absence of menstruation and can be categorized as normal, primary, or secondary. **Normal amenorrhea** follows menopause, occurs during pregnancy, and in some women may occur during lactation. **Primary amenorrhea** is when a woman has never menstruated although she has passed the age when menstruation

---

bicornuate: L. *bi,* two; *cornu,* horn
neoplasm: Gk. *neos,* new; *plasma,* something formed

normally begins. Primary amenorrhea is generally accompanied by failure of the secondary sexual characteristics to develop. Endocrine disorders may cause primary amenorrhea and abnormal development of the ovaries or uterus.

Secondary amenorrhea is the cessation of menstruation in women who previously have had normal menstrual periods and are not pregnant and have not gone through menopause. Various endocrine disturbances as well as psychological factors cause secondary amenorrhea. It is not uncommon, for example, for young women who are in the process of making major changes or adjustments in their lives to miss menstrual periods. Secondary amenorrhea is also frequent in women athletes during periods of intense training. A low percentage of body fat may be a contributing factor. Sickness, fatigue, poor nutrition, or emotional stress also cause secondary amenorrhea.

Dysmenorrhea is painful or difficult menstruation accompanied by severe menstrual cramps. The causes of dysmenorrhea are not totally understood but may include endocrine disturbances (inadequate progesterone levels), a faulty position of the uterus, emotional stress, or some type of obstruction that prohibits menstrual discharge.

Abnormal uterine bleeding includes **menorrhagia** *(men″o-ra′je-ah)*, or excessive bleeding during the menstrual period, and **metrorrhagia,** or spotting between menstrual periods. Other types of abnormal uterine bleeding are menstruations of excessive duration, too frequent menstruations, and postmenopausal bleeding. These abnormalities may be caused by hormonal irregularities, emotional factors, or various diseases and physical conditions.

Uterine neoplasms are an extremely common problem of the female reproductive tract. Most of the neoplasms are benign and include cysts, polyps, and smooth muscle tumors (leiomyoma). Any of these conditions may provoke irregular menstruations and may cause infertility if the neoplasms are massive.

Cancer of the uterus is the most common malignancy of the female reproductive tract. The most common site of uterine cancer is in the cervix (fig. 21.24). Cervical cancer is second only to cancer of the breast in frequency of occurrence and is a disease of young women (ages thirty through fifty), especially those who have had frequent coitus with multiple partners during their teens and onward. If detected early through regular Pap smears, the disease can be cured before it metastasizes. The treatment of cervical cancer depends on the stage of the malignancy and the age and health of the woman. In the case of women who are through having children, a **hysterectomy** *(his″tĕ-rek′to-me)* (surgical removal of the uterus) is usually performed.

Endometriosis is a condition characterized by the presence of endometrial tissues at sites other than the inner lining of the uterus. Frequent sites of ectopic endometrial cells are on the ovaries, outer layer of the uterus, abdominal wall, and urinary bladder. Although it is not certain how endometrial cells become established outside the uterus, it is speculated that some discharged endometrial tissue might be flushed backward from the uterus and through the uterine tubes during menstruation. Women with endometriosis will bleed internally with each menstrual period because the ectopic endometrial cells are stimulated along with the normal endometrium by ovarian hormones. The most common symptoms of endometriosis are extreme dysmenorrhea and a feeling of fullness during each menstruation period. Endometriosis can cause infertility. It is treated by suppressing the endometrial tissues with oral contraceptive pills or by surgery. An oophorectomy, or removal of the ovaries, may be necessary in extreme cases.

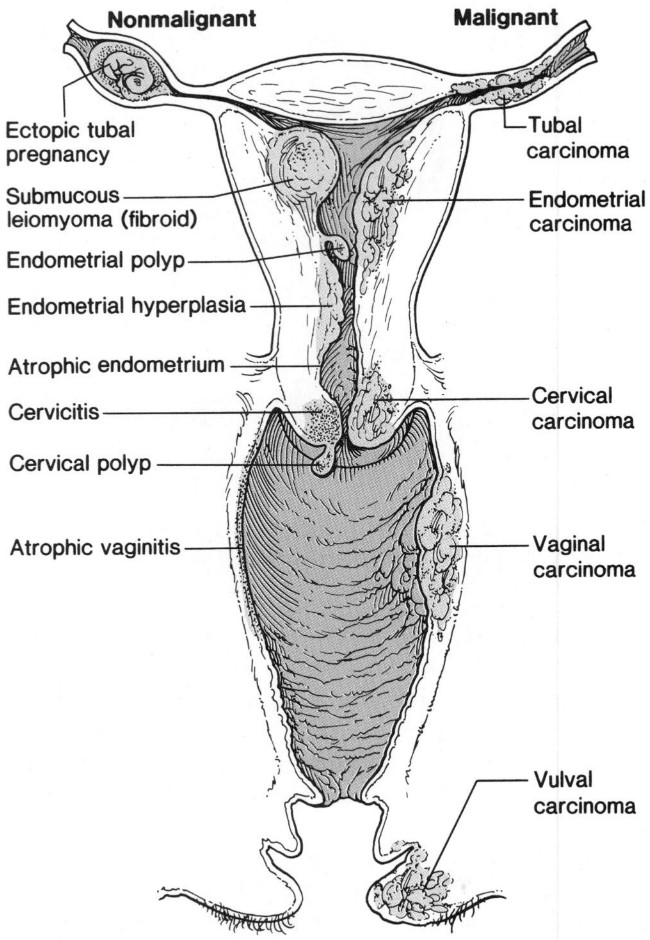

Figure 21.24 Sites of various conditions and diseases of the female reproductive tract, each of which could cause an abnormal discharge of blood.

Uterine displacements are relatively common in elderly women. When uterine displacements occur in younger women, they are important because of dysmenorrhea, fertility, or parturition problems. **Retroversion,** or **retroflexion,** is a displacement backward; **anteversion,** or **anteflexion,** is displacement forward. **Prolapse of the uterus** is a marked downward displacement into the vagina.

An **abortion** is defined as the termination of a pregnancy before the twenty-eighth week of gestation. A **spontaneous abortion,** or **miscarriage,** occurs without mechanical aid or medicinal intervention and may happen in as many as 10% of all pregnancies. Spontaneous abortions usually occur when there is abnormal development of the fetus or disease of the maternal reproductive system. An **induced abortion** is removal of the

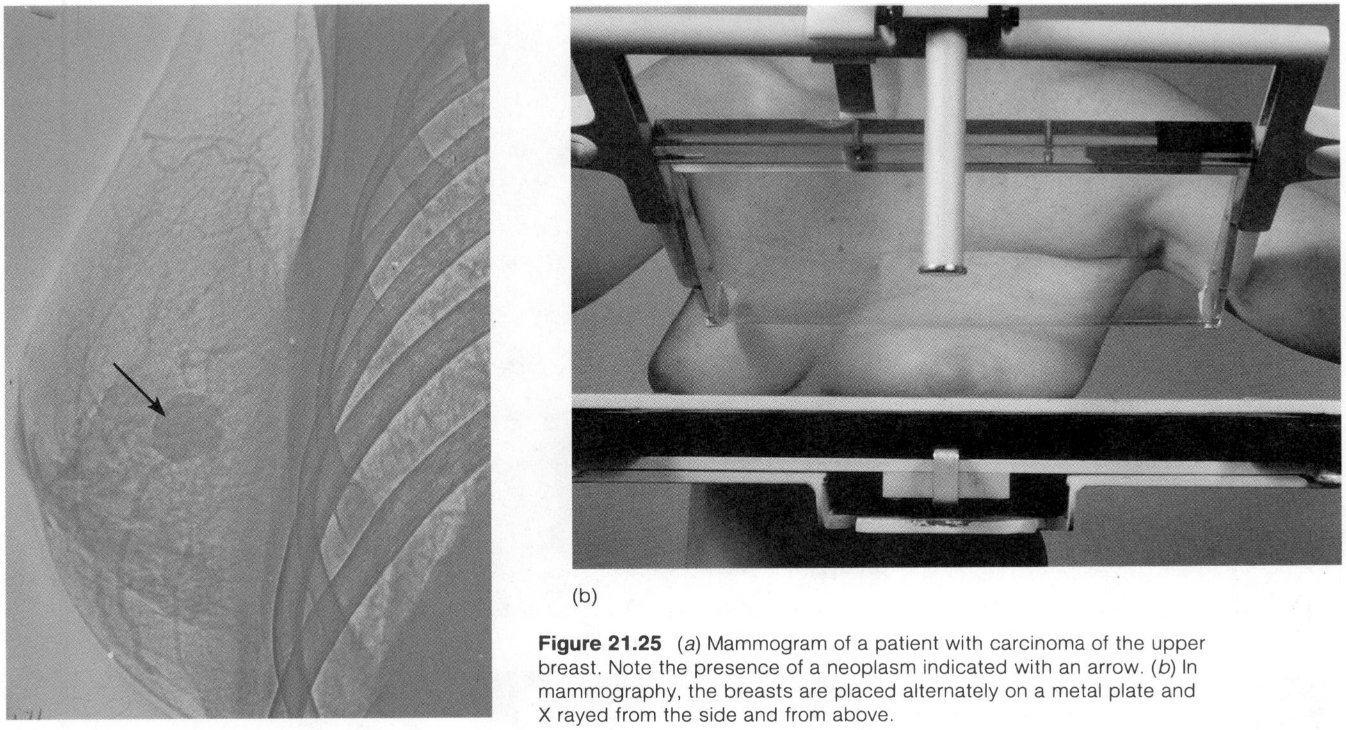

(a)

(b)

Figure 21.25 (*a*) Mammogram of a patient with carcinoma of the upper breast. Note the presence of a neoplasm indicated with an arrow. (*b*) In mammography, the breasts are placed alternately on a metal plate and X rayed from the side and from above.

fetus from the uterus by mechanical means or drugs. Induced abortions are a major issue of controversy because of questions regarding individual rights—those of the mother and those of the fetus—the definition of life, and moral concerns.

Diseases of the Vagina and Vulva

Pelvic inflammatory disease (PID) is a general term for inflammation of the female reproductive organs within the pelvis. The infection may be confined to a single organ, or it may involve all of the internal reproductive organs. The pathogens generally enter through the vagina during coitus, induced abortion, childbirth, or postpartum. Inflammation of the ovaries is called **oophoritis,** and inflammation of the uterine tube is **salpingitis** *(sal''pin-ji'tis).*

The vagina and vulva are generally resistant to infection because of the acidity of the vaginal secretions. Occasionally, however, localized infections and inflammations do occur; these are termed **vaginitis,** if confined to the vagina, or **vulvovaginitis,** if both the vagina and external genitalia are affected. The symptoms of vaginitis are a discharge of pus *(leukorrhea)* and itching *(pruritus).* The two most common organisms that cause vaginitis are the protozoan *Trichomonas vaginalis* and the fungus *Candida albicans.*

Diseases of the Breasts and Mammary Glands

The breasts and mammary glands of females are highly susceptible to infections, cysts, and tumors. Infections involving the mammary glands usually follow the development of a dry and cracked nipple during lactation. Bacteria enter the wound and establish an infection within the lobules of the gland. During an infection of the mammary gland, a blocked duct frequently causes a lobe to become engorged with milk. This localized swelling is usually accompanied by redness, pain, and an elevation of temperature. Administering specific antibiotics and applying heat are the usual treatments.

Nonmalignant cysts are the most frequent diseases of the breast. These masses are generally of two types, neither of which is life threatening.

Dysplasia (fibrocystic disease) is a broad condition involving several nonmalignant diseases of the breast. All dysplasias are benign neoplasms of various sizes that may become painful during or prior to menstruation. Most of the masses are small and remain undetected. Dysplasia affects nearly one-half of women over thirty years of age prior to menopause.

Fibroadenoma *(fi''bro-ad''ĕ-no'mah)* is a benign tumor of the breast that frequently occurs in women under the age of thirty-five. Fibroadenomas are nontender, rubbery masses, which are easily moved about in the mammary tissue. A fibroadenoma can be excised in a physician's office under local anesthetics.

Carcinoma of the breast is the most common malignancy in women (see fig. 24.13). One in ten women will develop breast cancer and one-third of these will die from the disease. Breast cancer is the leading cause of death in women between forty and fifty years of age. Men are also susceptible to breast cancer, but it is one hundred times more frequent in women. Breast cancer in men is usually fatal.

The causes of breast cancer are not known, but women who are most susceptible are those who are over age thirty-five, who have a family history of breast cancer, and who are nulliparous (no children). The early detection of breast cancer is important because the progressed state of the disease will determine the treatment and prognosis.

Confirming suspected breast cancer generally requires *mammography* (fig. 21.25). If the mammograph suggests breast cancer, surgery is performed so that a biopsy can be obtained

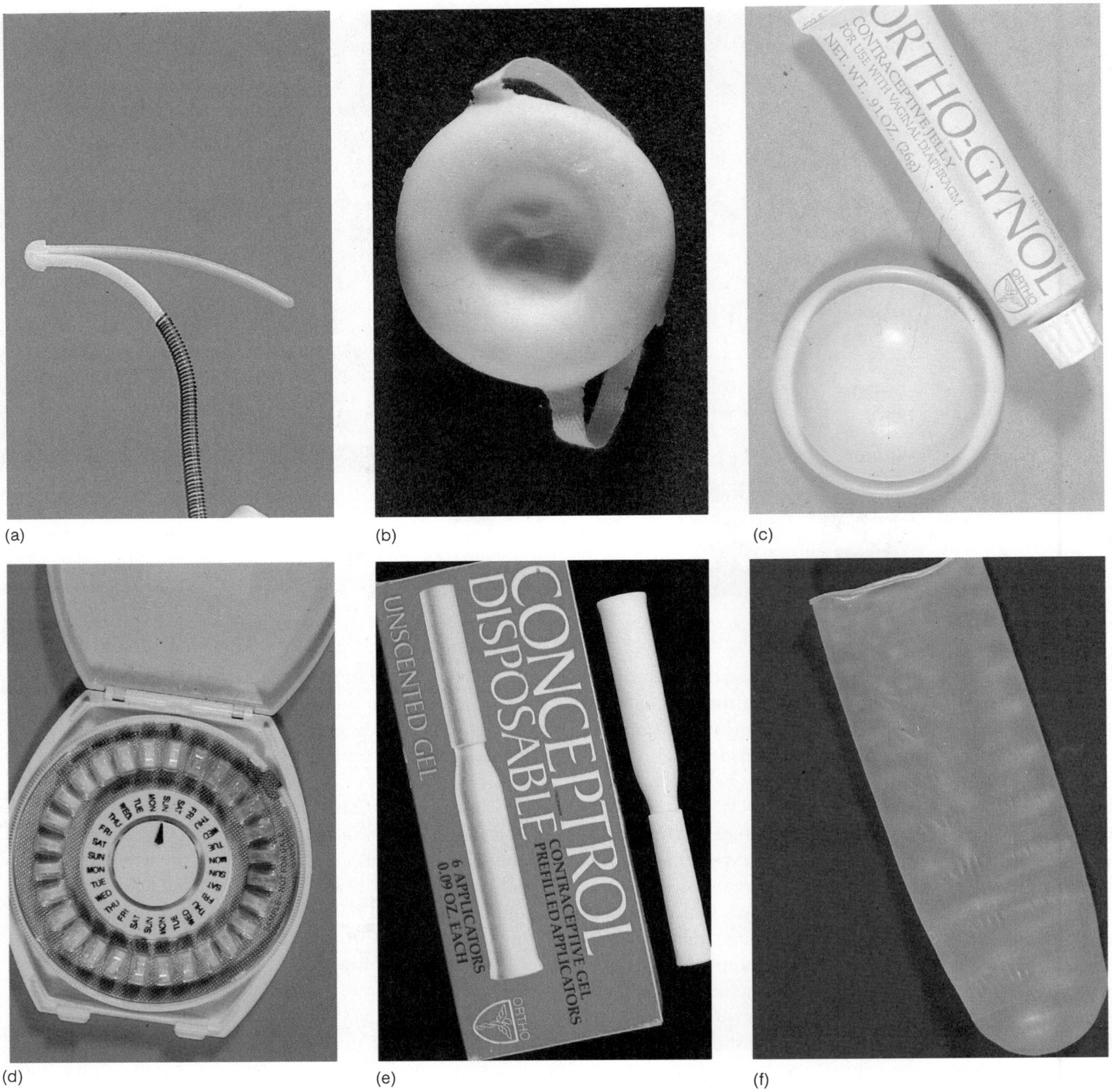

Figure 21.26 Various types of birth control devices. (*a*) IUD, (*b*) contraceptive sponge, (*c*) diaphragm, (*d*) birth control pills, (*e*) vaginal spermicide, and (*f*) condom.

and the tumor assessed. If the tumor is found to be malignant, extensive surgery is performed depending on the size of the tumor and whether metastasis has occurred. The surgical treatment for breast cancer is generally some degree of *mastectomy*. A *simple mastectomy* is removal of the entire breast but not the underlying lymph nodes. A *modified radical mastectomy* is the complete removal of the breast, the lymphatic drainage, and perhaps the pectoralis major muscle. A *radical mastectomy* is similar to a modified except that the pectoralis major muscle is always removed as well as the axillary lymph nodes and adjacent connective tissue.

Methods of Contraception

The *rhythm method* of birth control is used by many people, but the popularity of this technique has declined as more successful methods of contraception have been introduced (fig. 21.26). In the rhythm method of birth control, the woman attempts to predict the day of her ovulation and restricts coitus to safe times of the cycle when ovulation does not occur. The day of ovulation can be determined by a drop in basal body temperature or by a change in mucus discharge from the vagina. This technique is subject to great error, however, because most women don't ovulate on precisely the same day of the cycle each month.

More popular methods of birth control include (1) sterilization, (2) oral contraceptives, (3) intrauterine devices (IUDs), and (4) barrier methods—including condoms for the male and diaphragms for the female. All of these techniques are effective, but the safety, side effects, and efficacy of each technique are different.

Sterilization techniques include *vasectomy* for the male and *tubal ligation* for the female. In the latter technique (which accounts for over 60% of sterilization procedures performed in the United States), the uterine tubes are cut and tied. This is analogous to the procedure performed on the ductus (vas) deferens in a vasectomy and prevents fertilization of the ovulated ovum.

About 10 million women in the United States and 60 million women in the world are currently using **oral steroid contraceptives.** These contraceptives usually consist of a synthetic estrogen combined with a synthetic progesterone in the form of pills, which are taken once each day for three weeks after the last day of a menstrual period. This procedure causes an immediate increase in blood levels of ovarian steroids (from the pill), which is maintained for the normal duration of a monthly cycle. As a result of *negative feedback inhibition* of gonadotrophin secretion, *ovulation never occurs.* The entire cycle is like a false luteal phase, with high levels of progesterone and estrogen and low levels of gonadotrophins.

Since the contraceptive pills contain ovarian steroid hormones, the endometrium proliferates and becomes secretory just as it does during a normal cycle. In order to prevent an abnormal growth of the endometrium, women stop taking the pill after three weeks. This causes estrogen and progesterone levels to fall, and permits menstruation to occur. The contraceptive pill is an extremely effective method of birth control, but it does have potentially serious side effects—including an increased incidence of thromboembolism and cardiovascular disorders. It has been pointed out, however, that the mortality risk of contraceptive pills is still much lower than the risk of death from the complications of pregnancy—or from automobile accidents.

Intrauterine devices (IUDs) don't prevent ovulation, but instead prevent implantation of the blastocyst into the uterus in the event that fertilization occurs. The mechanisms by which these contraceptive effects are produced are not well understood, but the efficiency of different IUDs appears to be related to their ability to cause inflammatory reactions in the uterus. Uterine perforations are the foremost complication of the use of IUDs. Because of the potential problems of IUDs, their use has diminished.

Barrier contraceptives—condoms, diaphragms, and cervical caps—are slightly less effective than the other methods of contraception, but they do not cause serious side effects. Barrier contraceptives are quite effective, especially when they are used in conjunction with spermicidal foams, gels, and sponges. Many couples avoid them, however, because they detract from the spontaneity of sexual intercourse. Latex condoms offer an additional benefit; they provide some protection against sexually transmitted diseases, including AIDS.

CLINICAL CASE STUDY ANSWER

An ectopic pregnancy is any pregnancy that implants outside of the uterine cavity. This is most likely to occur in the uterine tube. The events leading up to our patient's problem are as follows: Ovulation occurs wherein an oocyte is extruded from the ovary and received into the uterine tube. Soon after arrival into the tube, the oocyte is fertilized, creating a zygote. The zygote is transported inside the tube toward the uterine cavity. Up to this point, the process is no different than that which occurs in a normal pregnancy. In the case of tubal pregnancy, the transporting ability of the uterine tube fails, causing the conceptus to be retained in the tube. Implantation then occurs within tissues that are not well-suited for that purpose. For example, the uterine tube does not expand well to accommodate a growing embryo, nor does it possess the necessary epithelium and glandular structures as does the endometrium. The result is over-expansion and erosion of the tubal wall, leading to rupture and hemorrhage. Because the uterine tube is basically exposed to the peritoneal cavity, blood from the site of rupture can flow into and collect in the area known as the *culdesae,* which is posterior to the dome of the uterus and vaginal fornix. There it is easily aspirated by a needle placed through the vaginal wall. ■

Chapter Summary

I. Introduction to the Female Reproductive System
   A. The reproductive period of a female is between puberty (about age twelve) to menopause (about age forty-seven). During this age span, cyclic ovulation and menstruation patterns occur when a female is not pregnant.
   B. The functions of the female reproductive system are to produce ova, secrete sex hormones, receive the sperm from the male, provide sites for fertilization and implantation, facilitate parturition, and secrete milk from the mammary glands.
   C. The female reproductive system is divided into (1) primary sex organs—the ovaries, (2) secondary sex organs—those that are essential for sexual reproduction, and (3) secondary sex characteristics—features that are sexually attractive and are expressed after puberty.

II. Development of the Female Reproductive System
   A. Oogonia contained within follicles appear in the ovaries, and the uterus and uterine tubes develop from paramesonephric ducts.
   B. The female external genitalia develop from the same indifferent structures that give rise to the external genitalia of a male.

III. Structure and Function of the Ovaries
   A. The ovaries are supported by the mesovarium from the broad ligament, the ovarian ligament, and the suspensory ligament.
   B. The ovarian follicles within the cortex of the ovary undergo cyclic changes.
      1. Primary oocytes, arrested at prophase I of meiosis, are contained within primordial follicles.

2. Due to gonadotrophin stimulation, some of the primordial follicles enlarge to become primary follicles.
3. When a follicle develops a fluid-filled antrum, it is called a secondary follicle.
4. One follicle continues to grow and becomes a vesicular ovarian follicle.
5. The vesicular ovarian follicle contains a secondary oocyte arrested at metaphase II of meiosis.
6. In the process of ovulation, the vesicular ovarian follicle ruptures and releases its secondary oocyte, which becomes an ovum upon fertilization.
7. After ovulation the empty follicle becomes a corpus luteum.

IV. Secondary Sex Organs
  A. The uterine tubes, which convey ova from the ovaries to the uterus, are the sites for fertilization.
    1. The open-ended portion of each tube is expanded, and its margin bears fimbriae, which extend over the lateral surface of the ovary.
    2. Movement of an ovum is aided by ciliated cells that line the lumen and by peristaltic contractions in the wall of the tube.
  B. The uterus is supported by the broad ligaments, uterosacral ligaments, cardinal ligaments, and round ligaments.
    1. The endometrium consists of a stratum basale and a stratum functionale, which is shed in menstruation.
    2. The myometrium produces the muscular contractions needed for labor and parturition.
  C. The vagina serves to receive the erect penis, to convey menses to the outside, and to transport the fetus during parturition.
    1. The vaginal wall is composed of an inner mucosa, a middle muscularis, and an outer fibrous layer.
    2. The vaginal orifice may be partially covered by a thin membranous hymen.
  D. The external genitalia, or vulva, include the labia majora and minora, clitoris, vestibule, and vestibular glands.
  E. Impulses through parasympathetic nerves stimulate erectile tissues in the clitoris and vestibular bulbs; in orgasm, muscular contraction occurs in the perineum, uterus, and uterine tubes.

V. Mammary Glands
  A. Mammary glands are located within the breasts and are modified sweat glands.
    1. Each mammary gland is composed of fifteen to twenty lobes, which are subdivided into lobules that contain alveoli.
    2. During lactation, the alveoli secrete milk, which passes through secondary tubules, a lactiferous duct, and an ampulla, and is discharged through the nipple.
  B. The nipple is a cylindrical projection from the breast surrounded by the circular pigmented areola.

VI. Ovulation and Menstruation
  A. Ovulation and menstruation are reproductive cyclic events that are regulated by hormones secreted by the hypothalamus, the anterior pituitary gland, and the ovaries.
  B. Menstruation is divided into menstrual, proliferative, and secretory phases.
  C. The principal hormones that regulate ovulation and menstruation are estrogen, progesterone, follicle-stimulating hormone (FSH), and luteinizing hormone (LH).

Review Activities

Objective Questions

1. The paramesonephric (müllerian) ducts give rise to the
   (a) uterine tubes.
   (b) uterus.
   (c) pudendum.
   (d) both *(a)* and *(b)*.
   (e) both *(b)* and *(c)*.
2. In a female, the homologue of the male scrotum is the
   (a) labia majora. (c) clitoris.
   (b) labia minora. (d) vestibule.
3. The cervix is a portion of the
   (a) vulva. (c) uterus.
   (b) vagina. (d) uterine tubes.
4. Fertilization normally occurs in the
   (a) ovary. (c) uterus.
   (b) uterine tube. (d) vagina.
5. The secretory phase of the endometrium corresponds to which of the following ovarian phases?
   (a) the follicular phase
   (b) ovulation
   (c) the luteal phase
   (d) the menstrual phase
6. Which of the following statements about oogenesis is *true?*
   (a) Oogonia, like spermatogonia, form continuously during postnatal life.
   (b) Primary oocytes are haploid.
   (c) Meiosis is completed prior to ovulation.
   (d) In ovulation, secondary oocytes are released from vesicular ovarian follicles.
7. The function of the mesovarium is
   (a) movement of the sperm to the ova.
   (b) nourishment of the ovarian walls.
   (c) muscular contraction of the uterus.
   (d) suspension of the ovary.
8. Which of the following layers is shed as menses? The
   (a) perimetrial layer.
   (b) fibrous layer.
   (c) functionalis layer.
   (d) menstrual layer.
9. The transverse folds in the mucosal layer of the vagina are called the
   (a) perineal folds.
   (b) vaginal rugae.
   (c) fornices.
   (d) labia gyri.
10. Which of the following is *not* a part of the vulva? The
    (a) mons pubis. (d) vagina.
    (b) clitoris. (e) labia minora.
    (c) vestibule.

Essay Questions

1. Define *puberty, ovulation, menstruation,* and *menopause.*
2. List the homologous reproductive organs of the male and female reproductive systems, and state the undifferentiated structures from which they develop.
3. Describe the follicular changes within the cortex of an ovary during the events that precede and follow ovulation.

4. Define *oogenesis* and explain when the process is initiated and when it is completed.
5. Describe the gross and histologic structure of the uterus, and explain the significance of the strata functionale and basale of the endometrium.
6. Identify the secondary sex organs and explain their functions.
7. Summarize the events in a menstrual cycle, and explain the roles of estrogen and progesterone.
8. List the functions of the vagina and describe its structure.
9. Distinguish between the labia majora and labia minora, between the pudendal cleft and vestibule, and between the vestibular bulbs and vestibular glands.
10. List the events that cause an erection of the female genitalia, and explain the process of orgasm.
11. Describe the structure and position of the mammary glands in the breasts.
12. Define *gynecology, obstetrics, speculum, laparoscopy, breast self-examination,* and *Pap smear.*
13. Distinguish between normal, primary, and secondary amenorrhea. What causes each?
14. List the various kinds of uterine neoplasms. Why is it important to know about cancer of the uterus? How is it generally detected?
15. Distinguish between dysplasia, fibroadenoma, and carcinoma of the breast.

CHAPTER 22

Developmental Anatomy and Inheritance

Outline and Concepts

Clinical Case Study

Fertilization

The fertilization of a secondary oocyte by a sperm in the uterine tube promotes the completion of meiotic development and the formation of a diploid zygote.

Pre-Embryonic Stage

The events of the two-week pre-embryonic stage include fertilization, transportation of the zygote through the uterine tube, mitotic divisions, implantation, and the formation of primordial embryonic tissue.

- Cleavage and Formation of the Blastocyst
- Implantation
- Formation of Germ Layers

Embryonic Stage

The events of the six-week embryonic stage include the differentiation of the germ layers into specific body organs and the formation of the placenta, the umbilical cord, and the extraembryonic membranes, which provide sustenance and protection to the embryo.

- Extraembryonic Membranes
- Placenta
- Umbilical Cord
- Structural Changes of the Embryo by Weeks

Fetal Stage

The fetal stage lasts from nine weeks until birth and is characterized by tremendous growth and the specialization of body structures.

Labor and Parturition

Labor and parturition are the culmination of gestation and require the action of oxytocin secreted by the posterior pituitary and prostaglandins, produced in the uterus.

Inheritance

Inheritance is the passage of hereditary traits carried by the genes on chromosomes from one generation to another.

Clinical Considerations

- Abnormal Implantation Sites
- *In Vitro* Fertilization and Artificial Implantation
- Multiple Pregnancy
- Fetal Monitoring
- Congenital Defects
- Some Genetic Disorders of Clinical Importance

Clinical Case Study Answer

Chapter Summary

Review Activities

CLINICAL CASE STUDY

A twenty-seven-year-old woman gives birth to twin boys, followed by an apparently single placenta. After examining the two infants the pediatrician informs the mother that one of the babies has a cleft palate but that the other is normal. She then adds that in some instances cleft palate is a genetic disorder and asks if the problem has existed in any relatives. The mother replies that it has not. Further examination of the placenta reveals two amnions and only one chorion.

Does the presence of two amnions and one chorion indicate monozygotic or dizygotic twins? Can any conclusions, therefore, be drawn regarding genetic similarities between the two infants? Does it seem unusual that one baby would exhibit a cleft palate and the other not? Explain. ■

Fertilization

The fertilization of a secondary oocyte by a sperm in the uterine tube promotes the completion of meiotic development and the formation of a diploid zygote.

Objective 1. Define *capacitation, fertilization,* and *morphogenesis.*
Objective 2. Describe the changes that occur in the sperm and ovum during fertilization.

The structure of the human body forms before birth, or *prenatally,* through a process called **morphogenesis** *(mor"fo-jen'e-sis).* Through morphogenic events the organs and systems of the body are established in a functional relationship. Understanding morphogenesis is important for a complete perspective of the structure and function of each body system. There are sensitive stages of morphogenesis for each organ and system, during which genetic or environmental conditions (nutrition, smoking, or drugs taken by the mother) may affect the normal development of the baby. Many clinical problems are congenital in nature and originate during morphogenic development.

The prenatal development of a human is a fascinating and awesome event. It begins with a single fertilized egg and culminates some thirty-eight weeks later with a complex organization composed of billions of cells. Prenatal development can be divided into a *pre-embryonic stage,* which is initiated by the fertilization of a secondary oocyte, an *embryonic stage,* during which the body's organ systems are formed; and a *fetal stage,* which culminates in parturition, or the birth of the baby.

During coitus (sexual intercourse), a male ejaculates between 100 million and 500 million sperm into the vagina. This tremendous number is needed because of the high rate of sperm fatality—only about 100 survive to contact the secondary oocyte in the uterine tube. During passage through the acidic female tract, sperm gain the ability to fertilize a secondary oocyte through a process called **capacitation.** The changes that occur in capacitation are not fully understood. Experiments have shown, however, that freshly ejaculated sperm are infertile and must be in the female tract for at least seven hours before they can fertilize a secondary oocyte.

A woman usually ovulates one secondary oocyte a month, totaling approximately four hundred during her reproductive years. Each ovulation releases a secondary oocyte arrested at metaphase of the second meiotic division. As the secondary oocyte enters the uterine tube, it is surrounded by a thin transparent layer of protein and polysaccharides, called the **zona pellucida,** and a layer of granulosa cells, called the **corona radiata** (fig. 22.1).

The head of each sperm is capped by an organelle called an **acrosome** (see figs. 20.13, 22.1, 22.2). The acrosome contains a trypsinlike protein-digesting enzyme and *hyaluronidase (hi"ah-lu-ron'ĭ-das),* which digests hyaluronic acid, an important constituent of connective tissue. When a sperm meets a secondary oocyte in the uterine tube, an *acrosomal reaction* occurs that exposes the acrosome's digestive enzymes and allows a sperm to penetrate through the corona radiata and the zona pellucida.

As the first sperm penetrates the zona pellucida, a chemical change in the zona occurs that prevents other sperm from entering. Only one sperm, therefore, is allowed to fertilize a secondary oocyte. When the cell membranes of the sperm and the secondary oocyte merge, the secondary oocyte becomes an ovum. As fertilization occurs, the ovum is stimulated to complete its second meiotic division (fig. 22.3). Like the first meiotic division, the second produces one cell that contains all of the cytoplasm and one polar body. The healthy cell is the mature ovum, and the second polar body, like the first, ultimately fragments and disintegrates.

At fertilization, the entire sperm enters the cytoplasm of the much larger ovum (fig. 22.4). Within twelve hours the nuclear membrane in the ovum disappears, and the *haploid number* of chromosomes (twenty-three) in the ovum is joined by the haploid number of chromosomes from the sperm. A fertilized egg, or zygote *(zi'gōt),* containing the *diploid number* of chromosomes (forty-six) is thus formed.

A secondary oocyte that is ovulated but not fertilized does not complete its second meiotic division, but instead it disintegrates twelve to twenty-four hours after ovulation. Fertilization therefore cannot occur if coitus takes place beyond one day following ovulation. Sperm, in contrast, can survive up to three days in the female reproductive tract. Fertilization therefore can occur if coitus is performed within three days prior to the day of ovulation.

---

1. Explain why capacitation and the acrosomal reaction of sperm are necessary to accomplish fertilization of a secondary oocyte.
2. Define the term *morphogenesis,* and discuss its importance in the development of a healthy baby.
3. Discuss the changes that occur in a sperm from the time of ejaculation to the time of fertilization. What changes occur in a secondary oocyte following ovulation to the time of fertilization?

---

capacitation: L. *capacitas,* capable of

zona pellucida: L. *zone,* a girdle; *pellis,* skin
corona radiata: Gk. *korone,* crown; *radiata,* radiate
acrosome: Gk. *akron,* extremity; *soma,* body
haploid: Gk. *haplous,* single; L. *ploideus,* multiple in form
diploid: Gk. *diplous,* double; L. *ploideus,* multiple in form

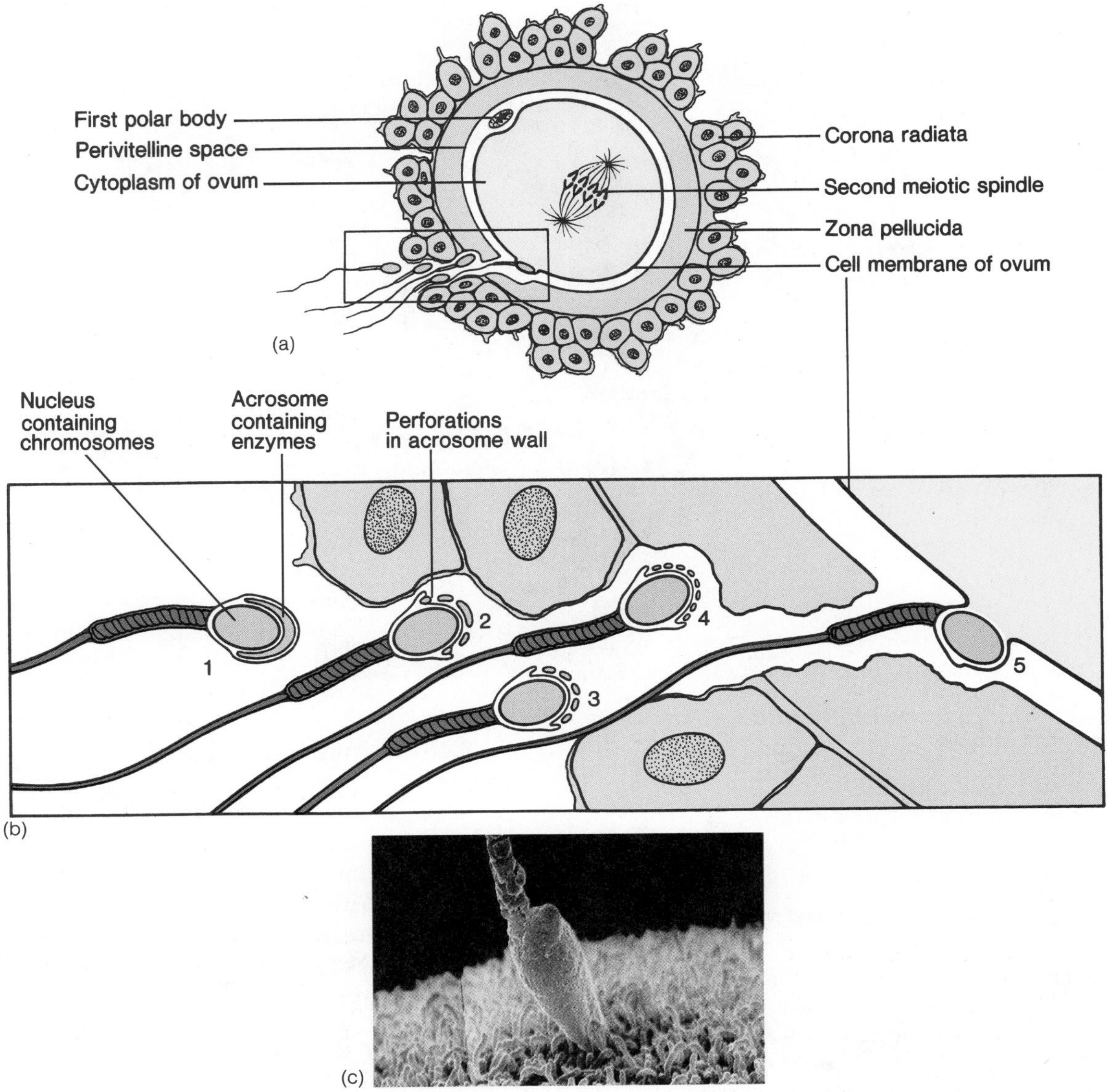

Figure 22.1 The process of fertilization. (*a, b*) Diagrammatic representations. As the head of the sperm encounters the gelatinous corona radiata of the egg (*2*), the acrosomal vesicle ruptures and the sperm digests a path for itself by the action of enzymes released from the acrosome (*3, 4*). When the cell membrane of the sperm contacts the cell membrane of the egg (*5*), they become continuous, and the sperm nucleus and other contents move into the egg cytoplasm. (*c*) A scanning electron micrograph of sperm bound to the egg surface.

Pre-Embryonic Stage

The events of the two-week pre-embryonic stage include fertilization, transportation of the zygote through the uterine tube, mitotic divisions, implantation, and the formation of primordial embryonic tissue.

Objective 3. Describe the events of pre-embryonic development that result in the formation of the blastocyst.

Objective 4. Discuss the role of the trophoblast in the implantation and development of the placenta.

Objective 5. Explain how the primary germ layers develop, and list the structures produced by each layer.

Objective 6. Define *gestation,* and explain how the parturition date is determined.

Cleavage and Formation of the Blastocyst

Fertilization occurs within the uterine tube, usually about twelve to twenty-four hours following ovulation. The fertilized egg (ovum) is referred to as a **zygote.** Within thirty hours the **cleavage** process begins with a mitotic division that results in the formation of two identical daughter cells called *blastomeres* (fig. 22.5). Several more cleavages occur as the structure passes down the uterine tube and enters the uterus on about the third day. It is now composed of a ball of sixteen or more cells called a **morula.** Although the morula has undergone several mitotic divisions, it is not much larger than the zygote because there have not been additional nutrients necessary for growth entering the cells.

The developing structure remains unattached in the uterine cavity for about three days. During this time, the center of the morula fills with fluid passing in from the uterine cavity. As

zygote: Gk. *zygotos,* yolked, joined

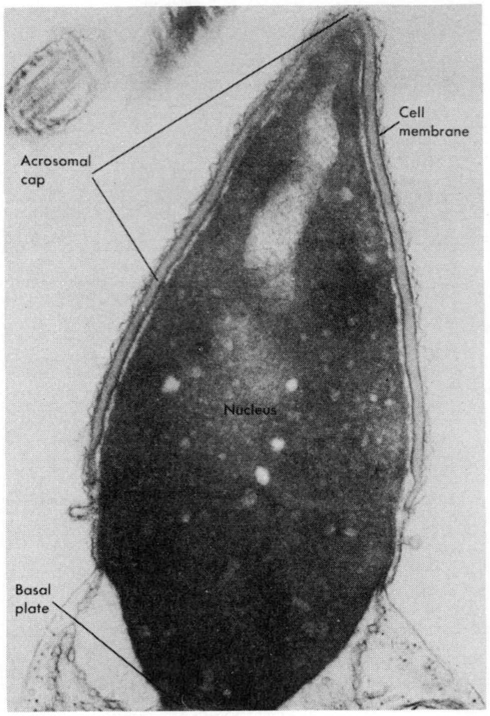

Figure 22.2 An electron micrograph showing the head of a human sperm with its nucleus and acrosomal cap.

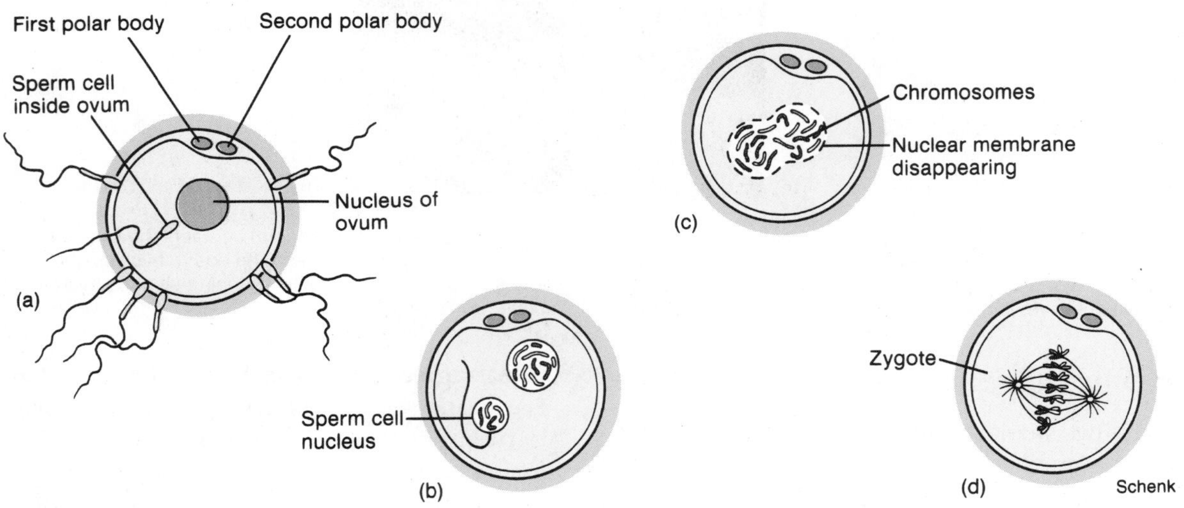

Figure 22.3 Fertilization and the union of chromosomes from the sperm and ovum to form the zygote. (*a*) Sperm penetration, (*b*) haploid number of chromosomes within the nucleus of each sex cell, (*c*) degeneration of nuclear membranes, and (*d*) matching and alignment of chromosomes.

the fluid-filled space develops inside the morula, two distinct groups of cells form and the structure becomes known as a **blastocyst.** The single outer layer forming the wall of the blastocyst is known as the **trophoblast,** whereas the small, inner aggregation of cells is called the **embryoblast,** or *inner cell mass.* With further development the trophoblast differentiates into a structure called the **chorion,** which will become a portion of the placenta, and the embryoblast will become the embryo. The hollow, fluid-filled center of the blastocyst is called the **blastocyst cavity.** A diagrammatic summary of the ovarian cycle, fertilization, and the morphogenic events of the first week is presented in figure 22.6.

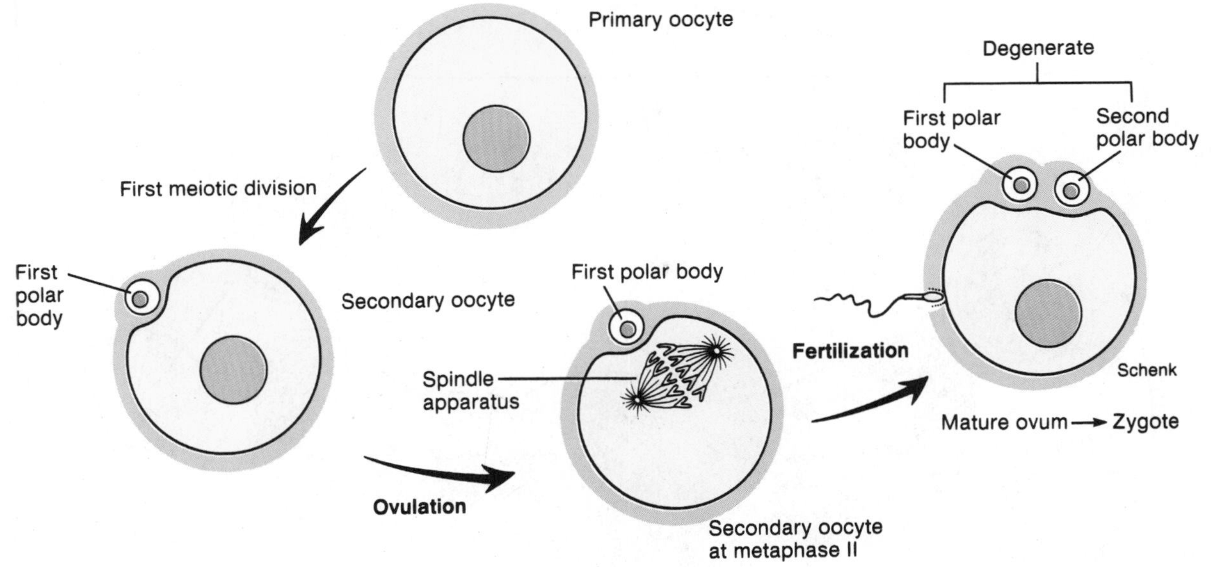

Figure 22.4 A secondary oocyte, arrested at metaphase II of meiosis, is released at ovulation. If this cell is fertilized, it becomes an ovum, completes its second meiotic division, and produces a second polar body.

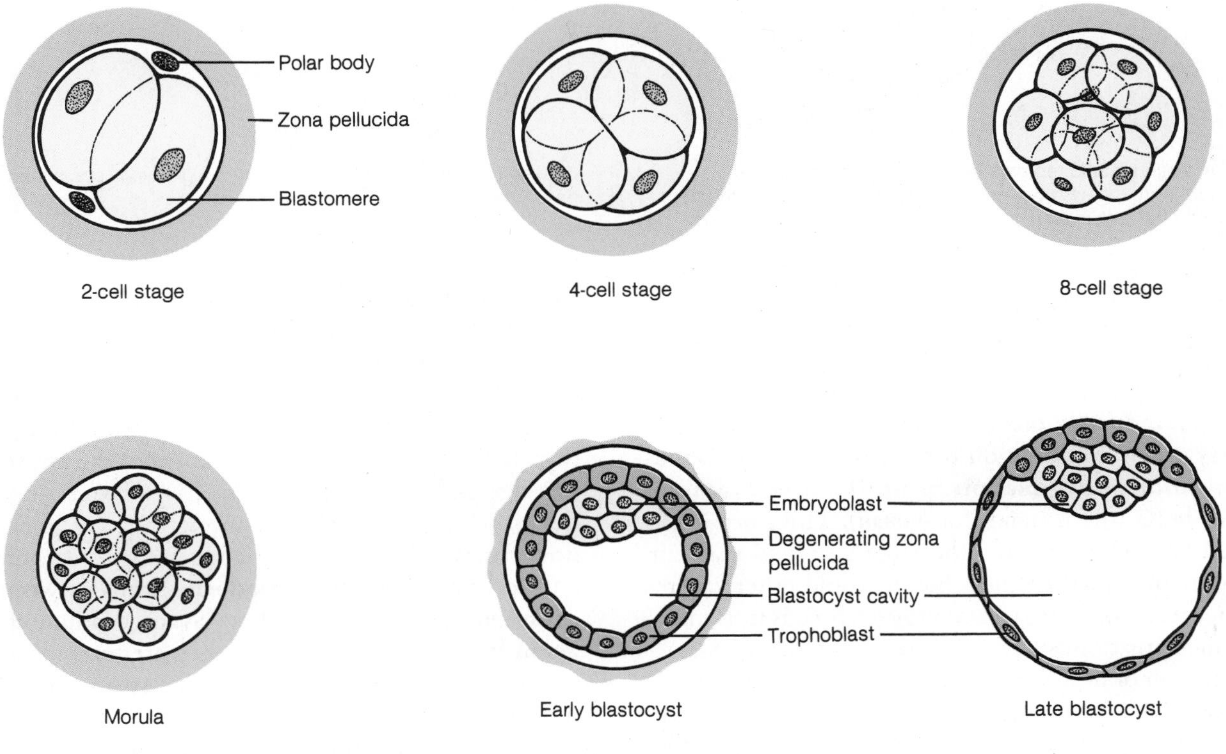

Figure 22.5 Sequential illustrations from the first cleavage of the zygote to the formation of the blastocyst. (Note the deterioration of the zona pellucida in the early blastocyst.)

Implantation

The process of *implantation,* or *nidation,* begins between the fifth and seventh day. Attachment is usually upon the posterior wall of the body of the uterus, with the side containing the embryoblast against the endometrium (fig. 22.7). Implantation is made possible by the secretion of *proteolytic enzymes* by the trophoblast, which digest a portion of the endometrium. The blastula sinks into the depression, and endometrial cells move back to cover the defect in the wall. At the same time, the part of the uterine wall below the implanting blastocyst thickens, and specialized cells of the trophoblast produce fingerlike projections, called **syncytiotrophoblasts** *(sin-sit''e-o-trof'o-blast),* into

implantation: L. *im,* in; *planto,* to plant
nidation: L. *nidus,* nest

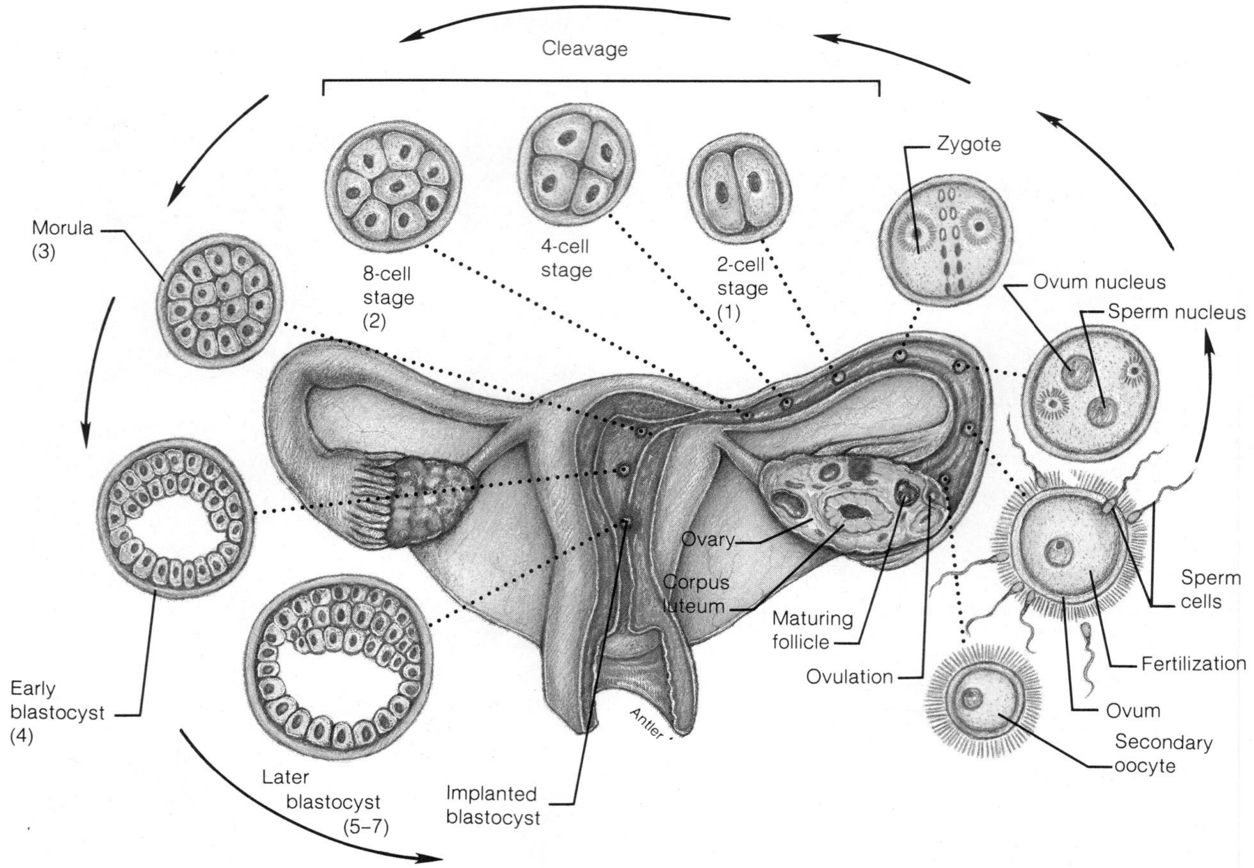

Figure 22.6 A diagrammatic representation of the ovarian cycle, fertilization, and the morphogenic events of the first week. The numbers indicate the days after fertilization. Implantation of the blastocyst begins between the fifth and seventh day and is generally completed by the tenth day.

the thickened area. The syncytiotrophoblasts arise from a specific portion of the trophoblast, at the embryonic pole, called the **cytotrophoblast.**

The blastocyst saves itself from being aborted by secreting a hormone that indirectly prevents menstruation. Even before the sixth day when implantation begins, the syncytiotrophoblasts secrete **chorionic gonadotrophin** *(ko''re-on'ik gon''ah-do-tro'fin),* or **hCG** (the h stands for *human*). This hormone is identical to LH in its effects and therefore is able to maintain the corpus luteum past the time when it would otherwise regress. The secretion of estrogen and progesterone is thus maintained and menstruation is normally prevented (fig. 22.8).

The secretion of hCG declines by the tenth week of pregnancy. Actually, this hormone is only required for the first five to six weeks of pregnancy, because the placenta itself becomes an active steroid-secreting gland. At the fifth to sixth week, the mother's corpus luteum begins to regress (even in the presence of hCG), but the placenta secretes more than sufficient amounts of steroids to maintain the endometrium and prevent menstruation.

All *pregnancy tests* assay for the presence of hCG in the blood or urine, because this is the only hormone that is secreted by the blastocyst but not by the mother's endocrine glands. Modern pregnancy tests detect the presence of hCG by use of antibodies against hCG or by the use of cellular receptor proteins for hCG.

Formation of Germ Layers

As the blastocyst completes implantation during the second week of development, the inner cell mass undergoes marked differentiation. A slitlike space, called the **amniotic cavity,** forms between the embryoblast and the invading trophoblast (fig. 22.9). The embryoblast flattens into the **embryonic disc** (see fig. 22.11), which consists of two layers: an upper **ectoderm,** which is closer to the amniotic cavity, and a lower **endoderm,** which borders the blastocyst cavity. A short time later, a third layer called the **mesoderm** forms between the endoderm and ectoderm. These three layers constitute the **primary germ layers** (fig. 22.9). Once they are formed, at the end of the second week, the pre-embryonic stage is completed, and the embryonic stage begins.

The primary germ layers are important because various cells and tissues of the body are derived from them. Ectodermal cells form the nervous system; the outer layer of skin (epidermis), including hair, nails, and skin glands; and portions of the sensory organs. Mesodermal cells form the skeleton, muscles, blood, reproductive organs, dermis of the skin, and connective tissue. Endodermal cells produce the lining of the digestive tract, the digestive organs, the respiratory tract and lungs, and the urinary bladder and urethra.

The events of the pre-embryonic stage are summarized in table 22.1 and the derivatives of the three primary germ layers

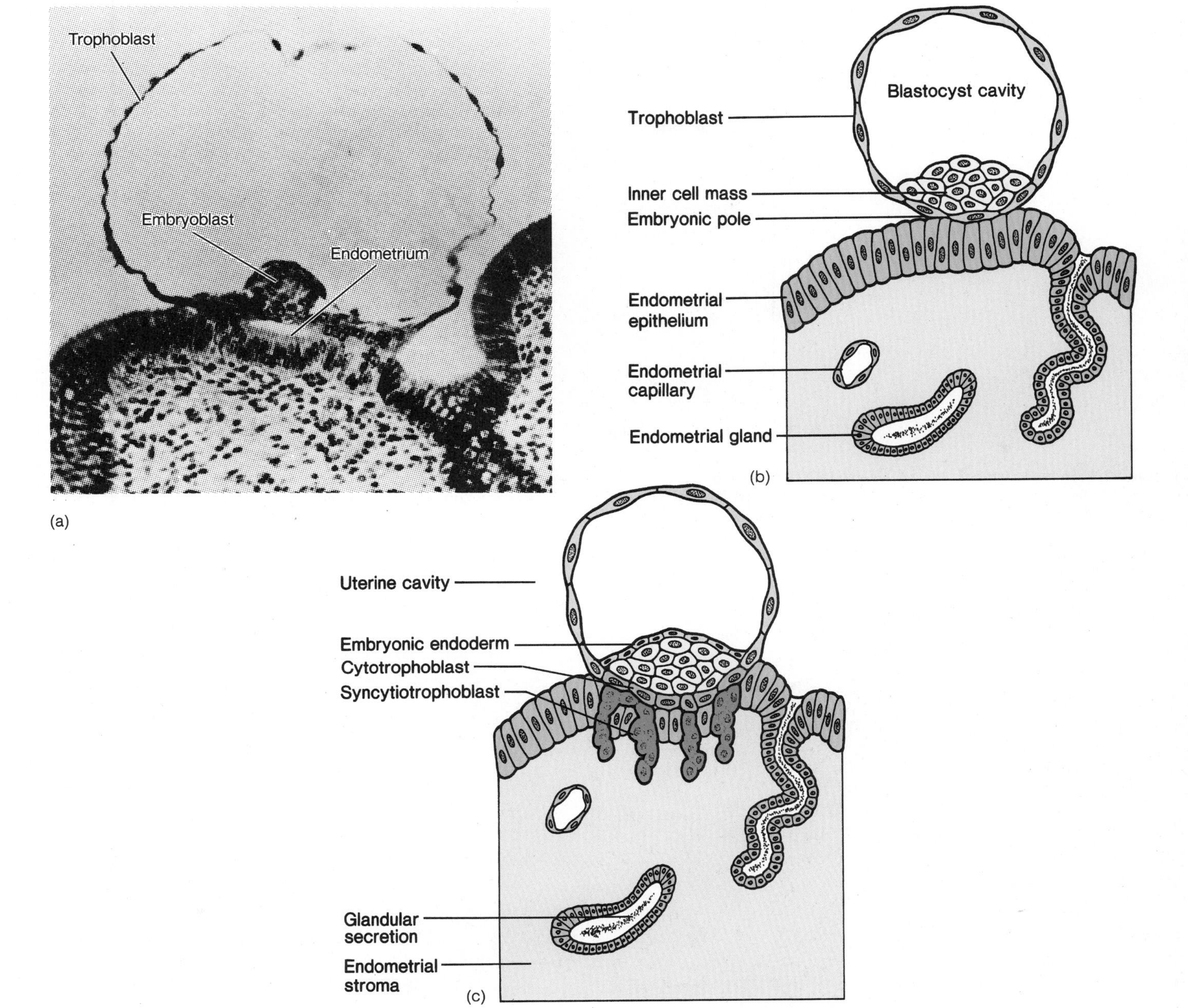

Figure 22.7 The blastocyst adheres to the endometrium on about the sixth day as seen in (*a*) a photomicrograph and (*b*) a diagram of the specific cells of the blastocyst. By the seventh day (*c*), specialized syncytiotrophoblasts from the trophoblast invade the endometrium. The syncytiotrophoblasts secrete hCG to sustain pregnancy and will eventually participate in the formation of the placenta for embryonic and fetal sustenance.

are listed in table 22.2. Refer to figure 22.10 for an illustration of the organs and body systems that derive from each of the primary germ layers.

The period of prenatal development is referred to as *gestation.* Normal gestation for humans is nine months. Knowing this and the pattern of menstruation make it possible to determine the delivery date of a baby. In a typical reproductive cycle, a woman ovulates fourteen days prior to the onset of the next menstruation and is fertile for approximately twenty to twenty-four hours following ovulation. Adding nine months, or thirty-eight weeks, to the time of ovulation gives one the estimated delivery date, or parturition (childbirth) date.

1. List the structural characteristics of a zygote, morula, and blastocyst. What is the general time of each of these stages of the pre-embryonic period of development?
2. Discuss the process of implantation, and describe the physiological events that ensure pregnancy.
3. Describe the development of the placenta.
4. List the major structures or organs that derive from each germ layer.
5. What is the length of time of a typical pregnancy? Define *gestation.* How is the parturition date determined?

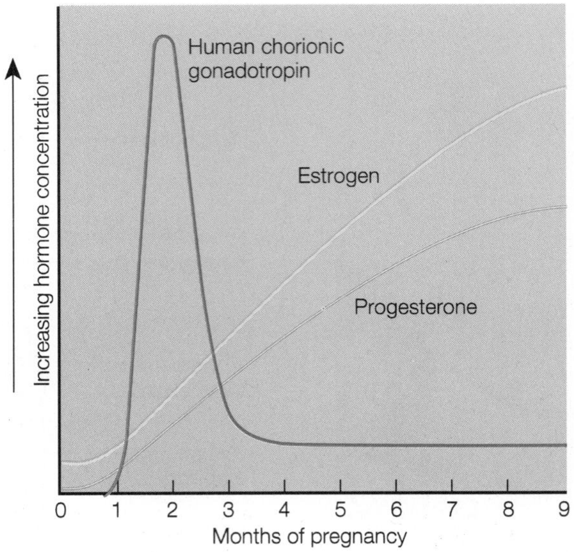

Figure 22.8 Human chorionic gonadotrophin (hCG) is secreted by syncytiotrophoblasts during the first trimester of pregnancy. This hormone maintains the mother's corpus luteum for the first five and one-half weeks. After that time the placenta becomes the major sex-hormone-producing gland, secreting increasing amounts of estrogen and progesterone throughout pregnancy.

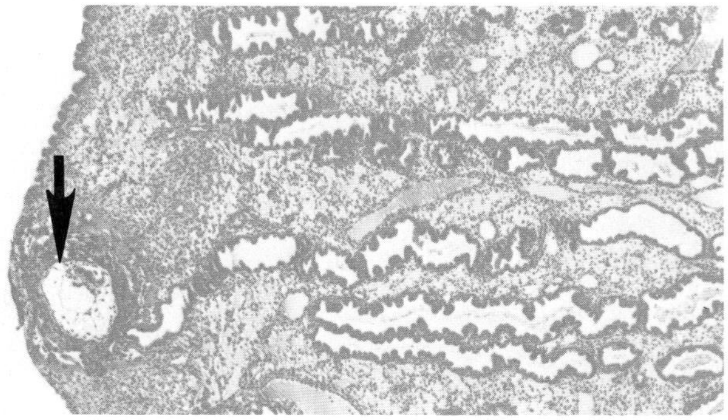

(a)

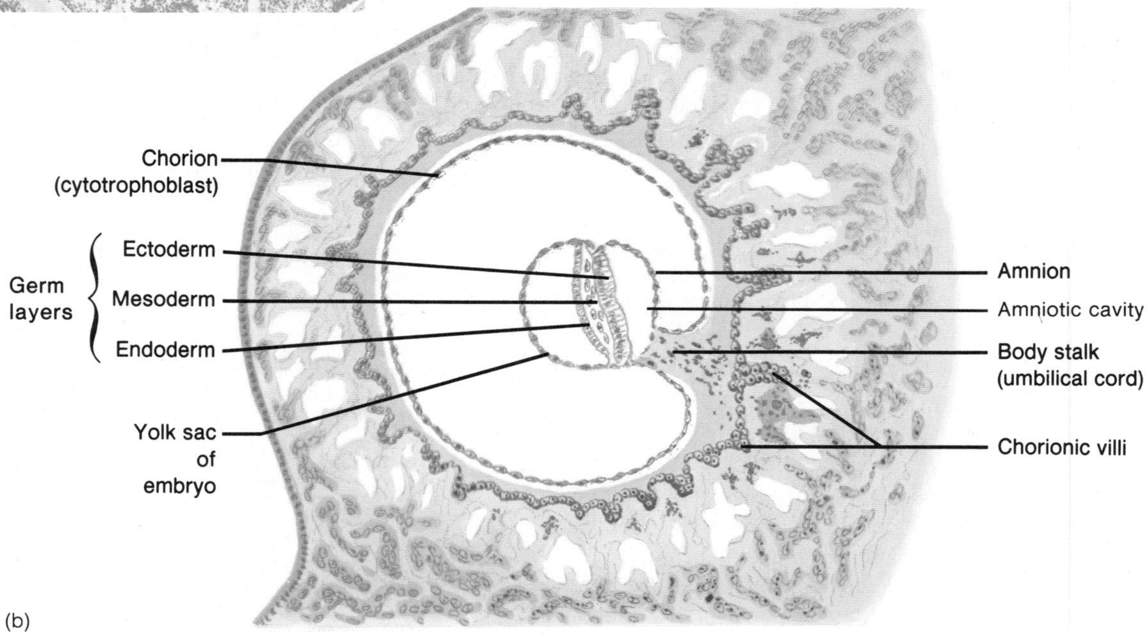

(b)

Figure 22.9 The completion of implantation. (*a*) A photomicrograph of an implanted blastocyst in the uterine wall. (*b*) The formation of the primary germ layers at the end of two weeks demarks the end of the pre-embryonic stage and the beginning of the embryonic stage. (Note the formation of extraembryonic membranes at this early period of development.)

Table 22.1 Morphogenic stages and principal events during pre-embryonic development

Stage	Time period	Principal events
Zygote	Twelve to twenty-four hours following ovulation	Egg is fertilized; zygote has twenty-three pairs of chromosomes (diploid) from haploid sperm and haploid egg; genetically unique
Cleavage	Thirty hours to third day	Mitotic divisions produce increased number of cells
Morula	Third to fourth day	Solid ball-like structure forms
Blastocyst	Fifth day to end of second week	Embryoblast and trophoblast form; hollow center; implantation occurs; embryonic disc forms followed by primary germ layers

Table 22.2 Derivatives of germ layers

Ectoderm	Mesoderm	Endoderm
Epidermis of skin and epidermal derivatives: hair, nails, glands of the skin; linings of oral, nasal, anal, and vaginal cavities	Muscle: smooth, cardiac, and skeletal	Epithelium of pharynx, auditory canal, tonsils, thyroid, parathyroid, thymus, larynx, trachea, lungs, digestive tract, urinary bladder and urethra, and vagina
Nervous tissue; sense organs	Connective tissue: embryonic, connective tissue proper, cartilage, bone, blood	Liver and pancreas
Lens of eye; enamel of teeth	Dermis of skin; dentin of teeth	
Pituitary gland	Epithelium of blood vessels, lymphatic vessels, body cavities, joint cavities	
Adrenal medulla	Internal reproductive organs	
	Kidneys and ureters	
	Adrenal cortex	

Embryonic Stage

The events of the six-week embryonic stage include the differentiation of the germ layers into specific body organs and the formation of the placenta, the umbilical cord, and the extraembryonic membranes, which provide sustenance and protection to the embryo.

Objective 7. Define the term *embryo,* give the time duration of the embryonic stage, and describe the major events of this period of development.

Objective 8. List the embryonic needs that must be met to avoid a spontaneous abortion.

Objective 9. Describe the structure and function of each of the extraembryonic membranes.

Objective 10. Describe the development and function of the placenta and the umbilical cord.

The embryonic stage lasts from the beginning of the third week to the end of the eighth week. At this stage, the developing organism can correctly be called an **embryo.** During the embryonic stage all of the body organs form, as well as the placenta, umbilical cord, and extraembryonic membranes. The term **conceptus** refers to the embryo, or fetus, and all of the extraembryonic structures—the products of conception. During parturition the baby and the remaining portion of the conceptus are expelled from the uterus.

Embryology is the study of the sequential changes in an organism as the various tissues, organs, and systems develop. Chick embryos are frequently studied because of the easy access through the shell and their rapid development. Mice and pig embryos are also extensively studied as mammalian models. Genetic manipulation, induction of drugs, exposure to disease, radioactive tagging or dyeing of developing tissues, and x-ray treatments are some of the commonly conducted experiments that provide information that can be applied to human development and birth defects.

During the pre-embryonic stage of cell divisions and differentiation, the developing structure is self-sustaining. The embryo, however, is not self-supporting and must derive sustenance from the mother. For morphogenesis to continue, certain immediate needs must be met, which include the (1) formation of a vascular connection between the uterus of the mother and the embryo so that nutrients and oxygen can be provided and metabolic wastes and carbon dioxide can be removed; (2) establishment of a constant, protective environment around the embryo, which is conducive to development; (3) establishment of a structural organization for embryonic morphogenesis along a longitudinal axis; (4) ensurance of structural support for the embryo both internally and externally; and (5) coordination of the morphogenic events through genetic expression. If these needs are not met, a spontaneous abortion will generally occur.

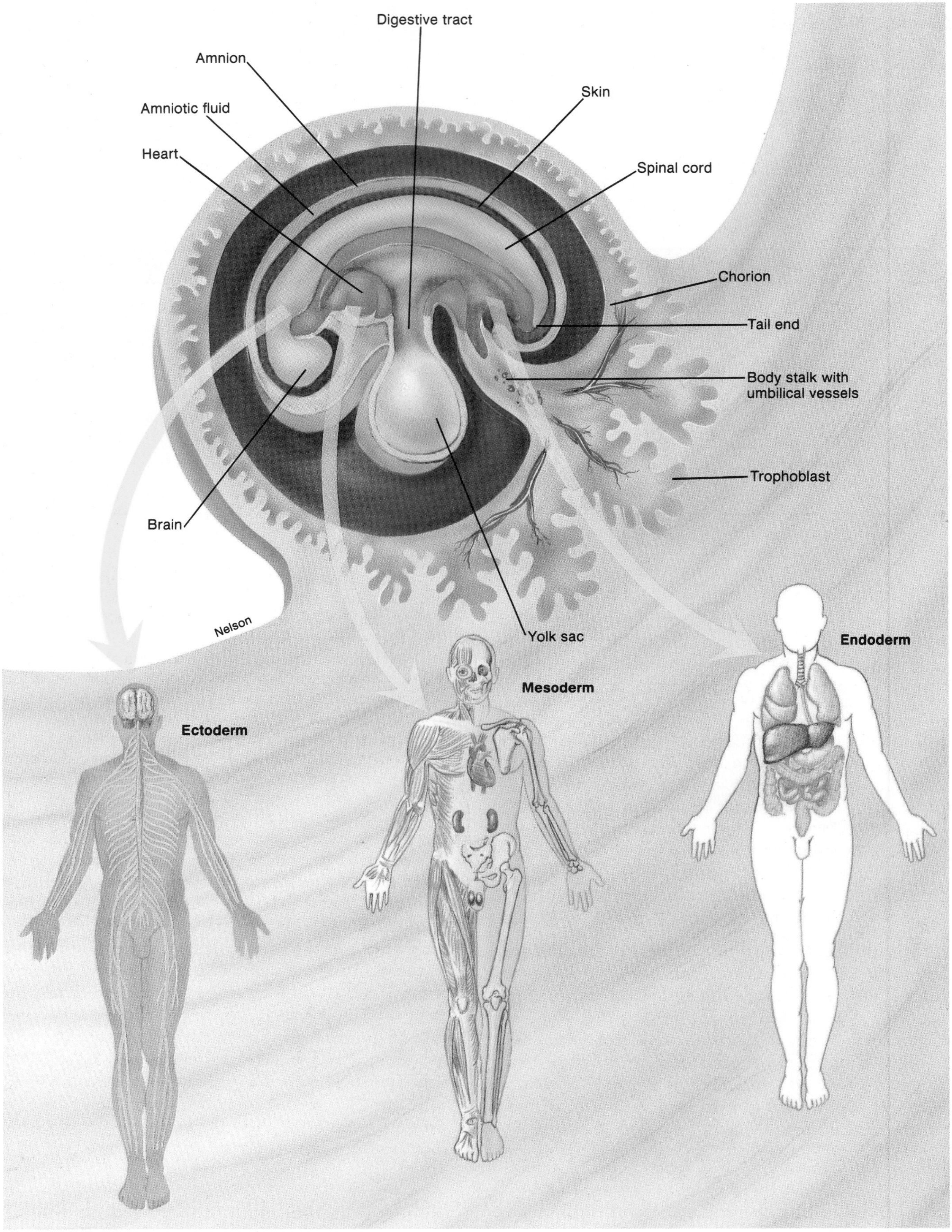

Figure 22.10 The body systems and the primary germ layers from which they develop.

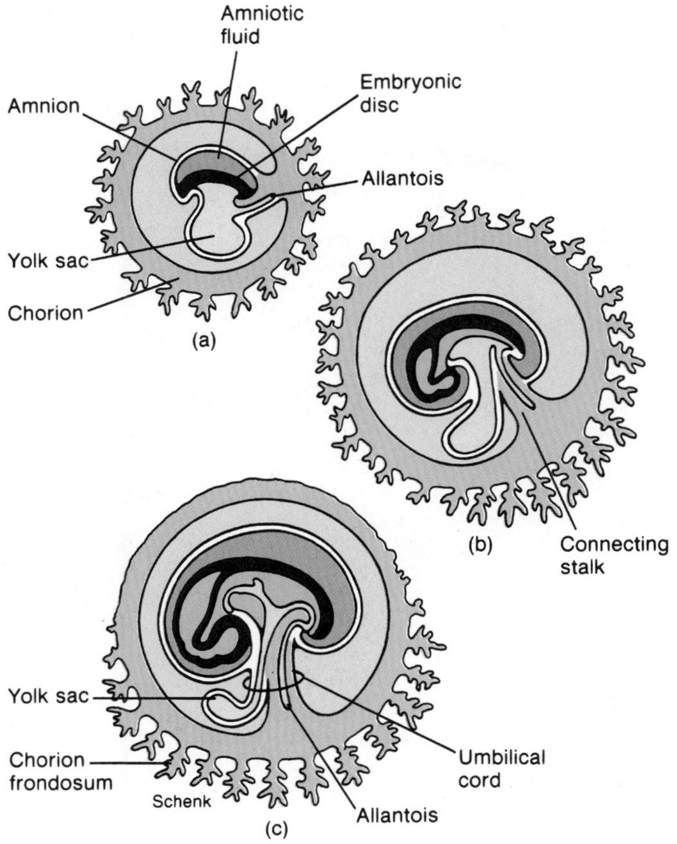

Figure 22.11 The formation of the extraembryonic membranes during a single week of rapid embryonic development: (*a*) three weeks, (*b*) three and one-half weeks, and (*c*) four weeks.

The first two of the aforementioned needs are provided by extraembryonic structures, whereas the other three are provided intraembryonically. The extraembryonic membranes, the placenta, and the umbilical cord will be discussed separately from the development of the embryo.

Most serious developmental defects cause the embryo to be naturally aborted. About 25% of early aborted embryos have chromosomal abnormalities. Other abortions may be caused by environmental factors such as infectious agents or teratogenic drugs (drugs that cause birth defects). In addition, an implanting, developing embryo is regarded as foreign tissue by the mother and is rejected and aborted unless maternal immune responses are suppressed.

Extraembryonic Membranes

While the many intraembryonic events are forming the body organs, a complex system of extraembryonic membranes is developing as well (fig. 22.11). The **extraembryonic membranes** are the amnion, yolk sac, allantois, and chorion. These membranes are responsible for the protection, respiration, excretion, and nutrition of the embryo and subsequent fetus. At parturition, the placenta, umbilical cord, and extraembryonic membranes separate from the fetus and are expelled from the uterus as the *afterbirth.*

Amnion The amnion *(am'ne-on)* is a thin extraembryonic membrane, derived from ectoderm and mesoderm, which loosely envelops the embryo, forming an **amniotic sac,** filled with *amniotic fluid* (fig. 22.12). In later stages of fetal development, the amnion expands to come in contact with the chorion. Amniotic development is initiated during early embryonic development, at which time its margin is attached around the free edge of the embryonic disc (fig. 22.11). As the amniotic sac enlarges during the late embryonic period (about eight weeks), the amnion gradually sheaths the developing umbilical cord with an epithelial covering (fig. 22.13).

As a buoyant medium, amniotic fluid performs four functions for the embryo and subsequent fetus: (1) it permits symmetrical structural development and growth; (2) it cushions and protects by absorbing jolts that the mother may receive; (3) it helps to maintain consistent pressure and temperature; and (4) it enables freedom of fetal development, which is important for musculoskeletal development and blood flow.

Amniotic fluid is formed initially as an isotonic fluid absorbed from the maternal blood in the endometrium surrounding the developing embryo. Later, the volume is increased and the concentration changed by urine excreted from the fetus into the amniotic sac. Amniotic fluid also contains cells that are sloughed off from the fetus, placenta, and amniotic sac. Since all of these cells are derived from the same fertilized egg, all have the same genetic composition. Many genetic abnormalities can be detected by aspirating this fluid and examining the cells thus obtained in a procedure called *amniocentesis (am''ne-o-sen-te'sis).*

Amniotic fluid is normally swallowed by the fetus and absorbed in the GI tract. Prior to delivery, the amnion is naturally or surgically ruptured, and the amniotic fluid (bag of waters) is released.

Amniocentesis (fig. 22.14) is usually performed at the fourteenth or fifteenth weeks of pregnancy, when the amniotic sac contains 175–225 ml of fluid. Genetic diseases such as *Down's syndrome* (where there are three instead of two number-21 chromosomes) can be detected by examining chromosomes; diseases such as *Tay-Sachs disease,* in which there is a defective enzyme involved in formation of myelin sheaths, can be detected by biochemical techniques.

Yolk Sac The yolk sac is established during the end of the second week as cells from the trophoblast form a thin *exocoelomic (ek''so-se-lo'mik) membrane.* Unlike many vertebrates, the human yolk sac contains no nutritive yolk but is an essential structure during early embryonic development. It is attached to the underside of the embryonic disc (figs. 22.11, 22.12), where it produces blood for the embryo until the liver forms during the sixth week. The dorsal portion of the yolk sac is involved in the formation of the primitive gut. In addition, primordial germ cells form in the wall of the yolk sac and migrate during the fourth week to the developing gonads where they become the primitive germ cells (spermatogonia or oogonia).

Down's syndrome: from John L. H. Down, English physician, 1828–96

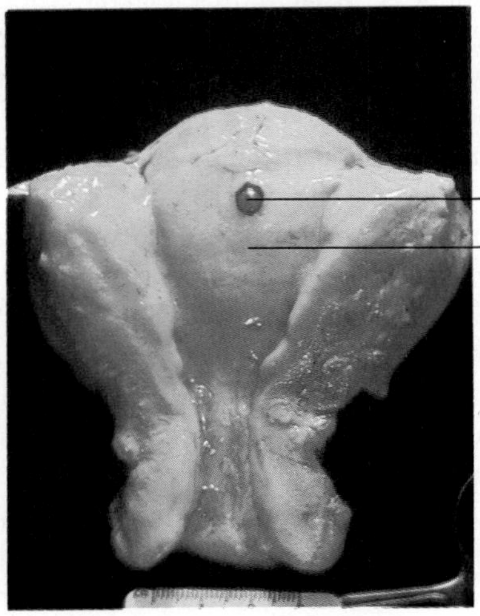

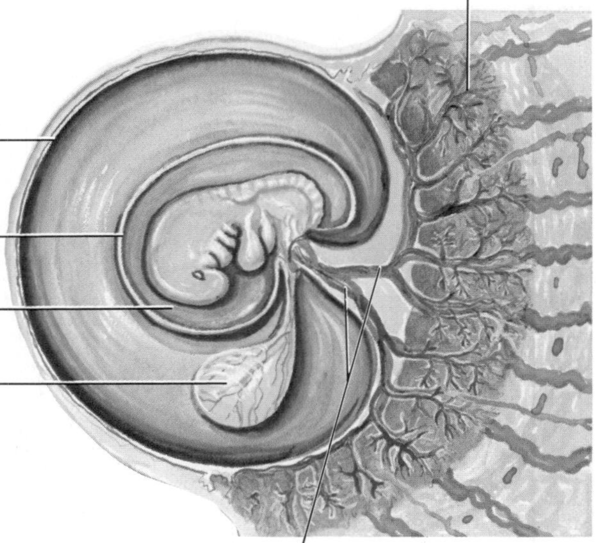

Figure 22.12 An implanted embryo at approximately four and one-half weeks. (*a*) The interior of a uterus showing the implantation site and elevated *decidua capsularis* caused by the expanded chorion. (*b*) The developing embryo, extraembryonic membranes, and the formation of the placenta.

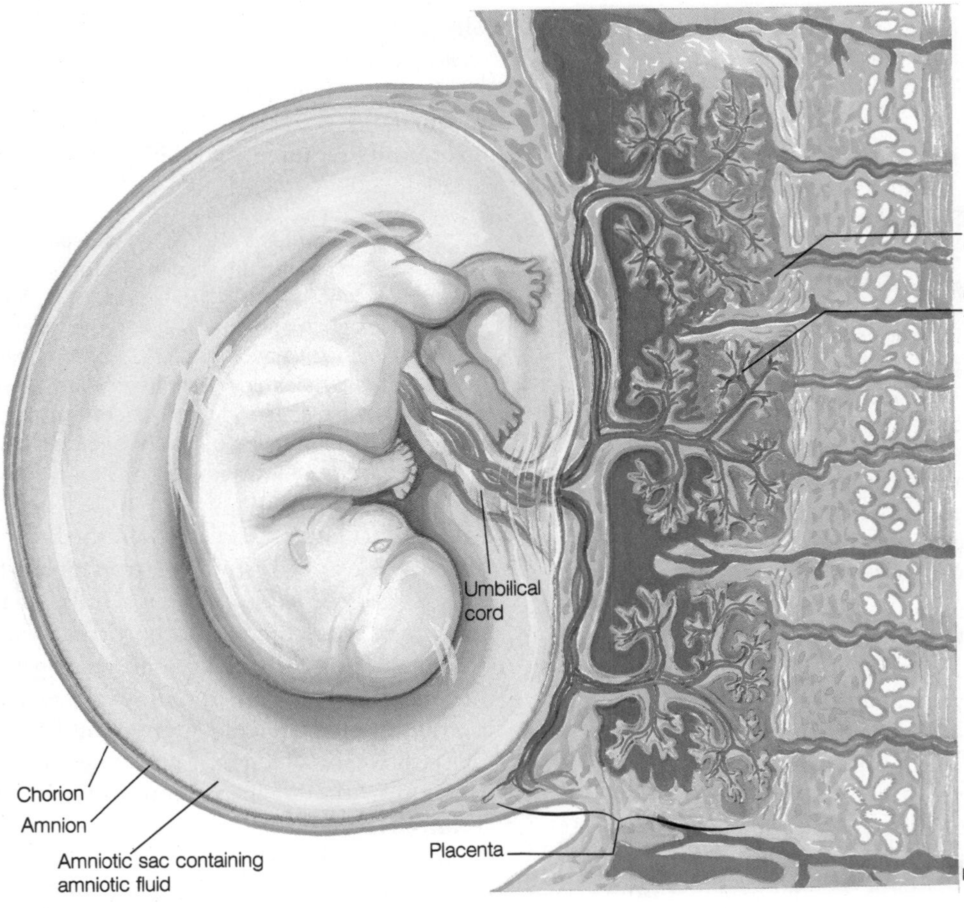

Figure 22.13 The embryo, extraembryonic membranes, and placenta at approximately seven weeks of development. Note that at this time, the amnion and chorion are adherent and are frequently referred to as the *amniochorionic membrane.* Blood from the embryo is carried to and from the chorion frondosum by the umbilical arteries and vein. The maternal tissue between the chorionic villi is known as the decidua basalis, and this tissue, together with the villi, forms the functioning placenta.

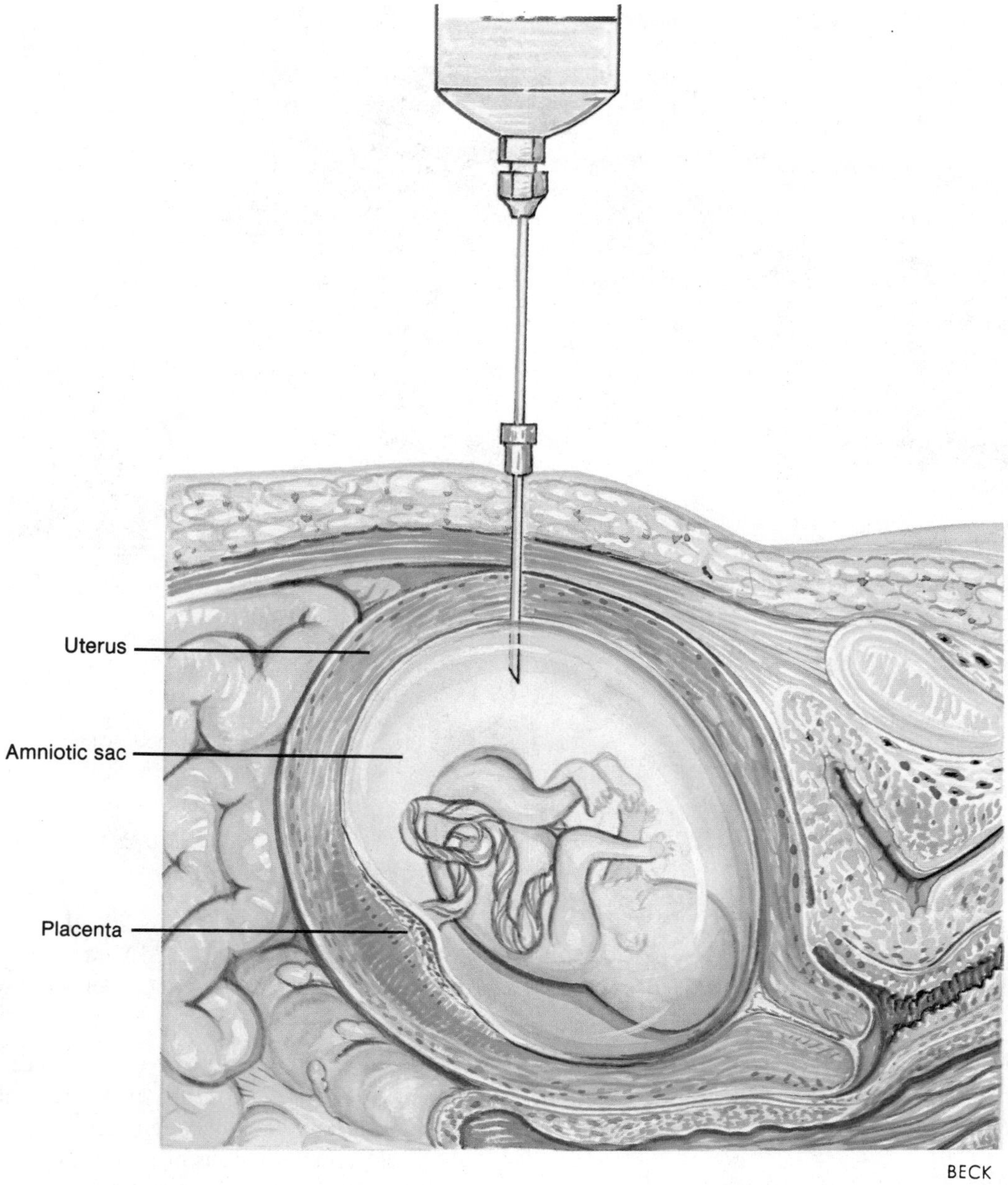

Figure 22.14 Amniocentesis. In this procedure amniotic fluid, together with suspended cells, is withdrawn for examination. Various genetic diseases can be prenatally detected by this means.

The stalk of the yolk sac usually detaches from the gut by the sixth week. Following this, the yolk sac gradually shrinks as pregnancy advances. Eventually it becomes very small and serves no additional developmental functions.

Allantois The allantois forms during the third week as a small outpouching, or diverticulum, from the caudal wall of the yolk sac (see fig. 22.11). The allantois remains small but is involved in the formation of blood cells and gives rise to the fetal umbilical arteries and vein. It also contributes to the development of the urinary bladder.

The extraembryonic portion of the allantois degenerates during the second month. The intraembryonic portion involutes to form a thick urinary tube called the **urachus.** After birth, the urachus becomes a fibrous cord, called the median umbilical ligament, which attaches to the urinary bladder.

Chorion The chorion is a highly specialized extraembryonic membrane that participates in the formation of the placenta (see fig. 22.11). It is the outermost membrane and originates from the trophoblast of the blastocyst. Numerous, small, fingerlike extensions, called **villi** (see fig. 22.12), form from the chorion and penetrate deeply into the uterine tissue. Initially, the entire surface of the chorion is covered with villi. But those villi on the surface toward the uterine cavity gradually degenerate and produce a smooth, bare area known as the **smooth chorion.** As this occurs, the villi associated with the uterine wall rapidly increase in number and branch out. This portion of the chorion

allantois: Gk. *allanto,* sausage; *iodos,* resemblance

chorion: Gk. *chorion,* external fetal membrane

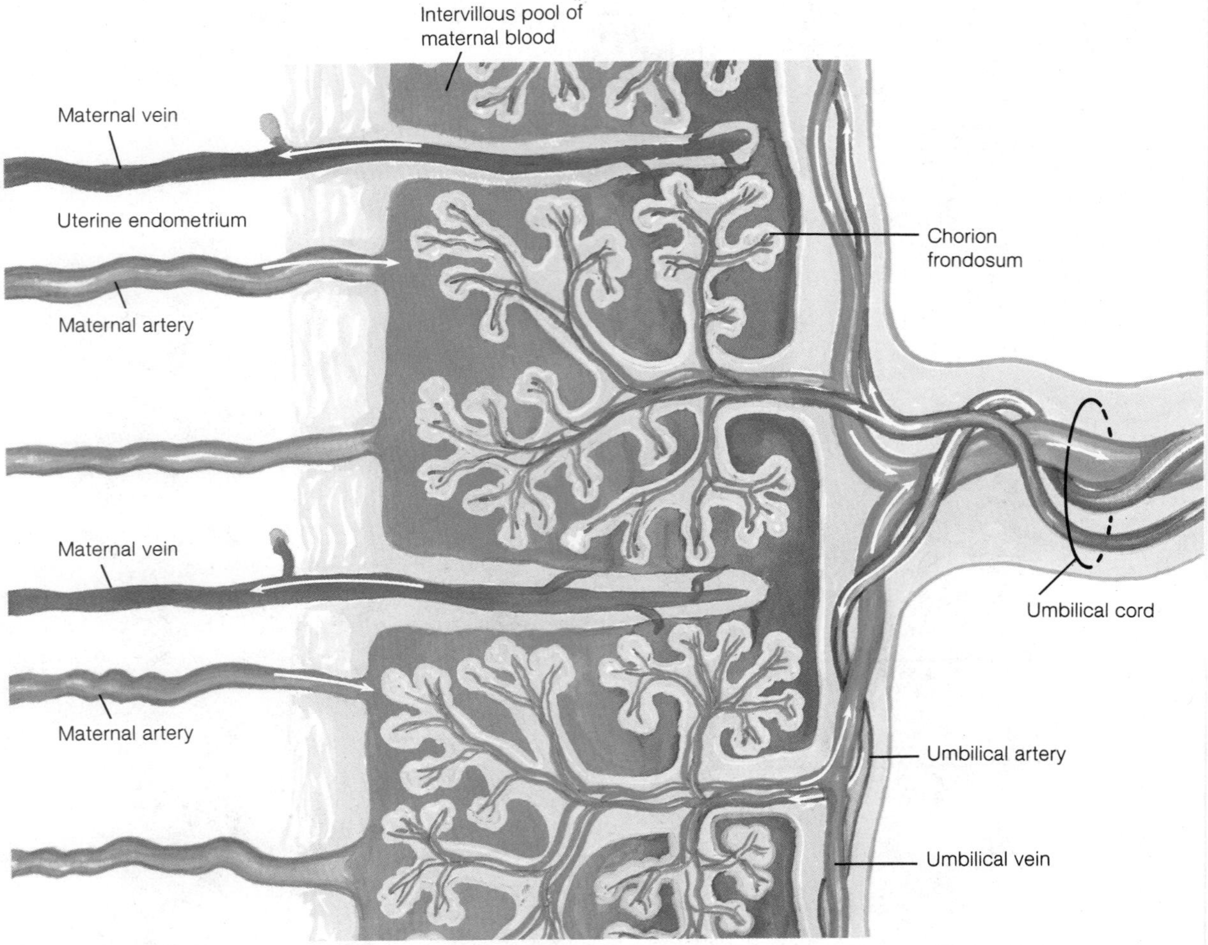

Figure 22.15 The circulation of blood within the placenta. Maternal blood is delivered to and drained from the spaces between the chorionic villi. Fetal blood is brought to blood vessels within the villi by branches of the umbilical artery and is drained by branches of the umbilical vein.

is known as the **villous chorion.** The villous chorion becomes highly vascular, and as the embryonic heart begins to function, blood is pumped in close proximity to the uterine wall.

Chorionic villus biopsy, is a technique used to detect genetic disorders much earlier than permitted by amniocentesis. In chorionic villus biopsy, a catheter is inserted through the cervix to the chorion, and a sample of a chorionic villus is obtained by suction or cutting. Genetic tests can be performed directly on the villus sample, since this sample contains much larger numbers of fetal cells than does a sample of amniotic fluid. Chorionic villus biopsy can provide genetic information at ten to twelve weeks gestation.

Placenta

The **placenta** is a vascular structure by which an unborn child is attached to its mother's uterine wall and through which metabolic exchange occurs (fig. 22.15). The placenta is formed in part from maternal tissue and in part from embryonic tissue. The embryonic portion of the placenta consists of the villi of the **chorion frondosum,** whereas the maternal portion is composed of the area of the uterine wall called the **decidua basalis** (see fig. 22.13), into which the villi penetrate. Blood does not flow directly between these two portions, but because of the close membranous proximity, certain substances diffuse readily.

When fully formed, the placenta is a reddish brown oval disc with a diameter of 15–20 cm (8 in.) and a thickness of 2.5 cm (1 in.). It weighs 500–600 gm, being about one-sixth the weight of the fetus.

Exchange of Molecules across the Placenta The two **umbilical arteries** deliver fetal blood to vessels within the villi of the chorion frondosum of the placenta. This blood circulates within the villi and returns to the fetus via the **umbilical vein.** Maternal blood is delivered to and drained from the cavities within the decidua basalis, which are located between the chorionic villi. In this way, maternal and fetal blood are brought close together but never mix within the placenta.

The placenta serves as a site for the exchange of gases and other molecules between the maternal and fetal blood. Oxygen diffuses from mother to fetus, and carbon dioxide diffuses in the

villous: L. *villus,* tuft of hair
placenta: L. *placenta,* a flat cake

Table 22.3 Hormones secreted by placenta

Hormones	Effects
Pituitary-like hormones	
Chorionic gonadotrophin (hCG)	Similar to LH; maintains mother's corpus luteum for first 5½ weeks of pregnancy; may be involved in suppressing immunological rejection of embryo
Placental lactogen (hPL)	Similar to prolactin; synergizes with growth hormone to promote, in the mother, increased fat breakdown and fatty acid release from adipose tissue, and to promote the sparing of glucose use by maternal tissues (diabetic-like effects)
Chorionic thyrotrophin (hCT)	Similar to TSH; physiological significance uncertain
Sex steroids	
Progesterone	Helps maintain endometrium during pregnancy; helps suppress gonadotrophin secretion; promotes uterine sensitivity to oxytocin; helps stimulate mammary gland development
Estrogens	Help maintain endometrium during pregnancy; help suppress gonadotrophin secretion; help stimulate mammary gland development; inhibit prolactin secretion

From Stuart Ira Fox, *Human Physiology,* 3d ed. Copyright © 1990 Wm. C. Brown Publishers, Dubuque, Iowa. All Rights Reserved. Reprinted by permission.

opposite direction. Nutrient molecules and waste products likewise pass between maternal and fetal blood.

The placenta is not merely a passive conduit for exchange between maternal and fetal blood, however. It has a very high metabolic rate, utilizing about a third of all the oxygen and glucose supplied by the maternal blood. The rate of protein synthesis is, in fact, higher in the placenta than it is in the liver. Like the liver, the placenta produces a great variety of enzymes capable of converting biologically active molecules (such as hormones and drugs) into less active, more water-soluble forms. In this way, potentially dangerous molecules in the maternal blood are often prevented from harming the fetus.

Most drugs ingested by a pregnant woman can readily pass through the placenta to the developing baby. Many drugs that are not particularly harmful to the mother may be deleterious to her child. For example, the nicotine taken in by a chain-smoking pregnant woman will stunt the growth of her fetus. Hard drugs, such as heroin, can lead to fetal drug addiction. Depressant drugs given to a mother during labor can readily cross the placenta and, if given in high dosages, can cause respiratory depression in the newborn infant.

Although the placenta is an effective barrier against diseases of bacterial origin, viruses such as rubella and certain blood-borne diseases such as syphilis can diffuse through the placenta and affect the fetus. During parturition, small amounts of fetal blood may pass across the placenta to the mother. If the fetus is Rh positive and the mother Rh negative, the antigens of the fetal red blood cells elicit an antibody response in the mother. In a subsequent pregnancy, the maternal antibodies then cross the placenta and cause a breakdown of fetal red blood cells, a condition called *erythroblastosis fetalis.*

Endocrine Functions of the Placenta The placenta functions as an endocrine gland in secreting both steroid and glycoprotein hormones. The glycoprotein hormones perform actions similar to those of some anterior pituitary hormones. The importance of these hormones is to maintain pregnancy and to insure optimal nutrition to the fetus.

The placenta converts androgens, secreted from the mother's adrenal glands, into estrogen to help protect the female embryo from becoming masculinized. In addition, the placenta secretes large amounts of a weak estrogen, called *estriol,* which helps to maintain the endometrium and stimulate the development of the mother's mammary glands. The production of estriol increases tenfold during pregnancy, so that by the third trimester estriol accounts for about 90% of the estrogens excreted in the mother's urine. Since almost all of this estriol comes from the placenta (rather than from maternal tissues), measurements of urinary estriol can be used clinically to assess the health of the placenta.

In the later stages of gestation, the steroid hormones stimulate the development of the mammary glands. The hormones secreted by the placenta and the effect they have upon their target tissues are summarized in table 22.3.

Umbilical Cord

The **umbilical cord** forms as the yolk sac shrinks and the amnion expands to envelop the tissues on the underside of the embryo. Refer to figure 22.16 for an illustration of the formation of the umbilical cord.

The umbilical cord usually attaches near the center of the placenta. When fully formed, the umbilical cord is about 1–2 cm (0.5–1 in.) in diameter and approximately 55 cm (2 ft) in length. The umbilical cord contains two **umbilical arteries,** which carry deoxygenated blood toward the placenta, and one **umbilical vein,** which carries oxygenated blood from the placenta to the embryo. These vessels are surrounded by embryonic connective tissue called *Wharton's jelly.*

The umbilical cord has natural twists because the umbilical vein is longer than the arteries. In about one-fifth of all deliveries, the cord is looped once around the baby's neck. If drawn tightly, the cord could cause death or serious perinatal problems.

Wharton's jelly: from Thomas Wharton, English anatomist, 1614–73

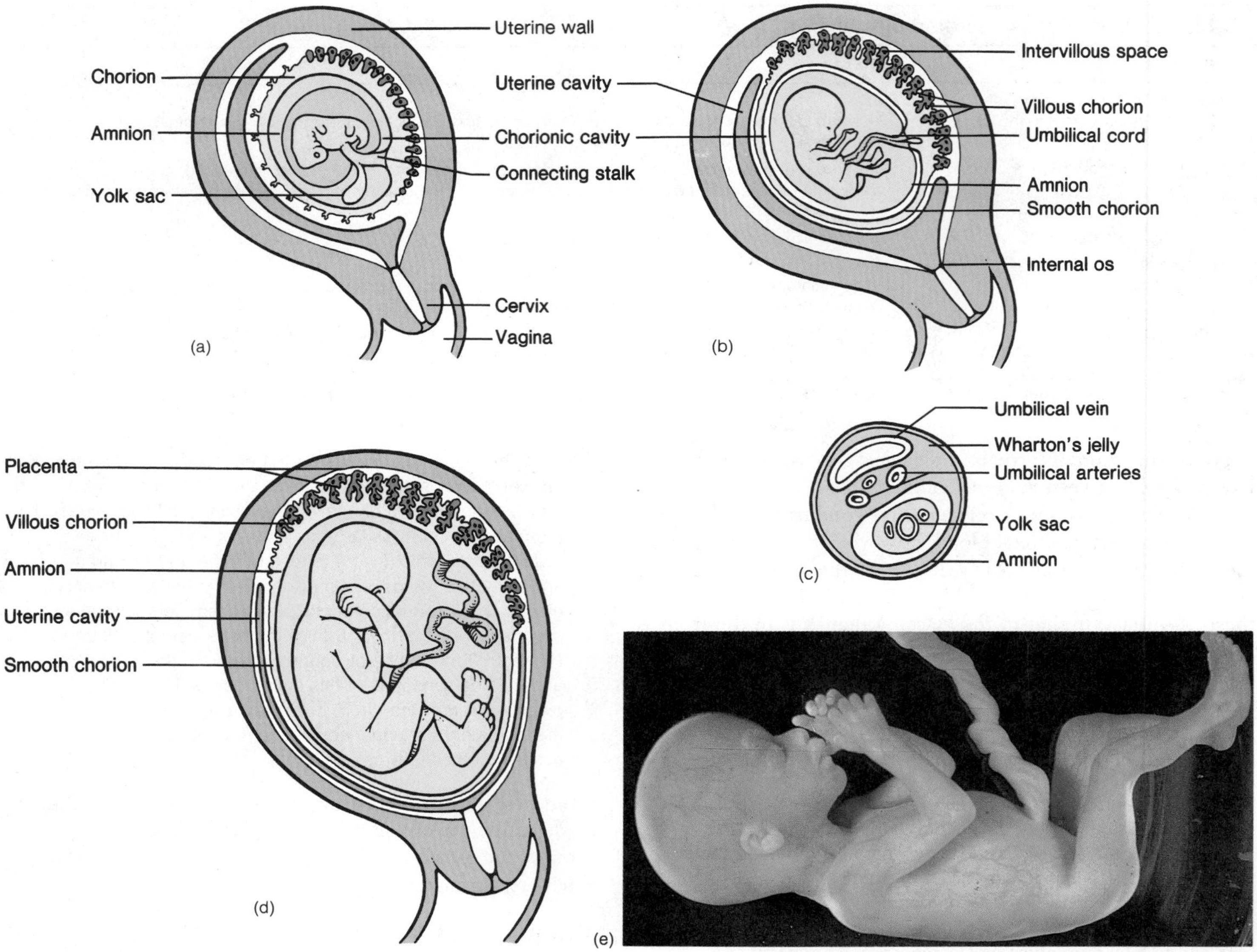

Figure 22.16 The formation of the umbilical cord and other extraembryonic structures as seen in sagittal sections of the gravid pregnant uterus from the fourth to the twenty-second week. (*a*) A connecting stalk forms as the developing amnion expands around the embryo and finally meets ventrally. (*b*) The umbilical cord begins to take form as the amnion ensheathes the yolk sac. (*c*) A transverse section of the umbilical cord showing the embryonic vessels, Wharton's jelly, and the tubular connection to the yolk sac. (*d*) By the twenty-second week, the amnion and chorion are fused, and the umbilical cord and placenta are well-developed structures. (*e*) A sixteen-week-old fetus.

Structural Changes of the Embryo by Weeks

Third Week Early in the third week a thick linear band, known as the **primitive streak,** appears along the dorsal midline of the embryonic disc (see fig. 22.11). The primitive streak is derived from the mesodermal cells, which are specialized from the ectodermal layer of the embryonic mass. The primitive streak establishes a structural foundation for embryonic morphogenesis along a longitudinal axis. As the primitive streak elongates, a prominent thickening known as the **primitive knot** appears at its cranial end (fig. 22.17). The primitive knot later gives rise to the mesodermal structures of the head and the **notochord.** To give support to the embryo the notochord forms a midline axis, which is the basis of the embryonic skeleton. The primitive streak also gives rise to loose embryonic connective tissue called **mesenchyme,** which differentiates into all the various kinds of connective tissue found in the adult. One of the earliest formed organs is the skin, which develops to support and maintain homeostasis within the embryo.

A tremendous amount of change and specialization occurs during the embryonic stage. The factors that cause precise, sequential change from one cell or tissue type to another are not fully understood. It is known, however, that the potential for change is programmed into the genetics of each cell and that under conducive environmental conditions these characteristics become expressed. The process of developmental change is referred to as **induction.** Induction occurs when one tissue, called the **inductor tissue,** has a marked effect on an adjacent tissue causing it to become **induced tissue** and stimulating it to differentiate.

induction: L. *inductus,* to lead in

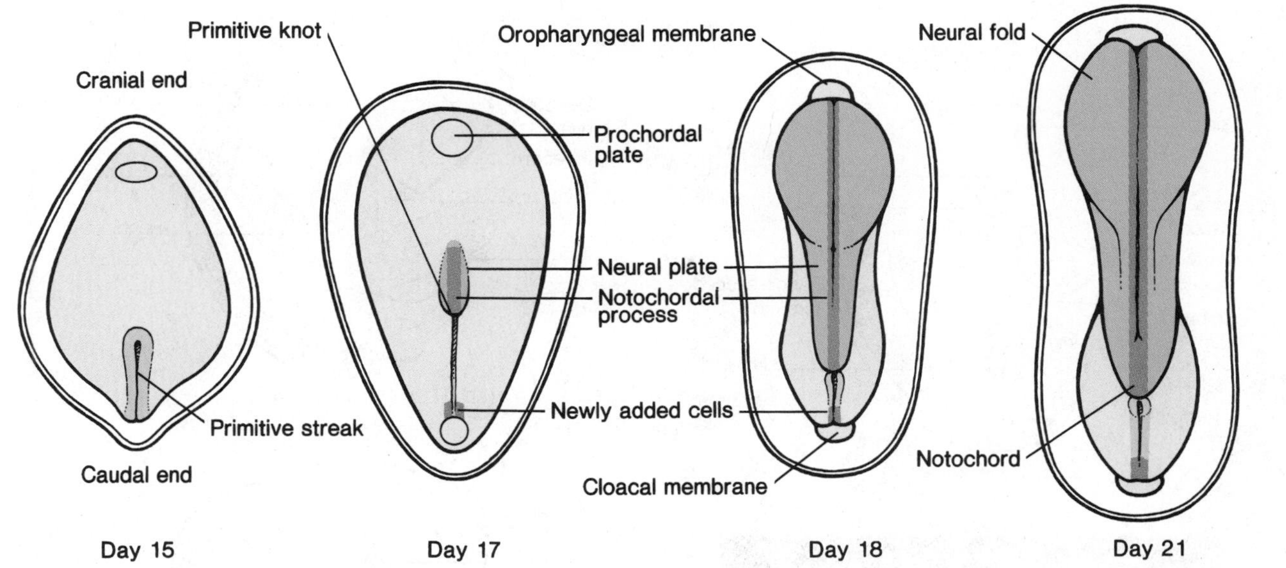

Figure 22.17 The appearance of the primitive streak and primitive knot along the embryonic disc. These progressive changes occur through the process of induction.

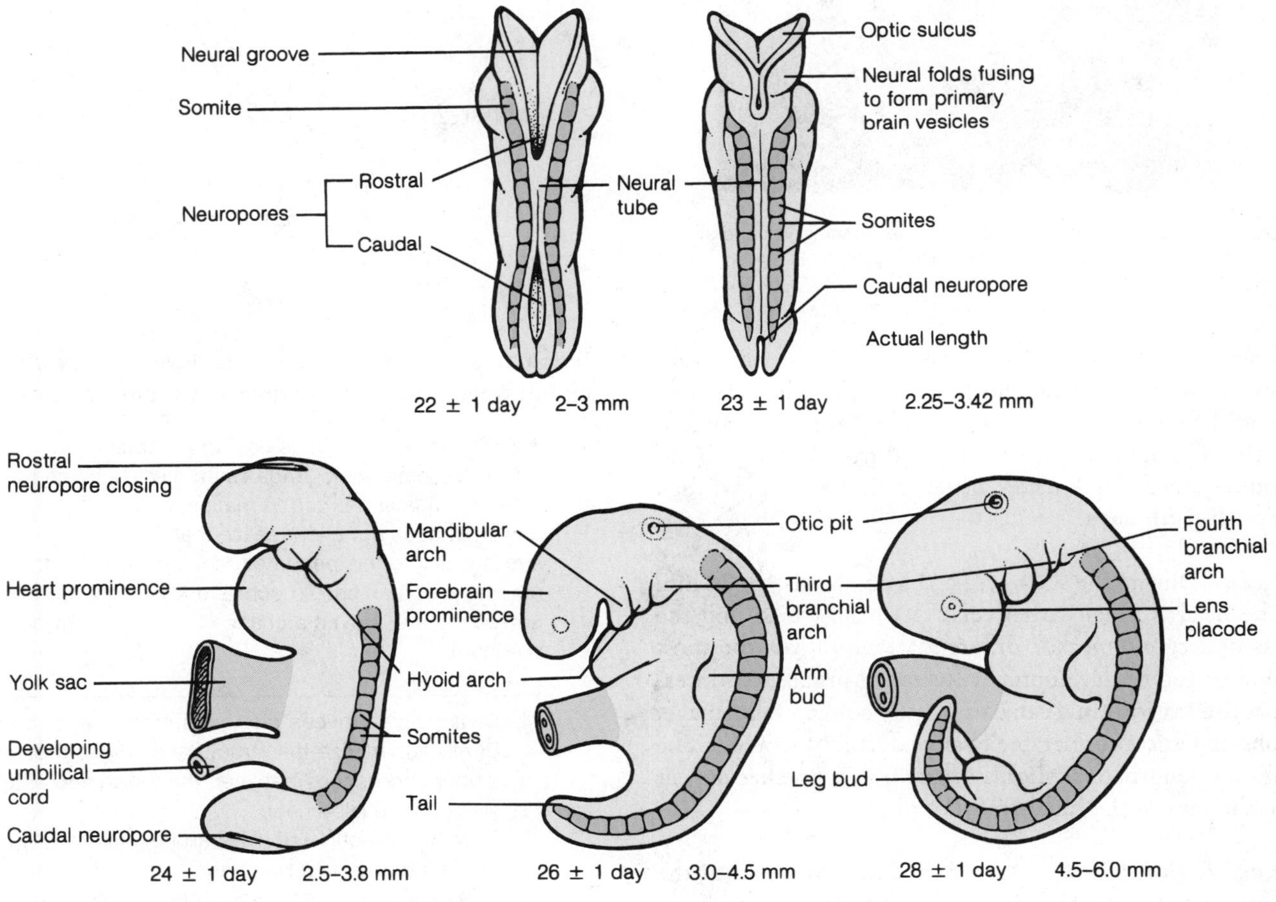

Figure 22.18 Progressive illustrations of four-week-old embryos.

Fourth Week During the fourth week of development, the embryo increases about 4 mm in length. A **connecting stalk,** which is later involved in the formation of the umbilical cord, is established from the body of the embryo to the developing placenta (fig 22.18). By this time, the heart is beating firmly to pump blood to all parts of the embryo. The head and jaws are apparent, and the primordial tissue that will form the eyes, brain, spinal cord, lungs, and digestive organs has developed. The **arm** and **leg buds** are recognizable as small swellings on the lateral body walls.

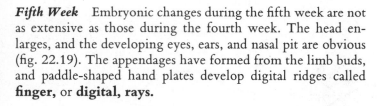

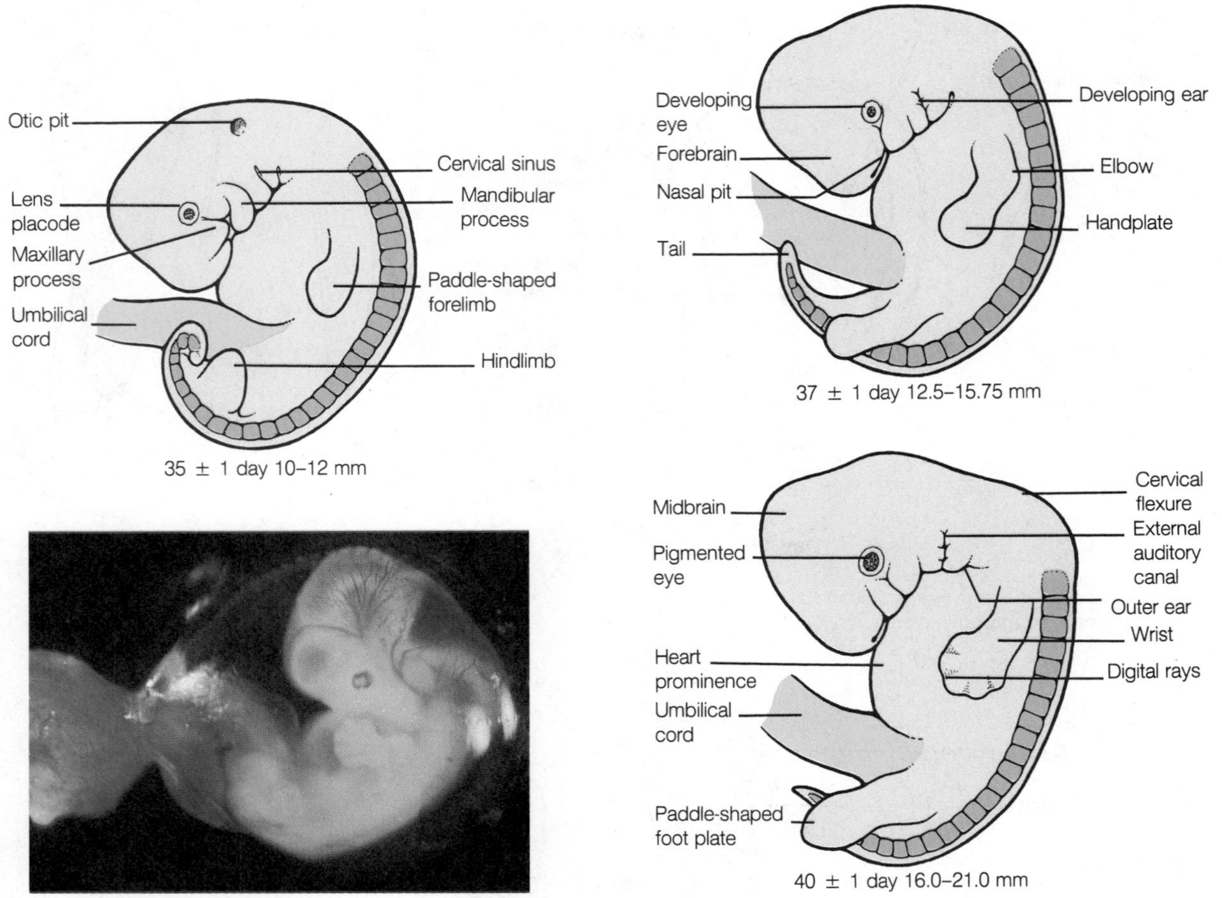

Figure 22.19 Progressive illustrations of five-week-old embryos and a photograph of a five-week-old embryo. (Note the digital rays and the developing eyes and ears.)

Fifth Week Embryonic changes during the fifth week are not as extensive as those during the fourth week. The head enlarges, and the developing eyes, ears, and nasal pit are obvious (fig. 22.19). The appendages have formed from the limb buds, and paddle-shaped hand plates develop digital ridges called **finger,** or **digital, rays.**

Sixth Week During the sixth week, the embryo is 16–24 mm long. The head is much larger relative to the trunk, and the brain has undergone marked differentiation. This is the most vulnerable period of development for many organs. An interruption at this critical time can easily cause congenital damage. The limbs undergo considerable change during this week. The forelimbs are lengthened, slightly flexed, and notches appear between the rays in the hand and foot plates.

Seventh and Eighth Weeks During the last two weeks of the embryonic stage, the embryo, which is now 28–40 mm long, has distinct human characteristics (figs. 22.20, 22.21). The body organs are formed, and the nervous system begins coordinating body activity. The neck region is apparent, and the abdomen is less protuberant. The eyes are well developed, but the lids are stuck together to protect against probing fingers during muscular movement. The nostrils are developed but plugged with mucus. The external genitalia are forming but are still undifferentiated. The body systems are developed by the end of the eighth week, and from this time on the embryo is called a **fetus.**

The most precarious time of prenatal development is during the embryonic stage when there is much tissue differentiation and organ formation. Frequently, however, a woman does not even realize that she is pregnant until she is well into this period of development. For this reason, a woman should consistently take good care of herself if there is even a chance that she might become pregnant.

---

1. Distinguish between the terms *embryo* and *fetus.* Briefly summarize the structural changes that an embryo undergoes between the fourth and eighth weeks of development.
2. Describe the origin of hCG, and explain why it is needed to maintain pregnancy for the first ten weeks.
3. What are the five embryonic needs that must be met in order to avoid spontaneous abortion? Which of these needs are met by the extraembryonic membranes?
4. Name the fetal and maternal components of the placenta, and describe the circulation in these two components. Explain how fetal and maternal gas exchange occurs.

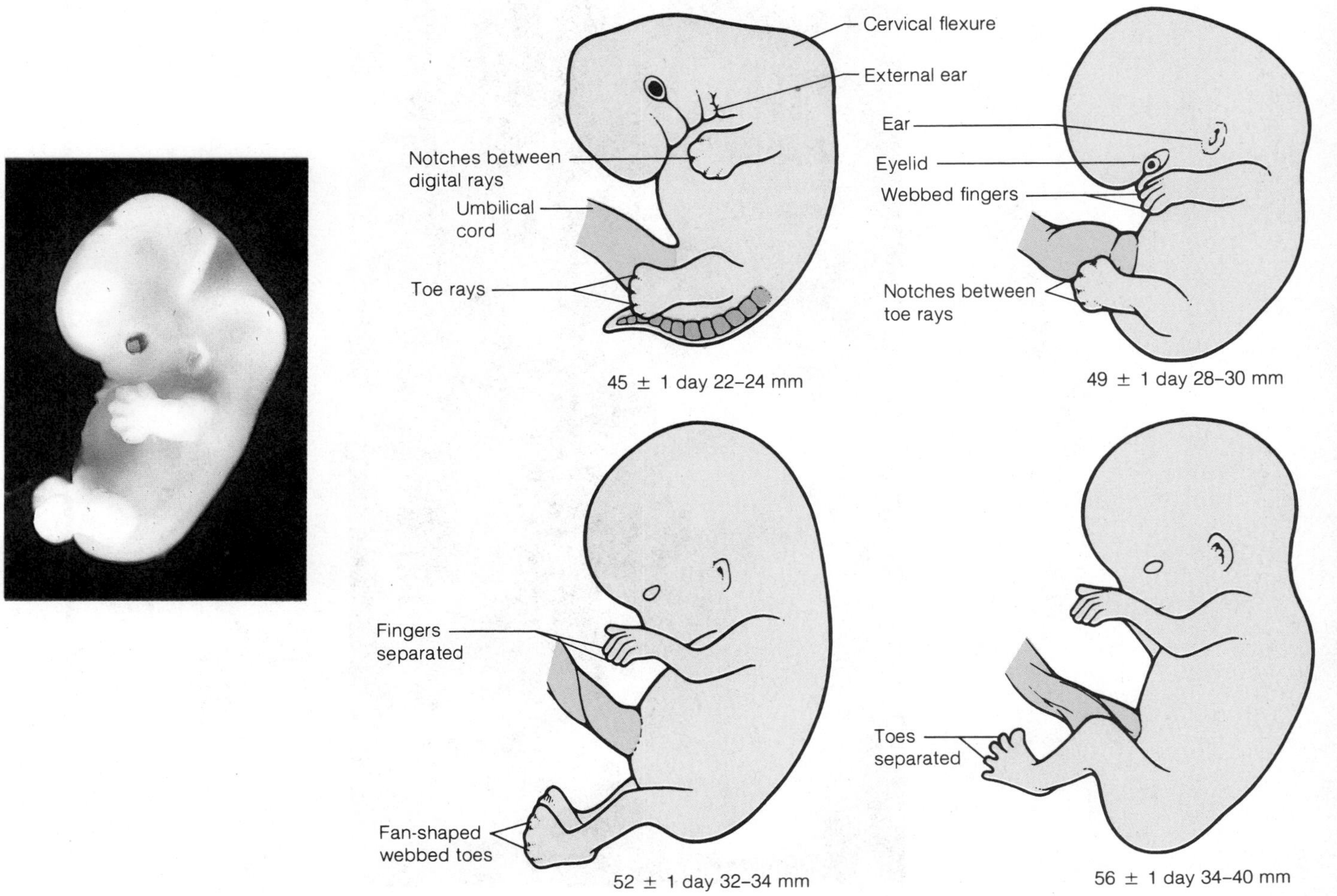

Figure 22.20 A photograph of a six-week-old embryo; progressive illustrations of six- and seven-week-old embryos.

Fetal Stage

The fetal stage lasts from nine weeks until birth and is characterized by tremendous growth and the specialization of body structures.

Objective 11. Define the term *fetus,* and discuss the major events of the fetal stage of development.

Objective 12. Describe the various techniques available for fetal examination or for monitoring fetal activity.

Since most of the tissues and organs of the body appear during the embryonic period, the **fetus** is recognizable as a human being at nine weeks and is far less vulnerable than the embryo to deformation from viruses, drugs, and radiation. A small amount of tissue differentiation and organ development still occurs during the fetal stage, but for the most part fetal development is primarily limited to body growth. Changes in external appearance of the fetus from the ninth through the thirty-eighth week are depicted in figure 22.22. The following is a discussion of the weekly structural changes of the fetus.

Nine to Twelve Weeks At the beginning of the ninth week, the head is as large as the rest of the body. The eyes are widely spaced, and the ears are set low. Head growth slows during the next three weeks, whereas growth in body length accelerates. Ossification centers appear in most bones during the ninth week. Differentiation of the external genitalia becomes apparent at the end of the ninth week, but the genitalia are not developed to the point of sex determination until the twelfth week. By the end of the twelfth week, the fetus is 87 mm (3.5 in.) long and weighs about 45 g (1.6 oz). It can swallow, digest the fluid that passes through its system, and defecate and urinate into the amniotic fluid. The nervous system and muscle coordination are developed enough so that the fetus will withdraw its leg if tickled. The fetus begins inhaling through its nose but can take in only amniotic fluid. The external appearance of fetuses at ten and twelve weeks is depicted in figure 22.23.

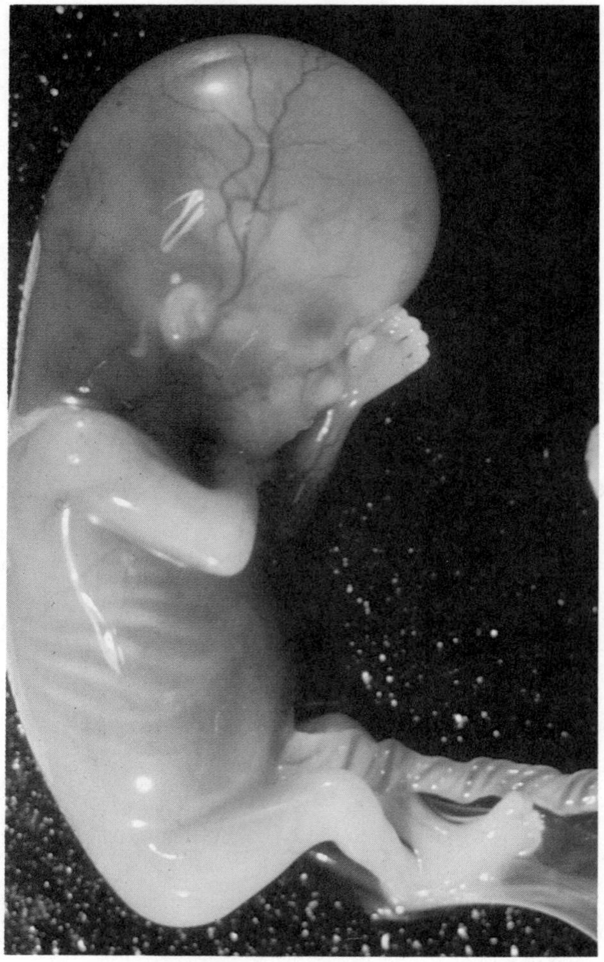

Figure 22.21 Photograph of an eight-week-old embryo. The body systems are developed by the end of the eighth week and the embryo is recognizable as human.

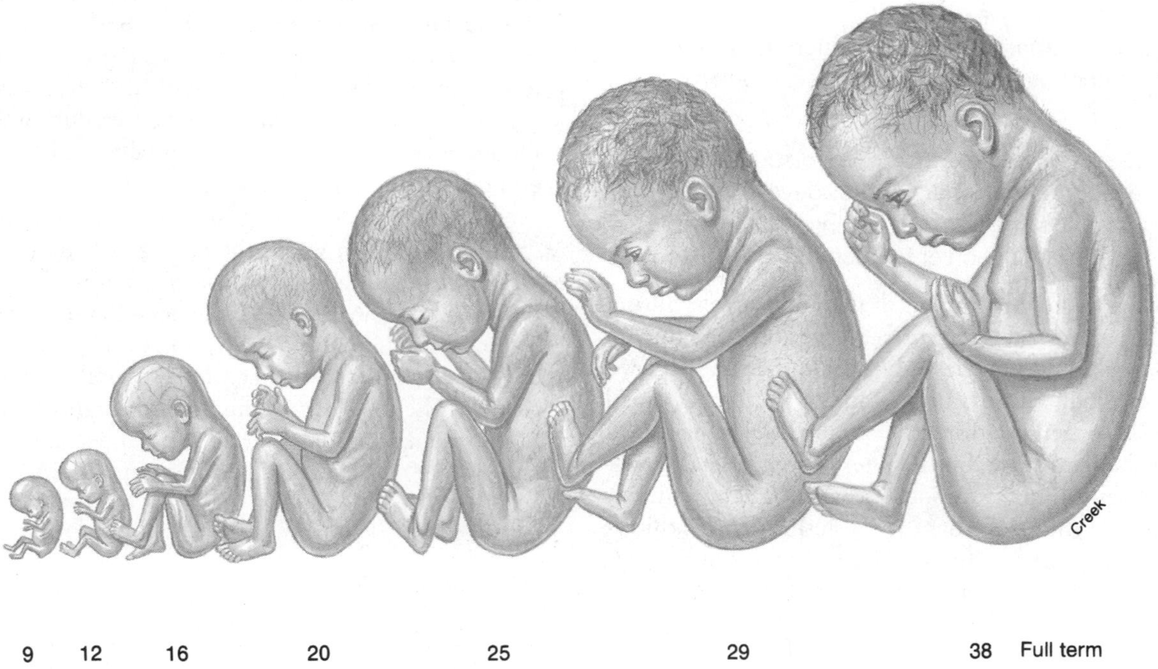

Figure 22.22 Changes in the external appearance of the fetus from the eleventh through the thirty-eighth week.

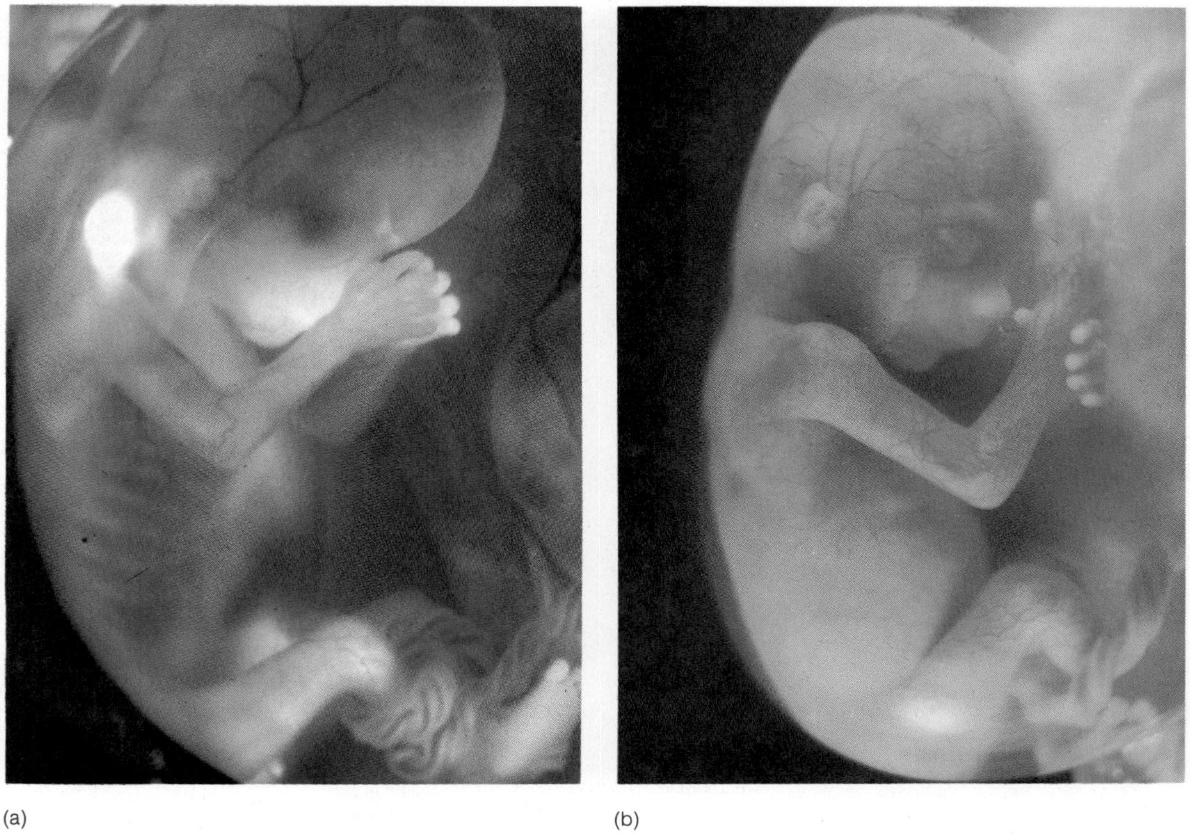

Figure 22.23 External appearances of fetuses at (*a*) ten weeks, and (*b*) twelve weeks.

Major structural abnormalities, which may not be predictable from genetic analysis, can often be detected by *ultrasonography* (fig. 22.24). Sound wave vibrations are reflected from the interface of tissues with different densities—such as the interface between the fetus and amniotic fluid—and used to produce an image. This technique is so sensitive that it can be used to detect a fetal heartbeat several weeks before it can be detected by a stethoscope.

Thirteen to Sixteen Weeks During the period from thirteen through sixteen weeks, the facial features are well formed, and epidermal structures such as eyelashes, eyebrows, hair on the head, fingernails, and nipples begin to develop. The appendages lengthen, and by the sixteenth week the skeleton is sufficiently developed so that it shows clearly on x-ray films. During the sixteenth week, the fetal heartbeat can be heard by applying a stethoscope to the mother's abdomen. By the end of the sixteenth week, the fetus is 140 mm in length (5.5 in.) and weighs about 200 g (7 oz).

After the sixteenth week, fetal length can be determined using ultrasonography. The reported length of a fetus is generally derived from a straight line measurement from the crown of the head to the developing ischium (crown-rump length). Measurements made on an embryo prior to the fetal stage, however, are not reported as crown-rump measurements but as total length.

Seventeen to Twenty Weeks During the period from seventeen to twenty weeks, the legs achieve their final relative proportions, and fetal movements, known as **quickening,** are commonly felt by the mother. The skin is covered with a white, cheeselike material known as **vernix caseosa.** It consists of fatty secretions from the sebaceous glands and dead epidermal cells. The function of vernix caseosa is to protect the fetus while it is bathed in amniotic fluid. Twenty-week-old fetuses usually have fine, silklike fetal hair, called **lanugo** *(lah-nu'go),* covering the skin (see chap. 5). Lanugo is thought to hold the vernix caseosa on the skin and produce a ciliarylike motion that moves amniotic fluid. The length of a twenty-week-old fetus is about 190 mm (7.5 in.) and the weight is 460 gm (16 oz). Because of cramped space the fetus develops a marked spinal flexure and is in what is commonly called the fetal posture with the head bent down in contact with the flexed knees.

Twenty-one to Twenty-five Weeks During the period from twenty-one to twenty-five weeks, the fetus increases its weight substantially to about 900 gm (32 oz). Body length increases only moderately (240 mm), however, so the weight is evenly proportioned. The skin is quite wrinkled and is translucent pinkish in color because the blood flowing in the capillaries is now visible.

---

vernix caseosa: L. *vernix, varnish;* L. *caseus,* cheese

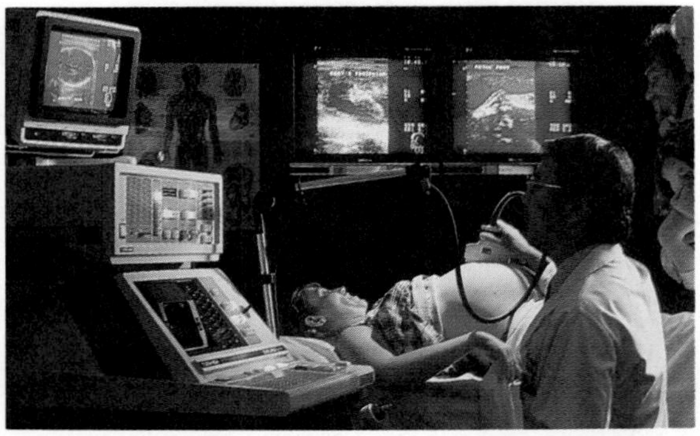

Figure 22.24 Ultrasonography. (*a*) Sound wave vibrations are reflected from the internal tissues of a person's body. (*b*) Structures of the human fetus observed through an ultrasound scan.

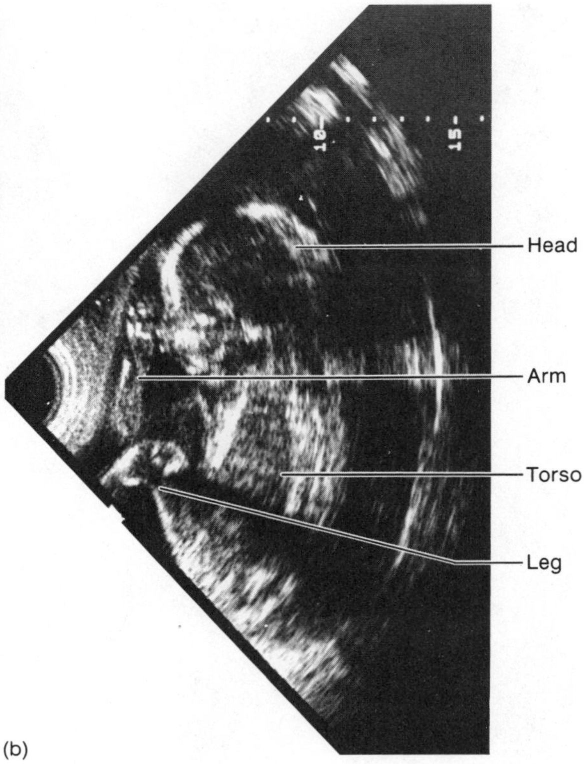

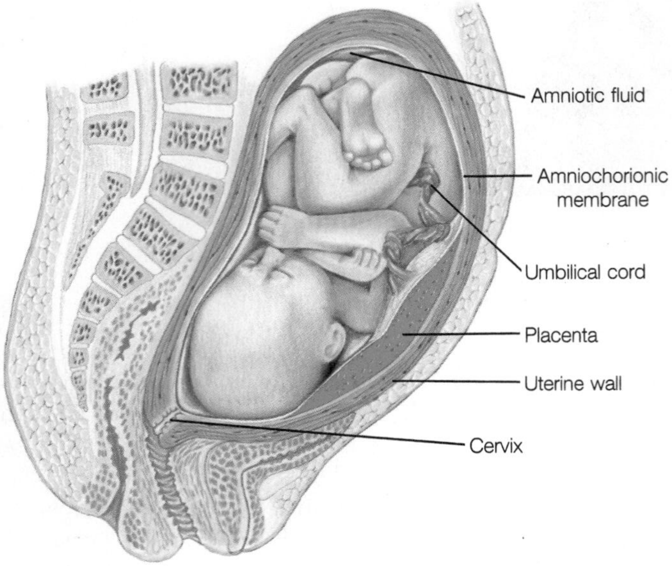

Figure 22.25 A fetus in vertex position. Towards the end of most pregnancies, the weight of the fetal head causes a rotation of its entire body such that the head is positioned in contact with the cervix of the uterus.

Twenty-six to Twenty-nine Weeks Toward the end of the period from twenty-six to twenty-nine weeks, the fetus will be 275 mm (about 11 in.) in length and will weigh 1,300 gm (46 oz). A fetus might now survive if born prematurely, but the mortality rate is high. Its body metabolism cannot yet maintain a constant temperature, and the respiratory muscles have not matured enough to provide a regular respiratory rate. If, however, the premature infant is put in an incubator and a respirator is used to maintain its breathing, it may survive. The eyes open during this period, and the body is well covered with lanugo. If the fetus is a male, the testes should have begun descent into the scrotum (see fig. 20.7). As the time of birth approaches, the fetus rotates to a **vertex position** (fig. 22.25). The head repositions toward the cervix because of the shape of the uterus and because the head is the heaviest part of the body.

vertex: L. *vertex*, summit

Thirty to Thirty-eight Weeks By the end of thirty-eight weeks, the fetus is considered full-term. It has reached a crown-rump length of 360 mm (14 in.) and weighs 3,400 gm (7.5 lbs). The average total length from crown to heel is 50 cm (20 in.). Most fetuses are plump with smooth skin because of the accumulation of subcutaneous fat. The skin is pinkish blue in color even on fetuses of dark-skinned parents, because melanocytes do not produce melanin until the skin is exposed to sunlight. Lanugo hair is sparse and is generally found on the head and back. The chest is prominent, and the mammary area protrudes in both sexes. The external genitalia are somewhat swollen.

1. Explain why the ninth week is designated as the beginning of the fetal stage of development.
2. List the approximate fetal age that each of the following occur: first detection of fetal heartbeat; presence of vernix caseosa and lanugo; fetal rotation into vertex position.

Labor and Parturition

Labor and parturition are the culmination of gestation and require the action of oxytocin, secreted by the posterior pituitary and prostaglandins, produced in the uterus.

Objective 13. Describe the hormonal action that controls labor and parturition.

Objective 14. Describe the three stages of labor.

The time of prenatal development, or the time of pregnancy, is called **gestation.** The human gestational period is usually 266 days or about 280 days from the beginning of the last men-

gestation: L. *gestatus*, to bear

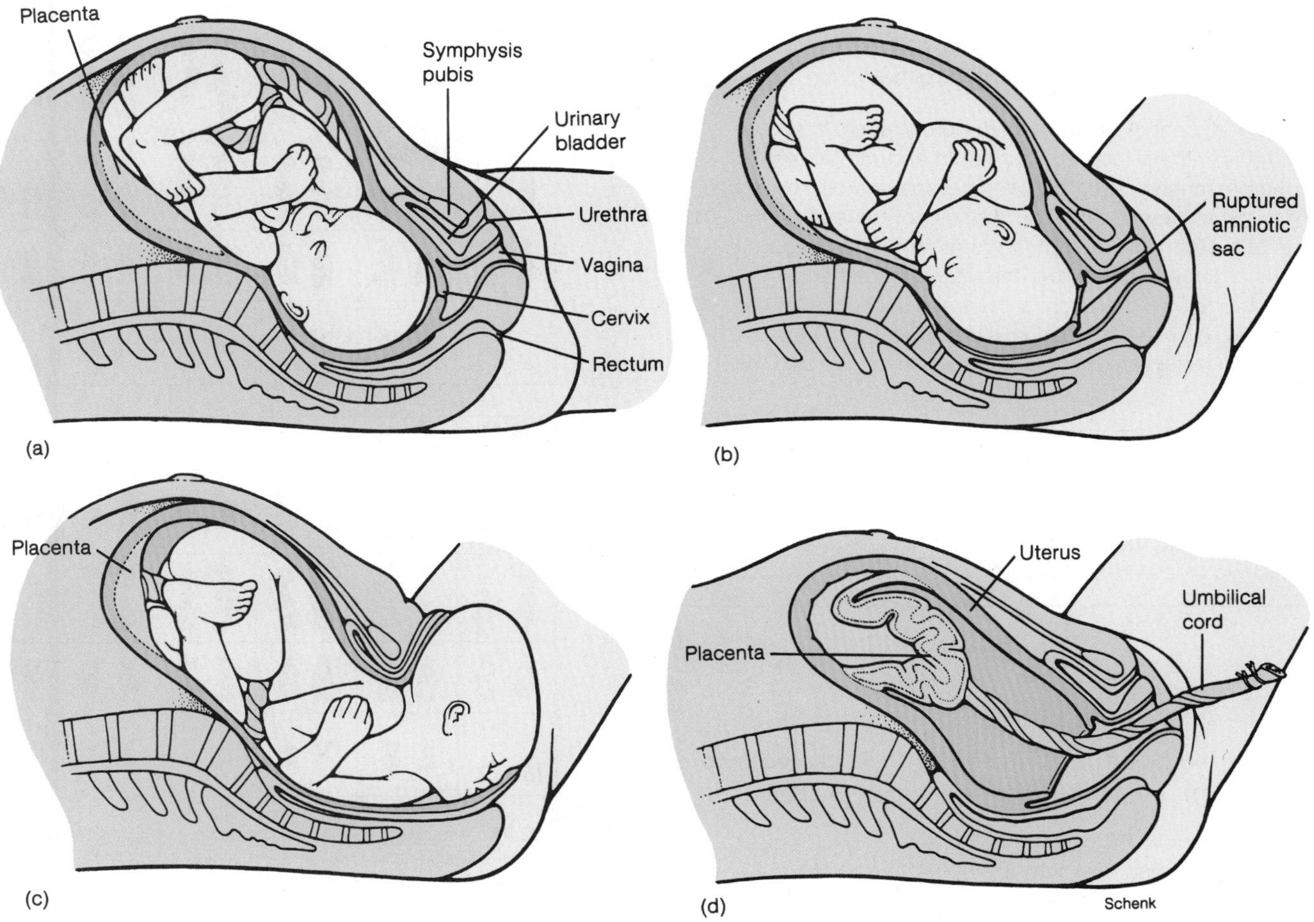

Figure 22.26 The stages of labor and parturition. (*a*) The position of the fetus prior to labor. (*b*) The ruptured amniotic sac and early dilation of the cervix; (*c*) expulsion stage, or the period of parturition; and (*d*) the placental stage.

strual period to **parturition,** or birth. Most fetuses are born within 10 to 15 days before or after this time. Parturition is accompanied by a sequence of physiological and physical events called **labor.**

The *onset of labor* is denoted by rhythmic and forceful contractions of the myometrial layer of the uterus (see fig. 21.12). In *true labor,* the pains from uterine contractions occur at regular intervals and intensify as the time between contractions shorten. A reliable indication of true labor is dilation of the cervix and a show, or discharge of blood-containing mucus in the cervical canal and vagina. In *false labor,* abdominal pain is felt at irregular intervals, and there is a lack of cervical dilations and show.

The uterine contractions of labor are stimulated by two agents: (1) **oxytocin** *(ok''si-to'sin),* a polypeptide hormone produced in the hypothalamus and secreted by the posterior pituitary, and (2) **prostaglandins** *(pros''tah-glan'dins),* a class of fatty acids produced within the uterus itself. Labor can indeed be induced artificially by injections of oxytocin or by the insertion of prostaglandins into the vagina as a suppository.

The hormone *relaxin,* produced by the corpus luteum, may also be involved in labor and parturition. Relaxin is known to soften the symphysis pubis in preparation for parturition and is thought to also soften the cervix in preparation for dilation. It may be, however, that relaxin does not affect the uterus, but rather progesterone and estradiol may be responsible for this effect. Further research is necessary to understand the total physiological effect of these hormones.

Labor is divided into three stages (fig. 22.26).

1. **Dilation stage.** In this period the cervix dilates to a diameter of approximately 10 cm. There are regular contractions during this stage and usually a rupturing of the amniotic sac (bag of waters). If the amniotic sac does not rupture spontaneously, it is done surgically. The dilation stage may last eight to twenty-four hours depending on whether it is occurring in the first or subsequent pregnancies.
2. **Expulsion stage.** This is the period of parturition, or actual childbirth. It consists of forceful uterine contractions and abdominal compressions to expel the fetus from the uterus and through the vagina. This stage may require thirty minutes in a first pregnancy, but only a few minutes in subsequent pregnancies.
3. **Placental stage.** Generally within ten to fifteen minutes after parturition, the placenta is separated from the uterine wall and expelled as the *afterbirth.* Forceful uterine contractions characterize this stage, constricting uterine blood vessels to prevent hemorrhage. In a normal delivery, blood loss does not exceed 350 milliliters.

A *pudendal nerve block* may be administered during the early part of the expulsion stage to ease the trauma of delivery for the mother and make an episiotomy possible.

Five percent of newborns are born *breech.* In a breech birth, the fetus has not rotated and the buttocks are the presenting part. The principal concern of a breech birth is the increased time and difficulty of the expulsion stage of parturition. Attempts to rotate the fetus through the use of forceps may injure the infant. If an infant cannot be delivered breech, a *cesarean section* must be performed. A cesarean section is delivery of the fetus through an incision made into the abdominal wall and the uterus.

---

1. Distinguish between labor and parturition.
2. Describe the hormonal mechanisms responsible for labor, and explain two techniques for inducing labor.
3. Describe the three stages of labor, and list the relative time duration of each.

Inheritance

Inheritance is the passage of hereditary traits carried by the genes on chromosomes from one generation to another.

Objective 15. Define the term *genetics.*
Objective 16. Discuss the variables that account for a person's phenotype.
Objective 17. Explain how probability is involved in predicting inheritance, and use a Punnett square to illustrate selected probabilities.

Genetics is the branch of biology that deals with inheritance. Genetics and inheritance are important in anatomy because of the numerous developmental and functional disorders that have a genetic basis. Genetic counseling is the practical application of knowing which disorders and diseases are inherited. The genetic inheritance of an individual begins with conception.

Each zygote inherits twenty-three chromosomes from its mother and twenty-three chromosomes from its father. This does not produce forty-six different chromosomes, but rather, twenty-three pairs of *homologous chromosomes.* Each member of a homologous pair, with the important exception of the sex chromosomes, looks like the other and contains similar genes (such as those coding for eye color, height, and so on). These homologous pairs of chromosomes can be **karyotyped** (photographed or illustrated) and identified (as shown in fig. 22.27). Each cell that contains forty-six chromosomes (that is *diploid*) has two chromosomes number 1, two chromosomes number 2, and so on through chromosomes number 22. The first twenty-two pairs of chromosomes are called **autosomal** *(aw″to-so′-mal)* **chromosomes.** The twenty-third pair of chromosomes are the **sex chromosomes,** which may look different and may carry different genes. In a female these consist of two X chromosomes, whereas in a male there is only one X chromosome and one Y chromosome.

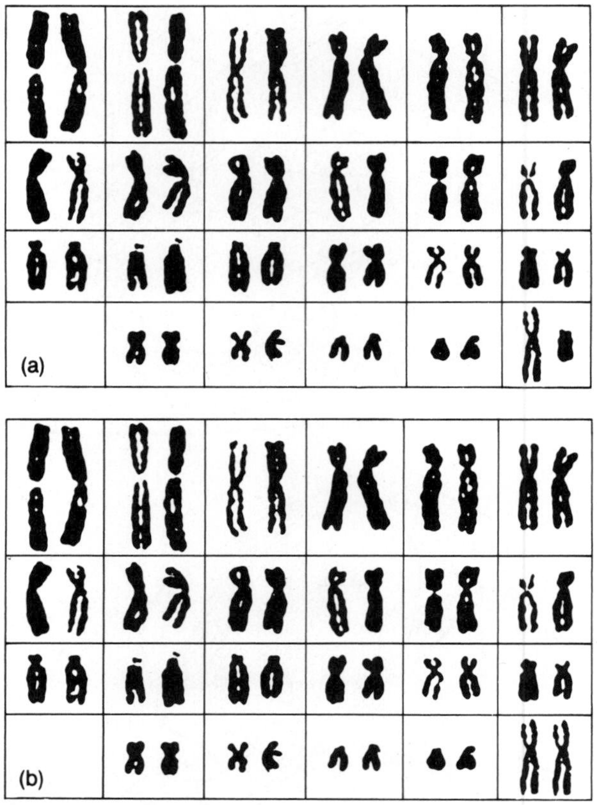

Figure 22.27 A karyotype of homologous pairs of chromosomes obtained from a human diploid cell. The first twenty-two pairs of chromosomes are called the autosomal chromosomes. The sex chromosomes are (*a*) XY for a male and (*b*) XX for a female.

Genes and Alleles A **gene** is a portion of the DNA of a chromosome that contains the information needed to synthesize a particular protein molecule. Although each diploid cell has a pair of gene locations for each characteristic, there may be a number of alternate forms of each gene. Those alternative forms of a gene that affect the same characteristic, but produce different expressions of that characteristic, are called **alleles** *(ah-lēls′).* One allele of each pair originates from the female parent and the other from the male. The shape of a person's ears, for example, is determined by the kind of allele received from each parent and how the alleles interact with one another. Alleles are always located on the same spot, called a **locus,** on homologous chromosomes (fig. 22.28).

For any particular pair of alleles in a person, the two alleles are either identical or not identical. If the alleles are identical, the person is said to be **homozygous** *(ho″mo-zi′gus)* for that particular characteristic. But if the two alleles are different, the person is **heterozygous** *(het″er-o-zi′gus)* for that particular trait.

Genotype and Phenotype A person's DNA contains a catalog of genes known as the **genotype** of that person. The expression of those genes results in certain observable characteristics referred to as the **phenotype.**

If the alleles for a particular trait are homozygous, the characteristic expresses itself in a specific manner (two alleles for attached earlobes, for example, results in a person with attached

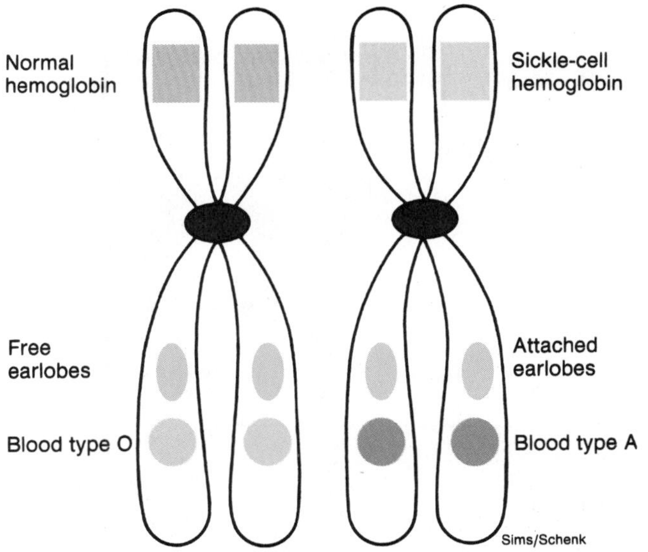

Figure 22.28 A pair of homologous chromosomes. Homologous chromosomes contain genes for the same characteristic at the same locus.

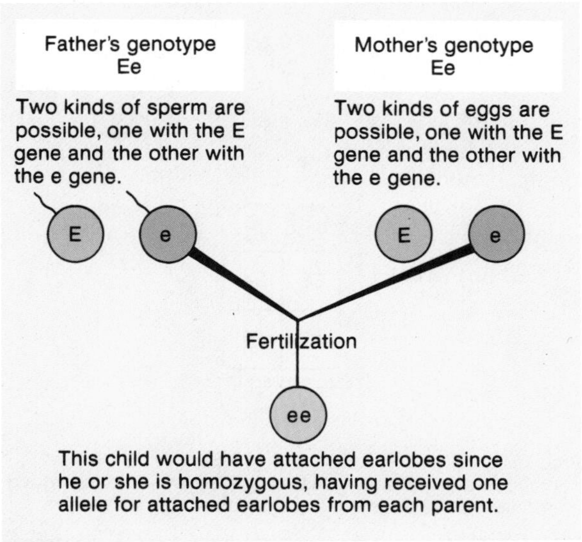

Figure 22.29 Inheritance of ear shape. Two heterozygous parents with free earlobes can have a child with attached earlobes.

Table 22.4 Some hereditary traits in humans determined by single pairs of dominant and recessive alleles

Dominant	Recessive	Dominant	Recessive
Free earlobes	Attached earlobes	Color vision	Color blindness
Dark brown hair	All other colors	Broad lips	Thin lips
Curly hair	Straight hair	Ability to roll tongue	Lack of this ability
Pattern baldness (♂♂)	Baldness (♀♀)	Arched feet	Flatfeet
Pigmented skin	Albinism	A or B blood factor	O blood factor
Brown eyes	Blue or green eyes	Rh blood factor	No Rh blood factor

earlobes). If the alleles for a particular trait are heterozygous, however, which of the alleles expresses itself and how the genes for that trait interact will determine the phenotype. Often one of the alleles expresses itself as the **dominant allele,** whereas the other does not and is the **recessive allele.** The combinations of dominant and recessive alleles are responsible for a person's hereditary traits (table 22.4).

In describing genotypes, it is traditional to use letter symbols to refer to the alleles of an organism. The dominant alleles are symbolized by uppercase letters and the recessive alleles are symbolized by lowercase. Thus, the *genotype* of a person who is homozygous for free earlobes due to a dominant allele is symbolized *EE;* a heterozygous pair is symbolized *Ee.* In both of these instances, the *phenotypes* of the individuals would be free earlobes, since a dominant allele is present in each genotype. A person who inherited two recessive alleles for earlobes has the genotype *ee* and will have attached earlobes.

Thus, three genotypes are possible when gene pairing involves dominant and recessive alleles. They are *homozygous dominant (EE), heterozygous (Ee),* and *homozygous recessive (ee).* Only two phenotypes are possible, however, since the dominant allele is expressed in both the homozygous dominant (*EE*) and the heterozygous (*Ee*) individuals. The recessive allele is expressed only in the homozygous recessive (*ee*) condition. Refer to figure 22.29 for an illustration of how a homozygous recessive trait may be expressed in a child of parents who are heterozygous dominant.

Probability A **Punnett square** is a convenient way to express the probabilities of allele combinations for a particular inheritable trait. In constructing a Punnett square, the male gametes (spermatozoa) carrying a particular trait are placed at the side of the chart, and the female gametes (ova) at the top (as in fig. 22.30). The four spaces on the chart represent the possible combinations of male and female gametes that could form zygotes. The probability of an offspring having a particular genotype is one in four (.25) for homozygous dominant and homozygous recessive, and one in two (.50) for heterozygous dominant.

A genetic study in which a single characteristic (e.g., ear shape) is followed from parents to offspring is referred to as a **monohybrid cross.** A genetic study in which two characteristics are followed from parents to offspring is referred to as a **dihybrid cross** (fig. 22.31). The term **hybrid** refers to an offspring descended from parents that have different genotypes.

Sex-linked Inheritance Certain inherited traits are located on a sex-determining chromosome and are called **sex-linked** characteristics. The allele for red-green *color blindness,* for example, is determined by a recessive allele (designated c) found on the X chromosome but not the Y chromosome. Normal

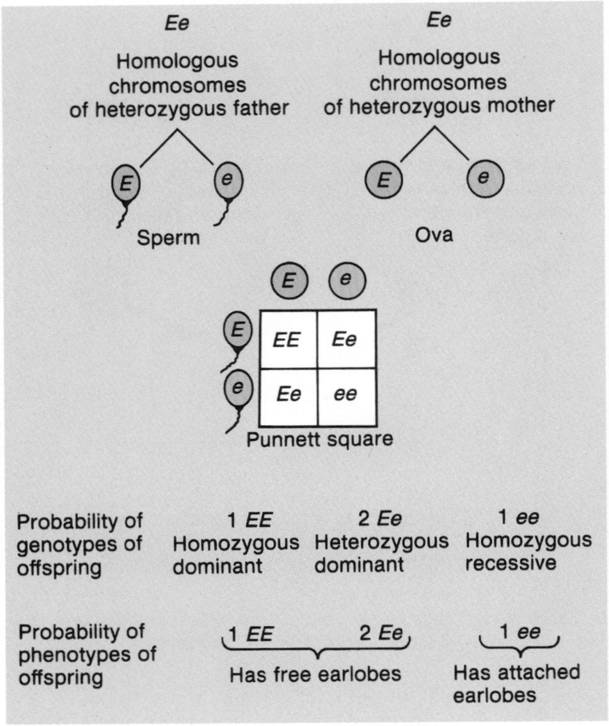

Figure 22.30 Inheritance of the shape of earlobes.

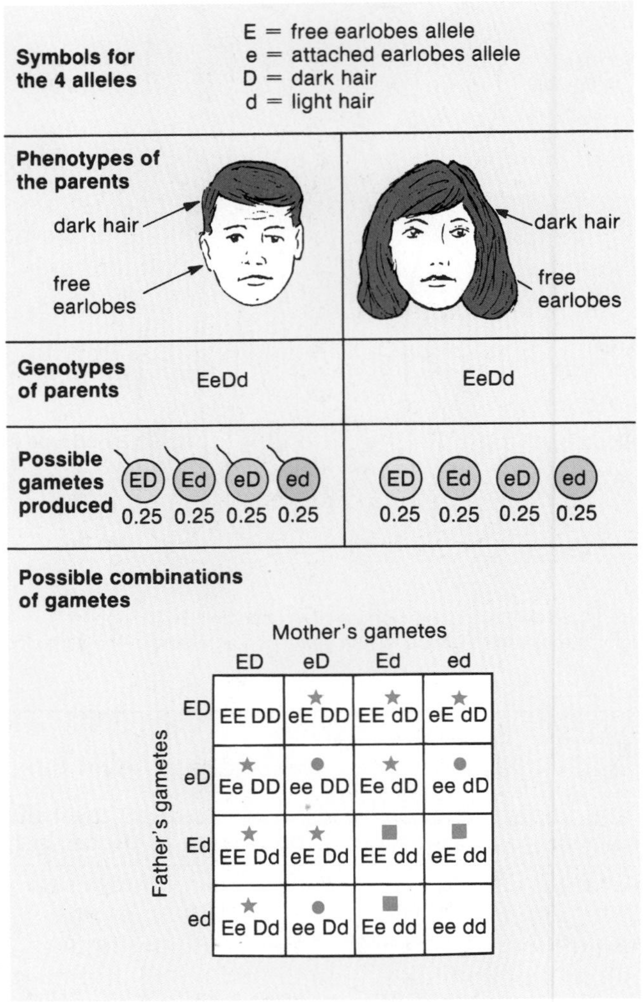

Figure 22.31 A dihybrid cross studies the probability of two characteristics at the same time. Any of the combinations of genes that have a *D* and an *E* (nine possibilities) will have free earlobes and dark hair. These are indicated with an asterisk (*). Three of the possible combinations have two alleles for attached earlobes (*ee*) and at least one allele for dark hair. They are indicated with a dot (•). Three of the combinations have free earlobes and light hair. These are indicated with a square (□). The remaining possibility has the genotype (*eedd*) for attached earlobes and light hair.

color vision (designated C) dominates. The ability to discern red-green colors, therefore, depends entirely on the X chromosomes. The genotype possibilities are

X^CY	Normal male
X^cY	Color-blind male
X^CX^C	Normal female
X^CX^c	Normal female carrying the recessive allele
X^cX^c	Color-blind female

In order for a female to be red-green color-blind, she must have the recessive allele on both of her X chromosomes. A male with only one such allele on his X chromosome, however, will show the characteristic.

Hemophilia is a sex-linked condition caused by a recessive allele. The blood in a person with hemophilia fails to clot or clots very slowly after an injury. If H represents normal clotting and h represents abnormal clotting, then males with X^HY will be normal and males with X^hY will be hemophiliac. Females with X^hX^h will have the disorder.

1. Define the following terms: *genetics, genotype, phenotype, allele, dominant, recessive, homozygous,* and *heterozygous.*
2. List several dominant and recessive traits inherited in humans. What are some variables that determine a person's phenotype?
3. Construct a Punnett square to show the possible genotypes for color blindness of an X^CY male and an X^CX^c female.

Clinical Considerations

Pregnancy and childbirth are natural events in human biology and generally progress smoothly without complications. Prenatal development is amazingly precise, and although traumatic, childbirth for most women in the world takes place without the aid of a physician. Occasionally, however, serious complications arise, and the knowledge of an obstetrician is required. The physician's knowledge of normal development and the causes of congenital malformations ensures the embryo every possible chance to develop normally. Many of the clinical aspects of prenatal development involve what might be referred to as applied developmental biology.

In clinical terms, gestation is frequently divided into three phases, or **trimesters,** each lasting three calendar months. By the end of the **first trimester** all of the major body systems are formed, the fetal heart can be detected, the external genitalia are developed, and the fetus is about the width of the palm of

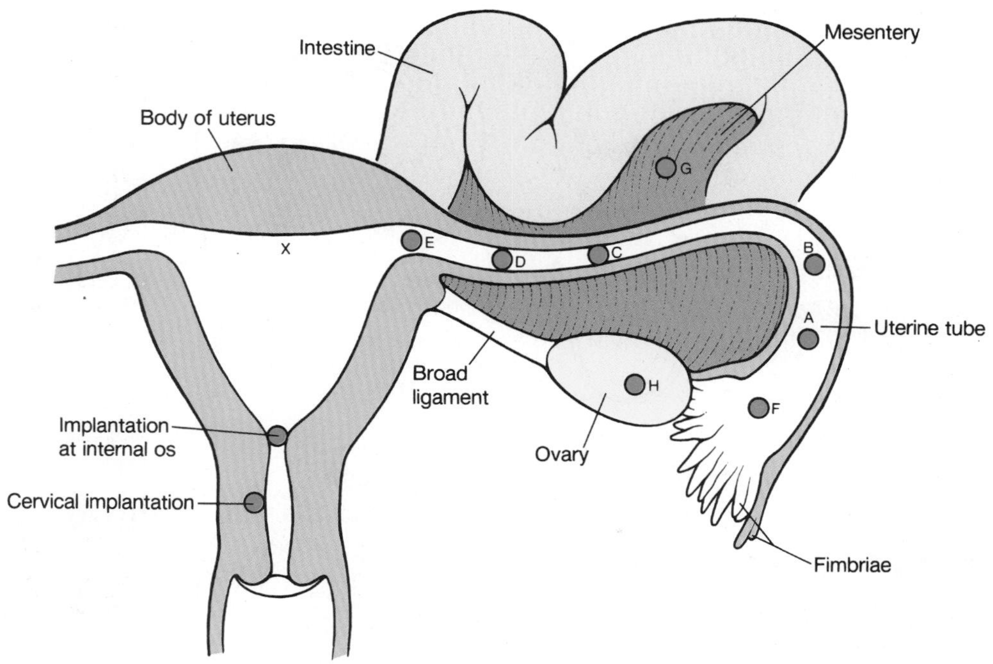

Figure 22.32 The various sites of ectopic pregnancies. The normal site is indicated by an X, and the abnormal sites are indicated by letters in order of frequency of occurrence.

an adult's hand. During the **second trimester,** fetal quickening can be detected, epidermal features are formed, and the vital body systems are functioning. The fetus, however, is still unlikely to survive if birth were to occur. At the end of the second trimester, fetal length is about equal to the length of an adult's hand. The fetus experiences a tremendous amount of growth and refinement in system functioning during the **third trimester.** A fetus of this age may survive if born prematurely, and of course, the chances of survival improve as the length of pregnancy approaches the natural delivery date.

There are many clinical considerations associated with prenatal development, some of which relate directly to the female reproductive system. Other developmental problems are genetically related and will be mentioned only briefly. Of clinical concern for developmental anatomy are topics such as implantation sites, test-tube development, multiple pregnancy, fetal monitoring, and congenital defects.

Abnormal Implantation Sites

In an **ectopic pregnancy** the blastocyst implants outside the uterus or in an abnormal site within the uterus (fig. 22.32). The most common ectopic location (about 95%) is within the uterine tube and is referred to as a **tubal pregnancy.** Occasionally, implantation occurs near the cervix where development of the placenta blocks the cervical opening. This condition, called **placenta previa,** causes serious bleeding. Ectopic pregnancies will not develop normally in unfavorable locations, and the fetus seldom survives beyond the first trimester. Tubular pregnancies are terminated through medical intervention. If a tubular pregnancy is permitted to progress, however, the uterine tube generally ruptures, followed by hemorrhaging. Depending on the location and the stage of development (hence vascularity) of tubal pregnancies, they may not be serious or the hemorrhaging and shock may cause the death of the woman.

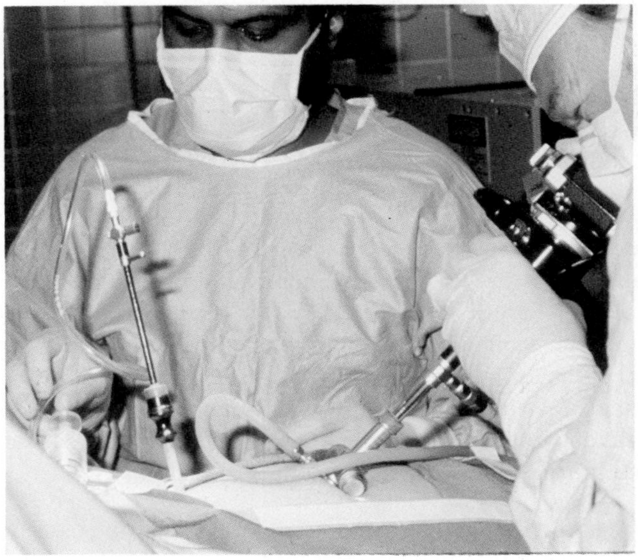

Figure 22.33 A laparoscope, used for various abdominal operations, including the extraction of a preovulatory oocyte.

In Vitro Fertilization and Artificial Implantation

Reproductive biologists have been able to fertilize a human oocyte *in vitro* (outside the body), culture it to the blastocyst stage, and then perform artificial implantation, leading to a full-term development and delivery. This is the so-called test-tube baby. To obtain the oocyte, a specialized laparoscope (fig. 22.33)

previa: L. *previa,* appearing before or in front of

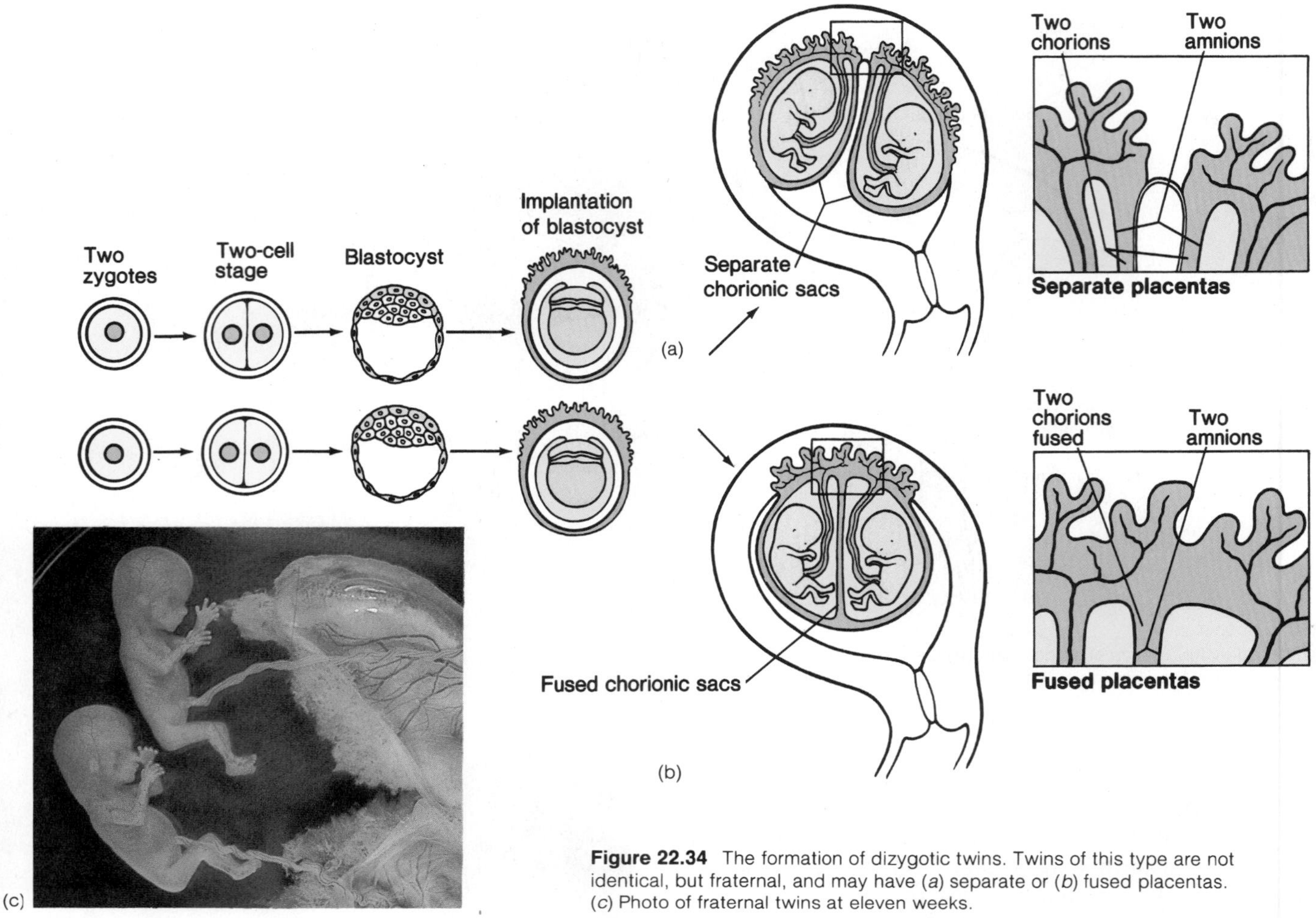

Figure 22.34 The formation of dizygotic twins. Twins of this type are not identical, but fraternal, and may have (*a*) separate or (*b*) fused placentas. (*c*) Photo of fraternal twins at eleven weeks.

is used to aspirate the preovulatory egg from a mature vesicular ovarian follicle. The oocyte is then placed in a suitable culture medium where it is fertilized with sperm. After the zygote forms, the sequential pre-embryonic development continues until the blastocyst stage, at which time implantation is performed. *In vitro* fertilization with artificial implantation is a means of overcoming infertility problems due to blocked uterine tubes in females or low sperm counts in males.

Multiple Pregnancy

Twins occur about once in eighty-five pregnancies. They can develop in two ways. **Dizygotic** (fraternal) **twins** develop from two zygotes resulting from two spermatozoa fertilizing two oocytes in the same ovulatory cycle (fig. 22.34). **Monozygotic** (identical) **twins** form from a single zygote (fig. 22.35). Approximately one-third of twins are monozygotic.

Dizygotic twins may be of the same sex or different sexes and are not any more alike than brothers or sisters born at different times. Dizygotic twins always have two chorions and two amnions, but the chorions and the placentas may be fused.

Monozygotic twins are of the same sex and are genetically identical. Any physical differences in monozygotic twins are caused by environmental factors during morphogenic development (e.g., there might be a differential vascular supply that causes slight differences to be expressed). Monozygotic twinning is usually initiated toward the end of the first week when the embryoblast divides to form two embryonic primordia. Monozygotic twins have two amnions but only one chorion and a common placenta. If the embryoblast fails to completely divide, **conjoined twins** (Siamese twins) may form.

Triplets occur about once in 7,600 pregnancies and may be (1) all from the same ovum and identical, (2) two identical and the third from another ovum, or (3) three zygotes from three different ova. Similar combinations occur in quadruplets, quintuplets, and so on.

Fetal Monitoring

Obstetrics has benefited greatly from the advancements made in fetal monitoring in the last two decades. Before these techniques became available, physicians could determine the welfare of the unborn child only by auscultation of the fetal heart and palpation of the fetus. Currently, there are several tests that provide much information about the fetus during any stage of development. Fetal conditions that can now be diagnosed and evaluated include genetic disorders, hypoxia, blood disorders, growth retardation, placental functioning, prematurity, postmaturity, and intrauterine infections. These tests also help determine the advisability of an abortion.

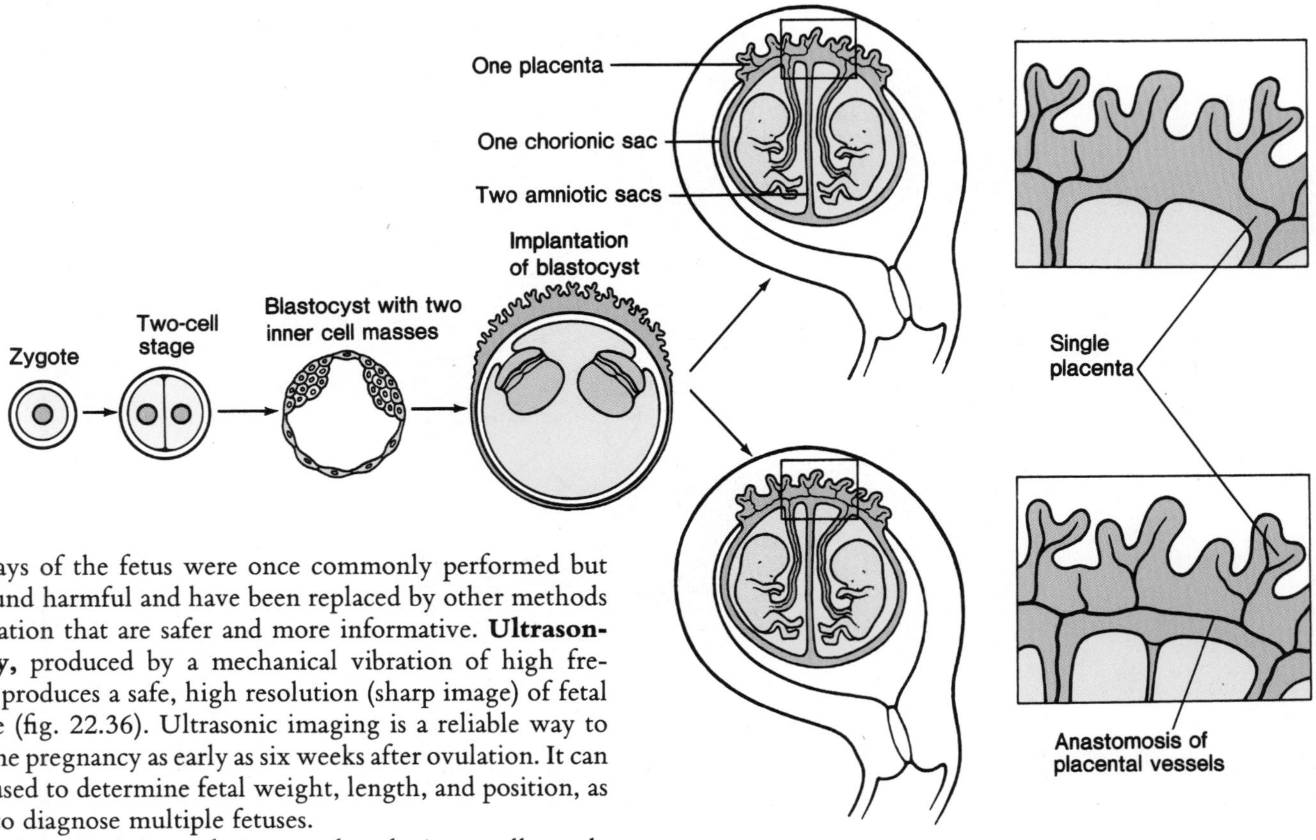

Figure 22.35 The formation of monozygotic twins. Twins of this type develop from a single zygote and are identical. Such twins have two amnions but one chorion and a common placenta.

X rays of the fetus were once commonly performed but were found harmful and have been replaced by other methods of evaluation that are safer and more informative. **Ultrasonography,** produced by a mechanical vibration of high frequency, produces a safe, high resolution (sharp image) of fetal structure (fig. 22.36). Ultrasonic imaging is a reliable way to determine pregnancy as early as six weeks after ovulation. It can also be used to determine fetal weight, length, and position, as well as to diagnose multiple fetuses.

Amniocentesis is a technique used to obtain a small sample (5–10 ml) of amniotic fluid with a syringe so that the fluid can be assessed (see fig. 22.14). Amniocentesis is most often performed to determine fetal maturity, but it can also help predict serious disorders like *Down's syndrome* and *Gaucher's disease* (a metabolic disorder).

The technique of **fetoscopy** (fig. 22.37) goes beyond amniocentesis by allowing direct examination of the fetus. Using fetoscopy, physicians scan the uterus with pulsed sound waves to locate fetal structures, the umbilical cord, and the placenta. Skin samples are taken from the head of the fetus and blood samples extracted from the placenta. The principal advantage of fetoscopy is that external features of the fetus (such as fingers, eyes, ears, mouth, and genitals) can be carefully observed. Fetoscopy is also used to determine several diseases, including hemophilia, thalassemia, and the 40% of sickle-cell anemia cases missed by amniocentesis.

Most hospitals are now equipped with instruments that monitor fetal heart rate and uterine contractions during labor and can detect any complication that arises during the delivery. This procedure is called Electronic Monitoring of Fetal Heart Rate and Uterine Contractions (FHR-UC Monitoring). The stress to the fetus from uterine contractions can be determined through monitoring (fig. 22.38). Long, arduous deliveries can be taxing to both the mother and fetus. If the baby's health and vitality are diagnosed to be in danger because of a difficult delivery, the physician may decide to perform a cesarean section.

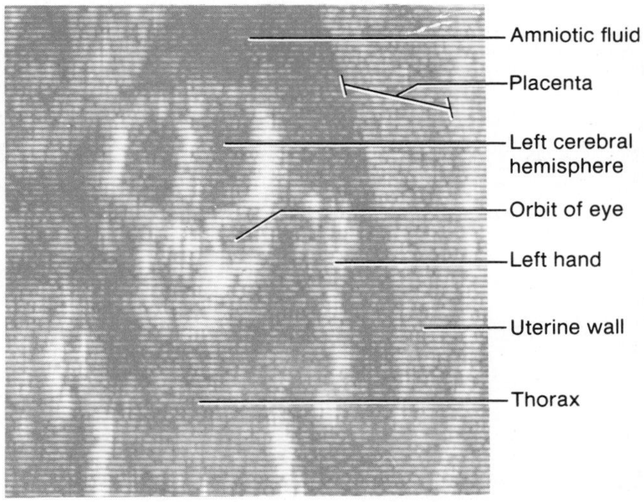

Figure 22.36 Color-enhanced ultrasonogram of a fetus during the third trimester. The left hand is raised, as if waving to the viewer.

amniocentesis: Gk. *amnion,* lamb (fetal membrane); *kentesis,* puncture
fetoscopy: L. *fetus,* offspring, *skopein,* to view

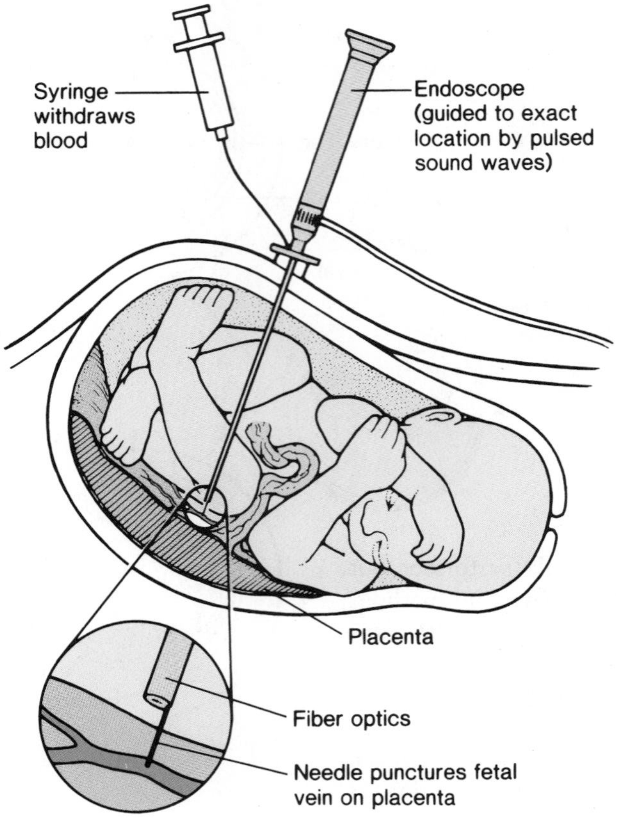

Figure 22.37 Fetoscopy.

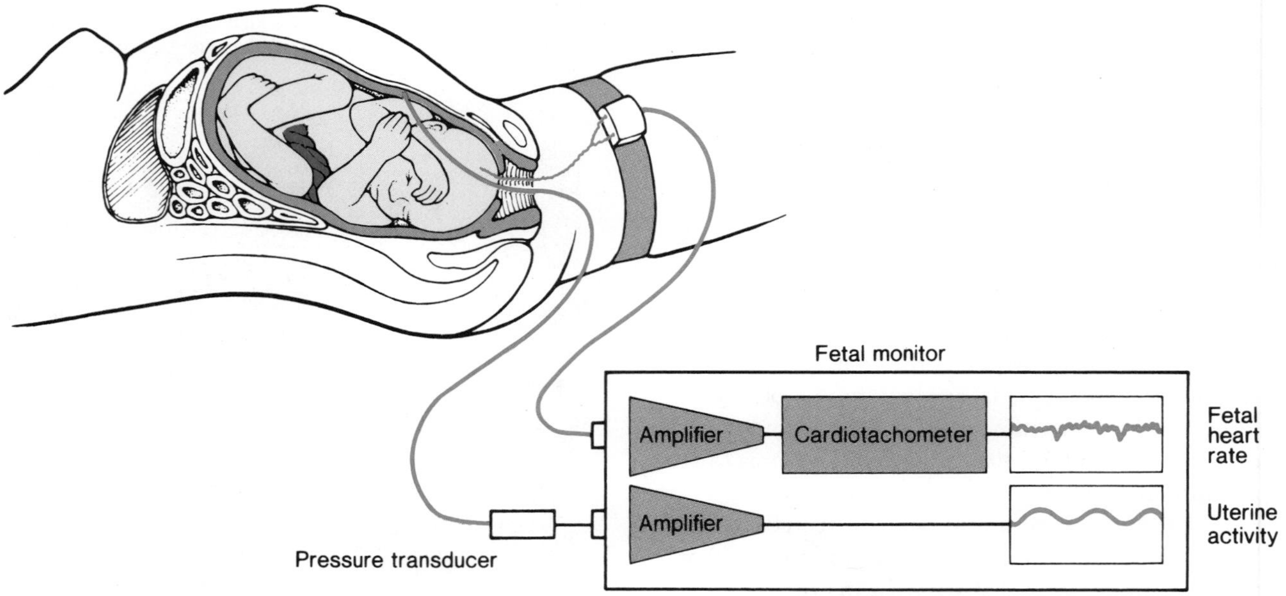

Figure 22.38 Monitoring fetal heart rate and uterine contractions using a FHR-UC device.

Congenital Defects

Major developmental problems called **congenital malformations** occur in approximately 2% of all newborn infants. The causes of congenital conditions include genetic inheritance, mutation (genetic change), and environment. About 15% of neonatal deaths are attributed to congenital malformations. The branch of developmental biology concerned with abnormal development and congenital malformations is called *teratology*. Many congenital problems have been discussed in previous chapters with the body system in which they occur.

congenital: L. *congenitus*, born with

teratology: Gk. *teras*, monster; *logos*, study of

Some Genetic Disorders of Clinical Importance

cystic fibrosis An autosomal recessive disorder that is characterized by the formation of thick mucus in the lungs and pancreas, which interferes with normal breathing and digestion.

familial cretinism An autosomal recessive disorder that is characterized by a lack of thyroid secretion, due to a defect in the iodine transport mechanism. Untreated children are dwarfed, sterile, and may be mentally retarded.

galactosemia An autosomal recessive disorder that is characterized by an inability to metabolize galactose, a component of milk sugar. Patients with this disorder have cataracts, damaged livers, and mental retardation.

gout An autosomal dominant disorder that is characterized by an accumulation of uric acid in the blood and tissue due to an abnormal metabolism of purines.

hepatic porphyria An autosomal dominant disorder that is characterized by painful gastrointestinal disorders and neurologic disturbances due to an abnormal metabolism of porphyrins.

hereditary hemochromatosis A sex-influenced, autosomal dominant disorder that is characterized by an accumulation of iron in the pancreas, liver, and heart, resulting in diabetes, cirrhosis, and heart failure.

hereditary leukomelanopathy An autosomal recessive disorder that is characterized by decreased pigmentation in the skin, hair, and eyes, and abnormal white blood cells. Patients with this condition are generally susceptible to infections and early deaths.

Huntington's chorea An autosomal dominant disorder that is characterized by uncontrolled twitching of skeletal muscles and the deterioration of mental capacities. A latent expression of this disorder allows the mutant gene to be passed to children before the symptoms develop.

Marfan's syndrome An autosomal dominant disorder that is characterized by tremendous growth of the extremities, dislocation of the lenses, and congenital cardiovascular defects.

phenylketonuria *(fen''il-ke''to-nu're-ah)* **(PKU)** An autosomal recessive disorder that is characterized by an inability to metabolize the amino acid phenylalanine. This is accompanied by brain and nerve damage and mental retardation.

pseudohypertrophic muscular dystrophy A sex-linked recessive disorder that is characterized by progressive muscle atrophy. It usually begins during childhood and causes death in adolescence.

retinitis pigmentosa A sex-linked recessive disorder that is characterized by progressive atrophy of the retina and eventual blindness.

Tay-Sachs disease An autosomal recessive disorder that is characterized by a deterioration of physical and mental abilities, early blindness, and death.

Huntington's chorea: from George Huntington, U.S. physician, 1850–1916

Marfan's syndrome: from Antoine Bernard-Jean Marfan, French physician, 1858–1942

Tay-Sachs disease: from Warren Tay, English physician, 1843–1927, and Bernard Sachs, U.S. neurologist, 1858–1944

CLINICAL CASE STUDY ANSWER

Two amnions and one chorion in all but very unusual cases prove the twins to be monozygotic. The two infants, therefore, are genetically identical. Thus, when considering genetic disorders such as cleft palate, one would expect a high degree of concordance (both twins of a monozygotic pair exhibit a particular anomaly). Many such defects however can be present in only one twin, a consequence of nongenetic factors such as intrauterine environment. An example would be inadequate blood supply to only one twin resulting in a defect, while the other twin is spared. ■

Chapter Summary

I. Fertilization
   A. The fertilization of a secondary oocyte by a sperm in the uterine tube promotes the completion of meiotic development and the formation of a diploid zygote.
   B. Morphogenesis is the sequential formation of body structures during the prenatal period of human life. The prenatal period lasts thirty-eight weeks and is divided into a pre-embryonic, an embryonic, and a fetal stage.
   C. A capacitated sperm digests its way through the zona pellucida and corona radiata layers of the secondary oocyte to complete the fertilization process and formation of a zygote.

II. Pre-Embryonic Stage
   A. Cleavage of the zygote is initiated within thirty hours and continues until a morula forms, which enters the uterine cavity on about the third day.
   B. A hollow, fluid-filled space forms within the morula, and it is then called a blastocyst.
   C. Implantation begins between the fifth and seventh day and is made possible by the secretion of enzymes that digest a portion of the endometrium.
      1. During implantation, the trophoblast cells secrete human chorionic gonadotrophin (hCG), which prevents the breakdown of the endometrium and menstruation.
      2. The secretion of hCG declines by the tenth week as the developed placenta secretes steroids which maintain the endometrium.
   D. The embryoblast of the implanted blastocyst flattens into the embryonic disc, from which the primary germ layers of the embryo develop.

1. Ectoderm gives rise to the nervous system, the epidermis of the skin and epidermal derivatives, and portions of sensory organs.
2. Mesoderm gives rise to bones, muscles, blood, reproductive organs, the dermis of the skin, and connective tissue.
3. Endoderm gives rise to linings of the GI tract, digestive organs, the respiratory tract and lungs, and the urinary bladder and urethra.

III. Embryonic Stage
A. The events of the six-week embryonic stage include the differentiation of the germ layers into specific body organs and the formation of the placenta, the umbilical cord, and the extraembryonic membranes, which provide sustenance and protection to the embryo.
B. The extraembryonic membranes include the amnion, yolk sac, allantois, and chorion.
1. The amnion is a thin membrane surrounding the embryo and contains amniotic fluid, which cushions and protects the embryo.
2. The yolk sac produces blood for the embryo.
3. The allantois also produces blood for the embryo and gives rise to the umbilical arteries and vein.
4. The chorion participates in the formation of the placenta.

C. The placenta, formed from both maternal and embryonic tissue, has a transport role in providing for the metabolic needs of the fetus and in removing its wastes.
1. The placenta produces steroids and hormones.
2. Nicotine, drugs, alcohol, and viruses can cross the placenta to the fetus.

D. The umbilical cord, containing two umbilical arteries and one umbilical vein, is formed as the amnion envelops the tissues on the underside of the embryo.
E. During the third to the eighth week, the structure of all the body organs, except the genitalia, becomes apparent.
1. During the third week, the primitive knot forms from the primitive streak, which later gives rise to the notochord and mesenchyme.
2. By the end of the fourth week, the heart is beating, the primordial tissue of the eyes, brain, spinal cord, lungs, and digestive organs are in their proper place, and the arm and leg buds are recognizable.
3. At the end of the fifth week, the sense organs are formed in the enlarged head and the appendages have developed with finger rays present.
4. The brain is well developed by the end of the sixth week, and the digits are separate on elongated appendages.
5. During the seventh and eighth weeks, the body organs, except the genitalia, are formed and the embryo appears distinctly human.

IV. Fetal Stage
A. A small amount of tissue differentiation and organ development occurs during the fetal stage, but for the most part fetal development is primarily limited to body growth.
B. Between weeks nine and twelve, ossification centers appear, the genitalia are formed, and the digestive, urinary, respiratory, and muscle systems show functional activity.
C. Between weeks thirteen and sixteen, facial features are formed and the fetal heart beat can be detected with a stethoscope.
D. During the period from seventeen to twenty weeks, quickening can be felt by the mother, and vernix caseosa and lanugo cover the skin.
E. During the period from twenty-one to twenty-five weeks, substantial weight gain occurs, and the fetal skin becomes wrinkled and pinkish.
F. Toward the end of the period from twenty-six to twenty-nine weeks, the eyes have opened, the gonads have descended in a male, and the weight and development of the fetus is such that it may survive if born prematurely.
G. By thirty-eight weeks, the fetus is full-term; the normal gestation is 266 days.

V. Labor and Parturition
A. Labor and parturition are the culmination of gestation and require the action of oxytocin, secreted by the posterior pituitary, and prostaglandins, produced in the uterus.
B. Labor is divided into the dilation, expulsion, and placental stages.

VI. Inheritance
A. Inheritance is the passage of hereditary traits carried on the genes of chromosomes from one generation to another.
B. Each zygote contains twenty-two pairs of autosomal chromosomes and one pair of sex chromosomes; XX in a female and XY in a male.
C. A gene is a portion of a DNA molecule that contains information for the production of one kind of protein molecule; alleles are different forms of genes that occupy corresponding positions on homologous chromosomes.
D. The combination of genes present in a person's cells constitutes a genotype; the appearance of a person is a phenotype.
1. Dominant alleles are symbolized by uppercase letters and recessive alleles are symbolized by lowercase letters.
2. The three possible genotypes are homozygous dominant, heterozygous, and homozygous recessive.

E. A Punnett square is a convenient way to express probability.
1. The probability of a particular genotype is one in four (.25) for homozygous dominant and homozygous recessive, and one in two (.50) for heterozygous dominant.
2. A single trait is studied in a monohybrid cross, and two traits are studied in a dihybrid cross.

F. Sex-linked traits, such as color blindness or hemophilia, are carried on the sex-determining chromosome.

Review Activities

Objective Questions

1. The pre-embryonic stage is completed when the
(a) blastocyst implants.
(b) placenta forms.
(c) blastocyst reaches the uterus.
(d) primary germ layers form.
2. The yolk sac produces blood for the embryo until the
(a) heart is functional.
(b) kidneys are functional.
(c) liver is functional.
(d) baby is delivered.

3. Which of the following is *not* a function of the placenta?
   (a) production of steroids and hormones
   (b) diffusion of nutrients and oxygen
   (c) removal of metabolic wastes
   (d) production of enzymes
4. The decidua basalis is
   (a) a component of the umbilical cord.
   (b) the embryonic portion of the villous chorion.
   (c) a contributor to the formation of the placenta.
   (d) a vascular membrane derived from the trophoblast.
5. Which of the following could diffuse across the placenta?
   (a) nicotine (c) heroin
   (b) alcohol (d) All of the above.
6. During which week following conception does the embryonic heart begin pumping blood?
   (a) fourth (c) sixth
   (b) fifth (d) eighth
7. Twins that develop from two zygotes resulting from the fertilization of two oocytes by two sperm in the same ovulatory cycle are referred to as
   (a) monozygotic. (c) dizygotic.
   (b) conjoined. (d) identical.
8. Match the genotype descriptions with the correct symbols:
   homozygous recessive Bb
   heterozygous bb
   homozygous dominant BB
9. An allele that is *not* expressed in a heterozygous genotype is called
   (a) recessive. (c) genotypic.
   (b) dominant. (d) phenotypic.
10. If the genotypes of both parents are Aa and Aa, the offspring probably will be
   (a) ½ AA and ½ aa.
   (b) all Aa.
   (c) ¼ AA, ½ Aa, ¼ aa.
   (d) ¾ AA and ¼ aa.

Essay Questions

1. Define *implantation, morphogenesis, gestation,* and *parturition.* Explain how the delivery date is determined.
2. List in sequence the major events of the pre-embryonic stage and describe the locations within the female reproductive tract where each of these events take place.
3. Discuss the implantation of the trophoblast into the uterine wall and its involvement in the formation of the placenta.
4. Explain how the primary germ layers form. What major structures does each germ layer give rise to?
5. Explain why development during the embryonic stage is so critical, and list the embryonic needs that must be met during the embryonic stage for morphogenesis to continue.
6. List the extraembryonic membranes and discuss the functions of each.
7. Identify the approximate time period (in weeks) for the following occurrences:
   (a) The arm and leg buds appear.
   (b) The external genitalia differentiate.
   (c) Quickening is perceived by the mother.
   (d) The embryonic heart is functioning.
   (e) The ossification of bone is initiated.
   (f) Lanugo and vernix caseosa appear.
   (g) The fetus has a chance of survival if born prematurely.
   (h) All major body organs are formed.
8. Describe the various possible genetic combinations of multiple pregnancies.
9. List the various techniques of fetal monitoring and examination and explain the advantages and limitations of each.
10. State the features of a genetic disorder that would lead one to believe that it is a form of sex-linked inheritance.

CHAPTER 23

Postnatal Growth, Development, and Aging

Outline and Concepts

CLINICAL CASE STUDY

A fourteen-year-old girl comes to her doctor complaining of lower abdominal pain. On initial examination, the doctor notes a palpable lower abdominal mass that is tender. Mature breasts and pubic hair are present, which leads the doctor to ask if the patient is having trouble with her periods. The patient responds, "I've never had one." The doctor knowingly looks at his nurse and asks her to help prepare for a vaginal pelvic examination.

Is it within the realm of normalcy for a fourteen-year-old girl to not have experienced menarche? What about the case of this particular patient? Explain. What do you suppose the doctor will find when he examines the vulva? ■

Introduction to Postnatal Growth, Development, and Aging

Mitotic potential is an important aspect of cellular specialization and body growth. Growth is not a linear process, and growth rates vary for different organs and structures of the body.

Objective 1. List the factors that presumably influence mitosis.

Objective 2. Discuss how an organism increases in body size.

Although the body organs are formed prenatally, the body undergoes continual growth and development into postnatal life. Body growth, developmental changes, and death are fundamental principles of life. Understanding the biological importance of these principles will make the material in this chapter more meaningful. Growth and physical development are necessary for an organism to sustain itself and contribute to its own species population. The principal contribution of an individual is reproduction, which ensures the continuance of the species. In a biological sense, once an individual has completed its reproductive potential, it is a liability to the species population because of competition for sustaining resources. Death of the aged is beneficial to a population because it eliminates those members who can no longer reproduce and provides room for offspring with genetic differences to add to the gene pool of the population.

Although humans have the capability and opportunity long beyond the natural reproductive years to contribute far more to the population than simply producing and protecting offspring, humans are still subject to the biological phenomena of growth, development, aging, and death. Clinical aspects of anatomy have been stressed throughout this book, and many of these conditions are, in part, the result of changes in the body during postnatal growth, development, and aging.

Cell Division

A zygote is a single fertilized egg cell, which has the potential to divide repeatedly, differentiate, and grow into a complex individual with a body that contains trillions of cells. As specialized cell types are formed, further mitotic divisions occur until distinct organ systems become functional. Embryonic development and early fetal development are characterized by phenomenal cellular mitotic activity, differentiation and specialization, and growth.

Apparently included within cellular specialization of structure and function is mitotic potential. Certain cells do not require further division once the organ to which they contribute becomes functional. Others, as part of their specialization, require continuous mitosis to keep an organ healthy. Thus, in the adult, it is found that some cells divide continually, some occasionally, and some not at all. For example, epidermal cells, hemopoietic cells, and cells that line the lumen of the GI tract divide continually throughout life. Cells within specialized organs, such as the liver or kidneys, divide as the need becomes apparent. Naturally occurring cellular death, disease, or trauma from surgery or injury may necessitate mitosis in these organs. Still other cells, such as muscle or nerve cells, lose their mitotic ability as they become differentiated. Trauma to these cells frequently causes a permanent loss of function.

The factors regulating mitosis are unclear. Evidence suggests that mitotic ability is genetically controlled and, of those cells that do divide, even the number of divisions is predetermined. If this is true, it would certainly account for the aging process. Physical stress, nutrition, and hormones definitely have an effect on mitotic activity. It is thought that the reproductive activity of cells might be controlled through a feedback mechanism involving the release of a *growth-inhibiting substance.* Such a substance may slow or inhibit the divisions and growth of particular organs once the organ has a certain number of cells or reaches a certain size.

Growth

Growth is a normal process by which an organism increases in size as a result of the accretion of tissues similar to those already present. Growth is an integral part of development that continues until adulthood. There are three aspects to the growth process: (1) an increase in the number of constituent cells, (2) an increase in the size of the existing cells through the addition of protoplasmic substance, and (3) an increase in the amount of intercellular substance.

Human growth is not a steady, linear process. It varies with age and the individual and is somewhat sex-dependent (fig. 23.1). At birth, an average full-term baby has a weight of 3,400 g (7.5 lb) and a length of about 50 cm (20 in.). The head circumference of newborns averages 35.5 cm (14 in.). In the first days of life, most newborns lose nearly 100 g (about 4 oz) before their bodies adapt to ingested food. The head of the newborn is disproportionately large compared to the trunk and appendages.

Growth rates vary for different organs and structures of the body. For example, the brain at birth is about 24% of its adult weight, whereas the neonatal body is only 6% of its adult weight. Phenomenally rapid growth of the brain continues, so that by the time a child is four years old the brain has reached 90% of its adult weight, whereas its body weight is only 25% of its eventual adult weight. The reproductive organs experience latent development because they are under hormonal control. These organs remain at less than 10% of their final weight until the onset of puberty.

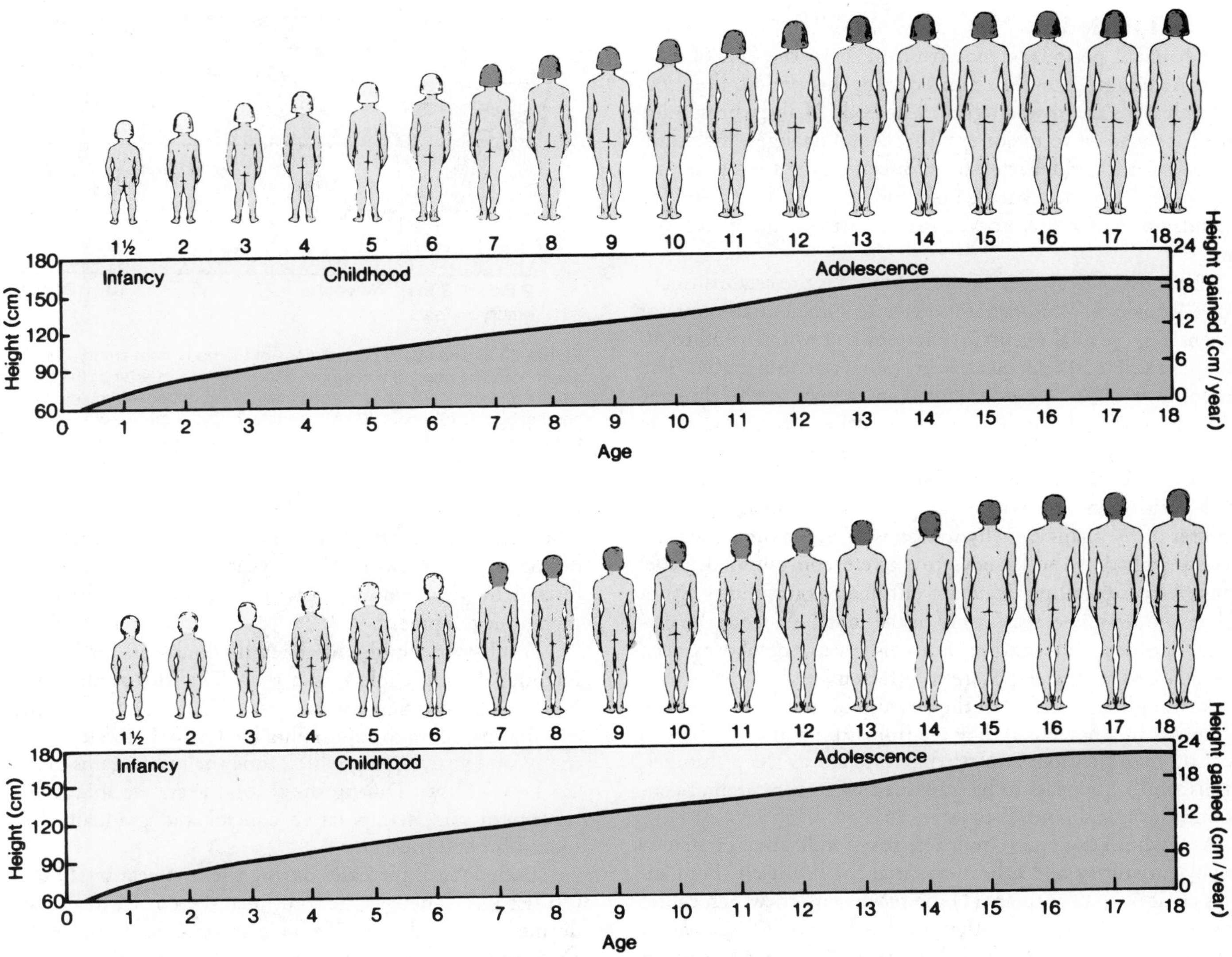

Figure 23.1 A chronological sequence of heights, growth rates, and physical changes from infancy through adolescence. Each chart (females above and males below) displays a growth curve of total average heights. The sharp peaks in the growth-rate curves are the adolescent growth spurts. (Note that the adolescent growth spurt in males follows that of females by about two years.) (From "Growing Up," by J. M. Tanner. Copyright © 1973 by Scientific American, Inc. All rights reserved.)

Normal growth depends not only on proper nutrition but on the concerted effect of several hormones, including insulin, growth hormone, thyroid hormone, and (during adolescence) the androgens. The hypothalamus, through the action of hormone-releasing factors, seems to play the governing role in growth, including the initiation of puberty and the adolescent growth spurt. Both hyper- and hypoactivity of the hypothalamus may have profound effects on the growth process.

1. List examples of cells that divide continuously, those that divide occasionally, and those that never divide in an adult person.
2. Diagram a cell, and construct a model that depicts the factors that may influence mitosis. Which factor is intrinsic to the cell, and which factors are extrinsic?
3. List the three aspects of the growth process that occur at the cellular level.

Stages of Postnatal Growth

The course of human life after birth is seen in terms of physical and physiological changes and the attainment of maturity in the neonatal, infant, childhood, adolescent, and adulthood periods.

Objective 3. Describe the growth and developmental events of the neonatal, infant, childhood, and adolescent periods.

Objective 4. Define *puberty,* and discuss what determines its onset in males and females.

Objective 5. Define the term *adulthood,* and discuss sexual dimorphism in adult humans.

Neonatal Period

The **neonatal** period extends from birth to the end of four weeks. Although growth is rapid during this period, the most drastic changes are physiological. The body of a newborn must immediately adapt to major environmental changes, including thermal stress; rapid bacterial colonization of the skin, oral cavity, and GI tract; a barrage of sensory stimuli; and sudden demands on cardiorespiratory, gastrointestinal, and renal functions.

The most critical need of the newborn is the establishment of an adequate respiratory rate to ensure sufficient amounts of oxygen. The normal respiratory rate of a newborn is 30 to 40 times per minute. An adequate heart rate is also imperative. The heart of a newborn seems enlarged in respect to the thoracic cavity (compared to the heart of an adult) and has a rapid rate that ranges from 120 to 160 beats per minute.

Most full-term newborn babies appear chubby because of the deposition of fat within adipose tissue during the last trimester of pregnancy. Dehydration is a serious threat because of the inability of the kidneys to excrete concentrated urine; large volumes of dilute urine are eliminated. Immunity is not well developed and consists only of that obtained from the placental transfer of the mother. For this reason, newborns need to be guarded against exposure to infected persons.

Although virtually all of the neurons of the nervous system are present in a newborn, they are immature and the newborn has little coordination. Most behavior such as sleep, hunger, and discomfort appears to be governed by lower cerebral centers and the spinal cord.

A newborn has many reflexes, some indicative of neuromuscular maturity and others essential for life itself. Four reflexes critical to survival are (1) the *suckling reflex,* which causes a newborn to suck anything that touches the lips; (2) the *rooting reflex,* which helps a baby find a nipple by causing it to turn its head and start suckling whenever something brushes its cheek; (3) the *crying reflex,* when its stomach is empty or it is experiencing other discomforts; and (4) the *breathing reflex,* which is apparent in a normal newborn even before the umbilical cord, with its supply of oxygen, is cut.

Babies born more than three weeks before the due date are generally considered *premature,* but because errors are commonly made in calculating the conception date, prematurity is defined by neonatal body weight rather than due date. Newborns weighing less than 2,500 grams (5.5 lbs) are considered premature. By this definition, approximately 8% of newborns in the United States are premature.

Postmature babies are those born two or more weeks after the due date. They frequently weigh less than they would have if they had been born at term because the placenta often becomes less efficient after a full-term pregnancy. Approximately 10% of newborns in the United States are postmature.

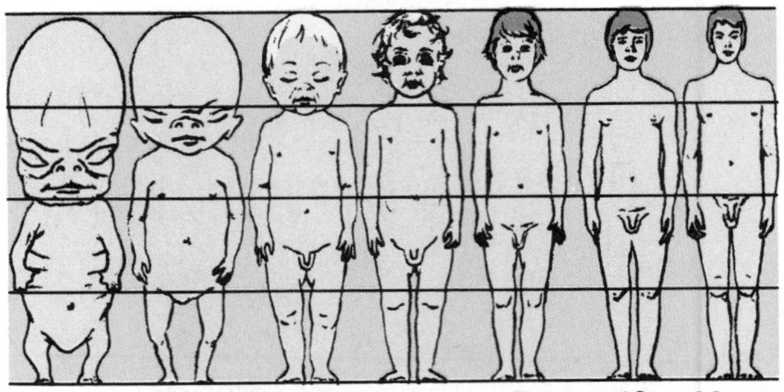

Figure 23.2 The relative proportions of the body from embryo to adulthood. The head of a newborn accounts for a quarter of the total body length, and the lower appendages make up about one-third. In an adult, the head accounts for about 13% of the total body length, whereas the length of the lower appendages constitutes approximately one-half.

Infancy

The period of **infancy** extends from the end of the neonatal period at four weeks until two years of age. Infancy is characterized by tremendous growth, increased coordination, and mental development.

A full-term child will generally double its birth weight by five months and triple it in a year. The formation of subcutaneous adipose tissue reaches its peak at about nine months, causing the infant to appear chubby. Growth decelerates during the second year, during which time the infant gains only about 2.5 kg (5–6 lbs). During the second year, the infant develops locomotor and manipulative control and gradually becomes more lean and muscular.

Body length increases during the first year by 25 to 30 cm (10–12 in.). There is an additional 12 cm (5 in.) of growth during the second year. The brain and circumference of the head also grow rapidly during the first year and only moderately during the second. Head circumference increases approximately 12 cm (5 in.) during the first year and only an additional 2 cm during the second. The anterior fontanel gradually diminishes in size after six months and becomes effectively closed at any time from nine to eighteen months. It is the last of the fontanels to close. The brain is two-thirds of its adult size at the end of the first year and four-fifths of its adult size by the end of the second year.

By two years, most infants weigh approximately four times their birth weight and average between 81 and 91 cm (32–36 in.) in length. The body proportions of a two-year-old are certainly not the same as an adult (fig. 23.2). Growth is a differential process, resulting in gradual changes in body proportions.

Deciduous teeth begin to erupt in most infants between five and nine months. By one year of age, most infants have six to eight teeth. Eight more teeth erupt during the second year, making a total of fourteen to sixteen, including the first deciduous molars and canine teeth.

neonatal: Gk. *neos,* new; *natus,* born
premature: L. *prae,* before; *maturus,* ripe

infancy: L. *in,* not; *fans,* speaking

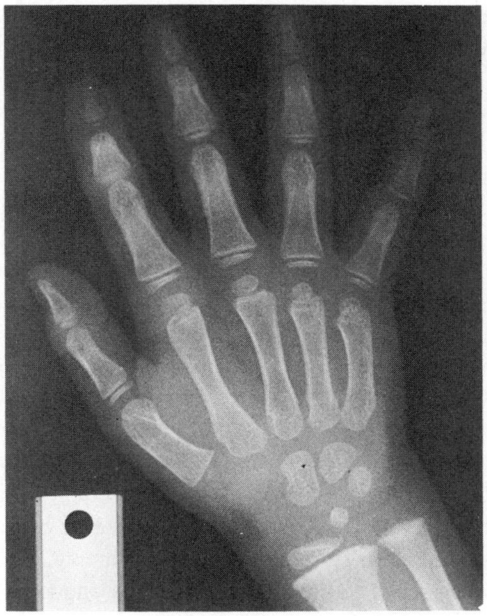

(a)

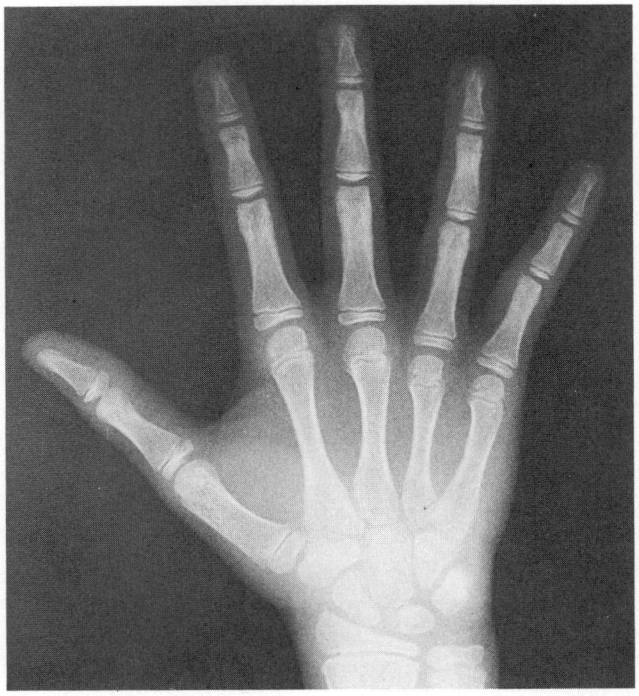

(b)

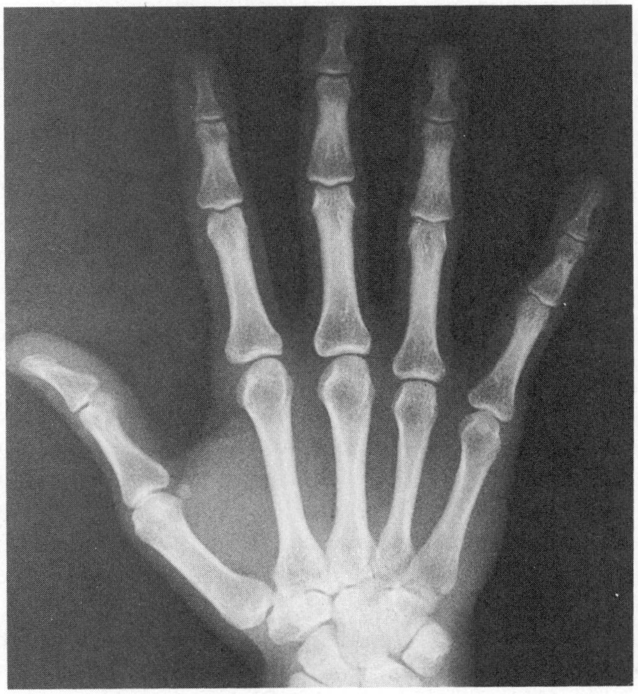

(c)

Figure 23.3 X rays of the right hand of (*a*) a child, (*b*) an adolescent, and (*c*) an adult.

The growth rates of children vary tremendously. Body lengths and weights are not always reliable indicators of normal growth and development. A more objective evaluation of a child's physical developmental progress is determined through x-ray analysis of skeletal ossification in the carpal region (fig. 23.3).

Childhood

Childhood is the period of growth and development extending from infancy to adolescence, at which time puberty begins. The chronological duration of childhood varies because puberty begins at different ages for different people.

Childhood years are a period of relatively steady growth until preadolescence, when there is a growth spurt. The average weight gain during childhood is about 3 to 3.5 kg (7 lb) per year. There is an average increase in height of 6 cm (2.5 in.) per year. The circumference of the head increases by only about 3 to 4 cm (1.5 in.) during childhood, and by adolescence the head and brain are virtually adult size.

The facial bones continue to develop during childhood (fig. 23.4). Especially significant is the enlargement of the sinuses. The first permanent teeth generally erupt during the seventh year, and then the deciduous teeth are shed approximately in the same sequence as they were acquired. Deciduous teeth are replaced at a rate of about four per year over the next seven years.

Although there is an average rate of growth during childhood due to genetics, there is a tremendous range in what is considered normal growth. If, for example, the eight-year-olds who are within the tallest and heaviest 10% of their age group were to stop growing for a year while their classmates grew normally, they would still be taller than half of their contemporaries and heavier than three-quarters of them.

During childhood, the average child becomes thinner and stronger each year as he or she grows taller. The average ten-year-old, for example, can throw a ball twice as far as the average six-year-old. Visceral organs, particularly the heart and lungs, develop tremendously during this period, enabling a child to run faster and exercise longer.

Lymphatic tissue is at its peak of development during mid-childhood and generally exceeds the amount of such tissue in the normal adult. Children need the extra lymphatic tissue to

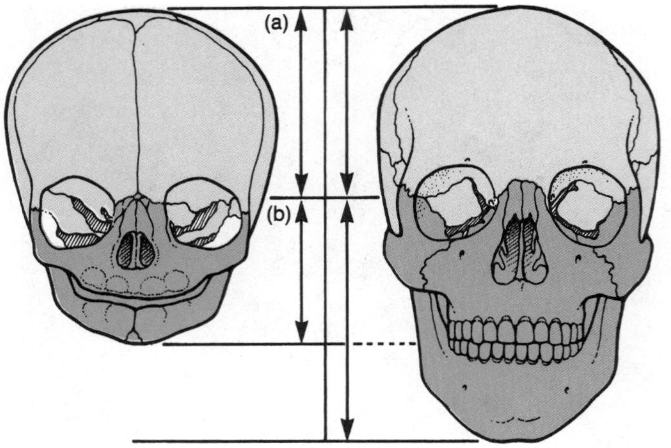

Figure 23.4 Growth of the skull. The height of the cranial vault (distance between planes a and b) is drawn the same in both the infant and adult skulls. Growth of the skull occurs almost exclusively within the bones of the facial region.

combat childhood diseases, which take a tremendous toll, particularly in countries where nutrition is poor and health care is minimal.

Childhood *obesity* can become a serious physical and psychological problem if not corrected. Overweight children usually exercise less and run a greater risk of serious illnesses. Frequently, they are teased and rejected by classmates, which causes psychological stress and learning impairments. At least 5% of children in the United States can be classified as obese. There is generally a correlation between childhood obesity and adult obesity. Obesity in adults is a major health problem, in light of the fact that one of five adults are 30% or more over their ideal, healthy weight. The obvious help for obese children and adults is a controlled diet and regular exercise.

Adolescence

Adolescence *(ad″o-les′ens)* is the period of growth and development between childhood and adulthood. It begins around the age of ten years in girls and the age of twelve years in boys. The end of adolescence is frequently said to be the age of twenty years, but it is not clearly delineated and varies with the developmental, physical, emotional, mental, or cultural criteria that define an adult.

Puberty *(pu′ber-te)* is the stage of early adolescence when the secondary sexual characteristics become expressed and the sexual organs become functional. **Pubescence** *(pu-bes′ens)* refers to the continuum of physical changes during puberty, particularly in regard to body hair.

For both sexes there is wide individual variation in the onset and duration of puberty. Although puberty is under hormonal control, a complex interaction of other factors, including nutrition and socioeconomic forces, has a decisive influence on the onset and duration of puberty. The end result of puberty is that the sexes are **sexually dimorphic** *(di-mor′fik);* that is, they are structurally different (fig. 23.5). The average adult male, for example, has a deeper voice and more body hair and is taller than the average adult female. Prior to puberty, male and female children have few major structural differences aside from the general appearance of the external genitalia.

Puberty actually begins before it is physically expressed. In most instances, significant amounts of sex hormones appear in the blood of females by the age of ten years and in males by the age of eleven years. Sexual changes are usually thought of as the only features of puberty, but major musculoskeletal changes take place as well. During late childhood, the body proportions of the sexes are similar, males being slightly taller. Under the influence of hormones, females experience a growth spurt in early adolescence that precedes that of males by nearly two years (see fig. 23.1). During this time, females are temporarily taller. Once puberty begins in males, the heights of both sexes are soon the same, and at the culmination of puberty, males average about 10 cm (4 in.) taller than females. By the time growth is completed at the end of adolescence, males average 13 cm (5 in.) taller than females.

Other dimorphic differences involving skeletal structures include a broadening of the pelvic girdle in females. The muscles of males become more massive and stronger than those of females. Females acquire a thicker subcutaneous layer of the skin during adolescence, which gives them a softer appearance.

Sexual maturation during adolescence includes not only the development of the reproductive organs but the appearance of secondary sexual characteristics. The sequence of average sexual maturation and the expression of secondary sexual characteristics for both males and females is presented in table 23.1.

In females, the first physical indication of puberty is the appearance of **breast buds,** which are swellings of the breasts and slight enlargement and pigmentation of the areolar areas. The average age for breast buds to appear in healthy girls is about eleven years, but it ranges from nine to thirteen years. Approximately three years are required after the appearance of breast buds for the maturation of the breasts. Usually pubic hair begins to appear shortly after the breast buds, but in about one-third of all girls, sparse pubic hair appears before the breast buds. Axillary hair appears a year or two after pubic hair.

The first menstrual period, referred to as **menarche** *(mĕ-nar′ke),* is generally at the age of thirteen years but may range as much as from nine to seventeen years of age. During puberty, the vaginal secretions change from alkaline to acidic.

The onset of puberty in males varies just as much as in females but generally lags behind by about one and a half years. The first indication of puberty in boys is growth of the testes and the appearance of sparse pubic hair at the age of twelve years on the average. This is followed by growth of the penis, which continues for about two years, and the appearance of axillary hair. Vocal changes generally begin during early puberty but are not completed until midpuberty. Facial hair and chest hair (which may or may not be present) come toward the end

adolescence: L. *adolescere,* to grow up
puberty: L. *pubertas,* adult form
dimorphic: Gk. *di,* two; *morphe,* form

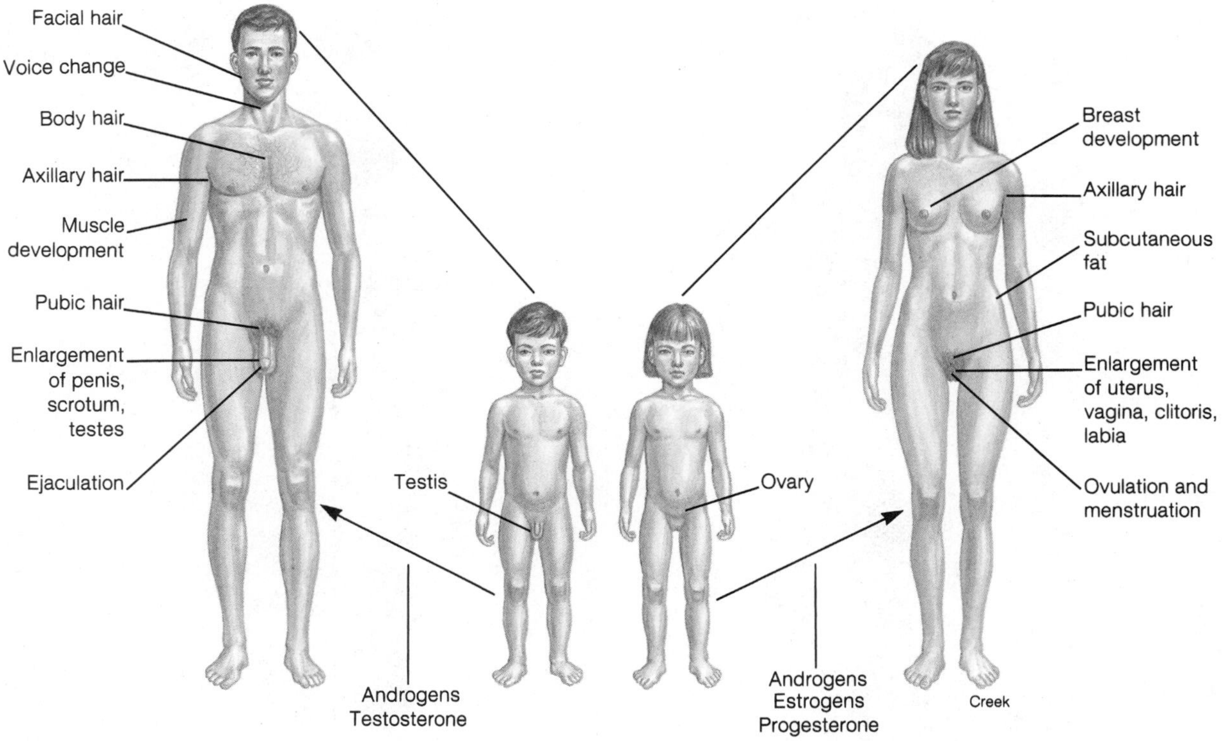

Figure 23.5 Developmental changes from childhood to adolescence. Puberty is hormonally controlled and is the stage of early adolescence when the secondary sexual characteristics are expressed and the sexual organs become functional.

Table 23.1 Sequence of adolescent physical development

Females	Age span		Males
Growth spurt begins; breast buds appear; sparse pubic hair	10–11 yrs	11.5–13 yrs	Growth of testes and scrotum; sparse pubic hair; growth spurt begins; penis growth begins
Appearance of straight, pigmented pubic hair; some deepening of voice; rapid growth of ovaries, uterus, and vagina; acidic vaginal secretion; menarche; further enlargement of breasts; kinky pubic hair; age of maximum growth	11–14 yrs	13–16 yrs	Appearance of straight, pigmented pubic hair; deepening of voice; maturation of penis, testes, scrotum, and accessory reproductive glands; ejaculation of semen; axillary hair; kinky pubic hair; sparse facial hair; age of maximum growth
Appearance of axillary hair; breasts are adult size and shape; culmination of physical growth	14–16 yrs	16–18 yrs	Increased body hair; marked vocal change; culmination of physical growth

of puberty. The mean age when semen can be ejaculated is 13.7 years, but sufficient mature sperm for fertility are generally not produced until fourteen to sixteen years of age.

Acne is an inflammatory disease of the integument, which is common during adolescence. The increase in hormonal activity that is responsible for the physical changes taking place during puberty also affects the activity of the sebaceous glands and promotes the formation of inflamed superficial pustules and comedones. Tension and emotional stress may also promote acne. Acne in teenagers may cause serious psychological problems related to a concern about and need for peer acceptance.

Adulthood

Adulthood is the final stage of human physical change. It is the period of life beyond adolescence. An adult has reached maximum physical stature as determined by genetic, nutritional, and environmental factors. Although skeletal maturity is reached in early adulthood, anatomical and physiological changes continue throughout adulthood and are part of the aging process.

Sexual dimorphism exists in adults apart from the obvious primary and secondary sex differences. The sexes differ anatomically, physiologically, metabolically, and behaviorally (psychologically or socially). Some of these differences manifest themselves prenatally and during childhood. Others are

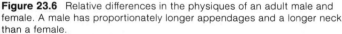

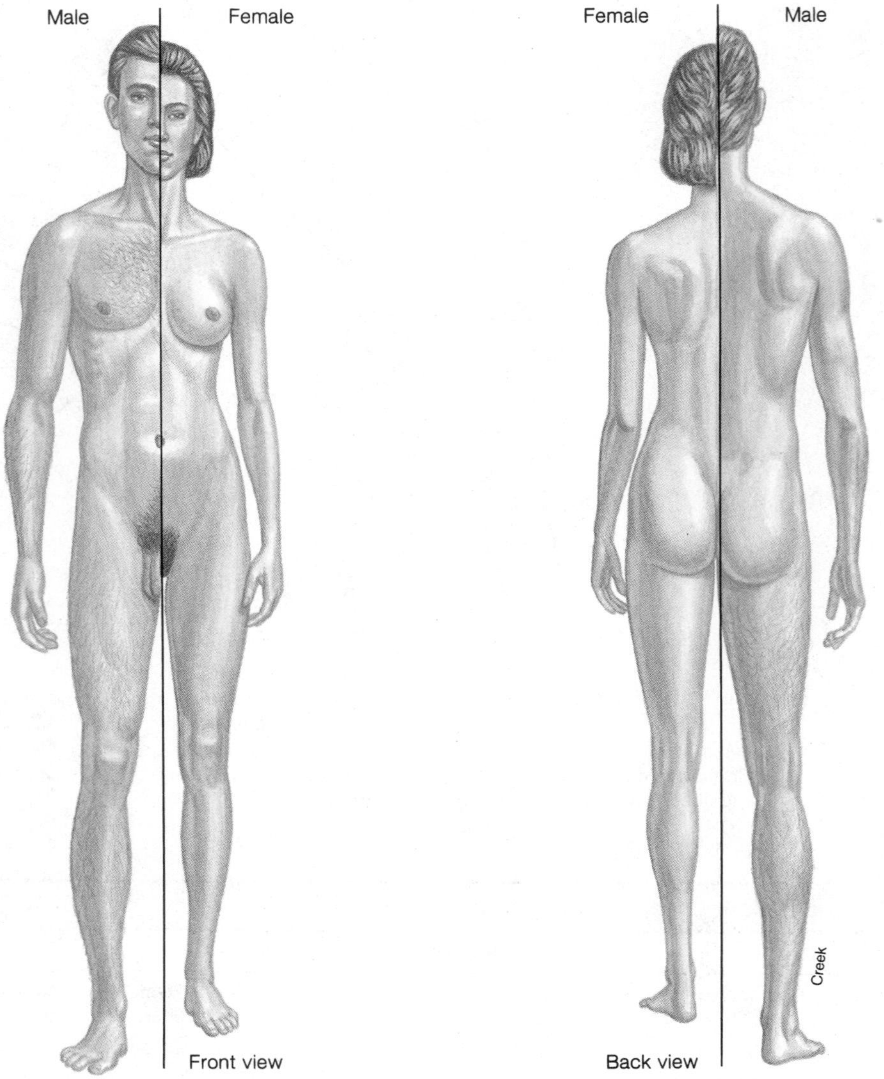

Figure 23.6 Relative differences in the physiques of an adult male and female. A male has proportionately longer appendages and a longer neck than a female.

characteristics of adolescence and adulthood. It is uncertain to what extent specific dimorphisms of the sexes are genetically determined through hormonal action or influenced by environmental (including cultural) factors. It is also unclear how these governing factors are expressed in observed physical characteristics.

The shape of the adult body is determined primarily by the skeleton and attached muscles as well as the subcutaneous connective tissue (especially adipose tissue) and extracellular body fluids. Although there is considerable variation in body proportions between adult males and females (fig. 23.6), in general, adult males have relatively longer appendages than females, their shoulders are relatively broader, and their pelvises are narrower. A male also has a relatively longer neck.

The general body composition of males and females can also be compared numerically. Mean data of body composition are summarized in table 23.2. These data show that total body fluid and skeletal weight is lower in adult females than in adult males. Females, however, have a greater percentage of body fat.

Other differences not shown in table 23.2 are that adult females have lower blood pressures, erythrocyte counts (hematocrits), basal metabolic rates, and respiratory rates than adult males. Females, however, have higher heart rates and oral temperatures.

Some of the physical and behavioral characteristics of each stage of human growth and development are summarized in table 23.3.

Physical anthropologists have for a long time been interested in the body proportions of racial groups. *Anthropometry* is the study of the physical differences, particularly skeletal, among racial groups. Proportions in anthropometric studies are expressed in indices; that is, one measurement reckoned as a per-

---

anthropometry: Gk. *anthropos,* human; *metron,* measure

Table 23.2 Body composition in average adult males and females at age 25 and 65

	Males				Females			
	Absolute		Relative (% body weight)		Absolute		Relative (% body weight)	
	Age 25	Age 65	Age 25	Age 65	Age 25	Age 65	Age 25	Age 65
Body weight	70.0 kg	70.0 kg			69.0 kg	60.0 kg		
Body fluids	41.0 liters	37.0 liters	58.9%	52.9%	30.8 liters	28.0 liters	51.3%	46.7%
Intracellular	24.0 liters	19.2 liters	34.3%	27.4%	16.6 liters	14.3 liters	27.7%	23.8%
Extracellular	17.2 liters	17.8 liters	24.6%	25.5%	14.2 liters	13.7 liters	23.6%	22.9%
Plasma volume	3,302 ml	2,940 ml	4.7%	4.2%	2,760 ml	2,462 ml	4.6%	4.1%
Lean body weight	56.3 kg	50.5 kg	80.4%	72.1%	42.0 kg	38.2 kg	70.2%	63.7%
Body fat	13.7 kg	19.5 kg	19.6%	27.9%	17.9 kg	21.8 kg	29.8%	36.3%
Skeletal weight	5.8 kg	5.7 kg	8.3%	8.1%	4.4 kg	4.2 kg	7.3%	7.0%

From K. H. Oleson, "Body Composition in Normal Adults" in *Human Body Composition*, Vol. 7:177–190. Copyright © 1965 Pergamon Press Ltd., Oxford, England. Reprinted by permission.

Table 23.3 Summary of stages of postnatal periods

Stage	Time period	Some physical and behavioral characteristics
Neonatal period	Birth to end of fourth week	Stabilizing of body systems necessary to carry on respiration, obtain nutrients, digest nutrients, excrete wastes, regulate body temperature, and circulate blood
Infancy	End of fourth week to two years	Tremendous growth; teeth begin to erupt; muscular and nervous systems develop so that motor activities are possible; verbal communication begins
Childhood	One year to puberty	Consistent growth; deciduous teeth erupt and are replaced by permanent teeth; good motor control; urinary bladder and bowel controls are established; intellect greatly improved
Adolescence	Puberty to adulthood	Reproductive system matures; growth spurts in skeletal and muscular systems; intellect and emotional maturity increase
Adulthood	Adolescence to old age	Maximum physical stature and strength obtained; anatomical and physiological degenerative changes begin
Senescence	Old age to death	Senescence continues; body becomes less able to cope with diseases and physical demands; death—usually from physical disturbances in the cardiovascular system or disease processes in vital organs

centage of another measurement. The cranial index, for example, is the breadth of the skull expressed as a percentage of its length.

There are some racial differences in the physical stature of both males and females. A particularly standardized expression of racial differences are the indices between the appendage lengths and sitting height. Negroes, for example, have comparatively long appendages to sitting height; moreover, the forearm and leg are long relative to the brachium and thigh (fig. 23.7). Australian Aborigines have even longer legs proportionately than do Negroes. Amongst all racial groups, the Negro has the narrowest pelvic girdle for a given shoulder width. Mongoloids possess relatively short appendage lengths to sitting heights. These differences provide distinct advantages and disadvantages in certain sports. Negroes have an advantage in many track events, particularly the sprints and high hurdles. Caucasians are generally adapted to distance running. Mongoloids prove apt in gymnastics and weight lifting.

1. Construct a table that lists the stages from infancy through adolescence of postnatal growth, and indicate the events or characteristics of each.
2. List four reflexes in a newborn that are critical for survival.
3. Define the terms *puberty, pubescence, sexual dimorphism,* and *menarche.* What is the mean age of puberty, and what causes its occurrence?
4. Describe the physical characteristics of adulthood. Compare the body structure of an adult female to an adult male.

Aging and Senescence

Every living organism experiences chronological and biological aging. Senescence accompanies biological aging as each body system experiences a decrease in viability and an increase in susceptibility to injury and disease.

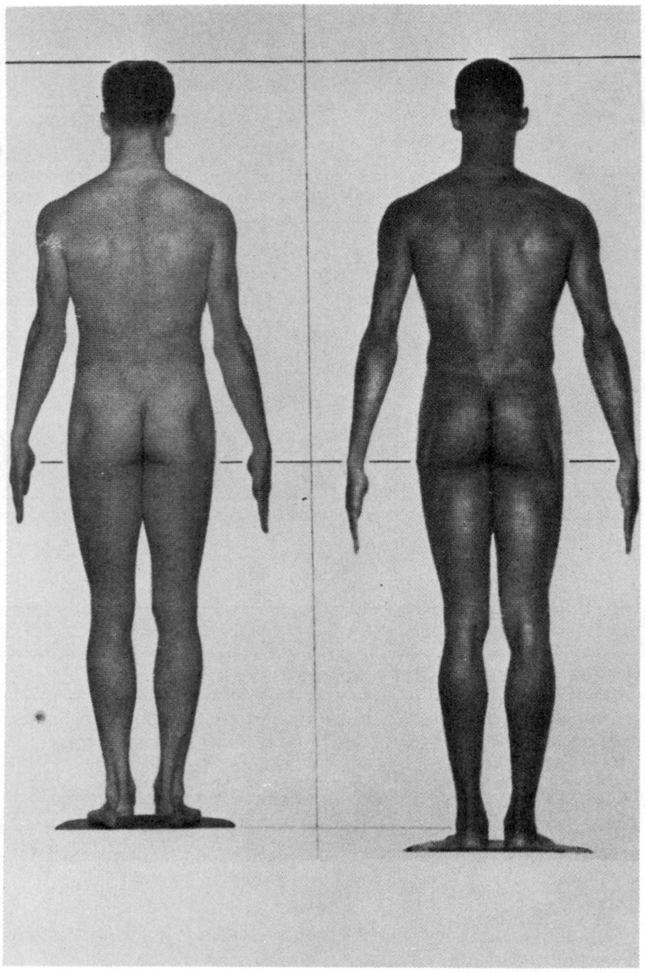

Figure 23.7 Comparison of Caucasian and Negro physiques. Photographs are of two Olympic 400 meter runners and have been scaled so that both have the same sitting height.

Objective 6. Define the terms *aging* and *senescence.*
Objective 7. Discuss the senescent changes that occur in each body system.

Aging

Aging is a term that has a broad definition. *Chronological aging* is an expression of time that can apply to prenatal or postnatal life. The age of an embryo is given chronologically in days or weeks from the date of conception, whereas the age of the fetus is expressed in weeks or months. The age during the first year is given in weeks and months from the date of birth and thereafter is indicated in years. *Biological aging,* which begins with conception and ends at death, is a process characterized by changes in the structure and function of organ systems. These changes are generally predictable and, at least during certain age periods, not always detrimental. The changes in biological aging that take place during adolescence, for example, permit humans to be at peak performance as they enter early adulthood. Aging during midadulthood, on the other hand, results in a gradual reduction of capabilities.

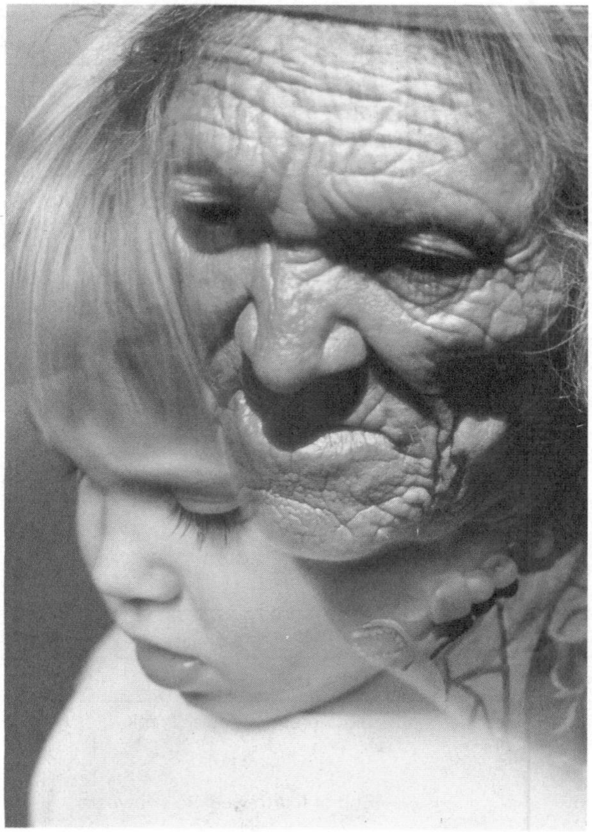

Figure 23.8 Senescence of the skin results in a loss of elasticity and the appearance of wrinkles.

Senescence of Body Systems

Senescence *(se-nes'ens)* refers to biological aging, characterized by a gradual deterioration of body structure and function. As senescence progresses through adulthood and old age, the viability of the body decreases and vulnerability increases. Body organs become more susceptible to injury and disease.

The following paragraphs of this section briefly discuss senescence of the body systems that have been discussed in the previous chapters of the text.

Integumentary System With senescence, the skin becomes thin, dry, and inelastic. Collagenous fibers in the dermis become thicker and less elastic, and the amount of adipose tissue in the hypodermis diminishes, making it thinner. Skinfold measurements indicate that the diminution of the hypodermis begins with regularity at the age of forty-five years. With a loss of elasticity and a reduction in the thickness of the hypodermis, wrinkling, or the permanent infolding of the skin, becomes apparent (fig. 23.8).

During senescence of the integument, the number of hair follicles, sweat glands, and sebaceous glands also diminishes, as well as their activity. Consequently, there is a marked thinning of scalp hair and hair on the extremities, reduced sweating, and decreased sebum production. Since elderly people cannot perspire as freely, they are more likely to complain of heat and are

senescence: L. *senis,* old

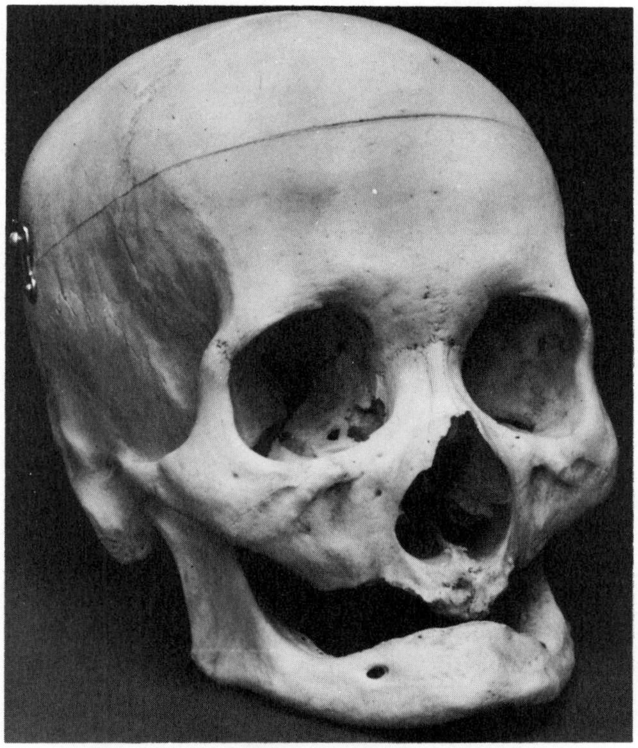

Figure 23.9 The geriatric skull. Note the loss of teeth and the degeneration of bone, particularly in the facial region.

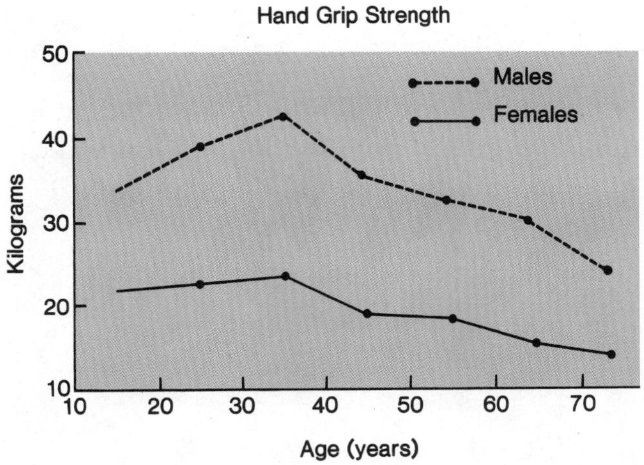

Figure 23.10 A gradual diminishing of muscle strength occurs after the age of thirty-five, as shown with a graph of hand-grip strength.

more subject to heat exhaustion. They also become more sensitive to cold because of the loss of insulating adipose and diminished circulation. A decrease in the production of sebum causes the skin to dry and crack frequently.

The integument is not as well protected from the sun because of thinning, and melanocytes that produce the brown pigment melanin gradually atrophy. The loss of melanocytes accounts for graying of the hair and pallor of the skin. After the age of fifty, brown, plaquelike growths, called *seborrheic (seb″o-re′ik) hyperkeratoses,* appear within the skin, particularly on exposed portions. Skin that has been exposed to excessive sunlight tends to develop more cutaneous carcinomas than less exposed skin. Caucasians are more susceptible to skin cancer than more darkly pigmented races.

Whereas a general loss of hair is characteristic of aging, females who have gone through menopause may develop more facial hair, particularly around the lips and on the chin. Changes in the androgen/estrogen ratio are responsible for the growth of this coarse, darkly pigmented hair. It can be removed through a procedure referred to as *depilation.* For reasons that are not understood, elderly males frequently grow coarse hair in the openings of their ears and nares, and their eyebrows grow more bushy.

Skeletal System Senescence affects the skeletal system by decreasing skeletal mass and density and increasing porosity and erosion (fig. 23.9). Bones become more brittle and susceptible to fracture. Articulating surfaces also deteriorate, contributing to arthritic conditions. Arthritic diseases are second to heart disease as the most common debilitation in elderly persons. *Osteoporosis (os″te-o-po-ro′sis)* is the most prevalent metabolic disorder of bone and develops in the aged. It is characterized by a decrease in skeletal mass and density. Postmenopausal women are most susceptible to this condition and because of it frequently sustain fractures of the hip, vertebra, or wrist. Osteoporosis is attributed to immobilization, decreased estrogen levels, high steroid levels, and environmental factors.

Distinct losses in height occur during middle and old age. Between the ages of fifty and fifty-five years, there is a decrease of 0.5–2 cm (0.25–0.75 in.) because of compression and shrinkage of the intervertebral discs. Elderly persons may suffer a further major loss of height because of osteoporosis.

Muscular System Although the aged experience a general decrease in the strength, endurance, and agility of skeletal muscle (fig. 23.10), the extent of senescent changes varies considerably among individuals. Apparently the muscular system is one of the body systems in which a person may actively slow senescent changes. A decrease in muscle mass, in part, is due to changes in connective and circulatory tissues. Atrophy of the muscles of the appendages causes the arms and legs to appear thin and bony. Degenerative changes in the nervous system decrease the effectiveness of motor activity. Muscle reflexes become less efficient, causing a marked reduction in physical capabilities.

Diminished muscular capabilities may affect the functioning of other body systems. A decrease in the strength of the respiratory muscles may limit the ability of the lungs to ventilate. Reduced muscularity of the urinary bladder causes difficult micturition and may cause urinary infections.

Exercise is important at all stages of life but is especially beneficial as one approaches old age (fig. 23.11). Exercise not only strengthens bones and muscles, but it also contributes to a healthy circulatory system and thus ensures an adequate blood supply to all body tissues. If an elderly person does not maintain muscular strength through exercise, he or she will be more prone to debility. Learning to use a cane, crutches, or a walker would be extremely difficult for such a person.

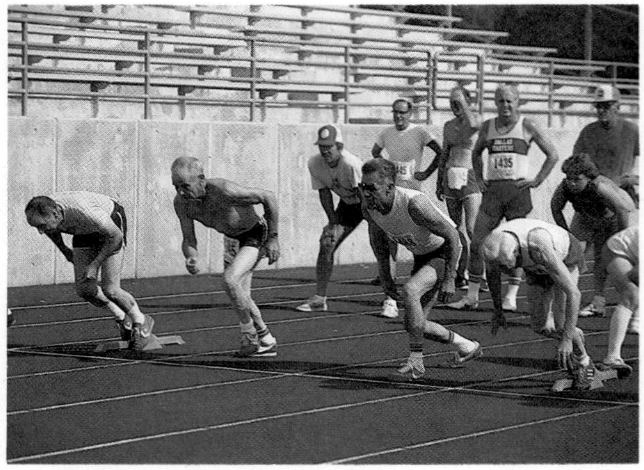

Figure 23.11 Competitive runners in late adulthood, recognizing the benefits of exercise.

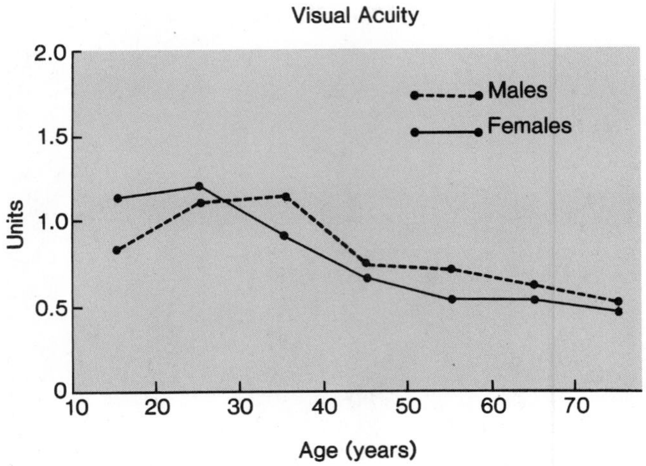

Figure 23.12 The decline in visual acuity in males and females due to senescence.

Nervous System The extent of senescent changes within the brain is not known. Previous textbooks have reported that perhaps 100,000 neurons die each day of our adult life. Other studies, however, show that such claims are unfounded. It is believed that relatively few neural cells are lost during the normal aging process. Neurons are, however, extremely sensitive and susceptible to various drugs or interruptions of vascular supply such as caused by a stroke or other cardiovascular diseases.

There is evidence that senescence alters neurotransmitters. Age-related conditions such as depression or specific diseases such as Parkinson's disease may be caused by an imbalance of neurotransmitter chemicals. Changes in sleeping patterns in aged persons also probably result from neurotransmitter problems.

The slowing of the nervous system with age is most apparent in tests of reaction time. It is not certain whether this is a result of a slowing in the transmission of impulses along neurons or in neurotransmitter relays at the synapses.

Although the nervous system does deteriorate with age (in ways that are not well understood), in most people it functions effectively throughout life. Brain dysfunction is not a common characteristic of senescence.

Sensory Organs Senescence is undoubtedly characterized by sensory impairments. An elderly person may require more intense levels of stimulation in vision, hearing, taste, and smell in order for the sensory receptors to perform with precise acuity. Such declines in sensory capabilities, especially in vision and hearing, are important not only because they affect a person's ability to function, but because they may socially isolate the person and cause serious psychological problems.

Many conditions of the eye are related to age, and some cause visual impairments. Most elderly persons develop lipid infiltrates of the cornea in a condition called *arcus senilis.* Visual acuity remains fairly constant until the ages of twenty-five to thirty in females and thirty-five to forty in males and then gradually declines (fig. 23.12). By the age of sixty-five, 40% of males and 60% of females have vision poorer than 20/70. The pupil of an elderly person's eye cannot dilate fully, and the amount of light reaching the retina may be only 50% of the amount that reaches the retina of a youth. In addition, the depth of the anterior chamber diminishes, creating poor drainage of aqueous humor, which results in increased intraocular pressure and a greater likelihood of glaucoma. Ptosis, or drooping, of the lids and diminished lacrimal secretion, contributing to dryness and irritation of the eyes, is typical in elderly persons.

The lens gradually loses its elasticity during senescence, and its ability to accommodate for viewing near objects diminishes. This change, called *presbyopia (pres'' be-o'pe-ah),* makes it necessary to hold reading material farther from the eye for the print to be focused on the retina. A person with presbyopia may require bifocals to restore visual abilities.

Cataract is the major cause of visual disability in elderly persons. In this condition, the transparent lens becomes opaque (cataractal), restricting the passage of light waves.

Retinal blindness is caused by a slow degeneration of the macula of the retina. Focusing for reading is gradually impaired, whereas perimeter retinal vision remains unaffected.

Actually, the number of legally blind persons in the United States is relatively low (about 225 per 100,000 population), but 60% of blind people are older than seventy years. Most cases of senescent blindness are preventable with early detection and treatment.

Senescence of hearing is a relatively frequent occurrence. About 5% of the population have hearing impairments by the age of fifty; after the age of sixty-five, 30% have hearing difficulties. Senescent hearing loss is referred to as *presbycusis (pres''be-ku-sis)* and is generally a loss of the ability to perceive high-frequency sounds. Changes that accompany presbycusis are so gradual that many elderly people fail to recognize them until the disability is extreme.

The discernment of taste and smell declines with age. About 30% of people over the age of eighty have difficulty identifying common substances by smell. The ability to recognize common foods by taste likewise diminishes with age. The loss in the sensory perception of taste and smell may perhaps be responsible for the clinically noted frequency of complaints about food among elderly people.

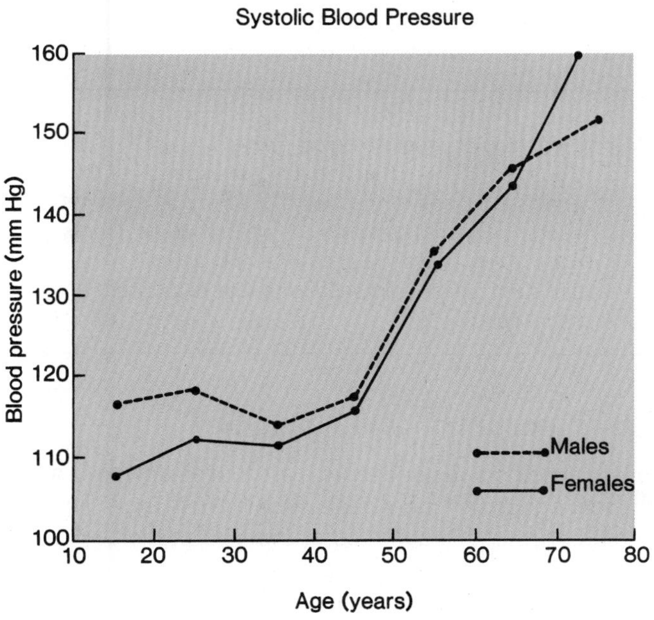

Figure 23.13 There is a marked increase in systolic blood pressure after the age of thirty-five as the heart and blood vessels undergo senescent changes.

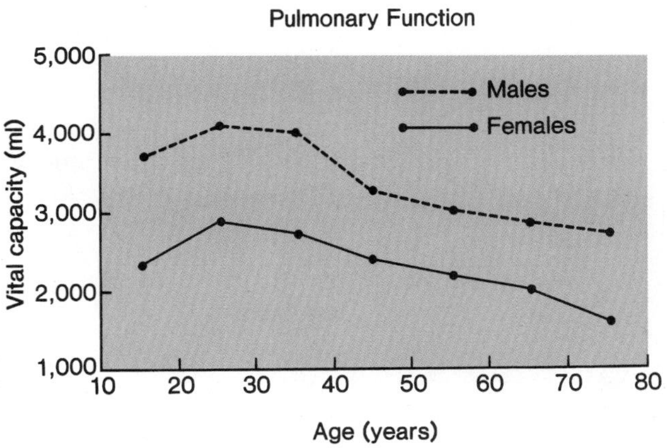

Figure 23.14 Senescence of the lungs and muscles of respiration reduces vital capacity. The graph represents data accumulated from healthy nonsmokers.

Endocrine System Although the organs of the endocrine system do undergo senescence, the consequences on overall body function are not well understood. The alteration in responsiveness to hormones is likely a major factor in the deterioration of biochemical activity in aging. The ability to secrete insulin and glucocorticoids is impaired in aged persons, and there seem to be changes in hormone transport, action, and feedback systems.

The organs of the endocrine system are susceptible to disorders of the circulatory system and to the diseases of other body organs that have feedback mechanisms with specific endocrine glands.

Circulatory System Cardiovascular senescence is the leading cause of death in the aged. As the heart and blood vessels age, a person develops or becomes more susceptible to life-threatening diseases. The pulse and electrocardiogram give some insight into general circulatory changes in aging. Changes in blood vessels include an increased stiffness of the arterial walls (*arteriosclerosis*) and the deposition of fatty, plaquelike material (*atherosclerosis*) along the lumina of arteries.

Falling is a major fear for the elderly. Arteriosclerosis of the carotid and vertebral arteries resulting in ishemia (dimished blood flow) to the brain may cause lightheadedness and dizziness and an increased risk of falling. Individuals who have had a stroke may have some loss of motor control, which creates difficulty in postural changes and walking. This too puts some elderly persons at risk for falling. Because of porous and weakened bones, the chance for sustaining a serious injury from a fall is high in elderly people. Furthermore, for many, the subsequent hospitalization and surgery may also carry a high risk.

The senescent heart may exhibit three principal characteristics: (1) structural changes in the heart and valves; (2) a diminished blood flow through the heart; and (3) an altered conduction system. The heart generally shrinks with age and accumulates adipose tissue along the coronary vessels. The endocardium increases 25% in thickness from the age of thirty to the age of eighty. After the age of thirty, the valves gradually become more rigid, thickened, and distorted because of sclerosis. The heart in an aged person reacts poorly to stress and cannot utilize oxygen as well. The maximum blood flow through the coronary arteries at the age of sixty is about 35% lower than in a person at the age of thirty. Blood pressure in elderly people may vary from 120 to 160 systolic (fig. 23.13) and from 70 to 90 diastolic. Electrocardiograms indicate slower and more erratic depolarization within the conduction system of the senescent heart.

Respiratory System Senescence significantly changes the structure of the lung and affects the capacity for ventilation. After the age of twenty-five, there is a gradual loss of elasticity within the alveoli and in the pulmonary capillaries, so that by the age of seventy vital capacity has decreased by about 40% (fig. 23.14). Elderly persons thus require increased muscular exertion to maintain an adequate level of ventilation. Vital lung capacity decreases at a much faster rate if the person is a cigarette smoker.

In addition to a loss of elasticity, pulmonary aging causes a change in gaseous exchange across pulmonary membranes. By the age of sixty-five, there is about a 48% decrease in diffusion capacity from that of the age of thirty.

Digestive System It is estimated that 60% of persons over fifty years of age have senescent problems of the digestive system that are generally more of an irritation than an impairment of health. Common problems include constipation, difficulties in swallowing (dysphagia), increased amounts of intestinal gas, heartburn, and the formation of diverticula along the large intestine. In addition, persons over the age of thirty-five are more susceptible to periodontal disease than those under the age of thirty-five.

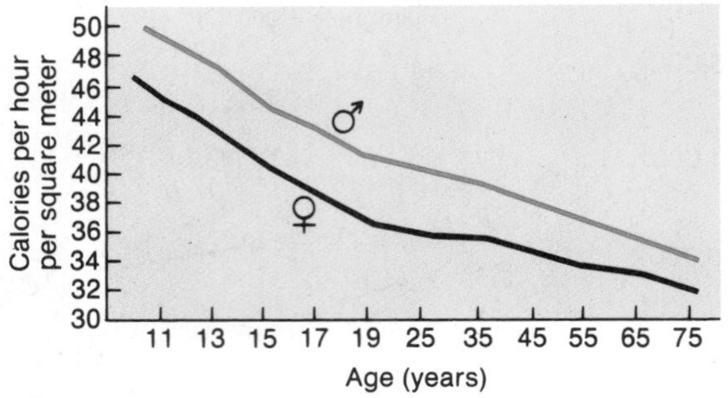

Figure 23.15 The basal metabolism rate.

Although GI problems such as those listed above are normal in elderly persons, they may be symptomatic of more serious conditions or diseases such as hypertension, ulcers, or cancer.

The *basal metabolism rate (BMR)* is the minimum amount of energy a person utilizes in a state of rest and is an index of caloric usage and requirements. The BMR rates are usually higher for males than for females (fig. 23.15) and decline proportionally with age for both sexes. Young adults with a high BMR can eat almost anything and not get fat, whereas middle-aged people with a lower BMR must constantly control food intake to avoid obesity.

Urinary System A progressive decline in kidney function accompanies senescence, due to general atrophy of the kidney and sclerosis of glomeruli. At the age of seventy the rate of glomerular filtration is only about 50% of the rate at the age of thirty. Renal blood flow diminishes from approximately 1,100 ml per minute at the age of thirty to only about 475 ml per minute at the age of eighty. The kidneys are able to regulate acid-base balance during senescence, although they respond more slowly to a sudden, large acid load.

Reproductive System The senescent changes in the reproductive system in both males and females are generally more important psychologically than pathologically. A decrease in sex hormones in elderly persons may decrease sexual desires or abilities. There is a decreased sensitivity of the genitalia, which may also affect sexual responsiveness.

A number of physical changes occur in aging males. The testes atrophy and the seminiferous tubules thicken and decrease in diameter. The prostate gland hypertrophies, which may cause pathological conditions. Although healthy males maintain the ability to ejaculate throughout old age, there is a reduction in the force of ejaculate, the volume, and the general quality of the seminal fluid.

Females experience major physiological changes at menopause, which generally occurs near the age of fifty. Hormonal changes bring about a cessation of menstruation and ovulation. Physical changes following menopause include atrophy of the uterus (fig. 23.16), with a reduction in the size of the cervix and a thinning of the vaginal wall and walls of the uterine tubes. The ovaries become irregular in shape. The vulva develops less pronounced folds as the skin becomes thinner. Vascularity and elasticity decrease, causing the vulva to become more susceptible to tissue trauma and pruritus (itching).

The principal senescent changes that occur within each body system are summarized in table 23.4.

---

1. Using the term *senescence,* distinguish between chronological and biological aging.
2. Construct a table that lists the senescence of the principal organs of the body. Which body systems seem to age most rapidly? In which body systems does aging present a greater threat to life?
3. List three sensory impairments that result from senescence.

Life Expectancy and Life Span

Biological aging and senescence are just as universal as chronological aging, but individuals, within limits, age biologically at different rates. Regardless of the rate of biological aging, each person has a life expectancy, and the individuals within the human population have a life span.

Objective 8. Define the terms *life expectancy, life span,* and *gerontology.*

Life expectancy is best defined as the average number of years lived by persons born in a particular location and period of time. A person born today in the United States has a greater life expectancy than a person born a century ago or a person born today in a Third World country. Life expectancy is relative to age. For example, life expectancy at birth is greater than it is at adolescence.

Life span is the maximum length of life possible in a particular culture. The average life span in the United States has increased twenty-six years in this century, from forty-seven to seventy-three years. Owing to cultural and environmental factors, the current average life span of males is about six years less (seventy years) than females (seventy-six years). The potential human life span is between 120 and 150 years if unaffected by accidents or diseases.

Gerontology is the study of senescence. This science is concerned primarily with the physical and psychological changes that occur between middle and old age until death. Gerontological research is concentrated in two principal areas: (1) studying the process of aging to determine if and how rates of senescence can be altered; and (2) examining and improving the quality of old age. Because of the increasing percentage of elderly people in our population, gerontology is becoming a more popular and vital science.

---

1. Discuss what determines life expectancy and what determines life span. Can either life expectancy or life span be increased? If so, how?
2. Define *gerontology.* List the kinds of classes probably taken by a person studying to be a gereontologist.

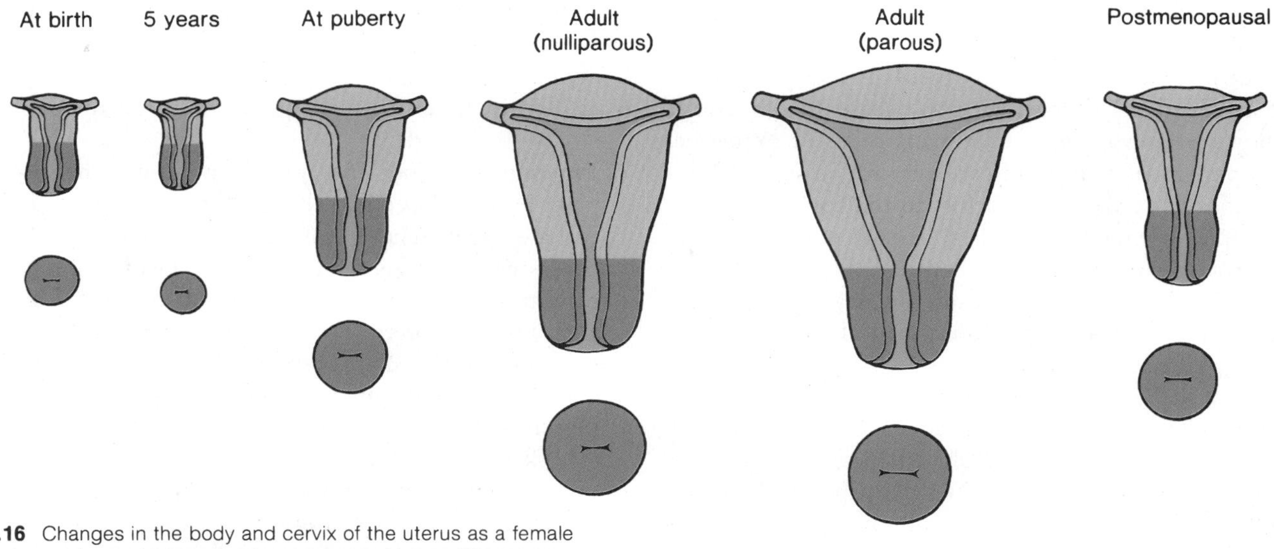

Figure 23.16 Changes in the body and cervix of the uterus as a female ages. The uterus of a postmenopausal woman atrophies and becomes fibrous.

Table 23.4 Summary of aging in body systems

Organ system	Principal senescent changes
Integumentary system	Degenerative change in collagenous and elastic fibers in dermis; decreased production of pigment in skin and hair follicles; reduced activity of sweat and sebaceous glands Skin tends to become thinner, wrinkled, and dry with pigment spots; hair becomes gray and then white
Skeletal system	Degenerative loss of bone matrix; deteriorating articulations Bones become thinner, more brittle, and more likely to fracture; stature may shorten due to compression of intervertebral discs; susceptibility to joint diseases increases
Muscular system	Loss of skeletal muscle mass; degenerative changes in neuromuscular junctions Loss of muscular strength and motor response
Nervous system	Degenerative changes in neurons; loss of dendrites and synapses; decrease in sensory sensitivity Decreased efficiency in processing and recalling information; decreased ability to communicate; diminished senses of smell, taste, sight, hearing, and touch
Endocrine system	Slightly reduced hormonal secretions Decreased responsiveness to hormones; decreased metabolic rate; reduced ability to maintain homeostasis
Circulatory system	Degenerative changes in cardiac muscle; decreased diameters of lumina of arteries and arterioles; decreased efficiency of immune system Decreased cardiac output; increased resistance to blood flow; increased blood pressure; increased incidence of autoimmune diseases
Respiratory system	Degenerative loss of elastic fibers in lungs; reduced number of functional alveoli Reduced vital capacity; increased dead air space; reduced ability to clear airways by coughing
Digestive system	Decreased motility in GI tract; reduced secretion of digestive enzymes; increased occurrence of periodontal disease Reduced efficiency of digestion
Urinary system	Degenerative changes in kidneys; reduced number of functional nephrons Reduced filtration rate, tubular secretion, and reabsorption
Reproductive system	
Male	Reduced secretion of sex hormones; reduced production of spermatozoa; enlargement of prostate gland Decreased sexual capabilities
Female	Degenerative changes in ovaries; decreased secretion of sex hormones Menopause; regressed secondary sex organs and characteristics

Theories of Aging

The many theories of aging are evidence that it is a complex, multifaceted process, which is not well understood. Evidence suggests that there is some genetic control of the aging process.

Objective 9. Discuss the five principal theories of aging.

Current literature in gerontology presents nearly two hundred theories of the mechanism of aging. These theories can be synthesized, however, into several principal types. Most of these theories continue to be scrutinized scientifically and are subject to continued modification.

Genetic Mutation Theory The foundation for the mutation theory is that cells with unusual or different characteristics are more frequently noted as a person ages. It is thought that senescence is related to the accumulation of mutational damage within the genetic mechanism of somatic cells. The processes that DNA undergoes during mitosis and protein synthesis increase the chance of damage or defect to the copying mechanism. As DNA becomes a modified or inaccurate blueprint, defective proteins are synthesized. Some of these proteins are essential enzymes whose loss will cause the death of a cell. Senescence occurs as cells mutate. When a critical number of cells are altered or killed, the organism dies.

Autoimmunity Theory The autoimmunity *(au''to-im-mu'nĭ-te)* theory suggests that senescence results when various body systems begin to reject their own tissues. As a person ages, the production of autoantibodies in the body increases. The antibodies are produced in response to somatic mutations or post-DNA errors. This causes organs to behave self-destructively, and the immune mechanism operates against its own body cells as if they were foreign bodies. Autoimmune reactions are known to occur in various cardiovascular diseases (e.g., giant-cell arteritis and amyloidosis), diabetes, rheumatoid arthritis, and certain types of cancer. It is also suggested that immunological surveillance becomes impaired as a person ages, so that cells that might give rise to neoplasms are no longer suppressed.

Cross-Linking of Molecules Theory Homogeneity is a characteristic of some organic molecules. That is, different molecules tend to attract each other and become uniform. Chemical substances adhere to, or cross-link with, adjacent protein molecules so that each loses its functional identity. During senescence, tendons, skin, and even blood vessels lose elasticity. As the functional characteristics of collagenous and elastic fibers change, the skin wrinkles, muscles sag, and wounds in the skin heal more slowly.

Cellular Aging Theory The cellular aging theory suggests that there is a genetically determined limit to the doubling potential of each normal mitotically active cell. This means that cells have an intrinsic finite life span. As cells approach their final divisions, the time interval for mitosis progressively increases, causing a gradual cessation of activity and an accumulation of cellular debris. Eventually there is a total loss of vigor, and cellular death occurs. Degenerative phenomena are manifest after about fifty cell population doublings.

Another characteristic of aging cells is the increased abundance of *free radicals.* Free radicals are parts of altered molecules or molecules that have had an electron stripped from their structure. Free radicals frequently alter other molecules by attaching to them and disrupting their function. Environmental factors, such as pollutants, accelerate free radical activity. In the aging process, the number of free radical compounds are thought to increase faster than body cells can repair the damage.

Programmed Aging Theory The preceding theories fail to explain why the specific life span for each species (even species that are closely related biologically) is so consistently fixed. Another unexplained observation is that the offspring of long-lived parents tend to live longer than the offspring of short-lived parents. Furthermore, the various cells constituting an organism have different fixed life spans. Epidermal cells, for example, live only a relatively short time before they are replaced, whereas muscle and nerve cells endure the lifetime of the body.

Clearly, there must be some genetic control of the aging process that is not only species and individual specific but also cell specific. If there is an aging gene responsible for senescence, then someday it should be possible to identify and perhaps eventually to manipulate that gene.

Senescence is a complex process, and all of the theories are probably valid to some extent. The process almost certainly has numerous factors. A major challenge for gerontology is to explain how events that occur at the cellular level are transcribed into events at the organismic level.

---

1. Using hypothetical situations, describe how laboratory experiments could be conducted to test each of the five current theories of aging.
2. Which of the five theories of aging accounts for closely related species having markedly different life spans?

CLINICAL CASE STUDY ANSWER

It is within normal limits for a fourteen-year-old girl to not have yet experienced menarche. By age fifteen or sixteen, however, one would suspect an abnormality. For our patient to actually not have begun having periods would be inconsistent with the physical findings of mature breasts and pubic hair. Menarche usually precedes these structural changes. The doctor, upon performing a vaginal examination, will most likely discover an *imperforate hymen* (see page 650). The hymen normally exists as a ring of tissue that in some cases is nearly closed but still allows egress of menstrual flow. This would explain our patient's history of no menstruation and a tender lower abdominal mass, which is likely an accumulation of menstrual fluid within the vagina, a condition called *hematocolpos.* Hematocolpos is treated by excision of the hymen allowing menstrual discharge. ■

Chapter Summary

I. Introduction to Postnatal Growth, Development, and Aging
   A. Growth and development are necessary so that an individual organism can sustain itself and contribute to its own species population.
   B. Mitotic potential is apparently just as much of a cellular specialization as is any structural or functional feature of a cell.
   C. Growth is not a linear process, and growth rates vary for different organs and structures of the body.
   D. A growth-inhibiting substance released via a feedback mechanism may influence cellular reproductive activities to limit the number of cells or the size of specific organs.
   E. Growth is a normal process by which an organism increases in size as a result of the accretion of tissues similar to those already present.

II. Stages of Postnatal Growth
   A. The course of human life after birth is seen in terms of physical and physiological changes and the attainment of maturity in the neonatal, infant, childhood, adolescent, and adulthood periods.
   B. The neonatal period, extending from birth to the end of the fourth week, is characterized by major physiological changes.
      1. The most critical need of the newborn is to establish adequate respiratory and heart rates. A normal respiratory rate is 30 to 40 times per minute, and a normal heart rate ranges from 120 to 160 beats per minute.
      2. The four reflexes in the newborn critical to survival are the suckling reflex, the rooting reflex, the crying reflex, and the breathing reflex.
   C. Infancy, extending from four weeks until the age of two years, is characterized by tremendous growth, increased coordination, and mental development.
      1. By two years, most infants weigh about four times their birth weight and average between 32 and 36 inches in length.
      2. Growth is a differential process resulting in gradual changes from infant to adult body proportions.
   D. Childhood, extending from infancy to adolescence, is characterized by steady growth, until preadolescence when there is a marked growth spurt.
      1. During childhood, the average child becomes thinner and stronger each year as he or she grows taller.
      2. The fact that disease and death are more rare during childhood may be because lymphatic tissue is then at its peak of development and is present in greater amounts in children than in adults.
   E. Adolescence is the period of growth and development between childhood and adulthood.
      1. Puberty is the stage of early adolescence when the secondary sexual characteristics become expressed and the sexual organs become functional.
      2. The end result of puberty is the structural expression of gender, or sexual dimorphism.
      3. The onset of menarche is generally at the age of thirteen years but ranges from nine to seventeen years. At this time, vaginal secretions change from alkaline to acid.
      4. The first physical indications of puberty are the appearance of breast buds in females and the growth of the testes and the appearance of sparse pubic hair in males.
      5. Although semen may be ejaculated at the ages of thirteen or fourteen, sufficient mature sperm for fertility are not produced until fourteen to sixteen years of age.
   F. Adulthood, the final stage of human physical change, is characterized by gradual senescence as a person ages.
      1. Although skeletal maturity is reached in early adulthood, anatomical and physiological changes continue throughout adulthood and are part of the aging process.
      2. Sexual dimorphism exists anatomically, physiologically, metabolically, and behaviorally between males and females.
      3. Differences in body statures, proportions, and compositions between males and females may become more expressed with age.

III. Aging and Senescence
   A. Aging can be expressed as chronological or as biological events. Senescence refers to the biological aging process, characterized by a gradual deterioration of body structure and function.
   B. Senescence affects each body system in a predictable but somewhat individual sequence.
      1. In the integumentary system, the skin thins, becomes dry, is less elastic, and is more fragile. There is an increased sensitivity to temperature extremes and an increased occurrence of cutaneous carcinomas in less pigmented races.
      2. In the skeletal system, the skeletal mass decreases in density and increases in porosity and erosion, predisposing humans to osteoporosis and arthritic conditions. A decrease in height during middle and old age results from the compression or shrinkage of intervertebral discs.
      3. In the muscular system, there is a general decrease in strength, endurance, and agility that varies considerably among individuals. Senescent changes may be slowed through regular exercise.
      4. In the nervous system, there is evidence that senescence causes an alteration in

neurotransmitters, which may result in depression or conditions such as Parkinson's disease. The nervous system slows with age but functions effectively in most people throughout life.

5. Sense organs are impaired and become less acute in response to stimuli with aging. Vision may be decreased by corneal lipid infiltrates (arcus senilis), glaucoma, ptosis of the eyelids, presbyopia, cataracts, or retinal blindness. Presbycusis refers to the loss of high-frequency sound perception. The discernment of taste and smell also decline with age.
6. In the endocrine system, the consequences of senescence on overall body functions are not well understood. Organs of this system are susceptible to circulatory disorders and to the diseases of other body organs through feedback mechanisms.
7. Cardiovascular senescence is the leading cause of death in the aged. Blood vessels may develop arteriosclerosis or atherosclerosis, and the heart may develop structural changes in the valves, a decreased blood flow, and an altered conduction system.
8. In the respiratory system, the structure of the lung changes significantly with senescence, decreasing its capacity for ventilation. Elasticity decreases as does the effectiveness of gaseous diffusion in lung tissue.
9. In the digestive system, senescence irritates rather than impairs health in most cases. Common problems include constipation, dysphagia, increased flatulence, heartburn, diverticuli formation along the colon, and an increased susceptibility to periodontal disease. These problems may be symptomatic of more serious diseases such as hypertension, ulcers, or cancer.
10. In the urinary system, kidney function progressively declines. The kidneys also respond slower in regulating acid-base balance.
11. In the reproductive system, senescent changes are more important psychologically in regard to maintaining sexuality than they are pathological. Although able to ejaculate, males experience a decrease in the force, volume, and general quality of the ejaculate. Females experience major physiological changes at menopause.

IV. Life Expectancy and Life Span
  A. Biological aging and senescence are just as universal as chronological aging, but individuals, within limits, age biologically at different rates.
  B. Life expectancy is the average number of years lived by persons born during a particular time period in a certain location.
  C. Life span is the maximum length of life attainable in a culture.

V. Theories of Aging
  A. The many theories of aging are evidence that it is a complex, multifaceted process, which is not well understood.
  B. There are five principal theories of aging that are probably not mutually exclusive.
    1. Genetic mutation theory—with age, somatic cells with mutational damage in the genetic mechanism accumulate.
    2. Autoimmunity theory—senescence results when various body systems reject their own tissue.
    3. Cross-linking of molecules theory—chemical substances adhere to, or cross-link with, adjacent protein molecules so that each loses its functional identity.
    4. Cellular aging theory—cells have an intrinsic, finite life span due to the genetically determined, limited doubling potential characteristic of each normal mitotically active cell.
    5. Programmed aging theory—evidence indicates that the aging process is genetically species specific, individual specific, and cell specific.

Review Activities

Objective Questions

1. Which of the following is the period of growth from birth to the end of the fourth week? The
   (a) neonatal. (c) infant.
   (b) fetal. (d) suckling.
2. The most drastic changes in the first four weeks of life are
   (a) anatomical.
   (b) environmental.
   (c) psychological.
   (d) physiological.
3. The normal newborn heart rate is
   (a) 70–80 beats/min.
   (b) 120–160 beats/min.
   (c) 100–120 beats/min.
   (d) 180–200 beats/min.
4. The continuum of physical change occurring in adolescence that regulates the growth of body hair is known as
   (a) puberty.
   (b) pubal progression.
   (c) pubescence.
   (d) dimorphism.
5. At birth, a baby is regarded as premature if it weighs less than
   (a) 2,500 grams (5.5 lbs).
   (b) 1,350 grams (3.0 lbs).
   (c) 1,050 grams (2.5 lbs).
   (d) 3,200 grams (7.0 lbs).
6. Which is *not* characteristic of the female when compared to the male?
   (a) lower blood pressure
   (b) higher basal metabolic rate
   (c) lower red blood cell count
   (d) faster heart rate
7. The first physical indication of puberty in females is generally
   (a) alkaline vaginal secretions.
   (b) a widening pelvis.
   (c) breast buds.
   (d) axillary hair.
8. Which statement is *not* true of a senescent heart of an elderly person?
   (a) The heart enlarges to pump additional blood.
   (b) The endocardium increases in thickness.
   (c) Adipose tissue accumulates along the coronary vessels.
   (d) Slower and more erratic depolarization exists within the conduction system.
9. Which statement is *not* true of life expectancy?
   (a) It can be changed with improved diet and health care.
   (b) It has an inverse ratio to chronological age.
   (c) It is the maximum length of life possible in a particular culture.
   (d) It varies from country to country.

10. The trend that the offspring of long-lived parents are often long-lived themselves supports which theory of aging? The
    (a) cross-linked.
    (b) autoimmunity.
    (c) programmed aging.
    (d) genetic mutation.

Essay Questions

1. Explain the biological importance of growth, development, aging, and death.
2. Give examples of cells within the body that divide continuously throughout a person's life, those that divide occasionally, and those that never divide after a person is born.
3. Discuss the growth and developmental events characteristic of the neonatal period.
4. Define the term *infancy,* and discuss the growth and developmental events that are characteristic of this stage of life.
5. Define the term *childhood,* and discuss the growth and developmental events that are characteristic of this stage of life.
6. Define the term *adolescence,* and discuss the role that puberty has in this stage of life for both males and females.
7. Distinguish between puberty and pubescence.
8. Define the term *adulthood,* and describe the characteristics of this stage of life.
9. List the sexual dimorphic structural features of adult males and females.
10. Define *anthropometry* and give examples of anthropometric characteristics.
11. For each body system, list two senescent changes.
12. Distinguish between chronological aging, biological aging, and senescence.
13. In which body systems are senescent changes life-threatening?
14. Distinguish between life expectancy and life span. What factors determine each?
15. Discuss gerontology as a contemporary science.

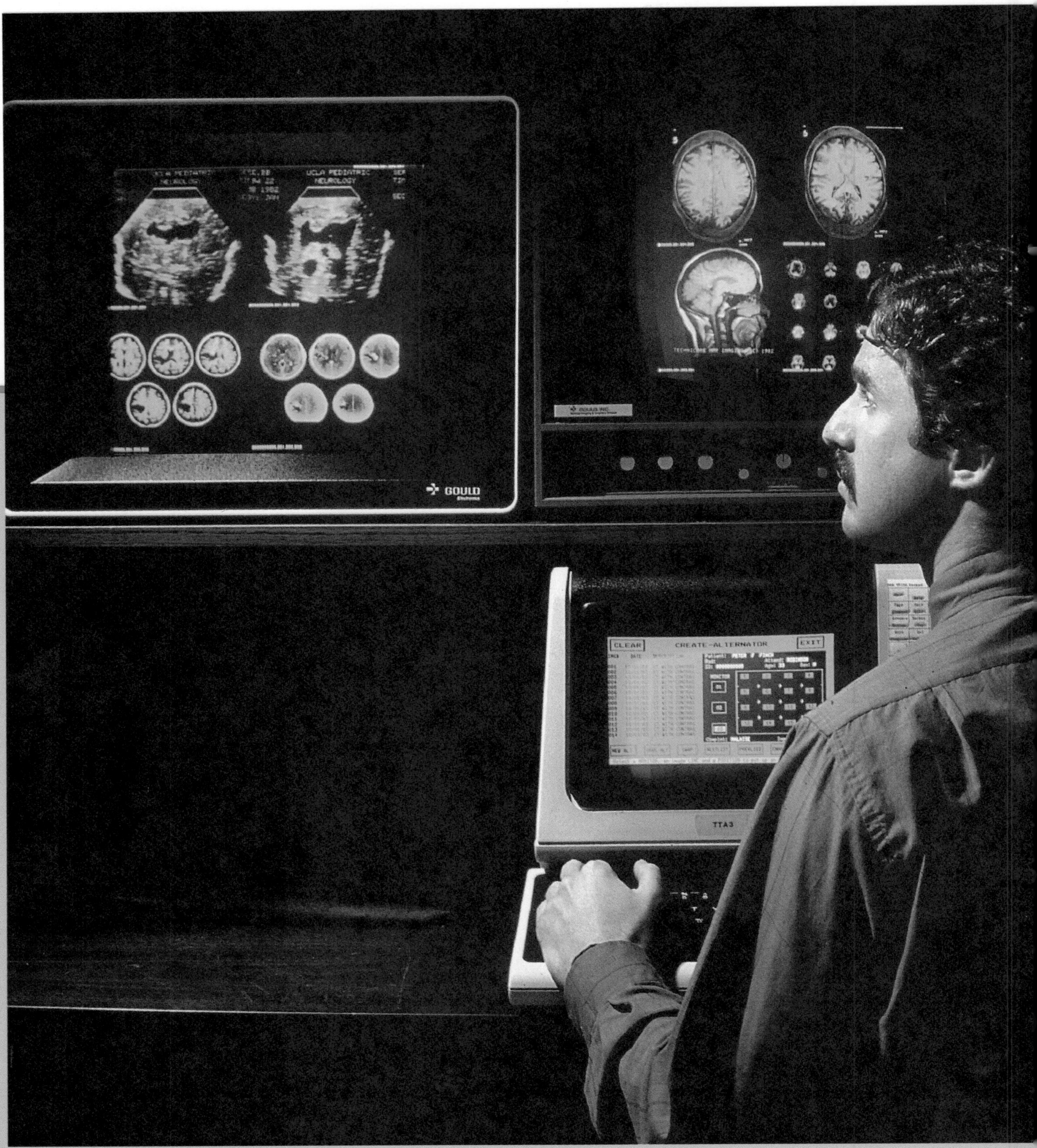

New medical equipment and radiographic techniques have greatly enhanced the observation of body structures and the diagnosing of diseases. Ultrasound scans, magnetic resonance imaging (MRI), and computerized axial tomographic scans (CT scans) as shown in this photograph, are noninvasive techniques for making detailed observations of specific body regions or organs in an array of visual sections and perspectives.

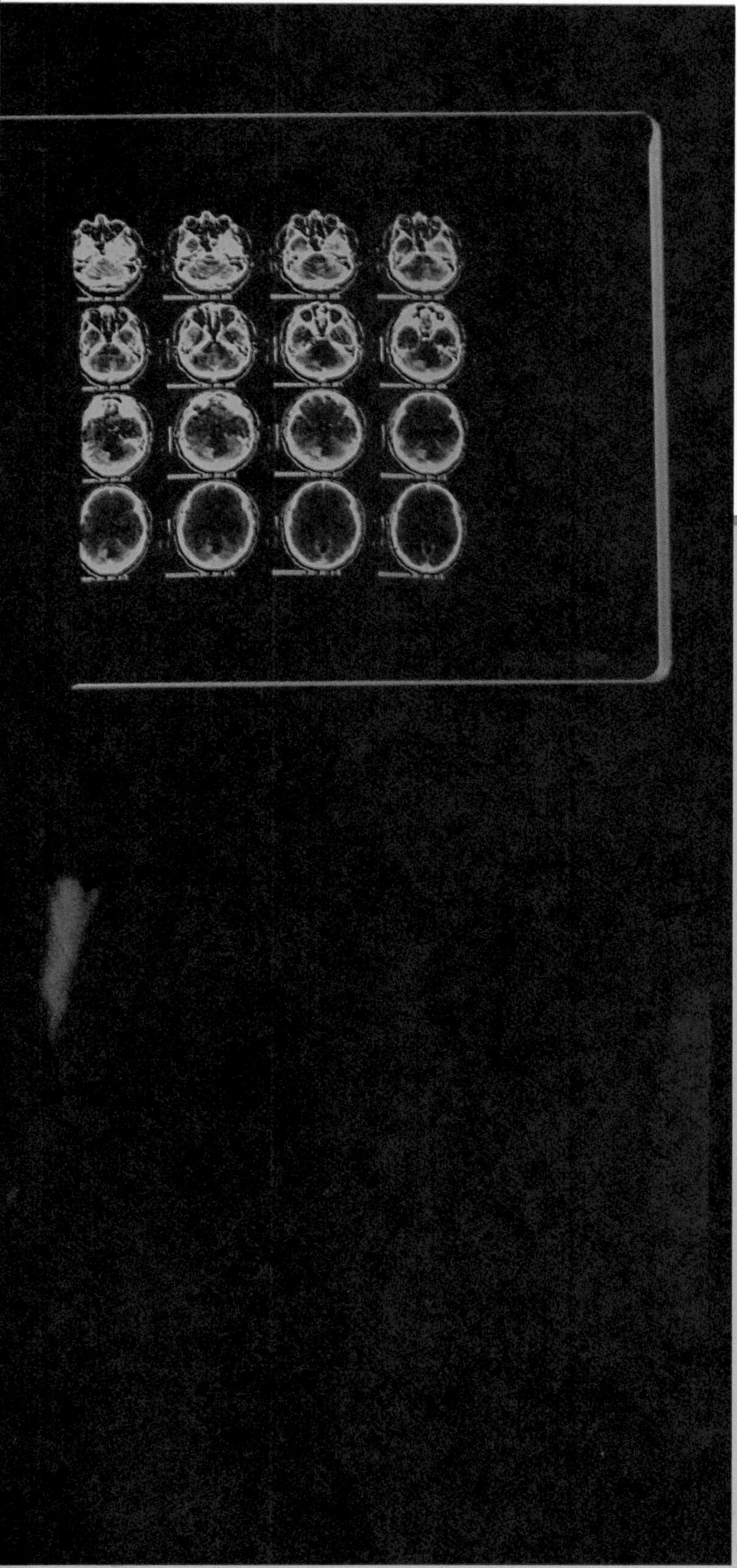

UNIT

8

Regional Anatomy

Through photographs of dissected human cadavers, the chapter in unit 8 depicts the integration of the body systems in the principal body regions. The text portion of this chapter provides a synopsis of the developmental, traumatic, and disease conditions that commonly afflict the organ systems within each of the body regions.

This unit includes:

CHAPTER 24

Clinical Synopsis of Principal Body Regions

Outline and Concepts

Head and Neck Regions

The head and neck are the most highly integrated regions of the body as they communicate with and control all of the systems of the body. Congenital conditions, trauma, and disease are common to these regions and vary in severity from minor discomfort to life threatening.

Thoracic Region

The thorax is that portion of the torso positioned between the neck and the diaphragm. The cavity of the thorax contains the heart, lungs, several large blood vessels, the thoracic duct, respiratory passages, the esophagus, and several autonomic nerves. Congenital conditions, trauma, and disease are common to this body region and are generally serious with the potential of being life threatening.

Abdominal Region

The abdomen is that portion of the torso positioned between the diaphragm and the pelvis. Because of the domed shape of the diaphragm, some of the abdominal viscera lie within the protection of the rib cage. The cavity of the abdomen contains the stomach and intestines, the liver and gallbladder, the spleen, the internal genitalia, and major vessels and nerves. Congenital conditions, trauma, and disease are common to this body region and are generally serious with the potential of being life threatening.

Shoulder and Upper Extremity

The shoulder and upper extremity include the pectoral girdle, the brachium, the antebrachium, and the hand. There are only a few serious congenital conditions of these regions, and there are no serious endemic diseases. Trauma to the skin, bones, muscles, vessels, and nerves, however, is common and ranges from annoying to extremely serious.

Hip and Lower Extremity

The hip and lower extremity include the pelvic girdle, the thigh, the leg, and the foot. There are only a few serious congenital conditions of these regions and only a few endemic diseases. Trauma to the skin, bones, muscles, vessels, and nerves of these regions is common and ranges from annoying to extremely serious.

Chapter Summary

Review Activities

Head and Neck Regions

The head and neck regions are the most highly integrated regions of the body as they communicate with and control all of the systems of the body. Congenital conditions, trauma, and disease are common to these regions and vary in severity from minor discomfort to life threatening.

Objective 1. Describe the common congenital conditions that may afflict the head and neck and are of clinical concern.

Objective 2. Explain why the head and neck are especially susceptible to trauma, and describe the common types of head and neck injuries.

Objective 3. List some common diseases of the head and neck, and explain how the appearance of the head is of clinical importance in the diagnosis of several diseases or body conditions.

The highly specialized head and neck regions (figs. 24.1, 24.2, 24.3, 24.4, and 24.5) are of tremendous clinical importance. These regions are extremely vulnerable to trauma and disease. Furthermore, because of the incredible complexity of these body regions, they are susceptible to numerous congenital conditions occurring during prenatal development.

The aggressive nature of humans, partially exhibited in contact sports and a fast-moving life-style, as evidenced by their use of speeding vehicles, puts the human head and neck in constant danger of injury. Pathogens readily gain access to the internal structures of the head and neck through the several openings into the head. Also, the risk for contracting certain diseases is greatly increased by the social nature of humans.

Developmental Conditions of the Head and Neck

Congenital malformations of the head and neck regions result from genetic or environmental (teratogenic) causes and are generally very serious. The less severe malformations may result in functional disability, whereas the more severe malformations usually make life impossible.

Anencephaly, *(an''en-sef'-ah-le)* a severe underdevelopment of the brain and surrounding cranial bones, is always fatal. The cause of anencephalia is unknown, but there does seem to be some hereditary and geographic influences. South Wales, for example, reports incidences of anencephalia as high as 1 in every 105 births. It occurs more frequently in females than males, and the damage to the developing embryo occurs between the sixteenth and twenty-sixth day after conception.

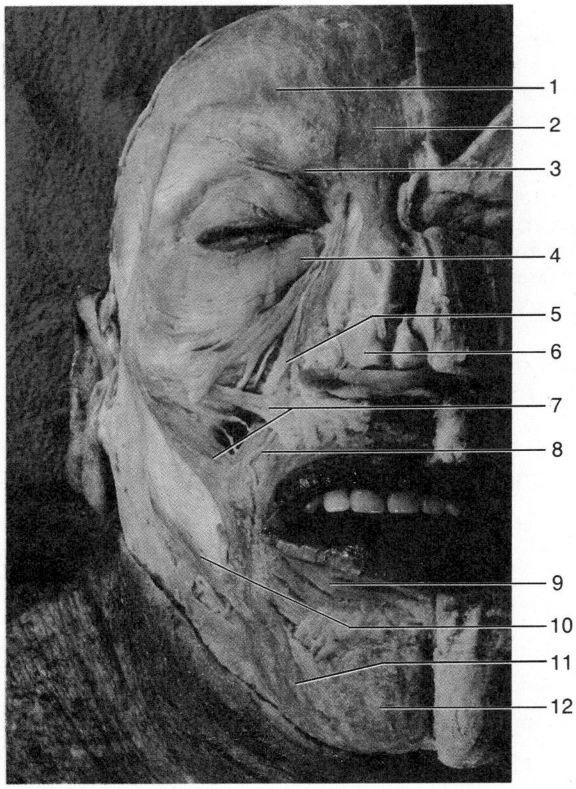

1 Frontalis m.
2 Supratrochlear a.
3 Corrugator m.
4 Orbicularis oculi m.
5 Levator labii superioris m.
6 Alar cartilage
7 Zygomaticus m.
8 Facial a.
9 Orbicularis oris m.
10 Risorius m.
11 Triangularis m.
12 Mentalis m.

Figure 24.1 Anterior view of the muscles of the head.

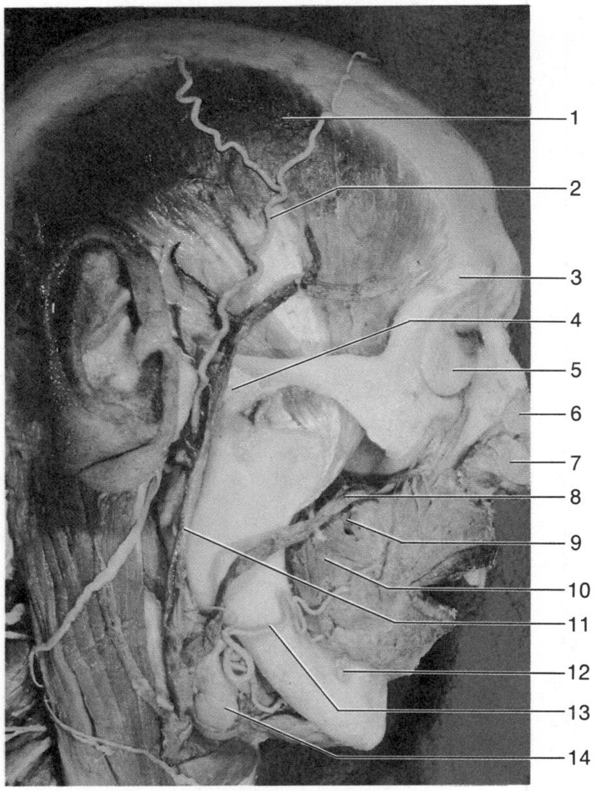

1 Temporalis m.
2 Superficial temporal a.
3 Supraorbital ridge
4 Temporomandibular joint
5 Orbital fat
6 Greater alar cartilage
7 Lateral alar cartilage
8 Facial v.
9 Parotid duct
10 Buccinator m.
11 Retromandibular v.
12 Mental a., v., n.
13 Facial a.
14 Submandibular gland

Figure 24.2 Lateral view of the deep muscles of the head.

Altered cranial bones and sutures, resulting in pressure on the brain, accompany several kinds of congenital conditions. **Microcephaly** is characterized by premature closure of the sutures of the skull. If the child is untreated, there will be underdevelopment of the brain and mental retardation. **Cranial encephalocele** is a condition in which the skull does not develop properly and portions of the brain often protrude through it. In **hydrocephalus** there is an excessive accumulation of cerebrospinal fluid dilating the ventricles of the brain and causing a separation of the cranial bones.

A **cleft palate** and **lip** is a common congenital condition of varying degrees of severity. A vertical split on one side, where the maxillary and median nasal processes fail to unite, is referred to as a *unilateral cleft.* A *bilateral,* or *double, cleft* occurs when the maxillary and median nasal processes on both sides fail to unite.

Normally by the age of thirty the sutures of the skull synostose and cranial bone growth ceases. Premature **synostosis (microcephaly)** is an early union of the cranial sutures before the brain has reached its normal size. **Scaphocephaly** is a malformation in which the sagittal suture prematurely closes. **Oxycephaly** is a condition in which the coronal suture closes prematurely. The skull will be noticeably crooked in a condition called **plagiocephaly.**

Torticollis (wryneck) is a congenital or acquired condition of the neck in which the head is drawn to one side and rotated so that the face is directed upward. It is not known what causes the congenital condition, but trauma to the neck, or certain diseases such as rheumatism, may cause its occurrence in newborns. The muscles that are involved in the spasmodic contractions are those innervated by the accessory nerve, particularly the sternocleidomastoid muscle.

Trauma to the Head and Neck

The head and neck are extremely susceptible to trauma and blows, which are frequently physically debilitating if not fatal. Striking the head from the front or back often causes **subdural hemorrhage,** resulting from the tearing of the superior cerebral veins at their points of entrance to the superior sagittal sinuses. Blows to the side of the head tend to be less severe because the falx cerebelli and tentorium cerebelli restrict the sideways displacement of the brain. With a sudden violent lateral movement of the head such as in a serious automobile accident, the serrated edge of the lesser wing of the sphenoid may severely damage the brain and frequently tear cranial nerves.

synostose: Gk. *syn,* together; *osteon,* bone
scaphocephalus: Gk. *skaphe,* boat; *kephale,* head
oxycephalus: Gk. *oxys,* pointed; *kephale,* head

plagiocephaly: Gk. *plagios,* oblique; *kephale,* head
torticollis: L. *tortus,* twisted; *collum,* neck

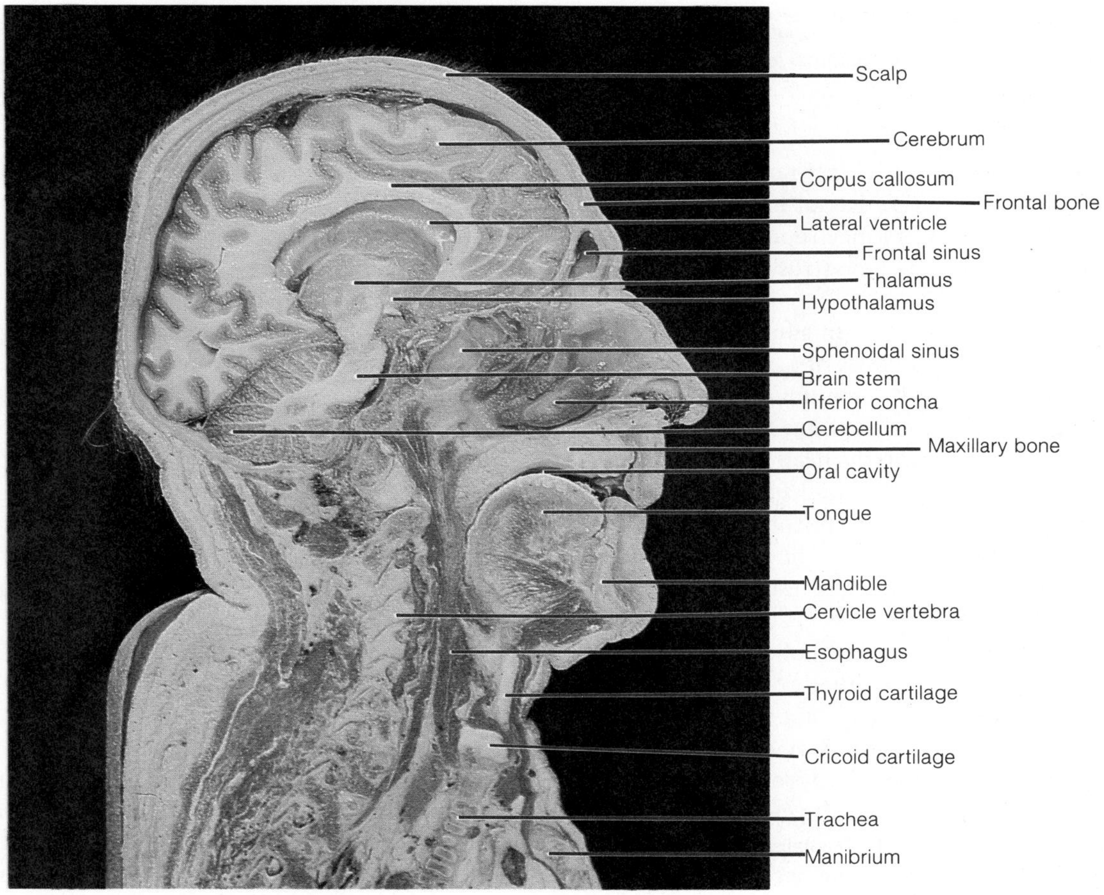

Figure 24.3 Sagittal section of the head and neck.

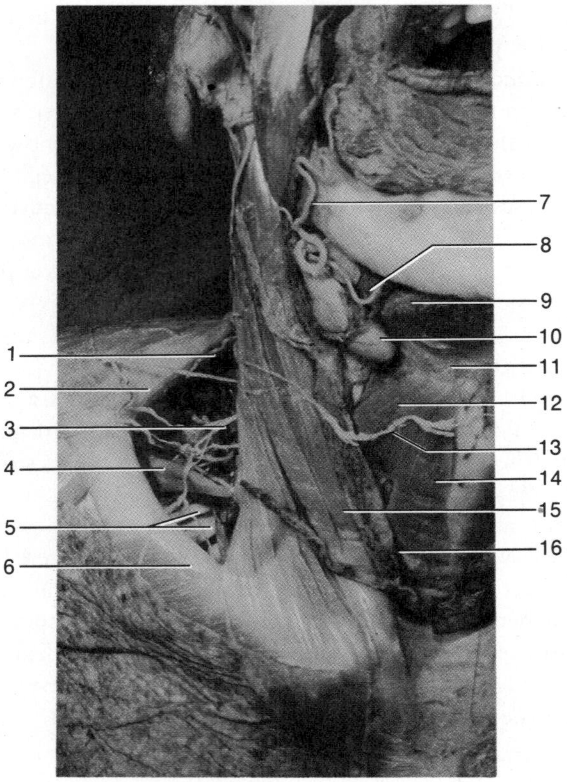

1 Accessory n.
2 Trapezius m.
3 Supraclavicular n.
4 Omohyoid m.
5 Brachial plexus
6 Clavicle
7 Facial a.
8 Mylohyoid m.
9 Digastric m.
10 Submandibular gland
11 Hyoid bone
12 Omohyoid m.
13 Transverse cervical n.
14 Sternohyoid m.
15 Sternocleidomastoid m.
16 External jugular v.

Figure 24.4 Anterior view of the right neck region.

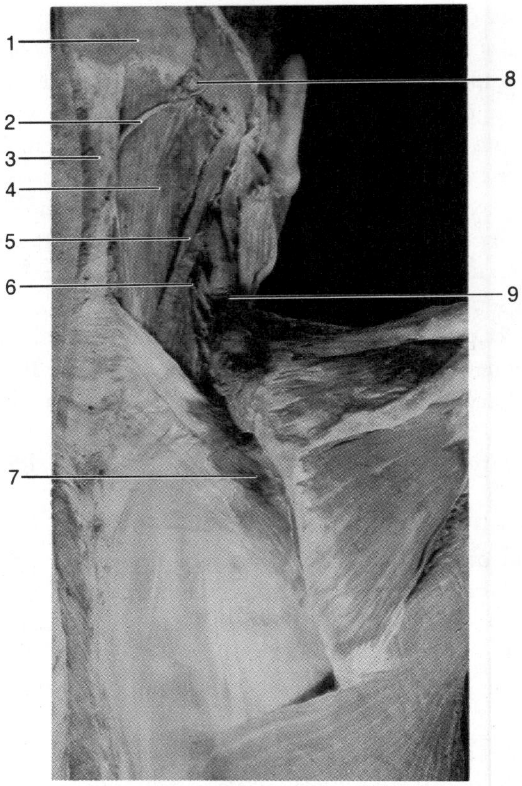

1 Occipital bone
2 Greater occipital n.
3 Ligamentum nuchae
4 Semispinalis capitis m.
5 Longissimus capitis m.
6 Longissimus cervicis m.
7 Serratus posterior m.
8 Occipital a.
9 Levator scapulae m.

Figure 24.5 Posterior view of the deep muscles of the neck.

The arteries of the head and neck are rarely damaged because of their elasticity. In a severe lateral blow to the head, however, the internal carotid artery may rupture resulting in a roaring sound being perceived as blood rushes into the cavernous sinuses of the temporal bone.

Skull fractures are fairly common in adults but much less common in children. The cranial bones of a child are resilient, and the sutures are not ossified. The cranium of an adult, however, is much like an eggshell in that it has limited resilience and tends to splinter. A hard blow to the head frequently breaks the bone on the opposite side of the skull in what is called a **contrecoup fracture.** Actually, the sphenoid, with its numerous foramina, is the weakest bone of the cranium and frequently sustains a contrecoup fracture as a result of a hard blow to the top of the head.

The two most commonly fractured bones of the face are the nasal and mandible. Trauma to either of these bones generally results in simple fracture, which is not usually serious. If the nasal septum or cribriform plate is fractured, however, careful treatment is required. If the cribriform plate is severely fractured, there may be a tear in the meninges causing a sudden loss of cerebrospinal fluid and death.

Whiplash is a common injury to the neck due to a sudden and forceful displacement of the head (see fig. 11.54). Injury to the muscle, bone, or ligaments may be involved as well as to the spinal cord and cervical nerves. A whiplash is usually extremely painful and difficult to treat because of the complexity in diagnosing the extent of the injury.

The sensory organs within the head are unfortunately very prone to trauma. The eyes and middle ears are extremely vulnerable to protruding objects. Sudden bright light and loud noise can also damage structures in each of these respective organs. A nonpenetrating blow to the eye may result in a herniation of the orbital contents through a fracture created in the floor of the orbit. The nerves that control the eye may also be damaged.

Diseases of the Head and Neck

The head and neck are extremely susceptible to infection, especially along the mucous membranes lining body openings. Sinusitis, tonsillitis, laryngitis, pharyngitis, esophagitis, and colds are common, periodically recurring ailments of the mucous-lined digestive and respiratory tracts of the head and neck.

The cutaneous area of the head most susceptible to infections extends from the upper lip to the mid-portion of the scalp. An infection of the scalp may spread via the circulatory system to the bones of the skull causing **osteomyelitis.** The infection may even spread into the sagittal venous sinus causing **venous sinus thrombosis.** A **boil** in the facial region may secondarily cause thrombosis of the facial vein or the spread of the infection to the sinuses of the skull. Before antibiotics, such sinus infections had a mortality rate of 90%.

Close observation of the head by a physician can be helpful in diagnosing several diseases and body conditions. The nose becomes greatly enlarged in a person with **acromegaly** and very wide in a person with **hypothyroidism.** The bridge of the nose is depressed in a person with **congenital syphilis.**

The color of the mucous membranes of the mouth may be important in diagnoses. Pale lips generally indicate *anemia,* yellow lips indicate *pernicious anemia,* and blue lips are characteristic of *cyanosis,* or cardiovascular problems. In *Addison's disease,* the normally pinkish mucous membranes of the cheeks have brownish areas of pigmentation.

---

1. List the common congenital conditions that involve the cranial bones and their sutures. Discuss the similarities and differences of each.
2. Describe the condition of *torticollis* and the structures involved.
3. Why are the head and neck so vulnerable to trauma? What precautions might be taken to minimize the risk of injury to these body regions?
4. Explain why certain portions of the head and neck are especially susceptible to infections.
5. Give examples of how the condition or appearance of certain portions of the head are of diagnostic value.

Thoracic Region

The thorax is that portion of the torso positioned between the neck and the diaphragm. The cavity of the thorax contains the heart, lungs, several large blood vessels, the thoracic duct, respiratory passages, the esophagus, and several autonomic nerves. Congenital conditions, trauma, and disease are common to this body region and are generally serious with the potential of being life threatening.

Objective 4. List the body systems that have organs within the thorax.

Objective 5. Describe the common congenital conditions that may afflict the thorax and are of clinical concern.

Objective 6. Explain how the heart and lungs are well protected and why trauma to the thorax is generally serious.

Objective 7. List some common diseases that afflict the organs of the thorax, and discuss how the anatomical surface landmarks are of clinical importance in diagnosing diseased organs within the thoracic cavity.

The thorax (figs. 24.6 through 24.12) includes the rib cage and its contents, the thoracic musculature, and the mammary glands and breasts of a female. The rib cage is formed by the sternum, the costal cartilages, and the ribs attached to the thoracic vertebrae. It gives protection to the lungs, several large vessels, and the heart. It also affords a site of attachment for the muscles of the thorax, upper extremities, back, and diaphragm. The prin-

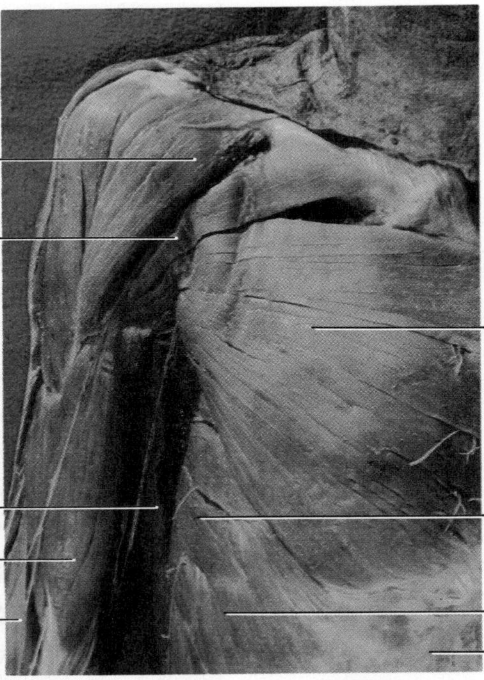

1 Deltoid m.
2 Cephalic v.
3 Latissimus dorsi m.
4 Biceps brachii m.
5 Brachioradialis m.
6 Pectoralis major m.
7 Serratus anterior m.
8 External oblique m.
9 Rectus sheath

Figure 24.6 Anterior view of the right thorax, shoulder, and brachium.

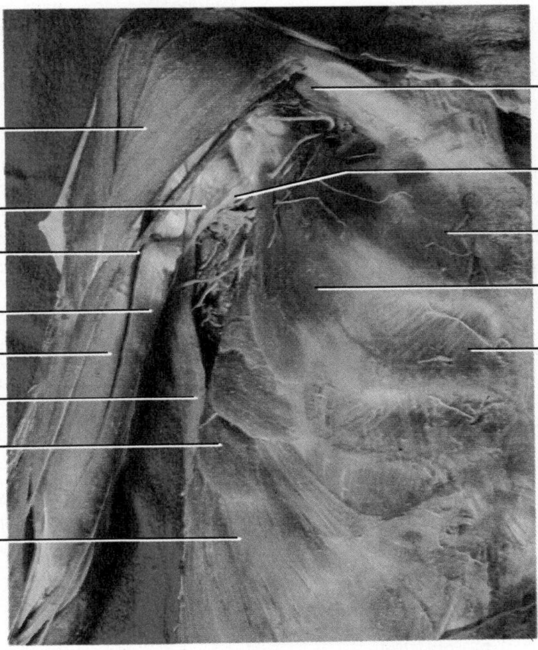

1 Deltoid m.
2 Coracobrachialis m.
3 Cephalic v.
4 Biceps brachii m:
4a Short head
4b Long head
5 Latissimus dorsi m.
6 Serratus anterior m.
7 External oblique m.
8 Subclavius m.
9 Brachial plexus
10 Internal intercostal m.
11 Pectoralis minor m.
12 External intercostal m.

Figure 24.7 Anterior view of the deep muscles of the right thorax, shoulder, and brachium.

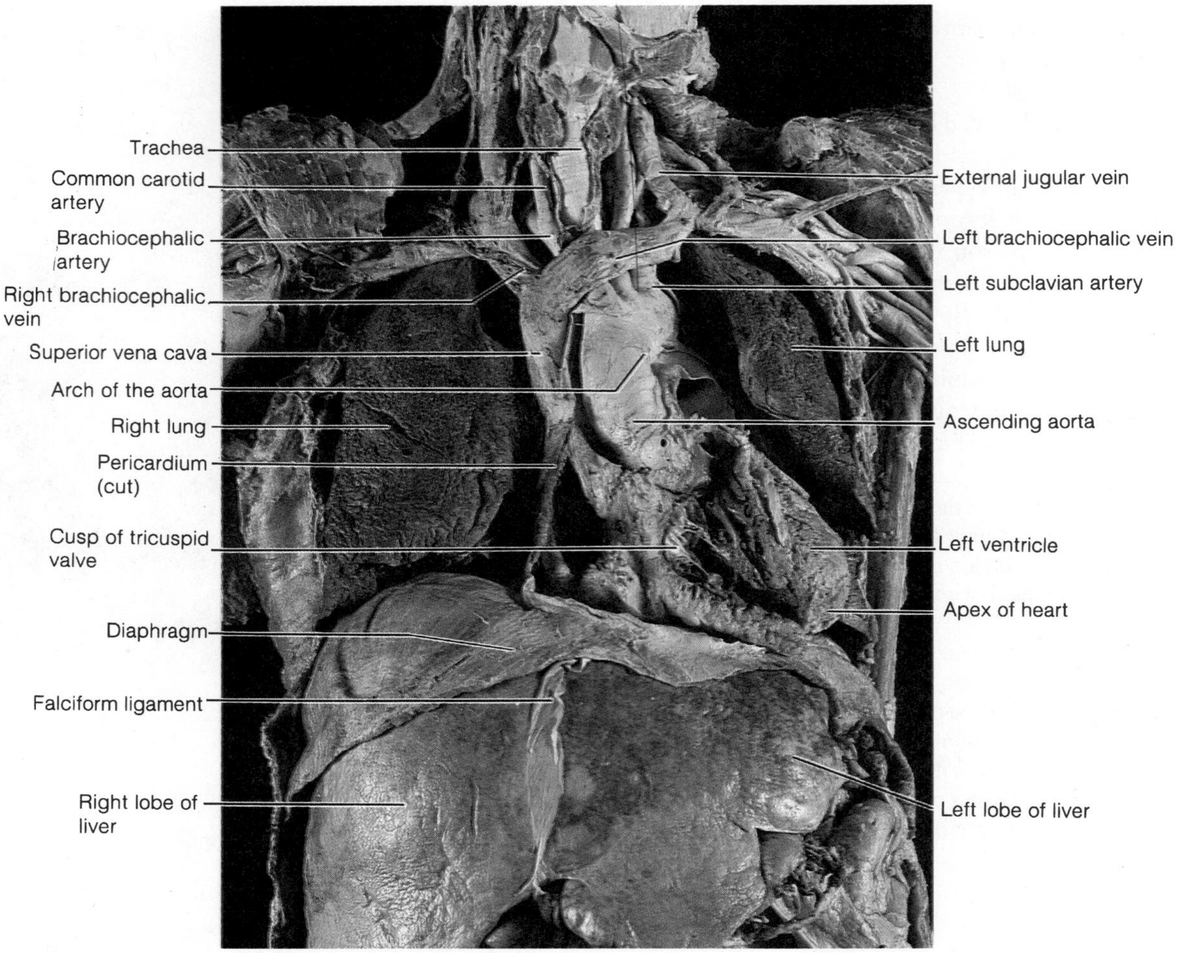

Figure 24.8 Viscera of the thorax. The heart has been coronally sectioned to expose the chambers.

cipal organs of the respiratory and circulatory systems are located within the thorax, and the esophagus of the digestive system passes through the thorax. Because the viscera of the thoracic cavity are vital organs, the thorax is of immense clinical importance.

Developmental Conditions of the Thorax

Although the rib cage manifests tremendous variation between individuals, its development is usually complete. When serious deformities of the chest do occur, they are almost always due to an overgrowth of the ribs. In **pigeon breast** (*pectus carinatum*), the sternum is pushed forward and downward like the keel of a boat. In **funnel chest** (*pectus excavatum*), the sternum is pushed posteriorly, causing an anterior concavity in the thorax. Rarely, there may be a congenital absence of a pair, or pairs, of ribs. The absence of ribs is due to incomplete development of the thoracic vertebrae, a condition termed **hemivertebrae,** and may result in impaired respiratory functions. There is a 0.5% occurrence of **cervical rib,** and in half of these it is bilateral. A cervical rib is attached to the transverse process of the seventh cervical vertebra and it either has a free anterior portion or is attached to the first (thoracic) rib. Pressure of a cervical rib onto the brachial plexus may produce paresthesia (burning, prickling) along the ulnar border of the forearm and atrophy of the medial (hypothenar) muscles of the hand.

The pectoralis major muscle may be congenitally absent, either partially or wholly. A person with this anomaly will have a sunken appearance in the chest area and have to rely upon the contractions of synergistic muscles to the pectoralis major for flexion at the shoulder joint.

The rapid and complex development of the heart and major thoracic vessels accounts for the numerous congenital abnormalities that may affect these organs (see chap. 16). Congenital heart problems occur in approximately 3 of every 100 births and account for about 50% of early childhood deaths. Cardiac malformations usually arise from developmental defects in the heart valves, septa (atrial and/or ventricular), or both. A **patent foramen ovale** is an example of a septal defect. Such a malformation permits venous blood from the right atrium to mix freely with the oxygenated blood in the left atrium. A **ventricular septal defect** usually occurs in the upper portion of the interventricular septum and is generally more serious than an atrial septal defect because of the greater fluid pressures in the ventricles and the greater chance of heart failure.

Congenital valvular problems are either an **incompetence** (leakage) or a **stenosis** (constriction) of valves. Improper closure of a valve permits some backflow of blood, causing a sound referred to as a **murmur.** Murmurs are common and generally do not affect a person's health.

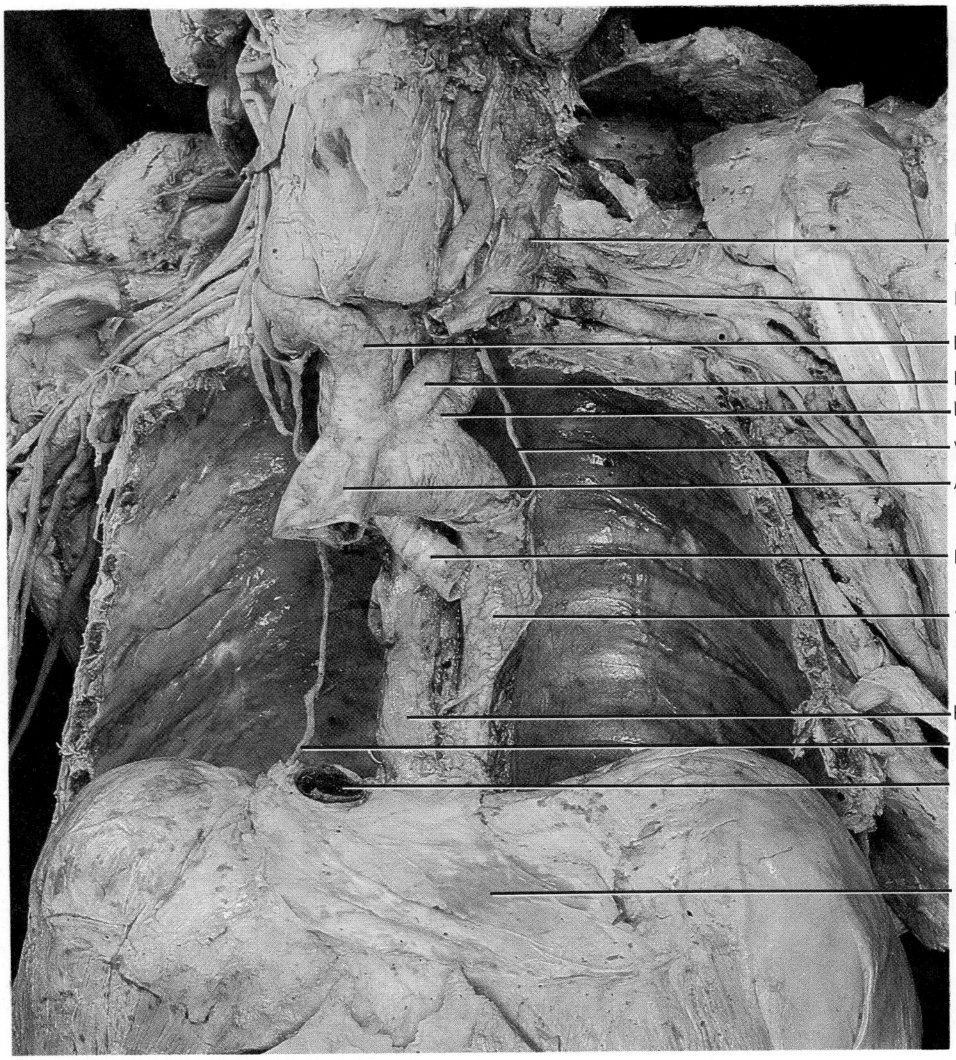

Figure 24.9 Thoracic cavity with the heart and lungs removed.

A **tetralogy of Fallot** is a combination of four defects within a newborn and immediately causes a cyanotic condition (blue baby). The four characteristics of this anomaly are (1) a ventricular septal defect, (2) an overriding aorta, (3) pulmonary stenosis, and (4) right ventricular hypertrophy. Although tetralogy of Fallot is one of the most common cardiac defects, it is one of the simplest to correct surgically.

Abnormal development of the primitive aortic arches may result in a functional right aortic arch and occasionally both a right and a left aortic arch. When these conditions occur, there are generally anomalies of other vessels as well.

Trauma to the Thorax

Because of its resilience, the rib cage generally provides considerable protection to the thoracic viscera. The ribs of children are highly elastic and fractures are rare. Although the ribs of adults are strong, they are frequently fractured by direct trauma or indirectly by crushing injuries. Ribs three through eight are the most commonly fractured. The first and second ribs are somewhat protected by the clavicle and the last four ribs are more flexible to blows. The costal cartilages in elderly people may undergo some ossification, reducing the flexibility of the rib cage and causing some confusion when examining a chest X ray.

A common complication of a rib fracture is a puncture of the lung or the protrusion of a bone fragment through the skin. In either case, pleural membranes will likely be ruptured resulting in a **pneumothorax** (air into the pleural cavity) or a **hemothorax** (blood into the pleural cavity).

Any puncture wound to the thorax—from a bullet or a knife, for example—may cause a pneumothorax. A **collapsed lung** generally results from a pneumothorax, making respiration extremely difficult. Because each lung is contained within its own pleural cavity, trauma to one lung does not usually directly affect the other lung.

The heart, the ascending aorta, and the pulmonary trunk are enclosed by the fibrous pericardium. A severe blow to the chest, such as hitting the steering wheel during an automobile accident, may cause a sudden surge of blood from the ventricles sufficient to rupture the ascending aorta. Such an injury will flood the pericardial sac with blood, causing **cardiac tamponade** (fluid compression) and immediate death.

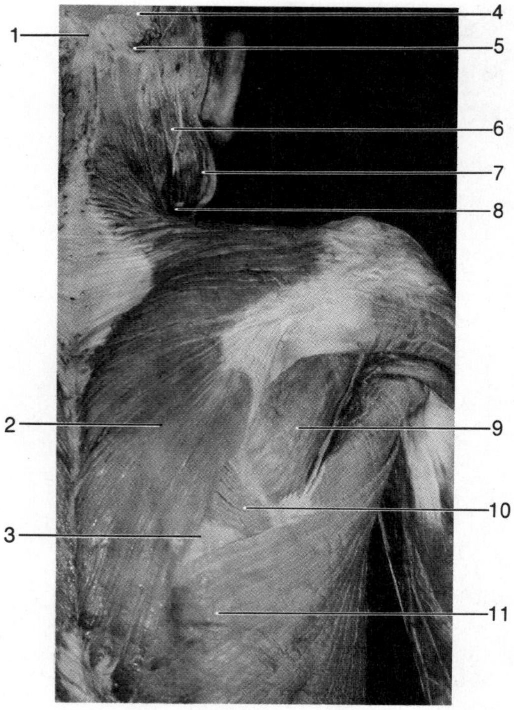

1 External occipital protuberance
2 Trapezius m.
3 Triangle of auscultation
4 Occipital a.
5 Greater occipital n.
6 Lesser occipital n.
7 Sternocleidomastoid m.
8 Greater auricular n.
9 Infraspinatus m.
10 Rhomboideus major m.
11 Latissimus dorsi m.

Figure 24.10 Posterior view of the right thorax and neck.

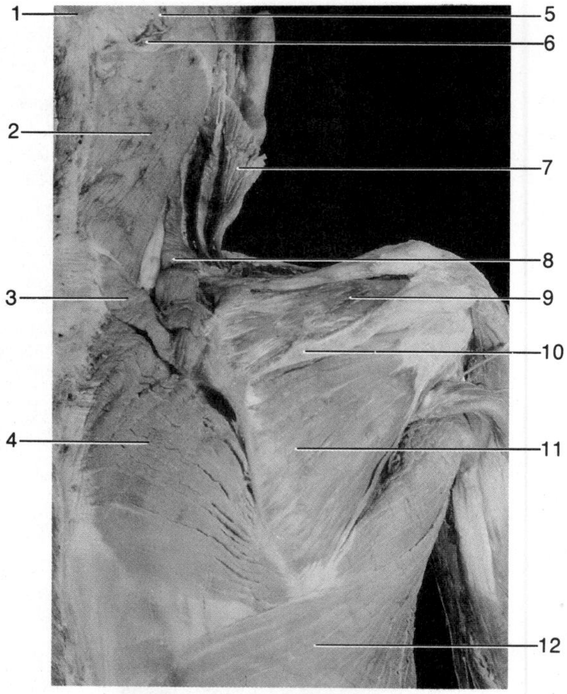

1 External occipital protuberance
2 Splenius capitis m.
3 Rhomboideus minor m.
4 Rhomboideus major m.
5 Occipital a.
6 Greater occipital n.
7 Sternocleidomastoid m. (cut)
8 Levator scapulae m.
9 Supraspinatus m.
10 Spine of scapula
11 Infraspinatus m.
12 Latissimus dorsi m.

Figure 24.11 Posterior view of the deep muscles of the right thorax and neck.

Diseases of the Thorax

The leading causes of death in the United States are due to disease or dysfunction of the thoracic organs. Consequently, the surface features of the thorax are extremely important to a physician in providing reference locations for performing *inspections* (visual observation), *palpations* (feeling with firm pressure), *percussions* (detecting densities through tapping), and *auscultations* (listening with a stethoscope). Many of the ribs can be seen on a thin person, and all of the ribs, except the first and at times the twelfth, can be palpated. The sternum, clavicles, and scapulae also provide important bony landmarks (see chapter 10) for doing a physical examination. The nipples in the male and prepubescent female are located at the fourth intercostal spaces. The position of the left nipple in males is important in knowing where to listen to various heart sounds and for determining if the heart is enlarged.

The clinical importance of the female breasts lies in their periodic phases of activity (during pregnancy and lactation) and their susceptibility to neoplastic change. The breasts and mammary glands are highly susceptible to infections, cysts, and tumors. The superficial position of the breasts permits a chance for effective treatment by surgery and radiotherapy if tumors are recognized early. The importance of *breast self-examination (BSE)*, therefore, cannot be overemphasized (see chapter 21). If untreated, breast cancer is soon fatal (fig. 24.13). Breast cancer, or **carcinoma of the breast,** is the most common malignancy in women. One in ten women will develop breast cancer, and one-third of these will die from the disease.

The lungs can be examined through percussion, auscultation, or observation. *Bronchoscopy* enables a physician to examine the trachea, carina, primary bronchi, and secondary bronchi. Such an instrument also enables a physician to remove foreign objects from these passageways. Swallowed objects that are aspirated beyond the glottis usually lodge (in 90% of the cases) in the right primary bronchus because of its near vertical position to the trachea. *Bronchiograms,* or chest X rays, are perhaps the best means of inspecting the lungs.

The lungs are a common site for cancer. Fortunately, each lung is divided into different segments allowing a surgeon to remove a diseased portion and leave the remainder of the lung intact. Pneumonia, tuberculosis, asthma, pleurisy, and emphysema are all other common diseases that directly or indirectly afflict the lungs.

Cardiovascular diseases are the leading cause of death in the United States. Included among these diseases are heart attacks, which are caused by an insufficient blood supply to the myocardium (called **myocardial ischemia**). Poor cardiac circulation is due to an accumulation of artherosclerotic plaques or the presence of a thrombus (clot). A heart attack is generally

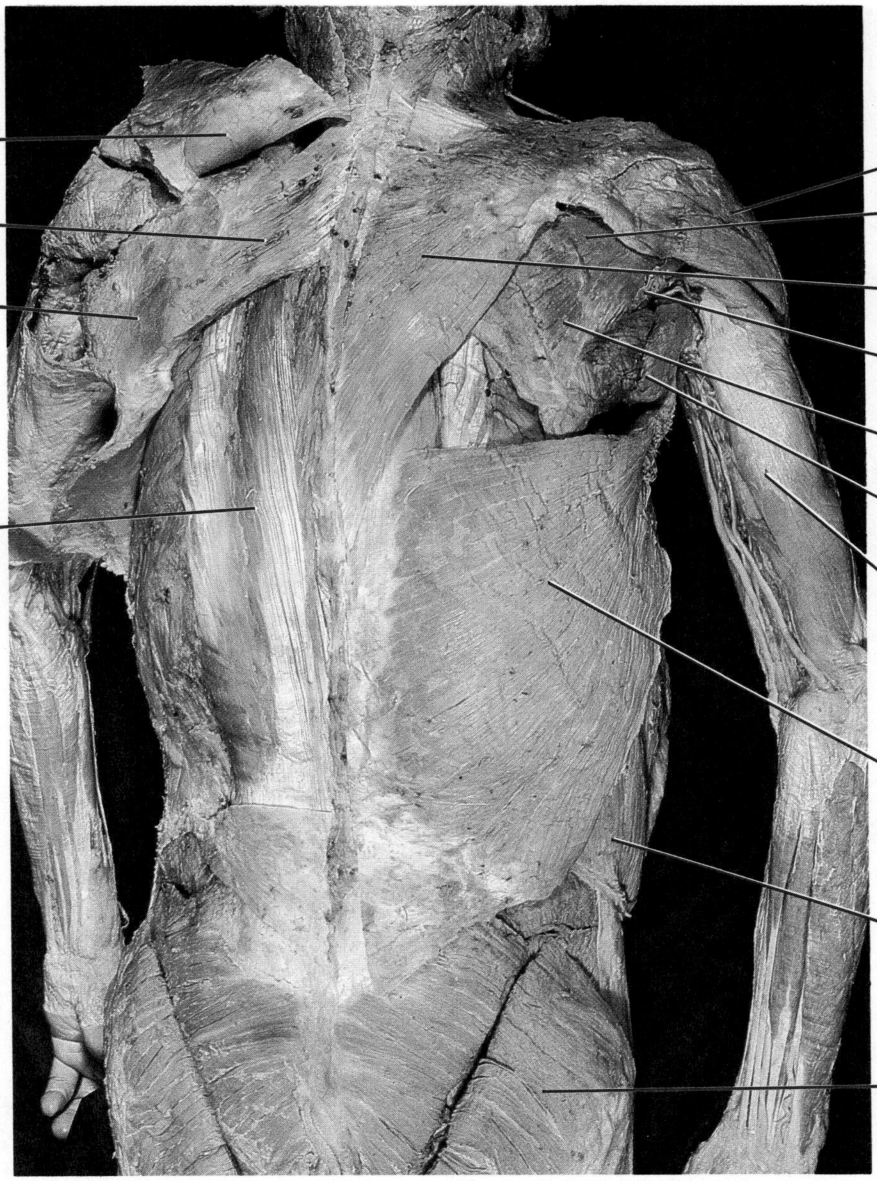

Figure 24.12 Posterior view of the torso; deep muscles are exposed on left.

accompanied by severe chest pains (**angina pectoris**) and usually by referred pain, perceived as arising from the left arm and shoulder. If the coronary deficiency is continuous, local tissue necrosis results, causing a permanent loss of cardiac muscle fibers (**myocardial infarction**). Extensive cardiac necrosis results in cardiac arrest.

Ventricular fibrillation follows a functional breakdown of the conduction system of the heart. Such a loss of coordinated ventricular contraction causes a loss of coronary circulation and low blood pressure (called **hypotension**). If untreated, continuous ventricular fibrillation results in death.

Various other heart diseases include infections of the serous membrane (**pericarditis**), infection of the lining of the heart chambers (**endocarditis**), and infection of the valves (a complication of **rheumatic fever**). Valvular disease may cause the cusps to function poorly and result in an enlarged heart.

1. Except for a passageway in the head and neck, which body system is predominately located in the thorax? Which body system is centrally located in the thorax? Which body system has a tubular passageway through the thorax?
2. Describe three common congenital conditions of the rib cage, and describe three congenital conditions that may afflict the heart. Which of these are of major clinical concern because of their incompatibility to life?
3. Explain how the rib cage is protective of the thoracic viscera. Discuss the various kinds of thoracic trauma that may impair the functioning of the heart or lungs.
4. List the anatomical structures of the thorax that may serve as landmarks for surface anatomy. Describe some of the techniques for assessing the condition of the thoracic viscera.

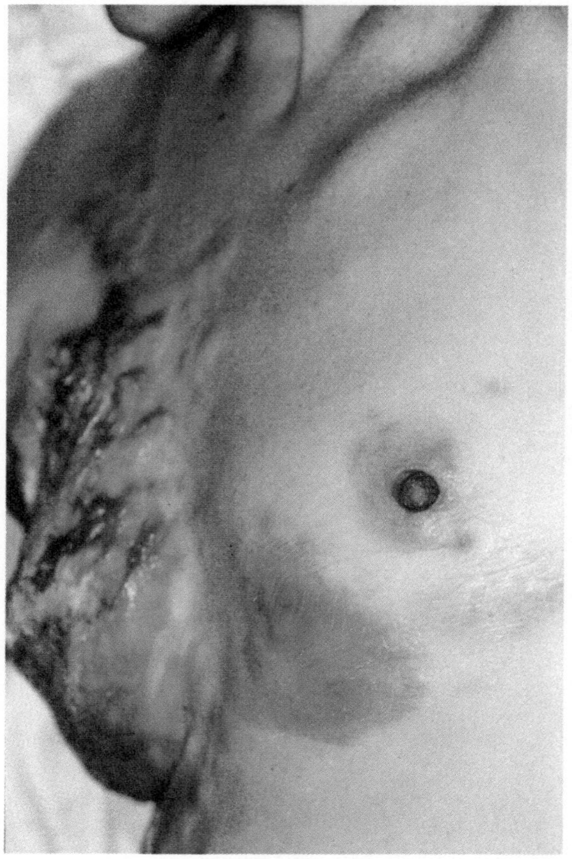

Figure 24.13 Advanced stages of untreated breast cancer that has metastasized to afflict both breasts.

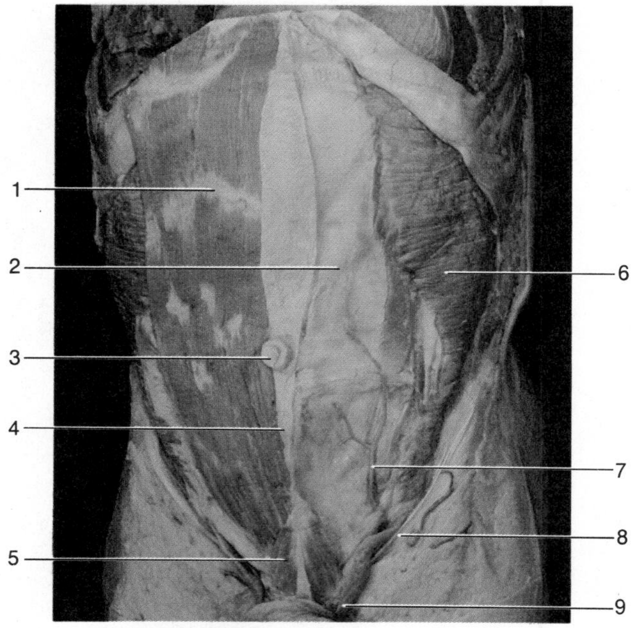

1 Rectus abdominis m.
2 Rectus sheath
3 Umbilicus
4 Linea alba
5 Pyramidalis m.
6 Transverse abdominis m.
7 Inferior epigastric a.
8 Inguinal ligament
9 Spermatic cord

Figure 24.14 Anterior view of the muscles of the abdominal wall.

Abdominal Region

The abdomen is that portion of the torso positioned between the diaphragm and the pelvis. Because of the domed shape of the diaphragm, some of the abdominal viscera lie within the protection of the rib cage. The cavity of the abdomen contains the stomach and intestines, the liver and gallbladder, the spleen, the internal genitalia, and major vessels and nerves. Congenital conditions, trauma, and disease are common to this body region and are generally serious with the potential of being life threatening.

Objective 8. List the body systems that have organs within the abdomen.

Objective 9. Describe the common congenital conditions that may afflict the abdomen and are of clinical concern.

Objective 10. Describe the arrangement of the abdominal serous membranes, and discuss the common clinical conditions that may afflict these membranes.

Objective 11. List some common diseases that afflict the organs of the abdomen, and discuss how the anatomical surface landmarks are of clinical importance in diagnosing diseased organs within the abdominal cavity.

The abdominal region is shown in photographs of cadavers in figures 24.14, 24.15, and 24.16.

Developmental Conditions of the Abdomen

The development of the abdominal viscera from endoderm and mesoderm is a highly integrated, complex, and rapidly occurring process and is therefore susceptible to a wide range of congenital malformations.

The diaphragm develops in four directions simultaneously as skeletal muscular tissue coalesce toward the posterior center at the hiatus where the esophagus, abdominal aorta, and inferior vena cava traverse. Failure of these four sections to fuse results in a **congenital diaphragmatic hernia.** This anomalous opening may permit certain abdominal viscera to project into the thoracic cavity.

The developmental role of the umbilicus with the fetal urinary, circulatory, and digestive systems may present some interesting congenital defects. A **patent urachus** is an opening from the urinary bladder to the outside through the umbilicus. For a short time during development, this opening is natural as urine passes from the fetus into the allantois. Closure of the urachus occurs in most fetuses during further development of the urinary system. A patent urachus will generally go undetected unless there is extreme difficulty with urination, such as that caused by an enlarged prostate gland. In such a case, some urine may be forced through the patent urachus and out the umbilicus.

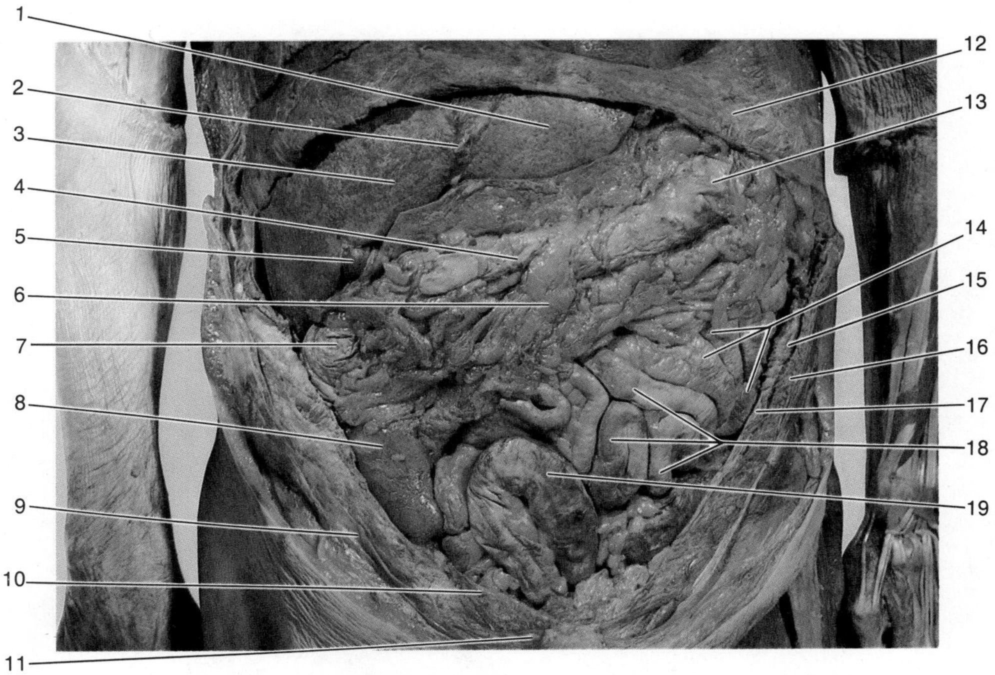

1 Left lobe of liver
2 Falciform ligament
3 Right lobe of liver
4 Transverse colon
5 Gallbladder
6 Greater omentum
7 Hepatic flexure of colon
8 Fat globule on greater omentum
9 Aponeurosis of internal oblique m.
10 Rectus abdominis m. (cut)
11 Rectus sheath (cut)
12 Diaphragm
13 Splenic flexure of colon
14 Jejunum
15 Transversus abdominis m. (cut)
16 Internal and external oblique mm. (cut)
17 Parietal peritoneum (cut)
18 Ileum
19 Anterior sigmoid colon (normal variation)

Figure 24.15 Viscera of the abdomen.

Being present in approximately 3% of the population, **Meckel's diverticulum** is the most common anomaly of the small intestine. It is the result of failure of the embryonic yolk sac to atrophy. When present, Meckel's diverticulum consists of a pouch approximately 6.5 cm (2.5 in.) in length, resembling the appendix and arising near the center of the ileum. It may terminate freely or be attached to the anterior abdominal wall near the umbilicus. Like the appendix, a Meckel's diverticulum is prone to infections and may become inflamed, producing symptoms similar to appendicitis. Because of its clinical importance, it is usually removed as a precautionary measure when discovered during abdominal surgery.

Sometimes, the connection from the ileum of the small intestine to the outside is patent at the time of birth; this condition is called a *fecal fistula.* This anomalous opening permits the passage of fecal material through the umbilicus and must be surgically corrected in a newborn.

Other parts of the GI tract are also common sites for congenital problems. **Pyloric stenosis** is a condition in which there is a narrowing of the pyloric orifice of the stomach due to hypertrophy of the muscular layer of the pyloric sphincter. This anomaly is more common in males than in females, and the symptoms usually appear early in infancy. The constricted opening interferes with the passage of food into the duodenum and therefore causes dilation of the stomach, vomiting, and loss of weight. Treatment involves a surgical incision of the pyloric sphincter.

Congenital megacolon (Hirschsprung's disease) is a condition in which there is failure of ganglia to develop in the submucosal (Meissner's) and the myenteric (Auerbach's) plexuses in a portion of the colon. The absence of these ganglia results in enlargement of the affected portion of the colon due to a lack of innervation and muscle tone. With the loss of peristalsis there is severe constipation. Treatment involves surgical resection of the affected portion of the colon.

Congenital malformations may develop in any of the abdominal viscera, most of which are inconsequential. Accessory spleens, for example, occur in about 10% of the population. Located near the hilum of the spleen, these anomalous organs are small (about 1 cm. in diameter), number from two to five, and are only moderately functional. Generally they atrophy within a few years after birth.

There can be tremendous variation in the formation of the kidneys. Frequently they are multilobed, fused, malpositioned, or they differ in number from the normal two (see figure 19.16). When there is an anomalous kidney, there is usually an accompanying variation in the vascular supply. It is not uncommon to have multiple renal arteries serving a kidney. Most renal anomalies do not pose serious problems.

Because of the effects of sex hormones upon the developing genitalia, there may be considerable malformation of these organs. These anomalies may range from being of cosmetic

Hirschsprung's disease: from Harold Hirschsprung, Danish physician, 1830–1916.

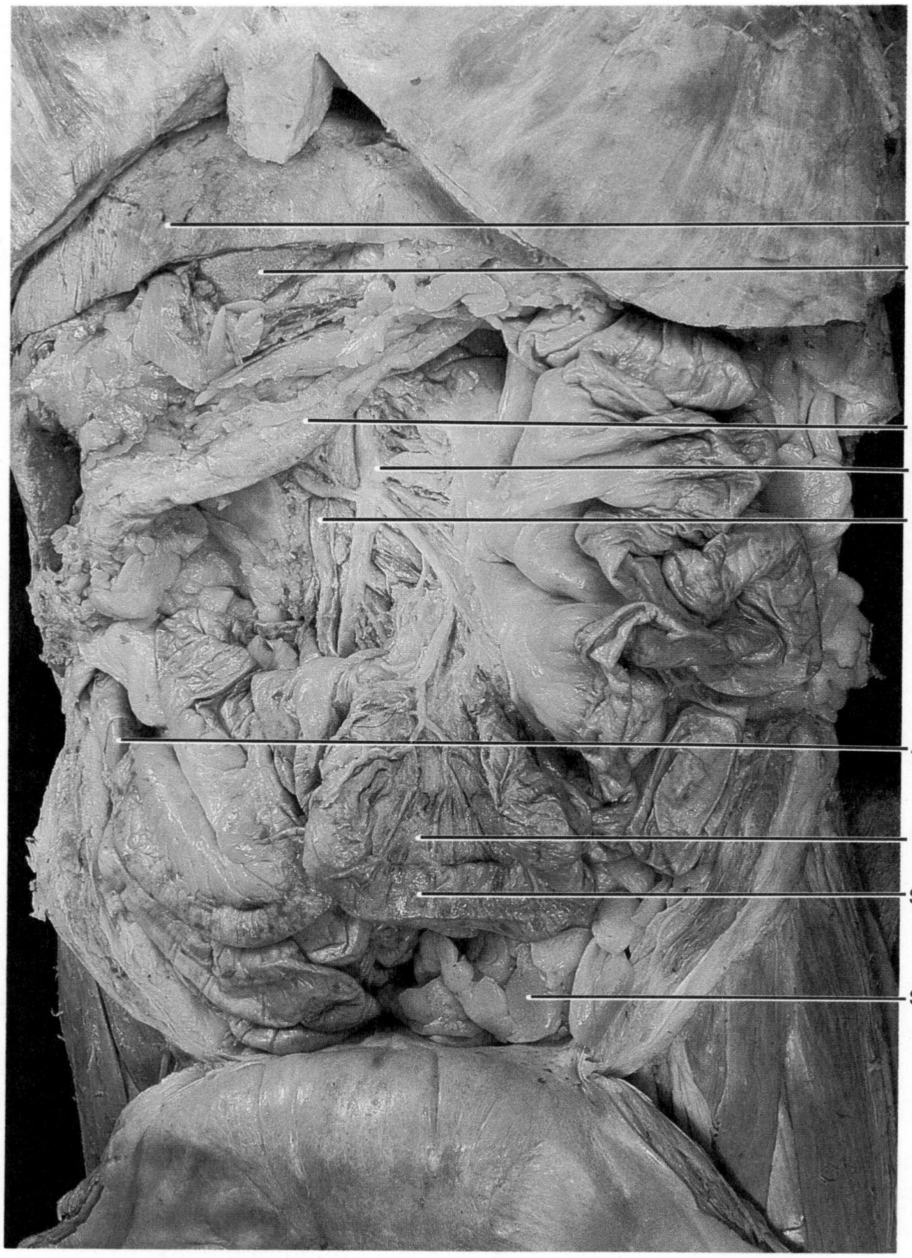

Figure 24.16 Viscera of the abdomen with greater omentum removed and small intestine displaced to the left.

concern to being totally nonfunctional; some may even be so severe as to prohibit the determination of an individual's sex based upon external appearance. Most of these conditions can be surgically corrected. Also of clinical concern and treatable are the various problems that may occur during descent of the testes into the scrotum. In the normal development of the male fetus, the testes are in scrotal position by twenty-eight weeks of gestation. If they are undescended at birth, medical intervention may be necessary.

Trauma to the Abdomen

The rib cage, the omentum, and the pendant support of the abdominal viscera offer some protection from trauma. Puncture wounds, compression, and severe blows to the abdomen do result, however, in serious abdominal injury.

The large and dense liver, located in the upper right quadrant of the abdomen, is quite vulnerable to blunt trauma, stab wounds, or puncture wounds by fractured ribs. A lacerated liver is extremely serious because of the possibility of internal hemorrhage from such a vascular organ.

The spleen is another highly vascular organ that is frequently injured, especially from blunt abdominal trauma. A ruptured spleen causes severe internal hemorrhage and shock. A prompt splenectomy is necessary to keep the patient from bleeding to death. The spleen may also rupture spontaneously due to infectious diseases that cause it to hypertrophy.

Rupture of the pancreas is not nearly as common as rupture of the spleen, but it could occur if a strong compression of the upper abdomen would force the pancreas against the vertebral column. The danger of a ruptured pancreas is the flow of pancreatic juice into the peritoneal cavity, the subsequent digestive action, and peritonitis.

The kidneys are vulnerable to trauma in the lumbar region, such as with a kidney punch. Because the kidney is fluid-filled, a blow to one side propagates through the kidney and may possibly rupture the renal pelvis or the proximal portion of the ureter. Blood in the urine is symptomatic of kidney damage. The medical treatment of a traumatized kidney is dependent upon the severity of the injury.

Trauma to the external genitalia of both the male and female is a relatively common occurrence. The pendant position of the penis and scrotum makes them vulnerable to compression forces—such as would be experienced by a construction worker slipping and straddling a steel beam, which would compress the external genitalia between the beam and his pubic bone. In such an injury the penis, including the urethra, may split open, and one or both testes may be crushed.

Trauma to the female genitalia usually results from sexual abuse. Vaginal tearing and a displaced uterus are common in molested girls. The physical and mental consequences are generally severe.

Diseases of the Abdomen

Each of the abdominal organs may be afflicted by an array of diseases. It is beyond the scope of this text to discuss any of these diseases in depth, so only an overview of some general conditions will be presented.

As with the thorax, surface features of the abdomen provide the physician with reference points for doing a physical exam. Knowledge of the clinical regions of the abdomen (see figs. 2.13, 2.14), and the organs within these regions (see table 2.3) provides the physician an important basis for performing a physical examination. Also important are the locations of the linea alba extending from the xiphoid process to the symphysis pubis, the umbilicus, the inguinal ligament, and the bones and processes, which can be palpated on the rib cage, and the pelvic girdle.

Peritonitis is of major clinical concern. The peritoneum is the serous membrane of the abdominal cavity that lines the abdominal wall, as the parietal peritoneum, and covers the visceral organs, as the visceral peritoneum. The peritoneal cavity is the moistened space betwen the parietal and visceral portions of the peritoneum. Peritonitis results from any type of contamination of the peritoneal cavity, such as from a puncture wound, blood-borne diseases, or a ruptured visceral organ. In females, peritonitis is frequently a complication of infections of the reproductive tract, which have entered the peritoneal cavity via the uterine tubes. Without medical treatment, peritonitis is generally fatal.

Ulcers may occur throughout the GI tract. *Peptic ulcers* are erosions of the mucous membranes of the stomach or duodenum, produced by the action of HCl. Agents that weaken the mucosal lining of the stomach, such as alcohol and aspirin, and abnormally high secretions of HCl increase the likelihood of developing peptic ulcers. Many people subject to chronic stress produce too much gastric acid and develop a peptic ulcer as a result.

Enteritis, or inflammation of the intestinal mucosa, is frequently referred to as intestinal flu. Causes of enteritis include bacterial or viral infections, irritating foods or fluids, and emotional stress. The symptoms are abdominal pain, nausea, and diarrhea. *Diarrhea* is symptomatic of inflammation, stress, and other body dysfunctions. Diarrhea in children is of immense clinical importance because of the rapid loss of body fluids.

---

1. Construct a table listing the principal organs of the abdomen according to the body system to which each belongs.
2. Describe three congenital conditions involving the umbilicus. Discuss two congenital anomalies of the abdomen, which, if untreated, would cause death.
3. Describe the locations of the abdominal serous membranes, and list some of the causes of peritonitis.
4. List the anatomical structures of the abdomen that serve as landmarks of surface anatomy.
5. Discuss the extent and seriousness of three diseases of your choice that afflict organs of the abdominal cavity.

Shoulder and Upper Extremity

The shoulder and upper extremity include the pectoral girdle, the brachium, the antebrachium, and the hand. There are only a few serious congenital conditions of these regions, and there are no serious endemic diseases. Trauma to the skin, bones, muscles, vessels, and nerves, however, is common and ranges from annoying to extremely serious.

Objective 12. Describe the common congenital conditions of the shoulder and/or upper extremity that are of clinical concern.

Objective 13. Discuss the frequent kinds of trauma to the shoulder and/or upper extremity.

Objective 14. List the common diseases that may afflict the shoulder and/or upper extremity regions.

The shoulder and upper extremity are shown in photographs of cadavers in figures 24.17 through 24.21.

Developmental Conditions of the Shoulder and Upper Extremity

Twenty-eight days after conception, a limb bud appears on the upper lateral side of the embryo, which eventually becomes a shoulder and an upper extremity. Three weeks later (seven weeks after conception) the shoulder and upper extremity are present in the form of mesenchymal primordium of bone and muscle. It is during this crucial three weeks of development that malformations of the extremities can occur due to genetic expression or to the effect of teratogens.

If a pregnant woman uses certain teratogenic drugs or is exposed to certain diseases (*Rubella* virus, for example) during development of the embryo, there is a strong likelihood that the appendage will be incompletely developed. A large number of limb deformities occurred between 1957 and 1962 as a result

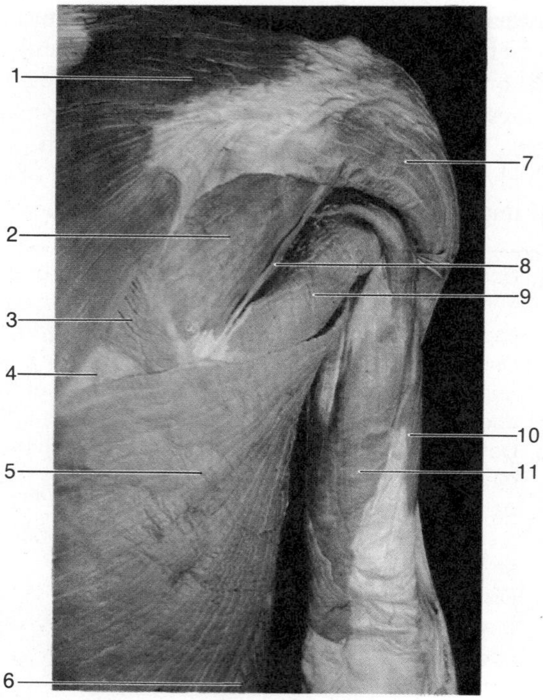

1 Trapezius m.
2 Infraspinatus m.
3 Rhomboideus major m.
4 Triangle of auscultation
5 Latissimus dorsi m.
6 External oblique m.
7 Deltoid m.
8 Teres minor m.
9 Teres major m.
10 Lateral head of triceps brachii m.
11 Long head of triceps brachii m.

Figure 24.17 Posterior view of the right shoulder and brachium.

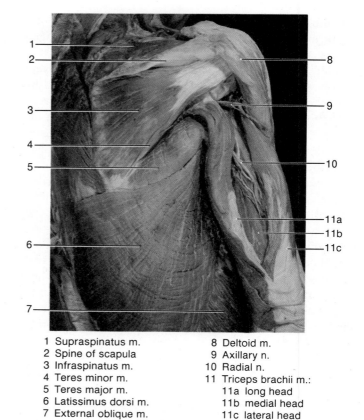

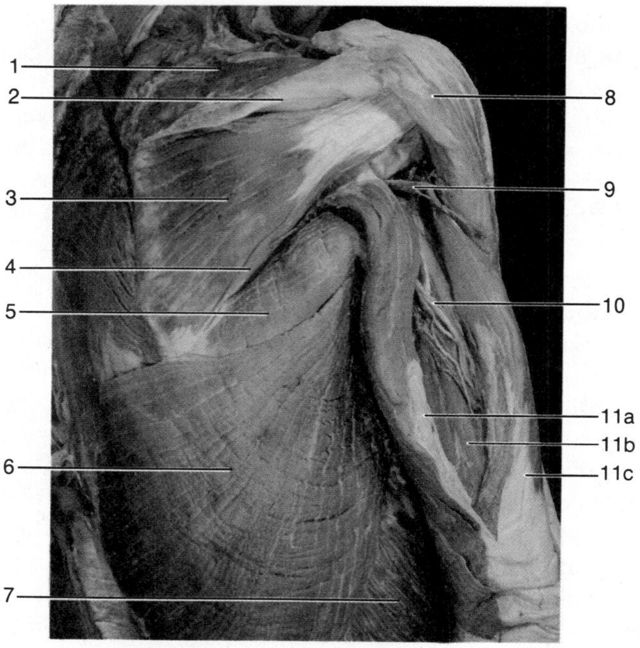

1 Supraspinatus m.
2 Spine of scapula
3 Infraspinatus m.
4 Teres minor m.
5 Teres major m.
6 Latissimus dorsi m.
7 External oblique m.
8 Deltoid m.
9 Axillary n.
10 Radial n.
11 Triceps brachii m.:
11a long head
11b medial head
11c lateral head

Figure 24.18 Posterior view of the deep muscles of the right shoulder and brachium.

of women ingesting thalidomide during early pregnancy to relieve morning sickness. It is estimated that seven thousand infants were malformed by thalidomide. The malformations ranged from *micromelia* (short limbs) to *amelia* (the absence of limbs).

Although there are a large number of genetic deformities of the shoulder and upper extremity, there are only a few that are relatively common. **Sprengel's deformity** affects the development of one or both scapulae. In this condition, the scapula is smaller than normal and is positioned at an elevated level. As a result, abduction of the arm is not possible beyond a right angle to the plane of the body.

Minor defects of the extremities are relatively common malformations. Extra digits, a condition called **polydactyly,** is the most common limb deformity. Usually an extra digit is incompletely formed and does not function. **Syndactyly,** or webbed digits, is likewise a relatively common limb malformation. Polydactyly is inherited as a dominant trait, whereas syndactyly is a recessive trait.

Trauma to the Shoulder and Upper Extremity

There is a variety of types of trauma to the shoulder and upper extremity, ranging from injury to the bones and surrounding muscles, tendons, vessels, and nerves to damage of the joints in the form of sprains or dislocations.

It is not uncommon to traumatize the shoulder and upper extremity of a newborn during a difficult delivery. **Upper arm birth palsy** (*Erb-Duchenne palsy*) is the most common type of birthing injury, caused by a forcible widening of the angle between the head and shoulder. Using forceps to rotate the fetus *in utero* or pulling on the head during delivery may cause this injury. The site of injury is at the junction of C5 and C6 (*Erb's point*), as they form the upper trunk of the brachial plexus. The expression of the injury is paralysis of the abductors and lateral rotators of the shoulder and the flexors of the elbow. The arm will permanently hang at the side in medial rotation.

The stability of the shoulder is largely dependent on the support of the clavicle and acromion process superiorly and the tendons forming the rotator cuff anteriorly. Because support is weak along the inferior aspect of the shoulder joint, dislocations are frequent in this direction. Injuries of this sort are common in athletes engaging in contact sports. Sudden jerks of the arm are also likely to dislocate the shoulder, especially in children who have weak muscles spanning this area.

Fractures are common in any location of the shoulder and upper extremity. Many fractures are the result of extending the arm to break a fall. The clavicle is the most frequently broken bone in the body. Fractures of the humerus are also common and often serious because of injury to the nerves and vessels that

Sprengel's deformity: from Otto G. K. Sprengel, German surgeon, 1852–1915

Erb-Duchenne palsy: from Wilhelm H. Erb, German neurologist, 1840–1921, and Guillaume G. A. Duchenne, French neurologist, 1806–1875

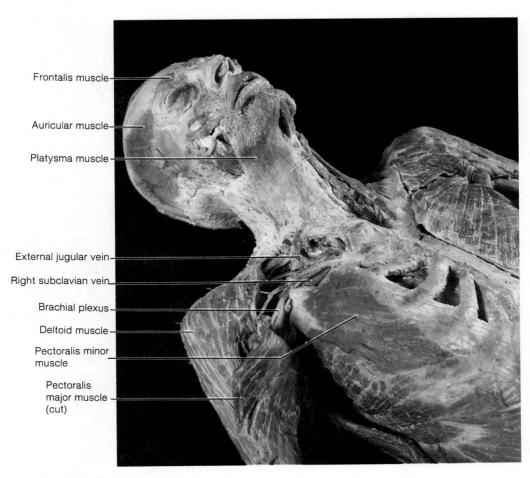

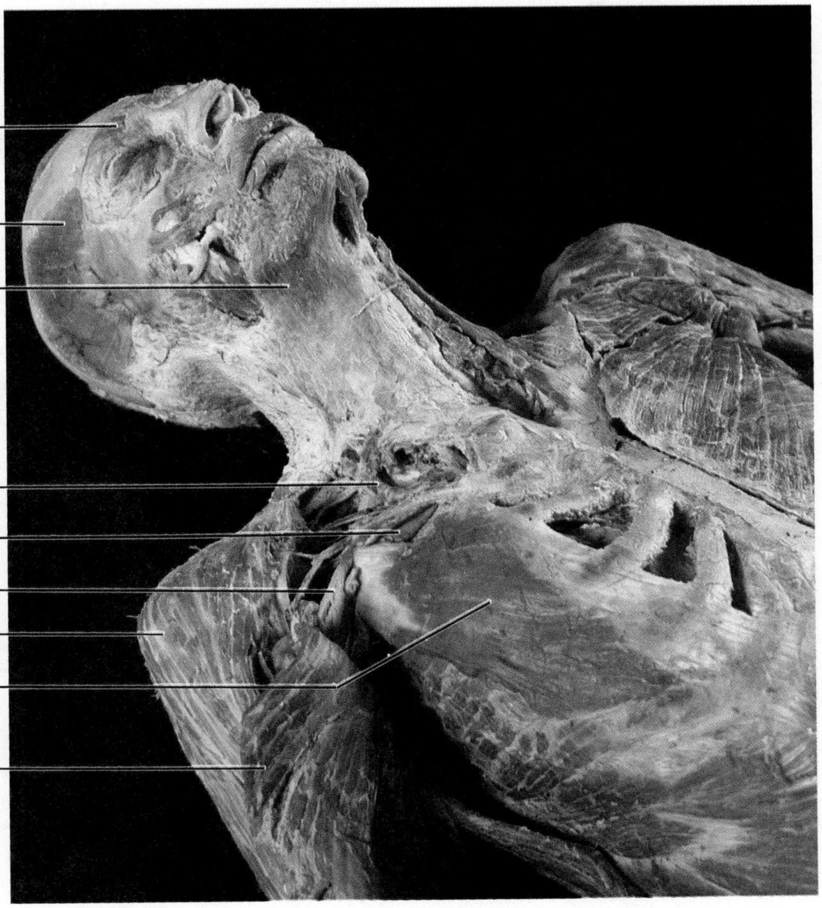

Figure 24.19 Anteriolateral view of the axilla and thorax.

parallel the bone. The surgical neck of the humerus is a common fracture site, which often damages the axillary nerve, thus limiting abduction of the arm. A fracture in the middle third of the humerus may damage the radial nerve, causing paralysis of the extensor muscles of the hand (wristdrop). A fracture of the olecranon process of the ulna often damages the ulnar nerve, resulting in paralysis of the flexor muscles of the hand and the adductor muscles of the thumb. The distal portion of the radius is frequently fractured (*Colles' fracture*) when one falls on the outstretched arm. In this fracture, the hand is displaced backward and upward.

Sports injuries frequently involve the upper extremity. Repeated extension of the wrist against a force, such as occurs during the backhand stroke in tennis, may cause **lateral epicondylitis** (tennis elbow). Wearing an elbow brace or a compression band may help reduce the pain, but only abstaining from tennis for a while will heal the area.

Athletes frequently jam a finger. This occurs when a ball forcefully strikes a distal phalanx as the fingers are extended, causing a sharp flexion at the joint between the middle and distal phalanges. Treatment involves splinting the finger for a period of time. If splinting is not effective, however, surgery is necessary or the person will suffer a permanent crook in the finger.

Diseases of the Shoulder and Upper Extremity

Infections in specific locations of the shoulder or upper extremity are the only common diseases that are endemic to these regions. **Bursitis,** for example, may specifically afflict any of the numerous bursae of the shoulder, elbow, or wrist joints. There are several types of **arthritis,** but generally they involve diarthrotic joints throughout the body rather than just in the hands and fingers.

Carpal tunnel syndrome is caused by compression of the median nerve by the transverse carpal ligament, forming the palmar aspect of the carpal tunnel. The nerve compression results in a painful burning sensation or numbness of the first three fingers and some muscle atrophy. The compression is due to an inflammation of the transverse carpal ligament. Surgery eases the pressure of the ligament and provides the entry for antibiotic treatment.

Tenosynovitis is an infection of the synovial tendon sheath in the wrist or hand. Digital sheath infections are quite common following a puncture wound in which pathogens enter the closed synovial sheath. The increased pressure from the swollen, infected sheath may cause severe pain and eventually result in necrosis of the flexor tendons. The loss of hand function can be prevented by draining the synovial sheath and providing antibiotic treatment.

Colles' fracture: from Abraham Colles, Irish surgeon, 1773–1843

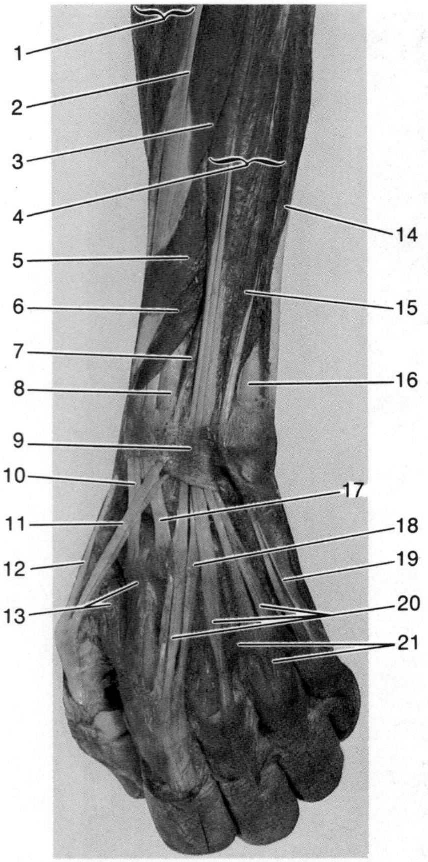

1 Brachioradialis m.
2 Extensor carpi radialis longus tendon
3 Extensor carpi radialis brevis m.
4 Extensor digitorum m.
5 Abductor pollicis longus m.
6 Extensor pollicis brevis m.
7 Extensor pollicis longus m.
8 Radius
9 Extensor retinaculum
10 Extensor carpi radialis longus tendon
11 Extensor pollicis longus tendon
12 Extensor pollicis brevis tendon
13 First dorsal interosseous m.
14 Extensor carpi ulnaris m.
15 Extensor digiti minimi m.
16 Ulna
17 Extensor carpi radialis brevis tendon
18 Extensor indicis tendon
19 Extensor digiti minimi tendon
20 Extensor digitorum tendons
21 Intertendinous connections

Figure 24.20 Posterior view of the left forearm and hand.

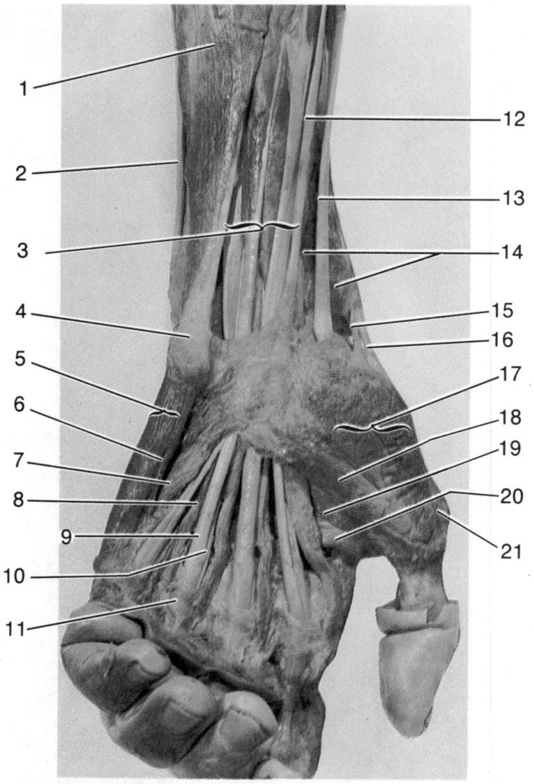

1 Flexor carpi ulnaris m.
2 Extensor carpi ulnaris m.
3 Flexor digitorum superficialis m.
4 Pisiform bone
5 Abductor digiti minimi m.
6 Flexor digiti minimi m.
7 Opponens digiti minimi m.
8 Lumbrical m.
9 Flexor digitorum superficialis tendon
10 Flexor digitorum profundus tendon
11 Fibrous digital sheath
12 Palmaris longus tendon
13 Flexor carpi radialis tendon
14 Pronator quadratus m.
15 Extensor pollicis brevis tendon
16 Extensor pollicis longus tendon
17 Abductor pollicis brevis m.
18 Flexor pollicis brevis m.
19 Adductor pollicis m. (oblique head)
20 Adductor pollicis m. (transverse head)
21 Opponens pollicis m.

Figure 24.21 Anterior view of the left forearm and hand.

Dupuytren's contracture is a condition in which digits four and five are progressively flexed into an immovable position. The condition is due to abnormal fibrous growth on the palmar fascia. Surgical excision of this tissue is necessary to restore the hand to normal function.

---

1. List the common congenital conditions of the shoulder and upper extremity, and state the point of development when these anomalies most commonly occur.
2. Construct a table listing the most common kinds of trauma to the shoulder and upper extremity. For each trauma indicate the treatment that may be required and the possible complications.
3. Discuss the extent and seriousness of three diseases of your choice that afflict the shoulder or portions of the upper extremity.

Hip and Lower Extremity

The hip and lower extremity include the pelvic girdle, the thigh, the leg, and the foot. There are only a few serious congenital conditions of these regions and only a few endemic diseases. Trauma to the skin, bones, muscles, vessels, and nerves of these regions is common and ranges from annoying to extremely serious.

Objective 15. Describe the common congenital conditions of the hip and/or the lower extremity that are of clinical concern.

Objective 16. Discuss the frequent kinds of trauma to the hip and/or lower extremity.

Objective 17. List the common diseases that may afflict the hip and/or lower extremity regions.

The hip and lower extremity are shown in photographs of cadavers in figures 24.22 through 24.28.

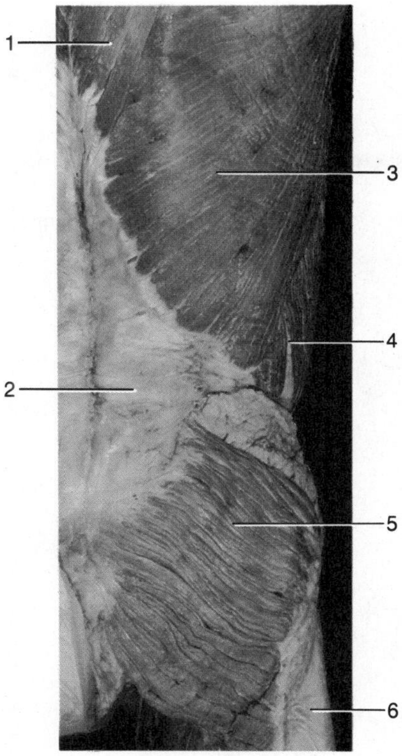

1 Trapezius m.
2 Lumbar aponeurosis
3 Latissimus dorsi m.
4 External oblique m.
5 Gluteus maximus m.
6 Fascia lata

Figure 24.22 Posterior view of the right abdominal and gluteal regions.

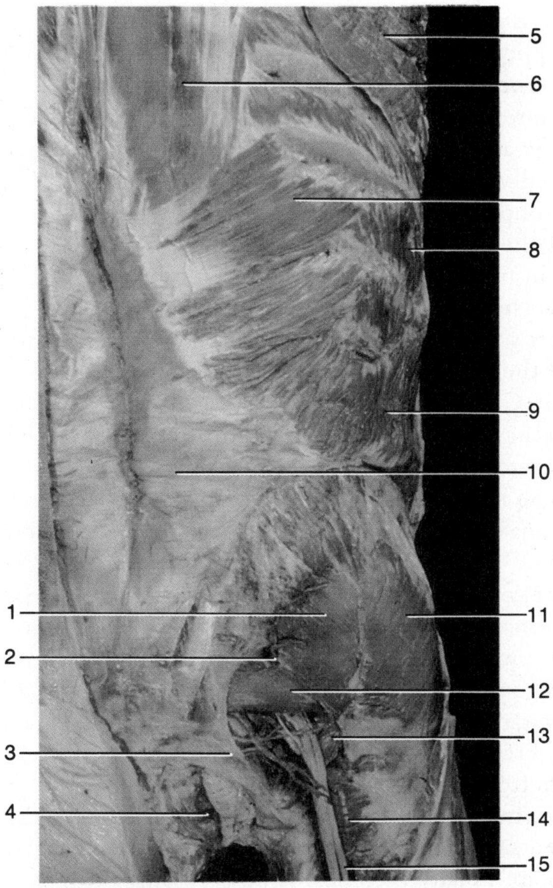

1 Superior gluteal vessels
2 Inferior gluteal vessels
3 Sacrotuberous ligament
4 Levator ani m.
5 Serratus anterior m.
6 Erector spinae m.
7 Serratus posterior m.
8 External intercostal m.
9 Internal oblique m.
10 Lumbar aponeurosis
11 Gluteus medius m.
12 Piriformis m.
13 Obturator internus m.
14 Quadratus femoris m.
15 Sciatic n.

Figure 24.23 Posterior view of the deep muscles of the right abdominal and gluteal regions.

Developmental Conditions of the Hip and Lower Extremity

The embryonic development of the hip and lower extremity is the same as that of the shoulder and upper extremity: the appearance of the limb bud is followed by the formation of the mesenchymal primordium of bone and muscle in the shape of an appendage. Development of the lower extremity, however, lags behind that of the upper extremity by three or four days.

There is minimal chance for congenital deformities of the hips and lower extremities in a newborn if the pregnant mother has been healthy and well nourished, especially prior to and during embryonic development. The few congenital malformations that exist generally have a genetic basis.

In **congenital dislocation of the hip,** the acetabulum fails to develop adequately, and the head of the femur slides out of the acetabulum onto the gluteal surface of the ilium. If untreated, the infant would never be able to walk normally.

Polydactyly and syndactyly occur in the feet as well as in the hands. Treatment of the feet is the same as treatment of the hands.

Talipes, or **clubfoot,** is a congenital malformation in which the sole of the foot is twisted medially. It is uncertain if abnormal positioning or restricted movement *in utero* causes this anomaly, but both genetical and environmental conditions are involved in most cases.

Trauma to the Hip and Lower Extremity

As with the shoulder and upper extremity, there is a variety of traumas to the hip and lower extremity, ranging from injury to the bones and surrounding muscles, tendons, vessels, and nerves to damage of the joints in the form of sprains or dislocations.

Dislocation of the hip is a common and severe result of an automobile accident when a seat belt is not worn. When the hip is in the flexed position, as in sitting in a seat, a sudden force onto the distal end of the femur will drive the head of the femur out of the acetabular socket, fracturing the posterior acetabular lip. In this kind of injury, there is usually damage to the sciatic nerve.

Trauma to the nerve roots that form the sciatic nerve may also occur from a herniated disc or pressure from the uterus during pregnancy. An improperly administered injection into the buttock may damage the sciatic nerve itself. Sciatic nerve damage is usually very painful and is expressed throughout the posterior length of the lower extremity.

Fractures are common in any location of the hip and lower extremity. Athletes (such as skiers) and elderly persons seemingly are the most vulnerable. *Osteoporosis* markedly weakens the bones of the hip and thigh regions making them vulnerable for fracture. A common fracture site, especially in elderly women, is across the femoral neck. A fracture of this kind may be complicated with vascular and nerve interruption. A direct blow to the knee will frequently fracture the patella. A potentially more serious knee trauma, however, is a clipping injury, caused by a blow to the lateral side. In this type of injury there is generally damage to the cruciate ligament and menisci. Serious complications arise if the common peroneal nerve, traversing the popliteal fossa, is damaged. Damage to this nerve results in paralysis of the ankle and foot extensors (footdrop) and inversion of the foot.

Pott's fracture is a common and serious break to either or both of the distal ends of the tibia and fibula at the level of the malleoli. Complications of a Pott's fracture are restricted blood flow to the foot and necrosis. A Pott's fracture is always of immense concern if the injury occurs to an adolescent or a young child who is still growing. If the epiphyseal plates are involved, the trauma could mean one leg would be somewhat shorter than the other.

Stress fractures of the long bones of the lower extremity are common to athletes. *Shinsplints,* which are trauma to the anterior muscles of the leg and their periosteal attachments, are often accompanied by tibial stress fractures. Stress fractures of the metatarsals may be very painful and yet not even appear on X rays. Frequently, the only way to heal stress fractures is abstinence from exercise.

Sprains are common in the joints of the lower extremity. Ligaments and tendons are torn to varying degrees in sprains. Sprains are usually accompanied by *synovitis,* an inflammation of the joint capsule.

Diseases of the Hip and Lower Extremity

As in the shoulder and upper extremity, infections in the hip and lower extremity—such as bursitis and tendonitis—can be localized in any part of the hip or lower extremity. Likewise, several types of arthritis may affect joints in these regions.

There are numerous skin diseases that afflict the foot, including athlete's foot, plantar warts, and dyshidrosis. Most of the diseases of the feet can be prevented, or if they do occur, they can be treated effectively.

Because arterial occlusive disease is common in elderly people, palpation of the posterior tibial artery is clinically important in general physical assessment. This can be accomplished by gently palpating between the medial malleolus and the gastrocnemius tendon.

Many neuromuscular diseases have a direct effect upon the functional capabilities of the lower extremities. Muscular dystrophy and poliomyelitis are both serious immobilizing diseases because of muscle paralysis.

1. List three congenital conditions of the hip or lower extremity, and describe the anatomical appearance of these conditions.
2. Describe the common kinds of trauma to the hip or lower extremity.
3. Discuss the types of diseases that may afflict the hip or lower extremity—including infections, skin diseases, skeletal diseases, and neuromuscular diseases.

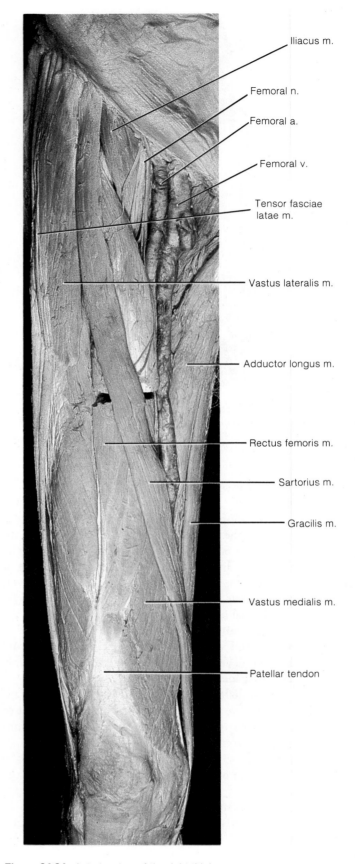

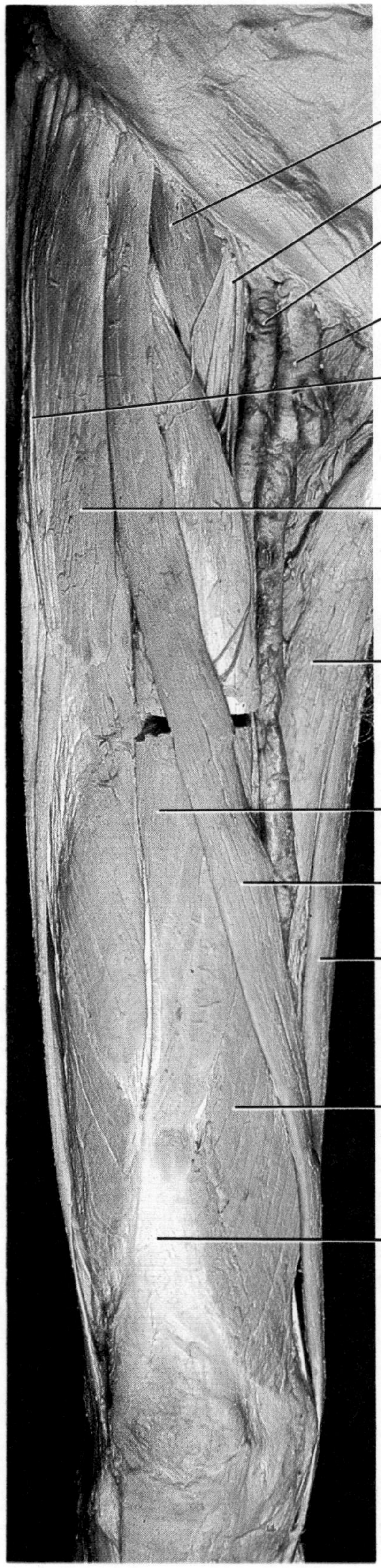

Figure 24.24 Anterior view of the right thigh.

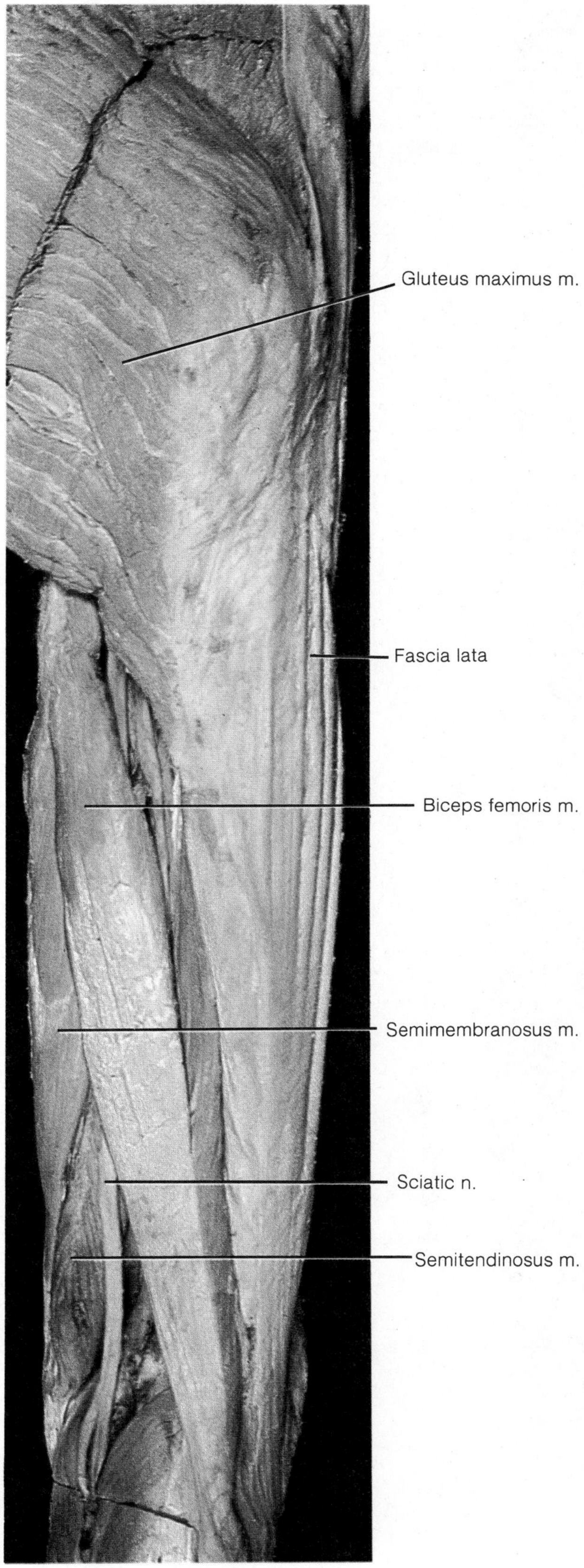

Figure 24.25 Posterior view of right hip and thigh.

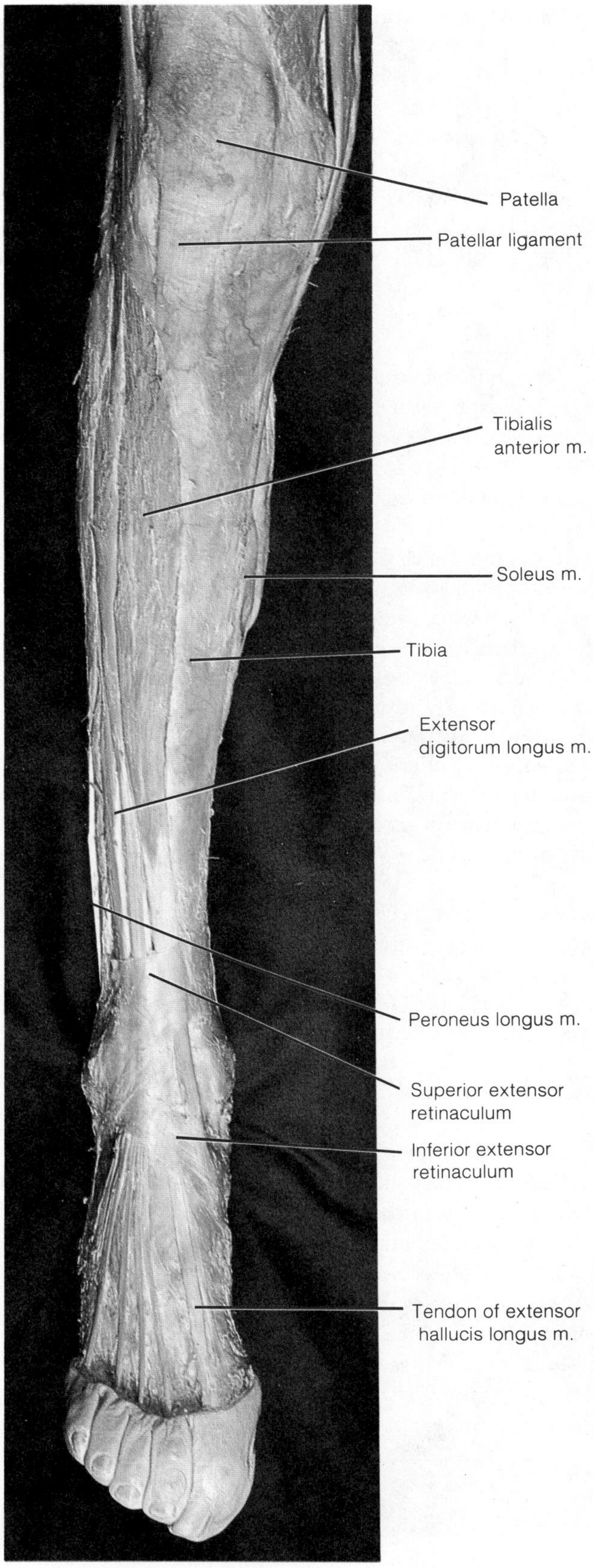

Figure 24.26 Anterior view of right leg.

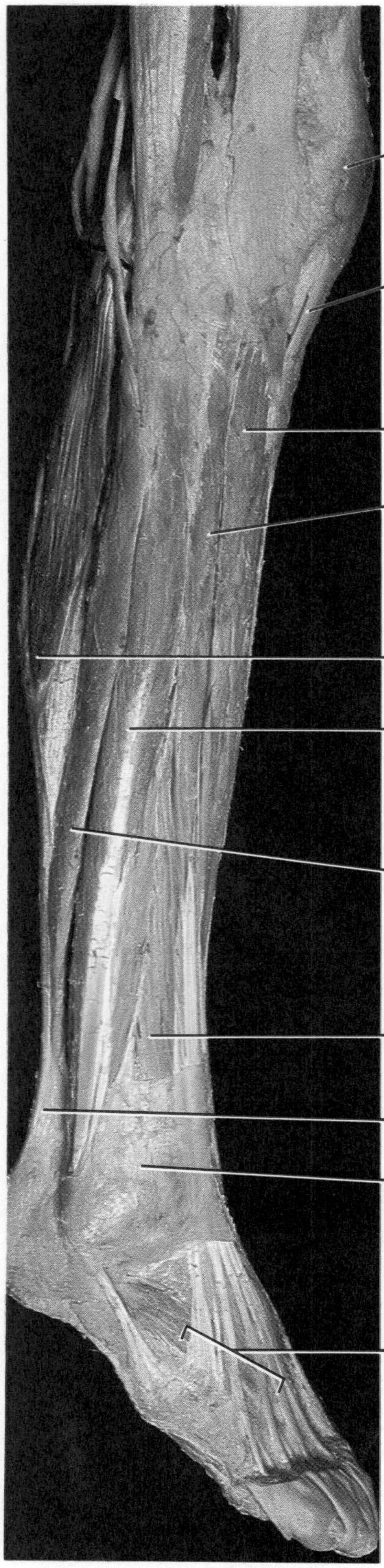

Figure 24.27 Lateral view of right leg.

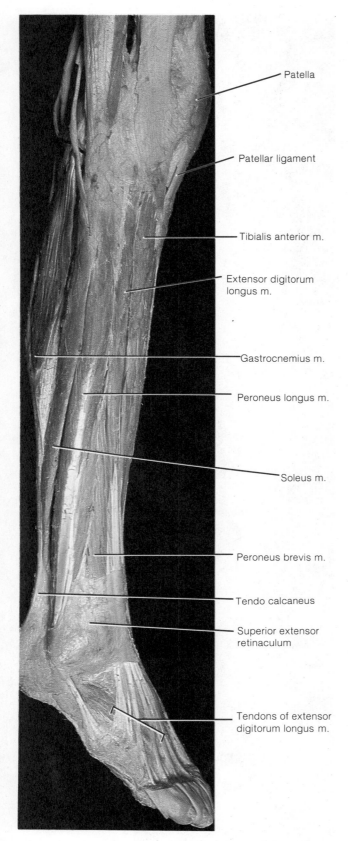

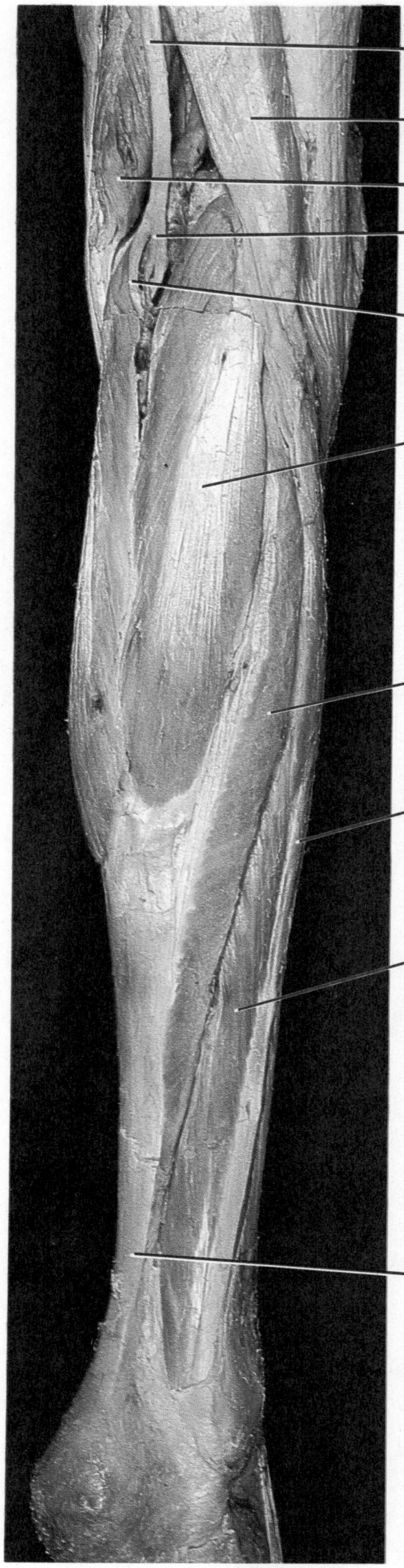

Figure 24.28 Posterior view of right leg.

Chapter Summary

I. Head and Neck Regions
   A. The head and neck regions are the most highly integrated regions of the body.
      1. These regions communicate with or control all of the systems of the body.
      2. Pathogens have easy access to the head and neck regions.
   B. Congenital malformations of the head include anencephaly, microcephaly, hydrocephaly, and cleft palate.
   C. Blows to the head or neck are frequently debilitating, if not fatal.
      1. Subdural hemorrhage is a frequent consequence of head injuries.
      2. Whiplash is usually painful and difficult to diagnose because of the innervation and complexity of the neck.
   D. Infections of the mucous membranes of the head and neck are common and can spread from one location to another.

II. Thoracic Region
   A. The thorax includes the rib cage and its contents, the thoracic musculature, and the mammary glands and breasts of a female.
   B. Deformities of the rib cage are common, but they are generally not life threatening.
      1. Pigeon breast and funnel chest are anomalies caused by an overgrowth of the ribs.
      2. Hemivertebrae are the developmental absence of some ribs, whereas cervical ribs are extra ribs in the lower neck region.
   C. Heart development is rapid and complex, resulting in 3% of newborns having some type of congenital cardiac defect.
      1. The cardiac malformations usually involve the heart valves, septa (atrial and/or ventricular), or both.
      2. A tetralogy of Fallot is a common heart defect characterized by four defects.
   D. Fractured ribs are common and may tear a pleural membrane, resulting in a pneumothorax and a collapsed lung.
   E. A cardiac tamponade results from a severe blow to the chest and is fatal because blood floods the peridcardial cavity compressing the heart.
   F. Because of the high occurrence of breast cancer, breast self-examinations are important as well as follow-up treatment if suspected neoplasms are detected.
   G. Cardiovascular diseases include an array of conditions and are the leading cause of death in the United States.

III. Abdominal Region
   A. The umbilicus is susceptible to several congenital anomalies, including patent urachus, Meckel's diverticulum, and fecal fistula.
   B. Anomalies of the GI tract include pyloric stenosis and megacolon. Likewise, the kidneys, spleen, and genitalia are susceptible to congenital problems.
   C. The liver, spleen, pancreas, and kidneys are highly vascular organs which are frequently damaged by abdominal trauma.
   D. Trauma to the external genitalia is common in males due to compression and in females due to sexual abuse.
   E. Peritonitis results from contamination of the peritoneal cavity.
   F. Ulcers may occur throughout the GI tract as a result of a breakdown of the mucosal lining.

IV. Shoulder and Upper Extremity
   A. Congenital abnormalities are most likely to occur in the extremities during the sensitive three weeks (third to seventh week) of embryonic development.
      1. Sprengel's deformity affects the development of one or both scapulae.
      2. Polydactyly and syndactyly are common anomalies of the digits.
   B. Upper arm birth palsy results from serious trauma to a baby during a difficult delivery.
   C. Fractures are common in each of the bones of the shoulder and upper extremity, and most frequently result from falling on an outstretched arm.
   D. Localized infections caused by trauma and introduced pathogens are the most common diseases of the shoulder and upper extremity.

V. Hip and Lower Extremity
   A. Congenital malformations of the hip and lower extremity include dislocation of the hip, polydactyly, syndactyly, and talipes; each of these can be corrected surgically.
   B. Trauma to the sciatic nerve may result from a herniated disc, pressure during pregnancy, or an improperly administered injection into the buttock.
   C. Fractures are common in each of the bones of the hip and lower extremity and frequently cause serious complications in the surrounding vessels and nerves.
   D. Diseases of the hip and lower extremity may be in the form of localized infections or as specific endemic diseases of body parts, such as the foot.

Review Activities

Objective Questions

1. Premature closure of the cranial sutures and mental retardation are characteristic of
   (a) anencephaly.
   (b) plagiocephaly.
   (c) encephalocele.
   (d) microcephaly.
2. Which of the following anomalies does *not* involve the digestive system?
   (a) patent urachus
   (b) cleft palate
   (c) megacolon
   (d) Meckel's diverticulum
3. Congenital limb deformities usually occur during weeks
   (a) 2–4. (c) 8–10.
   (b) 3–7. (d) 10–12.
4. Sprengel's deformity is a congenital anomaly of the
   (a) scapula.
   (b) foot.
   (c) small intestine.
   (d) diaphragm.
5. Micromelia and amelia are congenital conditions of the
   (a) rib cage.
   (b) upper extremities.
   (c) os coxae.
   (d) lower extremities.

Essay Questions

1. Construct a table of the congenital malformations of the GI tract in the head and neck, thorax, and abdomen. Indicate the possible complications of each of these anomalies.
2. Discuss the clinical problems that can afflict the serous membranes of the thoracic and abdominal regions.
3. Compare the kinds of trauma that are common in the shoulder and hip joints.
4. List the important surface landmarks for doing physical assessment of the thoracic and abdominal regions.
5. Discuss the kinds of trauma that are common to the genitalia.

Appendixes

Appendix A
Answers to Objective Questions

Chapter 1
1. (a) 2. (a) 3. (c) 4. (d) 5. (a) 6. (c) 7. (c) 8. (d) 9. (b) 10. (a)

Chapter 2
1. (b) 2. (a) 3. (b) 4. (a) 5. (b) 6. (c) 7. (c) 8. (b) 9. (b) 10. (b) 11. (c) 12. (c) 13. (c) 14. (d)

Chapter 3
1. (c) 2. (d) 3. (a) 4. (e) 5. (b) 6. (c) 7. (d) 8. (a) 9. (b) 10. (b)

Chapter 4
1. (b) 2. (c) 3. (a) 4. (b) 5. (d) 6. (a) 7. (c) 8. (a) 9. (b) 10. (d)

Chapter 5
1. (a) 2. (b) 3. (b) 4. (a) 5. (d) 6. (a) 7. (c) 8. (b) 9. (b) 10. (d)

Chapter 6
1. (c) 2. (a) 3. (b) 4. (b) 5. (a) 6. (d) 7. (a) 8. (b) 9. (d) 10. (c)

Chapter 7
1. (a) 2. (d) 3. (d) 4. (b) 5. (b) 6. (d) 7. (e) 8. (b) 9. (d) 10. (a)

Chapter 8
1. (b) 2. (a) 3. (d) 4. (d) 5. (a) 6. (d) 7. (c) 8. (b) 9. (b) 10. (d)

Chapter 9
1. (d) 2. (c) 3. (e) 4. (d) 5. (c) 6. (a) 7. (a) 8. (b) 9. (c) 10. (b)

Chapter 10
1. (e) 2. (b) 3. (d) 4. (d) 5. (d) 6. (a) 7. (a) 8. (b) 9. (c) 10. (a)

Chapter 11
1. (d) 2. (a) 3. (e) 4. (c) 5. (d) 6. (a) 7. (a) 8. (a) 9. (d) 10. (b) 11. (c) 12. (d)

Chapter 12
1. (a) 2. (b) 3. (b) 4. (c) 5. (a) 6. (b) 7. (c) 8. (d) 9. (a) 10. (c)

Chapter 13
1. (d) 2. (d) 3. (c) 4. (c) 5. (a) 6. (c) 7. (b) 8. (d) 9. (e) 10. (c)

Chapter 14
1. (e) 2. (b) 3. (d) 4. (c) 5. (a) 6. (c) 7. (d) 8. (e) 9. (e) 10. (d) 11. (a) 12. (b) 13. (e) 14. (d) 15. (a)

Chapter 15
1. (d) 2. (b) 3. (d) 4. (c) 5. (d) 6. (b) 7. (b) 8. (b) 9. (d) 10. (b)

Chapter 16
1. (b) 2. (c) 3. (a) 4. (b) 5. (c) 6. (f) 7. (c) 8. (b) 9. (a) 10. (a) 11. (a) 12. (c)

Chapter 17
1. (e) 2. (a) 3. (b) 4. (a) 5. (e) 6. (a) 7. (d) 8. (d) 9. (c) 10. (a)

Chapter 18
1. (a) 2. (c) 3. (b) 4. (c) 5. (b) 6. (d) 7. (c) 8. (a) 9. (a) 10. (a)

Chapter 19
1. (d) 2. (b) 3. (d) 4. (c) 5. (a) 6. (d) 7. (d) 8. (e) 9. (c) 10. (b)

Chapter 20
1. (a) 2. (b) 3. (d) 4. (b) 5. (e) 6. (b) 7. (a) 8. (a) 9. (c) 10. (a)

Chapter 21
1. (d) 2. (a) 3. (c) 4. (b) 5. (c) 6. (d) 7. (d) 8. (c) 9. (b) 10. (d)

Chapter 22
1. (d) 2. (c) 3. (a) 4. (c) 5. (d) 6. (a) 7. (c) 8. homozygous recessive (bb); heterozygous (Bb); homozygous dominant (BB) 9. (a) 10. (c)

Chapter 23
1. (a) 2. (c) 3. (b) 4. (c) 5. (a) 6. (b) 7. (c) 8. (a) 9. (b) 10. (c)

Chapter 24
1. (d) 2. (a) 3. (b) 4. (a) 5. (b)

Appendix B
Scientific Journals of Anatomy

Acta Anatomica (Basel)
Acta Biologica et Medica Germanica (Berlin)
Acta Cytologica (St. Louis)
Acta Embryologiae Experimentalis (Palermo)
Acta Morphologica Academiae Scientiarum Hungaricae (Budapest)
Acta Morphologica Academiae Scientiarum Hungaricae Supplementum (Budapest)
Acta Morphologica Neerlando-Scandinavica (Utrecht)
Activitas Nervosa Superior (Prague)
Age and Ageing (London)
Agressologie (Paris)
American Journal of Anatomy (New York)
Anatomical Record (New York)
Anatomischer Anzeiger (Jena)
Anatomy and Embryology (Berlin)
Andrologia (Berlin)
Archives d'Anatomie, d'Histologie et d'Embryologie (Strasbourg)
Archives d'Anatomie et de Cytologie Pathologiques (Paris)
Archives d'Anatomie Microscopique et de Morphologie Experimentale (Paris)
Archivio di Fisiologia (Florence)
Archivio Italiano di Anatomia e di Embriologia (Florence)
Archivum Histologicum Japonicum. Nihon Soshikigaku Kiroku (Niigata)
Arkhiv Anatomii, Gistologii i Embriologii (Moscow)
Biology of the Neonate (Basel)
Brain, Behavior, and Evolution (Basel)
Calcified Tissue International (New York)
Canadian Journal of Genetics and Cytology (Ottawa)
Cell (Cambridge MA)
Cell Biology International Reports (London)
Cell Differentiation (Limerick)
Cell and Tissue Kinetics (Oxford)
Cell and Tissue Research (Berlin)
Cellule (Brussels)
Chronobiologia (Milan)
Computers and Biomedical Research (New York)
Connective Tissue Research (London)
Contraception (Los Altos CA)
Cytobiologie (Stuttgart)
Cytobios (Cambridge ENG)
Cytogenetics and Cell Genetics (Basel)
Cytologia (Tokyo)
Developmental Biology (New York)
Differentiation (New York)
EEG/EMG (Stuttgart)
Electroencephalography and Clinical Neurophysiology (Limerick)
Experimental Cell Research (New York)
Folia Morphologica (Prague)
Growth (Lakeland FL)
Human Development (Basel)
International Journal of Aging and Human Development (Farmingdale NY)
In Vitro (Gaithersburg MD)

Journal of Anatomy (Cambridge ENG)
Journal of Cell Biology (New York)
Journal of Cell Science (London)
Journal of Electron Microscopy (Tokyo)
Journal of Embryology and Experimental Morphology (Colchester)
Journal of Human Ergology (Tokyo)
Journal of Membrane Biology (New York)
Journal of Microscopy (Oxford)
Journal of Molecular Evolution (Berlin)
Journal of Morphology (New York)
Journal of Neurocytology (London)
Journal of Physiology (London)
Journal of Ultrastructure Research (New York)
Kaibogaku Zasshi. Journal of Anatomy (Tokyo)
Life Sciences (Oxford)
Mechanisms of Ageing and Development (Limerick)
Okajima's Folia Anatomica Japonica (Tokyo)
Prostaglandins, Leukotrienes, and Medicine (Edinburgh)
Scanning Electron Microscopy (Chicago)
Stain Technology (Baltimore)
Teratology (New York)
Tissue and Cell (Harlow)
Tsitologiya (Leningrad)
Tsitologiya i Genetika (Kiev)
Ultramicroscopy (Amsterdam)
Undersea Biomedical Research (Bethesda)
Virchows Archiv. A. Pathological Anatomy and Histology (New York)
Zeitschrift fuer Mikroskopisch-Anatomische Forschung (Leipzig)
Zeitschrift fuer Morphologie und Anthropologie (Stuttgart)
Zeitschrift fuer Tierphysiologie Tierernaehrung und Futtermittelkunde (Hamburg)

Appendix C
Laboratory Demonstrations in Anatomy

The study of human anatomy can be augmented by obtaining various vertebrate organs such as the brain, heart, eye, trachea, lung, and kidney from a local slaughterhouse.

Fresh organs can be preserved by placing them in an embalming solution made according to the following formula:

Carbolic acid (melted crystals)	5 parts
Formalin (40%)	5 parts
Glycerin	5 parts
Water	85 parts

Carbolic acid is a disinfectant and helps maintain natural tissue color. Formalin is the main preservative. Glycerin prevents the anatomical structures from drying out too rapidly. Once an organ is preserved, it can be stored for further, more detailed dissection at a later date.

Using human cadavers benefits learning human anatomy. Cadavers have many advantages over anatomical charts, models, and illustrations when demonstrating size, position, and regional relationships within the human body. Furthermore, advanced anatomy students may be able to dissect the cadaver for use in an elementary human anatomy course.

Appropriate X rays displayed on a viewing screen give an added dimension to the anatomy course. Outdated X rays are usually available from doctors' offices, clinics, or hospitals.

Directions for preparing many interesting and valuable anatomical demonstrations can be found in the following books:

Hildebrand, M. 1968. *Anatomical preparations.* Berkeley: Univ. of California Press.

Tompsett, D. H. 1970. *Anatomical techniques.* 2d ed. Edinburgh: Livingstone.

Appendix D
Medical and Pharmacological Abbreviations

aa	of each
a.c.	before meals
A/G	albumin globulin ratio
ANS	autonomic nervous system
Bib.	drink
b.i.d.	twice a day
bihor	during two hours
B.M.R.	basal metabolic rate
B.P.	blood pressure
BUN	blood urea nitrogen
b.v.	vapor bath
c̄	with
caps.	capsule
c.b.c.	complete blood count
cc.	cubic centimeter(s)
cm.	centimeter(s)
CNS	central nervous system
Co., Comp.	compound
cr	tomorrow
C.S.F.	cerebrospinal fluid
CVP	central venous pressure
d.	a day
D & C	dilatation and curettage
D.C.	discontinue
D, Det.	give
de d. in d.	from day to day
Dieb. secund	every second day
Dieb. tert.	every third day
dim.	one-half
d. in dup.	give twice as much
D. in p. aeq.	divide into equal parts
dr.	dram
D.T.D.	give of such doses
Dur. dolor.	while pain lasts

e	out of, with
ECG, EKG	electrocardiogram
EEG	electroencephalogram
e.m.p.	in the manner prescribed
feb	fever
G.I.	gastrointestinal
gm.	gram
gr.	grain
Grad.	gradually
gtt.	drop(s)
h.	hour
HCT	hematocrit
Hg.	mercury
Hgb	hemoglobin
h.s.	at bedtime
ibid.	in the same place
I.M.	intramuscular
incid.	cut
in d.	in a day
inj.	an injection
int. cib.	between meals
int. noct.	during the night
IPPB	intermittent positive pressure breathing
I.V.	intravenous
kg.	kilogram
Lat. dol.	to the painful side
M.	mix
man.	in the morning
mEq.	milliequivalent
mg.	milligram
ml.	milliliter
Noct.	at night
Noct. maneq.	night and morning
N.P.O.	nothing by mouth
O.D.	in the right eye
o.d.	every day
Omn. hor.	every hour
Omn. man.	every morning
Omn. noct.	every night
O.S.	in the left eye
O.U.	in each eye
oz.	ounce
Part. aeq.	equal parts
PBI	protein-bound iodine
p.c.	after meals
pCO_2	partial pressure of carbon dioxide
PNS	peripheral nervous system
P.O.	by mouth
pO_2	partial pressure of oxygen
p.p.a.	having first shaken the bottle
p.r.n.	as needed
pro. us. ext.	for external use
pt.	let it be continued
q.	each; every
q.d.	every day
q.h.	every hour
q. ____ h.	every ____ hours
q.i.d.	four times a day
q. noct.	every night
q.o.d.	every other day
q.q.	also
q.s.	sufficient quantity
RBC	red blood cell
s̄	without
Semih.	half an hour
Sig.	write, label
S.O.S.	if needed
sp. gr.	specific gravity
ss., s̄s̄	one-half
s.s.s.	layer on layer
stat.	immediately
sum.	take
s.v.r.	alcohol
t.	three times
tab.	tablet
t.i.d.	three times a day
ung.	ointment
Ut. dict.	as directed
vic.	times
WBC	white blood cell

Appendix E

Units of Measurement and Their Equivalents

Apothecaries' Weights and Their Metric Equivalents

1 grain (gr) =
0.05 scruple (s)
0.017 dram (dr)
0.002 ounce (oz)
0.0002 pound (lb)
0.065 gram (g)
65. milligrams (mg)

1 scruple (s) =
20. grains (gr)
0.33 dram (dr)
0.042 ounce (oz)
0.004 pound (lb)
1.3 grams (g)
1,300. milligrams (mg)

1 dram (dr) =
60. grains (gr)
3. scruples (s)
0.13 ounce (oz)
0.010 pound (lb)
3.9 grams (g)
3,900. milligrams (mg)

1 ounce (oz) =
480. grains (gr)
24. scruples (s)
8. drams (dr)
0.08 pound (lb)
31.1 grams (g)
31,100. milligrams (mg)

1 pound (lb) =
5,760. grains (gr)
288. scruples (s)
96. drams (dr)
12. ounces (oz)
373. grams (g)
373,000. milligrams (mg)

Apothecaries' Volumes and Their Metric Equivalents

1 minim (min) =
0.017 fluid dram (fl dr)
0.002 fluid ounce (fl oz)
0.0001 pint (pt)
0.06 milliliter (ml)
0.06 cubic centimeter (cc)

1 fluid dram (fl dr) =
60. minims (min)
0.13 fluid ounce (fl oz)
0.008 pint (pt)
3.70 milliliters (ml)
3.70 cubic centimeters (cc)

1 fluid ounce (fl oz) =
480. minims (min)
8. fluid drams (fl dr)
0.06 pint (pt)
29.6 milliliters (ml)
29.6 cubic centimeters (cc)

1 pint (pt) =
7,680. minims (min)
128. fluid drams (fl dr)
16. fluid ounces (fl oz)
473. milliliters (ml)
473. cubic centimeters (cc)

1 milligram (mg) =
0.000001 kilogram (kg)
0.001 gram (g)
1,000. micrograms (μg)
0.0154 grains (gr)
0.000032 ounce (oz)

Metric Weights and Their Apothecaries' Equivalents

1 gram (g) =
0.001 kilogram (kg)
1,000. milligrams (mg)
1,000,000. micrograms (μg)
15.4 grains (gr)
0.032 ounce (oz)

1 kilogram (kg) =
1,000. grams (g)
1,000,000. milligrams (mg)
1,000,000,000. micrograms (μg)
32. ounces (oz)
2.7 pounds (lb)

Metric Volumes and Their Apothecaries' Equivalents

1 liter (l) =
1,000. milliliters (ml)
1,000. cubic centimeters (cc)
2.1 pints (pt)
270. fluid drams (fl dr)
34. fluid ounces (fl oz)

1 milliliter (ml) =
0.001 liter (l)
1. cubic centimeter (cc)
16.2 minims (min)
0.27 fluid dram (fl dr)
0.034 fluid ounce (fl oz)

Approximate Equivalents of Household Measures

1 teaspoon (tsp) =
4. milliliters (ml)
4. cubic centimeters (cc)
1. fluid dram (fl dr)

1 tablespoon (tbsp) =
15. milliliters (ml)
15. cubic centimeters (cc)
0.5 fluid ounce (fl oz)
3.7 teaspoons (tsp)

1 cup (c) =
240. milliliters (ml)
240. cubic centimeters (cc)
8. fluid ounces (fl oz)
0.5 pint (pt)
16. tablespoons (tbsp)

1 quart (qt) =
960. milliliters (ml)
960. cubic centimeters (cc)
2. pints (pt)
4. cups (c)
32. fluid ounces (fl oz)

Conversion of Units from One Form to Another

Refer to the preceding equivalency lists when converting one unit to another equivalent unit.

To convert a unit shown in bold type to one of the equivalent units listed immediately below it, multiply the first number (bold type unit) by the appropriate equivalent unit listed below it.

Sample Problems:

1. Convert 320 grains into scruples (1 gr = 0.05 s).

$$320\text{ gr} \times \frac{0.05\text{ s}}{1\text{ gr}} = 16.0\text{ s}$$

2. Convert 320 grains into drams (1 gr = 0.017 dr).

$$320\text{ gr} \times \frac{0.017\text{ dr}}{1\text{ gr}} = 5.44\text{ dr}$$

3. Convert 320 grains into grams (1 gr = 0.065 g).

$$320\text{ gr} \times \frac{0.065\text{ g}}{1\text{ gr}} = 20.8\text{ g}$$

Body Temperatures in ° Fahrenheit and ° Celsius

°F	°C	°F	°C
95.0	**35.0**	**100.0**	**37.8**
95.2	**35.1**	**100.2**	**37.9**
95.4	**35.2**	**100.4**	**38.0**
95.6	**35.3**	**100.6**	**38.1**
95.8	**35.4**	**100.8**	**38.2**
96.0	**35.5**	**101.0**	**38.3**
96.2	**35.7**	**101.2**	**38.4**
96.4	**35.8**	**101.4**	**38.6**
96.6	**35.9**	**101.6**	**38.7**
96.8	**36.0**	**101.8**	**38.8**
97.0	**36.1**	**102.0**	**38.9**
97.2	**36.2**	**102.2**	**39.0**
97.4	**36.3**	**102.4**	**39.1**
97.6	**36.4**	**102.6**	**39.2**
97.8	**36.6**	**102.8**	**39.3**
98.0	**36.7**	**103.0**	**39.4**
98.2	**36.8**	**103.2**	**39.6**
98.4	**36.9**	**103.4**	**39.7**
98.6	**37.0**	**103.6**	**39.8**
98.8	**37.1**	**103.8**	**39.9**
99.0	**37.2**	**104.0**	**40.0**
99.2	**37.3**	**104.2**	**40.1**
99.4	**37.4**	**104.4**	**40.2**
99.6	**37.6**	**104.6**	**40.3**
99.8	**37.7**	**104.8**	**40.4**
		105.0	**40.6**

To convert ° F to ° C

Subtract 32 from ° F and multiply by 5/9.
——° F − 32 × 5/9 = —— ° C

To convert ° C to ° F

Multiply ° C by 9/5 and add 32.
——° C × 9/5 + 32 = ——° F

Appendix F
Suggested Additional Resources

Chapter 1

Chewning, E. B. 1979. *Anatomy illustrated.* New York: Simon and Schuster.
Corner, G. W. 1977. *Anatomical texts of the earlier Middle Ages.* Carnegie Institution of Washington, Publication No. 364. Washington, D.C.: National Publishing Co.
Geller, S. A. March 1983. Autopsy. *Scientific American.*
Kevorkian, J. 1959. *The story of dissection.* New York: Philosophical Library.
Lyons, A. S., and R. J. Petrucelli. 1978. *Medicine, an illustrated history.* New York: Harry N. Abrams.
Persaud, T. V. N. 1984. *Early history of human anatomy from antiquity to the beginning of the modern era.* Springfield, Ill.: Charles C. Thomas, Publisher.
Singer, C. 1957. *A short history of anatomy and physiology from the Greeks to Harvey.* 2d ed. New York: Dover Publications.
Youngson, A. J. 1979. *The scientific revolution in victorian medicine.* London: Croom Helm.

Chapter 2

Bronowski, J. 1971. *The identity of man.* rev. ed. New York: Doubleday and Co.
Clemente, C. D. 1984. *Gray's anatomy of the human body.* 30th ed. Philadelphia: Lea and Febiger.
Clemente, C. D. 1987. *Anatomy: a regional atlas of the human body.* 3d ed. Baltimore: Urban and Schwarzenberg.
Garn, S. M. 1971. *Human races.* 3d ed. Springfield, Ill.: Charles C. Thomas, Publisher.
Hall-Craggs, E. C. B. 1990. *Anatomy as a basis for clinical medicine.* 2d ed. Baltimore-Munich: Urban and Schwarzenberg.
Kennedy, D. 1967. Introduction to *From cell to organism: readings from* Scientific American. San Francisco: W. H. Freeman and Co.
Last, R. J. 1984. *Anatomy, regional and applied.* 7th ed. Boston: Little, Brown and Co.
Lockhart, R. D. 1974. *Living anatomy.* London: Faber and Faber.
Lockhart, R. D. et al. 1981. *Anatomy of the human body.* Philadelphia: Lea and Febiger.
Morris, D. 1984. *The naked ape.* New York: Dell.
Romanes, G. J. 1981. *Cunningham's textbook of anatomy.* 12th ed. New York: Oxford University Press.
Royce, J. 1973. *Surface anatomy.* Philadelphia: F. A. Davis Co.
Sobotta, J., and F. Figge. 1983. *Atlas of human anatomy.* 10th ed. New York: Hafner Press.
Van De Graaff, K. M., and S. I. Fox. 1989. *Concepts of human anatomy and physiology.* 2d ed. Dubuque: Wm. C. Brown Publishers.
Woodburne, R. T. 1988. *Essentials of human anatomy.* 8th ed. New York: Oxford University Press.

Chapter 3

Avers, C. J. 1988. *Cell biology.* 3d ed. New York: D. Van Nostrand Co.
Bretscher, M. S. October 1985. The molecules of the cell membrane. *Scientific American.*
Corbett, T. H. 1977. *Cancer and chemicals.* Chicago: Nelson-Hall.
Dautry-Varsat, A., and H. Lodish. May 1984. How receptors bring proteins and particles into cells. *Scientific American.*
DeDuve, C. May 1983. Microbodies in the living cell. *Scientific American.*
DeWitt, W., and E. R. Brown. 1977. *Biology of the cell.* Philadelphia: W. B. Saunders.
Dustin, P. August 1980. Microtubules. *Scientific American.*
Grivell, L. A. March 1983. Mitochondrial DNA. *Scientific American.*
Kornberg, R. D., and A. Klug. February 1981. The nucleosome. *Scientific American.*
Lake, J. A. August 1981. The ribosome. *Scientific American.*
Lodish, H. F., and J. E. Rothman. January 1979. The assembly of cell membranes. *Scientific American.*
Lyon, J. L. July 1984. Radiation exposure and cancer. *Hosp. Prac.*
Mirsky, A. E. June 1968. The discovery of DNA. *Scientific American.*
Nomura, M. January 1984. The control of ribosome synthesis. *Scientific American.*
Reif, A. E. July–August 1981. The causes of cancer. *Amer. Scientist.*
Rothman, J. E. 1981. The Golgi apparatus. *Science* 213:1212.
Sloboda, R. D. May–June 1980. The role of microtubules in cell structure and cell division. *Amer. Scientist.*
Staehelin, L. A., and B. E. Hull. May 1978. Junctions between living cells. *Scientific American.*
Unwin, N., and R. Henderson. February 1984. The structure of proteins in biological membranes. *Scientific American.*
Weber, K., and M. Osborn. October 1985. The molecules of the cell matrix. *Scientific American.*
Weinberg, R. A. October 1985. The molecules of life. *Scientific American.*
Wolfe, S. L. 1981. *Biology of the cell.* 2d ed. Belmont, Calif.: Wadsworth.

Chapter 4

Bevelander, G., and J. A. Ramaley. 1979. *Essentials of histology.* 8th ed. St. Louis: C. V. Mosby Co.
Bloom, W. B., and D. W. Fawcett. 1986. *A textbook of histology.* 11th ed. Philadelphia: W. B. Saunders Co.
Di Fiore, M. S. H. 1988. *The atlas of normal histology.* 6th ed. Philadelphia: Lea and Febiger.
Kessel, R. G., and R. H. Kardon. 1979. *Tissues and organs: a text-atlas of scanning electron microscopy.* San Francisco: W. H. Freeman and Co.
Porter, K. R., and M. A. Bonneville. 1973. *An introduction to the fine structure of cells and tissues.* 4th ed. Philadelphia: Lea and Febiger.
Weiss, L., and R. O. Greep. 1983. *Histology.* 5th ed. New York: McGraw-Hill.
Wheater, P. R., H. G. Burkitt, and V. G. Daniels. 1987. *Functional histology: a text and colour atlas.* 2d ed. London: Churchill Livingstone.
Windle, W. F. 1976. *Textbook of histology.* 5th ed. New York: McGraw-Hill.

Chapter 5

Bluefarb, S. M. 1978. *Dermatology.* Kalamazoo, Mich.: The Upjohn Co.
Elden, H. R., ed. 1971. *Biophysical properties of the skin.* New York: John Wiley and Sons.
Helwig, E. B., and F. K. Mostofi, eds. 1980. *The skin.* New York: R. E. Krieger Pub. Co.
Loomis, W. F. December 1970. Rickets. *Scientific American.*
Marback, H. I., and H. S. Gavin. *Skin bacteria and their role in infection.* New York: McGraw-Hill.
Marples, M. J. 1965. *The ecology of the human skin.* Springfield, Ill.: Charles C. Thomas, Publisher.
Marples, M. J. January 1979. Life on the human skin. *Scientific American.*
Moncrief, J. A. 1973. Burns. *New Engl. J. Med.* 228:444.
Montagna, W. June 1969. The skin. *Scientific American.*
Montagna, W., and P. F. Parakkal. 1974. *The structure and function of the skin.* 3rd ed. New York: Academic Press.
Penrose, L. S. December 1969. Dermatoglyphics. *Scientific American.*
Reith, E. J., and M. N. Ross. 1985. *Histology: A text and atlas.* New York: Harper & Row.
Ross, R. June 1969. Wound healing. *Scientific American.*
Rushmer, R. L., et al. 1966. The skin. *Science* 154:343.
Tregear, R. T. 1966. *Physical functions of skin.* New York: Oxford Univ. Press.

Chapter 6

Burnstein, A. H. 1976. Aging of bone tissue: mechanical properties. *J. Bone Joint Surgery.* 58:82.
Evans, F. G., ed. 1966. *Studies on the anatomy and function of bones and joints.* New York: Springer-Verlag.
Hall, B. K. 1970. Cellular differentiation in skeletal tissue. *Biol. Rev.* 45:455.
Harris, W. H., and R. P. Heaney. 1970. *Skeletal renewal and metabolic bone disease.* Boston: Little, Brown, and Co.
Loomis, W. F. December 1970. Rickets. *Scientific American.*
Moore, K. L. 1988. *The developing human: clinically oriented embryology.* 4th ed. Philadelphia: W. B. Saunders Co.
Napier, J. April 1967. The antiquity of human walking. *Scientific American.*
Platzer, W. 1986. *Locomotor system.* Vol. 1 of *Color atlas and textbook of human anatomy.* 3d rev. ed. Chicago: Year Book Medical Publishers.
Prichard, J. J. 1977. Skeletal development. *Postgrad. Med. J.* 53:429.
Sharpe, W. D. 1979. Age changes in human bones: an overview. *Bull. N.Y. Acad. Med.* 55:757.
Snell, R. S. 1978. *Atlas of clinical anatomy.* Boston: Little, Brown and Co.
Trueta, J. 1968. *Studies in the development and decay of the human frame.* London: Heinemann Medical.
Vaughan, J. M. 1981. *The physiology of bone.* 3d ed. New York: Oxford University Press.

Chapter 7

Burnstein, A. H. 1976. Aging of bone tissue: mechanical properties. *J. Bone Joint Surgery.* 58:82.
Evans, F. G., ed. 1966. *Studies on the anatomy and function of bones and joints.* New York: Springer-Verlag.
Hall, B. K. 1970. Cellular differentiation in skeletal tissue. *Biol. Rev.* 45:455.
Harris, W. H., and R. P. Heaney. 1970. *Skeletal renewal and metabolic bone disease.* Boston: Little, Brown, and Co.
Loomis, W. F. December 1970. Rickets. *Scientific American.*
Moore, K. L. 1983. *Before we are born: basic embryology and birth defects.* 2d ed. Philadelphia: W. B. Saunders Co.

Napier, J. April 1967. The antiquity of human walking. *Scientific American.*
Platzer, W. 1986. *Locomotor system.* Vol. 1 of *Color atlas and textbook of human anatomy.* 3d rev. ed. Chicago: Year Book Medical Publishers.
Prichard, J. J. 1977. Skeletal development. *Postgrad. Med. J.* 53:429.
Sharpe, W. D. 1979. Age changes in human bones: an overview. *Bull. N.Y. Acad. Med.* 55:757.
Snell, R. S. 1978. *Atlas of clinical anatomy.* Boston: Little, Brown and Co.
Trueta, J. 1968. *Studies in the development and decay of the human frame.* London: Heinemann Medical.
Vaughan, J. M. 1981. *The physiology of bone.* 3d ed. New York: Oxford University Press.

Chapter 8

Aufranco, D. E., and R. H. Turner. October 1971. Total replacement of the arthritic hip. *Hosp. Prac.*
Evans, F. G., ed. 1966. *Studies on the anatomy and function of bones and joints.* New York: Springer-Verlag.
Napier, J. April 1967. The antiquity of human walking. *Scientific American.*
Platzer, W. 1978. *Locomotor system.* Vol. 1 of Kahle, W., H. Leonhardt, and W. Platzer. *Color atlas and textbook of human anatomy.* Chicago: Year Book Medical Publishers.
Rasch, P. J. 1989. *Kinesiology and applied anatomy: The science of human movement.* 7th ed. Philadelphia: Lea and Febiger.
Rosse, C., and D. K. Clawson. 1980. *The musculoskeletal system in health and disease.* Philadelphia: Harper and Row Publishers.
Simon, W. H., ed. 1978. *The human joint in health and disease.* Philadelphia: Univ. of Pennsylvania Press.
Sonstegard, D. A., L. S. Matthews, and H. Kaufer. January 1978. The surgical replacement of the human knee joint. *Scientific American.*

Chapter 9

Anderson, J. E. 1983. *Grant's atlas of anatomy.* 8th ed. Baltimore: The Williams and Wilkins Co.
Basmajian, J. V. 1989. *Grant's method of anatomy.* 11th ed. Baltimore: The Williams and Wilkins Co.
———. 1985. *Muscles alive.* 5th ed. Baltimore: The Williams and Wilkins Co.
Bendall, J. R. 1969. *Muscles, molecules, and movement.* New York: Elsevier Publishers.
Chapman, C. B., and J. H. Mitchell. May 1965. The physiology of exercise. *Scientific American.*
Clemente, C. D. 1987. *Anatomy: A regional atlas of the human body.* 3rd ed. Philadelphia: Lea and Febiger.
Close, R. I. 1972. Dynamic properties of mammalian skeletal muscles. *Physiol. Rev.* 52:129.
Cohen, C. November 1975. The protein switch of muscle contraction. *Scientific American.*
Eisenberg, E., and L. E. Greene. 1980. The relation of muscle biochemistry to muscle physiology. *Ann. Rev. Physiol.* 42:293.
Ferner, H., and J. Staubesand, ed. 1983. *Sobotta atlas of human anatomy.* 10th ed. Baltimore: Urban and Schwarzenberg.
Grinnell, A. D., and M. A. B. Brazier, eds. 1981. *Regulation of muscle contraction: Excitation-contraction coupling.* New York: Academic Press.
Hinson, M. M. 1981. *Kinesiology.* 2d ed. Dubuque, IA: Wm. C. Brown Company Publishers.
Hoyle, G. April 1970. How a muscle is turned off and on. *Scientific American.*
Huxley, H. E. November 1968. The contraction of muscle. *Scientific American.*
———. 1969. The mechanism of muscular contraction. *Science* 164:1356.
Junge, D. 1981. *Nerve and muscle excitation.* 2d ed. Sunderland, MA.: Sinauer Associates, Inc.
Last, R. J. 1984. *Anatomy, regional and applied.* 7th ed. Boston: Churchill Livingstone.
Margaria, R. March 1972. The sources of muscular energy. *Scientific American.*
Merton, P. A. May 1972. How we control the contraction of our muscles. *Scientific American.*
Murray, J. H., and A. Weber. February 1974. The cooperative action of muscle proteins. *Scientific American.*
Nadel, E. R. 1985. Physiological adaptations to aerobic training. *Amer. Scientist* 73:334.
Peterson, B. W. 1979. Reticulospinal projections to spinal motor nuclei. *Ann. Rev. Physiol.* 41:127.
Porter, K. R., and C. Franzini-Armstrong. March 1965. The sarcoplasmic reticulum. *Scientific American.*
Rasch, P. J. 1989. *Kinesiology and applied anatomy: The science of human movement.* 7th ed. Philadelphia: Lea and Febiger.
Rosse, C., and D. K. Clawson. 1980. *The musculoskeletal system in health and disease.* Philadelphia: Harper and Row Publishers.
Sandow, A. 1970. Skeletal muscle. *Physiol. Rev.* 32:87.
Snell, R. S. 1978. *Atlas of clinical anatomy.* Boston: Little, Brown, and Co.

Chapter 10

Anderson, J. E. 1983. *Grant's atlas of anatomy.* 8th ed. Baltimore: The Williams and Wilkins Co.
Basmajian, J. V. 1989. *Grant's method of anatomy.* 11th ed. Baltimore: The Williams and Wilkins Co.
———. 1983. *Surface anatomy: An instruction manual.* 2d ed. Baltimore: The Williams and Wilkins Co.
Clemente, C. D. 1987. *Anatomy: A regional atlas of the human body.* 3d ed. Baltimore: Urban and Schwarzenberg.
Clemente, C. D., ed. 1984. *Gray's anatomy of the human body.* 30th American ed. Philadelphia: Lea and Febiger.
Ferner, H., and J. Staubesand, ed. 1983. *Sobotta atlas of human anatomy.* 10th ed. Baltimore: Urban and Schwarzenberg.
Hall-Craggs, E. C. B. 1990. *Anatomy as a basis for clinical medicine.* 2d ed. Baltimore: Urban and Schwarzenberg.
Last, R. J. 1984. *Anatomy, regional and applied.* 7th ed. Boston: Little, Brown and Co.
Royce, J. 1973. *Surface anatomy.* Philadelphia: F. A. Davis Co.
Snell, R. S. 1978. *Atlas of clinical anatomy.* Boston: Little, Brown and Co.
Woodburne, R. T., and W. E. Burkel. 1988. *Essentials of human anatomy.* 8th ed. New York: Oxford University Press.

Chapter 11

Angevine, J. R., Jr., and C. W. Cotman. 1981. *Principles of neuroanatomy.* New York: Oxford Univ. Press.
Axelrod, J. June 1974. Neurotransmitters. *Scientific American.*
Block, I. B. et al. 1984. Neurotransmitter plasticity at the molecular level. *Science* 225:1266.
Bloom, F. E. October 1981. Neuropeptides. *Scientific American.*
Cottman, C. W., and M. Nieto-Sampedro. 1984. Cell biology of synaptic plasticity. *Science* 225:1287.
Cowen, W. M. September 1979. The development of the brain. *Scientific American.*
Coyle, J. T., D. L. Prince, and M. R. DeLong. 1983. Alzheimer's disease: A disorder of cortical cholinergic innervation. *Science* 2B 219:1184.
Decara, L. V. January 1970. Learning in the autonomic nervous system. *Scientific American.*
Evarts, E. V. September 1979. Brain mechanisms of movement. *Scientific American.*
Ferstrom, J. D., and R. J. Wurtman. February 1974. Nutrition and the brain. *Scientific American.*
Fincher, J. 1981. *The brain: mystery of matter and mind.* Washington, D.C.: U.S. News Books.
Geschwind, N. September 1979. Specializations of the human brain. *Scientific American.*
Goldstein, G. W., and A. L. Betz. September 1986. The blood-brain barrier. *Scientific American.*
Hubel, D. H. 1984. The brain. *Scientific American Offprint.*
Hubel, D. H. September 1979. The brain. *Scientific American.*
Hubel, D. H., and T. N. Wiesel. September 1979. Brain mechanisms of vision. *Scientific American.*
Iversen, L. L. September 1979. The chemistry of the brain. *Scientific American.*
Jacobs, B. L., and M. E. Trulson. 1979. Mechanisms of action of LSD. *American Scientist* 67:396.
Kandel, E. R. September 1979. Small systems of neurons. *Scientific American.*
Kety, S. S. September 1979. Disorders of the human brain. *Scientific American.*
Kimura, D. March 1973. The asymmetry of the human brain. *Scientific American.*
Krieger, D. T. 1983. Brain peptides: What, where, and why? *Science* 222:975.
Lester, H. A. February 1977. The response to acetylcholine. *Scientific American.*
Llinas, R. R. January 1975. The cortex of the cerebellum. *Scientific American.*
Nathanson, J. A., and P. Greengard. August 1977. Second messenger in the brain. *Scientific American.*
Nauta, W. J. H., and M. Feirtag. September 1979. The organization of the brain. *Scientific American.*
Norman, D. A. 1982. *Learning and memory.* San Francisco: W. H. Freeman.
Norton, W. T., and P. Morell. May 1980. Myelin. *Scientific American.*
Schwartz, J. H. April 1980. The transport of substances in nerve cells. *Scientific American.*
Shashoua, V. E. July–August 1985. The role of extracellular proteins in learning and memory. *American Scientist.*
Shepherd, G. M. February 1978. Microcircuits in the nervous system. *Scientific American.*
Siegal, R. K. October 1977. Hallucinations. *Scientific American.*
Snyder, S. H. 1984. Drugs and neurotransmitter receptors in the brain. *Science* 224:22.
Stevens, C. F. September 1979. The neuron. *Scientific American.*
Vellutino, F. R. March 1987. Dyslexia. *Scientific American.*
Wurtman, R. J. April 1982. Nutrients that modify brain function. *Scientific American.*

Chapter 12

Barr, M. L. 1988. *The human nervous system, an anatomical viewpoint.* 5th ed. Philadelphia: Lippincott.
Carpenter, M. B. 1985. *Core text of neuroanatomy.* 3d ed. Baltimore: Williams and Wilkins Co.
Clark, R. G. 1979. *Essentials of clinical neuroanatomy and neurophysiology.* Philadelphia: F. A. Davis.
Kandel, E. R., and E. R. Schwartz, eds. 1985. *Principles of neural science.* 2d ed. New York: Elsevier North-Holland.
Morell, P., and W. T. Norton. May 1980. Myelin. *Scientific American.*
Waxman, S. G. 1982. Membrane, myelin, and the pathophysiology of multiple sclerosis. *N. Engl. J. Med.* 306:1529.

Chapter 13

Angevine, J. B., Jr., and C. W. Cottman. 1981. *Principles of neuroanatomy.* New York: Oxford University Press.
Decara, L. V. January 1970. Learning in the autonomic nervous system. *Scientific American.*
Hoffman, B. B., and R. J. Lefkowitz. 1980. Alpha-adrenergic receptor subtypes. *New Engl. J. Med.* 302:1390.
Lefkowitz, B. B. 1976. Beta-adrenergic receptors: Recognition and regulation. *New Engl. J. Med.* 295:323.
Motulsky, J. H., and P. A. Insel. 1982. Adrenergic receptors in man. *New Engl. J. Med.* 307:18.
Noback, C. E., and R. J. Demerest. 1981. *The human nervous system: Basic principles of neurobiology.* 3d ed. New York: McGraw-Hill Book Company.

Chapter 14

Austin, L. A., and H. Heath, III. 1981. Calcitonin: Physiology and pathophysiology. *New Engl. J. Med.* 304:269.
Axelrod, J., and T. D. Reisine. 1984. Stress hormones: Their interaction and regulation. *Science* 224:452.
Baxter, J. D., and W. J. Funder. 1979. Hormone receptors. *New Engl. J. Med.* 300:117.
Brownstein, M. J. et. al. 1980. Synthesis, transport and release of posterior pituitary hormones. *Science* 207:373.
Carmichael, S. W., and H. Winkler. August, 1985. The adrenal chromaffin cell. *Scientific American.*
Frohman, L. A. 1975. Neurotransmitters as regulators of endocrine functions. *Hosp. Prac.* 10:54.
Ganong, W. F., L. C. Alpert, and T. C. Lee. 1974. ACTH and the regulation of adrenocortical secretion. *New Engl. J. Med.* 290:1006.
Gillie, R. B. June 1971. Endemic goiter. *Scientific American.*
Goldsmith, R. S. 1969. Hyperparathyroidism. *New Engl. J. Med.* 281:367.
Guillemin, R., and R. Burgus. November 1972. The hormones of the hypothalamus. *Scientific American.*
Katzenellenbogen, B. S. 1980. Dynamics of steroid hormone receptor action. *Ann. Rev. Physiol.* 42:17.
Krieger, D. T. 1984. Brain peptides: What, where, and why? *Science* 222:975.
McEwen, B. S. July 1976. Interactions between hormones and nerve tissue. *Scientific American.*
O'Malley, B., and W. T. Shrader. February 1976. The receptors of steroid hormones. *Scientific American.*
Rasmussen, H., and M. M. Pechet. October 1970. Calcitonin. *Scientific American.*
Reichlin, S. et. al. 1976. Hypothalamic hormones. *Ann. Rev. Physiol.* 39:389.
Roth, J. 1980. Insulin receptors in diabetes. *Hosp. Prac.* 15:98.
Roth, J., and S. I. Taylor. 1982. Receptors for peptide hormones: Alterations in diseases of humans. *Ann. Rev. of Physiol.* 44:639.
Schally, A. V. 1978. Aspects of the hypothalamic control of the pituitary gland. *Science* 202:18.
Selye, H. 1973. The evolution of the stress concept. *American Scientist* 61:692.
Wilson, J. D., and D. W. Foster, eds. 1985. *Textbook of endocrinology.* 7th ed. Philadelphia: W. B. Saunders Co.

Chapter 15

Brindley, G. S. 1970. Central pathways of vision. *Ann. Rev. Physiol.* 32:259.
Casey, K. L. 1973. Pain: a current view of neural mechanisms. *American Scientist* 61:194.
Daw, N. W. 1973. Neurophysiology of color vision. *Physiol. Rev.* 53:571.
Durrant, J. D., and J. H. Lovrinic. 1984. *Bases of hearing science.* 2d ed. Baltimore: The Williams and Wilkins Co.
Freese, A. J. 1977. *The miracle of vision.* New York: Harper and Row Publishers.
Gombrich, E. H. September 1972. The visual image. *Scientific American.*
Green, D. M. 1976. *An introduction to hearing.* New York: L. Erlbaum Assoc.
Harris, J. D. 1972. Audition. *Ann. Rev. Psych.* 23:313.
Hubel, D. H., and T. N. Weisel. September 1979. Brain mechanisms of vision. *Scientific American.*
Hudspeth, A. J. January 1983. The hair cells of the inner ear. *Scientific American.*
Kaufman, H. E. July 1973. Corneal transplantation, a progress report. *Hosp. Prac.*
Lim, R. K. S. 1970. Pain. *Ann. Rev. Physiol.* 32:269.
Loeb, G. E. February 1985. The functional replacement of the ear. *Scientific American.*
Pettigrew, J. D. August 1972. The neurophysiology of binocular vision. *Scientific American.*
Ross, J. March 1976. The resources of binocular perception. *Scientific American.*
Rubenstein, E. March 1980. Diseases caused by impaired communication among cells. *Scientific American.*
Rushton, W. A. H. March 1975. Visual pigments and color blindness. *Scientific American.*
Snyder, S. H. March 1977. Opiate receptors and internal opiates. *Scientific American.*
Toates, F. M. 1972. Accommodation function of the human eye. *Physiol. Rev.* 52:828.
Van Heyninger, R. December 1975. What happens to the human lens in cataract? *Scientific American.*
Von Bekesy, G. August 1975. The ear. *Scientific American.*
Werblin, F. S. January 1973. The control of sensitivity in the retina. *Scientific American.*
Wertenbaker, L. 1981. *The eye: Window to the world.* Washington, D.C.: U.S. New Books.

Chapter 16

Adolph, A. F. March 1967. The heart's pacemaker. *Scientific American.*
Benditt, E. P. February 1977. The origin of atherosclerosis. *Scientific American.*
Dubin, D. 1988. *Rapid interpretation of EKGs.* 4th ed. Tampa, FL: Cover Publishing Co.
Eder, H. A. May 1983. Lipoproteins and coronary artery disease. *Hosp. Prac.* 18:215.
Ferrans, V. J. July 1983. Morphology of the heart in hypertrophy. *Hosp. Prac.* 18:67.
Folkow, B., and E. Neill. 1971. *Circulation.* New York: Oxford Univ. Press.
Fuchs, R., and S. S. Scheidt. May 1983. Prevention of coronary atherosclerosis: part 1. *Cardiovascular Reviews and Reports* 4:671.
Lehman, J. 1972. Auscultation of heart sounds. *Amer. J. Nurs.* 72:1242.
Levy, M. N., and P. J. Martin. 1981. Neural regulation of the heartbeat. *Ann. Rev. Physiol.* 43:443.
Rushmer, R. F. 1976. *Structure and function of the cardiovascular system.* 2d ed. Philadelphia: W. B. Saunders Co.
Spear, J. F., and E. N. Moore. 1982. Mechanisms of cardiac arrhythmias. *Ann. Rev. Physiol.* 44:485.
Whittemore, R. December 1983. Congenital heart disease: Its impact on pregnancy. *Hosp. Prac.* 18:65.
Wood, J. E. January 1968. The venous system. *Scientific American.*

Chapter 17

Avery, M. E., N. S. Wang, and H. W. Taeusch, Jr. March 1975. The lung of the newborn infant. *Scientific American.*
Berger, A. J., R. A. Mitchel, and J. W. Severinghaus. 1977. Regulation of respiration. *New Engl. J. Med.* 297: first part, p. 92; second part, p. 138; third part, p. 194.
Brample, D. M., and D. R. Carrier. 1983. Running and breathing in mammals. *Science* 219:251.
Browning, R. J. 1982. Pulmonary disease: Back to basics (part 1); Putting blood gases to work (part 2). *Diagnostic Med.* First part: Jan/Feb, p. 39; second part: March/April, p. 59.
Cherniak, N. S. 1986. Breathing disorders during sleep. *Hosp. Pract.* 21:81.
Fielding, J. E. 1985. Smoking: Health effects and control. *New Engl. J. Med.* 313:491.
Fraser, R. G., and J. A. P. Pare. 1977. *Structure and function of the lung.* 2d ed. Philadelphia: W. B. Saunders Co.
Guz, A. 1975. Regulation of respiration in man. *Ann. Rev. Physiol.* 37:303.
Haddad, G. G., and R. B. Mellins. 1984. Hypoxia and respiratory control in early life. *Ann. Rev. Physiol.* 46:629.
Irsigler, G. B., and J. W. Severinghaus. 1980. Clinical problems of ventilatory control. *Ann. Rev. Med.* 31:109.
Naeye, R. L. April 1980. Sudden infant death. *Scientific American.*
Perutz, M. F. December 1978. Hemoglobin structure and respiratory transport. *Scientific American.*
Rigatto, H. 1984. Control of ventilation in the newborn. *Ann. Rev. Physiol.* 46:661.
Roussos, C., and P. T. Macklem. 1982. The respiratory muscles. *New Engl. J. Med.* 307:786.
Shannon, D. C., and D. H. Kelly. 1982. SIDS and near-SIDS. *New Engl. J. Med.* 306: first part, p. 959; second part, p. 1022.
Walker, D. W. 1984. Peripheral and central chemoreceptors in the fetus and newborn. *Ann. Rev. Physiol.* 46:687.
West, J. B. 1984. Human physiology at extreme altitudes on Mount Everest. *Science* 223:784.
Whipp, B. J. 1983. Ventilatory control during exercise in humans. *Ann. Rev. Physiol.* 45:393.

Chapter 18

Binder, H. J. 1984. The pathophysiology of diarrhea. *Hosp. Prac.* 19:107.
Bortoff, A. 1972. Digestion. *Ann. Rev. Physiol.* 28:201.
Carey, M. C., D. M. Small, and C. M. Bliss. 1983. Lipid digestion and absorption. *Ann. Rev. Physiol.* 45:651.
Chou, C. C. 1982. Relationship between intestinal blood flow and motility. *Ann. Rev. Physiol.* 44:29.
Christensen, R. R. 1971. The controls of gastrointestinal movements; Some old and new views. *New Engl. J. Med.* 285:483.
Cohen, S. 1983. Neuromuscular disorders of the gastrointestinal tract. *Hosp. Prac.* 18:121.
Davenport, H. W. 1982. *Physiology of the digestive tract.* 5th ed. Chicago: Year Book Medical Publishers.
Freeman, H. J., and Y. S. Kim. 1978. Digestion and absorption of proteins. *Ann. Rev. Med.* 29:99.
Gray, G. M. 1975. Carbohydrate digestion and absorption: Role of the small intestine. *New Engl. J. Med.* 292:1225.
Grossman, M. I. 1979. Neural and hormonal regulation of gastrointestinal function: An overview. *Ann. Rev. Physiol.* 41:27.
Guth, P. H. 1982. Stomach blood flow and acid secretion. *Ann. Rev. Physiol.* 44:3.
Holt, K. M., and J. I. Isenberg. 1985. Peptic ulcer disease: Physiology and pathophysiology. *Hosp. Prac.* 20:89.
Moog, F. November 1981. The lining of the small intestine. *Scientific American.*
Richardson, I. and P. G. Withrington. 1982. Physiological regulation of the hepatic circulation. *Ann. Rev. Physiol.* 44:57.
Salen, G. and S. Shefer. 1983. Bile acid synthesis. *Ann. Rev. Physiol.* 45:679.
Soll, A. and J. H. Walsh. 1979. Regulation of gastric acid secretion. *Ann. Rev. Physiol.* 41:35.
Van De Graaff, K. M. 1986. Anatomy and physiology of the gastrointestinal tract. *Pediat. Infect. Dis.* 5:S11.
Weisbrodt, N. W. 1981. Patterns of intestinal motility. *Ann. Rev. Physiol.* 43:21.
Williams, J. A. 1984. Regulatory mechanisms in pancreas and salivary acini. *Ann. Rev. Physiol.* 46:361.
Wood, J. D. 1981. Intrinsic neural control of intestinal motility. *Ann. Rev. Physiol.* 43:33.

Chapter 19

Alexander, E. 1986. Metabolic acidosis: Recognition and etiologic diagnosis. *Hosp. Prac.* 21:100E.
Anderson, B. 1977. Regulation of body fluids. *Ann. Rev. Physiol.* 39:185.
Bauman, J. W., and F. P. Chinard. 1975. *Renal function: Physiological and medical aspects.* St. Louis: C. V. Mosby Co.
Beeuwkes, R., III. 1980. The vascular organization of the kidney. *Ann. Rev. Physiol.* 42:531.
Brenner, B. M., and R. Beeuwkes, III. 1978. The renal circulation. *Hosp. Prac.* 13:35.
deBold, A. 1985. Atrial natriuretic factor: A hormone produced by the heart. *Science* 230:767.
Giebisch, G. H., and B. Stanton. 1979. Potassium transport in the nephron. *Ann. Rev. Physiol.* 41:241.
Hays, R. M. 1978. Principles of ion and water transport in the kidneys. *Hosp. Prac.* 13:79.

Hollenberg, N. K. 1986. The kidney in heart failure. *Hosp. Prac.* 21:81.
Kokko, J. S. 1979. Renal concentrating and diluting mechanisms. *Hosp. Prac.* 14:110.
Rector, F. C., Jr., and M. G. Cogan. 1980. The renal acidoses. *Hosp. Prac.* 15:99.
Reid, I. A., B. J. Morris, and W. F. Ganong. 1978. The renin-angiotensin system. *Ann. Rev. Physiol.* 40:377.
Renkin, E. M., and R. R. Robinson. 1974. Glomerular filtration. *New Engl. J. Med.* 290:79.
Steinmetz, P. R., and B. M. Koeppen. 1984. Cellular mechanisms of diuretic action along the nephron. *Hosp. Prac.* 19:125.
Tanner, R. L. 1980. Control of acid excretion by the kidney. *Ann. Rev. Med.* 31:35.
Van De Graaff, K. M., and S. I. Fox. 1989. *Concepts of Human Anatomy and Physiology.* 2d ed. Dubuque: Wm. C. Brown Publishers.
Vander, A. J. 1985. *Renal physiology.* 3d ed. New York: McGraw-Hill Book Company.
Walker, L. A., and H. Valtin. 1982. Biological importance of nephron heterogeneity. *Ann. Rev. Physiol.* 44:203.
Warnock, D. G., and F. C. Rector, Jr. 1979. Proton secretion by the kidney. *Ann. Rev. Physiol.* 41:197.

Chapter 20

Bardin, C. W. 1979. The neuroendocrinology of male reproduction. *Hosp. Prac.* 14:65.
Bartke, A., A. A. Hafiez, F. J. Bex, and S. Dalterio. 1978. Hormonal interaction in the regulation of androgen secretion. *Biology of Reproduction* 18:44.
Chan, L., and B. W. O'Malley. 1976. Recent studies on the mechanisms of action of the steroid hormones. *New Engl. J. Med.* 294:1322.
Epel, D. November 1977. The program of fertilization. *Scientific American.*
Friedman, T. November 1971. Prenatal diagnosis of genetic disease. *Scientific American.*
Goldstein, B. 1976. *Human sexuality.* New York: McGraw-Hill.
Jones, K. L. et al. 1973. *Sex.* 2d ed. New York: Harper and Row Publishers.
Katchadourian, H. A. 1989. *Fundamentals of human sexuality.* 5th ed. New York: Holt, Rinehart and Winston.
Mader, S. S. 1980. *Human reproductive biology.* Dubuque, Iowa: Wm. C. Brown Co.
Marx, J. L. 1973. Birth control: Current technology, future prospects. *Science* 179:1222.
Masters, W. H. 1986. Sex and aging—expectations and reality. *Hosp. Pract.* 21:175.
Oaks, W. W. et al. 1976. *Sex and the life cycle.* New York: Grune and Stratton.
Odell, W. D., and D. L. Moyer. 1971. *Physiology of reproduction.* St. Louis: C. V. Mosby Co.
Page, E. W. et al. 1981. *Human reproduction: The core content of obstetrics, gynecology, and perinatal medicine.* 3d ed. Philadelphia: W. B. Saunders Co.
Parkes, A. 1976. *Patterns of sexuality and reproduction.* New York: Oxford Univ. Press.
Pengelley, E. T. 1978. *Sex and human life.* 2d ed. Reading, Mass.: Addison-Wesley.
Reiter, E. O., and M. M. Grumback. 1982. Neuroendocrine control mechanisms and the onset of puberty. *Ann. Rev. of Physiol.* 44:595.
Segal, S. J. September 1974. The physiology of human reproduction. *Scientific American.*
Swanson, H. D. 1974. *Human reproduction: Biology and social change.* New York: Oxford Univ. Press.
Witters, W. L., and P. Jones-Witters. 1975. *Drugs and sex.* New York: Macmillan.

Chapter 21

Boyar, R. M. 1978. Control of the onset of puberty. *Ann. Rev. Med.* 31:329.
Epel, D. November 1977. The program of fertilization. *Scientific American.*
Goldstein, B. 1976. *Introduction to human sexuality.* New York: McGraw-Hill.
Grabowski, C. T. 1983. *Human reproduction and development.* Philadelphia: W. B. Saunders.
Hatcher, R. A., and A. K. Stewart. 1987. *Contraceptive technology.* 13th rev. ed. New York: Wiley.
Jones, K. L. et al. 1985. *Dimensions of human sexuality.* Dubuque, Ia.: Wm. C. Brown.
Jones, R. E. 1984. *Human reproduction and sexual behavior.* Englewood Cliffs, N.J.: Prentice-Hall.
Katchadourian, H. A. 1989. *Fundamentals of human sexuality.* 5th ed. New York: Holt, Rinehart and Winston.
Lein, A. 1979. *The cycling female.* San Francisco: W. H. Freeman.
Mader, S. S. 1980. *Human reproductive biology.* Dubuque, Ia.: Wm. C. Brown.
Marx, J. L. 1978. The mating game: What happens when sperm meets egg. *Science* 200:1256.
Masters, W. H. 1986. Sex and aging—expectations and reality. *Hosp. Prac.* 21:175.
Parkes, A. 1976. *Patterns of sexuality and reproduction.* New York: Oxford Univ. Press.
Pengelley, E. T. 1978. *Sex and human life.* Reading, Mass.: Addison-Wesley.
Shiu, P. C., and H. G. Friesen. 1980. Mechanisms of action of prolactin in the control of mammary gland functions. *Ann. Rev. Physiol.* 42:83.
Short, R. V. April 1984. Breast feeding. *Scientific American.*
Simpson, E. R., and P. C. MacDonald. 1981. Endocrine physiology of the placenta. *Ann. Rev. Physiol.* 43:163.
Soloff, M. S. et al. 1979. Oxytocin receptors. *Science* 204:131.
Swanson, H. D. 1974. *Human reproduction: Biology and social change.* New York: Oxford Univ. Press.
Wilson, J. D., F. W. George, and J. E. Griffin. March 1981. The hormonal control of sexual development. *Science* 211(No. 4488):1278.

Chapter 22

Annis, L. F. 1978. *The child before birth.* New York: Cornell Univ. Press.
Balinsky, B. I. 1981. *An introduction to embryology.* 5th ed. Philadelphia: W. B. Saunders.
Beaconsfield, P., G. Birdwood, and R. Beaconsfield. July 1980. The placenta. *Scientific American.*
Beer, A. E., and R. E. Billingham. April 1974. The embryo as a transplant. *Scientific American.*
Birnholz, J. C., and E. E. Farrell. 1984. Ultrasound images of human fetal development. *American Scientist,* 72:608.
Chervenak, F. A., G. Isaacson, and M. J. Mahoney. 1986. Advances in the diagnosis of fetal defects. *New Engl. J. Med.* 315:305.
Danforth, D. N., ed. 1986. *Obstetrics and gynecology.* 5th ed. New York: Harper and Row Publishers.
England, M. A. 1983. *Color atlas of life before birth: Normal fetal development.* Chicago: Year Book Medical Publishers, Inc.
Gellis, S. S., ed. 1973. *Year book of pediatrics.* Chicago: Year Book Medical Publishers, Inc.
Grobstein, C. November 1977. External human fertilization. *Scientific American.*
Jackson, L. G. 1985. First-trimester diagnosis of fetal genetic disorders. *Hospital Practice.* 20:39.
Lagerkrantz, H., and T. A. Slotkin. April 1986. The "stress" of being born. *Scientific American.*
Moore, K. L. 1982. *The developing human.* 3d ed. Philadelphia: W. B. Saunders.
Moore, K. L. 1988. *The developing human: clinically oriented embryology.* 4th ed. Philadelphia: W. B. Saunders Co.
Nilsson, L., A. Ingelman-Sundbert, and C. Wirsen. 1977. *A child is born.* Rev. ed. New York: Dell Publishing Co.
Oppenheimer, S. B., and G. Lefevere. 1989. *Introduction to embryonic development.* 3rd ed. Boston: Allyn and Bacon.
Rugh, R., and L. B. Shettles. 1971. *From conception to birth.* New York: Harper and Row Publishers.
Volpe, E. P. 1983. *Biology and human concerns.* 3d ed. Dubuque: Wm. C. Brown Co.
Warkentin, D. L. April 1984. From A to hCG in pregnancy testing. *Diagnostic Med.* p. 35.

Chapter 23

Bower, T. G. 1982. *Development in infancy,* 2d ed. San Francisco: W. H. Freeman.
Carter, N., ed. 1980. *Development, growth and aging.* London: Croon Helm.
Comfort, A. 1979. *The biology of senescence.* 3d ed. London: Churchill Livingstone.
Diamond, M. C. January–February 1978. The aging brain. *American Scientist.*
Fischer, K., and A. Lazerson. 1984. *Human development: From conception through adolescence.* San Francisco: W. H. Freeman.
Fries, J. F., and L. M. Crapo. 1981. *Vitality and aging.* San Francisco: W. H. Freeman.
Gersh, E. S., and I. Gersh. 1981. *Biology of women.* Baltimore: Univ. Park Press.
Hayflick, L. January 1980. The cell biology of human aging. *Scientific American.*
Heymann, M. A. et al. 1981. Factors affecting changes in the neonatal systemic circulation. *Ann. Rev. Physiol.* 43:371.
Katchadourian, H. 1977. *The biology of adolescence.* San Francisco: W. H. Freeman.
Lazarus, R. S., and A. DeLongis. 1983. Psychological stress and coping in aging. *Amer. Psychologist* 38:245.
Lehman, H. C. 1978. *Age and achievement.* Princeton, N.J.: Princeton Univ. Press.
Miller, C. A. July 1985. Infant mortality in the U.S. *Scientific American.*
Neuhaus, R. H., and R. H. Neuhaus. 1982. *Successful aging.* New York: John Wiley and Sons.
Oleson, K. H. 1965. Body composition in normal adults. In *Human body composition,* edited by J. Brozek. Symposia of the Society for the Study of Human Body Composition 7:177. New York: Pergamon Press.
Palmore, E. 1970. *Normal aging.* Durham, N.C.: Duke Univ. Press.
Palmore, E., and E. Jeffers, eds. 1971. *Prediction of life span.* Lexington, Ky.: Heath Lexington Books.
Riley, M. W. et al. 1968. *Aging and society.* St. Louis: C. V. Mosby.
Rockstein, M. 1968. The biological aspects of aging. *The Gerontologist* 8:124.
Santrock, J. W. 1985. *Adult development and aging.* Dubuque: Wm. C. Brown.
Selye, H. 1970. Stress and aging. *J. Am. Geriatr. Soc.* 18:669–80.
Tanner, J. M. September 1973. Growing up. *Scientific American.*
Tanner, J. M. 1978. *Foetus into man.* Cambridge, Mass.: Harvard Univ. Press.
Wantz, M. S., and J. E. Gay. 1981. *The aging process: A health perspective.* Cambridge, Mass.: Winthrop.

Chapter 24

Anderson, J. E. 1983. *Grant's atlas of anatomy.* 8th ed. Baltimore: The Williams and Wilkins Co.
Basmajian, J. V. 1989. *Grant's method of anatomy.* 11th ed. Baltimore: The Williams and Wilkins Co.
———. 1983. *Surface anatomy: An instruction manual.* 2d ed. Baltimore: The Williams and Wilkins Co.
Clemente, C. D. 1987. *Anatomy: A regional atlas of the human body.* 3d ed. Baltimore: Urban and Schwarzenberg.
———. 1984. *Gray's anatomy of the human body.* 30th American ed. Philadelphia: Lea and Febiger.
Hall-Craggs, E. C. B. 1990. *Anatomy as a basis for clinical medicine.* 2nd ed. Baltimore: Urban and Schwarzenberg.
Last, R. J. 1984. *Anatomy, regional and applied.* 7th ed. Boston: Little, Brown and Co.
Royce, J. 1973. *Surface anatomy.* Philadelphia: F. A. Davis Co.
Snell, R. S. 1978. *Atlas of clinical anatomy.* Boston: Little, Brown and Co.
Sobotta, J. and F. Figge. 1983. *Atlas of human anatomy.* 10th ed. New York: Hafner Press.
Woodburne, R. T., and W. E. Burkel. 1988. *Essentials of human anatomy.* 8th ed. New York: Oxford University Press.

Glossary

The words in this glossary are followed by a phonetic guide to pronunciation. This is a simplified system that is standard in medical usage and terminology.

Any unmarked vowel that ends a syllable or stands alone as a syllable is long. Any unmarked vowel that is followed by a consonant has the short sound.

If a long vowel appears in the middle of a syllable (followed by a consonant), it is marked with a macron (¯). Similarly, if a vowel stands alone or ends a syllable, but should have a short sound, it is marked with a breve (˘).

abdomen (ab-do'men) A region of the body between the diaphragm and pelvis.

abduction (ab-duk'shun) The movement of a body part away from the axis or midline of the body.

accessory organs (ak-ses'o-re) Organs that assist the functioning of other organs within a system.

acetabulum (as"ĕ-tab'u-lum) A socket in the lateral surface of the hipbone (os coxa) into which the head of the femur articulates.

Achilles tendon (ah-kil'ēz) (see tendo calcaneus)

actin (ak'tin) A protein in muscle fibers that together with myosin is responsible for contraction.

adduction (ah'duk'shun) The movement of a body part toward the axis or midline of the body.

adrenal cortex (ah-dre'nal kor'teks) The outer layer of the adrenal gland.

adrenal glands Endocrine glands; two glands, one superior to each kidney.

adventitia (ad"ven-tish'e-ah) The outermost epithelial layer of a visceral organ; also called *serosa.*

afferent (af'er-ent) Conveying or transmitting to.

afferent arteriole (ar-te're-ōl) A blood vessel within the kidney that supplies blood to the glomerulus.

afferent neuron (nu'ron) A sensory nerve cell that transmits an impulse toward the central nervous system.

agonist (ag'o-nist) The prime mover muscle, which is directly engaged in the contraction that produces the desired movement.

alimentary canal (al"ĕ-men'tar-e) The tubular portion of the digestive tract.

allantois (ah-lan'to-is) An extraembryonic membranous sac that is involved in the formation of blood cells and gives rise to the fetal umbilical arteries and vein. It also contributes to the formation of the urinary bladder.

all-or-none principle The statement of the fact that muscle fibers of a motor unit contract to their maximum extent when exposed to a stimulus of threshold strength.

alveolar sacs (al-ve'o-lar) A cluster of alveoli that share a common chamber or central atrium.

alveolus (al-ve'o-lus) An individual air capsule within the lung. The alveoli are the basic functional units of respiration.

amniocentesis (am"ne-o-sen-te'sis) A procedure for removal of a sample of amniotic fluid and suspended cells to examine for various genetic diseases.

amnion (am'ne-on) A developmental membrane that surrounds the fetus and contains amniotic fluid.

amphiarthrosis (am"fe-ar-thro'sis) A slightly movable articulation.

ampulla (am-pul'lah) A saclike enlargement of a duct or tube.

ampulla of Vater (fah'ter) (see hepatopancreatic ampulla)

anal canal (a'nal) The terminal tubular portion of the rectum that opens through the anus of the alimentary canal.

anal glands Enlarged and modified sweat glands that empty into the anal opening.

anastomosis (ah-nas"to-mo'sis) An interconnecting aggregation of blood vessels or nerves that form a network plexus.

anatomical position (an"ah-tom'e-kal) An erect body stance with the eyes directed forward, the arms at the sides, and the palms of the hands facing forward.

anatomy (ah-nat'o-me) The branch of science concerned with the structure of the body and the relationship of its organs.

antagonist (an-tag'-o-nist) A muscle that acts in opposition to a prime mover, or an agonist.

antebrachium (an"te-bra'ke-um) The forearm.

anterior (ventral) (an-te're-or) Toward the front; the opposite of *posterior (dorsal).*

anterior root The anterior projection of the spinal cord, which is composed of axons of motor, or efferent, fibers.

anus (a-nus) The terminal portion and outlet of the alimentary canal.

aorta (a-or'tah) The major systemic vessel of the arterial system of the body, emerging from the left ventricle.

apex (a'peks) The tip or pointed end of a conical structure.

apocrine gland (ap'o-krin) A type of sweat gland that functions in evaporative cooling. It may respond during periods of emotional stress.

aponeurosis (ap"o-nu-ro'sis) A fibrous or membranous sheetlike tendon.

appendix (ah-pen'diks) A short pouch that attaches to the cecum; also called the *vermiform appendix.*

aqueous humor (a'-kwe-us hum'or) The watery fluid that fills the anterior and posterior chambers of the eye.

arachnoid (ah-rak'noid) The weblike middle covering (meninge) of the central nervous system.

arbor vitae (ar'bor vi'tah) The branching arrangement of white matter within the cerebellum.

arch of aorta The superior left bend of the aorta between the ascending and descending portions.

arm (brachium) The portion of the upper extremity from the shoulder to the elbow.

arrector pili (ah-rek'tor pih'le) The smooth muscle attached to a hair follicle, which upon contraction, pulls the hair vertical, resulting in goose bumps.

arterial circle (ar-te're-al) An arterial vessel located on the ventral surface of the brain around the pituitary gland; also called the *circle of Willis.*

arteriole (ar-te're-ōl) A minute arterial branch.

artery (ar'ter-e) A blood vessel that carries blood away from the heart.

arthrology (ar-throl'o-je) The scientific study of the structure and function of joints.

articular cartilage (ar-tik'u-lar kar'ti-lij) A hyaline cartilaginous covering over the articulating surface of bones of synovial joints.

articulation (ar-tik"u-la'shun) A joint.

arytenoid cartilages (ar"ĕ-te'noid) A pair of small cartilages located on the superior aspect of the larynx.

ascending colon (ko'lon) The portion of the large intestine between the cecum and the hepatic flexure.

association neuron (nu'ron) A nerve cell located completely within the central nervous system. It conveys impulses in an arc from afferent to efferent neurons.

atrioventricular bundle (a"tre-o-ven-trik'u-lar) A group of specialized cardiac fibers that conducts impulses from the atrioventricular node to the ventricular muscles of the heart; also called the *bundle of His* or *AV bundle.*

atrioventricular node A microscopic aggregation of specialized cardiac fibers located in the interatrial septum of the heart that are a part of the conduction system of the heart; *AV node.*

atrioventricular valve A cardiac valve located between an atrium and a ventricle of the heart; *AV valve.*

atrium (a'tre-um) Either of the two superior chambers of the heart that receive venous blood.

atrophy (at′ro-fe) A gradual wasting away or decrease in the size of a tissue or organ.
auditory (aw′dĭ-to″re) Pertaining to the structures of the ear that are associated with hearing.
auditory tube A narrow canal that connects the middle ear chamber to the pharynx; also called the *eustachian canal.*
auricle (aw′rĕ-kl) 1. The fleshy pinna of the ear. 2. An ear-shaped appendage of each atrium of the heart.
autonomic nervous system (aw″to-nom′ik) The sympathetic and parasympathetic portions of the nervous system that function to control the actions of the visceral organs and skin.
axilla (ak-sil′ah) Pertaining to the depressed hollow commonly called the armpit.
axon (ak′son) The elongated process of a nerve cell that transmits an impulse away from the cell body.

ball-and-socket joint The most freely movable type of diarthrosis (e.g., the shoulder or hip joint).
baroreceptor (bar″o-re-sep′tor) A cluster of neuroreceptors that are stimulated by pressure changes and monitor blood pressure.
basal ganglion (ba′sal gang′gle-on) A mass of nerve cell bodies located deep within a cerebral hemisphere of the brain.
basement membrane A thin sheet of extracellular substance to which the basal surfaces of epithelial cells are attached; also called the *basal lamina.*
basophil (ba′so-fil) A granular leukocyte that readily stains with basophilic dye.
belly The thickest circumference of a skeletal muscle.
benign (be-nīn′) Not malignant.
bifurcation (bi″fur-ka′shun) Forked; divided into two branches.
bile (bīl) An excreted fluid from the liver that is stored and concentrated in the gallbladder, released through the common bile duct into the duodenum, and is essential for the absorption of fats.
bipennate (bi-pen′āt) Denoting muscles that have a fiber architecture coursing obliquely on both sides of a tendon.
blastula (blas′tu-lah) An early stage of prenatal development between the morula and embryonic stages.
blood The fluid connective tissue that circulates through the cardiovascular system to transport substances throughout the body.
blood-brain barrier A specialized mechanism that inhibits the passage of certain materials from the blood into brain tissue and cerebrospinal fluid.
bolus (bo′lus) A moistened mass of food that is swallowed from the oral cavity into the pharynx.
bone A solid, rigid, ossified connective tissue forming the skeletal system.
bony labyrinth (lab′ĭ-rinth) A series of chambers within the petrous portion of the temporal bone associated with the vestibular organs and the cochlea.
Bowman's capsule (bo′manz kap′sūl) (see glomerular capsule)
brachial plexus (bra′ke-al plek′sus) A network of nerve fibers that arise from C5–C8 and T1. Nerves arising from the brachial plexuses supply the upper extremity.
brain The enlarged superior portion of the central nervous system, located in the cranial cavity of the skull.
brain stem The portion of the brain consisting of the medulla oblongata, pons, and midbrain.
bronchial tree (brong′ke-al) The bronchi and their branching bronchioles.
bronchiole (brong′ke-ōl) A small division of a bronchus within the lung.
bronchus (brong′kus) A branch of the trachea that leads to a lung.
buccal cavity (buk′al) The mouth, or oral cavity.
bulbourethral glands (bul″bo-u-re′thral) A pair of glands that secrete a viscous fluid into the male urethra during sexual excitement; also called *Cowper's glands.*
bundle of His (see atrioventricular bundle)
bursa (ber′sah) A saclike structure filled with synovial fluid, which occurs around joints and over which tendons can slide without contacting bone.
buttock (but′ok) The rump or fleshy mass on the posterior aspect of the lower trunk, formed primarily by the gluteal muscles.

calyx (ka′liks) A cup-shaped portion of the renal pelvis that encircles renal papillae.
canaliculus (kan″ah-lik′u-lus) A microscopic channel in bone tissue that connects lacunae.
canal of Schlemm (shlem) (see venous sinus)
cancellous bone (kan′sĕ-lus) Spongy bone; bone tissue with a latticelike structure.
capillary (kap′ĭ-lar″e) A microscopic blood vessel that connects an arteriole and a venule; the functional unit of the circulatory system.
carotid sinus (kar-rot′id) An expanded portion of the internal carotid artery immediately above the point of branching from the external carotid artery; contains baroreceptors that monitor blood pressure.
carpus (kar′pus) Pertaining to the wrist; collectively, the eight wrist bones.
cartilage (kar′ti-lij) A type of connective tissue with a solid elastic matrix.
cartilaginous joint (kar″ti-laj′ĭ-nus) A joint that lacks a joint cavity and permits little movement between the bones held together by cartilage.
cauda equina (kaw′dah e-kwi′nah) The lower end of the spinal cord where the roots of spinal nerves have a tail-like appearance.
caudal (kaw′dal) Referring to a position more toward the tail.
cecum (se′kum) The pouchlike portion of the large intestine to which the ileum of the small intestine is attached.
cell The structural and functional unit of an organism; the smallest structure capable of performing all of the functions necessary for life.
cementum (se-men′tum) Bonelike material that binds the root of a tooth into its bony socket.
central canal An elongated, longitudinal channel in the center of an osteon in bone tissue, containing branches of the nutrient vessels and nerve; also called an *haversian canal.*
central nervous system (CNS) The brain and the spinal cord.
centrosome (sen′tro-sōm) A dense body near the nucleus of a cell that contains a pair of centrioles.
cerebellar peduncle (ser″ĕ-bel′ar pe-dung′-k′l) An aggregation of nerve fibers that connect the cerebellum with the brain stem.
cerebellum (ser″ĕ-bel′um) The portion of the brain concerned with the coordination of movements. Part of the metencephalon, it consists of two hemispheres and a central vermis.
cerebral aqueduct (ser′ĕ-bral ak′we-dukt″) The channel that connects the third and fourth ventricles of the brain; also called the *aqueduct of Sylvius.*
cerebral peduncles A paired bundle of nerve fibers along the ventral surface of the midbrain, conducting impulses between the pons and the cerebral hemispheres.
cerebrospinal fluid (ser″ĕ-bro-spi′nal) A fluid produced by the choroid plexus of the ventricles of the brain. It fills the ventricles and surrounds the central nervous system in association with the meninges.
cerebrum (ser′ĕ-brum) The largest portion of the brain, composed of the right and left hemispheres.
ceruminous (sĕ-roo′mĭ-nus) **gland** Specialized integumentary gland that secretes cerumen, or earwax, into the external auditory canal.
cervical (ser′vĭ-kal) Pertaining to the neck or a necklike portion of an organ.
cervical ganglion (gang′gle-on) A cluster of postganglionic sympathetic nerve cell bodies located in the neck, near the cervical vertebrae.
cervical plexus (plek′sus) A network of spinal nerves formed by the anterior branches of the first four cervical nerves.
cervix (ser′viks) 1. The narrow necklike portion of an organ. 2. The inferior end of the uterus that adjoins the vagina.
chemoreceptor (ke″mo-re-sep′tor) A neuroreceptor that is stimulated by the presence of chemical molecules.
chiasma (ki-as′mah) A crossing of nerve tracts from one side of the CNS to the other.
chondrocranium (kon″dro-kra′ne-um) The portion of the skull that supports the brain and is derived from endochondral bone.
chondrocyte (kon′dro-sīt) A cartilage cell.
chordae tendineae (kor′de ten-din′e-e) Chordlike tendinous bands that connect papillary muscles to the atrioventricular valves within the ventricles of the heart.
chorion (ko′re-on) An extraembryonic membrane that participates in the formation of the placenta.
choroid (ko′roid) The vascular, pigmented middle layer of the wall of the eye.
choroid plexus (ko′roid plek′sus) A mass of vascular capillaries from which cerebrospinal fluid is secreted into the ventricles of the brain.

chromatophilic substances (kro″mah-to-fil′ik) Clumps of rough endoplasmic reticulum in the cell's bodies of neurons; also called *Nissl bodies.*
chromosome (kro′mo-sōm) Structures in the nucleus that contain the genes for genetic expression.
chyme (kīm) The mass of partially digested food that passes from the pylorus of the stomach into the duodenum of the small intestine.
cilia (sil′e-ah) Microscopic, hairlike processes that move in a wavelike manner on the exposed surfaces of certain epithelial cells.
ciliary body (sil′e-er″e) A portion of the choroid layer of the eye that secretes aqueous humor and contains the ciliary muscle.
circle of Willis (see arterial circle)
circumduction (ser″kum-duk′shun) A conelike movement of a body part, such that the distal end moves in a circle while the proximal portion remains relatively stable.
clitoris (kli′to-ris) A small, erectile structure in the vulva of the female, homologous to the glans penis in the male.
coccygeal (kok-sij′e-al) Pertaining to the region of the coccyx; the caudal termination of the vertebral column.
cochlea (kok′le-ah) The spiral portion of the inner ear that contains the organ of Corti.
coelom (se′lom) The abdominal cavity.
collateral (kŏ-lat′er-al) A small branch of a blood vessel or nerve fiber.
colon (ko′lon) The large intestine.
common bile duct A tube that is formed by the union of the hepatic duct and cystic duct and that transports bile to the duodenum.
compact (dense) bone Tightly packed bone that is superficial to spongy bone and covered by the periosteum.
condyle (kon′dīl) A rounded process at the end of a long bone that forms an articulation.
cone (kōn) A color receptor cell in the retina of the eye.
congenital (kon-jen′ĭ-tal) Present at the time of birth.
conjunctiva (kon″junk-ti′vah) The thin membranous covering on the anterior surface of the eyeball and lining the eyelids.
connective tissue One of the four basic tissue types within the body. It is a binding and supportive tissue with abundant matrix.
conus medullaris (ko′nus med″u-lăr′is) The caudal, tapering portion of the spinal cord.
convolution (kon-vo-lu′shun) An elevation on the surface of a structure and an infolding of the tissue upon itself.
cornea (kor′ne-ah) The transparent convex, anterior portion of the outer layer of the eyeball.
coronary circulation (kor′ŏ-na-re) The arterial and venous blood circulation to the wall of the heart.
coronary sinus A large venous channel on the posterior surface of the heart into which the cardiac veins drain.
corpora quadrigemina (kor′po-rah kwod″rĭ-jem′ĭ-nah) Four dorsal lobes of the midbrain concerned with visual and auditory functions.
corpus callosum (kor′pus kah-lo′sum) A major tract within the brain that is composed of white matter and connects the right and left cerebral hemispheres.
cortex (kor′teks) The outer layer of an organ such as the convoluted cerebrum, adrenal gland, or kidney.
costal cartilage (kos′tal) The cartilage that connects the ribs to the sternum.
cranial (kra′ne-al) Pertaining to the cranium.
cranial nerve One of twelve pairs of nerves that arise from the ventral surface of the brain.
cranium (kra′ne-um) The endochondral bones of the skull that enclose or support the brain and the organs of sight, hearing, and balance.
crest A thickened ridge of bone for the attachment of muscle.
cricoid cartilage (kri′koid) A ring-shaped cartilage that forms the inferior end of the larynx.
crista (kris′tă) A crest, such as the crista galli extending superiorly from the cribriform plate.
cubital (ku′bĭ-tal) Pertaining to the forearm. The cubital fossa is the anterior aspect of the elbow joint.
cystic duct (sis′tik dukt) The tube that transports bile from the gallbladder to the common bile duct.
cytology (si-tol′o-je) The science dealing with the study of cells.
cytoplasm (si′to-plazm″) In a cell, the protoplasm located outside of the nucleus.

deciduous (de-sid′u-us) Not permanent. Deciduous teeth are shed and replaced by permanent teeth during development.
decussation (de″kŭ-sa′shun) A crossing of nerve fibers from one side of the CNS to the other.
defecation (def″ĕ-ka′shun) The elimination of feces from the rectum through the anus.
deglutition (deg″loo-tish′un) The process of swallowing.
dendrite (den′drīt) A nerve cell process that transmits impulses toward a neuron cell body.
dentin (den′tĭn) The main substance of a tooth, covered by enamel over the crown of the tooth and by cementum on the root.
dentition (den-tish′un) The number, arrangement, and shape of teeth.
dermal papilla (der′mal pah-pil′ah) A projection of the dermis into the epidermis.
dermis (der′mis) The second, or deep, layer of skin beneath the epidermis.
descending colon The segment of the large intestine that descends on the left side from the level of the spleen to the level of the left iliac crest.
diaphragm (di′ah-fram) A sheetlike dome of muscle and connective tissue that separates the thoracic and abdominal cavities.
diaphysis (di-af′ĭ-sis) The shaft of a long bone.
diarthrosis (di″ar-thro′sis) A type of joint in which the articulating bones are freely movable; also called a *synovial joint.*
diastole (di-as′to-le) The sequence of the cardiac cycle during which a heart chamber wall is relaxed.
diencephalon (di″en-sef′ah-lon) A major region of the brain that includes the third ventricle, thalamus, hypothalamus, and pituitary.
digestion The process by which larger molecules of food substance are broken down mechanically and chemically into smaller molecules that can be absorbed.
diploe (dip′lo-e) The spongy layer of bone positioned between the inner and outer layers of compact bone.
distal (dis′tal) Away from the midline or origin; the opposite of *proximal.*
dorsal (dor′sal) Pertaining to the back or posterior portion of a body part; the opposite of *ventral.*
dorsiflexion (dor″sĭ-flek′shun) Movement at the ankle or wrist as the dorsum of the foot or hand is elevated.
ductus arteriosus (duk′tus ar-te″re-o′sus) The blood vessel that connects the pulmonary trunk and the aorta in a fetus.
ductus deferens (def′er-enz) A tube that carries spermatozoa from the epididymis to the ejaculatory duct; also called the *vas deferens* or *seminal duct.*
ductus venosus (ven-o′sus) A fetal blood vessel that connects the umbilical vein and the inferior vena cava.
duodenum (du″o-de′num) The first portion of the small intestine that leads from the pyloric sphincter of the stomach to the jejunum.
dura mater (du′rah ma′ter) The outermost meninx.

eccrine gland (ek′rin) A sweat gland that functions in thermoregulation.
ECG Electrocardiogram; EKG.
ectoderm (ek′to-derm) The outermost of the three primary germ layers of an embryo.
edema (ĕ-de′mah) An excessive accumulation of fluid in the body tissues.
effector (ef-fek′tor) An organ such as a gland or muscle that responds to a motor stimulation.
efferent (ef′er-ent) Conveying away from the center of an organ or structure.
efferent arteriole (ar-te′re-ōl) An arteriole of the renal vascular system that conducts blood away from the glomerulus of a nephron.
efferent ductules (dukt′ūls) A series of coiled tubules that convey spermatozoa from the rete testis to the epididymis.
efferent neuron (nu′ron) A motor nerve cell that conducts impulses from the central nervous system to effector organs such as muscles or glands.
ejaculation (e-jak″u-la′shun) The discharge of semen from the male urethra during climax.
ejaculatory duct (e-jak′u-lah-to″re) A tube that transports spermatozoa from the ductus deferens to the prostatic urethra.
elastic fibers (e-las′tik) Protein strands that are found in certain connective tissue and have contractile properties.
elbow The diarthrotic joint between the brachium and the forearm.
electrocardiogram (e-lek″tro-kar′de-o-gram″) A recording of the electrical activity that accompanies the cardiac cycle; ECG or EKG.
electroencephalogram (e-lek″tro-en-sef′ah-lo-gram″) A recording of the brain wave patterns or electrical impulses of the brain; EEG.

electromyogram (e-lek″tro-mi′o-gram″) A recording of the electrical impulses or activity of a muscle; EMG.
embryology (em′bre-ol″o-je) The study of prenatal development from conception through the eighth week in utero.
enamel (en-am′el) The outer, dense substance covering the crown of a tooth.
endocardium (en″do-kar′de-um) The endothelial lining of the heart chambers and valves.
endochondral bone (en″do-kon′dral) Bones that develop as hyaline cartilage models first and then are ossified.
endocrine gland (en′do-krīn) Ductless, hormone producing gland that is part of the endocrine system.
endoderm (en′do-derm) The innermost of the three primary germ layers of an embryo.
endolymph (en′do-limf) A fluid within the membranous labyrinth and cochlear duct of the inner ear that aids in the conduction of vibrations involved in hearing and in the maintenance of equilibrium.
endometrium (en″do-me′tre-um) The inner lining of the uterus.
endomysium (en″do-mis′e-um) The connective tissue sheath surrounding each skeletal muscle fiber, separating the muscle cells from one another.
endoneurium (en″do-nu′re-um) The connective tissue sheath surrounding each nerve fiber, separating the nerve fibers one from another within a nerve.
endoplasmic reticulum (en-do-plas′mik rē-tik′u-lum) A cytoplasmic organelle composed of a network of canals running through the cytoplasm of a cell.
endothelium (en″do-the′le-um) The layer of epithelial tissue that forms the thin inner lining of blood vessels and heart chambers.
eosinophil (e″o-sin′o-fil) A type of white blood cell characterized by the presence of cytoplasmic granules that become stained by acidic eosin dye; eosinophils normally constitute about 2%–4% of the white blood cells.
epicardium (ep″ĭ-kar′de-um) A thin, outer layer of the heart; also called the *visceral pericardium.*
epicondyle (ep″ĭ-kon′dīl) A projection of bone above a condyle.
epidermis (ep″ĭ-der′mis) The outermost layer of the skin, composed of several stratified squamous epithelial layers.
epididymis (ep″ĭ-did′ĭ-mis) A highly coiled tube located along the posterior border of the testis. It stores spermatozoa and transports them from the seminiferous tubules of the testis to the ductus deferens.
epidural space (ep″ĭ-du′ral) A space between the spinal dura mater and the bone of the vertebral canal.
epiglottis (ep″ĭ-glot′is) A cartilaginous leaflike structure positioned on top of the larynx that covers the glottis during swallowing.
epimysium (ep″ĭ-mis′e-um) A fibrous, outer sheath of connective tissue surrounding a skeletal muscle.
epinephrine (ep″ĭ-nef′rin) A hormone secreted from the adrenal medulla resulting in actions similar to those from sympathetic nervous system stimulation; also called *adrenaline.*
epineurium (ep″ĭ-nu′re-um) A fibrous, outer sheath of connective tissue surrounding a nerve.
epiphyseal plate (ep″ĭ-fiz′e-al) A cartilaginous layer that is located between the epiphysis and diaphysis of a long bone and functions as a longitudinal growing region.
epiphysis (ē-pif′ĭ-sis) The end segment of a long bone, separated from the diaphysis early in life by an epiphyseal plate but later becoming part of the larger bone.
episiotomy (ē-piz″e-ot′o-me) An incision of the perineum at the end of the second stage of labor to facilitate delivery and avoid tearing the perineum.
epithelial tissue (ep″ĭ-the′le-al) One of the four basic tissue types; the type of tissue that covers or lines all exposed body surfaces.
eponychium (ep″o-nik′e-um) The thin layer of stratum corneum of the epidermis of the skin that overlaps and protects the lunula of the nail.
erythrocyte (ē-rith′ro-sīt) A red blood cell.
esophagus (ē-sof′ah-gus) A tubular portion of the GI tract that leads from the pharynx to the stomach as it passes through the thoracic cavity.
estrogen (es′tro-jen) Female sex hormone secreted from the ovarian (Graafian) follicle.
etiology (e″te-ol′o-je) The study of cause, especially of disease, including the origin and what pathogens, if any, are involved.
eustachian canal (u-sta′ke-an) (see auditory tube)
eversion (e-ver′zhun) A movement of the foot in which the sole is turned outward.
exocrine gland (ek′so-krin) A gland that secretes its product to an epithelial surface, directly or through ducts.
expiration (ek″spĭ-ra′shun) The process of expelling air from the lungs through breathing out; also called *exhalation.*
extension (ek-sten′shun) A movement that increases the angle between parts of a joint.
external (superficial) Located on or toward the surface.
external auditory meatus (aw′dĭ-to″re me-a′tus) An opening through the temporal bone that connects with the tympanum and the middle ear chamber and through which sound vibrations pass.
external ear The outer portion of the ear, consisting of the auricle (pinna), external auditory canal, and tympanum.
external nares (na′rēz) The openings into the nasal cavity; also called the *nostrils.*
extrinsic (eks-trin′sik) Pertaining to an outside or external origin.

face 1. The anterior aspect of the head not supporting or covering the brain. 2. The exposed surface of a structure.
facet (fas′et) A small, smooth surface of a bone where articulation occurs.
falciform ligament (fal′si-form lig′ah-ment) An extension of parietal peritoneum that separates the two major lobes of the liver.
fallopian tube (fal-lo′pe-an) (see uterine tube)
false vocal cords The supporting folds of tissue for the true vocal cords within the larynx.
falx cerebelli (falks) A fold of the dura mater that is anchored to the occipital bone and projects inward between the cerebellar hemispheres.
falx cerebri A fold of dura mater that is anchored to the crista galli and extends between the right and left cerebral hemispheres.
fascia (fash′e-ah) A tough sheet of fibrous tissue binding the skin to underlying muscles or supporting and separating muscles.
fasciculus (fah-sik′u-lus) A small bundle of muscle or nerve fibers.
fauces (faw′sēz) The passageway between the mouth and the pharynx.
feces (fe′sēz) Material expelled from the digestive tract during defecation, composed of food residue, bacteria, and secretions; also called *stool.*
fetus (fe′tus) A prenatal human after eight weeks of development.
fibroblast (fi′bro-blast) An elongated cell with cytoplasmic extensions present in connective tissue where it is capable of forming collagenous fibers or elastic fibers.
fibrous joint (fi′brus) A type of articulation that allows little or no movement (e.g., syndesmosis).
filiform papillae (fil′ĭ-form pah-pil′e) The numerous small projections over the entire dorsal surface of the tongue that contain no taste buds.
filum terminale (fi′lum ter-mĭ-nal′e) A fibrous, threadlike continuation of the pia mater, extending inferiorly from the terminal end of the spinal cord to the coccyx.
fimbriae (fim′bre-e) Fringelike extensions from the open end of the uterine tube.
fissure (fish′ūr) A groove or narrow cleft that separates two parts, such as the cerebral hemispheres of the brain.
flexion (flek′shun) A movement that decreases the angle between parts of a joint.
fontanel (fon″tah-nel) A membranous-covered region on the skull of a fetus or baby where ossification has not yet occurred; commonly called a *soft spot.*
foot The terminal portion of the lower extremity, consisting of the tarsus, metatarsus, and phalanges.
foramen (fo-ra′men), pl. *foramina* An opening, usually in a bone, for the passage of a blood vessel or a nerve.
foramen ovale (o-val′e) An opening through the interatrial septum of the fetal heart.
forearm (fōr′arm) The portion of the upper extremity between the elbow and the wrist; also called the *antebrachium.*
fornix (for′niks) 1. A recess around the cervix of the uterus where it protrudes into the vagina. 2. A tract within the brain connecting the hippocampus with the mammillary bodies.
fossa (fos′ah) A depressed area, usually on a bone.
fourth ventricle (ven′trĭ-k′l) A cavity within the brain, between the cerebellum and the medulla and pons, containing cerebrospinal fluid.
fovea centralis (fo′ve-ah sen-tra′lis) A depression on the macula lutea of the eye where only cones are located, which is the area of keenest vision.
frenulum (fren′u-lum) A membranous tissue that serves to anchor and limit the movement of a body part.

frontal 1. Pertaining to the region of the forehead. 2. A plane through the body, dividing the body into anterior and posterior portions; also called the *coronal plane.*

fungiform papillae (fun'jĭ-form pah-pil'e) Flattened, mushroom-shaped projections that are interspersed over the dorsal surface of the tongue and contain taste buds.

gallbladder A pouchlike organ, attached to the underside of the liver, which stores and concentrates bile.

gamete (gam'ēt) A haploid sex cell; either an egg cell or a sperm cell.

ganglion (gang'gle-on) An aggregation of nerve cell bodies occurring outside the central nervous system.

gastrointestinal tract (GI tract) (gas"tro-in-tes'tĭ-nal) The portion of the digestive tract that includes the stomach and the small and large intestines.

gingiva (jin-jī'vah) The fleshy covering over the mandible and maxilla through which the teeth protrude within the mouth; also called *gum.*

gland An organ that produces a specific substance or secretion.

glans penis (glanz pe'nis) The enlarged, sensitive, distal end of the penis.

gliding joint A type of diarthrotic joint in which the articular surfaces are flat, permitting only side-to-side and back-and-forth movements.

glomerular capsule (glo-mer'u-lar) The double-walled proximal portion of a renal tubule that encloses the glomerulus of a nephron; also called *Bowman's capsule.*

glomerulus (glo-mer'u-lus) A coiled tuft of capillaries that is surrounded by the glomerular capsule and filtrates urine from the blood.

glottis (glot'is) A slitlike opening into the larynx, positioned between the true vocal cords.

goblet cell A unicellular gland that secretes mucus and is associated with columnar epithelia.

Golgi apparatus (gol'je) A network of fibrils or a series of four to eight channels forming an organelle in the cytoplasm of a cell.

Golgi tendon organ An afferent receptor found near the junction of tendons and muscles.

gonad (go'nad) A reproductive organ, testis or ovary, that produces gametes and sex hormones.

gray matter The region of the central nervous system that is composed of nonmyelinated nerve tissue.

greater omentum (o-men'tum) A double-layered serosa membrane that originates on the greater curvature of the stomach and hangs inferiorly like an apron over the contents of the abdominal cavity.

gross anatomy A branch of anatomy concerned with structures of the body that can be studied without a microscope.

gustatory (gus'tah-to"re) Pertaining to the sense of taste.

gut (gut) Pertaining to the intestines; generally a developmental term.

gyrus (jī'rus) A convoluted elevation or ridge.

hair A threadlike appendage of the epidermis consisting of keratinized dead cells that have been pushed up from a dividing basal layer.

hair cells Specialized receptor nerve endings for detecting sensations, such as in the organ of Corti.

hair follicle (fol'lĭ-k'l) A tubular depression in the dermis of the skin in which a hair develops.

hand The terminal portion of the upper extremity, consisting of the carpus, metacarpus, and phalanges.

hard palate (pal'at) The bony partition between the oral and nasal cavities, formed by the maxillae and palatine bones and lined by mucous membrane.

haustra (haws'trah) Sacculations or pouches of the colon.

haversian canal (ha-ver'shan) (see central canal)

haversian system (see osteon)

head The superior portion of a human that contains the brain and major sense organs.

heart A four-chambered, muscular, pumping organ positioned in the thoracic cavity slightly to the left of midline.

hemoglobin (he"mo-glo'bin) The pigment of red blood cells that constitutes about 33% of the cell volume and transports oxygen and carbon dioxide.

hemopoiesis (hem"ah-poi-e'sis) The production of red blood cells.

hepatic duct (hĕ-pat'ik) Tubules that drain bile from the liver and merge with the cystic duct from the gallbladder to form the common bile duct.

hepatic portal circulation (por'tal) The return of venous blood from the digestive organs through a capillary network within the liver before draining into the heart.

hepatopancreatic ampulla A small, elevated area within the duodenum where the combined pancreatic and common bile duct empty.

hiatus (hi-a'tus) An opening or fissure; a foramen.

hilus (hi'lus) A concave or depressed area where vessels or nerves enter or exit an organ.

hinge joint A type of diarthrotic articulation characterized by the convex surface of one bone fitting into the concave surface of another so that movement is confined to one plane, such as in the knee or interphalangeal joint.

histology (his-tol'o-je) Microscopic anatomy of the structure and function of tissues.

horizontal (transverse) A directional plane that divides the body, organ, or appendage into superior and inferior or proximal and distal portions.

hormone (hor'mōn) A chemical substance that is produced in an endocrine gland and secreted into the bloodstream to cause an effect in a specific target organ.

hyaline cartilage (hi'ah-līn) A cartilage with a homogeneous matrix. It is the most common type, occurring at the articular ends of bones, in the trachea, and within the nose, and forms the precursor to most of the bones in the body.

hymen (hi'men) A developmental remnant of membranous tissue that partially covers the vaginal opening.

hyperextension (hi"per-ek-sten'shun) Extension beyond the normal anatomical position or 180°.

hypertension (hi"per-ten'shun) Elevated or excessive blood pressure.

hyponychium (hi"po-nik'e-um) A thickened, supportive layer of the stratum corneum at the distal end of a digit under the free edge of the nail.

hypothalamus (hi"po-thal'ah-mus) A structure within the diencephalon and below the thalamus, which functions as an autonomic center and regulates the pituitary gland.

ileocecal valve (il"e-o-se'kal) A modification of the mucosa at the junction of the small and large intestine that forms a one-way passage and prevents the backflow of food materials.

ileum (il'e-um) The terminal portion of the small intestine between the jejunum and cecum.

incus (ing'kus) The middle of three ear ossicles within the middle ear chamber; commonly called the *anvil.*

inferior vena cava (in-fēr'e-or ve'nah ka'vah) A large systemic vein that collects blood from the body regions inferior to the level of the heart and returns it to the right atrium.

infundibulum (in"fun-dib'u-lum) The flesh stalk that attaches the pituitary gland to the hypothalamus of the brain.

ingestion (in-jes'chun) The process of taking food or liquid into the body by way of the oral cavity.

inguinal (ing'gwi-nal) Pertaining to the groin region.

inguinal canal The circular passage through which a testis descends into the scrotum.

insertion The more movable attachment of a muscle, usually more distal.

inspiration (in"spi-ra'shun) The act of breathing air into the alveoli of the lungs; also called *inhalation.*

insula (in'su-lah) A deep, paired cerebral lobe.

integument (in-teg'u-ment) Pertaining to the skin.

intercalated disc (in-ter'kah-lāt-ed) A thickened portion of the sarcolemma that extends across a cardiac muscle fiber and indicates the boundary between cells.

intercellular substance (in"ter-sel'u-lar) The matrix or material between cells that largely determines tissue types.

internal (deep) Toward the center, away from the surface of the body.

internal ear The innermost portion or chamber of the ear, containing the cochlea and the vestibular organs.

internal nares (na'rēz) The two posterior openings from the nasal cavity into the nasopharynx; also called the *choanae.*

intervertebral disc (in"ter-ver'tĕ-bral) A pad of fibrocartilage located between the bodies of adjacent vertebrae.

intestinal gland (in-tes'tĭ-nal) A simple tubular digestive gland that opens onto the surface of the intestinal mucosa and secretes digestive enzymes; also called *crypt of Lieberkuhn.*

intramembranous ossification (see membraneous bone)

intrinsic (in-trin'sik) Situated in or pertaining to internal origin.

inversion (in-ver'zhun) A movement of the foot in which the sole is turned inward.
iris (i'ris) The pigmented-muscular portion of the eye that surrounds the pupil and regulates its diameter.
islets of Langerhans (i'lets of lahng'er-hanz) (see pancreatic islets)
isthmus (is'mus) A narrow neck or portion of tissue connecting two structures.

jejunum (je-joo'num) The middle portion of the small intestine, located between the duodenum and the ileum.
joint capsule (kap'sūl) A fibrous tissue cuff surrounding a diarthrotic joint.

keratin (ker'ah-tin) An insoluble protein present in the epidermis and in epidermal derivatives such as hair and nails.
kidney (kid'ne) One of the paired organs of the urinary system that contains nephrons and filters urine from the blood.
kinesiology (ki-ne"se-ol'o-je) The study of body movement.
knee A region in the lower extremity, between the thigh and the leg, containing a diarthrotic hinge joint.

labial frenulum (la'be-al fren'u-lum) A longitudinal fold of mucous membrane that attaches the lips to the gum along the midline of both the upper and lower lip.
labia majora (la'be-ah ma-jor'ah) A portion of the external genitalia of a female, consisting of two longitudinal folds of skin extending downward and backward from the mons pubis.
labia minora (ma-nor'-ah) Two small folds of skin, devoid of hair and sweat glands, lying between the labia majora of the external genitalia of a female.
labyrinth (lab'i-rinth) The complex system of interconnecting tubes within the inner ear, which includes the semicircular canals, cochlea, and vestibule.
lacrimal canal (lak'ri-mal) A drainage duct for tears, located at the medial corner of an eyelid and conveying the tears medially into the nasolacrimal sac.
lacrimal gland A tear-secreting gland, located on the superior lateral portion of the eyeball underneath the upper eyelid.
lactation (lak-ta'shun) The production and secretion of milk by the mammary glands.
lacteal (lak'te-al) A small lymphatic duct associated with a villus of the small intestine.
lacuna (lah-ku'nah) A small, hollow chamber that houses an osteocyte in mature bone tissue or a chondrocyte in cartilage tissue.
lambdoidal suture (lam'doid-al su'chur) The immovable joint in the skull between the parietal bones and the occipital bone.
lamella (lah-mel'ah) A concentric ring of matrix surrounding the central canal in an osteon of mature bone tissue.
lamellated corpuscle (lah-mel'a-ted kor'pus'l) A sensory receptor for pressure, found in tendons, around joints, and in visceral organs; also called *pacinian corpuscle.*
lamina (lam'i-nah) A thin plate of bone that extends superiorly from the body of a vertebra to form both sides of the arch of a vertebra.
lanugo (lah-nu'go) Short, silky fetal hair, which may be present for a short time on a premature infant.
large intestine The last major portion of the GI tract, consisting of the cecum, colon, rectum, and anal canal.
laryngopharynx (lah-ring"go-far'inks) The inferior or lower portion of the pharynx in contact with the larynx.
larynx (lar'inks) The structure located between the pharynx and trachea that houses the vocal cords; commonly called the *voice box.*
lateral (lat'er-al) Pertaining to the side; farther from the midline.
lateral ventricle (ven'tri-k'l) A cavity located in the cerebral hemisphere of the brain and filled with cerebrospinal fluid.
leg The portion of the lower extremity between the knee and ankle.
lens (lenz) A transparent refractive organ of the eye, derived from ectoderm and positioned posterior to the pupil and iris.
lesser omentum (o-men'tum) A peritoneal fold of tissue extending from the lesser curvature of the stomach to the liver.
leukocyte (lu'ko-sīt) A white blood cell; also spelled leucocyte.
ligament (lig'ah-ment) A tough chord or fibrous band of connective tissue that binds bone to bone to strengthen and provide flexibility to a joint; it also may support viscera.
limbic system (lim'bik) A portion of the brain concerned with emotions and autonomic activity.
linea alba (lin'e-ah al'bah) A vertical fibrous band extending down the anterior medial portion of the abdominal wall.
lingual frenulum (ling'gwal fren'u-lum) A longitudinal fold of mucous membrane that attaches the tongue to the floor of the mouth.
lower extremity A lower appendage, including the thigh, leg, and foot.
lumbar (lum'ber) Pertaining to the region of the loins.
lumbar plexus (plek'sus) A network of nerves formed by the anterior branches of spinal nerves L1 through L4.
lumen (lu'men) The space within a tubular structure through which a substance passes.
lung One of the two major organs of respiration positioned within the thoracic cavity on both sides of the mediastinum.
lunula (lu'nu-lah) The half-moon-shaped whitish area at the proximal portion of a nail.
luteinizing hormone (LH) (loo"te-in"i-zing) A hormone secreted by the adenohypophysis (anterior lobe) of the pituitary that stimulates ovulation and progesterone secretion by the corpus luteum, influences the mammary glands of milk secretion in females, and stimulates testosterone secretion by the testes in males.
lymph (limf) A clear, plasmalike fluid that flows through lymphatic vessels.
lymph node A small, oval mass of reticular tissue located along the course of lymph vessels.
lymphocyte (lim'fo-sīt) A type of white blood cell characterized by a granular cytoplasm. Lymphocytes usually constitute about 20%–25% of the white cell count.
macrophage (mak'ro-fāj) A wandering phagocytic cell.
macula lutea (mak'u-lah lu'te-ah) A yellowish depression in the retina of the eye that contains the fovea centralis, the area of keenest vision.
malignant (mah-lig'nant) Tending to become worse and end in death.
malleus (mal'e-us) The first of three ear ossicles attached to the tympanum; commonly called the *hammer.*
mammary gland (mam'er-e) The gland of the female breast responsible for lactation and nourishment of the young.
marrow (mar'o) The soft connective tissue that occupies the inner cavity of certain bones and produces red blood cells.
mastication (mas"ti-ka'shun) Pertaining to the chewing of food.
matrix (ma'triks) The intercellular substance of a tissue.
meatus (me-a'tus) A passageway or opening into a structure.
mechanoreceptor (mek"ah-no-re-sep'tor) A sensory receptor that responds to a mechanical stimulus.
medial (me'de-al) Toward or nearer the midline of the body.
mediastinum (me"de-as-ti'num) The space in the center of the thorax between the two pleural cavities.
medulla (mē-dul'ah) The center portion of an organ.
medulla oblongata (ob"long-ga'tah) A portion of the brain stem located between the spinal cord and the pons.
medullary (marrow) cavity (med'u-lār"e) The hollow core of the diaphysis of a long bone, occupied by marrow.
meiosis (mi-o'sis) A specialized type of cell division by which gametes, or haploid sex cells, are formed.
Meissner's corpuscle (mīs'nerz kor'pus'l) A touch sensory receptor found in the papillary layer of the dermis of the skin.
melanin (mel'ah-nin) A dark pigment found within the epidermis or epidermal derivatives of the skin.
melanocyte (mel'ah-no-sīt) A specialized melanin-producing cell found in the deepest layer of epidermis.
melanoma (mel"ah-no'mah) A dark, malignant tumor of the skin; frequently forms in moles.
membranous bone (mem'brah-nus) Bone that forms from membranous connective tissue rather than from cartilage; also called *intramembranous bone.*
menarche (mē-nar'ke) The first menstrual discharge.
meninges (mē-nin'jēz) A group of three fibrous membranes that cover the central nervous system, composed of the dura mater, arachnoid membrane, and pia mater.
menisci (mĕn-is'si) Wedge-shaped fibrocartilages in certain movable joints; also called *semilunar cartilages.*
menopause (men'o-pawz) The cessation of menstrual periods in the human female.
menstrual cycle (men'stru-al) The rhythmic female reproductive cycle, which is characterized by changes in hormone levels and physical changes in the uterine lining.
menstruation (men"stru-a'shun) The discharge of blood and tissue from the uterus at the end of the female reproductive cycle.

mesenchyme (mes'eng-kīm) An embryonic connective tissue that can migrate, and from which all connective tissues arise.
mesenteric patches (mes''en-ter'ik) Clusters of lymph nodes on the walls of the small intestine; also called *Peyer's patches.*
mesentery (mes'en-ter''e) A fold of peritoneal membrane that attaches an abdominal organ to the abdominal wall.
mesoderm (mes'o-derm) The middle of the three germ layers.
mesothelium (mes''o-the'leum) A simple squamous epithelial tissue that lines body cavities and covers visceral organs; also called *serosa.*
mesovarium (mes''o-va're-um) The peritoneal fold that attaches an ovary to the broad ligament of the uterus.
metabolism (mĕ-tab'o-lizm) The sum total of the chemical changes that occur within a cell.
metacarpus (met''ah-kar'pus) The region of the hand between the wrist and the phalanges, including the five bones that compose the palm of the hand.
metarteriole (met''ar-te're-ōl) A small blood vessel that emerges from an arteriole, passes through a capillary network, and empties into a venule.
metastasis (mĕ-tas'tah-sis) The spread of a disease from one organ or body part to another.
metatarsus (met''ah-tar'sus) The region of the foot between the ankle and the phalanges, comprised of five bones.
metencephalon (met''en-sef'ah-lon) One of the five regions of the brain that contains the cerebellum and the pons.
microglia (mi-krog'le-ah) Small phagocytic cells found in the central nervous system.
microvilli (mi''kro-vil'i) Microscopic, hairlike projections of cell membranes on certain epithelial cells.
micturition (mik''tu-rish'un) The process of voiding urine; also called *urination.*
midbrain The portion of the brain between the pons and the forebrain.
middle ear The middle of the three ear chambers, which contains the three ear ossicles.
midsagittal (mid-saj'ĭ-tal) A plane that divides the body or organ into right and left halves.
mitosis (mi-to'sis) The process of cell division, in which the two daughter cells are identical and contain the same number of chromosomes.
mitral valve (mi'tral) The left atrioventricular heart valve; also called the *bicuspid valve.*
mixed nerve A nerve that contains both motor and sensory nerve fibers.
monocyte (mon'o-sīt) A phagocytic type of white blood cell, normally constituting about 3%–8% of the white blood cell count.
mons pubis (monz pu'bis) A fatty tissue pad covering the symphysis pubis and covered by pubic hair in the female.
morula (mor'u-lah) An early stage of embryonic development characterized by a solid ball of cells.
motor area A region of the cerebral cortex from which originate motor impulses to muscles or glands.
motor nerve A nerve composed of motor nerve fibers.
motor neuron (nu'ron) An efferent neuron that conducts action potentials away from the central nervous system and innervates effector organs (muscles and glands). Motor neurons form the ventral roots of the spinal nerves.
motor unit A single motor neuron and the muscle fibers it innervates.
mucosa (mu-ko'sah) A mucous membrane that lines cavities and tracts opening to the exterior.
mucous cell (mu'kus) A specialized unicellular gland that produces and secretes mucus; also called a *goblet cell.*
multipolar neuron (mul''tĭ-po'lar nu'ron) A nerve cell with many processes originating from the cell body.
muscle (mus'el) A major type of tissue that is adapted to contract. The three kinds of muscle are cardiac, smooth, and skeletal.
muscularis (mus''ku-la'ris) A muscular layer or tunic of an organ, composed of smooth muscle tissue.
myelencephalon (mi''ĕ-len-sef'ah-lon) One of the five regions of the brain that contains the medulla oblongata.
myelin (mi'ĕ-lin) A lipoprotein material that forms a sheathlike covering around nerve fibers.
myenteric plexus (mi''en-ter'ik plek'sus) A network of sympathetic and parasympathetic nerve fibers located in the muscularis tunic of the small intestine; also called the *plexus of Auerbach.*
myocardium (mi''o-kar'de-um) The cardiac muscle layer of the heart.
myofibril (mi''o-fi'bril) A bundle of contractile fibers within muscle cells.
myogram (mi'o-gram) A recording of electrical activity within a muscle.
myology (mi-ol'o-je) The science or study of muscle structure function.
myometrium (mi''o-me'tre-um) The layer or tunic of smooth muscle within the uterine wall.
myoneural junction (mi''o-nu'ral) The site of contact between an axon of a motor neuron and a muscle fiber.
myopia (mi-o'pe-ah) A visual defect in which objects may be seen distinctly only when very close to the eyes; also called *nearsightedness.*.
myosin (mi'o-sin) A thick filament protein that together with actin causes muscle contraction.

nail A hardened, keratinized plate that develops from the epidermis and forms a protective covering on the dorsal surface of the distal phalanges of fingers and toes.
nasal cavity (na'zal) A mucosa-lined space above the oral cavity, which is divided by a nasal septum and is the first chamber of the respiratory system.
nasal concha (kong'kah) A scroll-like bone extending medially from the lateral wall of the nasal cavity; also called a *turbinate bone.*
nasal septum (sep'tum) A bony and cartilaginous partition that separates the nasal cavity into two portions.
nasopharynx (na''zo-far'inks) The first or uppermost chamber of the pharynx, positioned posterior to the nasal cavity and extending down to the soft palate.
neck 1. Any constricted portion, such as the neck of an organ. 2. The cervical region of the body between the head and thorax.
necrosis (nĕ-kro'sis) Cell death or tissue death due to disease or trauma.
neonatal (ne''o-na'tal) The stage of life from birth to the end of four weeks.
nephron (nef'ron) The functional unit of the kidney, consisting of a glomerulus, glomerular capsule, convoluted tubules, and the loop of the nephron.
nerve A bundle of nerve fibers outside the central nervous system.
neurilemma (nu''rĭ-lem'ah) A thin, membranous covering surrounding the myelin sheath of a nerve fiber.
neurofibril (nu''ro-fi'bril) One of many delicate threadlike structures within the cytoplasm of a cell body and the axon hillock of a neuron.
neurofibril node A gap in the myelin sheath of a nerve fiber; also called *node of Ranvier.*
neuroglia (nu-rog'le-ah) Specialized supportive cells of the central nervous system.
neurohypophysis (nu''ro-hi-pof'i-sis) The posterior lobe of the pituitary gland.
neurolemmocyte A specialized neuroglia cell that surrounds an axon fiber of a peripheral nerve and forms the neurilemmal sheath.
neuron (nu'ron) The structural and functional unit of the nervous system, composed of a cell body, dendrites, and an axon; also called a *nerve cell.*
neutrophil (nu'tro-fil) A type of phagocytic white blood cell, normally constituting about 60%–70% of the white blood cell count.
nipple A dark pigmented, rounded projection at the tip of the breast.
Nissl bodies (nis'l) (see chromatophilic substances)
node of Ranvier (rah-ve-a') (see neurofibril node)
notochord (no'to-kord) A flexible rod of tissue that extends the length of the back of an embryo.
nucleoplasm (nu'kle-o-plazm'') The protoplasmic contents of the nucleus of a cell.
nucleus (nu'kle-us) A spheroid body within a cell that contains the genetic factors of the cell.
nucleus pulposus (pul-po'sus) The soft, pulpy core of an intervertebral disc; a remnant of the notochord.
nurse cells Specialized cells within the testes that supply nutrients to developing spermatozoa; also called *Sertoli cells.*

olfactory (ol-fak'to-re) Pertaining to the sense of smell.
olfactory bulb An aggregation of sensory neurons of an olfactory nerve, lying inferior to the frontal lobe of the cerebrum on both sides of the crista galli of the ethmoid bone.
olfactory tract The olfactory sensory tract of axons, which conveys impulses from the olfactory bulb to the olfactory portion of the cerebral cortex.
oligodendrocyte (ol''ĭ-go-den'dro-sīt) A type of neuroglial cell concerned with the formation of the myelin of nerve fibers within the central nervous system.

oocyte (o'o-sīt) A developing egg cell.
oogenesis (o''o-jen'ĕ-sis) The process of female gamete formation.
optic (op'tik) Pertaining to the eye.
optic chiasma (ki-as'mah) An X-shaped structure on the inferior aspect of the brain, anterior to the pituitary gland, where there is a partial crossing over of fibers in the optic nerves.
optic disc A small region of the retina where the fibers of the ganglion neurons exit from the eyeball to form the optic nerve; also called the *blind spot.*
optic tract A bundle of sensory axons located between the optic chiasma and the thalamus that functions to convey visual impulses from the photoreceptors within the eye.
oral Pertaining to the mouth; also called *buccal.*
ora serrata (o'rah) The jagged peripheral margin of the retina.
organ A structure consisting of two or more tissues, which performs a specific function.
organ of Corti (kor'te) The spiral organ, or functional unit of hearing, consisting of a basilar membrane supporting receptor hair cells and a tectorial membrane within the endolymph of the cochlear duct.
organelle (or''gan-el') A minute living structure of a cell with a specific function.
organism (or'gah-nizm) An individual living creature.
orifice (or'ĭ-fis) An opening into a body cavity or tube.
origin (or'ĭ-jin) The place of muscle attachment—usually the more stationary point or proximal bone; opposite the insertion.
oropharynx (o''ro-far'inks) The second portion of the pharynx, located in a posterior position to the oral cavity and extending from the soft palate to the hyoid bone.
osseous tissue (os'e-us) Bone tissue.
ossicle (os'sĭ-k'l) One of the three bones of the middle ear.
ossification (os''ĭ-fĭ-ka'shun) The process of bone tissue formation.
osteoblast (os'te-o-blast'') A bone-forming cell.
osteoclast (os'te-o-klast'') A cell that causes erosion and resorption of bone tissue.
osteocyte (os''te-o-sīt'') A mature bone cell.
osteology (os''te-ol'o-je) The study of the structure and function of bone and the entire skeleton.
osteon (os'te-on) A group of osteocytes and concentric lamellae surrounding a central canal, constituting the basic unit of structure in osseous tissue; also called a *haversian system.*
otoliths (o'to-liths) Small, hardened particles of calcium carbonate in the saccule and utricle of the inner ear, associated with the receptors of equilibrium.
oval window An oval opening in the bony wall between the middle and inner ear, into which the footplate of the stapes fits.
ovarian follicle (o-va're-an fol'lĭ-k'l) A developing ovum and its surrounding epithelial cells.
ovarian ligament (lig'ah-ment) A cordlike connective tissue that attaches the ovary to the uterus.
ovary (o'vah-re) The female gonad in which ova and certain sexual hormones are produced.
oviduct (o'vĭ-dukt) The tube that transports ova from the ovary to the uterus; also called the *uterine tube* or *fallopian tube.*
ovulation (o''vu-la'shun) The rupture of an ovarian follicle with the release of an ovum.
ovum (o'vum) A secondary oocyte after ovulation but before fertilization.

pacinian corpuscle (pah-sin'e-an kor'pusl) (see lamellated corpuscle)
palate (pal'at) The roof of the mouth or oral cavity.
palatine (pal'ah-tin) Pertaining to the palate.
palmar (pal'mar) Pertaining to the palm of the hand.
palpebra (pal'pĕ-brah) An eyelid.
pancreas (pan'kre-as) A mixed organ in the abdominal cavity that secretes gastric juices into the digestive tract and insulin and glucagon into the blood.
pancreatic duct (pan''kre-at'ik) A drainage tube that carries pancreatic juice from the pancreas into the duodenum of the hepatopancreatic ampulla.
pancreatic islets A cluster of cells within the pancreas that forms the endocrine portion and secretes insulin and glucagon; also called *islets of Langerhans.*
papillae (pah-pil'e) Small, nipplelike projections.
papillary muscle (pap'ĭ-ler''e) Muscular projections from the ventricular walls of the heart to which the chordae tendineae are attached.
paranasal sinus (par''ah-na'zal si'nus) A mucous-lined air chamber that communicates with the nasal cavity.
parasympathetic (par''ah-sim''pah-thet'ik) Pertaining to the division of the autonomic nervous system concerned with activities that restore and conserve metabolic energy.
parathyroids (par''ah-thi'roids) Small endocrine glands that are embedded on the posterior surface of the thyroid glands and are concerned with calcium metabolism.
parietal (pah-ri'ĕ-tal) Pertaining to a wall of an organ or cavity.
parietal pleura (ploor'ah) The thin serous membrane attached to the thoracic walls of the pleural cavity.
parotid gland (pah-rot'id) One of the paired salivary glands located on the sides of the face over the masseter muscle just anterior to the ear and connected to the oral cavity through a salivary duct.
parturition (par''tu-rish'un) The process of childbirth.
pectoral (pek'to-ral) Pertaining to the chest region.
pectoral girdle The portion of the skeleton that supports the upper extremities.
pedicle (ped'ĭ-k'l) The portion of a vertebra that connects and attaches the lamina to the body.
pelvic (pel'vik) Pertaining to the pelvis.
pelvic girdle (ger'd'l) The portion of the skeleton to which the lower extremities are attached.
pelvis (pel'vis) A basinlike bony structure formed by the sacrum and os coxae.
penis (pe'nis) The external male genital organ, through which urine passes during urination and which transports spermatozoa to the female during sexual intercourse.
pennate (pen'āt) Skeletal muscle fiber arrangement in which the fibers are attached to tendinous slips in a feather-like pattern.
perforating canal A minute duct through compact bone by which blood vessels and nerves penetrate to the central canal of an osteon.
pericardium (per''ĭ-kar'de-um) A protective serous membrane that surrounds the heart.
perichondrium (per''ĭ-kon'dre-um) A toughened connective sheet that covers cartilage.
perikaryon (per''ĭ-kar'e-on) The cell body of a neuron.
perilymph (per'ĭ-limf) A fluid, secreted by cells lining the bony canals of the inner ear, that provides a liquid-conducting medium for the vibrations involved in hearing and the maintenance of equilibrium.
perimysium (per''ĭ-mis'e-um) Fascia or connective tissue surrounding a bundle (fascicle) of muscle fibers.
perineum (per''ĭ-ne'um) The floor of the pelvis, which is the region between the anus and the scrotum in the male and between the anus and the vulva in the female.
perineurium (per''ĭ-nu're-um) Connective tissue surrounding a bundle (fascicle) of nerve fibers.
periodontal membrane (per''e-o-don'tal) A fibrous connective tissue lining the sockets of teeth.
periosteum (per''e-os'te-um) A fibrous connective tissue covering the surface of bone.
peripheral nervous system (pĕ-rif'er-al) The nerves and ganglia of the nervous system that lie outside of the brain and spinal cord.
peristalsis (per''ĭ-stal'sis) Rhythmic contractions of smooth muscle in the walls of various tubular organs, which move the contents along.
peritoneum (per''ĭ-to-ne'um) The serous membrane that lines the abdominal cavity and covers the abdominal visceral organs.
Peyer's patches (pi'erz) (see mesenteric patches)
phalanx (fa'lanks), pl. *phalanges* A bone of a finger or toe.
pharynx (far'inks) The region of the digestive system and respiratory system located at the back of the oral and nasal cavities and extending to the larynx anteriorly and the esophagus posteriorly; also called the *throat.*
photoreceptor (fo''to-re-sep'tor) A sensory nerve ending that responds to the stimulation of light.
physiology (fiz''e-ol'o-je) The science that deals with the study of body functions.
pia mater (pi'ah ma'ter) The innermost meninx that is in direct contact with the brain and spinal cord.
pineal gland (pin'e-al) A small cone-shaped gland located in the roof of the third ventricle.
pinna (pin'nah) The outer, fleshy portion of the external ear; also called the *auricle.*

pituitary gland (pĭ-tu′ĭ-tār″e) A small, pea-shaped endocrine gland situated on the interior surface of the diencephalonic region of the brain, consisting of anterior and posterior lobes; also called the *hypophysis* and commonly called the "master gland."

pivot joint (piv′ut) A diarthrotic joint in which the rounded head of one bone articulates with the depressed cup of another to permit a rotational type of movement.

placenta (plah-sen′tah) The organ of metabolic exchange between the mother and the fetus.

plantar (plan′tar) Pertaining to the sole of the foot.

plasma (plaz′mah) The fluid, extracellular portion of circulating blood.

platelets (plāt′lets) Small fragments of specific bone marrow cells that function in blood coagulation; also called *thrombocytes.*

pleural (ploor′al) Pertaining to the serous membranes associated with the lungs.

pleural cavity The potential space between the visceral pleural and parietal pleural membranes.

pleural membranes Serous membranes that surround the lungs and provide protection and compartmentalization.

plexus (plek′sus) A network of interlaced nerves or vessels.

plexus of Auerbach (ow′er-bahk) (see myenteric plexus)

plexus of Meissner (mīs′ner) (see submucosal plexus)

plica circulares (pli′kah ser-ku-lar′is) A deep fold within the wall of the small intestine that increases the absorptive surface area.

pneumotaxic area (nu″mo-tak′sik) The region of the respiratory control center, located in the pons of the brain.

pons (ponz) The portion of the brain stem just above the medulla oblongata and anterior to the cerebellum.

popliteal (pop-lit′e-al, pop″lĭ-te′al) Pertaining to the concave region on the posterior aspect of the knee joint.

posterior (pos-tēr′e-or) Toward the back; also called *dorsal.*

posterior root An aggregation of sensory neuron fibers lying between a spinal nerve and the dorsolateral aspect of the spinal cord; also called the *dorsal root* or *sensory root.*

posterior root ganglion (gang′gle-on) A cluster of cell bodies of sensory neurons located along the posterior root of a spinal nerve.

postganglionic neuron (pōst″gang-gle-on′ik) The second neuron in an autonomic efferent pathway. Its cell body is outside the central nervous system, and it terminates at an effector organ.

postnatal (pōst-na′tal) After birth.

preganglionic neuron (pre″gang-gle-on′ik) The first neuron in an autonomic efferent pathway. Its cell body is inside the central nervous system, and it terminates on a postganglionic neuron.

pregnancy A condition where a female has a developing offspring in the uterus.

prenatal (pre-na′tal) The period of offspring development during pregnancy; before birth.

prepuce (pre′pūs) A fold of loose, retractable skin covering the glans of the penis or clitoris; also called the *foreskin.*

prime mover The muscle most directly responsible for a particular movement.

pronation (pro-na′shun) A rotational movement of the forearm in which the palm of the hand is turned posteriorly.

proprioceptor (pro″pre-o-sep′tor) A sensory nerve ending that responds to changes in tension in a muscle or tendon.

prostate gland (pros′tāt) A walnut-shaped gland that surrounds the male urethra just below the urinary bladder and secretes an additive to seminal fluid during ejaculation.

prosthesis (pros-the′sis) An artificial device to replace a diseased or worn body part.

protraction (pro-trak′shun) The movement of a body part, such as the mandible, forward on a plane parallel with the ground; the opposite of *retraction.*

proximal (prok′si-mal) Closer to the midline of the body or origin of an appendage; the opposite of *distal.*

puberty (pu′ber-te) The period of development in which the reproductive organs become functional.

pulmonary (pul′mo-ner″e) Pertaining to the lungs.

pulmonary circulation The system of blood vessels from the right ventricle of the heart to the lungs, transporting deoxygenated blood and returning oxygenated blood from the lungs to the left atrium of the heart.

pulp cavity A cavity within the center of a tooth, containing blood vessels, nerves, and lymphatics.

pupil The opening through the iris that permits light to enter the posterior cavity of the eyeball and be refracted by the lens.

Purkinje fibers (pur-kin′je) Specialized cardiac muscle fibers that conduct electrical impulses from the AV bundle into the ventricular walls.

pyloric sphincter (pi-lor′ik sfingk′ter) A modification of the muscularis tunica between the stomach and the duodenum, which functions to regulate the food material leaving the stomach.

pyramid (pir′ah-mid) Any of several structures that have a pyramidal shape, including the renal pyramids in the kidney and the medullary pyramids on the ventral surface of the brain.

ramus (ra′mus) A branch of a bone, artery, or nerve.

raphe (ra′fe) A ridge or a seamlike structure.

receptor (re-sep′tor) A sense organ or a specialized distal end of a sensory neuron that receives stimuli from the environment.

rectouterine pouch (rek″to-u′ter-in) A pocket of parietal peritoneal membrane between the uterus and the rectum, forming the lowest point in the pelvic cavity; also called the *pouch of Douglas.*

rectum (rek′tum) The terminal portion of the GI tract, from the sigmoid colon to the anus.

red marrow (mar′o) A tissue that forms blood cells, located in the medullary cavity of certain bones.

red nucleus (nu′kle-us) An aggregation of gray matter of a reddish color that is located in the upper portion of the midbrain and sends fibers to certain brain tracts.

reflex (re′fleks) A rapid involuntary response to a stimulus.

reflex arc The basic conduction pathway through the nervous system, consisting of a sensory neuron, interneuron, and a motor neuron.

regional anatomy The division of anatomy concerned with structural arrangement in specific areas of the body such as the head, neck, thorax, or abdomen.

renal (re′nal) Pertaining to the kidney.

renal corpuscle (kor′pus′l) The portion of the nephron consisting of the glomerulus and a glomerular capsule; also called the *Malpighian corpuscle.*

renal cortex The outer portion of the kidney, primarily vascular.

renal medulla (mĕ-dul′ah) The inner portion of the kidney, including the renal pyramids and renal columns.

renal pelvis The inner cavity of the kidney formed by the expanded ureter and into which the calyces open.

renal pyramid A triangular structure within the renal medulla, composed of the loops of the nephrons and the collecting ducts.

respiration (res″pĭ-ra′shun) The exchange of gases between the external environment and the cells of an organism.

respiratory center The structure or portion of the brain stem that regulates the depth and rate of breathing.

respiratory membrane A thin, moistened membrane within the lungs, composed of an alveolar portion and a capillary portion, through which gaseous exchange occurs.

rete testis (re′te tes′tis) A network of ducts in the center of the testis, site of spermatozoa production.

reticular formation (rĕ-tik′u-lar) A network of nervous tissue fibers in the brain stem that arouses the higher brain centers.

retina (ret′ĭ-nah) The inner layer of the eyeball that contains the photoreceptors.

retraction (re-trak′shun) The movement of a body part, such as the mandible, backward on a plane parallel with the ground; the opposite of *protraction.*

retroperitoneal (ret″ro-per″ĭ-to-ne′al) Positioned behind the parietal peritoneum.

rhythmicity area (rith-mis′ĭ-te) A portion of the respiratory control center that is located in the medulla and controls inspiratory and expiratory phases.

ribosome (ri′bo-sōm) A cytoplasmic organelle composed of protein and RNA, which is the site of protein synthesis.

right lymphatic duct (lim-fat′ik) A major vessel of the lymphatic system that drains lymph from the upper right portion of the body into the right subclavian vein.

rod A photoreceptor in the retina of the eye that is specialized for colorless, dim light vision.

root canal The hollow, tubular extension of the pulp cavity into the root of the tooth, containing vessels and nerves.

rotation (ro-ta′shun) The movement of a bone around its own longitudinal axis.

round window A round, membrane-covered opening between the middle and inner ear, directly below the oval window.
rugae (ru'je) The folds or ridges of the mucosa of an organ.

saccule (sak'ūl) A saclike cavity in the membranous labyrinth inside the vestibule of the inner ear, containing a vestibular organ for equilibrium.
sacral (sa'kral) Pertaining to the sacrum.
sacral plexus (plek'sus) A network of nerve fibers that arise from spinal nerves L4 through S3. Nerves arising from the sacral plexus merge with those from the lumbar plexus to form the lumbosacral plexus and supply the lower extremity.
saddle joint A diarthrotic joint in which the articular surfaces of both bones are concave in one plane and convex, or saddle shaped, in the other plane, such as in the distal carpal and proximal metacarpal joint of the thumb.
sagittal (saj'ĭ-tal) A vertical plane through the body that divides it into right and left portions.
salivary gland (sal'ĭ-ver-e) An accessory digestive gland that secretes saliva into the oral cavity.
sarcolemma (sar''ko-lem'ah) The cell membrane of a muscle fiber.
sarcomere (sar'ko-mēr) The portion of a striated muscle fiber between the two adjacent Z lines that is considered the functional unit of a myofibril.
sarcoplasm (sar'ko-plazm) The cytoplasm within a muscle fiber.
scala tympani (ska'lah tim'pah-ne) The lower channel of a cochlea that is filled with perilymph.
scala vestibuli (ves-tib'u-le) The upper channel of the cochlea that is filled with perilymph.
Schwann cell (shwahn) (see neurolemmocyte)
sclera (skle'rah) The outer white layer of fibrous connective tissue that forms the protective covering of the eyeball.
scrotum (skro'tum) A pouch of skin that contains the testes and their accessory organs.
sebaceous gland (se-ba'shus) An exocrine gland of the skin that secretes sebum.
sebum (se'bum) An oily, waterproofing secretion of sebaceous glands.
semen (se'men) The thick, whitish fluid secretion of the reproductive organs of the male, consisting of spermatozoa and additives from the prostate gland and seminal vesicles.
semicircular canals Tubular channels within the inner ear that contain the receptors for equilibrium.
semilunar valve (sem''e-lu'nar) Crescent or half-moon-shaped heart valves, positioned at the entrances to the aorta and the pulmonary trunk.
seminal vesicles (sem'ĭ-nal ves'ĭ-k'lz) A pair of accessory male reproductive organs lying posterior and inferior to the urinary bladder, which secrete additives to spermatozoa into the ejaculatory ducts.
seminiferous tubules (se''mĭ-nif'er-us tu'būls) Numerous small ducts in the testes where spermatozoa are produced.
senescence (sĕ-nes'ens) The process of aging.
sensory area A region of the cerebral cortex that receives and interprets sensory nerve impulses.
sensory neuron (nu'ron) A nerve cell that conducts an impulse from a receptor organ to the central nervous system.
septum (sep'tum) A membranous or fleshy wall dividing two cavities.
serous membrane (se'rus) An epithelial and connective tissue membrane that lines body cavities and covers visceral organs within these cavities; also called *serosa.*
Sertoli cells (ser-to'le) (see nurse cells)
serum (se'rum) Blood plasma with the clotting elements removed.
sesamoid bone (ses'ah-moid) A membranous bone formed in a tendon in response to joint stress (e.g., the patella).
shoulder The region of the body where the humerus articulates with the scapula; also called the *omos.*
sigmoid colon (sig'moid ko'lon) The S-shaped portion of the large intestine between the descending colon and the rectum.
sinoatrial node (sin''o-a'tre-al) A mass of specialized cardiac tissue in the wall of the right atrium that initiates the cardiac cycle; the SA node; also called the *pacemaker.*
sinus (si'nus) A cavity or hollow space within a body organ such as a bone.
sinusoid (si'nŭ-soid) A small, blood-filled space in certain organs such as the spleen or liver.
skeletal muscle A specialized type of muscle tissue that is multinucleated, occurs in bundles, has crossbands of proteins, and contracts in either a voluntary or involuntary fashion.
small intestine The portion of the GI tract between the stomach and the cecum whose function is the absorption of food nutrients.
smooth muscle A specialized type of muscle tissue that is nonstriated, composed of fusiform, single-nucleated fibers, and contracts in an involuntary, rhythmic fashion within the walls of visceral organs.
soft palate (pal'at) The fleshy, posterior portion of the roof of the mouth from the palatine bones to the uvula.
somatic (so-mat'ik) Pertaining to the nonvisceral parts of the body.
spermatic cord (sper-mat'ik) The structure of the male reproductive system composed of the ductus deferens, spermatic vessels, nerves, cremasteric muscle, and connective tissue. The spermatic cord extends from a testis to the inguinal ring.
spermatogenesis (sper''mah-to-jen'ĕ-sis) The production of male sex gametes, or spermatozoa.
spermatozoon (sper''mah-to-zo'on) A mature male sperm cell or gamete.
sphincter (sfingk'ter) A circular muscle that functions to constrict a body opening or the lumen of a tubular structure.
sphincter of ampulla The muscular constriction at the opening of the common bile and pancreatic ducts; also called the *sphincter of Oddi.*
sphincter of Oddi (o'de) (see sphincter of ampulla)
spinal cord (spi'nal) The portion of the central nervous system that extends downward from the brain stem through the vertebral canal.
spinal ganglion A cluster of nerve cell bodies on the dorsal root of a spinal nerve.
spinal nerve One of the thirty-one pairs of nerves that arise from the spinal cord.
spinous process (spi'nus) A sharp projection of bone or a ridge of bone, such as on the scapula.
spleen (splēn) A large, blood-filled, glandular organ located in the upper left of the abdomen and attached by mesenteries to the stomach.
spongy bone (spun'je) A type of bone that contains many porous spaces; also called *cancellous bone.*
squamous (skwa'mus) Flat or scalelike.
stapes (sta'pēz) The innermost of the ossicles of the ear, which fits against the oval window of the inner ear; also called the *stirrup.*
stomach A pouchlike digestive organ located between the esophagus and the duodenum.
stratified (strat'ĭ-fīd) Arranged in layers, or strata.
stratum basale (ba'sal) The deepest epidermal layer, where mitotic activity occurs.
stratum corneum (stra'tum kor'ne-um) The outer, cornified layer of the epidermis of the skin.
stroma (stro'mah) A connective tissue framework in an organ, gland, or other tissue.
subarachnoid space (sub''ah-rak'noid) The space within the meninges, between the arachnoid and pia mater, where cerebrospinal fluid flows.
subdural space (sub-du'ral) A space within the meninges between the dura mater and arachnoid of the brain and spinal cord.
sublingual gland (sub-ling'gwal) One of the three pairs of salivary glands; it is located below the tongue and secretes to the side of the lingual frenulum.
submandibular gland (sub-man-dib'u-lar) One of the three pairs of salivary glands; it is located below the mandible and secretes to the side of the lingual frenulum.
submucosa (sub''mu-ko'sah) A layer of supportive connective tissue that underlies a mucous membrane.
sulcus (sul'kus) A shallow impression or groove.
superficial (su''per-fish'al) Toward or near the surface.
superficial fascia (fash'e-ah) A binding layer of connective tissue between the dermis of the skin and the underlying muscle.
superior Toward the upper part of a structure or toward the head; also called *cephalic.*
superior vena cava (ve'nah ka'vah) A large systemic vein that collects blood from regions of the body superior to the heart and returns it to the right atrium.
supination (su''pi-na'shun) Rotation of the arm so that the palm is directed forward or anteriorly; the opposite of *pronation.*
surface anatomy The division of anatomy concerned with the structures that can be identified from the outside of the body.
surfactant (ser-fak'tant) A substance produced by the lungs that decreases the surface tension within the alveoli.

suspensory ligament (sus-pen'so-re) A portion of the peritoneum that extends laterally from the surface of the ovary to the wall of the pelvic cavity; also a ligament supporting the breast and a ligament supporting the lens of the eye.
sutural bone A small bone positioned within a suture of certain cranial bones; also called a *wormian bone.*
suture (su'chur) A type of immovable joint articulation found between bones of the skull.
sweat gland A skin gland that secretes a fluid substance for evaporative cooling.
sympathetic (sim"pah-thet'ik) Pertaining to that part of the autonomic nervous system concerned with processes involving the utilization of energy; also called the *thoracolumbar division.*
symphysis (sim'fi-sis) A type of articulation characterized by a fibrocartilaginous pad between the articulating bones, which provides slight movement.
symphysis pubis (pu'bis) A slightly movable joint anteriorly located between the two pubic bones of the pelvic girdle.
synapse (sin'aps) A minute space between the synaptic knob of a presynaptic neuron and a dendrite of a postsynaptic neuron.
synarthrosis (sin"ar-thro'sis) An immovable joint, such as a synchondrosis or a suture.
synchondrosis (sin"kon-dro'sis) An immovable cartilaginous joint in which the articulating bones are separated by hyaline cartilage.
syndesmosis (sin"des-mo'sis) A type of slightly movable, fibrous joint in which two bones are united by an interosseous ligament.
synergist (sin'er-jist) A muscle that assists the action of the prime mover.
synovial cavity (sĭ-no've-al) A space between the two bones of a diarthrotic joint, filled with synovial fluid.
synovial joint A freely movable joint in which there is a synovial cavity between the articulating bones; also called a *diarthrotic joint.*
synovial membrane The inner membrane of a synovial capsule, which secretes synovial fluid into the joint cavity.
system A group of body organs that function together.
systemic (sis-tem'ik) Relating to the entire organism rather than individual parts.
systemic anatomy The division of anatomy concerned with the structure and function of the various systems.
systemic circulation The portion of the circulatory system concerned with blood flow from the left ventricle of the heart to the entire body and back to the heart via the right atrium; as opposed to the pulmonary system, which involves the lungs.
systole (sis'to-le) The muscular contraction of a heart chamber during the cardiac cycle.
systolic pressure (sis'tol'ik) Arterial blood pressure during the ventricular systolic phase of the cardiac cycle.

tactile (tak'til) Pertaining to the sense of touch.
taeniae coli (te'ne-e ko'li) The three longitudinal bands of muscle in the wall of the large intestine.
target organ The specific body organ that a particular hormone affects.
tarsal gland An oil secreting gland that opens on the exposed edge of each eyelid; also called *Meibomian gland.*
tarsus (tahr'sus) The seven bones that form the ankle.
taste bud An organ containing the chemoreceptors associated with the sense of taste.
tectorial membrane (tek-to're-al) A gelatinous membrane positioned over the hair cells of the organ of Corti in the cochlea.
tendo calcaneus (ten'do kal-ka'ne-us) The tendon that attaches the calf muscles to the calcaneus bone.
tendon (ten'dun) A band of dense fibrous connective tissue that attaches muscle to bone.
tendon sheath A covering of synovial membrane surrounding certain tendons.
tentorium cerebelli (ten-to're-um ser"ĕ-bel'ē) An extension of dura mater that forms a partition between the cerebral hemispheres and the cerebellum and covers the cerebellum.
teratogen (ter'ah-to-jen) Any agent or factor that causes a physical defect in a developing embryo or fetus.
testis (tes'tis) The primary reproductive organ of a male, which produces spermatozoa and male sex hormones.
thalamus (thal'ah-mus) An oval mass of gray matter within the diencephalon that serves as a sensory relay area.
thigh The proximal portion of the lower extremity between the hip and the knee, containing the femur bone.
third ventricle (ven'trĭ-k'l) A narrow cavity between the right and left halves of the thalamus and between the lateral ventricles, containing cerebrospinal fluid.
thoracic (tho-ras'ik) Pertaining to the chest region.
thoracic duct The major lymphatic vessel of the body, which drains lymph from the entire body except the upper right quadrant and returns it to the left subclavian vein.
thorax (tho'raks) The chest.
thrombocyte (throm'bo-sīt) A blood platelet formed from a fragmented megakaryocyte.
thymus gland (thi'mus) A bilobed lymphoid organ positioned in the upper mediastinum, posterior to the sternum and between the lungs.
thyroid cartilage (thi'roid kar'tĭ-lij) The largest cartilage in the larynx, which supports and protects the vocal cords; also called the *Adam's apple.*
tissue An aggregation of similar cells and their binding intercellular substance, joined to perform a specific function.
tongue A protrusible muscular organ on the floor of the oral cavity.
tonsil (ton'sil) A node of lymphoid tissue located in the mucous membrane of the pharynx.
trabeculae (trah-bek'u-le) A supporting framework of fibers crossing the substance of a structure, as in the lamellae of spongy bone.
trachea (tra'ke-ah) The airway leading from the larynx to the bronchi, composed of cartilaginous rings and a ciliated mucosal lining of the lumen; also called the *windpipe.*
tract A bundle of nerve fibers within the central nervous system.
transection (tran-sek'shun) A cross-sectional cut.
transverse colon (trans-vers' ko'lon) A portion of the large intestine that extends from right to left across the abdomen between the hepatic and splenic flexures.
transverse fissure (fish'ūr) The prominent cleft that horizontally separates the cerebrum from the cerebellum.
tricuspid valve (tri-kus'pid) The heart valve located between the right atrium and the right ventricle.
trigone (tri'gōn) A triangular area in the urinary bladder between the openings of the ureters and the urethra.
trochanter (tro-kan'ter) A broad, prominent process on a bone, specifically on the femur.
trochlea (trok'le-ah) A pulley-shaped structure.
true vocal cords Folds of the mucous membrane in the larynx that produce sound as they are pulled taut and vibrated.
trunk The thorax and abdomen together.
tubercle (tu'ber-k'l) A small, elevated process on a bone.
tuberosity (tu"bĕ-ros'ĭ-te) An elevation or protuberance on a bone.
tunica albuginea (tu'nĭ-kah al"bu-jin'e-ah) A tough, fibrous tissue surrounding the testis.
tympanic membrane (tim-pan'ik) The membranous eardrum positioned between the outer and middle ear.

umbilical cord (um-bil'ĭ-kal) A cordlike structure containing the umbilical arteries and vein, which connects the fetus with the placenta.
umbilicus (um-bil'ĭ-kus) The site where the umbilical cord was attached to the fetus; commonly called the *navel.*
unipolar neuron (u"nĭ-po'lar nu'ron) A nerve cell that has a single nerve fiber extending from its cell body.
upper extremity The appendage attached to the pectoral girdle, consisting of the brachium, forearm, and hand.
ureter (u-re'ter) A tube that transports urine from the kidney to the urinary bladder.
urethra (u-re'thrah) A tube that transports urine from the urinary bladder to the outside of the body.
urinary bladder (u'rĭ-ner"e) A distensible sac situated in the pelvic cavity posterior to the symphysis pubis, which stores urine.
urogenital triangle (u"ro-jen'ĭ-tal) The region of the pelvic floor containing the external genitalia.
uterine tube (u'ter-in) The tube through which the ovum is transported to the uterus; the site of fertilization; also called the *oviduct* or *fallopian tube.*

uterus (u″ter-us) A hollow, muscular organ in which a fetus develops. It is located within the female pelvis between the urinary bladder and the rectum; commonly called the *womb.*

utricle (u′tre-k′l) An enlarged portion of the membranous labyrinth, located within the vestibule of the inner ear.

uvula (u′vu-lah) A fleshy, pendulous portion of the soft palate that blocks the nasopharynx during swallowing.

vacuole (vak′u-ōl) Small spaces or cavities within the cytoplasm of a cell.

vagina (vah-ji′nah) A tubular organ that leads from the uterus to the vestibule of the female reproductive tract and receives the male penis during coitus.

vallate papillae (sir″kum-val′āt pah-pil′e) The largest of the papillae on the dorsal surface of the tongue. They are arranged in an inverted V-shaped pattern at the posterior portion of the tongue.

vasa vasorum (va′sah va-so′rum) Tiny blood vessels that form a network through the tunica of larger blood vessels to supply them with nutrients.

vasomotor center (vas-o-mo′tor) A cluster of nerve cell bodies in the medulla oblongata that controls the diameter of blood vessels and is therefore important in regulating blood pressure.

vein A blood vessel that conveys blood toward the heart.

vena cava (ve′nah ka′vah) One of two large vessels that return deoxygenated blood to the right atrium of the heart.

venous sinus (ve′nus si′nus) A circular venous drainage for the aqueous humor from the anterior chamber; located at the junction of the sclera and the cornea; also called *canal of Schlemm.*

ventral (ven′tral) Toward the front or belly surface; the opposite of *dorsal.*

ventricle (ven′tri-k′l) A cavity within an organ; especially those cavities in the brain that contain cerebrospinal fluid and those in the heart that contain blood to be pumped from the heart.

venule (ven′ūl) A small vessel that carries venous blood from capillaries to a vein.

vermiform appendix (ver′mi-form ah-pen′diks) (see appendix)

vermis (ver′mis) The coiled, middle lobular structure that separates the two cerebellar hemispheres.

vertebral canal (ver′te-bral) The tubelike cavity that extends through the vertebral column and contains the spinal cord; also called the *spinal canal.*

vestibule (ves′ti-būl) A space or cavity at the entrance to a canal, especially that of the nose, inner ear, and vagina.

villus (vil′lus) A minute projection that extends outward into the lumen from the mucosal layer of the small intestine.

viscera (vis′er-ah) The organs within the abdominal or thoracic cavities.

visceral (vis′er-al) Pertaining to the membranous covering of the viscera.

visceral peritoneum (per″i-to-ne′um) A serous membrane that covers the surfaces of abdominal viscera.

visceral pleura (ploor′ah) A serous membrane that covers the surfaces of the lungs.

visceroceptor (vis″er-o-sep′tor) A sensory receptor that is located within body organs and responds to information concerning the internal environment.

vitreous humor (vit′re-us hu′mor) The transparent gel that occupies the space between the lens and retina of the eyeball.

Volkmann's canal (fōlk′ mahnz) (see perforating canal)

vulva (vul′vah) The external genitalia of the female that surround the opening of the vagina; also called the *pudendum.*

white matter Bundles of myelinated axons located in the central nervous system.

wormian bone (wer′me-an) (see sutural bone)

yellow marrow (mar′o) Specialized lipid storage tissue within bone cavities.

zygote (zi′gōt) A fertilized egg cell formed by the union of a sperm and an ovulated secondary oocyte (ovum).

Credits

Illustrations

Laurel Antler

Figures: 3.11b, 4.8a, 7.23a-d, 11.13, 19.6 a-c, 22.6

Ayres Assoc.

Figures: 11.52, 14.4, 14.5

Ernest Beck

Figures: 16.52, 22.9b, 22.12b, 22.13, 22.14, 22.15

Samuel Collins

Figures: 6.9a-f, 11.41, 12.19, 13.1, 15.24, 20.23

Chris Creek

Figures: 2.8, 2.9a, 2.10a,b, 2.17, 4.9a, 4.14, 4.15, 5.5, 5.10, 5.22, 5.27, 6.3, 6.8, 6.32, 7.19, 9.4, 9.6, 9.7, 9.8, 9.15, 9.16, 9.18, 9.19b, 9.20, 9.21, 9.22, 9.23, 9.24, 9.25, 9.26, 9.27, 9.28, 9.29, 9.30, 9.31, 9.32, 9.33, 9.34, 9.35, 9.36, 9.37, 9.38, 9.39, 9.40, 9.41, 9.42, 9.43, 9.44, 9.45, 9.46, 9.47, 9.48, 9.49, 9.50, 9.51, 9.52, 12.14, 12.15, 12.16, 12.17, 14.1, 14.2a, 14.3, 14.7, 15.2, 16.21, 16.23, 16.24, 17.23, 17.28, 17.29, 17.36, 19.1, 19.7a, 19.8, 21.15, 22.22, 23.5, 23.6

Fineline Illustrations, Inc.

Figures: 3.4, 3.12, 11.19, 11.20, 11.21, 14.25, 16.17b, 16.25, 17.32, 20.20, 20.21, 22.8, 22.28, 22.29, 22.30, 22.31

Rob Gordon

Figures: 3.2, 8.26, 8.27, 8.29, 8.30, 8.31, 8.33, 9.11, 11.23, 11.25, 11.27, 11.30, 11.31a, 11.34, 11.35, 12.11, 16.8, 16.9, 18.23, 21.12

Rob Gordon/Tom Waldrop

Figures: 12.1, 15.36, 21.11

Ruth Krabach

Figures: 4.26a, 6.7, 6.31, 22.25

Bill Loechel

Figures: 3.1, 5.23, 11.16, 18.8, 18.20, 18.22, 18.29, 18.32, 18.37, 18.43, 19.5, 19.9, 19.10, 20.8, 20.9

Bill Loechel/Tom Waldrop

Figure: 18.34

Robert Margulies/Tom Waldrop

Figures: 9.1, 16.26, 16.29, 16.30a, 16.31, 16.32, 16.33, 16.35, 16.38, 16.39, 16.42, 18.9

Nancy Marshburn

Figures: 12.3, 16.6

Nancy Marshburn/Tom Waldrop

Figures: 5.17c, 5.28, 8.28, 8.32

Steve Moon

Figures: 2.9b, 5.19, 11.1, 11.9, 12.4, 12.6, 12.9, 12.10, 12.12, 13.2, 13.4, 13.5, 14.19, 15.4, 15.6, 15.7a, 15.7b, 15.9, 15.11, 15.14, 15.18a,b, 15.33, 15.39b, 16.11, 16.20, 16.27, 17.34, 18.10, 18.11, 18.15, 18.17, 21.16
Reference Plates: 1, 2, 3, 4, 5, 6, 7

Robert Margulies/Steve Moon

Figure: 16.40

Diane Nelson

Figures: 2.1, 4.5b, 4.20a, 4.23a, 4.24a, 4.28, 6.4, 6.5, 6.12, 6.13, 6.14, 6.15, 6.17, 6.18, 6.19, 6.29, 7.22, 10.2, 11.7, 11.10, 11.22, 11.40, 11.43, 11.48, 11.49, 11.53, 11.54, 14.16a, 14.21, 14.24, 14.26, 15.8, 15.22, 15.45, 15.46a,b, 16.2, 16.7a, 16.13, 16.16, 16.17a, 16.19, 16.28a, 16.34, 16.36, 16.37, 16.41, 16.43, 16.44, 16.50, 16.55, 16.57d, 17.12, 17.14, 17.15, 17.25, 18.19a-d, 20.13a, 22.10

Precision Graphics

Figures: 15.47, 21.8

Mildred Rinehart

Figures: 11.50, 12.20, 16.47a, 16.48

Nancy Sally

Figures: 3.16b, 15.37

Mike Schenk

Figures: 2.2, 2.3a, 2.5, 2.6, 2.11b, 2.12b, 2.13a,b, 2.14, 2.15, 2.16, 3.10b, 3.14, 4.2, 4.3, 4.4, 4.5a, 4.6a, 4.8c, 4.9c, 4.10c, 4.11b, 4.12c, 4.13b, 4.18a, 4.22a, 4.31, 5.4, 5.18a, 5.21, 6.1, 6.6, 6.11, 6.21, 6.22, 6.23, 6.24, 6.25, 6.26, 6.27, 6.28, 6.30, 6.33b,c, 6.34, 6.35b,c, 6.36, 6.37, 6.38, 6.39, 7.1, 7.2, 7.3, 7.5, 7.6a,b, 7.7, 7.8a, 7.9, 7.11, 7.13, 7.14, 7.15, 7.17, 7.18c,d, 8.2, 8.4, 8.5, 8.7, 8.8, 8.9, 8.10, 8.11, 8.12, 8.13, 8.14, 8.15, 8.16, 8.17, 8.18, 8.19, 8.20, 8.24, 8.25, 9.13b, 9.19a, 10.12, 11.3, 11.4, 11.6, 11.15, 11.17, 11.18, 11.32, 11.36, 11.37, 11.38, 11.39, 11.42, 11.46, 12.7, 12.8, 12.13, 13.3, 14.6, 14.10b, 14.13, 14.18, 14.22, 15.1, 15.13, 15.17, 15.21a, 15.27, 15.29, 15.35, 16.12, 16.18, 16.46, 16.49, 16.53, 16.58a, 17.1, 17.3, 17.4, 17.5, 17.6, 17.7, 17.9, 17.11, 17.21, 17.27, 18.1, 18.6, 18.14, 18.27a,b, 18.28c, 18.35b, 18.36a,b, 18.38, 18.40, 19.13, 19.15, 19.16, 19.17, 19.18, 19.20, 20.11, 20.12b, 21.9, 22.1a,b, 22.3, 22.4, 22.5, 22.7b,c, 22.11, 22.16a-d, 22.17, 22.18, 22.19b, 22.20b, 22.26, 22.32, 22.34a,b, 22.35, 22.37, 22.38

Tom Sims

Figures: 3.21a–e, 4.12a, 10.3, 12.5, 15.3, 15.20, 15.23b, 17.17, 17.19, 17.30, 17.33, 18.24, 19.14, 19.19, 20.7, 21.18, 21.23, 21.24

Tom Sims/Mike Schenk

Figures: 5.24, 11.26, 11.51, 15.19, 15.26, 17.10, 17.26, 18.7, 20.22, 21.2, 21.13, 23.4, 23.16

Catherine Twomey

Figures: 3.3, 20.10

Tom Waldrop

Figures: 4.10a, 5.1, 5.2, 5.3, 6.2a, 8.1, 9.2, 9.3, 9.5, 9.9a, 9.10, 9.14a, 11.2, 11.5, 11.33, 11.45, 13.9b, 13.10, 14.8, 14.9, 14.10a, 14.14, 14.15, 15.5, 15.10, 15.30, 15.31, 15.32, 16.3, 16.4, 16.5, 17.2, 17.22, 18.2, 18.3, 18.4, 18.5, 18.26, 19.2, 19.3, 19.12, 20.1, 20.2, 20.3, 20.6a–f, 20.19, 21.1, 21.3, 21.19, 21.21

Yvonne Walston

Figures: 3.7b, 3.8b, 3.9b, 3.13b, 3.18a, 3.19, 3.20, 4.7a, 4.7c, 4.16a,b, 4.17b, 4.18c, 4.19a,c, 4.20c, 4.21a, 4.21c, 4.22c, 4.23c, 4.24c, 4.25b, 4.26c, 4.29a2, 4.29b2, 4.29c2, 4.30b

John Walters & Assoc.

Figures: 11.11, 11.47a, 13.7, 13.8, 15.41, 15.42, 15.44b–d, 16.1, 16.14, 16.22, 18.12a,b, 21.4

Bill Wood

Figures: 12.21, 12.22, 13.6

Chapter 1

Figure 1.2: From *A Short History of Anatomy and Physiology from the Greeks to Harvey* by C. Singer, 1957, Dover, NY. Reprinted by permission.

Figure 1.4: From J. Kevorkian, *The Story of Dissection.* Copyright © Philosophical Library, Inc., New York, NY.

Chapter 2

Figures 2.8, 2.16: From John W. Hole, Jr., *Human Anatomy and Physiology,* 5th ed. Copyright © 1990 Wm. C. Brown Publishers, Dubuque, Iowa. All Rights Reserved. Reprinted by permission.

Chapter 3

Figure 3.3: From Stuart Ira Fox, *Human Physiology,* 3d ed. Copyright © 1990 Wm. C. Brown Publishers, Dubuque, Iowa. All Rights Reserved. Reprinted by permission.
Figures 3.4, 3.11b, 3.12, 3.16b: From John W. Hole, Jr., *Human Anatomy and Physiology,* 5th ed. Copyright © 1990 Wm. C. Brown Publishers, Dubuque, Iowa. All Rights Reserved. Reprinted by permission.

Reference Figures

Reference Figures 1–7: From John W. Hole, Jr., *Human Anatomy and Physiology,* 5th ed. Copyright © 1990 Wm. C. Brown Publishers, Dubuque, Iowa. All Rights Reserved. Reprinted by permission.

Chapter 4

Figures 4.8a, 4.10a, 4.26a: From John W. Hole, Jr., *Human Anatomy and Physiology,* 5th ed. Copyright © 1990 Wm. C. Brown Publishers, Dubuque, Iowa. All Rights Reserved. Reprinted by permission.
Figure 4.20a, 4.23a, 4.24a, 4.28: From John W. Hole, Jr., *Human Anatomy and Physiology,* 4th ed. Copyright © 1987 Wm. C. Brown Publishers, Dubuque, Iowa. All Rights Reserved. Reprinted by permission.

Chapter 5

Figure 5.19: From John W. Hole, Jr., *Human Anatomy and Physiology,* 5th ed. Copyright © 1990 Wm. C. Brown Publishers, Dubuque, Iowa. All Rights Reserved. Reprinted by permission.

Chapter 6

Figures 6.4, 6.5, 6.18, 6.19: From John W. Hole, Jr., *Human Anatomy and Physiology,* 5th ed. Copyright © 1990 Wm. C. Brown Publishers, Dubuque, Iowa. All Rights Reserved. Reprinted by permission.

Figure 6.29: From John W. Hole, Jr., *Human Anatomy and Physiology,* 4th ed. Copyright © 1987 Wm. C. Brown Publishers, Dubuque, Iowa. All Rights Reserved. Reprinted by permission.

Chapter 7

Figures 7.22, 7.23a-d: From John W. Hole, Jr., *Human Anatomy and Physiology,* 5th ed. Copyright © 1990 Wm. C. Brown Publishers, Dubuque, Iowa. All Rights Reserved. Reprinted by permission.

Chapter 8

Figure 8.9b: From John W. Hole, Jr., *Human Anatomy and Physiology,* 5th ed. Copyright © 1990 Wm. C. Brown Publishers, Dubuque, Iowa. All Rights Reserved. Reprinted by permission.

Chapter 9

Figures 9.5, 9.14a: From John W. Hole, Jr., *Human Anatomy and Physiology,* 5th ed. Copyright © 1990 Wm. C. Brown Publishers, Dubuque, Iowa. All Rights Reserved. Reprinted by permission.
Figures 9.9a, 9.10: From Stuart Ira Fox, *Human Physiology,* 3d ed. Copyright © 1990 Wm. C. Brown Publishers, Dubuque, Iowa. All Rights Reserved. Reprinted by permission.
Figure 9.19a: From *The Color Atlas of Human Anatomy,* Dr. Richard T. Jolly, translator. Copyright © 1980 Fabbri Representative Offices, Inc., New York, NY. Reprinted by permission.

Chapter 11

Figures 11.7, 11.48, 11.49: From John W. Hole, Jr., *Human Anatomy and Physiology,* 3d ed. Copyright © 1984 Wm. C. Brown Publishers, Dubuque, Iowa. All Rights Reserved. Reprinted by permission.
Figures 11.9, 11.19, 11.20, 11.33, 11.45, 11.50, 11.54: From John W. Hole, Jr., *Human Anatomy and Physiology,* 5th ed. Copyright © 1990 Wm. C. Brown Publishers, Dubuque, Iowa. All Rights Reserved. Reprinted by permission.
Figures 11.10, 11.13, 11.22, 11.43, 11.53: From John W. Hole, Jr., *Human Anatomy and Physiology,* 4th ed. Copyright © 1987 Wm. C. Brown Publishers, Dubuque, Iowa. All Rights Reserved. Reprinted by permission.
Figures 11.11, 11.32, 11.52: From Stuart Ira Fox, *Human Physiology,* 3d ed. Copyright © 1990 Wm. C. Brown Publishers, Dubuque, Iowa. All Rights Reserved. Reprinted by permission.
Figures 11.21, 11.40: From John W. Hole, Jr., *Human Anatomy and Physiology,* 2d ed. Copyright © 1981 Wm. C. Brown Publishers, Dubuque, Iowa. All Rights Reserved. Reprinted by permission.
Figures 11.28: From T. L. Peele, *Neuroanatomical Basis for Clinical Neurology,* 2d ed. Copyright © 1961 McGraw-Hill, Inc. Reprinted by permission.

Chapter 12

Figures 12.3, 12.9, 12.20, 12.21, 12.22: From John W. Hole, Jr., *Human Anatomy and Physiology,* 5th ed. Copyright © 1990 Wm. C. Brown Publishers, Dubuque, Iowa. All Rights Reserved. Reprinted by permission.

Chapter 13

Figures 13.1, 13.7, 13.8: From Stuart Ira Fox, *Human Physiology,* 3d ed. Copyright © 1990 Wm. C. Brown Publishers, Dubuque, Iowa. All Rights Reserved. Reprinted by permission.

Chapter 14

Figures 14.4, 14.5, 14.15: From Stuart Ira Fox, *Human Physiology,* 3d ed. Copyright © 1990 Wm. C. Brown Publishers, Dubuque, Iowa. All Rights Reserved. Reprinted by permission.
Figure 14.10a, 14.19, 14.24, 14.25: From John W. Hole, Jr., *Human Anatomy and Physiology,* 5th ed. Copyright © 1990 Wm. C. Brown Publishers, Dubuque, Iowa. All Rights Reserved. Reprinted by permission.
Figure 14.16a: From Eldon D. Enger, et al., *Concepts in Biology,* 6th ed. Copyright © 1991 Wm. C. Brown Publishers, Dubuque, Iowa. All Rights Reserved. Reprinted by permission.
Figure 14.21: From John W. Hole, Jr., *Essentials of Human Anatomy and Physiology,* 3d ed. Copyright © 1989 Wm. C. Brown Publishers, Dubuque, Iowa. All Rights Reserved. Reprinted by permission.
Figure 14.26: From John W. Hole, Jr., *Human Anatomy and Physiology,* 4th ed. Copyright © 1987 Wm. C. Brown Publishers, Dubuque, Iowa. All Rights Reserved. Reprinted by permission.

Chapter 15

Figures 15.7a,b, 15.8, 15.11, 15.14, 15.22 (left), 15.33, 15.37: From John W. Hole, Jr., *Human Anatomy and Physiology,* 5th ed. Copyright © 1990 Wm. C. Brown Publishers, Dubuque, Iowa. All Rights Reserved. Reprinted by permission.
Figure 15.22 (right): From John W. Hole, Jr., *Human Anatomy and Physiology,* 3d ed. Copyright © 1984 Wm. C. Brown Publishers, Dubuque, Iowa. All Rights Reserved. Reprinted by permission.
Figure 15.25: From Stuart Ira Fox, *A Lab Guide to Human Physiology,* 4th ed. Copyright © 1987 Wm. C. Brown Publishers, Dubuque, Iowa. All Rights Reserved. Reprinted by permission.
Figures 15.27, 15.38, 15.42 (left), 15.43, 15.44c,d, 15.47: From Stuart Ira Fox, *Human Physiology,* 3d ed. Copyright © 1990 Wm. C. Brown Publishers, Dubuque, Iowa. All Rights Reserved. Reprinted by permission.
Figure 15.39a, 15.42 (right): From J. R. McClintic, *Physiology of the Human Body,* 2d ed. Copyright © 1978 John Wiley & Sons, Inc., New York, NY. Reprinted by permission of John Wiley & Sons, Inc.
Figure 15.41: From Stuart Ira Fox, *Human Physiology,* 2d ed. Copyright © 1987 Wm. C. Brown Publishers, Dubuque, Iowa. All Rights Reserved. Reprinted by permission.
Figures 15.45, 15.46a: From John W. Hole, Jr., *Human Anatomy and Physiology,* 4th ed. Copyright © 1987 Wm. C. Brown Publishers, Dubuque, Iowa. All Rights Reserved. Reprinted by permission.
Figure 15.46b: From John W. Hole, Jr., *Human Anatomy and Physiology,* 2d ed. Copyright © 1981 Wm. C. Brown Publishers, Dubuque, Iowa. All Rights Reserved. Reprinted by permission.

Chapter 16

Figures 16.2, 16.6, 16.9, 16.14b, 16.16, 16.25, 16.28a, 16.36, 16.37, 16.48: From John W. Hole, Jr., *Human Anatomy and Physiology,* 5th ed. Copyright © 1990 Wm. C. Brown Publishers, Dubuque, Iowa. All Rights Reserved. Reprinted by permission.
Figures 16.7a, 16.43, 16.50: From John W. Hole, Jr., *Human Anatomy and Physiology,* 4th ed. Copyright © 1987 Wm. C. Brown Publishers, Dubuque, Iowa. All Rights Reserved. Reprinted by permission.
Figures 16.22, 16.52, 16.55, 16.56: From Stuart Ira Fox, *Human Physiology,* 3d ed. Copyright © 1990 Wm. C. Brown Publishers, Dubuque, Iowa. All Rights Reserved. Reprinted by permission.

Chapter 17

Figure 17.12: From John W. Hole, Jr., *Human Anatomy and Physiology,* 3d ed. Copyright © 1984 Wm. C. Brown Publishers, Dubuque, Iowa. All Rights Reserved. Reprinted by permission.
Figures 17.14, 17.15, 17.25: From John W. Hole, Jr., *Human Anatomy and Physiology,* 4th ed. Copyright © 1987 Wm. C. Brown Publishers, Dubuque, Iowa. All Rights Reserved. Reprinted by permission.
Figure 17.22: From Stuart Ira Fox, *Human Physiology,* 3d ed. Copyright © 1990 Wm. C. Brown Publishers, Dubuque, Iowa. All Rights Reserved. Reprinted by permission.
Figure 17.32: From John W. Hole, Jr., *Human Anatomy and Physiology,* 5th ed. Copyright © 1990 Wm. C. Brown Publishers, Dubuque, Iowa. All Rights Reserved. Reprinted by permission.

Chapter 18

Figures 18.10, 18.11, 18.15, 18.17, 18.19a-d, 18.22, 18.34: From John W. Hole, Jr., *Human Anatomy and Physiology,* 5th ed. Copyright © 1990 Wm. C. Brown Publishers, Dubuque, Iowa. All Rights Reserved. Reprinted by permission.
Figures 18.26, 18.27, 18.38: From Stuart Ira Fox, *Human Physiology,* 3d ed. Copyright © 1990 Wm. C. Brown Publishers, Dubuque, Iowa. All Rights Reserved. Reprinted by permission.

Chapter 19

Figures 19.5, 19.6, 19.9, 19.10, 19.12: From John W. Hole, Jr., *Human Anatomy and Physiology,* 5th ed. Copyright © 1990 Wm. C. Brown Publishers, Dubuque, Iowa. All Rights Reserved. Reprinted by permission.
Figure 19.20: Courtesy American Medical Systems, Inc., Minnetonka, MN.

Chapter 20

Figures 20.1, 20.10, 20.13a: From John W. Hole, Jr., *Human Anatomy and Physiology,* 5th ed. Copyright © 1990 Wm. C. Brown Publishers, Dubuque, Iowa. All Rights Reserved. Reprinted by permission.
Figures 20.4, 20.5, 20.23: From Stuart Ira Fox, *Human Physiology,* 3d ed. Copyright © 1990 Wm. C. Brown Publishers, Dubuque, Iowa. All Rights Reserved. Reprinted by permission.

Chapter 21

Figures 21.1, 21.21: From John W. Hole, Jr., *Human Anatomy and Physiology,* 5th ed. Copyright © 1990 Wm. C. Brown Publishers, Dubuque, Iowa. All Rights Reserved. Reprinted by permission.
Figures 21.4b, 21.8, 21.19: From Stuart Ira Fox, *Human Physiology,* 3d ed. Copyright © 1990 Wm. C. Brown Publishers, Dubuque, Iowa. All Rights Reserved. Reprinted by permission.

Chapter 22

Figures 22.6, 22.8, 22.25: From John W. Hole, Jr., *Human Anatomy and Physiology,* 5th ed. Copyright © 1990 Wm. C. Brown Publishers, Dubuque, Iowa. All Rights Reserved. Reprinted by permission.
Figures 22.9b, 22.12b, 22.13, 22.14, 22.15: From Stuart Ira Fox, *Human Physiology,* 3d ed. Copyright © 1990 Wm. C. Brown Publishers, Dubuque, Iowa. All Rights Reserved. Reprinted by permission.
Figure 22.22: From K. L. Moore, *Before We Are Born: Basic Embryology and Birth Defects.* Copyright © 1974 W. B. Saunders, Philadelphia, PA. Reprinted by permission.

Chapter 23

Figure 23.2: From L. Patten, *Human Embryology.* Copyright © 1933 McGraw-Hill, Inc. Reprinted by permission.
Figure 23.10, 23.12, 23.13, 23.14: Hollingsworth JW, Hashizume A, Jablon S: "Correlations between Tests of Aging in Hiroshima Subjects: An Attempt to Define Physiologic Age" in *Yale J Biol Med* 38:11–26, 1965. Reprinted by permission.

Photos

Unit Openers

One: © Dr. R. Clark and M. Goff/SPL/Photo Researchers, Inc.; **Two:** © Bill Longcore/Science Source/Photo Researchers, Inc.; **Three:** © CNRI/SPL/Photo Researchers, Inc.; **Four:** © Ed Webber/Visuals Unlimited; **Five:** © Manfred Kage/Peter Arnold, Inc.; **Six:** © CNRI/SPL/Photo Researchers, Inc.; **Seven:** © SIU/Visuals Unlimited; **Eight:** © Gould/De Anza/Peter Arnold, Inc.

Chapter 1

Figure 1.1: Academy, Florence, Art Resources, New York; **Figure 1.3:** Published by permission of the Danish National Museum; **Figure 1.5:** The Trustees of the British Museum; **Figure 1.6:** Egyptian National Museum; **Figures 1.7, 1.8:** Bioliotheque Nationale; **Figures 1.9, 1.10:** Fratelli Alinari; **Figure 1.11:** Studio Fotofast; **Figures 1.12, 1.13:** Reproduced by gracious permissions of Her Majesty Queen Elizabeth, Royal Art Collection, Great Britain; **Figure 1.14:** New York Academy of Medicine Library; **Figure 1.15:** Johan Maurita Van Nassau Mawitsuia: The Hague; **Figure 1.16:** Historical Pictures Service, Inc. Chicago; **Figure 1.17a1:** Photography by Debi Stansbaugh; **Figure 1.17b:** © George Mosil/Visuals Unlimited; **Figure 1.17c:** © CNRI/

SPL/Photo Researchers, Inc.; **Figure 1.18a:** Courtesy of Kodak; **Figure 1.18b:** © Carroll H. Weiss, 1973; **Figure 1.18c:** Courtesy of Utah Valley Regional Medical Center, Department of Radiology; **Figure 1.19a:** © Lester V. Bergman and Associates; **Figure 1.19b:** Robb, R. A: Three-Dimensional Biomedical Imaging, Vol. 1, Ch. 5, p. 156, Copyright CRC Press, Inc. Boca Raton, FL, 1985; **Figure 1.19c:** © Hank Morgan/Science Source/Photo Researchers, Inc.; **Figure 1.19d:** © Monte S. Buchsbaum, M.D.

Chapter 2

Figure 2.3b: B. Haagensen, C.D, *Diseases of the Breast,* 2nd ed., 1974, W.B. Saunders Company; **Figure 2.4:** Dr. Sheril D. Burton; **Figure 2.7a:** © Joan Lebold Cohen/Photo Researchers, Inc.; **Figure 2.7b:** © John D. Cunningham/Visuals Unlimited; **Figure 2.7c:** © M. Courtney-Clarke/Photo Researchers, Inc.; **Figures 2.7d, 2.7e:** © Leonard Lee Rue III/Photo Researchers, Inc.; **Figure 2.7f:** © Tom Hollyman/Photo Researchers, Inc.; **Figures 2.11a, 2.12a, 2.18a, 2.19:** Dr. Sheril D. Burton.

Chapter 3

Figure 3.5a: © Keith Porter; **Figure 3.5b:** © Per H. Kjeldson, University of Michigan at Ann Arbor; **Figure 3.5c:** © John D. Cunningham; **Figure 3.6:** © W. Orme Rod/Visuals Unlimited; **Figure 3.7a:** © Keith R. Porter; **Figure 3.8a:** © Dr. Gordon Leedale/Biophoto Associates; **Figure 3.9a:** © Keith Porter/Science Source/Photo Researchers, Inc.; **Figure 3.10a:** © Keith Porter; **Figure 3.11a:** © David M. Phillips/Visuals Unlimited; **Figure 3.13a:** Dr. Joseph Gall, Reproduced from *Journal of Cell Biology* 31(1966). Copyrighted permission of Rockefeller University Press; **Figure 3.15:** © Dr. Don Fawcett/Penelope Goddum-Rosse/Photo Researchers, Inc; **Figure 3.16a:** © Stephen L. Wolfe; **Figure 3.17:** Runk/Schoenberger/Grant Heilman, Inc.; **Figure 3.18b:** Courtesy of Ealing Corporation, South Natick; **Figures 3.21a-1, 3.21b-1, 3.21c-1, 3.21d-1, 3.21e-1:** © Edwin A. Reschke; **Figure 3.22a:** © Visuals Unlimited/Copyright 1986 by SIU; **Figure 3.22b:** © John D. Cunningham/Visuals Unlimited.

Chapter 4

Figures 4.1a, 4.1b, 4.1c: © Edwin A. Reschke; **Figure 4.1d:** © Dr. Kerry L. Openshaw; **Figures 4.6b, 4.6c:** © Edwin A. Reschke; **Figures 4.7b, 4.8b, 4.9b:** © Edwin A. Reschke; **Figure 4.10b:** © Biological Photo Service Set 36 #020; **Figure 4.11a:** Edwin A. Reschke/Peter Arnold, Inc.; **Figures 4.12b, 4.13a:** © Edwin A. Reschke; **Figure 4.16c:** © Biophoto Associates/Science Source/Photo Researchers, Inc.; **Figures 4.17a, 4.18b:** © Edwin A. Reschke; **Figure 4.19b:** © Bruce Iverson, BSC.; **Figure 4.20b:** © Edwin A. Reschke; **Figures 4.21b, 4.22b, 4.23b, 4.24b, 4.25a, 4.26b:** © Edwin A. Reschke; **Figure 4.27:** From: *Tissues and Organs: A Text Atlas of Scanning Electron Microscopy* by R. G. Kessel and R. H. Kardon. W. H. Freeman and Company © 1979; **Figures 4.29a1, 4.29b1, 4.29c1, 4.30a:** © Edwin A. Reschke.

Chapter 5

Figure 5.6: © Edwin Reschke/Peter Arnold, Inc.; **Figure 5.7:** © Dr. Jeremy Burgess/Science Photo Library/Photo Researchers, Inc.; **Figure 5.8:** © Victor B. Eichler; **Figure 5.11:** © Dr. Sheril D. Burton; **Figure 5.12:** © James Clayton; **Figure 5.13:** © Martin M. Rotker/Taurus Photos, Inc.; **Figure 5.14:** © Lester V. Bergman and Associates; **Figure 5.15a:** World Health Organization; **Figure 5.15b:** © George P. Bogumil; **Figure 5.16:** © Dr. Sheril D. Burton; **Figure 5.17a:** © Michael Abbey/Science Source/Photo Researchers, Inc.; **Figure 5.17b:** © Dr. Kerry L. Openshaw; **Figure 5.18b:** © John D. Cunningham/Visuals Unlimited; **Figure 5.20:** © Dr. Kerry L. Openshaw; **Figure 5.23a2:** © Dorte Gronis/OKAPIA 1988/Photo Researchers, Inc.; **Figure 5.23b2:** © James Stevenson/SPL/Photo Researchers, Inc.; **Figure 5.23c2:** © John Radcliffe/SPL/Photo Researchers, Inc.; **Figure 5.25a,b:** © SIU/Visuals Unlimited; **Figure 5.26:** © Dan McCoy/Rainbow; **Figure 5.29:** © Jacques Jangoux/Peter Arnold, Inc.; **Figure 5.30:** © Normal R. Lightfoot/Photo Researchers, Inc.

Chapter 6

Figure 6.2b: © Ted Conde; **Figure 6.9g:** © Edwin A. Reschke; **Figure 6.10:** Courtesy of Utah Valley Regional Medical Center, Department of Radiology; **Figure 6.16:** Courtesy of Kent M. Van De Graaff; **Figure 6.20, 6.33a, 6.35a:** Courtesy of Utah Valley Regional Medical Center, Department of Radiology; **Figure 6.40:** © Blayne L. Hirshche; **Figure 6.41:** © CNRI/SPL/Photo Researchers, Inc.

Chapter 7

Figure 7.4: Courtesy of Kent Van De Graaff; **Figure 7.8b:** Dr. Sheril D. Burton; **Figures 7.10, 7.12, 7.16:** Courtsy of Utah Valley Regional Medical Center, Department of Radiology; **Figure 7.18a:** © Dr. Sheril D. Burton; **Figure 7.18b:** Courtesy of Utah Valley Regional Medical Center, Department of Radiology; **Figure 7.20:** © Blayne L. Hirshche; **Figure 7.21:** Courtesy of Kent Van De Graaff; **Figure 7.23e:** Courtesy of Eastman Kodak.

Chapter 8

Figure 8.3: Courtesy of Utah Valley Regional Medical Center, Department of Radiology; **Figure 8.6:** © Paolo Koch/Photo Researchers, Inc.; **Figures 8.21, 8.22, 8.23:** © Dr. Sheril D. Burton; **Figure 8.34:** Courtesy of Kent M. Van De Graaff; **Figure 8.35:** © SIU, School of Medicine; **Figure 8.36a:** © SIU, School of Medicine; **Figure 8.36b,c:** Courtesy of Kent Van De Graaff; **Figure 8.36d:** Dr. Nemmer's office.

Chapter 9

Figure 9.9b: © John D. Cunningham/Visuals Unlimited; **Figure 9.12a,b:** Dr. H. E. Huxley; **Figure 9.12c:** From: *Tissues and Organs: A Text-Atlas of Scanning Electron Microscopy* by R. G. Kessel and R. H. Kardon. W.H. Freeman and Company © 1979; **Figure 9.14b:** © Biophoto Associates/Photo Researchers, Inc.; **Figure 9.17:** © Dr. Sheril D. Burton.

Chapter 10

Figure 10.1: Historical Pictures Service, Chicago; **Figures 10.4, 10.5, 10.6, 10.7, 10.8:** © Dr. Sheril D. Burton; **Figure 10.9a:** Courtesy of Kent M. Van De Graaff; **Figures 10.9b, 10.10:** Dr. Sheril D. Burton; **Figures 10.11, 10.13, 10.14, 10.15, 10.16, 10.17, 10.18, 10.19, 10.20, 10.21, 10.22, 10.23, 10.24, 10.25, 10.26, 10.27, 10.28, 10.29, 10.30:** Courtesy of Kent M. Van De Graaff; **Figure 10.31a,b:** © Dr. Kenneth A. Johnson; **Figure 10.31c:** © Lester V. Bergman and Associates.

Chapter 11

Figure 11.8: © John D. Cunningham/Visuals Unlimited; **Figure 11.12:** H. Webster, from Hubbard, John, *The Vertebrate Peripheral Nervous System,* Plenum Press, 1974; **Figure 11.14:** © Edwin A. Reschke; **Figure 11.24:** © Monte S. Buchebaum, M.D.; **Figure 11.29a:** © Martin Rotker/Taurus Photos; **Figure 11.31b:** © William C. Brown Publishers/Karl Rubin Photographer; **Figure 11.44:** Courtesy of Utah Valley Regional Medical Center, Department of Radiology; **Figure 11.47a:** © Per H. Kjeldson, University of Michigan at Ann Arbor; **Figure 11.47b:** © Per H. Kjeldson, University of Michigan at Ann Arbor.

Chapter 12

Figure 12.2: From: *Tissues and Organs: A Text-Atlas of Scanning Electron Microscopy* by R. G. Kessel and R. H. Kardon. W. H. Freeman and Company © 1979; **Figure 12.18:** Courtesy of Kent M. Van De Graaff; **Figure 12.23:** Photographed by Dr. Sheril D. Burton, assisted by Douglas W. Hacking.

Chapter 13

Figure 13.9a: © Wm. C. Brown Publishers/Karl Rubin Photographer.

Chapter 14

Figure 14.2b: © Edwin A. Reschke; **Figure 14.11:** © John D. Cunningham/Visuals Unlimited; **Figure 14.12:** © Lester V. Bergman and Associates; **Figure 14.16b:** © Journalism Services; **Figure 14.17:** © Fred Hossler/Visuals Unlimited; **Figure 14.20:** © Bio Photo Associates/Photo Researchers, Inc.; **Figure 14.23:** © Edwin A. Reschke; **Figures 14.27, 14.28:** Lester V. Bergman and Associates.

Chapter 15

Figure 15.7c: © Victor B. Eichler, BIO—Art; **Figures 15.12, 15.15a-i:** © Dr. Sheril D. Burton; **Figure 15.16a,b:** © Camera M.D. Studios; **Figure 15.18c:** P. N. Farnsworth, University of Medicine & Dentistry, New Jersey Medical School; **Figure 15.21b:** © Per H. Kjeldson, University of Michigan at Ann Arbor; **Figure 15.23a:** © Thomas Sims; **Figure 15.34:** © Dr. Sheril D. Burton; **Figure 15.40:** From Kessel, R. G. and Shih, C. Y. SCANNING ELECTRON MICROSCOPY IN BIOLOGY, © 1976, Springer-Verlag; **Figure 15.44a:** © Dean E. Hillman; **Figure 15.48:** © Dr. Stephen Clark.

Chapter 16

Figure 16.7b: © Bill Longcore/Science Source/Photo Researchers, Inc.; **Figure 16.10:** Courtesy of Utah Valley Hospital Regional Medical Center, Department of Radiology; **Figure 16.14a:** © Wm. C. Brown Publishers/Karl Rubin Photographer; **Figure 16.15:** © Carroll Weiss RBP, 1973; **Figures 16.28b, 16.30b:** From: *Practische Ontleedkunde* from J. Dankmeijer, H. G. Lambers & JMF Landsmeer, Bohn, Scheltema & Holkema; **Figure 16.45:** © Edwin Reschke; **Figure 16.47b:** © Dr. John Cunningham/Visuals Unlimited; **Figure 16.47c:** © Igaku-Shoin, Ltd; **Figure 16.51a,b:** © Richard Menard; **Figure 16.54a:** Donald S. Baim, from Hurst et al., *The Heart,* 5th ed., McGraw-Hill Book Company, 1982; **Figure 16.54b:** Zamboni from Weiss/Greep, *Histology,* 3d ed., Elsevier Science Publishing Company, Inc.; **Figure 16.57a,b:** American Heart Association; **Figure 16.57c:** © Lewis Lainey; **Figure 16.58b:** Courtesy of Utah Valley Regional Medical Center, Department of Radiology.

Chapter 17

Figure 17.8: American Lung Association; **Figure 17.13:** Yokochi, C., and Rohen, J. W.: *Photographic Anatomy of The Human Body,* 2nd ed., © Igaku-Shoin, Ltd.; **Figure 17.16a:** © John D. Cunningham/Visuals Unlimited; **Figure 17.16b:** © Edwin A. Reschke; **Figure 17.18:** © John Watney Photo Library; **Figure 17.20:** Courtesy of Kodak; **Figure 17.24:** © David M. Phillips, The Population Council/Visuals Unlimited; **Figure 17.31:** © Edward Lettau/Photo Researchers, Inc.; **Figure 17.35:** © Edward C. Vasquez, R. T., CRT., Dept. of Radiologic Technology, Los Angeles City College; **Figure 17.37a,b:** © Martin Rotker/Taurus Photos, Inc.

Chapter 18

Figure 18.13: Courtesy of Kent M. Van De Graaff; **Figure 18.16a–c:** © Biophoto Associates, Science Source/Photo Researchers, Inc.; **Figure 18.18:** © Edwin A. Reschke; **Figure 18.19e:** Courtesy of Utah Valley Regional Medical Center, Department of Radiology; **Figure 18.21:** © Edwin A. Reschke; **Figure 18.25:** © Manfred Kage/Peter Arnold, Inc.; **Figure 18.28a:** © Keith R. Porter; **Figure 18.28b:** Alpers, D. H. & Seetharam, D., *New England Journal of Medicine,* 296(1977)1047; **Figure 18.30:** Sodeman, W. A. & Watson, T. M. *Pathologic Physiology,* 6th ed., W.B. Saunders Company; **Figure 18.31:** © Lester U. Bergman and Associates, Inc.; **Figure 18.33a:** © Edwin A. Reschke; **Figures 18.33b, 18.35b:** From *Tissues and Organs: A Text-Atlas of*

Scanning Electron Microscopy, R. G. Kessel and R. H. Kardon, W. H. Freeman and Company © 1979; **Figure 18.36c:** © Victor B. Eichler; **Figure 18.39A:** © Carroll H. Weiss, RBP; **Figure 18.39b:** © Dr. Sheril D. Burton; **Figure 18.41:** Courtesy of Dr. Ding Zhoa Zun; **Figure 18.42:** Courtesy of John Van De Graaff, DDS, and Larry S. Pierce, DDS, both at Northwestern University, School of Dentistry, Chicago.

Chapter 19

Figure 19.4: © Science Photo Library/Photo Researchers, Inc.; **Figure 19.7b:** From *Tissues and Organs: A Text-Atlas of Scanning Electron Microscopy* by R. G. Kessel and R. H. Kardon, W. H. Freeman and Company © 1979; **Figure 19.11:** © Per H. Kjeldson, University of Michigan at Ann Arbor.

Chapter 20

Figure 20.6g,h: © Dr. Landrum Shettles; **Figure 20.12a:** © Biophoto Associates Science Source/Photo Researchers, Inc.; **Figure 20.13b:** © Francis Leroy, Biocosmos/SPL/Photo Researchers, Inc.; **Figures 20.14, 20.15:** © Edwin A. Reschke; **Figures 20.16, 20.17:** Manfred Kage/Peter Arnold, Inc.; **Figure 20.18:** © Edwin A. Reschke; **Figure 20.24:** Center for Disease Control, Atlanta, GA.

Chapter 21

Figure 21.4a: © Edwin A. Reschke; **Figure 21.5a:** Dr. Landrum Shettles; **Figure 21.5b:** © R. J. Blandau, A TEXTBOOK OF HISTOLOGY 10th ed., W. B. Saunders Company; **Figure 21.6:** Bloom/Fawcett, *Histology,* 10th ed., p. 859, W. B. Saunders Company; **Figure 21.7:** © Dr. Landrum Shettles; **Figures 21.10, 21.14:** © Edwin A. Reschke; **Figure 21.17a,b:** © BioPhoto Associates/Photo Researchers, Inc.; **Figure 21.20:** Courtesy of Utah Valley Regional Medical Center, Department of Radiology; **Figure 21.25a,b:** © SIU School of Medicine; **Figure 21.26a:** © 1986 SIU/Visuals Unlimited; **Figure 21.26b-f:** Bob Coyle.

Chapter 22

Figure 22.1c: © David Phillips/Visuals Unlimited; **Figure 22.2:** Lucian Zamboni from Greep, Roy and Weiss, *Histology,* 3d ed., McGraw-Hill Book Company, 1973; **Figure 22.7a:** Ronan O'Rahilly/Carnegie Laboratories of Embryology, University of California; **Figure 22.9a:** Carnegie Institute of Washington, Dept. of Embryology; **Figure 22.12a:** © Donald Yeager/Camera M.D. Studios; **Figure 22.16e:** © Dr. Landrum Shettles; **Figure 22.19a:** © Donald Yeager/Camera M.D. Studios; **Figure 22.20a:** © Dr. Landrum Shettles; **Figure 22.21:** © Donald Yeager/Camera M.D. Studios; **Figure 22.23a,b:** © Donald Yeager/Camera M.D. Studios; **Figure 22.24a:** © Alexander Tseares/Science Source/Photo Researchers, Inc.; **Figure 22.24b:** Courtesy of Kent M. Van De Graaff; **Figure 22.27:** March of Dimes; **Figure 22.33:** © Lester V. Bergman and Associates; **Figure 22.34c:** © Dr. Landrum Shettles; **Figure 22.36:** © Gregory Dellore, M.D. and Steven L. Clark, M.D.

Chapter 23

Figure 23.3: Courtesy of Utah Valley Regional Medical Center, Department of Radiology; **Figure 23.7:** From *The Physique of the Olympic Athlete* by J. M. Tanner, 1964; **Figure 23.9:** © Carolina Biological Supply Company; **Figure 23.11:** John Bird

Chapter 24

Figure 24.1: Vidic S. D., F. R. Suarez, 1984 *Photographic Atlas of the Human Body* (pl 27 p. 34) St. Louis, The C. V. Mosby Co.; **Figure 24.2:** Vidic S. D., F. R. Suarez, 1984 *Photographic Atlas of the Human Body* (pl 29 p. 38) St. Louis, The C. V. Mosby Co.; **Figure 24.3:** © Wm. C. Brown Publishers/Karl Rubin Photographer; **Figure 24.4:** Vidic S. D., F. R. Suarez, 1984 *Photographic Atlas of the Human Body* (pl 34, p. 48) St. Louis, The C. V. Mosby Co.; **Figure 24.5:** Vidic S. D., F. R. Suarez, 1984 *Photographic Atlas of the Human Body* (pl 71, p. 62) St. Louis, The C.V. Mosby Co.; **Figure 24.6:** Vidic S. D., F. R. Suarez, 1984 *Photographic Atlas of the Human Body* (pl 113, p. 182) St. Louis, The C.V. Mosby Co.; **Figure 24.7:** Vidic S. D. F. R. Suarez, 1984 *Photographic Atlas of the Human Body* (pl 115, p. 184) St. Louis, The C. V. Mosby Co.; **Figure 24.8:** Courtesy of Kent M. Van De Graaff; **Figure 24.9:** © Wm. C. Brown Publishers/Karl Rubin Photographer; **Figure 24.10:** Vidic S. D., F. R. Suarez, 1984 *Photographic Atlas of the Human Body* (plate 39, p. 58) St. Louis, The C.V. Mosby Co.; **Figure 24.11:** Vidic S. D., F. R. Suarez, 1984 *Photographic Atlas of the Human Body* (plate 40, p. 60) St. Louis, The C. V. Mosby Co.; **Figure 24.12:** © Wm. C. Brown Publishers/Karl Rubin Photographer; **Figure 24.13:** © Dr. Blayne L. Hirsche; **Figure 24.14:** Vidic S. D., F. R. Suarez, 1984 *Photographic Atlas of the Human Body* (pl 148, p. 240) St. Louis, The C.V. Mosby Co.; **Figure 24.15:** Dr. Sheril D. Burton; **Figure 24.16:** © Wm. C. Brown Publishers/Karl Rubin Photographer; **Figure 24.17:** Vidic S. D. F. R. Suarez, 1984 *Photographic Atlas of the Human Body* (129, p. 210) St. Louis, The C. V. Mosby Co.; **Figure 24.18:** Vidic S. D., F. R. Suarez, 1984 *Photographic Atlas of the Human Body* (130, 212) St. Louis, The C. V. Mosby Co.; **Figure 24.19:** Courtesy of Kent M. Van De Graaff; **Figures 24.20, 24.21:** Dr. Sheril D. Burton; **Figure 24.22:** Vidic S. D., F. R. Suarez, 1984 *Photographic Atlas of the Human Body* (pl 205, p. 336) St. Louis, The C. V. Mosby Co.; **Figure 24.23:** Vidic S. D., F. R. Suarez, 1984 *Photographic Atlas of the Human Body* (pl 206, p. 338) St. Louis, The C.V. Mosby Co.; **Figures 24.24, 24.25, 24.26, 24.27, 24.28:** © Wm. C. Brown Publishers/Karl Rubin Photographer

Index

Prefixes and Suffixes of Anatomical and Medical Terminology

Element	Definition and Example
a-	absent, deficient, without: *atrophy*
ab-	off, away from: *abduct*
abdomin-	relating to the abdomen: *abdominal*
-able	capable of: *viable*
ac-	toward, to: *actin*
acou-	hear: *acoustic*
ad-	toward, to: *adduct*
af-	movement toward a central point: *afferent artery*
alba	pale or whilte: *linea alba*
-alg	pain: *neuralgia*
ambi-	both: *ambidextrous*
angi-	pertaining to the vessels: *angiology*
ante-	before, in front of: *antebrachium*
anti-	against: *anticoagulant*
aqua-	water: *aqueous*
archi-	first: *archeteron*
arthr-	joint: *arthritis*
-asis	condition or state of: *homeostasis*
aud-	pertaining to the ear: *auditory*
auto-	self: *autolysis*
bi-	two: *bipedal*
bio-	life: *biopsy*
blast-	generative or germ bud: *blastocyst*
brachi-	arm: *brachialis*
brachy-	short: *brachydont*
brady-	slow: *bradycardia*
bucc-	cheek: *buccal cavity*
cac-	bad, ill: *cachexia*
calc-	stone: *calculus*
capit-	head: *capitis*
carcin-	cancer: *carcinogenic*
cardi-	heart: *cardiac*
cata-	lower, under, against: *catabolism*
caud-	tail: *cauda equina*
cephal-	head: *cephalic*
cerebro-	brain: *cerebrospinal fluid*
chol-	bile: *cholic*
chondr-	cartilage: *chondrocyte*
chrom-	color: *chromocyte*
-cide	destroy: *germicide*
circum-	around: *circumduct*
co-	together: *copulation*
coel-	hollow cavity: *coelom*
-coele	swelling, an enlarged space or cavity: *blastocoele*
con-	with, together: *congenital*
contra-	against, opposite: *contraception*
corn-	denoting hardness: *cornified*
corp-	body: *corpus*
crypt-	hidden: *cryptorchidism*
cyan-	blue color: *cyanosis*
cysti-	sac or bladder: *cystoscope*
cyto-	cell: *cytology*

Element	Definition and Example
de-	down, from: *descent*
derm-	skin: *dermatology*
di-	two: *diarthrotic*
dipl-	double: *diploid*
dis-	apart, away from: *disarticulate*
duct-	lead, conduct: *ductus deferens*
dur-	hard: *dura mater*
dys-	bad, difficult, painful: *dysentery*
e-	out, from: *eccrine*
ecto-	outside, outer, external: *ectoderm*
-ectomy	surgical removal: *tonsillectomy*
ede-	swelling: *edema*
-emia	pertaining to a condition of the blood: *lipemia*
end-	within: *endoderm*
entero-	intestine: *enteritis*
epi-	upon, after, in addition: *epigenesis*
erythro-	red: *erythrocyte*
ex-	out of: *excise*
exo-	outside: *exocrine*
extra-	outside of, beyond, in addition: *extracellular*
fasci-	band: *fascia*
febr-	fever: *febrile*
-ferent	bear, carry: *efferent*
fiss-	split: *fissure*
for-	opening: *foramen*
-form	shape: *fusiform*
gastro-	relating to the stomach: *gastrointestinal*
-gen	an agent that produces or originates: *pathogen*
-genic	produced from, producing: *carcinogenic*
gloss-	tongue: *glossopharyngeal*
glyco-	sugar: *glycosuria*
-gram	a record, recording: *electroencephalogram*
gran-	grain, particle: *agranulocyte*
-graph	instrument for recording: *electrocardiograph*
gravi-	heavy: *gravid*
gyn-	female sex: *gynecology*
haplo-	simple or single: *haploid*
hema(o)-	blood: *hematology*
hemi-	half: *hemiplagia*
hepat-	liver: *hepatic portal*
hetero-	other, different: *heterosexual*
histo-	web, tissue: *histology*
holo-	whole, entire: *holocrine*
homo-	same, alike: *homologous*
hydro-	water: *hydrocoel*
hyper-	beyond, above, excessive: *hypertension*
hypo-	under, below: *hypoglycemia*
-ia	state or condition: *hypoglycemia*
-iatrics	medical specialties: *pediatrics*
idio-	self, separate, distinct: *idiopathic*
ilio-	ilium: *iliosacral*
infra-	beneath: *infraspinatus*
inter-	among, between: *interosseous*
intra-	inside, within: *intracellular*
-ion	process: *acromion*
-ism	condition or state: *rheumatism*
iso-	equal, like: *isotonic*
-itis	inflammation: *meningitis*